MW01075932

HANDBOOK OF POLYOLEFINS

PLASTICS ENGINEERING

Founding Editor

Donald E. Hudgin

Professor
Clemson University
Clemson, South Carolina

1. Plastics Waste: Recovery of Economic Value, *Jacob Leidner*
2. Polyester Molding Compounds, *Robert Burns*
3. Carbon Black-Polymer Composites: The Physics of Electrically Conducting Composites, *edited by Enid Keil Sichel*
4. The Strength and Stiffness of Polymers, *edited by Anagnostis E. Zachariades and Roger S. Porter*
5. Selecting Thermoplastics for Engineering Applications, *Charles P. MacDermott*
6. Engineering with Rigid PVC: Processability and Applications, *edited by I. Luis Gomez*
7. Computer-Aided Design of Polymers and Composites, *D. H. Kaelble*
8. Engineering Thermoplastics: Properties and Applications, *edited by James M. Margolis*
9. Structural Foam: A Purchasing and Design Guide, *Bruce C. Wendle*
10. Plastics in Architecture: A Guide to Acrylic and Polycarbonate, *Ralph Montella*
11. Metal-Filled Polymers: Properties and Applications, *edited by Swapan K. Bhattacharya*
12. Plastics Technology Handbook, *Manas Chanda and Salil K. Roy*
13. Reaction Injection Molding Machinery and Processes, *F. Melvin Sweeney*
14. Practical Thermoforming: Principles and Applications, *John Florian*
15. Injection and Compression Molding Fundamentals, *edited by Avraam I. Isayev*
16. Polymer Mixing and Extrusion Technology, *Nicholas P. Cheremisinoff*
17. High Modulus Polymers: Approaches to Design and Development, *edited by Anagnostis E. Zachariades and Roger S. Porter*
18. Corrosion-Resistant Plastic Composites in Chemical Plant Design, *John H. Mallinson*
19. Handbook of Elastomers: New Developments and Technology, *edited by Anil K. Bhowmick and Howard L. Stephens*
20. Rubber Compounding: Principles, Materials, and Techniques, *Fred W. Barlow*
21. Thermoplastic Polymer Additives: Theory and Practice, *edited by John T. Lutz, Jr.*
22. Emulsion Polymer Technology, *Robert D. Athey, Jr.*
23. Mixing in Polymer Processing, *edited by Chris Rauwendaal*
24. Handbook of Polymer Synthesis, Parts A and B, *edited by Hans R. Kricheldorf*
25. Computational Modeling of Polymers, *edited by Jozef Bicerano*
26. Plastics Technology Handbook: Second Edition, Revised and Expanded, *Manas Chanda and Salil K. Roy*
27. Prediction of Polymer Properties, *Jozef Bicerano*
28. Ferroelectric Polymers: Chemistry, Physics, and Applications, *edited by Hari Singh Nalwa*
29. Degradable Polymers, Recycling, and Plastics Waste Management, *edited by Ann-Christine Albertsson and Samuel J. Huang*
30. Polymer Toughening, *edited by Charles B. Arends*
31. Handbook of Applied Polymer Processing Technology, *edited by Nicholas P. Cheremisinoff and Paul N. Cheremisinoff*
32. Diffusion in Polymers, *edited by P. Neogi*

33. Polymer Devolatilization, *edited by Ramon J. Albalak*
34. Anionic Polymerization: Principles and Practical Applications, *Henry L. Hsieh and Roderic P. Quirk*
35. Cationic Polymerizations: Mechanisms, Synthesis, and Applications, *edited by Krzysztof Matyjaszewski*
36. Polyimides: Fundamentals and Applications, *edited by Malay K. Ghosh and K. L. Mittal*
37. Thermoplastic Melt Rheology and Processing, *A. V. Shenoy and D. R. Saini*
38. Prediction of Polymer Properties: Second Edition, Revised and Expanded, *Jozef Bicerano*
39. Practical Thermoforming: Principles and Applications, Second Edition, Revised and Expanded, *John Florian*
40. Macromolecular Design of Polymeric Materials, *edited by Koichi Hatada, Tatsuki Kitayama, and Otto Vogl*
41. Handbook of Thermoplastics, *edited by Olagoke Olabisi*
42. Selecting Thermoplastics for Engineering Applications: Second Edition, Revised and Expanded, *Charles P. MacDermott and Aroon V. Shenoy*
43. Metallized Plastics, *edited by K. L. Mittal*
44. Oligomer Technology and Applications, *Constantin V. Uglea*
45. Electrical and Optical Polymer Systems: Fundamentals, Methods, and Applications, *edited by Donald L. Wise, Gary E. Wnek, Debra J. Trantolo, Thomas M. Cooper, and Joseph D. Gresser*
46. Structure and Properties of Multiphase Polymeric Materials, *edited by Takeo Araki, Qui Tran-Cong, and Mitsuhiro Shibayama*
47. Plastics Technology Handbook: Third Edition, Revised and Expanded, *Manas Chanda and Salil K. Roy*
48. Handbook of Radical Vinyl Polymerization, *edited by Munmaya K. Mishra and Yusef Yagci*
49. Photonic Polymer Systems: Fundamentals, Methods, and Applications, *edited by Donald L. Wise, Gary E. Wnek, Debra J. Trantolo, Thomas M. Cooper, and Joseph D. Gresser*
50. Handbook of Polymer Testing: Physical Methods, *edited by Roger Brown*
51. Handbook of Polypropylene and Polypropylene Composites, *edited by Harutun G. Karian*
52. Polymer Blends and Alloys, *edited by Gabriel O. Shonaike and George P. Simon*
53. Star and Hyperbranched Polymers, *edited by Munmaya K. Mishra and Shiro Kobayashi*
54. Practical Extrusion Blow Molding, *edited by Samuel L. Belcher*
55. Polymer Viscoelasticity: Stress and Strain in Practice, *Evaristo Riande, Ricardo Díaz-Calleja, Margarita G. Prolongo, Rosa M. Masegosa, and Catalina Salom*
56. Handbook of Polycarbonate Science and Technology, *edited by Donald G. LeGrand and John T. Bendler*
57. Handbook of Polyethylene: Structures, Properties, and Applications, *Andrew J. Peacock*
58. Polymer and Composite Rheology: Second Edition, Revised and Expanded, *Rakesh K. Gupta*
59. Handbook of Polyolefins: Second Edition, Revised and Expanded, *edited by Cornelia Vasile*
60. Polymer Modification: Principles, Techniques, and Applications, *edited by John J. Meister*

Additional Volumes in Preparation

Polymer Modifiers and Additives, *edited by John T. Lutz, Jr. and Richard F. Grossman*

Handbook of Elastomers: Second Edition, Revised and Expanded, *edited by Anil K. Bhowmick and Howard L. Stephens*

HANDBOOK OF POLYOLEFINS

Second Edition, Revised and Expanded

edited by

CORNELIA VASILE

Romanian Academy
"P. Poni" Institute of Macromolecular Chemistry
Iasi, Romania

MARCEL DEKKER, INC. NEW YORK • BASEL

ISBN: 0-8247-8603-3

This book is printed on acid-free paper.

Headquarters
Marcel Dekker, Inc.
270 Madison Avenue, New York, NY 10016
tel: 212-696-9000; fax: 212-685-4540

Eastern Hemisphere Distribution
Marcel Dekker AG
Hutgasse 4, Postfach 812, CH-4001 Basel, Switzerland
tel: 41-61-261-8482; fax: 41-61-261-8896

World Wide Web
http://www.dekker.com

The publisher offers discounts on this book when ordered in bulk quantities. For more information, write to Special Sales/Professional Marketing at the headquarters address above.

Current printing (last digit):
10 9 8 7 6 5 4 3 2 1

PRINTED IN THE UNITED STATES OF AMERICA

I gratefully dedicate this book to Professor Raymond B. Seymour's memory, who devoted his entire life to the progress of science.

Foreword to the First Edition

This *Handbook of Polyolefins* is a highly welcome and, in fact, necessary addition to the existing literature on these extremely important and useful polymeric materials. The first edition was coedited by Dr. Cornelia Vasile of Romania and the late Professor Raymond B. Seymour of the University of Southern Mississippi. Professor Seymour was known as a distinguished expert who also was able to persuade leading scientists and engineers to give comprehensive descriptions of their fields of interest and activity.

In this new handbook, Professor Seymour again confirmed his leading role as an editor and organizer. The well-known contributors cover the important additions to the science and technology of polyolefins since these polymers came to the forefront of synthetic materials in the mid-1950s.

From the history of polyolefins to their various important applications, every significant progressive development and invention is covered in this book. There is little doubt that the *Handbook of Polyolefins* will become a major addition to the literature on these important and increasingly necessary materials.

Herman F. Mark

Preface

The continuing interest and significant advances in polyolefins (PO) make it necessary to present this second edition. Recent developments in the field of polyolefins synthesis and additives are worthwhile reporting and in fact many new materials and processes are already in use. This is also true of the rapidly increasing number of new specialized POs, additives (such as stabilizers, nucleating and clarifying agents, fillers and reinforcements, and coupling agents), compounding procedures with various additives, and so on.

Since nothing is as constant as change, the reader of this book will note many improvements and revisions compared to the previous edition. Several authors from the previous edition have changed jobs or retired. Qualified authors were acquired to write new chapters or revise previous ones. Chapters based on the first edition have been thoroughly updated or in some instances rewritten. Some topics that appeared in the first edition are not included in this edition to allow more coverage of major topics.

The production of most manufactured goods involves heterogeneous and homogeneous catalysis and related processes of some kind. Recently, important progress has been made in the modeling of catalytically active surfaces and catalysts, particularly in the field of metallocenes. The polyolefin industry is now at a crossroads with a revolution directly attributable to these metallocenes. The metallocene constrained geometry catalysts offer the control of the polyolefin structure and molecular weight as never before witnessed. Plastics and ultra-high-strength composites made of polyolefin materials are continually improved and find increased applications. Each year diversification of all generic PO by copolymerization, alloying, and special compounding emphasizing less-polluting technologies and processes, the technical and commercial development of advanced materials, as well as high-tech methods of processing and fields of applications become more and more significant. The latest trends and comprehensive knowledge have been dealt with in all chapters of the book, progressing from general concepts and literature references through PO synthesis, properties, compounding, processing, and manufacture to detailed description of such aspects as boundary areas. Optimization of engineering parameters requires a theoretical framework to translate from the language of design to a particular variable or the phenomenological parameters of shape, size, and history. If a material is to be designed to satisfy specific engineering requirements, the theoretical framework should include not only the connection between engineering and phenomenological parameters but also an understanding of how chemical composition, structure, and bonding of material are related to the particular properties of interest.

Important classes of advanced materials that are of great interest include: functional polyolefins, catalysts, blends and composites. The variety of PO-based materials is truly astounding. Not all materials are crystalline, monophasic, and composed of a single component. Some are amorphous and some are in the form of films, while others are complex mixtures of several components and phases. Today, composites and polymer blends occupy a prime position as high-performance PO materials. Therefore, recent advances in the following topics have been covered: main industrial and novel routes of synthesis, new materials, thermodynamic properties of PO solutions, surface treatments, interactions in composites materials, functionalized PO, controlled lifetime of PO materials, biocompatibility, and biodeterioration.

In their efforts to improve and characterize materials, specialists have made use of a large variety of physical techniques that yield precise information about not only bulk properties but also solid surfaces. Such information coupled with the use of computer graphics and simulation can lead to new classes of materials, with novel structures as well as chemical, electrical, magnetic, or mechanical properties. Techniques of materials characterization have undergone a dramatic change in the last few years. Present-day electron microscopes have atomic resolution enabling the study of materials at the atomic level under real conditions (in air, with a liquid interface, in a vacuum, etc.). These are reasons to reconsider the structure–morphology–properties relationships.

Over the last few years our fundamental understanding of polymer surfaces has indeed become advanced and sophisticated owing to cross-disciplinary efforts. Since the problems of interfacial and surface phenomena are of increasing concern for researchers and engineers, two chapters on these topics were added.

We are indebted to the contributing authors for their worthy contributions. Without their dedication, expertise, and hard work, timely publication of this book would not have been possible. We would like also to take this opportunity to acknowledge the technical support received from the entire staff at Marcel Dekker, Inc.

Cornelia Vasile

Contents

Contributors

A. A. Adamyan A. A. Vishnevskii Institute of Surgery of the Russian Academy of Chemical Sciences, Moscow, Russia

M. Avella Istituto di Ricerca e Tecnologia delle Materie Plastiche, CNR, Arco Felice (NA), Italy

Vasile Blaşcu Department of Knittings and Ready-Made Clothing, "Gh. Asachi" Technical University, Iasi, Romania

Bogdan Georgiev Bogdanov Bourgas University of Technology, Bourgas, Bulgaria

Mihaela-Emilia Chiriac Department of Research, SC INCERPLAST SA–Research Institute for Plastics Processing, Bucharest, Romania

Cheng-Hsiang Chuang Department of Plastics Engineering, University of Massachusetts at Lowell, Lowell, Massachusetts

Sossio Cimmino Istituto di Ricerca e Tecnologia delle Materie Plastiche, CNR, Arco Felice (NA), Italy

Rudolph D. Deanin Department of Plastics Engineering, University of Massachusetts at Lowell, Lowell, Massachusetts

Maria Laura Di Lorenzo Istituto di Ricerca e Tecnologia delle Materie Plastiche, CNR, Arco Felice (NA), Italy

Emilia Di Pace Istituto di Ricerca e Tecnologia delle Materie Plastiche, CNR, Arco Felice (NA), Italy

Danica Doskočilová Department of Vibrational and NMR Spectroscopy, Institute of Macromolecular Chemistry, Academy of Sciences of the Czech Republic, Prague, Czech Republic

Ileana Drăguţan Romanian Academy, Institute of Organic Chemistry, Bucharest, Romania

Valerian Drăguţan Romanian Academy, Institute of Organic Chemistry, Bucharest, Romania

Charles H. Fisher Department of Chemistry, Roanoke College, Salem, Virginia

Mariana Gheorghiu Faculty of Physics, "Al. I. Cuza" University, Iasi, Romania

Aurelia Grigoriu Department of Chemical Technology of Textiles, "Gh. Asachi" Technical University, Iasi, Romania

George Ervant Grigoriu Department of Physics and Structures of Polymers, Romanian Academy, "P. Poni" Institute of Macromolecular Chemistry, Iasi, Romania

K. Z. Gumargalieva N. N. Semenov Institute of Chemical Physics of the Russian Academy of Sciences, Moscow, Russia

Jun-ichi Imuta Mitsui Chemicals, Inc., Tokyo, Japan

Gheorghe Ivan Tofan Group, Bucharest, Romania

Silviu Jipa Advanced Research Center, Institute for Electrical Engineering, Bucharest, Romania

Norio Kashiwa Mitsui Chemicals, Inc., Tokyo, Japan

Ronald Koningsveld Max Planck Institute for Polymer Research, Mainz, Germany

Anand Kumar Kulshreshtha Research Centre, Indian Petrochemicals Corporation Limited, Vadodara, Gujarat, India

P. Laurienzo Istituto di Ricerca e Tecnologia delle Materie Plastiche, CNR, Arco Felice (NA), Italy

M. Malinconico Istituto di Ricerca e Tecnologia delle Materie Plastiche, CNR, Arco Felice (NA), Italy

Margaret A. Manion Library, University of Massachusetts at Lowell, Lowell, Massachusetts

E. Martuscelli Istituto di Ricerca e Tecnologia delle Materie Plastiche, CNR, Arco Felice (NA), Italy

Jimmy W. Mays Department of Chemistry, University of Alabama at Birmingham, Birmingham, Alabama

Marin Michailov† Bulgarian Academy of Sciences, Sofia, Bulgaria

Mihaela Mihaies Department of Research, Research Center for Designing and Production for Plastic Processing, SC Ceproplast SA, Iasi, Romania

Erik Nies Center for Polymers and Composites, University of Technology, Eindhoven, The Netherlands

Anton Olaru Department of Research, Research Center for Designing and Production for Plastic Processing, SC Ceproplast SA, Iasi, Romania

Mihaela Cristina Pascu Department of Biophysics, "Gr. T. Popa" Medicine and Pharmacy University, Iasi, Romania

A. Ya. Polishchuk Department of Chemical and Biological Kinetics, N. M. Emanuel Institute of Biochemical Physics of the Russian Academy of Sciences, Moscow, Russia

Gheorghe Popa Faculty of Physics, "Al. I. Cuza" University, Iasi, Romania

Aaron D. Puckett School of Dentistry/Biomaterials, University of Mississippi Medical Center, Jackson, Mississippi

Béla Pukánszky Department of Plastics and Rubber Technology, Technical University of Budapest, Budapest, and Department of Applied Polymer Chemistry and Physics, Institute of Chemistry, Chemical Research Center, Hungarian Academy of Sciences, Budapest, Hungary

Mihai Rusu Department of Macromolecules, "Gh. Asachi" Technical University, Iasi, Romania

---

† Deceased.

Bohdan Schneider Department of Vibrational and NMR Spectroscopy, Institute of Macromolecular Chemistry, Academy of Sciences of the Czech Republic, Prague, Czech Republic

Hans Adam Schneider FMF Freiburger Materialforschungszentrum, "Albert-Ludwigs-Universität" Freiburg, Freiburg, Germany

Radu Setnescu Advanced Research Center, Institute for Electrical Engineering, Bucharest, Romania

Clara Silvestre Istituto di Ricerca e Tecnologia delle Materie Plastiche, CNR, Arco Felice (NA), Italy

Roland Streck* Hüls AG, Marl, Germany

Subhasis Talapatra Research Centre, Indian Petrochemicals Corporation Limited, Vadodara, Gujarat, India

Krishnaraj N. Tejeswi Department of Plastics Engineering, University of Massachusetts at Lowell, Lowell, Massachusetts

Luc van Opstal BASF Antwerpen, Antwerpen, Belgium

Cornelia Vasile Romanian Academy, "P. Poni" Institute of Macromolecular Chemistry, Iasi, Romania

T. I. Vinokurova A. A. Vishnevskii Institute of Surgery of the Russian Academy of Chemical Sciences, Moscow, Russia

M. G. Volpe Istituto di Ricerca e Tecnologia delle Materie Plastiche, CNR, Arco Felice (NA), Italy

Traian Zaharescu Advanced Research Center, Institute for Electrical Engineering, Bucharest, Romania

G. E. Zaikov Department of Chemical and Biological Kinetics, N. M. Emanuel Institute of Biochemical Physics of the Russian Academy of Sciences, Moscow, Russia

* Retired.

1

Competitive New Technologies in Polyolefin Synthesis and Materials

Anand Kumar Kulshreshtha and Subhasis Talapatra
Research Centre, Indian Petrochemicals Corporation Limited, Vadodara, Gujarat, India

I. HISTORY OF POLYOLEFIN POLYMERIZATION

In the early 1950s Karl Ziegler, investigating his new triethyl-aluminum catalyzed synthesis of higher olefins, discovered the "nickel effects" which was caused by a colloidal nickel contaminant. In search of other displacement catalysts Ziegler and coworkers then tested $AlEt_3$ in combination with a wide range of transition metal compounds. A new breakthrough was made when zirconium acetylacetonate was found to catalyze the formation of high molecular weight polyethylene at low pressure. Another very active transition metal salt was found to be $TiCl_4$. The catalyst ($TiCl_4$ + $AlEt_3$) was developed for the manufacture of HDPE in 1953.

Parallel independent experiments on polyolefins were being conducted by Giulio Natta's group in Italy. In early 1954 Natta succeeded in isolating crystalline *polypropylene* comprising extended sequences of monomer units with the same configuration. Natta also recognized the importance of the catalyst crystal surface in determining polymer isotacticity [2].

Organometallic compounds known as Ziegler catalysts were produced industrially at rates measuring thousands of tonnes annually. Stereospecific polymerization of propylene, butylene and styrene was done by Natta in March 1954. Natta's discovery broke nature's monopoly of building macromolecules of regular structure using enzymes.

High pressure processes are the "grandparents" of the polyolefin industry and can make conventional LDPE. These free radical technologies cannot make HDPE. High pressure Ziegler-catalyzed processes are used only in a handful of plants around the world. The Ziegler catalysts employed were multi-site and produced low MW oligomer and dimer by-products which accumulated in recycle streams. These by-products were reincorporated into the polymer chains as short chain branches, reducing the density of the product. The Ziegler-catalyzed process is used in the production of soft VLDPE resins.

The first commercial production of crystalline PP was done by the Montedison company at Ferrara, Italy in 1957 using the MOPLEN batch slurry process.

The big breakthrough in catalyst development came in the 1970s when several manufacturers developed third generation catalysts supported on activated $MgCl_2$ which could control the atactic content of the resulting polymer. The process does not require costly catalyst removal or polymer extraction.

Fourth generation catalysts produce 25,000 pounds of product per pound of catalyst. This saves money not only for the catalyst, it also means there is so little catalyst left in the polymer that it remains as such in a dechirated state. The atactic content is also so low that it need not be removed. The product is in the form of spherical particles and does not require post-reactor extrusion.

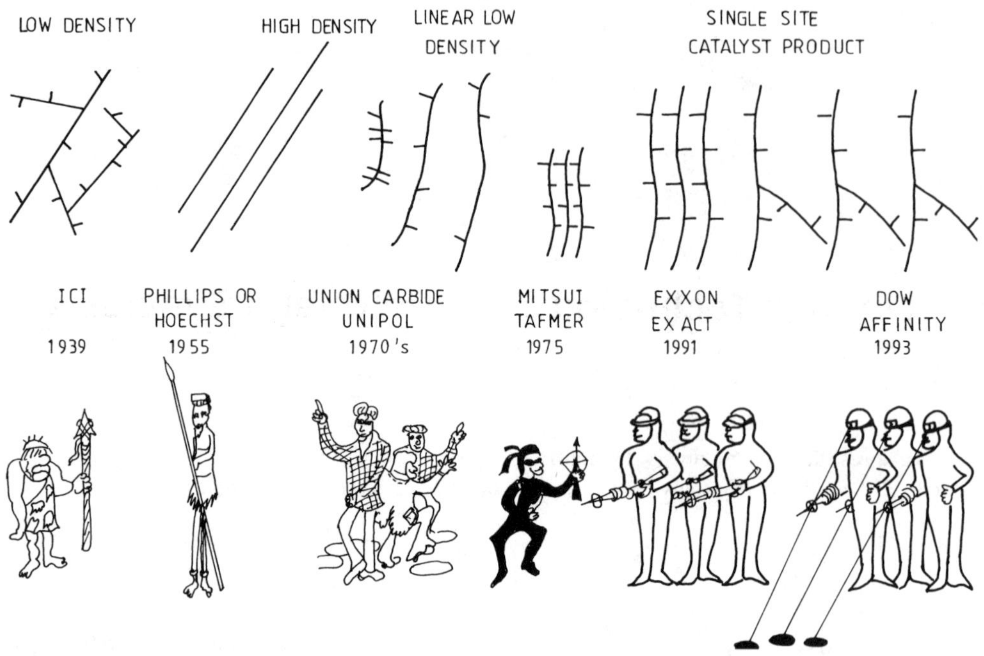

FIG. 1 Polyethylene/Ethylene Copolymer Evolution. (From Ref. 1.)

Increasing competition among various manufacturers resulted in a need to reduce the manufacturing cost of PP. This activity focussed primarily on improving the catalyst since catalyst activity, stereoselectivity and morphology affect the manufacturing economics and polymer properties. The progression in catalyst systems is shown in Table 1.

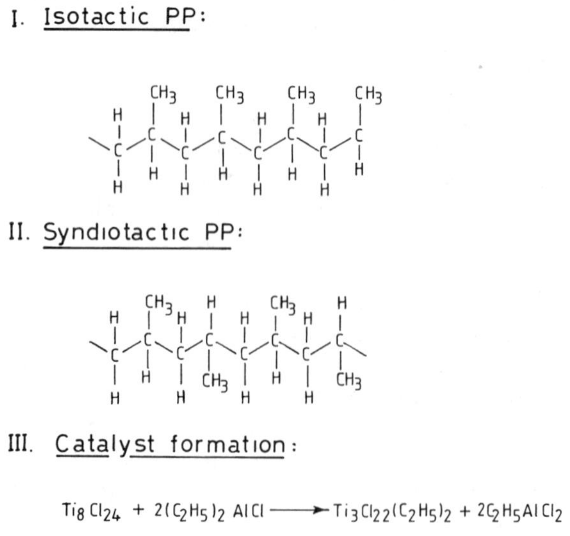

I. Isotactic PP:

II. Syndiotactic PP:

III. Catalyst formation:

$$Ti_8Cl_{24} + 2(C_2H_5)_2AlCl \longrightarrow Ti_3Cl_{22}(C_2H_5)_2 + 2C_2H_5AlCl_2$$

FIG. 2 Polymerization to Polypropylene.

Since its first use in 1954, the ZN catalyst has experienced a dramatic improvement in both activity and selectivity. The activity increased from a mere 0.2 kg of polymer per g of catalyst to 30 kg of polymer per g of catalyst, with selectivity improving from 86% to better than 96% isotactic index.

The first commercial processes for making PP used alkanes as diluents in the reactor. Catalyst stereoselectivity was such as to make PP with a 90% isotactic index. Catalyst efficiency was low

TABLE 1 Catalysts for Polypropylene

Catalyst type	Yield (kg/g)	Isotactic index (%)	Catalyst residue (ppm)
$TiCl_4 + AlEt_3$	0.2	40	10,000
First generation	0.8–1.5	90	4000
Second generation	1.5–5.0	93+	1000
Third generation	10–25	96+	100–300
Fourth generation	25–30	97	100–300

and washing vessels were needed for the deashing removal of catalyst residues from the polymer. With the advent of third generation high activity catalysts, the deashing step was no longer needed in the manufacture of PP. Figure 3 provides block diagrams of the early and modern processes for the manufacture of PP homopolymers.

Intensive R&D efforts in ZN catalysts have led to outstanding improvements in the technology of polyolefin manufacture. In this context the following advances can be mentioned:

new generations of catalysts,
polymerization processes which maximize catalyst performance,
broader range of products.

A polymerization process must be able to synthesize many polyethylene grades for various applications like injection moulding, blow moulding and extrusion. Figure 4 gives the methodology of evolution of polyolefins.

A. High Activity Catalysts

Due to discovery of these, the onerous step of polymer deashing is no longer necessary. These compounds have to be used in combination with a triether aluminum. They have large surface area, small crystallite size, and disordered crystal structure. Due to this many Mg^{2+} ions, coordinatively unsaturated, become available. On these sites $TiCl_4$ can be easily adsorbed and interact. $MgCl_2$ and $TiCl_4$ undergo solid solution due to a similarity of their layer lattice structure and ionic radii. This greatly increases catalytic activity: electron donation by Mg to active Ti species promotes the polymerization propagation rate constant.

On the left of Fig. 5 are shown typical $MgCl_2$ intermediates which, after contacting $TiCl_4$, give rise to high yield catalysts [4]. On the right of Fig. 5 the interaction between $MgCl_2$, $TiCl_4$, and an electron donor allowed the synthesis of bimetallic halide complexes. These model complexes are only "precursors" of the real catalysts being very active. These complexes could be considered as models of the interaction between Mg and Ti atoms on the surface of $MgCl_2$.

B. Morphology of $MgCl_2$

$MgCl_2$ has been found to be the ideal support; its host lattice can support chiral titanium-active sites. This can be attributed to the near isomorphism of $MgCl_2$ and $TiCl_3$ crystals. Crystals of both $TiCl_3$ and $MgCl_2$ consist of two planes of chlorine atoms sandwiching a plane of metal atoms; titanium atoms occupy two thirds of the octahedral sites in $TiCl_3$ while magnesium atoms occupy all the sites in $MgCl_2$. $MgCl_2$ possesses desirable morphology as a support: it is a highly porous polycrystalline material, resistant to fracture. It is also chemically inert. The porous morphology permits the diffusion of the monomer into the interior of the catalyst particles and serves to determine the morphology of the polymer as well. $MgCl_2$, which has a lower electronegativity than $TiCl_3$, increases the productivity of olefin polymerization. It is not known whether $TiCl_3$ is merely epitactically placed on the $MgCl_2$ lateral surface or whether it enters the $MgCl_2$ crystallite to form a solid solution in the catalyst. The smaller crystallites of $MgCl_2$ are held together in the form of layer aggregates by H-bonding with alcohols. Treatment of the $MgCl_2 \cdot$ alcohol complex with ethyl benzoate causes further disruption of the $MgCl_2$ structure.

C. Catalyst Morphology

ZN catalysts consist of primary crystallites (5–20 nm) which form agglomerates having 1–50 μm size. The internal structure of a catalyst particle is porous — it is through these pores that monomer and cocatalyst can reach the innermost crystallites (Fig. 6). It is on the surface of these crystallites that polymerization occurs. Heterogeneous ZN catalysts are capable of replicating their morphology in the morphology of the resulting polymer particles. The morphology of catalyst particles is a critical factor in determining process economics. The spheripol process of Himont produces polymer particles of 0–1 mm which does away with energy-expensive extrusion. Those polymer particles formed are enlarged replicas of the catalyst particles used.

D. Supported Catalysts

These catalysts comprise a $TiCl_4$ catalyst on a specially prepared $MgCl_2$ support. Electron donors of various type are incorporated into the $MgCl_2$ support and are also used as a "selectivity control agent". Electron donors are most frequently aromatic esters. The first supported $TiCl_4$ catalyst was made as early as 1955.

E. Activation of $MgCl_2$

Montedison issued patents in which $MgCl_2$ in an "active form" was used as catalyst support. Kaminsky [10] describes various routes to "activate $MgCl_2$":

TYPICAL 1960S PLANT

SOLVENT
C_2
ALCOHOL
H_2O
POLYMER-ZATION
C_3 REMOVAL
CATALYST DECOMPO-SITION
SLURRY WASHING
SOLVENT REMOVAL
DRYING
EXTRACTION
RECYCLE ALCOHOL
ALCOHOL RECOVERY
ATACTIC REMOVAL
SOLVENT PURIFICATION
RECYCLE ALCOHOL
RECYCLE SOLVENT
CATALYST RESIDUES
ATACTIC POLYMER
HEAVY ENDS

TYPICAL LATE 1980S PLANT

C_3
C_2 CATALYST
POLYMERIZATION
C_3 REMOVAL
POLYMER WITH SPHERICAL PARTICLES

a

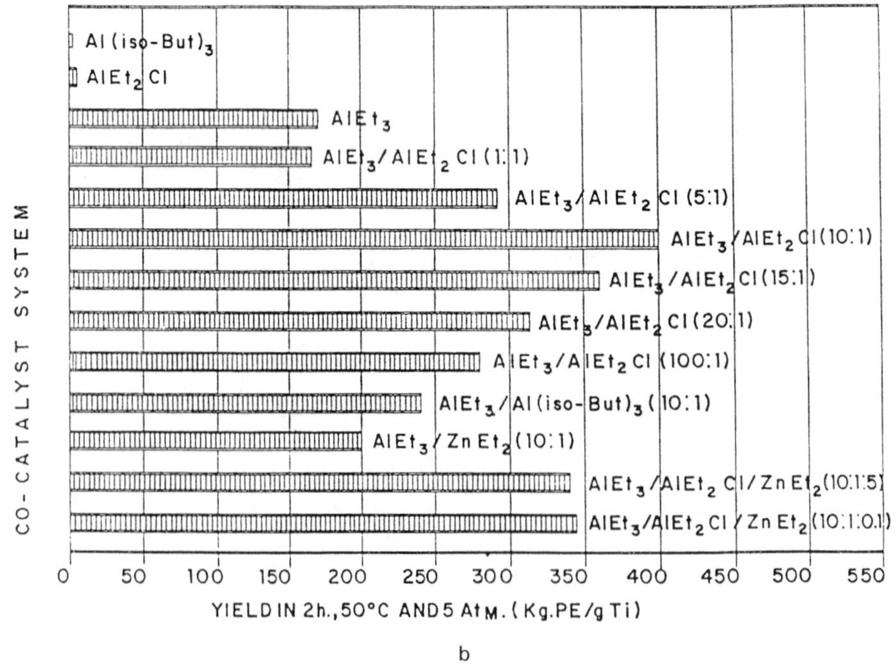

b

FIG. 3 (a) How the fourth generation catalyst changes a polypropylene plant. (b) Yields obtained after 2 hr. polymerization time, 50°C and 5 atm. with different cocatalytic systems.

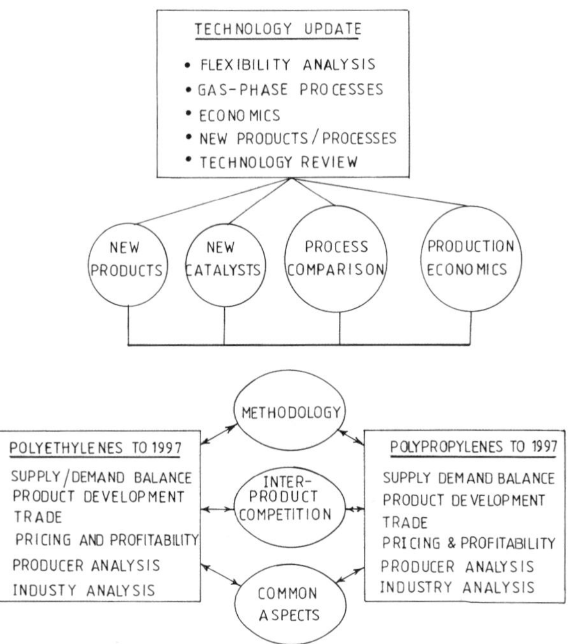

FIG. 4 Structure of the continuing evolution series. (From Ref. 3.)

treating $MgCl_2$ with electron donors (ethers, alcohols),
ball-milling commercial $MgCl_2$,
$MgCl_2$ derived by chlorination of Grignard reagents,
milling $MgCl_2$ with $TiCl_4$–electron donor complexes such as $TiCl_4$ ethylbenzoate.

$MgCl_2$ crystallites are broken down during milling and freshly cleaved surfaces are complexed by ethyl benzoate; thus inhibiting reaggregation shear forces during milling cause the Cl–Mg–Cl double layer to slide over each other, producing hexagonal $MgCl_2$ primary crystallites of only a few layers in thickness.

Commercially available anhydrous $MgCl_2$ and ball-milled (with ethyl benzoate) $MgCl_2$ crystallites are shown in Fig. 7.

The next step of catalyst preparation involves adding $TiCl_4$ to the ball-milled support. This is achieved by:

further ball-milling the catalyst support in the presence of $TiCl_4$, or
suspending the catalyst support in hot, undiluted $TiCl_4$. $TiCl_4$ becomes adsorbed onto free $MgCl_2$ surface vacancies.

This is structurally shown in Fig. 8a and 8b. The lattice structure and layers of $MgCl_2$ and $TiCl_3$ are shown in the (110) direction.

The creation of surface Mg ions which form the adsorption sites for ethyl benzoate and $TiCl_4$ can be envisaged. Three different types of surface Mg ions are present in the $MgCl_2$ crystal:

Type 1: On the lateral faces, single vacancy Mg ions.
Type 2: On the corners, single vacancy Mg ions.
Type 3: On the corners, double vacancy Mg ions.

Infrared spectroscopy (IRS) of $MgCl_2$/ethyl benzoate ball-milled samples indicates that ethyl benzoate is coordinated to the $MgCl_2$ surface via the carbonyl

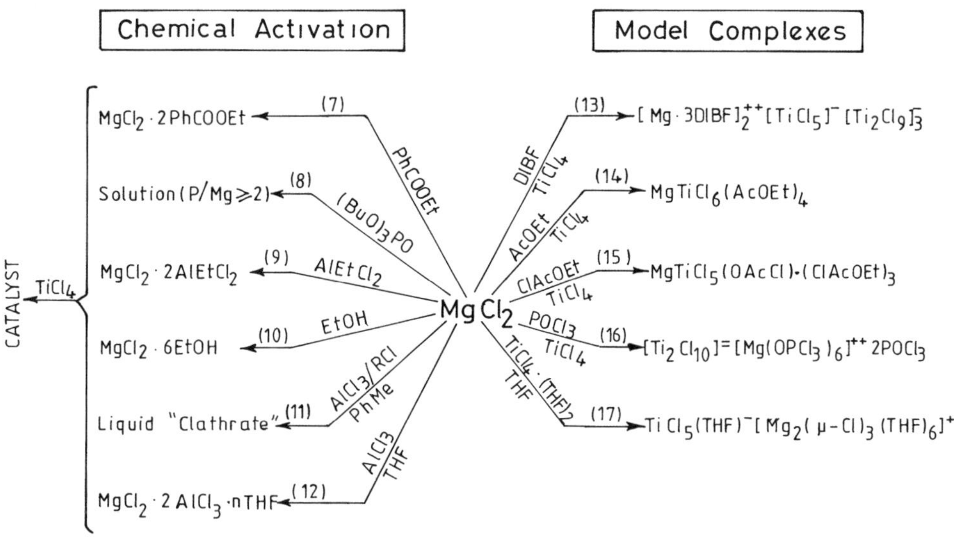

FIG. 5 Some $MgCl_2$ activation reactions and syntheses of model complexes. (From Ref. 4.)

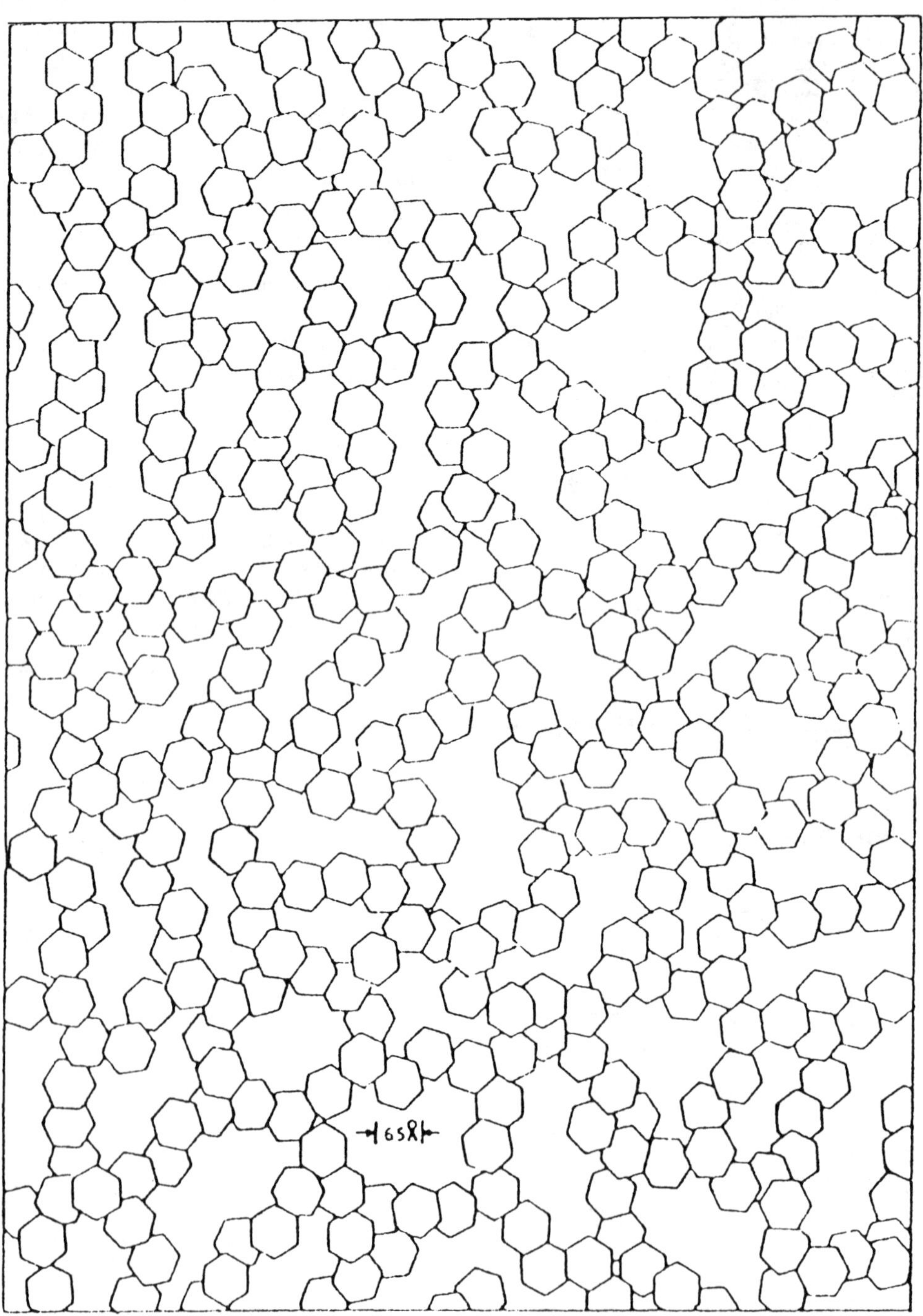

FIG. 6 Internal structure of a catalyst (Permission of Harwood Academic Publishers). (From Ref. 2.)

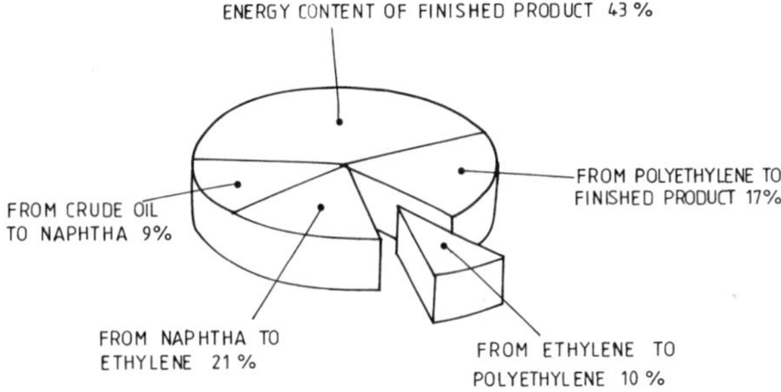

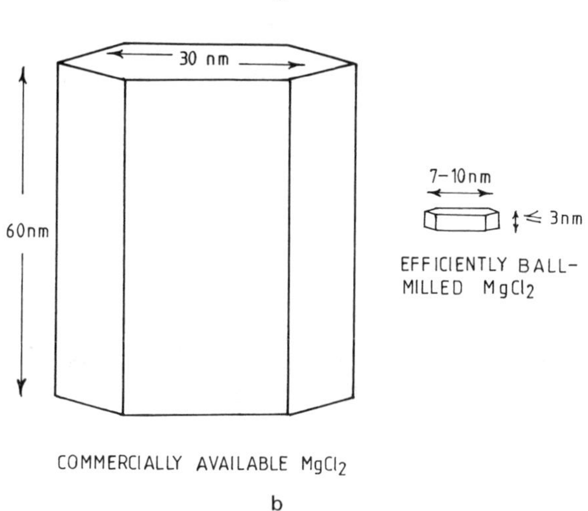

FIG. 7 (a) From crude oil to finished product energy requirement. (From Ref. 5a.) (b) $MgCl_2$ primary crystallite dimensions. (Permission from Harwood Academic Publishers.) (From Ref. 2.)

function. Ester molecules bound to the basal edges and corners of crystallite absorb near 1680 cm^{-1}. The broadness of the absorption demonstrates the inhomogeneity of the surface sites (Fig. 9).

In a ball-milled $MgCl_2$/ethyl benzoate sample, ethyl benzoates remains bound to almost all the surface sites. $TiCl_4$ coordinates to these sites by replacement of the ethyl benzoate molecules. Zirconium catalysts are three orders of magnitude more active than their titanium counterparts. These catalysts are contenders for future processes.

Doi [5c] has proposed an active center in which titanium is complexed by an aluminum species. This bimetalic active center is shown in Fig. 10. This model can explain many of the phenomena observed during polymerization reactions (without taking electron donors into account).

II. ROLE OF ESTER

The high level of stereoselectivity associated with $MgCl_2$-based catalysts is realized in the presence of an aromatic ester. The effect of an ester is to reduce activity while enhancing stereoselectivity. Ester is present both in the preparation of the procatalyst ($MgCl_2$–$TiCl_4$–ester) as well as in the cocatalyst. During the preparation of the procatalyst, ester plays a critical role in dispersing the $TiCl_4$ on small hexagonal crystallites of $MgCl_2$. Analysis shows that the ester is bonded through the carbonyl oxygen to magnesium rather than to titanium. In order to achieve high isotacticity, the ester has to be included with the cocatalyst. Triethylaluminum (TEA) forms a complex with ethyl benzoate ester to act as the selectivity control agent. Bridged Ti–Al species are generated on the surface.

The first two carbon atoms of the polymeric chain lie in the same plane with Ti and three of its attendant ligends [5]. The propylene bonds to Ti lie parallel to this plane with the double bond parallel to the titanium–chain bond. The most interesting feature of this depiction is that the propylene uses the same enantiometric face independent of the site location of the chain.

Only polymerization needs to be titanium while the second member of the dimer could be aluminum. The Ti(III) dimer has a vacant coordination site [5] which must be blocked to achieve high stereoselectivity. It is possible that this site is blocked by the aromatic portion of ethyl benzoate which is bonded through its carbonyl oxygen to a neighboring magnesium atom.

A good catalyst for olefin polymerization should exhibit good morphological control of the resulting polymers, high activity, and control of MW and MWD.

High activity catalysts for olefin polymerization can be prepared by supporting $TiCl_4$ on $MgCl_2$ in the presence of an electron donor. However, due to the low surface area of $MgCl_2$ and also to the poor interaction with $TiCl_4$, both components have to be intensively ball-milled in order to activate them and introduce sufficient amounts of titanium [5] into the crystal lattice of $MgCl_2$. The result is a highly active catalyst.

Catalysts with better ability to control polymer morphology can be prepared in many different ways. One of these uses the well defined morphology of silica, silica–alumina, or alumina which is loaded with $TiCl_4$ or other active transition metal compound. The best silicas used as supports have good mechanical strength, high porosity, and surface area. Thus, they do not break during the early steps of polymerization.

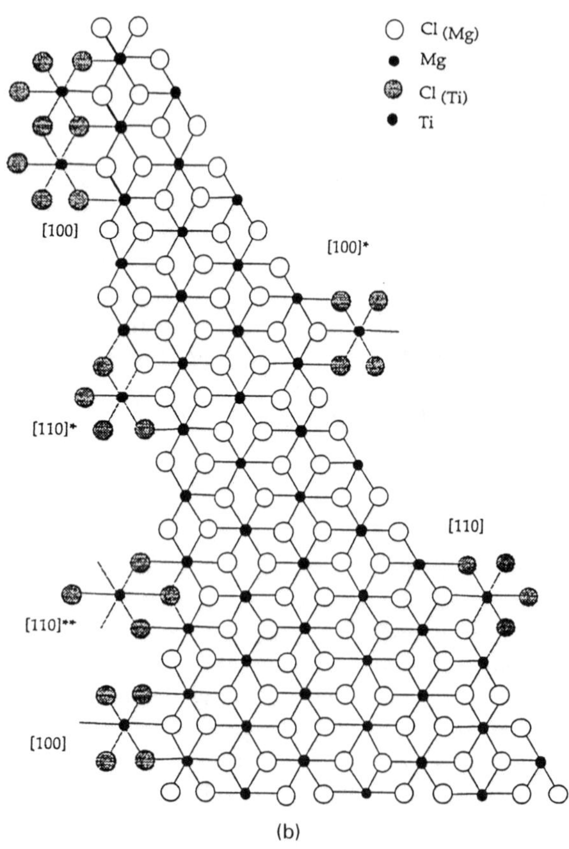

FIG. 8 (a) $TiCl_4$ dimer on $MgCl_2$ monolayer edge. (From Ref. 5.) (b) The mode of interaction of $TiCl_4$ with $MgCl_2$ of different surfaces. (From Ref. 5b.)

A. Mechanism of Catalyst Function (Fig. 11a and 11b)

Titanium species having different oxidation states, single and double-bonded to the surface of silica as well as strongly chemically and physically absorbed, can be formed. The role of the ester is to moderate the alkylating divalent titanium species. The bimetallic complex may be formed in the preparation of SiO_2 supported catalysts by impregnation [5] with $TiCl_4$–$AlEt_3$.

In the case of catalysts prepared by coimpregnation of $TiCl_4$–C_4H_9 MgCl mixtures on silica 951 the following intermediate complex may be formed [5] after heating at 450°C under vacuum.

Supported ZN catalysts, using SiO_2 as a carrier, control the polymer morphology more easily than $TiCl_3$–aluminum alkyl.

The yield using different catalyst systems is summarized [5]. The yield varies with the molar ratio of cocatalyst employed and compares favorably with commercial catalysts based on $MgCl_2$ as carrier.

B. Preparation of Supported ZN Catalysts for Propylene Polymerization

The solid precatalyst can be described by a common formula:

$$MgCl_2\text{–ED–}TiCl_4$$

in which the electron donor (ED) is always an aromatic ester (AE), generally ethyl benzoate (EB). The three components are reacted together by milling or impregnation, temperature, duration, and the ratios used. The high performance precatalyst is obtained by comilling $MgCl_2$ and EB followed by excess $TiCl_4$ impregnation and drying in an improved way. Precatalyst used shows a fast decrease in polymerization rate when used with cocatalysts ($AlEt_3$ or $AlBu_2$ mixed with dialkyl-aluminum chloride). The active center is a bimetallic titanium–aluminum complex in reversible equilibrium with a free aluminum complex in reversible equilibrium with a free aluminum alkyl (AA) in the solution. With $AlEt_3$, the exchange between free and complexed aluminum is very fast. The complex is itself in equilibrium [5] with AA–AE complex formation.

The protocol for the preparation of a catalyst from the patent literature is $MgCl_2$/EB/PC/$AlEt_3$/$TiCl_4$–$AlEt_3$/MPT where EB, PC, and MPT are abbreviations for ethyl benzoate, *p*-cresol and methyl-*p*-toluate, respectively.

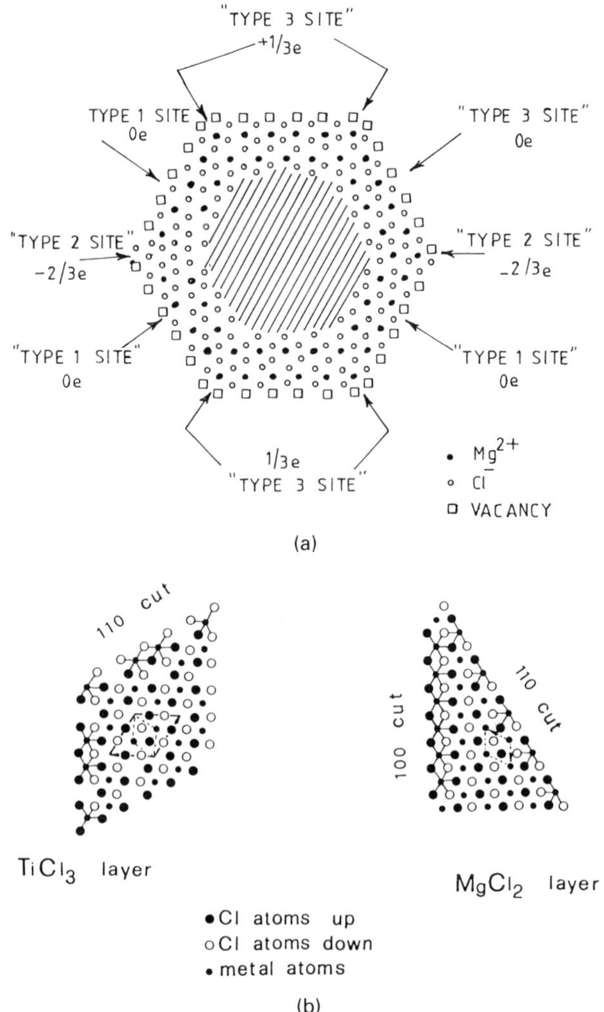

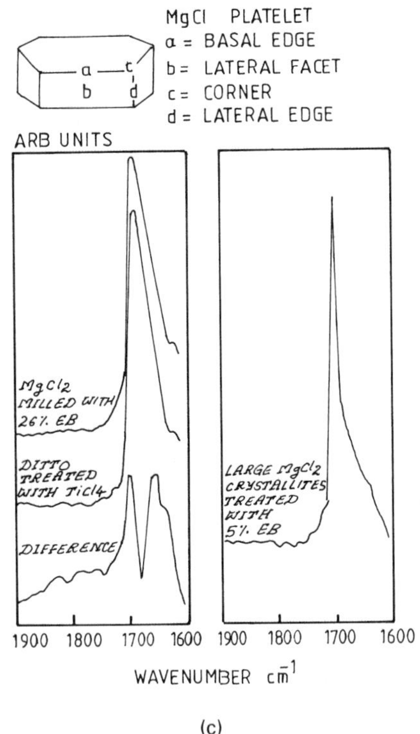

FIG. 9 (a) Surface structure and sites of hexagonal $MgCl_2$ crystallites (two-dimensional). (Permission from Harwood Academic Publishers.) (From Ref. 2.). (b) Schematic drawings of the unitary subcells and structural layers of violet $TiCl_3$ and $MgCl_2$. The bonds of the metal atoms belonging to the lateral cuts are also drawn to point out the different coordination numbers. (c) Carbonyl regions in the infrared absorption spectra of catalyst treated with ethyl benzoate. (From Harwood Academic Publishers.) (From Ref. 2.)

C. Catalyst Design

Reaction kinetics and polymer particle growth are determined by cocatalyst, comonomer, chain transfer agent, temperature, pressure, process, etc. (Fig. 12). Complete success in the catalyst design is reached only when a broad set of requirements is met (Table 2).

D. Single-Site Catalysts (SSCs)

The availability of SSCs for solution processes has broadened the product range considerably. It is possible to produce virtually any ethylene copolymer compostion ranging from completely amorphous copolymers to highly crystalline homopolymers. Operating conditions and production costs are similar across the entire product range. Thus, one of the leading low cost polyolefin processes can now be applied to

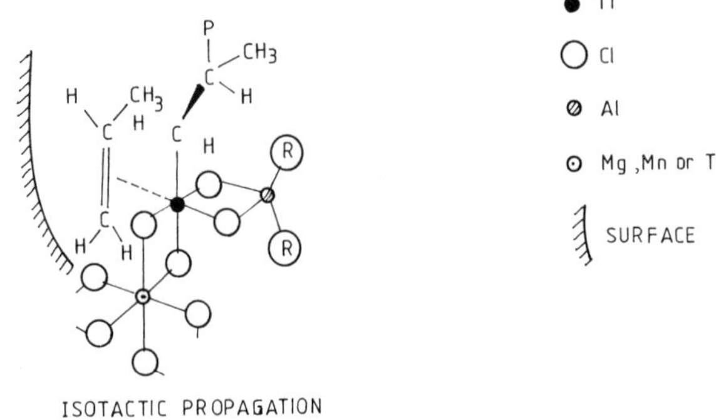

FIG. 10 Isotactic site model of Doi. (From Ref. 5c, with permission from Harwood Academic Publishers.) (From Ref. 2.)

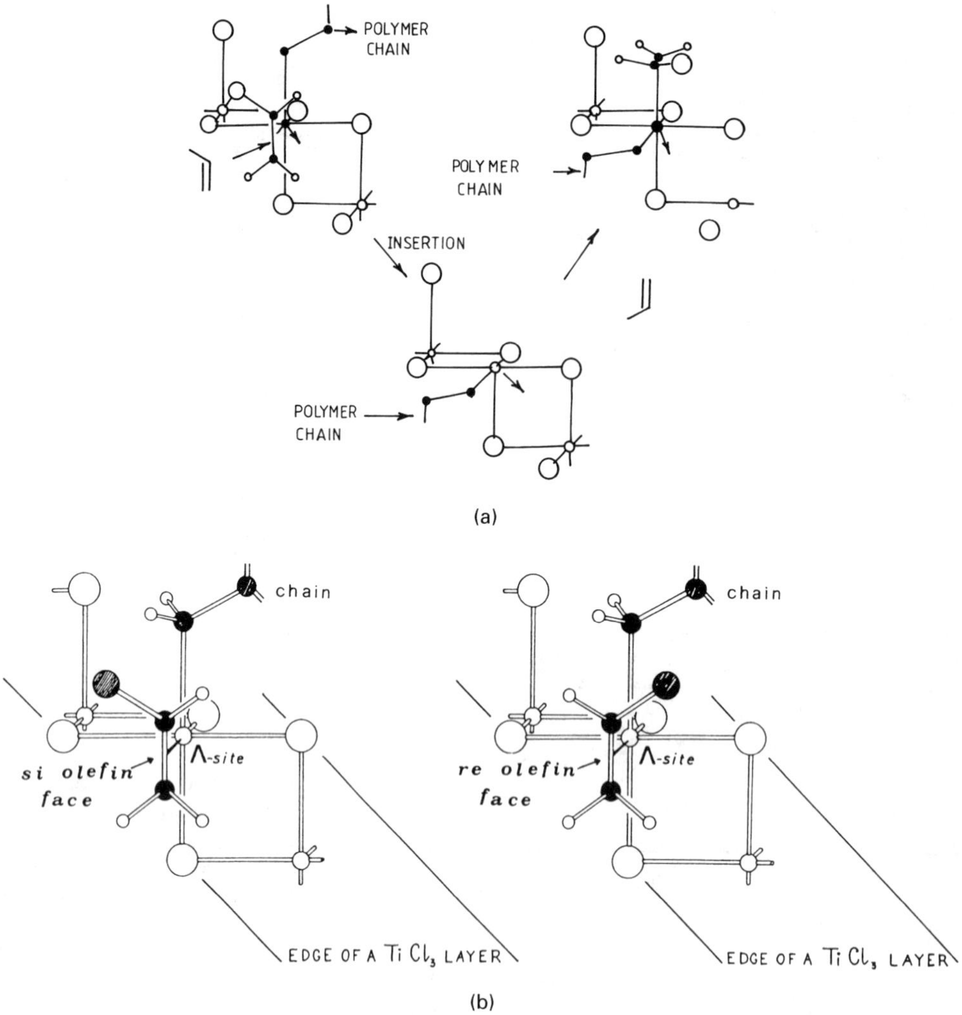

FIG. 11 (a) Isotactic specific site. (From Ref. 5.) (b) Schematic drawing of the local environment of a possible catalytic site located at a relief of the lateral surface of $TiCl_3$.

the production of a very broad range of elastomeric and soft specialities. The process limitation on product domain is the lower melt index limit, around 0.2 dg/min, set by the viscosity of the polymer solution as it exits the reactor. This limits the usefulness of the process in serving markets such as blow moulding and thin film which typically employ higher molecular weight polymers with melt indices as low as 0.01 dg/min.

E. Controlled Polymerization of Polyolefins [6]

In contrast to early generations of multi-site Ziegler–Natta catalysts, developed on the basis of "trial-and-error" research, modern metallocene catalysts contain essentially one type of catalytically active center, which can readily be fine-tuned to produce extraordinarily uniform homo- and copolymers. At present, metallocene structures, especially ligand substitution patterns, are tailored to control polymer microstructures, molecular weights, end groups, and morphology development with unprecedented precision [7–12].

Basic correlations between metallocene structures and polymer properties have been identified [13], eliminating the drawbacks of early metallocene catalyst generations. For instance, substantially increased molecular weights, that are of interest for commercial applications of polypropylenes, are obtained when a methyl substituent is introduced in the 2-position of

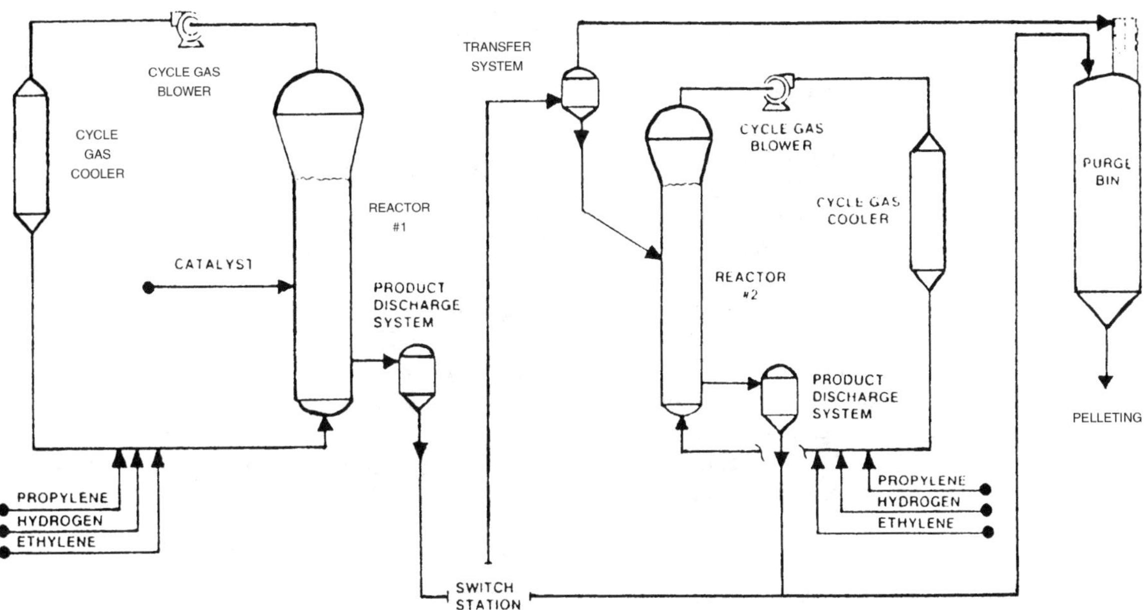

FIG. 12 Fourth generation PP plant.

metallocenes containing silylene-bridged bisindenyl ligands, while substituents in the 4-position as well as benzannelation promote higher catalyst activities. It is possible to improve polymer properties, e.g., stiffness and optical transparency, as a function of metallocene structures and to reduce the contents of low molecular weight fractions [14, 15]. Novel heterogenized metallocene-based catalysts were introduced successfully in gas phase polymerization [14]. In 1995 the focus in metallocene catalyst development was on new catalysts and processes for controlled olefin copolymerization such as ethylene/cycloolefin [16, 17], ethylene/styrene [18, 19], and ethene/l-olefin including long-chain l-olefins, which are known to give poor incorporation into polyethylene with conventional Ziegler–Natta catalysts [20–22]. Moreover, metallocene catalysts in situ copolymerize vinyl-terminated polyethylene — resulting from chain termination via hydride elimination — with ethene to produce long-chain branched polyethylenes containing 1–3 CH branching groups per 10,000 CH_2 groups. Such long-chain-branched high and linear low density polyethylenes exhibit much improved melt processability, reflected by lower melt viscosities at higher shear rates. With this technology novel poly (ethene-co-1-octene) products with high 1-octene content are available commercially. High-octene-LLDPE is useful as an elastomer, flexibilizer and blend component, e.g., for producing toughened polyolefins [23]. Progress has also been made in molecular modeling that supports the design of novel metallocene catalysts,

TABLE 2 Main Requirements of a Commercial Catalyst

Preparation	Polymerization reaction	Polymer properties
Easy feasibility	Easy feeding	MMD tailoring
No handling safety storage & aging problems	Hydrogen response	Chain branching and unsaturation control
Low cost	High productivity	Comonomer distribution control
	Proper kinetic	Influence on melt flow properties
	Control of polymer particle morphology	
	Efficient comonomer incorporation	

elucidation of structure/property correlations, and process control [24–27].

The most recent expansion of the metallocene catalyst families is represented by "oscillating" metallocenes such as bis(2-phenyl-indenyl)$_2$$ZrCl_2$, developed by Coates and Waymouth [28]. Rotation of the phenyl-substituted indenyl ligand accounts for formation of isoselective racemic and non-stereoselective meso-stereoisomers of the metallocene. Provided the isomerization rate is rapid in comparison to the chain propagation rate, segmented polypropylene is obtained, where crystalline isotactic and amorphous atactic segments alternate. The formation of such stereoblock polypropylenes is displayed in Fig. 13a and 13b. The ratio of amorphous and crystalline segments is controlled by polymerization temperature and monomer concentration. This allows polymer properties to be tailored and leads to the synthesis of novel thermoplastic elastomers based solely on propene.

F. Olefin Copolymerization and Branching

Recent progress has also been made in the area of homo- and copolymerization of polar monomers. For instance, metallocene catalysts copolymerize ethene with 11-undecene-1-ol to yield hydroxy-functional LLDPE exhibiting improved adhesion [29]. Metallocenes based upon lanthanoids, as developed by Yasuda *et al*, copolymerize non-polar and polar monomers to produce new families of block copolymers, e.g., polypropylene–block–polymethylmethacrylate [30].

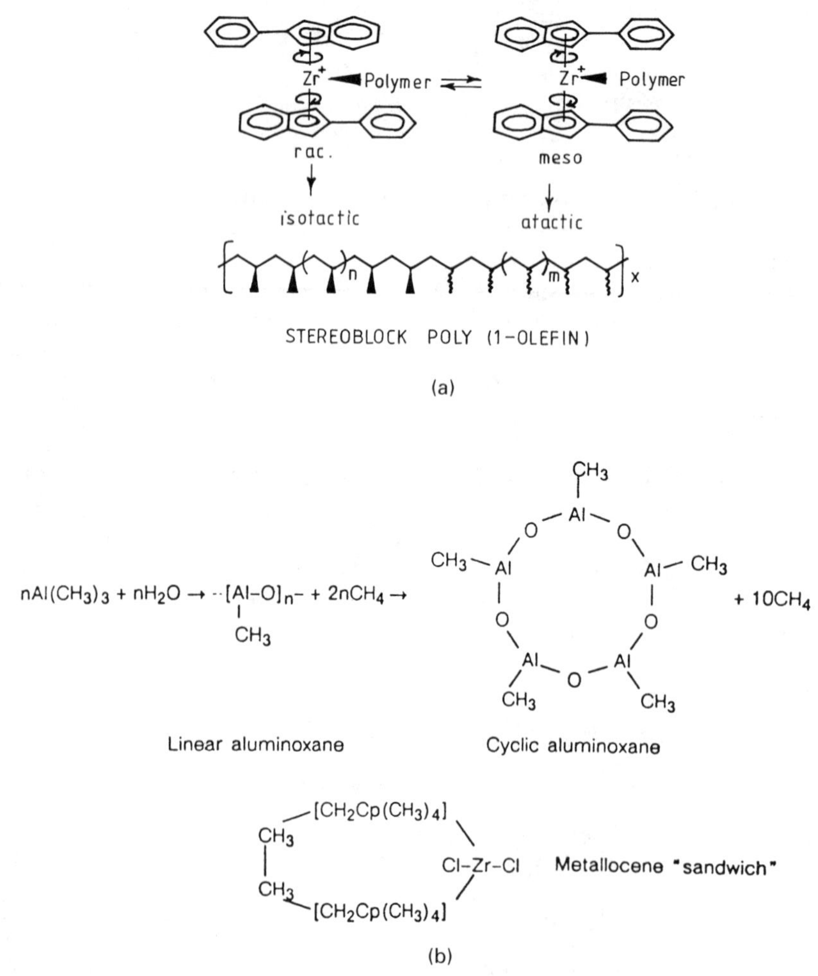

FIG. 13 (a) Oscillating metallocene catalysts. (From Ref. 6.) (b) Ethylene bis(pentamethylcyclopentadienyl) zirconium dichloride.

As a result of extensive development efforts [31, 32], alternating ethene/carbon monoxide copolymers, containing a small fraction of 1-olefin comonomer to enhance processability by lowering the melting temperature, are now becoming available commercially [22]. This attractive class of engineering polymers based upon cheap feedstocks exhibit high heat distortion temperatures and is starting to compete with polyamides, e.g., in automotive applications. A key development of this technology are special Pd-based catalysts which were developed by Drent and coworkers [33–35].

Takaya *et al* [36] reported the enantioselective alternatig copolymerization of propene and carbon monoxide by means of chiral phosphine–phosphite complexes of Pd II [36, 37]. According to Sen *et al* an equimolar mixture of the enantiomers of optically active poly(propene-co-carbon monoxide), which melts at 171°C, forms stereocomplexes melting at a considerably higher temperature (239°C) [38]. Brookhart [37] used Pd III and Ni II complexes with sterically hindered diimine ligands, e.g., $(Arn{=}C(R)C(R){=}N(Ar)M(CH_3(CF_3)(OE_2))^+BAr_4^-$ with M = Pd or Ni, Ar = 3.5-$C_6H_3(CF_3)_2$, to polymerize ethene [39] to highly branched polyethylene. Also propene is polymerized with such non-chiral Pd complexes to produce atactic polypropylene. Using Co III complexes, Brookhart obtained monofunctional polyethylenes with a narrow molecular weight distribution [40]. Brookhart's catalysts offer an intriguing potential for olefin copolymerization because Group VIII complexes tolerate Lewis bases, which are strong catalyst poisons for Group IV Ziegler–Natta catalysts. With chiral ligands it should be possible to achieve stereoselective homo- and copolymerization of 1-olefins, including 1-olefins with polar functional groups. This will lead to a large variety of novel polymer families.

Low-density polyethylene (LDPE) is manufactured by polymerization in reactors that operate at pressures of 1000–3000 atm and at temperatures of 200–275°C. Under these extreme conditions, ethylene is a supercritical fluid having a density of 0.4–0.5 g/mL. In the higher ranges of temperature and pressure the polymer remains dissolved in the ethylene phase, but in the lower ranges the polymer separates as a liquid. The polymerization is a free radical reaction. The reaction mechanism is summarized in Fig. 14. Initiation can either be by oxygen or peroxides. The termination and branching reactions shown in Fig. 14 are not the only ones taking place, but they are among the most important ones. The particular branching reaction shown is known as the backbiting reaction, and it controls the density of high-pressure polyethylene. The greater the amount of branching, the lower the density. As the pressure increases, the rates of the propagation steps increase faster than the rates for the termination and branching steps. High pressure thus favors higher densities and less branching (and higher molecular weights).

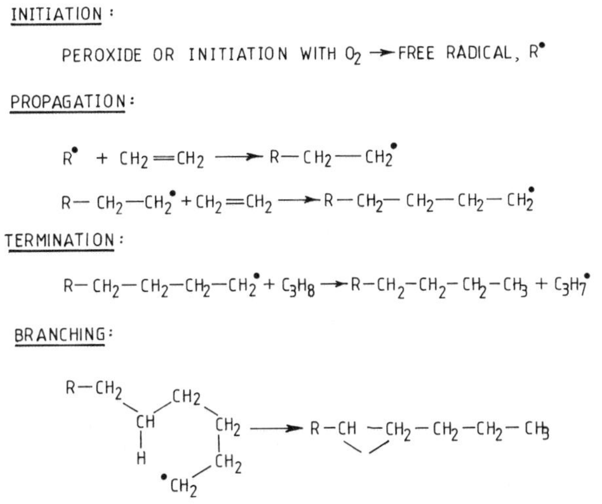

INITIATION:

PEROXIDE OR INITIATION WITH O_2 → FREE RADICAL, R•

PROPAGATION:

$R^{\bullet} + CH_2{=}CH_2 \rightarrow R{-}CH_2{-}CH_2^{\bullet}$

$R{-}CH_2{-}CH_2^{\bullet} + CH_2{=}CH_2 \rightarrow R{-}CH_2{-}CH_2{-}CH_2{-}CH_2^{\bullet}$

TERMINATION:

$R{-}CH_2{-}CH_2{-}CH_2{-}CH_2^{\bullet} + C_3H_8 \rightarrow R{-}CH_2{-}CH_2{-}CH_2{-}CH_3 + C_3H_7^{\bullet}$

BRANCHING:

R—CH2, CH2, CH, H, CH2, CH2, •CH2 → R—CH —CH2—CH2—CH2—CH3

FIG. 14 Polymerization to LDPE.

The polymerization of ethylene is a highly exothermic reaction. The heat of polymerization of ethylene was calculated to be 25.88 kcal/mol; heat liberated under reaction conditions is somewhat less, i.e. about 24 kcal/mol or 1540 Btu/lb.

Figure 15 shows some commercial PE production technologies.

The chemistry of polymerization of ethylene to high-density polyethylene is indicated in Fig. 16, showing a Ziegler-type catalyst. The first step is the reaction between titanium tetrachloride and the aluminum alkyl to form an alkylated complex. Propagation then proceeds as shown in step 2. The metal–alkyl bond MR is a weak bond, but it can be maintained in the absence of unsaturated hydrocarbons. As the olefin becomes π-bonded, the alkyl group is expelled as a radical which attaches itself to the nearest C-atom of the olefin. At the same time, the olefin connects itself to the metal. The configuration at the end of this propagation step is essentially the same as the beginning, except that the alkyl group and the chlorine vacancy change places. The process can now repeat itself by inserting the next monomer molecule into the new chlorine vacancy.

The structure of isotactic polypropylene is given in Fig. 2. The difference between polyethylene and polypropylene is that every other carbon atom along the

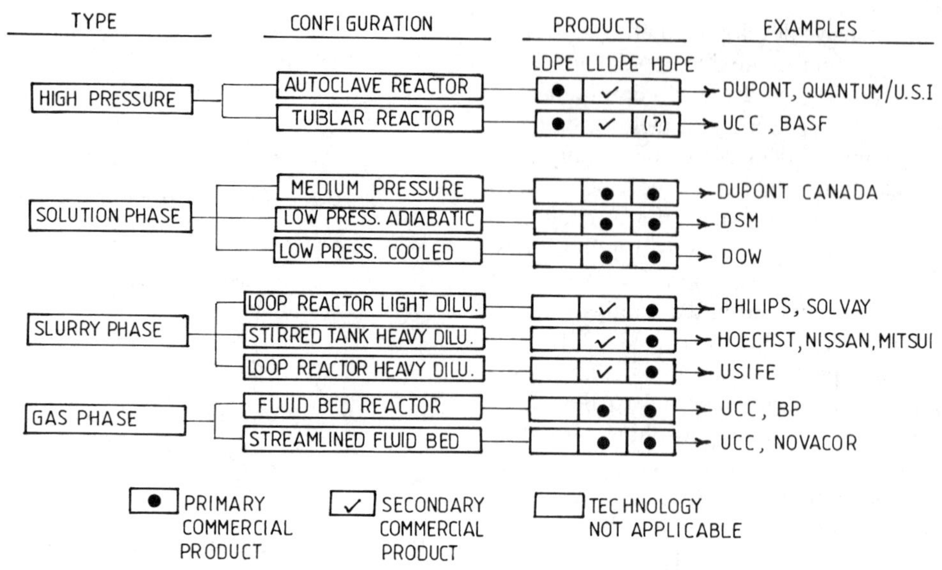

FIG. 15 Commercial PE production technologies.

chain in the polypropylene molecule has a pendant methyl group. The carbon atom to which the pendant methyl group is attached can be considered as the center of a tetrahedron. As one moves along the chain and approaches the apex of this carbon atom, the other three apexes are occupied by a hydrogen group, a methyl group, and a continuing chain. One can then continue down the chain, and as one approaches the next carbon atom to which a methyl group is attached, again one will see that the three apexes beyond the carbon atom have a hydrogen group, a methyl group, and the continuing chain. If these three groups are in the same clockwise configuration as the previous groups, and if this pattern is repeated all the way down the chain, then the structure is considered to be isotactic. On the other hand, if as one proceeds down the chain one finds that the positions of the methyl and hydrogen groups are alternately reversed, then one has syndiotactic polypropylene (Fig. 2). Both isotactic and syndiotactic polypropylene, because of their very regular structure, are highly crystalline in the solid state. On the other hand, if the configurations of the methyl group in the continuing polymer chain are rather random as one moves along the chain, then the material has little crystallinity and is considered to be atactic. It is the isotactic stereoregular polymer of polypropylene that is commercially important (Table 3 and Fig. 17). The syndiotactic polypropylene has been prepared but so far, at least, is of only academic interest.

I $TiCl_4 + Al(C_2H_5)_3 \longrightarrow$ complex

II Propagation

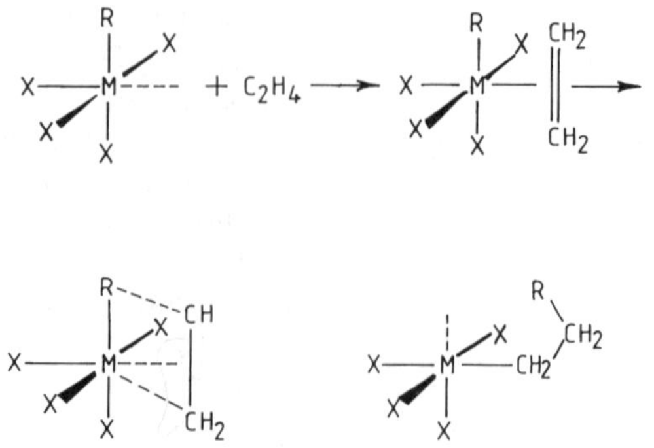

$$X_3M{-}R + C_2H_4 \longrightarrow X_3M(R)(CH_2{=}CH_2) \longrightarrow X_3M{-}CH_2{-}CH_2{-}R$$

FIG. 16 Polymerization to HDPE.

The basic reason for this is the considerable reduction in fixed capital requirements for the new high-density polyethylene processes, since all of the HDPE processes shown in Table 4 use highly active catalysts and do not require removal of catalyst residues or non-crystalline polymer. Polypropylene, however, is still manufactured primarily by processes which require the removal of catalyst residues and atactic material, and this fact is reflected both in higher capital costs and higher product values, even though

TABLE 3 Polypropylene Processes (Fig. 17), % of World Capacity

Liquid phase, slurry				% Total
Montecatini licensees	(%)	Self-developed processes	(%)	
Montedison	20	Amoco	12	
Hercules	15	Dart (Rexall)	9	
Shell	8	Phillips	5	
Hoechst	7	Exxon	5	
Mitsubishi Petrochemical	4	Tokuyama Soda	2	
Mitsui Toatsu	3	Idemitsu Kosan	1	
Mitsui Petrochemical	4	Solvay	1	
	59		35	94
Liquid phase, solution	Eastman			5
Gas phase	BASF			1
				100

propylene raw material is taken at 10 ¢/lb. whereas ethylene raw material is taken at 13 ¢/lb. A progressive evolution of PP technology from 1955 onward is shown in Fig. 18.

G. Spheripol Process

Figure 19 depicts a modern state-of-the-art plant based on Himont technology for making homopolymer and impact copolymers with supported catalysts. The latter are sufficiently active not to require any catalyst extraction, nor removal of atactic polymer because of the high stereospecificity. Homopolymerization takes place at ca. 70°C and 4 MPa in liquid propene circulating round one or more loop reactors. A single axial flow agitator in each loop maintains high flow rates to ensure good heat transfer to the water-cooled jackets, whilst also preventing any polymer particles settling from the slurry. Typically the PP concentration is ca. 40 wt%.

Continuously metered catalyst, triethylaluminum and a Lewis base stereoregulator such as a dialkyldi-

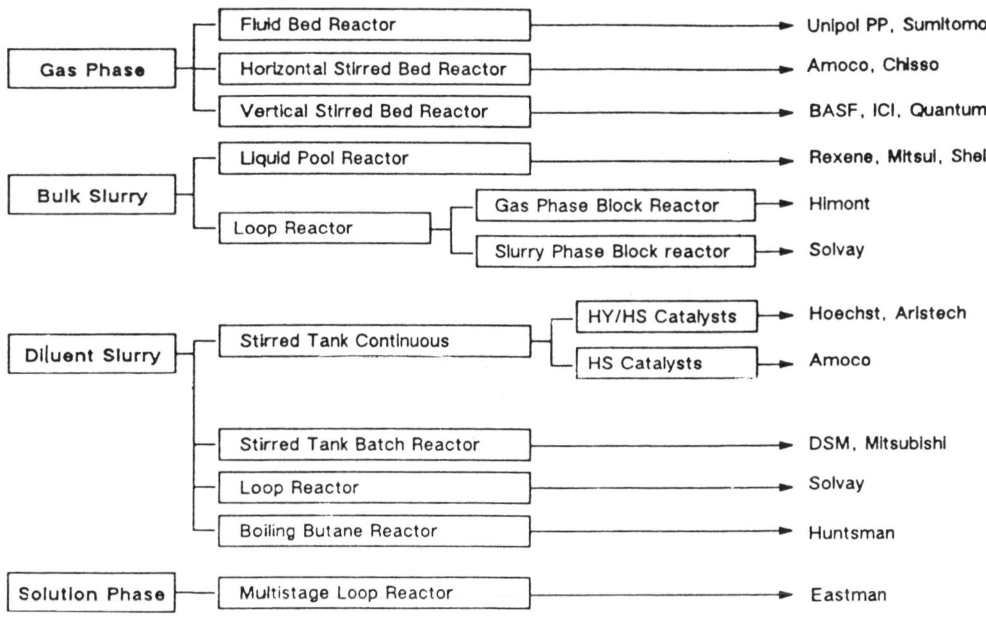

FIG. 17 Generic polypropylene processes. (From Ref. 3.)

TABLE 4 Comparative Economics of Polyolefin Processes, 200–220 million lb/yr

	Fixed capital ($ million)	Product values* (¢/lb)
LDPE		
Tubular (Exxon patents)	65.0	31.9
Autoclave (National Distillers patents)	72.3	33.4
HDPE		
Slurry (Phillips patents)	29.9	25.4
Slurry (Hoechst patents)	32.7	26.5
Slurry (Solvay patents)	37.2	28.1
Slurry (Montedison patents)	38.6	28.4
Solution (Stamicarbon patents)	31.3	27.3
Gas Phase (Union Carbide patents)	29.1	24.9
PP		
Slurry (Dart patents)	58.1	30.4
Gas phase (BASF patents)	48.7	26.0

*Product values assume ethylene at 13.0 ¢/lb, propylene at 10.0 ¢/lb, and include 25% return on fixed capital and 10% depreciation.

1955	Initial work on stereospecific polymerization
1960	Commercial solution and slurry processes
1967	Commercial gas phase process
1970–75	Second generation $TiCl_3 \cdot AlEt_2Cl$ catalysts
1975–80	Third generation supported catalysts
1980–85	Super-active third generation catalysts
1990–2000	"Fourth generation" catalysts?

FIG. 18 Milestones in PP technology development.

methoxysilane are fed into the reactor to maintain polymerization and stereocontrol. The initial few seconds of polymerization with a new high-activity catalyst particle are quite critical to secure good performance. For this reason, some processes have a prepolymerization stage in which the catalyst components react at lower temperature and monomer concentration. This can be either a batch or continuous pretreatment which produces only small amounts of polymer (~100 g/g) in the catalyst. This prepolymerized catalyst is then fed into the loop reactor as usual. Mean residence time in a single polymerizer is 1–2 h. Two loop reactors can be operated in series to narrow residence time distributions, modify the polymer, and increase output.

A continuous stream of polymer slurry discharges through a heated zone for the first stage of pressure letdown in cyclone (b). For homopolymers, this connects directly to the secondary cyclone (d), bypassing the

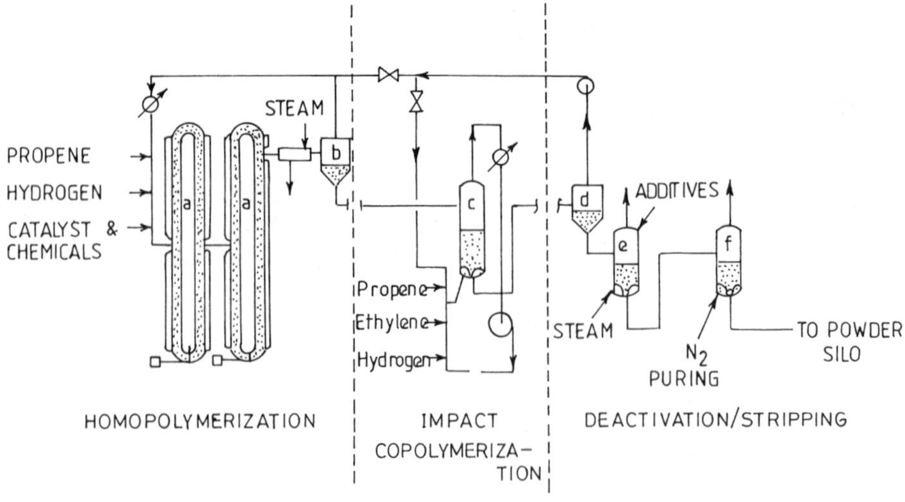

FIG. 19 Spherical process. (a) Loop reactors; (b) primary cyclone; (c) copolymer fluidized bed; (d) secondary and copolymer cyclone; (e) deactivactivation; (f) puring.

copolymerization unit. Unreacted propene flashes off from the first cyclone and is condensed with cooling water and recycled into the reactor. A compressor is required for gas from cyclone (d). Polymer powder from the cyclone is fed into vessel (e) for deactivation with small amounts of steam and undisclosed additives. Residual moisture and volatiles are removed by a hot nitrogen purge in vessel (f) before conveying polymer to storage silos for conventional finishing as stabilized powder or extruded pellets.

H. Loop Reactor Process

A simplified flowsheet for the Phillips particle form process is shown in Fig. 20. The novel double-loop reactor constructed from wide-bore jacketed pipe was developed by Phillips engineers to avoid deposits, which had been troublesome in a stirred autoclave. It also has a high surface-to-volume ratio, facilitating heat removal and allowing short residence times. The impeller forces the reaction mixture through the pipework in a turbulent flow with a velocity of 5–10 m/s. The reaction conditions of 100°C and 3–4 MPa correspond to the needs of the chromium-based Phillips catalyst and the required productivity. The diluent used is isobutane which facilitates the subsequent flash separation and, being a poor solvent for polyethylene, permits higher operating temperatures than the higher alkanes. The catalyst is flushed into the reactor with diluent from the metering device at the base of the catalyst slurry tank. The polymer is taken off from a sedimentation leg which enables the slurry to be passed to the flash tank at a concentration of 55–65% instead of the 30–35% circulating in the loop reactor. The isobutane diluent evaporates in the flash

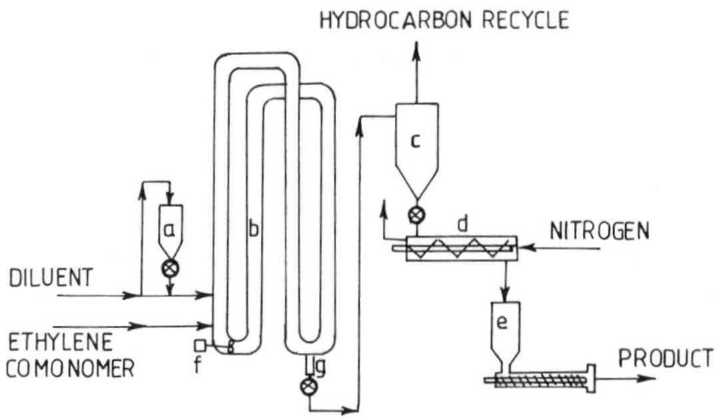

FIG. 20 Flowsheet of the Phillips particle form process. (a) Catalyst hopper and feed valve; (b) double loop reactor; (c) flash tank; (d) purge drier; (e) powder-fed extruder; (f) impeller; (g) sedimentation leg. (From Ref. 58.)

tank and is then condensed and recycled. Residual isobutane is removed in a nitrogen-flushed conveyor. Pelletization is carried out in a powder-fed extruder in a similar way to the Hoechst process.

I. New Polyolefin Alloys

Using $MgCl_2$-based catalysts for PP developed by Montedison it is possible to achieve control of the 3D structure of the catalyst particle and, as a result, of the obtained polymers. Such a polymer architecture control allows synthesis of new materials with improved properties like heteroplastic olefin copolymers, polyolefin alloys (Catalloy), and polyolefin alloys with non-olefinic polymers, which Montell has called Hivalloy.

1. Catalloy Process [40]

The research on reactor granule technology has opened up new possibilities in the field of polyolefin alloys. The Catalloy process allows unique reactor-made polymeric compositions endowed with properties no longer limited by mechanical considerations.

Such a process is a new and highly sophisticated technology, based on three mutually independent gas-phase reactors in series. Its versatility is suitable for designing special polymeric compositions able to meet today's existing polyolefin applications and to face tomorrow's developments. Broader MWD, higher stereoregularity, random copolymers containing up to 15% of comonomers, and heterophasic alloys containing up to 70% of multimonomer copolymers are achievable. No by-products are produced and no operating fluids are needed, with significative economical and ecological advantages.

This process, which is a maximization of the reactor granule technology, produces a broad range of multiphase polymer alloys directly in the reactor. These polyolefins can compete with other engineering resins such as nylon, PET, ABS, PVC, and even with many high-performance materials such as steel in the automotive sector.

2. Hivalloy Technology [40]

The reactor granule technology also makes possible an exciting new technical frontier, allowing the incorporation and polymerization of non-olefinic monomers in a polyolefin matrix. The porous polyolefin granule gives a very high specific surface area and a very high reactivity substrate suitable for easy reaction with non-olefinic monomers at a level greater than 50% wt via free radical graft copolymerization. Himont called this emerging technology Hivalloy and the combining of non-olefinic monomers with an olefinic substrate makes possible a family of materials not commercially achievable previously.

These resins are expected to bridge the performance gap between advanced polyolefin resins and engineering plastics, and are therefore, truly "specialty polyolefins". A first target of the Hivalloy family of products wil be those applications currently served by ABS. Possessing both olefinic and non-olefinic characteristics, Hivalloy products are designed to combine the most desirable properties of PP, such as processability, chemical resistance, and low density, with many of those desirable features of engineering resins which cannot be achieved with currently available polyolefins, such as improvements in the material's stiffness/impact balance, improved mar and scratch resistance, reduced molding cycle time, and improved creep resistance. Because of their olefinic base, Hivalloy polymers readily accept minerals and reinforcing agents, providing added flexibility and control over their properties and further expanding the PP property envelope into the specialty area.

By this technology it is possible to produce engineering polyolefinic based materials modified with polar and non-polar monomers. The ability to disperse and functionalize strongly hydrogen-bonding polymers within a polyolefinic matrix is unique and can produce polymers having engineering properties not possible by conventional compounding.

3. Synergism of Metallocene and $MgCl_2$-Based Catalysts [99]

The concept is to combine the traditional $MgCl_2$-based catalysts and metallocene-based catalysts. The resulting reactor blend constitutes a new composition of matter endowed with very interesting properties not achievable previously via mechanical blending.

III. METALLOCENE SCENE

The polyolefin business has experienced phenomenal growth through product innovation and market expansion in the last few decades. This growth could be possible due to the strong commitment among global companies for innovative technology development, as reflected by the estimated expenditure of US$ 2 billion on R&D and market development of new generation technologies. As a result, several types of new generation technologies have emerged. The most promising among them is the metallocene-based polyolefin technology.

Current developments on metallocene-based polyolefin technology [41–57] have far-reaching implications for the polymer industry. The present generation of metallocene catalysts comprises non-stereorigid, stereorigid, supported, and cationic metallocene. Metallocene in combination with alumoxane or alkyl–aluminum compounds affords a wide range of polyolefins with controlled molecular weight, composition, and stereo- and regio-structures. The compatibility of metallocene systems with existing solution, slurry, and gas phase processes has led to faster commercialization of metallocene-based polyolefins.

The major developments are presently targeted for polyethylene and new polymeric structures that command high prices. Exxon predicts that metallocene polyethylenes will take 12–15% of the total polyethylene industry by 2005 and will jump to 25% by 2010. It is expected that the price premium for metallocene polyethylene will be 4 ¢/lb in 2000, falling to 2 ¢/lb by 2005. This premium will be justified on the grounds of property advantages, higher catalyst cost, and the need to recover the huge R&D expenditure.

Metallocene-based polyethylene products are expanding the market for polyethylene by taking the polymer into both existing and new applications. Metallocene–LLDPE has been targeted for film and packaging applications while HDPE and MDPE grades are used for injection moulding and rotational moulding respectively. Full commercial applications of LLDPE is notably in blown and cast film for uses such as stretch film, can liner, and heavy duty sacks. Film applications for metallocene polyethylene are intended to maximize physical, optical, and heat sealing attributes.

Important properties of ethylene–cyclic olefin copolymers such as low melting point, high transparency, better optical characteristics, good heat stability, and high chemical resistance, have opened up its applications into areas like optoelectronic data transmission and data storage on a new generation of optical disks. These copolymers can effectively compete with existing acrylonitrile–butadiene–styrene polymers, polycarbonate and acrylics products. This will push polyolefins into the areas of speciality applications.

Many companies are involved in the commercialization of polyolefin technology based on metallocene catalysts (Figs. 21–24) using solution, slurry and gas phase processes. Exxon Chemical developed the "Exxpol process" to produce a wide range of polyethylene with different densities. Dow Plastics has focussed

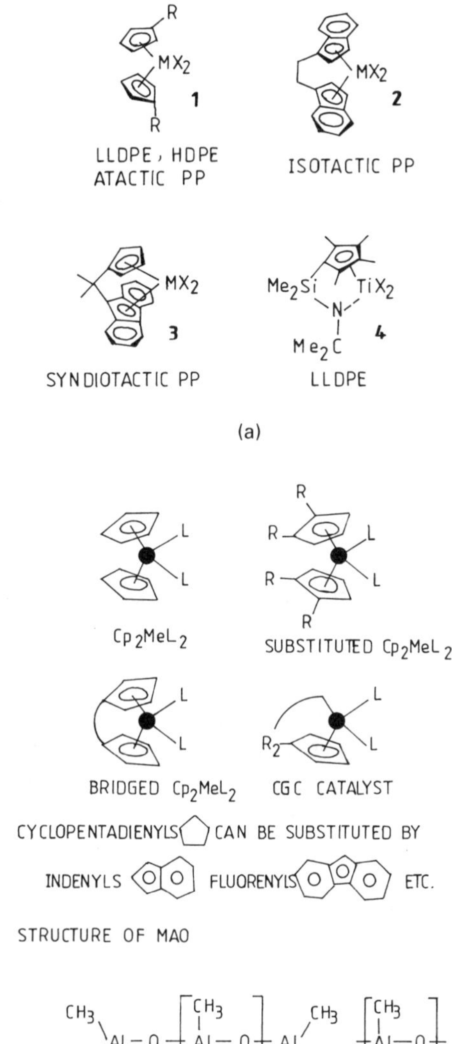

FIG. 21 (a) Representative examples of metallocenes for polyolefin synthesis M = Zr, Hf or Ti, X = (Cl or Me). (From Ref. 55.) (b) Main metallocene catalyst families and structural formula of methylalumoxane (MAO).

its attention on "Insite technology" for the production of ethylene–octene copolymers. Many other polyolefin technology licensors such as Mitsubishi, BP Chemicals, Nippon Petrochemicals, Ube Industries, BASF (Fig. 25), Chisso, Hoechst, Mitsui Toatsu, Mitsui Petrochemicals, Montell, Union Carbide, Shell, Mobil, etc., have, independently and/or jointly, progressed significantly towards metallocene-based processes and product developments.

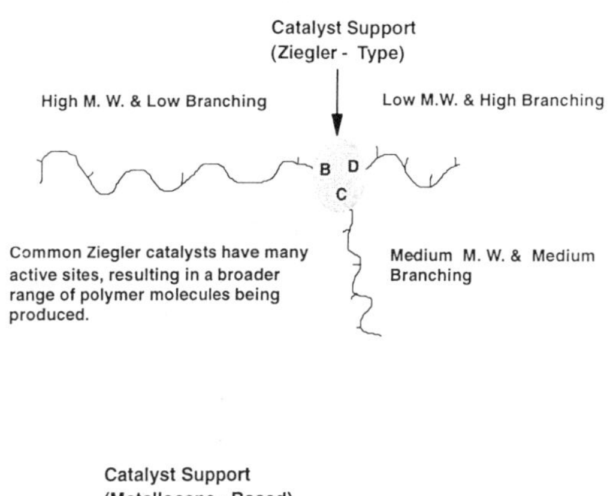

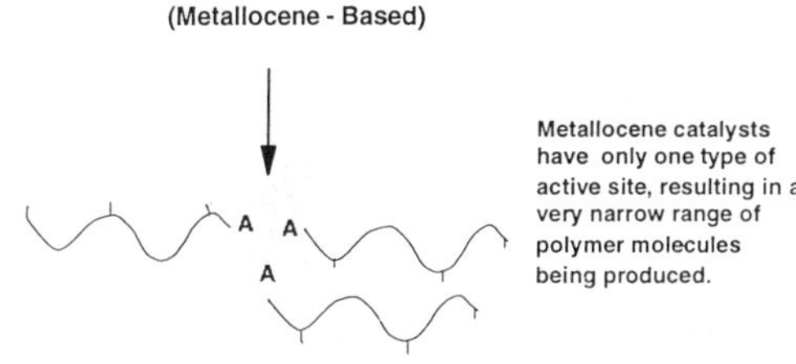

FIG. 22 Narrow distribution of polymer composition and molecular weight produced by metallocene catalyst. (From Ref. 1.)

A major rationalization of metallocene-based polyolefin technology is in progress following the announcement of joint ventures by four of the world-leaders in polyethylene technology. Dow and BP are to pool their respective work and Exxon and Union Carbide also plan to merge their knowhow. BP and Dow are bringing together their respective Innovene (gas phase) and Insite (metallocene) technologies. Dow moved rapidly to commercialize its technology into their plant in the USA and Europe and to form joint ventures with DuPont (in elastomers) and Montell (in polypropylene). Exxon similarly introduced its Exxpol catalysts in polyethylene plant in the USA and Europe and has gone into joint ventures with Hoechst for further development.

A. Metallocene-Related Transition Metal Catalysts [41]

Even as the metallocenes hit the commodity markets (Table 5), a new generation of polyolefin catalysts has emerged that could eventually rival the commercial impact of metallocenes. Most notably, DuPont has detailed work on late-transition-metal catalyst technology that it calls Versipol. While initial applications for the nickel and palladium-based catalysts will probably

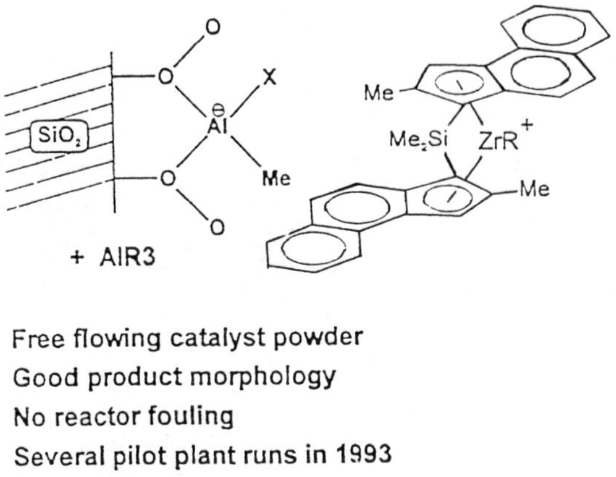

- Free flowing catalyst powder
- Good product morphology
- No reactor fouling
- Several pilot plant runs in 1993

FIG. 23 Supported metallocene catalyst for the Novolen® [BASF] process. (From Ref. 1.)

be in specialty polyolefins, such as ethylene copolymers, DuPont says the catalysts could also be used in making high-volume PE that combines the processibilities of LDPE with the properties of LLDPE.

In 1996 Mark Mack, Director of polyolefins technology, said it "aims to develop unique, low-cost catalyst systems" that are suitable for Lyondell's commodity PE and PP businesses. The company has developed a series of such single-site catalysts, including one that substitutes a nitrogen atom and one that substitutes a boron atom in the characteristic metallocene ring systems.

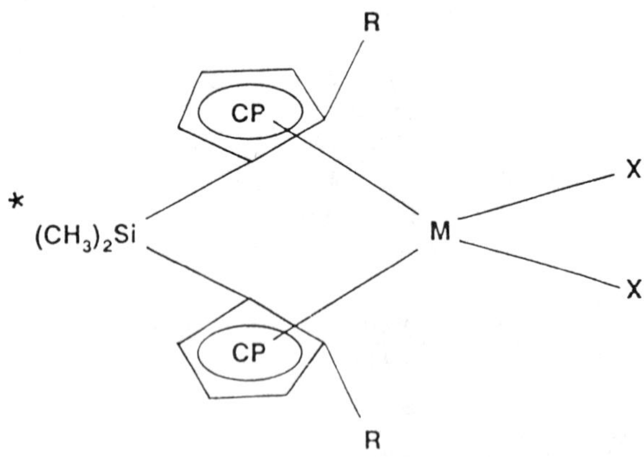

FIG. 24 The general formula for metallocene catalyst $(CP)_mMR_nX_q$, where M is a Group IV or V transition metal (Ti, Zr, Hf, V), CP is a cyclopentadienyl ring, bis(cyclopentadienyl) rings are joined together through a silyl bridge $(CH_3)_2Si$, R is a hydrocarboxy group (1–20 carbon atoms), X is a halogen (Cl), and m is an integer (1–3), n is an integer (0–3), q is an integer (0–3), ($m + n + q$ is equal to the valence of the transition metal M). (From Ref. 1.)

Metallocenes will probably expand the market (Table 6) by allowing new versions of EPDM and new applications. The ability of metallocenes is to expand and tailor the composition of polymers. In particular, he notes the ability of metallocenes to incorporate different comonomers, including higher α-olefins, into the polymer. It allows the molecular architects to go to work.

That, say observers, could be the most dramatic impact of metallocenes and related transition metal catalysts on the commodity polyolefins markets. The single-site catalysts allow greater control over molecular structure and composition, providing the once moribund polyolefin technology field with a range of new possibilities (see Fig. 31).

That prospect has clearly caught the attention of most of the polyolefins industry. But with the gold rush on in metallocenes, a number of producers have turned to related catalyst technologies, including those based on late-transition-metal catalysts, to find unique business advantages and avoid metallocene-related patent disputes. Like their metallocene cousins, the new-generation catalysts are single-site, homogeneous catalysts that allow the precise and predictable tailoring of polymer chains.

The late-transition-metal catalysts have two key advantages over metallocenes; first, they allow oxygen and other functional groups, including polar monomers such as acrylates, to be tailored onto an ethylene backbone; as a consequence, novel ethylene copolymers can be synthesized. Second, these catalysts can

- Metallocene Screening in BASF High Pressure Tubular Pilot Plant Reactor
- Methylalumoxane or Cationic Activator can be Used
- No Additional Al-Compound as Cocatalyst Required
- No Modifications of Process Necessary
- Drop-In Technology

FIG. 25 Metallocene development for BASF high pressure tubular process.

produce a high level of branching in the PE chain, resulting in low-density material that does not require hexene or other comonomers. By avoiding the use of comonomers, the highly branched PE could have significant cost advantages over LLDPE in large volume applications.

Most recently, DuPont scientists have also presented details on the development of a metallocene-based polycyclopentene with novel properties that could make it an attractive material for engineering plastic applications. Conventional polycyclopentene is isotactic, but analysis suggests the DuPont material is atactic.

B. Catalysts in Transition [56]

During the second half of 1996 the patent literature indicated a noticeable diversion of research activity away from strictly "*metallocene*"-*catalyzed olefin polymerization*. This movement was led by DuPont, but with noteworthy contributions from several other companies.

The search to find further structural variants on *metallocene catalysts* can be extended ad infinitum, but the potential benefits balanced agaist the high cost of this type of research mitigates against its pursuit. Increasingly *complex metallocene structures* imply

TABLE 5 History of Metallocene Development (From *Chemical Week*, May 21, 1977)

	Spotlight on polyolefin catalysts*
January	Borealis announces plans to commercialize metallocene PE using its slurry loop process. The company forms a license agreement with Exxon to gain rights to use metallocenes in a slurry process.
February	Exxon Chemical announces LLDPE capacity increase at Mont Belvieu, TX; about half of the capacity will use metallocenes. Startup of a 120,000 Mt/yr DSM–Exxon joint venture plastomer plant at Geleen, The Netherlands.
	Dow Chemical enters commodity metallocene LLDPE with its Elite line.
	Mitsui Sekka says it will build a metallocene EPDM plant in Southeast Asia.
March	Phillips Chemical makes metallocene LLDPE on a commercial-scale slurry reactor.
	Asahi Chemical forms agreement with Dow covering metallocene PE made in a slurry process. Asahi to restart 40,000 Mt/yr high-density PE unit using metallocenes.
April	Union Carbide and Exxon start up metallocene PE licensing venture, Univation Technologies.
	Lyondell Petrochemical details novel, next-generation metallocene catalysts.
May	Exxon declares force majeure on its Exceed line of metallocene LLDPE at its Mont Belvieu plant. Production is scheduled to resume in June.
	DuPont Dow Elastomers introduces metallocene EPDM and announces completion of its 90,000 Mt/yr plant at Plaquemine, LA.
	Chisso discloses plans for production of isotactic metallocene PP.

*Selected projects announced since January 1977. From company reports and CW estimates.

TABLE 6 Metallocene Plastics

	Growth spurt* (in thousands of Mt/yr) 1997	2005
Polyethylene		
LLDPE	120	5900
Plastomers	117	325
Elastomers	40	275
Polypropylene		
Isotactic PP	30	2000
Syndiotactic PP	10	100
Others		
Syndiotactic PS	—	40
Cyclo-olefins	—	30
EPDM	40	250

*Worldwide metallocene plastic demand. From The Catalyst Group Spring House, PA.

the use and eventual commercial availability of the correspondingly expensive ligands. The higher the metallocene cost, the greater the price constraints on the total catalyst systems required to meet the economic demands of commodity olefin producers. Also, there are fundamental inadequacies residing in the use of metallocene catalysts for making commodity polymers, i.e. their lack of thermal stability, robustness to contact with air and moisture, product versatility, and tolerance to polar functionality in the monomers. Emerging catalyst systems, although still mainly, but not exclusively, transition-metal complexes with π-bonded ligands, tolerate greater levels of impurities in the monomer feeds, and in some cases are capable of incorporating polar comonomers into a polyolefin backbone.

We have been well immersed in the world of polymerization by Ni(0) and Pd(0) α-diimine complexes, starting with Keim's work in 1981, and continuing with the famous DuPont blockbuster WO 9623010 A2 (1 Aug 1996), but what about some of the less heralded catalyst systems? One patent which surely must rate alongside the DuPont patent, and which provides a lucid demonstration of the fragility of the metallocene concept as an answer to all polymerization catalyst problems, is Occidental's WO 9633202 (24 Oct 1996). Here the inventors explicitly state that they have encountered high polymerization activity for ethylene and -olefin monomers in complexes not heretofore expected to show any catalytic activity whatsoever, namely a range of simple, easily prepared bidentate pyridine and quinoline complexes of Group IVB metals. To suggest that the complexes are variations on the Ziegler–Natta theme does not fit because, in common with metallocenes and the DuPont catalysts, the Occidental complexes require activation by alumoxane, or cationization with a Lewis acid, forming a non-coordinating anionic species. These systems used in solution or supported for gas-phase processes, have activities comparable with some of the best metallocene systems; and, interestingly, incorporation of a cyclopentadienyl ligand in these complexes actually reduces their catalytic activity.

A further interesting example is WO 96 16959 A1 (6 Jun 1996) granted to Borealis A/S. This describes a polymerization system based on a new type of platinum metal complex containing cyclic thioether ligands, activated by alumoxanes. High molar ratios of alumoxane to platinum are, however, still required and the activities are unremarkable.

BF Goodrich's patents US 5,569,730 (29 Oct 1996) and US 5,571,881 (5 Nov 1996) further erode the mystique of metallocene catalysts in making specially polyolefins, specifically for norbornene-functional monomers. Unique catalyst systems that catalyze the insertion of chain-transfer agents exclusively at a terminal end of a polymer chain are used to produce addition polymers from norbornene-functional monomers. These catalyst systems, in common with DuPont's, are based on nickel or palladium. There are two noteworthy factors: firstly, the nickel or palladium can be present as simple compounds such as the bromides, oxides, acetylacetonates, or 2-ethylhexanoates; secondly, and perhaps more surprisingly, these catalysts are activated by the same activators as are the metallocenes.

It remains to be seen whether Nova's new catalysts announced in December 1997, will also break the metallocene mold.

Why don't the other polymer producers follow the examples of Occidental, Goodrich, etc., and seek simpler, rather than more complex, catalytic systems?

C. Borealis Online with Metallocene PE

Borealis has started up its metallocene-based slurry loop polyethylene (PE) plant at Ruenningen, Norway. Borealis will produce 30,000 t/yr of metallocene-based products at the plant in 1997. Borealis has agreed a licensing deal with Exxon Chemical which should avoid future patent conflicts over the companies' respective metallocene technologies. In 1996

Exxon Chemical and Union Carbide formed a joint venture to develop metallocene technologies. Exxon and DSM also set up a metallocene joint venture in 1996. Dow began production of 65,000 t/yr of metallocene products at Tarragona, Spain in 1996.

D. Metallocene PP Challenge Tradition [57]

At the Metallocenes Europe '97 conference in April, John Grasmeder, BASF product technical manager, planned to unveil the results of injection moulding trials (Table 7) comparing a metallocene-catalyzed PP (mPP) with a random copolymer PP produced via Ziegler–Natta catalysts (ZN PP). Grasmeder concluded that mPP is an evolutionary step in propylene development, but estimated it will not eclipse traditional PP in the short term. He believed the best applications for mPP will be in thin-wall food packaging where good visibility is required.

The basic questions BASF attempted to answer with the study are valid ones. Do mPPs offer a better impact and stiffness balance? Are claims of higher crystallinity justified? To get some answers, Grasmeder tested Novolen M(mPP) against other Novolen ZN PP grades.

For the first test, samples were injection moulded in an open-ended spiral flow tool of constant cross section. For both metallocene and ZN PP, melt flows studied were in the range of 4–120 g/10 min. Grasmeder found that, although higher melt flow rates are required for mPP to produce the same flow lengths as ZN PP, there is no length unit as in ZN PP, and there is no compromise on impact strength. Traditional PPs lose impact strength proportionately as melt flow rises, but this is not the case with mPP, he observed.

A second trial involved moulding a ZN random copolymer PP, an mPP random copolymer, and an mPP homopolymer into large household articles. Easier flow of the mPPs in the barrel at lower shear rates gave a 10% reduction in plastication time and a 50% reduction in injection pressure. Optical properties of both mPPs were also noticeably better, improviing the aesthetics of the finished part.

Using a small two-impression thin-wall food packaging tool, a third trial showed no processing differences among any of the materials. The mPP packages appeared to have better gloss and transparency, as well as reduced odor. A single-impression tool, run with the latest fast cycling, high-transparency, and high-crystallinity grades of ZN PP, was then tested. After substituting an mPP, cycle times were 10% faster with parts of equal transparency, better gloss, and greater stiffness.

TABLE 7 PP Property Comparisons

Property	High crystallinity ZN PP	High transparency mPP	High transparency ZN PP
MFR (g/10 min)	23	60	48
Melting point (°C)	165	148	150
Shear modulus (MPa) at 23°C)	1050	860	650
Tensile yield stress (MPa)	40	35	29
Strain at break (MPa)	20	100	100
Tensile modulus (MPa)	2000	1710	1150
Ball indentation hardness (MPa)	93	77	60
HDTA (°C)	62	55	50
HDTB (°C)	105	86	70
Charpy impact strength (23°C, kJ/m^2)	90	91	180
Gloss (20°C, %)	41	100	100
Haze (%)	53	7	7

From these and other trials, Grasmeder concluded that homopolymer mPP offers a unique balance of stiffness, transparency, and organoleptic properties, a combination not currently achievable with ZN PP. However, he admitted, mPP is not yet as rigid as the latest generation of high-crystallinity ZN PP homopolymers. Excellent transparency for mPP lets it complete effectively against the new transparent ZN random copolymers, offering a 10% cycle time advantage as well. For random mPP copolymer, moulded parts achieve 96% transparency with the same mechanical properties as the new generation ZN transparent random copolymer.

As for processing, Grasmeder found no difference between mPP and the latest ZN PPs. Grasmeder noted, however, that mPPs of higher nominal melt flow rate are sometimes required. This does not seem to present any tradeoffs with impact strength, thanks to the narrow molecular weight distribution of the mPP.

E. Review of Metallocenes

For more than a decade, there has been a growing body of research and publications that describes the amazingly high activity of metallocene-based catalysts systems to polymerize α-olefins. The productivity of these metallocene systems now easily exceeds the range of 10–30 kg isotactic polypropylene (PP) per g of catalyst, typical of the Ziegler–Natta (ZN) catalyst systems presently in commercial use — under some conditions, productivity of metallocenes exceeds 1000 kg/g. Furthermore, metallocene catalysts have the capability to permit the user to better control polymer tacticity, molecular weight, and molecular weight distribution (MWD) — sometimes by supplementing with a second catalyst (Figs. 26 and 27). For olefin copolymers, better control may be exerted on comonomer content, as well as on distribution statistics. Another prominent feature of the new metallocene systems is the very high cost to manufacture the required ingredients. The purpose of this study is to develop some estimates of the capital required to build the production facilities and the costs to operate them. The accuracy of the estimates from a preliminary study of this sort is not expected to be better than ±50%.

The use of metallocene catalysts to make polyolefins is driven in part by enhanced properties, such as improved stiffness, higher impact strength, and, in certain cases, higher clarity and lower density. The hindrance has been the very high cost of these catalyst systems. Thus far metallocenes have seen successful commercial application in making selected ethylene-based copolyolefins that have less-stringent stereochemical requirements than PP. A metallocene system to make ethylene copolymer at high productivity is believed to have a cost distribution of 1–2 ¢/lb in the finished product. By way of comparison, the cost contribution of a conventional catalyst system is approximately 1 ¢/lb. On the other hand, the developmental metallocene systems tailored to make isotactic PP and ethylene–propylene (EP) impact copolymers have so

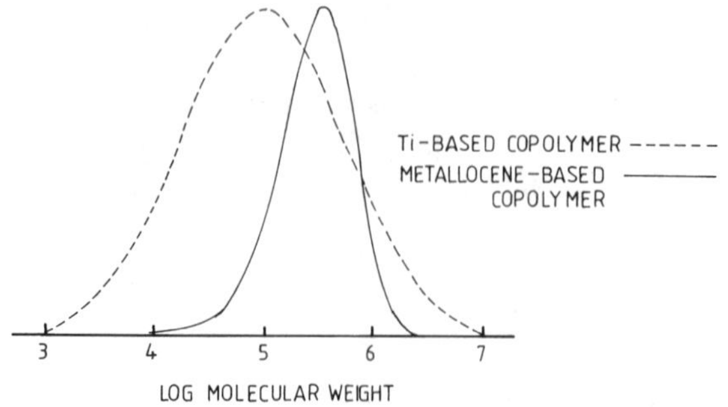

FIG. 26 Distribution of molecular masses for ethylene/propylenecopolymers from isospecific metallocenes. (From Ref. 1.)

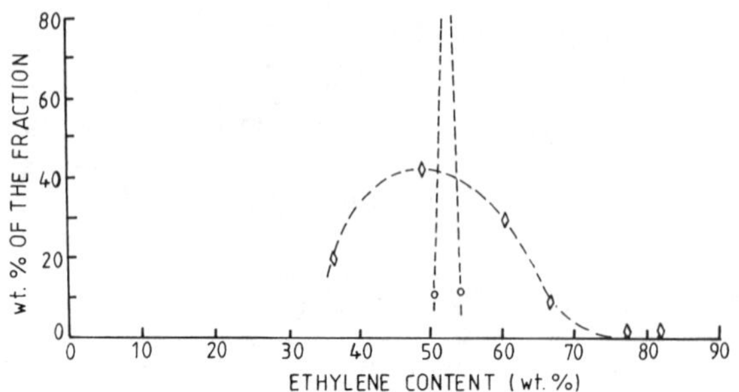

FIG. 27 Distribution of chemical composition for ethylene propylene copolymers from isospecific metallocenes. (From Ref. 1.)

far been much more costly. Yet several reviewers anticipate that, in a few years, metallocenes will be used to make propylene-based polymers having a cost differential of from 0–4 ¢/lb over the conventional ZN method. Consider the historical trend of catalyst price per productivity of commerical PP. From the early ZN catalysts to the latest supported high-activity ZN systems, this price has risen over a period of three decades from $5 to $10 per 1000 g/g. This price is estimated in this present study to reach $30 per 1000 g PP/g for metallocene systems used to make high-value specialty products. Metallocene polymerization conditions of temperature and reaction time will be similar enough to present ZN conditions that the new catalysts will be practically "drop-in" substitutes for the old. Relative ease of substitution is vital to control the high cost of the new catalyst.

The assumptions used in these cost estimates are strongly influenced by the discoveries of Kaminsky [12, 17, 43] and Brintzinger [7] in making isotactic PP using a particular zirconocene activated with considerable amounts of methylaluminoxane (MAO). Spaleck [14] and Razavi [53, 54] have studied the stereochemistry of propylene polymerization with metallocenes, particularly the influence of substituents and the role of an intramolecular bridge to enhance the isotactic quality of the polymer while at the same time giving high activity. Razavi [53, 54] has extended this research to reveal the interrelation between the catalyst ligand structure and the micro- and macrostructure of the polymer produced. This knowledge will enable chemists to tailor the catalysts to provide properties that are highly desired in the marketplace.

Figures 26–28 clearly show that copolymers from isospecific metallocenes have much narrower distributions of molecular masses and chemical composition.

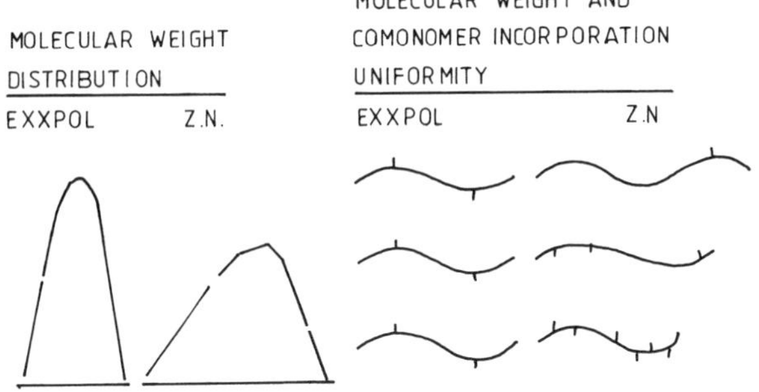

FIG. 28 Polymer structure of metallocene-based products vs. Ziegler–Natta products. (From Ref. 1.)

F. Metallocene Catalysts Can Retrofit Existing Plants

Much interest has been focussed in the field of metallocene catalyst and polymers on the catalyst which had first been reported by Kaminsky *et al* [43]. Kaminsky catalysts are a superactive catalyst system containing a soluble metallocene compound and methylaluminoxane (MAO). This catalyst system is very interesting from the viewpoint of high activity and reactivity with a wide variety of monomers, producing new types of copolymers. After this work, much effort has been spent to modify the catalyst. Brintzinger [7] and Ewen [47] have succeeded in preparing ansa – metallocene compounds and produced isotactic and syndiotactic polypropylene respectively (Fig. 29). Recently Dow Chemical Company has discovered a family of constrained geometry catalysts that could produce unique polyolefins. Catalysts are monocyclopentadienyl Group IV complexes with a covalently attached donor ligand.

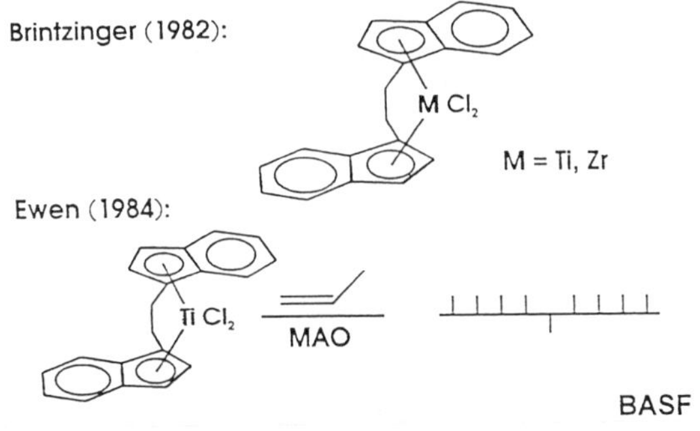

FIG. 29 The first isospecific *ansa*-metallocenes for propene polymerization. (From Ref. 1.)

On the other hand, much research has been carried out not only on finding excellent second components but also on modifying MAO as a cocatalyst. Exxon has discovered that anionic reagents like $B(C_6F_5)_4$ could act on metallocene as a catalyst and form real active cationic sites.

Many companies are challenging for the commercialization of new metallocene polymers, which has taken place in the 1990s (Fig. 30). We consider that the probability of commercialization mainly depends on the retrofittability of this catalyst to an existing plant without significant changes in plant operation and plant production rate. LLDPE are usually produced in gas phase, solution phase, slurry phase and high-pressure process (Fig. 8a). In general, gas phase process are preferred to produce LLDPE at low cost. The solution process is also suitable for LLDPE production, especially VLDPE production. High-pressure processes have essentially the same feature as the solution process. Metallocene catalysts are inherently homogeneous and are suitable for the solution and high-pressure process. To utilize the gas or slurry process, the preparation of supported catalysts (Fig. 23) is needed.

G. Manufacture of Specialty Polymers Using Single-Site Catalysts (SSC)

Shell is already producing ethylene/carbon monoxide polymers using a palladium-based SSC by a slurry phase process. BP and GE are working on similar technologies for the production of E/CO polymers. Norbornene and other cyclic olefins are very interesting and potentially low cost monomers that could be

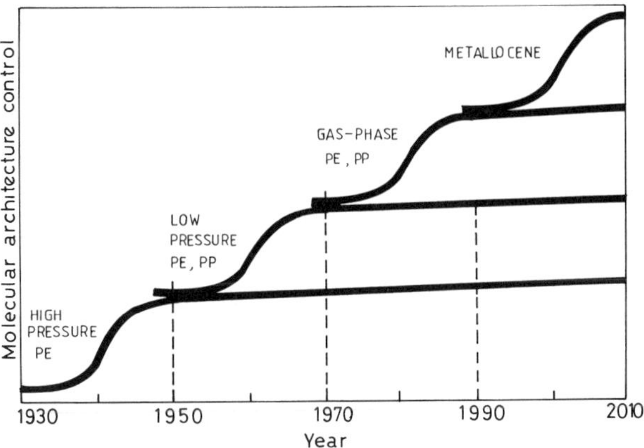

FIG. 30 Generations of polyolefin synthesis. (From Ref. 1.)

polymerized in standard polyolefin production processes to make anything from clear elastomers to engineering thermoplastics with glass transition temperatures above 370°C. Mitsui Petrochemical and Hoechst are already producing ethylene/cyclic olefin copolymers for optical markets using SSCs.

Goodrich and DuPont are developing nickel- and palladium-based SSCs with capabilities to produce anything from very high MW liquids (MW ~ 100,000) to engineering thermoplastics with MW above one million.

At the other end of the scale, Uniroyal is producing SSC-based low MW functionalized EP copolymers, and Exxon appears to be developing MPP (Table 8) and SSC-based copolymers for the next generation of lube oil base stocks.

A. METALLOCENE : SINGLE SITE

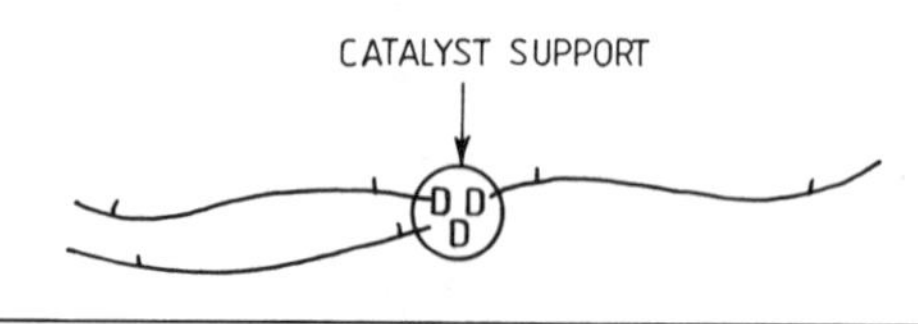

B. ZIEGLER-NATTA : MULTI-SITE

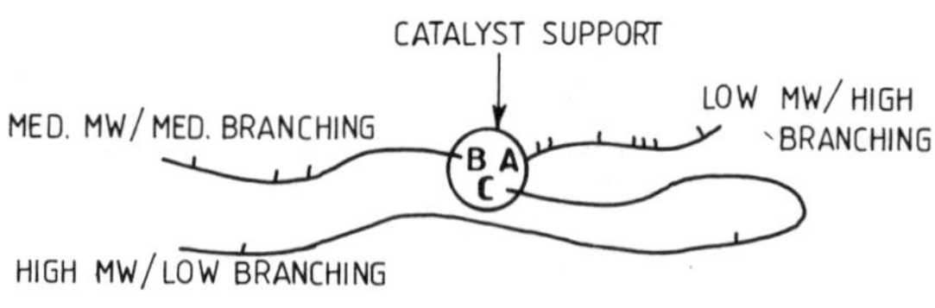

FIG. 31 Metallocene: single-site versus Ziegler–Natta multi-site catalyst. (From Ref. 55.)

H. Metallocene Catalyst

The introduction of metallocene and other single-site technologies (Fig. 31) made possible new process/comonomer combinations, use of novel comonomers such as styrene, norbornene, and carbon monoxide, and seemingly impossible property combinations. Potential adaptation of polyolefin manufacturing technologies to the production of engineering thermoplastics is possible.

The four primary process type — high pressure, solution phase, slurry phase and gas phase processes — have been developed continuously over 30–40 years.

For polymer manufacturers and their customers alike the metallocene/polyolefin process combination is a marriage made in heaven. Metallocene can be utilized in existing polyolefin plants.

TABLE 8 Potential Applications of Exxpol™ Metallocene-Based Polypropylene (From Ref. 1)

Exxpol PP shows exceptional potential in several important applications
- Fibrous products where stronger, finer fibers are valued.
- Specialty films where exceptional heat sealing performance is required.
- Applications where the unique combination of high modulus, high service temperature, and low processing temperature are important: OPP film, fast cycle molding are examples.
- Tough films, molded articles and soft, strong fibers from propylene–HAO copolymers.
- Any application where resistance to free radical attack is prized.

These favorable findings coupled with good prorgress in catalyst and process technology led Exxon Chemical to project commercialization of Exxpol PP in 1995.

I. Competitive Polyolefin Technologies (Table 8a)

1. High-Pressure Technology

These use pressures of 1000–3000 atm using free radical initiators to yield homopolymers containing both short and long chain branching. LDPEs have density ranges of 0.915–0.932 g/cm^3 and melt indices in the range of 0.1–100 dg/min. The process is flexible to grade change. However it cannot make HDPE.

High-pressure processes when used with ZN catalysts, greatly expand the product range and provide control over composition, chain linearity, and short chain branching. It became the most versatile technology through the 1980s.

2. Solution Phase Polyethylene Processes

Advantages of the process include the ability to process high melt index grades for injection moulding as well

TABLE 8a Suitability of Single-Site Metallocene Catalysts with Respect to Main Process Technologies

	Products		
	HDPE	LDPE	LLDPE
Slurry	X		
Solution	X	X	
High pressure		X	X
Gas phase	X		X

as a broad range of copolymers with densities down to 0.90 g/cm^3 using conventional ZN catalysts. The solution process covers the full LLDPE/HDPE density range from 0.915 g/cm^3 to above 0.965 g/cm^3. Besides, the process can use heavier comonomers such as 1-octene, at high concentration in the reactor. This process cannot make very low density material and polymers with blow moulding and thin film grades which require clarity.

3. Slurry Phase Polymerization Processes

This technology accounts for an annual production of 4–5 million tons of HDPE. The process involves polymerization of ethylene at temperatures below the melting point of the polymer using a solid catalyst to form solid polymer particles suspended in an inert hydrocarbon diluent. Recovery of polymer (by filtration, centrifugation, or flashing) is economic. The chromium oxide-on-silica catalyst developed by Phillips yields polymers which can be easily extruded and blow-moulded. The process is unable to produce copolymers of density below 0.937 g/cm^3. For a density of 0.92 g/cm^3 or below, the polymer swells, becomes sticky, and starts to dissolve in the reaction diluent.

4. Gas Phase Polyethylene Processes

The lower density limit for gas phase processes has been about 0.885 g/cm^3. This process has become the most widely licensed technology in the polyolefin industry. They cover the complete range of interesting property combinations used in both the LLDPE and HDPE markets. Their primary advantages are simplicity, low cost, and a broad product (copolymer) range capability. Introduction of SSCs removes the limitations of the gas phase process and extends their capabilities well beyond their previous limits. Gas phase SSC-based polymers have less tendency to be sticky as density is decreased to 0.885 g/cm^3 and below for safe production of polymers. Low MW fraction of polymers is eliminated and the polymer has a reduced tendency to swell and become sticky.

J. Product Developments

We have reviewed a number of developments in PP products which are at present only described in the patent and technical literature, or which are in the very early stages of commercialization. The more important of these developments are as follows.

The development of broad molecular weight distribution PP grades. A great deal of work has been done recently in broadening the MWD of PP, both by manipulation of catalysts, by production in multiple reactors, and by post-reactor treatment. The broad MWD gives PP much higher melt strength, which makes it more suitable for applications such as blow moulding, thermoforming and extrusion coating. These developments, combined with the dynamic growth of the packaging industries in the industrialized regions, are expected to provide significant new markets for PP.

Blends of PP with elastomers and other polymeric modifiers. A number of producers are making extensive use of compounding technologies to modify the physical properties and behavior of PP. By the use of carefully selected modifiers, reinforcements, and fillers, it is possible to specifically tailor the characteristics of specialization and differentiation. These activities are particularly prevalent in Japan and Western Europe, and are beginning to be used in a comparatively small way in the United States. The most common polymeric modifiers are EP rubber, EPDM, LLDPE, VLDPE, and amorphous poly-α-olefins (Fig. 32). The latter modifiers are possibly the most interesting, because they dramatically improve the low temperature impact properties of PP without too great a detriment to stiffness. Significant catalyst and process work is being done to control the molecular weight distribution, crystallinity, and molecular uniformity of these modifiers to obtain blend characteristics never before achieved.

The use of styrene graft copolymers to improve surface gloss [3]. The styrene component in ABS is the one responsible for the very good surface properties of these polymers. We have found examples of the use of styrene grafted onto an elastomeric polymer as a means to improve surface properties of PP compounds. With these new compounds it may be possible to increase interchangeability between PP and ABS.

Highly crystalline polypropylene. Modified catalyst systems have been discovered in the last few years which are capable of yielding almost perfectly stereoregular polypropylene. These polymers are very highly crystalline, have a significantly higher density than conventional PP (up to 0.936 g/cm^3), have a higher melting point and heat deflection temperature, and exhibit extraordinary stiffness properties. The properties of HCPP are so different from those of conventional PP that is in fact a completely new polymer and not just an extension of currently available grades. Because it results from a catalyst

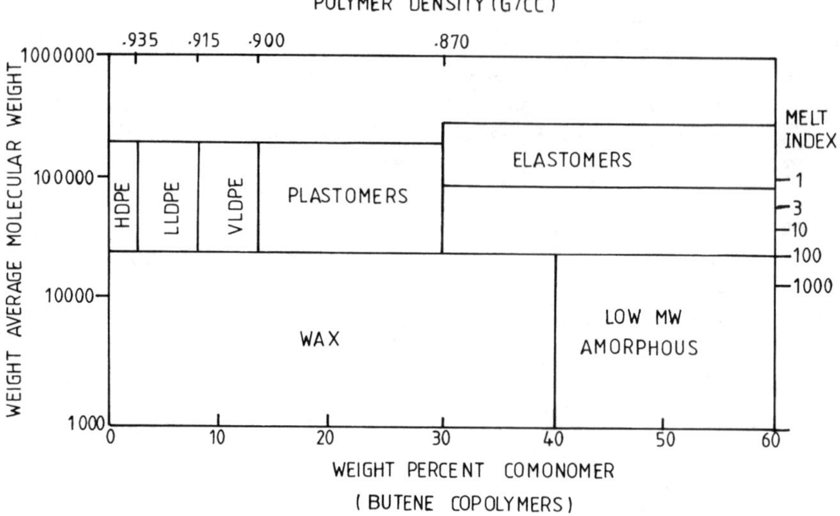

FIG. 32 Ethylene-based polymers product regions. (From Ref. 1.)

development, however, production costs are the same as for conventional PP.

Of these developments, the discovery of highly crystalline PP is probably the most significant, in that it immediately opens up a whole range of new applications for PP in injection molding (Fig. 33). HCPP is very much more competitive with ABS in terms of performance, and in many applications it will be superior. HCPP is already being commercialized by Nissan Motors.

Other developments in PP products demonstrate that it is technologically still a very young and versatile polymer. New catalyst systems are continually being discovered which yield polymer structures never before attainable. Blending and compounding technologies are expanding the range of property variations available to cover the complete spectrum from soft thermo-

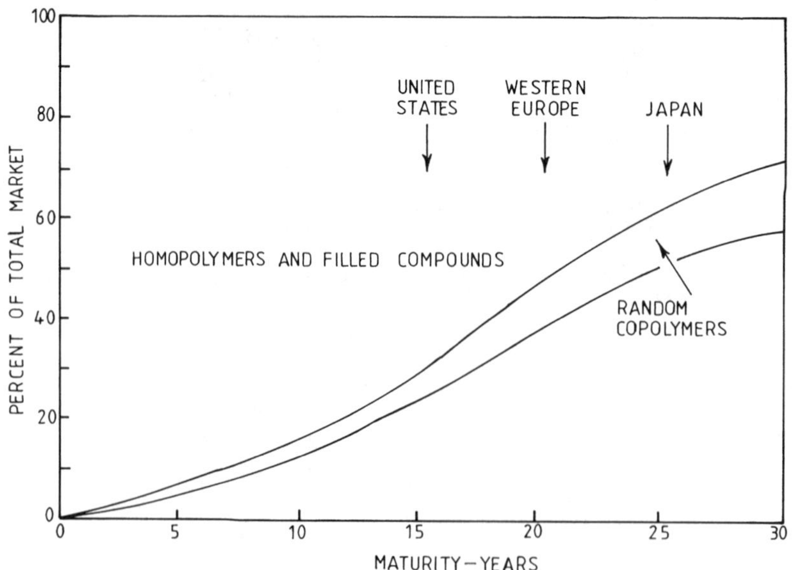

FIG. 33 Polypropylene injection molding market development. (From Ref. 3.)

plastic elastomers to rigid and hard resins for engineering applications.

K. Process Economics

In terms of cash production costs, there is virtually no difference between gas-phase and bulk slurry processes. In terms of product value, however (including capital charges), the somewhat higher investment for bulk slurry processes gives them a disadvantage of around 0.7 ¢/lb on a mid-1984 cost basis. Slurry processes based on a heavy diluent technology are significantly more expensive than the gas-phase and bulk slurry processes with a product value disadvantage of 2.5 ¢/lb, even with the use of high performance catalysts. Heavy diluent processes using a highly selective but lower yield catalyst have a disadvantage of about 4.5 ¢/lb, in terms of product value compared to gas-phase processes.

For production of random copolymers it was found that gas-phase and bulk slurry processes had similar product values for standard copolymer grades containing up to about 3.5 wt% ethylene. Heavy diluent processes were significantly more costly, having a net disadvantage of 2.3 ¢/lb in terms of product value [3]. For the production of high ethylene random copolymers, however, the gas-phase processes are the clear leader, to the extent that production of these grades by slurry processes may no longer be competitive in the future.

For the production of standard high-impact block copolymers, similar differences were found in overall economics between processes as found with homopolymers: gas-phase and bulk slurry processes with three or more continuous reaction stages in series. Such processes are significantly more costly than our standard two-reactor, gas-phase and bulk slurry processes, but the products they produce are very high performance and cannot yet be matched in a two-stage reaction system. The products command significantly higher prices, and these complex slurry processes are, therefore, expected to remain in use (at least in Japan and Western Europe) beyond the 1993 timeframe.

The share of competing technologies in production of polypropylenes in the three industrialized regions is undergoing dramatic change at the present time. Many of the existing heavy diluent slurry process plants are being revamped to bulk slurry processes, and many more new bulk slurry plants are presently under construction.

Typical penetration patterns for copolymers into homopolymer markets as a function of market maturity are shown in Fig. 33. Market maturity is expressed in the Japanese PP market, since PP was first commercialized in 1959. Copolymer penetration in Japan in 1984 is indicated in Fig. 33 by a maturity of 25 yr. The Japanese market has been selected as the basis for scaling, because it is the most advanced PP market worldwide and has the highest copolymer penetrations. Consumption patterns for Western Europe indicate a relative maturity of about 20 yr, whereas those found in the United States correspond to about 15 yr or less.

L. Market for Metallocene-Catalyzed Polyolefins

Metallocene-catalyzed polyolefins represent the latest wave of developments in the history of the polyolefins industry. Developments began in 1942 with the introduction of LDPE, followed by HDPE and PP in 1957, and LLDPE in 1978. Today's embryonic metallocene products represent a basic renewal for polyolefin technology in general.

Metallocene catalysts have been applied to a number of key product families across the chemical industry. Today, polyolefins and styrenics are either commercial or in development with additional polymers and specialty chemicals in the R&D pipeline. The work in polyolefins has focussed on polyethylene, polypropylene (isotactic, syndiotactic, and atactic), and cyclic olefins.

Figure 34 lays out the playing field in metallocene polyolefin development [59]. Activity in the various metallocene technologies is proceeding on a global basis, with many participants already involved in alliance arrangements. These materials have key properties which provide opportunities in many existing markets.

The current addressable market is huge:

approximately 21.4 Mt for metallocene polyethylene including markets in packaging films, electrical, automotive, medical devices and textiles targets. Targets include replacement of PVC, PP, LDPE/LLDPE/HDPE (enhancements), EP/EPDM, and EVA;

approximately 21.7 Mt for metallocene polypropylene including enhancements to current PP products, and replacement of PVC, HDPE, and EP/EPDM. Target markets include pipe/sheet, films, appliances, automotive, rigid packaging, fibers and filaments, blow molding and coating.

KEY PLAYERS BY REGION		KEY METALLOCENE-BASED PRODUCTS					
		POLY-ETHYLENE	POLYPROPYLENE			Syndiotactic Polystyrene	CYCLIC OLEFINS
			Syndiotactic	Isotactic	Other		
UNITED STATES	DOW[1]	✓				✓3	
	EXXON[2]	✓		✓4			
	MOBIL	✓					
	PHILLIPS	✓6					
WESTERN EUROPE	FINA		✓	✓			
	HOECHST		✓	✓4			✓5
	BP	✓					
	MONTECATINI				✓		
	BASF	✓		✓			
JAPAN	MITSUI TOATSU		✓				
	MITSUBISHI PETRO	✓					
	CHISSO		✓				
	IDEMITSU					✓3	✓
	MITSUI PETROCHEMICAL	✓					✓5
	NIPPON ZEON						✓
	JSR						✓
	SUMITOMO		✓7				
	TOSOH	✓8					

COMMENTS

1 ALSO DEVELOPING AN ETHYLENE/STYRENE INTER-POLYMER; WORKING WITH DuPONT IN ELASTOMERS; PE WORK IN SOLUTION
2 EXXON P.E. WORK (EXXPOL) DONE IN SLURRY, HIGH PRESSURE; GAS PHASE (PE) WITH MITSUI
3 DOW/IDEMITSU PARTNERSHIP IN SYNDIOTACTIC POLYSTYRENE
4 TONEN/EXXON/HOECHST PARTNERSHIP IN ISOTACTIC POLYPROPYLENE
5 HOECHST/MITSUI PETROCHEM. PARTNERSHIP IN CYCLIC-OLEFINS
6 DEVELOPMENT PROGRAM
7 NOW IN A JOINT VENTURE WITH PHILLIPS
8 IN DEVELOPMENT

FIG. 34 Activity in the various metallocene technologies is proceeding on a global basis, with many participants already involved in alliance arrangements.

By the end of the decade, the cost of metallocene-catalyzed polyolefins is expected to be competitive with conventional products.

In the area of metallocene technology, the industry has invested more than $3 billion in R&D. In the pursuit of leadership in this area, several new leaders have emerged:

Company	Product focus
Exxon	Polyethylene Isotactic polypropylene
Hoechst	Syndiotactic polypropylene Isotactic polypropylene Cyclic-olefins
Dow	Polyethylene Syndiotactic polystyrene

Figure 35 shows which companies hold the patents in the metallocene catalyst area. An interesting contrast is that today's apparent leaders in metallocene catalysis are different from the historical leaders in licensing. It is also clear that many companies are pursuing this area across all relevant products. At the same time as metallocene catalysis efforts intensify, incumbent producers continue to refine both process and catalyst systems in search of new/niche markets. Figure 36 describes some of these refinements to both process and catalyst being developed by incumbent producers. Clearly, developers of technology have already begun to separate catalyst and process; many of these aproaches are retrofittable on existing equipment/processes.

At the same time as technology developers pursue catalyst and processes enhancements, the distinction between traditional PE and PP is becoming blurred as these new catalyst and process technologies make the production of "polyolefins" possible (Fig. 37). This suggests that licensees will have many catalyst/process alternatives in the future to address the "polyolefin" market. It also suggests that, based on the above, the business of polyolefin technology licensing will become increasingly competitive [59].

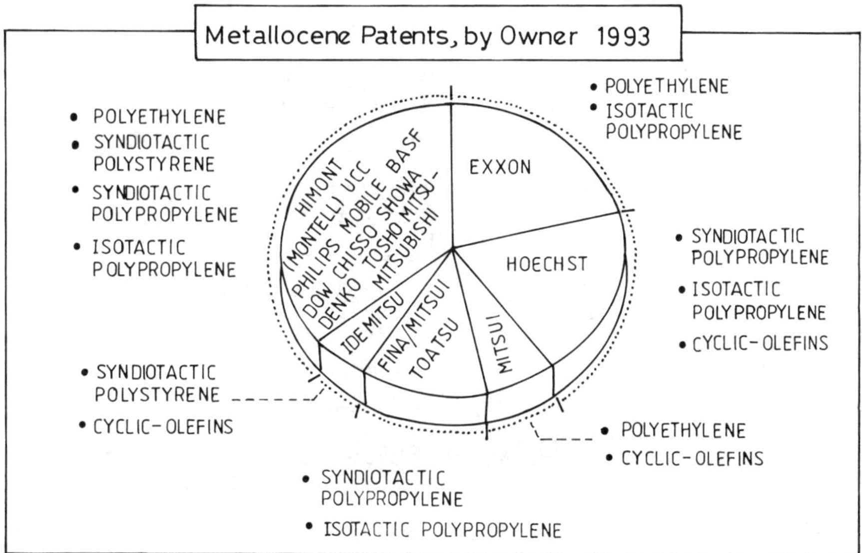

FIG. 35 The industry has invested more than $3 billion in R&D for metallocene technology, and several "new" leaders have emerged.

IV. NEW DEVELOPMENTS IN METALLOCENE-BASED POLYOLEFIN ALLOYS, BLENDS AND COMPOUNDS

A. What Makes Metallocenes Unique [60]?

The traditional Ziegler–Natta type catalyst, which contains several reactive sites, produces olefin polymer molecules with a broad range of molecular weights, molecular weight distributions, and comonomer distributions, while the newly developed single site metallocene-based catalyst produces highly uniform polymer molecular chains and evenly spaced comonomer. The novel "molecular engineering or architecture" of the homogeneous polymer molecules

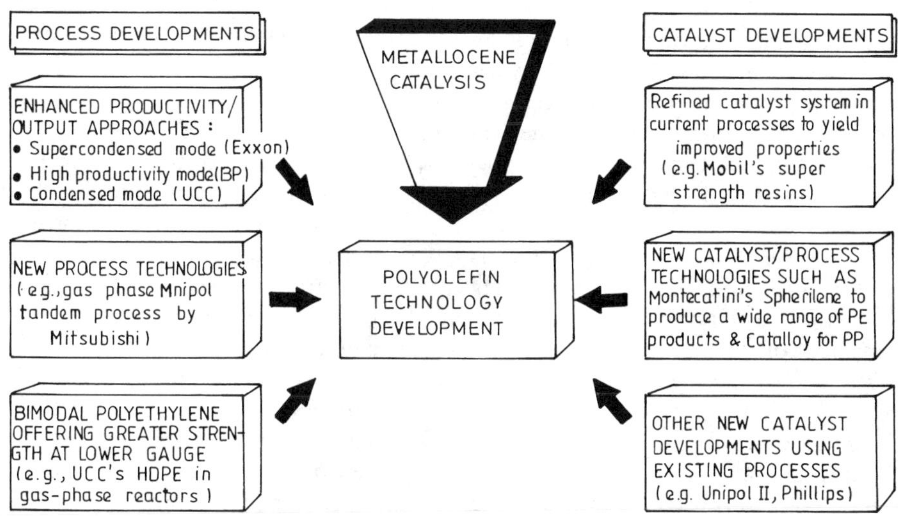

FIG. 36 Metallocene catalysis is just one body of activity, albeit a very large one, that is currently influencing the market. Developers of technology have already begun to separate catalyst and process.

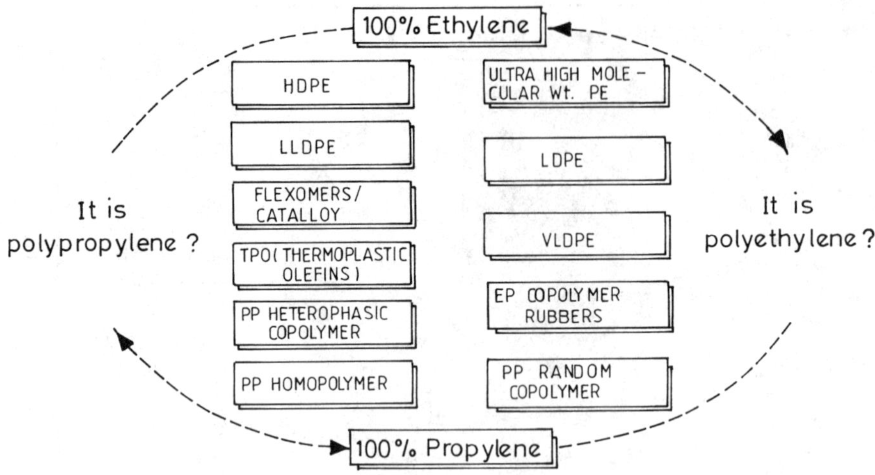

FIG. 37 The distinction between PE and PP is becoming blurred as new catalyst and process technologies make the production of polyolefins possible. Licensees will have many catalyst/process alternatives in the future to address the "polyolefin" market.

produced through the single site catalyst system provides a unique combination of performance properties for polyolefins, including polypropylenes, and offers great promise with new applications in various end-use markets.

Metallocene-based polyolefin alloys, blends and compounds, an emerging class of novel plastic products, offer an unique combination of performance properties. In polymer blending and/or alloying, often dissimilar polymers are physically mixed (compatibilized) and/or chemically reacted (reactive extrusion processing).

Some recent commercial developments are custom designed with the available metallocene-based polyolefin resins. These include process-improved halogen-free polyolefin alloys specially formulated for medical applications, newly developed highly flexible foam products from polypropylene/polyethylene blend systems, and value-added filled and reinforced polypropylene compounds to serve numerous market segments including the automotive, appliance and packaging areas. The major performance attributes of polyolefin alloy products include soft and flexible, halogen-free and plasticizer-free, good contact clarity, sterilizable, heat or radiofrequency sealable, good barrier properties, excellent chemical or solvent resistance, environmentally safe for disposition, ease of fabrication, and cost effective. The film (heat and RF sealable) tubing and injection molding grades and their performance properties have identified applications in the medical markets (Fig. 38).

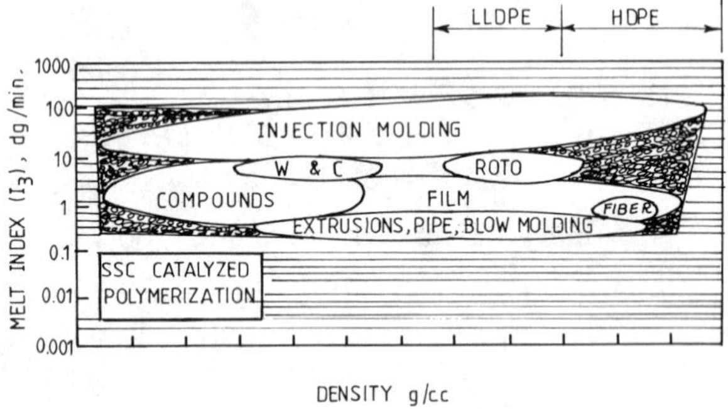

FIG. 38 Solution phase PE process product domains in the 1990s.

The newly developed foamed metallocene polypropylene (FMP) products are very flexible, processable, dimensionally stable, and heat resistant. Due to inherent chemical resistance, moisture barrier and insulating properties, FMP products offer alternatives to their more expensive polyurethane foam counterparts.

Metallocene-based impact modifiers offer improved stiffness and impact balance in PP compounds. This might suggest improved compatibility of the new developed impact modifiers with PP.

The performance property "envelopes" of the above novel and cost effective PP compounds are approaching the realm of more expensive engineering resin "specialties."

B. Is the Metallocene Catalytic System Worthy of Commercialization?

The Ziegler–Natta (ZN) catalytic system developed so far are heterogeneous in nature. The reactive metal ion exists on the catalytic surface in different electronic environments arising out of the crystal growth and the compound components used during its own preparative procedure. Consequently, in a particular catalyst sample, the reactive sites are different and thus produce olefin polymer molecules with a broad range of molecular weights, molecular weight distributions, and comonomer distributions. The newly developed metallocene catalyst possesses a single site character which is the representation of a particular electronic environment around the active metal-ion species. This is conducive to produce highly uniform polymer molecular chains. The novel "molecular engineering or architecture" of the homogeneous polymer molecules produced through the single site catalyst system provides an unique combination of performance properties for polyolefins, including polypropylenes, and offers great promise for new applications in the various end-use markets. Another area of existing development will be metallocene-based polyolefins alloys, blends and compounds.

Metallocene-based polyolefins blends can be developed which combine the performance characteristics of rubber, while maintaining a plastic's inherent ease of processibility. The extremely narrow molecular weight distributions and uniform comonomer composition result in improved and consistent physical properties.

The foamed metallocene polypropylene (FMP) has been developed by Sentinal Products Corporation in collaboration with Ferro. FMP products have advantages over competitive materials such as polyurethane foam, integral skin PVC foam and non-metallocene-based polyolefin foams. Due to the moisture barrier character, foamed metallocene polypropylene (FMP) offer alternatives to more expensive polyurethane foams in the medical packaging arena.

The speciality polyolefin alloys are soft and flexible, halogen-free and plasticizer-free, good contact clarity, sterilizable, heat or radiofrequency sealable, good barrier properties, excellent chemical or solvent resistance, comply wit USP class VI, environmentally safe for disposition, ease of fabrication, and cost effective. Rxloy™ products, developed by Ferro Corporation, find medical applications in fluid administration, prefilled containers, blood bags, flexible containers, urine collection bags, and dry bags. These applications are from Rxloy™ FFS (form/fill/seal) and RF (radio frequency) film grades. Rxloy™ TC (tubing compounds) have potential applications in IV infusion, drainage tubing corrugated respiratory systems. Rxloy™ IM (injection molding) grades are suitable candidates for IV administration components, cathether systems and prefilled syringes.

C. Can We Look Forward?

The analysis and understanding of the fundamentals at the moment support the belief that metallocene-catalyzed polyolefins will become a key technology of the 21st century and the business of polyolefin technology licensing will become increasingly competitive. The technology developers are pursuing vigorously both the catalyst developments and process enhancements. The distinction between traditional PE and PP is becoming blurred as these new catalyst and process technologies make the production of "polyolefins" possible. The metallocene-catalyzed polyolefins will have broad property attributes arising out of the possibility of new monomer combinations. These will address them to the newer needs and will have flexibility in the large existing market. The present activity is focussing on shaping the existing reactors of the major polyolefin processes (slurry, solution, gas phase, etc.) to accommodate metallocene catalysts. Coming of age, these diversified developmental attempts will facilitate many catalyst/process combinations in the future to sharpen the "polyolefin" markets to value-based products.

D. Metallocene-Based Technology: Some Determining Factors

1. The metallocene technology is set to make commercial impact.
2. Today's licensing leaders of polyolefins are not today's leaders of metallocene-based polyolefins.
3. The metallocene catalyst dominates product differentiation (vs process), since product attributes are created by catalyst.
4. In order to speed up the commercialization of metallocene technology and to lower legal barriers, rational alliances (among licensors) were created early to achieve leadership.
5. Incumbent leaders achieved new/additional bases of competitive advantage. Catalyst development remained the focus of development efforts.
6. Converters resisted the penetration of metallocene until the value was demonstrated to them. Some of the leading converters embraced metallocene either in suport of programs with end-users or as a basis

of competitive advantage for small volume applications.

7. The customized (precision) nature of metallocene-catalyzed products has been planned to control product cost.

E. Polycycloalkenes and Copolymers [61]

While it is very difficult to polymerize cyclic olefins such as cyclopentene or norbornene using heterogeneous catalysts without ring opening, a metallocene/alumoxane catalyst polymerizes them exclusively by double-bond opening.

Table 9 shows polymerization conditions and properties of crystalline polymers of cyclobutene, cyclopentene, norbornene, and tetracyclodocene produced by zirconocenes. The activities for the polymerization of cycloalkenes are significantly lower than for ethylene. The melting points are suprisingly high: they were found to be 395°C for polycyclopentene and over 400°C for the others: the decomposition temperatures lie in the same range.

Since the homopolymers of cyclic alkenes are insoluble in hydrocarbons, it is difficult to study their microstructure; therefore oligomers are produced. In the case of poly(cyclopentene), the configurational base units are *cis*- and *trans*-1,3-enchained (27) while poly(norbornene) shows *cis–exo* insertion.

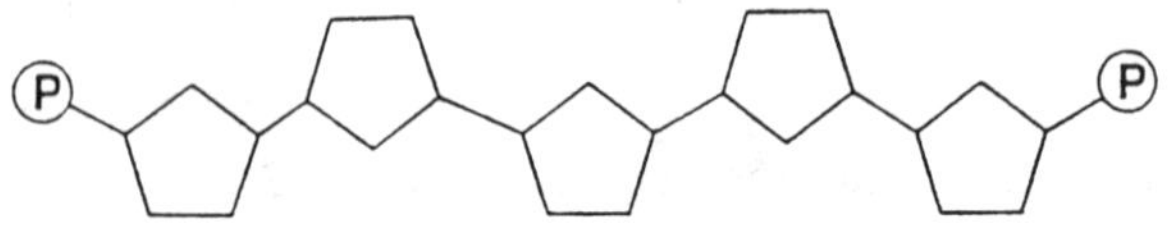

27

Comparisons of the activity, incorporation, molecular weight, and glass transition temperature for different norbornene/ethylene molar ratios for the ethylene/norbornene copolymerization are known. The bulky cycloolefin is incorporated into the growing polymer chain only two to three times more slowly than ethylene. The activity of copolymerization exceeds that of ethylene homopolymerization by a factor of 4–5. By varying the metallocene and the reaction conditions the molecular weight, the molecular distribution, and the microstructure of the COC are tailored.

Polymers with ring structures, interspaced with CH_2 groups, can be obtained by polymerization of 1,5-dienes. 1,2-insertion of the terminal double bond into the zirconium–carbon bond is followed by an intramolecular cyclization forming a ring. Waymouth describes the cyclopolymerization of 1,5-hexadiene to poly(methylene-1,3-cyclopentene) [62, 63] of the four possible microstructures. The optically active *trans*-, isotactic structure (Fig. 39) is predominant (68%) when using a chiral pure enantiomer of $[En(IndH_4)_2Zr](BINAP)_2$ and MAO.

F. Supported Metallocene Catalysts

The development of new techniques for supporting metallocene catalysts on particulate supports has been very rapid in the last decade. After the first patents focussed mainly on the adsorption of alumoxane and the metallocene on silica, we assisted in the development of other techniques such as the creation of a covalent bond between the metallocene and the support, or between the support and the cocatalyst. Another field of activity has been the development of new supports such as polymeric supports, magnesium-based supports and ion-exchange resins.

G. BASF's Proprietory "Novolen M" Metallocene Catalyst

The core for the production of metallocene-based polymers is the Novolen Gas Phase Process [64] which combines metallocene "drop-in" technology (Fig. 40). The core of the process is one or two vertical stirred reactors in series. The process can manufacture

TABLE 9 Polymerization of Cycloolefins using Metallocene Catalysts

Monomer	Metallocene structure	Temperature (°C)	Activity [kg PP/(mol Zr h)]	Melting point
Cyclobutene	11	−10	50	485
	11	0	149	485
Cyclopentene	11	0	32	395
	11	22	195	395
Norbornene	11	20	40	500
	18	23	458	500
Tetracyclododecene	9	25	35	500

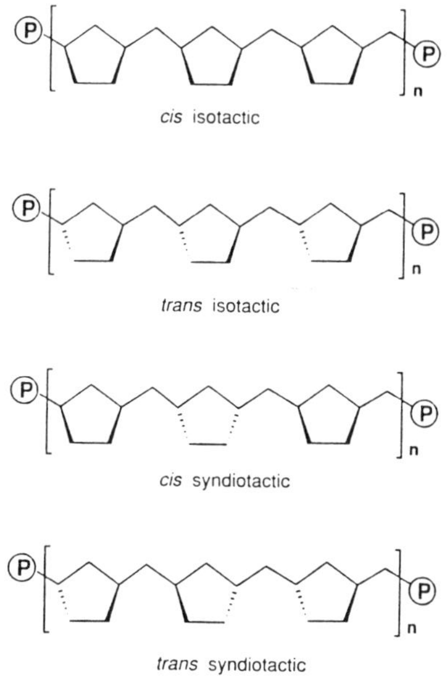

cis isotactic

trans isotactic

cis syndiotactic

trans syndiotactic

FIG. 39 Microstructures of poly(methylenecyclopentane). (From Ref. 61.)

Ziegler–Natta or metallocene-produced homopolymers, random copolymers and terpolymers. BASF have optimized the supporting concept and the support morphology of the Novolen M catalyst (Fig. 41). Cocatalyst MAO is chemically bound to silanol groups between the MAO-coated surface and the metallocene to prevent catalyst leaching. Novolen M is characterized by a homogeneous distribution of active centers throughout the silica support and fully preserves its single-site catalyst.

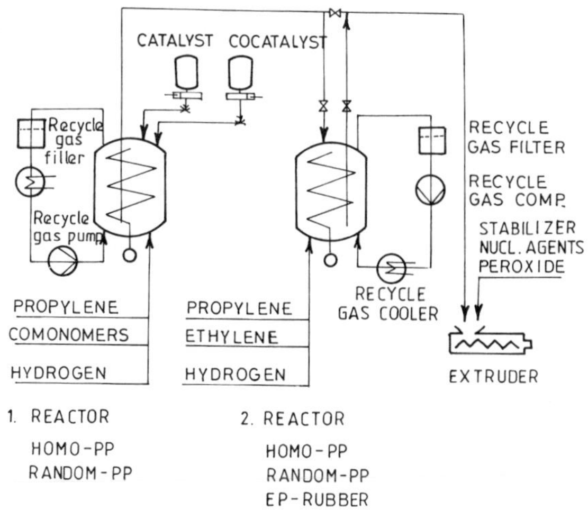

FIG. 40 The Novolen gas-phase process.

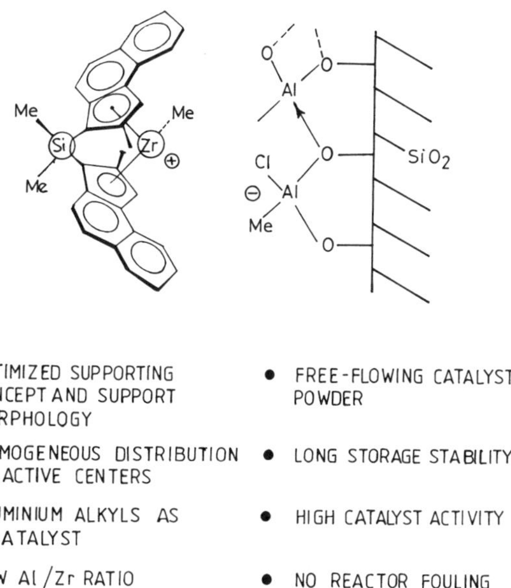

FIG. 41 BASF's MAO-supported metallocene–PP catalyst.

The melt flow rate (MFR) of the Novolen M polymers can be adjusted over a wide range by using a controlled feed of hydrogen into the reactor. The pronounced sensitivity to hydrogen enables the production of high melt flow rate polymers directly in the reactor. This avoids the disadvantages such as oligomer production, yellowing and odous, commonly found in peroxide-clipped extruder grades. Monomer mis-insertions (Fig. 42) cause melting point depressions in the resulting polymer.

H. Metallocene-Based Homopolypropylene [65]

Metallocene homopolypropylenes with a slightly decreased melting point are especially suitable for processes involving the simultaneous crystallization and stretching of the polymer. The first example of such a process is the production of very thin BOPP films (Table 10).

An extruded polypropylene film is stretched first in the machine direction, followed by a second stretching in the transverse direction. The resulting material shows highly orientated crystallites and no, or very little, spherulitic superstructure, which results in the film having both a higher Young's modulus and transparency.

Replacing conventional homopolypropylene with our metallocene homopolypropylene allows the pro-

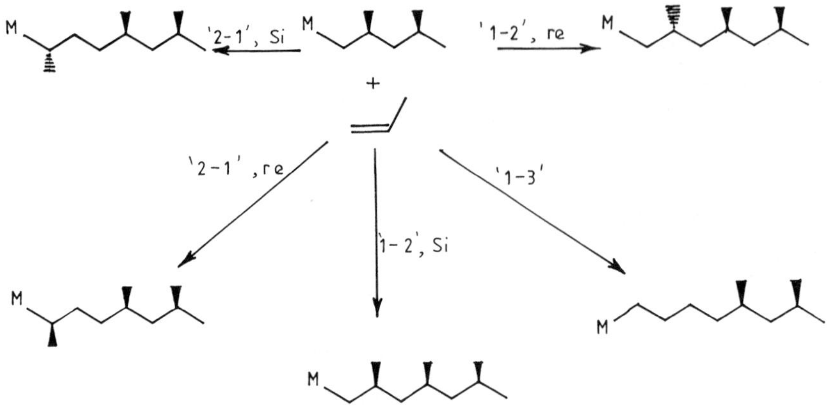

FIG. 42 Monomer mis–insertions using stereorigid, C_2-symmetric, *ansa*-metallocenes.

duction of thinner BOPP films, with enhanced tensile strength.

The slightly reduced elongation at break shown by the metallocene product indicates a higher degree of orientation for the crystalline portion. Furthermore, the polymer can be processed using lower temperatures, giving significant cost savings and the resulting films exhibit superior transparency and gloss.

I. Homopolymers for High Speed Spinning and Fine Fiber Applications

Another process involving simultaneous crystallization and stretching is the melt spinning process for fiber production. Here, the crystallizing polymer is stretched in only the machine direction.

Using metallocene homopolypropylene, spinning rates can be enhanced significantly, reaching up to 4000 m/min, a range previously achieved only by polyamides. The resulting fibers are much stronger than conventional CR grade PP fibers and show reduced elongation at break (Table 11).

The narrow molecular weight distribution of metallocene homopolypropylene directly translates into narrow stress relaxation time spectra of the molten polymer. In a spinning process, where the polymer melt is stretched and subsequently crystallized, this means a more rapid decay of the stress in the fibers, and as a consequence, allows higher spinning speeds and finer fibers. A fully viscoelastic simulation of the spinning process indicates that metallocene polymers exhibit a more pronounced spinline profile, allowing the production of thinner filaments than can be achieved using conventional polypropylene (Fig. 43).

J. High Clarity Random Copolymer [64]

Figure 44 shows the transparency of metallocene and ZN polypropylenes. Novolen 3248TC reaches a similar level of transparency to the metallocene homopolypro-

TABLE 10 Properties of BOPP Films of Homopolypropylene from a Metallocene Catalyst and from a Conventional Ziegler–Natta Catalyst (From Ref. 65)

		Metallocene Homo-PP (MFR: 8)	Conventional PP (MFR: 2.5)
Thickness (μm)		12	18
Processing temp. (°C)		140–150	165–170
Tensile strength (N/mm^2):	MD	134	120
	TD	362	260
Elongation at break (%):	MD	210	200–240
	TD	37	40–50
E modulus (N/mm^2):	MD	1700	2000
	TD	3110	2800
Haze (%)		0.1	2.5
Gloss		116	85–90

TABLE 11 Properties of Homopolypropylene PP-Fibers from a Metallocene Catalyst and a Conventional Ziegler–Natta Catalyst (From Ref. 65)

		Conventional Homo-PP (CR grade)		Metallocene Homo-PP (MFR: 18)	
Spinning rate (m/min)	Fiber diameter (μm)	Tensile strength (cN/decitex)	Elongation at break (%)	Tensile strength (cN/decitex)	Elongation at break (%)
2500	26	2.23	265	3.63	154
3000	26	2.20	255	3.69	142
3500	26	Cannot be achieved!		3.65	145
4000*	26			3.64	137

*Typical spinning rate of polyamides.

pylene but is inferior in stiffness. The metallocene random copolymer Novolen M NX70084 even reaches a transparency of up to 96%, the highest value seen so far for any PP (Fig. 44). It has a melting point of 135°C, xylene solubles of less than 0.6%, and mechanical properties comparable to Novolen 4348 TC.

K. Lower Xylene Extractables in Metallocenes vs. Ziegler–Natta Catalysts

The key feature of metallocene-based random copolymers is their homogeneity resulting in their low extractibles content compared with conventional polymers. For polymers with a melting point of 120°C, the metallocene copolymer beats the conventional random copolymer [64] with respect to xylene solubles by more than one order of magnitude (Fig. 45).

L. Statistics of Patents Filed on Metallocene Polymers [66]

The first consideration is related to the amount of European patent applications filed from 1984 to date by the most active companies in the metallocene field.

Figure 46 shows that Hoechst, Exxon and Mitsui have been by far the most active companies. However the amount of applications filed does not say anything about the relevance of the patents. Figure 47 shows the citations and number of metallocene patents.

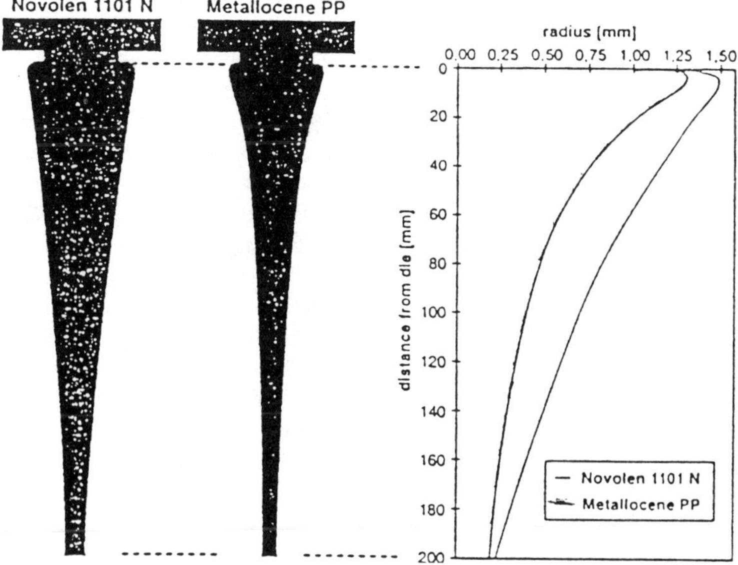

FIG. 43 Spinline profile simulations of metallocene and Ziegler–Natta catalyzed polypropylene. (From Ref. 64.)

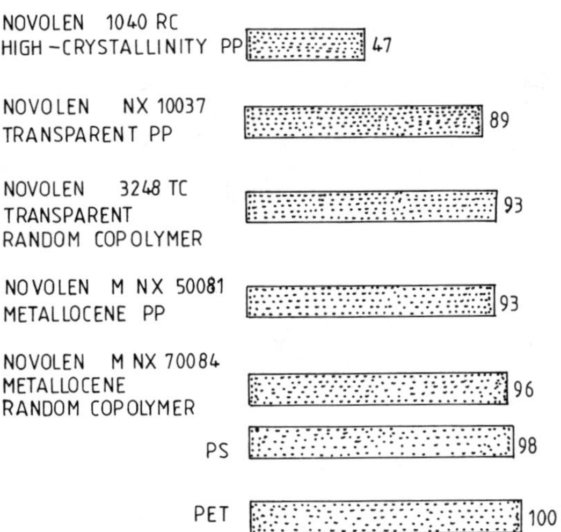

FIG. 44 Transparency of metallocene and Ziegler–Natta polypropylenes (1mm thick samples). (From Ref. 64.)

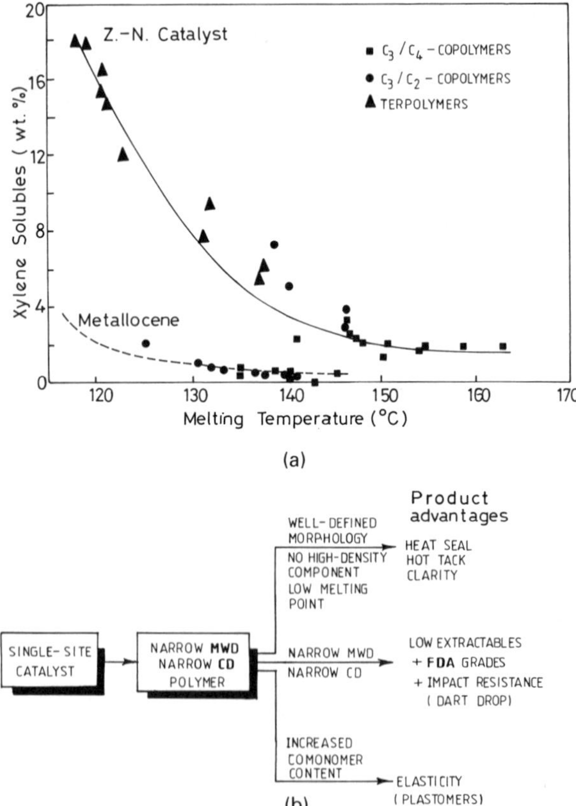

FIG. 45 (a) Superior organoleptic properties of metallocene polypropylenes. (From Ref. 64.) (b) Product advantages of single-site vs. Ziegler–Natta catalyzed. (Reprinted by permission from Hydrocarbon Processing, March 1994, copyright 1994 by Gulf Publishing Co., all rights reserved.)

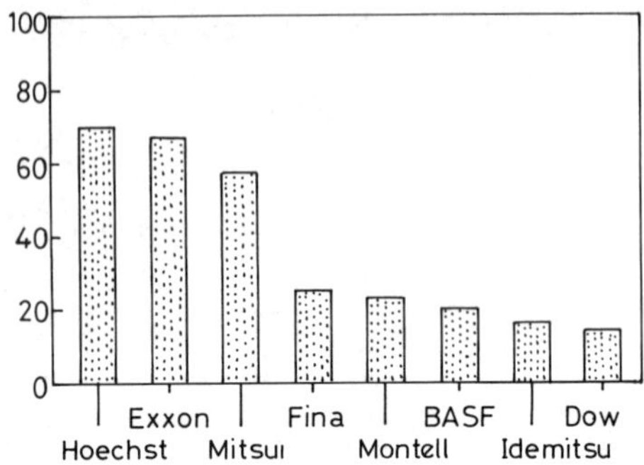

FIG. 46 European patent applications.

From these data it appears evident that conventional Ziegler–Natta catalysts can be considered as a mature technology and less research is now done in this field. However they still represent a very important percentage of the total amount of filed patent applications, confirming the opinion that the decline of polyolefins produced by conventional ZN catalysts will be slow and probably not complete.

The second conclusion is that the amount of research in the metallocene field (and thus the amount of money invested by industry) should be comparable with the amount of research done previously on the conventional Ziegler–Natta, since the amount of patent applications filed is comparable.

M. In Retrospect

The history of the breakthrough in metallocene catalysts from 1955 till now is given (Table 12). Natta pioneered the use of metallocene for olefin polymerization, but because their activity was low they were put aside. Twenty years later, much interest and attention was focussed on metallocenes with the discovery of a new activator, alumoxane. Since then innovations in this catalyst [67] have spurred the growth of business activity in this area.

Structural features depicting the evolution of metallocene catalysts from the 1950s to 1990s are illustrated [68] in Table 13. Bis-Cp structure when activated with methyl alumoxane could be commercialized in a limited way. Activity and MW capability of substituted bis-Cp metallocenes were substantially higher than that of the unsubstituted structure. The mono-Cp structure was further developed [68] which produced high MW ethylene copolymers with higher comonomer content. Bridged, substituted bis-Cp structures could polymerize propylene to an isotactic or syndio-

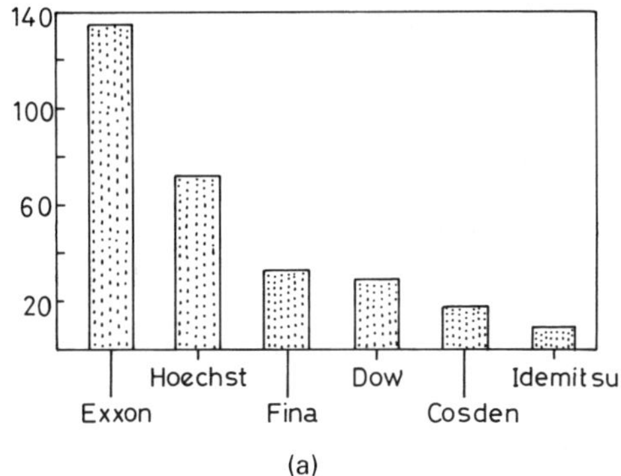

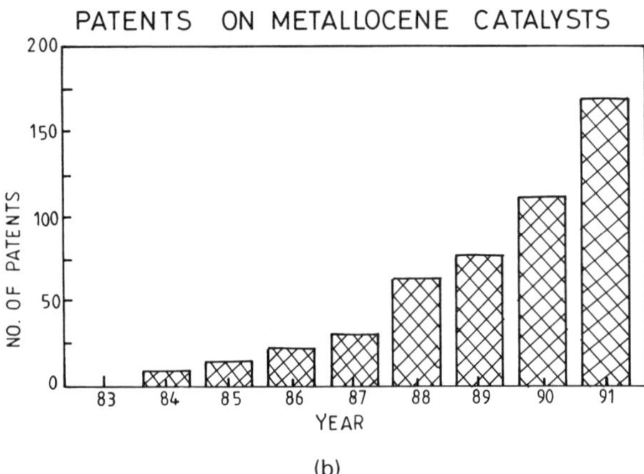

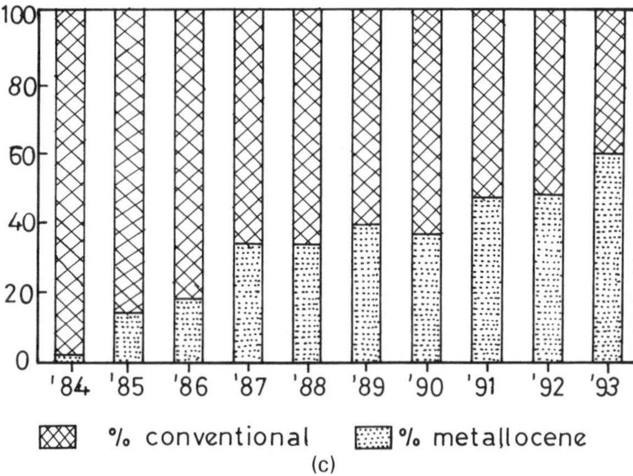

FIG. 47 (a) Amount of citations. (b) Patent production on metallocene catalysts for polyolefins. (c) Distribution of filings.

tactic polymer [68], enabling a high degree of stereoregulation, but low melting points. Spaleck *et al* [14] at Hoechst led to a breakthrough in the development of commercial i-PP technology by achieving "improved" stereoregulation.

The journey in time of the progress in activator technology is parallel [68] to that of metallocenes (Table 14). ZN catalysts rely on aluminum alkyls for activation which are ineffective in activating metallocenes. The discovery by Sinn, Kaminsky and coworkers [10, 42] that MAO was a more effective activator was pivotal in the development of metallocene catalysis. Other molecules, coordinating anions, were as effective as MAO in activating metallocenes.

N. Prospects

Key processes, in use today, can be retrofitted with metallocenes to produce designed polymers. The catalyst can be used in solution or slurry processes. It can also be used in gas-phase processes. High catalyst productivity minimizes catalyst removal needs and attendant waste streams. Metallocene catalysts offer process optimization opportunities that benefit both the resin supplier and the convertor. In Table 14a, the demand for metallocene-catalyzed polyolefins, polystyrene, and copolymers has been projected up to the year 2010.

With reference to Fig. 48, it was forecast that the application of single-site metallocene technology will define the first in a series of new S-curves [68] for the polyolefin industry. The second in the series will be driven by the emerging nonmetallocene catalysts. In this S-curve, product design capability will expand to include polar monomer incorporation and control of intermolecular distributions [68]. After this second technology wave the direction of development will shift to control of intramolecular architecture.

O. To Recapitulate

Ziegler–Natta catalysts for olefin polymerization were first developed in 1950, modified as high mileage catalysts, further improved for morphology control and are still in use [73] after a span of 40 years (Table 15).

Metallocenes were developed by Kaminsky and Sinn [42] by developing a new activator. Since then, metallocene technology has progressively undergone changes and has removed the carpet from under ZN catalysts. It is a mature technology now and more than 4 billion dollars has been spent in their development (Table 15). Syndiotactic PP, cyclo-olefin copolymers, and polar monomer incorporation are possible with metallocenes.

Figure 49 illustrates that metallocene catalysts are single-site catalysts with narrow MWD and constant composition. Table 16 elaborates this further.

TABLE 12 Key Milestones in Metallocene-Based Single-Site Catalysts

Early academic milestones	
1955–56	First metallocene polymerization catalysts (Natta, Breslow)
1975–76	First metallocene/alumoxane catalysts (Breslow, Sinn and Kaminsky)
Milestones in metallocene research	
1983	First practical catalysts based on substituted metallocene/alumoxane
1983	First controlled tacticity metallocenes
1983	Use of mixed metallocenes to control broadness of MWD
1986	Supported metallocene catalysts
1987	Composition of stable and active adducts of metallocene and alumoxane
1987	Discrete metallocene catalysis based on non-coordinating anions
1989	Mono-cyclopentadenyl metallocene

Source: Ref. 67.

TABLE 13 The Evolution of Metallocene Structure

Date	Metallocene	Stereoregulation	Performance
1950s	M, X, X	None	Moderate molecular weight PE with reasonable comonomer incorporation
Early 1980s	M, X, X	None	Higher molecular weight PE with better comonomer incorporation and activity
Early 1980s	Ti, Cl, Cl, Cl	Syndiotactic	Practical catalyst for commercial syndiotactic polystyrene
Late 1980s	R, R, N, R′, M, X, X	Slight	Very high molecular weight PE with excellent comonomer incorporation and activity
Late 1980s	R, R, M, X, X	Highly syndiotactic	Used commercially for syndiotactic PP
Early 1990s	R, R, M, X, X, R, R	Highly isotactic	Used commercially for isotactic PP

PE, Polyethylene; PP, polypropylene.
Source: Ref. 68. Reprinted with permission from Ref. 68. Copyright 1997 American Chemical Society.

TABLE 14 The Evolution of Activators for Transition Metal Catalysts

Date	Activator	Performance
Mid-1950s	Metal alkyls such as $(CH_3CH_2)_2AlCl$	Critical to the success of Ziegler–Natta catalysts
Mid-1970s	MAO	Higher activity and molecular weight from metallocenes; one of the keys to commercial viability of metallocene catalysts
1987	Noncoordinating anions	Performance at least as good as MAO; well-defined, discrete activation center

MAO, Methyl alumoxane.
Source: Ref. 68. Reprinted with permission from Ref. 68. Copyright 1997 American Chemical Society.

TABLE 14a Projected Demand for Metallocene-Catalyzed Polymers

	Demand, 10^3 t/yr		
Polymer	2000	2005	2010
Isotactic and syndiotactic polypropylene	1500	7000	20,000
EPDM	150	250	500
Ethylene–styrene copolymer	100	200	400
Syndiotactic polystyrene	80	150	300
Cyclic olefin copolymers	30	60	100

Source: Ref. 68.

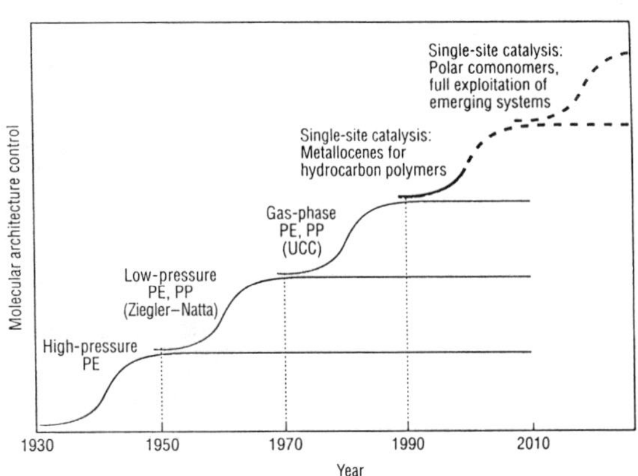

FIG. 48 Technology S-curves in the production of polyolefins. PE, polyethlene; PP, polypropylene; UCC, union carbide catalyst. (From Ref. 68.) Reprinted with permission from Ref. 68. Copyright 1997 American Chemical Society.

TABLE 15 Technology Advances. The Importance of Metallocene Technology: Measurable in Dollars and Time, yet Metallocenes may be Overshadowed by S-S Organometallics

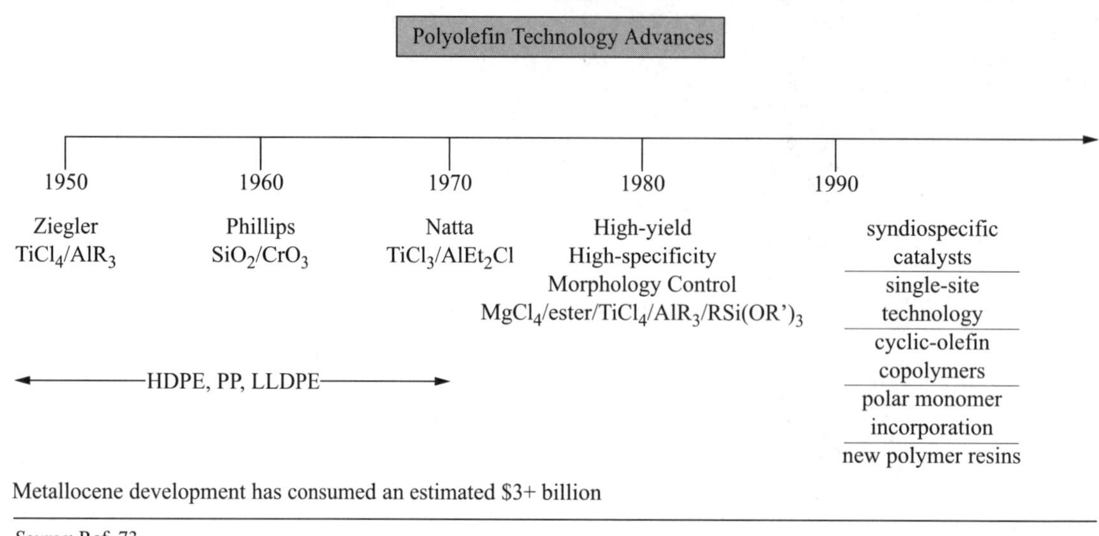

Source: Ref. 73

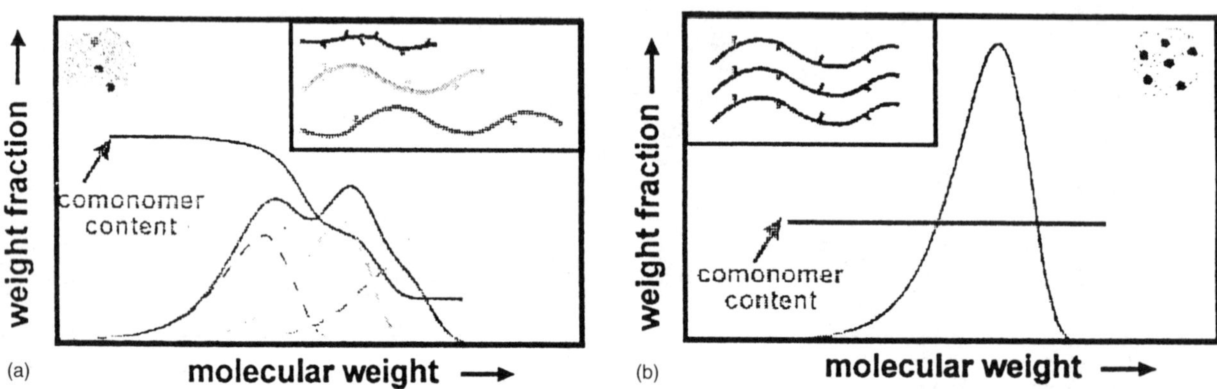

FIG. 49 Comparison of the polymers produced by (a) multi-site and (b) single-site catalysts. (From Ref. 64.)

Insite catalysts (Dow) and Exxpol catalysts (Exxon) [73] are successful commercial metallocene catalysts (Fig. 50).

Metallocenes are available [72] from BASF, BP, Dow, Exxon, Hoechst, Mitsubishi and Mobil (Table 17).

Number of monomers polymerized with metallocenes [73] is large (Table 18).

Metallocenes produce better products after processing and by 2000 their price is expected to come down at par (Table 19).

By 2005 approximately 2 billion 1b of metallocene plastics (70–80% film) are projected to be manufactured (Fig. 51).

Various products of metallocenes are listed in Table 20.

Metallocene catalysts are remarkably versatile for stereochemical control; they can be used to polymerize [71] propylene to atactic, isotactic, syndiotactic, isotactic stereoblock and hemi-isotactic polypropylene (Fig. 52).

TABLE 16 Narrow MW Distribution is a Prime Metallocene Product Attribute (Product Grades Tailored to Meet Application Needs)

Absence of low MW	*Absence of high MW*
– Low extractables	– Excellent clarity
– Non-sticky pellets	– Controllable peak melting point
Excellent comonomer incorporation	*High effective catalyst system*
– Low density, flexible without plasticizer	– Clean, low total ash
– Improved toughness properties	– Very stable
– Excellent filler acceptance	

Source: Ref. 73.

Finally, the adaptation of existing polyolefin manufacturing process to metallocene catalysts requires only minor changes, which makes use of these catalysts economically viable.

Key features of metallocene/MAO catalysts have been demonstrated in Table 20.

Mitsui has been operating a 300 t/yr capacity plant in which traditional catalyst technology is complemented by metallocene technology [74].

Table 21 compares the effect of SSC and ZN catalysts on polyethylene manufactured by different process routes. Applications of metallocene PE are given in Table 22.

P. Beyond Metallocenes

Several new catalysts have emerged on the horizon which are different from the metallocenes group. Figure 53 shows the structure and characteristics of DSM Lovacat and DuPont organometallic catalysts. The former produces higher MW EPDM and HDPE/LLDPE by higher temperature solution polymerization. The DuPont catalyst has been developed by collaboration with Brookhart (University of North Carolina) and can be used for copolymerization of nonpolar and polar monomers. Other nonmetallocene single-site catalysts are shown in Fig. 54. The Brookhart catalyst can produce branched ethylene polymers from an ethylene-only feed [68] and can incorporate polar monomers. The system is more tolerant of poisons than are ZN and metallocene catalysts.

The McConville catalyst (Fig. 54) is a zirconium complex and has the ability to polymerize 1-hexene and block copolymers [68]. The Goodall catalyst

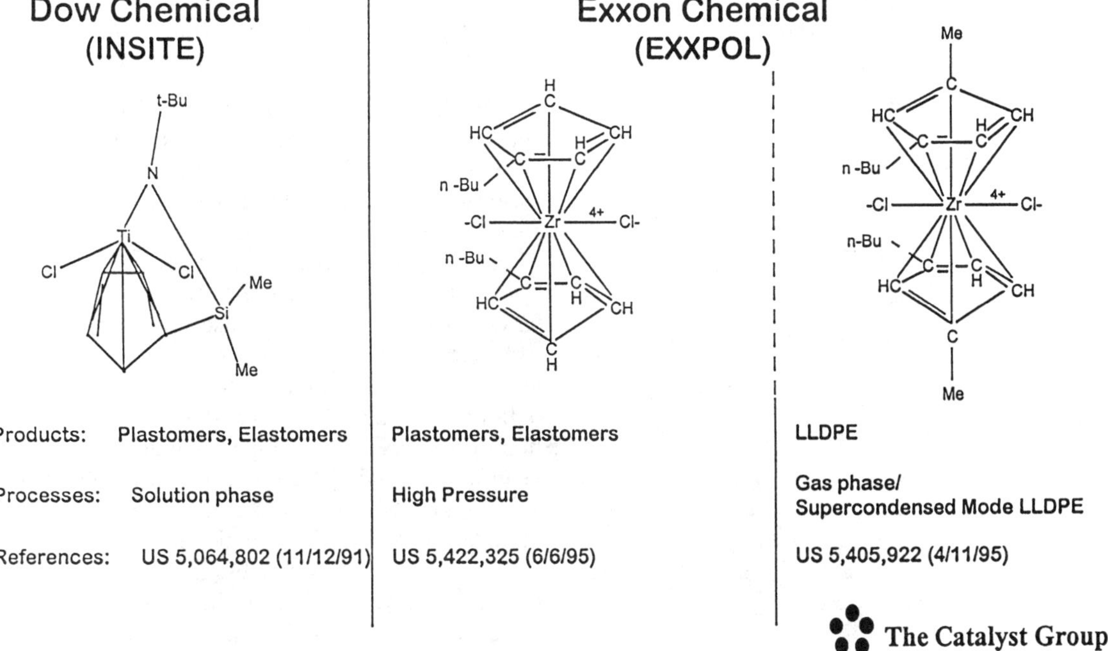

FIG. 50 Technology advances. Only a few metallocene catalyst/process systems have proven worthy at the commercial level. (From Ref. 73.)

TABLE 17 Typical Metallocenes Available

BASF
BP
Dow — affinity series
Exxon — Exact, Exceed
Hoechst
Mitsubishi
Mobil

Source: Ref. 72.

TABLE 19 Why Must Machine Manufacturers React?

Metallocenes offer processors
greater film strength
better dart drop
clarity, low haze, high gloss
better mechanical properties
better heat seal
Metallocene prices expected to reach parity by 2000

Source: Ref. 72.

TABLE 18 Some of the Monomers Polymerized with Metallocene Catalysts

Olefins	Cyclic olefins	Aromatics and others
Ethylene	Norbornene	Styrenes
Alpha Olefins	5-Vinyl-2-norbornene	Acenaphthalene
Propylene	Cyclopentene	Acrylates
1-Butene	Dicyclopentadiene	Acrylonitrile
1-Pentene		Acrylic acid
1-Hexene		Vinyl silane
Dienes, 1,3-butadiene		Indene
		Vinyl pyrene
		4-trimethyl siloxy-1,6-heptadiene
		5-N,N′-diisopropyl amino-1-pentene

Source: Ref. 73.

group consists of cationic nickel and palladium catalysts which can polymerize norbornenes at very high rates [68]. Thus, these "future catalysts" can be programmed to produce complex, multiphasic products in one simple, single-reactor process.

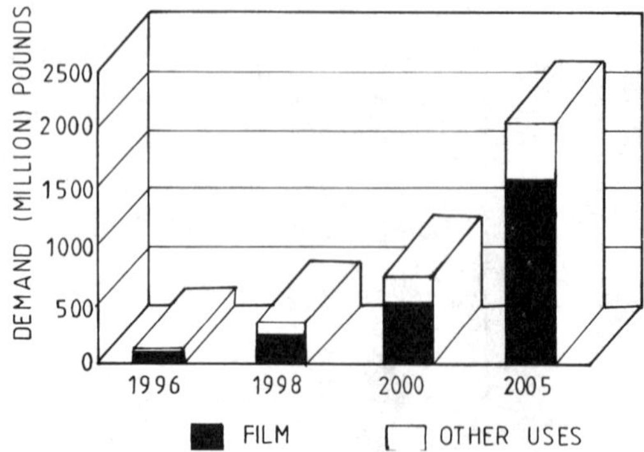

FIG. 51 Estimated growth of metallocenes. (From Ref. 72.)

Q. Control of Tacticity Distribution in Polyolefins Made from Dual Metallocene Catalysts

One of the defining characteristics of resins made with single-site catalysts is the narrow molecular weight distribution and uniform composition. In the case of PP, this leads to correspondingly narrow melt temperature ranges and, at times, dramatic processing limitations. Thus, early commercial uses of single-site PP were in applications where its narrow melt temperature range was not a problem, such as case film, or in compounds.

What Exxon Chemical Co. has developed is a new level of molecular control that can solve all of these limitations. Exxon has developed unique isotactic PP homopolymers and copolymers made in one reactor using dual metallocene catalysts on a single support. The copolymer is bimodal and probably the first true bimodal resin made in a single reactor.

Bimodal composition distribution, which applies only to copolymers, can produce resins having different concentrations of copolymer, i.e. different melting points in polymer chains that are the same length.

But the newest feature is Exxon's claim to control tacticity distribution. This means the ability to arrange

TABLE 20 Features of Metallocene/MAO Catalysts

Process, products	Remarks
Ethylene homopolymerization	High activity, highly linear, Mw/Mn = 2.
Ethylene copolymerization	Random comonomer distribution, LLDPE comonomers: propene, higher α-olefins, cycloolefins (COC), dienes.
EPDM elastomers	Low transition metal concentration in the polymer, Mw/Mn = 2.
Propene (α-olefin) polymerization to polymers of various microstructures.	
Syndiotactic polystyrene	
Cycloolefin polymerization to polymers with high melting points.	
Polymerization in the presence of fillers.	
Oligomerization of propene to optically active hydrocarbons.	
Cyclopolymerization of α,ω-dienes to optically active polymers.	

the propylene monomer in any of three variations — isotactic, atactic and syndiotactic — on one polymer chain.

Exxon's work shows it incorporating and using the aPP fractions in the polymer chains. Incidentally, this is similar to the result achieved with the addition of ethylene. The ability to insert and control blocks of different tacticity could theoretically lead eventually to TPOs made in one reactor with one monomer that would have property advantages that now require two monomers (ethylene and propylene) and 3–4 reactors! That is the potential down the road.

Exxon's paper describes a dual-metallocene catalyst on one support referred to as catalysts 1 and 3 making two different narrow MWD polymers. Catalyst 1 makes long isotactic polymer chains with a 150°C melting point; catalyst 3 makes shorter isotactic polymer chains with a 135°C melting point. The combination,

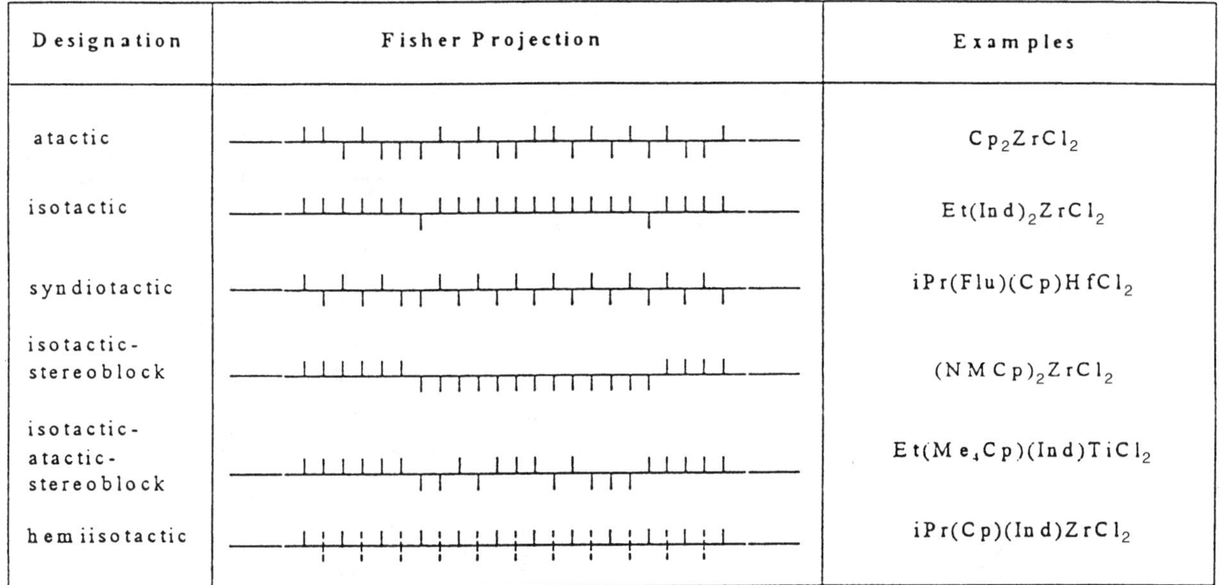

FIG. 52 Types of metallocene-made polypropylenes (Cp: cyclopentadienyl, Et: ethylidene, Ind: indenyl, iPr: isopropyl, Flu: fluorenyl, NM: neomenthyl, Me: menthyl). (From Ref. 71.)

TABLE 21 Single Site Expands the Property Range of PE

	Density range (g/cm^3)	MFI range
High pressure reactor		
Free radical polymerization	0.915–0.935	0.1–100
Ziegler–Natta	0.870–0.910	0.1–100
Single-site	0.865–0.950	0.1–1000
Solution reactor		
Ziegler–Natta	0.900–0.970	0.3–100
Single-site	0.855–0.970	0.3–12
Slurry loop reactor		
Phillips chromium catalyst	0.930–0.960	0.007–80
Single-site	0.880–0.970	0.001–100
Gas phase reactor		
Ziegler–Natta	0.885–0.970	0.001–1000
Single-site	0.855–0.970	0.002–1000

TABLE 22 Latest Field of Application of Metallocene PE by Replacing Conventional Polymer

Resin	Density (g/cm^3)	Application	Replacement
m-PE	0.900–0.915	Coextrusion heat seal layers	EVA (90%VA)
			ULDPE/VLDPE
m-PE elastomer	0.854–0.900	Stretch cling film	EMA
m-LLDPE	0.916–0.925	Stretch film	C4LLDPE, LDPE
		Industrial sacks, FFS application	LDPE–LLDPE blends
		Shopping bags	EVA.ULDPE
			Ionomer layers

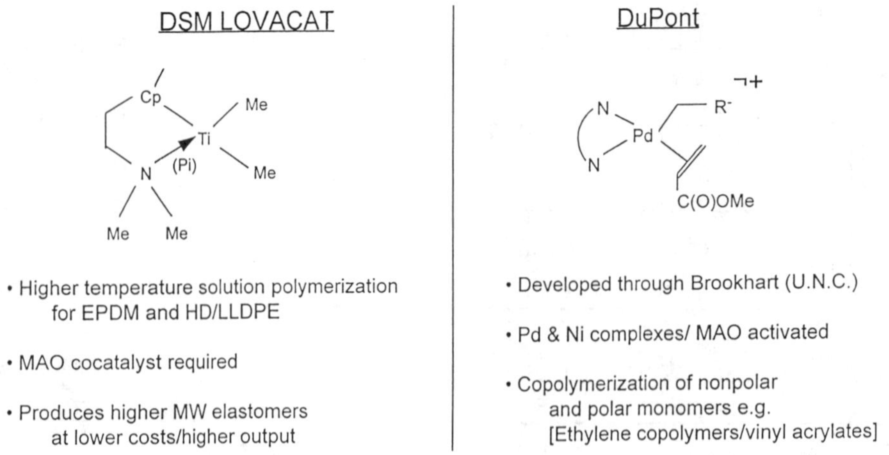

DSM LOVACAT

Cp, Ti, Me, Me, N, (Pi), Me, Me

- Higher temperature solution polymerization for EPDM and HD/LLDPE
- MAO cocatalyst required
- Produces higher MW elastomers at lower costs/higher output

DuPont

N, Pd, N, R⁻, +, C(O)OMe

- Developed through Brookhart (U.N.C.)
- Pd & Ni complexes/ MAO activated
- Copolymerization of nonpolar and polar monomers e.g. [Ethylene copolymers/vinyl acrylates]

FIG. 53 Beyond metallocenes. Beyond classic metallocenes, new organometallic catalysts are being developed. (From Ref. 73.)

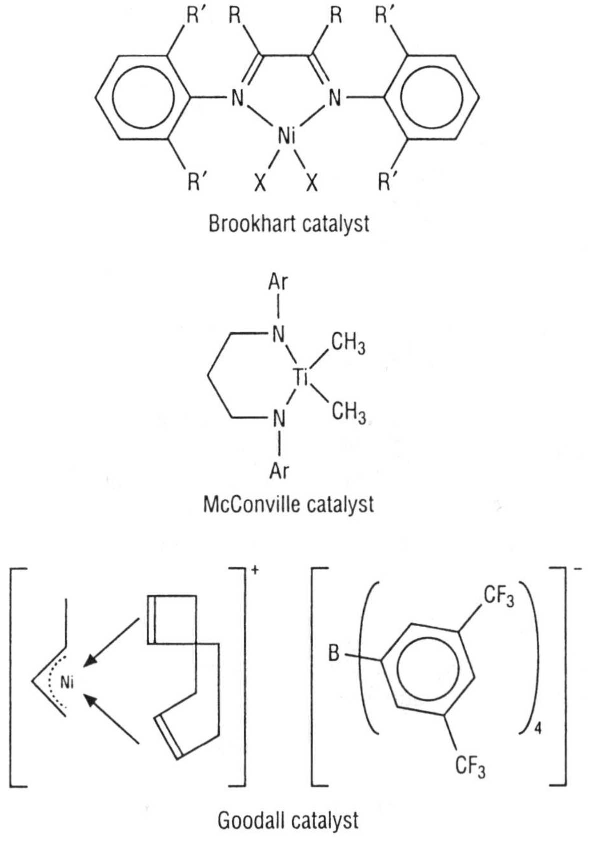

FIG. 54 Examples of nonmetallocene single-site catalysts. (From Ref. 68.) (Reprinted with permission from Ref. 68. Copyright 1997 American Chemical Society.)

however, is not a predictable linear blend. Catalyst 3 has an unexpected secondary effect on the long chain population made by catalyst 1, reducing them, broadening the MWD, and creating a higher concentration of short chains. This secondary effect also broadens the melt temperature range. The shorter polymer chains from catalyst 3 also have lower isotacticity or crystallinity, Exxon said.

In this early work, Exxon said the copolymers are more remarkable than the homopolymers. When ethylene is added, the initial chain-breaking reaction reverses and longer chains form again. The copolymers combine all three "tailored molecular distributions" — molecular weight, composition, and tacticity — whereas the broad MWD homopolymers display control of only molecular weight and tacticity distribution. These dual-metallocene PPs are such a recent development that Exxon appears not to have supplied them yet to customers.

The new bimodal copolymers can make good BOPP films at as much as 15°C below the processing temperatures of current ZN PPs.

V. REVOLUTIONARY NEW HIGH PERFORMANCE HDPE RESINS DEVELOPED

Asahi Chemical Industry Co., Ltd. and the Dow Chemical Company have succeeded in producing commercial HDPE resin products never before realized with conventional catalyst technology. The combination of Dow's INSITE* constrained geometry catalyst with Asahi's slurry process has made possible the production of HDPE resins with significantly improved mechanical strength and environmental stress crack resistance (ESCR), compared with HDPE produced from Ziegler–Natta (ZN) catalysts and conventional metallocene catalysts.

Researchers found that HDPE made with the Dow INSITE* catalysts can contain a significantly larger amount of comonomer than metallocene-catalyzed HDPE at the same density. The Dow catalysts also produce slightly broader molecular weight distribution (MWD) than can be made with metallocene catalysts. Still another plus is the surprising capability of the technology to produce HDPEs with a reversed comonomer distribution (that is, the higher the molecular weight, the higher the comonomer content). The high comonomer content in the high molecular weight portion helps to maximize tie molecules responsible for long life properties (e.g. ESCR). The slightly broader MWD resins enable their use in conventional extruders without losing the good impact properties of resins with narrower MWD.

A HDPE resin in the 0.9555 density and fractional melt flow range with bimodal MWD was prepared with a 50/50 blend of high and low molecular weight resins from the slurry process using the INSITE* technology. When compared with a bimodal blend of the same density and melt flow rate made with a conventional ZN catalyst, ESCR of the new resin was found to be several times higher and Charpy impact strength at −20°C improved by 60%.

VI. NEW METALLOCENE PLASTOMERS

Exxon Chemical has added five grades of exact plastomer to its FDA-compliant metallocene specialty portfolio. The addition is possible because of the US Food and Drug Administration's food additive amendment, specifically requested by Exxon, to increase the allowed hexene comonomer content from 10 wt% to 20 wt% in film applications intended for direct food contact. This FDA approval of increased hexene content broadens the food-compliance window of higher alphaolefin (HAO) metallocene resins, and is part of the company's program to develop other, even lower density and lower melting point packaging film resins.

Increased hexene content provides lower densities that reportedly yield unparalleled performance from metallocene plastomers. Food-compliant Exact plastomers are now offered with densities less than 0.895, and melting points 5–10°C lower than metallocene-based 20 wt% octene copolymers.

Physical properties, specifically toughness, breathability, and optical, combined with characteristics of enhanced sealability and broad FDA approval, are said to make Exact an ideal choice for film packaging. The five new grades of Exact are 3030, 3131, 3132, 4150 and 4151.

VII. FUNCTIONALIZED POLYOLEFINS

Metallocene-based catalysts have broadened the scope of synthesizing polyolefins with different functional groups. The polar monomers could now be polymerized with metallocene catalysts of cationic nature. Catalysts of the type $[Cp_2^* ZrMe^+ \ldots X^-]$ $(X^- = B(C_6F_5)_4^-$ or $H_3CB(C_6F_5)_3^-)$ are indeed capable of polymerizing a variety of functionalized α-olefins and dienes such as 4-trimethyl siloxy-1,6-heptadiene, 4-tert-butyldimethyl siloxy-1-pentene, and 5-N,N^1-diisopropylamino-1-pentene.

OSiMe3 Cat. n OSiMe3 OSiMe3 OSiMe3

Cat. n HCl R = N(iPr)2 n R R R R NH+ Cl−

SCHEME 1 Formation of polymers with O- or N-functional groups by use of borane-activated zircononcene catalysts reported by Waymouth and co-workers (MR Kesti, GW Coates, RM Waymouth, J. Am. Chem. Soc. 114: 9679, 1992). $R = NiPr_2$.

In contrast to heterogeneous polymerization catalysts, which afford shorter propene oligomers in the presence of high amount of hydrogen, only with saturated chain ends, metallocene catalysts give easy access to propene oligomers with olefinic end groups, which can be converted to various other functional groups [75, 76]. Oligopropenes with thiol end groups have been used as chain transfer reagents in methylacrylate polymerization to form poly(propene–block–methylmethacrylate). A new class of polymers containing pendant polypropene chains such as polymethylmethacrylate–graft–polypropene is derived from methacrylate terminated oligopropene macromonomers, which are copolymerized with various acrylic esters, acrylonitrile, or styrene [77, 78].

In order to achieve higher-performing engineering thermoplastic properties, significant modification of the polymer backbone is required. The use of carbon monoxide as a comonomer has been of interest based on its abundance and its ability to confer functionality. However, conventional "early" transition metal polyolefin catalysts are ineffective at copolymerizing olefins with carbon monoxide.

$$-\!\!\left(CH_2CH_3 = \overset{O}{\overset{\|}{C}} \diagup CH_2CH - \overset{O}{\overset{\|}{C}} \right)_{\!n}\!-$$

R=H, CH_3, alkyl, aryl, etc.

The first commercial polyketone polymer is a crystalline terpolymer based on carbon monoxide, ethylene and a minor amount of propylene ($T_m = 220$°C). The novel polymer architecture of polyketone polymers containing 50 mol% carbon monoxide deliver an excellent range of performance properties combining strength, stiffness, and toughness with thermal, chemical, abrasion/wear, and permeability resistance [79]. The breadth of properties will permit their use in a wide variety of applications including automotive, industrial, electrical/electronic, and customer markets. CARILOW thermoplastic polymers are melt processable using conventional methods including injection molding, extrusion, blow molding, powder coating, etc.

Colquhoun *et al* have studied insertion of CO into a transition metal–carbon σ-bond in homogeneous catalytic processes [80]. The recent works [81–84] are stressing unsymmetrical bidenate ligand systems compared to the previously studied C_2 symmetrical *cis*-bidentate ligands [85–87].

The alternating copolymerization of olefins with carbon monoxide is of great interest due to the potential use of the resulting polymer as a new material. The process includes two of the representative reactions of a palladium complex: (1) CO insertion into an alkyl–Pd bond, and (2) an olefin insertion into an acryl–Pd bond. Vrieze and his coworkers [86] investigated the CO insertion into a methyl palladium complex possessing a phosphorus–nitrogen ligand, 1-dimethyl amino-8-(diphenylphosphino) naphthalene (PAN). It has been revealed that both the methyl group in $Pd(CH_3)(CF_3SO_3)(PAN)$ and the acetyl group in $Pd(COCH_3)(CF_3SO_3)(PAN)$ are located *trans* to the nitrogen site. Two years later, Van Leewen and his coworkers [81] prepared an unsymmetrical biphosphine ligand which consists of a diarylmonoalkenyl phosphine site and a trialkyl phosphine site. With Pt(II) and Pd(II) complexes of this unsymmetrical bidentate ligand, they investigated the insertion of CO into phenyl platinum and methyl palladium complexes. They reported a possible migration of the phenyl and methyl groups from *trans* to the trialkyl phosphine site to *cis* and the higher activity of the former complex over the later in migratory insertion of CO. Theoretical studies indicated that the alkene insertion into the Pd–aryl bond *trans* to a phosphine is more favorable than that into the Pd–alkyl bond *trans* to a phosphite [88].

Compared to the CO insertion, fewer reports have appeared on the olefin insertion into an acyl palladium. Direct observation of this process has been reported very recently by Rix and Brookhart [89–92] using cationic 1,10-phenanthroline-Pd(II) complexes. They have investigated the microscopic steps responsible for the alternating copolymerization of ethene with CO using the same 1,10-phenanthroline system. On the basis of the kinetic and thermodynamic data, they proposed an accurate model for the polymer chain growth. In support stepwise isolation of the intermediates have been accomplished by norbornene as a substrate where symmetrical bidentate nitrogen ligands were used [93–95].

When α-olefins such as propene and styrene are used in place of ethene or norbornene for this copolymerization, regio- and enantio-selectivities of the olefin insertion arise and the control of these becomes a difficult aspect for obtaining stereo-regular polyketones. In the head-to-tail copolymer, a chirotopic center exists per monomer unit. If the same enantioface of each α-olefin is selected by a catalyst, the resulting copolymer is isotactic in which all the chirotopic carbons in a polymer backbone possess the same absolute configuration. Thus, asymmetric copolymerization using a chiral catalyst is now attracting much attention.

In recent publications polyketones have been synthesized by extending the copolymerization of carbon monoxide to dienes [95–98].

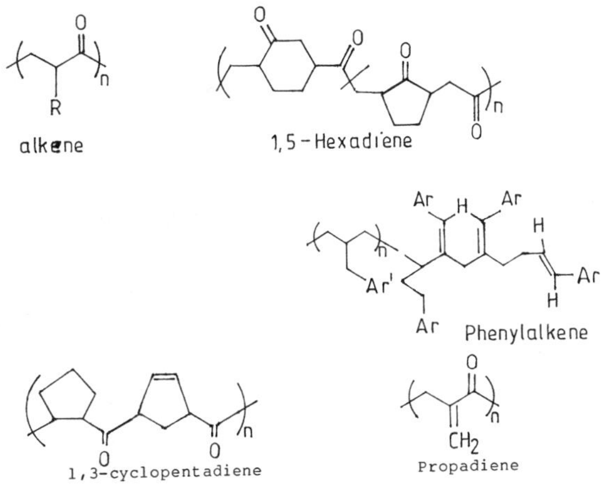

alkene 1,5-Hexadiene Phenylalkene 1,3-cyclopentadiene Propadiene

Control of structure and molecular weights in preparation of the polymers derived from 1,2-dienes and CO is of significant interest since it has a unique enone structure in the repeating unit and can be converted into further functionalized derivatives.

VIII. INDUSTRIAL EXPLOITATION OF METALLOCENE-BASED CATALYSTS (MBC)

These results are just an example of the potentiality of MBC and allow us also to envisage those possible industrial applications of great interest. An enormous research activity began after the first discoveries, both in the academic and in the industrial world. According to the Stanford Research Institute, the global expense for the research activities in the field of metallocenes is so far about 4 billion dollars.

All important companies active in the field of polyolefins are involved in this effort: Exxon, Hoechst, Fina, BASF, Dow, Idemitsu, Mitsui Toatsu, Mitsui Petrochemical and Montell. The importance of MBC, in view of industrial exploitation, can be summarized as follows.

1. All the traditional polyolefins can be prepared, often endowed with improved properties.
2. New polymers were already prepared. The most important are the following: (a) highly syndiotactic poly-1-olefins: poly-olefins: polypropylene, poly-1-butene, poly-1-pentene, poly-4-methyl-1-pentent, polyallyltrimethylsilane; (b) atactic polypropylene

with high molecular mass; (c) ethylene and propylene copolymers with 1-olefins, crystalline and amorphous, with new microstructure and physical–mechanical properties.
3. A more fruitful exploitation and even a simplification of the existing industrial processes can be envisaged.

Drawbacks of MBC are:

1. the use, in large amount, of an expensive cocatalyst as MAO,
2. the "homogeneous" nature of metallocenes that does not allow their use for bulk and gas phase polymerizations.

To solve the first problem, there have been developed in Montell, alternative MAO-free cocatalytic systems, based on branched aluminum alkyls. To have the chance of using the metallocenes in the above mentioned processes and to control the morphology of the obtained polymers, the way followed was the supporting of MBC on a carrier.

A. Support of Metallocene-Based Catalysts

Many scientific papers and patents report attempts to look at the support of the components of the catalyst system on different carriers: SiO_2, Al_2O_3, $MgCl_2$, and polymers were the ones most investigated. In general, the transition metal compound, MAO, and the carrier were reacted in different combinations and the resulting catalyst is employed in propylene polymerization with a common trialkylaluminum as cocatalyst. In a recent paper, Soga [76] has demonstrated that, when the metallocene is immobilized on a chemically modified silica, it can be activated by ordinary triakylaluminum to give isotactic polypropylene with a totally MAO-free catalyst system. In particular, a$=SiCl_2$ functionalized silica is further reacted with a Li salt of a cyclopentadienyl-type ligand to give a metallocene linked through its bridge.

In Montell they have developed a technology for the support of polymers on crosslinked polystyrene and porous polyethylene and on $MgCl_2$.

The catalytic activity and the investigated copolymer properties were found to be unchanged after supporting the metallocene on the above mentioned carriers.

IX. RECENT PROGRESS ON THE ADVANTAGES OF METALLOCENE CATALYSTS

Much of our knowledge in this area has been acquired by application of rational conceptional models to the design of new metallocene structures and catalyst activators. Some traits of these catalysts, in particular their stereoselectivity, are now close to being predictable, based on our understanding of the essential elementary reaction steps in these homogeneous reaction systems. Other traits, such as the formation of cyclic or functionalized polyolefins, have been and continue to be discovered by testing our understanding of the basic reaction mechanisms of these catalysts against increasingly demanding tasks. This approach derives substantial support from the rapidly advancing methods for modeling even large metallocene reaction systems by ab initio and density functional methods. These methods can also provide a calibration for useful molecular mechanics models for these catalysts, which may eventually allow estimates of steric effects on the course of essential reaction steps, for example on competing insertion and chain termination reactions.

The evolution of advanced catalysts and catalytic processes based on rational model hypotheses is now beginning to carry over to heterogeneous Ziegler–Natta catalysis as well. Practical application of metallocene catalysts requires their preadsorption on solid supports such as alumina or silica gels. Instead of the polymer dust produced by a homogeneously dissolved metallocene catalyst, the solid catalyst particles generate coherent polymer grains. As with classical heterogeneous catalysts, these grains appear to be enlarged replicas of the catalyst particles. Metallocene catalysts that are heterogenized, for example on a silica gel support, can thus be readily used in existing Ziegler–Natta production facilities, for instance in solvent-free slurry or gas phase reaction systems.

Detailed guidelines have been developed for the selection of supports with optimal composition, particle size, pore size distribution, and surface OH group density, and for their treatment with various alkyl aluminum and aluminoxane activators prior or subsequent to adsorption of the metallocene complex. The more advanced of these protocols require only limited excess ratios of alkylaluminum activators per zirconocene unit; Al:Zr ratios of roughly 100–400:1, much below those used for homogeneous catalyst systems, appear to give entirely satisfactory activities for these surface-supported metallocene catalysts. In contrast, to modern $MgCl_2$-supported heterogeneous Ziegler–

Natta catalysts, silica-gel-supported metallocene catalysts are practically free of chlorine. *This could be an advantage from an ecological standpoint when the chemical or thermal recycling of polymer products reaches a larger scale.*

Despite the practical advantages of supported catalysts, interactions between support materials and catalyst complexes are only partly understood on a molecular level. Based on the generally close resemblance of the polymer microstructures produced by a metallocene catalyst in homogeneous solutions and on solid supports, even in solvent-free gas phase systems, it appears likely that the active catalysts are quite similar, in other words that the (presumably cationic) metallocene catalyst is only physisorbed on the alkylaluminum-pretreated (possibly anionic) catalyst surface.

Based on this model, "microreactors" might be fabricated by immobilizing different types of single-site metallocene catalysts — or even catalyst cascades — on suitable supports for in situ production of novel polyolefin blends and other environmentally friendly polyolefin materials.

Polymers with properties distinctly different from those produced in homogeneous solution — with unusually high molecular weights — have recently been obtained by fixing a metallocene catalyst on unpretreated supports and by synthesizing covalently linked ansa-metallocenes directly on a SiO_2 support. These observations are probably connected with the site-isolation effects, i.e. with the strict suppression of all binuclear reaction intermediates, known to arise from linking catalyst centers covalently to a solid support. If methods become available for a controlled synthesis of such covalently supported metallocenes and for their structural characterization, one could imagine another round of developments towards novel metallocene-based Ziegler–Natta catalysts that are heterogeneous, like their predecessors discovered forty years ago, yet endowed with wider process and product variability and with *superior environmental properties.*

X. TYPES OF METALLOCENE RESINS

Because metallocene catalysis applies to different polymerization techniques, and because the resulting resins can be performance tailored, their processing characteristics differ by grade manufacturer. Current commercial metallocenes appear to be categorized as follows.

1. The first category targets the commodity resin market, designed to replace traditional LLDPE with a typical density of 0.915 g/cm^3, referred to as mLLDPEs, where m is metallocene catalyzed.
2. The second and third categories are lower in density (0.865–0.915 g/cm^3) and the resins can be classified as plastomers because of their elastic behavior; plastomers are divided at the 20% comonomer level, the principal monomer being ethylene: polyalkene (polyolefin) plastomers (POPs) are 0.895–0.915 g/cm^3 density and contain 20% comonomer; polyalkene (polyolefin) elastomers (POEs) are 0.986–0.895 g/cm^3 density and contain 20% comonomer.
3. The fourth type is metallocene catalyzed polypropylene and primarily aimed at the biaxially oriented polypropylene market.

Applications of the new generation resins and their pylace in the market are given in Table 23.

Figure 55 shows the effects of chain structure on the processing characteristics of ethylene polymers and copolymers and Table 24 attempts to simplify the correlation between resin properties and processing.

Table 25 shows typically worst-case torque requirements for four extruder sizes. The range of screw speed represents plastomers at the low end and LDPE at the high end. The specific torque value can be simply multiplied by the maximum screw speed to determine the recommended drive size. The transition from LDPE to plastomers is the most severe. When running LDPE and plastomers on the same machine, field weakening must be used to satisfy the increased screw speed and lower torque requirements of LDPE. The processing performance of narrow MWD, branched, unimodal and bimodal metallocene polymers is illustrated in Fig. 56.

A. Dual Metallocene PPs for Improved Processability

One of the main characteristics of resins made with single-site catalysts is narrow molecular weight distribution and uniform composition. In the case of PP, this leads to correspondingly narrow melt temperature ranges and at times dramatic processing limitations. Thus, early commercial uses of single-site PP were in applications where its narrow melt temperature range was not a problem, such as cast film, or in compounds.

Exxon has found a new level of molecular control that can solve all of these limitations. Exxon has developed unique isotactic PP homopolymers and copolymers made in one reactor using dual metallocene catalysts on a single support. The copolymer is a bi-

TABLE 23 New Generation Resins in the Marketplace

Resin	Density (g/cm^3)	Application	Replacement for
Metallocene polyethylene (mPE) plastomers	0.900–0.915	Coextrusion, heat seal layers	Ethylene–vinyl acetate (EVA) resins with 9% vinyl acetate, ultra low density polyethylene (ULDPE), very low density polyethylene (VLDPE)
mPE elastomers	0.855–0.900	Stretch-cling films	Ethylene methylacrylate
Metallocene linear low density polyethylene (mLLDPE)	0.916–0.925	Stretch films, industrial sacks, shopping bags, form/fill/seal applications, frozen food packaging	Butene coppolymer C4 LLDPE, LDPE–LLDPE blends, EVA, ULDPE, ionomer layers
Metallocene medium density polyethylene (mMDPE)	0.932	Good optics food packaging films, personal care, agriculture	Low density polyethylene (LDPE)
High performance, high clarity LLDPE	0.909–0.912	Liner stretch film, lamination, highly transparent food packaging	Higher α-alkene octene C8 LLDPE, LDPE–LLDPE blends
Ethylene-propylene resins, random heterophasic polypropylene (PP) copolymer	0.890	Coextruded form/fill/seal film, hot fill, heavy duty sacks, medical and personal care films, blends with other PP or PE grades	Soft poly(vinyl chloride), non-polymer packaging, LDPE, LLDPE

modal and probably the first true bimodal resin made in a single reactor.

Bimodal composition distribution, which applies only to copolymers, can produce resins having different concentrations of copolymer, i.e. different melting points in polymer chains that are the same length.

B. Tacticity Control

But the newest feature is Exxon's claim to control tacticity distribution. This means the ability to arrange propylene monomer in any of three variations — isotactic, atactic, and syndiotactic — on one polymer chain.

Pure sPP and aPP can be made, but only using specialized single-site catalysts. Exxon's work shows it incorporating and using the aPP fractions in the polymer chains. This is similar to the result achieved with the addition of ethylene. The ability to insert and control blocks of different tacticity could theoretically lead eventually to TPOs made in one reactor with one monomer that would have property

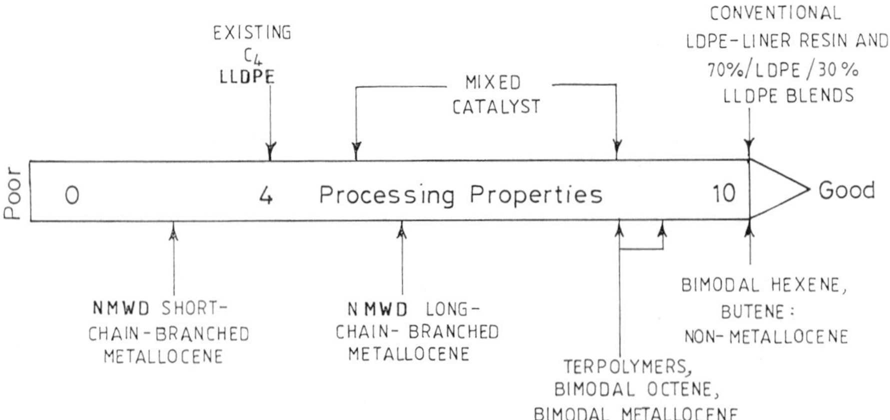

FIG. 55 Effects of chain branching on processing characteristics of ethylene polymers and copolymers (NMWD: Narrow Molecular Weight Distribution).

TABLE 24 Impact of Metallocene Polymer Properties on Processing

Polymer property	Interpretation	Implication	Process impact
Narrow molecular weight (MW) distribution with little or no long chain branching	Fewer low MW "lubricants"	Higher process pressures	Higher melt temperature
	Fewer low MW "extractables"	Higher torque	Change motor size or screw speed
		Low film blocking	Easier winding/unwinding
		Less chill roll plateout	Reduced housekeeping
	Fewer high MW "stiffeners"	Lower melt strength	Decreased bubble stability
			Easier drawdown
		Higher clarity	
	Fewer entanglements	Faster melt relaxation	Easier drawdown
Increased long chain branching	More chain entanglement	Increased melt tension	Increased bubble stability
			Less draw resonance
		Increased shear sensitivity	Less torque increase
			Less die pressure increase
Lower density	Lower softening point	Soft, tacky pellets	Decreased specific rate in grooved feed machines
			Increased specific rate in smooth bore machines
		Soft, tacky film	Increased collapser friction
			Increased wrinkling
	More elastic behavior	Sensitive to tension variation	Harder to wind

advantages that now require two monomers (ethylene and propylene) and 3–4 reactors! That is the potential down the road.

Exxon describes a dual-metallocene catalyst on one support, referred to as catalysts 1 and 3, making two different narrow MWD polymers. Catalyst 1 makes long isotactic polymer chains with a 150°C melting point; catalyst 3 makes shorter isotactic polymer chains with a 135°C melting point. The combination, however, is not a predictable linear blend. Catalyst 3 has an unexpected secondary effect on the long chain population made by catalyst 1, reducing them, broadening the MWD, and creating a higher concentration of short chains. This secondary effect also broadens the melt temperature range. The shorter polymer chains from catalyst 3 also have lower isotacticity or crystallinity.

Exxon say the copolymers are more remarkable than the homopolymers. When ethylene is added, the initial chain-breaking reaction reverses and longer chains form again. The copolymers combine all three "tailored molecular distributions" — molecular weight, composition, and tacticity — whereas the broad MWD homopolymers display control of only molecular weight and tacticity distribution. These dual-metallocene PPs are a very recent development.

The new three-way modality control can broaden melt temperature ranges and make single-site PP that may be used in high speed biaxially oriented film lines.

Biaxial orientation is commercially by far the most important PP film making process. It starts by forming a thick PP sheet, then stretching it in perpendicular directions at close to its melting point. BOPP resins need a low softening point and a broad range of stretch

TABLE 25 Typical Extruder Torque Requirements

Extruder	Typical screw speed (rev/min)		LDPE (kW/rpm)	LLDPE (kW/rpm)	mLLDPE (kW/rpm)	Plastomer (kW/rpm)
	Blown	Cast				
65 mm × 24:1	105–125	125–140	0.181	0.260	0.286	0.307
90 mm × 24:1	90–115	110–125	0.487	0.753	0.828	0.902
115 mm × 24:1	75–105	105–115	1.029	1.595	1.754	1.913
150 mm × 24:1	60–85	90–100	2.443	3.781	4.159	4.537

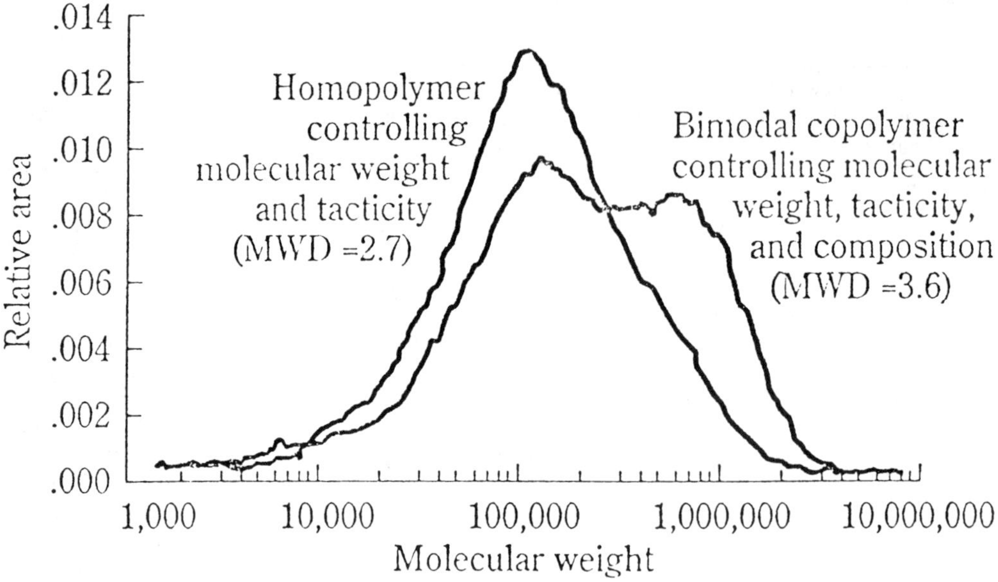

FIG. 56 These molecular weight profiles show Exxon's new developmental iPPs made with dual metallocene catalysts. The bimodal copolymer may be the first true bimodal made in one reactor; the broad MWD homopolymer may be the first pp with control of multitacticity. (From Exxon paper at Polyolefins X.)

temperatures. If the processing temperature window is too narrow, stretching is uneven and the web breaks.

The new bimodal copolymers can make good BOPP films at as much as 10°C below the processing temperatures of current ZN PPs. This suggests the possibility of a resin with "step-out" processability, allowing higher line speed and improved manufacturing economics to deliver levels of shrink performance previously unattainable. The melt temperature window on Exxon's first single-site PPs, commercialized as Achieve resins in 1995, for example, was only a 5°C window (152–157°C), too narrow for BOPP lines. Ziegler-made PPs have a 10°C window.

C. On the Horizon

Montell Polyolefins, Wilmington, DL, gave the latest update on its HMW-aPP, with molecular weights reported to range from 100,000 to 500,000 and able to make polymers from clear elastomers to liquids. Montell described new copolymers of aPP formed with the addition of ethylene. Like Exxon's findings with single-site iPP, Montell found ethylene to be a molecular weight regulator when used with a single-site catalyst. In aPP the copolymer's ethylene content lowers the glass transition temperature, in direct proportion to the amount of ethylene (Table 26).

Molecular control of experimental PPs, like these developmental atactic polymers, is enhanced by adding ethylene, here to control the glass transition temperature.

Albemarle Corp., a maker of metallocene catalysts and cocatalysts, and Fina gave new data on the processability of sPP in compounds, using Fina's sPP. Syndiotactic PPs well documented processing problem is a crystallization rate so slow that the first injection molded lab samples took several hours to set up.

Fina reported that it had made a third full-scale reactor run of sPP in 1996, following earlier runs in 93 and 95, all working to improve the crystallization rate and processability. With the 96 run crystallization time has apparently improved to where sPP can be used to replace ethylene as an impact modifier in 4% ethylene random copolymers. Using sPP as an additive improves resin clarity, Fina said.

Fina estimates sPP will represent as much as 5% of PP film markets in blends by the year 2010!

Albemarle studied the behavior of blends of 20% isotactic PP, made with a Ziegler titanium catalyst, and 80% sPP made with metallocene catalyst, which

TABLE 26 Ethylene Content Controls Glass Transition Temperature in Developmental Atactic PP Copolymers

Test batch	Intrinsic viscosity (cm^3/g)	Molecular weight	Ethylene content (%mol)	T_g (K)
1	0.67	67,400	40.8	223
2	0.65	64,700	45.1	218
4	0.6	58,000	48.6	216
5	1.32	169,200	5.5	260
6	1.36	176,200	13.6	255
10	0.96	109,800	23.7	247
12	1.43	188,600	42.3	227

uses zirconium as the active metal. Ablemarle wanted to see if the presence of two different active metals in combination in a compound would require different stabilization, even though only minute amounts of these very high activity catalysts stay in the resins.

The 80/20 compound was stabilized using several standard antioxidant packages and extruded five times to test for yellowing and long-term aging. The strands were fed first into a 60°C hot water bath and then into an icewater bath before strand cutting, a process Albemarle developed for the sPP material. Albemarle found melt stability and color were better for the sPP compound than for iPP alone, but long-term aging was poorer.

D. Industrial Activity Using Metallocenes

Industrial production of polyolefins using the new catalysts began in the early 1990s. The companies most present in the patent literature are Mitsui Petrochemical, Idemitsu, Exxon, Mitsui Toatsu, and Hoechst. In September 1989, Exxon announced its plans to commercialize PE from high pressure process (EXXPOL) and metallocene catalysts, produced in a 15,000 t/yr plant in Baton Rouge, LA. This was followed in March 1990 by an announcement by Chisso to produce highly stereoregular PP and in April 1991 by Mitsui Petrochemical for a process to produce SuperPE, i.e. a polymer, the properties of which can be regulated in order fit HDPE, LDPE and LLDPE. Also Idemitsu, Dow Chemical, Fina, Mitsui Toatsu, Tosoh, and Hoechst made analogous statements. However, in early 1993 and limited to PE resins, only Exxon came out with its resin EXACT and Mitsui Petrochemical with a 4000 t/yr plant to produce INSITE resin and is reported to have commissioned a 57,000 t/yr plant at Freeport, TX, based on solution technology.

In 1995 a capacity of PE production from metallocene catalysis of around 350,000 t/yr was installed.

The current status of some metallocene resin production is given in Table 27. Table 28 cites the global capacity for metallocene-based polyolefins. Figure 57 gives various metallocene structures used in olefin polymerization.

E. Metallocene Polypropylenes and Copolymers

m-PP homopolymer and copolymer have higher melting points (157–183°C), low melt strength, high MFR, lower extractables, and less blooming than conventional PP. The impact copolymer of PP has higher HDT, flexural modulus, and similar impact strength to its conventional PP and it also has lower comonomer content for making the impact grade. The film grade polymer has lower heat sealing temperature, and lower heat shrink than the conventional one. The random copolymer grade of PP also has sealing temperature 15°C lower than the conventional one. A BOPP film with low heat shrink (1% at 140°C) and high Young modulus (over 5000 MPa) for 23 μm film with high gloss and tensile strength has also been made possible.

F. Metallocene Syndiotactic Polypropylene (m-sPP)

It is a new molecular form of PP. Its film grade has unusually high clarity, gloss, toughness, softness, impact, tear resistance, and low seal temperature and is found to be better than random copolymer and impact modified grades. It crystallizes much slower than i-PP and blending with the latter is also being tried. The grades with broader molecular weight distribution and improved toughness at −25°C and better

TABLE 27 Current Status of Some Metallocene Resin Production

Metallocene product	Leading producers	Commercial status	Competitive materials
Linear low density polyethylene (LLDPE)	Exxon Chemical Dow Chemical Mitsui Seekha Mitsubishi Kagaku	1993[a] Acetate (EVA)	Conventional low density PE Ethylene vinyl Acetate (EVA)
PE-based elastomer	Dow Chemical Exxon	1991	EVA Ionomers LLDPE
Polypropylene	Exxon Hoechst Fina Dow BASF	Commercial scale trial runs[b]	Conventional PP engineering plastics
Polystyrene	Dow Idemitsu	Market development unit due on-line by March 1997[c]	Polyester nylon liquid crystal polymers
Ethylene–propylene–diene monomer (EPDM)	Du Pont–Dow Exxon	Commercial scale unit due on-line by March 1997[d]	Conventional EPDM
PE-based elastomers	Du Pont–Dow	1994	EVA Flexible PVC EPDM EP rubber
Cycloolefin polymers	Hoechst Mitsui Sekka	1997[e]	Polycarbonate Glass

[a]Exxon has 250,000 Mt/yr capacity in Mont Belvieu, TX; Dow has 227,000 Mt/yr capacity in Plaquemine, LA.
[b]Exxon has launched nonwoven spunbound products.
[c]Dow-Idemitsu unit in Chiba Japan.
[d]Du Pont–Dow's 90,000 Mt/yr plant in Seadrift, TX.
[e]Initial quantities supplied by Hoechst–Mitsui 3000 Mt/yr pilot plant in Iwakumi, Japan.

clarity than standard grades of impact and random copolymers.

G. Single-Site Constrained Geometry Catalyst (Table 29)

They have unusually high activities and thus can be used in small quantities. They also have high activity at higher temperature than metallocene. It is a constrained geometry catalyst and allows the polymer structure to be precisely controlled, thereby improving the properties. It is different from metallocene because of its ability to incorporate long chain branching into the polymer structure as a result of greater exposure of the active metal site. Several combinations of comonomer addition have been reported. They are styrene/α olefins/mixture of ethylene with unsaturated monomers (1-butene, 1-hexene and 1-octene, also diolefins 1,3-butadiene/1,4-hexadiene/1,5-hexadiene, and also norbornene/ethylidene norbornene/vinyl norbornene).

It can be produced in slurry, solution, high pressure or gas phase reactors.

H. Application of Metallocene Polyolefin Films

Until price compatibility comes, the trend is to coextrude m-PE as blending agents to get better properties, e.g. for 24 μm monolayer LLDPE film with 30% m-LLDPE, downgauging by 25% is possible to produce a tough bag to compete against woven PP.

In FFS, downgauging is possible from 180 μm (70% LDPE, 30% LLDPE) to 140 μm and replacing the LLDPE with the same percentage of m-PE. Thus film blend with metallocene PE will grow rapidly.

Supertough m-LLDPE substitutes the present blend of 75/25 LDPE/LLDPE, shear sensitivity is said to be similar to the blended one, and is also superior to butene LLDPE.

TABLE 28 Global Capacity for Metallocene-Based Polyolefins

Product and producer	Location	Capacity (Mt/yr)
Polyethylene		
Dow Plastics	US	113
Dow Plastics	Spain	57
Exxon Chemical	US	115
Mitsubishi	Japan	100
Nippon Petrochemicals	Japan	50
Ube Industries	Japan	20
Total		455
Polypropylene		
BASF	Germany	12
Chisso	Japan	20
Exxon Chemical	US	100
Hoechst	Germany	100
Mitsui Toatsu	Japan	75
Total		305
Polycyclic olefins		
Dow Plastics	US	Pilot
Hoechst	Germany	Pilot
Mitsui Petrochemicals	Japan	3

VLDPE is becoming a material of choice in packaging (of meat) as the precise control of molecular weight distribution virtually eliminates the low molecular weight fraction.

m-LLDPE with improved toughness, barrier properties, and heat sealing is becoming the ideal choice nowadays for packaging and applications are in the fields of sealant layer, overwrap film, stretch film, shrink package, coating, laminate wrap, foam, and label.

Low oligomer content in metallocene catalyzed copolymer offers some salient features. These are optical properties, above conventional clarity, and softness. Low oligomer content also eliminates undesirable odor/taste, reduces extractables, and imprvoes organoleptic properties which is superior to conventional polymers. It leads to potential application in packing, e.g. blood tubing, corrugated respiratory tubing, and food packaging.

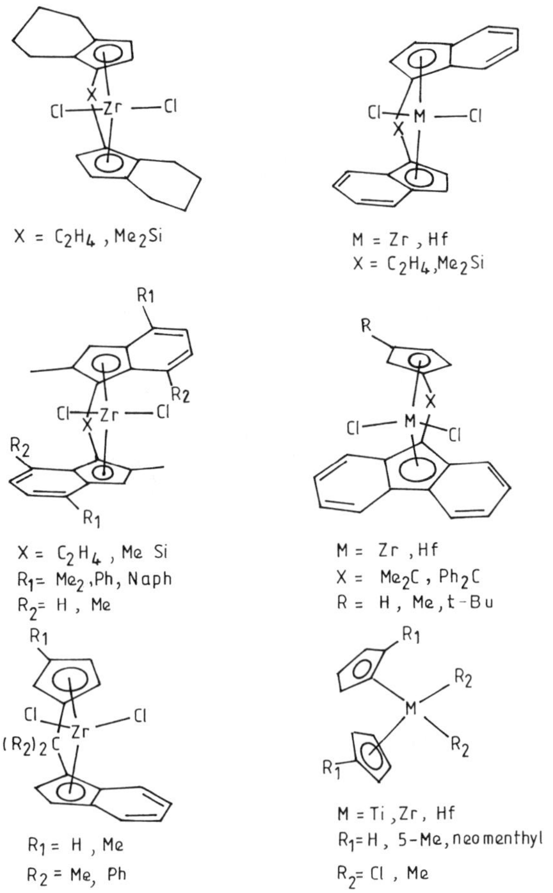

A recently patented use in flexible packaging is a three-layer breathable film in which a two-layer blown extrusion of K resin KR 10 and inner metallocene

FIG. 57 Structures of metallocenes that are used in the polymerization of olefins.

TABLE 29 Constrained Geometry Offers Salient Features Over Conventional Polyethylene

	Density (g/cm^3)	MWD (Mw/Mn)	M FR (g/10 min)	Melt strength (CN)
LLDPE				
single site	0.92	1.97	9.5	1.89
conventional	0.92	3.80	8.0	1.21
VLDPE				
single site	0.91	1.90	7.9	1.68
conventional	0.912	3.80	8.2	1.20

PE is laminated to a metallocene PE sealant layer. It has double the shelf life of LDPE, with improved oxygen and carbon dioxide exchange.

Coextruded film with m-PE offers high puncture resistance for frozen food packaging.

Two layer of LLDPE and LLDPE blend with m-PE gives detergent package strength.

m-PE is also replacing ionomer and high VA content EVA as heat seal resins in coextruded film. It would also replace plasticized PVC (e.g. blood bags/food wrap) and would grab the gamut of the PVC market.

Cost/performance balance also going to replace some engineering thermoplastics (ABS/PET).

m-sPP is suitable for food packaging applications. The market domain is in cosmetic packaging (cups) where transparency and chemical resistance are exploited.

Metallocene PE film grade resins are made by BP, Dow, Exxon and Mobil (Table 30). Metallocene processes for ethylene polymerization are summarized in Table 31.

I. Metallocene Films: The Latest Material for Packaging

Narrow molecular weight distribution (MWD) of metallocene polyolefin films effects: improved hot tack, lower heat seal temperature, reduces leak rates, increases packaging line speeds, improved toughness and puncture resistance, fewer failures, product protection, and downgauging.

All these features are of immense importance to improve packaging.

Other features in film production are: higher stiffness, higher or lower oxygen transmission, improved clarity, sealability, and crease resistance. Raffia: increased stiffness, resilience, recovery, dimensional stability, and purity. Sheet: better processability, clarity, thermoformability, and calenderability. Blowmoulding: higher clarity, melt strength, and controlled stiffness. Injection moulding: better impact stiffness balance, low temperature impact, and better flow.

J. New Metallocene Polymers

Metallocene catalysts have enabled the synthesis of novel polymers not possible with Ziegler–Natta catalysts (e.g. syndiotactic polystyrene, a material with a high melting point that promises to compete effectively with polyamides and other engineering thermoplastics). Dow and Idemitsu are leaders in this development. Dow has extended the styrene polymer family to a new class of ethylene–styrene copolymers with a

TABLE 30 Metallocene-Based Polyethylene Processes

Process	MI range (g/min)	Density (g/cm^3)	
Comonomer	Solution (Dow)	0.5–30.0	0.880–0.950
Octene (0–20%)	Solution (Exxon)	1.0–100.0	0.865–0.940
Propylene, butene, hexene	Gas phase (Mobil)	1.0	0.918
Hexene	Gas phase (BP Chem)	2.3	0.916

TABLE 31 Single-Site Metallocene Catalysts for Ethylene Polymerization

Year	Process	Licensor	Remarks
1991	High pressure	Exxon, Mitsubishi, BASF, Tosoh	Controlled polymerization, copolymerization possible to produce elastomers
1993	Solution	Dow, Mitsui Dex Plastomer	Low cost, wide product range, 0.2 MI material
1995	Loop reactor	BASF, Exxon, Fina	Very low MI product and wide product range
1995	Gas phase	Union Carbide, Nippon Oil	Production of elastomers, EPDM and EP rubber
Future trends			
Production of ethylene copolymer with comonomers, such as styrene, norbornene, carbon monoxide, etc., possible.			

wide range of properties that can be controlled, from flexible, leathery products to stiff plastics. The copolymerization of ethylene and cyclics such as norbornene has been commercially practised for some time, but the new metallocene catalysts open up new cyclic choices and composition ranges. Heat-resistant products with high stiffness, hardness, and optical clarity are possible. Hoechst and Mitsui are among companies pursuing these new products. Their use in CD-ROMs (competing against polycarbonate CD-ROMs) is an example of their potential applications.

The "fluxional" catalyst, a phenyl-substituted unbridged bis-Cp structure, periodically changes its conformation from one that makes iPP to one that produces aPP. The conformation cycle is short with respect to chain lifetimes. Thus true stereoblock copolymers are possible. A new generation of flexible propylene polymers is expected to result from applications of this catalyst.

K. Future Trends in Metallocene Polyolefins

Metallocene catalysts gave the product designer the building blocks from which to tailor MWD, composition distribution, and, in the 1990s when PP metallocenes matured, tacticity distribution.

Future catalysts will be programmed to produce complex, perhaps multiphasic, products in simple, single-reactor processes.

Within 10 years, 20% of the world's PE will be made by metallocene catalysis. The rapid market penetration is based on the high value attributes delivered by metallocene-based PE such as greater stiffness and impact strength.

XI. FUTURE 2025 [59]

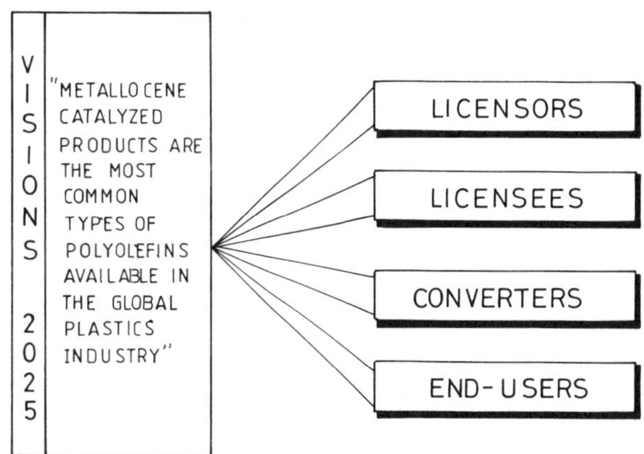

FIG. 58 Metallocene polyolefins will replace conventional polyolefins by the year 2025. (From Ref. 59.)

Today's licensing leaders will not be the future metallocene leaders.

"Catalyst" will dominate over "process", since product attributes are created by catalyst.

Technology owners will demonstrate the value of their technology to secure a return on investment before patent expiration.

To achieve commercialization of metallocene technology and to lower legal barriers, alliances will be formed between catalyst and process owners (Fig. 58).

Manufacturing will be modified to accompany the customized nature of metallocene-catalyzed products to control product cost.

Production planning will be re-engineered on the basis of new metallocene grade slates.

End-user and convertor alliances will be formed to exploit the technologies for specialties and for profitable market penetration (Fig. 58).

Influential end-users will drive the business.

Convertors will resist the penetration of metallocenes until the value to them is demonstrated. They will do so because it will be expensive for them to modify equipment without economic incentives.

The precision and wide property range will eliminate the need for some compounding.

The new behavior of metallocenes will change the types and reduce the quantities of additives required in polyolefins.

Equipment suppliers will supply machinery to suit the processing of metallocenes (e.g. tighter tolerances).

Identification of attractive markets for metallocene-catalyzed polyolefins will be done to fit them with suppliers' incumbent businesses.

Many end-users will specify metallocenes as the preferred polyolefin material and will be able to significantly enhance their business.

APPENDIX 1. RECENT METALLOCENE CATALYST PATENTS (1995–1996)

Albemarle

Patent Number	Title
5,391,529	Siloxy-aluminoxane compositions, and catalysts which include such compositions with a metallocene
5,412,131	Tertiary amino-aluminoxane halides
5,455,333	Preparation of metallocenes
5,466,647	Tertiary amino-aluminoxane halides
5,527,930	Aluminoxanes having increased catalytic activity

BASF

Patent Number	Title
5,453,475	Process for preparing low density ethylene copolymers
5,457,171	Catalyst systems for the polymerization of C_2–C_{10} alkenes
5,474,961	Deactivated and reactivated metallocene catalyst systems
5,491,205	Preparation of polymer of C_2–C_{10} alk-1-enes using racemic metallocene complexes as catalysts
5,496,902	Catalyst systems for the polymerization of C_2–C_{10} alk-1-enes
5,514,760	Soluble catalyst systems for the preparation of polyalk-1-enes having high molecular weights
5,527,868	Catalyst systems for the polymerization of C_2–C_{10} alk-1-enes

BP Chemicals

Patent Number	Title
5,439,995	Catalyst and prepolymer used for the preparation of polyolefins

Dow

Patent Number	Title
5,380,810	Elastic substantially linear olefin polymers
5,399,635	Process for the preparation of monocyclopentadienyl metal complex compounds and method of use

Enichem

Patent Number	Title
5,529,966	Catalyst and process for (co)polymerizing α-olefins

Exxon

Patent Number	Title
5,391,629	Block copolymers from ionic catalysts
5,422,325	Supported polymerization catalysts, their production and use
5,427,991	Polyonic transition metal catalyst composition
5,432,242	HP catalyst killer
5,441,920	Silicon-bridged transition metal compounds
5,442,019	Process for the transitioning between incompatible polymerization catalysts
5,444,145	Ethylene/branched olefin copolymers
5,446,221	Oleaginous compositions containing novel ethylene α-olefin polymer viscosity index improver additive
5,451,450	Elastic articles and a process for their production
5,462,807	Heat sealable films and articles
5,468,440	Process of making oriented film or structure
5,462,999	Process for polymerizing monomers in fluidized beds
5,470,811	Polymerization catalysts, their production and use
5,470,927	Ionic metallocene catalyst composition
5,475,075	Ethylene/longer α-olefin copolymers

5,491,207	Process of producing high molecular weight ethylene α-olefin elastomers wtih an indenyl metallocene catalyst system
5,525,128	Fuel oil additives and compositions
5,529,965	Polymerization catalyst systems, their production and use
5,530,054	Elastomeric ethylene copolymers for hot melt adhesives

Fina

Patent Number	Title
5,387,568	Preparation of metallocene catalysts for polymerization of olefins
5,393,851	Process for using metallocene catalyst in a continuous reactor system
5,395,810	Method of making a homogeneous–heterogeneous catalyst system for olefin polymerization
5,416,228	Process and catalyst for producing isotactic polyolefins
5,449,651	Metallocene compound for a catalyst component with good catalyst efficiency after aging
5,476,914	Syndiotactic polypropylene
5,519,100	Addition of aluminum alkyl for improved metallocene catalyst

Hoechst

Patent Number	Title
5,391,789	Bridged, chiral metallocenes, processes for their preparation and their use as catalysts
5,416,178	Process for the preparation of 1-olefin polymers
5,422,409	Cycloolefin (co)polymer with a narrow molecular weight distribution and a process for the preparation thereof
5,455,365	Process for the preparation of an olefin polymer using metallocenes containing specifically substituted indenyl ligands
5,455,366	Metallocenes having benzo-fused indenyl derivatives as ligands, processes for their preparation and their use as catalysts
5,475,060	Cycloolefin block copolymers and a process for their preparation
5,498,677	Process for the preparation and purification of material of a cycloolefin copolymer
5,504,232	Process for the preparation of an olefin polymer using specific metallocenes

Mitsubishi

Patent Number	Title
5,444,125	Aminated olefin polymers
5,468,781	Polypropylene resin expanded particles

Mitsui

Patent Number	Title
5,444,125	Aminated olefin polymers
5,464,905	Ethylene/α-olefin copolymer composition, graft modified, ethylene/α-olefin copolymer composition, ethylene copolymer composition, and multi-stage olefin polymerization process
5,468,781	Polypropylene resin expanded particles

Mobil

Patent Number	Title
5,397,757	Cocatalysts for metallocene-based olefin polymerization catalyst systems
5,455,214	Metallocenes supported on ion exchange resins
5,461,017	Olefin polymerization catalysts
5,473,028	Process and a catalyst for preventing reactor fouling
5,498,582	Supported metallocene catalysts for the production of polyolefins

Phillips Petroleum	
Patent Number	Title
5,399,636	Metallocenes and processes therefor and therewith
5,401,817	Olefin polymerization using silyl-bridged metallocenes
5,416,179	Catalyst composition and olefin polymerization
5,420,320	Method for preparing cyclopentadienyl-type ligands and metallocene compounds
5,436,305	Organometallic fluorenyl compounds, preparation, and use
5,451,649	Organometallic fluorenyl compounds, preparation, and use
5,459,218	Syndiotactic polypropylene prepared using silyl bridged metallocenes
5,466,766	Metallocenes and processes therefore and therewith
5,473,020	Polymer bound ligands, polymer bound metallocenes, catalyst systems, preparation, and use
5,492,973	Polymer bound ligands
5,492,974	Process for preparing polymer bound metallocenes
5,492,975	Polymer bound metallocenes
5,492,978	Process for preparing catalyst system
5,492,985	Polymerization process
5,496,781	Metallocene catalyst systems, preparation, and use
5,498,581	Method for making and using a supported metallocene catalyst system

Solvay	
Patent Number	Title
5,496,782	Catalyst system, use of this catalyst system for the (co)polymerization of olefins, process for preparing this catalyst system and olefin (co)polymerization process
5,525,690	Process for the preparation of a polyolefin and syndiotactic polypropylene

Tosoh	
Patent Number	Title
5,407,882	Olefin polymerization catalyst and olefin polymerization process
5,434,115	Process for producing olefin polymer

Source: Ref. 71.

APPENDIX 2. A DISCUSSION ON REACTION PATHWAYS FOR α-OLEFIN POLYMERIZATION WITH METALLOCENES

In 1957, Natta, Pino and their coworkers reported that homogeneous reaction mixtures of dicyclopentadienyl titanium dichloride (Cp_2TiCl_2) and diethyl aluminum chloride (El_2AlCl) catalyze the formation of polyethylene under conditions similar to those used with heterogeneous Ziegler catalysts. These metallocene studies, together with those done at the Hercules Laboratories by Breslow, Newburg, and Chien, helped to understand the mechanism of heterogeneous Ziegler–Natta catalysis through identification of reaction intermediates.

Early studies [103–110] on spectroscopies and kinetics predicted the formation of (alkyl) titanocene complexes ($Cp_2Ti(R)Cl$) (R=Me or Et) by ligand exchange reaction of alkylaluminum cocatalyst through polarization of Cp_2Ti–Cl bond by Lewis-acidic aluminum centers in an adduct of the type $Cp_2Ti(R)Cl{\cdot}AlRCl_2$. Insertion of the olefin take place into the Cp_2Ti–R bond of this (or some closely related) electron-deficient species [100–102].

In early research work in the 1960s, the polymerization of ethylene by metallocene systems raised an interesting question: does olefin insertion occur in a bimetallic species in which an alkyl group or a halogen atom bridges the titanium and aluminum centers as was thought by Natta and others [103, 104]. Or, does it require the firm formation of a cationic species $[Cp_2TiR]^+$ by abstraction of a halide anion and its incorporation into an anion $R_xCl_{4-x}Al$, as proposed by Dyachkovoskii's group [110]?

The crystal structures of complexes obtained from reaction mixtures of Cp_2TiCl_2 and an alkyl aluminum chloride, however, could not lead to any conclusion due to degradation of the product. Jordon and coworkers [111–116] showed that tetraphenyl borate salts of cations such as $[Cp_2Zr{\cdot}CH_3{\cdot}THF]^+$ and $[Cp_2ZrCH_2Ph{\cdot}THF]^+$ are capable of polymerizing ethylene without addition of any activator.

The functions in the alkylaluminum activated metallocene catalyst systems have been described by Reichert and coworkers [117–119]. The conceptualized point is that each metal–polymer species appears to alternate between a "dormant" and "active" state in which the polymer chain grows. This "intermittent-growth" model (Fig. 59) was further elaborated by Fink [120, 121] and Eisch [122, 123] and their coworkers in extensive kinetic and reactivity studies (Fig. 59). Consecutive equilibria appear to convert alkyl aluminum and (alkyl) metallocene halides first into Lewis acid–base adducts equivalent to an inner (or contact) ion pair and then into a dissociated (or separated) ion pair. In these dynamic equilibria, only the cation of a separated ion pair appears to be capable of interacting with an olefin molecule and thus to dominate in these equilibria, and can then be termed "dormant" in this regard [124].

This model has been used to explain the inability of metallocenes activated by alkylaluminum halides to catalyze the polymerization of propene and higher olefins due to the insufficient capability of the more weakly coordinating substituted α-olefins to form reactive, olefin-separated in pairs by displacement of an aluminate anion from the metal center [125, 126].

Sinn, Kaminsky and coworkers [127–129] observed that methyl alumoxane (MAO)-activated homogeneous metallocene catalysts, together with aluminum halides, are capable of polymerizing propene and higher olefins. Although the achiral metallocene catalysts were still lacking the stereoselectivity of heterogeneous Ziegler–Natta systems, aluminoxane-activated metallocene catalysts now come to be most promising as model systems. The hydrolysis of $AlMe_3$, a highly exothermic reaction, produces oligomers Me_2Al–[O–AlMe]$_n$–$OAlMe_2$ with $n = 5$–20 with a certain amount of residual $AlMe_3$ still existing in equilibrium. Spectroscopic studies predicts a cross-linking by methyl free-oxoaluminum centers to generate a microphase with an Al_xO_y core. Aluminoxane clusters $[RAl(\mu_3\text{-}O)]_n$, with R = tertbutyl and $n = 4$, 6, or 9 has been isolated and structurally characterized by Barron and his group [130, 131]. Although complexes with four-coordinate Al centers seem to predominate in MAO solutions, the presence of three-coordinate Al centers has also been deduced by Siedle and co-workers [132, 133] from ^{27}Al NMR data. ^{91}Zr and ^{13}C NMR spectra of Cp_2ZrMe_2/MAO solutions [134] and solid state XPS [135] studies and ^{13}C NMR studies [136] provide an indication for the formation of a cation $[CpZrR]^+$, and which is most likely stabilized by coordinative contact with its CH_3–MAO^-.

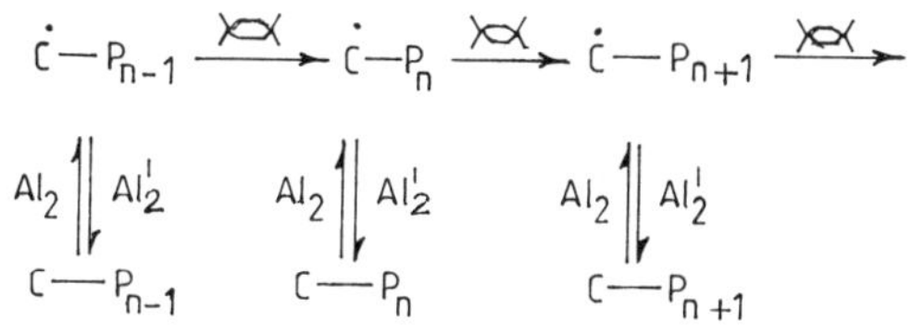

$$\dot{C}-P_{n-1} \xrightarrow{\asymp} \dot{C}-P_n \xrightarrow{\asymp} \dot{C}-P_{n+1} \xrightarrow{\asymp}$$

$$Al_2 \rightleftharpoons Al_2' \qquad Al_2 \rightleftharpoons Al_2' \qquad Al_2 \rightleftharpoons Al_2'$$

$$C-P_{n-1} \qquad C-P_n \qquad C-P_{n+1}$$

FIG. 59 "Intermittent-growth" model involving equilibria between polymer-bearing, but inactive, primary complexes (C-Pm) and active catalyst species (C*-Pn), generated by excess alkyl aluminum halide.

These contacts appear to give way, in the presence of even substituted olefins, to olefin-separated ion pairs $[Cp_2Zr(olefin)]^+CH_3$–MAO^-, the presumed prerequisite for olefin insertion into the Zr–R bond. This hypothesis — that the unusually low coordinating capability of the anion A^- is crucial for catalytic activity [137] — led to the discovery of a series of highly active cationic metallocene catalysts for the polymerization of propene and higher α-olefins.

When toluene solutions of Cp_2ZrMe_2 and MAO are reacted, these systems become catalytically active when the concentration of excess MAO is raised to Al:Zr ratios of about 200:1 or higher [138–140]. Initial ligand exchange reaction produces Cp_2ZrMe_2. It has been assumed that some of the Al centers in MAO have an exceptionally high propensity to abstract a CH_3 ion from Cp_3ZrMe_2 and to sequester it in a weakly coordinating ion CH_3–MAO^-. A fast, reversible transfer of $^{13}CH_3$ groups from Cp_2ZrMe_2 to the Al centers of a MAO activator was observed by Siedle *et al.* It has been suggested [141] from NMR spectroscopic evidence that alumoxane clusters like $(\mu_3\text{-}O)_6Al_6t\text{-}Bu_6)$ form complexes of the type $[Cp_2ZrMe^+ \ldots (\mu_3\text{-}O)_6Al_6(t\text{-}Bu)_6Me^-]$ from the reaction with Cp_2ZrMe_2 in $[D_8]$ toluene solution and this species polymerizes ethylene. This tendency of four-coordinate Al centers in these aluminoxane clusters to abstract a methyl anion is ascribed by these authors to the relief of ring strain upon formation of the methyl complex.

In the early work the $[Cp_2TiCH_3]^+$ cation has been trapped by insertion of the alkyne PhC=CSi $(CH_3)_3$ into the titanium–methyl bond with the corresponding cationic vinyl titanium complex as the tetrachloroaluminate in the $Cp_2TiCl_2/Al(CH_3)Cl_2$ system [122]. The importance of the anions in the coordinating sphere of the metallocene catalytic species for propylene polymerization has been shown from the comparison of the studies on weakly coordinating anion species like $(C_6H_5)_4B^-$ and $C_2B_9H_{12}^-$ with $[Cp_2ZrMe]^+$, producing polypropylene at low rates [142–149]. While perfluorinated tetraphenyl borate forms an anion pair with the similar cationic species and exists as $[Cp_2^xZrMe]^+(C_6F_5)_4B^-$ (where Cp^x is the indenyl ligand and is considered as substituted Cp), it polymerizes propylene at high rates without addition of further activator. This ion-pair species is observed to be capable of polymerizing higher α-olefins also (other than propylene) without needing another activator. This catalytic species was studied by several groups of researchers [150–152].

This single-component "cationic" metallocene catalysis is undoubtedly an extraordinary new development which eliminates the problems associated with the methyl alumoxane cocatalyst usage and has been represented as follows for its synthesis:

$$Cp_2^xZrMe_2 + [PhNHMe_2]^+[B(C_6F_5)_4]^- \xrightarrow{-CH_4,\ PhNMe_2} [Cp_2^xZrMe]^+[B(C_6F_5)_4]^-$$

$$Cp_2^xZrMe_2 + [Ph_3C]^+[B(C_6F_5)_4]^- \xrightarrow{-Ph_3CMe} [Cp_2^xZrMe]^+[B(C_6F_5)_4]^-$$

The ion-pair $[(Me_2C_5H_3)_2ZrCH_3^+ \ldots . H_3C{-}B(C_6F_5)_3^-]$ obtained by an abstraction of CH_3^- from a (dimethyl) zirconocene complex by the powerful Lewis acid $B(C_6F_5)_3$ was found to be a highly active α-olefin polymerization catalyst. The crystal structure of this species has been discussed by Marks and coworkers [153, 154] where coordinative contacts between the cationic Zr center and its counterion have been postulated.

Cationic metallocene complexes, particularly those arising by in situ activation of a stable zirconocene precursor, yield catalysts with very high activities. The easy deactivation of these catalysts is probably brought about by minute traces of impurities. These cationic metallocene catalysts are observed to be stabilized by addition of $AlMe_3$ or $AlEt_3$ through the formation of alkyl aluminum adducts [155].

The many questions of the role of methyl alumoxane (MAO) to activate metallocene complexes and increase the magnitude of their ability to increase the polymerization of olefins by several orders have remained unanswered [156]. The amorphous structure of MAO and its chemical fluxionality make it, and the chemical reactions occurring when it is mixed with metallocenes to form catalytically active systems, extremely difficult to study [157]. It is believed that methyl alumoxane is involved in several reactions: (i) alkylation of metallocene (in the case of dihalogenated precursors; (ii) formation of "cation-like" active species; (iii) deactivation of catalyst poisons.

Concerning step (ii), it is generally assumed that the main role played by aluminum-based cocatalysts should consist of the extraction of the X^- anion (X = halogen or alkyl group) from a neutral alkylated metallocene precursor L_2MtRX (L = cyclopentadienyl ligand, R = alkyl group) to form ion pairs [158–161] in which alkylated cationic metallocenes L_2MtR^+ should be responsible for olefin polymerization [135, 162–164]. In this framework, the high efficiency of MAO is generally ascribed [165] both to its strong Lewis acidity, making it a good anion extractor and its ability to form a weakly coordinating $X{\cdot}MAO^-$ macroanion.

The calculation of energetics [166] brought an idea that ion pair separation could play a critical role in determining the activity of metallocene-based catalytic systems. With the $Cp_2TiMeCl/AlMe_2$ ($Cp = n\text{-}C_5H_5$, $Me = CH_3$) system, the high energy (> 100 kcal/mol) required to separate Cp_2TiMe^+ from $AlMe_2Cl_2^-$ in vacuum can hardly be compensated for by the solvation energy in the weakly polar solvents usually used in polymerization (15–30 kcal/mol) [167]. Fusco *et al* [157] suggested that olefin-separated ion pairs (OSIP) $Cp_2TiMe^+/C_2H_4/AlMe_2Cl_2^-$, where an ethylene molecule is sandwiched between the cation and the anion, represent, in that case, the least unfavorable way to allow olefin coordination following insertion into the metal–carbon bond.

It has been well established that the presence of "free trimethylaluminum" in MAO is crucial for the exhibition of a high catalytic activity, and it is assumed to be partly coordinated to the alumoxane chain through methyl bridges [168, 169]. Although the real coordination situation of Al and O atoms in MAO is controversial, different theoretical considerations led one to assume that the $[-Al(Me)O-]_n$ chains can aggregate with each other. $MeAl[OAl_2Me_4]_2O$ has been proppposed as a minimal model of the cocatalytic site satisfying a model which takes into account the experimental observations and theoretical considerations, assuming that (a) aluminum atoms are able to interact with metallocene neutral precursors only in the metastable tricoordinated situations; (b) aluminum atoms in weakly coordinating $AlMe_3$ molecules have a higher probability to produce such a situation rather than those involved in inter- or intrachain O→Al dative bonds for both thermodynamic (the higher energy requirements expected to break O→Al bridges compared to that needing to break AlMe. . . Al (bridges) and kinetic factors (dissociation processes involving small molecules like $AlMe_3$ are faster than those implying cooperative rearrangements of macromolecular chains); (c) tricoordinated aluminum atoms bridging two tricoordinated oxygen atoms are expected to be strongly acidic because the p-electron back-donation from oxygen to aluminum should be at least partially inhibited. The minimal model of alumoxane could be considered as the aggregation product of trimeric $Me_2AlOAl(Me)OAlMe_2$ and dimeric $Me_2AlOAlMe_2$ methyl alumoxane chain fragments as shown in Fig. 60 [157].

The formation of two dative O→Al bonds between the two oxygen atoms of the trimer and the two aluminum atoms of the dimer gives a six-membered ring structure. According to point (c), the Al* aluminum atom is symmetrically coordinated to two tricoordinated oxygen atoms. To further reduce the complexity of the model, the terminal methyl groups were replaced with hydrogen atoms; this model molecule has been referred to as $MeAl(OAl_2H_4)_2O$ or MaO^H as given in Fig. 61.

The proposed reactions [169] involved in active species formation have been indicated as the following:

1. reaction between chloroalkylated metallocene and cocatalytic species with formation of Mt–Cl–Al chlorine-bridged adducts;

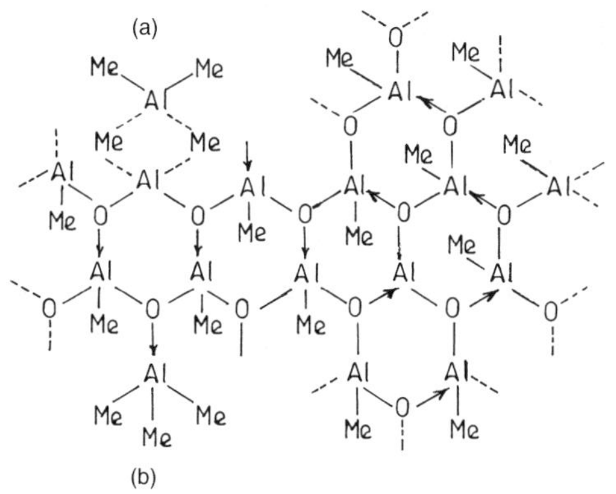

FIG. 60 The arrows denote O→Al dative bonds between different aluminoxanic chains. Two possible ways of coordination of $AlMe_3$ to methyl aluminoxane are shown through (a) a double methyl bridge, and (b) oxygen coordination.

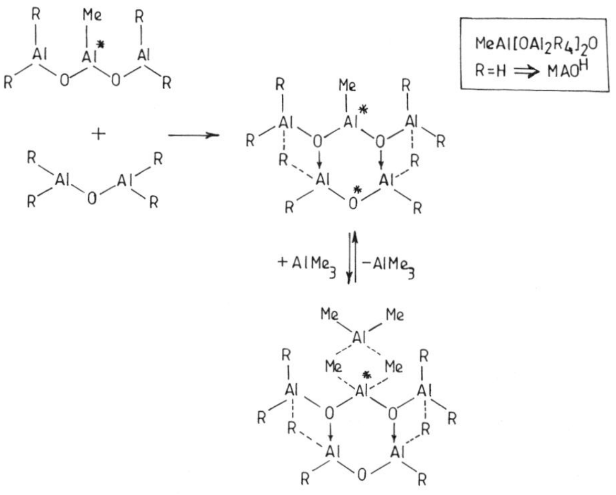

FIG. 61 The mechanism of cocatalysis with MAO. The tricoordinated Al* atom in the central structure represents the cocatalytic site.

2. dissociation of these adducts into free ion pairs;
3. formation of oxygen coordinated metallocene/MAO^H adducts;
4. formation of ion pairs from dissociation of these adducts;
5. stabilization pairs through the formation of metallocene/monomer cationic complexes and counter-ion/$AlMe_3$ adducts.

The potential use of metallocene catalysts to obtain diversified polymeric products have drawn the attention of chemists worldwide to understand the reaction pathways for edging product varieties. This endeavor will move rapidly in the coming years in order for competitive technologies to grow.

ABBREVIATIONS

BOPP film	Biaxially oriented polypropylene film
DEAC	Diethyl aluminum chloride
ED	Electron donor
EPDM	Ethylene–propylene–diene terpolymer
FMP	Foamed metallocene polypropylene
LDPE	Low density polyethylene
LLDPE	Linear low density polyethylene
MAO	Methyl alumoxane
MDPE	Medium density polyethylenes (from metallocenes)
$MgCl_2$	δ-MCl_2
$\overline{Mn}$	Number-average molecular weight
m-PP	Polypropylene obtained from metallocene catalysis

MSC	Multisite catalyst
$\overline{M}w$	Weight-average molecular weight
$\overline{M}w/\overline{M}n$	Molecular weight distribution (MWD)
PP	Polypropylene
SSC	Single-site catalyst
TEA	Triethyl aluminum
VLDPE	Very low density polyethylene
ZN catalyst	Ziegler–Natta catalyst

REFERENCES

1. Worldwide Metallocene Conference. METCON 94, Houston, TX, USA, May 25–27, 1994.
2. S Van der Ven. Polypropylene and Other Polyolefins. Amsterdam: Elsevier, 1990.
3. Polyolefins Through the 80's – The Continuing Evolution. SRI International Project No. 7892, Dec. 1990.
4. U Zucchini, T Dall'Occ. In: S Sivaram, ed. Polymer Science, Vol. I. Tata: McGraw-Hill, 1991, pp. 221–228.
5. RB Seymour, T Cheng, eds. Advances in Polyolefins. New York: Plenum Press, 1987.
5a. I Bousted, G Hancock. Handbook of Industrial Energy Analysis, Ellis Horwood Ltd., England, 1979.
5b. A Razavi. METCON 96, (1996).
5c. Doi. In: R Quirk, ed. Transition Metal Catalyzed Polymerizations. Alkenes and Dienes. New York: Harwood, 1983.
6. H Frey, J Kressler, W Richtering, R Mulhaupt. Acta Polymer 47: 131, 1996.
7. H-H Brintzinger, D Fischer, R Mulhaupt, B Rieger, R Waymouth. Angew Chem Int Ed Engl 34: 1143, 1995; Angew Chem 107: 1255, 1995.
8. G Fink, R Mulhaupt, HH Brintzinger, eds. Ziegler Catalysts. Berlin: Springer, 1995.
9. I Tritto, U Giannini, eds. Stereospecific Polymerization. Macromol Symp Vol. 89, Heidelberg: Huthig, 1995.
10. H Sinn, W Kaminsky. Alumoxanes. Macromol Symp Vol. 97, Heidelberg: Huthig, 1995.
11. SS Reddy, S Sivaram. Prog Polym Sci 20: 309, 1995.
12. W Kaminsky, Angew Makromol Chem 223: 101, 1994.
13. S Jungling, R Mulhaupt, U Stehling, HH Brintzinger, D Fischer, F Langhauser. J Polym Sci: Part A: Polym Chem 33: 1305, 1995.
14. W Spaleck, M Antberg, M Aulbach, B Bachmann, V Dolle, S Haftka, F Kuber, J Rohrmann, A Winter. In: G Fink, R Mulhaupt, HH Brintzinger, eds. Ziegler Catalysts. Berlin: Springer, 1995, p. 83.
15. KD Hungenberg, J Kerth, F Langhauser, B Marcinke, R Schlund. In: G Fink, R Mulhaupt, HH Brintzinger, eds. Ziegler Catalysts. Berlin: Springer, 1995, p. 362.
16. H Cherdron, M-J Brekner, F Osan. Angew Makromol Chem 223: 121, 1994.
17. W Kaminsky, A Noll. In: G Fink, R Mulhaupt, HH Brintzinger, eds. Ziegler Catalysts. Berling: Springer, 1995, p. 149.
18. J Ren, GR Hatfield. Macromolecules 28: 2588, 1995.
19. L Oliva, L Caporaso, C Pellecchia, A Zambelli. Macromolecules 28: 4665, 1995.
20. PH Muhlenbrock, G Fink. Z Naturforsch 50b: 423, 1995.
21. J Koivumaki, G Fink, JV Seppala. Macromolecules 27: 6254, 1994.
22. I Tritto, Z-Q Fan, P Locatelli, MC Sacchi. Macromolecules 28: 3342, 1995.
23. A Batistini. Macromol Symp 100: 137, 1995.
24. P Corradini, G Guerra, L Cavallo, G Moscardi, M Vacatello. In: G Fink, R Mulhaupt, HH Brintzinger, eds. Ziegler Catalysts. Berlin: Springer, 1995, p. 237.
25. K Angermund, A Hanuschik, M Nolte. In: G Fink, R Mulhaupt, HH Brintzinger, eds. Ziegler Catalysts. Berlin: Springer, 1995, p. 251.
26. N Koga, T Yoshida, K Morokuma. In: G Fink, R Mulhaupt, HH Brintzinger, eds. Ziegler Catalysts. Berlin: Springer, 1995, p. 275.
27. TK Woo, L Fan, T Ziegler. In: G Fink, R Mulhaupt, HH Brintzinger, eds. Ziegler Catalysts. Berlin: Springer, 1995, p. 291.
28. GW Coates, RM Waymouth. Science 267: 217, 1900; E Hauptman, RM Waymouth, JW Ziller. J Am Chem Soc 117: 11586, 1995.
29. P Aaltonen, B Lofgren. Macromolecules 28: 5353, 1995.
30. H Yasuda, E Ihara. Macromol Chem Phys 196: 2417, 1995.
31. S Kacker, A Sen. J Am Chem Soc 117: 10591, 1995.
32. A Wakker, HG Kormelink, P Verbeke, JCM Jordaan. Kunststoffe 85: 1056, 1995.
33. E Drent, JAM van Broekhoven, MJ Dovle, PK Wong. In: G Fink, R Mulhaupt, HH Brintzinger, eds. Ziegler Catalysts. Berlin: Springer, 1995, p. 481.
34. FC Rix, M Brookhart. Macromol Symp 98: 219, 1995.
35. FC Rix, M Brookhart. J Am Chem Soc 117: 1137, 1995.
36. K Nozaki, N Sato, H Takaya. J Am Chem Soc 117: 9911, 1995.
37. F Rix, J Barborak, M Wanger, S Tahliani, J DeSimone, M Brookhart, D Elder. Macromol Symp 98: 219, 1995.
38. Z Jiang, MT Boyer, A Sen. J Am Chem Soc 117: 7037, 1995.
39. LK Johnson, CM Killian, M Brookhart. J Am Chem Soc 117: 6414, 1995.
40. M Brookhart, JM DeSimone, DE Grant, MJ Tanner. Macromolecules 28: 5378, 1995.

E Alibizzati, M Galimberti. Chim Indust 79: 1053, 1997.
41. Metallocene Polyolefins, Chemical Week, 23–26 (21.5.1997).
42. H Sinn, W Kaminsky. Adv Organomet Chem 18: 99, 1980.
43. W Kaminsky, K Kulper, H Brintzinger, FRWP Wild. Angew Chem Int Ed Eng 97: 507, 1985.
44. JA Ewen, RI Jones, A Razuvi. J Am Chem Soc 110: 6255, 1988.
45. GG Hlatky, HW Turner, RR Eckman. J Am Chem Soc 111: 2728, 1989.
46. FRWP Wild, L Zsolnai, G Huttner, HH Brintzinger. J Organomet Chem 232: 233, 1982.
47. JA Ewen. J Am Chem Soc 106: 6355, 1984.
48. DF Bari. Paper presented at SPO'93, Houston, TX, USA, September 21, 1993.
49. NF Brockmeier. Paper presented at SPO'93, Houston, TX, USA, September 22, 1993.
50. KB Sinclair. Paper presented at SPO'93, Houston, TX, USA, September 22, 1993.
51. CF Payn. Paper presented at AIChE Spring National Meeting, Session No. 92, Olefin Polymerization, March 29, 1992.
52. W Splacek, A Winter, B Bachmann, V Dolle, F Kueber, J Rohrmann. Paper presented at METCON 93, Houston, TX, USA, May 27, 1993.
53. A Razavi. Paper presented at SPO'92 (1992).
54. A Razavi. Paper presented at SPO'93, Houston, TX, USA, September 22, 1993.
55. A Status Report on Metallocene-Based Polyolefins, Research Centre, IPCL, December 1996.
56. Catalysts in Transition, Focus on Catalysts. Royal Society of Chemistry, February 1997.
57. Metallocene PP's Challenge Tradition, Injection Moulding Int., April/May, 1997.
58. Polyolefins. Ullmann's Encyclopedia, Vol. A21, pp. 487–577, 1992.
59. BP Gersh. SPO'95 Conference, Houston, Texas, USA, September 20–22, 1995.
60. D Chundry, S Edge, B MacIver, J Vaughn. Plastics Formul Compound, March/April: 18–24, 1996.
61. W Kaminsky, M Arndt. In: B Cornils, WA Herrmann, eds. Applied Homogeneous Catalysis with Organometallic Compounds Vol. 1: Applications. VCH Publishers, pp. 000–000, 1900.
62. GW Coates, RM Waymouth. J Am Chem Soc 115: 91, 1993.
63. SA Miller, RM Waymouth. In G Fink, R Mulhaupt, HH Brintzinger, eds. Ziegler Catalysts. Berlin: Springer, 1994, p. 441.
64. W Bidell, D Fischer, J Grasmeder, H Gregorius, R Hingmann, P Jones, F Langhauser, B Marczinke, U Moll, V Rauschenberger, G Schweier. Paper presented at METCON '97, June 4–5, 1997.
65. W Bidell, D Fischer, R Hingmann, P Jones, F Langhauser, H Gregorius, B Marczinke. Paper presented at METCON '96, June 13, 1996.
66. M Serravalle. Paper presented at METALLOCENES '96, 3–10, 1996.
67. AA Montagna, JC Floyd. Hydrocarbon Processing, 57–62, March 1994.
68. AA Montagna, RM Burkhart, AH Dekmezian. Chemtech, 26–31, December 1997.
69. M Serravalle. Patent situation for supported metallocene catalysts. MetCon June '97, Houston, TX, USA.
70. MB Welch, HL Hsieh. Olefin Polymerization Catalyst Technology. Phillips Petroleum Co., pp. 21–38, Chapter 3.
71. JBP Soares, AE Hamielec, Molecular Structure of Metallocene Polyolefins: The Key to Understanding the Patent Literature.
72. AA Wheeler, MetCon '96, June 12–13, 1996.
73. HR Blum, MetCon '97, June 4–5, 1997.
74. V Wigotsky. Plastics Engineering, 18–23, August 1997.
75. R Mulhaupt, T Duschek, B Reiger. Macromol Chem Macromol Symp 48/49: 317, 1991.
76. T Shiono, K Soga. Makromol Chem Rapid Commun 13: 371, 1992.
77. T Duschek, R Mulhaupt. Polym Prep 33: 170, 1992.
78. R Mulhaupt, T Duschek, D Fischeu, S Setz. Polym Adv Technol 4: 439, 1993.
79. CE Ash. In MetCon '97.
80. HM Colquhoun, DJ Thompson, MV Twigg. Carlonylation, Direct Synthesis of Carbonyl Compounds. New York: Plenum Press, 1991.
81. PWNM Van Leewen, CF Roobek, HJ Vander Heijden. J Am Chem Soc 116: 12117, 1994.
82. RE Rulke, VE Kassjager, P Wehman, CJ Elsevier, PWNM Van Leeuwen, K Vrieze, J Fraanje, K Goubitz, AL Spek, YF Wang, CH Stam. Organometallics 11: 1937, 1992.
83. HA Ankersmit, N Veldman, AL Spek, K Eriksen, K Goutibz, K Vrieze, G Van Koten. Inorg Chem Acta 252: 203, 1996.
84. KJ Cavell, H Jin, BW Skelton, AH White, J Chem Soc, Dalton Trans: 2923, 1992.
85. A Yamamoto, F Ozawa, K Osakada, L Huang, TI Son, N Kawasaki, M-K Doh. Pure Appl Chem 63: 687, 1991.
86. GPCM Dekker, CJ Elsevier, K Vrieze, PWNM Van Leeuwen. Organometallics 11: 1598, 1992.
87. GM Kapteijn, A Dervisi, MJ Verhoef, MAFH Vanden Brock, DM Grove, G Van Koten. J Organomet Chem 517: 123, 1996.
88 K Nozaki, N Sato, Y Tonomura, M Yasutomi, H Takaya, T Hiyama, T Matsubara, N Koga. J Am Chem Soc 119: 12779, 1997.
89. FC Rix, M Brookhart. J Am Chem Soc 117: 1137, 1995.

90. M Brookhart, FC Rix, J M DeSimone. J Am Chem Soc 114: 5894, 1992.
91. FC Rix, M Brookhart, PS White. J Am Chem Soc 118: 4746, 1996.
92. R Van Asselt, EECG Gieleus, RE Rulke, CJ Elsevier. J Chem Soc, Chem Commun: 1203, 1993.
93. BA Markies, KAN Verkerk, MHP Rietveld, J Boersma, H Kooijman, AL Spek, G Van Koten. J Chem Soc, Chem Commun: 1317, 1993.
94. M Svensson, T Matsubara, K Morokuma. Organometallics 15: 5568, 1996.
95. SL Borkowsky, RM Waymouth. Macromolecules 29: 6377, 1996.
96. E Drent. Eur Pat Appl 504,985, 1992; Chem Abstr 188: 103023, 1993.
97. K Nozaki, N Sato, K Nakamoto, H Tokaya. Bull Chem Soc Japan 70: 659, 1997.
98. K Osakada, J-C Choi, T Yamamoto. Chem Commun 119: 12390, 1997.
99. P Galli. ANTEC'96, pp. 1620–23, 1996.
100. DS Breslow, NR Newburg. J Am Chem Soc 81: 3789, 1959.
101. WP Long. J Am Chem Soc 81: 81, 1959.
102. WP Long, DS Breslow. J Am Chem Soc 82: 1953, 1960.
103. G Natta, G Mazzauti. Tetrahedron 8: 96, 1960.
104. H Sinn, F Patat. Angew Chem 75: 805, 1963.
105. KH Reichert, E Schubert. Macromol Chem 123: 58, 1969.
106. KH Reichert, J Bethhold, V Dornow. Makromol Chem 121: 258, 1969.
107. K Meyer, KH Reichert. Angew Makromol Chem 12: 175, 1970.
108. KH Reichert. Angew Makromol Chem 13: 177, 1970.
109. G Henrici-Olive, S Olive. Angew Chem 79: 764, 1967.
110. FS Dyachkovoskii, AK Shilova, A Shilov. J Polym Sci, Part C 16: 2333, 1967.
111. RF Jordan, WE Dasher, SF Echols. J Am Chem Soc 108: 171, 1986.
112. RF Jordan, CS Bajgur, R Willett, B Scott. J Am Chem Soc 108: 7410, 1986.
113. RF Jordan, CS Bajgur, WE Desher, AL Rheingo. Organometallics 6: 1041, 1987.
114. RF Jordan, RE LaPointe, N Baenziger, GD Hinch. Organometallics 9: 1539, 1990.
115. DJ Crowther, S Borkowsky, D Swensow, TY Meyer, RF Jordan. Organometallics 12: 2897, 1997.
116. HH Brintzinger, D Fisher, R Mulhaupt, R Ruger, RM Waymouth. Angew Chem Int Ed Engl 34: 1143, 1995.
117. KH Reichert, E Schubert. Makromol Chem 58: 123, 1969.
118. KH Reichert, J Berthhold, V Dornow. Makromole Chem 121: 258, 1969.
119. KH Reichert. Macromol Chem 13: 177, 1970.
120. G Fink, R Rottler. Angew Makromol Chem 94: 25, 1981.
121. R Mynott, G Fink, W Fenzl. Angew Makromol Chem 154: 1, 1987.
122. JJ Eisch, AM Piotrowski, SK Brownstein, EJ Gabe, FL Lee. J Am Chem Soc 107: 7219, 1985.
123. JJ Eisch, KR Caldwell, S Werneu, C Kuiger. Organometallics 10: 3417, 1991.
124. PJT Tait, BL Booth, MO Jejelowo. Makromol Chem Rapid Commun 2: 393, 1988.
125. CG Overberger, FS Dyachkovsky, PA Jarovitzsky. J Polym Sci A2: 4113, 1964.
126. CG Overberger, P A Jarovitzsky. J Polym Sci A3: 1483, 1965.
127. W Kaminsky. MMI Press Symp Ser 4: 225, 1983.
128. W Kaminsky, M Miri, H Sin, R Woldt. Makromol Chem Rapid Commun 4: 417, 1983.
129. J Herwig, W Kaminsky. Polymre Bull 9: 464, 1983.
130. MR Mason, JM Smith, SG Bott, AR Barron. J Am Chem Soc 115: 4971, 1993.
131. CJ Harlan, MR Mason, AR Barron. Organometallics 13: 2957, 1994.
132. AR Siedle, RA Newmark, WM Lamanna, JN Schroepfer. Polyhedron 9: 301, 1990.
133. AR Siedle, WM Lamanna, JM Olofson, BA Nerad, RA Newmark. In: ME Davis, SL Sujb, eds. Selectivity in Catalysis (ACS Symp Ser 517: 156, 1993).
134. AR Siedle, WM Lamanna, RA Newmark, J Stevens, DE Richardson, M Ryan. In: A Guyot, R Spitz, eds. Advances in Olefin, Cycloolefin and Diolefin Polymerization. Makromol Chem, Macromol Symp, 1993.
135. PG Gassman, MR Callstrom. J Am Chem Soc 109: 7875, 1987.
136. C Sistha, RM Hathorn, TJ Marks. J Am Chem Soc 114: 1112, 1992.
137. AR Siedle, RA Newmark. J Organomet Chem 497: 119, 1995.
138. A Anderson, HG Cordes, J Herwig, W Kaminsky, A Merck, R Mottweiler, J Pein, H Sinn, HJ Vollmer. Angew Chem Int Ed Engl 15: 630, 1976.
139. H Sinn, W Kaminsky, HJ Vollmer, R Woldt. Angew Chem Int Ed Engl 92: 396, 1980.
140. W Kaminsky, A Bark, R Steiger. J Molec Catal 74: 109, 1992.
141. CJ Harlan, SG Bott, AR Barron. J Am Chem Soc 117: 6465, 1995.
142. M Bochmann, LM Wilson. J Chem Soc, Chem Commun: 1610, 1986.
143. M Bochmann, LM Wilson, MB Hursthouse, RL Short. Organometallics 6: 2556, 1987.
144. M Bochmann, AJ Jaggar, JC Nicholls. Angew Chem Int Ed Engl 29: 780, 1990.
145. RF Jordon, RE LaPointe, PK Bradley, N Baenziger. Organometallics 8: 2892, 1989.
146. RF Jordon, PK Bradley, NC Baenziger, RE LaPointe. J Am Chem Soc 112: 1289, 1990.

147. JJW Eshuis, YY Tan, A Meetsma, JH Teuleen, J Renkema, GG Evens. Organometallics 11: 362, 1992.
148. R Taube, L Krukowka. J Organomet Chem 347: C9, 1988.
149. HW Turner, GG Hlatky. Eur Pat Appl 277003, 1988; Chem Abstr 110: 58291b, 1989.
150. X Yang, CL Stern, TJ Marks. Organometallics 10: 840, 1991.
151. JA Ewen, MJ Elder. Eur Pat Appl 426638, 1991; Chem Abstr 115: 136987c, 1991; Eur Pat Appl 426637, 1991; Chem Abstr 115: 136988d, 1991.
152. GG Hlatky, DJ Upton, HW Turner. US Pat Appl 459921, 1990; Chem Abstr 115: 256897, 1991.
153. X Yang, CL Stern, TJ Marks. J Am Chem Soc 113: 3623, 1991.
154. X Yang, CL Stern, TJ Marks. Angew Chem Int Ed Engl 31: 1375, 1992.
155. M Bochman, SJ Lancaster. Angew Chem Int Ed Engl 33: 1634, 1994.
156. H Sinn, W Kaminsky. Adv Organomet Chem 18: 99, 1980.
157. R Fusco, L Longo, F Masi, F Garbassi. Macromolecules 30: 7673, 1997 and references therein.
158. D Cam, U Giannini. Makromol Chem 193: 1049, 1992.
159. C Sishta, R M Hathorn, TJ Marks, J Am Chem Soc 114: 1112, 1992.
160. I Tritto, S Li, MC Sacchi, S Sacchi, G Zannoni. Macromolecules 26: 7111, 1993.
161. I Trito, MC Sacchi, S Li. Macromol Rapid Commun 15: 217, 1994.
162. FS Dyachkovskii, AK Shilova, AE Shilov. J Polym Sci Part C 16: 2333, 1967.
163. X Yang, CL Stern, TJ Marks. J Am Chem Soc 113: 3623, 1991.
164. M Bochmann, SJ Lancaster. J Organomet Chem C1, 419: 434, 1992.
165. R Fusco, L Longo, F Masi, F Bargassi. Macromol Rapid Commun 18: 433, 1997.
166. DE Richardson, NG Alameddin, MF Rayan, T Hayes, JR Eyler, AR Siedle. J Am Chem Soc 118: 11244, 1996.
167. L Resconi, S Bossi, L Abis. Macromolecules 23: 4489, 1990.
168. AR Baron. Macromol Symp 97: 15, 1995.
169. R Fusco, L Longo, F Masi, G Fabio. Macromolecules 30: 7673, 1997.

2

Recent Progress on Single-Site Catalysts

Jun-ichi Imuta and Norio Kashiwa
Mitsui Chemicals, Inc., Tokyo, Japan

I. INTRODUCTION

Nowadays, production of polyolefins has grown into a huge industry, and the amount of polyolefins produced is about 46 million metric tons per year throughout the world. These developments are supported by enormous research activity in catalysts and process technologies, which have followed an "S" curve over the last 60 years. This simple pattern shows their introduction, growth and stabilization or maturity [1].

High-pressure polyethylene (PE) was found in 1933. The wide structural distribution of the obtained polymer showed that more than one reaction was occurring during polymerization. High-pressure PEs polymerize off an existing chain due to a non-specific free radical mechanism.

Ziegler–Natta PE was produced at low temperature and low pressure conditions in the early 1950s. Since the finding of Ziegler–Natta catalysts, catalysts have been developed aimed at the simplification of the production process from the viewpoint of resources and energy savings. Of them all, development of a series of highly active $MgCl_2$ supported Ti catalysts accompanied excellent catalyst performance (high reactivity, high stereospecificity), and brought about the simplification of the production process (non-deashing process, removal of atactic PP) to make it nearly perfect in the 1980s. However, Ziegler–Natta catalysts have the property of providing polymers with a wide molecular weight distribution (MWD) and composition distributions due to the multiple active sites (multi-site catalyst; MSC).

In contrast, metallocene catalysts, which can control (1) composition distribution, (2) MWD, (3) incorporation of various comonomers, and (4) stereoregularity are the next S curve in the polyolefin development cycle.

As early as 1953, the Hercules group (Breslow) and the Natta group studied homogeneous Cp_2TiCl_2/AlR_2Cl as a catalyst model for mechanistic investigation. Spectroscopic investigation in Wilmington and electrodialysis experiments in Moscow implicated a cation species [2] as the catalytic intermediate as shown in Eqs. (1) and (2).

$$Cp_2TiCl_2 + AlR_2Cl \rightarrow [Cp_2Ti^+R \cdot Al^-RCl_3] \qquad (1)$$

$$[Cp_2Ti^+R \cdot Al^-RCl_3] \rightarrow Cp_2Ti^+R/Al^-RCl_3 \qquad (2)$$

This catalyst system shows low ethylene polymerization activity, viz., 1×10^3 to 4.2×10^3 g PE/g Ti h atm, due to strong ion coupling [reverse of Eq. 2] and nucleophilic attack of the anion on the cation [reverse of Eq. (1)] [3]. However, the addition of small amounts of water was found to increase the activity of these catalyst systems significantly.

In the 1980s Sinn, Kaminsky, and coworkers found that metallocenes such as Cp_2ZrCl_2 and Cp_2ZrMe_2 in conjugation with methyl aluminoxane, which is obtained from trimethyl aluminum and water, polymerize ethylene with activities exceeding 25×10^6 g PE/g Zr h atm [4]. A polymer thus obtained showed controlled MWD and comonomer incorporation, because all the active sites behave in the same way

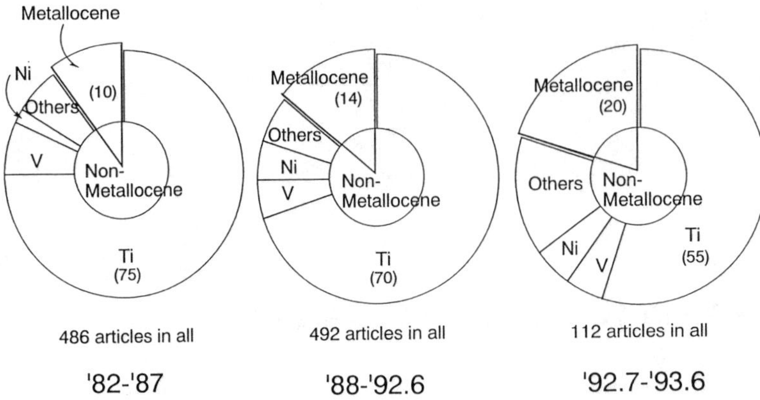

FIG. 1 Change of the object of study in the field of Ziegler–Natta-related catalyst.

during homopolymerization and copolymerization. Consequently, metallocene catalysts are called single-site catalysts (SSC).

Additionally, metallocene catalysts enable the design of catalysts for tailored polyolefins due to the intrinsic nature of the single site. Actually, new polymers which could never have been produced by conventional Ziegler–Natta catalysts, i.e., syndiotactic polypropylene, syndiotactic polystyrene, long chain branched polyolefins, cyclo-olefin polymer, and styrene copolymers, can be obtained by metallocene catalysts. This upcoming new S curve in the polyolefin development cycle means the evolution of new type of polyolefins.

Here, we review (1) research activities in metallocene catalysts, (2) polymerization performances of metallocene catalysts and other single-site catalyst technologies, with examples for polyethylene (PE), cyclo-olefin copolymer (COC), polypropylene (PP), syndiotactic polystyrene (SPS), and cyclo-olefin polymers, and (3) the computational design of metallocene catalysts.

II. RESEARCH ACTIVITIES IN METALLOCENE CATALYSTS

A. The Rapid Progress of Metallocene Catalysts Research

In the research field of Ziegler–Natta catalysts, *ca.* 600 articles have been contributed to the main journals and *ca.* 1700 patents have been applied for in Japan alone during the period from 1988 to June 1993. Figure 1 shows the classification of the studies, which have been contributed since 1982, into the titanium catalyst, metallocene catalyst, vanadium catalyst, and nickel catalyst systems.

The research into titanium catalyst systems went up to 75% in the period 1982–1987. Recently, however, they have been reduced to 70% (1988–June 1992) and then to 55% (July 1992–June 1993). In contrast, while studies in metallocene catalyst systems comprised 10% in 1982–1987, they have increased to 14 and 20% in the later two periods, respectively. These data show that studies into metallocene catalyst systems are growing rapidly these days.

B. Metallocene Catalysts—What are They?

1. Metallocene

Metallocenes are a class of compounds in which cyclopentadienyl or substituted cyclopentadienyl ligands are π-bonded to the metal atom. The stereochemistry of biscyclopentadienyl (or substituted cyclopentadienyl)–metal bis(unidentate ligand) complexes $[MX_2(\eta^5 - L)_2]$ $(M = Zr, Hf, Ti)$ can be most simply described as distorted tetrahedra, with each $\eta^5 - L$ group occupying a single coordination position, as in Fig. 2 [5].

The spatial arrangement and electronic properties can be altered by changing the structure of the metallocene: different substituents can be placed on each C atom of the cyclopentadienyl ligands, and the bridge structure as well as the metal center can also be changed, etc. Sc, Y, Yb, Sm, Lu, and some with U and Th as a different kind of metal center, have been studied so far.

Each arrangement causes various catalytic properties to be altered such as activity, relative reactivity toward different monomers, and stereospecificity. This potential to control catalyst performance through

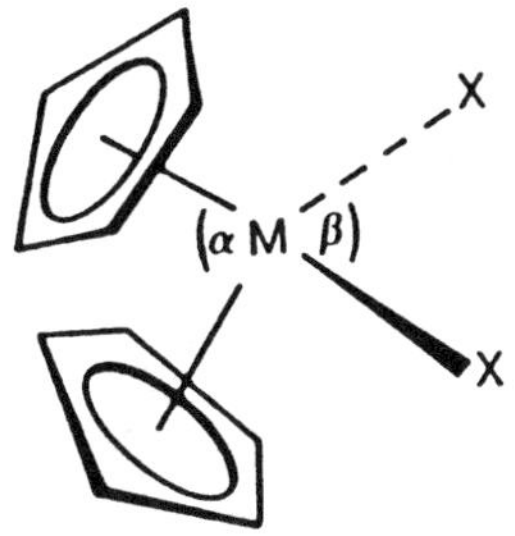

FIG. 2 Molecular Structure of Metallocene.

the chemical modification of the metallocene is its greatest advantage. Representative examples of each category of metallocenes are shown in Fig. 3.

2. Cocatalyst

Active species are formed by combination of metallocene and cocatalyst. Examples of cocatalysts are aluminoxane and boron compounds.

Simple synthetic routes to the methyl aluminoxane $[-O-Al(Me)-]_n$ are direct reaction between $Al(CH_3)_3$(TMA) and H_2O in a 1:1 molar ratio in a toluene solution. This reaction is extremely vigorous [6].

The formation of insoluble compounds was always observed even if hydrolysis conditions were carefully controlled as reported by Storr [7]. More reliable results could be obtained by the following synthetic procedure given by Sinn, Kaminsky, and Vollmer [8]. Many inorganic hydrated compounds such as $CuSO_4 \cdot 5H_2O$ [9], $Al(SO_4)_3 \cdot 6H_2O$ [10], $FeSO_4 \cdot 7H_2O$ [11],

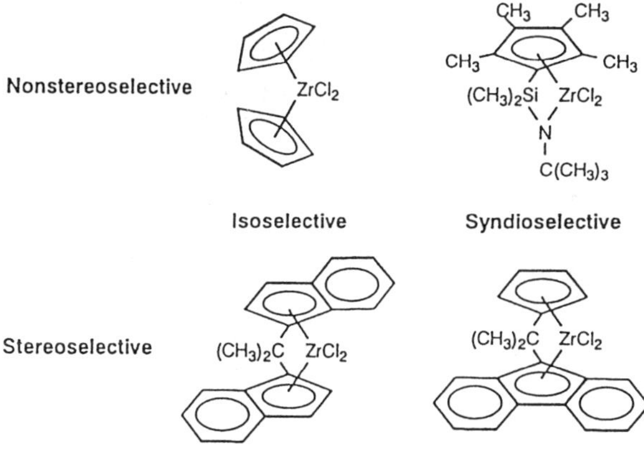

FIG. 3 Representative examples of metallocenes with classification by stereoselectivity.

and $MgCl_2 \cdot 6H_2O$ [12] are used as sources of water for preparing aluminoxane from alkylaluminum.

The structure of methylaluminoxane (MAO) was generally recognized as an oligomeric (cyclic or linear) chain with $[-O-Al(Me)-]$ units as shown in Fig. 4 [13], although the structure of MAO has not yet been clearly determined.

MAO is considered as an equilibrium mixture of the very strong Lewis acid (TMA) and oligomeric oxygen-containing species. Higher oligomers can, therefore, be obtained only by an appropriate shift of the dynamic equilibrium involved in Fig. 5 [11].

Spectroscopic data of MAO, such as NMR [14], mass spectra [4], UV [11], and IR, have already been reported. Their amorphous nature was confirmed by X-ray powder diffraction [11].

Recently, Sinn *et al.* proposed another possible structure of MAO. They found that the analytical data of purified MAO correspond to Eq. (3) [15]. So they proposed a structure like a soccer ball, in which trimethyl aluminum was held (Fig. 6) [16].

$$(CH_3)_2Al-(OAlCH_3)_5-O\cdot(CH_3)_2Al-(OAlCH_3)_5-OAl(CH_3)_2\cdot 3Al(CH_3)_3 \quad (3)$$

In the case of another kind of aluminoxane ($R = t-Bu$), Barron proposed a hexamer structure on the basis of X-ray structural analysis and he postulated the active species shown in Fig. 7 [17] from the experimental fact that ethylene polymerization proceeds by adding Cp_2ZrMe_2.

The cationic alkyl complexes of Group 4 metallocenes have been recognized as catalytically active species in polymerization reactions. Based on the chemistry found for the $Cp_2TiCl_2-AlR_2Cl$ catalyst systems [18], Chien [10] has assumed that MAO acts in the following sequence. The first step is to complex with metallocene [Eq. (4)], and the next step is to alky-

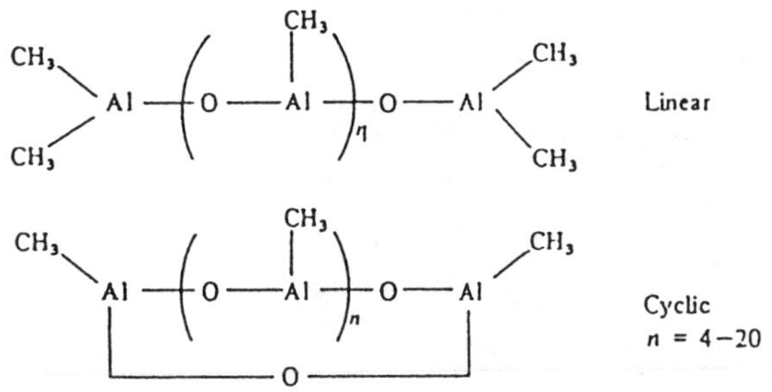

Linear

Cyclic $n = 4-20$

FIG. 4 Structures of methylaluminoxane.

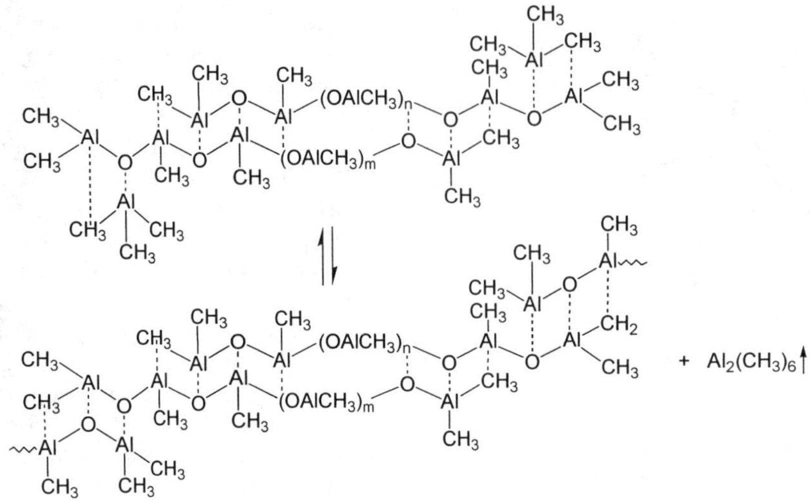

FIG. 5 Equilibrium complexes of the methylaluminoxane.

late the metallocene [Eq. (5)], resulting in the formation of catalytically active species:

$$Cp_2ZrCl_2 + MAO \rightarrow \underset{\textbf{(I)}}{Cp_2ZrCl_2 \cdot MAO} \quad (4)$$

$$Cp_2ZrCl_2 \cdot MAO \rightarrow \underset{\textbf{(II)}}{Cp_2ZrClMe \cdot MAO} \quad (5)$$

Propagation proceeds with repeated insertion of ethylene into the metal–alkyl bond as shown in Eq. (6):

$$Cp_2ZrClMe \cdot MAO + nC_2H_4 \rightarrow Cp_2ZrCl(C_2H_4)_n Me \cdot MAO \quad (6)$$

The second role of MAO is to act as a chain transfer agent.

$$Cp_2ZrCl(C_2H_4)_nMe \cdot MAO + MAO \rightarrow Cp_2ZrClMe{\cdot}MAO + MAO(C_2H_4)_nMe \quad (7)$$

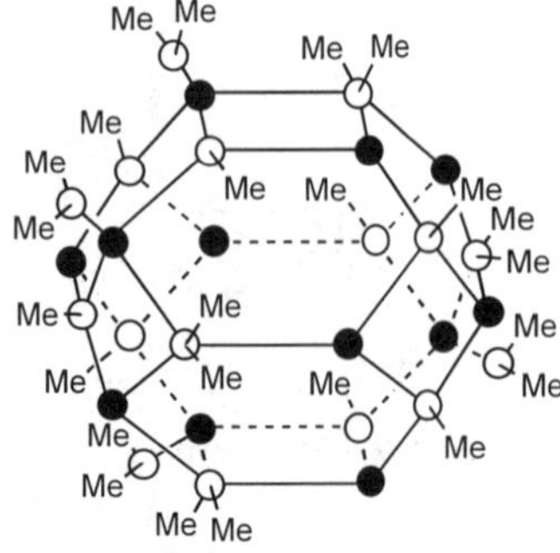

FIG. 6 Structure of methylaluminoxane proposed by H Sinn. ○: Al; ●: O.

Since TMA can be active in both the alkylation [Eq. (5)] and chain transfer [Eq. (7)], it seems plausible that MAO can be replaced in part by TMA.

Actually, many cationic metallocene catalysts have been prepared, and counter-anions of these ionic metallocenes are PF_6^-, BF_4^-, and $AlCl_4^-$ [19]. Because these counter-anions are tightly bound to metallocene cations, these complexes show no olefin polymerization activities.

However use of some kind of counter-anions such as tetraphenyl borate [20], tetrakis(pentafluoro)borate [21], carborane [22], and metallocarborane [23] gave high olefin polymerization activities. The representatives of these non-coordinated cationic metallocene catalysts are shown in Fig. 8 [24].

Recently, a new type of fluoroarylborane, tris(2,2′,2″-nonafluoro biphenyl)fluoroaluminate, was found. The steric effect of a substituted fluoroaryl borane plays an important role, and exhibit high polymerization activity at high temperature [25, 26].

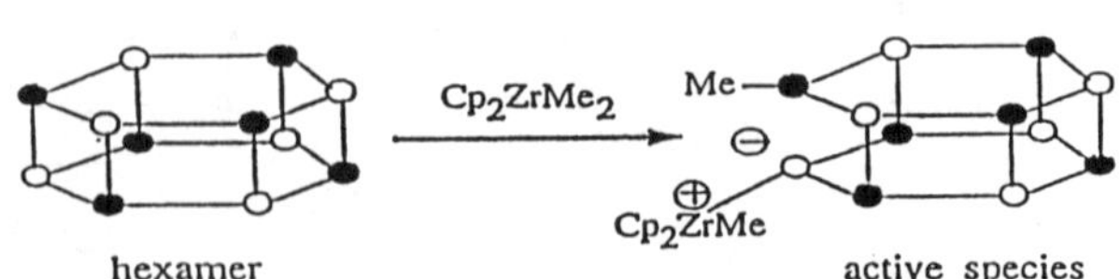

FIG. 7 Model for the formation of active species. ●: Al; ○: O: each Al has one *t*-Bu group.

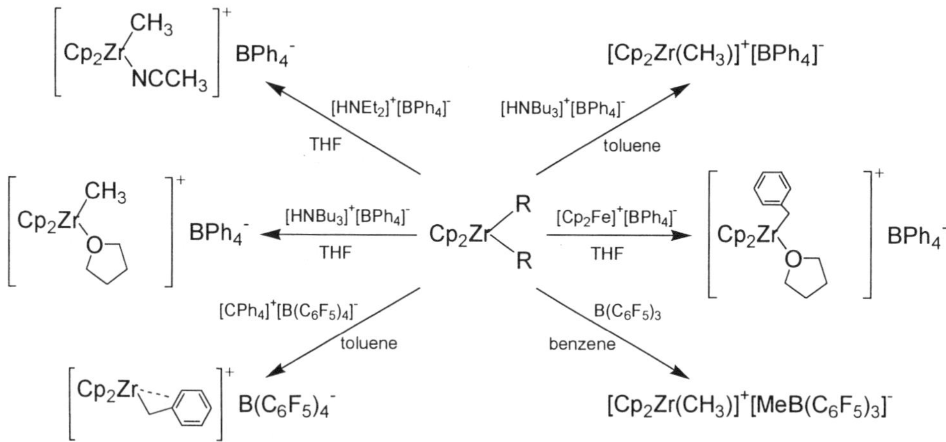

FIG. 8 Synthetic routes for cationic metallocenes.

3. Active Species

The first isolation and direct observation of highly active cationic metallocene catalysts was reported by Gassmann *et al.* [27], who used X-ray photoelectron spectroscopy (XPS) for characterization. Treatment of dichlorozirconocene, methylchlorozirconocene, or dimethylzirconocene with MAO in toluene gave a solution of a catalyst in which each zirconocene was all converted into a new zirconium derivative, having the same zirconium (IV) binding energy (182.4 eV). The results indicate the formation of a cationic metallocene with MAO as the counter-anion. Surface chemical [28], model synthetic [29], and theoretical [30] studies also suggest that the primary role of the Lewis acidic [31] aluminoxane may be to afford "cation-like" catalytic centers [(**III**) in Eq. (8)] [32].

$$Cp_2Zr(CH_3)_2 + [Al(CH_3)O]_n \rightarrow Cp_2ZrCH_3^+\{CH_3[Al(CH_3)O]_n\}^- \quad \textbf{(III)} \qquad (8)$$

In "cation-like" catalytic centers (**III**), the counter-anion is MAO and methyl anion derived from metallocene. Imuta *et al.* have pointed out that a stable ionic compound like (**V**) is the key structure for high polymerization activities, and proposed that the sulfonate anion from zirconium sulfonate (**IV**) forms stable "cation-like" catalytic centers (**V**) [Eq. (9)] [33]. This zirconium sulfonate (**IV**) and MAO shows higher activities compared to the corresponding chloride system:

$$\underset{\textbf{(IV)}}{Cp_2ZrCH_3(OSO_2R)} + [Al(CH_3)O]_n \rightarrow \underset{\textbf{(V)}}{Cp_2ZrCH_3^+\{(OSO_2R)[Al(CH_3)O]_n\}^-} \qquad (9)$$

Spectroscopic evidence for the formation of cationic species is reported by Marks *et al.* who applied solid-state ^{13}C CPMAS–NMR techniques to the solid (immobilized) catalytic species $\{Cp_2Zr(^{13}CH_3)_2$ with MAO$\}$ [32]. The study at different Al/Zr ratios provided the direct observation in the reaction between a metallocene dialkyl and MAO of a "cation-like" $Cp_2Zr^+CH_3$ species, the entirety of which undergoes facile ethylene insersion. The data also indicate that the Al:Zr stoichiometry required to form this cation is considerably lower than that employed in a typical catalytic reaction.

III. POLYMERIZATION PERFORMANCES OF METALLOCENE CATALYSTS

A. Ethylene Polymerization

PE is classified into three groups: low density PE (LDPE), high density PE (HDPE), and linear low density PE (LLDPE). The main difference between these PEs is their stiffness as shown in Fig. 9. Recently, very low density PE (VLDPE) and ultra low density PE (ULDPE) have joined the LLDPE family. So the PE family seems to cover a wide stiffness range from HDPE to elastomer. Here, we describe PE in general, and LLDPE and elastomer in further detail.

Kaminksy and coworkers found that a soluble catalyst styem comprising a bis(cyclopentadienyl) zirconium compound and methylaluminoxane (MAO) showed very high activity (10^8 g/g Zr h) for ethylene polymerization in 1980 [34]. On the basis of this catalyst performance, the alkyl substituted zirconocenes were prepared, and the effects of substituents on poly-

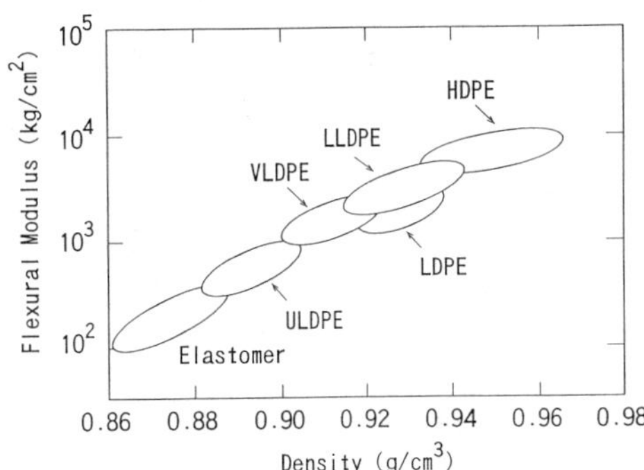

FIG. 9 Stiffness of various types of polyethylene.

merization activities and molecular weights were examined (Table 1) [35–37].

The importance of a balance between steric and electronic properties of the Cp ligand is apparent from the comparisons of Cp_2ZrCl_2 with the complexes having alkyl Cp substituents.

The bulky pentamethylcyclopentadienyl (Cp*) ligands decrease the ethylene polymerization rate and molecular weight. The decrease of ethylene polymerization rate is brought about by reducing both K_c and K_m (Scheme 1) [38].

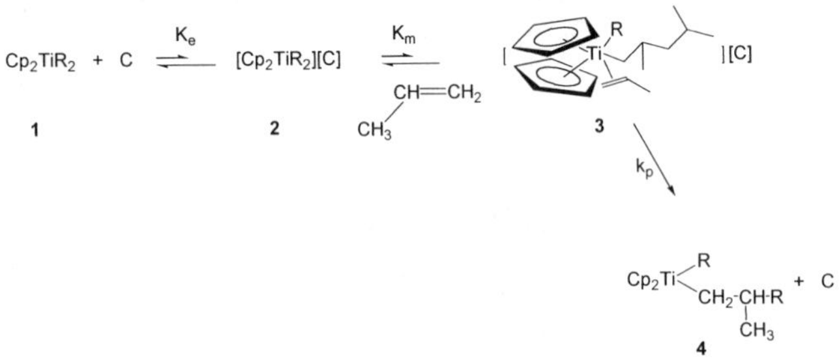

$$\underset{\mathbf{1}}{Cp_2TiR_2} + C \underset{}{\overset{K_e}{\rightleftharpoons}} \underset{\mathbf{2}}{[Cp_2TiR_2][C]} \overset{K_m}{\rightleftharpoons} \underset{\mathbf{3}}{[\ldots]}[C] \xrightarrow{k_p} \underset{\mathbf{4}}{Cp_2Ti(R)(CH_2\text{-}CHR\text{-}CH_3)} + C$$

SCHEME 1 Propylene polymerization with catalysts derived from $Cp_2Ti(Ph)_2$ [44].

TABLE 1 Effect of Substituted Cp Ligand

Catalyst	M_w	M_n	MWD	Pol. Act. (kg/g-M h atm)
Cp_2ZrCl_2	212,000	39,500	3.5	252
$(MeCp)_2ZrCl_2$	212,000	55,900	3.8	467
$(EtCp)_2ZrCl_2$	171,000	44,700	3.8	306
$(Cp^*)_2ZrCl_2$	63,000	13,200	4.7	71

Reaction conditions: 80°C, 60 psi, Al/Zr = 24,000, 30 min.

The reduction in molecular weight is not a consequence of increased termination rates. This conclusion is supported by the fact that the higher basicity of Cp* decreases k_t (termination by β-hydride elimination) relative to Cp in propylene polymerizations by lowering the acidity of the metal [39].

The monoalkyl substituted Cp ligands show an interesting result in that the catalyst with MeCp is most active and gives a higher molecular weight than the catalysts with EtCp and Cp ligands. The MeCp ligand receives the benefit of increased electron density at the metal (increasing k_p/k_t) with relatively little opposing steric effects on ethylene coordination.

The narrow molecular weight distributions obtained with individual metallocenes can be broadened considerably by adding controlled amounts of another kind of catalyst to the polymerization system. Further, the higher molecular weights obtained with Ti metallocene relative to Zr metallocene under identical polymerization conditions have permitted the use of mixed metal catalysts for the syntheses of materials with GPC elution curves having resolved bimodality (Fig. 10). Simply varying the Ti/Zr ratio permits tailoring of molecular weight distribution (Mn between 16,000 and 63,000 [38]).

Organolanthanide compounds are a new class of catalysts which are capable of polymerizing olefins. In 1990, Yasuda *et al.* discovered that Cp* organolanthanide compounds are excellent catalysts for the living polymerization of ethylene. A variety of isolated and nonisolated lanthanide complexes have been used in the polymerization of ethylene. Typical examples are shown in Table 2 [40]. Surprisingly, organolanthanide

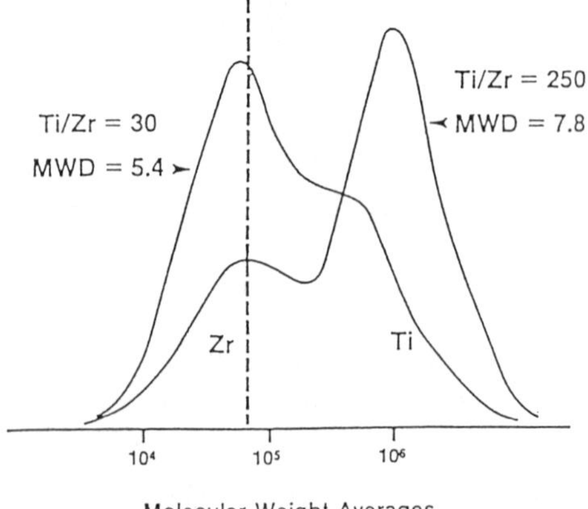

FIG. 10 GPC elution curves for HDPE produced with Cp_2TiPh_2 and Cp_2ZrCl_2 mixtures [44].

TABLE 2 Ethylene Polymerization by Organolanthanide Initiators

Reference	Initiators
$Ln(C_5Me_5)_2(Ln = Eu, Sm, Yb)$	
$LnCl_3/PBu_3/TiCl_4$	$M_w/M_n = 4$
$[LnMe(C_5H_4R)_2]_2 (M = Yb, Er. R = H, Me, SiMe_3)$	$M_w/M_n = 1.5–2.5$
$Ln(C_5H_4R)_2Me_2AlMe_2$ (M = Y, Er, Ho, Yb. R = H, Me)	$M_n = 1.7–14 \times 10^3$
$LnH(C_5Me_5)_2$ (Ln = La, Nd, Lu)	$M_w/M_n = 1.37–6.42$, $M_n = 96–676 \times 10^3$
$YbCl_3/AlEt_2Cl$	Alkylcyclopentanes

compounds can promote not only homopolymerization but also copolymerization with MMA and can yield high molecular weight copolymers [40].

1. LLDPE

LLDPE can be produced by copolymerization of ethylene with a comonomer, i.e., 1-butene, 1-hexene, 1-octene, and has excellent properties (impact strength and heat resistance, etc.).

In the case of Ziegler–Natta catalysts (MSC), the studies to control molecular structure such as Mw, MWD, and composition distribution (CD) have been extensively carried out because this molecular control was necessary to develop VLDPE and ULDPE. To produce the narrow CD LLDPE, a great deal of effort went into research to improve the traditional Ziegler–Natta catalyst (MSC), but this technology still has its limit.

On the other hand, the metallocene catalyst is a single-site catalyst (SSC), so the precise control of molecular structure unattainable via existing MSC becomes possible. Here are shown the characteristics of ethylene copolymers with 1-hexene produced by gas phase polymerization. The comonomer content of MSC's LLDPE varies according to the length of the molecular chains. The short chains tend to have more comonomers than the long chains have (Fig. 11) [41].

These short chains are undesirable fractions that cause low performance such as high blocking characteristics and low clarity of the film, etc. In the case of SSC's LLDPE, MWD and CD are very narrow compared with MSC's LLDPE (Fig. 12) and therefore it contains a much smaller amount of the n-decane soluble fraction (Fig. 13) [42].

2. Elastomer

Many researches on random copolymers in the presence of metallocene catalyst have been undertaken. Reactivity ratios r_1 and r_2 in the copolymerization of ethylene (1) or with propylene (2) performed in the presence of different metallocene/MAO catalytic systems are shown in Table 3.

The numerical values of the obtained elastomer, $r_1 \cdot r_2$, are less than 1, and r_1 varies from 250.0 to 1.3 while the symmetry of the metallocene is from C_{2v} to C_2 to C_s. According to the literature, for the metallocene catalysts the chain propagation rate increases in the same order [43, 44]. These phenomena should be commonly rationalized in terms of nonbonded interactions of the incoming monomer with some carbon atoms of the growing chain, e.g., the carbon at the β- or even γ-position from the metal [45].

With stereorigid C_2 or C_s symmetry of the metallocenes the stereospecificity of the incoming propylene could be strictly regulated, while for C_{2v} symmetry of the metallocenes the favored conformation of the growing chain could be such that the corresponding carbon atom protrudes in the direction of the incoming monomer. Such a conformation at the end of the growing chain should slow down the insertion rate, especially for the bulkier monomers (propylene) and consequently would lead to a large r_1. Metallocene catalysts incorporate higher α-olefins, which are difficult to polymerize with conventional catalysts. The

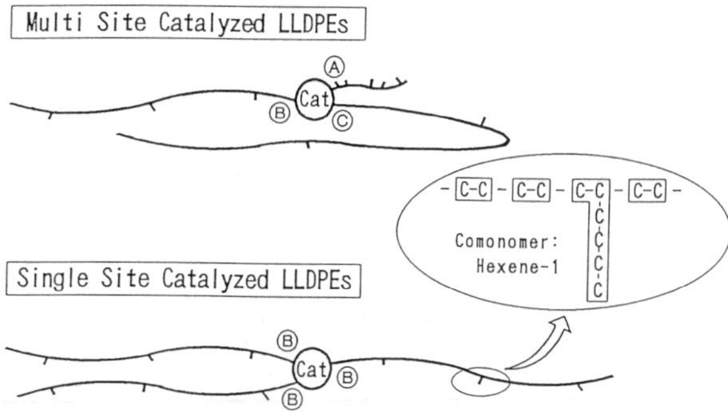

FIG. 11 Comparison between multi-site and single-site catalysts [47].

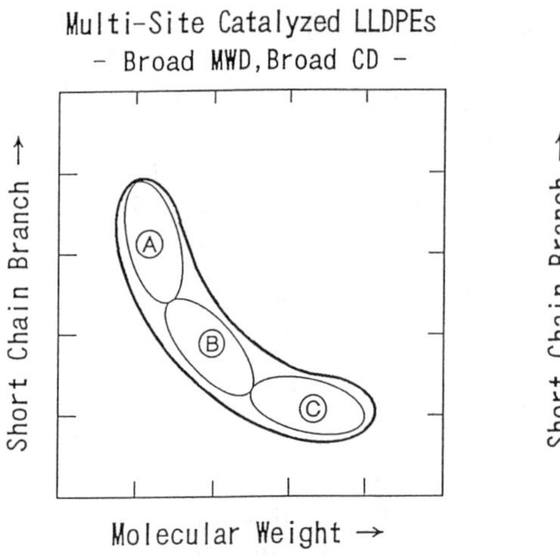

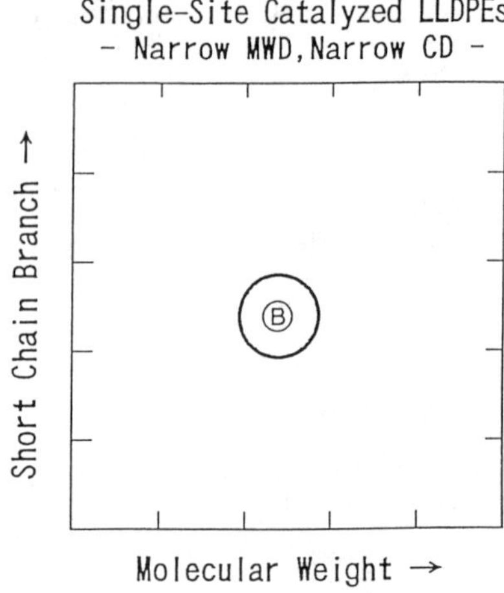

FIG. 12 MWD and CD characteristics for multi- and single-site catalysts.

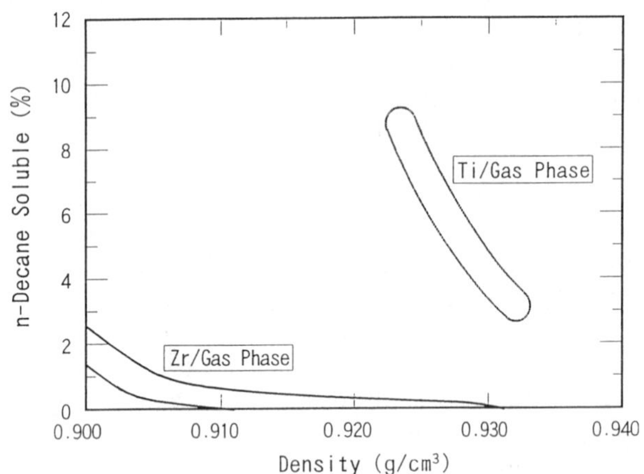

FIG. 13 *n*-Decane soluble fractions of LLDPEs obtained by Zr and Ti catalysts.

metallocene structure is the key for controlled comonomer incorporation, especially for benzannelation and 2-methyl substitution for indene is excellent for the incorporation of 1-octene [46].

Metallocene catalysts have a unique feature of polymerizing cyclo-olefin monomers (i.e., cyclopentene, norbornene) selectively without ring opening, and also enable copolymerization with ethylene [47, 48]. Application of metallocene catalysts to cyclo-olefin copolymer (COC) will be discussed in Section IV.

The terpolymerization, ethylene–propylene–diene, with high polymerization activity (100–1000 kg/mol Zr h) and high incorporation of diene (0.5–8%) is also reported [49, 50].

A common type of metallocene includes two Cp ligands; however, another type of metallocene which includes one Cp ligand is known (Fig. 14). Catalyst 1a

TABLE 3 Reactivity Ratio r_1, r_2 in Copolymerization of Ethylene (1) with Propene (2) in the Presence of Metallocene/MAO

Catalyst	Symmetry	Temp.	r_1	r_2	r_1r_2	Ref.
$(Cp^*)_2ZrCl_2$	C_{2v}	50	250	0.002	0.50	[44]
$(MeCp)_2ZrCl_2$	C_{2v}	50	60.0	—	—	[44]
Cp_2ZrCl_2	C_{2v}	50	48.0	0.015	0.72	[44]
Cp_2ZrCl_2	C_{2v}	40	16.5	0.031	0.51	[49]
Cp_2ZrMe_2	C_{2v}	60	31.5	0.005	0.25	[50]
$Me_2SiCp_2ZrCl_2$	C_{2v}	50	24.0	0.029	0.70	[50]
$Et(Ind)_2ZrCl_2$	C_2	50	6.61	0.060	0.40	[53]
$Et(Ind)_2ZrCl_2$	C_2	25	6.26	0.110	0.69	[53]
$Et(Ind)_2ZrCl_2$	C_2	0	5.20	0.140	0.73	[61]
$Et(Ind)_2ZrCl_2$	C_2	50	2.57	0.390	1.00	[52]
$Me_2C(Cp)(Flu)ZrCl_2$	C_s	25	1.30	0.200	0.26	[61]

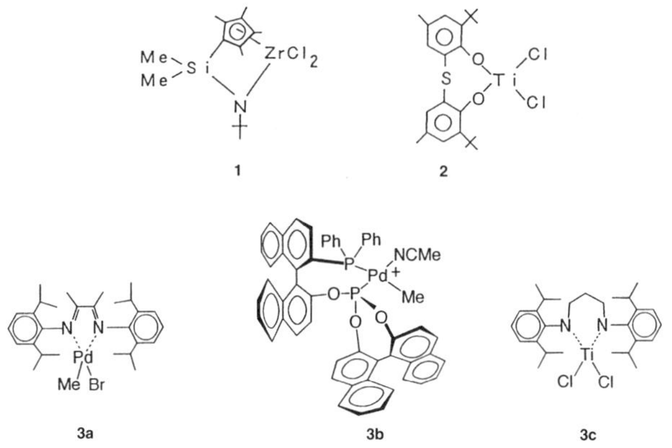

FIG. 14 New type of catalysts including one or no Cp ligand.

is a mono Cp type metallocene, which is useful for LLDPE and elastomers, and is now available in the market as AFFINITY and ENGAGE grades [51]. Another mono Cp type metallocene 1b using a lower-valency metal system is also announced. Catalysts 2 [52] and 3 [53] are new type catalysts, which have no Cp ligand, and produce an ethylene–styrene copolymer. In particular, 3a, 3b, and 3c have unique chemical structures. 3a produces a highly branched and methyl acrylate copolymer [54, 55], 3b enables highly enantioselective synthesis of ethylene–CO alternative copolymer [56], and 3c enables living polymerization of 1-hexene [57].

Recently, research into new organometallic compounds has accelerated. Group 4 metal–alkyl complexes incorporating 8-quinolino ligands [58] and bidentate pyridine-alkoxy ligands [59] are reported by Jordan. Moreover, it was reported that a very simplified catalyst, Ni(II)(dppe) or Pd(II)(dppp), produced hyperbranched PE by Sen [60].

B. Propylene Polymerization

1. Isotactic PP

A heterogeneous catalyst, e.g., $MgCl_2$ supported Ti catalyts, can produce highly isotactic PP, whose isotacticity is sufficient for commercial use. On the other hand, a homogeneous catalyst, Cp_2ZrCl_2 and MAO, gave atactic PP. However, some stereorigid chiral compounds consisting of zirconium or hafnium matallocenes in the presence of MAO are good catalysts for isospecific or syndiospecific polymerization. Here we describe the metallocene catalysts which produce isotactic PP in detail. There are two categories of catalyst to produce isotactic PP; C_2-symmetric metallocenes and C_1-symmetric metallocenes.

Ewen and coworkers have reported that a mixture of atactic PP and isotactic PP (37:63) was obtained with ethylene-bridged idenyl Ti complexes in 1984 [61]. They have supposed that isotactic PP could be derived from the chiral structure of Ti complexes, because Ti metallocenes are composed of mesoisomers (achiral structure) and racemic enantiomers (chiral structure) (44:56). On the other hand, Brintzinger and coworkers have succeeded in the isolation of racemic enantiomers of Zr metallocenes with the same ligand structure (catalyst 4 in Table 4) [62], and Kaminsky and coworkers have confirmed that isotactic PP is obtainable with this racemic enantiomer of Zr metallocenes [63].

Isospecificity derived from racemic enantiomers can be understood by considering that the approach of the incoming prochiral propylene to the reactive metal–carbon bond must be controlled by the chiral structure of reactive centers (enantiomorphic site control). Pino and coworkers obtained (S)-2,4-dimethylheptane with high optical purity by trimerizing propene in the presence of (−)(R) -bis(1-tetrahydroindenyl)ethane zirconium dimethyl/MAO in the presence of H_2, which causes very fast chain transfer. As insertion occurs with *cis* stereochemistry, they supposed the prevailing chirality of the hydro-oligomers is supposed to indicate that the Re enantioface of propylene preferentially approached Zr—carbon bond (Fig. 15) [64, 65].

Isotactic polypropylene and poly(1-butene) were prepared in the presence of the same catalyst system (M = Ti) and ^{13}C-enriched triethylaluminum [66] by Zambelli and coworkers. The result clearly shows propylene is inserted to reactive Ti — ^{13}C — C bond, which supports the model (Fig. 15) or the mechanism of the steric control proposed by Corradini and coworkers [67, 68]. According to these authors the chiral structure of the stereorigid transition metal cation imposes the conformation of the alkyl ligand (growing chain during the propagation). The stereospecific nonbonded interactions then occur between the β-carbon atom of the alkyl ligand and the substituent of the incoming monomer rather than the indenyl ligands. Another study using an ab initio MO and MM calculation also supports these results [69].

Many metallocenes carrying alkyl substituted Cp ligands bridged with a Me_2Si group were prepared, whose prochiral face selectivity would be superior to catalyst 4. It was found that the C_2-symmetric catalyst 5 produces highly isotactic PP (T_m = 163°C) under a moderate polymerization condition (30°C) [70].

TABLE 4 Performance-Relevant Structural Features of C_2-Symmetric Metallocenes

No.	Metallocene	Temp. (°C)	M_w (10^4)	mmmm (%)	T_m (°C)
4	$Et(Ind)_2ZrCl_2$	30	3.6	81.3	142
5	$Me_2Si(2, 3, 5\text{-}Me_3Cp)_2ZrCl_2$	30	13.4	98.7	162
6	$Me_2Si(2\text{-}Me,4\text{-}t\text{-}BuCp)_2ZrCl_2$	50	0.9	94	149
7	$Me_2Si(Ind)_2ZrCl_2$	70	3.6	81.7	137
8	$Me_2Si(2\text{-}MeInd)_2ZrCl_2$	70	19.5	88.5	148
9	$Me_2Si(2\text{-}Me,4\text{-}iPrInd)_2ZrCl_2$	50	46.0	91.6	152
10	$Me_2Si(2\text{-}Me,4\text{-}NaphInd)_2ZrCl_2$	70	92.0	99.1	161
11	$Me_2Si(2\text{-}Me,4,5\text{-}benzoInd)_2ZrCl_2$	70	33.0	88.7	146

4

5: $R^1=R^2=Me$
6: $R^1=t\text{-}Bu$
$R^2=H$

7: $R3=R4=H$
8: $R^3=Me$, $R^4=H$
9: $R^3=Me$, $R^4=iPr$
10: $R^3=Me$, $R^4=Naph$

11

Brintzinger and coworkers reported highly isotactic PP was obtained with the similar Si-bridged catalyst 6, which contains a *t*-Bu group at the 4-(β-) position and a methyl group at the 2-(α-) position in each ring ligand [71]. They explained this observation based on structural models for the reactive olefin-coordinated zirconocene alkyl cation. An effect of 2-(α-)methyl substituent is suppression of 2,1 and 1,3 insertions. On the other hand, the β-agostic binding required for chain termination by β-H transfer places the polymer chain (P1) with a stereoregular unit into that ligand sector which is occupied by the substituent R^1 at the 4-(β-) position (Fig. 16a). Stereoregular olefin insertion might proceed via agostic binding of that one of the two α-H atoms which places the polymer chain in an open sector of the ligand framework (Fig. 16b).

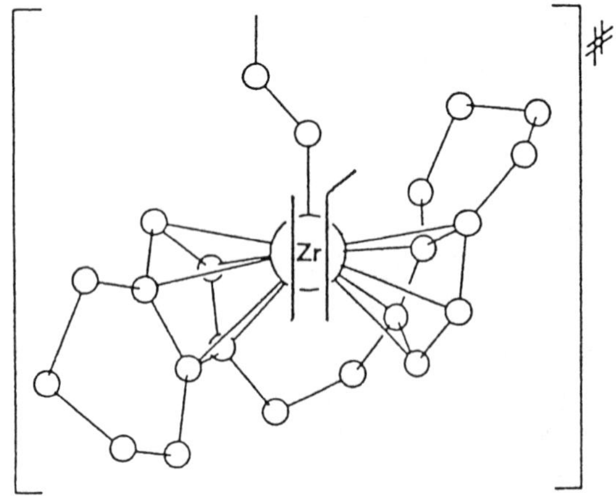

FIG. 15 A simplified stereochemical model of a possible transition state for the olefin insertion step [76].

There are a number of theoretical and some experimental supports [72] of this α-agostic Zr—H interaction, e.g., the fact that hydrocyclization of trans, trans-1,6-d_2-1,5-hexadiene affords trans-d_2-methylcyclopentane rather than the *cis* isomer [73].

The one-membered silicon bridge in catalyst 7 seems to be preferable to the two-membered unsubstituted ethylene bridge in catalyst 4, thus imposes higher rigidity and favorable electronic characteristics to the metallocene, and thus inducing higher isotacticity and molecular weight of the polymer [74]. The stereochemistry of PP produced by C_2-symmetric metallocene catalysts results from the interplay of two competing reactions, namely monomer polyinsertion and an epimerization of the polymer chain. Preliminary evidence for the existence of epimerization was reported in catalyst rac-Et(4,5,6,7-tetrahydro-Ind)$_2$ZrCl$_2$ [75]. Based on this assumption, it was demonstrated that the effectiveness of substituents strongly depends on their position and on a nonincremental synergism with alkyl substituents on the ligand frame, e.g., methyl substitu-

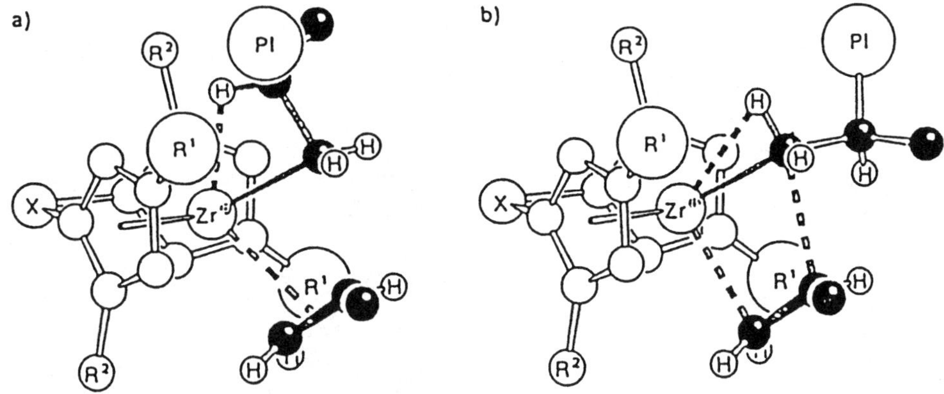

FIG. 16 Plausible model of (a) β- and (b) α-agostic interaction [86].

ent in the 2-position (catalyst 8) improves catalyst performance, especially molecular weight and stereospecificity [76], and isopropyl substituent in the 4-position (catalyst 9) is further effective, especially in its polymerization activity, and the molecular weight of the polymers [77]. The most striking effect was observed by introducing aromatic substituents, phenyl or naphthyl group, into the 4-position. The performance of catalyst 10 is nearly equal to the heterogeneous Ti catalysts [78].

It is also proposed that a combination of substituents at 2- and 4-positions is important for higher isotacticity. An especially favorable number of alkyl substituents at the 2-position is more than 2, and a favorable carbon number of the aryl substituent at the 4-position is more than 11.

The 4,5-benzoindenyl structure (catalyst 11) would be a logical candidate, combining aromatic annealation with 4-substitution of the indenyl six-membered rings, and was proposed by Spaleck [78] or Brintzinger [79]. The corresponding performance-relevant structural features are tabulated in Table 5.

Also, Zr complexes with modified benzindenyl ligand, Me_2Si(2-methylcyclopenta[1]phenanthry)$_2$-$ZrCl_2$, give highly active catalysts for the polymerization of propene to polymers with high isotacticities and molecular weights [75]. Stereochemistry of the first monomer insertion into the metal–methyl bond in

TABLE 5 Performance-Relevant Structural Features of C_s,C_1-Symmetric Metallocenes

No.	Metallocene	Temp. (°C)	M_w (10^4)	rrrr (%)	mmmm (%)	T_m (°C)
12	*i*Pr(Cp)(Flu)$_2$ZrCl$_2$	60	0.32	82.4	—	142
13	*i*Pr(3-MeCp)(Flu)$_2$ZrCl$_2$	60	0.50	19.2	19.2	—
14	*i*Pr(3-*t*-BuCp)(Flu)ZrCl$_2$	60	0.75	0.3	79.4	129
15	*i*Pr(3-*t*-BuCp)(3-*t*-BuInd)ZrCl$_2$	3	16.0	—	99.6 (mm)	162

Me, Me, R^2, R^1, $ZrCl_2$ Me, Me, t-Bu, $ZrCl_2$, t-Bu

12: R1=R2=H   15

13: R1=H, R2=Me

14: R1=H, R2=t-Bu

the benzoindenyl structure is reported [80]. The bulky benzannelated substituents lead to an almost total lack of steric control of the first monomer insertion. The methyl substitution in the 2-position of cyclopentadienyl ring does not seem to influence the choice of the face [81]. The superior performance effect of 2- or 4-substituent has been investigated theoretically with the ab initio molecular orbital (MO) method, where the structure and energetics of the reactants, the p-complex, the transition state, and the product were determined [82].

Recently, a convenient synthetic method for catalyst 7 is reported. The elimination reaction of $Me_2Si(Ind)_2H_2$ and $Zr(NMe)_4$ affords rac-$Me_2Si(Ind)_2Zr(NMe)_2$ in 65% isolated yield. This rac-$Me_2Si(Ind)_2Zr(NMe)_2$ is cleanly converted to catalyst 7 (100% NMR yield) [83].

Substituents on the Cp ring also affect catalyst performance. For example, introduction of the bulkier *t*-butyl group into the distal position (b-) of the Cp group, *i*Pr (3-*t*-BuCp-Flu)$ZrCl_2$ (catalyst 14), changes the character of the catalyst from syndiospecific to isospecific, and produces isotactic PP (mmmm:79%) under a mild polymerization condition (20°C) [84].

A different type of C_1 symmetric catalyst (catalyst 15) produces highly isotactic PP (mmmm:99%) also under a mild polymerization condition (0°C) [85]. Their performance-relevant structural features are illustrated in Table 5.

Recently, Kaminsky and Soga have reported that immobilization of metallocene on silica supports enables production of PP with high isotacticity, although polymerization activity is very low. The preparation method of immobilized catalysts is shown in Fig. 17 [86, 87].

In spite of the high activities of rare earth metal complexes for ethylene polymerization, they are inert to propylene due to the formation of η^3-allyl complexes as revealed by NMR spectra. Recently, an isospecific single component C_2 symmetric catalyst, [rac-$Me_2Si(2\text{-}SiMe_3\text{-}4\text{-}CMe_3C_5H_2)_2YH]_2$ (Fig. 18), was found, with which the polymerization of 1-pentene was performed over a period of several days, affording polymers ($M_n = 13{,}500$) with $M_w/M_n = 2.31$ [40].

Chain end analysis of the polypentene by ^1H NMR shows the presence of geminally disubstituted olefin resonances, indicating chain termination by β-H elimination. ^{13}C NMR analysis of the polymer at the pentad level shows a remarkably high degree of isotacticity. This catalyst also initiates the polymerization of 1-hexene and afford polymers ($M_n = 24{,}000$) with $M_w/M_n = 1.75$. A similar ^{13}C NMR analysis showed the tacticity of more than 95%.

2. Syndiotactic PP

Highly isotactic PP can be produced with C_2-symmetric metallocene catalysts. In such isotactic propagation, monomer insertion is to take place on two active sites on the metal center alternately, and the stereochemistry on one site must be the same as that on the other site. Highly syndiotactic PP has been

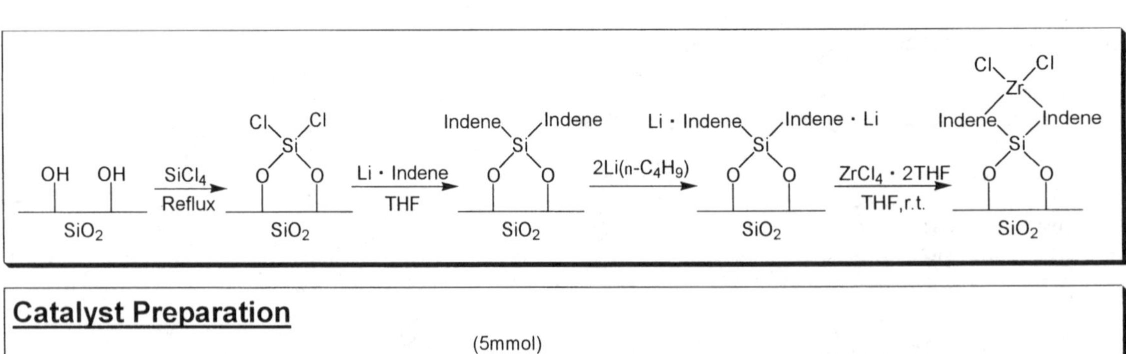

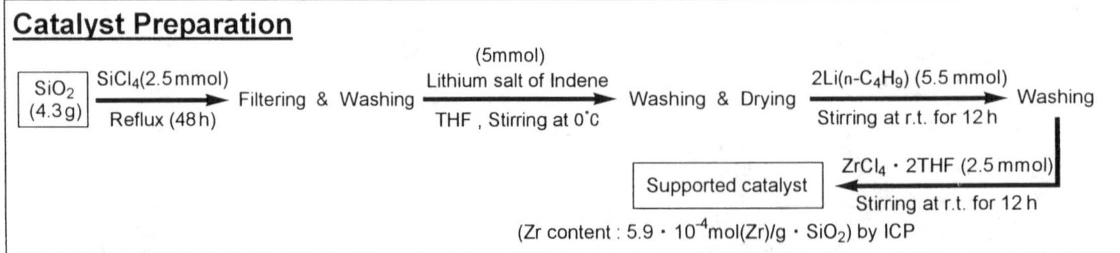

FIG. 17 Preparation method of immobilized catalysts.

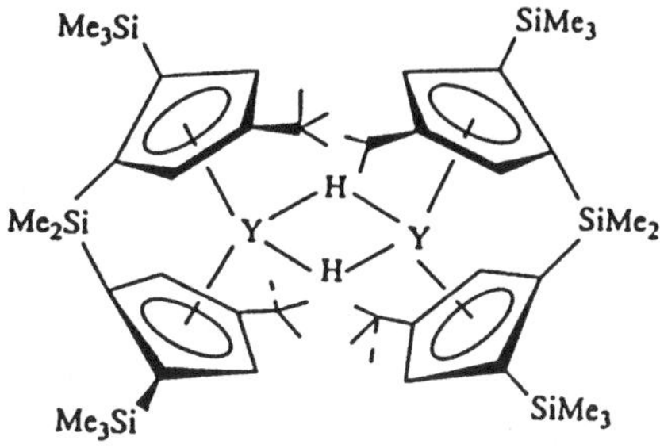

FIG. 18 Molecular structure of [rac-$Me_2Si(2\text{-}SiMe_3\text{-}4\text{-}CMe_3C_5H_2)_2YH]_2$.

obtained with the isopropylydene(cyclopentadienyl-1-fluorenyl) metallocene (catalyst 12 in Table 5), C_s-symmetric catalyst (catalyst 12 in Table 5) by Ewen and coworkers [88]. In syndiotactic polymerization, the stereochemistry of monomer insertion has to change from one site to the other.

Recently, a new type of C_s-symmetric catalyst to produce syndiotactic PP was reported [89]. The doubly-bridged zirconocene catalysts show very high syndiospecificity, especially under conditions of low temperature and high propylene concentration [90].

C_1 symmetric metallocene with β-Me introduced into the Cp ligand of catalyst 12 (catalyst 13) behaves differently with differing properties of each coordination site, and produces hemi isotactic PP [91].

Thus, the principle of manipulating the ligand structure is extremely versatile, and rational tailoring of the polymer structure (e.g., atactic PP, isotactic PP, hemi isotactic PP, syndiotactic PP) would be applied more widely in the future.

3. Microtacticity and Properties of Isotactic PP

Here we look into the differences between isotactic PP prepared by metallocene catalysts and that prepared by traditional supported Ti catalysts.

We have reported homopolymerization of propylene and its copolymerization with ethylene using catalyst 4/MAO systems [92]. Isotactic polypropylene obtained with catalyst 4/MAO catalyst systems in the range of reaction temperature from −30°C to 50°C was successively fractionated by boiling pentane, hexane, heptane, and trichloroethylene. It was found that the hexane-insoluble/heptane-soluble portion was the major fraction for all samples. The whole and fractionated polymers were characterized in comparison to those obtained with $MgCl_2/TiCl_4\text{–}Et_3Al$ catalyst system by ^{13}C NMR, GPC, and DSC. The ^{13}C NMR spectrum of the Zr system exhibited a number of small irregular peaks, which were not observed in the Ti system, showing a significant difference in microstructure between the two catalyst systems. They were assigned to the peaks arising from 2,1-insertion of propylene monomer (formation of head-to-head or tail-to-tail enchainment) and 1,3-insertion of propylene monomer (formation of $(CH_2)_4$- unit), based on (1) the additive rules for chemical-shift calculation, (2) DEPT (distortionless enhancement by polarization transfer) measurements for the determination of carbon species, and (3) two-dimensional NMR in ^{13}C-H correlation.

Furthermore, the relation between T_m and mm values was apparently different between Zr metallocene and $MgCl_2$ supported Ti catalyst, as seen in Fig. 19; for PP of the Ti system, two T_m values were plotted owing to the presence of two peaks in the DSC curves, while PP of the Zr system showed a single peak. T_m values of the Zr system are always more than 10°C or even 20°C lower than those of the Ti system having the same mm values. In other words, for example, the mm value to give $T_m = 150°C$ is 97% for the Zr system but only 90% for the Ti system.

These properties of the Zr-catalyzed polymers would be attributable to their characteristic microstructure of, i.e., the presence of head-to-head or tail-to-tail enchainment and the -$(CH_2)_4$- units in a polymer chain.

A blown film of isotactic homo-PP obtained with catalyst 4 (M=Hf) in conjunction with methylaluminoxane was characterized by its mechanical and thermal properties in comparison with those of commercially available homo- and random-PP obtained with a Ti catalyst system [93]. Table 6 shows data on the mechanical and heat sealable properties of the films, e.g., stiffness by tensile yield stress (YS), tensile strength (TS), toughness by film impact strength, and heat sealability.

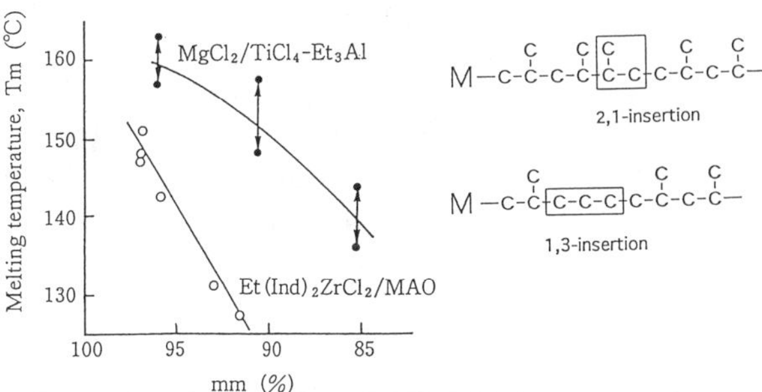

FIG. 19 Relationship between Tm and mm values of PP obtained by Zr and Ti catalysts PP: (○) Zr; (●) Ti.

TABLE 6 Properties of Hf and Ti Homo-Films

		Hf	Ti
Melting point	(°C)	133.7	164.1
Crystallinity	(%)	43.7	50.5
Thickness of lamella	(Å)	85	95
YS (I)/(II)	(kg/cm^2)	255/245	415/395
TS (I)/(II)	(kg/cm^2)	418/325	740/565
EL (I)/(II)	(%)	505/220	575/615
Impact strength	(kg cm/cm)	445	207
HST	(°C)	140	160

YS: Tensile yield stress.
TS: Tensile strength.
EL: Elongation.
(I): Measurement value for direction of tensile.
(II): Measurement value for transversal direction to (I).
HST: Lowest heat sealable temperature at which sealed strength is more than 800 g.

The values of YS and TS for the homo-PP film obtained by Hf metallocene catalyst were lower than thsoe for the homo-PP film obtained by $MgCl_2$ supported Ti catalyst, indicating that the former film is stiffer. This may be ascribed to the lower crystallinity of the Hf homo-film. The impact strength of the Hf homo-film was higher than that of the Ti homo-film. From the above features, the Hf homo-film is concluded to have properties clearly different from those of the Ti homo-film. Consequently, the film with the Hf system was found to show considerably different properties from homo PP film with the Ti system and relatively similar properties to random PP film with the Ti system. This is probably due to the differences and similarities in the microstructure of polymers.

In determining the film properties, the nonregiospecific units shown above play a role similar to that of ethylene units in the random copolymer, but the efficiency in changing the properties would be significantly different.

Performances of isotactic PP produced with catalyst 7 are different from those of isotactic PP obtained by $MgCl_2$ supported Ti catalyst [94]. Additionally, it is reported that the isotactic PP with a high melting point (T_m:160°C) produced with metallocene has a 35% higher modulus than conventional isotactic PP, which gives a higher indentation hardness [95].

4. Elastomeric PP

Propylene polymers are used as structural materials in the polyolefin industry. However, they can be used as elastomers by controlling catalyst stereospecificity. Some elastomers produced by characteristic catalysts are described below.

Catalyst 16, bridge Cp–Ind metallocene complexes, produce alternating (isoblock/atablock) stereoblock PP, because the two coordinating sites are chemically different [96]. Waymouth has investigated a closely related system and proposed a different mechanism to account for the stereoblock structures with a non-bridged 2-arylindenyl metallocene complex 17 [97]. This catalyst is designed to switch its coordination geometry from aspecific to isospecific during the course of polymerization in order to generate atactic and isotactic blocks. The influence of the structures and properties of the polymers by manipulation of the catalyst structure have been investigated, 1-substitution [98], and 1,3,4-substitution on the phenyl ring of bis(2-arylindene) metallocenes [99]. Moreover, replacement of one 2-aryl indenyl ligand to substituted Cp ring was carried (catalyst 18) [100].

C. Styrene Polymerization

Polystyrene (PS) now on the market is atactic PS (APS), but there is a problem of low heat resistance. Isotactic PS (IPS) is also known, but there is a problem of low crystallization rate. A homogeneous Ti/metallocene and MAO system is an effective catalyst for syndiotactic polystyrene (SPS). Advantages of SPS are heat resistance (T_m:270°C) and chemical resistance like engineering plastics, which are derived from its high crystallinity compared with APS produced by radical polymerization. Furthermore, the crystallization rate of SPS is faster than that of APS or IPS.

A comparative study was made for the stereoregularities of the polypropylene and the polystyrene formed by various metallocene catalysts is studied (Table 7) [101]. When the chiral metallocene was used, stereoregular polymers, IPP and SPS, were produced. In the syndiotactic polymerization of styrene, the secondary insertion occurred. On the other hand, in the case of isospecific polymerization of propylene,

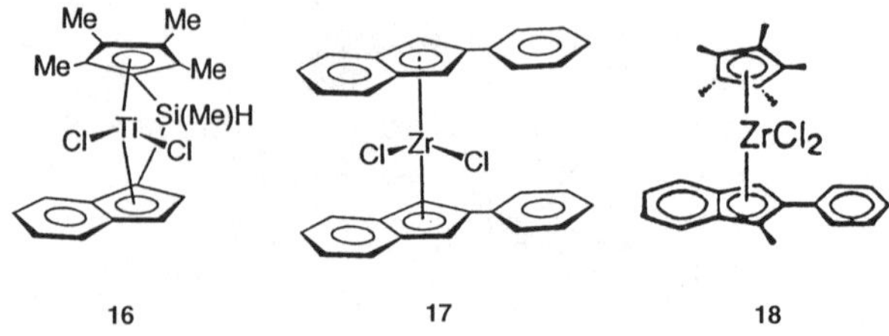

FIG. 20 Catalysts for elastic PP.

TABLE 7 Relationships Between the Catalysts and the Stereoregularities of the Polypropylene and Polystyrene Products

	Stereoregularity	
Catalysts	PP	PS
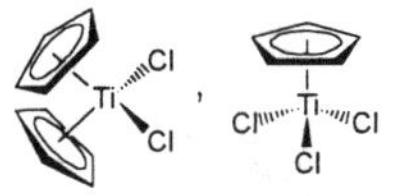	Atactic	Syndiotactic [rrrr] = 1.0
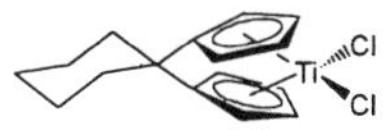	Atactic	Syndiotactic [rrrr] = 1.0
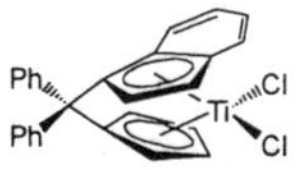	Isotactic [mm] = 0.57	Syndiotactic [rrrr] = 1.0
	Atactic [mm] = 0.19	Atactic
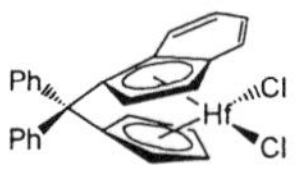	Isotactic [mm] = 0.52	Atactic
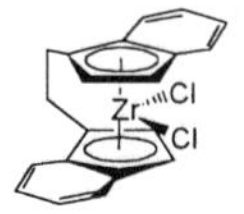	Isotactic	Atactic
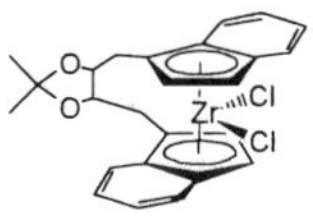	Isotactic [m] = 0.96	Atactic

the primary insertion could occur with the same catalyst. From these reults, the mechanism of syndiospecific polymerization of styrene should be different from that of isospecific polymerization of propylene described in the previous part.

IV. OTHER POLYMERS

A. Alternating Copolymers

Metallocene catalysts also offer the possibility of designing catalyst systems for nonrandom olefin copolymers. In particular, substituted Cp-fluorenyl complexes produce highly alternating ethylene/propylene copolymers [102]. A C_2-symmetric catalyst also produced the alternating copolymers, for examples, meso-$[Me_2Si)(1\text{-}MeInd)ZrCl_2]$/MAO systems give alternating ethylene/1-octene copolymers [103]. Another C_2-symmetric catalyst to produce ethylene/1-octene copolymers is rac-[isopropylidenebis(bnenzindenyl)] zirconium catalyst. This is the first example to produce a copolymer that consists of only isotactic alternating structure [104].

B. Cyclopentene (Co)polymers

Cyclopentene and other cyclic olefins can be polymerized using catalyst 4 in the presence of MAO catalyst to give polymers without detectable ring opening [47, 48]. In this system, cyclopentene is incorporated in a *cis*-1,3 manner in which the stereochemical relationship between cyclopentene rings in the polymer and hydrooligomers is isotactic [105]. Analogous reactions using a tetrahydroindenyl type of catalyst 4 leads to the production of oligomers in which cyclopentene is incorporated in a *cis*- or *trans*-1,3 manner [106].

C. MMA (Co)polymers

Yasuda and coworkers reported that MMA could be polymerized using lanthanocene initiators. They demonstrated a new enolate complex, formed in situ from the initiator and MMA, was responsible for propagation [107].

The homo- and copolymerization of methacrylates and acrylates and the synthesis of block copolymers is also reported [108]. Recently, it is reported that new organometallic compounds (late metals) also produce ethylene/MMA copolymers [54, 55]. The development of polymerization catalysts incorporating late transition metals is a promising area of research, since late metals are typically less oxophilic, and thus more functional group tolerant, than early metals. Examples of the functional group tolerance of late metals in insertion-type reactions include reports on Ru, Rh, Ni, and Pd catalysts.

Bimetallic complexes of Sm were used for the living bis-initiated polymerization of MMA and ε-caprolactone, giving polymers with discrete functionalities at the center of the backbone ("link-functionalized") [109]. When the bis-allyl complex $[Cp^*_2Sm(\mu\text{-}\eta^3\text{-}CH_2\text{-}CH\text{-}CH\text{-})]_2$ [110], depending on the mode of attack during initiation, the link functionality resulting from this process should comprise either two pendant vinyl groups or a backbone 1,3-diene unit.

V. COMPUTATIONAL DESIGN OF METALLOCENE CATALYSTS

Metallocenes as homogeneous catalysts have been studied theoretically compared to heterogeneous catalysts. Most of the theoretical studies are related to the mechanism of stereoregulation in α-olefin polymerization and the prediction of stereoselectivity for the design of new catalysts [111].

In the study on the olefin insertion, the most important step in the polymerization reaction, various methods are discussed and all of them support the Cossee mechanism and the presence of α-agostic effects [112, 113].

In particular, Sc metallocenes were studied recently by the use of the GVB method [114] to examine the prediction of stereoselectivity and the possibility of olefin insertion based on the calculated activation energies [115].

There are three methods to predict the stereoselectivity of homogeneous Ziegler–Natta catalysts.

1. From the coordination structure of the olefin. This method has been applied to the stereoselectivity prediction of various types of metallocenes by using the calculation method used in heterogeneous catalyst systems by Corradini [116].
2. From the structure of the activated complex. This method has been reported by Rappe *et al.*, who calculated steric energies of the activated complex using moelcular mechanics [117].
3. From the structure of the transition state. This method has been adopted by Kuribayashi *et al.*, wherer the transition state of olefin insertion is estimated by the use of simplified catalyst models, and then model structures are constructed to evaluate steric energies using molecular mechanics [118].

At the present stage of theoretical study, it seems to be difficult to decide which step is the most important in determining the stereoselectivity in the olefin polymerization [119, 120]. In order to predict the stereoselectivity more precisely, it is necessary to develop a new method that can reliably reproduce the coordination structure of an olefin.

VI. CONCLUDING REMARKS

As discussed in this chapter, metallocene catalysts have unique advantages over conventional Ziegler–Natta catalysts. This new type of catalyst brought us tailor-made polymers which have not been produced by conventional catalysts. The innovation of single-site catalyst by metallocene catalyst and other single-site catalyst technologies (new organometallic catalysts) from multi-site catalysts was a remarkable event in polyolefin industry, as shown in Fig. 21.

However, a new concept of a tailored multi-site catalyst is indispensable in the next generation, because highly functionalized homopolymers and copolymers can be produced by the combined single-site catalysts. I conclude that this catalyst will open up new regions of the polyolefin industry in the future.

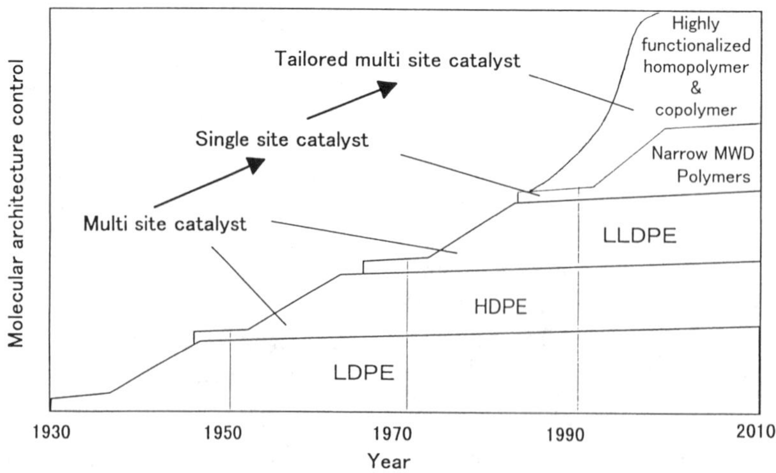

FIG. 21 Generations of innovation in polymerization catalyst.

REFERENCES

1. AA Montagna, JC Floyd. Hydrocarbon Processing/ March, 57, 1994.
2. CW Chien. In: Fink, Mulhaupt, Brintzinger, eds. Ziegler Catalysts. p. 199, 1995.
3. H Sinn and W Kaminsky. Adv Organomet Chem 18: 99, 1980.
4. H Sinn, W Kaminsky, HJ Vollmer, R Woldt. Angew Chem Int Ed Engl 19: 390, 1980.
5. DJ Cardin, MF Lappert, CL Raston. Chemistry of Organo-zirconium and -hafnium Compounds. London: Ellis Horwood, 1986.
6. A Anderson *et al.* Angew Chem Int Ed Engl 15: 630, 1976.
7. A Storr, J Jones, AW Laubengayer. J Am Chem Soc 90: 3173, 1968.
8. H Sinn, W Kaminsky, HJ Vollmer. Eur Pat 35242, 1980; Chem Abstr 95: 187927j, 1980.
9. K Peng, S Xiao. Maklomol Chem Rapid Comm 14: 633, 1993.
10. CW Chien, B-P Wang. J Polym Sci: Part A: Polym Chem 26: 3089, 1988.
11. E Giannetti, GM Nicoletti, R Mazzocchi. J Polym Sci Polym Chem Ed 23: 2117, 1985.
12. K Peng, S Xiao. J Molec Catal 90: 201, 1994.
13. W Kaminsky, R Steiger. Polyhedron 7: 2375, 1988.
14. N Veyama, T Arai, H Tani. Inorg Chem 12: 2218, 1988.
15. H Sinn *et al.* In: W Kaminsky, H Sinn, eds. Transition Metals and Organometallics as Catalysts for Olefin Polymerization. Berlin: Springer-Verlag, p. 257, 1988.
16. H Sinn. Presented at Hamburg Macromolecules Colloquium, Sept. 22–23, 1994.
17. MR Maison, JM Smith, SG Bott, AR Barron. J Am Chem Soc 115: 4971, 1993.
18. (a) ACW Chien. J Am Chem Soc 81: 86, 1959. (b) WP Long, DS Breslow. J Am Chem Soc 82: 1953, 1960.
19. RF Jordan. Adv Organomet Chem 32: 325, 1991.
20. M Bochmann, AJ Jagger. J Organomet Chem 424: C5, 1992.
21. X Yang, CL Stern, TJ Marks. J Am Chem Soc 113: 3623, 1991.
22. a. GG Hlatky, HW Turner, RR Eckman. J Am Chem Soc 111: 2728, 1989.
    b. DJ Crowther, RF Jordan. Macromol Chem Macromol Symp 66: 121, 1993.
23. GG Hlatky, RR Eckman, HW Turner. Organometallics 11: 1413, 1992.
24. K Mashima, K. Nakamura. Catalyst 36: 572, 1994.
25. TJ Marks *et al.* J Am Chem Soc 120: 1772, 1998.
26. TJ Marks *et al.* J Am Chem Soc 119: 2582, 1997.
27. PG Gassmann, MR Castrom. J Am Chem Soc 109: 7875, 1987.
28. KH Dahmen, TJ Marks. Langmuir 4: 1212, 1988.
29. RF Jordan, DF Taylor, NC Baenziger. Organometallics 9: 1546, 1990.
30. CA Jolly, DS Marynick. J Am Chem Soc 113: 3623, 1991.
31. SG Bott, AW Coleman, JL Atwood. J Am Chem Soc 108: 1709, 1986.
32. C Sishta, RM Hathorn, TJ Marks. J Am Chem Soc 114: 1112, 1992.
33. J Imuta, J Saitoh. Science and Technology in Catalysis 1994. Tokyo: Kodansha, p 295, 1995.
34. H Sinn, W Kaminsky, HJ Vollmer, R Woldt. Angew Chem Int Ed Engl 19: 390, 1980.
35. JA Ewen *et al.* Japan Pat 60-35006 (Exxon Chemical Company).
36. JA Ewen *et al.* Japan Pat 60-35007 (Exxon Chemical Company).
37. JA Ewen *et al.* Japan Pat 60-35008 (Exxon Chemical Company).
38. JA Ewen. In: T Keyi, K Soga, eds. Catalytic Polymerization of Olefins. Tokyo: p 271, 1986.
39. JA Ewen. J Am Chem Soc 106: 6355, 1984.
40. H Yasuda, H Tamai. Prog Polym Sci 18: 1097, 1993.
41. DM Selman. SPO'92, Houston, Texas, USA, p 22, 1992.
42. N Kashiwa. MetCon'93, Houston, Texas, USA, p 236, 1993.
43. W Kaninslay. Macromol Chem Macromol Symp 3: 377, 1986.
44. JA Ewen, RL Jones, A Razavi, JD Ferrara. J Am Chem Soc 110: 6255, 1988.
45. A Zambelli, A Grassi. Macromol Chem Rapid Commun 12: 523, 1991.
46. HH Brintzinger *et al.* Macromolecules 30: 3164, 1997.
47. Mitsui Petro Chem Ind. Japan Pat 61-221, 206, Japan Pat 64-106.
48. W Kaminsky, R Spiehl. Macromol Chem 190: 515, 1989.
49. W Kaminsky *et al.* J Polym Sci Part A Polym Chem Ed 23: 2151, 1985.
50. W Kaminsky *et al.* Makromol Chem Rapid Commun 11: 89, 1990.
51. JC Stevens. Int Symp on Catalyst Design for Tailor-Made Polyolefins, Kanazawa, Japan, 1994.
52. M Kakugo, K Mizunuma, T Miyatake. Macromol Chem Macromol Symp 66: 203, 1993.
53. RF Jordan. Int Symp on Catalyst Design for Tailor-Made Polyolefins, Kanazawa, Japan, 1994.
54. M Brookhart *et al.* J Am Chem Soc 117: 6414, 1995.
55. M Brookhart *et al.* J Am Chem Soc 120: 889, 1998.
56. K Nozaki *et al.* J Am Chem Soc 117: 9911, 1995.
57. DH McConville *et al.* J Am Chem Soc 118: 10008, 1996.
58. RF Jordan *et al.* Organometallics 16: 3282, 1997.
59. RF Jordan *et al.* Organometallics 16: 3303, 1997.
60. A Sen *et al.* J Am Chem Soc 120: 1932, 1998.
61. JA Ewen. J Am Chem Soc 106: 6355, 1984.

62. FRWP Wild, M Wasincionek, G Huttner, HH Brintzinger. J Orgonomet Chem 288: 63, 1985.
63. W Kaminsky, K Kulper, HH Brintzinger, FRWP Wild. Angew Chem Int Ed Engl 24: 507, 1985.
64. P Pino, P Cioni, J Wei. J Am Chem Soc 109: 6189, 1987.
65. P Pino, M Galimberti. J Organomet Chem 370: 1, 1989.
66. P Longo, A Grassi, C Pellecchia, A Zambelli. Macromolecules 20: 1015, 1987.
67. L Cavalloy, G Guerra, L Olivia, M Vacatello, P Corradini. Polym Commun 30: 16, 1989.
68. L Cavallo, P. Corradini, G Guerra, M Vacatello. Polymer 32: 1329, 1991.
69. HK Kuribayashi, N Koga, K Morokuma. J Am Chem Soc 114: 8687, 1992.
70. T Mise, S Miya, H Yamazaki. Chem Lett 1853, 1989.
71. W Roll, HH Brintzinger, B Rieger, R Zolk. Angew Chem Int Ed Engl 29: 279, 1990.
72. MH Prosenc, C Janiak, HH Brintzinger. Organometallics 11: 4036, 1992.
73. WE Piers, JE Bercaw. J Am Chem Soc 112: 9406, 1990.
74. WA Herrmann, J Rohrmann, E Herdweck, W Spaleck, A Winter. Angew Chem Int Ed Engl 28: 1511, 1989.
75. V Busico *et al.* Macromolecules 30: 3971, 1997.
76. A Winter, J Rohrmann, M Antberg, V Dolle, W Spaleck. Eur Pat 387690, 1989.
77. W Spaleck, M Antberg, J Rohrmann, A Winter, B Bachmann. Angew Chem Int Ed Engl 31: 1347, 1992.
78. W Spaleck, F Kuber, A Winter, J Rohrmann, B Bachmann, M Antberg, V Dolle, EF Paulus. Organometallics 13: 954, 1994.
79. U Stehling, J Diebold, R Kirsten, W Roll, HH Brintzinger, S Jungling, R Mulhaupt, F Langhauser. Organometallics 13: 964, 1994.
80. MC Sacchi *et al.* Macromolecules 30: 3955, 1997.
81. HH Brintzinger *et al.* Organometallics 16: 3413, 1997.
82. N Koga, K Morokuma. Int Symp on 40 years of Ziegler Catalysts, Freiburg, 1993.
83. RF Jordan *et al.* Organometallics 15: 4038, 1996.
84. JA Ewen. Eur Pat 0537130A, Fina Technology, Inc., 1992.
85. S Miyake. Japan Pat H5-209013, Shouwa Denkou, 1992.
86. W Kaminsky, F Renner. Makromol Chem Rapid Commun 14: 239, 1993.
87. K Soga. Makromol Chem Rapid Commun 12: 367, 1991.
88. JA Ewen, RL Jones, A Razavi. J Am Chem Soc 110: 6255, 1988.
89. D Veghini, JE Bercaw. Polymer Preprints 39 (No 1): 210, 1998.
90. JE Bercaw *et al.* J Am Chem Soc 118: 11988, 1996.
91. JA Ewen. Int Symp on 40 Years of Ziegler Catalysts, Freiburg, 1993.
92. T Tsutsui, N Ishimaru, A Mizuno, A Toyota, N Kashiwa. Polymer 30: 1350, 1989.
93. T Tsutsui, M Kioka, A Toyota, N Kashiwa. In: T Keiji, K Soga, eds. Catalytic Olefin Polymerization. Tokyo: Kodansha-Elsevier, p 493, 1990.
94. M Antberg, V Dolle, S Haftka, R Rohrmann, W Spaleck, S Winter, HJ Zimmermann. Makromol Chem Macromol Symp 48/49: 333, 1991.
95. W Spaleck. Int Symp on 40 Years of Ziegler Catalysts, Freiburg, 1993.
96. S Collins *et al.* Macromolecules 28: 3771, 1995.
97. RM Waymouth *et al.* J Am Chem Soc 117: 11586, 1995.
98. RM Waymouth *et al.* Organometallics 16: 3635, 1997.
99. RM Waymouth *et al.* Organometallics 16: 5909, 1997.
100. RM Waymouth *et al.* J Am Chem Soc 120: 2309, 1998.
101. N Ishihara, M Kuramoto. Int Symp on Catalyst Design for Tailor-Made Polyolefins, Kanazawa, Japan, p 339, 1994.
102. RM Waymouth *et al.* Angew Chem Int Ed Engl 37: 922, 1998.
103. K Soga *et al.* Makromol Chem Rapid Commun 18: 883, 1997.
104. T Arai, T Ohtsu, S Suzuki. Polymer Preprints 39 (No 1): 220, 1998.
105. S Collins *et al.* Macromolecules 27: 4477, 1994.
106. S Collins *et al.* Macromolecules 30: 3151, 1997.
107. H Yasuda *et al.* J Am Chem Soc 114: 4908, 1992.
108. H Yasuda *et al.* Prog Polym Sci 18: 1097, 1993.
109. BM Novak *et al.* Macromolecules 30: 3494, 1997.
110. BM Novak *et al.* Polymer Preprints 35 (No 1): 682, 1994.
111. K Ohshima, E Tanaka. Koubunshi High Polymers Japan 43: 784, 1994.
112. CA Jolly *et al.* J Am Chem Soc 111: 7968, 1989.
113. C Janiak. J Organomet Chem 452: 63, 1993.
114. WA Coddard *et al.* J Am Chem Soc 116: 1481, 1994.
115. T Ziegler *et al.* Organometallics 13: 432, 1994.
116. P Corradini *et al.* Macromol Chem Macromol Symp 69: 237, 1993.
117. AK Rappe *et al.* J Am Chem Soc 115: 6159, 1993.
118. H Kuribayashi *et al.* J Am Chem Soc 114: 8687, 1992.
119. P Corradini *et al.* J Am Chem Soc 116: 2988, 1994.
120. T Ziegler *et al.* Organometallics 13: 2257, 1994.

3

Synthesis of Monomers by Olefin Metathesis

Ileana Drăguţan and Valerian Drăguţan
Romanian Academy, Institute of Organic Chemistry, Bucharest, Romania

Olefin metathesis [1], having recently experienced precipitate development [2], is being applied increasingly for the preparation of a significant number of monomers for the plastic and rubber industries, either in a single step or in conjunction with other traditional hydrocarbon processes. Of these monomers, simple linear and branched olefins are of importance, the latter offering a new source for synthesis of common dienes used in the rubber industry. In addition, the metathesis reaction opens nonconventional ways for manufacturing certain particular aromatic or cyclic olefins.

I. SYNTHESIS OF LINEAR ACYCLIC OLEFINS

Extensively used in the polyethylene industry, ethene can be obtained, along with 2-butene, by metathesis disproportionation* of propene in the presence of alumina- or silica-supported molybdenum or tungsten oxide catalysts [1]:

$$2CH_2{=}CH{-}CH_3 \leftrightarrows CH_2{=}CH_2 + CH_3{-}CH{=}CH{-}CH_3 \quad (1)$$

Data on product composition, conversion, and selectivity of propene disproportionation by metathesis over supported molybdenum oxide catalysts at various temperatures are given in Table 1 [1a, 3].

Applied on an industrial scale [4], this reaction, the Phillips triolefin process, is carried out in a tubular reactor with a fixed bed of catalyst containing cobalt molybdate. Separation of the reaction products is effected by fractionation and subsequent distillation (Fig. 1).

Metathesis disproportionation of propene has been reported on a wide range of heterogeneous catalytic systems based mainly on carbonyl compounds, as well as oxides or sulfides of molbydenum, tungsten, or rhenium supported on alumina, silica, or mixtures of oxides or phosphates. Among these catalytic systems, only a restricted number present good activity and sufficient selectivity to allow reasonable yields of ethene and 2-butene. Examples of the most efficient heterogeneous catalysts appear in Table 2 [1, 3, 5–11].

An improved process for manufacturing ethene from propene involves a series of successive disproportionation, isomerization and hydrogenation steps, directly connected to the petroleum cracking unit. In this process [12], propene is converted to ethene and 2-butene; then 2-butene is isomerized to 1-butene and transformed (in the presence of propene and over disproportionation catalysts) into ethene, pentene and hexene (Fig. 2). Pentene and hexene are hydrogenated to pentane and hexane, respectively, which are recirculated to the cracking unit to again produce propene for disproportionation. This process

*The metathesis disproportionation of olefins that occurs by transalkylidenation, initially named disproportionation [1a] and later called metathesis [1b], is to be differentiated from other types of disproportionation by transalkylation, hydrogen transfer, or cracking.

TABLE 1 Propene Metathesis Disproportionation Over Supported Molybdenum Oxide Catalysts

Parameters	Catalytic system			
	$MoO_3 \cdot CoO/Al_2O_3$	MoO_3/Al_2O_3	MoO_3/SiO_2	$MoO_3/AlPO_4$
Temperature, °C	205	50	538	538
Conversion, %	42.6	11–12	28	5
Product composition, mol %				
ethene	27.4		38.7	40.0
1-butene	8.0		19.1	22.0
trans-2-butene	32.7		24.4	22.0
cis-2-butene	18.4		17.8	16.0
C_5^+	13.4		Traces	Traces
Selectivity in ethene and butenes, %	94	100	95	95

Source: Refs. 1a and 3.

is one of the most economical routes for manufacturing ethene.

When highly pure propene is not available commercially, it can be prepared by the reverse metathesis reaction of ethene and 2-butene [Eq. (1)]. The process is performed either at high temperatures (150–350°C) in the gas phase, over molybdenum or tungsten catalysts (Phillips triolefin process) [4], or at low temperatures (50°C) in the liquid phase, in the presence of rhenium-based catalysts (IFP-CPC process) [13]. The raw material may be either ethene and the C_4 fraction available from the hydroisomerization unit (previously submitted to an isomerization step to maximize its 2-butene content) or ethene alone, which, before admission to the metathesis unit, is partly dimerized to 1-butene, then isomerized to 2-butene in separate units. The process is useful in the event of a high demand for propene, since the C_4 fraction is readily available from a cracking unit.

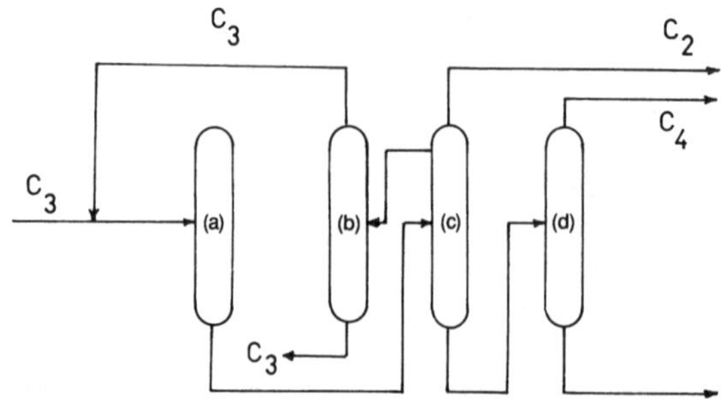

FIG. 1 Schematic diagram of the Phillips triolefin process for propene disproportionation: (a) disproportionation; (b) fractionation; (c) fractionation; (d) distillation. (From Ref. 4.)

Polymerization grade 1-butene, used as comonomer in polyethylene production, is obtained by metathesis disproportionation of propene according to the triolefin process, followed by 2-butene isomerization on specific catalysts [14]. When propene is available, the method is an alternative to the above-mentioned process of ethene dimerization to 1-butene. 1-Butene and 2-butene produced by propene disproportionation constitute a valuable source for manufacturing high purity butadiene via dehydrogenation, since the product contains only trace amounts of branched hydrocarbons.

According to another process, ethene and butadiene are economically obtained through a propene metathesis reaction integrated into the cracking and fractionation units of the petroleum plant [15]. Thus, ethene resulting from the disproportionation step enriches the ethene content of the C_2 fraction from the cracking unit, whereas butadiene is formed by subsequent dehydrogenation of the butene fraction.

Mixtures of butene, pentene, hexene, and higher olefins are commonly obtained by α-olefin disproportionation over supported heterogeneous catalysts [1]. Examples of disproportionation of propene, 1-butene, 1-pentene, and 1-hexene on alumina-supported molybdenum hexacarbonyl are included in Table 3. Adequate separation units may provide fractions of C_4, C_5, C_6, and higher olefins for petrochemical uses.

An alternative to the triolefin process produces hexenes, heptenes, and octenes along with ethene from propene by a three-step metathesis disproportionation. The product, containing hexene, heptene, and octene, has 95% linearity and will provide C_6, C_7, or C_8 cuts by fractionation or may be used as such in oxosynthesis reactions for the production of polyvinyl chloride (PVC) plasticizers. Higher internal olefins for use in

TABLE 2 Heterogeneous Catalytic Systems for Metathesis Disproportionation of Propene

Catalytic system	Propene conversion (%)	Product selectivity (%)	Ref.
$CoO \cdot MoO_3/Al_2O_3$	42.0	94.1	1
$Mo(CO)_6/Al_2O_3$	25.0	97.0	1
MoO_3/Al_2O_3	11.0	100.0	1
MoO_3/SiO_2	28.0	95.0	3
$MoO_3/AlPO_4$	5.0	95.0	3
$MoO_3 \cdot Cr_2O_3/Al_2O_3$	36.0	97.0	5
MoS_2/Al_2O_3	1.3	100.0	3
MoS_2/SiO_2	9.1	100.0	3
WO_3/Al_2O_3	7.4	100.0	6
WO_3/SiO_2	44.8	97.8	3
$WO_3/AlPO_4$	34.0	82.0	3
WS_2/Al_2O_3	1.0	100.0	3
WS_2/SiO_2	18.3	100.0	3
$Re(CO)_{10}/Al_2O_3$	20.4	100.0	7
Re_2O_7/Al_2O_3	19.2	100.0	8
Re_2O_7/SiO_2	4.0	100.0	3,9
Re_2O_7/ZrO_2	2.1	100.0	9
Re_2O_7/ThO_2	12.0	100.0	9
Re_2O_7/SnO_2	15.0	100.0	9
Re_2O_7/TiO_2	13.0	100.0	10
Re_2O_7/Fe_2O_3	0.4	100.0	10
Re_2O_7/MoO_3	3.9	100.0	10
Re_2O_7/WO_3	0.75	100.0	10
V_2O_5/SiO_2	11.2	44.0	11
Nb_2O_5/SiO_2	3.7	90.0	11
Ta_2O_5/SiO_2	8.3	56.0	11

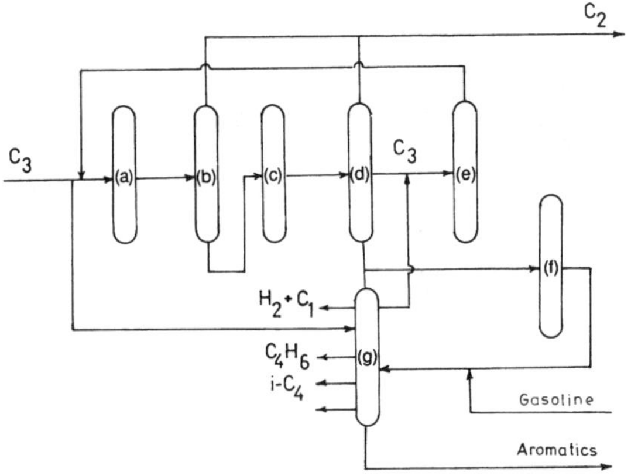

FIG. 2 Schematic representation of the process for manufacturing ethene from propene: (a) disproportionation; (b) fractionation; (c) isomerization; (d) fractionation; (e) disproportionation; (f) hydrogenation; (g) cracking. (From Ref. 12.)

TABLE 3 Metathesis Disproportionation of α-Olefins in the Presence of $Mo(CO)_6/Al_2O_3$

α-Olefin	Reaction products (mol %)			
	Propene	1-Butene	1-Pentene	1-Hexene
Ethene	42	8	2	5
Propene		34	21	13
Butene	55		27	12
Pentene	2	18		15
Hexene	1	32	27	
C_7^+		8	23	55
α-Olefin conversion, %	25	10	60	54

Source: Ref. 1.

detergent production can readily be obtained from ethene through a complex process involving oligomerization, isomerization, and disproportionation (SHOP process) [16]:

$$n\mathrm{CH_2{=}CH_2 \rightarrow R{-}CH{=}CH_2} \tag{2}$$

$$\mathrm{R{-}CH{=}CH_2 \rightarrow R_1{-}CH{=}CH{-}R_2} \tag{3}$$

$$\mathrm{3R_1{-}CH{=}CH{-}R_2 \rightarrow R_1CH{=}CHR_1 + R_1CH{=}CHR_2 + R_2CH{=}CHR_2} \tag{4}$$

where n is the degree of oligomerization and R_1, R_2, and R are alkyl radicals. In this process ethene is first oligomerized to a mixture of C_4–C_{20} olefins, then isomerized to α-olefins. These α-olefins are disproportionated into internal olefins, which are further fractionated into the required fractions.

Propene may be converted into a large number of higher acyclic olefins by employing variants of metathesis disproportionation and/or codisproportionation of the reaction products formed in successive stages of disproportionation. For example, a process for manufacturing 5-decene from propene comprises three successive steps of metathesis disproportionation combined with isomerization [17]. Thus, propene is first disproportionated to ethene and 2-butene; then the butene fraction is isomerized to 1-butene, which is disproportionated to ethene and 3-hexene. The latter is subsequently isomerized to 1-hexene and disproportionated to 5-decene and ethene (Fig. 3). Ethene produced in the first stage is further dimerized to butene, which is introduced into the butene fraction prior to the isomerization step. Disproportionation occurring between the α- and/or internal olefins encountered in various stages of this process, combined with isomerization, will provide an unlimited number of linear internal and α-olefins.

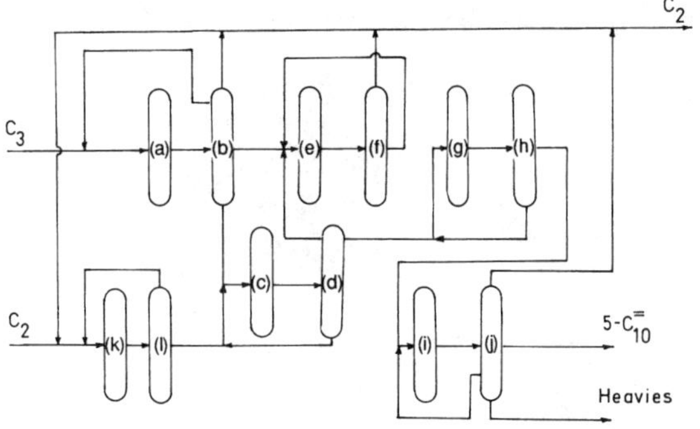

FIG. 3 Schematic diagram of the process for manufacturing 5-decene from propene: (a) disproportionation; (b) fractionation; (c) isomerization; (d) fractionation; (e) disproportionation; (f) fractionation; (g) isomerization; (h) fractionation; (i) disproportionation; (j) fractionation; (k) dimerization; (l) fractionation. (From Ref. 17.)

II. SYNTHESIS OF BRANCHED ACYCLIC OLEFINS

2-Methyl-1-butene and 3-methyl-1-butene (isoamylene), a valuable raw material for production of isoprene, have been efficiently obtained from propene via disproportionation. In fact, the process occurs by a two-step pathway involving dimerization and disproportionation [18] (Fig. 4). In a first reactor propene is dimerized to a branched product, which is further codisproportionated with propene to a fraction rich in 2-methyl-2-butene. After separation, 2-methyl-2-butene may be used as a monomer, whereas the remainder, consisting of 2-methyl-1-butene and 3-methyl-1-butene (isoamylene), is further dehydrogenated to isoprene.

3,3-Dimethyl-1-butene (neohexene), an important intermediate for the synthesis of musk perfumes, is produced on a commercial scale either from diisobutene (a mixture of α and β isomers), by codisproportionation of the β isomer with ethene, or from isobutene, which is first dimerized and then codisproportionated with ethene [19] (Fig. 5). An additional isomerization unit is provided for the continuous conversion of the α-diisobutene into the β isomer, the latter being consumed in the disproportionation stage.

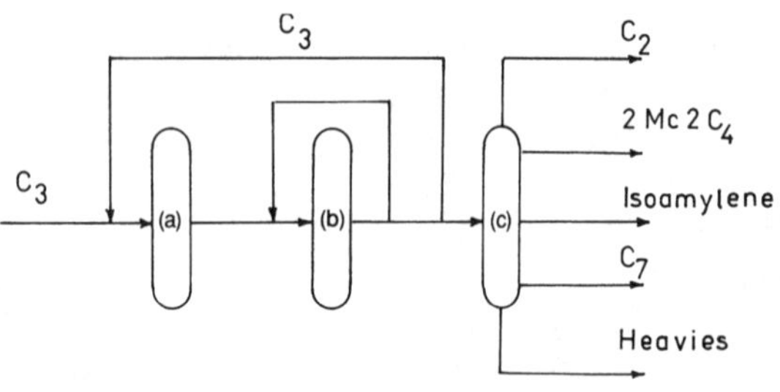

FIG. 4 Schematic representation of the process for manufacturing isoamylene from propene: (a) dimerization; (b) disproportionation; (c) fractionation. (From Ref. 18.)

1,2-Di-*t*-butylethene, a component also used as an additive for synthetic gasoline, is obtained from isobutene and ethene with a dual isomerization–disproportionation catalyst [20]. The process begins with dimerization in a reactor for conversion of isobutene to 2,4,4-trimethyl-1-pentene; this is followed by isomerization–disproportionation in a bifunctional unit (isomerization of 2,4,4-trimethyl-1-pentene to 2,4,4-trimethyl-2-pentene and conversion of the latter into di-*t*-butylethene) (Fig. 6). A by-product of the process, 2,3-dimethyl-2-butene, is recirculated to the disproportionation unit to be cleaved with ethene to isobutene, which is reintroduced into the process. When neohexene (3,3-dimethyl-1-butene) is employed as the starting material in this process, the installation consists solely of the disproportionation and fractionation units.

III. SYNTHESIS OF LINEAR DIENES AND POLYENES

α, ω-Linear dienes are prepared conveniently by metathesis cleavage of cycloolefins with lower alkenes such as ethene (ethenolysis) [2]. If cyclic dienes or polyenes are used instead of cycloolefins, linear trienes or polyenes can be obtained.

Thus, 1,6-heptadiene is readily produced by cyclopentene ethenolysis in the presence of molybedenum or tungsten oxide catalysts:

$$\text{cyclopentene} + \text{CH}_2{=}\text{CH}_2 \xrightarrow{\text{W-cat}} \text{CH}_2{=}\text{CH(CH}_2)_3\text{CH}{=}\text{CH}_2 \quad (5)$$

whereas 1,7-octadiene, 1,9-decadiene, 1,10-undecadiene, and 1,11-dodecadiene are similarly formed from cyclohexene, cyclooctene, cyclononene, and cyclodecene [21]. In another process [22], 1,5-hexadiene results from 1,5-cyclooctadiene and a twofold molar excess of ethene; the starting material is provided by cyclodimerization of butadiene (Fig. 7). Upon varying the molar ratio of 1,5-cyclooctadiene to ethene within the range 1–2, it is possible to obtain either 1,5,9-decatriene or 1,5-hexadiene. A different process yields the latter diene, simultaneously with 1,9-decadiene, by ethenolysis of a mixture of cyclooctene and 1,5-cyclooctadiene [21b].

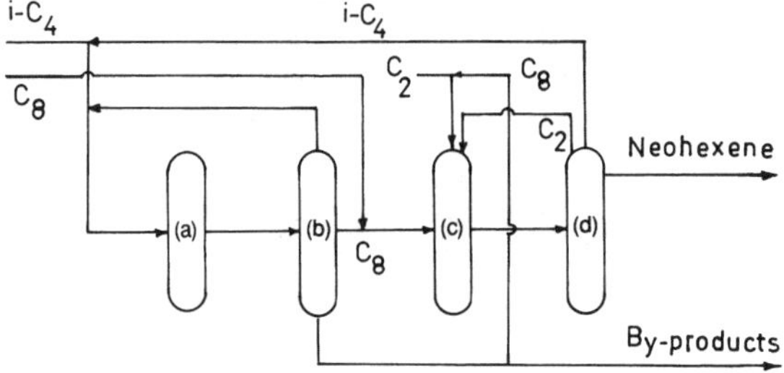

FIG. 5 Schematic diagram of the process for neohexene synthesis from isobutene: (a) dimerization; (b) separation; (c) disproportionation; (d) separation. (From Ref. 19.)

IV. SYNTHESIS OF CYCLIC MONOMERS

The metathesis reaction offers some opportunities for the synthesis of highly pure cycloolefins, as well. A first efficient method consists of the metathesis cyclization of linear α, ω-dienes in the presence of tungsten- or molybdenum-based catalysts. Ethene or other low alkenes formed as by-products are easily removed from the reaction mixture, enabling convenient separation of the cycloalkene. A pertinent example is the synthesis of cyclohexene, in high yield, from 1,7-octadiene with molybdenum nitrosyl complexes [23]:

$$\xrightarrow[\eta = 91\%]{\text{Mo-cat}} \quad + \quad \| \tag{6}$$

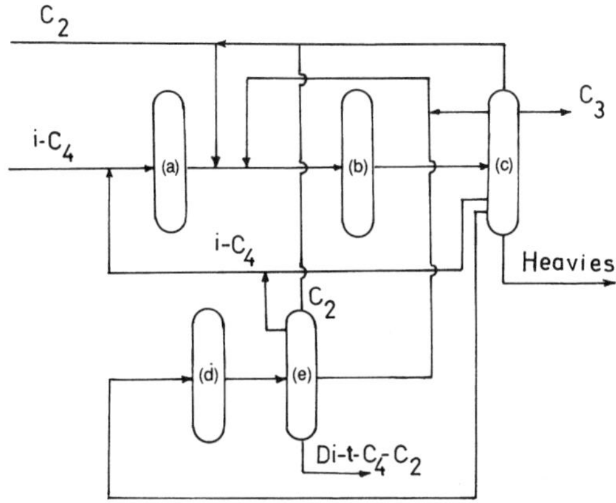

FIG. 6 Schematic representation of the process for manufacturing di-*t*-butylethene from ethene and isobutene: (a) dimerization; (b) disproportionation; (c) separation; (d) disproportionation; (e) separation. (From Ref. 20.)

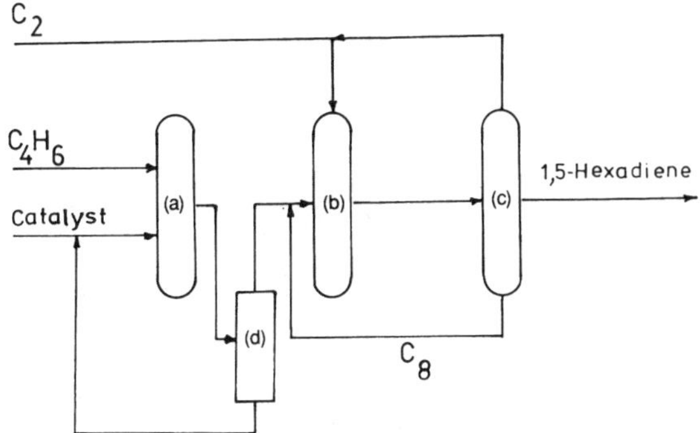

FIG. 7 Schematic diagram of the process for manufacturing 1,5-hexadiene: (a) dimerization; (b) disproportionation; (c) separation; (d) catalyst recovery. (From Ref. 22.)

Other related cyclic olefins, 1,3-cyclohexadiene and vinylcyclohexene, are reported to be formed as main products from butadiene over tungsten oxide catalysts [24]. Under the reaction conditions, hexatriene, resulting from the intermolecular metathesis of butadiene, underwent subsequent cyclization to 1,3-cyclohexadiene:

$$\xrightarrow{WO_3} \tag{7}$$

Another interesting procedure for the synthesis of pure cycloolefins, developed by Küpper and Streck [25], starts from a cyclodiene. In this method, cyclopentene is obtained through metathesis scission of 1,6-cyclodecadiene under the influence of WCl_6-based catalysts:

$$\xrightarrow{\text{W-cat}} 2 \tag{8}$$

The procedure can be extended to other cycloolefins (e.g., cyclohexene, cycloheptene, cyclooctene).

Conversely, cyclodienes can be produced by metathesis dimerization and cometathesis of cycloolefins [26]. Good yields of 1,6-cyclodecadiene, 1,9-cyclohexadecadiene, 1,10-cyclooctadecadiene, and 1,11-cycloeicosadiene are reported for the metathesis of cyclopentene, cyclooctene, cyclononene, and cyclodecene, respectively, in the presence of tungsten hexachloride or rhenium heptoxide catalysts:

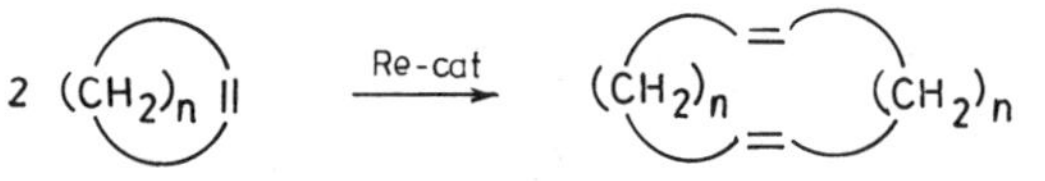

$$2\,(CH_2)_n \xrightarrow{\text{Re-cat}} (CH_2)_n \quad (CH_2)_n \tag{9}$$

where $n = 3$, 6, 7, or 8. Analogously, 1,7-cyclpentadecadiene arises from the cometathesis of cycloheptene with cyclooctene, as induced by rhenium heptoxide catalysts [26].

V. SYNTHESIS OF AROMATIC OLEFINS

Styrene, the best known representatives of the aromatic olefin class, is readily obtained by cometathesis reaction of stilbene with ethene in the presence of alumina-supported molybdenum oxide catalysts. In a two-step process [27] stilbene is first produced from toluene by catalytic dehydrogenative coupling over PbO/Al_2O_3, followed by codisproportionation to styrene, with ethene, over MoO_3/Al_2O_3:

$$2C_6H_5{-}CH_3 \rightarrow C_6H_5{-}CH{=}CH{-}C_6H_5 + 4H \tag{10}$$

$$C_6H_5{-}CH{=}CH{-}C_6H_5 + CH_2{=}CH_2 \rightarrow 2C_6H_5{-}CH{=}CH_2 \tag{11}$$

The process affords highly pure styrene by an unusually economic method, also facilitating the separation of the reaction products. If propene or 2-butene were used as the codisproportionation agent instead of ethene, β-methylstyrene could be recovered by this procedure.

VI. SYNTHESIS OF MACROCYCLIC OLEFINS

A considerable number of large size cycloolefins of macrocyclic nature have become accessible by metathesis cyclooligomerization reaction of small and medium size cycloolefins in the presence of common metathesis catalytic systems [2]. Thus, starting from cyclooctene, Wasserman and coworkers [28] manufactured unsatured carbocycles with up to 120 carbon atoms (degree of oligomerization up to 15) [Eq. (12)].

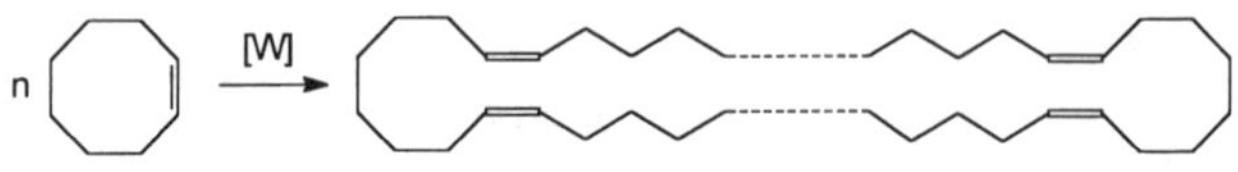

n [cyclooctene] $\xrightarrow{[W]}$ [macrocyclic polyene] (12)

The reaction was conducted under mild metathesis conditions, with the homogeneous catalyst $WCl_6/EtOH/EtAlCl_2$, at 5–20°C in benzene as a solvent. The products were conveniently separated by gas chromatography and were further analyzed by mass spectrometry. The corresponding saturated carbocycles were also obtained from unsaturated products by total catalytic hydrogenation [Eq. (13)].

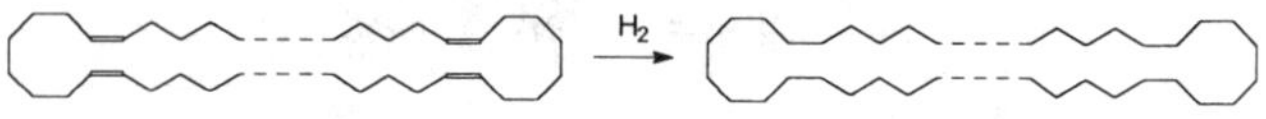

[macrocyclic polyene] $\xrightarrow{H_2}$ [saturated macrocycle] (13)

The nature of the metathesis oligomerization products obtained from cyclooctene was also investigated by Scott [29], Calderon [30, 31] and Hoecker and coworkers [32–35]. Thus, Scott and Calderon [29] determined the macrocyclic nature of the oligomeric products formed by metathesis of cyclooctene in the presence of $WCl_6/EtOH/EtAlCl_2$. In addition, Hoecker and coworkers [32–35] synthesized cyclohexadecadiene, tetraeicosatriene and dotriacontatetraene by metathesis dimerization, trimerization and tetramerization of cyclooctene with $WCl_6/EtOH/EtAlCl_2$ [Eqs. (14)–(16)].

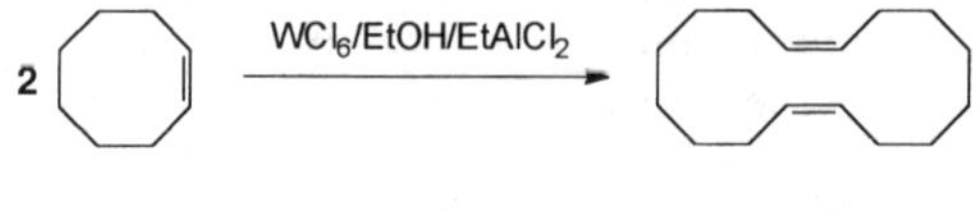

2 [cyclooctene] $\xrightarrow{WCl_6/EtOH/EtAlCl_2}$ [cyclohexadecadiene] (14)

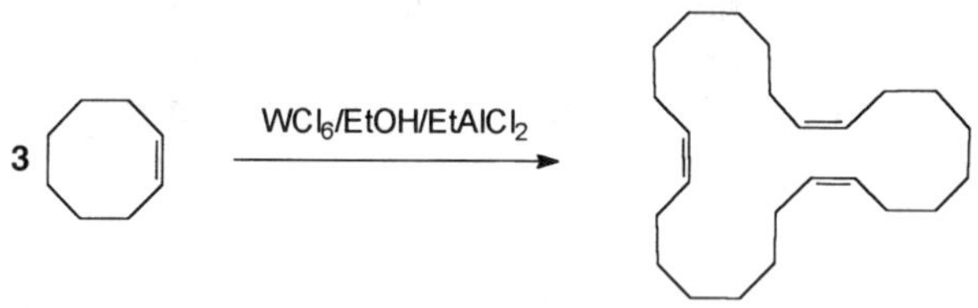

3 [cyclooctene] $\xrightarrow{WCl_6/EtOH/EtAlCl_2}$ [tetraeicosatriene] (15)

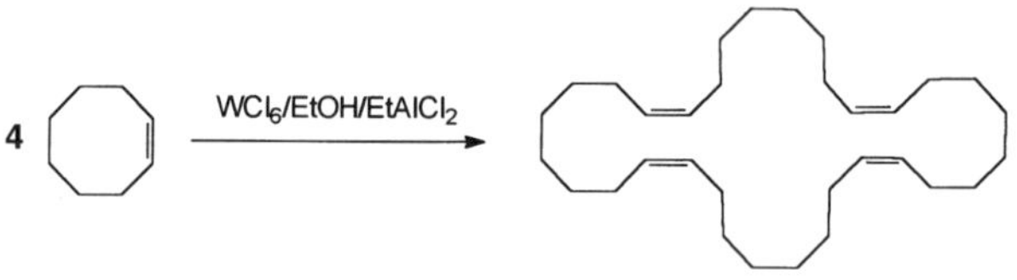

4 [cyclooctene] $\xrightarrow{WCl_6/EtOH/EtAlCl_2}$ [dotriacontatetraene] (16)

Hoecker and Musch [32] elaborated in more detail the methods of chromatographic and spectrometric analysis of these macrocyclic compounds and studied the influence of the reaction parameters on their formation. Interestingly, through metathesis reaction of some cyclic polyenes such as 1,5-cyclooctadiene and 1,5,9-cyclododecatriene, Scott and Calderon [29] obtained macrocycles with similar structures. It is noteworthy that the higher degree of unsaturation led in this case to a greater fragmentation of the reaction products.

A large number of unsaturated macrocycles were synthesized by Wolovsky and Nir [36] by metathesis oligomerization of cyclododecene in the presence of $WCl_6/EtOH/EtAlCl_2$ catalyst [Eq. (17)].

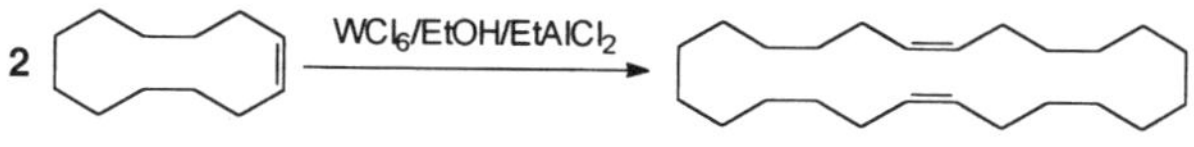

(17)

The unsaturated oligomers resulted in the first stage were further reduced by partial hydrogenation to the corresponding cycloolefins [Eq. (18)].

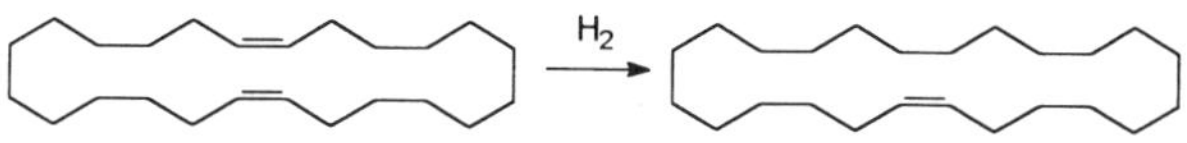

(18)

These products were further subjected to oligomerization in the presence of the same metathesis catalyst. By this procedure, unsaturated carbocycles with 24, 36 and 48 carbon atoms have been manufactured.

VII. SYNTHESIS OF HETEROCYCLIC OLEFINS

Olefin metathesis has recently become a convenient way for the synthesis of various heterocyclic olefins by a versatile intramolecular reaction of unsaturated compounds, ring-closing metathesis (RCM) reaction [37]. The ring-closing metathesis reaction takes advantage of the very active and selective transition metal metathesis catalysts that are tolerant toward functionalities [38]. Ruthenium carbene complexes such as $(Cy_3P)_2RuCl_2(=CHPh)$ (Cy = cyclohexyl, Ph = phenyl) effectively catalyze the ring closing metathesis of dienes to yield unsaturated carbocycles and heterocycles [39] [Eq. (19)].

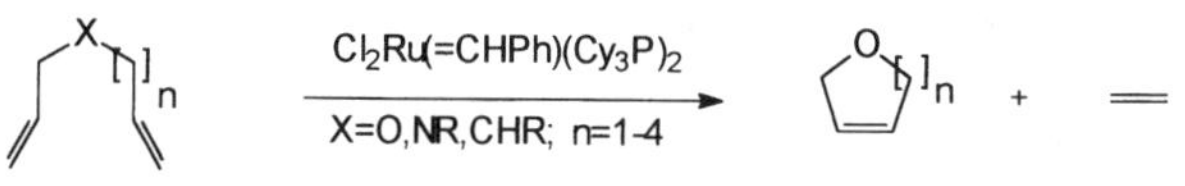

(19)

By this route, 2,5-dihydrofuran and 2-phenyl-2,5-dihydrofuran could be easily prepared from the corresponding unsaturated ethers in the presence of the above-mentioned ruthenium catalyst [40] [Eqs. (20) and (21)].

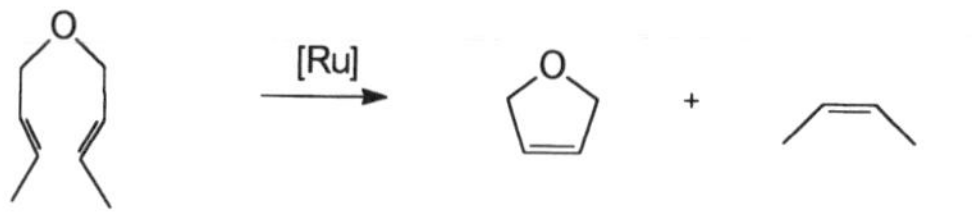

(20)

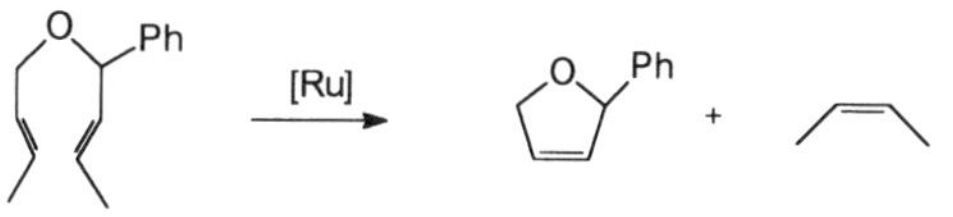

(21)

Similarly, oxygen-containing six- and seven-membered heterocycles have been conveniently prepared from a series of unsaturated ethers by ring closing metathesis reaction [Eqs. (22)–(25)].

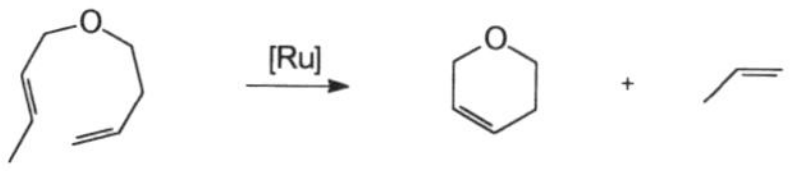

(22)

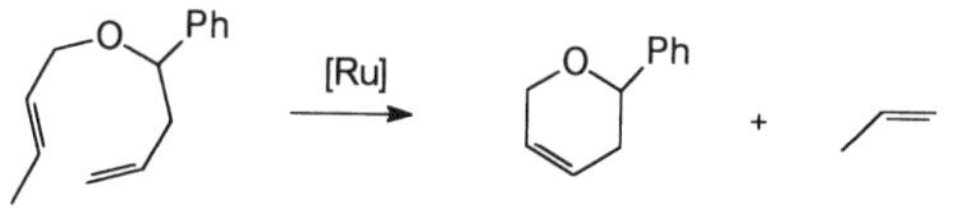

(23)

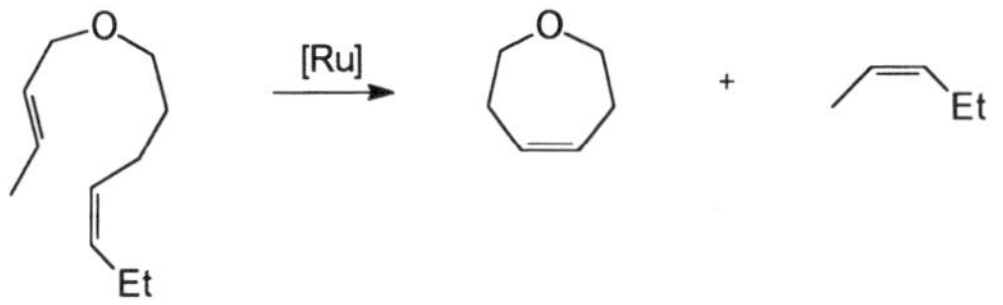

(24)

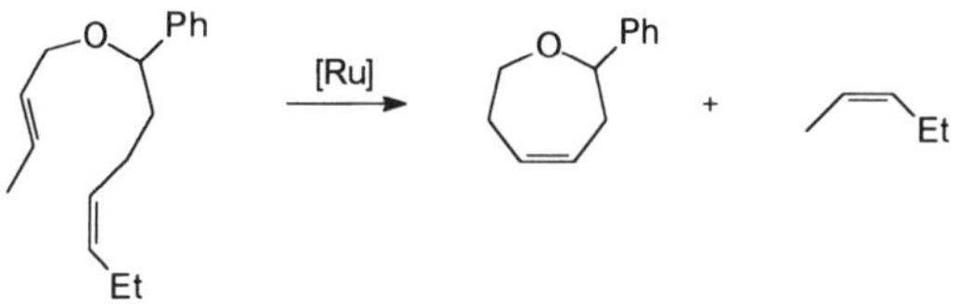

(25)

Under the same conditions, unsaturated seven-membered diethers have been prepared from the corresponding linear unsaturated diethers [41] [Eqs. (26) and (27)].

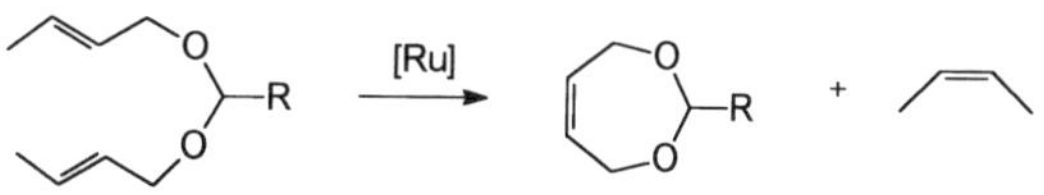

(26)

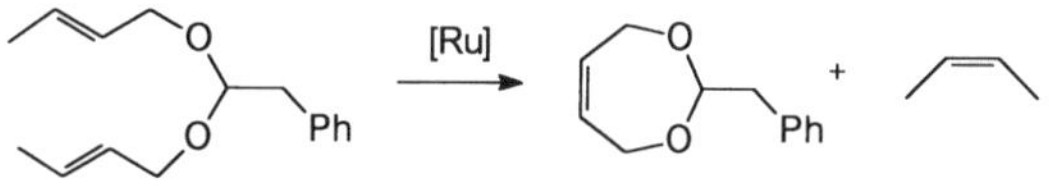

(27)

Remarkably, on applying tandem ring opening–ring closing metathesis reactions, unsaturated five- and six-membered diheterocyclic compounds could be prepared from unsaturated alicyclic diethers [39] [Eqs. (28)–(31)].

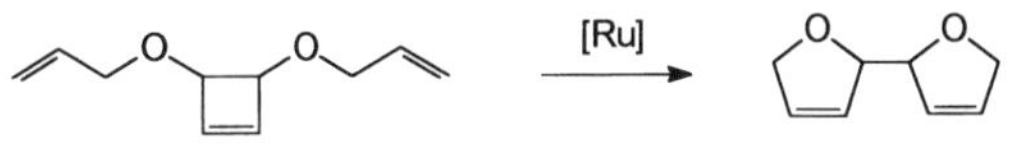

(28)

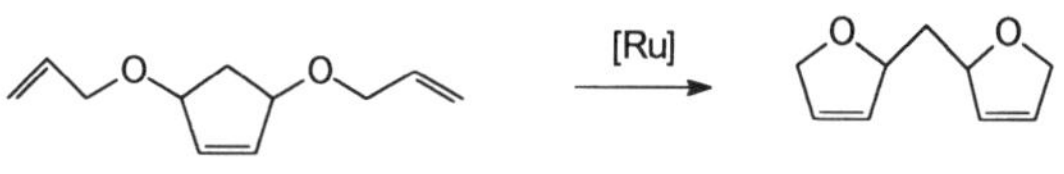

(29)

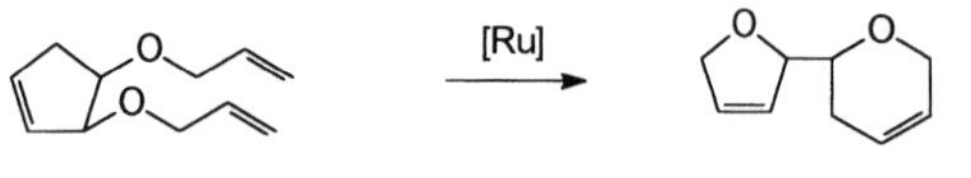

(30)

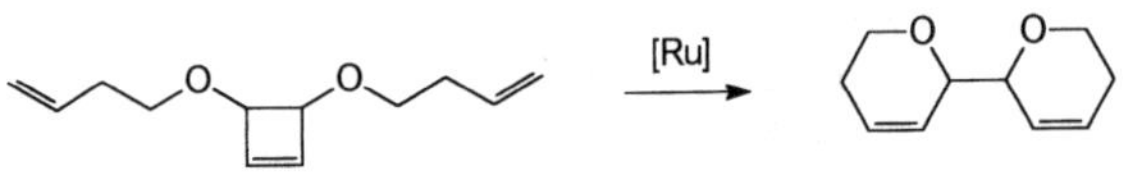

(31)

Nitrogen [42] and sulfur [43] heterocycles of various sizes have been prepared from the corresponding linear heteroatom-containing dienes by applying ring closing metathesis under appropriate conditions [Eqs. (32)–(34)].

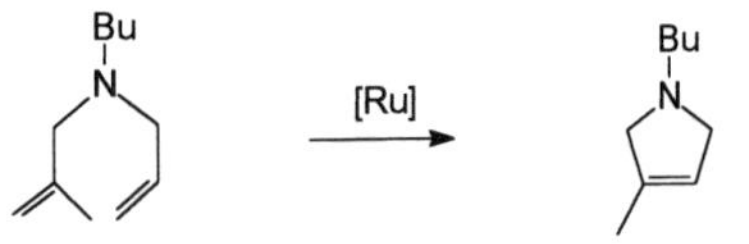

(32)

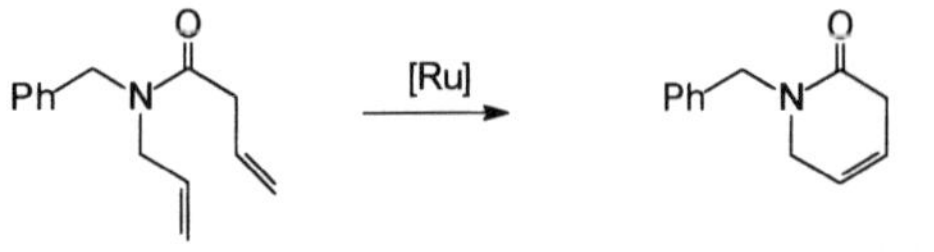

(33)

(34)

Of great interest, diheteroatom-containing cyclic olefins, e.g. incorporating oxygen and nitrogen [44] or oxygen and silicon [41] into the cycle, have been also prepared under ring closing metathesis conditions [Eqs. (35) and (36)].

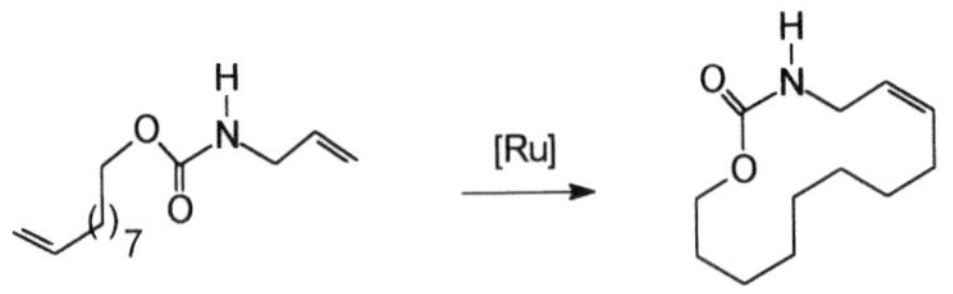

(35)

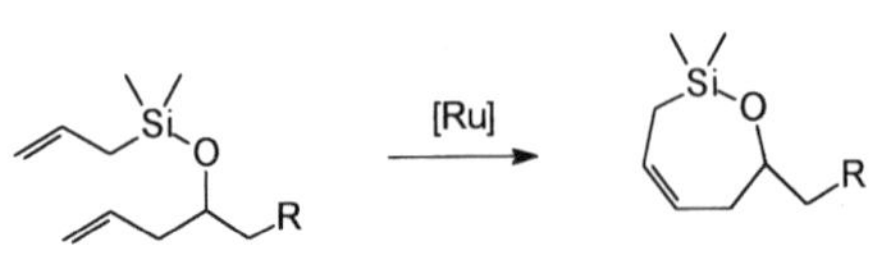

(36)

Finally, the application of ring closing metathesis for the synthesis of large size unsaturated heteroatom macrocycles such as crown ethers [45] or various unsaturated macrolides [46] is further illustrated in Eqs. (37) and (38).

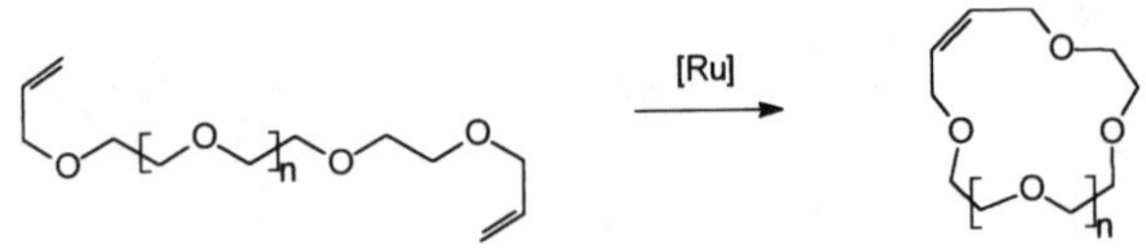

(37)

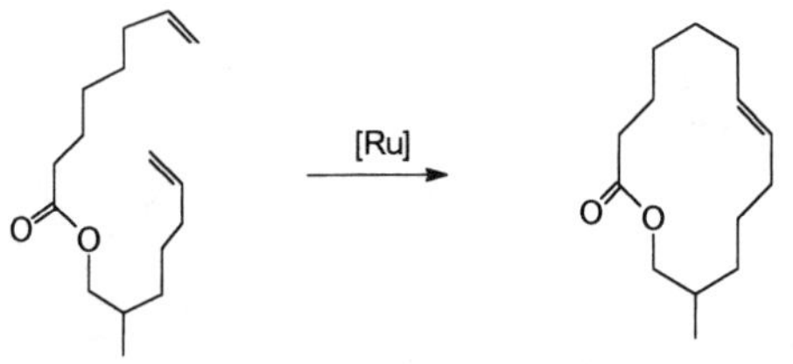

(38)

REFERENCES

1. (a) RL Banks, GC Bailey. Ind Eng Chem Prod Res Develop 3: 170–173, 1964. (b) N Calderon, HY Chen, KW Scott. Tetrahedron Lett 1967: 3327–3329.
2. (a) V Dragutan, AT Balaban, M Dimonie. Olefin Metathesis and Ring-Opening Polymerization of Cycloolefins, Bucharest: Ed. Academiei, 1981, rev. ed., Chichester: John Wiley & Sons, 1985. (b) KJ Ivin. Olefin Metathesis, London: Academic Press, 1983. (c) KJ Ivin, JC Mol. Olefin Metathesis and Metathesis Polymerization, London: Academic Press, 1997. (d) Y Imamoglu, B Zumreoglu-Karan, AJ Amass (eds). Olefin Metathesis and Polymerization Catalysts, Dordrecht: Kluwer, 1990. (e) Y Imamoglu (ed). Metathesis Polymerization of Olefins and Polymerization of Alkynes, Dordrecht: Kluwer, 1998.
3. LF Heckelsberg, RL Banks, GC Bailey. Ind Eng Chem Prod Res Develop 8: 259–261, 1969.
4. Phillips Petroleum Company. Hydrocarbon Process 46: 232, 1967.
5. VV Atlas, II Pis'man, AM Bakhshi-Zade. Khim Prom 45: 734, 1969.
6. RL Banks. U.S. Patent 3,261,879, 1966.
7. KV Williams, L Turner. Brit. Patent 1,116,243, 1968.
8. British Petroleum Company. Brit. Patent 1,105,564, 1965.
9. British Petroleum Company. Brit. Patent 1,093,784, 1965.
10. L Turner, KV Williams. Brit. Patent 1,096,250, 1965.
11. RL Banks. U.S. Patent 3,443,969, 1969.
12. RE Dixon. U.S. Patent 3,485,890, 1969.
13. P Amigues, Y Chauvin, D Commereuc, CC Lai, YH Liu, JP Pam. Hydrocarbon Process 1990: 79.
14. RL Banks, DS Banasiak, PS Hudson, JR Norell. J Mol Catal 15: 21, 1982.
15. PH Johnson. Hydrocarbon Process 46: 149–151, 1967.
16. ER Freitas, CR Gum. Chem Eng Prog 75(1): 73–76, 1979.
17. LF Heckeslberg. U.S. Patent 3,485,891, 1967.

18. RL Banks. U.S. Patent 3,538, 181, 1970.
19. RL Banks, RB Regier. Ind Eng Chem Prod Res Develop 10: 46–51, 1971.
20. RE Reusser, SD Turk. U.S. Patent 3,760,026, 1974.
21. (a) GC Ray, DL Crain. Fr. Patent 1,511,381, 1968. (b) VC Vives, RE Reusser. U.S. Patent 3,792,102, 1974.
22. VC Vives. U.S. Patent 3,786,111, 1974.
23. EA Zuech, WB Hughes, DH Kubicek, ET Kittleman. J Am Chem Soc 92: 528–531, 1970.
24. LF Heckelsberg, RL Banks, GC Bailey. J Catal 13: 99–100, 1969.
25. FW Kuepper, R Streck. Makromol Chem 175: 2055, 1974.
26. HJ Katker. PhD Thesis, RWTH Aachen, 1988.
27. DP Montgomery, RN Moore, RW Knox. U.S. Patent 3,965,206, 1976.
28. E Wasserman, DA Ben-Efraim, R Wolovsky. J Am Chem Soc 90: 3286–3287, 1968.
29. KW Scott, N Calderon, EA Ofstead, WA Judy, JP Ward. Adv Chem Ser 91: 399–418, 1969.
30. N Calderon. U.S. Patent 3,439,056, 1969; Chem Abstr 71: 38438c, 1969.
31. N Calderon. U.S. Patent 3,439,057, 1969; Chem Abstr 71: 80807x, 1969.
32. H Hoecker, R Musch. Makromol Chem 157: 201–218, 1972.
33. H Hoecker, W Reimann, K Riebel, Zs. Szentivanyi. Makromol Chem 177: 1707, 1976.
34. H Hoecker, L Reif, W Reimann, K Riebel. Recl. Trav. Chim 96: M47, 1977.
35. H Hoecker, FR Jones. Makromol Chem 161: 251–266, 1972.
36. R Wolovsky, Z Nir. Synthesis 1972: 134.
37. RH Grubbs, ST Nguyen, GC Fu. J Am Chem Soc 115: 9856–9857, 1993.
38. RH Grubbs. JMS Pure Appl Chem A31: 1829–1833, 1994.
39. WJ Zuercher, M Hashimoto, RH Grubbs. J Am Chem Soc 118: 6634–6640, 1996.
40. GC Fu, RH Grubbs. J Am Chem Soc 114: 5426–5427, 1992.
41. RH Grubbs, SJ Miller, GC Fu. Acc Chem Res 28: 446–452, 1995.
42. GC Fu, RH Grubbs. J Am Chem Soc 114: 7324–7325, 1992.
43. YS Shon, TR Lee. Tetrahedron Lett 38: 1283–1286, 1997.
44. AK Ghosh, KA Hussain. Tetrahedron Lett 39: 1881–1884, 1998.
45. RH Grubbs, S Chang. Tetrahedron 54: 4413–4450, 1998.
46. J Tsuji, S Hashiguchi. Tetrahedron Lett 1980: 2955.

4

Advances in Cycloolefin Polymerization

Valerian Drăguţan
Romanian Academy, Institute of Organic Chemistry, Bucharest, Romania

Roland Streck*
Hüls AG, Marl, Germany

I. INTRODUCTION

Under the influence of a great variety of cationic, Ziegler–Natta and metathesis catalysts, cycloolefins (also named cycloalkenes) undergo addition or ring-opening metathesis polymerization (ROMP) yielding vinylic and/or ring-opened polymers [Eq. (1)].

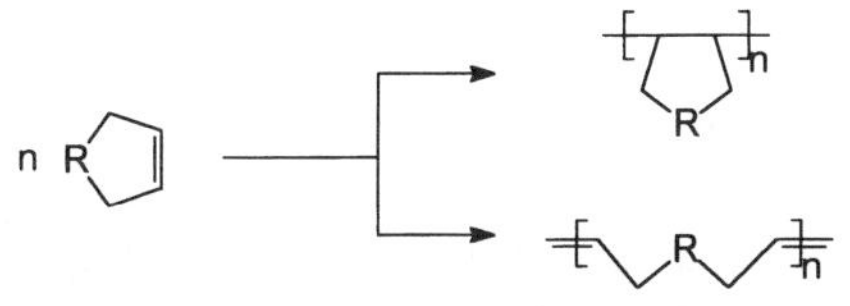

(1)

These two types of polymerization reactions are well documented for a large number of monomers and catalytic systems [1–16].

At present, the main theoretical and experimental aspects such as reaction mechanism and stereochemistry, reaction kinetics and thermodynamics are extensively treated by various research groups [6, 7, 14, 16]. On the other hand, whereas addition polymerization found earlier practical applications in the synthesis of hydrocarbon resins [2], ring-opening metathesis polymerization has recently become an important procedure for the manufacture of polyalkenamers having outstanding plastic and elastomeric properties [6, 7, 14, 15].

By addition polymerization, the carbon–carbon double bond can be opened via a *trans* mode resulting in *threo*-polymers of diisotactic, disyndiotactic or atactic structure or via a *cis* mode to form *erythro*-polymers of diisotactic, disyndiotactic or atactic configuration [5] [Eq. (2)].

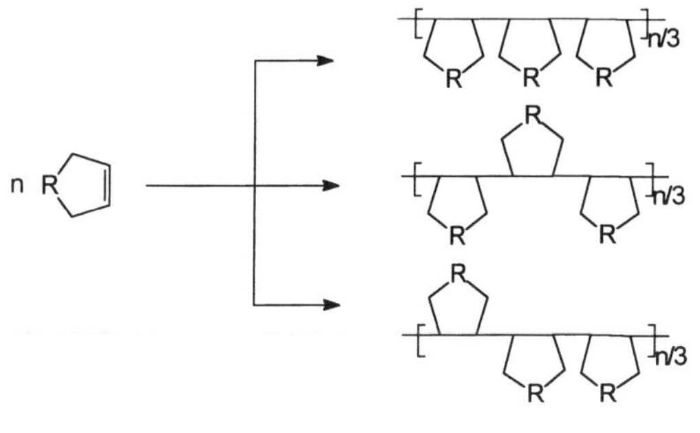

(2)

*Retired.

Analogously, the ring-opening polymerization of cycloolefins can result in *trans*- or *cis*-polyalkenamers via a *trans* or *cis* route [5, 7, 14] [Eq. (3)].

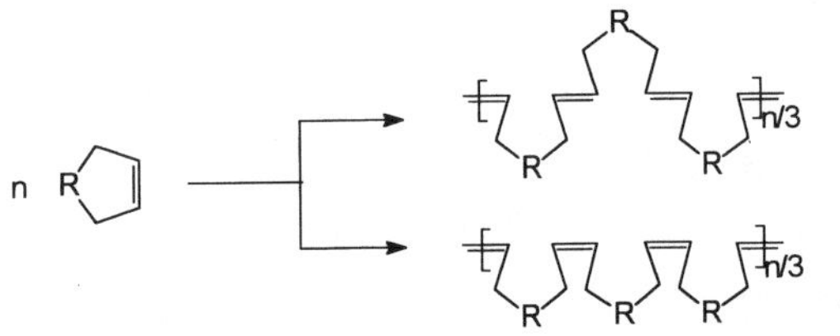

(3)

In the case of polycyclic olefins, the cycles incorporated into the polymer units give rise to structures in which the successive rings stand in an isotactic (m) or syndiotactic (r) relationship, respectively [14, 16] [Eq. (4)].

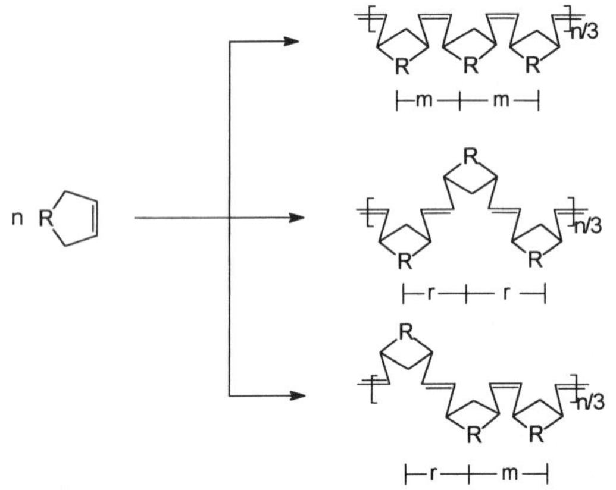

(4)

Moreover, when the cycloolefin has substituents in certain positions or is chiral, polymers with head–head (HH), head–tail (HT), and tail–tail (TT) structures may arise, each one having m or r configurations, containing *cis* or *trans* double bonds along the polymer chain [16].

II. CATALYTIC SYSTEMS

There are several types of catalytic systems active for addition and ring-opening metathesis polymerization of cycloolefins [1, 4–7]. The main catalytic systems could be grouped in cationic and coordination catalysts. The last group refers to Ziegler–Natta and ring-opening metathesis polymerization (ROMP) catalysts used largely for addition and ring-opening polymerization of cycloolefins, respectively.

A. Cationic Catalysts

Polymerization of cycloolefins is initiated under the action of both Broensted and Lewis acids [1]. The first type of catalyst has been used for a limited number of cycloolefins, whereas the second one has been largely employed, especially for polymerization of monocyclic and bicyclic terpene hydrocarbons in the manufacture of synthetic resins [2].

1. Bronsted Acids

Various Bronsted acids such as H_2SO_4, H_3PO_4, H_3BO_3, HF, HCl, HBr, HI, benzenesulphonic acid, etc., have been employed to polymerize different cycloolefins [1, 2]. The reactions were in the major part homogeneous and the process yield and product composition depended essentially on the nature and concentration of the acid used to promote the reaction. Generally, oligomers and low molecular weight polymers have been produced whose structures have not been completely elucidated.

Concentrated sulphuric acid has been initially used for dimerization and oligomerization of cyclic monoolefins such as cyclohexene and diolefins such as cyclopentadiene, and later on for indene-coumarone fractions. Diluted sulphuric acid and benezenesulphonic acid have been further employed for the polymerization of more active cycloolefins like norbornene and dicyclopentadiene. In these reactions, monomer conversion, product yield and molecular mass depended largely on the acid concentration and monomer nature as well as on the other reaction parameters. Various compositions of initiators containing sulphuric acid in association with phosphoric acid, boric acid, sulphonic acids or inorganic sulphates of the type $M_x(SO_4)_y$ (M = Al, Cr, Mg, Co, V) have also been reported for the polymerization of unsaturated alicyclic and cyclic fractions and for reactions with heavy aromatic fractions in hydrocarbon resin synthesis [2].

Heterogeneous catalytic systems consisting of organic hydrosulphates and formic acid deposited on bentonite or various aluminosilicates have also been found to be very active in the polymerization reaction of hydrocarbon fractions [1–3]. The product composition and reaction yield were largely dependent on the concentration of the acid and the nature of the support ranging from low oligomers to high molecular weight polymers of ill-defined structures. Related heterogeneous catalysts can also be formed by acid treatment of various solid supports such as inorganic sulphates, phosphates or borates, silica, alumina, aluminosilicates, etc. The nature of the support is very important in producing very active systems and allowing operation conditions that will direct the reaction toward the formation of high molecular mass products.

2. Lewis Acids

Due to their ready availability and excellent catalytic properties, this type of catalyst has been extensively used in cycloolefin polymerization for several decades, resulting in important industrial applications such as manufacture of hydrocarbon resins [3, 17, 18]. They are used mainly in homogeneous systems with adequate solvents but also heterogeneous catalysts are very active and promote cycloolefin polymerization to different reaction products, depending on the operation conditions. Generally, they are unicomponent, binary, ternary and multi-component catalytic systems and their final composition is strongly dependent on the nature and quality of the solvent, the reaction conditions and the monomer type and structure [1].

(a) Unicomponent Lewis Acid Catalysts. Traditional Lewis acids have been employed extensively as one-component catalyst to polymerize cycloolefins by addition reaction in adequate solvents that ensure a homogeneous medium and good reaction conditions [1]. These catalysts are inorganic halides such as $AlCl_3$, $AlBr_3$, BF_3, $SnCl_4$, $TiCl_4$, $FeCl_3$, $SbCl_5$, $ZnCl_2$, organometallic compounds like $AlEt_3$, Et_2AlCl, $EtAlCl_2$, and several metallic salts having Lewis acid character. Of these, the most frequently encountered for the polymerization of simple and substituted cycloolefins are the following: $AlCl_3$, BF_3, $SnCl_4$, $TiCl_4$. It has been documented that $AlBr_3$ is a more active catalyst than $AlCl_3$ and whereas the chloride is substantially insoluble in hydrocarbons, aluminum bromide is quite soluble. This difference in solubility accounts for the fact that in physicochemical studies, aluminum bromide is necessarily used as a catalyst. Several heterogeneous unicomponent Lewis acid catalysts for cycloolefin polymerization are also generated from different metallic oxides or salts having a significant acid character such as Al_2O_3, SiO_2, SnO_2, TiO_2, aluminosilicates, etc., but they work under special conditions and the control of the process is severely difficult [17, 18].

(b) Binary Lewis Acid Catalysts. The activation of the Lewis acid by protogenic or cationogenic compounds such as water, hydrohalides (HX, X = Cl, Br, I), alkyl halides, alcohols, phenols, organic acids, halohydrins is a convenient way to obtain highly efficient two-component cationic catalysts of a wide utility in cycloolefin polymerization [1, 2]. These "cocatalysts", or more properly named "coinitiators", play an important role in generating the actually active species by providing a proton or a carbenium ion in the catalytic system which is able to electrophilically attack, in a complexed or free state, the carbon–carbon double bond of the cycloolefin and start the initiation step of the polymerization process. Such systems are combinations of the following type: $AlCl_3/H_2O$, BF_3/HF, Et_2AlCl/tBuCl, $SnCl_4/$ CCl_3COOH, etc.

Ethers and esters are frequently employed in conjunction with Lewis acids to promote cationic polymerization of different cycloolefins. The traditional catalytic system of this type, BF_3/Et_2O, has been extensively used in reactions of various cycloolefins such as cyclopentene, cyclohexene, norbornene, norbornadiene, cyclopentadiene, dicyclopentadiene, pinene, etc., to produce polymers in high yield and having large molecular mass [1]. These strongly complexing agents impart a higher stability and selectivity to the catalytic system as compared to the uncomplexed Lewis acid.

In addition to the above catalytic systems, efficient cationic catalysts for cycloolefin polymerization have been derived from binuclear compounds consisting of Lewis acids associated with transition metal salts, essentially metal halides. Examples are found in the polymerization of cyclic dienes such as cyclopentadiene, cyclohexadiene and cyclooctadiene [1]. Similarly, cationic complexes of transition metal derivatives with Lewis acids, e.g. $Pd(MeCN)_4(BF_4)_2$ and $Pd(EtCN)_4(BF_4)_2$ have been used in norbornene polymerization [19]. Finally, organometallic compounds including organoaluminums (e.g. Et_3Al, $EtAlCl_2$, Et_2AlCl) were employed as such or in association with activators such as alkyl halides (tBuCl) for the polymerization of some cycloolefins [1, 20]. Relevant examples are the polymerizations of dicyclopentadiene initiated by $EtAlCl_2/t$BuCl and Et_2AlCl/tBuCl [20].

B. Coordination Catalysts

There are several types of Ziegler–Natta and ROMP catalysts employed for cycloolefin polymerization, the majority of them being derived from transition metal salts and organometallic compounds [4–7]. These types are grouped into unicomponent, binary, ternary, and multicomponent catalytic systems as a function of the presence or absence of the organometallic cocatalyst or other additives, each of them being differentiated on the catalyst composition and selectivity toward vinylic or ring-opened polymerization.

1. Unicomponent Coordination Catalytic Systems

Tungsten halides and oxyhalides such as WCl_6, WBr_6 or $WOCl_4$ were the first unicomponent ROMP catalysts reported to polymerize cyclopentene, cyclooctene, cyclodecene and cyclododecene to the respective polyalkenamers in hydrocarbon media without any cocatalyst. Later, $MoCl_5$, $ReCl_5$ and WCl_6 were found to be active in CCl_4 or CS_2 for the polymerization of norbornene, *exo*-trimethylenenorbornene and *endo*-dicyclopentadiene [7].

A very efficient group of metathesis polymerization catalysts active in polar media such as ethanol or water is represented by halides of ruthenium, osmium and iridium. Thus, $RuCl_3$ in polar medium has been employed by Natta and Dall'Asta [21, 22] for the polymerization of cyclobutene and 3-methylcyclobutene. Michelotti and Keaveney [23] used hydrates of $RuCl_3$, $OsCl_3$ and $IrCl_3$ in ethanol for polymerization of norbornene. Catalyst activity decreased in the order Ir > Os > Ru. Good yields in norbornene polymerization were obtained by Porri and coworkers [24] using η-chloro(cyclooctene)iridium. More recently, Ivin and coworkers [25, 26] investigated the reaction of norbornene and substituted norbornene in the presence of $ReCl_5$, $RuCl_3$, $RuCl_3$(1,5-cyclooctadiene), $OsCl_3$ or $IrCl_3$ obtaining polyalkenamers with variable stereoconfiguration at the double bonds in the chain, depending on the nature of the transition metal. Furthermore, Grubbs and coworkers [27] prepared polymers of a high thermal stability and good mechanical properties with $RuCl_3$, $OsCl_3$, $Rh(H_2O)_6$(tosylate)$_2$ and $RuCl_3$(1,5-cyclooctadiene) from 7-oxabicyclo[2.2.1]hept-5-ene. The one-component catalyst $PhWCl_3$ was also active in the polymerization of cyclopentene providing high yields in polypentenamer.

Carbene complexes of tungsten, molybdenum or rhenium initiate the ring-opening polymerization of cycloolefins to polyalkenamers. Thus, $Ph_2C{=}W(CO)_5$ (**1**) polymerizes cyclobutene, 1-methylcyclobutene, cyclopentene, cycloheptene, cyclooctene and norbornene while $Ph(MeO)C{=}W(CO)_5$ (**2**) polymerizes cyclobutene and norbornene to highly stereospecific polyalkenamers [28].

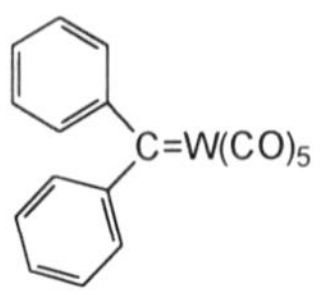

(1) 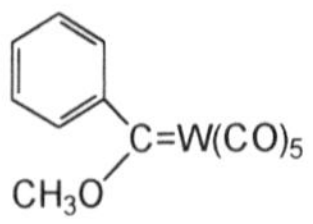 (2)

Unicomponent transition metal carbenes of the type $M({=}CHR)(NAr)(OR)_2$ (M = Mo, W or Re, R = *t*Bu, CMe_2CF_3, $CMe(CF_3)_2$) (**3**) or $(THF)M({=}CHR)X_3$ (M = Ta, THF = tetrahydrofuran, X = Cl)) (**4**) give living oligomers or polymers from norbornene and its derivatives [29].

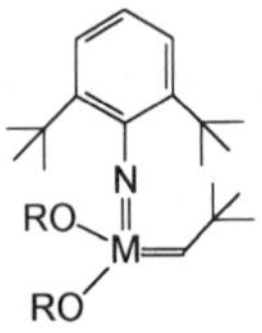

(3) 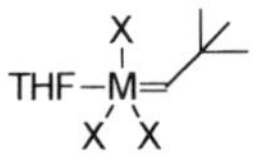 (4)

Related rhenium carbenes, $Re(CtBu)(CHtBu)(OR)_2$, (R = $CMe(CF_3)_2$ (**5**), more tolerant toward functionalities, have also been reported [30].

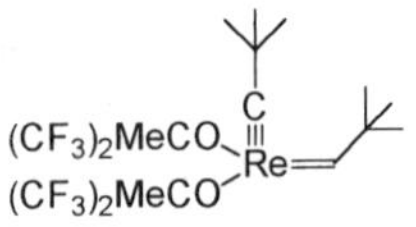

(5)

Titanacyclobutanes (**6** and **7**) and tantallacyclobutanes (**8** and **9**) showed to be a new class of active catalysts for living ring-opening polymerization of cycloolefins [31, 32].

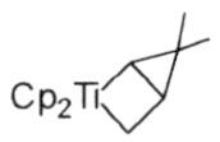

(6) 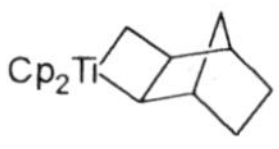 (7)

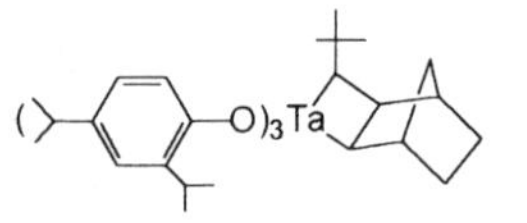

(8) 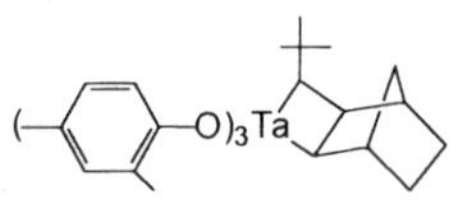 (9)

The ready availability of the Tebbe reagent as the starting material allowed Grubbs and coworkers [27] to prepare several titanacylobutane catalysts used for living ring-opening polymerization of norbornene and dicyclopentadiene. Unicomponent ruthenium–carbene complexes of formula $(Cp_3P)_2RuCl_2({=}CHR')$ or $(Cy_3P)_2RuCl_2({=}CHR')$ (R′ = Ph (**10**), $CHCPh_2$ (**11**)) (Cp = cyclopentyl, Cy = cyclohexyl) have been prepared and used by Grubbs and coworkers [33, 34] successfully in the ring-opening metathesis polymerization of norbornene and norbornene derivatives in water and polar media.

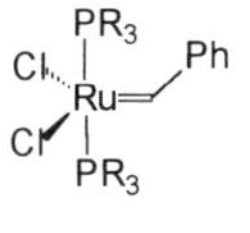

(10) 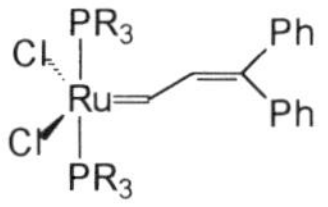 (11)

where R is Cp or Cy.

Well-defined bimetallic ruthenium catalysts $(R_3P)_2Cl_2Ru(=CHp\text{-}C_6H_4CH=)RuCl_2(PR_3)_2$ [R = Ph, Cy (cyclohexyl), Cp (cyclopentyl)) (**12**) have been employed successfully by Grubbs and coworkers for ring-opening polymerization of norbornene and norbornene derivatives [35].

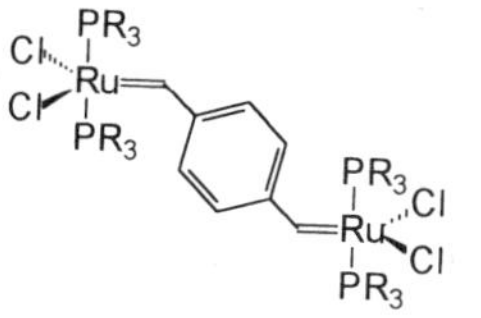

(12)

Various unicomponent catalysts based on π complexes of transition metals from groups IV–VII of the Periodic Table, e.g., $(\pi\text{-allyl})_4Zr$, $(\pi\text{-allyl})_3Cr$, $(\pi\text{-allyl})_2Ni$, $(\pi\text{-allyl})_3Co$, $(\pi\text{-allylPdX})_2$, $(\pi\text{-allylRhX})_2$, have been found to be very active in polymerization of a large number of cycloolefins like cyclobutene, cyclopentene, cyclooctene, cyclooctadiene and norbornene [36]. Some of these catalysts induce polymerization of the cycloolefin totally to vinyl polymers while other catalysts of this class give preferentially vinylic polymers accompanied in a large extension by ring-opened polymers.

2. Binary Coordination Catalytic System

A wide variety of binary catalysts of both Ziegler–Natta and ROMP type, consisting of group IV–VII transition metal salts associated with organometallic compounds, have been used in cycloolefin polymerization to manufacture high molecular weight vinylic and ring-opened polymers [4–7]. The activity and selectivity of these catalytic systems depend mainly on the nature of the transition metal and the structure of cycloolefin.

For the polymerization of cyclobutene, Natta and coworkers [37] reported binary catalytic systems containing titanium, vanadium, chromium, and tungsten to be the most active, those of molybdenum less active, and systems derived from cobalt, iron, manganese, and uranium totally inactive. Catalysts based on vanadium and chromium yield preferentially polycyclobutylene by addition polymerization, those with molybdenum and tungsten give polybutenamer by ring-opening polymerization, while titanium catalysts give polymers of both types. Similarly, in the polymerization of cyclopentene, vanadium and chromium lead to polycyclopentylene by addition polymerization, molybdenum and tungsten yield polypentenamer by ring-opening polymerization, while titanium gives both types of polymer. The polymerization of cycloolefins containing seven or more carbon atoms in the ring with the catalysts based on WCl_6, $MoCl_5$ or $ReCl_5$ and organometallic compounds or hydrides of aluminum, zinc, magnesium, berilium or cadmium has also been reported [38]. Of these catalysts, those based on tungsten proved to be more active than those based on molybdenum, the catalytic activity decreasing with the number of carbon atoms in the cycloolefin substrate.

Many authors have extensively used the binary catalytic systems to polymerize cyclopentene and higher unsubstituted or substituted cycloolefins like norbornene, norbornadiene, dicyclopentadiene and other norbornene-like derivatives to high molecular weight ring-opened polymers [6, 7]. Thus, catalysts based on tungsten, tantalum or niobium salts were used by Guenther *et al.* [39] in combination with organoaluminum, organosilicon, organotin or other organometallic compounds to obtain high yields of *trans*-polypentenamer. Interestingly, substitution of WF_6 or $ReCl_5$ for WCl_6 in such catalysts changed the stereoselectivity of the reaction to *cis*-polypentenamer. The binary system $WCl_6.EtAlCl_2$ has been employed frequently by Calderon *et al.* [40] for the polymerization of cyclooctene, 3-methylcyclooctene, 3-phenylcyclooctene, 1,5-cyclooctadiene, cyclodecene and 1,5,9-cyclcododecatriene to obtain *trans*-polyalkenamers; conversions over 75% were attained working under mild conditions at normal pressure and temperature. Systems consisting of WCl_6 and R_4Sn (R = Me, Et, Bu, Ph) have also been employed for the polymerization of a wide range of cycloolefins [6]. Other tungsten compounds, e.g., alkoxides or phenoxides of formula $W(OR)_xCl_{6-x}$, R = Ph, Et, or transition metal salts, e.g., of Ti, V, Cr, Mo, Re. Nb, Ta, Zr, associated with organoaluminum or organotin compounds have been reported in several patents as active catalysts for polymerization of cyclopentene, dicyclopentadiene, norbornene, norbornadiene and their derivatives [41]. Recently, chiral metallocenes (e.g., zirconocenes such as ethylene[bis(η^5-indenyl)zirconium] dichloride (**13**) and ethylene[bis(η^5-tetrahydroindenyl)zirconium] dichloride (**14**)) in conjunction with aluminoxanes (e.g., methylaluminoxane) were shown to be very active and stereospecific for the addition polymerization of cycloolefins such as cyclobutene, cyclopentene, norbornene, and dicyclopentadiene [5].

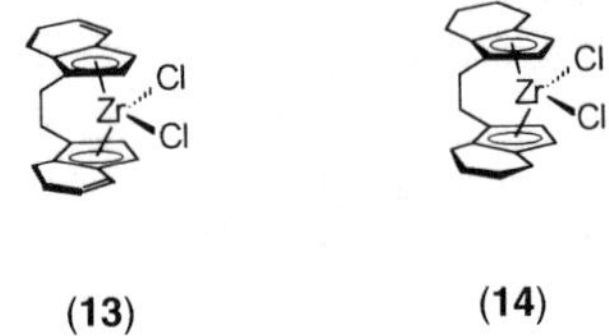

(13) (14)

A wide range of binary catalysts consisting of WCl_6, $WOCl_4$, WCl_4 or WBr_5 and $AlCl_3$ or $AlBr_3$ have been used for polymerization of cyclopentene, cyclooctene, 1,5-cyclooctadiene, cyclododecene and 1,5,9-cyclododecatriene to polyalkenamers of predominantly *trans* configuration. Of this class, the binary systems formed from WCl_6 or $MoCl_5$ and $AlBr_3$ were very active and stereoselective for the polymerization of cyclopentene, higher cycloolefin homologues, 1,5-cyclopentadiene and 1,5,9-cyclododecatriene. Catalysts obtained from WCl_6 or $MoCl_5$, $AlCl_3$ and Al powder have been active in the polymerization or copolymerization of cycloolefins such as cyclopentene, cyclooctene, 1,5-cyclododecene, and 1,5,9-cyclododecatriene [42, 43]. Instead of WCl_6 or $MoCl_5$, WF_6, WOF_4, WCl_4, MoF_6 or $MoCl_4$ have also been used as the active component. Remarkably, high activity was exhibited by the catalysts consisting of WCl_6 and $AlCl_3$ or $AlBr_3$ in the polymerization of cyclooctene. Similarly, W(*o*-phenanthroline)$(CO)_4$ with $AlBr_3$ afforded high yields in the synthesis of polyoctenamer [42]. Active systems prepared from inorganic compounds of tungsten such as tungsten trioxide, tungstic acid, isopolyacids, heteropolyacids or salts of these acids and $AlCl_3$ or $AlBr_3$ have been employed for polymerization of cyclopentene [39]. Although the yields in polpentenamer were considerable, the products obtained had complex structures, probably because of the accompanying Friedel–Crafts reactions. Good results have also been recorded using organic derivatives of group IV–VI transition metals associated with Lewis acids. In this context, the use of aryltungsten compounds in conjunction with halides of boron or tin resulted in high *trans* stereoconfiguration of the product in the polymerization of cyclopentene [44].

π-Complexes of tungsten and molybdenum associated with AlX_3 or GaX_3 (X = Cl or Br) provided very active binary catalysts for a number of cycloolefins, for instance, π-allyl$_4$W/$GaBr_3$ for polymerization of 1,5-cyclooctadiene [45, 46]. Surprisingly, π-allyl$_3$Cr and π-allyl$_4$W$_2$ in conjunction with WCl_6 afforded low conversions of cyclopentene, whereas the catalyst π-allyl$_4$W$_2$ with aluminosilicate and trichloroacetic acid exhibited particularly high stereospecificity in the polymerization of 1,5,9-cyclododecatriene [47].

Binary catalysts derived from metal carbenes and Lewis acids have been employed successfully to initiate the ring-opening polymerization of several substituted and unsubstituted cycloolefins [48]. To such catalytic systems pertain complexes of formula $(CO)_5M{=}CRR'$ (R = MeO, EtO; R′ = Me, Et; M = W, Mo) associated with $AlCl_3$, $EtAlCl_2$ or $TiCl_4$; these complexes were used for polymerization of cyclopentene. The tungsten complex $(Ph_3P)(CO)_4W{=}C(OMe)Ph$ with $TiCl_4$ was applied for the polymerization of cyclic trimers, tetramers and pentamers of cyclobutene, cyclooctene, and 1,5-cyclooctadiene. Another active tungsten–carbene complex used in the polymerization of norbornene and its derivatives is $Br_2(RO)_2W{=}CR'R''$ (R = neopentyl, R′ = H and R″ = *t*Bu) associated with $GaBr_3$. Addition of Et_3Al and particularly $AlCl_3$ to the carbene complex $(CO)_4W{=}C(OR)R'$ (R = CH_3, R′ = 3-buten-1-yl) enhanced considerably the activity of the resulting binary catalytic system in norbornene polymerization as compared to the initial tungsten–carbene complex [49].

3. Ternary Coordination Catalytic Systems

The activity and stability of binary catalysts for cycloolefin polymerization have been considerably improved by adding a third component containing essentially oxygen, halogen, sulfur or nitrogen [6]. These new catalytic systems allowed synthesis of a wide range of polyalkenamers, particularly *trans*-polypentenamer and *trans*-polyoctenamer, to be obtained. Moreover, their application in polymerization reactions permitted synthesis of highly performant copolymers, block and graft copolymers, interpolymers and polymers of special architecture to be manufactured [41].

A versatile catalytic system of this type which has been applied largely in the polymerization reactions of cyclopentene, cycloheptene, cyclooctene, and cyclodecene consists of WCl_6, $WOCl_4$ or $MoCl_5$ associated with organoaluminum (Et_3Al, Et_2AlCl, $EtAlCl_2$, *i*Bu_3Al), organozinc (Et_2Zn), or organoberyllium (Et_2Be) compounds and an oxygen-containing derivative such as a peroxide, hydroxyperoxide, alcohol, phenol, molecular oxygen or water. Examples of such derivatives are benzoyl peroxide, ethanol, phenol, hydrogen peroxide, *t*-butyl peroxide, and cumyl peroxide [6, 41]. Thus, the catalytic system WCl_6/EtOH/$EtAlCl_2$ has been extensively employed in the polymerization of cyclopentene, cyclooctene, 3-methylcyclooctene, 3-phenylcyclooctene, 1,5-cyclooctadiene, and 1,5,9-cyclododecatriene [40, 41]. Instead of ethanol, other oxygen-containing compounds such as methanol, allyl alcohol, cumyl alcohol, glycol, phenol, thiophenol, or cumyl peroxide have been employed. Cyanhydrins, chlorosilanes, nitrogen derivatives such as amines, amides, or nitro compounds, halogenated derivatives such as epichlorohydrin or choroethane, or inorganic peroxides such as sodium or barium peroxide have also been employed [39]. Numerous other tungsten-based catalytic systems containing oxygen and halogen, e.g., some halogen alcohols (2-chloroethanol, 2-bromoethanol, 1,3-dichloroisopropanol, 2-chlorocyclohexanol, 2-iodocyclohexanol), 2- chlorophenol, acetals ($CH_2(OCH_2CH_2)_2$, $CH_3CH(OC_2H_5)_2$, $Cl_3CCH(OCH_3)_2$ or $C_6H_5CH(OC_2H_5)_2$) and epoxides (ethylene oxide, butylene oxide) have been reported for the polymerization of various cycloolefins [6, 41].

4. Multicomponent Coordination Catalysts

In order to improve the catalytic activity and selectivity as well as the stability and reproducibility of the binary and ternary coordination catalysts used in cycloolefin polymerization several combinations of two or more catalyst components have been added to the parent system based on transition metal salts and organometallic compounds. Such catalytic systems were developed either by adding an active Lewis acid component to increase the catalytic activity, together with an oxygen-, nitrogen- or halogen-containing compound able to enhance the catalyst stability or by adding two new components, each directed to impart a high performance to the catalytic system and to confer special properties to the resulted polymer. For instance, particularly active quaternary catalytic systems derived from WCl_6 or $Mo(CO)_4L_2$ ($L = 1,5$-cyclooctadiene, norbornene, norbornadiene) and organoaluminum compounds containing Lewis acids and oxygen- or halogen-carrying components have been reported for the polymerization of cyclopentene, cyclooctene, norbornene, norbornadiene, dicylopentadiene and their derivatives [6, 7].

III. POLYMERIZATION OF MONOCYCLIC OLEFINS

Polymerization of monocyclic olefins has been carried out with the most important representatives of this series, from cyclobutene to cyclododecene and higher homologues [1–3]. Lower cycloolefins such as cyclobutene and cyclopentene proved to be rather reactive in both addition and ring-opening polymerizations whereas the larger members of the series, except cyclohexene, polymerized readily by ring-opening polymerization. In the cycloolefin series, generally, the reactivity decreases as the number of carbon atoms in the ring increases. Of the substituted cycloolefins, those monomers bearing substituents at the double bonds polymerize with difficulty because of the steric hindrance at the reaction center, while those having substituents in other positions yield easily polymers bearing the substituents along the chain.

A. Four-Membered Ring Monomers

1. Cyclobutene

Polymerization of cyclobutene has been carried out using a wide range of catalytic systems based on transition metal salts and organometallic compounds. Depending on the catalytic system employed, polycyclobutylene or polybutenamer or both types of polymers have been obtained [Eq. (5)].

n □ → [polycyclobutylene]$_n$; → [polybutenamer]$_n$ (5)

It is of interest that the polybutenamer resulting by ring-opening polymerization of cyclobutene is identical in structure with 1,4-polybutadiene, an elastomer manufactured so far by conventional polymerization of 1,3-butadiene. In their earlier publications on cyclobutene polymerization, Natta and coworkers [37, 38] described the polymerization of cyclobutene using several catalytic systems consisting of group IV–VI transition metal salts and various organometallic compounds; the systems containing chlorides or acetylacltonates of titanium, vanadium, chromium, molybdenum, and tungsten were the most active in these reactions. Interestingly, of these catalysts, those based on vanadium and chromium led to polycyclobutylene and those based on titanium to polybutenamer, while the other catalysts yielded polymers of both types. In all cases, the polybutenamer obtained was a mixture of *cis* and *trans* stereoisomers. Salts of group VIII transition metals in polar protic solvents such as water or ethanol have also been used. With nickel and rhodium salts, vinylic polymers were preferred, whereas ruthenium salts gave mainly polybutenamer. In the latter case, a strong effect of the solvent on the steric configuration at the double bond has been observed. It was found that when ethanol was used as the solvent, the amount of *trans* configuration in the polymer increased significantly.

2. Substituted Cyclobutene

(a) 1-Methylcyclobutene. In the presence of appropriate catalytic systems, 1-methylcyclobutene will undergo addition or ring-opening metathesis polymerization. In the first case poly(1-methylcyclobutene) will be formed whereas in the second case poly(1-methylbutenamer) [Eq. (6)].

(6)

It is noteworthy that poly(1-methylbutenamer) obtained by ring-opening metathesis polymerization has the structure of 1,4-polyisoprene, a well known elastomer of high commercial value.

Polymerization of 1-methylcyclobutene was first carried out with catalytic systems based on WCl_6 and organoaluminum compounds (e.g., WCl_6/Et_3Al and WCl_6/Et_2AlCl), leading to products having a polyisoprene skeleton but with a saturated structure. To explain this result, it was supposed that intermolecular cyclizations of the polyisoprene fragments occurred under the influence of the foregoing catalytic systems, accompanied by the formation of saturated cyclohexane structures [50].

In contrast, Katz and coworkers [51] obtained good results using molybdenum- and tungsten-based catalytic systems. Thus, with $MoCl_2(NO)_2(Ph_3P)_2/Me_3Al/AlCl_3$, conversions of about 67% were attained and the product had 20% polyisoprene structure. The system WCl_6/Ph_3SnEt and WCl_6/nBuLi proved to be rather reactive, but the polymer obtained had a high content of saturated structures. Remarkable results, however, were recorded with the tungsten carbene catalyst $Ph_2C{=}W(CO)_5$: the product was polyisoprene with 84–87% *cis* stereoconfiguration. It is of interest that the polymer had, along with 2-methyl-2-butene units, 2-butene and 2,3-dimethyl-2-butene units, the latter probably resulting by secondary metathesis reactions under the influence of the catalytic system after the primary ring-opening metathesis polymerization had occurred.

(b) 3-Methylenecyclobutene. 3-Methylenecyclobutene polymerizes rapidly in the presence of BF_3, BF_3/Et_2O and Et_2AlCl, leading to polymers with a predominant 1,5-enchainment [Eq. (7)].

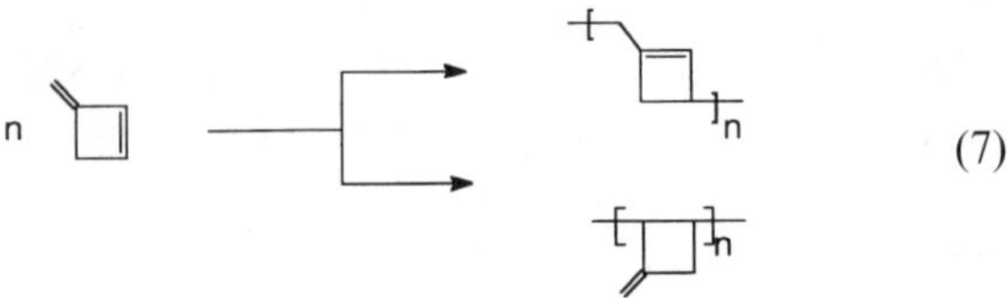

(7)

In contrast, by anionic polymerization in the presence of *n*BuLi catalyst, 1,2-addition product has also been obtained [52].

(c) 3-Methylcyclobutene. Polymerization of 3-methylcyclobutene was first carried out by Natta and coworkers with $RuCl_3$ in polar media and later with vanadium catalysts in nonpolar solvents [21]. In both cases the polymer had mainly a polybutenamer structure with the methyl groups distributed in each monomer unit [Eq. (8)].

(8)

Interestingly, when $RuCl_3$ served as a catalyst, the structure of the resulting polymer was strongly dependent on the nature of the polar solvent: ethanol led to an increase in the *trans* configuration while water increased the *cis* configuration of the polymer. In all cases, the polymer had an amorphous structure that was attributed to the stereoirregularity of the tertiary carbon atom in the chain.

The influence on the conversion of the monomer and the structure of the polymer of various organometallic compounds in vanadium-based catalysts has also been explored. Of these, the most active proved to be those catalysts containing Et_3Al as a cocatalyst attaining conversions of *ca.* 37%. The polymer thus obtained had largely polybutenamer structures (90%). On the other hand, the catalytic systems containing *n*BuLi as a cocatalyst, although exhibiting a lower activity, led to polybutenamer structures with 55% *trans* and 45% *cis* configuration.

(d) 3,4-Bis(dimethylmethylene)cyclobutene. The ring-opening metathesis polymerization of this substituted cyclobutene monomer catalyzed by titanacyclobutane [53] provides a conjugated polymer [Eq. (9)].

(9)

The living polymer obtained under these conditions may be blocked with other polyalkenamers (e.g., polynorbornene) to produce block copolymers with interesting electrical and mechanical properties.

B. Five-Membered Ring Monomers

1. Cyclopentene

Cyclopentene has been polymerized in the presence of numerous catalytic systems leading either to poly(cyclopentenylene) by addition polymerization or to polypentenylene or polypentenamer by ring-opening metathesis polymerization. Thus, in an early report, Hoffman [54] observed that the oligomerization of cyclopentene in the presence of BF_3/HF proceeded to dimers, trimers, and tetramers as well as to solid resins. The structure probably corresponded to a 1,2-addition polymer, poly(1,2-cyclopentenylene) [Eq. (10)].

(10)

More recently, Kaminsky and coworkers [5] polymerized cyclopentene with chiral zirconocene/aluminoxane catalysts (e.g., $Et(Ind)_2ZrCl_2$/methylaluminoxane and $Et(IndH_4)_2ZrCl_2$/methylaluminoxane) obtaining highly isotactic polycyclopentenylene with probably 1,2-addition structure. By contrast, Collins *et al.* [55, 56] evidenced the formation of poly(1,3-cyclopentenylene) in the polymerization of cyclopentene with zirconocene/methylaluminoxane catalysts [Eq. (11)].

(11)

The ring-opening polymerization of cyclopentene to polypentenamer was first carried out by Eleuterio [57] who used heterogeneous catalysts based on oxides of chromium, molybdenum, tungsten, or uranium [Eq. (12)].

(12)

The polymer was an amorphous elastomer, which developed crystallinity upon stretching. Later, Natta and coworkers [58] studied the polymerization of cyclopentene using homogeneous catalysts derived from various transition metal salts (e.g. Ti, Zr, W, Mo, V) and organometallic compounds. They observed that the systems based on vanadium led to poly(cyclopentenylene) while those based on titanium, zirconium, tungsten, and molybdenum to polypentenamer. Of the latter systems, tungsten and molybdenum were the most active ones. Experiments with catalytic systems of manganese, iron, cobalt, and uranium indicated polymer formation. It is of interest that tungsten-based catalysts formed predominantly *trans*-polypentenamer and molybdenum catalysts *cis*-polypentenamer.

A great variety of one-component, two-component, and three-component catalytic systems have been developed by Dall'Asta [41], Guenther [39], Nuetzel [59], Pampus [60], Streck [61], Uraneck [62], Witte [63], and many other investigators [6]. Guenther and Nuetzel [39] used several binary catalysts derived from tungsten and molybdenum containing alkali or alkaline earth metals, which exhibited a high activity and stereoselectivity. Highly active ternary systems based on tungsten and molybdenum salts associated with organometallic compounds and a third component were used by Dall'Asta and Carella [64] to obtain polypentenamers having a predominantly *trans* configuration. The stability of these catalysts has been improved by addition of oxygen-containing compounds (e.g., benxoyl peroxide, *t*-butyl peroxide, cumyl hydroperoxide, hydrogen peroxide, ethanol, phenol, oxygen, water). Pampus and coworkers [60] used epichlorohydrin and chloroethanol in WCl_6-based systems to increase catalyst stability in the synthesis of *trans* polypentenamer.

High conversions of cyclopentene were obtained by Nuetzel *et al.* [59] with hydroperoxides, inorganic peroxides, or aromatic nitroderivatives. To increase the stability of some tungsten-based catalytic systems, Witte *et al.* [63] employed α-halogenated alcohols such as chloroethanol, 2-chlorocyclohexanol, bromoethanol, 1,3-dichloro-2-isopropanol, o-chlorophenol, and 2-iodocyclohexanol. It was observed that methyl and ethyl acetals of formaldehyde, acetaldehyde, chloroform, and benzaldehyde impart good staiblity to the binary systems of tungsten, whereas epoxides such as ethylene oxide and butylene oxide lead at the same time to an increase in activity and stability.

Several unsaturated halogenated compounds added as activators to the tungsten systems allow an appropriate control of the molecular weight and impart useful properties for polymer processing. For instance, Oberkirch *et al.* [65] used vinyl chloride whereas Streck and Weber [61] employed vinyl fluoride, vinyl chloride, and vinyl bromide. Furthermore, Streck and Weber [61] obtained good results in controlling the molecular weight in cyclopentene polymerization on using unsaturated esters or ethers such as vinyl acetate or allyl phenyl ether. Catalytic systems based on niobium and tantalum compounds were reported by Uraneck and Trepka [62] to yield polypentenamers having special properties (e.g., low gel content). Further examples for cyclopentene polymerization by the addition or ring-opening metathesis mechanism employing various other transition metal catalytic systems such as carbene complexes, π-allyl complexes, cycloalkene and cyclodiene coordination compounds have been described in several extensive reviews [6,7].

2. Substituted Cyclopentene

(a) 1-Methylcyclopentene. Polymerization of 1-methylcyclopentene was carried out by Schmidt and Schuerlich [66] in the presence of BF_3 or BF_3/camphor complexes giving a viscous oily product whose structure was not fully elucidated. Probably, 1,2- and 1,3-addition reactions took place with formation of 1,2- and 1,3-recurring units [Eq. (13)].

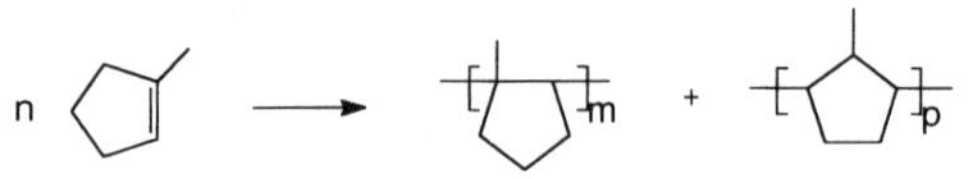

(13)

The reaction of this monomer has also been effected by Katz and coworkers [51] with $MoCl_2(NO)_2(Ph_3P)_2/Me_3Al_2Cl_3$, but undefined polymer at low conversions was obtained.

(b) 3-Methylcyclopentene. In the presence of $AlCl_3$ or BF_3/HF, the cationic polymerization of 3-methylcyclopentene has been reported [67] to give 1,3-poly-3-methylcyclopentylene, while metathesis polymerization in the presence of WCl_6-based catalysts [68] led to poly-3-methylpentenamer [Eq. (14)].

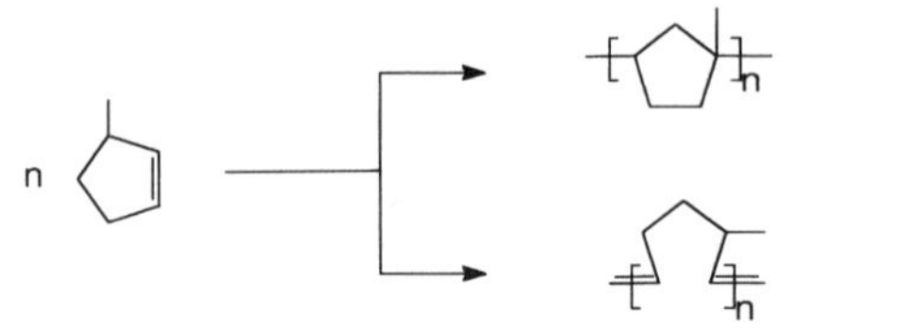

(14)

It is noteworthy that the poly-3-methylpentenamer thus obtained had more than 90% *trans* configuration and was an amorphous substance with properties totally different from the unsubstituted polypentenamer.

(c) 3-Vinylcyclopentene. Polymerization of 3-vinylcyclopentene in the presence of cationic initiators (e.g., $EtAlCl_2$) led to partially soluble polymers containing large amounts of bicyclic structures [1].

(d) 3-Allylcyclopentene. This monomer was reported to polymerize in the presence of cationic systems such as $EtAlCl_2$ and $EtAlCl_2/PhCH_2Cl$ to yield soluble polymers having essentially cyclic structures [1].

3. Cyclopentadiene

Cyclopentadiene readily polymerizes under the influence of a wide range of cationic initiators ($AlCl_3$, $AlBr_3$, BCl_3, BF_3/Et_2O, $SnCl_4$, $TiCl_3(O\text{-}n\text{Bu})$, etc.). In early studies on the polymerization of cyclopentadiene with various Lewis acids, Staudinger and Burson [69] prepared poly(cyclopentadiene) having 1,2 and 1,4 enchainments [Eq. (15)].

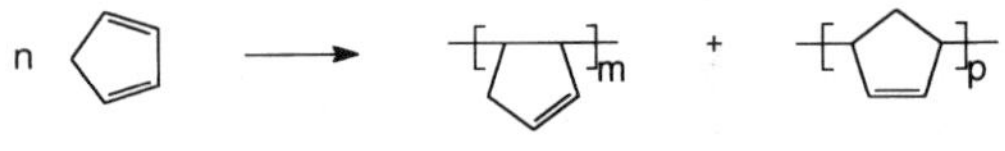

(15)

The molecular weight of the polymer was greatly affected by the nature of the Lewis acid. Later, Vairon and Sigwalt [70] succeeded in obtaining high molecular weight poly(cyclopentadiene)s using such mild cationic initiators like $TiCl_3(O\text{-}n\text{Bu})$. Further initiation with stable carbocation salts (e.g. $Ph_3C^+SbCl_6^-$) indicated direct addition of the trityl cation at the monomer in the initial reaction step. Extensive studies by Aso *et al.* [71] on the microstructure of the poly(cyclopentadiene) prepared with strong Lewis acids such as $AlBr_3$ revealed new isomer structures of the above repeat units in the polymer chain. Interesting kinetic data resulting from experiments with the catalysts $TiCl_4/CCl_3COOH$ and $SnCl_4/CCl_3COOH$ or BF_3/Et_2O have been reported by Higashimura *et al.* [72].

4. Substituted Cyclopentadiene

(a) 1- and 2-Methylcyclopentadiene. Studies on the isomer mixtures of 1- and 2-methylcyclopentadiene have been carried out using BF_3/Et_2O, $SnCl_4$, and $TiCl_4$ catalysts under various conditions (solvents, monomer concentration, temperature) [71]. The polymer microstructure corresponds to the several repeat units in the chain, with structure 1,4 prevailing [Eq. (16)].

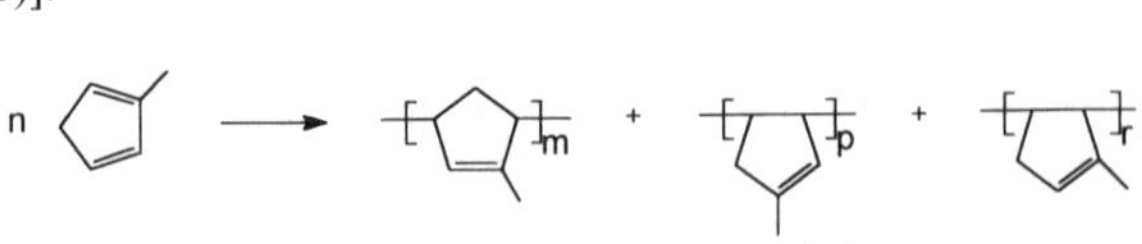

(16)

It is of interest that the 1,4-enchainment is preferentially formed with $SnCl_4$ or $TiCl_4$ as the catalysts and much less often with BF_3/Et_2O system.

(b) 1,2-Dimethylcyclopentadiene. In the presence of cationic initiators (e.g., $AlCl_3$, BF_3/Et_2O) 1,2-dimethylcyclopentadiene gives poly(1,2-dimethylcyclopentadiene) [71] having mainly 3,4 enchainment [Eq. (17)].

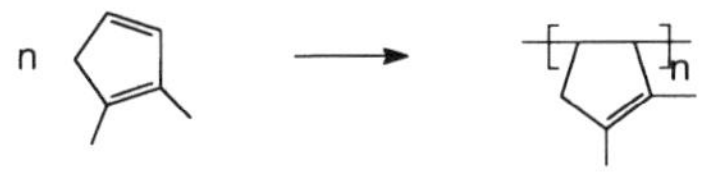

(17)

(c) 1,3-Dimethylcyclopentadiene. Polymerization of this monomer has been undertaken with BF_3/Et_2O, $SnCl_4$, and $TiCl_4$ leading to polymers with 1,4 and 4,3 enchainments [Eq. (18)].

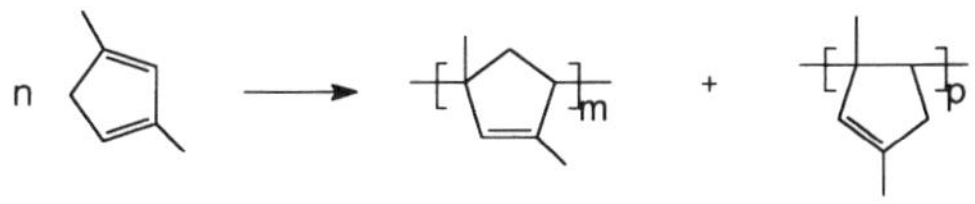

(18)

The microstructure of poly(1,3-dimethylcyclopentadiene) depended strongly on the reaction conditions (catalyst, solvent, temperature).

(d) 2,3-Dimethylcyclopentadiene. In the presence of various catalytic systems (e.g. $AlCl_3$ or BF_3/Et_2O), 2,3-dimethylcyclopentadiene gives rise to polymers having 1,4 enchainment [71] [Eq. (19)].

(19)

(e) Allylcyclopentadiene. Mixtures of isomers of allylcyclopentadiene have been polymerized with $AlBr_3$, BF_3/Et_2O, $SnCl_4$, and $TiCl_4$ to yield poly(allylcyclopentadiene) having about equal amounts of 1,4 and 3,4 structures in the polymer chain [Eq. (20)].

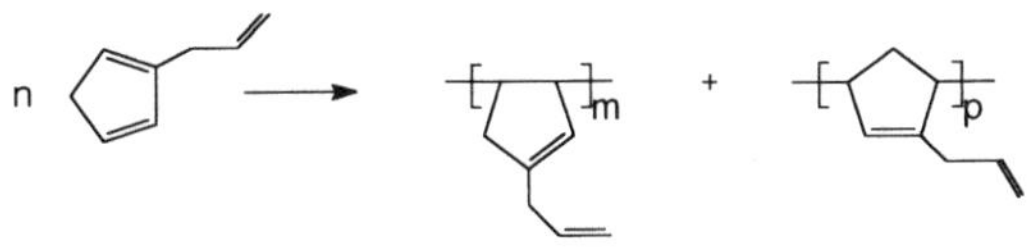

(20)

In the presence of the catalytic system BF_3/Et_2O, microstructures containing 1,4 along with 1,2 enchainments have also been reported [71].

(f) Methallylcyclopentadiene. Polymerization of mixtures of methallylcyclopentadiene [73] isomers induced by BF_3/Et_2O gives rise to poly(methallylcyclopentadiene) with 1,2 and 1,4 structures in the polymer chain [Eq. (21)].

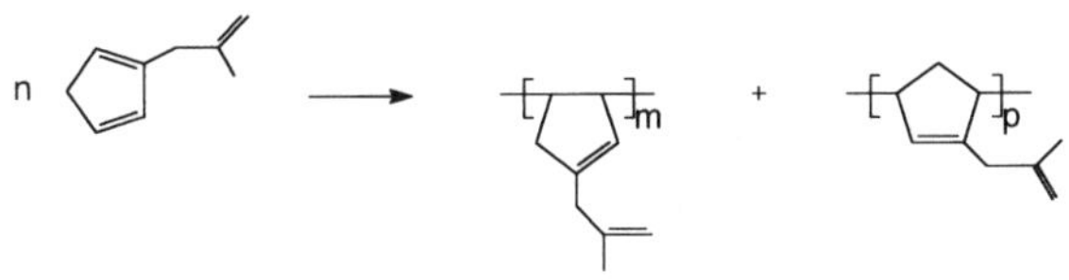

(21)

(g) Allylmethylcyclopentadiene. Under the influence of BF_3/Et_2O, isomers of allylmethylcyclopentadiene lead to poly(allylmethylcyclopentadiene) having predominantly 1,4 linkages along with 1,2 linkages in the polymer chain [Eq. (22)].

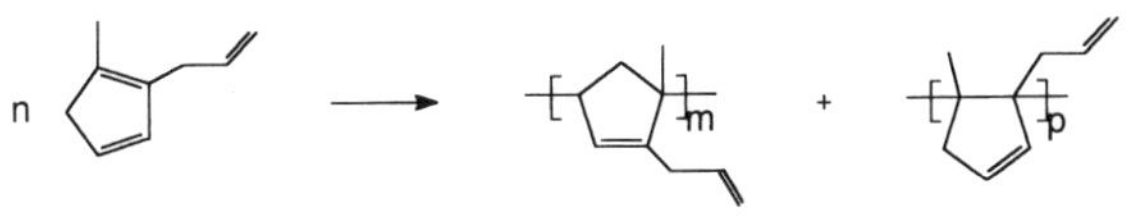

(22)

5. Substituted Fulvenes

(a) 6,6-Dimethylfulvene. Polymerization of 6,6-dimethylfulvene in the presence of cationic initiators such as $AlCl_3$, $SbCl_3$, $FeCl_3$, and $SnCl_4.5H_2O$ yielded powdery polymers, soluble in common solvents and displaying a pronounced unsaturated character [74].

(b) 6-Methyl-6-ethylfulvene. Polymers of 6-methyl-6-ethylfulvene have been obtained by cationic polymerization in the presence of several Lewis acids [74]. The structure of these polymers has not been fully elucidated.

C. Six-Membered Ring Monomers

1. Cyclohexene

Several reports have been published on the polymerization of cyclohexene by cationic or coordination catalysts. First, Hoffman [54] showed that cyclohexene can be readily oligomerized with BF_3 and HF to dimer, trimer, and tetramer. Eleuterio [57] in his studies on cycloolefin polymerization reported that cyclohexene reaction over heterogeneous catalysts consisting of oxides of chromium, molybdenum, tungsten, and uranium gives only traces of oligomers. Later, Amass *et al.* [75] carried out cyclohexene polymerization with WCl_6 but only ill-defined polymers were obtained. Better results have been published by Farona and Tsonis [76] who used the binary system $Re(CO)_5Cl/EtAlCl_2$. These investigators obtained poly(cyclohexylene) in 55% yield and not polyhexenamer [Eq. (23)].

(23)

Subsequently, Patton and McCarthy [77] described the metathesis polymerization of cyclohexene to hexenamer oligomers in the presence of WCl_6/Me_4Sn at very low temperatures but with poor evidence for the structure of the products claimed.

2. Substituted Cyclohexene

(a) 1-Methylcyclohexene. In the presence of $AlCl_3$, in a benzene solution at 40–45°C, 1-methylcyclohexene gives rise to a mixture of dimers and oligomers [78]. The structure of the reaction products was not well characterized.

(b) 3-Methylcyclohexene. 3-Methylcyclohexene has been polymerized using cationic initiators (e.g. $AlCl_3$ in ethyl chloride) at various temperatures [79]. At −78°C a solid poly-1,3-(3-methylcyclohexene) resulted in low yields whereas at −20°C only dimers and trimers were formed [Eq. (24)].

(24)

(c) 3-Methylenecyclohexene. 3-Methylenecyclohexene readily polymerized under the influence of various cationic initiators (e.g. $AlCl_3$, $TiCl_4$, VCl_4, BF_3/Et_2O, Et_2AlCl) giving soluble amorphous polymers whose structures corresponded mainly to 1,4 linkages in the polymer chain [80] [Eq. (25)].

(25)

(d) 1-Vinylcyclohexene. Cationic polymerization of 1-vinylcyclohexene induced by several Lewis acids (e.g. BF_3/Et_2O, $SnCl_4/CCl_3COOH$) gave rise to products with a predominantly 1,4 enchainment accompanied by some 1,2 linkages [81] [Eq. (26)].

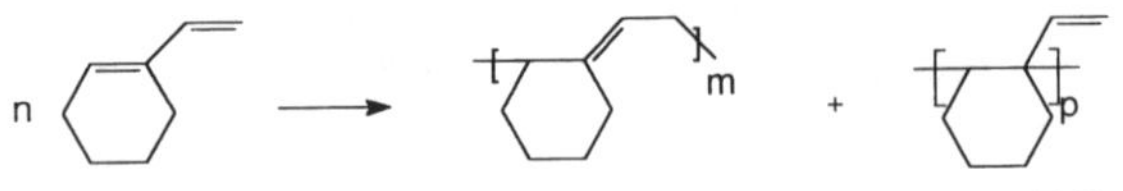

(26)

It has been suggested that under particular conditions (e.g., using BF_3/Et_2O) other possible structures could explain the presence of the methyl groups in the polymer chain.

(e) 4-Vinylcyclohexene. In the presence of cationic initiators (e.g. BF_3, BF_3/Et_2O, $TiCl_4$) 4-vinylcyclohexene forms low molecular weight polymers having predominantly bicyclic structures as a result of intramolecular migration reactions in the intermediate propagating species [81] [Eq. (27)].

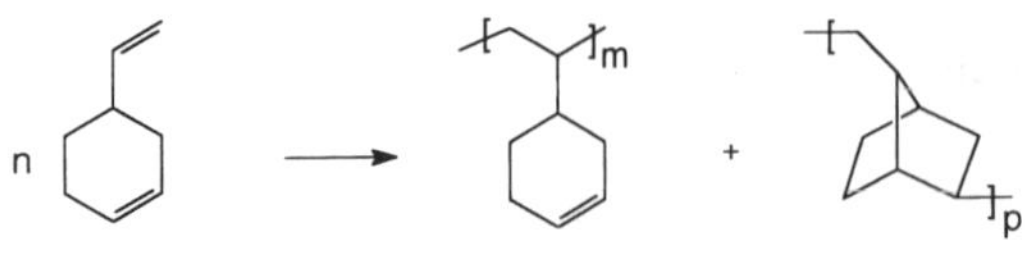

(27)

(f) Limonene. d-Limonene (1-methyl-4-isopropenylcyclohexene) or sulfate dipentene [2] was polymerized by Lewis acids (e.g. $AlCl_3$) to give polymers, mainly trimers and other oligomers, having at least 50% bicyclic structures along with the expected vinylic enchainment [Eq. (28)].

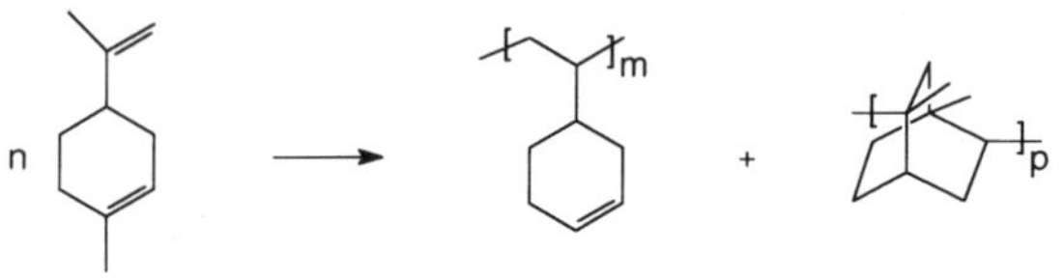

(28)

3. Cyclohexadiene

(a) 1,3-Cyclohexadiene. Several investigators have studied the polymerization of 1,3-cyclohexadiene in the presence of numerous cationic initiators such as BF_3, BF_3/Et_2O, PF_5, $TiCl_4$, $SnCl_4$, and $SnCl_4/CCl_3COOH$. Under these conditions 1,2 and 1,4 enchainments were formed in the polymer chain but the actual microstructure of the polymer was not always well characterized [82]. Probably, it corresponded to unsaturated cyclic repeat units [Eq. (29)].

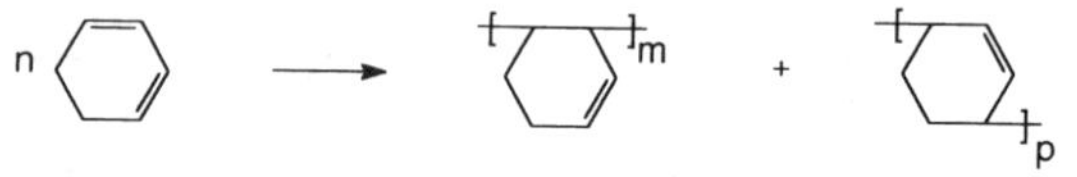

(29)

Chain branching has also been involved in the polymer structures obtained with several catalytic systems like $SnCl_4/CCl_3COOH$ and BF_3/Et_2O. Moreover, some peculiar kinetic features have been observed [83] in reactions with $SnCl_4$ in contrast to the complexed catalysts $SnCl_4/CCl_3COOH$ and BF_3/Et_2O. The polymerization of 1,3-cyclohexadiene has also been reported in the presence of iridium complexes with cyclooctene and 1,3-cyclooctadiene in polar media. When such a catalyst consisting of $IrCl_3.xH_2O$(cyclooctene) was used in ethanol or water, low polymer yields were obtained with ill-defined structures [84].

4. Substituted Cyclohexadiene

(a) α- and β-Phellandrene. The reaction of the terpene hydrocarbons α-phellandrene (1-methyl-4-isopropyl-1,3-cyclohexadiene) and β-phellandrene (1-methyl-4-isopropyl-2-cyclohexene), probably a mixture of the two isomers, has been reported to proceed in the presence of CCl_3COOH as a cationic initiator [2]. Under these conditions, polymers having a degree

of polymerization of *ca.* 10 were obtained, although details on the structure of the product were not given.

(b) p,p′-Dimethylene-1,4-cyclohexadiene. The polymerization of p,p'-dimethylene-1,4-cyclohexadiene (xylylene) occurred readily in the presence of several initiators (e.g. BF_3, $AlCl_3$, $TiCl_4$, $SbCl_5$, H_2SO_4, CCl_3COOH) in inert solvents, at low temperatures [85]. Products of low molecular weight having aromatic moieties in the polymer chain have been obtained [Eq. (30)].

(30)

D. Seven-Membered Ring Monomers

1. Cycloheptene

Cycloheptene polymerization in the presence of tungsten or molybdenum catalysts occurred via ring opening to polyheptenamer [Eq. (31)].

(31)

Natta and coworkers [86] carried out cycloheptene polymerization using homogeneous catalysts based on WCl_6 or $MoCl_5$ and Et_3Al or Et_2AlCl. Highly crystalline polyheptenamers with predominantly *trans* configuration have been obtained. Polymers with high *trans* content were also obtained by Porri [24] using the $Ir(CF_3COOH)$ (Cycloocetene) complex as initiator. By contrast, Katz and coworkers [28] prepared a highly *cis* polyheptenamer using the tungsten carbene $Ph_2C{=}W(CO)_5$. Heterogeneous catalysts like Re_2O_5/Al_2O_3 and Me_4Sn, giving mainly dimers and trimers of cycloheptene, have also been reported [87].

2. Cycloheptadiene

1,3-Cycloheptadiene under the influence of anionic initiators gives polymers with 1,2 and 1,4 enchainments [Eq. (32)].

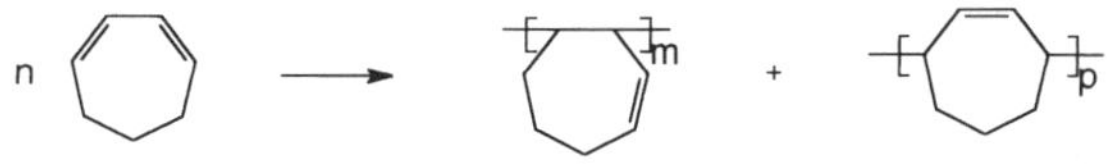

(32)

whereas in the presence of cationic initiators two new rearranged structures were also formed [88] [Eq. (33)].

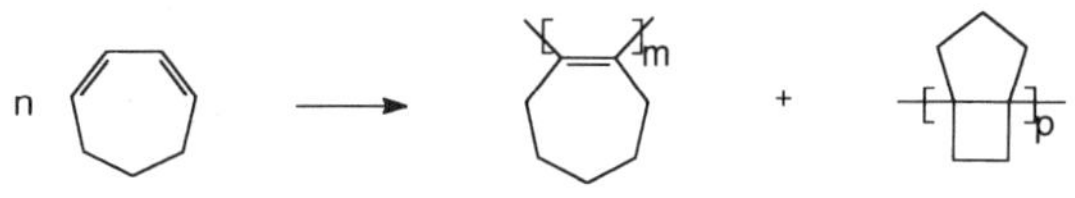

(33)

E. Eight-Membered Ring Monomers

1. Cyclooctene

Cyclooctene has been extensively polymerized in the presence of numerous binary and ternary catalysts, leading by ring-opening polymerization to polyoctenamer, a product of commercial interest [Eq. (34)].

(34)

Carbocycles of varying degree of polymerization have also been reported in this reaction.

Ring-opening polymerization of cyclooctene was first performed by Eleuterio [57] using heterogeneous catalysts based on oxides of chromium, tungsten, molybdenum, and uranium. Later, Natta and coworkers [86] obtained polyoctenamer having predominantly *trans* configuration and high crystallinity using WCl_6 or $MoCl_5$ in conjunction with Et_3Al or Et_2AlCl. Similar results were obtained by Calderon *et al.* [89] with the catalyst $WCl_6/EtOH/EtAlCl_2$ and $WCl_6/EtAlCl_2$. In addition, these authors observed the formation of carbocycles of different molecular weights which proved to be cyclic oligomers of cyclooctene. Such carbocyclic compounds were reported independently by Wassermann *et al.* [90] using the catalyst $WCl_6/EtOH/EtAlCl_2$. The *trans* content of the polyoctenamer was varied by Calderon and Morris [91] upon changing the polymerization time with the catalytic system $WCl_6/i Bu_3Al$. Ofstead [92] successfully employed highly active catalysts based on molybdenum and tungsten complexes with 1,5-cyclooctadiene and norbornadiene associated with alkylaluminum halides and containing oxygen, chlorine, bromine, iodine, or cyanogen halides. Binary heterogenous systems were reported by Crain [93] consisting of cobalt molybdate supported on SiO_2, ThO_2, Al_2O_3, or phosphates of aluminum, zirconium, calcium, magnesium or titanium to produce low molecular weight polymers. Similarly, catalysts based on oxides of molybdenum, tungsten or rhenium in conjunction with oxides of potassium, rubidium, cesium or calcium led predominantly to cyclooligomers. Homogeneous catalysts consisting of WCl_6 and Me_4Sn or Bu_4Sn were used by Bradshaw [94] to obtain a high molecular weight polyoctenamer with good elastomeric properties. Marshall and Ridgewell [95] used organometallic-free catalysts from WCl_6 and $AlBr_3$ to obtain high activity in cyclooctene polymerization. Two- and three-component catalysts derived from $WF_6/EtAlCl_2$ containing O_2, CO_2 or water as activators were reported by Dall'Asta and Manetti [96] to produce a highly crystalline *cis*-polyoctenamer (93%) [96]. Kuepper, Streck and coworkers [97, 98] obtained efficient ternary catalysts from WCl_6 $EtAlCl_2$, and organic acids (e.g. acetic acid), mineral salts of organic acids (e.g. lithium palmitate), or alkoxides (e.g. aluminum sec-butoxide). One-component tungsten carbene $Ph_2C{=}W(CO)_5$ allowed Katz and coworkers [99] to obtain a highly *cis* polyoctenamer by ring opening in a high yield.

Numerous π-complexes of transition metals were reported by Kormer *et al.* [100] to induce the conversion of cyclooctene to polyoctenamer or poly(octenylene) by ring-opening polymerization and to poly(cycloctenylene) by addition polymerization [Eq. (35)].

(35)

Thus, one component π-complexes of tungsten and molybdenum gave predominantly polyoctenamer by ring-opening reaction, catalysts of nickel and cobalt yielded poly(cyclooctylene) by addition polymerization whereas catalysts of zirconium and chromium afforded polymers of cyclooctene of both types. Furthermore, two-component π-complexes of group IV–VIII transition metals associated with metal halides (e.g. $AlBr_3$, $TiCl_4$, WF_6, and $MoCl_5$) showed to be active in cyclooctene polymerization [6].

2. Substituted Cyclooctene

It is of interest that 1-methyl-*trans*-cyclooctene reacts in the presence of the tungsten–carbene complex $Ph_2C{=}W(CO)_5$ to lead via head–tail enchainment to a perfectly alternating poly(1-methyloctenylene) of predominantly *trans* configuration while 1-methyl-*cis*-cyclooctene does not polymerize under the same conditions [101] [Eq. (36)].

(36)

3-Methylcyclooctene and 3-phenylcyclooctene polymerized readily with the two-component catalyst $WCl_6/EtAlCl_2$ giving the corresponding substituted polyalkenamer [90]. Studies on polyoctenamer microstructures indicated variable contents of *cis* and *trans* double bonds in the chain [Eqs. (37) and (38)].

(37)

(38)

5-Methylcyclooctene and 5-phenylcyclooctene have also been reported to polymerize in the presence of binary catalysts derived from WCl_6 or $MoCl_5$ and $EtAlCl_2$ or $Et_3Al_2Cl_3$ to form the corresponding substituted polyoctenamers having the structures of butadiene–ethene–propene and butadiene–ethene–styrene terpolymers, respectively [44] [Eqs. (39) and (40)].

(39)

(40)

3. Cyclooctadiene

(a) 1,3-Cyclooctadiene. Cationic polymerization of *cis, cis*-1,3-cyclooctadiene has been investigated extensively by Imanishi *et al.* [83] in the presence of BF_3, $TiCl_4$, $TiCl_4/CCl_3COOH$, and $SnCl_4$, CCl_3COOH at low temperatures, using methylene chloride or toluene as the solvents. Polymers with 1,4 enchainment having number average molecular weight, M_n, of 1600–1800 have been obtained [Eq. (41)].

(41)

Interestingly, these authors observed a different kinetic behavior of this system in the two solvents; this phenomenon was explained by a two-step mechanism involving distinct kinetic species as a function of the solvent. Mondal and Young [102] have studied the polymerization of this monomer with the $TiCl_4/H_2O$ system yielding polymers of M_n 10,000 having some branched structures. An alternate mechanism involving allylic initiation and termination was suggested to account for the formation of these new structures. Metathesis polymerization of 1,3-cyclooctadiene has been reported by Korshak [103] in the presence of WCl_6-based catalysts to obtain cross-linked, rubber-like polymers with conjugated double bonds in the chain.

(b) 1,4-Cyclooctadiene. Amass [104] tried to polymerize 1,4-cyclooctadiene with WCl_6 catalyst but undefined products were obtained. Some time later, Streck and coworkers [105] carried out the polymerization of 1,4-cyclooctadiene in the presence of the ternary catalytic system $WCl_6/EtAlCl_2/EtOH$, sometimes activated by allyl 2,4,6-tribromophenyl ether. The result was a high molecular weight polybutadiene-like rubber having the structure of an alternating propene–pentenamer. When acyclic olefins were used as the molecular weight controlling agents, low molecular weight products in the form of oxidatively drying oils were obtained.

(c) 1,5-Cyclooctadiene. 1,5-Cyclooctadiene readily polymerizes in the presence of binary and ternary metathesis catalysts leading to polyoctadienamer identical to polybutenamer or polybutadiene elastomers [Eq. (42)].

(42)

Thus, Natta and coworkers [106] used the ternary catalyst $WCl_6/i\text{Bu}_3\text{Al}$/cyclopentene hydroperoxide while other investigators [6] used binary and ternary catalysts based on WCl_6 (e.g., $WCl_6/i\text{Bu}_3\text{Al}$, $WCl_6/EtOH/EtAlCl_2$, $WCl_6/t\text{BuOH}/i\text{Bu}_3\text{Al}$) to polymerize 1,5-cyclooctene to polyoctadienamer. Organotin, organosilicon, and organogermanium compounds have been substituted for organoaluminum compounds to control the stereoselectivity of the reaction, but the loss in activity was substantial in this case. Binary catalysts such as $WCl_6/AlCl_3$ and $WCl_6/AlBr_3$ or catalysts with various transition metals, e.g., $MoCl_5/AlCl_3$ or $TaCl_5/EtOH/i\text{Bu}_3\text{Al}$, have also been reported [6]. Heterogeneous systems such as $CoO/MoO_3/Al_2O_3$ or Re_2O_7/Al_2O_3 were among the few metal oxides applied in 1,5-cyclooctadiene polymerization.

Generally, a high *cis* content of polymer was obtained with the above catalyst systems as a result of the already existing *cis* double bonds in the monomer, but the *cis* content could be altered by several factors including reaction temperature, nature of the cocatalyst, additives and ligands, or the molar ratio of the catalytic components. Microstructures of the polymers obtained with several WCl_6-based catalysts have been thoroughly examined by ^{13}C NMR spectroscopy, affording relevant evidence for secondary *cis/trans* isomerization and double bond migration in the polymer chain.

4. Substituted Cyclooctadiene

1-Methyl-1,5-cyclooctadiene and 1-ethyl-1,5-cyclooctadiene polymerized in the presence of the ternary catalyst $WCl_6/EtOH/EtAlCl_2$ affording an alternating copolymer with good elastomeric properties [107]. [Eqs. (43) and (44).]

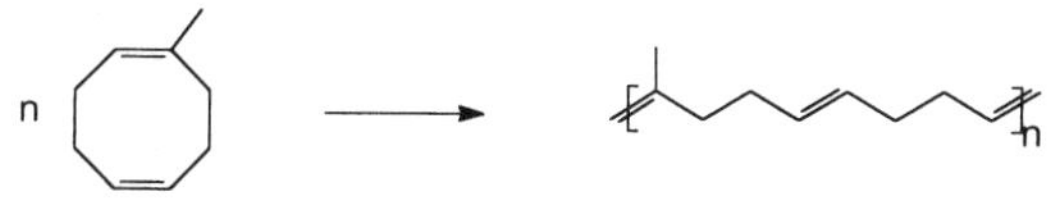

(43)

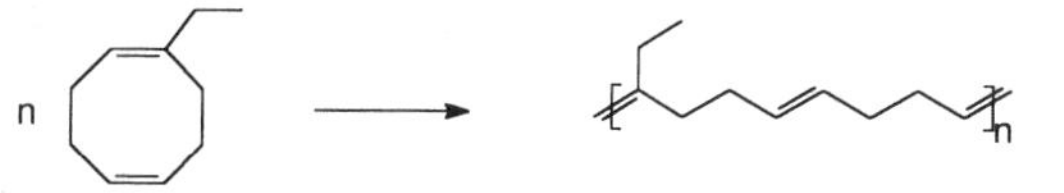

(44)

In the presence of WCl_6-based catalysts, 3-methyl-1,5-cyclooctadiene formed a randomly substituted polyalkenamer because the two double bonds in the ring-opening reaction had different reactivities [Eq. (45)].

(45)

1,2-Dimethyl-1,5-cyclooctadiene gave rise to a substituted polybutadienamer whose microstructure corresponded to the ring opening of one double bond [Eq. (46)].

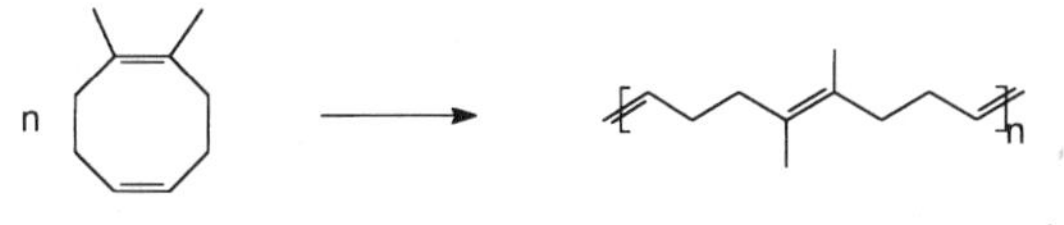

(46)

3,7-Dimethyl-1,5-cyclooctadiene reacted in the presence of WCl_6-based catalysts to form a uniform head-tail structure as a result of a comparable reactivity of the two double bonds [Eq. (47)].

(47)

5. Cyclooctatetraene

Ring-opening polymerization of cyclooctatetraene has been effected using either WCl_6 associated with Et_2AlCl or tungsten–carbene complexes W(=CH*t*Bu)$(OR)_2$(NHR) and W(=CH*t*Bu)$(OR)_2Br_2/GaBr_3$ as catalysts [108, 109]. Polyacetylene can be readily obtained in various yields and having special electrical and mechanical properties, depending on the catalyst employed and the polymerization technique applied [Eq. (48)].

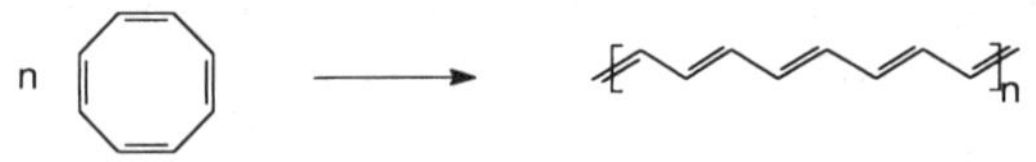

(48)

6. Substituted Cyclooctatetraene

Ring-opening polymerizations of 1-substituted and 1,5-disubstituted cyclooctatetraene have been effected also using either WCl_6 associated with Et_2AlCl or tungsten–carbene complexes W(=CH*t*Bu)$(OR)_2$(NHR) and W(=CH*t*Bu)$(OR)_2Br_2/GaBr_3$ as catalysts [109]. Substituted polyacetylene has been obtained in various yields and having special physical and mechanical properties, depending on the catalyst employed [Eqs. (49) and (50)].

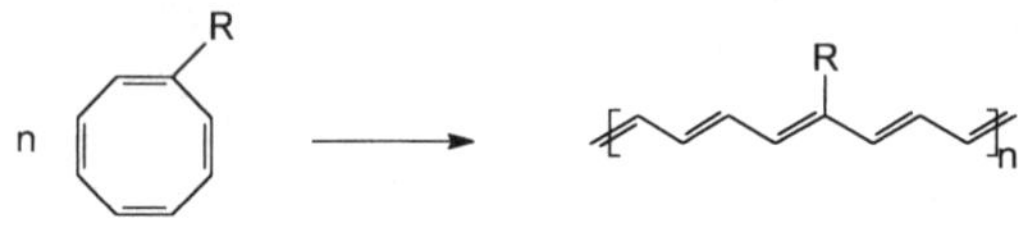

(49)

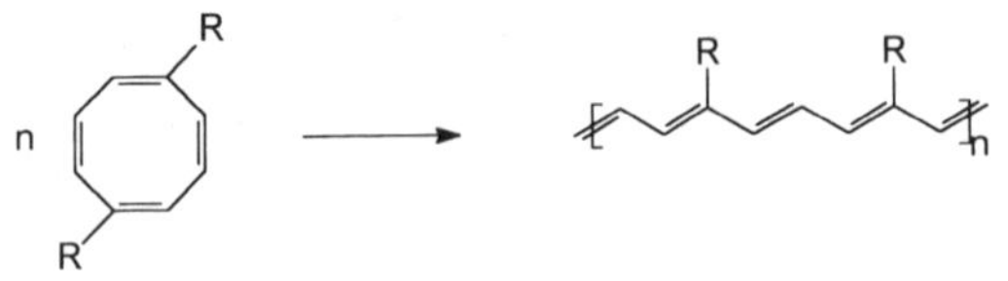

(50)

F. Nine-Membered Ring Monomers

1. Cyclononene

In 1966, Natta and coworkers [86] reported on the polymerization of *cis*-cyclononene with catalysts derived from WCl_6 and organoaluminum compounds, yielding highly *trans* polynonenamer [Eq. (51)].

(51)

Later, Rossi and Giorgi [110] described the reaction of *cis*-cyclononene in the presence of $WCl_6/LiAlH_4$. This process, however, seemed to involve both addition and ring-opening polymerization. More recently, metathesis dimerization of *cis*-cyclononene to 1,10-cyclooctadiene over Re_2O_7/Al_2O_3 accompanied by formation of cyclic trimer and oligomers has been also reported [87].

2. Cyclononadiene

cis, cis-1,5-Cyclononadiene has been polymerized by Rossi and Giorgi [110] in the presence of the binary catalyst $WCl_6/LiAlH_4$. Analysis of the product structure indicated a higher content of the butadiene units compared to the pentenamer and a small proportion of saturated structures [Eq. (52)].

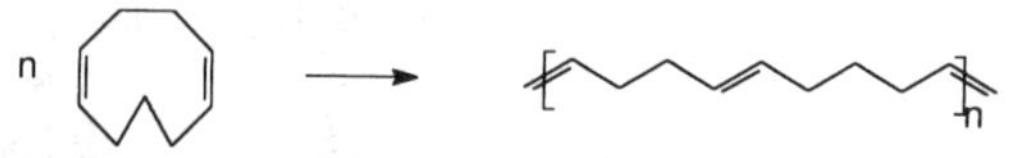

(52)

Cyclopentene and divinylcyclopentene, formed as byproducts, were explained in terms of intramolecular metathesis reactions under the influence of the catalytic system.

G. Ten-Membered Ring Monomers

1. Cyclodecene

Dall'Asta and Manetti [96] reported on the polymerization of *cis*-cyclodecene in the presence of the system WCl_6/Et_2AlCl and $WOCl_4/Et_2AlCl$ to obtain a *trans* polydecenamer of high crystallinity [Eq. (53)].

(53)

Both the activity and stability of the catalyst have been improved by using oxygen-containing compounds such as benzoyl and cumyl peroxide. Working with the heterogenous system Re_2O_7/Al_2O_3, cyclodecene formed preferentially a dimer product, 1,11-cycloeicosadiene [87].

2. 1,5-Cyclodecadiene

cis, cis-1,5-Cyclodecadiene has been polymerized with WCl_6 alone or in association with Et_3Al, Et_2AlCl, and

$EtAlCl_2$. Instead of WCl_6, $MoCl_5$ has also been employed. Polymers of high *trans* content have been obtained [Eq. (54)].

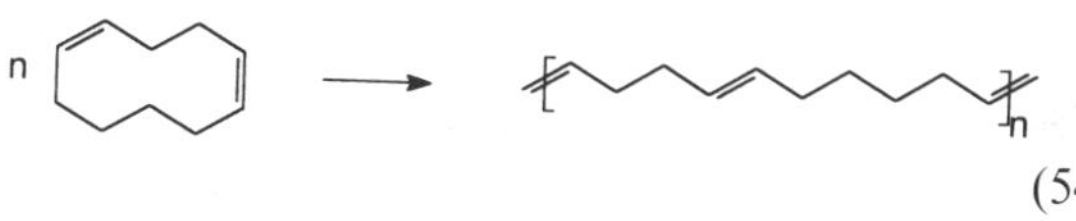

(54)

Cyclohexene formed as a by-product, as expected on thermodynamic grounds. Under the influence of WCl_6/Et_2AlCl, *cis,trans*-1,5-cyclodecadiene failed to give the corresponding polyalkenamer. However, *cis,trans*-1,5-cyclodecadiene was reported to act as a cocatalyst along with WCl_6 for the polymerization of several cycloolefins [6].

3. 1,6-Cyclodecadiene

The ring-opening polymerization of *cis,cis*-1,6-cyclodecadiene would be an alternative way for the production of polypentenamer. However, instead, the reaction of *cis,cis*-1,6-cyclodecadiene in the presence of $WCl_6/EtAlCl_2/EtOH$ yielded pure cyclopentene by an intramolecular metathesis reaction involving the two double bonds of the molecule [111] [Eq. (55)].

(55)

H. Twelve-Membered Ring Monomers

1. Cyclododecene

Several authors [112–114] reported on the polymerization of *cis* and *trans* cyclododecene leading either to polydodecenamer or to cyclic oligomers, depending on the catalyst employed and the reaction conditions. Highly *trans* polydodecenamer has been obtained using binary catalysts derived from WCl_6 or $MoCl_5$ and Et_3Al, Et_2AlCl or $EtAlCl_2$ [Eq. (56)].

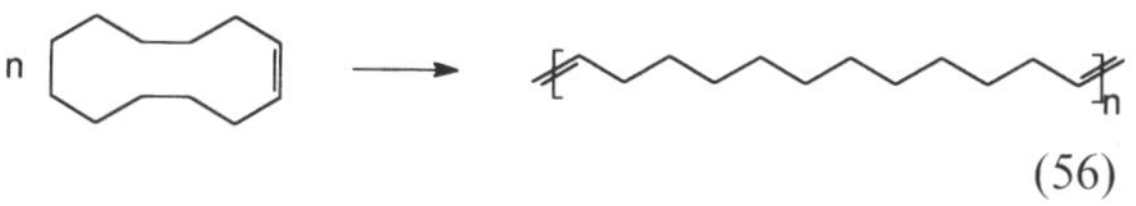

(56)

The systems WCl_6/1,5-cyclooctadiene and π-allyl$_4$W/$AlBr_3$ gave predominantly a *cis* polydodecenamer [Eq. (57)].

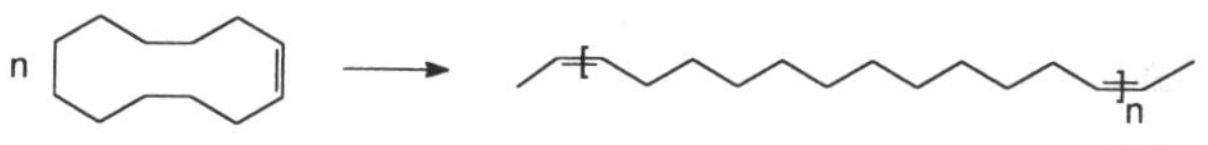

(57)

The polymerization of cyclododecene with $WCl_6/EtOH/EtAlCl_2$ yielding cyclic oligomers, including catennanes and interlocked ring systems, has been reported by Wasserman [90] and Wolovsky [114].

2. Substituted Cyclododecene

Dall'Asta [115] polymerized 3-methyl-*trans*-cyclododecene with the system $WCl_6/EtAlCl_2/(PhCOO)_2$ to yield mainly *trans*-poly(3-methyldodecenamer [Eq. (58)].

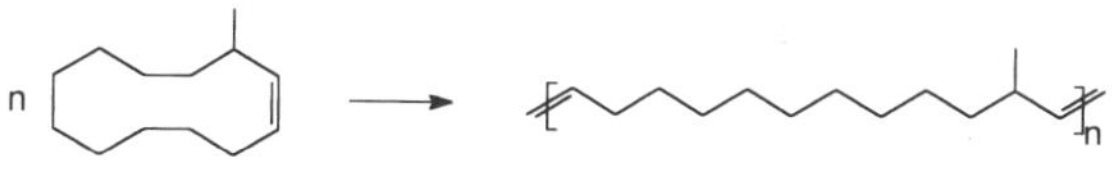

(58)

Microstructure studies indicated that small proportions of head–head enchainments were formed.

3. Cyclododecatriene

1,5,9-Cyclododecatriene formed in high yields 1,4-polybutadiene in the presence of $WCl_6/EtAlCl_2$ as a catalyst [Eq. (59)].

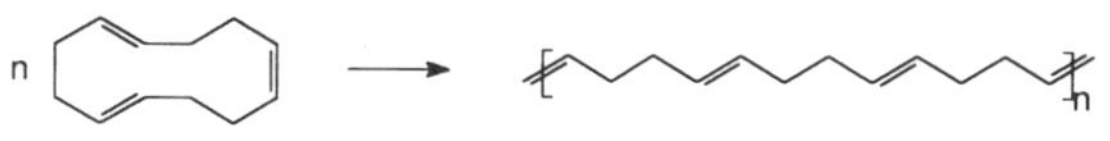

(59)

Starting from a mixture of *cis,trans,trans*- and *trans,trans,trans*-1,5,9-cyclododecatriene (40:60), Calderon and coworkers [89] obtained with the same catalytic system 1,4-polybutadiene with steric configuration closely related to that of the initial monomer. Several other binary and ternary catalysts have also been reported for the polymerization of 1,5,9-cyclododecatriene such as $WCl_6/AlBr_3$, $WCl_6/EtAlCl_2/EtOH$, $WCl_6/EtAlCl_2/PhOH$, π-allyl$_4$W/CCl_3COOH, π-allyl$_4$W/Al_2O_3/SiO_2, and Re_2O_7/Al_2O_3.

I. Higher-Membered Ring Monomers

1. Cyclopentadecene

Hoecker and coworkers [116] reported on the polymerization of *cis/trans*-cyclopentadecene (18:82) using WCl_6/Me_4Sn and $WCl_6/EtAlCl_2/EtOH$ as catalysts to produce cyclic oligomers ($n = 2$–7).

2. Cyclohexadecatetraene

Ring-opening polymerization of 1,5,9,13-cyclohexadecatetraene, the cyclic tetramer of butadiene, to poly-(hexadecatetraenamer) is a new way to manufacture 1,4-polybutadiene [Eq. (60)].

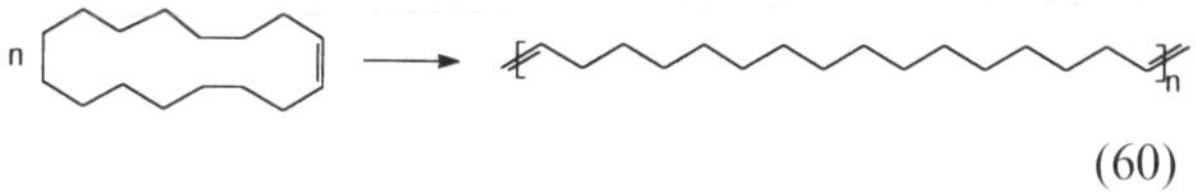

(60)

The reaction was carried out by Chauvin [117] using the tungsten carbene complex $Ph(MeO)C{=}W(CO)_4(Ph_3P)/TiCl_4$ and by Saito [118] with the catalytic system Re_2O_7/Al_2O_3. The reactivity of the monomer proved to be very high but the polymer structure was not fully characterized.

IV. POLYMERIZATION OF BICYCLIC OLEFINS

A. Norbornene (Bicyclo[2.2.1]hept-2-ene)

The cationic polymerization of norbornene with $AlCl_3$, $TiCl_4$ and BF_3/Et_2O was reported by Saegusa *et al.* [119] to give low molecular weight polymers of the 1,2-addition type, polynorbornylene. Later on, Kennedy and Makowski [120] carried out the reaction in the presence of $EtAlCl_2$, obtaining amorphous polymers whose structures were suggested to contain rearranged repeat units [Eq. (61)].

(61)

Ziegler–Natta systems based on $TiCl_4$ and several organometallic cocatalysts, including EtMgBr, $LiAlHept_4$, alkyl and aryl compounds, metal hydrides, and even alkali metals and earth alkaline metals, were employed in early investigations by Anderson and Merckling [121] who obtained polymers having structures that were not fully elucidated at that time. Subsequently, Truett *et al.* [122] proved that the polymerization of norbornene with $TiCl_4/LiAlHept_4$ proceeds to polynorbornylene by addition polymerization and to polynorbornenamer by ring-opening [Eq. (62)].

(62)

The polymerization route by addition or ring-opening mechanism is influenced by the molar ratio Al:Ti; thus, ratios Al:Ti < 1 favor addition polymerization while ratios Al:Ti > 1 favor ring-opening polymerization. Similar results have been obtained with other Ziegler–Natta systems [123], e.g. $TiCl_4/i Bu_3Al$, $TiCl_4/Et_3Al$.

Eleuterio [57], using several heterogeneous catalysts formed from oxides of chromium, molybdenum, tungsten or uranium on alumina, titania or zirconia showed that norbornene polymerizes by ring opening with $MoO_3/Al_2O_3/LiAlH_4$ and polynorbornene of both *cis* and *trans* configuration at the double bond is obtained. A wide range of homogeneous molybdenum catalysts were further developed for norbornene polymerization leading to high yields and stereoselectivities. $MoCl_5$ alone or associated with $i Bu_3Al$, $EtAlCl_2$ or Et_2AlI produced highly *trans* or *cis* polynorbornene depending on the nature of organometallic compound [7, 10]. The *cis* content of polynorbornene has been increased by adding Michael acceptors [16] such as ethyl acrylate, diethyl maleate or diethyl fumarate to the system $MoCl_5/EtAlCl_2$. Similarly, tungsten systems consisting of WCl_6 alone or associated with organoaluminum compounds ($EtAlCl_2$, Et_2AlCl, Et_3Al), organotin compounds (Me_4Sn, Bu_4Sn, Ph_4Sn) or organolithium compounds (*n*BuLi) allowed high yields in polynorbornene to be obtained. In many cases, the stability and stereoselectivity of such catalysts have been considerably improved by using additional compounds containing oxygen, nitrogen or sulfur. π-Allyl complexes of zirconium, molybdenum, tungsten, chromium, nickel, cobalt, and lead, by themselves or in conjunction with Lewis acids (e.g., $AlCl_3$, $AlBr_3$, $TiCl_4$) impart increased activity and selectivity to both types of polymerization. Thus, π-allylNiI_2 or π-allyl$_2$Pd complexes selectively give addition polymers while π-allyl$_4$Mo and π-allyl$_4$W lead primarily to polynorbornenamer. Similarly, metal carbenes of tungsten, molybdenum or tantalum, as such or with Lewis acids, are very active and stereoselective for ring-opening metathesis polymerization of norbornene. With certain catalysts of this type, e.g., $Ph_2C{=}W(CO)_5$, a highly *cis* form of polynorbornene has been obtained [124]. Well defined tungsten and molybdenum complexes ($M({=}CH t Bu)(NAr)(O t Bu)_2$)(M = W or Mo) as well as tantalum complexes ($Ta({=}CH t Bu)(OR)_3(THF)$) produced living monodisperse polynorbornene [125, 126]. Rhenium catalyst, $ReCl_5$, has been used to prepare high *cis* polynorbornene [127]; associated with organoaluminum compounds (e.g. $EtAlCl_2$) led to an increase in *trans* configuration. Interestingly, a new class of titanacyclobutanes has been employed successfully to obtain living polynorbornene and its derivatives [128].

Polymerization of norbornene has been also carried out with ruthenium, osmium, and iridium compounds in water, weakly polar or nonpolar media [23, 129]. In ethanol as a solvent, the hydrates $MCl_3 \cdot 3H_2O$ (M = Ru, Os, Ir) showed the following order of activity: Ir > Os > Ru in ring-opening polymerization [23]. Interestingly, when the solvent for $RuCl_3 \cdot 3H_2O$ was varied in the alcohol series, the following order of activity has been obtained: *i*PrOH > *n*BuOH > EtOH. Various other catalysts of this class consisting of iridium salts ($IrCl_3 \cdot 4H_2O$, $(NH_4)_2IrCl_6$, K_2IrCl_6) associated with cycloalkenes or cyclodienes (e.g., $IrCl(cyclooctene)_2$, $IrCl(cyclooctadiene)_2$, $IrCl(cyclohexadiene)_2$ and $Ir(cyclohexadiene)_2SnCl_2$) without any cocatalyst have also been employed in polar or nonpolar media. Numerous ruthenium complexes bearing various ligands such as $RuCl_2(Ph_3P)_2$, $RuCl_2(Ph_3P)_4$, $RuCl_2$(2,7-dimethyl-2,6-octadiene), $Ru(CF_3COO)_2$ (2,7-dimethyl-2,6-octadiene), $RuCl_3$(1,6,11-dodecatriene), $RuCl_3$(1,5-cyclooctadiene) and $RuCl_3$(norbornadiene) have been widely employed [12]. More recently, well defined ruthenium carbene complexes such as $RuCl_2(=CHPh)(R_3P)_2$ or $RuCl_2(=CHCH=CPh_2)(R_3P)_2$, have been also successfully employed [130, 131]. On the other hand, cationic complexes of palladium, e.g., $Pd(CH_3CN)_4(BF_4)_2$ and $Pd(C_2H_5CN)_4(BF_4)_2$, have been largely applied for addition polymerization of norbornene [132].

B. Monosubstituted Norbornene

1. Methylnorbornene

All methyl isomers of norbornene, 1-, 2-, 5-, and 7-methylnorbornene have been reacted in the presence of catalysts based on tungsten, rhenium, ruthenium, osmium and iridium compounds [7]. The polymers corresponded to ring-opened products having various microstructures. Racemic mixtures or pure enantiomers have been used as starting materials. Differences in reactivities as a function of the methyl position (1, 2, 5 or 7) and steric configuration (*endo–exo* and *syn–anti*) have been reported.

2. Methylenenorbornene

Cationic polymerization of 5-methylenenorbornene [133] in the presence of $AlBr_3$, VCl_4 or $EtAlCl_2$ gave polymers with nortricyclic structure [Eq. (63)].

n → n (63)

In contrast, metathesis polymerization of 5-methylenenorbornene in the presence of WCl_6/Et_3Al, $MoCl_5/Et_3Al$ or $IrCl_3$ led preferentially to ring-opened polymers [134] [Eq. (64)].

n → n (64)

3. Ethylnorbornene

On using $RuCl_3$ or $IrCl_3$ as a catalyst, 5-ethylnorbornene led to highly *trans* ring-opened polymer [7]. Interestingly, microstructure determination revealed overall head–tail enchainment in the case of the $IrCl_3$ catalyst [Eq. (65)].

n → n (65)

4. Vinylnorbornene

Reaction of vinylnorbornene in the presence of $EtAlCl_2$ leads to polymers having various microstructures depending on the reaction temperature [120]. At low temperatures, mainly structures corresponding to a 2,3 enchainment were obtained, whereas at higher temperatures, cross-linked polymers were formed, probably as a result of the participation of the less reactive vinyl group in this case [Eq. (66)].

n → n (66)

5. Isopropylnorbornene

5-Isopropylnorbornene polymerizes readily in the presence of $MoCl_5/Et_2AlI$ to the corresponding ring-opened polymers [135]. No migration of the isopropyl group has been observed [Eq. (67)].

n → n (67)

6. Isopropenylnorbornene

Polymerization of 5-isopropenylnorbornene in the presence of cationic initiators (e.g., $EtAlCl_2$) gives cross-linked polymers in which the two kinds of double bonds participate in the reaction [120].

7. Butenylnorbornene

5(1-Buten-4-yl)norbornene in the presence of a Ziegler–Natta system $TiCl_4/LiAl(C_{10}H_{21})_4$ affords two types of polymer arising from addition reaction at the exocyclic double bond and ring-opening reaction at the endocyclic double bond in the proportions of 59% and 41%, respectively [136].

8. Arylnorbornene

5-Arylnorbornenes having as the aryl group phenyl, *o*-, *m*- or *p*-tolyl, *o*-, *m*- or *p*-ethylphenyl, *p*-isopropylphenyl or 1,4-methano-1,4,5,8-tetrahydrofluorene have been copolymerized with various olefins in the presence of vanadium salts (e.g., VCl_4, VBr_4, $VOCl_3$, $VOBr_3$, $VO(OR)_3$) and organoaluminum compounds (e.g. Me_3Al, Et_2AlCl, $Et_3Al_2Cl_3$) yielding copolymers which are of interest for applications in optical disks and optical fibers [137].

C. Disubstituted Norbornene

1. 5,5-Dimethylnorbornene

The polymerization of 5,5-dimethylnorbornene in the presence of various metathesis catalysts based on molybdenum, tungsten, rhenium, osmium, ruthenium, and iridium compounds to ring-opened polymers having 0–100% *cis* content has been reported [138] [Eq. (68)].

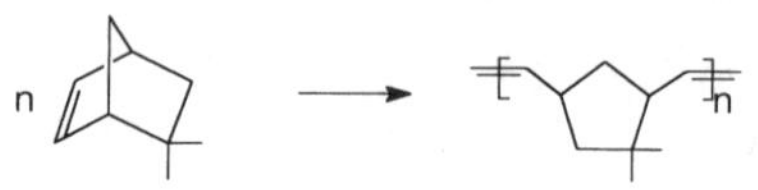

(68)

There was no correlation between tacticity and *cis* content, but all-*cis* polymers were mainly highly syndiotactic and all-*trans* polymers slightly isotactic when prepared from 3 molar monomer concentration at 20°C.

2. 7,7-Dimethylnorbornene

Polymerization of this monomer with the above mentioned catalysts led to ring-opened polymers of various configurations and tacticities depending on the catalyst employed and the reaction parameters [7] [Eq. (69)].

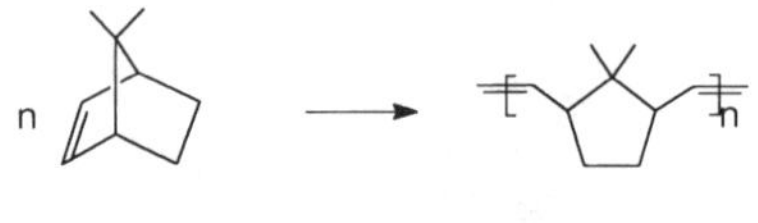

(69)

D. Norbornadiene (Bicyclo[2.2.1]hepta-2.5-diene)

Polymerization of norbornadiene with cationic catalysts (e.g., $AlCl_3$) occurs readily at various temperatures leading to saturated polymers [139]. Thus, at low temperatures, soluble polymers having 2,6-disubstituted nortricyclene structure were primarily obtained while at higher temperatures insoluble polymers with cross-linked structures, as a result of addition polymerization of the two double bonds, were formed [Eq. (70)].

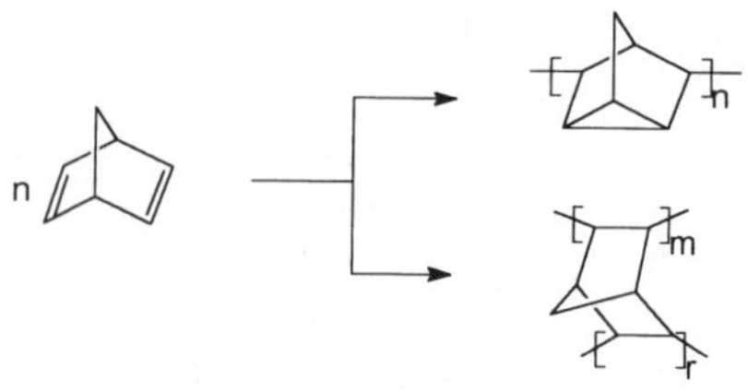

(70)

Ziegler–Natta catalysts based on $TiCl_4$ or VCl_4 and organometallic compounds gave addition and ring-opened polymers [Eq. (71)].

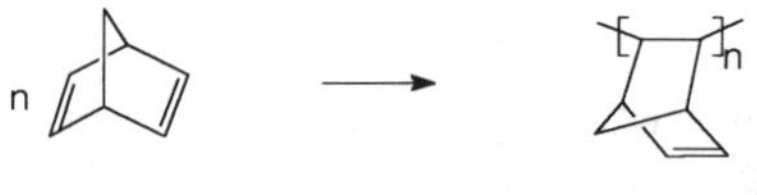

(71)

On the other side, WCl_6-based catalytic systems gave preferentially ring-opened polymers [140] [Eq. (72)].

(72)

E. Terpene Monomers

Polymerization of α-pinene and 3-carene is a reaction occurring in the synthesis of hydrocarbon resins from terpene monomers under the action of Lewis acids [2].

1. α-Pinene

The reaction of α-pinene has been carried out with a series of Friedel–Crafts catalysts (e.g., $AlCl_3$, $AlBr_3$, BF_3, $ZrCl_4$) in toluene at 40–45°C. The low molecular weight products obtained were similar to those produced from limonene [141]. The polymer structure could not be explained by a normal addition polymerization reaction; probably, rearrangements occur during the propagation step resulting in the incorporation of rearranged structures into the polymer chain.

2. 3-Carene

3-Carene has been polymerized in the presence of Friedel–Crafts systems (e.g., $AlCl_3$) to low molecular weight polymers of the hydrocarbon resin type [2,17].

F. Bicyclo[3.2.0]hept-2-ene

Reaction of bicyclo[3.2.0]hept-2-ene with $TiCl_4/Et_3Al$ gives undefined products, probably as a result of both addition and ring-opening polymerization [Eq. (73)].

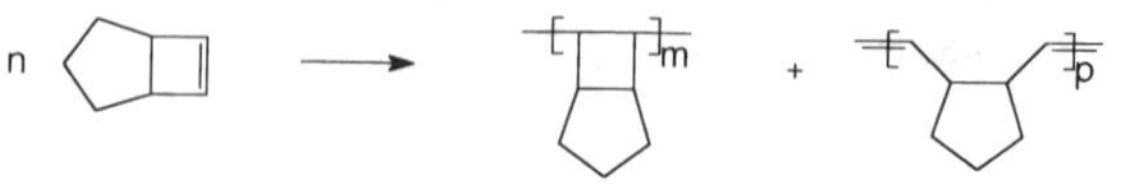

(73)

By contrast, WCl_6/Me_4Sn and $Ph_2C{=}W(CO)_5$ give polyalkenamers via opening of the cyclobutene ring [142, 143] [Eq. (74)].

(74)

G. Bicyclo[3.2.0]hepta-2,6-diene

Polymerization of bicyclo[3.2.0]hepta-2,6-diene has been carried out with Ziegler–Natta catalysts based on titanium and vanadium salts [140]. Structure investigation showed that titanium catalysts gave mainly ring-opened polymers [Eq. (75)]

(75)

while vanadium catalysts formed both addition and ring-opened polymers [Eq. (76)].

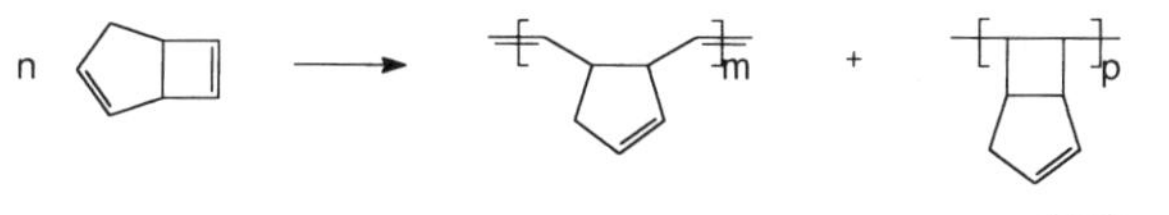

(76)

On the other hand, clean ring-opened polymers were obtained with $Ph_2C{=}W(CO)_5$ and WCl_6/Me_4Sn. In contrast, $RuCl_3$ in polar media produced polymers having transannular bonds whose structures have not been elucidated [143].

H. Bicyclo[4.2.0]oct-7-ene

Several binary Ziegler–Natta catalysts based on halides or acetylacetonates of titanium, vanadium, chromium, and tungsten associated with organoaluminum compounds have been employed to polymerize bicyclo[4.2.0]octene [140]. The catalysts based on titatanium gave primarily ring-opened polymers through cleavage of the cyclobutene ring [Eq. (77)].

(77)

The catalysts derived from vanadium and chromium led to addition polymers [Eq. (78)].

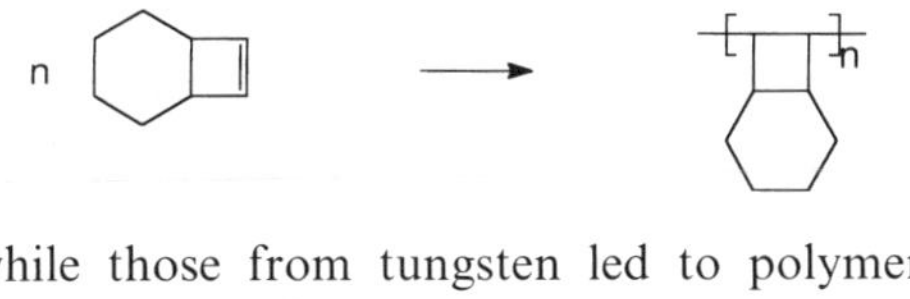

(78)

while those from tungsten led to polymers of both types [Eq. (79)].

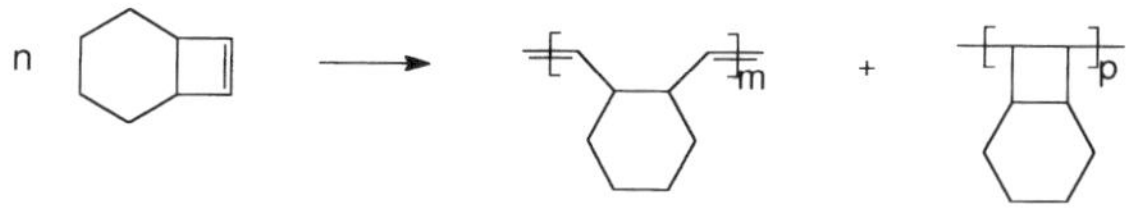

(79)

Interestingly, one-component catalysts consisting of salts of ruthenium, rhodium, iridium, nickel, and palladium in polar solvents such as ethanol, water, and dimethyl sulfoxide gave different polymers as a function of the metal salt: ruthenium and iridium catalysts yielded ring-opened polymers, while rhodium, nickel and palladium catalysts led to addition polymers.

I. Bicyclo[5.1.0]oct-2-ene

The tungsten carbene system, $Ph_2C{=}W(CO)_5/TiCl_4$ induced polymerization of bicyclo[5.1.0]oct-2-ene to polyalkenamers having cyclopropane rings along the polymer chain [144] [Eq. (80)].

(80)

J. Bicyclo[2.2.2]octene

Bicyclo[2.2.2]octene [145] has been polymerized in the presence of the ternary catalyst $WCl_6/EtOH/EtAlCl_2$ to ring-opened polymer having cyclohexane structures along the polymer chain [Eq. (81)].

(81)

The polymer had primarily *trans* configuration and contained the cyclohexane ring in the boat form. On the other side, polymerization of 5-methyl-substituted bicyclo[2.2.2]octene has been carried out with cationic and Ziegler–Natta systems [146] leading to low yields in polymers, probably of mainly vinylic structures.

K. Bicyclo[4.3.0]nona-3,7-diene (Tetrahydroindene)

Tetrahydroindene has been polymerized with Ziegler–Natta catalysts [140] derived from salts of titanium, vanadium, chromium, molybdenum, and tungsten associated with organoaluminum compounds. The catalysts based on tungsten and molybdenum (e.g., WCl_6/Et_2AlCl and $MoCl_5/Et_3Al$) proved to be very active and produced mainly *trans*-polyalkenamer as a result of the opening of the cyclopentene ring [Eq. (82)].

(82)

V. POLYMERIZATION OF POLYCYCLIC OLEFINS

A. Benzvalene

Benzvalene, a valence isomer of benzene, has been polymerized by ring-opening metathesis polymerization in the presence of tungsten-carbene complexes to form polybenzvalene [Eq. 83)].

(83)

This polymer was further isomerized to polyacetylene by catalytic treatment with salts of mercury or silver [147] [Eq. (84)].

(84)

B. Benzonorbornadiene

Polymerization of benzonorbornadiene has been performed in the presence of binary catalysts WCl_6/Ph_4Sn and WCl_6/Me_4Sn to lead preferentially to ring-opened polymers [Eq. (85)].

(85)

Metal-carbenes of Schrock type [125, 126], e.g., Mo(=CH*t*Bu)(NAr)(O*t*Bu)$_2$ led in a living process to ring-opened polymers of benzonorbornene.

C. Benzo[3.4]buta[1.2]norbornene

Both *exo* and *endo* isomers of benzo[3.4]buta[1.2]norbornene have been polymerized in the presence of WCl_6/Ph_4Sn and WCl_6/Me_4Sn to produce ring-opened polymers, having *cis* and *trans* configurations of double bonds in the polymer chain [148] [Eq. (86)].

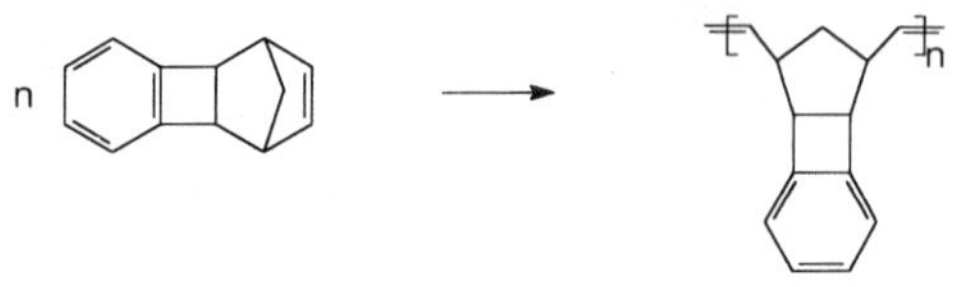

(86)

D. Benzobicyclo[4.2.0]octa-3,7-diene

Benzobicyclo[4.2.0]octa-3,7-diene reacts in the presence of tungsten carbene and tungstacyclobutane catalysts [27] to give a living ring-opened polymer [Eq. (87)].

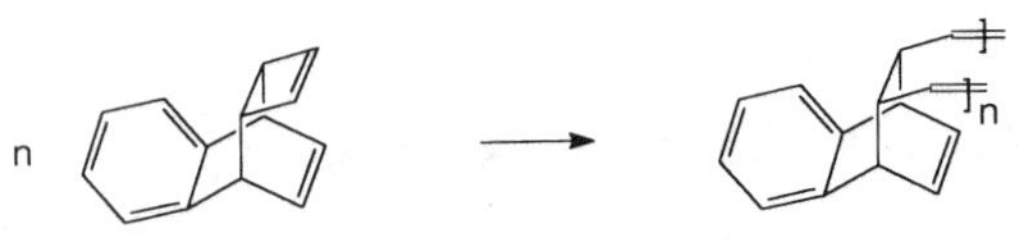

(87)

This product goes into polyacetylene with elimination of naphthalene upon thermal treatment [Eq. (88)].

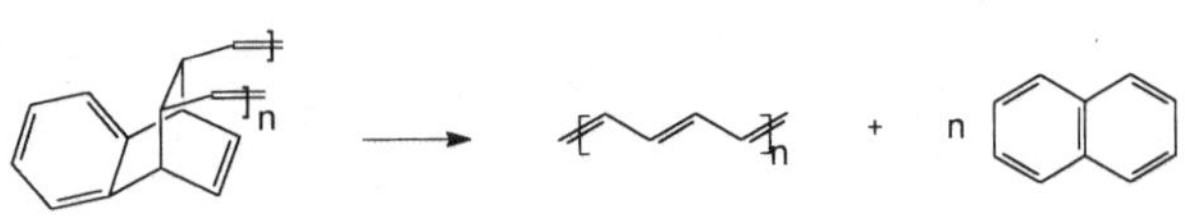

(88)

E. Acenaphthonorbornadiene

Acenaphthonorbornadiene, both *exo* and *endo* isomers, will undergo ring-opening metathesis polymerization [148] under the influence of WCl_6/Me_4Sn and WCl_6/Ph_4Sn [Eq. (89)].

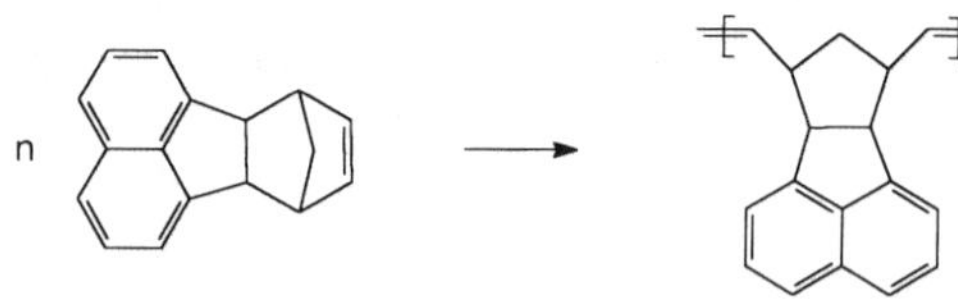

(89)

F. Deltacyclene

Deltacyclene and several substituted deltacyclenes have been polymerized in the presence of the catalytic systems $ReCl_5$, $RuCl_3.3H_2O$ and WCl_6/Ph_4Sn. Ring-opened polymers of different stereochemistries have been obtained depending essentially on the catalyst, solvent and reaction conditions employed [149] [Eq. (90)].

(90)

Using the above catalytic systems, butyl and phenyl substituted deltacyclenes have been polymerized to produce ring-opened polymers in various yields as a function of the catalyst and nature of the substituent [Eq. (91)].

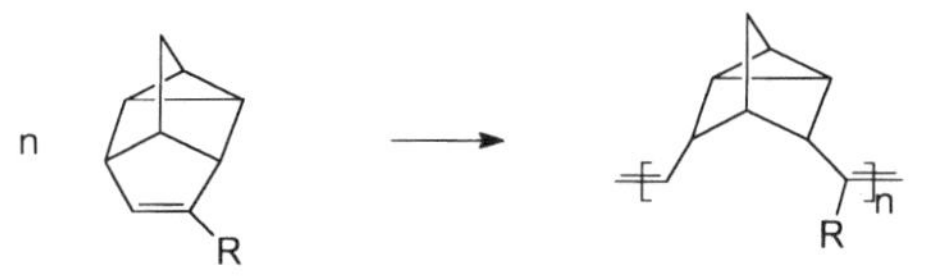

(91)

G. Dicyclopentadiene

Due to its ready accessibility, polymerization of *exo*- and *endo*-dicyclopentadiene has been effected with a wide range of cationic, Ziegler–Natta and ROMP catalysts [1, 6, 7]. The structure of the polymers obtained depends largely on the nature of the catalyst employed and reaction conditions.

Using cationic catalysts such as $AlCl_3$, BF_3, BF_3/Et_2O, $TiCl_4$, $SnCl_4$, $EtAlCl_2$, $EtAlCl_2/t$BuCl and Et_2AlCl/tBuCl, polymers with vinylic and rearranged structures in the polymer chain have been obtained from both *exo*- and *endo*-dicyclopentadiene [150] [Eq. (92)].

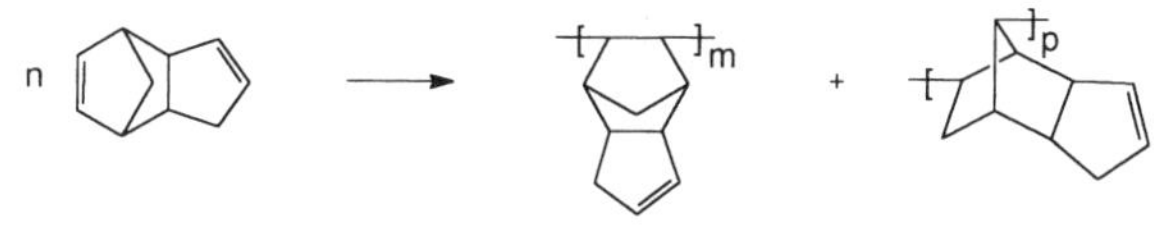

(92)

Under the influence of several cationic initiators, nortricyclene structures have been obtained starting with *endo*-dicyclopentadiene [Eq. (93)].

(93)

Various Ziegler–Natta and ROMP caalysts based on group IV–VIII transition metal salts showed to be very active and selective in the polymerization of both *exo*- and *endo*-dicyclopentadiene [151–155]. Thus, binary Ziegler–Natta systems derived from chromium, molybdenum, and tungsten halides associated with organoaluminum compounds form addition and ring-opened polymers [Eq. (94)].

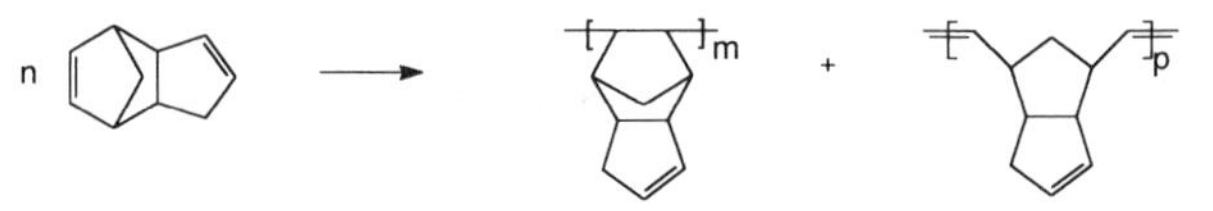

(94)

It is of particular interest that *endo*-dicyclopentadiene produced by ring-opening polymerization highly stereospecific polyalkenamer, with one-component catalysts consisting of transition metal salts of groups VI and VII. For instance, $MoCl_5$ yields highly *trans* polymer, $ReCl_5$ or $RuCl_3 \cdot 3H_2O$ give highly *cis* polymer, whereas WCl_6, $OsCl_3$, and $IrCl_3$ lead to both stereo-configurations [156]. By contrast, *exo*-dicyclopentadiene yields mainly *trans* polymer with $RuCl_3$ and *cis* polymer with $ReCl_5$. Numerous other binary and ternary catalytic systems derived from group V–VIII transition metal salts and organometallic compounds have been used for ring-opening polymerization of dicyclopentadiene and particularly for application in reaction injection moulding (RIM) and reaction transfer moulding (RTM) processes [157].

H. Dihydrodicyclopentadiene

Both 2,3- and 7,8-dihydrodicyclopentadiene, in their *exo* and *endo* conformations, easily yielded polymers in the presence of various cationic initiators, Ziegler–Natta catalysts and ROMP catalysts [1, 6, 7]. Generally, addition and ring-opened polymers are formed, often with rearranged structures, depending essentially on the catalyst and substrate. Eleuterio [57] polymerized 7,8-dihydrodicyclopentadiene over heterogeneous catalysts based on molybdenum, tungsten and chromium compounds yielding unsaturated polymers [Eq. (95)].

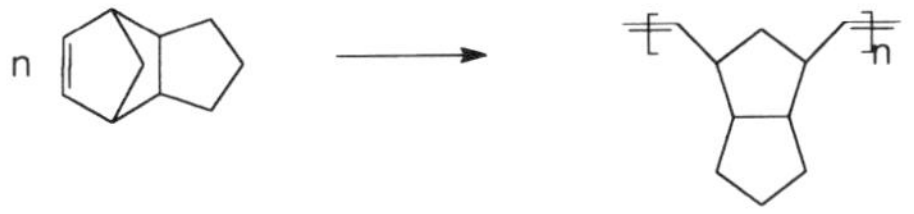

(95)

By contrast, using cationic initiators, e.g. $EtAlCl_2/t$BuCl, Cesca and coworkers [150] obtained saturated polymers by 1,2-addition reaction [Eq. (96)].

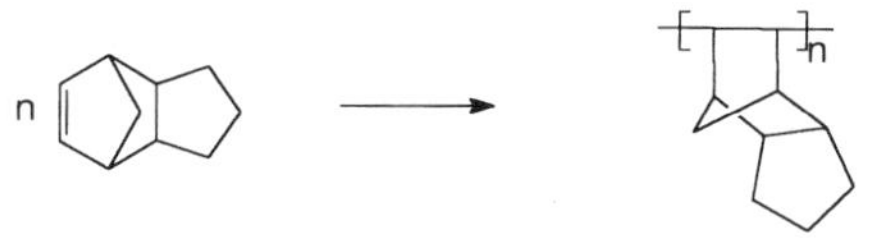

(96)

On the other side, 2,3-dihydrodicyclopentadiene produces low molecular weight oligomers having mainly rearranged structures with $EtAlCl_2/t$BuCl [Eq. (97)].

(97)

Probably, hydride migration reactions in the intermediate propagating carbenium species occurred under these conditions.

Numerous other Ziegler–Natta and ROMP catalysts have been used in the reaction of 7,8-dihydrodicyclopentadiene [151–152, 156]. For instance, with $TiCl_4/Et_3Al$ both addition and ring-opened polymers have been obtained as a function of the molar ratio Al:Ti. At low molar ratios, ring-opened polymers are preferred [152]. The yield and structure of the polyalkenamer depend mainly upon the nature of the catalyst. Thus, the catalyst WCl_6/Me_4Sn affords polymers of both *cis* and *trans* structures from *endo*- and *exo*-7,8-dihydrodicyclopentadiene, $ReCl_5$ forms predominantly the *cis* stereoconfiguration while $MoCl_5$ and $RuCl_3 \cdot 3H_2O$ give the *trans* configuration of the polymer [156].

I. Dimethylenehexahydronaphthalene

Polymerization of di-*endo*-methylenehexahydronaphthalene in the presence of cationic initiators, e.g., BF_3/Et_2O, gives rise to saturated products whose structures probably contain half-cage recurring units as a result of transannular reactions [158] [Eq. (98)].

(98)

J. Dimethyleneoctahydronaphthalene

Dimethyleneoctahydronaphthalene has been polymerized by a great variety of Ziegler–Natta and ROMP catalysts based on transition metal salts of titanium, zirconium, vanadium, molybdenum, tungsten, ruthenium, iridium, osmium, platinum or palladium and organometallic compounds [159]. Depending on the catalyst employed, addition or ring-opened polymers were preferentially formed [Eqs. (99) and (100)].

(99)

(100)

Furthermore, monosubstituted and disubstituted derivatives in the positions 9 and 10 of dimethyleneoctahydronaphthalene have been polymerized using the above catalytic systems to vinylic or ring-opened polymers [Eqs. (101) and (102), R_1 and R_2 = alkyl and aryl groups].

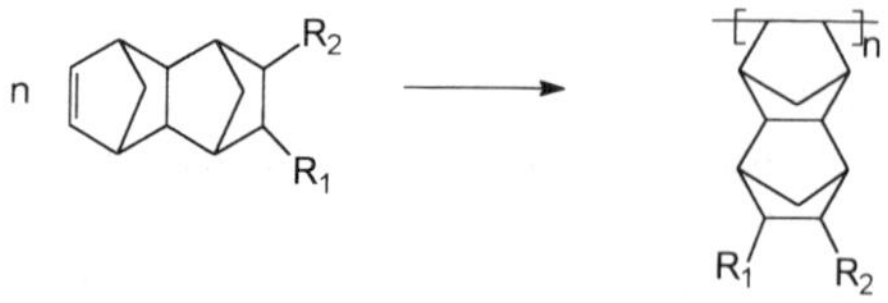

(101)

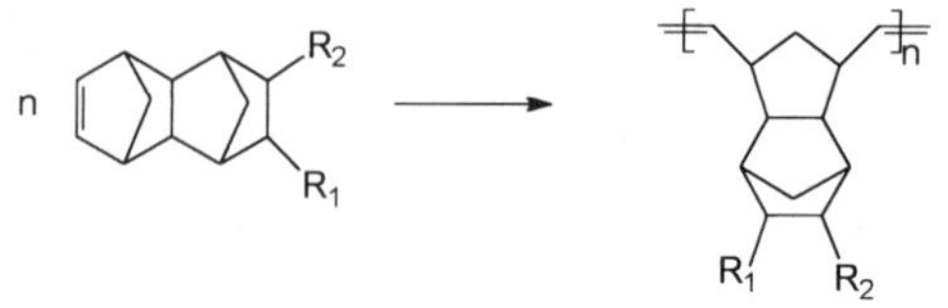

(102)

It is of interest that products obtained in these reactions display good mechanical and optical properties.

K. Pentacyclopentadecene

Pentacyclopentadecene undergoes addition and ring-opening polymerization [160] in the presence of several catalysts derived from titanium, vanadium, molybdenum, and tungsten associated with organometallic compounds of aluminum, tin or boron. Late transition metal salts such as rhodium, palladium, osmium, and ruthenium have been also employed. The polymers thus obtained have good physical, mechanical and electrical properties, high resistance to light, organic solvents and chemicals as well as improved heat resistance and transparency.

L. Pentacyclohexadecene

Pentacyclohexadecene has been reported to give addition and ring-opened polymers in the presence of Ziegler–Natta and ROMP catalysts [161]. The products obtained have valuable physical–mechanical properties, impart good transparency, low double refraction and high glass transition temperatures. They show further good stability to heat, water, moisture, light and chemicals and display good dimensional stability and thermal moldability.

M. Higher Norbornene-Like Monomers

A great number of norbornene-like monomers [e.g., $m = 1–3$, R_1 and R_2 = alkyl and aryl groups, Eqs. (103) and (104)] with or without substituents have been employed in polymerization reactions induced by Ziegler–Natta and ROMP catalysts derived from ruthenium, osmium, iridium, palladium, platinum, molybdenum, and tungsten halides or vanadium and zirconium halides or acetylacetonate associated with organometallic compounds [162, 163]. Both addition and ring-opened polymers have been obtained by this way depending on the catalyst employed [Eqs. (103) and (104)].

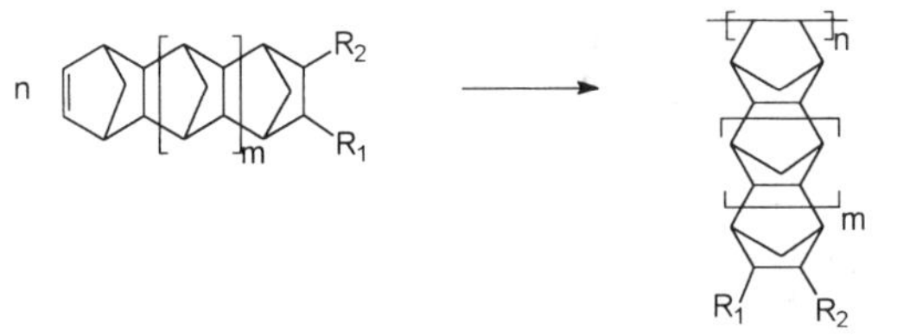

(103)

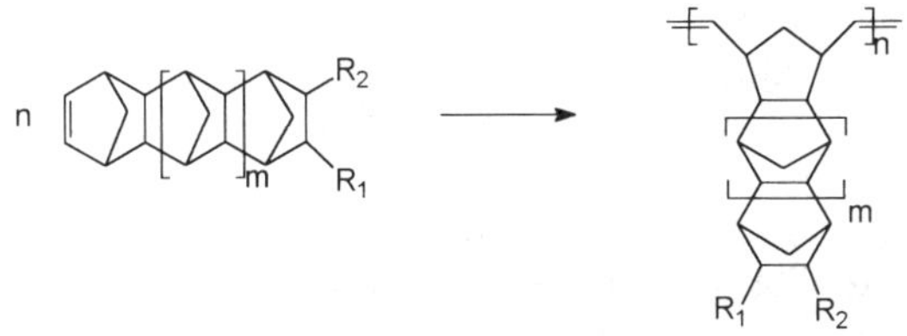

(104)

Polymers with different structures and physical–mechanical properties have been manufactured in these ways. Using a wide range of monomers bearing alkyl, aryl or various functional groups as substituents, the physical–mechanical properties of the polymers could be varied for desired practical applications. Numerous other examples of polymerization reactions of various substituted and unsubstituted polycyclic olefins are found in the open and patent literature [164–169].

VI. THERMODYNAMIC CONSIDERATIONS

The polymerization of cycloolefins presents several thermodynamic anomalies depending greatly on the vinylic or metathetic type of reaction and on the nature of the cycloolefin. It is noteworthy that the thermodynamic stability of cycloolefins makes an important contribution to the reaction enthalpy, while the nature of the cycloolefin influences the free energy by the entropic factor.

The thermodynamic parameters of cycloolefin polymerization are distinct during the passage from vinylic to ring-opening reaction. This difference arises because in a vinylic reaction a double bond of the monomer is converted into single bonds of the polymer, whereas in a ring-opening reaction the overall number of bonds and their type are the same in monomer and polymer. As a consequence, vinylic polymerization is strongly favored by enthalpy and weakly opposed by entropy, while ring-opening polymerization is favored also by enthalpy and in some cases by entropy.

The thermodynamic stability of cycloolefins will be a major determinant of the enthalpy and the free energy of the reaction, and even of the polymerizability of the monomers. Taking into account that the contribution of the double bond to the stability of the ring is a minor factor, at least for medium and large cycloolefins, thermodynamic stability can easily be followed along the entire range of cycloolefins by comparison with this property as it is known for cyclolkanes [170]. The steric strain of the cycloolefins will be slightly higher than that of cycloalkanes because of the particular sp^2 hybridization of the two olefinic carbon atoms.

In the polymerization reaction, the cycloolefins proved to have some peculiarities, that were closely related to the stability of the corresponding ring [6]. Thus, small strained rings, such as those in cyclopropene, cyclobutene, and cyclopentene, are easily polymerizable, even under mild conditions. Similar thermodynamic behavior is encountered in strained polycyclic olefins such as norbornene, dicyclopentadiene, and norbornene-like monomers. In contrast, cyclohexene is not easily polymerized, and the polymerizability of larger rings decreases as the ring size increases [171].

The low polymerizability of cyclohexene is attributed to the high stability of this six-membered ring. The cyclohexene is relatively unstrained, particularly in its generally occurring chair form, with the result that its free energy of polymerization ΔG is likely to be positive under readily accessible conditions. Early attempts to polymerize cyclohexene with cationic [172] or Ziegler–Natta [173] initiators led mainly to oligomers or low molecular weight polymers of the vinylic type. Recent studies indicate that cyclohexene can be polymerized or even copolymerized under controlled conditions [174].

The dependence of the polymerizability of small and medium cycloolefins on ring strain has been discussed extensively by Natta and Dall'Asta [175]. These investigators showed that for small-ring cycloolefins, polymerizability is directly correlated with ring strain, while for medium rings, when the strain energy is low, polymerizability is favored by other thermodynamic factors, such as the entropy of the reaction.

A. Thermodynamic Parameters of Cycloolefin Polymerization

The basic trend of the main thermodynamic parameters for cycloolefin polymerization (e.g., reaction enthalpy, entropy, free energy) is close to that of the hypothetical reaction of the corresponding cycloalkanes [176]. The use of an empirical method analogous to that of the hypothetical reaction of cycloalkanes has permitted the calculation of the main thermodynamic parameters of a series of cycloolefins in which the monomers are in the liquid state and the polymers in the condensed state [177]. For small and highly strained rings, polymerization is allowed because the value of the reaction enthalpy is high. For these rings, the change in enthalpy is negative, and it determines to a great extent the free energy of polymerization. Natta and Dall'Asta [175] showed that the reaction enthalpy for small rings depends mainly on ring strain, which is the driving force of the polymerization of these cycloolefins. For instance, cyclobutene, a cycloolefin with a large ring strain, shows a high negative enthalpy, favoring ring-opening polymerization, while its negative entropy brings a positive contribution to the free energy.

Several semiempirical thermodynamic parameters for polymerization of cycloolefins are given in Table 1 [178]. It is known that cyclopentene readily polymerizes even though the reaction enthalpy is lower than that of cyclobutene and the reaction entropy is negative. Experimental determinations for the reaction indicate that enthalpy values are −18.4, −18.8, or −20.5 kJ/mol; for entropy, we have −60.25 J/mol·K [175]. As a result of these findings, the average value of the enthalpy of the reaction is very close to the strain energy of cyclopentane (20.5 kJ/mol) [179]. The high polymerizability of cyclopentene is favored by the release of the ring strain and by the energy that is released during passage from the high energy eclipsed conformation of the monomer to a low energy staggered conformation adopted by the polymer. Although the reaction entropy is negative and has a relatively high absolute value at temperatures between 0 and 40°C, it does not exert a significant influence on the course of polymerization.

As mentioned above, cyclohexene does not easily polymerize under the usual conditions because this cycloolefin is highly stable in the chair conformation. It is also known that the hypothetical polymerization of cyclohexene is thermodynamically impossible, this reaction having positive values for the free energy and enthalpy. The presence of the double bond in the ring brings about only minor changes in the free energy of cyclohexene. The rection enthalpy for cyclohexene will have a value close to zero and the reaction entropy, very likely negative, will be low. These values will lead to a low free energy of reaction for this monomer. Even if the reaction is thermodynamically possible, the equilibrium will favor the reverse course as a consequence of the very high stability of cyclohexene.

For monomers possessing low ring strain energy, entropy is often the factor determining the free energy of polymerization [180]. For example, for five- and seven-membered cycloolefins, the enthalpy of the reaction for polymerization is lower than that of small rings and contributes only slightly to the free energy, whereas the entropy of the reaction may be a determining factor. For still larger cycloolefins (eight- to twelve-membered rings), strain factors again become important.

There is another feature of the process of ring-opening polymerization that depends on ring size and involves the significantly different contributions of translational or torsional and vibrational entropies. The negative translational entropy involved in polymerization, very high for small rings, becomes less negative for larger rings. In contrast, the torsional and vibrational entropies, which are always positive, decrease to a much smaller extent with increasing ring size. Accordingly, the negative translational entropy

TABLE 1 Semiempirical Values of Thermodynamic Parameters for the Polymerization of Cycloolefins at 25°C

Monomer	Polymer configuration	$-\Delta G^\circ$ (kJ/mol)	$-\Delta H^\circ$ (kJ/mol)	$-\Delta S^\circ$ (J/K·mol)
Cyclopentene	cis	2.3	16	46
	trans	6.3	20	46
Cyclohexene	cis	−6.2	−2	31
	trans	−7.3	2	28
Cycloheptene	cis	8.0	16	20
	trans	14.0	20	17
cis-Cyclooctene	cis	19.0	20	2
	trans	20.0	22	2

Source: Adapted from Ref. 178.

prevails over the other two types of entropy in small rings up to cyclohexene, and the positive torsional and vibrational entropies prevail over the translational one in larger rings.

B. Monomer–Polymer Equilibria

The existence of monomer–polymer equilibria in the polymerization reaction of cycloolefins under various conditions is based on both experimental and theoretical determinations. These equilibria are governed by the stability of the monomer and polymer, by such characteristic thermodynamic parameters as the enthalpy and entropy of the reaction, and by the working conditions. The equilibrium concentration of the monomer is given by:

$$\ln[\mathrm{M}]_{\mathrm{e}} = \frac{\Delta H_{\mathrm{p}}}{RT} - \frac{\Delta S_{\mathrm{p}}^{\mathrm{o}}}{R} \tag{105}$$

where $[\mathrm{M}]_{\mathrm{e}}$ represents the equilibrium concentration of the monomer, ΔH_{p} the variation in enthalpy during polymerization, $\Delta S_{\mathrm{p}}^{\mathrm{o}}$ the standard entropy during polymerization, T the reaction temperature, and R the ideal gas constant.

The monomer concentration at equilibrium is a measure of the polymerizability of cycloolefins. Depending on the nature of the ring, the terms of Eq. (105) make different contributions to the value of the monomer equilibrium with respect to polymerizability. Thus, Natta and Dall'Asta [175] showed that for small-ring cycloolefins, polymerizability is determined mainly by the ring strain of the cycle. On the other hand, the polymerizability of larger rings is determined primarily by the entropy of the reaction. As a consequence, the concentration of the monomer at equilibrium decreases with each increase in the ring size.

Ofstead and Calderon [181] determined quantitatively the monomer concentration at equilibrium during the polymerization of cyclopentene at various temperatures in the presence of the ternary catalyst $WCl_6/EtOH/EtAlCl_2$. Upon carrying out this reaction at 0, 10, 20, and 30°C, these researchers found significant dependence of cyclyopentene concentration on temperature (Table 2). It is noteworthy that the minimum equilibrium concentration of 0.51 mol/L was obtained at 0°C, which corresponds to a maximum cyclopentene conversion of 78%. The maximum equilibrium concentration, 1.19 mol/L, was reached at 30°C, corresponding to a conversion below 50%. From the equilibrium concentration of the monomer as a function of temperature, the thermodynamic parameters of this reaction can be determined.

TABLE 2 Equilibrium Concentration of Cyclopentene Polymerization with $WCl_6/EtOH/EtAlCl_2$ at Various Temperatures

Temperature (°C)	Equilibrium concentration, $[\mathrm{M}]_{\mathrm{e}}$ (mol/L)
0	0.51 ± 0.01
10	0.70 ± 0.01
20	0.88 ± 0.01
30	1.19 ± 0.04

Source: Adapted from Ref. 181.

C. Ring–Chain Equilibria

Numerous studies [182–184] on the polymerization of cycloolefins indicate that a ring–chain equilibrium exists between the linear high molecular weight polymer and the fractions composed of cyclic oligomers. The overall cyclic oligomer concentration in equilibrium with linear polymer is about 0.3–1 mol of repeat units per liter. As a consequence, when the starting concentration of the monomer is less than the equilibrium value, no linear polymer forms, and the products consist of cyclic oligomers. This dilution effect can be applied when cyclic oligomers are desired through reaction of cycloolefin.

Detailed investigation on ring–chain equilibria for the polymerization of cyclooctene [183] showed that the polymer obtained in the absence of acyclic olefins consists of 15–20% cyclic fraction and 80–85% linear fraction. Höcker *et al.* [184] correlated the product distribution as a function of initial cyclooctene concentration. They found that below a threshold monomer concentration of 0.1 mol, almost no polymer was present; at higher monomer concentrations, however, formation of cyclic oligomers and linear polymers was observed.

The ring–chain equilibrium for cyclooctene, cyclododecene, and cyclopentadecene has been treated in the limits of the Jacobson–Stockmayer theory [185]. According to Jacobson and Stockmayer, the equilibrium constant, K_n, of the resulting cyclooligomer is inversely proportional to the power 2.5 of the degree of polymerization n:

$$K_n = \left(\frac{3}{2\pi}\right)^{3/2} \frac{1}{N_{\mathrm{A}}} \left(\frac{1}{\gamma l^2 C_n}\right)^{3/2} n^{-5/2} \tag{106}$$

where N_A is Avogadro's number, γ is the number of bonds in the monomer unit, l is the average bond length, and C_n is the constant characteristic ratio. When the logarithms of the equilibrium constant K_n of the oligomers of cyclooctene, cyclododecene, and cyclopentadecene were plotted versus the logarithms of the degree of polymerization, straight lines with a slope of $-5/2$ were obtained [184]. Deviations were found for oligomers with $n > 4$. It was observed that the larger the ring, the lower the position of the respective straight line.

An important thermodynamic feature of the polymerization of cycloolefins is constant oligomer distribution, which can be obtained starting from monomers, from a fraction of higher oligomers, from the polymer, or from a single oligomer [185]. Above a minimum initial concentration, and independent of catalyst concentration or temperature in a range between 0 and 100°C, the absolute concentrations of cyclic oligomers were found to be constant. Up to a minimum monomer concentration, which is called the cutoff point (threshold value), the monomer is generally converted into cyclic oligomers, while above this concentration linear polymer is formed. It is important to note that higher polymer–oligomer ratios result when higher monomer concentrations are employed.

The equilibrium oligomer concentration $[M]_e$ is related to the entropy of formation ΔS^o by

$$R \ln[M]_e = \Delta S^o \quad (107)$$

It is significant that the plot of the entropy per carbon atom versus the number of carbon atoms for cyclobutene, cyclooctene, cyclododecene, and cyclopentadecene resulted in a common curve. Accordingly, the number of the double bonds has no effect on the entropy of formation or polymerization.

VII. KINETICS AND MECHANISM

A. Reaction Kinetics

Whereas interesting studies have been published on the kinetics of the polymerization of cycloolefins with cationic [1] and metathesis [6, 7] initiators, the main kinetic features of cycloolefin reactions with Ziegler–Natta catalysts have been inferred from the vast existing work on linear olefins [186].

Kinetic investigation on the polymerization of cyclopentadiene in the presence of cationic initiators such as BF_3, $TiCl_4$, or $TiCl_3 \cdot (On\text{-}Bu)$ pointed out a correlation of the reaction rate with monomer and initiator concentrations [187]. The initial rate was found to be first order in monomer, catalyst, and cocatalyst. Initiation was, however, slower than propagation and was independent on the initial monomer concentration. The rate constant for the initiation with $TiCl_3 \cdot (On\text{-}Bu)$ was $k_i = 11.1\ mol^{-1}\ s^{-1}$ at −70°C. A quasi-stationary concentration for the active centers resulted from theoretical calculations. Additional data obtained for the polymerization of cyclopentadiene with the catalysts $TiCl_4 \cdot CCl_3COOH$, $SnCl_4 \cdot CCl_3COOH$, and $BF_3 \cdot OEt_2$ allowed the correlation of the reaction rates with monomer concentration and the catalytic systems and the development of a kinetic theory of the process [188]. Valuable kinetic aspects have also been published on the polymerization of 1,3-cyclohexadiene [189] with $BF_3 \cdot OEt_2$ and $SnCl_4 \cdot CCl_3COOH$. Interestingly, the polymerizations with the latter catalyst were nonstationary at early stages, and high conversions have been obtained at low monomer concentrations. Similar kinetic behavior was found for the polymerization of *cis,cis*-1,3-cyclooctadiene [190] with $TiCl_4 \cdot CCl_3COOH$; in this case, however, strong dependence on the solvent was observed.

Kinetic studies on the polymerization of 1,3-cyclooctadiene with the $TiCl_4 \cdot H_2O$ system [191] confirmed that the reaction consists of two phases: a slow, stationary phase and a rapid, nonstationary phase, whose initial rate follows the law:

$$\text{Rate} = k\,[TiCl_4]\,[H_2O]\,[M] \quad (108)$$

where M is monomer. It was found that the yields were directly proportional to the concentration of the active centers but inversely proportional to the monomer concentration:

$$\text{Yield} = \frac{k_p}{k_{mt}} \frac{[P^*]_0}{[M]_0} \quad (109)$$

where k_p and k_{mt} are the rate constants for propagation and chain transfer with monomer, and $[P^*]_0$ and $[M]_0$ are the initial concentrations of the active centers or polymer ends and monomer, respectively. Additional kinetic data concerning the polymerization of other monomers (e.g., cyclohexene, cyclooctene, norbornene, norbornadiene, dicyclopentadiene) may be found in the literature [1].

A different kinetic behavior is encountered in the ring-opening polymerization of cycloolefins using ROMP catalysts [6]. This process must be complex to accommodate the well-known difficulties stemming from the treatment of Ziegler–Natta systems [186]. The main kinetic data come from homogeneous catalytic systems involving the metathesis polymerization of cyclopentene, cyclooctene, cyclooctadiene, and norbornene.

Early studies on cyclopentene polymerization with WCl_6-based catalysts [192] showed that the reaction kinetics correspond to a first-order equation. The following expression is found for the rate of monomer consumption:

$$\ln \frac{[M] - [M]_0}{[M] - [M]_e} = k_{cr} t \tag{110}$$

where [M], $[M]_0$ and $[M]_e$ represent the concentrations of the monomer at any moment, at the initial moment, and at equilibrium, respectively, k_{cr} is the rate constant, and t is the time.

More detailed studies on the polymerization of cyclopentene with $WCl_6/i\text{-}Bu_3Al$ evidenced more complex behavior on the part of this process [193a]. A first-order dependence of the rate on catalyst concentration was found, but the order with respect to the monomer was rather complicated. The rate of polymerization, which was initially first order with respect to cyclopentene, decreased rapidly as the reaction progressed. Based on these findings, two kinetic species have been proposed. The first species is formed by the reaction of a complex between cyclopentene and WCl_6 with 2 mol of organoaluminum compound, which is partially converted by further reaction with cyclopentene into a new species having a much lower activity. A simple kinetic scheme illustrating the generation of these two distinct active species was proposed:

$$WCl_6 + C_5H_8 \xrightarrow{k_1} \underset{\text{I}}{W^*} \xrightarrow{i\text{-}Bu_3Al} W^*/Al \tag{111}$$

$$\underset{\text{I}}{W^*} + C_5H_8 \xrightarrow{k_2} \underset{\text{II}}{W^{**}} \xrightarrow{i\text{-}Bu_3Al} W^{**}/Al \tag{112}$$

Species II displays complex behavior. A significant observation was that the rate of polymerization under a given set of conditions depended on the time between addition to the monomer of the two catalyst components (WCl_6 and $i\text{-}Bu_3Al$); the premixing time for WCl_6, before addition of $i\text{-}Bu_3Al$, that gave the highest polymerization rate was defined by

$$t_{max} = \frac{1}{(k_1 - k_2)[M]_0} \ln \frac{k_1}{k_2} \tag{113}$$

where $[M]_0$ is the initial cyclopentene concentration. The optimum premixing time, t_{max}, was inversely proportional to the monomer concentration. Accordingly, in the foregoing kinetic scheme, the initial rate of polymerization R_{p0} was represented by

$$R_{p0} = k_p[W^*/Al]_0[M]_0 \tag{114}$$

where $[W^*/Al]_0$ was estimated to be equal to [I]. In this way, the initial rate of polymerization can be used to measure the concentration of the active species $[W^*/Al]$, at any time. Further investigation of this system [193b] indicated that during the polymerization of cyclopentene, the rate of propagation decreases faster than can be accounted for by the consumption of the monomer alone. This observation suggested that the decrease in the rate of polymerization is a second-order reaction with respect to the concentration of the active species. Accordingly, a second-order termination reaction with respect to the active species has been proposed:

$$W^* + W^* \rightarrow \text{polymer} \tag{115}$$

similar to the bimolecular termination reaction that occurs in the Ziegler–Natta polymerization.

Further kinetic studies have been carried out for the elucidation of initiation, propagation, and termination mechanisms in the polymerization reactions of cyclopentene, cyclooctene, cyclooctadiene, norbornene, norbornadiene, and their derivatives [194–196]. Some interesting observations were collected in connection with the polymerization of cyclooctene [195], using the following catalytic systems: $WCl_6/EtOH/EtAlCl_2$, WCl_6/Me_4Sn, and $MoCl_2(NO)_2(PPh_3)_2/ EtAlCl_2$. The reaction proved to be first order with respect to the catalyst concentration and second order with respect to the monomer when $WCl_6/EtOH/EtAlCl_2$ was used in the range of 1 to 5×10^{-4} mol. In contrast, the other two catalysts gave first-order kinetics with respect to the monomer concentration. Pronounced kinetic control of the reaction products in the early stages of the reaction was observed: at high initial monomer concentration, there is immediate formation of polymer, while at low concentration the homologous series of oligomers is formed first, and eventually the polymer.

Studies on the polymerization of 1,5-cyclooctadiene [196] with $WCl_6/i\text{-}Bu_3Al$ showed significant similarities to the reaction of cyclopentene, although the rate of the former reaction was slower.

B. Reaction Mechanisms

As a function of the catalytic system used in the polymerization of cycloolefins, the reaction may proceed by a cationic, Ziegler–Natta, or ring-opening metathesis mechanism. The three basic mechanisms for each type of polymerization reaction are briefly outlined.

1. Cationic Polymerization

In the presence of cationic initiators, the polymerization of cycloolefins proceeds via a carbocationic pathway in which the initiation, propagation, chain transfer, and termination steps involve the well-documented processes related to carbocation chemistry [1]:

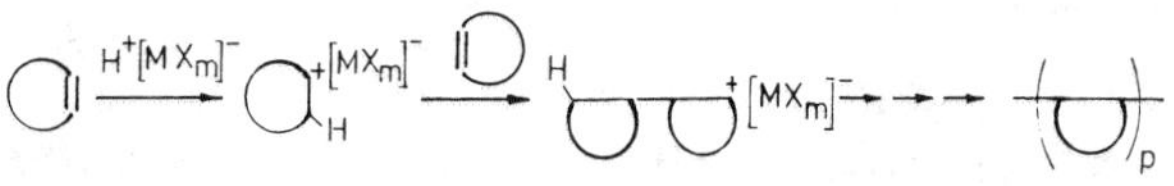

(116)

Initiation occurs by the addition of a proton or a cationic species to the carbon–carbon double bond of the cycloolefin, generating a carbenium ion as a free ion or an ion pair with the counteranion from the catalytic complex [197]. Propagation proceeds by successive additions of newly formed carbocations to the monomer, while chain transfer and termination take place by the usual deactivating reactions known for carbocations [1]. Intensive recent studies on the carbocationic mechanism revealed interesting aspects concerning the mode of generating the active species, the identification and characterization of the initiating and propagating species, the nature of breaking steps (chain transfer and termination), and other reaction parameters [197].

2. Ziegler–Natta Polymerization

The polymerization of cycloolefins in the presence of Ziegler–Natta catalysts generally involves the main steps known for this type of reaction from work with acyclic olefins [198] (e.g., cycloolefin coordination to the metal center, monomer insertion into the metal–carbon bond, chain termination, and reaction transfer):

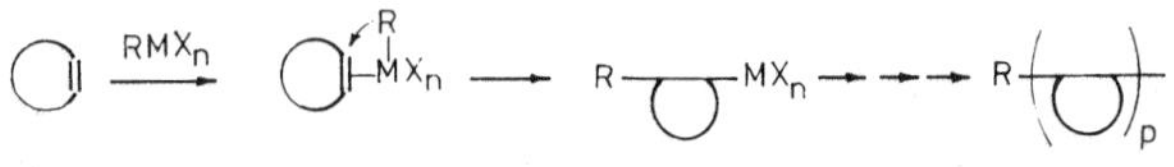

(117)

Coordination of olefins at the transition metal compounds with formation of π complexes is well documented experimentally and theoretically [199]. Whether a particular cycloolefin will coordinate first and then insert into the metal–carbon bond, or will insert directly, depends on the steric environment imposed on the cycloolefin by complexation; in addition, the reactivity of the cycloolefin plays an important part. This latter characteristic also influences double-bond opening during the insertion process. Since both the structure and the reactivity of the monomer extend over a wide range (from small to large and from mono- to polycyclic olefins), the borderline between complexation and direct insertion is difficult to define. Furthermore, because of marked differences in polymer structures compared to those obtained from simple α- or linear olefins, the polymerization of cycloolefins may involve specific chain transfer or termination processes.

3. ROMP Reaction

It is generally accepted today that the ring-opening metathesis polymerization (ROMP) of cycloolefins occurs by a carbene mechanism involving metallacarbene cycloaddition to form metallacyclobutane intermediates, which generate new metallacarbene to propagate the reaction chain [6, 7, 200].

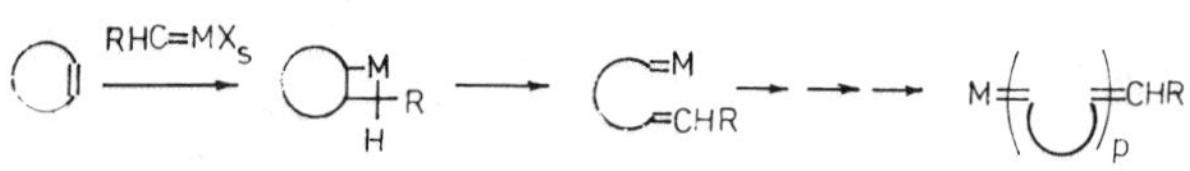

(118)

The main steps of this process (viz., generation of the initiating metallacarbene species from metathesis catalysts of various types with or without cycloolefins, cycloaddition of the metallacarbene to the carbon–carbon double bond to form the intermediate metallacyclobutane, propagation of the reaction chain by the newly generated metallacarbenes, further addition of the monomer) are well documented by physicochemical methods [200]. In this regard, it is important to note that the well-defined metallacarbenes [201] and metallacyclobutanes [202] initiate the ring-opening polymerization of cycloolefins. Moverover, the newly formed metallacarbenes and metallacyclobutanes can be observed and followed by, for example, proton and ^{13}C NMR spectroscopy during the polymerization process [200], and the microstructure of the polymers obtained from unsubstituted or substituted bicyclic and/or polycylic olefins (e.g., norbornene and its derivatives) can be readily correlated with the metallacarbene–metallacyclobutane nature of the intermediates [203]. In addition, living polymerization systems for cycloolefins using tungsten– or molybdenum–alkylidene complexes [201] or titanacyclobutanes [202] in which termination or transfer processes are completely eliminated allow the production of polymers in a narrow range of molecular weights, having attractive applications in block or graft copolymer area.

VIII. REACTION STEREOCHEMISTRY

By employing highly stereospecific Ziegler–Natta or metathesis catalysts for the polymerization of cycloolefins, stereoregular vinylic or ring-opened polymers can be readily obtained. The degree of stereoselectivity varies with the nature of the catalytic system and of the cycloolefin. Furthermore, the stereoselectivity of the reaction products depends substantially on the reaction conditions, especially the reaction temperature and the conversion rate. However, since vinylic and metathesis polymerization occur by distinct mechanisms, the origin of stereoselectivity will be different in the two reactions.

A. Steric Configuration of Vinylic Polymers

In general, four different isomeric structures, may arise by addition polymerization of cycloolefins:

I II (119)

III IV (120)

where **I** and **II** are *erythro*-diisotactic and -disyndiotactic and **III** and **IV** are *threo*-diisotactic and disyndiotactic, respectively. The opening of the double bond can take place in a cis or a trans fashion, thus forming two erythro (**I** and **II**) and two threo (**III** and **IV**) isomers. Some of these forms (e.g., *threo*-disyndiotactic) are expected to show optical activity [204].

B. Steric Configuration of Polyalkenamers (Poly-1-alkenylenes)

The configuration of the double bonds in polyalkenamers may have cis (or Z) and trans (or *E*) geometries. These configurations may be predominant or exclusive in all-cis (**I**) or all-trans (**II**) forms or may occur in random or blocky distribution (**III**):

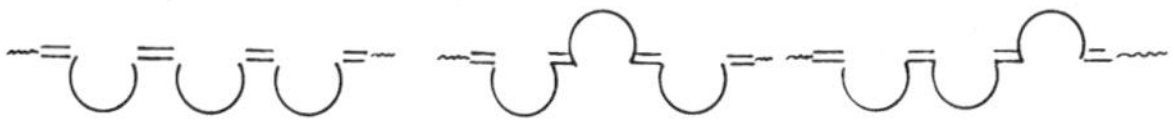

I II III

(121)

The double-bond pair sequence (or dyad configuration) in polyalkenamers may be of the type cc, ct, tc, or tt (c = cis and t = trans).

If the polyalkenamers have substituents and possess chiral centers, the monomer units may have isotactic, syndiotactic, or atactic relationships. In fully isotactic and syndiotactic structures the dyad relationship can be of the HT or TH type (H = head, T = tail) (**I** and **II**), while atactic structures have dyad sequences of the HH, TH, HT, and TT types (**III**):

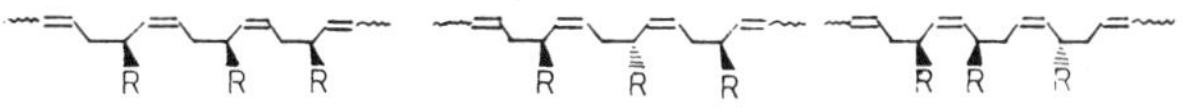

R R R R R R R R R

I II III

(122)

In polyalkenamers derived from bicyclic or polycyclic olefins, adjacent rings may have an isotactic (m), syndiotactic (r), or atactic (m/r) relationship. For polynorbornenes, these structures are represented as follows:

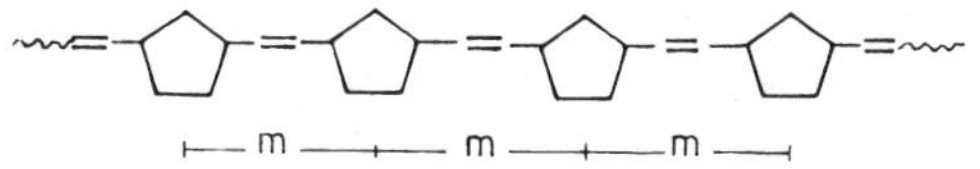

m m m (123)

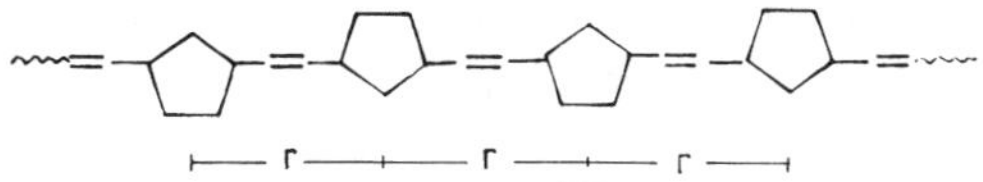

r r r (124)

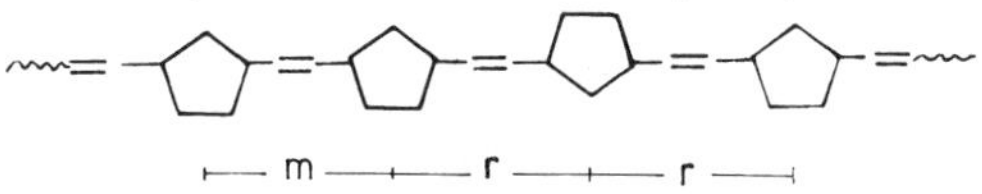

m r r (125)

Polyalkenamers of nonsymetrically substituted bicycloolefins, such as 5-alkylnorbornene, may have the monomer dyad in an HH, TT, or TH sequence:

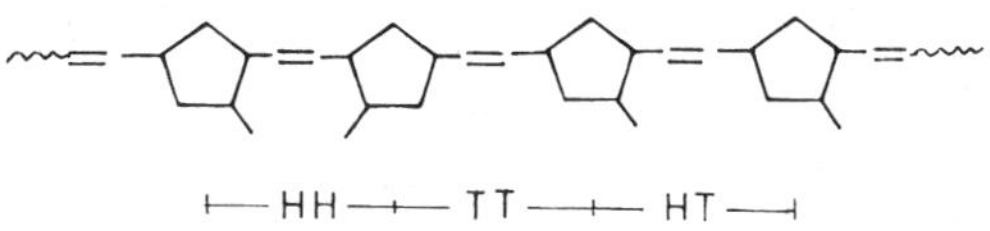

HH TT HT (126)

C. Stereoselectivity in Cycloolefin Polymerization

Ziegler–Natta catalysts exhibit varying degrees of selectivity in the vinylic polymerization of cycloolefins, strongly depending on the nature of the catalyst and reaction conditions [198]. When chiral catalysts (e.g., chiral metallocenes in combination with aluminoxanes) are used, the synthesis of stereoregular polycycloolefins is possible [205].

There is strong evidence that the ring-opening polymerization of cycloolefins provides a convenient approach for the synthesis of polyalkenamers with high trans or cis content [2, 3]. Except for polyhexenamer, polyalkenamers of predominantly trans structure have been obtained from cyclobutene to cyclododecene, and polyalkenamers of predominantly cis structure from cyclobutene to cyclooctene. In addition, polyalkenamers of high steric purity have been obtained from norbornene and its derivatives or from other bicyclic and polycyclic olefins.

Abundant data indicate that the nature of the catalyst is crucial in directing the polymerization stereoselectivity towards cis or trans polymers; definite classes of cis-specific and trans-specific catalysts are known [6, 7]. Within the same binary or ternary catalyst system, the cis or trans proportion of the polymer can be varied by appropriate changes in the molar ratio of the components. Other reaction conditions, such as temperature and solvent, also make important contributions during the steric course of the reaction.

The polymerization of cyclobutene with catalytic systems consisting of $MoCl_5$ and Et_3Al or of $RuCl_3$ without any cocatalyst leads to a polybutenamer having mainly cis configuration at the double bond [206]. By contrast, catalytic systems based on $TiCl_3$ or $TiCl_4$ and Et_3Al provide polybutenamers having a predominantly trans configuration. When catalytic systems derived from WCl_6 and Et_3Al or $MoO_2(acac)_2$ and Et_2AlCl are used, polybutenamers with random distribution of cis and trans configurations are obtained. It is noteworthy that a cis polybutenamer of high steric purity (93% cis content) was formed by polymerizing cyclobutene with the carbenic system $Ph_2C{=}W(CO)_5$ without cocatalyst [207].

The stereoselective polymerization of cyclopentene to cis or trans polypentenamer may readily be achieved with various binary and ternary catalysts. For instance, polypentenamer with up to 95% trans configuration can be obtained using a wide range of catalysts derived from WCl_6, $TiCl_4$, BCl_3, and organoaluminum halides with or without the addition of a ternary component such as alcohol, peroxide, hydroperoxide, or chlorinated or nitro compounds [208]. On the other hand, polypentenamers with a cis structure exceeding 90% are commonly synthesized with catalytic systems consisting of WF_6, $MoCl_5$, or $ReCl_5$ and organoaluminum compounds [209] or with the carbenic system $Ph_2C{=}W(CO)_5$ mentioned above [207]. Remarkably, polypentenamers of more than 97% cis configuration have been obtained using catalyst systems such as WCl_6/allyl$_4$Si, WCl_6/Bu_4Sn, WCl_6/aluminoxane [210] or $MoCl_5/Et_3Al$ [211]. The steric configuration of a polypentenamer can be substantially modified by changing the composition of the catalyst or the reaction temprature [211]. The effects of reaction temperature on the cis content of polypentenamers manufactured with $MoCl_5/Et_3Al$ are illustrated in Table 3.

TABLE 3 Effect of Temperature on the Polymer Yield and Stereostructure in the Polymerization of Cyclopentene with $MoCl_5/Et_3Al$

Temperature (°C)	Polymer yield (%)	Stereostructure	
		cis (%)	trans (%)
−80	3.3	99.4	0.6
−55	14.6	98.1	1.9
−40	40.8	99.3	0.7
−30	27.0	98.8	1.2
−10	13.3	97.7	2.3
+10	2.0	91.9	8.1
+30	0.3	85.4	14.6

Source: Adapted from Ref. 211.

The polymerization of cyclooctene induced by catalytic systems derived from WCl_6 and Et_3Al, $EtAlCl_2$, or other organoaluminum compounds produces mainly trans polyoctenamer [212]. By contrast, the use of WF_6 in conjunction with $EtAlCl_2$ or the cis-specific carbene system $Ph_2C{=}W(CO)_5$ has yielded polyoctenamers of 93 and 97%, respectively [207]. It is noteworthy that in these reactions only *cis*-cyclooctene is reactive, the trans isomer remains inert under the same conditions.

The trans polydecenamer is readily formed by the polymerization of *cis*-cyclodecene in the presence of such tungsten-based binary and ternary catalysts as WCl_6/Et_2AlCl, $WCl_6/EtAlCl_2$, $WCl_6/EtOH/EtAlCl_2$, or $WOCl_4/Cum_2O_2/Et_2AlCl$. When the reaction of *cis* – /*trans*-cyclodecene is carried out with WCl_6/Et_2AlCl, a polydecenamer having predominantly trans configuration is produced, but the residual monomer contains less cis isomer than the initial ratio. This result indicates that *cis*-cyclodecene manifests a higher reactivity than *trans*-cyclodecene in this reaction [213].

Of interest is the polymerization of cycloolefins containing several double bonds, such as *cis,cis*-1,5-cyclooctadiene, *cis,trans,trans*-, and *trans,trans,trans*-1,5,9-cyclododecatriene [214]. These monomers yield polybutenamers having prevailingly trans structures when WCl_6-based catalysts (e.g., $WCl_6/EtAlCl_2$, WCl_6/Et_2AlCl, and $WCl_6/EtOH/EtAlCl_2$) are employed.

High stereoselectivity has been reported in the polymerization of bicyclic and polycyclic olefins using various catalytic systems [207, 215, 216]. Thus, norbornene yields a trans polynorbornenamer in the presence of $MoCl_5$ or $RuCl_3$ and a cis polynorbornenamer with $ReCl_5$ or $Ph_2C{=}W(CO)_5$, while mixtures of cis and trans configurations are obtained when binary and ternary systems of WCl_6 or other transition metal (Ir, Rh, Os) catalysts are employed [216]. Polynorbornene with a moderately high cis double-bond content generally exhibits a blocky distribution of cis and trans double bonds. As the cis content decreases from 100% to 35%, a trend from fully syndiotactic to atactic ring dyads has been observed [203].

Different stereoselectivities are recorded in substituted cycloolefins [217]. With few exceptions, the polymerization of substituted cycloolefins is less stereospecific [26]. However, when the substituent is located at the double bond, translationally invariant polymers may be obtained in the presence of $Ph_2C{=}W(CO)_5$. For instance, with this catalytic system 1-methylcyclobutene formed polyisoprene having *Z* stereostructure [218], while 1-methylcyclooctene stereospecifically led to a perfectly alternating polymer with *E* stereostructure [219].

IX. INDUSTRIAL DEVELOPMENTS

A. Synthesis of Hydrocarbon Resins

Hydrocarbon resins obtained from terpene monomers find application in areas including adhesives, printing inks, hot-melt coatings, calks and sealants, rubber articles, textiles, paints, varnishes, and plastics [2, 220, 221]. They are generally compounded with elastomers, plastics, waxes, and oils to confer special properties for a particular use. Industrial scale production originated in the mid-1930s with α-pinene, β-pinene, and *d*-limonene or dipentene made with Friedel–Crafts catalysts, particularly $AlCl_3$. The physicomechanical properties range widely as a function of monomer, catalyst, and process parameters. β-Pinene and *d*-limonene give better yields of solid resins than α-pinene in solution polymerization with $AlCl_3$. Several cocatalysts (e.g., Bu_2SnX_2, R_2GeX_2, R_3SiX, SbX_3) considerably improve the yield of hard resins from α-pinene. α-Pinene resins display a narrower molecular weight distribution than β-pinene resins obtained under the conditions above. Differences in solubility range and tackifying efficiencies for natural rubber and styrene–butadiene rubber are also significant. Production of copolymers of terpenes with other terpene or nonterpene monomers gains importance in special areas of application—for example, the production of terpene phenol resin and the synthesis of terpene copolymers with styrene, isobutylene, or isoprene. Such resins as styrene–isobutylene–β-pinene terpolymer with softening points of 100°C are useful as tackifiers for pressure-sensitive adhesives and components of hot-melt coatings.

B. Synthesis of Polypentenamer (Poly-1-pentenylene)

The particular properties of trans polypentenamer as a general-purpose rubber drew the attention of industrial research groups as long ago as the early period of its discovery [6, 7]. However, aspects of the economical extraction of cyclopentadiene from the C_5 steam cracker cuts and subsequent selective hydrogenation to cyclopentene have delayed production of the trans polypentenamer on an industrial scale [222]. Despite the existence of pilot plants (e.g., Bayer, Goodyear, Nippon Zeon, Japan Synthetic Rubber), commercialization is not yet a reality.

The polymer has a fairly low melting point (18°C) and glass transition temperature (−97°C). These characteristics, very close to those of natural rubber, impart to the polymer good processability and elastomeric qualities [41]. It is noteworthy that the polymer hardly crystallizes at room temperature in a finite period, but under strain it readily crystallizes. Studies of compounding and processing properties reveal that it behaves well during mixing, dispersing ingredients and fillers, extrusion, calendering, tire assembly, and other operations on materials with a Mooney viscosity between 30 and 150. This behavior is favored by a wide range of molecular mass that is not encountered in conventional types of polymerization. The polymer can be compounded in open or Bambury mills without premastication. The change in energy absorption and the temperature profiles during mixing with HAF carbon black are close to those of natural rubber. Poly-1-pentenylene can be loaded with large amounts of carbon black and oil, preserving its good mechanical properties. A remarkable property of trans polypentenamer is its high stress crystallization in the unvulcanized state, which entails an increase in resistance during processing. This quality, even superior to that of natural rubber, allows easy processability during the operation of building up tire bodies. The polymer can be readily vulcanized with sulfur. Vulcanizates with high crosslinking require low consumption of sulfur, vulcanizing agent, and accelerator. The vulcanized products possess high tensile strength and modulus and high abrasion resistance. The polymer has very low air permeability and good aging resistance as well as stability to UV irradiation, ozone, and weathering. The trans polypentenamer can be used in compositions with various diene rubbers because it has good compatibility and covulcanizability with them. It is largely compatible with natural rubber, isoprene, butadiene, and styrene–butadiene rubbers and with ethylene–propylene–diene terpolymers, yielding compositions suitable for tires and rubber articles. The polymer confers to these blends high abrasion and aging resistance, good elasticity, and low air permeability.

The cis polypentenamer is a special rubber for low temperature applications [41]. It has a very low melting point (−41°C), but the glass transition temperature is also very low (−114°C), which imparts poor elastomeric properties to the polymer. The cis polypentenamer is compounded with difficulty from ingredients and fillers at room temperature, and its physicochemical properties are inferior to those of natural rubber or the trans polypentenamer. For low temperature uses, however, the polymer will compete with such special rubbers as polypropyleneoxide or *cis*-polybutadiene.

C. Synthesis of Polyoctenamer (Poly-1-octenylene)

The commercialization of polyoctenamer under the trade name of Vestenamer by Chemische Werke Hüls (now Hüls AG) has been possible because there exists an economical source of the monomer from butadiene, which is available in two highly selective steps (dimerization and partial hydrogenation of the resulting 1,5-cyclooctadiene) [15]. The metathesis polymerization of cyclooctene to polyoctenamer with specific catalysts under controlled conditions leads to polymers of several types as mixtures of linear and cyclic macromolecules of differing trans content [223]. The first product, Vestenamer 8012 (trans polyoctenamer or TOR), has 80% trans double bonds and a viscosity number of $120\,cm^3/g$ when measured in 0.1% solution in toluene at 25°C. It is highly crystalline (≈ 30%) and has a melting point of 54°C, which makes this polymer appropriate for use in various technical articles. The second type of polymer, Vestenamer 6213, with a lower trans content (only 62%) and lower crystallinity and melting point (30°C), is used for low temperature applications where the admixture of the standard type would lead to stiffening. The third polyoctenamer, recently introduced, is Vestenamer L, a low molecular weight material that promises to have good properties for coatings compositions after chemical modification.

The trans polyoctenamer possesses good processing characteristics and acquires good physical properties after vulcanization [224, 225]. It can be used in blends with other rubbers, including natural, butadiene, isoprene, and styrene–butadiene rubbers, leading to ease of incorporation and distribution of filler, higher green strength and rigidity of uncured compound at room temperature, lower viscosity at processing temperatures, and improved flow behavior during calendering, extrusion, injection, molding, and pressing. It can be vulcanized with various rubbers using sulfur, peroxides, or other crosslinking agents. After vulcanization, it confers good tensile strength and elasticity, high modulus and hardness, good abrasion resistance, and resistance to aging and ozone or to swelling in organic media and water, as well as good low temperature characteristics. As a result of its excellent properties, trans polyoctenamer offers versatile possibilities for applications in rubber processing for injection, molding, extrusion, calendering, fabric coating, coating of rollers, tire production (e.g., rim strips, belt strips, sidewalls or apex compounds for truck tires and HR- or VR-passenger vehicle tires), production of plasticizer-free compounds, and so on.

D. Synthesis of Polynorbornene and Its Derivatives

Polynorbornene is produced under the trade name of Norsorex using conventional metathesis catalysts [226]. The polymer has a high trans content (≈ 90%) and imparts to rubber compositions special qualities that lead to easy processability and better mechanical characteristics [227, 228]. The polymer is a white, freely flowing powder of extremely high molecular weight; it is able to absorb large amounts of extending oil and plasticizer or filler. The plasticized material has good green strength and is largely compatible with fillers. It can be molded and cured normally at temperatures up to 185°C. The glass transition temperature of 35°C can be lowered to −60°C if sufficient plasticizer is used. The vulcanizates acquire good resistance to ozone when suitable antiozonants are added in combination with wax. The vulcanized rubber has remarkable properties, particularly for application in engine mounts, vibration dampers, impact and sound absorption materials, flexible couplings, and so on. Various soft rubbers can be produced for utilization in arm rests, sealing rings, and printing rollers. Other special applications would be possible, using suitable compositions and additives.

Polymers from norbornene derivatives also offer real promise for industrial development [6, 7]. For instance, poly-5-cyanonorbornene produced with ROMP catalysts is reported to possess high impact and tensile strength, a wide temperature range for useful applications, good creep resistance and transparency, good blending properties, and good workability by extrusion, molding, or film. A drawback lies in the material's pronounced sensitivity to autoxidation.

Polycyclic monomers containing norbornene moieties have recently been employed to yield a wide range of specialty ring-opened polymers and copolymers characterized by high dimensional stability; unchanged optical qualities due to strain; high transparency; high resistance to heat, solvents, chemicals, water, and impact; high thermoplasticity; and low glass transition temperatures [159–161]. These products can be conveniently used for molding plastic lenses (Fresnel lens), optical fibers, filters, windows, and so on [162–164].

E. Synthesis of Polydicyclopentadiene

Polymers of dicyclopentadiene are produced commercially by different industrial procedures. The reaction injection molding technique is applied by Hercules to make polydicyclopentadiene (Metton) having outstanding physicomechanical properties (e.g., high modulus and high impact strength, excellent creep resistance, resistance to antioxidants) [229]. The polymer is used mainly in recreational equipment such as golf carts and snowmobiles, for satellite antenna dishes and already in the automotive industry. Telene polymers are being developed by Goodrich, via the RIM and resin transfer molding techniques, from norbornene monomers in the presence of ROMP catalysts [230–234]. The product provides an excellent combination of stiffening, impact strength, and heat deflection temperature with low specific gravity. Such polymers have superior hydrolytic stability and electrical insulating properties.

REFERENCES

1. JP Kennedy. Cationic Polymerization of Cycloolefins: A Critical Inventory. New York: John Wiley & Sons, 1975.
2. W Vredenburgh, KF Foley, AN Scarlatti. In: Encyclopedia of Polymer Science and Engineering, vol. 7. New York: John Wiley & Sons, 1987, p 758.
3. PH Plesch (ed). The Chemistry of Cationic Polymerization. New York: Macmillan, 1963.
4. (a) J Boor, Jr. Ziegler–Natta Catalysts and Polymerization. New York: Academic Press, 1979. (b) G Fink, R Mulhaupt, HH Brintzinger, eds. Ziegler Catalysts, Recent Scientific Innovations and Technological Improvements. Berlin: Springer, 1995.
5. W Kaminsky, A Bark, I Dake. Polymerization of cyclic olefins with homogeneous catalysts. In: T Keii, K Soga, eds. Catalytic Olefin Polymerization. Amsterdam: Elsevier, 1990.
6. V Dragutan, AT Balaban, M Dimonie. Olefin Metathesis and Ring-Opening Polymerization of Cycloolefins. New York: John Wiley & Sons, 1985.
7. KJ Ivin, JC Mol. Olefin Metathesis and Metathesis Polymerization. London: Academic Press, 1997.
8. (a) Y. Imamoglu, B Zumreoglu-Karan, AJ Amass, eds. Olefin Metathesis and Polymerization Catalysts. Dordrecht: Kluwer, 1990. (b) Y Imamoglu, ed. Metathesis Polymerization of Olefins and Polymerization of Alkynes. Dordrecht: Kluwer, 1998.
9. M Biermann. In: Houben-Weyl Methoden der Organischen Chemie, 4th edn, vol. E20, Stuttgart: Georg Thieme-Verlag, 1989, p 830.
10. KJ Ivin. In: Encyclopedia of Polymer Science and Engineering, vol. 9. New York: John Wiley & Sons, 1987, p 634.
11. EA Ofstead. In: Encyclopedia of Polymer Science and Engineering, vol. 11. New York: John Wiley & Sons, 1987, p 287.
12. I Pasquon, L Porri, U Giannini. In: Encyclopedia of Polymer Science and Engineering, vol. 15. New York: John Wiley & Sons, 1987, p 662.
13. J Witte. In: Houben-Weyl Methoden der Organischen Chemie, 4th edn, vol. E20. Stuttgart: Georg Thieme-Verlag, 1989, p 134.
14. KJ Ivin. Tacticity in polymers initiated by metathesis polymerization. In: Y Imamoglu, B Zumreoglu-Karan, AJ Amass, eds. Olefin Metathesis and Polymerization Catalysts. Dordrecht: Kluwer, 1990.
15. R Streck. Industrial aspects of olefin metathesis/polymerization catalysts. In: Y Imamoglu, B Zumreoglu-Karan, AJ Amass, eds. Olefin Metathesis and Polymerization Catalysts. Dordrecht: Kluwer, 1990.
16. KJ Ivin. Olefin Metathesis. London: Academic Press, 1983.
17. JF Holohan, IY Pean, WA Vredenburgh. In: Kirk-Othmer Encyclopedia of Chemical Technology, vol. 12. New York: John Wiley & Sons, 1980, p 852.
18. W Barendrecht *et al.* In: Ullmanns Encyklopaedie der Technischen Chemie, 4th edn, vol. 12. Weinheim: Verlag Chemie, 1976, p 539.
19. W Risse, C Mehler. Paper presented at the 7th International Symposium on Homogeneous Catalysis, Lyon, France, 1990, Abstracts, p 168.
20. S Cesca *et al.* Polymer Lett 8: 573, 1970.
21. G Natta *et al.* Makromol Chem 81: 253, 1964.
22. G Dall'Asta. J Polymer Sci A1, 6: 2397, 1968.
23. FW Michelotti, WP Keaveney. J Polymer Sci A3: 895, 1965.
24. L Porri *et al.* Chim Ind (Milan) 46: 428, 1965.
25. KJ Ivin *et al.* Makromol Chem 180: 1989, 1979.

26. (a) KJ Ivin *et al.* J Mol Catal 15: 245, 1982. (b) KJ Ivin *et al.* J Mol Catal 28: 255, 1985.
27. RH Grubbs, W Tumas. Science 243: 907, 1989.
28. (a) TJ Katz *et al.* Tetrahedron Lett 47: 4247, 1976. (b) TJ Katz *et al.* Tetrahedron Lett 47: 4251, 1976.
29. RR Schrock. Acc Chem Res 23: 158, 1990.
30. J Feldman, WM Davis, RR Schrock. Organometallics 8: 2266, 1989.
31. LR Gilliom, RH Grubbs. J Am Chem Soc 108: 733, 1986.
32. RR Schrock, J Feldman, LF Cannizzo, RH Grubbs. Macromolecules 20: 1169, 1987.
33. ST Nguyen, LK Johnson, RH Grubbs, JW Ziller. J Am Chem Soc 114: 3974–3975, 1992.
34. PE Schwab, RH Grubbs, JW Ziller. J Am Chem Soc 118: 100, 1996.
35. M Weck, P Schwab, RH Grubbs. Macromolecules 29: 1789–1793, 1996.
36. WA Kormer, IA Poleyatova, TL Yufa. J Polymer Sci A1, 10: 251, 1972.
37. G Natta *et al.* Makromol Chem 69: 163, 1963.
38. G Natta *et al.* Makromol Chem 56: 224, 1962.
39. P Guenther *et al.* Angew Makromol Chem 14: 87, 1970.
40. (a) N Calderon *et al.* J Macromol Rev C7: 105, 1972. (b) N Calderon *et al.* J Polymer Sci A2, 5: 1283, 1967.
41. G Dall'Asta. Rubber Chem Technol 47: 515, 1974.
42. (a) WA Judy. Fr. Patent 2,016,360, 1970. (b) WA Judy. Fr. Patent 2,106,364, 1970. (c) WA Judy. Fr. Patent 2,029,747, 1970.
43. WA Judy, U.S. Patent 3,746,696, 1973.
44. EA Ofstead. In: Kirk-Othmer Encyclopedia of Chemical Technology, vol. 8, 3rd edn. New York: John Wiley & Sons, 1979, p 592.
45. P Guenther, W Oberkirch, G Pampus. Fr. Patent 2,065,300, 1969.
46. TL Yufa *et al.* Belg. Patent 770,823, 1972.
47. IA Oreshkin *et al.* Izv Akad Nauk SSSR, Ser Khim 1971: 1123.
48. (a) D Commereuc, Y Chauvin, D Cruypelnick. Makromol Chem 177: 2637, 1976. (b) Y Chauvin *et al.* Ger Offen 2,151,662, 1972.
49. J Kress *et al.* J Mol Catal 36: 1, 1986.
50. G Dall'Asta *et al.* Atti Acad Naz Lincei, Rend 41: 351, 1966.
51. TJ Katz *et al.* J Am Chem Soc 98: 606, 1966.
52. CC Wu, RW Lenz. Polymer Preprints (Am Chem Soc, Div Polym Chem) 12: 209, 1987.
53. TM Swager, RH Grubbs. J Am Chem Soc 109: 894, 1987.
54. F Hoffman. Chem Ztg 57: 5, 1933.
55. WM Kelley, NJ Taylor, S Collins. Macromolecules 27: 4477–4485, 1994.
56. WM Kelley, S Wang, S Collins. Macromolecules 30: 3151–3158, 1997.
57. HS Eleuterio. U.S. Patent 3,074,918, 1957.
58. G Natta, G Dall'Asta, G Mazzanti. Angew Chem 76: 765, 1964.
59. K Nuetzel *et al.* Ger. Patent 1,720,971, 1968.
60. G Pampus *et al.* Ger. Patent 1,954,092, 1969.
61. (a) R Streck, H Weber. Ger. Patent 2,028,716, 1971. (b) R Streck, H. Weber. Ger. Patent 2,028,935, 1971.
62. CA Uraneck, WJ Trepka. Fr. Patent 1,542,040, 1968.
63. J Witte *et al.* Ger. Patent, 1,957,026, 1971.
64. G Dall'Asta, G Carella. Br Patent 1,062,367, 1965.
65. W Oberkirch *et al.* Ger. Patent 1,811,653, 1970.
66. GI Schmidt, G Scheuerlich. J Polym Sci 49: 287, 1961.
67. J Boor *et al.* Makromol Chem 90: 26, 1966.
68. K Neutzel *et al.* Ger. Patent 1,720,797, 1968.
69. H Staudinger, HA Burson. Liebigs Ann Chem 447: 110, 1926.
70. JP Vairon, P Sigwalt. Bull Soc Chim Fr. 1964: 482.
71. C Aso *et al.* Chem High Polym 19: 734, 1962.
72. T Higashimura *et al.* Kobunshi Kogaku 23: 56, 1966.
73. RS Mitchell *et al.* Macromol Symp 1: 417, 1968.
74. H Maines, JH Day. Polymer Lett 1: 347, 1963.
75. AJ Amass *et al.* Eur Polymer J 12: 93, 1976.
76. M Farona, C Tsonis. J Chem Soc, Chem Commun 1977: 363.
77. PA Patton, TJ McCarthy. Macromolecules 19: 1266, 1986.
78. WJ Roberts, AR Day. J Am Chem Soc 72: 1266, 1950.
79. M Marek. Czeck Patent 88,879, 1959.
80. WJ Bailey, JC Grossens. Cited after Ref. 1.
81. W Marconi *et al.* Polymer Lett 2: 301, 1964.
82. G Lefebvre, F Dawans. J Polym Sci A2: 3297, 1964.
83. Y Imanishi *et al.* J Macromol Sci, Chem A3: 237, 1968.
84. Uniroyal Inc. Br. Patent 1,131,160, 1968.
85. LA Errede *et al.* J Am Chem Soc 82: 5224, 1960.
86. G Natta *et al.* Makromol Chem 91: 87, 1966.
87. HJ Katker. PhD Thesis. Rheinisch-Westfalische Technische Hochschule (RWTH), Aachen, 1988.
88. S Kohjiya *et al.* Makromol Chem 182: 215, 1981.
89. N Calderon *et al.* J Polymer Sci A1, 5: 2209, 1967.
90. E Wasserman *et al.* J Am Chem Soc 90: 3286, 1968.
91. N Calderon, M Morris. J Polym Sci A2, 5: 1238, 1967.
92. EA Ofstead. U.S. Patent 3,597,403, 1971.
93. DL Crain. U.S. Patent 3,575,947, 1971.
94. CPC Bradshaw. Br. Patent 1,252,799, 1971.
95. PR Marshall, BJ Ridgewell. Eur Polym J 5: 29, 1969.
96. G Dall'Asta, R Manetti. Eur Polym J 4: 154, 1968.
97. FW Kuepper, R Streck, K Hummel. Ger. Patent 2,051,798, 1972.
98. FW Kuepper, R Streck, HT Heims, H Weber. Ger. Patent 2,051,799, 1972.
99. TJ Katz *et al.* J Am Chem Soc 97: 1592, 1975.
100. VA Kormer *et al.* Ger. Patent 1,944,753, 1970.
101. TJ Katz *et al.* J Am Chem Soc 98: 7818, 1976.
102. M Mandel, RH Young. Eur Polym J 7: 523, 1971.
103. Yu V. Korshak *et al.* J Mol Catal 15: 207, 1982.
104. AJ Amass *et al.* Eur Polym J 12: 93, 1976.

105. W Holtrup, R Streck, W Zaar, D Zerpner. J Mol Catal 36: 127, 1986.
106. G Natta *et al.* Fr. Patent 1,425,601, 1966.
107. KW Scott *et al.* Rubber Chem Technol 44: 1341, 1971.
108. Yu V Korshak *et al.* Makromol Chem Rapid Commun 6: 685, 1985.
109. FL Klavetter, RH Grubbs. J Am Chem Soc 110: 7807, 1988.
110. R Rossi, R Giorgi. Chim Ind (Milan) 58: 517, 1976.
111. FW Kuepper, R Streck. Makromol Chem 175: 2025, 1974.
112. G Natta *et al.* Ital. Patent 733,857, 1964.
113. LM Vardanyan *et al.* Dokl Akad Nauk SSSR 207: 345, 1972.
114. R Wolovsky. J Am Chem Soc 92: 2132, 1970.
115. G Dall'Asta. Makromol Chem 154: 1, 1972.
116. H Hoecker *et al.* J Mol Catal 8: 191, 1980.
117. Y Chauvin *et al.* Recl Trav Chim Pays Bas 36: M131, 1977.
118. K Saito *et al.* Bull Chem Soc Japan 52: 3192, 1979.
119. T Saegusa *et al.* Kogyo Kagaku Zasshi 67: 1961, 1964.
120. JP Kennedy, HS Makowsky. J Macromol Sci A1: 345, 1967.
121. A Anderson, NC Merckling. U.S. Patent 2,721,189, 1955.
122. WL Truett, DR Johnson, IM Robinson, BA Montague. J Am Chem Soc 82: 2337, 1960.
123. G Sartori *et al.* Chim Ind (Milan) 45: 1478, 1963.
124. TJ Katz, SJ Lee, MA Shippey. J Mol Catal 8: 219, 1980.
125. RR Schrock. In: DJ Brunelle, ed. Ring-Opening Polymerization. New York: Hanser Publishers, 1993, pp 129–156.
126. (a) RR Schrock *et al.* J Am Chem Soc 112: 8378, 1990. (b) RR Schrock *et al.* J Am Chem Soc 114: 5426, 1992. (c) RR Schrock *et al.* J Am Chem Soc 114: 7324, 1992. (d) RR Schrock *et al.* J Am Chem Soc 115: 4413, 1993. (e) RR Schrock *et al.* J Am Chem Soc 116: 3414, 1994.
127. T Oshika, H Tabuchi. Bull Chem Soc Japan 41: 211, 1968.
128. RH Grubbs, W Risse, BM Novak. Adv Polymer Sci 102: 47, 1992.
129. RE Rinehart, HP Smidt. J Polymer Sci A3: 1049, 1965.
130. PE Schwab, MB France, JW Ziller, RH Grubbs. Angew Chem Int Ed Engl 34: 2039, 1995.
131. ST Nguyen, RH Grubbs, JW Ziller. J Am Chem Soc 115: 9858–9859, 1993.
132. C Mehler, W Risse. Makromol Chem Rapid Commun 12: 255–259, 1991.
133. G Sartori *et al.* Chim Ind (Milan) 45: 1529, 1963.
134. KJ Ivin *et al.* Makromol Chem Rapid Commun 1: 467, 1980.
135. LP Tenney, PC Lane. U.S. Patent 4,136,247, 1979.
136. HGG Dekking. J Polym Sci 55: 525, 1961.
137. Nippon Oil KK. Eur. Patent 325,260A, 1988.
138. HT Ho, KJ Ivin, JJ Rooney. J Mol Catal 15: 245, 1982.
139. JP Kennedy, JA Hinlicky. Polymer 6: 133, 1965.
140. G Dall'Asta, G Motroni. J Polym Sci A1, 6: 2405, 1968.
141. WJ Roberts, AR Day. J Am Chem Soc 72: 1266, 1950.
142. KJ Ivin *et al.* Pure Appl Chem 54: 447, 1982.
143. CT Thu *et al.* Makromol Chem Rapid Commun 2: 7, 1981.
144. SP Kolesnikov, NI Povarova. Izv Akad Nauk SSSR, Ser Khim 1979: 2398.
145. EA Ofstead, N Calderon. Makromol Chem 154: 21, 1972.
146. A Takada *et al.* J Chem Soc Japan 69: 715, 1966.
147. TM Swager, RH Grubbs, DA Dougherty. J Am Chem Soc 110: 2973, 1988.
148. IFAF El-Saafin, WJ Feast. J Mol Catal 15: 61, 1982.
149. (a) M Lautens, AS Abd-El-Aziz, J Reibel. Macromolecules 22: 4134–4136, 1989. (b) M Lautens, AS Abd-El-Aziz, G Schmidt. Macromolecules 23: 2821–2823, 1990.
150. S Cesca *et al.* Polymer Lett 8: 573, 1970.
151. T Oshika, H Tabuchi. Bull Chem Soc Japan 41: 211, 1968.
152. G Dall'Asta, G Motroni, R Manetti, C. Tossi. Makromol Chem 130: 153, 1969.
153. (a) A Bell. In: Polymeric Materials Science and Engineering, vol. 64. New York: American Chemical Society, 1991, p 102. (b) A Bell. J Mol Catal 76: 165–180, 1992.
154. RA Fisher, RH Grubbs. Makromol Chem, Macromol Symp 63: 271–277, 1992.
155. A Pacreau, M Fontanille. Macromol Chem 188: 2585, 1987.
156. J Hamilton, KJ Ivin, JJ Rooney. J Mol Catal 36: 115–125, 1986.
157. (a) BL Goodall, WJ Kroenke, RJ Minchak, LF Rhodes. J Appl Polymer Sci 47: 607, 1993. (b) LB Goodall, LH McIntosh, LF Rhodes. Macromol Symp 89: 421, 1995.
158. RG Foster, P Hepworth, cited after Ref. 1.
159. Mitsui Petrochemical Industries KK. Japan. Patent 253, 192, 1988.
160. Mitsui Petrochemical Industries KK. Japan. Patent 246,559, 1986.
161. Showa Denko KK. Japan. Patent 252,965, 1988.
162. (a) Japan. Synthetic Rubber. Japan. Patent 288,528, 1988. (b) Japan. Synthetic Rubber. Japan. Patent 288,527, 1988.
163. (a) Mitsui Petrochemical Industries KK. Japan. Patent 085,212, 1990. (b) Mitsui Petrochemical Industries KK. Japan. Patent 186,242, 1991.
164. L Mateika, C Houtman, W Makosko. J Appl Polym Sci 30: 2787, 1985.
165. (a) DS Breslow. Polym Prep (Am Chem Soc, Div Polymer Chem) 31: 410, 1990. (b) DS Breslow.

Chemtech 1980: 540. (c) DS Breslow. Polym Mater Sci Eng (Am Chem Soc, Div Polymer Chem) 58: 223, 1998.
166. (a) RJ Minchak (to BF Goodrich). U.S. Patent 4,002,815, 1977. (b) RJ Minchak, TJ Kittering, WJ Kroenke (to BF Goodrich). U.S. Patent 4,380,617, 1987. (c) RJ Minchak (to BF Goodrich). U.S. Patent, 4,426,502, 1984. (d) RJ Minchak, PC Lane (to BF Goodrich). U.S. Patent 4,701,510, 1987.
167. (a) DW Klosiewicz (to Hercules). U.S. Patent No 4,400,340, 1983. (b) DW Klosiewicz (to Hercules). U.S. Patent No 4,469,809, 1984. (c) DW Klosiewicz (to Hercules). U.S. Patent 4,657,981, 1987. (d) DW Klosiewicz. U.S. Patent 4,520,181, to Hercules, 1987.
168. (a) W Sjardijn (to Shell Oil Compnay), AH Kramer. U.S. Patent 4,729,976, 1986. (b) W Sjardijn, AH Kramer (to Shell Oil Company). U.S. Patent 4,810,762, 1987.
169. (a) PA Devlin, EF Lutz, RJ Pattan (to Shell Oil Company). U.S. Patent 3,627,739, 1971. (b) GM Tom (to Hercules). U.S. Patent 4,507,453, 1985. (c) A Bell (to Hercules). U.S. Patent 5,982,909, 1992.
170. EL Eliel. Stereochemistry of Carbon Compounds. New York: McGraw-Hill, 1962.
171. G Natta, G Dall'Asta, IW Bassi, G Carella. Makromol Chem 91: 87, 1966.
172. F Hoffman. Chem Ztg 57: 5, 1933.
173. MF Farona, C Tsonis. J Chem Soc, Chem Commun 1977: 363.
174. (a) KJ Ivin, G Lapienis, JJ Rooney, CD Stewart. Polymer 20: 1308, 1979. (b) PA Patton. PhD Thesis, University of Massachusetts, Amherst, 1988.
175. G Natta, G Dall'Asta. In: JP Kennedy, E Thornquist, eds. Polymer Chemistry of Synthetic Elastomers. New York: Wiley-Interscience, 1969, p 703.
176. FS Dainton *et al.* Trans Faraday Soc 51: 1710, 1955.
177. L Hocks, D Berek, AJ Hubert, P Teyssie. J Polym Sci, Polym Lett 13: 391, 1975.
178. (a) VM Cherednichenko. Polym Sci USSR A20: 1225, 1979. (b) L Hocks *et al.* Bull Soc Chim Fr. 1975: 1893. (c) A Drapeau, J Leonard. Macromolecules 18: 144, 1985.
179. FD Rossini *et al.* In: Selected Values of Physical and Thermodynamic Properties of Hydrocarbons and Related Compounds. Pittsburgh: Carnegie Press, 1953.
180. FS Dainton, KJ Ivin. Q Rev 12: 61, 1958.
181. EA Ofstead, N Calderon. Makromol Chem 154: 21, 1972.
182. UW Suter, H Hoecker. Makromol Chem 189: 1603, 1988.
183. KW Scott *et al.* Rubber Chem Technol 44: 1341, 1971.
184. H Hoecker *et al.* Recl Trav Chim Pays-Bas 96: M47, 1977.
185. H Jacobson, WH Stockmayer. J Chem Phys 18: 1600, 1950.
186. T Keii. Kinetics of Ziegler–Natta Polymerization. London: Chapman & Hall, 1972.
187. JP Vairon, P Sigwalt. Bull Soc Chim Fr. 1971: 559, 569.
188. Y Imanishi *et al.* J Macromol Sci, Chem A2: 471, 1968.
189. Y Imanishi *et al.* J Macromol Sci, Chem A: 223, 1969.
190. Y Imanishi *et al.* J Macromol Sci, Chem A3: 237, 1969.
191. MAS Mondal, RN Young. Eur Polym J 7: 523, 1971.
192. VA Hodjemirov *et al.* Vysokomol Soedin B14: 727, 1972.
193. (a) AJ Amass, CN Tuck. Eur Polym J 14: 817, 1978. (b) AJ Amass, JA Zurimendi. J Mol Catal 8: 243, 1980. (c) AJ Amass *et al.* In: Y Imamoglu, B Zumreoglu-Karan, AJ Amass, eds. Olefin Metathesis and Polymerization Catalysts. Dordrecht: Kluwer, 1990.
194. (a) KJ Ivin *et al.* Makromol Chem 180: 1989, 1979. (b) KJ Ivin *et al.* J Mol Catal 8: 203, 1980.
195. H Hoecker *et al.* J Mol Catal 8: 191, 1980.
196. Yu V Korshak *et al.* Recl Trav Chim Pays-Bas 96: M64, 1977.
197. JP Kennedy. J Macromol Sci, Polym Symp 56, 1976.
198. J Boor Jr. Ziegler–Natta Catalysts and Polymerizations. New York: Academic Press, 1979.
199. P Pino, R Mulhaupt. Angew Chem Int Ed Engl 19: 857, 1980.
200. KJ Ivin. In: Y Imamoglu, B Zumreoglu-Karan, AJ Amass, eds. Olefin Metathesis and Polymerization Catalysts. Dordrecht: Kluwer, 1990, p 192.
201. J Feldman, RR Shrock. In: SJ Lippard, ed. Progress in Inorganic Chemistry, vol. 39. New York: John Wiley & Sons, 1991, pp 1–66.
202. RH Grubbs, W Tumas. Science 243: 907, 1989.
203. KJ Ivin. In: Y Imamoglu, B Zumreoglu-Karan, AJ Amass, eds. Olefin Metathesis and Polymerization Catalysts. Dordrecht: Kluwer, 1990, p 197.
204. G Natta. Pure Appl Chem 12: 165, 1966.
205. (a) W Kaminsky. Angew Makromol Chem 223: 101, 1994. (b) W Kaminsky, A Noll. In: G Fink, R Mulhaupt, HH Brintzinger, eds. Ziegler Catalysts, Recent Scientific Innovations and Technological Improvements. Berlin: Springer, 1995. (c) W Kaminsky, A Bark, I Dake. In: T Keii, K Soga, eds. Catalytic Olefin Polymerization. Amsterdam: Elsevier, 1990, pp 425–438.
206. G Natta, G Dall'Asta, G Mazzanti, G Motroni. Makromol Chem 69: 163, 1963.
207. TJ Katz, J McGinnis, SJ Lee, N Acton. Tetrahedron Lett 1976: 4246.
208. P Guenther *et al.* Angew Makromol Chem 16/17: 27, 1971.
209. MA Golub, SA Fuqua, NS Bhacca. J Am Chem Soc 84: 4981, 1962.

210. (a) V Dragutan *et al.* Paper presented at the 9th International Symposium on Olefin Metathesis and Polymerization (ISOM 9), Collegeville, Pennsylvania, USA, July 21–26, 1991, Abstracts, p 19. (b) M Dimonie, S Coca, V Dragutan. J Mol Catal 76: 79–91, 1992. (c) V Dragutan, M Dimonie, S Coca. Polymer Preprints (Am Chem Soc, Div Polym Chem) 35: 698–699, 1994. (d) V Dragutan, S Coca, M Dimonie. In: Y Imamoglu, ed. Metathesis Polymerization of Olefins and Polymerization of Alkynes. Dordrecht: Kluwer, 1998, pp 89–102.
211. G Natta, G Motroni. Angew Makromol Chem 16/17: 51, 1971.
212. G Natta, G Dall'Asta, G Mazzanti. Angew Chem 76: 765, 1964.
213. G Dall'Asta, R Manetti. Eur Polym J 4: 145, 1968.
214. N Calderon, EA Ofstead, WA Judy. J Polym Sci A1, 5: 2209, 1967.
215. KJ Ivin, DT Laverty, JJ Rooney. Makromol Chem 178: 1545, 1977.
216. T Oshika, H Tabuchi. Bull Chem Soc Japan 411: 211, 1968.
217. G Dall'Asta. Makromol Chem 154: 1, 1972.
218. TJ Katz, J McGinnis, C Altus. J Am Chem Soc 98: 606, 1970.
219. TJ Katz, SJ Lee, MA Shippey. J Mol Catal 8: 219, 1980.
220. JF Holohan, IY Pean, WA Vredenburgh. In: Kirk-Othmer Encyclopedia of Chemical Technology, vol. 12. New York: John Wiley & Sons, 1980, p 861.
221. W Barendrecht *et al.* In: Ullmanns Encyklopaedie der Technischen Chemie, 4th edn, vol. 12. Weinheim: Verlag Chemie, 1976, p 544.
222. EA Ofstead. In: Kirk-Othmer Encyclopedia of Chemical Technology, vol. 8, 3rd edn. New York: John Wiley & Sons, 1979, p 602.
223. A Draxler. Kaut Gummi Kunst 34: 185, 1981.
224. (a) KM Diedrich, G Huhn, K Zur Nedden. Paper presented at the Elastomery '90 Scientific Conference, Piastow, Poland, October 15–17, 1990. (b) K Zur Nedden, KM Diedrich, G Huhn. Paper presented at Rubber Division, American Chemical Society, Detroit, Michigan, October 17–20, 1989. (c) Huls AG, An Unusual Rubber with Versatile Possibilities-Vestenamer, 1st edn. Marl, Germany, August 1989.
225. A Draxler. Elastomerics 16, Feb. 1983.
226. Plast Rubber Wkly 20: 1030, 1985.
227. P Le Delliou. Hule Max Plast 33: 17, 20, 1977.
228. P Le Delliou. Int Rubber Conf 1977: 1.
229. (a) Chem Week Oct. 9, 1985. (b) Plast World, May 1985.
230. LP Tenney. Paper presented at the ANTEC '82, San Francisco, California, USA.
231. R Streck. Industrial aspects of olefin metathesis/polymerization catalysts. In: Y Imamoglu, B Zumreoglu-Karan, AJ Amass, eds. Olefin Metathesis and Polymerization Catalysts. Dordrecht: Kluwer, 1990.
232. A Bell. In: Polym Mater Sci Eng. 64: 102, 1991.
233. DS Breslow. Polym Mater Sci Eng. (Am Chem Soc, Div Polymer Chem) 58: 223, 1998.
234. (a) A Bell. Paper presented at the 9th International Symposium on Olefin Metathesis and Polymerization, Collegeville, Pennsylvania, USA, July 21–26 (1991), Abstracts, p 40. (b) A Bell. J Mol Catal 76: 165–180, 1992.

5

Synthesis of Copolymers from Cycloolefins

Valerian Drăguţan
Romanian Academy, Institute of Organic Chemistry, Bucharest, Romania

Roland Streck*
Hüls AG, Marl, Germany

I. INTRODUCTION: REACTION TYPES

Copolymers of cycloolefins are readily obtained by catalytic copolyermization reactions under appropriate conditions. They can be random, block, or graft copolymers depending on the monomer reactivity, catalyst nature, polymerization procedure, or reaction parameters and they may display physical and mechanical properties totally different from those of the corresponding homopolymers.

Two major types of copolymer arise from the copolymerization of two or more cycloolefins: vinylic and ring-opened copolymers, essentially as a function of the catalyst employed. Vinylic copolymers result via addition polymerization in the presence of cationic initiators [1] or anionic–coordinative catalysts of the Ziegler–Natta type [2]. In this case, mainly saturated repeat units (**1**) prevail in the polymer chain:

(1)

(1)

Ring-opened copolymers, called copolyalkenamers or copolyalkenylenes, are formed by the metathetic copolymerization of cycloolefins in the presence of metathesis catalysts [3, 4]. This process involves cleavage of the carbon–carbon double bond in the ring, and unsaturation is preserved in the copolymer chain, (**2**):

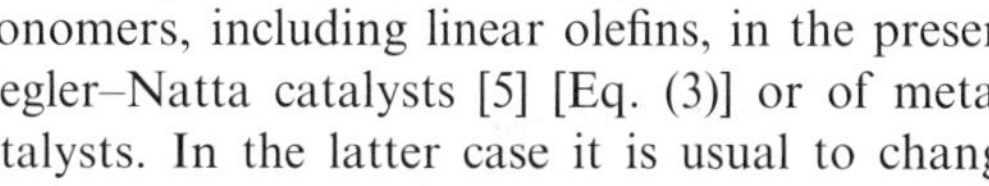

(2)

(2)

Block copolymers can be obtained by copolymerization of cycloolefins of entirely different reactivities or by applying adequate sequential addition of the monomer. They also arise from cycloolefins and vinylic monomers, including linear olefins, in the presence of Ziegler–Natta catalysts [5] [Eq. (3)] or of metathesis catalysts. In the latter case it is usual to change the reaction mechanism to Ziegler–Natta [6] and group transfer polymerization [7] or from anionic–coordinative to metathesis polymerization [8] [Eq. (4)].

(3)

(3)

*Retired.

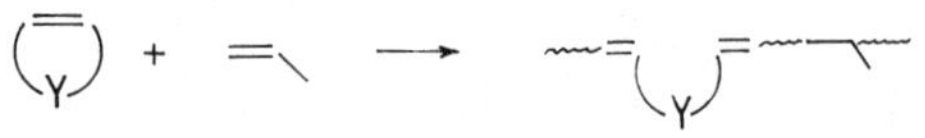

(4)

(4)

In addition, block copolymers of cycloolefins and unsaturated polymers are obtained when cycloolefins are polymerized in the presence of these polymers by a vinylic or metathetic route:

(5)

(5)

(6)

(6)

If unsaturation occurs in the branched arm of the polymer that is to be blocked (e.g., 1,2-polybutadiene), graft copolymers (**7**) and (**8**) will arise by vinylic and metathesis pathways, respectively:

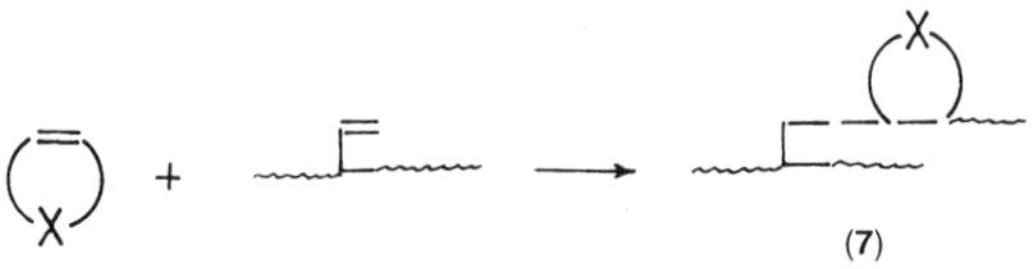

(7)

(7)

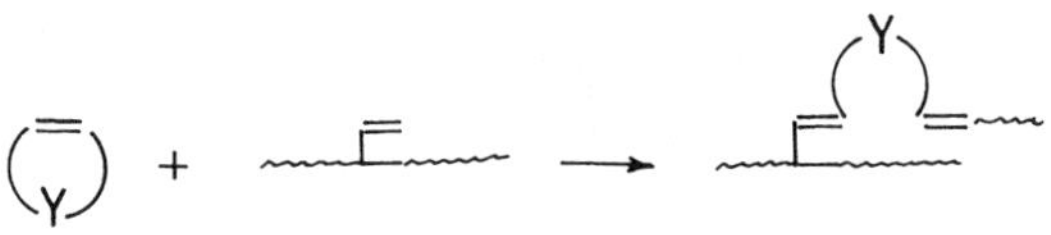

(8)

(8)

II. COPOLYMERS BY VINYLIC POLYMERIZATION

Vinylic copolymers derived both from cycloolfins and from a cycloolefin and an acyclic olefin are known. These are obtained by using cationic initiators or anionic–coordinative catalysts of the Ziegler–Natta type.

A. Cationic Polymerization

Relatively few data are published concerning the polymerization of cyclobutene, cyclopentene, cyclohexene, or cyclooctene and their substituted derivatives in the presence of cationic initiators [1]. Cyclopentene and cyclohexene proved to be unreactive toward $AlCl_3$ at low temperatures (−20°C) in ethyl chloride, but they give dimers, trimers, and tetramers as well as higher polymers with BF_3 and hydrogen fluoride [9, 10]. On the other hand, copolymerization of cyclohexene with styrene and α-methylstyrene in the presence of $AlCl_3$ or an $AlCl_3$/*t*-BuCl complex [11] gives low molecular weight copolymers, probably having structures of type (**9**) and (**10**), in high yield. In the case of the $AlCl_3$/*t*-BuCl complex, the yield and molecular weight are significantly dependent on the molar ratio of the catalytic components. When the proportion of the *t*-BuCl is increased, the copolymer molecular weight decreases considerably, in contrast to the case of the homopolymerization of cyclohexene.

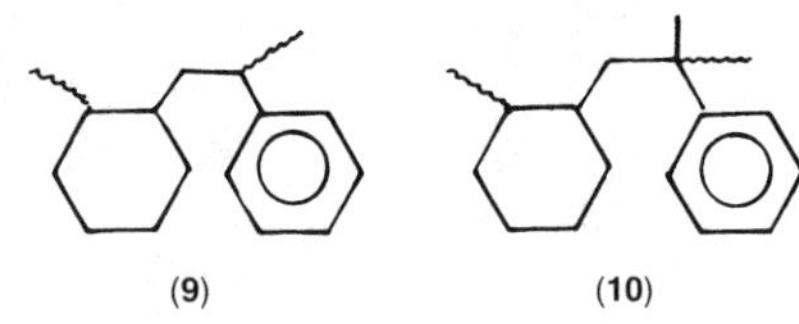

(9) (10)

Cyclopentadiene proved to be highly reactive in cationic copolymerization with acyclic olefins. The reaction with isobutene and α-methylstyrene in the presence of different cationic initiators, such as Et_3Al/*t*-BuCl, Et_2AlCl/*t*-BuCl, $TiCl_4/AlBr_3$, $AlCl_3$, and $AlBr_3$, has been reported to yield copolymers containing repeat units (**11**) and (**12**), respectively [12, 13].

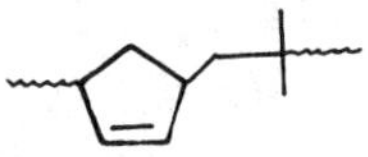

(11) 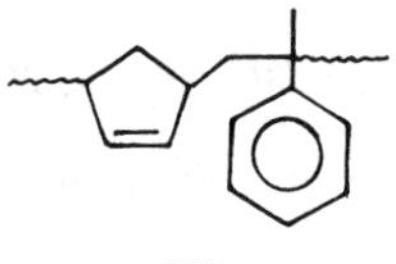 (12)

For instance, random copolymers have been obtained in high yields from cyclopentadiene and α-methylstyrene in toluene and methylene chloride with $TiCl_4/AlBr_3$ (molar ratio 60:1). The random distribution of the repeat units in the copolymer chain indicates that the two monomers are similar in reactivity. In this reaction, the author found the catalyst activity to be in the following order: Et_2AlCl/*t*-BuCl > $AlCl_3$ > $AlBr_3$ > $TiCl_4/AlBr_3$.

The copolymerization of cyclopentadiene with 1- and 2-methylcyclopentadiene has been reported [14] in toluene and methylene chloride in the presence of BF_3/OEt_2, $SnCl_4$, and $TiCl_4$; products having structures (**13**) and (**14**) are formed in the polymer chain.

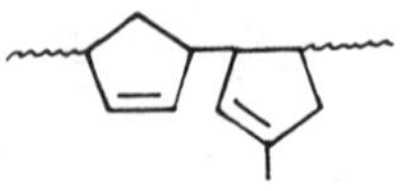

(13) 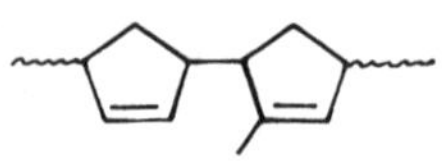 (14)

The reactivity ratios, r_{Me}/r were markedly in favor of the methyl-substituted compound, depending on the solvent employed (e.g., under certain conditions r_{Me}/r was 8.5:0.36 in toluene and 14.9:0.42 in methylene chloride). However, the nature of the cationic initiator had little effect on copolymer composition. The enhanced reactivity of the methylcyclopentadiene over that of the unsubstituted monomer was attributed to the inductive effect of the methyl group, particularly when the substitutent is in the position 2.

The copolymerization of cyclopentadiene with 1,3-dimethylcyclopentadiene [15] in toluene at −78°C, induced by BF_3/Et_2O, gave products probably having repeat units (**15**) and (**16**).

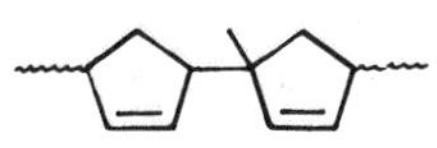

(15) 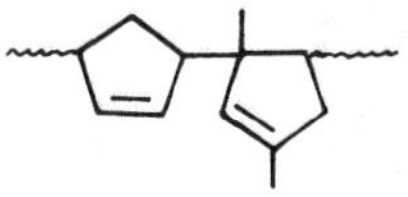 (16)

The reactivity ratio r_{13}/r of 6.85:0.30 was heavily in favor of the 1,3-disubstituted monomer; it is obvious that 1,3-disubstitution enhances the stability of the cyclopentadienyl cation. These products displayed remarkable properties, totally different from those of the homopolymers. For instance, (1) they can be dissolved in benzene and the solution can be cast to yield transparent films, (2) the softening points of 150–155°C are somewhat higher than those of the homopolymers, and (3) the rate of oxidation and oxygen absorption upon exposure to air are found to be slower. These particular properties were attributed to the decreasing number of tertiary allylic structures in the copolymer chain as compared to the respective homopolymers.

Reaction of 1-methylcyclopentadiene with 2-methyl isomer [14] in the presence of the catalyst above led to copolymers having repeat units (**17**)–(**20**).

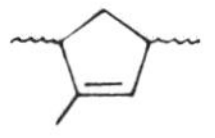

(17) 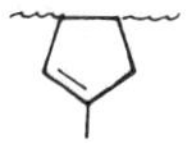 (18)  (19) 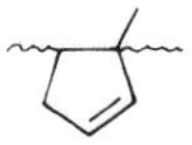 (20)

The reactivity ratios r_1/r_2 of 0.01:1.1 were strongly in favor of the 2-methylcyclopentadiene isomer. Furthermore, examination of the product microstructure by IR and NMR spectroscopy indicated that the most important contributing repeat unit was (**17**), resulting from 2-methylcyclopentadiene by 1,4 addition. The repeat units (**18**) and (**19**) resulted from 1- and 2-methylcyclopentadiene, respectively, by 3,4 addition, and structure (**20**) was virtually absent.

Ternary copolymers starting from a mixture of 1-methyl-, 2-methyl-, and 5-methyl-cyclopentadiene (1-/2-/5-isomer = 45:52:3) were obtained using BF_3/Et_2O, $SnCl_4/CCl_3COOH$ (TCA), and $TiCl_4/CCl_3COOH$ as the initiators [16]. With $TiCl_4$/TCA and $SnCl_4$/TCA, the reactions were typically nonstationary: rapid termination soon followed strong initiation. Fresh initiator was needed to attain high conversions of the monomers. In contrast, when BF_3/Et_2O was used as initiator, after a rapid initiation, a stationary phase was reached that slowly progressed until the monomers were consumed. Evidently, the reactivity ratios of these monomers differed essentially with the position of the substituent, the most reactive being the 2-substituted isomer.

The formation of random copolymers by the reaction of cyclopentadiene with 2-chloroethyl vinyl ether at −78°C in various solvents using BF_3/Et_2O as the catalyst has been reported [17]. Examination of the copolymer microstructure was carried out by NMR spectroscopy, but details on its structure were not given. Apparently because of the difference in reactivity of the two monomers, repeat units (**21**) do not form a real random copolymer under the conditions studied. Data concerning the reactivity ratios and physical properties of the sulfur vulcanizates can also be found for this copolymer.

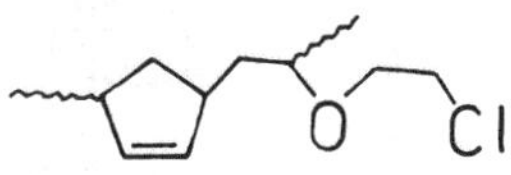

O Cl

(21)

Copolymer of cyclopentene and cyclopentadiene with α-methylstyrene and piperylene are frequently encountered in the synthesis of C_5 aliphatic petroleum resins using the following catalysts: $AlCl_3$, $AlBr_3$, BF_3/Et_2O, $SnCl_4$, and $TiCl_4$ [18]. In addition, mixtures of C_5 streams containing cyclopentene and cyclopentadiene are copolymerized with styrene and vinyltoluene to produce resins that have variable properties, depending on the reaction conditions: Et_2AlCl, $Et_3Al_2Cl_3$, and $EtAlCl_2$ in the presence of *t*-BuCl were used as initiators. Furthermore, terpene hydrocarbons such as α-pinene and dipentene are currently copolymerized with isobutene, isoprene, styrene, and terpene oligomers in the presence of $AlCl_3$ or $EtAlCl_2$ to manufacture useful resins with special properties.

Dicyclopentadiene, a monomer with a particularly high reactivity, has been copolymerized both thermally and catalytically (e.g., $AlCl_3$) with styrene, indene, α-pinene, and various unsaturated hydrocarbon fractions to yield copolymers of hydrocarbon resins having improved properties [18, 19]. The copolymers obtained from cyclopentadiene and styrene or indene probably contain repeat units of the types (**22**) and (**23**):

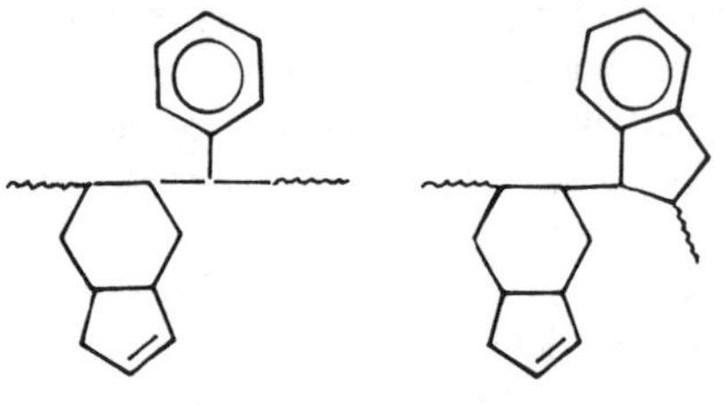

(22) (23)

but other structures may also occur as a result of the possibilities for isomerization reactions of the cyclopentadiene skeleton under the influence of the cationic initiators.

B. Ziegler–Natta Polymerization

Whereas copolymers of two or more cycloolefins via Ziegler–Natta polymerization are seldom reported [2], an abundant literature covers the copolymerization of cycloolefins with linear olefins using this type of catalyst. This mode of introducing one or more acyclic repeat units into the polymer chain formed from cycloolefins will effect drastic changes in the physicomechanical properties of these polymers, enabling them to be applied on a large scale in various areas.

Early reports [20] refer to the copolymerization of several cycloolefins such as cyclobutene, cyclopentene, cyclohexene, cycloheptene, and cyclooctene with ethylene, propylene, isobutene, and other lower linear olefins, using catalytic systems prepared from transition metal salts of groups IV–VI and organometallic compounds of groups I–III of the periodic table. Linear thermoplastic polymers containing up to 50 mol% cycloolefin alternating with polyolefin units were obtained and characterized by fractionated extraction, IR and X-ray spectroscopy, and other methods. An illustrative example is the copolymerization of cyclobutene with [^{14}C]ethylene in the presence of $V(acac)_3$ and Et_2AlCl. When the reaction was conducted at −60°C, a powdery, white thermoplastic copolymer was obtained in 88% yield, bearing structural units (**24**):

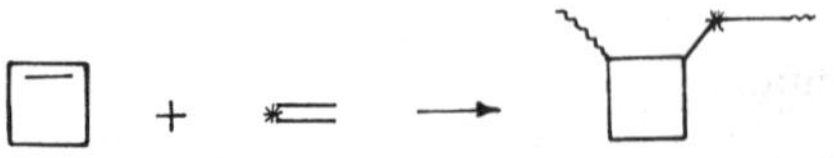

(9)

(24)

The reactivity of various cycloolefins in copolymerization reactions with ethylene and 2-butene has been examined using $V(acac)_3/Et_2AlCl$ and VCl_4/Hex_3Al as the catalysts [21]. It was observed that, while the catalyst system has little effect on the relative reactivity of the cycloolefin, the nature of the cyclic and acyclic monomer proved to be quite determinant for reaction kinetics and stereospecificity. In this respect, cyclopentene and cycloheptene displayed high reactivity compared to cyclohexene and cyclooctene, while *cis*-2-butene was more reactive than *trans*-2-butene. These observations were rationalized by considering steric factors induced by monomers rather than catalyst activity and specificity.

Copolymers of norbornene with ethylene and/or lower linear olefins having repeat units (**25**):

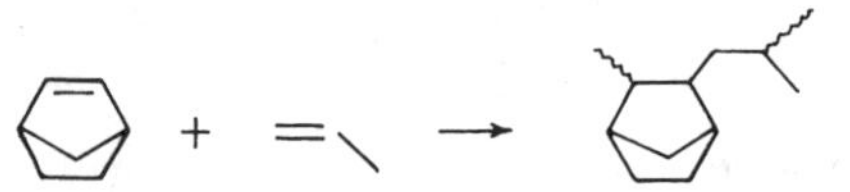

(10)

(25)

which were obtained in high yield with Ziegler–Natta catalysts, exhibit improved chemical, mechanical, optical, and electric properties and find wide application as such or in blends with polyolefins or elastomers, especially in the electrical, electronic, fine mechanical, and optical fields. Numerous catalysts based on salts of titanium and vanadium and organometallic compounds were claimed for this reaction, but best results were recorded with $TiCl_4$, $TiCl_{4-n}(OR)_n$, or $TiCl_3$ in conjunction with organoaluminum compounds of the R'_3Al or $R'_{3-m}AlCl_m$ type. According to some procedures using these systems [22–24], it is preferable to add the catalytic components at −30° to +60°C; the norbornene concentration during catalyst formation is between 0.25 and 5.00 mol/L, and the catalyst is aged at 20–70°C for 15–180 min. The process takes place in a liquid inert medium (e.g., aliphatic hydrocarbons or benzines poor in aromatics and cycloaliphatics) at temperatures and pressures required to control the desired monomer ratio in the copolymers and the physicomechanical properties of the product. As dielectric materials for capacitors in electrical and electronic engineering, ethylene–norbornene copolymers impart better resistance to aging, better stability of electrical parameters, improved dielectric strength and low dielectric loss at high frequencies, and high voltage resistance over a range of times and temperatures [25]. It is of interest that blends of norbornene copolymers with a great number of elastomers form high impact compositions with high heat resistance, a property that is improved by fiber or particulate reinforcement and crosslinking. Multiple compositions having improved mechanical properties are also obtained from norbornene copolymers with aliphatic and/or aromatic polyamides, elastomers, and various reinforcing agents [26]. Novel copolymer compositions contain ethylene–norbornene polymers and thermoplastics (e.g., polyethylene, polyamides, polyethers) with particulate reinforcing materials (PRM), bonding agents, crosslinking agents, glass fibers, lubricants, stabilizers, and pigments [27].

Saturated terpolymers of structure (**26**), obtained from norbornene or other bicyclic and polycyclic olefins with ethylene and higher α- or internal olefins, display variable properties depending on the nature and ratio of the monomers.

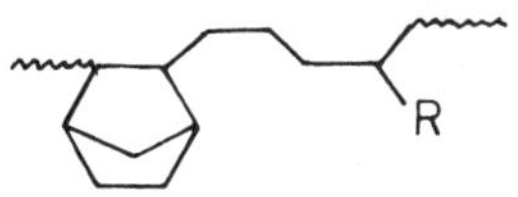

R

(**26**)

Efficient procedures yield high molecular weight, partially crystalline copolymers with improved green strength from norbornene and ethylene with C_4–C_{10} α-olefins in the presence of $VOCl_3$, $Et_3Al_2Cl_3$/methylperchlorocrotonate catalyst [28]. In one example, a copolymer containing 3.35% norbornene, 3.35% propylene, and 93.3% ethylene had a green strength of 138 kg/cm^2, elongation of 586%, 100% modulus at 90 kg/cm^2, and Shore hardness of 92 at 22°C, compared to the values of 87 kg/cm^2, 507%, 63 kg/cm^2, and 93, respectively, for a control ethylene–propylene copolymer containing 10% propylene.

Copolymers of norbornene and 1-butene, isobutene, or 2-butene (**27**–**29**) are preferably obtained using $TiCl_3$, $TiCl_4$, $Ti(OR)_2X_2$, or $Ti(OR)_3X$ associated with Et_2AlCl, Et_2AlBr, or Me_2AlCl in the presence of electron-donating agents such as amines, ethers, and carbonyl-, phosphor-, or sulfur-containing compounds [29a].

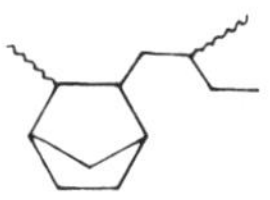

(**27**) 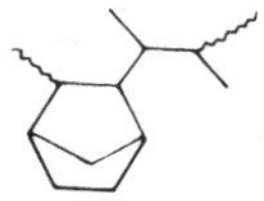 (**28**) 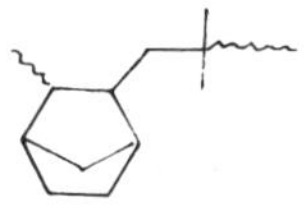 (**29**)

The role of the electron-donating component is essential in determining both the monomer reactivity and the copolymer structure. The physical and mechanical properties of these products may be altered gradually by modifying the catalyst composition within desired limits. Other norbornene derivatives, such as compounds (**30**–**32**), find valuable applications in copolymers with ethylene and/or C_3–C_8 α-olefins to produce materials with interesting properties.

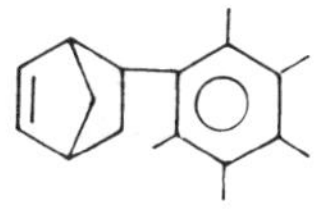

R'X R''X R'X R''X

(**30**) 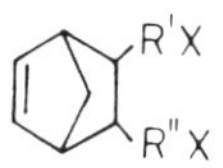 (**31**) 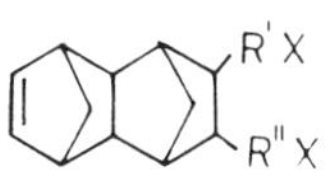 (**32**)

For instance, compound (**30**) enables the production of plastics with improved light transmissibility and dielectric parameters, thermochemical resistance, and mechanical workability [29b]. The comonomers for these products are selected from 1-butene, 3-methyl-1-butene, 1-pentene, 1-hexene, 1-decene, styrene, and α-methylstyrene, and the catalysts consist of organoaluminum compounds (e.g., Me_3Al, Et_2AlCl, $Et_3Al_2Cl_3$) and vanadium salts (e.g., VCl_4, VBr_4, $VOCl_3$, $VOBr_3$, $VO(OR)_3$).

Various monomers containing norbornene and norbornadiene structure have been employed in copolymerization reactions to obtain new products with improved physico-chemical and mechanical properties [30]. Synthetic rubber with unsaturated units (**33**) is manufactured by copolymerizing ethylene and propylene with dicyclopentadiene in inert hydrocarbon solvents using VCl_4 or $VOCl_3$ associated with R'_2AlCl or $R'AlCl_2$ as the catalysts [30a,b].

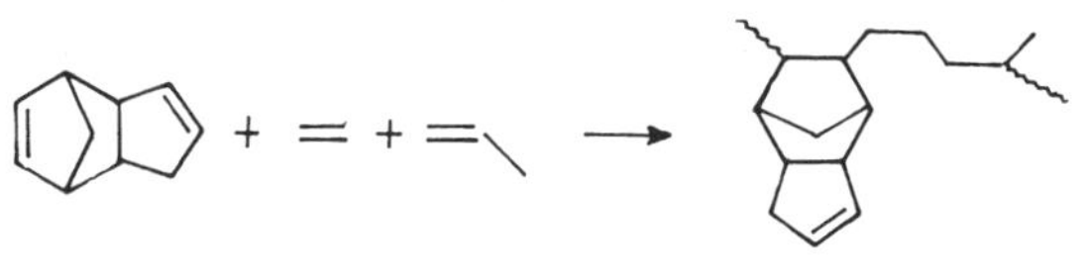

(**33**)

(11)

Extending the α-olefin to $RCH{=}CH_2$, where R is a C_2–C_6 alkyl group, one obtains copolymers with superior processibility and physical properties, including high creep resistance and low cold flow [30c,d]. Many other cyclic and polycyclic dienes have been copolymerized with ethylene, propylene, and various α-olefins using conventional Ziegler–Natta catalysts. Of these, compounds (**34**–**40**) yield products displaying attractive physico-mechanical parameters [30e–l].

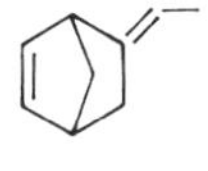

(**34**) 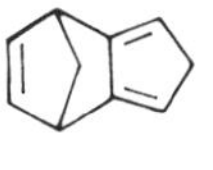 (**35**)  (**36**) 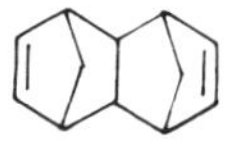 (**37**)

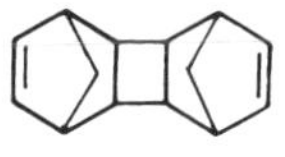

CH_2 $(CH_2)_n$

(**38**) 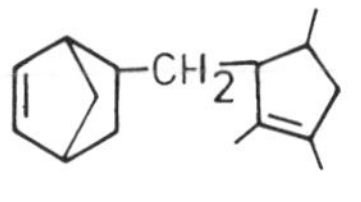 (**39**) 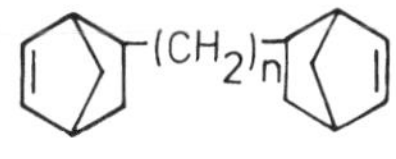 (**40**)

Remarkably, very active catalysts for the copolymerization of bicyclic and polycyclic olefins with α-olefins are derived from transition metal metallocenes and cyclic or linear aluminoxanes [31]. Cycloolefins (**41–45**) are reported to undergo high yield polymerization with linear olefins using metallocenes of titanium, zirconium, vanadium, or chromium associated with aluminoxanes [31a].

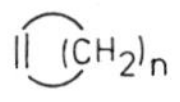

(41) 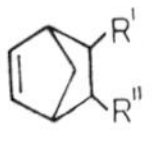 (42) 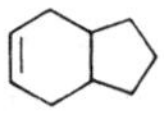 (43) 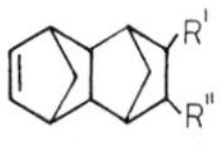 (44) 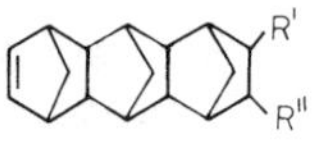 (45)

The literature describes several cases of the copolymerization of ethylene with cyclopentene [31b], cycloheptene [31b], norbornene [31c], 2-methylnorbornene [31c], dicyclopentadiene [31d], dimethanooctahydronaphthalene [31a,e], and 2-methyldimethanooctahydronaphthalene [31a]. When chiral catalysts were employed (e.g., ethylenebis(indenyl)zirconium dichloride associated with methylaluminoxane), the activity was much higher than that of the achiral catalyst. In addition, the chiral catalyst increased the polymer stereoselectivity and the amount of the cycloolefin incorporated into the copolymer. Materials having excellent transparency, thermal stability, and chemical resistance—properties that are useful for optical disks—may be produced efficiently by this method.

A significant number of patents describe procedures for the synthesis of various copolymers of polycyclic olefins of the general formulas (**46–49**); in each case, ethylene or α-olefins are used in the presence of different Ziegler–Natta catalysts [32].

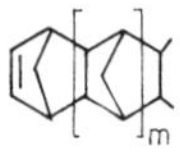

(46) 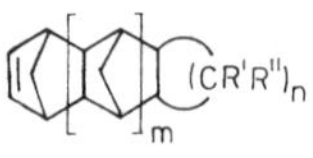 (47) 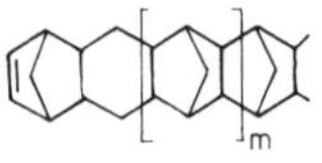 (48) 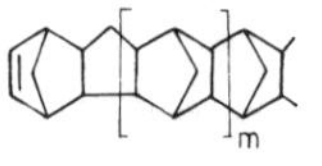 (49)

These norbornenelike monomers, easily obtained by the Diels–Alder route, display a fairly high reactivity, which leads to copolymers exhibiting random to blocky distribution of the structural units of type (**50**) or (**51**), depending on the particular structure and the substituent on the monomer.

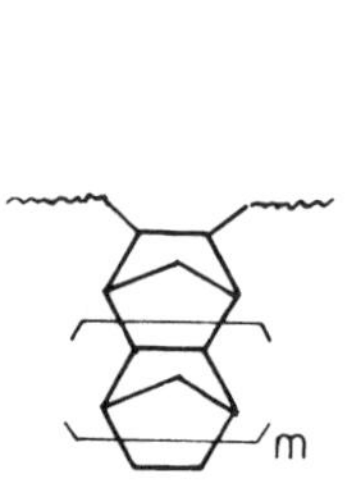

(50) 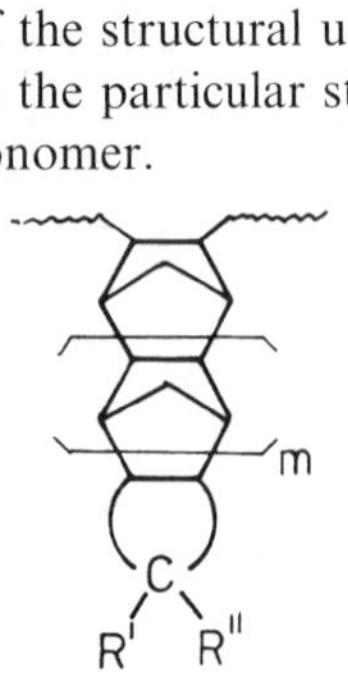 (51)

It is noteworthy that the products thus prepared possess remarkable physico-chemical properties (glass transition point; mechanical, electrical, and optical characteristics), as evidenced by excellent transparency, good moldability, and high resistance to heat, aging, solvents, and weathering. Furthermore, they are compatible with many polymers, including polyesters, polycarbonates, polyamides, and polyolefins, and find broad application in the manufacture of optical devices, photo disks, circuit boards for crystalline liquids, printed circuit boards, special electrical and electronic devices, and so on.

Copolymers containing functional groups [33] are produced by copolymerizing ethylene and C_3–C_{18} α-olefins with compounds of the general formulas (**52**) and (**53**), where X = OR, COOR, CRO, NH_2, NR_2, $CONR_2$, CN, OH.

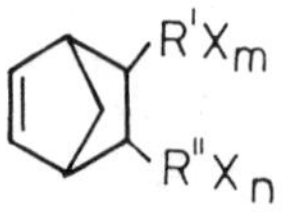

(52) 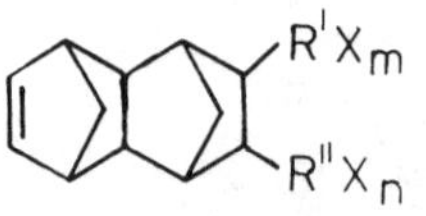 (53)

In one process [33a], the monomer is preferentially contacted first with at least an equimolar amount of an organometallic compound and/or a halogen-containing metal salt; then it is fed into the reaction system comprising a soluble vanadium compound (e.g., $VO(OEt)Cl_2$) and an organoaluminum compound (e.g., $Et_3Al_2Cl_3$). The products thus obtained have excellent resistance against ozone, weathering, and heat, as well as high adhesion to metals and good compatibility with other resins and elastomers. They are usable as modifiers for resins and adhesives, as additives to lubricating oils, and in the manufacture of paint, inks, and so on. Interesting copolymers are prepared from norbornene derivatives containing silicon groups by reacting with ethylene and/or C_3–C_{20} monoolefins or polyolefins [34]. For instance, with a catalyst consisting of $VOCl_3$ and $Et_3Al_2Cl_3$, a terpolymer of 7-isopropylidenyl-5-trichlorosilyl-2-norbornene with ethylene and propylene, (**54**), was prepared in high yield, working in hexane by continuous addition of the monomers:

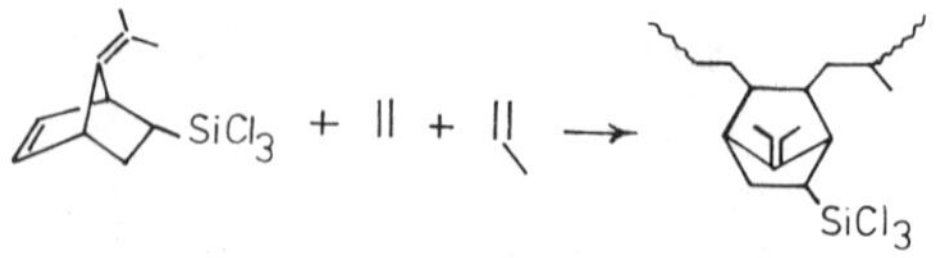

(12)

(54)

Such a product can be cured or subjected to sulfur vulcanization, and it displays special elastomeric properties.

III. COPOLYMERS FORMED BY RING-OPENING METATHESIS POLYMERIZATION

A. Random Copolymers

Unless special procedures are employed, cycloolefins formed by ring-opening copolymerization in the presence of classical metathesis catalysts derived from WCl_6 or $MoCl_5$ and organometallic compounds lead mainly to random copolyalkenamers. The composition and structure of the copolymers, as well as their elastomeric properties, depend greatly on the nature of the cycloolefin, as well as its reactivity, the reaction parameters, and the nature of the catalyst [3, 4]. In many cases, the catalytic systems exhibit significant differences in their activity toward cycloolefins, regardless the reactivity of the latter. On the other hand, the cycloolefins show quite different reactivities with a particular catalyst at a given temperature, which results in pronounced inhomogeneities in the composition or structure of the copolymer. Since differences always exist between the reactivities of the cyclic olefins as a result of their distinct electronic and steric environment, it is difficult in practice to obtain copolymers with ideal random distribution.

Cyclopentene was reported to copolymerize with cycloheptene in the presence of WCl_6- or $WOCl_4$-based catalysts [35], yielding copolymers with structure (**55**), having a high trans content of the double bonds in the chain:

(**55**)

(13)

Similar copolymers (**56**) and (**57**) have been reported in the copolymerization reactions of cyclopentene with cyclooctene [35] and of cyclooctene with cyclododecene [36] in the presence of the foregoing catalytic systems:

(**56**) (14)

(**57**)

(15)

Copolymer composition, cis/trans ratio of the double bonds, and inherent viscosities of copolymers obtained at varying conversions using different monomer mixtures [37] are presented in Table 1. It is evident from a comparison of the molar ratios of the monomers and the monomeric units in the chain that the reactivity of lower cycloolefins is considerably higher. Although the trans content of the copolymers is not in all cases as high as in homopolymers made with the same catalyst, all copolymers, even those that are

TABLE 1 Copolymerization of Cycloolefins with the Catalytic System $WOCl_4$/Benzoyl Peroxide/Et_2AlCl*

Monomers	Monomer composition (molar ratio)	Copolymer conversion (%)	Copolymer composition (molar ratio)	cis:trans ratio	Inherent viscosity (dL/g)
Cyclopentene–cycloheptene	C_5:C_7				
	50:50	45	72:28	25:75	1.4
	25:75	37	44:56	20:80	1.0
	10:90	44	18:82	20:80	1.1
Cyclopentene–cyclooctene	C_5:C_8				
	50:50	13	97:3	30:70	2.5
	25:75	10	53:47	35:65	3.1
	10:90	13	13:87	25:75	3.0
Cyclooctene–cyclododecene	C_8:C_{12}				
	80:20	23	85:15	20:80	1.4
	50:50	12	69:31	20:80	1.3
	5:95	17	14:86	30:70	1.6

*Molar ratios: monomer/$WOCl_4$/benzoyl peroxide/Et_2AlCl = 500:1:1:2.5.
Source: Data from Ref. 37.

amorphous in the unstretched state at room temperature, are crystallizable upon stretching and annealing. The molecular weights of these copolymers are as high as those of the corresponding homopolymers, which would not significantly influence the melting temperature or crystal structure. The crystallization temperatures are generally 20–30°C below the melting temperature and, consequently, the rate of crystallization is low compared to that of the corresponding homopolymers. This phenomenon indicates that the copolymerization technique is of interest for obtaining elastomeric copolyalkenamers from monomers, the homopolymers of which are thermoplastic materials.

Structural investigation by X-ray methods of the foregoing copolymers [37] showed that, for the range of composition in which one component is predominant, the structure of the copolymer is the same as in the corresponding homopolymer in its more stable modification. In C_5–C_7 copolymers, orthohombic geometry characterizes all compositions, while in the C_5–C_8 copolymers the product has an orthorhombic symmetry in the range of prevalence of the C_5 units and a triclinic symmetry when C_8 units prevail. In the C_8–C_{12} copolymers, for the region in which C_8 units predominate, the product has triclinic symmetry, whereas when C_{12} units prevail the structure is monoclinic. In the intermediate composition range of 1:1, the degree of crystallinity is comparatively low and the X-ray pattern resembles that of orthorhombic polyethylene.

The results of studies on the copolymerization of labeled cycloolefins have been successfully applied to the elucidation of the basic mechanism of metathesis polymerization [38]. In this regard, reaction of cyclooctene with cyclopentene, isotopically labeled at one carbon atom of the double bond, in the presence of the catalytic system $WOCl_4/Et_2AlCl$/benzoyl peroxide, proved that the process occurred by the cleavage of the carbon–carbon double bond:

(58)

(16)

Radio–gas chromtographic analyses of the acetylated diols, resulting in degradation of the copolymer (**58**), indicated that all the radioactivity was found in pentandiol, whereas octandiol was nonradioactive. Analogously, copolymerization of isotopically labeled cyclobutene with 3-methylcyclobutene, in the presence of tungsten, molybdenum, titanium, and ruthenium-based catalysts, proved that the cleavage occurs at the double bond in the ring [39]. Copolymers with structures (**59**), bearing the methyl group in the initial position, were produced by this route:

(59)

(17)

The yield and structure of the resulting copolyalkenamer depended markedly on the catalytic system employed (Table 2). The $TiCl_4/Et_3Al$ catalyst was shown to be the most active and the least stereoselective. The ratios of the structural units butenamer to 3-methylbutenamer were varied as a function of the catalyst; thus, while the system WCl_6/Et_3Al afforded copolymers with equal proportions of the two monomer units, the other catalysts gave predominantly products having 3-methylbutenamer units.

Cyclohexene, a monomer of pronounced stability that does not at all polymerize under usual conditions, was reported to copolymerize with cyclopentene at room temperature in the catalytic presence of $WCl_6/EtAlCl_2$ (molar ratio 1:13) [40]:

TABLE 2 Copolymerization of Cyclobutene (CB) with 3-Methylcyclobutene (3MeCB) with Different Catalytic Systems

Catalytic system	CB/3MeCb (molar ratio)	Yield %	Copolymer composition (CB/3MeCB)	Copolymer configuration (%)		
				cis	trans	saturated
WCl_6/Et_3Al	30:70	38	50:50	60	25	15
$MoO_2(acac)_2/Et_2AlCl$	35:65	47	40:60	70	25	5
$TiCl_4/Et_3Al$	26:74	100	25:60	30	55	15
$RuCl_3/H_2O/EtOH$	35:65	20	45:55	Traces	80	20

Source: Data adapted from Ref. 39.

(60)

(18)

The maximum proportion of hexenamer units introduced into the polymer chain, (**60**), was about 25%. The ratio of catalyst to cocatalyst seemed to be crucial for copolymer composition; at Al/W = 4, homopolymers of cyclopentene are already formed. By contrast, copolymers of cycloheptene and cyclopentene obtained with the same catalyst had mainly a random distribution of the two monomer units, with a slight tendency in favor of the cyclopentene monomer under certain conditions.

Copolymerization of 1,5-cyclooctadiene with cyclooctatetraene will form random copolymers (**61**), displaying interesting electrical and optical properties:

(61)

(19)

By controlling the ratios of 1,5-cyclooctadiene to cyclooctatetraene, products with varying conjugation length and electrical properties will be produced. When 1,5-cyclooctadiene is copolymerized with hexachlorocyclopentadiene [41b], elastomeric copolymers (**62**), having improved chemical resistance, can be obtained:

(62)

(20)

Both $WCl_6/2Et_2O$ and $Ph_3SiOWCl_5/2Et_2O$ complexes have been employed in conjunction with Bu_4Sn to attain high yields in copolymers containing 30% chlorine. These catalytic systems were tolerant of the monomer functionality, and both monomers displayed substantial reactivity at room temperature.

Copolymers of several cycloolefins with norbornene and/or norbornadiene have been widely explored with catalytic systems based on tungsten and molybdenum compounds. Products of varying compositions, of interest for their plastic or elastomeric properties, have been prepared by this way. Valuable information concerning reaction mechanisms and stereochemistry has been obtained by selecting particular substituted or nonsubstituted cycloolefins and/or norbornene or norbornadiene monomer [42].

Cyclopentene copolymerized with norbornene in the presence of a wide variety of transition metal catalysts yields copolymalkenamers with structures (**63**), having rubber-like properties [42]:

(63)

(21)

It is of great interest to model the product properties by changing the copolymer composition under different reaction conditions (e.g., feed ratio, nature of catalyst, monomer concentration, reaction temperature, sovlent). Studies with this and other monomer systems have shown that the copolymer structure from a given feed, hence the reactivity ratios of a pair of monomers, are markedly dependent not only on the catalyst or reaction parameters but also, significantly, on the method of mixing the reaction components [4]. The reactivity ratios for the copolymerization of cyclopentene–norbornene in the presence of several catalytic systems are given in Table 3.

It is obvious that in all cases, the catalytic systems discriminate in favor of norbornene, a strained monomer with particularly high reactivity. Copolymer microstructure is well documented from detailed ^{13}NMR spectroscopy studies, from which valuable information concerning the reaction mechanism and stereochemistry or the nature of the initiating and propagating species can be inferred.

Linear and branched copolymers are obtained by copolymerizing cyclopentene with norbornadiene under appropriate conditions [43]:

TABLE 3 Reactivity Ratios for Cyclopentene–Norbornene Copolymerization (r_2/r_2) in the Presence of Several Catalysts

Catalytic system	Mole ratio	r_1	r_2
$WCl_6/EtAlCl_2$	1:4	0.32	13
$WCl_6/EtAlCl_2$	1:10	0.09	20
WCl_6/Bu_4Sn	2:1	0.27	12
WCl_6/Ph_4Sn	1:1	0.55	2.6
$WCl_6/Ph_4Sn/EA$[a]	1:1	0.62	2.2
$WCl_6/allyl_4Sn$	1:1		70
$MoCl_5/EtAlCl_2$	1:1		56

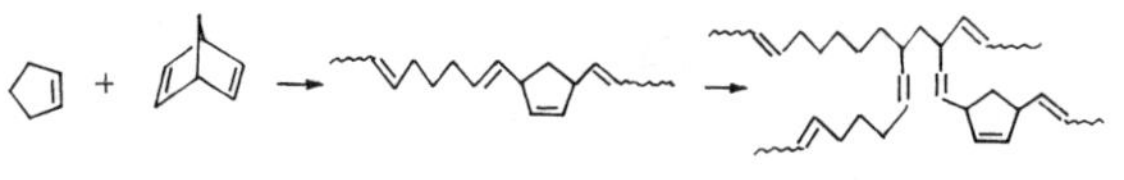

(64) (64a)

(22)

The products (**64**) and (**64a**), with predominantly trans configuration at the double bonds, have rubber-like properties and can be used in blends with synthetic rubber.

Of particular interest is the reaction of cyclopentene with cyclopentadiene [44] or dicyclopentadiene [45], the latter providing an efficient route for the synthesis of various products that have good processing properties. Both cyclopentene and dicyclopentadiene proved to be fairly reactive in the presence of several tungsten- and molybdenum-based catalytic systems to form copolymers of type (**65**) and (**65a**) with a high degree of unsaturation:

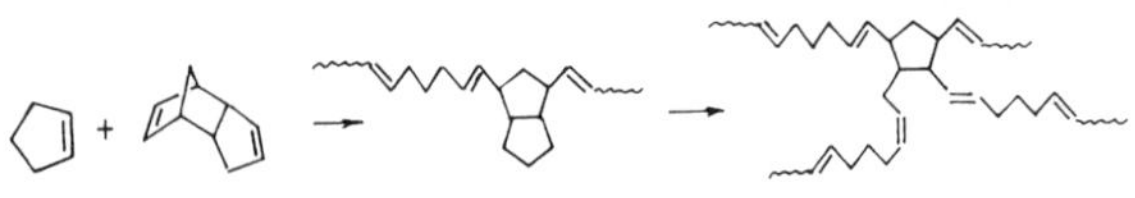

(65) (65a)

(23)

Grafted and crosslinked polymers can be manufactured depending essentially on feed composition, catalyst activity, and reaction parameters.

Numerous other cycloolefins have been coupled with norbornene, norbornadiene, dicyclopentadiene, or norbornenelike polycyclic olefins, affording a wide variety of copolymers with a large spectrum of physical and mechanical properties [3]. For such pairs of monomers, the reactivity ratios in the presence of different catalyst can be found in the literature [4].

It is significant that cyclopentene has been copolymerized with a large number of norbornene derivatives bearing functional groups X, where X is cyano-, ester-, amido-, amino-, or methyl-dichlorosilyl [46]:

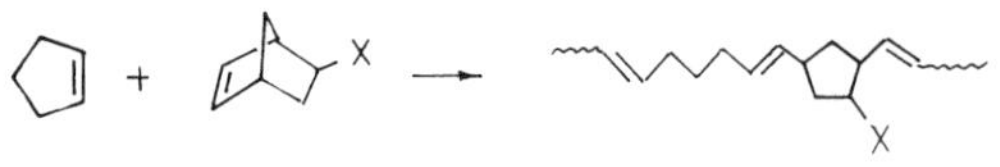

(66)

(24)

The products resulting from this reaction exhibit valuable physical and chemical properties depending on the nature of the substituent. The catalyst that must tolerate monomer functionality also strongly influences the copolymer composition, hence the polymer properties. Instead of cyclopentene, other easily polymerizable monomers can be employed (e.g., 1,5-cyclooctadiene) [47]:

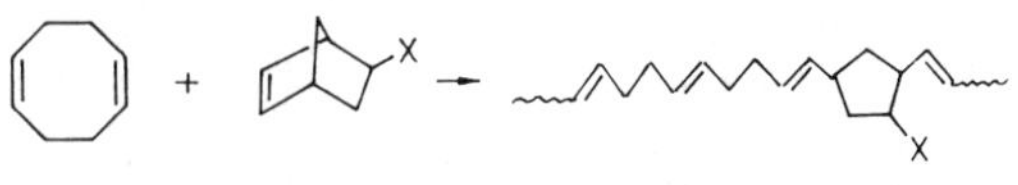

(67)

(25)

Copolymers of type (**67**) will possess good elastomeric properties and chemical resistance. Through proper choice of catalyst and monomer to be reacted with the substituted norbornene, new products having special characteristics may be produced by this method.

Copolymers of two or more norbornenelike polycyclic olefins [48a–c] [e.g., norbornene, alkylnorbornenes (where alkyl is a C_8–C_{12} group), dicyclopentadiene, dimethanooctahydronaphthalene, alkyldimethanooctadhydronaphthalene] will be obtained by appropriate procedures using molybdenum- or tungsten-based soluble ring-opening metathesis polymerization (ROMP) catalysts [48]:

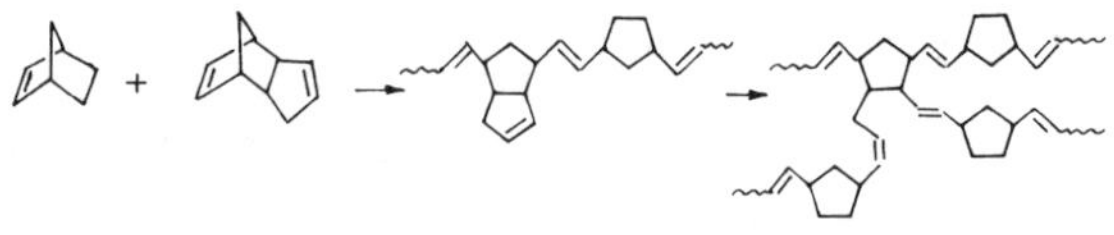

(68) (68a)

(26)

These products will have linear or branched structures of type (**68**) or (**68a**), respectively, depending on catalyst activity and monomer polymerizability. Furthermore, ternary copolymers (e.g., (**69**) or higher interpolymers), usable as thermoplastics or for special elastomer compositions, may be made by combining three or more norbornenelike monomers in such a procedure [48d].

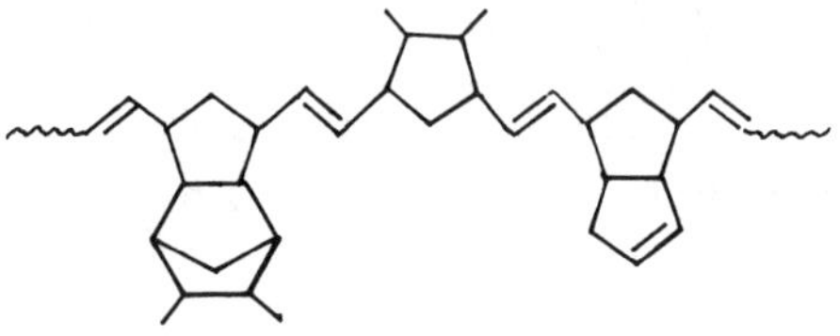

(69)

One or both monomers can be functionalized. Copolymers of norbornene with hexachlorocyclopentadiene **70** [Eq. (27)] will change the product properties considerably compared to the corresponding copolymer produced from norbornene and unsubstituted cyclopentadiene [49].

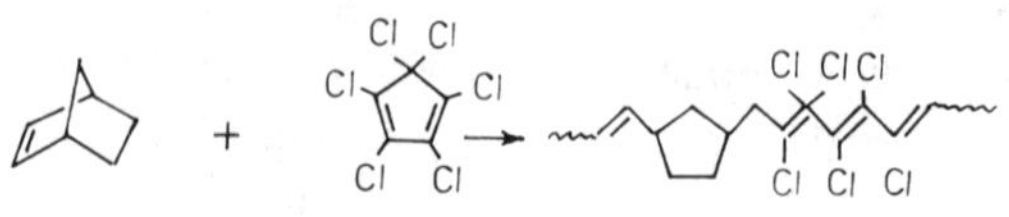

(70)

(27)

Similarly, copolymers of norbornene, with norbornene oxo derivatives such as those represented by **(71)** and **(72)**, obtained in the catalytic presence of WCl_6/Me_4Sn, display properties totally different from those of the homopolymers [50]:

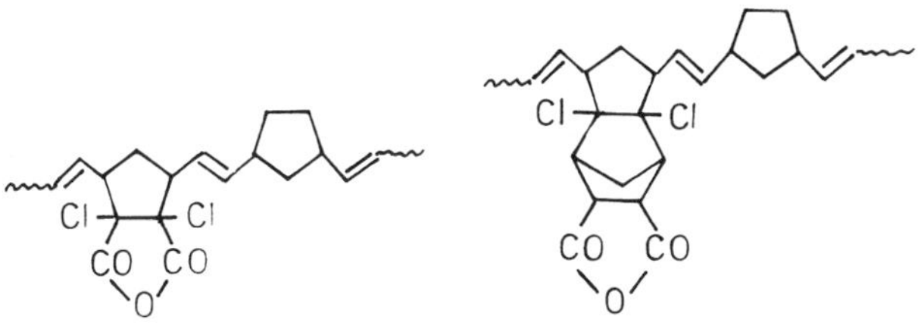

(71) (72)

Likewise, copolymers of 5-cyanorbornene with 5-phenylnorbornene, (**73**), prepared using WCl_6/Et_2AlCl as the catalyst [51a], will have a combination of interesting properties:

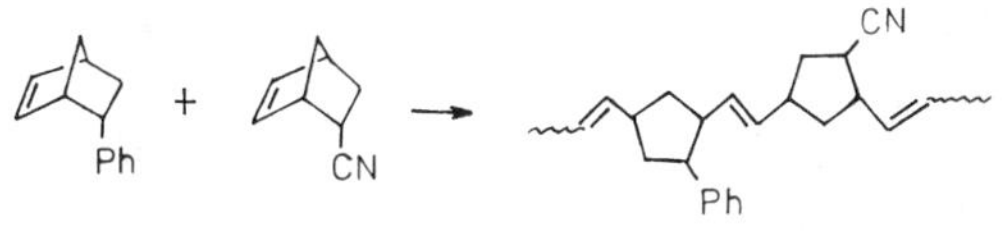

(73)

(28)

When the same substituted norbornene is reacted with methyltetracyclododecene [51b] (methyldimethanooctahydronaphthalene) in the presence of WCl_6/organoaluminum compounds and activators (diethylacetal of acetaldehyde), random copolymers of type **(74)** will result:

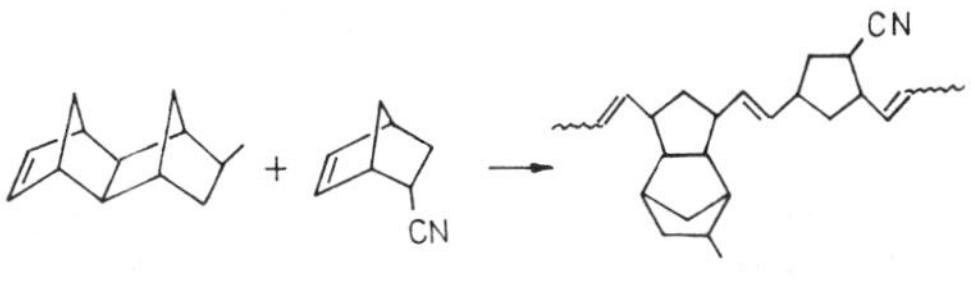

(29)

(74)

At high methyltetracyclododecene concentration, there appears to be phase separation, indicating block copolymer formation. It was observed that in this reaction the norbornene derivative controls the polymerization rate favoring copolymers with a high glass transition temperature. The choice of catalytic system and monomer allows application of the process in the reaction injection molding (RIM) procedure. Many other functionalized groups incorporated into the norbornene or norbornene-like hydrocarbons (e.g., ester, amido, imido, anhydride, halogen) have been used to yield derivatized copolymers with particular physico-chemical properties.

B. Block and Graft Copolymers

Block copolymers from cyclooolefins have been prepared by various experimental techniques [52]. Some interesting methods use living ROMP catalysts, which allow ready synthesis of new products having controllable structures and properties. Other methods apply cross-metathesis between unsaturated polymers and/or polyalkenamers [3], polymerization of cycloolefins in the presence of unsaturated polymers [4], polymerization of two or more cycloolefins of quite different reactivities with classical ROMP catalysts [4], and copolymerization of cycloolefins with other monomers, effected by changing the polymerization mechanism from ROMP to anionic, cationic, Ziegler–Natta, and group transfer, and vice versa [6–8, 52].

The formation of triblock copolymers by successive addition of monomers was first claimed for cyclopentene and 5-cyanorbornene, using the conventional catalyst WCl_6/Et_3Al in the presence of small amounts of 1-hexene and triethyl phosphite [53]. The reaction was carried out by adding batches of 5-cyanonorbornene, then cyclopentene, and finally 5-cyanonorbornene every 2 h. The yield in the copolymer **(75)*** was high (94%), and the product had considerably improved tensile and flexural strength compared to the random copolymer of the same composition.

(75)

Cyclopentene–norbornene block copolymers have been prepared with the living tungsten–carbene system ($W(CHt\text{-}Bu)(NAr)(Ot\text{-}Bu)_2$, but the structure of this product has not been studied in detail [54]. Because there is a marked difference in the reactivity of the two monomers, rigorous reaction conditions must be applied to obtain the virtual block copolymer **(76)**.

(76)

The more active catalyst $W(C(CH_2)_3CH_3)(OCH_2CH_3)_2Br_2$, used as such or complexed with $GaBr_3$, will form block copolymers from norbornene and norbornene derivatives, *syn-* and *anti-* 7-methylnorbornene, or 1,5-cyclooctadiene [55]. Determination of product structure by NMR spectroscopy indicated

*The unit structures in parentheses in (**75**–**104**) represent chain block segments arising from each of the monomers.

block units of types (**77**) (X = COOMe, CN), (**78**), and (**79**) when starting from norbornene. By reaction of 1,5-cyclooctadiene with norbornene, copolymers having structures (**80**) were produced; in this case, however, secondary metathesis reactions of double bonds in the polymer chain occur under the influence of the catalyst, changing the molecular weight of the polymer accordingly.

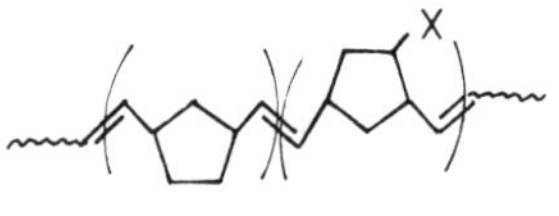

X

(**77**) 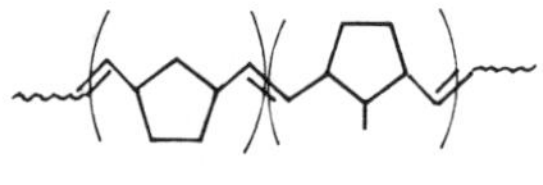 (**78**)

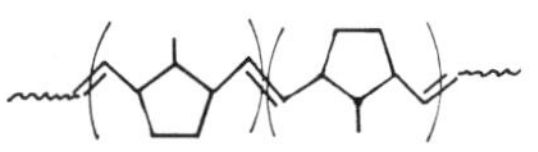

(**79**) 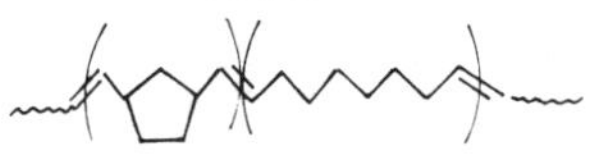 7 (**80**)

Cyclooctatetraene readily reacts with norbornene [41a] to yield the diblock copolymers (**81**), which have variable block length depending on the reaction conditions.

(**81**)

Block copolymers of the AB and ABA types, containing segments of narrow molecular weight distribution, are obtained from norbornene, benzonorbornene, 6-methylbenzonorbornene, and *endo*- or *exo*-dicyclopentadiene in the catalytic presence of titanocyclobutanes [56]. Illustrative examples are (**82**) and (**83**), the diblock and triblock copolymers from norbornene and dicyclopentadiene, having polydispersity indices of 1.08 and 1.14, respectively:

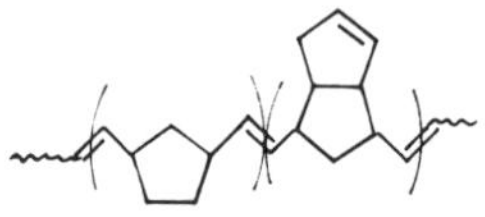

(**82**) 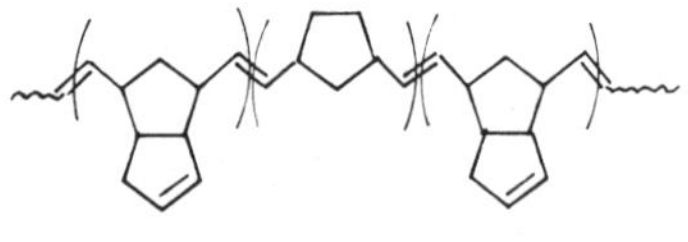 (**83**)

Analysis by differential scanning calorimetry (DSC) showed a single glass transition temperature for these block copolymers, indicating that they may be compatible.

When polydimethylenecyclobutene is blocked with polynorbornene as in (**84**), the physical properties of the initial polymer change substantially (e.g., from brittle and insoluble into a rubbery material with more desirable mechanical properties) [57]:

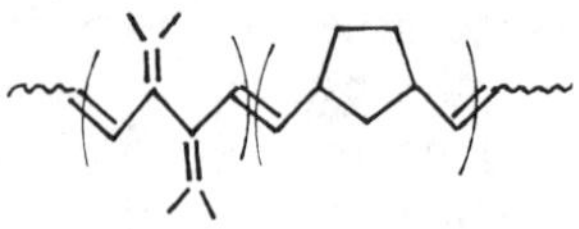

(**84**)

Doping this product with reducing or oxidizing agents results in conductivities comparable to that of polydimethylenecyclobutene but possessing better mechanical properties.

Living block copolymers of type (**85**) have been prepared via sequential addition of norbornene and *endo,endo*-dicarboxymethylnorbornene (and vice versa) to Mo(CH*t*-Bu)(NAr)(O*t*-Bu)$_2$ catalyst [58]:

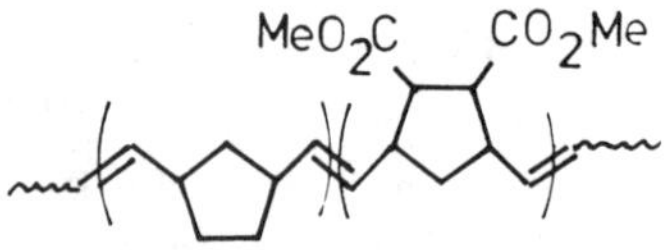

MeO_2C CO_2Me

(**85**)

The molybdenacarbene catalyst proved to be rather active toward both monomers and tolerant to the ester functionality on the time scale of the reaction. Low polydispersity indices of 1.06 and 1.09 have been obtained for the two copolymers, depending on the sequence of monomer addition.

Remarkably, the tolerance of the M(CH*t*-Bu)(NAr)(O*t*-Bu)$_2$ catalysts (M = W or Mo) for functionalities allowed them to be used to effect ring-opened polymerization of norbornenes containing metals. Such monomers have already been copolymerized with norbornene or methyltetracyclododecene, to obtain semiconductor clusters of controlled cluster dimensions. A first block copolymer containing lead (e.g., **86**) was prepared from norbornene and $(C_7H_9CH_2C_5H_4)_2Pb$ in the presence of Mo(CH*t*-Bu)(NAr)(O*t*-Bu)$_2$ (Ar = 2, 6-C_6H_3i-Pr_2) [59].

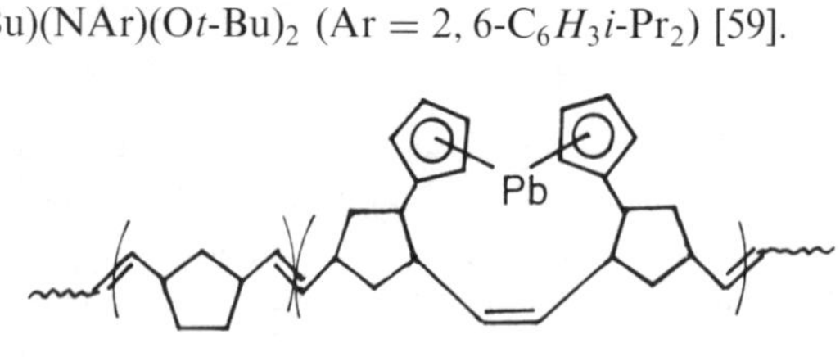

Pb

(**86**)

Treatment of this block copolymer with hydrogen sulfide resulted in the presence of small particles of lead sulfide in an essentially polynorbornene matrix. Semiconductors based on metal sulfides (PbS, CdS) have been generated by this method. Similar block copolymers have been prepared from norbornenes that contain tin, zinc, cadmium, palladium, and gold.

Further treatment of these copolymers with hydrogen afforded metal clusters [60].

Norbornene can be copolymerized with di-, tri-, and tetrasubstituted fluoroalkyl norbornenes in the presence of the above-mentioned living initiator $Mo(CHt\text{-}Bu)(NAr)(Ot\text{-}Bu)_2$ to form products of narrow molecular weight distribution **(87)**, having attractive physico-chemical properties [61].

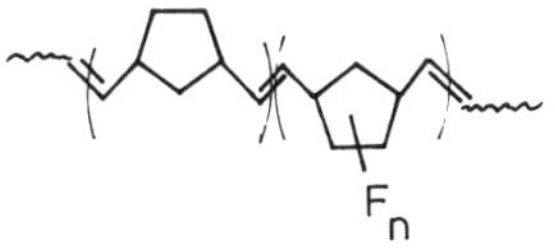

(87) 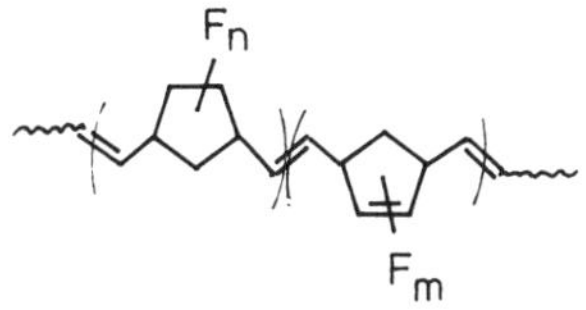 (88)

Analogously, block copolymers of fluorinated norbornenes and fluorinated norbornadienes of type **(88)** have been prepared by this procedure.

Graft and block copolymers containing polyacetylene segments have been synthesized on a large scale recently, in an effort to take advantage of their valuable conductive properties. Two of the most efficient techniques use living ROMP catalysts: direct copolymerization of acetylene with other monomers to form diblock and triblock copolymers with polyacetylene, and generation of block copolymers containing one precursor polymer, which converts to polyacetylene via a retro Diels–Alder reaction (the "Durham route" to polyacetylene). In the first process [62], acetylene is polymerized by $M(CHt\text{-}Bu)(NAr)(Ot\text{-}Bu)_2$ catalyst (M = Mo, W) to yield diblock and triblock copolymers with polynorbornene, **(89)** and **(90)**, respectively.

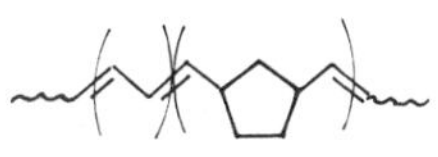

(89) 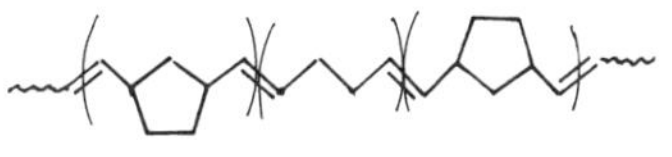 (90)

The products thus obtained have a narrow molecular weight distribution. These polymers, with a high trans content in the double bonds, exhibit a greater insolubility and a pronounced tendency to crosslink, even when the polyene is present at the central block in the triblock copolymers with polynorbornene.

Conveniently applied to obtain monodisperse block copolymers of polyenes, the second process, the so-called Durham route, uses presently several precursor monomers and catalysts [63, 64]. Diblocks and triblocks, each containing one or two sequences of polynorbornene and one or two sequences of a precursor polymer that converts to polyacetylenes (polyenes) upon heating, are formed with titanacycles [52] or with well-defined tungsten or molybdenum alkylidene complexes [62]. Such products are manufactured by one of the two processes. The first entails blocking a living polymer, **(91)**, prepared from the Feast monomer with polynorbornene, and subsequent thermolysis of copolymer **(91a)** to yield controllable block lengths of polyacetylene within a polymer matrix, **(91b)** [Eq. (30)]. The second approach consists of blocking a living polynorbornene, **(92)**, with a precursor polymer: the initial product is a copolymer, **(92a)**, which converts to a diblock of polynorbornene and polyacetylene, **(92b)** [Eq. (31)]:

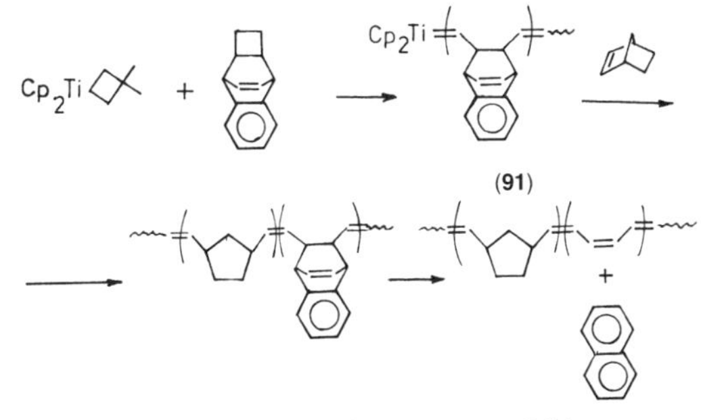

(91) (91a) (91b) (30)

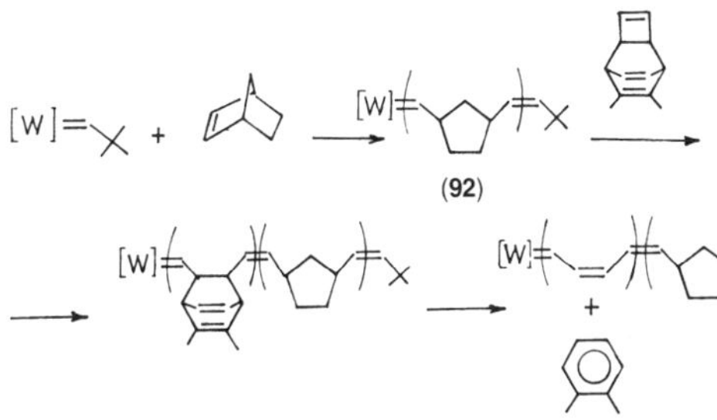

(92) (92a) (92b) (31)

Several precursor diblock copolymers have been prepared by the second method [64], including the following:

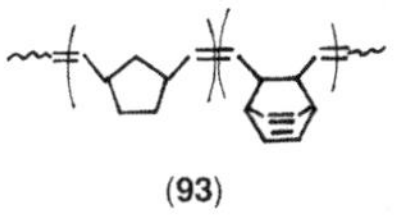

(93) 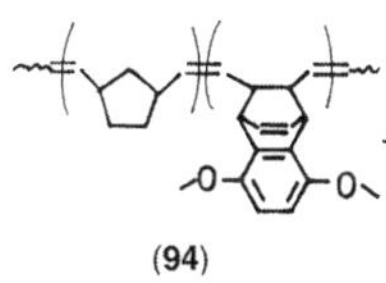 (94) 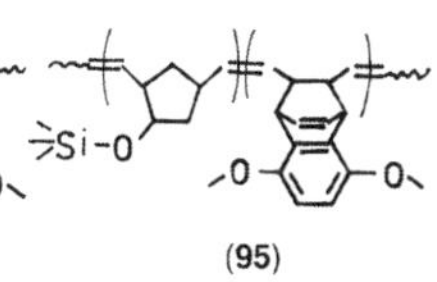 (95)

Difunctional dititanacyclobutane living catalysts [65], **(96)**, allowed synthesis of polyacetylene–polynorbornene–polyacetylene triblock copolymers by the following pathway:

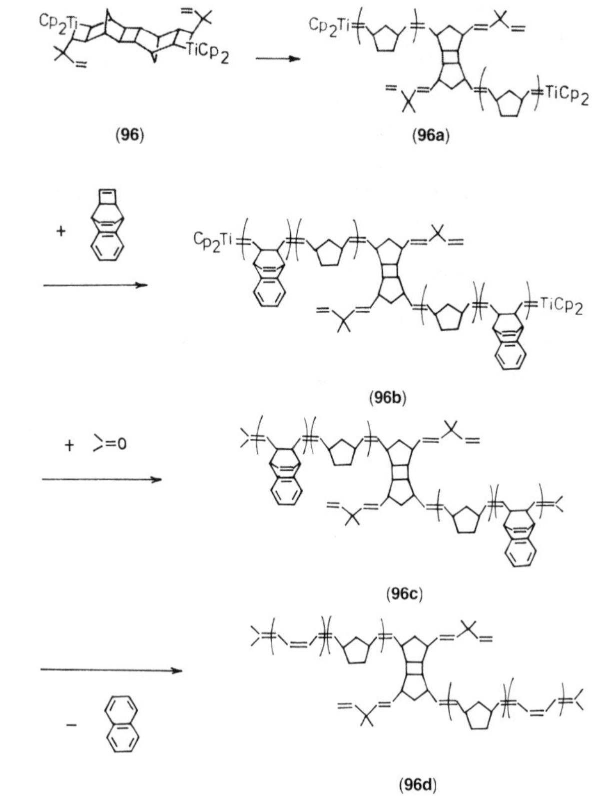

(96) (96a)

(96b)

(96c)

(96d)

(32)

Similarly, "star" and "heterostar" copolymers, **(97)** and **(98)**, have been reported to form from norbornene and 2,3-dicarbomethoxynorbornene or 5-cyanonorbornene and 2,3-bis(trifluoromethyl)norbornadiene.

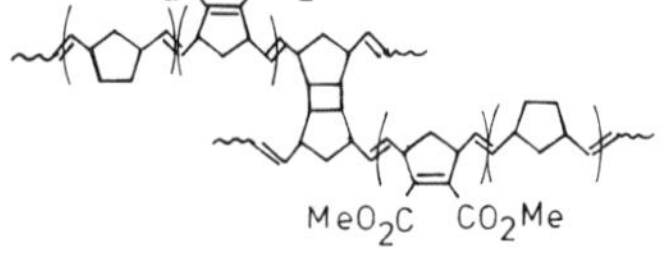

(97) 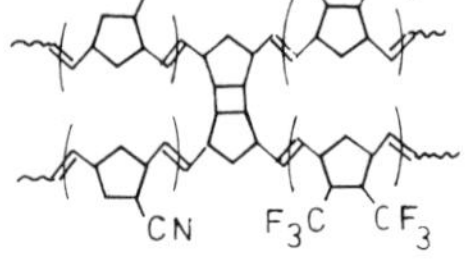 (98)

The norbornadiene dimer *exo-trans-exo*-pentacyclotetradeca-5,11-diene has been employed as a crosslinking agent for the controlled synthesis of these star-shaped copolymers in the catalytic presence of W(CH*t*-Bu)(NAr)(O*t*-Bu)$_2$ or Mo(CH*t*-Bu)(NAr)O*t*-Bu)$_2$ [66]. Star polymers in which the arms are block copolymers of different polarities can be prepared by this method, as well as star polymers in which the composition of the arms is different, as are the solubility characteristics and stereochemical regularity. "Heterostars" thus obtained, in which the number of the different arms will be approximately equal, may exhibit interesting intramolecular phase separations.

New block copolymers have been obtained by combining various polymerization techniques with the metathesis polymerization of cycloolefins [52]. For this purpose, the living polystyryllithium system has been changed into a metathesis catalyst by treatment with WCl_6. This living system initiates further cyclopentene ring-opening polymerization to produce the following diblock copolymer, consisting of polystyrene and polypentenamer blocks [8]:

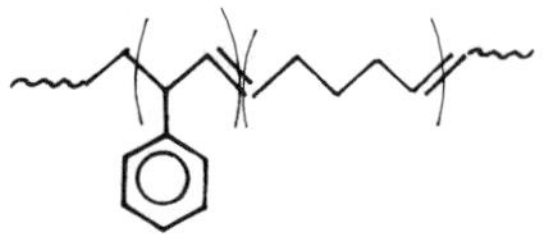

(99)

Similarly, grafting living polynorbornene onto polymers that contain carbonyl groups formed by conventional polycondensation reactions leads to triblock copolymers, **(100)**, having properties totally different from those of the homopolymers [67].

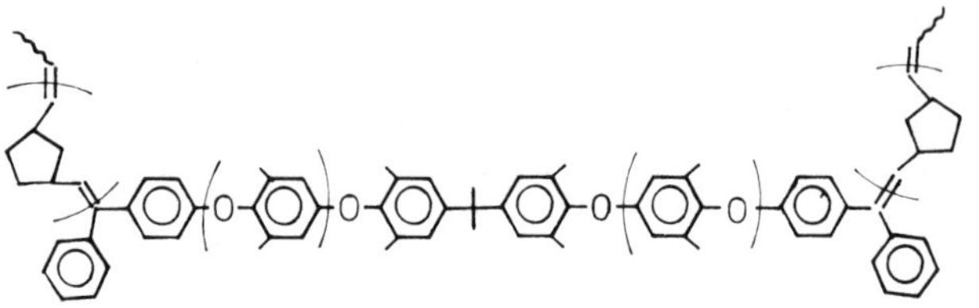

(100)

A second block can also be grown on a ring-opened polymer by the use of group transfer [68] or Ziegler–Natta polymerization [6]. In the first case, polynorbornene and poly(*exo*-dicyclopentadiene) with one aldehyde end group have been synthesized by metathesis polymerization and successive reaction of terephthaldehyde [68]. The aldehyde end group served as an initiator for the silyl aldol condensation polymerization of *t*-butyldimethyl vinyl ether to give diblock polynorbornene–poly(silyl vinyl ether) copolymers with narrow molecular weight distribution. The cleavage of the silyl groups led to hydrophobic–hydrophilic diblock copolymers having polyvinyl alcohol as the second segment. Further transformation of polyvinyl alcohol block into polyvinyl acetate is possible. The blocking of polynorbornene-bearing aldehyde end groups, **(101)**, with poly(silyl vinyl ether) and subsequently with polyvinyl alcohol and polyvinyl acetate to the respective diblock copolymers **(101a)**, **(101b)**, and **(101c)** proceeds as follows:

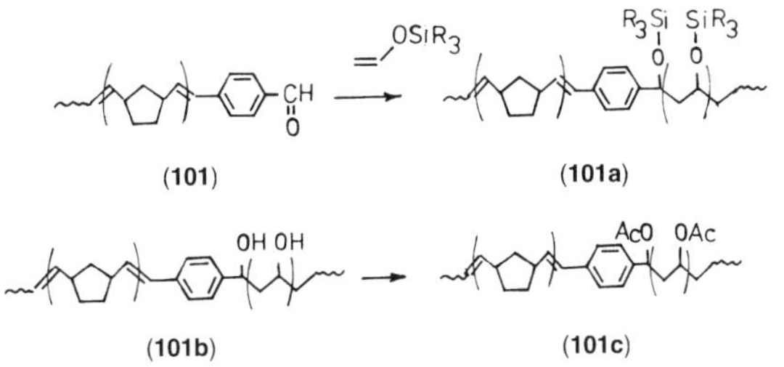

(101) (101a)

(101b) (101c)

(33)

Unsaturated polymers can be blocked and grafted with polyalkenamers by ring-opening copolymerization of cycloolefins in the presence of metathesis catalysts. Thus, diblock copolymers (**102**) consisting of the initial unsaturated chain and the newly grown polyalkenamer can be obtained:

(102) (34)

The products have superior mechanical properties compared with the random copolymers or blends of homopolymers of the same overall composition. The literature reports block copolymers of polybutadiene with cyclopentene [69a], cyclooctadiene [69b], cyclododecene [69c] and substituted norbornenes [69d], of polyisoprene, polychloroprene, polypentenamer, and butyl rubber with norbornene derivatives [69c] and styrene–butadiene copolymers with cyclopentene [69a] and norbornene derivatives [69c]. Graft copolymers of type (**103**) will arise when unsaturation occurs in branched arms of the polymer to be grafted (e.g., 1,2-polybutadiene with cycloolefins):

(103)

(35)

Cross-metathesis of unsaturated polymers with polyalkenamers affords another versatile method of forming block copolymers from cycloolefins:

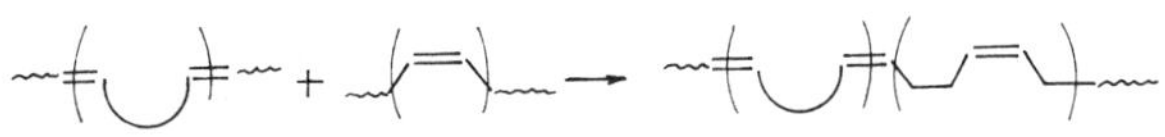

(104)

(36)

Such a block copolymer of type (**104**) has been prepared from 1,4-*cis*-polybutadiene and polydodecenamer in the catalytic presence of $WCl_6/Et_3Al_2Cl_3$ [70]. By careful control of the reaction conditions, the amount of simultaneous degradation of polybutadiene has been diminished to acceptable proportions. The technological demands of these block copolymers are comparable to those of blends from the corresponding polymers.

C. Copolymers with Heterocyclic Olefins

With the discovery of well defined functional group tolerant ROMP catalysts, synthesis of block copolymers from cycloolefins and heterocyclic olefins bearing oxygen or nitrogen atoms became readily available [71]. The synthesis of such copolymers is of interest because the heteroatoms incorporated into the polymer chain impart new physical and chemical properties to the products obtained. For instance, the copolymers (**105**) (R = H, Ph) prepared from 1,5-cyclooctadiene and 4,7-dihydro-1,3-dioxepin with ruthenium carbene initiators, $(Cy_3P)_2Cl_2Ru{=}CHR'$ (Cy = cyclohexyl, Ph = phenyl, R′ = Ph or $CHCPh_2$) [Eq. (37)] can be degraded at the acetal group to 1,4-hydroxytelechelic polybutadiene [72].

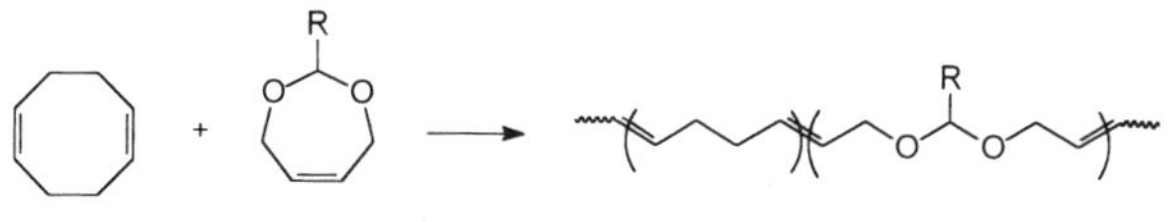

(105)

(37)

Well-defined block copolymers such as (**106**) and (**107**) may be efficiently prepared in aqueous media by copolymerization of functionalized norbornenes and 7-oxanorbornenes with monometallic ruthenium initiators, $(Cy_3P)_2(Cl_2Ru{=}CHR$ (R = Ph) [73] [Eq. (38)]

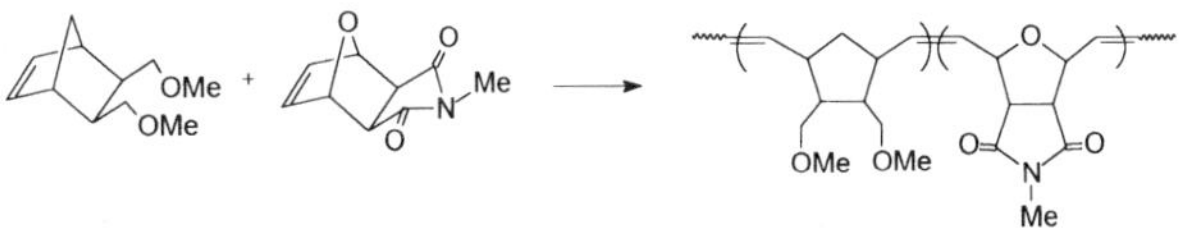

(106)

(38)

or bimetallic ruthenium initiators, $(PR_3)_3Cl_2Ru{=}CHpC_6H_4CH{=}RuCl_2(PR_3)_2$ (R = Ph, cyclohexyl, cyclopentyl) [74] [Eq. (39)].

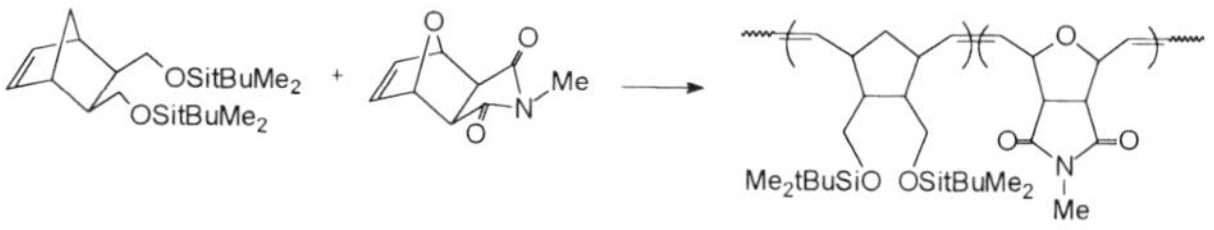

(107)

(39)

The association of such hydrophilic with hydrophobic monomers may result in copolymers with unique physical properties particularly in applications for semipermeable membranes [75].

The recent development of new water-soluble ruthenium metathesis catalysts such as $(Cy_2(CH_2CH_2N(Me)_3^+Cl^-)_2Cl_2Ru{=}CHPh$ and $(Cy_2(Me_2piperidinium^+Cl^-)_2Cl_2Ru{=}CHPh$ allowed synthesis of well defined water soluble copolymers (**108**), $R = CH_2CH_2N(Me)_3^+Cl^-$, by sequential monomer addition [76] [Eq. (40)].

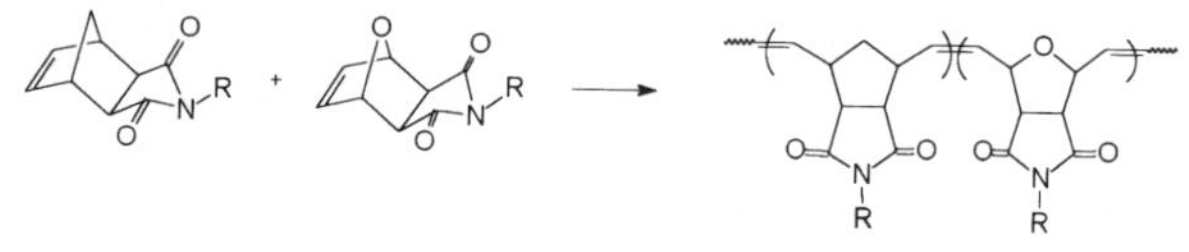

(**108**)

(40)

The synthesis of this type of copolymers with the above catalysts becomes important in the light of emerging biomedical applications for water-soluble products which demand precise control over the polymer length and polydispersity [77].

D. Copolymers from Cycloolefins and Acetylenes

After being recognized that olefin metathesis and acetylene polymerization reactions occur through a common metallacarbene mechanism [3, 4, 78, 79], it became obvious that cycloolefins and substituted acetylenes could polymerize with each other. Such copolymerization reactions occurred successfully in the presence of a wide range of W- and Mo-based metathesis catalysts and various substrates to provide a novel type of copolymers [80]. Thus, norbornene was found to copolymerize rather easily in the presence of tungsten catalysts, e.g., WCl_6/Ph_4Sn and $W(CO)_6$-$h\nu$ (CCl_4), with phenylacetylene to produce copolymers (**109**) having both repeat units, 1,3-cyclopentylenevinylene and phenylethylidene, in the polymer chain [Eq. (41)].

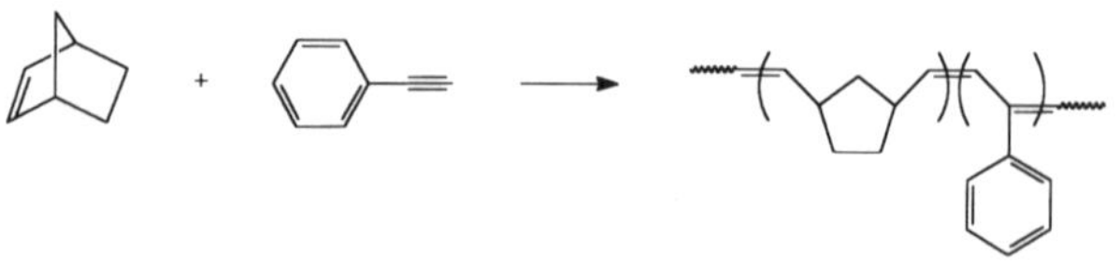

(**109**)

(41)

Under similar conditions, dimethyleneoctahydronaphthalene (tetracyclo[6.2.1$1^{3,6}$.$0^{2,6}$]dodec-4-ene) was reacted with phenylacetylene to produce copolymers (**110**) having polycyclic repeat units in the polymer chain [80] [Eq. (42)].

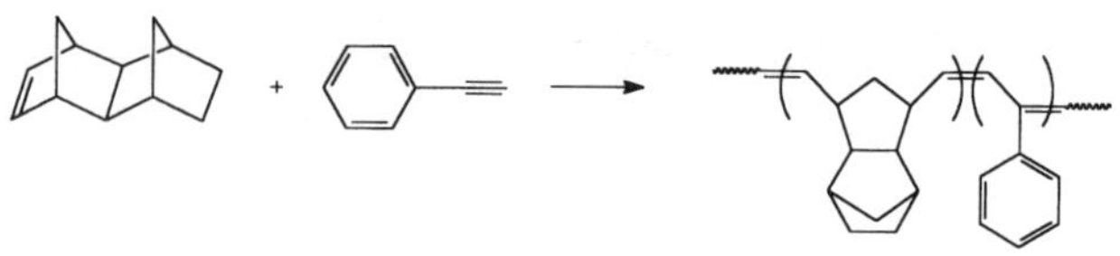

(**110**)

(42)

Furthermore, ring-substituted phenylacetylenes were employed in reactions with norbornene in the presence of WCl_6-based catalysts to prepare the corresponding copolymers (**111**–**113**) ($R = CH_3$, CF_3, $SiMe_3$) [81, 82] [Eqs. (43)–(45)].

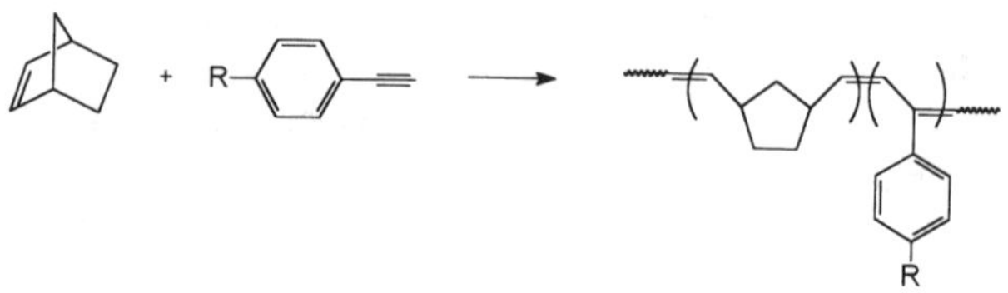

(**111**)

(43)

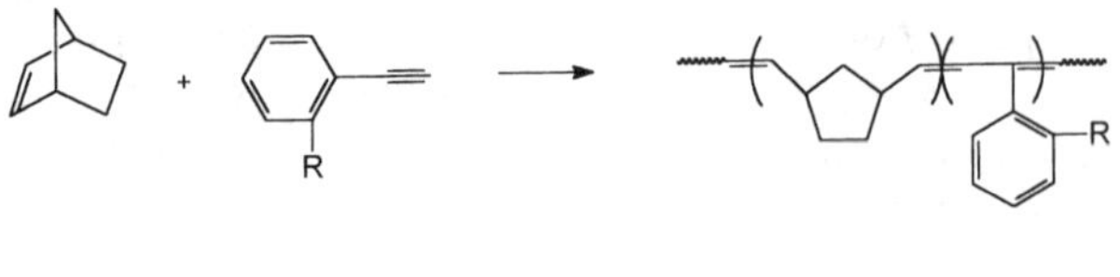

(**112**)

(44)

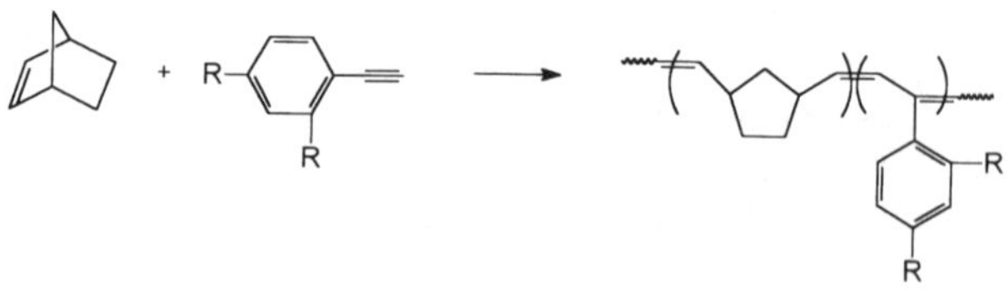

(**113**)

(45)

Of a particular interest is the fluorinated copolymer (**114**) prepared from norbornene and *o,m*-tetrafluoro-*p*-*n*-butylphenylacetylene with WCl_6-based catalysts which could exhibit attractive physical properties [Eq. (46)].

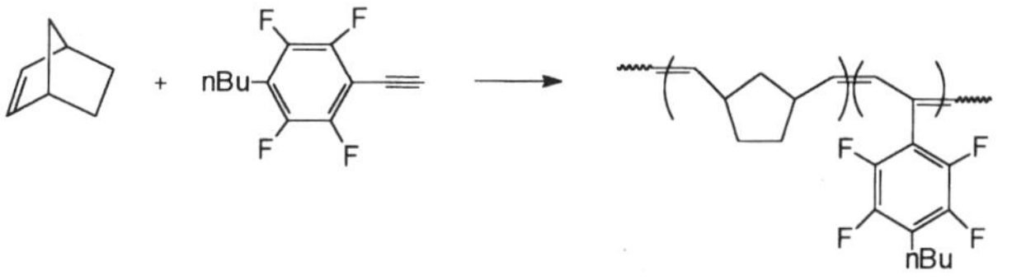

(114)

(46)

On the other hand, chlorine-containing acetylenes gave unimodal copolymers (**115** and **116**) with norbornene in the presence of $MoCl_5/n$BuSn and WCl_6/nBuSn [83] [Eqs. (47)–(48)].

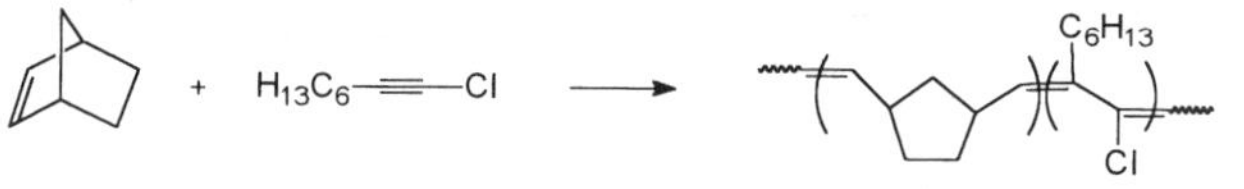

(115)

(47)

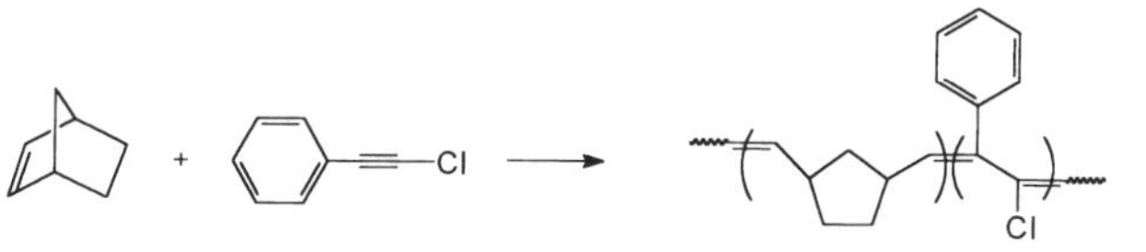

(116)

(48)

Films were fabricated from the copolymerization products and from mixtures of both homopolymers with the same composition by casting them from toluene solution. Interestingly, while films of the homopolymer mixtures were obviously inhomogeneous, the films from copolymerization products were uniform, indicating that the two monomer units were present in the same polymer molecule.

E. Copolymers from Macromonomers

Cycloolefin macromonomers have been recently used in ring-opening metathesis polymerization reactions to manufacture block and graft copolymers of novel macromolecular architectures [84]. For this purpose, α- and ω-norbornenyl–polybutadiene macromonomers, α-NBPB (R = CH_2) and ω-NBPB (R = COO), were reacted in the presence of molybdenum alkylidene complex, Mo(NAr)(CH*t*Bu)(O*t*Bu)$_2$, to form polynorbornene–polybutadiene diblock copolymers (**117**), with comb-like structure [85] [Eq. (49)].

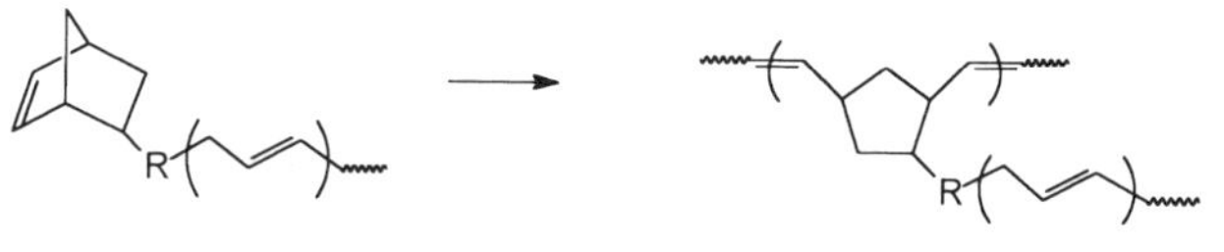

(117)

(49)

The polymerization reactions with the above catalyst were carried out in toluene at ambient temperatures to almost complete conversion, provided the vinyl content of the macromonomer is low.

Polymerization of α-norbornenyl–polystyrene macromonomer, α-NBPS, with the Schrock-type catalyst Mo(NAr)(CH*t*Bu)(OC(CH_3)(CF_3)$_2$)$_2$ in toluene at room temperature gave polynorbornene–polystyrene diblock copolymer (**118**) in a quantitative yield [86] [Eq. (50)].

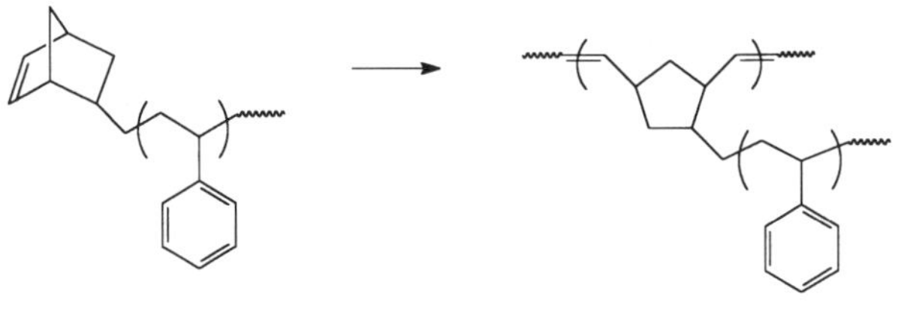

(118)

(50)

The polymacromonomers were characterized by size exclusion chromatography (SEC), equipped with a laser light scattering (LLS) detector at its outlet. Even though the M_ns of the polymer did not perfectly agree with the targeted values, there were enough reasons to conclude that the ring-opening metathesis reaction occurred under truly "living" conditions. On using a similar method, polymerization of norbornenyl–poly(ethylene oxide) macromonomers, NBPEO, in toluene at room temperature with Mo(NAr)(CH*t*Bu)(OC(CH_3)(CF_3)$_2$)$_2$, led to polynorbornene–poly(ethylene oxide) block copolymers (**119**) in high yields [87] [Eq. (51)].

(119)

(51)

The polymacromonomer, as determined by SEC technique, exhibited narrow molecular weight distribution and low polydispersities when working in optimum conditions at low catalyst and monomer concentrations.

As polymers that contain both hydrophilic and hydrophobic components aroused keen interest from theoretical and practical points of view over the past years, synthesis of amphiphilic branched copolymers by ring-opening metathesis polymerization of miscellaneous macromonomers is an important goal of the actual research. Thus, in order to obtain globular shape macromolecules that would present the same features as those exhibited by certain assemblies of molecules such as the micelles or the latices, with a bulk part different from the external surface, polymerization of norbornyl–polystyrene–poly(ethylene oxide) macromonomer has been conducted in the presence of Schrock-type catalyst $Mo(NAr)(CHtBu)(OC(CH_3)(CF_3)_2)_2$ in toluene at room temperature to produce poly-norbornene–polystyrene–poly(ethylene oxide) block copolymers (**120**) [88] [Eq. (52)].

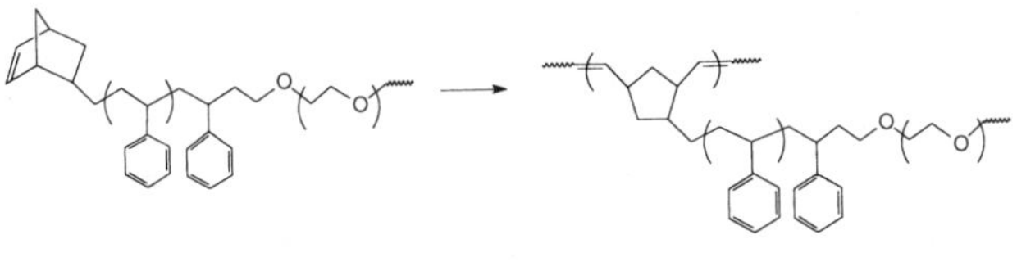

(120)

(52)

The topology of this product with a bulk hydrophobic structure inside and a hydrophilic part outside makes it particularly attractive for applications such as unimolecular micelles. On using the same procedure, norbornyl–poly(ethylene oxide)–polystyrene macromonomer was polymerized to polynorbornene–poly(ethylene oxide)–polystyrene block copolymer (**121**) [Eq. (53)].

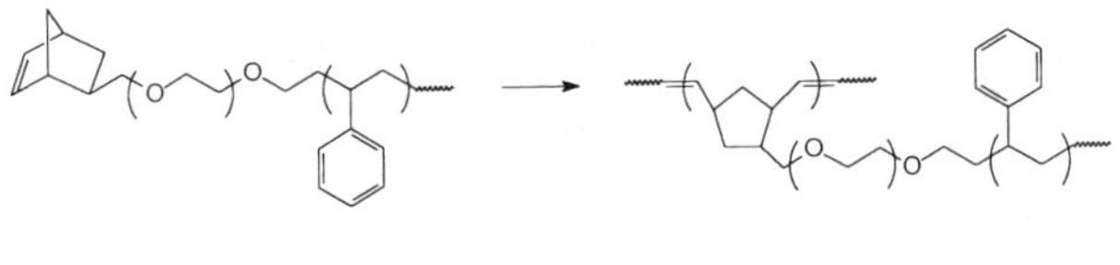

(121)

(53)

On controlling the molecular weight growth in macromonomer and polymer synthesis and the molecular weight polydispersity of the block copolymer by a proper choice of the reaction conditions, polymacromonomers with desirable properties could be produced. The amphiphilic block copolymer thus obtained having a combination of hydrophobic–hydrophilic–hydrophobic architectures is of interest for associative thickeners applications.

It is noteworthy that polymers of practical interest that exhibit a globular shape and unsymmetrical faces of the Janus-type architecture, not easy to make by conventional methods [89, 90], can be prepared by sequential ring-opening metathesis copolymerization of two different norbornenyl macromonomers. To this end, sequential polymerization of ω-norbornyl–polystyrene and α-norbornyl–poly(ethylene oxide) macromonomers in the presence of the Schrock initiator $Mo(NAr)(ChtBu)(OC(CH_3)(CF_3)_2)_2$ produced poly[polystyrene–oxycarbonyl–norbornene)-b-poly(ethylene oxide)–oxymethylene–norbornene)] block copolymer (**122**) in high yield [88] [Eq. (54)].

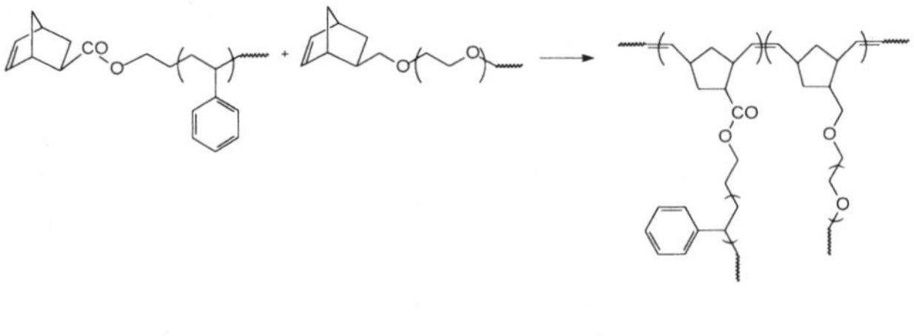

(122)

(54)

The products obtained under these conditions displayed high molecular weights and polydispersities indexes lower than 1.3. In addition, statistical poly[polystyrene–oxycarbonyl–norbornene)-b-poly(-ethylene oxide)–oxymethylene–polynorbornene)] copolymers were obtained from ω-norbornyl–polystyrene and α-norbornyl–poly(ethylene oxide) macromonomers by copolymerization under similar conditions. The distribution of PS and PEO grafts along the polymer backbone and the tendency of the copolymerization to blockiness in this case were determined by the reactivity ratios of the two macromonomers.

Comb graft copolymers with polystyryl grafts with average degrees of polymerization (DPs) of 4, 7, and 9 have also been successfully prepared from disubstituted norbornene–polystyrene macromonomers by living ring-opening metathesis polymerization using the well defined molybdenum carbene initiators of the Schrock type, $Mo(NAr)(CHR)(OR')_2$, where R is $C(CH_3)_3$ or $C(CH_3)_2C_6H_5$, Ar is 2,6-diisopropylphenyl and R′ is $C(CH_3)_3$. Thus, starting from bicyclo[2.2.2]-hept-5-ene-2,3-*trans*-bis(polystyrylcarboxilate), in the presence of the above molybdenum carbene initiators, the corresponding ring-opened block copolymer (**123**) has been obtained [91] [Eq. (55)].

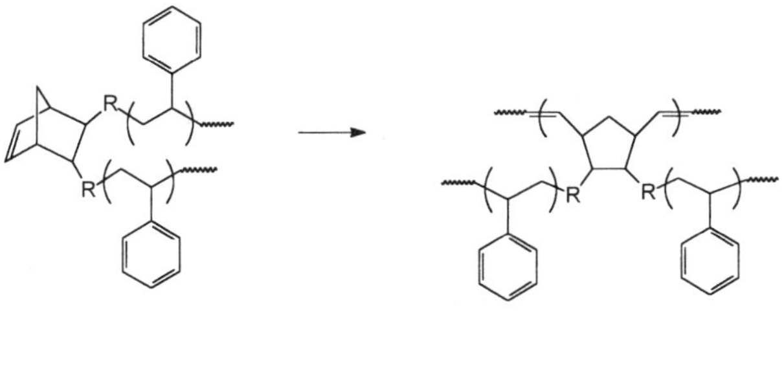

(123)

(55)

The graft copolymers exhibited single-mode molecular weight distributions and narrow polydispersities. Note that attempts to prepare similar copolymers with longer polystyryl grafts gave products which exhibited bimodal moelcular weight distributions in which one component of the distribution had the same retention time as that of the macromonomer.

Finally, copolymers containing polyethylene backbone and polystyryl grafts, (**124**), have been manufactured by ring-opening metathesis copolymerization of cycloalkenes like cyclooctene with ω-norbornenyl macromonomers in the presence of the molybdenum carbene initiator $Mo(NAr)(CH\textit{t}Bu)(OC(CH_3)(CF_3)_2)_2$ [92] [Eq. (56)].

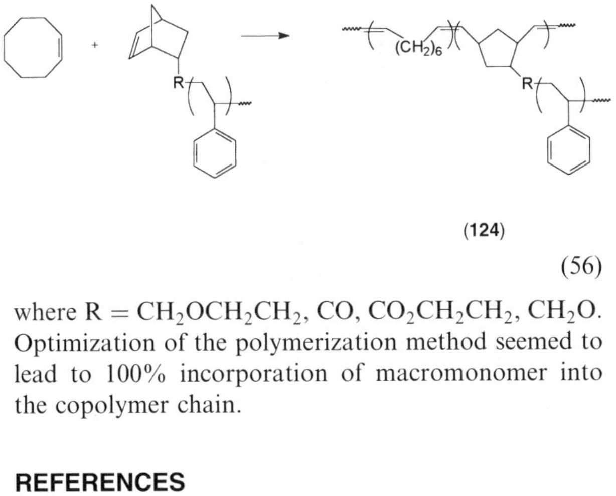

(124)

(56)

where R = $CH_2OCH_2CH_2$, CO, $CO_2CH_2CH_2$, CH_2O. Optimization of the polymerization method seemed to lead to 100% incorporation of macromonomer into the copolymer chain.

REFERENCES

1. (a) JP Kennedy. Cationic Polymerization of Olefins: A Critical Inventory. New York: John Wiley & Sons, 1975. (b) PH Plesch, ed. The Chemistry of Cationic Polymerization. New York: Macmillan, 1963.
2. (a) J Boor Jr. Ziegler–Natta Catalysts and Polymerization. New York: Academic Press, 1979. (b) G Fink, R Mulhaupt, HH Brinzinger, eds. Ziegler Catalysts, Recent Scientific Innovations and Technological Improvements. Berlin: Springer, 1995.
3. (a) V Dragutan, AT Balaban, M Dimonie. Olefin Methathesis and Ring-Opening Polymerization of Cycloolefins. New York: John Wiley & Sons, 1985. (b) Y Imamoglu, B Zumreoglu-Karan, AJ Amass, eds. Olefin Metathesis and Polymerization Catalysts. Dordrecht: Kluwer, 1990.
4. (a) KJ Ivin, JC Mol. Olefin Metathesis and Metathesis Polymerization. London: Academic Press, 1997. (b) KJ Ivin. Olefin Metathesis. London: Academic Press, 1983.
5. G Natta *et al.* Fr. Patent 1,392,142, 1965.
6. I Tritto, RH Grubbs. In: T Keii, K Soga, eds. Catalytic Olefin Polymerization. Amsterdam: Elsevier, 1990, pp 301–312.
7. W Risse, RH Grubbs. Macromolecules 22: 1558, 1989.
8. AJ Amass *et al.* In: Y Imamoglu, B Zumreoglu-Karan, AJ Amass, eds. Olefin Metathesis and Polymerization Catalysts. Dordrecht: Kluwer, 1990.
9. J Boor Jr., EA Joungman, M Dimbat. Makromol Chem 90: 26, 1966.
10. F Hoffman. Chem Z 57: 5, 1933.
11. E Ionescu *et al.* Mat Plast (Bucharest) 22: 84, 1985.
12. G Heublein *et al.* J Macromol Sci, Chem A21: 1355, 1984.
13. G Heublein, G Albrecht. Acta Polym 33: 505, 1982.
14. C Aso, O Ohara. Makromol Chem 109: 161, 1967.
15. C Aso, O Ohara. Makromol Chem 127: 78, 1969.
16. S Kohjiya, I Imanishi, S Okamura. J Polym Sci A1, 6: 809, 1968.
17. S Kohjiya, K Nakamura, S Yamashita. Angew Makromol Chem 27: 189, 1972.
18. W Vredenburgh, KF Foley, N Scarlatti. In: Kirk-Othmer Encyclopedia of Polymer Science and Engineering, vol. 7. New York: John Wiley & Sons, 1987, pp 758–782.
19. HB Wheeler, EP D'Amico (to Neville Chemical Co.). US. Patent 3,468,837, 1969.
20. G Natta *et al.* Belg. Patent 619,877, 1962.
21. G Natta, G Mazzanti. Makromol Chem 61: 178, 1963.
22. (a) Exxon Research & Engineering Co. US. Patent 4,306,041, 1980. (b) Exxon Research & Engineering Co. US. Patent 4,016,342, 1977.
23. (a) VEB Leuna-Werk Ulbricht. Ger. Patent (DDR) 211,612, 1986. (b) VEB Leuna-Werk Ulbricht. Ger. Patent (DDR) 233,043, 1983.
24. VEB Leuna-Werk Ulbricht. Ger. Patent (DDR) 232,659, 1987.
25. VEB Leuna-Werk Ulbricht. Ger. Patent (DDR) 230, 071, 1987.
26. (a) VEB Leuna-Werk Ulbricht. Ger. Patent (DDR) 235,391, 1983. (b) VEB Leuna-Werk Ulbricht. Ger. Patent (DDR) 235,392, 1983.
27. VEB Leuna-Werk Ulbricht. Ger. Patent (DDR) 235,394, 1983.
28. (a) Chemische Wrke Huels. Ger. Offen 2,221,982, 1972. (b) H Emde. Ger. Patent 2,532,115, 1977. (c) H Emde.

Angew Makromol Chem 60/61: 1, 1977. (d) Comb. Petrochim. Romanian Patent 102,489, 1982.

29. (a) G Heublein *et al.* (To Friedrich Schiller University, Jena). Ger. Patent (DDR) 153,880, 1980. (b) Nippon Oil KK. Japan. Patent 010,068, 1989.
30. (a) H Schenko *et al.* Angew Makromol Chem 20: 141, 1971. (b) Synthetic Rubber RE. Swiss Patent 950,838, 1978. (c) Sumimoto Chemical Industries KK. Japan. Patent 016,102, 1982. (d) Sumimoto Chemical Industries KK. Japan. Patent 290,970, 1987. (e) Daicel Chemical Industries KK. Japan. Patent 046,735, 1988. (f) S Cesca *et al.* Makromol Chem 175: 2539, 1974. (g) Sumitome Chemical Industries KK. Japan. Patent 41,604, 1971. (h) Nippon Oil KK. Japan. Patent 010,068, 1989. (i) Nippon Oil KK. Japan. Patent 210,205, 1989.
31. (a) Hoechst AG. Ger. Offen. 3,835,044, 1990. (b) W Kaminsky, R Spiehl. Makromol Chem 190: 515, 1989. (c) Mitsui Petrochemical Industries Ltd. Japan. Patent 45,152, 1987. (d) Mitsui Petrochemical Industries Ltd. Japan. Patent 312,112, 1987. (e) Mitsui Petrochemical Industries Ltd. Japan. Patent 315,206, 1987. (f) Mitsui Petrochemical Industries Ltd. Japan. Patent 019,622, 1988. (g) Mitsui Petrochemical Industries Ltd. Japan. Patent 110,486, 1989. (h) W Kaminsky *et al.* In: T Keii and K Soga, eds. Catalytic Olefin Polymerization. Amsterdam: Elsevier, 1990.
32. (a) Mitsui Petrochemical Industries Ltd. Japan Patent 016,839, 1986. (b) Mitsui Petrochemical Industries Ltd. Japan. Patent 242,336, 1986. (c) Mitsui Petrochemical Industries Ltd. Japan. Patent 014,761, 1987. (d) Mitsui Petrochemical Industries Ltd. Japan. Patent 016,839, 1987. (e) Mitsui Petrochemical Industries Ltd. Japan. Patent 057,006, 1987. (f) Mitsui Petrochemical Industries Ltd. Japan. Patent 084,302, 1987. (g) Mitsui Petrochemical Industries Ltd. Japan. Patent 095,906, 1987. (h) Mitsui Petrochemical Industries Ltd. Japan. Patent 110,545, 1987. (i) Mitsui Petrochemical Industries Ltd. Japan. Patent 248,113, 1989. (j) Mitsui Petrochemical Industries Ltd. Japan. Patent 296,992, 1982. (k) Mitsui Petrochemical Industries Ltd. Japan. Patent 199,477, 1990. (l) Mitsui Petrochemical Industries Ltd. Japan. Patent 201,994, 1990. (m) Mitsui Petrochemical Industries Ltd. Japan. Patent 223,777, 1990. (n) Mitsui Petrochemical Industries Ltd. Japan. Patent 276,739, 1990. (o) Mitsui Petrochemical Industries Ltd. Japan. Patent 300,357, 1990. (p) Mitsui Petrochemical Industries Ltd. Japan. Patent 187,250, 1987.
33. (a) Mitsui Petrochemical Industries Ltd. Japan. Patent 085,351, 1989. (b) Mitsui Petrochemical Industries Ltd. Japan. Patent 259,012, 1989.
34. Copolymer Rubber Chemicals. US. Patent 3,994,947, 1971.
35. G Dall'Asta, G Motroni. Ger. Patent 2,337,359, 1973.
36. KW Scott, N Calderon. Ger. Patent 2,058,198, 1971.
37. G Motroni, G Dall'Asta, IW Bassi. Eur Polym J 9: 257, 1973.
38. G Dall'Asta, G Motroni. Eur Polym J 7: 707, 1971.
39. G Dall'Asta, G Motroni, L Motta. J Polym Sci A1, 10: 1601, 1972.
40. KJ Ivin, G Lapienis, JJ Rooney, CD Stewart. Polymer 20: 1308, 1979.
41. (a) FL Klaweter, RH Grubbs. Synth. Met 28: D105, 1989. (b) AJ Bell. Ger. Patent 2,553,954, 1976.
42. (a) KJ Ivin *et al.* Pure Appl Chem 54: 447, 1982. (b) KJ Ivin, G Lapienis, JJ Rooney. Makromol Chem 183: 9, 1982.
43. G Pampus *et al.* Ger. Patent 1,961,865, 1969.
44. JD Donald, CA Uraneck, JE Burgleigh. Ger. Patent 2,137,524, 1972.
45. CA Uraneck, JE Burgleigh. US. Patent 3,681,300, 1972.
46. (a) H Ikeda, S Matsumoto, K Makino. Jpn Kokai Tokkyo Koho 77: 47900, 1977. (b) F Imaizumi, K Enyo, Y Nakada. Jpn Kokai Tokkyo Koho 77: 45697, 1977. (c) M Zimmermann, G Pampus, D Maertens. Ger. Patent 2,460,911, 1976.
47. LM Vardanyan, Y V Korshak, BA Dolgoplosk. Vysokomol Soedin B 15: 268, 1973.
48. (a) BF Goodrich Co. US. Patent 4,136,247, 1977. (b) BF Goodrich Co. US. Patent 4,136,248, 1977. (c) BF Goodrich Co. US. Patent 4,069,376, 1977. (d) BF Goodrich Co. Eur Patent 140,319, 1983.
49. RJ Minchak (to BF Goodrich Co.). US. Patent 4,138,448, 1979.
50. WJ Feast, K Harper. J Mol Catal 28: 293, 1985.
51. (a) Y Tanaka, T Ueshima, S Kurosawa. Jpn Koai Tokkyo Koho 75: 160400, 1975. (b) J Asrar, SA Caran. J Mol Catal 65: 1, 1991.
52. RH Grubbs, W Tumas. Science 243: 907, 1989.
53. F Imaizumi, S Matsumoto, T Kotani, K Enyo. Jpn Kokai Tokkyo Koho 77: 51500, 1977.
54. RR Schrock *et al.* J Mol Catal 46: 243, 1988.
55. (a) J Kress *et al.* Makromol Chem 191: 2237, 1990. (b) RME Greene *et al.* Makromol Chem 189: 2797, 1988.
56. LF Cannizzo, RH Grubbs. Macromolecules 21: 1961, 1988.
57. TM Swager, RH Grubbs. J Am Chem Soc 109: 894, 1987.
58. RR Schrock *et al.* Paper presented at the 7th International Symposium on Olefin Metathesis and Polymerization, Kingston Upon Hull, UK, August 24–28, 1987, abstracts, p 17.
59. V Sankaran *et al.* J Am Chem Soc 112: 6858, 1990.
60. RR Schrock *et al.* Paper presented at the 9th International Symposium on Olefin Metathesis and Polymerization, Collegeville, Pennsylvania, USA, July 21–26, 1991, abstracts, p 31.
61. E Khosravi, WJ Feast, VC Gibson, EL Marshall. Paper presented at the 9th International Symposium on Olefin Metathesis and Polymerization, Collegeville, Pennsylvania, USA, July 21–26, 1991, abstracts, p 13.

62. LY Park, RR Schrock, SG Stieglitz, WE Crowe. Macromolecules 24: 3489, 1991.
63. FW Klaweter, RH Grubbs. Synth Met 26: 311, 1988.
64. RS Saunders, RE Cohen, RR Schrock. Macromolecules 24: 5599, 1991.
65. F Stelzer, RH Grubbs, G Leising. Polymer 32: 1851, 1991.
66. GC Bazan, RR Schrock. Macromolecules 24: 817, 1991.
67. W Risse, RH Grubbs, cited after Ref. 52.
68. W Risse, RH Grubbs. Macromolecules 22: 1558, 1989.
69. (a) G Pampus, J Witte, M Hoffman. Rev Gen Caout Plast 47: 1343, 1970. (b) BA Dolgoplosk *et al.* Recl Trav Chim Pays-Bas 96: M35, 1977. (c) R Streck, K Nordsiek, H Weber, K Mayr. Ger. Offen. 2,131,355, 1973. (d) S Matsumoto, K Komatsu, K Enyo. US. Patent 4,039,491, 1977.
70. R Streck. J Mol Catal 15: 3, 1982.
71. RH Grubbs. JMS Pure Appl Chem A31: 1829–1833, 1994.
72. (a) C Fraser, MA Hillmyer, E Gutierrez, RH Grubbs. Macromolecules 28: 7256–7261, 1995. (b) C Fraser, MA Hillmyer, E Gutierrez, RH Grubbs. Polym Prepriats (Am Chem Soc Div Polym Chem) 36: 237–238, 1995.
73. DM Lynn, S Kanaoka, RH Grubbs. J Am Chem Soc 118: 784–790, 1996.
74. M Weck, P Schwab, RH Grubbs. Macromolecules 29: 1789–1793, 1996.
75. A Noshay, JE McGrath. Block Copolymers. New York: Academic Press, 1977.
76. DM Lynn, B Mohr, RH Grubbs. J Am Chem Soc 120: 1627–1628, 1998.
77. (a) MC Schuster, KH Mortell, AD Hegeman, LL Kiessling. J Mol Catal A: Chem 116: 209–216, 1997. (b) LL Kiessling, NL Pohl. Chem Biol 3: 71, 1996.
78. AJ Amass, WJ Feast. In: G Allen ed. Comprehensive Polymer Science. Oxford: Pergamon Press, 1989.
79. T Masuda, T Higashimura. Adv Polym Sci 81: 121, 1986.
80. T Masuda, Y Yoshida, H Makio, MZA Rahman, T Higashimura. J Chem Soc, Chem Commun 1991: 503–504.
81. H Makio, T Masuda, T Higashimura. Polymer 34: 1490–1495, 1993.
82. H Makio, T Masuda, T Higashimura. Polymer 34: 2218–2223, 1993.
83. T Ohgane, T Masuda, T Higashimura. Polym Bull 32: 517–524, 1994.
84. S Breunig, V Heroguez, Y Gnanou, M Fontanille. Macromol Symp 95: 151, 1995.
85. V Heroguez, J-L Six, Y Gnanou, M Fontanille. Macromol Chem Phys 199: 1405–1412, 1998.
86. V Heroguez, Y Gnanou, M Fontanille. Macromol Chem Rapid Commun 17: 137–142, 1996.
87. V Heroguez, S Breunig, Y Gnanou, M Fontanille. Macromolecules 29: 4459–4464, 1996.
88. V Heroguez, Y Gnanou, M Fontanille. Macromolecules 30: 4791–4798, 1997.
89. E Sanford, J Frechet, K Wooley, C Hawker. Polymer Preprints (Am Chem Soc, Div Polym Chem) 34: 654, 1993.
90. C Hawker, J Wooley, J Frechet. J Chem Soc, Perkin Trans 1: 1287, 1993.
91. WJ Feast, VC Gibson, AF Johnson, E Khosravi, MA Mohsin. Polymer 35: 3542, 1994.
92. S Breunig, V Heroguez, J-L Six, Y Gnanou, M Fontanille. Polymer Preprints (Am Chem Soc, Div Polym Chem) 35: 526–527, 1994.

6

Structure of Polyolefins

Bohdan Schneider and Danica Doskočilová
Institute of Macromolecular Chemistry, Academy of Sciences of the Czech Republic, Prague, Czech Republic

I. INTRODUCTION

The structure of polyolefins is governed by a number of intramolecular characteristics, which combine to produce a great variety of substances. Even polyethylene, the simplest polyolefin, can form various conformational structures of the macromolecular chains, which in the solid state assume various states of intermolecular order. Technical polyethylenes differ by the degree and quality of branching. Monomeric units of polypropylene and of the higher poly-1-olefins (with the exception of polyisobutylene) contain an asymmetric carbon atom, with the result that the macromolecules can differ in the ordering of these asymmetric centers in the chain—in stereoregularity. Together with conformational structure, stereoregularity determines the structure of the macromolecules in the liquid state, and it also affects the crystalline structure in the solid state. The structures of olefin copolymers are even more varied, inasmuch as all the foregoing characteristics are further combined with the effects of ordering chemically differing units in the chain.

II. CONFORMATIONAL STRUCTURE

In polyolefins, rotation around the backbone single bonds can generate an enormous number of possible geometrical structures of the chains. The elements of these chains can assume many positions with respect to one another, and various statistical quantities have been introduced to characterize the overall conformational structure of macromolecules. A quantity frequently used to describe the conformation of polymer chains in solution is the so-called end-to-end distance. The end-to-end distance corresponds to the vector **r** characterizing the distance and orientation of various chain elements. Thus $\mathbf{r} = \sum \mathbf{l}_i$ where each element is represented by one position vector $\mathbf{l}_i$. The quantity **r** cannot be experimentally determined, rather the scalar r^2:

$$r^2 = \mathbf{r} \cdot \mathbf{r} = \sum_{ij}^{n} \mathbf{l}_i \mathbf{l}_j$$

is determined, for example, by light scattering measurements. Another such quantity often applied to describe the overall conformation of polymer chains is the so-called radius of gyration s. This parameter follows from a description in which the polymer chain is represented by a collection of mass points at positions defined by the distance s_i with respect to the center of gravity. The radius of gyration is defined by the expression

$$s^2 = \frac{1}{n} \sum_{i=1}^{n} s_i^2$$

It is usually determined from light scattering or viscometric measurements.

The mutual positions of the segments in a polymer chain are determined by the conformational structures generated by rotation about the backbone chemical bonds. In general this rotation is not free, but is connected with a change in potential energy with the angle of rotation. This can be graphically represented by the dependence of potential energy on the angle between the neighboring bonds. For the simplest example, *n*-butane, in which the almost free rotation of the end methyl groups can be neglected, spectroscopic studies and quantum mechanical calculations have shown that the potential energy generated by rotation about the $C_2—C_3$ bond has the form shown in Fig. 1. The minima in this diagram correspond to stable conformers, and the difference in height between two non-equivalent minima is equivalent to ΔH_{ij}, defined as the enthalpy difference of these two conformers. The difference in height between a minimum and the neighboring maximum defines the activation enthalpy ΔH^*_{ij}. All parameters connected with conformational structure formation can be determined from the dependence of potential energy on the rotation angle. At dynamic equilibrium, which practically exists in solutions, the content of various conformational structures at temperature T is given by the relation

$$K_{ij} = \frac{c_i}{c_j} = e^{(-\Delta G_{ij}/RT)}$$

where c_i and c_j are the concentrations of the considered conformational forms and ΔG_{ij} is the free enthalpy difference (Gibbs energy). This is related to the enthalpy difference ΔH_{ij} by the relation

$$\Delta G_{ij} = \Delta H_{ij} - T\Delta S_{ij}$$

where ΔS_{ij} is the change in entropy. The rate at which conformational equilibrium is reached is a function of

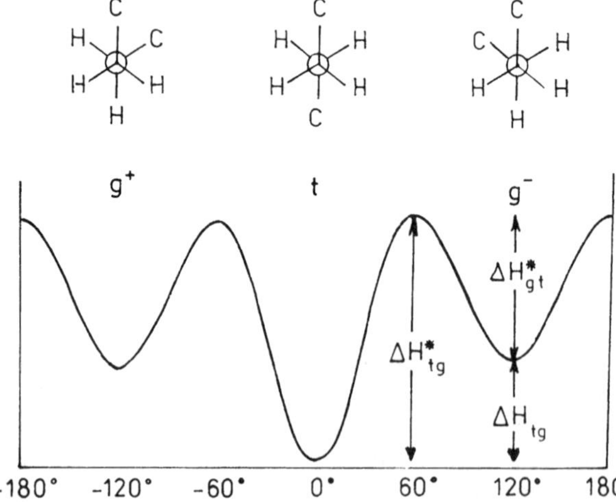

FIG. 1 Potential energy function generated by rotation about the $C_2—C_3$ bond in *n*-butane.

the rotation barrier hindering the rotation and is given by the Arrhenius equation

$$k_{ij} = A_{ij}e^{(-E_{ij}/RT)}$$

where k_{ij} is the rate constant, A_{ij} is the Arrhenius frequency factor, and E_{ij} is the activation energy, while $E_{ij} \cong \Delta H^*_{ij}$.

It is customary to designate the stable conformational forms with three coplanar C—C bonds as t, and those in which one bond is rotated from the plane of the other two by 120° as g. Figure 1 shows these forms for butane. These are the so-called staggered conformational forms, and they are important for describing conformational structures in polymer chains.

All the above-mentioned parameters can be determined by means of molecular spectra. In the liquid state, the lifetime of a conformational form in hydrocarbons is 10^{-9}–10^{-11} s. In vibrational (infrared (IR) and Raman) spectra, the timescale of measurement is about 10^{-13}, therefore vibrational spectra exhibit overlapping spectra of all conformational forms present, and thus are very suitable for determining the populations of various conformers. In NMR spectra, with a spread of chemical shifts on the order of kHz, a chance for discerning individual conformers arises for lifetimes on the order of milliseconds. Therefore in liquids the individual conformers are not directly manifested, and the spectrum appears as that of a single compound, with parameters (chemical shifts and coupling constants) corresponding to the weighted average of the conformers present. The situation is different in solids, where the lifetimes of various conformational forms are sufficiently long so that in high resolution solid state NMR spectra, all conformers present can be individually manifested.

In polymer chains the situation is much more complicated than that described above for butane. The generation of conformational structures depends not only on the rotational barriers due to neighboring groups, but also on the cooperative effects of additional bonds. Also the sense of rotation must be respected, because the two g forms, g^+ and g^- of Fig. 1, are no longer equivalent.

When the interactions of only three bonds are considered in the generation of conformational structures, as in the case of *n*-butane, three rotational isomers (t, g^+, g^-) can be generated by rotation about each single bond, and thus a polymer chain of length n can form 3^{n-2} conformational structures. A description considering the effect of four bonds in the generation of conformational structures is based on the potential curve for *n*-pentane, and the conformations thus generated can schematically be represented as shown in Fig. 2. The conformational structure of polymer chains can usually be fairly well described in this approximation. The structures g^+g^+ and g^-g^- are energetically unfavored and almost never occur in hydrocarbons.

III. STRUCTURE OF POLYETHYLENE: CONFORMATION AND CRYSTALLINITY

According to theoretical calculations and experimental measurements, in polyethylene (PE) solutions about 65% of monomeric units assume the t and 35% the g conformational form [1]. The conformational structure of polyethylene in melt is very similar to that in solution [1]. This is because in polymer melts, the polymer chains behave in the same way as in Θ solvents.

In solid polyethylene, the situation is different. Polyethylene is one of the so-called semicrystalline polymers. In these polymers, the presence of three phases usually is considered: the crystalline phase, the amorphous phase, and a third phase with various designations, which is best characterized as a phase with partial anisotropy of polymer chains (partially ordered phase).

According to X-ray studies, in the most frequent orthorhombic crystalline form of polyethylene, the chains assume the planar zigzag ttt conformational structure [2]. The unit cell contains two polyethylene chains, and this is also manifested in IR spectra, where many bands exhibit characteristic doublets caused by vibrational dipolar interactions of polymer chains [3]. Several other crystalline modifications of polyethylene

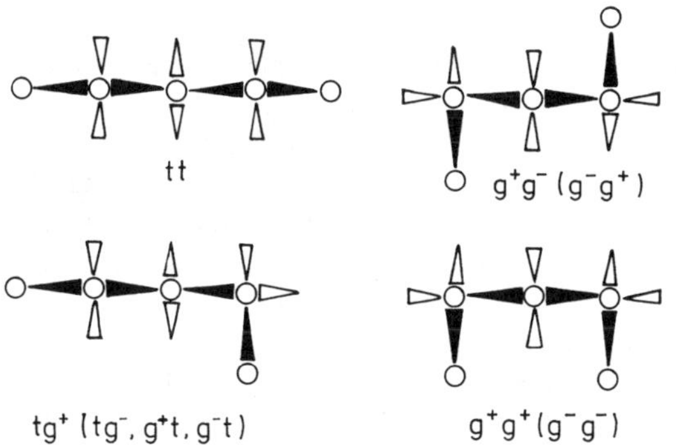

FIG. 2 Conformers generated by rotation about two neighboring bonds in a polymethylene chain. ○ Carbon atoms involved in γ-gauche interaction.

have been described, but they differ from the most stable orthorhombic form only by minor changes in the ordering of chains in the unit cell.

In the amorphous phase of solid polyethylene, we find the same conformational structures as in polyethylene solutions and melts [4]. The only difference rests in the circumstance that in solid amorphous polyethylene, conformer averaging is less rapid. Therefore bands characteristic of structures containing g forms are exhibited not only in vibrational spectra, but also in solid state, high resolution ^{13}C NMR spectra [5, 6] (Fig. 3). Thus in the range of CH_2 carbons, ^{13}C CP MAS (crosspolarization magic angle spinning) NMR spectra of all types of polyethylene exhibit a band assigned to long sequences of planar zigzag . . . ttt . . . conformational structures corresponding to the crystalline phase of PE (peak 1 in Fig. 3), and another band assigned to sequences containing g conformational structures (peak 2 in Fig. 3). The position of this band is affected by the orientation of CH_2 groups in the γ position (i.e., three bonds from the carbon of interest [7] and corresponds to the expected value of the so-called γ-gauche effect (the respective carbons are marked ⊗ in Fig. 2). The width and shape of this band are given by the sequence length distribution in segments containing g conformational structures [5].

Also Raman spectroscopy has greatly contributed to our knowledge of the conformational structure of polyethylene in the solid state. In the range below $250\,cm^{-1}$ there appear the so-called acoustic branches of chain vibrations; from their position, the length of uninterrupted sequences of t structure can be determined by means of the expression originally proposed by Schaufele and Shimanouchi [8]:

$$\nu = (m/2L)E/\rho)^{1/2}$$

where m is the acoustic mode order, L the length of the straight extended chain segment, E the elastic modulus, and ρ is density. Sequence distributions calculated from experimental Raman spectra by means of expressions analogous to the equation above are in good agreement with the results of X-ray methods [9]. It should be noted that Raman data measure the length of all-trans sequences in the crystal, whereas folds are included in the data obtained by SAXS. More recently, domain sizes are measured by ^1H spin-diffusion NMR experiments [6, 10, 11]. The agreement of various methods is demonstrated in Table 1.

Conformational structure in the partially ordered phase depends on the method and history of polyethylene crystallization: it has different structures at the

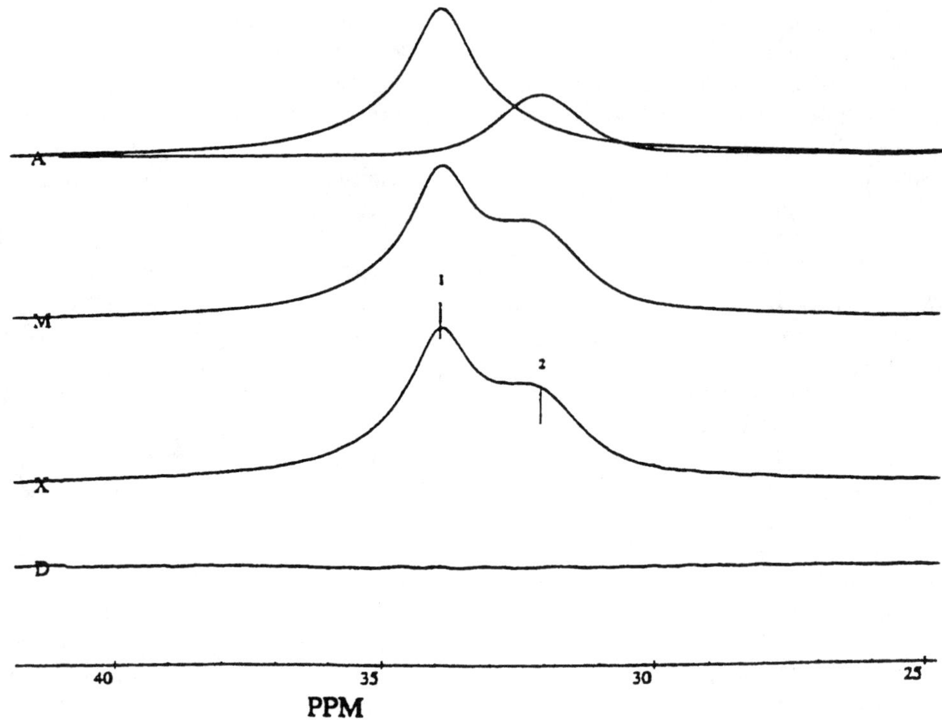

FIG. 3 ^{13}C CP-MAS NMR spectrum of semicrystalline polyethylene (X), with line separation and peak fitting (A,M,D). (From Ref. 6.)

surface of monocrystals prepared from high-density polyethylenes by solution crystallization and in semicrystalline polyethyelenes prepared by thermal crystallization [1]. This is because, in the partially ordered phase, the structural freedom of individual chains is limited to different extents when each chain passes from the crystalline into the amorphous phase.

Upon analysis it was found that IR, Raman, broad-line or solid-echo ^{1}H NMR, and solid-state ^{13}C NMR spectra usually cannot be fully described when only two phases are assumed to be present, but three phases are usually sufficient for a satisfactory description. The same holds for the analysis of NMR relaxation data [6, 10, 11]. In broad-line ^{1}H NMR spectra, the widest band corresponds to the crystalline state, the intermediate band to the partially ordered phase, and the narrow band to the amorphous phase of polyethylene. The content of these phases can be determined by numerical analysis of the spectra [12, 13] (Fig. 4). Also IR [14] and Raman [15, 16] bands can be subjected to a similar bandshape analysis, assuming the presence of three phases. Also in solid-state ^{13}C NMR [6], analysis of T_{1C} relaxation data for the crystalline peak reveals the presence of two components, the true crystalline and the intermediate one, differing in mobility. The spectral characteristics of the "intermediate" phase in both Raman and solid-state ^{13}C NMR spectra are those of longer tt sequences. The results of crystallinity determination in polyethylene by various methods are summarized in Table 2. The agreement of simple crystallinity data obtained by various methods is better than might have been expected, since the methods are based on quite different properties of the material: in DSC and X-ray analysis, the crystalline phase is understood to be that with regular mutual chain ordering, while vibrational spectroscopy

TABLE 1 Lamellar Thickness (Å) in Linear PE from Raman, X-ray (SAXS) and NMR Spin Diffusion Measurements

Ref.	Raman	SAXS	^{1}H NMR spin diffusion
	106	111	
[9]	155	167	
	238	249	
[11]	178		221

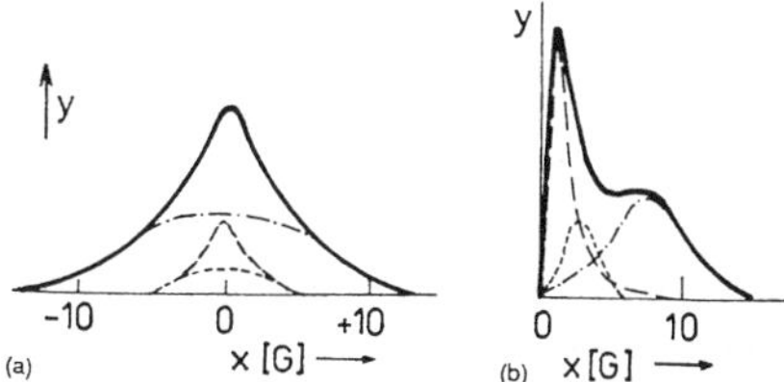

FIG. 4 Broad line ^{1}H NMR spectrum of solid polyethylene and its components obtained by digital decomposition: (a) absorption lineshape and (b) derivative lineshape. Solid curves, experimental; dash-dot, crystalline; short dashes, partially ordered; long dashes, amorphous phase. (Adapted from Ref. 10.)

essentially characterizes the conformational structure, and NMR spectroscopy is sensitive both to conformation and to segmental mobility in various phases. In a series of samples, differences in crystallinity may originate from differences in chain branching, or from different crystallization procedures applied to a single, essentially linear polymer, and this may also be reflected in the results of various methods. Moreover, particularly the content of the intermediate phase is affected by the method applied: it always strongly depends on the bandshapes introduced in the line-separation procedures, on the functions used in the analysis of relaxation data [20]: a continuous transition in order and a continuous distribution of mobilities in the amorphous phase adjacent to a crystalline surface is probably equally plausible [11, 21].

IV. CONFIGURATIONAL STRUCTURE

Contrary to the simplest polyolefin polyethylene, the monomeric units in polypropylene and the higher poly-1-olefins contain an asymmetric center that leads to the possibility of generating an even greater variety of molecular chain structures. First of all, under some polymerization conditions, monomer inversion can occur, creating tail-tail (**2**) or head-head (**3**) structures, among the more common head–tail (**1**) ones:

$$CH_2{-}CH(CH_3){-}CH_2{-}CH(CH_3){-}CH_2{-} \quad (1)$$

$$-CH(CH_3){-}CH_2{-}CH_2{-}CH(CH_3){-} \quad (2)$$

$$-CH_2{-}CH(CH_3){-}CH(CH_3){-}CH_2 \quad (3)$$

TABLE 2 Comparison of Crystallinity Values as Determined by Various Methods for Selected Polyethylene Samples

	Calorimetry	X-ray		Ra		^{1}H-NMR		^{13}C-NMR			
Density	DSC	WAXS	IR	c	i	c	i	c	Cryst	i	Ref.
67			67			62	?				[17]
74			80			82	?				
		57	53								[18]
		74	72								
63	66					69	temp				[12]
94	–					91	dep.				
64						64	18				[19]
94						94	5				
		77		77	6						
		65		67	6						[15]
		41		40	13						
		39		37	12						
	85	93							91		
								73		18	
	69	83							80		[6]
								59		21	
	67	75							79		
								58		21	
71	62			66	14	66					[11]
90	82			87	3	86					

Structures (**1**) to (**3**) can be regarded as variations in chemical structure.

Second, even in a purely head–tail polymer, the asymmetric centers may be placed in mutual meso (i) or racemic (s) orientation, thus forming isotactic or syndiotactic sequences of various lengths.

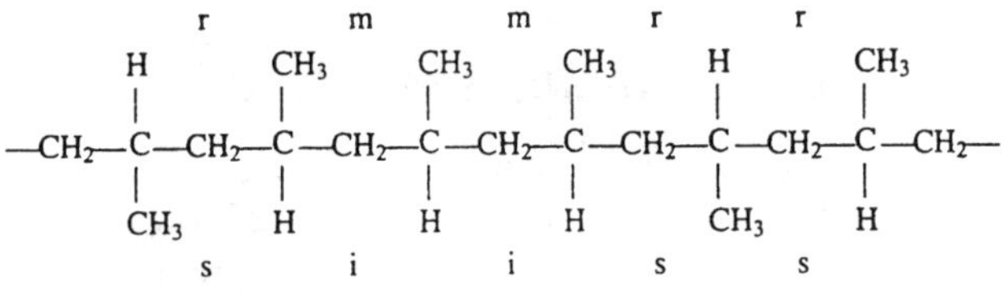

```
            r          m          m          r          r
      H         CH3        CH3        CH3        H          CH3
      |          |          |          |         |           |
—CH2—C—CH2—C—CH2—C—CH2—C—CH2—C—CH2—C—CH2—
      |          |          |          |         |           |
     CH3         H          H          H        CH3          H
            s          i          i          s          s
```

These sequences are of identical chemical, but different configurational structure.

It appears that such a polymer can be described as a copolymer composed of i and s units. The properties of the polymer are strongly dependent not only on the relative amounts of i and s units (composition, tacticity), but also on their distribution. Polymers with comparable fractions of randomly distributed i and s units are called atactic. Polymers with a predominance of either i or s units are called isotactic or syndiotactic, respectively. When the distribution of i and s units is random, we speak of a statistical polymer, in contrast to a block polymer, which may contain long sequences of only i or only s units.

Polymers with long regular sequences of like structures will exhibit a higher tendency to crystallize, while polymers with a more varied distribution of the basic units can present a wide range of elastic properties. It is therefore of considerable practical interest to characterize quantitatively the tacticity and sequence distribution (chain microstructure) in polymers of this kind. To this end, the methods of NMR spectroscopy, and particularly ^{13}C NMR in the case of polyolefins, are the methods of choice. The chemical shifts of ^{13}C nuclei are sensitive to the chemical nature and geometrical arrangement of the neighboring atoms over several bonds. Thus in sufficiently dilute solutions and at sufficiently elevated temperatures, separate signals of carbons in various surroundings can be resolved and, under appropriate experimental conditions, quantitatively measured. Various methods have been devised for assigning these signals to specific structural units, using specially synthesized model compounds and polymers, combined with evaluation of their structure by statistical methods. These methods are identical to those applied for the characterization of the microstructure of copolymer chains to be discussed later on.

The structure of a chain built from two types of structural units (e.g., i and s in a stereopolymer) can be described by the populations of various n-unit sequences. Two types of elementary unit can generate 2^n types of sequence of length n (where some cannot be distinguished), the populations of which are bound by a number of necessary relations (Table 3). These relations are very useful, because they are valid, irrespective of the mechanism by which the polymer has been generated and of the statistics it obeys. They are often applied to substitute for incomplete experimental data. Thus the populations of all six three-unit sequences can be established by the determination of only two of them (e.g., the easily assignable iii and sss) when the two-unit populations are known, etc.

When in polymerization the addition probability of a monomer unit is independent of the type of the last unit in the growing chain (as is the case in most radical polymerizations), then the chain growth is governed by the two probabilities, P_1 and P_2, of the addition of monomer 1 or 2. Since these two probabilities are bound by the relation $P_1 + P_2 = 1$, the polymer statistics is determined by a single parameter $p = P_1 = 1 - P_2$. This is the so-called Bernoulli statistics. When the probability of monomer addition depends on the type of the last monomer unit in the growing chain, the system is characterized by four transition probabilities, P_{11}, P_{12}, P_{21}, and P_{22}, bound by the conditions $P_{11} + P_{12} = 1$ and $P_{21} + P_{22} = 1$, defining two independent parameters (e.g., $P_{21} = 1 - P_{22}$ and $P_{12} = 1 - P_{11}$). This is the so-called first-order Markov statistics. Markov

TABLE 3 Relations Between the Populations of Sequences for a Chain Composed of i and s Units

n: 1–2
(i) = (ii) + $\frac{1}{2}$ (is)
(i) = (ss) + $\frac{1}{2}$ (is)
n: 2–3
(ii) = (iii) + $\frac{1}{2}$ (iis)
(is) = (iis) + 2(sis) = (iss) + 2(isi)
(ss) = (sss) + $\frac{1}{2}$ (iss)
n: 3–4
(iii) = (iiii) + $\frac{1}{2}$ (iiis)
(iis) = (iiis) + 2(siis) = (iiss) + 2(iiss)
(isi) = $\frac{1}{2}$ (iisi) + $\frac{1}{2}$ (isis)
(iss) = 2(issi) + (isss) = (iiss) + (siss)
(sis) = $\frac{1}{2}$ (isis) + $\frac{1}{2}$ (siss)
(sss) = (ssss) + $\frac{1}{2}$ (isss)

statistics of higher order can be derived on the same principles. By means of the independent statistical parameters, the polymer microstructure is fully characterized, and the populations of sequences of arbitrary length can be calculated as shown in Table 4.

The single parameter of Bernoulli statistics can be simply determined from one-unit "sequences": that is, from the tacticity of a stereopolymer (or the composition of a copolymer). To determine the two parameters of first-order Markov statistics, the populations of all three two-unit sequences must be known (use can be made of the necessary relations, Table 3). To verify whether the proposed order is actually valid, the populations of sequences longer by one unit must be checked by experiment (e.g., Bernoulli statistics is established by a check on the two-unit populations, etc.).

By means of the statistical parameters, the number-average sequence lengths can be calculated:

$$\bar{l}_1 = \frac{1}{P_{12}} \quad \text{and} \quad \bar{l}_2 = \frac{1}{P_{21}}$$

Being defined as the ratio $\bar{l}_1$ = (number of type 1 units)/(number of blocks of type 1 units), these can be directly calculated—for example, from the experimentally determined populations of two-unit sequences of i and s units as follows:

$$\bar{l}_i = \frac{2(\text{ii}) + (\text{is})}{(\text{is})} \qquad \bar{l}_s = \frac{2(\text{ss}) + (\text{is})}{(\text{is})}$$

In the sense of the definitions above, the number-average sequence length is independent on the order of the

TABLE 4 Relations Between i and s Sequence Populations and Statistical Parameters

n	$P(n)$	Bernoulli $P_i + P_s = 1$	First-order Markov $P_{ii} = P_{is} = 1, P_{si} + P_{ss} = 1$
1	(i)	P_i	$P_{is}/(P_{si} + P_{is})$
	(s)	$1 - P_i$	$P_{si}/(P_{si} + P_{is})$
	(ii)	P_i^2	$(1 - P_{si})P_{is}/(P_{si} + P_{is})$
2	(is)	$2P_i(1 - P_i)$	$2P_{si}P_{is}/(P_{si} + P_{is})$
	(ss)	$(1 - P_i)^2$	$P_{si}(1 - P_{si})/(P_{si} + P_{is})$
	(iii)	P_i^3	$(1 - P_{si})^2 P_{is}/(P_{si} + P_{is})$
	(iis)	$2P_i^2(1 - P_i)$	$2P_{si}(1 - P_{si})P_{is}/(P_{si} + P_{is})$
3	(isi)	$P_i^2(1 - P_i)$	$P_{si}P_{is}^2/(P_{si} + P_{is})$
	(iss)	$2P_i(1 - P_i)^2$	$2P_{si}P_{is}(1 - P_{is})/(P_{si} + P_{is})$
	(sis)	$P_i(1 - P_i)^2$	$P_{si}^2 P_{is}/(P_{si} + P_{is})$
	(sss)	$(1 - P_i)^3$	$P_{si}(1 - P_{is})^2/(P_{si} + P_{is})$

statistics. This is important because stereospecific polymerizations in particular sometimes do not obey simple statistics.

V. STRUCTURE OF POLYPROPYLENE AND OF THE HIGHER POLYOLEFINS: STEREOREGULARITY, CONFORMATION AND CRYSTALLINITY

Polypropylene is a typical polymer whose properties are profoundly affected by stereoregularity. Isotactic polypropylene is prepared with various modifications of Ziegler–Natta coordination catalysts, producing polymers with varying degrees of stereoregular order, with isotacticity reaching up to 98%. Syndiotactic polypropylene is prepared with soluble coordination catalysts and the stereoregularity attained generally is lower than that of the isotactic polymers. Atactic polypropylene can be obtained by extraction with boiling *n*-heptane from isotactic polypropylene of lower stereoregularity.

The NMR chemical shift of the carbon (or protons) of a methylene group in a head–tail polypropylene is sensitive to the s or i character of that group, in the first place. With sufficient resolution, the fine structure of the methylene bands can be assigned to i- or s-centered sequences with an odd number of units. The chemical shifts of the methine or methyl carbons (or protons) are sensitive to the structure of the two contiguous units, ii, is, or ss, and with sufficient resolution to the structure of the corresponding sequences with an even number of units. The extent to which bands of longer sequences can be resolved and the corresponding populations analyzed with the use of modern NMR instruments is demonstrated in Fig. 5, which shows the fine structure of the methyl-carbon band of an atactic polypropylene sample, together with band assignments.

The polymerization of isotactic polypropylene with Ziegler–Natta catalysts is an example of a process that does not obey simple statistics of the Bernoulli or Markov type, and a two-site mechanism has been proposed to describe the sequence distribution in the resulting polymers [22].

The type of error (I or II) in stereosequence propagation in the most highly isotactic polypropylenes has been subject of considerable attention. This is because error type has implications not only for the determination of mean isotactic sequence length, but also for the mechanism of stereosequence polymerization.

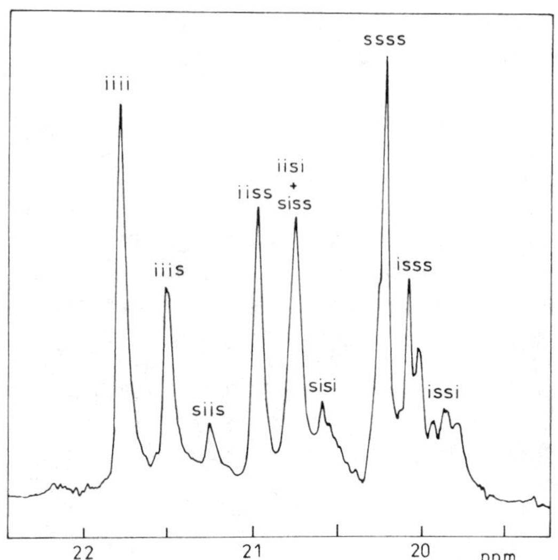

FIG. 5 The fine structure of the methyl band in the ^{13}C NMR spectrum of atactic polypropylene at 90 MHz (1,2,4-trichlorobenzene, 100°C). (Adapted from Ref. 23.)

```
        i       i       s       i       i
   CH3     CH3     CH3     H       H       H
   |       |       |       |       |       |
—CH2—C—CH2—C—CH2—C—CH2—C—CH2—C—CH2—C—CH2—
   |       |       |       |       |       |
   H       H       H       CH3     CH3     CH3
                   I

        i       i       s       s       i       i
   CH3     CH3     CH3     H       CH3     CH3     CH3
   |       |       |       |       |       |       |
—CH2—C—CH2—C—CH2—C—CH2—C—CH2—C—CH2—C—CH2—C—CH2—
   |       |       |       |       |       |       |
   H       H       H       CH3     H       H       H
                   II
```

^{13}C NMR analysis combined with elaborate model compound studies gives preference to the iss (**II**) over the isi (**I**) type of error [23].

With high-field (600 MHz) ^{13}C NMR analysis which permits microstructure characterization at the heptad level, the presence of chemical junctions between i- and s-rich sequences (stereoblocks) in single polypropylene molecules could be proved in predominantly isotactic polymers prepared with "high-yield" $MgCl_2$-supported catalysts [24].

The presence of highly stereoregular components, together with less stereoregular material of high molecular weight in some isotactic polypropylenes, can be utilized for the preparation of materials with a wide range of elastic properties (elastomeric polypropylene: ELPP) [25]. These specific properties are thought to result from cocrystallization of the high molecular weight elastomeric component with the more crystalline insoluble fraction, producing a physical network.

Configurational structure naturally affects the conformational structure and crystallinity of polypropylene. For isotactic polypropylene, the energetically most favored structure can, in the staggered approximation, be described as tg [26]. This is also the main conformational form of isotactic polypropylene in solution. In the crystalline state, two structures have been observed [27], designated α and β; they have the some conformational structure, tg, but they differ with respect to mutual chain orientation in the unit cell: α is monoclinic, β is hexagonal. In consequence of the tg conformation, in both these crystalline modifications, the chains form a 3_1 helix.

Isotactic polypropylene is, of course, semicrystalline. The crystalline and amorphous phases can be differentiated in solid state ^{13}C NMR spectra [28]. Crystallinities determined from NMR are in agreement with those from DSC, both being lower than crystallinities determined from density which probably include part of the amorphous phase. Similarly as in the case of polyethylene, the existence of an intermediate partially ordered phase is being considered, and evidence for it has been presented from line shape analysis in solid-state ^{13}C NMR spectra. Once more, quantitative data on the content of the intermediate phase have to be regarded with caution.

In syndiotactic polypropylene, the energetically most favored conformational structures [29] are ...tttt... and ...(tt)(gg)... . The presence of both these conformational structures was proved by infrared spectra in solution [29]. Syndiotactic polypropylene can crystallize in either of these two conformational forms. The most stable crystalline form generated during slow cooling from the melt contains the (tt)(gg) helix [30]. The linear zigzag ...(tttt)... form appears in oriented crystalline syndiotactic polypropylene, where its presence was proved by infrared and X-ray spectroscopy [31]. The structure of the stable crystalline form with (tt)(gg)... conformation, mainly established by solid state ^{13}C NMR studies [32, 33], features two types of CH_2 groups, two inside and two outside the helix (Fig. 6). In studies of highly syndiotactic polypropylene with crystallinity varied by thermal treatment [34], solid state ^{13}C NMR can also reveal fine details about the crystal packing in the crystalline phase, and about the distribution of conformations in the amorphous phase where trans sequences were found to dominate. The presence of

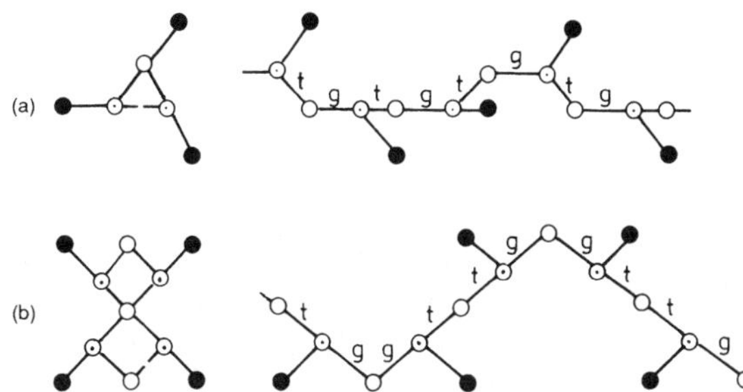

FIG. 6 Conformational structure of crystalline polypropylene: (a) isotactic and (b) syndiotactic. Solid circles, CH_3; open circles, CH_2; circles with dot, CH.

an interfacial phase was also indicated from relaxation measurements.

In the common forms of polypropylene, head–tail sequences strongly predominate. Forms with more abundant head–head and tail–tail sequences can be prepared by nonregiospecific catalysts [35]. Such structures, with tacticity taken into account, can be identified by 2D ^{13}C NMR methods [36]; isolated head–head units were found to always occur as meso structures.

The considerations discussed in detail in connection with the chemical, configurational, and conformational structure of polypropylene apply similarly to poly-1-butene and the higher poly-1-olefins. Also with poly-1-butene, the preferred backbone conformation is tgtg for the isotactic chain [37]. The various crystalline modifications correspond to greater or lesser deviations from these ideal chain conformations, coupled with variations in chain packing in the crystal lattice.

The configurational structure (stereoregularity) of 1-butene and of the higher polyolefins up to 1-nonene has been studied by ^{13}C NMR spectroscopy in solution [38, 39], interpreted with the aid of chemical shift calculations, consideration of the γ effect and of the rotational isomeric state model of Flory. The evaluation of the results favors the bicatalytic sites model of polymerization [40] over simple Markovian statistics. In contrast to polypropylene, side-chain conformation also has to be considered. Comparison with alkane model compounds indicates that in meso-units of poly-1-butene, trans conformation of backbone is less favored than in isotactic polypropylene because of contiguous ethyl group interactions. Introduction of racemic units in both poly-1-butene and polypropylene leads to a decrease of the gauche fraction.

Somewhat special is the position of polyisobutylene, which is due to the crowding of the two bulky methyl groups on the α carbon. Stereoregularity considerations are irrelevant in this case. Because of the crowding of the methyl groups, conformational states are separated by low barriers, and thus under normal conditions, at room temperature without stress, the polymer behaves like a rubber. When oriented by drawing, the chain shows a tendency to crystallize, with a chain conformation of an 8_3 helix, which is near to the tg conformation [41]. This behavior is a result of the regularity of the chain.

VI. COPOLYMERS OF ETHYLENE WITH HIGHER OLEFINS AND BRANCHING IN POLYETHYLENE

By copolymerization of ethylene with propylene and the higher olefins, the properties of the polymer can be varied at will in an extremely broad range. Moreover, combined with ^{13}C NMR spectroscopy, these copolymers have played a major role in elucidating the character of branching in both low density and high density polyethylene. Copolymers with a random distribution of the higher 1-olefin units tend to decrease the crystallizability of polyethylene and enhance the rubbery character of the polymer. Polymers containing blocks of isotactic polypropylene in combination with random ethylene–propylene or branched polyethylene sequences are likewise of commercial interest. In these the blocks of polypropylene (or higher poly-1-olefin) sequences are of course subject to stereospecific variations, and may also contain inverted sequences.

To characterize so wide a variety of structures, ^{13}C NMR spectroscopy is again the method of choice. In analyzing copolymer structure and branching in polyethylene, it has to be kept in mind that ethylene units are present in the copolymer in the form of sequences of methylene units, and in some of these the methylene units from ethylene cannot be differentiated from those coming from the higher 1-olefin comonomer. In copolymer structure analysis by ^{13}C NMR spectra (Fig. 7) it is therefore customary to designate the carbon atoms by their distance from the nearest branching point: that is, the nearest methine carbon of a propylene or higher 1-olefin unit, as shown in **III**, **IV**, and **V**. The customary way of describing a branch is included.

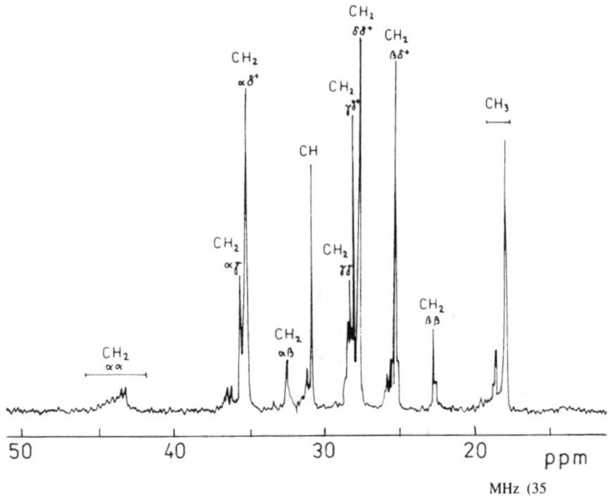

FIG. 7 The ^{13}C NMR spectrum of ethylene–propylene copolymer at 75 MHz (35 mol% propylene units, $CDCl_3$, 60°C).

αα: CH–CH_2–CH, each CH bearing R

III

αβ: CH(R)–CH_2–CH_2–CH(R), both CH_2 labelled αβ

IV

αδ⁺, βγ: CH(CH_2–CH_3; 2B_2, 1B_2)–CH_2 (αδ⁺)–CH_2 (βγ)–CH_2 (βγ)–CH_2 (αδ⁺)–CH(CH_2–CH_2–CH_3; 3B_3, 2B_3, 1B_3)

V

It is clear that structure **IV**, with R = Me, could arise as part of a PEP (propylene–ethylene–propylene) sequence, or an inverted tail–tail PP sequence. Methine, methyl, and other branch carbons can be defined by comonomer sequences (EPE, EPP, PPP, etc.). A statistical characterization can then be developed, similar to that described above for stereosequence analysis [42].

In crystalline propylene–coethylene copolymers with low contents of ethylene units, the type of defects in both intramolecular and intermolecular structure as characterized by a combination of ^{13}C NMR, IR, WAXS and DSC methods could be correlated with physical properties [43]. A special class of ethylene–propylene copolymers are the high impact materials of low crystallinity prepared by sequential polymerization connecting random ethylene–propylene sequences to a highly isotactic polypropylene core [44, 45]. By special fractionating procedures and ^{13}C NMR analysis, several compositional components, each obeying Bernoullian statistics, have been revealed in such material. By solid state ^{13}C NMR, no rigid copolymer phase was found, indicating that amorphous propylene forms the interface to the polypropylene matrix. Evidence for crystallizable runs of ethylene, of domain size > 30 Å, was also presented [46].

With respect to branching, polyethylenes fall into three broad categories: high density (HDPE), low density (LDPE), and linear low density polyethylene (LLDPE).

By ^{13}C NMR spectroscopy, and by comparison with copolymers containing 1-propene and 1-butene comonomer units (E/P and E/B copolymers), typical HDPE samples were shown to contain on the order of 0.5–2.5 methyl branches per thousand backbone carbon atoms: longer branches were not detected and are assumed to be absent [47]. Combined ^{13}C NMR and IR evidence indicates that the number of methyl branches in HDPE samples does not correlate with attainable crystallinity values, which implies that the methyl branches can be largely accommodated in the crystalline domains of polyethylene [47, 48].

In LDPE, prepared by radical polymerization in high pressure reactors, the content of branches is greater by an order of magnitude, and the branches are of varying lengths. Short-chain branching is assumed to affect morphological and solid state properties, while long-chain branching is mostly manifested in viscoelastic properties. Comparison with E/P, E/B, E/H (1-hexene and E/O (1-octene) model copolymers has shown [49] that a typical LDPE with a total of 6–20 methyl groups per thousand carbon atoms contains 2–3% methyl, 31–37% ethyl, ≈ 2% propyl, 34–37% butyl, 11–13% pentyl, and 14–16% longer branches. The ethyl branches are mostly present as 1,3 (predominantly racemic) ethyls, or 1,3-ethyls with one ethyl group on a quaternary carbon; isolated ethyls are rare, as are hexyl groups. This structure conforms with the backbiting polymerization mechanism.

LLDPEs are copolymers of ethylene with a few mol% of 1-butene, 1-hexene, or 1-octene, tailor-made to meet desired density and processing requirements. 1-Butene, the most frequently used comonomer, is known to have a tendency to form contiguous sequences [50].

With the use of such a well defined, ultrahigh molecular weight E/B copolymer, a very sophisticated solid state ^{13}C NMR method has been developed, suitable to determine the partitioning of end groups and branches among the crystalline and amorphous phases [51]. The method claims sensitivity of 2 per thousand carbon atoms. It was found that methyl and vinyl end groups are, to a considerable extent, incorporated in the crystal, whereas only a small fraction of methyl branches, and almost no ethyl branches, enter the crystalline phase. The ease of incorporation into the crystalline phase thus can be represented as follows:

saturated end groups > vinyl end groups > methyl branches > ethyl branches

In addition to appearing in the high pressure radical polymerization process, there exists evidence [42] that long branches may be shear-induced during severe mechanical processing, prolonged melting, or γ-irradiation of polyethylene, together with carbonyl and peroxide groups. This phenomenon is accompanied by a decrease in the content of vinyl end groups, while the content of saturated end groups and short-chain branches remains unchanged. The long, so-called Y branches are identified by the appearance of $\alpha\delta^+$, $\beta\delta^+$, and $\gamma\delta^+$ peaks in ^{13}C NMR spectra.

VII. COPOLYMERS OF OLEFINS WITH OTHER MONOMERS: FOREIGN STRUCTURES IN POLYOLEFINS

In copolymerization of ethylene with different class monomers, all the foregoing aspects are superimposed: various chemical comonomer sequences can be generated, with various mutual stereospecific placements. Chemical comonomer sequences can be analyzed by statistical methods completely analogous to those described in Section IV (Tables 3 and 4), but practical analysis may be complicated by superposition of stereosequence statistics. Short-chain branching in longer polyethylene sequences is an additional factor to be considered. All these aspects can be demonstrated on the example of the ethylene–vinyl alcohol E/V copolymer prepared by a high pressure, free radical process, which has been characterized in great detail by modern NMR methods.

This analysis takes advantage of the fact that, besides CH_2 and CH signals in ^1H NMR spectra, the OH peaks become analyzable in dimethyl sulfoxide solution. Using one-dimensional ^1H and ^{13}C NMR methods, Ketels *et al.* [52] found that at 50°C, comonomer effects dominate the shifts of methine protons, with the result that (with CH_2 decoupling) separate signals of VVV, EVV, and EVE sequences are well resolved. Steric effects are manifested by the fine structure on these peaks. By quantitative analysis of the peak intensities, the copolymerization was found to be governed by Bernoulli statistics, and the distribution of comonomer sequence lengths could be characterized from both ^1H and ^{13}C NMR spectra. From the fine structure of the OH peaks in ^1H NMR spectra, the stereoregularity of V sequences could be characterized by a Bernoullian placement probability P_i, very near to 0.5. From ^{13}C NMR spectra, short-chain branching in polyethylene sequences was found to vary from 2 to 6 CH_3 per thousand carbon atoms for ethylene content varying from 8 to 35 mol %, respectively. Anomalous 1,4- and 1,2-diol structures could also be determined from ^{13}C NMR spectra, amounting to $\approx$ 1% of total V units.

In an almost simultaneous publication by Bruch [53] concerning the same class of E/V copolymers, the chemical and steric sequence assignments were established by two-dimensional NMR methods. Also, the vicinal CH—OH proton coupling constants were found to be smaller in V units with a meso-V neighbor than with a racemic-V neighbor. This information could be utilized for characterizing some conformational aspects of the polymer chain. Likewise the chemical shifts in ^{13}C NMR spectra that are sensitive to the orientation of substituents in γ position (the γ-gauche effect) [7] could be considered for conformational characterization of these polymers.

An important class of ethylene copolymers consists of terpolymers with maleic acid anhydride and acrylic esters, of the following general structure [54]:

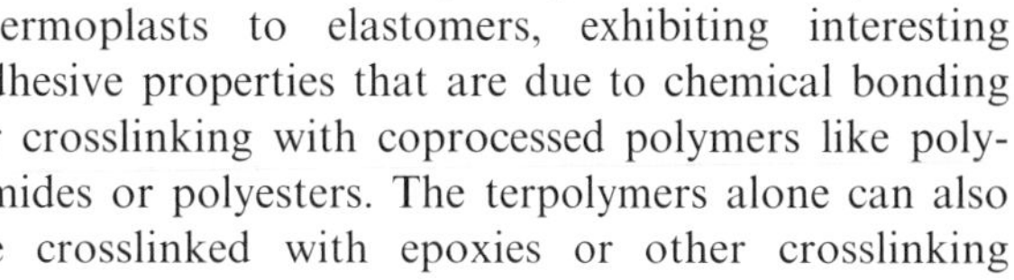

$$-\!\!\!+\!(CH_2\!-\!CH_2)_m\!-\!CH_2\!-\!\underset{\underset{\displaystyle OR}{|}}{\underset{C=O}{|}}{CH}\!-\!(CH_2\!-\!CH_2)_n\!-\!\underset{O=C\diagdown}{\overset{}{CH}}\!-\!\underset{\diagup C=O}{CH}\!-\!\!\!+ \quad (\text{anhydride ring via } O)$$

The distribution of comonomer units is mostly statistical, based on NMR analysis. Depending on composition, such materials may range from semicrystalline thermoplasts to elastomers, exhibiting interesting adhesive properties that are due to chemical bonding or crosslinking with coprocessed polymers like polyamides or polyesters. The terpolymers alone can also be crosslinked with epoxies or other crosslinking agents.

Grafting, another method of introducing foreign structures into polyolefins, is particularly useful for combining widely differing incompatible homopolymers. During the grafting of methacrylic acid onto polyethylene, it was found that grafting is rather inefficient in high density polyethylene because methacrylic acid groups were predominantly introduced at tertiary carbon atoms [55]. Homogeneous systems are obtained by grafting in solution [56]. Wide-angle X-ray scattering and small-angle light scattering measurements indicate that during grafting in suspension, methacrylic acid groups enter chains outside the spherulites only [57].

Various service properties required in protective films (adhesion, wettability, etc.) are obtained, for example, by oxidation of polypropylene films, introducing hydroxyl, carbonyl, carboxyl, or hydroperoxide groups and double bonds into the macromolcule. Similar groups are introduced into polyethylene by irradiation in the presence of oxygen, but they can also spontaneously appear during thermal degradation and natural aging. Vibrational spectra are particularly sensitive in detecting these groups, even at "natural" abundance. Since, however, modern NMR methods are reported to be able to analyze these "defects" at a level of 0.05% [58], combined spectroscopic evidence [59] can be utilized for detailed characterization of such structures.

A specific group of olefins with intentionally introduced foreign structures are the halogenated polyolefins, particularly the chlorinated ones. With increasing chlorine content in polyethylene, the crystallinity of the polymer is reduced; at about 30 wt% chlorine the crystalline phase disappears and the polymer behaves like an elastomer [60]. During photochlorination of polyethylene in solution, the chlorination of the polyethylene chain is random up to the molar ratio Cl:H = 1:1. Besides the original CH_2 groups, the product generated contains both CCl_2 and CHCl groups; CHCl-CHCl groups appear at higher degrees of chlorination [61]. During suspension chlorination, only noncrystalline domains are chlorinated at first [62, 63].

Chlorine enters polyethylene chains together with oxygen during chlorocarboxylation; structures of the type:

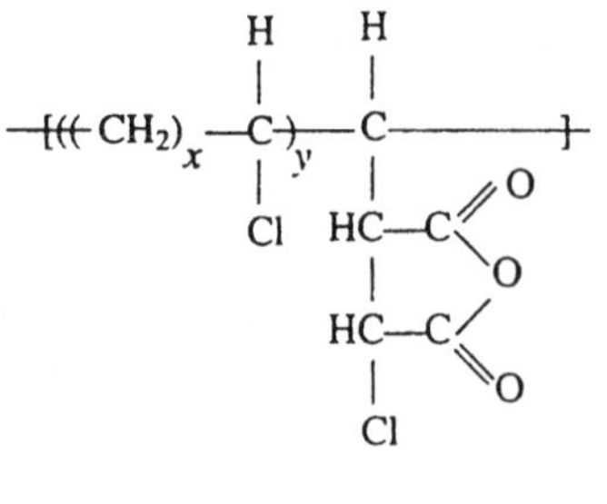

$$-\!\left[\left[\left(CH_2\right)_x-\underset{Cl}{\overset{H}{C}}\right]_y-\overset{H}{C}\left(\underset{\underset{Cl}{|}}{\overset{}{HC}}\right)-\right] \quad \text{with } \ HC-C(=O)-O-C(=O)-CH(Cl) \text{ ring}$$

are formed in a simultaneous reaction of chlorine and maleic anhydride with polyethylene [64].

In reactions of polyethylene with peroxides, branching and network formation are also observed [65]. Chemical crosslinking of polyethylene can be induced by photoinitiated reactions with aromatic ketones. Crosslinking efficiency was found to increase in the order benzoquinone < naphthoquinone < anthraquinone [66]. The form-stability of polyethylene above its normal melting point, as well as its environmental resistance, can be improved in this way. Crosslinking strongly affects the morphological and mechanical properties of polymers, even at very low degrees of crosslinking. At the same time, the conformational structure of polymer chains and the dynamical behavior of polymer segments both in the solid state and in swollen gels of weakly crosslinked polymers are equal to those of uncrosslinked solid polymers or of linear polymers in corresponding solutions [67–69].

VIII. POLYOLEFINS PREPARED BY METALLOCENE CATALYSTS

By these catalysts, polyolefins with a wide range of mechanical properties, narrow molecular weight distribution, specific melting behavior and other interesting properties can be prepared. The structural characteristics underlying these properties have mostly been determined by spectroscopic methods analogous to those described in the preceding paragraphs, together with temperature rising elution fractionation (TREF) and analysis of crystallization processes. Thus while in LLDPE prepared with conventional Ziegler–Natta catalysts, uneven distribution of branches produces semicrystalline material with greater lamellar thickness in the crystalline phase, in LLDPE prepared with Cp_2MCl_2 type metallocene catalysts, branch distribution is even, is independent of chain length, and lamellar thickness is reduced presumably to interbranch spacing [70, 71].

Various chiral metallocenes with methylaluminoxane cocatalysts (MAO) are specific for the preparation of isotactic, syndiotactic, atactic and hemiisotactic polypropylenes and higher polyolefins [72]. In hemiisotactic polypropylene (hit-PP) every other methyl is placed isotactically, the remaining methyls randomly. This type of polypropylene has served as a keypoint in the elucidation of the polymerization mechanism with metallocene catalysts [73]. The chiral metallocene catalysts are not as stereorigid as the conventional heterogeneous systems. Consequently under some polymerization conditions, stereoregular polymers with shorter homosteric sequences are produced. In polypropylene produced in this way, designated as anisotactic polypropylene (ani-PP), low homosteric sequence length favors the γ-crystalline form, rather than the α-crystalline form of conventional i-PP [74]. The relation between the symmetry type of the metallocene catalyst and stereochemical structure of PP and the higher polyolefins has been discussed, similarly as the regulation of the comonomer ordering in the higher olefin copolymers [75].

REFERENCES

1. FA Bovey, FH Winslow. Macomolecules. New York: Academic Press, 1979.
2. CW Bunn. Chemical Crystallography. London: Oxford University Press, 1946.
3. S Krimm, CY Liang, GBBH Sutherland. J Chem Phys 25: 549, 1956.
4. RG Snyder, NE Schlotter, R Alamo, L Mandelkern. Macromolecules 19: 621, 1986.
5. WL Earl, DL VanderHart. Macromolecules 12: 762, 1979.
6. J Cheng, M Fone, VN Reddy, KB Schwartz, HP Fisher, B Wunderlich. J Polymer Sci Polym Phys Ed 32: 2683, 1994.
7. AE Tonelli. J Am Chem Soc 102: 7635, 1980.
8. RF Schaufele, T Shimanouchi. J Chem Phys 47: 3605, 1967.
9. JL Koenig, DL Tabb. J Macromol Sci Phys B9(1): 141, 1974.
10. KJ Packer, JM Pope, RR Yeung. J Polym Sci Polym Phys Ed 22: 589, 1984.
11. RR Eckman, PH Henrichs, AJ Peacock. Macromolecules 30: 2474, 1997.
12. K Bergmann, K Nawotki. Kolloid-Z Z Polym 219: 132, 1967.
13. K Bergmann. J Polym Sci Polym Phys Ed 16: 1611, 1978.
14. G Kerestazy, E Foldes. Polym Test 9: 329, 1990.
15. RG Strobl, WJ Hagedorn. J Polym Sci Polym Phys Ed 16: 1181, 1978.
16. L Mandelkern, RG Alamo. Macromolecules 28: 2988, 1995.
17. B Schneider, J Jakeš, H Pivcová, D Doskočilová. Polymer 20: 9939, 1979.
18. H Hendus, G Schnell. Kunststoffe 51: 69, 1961.
19. R Kitamaru, F Horii, S-H Hyou. J Polym Sci Polym Phys Ed 15: 821, 1977.
20. R Kitamaru, F Horii, K Murayama. Macromolecules 19: 636, 1986.
21. D Doskočilová, B Schneider, J Jakeš, P Schmidt, J Baldrian, I Hernández-Fuentes, M Caceres Alonso. Polymer 27: 1658, 1986.
22. Y Inoue, Y Itabashi, R Chujo, Y Doi. Polymer 25: 1640, 1984; M Kakugo, T Miyatake, Y Naito, K Mizunuma. Macromolecules 21: 314, 1988.
23. FA Bovey, LW Jelinski. Chain Structure and Conformation of Macromolecules. New York: Academic Press, 1982.
24. V Busico, P Corradini, R DeBiasio, L Landriani, AL Segre. Macromolecules 27: 4521, 1994.
25. JW Collette, CW Tullock, RN MacDonald, WH Buck, ACL Su, JR Harrell, R Mulhaupt, BC Anderson. Macromolecules 22: 3851, 1989.
26. G Natta, P Corradini. Nuovo Cimento 15: 40, 1960.
27. A Bunn, MEA Cudby, RK Harris, KJ Packer, BJ Say. Polymer 23: 694, 1982.
28. S Saito, Y Moteki, M Nakagawa, F Horii, R Kitamaru. Macromolecules 23: 3256, 1990.
29. T Miyazawa, Y Ideguchi. J Polym Sci B3: 541, 1965.
30. G Natta, M Peraldo, G Allegra. Macromol Chem 75: 215, 1964.
31. M Peraldo, M Cambini. Spectrochim Acta 21: 1509, 1965.
32. A Bunn, MEA Cudby, RK Harris, KJ Packer, BJ Say. J Chem Soc Chem Commun 1981: 15.
33. RA Komoroski. High Resolution NMR Spectroscopy of Synthetic Polymers in Bulk. Weinheim: VCH Publishers, 1986.
34. P Sozzani, R Simonutti, M Galimberti. Macromolecules 26: 5782, 1993.
35. A Zambelli, C Tossi. Adv Polymer Sci 15: 31, 1974.
36. T Asakura, N Nakayama, M Demura, A Asano. Macromolecules 25: 4876, 1992.
37. LA Belfiore, FC Schilling, AE Tonelli, AJ Lovinger, FA Bovey. Macromolecules 17: 2561, 1984.
38. T Asakura, M Demura, K Ymamoto, R Chujo. Polymer 28: 1037, 1987.
39. T Asakura, M Demura, Y Nishiyama. Macromolecules 24: 2334, 1991.
40. P Pino, R Mullaupt. Angew Chemie Int Ed Engl 19: 875, 1980.
41. G Allegra, E Benedetti, D Pedone. Macromolecules 3: 727, 1970.
42. JC Randall. J Macromol Sci Rev Macromol Chem Phys C29: 202, 1989.
43. M Avella, E Martuscelli, GD Volpe, A Segre, E Rassi. Macromol Chem 187: 1927, 1986.
44. HN Cheng, M Kakugo. Macromolecules 24: 1724, 1991.
45. X Zhang, H Chen, Z Zhou, B Huang, Z Wang, M Jiang, Y Yang. Macromol Chem Phys 195: 1063, 1994.
46. NJ Clayden. Polymer 33: 3145, 1992.
47. J Spěváček. Polymer 19: 1149, 1978.
48. MA McRae, WF Maddams. Macromol Chem 177: 449, 1976.
49. T Usami, S Takayama. Macromolecules 17: 1756, 1984.
50. ET Hsieh, JC Randall. Macromolecules 15: 353, 1982.

51. DL VanderHart, E Pérez. Macromolecules 19: 1902, 1986; E Pérez, DL VanderHart. J Polym Sci Polym Phys Ed 25: 1637, 1987.
52. H Ketels, J Beulen, G van der Velden. Macromolecules 21: 2032, 1988.
53. MD Bruch. Macromolecules 21: 2707, 1988.
54. M Hert, L Guerdoux, J Lebez. Angew Makromol Chem 154: 111, 1987.
55. M Bryjak, W Trochimczuk. Angew Makromol Chem 118: 191, 1983.
56. M Bryjak, W Trochimczuk. Angew Makromol Chem 116: 221, 1983.
57. M Bryjak, W. Trochimczuk. Angew Makromol Chem 133: 37, 1985.
58. LW Jelinski, JJ Dumais, JP Luongo, AL Cholli. Macromolecules 17: 1650, 1984.
59. KW Lee, TJ McCarthy. Macromolecules 21: 309, 1988.
60. W Busch, F Kloos, J Brandrup. Angew Makromol Chem 105: 187, 1982.
61. C Zhikuan, S Lianghe, RN Sheppard. Polymer 25: 369, 1984.
62. VA Era, JJ Lindberg. J Polym Sci A-2, 10: 937, 1972.
63. VA Era. Makromol Chem 175: 2191, 2199, 1974.
64. SG Joshi, AA Natu. Angew Makromol Chem 140: 99, 1986.
65. J Bongardt, GH Michler, I Naumann, G Schulze. Angew Makromol Chem 153: 55, 1987.
66. PV Zamotaev, SV Luzgarev. Angew Makromol Chem 173: 47, 1989.
67. D Doskočilová, B Schneider. Pure Appl Chem 54: 575, 1982.
68. B Schneider, D Doskočilová, J Dybal. Polymer 26: 253, 1985.
69. C Schmid, JP Cohen-Adad. Macromolecules 22: 142, 1989.
70. JA Parker, DC Bassett, RH Olley, P Jaaskelainen. Polymer 35: 4140, 1994.
71. RG Alamo, EM Chan, L Mandelkern. Macromolecules 25: 6381, 1992.
72. JA Ewen, MJ Elder, RL Jones, L Haspelagh, JL Atwood, SG Bott, K Robinson. Makromol Chem Macromol Symp 48/49: 253, 1991.
73. M Farina, G DiSilvestro, P Sozzani. Macromolecules 26: 946, 1993.
74. J Huang, GL Rempel. Prog Polym Sci 20: 459, 1995.
75. O Olabisi, M Atiqullah, W Kaminsky. JMS Rev Macromol Chem Phys C37(3): 519, 1997.

7

Morphology of Polyolefins

Clara Silvestre, Sossio Cimmino, and Emilia Di Pace
Istituto di Ricerca e Tecnologia delle Materie Plastiche, CNR, Arco Felice (NA), Italy

I. INTRODUCTION

Polymer morphology is the study of order within solid macromolecules. Several reviews are available for this topic, so only general aspects of the polymer morphology are covered here [1–3].

After solidification two extreme states are possible: the amorphous state and the crystalline state. The conformations of the single chains and their relative organization define these two states.

The occurrence of one state or another is depending on the chemical structure of the polymers and on the conditions of preparation.

The amorphous state is the state of molecular disorder. In this state the chain assumes a statistical random coiled conformation. The crystallization for amorphous polymers cannot take place, due to their disordered structure.

In the crystallizable polymers, under appropriate conditions of temperature and pressure, an ordering of segments of the chain can occur.

Whether or not polymers are able to crystallize depends primarily on the chemical, geometrical and spatial regularity of the macromolecules. A familiar example is provided by the poly-α-olefins, schematically illustrated in Fig. 1. To permit crystallization the pendant R group must be systematically placed along the chain. Two regular arrangements (configurations) can occur, called isotactic and syndiotactic. In the isotactic configuration the substituent R groups have the same placement in space in relation to the atoms of the backbone. In the syndiotactic configuration the placement of the R group is alternated with respect to the atoms of the main chain. The other possible configurations are spatially disordered and they are known as atactic. Isotactic and syndiotactic polymers can crystallize, whereas for atactic polymers crystallization is not allowed.

The crystallization of long chain molecules will occur at large undercooling, giving rise to a very complex arrangement, with polycrystalline character and coexistence of crystalline and amorphous components. So a crystalline polymer is always characterized by the presence of ordered crystallized regions as well as amorphous zones. Crystalline polymers show ordering at a variety of dimensional levels from molecular spacing to macroscopic measures.

The molecular level of ordering is often called *crystal structure* and describes the relative position and orientation of the atoms in the molecular chains. The arrangement at a higher dimensional level is called *molecular morphology* or *supermolecular structure* and describes the organization of the crystalline and amorphous zones and their associated interfacial regions. It also describes the organization of the material into large entities.

Three distinct regions are of importance in crystalline polymers: the crystalline region, the interfacial region, and the amorphous region. They are schematically illustrated in Fig. 2. All together, these three regions form the crystallite.

The crystalline region represents the three-dimensional regular structure. It has a lamellar-like habit.

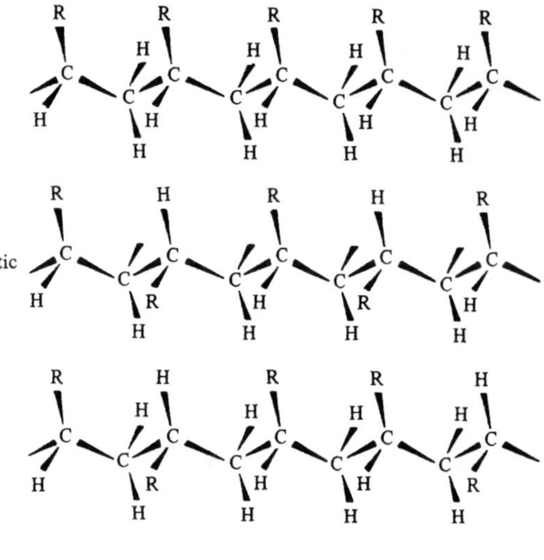

Isotactic

Syndiotactic

Atactic

FIG. 1 Configurations of the chains in the poly-α-olefins.

This habit is one of the morphological characteristics developed by the polymer during crystallization. The lamellar thickness depends on the crystallization conditions and can range from 50 to a few hundred Å. In the lamella the chains are preferentially oriented perpendicular to the plane of the lamellae. As the length of the macromolecules can be 10,000 Å, a single chain must traverse a crystallite many times, giving rise to chain folding. So some chains or parts of the chain will exit from the crystallite, and some will enter, forming a diffuse and almost disordered transition region. Finally the amorphous region, characterized by chains in random conformation, connects one crystallite to another, giving rise to the supermolecular structure, where the crystallites are organized and assembled.

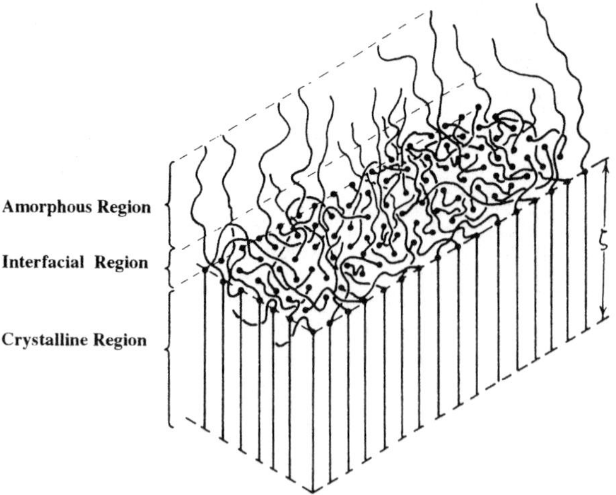

FIG. 2 Scheme illustrating a crystallite.

The dimension of the different regions changes with molecular mass and crystallization conditions and for blends with composition [4, 5]. Thicker crystallites are usually obtained by annealing the sample at a temperature higher than the crystallization temperature.

Insight into and information about structural details of polymer morphology was mainly based on small- and wide-angle X-ray scattering, optical, atomic force and electron microscopy.

In this review the crystal structure and the supermolecular structure of the most used polyolefins is discussed. In particular the latest papers on the morphology of polyethylene, isotactic and syndiotactic polypropylene, isotactic poly(1-butene), and finally isotactic poly(4-methylpentene-1) are summarized and integrated with the fundamental work on the topic. After a short general introduction, the first part of the chapter is dedicated to the analysis of the order at the molecular level (the crystal structure), and the second part deals with the supermolecular structures.

The aim of the chapter is to underline the dependence of the crystal and supermolecular structure on the molecular mass and crystallization conditions.

The authors wish to dedicate their work to the memory of Professor Fatou. He made a significant contribution to the study of the morphology and crystallization of polymers. An extensive review on the morphology, crystallization and melting of polyolefins was made by Professor Fatou and represents a fundamental part of the first edition of the Handbook of Polyolefins [6]. His work is the starting point of the present contribution. The authors had the honor and the fortune to meet Professor Fatou and to benefit from his cultural zeal and inspiration.

II. CRYSTAL STRUCTURE

A. Polyethylene

Crystals of polyethylene, PE, exist in three different modifications. The well known orthorhombic modification, a metastable monoclinic phase, and a hexagonal phase that appears at high pressure.

The lattice dimensions of the three different modifications of PE are given in Table 1. Crystallization from the melt and solution at atmospheric pressure gives rise to the stable orthorhombic modification. In this modification the angle between the zigzag plane and the *bc* plane (the setting angle in the orthorhombic form) was

TABLE 1 Crystallographic Data of PE

Form	Crystal system	Space group, lattice constants and number of chains per unit cell	Molecular conformation	Crystal density (g/cm^3)
I	Orthorhombic stable form	$Pnam\text{-}D_{2h}$ $a = 7.417$ Å, $b = 4.945$ Å, $c = 2.547$ Å, $N = 2$	Planar zigzag (2/1)	1.00
II	Monoclinic metastable form	$C2/m\text{-}C_{2h}$ $a = 8.09$ Å, b(f.a.) $= 2.53$ Å, $c = 4.79$ Å, $\beta = 107.9°$, $N = 2$	Planar zigzag (2/1)	0.998
III	Orthohexagonal (assumed) high pressure form	$a = 8.42$ Å, $b = 4.56$ Å, c has not been determined		

first reported to be about 41° by Bunn [7] and this value has been used for many years. Avitabile *et al.* [8] have reported this angle to be 49° as determined by neutron diffraction. According to their paper, the C—C bond distance is 1.574 Å, appreciably longer than the value 1.54 Å so far reported from X-ray results. X-ray analyses [9, 10] on high density PE have shown that the setting angle is about 45° for non-oriented samples and about 46° for oriented samples. In the case of low density PE, the angle changes to 49–51° with the increase of the a axis. Several authors report that branching affects the lattice constants [9, 11–13].

The monoclinic phase of PE has been obtained from the orthorhombic one by the application of stress. In particular the monoclinic form is obtained by rolling or biaxial stretching, but this form is metastable and transforms just below the melting point into the orthorhombic form on heating [14–16].

The monoclinic phase of PE has also been obtained under different growth conditions, in the absence of mechanical stress. Teare and Holmes [15] suggested that it can be produced by crystallization at low temperatures. Keller and coworkers [17] found traces of this phase in germ-free PE droplets crystallized at very high supercooling, which is indeed known to induce unstable crystal forms in many systems. In [18–20] it was reported that epitaxial crystallization of PE from dilute solutions on alkali halide substrates may give rise to the monoclinic phase that transforms to the orthorhombic phase. Transformation takes place within a few nanometers from the substrate surface, and growth proceeds thereafter in the orthorhombic phase.

The orthorhombic to monoclinic phase transformation has been much investigated, for example, by electron diffraction on deformed solution-grown PE single crystals [21] and by conformational energy analysis [22, 23]. In their early work, Seto *et al.* [14] suggested three different possible transformation pathways and favored the mode which requires only small chain displacements, a view supported by lateral conformational energy analysis results [22].

A high-pressure form was also found (5000 kg/cm^2, 240°C) with an orthohexagonal unit cell [24, 25].

B. Isotactic Polypropylene

Isotactic polypropylene, iPP, is a polymorphic material with five crystal modifications [26, 27], α, β, γ, δ, and a modification with intermediate crystalline order.

The monoclinic lattice of the α form in iPP was identified by Natta and Corradini [28]. Shortly afterwards, a polymorph with a hexagonal lattice was recognized, and designed as the β form [26, 28–30]. An even rarer third polymorph was found, based on a triclinic lattice; this was called the γ form [26, 28, 31–33]. There is mention of another structure, the δ form, in iPP with a high percentage of amorphous material, which may be associated with the nonisotactic portion of PP [34]. In addition to these crystal structures, a quenched crystal form, called the smectic form by Natta and Corradini [28], was also observed in iPP. In all of these structures, the chain conformation of each individual chain molecule is believed to be identical with $2 \times 3/1$ helix (tgtgtg conformation) [28, 35–38]. The polymorphs are distinct due to the different chain packing geometries of the helices. The unoriented powder wide-angle X-ray diffraction patterns for different forms of iPP crystals are shown in Fig. 3.

It was also reported that the α form in iPP can be recrystallized and/or annealed from a less ordered α_1 form with a random distribution of "up" and "down" chain packing of methyl groups to a more ordered α_2

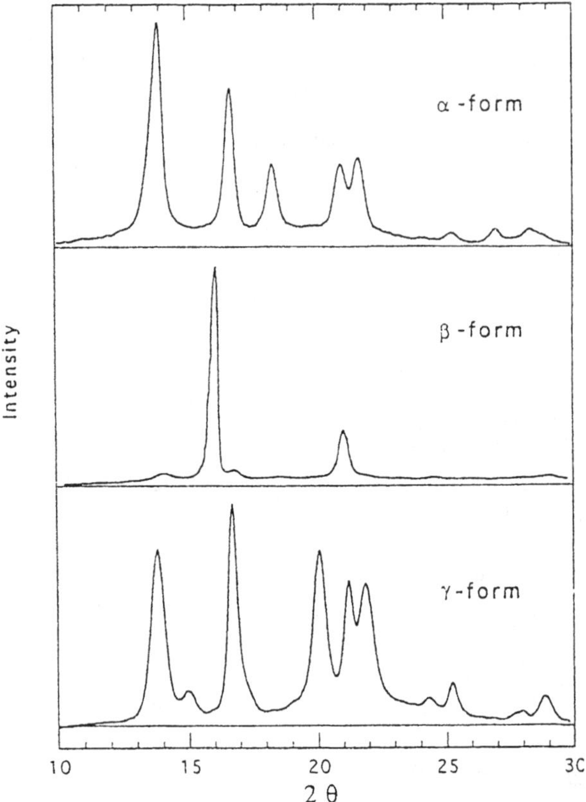

FIG. 3 WAXS of different crystalline forms of IPP.

form with a well defined deposition of "up" and "down" helices in the crystal unit cell [39–44].

The crystallographic data of iPP are reported in Table 2.

The appearance of these structures is critically dependent upon crystallization condition and pressure. The monoclinic α modification occurs most frequently [45, 46]. In the crystallization of conventional iPP grades, essentially the α modification is formed which may be accompanied by a lower or higher amount of the hexagonal β modification at higher undercooling [47]. Under special crystallization conditions, however, when the "temperature gradient method" [48, 49] is used or when selective β-nucleating agents are present, the product will be quite rich in β modifications [50, 51]. Varga *et al.* [52, 53] prepared the pure β modifications in the presence of selective β modification agents, by the appropriate selection of the thermal conditions of crystallization. The γ modification may form in degraded, low molecular mass PP ($M = 1000$–3000) or in samples crystallized under high pressure [54–57]. Also random copolymers of propene into 2.5–20 wt% of other 1-olefins may crystallize preferably in the γ modification [58–60]. More recently it has been shown that high molecular mass iPP samples, prepared by employing metallocene catalyst systems, may tend to crystallize in the γ modification [61, 62]. Similar to propene copolymers with 1-olefins, the steric irregularities of these samples have been proposed to be responsible for the formation of the γ modification [62]. The formation of the γ modification was associated with chemical heterogeneity in the polypropylene chain caused by atacticity or by copolymerization [59]. According to Turner-Jones [58] the presence of comonomers like ethylene or others enhanced the formation of the γ modification. Studies conducted by Phillips [63, 64], using variable amount of ethylene content confirmed the results of Turner-Jones, also indicating that the amount of γ modification was proportional to the ethylene content. It was also demonstrated that the γ modification is produced at elevated pressures from high molecular mass homopolymers and has the same diffraction patterns as the low molecular mass polymers crystallized at atmospheric pressure. This study has also confirmed that the γ modification is not the result of some unexpected degradation reaction at elevated pressures.

Since its description in 1989 the γ modification has been an enigma. Its structure, unique to polymer science, is composed of sheets of parallel chains just positioned next to one another so that nonparallel chains are generated normal to the sheets [65–67].

TABLE 2 Crystallographic Data of IPP

Form	Crystal system	Space group, lattice constants and number of chains per unit cell	Molecular conformation	Crystal density (g/cm^3)
α	Monoclinic	$P2_1/c$ $a = 0.666$ Å, $b = 2.078$ Å, $c = 6.495$ Å, $N = 12$	Helix (3/1)	0.936
β	Hexagonal	$a = 12.74$ Å, $c = 6.35$ Å	Helix (3/1)	0.921
γ	Triclinic	$a = 6.54$ Å, $b = 21.4$ Å, $c = 6.50$ Å	Helix (3/1)	0.954

Phillips reported that, as the pressure is increased, the proportion of γ modification increases from 0 at atmospheric pressure to close to 100% at 2 kbar. It appears that the lower the supercooling the higher the amount of γ modification produced at a specific pressure [65]. A model was proposed for the crystallization in which the two crystals deposit within the same lamellae through an epitaxial process. Theoretical and experimental prediction of the pure γ modification as a function of temperature and pressure can be found in the phase diagram proposed by Phillips [68], see Fig. 4.

Quenching (abrupt cooling) the molten polymer leads to an intermediate crystalline order, indicated by two diffuse X-ray diffraction peaks [28, 69], which are different from both the crystalline and the amorphous states. The interpretation of the structure formed is still disputed in the literature.

Natta described this structure as "smectic", and suggests that it is composed of parallel 2 × 3/1 helices but that disorder exists in the packing of the chains perpendicular to their axes [70, 71]. Miller proposed that the quenched form is "paracrystalline" [71], whereas Wunderlich classified it as a "conformationally disordered" (condis) crystal [72]. Gailey and Ralston [73], as well as Gezovich and Geil [74], found that the quenched form consists of very small hexagonal crystallites. Lately, McAllister *et al.* suggested that the quenched form represented approximately 60% of the amorphous fraction, with the rest of the sample having the helices arranged in a "square" array with a cubic or tetragonal symmetry [75]. Gomez *et al.* have reported that the local packing of 2 × 3/1 helices are very similar in the β and quenched smectic forms of iPP [76]. On the other hand, Corradini *et al.* indicated that in quenched iPP a fairly high correlation of distances is present within each chain and between neighboring chains to form small bundles. The local correlations between chains are probably nearer to those characterizing the crystal structure of the monoclinic form than to those characterizing the structure of the hexagonal form [77].

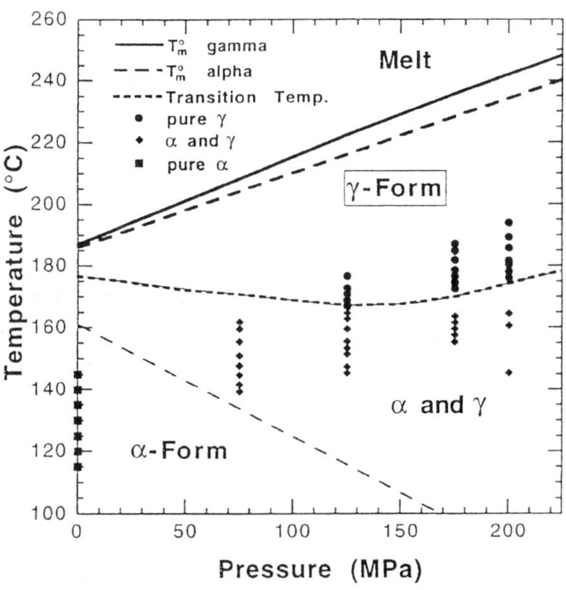

FIG. 4 Theoretical prediction of the pure γ form of IPP as a function of temperature and pressure and its comparison with the experimental data. (Reprinted from Ref. 68.)

C. Syndiotactic Polypropylene

Syndiotactic polypropylene, sPP, exists in three different conformations: a stable two-fold helix, a planar zigzag both with orthorhombic lattice and an intermediate conformation with a triclinic lattice. The crystallographic data of sPP are reported in Table 3. The stable two-fold helix structure is called the high-temperature orthorhombic form. Three different unit cells for this form are proposed [78–85].

For cell I the lattice is fully isochiral with all molecules of the same hand on a *bc* plane. Cell II, that has the same dimensions as cell I, is composed of helices of alternating handedness incorporated along the *a* axis, resulting in a unit cell face centered on the *ac* plane with a space group $Pca2_1$. For cell III the alternation of left and right-hand helices, known as antichiral packing, is found.

Lovinger *et al.* [84] investigated the temperature dependence on the sPP structure and morphology and its epitaxial relationship with iPP. Their results indicate that the origin of the different packing schemes may lie in the kinetics of crystallization. It is likely that packing helices of opposite hand along *a* and *b* are energetically more favorable as opposed to incorporation of helices of the same hand. As a result, isochiral chains may be preferentially rejected from the growth surface. In that case, adsorption and nucleation of molecules directly on top of chains already on the growth surface is preferred, in which case an antichiral packing results. This view is supported by the observation that at higher crystallization temperatures the antichiral packing is favored. At lower crystallization temperatures, with a much faster growth rate, the regularity found in the antichiral packing is reduced, indicating the dominant effect of kinetic factors. The authors conclude that cells I and II may exist as a

TABLE 3 Crystallographic Data of SPP

Form	Crystal system	Space group, lattice constants and number of chains per unit cell	Molecular conformation	Crystal density (g/cm^3)
I	Orthorhombic high temperature	C2221 $a = 14.5$ Å, $b = 5.8$ Å, $c = 7.40$ Å, $N = 8$ Ibca $a = 14.5$ Å, $b = 11.2$ Å, $c = 7.4$ Å	Helix (4/1)	0.90
II	Orthorhombic low temperature	$a = 5.22$ Å, $b = 11.17$ Å, $c = 5.06$ Å	Planar zigzag	0.945
III	Triclinic	P1 $a = 5.72$ Å, $b = 7.64$ Å, $c = 1.16$ Å, $N = 6$	Helix (3/1)	0.939

result of packing defects in the crystal structure as isochiral packing defects are incorporated in the crystal.

According to Corradini [86] annealing under stress produces some crystals with unit cell I whereas in samples obtained under isotropic conditions a mixture of unit cell II and III is predominant; unit cell III is present prevalently in single crystals grown at low supercooling.

The planar zigzag conformation, also known as the low-temperature orthorhombic form, was observed by Natta for samples obtained by cold drawing the polymer quenched from the melt. Upon annealing at about 100°C for a few hours, the samples were converted from zigzag planar to the more stable helical form.

Chatani *et al.* [87] observed the triclinic lattice in cold draw sPP samples quenched in ice water. The samples were subjected to vapor absorption for several days; the solvents used were benzene, toluene and *p*-xylene. The triclinic lattice transforms into the helical form by annealing above 50°C.

D. Isotactic Poly(butene-1)

Isotactic poly(butene-1), PB1, may exist in at least three distinct crystal modifications. The crystallographic data of PB1 are reported in Table 4. The different modifications are commonly referred to as I, II, and III.

Early work by Natta *et al.* [88] revealed the existence of two polymorphs, referred to as I and II. Cooling of molten polymer results in the growth of crystallites of modification II; on standing at room temperature they gradually transform to I. Some time thereafter it was disclosed that a third polymorph, known as III, could be obtained by precipitating the polymer from various solvents [89]. A low melting form of modification I (modification I′) may be prepared by polymerizing 1-butene in *n*-heptane with a Ziegler–Natta catalyst at −20°C [90]. Another form, II′, is formed by pressure crystallization under specific condition [91].

The X-ray diffraction diagram of poly(1-butene) is reported in Fig. 5. Modification I is characterized by a chain in the 3/1 helix conformation with axes $a = 17.7$ Å and $c = 6.50$ Å. The unit cell is rhombohedral. For modification II a tetragonal unit cell with axes $a = 7.49$ Å and $c = 8.5$ Å in the 11/3 helix conformation has been proposed. For modification III an orthorhombic cell has been suggested with $a = 12.49$, $b = 8.96$ and $c = 7.6$ Å.

E. Poly(4-methylpentene-1)

Poly(4-methylpentene-1), P4MP1, can crystallize in a variety of crystal lattices. Five different crystalline forms have been described (the nomenclature utilized

TABLE 4 Crystallographic Data of PB1

Form	Crystal system	Space group, lattice constants and number of chains per unit cell	Molecular conformation	Crystal density (g/cm^3)
I	Rhombohedral	R3c-D_{3d} $a = 17.70$ Å, $c = 6.50$ Å, $N = 6$	Helix (3/1)	0.95
II	Tetragonal	P4b2-D_{2d} $a = 7.49$ Å, $b = 8.96$ Å, $c = 7.6$ Å, $N = 4$	Helix (4/1)	0.998
III	Orthorhombic	$a = 8.42$ Å, $b = 4.56$ Å	Helex (4/1)	

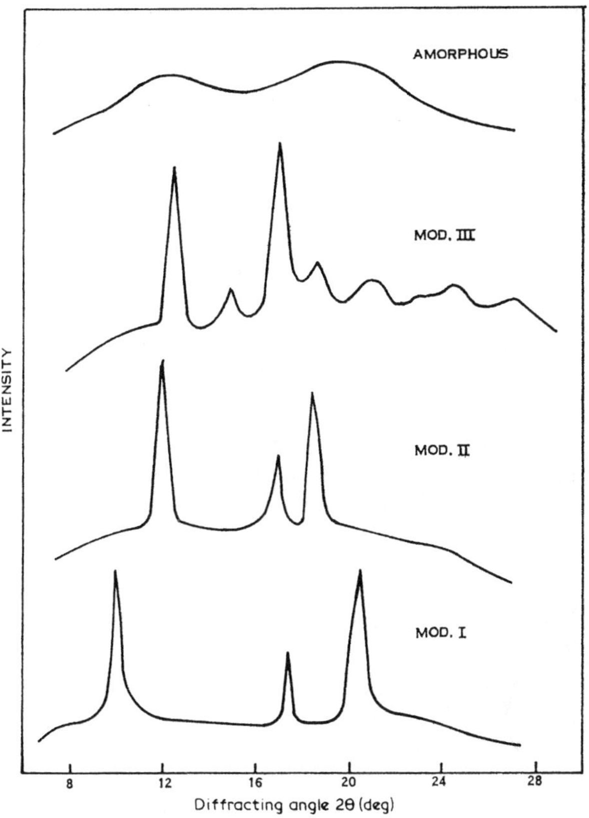

FIG. 5 WAXS of different crystalline forms of PB1.

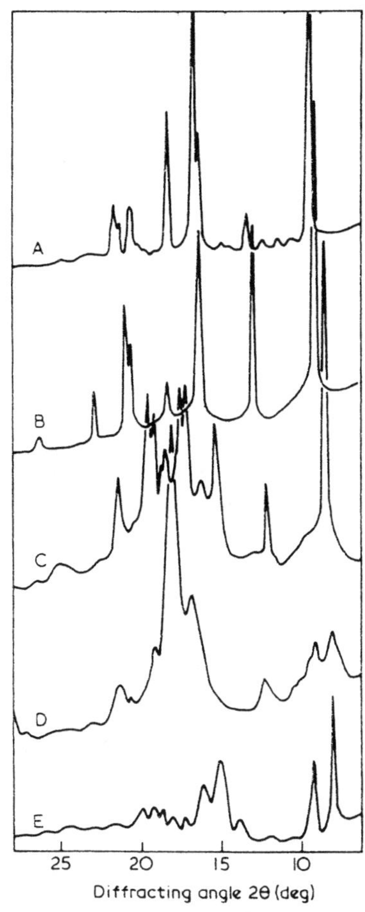

FIG. 6 WAXS of different crystalline forms of P4MP1; (A) I, (B) III, (C) (V), (D) IV, and (E) II.

by Delmas and coworkers [92] will be used throughout this discussion). The modifications of P4MP1 are denoted as I, II, III, IV, and V. The various forms differ from the conformation of the polymer chains and for the mode of packing of chains having the same conformation. The crystallographic data of P4MP1 are reported in Table 5.

Figure 6 presents the wide-angle X-ray diffraction patterns corresponding to the five modifications.

Modification I is characterized by chains in the 7/2 helical conformation packed in a tetragonal unit cell with axes $a = b = 18.66$ Å and $c = 13.80$ Å [93–97].

For modification II, Takayamagi *et al.* [96] proposed a tetragonal unit cell with axes 19.16 Å and $c = 7.12$ Å with the chain in the 4/1 helical conformation. Modification III is characterized according to Charlet *et al.* [97] by a tetragonal unit cell with parameters $a = 19.38$ and $c = 6.98$ Å and a chain in the 4/1 helical conformation. For modification IV a hexagonal unit cell with axes $a = 22.17$ and $c = 6.96$ Å. A hexagonal

TABLE 5 Crystallographic Data of P4MP1

Form	Crystal system	Lattice constants (Å)	Molecular conformation	Crystal density (g/cm^3)
I	Tetragonal	$a = 18.66, c = 13.80$	Helix (7/2)	0.812
II	Tetragonal	$a = 19.16, c = 7.12$	Helix (4/1)	—
III	Tetragonal	$a = 19.38, c = 6.98$	Helix (4/1)	—
IV	Hexagonal	$a = 22.17, c = 6.5$	Helix (3/1)	—
V	Hexagonal	$a = 22.17, c = 6.69$	—	—

unit cell has also been proposed for modification V. For this modification the cell parameters are $a = 22.17$ and $c = 6.69$ Å [98, 99].

The different modifications can be obtained by changing the crystallization conditions, nature of solvent, solution concentration, thermal history of the solution, and cooling rate, see Tables 6 and 7 [100–102].

Crystallization from the melt only generates modification I. Crystallization from solution, however, promotes the formation of various crystalline modifications.

Solid–solid transitions from a modification to others have been observed. In fact, P4MP1 is found [102] to exhibit changes in crystal modification on uniaxial draw. In particular He and Porter [103] report that the crystal structure at lower planar zigzag (2/1) draw ratio is a common tetragonal form and that at higher draw it is orthorhombic.

The effect of pressure and temperature on the phase diagram of P4MP1 was studied by Rastogi *et al.* [104], see Fig. 7. The polymer crystalline under ambient conditions becomes amorphous reversibly on increasing pressure in two widely separated temperature regimes (20°C and 200°C). The transformation occurs via a liquid crystalline state. The lower-temperature amorphous phase becomes crystalline on heating and reverts to the glass disordered phase on cooling.

TABLE 6 Effect of Solvent on the Structure of P4MP1 Crystallized from Solutions

Solvent	T_{max} (°C)*	Crystal modification
n-Hexadecane	190	I
n-Dodecane	165	I
Xylene/amyl acetate	165	I
n-Nonane	165	I
e-Octane	165	I + II
n-Heptane	135	III
n-Pentane	135	III
2,2,4-Trimethylpentane	165	III
3,4-Dimethylhexane	135	III
2,4-Dimethylpentane	165	III
2,3-Dimethylpentane	165	III
3-Ethylpentane	165	III
Cyclooctane	135, 165	III
Cycloheptane	85	III
Cyclohexane	85	III
Carbon tetrachloride	85	III
Cyclopentane	85	IV
Carbon disulfide	85, 135	II + III

*T_{max} is the maximum temperature used for complete dissolution. The solution concentrations were $0.02 = \phi = 0.08$.

TABLE 7 Effect of Thermal History on the Structure of P4MP1 Crystallized from Solution

Solvent	Crystal modification	
	Low supercooling Slow cooling	High supercooling Rapid cooling
n-Octane	I + III	III
Xylene	I	III
Toluene	I	III
Benzene	I	III
Cyclooctane	III	II + III
Cycloheptane	III	II + III
Carbon tetrachloride	III	V
Cyclohexane	III	V
Tetraethyltin	II + III	II
Tetramethyltin	II + III	II

III. SUPERMOLECULAR STRUCTURE

Several polymer superstructure morphologies can be found in crystallizing a polymer: lamellae, rods, sheet-like structures, hedrides or axialites, and spherulites. The formation of specific superstructures depends on molecular mass, crystallization condition, and structural regularity of the individual macromolecules. The polymer superstructures most likely to be met are lamellae, spherulites, and hedrites [1–3].

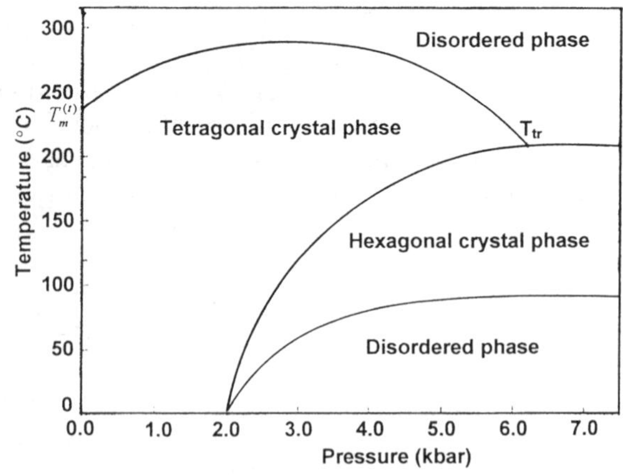

FIG. 7 Phase diagram of P4MP1. (Reprinted with permission from S Rastogi, M Newman, A Keller. Nature 353: 55–57, 1991. Copyright 1991 Macmillan Magazines Limited.)

A. Lamellae

Lamellae are the primary morphological entities observed after crystallization from solution. They can also be formed by isothermal crystallization from a quiescent supercooled melt. In the lamellae, the polymer chains are usually oriented perpendicular or nearly perpendicular to the lamellae face. If the molecular mass is low and the sample essentially monodisperse, completely extended chains packed side-by-side are obtained. However for most synthetic and naturally occurring polymers, the chain length is greater than the lamellar thickness, requiring the parts of the chain immediately preceding and following a particular crystal stem to either form folds connecting to other crystal stems or to protrude from the lamellae. The lamellae usually encountered a few tens of nm thick and a few nm wide, although these dimensions can vary greatly, increasing by two orders of magnitude under certain conditions. The crystals come in a great variety of size and habits, within the general rule that the most regular crystals are those which form most slowly, e.g., at the highest crystallization temperatures. Several geometries are reported: they include diamond, hexagonal, pyramidal, square, and lath shapes. Lamellae with rounded edges are also observed. The geometric shapes encountered are dependent on the type of crystal lattice adopted, crystallization solvent and kinetics factors. Some lamellae are composed of sectors: triangular shaped zones with apexes usually meeting at the center of the lamellae. Due to distortion of the subcell by fold packing, the sectored polyer lamellae can be nonplanar and therefore features of collapse, such as creasing, splitting, and buckling, can be evident in samples deposited from solution. The sectors can join by forming a hollow-pyramidal sheet. The pyramidal form is a familiar feature of lamellae grown from solution and is especially prominent at low or moderate concentration. Despite the great variation in habits, it is found that the thickness depends on temperature and type of solvent and is independent of concentration and molecular mass.

B. Spherulites

The spherulites are birefringent spherical units observable in a series of polymers crystallized from the melt.

The dimension of a spherulite of a given polymer is related to their relative nucleation and growth rates. The experimental lowest limit is a few μm, whereas the upper limit can only be estimated. Spherulites of 1 cm in diameter can be observed, but there is no theoretical limit to their growth to even larger units, if there is enough crystallizable material and no other growing spherulite near.

The word "spherulite" means "little sphere", but the spherical shape is maintained only in the initial stages of growth. In fact when neighboring spherulites impinge, they become polyhedral. In this case, if the spherulites are nucleated simultaneously, the common outline will be planar; otherwise the boundary is going to be a hyperboloid of revolution.

In the investigation of the morphology of spherulites, polarizing microscopy has been the predominant and most informative tool. Very often the samples have been observed between two microscope slides in thin films, giving rise to an almost two-dimensional image. This image resembles rather closely diametrical sections of spherulites grown in bulk. The microstructure of spherulites is a polycrystalline array of equivalent radiating units. Observed under crossed polars the spherulites look like birefringence bright circular areas, indicating crystalline entities. They often display a black Maltese cross, with the arms lying parallel and perpendicular to the extinction direction of the polars. When the specimen is rotated in its own plane, the cross remains stationary, implying that all the radiating units composing the spherulite are crystallographically equivalent.

From a birefringence point of view, we can have two different kinds of spherulites, called positive and negative. If the major refractive index is parallel to the extinction direction of the polar, we have positive spherulites. If it is perpendicular we have negative spherulites. The sign of birefringence depends on the kind of material studied and on the crystallization temperature [105]. In fact, it was reported that therer is a gradual change from positive to negative spherulites on increasing the crystallization temperature, indicating a change of orientation of the chain with temperature.

There are examples of more complicated spherulites, which in addition to the Maltese cross, also show concentric rings.

Two mechanisms for spherulite growth have been proposed: one by Keith and Padden [106], the fibrillation mechanism, and the other by Bassett [107], the splaying and branching mechanism.

According to Keith and Padden, the radiating units are multilamellar fibers. Once a crystalline unit is nucleated, a concentration instability of the interface will result. Due to the accumulation at the interface of noncrystallizable or slower growing molecular species, long units called fibrils develop. During the process of crystallization, a branching phenomenon is also supposed in order to give the spherulite its spherical shape.

For Bassett [107] the radiating units are lamellae. The explanation of Bassett supposes splaying and branching of individual lamellae. The splaying is the deviation of adjacent lamellae from their ideal natural crystallographic orientation, whereas the branching is the creation of new lamellae. The origins of the splaying forces were attributed to cilia of uncrystallized molecule ends or long loops, attached to a lamella. If these impurities are in sufficient concentration they could promote nucleation of new lamella.

C. Hedrites

As reported by Woodward [108] hedrites or axialites are stacks of centrally connected lamellae that grow from solution or melt [109, 110]. One or more chains with elongated conformations probably commonly nucleate the lamellae in a particular stack. The lamellae are usually somewhat longer than they are wide, with growth occurring principally at the ends. During growth, these lamellae can develop further connections due to the incorportion of a single chain or a group of chains in neighboring lamellae. One consequence is that two lamellae can be closely connected to each other at more than one place in the stack; repeated events of this sort give the hedrite a cellular appearance. A lamellar stack can exhibit considerable curvature, mainly to one side but also along the width and length of the lamellae, causing the stack to appear cup-shaped. Separation or splaying of the lamellar ribbons can occur. When crystallization is carried out at high solution concentrations, gels containing hedrite-like structures can be obtained [111, 112]. Crystallization from a supercritical fluid yields hedritres [113]. Branching, cellulation and splaying are observed in these structures.

Hedrites are usually birefringent when viewed through crossed polaroids. When viewed face-on in suspension, they show an oval of light, the central portion being dark due to insufficient thickness. When viewed from the edges of the stack in suspension, four thick, bright quadrants are usually seen and, in some cases, six bright parts are observed [114]. Rotation of the stack during observation brings about rotation of the bright (and dark) portions of the pattern.

The crystal stems in a hedrite are nearly perpendicular to the faces of the ribbons: these faces containing chain folds. The amorphous fraction in hedrites depends on the molecular mass and, at high molecular mass, is 20% larger than that in single lamellae [115]; this larger amorphous fraction is believed to be due mainly to exposed sections of chains connecting the crystal stems in two adjacent lamellae [116]. Hedrites containing over 100 lamellae have been obtained [117].

1. Polyethylene

The lamellar nature of polyethylene single crystals grown from dilute solution is well known. Extensive studies have been carried out on such crystals grown at temperatures up to 120°C [118–121], and they are known to exhibit well defined fold lengths and habits characteristic of the conditions of growth. The typical habits closely reflect the symmetry of the unit cell (see Fig. 8). Crystals are bounded by four {110} faces, with truncating {100} faces appearing as the concentration or crystallization temperature is raised. By using different solvents it is possible to extend the upper limit of crystallization temperature, whereupon further developments in morphology can be seen.

The major previous studies of this type have been those of Keith [120] and Khoury and Bolz [121]. Keith crystallized low and high molecular mass fractions of polyethylene at temperatures up to 110°C from thin films, using *n*-alkanes (mainly n-$C_{32}H_{66}$) as solvents. Khoury and Bolz, using mainly dodecanol and heptyl acetate as solvents to grow crystals at temperatures up to 120°C, confirm the findings of Keith with respect to crystallization temperature, concentration and molecular mass. Organ and Keller [122] also have grown polyethylene single crystals from solution in a variety of solvents at temperatures up to 120°C. The solvents and the ranges of dissolution and crystallization temperatures are reported in Table 8.

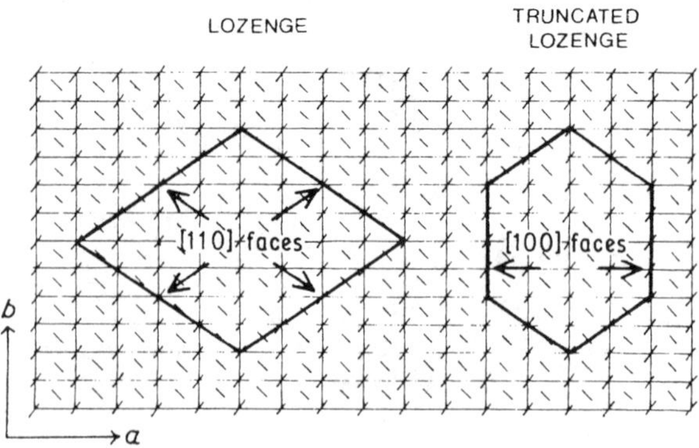

FIG. 8 Typical morphologies of PE single crystals and their relationship to the unit cell. (Reprinted from Ref. 122.)

TABLE 8 Solvents and Ranges of Crystallization and Dissolution Temperatures used by Organ and Keller [122] for the Study of PE Morphology

Solvent	T_s range (°C)	T_c range (°C)
Xylene	101–102	78–95
Octane	109–110	85–100
Dodecane	111–112	87–106
Hexadecane	115–116	89–108
Tetracosane	121–122	98–114
Hexatriacontaine	123–124	102–116
Hexyl acetate	119–120	100–112
Ethyl esters	120–121	100–112
1-Dodecanol	126–127	105–120
1-Tetradecanol	125–126	104–118

They found that the axial ratio of the crystals increased with crystallization temperature and with concentration, with the crystals becoming more elongated in the *b* direction. Crystals grown from low molecular mass polymer tended to have higher axial ratios than those grown under the same conditions from high molecular mass material. Keith also obtained radially arranged aggregates of crystals and compared these structures to the radial crystalline units in melt grown spherulites.

Organ and Keller, comparing crystals grown from different solvents, found that at high T_c the axial ratio was roughly related to the crystallization temperature, although there was some solvent effect. The {100} crystal faces were usually curved and the degree of curvature increased with crystallization temperature. {110} crystal faces were initially straight, but showed an increasing slight curvature with increasing crystallization temperature. At high temperatures the crystals tended to grow in clusters, with many crystals growing in all directions from a common center.

Electron microscopy provides experimental evidence of the nature of the fold surface. The observation of distinct sectors in lozenge-shaped polyethylene crystals provides strong evidence for folds that occur preferentially along the {110} [118]. Attempts to predict or model the possible fold conformations in polyethylene crystals have been made over the years [123]. Generally, results have been extensively dependent on the contributions to the potential energy included in any particular model [124].

Small-angle X-ray diffraction measurements also have a significant bearing on the question of fold surface structure of PE. Various methods have been used, including study of the variation of diffracted intensity with diffraction order, analysis of the shape of diffraction lines, and a comparison between the long period from X-ray diffraction and the chain length as determined by the Raman-active longitudinal acoustic mode. The results generally seem to favor a narrow region over which crystallographic register is lost, probably less than 2–3 nm [125]. NMR and X-ray analysis indicate that the boundary between the crystalline and amorphous regions is not sharp, but an intermediate interfacial region is present.

Wittman and Lotz have developed the method of the fold surface decoration of polyethylene crystals grown from solution and melt [126]. In both lozenge-shaped polyethylene crystals and those with rounded faces, they observed decorating lines perpendicular to the (local) growth faces, clearly revealing the different crystal sectors with their differing fold directions. A spherulitic polyethylene film, prepared by slow cooling, showed a preferential orientation of decoration normal to the lateral faces of the lamellae involved. This provides strong evidence against the possibility of random folding occurring in any of the polyethylene systems considered.

Neutron scattering [127, 128] and mixed crystal IR techniques [129, 130] strongly favor a model with a high percentage of adjacent re-entry.

The first atomic force microscopy (AFM) images of PE were obtained by Patil *et al.* [131]. This study was on dendritic crystals of linear polyethylene formed on cooling a hot solution in xylene. The suspended crystalline precipitate was dried down on mica and imaged in air in the constant-force mode. Images typical of the apices of dendritic polymer crystals [132] were obtained. Lozenge-shaped, chain-folded crystals with spiral terraces, familiar from TEM studies, were also found. In [131] is reported an AFM image of solution-grown single crystals of PE, see Fig. 9, showing multiple layers of overgrowth.

The nature of the PE single-crystal surface as a function of molecular mass has been the subject of a study using scanning force microscopy by Takahara *et al.* [133]. In particular, the following molecular masses were used: $M_w = 520 \times 10^3$, 10×10^3, and 45×10^3. Characteristic lozenge shaped lamellar crystals *ca.* 40 μm long and 10 nm thick were observed. The results indicated that, for the low molecular mass, the surface of the HDPE single crystals is formed of sharp and regular folds, whereas for the high M_w the surface has switchboard-like random loose loops. The surface structure of the intermediate molecular mass is also intermediate.

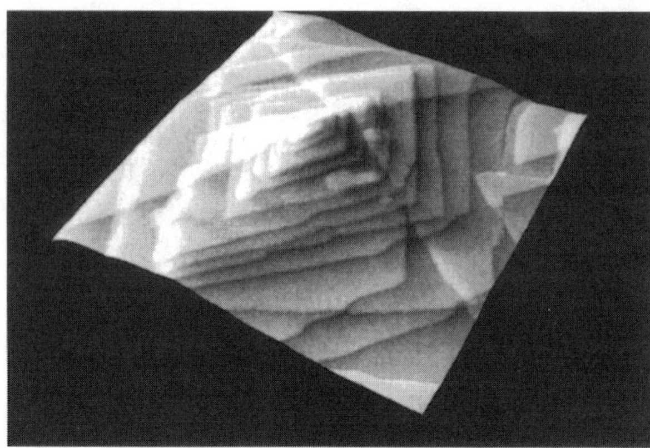

FIG. 9 Atomic force microscopy image of solution-grown PE crystals. (Reprinted from Ref. 130.)

Polyethylene crystallites are generally not planar; rather, they have a hollow-pyramidal shape [134, 135]. A pleat or a trapezoid-shaped structure is observed in the center of the lamellae. By precipitation of crystals from highly diluted solutions, mats are obtained which are very brittle because of the scarcity of interlamellar connections and high crystallinity.

The ridged lozenge is related to hollow-pyramdial conformation by reversals of obliquity in the major sectors. Ridging a lozenge modifies the mode of collapse of nonplanar lozenges. In the case of ridged polyethylene lozenges, each facet contains an ordered pattern of folding and is a microsector of the {110}{311} type. They do not have different growth faces for adjacent units, and all have the same {110} plane. Revol and Manley [136] using electron crystallography were able to directly visualize in the collapsed polyethylene lamellae crystalline regions corresponding to alternate bands of titled and untitled molecular chain stems.

Table 9 shows structural properties for solution-crystallized polyethylene. Of particular interest are the relationships between crystallite size, crystallization temperature, and the stability of the crystals formed.

Crystallite size depends very strongly on the temperature of crystallization and annealing. Its thickness can be determined from small-angle X-ray scattering [137–139], electron microscopy, and from the analysis of the Raman low frequency longitudinal mode [140–142]. These methods allow also the determination of the crystallite size distribution [139, 142, 143].

Figure 10 plots crystallite sizes, as a function of crystallization temperature, for polyethylene. The sizes range from about 80 to 200 Å, and the dependence on crystallization temperature is very evident. The size is independent of molecular mass in the range specified. It is important to note that both crystallite size and dissolution temperature remain invariant with time. Although solution-formed crystals thicken with time during annealing, thickening does not take place during isothermal crystallization from dilute solutions [143, 144]. This can be explained by the

TABLE 9 Structural Properties for PE Crystallized from *p*-Xylene Solution

M_w	M_w/M_n	T_c (°C)	L (Å)
37,800	1.11	89.5	
93,000	1.11	89.5	
188,500	1.11	89.5	
428,000	1.45	89.5	
37,800	1.11	23*	
188,500	1.11	23*	
52,600	1.06	87	127†
115,000	1.07	87	127†
161,000	1.05	87	136†
225,000	1.09	87	127†
800,000	1.11	87	134†
1500,000‡		87	127†

*Rapidly cooled to the temperature indicated.
†From (most probable) LAM (low acoustic mode).
‡M_η.

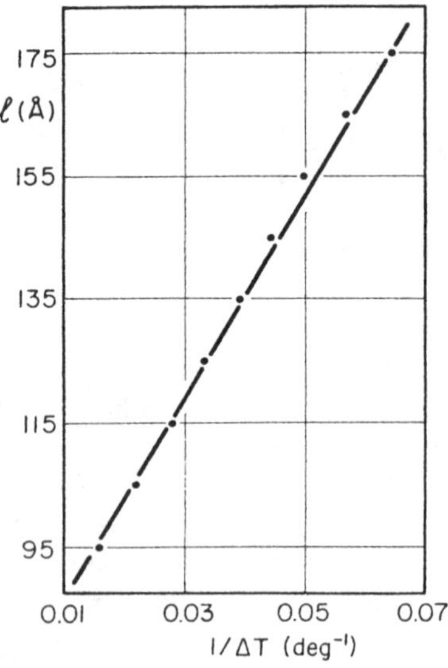

FIG. 10 Lamellar thickness of solution-grown PE as a function of crystallization temperature. (Reprinted from Ref. 110.)

high undercooling involved in the crystallization process for dilute systems. Zhou and Wilkes [139] report for polyethylene the lamellae thickness distribution histograms for films annealed from 120 and 130°C, see Fig. 11.

Crystal size increases with temperature and with time of annealing. This increase is faster with higher annealing temperatures. Explanations for the mechanism include a diffusion of the chains along their backbones and the effect of melting and recrystallization of thinner crystals.

In solution-crystallized polyethylene fractions, Raman spectra have demonstrated that crystalline structure is invariant with molecular mass and that crystallinity is far from complete. The interfacial region is relatively small, as expected from theoretical considerations. Densities of solution-grown crystals of linear polyethylene show that the crystals are 80–90% crystalline [145–147]. This conclusion is supported by measurements of the enthalpy of fusion, infrared and Raman spectroscopy, and other physical properties [148]. Consequently there is a small but appreciable amount of amorphous material, and disorder occurs at the lamellar surfaces.

The lamellar morphology and high crystallinity of polyethylene suggest that there is intimate contact between crystals, after both solution and melt crystallization. More specifically, the close proximity of fold surfaces from neighboring lamellae, and also the possibility of the molecules between them, indicates considerable interactions between lamellae.

With special methods such as using very thin films (~ 50 nm) or quenching and subsequent extraction in a solvent [149–151], Labaig [149], Keith *et al.* [150] and Bassett *et al.* [151] have obtained single crystals grown from the melt in the same temperature range where spherulites and axialites are formed. The shape of melt-grown single crystals is dependent on crystallization temperature and molecular mass.

They have obtained lenticular single crystals that have sharp tips in the direction of the *b* axis. Toda [152] obtained single crystals from polyethylene fractions isothermally crystallized from the melt at atmospheric pressure by using the same technique.

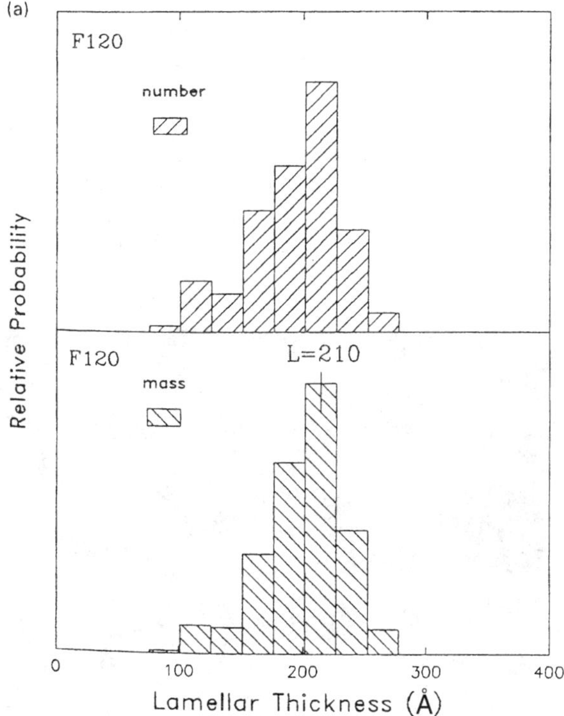

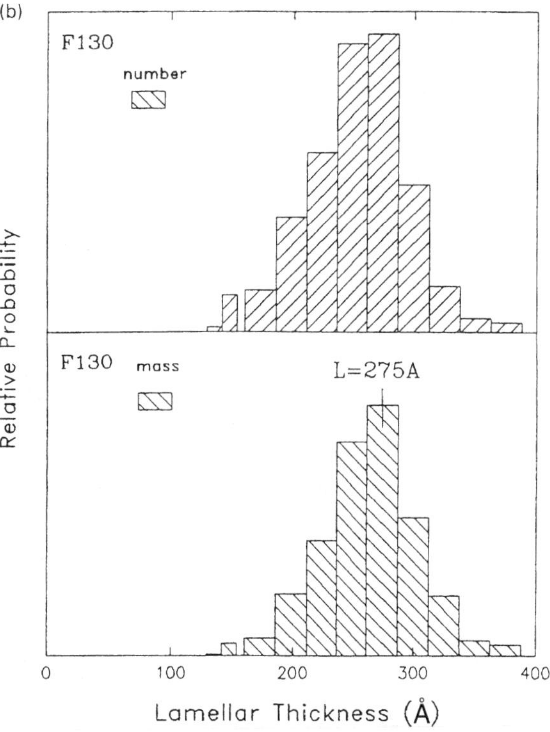

FIG. 11 Lamellar thickness distribution histograms based on transmission electron microscopy results for PE annealed film at (a) 120°C, and (b) 130°C. (Reprinted from Ref. 139.)

Three different fractions were used, see Table 10. He observed two types of crystals. It has been found that the lateral habit of single crystals changes in the vicinity of the transition temperature of the growth regime (regime I–II): a lenticular shape elongated in the direction of the b axis (type A) in the range of regime I and a truncated lozenge with curved edges of {200} and {110} growth faces (type B) in regime II. The transition of lateral habit causes a drastic change in the width of {110} growth faces; {110} growth faces are well developed in type B crystals whereas they cannot be observed and must be very small in type A crystals. It has been shown that the growth regime of small {110} growth face of type A crystals must be in regime I; hence, the regime I–II transition can be explained as the result of this change in lateral habit (with the {110} growth face).

Lateral habit of single crystals of 14×10^3 fraction remained lenticular within the examined range (125–130°C). The growth has been supposed to be in regime I [153].

Figure 12 shows the change in the lateral habit of single crystals of the 32×10^3 fraction with crystallization temperature. Lateral habit has changed from lenticular shape to truncated lozenge with curved edges of {200} and {110} growth faces with decreasing crystallization temperature (increasing supercooling).

Crystals of 120×10^3 fraction showed morphologically quite different shapes from those of the 14×10^3 and 32×10^3 fractions. The external shape is close to truncated lozenge; Bassett *et al.* [151] have reported similar shapes for a high molecular mass fraction. The growth front of {110} growth sectors is serrated and divided into small growth tips similar to the tips of lenticular crystals.

For two low molecular mass fractions ($M_n = 3.9 \times 10^3$ and 5.8×10^3) the morphological change from a truncated lozenge with curved plans to a lenticular crystal was observed by Cheng *et al.* [154] at a temperature higher than the regime transition as proposed by Toda [152] for PE having molecular mass ranging from 14×10^3 to 120×10^3. The morphological change is accompanied by a sharp long period increase. The morphological change may result from a sudden increase in the G_b coupled with a smaller change in the growth rate along the a axis with undercooling. This implies that, within this temperature region (2°C), the crystals may undergo substantial changes in the geometry of the {110} and {200} crystal growth fronts and chain-folding behavior.

TABLE 10 Morphological Characteristics of PE Fractions

Fraction	M_w	M_w/M_n	Morphological habits
14×10^3	13,600	1.24	Lenticular
32×10^3	32,100	1.11	Lenticular/truncated lozenges
120×10^3	119,600	1.19	Lozenges

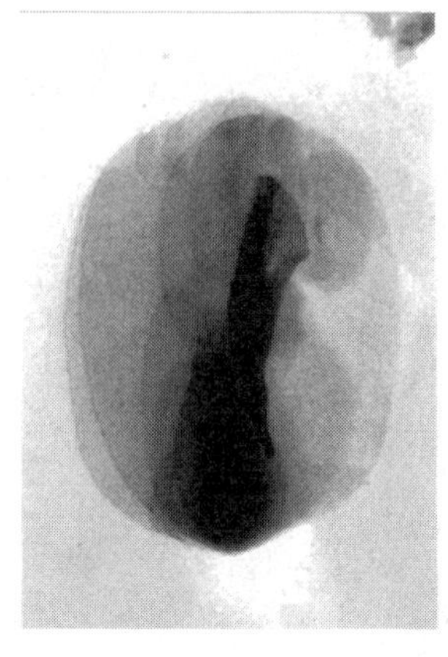

(a)

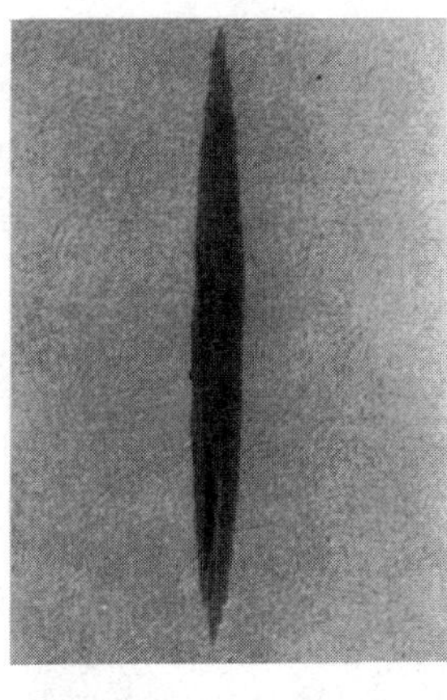

(b)

FIG. 12 Electron micrographs of the 32×10^3 PE fraction crystallized at: (a) 125°C, and (b) 131°C. (Reprinted from Ref. 152.)

Samples crystallized by air quenching were studied by Cheng [154]. Banded spherulites are found when the M_n is 5.8×10^3 (see Fig. 13). The lower molecular mass fraction ($M_n = 3.9 \times 10^3$) does not show the banded texture. When the fractions were isothermally crystallized at different temperatures, a spherulitic texture was observed for both fractions at relatively low tem-

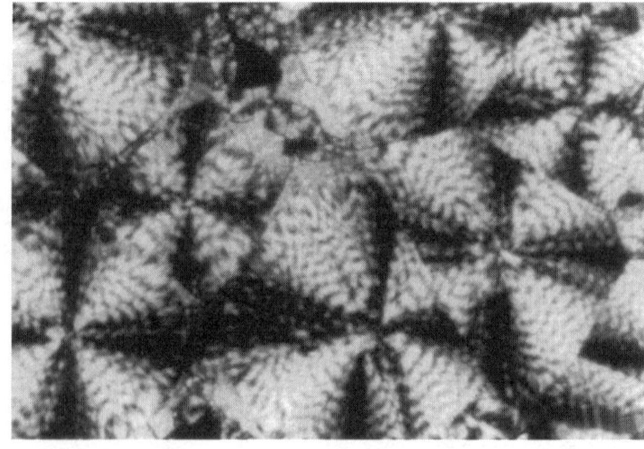

FIG. 13 Optical micrograph of ringed spherulites of PE. (Reprinted from Ref. 154.)

peratures. Also Bassett and Magill *et al.* [155, 156] obtained banded spherulites of PE by rapid quenching from the melt (see Fig. 14). According to Bassett this orientation is achieved by interleaving of curved lamellae units.

Polyethylene generally crystallizes from the melt into polycrystalline aggregates such as spherulites and axialites even if many other supermolecular structures have been described. Hoffman *et al.* [153] studying the kinetics have found a transition in the dependence of growth rate on crystallization temperature at a definite supercooling (Fig. 15). In spherulites and axialites the *b* axis of crystals is oriented in the radial direction [157], while in single crystals a pair of {110} faces grows in the direction of the *b* axis. Therefore, the transition of growth rate in these aggregates should be due to the change in the growth mode of the {110} growth faces. Hoffman *et al.* [1] have attributed the transition to the change of growth regime (regime I–II) of {110} growth faces. It was also reported that the transition is accompanied by a morphological change: spherulite (regime II) and axialite (regime I) [157].

A detailed analysis of the crystallite and supermolecular structure of polyethylene was performed by Mandelkern *et al.* [158, 159], who investigated the effect of molecular mass on the morphology using narrow fractions of polyethylene crystallized from the melt by quenching and by isothermal crystallization, using electron microscopy and light scattering techniques. A summary of the morphological characteristics of the crystallites is given in Table 11.

FIG. 14 Scanning electron micrographs of a ringed spherulite of PE. (Reprinted from Ref. 156.)

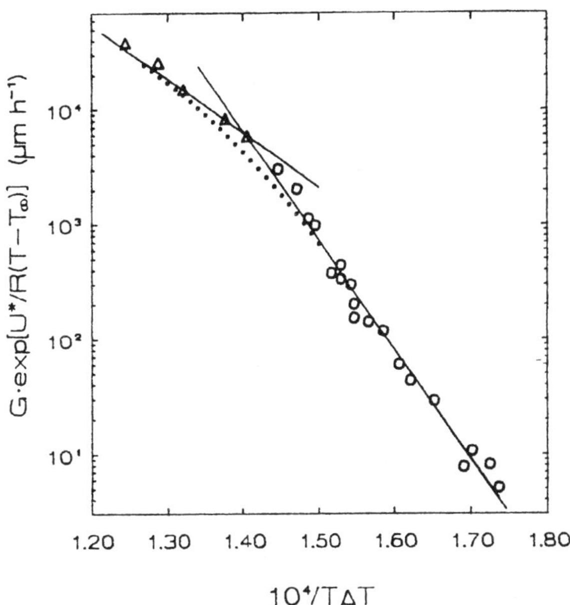

FIG. 15 Typical supercooling dependence of growth rate (○ axialite and Δ spherulite) of a PE fraction. (Reprinted from Ref. 153.)

It can be stated that the supermolecular structures that are formed depend on the molecular mass, the crystallization conditions (such as the isothermal crystallization temperature or the cooling rate), the molecular constitution and the molecular mass polydispersity.

A morphological map that depicts how the supermolecular structures depend on molecular mass and crystallization conditions was established from isothermal to nonisothermal crystallization. Figures 16 and 17 highlight that spherulitic structures are not always observed. In fact, as the map clearly indicates, superstructures do not always develop. In isothermal conditions, the low molecular mass polymers form thin rod-like structures. As the molecular mass is increased, a sheet-like type morphology is observed at the higher crystallization temperatures. Here the length and breadth of the rod-like structures are comparable to one another. If the isothermal crystallization temperature is lowered then, in this molecular mass range, spherulites will form. The spherulitic structure deteriorates as the chain length increases in this range. For molecular masses greater than about 2×10^6 no organized superstructures are observed at all, although the crystallinity level is of the order of 0.50–0.60 for these samples.

A morphological map for the nonisothermal crystallization of linear polyethylene is given in Fig. 17. In

TABLE 11 Morphological Characteristics of PE Fractions as Function of Temperatures

	Molecular weight		
	5.6×10^3	7.7×10^4–2.5×10^5	1.6×10^6–6×10^6
High T_c	Large roof shaped crystals	Large rod shaped crystals	Curved plus roof shaped crystals
Intermediate T_c	Large intermediate roof shaped crystals	Intermediate roof shaped and curved crystals	Short curved crystals
Quenched	Thin roof shaped crystals	Curved thin crystals	Small crystallites

these experiments the temperature listed is not the crystallization temperature but that of the quenching bath to which the sample is rapidly transferred from the melt. Although no new structural forms are observed, major differences can again be developed, depending on the molecular mass and quenching temperature. The random-type morphology can now be formed for molecular masses as low as 1×10^5 for crystallization after relatively rapid cooling. Hence this structure is not limited to very high molecular masses. In contrast, well developed spherulites can also be generated at low temperatures for very low molecular mass fractions.

By using the maps of Figs. 16 and 17 it is found that it is possible to prepare different supermolecular structures from the same molecular mass by choosing the appropriate crystallization conditions. In certain situations the different superstructures can be formed with the same molecular mass at the same level of crystallinity.

Cimmino *et al.* [160] observed a microspherulitic morphology ($M_w = 3.5 \times 10^4$, $M_n = 3.3 \times 10^4$, $M_z = 1.91 \times 10^6$) and a SALS photograph characterized by a four leaf pattern for the PE crystallized isothermally at $T_c = 125°C$.

Although several supermolecular structures are formed during the crystallization of PE, depending strongly on the molecular mass and crystallization conditions, excluding the very high and very low molecular mass samples, a correlation between growth regime and morphology can be schematically stated:

> Generally in regime I the habit of single crystals is lenticular, whereas the habit of the polycrystallite aggregates is axialitic. In regime II the lamellae have truncated lozenges shape, whereas the superstructure is spherulitic.

An internal surface of a cold-extruded rod of polyethylene was prepared for AFM by Magonov *et al.* [161] by ultramicrotomy. As expected, a fibrillar structure was observed. The authors do not report any evidence of stacked lamellae. At high magnification, ridges, running roughly parallel and in the extrusion direction, were seen, and measurements of their diameters gave values of 0.4–0.5 nm. This corresponds to the van der Waals diameter of a polyethylene molecule when the hydrogen atoms are included. In some regions, extended chains of PE in the all-*trans* conformation were seen with the expected repeat distance along the chain of 0.25 nm. For a drawn solution-cast film of ultra-high molecular mass polyethylene,

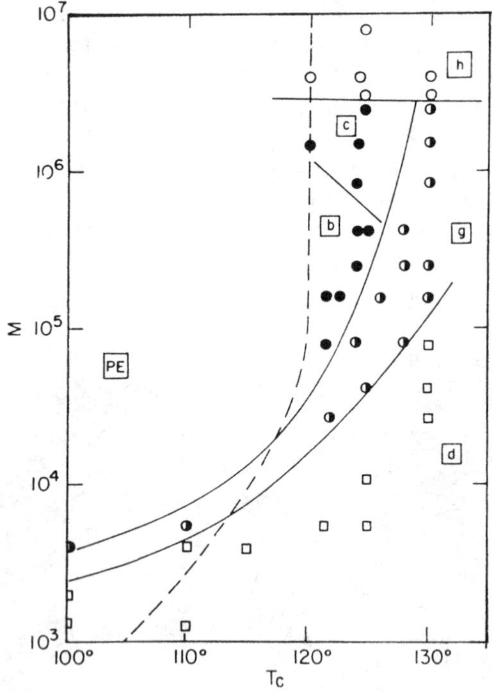

FIG. 16 Morphological map for isothermally crystallized molecular mass fractions of PE. Regions marked with b and c denote spherulites; region d, thin rods; region g, rods and sheet-like crystals; region h, randomly oriented lamella. (Reprinted with permission from Ref. 158. Copyright 1984 American Chemical Society.)

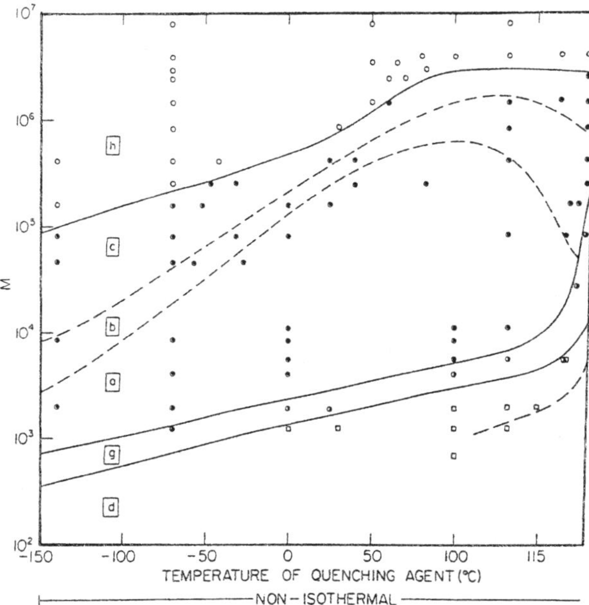

FIG. 17 Morphological map for nonisothermally crystallized molecular mass fractions of PE. Regions marked with a, b and c denote spherulites; region d, thin rods; region g, rods and sheet-like crystals; region h, randomly oriented lamella. (Reprinted with permission from Ref. 158. Copyright 1984 American Chemical Society.)

Snétivy *et al.* [162] imaged regions of ordered molecules at many areas on the surface of the specimen. The values of the lateral spacing of molecules in the different regions fell into two groups: 0.73 ± 0.018 nm and 0.49 ± 0.023 nm. These values are in good agreement with molecular spacings on the *a–c* and *b–c* planes, respectively.

2. Isotactic Polypropylene

Well defined lath-shaped lamellar crystals of isotactic polypropylene were grown from solution [163]. The thickness of these lamellae, as determined by small-angle X-ray diffraction of crystal aggregates, is 125 Å. The crystals were mainly monolayer with the shorter and larger dimensions, respectively, of about 0.2 and 1 μm. Similar lamellar crystals were observed by Sauer *et al.* [164], Morrow *et al.* [165], Kojima [166], and Lotz and Wittmann [167]. Fibrils parallel to the longitudinal direction of the lamella were often observed. According to Kojima [166] these fibrils connect the planes of fracture and cleavage that are parallel to the *a* axis. This is an indication that the {010} planes are the most probable fold planes. Lotz and Wittmann [167] observed microsectorization along the {110} crystallographic direction in the center of the lath.

Square lamellae and lath-shaped lamellae were also detected for iPP by Al-Raheil *et al.* [168]. They observed that the iPP spherulite developed from initials quare lamellae by branching probably from screw dislocations and spreading out from four sides to form sheaflike structures. This process continued, producing radiating individual lamellae, within which short lamellae (crosshatched) traversed the distance between two neighboring radiating lamellae, leading eventually to a spherical envelope.

All the observations of the morphology of solution-crystallized and melt-crystallized iPP indicate that the morphology of iPP is dominated by a highly characteristic lamellar branching (crosshatching), which has no counterpart in other crystalline polymers [47, 109, 169–178]. It has been recognized that this lamellar branching is characterized by a constant angle between daughter and parent lamellae (80° or 100°). This feature of the morphology is illustrated schematically in Fig. 18. As indicated the preferred growth direction of the dominant radial lamella has been determined to be associated with the a^* crystallographic direction, with the chain axis nearly perpendicular to the radial direction. An extensive study on this subject was made by Lotz and Wittmann [167]. They suggested that structurally the daughter lamellae in iPP grow expitaxially on the lateral {010} crystallographic plane of the parent lamellae by a satisfactory interdigitation of the

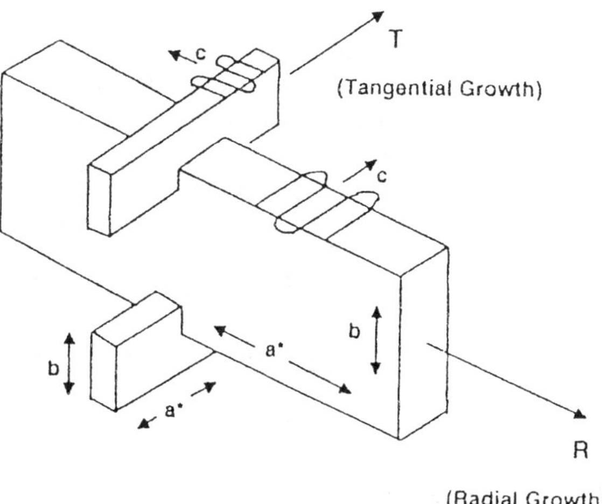

FIG. 18 Schematic of crosshatched lamellar morphology of iPP. The radial and tangential growth are shown together with the relative relationships of the chain axis in each.

methyl groups of facing planes. This condition arises from the near identity of the size of the *a* and *c* crystallographic axes in the unit cell of the monoclinic α form. From a molecular point of view, chains that deposit onto the {010} plane for the initiation of this epitaxy have the same helical handiness as chains in the [010] plane, but with a substantial angle of 80° or 100° in order to have favorable interactions of the methyl groups in the helices.

It has also been found that the degree of branching decreases with isotacticity in iPP crystals and is T_c-dependent. In particular the crosshatching increases as the isothermal crystallization temperature is reduced.

The crystallinity, apparent crystal size and lamellar thickness are found to be dependent on crystallization temperature and isotacticity [179]. Table 12 lists the SAXS results for PP fractions having similar molecular mass but different isotacticity, which were crystallized at different crystallization temperatures. The lamellar thickness of the crystals decreases with isotacticity for a given supercooling. On the other hand, the lamellar thickness for each fraction increases with decreasing supercooling. Nevertheless, the lamellar thickness of these fractions at a constant temperature shows an interesting tendency. Namely, with decreasing isotacticity, the thickness seems to increase, as shown in Table 12.

Furthermore, the width of the SAXS peaks for the sample with highest tacticity is about three times as narrow as that for the iPP fraction with lowest tacticity at the same ΔT, indicating different distributions of the long spacing sizes for these fractions with different isotacticities.

Early studies by Padden and Keith [180] suggested the formation of five different types of spherulites consisting of α and β modifications (α- and β-spherulites formed during the melt crystallization of iPP). They demonstrated that three types of α-spherulite might be formed depending on the temperature of crystallization: positive radial (α_I) below 134°C, negative radial above 137°C (α_{II}), and mixed-type spherulites in the intermediate range (α_{III}).

Two types of β-spherulite formed sporadically with the α modification [180]: β_{III}, highly birefringent, below 128°C, and negative ringed (β_{IV}) between 128–132°C. The different types of iPP spherulites were studied in detail by Varga and reported in Fig. 19 [27]. Several authors supported the observations of Padden and Keith with their subsequent investigations, but they report different temperature ranges for the formation of the particular types of spherulites, presumably due to the variability in the molecular characteristics of the iPP types used [181, 182].

TABLE 12 Lamellar Thickness of the iPP Fractions with Different Isotacticity Content Isothermally Crystallized from the Melt

Isotacticity	T_c (°C)	L_{exp} (nm)*	l_{exp} (nm)
0.988	112	19.2	8.4
	122	21.5	9.9
	132	22.4	10.8
	142	25.7	13.1
	152	30.4	20.1
0.978	107	20.7	8.3
	117	21.9	9.0
	127	22.8	9.8
	137	25.1	11.8
	147	30.1	18.1
0.953	107	23.1	8.3
	117	24.8	9.2
	127	26.9	10.8
	137	29.4	12.6
	147	32.5	18.2
0.882	87	20.2	6.3
	97	22.2	7.0
	107	25.5	8.4
	117	27.0	9.3
	127	29.5	10.9
0.787	62	24.6	4.4
	72	24.7	4.7
	82	29.0	5.8
	92	31.4	6.6
	102	35.0	7.7

*L_{exp} represents the long spacing observed via SAXS measurements.

The different optical characteristics of the α-spherulites have been linked to the lamellar morphology through the balance of crosshatched radial and tangential lamellae. As shown schematically in Fig. 18, the lamellae are composed of chains nearly normal to the lamellar surfaces. In this way, the negatively birefringent α_{II} spherulites are dominated by the radial lamellae (chain perpendicular to the spherulitic radius), whereas the α_I positively birefringent spherulite contains increased quantities of the tangential lamellae (chain parallel to spherulitic radius). During heating, a sign change of the birefringence for α_I spherulites is observed due to the premelting of the orthogonal overgrowths. This transition between negative and positive birefringence occurs when approximately one-third of the lamella is tangential [173]. At higher crystallization temperature [183] axialitic morphologies composed of sheaf-like structures have been observed, with little tendency for crosshatching [171, 178, 179, 184].

FIG. 19 Optical micrographs of the different α and β types of iPP spherulites: (a) α_{I}; (b) α_{II}; (c) α_{III}; (d) β_{III}; (e) β_{IV}. (Reprinted from Ref. 27.)

Norton and Keller [181] reviewed the experimental results on the structure of positive and mixed spherulites in iPP. It was demonstrated that α-spherulites contained fibrils with large-angle branches even at about 80° to the radius of the spherulite [173]. This was in good agreement with other observations for crystals crystallized in solution [185–187]. The structure of the positive and mixed spherulites was interpreted by the "crosshatched" model [178, 181, 186] where radial fibrils were accompanied by a large number of tangential ones. Assuming a quadratic arrangement of fibrils, the relative amount of radial and tangential fibrils in various types of α-spherulites was determined by Idrissi *et al.* [188] using optical microscopy, light scattering, and calorimetry.

The optical character of α-spherulites is controlled by the ratio of radial to tangential fibrils. Raising the temperature of crystallization leads to a reduction in the proportion of tangential ones [181] and, simultaneously, a positive to negative character transformation. No tangential fibrils form above 155°C [172]. It was also revealed that the thickness of tangential fibrils was lower than that of the radial ones.

Structural and optical characteristics of β-spherulites were studied comprehensively by Samuels [189, 190]. He found intrinsic refractive indices in the hexagonal (β) modification of 1.536 and 1.506 in the *c*- (chain direction) and *a*-axis directions respectively, resulting in a strong negative optical character. He also revealed that the tangential refractive index of β_{III}-spherulites was constant (1.507) while it changed periodically between 1.496 and 1.519 in β_{IV} spherulites, according to their ringed feature.

The arrangement of the fibrils in β-spherulites is radial, as is usual in polymeric spherulites, and no traces of a crosshatched structure can be detected [181]. The central region of β-spherulites comprises nohomogeneous and sheaf-like branched fibrils.

Varga [52, 53, 191] demonstrated that the formation of β-PP had a theoretical upper temperature limit ($T_{\beta\alpha} = 140$–141°C), whereas indirect experimental data [52] referred to a possible lower temperature limit ($T_{\beta\alpha} \approx 100$°C). The existence of $T_{\alpha\beta}$ was proved by Lotz *et al.* with direct experiments [192].

Formation of some new types of spherulite was also detected with a high-temperature crystallization of iPP ($T_c = 150$–160°C). Awaya [171] observed several new types of α-spherulites, designated pseudo-positive, pseudo-negative, neo-mixed, high-temperature positive, and flower-like ones. Varga observed for highly degraded iPP positive α-spherulites independently of the crystallization temperature, where monofibrils and small-angle branches of fibrils became visible even by optical microscopy. Moreover during a self-seeded high-temperature crystallization ($T_c = 150$–160°C), positive α-spherulites were found [172, 178], in contrast to the observations of Padden and Keith [180]. In some experiments, the random appearance of hexagonal hedrites of β modification was detected in addition to the α-spherulites, even at a low-resolution optical level. At a further stage of growth, they eventually transformed into a radial symmetry characteristic of the spherulitic structure.

Using electron microscopy, Geil [193] and Olley [194] detected β hedrites produced by melt crystallization. The micrographs suggest that hedrites are derived from a spiral growth of lamellae round a screw dislocation. The size of the hedrites is about 1 μm.

Few publications report on the morphology of the iPP γ modification. These studies were carried on mainly by Phillips [63, 195] and show that the γ modification exhibits both positive and negative birefringent spherulites, Fig. 20. However, when α and γ modifications are present in the same sample, the morphological features become complex. The optical studies on microtomed sections show no evidence of Maltese cross formation when less than 10% of the material is in the γ modification, whereas when more than 60% of γ modification is present, a clear Maltese cross exists. In addition, optical and electron microscope studies of etched specimens reveal no crosshatching in these samples. The birefringence of the pure γ modification crystallized at 200 MPa changes from positive to negative as supercooling is increased. It was also shown that spherulites grown at low supercooling take the form of large "feather-like" structures, apparently caused by massive self-epitaxy, see Fig. 21.

3. Syndiotactic Polypropylene

Few publications have made reference to the supermolecular structure of sPP. Marchetti and Martuscelli [196] and Lovinger *et al.* [82, 197], and Rodriguez-Arnold *et al.* [198, 199] published mostly papers in which the morphology was systematically studied with regard to chain stereoregularity and crystallization conditions.

Marchetti and Martuscelli in 1974 studied the morphology of solution-grown single crystals of sPP and the effects of thermal history and configurational chain defects that occur when there is an interruption in the chain's stereoregular sequence. Various solvents and

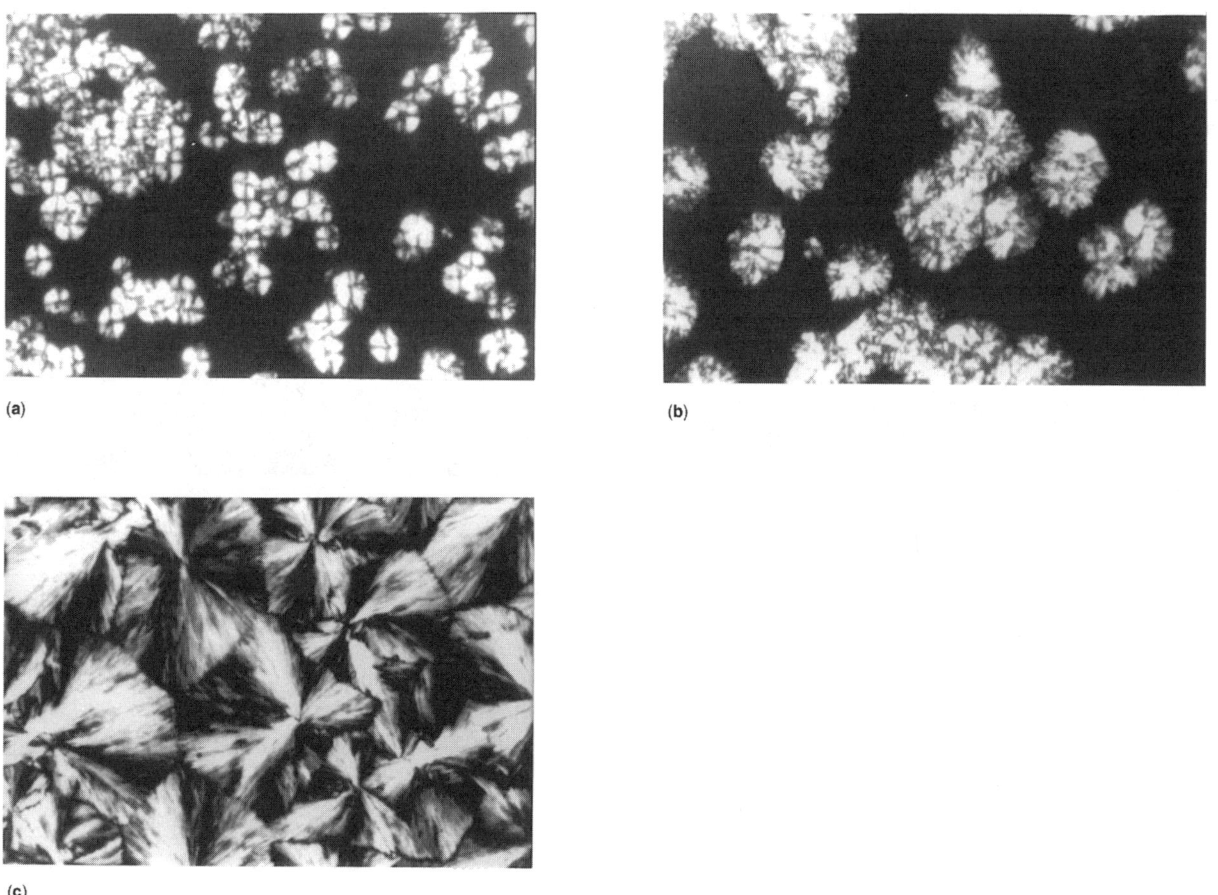

FIG. 20 Optical micrographs of the different γ types of iPP spherulites: (a) negative birefringence spherulites; (b) mixed birefringence spherulites; (c) positive birefringence spherulites. (Reprinted from Ref. 195.)

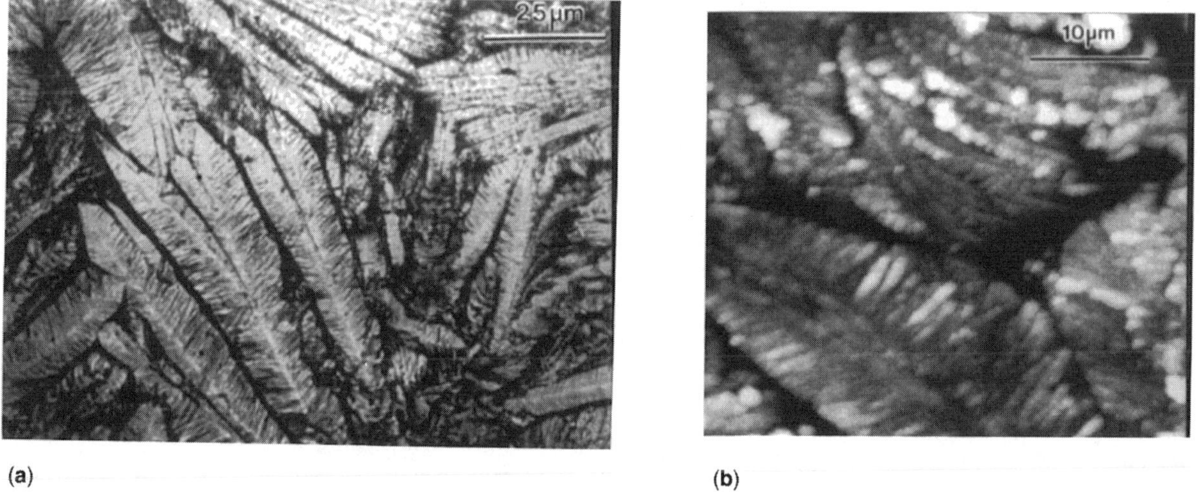

FIG. 21 Feather-like structure in iPP γ type spherulites, crystallized at 200 Mpa and 203°C. (a) Reflection otpical micrograph; (b) atomic force micrograph. (Reprinted from Ref. 195.)

crystallization conditions were used in an attempt to grow single crystals. The sample with the highest stereoregularity showed elongated single crystals with irregular edges. The crystallographic facet was proposed to be along the longitudinal direction of the single crystals. They also suggested that most likely the fold planes were the {100} and the {110} of cell I (cells II and III were not recognized at that time) and that the growth direction may be coincident with both the *a* and *b* axes. Twinned monolayer crystals were also found which were believed to develop from twinned nuclei [200]. Wide-angle X-ray diffraction patterns showed a crystal structure corresponding to the packing of helical chains as described by Corradini *et al.* [78]. From oriented WAXD patterns of single crystal mats, the authors concluded that the chain *c* axis is almost perpendicular to the mat plane and, therefore, to the fold surface of the lamellar crystals. Samples with lower syndioregularity exhibited dentritic-like structures.

The lamellar thicknesses after solution crystallization and annealing at different temperatures showed a decrease with decreasing syndiotacticity index and annealing temperatures. The long spacings are also critically dependent upon the solvent systems used in solution crystallization. It was postulated that defects might be included, at least to a certain degree, in the crystal lattice.

Lovinger *et al.* [82] studied the crystal morphology of sPP crystals grown from the melt in thin films and its dependence on supercooling. Samples possessed a racemic diad content of 0.769 and syndiotactic and isotactic triads of 0.698 and 0.159, respectively. At the lowest supercoolings ($T_c > 105°C$), large, rectangular, faceted lamellar crystals were observed, see Fig. 22. These crystals had the same unit cell and interchain packing consistent with cell III, as previously proposed by the same authors. At a high supercooling (a crystallization temperature of 90°C), twinned crystals seemed to be more common. As T_c was further lowered, at around 60°C, the morphology became axialitic and eventually spherulitic, see Table 13.

Electron diffraction patterns suggested that, with increasing supercooling, the intermolecular packing along the *a* axis systematically deviated from the regular antichiral form due to kinetic reasons during crystallization. More specifically, the packing disorder is probably introduced along the *b* axis with the incorporation of isochiral chains and thus a unit cell I type crystal structure results. Moreover Lovinger observed overgrowths on the sPP crystals grown from relatively low syndiotacticity samples identified as individual iPP crystals. These crystals were initiated by specific epitaxial relationships between iPP and lateral surfaces of sPP lamellae. The presence of iPP in the sPP samples was due to individual iPP chains in the samples, and not stereoblock copolymers.

FIG. 22 Electron micrograph of crystals of SPP. (Reprinted with permission from Ref. 197. Copyright 1994 American Chemical Society.)

Also Rodriguez-Arnold *et al.* pointed out the dependence of the morphology on the temperature, but comparing with Lovinger's observations, the temperatures where the different morphologies formed are shifted to higher temperatures [198, 199], see Table 13. At crystallization temperatures lower than 105°C spherulites and axialites are observed. With increasing temperature, the spherulitic texture becomes relatively coarse, with irregular Maltese cross extinction. Further temperature increases, at 125°C, lead to the growth of single large rectangular lamellar crystals. These crys-

TABLE 13 Morphological Characteristics of SPP as Function of Temperature According to Different Investigators

T_c (°C)	Morphology characteristics	Ref.
60	Spherulites and axialites	[84]
90	Twinned crystals	[84]
> 105	Large rectangular faced lamellar crystals	[84]
< 105	Spherulites	[85]
105–125	Spherulites with coarse structure and irregular Maltese cross	[85]
125	Single crystals	[85]

tals have a preferred crystallographic growth direction along the *b* axis. This behavior is independent of molecular mass. However, with increasing molecular mass lamellae crystals are slightly more difficult to grow. Other interesting observations reported by Rodriguez are the cracks are always perpendicular to the *b* axis found in single crystals. This unusual behavior has been attributed to the very large anisotropy of the thermal expansion coefficient of sPP with respect to the crystallographic direction, as reported by Lovinger [84]: (2.3×10^{-5} nm/°C *a* axis; 2.3×10^{-4} nm/°C *b* axis).

It is expected that these cracks may form when the lamellar crystals are cooled to room temperature after isothermal crystallization.

A similar anisotropy of the thermal expansion coefficient has been noted in the α form of iPP, although of somewhat reduced magnitude [44].

In the melt-crystallized thin films on selected substances, some tendency for homoepitaxy has been noted that could give rise to qualitatively similar cross-hatching behavior as in the α form of iPP [201].

A typical sPP rectangular, faceted, single lamellar crystal was also obtained by Rodriguez-Arnold *et al.* It was crystallized at 125°C from the isotropic melt. An electron diffraction pattern indicated that the long axis of this rectangular crystal is the *b* axis. Very recently, at high crystallization temperature, we also found that the single lamellar crystals possess microsectors for a sPP fraction with high syndiotacticity ([r] 94%, [rr] 92%, and [rrrr] 86%). This microsector may be an indication of chain folding domains present in the crystal. Under both polarized light microscopy and TEM, spherulitic morphology can also be observed in an intermediate supercooling range (Di Pace and Silvestre private communication, 1998).

4. Isotactic Poly(butene-1)

Well defined single crystals have been grown from solutions of the polymer in amyl acetate and its mixtures with *n*-butanol in the three modifications by varying the concentration and crystallization temperature by Sastry and Patel [202].

The morphology is linked to the crystal modification, Fig. 23. The crystals with orthorhombic modification are multilayers with square and lath shape. The crystals with tetragonal and hexagonal modification have square or hexagonal habits and present spiral ramps and terraces. In the morphology of the orthorhombic modification, twinning was frequently observed when the crystals were grown at the turbidity temperature from 1.5% solution in amyl acetate. In the tetragonal modification crystals rotation of the screw dislocations as well as interlacing multilayered growths have been observed. Reflection and supplementary twinning was commonly observed in crystal with orthorhombic and hexagonal modification. An interesting type of twinning was the transformation twin. In this twinning an orthorhombic basal lamella gave rise to hexagonal modification. The transformation started from the center of the orthohombic single crystal, yielding a hexagonal lamella with further overgrowth maintaining the same lattice orientation.

No evidence of twinning was observed in the tetragonal modification.

Holland and Miller [203] observed square crystals with tetragonal diffraction pattern within 10 min of their growth. During the lattice transformation from the tetragonal to the orthorhombic lattice, the morphology of the square crsytal did not change when there was no solvent. When the crystals were in contact with solvent a change in morphology along with the change in unit cell was found.

Mostly monolamellar and some multilamellar square crystals were observed by Kopp *et al.* [204] in their investigation on the phase II to phase I crystal transformation. The samples were obtained by slow evaporation of a hot octanol solution. Dark-field pictures further reveal the fine structure of the domains that appear to be striated. These striations mostly with the same orientation corresponded to the transformed domains from form II to form I.

Spirals on a poly(butene-1) lamellar structure grown from solution are seen in the TEM micrgraphs in Fig. 24 [205]. Spiral growths are frequently found when crystallization is carried out nonisothermally or at high degrees of supercooling.

The morphology of PB1 films drawn from the melt has been investigated by Fuchs *et al* [206] and Jandt et al. [207] by using the scanning tunneling microscopy analysis. Fuch *et al.* showed images of PB1 flakes extending over some hundred nanometers. Jandt *et al.* obtained structures similar to crystalline needles.

For samples isothermally crystallized from the melt, Silvestre *et al.* [208, 209] observed well defined spherulites of PB1 in form II modification, Fig. 25. At high crystallization temperature, $T_c > 100$°C, some crystallites lose part of the spherulitic character and square crystals were also observed, with hedritic morphology. For samples crystallized during the cooling, the morphology was characterized by large and small spherulites nucleated at different times and temperatures. The time lage between the two nucleation processes was

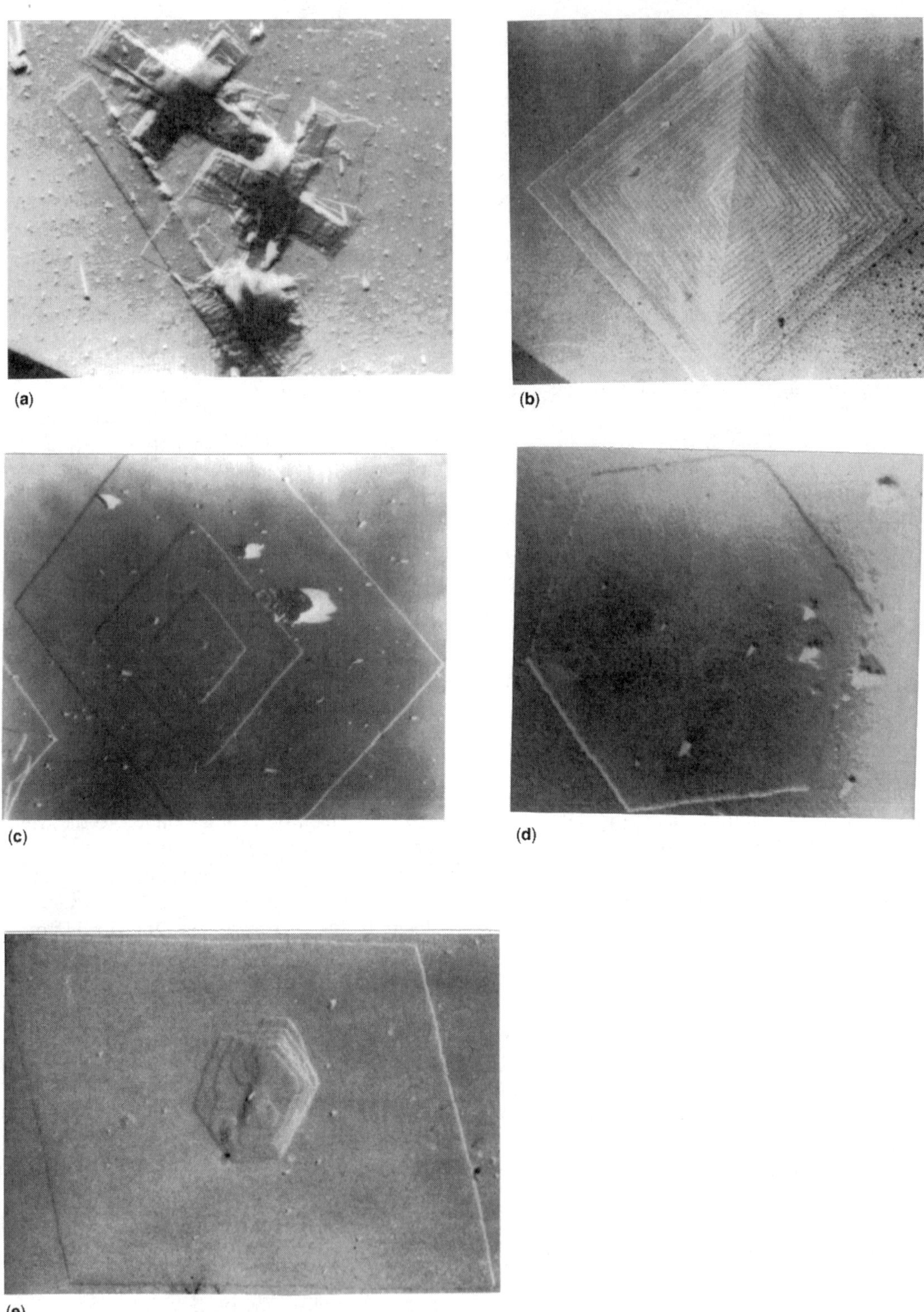

FIG. 23 Transmission electron micrographs of PB1 single crystals: (a) lath-shaped orthorhombic crystals; (b) terraced orthorhombic crystals; (c) terraced tetragonal crystals; (d) hexagonal crystals; (e) transformation twin from the orthorhombic to the hexagonal modification. (Reprinted from Ref. 202.)

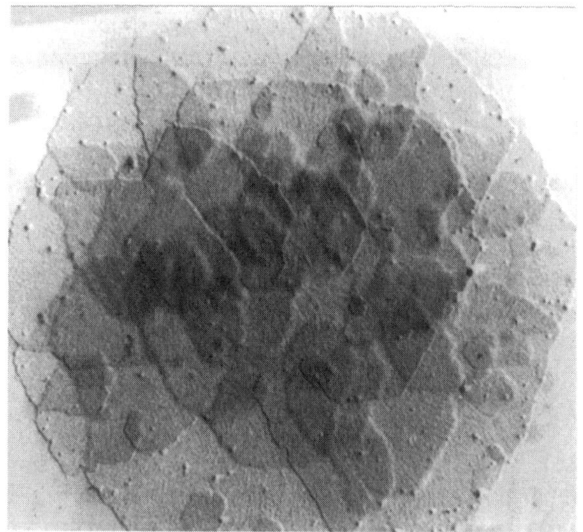

FIG. 24 Transmission electron micrographs of a PB1 single crystal, showing growth spirals. (Reprinted from Ref. 205.)

(a)

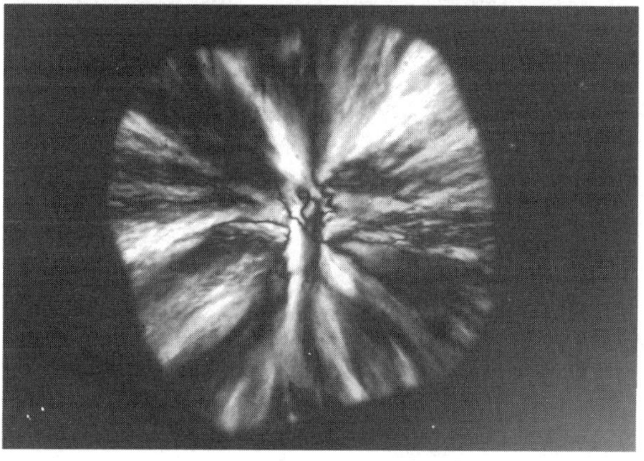

(b)

FIG. 25 Optical micrographs of thin films PB1 crystallized from the melt: (a) spherulite; (b) axialite.

indicated by the curvature of the common boundary between the different spherulites. The authors found that decreasing the temperature from the melting point, at a certain T that depended on composition, few nuclei started to grow, giving rise to large spherulites. At lower T, independently of the cooling rate used, suddenly many more nuclei appeared and grew. As their number was elevated, they impinged on each other very quickly and hence they could not grow into large spherulites. The number of the nuclei and the temperature at which both kinds of nuclei become active were independent of the premelting temperature. The two kinds of spherulites were attributed to the presence of heterogeneous nuclei having different activation energy.

5. Poly(4-methylpentene-1)

Several authors [102, 109, 210–213] reported square lamellae when P4MP1 is crystallized from dilute solution of several solvents (xylene, amyl acetate and toluene), Fig. 26. Bassett [212] reported that these lamellae were slightly dished because of distortion of the subcell by ordered arrangements of folds. Multilayer aggregates from concentrated solutions show bowl-shaped habits and develop in three dimensions around screw dislocation. Frank *et al.* [211] indicated that the appearance of the crystals was dependent on the cooling rate. Some crystals were pyramidal, some form polygons or spheres. Dentritic crystals containing numerous spiral growths were often observed. Solution of the polymer in tricloroethylene produced rows of parallel sheaves composed of fibrils, when a drop of the suspension was placed on slide and the solvent evaporated. The long period was 108 Å.

Pratt and Geil [214] studied the morphology of amorphous thin films of P4MP1 very rapidly quenched from the melt. Electron microscopy of the films presented a rather smooth surface with widely separated nodules of 200–250 Å in diameter. Cold electron diffraction gave no indication of crystallinity. Annealing these films at 47°C for 48 h produced a rough texture; upon longer annealing times, well defined structures with fibers or ribbons 100 Å thick were observed. Selected area diffraction showed continuous crystalline ring patterns of randomly oriented polycrystalline samples. Annealing for 94 h caused lathlike structures of 250 Å in width and 2000 Å in length, randomly placed on the film.

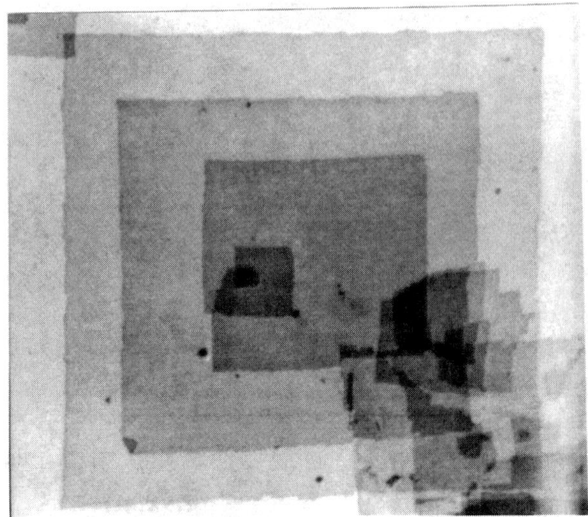

FIG. 26 Transmission electron micrograph of a single crystal of P4MP1 crystallized at 90°C. (Reprinted from Ref. 102.)

The morphology of nascent P4MP1 polymerized by Ziegler–Natta catalysis was studied by Blais and Manley [215, 216]. The P4MP1 possessed the modification I crystal structure. Optical microscopy showed well defined globular structures (spheroidal) with sizes between 0.5–1.0 mm. These globules failed to exhibit the characteristic Maltese cross patterns between crossed polarizers, indicating that they were not spherulitic in structure. Furthermore, the globules were found to be hollow, which upon drying, collapsed to flat sheets. The mean size and appearance of the globules were a function of the type of catalyst, its concentration, diluent used, reaction temperature, and mode of agitation. The globule walls were formed by a fibrillar structure. The fibrils were composed of platelets or lamellae about 100 Å thick, oriented at 45° with respect to the fiber axis obliquely stacked along the fibril axis that resembled shish-kebabs. Therefore, Blais and Manley proposed two models to account for the morphological observations. One model assumed that the fibrils consist of lamellae growing epitaxially on one another and held together by tie-molecules. The second one suggested that helicoidal crystals grow from a single screw dislocation. Crystalline growth would occur concurrently with synthesis. The growth rate was controlled by the rate of polymerization on the catalyst site. The globule formation was a secondary event due to collision of catalyst/fibril particles upon stirring.

Rybnikar and Geil [217] studied the lamellar structure of P4MP1 crystallized from the melt and solution. The lamellar thickness was independent of crystallization temperature for both crystallization procedures. The lamellar thickness was found to be 350 Å for samples crystallized from the melt in the 201–235°C temperature range, whereas for samples crystallized from 0.01% xylene, solution over the range of 25–90°C, the lamellar thickness was about 100 Å. In the melt-crystallized samples, the annealing did not change the mean lamellar thickness, indicating an upper limit. Solution-crystallized lamellae exhibited thickening upon annealing, to reach a mean lamellar thickness of 350 Å. This behavior was attributed to the low packing density of P4MP1 in the crystalline state, which induced high chain mobility in the crystal at high temperature. Electron microscopy of fracture surface replicas showed that lamellar thickness was not uniform. These authors reported a mosaic block structure in which the individual blocks in a given lamella vary from 200 to 500 Å in thickness. Such a mosaic structure was previously observed in single crystals of P4MP1 [96, 218].

Hase and Geil [219, 220] utilizing swelling techniques showed the presence of interlamellar fibrils between separated lamellae. The shape and density of interlamellar links were also revealed by electron microscopy of these preswollen samples. The density and thickness of the links were independent of crystallization temperature. These authors also indicated that swelling occurred heterogeneously in the interlamellar regions with only a small number of interlamellar links.

The morphological investigations mostly utilized transmission electron microscopy of sample replicas. This occurred because the density difference between crystalline and amorphous phases of P4MP1 at room temperature is very small ($\rho_a = 0.838\ \text{g/cm}^3$, $\rho_c = 0.828\ \text{g/cm}^3$) [221]. Consequently, the use of techniques such as small angle X-ray scattering to study the fine structure of P4MP1 has been limited. However SAXS measurements were used at high temperatures to study the effect of annealing on the long spacing of P4MP1 drawn samples [222, 223]. An increase of the long spacing as the annealing temperature increased was observed. No discrete SAXS peak in samples annealed below 180°C was observed. This was attributed to periodicity not being sufficient to be detected by SAXS although little further explanation was provided. One possible explanation may be the presence of random lamellae due to quenching of the samples from the melt, as has been observed in branched polyethylene upon quenching [224].

A technique was developed to observe the lamellar morphology directly with transmission electron microscopy without the need of replication. The technique is a modification of the staining method by Trent [225]. The lamellar thickness in these sections was 190 Å.

The morphology of melt-crystallized P4MP1 was studied by Patel and Bassett [226, 227] by X-rays and electron microscopy following permanganic etching for a wide range of crystallization temperatures (210–244°C). They observed a spherulitic morphology. The average diameter of the spherulites was 20 μm. The resulting spherulites grew along a common radial direction within a spherical envelope. These spherulites developed from initial square lamellae by iteration of branching and divergence of individual dominant lamellae to give a framework within which subsequent subsidiary lamellae developed. The average lamellar thickness, measured on electron micrographs, increased with the logarithm of annealing time and reached a value of about 600 Å when the sample was kept at 241°C for 1000 min [227]. The essential origin of spherulitic growth was identified as the lamellar divergence at spatially distributed arrays of branch points. A principal cause of this is suggested to be the pressure of noncrystallized molecular cilia attached to fold surfaces.

Polarized optical microscopy of P4MP1 films has brought out a very peculiar behaviour of P4MP1 with regard to its sphrulitic birefringence [228]. Saunders [229] found that the spherulitic birefringence was temperature dependent in P4MP1. At room temperature up to 50°C the spherulitic superstructure of P4MP1 showed no birefrigence under cross polarizers. Furthermore, a change of sign in the birefringence in the vicinity of 50°C was observed. Above 50°C the birefringence is negative, whereas it is positive below 50°C. Saunders interpreted the phenomenon as the resultant of different temperature dependence of the tangential and radial refractive indices. The morphological changes that may produce such changes in birefringence are not well understood, but was probably related to the fact that the densities of the crystalline and amorphous phases are equal at 50°C [223] and that the *a* axis of the unit cell increases at a greater rate with temperature than the *c* axis, probably contributing to the change of refractive indices. This birefringence dependence makes observation of spherulitic textures difficult under the polarized microscope at room temperature. Also Silvestre *et al.* [230] reported that the observation of the birefringence of all samples is possible only at temperatures higher than 60°C. In fact, from room temperature up to 60°C, even if the sample was crystalline, no birefringence was shown. For P4MP1 isothermally crystallized at $T = 220°C$ birefringent microspherulites was conversely observed, see Fig. 27. The authors reported that, during the crystallization, the microspherulites did not develop completely into mature spherulites, because of the large number of nuclei.

Scanning electron [228] and atomic force microscopy (Silvestre, private communication, 1998) of P4MP1 revealed moreover that the microspherulites were composed of stacks of lamellae that diverge during the growth, see Fig. 28.

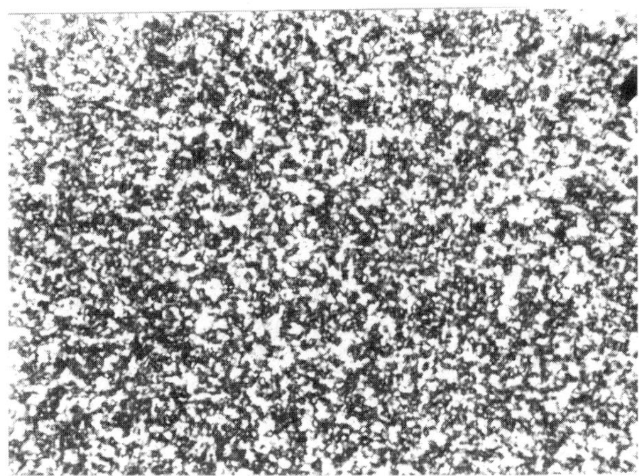

FIG. 27 Optical micrograph of a thin film of P4MP1 crystallized from the melt, showing birefringent microspherulites.

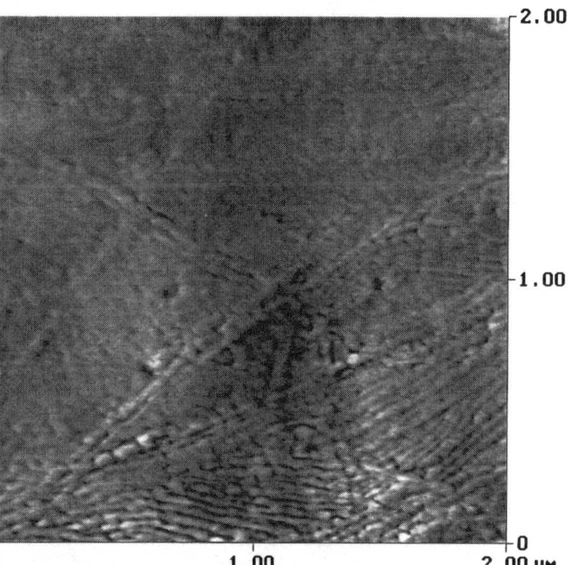

FIG. 28 Atomic force micrograph of the surface of a film of P4MP1 crystallized isothermally at 220°C.

REFERENCES

1. AE Woodward. Atlas of Polymer Morphology. Munich: Hanser Publishers, 1988.
2. DC Bassett. Principles of Polymer Morphology. Cambridge: Cambridge University Press, 1981.
3. L Mandelkern. Physical Properties of Polymers. Washington, DC: American Chemical Society, 1984, chap. 4.
4. EW Fisher and GF Schmidt. Angew Chem 74: 551, 1962.
5. C Silvestre, S Cimmino, E Martuscelli, FE Karasz, WJ MacKnight. Polymer 28: 1190, 1987.
6. JG Fatou. In: C Booth and C Price, eds. Handbook of Polyolefins, 1st edn, vol. 2. New York: Pergamon Press, 1989.
7. CW Bunn. Trans Faraday Soc 35: 482, 1939.
8. G Avitabile, R Napolitano, B Pirozzi, KD Rouse, MW Thomas, BTM Willis. J Polym Sci B 13: 351, 1975.
9. Y Chatani, Y Ueda, H Tadokoro. Annual Meeting of the Society of Polymer Science, Japan, Tokyo, Preprint, 1977, p 1326.
10. S Kavesh, JM Schultz. J Polym Sci A-2 8: 243, 1970.
11. PR Swan. J Polym Sci 56: 409, 1962.
12. ER Walter, FP Reding. J Polym Sci 21: 561, 1956.
13. EA Cole, DR Holmes. J Polym Sci 46: 245, 1960.
14. T Seto, T Hara, K Tanaka. Jpn J Appl Phys 7: 31, 1968.
15. PW Teare, DR Holmes. J Polym Sci 24: 486, 1957.
16. A Turner-Jones. J Polym Sci 62: S53, 1962.
17. PJ Barham, DA Jarvis, A Keller. J Polym Sci, Polym Phys Ed 20: 1733, 1982.
18. S Wellinghoff, F Rybnikar, E Baer. J Macromol Sci Phys B10: 1, 1974.
19. JC Wittmann, B Lotz. Polymer 30: 27–34, 1989.
20. PR Swan. J Polym Sci 56: 403, 1962.
21. H Kiho, A Peterlin, PH Geil. J Appl Phys 35: 1599, 1964.
22. T Yemni, L McCullough. J Polym Sci, Polym Phys Ed 11: 1385, 1973.
23. M Kobayashi, H Tadokoro. Macromolecules 8: 897, 1975.
24. DC Bassett, S Block, GJ Piermarini. J Appl Phys 45: 4146, 1974.
25. M Yasuniwa, Y Enoshita, T Takemura. Jpn J Appl Phys 15: 1421, 1976.
26. A Turner-Jones, JM Aizlewood, DR Beckett. Makrom Chem 75: 134, 1964.
27. J Varga. J Mat Sci 27: 2557–2579, 1992.
28. G Natta, P Corradini. Nuovo Cimento Suppl 15: 40, 1960.
29. HD Keith, FJ Padden Jr, NM Walker, HW Wyckoff. J Appl Phys 30: 1485, 1959.
30. A Turner-Jones, AJ Cobbold. J Polym Sci, Part B Polym Lett 6: 539, 1968.
31. DR Morrow, BA Newman. J Appl Phys 39: 4944, 1968.
32. JL Kardos, AW Christiansen, EJ Baer. J Polym Sci, Part A2 4: 777, 1966.
33. JA Sauer, KD Pae. J Appl Phys 30: 4950, 1968.
34. EJ Addink, J Bientema. Polymer 2: 185, 1961.
35. G Natta, P Corradini. J Polym Sci 39: 29, 1959.
36. G Natta, P Corradini, P Ganis. Makrom Chem 39: 238, 1960.
37. G Natta, P Corradini, P Ganis. J Polym Sci 58: 1191, 1962.
38. G Natta, G Dall'Asta, G Mazzanti. Makrom Chem 54: 95, 1962.
39. M Hikosaka, T Seto. Polym J 5: 111, 1973.
40. V Petraccone, C De Rosa, G Guerra, A Tuzi. Makrom Chem, Rapid Comm 5: 63, 1984.
41. G Guerra, V Petraccone, P Corradini. J Polym Sci, Poly Phys Ed 22: 1029, 1984.
42. C De Rosa, G Guerra, R Napolitano. J Europ Polym 20: 937, 1984.
43. V Petraccone, G Guerra, C De Rosa, A Tuzi. Macromolecules 18: 813, 1985.
44. R Napolitano, B Pirozzi, V Varriale. J Polym Sci, Polym Phys Ed 28: 139, 1990.
45. G Natta, P Corradini. Atti Accad Nazl Lincei 21: 365, 1956.
46. ZW Wilchinsky. J Appl Phys 31: 1969, 1960.
47. FJ Padden, HD Keith. J Appl Phys 30: 1479, 1959.
48. JM Crissman. J Polym Sci A2: 398, 1969.
49. AJ Lovinger, JO Chua, CC Gryte. J Polym Sci Polym Phys Ed 15: 641, 1977.
50. HJ Leugering. Makromol Chem 109: 204, 1967.
51. S Guanyi, H Bin, Z Jingyun, C Youhong. Sci Sinica, Ser B 30: 225, 1987.
52. J Varga. J Thermal Anal 31: 165, 1986.
53. J Varga. J Thermal Anal 35: 1891, 1989.
54. A Turner-Jones, JM Aizlewood, DR Beckett. Makromol Chem 75: 134, 1964.
55. DR Morrow. J Macromol Sci Phys B3: 53, 1969.
56. EJ Addink, J Beintema. Polymer 2: 185, 1961.
57. B Lotz, S Graff, JC Wittmann. J Polym Sci Polym Phys Ed 24: 2017, 1986.
58. A Turner-Jones. Polymer 12: 487, 1971.
59. K Mezghani, PJ Phillips. Polymer 35: 2407, 1995.
60. R Thomann, C Wang, J Kressler, R Mülhaupt. Macromolecules 29: 8425–8434, 1996.
61. D Fisher, R Mülhaupt. Makromol Chem Phys 195: 1143, 1994.
62. S Jüngling. PhD Thesis, Freiburg, 1995.
63. RA Campbell, PJ Phillips, JS Lin. Polymer 34: 4809, 1993.
64. K Mezghani, PJ Phillips. Polymer 36: 2407, 1995.
65. S Bruckner, SV Meille. Nature 340: 455, 1989.
66. SV Meille, S Bruckner, W Porzio. Macromolecules 23: 4114, 1990.

67. SV Meille, PJ Phillips, K Mezghani, S Bruckner. Macromolecules 29: 795, 1996.
68. K Mezghani, PJ Phillips. Polymer 39: 3735, 1998.
69. P Corradini, C De Rosa, G Guerra, V Petraccone. Polymer Commun 30: 281, 1989.
70. SZD Cheng, JJ Janimak, A Zhang, ET Hsieh. Polymer 32: 648–655, 1991.
71. RL Miller. Polymer 1: 135, 1960.
72. B Wunderlich, J Grebowicz. J Adv Polym Sci 60/61: 1, 1984.
73. JA Gailey, RH Ralston. SPE Trans 4: 29, 1964.
74. DM Gezovich, PH Geil. Polym Eng Sci 8: 202, 1968.
75. PB McAllister, TJ Carter, RM Hindle. J Polym Sci, Polym Phys Ed 16: 49, 1978.
76. MA Gomez, H Tanaka, AE Tonelli. Polymer 28: 2227, 1987.
77. P Corradini, V Petraccone, C De Rosa, G Guerra. Macromolecules 19: 2699, 1986.
78. P Corradini, G Natta, P Ganis, PA Temussi. J Polym Sci, Part C 16: 2477, 1967.
79. B Lotz, AJ Lovinger, RE Cais. Macromolecules 21: 2375, 1988.
80. AJ Lovinger, B Lotz, D Davis. Polymer Preprints ACS 33(1): 270, 1992.
81. AJ Lovinger, B Lotz, D Davis. Polymer 31: 2253, 1990.
82. AJ Lovinger, D Davis, B Lotz. Macromolecules 24: 552, 1991.
83. G Natta, M Peraldo, G Allegra. Makromol Chem 75: 215, 1964.
84. AJ Lovinger, B Lotz, D Davis, FJ Padden Jr. Macromolecules 26: 3494, 1993.
85. J Rodriguez-Arnold, Z Bu, SZD Cheng. JMS Rev Macromol Chem Phys, C35(1): 117–154, 1995.
86. C De Rosa, P Corradini. Macromolecules 26: 5711, 1993.
87. Y Chatani, H Maruyama, T Asanuma, T Shiomura. J Polym Sci, Polym Phys Ed 29: 1649, 1991.
88. G Natta, P Corradini, IW Bassi. Atti Accad Nazl Lincei, Rend, Classe Sci Fis, Mat Nat 19: 404, 1955.
89. R Zannetti, P Manaresi, GC Buzzoni. Chim Ind (Milan) 43: 735, 1961.
90. J Boor Jr, EA Youngman. J Polym Sci B2: 903, 1964.
91. C Nakafuku, T Miyaki. Polymer 24: 141, 1983.
92. G Charlet, G Delmas. Polymer 25: 1619, 1984.
93. G Natta, P Corradini, IW Bassi. Rend Fis Acc Lincei 19: 404, 1995.
94. W Bassi, O Bonsignori, GP Lorenzi, P Pino, P Corradini, PA Temussi. Polym Sci, Polym Phys Ed 9: 193, 1971.
95. C De Rosa, A Borriello, V Venditto, P Corradini. Macromolecules 27: 3864–3868, 1994.
96. M Takayanagi, N Kawasaki. J Macromol Sci Phys Ed B1: 741, 1967.
97. G Charlet, G Delmas, FJ Revol, R Manley. Polymer 25: 1613, 1984.
98. LA Novokshonova, GP Berseneva, VI Tsvetkov, NM Chirkov. Vysokomol Soedin (A) 9: 562, 1967.
99. YV Kissin. In: JI Kroschwitz, ed. Encyclopedia of Polymer Science and Engineering, vol. 9. New York: Wiley, 1987, p 707.
100. J Pradere, JF Revol, RStJ Manley. Macromolecules 21: 2747, 1988.
101. DR Burfield, ID McKenzie, PJT Tait. Polymer 17: 130, 1976.
102. Lopez. JMS-Rev Macromol Chem Phys C32: 407–519, 1992.
103. T He, RS Porter. Polymer 28: 1321–1325, 1987.
104. S Rastogi, M Newman, A Keller. Nature 353: 55–57, 1991.
105. HD Keith. In: TG Fox and A Weisseberg, eds. Physics and Chemistry of the Organic Solid State. New York: Wiley Interscience, 1963.
106. HD Keith, FJ Padden. J Appl Phys 34: 2409, 1963.
107. DC Bassett, AS Vaughan. Polymer 26: 717, 1985.
108. AE Woodward. Understanding Polymer Morphology. Munich: Unser Publishers, 1995, pp 40–43.
109. PH Geil. Polymer Single Crystals. New York: Wiley, 1963.
110. B Wunderlich. Macromolecular Physics, vol. I. New York: Academic Press, 1976.
111. L Mandelkern, CO Edwards, RC Domszy, MW Davidson. Microdomains in Polymer Solutions. New York: Plenum Press, 1985, pp 121–141.
112. PG Wang, AE Woodward. Macromolecules 20: 2718, 1987.
113. PJ Bush, D Pradhan, P Ehrlich. Macromolecules 24: 1439, 1991.
114. CC Kuo, AE Woodward. Macromolecules 17: 1034, 1984.
115. J Xu, AE Woodward. Macromolecules 21: 83, 1988.
116. HD Keith, FJ Padden Jr, RG Vadimsky. J Polym Sci A-2 4: 267, 1966.
117. GJ Rench, PJ Phillips, N Vatansever, A Gonzalez. J Polym Sci: Polym Phys 24: 1943, 1986.
118. DC Bassett, FC Frank, A Keller. Phil Mag 8: 1753, 1973.
119. T Kawai, A Keller. Phil Mag 11: 1165, 1965.
120. HD Keith. J Appl Phys 35: 3115, 1964.
121. F Khoury, LH Bolz. In: GW Bailey, ed. 38th Annual Proceeding of the Electron Microscopy Society of America San Francisco, California, 1980.
122. SJ Organ, A Keller. J Mat Sci 20: 1571–1585, 1985.
123. A Keller. Polymer 3: 393, 1962.
124. V Petraccone, G Allegra, P Corradini. J Polym Sci C 38: 419, 1972.
125. DM Slader. In: RA Pethrich and RW Richards, eds. Static and Dynamic Properties of the Polymeric Solid State. Dordrecht: D Reidel, 1981.
126. JC Wittmann, B Lotz. J Polym Sci Polym Phys Ed 23: 205, 1985.
127. DY Yoon, PJ Flory. Disc Faraday Soc 68: 288, 1979.

128. SJ Spells, DM Sadler. Polymer 25: 739, 1984.
129. SJ Spells. Polymer 26: 1921, 1985.
130. SJ Spells. Characterization of Solid Polymers. Cambridge: Chapman & Hall, 1994.
131. R Patil, SJ Kim, E Smith. Polymer Commun 31: 455, 1990.
132. PH Geil, DH Reneker. J Polym Sci 51: 569, 1961.
133. A Takahara, I Ohki, T Kajiyama. Polymer Preprints 37(2): 577, 1996.
134. L Mandelkern. In: G Allen and JC Bevington eds. Comprehensive Polymer Science, Crystallization and Melting. vol. 2. Oxford: Pergamon Press, 1987.
135. JG Fatou.. In: HF Mark, NH Bikales, CG Overberger, G Henges eds. Encyclopedia of Polymer Science and Engineering Crystallization Kinetics Suppl. New York: Wiley, 1989.
136. JF Revol, RStJ Manley. J Mater Sci Lett 5: 249–251, 1986.
137. IO Salyer, AS Kenyon. J Polym Sci A-1 9: 3083, 1971.
138. N Nakamae, M Kameyama, T Matsumoto. Polym Eng Sci 19: 572, 1979.
139. H Zhou, GL Wilkes. Polymer 38: 5735, 1997.
140. RC Domszy, R Alamo, PJM Mathieu, L Mandelkern. J Polym Sci Polym Phys Ed 2: 1727, 1984.
141. RG Alamo, L Mandelkern. Macromolecules 22: 1273, 1989.
142. R Alamo, RH Glaser, L Mandelkern. J Polym Sci Polym Phys Ed 26: 2169, 1988.
143. TM Krigas, JM Carella, MJ Struglinksi, B Crist, WW Graessly, FC Schilling. J Polym Sci, Polym Phys Ed 23: 509, 1985.
144. NS Murthy, S Chandrasekaran, HK Reimschuessel. Polymer 7: 23, 1966.
145. SH Ryn, CG Gogos, M Xanthos. Polymer 32: 2449, 1991.
146. E Martuscelli, M Pracella, L Crispino. Polymer 24: 693, 1983.
147. JG Fatou. Eur Polym J 7: 1057, 1971.
148. GM Stack, L Mandelkern, C Kröhnke, G Wegner. Macromolecules 22: 4351, 1989.
149. JJ Labaig. PhD Thesis. Faculty of Science, University of Strasbourg, 1978.
150. HD Keith, FJ Jr Padden, B Lotz, JC Wittmann. Macromolecules 22: 2230, 1989.
151. DC Bassett, RH Olley, IAM Al Raheil. Polymer 29: 1539, 1988.
152. A Toda. Colloid Polym Sci 270: 667, 1992.
153. JD Hoffman, LJ Frolen, GS Ross, JI Jr Lauritzen. J Res Nat Bur Stand 79A: 671, 1975.
154. SZD Cheng, ET Hsieh, CC Tso, BS Hsiao. J Macromol Sci, Phys B36(5): 553, 1997.
155. DC Bassett, AM Hodge. Polymer 19: 469, 1978.
156. MJ Shankernarayanan, DC Sun, M Kojima, JH Magill. J Int Processing Soc 1: 66, 1987.
157. A Keller. J Polym Sci 17: 351, 1955.
158. JE Mark, A Eisenberg, WW Graessley, L Mandelkern, JL Koenig. Physical Properties of Polymers. Washington DC: ACS, 1984, pp 192–199.
159. IG Voigt-Martin, L Mandelkern. J Polym Sci Polym Phys Ed 22: 1901, 1984.
160. S Cimmino, E Di Pace, E Maruscelli, LC Mendes, C Silvestre. J Polym Sci Polym Phys Ed 32: 2025, 1994.
161. SN Magonov, K Ovarnström, V Elings, HJ Cantow. Polym Bull 25: 689, 1991.
162. D Snétivy, H Yang, GJ Vancso. J Mater Chem 2: 891, 1992.
163. E Martuscelli, M Pracella, A Zambelli. J Polym Sci Polym Phys Ed 18: 619, 1980.
164. JA Sauer, DR Morrow, GC Richardson. J Appl PHys 36: 3017, 1965.
165. DR Morrow, JA Saver, AF Woodward. J Polym Sci B3: 463, 1965.
166. M Kojima. J Polym Sci A-2 5: 597, 1967.
167. B Lotz, JC Wittmann. J Polym Sci Polym Phys Ed 24: 1541, 1986.
168. A Al-Raheil, AM Qudah, M Al-Share. J Appl Polym Sci 67: 1259, 1998.
169. DR Morrow, BA Newman. J Appl Phys 39: 4944, 1968.
170. DR Norton, A Keller. Polymer 26: 704, 1985.
171. H Awaya. Polymer 29: 591, 1988.
172. RH Olley, DC Bassett. Polymer 30: 399, 1989.
173. FL Binsbergen, BGM De Lange. Polymer 9: 23, 1968.
174. F Khoury. J Res Nat Bur Stand, Part A 70: 29, 1966.
175. FJ Jr Padden, HD Keith. J Appl Phys 37: 4013, 1966.
176. FJ Jr Padden, HD Keith. J Appl Phys 44: 1217, 1973.
177. AJ Lovinger. J Polym Sci Polym Phys Ed 21: 97, 1983.
178. DC Bassett, RH Olley. Polymer 25: 935, 1984.
179. SZD Cheng, JJ Janimak, A Zhang, ET Hsieh. Polymer 32: 648–655, 1991.
180. FJ Padden, HD Keith. J Appl Phys 30: 1485, 1959.
181. DR Norton, A Keller. Polymer 26: 704, 1985.
182. MOB Idrisse, B Chalbert, J Guillet. Makromol Chem 186: 881, 1985.
183. T Asano, Y Fujiwara. Polymer 19: 99, 1978.
184. SZD Cheng, JJ Janimak, A Zhang, HN Cheng. Macromolecules 23: 298, 1990.
185. FJ Khoury. J Res Nat Bur Stand A70: 29, 1966.
186. JA Sauer, DR Morrow, GC Richardson. J Appl Phys 30: 3017, 1965.
187. B Lotz, JC Wittmann. J Polym Sci Polym Phys Ed 24: 1541, 1986.
188. MOB Idrissi, B Chabert, J Guillet. Makromol Chem 187: 2001, 1986.
189. RJ Samuels, RJ Yee. J Polym Sci Polym Phys Ed A2 10: 385, 1972.
190. RJ Samuels. J Polym Sci Polym Phys Ed A2 13: 1417, 1975.
191. J Varga. Angew Makromol Chem 104: 79, 1982.
192. B Lotz, B Fillon, A Therry, JC Wittmann. Polym Bull 25: 101, 1991.

193. PH Geil. J Appl Phys 33: 642, 1962.
194. RH Olley. Sci Prog Oxford 70: 17, 1986.
195. K Mezghani, PJ Phillips. Polymer 38: 5725, 1997.
196. A Marchetti, E Martuscelli. J Polym Sci Polym Phys Ed 12: 1649, 1974.
197. AJ Lovinger, B Lotz, DD Davis, M Schumacher. Macromolecules 27: 6603, 1994.
198. J Rodriguez-Arnold, A Zhang, SZD Cheng, A Lovinger, ET Hsieh, P Chu, TW Johnson, KG Honnell, RG Geerts, SJ Palackal, GR Hawley, MB Welch. Polymer 35: 1884, 1994.
199. J Rodriguez-Arnold, Z Bu, SZD Cheng, ET Hsieh, TW Johnson, RG Geerts, SJ Palackal, GR Hawley, MB Welch. Polymer 35: 5194, 1994.
200. RM Gohil, KG Patel, RD Patel. Makromol Chem 169: 291, 1973.
201. M Schumacher, AJ Lovinger, P Agarwal, JC Wittmann, B Lotz. Macromolecules 27: 6956, 1994.
202. KS Sastry, RO Patel. Europ Polym J 9: 177, 1973.
203. VF Holland, RL Miller. J Appl Phys 35: 3241, 1964.
204. S Kopp, JC Wittmann, B Lotz. J Mater Sci 29: 6159, 1994.
205. AE Woodward, DR Morrow, J Polym Sci A26: 1987, 1968.
206. H Fuchs, LM Eng, R Sander, J Petermann, KD Jandt, T Hoffman. Polym Bull 26: 95, 1991.
207. KD Jandt, TJ McMaster, MJ Miles, J Petermann. Macromolecules 26: 6552, 1993.
208. C Silvestre, S Cimmino, ML Di Lorenzo. J Appl Polym Sci 71: 1677, 1999.
209. S Cimmino, ML Di Lorenzo, E Di Pace, C Silvestre. J Appl Polym Sci 67: 1369, 1998.
210. SB Eng, AE Woodward. J Macromol Sci Phys, B 10: 627, 1974.
211. FC Frank, A Keller, A O'Connor. Phil Mag 4: 200, 1959.
212. DC Bassett. Phil Mag 10: 595–615, 1964.
213. F Khoury, JD Barnes. J Res Nat Bur Stand, A, Phys Chem 76A: 225, 1972.
214. CF Pratt, PH Geil. J Macromol Sci-Phys, B21: 617, 1982.
215. P Blais, RStJ Manley. Science 153: 539, 1966.
216. P Blais, RStJ Manley. J Polym Sci, Part A-1, 6: 291, 1968.
217. F Rybnikar, PH Geil. J Macromol Sci Phys, B 1:1, 1973.
218. M Takayanagi, N Kawasaki. J Macromol Sci Phys, B 1: 741, 1967.
219. Y Hase, PH Geil. Polym J 2: 560, 1971.
220. PH Geil, F Rybnikar, Y Hase. Mech Behav Mater, Proc Int Conf 3: 466, 1972.
221. M Litt. J Polym Sci, Part A, 1: 2219, 1963.
222. T Tanigami, K Miyasaka. J Polym Sci, Polym Phys Ed 19: 1865, 1981.
223. JH Griffith, BG Ranby. J Polym Sci 44: 369, 1960.
224. L Mandelkern, M Glotin, RA Benson. Macromolecules 14: 22, 1981.
225. JS Trent. Macromolecules 17: 2930, 1984.
226. D Patel, DC Bassett. Proc R Soc A 445: 577, 1994.
227. DC Bassett, D Patel. Polymer 35: 1855, 1994.
228. TW Campbell. J Appl Polym Sci 5: 184, 1960.
229. FL Saunders. J Polym Sci, Polym Lett Ed 2: 755, 1964.
230. C Silvestre, S Cimmino, E Di Pace, M Monaco. J Mater Sci, Pure Appl Chem a35: 1507, 1998.

8

The Glass Transition of Polyolefins

Hans Adam Schneider
"Albert-Ludwigs-Universität" Freiburg, Freiburg, Germany

I. INTRODUCTION

The glass transition is considered one of the distinguishing marks of polymeric materials. In structural terms, glasses are characterized by the absence of long range order. Experimentally the glass transition is not a precise point, but a relatively narrow temperature range, whereas the time dependence of the glass transition is an indication that the transition does not occur under equilibrium conditions. Due to this time dependence of the rate of transition, glasses will show not only relaxation phenomena, but it imposes additionally to define the timescale of the experiment. The relaxation time depends on entropy and "free volume", while the rate of change of both is controlled by the changing relaxation time, i.e., the relaxation process is controlled by the tendency of the metastable glass to approach thermodynamic equilibrium. The microscopic origin of the glass transition is a kinetic controlled freezing phenomenon which formally shows great similarity to a thermodynamic second-order transition.

Taking into account this dual aspect of the glass transition, two main models have been proposed for the theoretical description of the glass transition phenomenon: the kinetic based "free volume", and the thermodynamic based conformational entropy theory.

For a full specification of a glass transition, besides the specific changes of the measured temperature-dependent characteristics of a polymer, must be stated both the cooling and heating rate (respective the applied frequency), as well as the used mode of definition of the glass transition temperature. Additionally the related experimental temperature parameters, T_i and T_f, characterizing the glass temperature range of the glass transition are required. Unfortunately, the specification of all these characteristic parameters is mostly missing in the respective publications. Nevertheless, in Table 1 are presented the glass temperatures of polyolefins adopted by the 'Polymer Handbook' [1].

II. THEORETICAL BACKGROUND OF THE GLASS TRANSITION

The kinetic nature of the glass transition is supported by the observation that any liquid can be transformed into a glass with the condition that the cooling rate exceeds the crystallization rate. Below the glass transition temperature the structure of the glass remains that of the liquid at the glass transition temperature, due to the lack of large-amplitude mobility because the concentration of frozen holes (i.e., of "free volume") remains fixed at the value characteristic for T_g. Taking into account that the experimentally observed glass transition shows the characteristics of a second-order transition, the thermodynamic model assumes the existence of the glass as a real thermodynamic stable fourth state of matter, characterized by a true thermodynamic second-order transition situated, however, far below the experimentally accessible glass transition. This thermodynamic background of the glass

TABLE 1 Glass Temperatures of Polyolefins

Polymer	x	T_g (K)	μ/ρ[a]	Ref.
Polyethylene		148, 222–240	28/2	[1]
		252	(branch point transition)	
		195 ($\pm$ 10)[b], 200[c]		[2, 3]
Poly(methylene)		*155*	*14/1*	[1]
		Poly-α-Olefines		
Poly(alkyl-ethylene)s		$-(CH_2-CH)-$ with side chain $(CH_2)_x$–CH_3		
Poly(methylethylene) = Poly(propylene)				
atactic	0	~ 260, (258 to 270)	42/2	[1]
isotactic		266, 272 (238 to 260)		
syndiotactic		~ 265 (263–267)		
Poly(butene-1) = poly(ethylethylene)	1	249	56/3	[1]
Poly(propylethylene)	2	233	70/4	[1]
Poly(butylethylene)	3	223	84/5	[1]
Poly(pentylethylene)	4	242	98/6	[1]
Poly(hexylethylene)	5	208, 228	112/7	[1]
Poly(heptylethylene)	6	226	126/8	[1]
Poly(octylethylene)	7	232	140/9	[1]
Poly(nonylethylene)	8	236	154/10	[1]
Poly(decylethylene)	9	237	168/11	[1]
Poly(dodecylethylene)	11	241	196/13	[1]
Poly(tetradecylethylene)	13	246	220/15	[1]
Poly(iso-alkyl-ethylenes)		$-(CH_2-CH)-$ with side chain $(CH_2)_x$–$CH(CH_3)_2$		
Poly(iso-propylethylene)				
atactic	0	323	70/2.5[d]	[1]
isotactic		367		
Poly(iso-butylethylene)	1	302	84/3.5[d]	[1]
Poly(iso-pentylethylene)	2	259	98/4.5[d]	[1]
Poly(iso-hexylethylene)	3	239	112/5.5[d]	[1]
Poly[(methylene)$_x$-tert-butyl-ethylene]s		$-(CH_2-CH)-$ with side chain $(CH_2)_x$–$C(CH_3)_3$		
Poly(tert-butylethylene)	0	337	84/2.5[d]	[1, 4]
Poly(neopentylethylene)	1	332	98/3.5[d]	[1, 4]
Poly[(dimethylene)tert-butylethylene]	2	326	112/4.5[d]	[4]
Poly[(trimethylene)tert-butylethylene]	3	313	126/5.5[d]	[4]

TABLE 1 Continued

Polymer	x	T_g (K)	μ/ρ[a]	Ref.
Poly[(methylene)$_x$-cyclopentyl-ethylene]s $-\!\!\left(CH_2-CH\right)\!\!-$ with side group $(CH_2)_x$–cyclopentyl				
Poly(cyclopentylethylene)	0	348	96/3	[1]
Poly[(methylene)cyclopentylethylene]	1	333	110/4	[1]
Poly[(methylene)$_x$-cyclohexyl-ethylene]s $-\!\!\left(CH_2-CH\right)\!\!-$ with side group $(CH_2)_x$–cyclohexyl				
Poly(cyclohexylethylene)				
atactic	0	393	110/3	[1]
isotactic		406		
Poly[(methylene)cyclohexylethylene]	1	348	124/4	[1]
Poly[(dimethylene)cyclohexylethylene]	2	313	138/5	[4]
Poly[(trimethylene)cyclohexylethylene]	3	248	152/6	[4]
Poly[(methylene)$_x$-phenyl-ethylene]s $-\!\!\left(CH_2-CH\right)\!\!-$ with side group $(CH_2)_x$–phenyl				
Poly(phenylethylene) = Poly(styrene)				
atactic and isotactic	0	368, 371–377	104/3	[4, 1]
Poly[(methylene)phyenlethylene]	1	351	118/4	[4]
Poly[(dimethylene)phenylethylene]	2	313	132/5	[4]
Poly[(trimethylene)phenylethylene]	3	245	146/6	[4]
Poly(alkyl,alkyl-ethylene)s and poly(alkyl-n-methylene)s				
Poly(ethyl-2-propylene)		268	70/3.5[d]	[1]
Poly(propyl-2-propylene)		300	84/4	[1]
Poly(isobutylene), poly(1,1-dimethylethylene)		200	56/3.5[d]	[1]
Poly(1,1-dimethyltrimethylene)		263	70/3.5[d]	[1]
Poly(1,1-dimethyltetramethylene)		253	84/4.5[d]	[1]
Poly(1-ethyl-1-methyltetramethylene)		~ 250	98/5.5[d]	[1]
Poly(1,1,2-trimethyltrimethylene)		310	84/3.5[d]	[1]
Poly(1-methyloctamethylene)		215	126/8	[1]

[a] μ/ρ, mass/"flexible" bond.
[b] Data compiled by Boyer [2] for polyethylene.
[c] Value reported by Miller [3] obtained by extrapolation of glass temperatures of n-alkanes for $n = \infty$.
[d] Value of 0.5 attributed to the "flexible" bonds of two or three methyls attached to the same or adjacent C-atoms, assuming a reciprocal hindered rotation.

transition is supported by the so-called "Kauzmann paradox", which establishes that any thermodynamic property of a liquid extrapolated through and below the glass temperature results in absurd findings, such as smaller specific volumes and entropies (or even negative entropies) of the amorphous glasses compared with that of the ordered crystals [5].

The kinetic based "free volume" theory starts with the Doolitle viscosity equation [6]

$$\eta = A \exp(BV_o/V_f) \tag{1}$$

with V_o the occupied and V_f the "free volume" of the flowing liquid and A and B the material specific constants. The Doolitle equation provides the theoretical background for the (William–Landel–Ferry) WLF equation:

$$\log(a_T) = [-C_1^0(T - T_o)]/(C_2^0 + T - T_o) \tag{2}$$

which assumes that the fractional free volume at the glass temperature is for polymers an "universal" constant, $f_g = V_f/V \sim 0.022$–0.025, $V = V_o + V_f$ being the total volume [7]. a_T is the "shift factor" for time–temperature superposition of experimental viscoelastic properties measured at different temperatures, by shifting the measured property isomethermal at T along the logarithmic timescale to a reference temperature, T_o. The constants C_1^0 and C_2^0 have been considered for long time universal constants ($C_1^0 = 8.86$ and $C_2^0 = 101.6$ for $T_o = T_g + 50\,\mathrm{K}$). Accordingly it has been assumed that the glass transition characterizes an "iso-free volume state".

The kinetic aspect of the glass transition has been thoroughly emphasized by Eyring *et al.* [8, 9]. According to the kinetic theory the molecular mobility is controlled by the free volume and the glass is considered a *frozen state of matter* [10–12], described by the *P–V–T equation of state* and by an additional, kinetically controlled internal order parameter [13]. Due to this kinetic aspect of the glass transition, polymer glasses are thermodynamically metastable, being characterized by "relaxation" and "aging" phenomena [14].

The thermodynamic theory considers the glass to be a fourth state of matter, characterized by *zero conformational entropy*. The thermodynamic theory uses the *S–V–T equation of state* and the transition to the thermodynamic stable glassy state is characterized by a *second-order transition* situated at a temperature $T_2 \sim 50\,\mathrm{K}$ below the experimentally observed glass temperature, T_g [15, 16]. The second-order transition occurs for zero conformational entropy in the *S–V–T* state diagram. The experimentally observed relaxation properties of the glass-forming liquids are explained by the temperature variation of the different cooperatively rearranging regions [17].

Unfortunately, there is no way to verify experimentally the validity of these two theories. Although the zero conformational entropy can be calculated, the second-order transition temperature, T_2, is experimentally nonverifiable because of the foregoing kinetic freeze-in phenomenon. On the other hand, the "free volume", responsible for the experimentally observed glass transition, cannot be quantified because, depending on the definition used, the estimated values of the critical free volumes differ by several orders of magnitude [18].

Further details concerning glass formation, glassy behavior and glass temperature are given elsewhere [19–23].

III. MOLECULAR PARAMETERS AFFECTING THE GLASS TRANSITION

Beside the molecular weight ("chain end effect") and degree of crosslinking [24, 25], both diluents and plasticizer as well as pressure affect via the related "free volume" changes the glass transition polymers. Concerning the pressure influence it has been shown that the increase of the glass temperature decreases asymptotically with increasing pressure [26] and that the pressure influence is the higher the higher the glass temperature at atmospheric pressure [27]. Additionally a series of molecular parameters determine the glass temperature of polymers, the prevailing two factors being the *chain flexibility* and *intermolecular interactions* in polymers. The chain flexibility is related to the characteristic ratio [28], whereas the intermolecular forces are characterized by the cohesive energy density [21].

The chain flexibility (stiffness) is controlled by the barrier to rotation around the backbone C—C bonds and the larger the substituent the more hindered is the rotation, multiple anchoring (two or three substituents on the same C) increasing additionally this hindrance and thus the glass transition temperature. On the other hand, longer side chains lower the glass temperature due to an internal "plasticizing effect" of the main chain. Additionally the glass temperature of polymers is affected by tacticity effects as it has ben demonstrated for instance for poly(methylethylene) by Cowie [29] and Burfield and Doi [30], respectively (see Fig. 1).

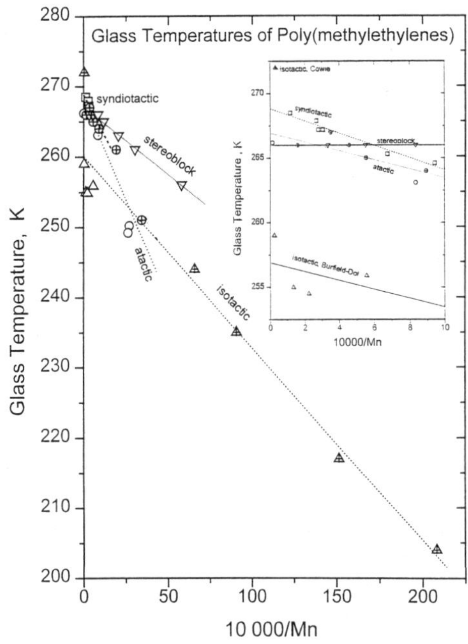

FIG. 1 Influence of the molecular weight, M_n, on the tacticity dependence of the glass temperature of poly(methylethylene)s. T_g by DSC, heating rate 10 K/min. Atactic polypropylene data of: ○, Burfield and Doi [30]; ⊕, Cowie [29]. Isotactic PP: Δ, [30]; ⟁ [29]. Syndiotactic PP: □, [30]; stereoblock PP: ∇, [29].

Taking into account the dependence of the glass transition on the different molecular parameters, a series of semiempirical and/or theoretical attempts at correlating T_g with various characteristics of the polymers are presented in the literature to predict the glass temperature of polymers.

One of these semiempirical methods scales the volume contributions to T_g of the different moieties existing in the chemical structure of the repeating unit of polymers [31, 32]. Schneider and DiMarzio [33] suggested the more simple correlation of T_g with the mass/"flexible" bond of the repeating unit.

$$T_g = A + C[m/r] \tag{3}$$

with m the mass and r the number of "flexible" bonds of the repeating unit. By linear trial and error fit the following values were obtained for the constants: $A = 135.04$ and $C = 5.74$, respectively. The only problem in this latter correlation is related to the difficulty of an exact counting of the number of "flexible" bonds contributing by rotation to conformational changes of the repeating unit because of unknown hindrances introduced by inter-/intramolecular interactions and/or steric effects. It has been shown that the T_g vs mass/"flexible" bond of monomeric unit correlation holds exactly for the different classes of polymers with class specific A and C constants, which is explained by the very similar interactions prevailing within a given class [34].

In a more sophisticated attempt, Koehler and Hopfinger [35] correlates the glass temperature with conformational entropies and mass moments to account for intermolecular interactions, assuming group additivity and applying molecular modeling. These characteristics are estimated in terms of torsion angle units composing the polymer. Conformational entropy contributions and mass moments associated with the different torsion angle units were, however, estimated by correlating calculated T_g data with experimental T_g data. For newer theoretical concepts concerning the glass transition of polymers see [36, 37] and the literature cited therein.

IV. ASPECTS OF THE GLASS TRANSITION OF POLYETHYLENE

One of the controversial problems concerning the glass transition of polymers refers to the glass transition temperature of polyethylene. In the literature glass temperatures for polyethylene were reported ranging between about 140 and 340 K, with the most frequent values centered around 150 and 250 K, respectively [38]. Wunderlich [39] attributed the observed large glass transition interval ranging between 250 and 120 K to the separate freezing in of the relatively fast *hole equilibrium* situated at about 237 K and the much slower freezing in of the *trans–gauche equilibrium*, which completely stops at 120 K.

Dynamic mechanical [40–43] and dynamic electrical [44] loss data suggested the existence of three loss peaks in the temperature range between 150 and 350 K. The dispersion regions characterized by the relevant loss peaks are customarily named, in decreasing temperatures as α, β, γ, and are situated near 360, 275 and 165 K, respectively. Generally the observed α and β loss peaks are the higher the higher the crystallinity, respective the higher the density and the smaller the branching, i.e., the number of methyl groups per 100 carbon atoms of the main chain [42, 44]. Illers has emphasized that an exact recognition of the glass temperature of polyethylene is impeded by the relatively high rate of crystallization which prevents any freezing in of the pure amorphous state. Additionally polyethylenes are normally "contaminated" by a more or less intense branching during syntheses, so that a completely amorphous polyethylene is unrealizable [45]. Using the Gordon–Taylor equation [46],

$$T_g = (w_1 T_{g_1} + K w_2 T_{g_2})/(w_1 + K w_2) \quad (4)$$

to evaluate by extrapolation the glass temperature of amorphous polymethylene by analyzing the composition dependence of the glass temperatures of different random copolymers of ethylene, Illers suggested ~ 200 K to be the glass temperature of polymethylene.

In the Gordon–Taylor equation (4) T_g is the glass temperature of the copolymer or of the miscible polymer blend, respectively, w_i and T_{g_i} are the weight fractions resp. glass temperatures of the components. K is a model specific parameter, i.e., $K_{GT} = (\rho_1/\rho_2)/(\Delta\alpha_2/\Delta\alpha_1)$ for volume additivity [46] and $K_{DM} = (m_1/r_1)/(m_2/r_2)$ for "flexible" bond additivity [47]. ρ_i are the densities and $\Delta\alpha_i = (\alpha_{melt} - \alpha_{glass})_i$ the increments at T_g of the expansion coefficients of the components. m_i and r_i are the masses and "flexible" bonds of the repeating units—see (3).

Illers has performed further extensive studies concerning the dependence on crystallinity and branching of the dispersion regions of polyethylene [48–50]. He has also shown that the low temperature γ process is, in fact, composed of at least two superposed relaxation processes γ_I and γ_{II}, which according to Pechhold [51] are explained by cooperative jumps or "kinks" in the amorphous and crystalline regions of polyethylene, respectively.

Due to this experimentally observed multiple relaxation processes, the attribution of the glass temperature of polyethylene is essentially dependent on the individual interpretation of the experimental data.

Based on experimentally evaluated specific heat data of polyethylenes, Wunderlich [39] assumed the value of the inflexion point at 237 K observed in the heat capacity, C_p vs T curve (estimated by extrapolation to zero crystallinity) as the true glass temperature of polyethylene. This value was considered to characterize the freeze in of the hole equilibrium, whereas the second inflexion point in C_p at about 150 K characterizes the freezing in of cooperative trans–gauche motions of at least three successive carbon atoms.

Stehling and Mandelkern [52], looking at the experimental detectibility of the different typical changes observed during expansion, calorimetric and mechanical dynamic investigation conclude, on the contrary, that the γ transition at about 143 K is the primary glass transition of linear polyethylene (LPE). Using the same criterion for the interpretation of the results of specific volume data, Swan [53] also preferred the low temperature value of 143 K for the glass temperature of LPE.

Based on small-angle X-ray scattering, Fischer and Kloss [54] attributed, on the contrary, the value of 148 K to a glass transition in the surface layer of PE single crystals. Kuzmin *et al.* [55], on the other hand, observed by analysis of the X-ray amorphous scattering curves of semicrystalline polyethylene two transitions at about 170 and 268 K, respectively. A low temperature T_g value is also preferred by Hendra *et al.* [56], based on the observation that crystallization of PE occurs even at temperatures below 190 K.

Danusso *et al.* [57], on the contrary, attributed the dilatometrically determined value of 252 K to the "apparent second-order " (i.e., glass) transition of polyethylene. In a further paper Natta *et al.* [58] emphasized for the first time that the glass transition temperatures of semicrystalline poly(α-olefin)s decrease from polyethylene to a minimum value for poly(propylethylene), increasing then again starting with poly(butylethylene), suggesting thus a plasticizing effect of the shorter side chains. Additionally the authors stated that this observation is in contrast to the believed constancy of the T_m/T_g ratio of polymers.

Miller [3], for his part, attempted to evaluate the glass temperature of polyethylene by extrapolation of the glass temperatures of amorphous n-alkanes for $(CH_2)_n$, $n = \infty$. He obtained in this way the value of 200 K for T_g and 160 K for the thermodynamic second order transition, T_2. Additionally the author showed that the T_g/T_2 ratio for n-alkanes, beginning with $C_{12}H_{26}$, reached the limiting value of 1.25. Finally, Braun and Guillet [59] obtained by inverse gas chromatography the values of 238, 248 and 255 K for the phase transitions of low-, medium-, and high-density polyethylene, respectively. Due to the fact that the chromatographic technique involves essentially a sensitive detection of the "free volume" in polymers, the authors suggest that the reported values represent the glass temperatures of the respective PE's, on condition that T_g is defined as an iso-free volume state of the amorphous polyethylene.

In his attempt to review the literature data concerning the glass temperature of polyethylenes, Boyer [2] concluded that the glass temperature of amorphous polyethylene is 195 ± 10 K. According to Boyer, the 145 ± 10 K process is a $T < T_g$ relaxation, commonly known as T_γ. Partially crystalline polyethylenes ($x \sim 0.5$–0.8) exhibit two glass transitions: T_g(L) at around 195 K and T_g(U) at around 240 K, both being the result of the chain-folded morphology of the melt crystallized polyethylene, and increasing with crystallinity (see Fig. 2). It is tentatively assumed that $T_g(L)$ is associated with dangling cilia and $T_g(U)$ with loose loops or intercrystalline tie molecules, suggesting that there exist two distinct types of amorphous material in PE. It is interesting to notice that both glass temperatures extrapolate for zero crystallinity to almost the same value of 195 K, which is considered to be the effective glass temperature of amorphous polyethylene.

The observation that the T_γ relaxation also occurs in ethylene copolymers, containing at least 3–5 consecutive —(CH_2)— sequences confirms both the above statement of Wunderlich [39] and the assumption of Reding *et al.* [60] that this relaxation characterizes "crankshaft" like motions during the development of the trans–gauche equilibrium.

V. SOME REGULARITIES CONCERNING THE GLASS TEMPERATURES OF POLYOLEFINS

Natta *et al.* [58] emphasized already since 1957 the plasticizer effect of short side chains on the glass temperature of poly(α-olefin)s, as well as the nonconstancy of the ratio between melting and glass temperature, T_m/T_g, and in Fig. 3a are presented the respective dependences on the number of side-chain C atoms of T_m, T_g and T_m/T_g, using the values reported in the

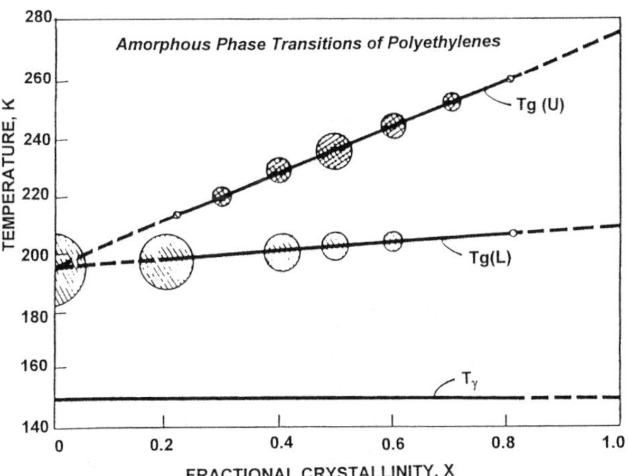

FIG. 2 The double glass transition, $T_g(L)$ and $T_g(U)$, and the T_γ amorphous relaxation in linear polyethylene. The size of the circles indicate the intensity of the two T_g. Intensity of the T_γ transition increases continuously with decreasing crystallinity. (Reproduced from the review paper of Boyer [2] with the permission of ACS Publication Division.)

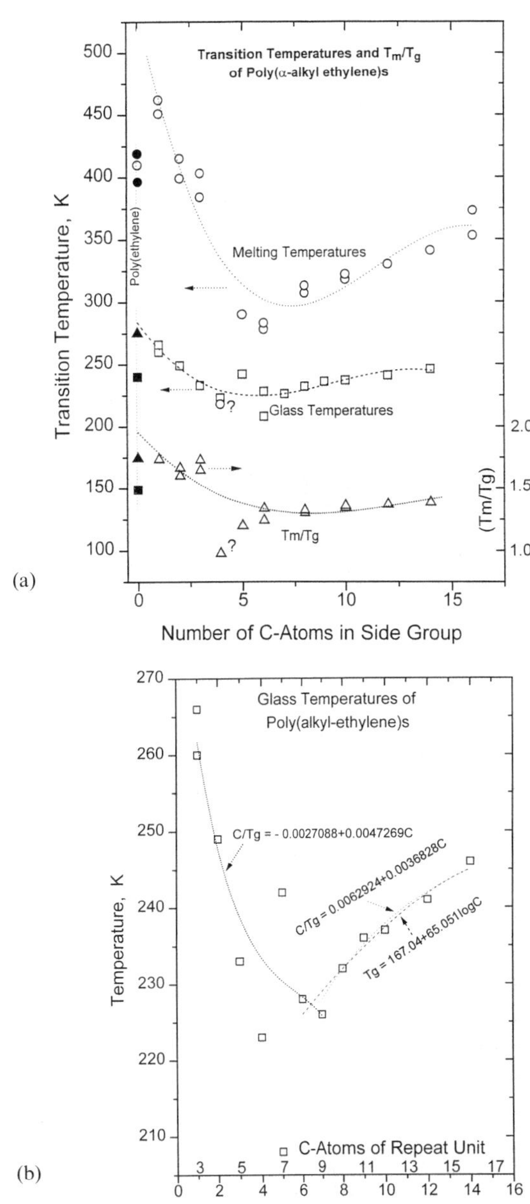

FIG. 3 (a) The dependence of the melting, T_m, and glass transition temperatures, T_g, as well of the T_m/T_g ratio of poly(α-alkyl ethylene)s on the number of C atoms in the linear side-chain. (b) Correlations of the glass transition temperature of poly(α-alkyl ethylene)s with the number of carbon atoms of the repeating unit.

Polymer Handbook [1, 61]. The experimentally observed "plasticizer" effect being more pronounced for the melting than for the glass temperature explains the stated nonconstancy of the T_m/T_g ratio. At the same time it is observed that the plasticizer effect ceases for more than 6 C atoms in the side chain, increasing both T_m and T_g. This is attributed to a denser packing due to a possible side-chain crystallization.

A series of equations have been proposed in the literature to correlate both the melting and glass temperatures of the lower and upper members of poly-(alkyl-ethylene)s with the respective number of C atoms of the repeating unit [62]. The agreement between these T_g vs number of C atoms of repeat unit equations and literature T_g data of poly(alkyl-ethylene)s is especially acceptable for the higher members (C_8–C_{16}), as demonstrated in Fig. 3b.

The plasticizer effect of short side chains is common in all poly(α-olefin)s, as it results from the data exhibited in Fig. 4 and Table 2.

On the other hand, it is known that the glass temperature of the mono-substituted polymers increases with the bulkiness of the side group but it is, however, surprising that the increse of the T_g of poly(α-olefin)s is directly proportional to the atomic mass of the side group, as results from the data shown in Fig. 5. Additionally, it is interesting to notice that the extrapolation of these T_g data supposing the H atom attached as a "side group" to the main chain α-C, results in a value of about 223 K for the glass temperature of the "amorphous linear polyethylene".

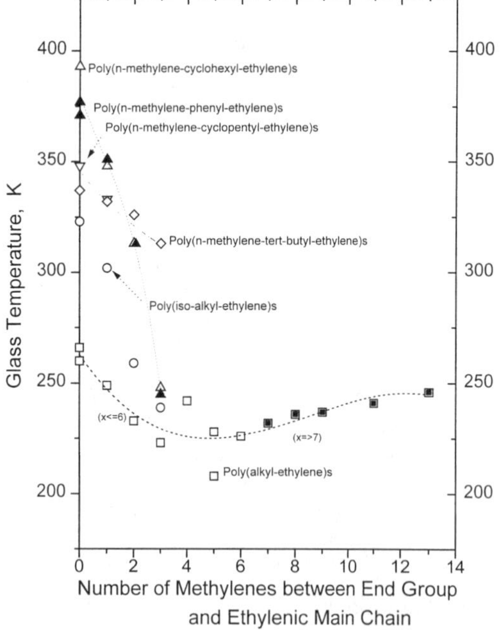

FIG. 4 The dependence of the glass temperature of poly(α-olefin)s on the number of methylenes between main chain and end group of linear side chain.

TABLE 2 Dependence of the Glass Temperature of Poly(α-olefin)s on the Nature of the Side Group

Side group	Glass temperature (K)	Mass/"flexible"/Bond
-n-Alkyls		
-Methyl (—CH_3)	260, 266	15
-(methylene)-Methyl	249	29/2
-(dimethylene)-Methyl	233	43/3
-(trimethylene)-Methyl	242	57/4
-i-Alkyls		
i-Propyl [—$CH(CH_3)_2$]	323	43
-(methylene)-iso-Propyl	302	57/1.5[a]
-(dimethylene)-iso-Propyl	259	71/2.5[a]
-(trimethylene)-iso-Propyl	239	85/3.5[a]
-t-Butyls		
-*t*-Butyl [—$C(CH_3)_3$]	337	57
-(methylene)-tert-Butyl	332	71/1.33[b]
-(dimethylene)-tert-Butyl	326	85/2.33[b]
-(trimethylene)-tert-Butyl	313	99/3.33[b]
-Cyclopentyls		
-Cyclopentyl (-cyclo-C_5H_9)	348	69
-(methylene)-Cyclopentyl	333	83/2
-Cyclohexyls		
-Cyclohexyl (-cyclo-C_6H_{11})	393	83
-(methylene)-Cyclohexyl	348	97/2
-(dimethylene)-Cyclohexyl	313	111/3
-(trimethylene)-Cyclohexyl	248	125/4
-Phenyls		
-Phenyl (-C_6H_5)	371, 377	77
-(methylene)-Phenyl	351	91/2
-(dimethylene)-Phenyl	313	105/3
-(trimethylene)-Phenyl	245	119/4

[a]The value of 0.5 was assumed for the number of "flexible" bonds in the final ~CH-$(CH_3)_2$ group, i.e., between ~CH and the two attached methyls, supposing reciprocal hindrance of the two methyl groups.

[b]The value of 0.33 was assumed for the number of "flexible" bonds in the final ~C-$(CH_3)_3$ group, supposing reciprocal hindrances of the three methyl groups substituted at the same ~C of the side group.

The plasticizer effect of short side chains on the T_g of the diverse poly(α-olefin)s is emphasized by the data exhibited in Fig. 6. From the results presented in Fig. 6a it seems that the plasticizer effect of the increasing number of methylenes inserted between the side end-group and the polymeric main chain is the higher the higher the molecular mass of the end-group. With exception of the poly[(methylene)$_x$-*t*-butyl-ethylene] series, a possible correlation between T_g and the overall molecular mass of the side group is additionally

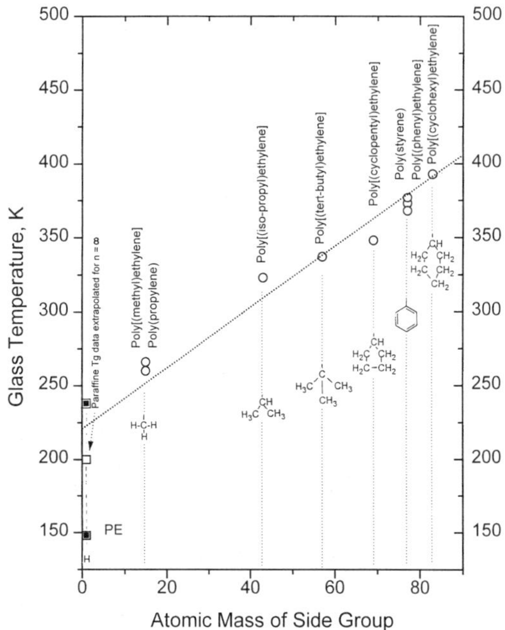

FIG. 5 The influence on the glass transition temperature of poly(α-olefin)s of the atomic mass of the side-group radical attached directly to the α-C atom in the ethylenic main chain.

roughly confirmed. An attempt of shifting these data according to an adapted T_g vs mass/"flexible" bond rule (3) for the side chains of the poly(α-olefin)s is shown in Fig. 6b. The assumed correlation seems to hold even for the side-chain dependence of the glass temperature, although the scatter of the data is relatively high. The large scatter of the data is mainly attributed to effective difficulties encountered in an exact counting of the number of "flexible" bonds responsible for conformational rearrangements, because of unknown reciprocal intra- and inter-molecular hindrances and steric effects of the bulky end groups of the side chains.

Finally, the validity of the T_g vs mass/"flexible" bond rule is demonstrated in Fig. 7 for the T_g data of the polyolefins shown in Table 1. The linear correlation is surprisingly good, taking into account the general difficulties encountered in the counting of an exact number of "flexible" bonds contributing to conformational change because of the unknown effects of intra- and interactions and steric hindrances as well. The experimentally observed increase of the T_g's of poly-(alkyl-ethylene)s for side chain $-(CH_2)_n-$ with $n < 6$, is explained by side-chain crystallization, which may reduce substantially the respective number of "flexible" bonds stipulated in Table 1. Extrapolation of the data for the value of 14, i.e., the number of

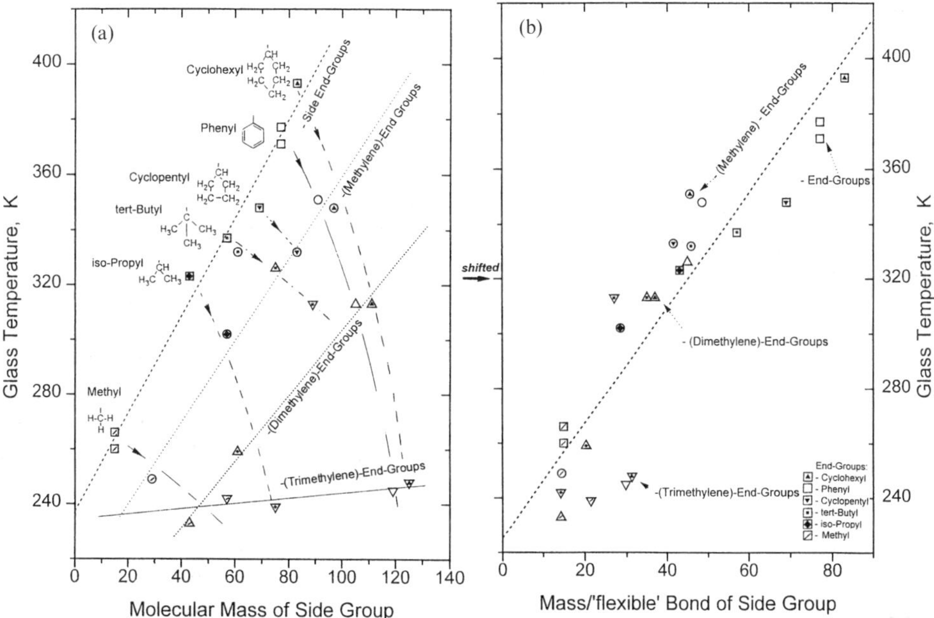

FIG. 6 Glass temperature of poly(α-olefin)s. (a) The dependence of the glass temperature on the molecular mass of the linear side groups. (b) The dependence of the glass temperature on the mass/"flexible" bond ratio of the linear side groups.

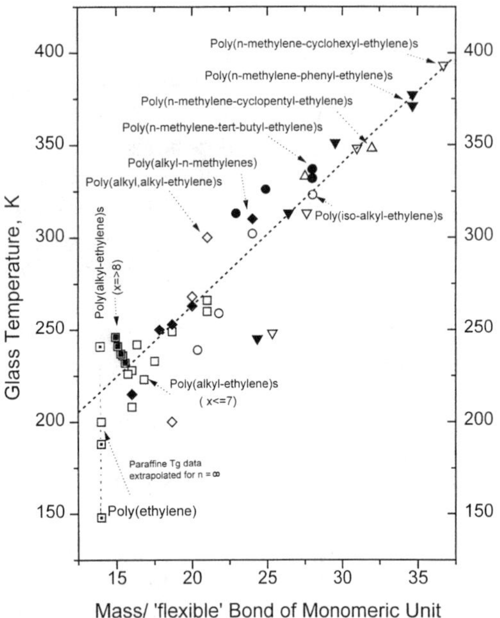

FIG. 7 The dependence of the glass temperature of poly-(olefin)s on the mass/"flexible" bond ratio of the monomeric units.

mass/"flexible" bonds of poly(ethylene), results this time in a value of 215 K for the glass temperature of poly(ethylene).

It may be interesting to notice that all the above presented extrapolation procedures used for estimating the glass temperature of "amorphous polyethylene" result in a value situated around or slightly above 200 K.

VI. ABOUT THE GLASS TEMPERATURES OF COPOLYMERS AND MISCIBLE POLYMER BLENDS

Taking into account the widespread applications of polyolefins and the increasing demands of speciality polymeric materials, from the beginning it was attempted to improve the given properties of the olefinic polymers either by copolymerization or by a corresponding blending. Thus, for instance, ethylene is copolymerized with different olefinic or nonolefinic comonomers in order to obtain more flexible and tougher materials, mainly by reduction of the degree of crystallinity and/or by the increase of polarity [63]. Propylene, on the other hand, is copolymerized for improving either the dyeability of polypropylene or the mechanical brittleness of isotactic polypropylene [64]. The overwhelming majority of contributions in this field are, however, found in the patent literature. We will restrict ourselves thus to stress only some general rules concerning the glass temperature of such polymeric systems, based exclusively on polyolefinic components.

It is well known that random copolymers as well as miscible polymer blends are characterized by a single, composition-dependent glass temperature and at the beginning, additivity rules were suggested in the literature. Both the free volume approach [46] and the thermodynamic model [47] were applied first to explain by additivity assumptions the composition dependence of the glass temperature of random copolymers. The equations have then been adapted for miscible polymer blends too. It has been shown that the two approaches are almost equivalent and that the most simplified equation for additivity of both the free volumes and the masses/"flexible" bonds of the repeating units of the components is reflected by the well known Fox equation [65],

$$1/T_g = w_1/T_{g_1} + w_2/T_{g_2} \tag{5}$$

which results by substitution of the K parameters in the "Gordon–Taylor" equation (4) by the ratio $K_F = T_{g_1}/T_{g_2}$. K_F results in a first approximation from K_{GT} for volume addivity by assuming the validity of the Simha–Boyer rule [66], $\Delta\alpha T_g = 0.133$ and substituting $\rho_1/\rho_2 = 1$ for the ratio of the mostly very similar densities of polymers. K_{DM} for flexible bond additivity is also reduced to K_F, taking into account the correlation between T_g and the m/r ratio—see Eq. (3). The significance of T_g, T_{g_i} and w_i in Eq. (5) is that of the Gordon–Taylor Eq. (4).

The experimentally demonstrated equivalence between the kinetic based "free volume" approach and the thermodynamic supported "mass/flexible bond" assumption [33] raises the *philosophical question* if, during heating of a polymeric glass, at first the accompanying increase of the "free volume" permits beginning with a critical value conformational changes by rotation around simple chemical bonds or the accumulated thermal energy enables first at all the rotation around chemical bonds, thus creating the additional "free volume" necessary for allowing the mobility characteristic for the glass transition. It is, however, interesting to notice that in the literature the kinetic "free volume" approach is accepted.

It has been demonstrated experimentally that the glass temperatures of both random copolymers and compatible polymer blends are mostly not respecting these additivity rules. Thus the corresponding T_g vs composition equations were extended to account for the contribution of binary and/or ternary heterosequences in copolymers [67–70], respective for the effect of interactions and conformational entropy changes in polymer blends [71–73]. Taking into account that generally the glass temperature vs composition dependences of neither random copolymers nor of compatible polymer blends obey additivity based rules, any attempt of extrapolation of the corresponding T_g data of such systems for evaluating the unknown glass temperature of one of the components by using T_g data of copolymers or miscible polymer blends and supposing the validity of additivity rules seems questionable at least.

Generally, it may be emphasized that experimentally observed positive deviations from the predicted additivity of the glass temperature are rather caused by overwhelming energetic favorable interactions between the respective contacting units of the components. On the other hand, negative deviations are characteristic of pronounced conformational entropy contributions due to a looser packing caused either by repulsions within the heterosequences of copolymers or by the mismatch and/or a nonrandom mixing of the interacting units of the blend components [34, 74]. Thus, any experimentally observed additivity suggests rather an overall compensation by interaction favored energetic effects with conformational entropy contributions, remembering the well known θ conditions in polymer solutions.

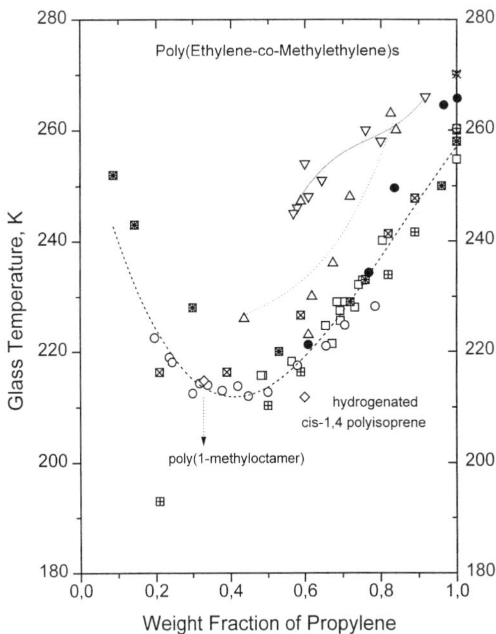

FIG. 8 The composition dependence of the glass temperatures of ethylene–propylene copolymers. Experimental data of: □ Manaresi and Gianelli [76] by capillary dilatometry, cooling rate 0.5 k/min; ⊞ Kontos [77] by capillary dilatometry and ⊠ mechanical dynamic resilience measurements; ○ adopted from Fig. 2 in Gianotti *et al.* [78]; ◇ model E/P copolymers by DSC [78]; ∇ De Candia *et al.* [79], tan δ by dynamic mechanical measurements at 110 Hz; △ E/P copolymers by syndiotactic regulating catalyst; ⊠ Privalko and Shmorgun [80] by DSC, heating rate 2 K/min; ● Suhm [81] by DSC, heating rate 10 K/min.

VII. THE GLASS TEMPERATURE OF RANDOM ETHYLENE/ALKYLETHYLENE COPOLYMERS

Several papers were published concerning the composition dependence of the glass temperature of ethylene/methylethylene (i.e., propene) copolymers. Although the respective T_g data were obtained by different methods, the scatter of the T_g vs composition data is acceptable (see Fig. 8), except for the data reported by de Candia *et al.* [79] obtained during mechanical dynamic measurements. A possible explanation could be that the "free volume" is partially squeezed out by pressmolding of the samples used for mechanical dynamic studies. At least results of PVT measurements support this explanation [27].

In Fig. 9 are illustrated the T_g vs composition data of ethylene copolymers with ethylethylene (i.e., 1-butene), butylethylene (i.e., 1-hexene), and hexylethylene (i.e., 1-octene). By the strong dashed line is shown, omitting the corresponding experimental data, the respective T_g vs composition dependence of the ethylene/methylethylene copolymers, obtained by fitting the respective T_g data (see Fig. 8).

By extrapolation of these T_g vs composition data of both the E/P and the respective E/1-alkene copolymers a value of about 260 ± 10 K is obtained for the glass temperature of poly(ethylene). Although this value is extremely high, it is still comparable with the high temperature glass temperature, T_g(u), reported by Boyer for highly crystalline polyethylene (see Fig. 2). It may thus be supposed that the crystallinity of E/1-alkylethylene copolymers increases rapidly with increasing length of the CH_2 sequences in the main chain of the copolymer, causing the experimentally observed increase of the T_g of poly(ethylene) rich copolymers.

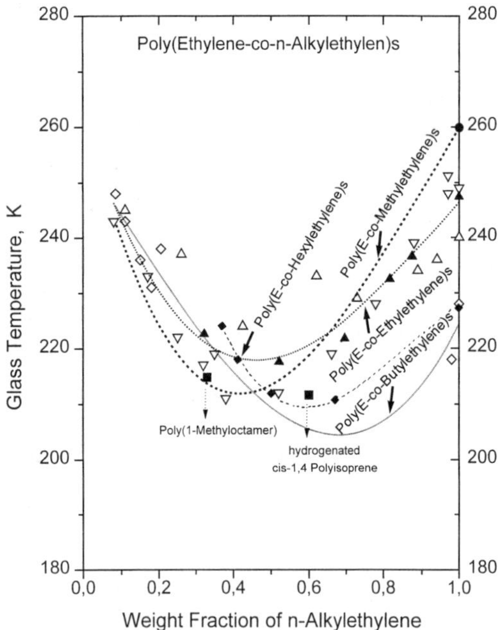

FIG. 9 The composition dependence of the glass temperature of ethylene/α-olefin copolymers. Ethylene/1-ethylethylene copolymers: △ data of Privalko and Shmorgun [80] by DSC, heating rate 2 K/min; ∇ data of Graessley *et al.* [83] by PVT. Ethylene/1-butylethylene copolymers: ◇ data of Privalko, Shmorgun [80]. Ethylene/1-hexylethylene copolymers: ◆ data of Suhm [81], by DSC, heating rate of 10 K/min.

It is, however, worth mentioning that strong negative deviations from additivity are exhibited by all ethylene copolymers, suggesting that the respective heterosequences are characterized by considerable repulsions between the repeating units of the copolymer components. This assumption is also supported by the value of the glass temperature (211.6 K [78]) of the strictly alternating model E/P copolymer, i.e., the hydrogenated cis-1,4 polyisoprene (see Fig. 8), which is situated well below the respective additivity T_g value of 229 K, estiamted according to the Fox equation and assuming for the glass transition temperature of PE the value of 195 K suggested by Boyer [2].

VIII. THE GLASS TEMPERATURE OF POLYOLEFIN BLENDS

The special combination of properties realizable by blending polyolefins opened up a number of new applications for the pipe, tubing and cable industry as well as for speciality flexible filters, sanitary wares, the shoe industry and last but not least in the automotive industry. The newer trends are mainly restricted for the development of tailor-made grade polyolefinic blends for materials having high impact strength, high elastic modulus and improved dimensional stability as well as materials with improved UV and weathering resistance [83]. Generally polyolefins are not miscible due to the lack of strongly interacting units. Thus mostly compatibilized polymer blends are recommended in commercial practice. These practical compatible polymer blends show at least the two glass transition temperatures, more or less identical with those of the blend components and will not be further analyzed in this context. As a general rule it may be emphasized that the shift of the glass temperatures of the components of compatibilized blends towards each other is the higher the larger the interphase, i.e. the stronger the adhesion effect of the compatibilizer.

Miscible polymer blends, on the contrary, are characterized by a single composition-dependent glass temperature and will be discussed in the following.

Traditionally it has been supposed in the framework of the Flory–Huggins theory [84] that, because of negligible combinatorial entropy contributions, polymer miscibility is conditional on favorable interactions between the blend components. Thus T_g of miscible polymer blends should always show positive deviations from additivity because of the denser packing resulting from the increased energetic interaction between the components [34]. Experimentally, however, mostly negative deviations from additivity of the blend T_g are observed. Thus, in light of the limitations of the classic F–H lattice theory as a zeroth-order approximation concerning the miscibility of polymers, new theoretical models were developed by introducing an effective interaction parameter to account for the experimentally observed entropy contributions. Due to the large number of theoretical papers, in the following are presented only some representative theoretical models related above all with the miscibility of polyolefins.

Taking into account that polyolefins can be blended in spite of only small van der Waals enthalpic interactions, in a series of papers theoretical models were developed to explain the miscibility in such nearly athermal and/or athermal polymeric mixtures. Thus Schweizer *et al.* [85] found by extending the RISM ("reference interaction side model") theory, that structural asymmetry between the polymer components leads to negative interaction parameters because of significant noncombinatorial mixing entropy contributions which stabilize the polymer blend by spatially nonrandom packing, which is enhanced by the structural differences. Although according to the Flory–Huggins theory polyolefins should not be miscible because of negligible enthalpic interaction, the RISM theory suggests that polyolefins should be miscible, even at low temperatures. Freed and Dodowicz [86], on the other hand, have demonstrated using the lattice cluster theory that, except for very immiscible polyolefin blends, both the "entropic" and the "enthalpic" portions of the interaction parameter are relevant and comparable to each other, however, with an increasing dominance of the entropic part when the blend miscibility improves. They concluded, however, contrary to Schweizer *et al.* [85], that blends with the higher asymmetry are less miscible, while blends whose side groups are more "similar" exhibit the better compatibility.

The favorable effect on polyolefin miscibility of statistical segment length asymmetry due to the entropy contributions required for conformational adjustments has also been emphasized by Bates *et al.* [87]. In a series of papers, Bates and Fredrickson [88] attributed the miscibility of athermal or nearly athermal polymer mixtures mainly to these "conformational asymmetries" which contribute substantially to a "nonlocal conformational excess entropy" of mixing. The effect is exemplified for the amorphous polyethylene/poly(ethylethylene) blend. Due to the fact that unperturbed PE and PEE molecules cannot be randomly interchanged, a positive excess free energy of mixing caused by nonlocal excess entropy contribution is anticipated by the authors. The effect of asymmetry on polymer miscibility is also supported by computer simulations, which suggest additional contributions due to entropy density differences of the pure polymeric phases [89].

Considering both the effect of enthalpic and conformational entropic contributions for polymer miscibility, Brekner *et al.* [73] derived a virial-like concentration third power equation for the composition dependence of the glass transition temperature of miscible polymer blends:

$$(T_g - T_{g_1})/(T_{g_2} - T_{g_1}) = (1 + K_1)w_{2c} - (K_1 + K_2)w_{2c}^2 + K_2 w_{2c}^3 \qquad (6)$$

where $w_{2c} = Kw_2/(w_1 + Kw_2)$ with $K = T_{g_1}/T_{g_2}$, is the weight fraction of the polymer with the higher T_g, corrected to account via the Simha–Boyer rule [56] for the different expansivity of the blend components. K_1 is essentially characteristic of the difference between the interaction energies of hetero- and homo-contacts, whereas K_2 depends on the conformational entropy changes during mixing. By adapting experimental T_g data of polymer blends the values of the obtained K_i parameters of the third power Eq. (6) are characteristic of the different shapes of T_g vs composition curves.

In a series of papers by Karasz *et al.* suggested an effect of both sequence distribution [90] and branching [91] on the miscibility of copolymers. Experimental results confirmed these suppositions. Cantow and Schulz [92], on the other hand, have emphasized that both chemical and configurational sequence distribution affects polymer miscibility.

Nitta *et al.* [64] have shown that isotactic poly(methylethylene), i-PP, is miscible with poly(ethylene-co-ethylethylene)s, EBR, for only a very restricted composition range of the copolymer. Thus i-PP is miscible with EBR 56 (i.e., a EBR copolymer containing 56 mol% 1-butene) and EBR 62, but not miscible with EBR 36 and EBR 45.

In a series of papers Graessley *et al.* were concerned with the glass transition of copolymers [82] and the miscibility of polyolefins [93–98]. They concluded that the miscibility of polyolefins with polyolefinic copolymers depended overall on the composition of the copolymers. Thus, for instance poly(1,1-dimethylethylene)—i.e., polyisobutylene—is miscible with poly(ethylene-co-ethylethylene) for the restricted composition range of the EB copolymer between 52–90 wt% ethylethylene (i.e., butene). The respective T_g vs composition data of the blend of polyisobutylene with an EB copolymer with 66 wt% butene [98] are shown in Fig. 10. Also included are the T_g vs composition data for the poly(ethylene)/poly(methylethylene) blends of Piloz *et al.* [99]. It shows that, except for the T_g data estimated by mechanical dynamic method, the glass transition temperatures of the polyolefin blends are characterized by negative deviations from additivity, suggesting large conformational entropy contributions to miscibility, as recently has been emphasized in a review paper concerning the glass temperatures of miscible polymers [100].

Graessly *et al.* [95] concluded that the thermodynamic interactions in blends of saturated hydrocarbon polymers as well of model polyolefins originate from induced-dipole forces, and that they differ in subtle but important ways, depending on the blend component structures and statistical segment mismatch, resulting in strongly anomalous mixing behavior of the polyolefins. Generally it seems there exists, however, a parallel between the segment length of a component and its "regular mixing" behavior. The magnitude and the temperature dependence of the interaction parameter vary from system to system and no clear trends with other specific parameters are apparent.

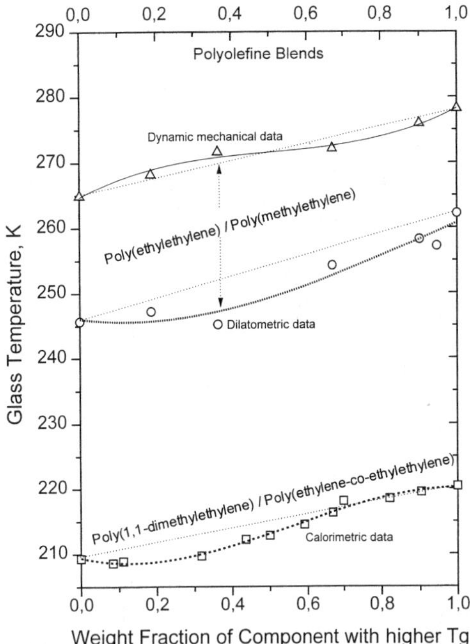

FIG. 10 The composition dependence of the glass temperature of miscible polyolefin blends. Poly(ethylene)/poly(methylethylene) blend; data of Piloz *et al.* [99] by: ○ dilatometry; △ mechanical dynamic measurements. Poly(1,1-dimethylethylene)/poly(33 ethylene-co-66 ethylethylene) blend: □ data of Krishnamoorti *et al.* [99] by DSC.

Based on the above presented experimental results it may be generally supposed that both polyolefinic copolymers and miscible blends are sooner characterized by important conformational entropy contributions, i.e., the glass temperature of these systems are situated well below by additivity assumption calculated values.

Before closing this chapter on the glass temperature of polyolefins, we will mention some remarks concerning the glass temperature of immiscible blends of polyolefins. The large spectrum of properties of polyolefins is broadened mainly for improving their processibility by blending the individual polyolefins with other polymers, even if the components are immiscible. Taking into account the thermodynamic immiscibility of these blends of great practical importance, they are separated into two or even more phases, each phase showing more or less accurately their own characteristic glass temperature. Generally the behavior of these blends is governed by the interfacial properties, i.e., interfacial tension or adhesion, whereas the morphology is characterized by the particle size and interparticle distance, which is related to the ratio of the blend components. To increase adhesion between the immiscible phases, in order to improve the properties of the blend, either "compatibilizers" are used as additional components, or interacting groups are introduced or grafted onto the components. The better the adhesion between the blend components, the stronger is the observed shift of the glass temperatures, sometimes even exhibiting additional glass temperatures characteristic of the "compatibilizer" and/or the interphase.

Impact modification of the brittle polyolefins, for instance, is usually realized by blending with rubbers in order to produce blends with improved toughness. This impact improvement is the result of superposition of local stresses induced in the soft dispersed rubber particles and is related to a critical particle diameter and a critical interparticle distance. The general condition for toughening is that the interparticle distance must be smaller than the critical value [101]. These stresses may be morphology-induced stresses, but also stresses arising from the thermal history of the blend, the latter being the result of differences in the thermal expansion coefficients of the components. It can be supposed that this thermal stress influences the mobility of the blend components and consequently also the glass temperatures.

Taking into account the much higher glass temperature of the brittle polyolefinic matrix, T_g^{matrix}, in the temperature between T_g^{matrix} and T_g^{rubber} the shrinkage during cooling of the already frozen, glassy polyolefinic matrix will be smaller then of the dispersed rubber particles existing still in the molten state. If the adhesion between the phases is strong enough to prevent debonding between the particles and matrix, a volume dilatation occurs, inducing radial stresses in the dispersed particles. This volume dilatation is equivalent to an apparent reduction of the density of network points as compared with the neat rubber, leading consequently to a depression of the experimentally observed T_g^{rubber}.

Thus Mäder *et al.* [102] have shown, for instance, by mechanical dynamic analysis of blends of poly(methylethylene)s of different tacticity with various rubbers, that, depending on the rubber content of the blends, the T_g depression of the dispersed rubber phase reaches up to about $-10°C$.

Generally, it may be assumed that the greater the observed shift of the glass temperatures of the components, the stronger the adhesion between the phases in immiscible polymer blends. Additionally the experimental evidence of a third glass temperature in a binary blend points at a well developed interphase due to "compatibilization" effects.

REFERENCES

1. P Peiser. In: J Brandrup, EH Immergut, eds. Polymer Handbook. New York: John Wiley & Sons, 1989, pp VI/209–VI/227.
2. RF Boyer. Macromolecules 6: 288–299, 1973.
3. AA Miller. J Polym Sci A-2, 6: 249–257, 1968.
4. KR Dunham, J Vandenberghe, JWH Faber, LE Contois. J Polym Sci A, 1: 751–762, 1963.
5. W Kauzmann. Chem Rev 43: 219–256, 1948.
6. AK Doolitle. J Appl Phys 22: 1471–1475, 1951.
7. JD Ferry. Viscoelastic Properties of Polymers. New York: John Wiley & Sons, 1980, pp 287–289.
8. N Hirai, H Eyring. J Polym Sci 37: 51–70, 1959.
9. AS Krausz, H Eyring. Deformation Kinetics. New York: Wiley Interscience, 1975, pp 36–45.
10. MH Cohen, D Turnbull. J Chem Phys 34: 120–125, 1961.
11. R Simha, T Smocynsky. Macromol 4: 342–350, 1969.
12. R Simha. Macromol 10: 1025–1030, 1977.
13. G Rehage. J Macromol Sci Phys B18: 423–443, 1980.
14. RF Robertson, R Simha, JG Curro. Macromol 17: 911–919, 1984.
15. JH Gibbs, EA DiMarzio. J Chem Phys 28: 373–383, 807–813, 1958.
16. EA DiMarzio. Ann NY Acad Sci 37: 1–21, 1981.
17. G Adam, JH Gibbs. J Chem Phys 43: 139–146, 1965.
18. RN Haward. J Macromol Sci Revs Macromol Chem C4: 191–242, 1970.
19. AJ Kovacs. Fortschr Hochpolym Forsch 3: 394–507, 1963.
20. G Rehage, W Borchard. In: RN Haward, ed. The Physics of Glassy Polymers. New York: John Wiley & Sons, 1973, pp 55–102.
21. A Eisenberg. In: Physical Properties of Polymers. Washington, DC: ACS, 1984, pp 57–95.
22. GB McKenna. In: C Booth, C Price, eds. Polymer Physics. Oxford: Pergamon, 1989, vol. II, pp 311–362.
23. HA Schneider. In: JC Salamone, ed. Polymeric Materials Encyclopedia. Boca Raton: CRC Press, 1996, vol. 4, pp 2777–2789.
24. TG Fox, S Loshaek. J Polym Sci 15: 371–390, 1955.
25. H Stutz, KH Illers, J Mertes. J Polym Sci Part B 28: 1483–1498, 1990.
26. VF Skorodumov, YuK Godowskii. Vysokomol Soed B 35: 214–226, 1993.
27. HA Schneider, B Rudolf, K Karlou, H-J Cantow. Polym Bull 32: 645–652, 1994.
28. KK Che. J Appl Polym Sci 43: 1205–1208, 1991.
29. JMG Cowie, IJ McEwen, MT Garay. Europ Polym J 9: 1041–1049, 1973.
30. DR Burfield, Y Doi. Macromol 16: 702–704, 1983.
31. DW van Krevelen, PJ Hoftyzer. Properties of Polymers. Elsevier: Amsterdam, 1976, pp 99–127.
32. DR Wiff, MS Altieri, IJ Goldfarb. J Polym Sci, Polym Phys Ed 23: 1165–1176, 1985.
33. HA Schneider, EA DiMarzio. Polymer 33: 3453–3461, 1992.
34. HA Schneider. J Res Nat Inst Stand Technol 102: 228–248, 1997.
35. MG Koehler, AJ Hopfinger. Polymer 30: 116–126, 1989.
36. X Lu, B Jiang. Polymer 32: 471–478, 1991.
37. K Binder. Ber Bunsenges Phys Chem 100: 1381–1387, 1996.
38. GT Davis, RK Eby. J Appl Phys 44: 4274–4281, 1973.
39. B Wunderlich. J Chem Phys 37: 2429–2432, 1962.
40. K Schmieder, K Wolf. Kolloid-Z Z Polym 134: 149–189, 1953.
41. JA Sauer, DE Kline. J Polym Sci 18: 491–495, 1955.
42. DE Kline, JA Sauer, AE Woodward, J Polym Sci 22: 455–462, 1956.
43. LE Nielsen. J Polym Sci 42: 357–366, 1960.
44. WG Oakes, DW Robinson. J Polym Sci 14: 505–506, 1954.
45. KH Illers. Kolloid-Z Z Polym 190: 16–34, 1963.
46. M Gordon, JS Taylor, J Appl Chem, USSR 2: 493–500, 1952.
47. EA DiMarzio. Polymer 31: 2294–2298, 1990.
48. KH Illers. Kolloid-Z Z Polym 250: 426–433, 1972.
49. KH Illers. Kolloid-Z Z Polym 251: 394–401, 1973.
50. KH Illers. Kolloid-Z Z Polym 252: 1–7, 1974.
51. W Pechhold. Kolloid-Z Z Polym 228: 1–34, 1968.
52. FC Stehling, L Mandelkern. Macromol 3: 242–252, 1970.
53. PR Swan. J Polym Sci 42: 525–534, 1960.
54. EW Fischer, F Kloos. J Polym Sci, Polym Lett 8: 685–693, 1970.
55. NN Kuzmin, EV Matukhina, NN Makarova, BM Polycarpov, AM Antipov. Makromol Chem, Makromol Symp 44: 155–164, 1991.
56. PJ Hendra, HP Jobic, K Holland-Moritz. J Polym Sci, Polym Lett 13: 365–368, 1975.
57. F Danusso, G Moraglio, G Talamini. J Polym Sci 21: 139–140, 1956.
58. G Natta, F Danusso, G Moraglio. J Polym Sci 25: 119–122, 1957.
59. J-M Braun, JE Guillet. J Polym Sci, Polym Chem Ed 14: 1073–1081, 1976.
60. FP Reding, JA Faucher, RD Whitman. J Polym Sci 57: 483–498, 1962.
61. RL Miller. In: J Brandrup, EH Immergut, eds. Polymer Handbook. New York: John Wiley & Sons, 1989, pp V1/96–VI/171.
62. CH Fisher. In: C Vasile, RB Seymour, eds. Handbook of Polyolefines. New York: Marcel Dekker, Inc., 1993, pp 11/471–478.
63. D Feldman, A Barbalata. Synthetic Polymers: Technology, Properties, Applications. London: Chapman & Hall, 1997, pp 30–33.
64. K-H Nitta, K Okamoto, M Yamaguchi. Polymer 39: 53–58, 1998.

65. TG Fox. Bull Am Phys Soc 1: 123, 1965.
66. R Simha, RF Boyer. J Chem Phys 37: 1003–1007, 1962.
67. JM Barton. J Polym Sci Part C 30: 573–597, 1970.
68. NW Johnston. J Macromol Sci Rev Macromol Chem C14: 215–250, 1976.
69. H Suzuki, VBF Mathot Macromol 22: 1380–1384, 1989.
70. HA Schneider, J Rieger, E Penzel. Polymer 38: 1323–1337, 1997.
71. TK Kwei. J Polym Sci Polymer Lett Ed 22: 307–313, 1984.
72. HA Schneider. Mackromol Chem 189: 1941–1955, 1988.
73. M-J Brekner, HA Schneider, H-J Cantow. Polymer 78: 78–85, 1988; Makromol Chem 189: 2085–2097, 1988.
74. PR Couchman. Macromolecules 24: 5772–5774, 1991.
75. HA Schneider. J Res Inst Stand Technol 102: 228–248, 1997
76. P Manaresi, V Giannella. J Appl Polym Sci 4: 251–252, 1960.
77. EG Kontos, WP Slichter. J Polym Sci 61: 61–68, 1962.
78. G Gianotti, G Dall'Asta, V Valvassori, V Zamboni. Makromol Chem 149: 117–125, 1971.
79. F De Candia, G Maglio, R Palumbo, G Romano, A Zambelli. Makromol Chem 179: 1609–1616, 1978.
80. VP Privalko, AV Shmorgun. J Thermal Anal 38: 1257–1270, 1992.
81. J Suhm, PhD Thesis, University Freiburg, 1998.
82. WW Graessley, R Krishnamoorti, NP Balsara, RJ Butera, LJ Fetters, DH Lohse, DN Schulz, JA Sissano. Macromol 27: 3896–3901, 1994.
83. A Mattiussi, F Forcucci. In: E Martuscelli, R Palumbo, M Kryszewski, eds. Polymer Blends. New York: Plenum Press, 1980, pp 469–484.
84. PJ Flory. Principles of Polymer Chemistry. Ithaca, New York: Cornell University Press, 1953, pp 495–519.
85. JC Corro, KS Schweizer. Macromol 23: 1402–1411, 1990; C Singh, KS Schweizer. Macromol 28: 8692–8695, 1995.
86. KF Freed, J Dudowicz. Macromol 29: 625–636, 1996.
87. FS Bates, MF Schulz, JH Rosedale. Macromol 25: 5547–5550, 1992.
88. FS Bates, GH Fredrickson. Macromol 27: 1065–1067, 1994; GH Fredrickson, AJ Liu, FS Bates. Macromal 27: 2503–2511, 1994.
89. M Müller, K Binder. Macromol 28: 1825–1834, 1995.
90. C Zhikuan, S Ruona, FE Karasz. Macromol 25: 6113–6118, 1992.
91. Z Chai, R Sun, S Li, FE Karasz. Macromol 28: 2297–2302, 1995.
92. H-J Cantow, O Schulz. Polym Bull 15: 449–453, 1986.
93. NP Balsara, LJ Fetters, N Hadjichristidis, DH Lohse, CC Han, WW Grasessley, R Krishnamoorti. Macromol 25: 6137–6147, 1992.
94. WW Grassley, R Krishnamoorti, NP Balsara, LJ Fetters, DJ Lohse, DN Schulz, JA Sissano. Macromol 27: 2574–2579, 1994.
95. R Krishnamoorti, WW Graessley, NP Balsara, DJ Lohse. Macromol 27: 3073–3081, 1994.
96. R Krishnamoorti, WW Graessley, NP Balsara, DJ Lohse. J Chem Phys 100: 3894–3904, 1994.
97. R Krishnamoorti, WW Graessley, LJ Fetters, RT Garner, DH Lohse. Macromol 28: 1252–1259, 1995.
98. WW Graessley, R Krishnamoorti, GC Reichart, NP Balsara, LJ Fetters, DJ Lohse. Macromol 28: 1260–1270, 1995.
99. A Piloz, J-Y Decroix, J-F May. Angew Makromol Chem 54: 77–90, 1976.
100. HA Schneider. Polym Bull 40: 321–328, 1998.
101. S Wu. Polymer 26: 1855–1863, 1985.
102. D Mäder, M Bruch, R-D Maier, F Striker, R Mülhaupt, Macromol 32: 1252–1259, 1999.

9

Crystallization of Polyolefins

Clara Silvestre, Maria Laura Di Lorenzo, and Emilia Di Pace
Istituto di Ricerca e Tecnologia delle Materie Plastiche, CNR, Arco Felice (NA), Italy

I. INTRODUCTION

Crystallization of polymers of sufficient structural regularity can occur in a range of temperature limited by the glass transition temperature, T_g, and the melting temperature, T_m. All studies show that isothermal crystallization rate varies with temperature, as shown in Fig. 1. This characteristic shape is a consequence of growth being slowed by increasing viscosity at temperatures close to T_g and by diminishing thermodynamic drive as the melting point is approached.

In the range between T_g and T_m, the extent over which crystallization can be measured depends on the size of the crystals, i.e., on the nucleation and growth rate. For a given T_c if nucleation is slow and growth fast, few large spherulites result, whereas rapid nucleation leads to a profusion of spherulites. When both nucleation and growth are very fast, the crystallization can start during the cooling from T_m to T_c, but in this case crystallization is not isothermal. For the polymers that behave in this way, only crystallization at temperatures close to T_m can be followed isothermally.

The process of crystallization of a polymer from the melt can be divided in three stages: (1) primary nucleation; (2) growth of the crystals (secondary nucleation); (3) secondary crystallization.

Primary nucleation is the process by which a crystalline nucleus is formed in the melt state. After the nucleus is formed, a new layer grows on the face of the existing one with a secondary nucleation, a process similar to primary nucleation. The crystallization does not stop with the growth of the crystals, but a process called secondary crystallization occurs, giving an increase of crystallinity and thickness of the already formed lamellar crystals. The secondary crystallization will not be considered in this chapter.

The study of crystallization of polymers is generally conducted in isothermal conditions, since the use of a constant temperature permits an easier theoretical treatment and limits the problems connected with thermal gradients within the samples. Nevertheless, the analysis of nonisothermal crystallization is also of great importance, since it permits us to better simulate the polymer processing conditions.

The study of crystallization kinetics is possible by dilatometric, microscopic, spectroscopic, and calorimetric techniques.

The isothermal and nonisothermal crystallization process of the most used polyolefins is discussed in this review. Major attention is directed to crystallization from the melt; some examples of crystallization from solution and cold state are also reported. The description of isothermal crystallization of polyolefins constitutes the first part of this review. In the second part the nonisothermal crystallization process is treated.

For both isothermal and nonisothermal crystallization, general theoretical aspects are mentioned. For details many excellent reports can be found in the literature [1–3]. Those papers, whose results are reported, were selected to underline the relationships among molecular characteristic, crystallization conditions, and kinetic and thermodynamic parameters.

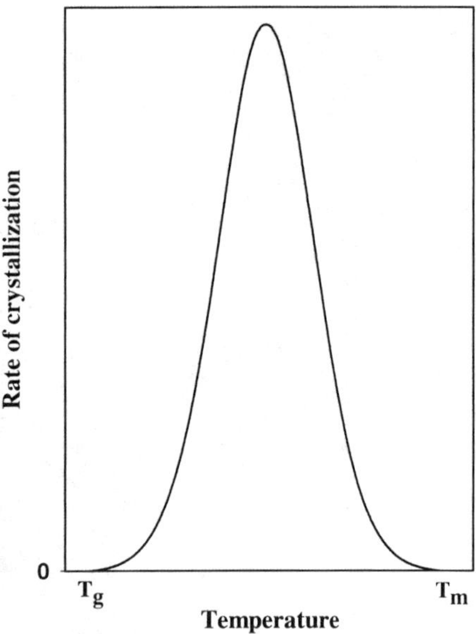

FIG. 1 Variation of the crystallization rate with temperature. Schematic.

To have a more complete pattern of the crystallization process of the polyolefins, the authors suggest you also consider the morphological aspects that develop during the crystallization of such important materials. These aspects are analyzed in details in the chapter dealing with the morphology of the polyolefins [4].

II. ISOTHERMAL CRYSTALLIZATION

A. Nucleation

Nucleation in polymers can be homogeneous or heterogeneous. Nuclei can be formed homogeneously, by means of statistical fluctuation in the melt phase, or heterogeneously, i.e., catalyzed by the presence of heterogeneities. In the latter process, nucleation starts on surfaces, cavities and cracks of insoluble impurities.

Very often the nucleation of polymers is heterogeneous and is favored by longer molecules. This is probably due to the fact that the longer a molecule is, the greater is the chance to adopt a conformation suitable for the crystallization.

The classical concept of crystal nucleation, based on the assumption that fluctuations in the undercooled phase can overcome an energy barrier at the surface of the crystal, was first developed by Gibbs and later by Kossel and Voirner, as reported in [5, 6]. the rate of nucleation, I^*, has been derived by Turnbull and Fisher [7] to be

$$I^* = \frac{NkT}{h} \mathrm{e}\left(-\frac{\Delta\Phi^* + \Delta E}{kT}\right) \tag{1}$$

where N is related to the number of crystallizable elements, $\Delta\Phi^*$ is the energy of formation of a nucleus of critical size, and ΔE is the activation energy for chain transport.

Generally in polymers, as the temperature is lowered from the melting temperature, a rapid decrease in $\Delta\Phi^*$ and a slow increase in ΔE occur, causing I^* to increase. As the temperature is lowered even further, the decrease in $\Delta\Phi^*$ becomes moderate but the increase in ΔE is more significant, resulting in a decrease in I^*. Therefore, a maximum in I^* exists and is related to the ease with which crystallizable elements can cross the phase boundary.

Expressions for critical sizes of nucleus can be obtained by zeroing the first derivative of $\Delta\Phi^*$ with respect to the dimensions of the nucleus.

Hoffman [8–11] presented extensive work in this area. He found that the typical homogeneous nucleus dimensions are about 10^3–10^5 Å^3, where a typical polymer chain volume is about 10^5–10^7 Å^3. Thus, only a small portion of the polymer chain is involved in forming a nucleus. This initial organization of a part of a flexible macromolecule on a crystal surface has been named *molecular nucleation* [12].

The heterogeneous nucleation was studied extensively by Binsbergen [13–18].

Formation of a nucleus on a foreign surface involves a creation of a new interface, similar to the case of homogeneous nucleation. However, a pre-existing foreign surface greatly reduces the free enthalpy of the formation of a critical nucleus, $\Delta\Phi^*$. This lowers the critical size of the nucleus and results in the formation of heterogeneous nuclei at lower undercooling.

The theories of nucleation describe the mechanism, but hardly predict the real habit of nucleation in polymers. The experimental methods of determination of nucleation are hence of particular importance. The knowledge of nucleation data is often essential for controlling physical properties of polymers: mechanical properties depend to a great extent on the spherulite average size and size distribution [19, 20], that are determined by the primary nucleation process. For some applications it is sufficient to determine only the total number of nuclei activated during the crystallization. The simplest way of obtaining this value is from the average spherulite size of samples filled with spherulites [21], on the basis of direct characterization of spherulite patterns, by using polarized light microscope, scanning and transmission electron microscope, small-angle light scattering [22–24] and light depolarization techniques [25, 26].

However, if the time dependence of activation of nuclei is required, other methods must be used. The data on time distribution of primary nucleation are usually obtained by direct microscopic observation of a crystallizing sample [27, 28]. The time lag between the nucleation of two neighboring spherulites can be found from the curvature of their common boundary.

In nucleation of polymer melts four distinct temperature regions [3] can be recognized. Immediately below the melting point there is a 10–30°C region where no crystal growth takes place. Even the addition of heterogeneous nuclei does not promote crystallization. Next, there is a region of about 20–40°C, where crystal growth occurs on heterogeneous nuclei, but no homogeneous nucleation is possible. The number of heterogeneous nuclei remains constant throughout the crystallization experiment. The third region of temperature, of perhaps as much as 10–30°C, is one where, in addition to the initial heterogeneous nuclei, some smaller or less perfect heterogeneous nuclei can grow to critical size. These latter nuclei are also limited in number. Homogeneous nucleation finally becomes possible at a relatively lower well defined temperature. The number of new nuclei in this region is usually so large that it does not lead to optically resolvable crystals, when observed in bulk samples after completion of crystallization.

1. Polyethylene

Homogeneous nucleation from dilute solution of polyethylene, PE, in toluene and xylene was concluded by Wunderlich and Metha to occur at about 32°C undercooling [29]. From the melt, experiments to gain information on homogeneous nucleation of PE were also performed by Vonnegut [30], dividing the melt into small droplets. The homogeneous nucleation was followed by observation of the crystallization of the droplets under the optical microscope. At the temperatures of homogeneous nucleation, the crystallization rate is usually so high that a single nucleus crystallizes the whole droplet. Crystallization from dilute solution yields well separated crystals so that morphological evidence for homogeneous and heterogeneous nucleation can be sought.

The four regions of nucleation in PE from the melt were clearly delineated by Cormia *et al.* [31]:

1. From the melt to 125°C: no nucleation.
2. From 125–100°C: heterogeneous nucleation.
3. From 100–85°C: second heterogeneous nucleation.
4. Below 85°C: homogeneous nucleation.

Figure 2 illustrates the nucleation activity in polyethylene as a function of undercooling.

2. Isotactic Polypropylene

Isotactic polypropylene, iPP, is a polymorphic material with five crystal modifications [32, 33], α, β, γ, δ and a modification with intermediate crystalline order. For details see the chapter on morphology of polyolefins in this book [4].

Nucleation of the α form of isotactic polypropylene was studied by several authors. The course of primary nucleation in isotactic polypropylene down to 70°C was first demonstrated by Burns and Turnbull [34] and Koutsky *et al.* [35] employing the droplet technique.

The four distinct regions in the nucleation of isotactic polypropylene melt were listed by Galeski [36] as follows:

1. Immediately below the DSC determined melting point (165–167°C) there is a gap where the crystal nucleation and growth hardly takes place. Neither the present heterogeneities nor introduced nucleating agent can promote the nucleation.
2. Most of the published nucleation data concern the region of temperature between 115–150°C, where regular spherulites are nucleated [37, 38]; this is the extended region of activity of heterogeneous nuclei. The number of these heterogeneous nuclei is limited.
3. Some of the heterogeneous nuclei become active at even lower temperatures which follow from their

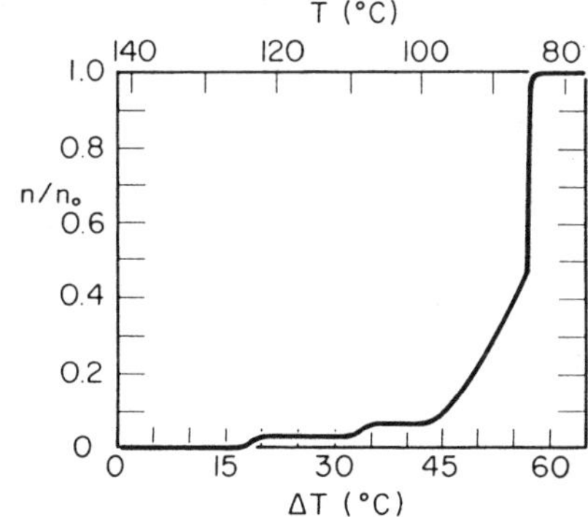

FIG. 2 Fraction of solidified polyethylene as a function of undercooling. (From Ref. 3.)

smaller size or lower perfection. These nuclei are also limited in number.

4. Finally, at approximately 80–85°C and below, there is the region of homogeneous nucleation. The number of nuclei in this region increases rapidly with the decrease of the temperature.

Figure 3 illustrates the nucleation activity of isotactic polypropylene, as a function of crystallization temperature. Bicerano [39], using literature data, elaborated some equations to obtain a general theoretical estimate of the density of the nucleation, ρ, in iPP crystallized from the glass state and from the melt. He also showed that ρ decreased rapidly with increasing T_c.

Bartczak *et al.* [40, 41] also showed that the average spherulite radius in iPP is a function of T_c in the regime at low T_c in which the dominant type of nucleation crosses over from heterogeneous to homogeneous. The data points extracted from their graphical representations are shown in Fig. 4, along with the curves that describe the empirical fits to these data. In the higher T_c region in which heterogeneous nucleation was dominant [42, 43], ρ decreased quite rapidly with increasing T_c, causing an increase of the average spherulite size. Very broad spherulite size distributions were found.

Annealing of the melt has a great influence on primary nucleation in isotactic polypropylene. First extensive study of the effects of thermal history on crystallization of isotactic polypropylene was conducted by Pae and Sauer [44, 45]. Annealing of poly-

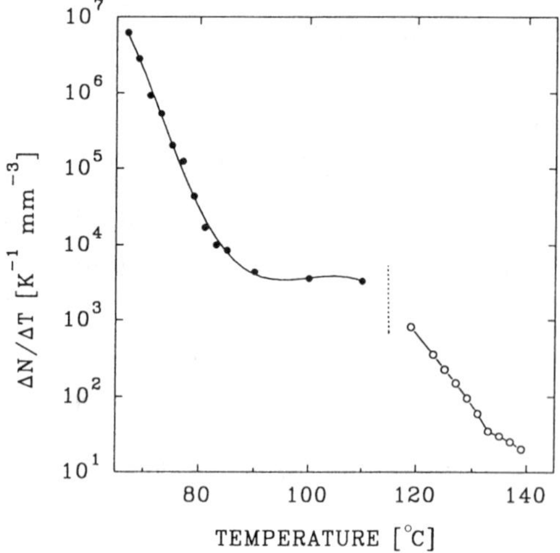

FIG. 3 Nucleation activity in isotactic polypropylene as a function of temperature. (From Ref. 36.)

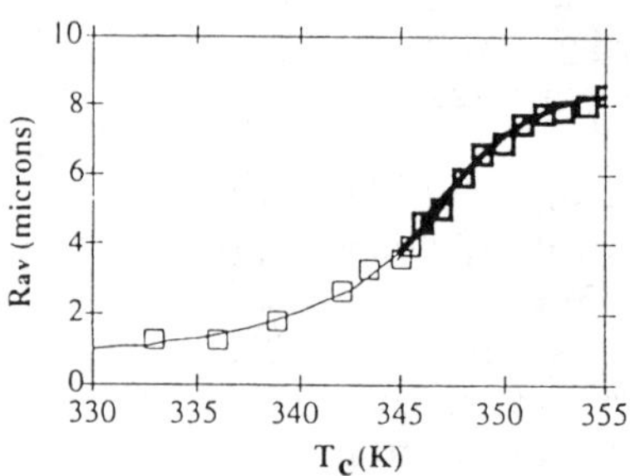

FIG. 4 Average spherulite radius of iPP as function of crystallization temperature. (From Ref. 39.)

propylene melt prior to crystallization decreased the active fraction of primary nuclei. The crucial factor was the temperature of melt annealing—below or above the equilibrium melting temperature, T_m^0. At 190°C a vast number of those thermally sensitive nuclei remained untouched, whereas even short exposure to temperature around 220°C decreased the number of active nuclei by orders of magnitude.

The structure of early stages of the growth of spherulites in polypropylene was studied by Bassett *et al.* [46–48]. The nucleus was built from more or less regular lamellae showing a multilayer arrangement. Early objects developed by branching, usually at rather large angles, and splaying apart the dominant lamellae. A sheaf-like center was usually seen for crystallization temperatures above 155°C or a crosshatched structure for the crystallization temperature 155°C and below. The change of morphology of spherulite centers at 155°C was apparently connected with the regime II–I transition. The multilayer arrangement of the center of a spherulite was explained by Bassett [47] as the result of a shish-kebab type of nucleation on straightened fragments of macromolecules which can always be found in a polymer melt. He suggested that the core of a nucleus is built of a single or a bundle of elongated macromolecules. The elongated fragments extend as far as several lamallae thickness, i.e., 500–1000 Å. The volume of such nuclei agrees with the estimation of the volume of homogeneous nuclei based on calculations of the critical nucleus size. However, such an elongated shape of the nucleus is inconsistent with the intuitive assumption that the nucleus is limited to a single lamella thickness which was made in most calculations and modeling concerning the primary nucleation.

The crystallization of isotactic polypropylene from melt could be enhanced in the region of the temperature where heterogeneous nucleation is observed by adding some extra heterogeneous nuclei. The interest in such experiments was stimulated by industrial efforts to decrease the size of spherulites for improvement of optical and mechanical properties. It was shown that solids, liquids, and even gas bubbles are able to nucleate polypropylene spherulites (for the list of patents see [3]). Binsbergen [15] has found, in contrast to some patent claims, that most inorganic salts and oxides were inactive in nucleating polypropylene. Beck and Ledbetter [49], Beck [50], and Binsbergen [13, 15] tested a large number of substances for their possible nucleating effect on the crystallization of polypropylene. The list of most active nuclei for isotactic polypropylene contains: sodium tertiary butylbenzoate, monohydroxyl aluminum *p*-tertiary butyl benzoate, sodium *p*-methylbenzoate, sodium benzoate, colloidal silver, colloidal gold, hydrazones, aluminum salts of: aromatic and cyclo aliphatic acids, aromatic phosphonic acids, phosphoric acid, phosphorous acid, several salts of Ca^{2+}, Ba^{2+}, Cu^{2+}, Co^{2+}, Ga^{3+}, In^{3+}, Ti^{4+} and V^{4+}. There are also other reports on nucleation activity of certain seeds: indigo [51], talc [52], and certain crystallographic planes of calcite [53]. Good nucleating agents are insoluble in the polymer or crystallized before crystallization of polypropylene. The important feature of a good nucleating agent appears to be the existence of alternating rows of polar and nonpolar groups at the seed surface. It should be mentioned here that the best cleavage planes of seed crystals are not necessarily the crystallographic planes exposing alternating polar–nonpolar rows if they exist.

Nucleation of the β form occurs much more rarely in the bulk samples than the predominant α form. Keith and Padden [37] observed sporadic formation of β spherulites during crystallization in the range 128–132°C. The characteristics of β phase formation were studied extensively by Varga and colleagues [54–59], who found polypropylene fractions, having high molecular mass, are more susceptible to β crystallization. It was established that the nucleation of β form was instantaneous, and that the density of nuclei decreased with increasing temperature of crystallization [59]. In contrast, Shi and Zhang [60] and Varga *et al.* [61] reported that amounts of the α phase could be suppressed by slowing the cooling rate to below 5°C/min. In this way β form of high purity can be obtained. Lovinger *et al.* [62] showed that a large amount of β phase could be obtained by crystallization in a temperature gradient. Although the primary nucleation of β phase spherulites is extremely rare in this case, the β phase is easily initiated by the growth transition along the growing front of α phase spherulites. Leugering [63] has demonstrated that a certain quinacridone dye, known as permanent red E3B, was very effective in generating spherulites of β form below 130°C. However, its effectiveness depended on nucleant concentration, and on the dispersion and cooling rate. Since then a series of other crystalline substances were found to nucleate the β form [64]: 2-mercaptobenzimidazole, phenothiazin, triphenodithiazine, anthracene and phenanthrene, and pimelic acid [65].

In the literature there are suggestions that the observation of the β form during slow crystallization results from the retardation of one of the substages of a multistage process leading to the α form [66]. The β–α transition observed at 145°C [67] seems to confirm that hypothesis.

Several observations [68–72] showed clearly that the β form generally occurs at the level of only a few per cent, unless certain heterogeneous nuclei are present [73], or the crystallization occurs in a temperature gradient [62], or in presence of shearing forces [74, 75].

3. Poly(butene-1)

Burns and Turnbull have performed droplets experiments on poly(butene-1), PB1, in order to study the crystal nucleation [76]. Microscopic and light-scattering observations indicate that PB1 can be cooled to 0°C and held at room temperature for very long periods (e.g., days) without homogeneous nucleation to occur. This represents an undercooling of approximately 115–140°C, from the melting point of the highest melting form.

The four distinct regions of nucleation for PB1 were identified by Silvestre *et al.* [77] cooling the samples from the melting point:

1. Below the experimental melting point there is a temperature gap (112–90°C) where the nucleation does not take place.
2. Below 90°C there is a temperature region of about 20°C, where some of the heterogeneous nuclei become active. These nuclei are limited in number and appear sporadically, giving rise to large spherulites.
3. At even lower temperatures ($T_c < 69$°C), an intense nucleation process takes place. This could be considered the region of activity of heterogeneous instantaneous nuclei. These nuclei, as their number is elevated, grow in small, but well defined spherulites.
4. Finally, at very low temperatures, reached by quenching the samples from the melt, there is probably the region of the homogeneous nuclea-

tion. A very high number of nuclei become active, from which very small crystals can be obtained, due to impingement of the crystals among each other.

A number of effective heterogeneous nucleating agents for the melt crystallization of poly(butene-1) homopolymer have been identified [78–83]. Some of the effective nucleants are graphite [81], 1-naphthylacetamide, *N*-stearoyl-*p*-aminophenol, mercapto-N-2-naphthylacetamide, *N*-stearoyl-*p*-aminophenol, mercapto-N-2-naphthylacetamide, malonamide, nicotinamide, isonicotinamide, benzamide, salicylamide, anthranilamide [80], *N,N'*-ethylenedistearamide, stereamide [82], adipic acid, *p*-amino benzoic acid [83], and aromatic sulfonic acids and their salts [79]. Hardness, tensile yield strength, stiffness, and heat distortion temperature of poly(1-butene) can be enhanced by nucleation with nonmetallic salts and complexes of organic carboyxlic acids [84].

4. Poly(4-methylpentene-1)

Few studies have been reported on the nucleation of poly(4-methylpentene-1), P4MP1.

Binsbergen [15] tested about 2000 substances for possible nucleating effect on the polypropylene and found that many nuclei active for iPP also had activity for P4MP1. The nucleating effect was judged by the decrease in size and increase in number of the spherulites when compared to the pure polymer, and by the possible increase in crystallization temperature at constant cooling rate. Silvestre *et al.* [85] reported that from the melt, the crystallization of P4MP1 can start at low undercooling ($\sim 10°C$). At these temperatures many nuclei became active and start to grow into microspherulites. Due to the impingement, microspherulites of very small dimension, hardly measured by optical microscopy, were obtained.

B. Growth of Crystals

Modern theories on polymer crystallization, elaborated in order to predict the temperature dependence of the growth rate, are kinetic theories. They assume that a particular stage of nucleation, in particular the secondary nucleation, is the rate-controlling step in the growth of polymer lamellae. These lamellae characterized by chain folding are those which grow faster than the most stable extended chain crystals. According to the Lauritzen and Hoffman theory [86, 87], formation of a secondary nucleus and growth of the crystals with folded chains can be regarded as a sequential addition of segments, see Fig. 5.

The steady-state rate of transfer of segment from the melt along the crystallizing unit can be found considering the crystallization of a molecular layer on a substrate as follows:

1. Attachment of a segment with volume *abl* (see Fig. 5). The free energy relative to this process is

$$\Delta\Phi = 2bl\sigma - abl\Delta F \tag{2}$$

   This attachment provides the creation of two new surfaces, whose energetic cost is $2bl\sigma$, and at the same time there is a gain in the energy ΔF due to the crystallization of the volume element *abl*.

2. Formation of the first folding with the attachment of the second element. The free energy relative to this process is

$$\Delta\Phi = 2ab\sigma_e - abl\Delta F \tag{3}$$

   With this second step two new folding surfaces are created, whose energetic cost is $2ab\sigma_e$ and as in

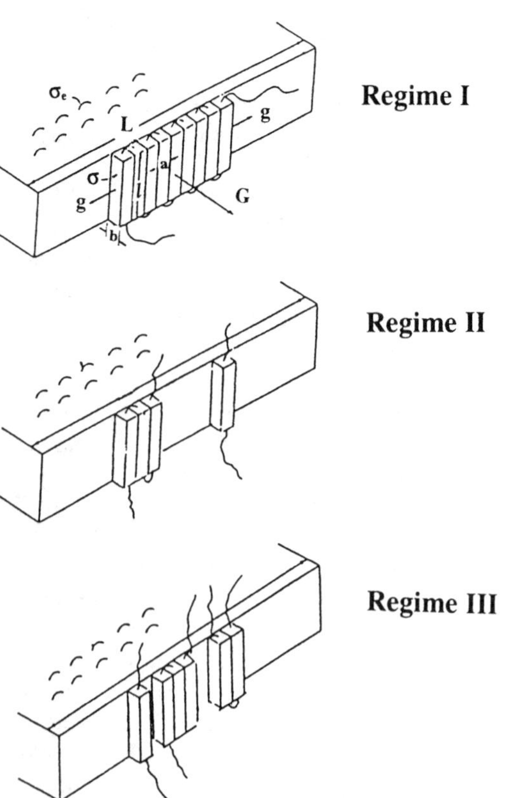

FIG. 5 Growth of polymer crystals according to regime I, regime II and regime III.

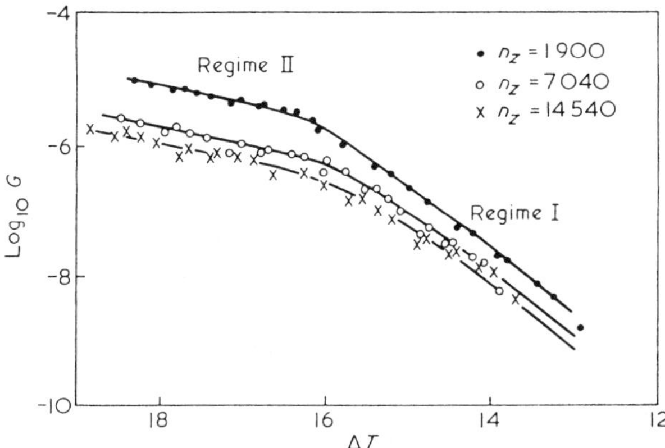

FIG. 6 Relationships of logarithmic linear growth rates with respect to ΔT for polyethylene fractions. (From Ref. 9.)

the first attachment there is again the energy balance due to the crystallization of the second element of volume *abl*.

The formation of a stable crystal is possible when l is larger than $2\sigma_e/\Delta F$. The free energy of formation of a secondary nucleus constituted by ν segments and $\nu - 1$ folds is hence the following:

$$\Delta\Phi = 2bl\sigma + 2(\nu - 1)ab\sigma_e - \nu bl\Delta F \tag{4}$$

The rate of crystallization, G, is related to the rate of attachment of the first segment, i, and the rate of completion of the layers, g. (Both these rates are also depending on the rate on which untransformed chain units are brought to the growing face crystals).

At lower temperatures i is higher than g, whereas at higher T the opposite occurs: so, in dependence on temperature, different crystallization regimes can occur, that define the mechanism of growth.

There are three extreme cases, which are known as regime I, regime II, and regime III [88].

At higher T_c, the nucleation rate is sufficiently low that one attachment is enough for the completion of the substrate, L, before another attachment happens. In this temperature region the rate of advancement G is proportional to i: $G = biL$ (regime I). At lower T_c (regime III), the rate of nucleation is so high that the segments attached on the substrate have little space for subsequent lateral growth, G is again proportional to i: $G = biL$.

Between these two ranges of temperature, the growth mechanism is intermediate between the two extremes. There is a multiple nucleation on the substrate, but the nuclei can spread laterally. The rate of crystallization will be proportional to the square root of ig: $G = bL(2ig)^{1/2}$ (regime II).

Also the resulting morphology depends on the growth mechanism. In fact, in regime III and regime II, the growing surface is rougher than in regime I, because of the multiple nucleation involved.

The occurrence of the three regimes has been demonstrated by experimental evidences. In all cases the linear growth rate takes the form:

$$G = G_0 e\left(-\frac{U^*}{R(T_c - T_\infty)}\right) e\left(-\frac{Kg}{T_c \Delta T f}\right) \tag{5}$$

where G_0 contains all the terms that are independent of temperature, U^* takes into account the energy for the transport of the macromolecules in the melt, Kg takes into account the energy required for the formation of a nucleus of critical size and T_∞ is equal to $T_g - C$, where C is a constant, f is a correction term that takes into account the dependence of ΔH^0 on T_c and is given empirically as $f = 2T_c/T_m^0 + T_c$.

A suitable graph to test this prediction is obtained by rewriting the equation in the form

$$\ln G + \frac{U^*}{R(T_c - T_\infty)} = \alpha = \ln G_0 - \frac{Kg}{T_c \Delta T f} \tag{6}$$

Plotting α against $\frac{1}{\sqrt{T_c \Delta T f}}$ two changes in the slope, always by a factor of 2, should occur, as the crystallization temperature increases.

The regimes do not depend only on the temperature, but also on molecular mass [89, 90]: it was found that, on increasing the molecular mass, the mechanisms of growth with multiple nucleations (regimes II and III) are active at higher temperature.

Regime I–II transitions have been observed in polyethylene [91, 92], poly(1,3-dioxalane) [93], *i*-polypropylene [94, 95], *cis*-1,4-polyisoprene [96], polyethylene oxide [97], and poly(3,3-dimethylthietane) [98].

Transitions from regime II to regime III at higher undercoolings have been observed in *cis*-1,4-polyisoprene [99], isotactic polypropylene [100, 101], polyoxymethylene [102], poly(1-butene) [103], poly(*p*-phenylene sulfide) [104], and poly(3,3-dimethylthietane) [98].

The existence of three regimes in the same polymer has been observed in: *cis*-polyisoprene [96], poly(3,3-dimethylthietane) [98], poly(1-butene) [103], and polypropylene [100].

1. Polyethylene

Polyethylene has been the most popular model compound in the study of polymer crystallization. A sizable number of publications exists on the topic.

Polyethylene crystallizes from melt at atmospheric pressure into crystalline aggregates, such as spherulites and axialites. Several authors measured the isothermal linear growth rate of such aggregates in the crystallization temperature ranging between 118–130°C. Fatou tabulated a large amount of data of the linear growth rate of PE in dependence on temperature and molecular mass. The growth rate is depending on temperature and molecular mass. For a given molecular mass the linear growth rate decreased by about four orders of magnitude on increasing the crystallization temperature. For a given T_c, the dependence on molecular mass is not so clear, probably depending on undercooling and molecular mass distribution.

Linear growth rates of PE fractions in a broad molecular mass range were examined by Hoffman *et al.* [10, 91] and Hoffman [9] as part of their investigation of the regime II–I transition in the polymer growth rate.

For a given undercooling, the growth rate is dependent on molecular mass, decreasing increasing the molecular mass. Moreover they found a regime behavior (regime I and II). The analysis of the growth rate data at a fixed undercooling near the center of regimes I and II show for all the fractions $G \propto 1/n_z^{1.3\pm0.3}$, where n_z is the number of chain units associated with the molecular mass M_z. The temperature of the regime I to regime II transition increases with molecular mass. For all the fraction the transition occurs at $\Delta T = 17°C$. The same regime behavior was observed by Labaig [105] and Toda [106]. Also these authors found a transition from regime I to regime II at the same undercooling $\sim 17°C$, see Fig. 6.

For polyethylene, the transition from regime II to regime III has been predicted to occur at an undercooling of $\Delta T = 23°C$ [91]. The existence of regime III has been inferred by estimates of growth rates from data obtained from the crystallization of droplets [107].

2. Isotactic Polypropylene

Since 1958, the isotactic polypropylene became commercially available, the linear growth rate behavior has been extensively studied. iPP is a favorable model substance, because its linear growth rate can be determined with high precision in a wide temperature range. In particular the growth rate of α and β spherulites were analyzed in details. A large amount of data of linear growth rate, G, of iPP α-spherulites was tabulated by Fatou [108]. Janeschitz-Kriegel *et al.* [109] also collected data on G from several sources. It can be seen that there are significant differences among the values of G reported by different workers at given values of T_c and molecular mass. No systematic molecular mass dependence can be discerned, indicating that the molecular mass had no influence on G. Bicerano [39], using these data, obtained an empirical equation relating G to T_c. The maximum in G is located at $T_c = 375°C$.

G may be decreased by orders of magnitude by the reduction of chain regularity. In particular Avella *et al.* [110], and Janimak *et al.* [111] provided several data for the dependence of G on the isotacticity, see Fig. 8. They showed that G decreases with decreasing the isotacticity content, whereas it is not affected significantly by the choice of catalyst used during synthesis. At a constant undercooling, the linear growth rate data differ by about 3 orders of magnitude because of the differences in isotacticity. Also Ibhadon reported that highly tactic samples crystallize faster than those of low tacticity [112]. Cheng *et al.* [100] measured the crystal growth rate of two molecular mass fractions ($M_w = 15{,}000$, $M_w = 300{,}000$) crystallized from the melt in an extensive undercooling range (20–70°C).

Results from kinetic studies on the growth of iPP spherulites have contributed towards checking the applicability and the validity of the growth regime theory proposed by Hoffman.

Clark and Hoffmann [11], utilizing several data sources [13, 62, 113–117], showed the existence of two crystal growth regimes. According to their studies, the regime III–II transition occurred near 137°C. A transition from regime II to regime I was reported by

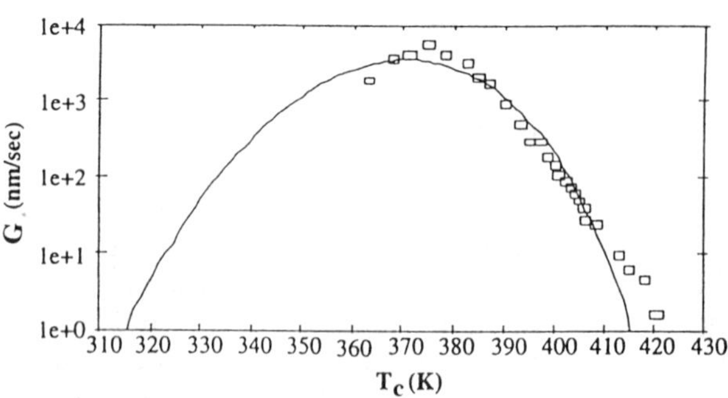

FIG. 7 Dependence of growth rate on T_c for iPP. (From Ref. 39.)

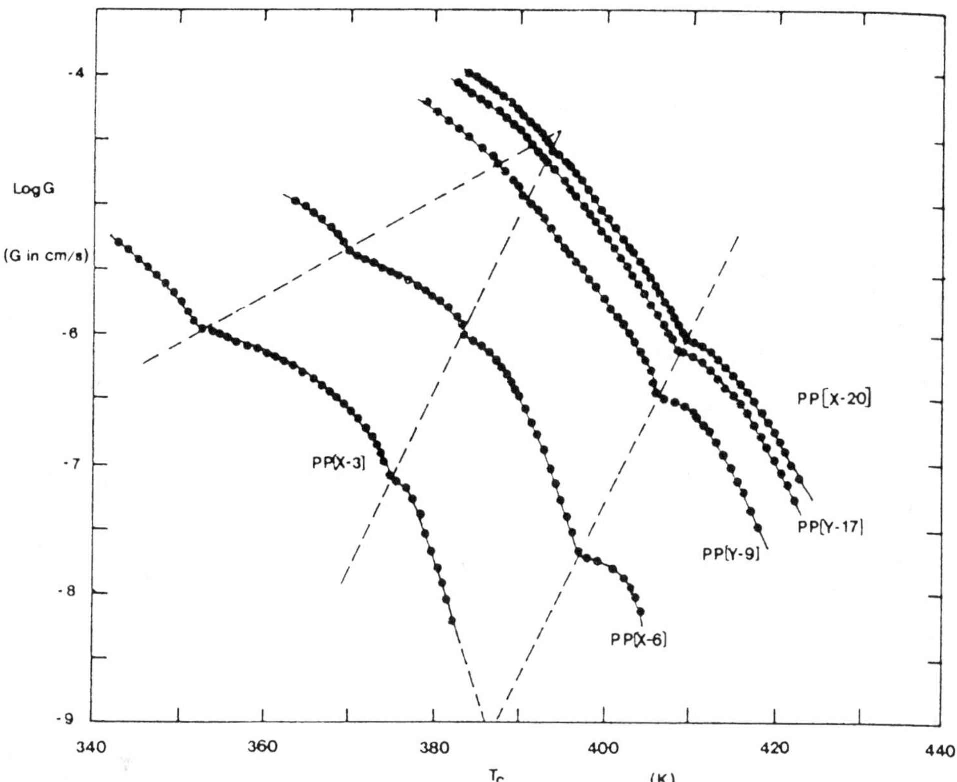

FIG. 8 Relationships of logarithmic linear growth rates with respect to crystallization temperatures for iPP fractions. (From Ref. 111.)

Allen and Mandelkern [99] at 148°C. Also Cheng *et al.* [100] observed, both by linear growth rate and overall crystallization measurements, the presence of a regime II to regime I transition for low molecular mass iPP. Detailed regime analysis indicates that a regime III exists above $\Delta T = 48$°C, a regime II is active between 48 and 37°C of undercooling and below $\Delta T = 37$°C a regime I crystal growth appears. In the case of high molecular mass only regime III and regime II was observed [100].

Crystal growth rate data at moderate ($\Delta T = 20$°C) to high undercoolings ($\Delta T = 70$°C) in an iPP fraction of different molecular mass were obtained by Ibhadon, who reported regime II–I and III–II transitions for a wide range of molecular masses, from 12,000 to 630,000 [118]. The regime II–I transition occurred at $\Delta T = 35$°C, whereas the III–II transition occurred at ΔT of about 47°C. The crystallization behavior of a set of polypropylene with similar molecular masses and distributions but different isotacticity ranging from 88–98.8% was studied by Janimak *et al.* [111]. A regime III–II transition was found for all samples. The transition temperatures decreased with isotacticity, but occurred at a constant undercooling (48°C) from their equilibrium melting temperatures. At low crystallization temperature, two other crystal growth regions were identified at an undercooling larger than 60°C. The change of a regular α crystal to a disordered α crystal form was suggested as a possible cause for such behavior. The calculated values of σ_e show a large scattering: between 40×10^{-3} J/m^2 and 230×10^{-3} J/m^2 [38, 117, 119, 120].

Kinetic studies of crystallization of the β form of iPP were performed by several authors [65, 121–124]. Padden and Keith [37] first pointed out that the growth rate of β-spherulites was higher than that of α-spherulites at the same undercooling. Lovinger *et al.* [125] demonstrated that the ratio of growth rates of the two modifications (G_β/G_α), decreased with elevating T_c and was always higher than 1 in the temperature range studied. Studies by Varga and coworkers [126, 127] established that:

1. The ratio G_β/G_α decreases with increasing T_c and below a critical temperature, denoted as $T(\beta\alpha) = 140$–$141°C$, $G_\beta > G_\alpha$, see Fig. 9.
2. At a temperature above $T(\beta\alpha)$, punctiform α nuclei formed on the surface of β-spherulites, growing into α spherulites segments which finally encompassed the basic β spherulite. A further increase in T_c enhanced the probability of $\beta\alpha$-transitions (while G_α and G_β decreased steadily), being predominant as approaching the melting temperature of the β form.
3. The critical temperature $T(\beta\alpha)$ was accepted as an upper limit temperature for the formation of β iPP, since above this temperature formation of β iPP was not possible. The formation of β iPP, however, also had a lower limit temperature, $T(\alpha\beta)$, as demonstrated by Lotz *et al.* [128].
4. The growth of β iPP proceeded in the Hoffman's regimes II and III [57, 65, 125]. The transition temperature, T(II–III), occurred at about 133°C ($\Delta T = 43°C$) according to Varga and colleagues [57, 58], whereas Shi *et al.* calculated T(II–III) = 133–137°C for various samples [65] ($\Delta T = 39$–$43°C$).

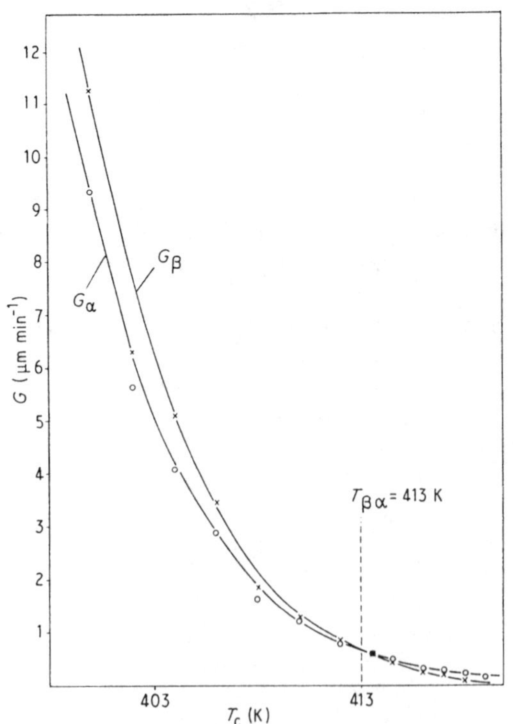

FIG. 9 Temperature dependence of the linear growth rates ($G_\alpha G_\beta$) of α and β modifications. (From Ref. 33.)

3. Syndiotactic Polypropylene

Miller and Seeley reported a study on the linear crystal growth rate kinetics of sPP [129]. They measured the spherulitic growth of a sPP sample of 72% syndiotacticity in an isothermal crystallization temperature range of 40°C (97.4°C $\leq T_c \leq$ 137.3°C).

Using values of $T_m^0 = 161°C$, $\Delta h_m = 3.14\,kJ/mol$, and assuming a regime II growth, the lateral and fold surface free energies were estimated to be 4.4 and 47 erg/cm^2, respectively. The work of chain folding was 23.4 kJ per mole of folds. It was noted that the growth rate data of sPP is of the same order of magnitude as iPP for similar degrees of undercooling.

Clark and Hoffman [11] fitted the growth rate data of Miller and Seeley with a straight line above temperatures of 110°C [130]. They suspected a break in the curve below this temperature which would lead to regime III (at about ΔT of 50°C). Because of the lack of data at the corresponding temperatures, this transition could not be confirmed. Clark and Hoffman estimated $\sigma_e = 49.9\,erg/cm^2$ and $q = 24.3$ kJ/mol of folds.

The effect of different molecular masses on the crystallization kinetics and regime analysis of sPP has been reported for a set of sPP fractions with the same syndiotacticity and molecular mass distribution but different molecular masses [130]. Linear crystal growth rates of these fractions have been measured via polarized light microscopy. At $\Delta T = 50°C$, a regime III–II transition has been clearly seen in original growth rate data [130]. An example of the regime analysis on two sPP fractions is shown in Fig. 10. Morphological changes corresponding to crystallization in a specific regime have been investigated for several polymers. For sPP no correlation between morphological changes and regime transitions was observed [130]. In electron diffraction experiments the growth direction of sPP spherulites and the lamellar morphology in the spherulites were checked. It was found that the radius direction of the spherulites is parallel to the b axis both below and above the regime transition temperature. The rectangular shaped lamellar crystals are also not changed in this temperature range.

4. Isotactic Poly(butene-1)

The dependence of linear growth rate of poly(1-butene), on temperature has been measured by optical microscopy by Monasse and Haudin [103], and Silvestre *et al.* [77]. Monasse and Haudin reported the growth rate, G, of two PB1 with different molecular mass, M_w = 174,000 and 491,000, but similar poly-

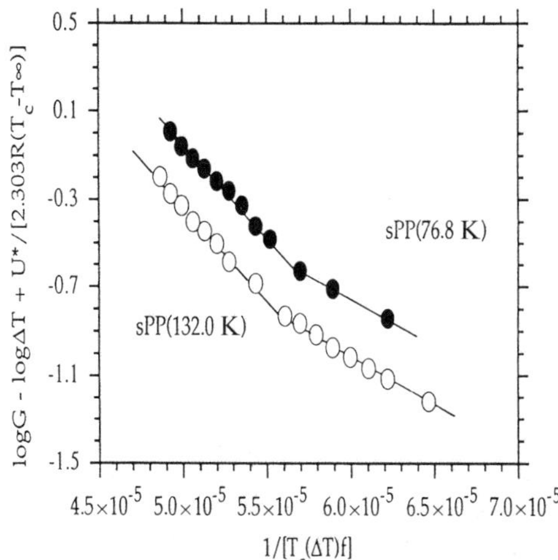

FIG. 10 Regime transition analysis of two sPP fractions crystallized from isotropic melt. (From Ref. 130.)

dispersity, see Fig. 11. Each sample was isothermally crystallized in the 40–113°C temperature range with a 1°C step between the different T_c. Both samples had a very similar temperature dependence of the growth rate, presenting a maximum at $T \cong 60°C$. The values of G were almost independent of molecular mass. The growth regime behavior was analyzed according to Hoffman theory. Plots of $\ln G + U^*/R(T_c - T_\infty)$ versus $1/T_c\Delta T$ then presented straight lines with different slopes (Fig. 10) related to the three grow regimes. The regime transition temperatures are reported on the plot and, like PE, increase with molecular mass. The regime III–II and II–I transitions of the lower molecular mass sample were located at 92 and 98°C, whereas those of the higher molecular mass sample were at 97 and 101°C, respectively.

Silvestre *et al.* investigated the growth rate of an unfractioned PB1 sample ($M_w = 185{,}000$) in a shorter range of temperature (85–120°C). They found, for a given T_c, values of the growth rate similar to those obtained by Monasse and Haudin.

5. Poly(4-methylpentene-1)

Only one publication by Patel and Bassett [131] was found in the literature on the crystal growth of P4MP1. Due to the very small dimensions of the spherulites the study was performed using electron microscopy.

Isothermal crystallization was conducted at temperatures within the range 244–210°C, corresponding to undercoolings of *ca.* 27–62°C with respect to the extrapolated melting point of 272°C. Figure 12 shows

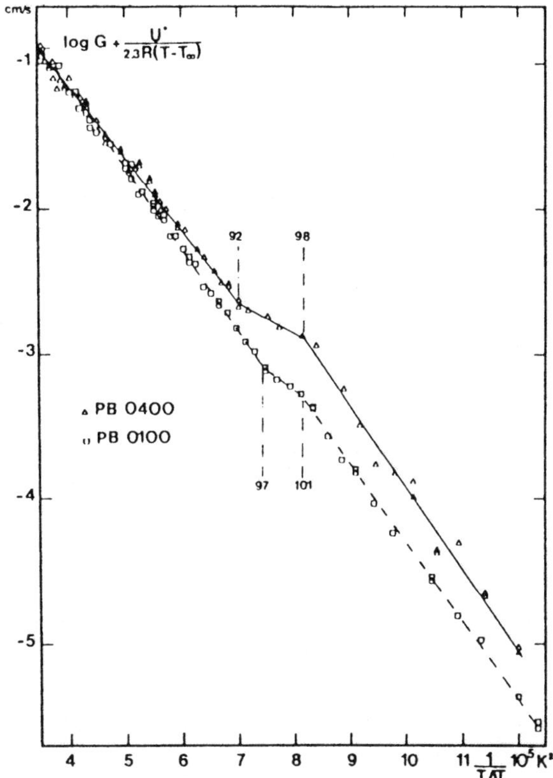

FIG. 11 $\log G + U^*/(T_c - T_\infty)$ versus $1/T_c\Delta T$ for two PB1 fractions. (From Ref. 103.)

that rates change by more than three orders of magnitude over this interval.

It is interesting to note in the plot that the data can be fitted by two straight lines. The ratio between the slopes of these two lines is about 2. This observation could be accounted by assuming that a regime II-regime I transition is present. The regime transition is located at $T \approx 235°C$.

C. Overall Crystallization

The overall crystallization rate is determined by the superposition of crystallite nucleation rate and growth rate.

Measurements of the overall crystallization rate involve often the macroscopic determination of crystallinity as a function of time. The first effort to describe quantitatively macroscopic development of crystallinity in term of nucleation and linear crystal growth was made by Kolmogoroff [132], Johnson and Mehl [133], and Avrami [134].

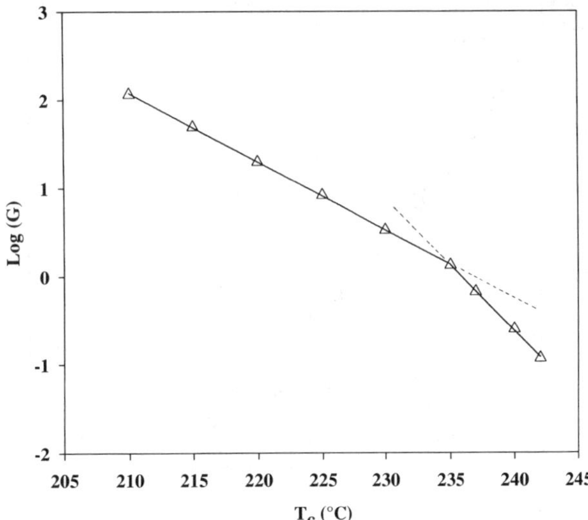

FIG. 12 Relationships of linear growth rates with respect to crystallization temperatures for P4MP1. (Data from Ref. 131.)

The overall crystallization kinetics is often described by the Avrami expression [134]:

$$1 - X(t) = \exp(-Kt^n) \tag{7}$$

where $X(t)$ is the crystallinity fraction in the crystallized material at time t. K and n are constants typical of a given crystalline morphology and type of nucleation. These constants contain information on nucleation, linear growth rate, and crystal geometry. K is the crystallization rate constant and is temperature dependent. n, an integral, is the Avrami index. The values of n for particular mechanisms of nucleation and crystal growth are reported in Table 1. Interpretation of the exponent n is not unique, and its determination may be complicated by factors such as volume changes on crystallization, incomplete crystallization, annealing, or different mechanisms involved during the process. These complications give rise to fractional values of n. In general, both K and n may be obtained from experimental data by construction of a plot of $\log[-\ln(1 - X(t))]$ versus $\log t$ from the expression

$$\log[-\ln(1 - X(t)] = \log K + n \log t \tag{8}$$

which is equivalent to Eq. (7). However, it should be noted that, in order to get a definite picture of the microscopic mechanism of crystal growth, morphological observations are needed, otherwise the Avrami treatment represents only a convenient overall data representation.

TABLE 1 Avrami Indices for Particular Mechanisms of Nucleation and Linear Growth

	Nucleation	
Linear growth	Homogeneous	Heterogeneous
Three-dimensional	4	3
Two-dimensional	3	2
Mono-dimensional	2	1

1. Polyethylene

The overall crystallization rate of PE has been extensively studied by several authors. The dependence of the overall crystallization rate in polyethylene on chain length was studied in detail [92, 135–152].

The most extensive work was carried out by Ergoz *et al.* [150] on 17 fractions of linear polyethylene covering a wide molecular mass range from 4200 to 8×10^6. For fractions up to 660,000 there is a good fit of the experimental data to the Avrami equation and a linear relation is obtained for a significant portion of the transformation. In the range of temperatures investigated, the isotherms are superimposable over the complete extent of the transformation for a given molecular mass. However, for crystallization temperatures of 130°C and above, the final portions of the isotherms are no longer superimposable with one another. As molecular mass increases, deviations from the Avrami theory occur at lower degree of transformation. The crystallinity reached at the crystallization temperature and after cooling to room temperature decreases with molecular mass. The Avrami exponents are independent of temperature and have an integral value, depending on molecular mass. For the lowest molecular mass ($M_n \leq 5800$) the slope is 4. As molecular mass increases, the slope is 3, and for higher molecular mass (1.2×10^6) the slope is 2.

Similar results have been found upon analysis of the crystallization kinetics by differential scanning calorimetry [151, 152] of polyethylene fractions encompassing a very wide range of molecular masses, from 2900 to 8×10^6. This is the widest range in molecular masses to have been reported for the crystallization kinetics of this polymer. The influence of molecular mass on regime formation has been explored.

Figure 13 reports $\tau_{0.01}$, the time required for 1% of the absolute amount of crystallinity to develop as a function of molecular mass. In the lower molecular mass range, crystallization time decreases over several orders of magnitude as molecular mass is increased. It

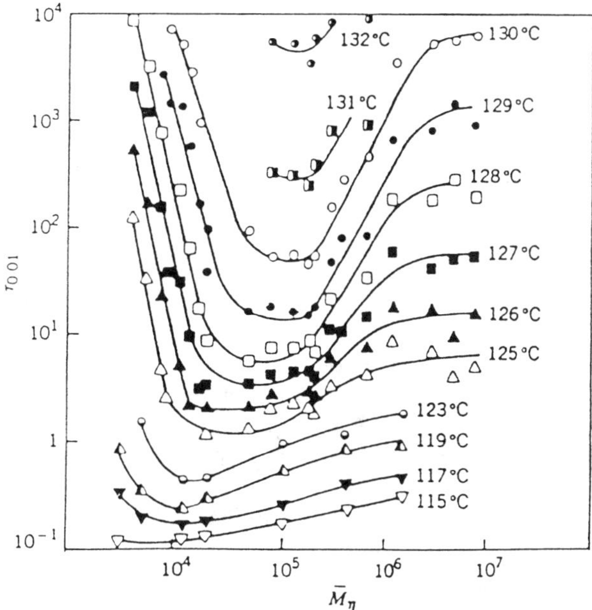

FIG. 13 Double logarithmic plot $\tau_{0.01}$ versus molecular mass for indicated crystallization temperatures. (From Ref. 108.)

was observed as a minimum in the timescale, a maximum in the crystallization rate. The molecular mass at which the crystallization rate achieves its maximum value depends on crystallization temperature. The maximum is shifted to lower molecular masses as the crystallization temperature is decreased. This relation is, however, clearly dependent on the crystallization temperature for molecular masses exceeding that for the maximum crystallization rate. Although the change of $\tau_{0.01}$ with molecular mass is small at the lower crystallization temperatures, this dependence becomes very steep at higher crystallization temperatures and change between the extremes.

By using the overall crystallization data the regime transition was analyzed. The data for high molecular mass ($M \geq 3 \times 10^6$) are represented, in the plots of ln $(\tau_{0.025})^{-1}$ against the temperature function by a single straight line over the whole range of T_c. The slope is the same as that of the lower molecular masses at higher T_c. For these fractions, crystallization can be assigned to regime I.

As T_c is lowered, the transition to regime II occurs with a change in the slope. The slopes for these fractions increase in this regime as the molecular mass increases, and this means that the ratio of slopes from regime I to II varies with molecular mass.

As T_c is further lowered, for all fractions except for those with $M \geq 3 \times 10^6$, there is another sharp break in the slope, indicating crystallization according to regime III.

2. Isotactic Polypropylene

Several authors studied the overall crystallization of isotactic polypropylene. In particular the dependence of $\tau_{1/2}$ on T_c, molecular mass, and content of isotacticity is well documented [42, 113, 153–159].

The measured $\tau_{1/2}$ values of many polypropylene samples of high isotacticity are shown in Fig. 14 as a function of T_c. It is seen that $\tau_{1/2}$ increases very rapidly with T_c, whereas at any given T_c, no statistically significant correlation with molecular mass is found. The independence of $\tau_{1/2}$ with molecular mass agrees with the fact that nucleation density and spherulite radial growth rate were found to have no identifiable dependence on molecular mass.

Martuscelli and coworkers studied the effects of isotactic fractions, f_{iso}, and catalyst type on the overall crystallization of iPP [154, 155]. For a given crystallization temperature and catalyst type, $\tau_{1/2}$ increased with increasing f_{iso}. However, the effect of the choice of the catalyst was strong enough to contravene the effect of even a large change in f_{iso}. The dependence of $\tau_{1/2}$ on catalyst type was probably due to a combination of two effects: the different impurity levels left in the resin with the different catalysts, that acted as heterogeneous nucleation sites, and the different distribution

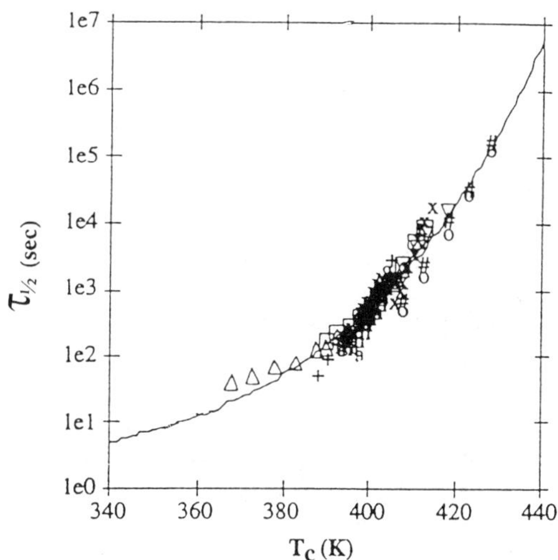

FIG. 14 Half-time as function of crystallization temperature for iPP. Data assembled by Bicerano from a variety of sources. (From Ref. 39.)

of stereosequences of repeat units at a fixed value of f_{iso}.

The crystallization kinetics of iPP has been extensively examined via the Avrami analysis, as reported by Janimak [111] and Hieber [160].

Exponents between $n = 3$ and $n = 4$ were found in most of the experiments.

The dependence of the Avrami index on the isotacticity was studied by Janimak *et al.* [111]. They found for these fractions that the Avrami exponents are between 1.8–3.2 over the crystallization temperature range studied. A tendency of increase in the values of n with decreasing undercooling appears. At constant undercooling, on the other hand, this value decreases with isotacticity.

3. Syndiotactic Polypropylene

The overall crystallization kinetics of syndiotactic polypropylene (sPP) has been studied by Balbontin *et al.* [161] in fractions of sPP with varying syndiotacticities and molecular masses, in the temperature range between 90–136°C. In this temperature range, for a given sample, $\tau_{1/2}$ increased with crystallization temperature. At constant undercooling, moreover, $\tau_{1/2}$ decreased with decreasing stereotacticity. The isothermal kinetic data of sPP were analyzed according to the Avrami equation. The n values were substantially invariant with the crystallization temperature. In particular they reported values for the Avrami exponent n between 1.8–3.7, and for the kinetic constant K between 0.16×10^{-3} and 4.77×10^{-3}, depending on percentage of racemic dyads and molecular mass.

Rodriguez-Arnold *et al.* [131, 162] reported the study of overall crystallization kinetics on a series of sPP with different molecular masses, but same syndiotacticity, in the temperature range 94.1–118.8°C. Since the sPP fraction used by the authors had a similar syndiotacticity, the overall crystallization rate difference among these fractions truly reflected the molecular mass dependence. From the low molecular mass samples, the crystallization decreased with increasing molecular mass up to about 50 K. Further increasing the molcular mass led to little change in the overall crystallization rate. A discontinuity of the slope was observed in the kinetic study based on a conventional nucleation theory plot and was associated with a regime III to regime II transition. The discontinuity temperature corresponded to a constant undercooling of about 50°C for all the fractions. The ratio between the two slopes was close to 2 as would be expected for a regime transition. The Avrami treatment led to values of n ranging from 1.9 to slightly over 3.

4. Isotactic Poly(butene-1)

The crystallization kinetics of an unfractionated sample of isotactic poly(butene-1), PB1, have been studied from the melt and from dilute solutions in amyl acetate by the dilatometric method by Sastry and Patel [163]. The isothermal crystallization was studied in the range of temperature between 95–105°C. The half-time of crystallization increased rapidly with T_c by passing from 125 min ($T_c = 95$°C) to 3160 min ($T_c = 105$°C). The kinetics of bulk crystallization followed the Avrami equation for most of the transformations with a deviation towards the end of the crystallization process. The Avrami exponent was found to be temperature dependent with the value of $n \approx 4$ at high undercooling (indicating a homogeneous nucleation process) and $n \approx 3$ at lower undercooling (indicating a heterogeneous nucleation process). The temperature coefficients of the rate constants indicated a nucleation controlled process of crystallization. In the case of solution crystallization, when amyl acetate was used as a diluent, the crystallization was found to follow the Avrami formulation only at a undercooling, whereas at the other temperatures it was applicable only to part of the transformations. Cortazar and Guzman [164] studied the overall isothermal crystallization and the crystal growth of isotactic poly (butene-1), PB1, by calorimetry and microscopy for a wide range of molecular masses, from 96,000 to 964,000.

The half-time of crystallization, $\tau_{1/2}$, increased with T_c for all the molecular masses studied. For a given T_c, $\tau_{1/2}$ decreased with the molecular mass, see Fig. 15.

The overall crystallization of PB1 was analyzed by means of the Avrami equation by several authors. The linear relationship between $\log(-\ln(1 - Xt)$ versus $\log t$ showed that the results were compatible with the Avrami kinetic model. Cortaraz and Guzman found a value of $n = 2$ for all the samples independently of crystallization temperature and molecular mass. For unfractionated PB1 Danusso [165] and Gordon and Hiller [166] found $n = 3$, whereas Silvestre *et al.* found values lying between 2 and 3.

The different values of n, due to the different nucleation mechanism and shapes of the crystals, were caused by the different crystallization conditions, molecular masses and molecular mass distributions.

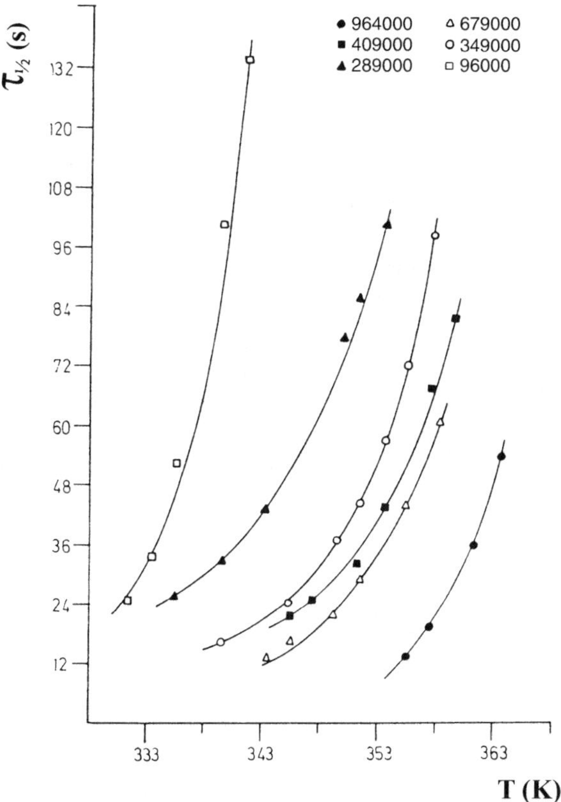

FIG. 15 Half-time of crystallization against crystallization temperatures for PB1 fractions. (From Ref. 83.)

5. Poly(4-methylypentene-1)

The study of the bulk crystallization behavior of P4MP1 has not received much attention [167]. However, a few studies describe the crystallization kinetics of PMP1 from the melt in the 214–236°C temperature range [168, 169]. Griffith and Ranby [168] used dilatometry to study the crystallization rates of P4MP1 from the melt. An Avrami equation [134] with exponent $n = 4$ seemed to describe the crystallization events rather well. However, Yadav *et al.* [169] reported a value of 3 for the Avrami exponent, whereas Silvestre and coworkers [85] calculated $n = 2$.

Figure 16 is a plot of the half-time of crystallization as a function of the degree of undercooling ΔT. A value of 247°C is used for the melting temperature T_m in this case.

III. NONISOTHERMAL CRYSTALLIZATION

The kinetic of crystallization in polymers has been studied mainly with regard to isothermal processes. In

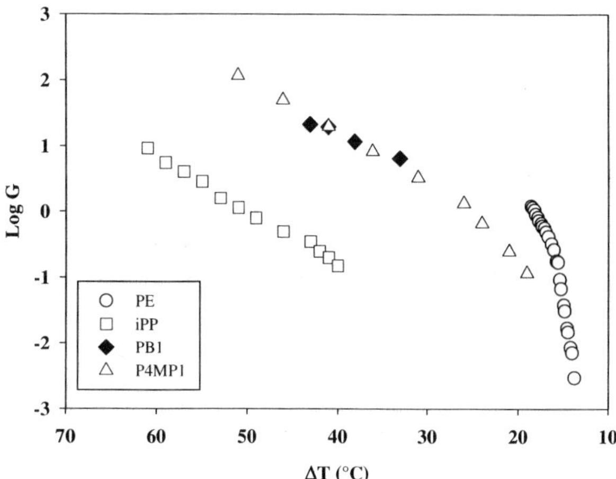

FIG. 16 Relationships of linear growth rates as a function of undercooling, ΔT, for PE, iPP, PB1 and P4MP1. (Data from Refs. 9, 33, 77, 131.)

isothermal conditions, the theoretical analysis is relatively easy and problems relative to cooling rate and thermal gradients within the specimens are avoided. The treatment of dynamic crystallization data is relatively more complex, but is of much greater interest, since industrial processes proceed under nonisothermal conditions. Moreover, from a scientific point of view, the study of crystallization in nonisothermal conditions may expand the general understanding of the crystallization behavior of polymers, since the isothermal methods are often restricted to narrow temperature ranges.

To study kinetic parameters of nonisothermal crystallization processes, several methods have been developed and the majority of the proposed formulations are based on the Avrami equation, which was developed for isothermal crystallization conditions.

Ziabicki proposed to analyze nonisothermal processes as a sequence of isothermal steps [170–172]. The proposed equation is a series expansion of the Avrami equation. In quasi-static conditions, provided that nucleation and growth of the crystals are governed by thermal mechanisms only, that their time dependence comes from a change in external conditions, and that the Avrami exponent is constant throughout the whole process, the nonisothermal crystallization kinetics can be expressed in terms of an observable half-time of crystallization, $\tau_{1/2}$, a function of time, and of the external conditions applied. The following equation was derived for the dependence of the total volume of the growing crystal, $E(t)$, with time:

$$E(t) = \ln 2\left(\int_0^t \frac{ds}{\tau_{1/2}}\right)^n\left[1 + a_1\int_0^t \frac{ds}{\tau_{1/2}} + a_2\left(\int_0^t \frac{ds}{\tau_{1/2}}\right)^2 + a_3\ldots\right] \tag{9}$$

where $a_1, a_2, \ldots, a_n$ are the coefficients of the series and s is the time required for the nucleation of the crystals. When $a_1 = a_2 = \ldots = a_n = 0$, Eq. (9) becomes:

$$E(t) = \ln 2\left(\int_0^t \frac{ds}{\tau_{1/2}}\right)^n \tag{10}$$

This equation can be used when the ratio between the nucleation rate and the growth rate is constant with time and, therefore, the athermal nucleation is negligible (isokinetic approximation). The drawback of Ziabicki's theory is that it can be applied only in the range of temperatures where isothermal crystallization data are available.

Another method was proposed by Nakamura and coworkers [173, 174] who, assuming isokinetic conditions, derived the following equation:

$$X(t) = 1 - \exp\left[-\left(\int_0^t K'(T)d\tau\right)^n\right] \tag{11}$$

where n is the Avrami index determined from isothermal crystallization data (see Section II.C) and K' is related to the Avrami constant K through the following equation:

$$K' = K^{1/n} = (\ln 2)^{1/n}(1/\tau_{1/2}) \tag{12}$$

where $\tau_{1/2}$ is the isothermal half-time of crystallization.

Therefore, the crystallinity–time relation in nonisothermal processes can be predicted from the isothermal crystallization data.

Patel and coworkers [175] also suggested a differential form for the Nakamura model:

$$\frac{dX(t)}{dt} = nK'(T)[1 - X(T)]\left\{\ln\left[\frac{1}{1 - X(t)}\right]\right\} \tag{13}$$

The principal drawback of the Nakamura model is that it does not consider the effect of the induction time. Chan and Isayev [176] proposed to calculate nonisothermal induction times, t_I from isothermal ones according to the equation:

$$t_I = \int_0^{t_{ni}} \frac{dt}{t_i(T)} \tag{14}$$

where $t_i(T)$ is the isothermal induction time as a function of temperature. When the value of the dimensionless induction time index t_I reaches unity, the upper limit of integration is taken as the nonisothermal induction time t_{ni}.

Another widely used method is the one proposed by Ozawa [177], which can be used when crystallization occurs at a constant cooling rate and the nuclei grow as spherulites. According to Ozawa's theory, the degree of conversion at temperature T, $X(T)$, can be calculated as:

$$-\ln[1 - X(T)] = \frac{K^*(T)}{\chi^n} \tag{15}$$

where χ is the cooling rate, n is the Avrami exponent, and K^* is the cooling crystallization function. K^* is related to the overall crystallization rate and indicates how fast crystallization occurs. The Avrami exponent calculated with the Ozawa method does not have the same physical meaning as in isothermal crystallization, since under nonisothermal conditions, the temperature changes continuously and both the spherulite growth rate and nuclei formations are temperature dependent. From Eq. (15) it follows:

$$\log\{-\ln[1 - X(T)]\} = \log K^*(T) - n\log\chi \tag{16}$$

By plotting the left term of Eq. (16) versus χ, a straight line should be obtained and the kinetic parameters K^* and n can be derived from the slope and the intercept, respectively.

The most interesting feature of Ozawa's method is the possibility to compare the continuous cooling results with the calculations obtained by means of the Avrami equation in isothermal conditions. However, this treatment requires values of relative crystallinity at a given temperature for different cooling rates, so it is not possible to include a wide range of cooling rates. The temperature range over which the analysis can be applied is therefore narrow. Moreover, the assumption of constant cooling rates may cause problems in modeling the development of crystallization in polymer processing.

In deriving Eq. (15), Ozawa ignored the secondary crystallization and the dependence of the fold length on temperature. However, López and Wilkens [178] argued that, during the cooling, the secondary crystallization would be virtually absent since the temperature is lowered continuously. The dependence of the fold length on temperature may also be neglected, since in most cases it appears to have no effect on the kinetic parameters [178].

As will be discussed later, the Ozawa method can be applied successfully for the analysis of dynamic solidification of some polymers, but for others it fails. In order to apply the Ozawa method, the $X(T)$ chosen at a given temperature include the values selected from the earliest stages of crystallization at high cooling rates and the values from the end stage at lower rates. At high conversion, the crystallization rate may be lowered by factors like spherulite impingement and secondary crystallization, and the values may not be comparable with those obtained at early stages of conversion, when nucleation may be the rate controlling step.

Another approach used in the literature is the use of the mere Avrami analysis also on data obtained from nonisothermal measurements [179, 180]. By plotting $\log\{-\ln[1 - X(t)]\}$ against $\log t$ for each cooling rate, one single line can be obtained. It must be taken into account that, in this treatment, the values of n and K do not have the same physical meaning as in isothermal processes, because under nonisothermal conditions the temperature changes continuously. In this case, K and n are two adjustable parameters to be fitted to the data.

Dietz [181] noted that none of the above relations account for the effects of secondary crystallization. He proposed the following equation to account for slower secondary crystallization:

$$\frac{\mathrm{d}X}{\mathrm{d}t} = nX(T)(1 - X)t^{(n-1)} \exp\left(-\frac{\gamma X}{1 - X}\right) \tag{17}$$

where the parameter γ lies between zero and one.

Phillips and Månson [182] described the nonisothermal crystallization as a linear combination of homogeneous and heterogeneous nucleation, including the growth process, by using two modified Avrami equations. The predicted evolution of the absolute crystallinity showed good agreement with experimental values obtained for a wide range of cooling rate.

Lim and coworkers [183] proposed the following modification of the Hoffman and Lauritzen Eq. (8) to measure spherulite radial growth rate, as a function of T and cooling rate:

$$\ln G + \frac{U^*}{R(T_b - \chi t - T_\infty)} = \ln G_0 - \frac{K_g}{(T_c - \chi t)[T_m^0 - (T_b - \chi t)]f} \tag{18}$$

where T_b is the temperature at which the first measurable value is recorded, and χ is the cooling rate. In this model, G_0 is the only adjustable parameter and accounts for the factors that affect the transport of macromolecular chains towards the growing site that are not temperature dependent.

To calculate the main parameters of nonisothermal crystallization and to compare the crystallization rate of different polymer systems, further methods were reported by several other authors. Khanna [184] introduced a "crystallization rate coefficient", CRC, defined as the change in cooling rate required to bring a 1°C change in the undercooling of the polymer melt. The CRC can be measured from the slope of the plot of cooling rate vs crystallization peak temperature and can be used as a guide for ranking the polymers on a single scale of crystallization rates. The CRC values are higher for faster crystallizing systems. A similar method was proposed by Zhang *et al.* [185]: crystallization rates of different polymers can be compared by plotting the reciprocal of the nonisothermal half-crystallization time divided by the cooling rate against the cooling rate. The resulting value was named "crystallization rate parameter" (CRP). Other procedures were also suggested by Jeziorny [186], Piorkovska [187], Ding and Spruiell [188], Cazé *et al.* [189], and Chuah *et al.* [190].

A. Polyethylene

The nonisothermal crystallization process of polyethylene has been investigated by several authors. Supaphol and Spruiell [191] analyzed the nonisothermal bulk crystallization kinetics of HDPE by a modified light-depolarizing microscopy technique, using cooling rates up to 2500°C/min. The crystallization kinetic was investigated by Avrami analysis applied to nonisothermal conditions. An Avrami exponent near 3 was found, suggesting spherical growth geometry and heterogeneous nucleation at predetermined sites. The analysis of the half-time of crystallization, based on Hoffman and Lauritzen secondary nucleation theory, showed a regime III–II transition at about 119°C, which corresponded to an undercooling of approximately 22°C [191], in agreement with the predictions obtained in isothermal conditions [91, 107].

Nakamura *et al.* [173, 174] studied nonisothermal kinetics of HDPE by detecting the changes in crystallinity by X-ray scattering and measuring the integrated intensity of the 2θ range in which the (110) reflection is located. The experimental data were analyzed by using the theoretical treatment discussed in the previous section, and the agreement with the theoretical values was satisfactory.

Eder and Wlochowicz [192] crystallized PE at constant cooling rates ranging from 0.5 to 10°C/min. Their experimental data did not conform to the theoretical treatment developed by Ozawa [177]. The authors attributed the deviation from the equation to factors such as secondary crystallization (for polyethylene it may be greater than 40% of the total [140]), dependence of the lamellar thickness on crystallization temperature, and occurrence of different mechanisms of nucleation. However, it is worth commenting that the occurrence of different kinds of nucleation would not affect the validity of Ozawa's equation, but only the value of the Avrami exponent.

Harnisch and Muschick [193] used the Avrami equation to determine the Avrami exponents for PE, low density PE and iPP. However, the values they found ($n = 2.9$ for PE, $n = 1.3$ for low density PE, and $n = 2.2$ for iPP) are not in good agreement with those reported for isothermal crystallization.

Mandelkern and coworkers [194] studied the morphology of linear and branched polyethylenes crystallized under controlled nonisothermal conditions. They proved that various morphological forms could develop by varying molecular mass, concentration of branch groups, and quenching temperature. A review of the supermolecular structures of polyethylene developed in nonisothermal crystallization conditions and the related morphological maps has been presented in a previous chapter of this handbook by Silvestre *et al.* [4].

B. Isotactic Polypropylene

Monasse and Haudin [195–197] studied the thermal dependence of nucleation and growth rate in isotactic polypropylene by performing nonisothermal crystallizations with differential scanning calorimetry. The experimental data, analyzed with the Ozawa theory, showed a transition between heterogeneous and homogeneous nucleation at about 122°C. From 107–116°C, the Avrami exponent, n, was close to 4, which corresponded to homogeneous nucleation and three-dimensional growth of the crystallites, whereas in the 122–129°C temperature range, n close to 3 was found. This value was attributed to heterogeneous nucleation and three-dimensional growth. Between 117–121°C, n varied between 4 and 3.

In the temperature range of heterogeneous nucleation, the surface free energies σ and σ_e were also calculated and the reported values, $\sigma = 9.2 \times 10^{-3}$ erg/cm^2 and $\sigma_e = 144$ erg/cm^2, agree quite well with those obtained with isothermal measurements by the same authors [196, 197].

Lim and Lloyd [198, 199] and Eder and Wlochowicz [192] analyzed the nonisothermal crystallization process of iPP with the Ozawa method. The calculated value for the Avrami exponent, close to 3, indicated that the crystallization proceeded through heterogeneous nucleation and three-dimensional growth of the crystallites and that at lower crystallization temperatures, homogeneous nucleation occurred. The values of the Avrami exponent as a function of temperature, calculated by Eder and Wlochowicz [192], are shown in Table 2. The analysis of the nonisothermal crystallization process with the Ziabicki theory, performed by Lim and Lloyd [198, 199], also indicated a three-dimensional growth of the crystallites with instantaneous nucleation.

Burfield and coworkers [200] studied the dependence of the thermal properties of polypropylene on tacticity and catalyst system used. The samples tacticity ranged from 0.35 to 1.00 *mm*, where *mm* is the triad tacticity. Crystallization studies at constant cooling rate showed that the crystallization onset temperature, T_b, decreased with reduced isotacticity and that, at a given tacticity, the catalyst used had a high influence on crystallization: samples prepared with a conventional catalyst ($TiCl_3/AlEt_2Cl$/diglyme) had $T_b = 119 \pm 2$°C, whereas samples synthesized with a supported catalyst ($TiCl_4/MgCl_2/AlEt_3$) showed $T_b = 111 \pm 1$°C, as reported in Table 3. Differences in crystallization behavior were suggested to be related to variation in the distribution of stereochemical defects: although the samples could have the same overall isotacticity, the nonisotactic units could be dis-

TABLE 2 Avrami Exponent, n, as a Function of Temperature for Isotactic Polypropylene. (Data from Ref. 192.)

Temperature (K)	n	Temperature (K)	n
406	3.29	396	2.55
405	3.13	395	2.57
404	2.96	394	2.66
403	2.97	393	2.59
402	2.90	392	2.71
401	2.77	391	2.74
400	2.74	390	2.78
399	2.60	389	2.80
398	2.55	388	2.82
397	2.60	387	2.80

tributed differently, leading to variation in crystallizability. Supported catalysts seemed to be characterized by a greater uniformity in active sites, thus producing a more uniform distribution of stereodefects in the polymers. Moreover, variation of molecular mass, M_n, in the range from 3.4×10^4 to 3.4×10^5, had no influence on the crystallization behavior, as shown in Table 3. Crystallization parameters for unfractioned and fractionated polypropylene samples are also presented in Table 4.

A correlation between isotacticity and enthalpy of crystallization (ΔH_c) was also found [200], with ΔH_c varying from about 25 cal/g for highly isotactic samples to 2 cal/g for samples with $mm = 0.35$.

A similar study was conducted by Paukkeri and Lehtinen [201], who proved that for molecular masses, M_w, ranging from 2.2×10^4 to 9.5×10^5, isotacticity was the main parameter determining crystallization peak temperature and crystallinity, whereas the molecular mass had a much less pronounced effect on the crystallization rate, and almost no influence on crystallinity. Moreover, in contrast with the findings of Burfield *et al.* [200], very small differences between the samples produced with different processes and catalysts were reported.

Also Lim and Lloyd [198, 199] studied the overall nonisothermal crystallization kinetics for nucleated iPP, by using adipic acid as a nucleating agent. The crystallization peak temperatures of nucleated samples were higher than those of nonfilled samples, indicating that adipic acid is a good nucleating agent for iPP.

TABLE 3 Influence of Catalysts and Molecular Mass on Crystallization Behavior of Polypropylene. (Data from Ref. 200.)

Sample	mm^a	M_n^b	T_b (°C)^c
Conventional Catalyst[c]			
Sample 1	0.95	142	116
Sample 2	0.95	140	117
Sample 3	0.95	93	121
Sample 4	0.95	34	120
Supported catalyst[d]			
Sample 1	0.94	338	112
Sample 2	0.93	162	112
Sample 3	0.94	108	109
Sample 4	0.92	89	111
Sample 5	0.94	65	112

[a]Measured by NMR.
[b]Determined by GPC.
[c]$TiCl_3/AlEt_2Cl$/Diglyme.
[d]$TiCl_4/MgCl_2/EB/AlEt_3$.

TABLE 4 Crystallization Parameters for a Fractionated Polypropylene Sample. (Data from Ref. 200.)

Sample	ω^*	*mm* (NMR)	T_b (°C)	ΔH_c (cal/g)
Unfractionated	1	0.84	114	−17.1
Heptane-insoluble fraction	0.32	1.00	120	−24.5
Heptane-soluble fraction	0.68	0.77	102	−13.5

*Mass fraction of the whole sample.

A different additive (dibenzylidene sorbitol, DBS), suggested by Feng *et al.* [202], could also satisfactorily increase the crystallization rate of iPP. The analysis of σ_e with the Hoffman and Lauritzen theory showed that σ_e decreased with the addition of DBS, and that the crystallization temperature for nucleated iPP was higher than that of the nonnucleated samples. The addition of DBS also reduced the Avrami exponent, suggesting a change in crystallite morphology.

The dynamic crystallization process of iPP in presence of different nucleating substances was also studied by several other authors, like Zhang and coworkers [185], Cazé and coworkers [189], and Bogoeva-Gaceva *et al.* [203].

C. Isotactic Poly(butene-1)

The nonisothermal crystallization process of isotactic poly(butene-1) as a function of cooling rate was investigated by Silvestre *et al.* [77]. The samples were melted at 160°C for ten minutes and cooled to room temperature at different rates: 0.5, 1, 2 and 4°C/min. The induction time and the time required for the completion of the phase transition were found to decrease with cooling rate. The calculated CRC parameter was $18\,h^{-1}$.

The experimental data were analyzed with the Ozawa and Ziabicki theories. The Ozawa equation was satisfactorily used to describe the dynamic solidification of PB1. The value of the Avrami exponent, calculated with the Ozawa method, was close to 3, as shown in Table 5, in quite good agreement with the value obtained in isothermal conditions (see Section II.C.4). Conversely, the use of Ziabicki theory was not in good agreement with the experimental results: it was found that the zero-order approximation did not describe the nonisothermal crystallization process of PB1, probably indicating that athermal nucleation is not negligible.

TABLE 5 Avrami Exponent, n, as a Function of Temperature for Isotactic Poly(butene-1). (Data from Ref. 204.)

Temperature (K)	n
86	3.4
85	3.4
84	3.3
83	3.6
82	3.6
81	3.5
80	3.2
79	3.2
78	3.2

TABLE 6 Basic Parameters for the Nucleation and Crystallization of Polyethylene

T_m^0	141 [205], 142–145°C [9]
Δh_m	280 J/cm^3, 4.11 kJ/mol [1]
Regime II–I transition temperature	127–128°C [9]
Regime III–II transition temperature	119°C [191], 118°C [91]
σ_e	90 erg/cm^2 [9]
q	20.5 kJ/mol [9]
σ	11.2 erg/cm^2 [9]

IV. CONCLUDING REMARKS AND FUTURE DEVELOPMENTS

To conclude this work, some tables, (namely Tables 6–10) and two figures (16 [p.237] and 17) are reported in order to supply basic parameters for nucleation and crystallization rate of PE, iPP, sPP, PB1 and P4MP1 and to compare the crystallization process of the four polyolefins object of this review. It is also interesting to underline the possible dependences of the parameters related to the crystallization process on the bulkiness of the side group. In fact, each of these polymers has a carbon–carbon backbone, but different side group. A bulkier side group on the backbone makes the chain stiffer and should give steric hindrance to chain folding. The rate of crystallization, hence, should decrease with the bulkiness of the side group.

Figures 16 [p.237] and 17 illustrate respectively the crystal growth rates and the crystallization half-time as a function of undercooling for the four polyolefins. Analyzing the figures, it is not possible to draw any general conclusion on the dependences of the crystallization rate on the bulkiness of the polyolefin side group. Other variables, like molecular mass, crystallization condition, melting, and glass transition temperatures, as well as viscosity of the melt, are involved and should be taken into account for defining the crystallization process of these important polymeric materials.

TABLE 7 Basic Parameters for the Nucleation and Crystallization of Isotactic Polypropylene

α form	
T_m^0	187 [206], 208°C [38, 207]
Δh_m	8.7 kJ/mol [208], 209 J/g [209]
Regime II-I transition temperature	148 [99], 155°C [38, 210]
Regime III–II transition temperature	137°C [11]
σ_e	62.5 [11], 72 [11], 122 [38], 230 [117, 120] erg/cm^2
q	27.6 kJ/mol [209]
σ	9.2–11.5 × 10 erg/cm^2 [38, 211]
β form	
T_m^0	176°C [65]
Δh_m	177 J/cm^3 [65]
Regime III–II transition temperature	123–129°C [65]
σ_e	48–55 erg/cm^2 [65]

TABLE 8 Basic Parameters for the Nucleation and Crystallization of Syndiotactic Polypropylene

T_m^0	151–161°C [130, 212, 213]
Δh_m	45–75 J/g [130, 214, 215], 197 J/g [212]
Regime III–II transition temperature	111°C [130]
σ_e	49.9 erg/cm^2 [11]
q	24.3 kJ/mol of folds [11]

TABLE 9 Basic Parameters for the Nucleation and Crystallization of Isotactic Poly(1-butene)

I form	
T_m^0	135 [216], 138 [217, 218], 141°C [219]
Δh_m	6.1 [215], 6.6, 7.0, 7.2 [220] kJ/mol 7.0 kJ/mol [218]
Entropy of fusion	15 J/mol °C [215]
II form	
T_m^0	124 [215], 128 [216], 130°C [218]
Δh_m	3.3 [221], 4.1 [219], 6.3 kJ/mol [215]
Entropy of fusion	15.9 J/mol K [215]
Regime II–I transition temperature	98–101°C [103]
Regime III–II transition temperature	92–97°C [103]
III form	
T_m^0	106.5 [13], 109°C [222]
Δh_m	6.5 kJ/mol [215]
Entropy of fusion 17 J/mol °C [215]	

TABLE 10 Basic Parameters for the Nucleation and Crystallization of Isotactic Poly(4-methylpentene-1)

T_m^0	250°C [217], 261°C [85], 272°C [131]
Δh_m	9.96 kJ/mol [223]
Regime II–I transition temperature	235°C

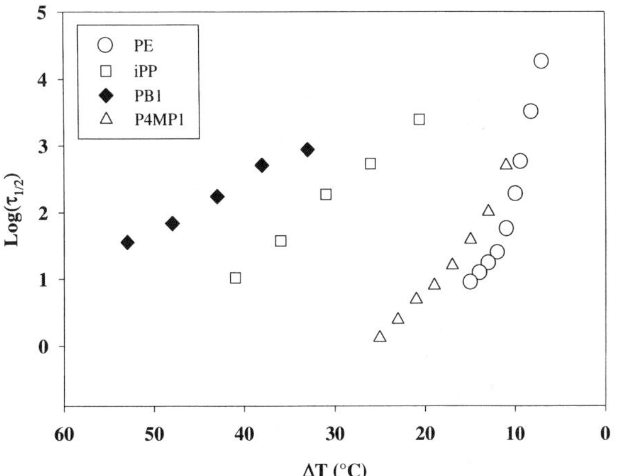

FIG. 17 Crystallization half-time as a function of undercooling, ΔT, for PE, iPP, PB1 and P4MP1. (Data from Refs. 77, 167.)

LIST OF ABBREVIATIONS

a	Width of macromolecular chain
b	Thickness of a macromolecular layer
b_0	Molecular thickness
f	Correction term ($= 2T_c/(T_m^0 + T_c)$)
f_{iso}	Isotactic fraction
g	Rate of completion of the layers
h	Planck constant
h_t	Height of mercury in a dilatometric column
i	Rate of attachment of first segment
iPP	Isotactic polypropylene
k	Boltzmann constant
l	Lamella thickness
mm	Triad tacticity
n	Avrami exponent
n_z	Number of chain units associated with M_z
q	Work of chain folding
s	Time required for the nucleation of the crystals
sPP	Syndiotactic polypropylene
t	Time
$t_i(T)$	Isothermal induction time
t_I	Nonisothermal induction time index
t_{ni}	Nonisothermal induction time
C	Constant
CRC	Crystallization rate coefficient
CRP	Crystallization rate parameter
DBS	Dibenzylidene sorbitol
$E(t)$	Total volume of growing crystals
G	Radial crystallization rate
G_0	Preexponential factor in growth rate
HDPE	High density polyethylene
I^*	Rate of nucleation
K	Avrami constant
Kg	Energy required for the formation of a nucleus of critical size
K'	Avrami constant, calculated with Nakamura method
K^*	Avrami constant, calculated with Ozawa method
L	Length of substrate
LDPE	Low density polyethylene
M	Molecular mass
M_n	Number average molecular mass
M_w	Weight average molecular mass
M_z	z-average molecular mass
N	Number of crystallizable elements
P4MP1	Isotactic poly(4-methylpentene-1)
PB1	Isotactic poly(1-butene)
PE	Polyethylene
R_{av}	Average spherulite radius
T	Temperature
T_b	Onset temperature of crystallization
T_g	Glass transition temperature
T_m^0	Equilibrium melting temperature
T_m	Nonequilibrium melting temperature
T_c	Crystallization temperature
T_∞	$T_g - C$
U^*	Energy required for the transport of crystallizable molecules in the melt
X	Crystallinity fraction
α	$\ln G_0 - (Kg/T_c\Delta Tf)$
γ	Constant of Dietz equation
ν	Number of segments attached on a substrate
ρ	Density of nucleation
σ	Lateral surface free enthalpy
σ_e	Specific free enthalpy of folding
$\tau_{0.01}$	Time required for 1% of the absolute crystallinity to develop
$\tau_{0.25}$	Time required for 25% of the absolute crystallinity to develop
$\tau_{1/2}$	Half-time of crystallization
χ	Constant cooling rate
Δh_m	Specific enthalpy of fusion
ΔE	Activation energy for chain transport
ΔF	Energy of crystallization of the volume element *abl*
ΔT	Undercooling ($= T_m^0 - T_c$)
$\Delta\sigma$	Specific free enthalpy at the interface nucleus-foreign surface
$\Delta\phi$	Specific free enthalpy of fusion
$\Delta\Phi$	Free enthalpy of fusion
$\Delta\Phi^*$	Energy of formation of a nucleus of critical size

REFERENCES

1. L Mandelkern. Crystallization and melting. In: Comprehensive Polymer Science, vol. 2. Oxford: Pergamon Press, 1987.
2. A Keller, G Goldbeck-Wood. Polymer crystallization: fundamentals of structure and crystal growth of flexible chains. In: Comprehensive Polymer Science, vol. 7. Oxford: Pergamon Press, 1987.
3. B Wunderlich. Macromolecular Physics, vol. 2, Crystal Nucleation, Growth, Annealing. New York: Academic Press, 1976.
4. C Silvestre, S Cimmino, E Di Pace. Morphology of Polyolefins. In: Handbook of Polyolefins: Second Edition, Revised and Expanded. Cornelai Vasile Editor, Marcel Dekker Inc, New York, same volume.
5. AC Zettiemoyer. Nucleation. New York: M Dekker, 1969.

6. FP Price. In: AC Zettiemoyer, ed. Nucleation in Polymer Crystallization. New York: M. Dekker, 1969.
7. D Turnbull, JC Fisher. J Chem Phys 17: 71–73, 1949.
8. JD Hoffman, GT Davis, JI Lauritzen Jr. The rate of crystallization of linear polymers with chain folding. In: NB Hannay, ed. Treatise on Solid State Chemistry. New York: Plenum Press, 1976, vol. III, Ch. 7, pp 497–614.
9. JD Hoffman. Polymer 23: 656–670, 1982.
10. JD Hoffman, RL Miller. Macromolecules 21: 3038–3051, 1988.
11. EJ Clark, JD Hoffman. Macromolecules 17: 878–885, 1984.
12. B Wunderlich, A Metha. J Polym Sci, Polym Phys Ed 12: 255–263, 1974.
13. FL Binsbergen, BGM deLange. Polymer 11: 309–332, 1970.
14. FL Binsbergen. Kolloid Z Z Polym 237: 289–297, 1970.
15. FL Binsbergen. Polymer 11: 253–267, 1970.
16. FL Binsbergen. Kolloid Z Z Polym 238: 389–395, 1970.
17. FL Binsbergen. J Cryst Growth 16: 249–258, 1972.
18. FL Binsbergen. J Polym Sci, Polym Phys Ed 11: 117–135, 1973.
19. A Galeski, E Piorkowska. J Polym Sci, Polym Phys Ed 21: 1299–1312, 1983.
20. A Galeski, E Piorkowska. J Polym Sci 21: 1313–1322, 1983.
21. GEW Schultze, R Willers. J Polym Sci, Polym Phys 25: 1311–1324, 1987.
22. RS Stein, MB Rhodes. J Appl Plys 31: 1873–1884, 1960.
23. S Clough, JJ van Aartsen, RS Stein. J Apply Phys 36: 3072–3085, 1965.
24. JV Champion, A Killey, GH Meeten. J Polym Sci, Polym Phys Ed 23: 1467–1476, 1985.
25. JH Magill. Nature 191: 1092–1093, 1961.
26. JH Magill. Polymer 3: 35–42, 1962.
27. T Pakula, A Galeski, M Kryszewski, E Piorkowska. Polym Bull 1: 275–279, 1979.
28. A Galeski, E Piorkowska. Polym Bull 2: 1–6, 1980.
29. B Wunderlich, A Metha. J Mater Sci 5: 248, 1970.
30. B Vonnegut. J Colloid Sci 3: 563, 1948.
31. RL Cormia, FP Prince, D Turnbull. J Chem Phys 37: 1333, 1962.
32. A Turner-Jones, JM Aizlewood, DR Beckett. Makrom Chem 75: 134, 1964.
33. J Varga. J Mat Sci 27: 2557–2579, 1992.
34. JR Burns, D Turnbull. J Appl Phys 37: 4021–4026, 1966.
35. JA Koutsky, AG Walton, E Baer. J Appl Phys 38: 1832–1839, 1967.
36. A Galeski. In: J Karger-Kocsis, ed. Polypropylene: Structure, Blends and Composites 1, Structure and Morphology. London: Chapman and Hall, 1995.
37. J Keith, HD Padden. J Appl Phys 30: 1479–1484, 1959.
38. B Monasse, JM Haudin. Colloid Poly Sci 263: 822–831, 1985.
39. J Bicerano. JMS Rev Macromol Chem Phys C38(3): 391–479, 1998.
40. Z Bartczak, E Martuscelli, A Galeski. In: J Karger-Kocsis, ed. Polypropylene: Structure, Blends and Composites, 2. London: Chapman and Hall, 1995, chap. 2.
41. Z Bartczak, A Galeski. Polymer 31: 2027–2038, 1990.
42. A Galeski, Z Bartczak, M Pracella. Polymer 25: 1323–1326, 1984.
43. Z Bartczak, A Galeski, M Pracella. Polymer 27: 537–543, 1986.
44. KD Pae, JA Sauer. J Appl Polym Sci 12: 1901–1919, 1968.
45. JA Sauer, KD Pae. J Appl Polym Sci 12: 1921–1938, 1968.
46. DC Bassett, RH Olley. Polymer 25: 935–943, 1984.
47. DC Bassett. In: DC Bassett, ed. Developments in Crystalline Polymers 2. London: Elsevier, 1988, p 102.
48. RH Olley, DC Bassett. Polymer 30: 399–409, 1989.
49. HN Beck, HD Ledbetter. J Appl Polym Sci 9: 2131–2142, 1965.
50. HN Beck. J Appl Polym Sci 10: 673–685, 1967.
51. VA Kargin, TT Soqolova, NYa Rapoport, II Kurbanova. J Polym Sci Part C 16: 1609–1617, 1967.
52. J Menczel, J Varga. J Thermal Anal 28: 161–174, 1983.
53. T Kowalewski, A Galeski. J Appl Polym Sci 32: 2919–2934, 1989.
54. J Varga, F Tòth. Makromol Chem, Macromol Symposia 5: 213–223, 1986.
55. J Varga. J Therm Anal 35: 1891–1912, 1989.
56. J Varga, G Garzò. Angew Makrom Chem 180: 15–24, 1990.
57. J Varga, Y Fujiwara, A Ille. Periodica Polytechnica, Chem Eng 34: 255–271, 1990.
58. J Varga, G Garzò. Acta Chimica Hungarica 128: 303–317, 1991.
59. J Varga, FS Tòth. Angew Makrom Chem 188: 11–25, 1991.
60. GY Shi, JY Zhang. Kexue Tongbao 27: 290–294, 1982.
61. J Varga, G Garzò, A Ille. Angew Makromol Chem 142: 171–181, 1986.
62. AJ Lovinger, JO Chua, CC Gryte. J Polym Sci, Polym Phys Ed 15: 641–656, 1977.
63. HJ Leugering. Makromol Chem 109: 204–216, 1967.
64. J Garbarczyk, D Paukszta. Colloid Polym Sci 263: 985–990, 1985.
65. GY Shi, XD Zhang, ZX Qiu. Makromol Chem 193: 583–591, 1992.
66. J Garbarczyk. Makromol Chem 186: 2145–2151, 1985.
67. P Forgacs, BP Tolochko, MA Sheromov. Polym Bull 6: 127–133, 1981.

68. A Turner Jones, JM Aizlewood, DR Beckett. Makrom Chem 75: 134–158, 1964.
69. Y Fujiwara. Colloid Polym Sci 253: 273–282, 1975.
70. W Ullmann, JH Wendorff. Progr Colloid Polym Sci 66: 25–33, 1979.
71. KH Moss, B Tilger. Angew Makrom Chem 94: 213–225, 1981.
72. GY Shi, B Huang, JY Zhang. Makrom Chem Rapid Comm 5: 573–578, 1984.
73. P Jacoby, BH Bersted, WJ Kissel, CE Smith. J Polym Sci, Polym Phys Ed 24: 461–491, 1986.
74. H Dragaun, H Hubeny, H Muschik. J Polym Sci, Polym Phys Ed 15: 1779–1799, 1977.
75. HJ Leugering, G Kirsch. Angew Makrom Chem 33: 17–23, 1973.
76. JR Burns, D Turnbull. J Polym Sci Part A-2, 6: 773–782, 1968.
77. C Silvestre, S Cimmino, ML Di Lorenzo. J Appl Polym Sci 71: 1677–1690, 1999.
78. Belg Pat 695,803 (Jan 9, 1967), C Geacintov (to Mobil Oil Corp).
79. US Pat 3,756,997 (Sept 4, 1973), U Eichers, O Hahmann, H Meyer, K Rombush, M Rossbach (to Chemische Werke Huels).
80. US Pat 4,320,209 (Mar 16, 1982). AM Chatterjee (to Shell Oil Co).
81. US Pat 4,321,334 (Mar 23, 1982). AM Chatterjee (to Shell Oil Co).
82. US Pat 4,322,503 (Mar 30, 1982). AM Chatterjee (to Shell Oil Co).
83. M Cortazar, GM Guzman. Polym Bull 5: 635, 1981.
84. US Pat 3,408,341 (Oct 29, 1968). SB Joyner, GO Cash Jr (to Eastman Kodak Co).
85. C Silvestre, S Cimmino, E Di Pace, M Monaco. J Macromol Sci A35: 1507–1525, 1998.
86. JD Hoffman, GT Davis, JI Lauritzen. Treatise on Solid State Chemistry, vol. 3. New York: Plenum Press, 1976, Chap 7.
87. L Mandelkern. Crystallization of Polymers. New York: McGraw-Hill, 1964.
88. JD Hoffman. Polymer 24: 3, 1983.
89. JC Fatou, C Marco, L Mandelkern. Polymer 31: 1685, 1990.
90. PJ Phillips, WS Lambert. Macromolecules 23: 2075, 1990.
91. JD Hoffman, LT Frolen, GS Ross, JI Lauritzen. J Res Nat Bur Stand (USA), Sec A 79: 671, 1975.
92. E Ergoz, JG Fatou, L Mandelkern. Macromolecules 5: 147, 1972.
93. R Alamo, JG Fatou, J Guzman. Polymer 23: 379, 1982.
94. RC Allen, PhD Dissertation, Virginia Polytechnic Institute, 1981.
95. SZD Cheng, JJ Janimak, A Zhang, HN Cheng. Macromolecules 23: 298, 1990.
96. PJ Phillips, N Vatansever. Macromolecules 20: 2138, 1987.
97. RC Allen, L Mandelkern. Macromolecules 5: 147, 1972.
98. S Lazcano, JG Fatou, C Marco, A Bello. Polymer 29: 2076, 1988.
99. RC Allen, L Mandelkern. Polym Bull 17: 473, 1987.
100. SZD Cheng, JJ Janimak, A Zhang, HN Cheng. Macromolecules 23: 298–303, 1990.
101. EJ Clark, JD Hoffman. Macromolecules 17: 878, 1984.
102. Z Pelzbauer, A Galeski. J Polym Sci C38: 23, 1972.
103. B Monasse, JM Haudin. Makromol Chem, Makromol Symp 20/21: 295, 1988.
104. AJ Lovinger, DD Davis, EJ Padden. Polymer 26: 1595, 1985.
105. JJ Labaig, PhD Thesis, Faculty of Science, University of Strasbourg, 1978.
106. A Toda. Colloid Polym Sci 270: 667, 1992.
107. PH Geil, FR Anderson, B Wunderlich, T Arakawa. J Polym Sci A2: 3707, 1964.
108. JG Fatou. Morphology and Crystallization in Polyolefins. In: C Vasile and RB Seymour, eds. Handbook of Polyolefins: Synthesis and Properties. New York: Marcel Dekker, 1993.
109. H Janeschitz-Kriegl, E Fleischmann, W Geymayer. In: I Karger-Kocsis, ed. Polypropylene: Structure, Blends, and Composites, 1, Structure and Morphology. London: Chapman and Hall, 1995.
110. M Avella, E Martuscelli, M Pracella. J Therm Anal 28: 237–248, 1983.
111. JJ Janimak, SZD Cheng, PA Giusti, ET Hsieh. Macromolecules 24: 2253–2260, 1991.
112. AO Ibhadon. PhD Thesis, University of Birmingham, 1989.
113. A Wlochowicz, M Eder. Polymer 22: 1285–1287, 1981.
114. B von Falkai, HA Stuart. Kolloid Z 162: 138, 1959.
115. HD Keith, FJ Padden. J Appl Pys 25: 1286, 1964.
116. L Goldfarb. Makromol Chem 179: 2297, 1978.
117. E Martuscelli, C Silvestre, G Abate. Polymer 23: 229–237, 1982.
118. AO Ibhadon. J Appl Polym Sci 71: 579–584, 1999.
119. EJ Clark, JD Hoffman. Macromolecules 17: 878–885, 1984.
120. C Silvestre, S Cimmino, E D'Alma, ML Di Lorenzo, E Di Pace. Polymer 40: 5119–5128, 1999.
121. KH Moos, BJ Jungnickel. Angew Makromol Chem 132: 135–160, 1985.
122. G Shi, B Huang, J Zhang, Y Cao. Sci Sin Series B 30: 225–233, 1987.
123. J Varga, G Garzó, A Ille. Angew Makrom Chem 190: 907–913, 1986.
124. J Varga, G Garzó. Acta Chimica Hungarica 128: 303–317, 1991.
125. AJ Lovinger, JO Chua, CC Gryte. J Polym Sci, Polym Phys Ed 15: 641–656, 1977.
126. J Varga. Angew Makrom Chem 104: 79–87, 1982.

127. J Varga, Y Fujiwara, A Ille. Period Polytech Chem Engng 34: 256–271, 1990.
128. B Lotz, B Fillon, A Therry, JC Wittman. Polym Bull 25: 101–105, 1991.
129. RL Miller, EG Seeley. J Polym Sci, Polym Phys Ed 20: 2297, 1982.
130. J Rodriguez-Arnold, Z Bu, SZD Cheng. JMS Rev Macromol Chem Phys C35(1): 117–154, 1995.
131. D Patel, DC Bassett. Proc R Soc A 445: 577–595, 1994.
132. AN Kolmogonoff. On the statistics of crystallization process in metals. Isv Akad Nauk SSSR Ser Math 1: 335, 1937.
133. WA Johnson, RT Mehl. Reaction kinetics in processes of nucleation and growth, Trans AIME 135: 416, 1939.
134. M Avrami. J Chem Phys 7: 1103, 1939; 8: 212, 1940; 9: 177, 1941.
135. B Buckser, LH Tung. J Phys Chem 63: 763, 1959.
136. S Matsouka. J Polym Sci 42: 511, 1960.
137. FP Price. J Phys Chem 64: 169, 1960.
138. F Rybnikar. J Polym Sci 44: 517, 1960.
139. F Rybnikar. J Polym Sci, Part A1: 2031, 1963.
140. WM Banks, RJ Gordon, A Sharples. Polymer 4: 61, 1963.
141. H Heber. J Polym Sci, Part A2: 1291, 1964.
142. HD Keith, FJ Padden. J Appl Phys 35: 1286, 1964.
143. PH Lindenmeyer, VF Holland. J Appl Phys 35: 55, 1964.
144. M Gordon, IH Hillier. Trans Faraday Soc 60: 763, 1964.
145. M Gordon, IH Hillier. Phil Mag 11: 31, 1965.
146. Yu K Godovsky, B Barsky. Vysokomol Soed 8: 395, 1966.
147. Yu K Godovsky, B Barsky. Polym Sci USSR 8: 431, 1966.
148. JL Stafford. J Polym Sci Symp 42: 837, 1973.
149. V Era, E Venäläinen. J Polym Sci Symp 42: 879, 1973.
150. E Ergoz, JG Fatou, L Mandelkern. Macromolecules 5: 147, 1972.
151. JG Fatou, C Marco, L Mandelkern. Polymer 31: 1685, 1990.
152. JG Fatou, C Marco, L Mandelkern. Polymer 31: 890, 1990.
153. J Menczel, J Varga. J Therm Anal 28: 161–174, 1983.
154. M Avella, E Martuscelli, M Pracella. J Therm Anal 28: 237–248, 1983.
155. E Martuscelli, M Pracella, L Crispino. Polymer 24: 693–699, 1983.
156. J Varga, J Menczel, A Solti. Polym Sci USSR 26: 2763–2773, 1984.
157. B von Falkai. Makromol Chem 41: 86–109, 1960.
158. E Martuscelli, M Pracella, G Della Volpe, P Greco. Makromol Chem 185: 1041–1061, 1984.
159. Yu K Godovsky, GL Slonimsky. J Polym Sci, Polym Phys Ed 12: 1053–1080, 1974.
160. CA Hieber. Polymer 36: 1455–1467, 1995.
161. G Balbontin, D Dainelli, M Galimberti, G Paganetto. Makromol Chem 193: 693, 1992.
162. J Rodriguez-Arnold, A Zhang and SZD Cheng. Polymer 35: 1884–1894, 1994.
163. KS Sastry, RD Patel. Eur Polym J 8: 63–74, 1972.
164. M Cortazar, GM Guzman. Makromol Chem 183: 721–729, 1982.
165. D Danusso, G Gianotti. Makromol Chem 88: 149, 1965.
166. M Gordon, IH Hiller. Polymer 6: 213, 1965.
167. LC Lopez, GL Wilkes. JMS Rev Macromol Chem Phys, C32: 301–406, 1992.
168. JH Griffith, BG Ranby. J Polym Sci 44: 369–381, 1960.
169. YS Yadav, PC Jain, VS Nanda. Thermochim Acta 71: 313, 1983.
170. A Ziabicki. Coll Polym Sci 252: 433, 1974.
171. A Ziabicki. J Chem Phys 48: 4368, 1968.
172. A Ziabicki. Atti della Scuola-Convegno su Cristallizzazione dei Polimeri, Gargnano, 1979.
173. K Nakamura, K Katayama, T Amano. J Appl Polym Sci 17: 1031–1041, 1973.
174. K Nakamura, T Watanabe, K Katayama, T Amano. J Appl Polym Sci 16: 1077–1091, 1972.
175. RM Patel, JH Bheda, JE Spruiell. J Appl Polym Sci 42: 1671, 1991.
176. TW Chan, AI Isayev. Polym Eng Sci 34: 461–471, 1994.
177. T Ozawa. Polymer 12: 150, 1971.
178. LC Lόpez, GL Wilkens. Polymer 30: 882–887, 1989.
179. CR Herrero, JL Acosta. Polym J 26: 786–796, 1994.
180. P Cebe. Polym Comp 9: 271, 1988.
181. W Dietz. Coll Polym Sci 259: 413, 1981.
182. R Phillips, JAE Månson. J Polym Sci Part B Polym Phys 35: 875–888, 1997.
183. BA Lim, KS McGuire, DR Lloyd. Polym Eng Sci 33: 537, 1993.
184. P Khanna. Polym Eng Sci 30: 1615–1619, 1990.
185. R Zhang, H Zheng, X Lou, D Ma. J Appl Polym Sci 51: 51–56, 1994.
186. A Jeziorny. Polymer 19: 1141–1144, 1978.
187. E Piorkovska. Coll Polym Sci 275: 1046–1059, 1997.
188. Z Ding, JE Spruiell. J Polym Sci Part B Polym Phys 35: 1077–1093, 1997.
189. C Cazé, E Devaux, A Crespy, JP Cavrot. Polymer 38: 497–502, 1997.
190. KP Chuah, SN Gan, KK Chee. Polymer 40: 253–259, 1998.
191. P Supaphol, JE Spruiell. J Pólym Sci: Part B: Polym Phys 36: 681–692, 1998.
192. M Eder, A Wlochowicz. Polymer 24: 1593–1595, 1983.

193. K Harnisch, H Muschick. Coll Polym Sci 261: 908–913, 1983.
194. L Mandelkern, M Glotin, RA Benson. Macromolecules 14: 22–34, 1981.
195. B Monasse, JM Haudin. Coll Polym Sci 264: 117–122, 1986.
196. B Monasse. Doct Ing Thesis, Ecole Nationale Supérieure des Mines de Paris, Paris, France, 1982.
197. B Monasse, JM Haudin. 29th International Symposium on Macromolecules, IUPAC Macro 83, Bucharest, 1983, p 146.
198. GBA Lim, DR Lloyd. Polym Eng Sci 33: 529–536, 1993.
199. GBA Lim, KS McGuire, DR Lloyd. Polym Eng Sci 33: 537–542, 1993.
200. DR Burfield, PST Loi, Y Doi, J Mejzík. J Appl Polym Sci 41: 1095–1114, 1990.
201. R Paukkeri, A Lehtinen. Polymer 34: 4075–4082, 1993.
202. Y Feng, X Jin, JN Hay. J Appl Polym Sci 69: 2089–2095, 1998.
203. G Bogoeva-Gaceva, A Janevski, A Grozdanov. J Appl Polym Sci 67: 395–404, 1998.
204. C Silvestre, S Cimmino, ML Di Lorenzo. Unpublished data.
205. B Wunderlich, G Czornyj. Macromolecules 10: 906, 1977.
206. U Gaur, B Wunderlich. J Phys Chem Ref Data 10: 1051, 1981.
207. JG Fatou. Eur Polym J 7: 1057–1064, 1971.
208. HS Bu, SZD Cheng, B Wunderlich. Makromol Chem, Rapid Commun 9: 75, 1988.
209. WR Krigbaum, I Uematsu. J Polym Sci, Part A 3: 767–776, 1965.
210. JI Lauritzen. J Appl Phys 44: 4353–4359, 1973.
211. JD Hoffman. Polymer 23: 656–670, 1982.
212. J Boor Jr, EA Youngman. Polym Lett 3: 577–580, 1965.
213. S Haftka, H Könnecke. J Macromol Sci, Phys B30: 319–334, 1991.
214. J Boor Jr, EA Youngman. J Polym Sci, Part A1 4: 1861–1884, 1966.
215. SL Aggarwal. Physical Constant of Poly(propylene). In: J Bandrup, EH Immergut, W McDowell, eds. Polymer Handbook, 2nd edn. New York: Wiley, 1975, pp V23–28.
216. F Danusso, G Gianotti. Makromol Chem 61: 139, 1963.
217. J Powers, JD Hoffman, JJ Weeks, FA Quinn Jr. J Res Nat Bur Stand 69A: 335, 1965.
218. U Gaur, BB Wunderlich, B Wunderlich. J Phys Chem Ref Data 12: 29, 1983.
219. F Danusso, G Gianotti. Makromol Chem 80: 1, 1964.
220. H Wilski, T Greewr. J Polym Sci C6: 33, 1964.
221. ID Rubin. J Polym Sci B2: 747, 1964.
222. HW Holden. J Polym Sci C6: 209, 1964.
223. B Wunderlich. Macrmolecular Physics, vol. 3, Crystal Melting. New York: Academic Press, 1980.

10

Estimation of Properties of Olefin Monomers and Polymers

Charles H. Fisher
Roanoke College, Salem, Virginia

I. OLEFIN MONOMERS

A. Introduction

Olefins played the principal role in the unprecedented, explosive burst of creativity, invention, and successful development that occurred in the latter half of the 20th century and gave the world a new class of polymers, a family of new plastics, a family of new synthetic rubbers a new class of catalysts (Ziegler–Natta coordination catalysts), and a new nomenclature (isotactic, syndiotactic, and atactic) [1]. This period of discovery also spawned two Nobel prizes, thousands of patents, thousands of scientific papers, not to mention dozens of lawsuits [1].

Olefins (C_nH_{2n}), or alkenes, have two fewer hydrogens than their corresponding alkanes (C_nH_{2n+2}). The "olefin" related to the corresponding alkane, methane (CH_4), is the unstable methylene radical ($-CH_2-$), which has been made by pyrolyzing diazomethane [2]. Pyrolysis of diazomethane under suitable conditions gives polymethylene or linear polyethylene, $(CH_2)_n$ [3, 4].

Ethylene (ethene, $CH_2{=}CH_2$), the first stable member of the alpha- or 1-olefin family, was discovered in 1795 by four Dutch chemists, Deimann, Van Troostwick, Bondt, and Louwrenburgh [2, 3]. Additional olefins, including propylene, isobutylene, and pentene, were prepared and studied in the 19th century [3, 4].

The number of olefin structural isomers increases greatly with increase in molecular weight. The number of such isomers for the lower olefins are: C_4, 3; C_5, 5; C_6, 13; C_7, 27; and C_8, 68 [2]. All the aliphatic mono olefins, however, have the same percentages of carbon and hydrogen: 85.63% carbon and 14.37% hydrogen.

The physical properties of the olefins resemble those of the corresponding alkanes. The chemical reactivity of the olefins, however, is much greater than the alkane's reactivity. Not only do the olefins react readily with a wide variety of reagents but they react also with each other to form polymers.

Because of their chemical reactivity, olefins generally are not found in natural products. However, they are formed by the thermal decomposition of almost all carbon compounds [2]. The thermal cracking of cottonseed oil, for example, gives about 30% of unsaturated compound [2]. The total amount of by-product olefins obtained in the cracking process of making gasoline is enormous [2].

About 130 years after the 1795 discovery of ethylene [2, 3], the lower olefins became important intermediates for manufacturing many industrial chemicals, including glycol, glycol mono- and di-esters, glycol mono- and di-ethers, glycol ether esters, polyethylene glycols, ethanolamines, and morpholine [2, 5].

Markets for olefins increased dramatically when the polyolefins were developed and became tremendously important. Today, some 200 years after the discovery of ethylene, olefins are produced in large quantities because of their value as intermediates in

TABLE 1 1-Olefins: Freezing Point, Normal Boiling Point, Refractive Index (n_D^{20}), Density (d_4^{20}), and Viscosity (n)*

C	F.p., °C	B.p., °C	n_D^{20}	d_4^{20}	n^{20}	n^{100}
5	−165.2	29.97	1.3715	0.6405		
6	−139.8	63.49	1.3879	0.6732	0.39	
7	−119.0	93.64	1.3998	0.6970	0.50	
8	−101.7	121.3	1.4087	0.7149	0.656	0.363
9	−81.4	146.9	1.4157	0.7292	0.851	0.427
10	−66.3	170.6	1.4215	0.7408	1.091	0.502
12	−35.2	213.4	1.4300	0.7584	1.72	0.678
14	−12.9	251.0	1.4364	0.7713	2.61	0.894
16	4.12	284.4	1.4412	0.7811	3.83	1.152
18	17.6	314.2	1.4450	0.7888	5.47	1.46
20	28.6	341.2	1.4481	0.7950		1.82

* Data from [12–14].

TABLE 2 1-Olefin Properties*

C	t_b, °C	P_{mm}	E_R^C	δ^{25}	D^{25}	d^{25}
4	−6.3	2229	13722	6.67	4.905	0.5878
5	30.0	637.7	4388	7.07	5.298	0.6349
6	63.5	186.0	1402	7.31	5.575	0.6681
7	93.6	56.3	458.7	7.47	5.777	0.6923
8	121.3	17.38	151.3	7.59	5.928	0.7104
9	146.9	5.34	49.29	7.71	6.042	0.7240
10	170.6	1.63	15.88	7.84	6.139	0.7357

* Data from [15].
Normal boiling point (t_b, °C), vapor pressure at 25°C (P, mm); evaporation rate (E_R); solubility parameter, Hildebrands (δ^{25}), density (lb gal^{-1}, 25°C), and density, g/cm^{-3}.
Relative evaporation rate (Bu acetate = 100).
Multiply Hildebrands by 2.046 to get solubility parameters in S.I. units.

manufacturing a wide variety of chemicals [2, 5] and olefin polymers and copolymers [5–8].

The higher linear 1-olefins first became available in commercial quantities in 1962 when Chevron began production based on cracking C_{20} and higher waxes. Gulf was the first to produce 1-olefins, based on ethylene, when operations were started in 1966. Oligomerization, or chain growth of ethylene, is the primary source of linear 1-olefins today [5]. Olefins, only laboratory chemicals a few decades ago, are now a major industry producing many types of plastics [3–5, 9–11] and amounting to billions of pounds per year [5].

B. Structure and Chemical Properties

Olefins can be categorized as linear alpha- or 1-olefins, $(H(CH_2)_n\,CH{=}CH_2$ (Tables 1–4); branched 1-olefins, $RCH{=}CH_2$ (R, branched alkyl) (Table 5); vinylidene olefins, $R_2C{=}CH_2$; internal olefins, $RCH{=}CHR$; cyclic olefins; and vinylcycloalkanes [7].

Olefins [12–23] are chemically reactive [2, 24–26], due to the carbon–carbon double bond or olefinic linkage (lacking in the alkanes). The olefin reactions have been classified [5] as:

1. Addition reactions
   a. Electrophilic
   b. Free radical
2. Substitution reactions

More than fifty olefin chemical reactions have been described [24–26]. The 1-olefins copolymerize with many monomers, including various olefins, vinyl esters, acrylic acid, acrylic acid esters, sulfur dioxide, and carbon monoxide [3, 27–29]. The principal industrial

TABLE 3 1-Olefin Properties*

C	T_b K	T_c, K	P_c, K	V_c	γ(25°C)	n(25°C)	ε(20°C)	Sp(25°C)
2	169.4	282.3	5.04	131				
3	225.5	364.9	4.60	181				
4	266.9	419.6	4.02	240				13.6
5	303.0	464.8	3.53	293	15.45	0.195	2.01	14.5
6	336.5	504.1	3.21	348	17.90	0.252	2.08	15.0
7	366.7	537.3	2.92	402	19.80	0.340	2.09	15.3
8	394.3	566.7	2.68	464	21.28	0.45	2.11	15.5
10	443.6	616.4	2.22	584	23.54	0.756	2.14	16.0
12	486.9	657.6	1.93	680	25.15	1.20	2.15	

* Data from [16–20].
Normal boiling points (T_b, K), surface tensions (γ, 25°C), viscosity at 25°C) (mPa s), dielectric constants (ε, 20°C), solubility parameters at 25°C (S.I. units, J$^{1/2}$ cm$^{-3/2}$).

TABLE 4 1-Olefins (RC=C) and *n*-Alkanes (*n*-RH): Comparison of Properties*

	Melting point, °C			Boiling point, °C			Density, d_4^{20}		
C	*n*-RH	RC=C	Diff.	*n*-RH	RC=C	Diff.	*n*-RH	RC=C	Diff.
5	−129.7	−165.2	35.5	36.0	29.9	6.10	0.6262	0.6405	0.143
6	−95.3	−139.8	44.5	68.74	63.49	5.25	0.6594	0.6371	0.014
8	−56.8	−101.7	44.9	125.7	121.3	4.4	0.7025	0.7149	0.012
10	−29.7	−66.3	36.6	174.1	170.6	3.5	0.7301	0.7408	0.011
12	−9.6	−35.2	25.6	216.3	213.4	2.9	0.7487	0.7584	0.010
14	5.9	−12.9	18.8	253.6	251.1	2.5	0.7628	0.7713	0.009
20	36.8	28.6	8.2	342.7	341.2	1.5	0.7887	0.7950	0.006
30	65.8	62.4	3.4	449.7	448	1.7	0.8097	0.8144	0.004
40	81.5	79.8	1.7	525	523	2.0	0.8205	0.8238	0.003
60	99.2			620			0.8315		
100	115.2			715			0.8404		
∞	141.2		0.0	917†		0.0	0.854		0.0

* Data from [13, 14, 18, 21, 23, 31].
Boiling points are normal.
The C_{60} and C_{100} data are for *n*-alkanes [21, 23].
† Dr Benjamin P Huddle, Jr., personal communication, 1988.

TABLE 5 Methyl-1-Olefins and Methylalkanes: Comparison of Properties*

C		t_m, °C	t_b, °C	d_4^{20}	n_D^{20}	t_c, °C	P_c, mm	V_c
4	2-Me-1-propene	−140.4	−6.90	0.5942		144.7	29,982	4.28
4	2-Me propane	−159.6	−11.73	0.5572		135.0	27,360	4.53
5	3-Me-1-butene	−168.5	20.06	0.6272	1.3643	170	23,294	4.57
5	2-Me butane	−159.9	27.9	0.6197	1.3537	187.8	25,004	4.27
6	4-Me-1-pentene	−153.6	53.88	0.6642	1.3828	212	22,185	4.30
6	2-Me pentane	−153.7	60.27	0.6532	1.3715	224.9	22,762	4.26
7	5-Me-1-hexene		85.31	0.6920	1.3966	249	20,522	4.20
7	2-Me hexane	−118.3	90.05	0.6786	1.3849	257.9	20,672	4.27
8	6-Me-1-heptene		113.2	0.712	1.4070	280	18,887	4.17
8	2-Me heptane	−109.0	117.6	0.6979	1.3949	288	18,848	4.27

* Data from Refs. 13 and 14.
Melting point (t_m, °C), normal boiling point (t_b, °C), density (d_4^{20}), refractive index (n_D^{20}), critical temperature (t_c, °C), critical pressure (P_{mm}), and critical volume (V_c).

olefin reactions are:

1. Polymerization (and oligomerization)
2. Hydroformylation (oxo reaction)
3. Miscellaneous simple addition reactions
4. Alkylation reactions
5. Sulfations and, sulfonations
6. Oxidations

The numerous industrial and consumer products made from olefins include many polymers and copolymers, synthetic rubbers [10, 11], plasticizers, surfactants, lube oil additives, synthetic lubricants, chemicals for oil field and fiber industry applications, amines, alcohols, fatty acids, mercaptans, and miscellaneous chemicals [2, 5].

C. Physical and Thermal Properties

1. Data Sources and Equations

References [12–23, 30–39] are useful sources of olefin and alkane properties. Properties of 1-olefins having two to forty carbons were published in 1953 [14] and 1959 [13]. Many of these properties are given in Tables 1–5. *n*-Alkane densities [36, 37, 40] and melting points (C_1 to C_∞) (amorphous polymethylene) [23, 31] and

n-alkane boiling points (some estimated) to C_{100} are available [21].

Alkane data are abundant, relevant, and useful because their physical properties are virtually identical with those of the higher olefins and generally similar to those of the lower olefins (Tables 4 and 5).

The olefins, alkanes, and other homologs (e.g., RCl, ROH, RCOOH, RNH_2 RSH, etc., where R is n-alkyl) of infinite chain length (M_∞ and C_∞) have the same structure [$(CH_2)_n$] and hence the same properties (P_∞) [41]. Some limiting properties (e.g., fluidity, dipole moment, iodine number, functional groups, water solubility, and vapor pressures) are zero or approximately zero [41].

Simple correlation equations were used extensively in the present work because of their advantages, which include ease of use, space economy (much information in little space), new data estimated by interpolation or prudent extrapolation, identification of grossly inaccurate (non-fitting) data, and evaluation of data quality (accurate data give high correlation coefficients).

The olefin physical properties are proportional to the content of the olefinic group ($-CH{=}CH_2$) in the molecule. For ethylene, the olefinic group is about 100% of the molecule; for propylene about 66%; for butene, about 50%; for pentene, about 40%; decene, 20%; C_{20}, 10%; C_{50}, 4%; C_{1000}, 0.2%; and C_∞, 0%. The C_5 and higher olefins are more paraffinic, by proportion of chain length, than olefinic. Some of the olefin properties can be correlated with the degree or proportion of alkanicity [$(M$-$27)/M$] [Eqs. (1)–(3)], where M is the olefin molecular weight and 27 is the weight of the ethenyl group. According to Eqs. (1)–(3), the limiting properties (for M_∞) are: $d_4^{20} = 0.8537$; $n_{D\infty}^{20} = 1.4766$; and $t_{f\infty}, = 130°C$.

$$d_4^{20} = 0.24061 + 0.61304[(M\text{-}27)/M] \quad (1)$$

(C_{20}–C_∞; $r = 0.99975$)

$$n_D^{20} = 1.1788 + 0.29781[(M\text{-}27)/M] \quad (2)$$

(C_{16}–C_∞; $r = 0.99975$)

$$t_f, °C = -919.40 + 1049.2[(M\text{-}27)/M] \quad (3)$$

(C_{16}–C_{40}; $r = 0.999915$)

Some properties (e.g., heat capacity and heats of formation, fusion, and vaporization) are usually reported [17–22, 30] and treated as molar properties (MP). Densities are sometimes reported as molar volumes, M/P (where $M/P = M^2/MP$). Molar properties of the higher homologs have the advantage of being linear with chain length. A second advantage is that molar properties are linear with other homolog properties and property functions that are linear with chain length.

2. Estimating Properties from Homolog Chain Length

Some, but limited, effort was given in the present work to estimating monolefin properties by comparison (Tables 4 and 5) and by group contribution methods [42–44]. Methods based on homology, using olefins as homologs, received most attention.

More than 150 years ago, Kopp used the principles of homology to correlate homolog molar volumes and boiling points with chain length [38, 45]. Subsequent workers [38, 41] have used homology advantageously to correlate properties of homologs (RG, where R is a normal alkyl group and G is a functional group) with M or the number of carbons, C.

1-Olefin boiling points (starting with the ethylene boiling at $-103.8°C$) increase regularly with increasing M and C until the limiting boiling point (for M_∞ and C_∞) of approximately 900°C is reached. Melting points (starting with ethylene at $-160°C$) increase with increasing M and C until the limiting melting point (approximately 141°C) is reached. The lower liquid olefins at room temperature have 20° and 25°C densities of approximately 0.64 to 0.74; these increase with increasing M and C until the limiting values of 0.854 (20°C) and 0.851 (25°C) are reached (Table 4).

Two-parameter equations correlating many 1-olefin molar properties with chain length (boiling points, melting points, critical properties, aniline points, dielectric constants, pounds/gallon, densities, refractivities, surface tensions, solubility parameters, cohesive energies, autoignition temperatures, flash points and thermal properties) are given in Tables 6 and 7.

Molar property Eqs. (4) and (5) can be used with Eq. (6) (b' is zero for olefins) to get property Eqs. (7) and (8). Dividing Eq. (6) by Eq. (4) gives Eq. (7). Dividing Eq. (5) by Eq. (6) gives Eq. (8).

$$M/P = b + mC \quad (4)$$

$$MP = b + mC \quad (5)$$

$$M = b' + 14.027C \quad (6)$$

$$P(C + b/m) = b'/m + 14.027C/m \quad (7)$$

$$P(C + b'/14.027) = b/14.027 + mC/14.027) \quad (8)$$

3. Estimating from Published Equations

Many 1-olefin properties (molar volume, V; melting pint, t_m, °C; boiling point, T_b, K; critical properties,

TABLE 6 1-Olefins: Equations ($y = b + mC$) Correlating Molar Properties (M/P or PM) with Number of Carbons (C)*

	Carbons	y	Intercept b	Slope, m	Corr. Coeff., r	Ref.
Boiling point, K	20–40	M/T_b, K	0.2097	0.01240	$\underline{4}$, 26	13
Critical temp., K	5–12	M/T_c	0.0771	0.01499	$\underline{3}$, 74	19
Critical density	5–10	M/d_c	−2.027	58.36	$\underline{3}$, 64	19
Melting point, K	20–40	M/T_m	0.2648	0.03306	$\underline{4}$, 49	13
Aniline point, K[c]	8–17	M/T_a	0.0900	0.03497	$\underline{3}$, 84	54
Dielectric constant, 20°C	6–13	M/ε	2.384	6.3463	$\underline{4}$, 15	19
Density, lb gal^{-1}	5–10	M/D^{25}	3.5649	1.9241	$\underline{4}$, 47	15
Density, g cm^{-3}	10–40	M/d_D^{20}	25.18	16.390	$\underline{6}$, 70	13, 14
Refractivity	8–40	M/N_D^{20}	39.21	29.366	$\underline{6}$, 70	13, 14
Surface tension dyn cm^{-1}	9–18	M/γ	2.081	0.37558	$\underline{4}$, 45	35
Solubility parameter, S.I.	5–10	M/δ	0.8952	0.78886	$\underline{4}$, 21	15, 16
Cohesive energy	8–16	E_cM	−1729	4978.6	$\underline{4}$, 36	16
Autoignition temp., K	6–16	M/AT	0.01430	0.02835	$\underline{3}$, 90	5
Flashpoint, K	8–18	M/FP	0.2317	0.02061	$\underline{3}$, 30	5

* $y = b + mC$, where y is M/P or PM, b is intercept, and m is slope.
See [5, 13–16, 18, 19, 23, 25, 54].
Correlation coefficient, r, of 0.99945 given as $\underline{3}$, 45.
Critical solution temperature, K.
Refractivity, $N_D^{20} = n_D^{20} - 1$.
Solpar, J$^{1/2}$ cm$^{-3/2}$.

TABLE 7 1-Olefins: Equations ($y = b + mC$)* Correlating Molar Thermal Properties† with Carbons (C)‡

Heat	y	Carbons	Intercept, b	Slope m	Corr. coeff., r
Capacity, gas	C_p, J mol^{-1}	6–20	−4.888	22.87	$\underline{7}$, 65
Capacity	C_p, J mol^{-1}	6–13	3.354	29.81	$\underline{4}$, 52
Combustion, gas	H_cM, kJ mol^{-1}	6–20	−82.51	−658.8	$\underline{7}$, 47
Combustion, liq.	H_cM, kJ mol−1	5–13	−81.27	−653.7	$\underline{7}$, 83
Formation, gas	H_fM, K mol−1	3–19	−82.1	−20.62	−$\underline{5}$, 75
Formation, liq.	H_fM, K mol^{-1}	5–13	81.13	−25.63	−$\underline{6}$, 72
Fusion	H_mM, kJ mol^{-1}	6–10	−8.216	2.946	$\underline{3}$, 47
Vaporization	H_vM, kJ mol^{-1}	5–20	0.9062	4.950	$\underline{5}$, 40
Vaporization§	H_vM, kJ mol^{-1}	6–16	−6.569	14.28	$\underline{4}$, 10
Vaporization¶	$(H_vM)^2$, kJ mol^{-1}	6–16	−249.8	174.6	$\underline{4}$, 86
Entropy	S, kJ mol^{-1}	6–20	150.9	38.59	$\underline{7}$, 80

* $y = b + mC$, except for vaporization heats at normal boiling points.
† See [14–20, 30, 32].
‡ Properties at 25°C, except for fusion and boiling point data.
§ Equation for data at boiling point, $y = b + mC^{1/2}$.
¶ Equation for data at boiling point, $(H_vM)^2 = b + mC$.
Correlation coefficient, r, of 0.9999999 65 given as $\underline{7}$, 65.

T_c, K; P_c, and V_c; surface tension, γ; and solubility parameter, δ) can be estimated by published Eqs. (9)–(18) [46–53]:

$$(C_{10}\text{–}C_{40})[46]:\ t_m,\ ^\circ\text{C (even carbons)} = 133.1 - 2192/(C\text{-}1) \quad (9)$$

$$(C_6\text{–}C_{20})[47]:\ (T_b, \text{K})^{1/2} = 8.695 + 12.39 \log C \quad (10)$$

$$(C_5\text{–}C_{16})[48]:\ \log(1078 - T_b, \text{K}) = 3.037 - 0.05057C^{2/3} \quad (11)$$

$$(C_5\text{–}C_{16})[49]:\ T_c, \text{K} = 113.0 + 502.6 \log C \quad (12)$$

$$(C_3\text{–}C_8)[50]:\ 1/P_c = 0.01165 + 0.003335C \quad (13)$$

$$(C_2\text{–}C_8)[50]:\ V_c = 16.78 + 55.88C \quad (14)$$

$$(C_7\text{–}C_{18}[51]\text{: } (\gamma)^3 = -22{,}508 + 36{,}403 \log C \quad (15)$$

$$(C_5\text{–}C_{10})[52]\text{: } \delta(M + 21) = 67.03 + 17.89M \quad (16)$$

$$(C_5\text{–}C_{10})[52]\text{: } M/\delta = 0.8939 + 0.05624M \quad (17)$$

$$[53]\text{: } V^{25} = 57.08 + 16.484C + 10.371/(C\text{-}1) - 5.332/(C\text{-}1)^2 \quad (18)$$

Equations (19)–(22) are also useful for estimating 1-olefin properties:

$$T_b, \text{K} = 1190 - 25{,}034/(C + 23.52) \quad (19)$$

$(C_8\text{–}C_{40};\ r = -0.999984)$

$$T_m, \text{K} = 414 = 2630.2/(C + 3.364) \quad (20)$$

$(C_{14}\text{–}C_{40};\ r = -0.99979)$

$$\eta^{1/2} = -3.322 - 162.31/(C - 48.7) \quad (21)$$

$(C_8\text{–}C_{19};\ r = -0.999983)(25°\text{C})$

$$\log p(\text{kPa}) = 4.523 - 0.52007C \quad (22)$$

$(C_5\text{–}C_{12};\ r = -0.999985[18])$

4. Temperature Dependence of Properties

Because properties frequently are available for only one or a few temperatures, it is advantageous to have property–temperature equations to calculate properties at any desired temperature.

Property–temperature equations are in Tables 8 and 9. The properties and property functions that are linear with temperatures are linear also with other temperature-linear properties and property functions.

TABLE 9 1-Olefins: Equations [$\log P_{mm} = A - B/(t + C)$] Correlating Vapor Pressure (P_{mm}) with Temperature (t, °C)*

C	A	B	C	Temp. range, °C
5	6.84650	1044.9	234	−39/73
6	6.86572	1152.97	226	−12/73
7	6.90069	1257.51	219.2	10/128
8	6.93263	1353.5	212.8	0/151
9	6.95387	1435.4	205.5	25/173
10	6.96034	1501.87	197.6	25/233
12	6.97522	1619.86	182.3	112/280
20	6.859	1807.9	113	223/405
30	7.61184	2859.4	154.5	313/545
40	7.66107	3174.1	141	376/625

* Equations from [13].
More p vs. t equations are in [13] and [14].

Charts or figures are available for the following 1-olefin property correlations [5]:

1. Vapor pressure vs temperature, high pressure range
2. Vapor pressure vs temperature, low pressure range
3. Liquid density vs temperature
4. Saturated liquid viscosity versus temperature
5. Ideal gas state enthalpy versus temperature
6. Liquid thermal conductivity versus temperature.

II. OLEFIN POLYMERS

A. Introduction

Investigations of polyolefin properties began more than a century ago. Polymethylene, or linear polyethylene,

TABLE 8 1-Olefins: Equations* ($y = b + mx$) Correlating Properties with Temperature (t, °C)†

C	Temperature °C	y	x	Intercept, b	Slope, m	Corr. coeff., r
8	0/100	d	t	0.7316	0.0008337	$-\underline{6}$, 65
8	0/100	$100/V$	t	6.5193	−0.007431	$-\underline{6}$, 47
8	20/30	n_D	t	1.4185	0.0004900	$-\underline{4}$, 31
8	10/100	γ	t	23.68	0.9581	
6	0/100	T/δ	t	17.308	0.09882	$\underline{3}$, 76
6		$T/\delta^{1/2}$	t	68.766	0.31981	$\underline{4}$, 50
7	0/75	$T \log \eta$	t	−97.050	−1.7025	$-\underline{6}$, 76
7	0/75	$T/\eta^{0.2}$	t	321.20	1.9600	$\underline{4}$, 73
6	−25/75	$T \log \eta$	t	−133.25	−1.8163	$-\underline{4}$, 80
6	−25/75	$T/\eta^{0.18}$	t	323.40	1.7360	$\underline{4}$, 66

* $y = b + mx$.
† See [13, 14, 16, 18, 35].
Density, d; molar volume, V; refractive index, n_D; surface tension, γ; solubility parameter, δ; and viscosity, η.
Correlation coefficient, r, of −0.999999 65 given as $\underline{6}$, 65.

made from diazomethane, was studied by several groups during the period 1887–1938 [55]. Chemists at the Phillips Petroleum Company did similar work on diazomethane-produced polymethylene and reported the following properties: crystallinity (X-ray), 90–92%; melting point, 134–137°C; tensile strength, 3720–4140 p.s.i.; and annealed density, 0.964–0.970 [55].

Polyisobutylene and its properties were studied early in the 20th century. The industrial importance of polyisobutylene (made by the polymerization of isobutylene) increased when Thomas and Sparks cross-linked the polymer to get the elastomer, butyl rubber [10].

In the middle of the 20th century, polymers similar to polyethylene (PE) were made by hydrogenating poly(butadiene) [56]. Properties of a hydrogenated poly (butadiene) are compared with those of a molding grade polyethylene in Table 10.

The estimation of polymer properties began as early as 1944, when the properties of homologous polymers were correlated with the dimensions (molecular weight, M, or number of carbons, C) of the repeat units [23, 57]. The resulting correlations and equations could be used, by interpolation, to estimate properties.

In 1953, Sperati and coworkers [58] published equations correlating polyethylene properties with each other and with molecular weight; Eqs. (23) and (24) are examples:

$$\log(\text{melt index}) = 5.09 - 0.000153M_n \tag{23}$$

(M_n from 15,000 to 50,000)

$$\log(\text{melt viscosity}) = 0.64 + 0.0274(M_n)^{1/2} \tag{24}$$

(M_n from 1800 to 52,000)

TABLE 10 Comparison of Properties of a Hydrogenated Poly(butadiene) with Molding Grade Polyethylene [56]

	Hydrogenated poly(butadiene)	Polyethylene
Tensile strength (77–80°F), p.s.i.	2350	1900
Elongation (77–80°F), %	750	600
Stiffness modulus (77°F), p.s.i.	15,000	20,000
Hardness (durometer)	78A-2	98 A
Brittle point, °F	<-100	>-100
Refractive index (77°F)	1.50	1.51

TABLE 11 Limiting Properties (P_∞) and Amorphous Polymethylene Properties (P_{PM})*

	P_∞	P_{PM}
Melting point, T_m, K	414.3	414.6
Density, d^{20}, g cm^{-3}	0.854	0.855
Density, d^{25}, g cm^{-3}	0.851	0.8519
Refractive index, n_D^{25}	1.475	1.476–1.49
Dielectric constant (25°C)	2.18†	2.3
Surface tension, dyn cm^{-1} (20°C)	35.4	35.7
Solubility parameter (25°C)‡	17.4	17.0

* See [16, 23, 31, 41, 53, 60, 66, 72].
† Square of refractive index ($n_D^{25} = 1.475)^2$.
‡ Solpar, Mpa$^{1/2}$ or J$^{1/2}$ cm$^{-3/2}$ [16, 44].

Van Krevelen [44] and Bicerano [59] published books describing methods for estimating polymer properties; they emphasized the use of molar properties, group contributions, and connectivity indices.

The method used most in the present work to estimate amorphous polyethylene properties consisted in using oligomer (n-alkane) properties to calculate limiting properties (P_∞, properties of an infinite-length homolog); such limiting properties resemble the corresponding properties of amorphous, linear polyethylene (or amorphous polymethylene) (Table 11).

The use of n-alkane properties (with suitable equations) to estimate amorphous polyethylene properties has the advantage the n-alkanes (unlike the polyolefins) can be prepared pure, and accurate properties can be determined. The use of n-alkane data has the limitations, however, that some properties (e.g., boiling points) are irrelevant for polymers and some polymer properties (e.g., tensile strength) are lacking in the lower alkanes. This method treats the main backbone polymer chain as the functional group and the pendant alkyl groups or repeat units as the homolog chain for correlation purposes.

The method, used with the higher poly(1-olefins) consisted in correlating poly(1-olefin) properties with the dimensions (M or number of carbons) of the repeat units.

A second method of estimating poly(1-olefin) properties consisted in creating equations correlating n-alkane properties with n-alkane densities, and then using the equations with poly(1-olefin) densities to estimate poly(1-olefin) properties.

B. Data Sources and Equations

n-Alkane properties [13, 14, 17–22] and amorphous isotactic polyolefin properties from several sources

[6, 16, 23, 34, 44, 60–63] were used in developing correlations. The work was hindered in some instances by lack of data or for lack of information about crystallinity or degree of branching. Selecting the proper properties was frequently difficult because of disagreement among the published data. The selected data, however, are reasonably consistent with each other and appear suitable for correlation purposes.

Molar properties (M/P and MP, where M is molecular weight and P is property) were frequently used [Eqs. (4) and (5)] to estimate properties. The number of carbons divided by properties (C/P) is also linear (except for a few lower homologs) with chain length (M or C).

Squared refractive indexes were used as dielectric constants; this is permissible for the nonpolar *n*-alkanes [44].

Equations (4), (5), (25) and (26) are among those [46, 64–65] that can be used conveniently to estimate limiting properties (P_∞) from *n*-alkane properties, and also to correlate amorphous poly (1-olefin) properties with the molecular weight or number of carbons in the repeat units:

$$M/P = b + mC \qquad (P_\infty \text{ is } 14.027/m) \qquad (4)$$

$$MP = b + mC \qquad (P_\infty \text{ is } m/14.027) \qquad (5)$$

$$P = b + m/(C + k) \qquad (P_\infty \text{ is } b) \qquad (25)$$

$$(C + k)/P = b + mC \qquad (P_\infty \text{ is } 1/m) \qquad (26)$$

where b is intercept, m is slope, and k is an adjustable parameter.

In developing equations [such as (4), (5), (25), and (26)] it is helpful to know limiting properties (P_∞) because they are equation parameters. Several limiting properties are given in Table 11; others can be calculated from Somayajulu's methylene increments [slopes in Eqs. (4) and (5)] [67]. Some of Somayajulu's methylene increments and the limiting properties calculated from them are: formation heat (liquid), −25.627 and −1.827; vaporization heat, 4.959 and 0.3535; heat capacity (liquid), 31.733 and 2.262; molar volume (20°C), 16.428 and 0.8539; molar volume (25°C), 16.486 and 0.8508; and critical volume, 59.622 and 0.2353 [67, 68].

Some limiting properties (and polyolefin properties) are zero or approximately zero, e.g., dipole moments, vapor pressures, fluidity, iodine numbers, saponification values, acid strength, functional groups, and water solubility.

C. Limiting and Polyethylene Properties

Some of the equations developed to correlate *n*-alkane properties with homolog chain length and to estimate limiting and amorphous polyethylene properties are given below.

Melting points: Broadhurst [31] used *n*-alkane melting points and Eq. (27) to estimate the melting point of amorphous polyethylene (PE). The function $[(T_m K(C+5)]$ in Eq. (28) [derived from Eq. (27)] is linear with homolog chain length; the indicated T_m^∞ (414.3 K) agrees with the PE melting point, Table 11:

$$T_m, \text{K} = 414.3(C - 1.5)/(C + 5) \qquad (27)$$

$$T_m, \text{K}(C + 5) = -621.45 + 414.3C \qquad (28)$$

Wunderlich [23] reviewed expressions generally similar to Eq. (28), and proposed Eq. (29) for correlating polyethylene melting points with molecular weights ranging from 3200 to 1300,000 (carbons from 228 to 92,678) [23, 60]

$$T_m, \text{K} = 414.1 - 2071/C \qquad (29)$$

Meyer [69] defined PE melting points by Eq. (30), which can be rearranged to Eq. (31). The indicated PE limiting melting point is 417.5 K. The melting point of the C_{100} *n*-alkane calculated by Eqs. (30) and (31) is 389.7 K; this is similar to the published C_{100} melting points: 388.4 to 289.5 K [23]:

$$1/T_m, \text{K} = 0.002395 + 0.0171/C \qquad (30)$$

$$C/T_m, \text{K} = 0.0171 + 0.002395C \qquad (31)$$

Densities: eqs. (32) and (33) are examples of expressions that have been used [39] to estimate limiting densities (d_∞^{20}) (where C is the number of *n*-alkane carbons):

$$(C - 0.528)/d_4^{20} = 1.2712 + 1.1709C \qquad (32)$$

$$(C_5\text{–}C_{40};\ d_\infty^{20} = 1/\text{slope} = 0.8540)$$

$$d_4^{25} = 0.8510 - 1.3874/(C + 1.12) \qquad (33)$$

$$(C_9\text{–}C_{40};\ d_\infty^{25} = \text{intercept} = 0.8510)$$

Equation (34) can be used [40] to develop equations indicating d_∞^t at many temperatures:

$$d(C + 1.215) = -0.29796 + 0.86555C - (0.0025302 + 0.00057592C)t, \text{°C.} \qquad (34)$$

where t, °C is zero ($d_\infty^0 = 0.8656$):

$$d(1.215) = -0.29796 + 0.86555C \qquad (35)$$

where t, °C is 20° ($d_{\infty}^{20} = 0.8540$):

$$d(C + 1.215) = 0.34856 + 0.85403C \quad (36)$$

Refractivities: eqs. (37) and (38) have been used to estimate limiting refractivities ($N_D = n_D - 1$) from n-alkane data [39]:

$$N_D^{20} = 0.47712 - 0.73460/(C + 1.25) \quad (37)$$

$$(C_9\text{–}C_{40};\ N_{\infty}^{20} = \text{intercept} = 0.477)$$

$$N_D^{25} = 0.47517 - 0.73089/(C + 1.12) \quad (38)$$

$$(C_6\text{–}C_{40};\ N_{\infty}^{25} = \text{intercept} = 0.475)$$

The limiting values indicated by Eqs. (32)–(38) agree with those (Table 11) published by other investigators [53, 66].

Dielectric constants: molar dielectric constants (M/ε) at 20°C [18] were used to create Eq. (39). The indicated limiting value of 2.155 is lower than the square (2.182) of the limiting refractive index (1.477) at 20°C [Eq. (37)]:

$$M/\varepsilon = 6.5819 + 6.5088C \quad (39)$$

$$(C_5\text{–}C_{13};\ r = 0.999989;\ \varepsilon_{\infty}^{20} = 2.155)$$

The squared refractive indexes of the n-alkanes were used as dielectic constants (ε, 25°C) in developing Eq. (40) which correlates molar dielectric constants with chain length. The indicated limiting dielectric constant ($\varepsilon_{\infty}^{25}$) is 2.173; this value is similar to the square (2.176) of the limiting refractive index ($n_{\infty}^{25} = 1.475$) estimated by Eq. (38) but lower than the dielectric constant (2.2–2.3) reported for PE (Table 11):

$$M/\varepsilon = 7.067 + 6.4542C \quad (40)$$

$$(C_6\text{–}C_{40};\ r = 0.9999996,\ \varepsilon_{\infty}^{25} = 2.173)$$

Surface tensions: Wu [70] preferred the use of n-alkane surface tensions [35] for estimating limiting surface tensions.

Jasper's n-alkane surface tensions at 20°C [35] were used to develop Eq. (41). Equation (42), based on Rossini's surface tensions at 25°C [14], was used to estimate the limiting surface tension at 25°C; the indicated limiting surface tension is 35.69 dyn/cm:

$$\log \gamma^{20} = 1.5441 - 1.6754/C \quad (41)$$

$$(C_6\text{–}C_{26};\ r = 0.999971;\ \gamma_{\infty}^{20} = 35.00)$$

$$\gamma^{25} = 35.690 - 167.13/(C + 3.6) \quad (42)$$

$$(C_8\text{–}C_{20};\ r = -0.999900;\ \gamma_{\infty}^{25} = 35.69)$$

Solubility parameters: the Hansen–Beerbower n-alkane solpar data, reported by Barton [16], were used to develop Eq. (43), which correlates molar solpars (M/δ) with carbons. The limiting solpar indicated by Eq. (43) is 17.4 (SI units, $\text{MPa}^{1/2}$, at 25°C):

$$M/\delta = 0.93999 + 0.80449C \quad (43)$$

$$(C_5\text{–}C_{20};\ r = 0.999968;\ \delta_{\infty}^{25} = 17.4)$$

Equations (44) and (45) correlate Hoy's n-alkane solpars at 25°C [15] with molecular weights. The indicated limiting values (17.41 and 17.68) are similar to the published PE solpars [16]:

$$\delta M = -220.4 + 17.41M \quad (44)$$

$$(C_5\text{–}C_{12};\ r = 0.99988)$$

$$M/\delta = 0.9276 + 0.05655M \quad (45)$$

$$(C_5\text{–}C_{12};\ r = 0.99985)$$

Molar thermal properties: the molar thermal properties (MP) of the n-alkanes are linear with the number of carbons, C (Eq. (5) and Table 12]. The slopes, m, of the n-alkane equations (Table 12) are similar to the 1-olefin slopes (Table 7). The limiting properties, P_{∞} can be estimated from the slope: $m/14.027 = P_{\infty}$.

D. Poly(1-olefin) Properties

1. Molar Volumes and Densities

Molar volumes ($V = M/d$) are important for various reasons, including their role in molar property expressions [44]: density, $d = M/V$; refractive index, $n = R_{GD}/V + 1$, where R_{GD} is the Gladstone–Dale molar refraction; dielectric constant, $\varepsilon = (1 + Q/V)(1 - Q/V)$, where R_{LL} is molar refraction, and Q is $R_{LL}V$; cohesive energy density, E_cM/V, where E_cM is cohesive energy; solubility parameter, $\delta = F/V$, where F is molar attraction constant; surface tension, $\gamma = (P_s/V)^4$, where P_s is parachor.

Equations (46) and (47) developed with Wunderlich [23] and Van Krevelen data [44] are similar; they give calculated molar volumes that agree reasonably well with the literature values (Table 13) and indicate a limiting density (d_{∞}^{25}) of approximately 0.851. Equation (48) was obtained by dividing Eq. (47) by $M = 14.027C$:

$$V^{25} = -0.1560 + 16.476C \quad (46)$$

$$(C_2\text{–}C_5;\ r = 0.99989;\ d_{\infty}^{25} = 0.8514)$$

TABLE 12 *n*-Alkanes: Equations* ($MP = b + mC$) Correlating Molar Thermal Properties (MP) with Number of Carbons (C)†

	C	Intercept, b	Slope m	Corr. coeff, r
C_p, J mol^{-1} (g)	6–20	5.892	22.871	$\underline{7}$, 67
C_p, J mol^{-1} (l)	6–16	13.55	30.178	$\underline{4}$, 55
H_cM, kJ mol^{-1} (g)	6–20	−240.0	−658.88	$-\underline{7}$, 64
H_cM, kJ mol^{-1} (l)	6–20	−240.4	−653.79	$\underline{8}$, 70
S, J mol^{-1} (g)	6–20	154.83	38.967	$\underline{6}$, 82
H_fM, kJ mol^{-1} (g)	6–20	−43.51	−20.614	$\underline{7}$, 6
H_fM, kJ mol^{-1} (l)	5–16	−44.61	−25.652	$-\underline{4}$, 60
H_mM, kJ mol^{-1}‡	8–20	−12.64	4.1377	$\underline{4}$, 33
H_mM, kJ mol^{-1}§	9–17	−12.34	3.1215	$\underline{3}$, 67
H_vM, kJ mol^{-1}	5–17	1.725	4.9692	$\underline{4}$, 84

* $MP = b + mC$; $P_\infty = m/14.027$; g, gas; l, liquid.
† Heat capacity, C_p; combustion heat, H_cM; entropy, S; formation heat, H_fM; fusion heat H_mM; vaporization heat, H_vM.
Data from [14, 17, 18, 20].
‡ Even carbons.
§ Odd carbons.
Correlation coefficients, r, of 0.9999999 67 given as $\underline{7}$, 67.

TABLE 13 Poly(1-olefins): Molar Volumes (V)* and Densities (d^{25})†

		Wunderlich [23]			Van Krevelen [44]		
C‡	M	Lit.	(V^{25})§	(d^{25})§	Lit.	(V^{25})¶	(d^{25})‖
2	28.05	32.9	32.80	0.8552	32.8	32.78	0.8558
3	42.08	49.34	49.27	0.8541	49.5	49.26	0.8542
4	56.11	65.3	65.75	0.8534	65.2	65.74	0.8535
6	84.16		98.70	0.8527	97.9	98.70	0.8527
8	112.22		131.65	0.8524		131.7	0.8523
12	168.32		197.6	0.8520		197.60	0.8519
18	252.49		296.4	0.8518	293.6	296.5	0.8517

* V, molar volume, cm^3 mol^{-1}.
† d, density, g cm^{-3}.
‡ Repeat unit carbons.
§ $V^{25} = -0.1560 + 16.476C$.
¶ $V^{25} = -0.1800 + 16.480C$.
‖ $C/d = -0.01283 + 1.1749C$.

$$V^{25} = -0.1800 + 16.480C \tag{47}$$

$$(C_2\text{–}C_5;\ r = 0.99984;\ d_\infty^{25} = 0.8512)$$

$$C/d = -0.01283 + 1.1749C \tag{48}$$

Densities (20–300°C) of the polyethylene Marlex 6050 [37] can be estimated by Eq. (49):

$$d^{1/2} = 0.93087 - 000032096t,\ ^\circ\mathrm{C} \tag{49}$$

2. Molar Refractions and Refractivities

The Lorenz–Lorentz (R_{LL}) and Gladstone–Dale (R_{GD}) molar refractions have been shown to be useful [44] for estimating the refractive indexes (n_D) and refractivities ($N_D = n_D - 1$) of many types of polymer. The R_{LL} and R_{GD} methylene values of 4.649 and 7.831, respectively, were used to estimate polyolefin refractive indexes and refractivities [Eqs. (50)–(57)]; the R_{LL} calculated refractivities are somewhat higher than the R_{GD} values (Table 14):

TABLE 14 Poly(1-olefins): Refractive Indexes Estimated from Molar Refractions*

C†	(d^{25})‡	Q^{d}§	(η_D^{25})¶	$N_D^{25})$‖	(η_D^{25})**	(N_D^{25})††
2	0.8558	0.2836	1.4791	0.4790	1.4791	0.4778
3	0.8542	0.2831	1.4781	0.4781	1.4781	0.4769
4	0.8535	0.2829	1.4777	0.4776	1.4776	0.4765
6	0.8527	0.2826	1.4771	0.4771	1.4771	0.4761
8	0.8523	0.2825	1.4769	0.4769	1.4769	0.4758
12	0.8519	0.2824	1.4766	0.4766	1.4766	0.4756
18	0.8517	0.2823	1.4765	0.4765	1.4765	0.4755

* Lorenz–Lorentz, R_{LL}; Gladstone–Dale, R_{GD} [44].
† Repeat unit carbons.
‡ $C/d^{25} = -0.01283 + 1.1749C$.
§ $Q = R_{LL}/V^{25}(V^{25}$, molar volume).
¶ $(\eta_D^{25})^2 = (1+2Q)/(1-Q)$.
‖ $M/N_D^{25} = -0.3826 + 29.466C$.
** $\eta_D^{25} = 0.9319 + 0.63944d^{25}(R_{LL}$ data).
†† $R_{GD}/V^{25} = N_D^{25} = 0.4752C/(C - 0.0109)$.

$$\text{molar refraction, } R_{LL} = 0.0 + 4.649C \quad (50)$$

$$R_{LL}/V^{25} = Q = 0.2821C/(C - 0.0109) \quad (51)$$

$$(n_D^{25})^2 = (1+2Q)/(1-Q), \text{ where } Q = R_{LL}/V \quad (52)$$

$$(n_D^{25})^2 = [3/(1-Q)] - 2 \quad (53)$$

$$\text{molar refractivities: } M/N_D^{25} = -0.3826 + 29.466C \quad (54)$$

$$\text{molar refraction, } R_{GD} = 0.0 + 7.831C \quad (55)$$

$$R_{GD}/V = N_D^{25} = 0.4752C/(C - 0.0109) \quad (56)$$

$$C/N_D^{25} = -0.0230 + 2.1045C \quad (57)$$

3. Dielectric Constants

Refractive indexes (Table 14) can be squared to get dielectric constants, $[\varepsilon = (n_D)^2]$.

Lorentz–Lorentz molar refractions, R_{LL}, and Gladstone–Dale molar refractions, R_{GD}, can be used with Eqs. (58) and (59) to estimate dielectric constants. The estimated dielectric constants (Table 15) are lower than the reported experimental values [60]:

$$\varepsilon = [3/(1-Q)] - 2, \text{ where } Q \text{ is } R_{LL}/V \quad (58)$$

$$\varepsilon = (1+W)^2, \text{ where } W \text{ is } R_{GD}/V \quad (59)$$

Equation (60), based on n-alkane C_{10}–C_{20} data and limiting values ($d_\infty^{20} = 0.854$ and $\varepsilon_\infty = 2.182$), was used with poly(1-olefin) densities [Eq. (48)] to estimate

TABLE 15 Poly(1-olefins): Estimated Dielectric Constants (ε)

C*	V^{25}	(d^{25})†	(ε^{25})‡	(ε^{25})§	(ε^{25})¶	(ε^{25})‖
2	32.78	0.8558	2.188	2.184	2.184	2.1836
3	49.26	0.8542	2.185	2.181	2.181	2.1812
4	65.74	0.8535	2.184	2.180	2.180	2.180
6	98.70	0.8527	2.182	2.179	2.179	2.1789
8	131.66	0.8523	2.181	2.178	2.178	2.1783
12	197.58	0.8519	2.181	2.177	2.177	2.1777
18	296.46	0.8517	2.180	2.177	2.177	2.1773

* Repeat unit carbons.
† $C/d^{25} = -0.01283 + 1.1749C$.
‡ Estimated from R_{LL} molar refractions [44]: $\varepsilon = [3/(1-Q)] - 2$, where Q is R_{LL}/V.
§ Estimated from R_{GD} molar refractions [44]: $\varepsilon = (1+W)^2$, where W is R_{GD}/V.

poly(1-olefin) dielectric constants. The ε values estimated in this manner are similar to those estimated from the R_{LL} and R_{GD} molar refractions (Table 14). Dielectric constants can be estimated conveniently by Eq. (61):

$$\varepsilon = 0.8824 + 1.5204d_4^{20} \quad (60)$$

$$(C_{10}\text{–}C_{20}, C_\infty, r = 0.999944)$$

$$C/d_4^{25} = -0.01283 + 1.1749C \quad (48)$$

$$M/\varepsilon = -0.0423 + 6.4448C \quad (61)$$

Van Krevelen's dielectric constants (ε) and solubility parameters (δ) [44] for 16 different polymers (ε ranged from 1.7–4.0 and δ ranged from 11.7–28) were used to develop Eq. (62), which was used to estimate the following ε values: $\delta = 15$, $\varepsilon = 2.14$; $\delta = 16$, $\varepsilon = 2.28$; and $\delta = 17$, $\varepsilon = 2.422$:

$$\varepsilon = -0.00701 + 0.1429\delta \quad (62)$$

4. Surface Tensions

Parachors (P_s) [Eq. (63)] and molar volumes [Eq. (47)] were used to estimate surface tensions (γ). Equation (63) divided by Eq. (47) gave an expression [Eq. (64)] that was used to estimate amorphous poly(1-olefin) surface tensions (Table 16):

$$P_s = 40.00C \quad (63)$$

$$V^{25} = -0.1800 + 16.480C \quad (47)$$

$$\gamma^{1/4} = 2.4272C/(C - 0.0109) \quad (64)$$

TABLE 16 Poly(1-olefins): Estimated Surface Tensions (γ^{25})

C*	d^{25}	γ†	γ‡	γ§	γ¶
2	0.8558	35.47	35.47	35.50	35.59
3	0.8542	35.22	35.20	35.23	35.42
4	0.8535	35.09	35.09	35.10	35.35
6	0.8527	34.96	34.96	34.97	35.27
8	0.8523	34.90	34.89	34.90	35,23
12	0.8519	34.83	34.83	34.84	35.18
18	0.8517	34.79	34.79	34.80	35.16
∞	0.8511	34.71	34.68	34.71	35.40

* Repeat unit carbons.
† $\gamma^{1/4} = 2.4272\ C/(C - 0.0109)$.
‡ $\gamma^{1/4} = 28516d^{25}$.
§ $C/\gamma = -0.001278 + 0.02881C$.
¶ $\gamma^{1/2} = -1.422 + 8.6321d_4^{20}$.

Surface tensions can be estimated from polyolefin densities, d^{25} [from Eq. (48)] and Eq. (66):

$$C/d^{25} = -0.01283 + 1.1749C \quad (48)$$

$$\gamma = (2.8516d^{25})^4 \quad (66)$$

Equation (67) [based on γ values calculated by Eq. (64)] can be used conveniently to estimate amorphous polyolefin surface tensions:

$$M/\gamma = -0.01793 + 0.40417C \quad (67)$$

Equation (68) was developed with n-alkane data [13, 14] and limiting properties ($d_\infty^{20} = 0.854$, $\gamma_\infty^{20} = 35.4$). Equation (68) was used with polyolefin densities (Eq. (48)] to estimate surface tensions, which are compared with other γ values in Table 16:

$$\gamma^{1/2} = -1.422 + 8.6321d_4^{20} \quad (68)$$

$$(C_{10}–C_{20}, C_\infty;\ r = 0.999919)$$

The PE surface tensions published by Wu [70] (20°C, 35.7; 140°C, 28.8: and 180°C, 26.5) are linear with temperature [Eq. (69)]. Wu [70] gives 53.7 as the γ value at −273.2°C; the γ value at −273.2°C calculated by Eq. (69) is 52.56:

$$\gamma = 36.85 - 0.057500t,\ ^\circ C \quad (69)$$

$$(20^\circ C/180^\circ C;\ r = 1.0000)$$

5. Solubility Parameters

Solubility parameters (δ, solpars, $J^{1/2}cm^{-3/2}$), molar cohesive energies (E_cM, J mol^{-1}), and vaporization heats (H_vM, J mol^{-1}) are related [Eq. (70)]:

$$\delta^2 = E_cM/V = (H_vM - RT)/V \quad (70)$$

where R is the gas constant (8.3144 J) and RT is 2479 at 298.15 K.

The Van Krevelen [44] molar attraction constant (F, $CH_2 = 280$) was used to develop Eq. (71). Equation (71) was divided by Eq. (47) to get Eq. (72), which was used to estimate polyolefin solubility parameters (δ at 25°C) (Table 17). Molar solubility parameters (M/δ) and C/δ are linear with repeat unit carbons [Eqs. (73) and (74)]:

$$F(\text{molar attraction constant}) = 280.0C \quad (71)$$

$$V^{25} = -0.1800 + 16.480C \quad (47)$$

$$\delta = 16.990C/(C - 0.0109) \quad (72)$$

$$M/\delta = -0.0090 + 0.8256C \quad (73)$$

$$C/\delta = -0.0006 + 0.05886C \quad (74)$$

The cohesive energy CH_2 value of 4940 [44] was used with molar volumes [Eq. (47)] to develop Eqs. (75) and (76):

$$E_cM = 4940C \quad (75)$$

$$\delta^2 = 299.76C/(C - 0.0109) \quad (76)$$

n-Alkane data [13, 14] were used to create an expression [Eq. (77)] correlating solpars with densities. Equation (77), when used with poly(1-olefin) densities, provided the calculated polyolefin solpars (Table 17):

$$\delta = 6.693 + 12.592d^{25} \quad (77)$$

$$(n\text{-alkanes}; C_5–C_{20}, C_\infty;\ r = 0.99968)$$

Equation (78) [16, 61] was used with poly(1-olefin) refractive indexes (calculated from R_{LL} molar

TABLE 17 1 Poly(1-olefins): Estimated Solubility Parameters (δ) at 25°C

C*	d^{25}	δ†	δ‡	δ§	δ¶
2	0.8558	17.08	17.08	17.36	17.47
3	0.8542	17.05	17.05	17.35	17.45
4	0.8535	17.04	17.04	17.34	17.44
6	0.8527	17.02	17.02	17.33	17.43
8	0.8523	17.01	17.01	17.33	17.43
12	0.8519	17.01	17.01	17.32	17.42
18	0.8517	17.00	17.00	17.32	17.42
∞		16.99	16.99	17.31	17.41

* Repeat unit carbons.
† $\delta = 16.990C/(C - 0.0109)$.
‡ $M/\delta = -0.0090 + 0.8256C$.
§ $\delta^2 = 299.76C/(C - 0.0109)$.
¶ $\delta = 6.693 + 12.592d^{25}$.

refractions, Table 14) as another method of estimating poly(1-olefin) solpars (Table 17)

$$\delta = -11.4 + 19.5\eta_D \quad (78)$$

The hypothetical polyolefin vaporization heats (H_vM) published in 1968 [71] were used to calculate molar vaporization heats [Eq. (79)] and energies, which were used with molar volumes to estimate solpars. The solpars estimated in this manner ranged from 19.4 for PE to 17.3 for poly(octadecene):

$$H_vM, \text{kJ mol}^{-1} = 0.5230 + 4.7747C \quad (79)$$

$(C_4–C_{16}; r = 0.999940)$

The calculated poly(1-olefin) solpars (Table 17) agree reasonably well with Bicerano's values [59] and the nonpolar poly(vinyl ether) molar volumes and solpars [Eq. (80) and (81)]:

$$V = 9.477 + 16.024C \quad (80)$$

$(C_4–C_{16}; r = 0.999986)$

$$C/\delta = -0.0290 + 0.062063C \quad (81)$$

$(C_6–C_{12}; r = 0.99983)$

The relatively low solubility parameters (solpars) of the amorphous poly(1-olefins) have been compared with the solpars at 25°C) of other polymers: poly(tetrafluoroethylene), 13.5; poly(dimethylsiloxane) 15.5; polypropylene, 16.5; polyisobutylene, 16.5; polyethylene, 17.0; poly(ethyl methacrylate); 18.5; polystyrene, 18.5; poly(vinyl acetate); 20; cellulose nitrate, 21; poly(ethylene oxide), 24; and cellulose acetate, 24 [16].

6. Glass Transition Temperatures

The poly(1-olefin) glass points (T_g, K) require different correlation equations for the $C_3–C_9$ polyolefins and the C_{10} and higher polyolefins [Eqs. (82)–(87) and Table 18]. Molar glass points are linear with repeat unit carbons (C) [Eqs. (82) and (83)]; the same is true for the products of T_g, K, and C [Eqs. (84) and (85)]:

$$M/T_g, \text{K} = -0.03816 + 0.06635C \quad (82)$$

$(C_3–C_9; r = 0.999972)$

$$M/T_g, \text{K} = 0.08830 + 0.05165C \quad (83)$$

$(C_{10}–C_{16}; r = 0.99977)$

TABLE 18 Poly(1-olefins): Glass Transition Temperatures (T_g, K)*

		Equation					
C†	Lit*	82	84	83	85	86	87
3	260	262	260				
4	249	247	247				
5	(233)	239	240				
6	233	234	234				
7	228/242	230	231				
8	228	228	228				
9	226	226	226			226	227
10	232			232	232	232	231
11	236			235	235		235
12	237			238	238	237	237
14	241			242	242	241	242
16	246			245	245	246	245
18	(328)			248	248	249	248

* Data from [60] and [68].
† Repeat unit carbons.

$$CT_g, \text{K} = 156.09 + 208.39C \quad (84)$$

$(C_3–C_9; r = 0.999947)$

$$CT_g, \text{K} = -360.85 + 267.85C \quad (85)$$

$(C_{10}–C_{16}; r = 0.99983)$

$$T_g, \text{K} = 59.249 + 0.55119T_m, \text{K} \quad (86)$$

$(C_9–C_{16}; r = 0.99954)$

$$T_g, \text{K} = 269.3 - 381.80/C \quad (87)$$

$(C_9–C_{16}; r = -0.98937)$

Equations (83), (85), and (87) suggest the limiting T_g, K values are 272 K, 268 K, and 269 K, respectively. When the T_m, K value of 415 K is used with Eq. (86), the estimated limiting value of T_g, K is 288 K.

7. Melting Points

Equations (88)–(92), which correlate poly(1-olefin) melting points (T_m, K) with repeat unit carbons (C) and with glass points (T_g, K), give calculated values that agree well with the selected literature melting points (Table 19). Molar melting points (M/T_m, K) and C/T_m, K are linear with C [Eqs. (89) and (92)]. The adjustable parameter (5.84) in Eq. (88) was selected to give the intercept of 414, the approximate T_m, K of amorphous polyethylene. Equation (90) is based on Mandelkern's claim [72] that T_m, K is linear

TABLE 19 Melting Points (T_m, K) of the Poly(1-olefins)*

C†	Lit.*	Equation 88	89	90	91	92
4						
7	290	287		288		
8	293	296	294	296		294
9	292‡	304	303	304	303	303
10	313	311	311	311	313	311
12	322	323	323	322	323	323
14	330	332	332	331	330	332
15	331‡	336	336	335		336
16	341	339	339	338	339	339
18	345	346	345	347		345
22	365‡	355	354	354		354

* Data from [23, 60].
† Carbons in repeating units.
‡ Not used in developing equations.

with ln C/C:

$$T_m, \text{K} = 414 - 1630.8/(C + 5.84) \quad (88)$$

$$M/T_m, \text{K} = 0.1021 + 0.034953C \quad (89)$$

$$T_m, \text{K} = 422.0 - 482.7 \ln C/C \quad (90)$$

$$T_m, \text{K} = -107.08 + 1.8126T_g, \text{K} \quad (91)$$

$$C/T_m, \text{K} = 0.00728 + 0.0024918C \quad (92)$$

Wunderlich's lower transition temperatures, reported by Brogdanov [60], can be estimated by Eq. (93):

$$M/T, \text{K} = 0.25533 + 0.033307C \quad (93)$$

(C_{16}–C_{44}; $r = 0.999960$)

8. Thermodynamics Properties

Equations (from [73]) for correlating thermodynamic properties and functions with repeat unit carbons are given in Table 20.

9. Simple Correlation Equations

The simple equations in Table 21 can be used conveniently to estimate many of the properties of amorphous, or largely amorphous, poly(1-olefins).

E. Copolymers: Ethylene with the Higher 1-Olefins

Equations (94)–(98) correlate the concentration (mol%) of 1-hexene (in ethene–hexene copolymers)

TABLE 20 Poly(1-olefins): Equations ($P = b + mC$)* Correlating Properties (P) with Repeat Unit Carbons (C)†

		Intercept, b	Slope, m
Heat capacity, C_p, J mol^{-1}	Cryst.	−8	25.5
Heat capacity, C_p, J mol^{-1}	RE‡	8	27.8
$H^0(T) - H^0(0)$, kJ mol^{-1}	Cryst.	−1.1	4.0
$H^0(T) - H^0(0)$, kJ mol^{-1}	RE‡	−2.42	4.68
S^0, J mol^{-1}	Cryst.	−21.2	30.2
S^0, J mol^{-1}	RE‡	−17.8	32.2
$G^0(T) - H^0(0)$, kJ mol^{-1}	Cryst.	6.1	−5.2
$G^0(T) - H^0(0)$, kJ mol^{-1}	RE‡	3.5	−5.1
Combustion heat, $\Delta_c H^0$, kJ mol^{-1}	Cryst.	−4.3	−648.95
Combustion heat, $\Delta_c H^0$, kJ mol^{-1}	Re‡	−7.97	−649.5
Formation heat, $\Delta_f H^0$, kJ mol^{-1}	Cryst.	4.4	−30.4
Formation heat, $\Delta_f H^0$, kJ mol^{-1}	RE‡	8.35	−29.9
Formation, $\Delta_f S^0$, J mol^{-1}	Cryst.	−6.1	−108
Formation, $\Delta_f S^0$, J mol^{-1}	RE‡	−3.8	−105.4

* $P = b + mC$, where b is intercept and m is slope.
† Equations from [73].
‡ Rubber-elastic.

TABLE 21 Simple Equations ($C/P = b + mC$)* for Estimating Poly(1-olefin) Properties (P)†

No.	y	Intercept, b	Slope, m	1/slope
1	C/d^{25}	−0.0128	1.1749	0.8511
2	C/n_D^{25}	−0.0027	0.6774	1.476
4	C/γ^{25}	−0.0013	0.02881	34.71
5	C/δ^{25}	−0.0006	0.05886	16.99
6‡	C/T_g, K	−0.0027	0.00473	211
7§	C/T_g, K	0.0063	0.00368	272
8¶	C/T_m, K	0.0073	0.00249	402

* $C/P = b + mC$, where C is repeat unit carbon, and 1/slope is the indicated limiting property (P_∞).
† Density, d^{25}; refractive index, n_D^{25}; dielectric constant, ε^{25}; surface tension, γ^{25}; solubility parameter, δ^{25}; glass point, T_g, K; and melting point, T_m, K.
‡ T_g, K, C_3–C_9.
§ T_g, K, C_{10}–C_{16}.
¶ T_m, K, C_8–C.

TABLE 22 Ethene-1-Hexene Copolymers*: Equations (91)–(95) Correlating Properties with 1-Hexene Concentration, mol %

	B		C		D		E	
A	d, lit.	Eq. (91)	η	Eq. (95)	t_m, °C	Eq. (96)	%	Eq. (97)
0.0	0.965		2.5		130		61	
1.5	0.958	0.957	2.3	2.38	127	127	55	54.8
3	0.954	0.955	1.15	1.10	122	122	52	52.3
6	0.950	0.950	0.85	0.863	117	117	48	47.8
9	0.946	0.946	0.8	0.806	114	114	44	44.1

* Data from [34].
A. 1-Hexene concentration, mol%.
B. Density, g cm^{-3}.
C. Viscosity, dL g^{-1}.
D. Melting point, °C.
E. Crystallinity.

with density (g cm^{-3}), viscosity (dL g^{-1}) melting point (t_m, °C), crystallinity (%), and branching (Me groups per 1000 carbons); the first four properties decrease in magnitude with increase in 1-hexene concentration [34], while branching increases (Table 22):

$$1/d \text{ (density)} = 1.042 - 0.0016965 \text{ mol\% 1-hexene} \quad (94)$$

(1.5–9 mol%; $r = 0.99026$)

$$1/\eta \text{ (viscosity)} = 1.405 - 1.4779/\text{mol\% 1-hexene} \quad (95)$$

(1.5–9 mol%; $r = -0.99701$)

$$t_m, {}^\circ\text{C (melting point)} = 130.0 - 16.686 \log \text{mol\% 1-hexene} \quad (96)$$

(1.5–9 mol%; $r = -0.999989$)

$$1/\text{crystallinity\%} = 0.01734 + 0.0005956 \text{ mol\% 1-hexene} \quad (97)$$

(1.5–9 mol%; $r = 0.99912$)

$$\text{Me groups per 1000 C} = 0.6222 + 4.2106 \text{ mol\% 1-hexene} \quad (98)$$

(0.66–5.25 mol%; $r = 0.99993$)

Equations (99) and (100) describe the effects of 1-octene on branching and densities of ethene–octene copolymers [34]:

$$\text{Me/1000 C} = -0.1194 + 5.0339 \text{ mol\% 1-octene} \quad (99)$$

(0.7–6.2 mol%; $r = 0.999985$)

$$d \text{ (density)} = 0.9622 - 0.011596 \text{ mol\% 1-octene} \quad (100)$$

(0.0–6.2 mol%; $r = -0.9989$)

The influence of 1-tetradecene (mol%) on the properties (Me groups per 1000 C and densities) of ethene–tetradecene copolymers is shown by Eqs. (101) and (102) [34]:

$$\text{Me/1000 C} = -0.02761 + 5.0127 \text{ mol\% 1-tetradecene} \quad (101)$$

(0.0–6.5 mol%; $r = 0.999937$)

$$d \text{ (density)} = 0.9444 - 0.008253 \text{ mol\% 1-tetradecene} \quad (102)$$

(0.8–6.5%; $r = -0.99107$)

F. Ionomers

Ionomers, or ionic polymers, have been called thermosets that can be processed as thermoplastics. They are members of a broad class of copolymers with inorganic salt groups attached to the polymer chain [74]. The combination of low ionic content and low polarity backbone results in a group of polymers that continue to be of scientific and commercial interest. The word ionomer was applied originally to olefin-based polymers containing a relatively low percentage of ionic groups. Over the years this definition was broadened to include other parent polymers [29].

The literature on ionomers has grown exponentially, with a doubling period of six years. There are more than 7500 papers and patents in the 1998 ionomer literature, which is growing at the rate of almost 600 documents per year [29].

The word ionomer dates back to 1965, but materials of this type had been synthesized and investigated long before. Examples of early work include the investigation of Littman and Marvel in 1930, synthesis of an elastomer based on butadiene and acrylic acid in the early 1930's, and the appearance in the 1950's of du Pont's Hypalon (sulfonated, chlorinated, and cross-linked polyethylene) [29].

Ionomer properties depend upon several variables, including the natures of the backbone and ion-containing monomers, the ionic content, degree of neutralization, and type of cation. With these several variables, the range of ionomer properties and potential applications is broad [74]. Specific properties, such as glass transition temperature, rubbery modules above the glass transition, dynamic mechanical behavior, melt rheology, relaxation behavior, dielectric properties, and solution behavior are modified by even small amounts of ionic groups.

ACKNOWLEDGMENTS

I am grateful to Dr. Benjamin P Huddle, Jr., Chair, Chemistry Department, and Margaret Anderson, Secretary, Chemistry Department, Roanoke College, for valuable assistance and to Dr. Vernon R Miller, Professor, Chemistry Department, Roanoke College, and Mr. Vernon C Miller for developing and providing the excellent computer program used in developing equations.

SYMBOLS AND ABBREVIATIONS

AT	Autoignition temperature
b	Intercept in equation $y = b + mx$
C	Homolog carbon or repeat unit carbon
C_p	Molar heat capacity (constant pressure), J mol^{-1}
E_cM	Molar cohesive energy, kJ mol^{-1}
E_cM/V	Cohesive energy density, where V is molar volume
ε	Dielectric constant
F	Molar attraction function
FP	Flash point
G	Functional group in RG, where R is alkyl
H_cM	Molar combustion heat, kJ mol^{-1}
H_fM	Molar formation heat, kJ mol^{-1}
H_mM	Molar fusion heat, kJ mol^{-1}
H_vM	Molar vaporization heat, kJ mol^{-1}
M_n	Average number molecular weight
P_c	Critical pressure
P_s	Molar parachor, used in calculating surface tension, γ
P_∞	Limiting property (property of infinite-length homolog)
R	Gas constant 8.31441 J (2749 at 298.15 K)
R_{GD}	Molar refraction, Gladston–Dale
R_{LL}	Molar refraction, Lorenz–Lorentz
r	Correlation coefficient
S	Molar entropy, J mol^{-1}
SI	International System of units
Solpar	Solubility parameter δ, Hildebrands (H), cal$^{1/2}$ cm$^{-3/2}$. Multiply by 2.046 to get solpars in S.I. units, MPa$^{1/2}$ or J$^{1/2}$ cm$^{-3/2}$
T_a	Aniline point, K (critical solution temperature in aniline)
T_b	Boiling temperature, K
T_c	Critical temperature, K
T_g	Glass transition temperature, K
T_m	Melting temperature, K
t_m	Melting point, °C
V	Molar volume (M/d)
V_c	Critical volume
ε	Dielectric constant (equal to $(\eta_D)^2$ for nonpolar materials)
γ	Surface tension (dyn cm^{-1})
δ	Solubility parameter (solpar) $= (E_cM/V)^{1/2}$
η	Viscosity, mPa s (equivalent to centipoise, cP)

REFERENCES

1. FM McMillan. In: RW Tess, GW Poehlien, eds., Applied Polymer Science, 2nd Ed. Washington, DC: American Chemical Society, 1985, pp 333–361.
2. FC Whitmore. Organic Chemistry, New York: Van Nostrand, 1937 p 30.
3. RB Seymour. In: C Vasile, RB Seymour, eds. Handbook of Polyolefins. New York: Marcel Dekker, 1993.
4. RB Seymour. In: RB Seymour, T Cheng, eds. Advances in Polyolefins. New York: Plenum Press, 1987, pp 3–14.
5. GR Lappin, JD Sauer. Alpha-Olefins Applications Handbook. New York: Marcel Dekker, 1989.
6. C Vasile, RB Seymour, eds. Handbook of Polyolefins. New York: Marcel Dekker, 1993.
7. V Dragutan. In: C Vasile, RB Seymour. Handbook of Polyolefins. New York: Marcel Dekker, 1993, pp 11–20.

8. RB Seymour, C Vasile, M. Rusu. In: C Vasile, RB Seymour, eds. Handbook of Polyolefins. New York: Marcel Dekker, 1993.
9. SC Stinson, Chem. Eng. News, Apr. 20, 1998, pp 63–66.
10. RM Thomas, WJ Sparks. In: GS Whitby, ed. Synthetic Rubber. New York: John Wiley & Sons, 1954, pp 838–891.
11. DL Hertz, Jr. In: AK Bhowmick, HL Stephens, eds. Handbook of Elastomers. New York: Marcel Dekker, 1988, pp 443–482.
12. DG Demianiw. In: M Grayson, ed. Concise Encyclopedia of Chemical Technology. New York: John Wiley & Sons, 1985, pp 818–819.
13. RR Dreisbach. Physical Properties of Chemical Compounds, Vol. 2. Washington, DC: American Chemical Society, 1959.
14. FD Rossini, KS Pitzer, RL Arnett, RM Braun, GC Pimentel. Selected Values of Physical and Thermodynamic Properties of Hydrocarbons and Related Compounds, Pittsburgh, PA: Carnegie Press, 1953.
15. KL Hoy, RA Martin. Tables of Solubility Parameters. 3rd ed. Tarrytown, NY: Union Carbide Corp., May 16, 1975.
16. AFM Barton. Handbook of Solubility Parameters and Other Cohesion Parameters, 2nd ed. Boca Raton: CRC Press, 1991.
17. DR Lide. Basic Laboratory and Industrial Chemicals. Boca Raton: CRC Press, 1993.
18. DR Lide. Handbook of Organic Solvents. Boca Raton: CRC Press, 1995.
19. DR Lide, ed. Handbook of Chemistry and Physics, 78th ed. Boca Raton: CRC press, 1997–1998.
20. JA Riddick, WB Bunger, JA Sakonao, eds. Organic Solvents, 4th ed. New York: John Wiley & Sons, 1986.
21. BJ Zwolinski, RC Wilhoit. Handbook of Vapor Pressures and Heats of Vaporization of Hydrocarbons and Related Compounds, AP144-TRC 101. College Station, TX: Thermodynamic Research Center, 1971, pp 276–281.
22. JA Dean, ed. Lange's Handbook of Chemistry, 14th ed. New York: McGraw-Hill, 1998.
23. B Wunderlich. Macromolecular Physics, Vol. 3. New York: Academic Press, 1980, p 27.
24. JD Sauer. In: GR Lappin, JD Sauer eds. Alpha Olefins Applications Handbook, New York: Marcel Dekker, 1989 pp 15–34.
25. S Patai. The Chemistry of the Alkenes. New York: Wiley-Interscience, 1965.
26. RC Fuson. Reactions of Organic Compounds. New York: John Wiley & Sons, 1962.
27. RL Wakeman. Chemistry of Commercial Plastics. New York: Reinhold Publishing Corp., 1947, pp 788-789.
28. Z Jiang, GM Dahlen, A Sen. In: TC Cheng, ed. New Advances in Polyolefins. New York: Plenum Press, 1993, pp 47–57.
29. A Eisenberg, JS Kim. Introduction to Ionomers. New York: John Wiley & Sons, 1998.
30. RC Reid, JM Prausnitz, BE Poling, 4th ed., Properties of Gases and Liquids, New York: McGraw-Hill, 1987.
31. MG Broadhurst. J Chem Phys 1962; 36: 2578; J Res NB S 1966; 70A: 481.
32. JA Dean. Handbook of Organic Chemistry. New York: McGraw-Hill, 1987.
33. V Majer, V Svoboda. Enthalpies of Vaporization of Organic Compounds. London: Blackwell Scientific Publications, 1985.
34. BA Krentsel, YV Kissin, VJ Kleiner, LL Stotskaya. Polymers and Copolymers of Higher Alpha - Olefins. New York: Hanser Publishers, 1997.
35. JJ Jasper. J Phys Chem Ref Data 1972; 1: 841–1009.
36. PJ Flory, RA Orwell, A Vrij. J Am Chem Soc 1964; 86: 3507–3514.
37. RA Orwoll, PJ Flory. J Am Chem Soc 1967; 89: 6814–6822.
38. JR Partington. Advanced Treatise on Physical Chemistry. Vol. II. London: Longman, Green & Co., 1951, pp 17–19.
39. CH Fisher. Chem Engng 1982; Sept 20: 111–113.
40. CH Fisher. Chem Engng 1989; Oct 1989, p 195.
41. CH Fisher. J Am Oil Chem Soc 1988; 65 1647–1651.
42. EJ Baum. Chemical Property Estimation. New York: Lewis Publishers, 1998.
43. WJ Lyman. Handbook of Chemical Property Estimation Methods. Washington, DC, American Chemical Society, 1990.
44. DW Van Krevelen, PJ Hoftyzer. Properties of Polymers. New York: Elsevier Scientific Publishing Co., 1976.
45. A Bernthsen. A Textbook of Organic Chemistry. New York: Van Nostrand, 1922, p 51.
46. CH Fisher, BP Huddle, Jr., Ind J Chem 1978; 16A: 1008–1010.
47. CH Fisher, J Coat Techn 1991, 63: 79–83.
48. A Kreglewski, BJ Zwolinski. J Phys Chem 1961; 65: 1050–1052.
49. CH Fisher. J Am Oil Chem Soc 1990; 67: 101–102.
50. CH Fisher. Chem Engng 1991, Jan: 110–112.
51. CH Fisher, J Colloid Interface Sci 1991; 141: 589–592.
52. CH Fisher. Proc Am Chem Soc Div PMSE 1984; 51: 588–592.
53. FD Rossini, K Li, RL Arnett, MB Epstein, RB Ries, LPBJM Lynch. J Phys Chem 1956, 60: 1400–1406.
54. AW Francis. Critical Solution Temperatures. Washington, DC: American Chemical Society, 1961.
55. HR Sailors, JP Hogan. In: RB Seymoru, ed. History of Polymer Science and Technology. New York: Marcel Dekker, 1982, pp 313–338.
56. RV Jones, CW Moberly, WB Reynolds, Rubber Chem Technol 1954; 27: 74–87.
57. CE Rehberg, CH Fisher. Ind Engng Chem 1948; 40: 1429.

58. CA Sperati, WA Franta, HW Stark, Weather, Jr. J Am Chem Soc 1953; 75: 6127.
59. J Bicerano. Prediction of Polymer Properties, 2nd ed. New York: Marcel Dekker, 1996.
60. BG Bogdanov, M Michailov. In: C Vasile, RB Seymour, eds. Handbook of Polyolefins. New York: Marcel Dekker, 1993, pp 295–469.
61. J Brandrup, EH Immergut, eds. Polymer Handbook, 3rd edn. New York: John Wiley & Sons, 1989.
62. V Gaur, B Wunderlich. J Phys Chem Ref Data 1981; 10: 119–152.
63. OG Lewis. Physical Constants of Linear Homopolymers. New York: Springer Verlag, 1986.
64. CH Fisher, BP Huddle, Jr. Chem Engng 1982, Jan: 99–101.
65. BP Huddle, Jr, Reciprocal Relationship Between Homolog Properties and Chain Length. Virginia Academy of Science, 56th Annual Meeting, 11 May, 1978, Blacksburg, VA.
66. TH Gouw, JC Vlugter. J Am Oil Chem Soc 1964; 41: 426, 675.
67. GR Somayajulu. Bulletin Thermodynamic Research Center; Sept. '84 - Mar. '85, Texas A&M University, College Station, TX 77843.
68. CH Fisher. In: RB Seymour, T Cheng. Advances in Polyolefins, New York: Plenum Press, 1987, pp 23–34.
69. KH Meyer. Natural and Synthetic High Polymers, 2nd ed. New York: Interscience Publishers, 1950.
70. S Wu. Polymer Interface and Adhesion. New York: Marcel Dekker, 1982 p 73.
71. RM Joshi, BJ Zwolinski, CW Hayes. Macromolecules 1968; 1: 30–36.
72. L Mandelkern, GM Stack. Macromolecules 1984; 17: 871–877.
73. BV Lebedev, NN Smirnova. In: BA Krentsel, YV Kissen, VJ Kleiner, LL Stotskaya, eds. Polymers and Copolymers of Higher Olefins, New York: Hanser Publishers, 1997, pp 336–363.
74. CE Carraher Jr. 1996 Polymer Chemistry, 4th ed. New York: Marcel Dekker.

11

Mechanical Properties and Parameters of Polyolefins

Mihaela Mihaies and Anton Olaru
SC Ceproplast SA, Iasi, Romania

I. INTRODUCTION

This chapter is an extended and updated review of the systematized data on the mechanical properties of polyolefins (PO). The mode of presentation from the first edition of *Handbook of Polyolefins* as used by Bogdanov [1] has been adopted. This kind of presentation was considered very useful for readers involved in PO manufacture, processing, application, and research.

The data presented refer to the most important POs with the general formulae:

$$—(—CH_2—\underset{\displaystyle R}{\underset{|}{CH}}—)_n—$$

(a) with R = H, polyethylene (PE):

High density polyethylene (HDPE) and ultrahigh molecular weight polymer (UHMWHDPE);
Low density polyethylene with branched carbon backbone (LDPE) or linear carbon backbone (LLDPE);
Random copolymer ethylene/α-olefins.

(b) with R = CH_3, polypropylene (PP):

Isotactic polypropylene (IPP);
Syndiotactic polypropylene (SPP);
Atactic polypropylene (APP);
Ethylene–propylene block copolymers with isotactic polypropylene segments (I-PEP).

New commercial grades of polyethylenes synthesized with both Ziegler–Natta and metallocene catalysts were included.

The selected mechanical properties refer to one-component polyolefin systems.

These properties as well as their dependences on various factors (F_i)—Pr(F_i)—are presented in Table 1 where the literature (articles, monographs, encyclopaedias, handbooks, etc.) containing information about them is classified. Only a limited number of data are presented here in tabular and graphical form. All properties and dependences presented in tabular and graphical form are accompanied by corresponding references which are underlined in Table 1.

The text, abbreviation, and symbols to the data from the English references in graphic and tabular form are not changed (original) or insignificantly changed; those from the references in other languages are translated using our own terms and abbreviations; data in tabular form are not given fully because of the limited number of pages.

II. MECHANICAL PROPERTIES

Only mechanical properties (behavior) (Mch. Pr.) of PO are chosen mainly at uniaxial strain.

Selected mechanical properties and parameters as well as their dependences on definite factors are presented in Tables 2–14 and Figs. 1–96.

(Text continues on p 271)

TABLE 1 Mechanical properties (Mch. Pr.) of PO and their relationship to different factors

			Static-mechanical properties				
No. 1	PO 2	State 3	$\sigma(\varepsilon)$ 4	$\sigma(\varepsilon, T)$ 5	$\sigma(\varepsilon, V_d)$ 6	$\sigma(\varepsilon, p)$ 7	$\varepsilon(t, \sigma)$ 8
1	HDPE UHMWHDPE	Isotropic	[2] p 298; [3] p 4694	[25] p 35		[34] (Fig. 19); [35]	[2] p 315 (Fig. 22)
		Oriented	[4, 5]		[30]		
2	LDPE LLDPE included metallocene obtained	Isotropic	[2] p 298; [6] p 274; [7] (Fig. 1); [8] (Fig. 2); [9] (Fig. 3);	[26]; [9] (Fig. 11)	[31] (Fig. 14); [9] (Fig. 15)	[36] p 125 (Fig. 20)	
		Oriented	[10]		[31]		
3	Different types of PE	Isotropic	[11]	[2] p 311 (Fig. 12); [26]	[26]		
		Oriented	[12]				
4	IPP APP SPP and I-PEP included metallocene obtained	Isotropic	[13] (Fig. 4); [14]; [2] p 299	[27] p 12	[32]; [33] (Fig. 16)	[37] (Fig. 21); [33]	[2] p 325 (Fig. 23)
		Oriented	[15] (Fig. 5); [16, 17]; [18] p 59; [19] (Fig. 6)	[28]			[38]
5	Different types of PO	Isotropic	[2] p 299				[39] [40]
		Oriented					
6	Random copolymers E-H, E-O, E-B included metallocene obtained	Isotropic	[9] (Figs. 7, 8); [20] p 84; [21] (Fig. 9); [22] (Fig. 10); [23] p 1410 [24] p 1115	[29] p 1161; [24] (Fig. 13)	[9] (Fig. 17); [29] (Fig. 18); [24] p 1116		

TABLE 1 (Continued)

No. 1	State 3	Static-mechanical properties E_{el}, σ_y, σ_B, ε_y, ε_B, σ_{imp} 9	Dynamic-mechanical properties E', E'', tan δ 10
1	Isotropic	E_{el} (ε, T), σ_B (ε, T): [30]; E_{el}, σ_B (V_d): [32]; E_{el} (of chain): [41] (Fig. 24); E_{el} (a, b, 110 in ab crystalline plane): [42, 43] (Table 2); E_{el} (α_c): [44]; E_{el} (T_c): [45] (Table 3); E_{el} (T, V_d): [46] (Fig. 25); E_{el} (p): [35]; E_{el} (T): [47] (Fig. 26); E_{tsh} (T): [2] p 312 (Fig. 27); E_{fx} (t): [2] (Fig. 28) [48]; E_{creep} (T, t): [25] p 38; E_{creep} (t, σ): [47] (Fig. 29); A_{tear} (V_{tear}, T): [49]; σ_y (T, M): [25] p 35 (Fig. 30); σ_y (T, V_d): [46] (Fig. 31); σ_B (T): [26]; σ_B (T, M): [25] p 35 (Fig. 32); ε_y (V_d): [50] (Fig. 33); ε_B (T, M): [25] p 35 (Fig. 34); ε_B (T, V_d): [46] (Fig. 35); ε_{creep} (σ_d, T, t): [25] p 37; ε_{creep} (t, σ); [51] p 1343, σ_{imp} (T): [25] p 33 (Fig. 36); σ_{imp} (I_m): [25] p 35 (Fig. 37); J_c (t, σ): [51] p 1341; J (t, σ): [51] p 1342; Mch. Pr.: [52] p 2910	E' (ν): [49]; tan $\delta(T)$: [105] pp 126–127; tan $\delta(\nu)$: [105] pp 128–131; tan $\delta(t)$: [106]; E', tan $\delta(t)$: [107] (Fig. 74)
	Oriented	E_{el} (ε): [53, 54]; E_{el} (ρ): [5]; E_{el}, σ_B (ε): [55, 56] (Fig. 38); E_{el}, σ_B (dimensions of the fibrils): [30, 57] (Fig. 39); E_{el}, σ_y (T, θ): [46]; E_{el} (ε_{or}, T): [58] p 3481; σ_B (θ): [59]; Mch. Pr. (ε, T_{or}): [60]; Mch. Pr.: [54]; Mch. Pr. (ε_{or}): [61] p 1385 (Table 4)	E' (T, ε), tan $\delta(T, \varepsilon)$: [64], [60]; E', E'' (T, ε): [97]; E'_{rel} (T, ε_{or}): [61] p 1389; tan $\delta(\nu^{-1}, T)$: [61] (Fig. 75); tan $\delta(a_T, \nu)$: [61] p 1391; E', tan δ (T): [108] (Fig. 76); E', tan $\delta(\varepsilon)$: [108] (Fig. 77); E', tan $\delta(t, T)$: [108] p 1033;
2	Isotropic	E_{el}: [6] p 299; E_{el} (p): [62] (Table 5); E_{el} (α_c, M): [8] (Fig. 40); E_{el} (L_c, M): [8] p 5307; E_{el} (L_a, M): [8] p 5307; σ_{tear} (film thickness): [63]; σ_B (T): [26]; σ_B (M): [8] (Fig. 41); σ_y (α_c): [8] (Fig. 42); ε_b (α_c, M): [8] p 5303; ε_b (L_a, M): [8] p 5303; σ_{imp} (film thickness): [63]; ε_B, σ_u (M, SCB): [7] (Fig. 43); $\sigma_{T,I}$ (M, SCB): [7] (Fig. 44)	E' (ν), tan $\delta(\nu)$: [2], [109]; (Fig. 78); E'', tan $\delta(T)$: [110] (Fig. 79); E', tan δ (T, ν): [111] (Fig. 80); E_{shear},
	Oriented	E_{el} (p): [64]; E_{el} (θ): [65] p 185; E_{el} (T_d, θ): [66]; σ_B (M): [67]; σ_B (ε_{or}): [67]	tan δ (T, ε_{or}): [111] (Fig. 81); tan $\delta(T, \varepsilon)$: [112]; E, tan δ (T): [111] (Fig. 82); $E(\varepsilon_{or}, T)$: [110] (Fig. 83); $E(\theta, T)$: [111] (Fig. 84)
3	Isotropic	E_{el} (T, ρ): [6] p 270; E_{fx} (T): [48]; σ_y (T): [47]; σ_h (T): [68] p 64, [69]; σ_h (T, ρ): [6] p 271 (Fig. 45); σ_y (T): [69]; σ_B (T): [2] p 312 (Fig. 46); σ_{imp} (I_m, ρ): [2] p 309 (Fig. 47); σ_{imp}: [69]; Mch. Pr.: [70]; Mch. Pr. (ρ, α_c): [2] p 308 [49]; Mch Pr. (T, M, branchiness): [69]; Mch. Pr.: [71] (Table 6); Mch. Pr.: [48]; Mch. Pr.: [72] p 24 (Table 7)	E', E'', tan δ (mol. structure, α_c): [113] (Table 14) E', tan $\delta(T, \alpha_c)$: [113], [114] (fig. 85); tan $\delta(T, \alpha_C)$: [115];
	Oriented	E_{el} (ε_{or}, M): [30] (Fig. 48); E_{el} (ε_{or}, T): [73] p 451; E_{el} (ε, T_{or}): [74]; E_{el} [75]; E_{el}, σ_B (M, M_W/M_n): [67] (Table 8); $\sigma_{h\parallel}$, $\sigma_{h\perp}(\varepsilon)$: [74]; Mch. Pr. ($\varepsilon$, σ_{or}): [76]; Mch. Pr.: [73] p 450	E', tan $\delta(T)$: [67]; tan $\delta(T, \nu)$, $E'(T)$: [116] (Fig. 86); E', $E''(T)$: [12]; tan δ (T, ε): [117] (Fig. 87);
4	Isotropic	E_{el} (of the chains): [41] (Figure 49); $E_{el}(T)$: [6] p 272; E_{el} (p, T): [37]; E_{el} (V_d, T): [46] (Fig. 50); E_{fx} (I_m): [77] p 653, [78] (Fig. 51); E_{fx} (M): [79] (Fig. 52); E_{fx}, σ_{imp} (tactic): [78] (Fig. 53); E_{fx} (MWD): [78] (Fig. 54); σ_y (p): [80] (Fig. 55); σ_y (T): [2] p 323 (Fig. 56); σ_y (p, T): [37]; σ_y (V_d): [32] (Fig. 57), [46]; σ_y (T, α_c): [68] p 64; σ_y, σ_B, ε_B (T_c): [27]; σ_B (T): [2] p 323 (Fig. 58); σ_B (V_d): [32]; σ_B, ε_B (M): [68] p 54; ε_y (V_d): [32] (Fig. 59); ε_B (V_d, T): [46] (Fig. 60); σ_h (T): [6] p 273, [81]; σ_h (distance from the center of a spherulite): [13]; σ_{imp} (T): [82] (Fig. 61), [2] p 323 (Fig. 62); [68] p 64; [46]; σ_{imp} (I_m): [68] p 52 (Fig. 63); σ_{imp} (I_m, IPP): [83] p 50 (Fig. 64); σ_{imp} (I_m): [77] p 654, 655; σ_{imp} (E_{fx}): [78] p 240; σ_{imp} (M): [78] p 242; σ (ε, NC): [84] (Fig. 65), σ (ε_m, NC): [84] p 1458; σ (t): [84] p 1459; ε_y (M): [79] p 1870; Mch. Pr.: (M, APP): [2] p 321; Mch. Pr.: [78] p 238; Mch. Pr.: [78] p 404; Mch. Pr.: [70], [68] p 44, 46; Mch. Pr.: [85, 86]; Mch. Pr. [87] (Table 9); Mch. Pr.: [78] p 406 (Table 10)	E', $E''(T)$: [81]; E', tan $\delta(T)$: [105] p 134, [2] p 323 (Figs 88, 89), [114] (Fig. 90); E', E'', tan $\delta(\nu)$: [118]

TABLE 1 (Continued)

No.	State	Static-mechanical properties	Dynamic-mechanical properties
		E_{el}, σ_y, σ_B, ε_y, ε_B, σ_{imp}	E', E'', tan δ
1	3	9	10
	Oriented	E_{el} (T, σ_{or}): [88]; E_{el} (ε, T_{or}): [88], [89]; E_{el} (ε): [88], [90], [18] (Fig. 66); E_{el} (θ, mol. or.): [91]; $\sigma_{y\parallel}$, $\sigma_{y\perp}$, $\varepsilon_{y\parallel}$, $\varepsilon_{y\perp}$ (T): [82], σ_B (V_d, mol. or): [92] (Fig. 67), [93]; σ_B (ε_{or}): [18] (Fig. 68); σ (ε, α_c): [19] p 5844; σ_h (ε_c): [94] (Fig. 69), [56] p 192; σ_{imp} (ε): [94] (Fig. 70); ε_B (V_d): [28]; $\varepsilon_{or,max}$ (T, α_c): [18] p 59; ε_b (ε_{or}): [18] (Fig. 71); ε_{or} (ε_c): [56] p 156; E_y (T_{or}): [56] (Fig. 72); Mch. Pr. (fiber): [78] p 250	E', $E''(T)$: [89]; E', tan $\delta(T, \varepsilon)$: [88]; $E'(T, \alpha_c)$: [19] (Fig. 91); E' (ε, α_c): [19] p 5848; $E''(T)$: [18] (Fig. 92); tan δ (T): [19] (Fig. 93)
5	Isotropic	E_{el}: [95] p VIII-1; E_{el}: [96–99] (Table 11); σ_B: [95] p VIII-2, 3; σ (ε, α_c): [19] (Fig. 73); Mch. Pr. (T): [100]; Mch. Pr.: [101] p 32 (Table 12); p 78, 133, 137; Mch. Pr.: [77] p 653; Mch. Pr. (film): [78] p 251 (Table 13)	$E'(T)$: [105] p 256; E', $E''(t)$: [119], [120]; E', tan $\delta(T)$: [109] (Fig. 94)
	Oriented	E_{el} (ε): [102] p 284; E_{el} (σ_B): [103]	
6	Isotropic	σ_y (α_c): [104] p 762; σ_y (L_c): [104] p 762; σ_y (V_d): [104] p 763; σ_y (T): [22] p 1277	$E'(T)$: [22] p 1277 (Fig. 95)

Source: Data from Refs. 1, 121.

TABLE 2 Comparison of elastic Moduli E_a, E_b, and $E(110)$ in the *ab* plane of crystalline polyethylene at 293 K measured by Sakurada *et al.* [42] by X-ray scattering and values calculated from measured elastic constants [43]

Source	E_a, GPa	E_b, GPa	$E(110)$, GPa
Sakurada *et al.*	3.2	3.9	4.3
Calculated from elastic constants	3.2 ± 0.5	3.9 ± 0.5	4.6 ± 0.5

TABLE 3 Effect of morphology–crystallinity on elastic modulus at constant time duration but at different crystallization temperatures of HDPE

Crystallization time, min	Crystallization temperatures, °C	Spherulite size, μm	Lamellar thickness, Å	Crystallinity by DSC, %	Elastic modulus, GPa
150	100	11.3	132.0	65.0	3.207
150	105	11.9	129.0	68.6	3.417
150	110	13.3	136.6	68.9	3.438
150	115	16.6	145.0	70.9	3.482
150	120	20.4	162.7	72.1	3.541
150	125	—	185.7	72.3	3.655
150	127	—	189.8	73.4	4.358
150	129	—	200.6	69.6	3.911

Source: Data from Ref. 45.

TABLE 4 Results of tensile testing at 20°C and some other characteristics of HDPE, $M_w = 175{,}000$, $\rho = 0.952\,\mathrm{g\,cm^{-3}}$, $I_m = 0.5\,\mathrm{g/10\,min}$ as a function of the draw ratio

Property	Draw ratio					
	1.0	5.5	7.5	9.1	10.9	12.2
Density, g cm^{-3}	0.9520	0.9575	0.9605	0.9615	0.9620	0.9620
Elastic modulus, $E_{el} \times 10^{-3}$ kJ cm^{-3}	1.07	3.6	4.6	5.7	6.2	7.1
Tensile strength, $\times 10^{-3}$, kJ cm^{-3}	—	0.20	0.25	0.30	0.32	0.37
Relative elongation at fracture, 100%	—	94	57	41	30	13
Degree of crystallinity, %	0.50	0.54	0.55	0.57	0.60	0.65
Degree of orientation in the crystalline phase	—	0.95	0.95	0.95	0.96	0.96
Degree of orientation in the amorphous phase	—	0.91	0.92	0.91	0.91	0.92
Specific heat capacity, J K^{-1} g^{-1}	170	175	—	178	181	183

Source: Data from Ref. 61.

TABLE 5 Young's modulus of PE at 298 K [62]

Polymer	*E*, GPa at *p* MPa		
	0.1	345	689
LDPE	0.24	1.5	2.7
MDPE	1.2	3.8	5.0

Source: Data from Ref. 102.

A. Static Mechanical Properties

Mechanical properties are summarized as follows:

1. Stress–strain dependences of PO at uniaxial strain in a static-mechanical field, i.e., $\sigma(\varepsilon)$ and its dependence on a number of factors, i.e., $\sigma(\varepsilon, F_i)$ where F_i is temperature (T), pressure (p), draw rate (V_d), molecular and supermolecular structural parameters of PO systems, molecular weight (M), and

TABLE 6 Properties of LLDPE, relative to LDPE and HDPE

Property	LDPE	HDPE	Relative to LDPE	Relative to HDPE
Tensile strength, MN m^{-2}	6.9–15.9	21.4–38	Higher	Lower
Elongation, %	90–650	50–800	Higher	Higher
Impact strength, J/12.7 mm	No break	1.02–8.15	Better	Similar
Environmental stress-cracking resistance	—	—	Better	Similar
Heat distortion temp., °C	40–50	60–82	15°C Higher	Lower
Stiffness, 4.5 MN m^{-2}	1.18–2.42 mode of elasticity	5.53–10.4 mode of elasticity	Less	Can be same
Warpage processibility	Excellent	Good	More difficult	Easier
Haze, %	40	—	Worse	Better
Gloss, 45° in %	83	—	Worse	Better
Clarity	Near transparent to opaque	Translucent to opaque	Worse	Better
Melt strength	—	—	Lower	Lower
Softening point range, °C	85–87	120–130	Narrower	Narrower
Permeability, mL cm^{-2} g^{-1} mL^{-1} cm Hg^{-1} at 25°C × 10^{3}				
(a) H_2O vap	420	55	Better	Worse
(b) CO_2	60	13	Better	Worse

Source: Data from Ref. 71.

TABLE 7 Some properties of different grades of polyethylene

Property	LLDPE	LDPE	HDPE	UHMWPE
Density, g cm^{-3}	0.910–0.925	0.915–0.935	0.941–0.967	0.93
Melting temperature, °C	125	106–112	130–133	132
Tensile strength, MPa	14–21	6.9–17.2	18–30	20–41
Elongation at break, %	200–1200	100–700	100–1000	300
Flexural modulus, MPa	248–365	415–795	689–1654	—
Izod impact strength, J m^{-1}	—	0.67–21	27–160	No break
Hardness, shore D	41–53	45–60	60–70	—

Source: From Ref. 72.

TABLE 8 Effect of number average molecular weight (M_n) on the tensile strength and tensile modulus of oriented polyethylenes at −55°C*

Sample	$M_n \times 10^{-3}$	$M_w \times 10^{-3}$	M_w/M_n	σ_B, GPa	E_{el}, GPa
Alathon 7050	22.0	59	2.7	0.92	38.5
Rigidex 50b	7.8	104	13.3	0.86	32.2
Rigidex 50a	12.3	101	8.2	0.86	31.8
BXP 10	16.8	94	5.6	0.94	31.8
Alathon 7030	28.0	115	4.1	1.12	30.1
NBS SRM 1484	110.0	120	1.1	1.23	31.4
BP 206	16.6	213	12.8	1.21	33.6
Unifos 2912	24.2	224	9.3	1.11	32.2
XGR 661	27.8	220	7.9	1.17	31.4
H020 54P	33.0	312	9.5	1.23	36.6

* Draw ratio 15.
Source: Data from Ref. 67.

TABLE 9 Properties of product from the catalloy process (1997)

Type	I_m	Flexural modulus, MPa	Elongation at break, %	Hardness, D scale
Adstif KC 732P	20	2000	—	88*
Hifax 7135	15	950	>150	—
Hifax CA 53 A	10	650	>500	52
Hifax CA 138 A	3	420	>200	39
Hifax CA 162 A	14	80	>500	32
Adflex Q 300 F	0.8	350	430	36
Adflex Q 100 F	0.6	80	800	30
Adflex C 200 F	6	230	—	41
Adflex X 101 H	8	80	—	—
Adsyl	6	700	—	—

* R scale.
Source: From Ref. 87.

TABLE 10 Properties of SPPs

Property	Polypropylene sample			
	SPP 1	SPP 2	SPP 3	Conv. IPP
Melt flow rate, g/10 min	5.3	8.9	2.9	
Melting point, °C	125	126	148	163
Density, g cm^{-3}	0.87	0.87	0.89	0.91
Crystallinity, %	21	22	29	55
CH_3 placement: racemic*, %	91.4	91.9	96.5	1.4
CH_3 placement meso*, %	8.6	8.1	3.5	98.6
M_w/M_n	2.6	2.6	1.7	8
Flexural modulus†, MPa	380	415	760	1170
Notched Izod, J m^{-1}	775	670	750	25
Haze†, %	20	27	48	—

* By ^{13}C NMR; total % in pentads.
† Comparative values.
Source: From Ref. 78.

TABLE 11 Tensile modulus of ideal crystal of polyolefins

Polymer	$E_{el\parallel}$, GPa	$E_{el\perp}$, GPa	Ref.
Polybutene-1	25	2.0	96, 97, 98
Poly-4-methylpentene-1	6.7	2.9	96, 97, 98
IPP	34	3.1	97, 98, 99
HDPE	235	5.2	97, 98, 99

degree of crystallinity (α_c), chemical nature of the environment, etc., are presented in Table 1, columns 4–7.

2. Deformation (ε) of PO vs time (t) at uniaxial strain in a static-mechanical field with tension σ, i.e., the dependence $\varepsilon(t, \sigma)$ is given in Table 1, column 8.
3. Mechanical parameters of PO determined by specific points and regions of the load–elongation curve $\sigma(\varepsilon, F_i)$ are given in paragraphs 1 and 2, or special methods used to determine their dependencies on the enumerated factors are given in paragraphs 1 and 2 and Table 1: static modulus of elasticity (E_{el}), yield stress (σ_y), yield strain (ε_y), tensile strength or tensile at break (σ_B), etc., of PO in isotropic and oriented state at an angle θ between the tensile axis and the orientation direction are given in Table 1, column 9.

B. Dynamic Mechanical Properties

The study of the dynamic-mechanical properties comprises the determination of the following values: dynamic storage modulus (real part of the complex modulus) E', loss modulus (imaginary part of the complex modulus) E'' and their ratio E''/E', called the factor of dynamic-mechanical loss, tan δ.

Dynamic-mechanical parameters and their dependences on particular factors F_i (temperature T, frequency ν, amplitude of tension of a varying

TABLE 12 Mechanical properties of PO

No.	Properties	LDPE	HDPE	IPP	PB
1	Yield stress, MPa	7–13	24–33	28–35	15–25
2	Tensile strength, MPa	10–17	20–35	26–43	15–40
3	Tensile elongation at break, %	200–600	300–1000	250–700	150–400
4	Bending strength, MPa	17–20	26–43	34–50	15–25
5	Tensile modulus, MPa	—	900–1200	1000–1500	500–900
6	Shore hardness	42–50	62–69	70–75	60–68
7	Izod impact strength with notch, kJ × m^{-1}	Without break	2–150	5–8	>40

Source: Data from Ref. 101.

TABLE 13 Properties of different polyolefin films

Property	ASTM	Units	LDPE	HDPE	Unoriented PP	Biaxially oriented PP
Tensile strength	D-822	MPa	17–24	34–69	40–60	140–240
		(kpsi)	(2.5–3.5)	(5–10)	(6–9)	(20–35)
Modulus	D-822	MPa	140–210	550–1250	690–960	1720–3100
		(kpsi)	(20–30)	(80–180)	(100–140)	(250–450)
Elongation	D-822	%	300–600	—	400–800	50–130
Tear strength	D-1922	N/mm	80–160	—	16–160	1.5–2
		(g/mil)	(200–400)		(40–400)	(4–6)
Haze	D-1003	%	5–8	High	1–4	1–4
MVTR	E-96	$\frac{\text{g mil}}{100\ \text{in}^2\ \text{d}}$	1.2	0.3	0.7	0.3
O_2 trans. rate	D-1434	$\frac{\text{cc mil}}{100\ \text{in}^2\ \text{d atm}}$	450	150	240	160

Source: Data from Ref. 78.

deformation mechanical field) are presented in Table 1, column 10.

C. Fatigue Behavior

Fatigue behavior is presented in Figs. 65, 74, 76, and 77.

The static- and dynamic-mechanical properties and parameters grouped in Table 1 refer to crystalline PO below the melting temperature, but some of them characterize the corresponding PO in the molten (viscoelastic) state, too.

Table 1 gives the following dependences of these properties and parameters as well as their references.

Mechanical behavior of polyolefins is firstly dependent on their semicrystalline nature. Generally, polyolefins exhibit three phases: a tridimensional ordered phase, an amorphous disordered phase, and an interfacial layer between the two phases. The relative content of these morphological forms influences all properties that depend on the response of each phase on the external tensions.

The most important tensile properties depend on the molecular characteristics such as structural regularity of chains, molecular weight distribution and also on the morphological characteristics: crystallinity degree, crystallite size, distribution of the crystallite size, etc.

Other properties as stress-cracking, impact or tear resistance are controlled by the topology of the amorphous phase. The interlamella layer constitution is responsible for the propensity of crack propagation.

For polyethylenes the complex of properties such as stiffness and hardness at moderate temperatures and high stability at low temperature are the most important and valuable among the mechanical properties.

One of the most important characteristics that predominantly determines the properties and the behavior of different grades of PE is their branching which influences the ability of the polymer to crystallize. The nature, size, and distribution of branches have a dominant influence on crystallinity, density and consequently on mechanical properties (Fig. 96).

Other properties depending on crystallinity, such as stiffness, hardness, tear strength, yield point, Young's modulus and chemical resistance, increase with increasing degree of crystallinity whereas flexibility and toughness decrease under the same conditions.

Long branches affect more pronouncedly the polydispersity. It is generally considered that, when the other structural factors are constant, a narrower MWD leads to an increase in impact strength, tensile strength, toughness, softening point and resistance to environmental stress cracking.

Another factor that influences the properties of the polyolefins is the weight average, M_w. Ultimate tensile strength, tear strength, low temperature toughness, softening temperature, impact strength and environmental stress cracking increase as the M_w increases.

In recent years new ethylene copolymers (LLDPE) arrived on the market with density in the 0.915–0.935 g cm^{-3} interval. LLDPE are copolymers of ethylene and small amount of another α-olefin, such as l-butene, l-hexene, 4-methyl pentene or l-octene. New generation of "super strength" grade have been prepared, most recently using higher olefins comonomer.

(Text continues on p 279)

TABLE 14 Thermal characterization of polyethylenes: dynamic-mechanical analysis

Sample identification			Sample characteristics					
			Thermal history					
No.	Type	Source	Slow cooled	Annealed 25 h, °C	Quenched	Density, g cm^{-3}	Mol. wt. $M_w \times 10^{-3}$	Crystallinity index, %
1	2	3	4	5	6	7	8	9
1	LDPE	Linear	+			0.920		49
		LL-1001 Exxon		70		0.920	142	
					+	0.916		36
			+			0.926	*	50
		Conventional		70		0.928		
2		HBS, 1476, NBS			+	0.921		43
		High mol weight	+			0.930		64
		Lot 90449		70		0.930		
3		Hercules		100		0.930	5400	
					+	0.924		55
4		6097 Union Carbide	+			0.950	248	63
5	HDPE	Lot 273908	+			0.951	195	69
		Paxon 4100; Allied			+	0.941		60
6		Lot 293922; Paxon 4100; Allied	+			0.951	246	67
		Milk Bottle grade	+			0.963	157	74
7		Allied			+	0.947		66
		Milk Bottle grade	+			0.962		75
		Allied		70		0.963	160	
8				100		0.964		
					+	0.949		66

DMA												
	log E' values, Pa			log E'' transitions, °C			log E'' values†			tan δ values		
No.	−120°C	25°C	75°C	γ	β	α	γ_{max}	β_{max}	α_{max}	−120°C	25°C	75°C
	10	11	12	13	14	15	16	17	18	19	20	21
1	9.41	8.60	7.83	−111	−22	29	0.35	0.20	0.14	0.024	0.119	0.299
	9.39	8.58	7.84	−110	−22	28	0.36	0.19	0.15	0.023	0.125	0.308
	9.40	8.43		−111	−18	26	0.36	0.27	0.12	0.027	0.168	
2	9.44	8.68	7.87	−114	−10	18	0.34	0.26	0.27	0.027	0.138	0.297
	9.42	8.66		−114	−14	19	0.33	0.35	0.29	0.027	0.128	0.301
	9.40	8.39	7.56	−114	−7		0.35	0.37		0.024	0.202	0.096
3	9.43	8.93	8.50	−107	−10	56	0.55	0.10	0.37	0.016	0.050	0.189
	9.39	8.90	8.47	−108	−8	58	0.55	0.11	0.37	0.016	0.048	0.194
	9.43			−106	−8		0.56	0.12		0.018		
	9.39	8.84	8.31	−106	−19	45	0.58	0.09	0.34	0.015	0.066	0.222
4	9.49	9.03	8.45	−108		41	0.41		0.51	0.017	0.084	0.217
5	9.50	9.03	8.46	−108	−24	41	0.40	0.15	0.52	0.016	0.089	0.220
	9.46	8.93	8.26	−109		35	0.45		0.52	0.018	0.112	0.245
6	9.51	9.05	8.46	−108	−25	41	0.40	0.13	0.50	0.016	0.086	0.222
7	9.51	9.18	8.70	−106	−27	51	0.36	0.07	0.54	0.015	0.059	0.210
	9.47	9.07	8.50	−107		43	0.47		0.61	0.016	0.082	0.248
8	9.52	9.18	8.71	−107	−29	52	0.37	0.06	0.54	0.015	0.059	0.218
	9.53	9.18	8.73	−107	−33	53	0.41	0.05	0.61	0.013	0.061	0.219
	9.53	9.15	8.70	−106	−28	53	0.37	0.10	0.57	0.013	0.061	0.220
	9.45	9.04	8.48	−107	−40	42	0.46	0.06	0.57	0.017	0.085	0.240

* Due to the excessive branching, no attempt was made to determine M_w.
† Measured as the peak height above the lowest point between the end peaks.
Source: Data from Ref. 113.

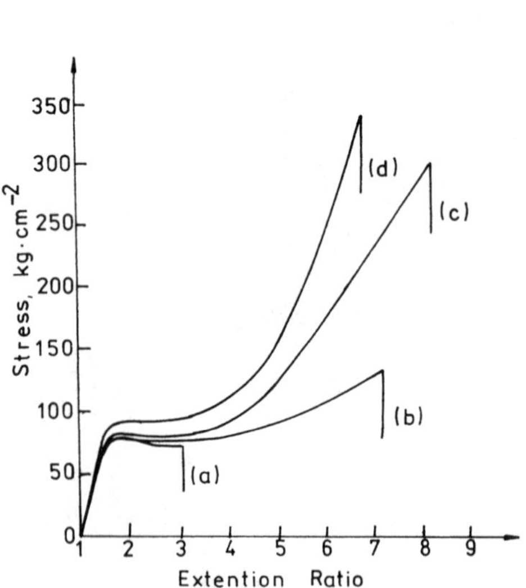

FIG. 1 Typical stress–strain curves at 25°C for LLDPE, $I_m = 0.8$ g/10 min, $\rho = 0.919$ g cm^{-3}, SCB = 17.4/1000*C*, fraction of molecular weight of (a) 3.0×10^4, (b) 5.4×10^4, (c) 9.9×10^4 and (d) 21.3×10^4. (From Ref. 7.)

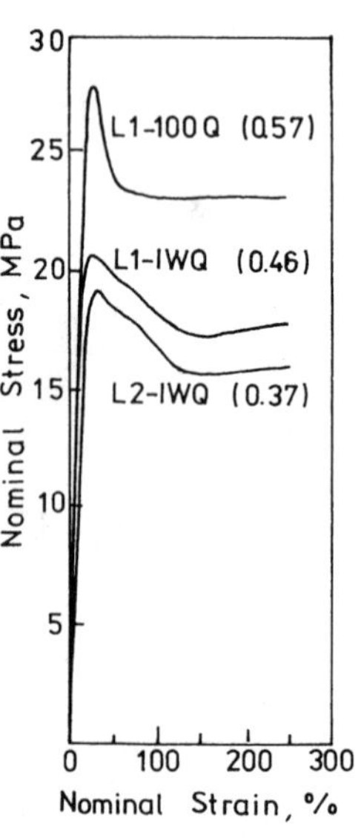

FIG. 3 Stress–strain curves in yield region for linear polyethylene: L2-IWQ – $M_w = 970{,}000$, $M_w/M_n = 4.42$, $\alpha_c = 0.37$; L1-IWQ – $M_w = 173{,}000$, $M_w/M_n = 2.0$, $\alpha_c = 0.46$; L1-100Q – $M_w = 173{,}000$, $M_w/M_n = 2.0$, $\alpha_c = 0.57$. IWQ ÷ quenched into ice-water mixture; 100Q ÷ quenched into water at 100°C. (From Ref. 9.)

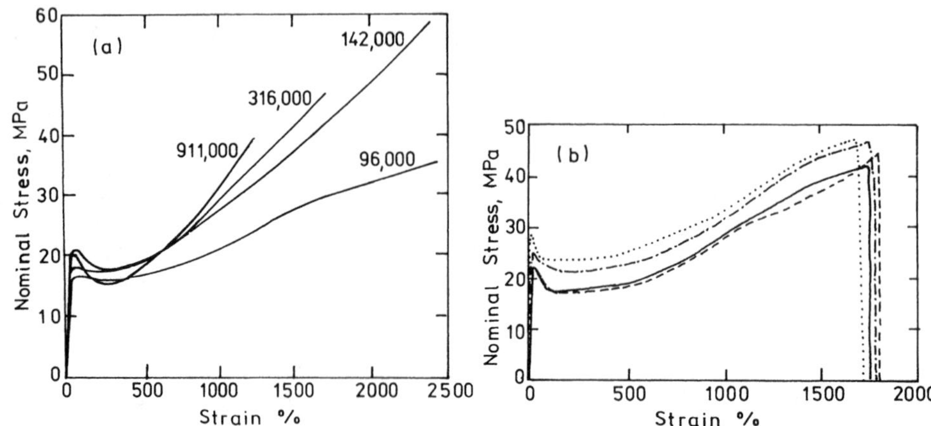

FIG. 2 Nominal stress–strain curves for linear polyethylene as a function of M_W (a) and different crystallinity levels (b) (···) $\alpha_c = 0.64$; (–·–·) $\alpha_c = 0.55$; (——) $\alpha_c = 0.44$; (- - - -) $\alpha_c = 0.46$. (From Ref. 8.)

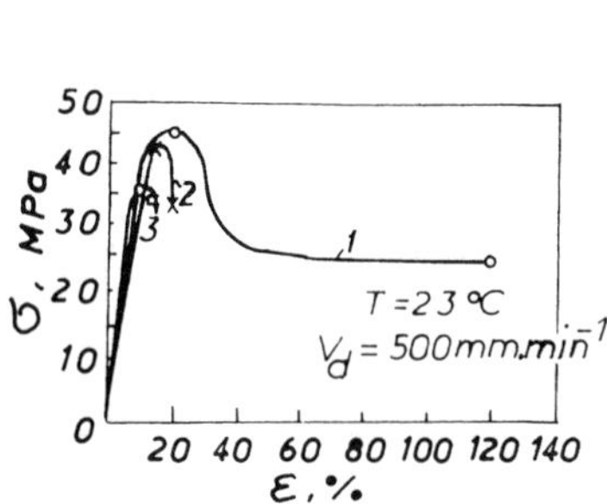

FIG. 4 Engineering stress–strain diagrams of polypropylene 1120 LX as a function of the structure: curve 1, fully quenched structure; curve 2, partially coarse spherulitic structure; curve 3, coarse spherulitic structure. (From Ref. 13.)

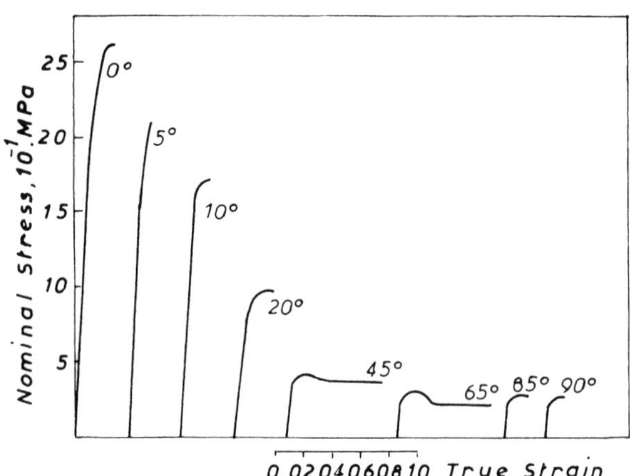

FIG. 5 Nominal stress–true strain curves for oriented PP pulled at various angles (marked on the curves) to the molecular direction. (From Ref. 15.)

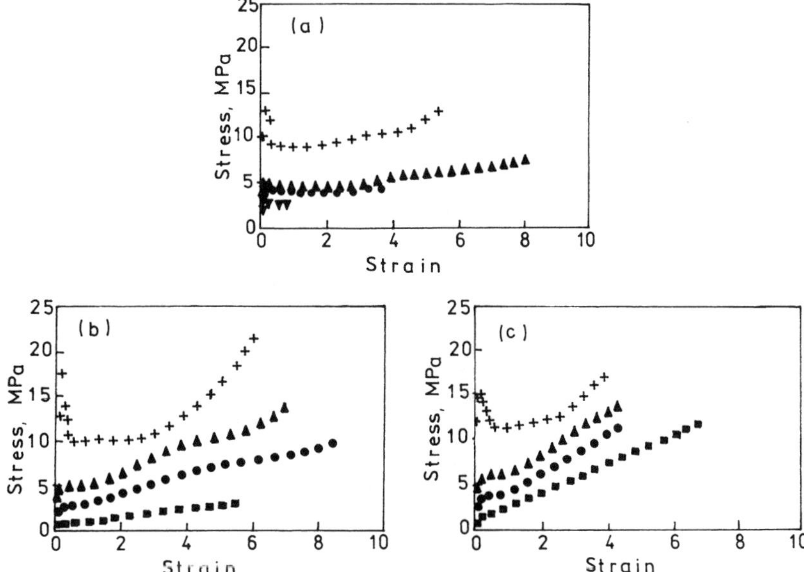

FIG. 6 Stress–strain curves of quenched sample of syndiotactic polypropylene as a function of drawing temperature: (a) $M_w = 69{,}800$, $M_w/M_n = 2.2$, $T_m = 140°C$, (b) $M_w = 152{,}000$, $M_w/M_n = 2.0$; $T_m = 139°C$; (c) $M_w = 490{,}000$, $M_w/M_n = 2.4$; $T_m = 138°C$. All samples drawn at 5 mm min^{-1}. Draw temperature: +, 20°C; ▲, 80°C; ●, 90°C; ▼, 110°C and ■, 130°C. (From Ref. 19.)

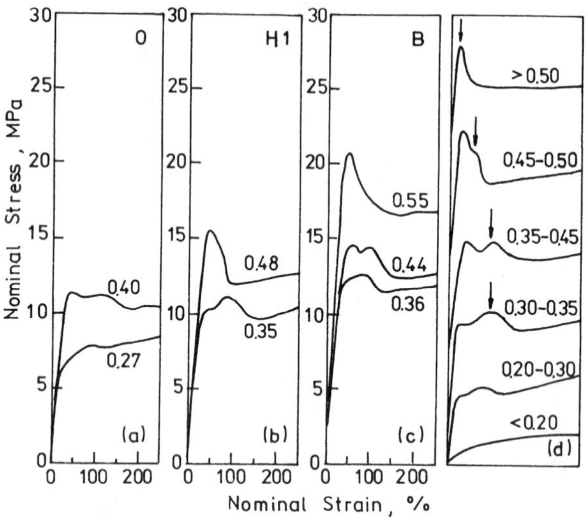

FIG. 7 Stress–strain curves in yield region for specific copolymers: (a) ethylene–octene copolymers; (b) ethylene–hexene copolymers; (c) ethylene–butene copolymers; (d) schematic representation. Core level of crystallinity indicated with each curve. (From Ref. 9.)

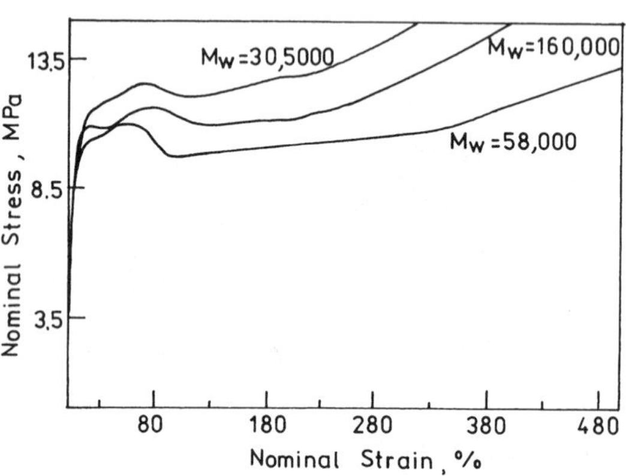

FIG. 8 Stress–strain curves for ethylene–hexene copolymers of indicated molecular weights. Core crystallinity level in range 0.30 to 0.35 for all samples. (From Ref. 9.)

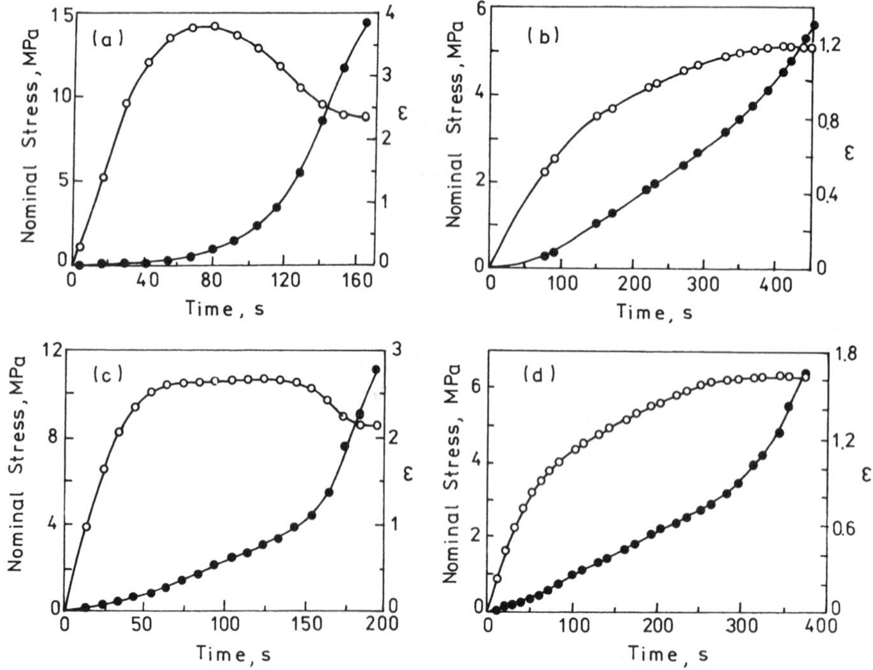

FIG. 9 Nominal stress (○) and true strain (●) versus draw time at the cross-head speed of 5 mm min^{-1}, at 60°C: for ethylene–butene copolymers, (a) $\rho = 0.943\,\mathrm{g\,cm^{-3}}$, $M_n = 30{,}000$, $M_w/M_n = 5.2$, co-unit. mol. = 1.2%, (b) $\rho = 0.910\,\mathrm{g\,cm^{-3}}$, $M_n = 27{,}000$, $M_w/M_n = 5.4$, co-unit. mol. = 7.6% and ethylene–octene copolymers: (c) $\rho = 0.941\,\mathrm{g\,cm^{-3}}$, $M_n = 35{,}000$, $M_w/M_n = 2.2$, co-unit. mol. = 0.9%; (d) $\rho = 0.908\,\mathrm{g\,cm^{-3}}$; $M_n = 37{,}000$, $M_w/M_n = 2.2$, co-unit. mol. = 3.9%. (From Ref. 21.)

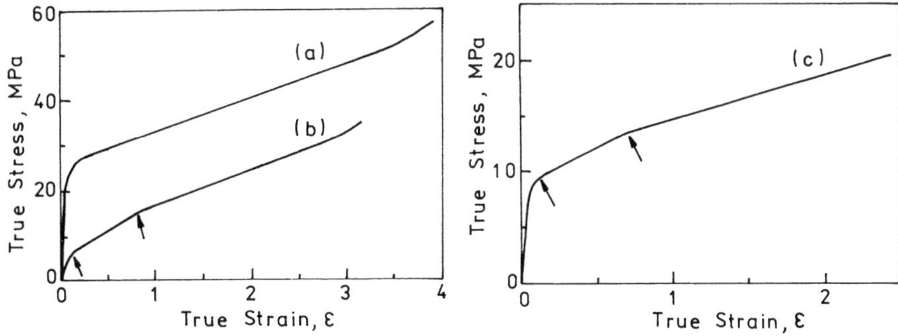

FIG. 10 True stress–strain curves of ethylene–butene copolymers, $M_w = 157{,}000$, $M_n = 30{,}000$, $\rho = 0.945\,\mathrm{g\,cm^{-3}}$, $\alpha_c = 0.67$, at 20°C (curve a) and 80°C (curve c) and ethylene–butene copolymers, $M_w = 146{,}000$, $M_n = 27{,}000$, $\rho = 0.910\,\mathrm{g\,cm^{-3}}$, $\alpha_c = 0.35$ at 20°C (curve b). (From Ref. 22.)

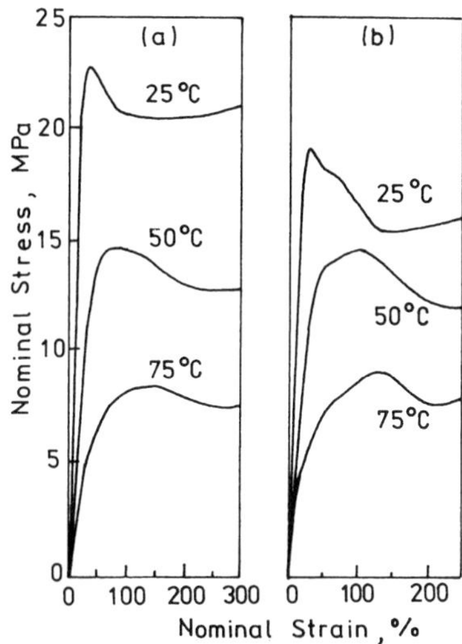

FIG. 11 Stress–strain curves for linear polyethylene as function of temperature. (a): $M_w = 970{,}000$, $M_w/M_n = 4.42$, $\alpha_c = 0.51$, draw rate 0.1 in min^{-1}; (b): $M_w = 970{,}000$, $M_w/M_n = 4.42$, $\alpha_c = 0.37$, draw rate 1 in min^{-1}. (From Ref. 9.)

Properties of copolymers are especially determined by the nature, amount and distribution of comonomer, and by the catalyst system used in their synthesis.

Thanks to the "single-site" nature of the metallocene catalysts a new class of copolymers named "homogeneous copolymers" has been obtained. They have a high molecular weight, narrow molecular weight distribution (MWD), and narrow distribution of chemical composition.

These materials have improved tensile strength and elongation characteristics, higher stiffness at a given density, and better heat and stress-crack resistance compared to conventional highly branched low density polyethylene.

The dominant feature in all of the stress–strain curves of ethylene/α-olefins, beyond the yield region is the development of significant strain hardening. While for homopolymers the strain hardening region only becomes dominant at very high molecular weight ($>10^6$), for the copolymers, beginning at relatively low molecular weights, the slope of the strain hardening region increases with chain length.

Homogeneous copolymers have a strong strain hardening rate which helps to reduce the stress concentration due to localized external effects and to improve resistance towards crack propagation. The crystal thickness is the main structural parameter that governs the occurrence of the homogeneous crystal slip in addition to the experimental parameters such as draw temperature and strain rate. Consequently, the reduced most probable crystal thickness of the homogeneous copolymers compared with the heterogeneous ones, at equivalent crystal content, is suggested to be one of the basic parameters of the improved-use properties of the former kind of materials.

The metallocene catalysts use makes possible the preparation of both the syndiotactic and isotactic polypropylene with low polydispersity (approximately 2–2.5 compared with 6–8 for conventional polypropylene). This characteristic confers the superior mechanical properties of a syndiotactic polymer in respect to polymers prepared by classical procedures.

(Text continues on p 282)

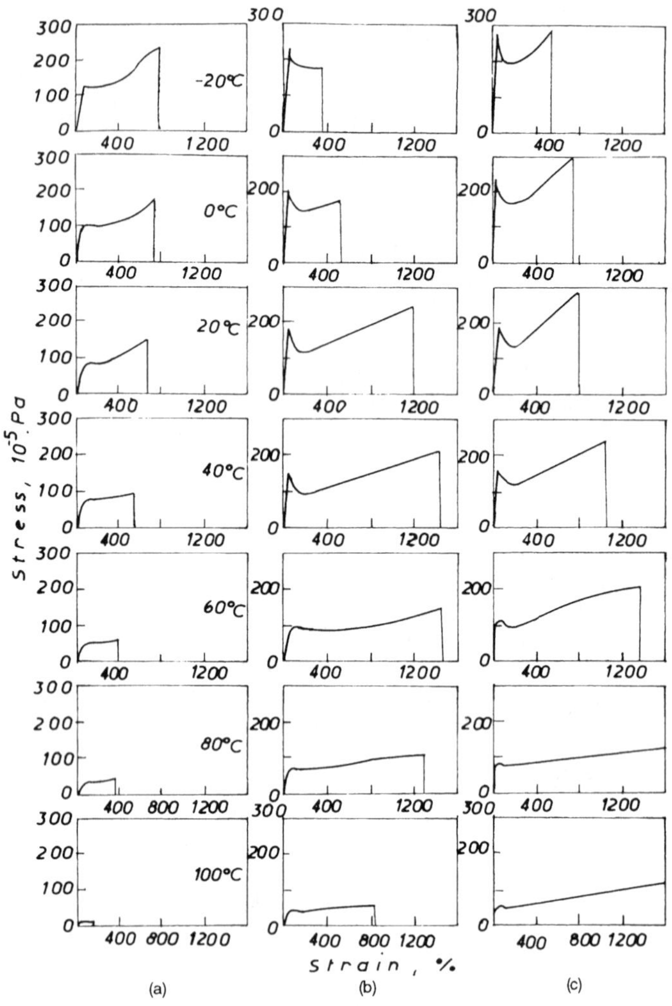

FIG. 12 Stress–strain diagrams of PE with different densities at various temperatures (from −20 to 100°C). Density, g cm^{-3}: (a) 0.918; (b, c) 0.950, Melt flow index, g/10 min: (a, b) 1; (c) < 0.1. (From Ref. 2.)

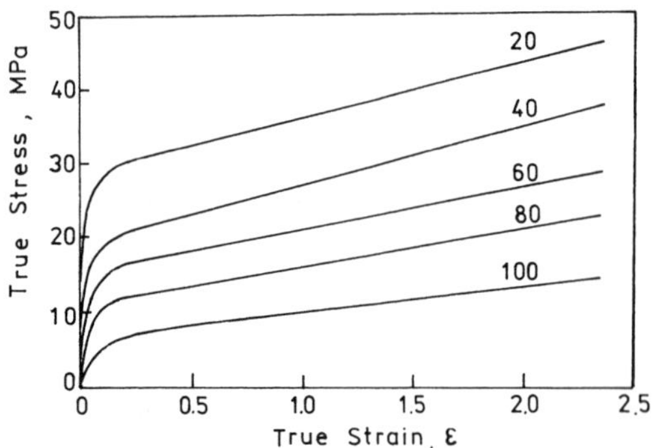

FIG. 13 True stress–strain curves for ethylene–butene random copolymer: $M_w = 157{,}000$, $\rho = 0.950$ g cm^{-3}, $\alpha_c = 0.70$ recorded at a crosshead speed of 5 mm min^{-1} for various draw temperature (T in °C). (From Ref. 24.)

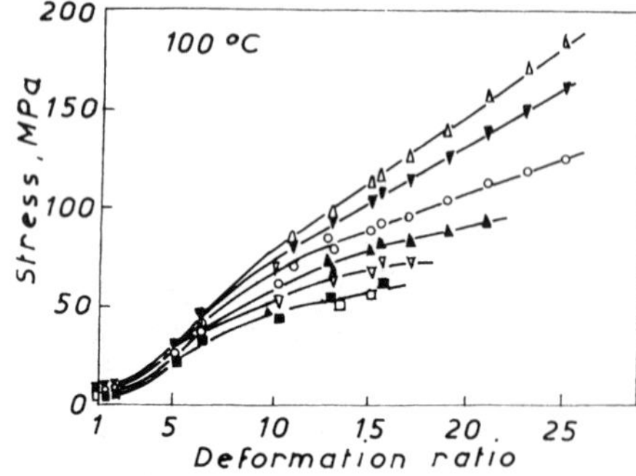

FIG. 14 True stress–deformation ratio–strain rate relationships for Rigidex LPE (slow-cooled) at 100°C. Strain rates s^{-1}: △, 5×10^{-2}; ▼, 2.5×10^{-2}; ○, 1×10^{-2}; ▲, 5×10^{-3}; ▽, 2.5×10^{-3}; ■, 1×10^{-3}; ●, 5×10^{-4}. (From Ref. 31.)

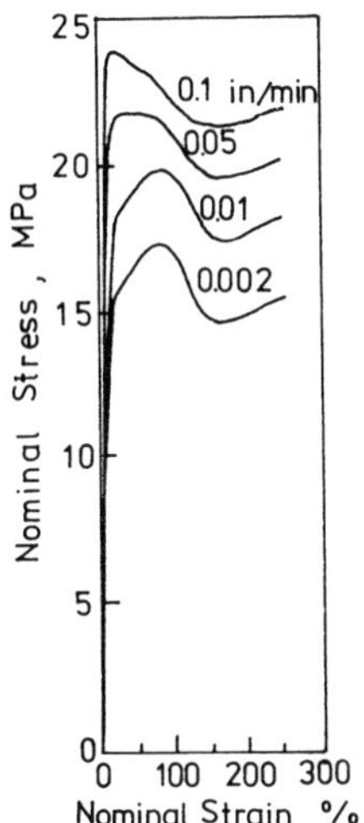

FIG. 15 Stress–strain curves in yield region as a function of indicated draw rates at 25°C for linear polyethylene: $M_w = 970{,}000$, $M_w/M_n = 4.42$, $\alpha_c = 0.51$. (From Ref. 9.)

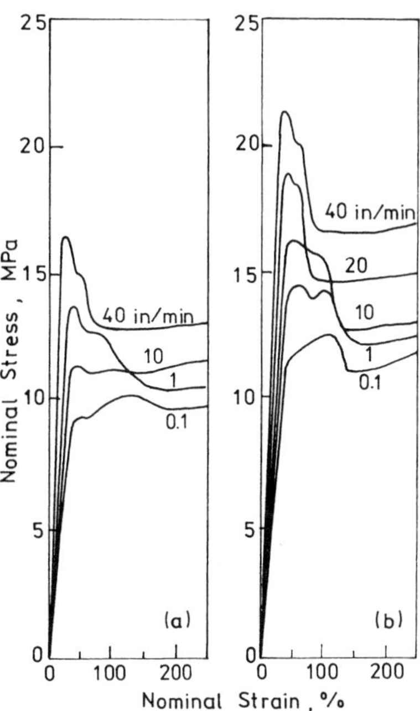

FIG. 17 Stress–strain curves in yield region as a function of indicated draw rates at 25°C. Several of the curves slightly displaced for purposes of clarity. (a) ethylene–octene copolymer: $M_w = 79{,}000$, $M_w/M_n = 2.0$, $\alpha_c = 0.4$; (b) ethylene–butene copolymer: $M_w = 108{,}000$, $M_w/M_n = 2.1$, $\alpha_c = 0.44$. (From Ref. 9.)

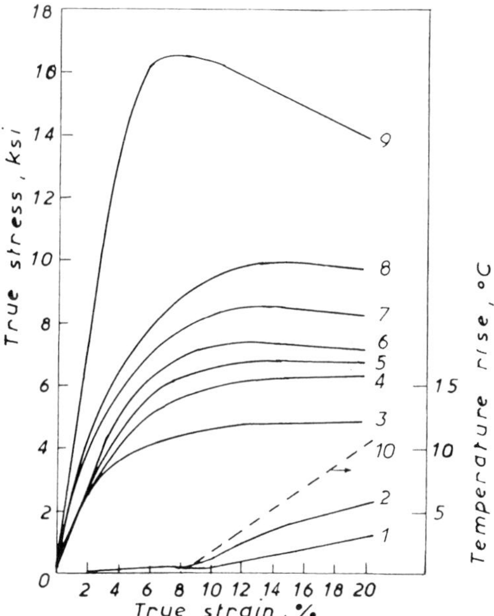

FIG. 16 Stress–strain (curves 1–9) and stress–temperature rise (curve 10) for PP at various strain rates in compression at 22°C. Deformation rate, s^{-1}: curve 1, $\sim 2 \times 10^{-3}$; curve 2, 50; curve 3, 2×10^{-4}; curve 4, 2×10^{-3}; curve 5, 2×10^{-2}; curve 6, 2×10^{-1}; curve 7, 4; curve 8, 50; curve 9 and 10, 1500. (From Ref. 33.)

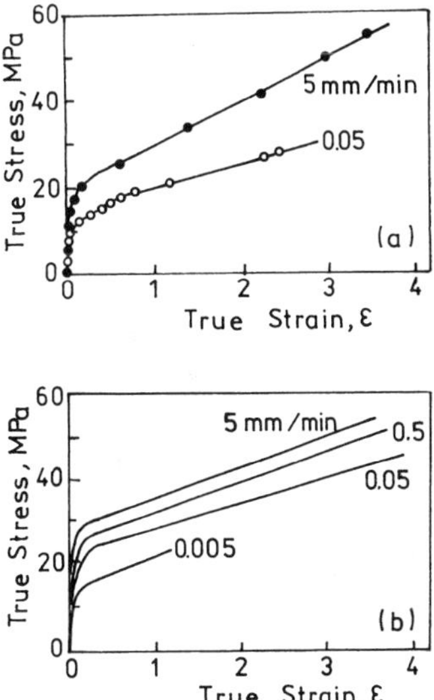

FIG. 18 True stress–strain curve of ethylene–butene copolymers at the draw temperature 20°C for various values of crosshead speed, (a): $M_w = 136{,}000$, $M_n = 31{,}000$, $\alpha_c = 0.55$; (b) $M_w = 157{,}000$, $M_n = 30{,}000$, $\alpha_c = 0.65$. (From Ref. 29.)

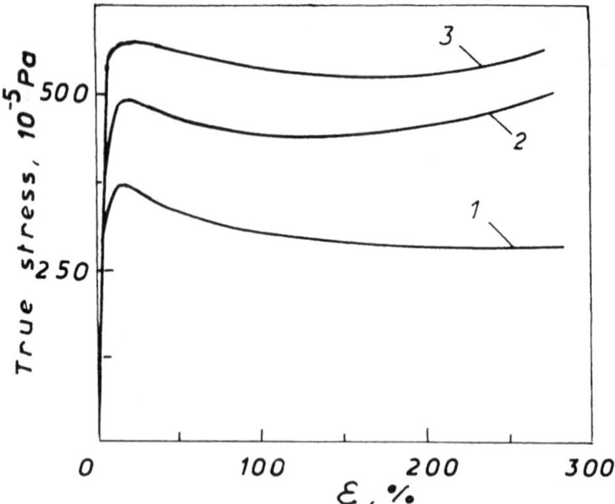

FIG. 19 True tensile stress–strain curves for HDPE at various pressures, kg cm^{-2}: curve 1, 1; curve 2, 1000; curve 3, 2000. (From Ref. 34.)

As a consequence, these copolymers offer remarkable mechanical properties such as superior tensile strength and elongation at break, high tear and puncture resistance, are stress-crack resistance compared with polyethylenes obtained by conventional procedures.

Recently special attention has been paid to the oriented polyolefins such as UHMWHDPE and LLDPE. These can be subjected to high drawing ratios resulting in materials with special properties as ultrahigh modulus and superior tensile strength. The orientation takes place both in the amorphous and crystalline states, so the material exhibits mechanical anisotropy in transversal direction. The elasticity modulus of these materials is very high and close to the theoretical one.

The principal mechanical properties of PP homopolymers are good rigidity and high thermal resistance, with limited impact resistance at low temperature. The main structural factors affecting these properties are isotacticity, molecular weight, and MW distribution, mostly through their influence on crystallinity.

(Text continues on p 293)

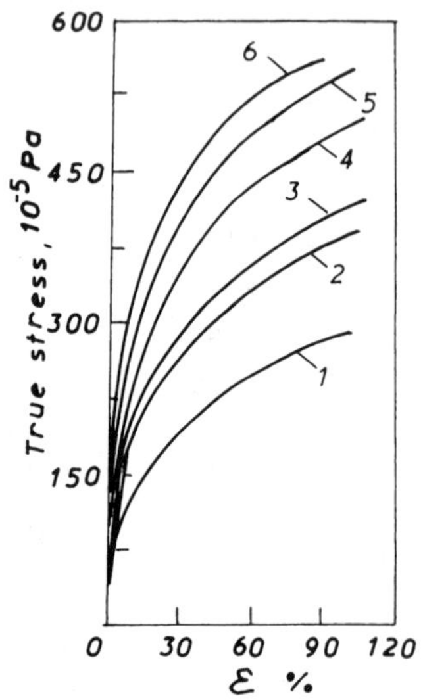

FIG. 20 True tensile stress–strain curves for LDPE at various pressures, kg cm^{-2}: curve 1, l: curve 2, 300; curve 3, 500; curve 4, 1000; curve 5, 1500; curve 6, 2000. (From Ref. 36.)

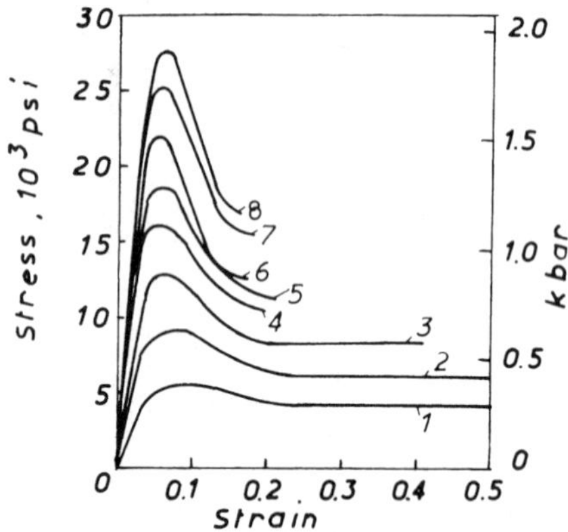

FIG. 21 Nominal tensile stress–strain curves of PP at various pressures obtained at 20°C. Pressure, kbar: curve 1, 1×10^{-3}; curve 2, 1; curve 3, 2; curve 4, 3; curve 5, 4; curve 6, 5; curve 7, 6; curve 8, 7. (From Ref. 37.)

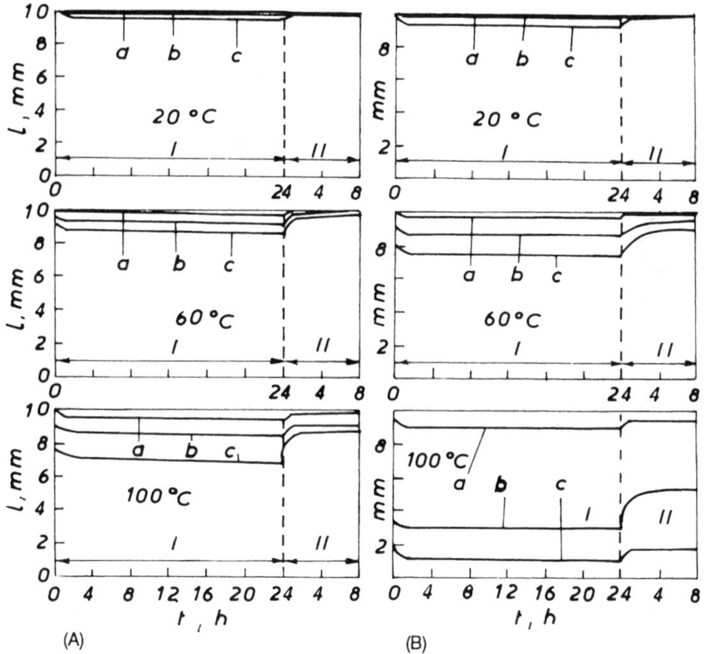

FIG. 22 Compression strain for polyethylene at various temperatures (20, 60 and 100°C). Density, g cm^{-3}: (A) 0.950: (B) 0.918. Melt flow index, g/10 min.: (A) 0.8; (B) 1.0. Pressures, MPa: a, 0.6; b, 1.5; c, 4.0. I, loading; II, unloading. (From Ref. 2.)

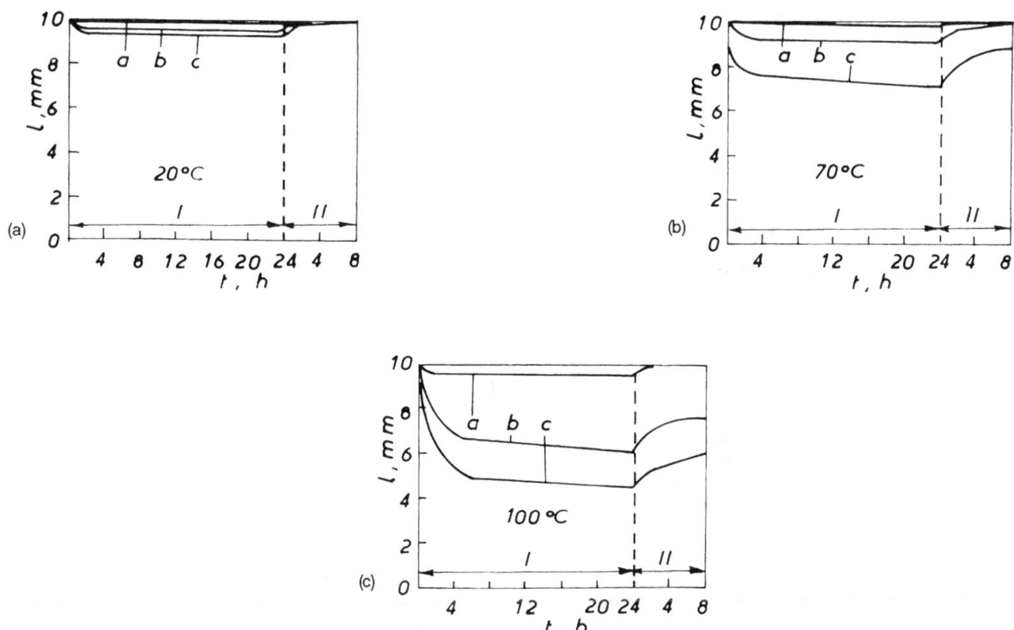

FIG. 23 Compression strain for IPP at various temperatures. Melt flow index $I_m = 0.3$ g/10 min. Density $\rho = 0.905\,\mathrm{g\,cm^{-3}}$. Pressure, MPa: a, 2.5; b, 5.0; c,10.0. I, loading; II, unloading. (From Ref. 2.)

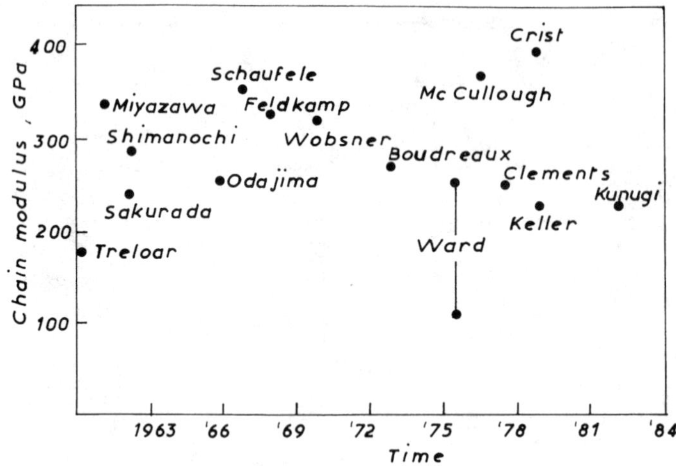

FIG. 24 Chain modulus values for polyethylene are shown as reported by many investigators. (From Refs. 41 and 122.)

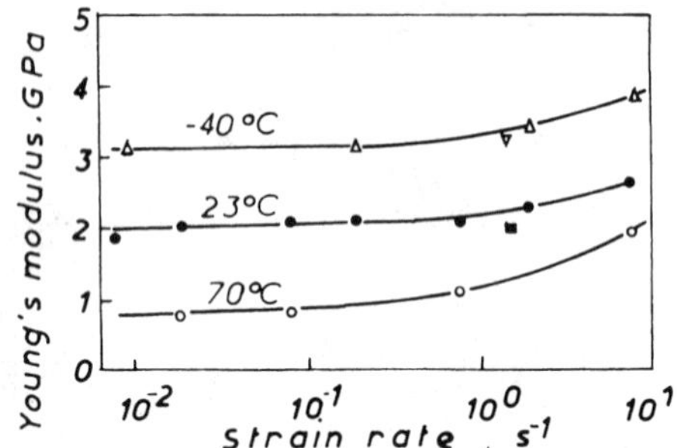

FIG. 25 Modulus of HDPE: (△, ●, ○) data of BASF AG (Germany); (▽, ■) data of Montepolimeri (Italy). (From Ref. 46.)

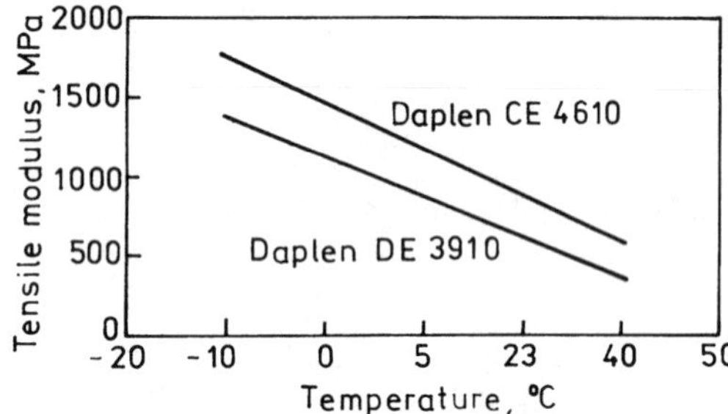

FIG. 26 The tensile modulus in relation to the temperature. Daplen DE 3910, $\rho = 0.94\,\mathrm{g\,cm^{-3}}$, $I_m(190°C/5\,kg)$, g/10 min = 0.8; Daplen CE 4610, $\rho = 0.945\,\mathrm{g\,cm^{-3}}$, $I_m = 0.5$ g/10 min. (From Ref. 47.)

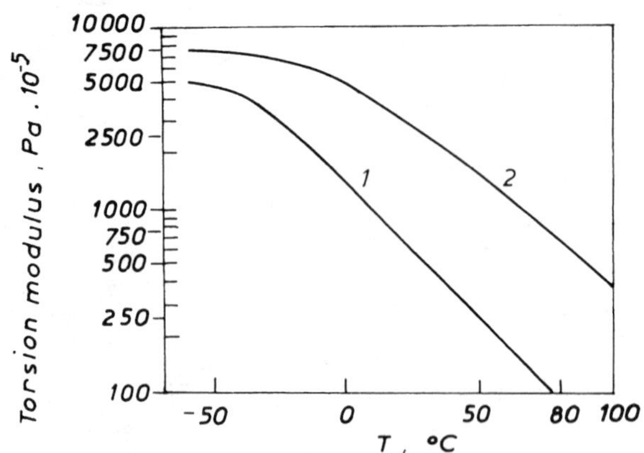

FIG. 27 Dependence of torsion modulus (ASTMD-1043-51) of PE (melt flow index $I_m = 0.3$ g/10 min) on temperature. Density, $\mathrm{g\,cm^{-3}}$: curve 1, 0.918; curve 2, 0.98. (From Ref. 2.)

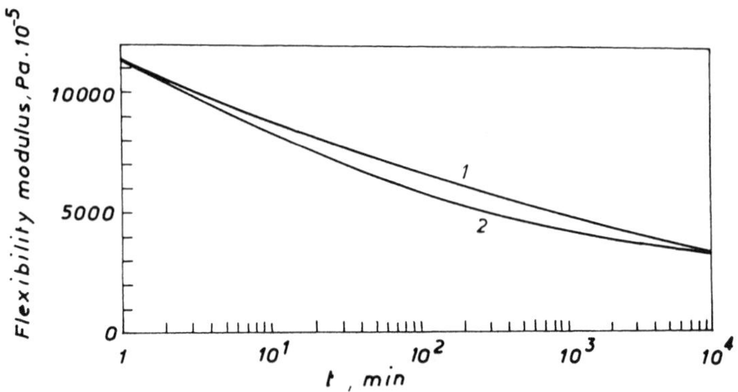

FIG. 28 Dependence of flexibility modulus of PE (density = 0.960 g cm^{-3}; melt flow index $I_m(5) = 10$ g/10 min) on time of deformation at 22°C. Stress, MPa: curve 1, 3; curve 2, 5. (From Ref. 2.)

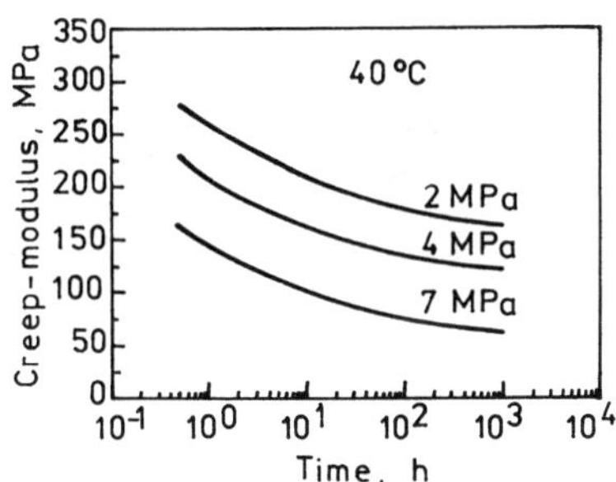

FIG. 29 Creep modulus as a function of time for HDPE, $\rho = 0.94$ g cm^{-3}, $I_m(190°C/5\,kg)$, g/10 min = 0.8. (From Ref. 47.)

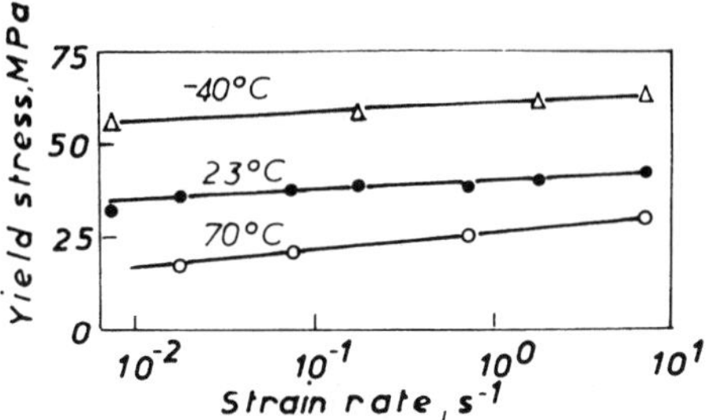

FIG. 31 Yield stress of HDPE vs strain rate. Data of BASF. (From Ref. 46.)

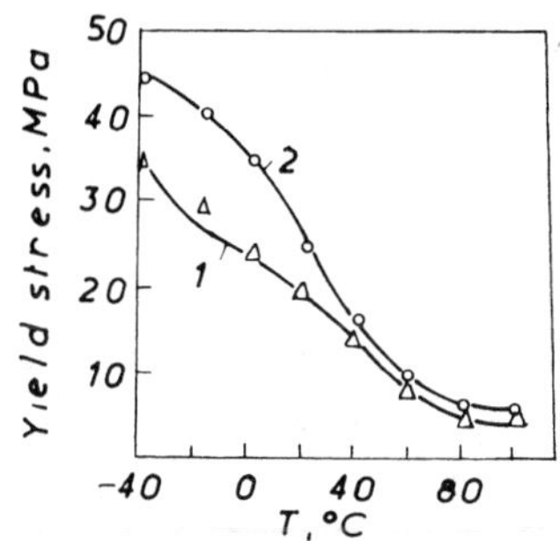

FIG. 30 Yield stress of PE vs temperature. Curve 1 UHMWHDPE, $M_w = 1.5 \times 10^6$, $\rho = 0.937$ g cm^{-3}; curve 2, HDPE, $M_w = 260{,}000$, $\rho = 0.950$ g cm^{-3}. (From Ref. 25.)

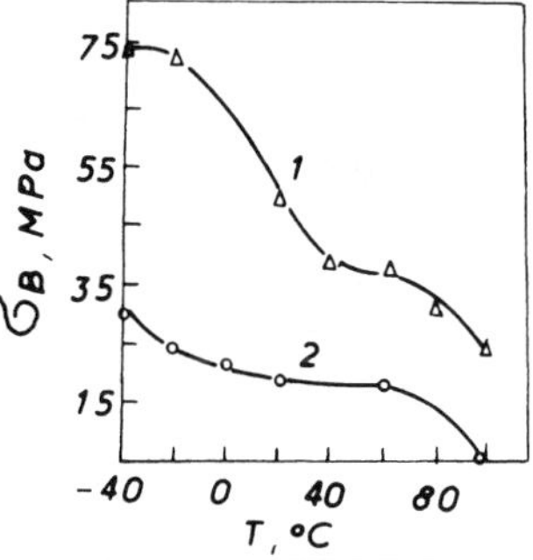

FIG. 32 Dependence of tensile strength of PE on temperature. Curve 1, UHMWHDPE: $M_w = 1.5 \times 10^6$; $\rho = 0.937$ g cm^{-3}. Curve 2, HDPE: $M_w = 260{,}000$, $\rho = 0.950$ g cm^{-3}. (From Ref. 25.)

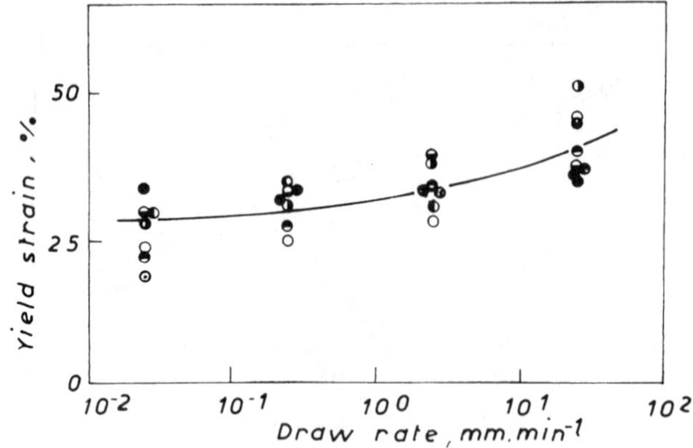

FIG. 33 Yield strain of HDPE plotted against draw rate for various temperature: (○) 80°C; (◑) 50°C; (◒) 20°C; (⊙) 0°C; (◐) −10°C; (●) −20°C; (⊗) −30°C; (◓) −40°C. (From Ref. 50.)

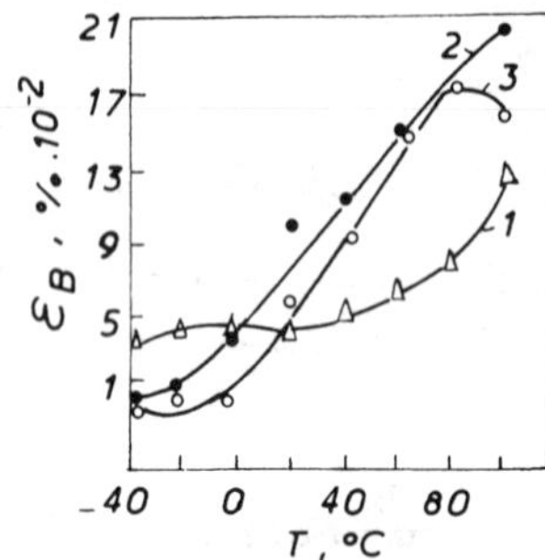

FIG. 34 Dependence of tensile elongation at break in PE specimens on temperature. Curve 1, UHMWHDPE: $M_w = 1.5 \times 10^6$; $\rho = 0.937\ \mathrm{g\,cm^{-3}}$. Curve 2, HDPE: $M_w = 260{,}000$, $\rho = 0.950\ \mathrm{g\ cm^{-3}}$. Curve 3, LDPE: $M_w = 190{,}000$; $\rho = 0.925\ \mathrm{g\,cm^{-3}}$. (From Ref. 25.)

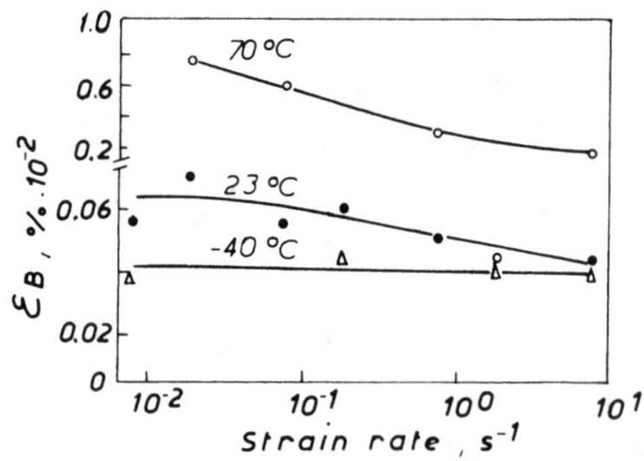

FIG. 35 Tensile elongation at break in HDPE specimens vs strain rate. Data of BASF AG. (From Ref. 46.)

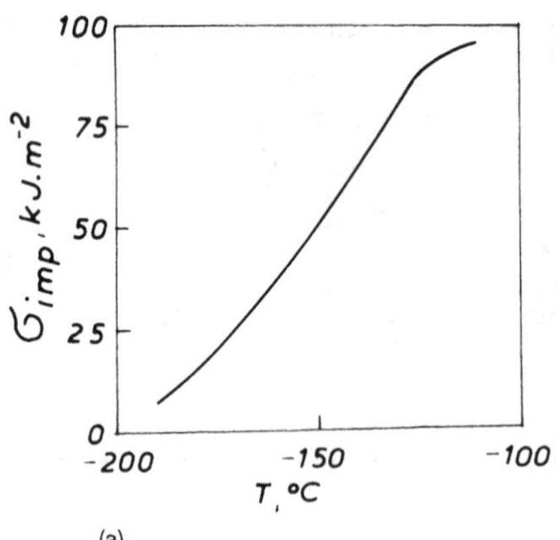

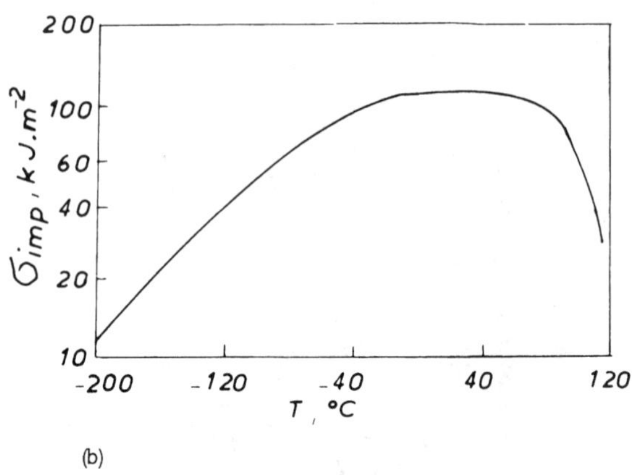

FIG. 36 Dependence of impact strength (DIN 53453) of UHMWPE on temperature: (a) without notch; (b) with sharp (15°) notch. (From Ref. 25.)

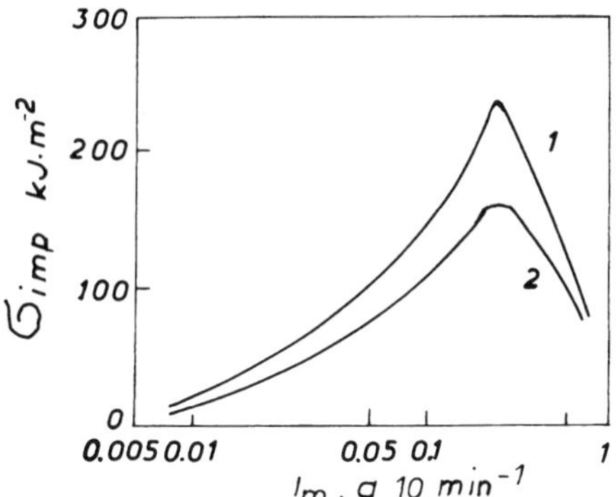

FIG. 37 Dependence of impact strength of UHMWPE at 23°C notched (curve 1) and double-edge notched (curve 2) specimens on melt flow index. (From Ref. 25.)

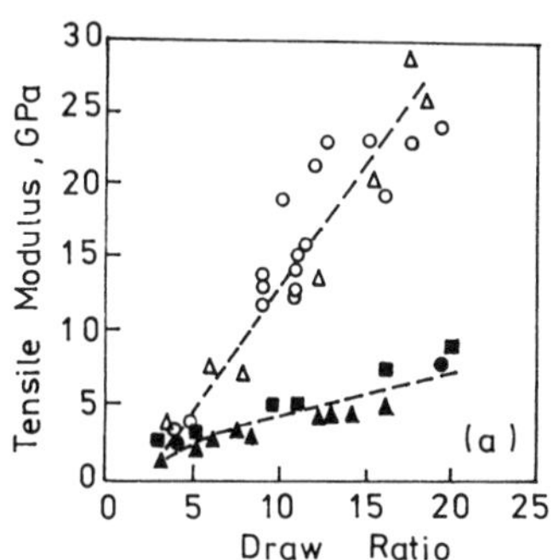

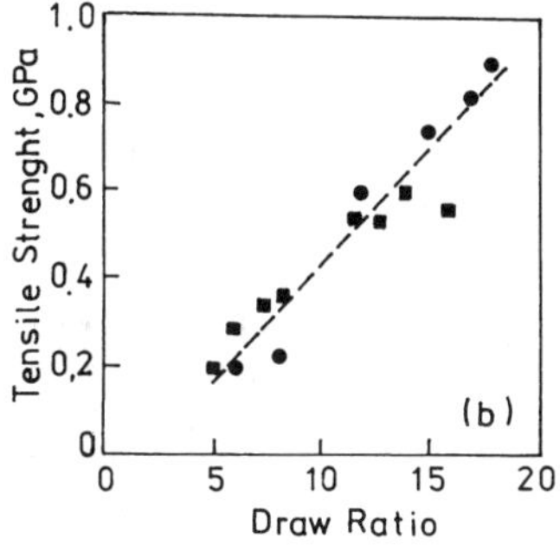

FIG. 38 Relationship between tensile modulus (a), tensile strength (b) and draw ratio for: (a) uniaxially (○, △) and biaxially (●, ▲, ■) oriented UHMWPE gel films; (b) uniaxially (▲) and biaxially (●) oriented UHMWPE gel films. (From Ref. 56.)

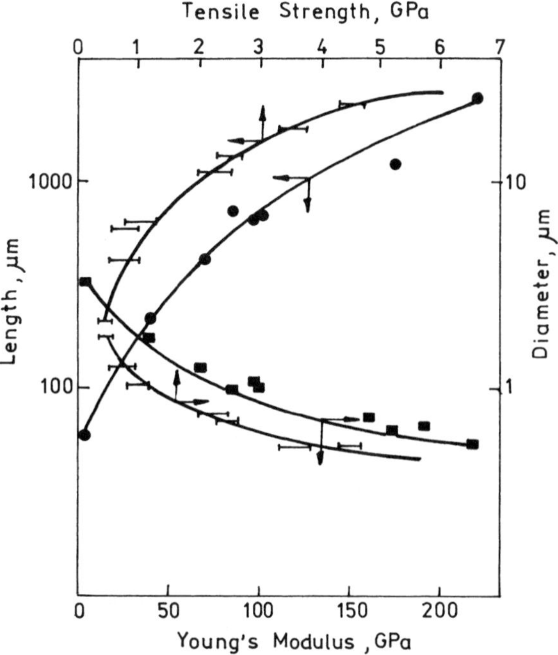

FIG. 39 Variation of the Young's modulus and tensile strength with the dimension of the fibrils of drawn UHMWPE crystals grown from solution. (From Ref. 57.)

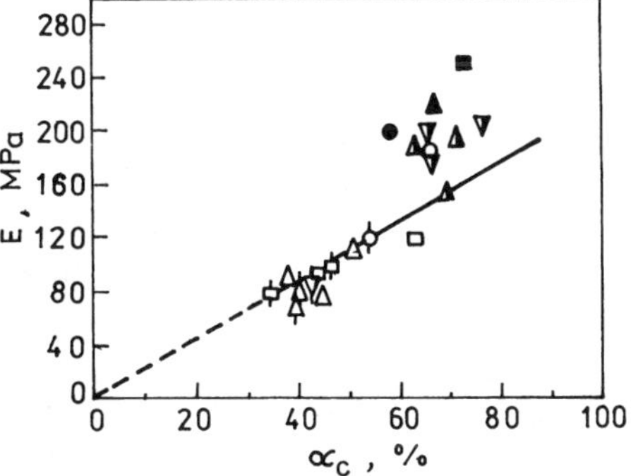

FIG. 40 Plot of initial modulus E against core crystallinity level α_c for molecular weight fractions of linear polyethylene: $M_w = 23{,}000$ (●); $M_w = 38{,}000$ (■); $M_w = 68{,}700$ (▲); $M_w = 96{,}000$ (▽); $M_w = 115{,}000$ (○); $M_w = 142{,}000$ (□); $M_w = 316{,}000$ (△); $M_w = 911{,}000$ (▽). Open symbols indicate ductile deformation, closed symbols brittle and half-filled symbols transitional. (From Ref. 8.)

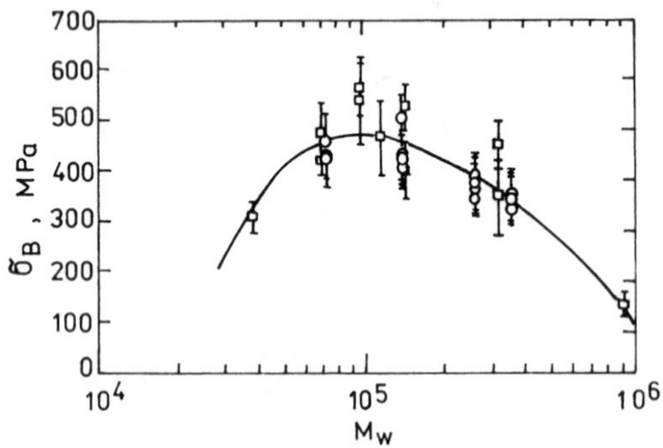

FIG. 41 Plot of the true ultimate tensile strength against log M_w (solid curve), unfractionated linear polyethylene, (□) fractions, (○) linear polyethylene with the most probable molecular weight distributions. (From Ref. 8.)

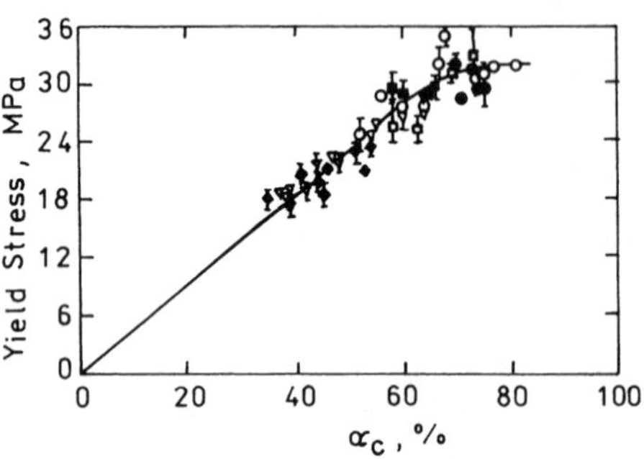

FIG. 42 Plot of yield stress against the crystallinity level as determined from the Raman internal modes for linear polyethylene. (From Ref. 8.)

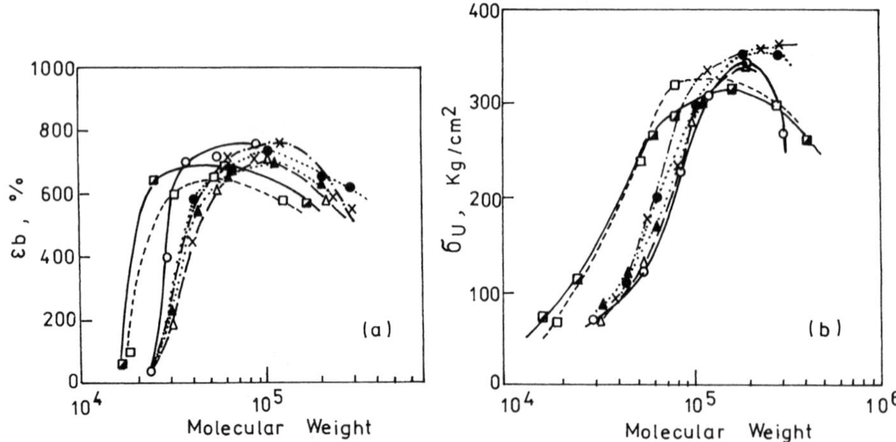

FIG. 43 Molecular weight dependencies of (a) elongation at break, (b) ultimate strength for various LLDPE fractions: (○) I_m (g/10 min) = 0.8, ρ(g cm^{-3}) = 0.920, SCB = 20.1/1000C; (●) I_m (g/10 min) = 0.5, ρ(g cm^{-3}) = 0.920, SCB = 16.4/1000C; (△) I_m (g/10 min) = 0.8, ρ(g cm^{-3}) = 0.919, SCB = 17.4/1000C; (▲) I_m (g/10 min) = 0.8, ρ (g cm^{-3}) = 0.920, SCB = 17.5/1000C; (**Δ**) I_m (g/10 min) = 0.9, ρ(g cm^{-3}) = 0.919, SCB = 16.1/1000C; (×) I_m (g 10 min = 0.8, ρ(g cm^{-3}) = 0.920, SCB = 14.3/1000C; (□) I_m (g/10 min) = 2. 1 ρ(g cm^{-3}) = 0.919, SCB = 14.1/1000C; (◩) I_m (g/10 min) = 1.0, ρ(g cm^{-3}) = 0.921, SCB = 11.0/1000C. (From Ref. 7.)

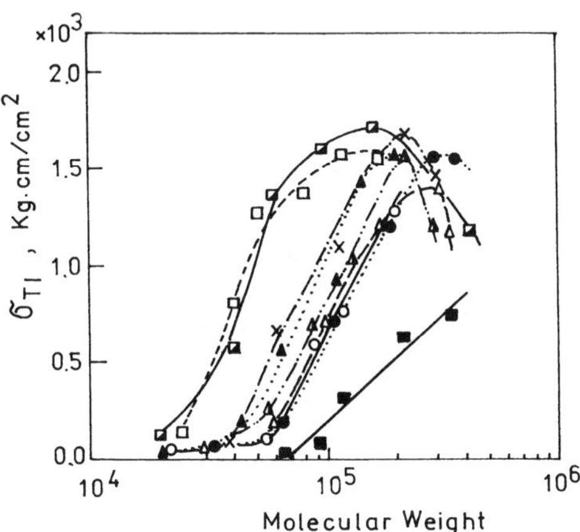

FIG. 44 Molecular weight dependence of tensile impact strength measured at 25°C for LLDPE, symbols are the same as in Fig. 43. (From Ref. 7.)

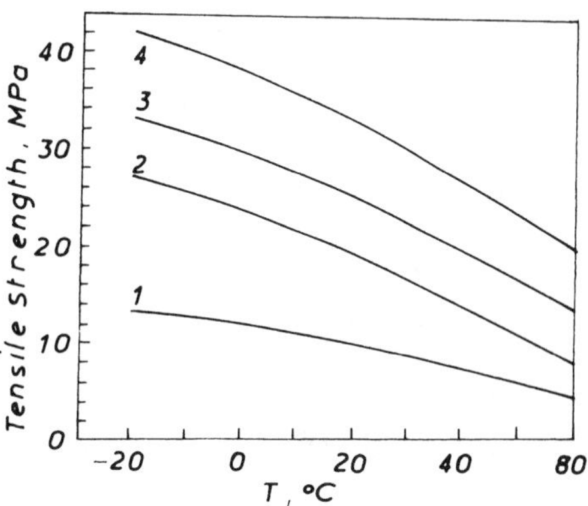

FIG. 46 Dependence of tensile strength of PE with various densities on temperature. Melt flow index $I_m = 18$ g/10 min. Density, g cm^{-3}: curve l, 0.918 ; curve 2, 0.940; curve 3, 0.950; curve 4, 0.965. (From Ref. 2.)

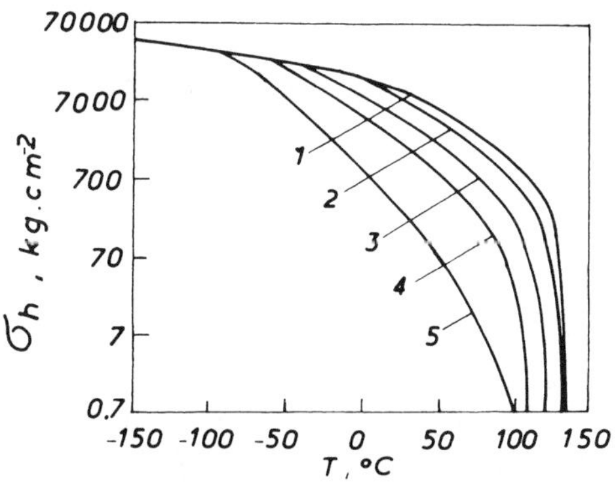

FIG. 45 Dependence of hardness of PE with various densities, on temperature. Density, g cm^{-3}: curve 1, 0.968; curve 2, 0.950; curve 3, 0.935; curve 4, 0.918; curve 5, 0.895. (From Ref. 6.)

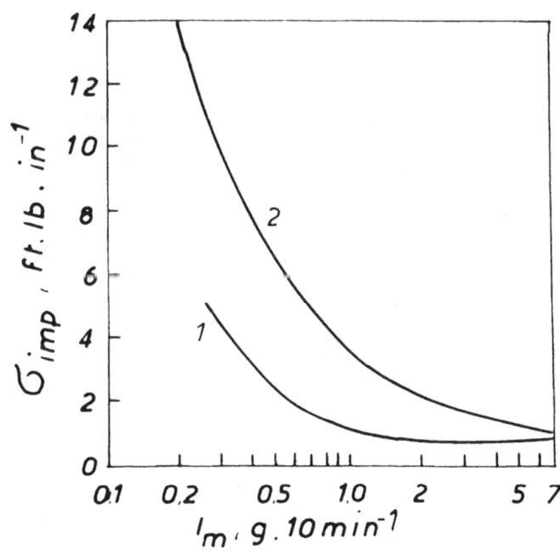

FIG. 47 Dependence of Izod impact strength, (ASTM D-256-56) of PE with different density on melt flow index. Density, g cm^{-3}: curve l, 0.950; curve 2, 0.960. (From Ref. 2.)

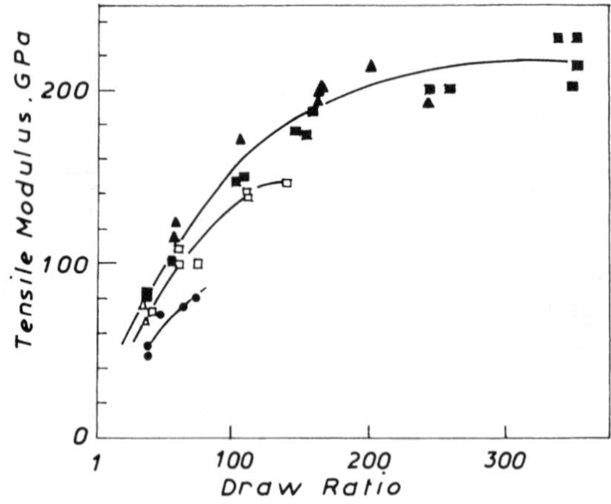

FIG. 48 Variation of the tensile modulus of superdrawn polyethylene crystalline morphologies of different molecular weight grown from solution with drawn ratio. $M_w \times 10^{-5}$: (■) 21; (▲) 15; (□) 5; (●) 2. (From Ref. 30.)

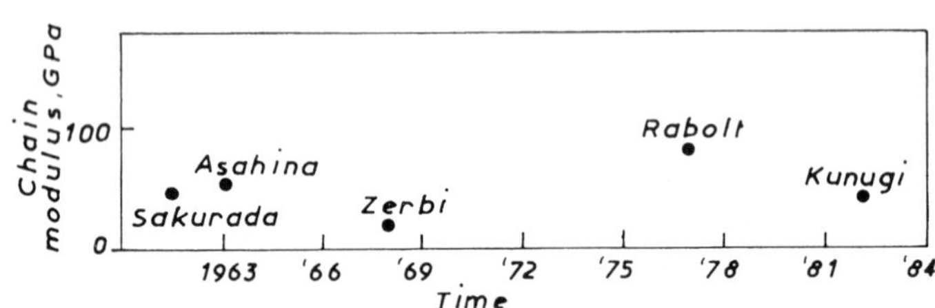

FIG. 49 Chain modulus values for polypropylene are shown as reported by many investigators. (Data from Refs. 41 and 122.)

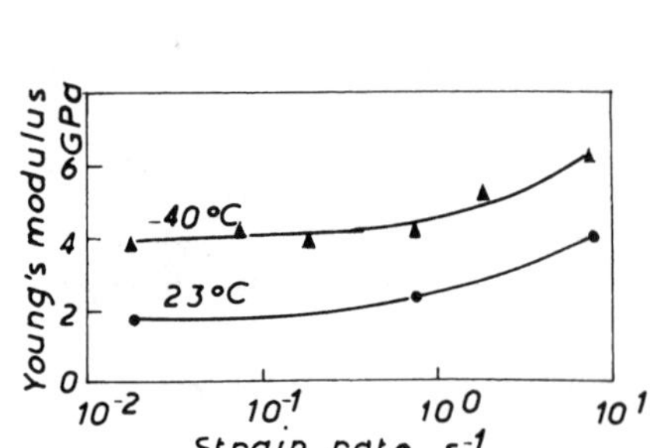

FIG. 50 Young's modulus of PP homopolymer vs strain rate. (Data of BASF AG. From Ref. 46.)

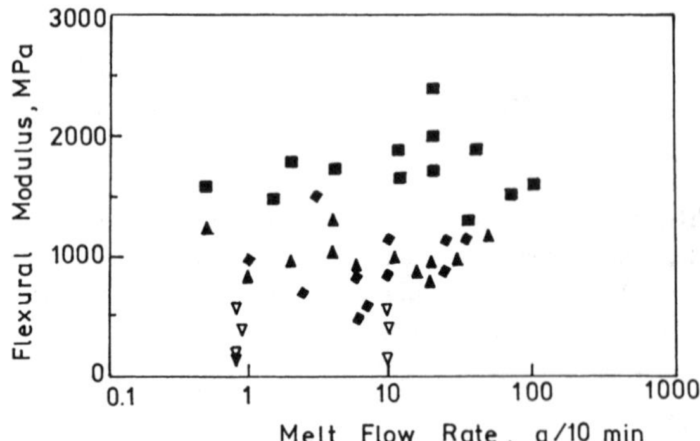

FIG. 51 Flexural modulus dependencies on melt flow rate for PPs: ■ homopolymer, ◆ random copolymer, ▲ impact copolymer, ▽ high alloy copolymer. (From Ref. 78.)

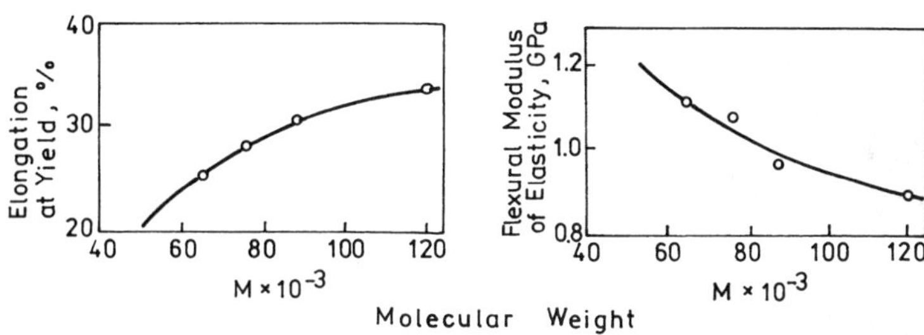

FIG. 52 Molecular weight dependence of elongation at yield and elastic modulus in flexural test of commercial polypropylenes. (From Ref. 79.)

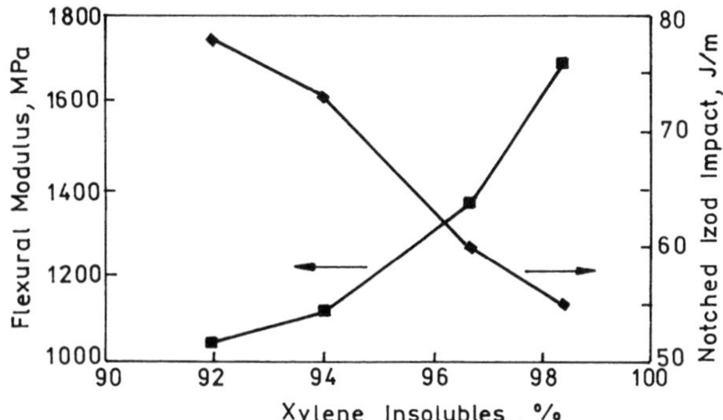

FIG. 53 Effects of tacticity (xylene insolubles) on stiffness and impact for polypropylene, $I_m = 2$ g/10 min. (From Ref. 78.)

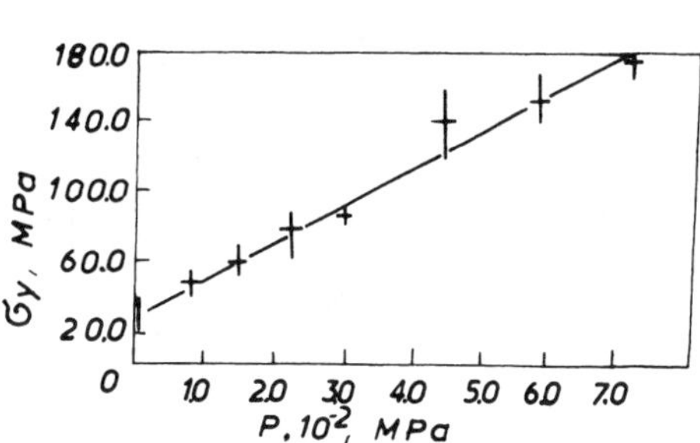

FIG. 55 Yield stress of IPP vs pressure. (From Ref. 80.)

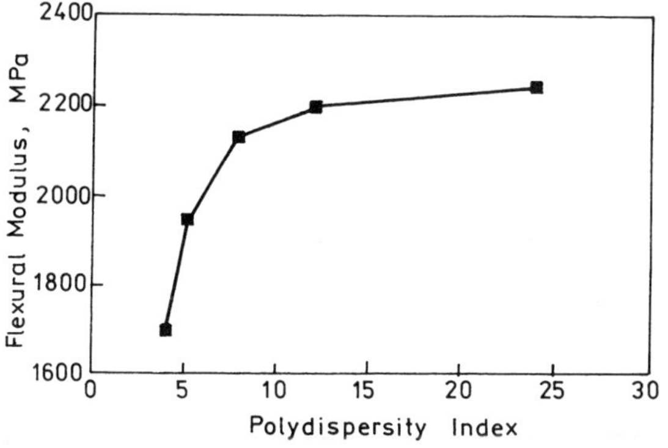

FIG. 54 Effect of MWD on stiffness for polypropylene homopolymer, $I_m = 25$ g/10 min, xylene insolubles = 98%. (From Ref. 78.)

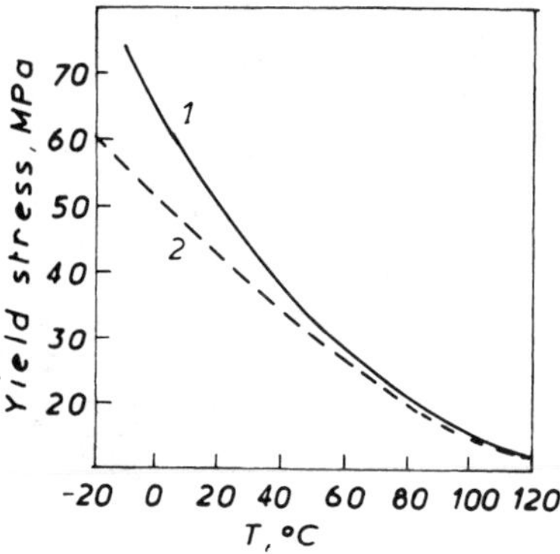

FIG. 56 Yield stress of IPP vs temperature. Melt flow index I_m, g/10 min; curve 1, 0.5; curve 2, 8. (From Ref. 2.)

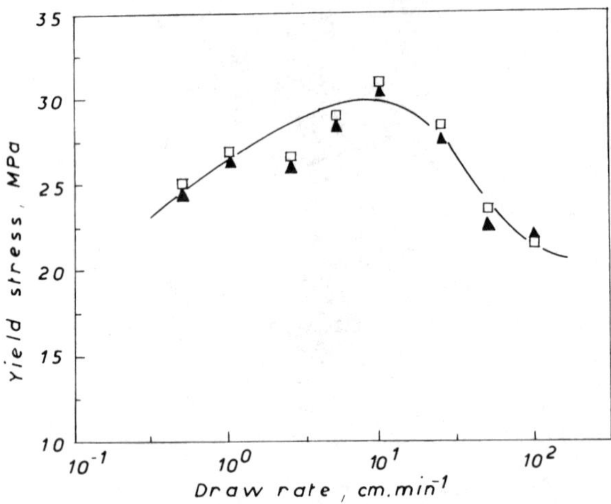

FIG. 57 Effect of draw rate on the yield stress of polypropylene drawn in air (▲) and under water (□). (From Ref. 32.)

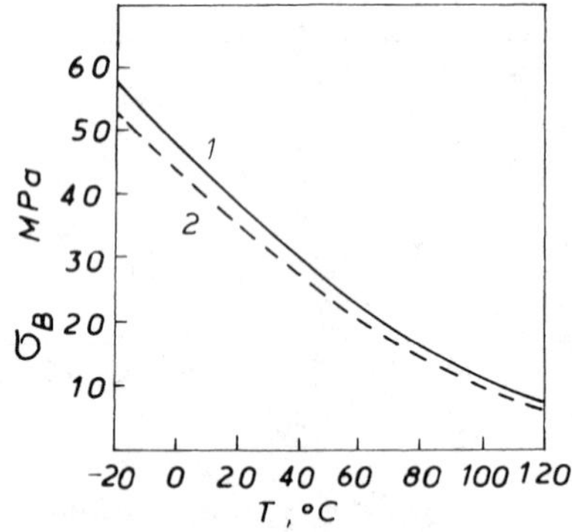

FIG. 58 Tensile strength of IPP vs temperature. Melt flow index I_m, g/10 min: curve 1, 0.5; curve 2, 8. (From Ref. 2.)

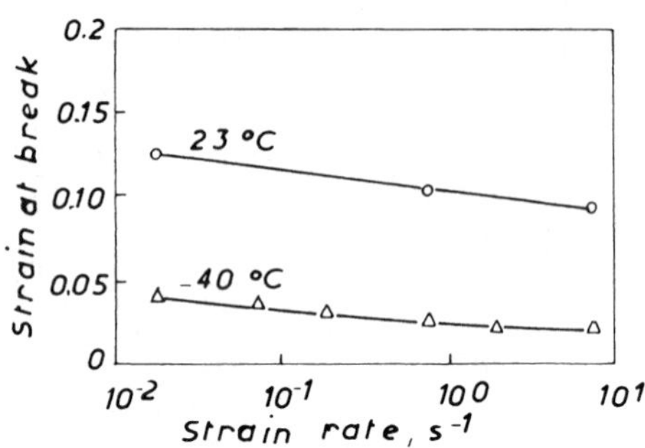

FIG. 60 Tensile elongation at break in IPP homopolymer specimens. (Data of BASF.) (From Ref. 46.)

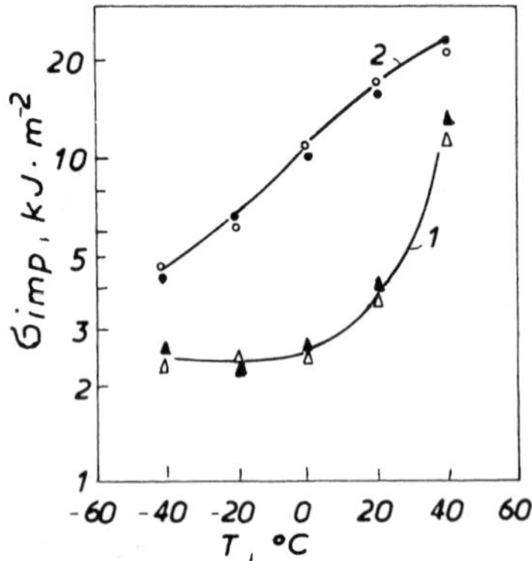

FIG. 61 Charpy impact energy of compression vs temperature. Molded polypropylene bars with 2 mm radius notch. Aging time after molding: (○, △) 1 week; (●, ▲) 10 weeks. Data of TNO (Netherlands). Curve 1, homopolymer Moplene T30S; Curve 2, copolymer (rubber-toughened), propathene GWM 101 (supplied by ICI). (From Ref. 82.)

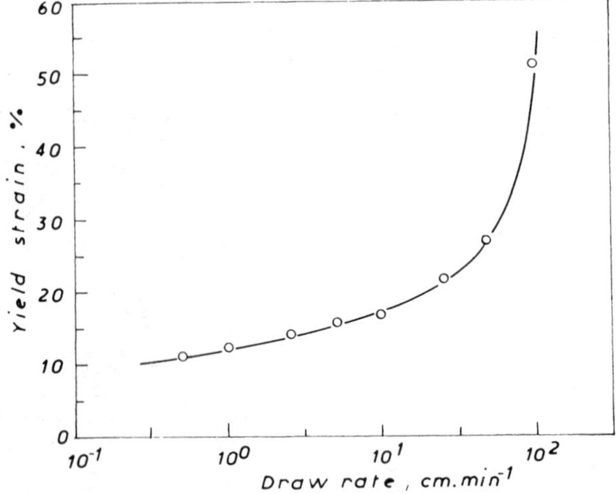

FIG. 59 The effect of draw rate on the yield strain of IPP. (From Ref. 32.)

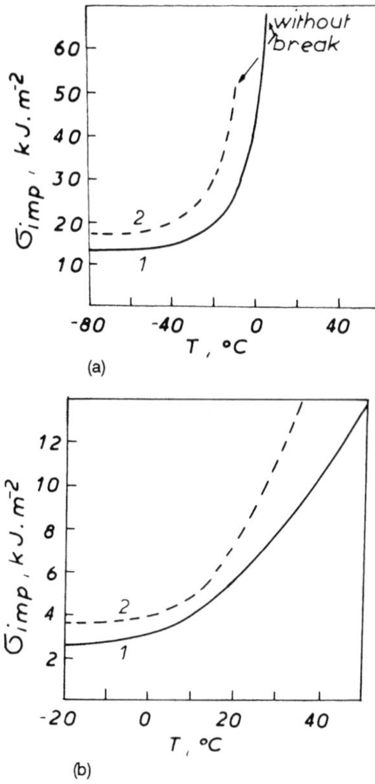

FIG. 62 Dependence of impact strength of IPP on temperature (a) without notch; (b) with notch. Melt flow index, I_m, g/10 min. Curve 1, 0.5; Curve 2, 8. (From Ref. 2.)

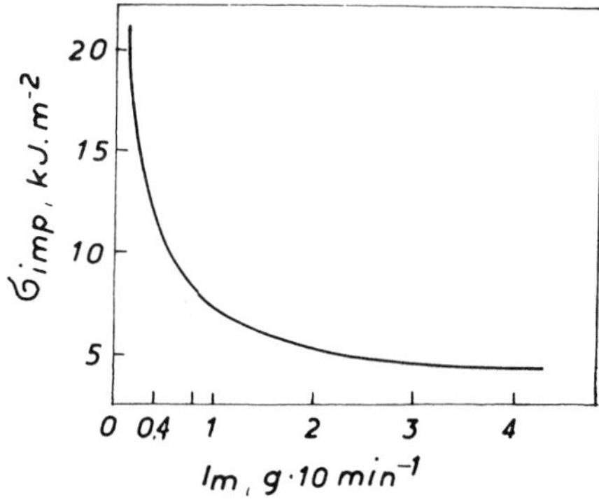

FIG. 63 Dependence of impact strength of IPP on melt flow index. (From Ref. 68.)

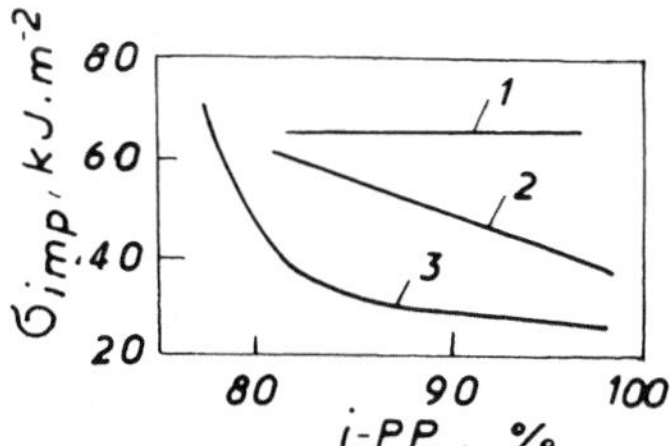

FIG. 64 Dependence of impact strength of IPP on tacticity. Melt flow index, I_m, g/10 min: curve 1, >1.0; curve 2, 0.5–0.6; curve 3, <0.3. (From Ref. 83.)

PP homopolymer and copolymer from metallocene catalysis exhibit narrow MWD, low melting temperature and atactic content and consequently they have superior mechanical properties: hot tack strength, toughness, better elastic recovery.

Both ethylene (e.g., LLDPE) and propylene copolymers, because of their superior mechanical properties, are widely applicable in polymer blends and composites with excellent results.

(Text continues on p 304)

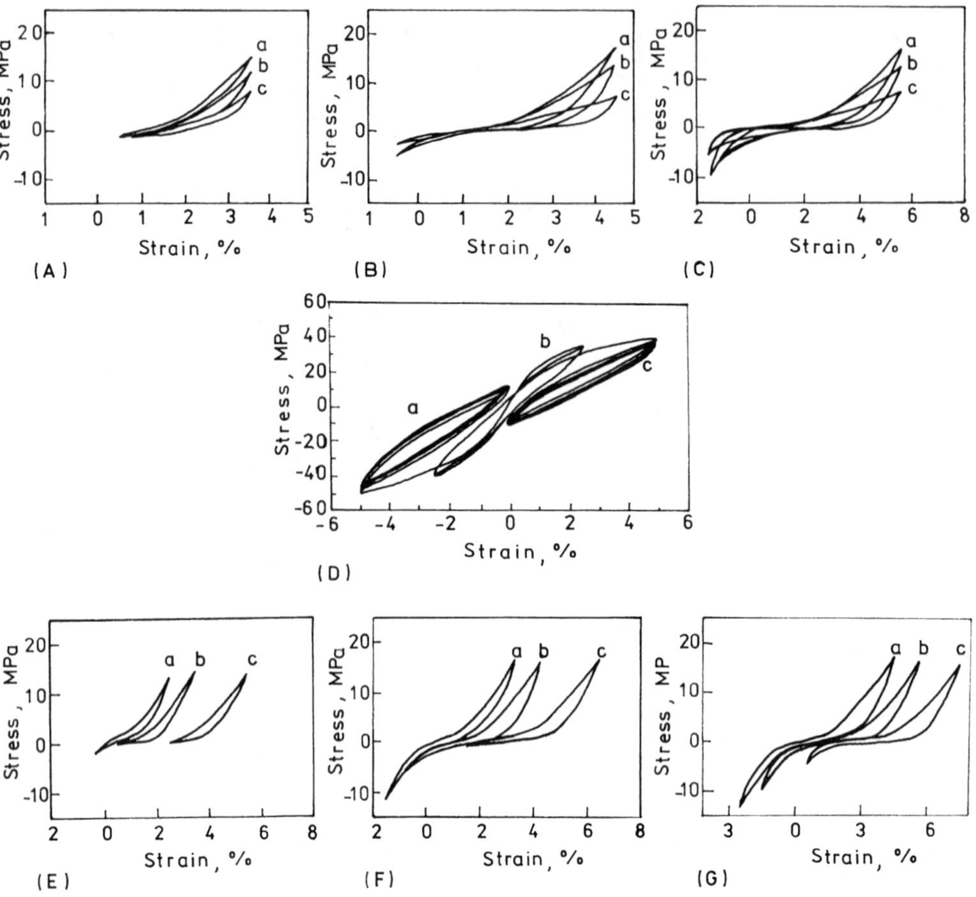

FIG. 65 Stress–strain curves for IPP, $\rho = 0.907$ g cm^{-3}, $\alpha_c = 0.56$, $M_w = 285{,}000$, $I_m = 5$ g/10 min. A: mean strain of 2% at a strain width of 3% at different number of cycles: (a) $N = 35$, (b) $N = 45$; (c) $N = 50$; B: as A at strain width of 5%; C: as A at strain width of 7%; D: strain width of 5%, at first ten number of cycles for three mean strains; (a) −2.5%; (b) 0%; (c) +2.5%; E: strain width $\Delta\varepsilon = 3\%$ after a number of cycles of $N = 35$ at three mean strains: (a) 1%; (b) 2% (c) 4%; F: as E at strain width $\Delta\varepsilon = 5\%$; G: as E at strain width $\Delta\varepsilon = 7\%$. (From Ref. 84.)

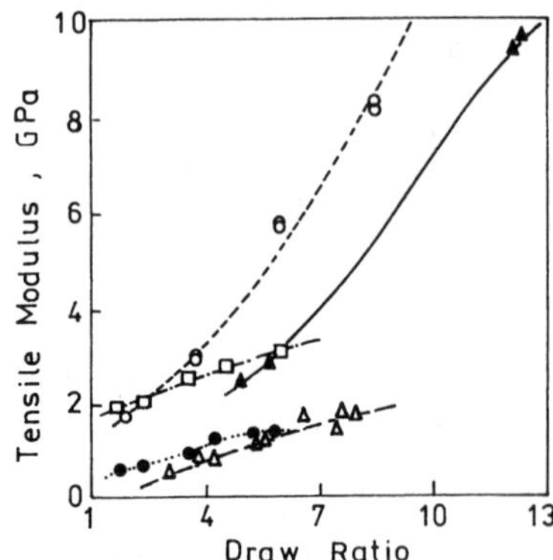

FIG. 66 Tensile modulus vs draw ratio for extrusion drawing of SPP ($\alpha_c = 0.63$) (□); SPP ($\alpha_c = 0.44$) (●) and IPP ($\alpha_c = 0.68$) (○) and for tensile drawing of SPP ($\alpha_c = 0.29$) (△) and IPP ($\alpha_c = 0.57$) (▲). (From Ref. 18.)

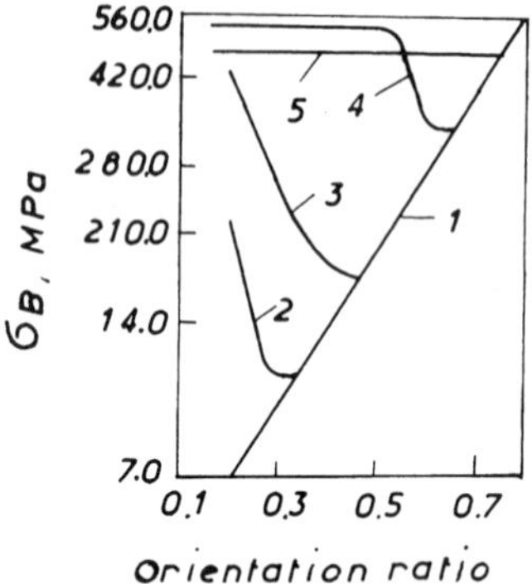

FIG. 67 Dependence of tensile strength (calculated with final cross-sectional area) of oriented IPP on the average ratio of chain orientation. Strain rate, % min^{-1}: curve 1, 10^6; curve 2, 10^5; curve 3, 10^3; curve 4, 10^2; curve 5, 1 and 50. (From Ref. 92.)

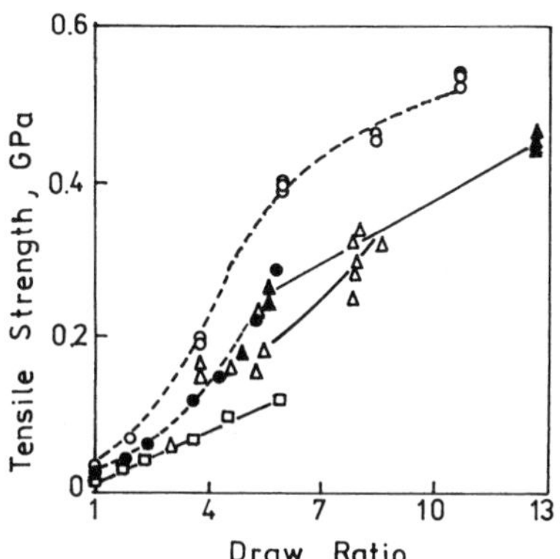

FIG. 68 Tensile strength vs draw ratio for extrusion drawing of SPP, symbols are the same as in Fig. 66. (From Ref. 18.)

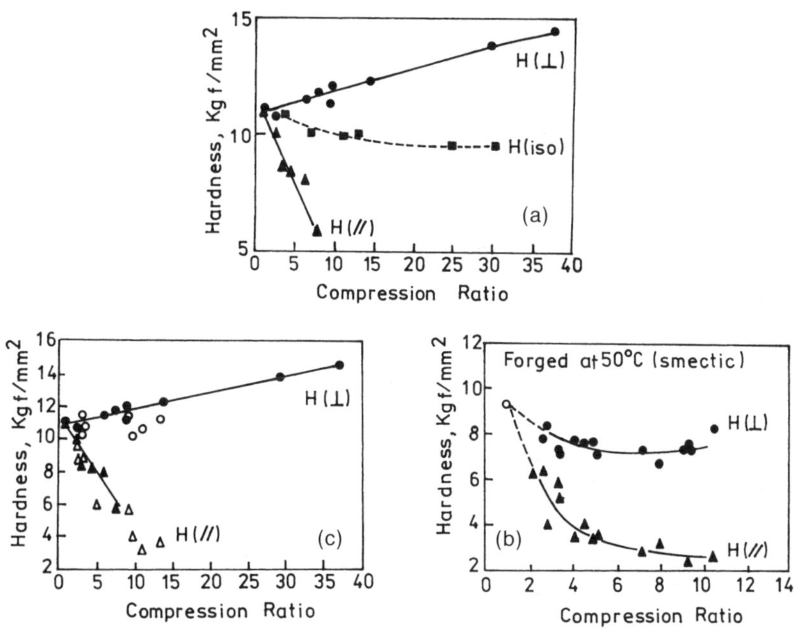

FIG. 69 The microhardness of the IPP, $M_w = 290{,}000$, $I_m = 4$ g/10 min as a function of compression ratio for perpendicular ($H_\perp$) and parallel ($H_\parallel$) to the plane direction: (a) sample forged at 140°C; (b) sample prepared at 50°C; (c) samples forged at 50°C for perpendicular (○) and parallel (△) subsequently heat-treated (140°C for 30 min) and forged at 140°C for perpendicular (●) and parallel (▲) to the plane direction. (From Ref. 94.)

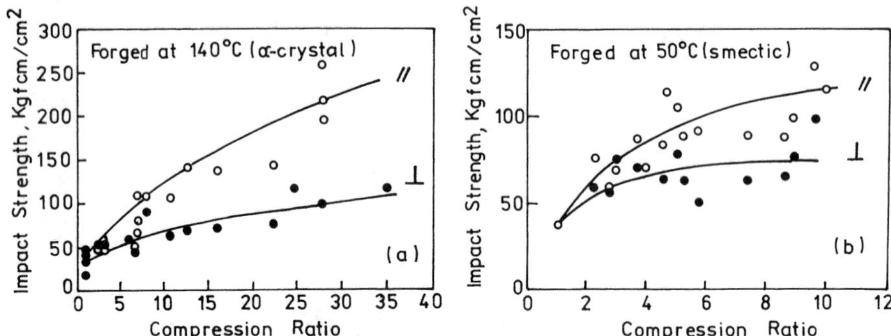

FIG. 70 The impact strength of the IPP vs compression draw ratio, $M_w = 290{,}000$, $I_m = 4$ g/10 min sample prepared at 140°C (α-crystal) (a) and the sample prepared at 50°C (with smectic) (b) tested parallel (○) and perpendicular (●) to the plane direction. (From Ref. 94.)

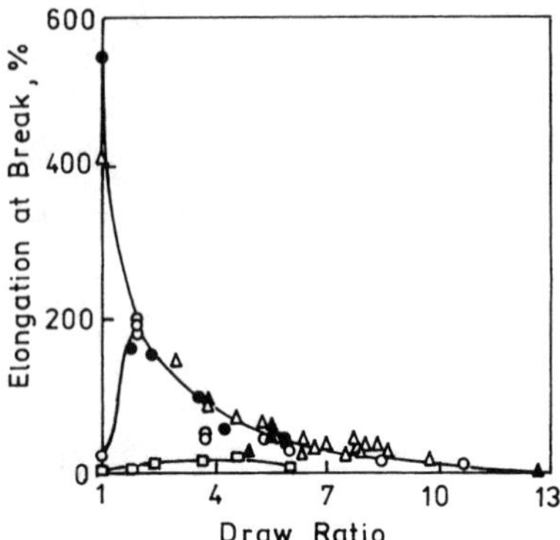

FIG. 71 Elongation at break vs draw ratio for polypropylene, the symbols are the same as in Figs. 67 and 68. (From Ref. 18.)

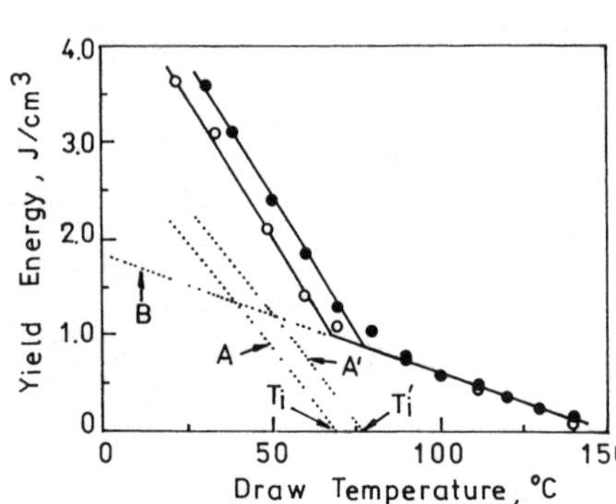

FIG. 72 The yield energy vs draw temperature for polypropylene for two compression speeds: 0.254 (●) and 0.0254 cm min^{-1} (○). The dotted line A (or A′) and B represent the components of amorphous and crystal phases, respectively A and A′ are for higher and lower compression speed, respectively. T_i and T_i' are the intercept temperatures at zero yield energy for A and A′. (From Ref. 56.)

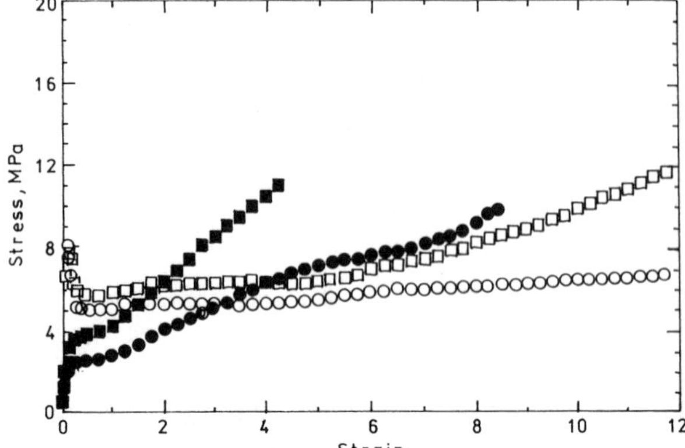

FIG. 73 Stress–strain curves of quenched samples of sindiotactic and isotactic polypropylene drawn at 5 mm min^{-1} and 110°C: (■), $M_w = 490{,}000$, $M_w/M_n = 2.4$, $\alpha_c = 0.21$, syndiotactic; (●), $M_w = 152{,}000$, $M_w/M_n = 2.0$, $\alpha_c = 0.2$, syndiotactic; (□), $M_w = 465{,}000$, $M_w/M_n = 2.1$, isotactic; (○), $M_w = 135{,}000$, $M_w/M_n = 1.9$, isotactic. (From Ref. 19.)

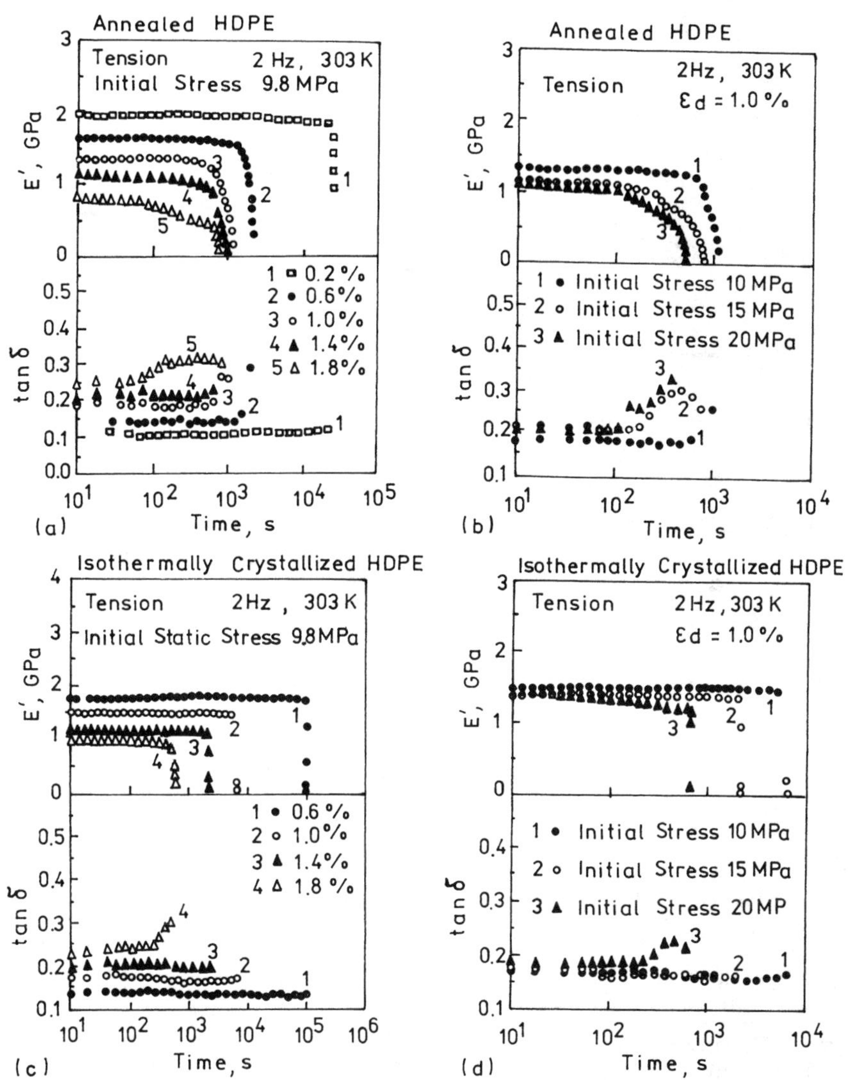

FIG. 74 Variations of E′ and tan δ with time during the fatigue process. (a) The annealed HDPE, the initial static stress of 9.8 MPa under various dynamic imposed strain amplitudes; (b) the annealed HDPE, the imposed dynamic strain amplitude of 1% under various initial static stresses; (c) the isothermally crystallized HDPE, the initial static stress of 9.8 MPa under various dynamic imposed strain amplitudes; (d) the isothermally crystallized HDPE, the dynamic strain amplitude of 1.0% under various initial static stresses. (From Ref. 107.)

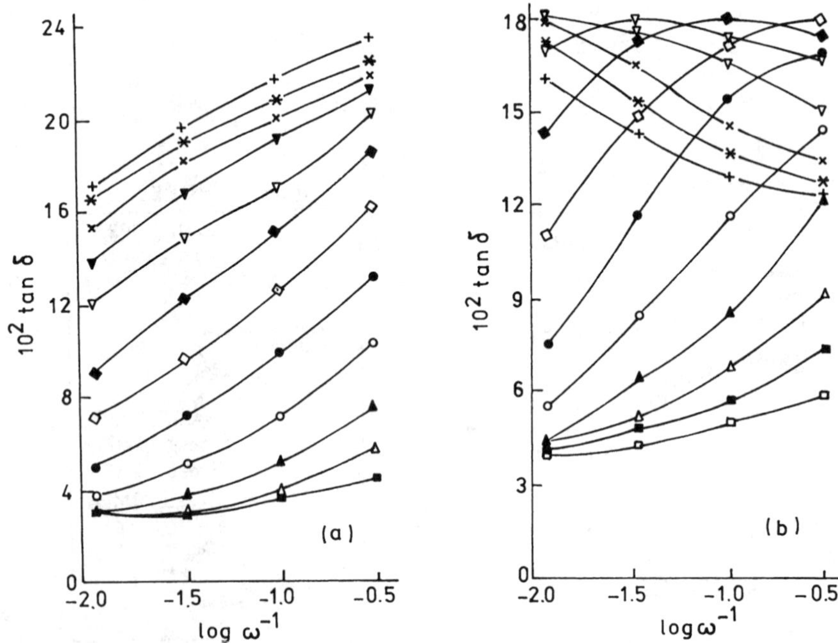

FIG. 75 Tan δ as a function of logarithmic reciprocal frequency for HDPE, $\rho = 0.952$ g cm^{-3}, $I_m = 0.5$ g/10 min (190°C), $M_w = 175{,}000$, $M_n = 1 \times 10^6$, for the draw ratio $\varepsilon_{or} = 1$ (a) and $\varepsilon_{or} = 12.2$ (b) at several temperatures: 0°C (□); 10°C (■); 20°C (△); 30°C (▲); 40°C (○); 50°C (●); 60°C (◇); 70°C (◆); 80°C (▽); 90°C (▼); 100°C (×); 110°C (∗); 120°C (+). (From Ref. 61.)

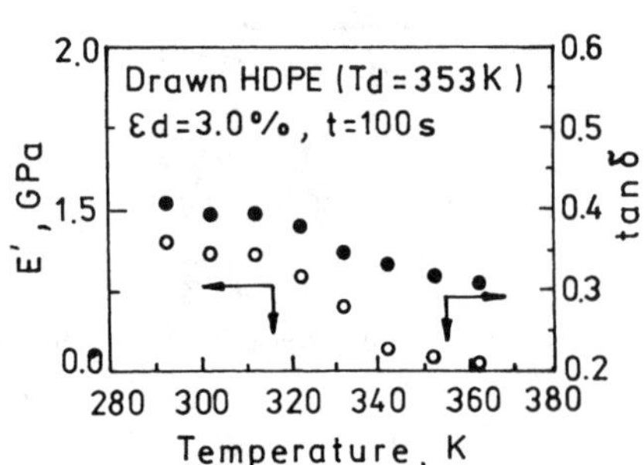

FIG. 76 Temperature dependencies of the magnitude of E′ and tan δ for the drawn HDPE at 353 K in the case of cyclic fatigue after 100 s from the start of fatigue test at $\varepsilon_d = 3.0\%$. (From Ref. 108.)

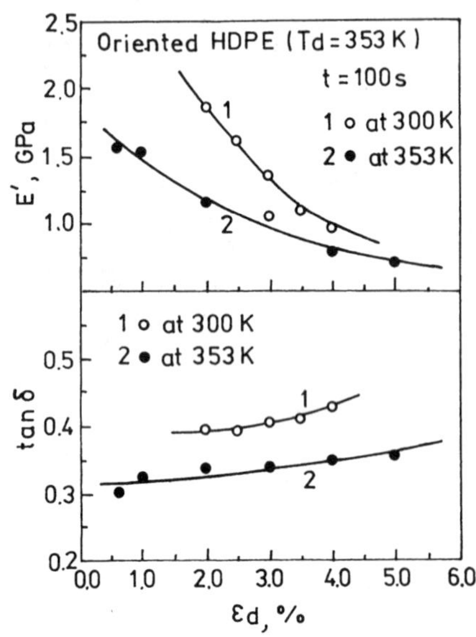

FIG. 77 Variation of E' and tan δ for HDPE with imposed strain amplitudes at the ambient temperatures of 300 K and 353 K. (From Ref. 108.)

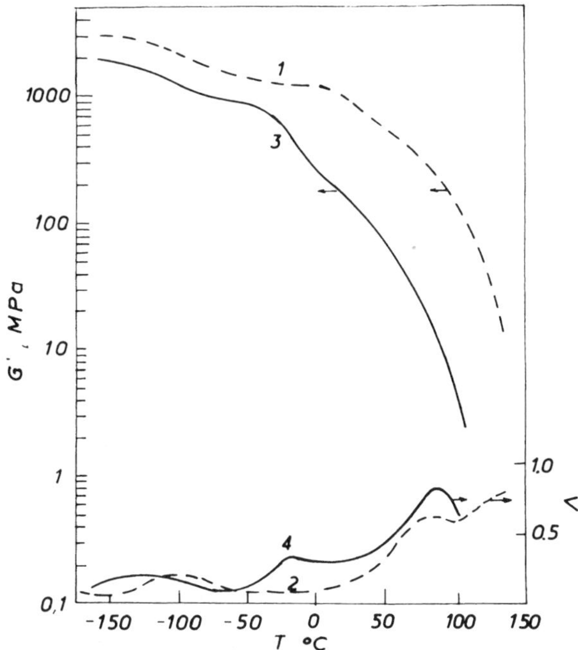

FIG. 78 Dependence of mechanical loss spectra and dynamic torsion G′ modulus of LLDPE at 1 Hz (by Schmieder-Wolf) [99] on temperature. $E' = 2G'(1 + \mu)$, where μ is the Poisson constant. Curves 1, 2, one CH_3 per 1000 carbons. Density $\rho = 0.960$ g cm^{-3}; curves 3, 4, 30 CH_3 per 1000 carbons. Density $\rho = 0.918$ g cm^{-3}. (From Ref. 2.)

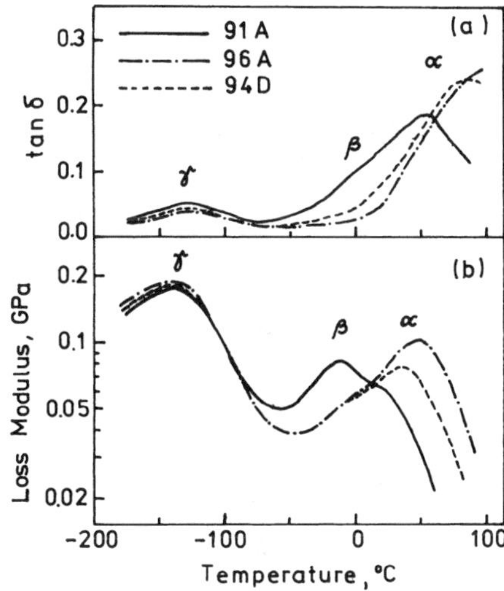

FIG. 79 Temperature dependence of the (a) loss tangent and (b) loss modulus for LLDPE Sclair 91A: $M_w = 123{,}000$, $M_n = 23{,}700$, SCB = 20/1000C; Sclair 94D: $M_w = 110{,}000$, $M_n = 26{,}800$, SCB = 6/1000C: Sclair 96A: $M_w = 193{,}000$, $M_n = 17{,}800$, SCB = 2.5/1000C. (From Ref. 110.)

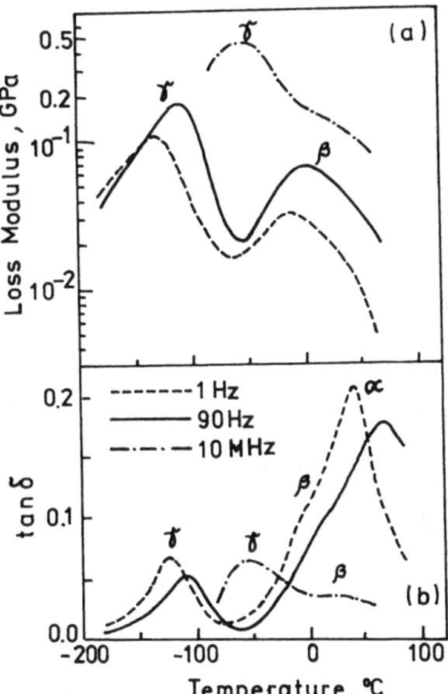

FIG. 80 Temperature dependence of the (a) loss modulus and (b) loss tangent of isotropic LLDPE at frequencies of 1 Hz, 90 Hz and 10 MHz. (From Ref. 111.)

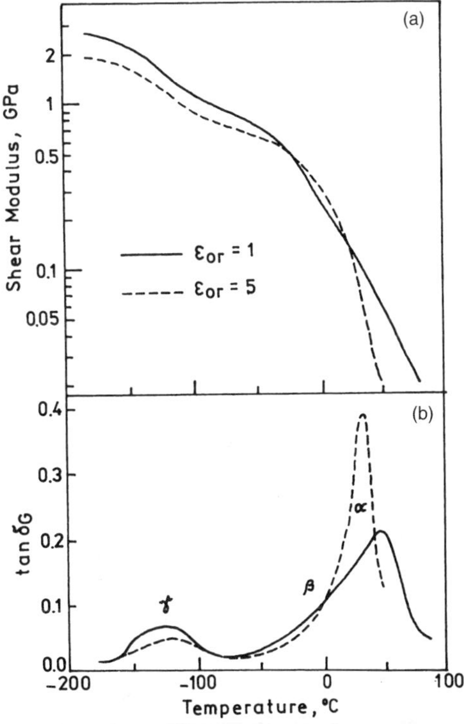

FIG. 81 Temperature dependencies of the (a) shear modulus and (b) loss tangent of LLDPE obtained from torsional measurement at 1 Hz for isotropic sample and oriented sample ($\varepsilon_{or} = 5$). (From Ref. 111.)

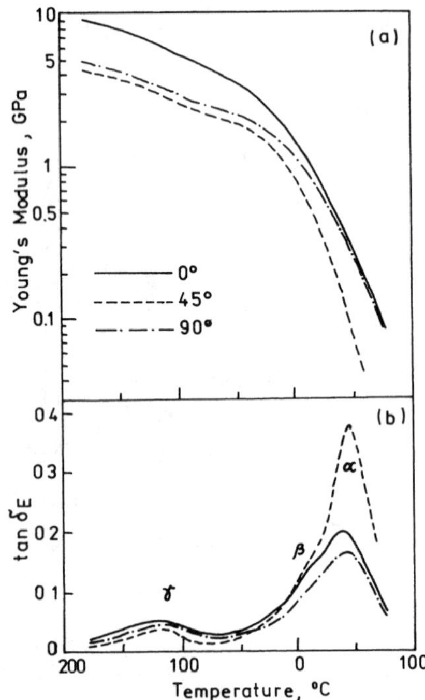

FIG. 82 Temperature dependence of the (a) Young's modulus and (b) loss tangent at 0°, 45° and 90° to the draw axis measured at 10 Hz for LLDPE, $M_w = 123{,}000$, $M_n = 23{,}000$, SCB = 20/1000*C*. (From Ref. 111.)

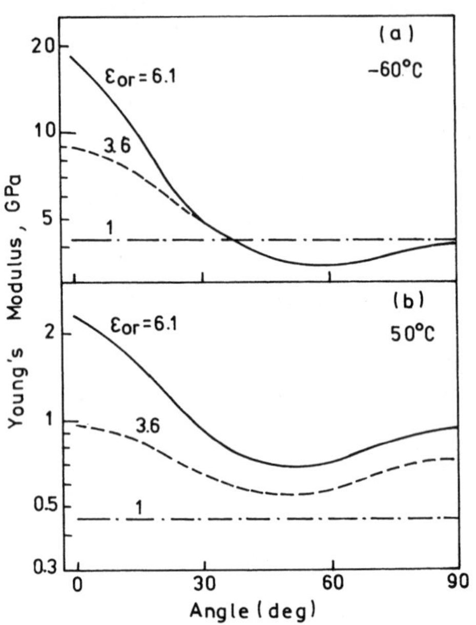

FIG. 84 Variation of the Young's modulus of oriented LLDPE as a function of the angle relative to the draw direction. (a) −60 and (b) 50°C. (From Ref. 111.)

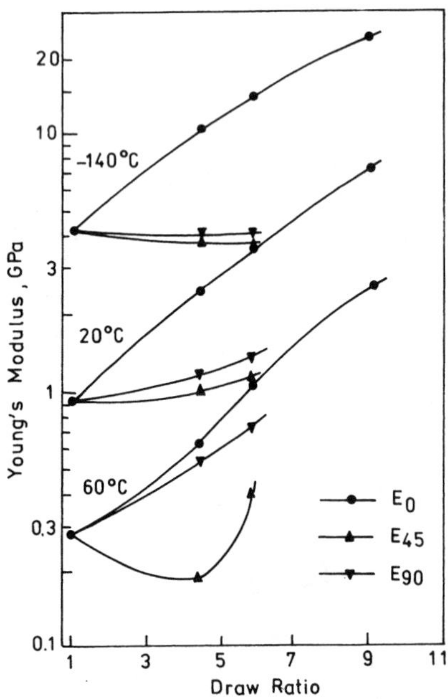

FIG. 83 Draw ratio dependence of the Young's modulus at 10 Hz and −140, 20 and 60°C for LLDPE, $M_w = 123{,}000$, $M_n = 23{,}700$, SCB = 20/1000*C*. E_0 (●), E_{45}, (▲), E_{90} (▼). (From Ref. 110.)

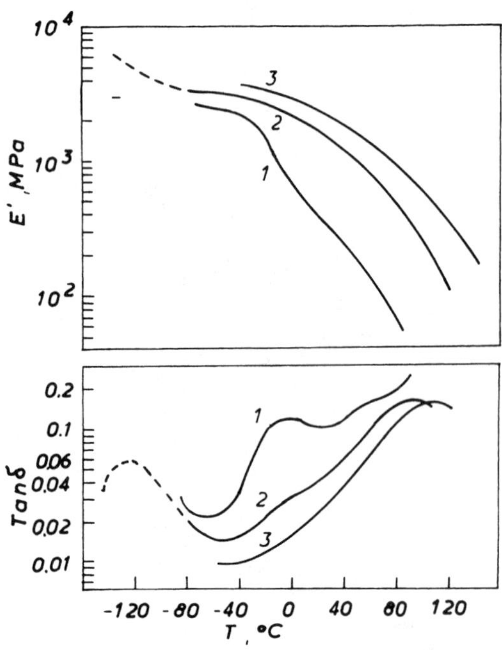

FIG. 85 Dynamic Young's modulus (E') and mechanical loss spectra (by Bohn-Oberst) [114] vs temperature for polyethylene. Density, g cm^{-3}: curve 1, 0.918; curve 2, 0.950; curve 3, 0.965. (From Ref. 2.)

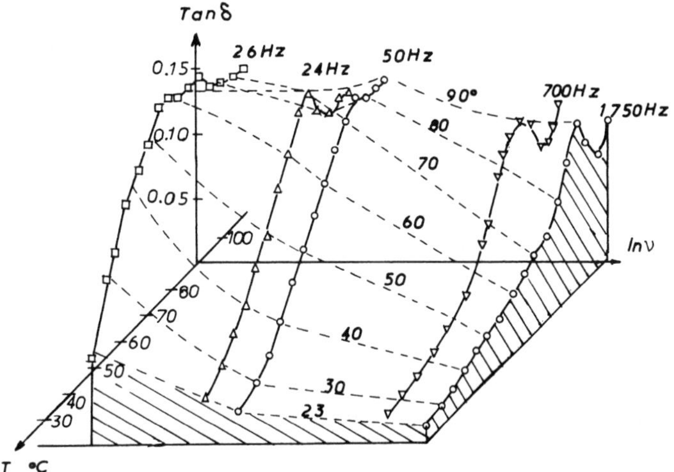

FIG. 86 High-temperature dynamic mechanical behavior of highly oriented PE rod in flexure. Three-dimensional plot showing tan δ, temperature, and natural logarithm of frequency for $T > 23°C$. (From Ref. 116.)

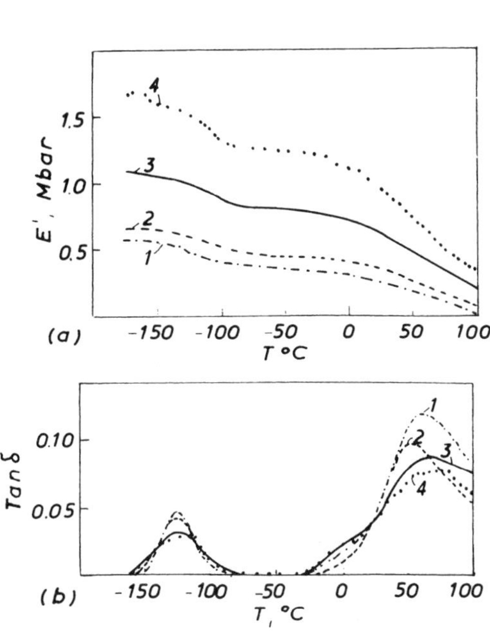

FIG. 87 Dynamic Young's modulus vs temperature for PE (Rigidex 140-60) samples at different draw ratios (measured at 20 Hz, experimental points shown for $\varepsilon_{or} = 35$). (b) tan δ vs temperature for PE (Rigidex 140-60) samples at different draw ratios. Draw ratio: curve 1, 11; curve 2,15; curve 3, 28; curve 4, 35. (From Ref. 117.)

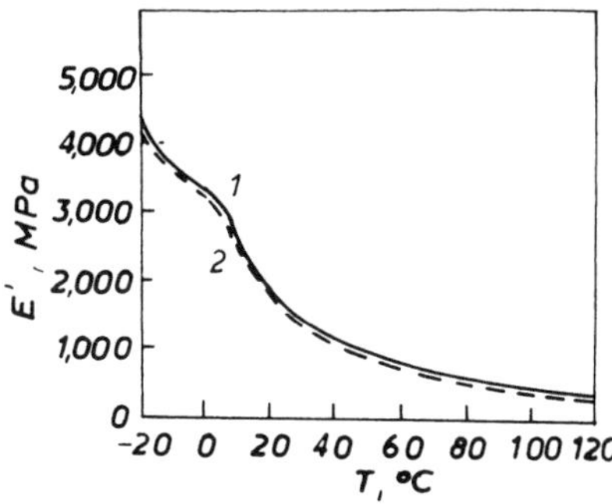

FIG. 88 Dependence of dynamic Young's modulus (E') at 100 Hz of IPP on temperature. Melt flow index, $I_m(5)$, g/10 min: curve 1, 0.5; curve 2, 8. (From Ref. 2.)

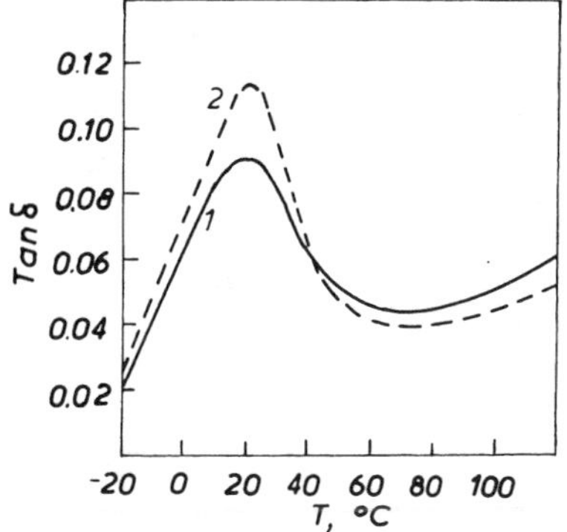

FIG. 89 Mechanical loss spectra (measured at 100 Hz) vs temperature for IPP. Melt flow index, $I_m(5)$, g/10 min: curve 1, 0.5; curve 2, 8. (From Ref. 2.)

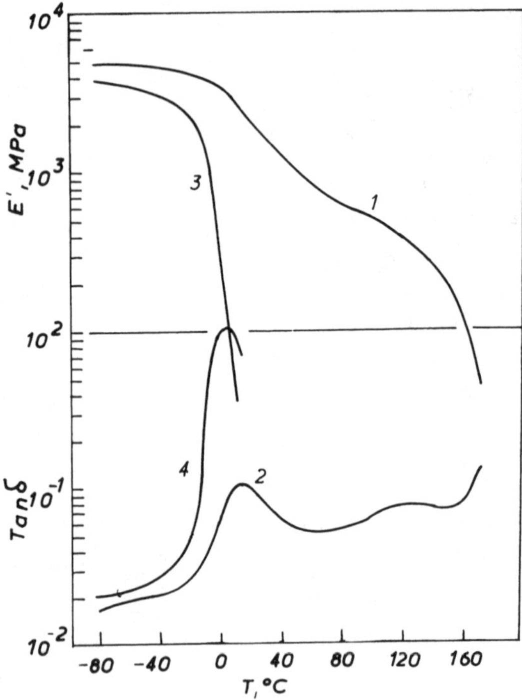

FIG. 90 Dynamic Young's modulus (E') (curve 1, 3) and mechanical loss spectra (curve 2, 4) (measured at 100 Hz) (by Bohn-Oberst) [114] vs temperature for IPP (curve 1, 2) and APP (curve 3, 4). (From Ref. 2.)

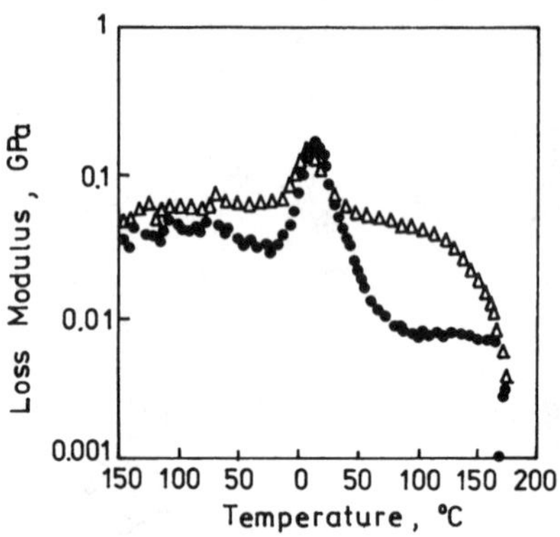

FIG. 92 Temperature dependence of the dynamic loss modulus for SPP: $M_w = 202{,}000$, $M_w/M_n = 1.8$, $\alpha_c = 0.44$ (●) and IPP: $M_w = 236{,}000$, $M_w/M_n = 2.3$, $\alpha_c = 0.68$ (△) (sample slow cooling to room temperature). (From Ref. 18.)

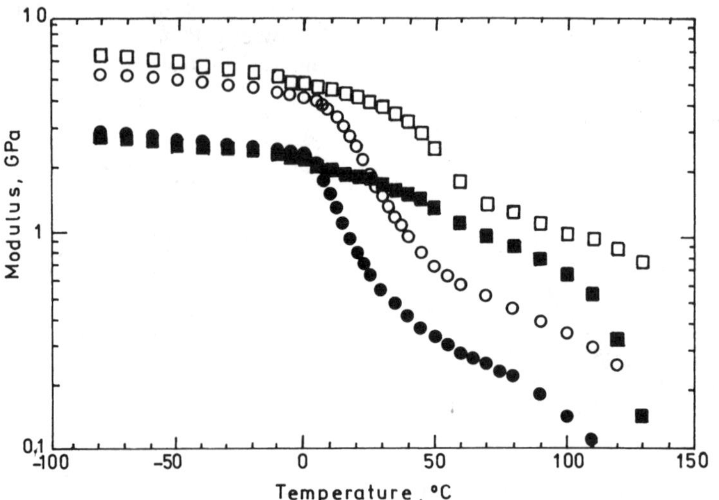

FIG. 91 Temperature dependence of the dynamic modulus of isotropic and drawn samples of syndiotactic polypropylene: $M_w = 152{,}000$, $M_w/M_n = 2.0$, $\alpha_c = 0.20$, $\varepsilon_{or} = 1$ (●) and $\varepsilon_{or} = 5.25$ (○); SPP: $M_w = 490{,}000$, $M_w/M_n = 2.4$, $\alpha_c = 0.43$, $\varepsilon_{or} = 1$ (■) and $\varepsilon_{or} = 10.5$ (□). (From Ref. 19.)

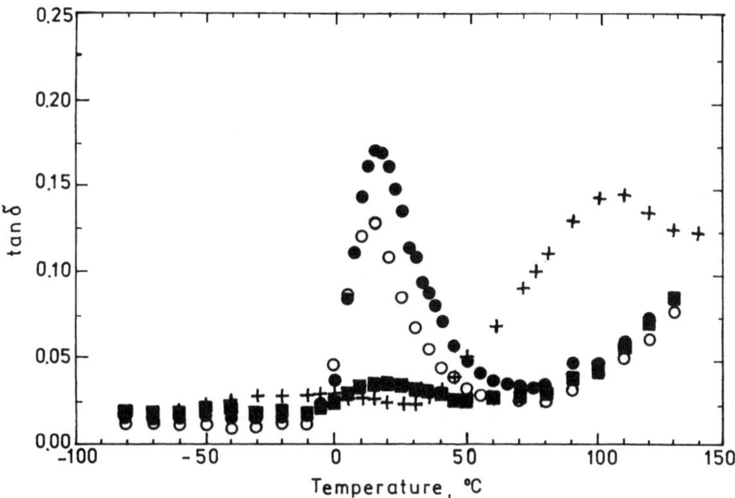

FIG. 93 Temperature dependence of the dynamic loss factor of syndiotactic and isotactic polypropylenes, SPP: $M_w = 152{,}000$, $M_w/M_n = 2.0$, $\alpha_c = 0.2$ (●) and $\alpha_c = 0.29$ (○), SPP: $M_w = 490{,}000$; $M_w/M_n = 2.4$, $\alpha_c = 0.43$ (■), IPP: $M_w = 135{,}000$, $M_w/M_n = 1.9$ (+). (From Ref. 19.)

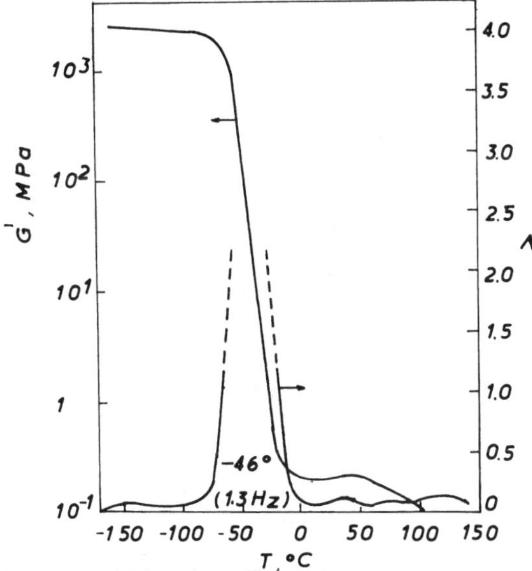

FIG. 94 Mechanical loss spectra and dynamic torsion G' modulus vs temperature for polyisobutylene (by Schmieder and Wolf) [109]. (From Ref. 2.)

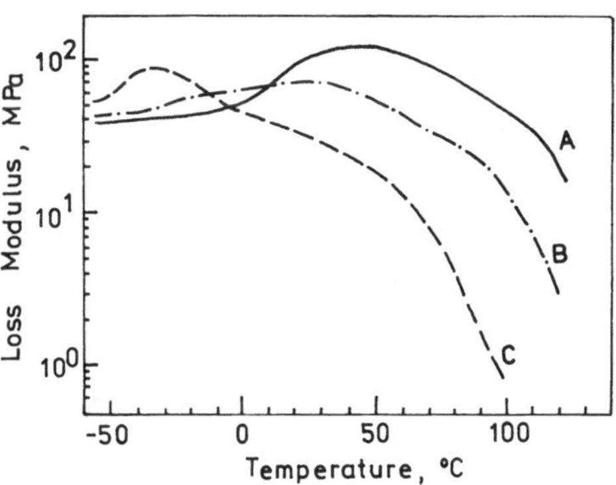

FIG. 95 Loss modulus vs temperature for ethylene–butene copolymers: (A) $M_w = 157{,}000$, $M_n = 30{,}000$, $\rho = 0.945$ g cm^{-3}, $\alpha_c = 0.67$, co-unit. mol. % = 1.2; (B) $M_w = 136{,}000$, $M_n = 31{,}000$, $\rho = 0.931$ g cm^{-3}, $\alpha_c = 0.55$, co-unit. mol. % = 2.7; (C) $M_w = 146{,}000$, $M_n = 27{,}000$, $\rho = 0.910\,g\,cm^{-3}$, $\alpha_c = 0.35$, co-unit. mol. % = 7.6. (From Ref. 22.)

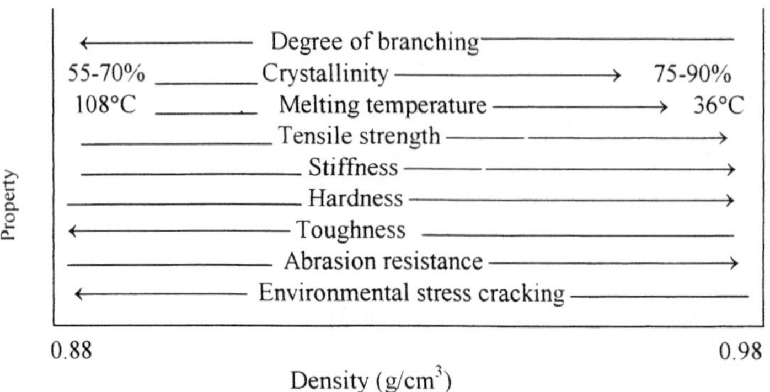

FIG. 96 Effect of density on polyethylene properties. (From Ref. 3.)

SYMBOLS AND ABBREVIATIONS

a_T	Empirical shift factor
A_{tear}	Tear energy
APP	Atactic polypropylene
BPE	Branched PE (LDPE)
E'	Storage modulus
E''	Loss modulus
E–B	Ethylene–butene copolymer
E_{creep}	Creep modulus
E_{el}	Static elastic modulus
$E_{el\parallel}$	Tensile modulus parallel to the molecular direction
$E_{el\perp}$	Tensile modulus normal to the molecular direction
E_{fx}	Flexibility modulus
E–H	Ethylene–hexene copolymer
E–O	Ethylene–octene copolymer
E_{tsh}	Torsion modulus
E_y	Yield energy
F	Strength
F_i	Factor
G', G''	Dynamic moduli
HDPE	High-density polyethylene
I_m	Melt index
I-PEP	isotactic polyethylene–polypropylene block copolymer
IPP	Isotactic polypropylene
J	Compliance
J_c	Creep compliance
L_a	Interlamellar thickness
L_c	Crystal thickness
LDPE	Low-density polyethylene
LLDPE	Linear low-density polyethylene
LPE	Linear PE (LLDPE)
M	Average molecular weight
M_n	Number average molecular weight
M_w	Weight average molecular weight
MDPE	Middle-density polyethylene
Mch. Pr.	Mechanical properties
MWD	Molecular weight distribution
n	Degree of polymerization
NC	Numbers of cycles
p	Pressure
PB	Polybutene-1
PE	Polyethylene
PIB	Polyisobutylene
PO	Polyolefins
PP	Polypropylene
Pr	Properties
SCB	Short chain branches
SPP	Syndiotactic polypropylene
t	Time
t_c	Crystallization time
T	Temperature
T_c	Crystallization temperature
T_g	Glass transition temperature
T_m	Melting temperatures
T_{or}	Orientation temperature
T_α, T_β, T_γ	α, β, γ are relaxation transition temperatures
UHMWHDPE	Ultrahigh molecular weight high-density polyethylene
V_d	Rate of deformation ($d\varepsilon/dt$); draw rate
V_{tear}	Rate of tear propagation
α_c	Crystallinity
tan δ	Loss (dynamic-mechanical) factor
ε	Elongation (draw) ratio; stretching strain
ε_{bl}	Blowing ratio
ε_B	Tensile strain

ε_b	Elongation at break
$\varepsilon_{B,b}$	Draw ratio after break
ε_c	Compression draw ratio
ε_{or}	Draw ratio
ε_y	Yield strain
θ	Angle between tensile axis and the orientation direction
ν	Frequency
ρ	Density
σ	Stress, tension
σ_B	Tensile strength
σ_d	Deformation stress
σ_h	Hardness
σ_{imp}	Impact strength
σ_{or}	Orientation stress
σ_{tear}	Tear strength
$\sigma_{T,imp}$	Tensile impact strength
σ_u	Ultimate strength
σ_y	Yield stress

REFERENCES

1. BG Bogdanov, M Michailov. In: C Vasile, ed. Handbook of Polyolefins. New York: Marcel Dekker, 1993, pp 295–466.
2. R Vieweg, A Schley, A Schwarz, eds. Kunststoff-Handbuch, Polyolefine. Munchen: Carl Hanser Verlag, 1969, vol. 4.
3. Z Bartczak, RE Cohen, AS Argon. Macromol 25: 4692–4704, 1992.
4. X Guofong, D Qianggou, LH Wang. Makromol Chem Rapid Commun 8: 539, 1987.
5. DC Bassett, DR Garder. Philos Mag 28: 535, 1973.
6. RAV Raff, KW Doak, eds. Crystalline Olefin Polymers (in Russian). Moskva: Khimiya, 1970, Part 2.
7. S Hosoda, A Uemura. Polym J 24: 939-949, 1992.
8. MA Kennedy, AJ Peacock, L Mandelkern. Macromol 27: 5297–5310, 1994.
9. JC Lucas, MD Failla. FL Smith, L Mandelkern. Polym Eng Sci 35: 1117–1123, 1995.
10. A Peterlin. J Mater Sci 6: 490, 1971.
11. MER Shanahan. J Mater Sci Lett 2: 28, 1983.
12. SJ Kang, M Matsuo. Polym J 21: 49, 1989.
13. I Wittkamp, K Friedrich. Prakt Metallog 15: 321, 1978.
14. EW Billington, C Brissenden. Int J Mech Sci l3: 531, 1971.
15. D Shinozaki, GW Groves. J Mater Sci 8: 71, 1973.
16. PM Pakhomov, MV Shablygin, ES Cobkallo, AS Chegolya. Vysokomol Soedin A-28-558, 1986.
17. SJ Pan, HI Tang, A Hiltner, E Baer. Polym Eng Sci 27: 869, 1987.
18. H Uehara, Y Yamazaki, T Kanamoto. Polym 37: 57, 1996.
19. Y Sakata, AP Unwin, IM Ward. J Mater Sci 30: 5841–5849, 1995.
20. S Elkoun, V Gaucher-Miri, R Seguela. Mater Sci Eng A 234–236: 83–86, 1997.
21. V Gaucher-Miri, S Elkoun, R Seguela. Polym Eng Sci 37: 1672-1683, l997.
22. R Seguela, V Gaucher-Miri, S Elkoun. J Mater Sci 33: 1273–1279, 1998.
23. MA Kennedy, AJ Peacock, MD Failla, JC Lucas, L. Mandelkern. Macromol 28: 1407–1421, 1995.
24. V Gaucher-Miri, P Francois, R Seguela. J Polym Sci: Part B: Polym Phys 34: 1113–1125, 1996.
25. IN Andreeva, EV Veselovskaya, EI Nalivaiiko, AD Pechenkin, VI Bukhgalter. Ultrahighmolecular High Density Polyethylene (in Russian). Leningrad: Khimiya, 1982.
26. E Kamel, N Brown. J Polym Sci Polym Phys Ed 22: 543, 1984.
27. JH Reinshagen, RW Dunlap. J Appl Polym Sci 20: 9, 1976.
28. HI Tang, A Hiltner, E Baer. Polym Eng Sci 27: 876, 1987.
29. V Gaucher-Miri, R Seguela. Macromol 30: 1158–1167, 1997.
30. AE Zachariades, T Kanamoto. J Appl Polym Sci 35: 1265, 1988.
31. PD Coates,. IM Ward. J Mater Sci 15: 2897, 1980.
32. T Lui, IR Harrison. Polym 29: 233, 1988.
33. SC Chou, KD Robertson, JH Rainey. Exp Mechanics 13: 422, 1973.
34. MG Laka, AA Dzhenis. Mekhanika Polimerov 6: 1043, 1967.
35. WA Spitzig, O Richmond. Polym Eng Sci 19: 1129, 1979.
36. SB Aiinbinder, KI Alksne, EL Tjunina, MG Laka. High Pressure Properties of Polymer (in Russian). Moskva: Khimiya, 1973.
37. HN Yoon, KD Pae, JA Sauer. J Polym Sci Polym Phys Ed 14: 1611, 1976.
38. J Duxbury, IM Ward. J Mater Sci 22: 1215, 1987.
39. FP La Mantia, D Acierno. Polym Eng Sci 19: 800, 1979.
40. WN Findley. Polym Eng Sci 27: 582, 1987.
41. RP Wool, RH Boyd. J Appl Phys 51: 5116, 1980.
42. I Sakurada, T Ito, K Nakamae. J Polym Sci C-15: 75, 1966.
43. D Heyer, U Buchenau, M Stamm. J Polym Sci Polym Phys Ed 22: 1515, 1984.
44. TT Wang. J Appl Phys 44: 2218, 1973.
45. F Daver, BW Cherry. J Appl Polym Sci 58: 2429–2432, 1995.
46. CB Bucknall. Pure and Appl Chem 58: 999, 1986.
47. Technical Brochure DAPLEN HDPE, PCD Polymere, 1995.

48. Technical Brochure FINATHENE, FINA Chemicals, 1993.
49. DS Chiu, AN Gent, JR White. J Mater Sci 19: 2622, 1984.
50. Y Wada, A Nakayama. J Appl Polym Sci 15:183, 1971.
51. J Lai, A Bakker. Polym Eng Sci 35: 1339–1347, 1995.
52. M Goldman, R Gronsky. Polym 37: 2909–2913, 1996.
53. T Kanamoto, RS Porter. J Polym Sci Polym Lett Ed 21: 1005, 1983.
54. AE Zachariades. J Appl Polym Sci 29: 867, 1984.
55. MR Mackley, S Solbai. Polym 28: 1115, 1987.
56. S Osawa, AE Zachariades, RF Saraf, RS Porter. J M S Rev Macromol Chem Phys C 37 (1): 149–198, 1997.
57. AE Zachariades, T Kanamoto. J Appl Sci 35: 1265–1281, 1988.
58. L Govaert, B Brown, P Smith. Macromol 25: 3480–3483, 1992.
59. LA Simpson, T Hinton. J Mater Sci 6: 558, 1971.
60. A Kaito, K Nakayama, H Kanetsuna. J Appl Polym Sci 30: 4591, 1985.
61. YM Boiko, W Brostow, AY Goldman, AC Ramamurthy. Polym 36: 1383–1392, 1995.
62. KD Pae, SK Bhateja. J M S - Rev Macromol Chem C-13:1, 1975.
63. V Dobrescu, G Andrei, A Cimpeanu, C Andrei. Rev Roum Chim 33: 399, 1988.
64. IM Ward. Plas Rubber Proc Appl 4: 77, 1984.
65. GP Andrianova. Polyolefin's Physical Chemistry (in Russian). Moskva: Khimiya, 1974
66. VB Gupta, IM Ward. J Macromol Sci B-2: 373, 1968.
67. MA Hallam, DLM Cansfield, IM Ward, G Pollard. J Mater Sci 21: 4199, 1986.
68. DV Ivanjukov, ML Fridman. Polypropylene (Properties and Application) (in Russian). Moskva: Khimija, 1974.
69. S Hashemi, JG Williams. Plastic Rubb Proc Appl 6: 363, 1986.
70. Mach Des Materials Reference Issue 52: 137, 1980.
71. SH Hamid, FS Qureshi, MB Amin, AG Maadhah. Polym - Plast Technol Eng 28: 475–492, 1989.
72. D Feldman, A Barbalata. Synthetic Polymers - Technology, Properties, Application. London: Chapman & Hall, 1996, pp 21–24.
73. T Ogita, H Yasuda, N Suzuki, M Minagawa, M Matsuo. Polym J 25: 445-452, 1993.
74. DR Rueda, FJ Balta Calleja, J Garcia Pena, IM Ward, A Richardson. J Mater Sci 19: 2615, 1984.
75. J Purvis, DI Bower. J Polym Sci Polym Phys Ed 14: 1461, 1976.
76. M Gilbert, DA Hemsley, SR Patel. Br Polym J 19: 9, 1987.
77. M Gahleitner, J Wolfschwenger, C Bachner, K Bernreitner, W Neißl. J Appl Polym Sci 61: 649–657, 1996.
78. D Del Duca, EP Moore. In: EP Moore, Jr., ed. Polypropylene Handbook. New York: Hanser Publishers, 1996.
79. T Ogawa. J Appl Polym Sci 44: 1869–1871, 1992.
80. DR Mears, KD Pae, JA Sauer. J Appl Phys 40: 4229, 1969.
81. B Martin, JM Perena, JM Pastor, JA De Saja. J Mater Sci Lett 5: 1027, 1986.
82. CB Bucknall. Pure Appl Chem 58: 985, 1986.
83. T Kresser. Polypropylene (in Russian). Moskva: Izdatinlit, 1963.
84. T Ariyama. Polym Eng Sci 35: 1455–1460, 1995.
85. Technical Brochure, MOPLEN, HIMONT, 1991.
86. Technical Brochure, Flexible packaging-new developments in polyolefin materials, HIMONT, 1993.
87. EP Moore, Jr. The Rebirth of Polypropylene: Supported Catalysts, How the People of the Montedison Laboratories Revolutionized the PP Industry. Munich: Hanser Publishers, 1998.
88. AK Taraiya, A Richardson, IM Ward. J Appl Polym Sci 33: 2559, 1987.
89. T Kunugi, T Ito, M Hashimoto, M Ooishi. J Appl Polym Sci 28: 179, 1983.
90. FM Mirabella, Jr. J Polym Sci Polym Phys B-25: 591, 1987.
91. JC Seferis, R J Samuels. Polym Eng Sci 19: 975, 1979.
92. RJ Samuels. J Macromol Sci B-4: 701, 1970.
93. T Kunugi, S Oomori, S Mikami. Polymer 29: 814, 1988.
94. S Osawa, RS Porter. Polymer 37: 2095–2101, 1996.
95. J Brandrup, EH Immergut, eds. Polymer Handbook. New York: Interscience, 1960.
96. L Holliday, JW White. Pure Appl Chem 26: 545, 1971.
97. I Sakurada, K Kaji. Makromol Chem l: Suppl 599, 1975.
98. I Sakurada, K Kaji. J Polym Sci C-31: 57, 1970.
99. H Tadokoro Structure and Elastic Moduli of Crystalline Polymers. Sen-I Gakkaishi, 1978, pp 34–89.
100. TI Sogolova, MI Demina. Mekhanika Polimerov 3: 387, 1977.
101. BA Krencel, VI Kleiner, LL Stockaya. α-Polyolefins (in Russian). Moskva: Khimiya, 1984.
102. VP Privalko. Handbook of Polymer Physical Chemistry, Properties of Polymers in Solid State (in Russian). Kiev: Naukova Dumka, 1984, Vol. 2.
103. HW Starkweather, Jr., TF Jordan, GB Dunnington. Am Chem Soc Polym 15: 143, 1974.
104. O Darras, R Seguela. J Polym Sci 31: 759-766, 1993.
105. II Perepechko. Low Temperature Properties of Polymers (in Russian). Moskva: Khimiya, 1977.
106. J Kubat, LA Nilsson, M Rigdahl. Rheol Acta 23: 40, 1984.
107. N-J Jo, A Takahara, T Kajiyama. Polym J 25: 721–729, 1993.
108. N-J Jo, A Takahara, T Kajiyama. Polym J 26: 1027–1036, 1994.

109. K Schmieder, K Wolf. Kolloid Z. 134: 149, 1953.
110. WP Leung, CL Choy. J Appl Polym Sci 36: 1305–1324, 1988.
111. CL Choy, WP Leung. J Appl Polym Sci 32: 5883–5901, 1986.
112. YS Papir, E Baer. J Appl Phys 42: 4667, 1971.
113. YP Khana, EA Turi, TJ Taylor, VV Vickroy, RF Abbott. Macromol 18: 1302, 1985.
114. H Oberst, L Bohn. Rheologica Acta 1: 608, 1961.
115. L Mandelkern, M Clatin, R Popli, RS Benson. J Polym Lett Ed 19: 435, 1981.
116. JM Crissman, LJ Zapas. J Polym Sci Polym Phys Ed 15: 1685, 1977.
117. JB Smith, GR Davies, G Capaccio, IM Ward. J Polym Sci Polym Phys Ed 13: 2331, 1975.
118. DL Plazek, DJ Plazek. Macromol 16:1469, 1983.
119. Y Isono, JD Ferry. J Rheol 29: 273, 1985.
120. U Gaur, B Wunderlich. Polymer Preprints 20: 429, 1979.
121. EA Turi, ed. Thermal Characterization of Polymers. New York: Academic Press, 1981.
122. T Kunugi, T Ikuta, M Hashimoto, K Matsuzaki. Polymer 23: 1983, 1982.

12

Electrical Properties

Silviu Jipa, Radu Setnescu, and Traian Zaharescu
Institute for Electrical Engineering, Bucharest, Romania

The extensive utilization of polyolefins for various applications such as electrical insulators involves a description of the influence of material behavior on the main electrical properties as well as their changes induced by outer stress factors.

The chemical structures of polyolefins, being somewhat similar, gather these materials into a large class of organic dielectrics that exhibit remarkable electrical properties; in this sense polyolefins have to prevent current flow between conductors at different potentials. Detailed reports on electrical properties have been presented by various authors, each of them touching on particular aspects of this topic [1–6].

One of the most important questions for long-term service is the lifetime prediction that involves the evaluation of the changes resulted during the application of stresses [7–12].

A particular interest in the space charge distribution can be explained by the consequences of charge accumulation in dielectrics on their durability. [13–17]. Irregular charge distribution caused by various local defects will lead to tree development [18–24] or a quick failure during degradation [25–33].

Differences in the molecular configurations emphasize the specific responses of materials to the applied stress; it is intended to offer suggested examples of a practical interest for electrical engineering.

The selected electrical properties presented in this chapter can be artificially divided into two kinds of characteristics that influence the electrical behavior of polyolefins:

1. *Intrinsic properties*: dielectric constant and dissipation factor. Their changes describe the modifications in the electronic density distribution in various macromolecules (Figs. 1–9).
2. *Extrinsic properties* revealed by the application of an outer electrical field: voltage–current dependences, changes in voltage or current values over certain time, electrical conductivity, electrical breakdown (Figs. 9–30).

Because of the irreversible process of degradation, especially oxidation, worsening in electrical properties has been largely assessed [34–37]. Some representative results obtained during thermal and/or radiochemical aging of various polyolefins are presented (Figs. 31–42), taking into account that the operation duration is strongly dependent on the chemical strength of materials.

SYMBOLS AND ABBREVIATIONS

A_{1720}	absorbance at 1720 cm^{-1}
BVD	breakdown voltage
C	black carbon
d	days
E	electrical field
EPDM	ethylene–propylene terpolymer
EPR	ethylene–propylene rubber
E_u	electrical field
EVA	ethylene vinyl acetate copolymer

(Text continues on p 318)

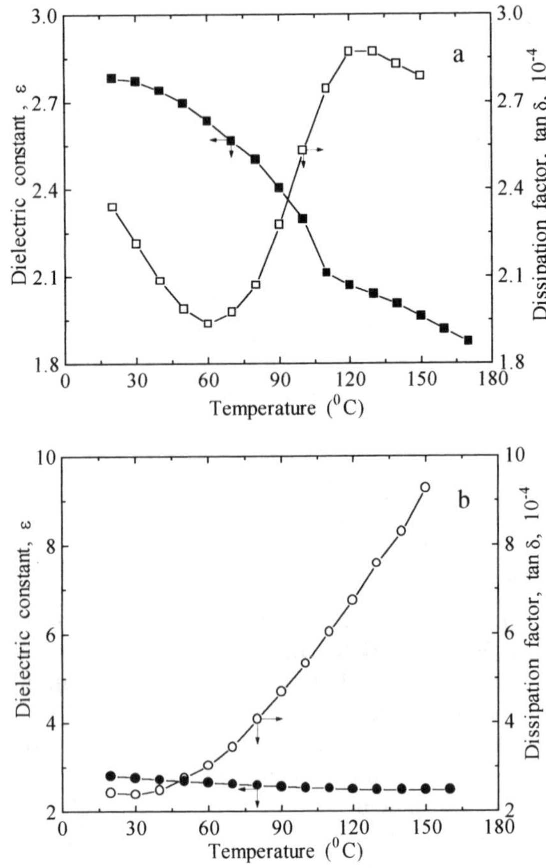

FIG. 1 Temperature dependencies of permittivity and tan δ for XLPE (a) and EPR (b). (Adapted from Ref. 38.)

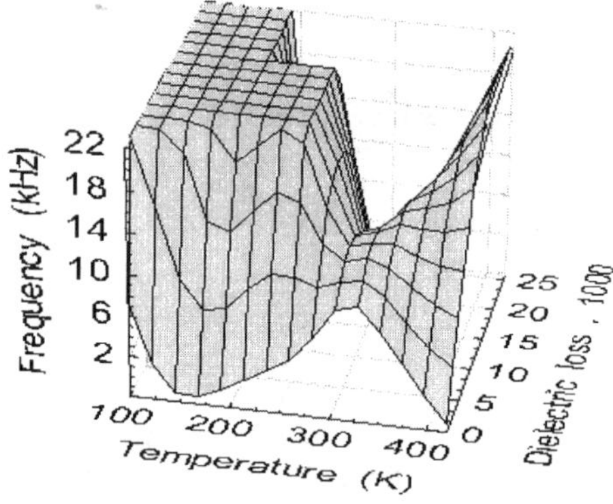

FIG. 2 Dependence of dielectric loss on the temperature and frequency for linear polyethylene, $\rho = 0.961$ g cm^{-3}. (Adapted from Ref. 39.)

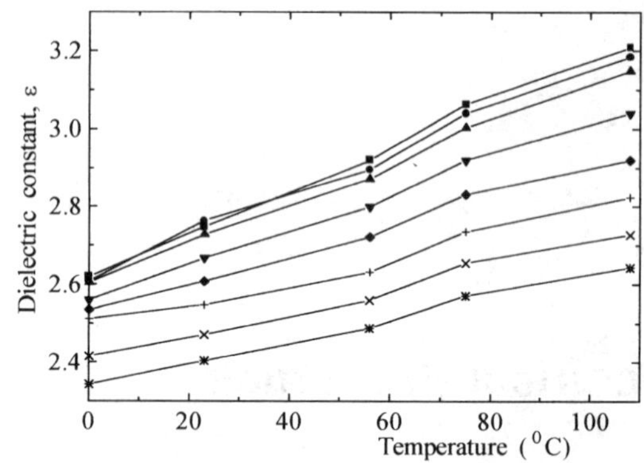

FIG. 3 Dependencies of dielectric constant of tetrafluoroethylene-propylene copolymer on temperature at various frequencies. (■) 20 Hz; (●) 50 Hz; (▲) 100 Hz; (▼) 300 Hz; (◆) 1 kHz; (+) 3 kHz; (×) 10 kHz; (∗) 30 kHz. (Adapted from Ref. 40.)

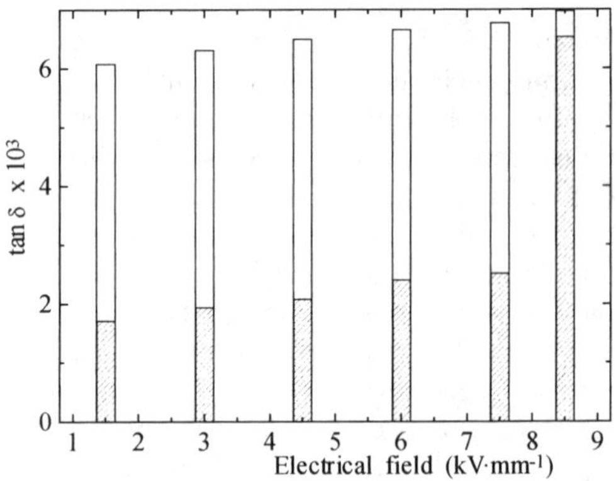

FIG. 4 Electrical field dependence of tan δ of EPR sheet. Hollow: 300 K; hatched: 77 K. (Adapted from Ref. 41.)

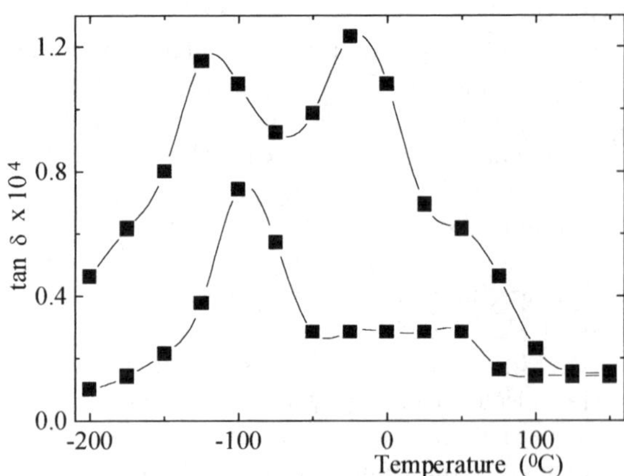

FIG. 5 Tan δ vs. temperature for HDPE (●) and (■) LDPE. Testing frequency: 10^5 Hz. (Adapted from Ref. 42.)

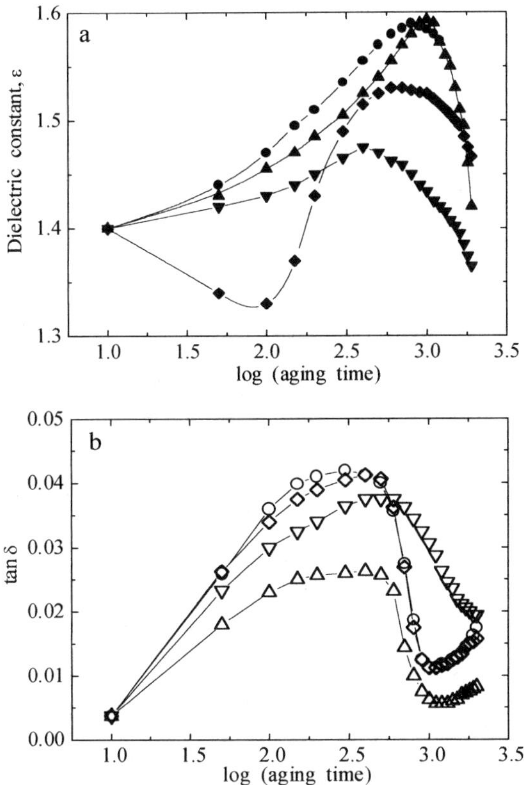

FIG. 6 Changes in dielectric constant (a) and dissipation factor (b) for LDPE aged in different conditions: in water at 40°C (●); in water at 100°C (▲); in humidity 100% at 40°C (▼); in humidity 100% at 100°C (◆). (Adapted from Ref. 43.)

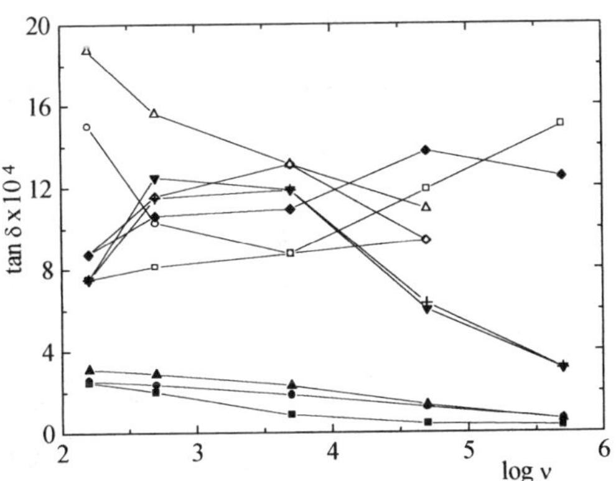

FIG. 7 Isothermal changes in tan δ vs. log ν for IPP measured at various temperatures. (■) −75°C; (●) −50°C; (▲) −25°C; (▼) 0°C; (◆) 25°C; (+) 50°C; (□) 75°C; (○) 100°C; (□) 125°C; (◇) 150°C. (Adapted from Ref. 44.)

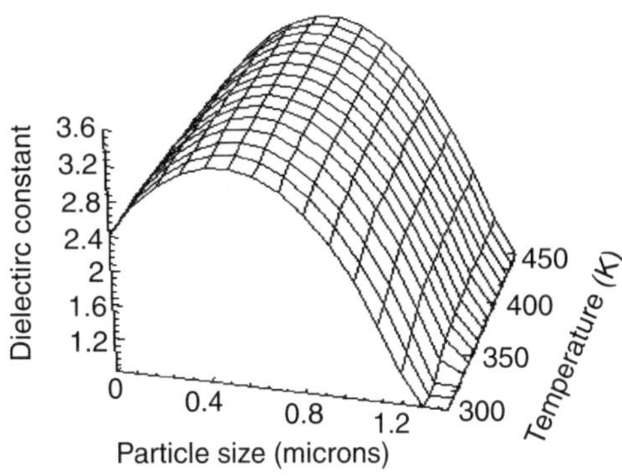

FIG. 8 Dependencies of dielectric constant on temperature and added Al_2O_3 particle size in EPDM. (Adapted from Ref. 45.)

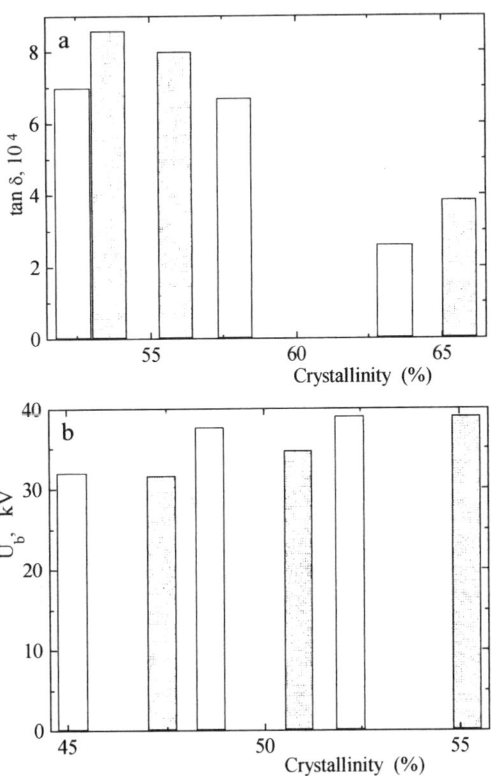

FIG. 9 Dependencies of dielectric loss factor (a) and dielectric strength (b) on the crystallinity degree of polyethylene. White: thermally aged at 80°C; gray: thermally aged at 90°C. (Adapted from Ref. 46.)

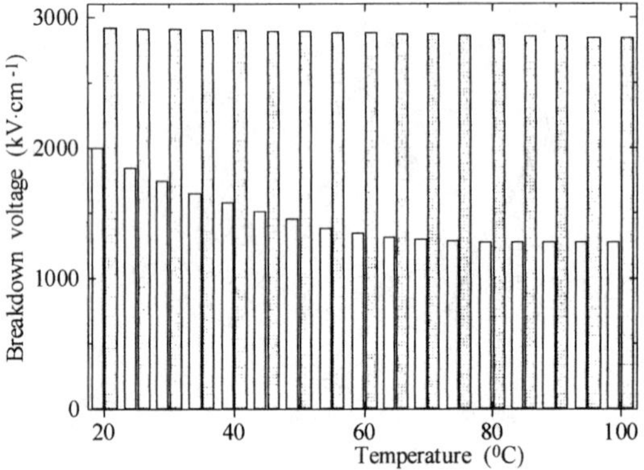

FIG. 10 Dependencies of breakdown voltage of IPP foil on temperature. White: extruded sheet; gray: biaxially oriented sheet. (Adapted from Ref. 42.)

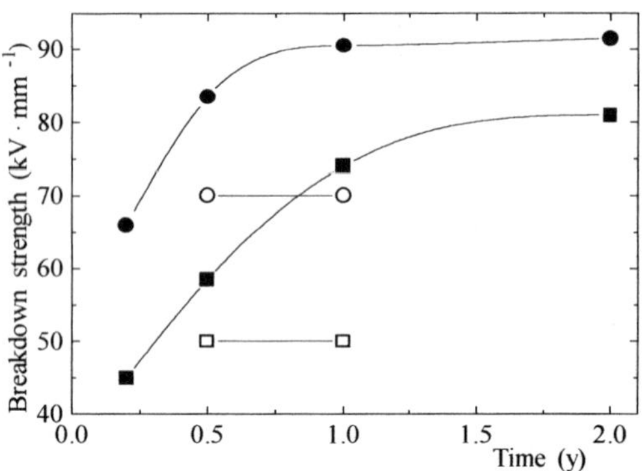

FIG. 11 Breakdown strength vs. time elapsed after vulcanization of EPR cables. Applied voltage: 10 kV. (□, ■) amorphous material; (●, ○) semi-crystalline material. (Adapted from Ref. 47.)

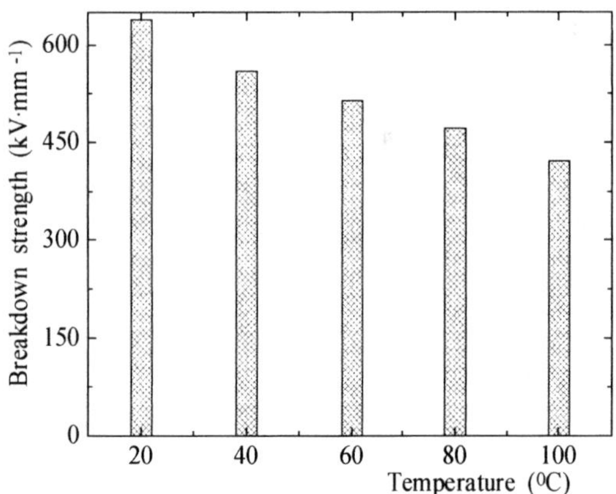

FIG. 12 Electrical d.c. breakdown strength versus temperature for two layer of the 25 μm thick PP film subjected to a linearly increasing voltage at a rate of 25 kV min^{-1}. (Adapted from Ref. 48.)

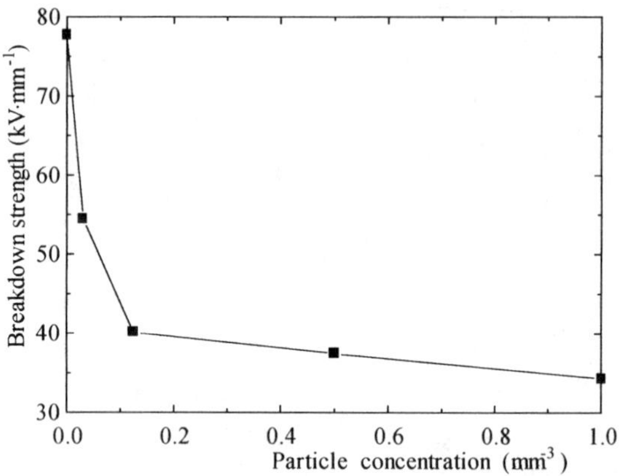

FIG. 13 Breakdown strength dependence on 165 μm Fe particle concentration in EPDM rubber. (Adapted from Ref. 49.)

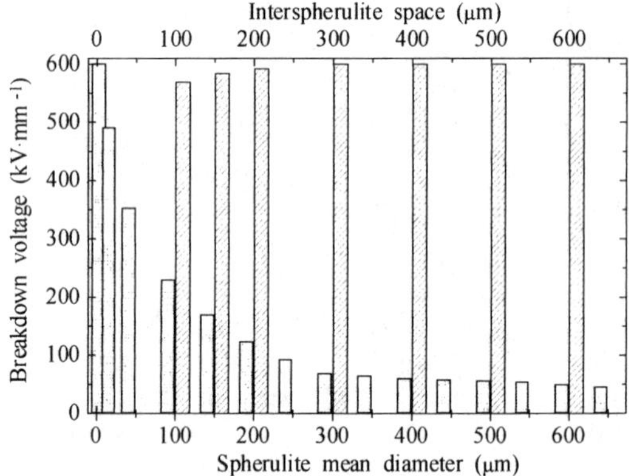

FIG. 14 Dependence of spherulite electrical strength on the spherulite mean diameter (hatched) and on the interspherulite space (gray) in PP specimens. (Adapted from Ref. 50.)

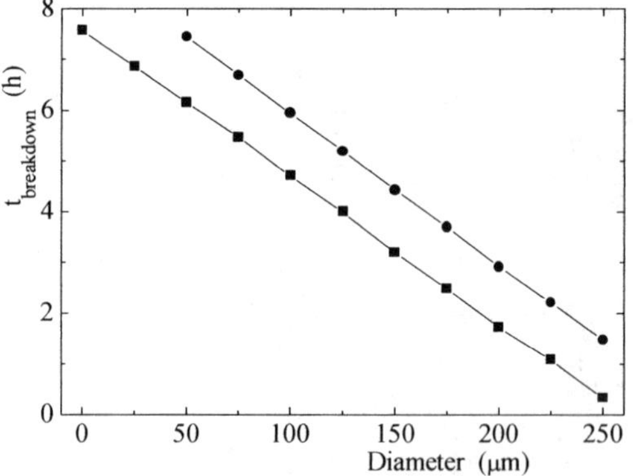

FIG. 15 Electrical breakdown time vs. spherulite diameter obtained for IPP crystallized at 130°C (●) and 140°C (■). Applied voltage 11.9 kV. (Adapted from Ref. 51.)

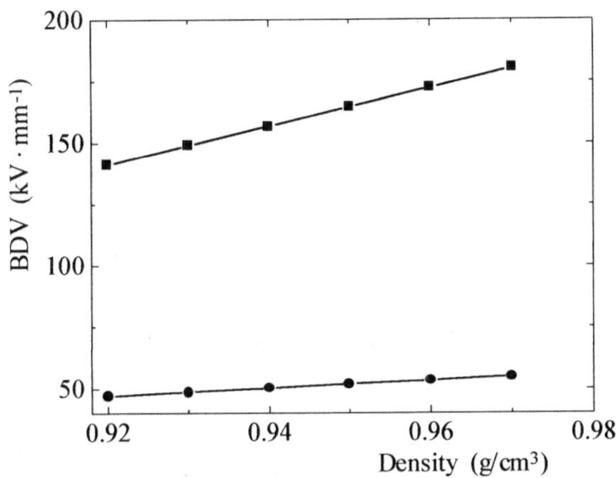

FIG. 16 Relationships between density and ac (■) or impulse (●) breakdown stresses of XLPE. (Adapted from Ref. 52.)

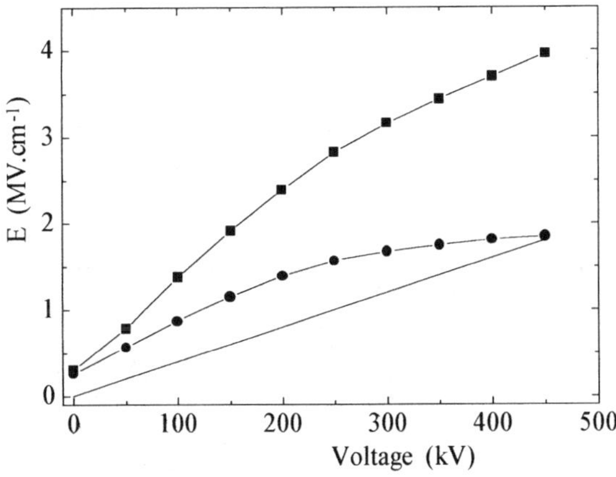

FIG. 17 Relation between the applied voltage and the peak value of the field at electrode interfaces when a negative voltage is applied to XLPE insulation. (■) anode; (●) cathode; (——) average voltage. (Adapted from Ref. 53.)

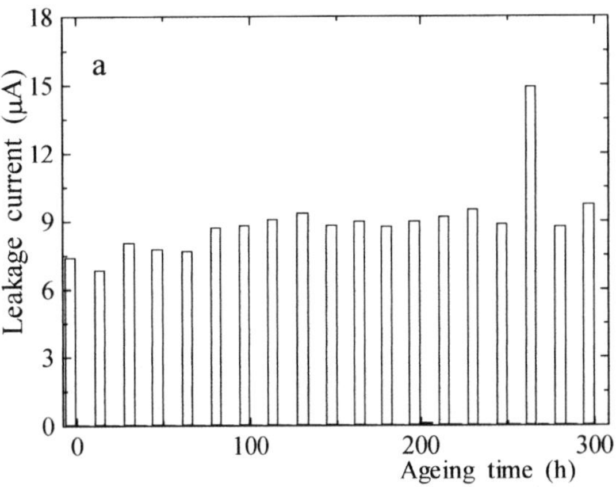

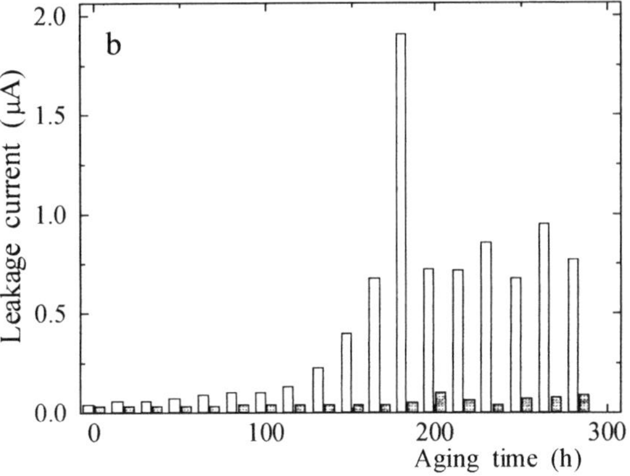

FIG. 18 Leakage current at different humidity. (a) EPDM; (b) HDPE; 87% (gray), 96% (hatched). (Adapted from Ref. 54.)

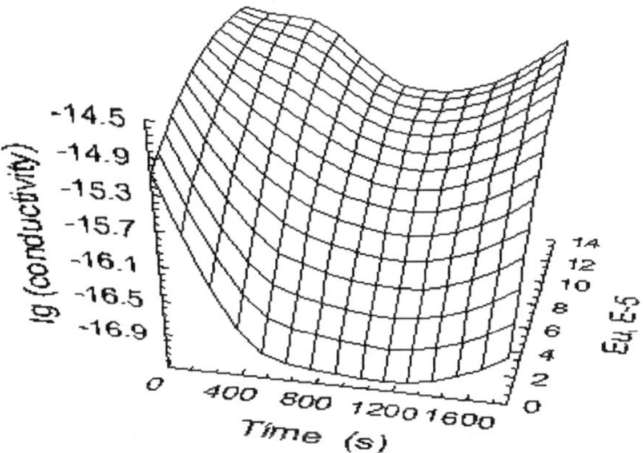

FIG. 19 Dependence of conductivity (Ω^{-1} cm^{-1}) on applied field for LDPE measured at 106°C at various measuring times. (Adapted from Ref. 55.)

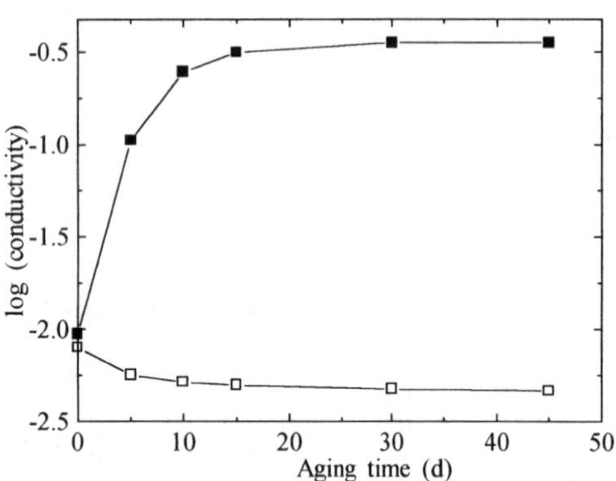

FIG. 20 Electrical conductivity (Ω^{-1} cm^{-1}) vs. aging time at the surface (■) and in the bulk (□) in EPDM rubber. (Adapted from Ref. 56.)

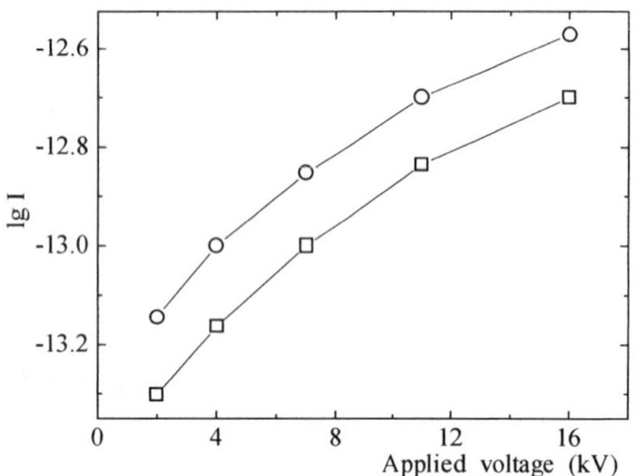

FIG. 21 Isochronal charging (○) and discharging (□) currents (A) vs. applied voltage in virgin XLPE. (Adapted from Ref. 57.)

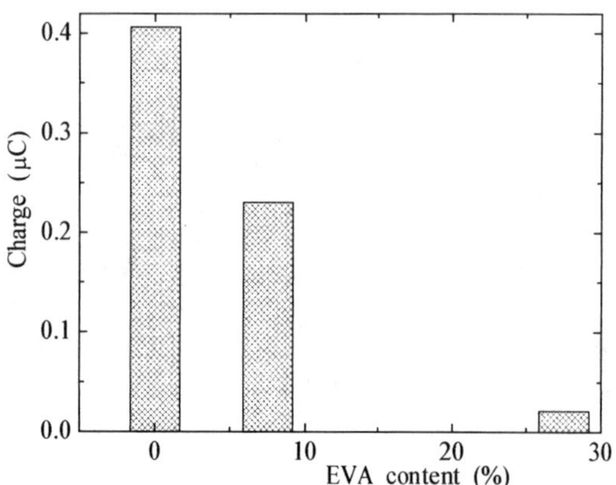

FIG. 22 Charge near the cathode of PE/EVA blends. Polyethylene was low density material ($\rho = 0.92$ g cm^{-3}); percentage of vinyl acetate in EVA was 15%. Amplitude of electric pulses: −2 kV. Width pulse: 30 ns. (Adapted from Ref. 58.)

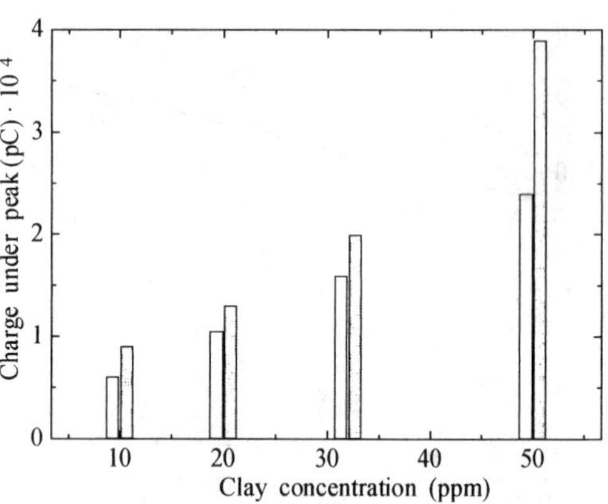

FIG. 23 Charge stored under the high temperature TSDC peak as a function of clay concentration for two particle size clays: white: 0.8 μm diameter; gray: 1.4 μm diameter. (Adapted from Ref. 59.)

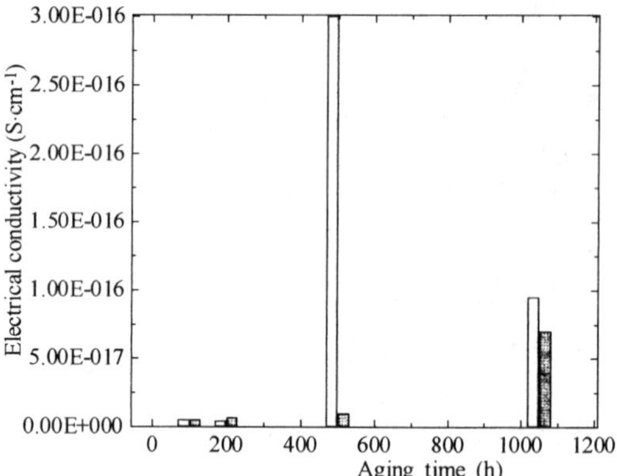

FIG. 24 Electrical conductivity vs. aging time for XLPE aged at 12 kV and 60°C. Gray: 1 min; hatched: 60 min. (Adapted from Ref. 60.)

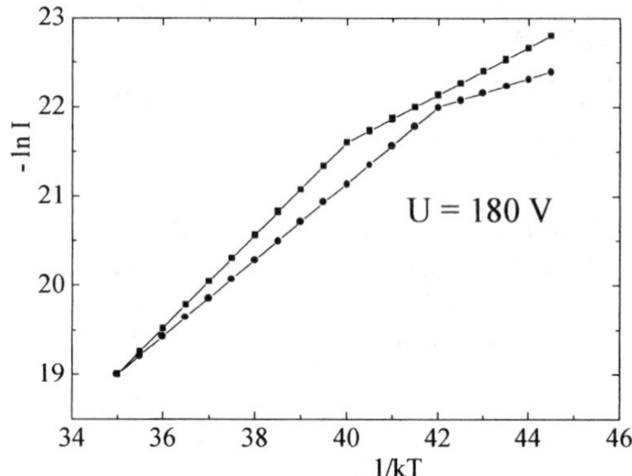

FIG. 25 Dependencies $\ln I = f(1/kT)$ for 1,4-*cis*-polybutadiene. (■) 96% *cis* form; (Au–Al, Au(+), $d = 6.35$ μm); (●) 100% *cis* form; (Au–Al, Au(−), $d = 6.70$ μm). (Adapted from Ref. 61.)

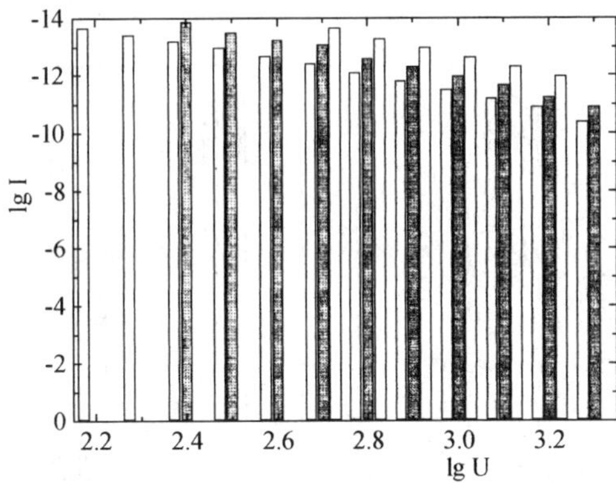

FIG. 26 $I = f(U)$ for HDPE at various temperatures. White: 85°C; hatched: 95°C; gray: 110°C. Current was expressed in ampères. (Adapted from Ref. 62.)

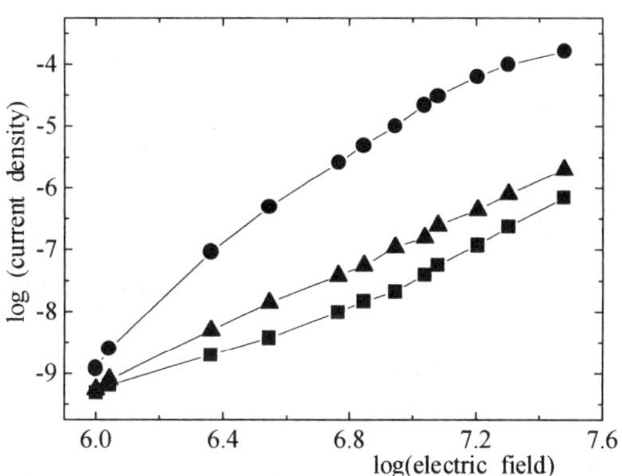

FIG. 27 Plots of current density (A m^{-2}) vs. electric field of LDPE extracted with *p*-xylene at 53°C and measured at various temperatures: 50°C (●); 70°C (▲); 90°C (■). (Adapted from Ref. 63.)

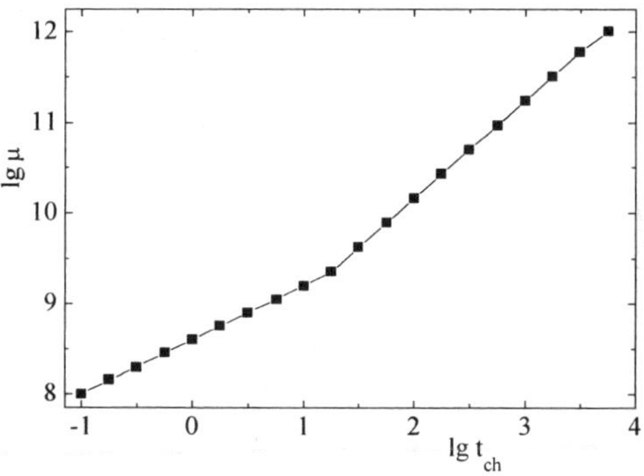

FIG. 28 Dependence of charge carrier mobility (μ, cm^2 V^{-1} s^{-1}) on the charging time (t_{ch}, s) in 50 μm sheet of polyethylene. Applied voltage. −1000 V. (Adapted from Ref. 64.)

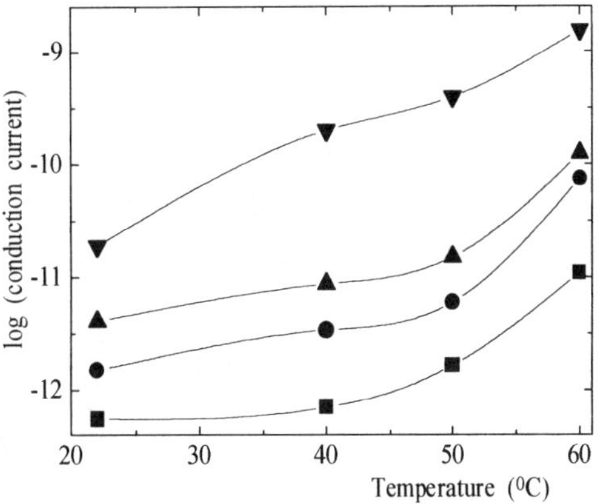

FIG. 29 Conduction current (A) after 16 h of charging as a function of temperature in polypropylene at various breakdown strengths: 40 kV min^{-1} (■); 80 kV mm^{-1} (●); 120 kV mm^{-1} (▲); 240 kV mm^{-1} (▼). (Adapted from Ref. 48.)

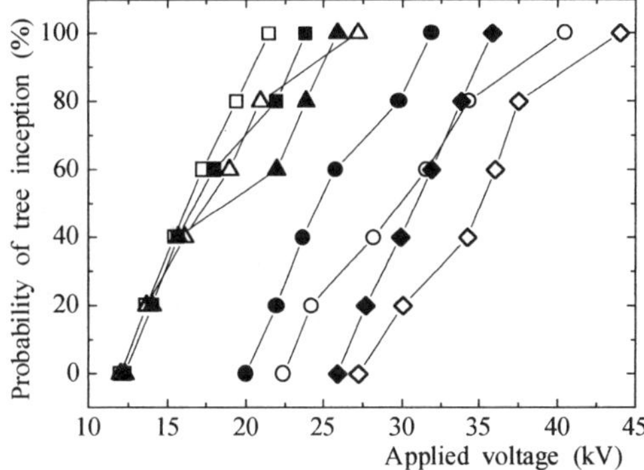

FIG. 30 Grounded tree initiation voltage for nondegassed XLPE. Solid marks regard XLPE free of antioxidant; hollow marks regards XLPE with antioxidant (0.16% phenolic antioxidant). (Adapted from Ref. 65.)

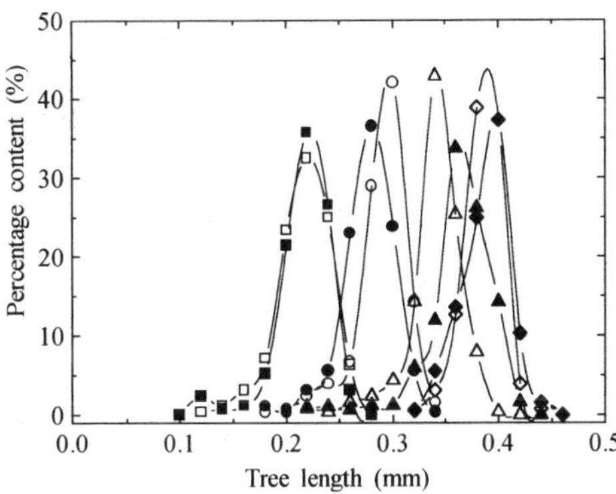

FIG. 31 Tree length distribution vs. percentage content of low molecular mass fraction in two formulations of linear polyethylene (LPE) at various aging times. 100 h (■, □); 240 h (●, ○); 500 h (▲, △); 750 h (◆, ◇). Solid marks regard 100% LPE; hollow marks regard 80% LPE + 20% LDPE. (Adapted from Ref. 66.)

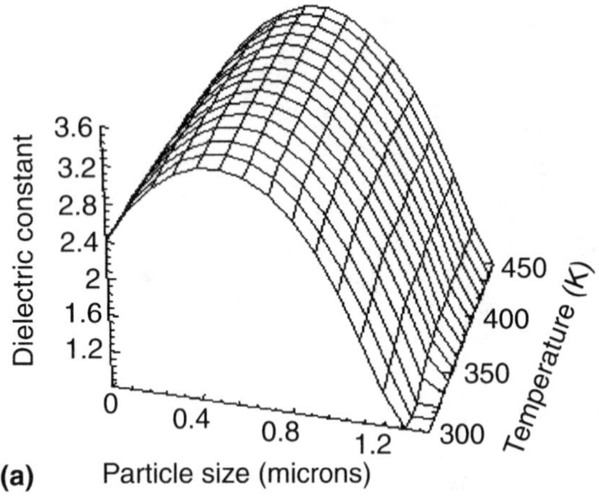

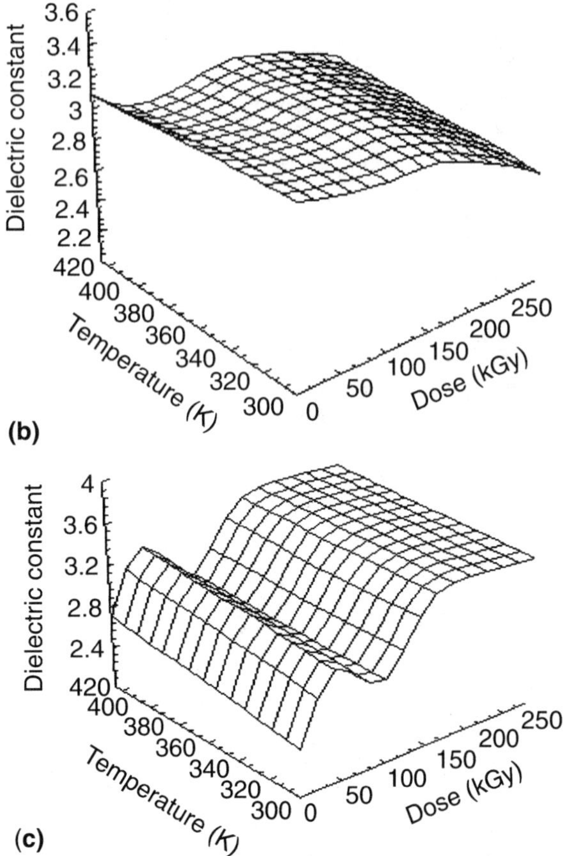

FIG. 32 Dependencies of dielectric constant on temperature and irradiation dose for EPDM filled with Al_2O_3 of various sizes. (a) 0.02 µm; (b) 0.3 µm; (c) 1 µm. (Adapted from Ref. 45.)

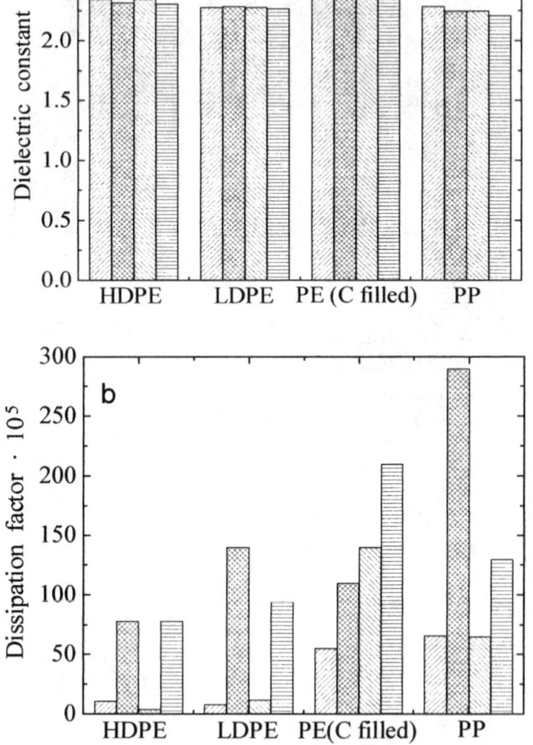

FIG. 33 Dependencies of electrical properties: dielectric constant (a) and dissipation factor (b) on irradiation aging measured at various frequencies (accelerated electrons: dose 10 MGy; dose rate 3 kGy/h). (Adapted from Ref. 67.) Column 1; 1 kHz, nonirradiated materials; column 2; 1 kHz, irradiated materials; column 3; 1 MHz, nonirradiated materials; column 4; 1 MHz, irradiated materials.

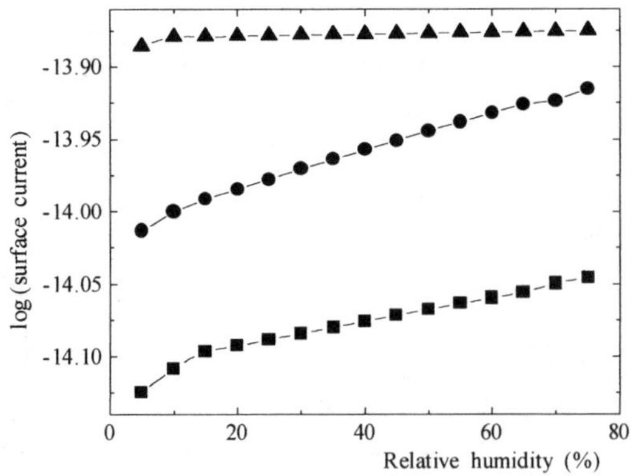

FIG. 34 Dependence of surface current (*A*) on relative humidity percentage for γ-irradiated LDPE (6 Mrad) in nitrogen. Applied voltage: 100 V (■); 200 V (●); 400 V (▲). (Adapted from Ref. 68.)

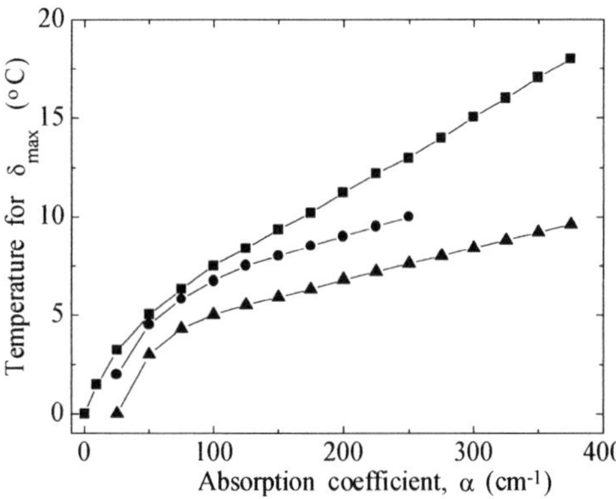

FIG. 35 Change in the temperatures corresponding to tan (δ_{max}) as a function of absorption coefficient of carbonyl groups for various treated LDPE samples. (Adapted from Ref. 69.) Measurements performed at 1 kHz. (■) γ irradiation at room temperature (5.02×10^2 Gy h^{-1}; (●) thermal aging (168 h at 90°C) after γ irradiation; (▲) simultaneous application of thermal aging and γ irradiation performed at 90°C.

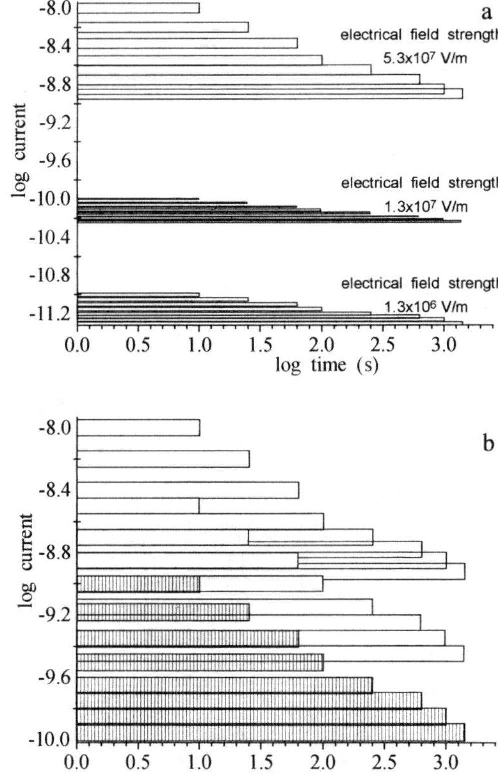

FIG. 36 Transient charging currents (A) at 70°C for γ-exposed LDPE in nitrogen. (a) 10^4 Gy; (b) 10^6 Gy. (Adapted from Ref. 70.)

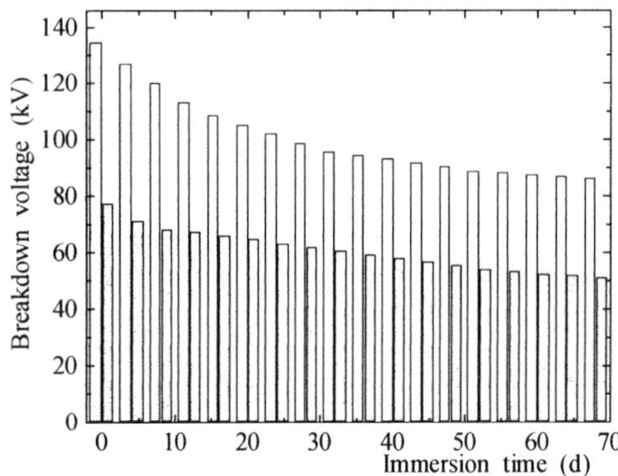

FIG. 37 A.C. breakdown voltage of 11 kV cable immersed in water at 70°C. White: XLPE; gray: EPR. (Adapted from Ref. 71.)

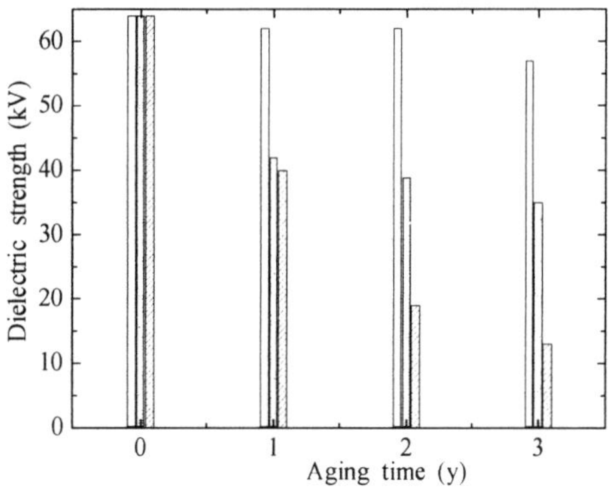

FIG. 38 Predicted dielectric strength for aged ethylene-propylene rubber insulation used for manufacture of high voltage cables: white: degraded in air at 50°C; gray: degraded in water at 50°C; hatched: degraded in water at 35°C. (Adapted from Ref. 72.)

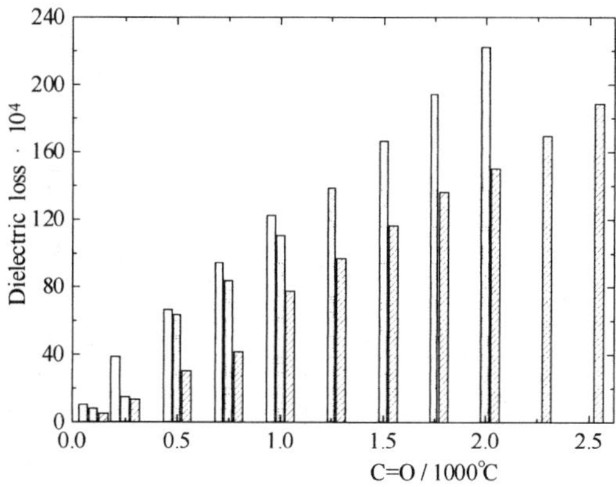

FIG. 39 The development of dielectric loss measured at room temperature as a function on carbonyl content during the oxidation of various polyethylene. White: LDPE, about 20 CH_3/1000C, $\rho = 0.925$ g cm^{-3}; gray: HDPE, about 4 CH_3/1000C, $\rho = 0.957$ g cm^{-3}; hatched: HDPE, about 2 $CH_3$1000C, $\rho = 0.974$ g cm^{-3}. (Adapted from Ref. 73.)

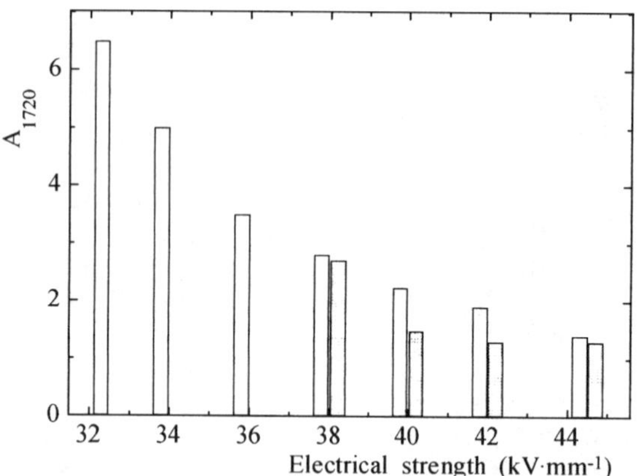

FIG. 40 Relationship between 1720 cm^{-1} band absorption and electrical strength of LDPE insulated cables. Aging conditions: (white) heated water at 70°C and 50 kV applied a.c. voltage; (gray) heated water at 70°C. (Adapted from Ref. 74.)

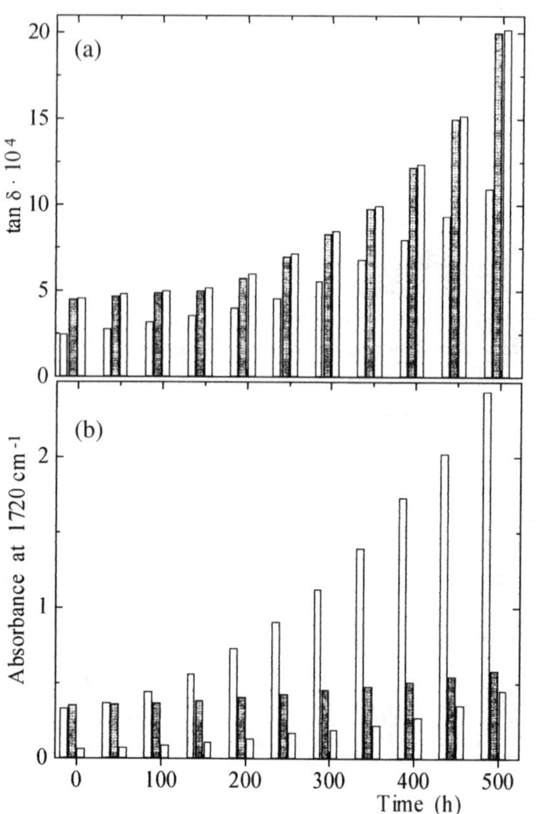

FIG. 41 Changes in dielectric loss factor (a) and (b) absorption in 1720 cm^{-1} band as functions of aging time of XLPE: gray: thermally degraded in air at 90°C; hatched: thermally degraded in air at 135°C; white: thermally degraded in water at 90°C. (Adapted from Ref. 75.)

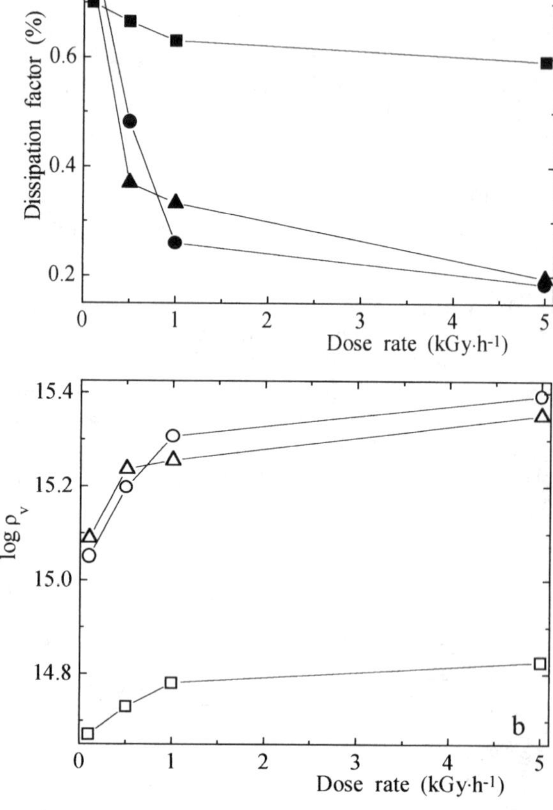

FIG. 42 Effects of dose rate on electrical dissipation factor (a) and volume resistivity (b) for some polyolefins irradiated in air at 1 MGy. Ethylene–propylene rubber with quinoline as antioxidant (■, □); XLPE free of antioxidant (●, ○); XLPE with thiobisalkylphenol as antioxidant (▲, △). (Adapted from Ref. 76.)

HDPE	high density polyethylene
I	current
IPP	isotactic polypropylene
k	Boltzmann's constant
LDPE	low density polyethylene
LPE	linear polyethylene (LLDPE)
XLPE	crosslinked polyethylene
PE	polyethylene
PP	polypropylene
T	temperature
$t_{breakdown}$	electrical breakdown time
t_{ch}	charging time
TSDC	Thermally Stimulated Depolarization Current
U	voltage

U_b	breakdown voltage
yr	year
γ	gamma radiation
ϵ	dielectric constant
tan δ	loss (dissipation) factor
ρ	density
ρv	volume resistivity
μ	charge carrier mobility
ν	frequency

REFERENCES

1. A Tager. Physical Chemistry of Polymers. English ed. Moscow: Mir Publishers, 1978, pp 306–333.
2. DE Seanor. Electrical Properties of Polymers. New York: Academic Press, 1982.
3. VK Agarwal. IEEE Trans Electr Insul EI-24: 741–764, 1989.
4. JR Laghari, A Hammoud. IEEE Trans Nucl Sci NS-37: 1076–1083, 1990.
5. LA Dissado, JC Fothergill. Electrical Degradation and Breakdown in Polymers. London: Peter Peregrinus Ltd., 1992.
6. BG Bogdanov, M Michailov. In: Handbook of Polyolefins. Synthesis and Properties. 1st ed. New York: Marcel Dekker, 1993, pp 421–438.
7. K Soma, M Aihara, Y Kataoka. IEEE Trans Electr Insul EI-21: 1027–1032, 1986.
8. H Hirose, IEEE Trans Electr Insul EI-22: 745–753, 1987.
9. W Khachen, JR Laghari. IEEE Trans Electr Insul EI-27: 1022–1025, 1992.
10. GC Montanari. IEEE Trans Electr Insul EI-27: 974–986, 1992.
11. LA Dissado, G Mazzanti, GC Montanari. IEEE Trans Dielectr Electr Insul 2: 1147–1158, 1995.
12. J-P Crine. IEEE Trans Dielectr Electr Insul 4: 487–495, 1994.
13. T Mizutani. IEEE Trans Dielectr Electr Insul 1: 923–933, 1994.
14. D Malec, R Essolbi, H The-Giam, Bui Ai, B Garros. Trans Dielectr Electr Insul 3: 64–69, 1996.
15. Y Zhang, J Lewiner, C Alquié, N Hampton. IEEE Trans Dielectr Electr Insul 3: 778–783, 1996.
16. NH Ahmed, NN Srinivas. IEEE Trans Dielectr Electr Insul 4: 644–656, 1997.
17. N Hazumi, T Takeda, H Suzuki, T Okamoto. IEEE Trans Dielectr Electr Insul 5: 82–90, 1998.
18. MC Michel, JC Bobo. J Appl Polym Sci 35: 581–587, 1979.
19. SS Bamji, AT Bulinski, Y Chen, RJ Densley. IEEE Trans Electr Insul EI-27: 402–404, 1992.
20. AT Bulinski, SS Bamji, RJ Densley. Factors affecting the transition from a water tree to an electrical tree. Proceedings of IEEE International Symposium on Electrical Insulation, Boston, 1988, pp 327–331.
21. E David, J-L Parpal, J-P Crine. IEEE Trans Dielectr Electr Insul 3: 248–257, 1996.
22. Ying Li, Y Ebinuma, Y Fujiwara, Y Ohki, Y Tanaka, T Tanaka. IEEE Trans Dielectr Electr Insul 4: 52–57, 1997.
23. A Garton, S Bamji, AT Bulinski, J Densley. IEEE Trans Electr Insul EI-22: 405–412, 1987.
24. T Zaharescu, I Mihalcea. Irradiation effects in synthetic polymers. Proceedings of IEEE International Conference on Conduction and Breakdown in Solid Dielectrics, Leicester (GB), pp 542–545, 1995.
25. PV Notinger, F Ciuprina, JC. Filipini, B Gosse, S Jipa, T Setnescu, R Setnescu, I Mihalcea. Studies of water treeing and chemiluminescence on irradiated polyethylene. Proceedings of IEEE International Symposium on Electrical Insulation, Montreal, 1996, pp 163–166.
26. J-P Crine. IEEE Trans Electr Insul EI-26: 811–818, 1991.
27. G Chen, RA. Fouracre, HM Bandford, DJ Tedford. Radiat Phys Chem Int. J. Radiat. Appl. Instrum Part C 37: 523–530, 1991.
28. A Gadoum, B Goose, J-P. Gosse. IEEE Trans Dielectr Electr Insul 2: 1075–1082, 1995.
29. G Mazzanti, GC Montanari, A Motori, P Anelli. IEEE Trans Dielectr Electr Insul 2: 1095–1099, 1995.
30. GC Montanari, A Motori, AT Bulinski, SS Bemji, J Densley. IEEE Trans Dielectr Electr Insul 3: 351–360, 1996.
31. J Jonsson, B Rånby, C Laurent, C Mayoux. IEEE Trans Dielectr Electr Insul 3: 148–152, 1996.
32. HM Bandford, RA Fouracre, A Faucitano, A Buttafava, F Martinotti. IEEE Trans Dielectr Electr Insul 3: 594–598, 1996.
33. RJ Densley, SS Bamji, AT Bulinski, J-P. Crine, Water teenig and polymer oxidation. Proceedings of IEEE International Symposium on Electrical Insulation, Toronto, 1990, pp 178–182.
34. AL Buchachenko. J Polym Sci 57: 299–310, 1976.
35. T Zaharescu. Rom J Phys 41: 771–778, 1996.
36. R Setnescu, S Jipa, T Setnescu, M Dumitru, I Mihalcea, C Podin, PV Notinger. Synergetic effects of temperature and radiation in degradation of LDPE and emphasized by chemiluminescence techniques. Proceedings of IEEE International Conference on Dielectrics and Insulation, Budapest, 1997, pp 329–332.
37. DJ Carlsson. In: G. Scott, ed. Atmosphere Oxidation and Antioxidants, vol II. Amsterdam: Elsevier, 1993, pp 495–310.
38. R Barticas. IEEE Trans Dielectr Electr Insul 4: 544–557, 1997.
39. MS Graff, RH Boyd. Polymer 35: 1797–1801, 1994.
40. M Ito, S Okada. J Appl Polym Sci 50: 233–243, 1992.
41. Z Mizumo, M Nagao, N Shimizu, K Horii. IEEE Trans Electr Insul EI-27: 1108–1111, 1992.
42. Kunstsoff Handbuch. In: Vol. 4, Polyolefine, eds: R.

Vieweg, A. Schley, A. Schwarz, Munchen: Carl Hanser Verlag, 1969.
43. V Maslow. Moisture and Water Resistance of Electrical Insulation, English ed. Moskow: Mir Publisher, 1975, pp 125–121.
44. DV Ivanjukov, ML Fridman (ed.). Propylene. Properties and Application, Russian ed. Moskow: Khimija, 1974, p 73.
45. MM Abdel-Aziz, SE Gwaily, M Madoni. Polym Degrad Stab 62: 587–597, 1998.
46. S Grzybowski, P Zubielik, ER Kuffel. Trans IEEE Power Delivery 4: 1507–1512, 1989.
47. F Selle, D Schultz. Electrical breakdown of EPR insulated cables at dc and impulse voltage. Proceedings of IEEE International Conference of Conduction and Breakdown in Solid Dielectrics, Erlangen, 1986, pp 61–64.
48. B Sanden, E Ildstad. DC electrical and mechanical characteristics of polypropylene film. Proceedings of IEEE International Conference on Conduction and Breakdown in Solid Dielectrics, Västerås, 1998, pp 210–213.
49. J Svahn, SM. Gubanski. Influence of metal inclusions on the AC breakdown strength in EPDM, Proceedings of IEEE International Conference on Conduction and Breakdown in Solid Dielectrics, Vasteras, 1998, pp 498–501.
50. SN Kolesov. IEEE Trans Electr Insul EI-15: 382–390, 1980.
51. BV Ceres, JM Schultz. J Appl Polym Sci 29: 4183–4197,1984.
52. H Kato, T Fukuda. IEEE Trans Electr Insul EI-21: 925–927, 1986.
53. N Hozumi, H Suzuki, T Okamoto, K Watanabe, A Watanabe. IEEE Trans. Dielectr Electr Insul 1: 1068–1076, 1994.
54. M Otsubo, Y Shimono, T Hikami, C Honda, K Ito. Influence of the humidity on leakage current under accelerated ageing of polymer insulating materials. Proceedings of IEEE International Symposium on Electrical Insulation, Montreal, 1966, pp. 267–270.
55. I Sazhin, VP Shuvaev, BS Skurikhina. Vysokomol Soedin Part A 12: 2728–2731, 1970.
56. B Mattson, B Stenberg. Rubb Chem Technol 65: 315–328, 1992.
57. DK Das-Gupta, K Doughty, DE. Cooper. Dielectric ageing of high voltage cable insulators. Proceedings of IEEE International Conference on Conduction and Breakdown in Solid Dielectrics, Erlangen, 1986, 82–85.
58. Fwang S Suh, Jae Young Kim, Chang Ryong Lee, T Takada. IEEE Trans Dielectr Electr Insul 3: 201–206, 1996.
59. A-M Jeffery, DH. Damon. IEEE Trans Dielectr Electr Insul 2: 394–407, 1995.
60. A Motori, F Sandrolini, GC Montanari. Degradation and electrical behavior of aged XLPE cable models. Proceedings of IEEE International Conference on Conduction and Breakdown in Solid Dielecterics, Trodheim, 1989, pp 352–358.
61. SW Traczyk, T Galiński, J Świątek. Adv Mater Optics Electron 7: 87–91, 1997.
62. R Danz, H van Berlepsch, L Brehmer, B Elling, D Greiss, M Kornelson, W Kunstler, M Pinrow, W Stark, A Wedel. Acta Polym 40: 206–210, 1989.
63. Seung Hyung Lee, Jung-Ki Park, Jae Hong Han, Kwang S. Suh. IEEE Trans Dielectr Electr Insul 2: 1132–1139, 1995.
64. K Yoshino. IEEE Trans Electr Insul EI-21: 999–1006, 1986.
65. Y Sekii, K Maruta, T Okada. The effect of inclusions in XLPE insulators on electrical trees generated by ac and grounded dc voltage. Proceedings of IEEE International Conference on Conduction and Breakdown in Solid Dielectrics, Västerås, 1998, pp 317-321.
66. AT Bulinski, S Bamji, J Densley, A Gustafsson, UW Gedde. IEEE Trans Dielectr Electr Insul 1: 949–962, 1994.
67. HM. van der Voorde, C Restat. Report CERN 72-2, p 94, 1972.
68. DK. Das-Gupta. IEEE Trans Electr Insul EI-27: 909–923, 1992.
69. S Nakamura, F Murabayashi, K Iida, G Sawa, M Ieda. IEEE Trans Electr Insul EI-22: 715–720, 1987.
70. HM Bandford, RA Fouracre, G Chen, DJ Tedford. Radiat Phys Chem Part C 40: 401–410, 1992.
71. RM Eichhorn. IEEE Trans Electr Insul EI-16: 469–482, 1981.
72. WD Wilkens. IEEE Trans Electr Insul EI-16: 521–527, 1981.
73. HC. Booij. J Polym Sci Part C 16: 1761–1766, 1967.
74. G Grzybowski A Rakowska, JE. Tompson. Examination of ageing process in polyethylene cable insulation. Proceedings of IEEE International Symposium on Electrical Insulation, Montreal, 1984, pp 262–266.
75. S Grzybowski, A Rakowska, K Siodla, P Zubielik. Influence of the water environment on the high voltage polyethylene insulation. Proceedings of 5th International Symposium on High Voltage Engineering, Braunschweig, 1987, paper 21.06.
76. I Kuriyama, N Hayakawa, Y Nakase, J Ogura, H Yagyu, K Kasai. IEEE Trans Electr Insul EI-14: 272–277, 1979.

13

Rheological and Optical Properties and Parameters

Bogdan Georgiev Bogdanov
Bourgas University of Technology, Bourgas, Bulgaria

Marin Michailov†
Bulgarian Academy of Sciences, Sofia, Bulgaria

I. RHEOLOGICAL PROPERTIES AND PARAMETERS

The rheology of polymer melts in shear flow is considered. It is included with limited data for rheological properties of some PO melts in uniaxial elongational (extensional) flow. Table 1 systematizes the following.

A. Steady Flow Rheological Properties

1. Shear stress–shear rate dependences of PO in the steady flow, i.e., $\tau(\dot{\gamma})$ and its dependence on a number of factors, i.e., $\tau(\dot{\gamma}, F_i)$ where F_i is T, $\bar{M}$, I_m, etc.
2. Rheological properties of PO determined in shear flow, i.e., zero shear (zero-limiting) viscosity (η_0), apparent (melt) shear viscosity (η), temperature-invariant dependence of melt viscosity (η/η_0), critical shear rate ($\dot{\gamma}_0$) and critical shear stress (τ_0, first normal stress difference (N_1), die swell (B), and activation energy of flow (E_a), melt index or flow rate (I_m or G_m, respectively), and their dependences on a different factors and viscosity in elongational flow (η_μ).

B. Dynamic Rheological Properties

Dynamic viscosity and dynamic moduli. Dynamic rheological parameters and their dependencies on particular factors (frequency, shear rate, time, and temperature).

Complex dynamic viscosity η^*, and its components—viscous (η') and elastic (η''), shear relaxation modulus G^* and its components—storage modulus G' and loss modulus G''.

The study of tan δ is of great interest with regard to the present processing problems, but it is not included here. This parameter represents the ratio of viscosity and elasticity, i.e., $\tan\delta = \eta'/\eta''$, and it is possible to calculate if necessary using the presented data.

Tables 1–11 and Figures 1–68 show the selected literature data concerning Rheol. Pr. (F_i) and Rheol. Pm. (F_i) of POL.

II. OPTICAL PROPERTIES AND PARAMETERS

Only optical behaviors of PO in the visible part of the wave spectrum (as an exception and in the UV spectrum for particular cases) are treated here. Table 12 includes:

1. Refractive index (n) and molecular refraction (R) of PO and their dependence on different factors (F_i).

(Text continues on p 345)

†Deceased.

Arranged for the second edition by Cornelia Vasile, "P. Poni" Institute of Macromolecular Chemistry, Iasi, Romania.

TABLE 1 Rheological properties of PO and their relationship from different factors [Rheol. Pr. (F_i)]

		Steady flow rheological properties			
N	PO	$\tau(\dot{\gamma})$; $(\dot{\gamma}(\tau))$	$\tau(\dot{\gamma}, T)$	$\tau(\dot{\gamma}, \bar{M})$ $\tau(\dot{\gamma}, \bar{M}_w/\bar{M}_n)$	$\tau(\dot{\gamma}, I_m)$
1	2	3	4	5	6
1.	HDPE UHMWHDPE LHDPE	[1], [2], [3]	[11] p 80 (Fig. 1), p 130, [12] p 294, [13] [14]		[12] p 293
2.	LDPE, LLDPE	[3], [4], [5], [6]	[15] (Fig. 2), [11] p 80, [16] p 62, [14], [6]		[18], [19]
3.	Different PEs	[4] [12] p 293			[20]
4.	IPP, APP, i-PEP	[7] [8]	[11] p 136 (Fig. 3), [17] p 176	[12] p 295 (Fig. 5), [7]	
5.	Different POs: PE, PP, α-PO	[9], [10]	[14] (Fig. 4)	[12] p 294 (Fig. 6)	

	Steady flow rheological properties	
N	Zero shear viscosity (η_0)	Apparent shear viscosity (η)
1	7	8
1.	$\eta_0(\bar{M})$: [21] p 182 (Fig. 7); $\eta_0(g.\bar{M})$: [22]; $\eta_0(1/T)$: [13]	$\eta(\dot{\gamma})$: [31] (Fig. 12); [32], [33], [10] p 68; $\eta.I_m(\dot{\gamma}/I_m)$: [34], [35], [36]; $\eta(\dot{\gamma}$, branching): [36]; $\eta(\dot{\gamma}, T)$: [37], [38], [11]; $\eta(\tau, T)$: [13] (Fig. 13), [39]; $\eta(\bar{M}, \dot{\gamma})$: [31]; [40] p 88 (Fig. 14), [41], [42]
2.	$\eta_0(\bar{M})$: [23] (Table 2); [24], [25]; $\eta_0(1/T)$: [5]	$\eta(\dot{\gamma})$: [43], [44], [15], [45], [24], [46], [32], [35], [6]; $\eta(\dot{\gamma}, I_m)$: [47], $\eta \cdot I_m(\dot{\gamma}/I_m)$: [34]; $\eta(\dot{\gamma}, T)$: [11] p 87, [37] (Fig. 15); $\eta(\tau, T)$: [15] (Fig. 16), [18]; $\eta(\bar{M}, T)$: [5]; $\eta(t)$: [48] (Fig. 17), [45], [39]
3.	$\eta_0(\bar{M})$: [4] (Fig. 8), [22]; $\eta_0((M_w/M_n \cdot \dot{\gamma}_o)$: [4]; $\eta_0(1/T)$: [26] (Fig. 9)	$\eta(\dot{\gamma})$: [41], [20], [23], [49]; $\eta(\dot{\gamma}, I_m)$: [47]; $\eta(\dot{\gamma}, T)$: [50]; p 165 $\eta(\tau)$: [25] (Fig. 18), [51], [20]; $\eta(T)$: [52] p 402 (Fig. 19)
4.	$\eta_0(\bar{M})$: [17] p 180 (Table 3); [27] (Fig. 10); $\eta_0(1/T)$: [17] p 182 (Fig. 12), [28]	$\eta(\dot{\gamma})$: [27] (Fig. 20), [32], [31], [10]; $\eta \cdot I_m(\dot{\gamma}/I_m)$: [34]; $\eta(\dot{\gamma}, T)$: [31], [11] p 137; $[1/\eta](\tau)$: [8]; $\eta(\tau, T)$: [17] (Fig. 21); $\eta(\tau, \dot{\gamma}, T)$: [53] p 55; $\eta(T)$: [54] (Fig. 22), [55]; $\eta(t)$: [54]

TABLE 1 (Continued)

N	Steady flow rheological properties Zero shear viscosity (η_0)	Apparent shear viscosity (η)
1	7	8
5.	$\eta_0(\bar{M})$: [29] p 273; $\eta_0(1/T)$: [22], p 132; $\eta_0(\bar{M}, T)$: [30] (Table 4); η_0(p): [29] p 281	η: [51], [56] (Table 5); $\eta(\dot{\gamma})$: [57], [10], [58]; $\eta(\dot{\gamma}, T)$: [59] p 82; $\eta(\tau, T)$: [22], p 179 (Fig. 23); $\eta(T/T_g)$: [60] p 268; $\eta(t)$: [39]

N	Steady flow rheological properties Elongation viscosity (η_μ)
1	9
1.	$\eta_\mu(t)$: [61] (Fig. 24)
2.	$\eta_\mu(V_d)$: [62], [24]; $\eta_\mu(\tau, T)$: [39] (Fig. 25); $\eta_\mu(\varepsilon)$: [39], [63], [64], [65], [66]; $\eta_\mu(t)$: [48] (Fig. 17), [61], [45], [67], [68]; $\eta_\mu(M)$: [64], [69]
3.	$\eta_\mu(\varepsilon)$: [39] (Fig. 26)
4.	$\eta_\mu(V_d)$: [27]; $\eta_\mu(t)$: [61] (Fig. 27), [27]
5.	$\eta_\mu(\tau)$: [26] p 163

N	Steady flow rheological properties η/η_0	Melt index (I_m), Melt flow rate (G_m)
1	10	11
1.	$\eta/\eta_0(\dot{\gamma} \cdot \eta_o)$: [13] (Fig. 28)	$G_m(p)$: [72] (Fig. 32); $G_m(p, \tau, \dot{\gamma})$: [72]
2.	$\eta/\eta_0(\dot{\gamma})$: [44]; $\eta/\eta_0(\dot{\gamma}, a_T)$: [45] (Fig. 29)	$I_m(T, t)$: [16], p 371; $G_m(t)$: [26] p 58; $G_m(p)$: [16] p 63 (Fig. 33); $I_m(G'')$: [47]
3.		$I_m(\bar{M})$: [51]; G_m (mol. composition): [73]
4.	$\eta/\eta_0(\tau)$: [70], [17] p 178; $\eta/\eta_0(\dot{\gamma} \cdot \tau)$: [17] p 178; $\eta/\eta_0(\dot{\gamma} \cdot \eta_o)$: [7] (Fig. 30), [17] p 178; $\eta/\eta_0(\tau \cdot \bar{M}_w/M_n)$: [71]	$I_m(p, T)$: [12] p 292 (Fig. 34) $I_m(\bar{M})$: [74] (Fig. 35)
5.	$\eta/\eta_0(\dot{\gamma} \cdot \eta_0)$: [22] p 228 (Fig. 31); $\eta/\eta_0(\tau/T)$: [50] p 168	$I_m(\bar{M})$: [51]; $I_m(\bar{M}$, short-chain branching); [51] (Fig. 36); $I_m(\dot{\gamma})$: [75] p 79

TABLE 1 (Continued)

N	Steady flow rheological properties: Critical shear rate ($\dot{\gamma}_o$) and critical stress (τ_o)	First normal stress difference (N_1)
1	12	13
1.	$\dot{\gamma}_o(T)$: [76]	$N_1(\dot{\gamma})$: [57] (Fig. 39); [10] p 68; $N_1(\dot{\gamma}, T)$: [80], [63]; $N_1(1/T)$: [80] (Fig. 40)
2.	$\dot{\gamma}_o(\bar{M})$: [24] (Table 6); $\dot{\gamma}_o(\bar{M}, \bar{M}_w/\bar{M}_n)$: [47]; $\dot{\gamma}_o(\eta)$: [77]; $\dot{\gamma}_o, \tau_o(I_m)$: [78]	$N_1(\dot{\gamma})$: [46] (Fig. 41) [45]; $N_1(t)$: [48] (Fig. 42); $N_1(\dot{\gamma}, t)$: [48]; $N_1(\tau, T)$: [45]; $N_1(\nu)$: [33]
3.	$\dot{\gamma}_o\tau_o(T)$: [79] (Fig. 37), [51]; $\dot{\gamma}_o(\bar{M}, \bar{M}_w/\bar{M}_n)$: [4]; $\dot{\gamma}_o$(side chain): [51]	$N_1(\dot{\gamma})$: [25]
4.	$\dot{\gamma}_o, \tau_o(M)$: [51]	$N_1(\dot{\gamma})$: [81], [10], [80]; $N_1(\dot{\gamma}, I_m)$: [27] (Fig. 43); $N_1(\tau)$: [55]; $N_1(\tau, I_m)$: [27]; $N_1(1/T)$: [80] (Fig. 44)
5.	$\tau_o(\bar{M}, \bar{M}_n/\bar{M}_w)$: [22] p 201 (Fig. 38)	$N_1(\dot{\gamma})$: [10], [82] p 348 (Fig. 45), [57], [58]; $N_1(\tau)$: [80] (Fig. 46), [22] p 350

N	Steady flow rheological properties: Die swell (B)	Activation energy for steady shear viscosity (E_a)
1	14	15
1.	$B(\dot{\gamma})$: [72] (Fig. 47); $B(t, \dot{\gamma})$: [10] p 19; [57] (Fig. 48); $B(\tau, p, \dot{\gamma})$: [72] (Table 7)	E_a (number of C atoms): [16] p 84 (Fig. 54); $E_a(t)$: [25] (Fig. 55)
2.	$B(\dot{\gamma})$: [46], [6]; $B(\dot{\gamma}, T)$: [83] p 65, [6]; [19]; $B(G_m, \dot{\gamma})$: [19]; $B(\tau)$ [6]; $B(p, T)$: [53] p 66, [18] (Fig. 49); $B(\dot{\gamma}, \bar{M}, \bar{M}_w/\bar{M}_n)$: [47] (Table 8); $B(\tau, \bar{M}, \bar{M}_w/\bar{M}_n)$: [47]; $B(L_d/D, \dot{\gamma})$: [18] (Fig. 50) [53] p 67; B(LCBF): [84] (Fig. 51), [23]	$E_a(\tau)$: [15]; $E_a(M)$: [5] (Fig. 56)
3.	$B(\dot{\gamma})$: [83]; $B(\tau)$: [37] p 570; $B(\bar{M}, \bar{M}_w/\bar{M}_n)$: [4]	E_a: [24] (Table 9); E_a (branched PE): [25]; $E_a(\bar{M}, \bar{M}_w/\bar{M}_n)$: [4]
4.	$B(\dot{\gamma})$: [57] p 78 (Fig. 52); $B(p)$: [55]	$E_a(\dot{\gamma})$: [24] p 10, [59] (Fig. 57) $E_a(M)$: [51], [85] E_a (wt% of APP): [70] (Fig. 58)

TABLE 1 (Continued)

	Steady flow rheological properties	
N	Die swell (B)	Activation energy for steady shear viscosity (E_a)
1	14	15
5.	$B(\dot{\gamma})$: [75] p 79, [22] p 397; [57]; $B(t)$: [22] p 396 (Fig. 53)	$E_a(\dot{\gamma})$: [24]; E_a (C atoms in side chain): [22] p 138 (Fig. 59), [51]; E_a: [86], [75] p 79, [29] p 227 (Table 10)

	Dynamic rheological properties
N	Dynamic viscosity (η^*, η', η''), and dynamic moduli (G^*, G', G'')
1	16
1.	$\eta^*(\nu, \bar{M})$: [42]; η', $G'(\dot{\gamma})$: [10] p 68; η', $G'(\dot{\gamma})$: [33]; η', G', $G''(\nu)$: [41] (Fig. 60); $G'(t)$: [87]
2.	$\eta'(\nu)$: [25]; η', $\eta''(\dot{\nu})$: [23, 88] (Fig. 61); $\eta''(\eta')$: [44]; G^*, G', $G''(t, 1/\nu)$: [48] (Fig. 62)
3.	$\eta'(\nu)$: [41]; η', $G'(\nu)$: [47]; η', G', $G''(\bar{M}, \bar{M}_w/\bar{M}_n)$: [47] (Table 11); $G'(\nu)$: [25] p 526
4.	η', $\eta''(\nu)$: [89] (Fig. 63, 64); $\eta'(\nu, T)$: [28]; $\eta'(T)$: [54]; η^*, $\eta'(\nu)$: [81]; G', $G''(\nu)$: [81] (Fig. 65); η', $G'(\nu)$): [10]
5.	$\eta^*(\nu)$: [30] (Fig. 66), [22]; η', $G'(\nu)$: [57]; G', $G(t)$: [90] (Fig. 67); $G'(\nu)$: [30] (Fig. 68); $G''/G'(\nu)$): [22] p 310

TABLE 2 Dependence of zero shear rate viscosity η_0 on the molecular weight and LCB LDPE

Author	Ref.	Relation	Observation
1	2	3	4
Tung	[91]	$\log \eta_0 = 3.4 \log M_w + \frac{3.16 \times 10^3}{T} - 19$	$T > 165°C$
Peticolas and Watkins	[92]	$\eta_0 = 3.01 \times 10^{-12} M_w^{3.4} e^{-2.35}\left(\frac{n_w}{12}\right)$	$T = 150°C$
Busse and Longworth	[93]	$\log \eta_0 = 12.3 + \frac{3.45}{a} \log (w_i M_i^a)$	a is related LCB
Schreiber and Bagley	[94]	$\log \eta_0 = 4.22 \log M_w - 15.78 - aN_L$	N_L is number of LCB; a is constant
Bontinck	[94]	$\log \eta_0 = 3.84 M_w^{3.38} e^{-2.35}\left(\frac{n_w}{12}\right)$	$T = 150°C$
Mendelson	[95]	$\log \eta_0 = 24.18 + 6.56 \log (gM_w)$	$T = 150°C$
Miltz and Ram	[96]	$\eta_0 = K(gM_w)^{4.73}$	
Locati and Garganti	[97]	$\log \frac{\eta_0}{[\eta]^{2.84} t_0^{0.23}} - \log k$	t_0 is relaxation time, k is a constant
Briedis and Faitelson	[98]	$\eta_0 = \frac{K_b(gM_w)^6}{\exp\left[6K_p\left(\frac{M_w}{M_n} - 1\right)\right]}$	For 36 commercial LDPE
Ram	[99]	$\eta_0 = K(gM_w)^{3.9}$	$T = 148°C$
Pederson and Ram	[100]	$\eta_0 = -12.75 + 4.25 \log (gM_w)$	$T = 148°C$
		$\eta_0 = -13.47 + 4.28 \log (gM_w)$	$T = 190°C$

Note: g, ratio of the intrinsic viscosities of branched and linear molecules at the same molecular weight; n_w, weight-average number of branch points per molecule.
Source: Ref. 23.

TABLE 3 Empirical equations of zero shear rate viscosity η_0 on molecular weight of polypropylene

Equation	Ref.
$\log \eta_0 = 3.5\bar{M}_w + \log(\bar{M}_z/\bar{M}_w) - 12.7$	[101]
$\log \eta_0 = -12.2 + 3.12(M_v)$	[102]
$d \log \eta_0/d \log(\bar{M}_w) = 3.5$	[103]
$\log \eta_0 = 5.15 - 0.926\, l_g (I_m))$	[104]
$\log \eta_0 = 4.61 \log[\eta] - 10.33$	[8, 105]
$\log \eta_0 = 3.69 \log(M_w) - 11.90$	[8, 105]

Source: Data from Ref. 17.

TABLE 4 Zero shear rate viscosity η_0 of polyisobutylenes

Sample	$M_w \times 10^{-5}$	$\bar{M}_w/\bar{M}_n$	Temp., °C	$\log \eta_0$, poise
I	0.847	2.99	50	5.17
			74	4.50
			99	3.94
			121	3.53
			140	3.19
II	8.48	2.54	142	7.37
			174	6.95
			197	6.65
			222	6.39
			242	6.12
IV	15.0	2.38	139	8.69
			179	7.79
			202	7.4
			222	7.13
			242	6.92

Source: Data From Ref. 30.

TABLE 5 Rheological properties of poly(4-methyl-1-pentene) and poly(4-methyl-1-pentene)hexene-1 copolymer

Content of hexene-1, mass %	0	10	10	12	15	22
Melt index I_m, g/10 min	5.60	0.28	14.00	2.80	2.70	0.20
Shear rate $\dot{\gamma}$ (250°C, $\tau = 0.06$ MPa), s^{-1}	57	15	50	60	50	49
Melt viscosity $\times 10^{-3}$ (at $\dot{\gamma} = 50\ s^{-1}$), Pa s	1.14	1.68	0.70	1.12	1.22	1.37

Source: Data from Ref. 56.

TABLE 6 Expressions relating shear rate $\dot{\gamma}_0$ at which non-Newtonian behavior begins to molecular weight and LCB for LDPE

Author	Ref.	Expression	Observation
Guillet *et al.*	[106]	$\log \dot{\gamma}_0 = 7.09 - \log \eta_0$	For linear and branched PE
Mendelson *et al.*	[95]	$\log \dot{\gamma}_0 = 26.81 = 6.09 \log (gM_w)$	$\dot{\gamma}_0$ defined as $\eta(\dot{\gamma}_0) = 95\%\ \eta_0$
Miltz and Ram	[96]	$\log \dot{\gamma}_0 = K - 4.3 \log (gM_w)$	
Graessley	[107]	$\eta_0 J_e^o \dot{\gamma}_0 = 0.6 \pm 0.2$	$\dot{\gamma}_0$ is the value at which $\eta(\dot{\gamma}_0) = 0.8\ \eta_0$

J_e^o, Shear compliance (steady-static elastic compliance).
Source: Ref. 24.

TABLE 7 Rheological data for constant pressure operation of LHDPE Marlex 6009, at 186°C*

Pressure, p.s.i.	Shear stress, (dyn cm^{-2}) $\times 10^{-6}$†	Shear rate, s^{-1}†	Average density, g cm^{-3}‡	Mass flow rate, g min^{-1}	Die swell ratio	Extrudate appearance
1	2	3	4	5	6	7
856	1.32	133.2	0.761	0.2854	1.74	Smooth
877	1.35	142.3	0.761	0.3052	1.75	Smooth
1022	1.58	196.4	0.768	0.4250	1.80	Smooth
1493	2.31	486.8	0.773	1.0596	—	Smooth
1950	3.01	907.1	0.780	1.9926	2.14	Smooth
2015	3.11	2467.6	0.780§	5.4233	—	Rough
2073	3.20	1053.7	0.779	2.3115	2.18	Smooth
2124	3.28	2643.4	0.7769§	5.7774	—	Rough
2146	3.31	1174.1	0.780	2.5811	—	Wavy
2146	3.31	1170.1	0.777	2.5592	2.20	Smooth
2189	3.38	1267.8	0.772	2.7571	—	Wavy
2233	3.44	1272.0	0.781	2.8010	—	Wavy
2320	3.58	3244.7	0.778§	7.1100	—	Rough
2147	3.31	1120.5	0.778	2.4645	2.17	Smooth
1987	3.06	945.5	0.776	2.0667	2.12	Smooth
1668	2.57	628.3	0.769	1.3608	2.01	Smooth
1428	2.20	425.7	0.764	0.9156	1.90	Smooth
1160	1.79	257.6	0.760	0.5514	1.83	Smooth
808	1.25	112.0	0.764	0.2408	1.71	Smooth
532	0.82	48.0	0.762	0.1029	1.62	Smooth

* Steel die, $R = 0.0391$ cm, $L_d/R = 22.30$.
† Apparent values.
‡ More than 24 samples were taken to obtain each average density value except those values in note § below.
§ Six sample measurements were taken to obtain the average value.
Source: Ref. 72.

TABLE 8 Swelling behavior of LDPE samples

N	Resin	$\bar{M}_w$	$\bar{M}_w/\bar{M}_n$	Die swell, % Shear rate $3.3\ s^{-1}$	$33\ s^{-1}$	$330\ s^{-1}$	$660\ s^{-1}$
1	2	3	4	5	6	7	8
1	A 30	408000	17.4	33.6	47.7		
2	A 35a	296300	14.9	41.1	58.7		
3	A 35b	200500	8.0	32.2	48.0		
4	A 50 a	170000	6.4	30.3	43.4	44.0	
5	A 50b	225000	8.4	35.7	52.3	52.2	
6	A 50c	193000	7.4	33.0	44.7	47.6	
7	A 70a	165600	6.4	30.4	45.9	50.7	
8	B 18a	127200	5.6			44.6	47.3
9	B 18b	150900	7.0			52.6	54.9
10	B 20	167000	9.5	19.1	42.5	54.4	
11	B 30a	456400	19.9	19.4	44.5	63.3	
12	B 30b	495000	25.5				
13	PTR 35	101800	5.1			37.6	43.0
14	B 80a	710000	35.5		53.8	68.3	
15	B 80b	533800	25.3				
16	B 80c	3330400	18.1		38.5	51.3	
17	PTR 43	631500	30.5		42.7	62.3	

Source: Data from Ref. 47.

TABLE 9 Activation energies for branched polyethylenes

Author	Ref.	Activation energy, kcal mol^{-1}	Observations
1	2	3	4
Tung	[91]	14.6 for LDPE; 7.5 for HDPE	
Schott and Kaghan	[108]	11–14.6	E_a decreases when shear rate is increased
Aggarwall	[109]	At 150°C 11.8–17; at 180°C 8.9–13	Eight samples of different degrees of LCB
Meissner	[110]	13	
Sabia	[109]	At $\tau = 2.2 \times 10$ dyn cm^{-2}; 13.2 for LDPE; 7.9 for HDPE	E_a increases with shear stress
Mendelson	[111]	11.3 for LDPE; 6.3 for HDPE	Uses the superposition method
Boghetich and Kratz	[112]	5.5–18.6	E_a increases with the number of branches
Porter and Johnson	[113]	Twice for branched than for linear PE	E_a decreases with shear stress
Porter *et al.*	[114]	18 for LDPE; 6–7 for HDPE	E_a decreases with stress
Mendelson	[83]	11.7 for LDPE; 6.3 for HDPE	Uses the superposition method for S_R*
Semionov	[115]	13–13.6 for LDPE; 6–7.5 for HDPE	Data taken from the literature
Blyler	[116]	8–11 for LDPE; 6.7 for HDPE	E_a is constant with stress for HDPE but decreases for LDPE
Miltz and Ram	[96]	12–17 for LDPE; 6.5 for HDPE	E_a decreases with shear rate and increases with LCB
Meissner	[63]	13.6 ± 0.2	IUPAC LDPE
Schroff *et al.*	[117]	11.7 for LDPE: 6 for HDPE	
Starck *et al.*	[147]	12–18	E_a increases with LCB
Jacovic *et al.*	[88]	8.7–12.5 for LDPE; 6.3 for HDPE	E_a varies with strain; remains constant for HDPE

* S_R recoverable shear strain ($= N_1/2\tau$).
Source: Ref. 24.

TABLE 10 Activation energies for PO

PO	Temp., K	E_a, kJ mol^{-1}	Ref.
Polybutene-1	420–470	46–52	[118, 119]
Polyhexadecene-1	355–500	32	[120]
Polyhexene-1	355–500	75	[121, 122, 120]
Polydecene-1	355–500	58	[120]
Polydodecene-1	370–450	31	[51]
Polyisobutylene	350–420	45	[21, 121]
Poly-4-methyl	513-553	7	[123]
pentene-1	553–583	170	[123]
Polyoctadecene-1	355–500	32	[120]
Polyoctene-1	355–500	78	[120]
Polypropylene			
Atactic	350–370	115	[70]
Isotactic	480–560	45–75	[70, 124, 115]
Polytridecene-1	355–500	33	[120]
Polyethylene			
HDPE	430–550	25–30	[21, 125]
LDPE	400–510	30–60	[21, 124, 115]

TABLE 11 Dynamic viscosity data of different samples of PE

N	Resin	$\bar{M}_z/\bar{M}_w$	$[\eta]/g^{0.5}$	G' 0.08 s^{-1}/1.25 s^{-1} (260°C)	η' (0.08 s^{-1}) (260°C)	G'' (0.08 s^{-1}) (260°C)
	LDPE					
2	A 35a	8.2	2.94	142/2013	6005	503
4	A 50a	4.1	2.28	28/1573	4128	345
5	A 50b	6.4	2.61	76/1510	4429	371
6	A 50c	5.7	2.40	25/1532	4504	377
7	A 70	5.3	2.18	76/708	4504	377
8	B 18a	3.9	1.82	25/340	1126	94
9	B 18b	4.6	2.00	32/431	1201	100
10	B 20	7.4	2.07	25/274	750	63
12	B 30b	12.6	4.50	19/233	750	232
13	PTR 35	3.3	1.54	22/120	676	57
14	B 80a	10.9	5.33	54/264	788	66
15	B 80b	11.4	4.20	38/176	826	69
16	B 80c	11.1	3.08	28/104	488	41
17	PTR 43	10.7	4.89	28/164	600	50
18	C 22	9.0	2.45	125	703	59
	HDPE					
19	Vestolen A	—	—	13/75.5	660	207
20	NBC 1475	—	—	19/63	300	25

Source: Data from Ref. 47.

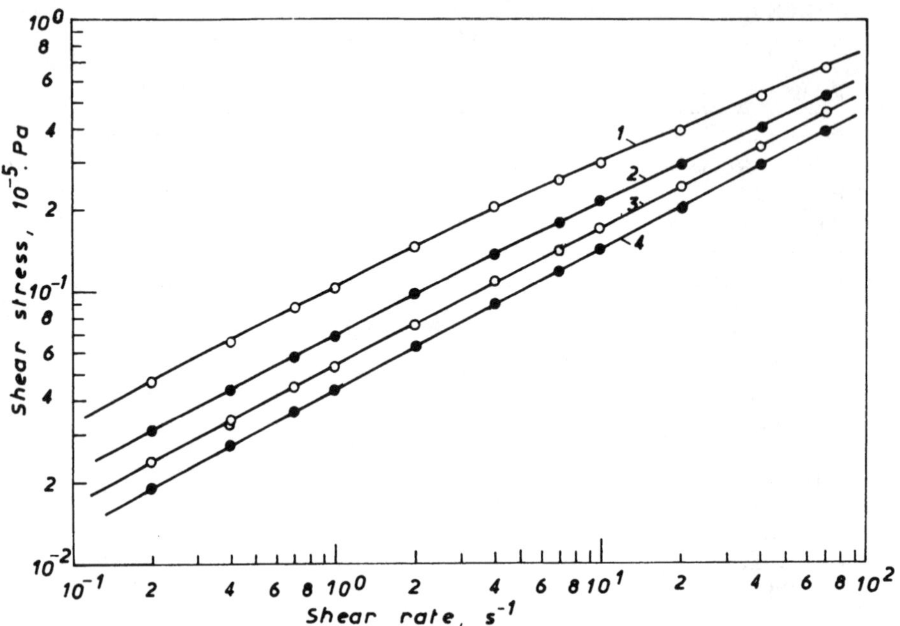

FIG. 1 Dependence of shear stress on shear rate for different temperatures of HDPE ($\rho = 0.942$ g cm^{-3}, I_m =2.21 g/10 min.) Temperature, °C: 1, 150; 2, 170; 3, 190; 4, 210. (From Ref. 11.)

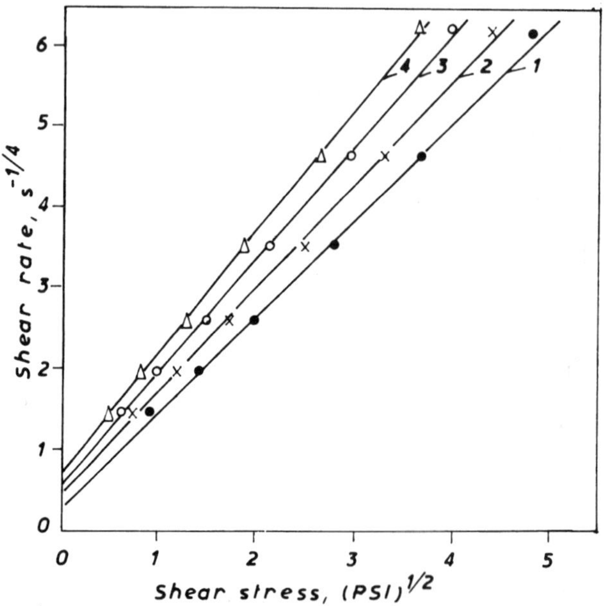

FIG. 2 Dependence of shear rate on shear stress for different temperatures of LDPE (Dow 748). Temperature, °C: 1, 190; 2, 210; 3, 230; 4, 250. (From Ref. 15.)

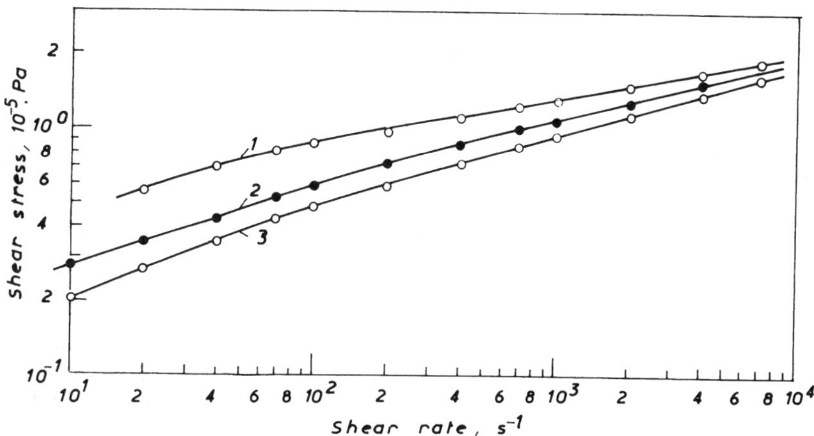

FIG. 3 Dependence of shear stress on shear rate for different temperatures of IPP. ($\rho = 0.910\,\text{g cm}^{-3}$, $I_m = 0.72\,\text{g}/10$ min). Temperature, °C: 1, 190; 2, 205; 3, 220. (From Ref. 13.)

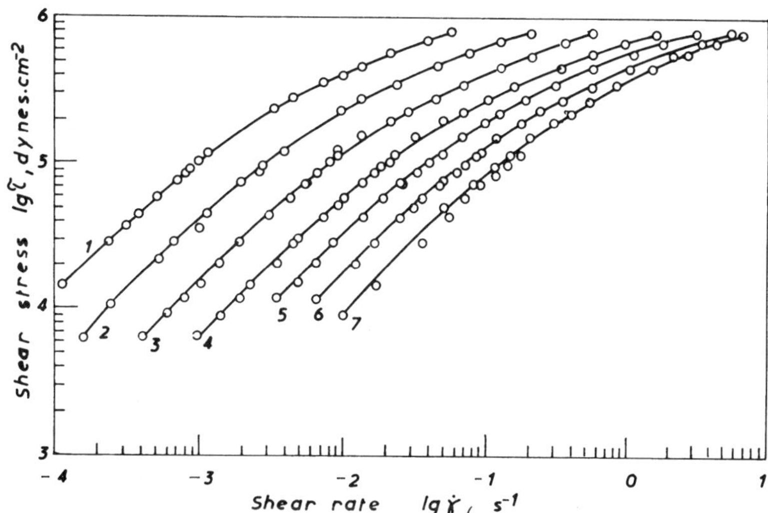

FIG. 4 Flow curves of low molecular weight polyisobutylene L-80. Temperature, °C: 1, 93.3; 2, 121; 3, 148.9; 4, 176.7; 5, 204.4; 6, 232.2; 7, 260. (From Ref. 14.)

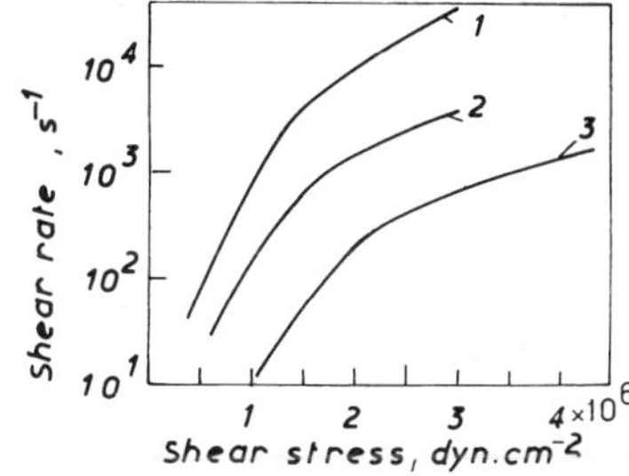

FIG. 5 Dependence of shear rate on shear stress of melt of IPP. Intrinsic viscosity dL/g: 1–2, 3; 2–3, 4; 3–12. (From Ref. 12.)

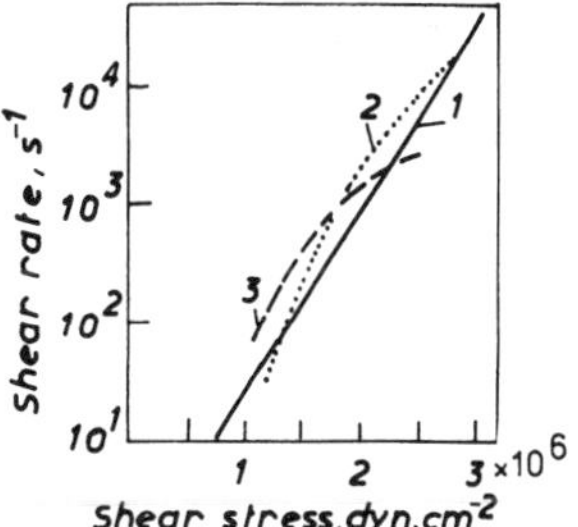

FIG. 6 Dependence of shear rate on shear stress of melt of PIB with different molecular weights $\bar{M}$: 1, B50; 2, B100; 3, B200 ($M_1 < M_2 < M_3$). (From Ref. 12.)

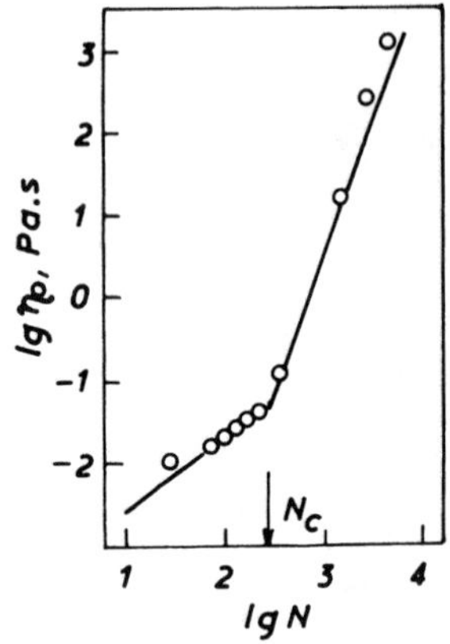

FIG. 7 Dependence of zero shear rate viscosity for mixture of polyethylene and paraffin on number of C atoms in backbone. (Data from Ref. 21.)

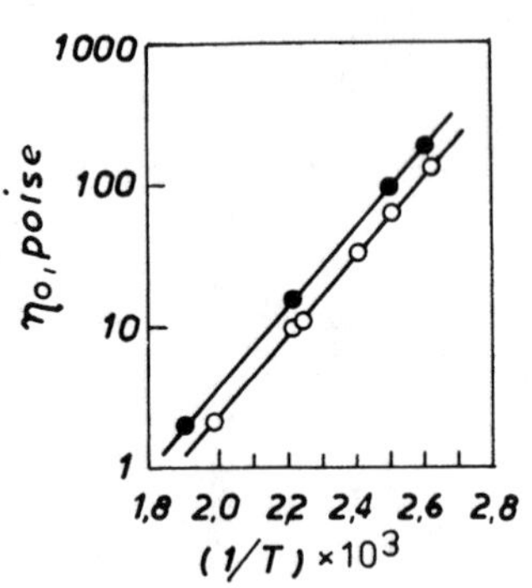

FIG. 9 Plots of log (viscosity at zero shear rate, η_0) vs reciprocal absolute temperature for two grades of polyethylene. (From Ref. 26.)

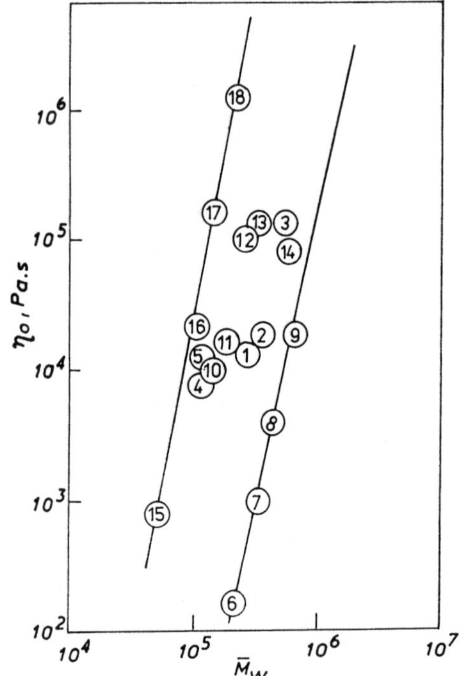

Sample of PE N	ρ, 23°C g cm^{-3}	I_m g/10 min	$\bar{M}_w$ 10^{-3}	$\bar{M}_w/\bar{M}_a$
LDPE				
1	0.9176	2.14	280	5.4
2	0.9172	1.57	360	5.7
3	0.9176	0.24	550	5.7
4	0.9205	3.12	120	4.9
5	0.9234	2.13	135	4.1
6	0.9128	84.0	210	6.6
7	0.9144	22.4	340	8.8
8	0.9156	6.58	420	9.6
9	0.9176	1.19	690	9.4
10	0.9198	1.75	140	4.5
11	0.9204	1.69	185	5.5
12	0.9211	0.25	275	4.7
13	0.9207	0.24	330	5.8
14	0.9217	0.24	600	6.6
HDPE				
15	0.9780	15.1	52	3.0
16	0.9500	0.38	110	5.3
17	0.9445	0.21	140	8.0
18	0.9440	0.04	220	9.1

FIG. 8 Melt Newtonian viscosity η_0 of PE at 180°C vs weight average molecular weight $\bar{M}_w$ of samples examined. (From Ref. 4.)

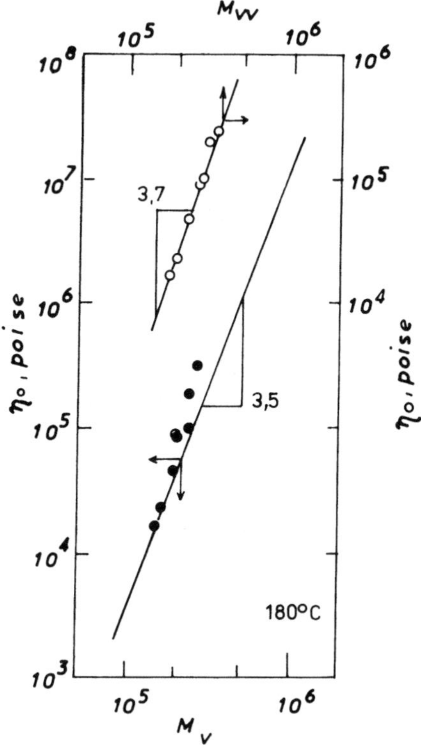

FIG. 10 Zero shear viscosity of IPP as a function of weight average molecular weight and viscosity average molecular weight. (From Ref. 27.)

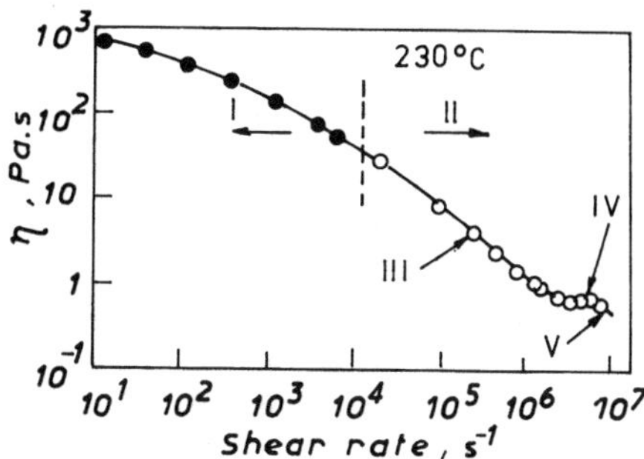

FIG. 12 Plot of apparent viscosity η vs apparent shear rate $\dot{\gamma}$ for HDPE at 230°C; $L_d/D = 40$. I, Instron capillary rheometer; II, specially designed capillary rheometer; III, first non-Newtonian region; IV, second Newtonian region; V, second non-Newtonian region. (From Ref. 31.)

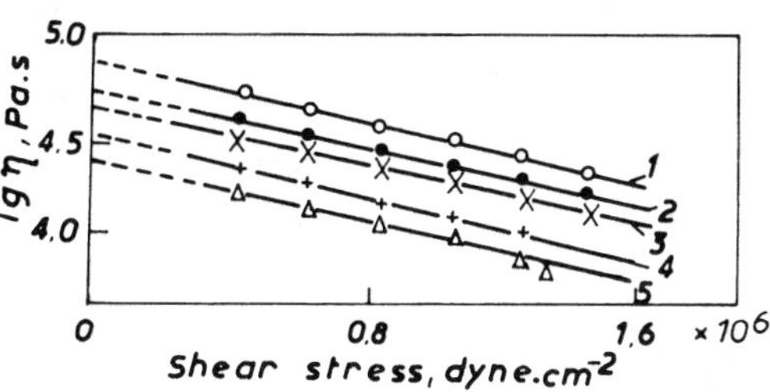

FIG. 13 Viscosity vs shear stress for HDPE. Temperature, °C: 1, 100; 2, 180; 3, 200; 4, 200; 5, 240. (From Ref. 13.)

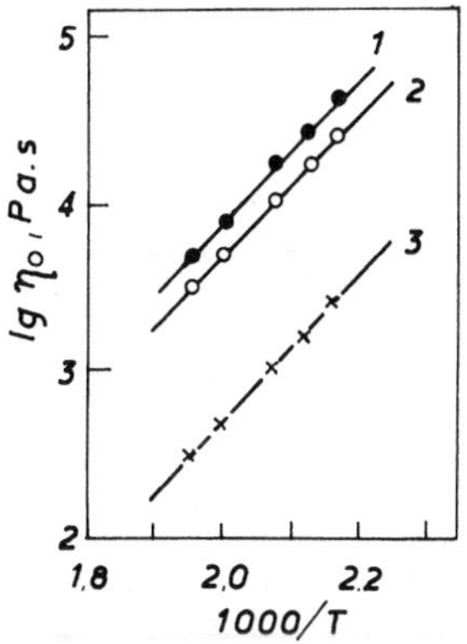

FIG. 11 Temperature dependence of zero shear rate viscosity IPP with different molecular weight. I_m, g/10 min: 1, 0.45; 2, 2.1; 3, 14. (From Ref. 17.)

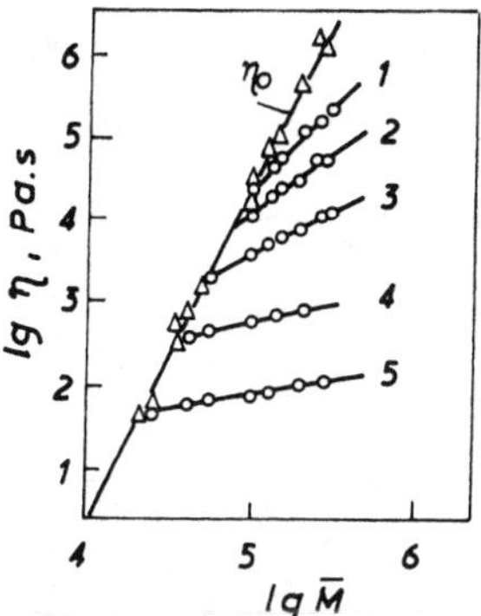

FIG. 14 Viscosity of LHDPE at 190°C vs molecular weight. Shear rate, s^{-1}: 1, 1; 2, 10: 3, 10^2; 4, 10^3; 5, 10^4. (From Ref. 40.)

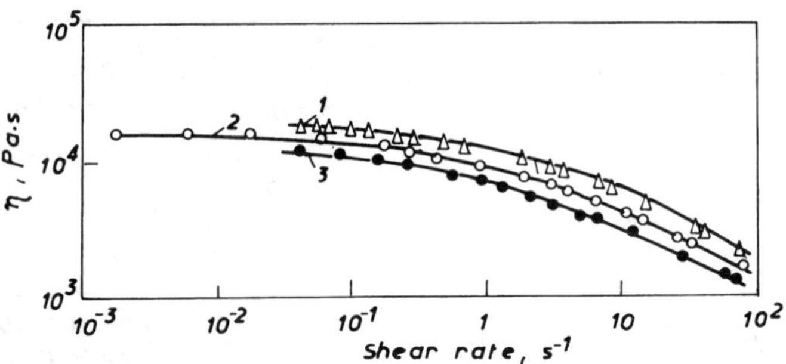

FIG. 15 Viscosity vs shear rate for sample of LLDPE. Temperature, °C: 1, 160; 2, 180; 3, 200. (From Ref. 37.)

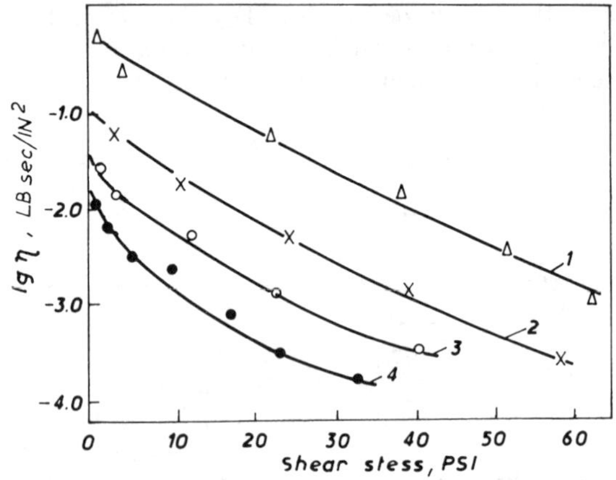

FIG. 16 Dependence of melt viscosity on shear stress for LLDPE (Dow 2047) at different temperatures. Temperature, °C: 1, 160; 2, 200; 3, 240; 4, 280. (η, LB. SEC/IN^2). (From Ref. 15.)

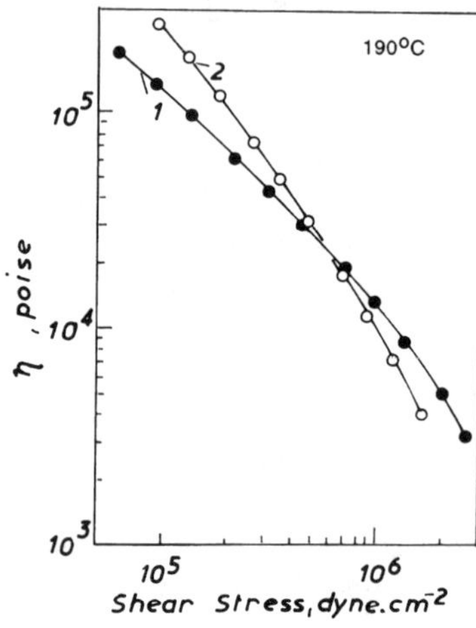

FIG. 18 Viscosity vs shear stress obtained by the Instron capillary rheometer for samples of LHDPE (curve 1) and LDPE (curve 2) at 190°C. (From Ref. 25.)

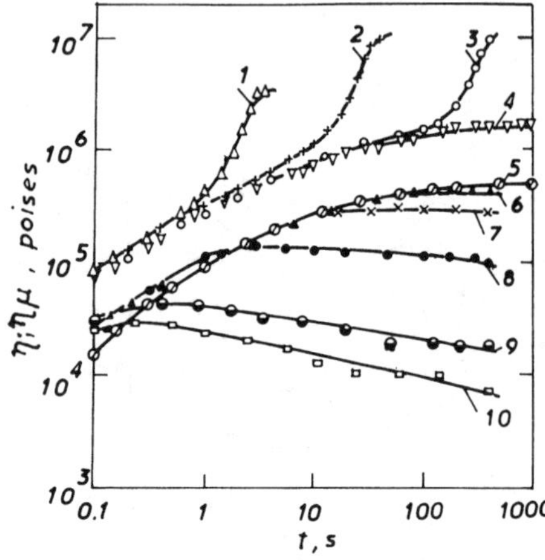

FIG. 17 Stress viscosity in shear η(t) (curves 5–10) and in extension η_μ(t) (curves 1–4) at different deformation rates. Materials, LDPE. Temperature 150°C. Shear rate, s^{-1}: 5, 0.001; 6, 0.01; 7, 0.1; 8, 1; 9, 10; 10, 20. Deformation rates, s^{-1}: 1, 1; 2, 0.1; 3, 0.01; 4, 0.001. (From Ref. 42.)

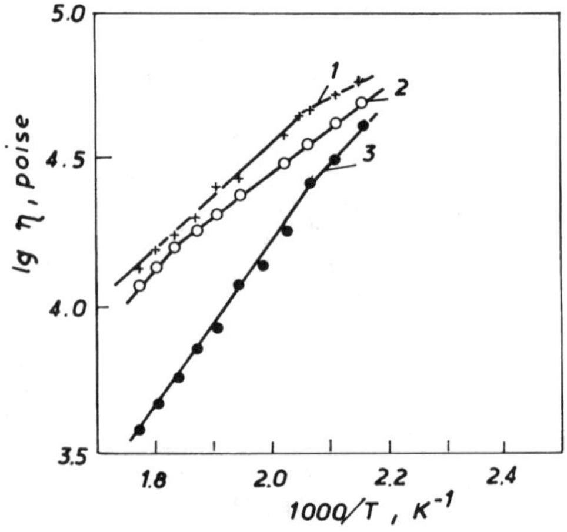

FIG. 19 Temperature dependence of viscosity of HDPE (1), LLDPE (2), and LDPE (3) at shear stress $\tau = 5.10^5$ dyn cm^{-2}. (From Ref. 47.)

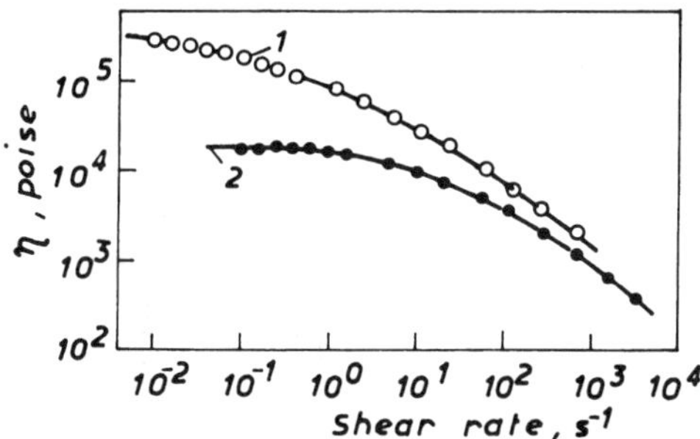

FIG. 20 Shear viscosity-shear rate plots for polypropylene at 180°C. I_m, g/10 min: 1, 3.7; 2, 25; $M_w \times 10^5$: 1, 3.39; 2, 1.79. (From Ref. 27.)

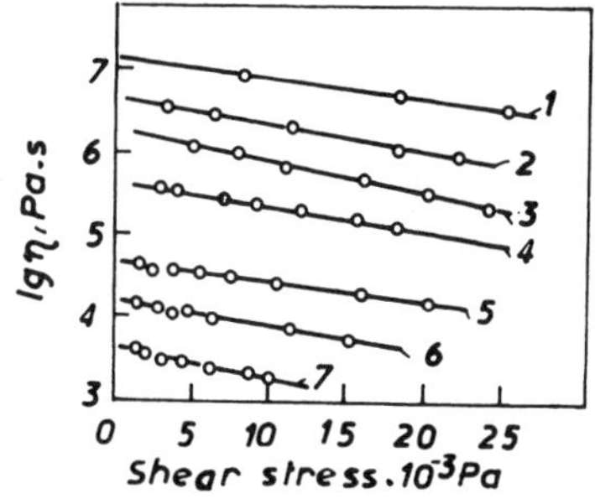

FIG. 23 Dependence of viscosity of PIB ($M_w = 1 \times 10^5$) on shear stress at different temperatures, °C: 1, (−20); 2, (−10); 3, 0; 4, 20; 5, 40; 6, 60; 7, 80. (From Ref. 22.)

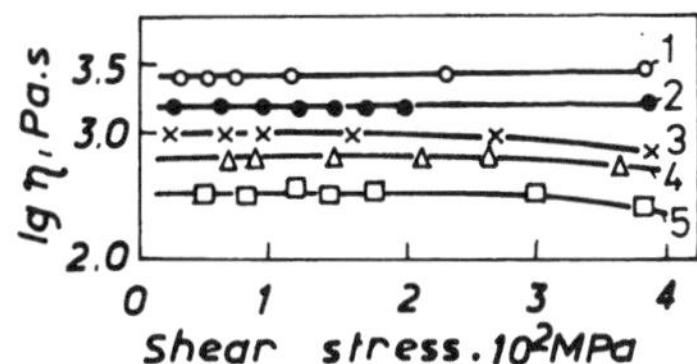

FIG. 21 Dependence of melt viscosity of IPP on shear stress (in Newtonian region) for different temperatures, °C: 1, 190; 2, 200; 3, 210; 4, 230; 5, 240. (From Ref. 17.)

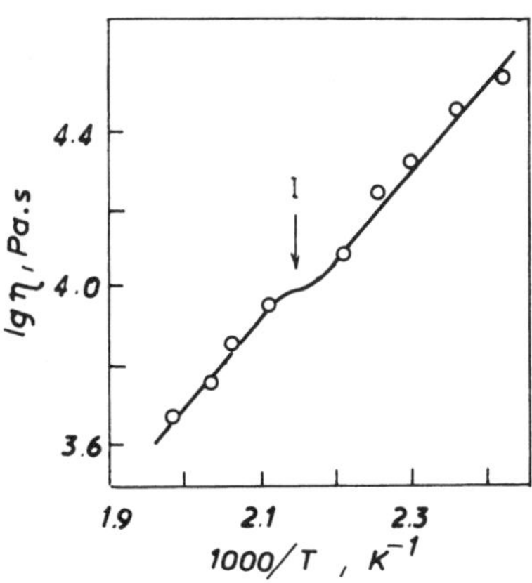

FIG. 22 Temperature dependence of Newtonian viscosity of IPP I, isotropic-liquid crystal state transformation (180–205°C, $E_a = 4\,\mathrm{kJ\,mol^{-1}}$). (From Ref. 54.)

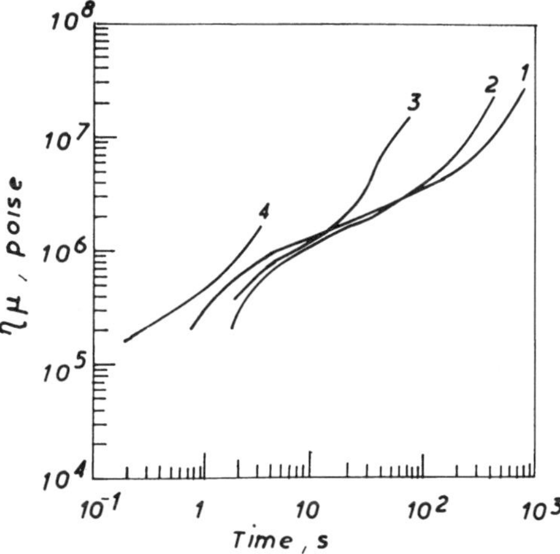

FIG. 24 Extensional viscosity growth function vs time for Phillips Marlex 5502, high-density polyethylene. Strain rate, s^{-1}: 1, 0.0022; 2, 0.005; 3, 0.055; 4, 1.10. (Data from Ref. 61.)

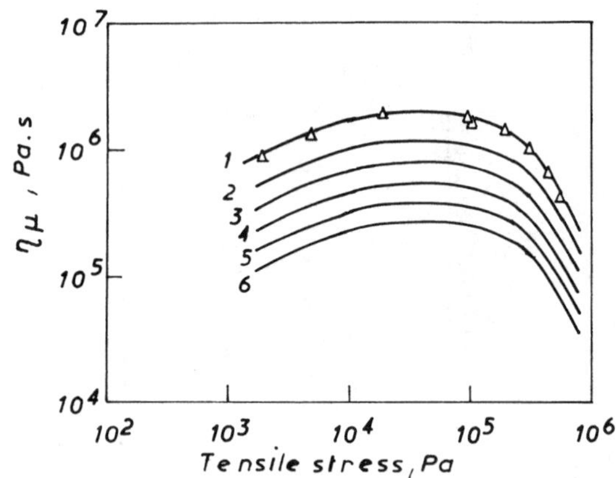

FIG. 25 Temperature dependencies of elongational viscosities of a low-density polyethylene melt. Unfilled triangles denote directly measured values at 170°C. Full curves for other temperatures have been shifted using $E_a = 40.2\,\mathrm{kJ\,mol^{-1}}$. Temperature, °C: 1, 170; 2, 190; 3, 210; 4, 230; 5, 250; 6, 270. LDPE $\bar{M}_w \cdot 10^{-3} = 353$; $M_w/M_n = 20$; $CH_3/1000C = 19.5$; $\rho = 0.924\,\mathrm{g\,cm^{-3}}$. (From Ref. 39.)

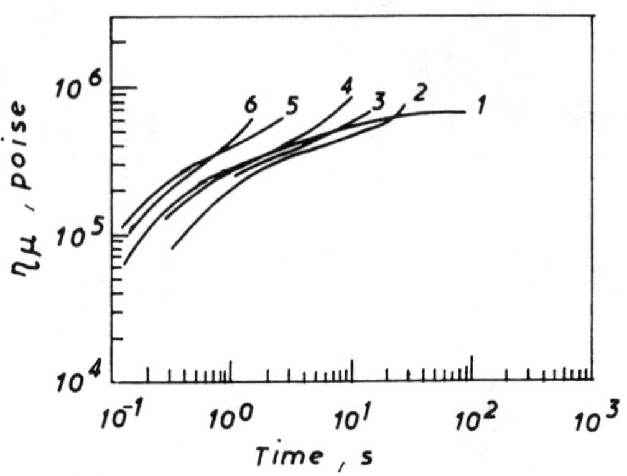

FIG. 27 Extensional viscosity growth function vs time for Exxon CD-263 unmodified polypropylene. Strain rates, s^{-1}: 1, 0.011; 2, 0.055; 3, 0.11; 4, 0.22; 5, 0.55; 6, 1.10. (From Ref. 61.)

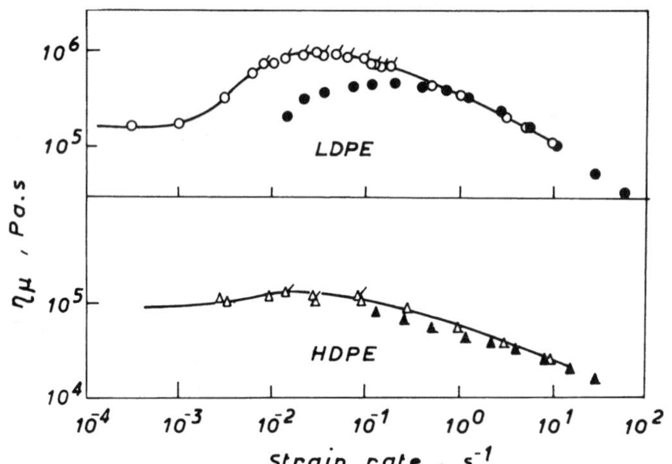

	$\bar{M}_w \times 10^{-3}$	$\bar{M}_w/\bar{M}_n$	$CH_3/1000C$	ρ g cm^{-3}
LDPE	467	25	30	0.918
HDPE	152	14	0	0.960

(From Ref. 39.)

FIG. 26 Comparison of steady-state or plateau elongational viscosities of polyethylene melts from homogeneous isothermal elongation tests (unfilled symbols) and transient tensile viscosities evaluated from converging flow (filled symbols). Unfilled symbols with tick denote tensile creep measurements. $T = 150$°C.

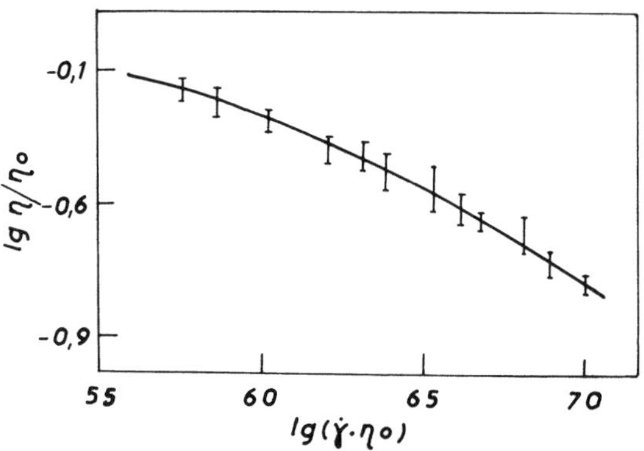

FIG. 28 Temperature-invariant dependence (Vinogradov function) of melt viscosity of HDPE (η/η_0 vs $\dot{\gamma} \cdot \eta_0$) at various melt temperatures from 140°C to 240°C. (From Ref. 13.)

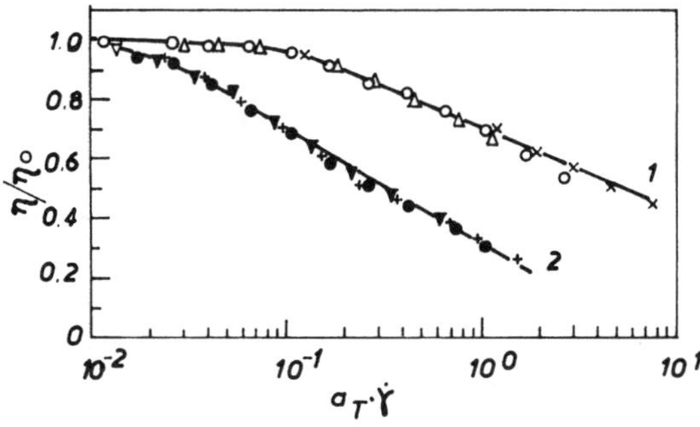

FIG. 29 η/η_0 vs $a_T \cdot \dot{\gamma}$ of LDPE: curve 1, Rexene PE-111 at various melt temperatures, °C: (○) 180; (△) 200; (×) 220; curve 2, Norchem PE-962 at various melt temperatures, °C: (●) 180; (▲) 200; (+) 220. a_T is an empirical shift factor obtained by shifting the viscosity data at various temperatures to the reference temperature (180°C in the present case). In other words, one obtains a temperature-independent correlation of viscosity curves in terms of η/η_0 and $a_T \cdot \gamma$ (Data from Ref. 45.)

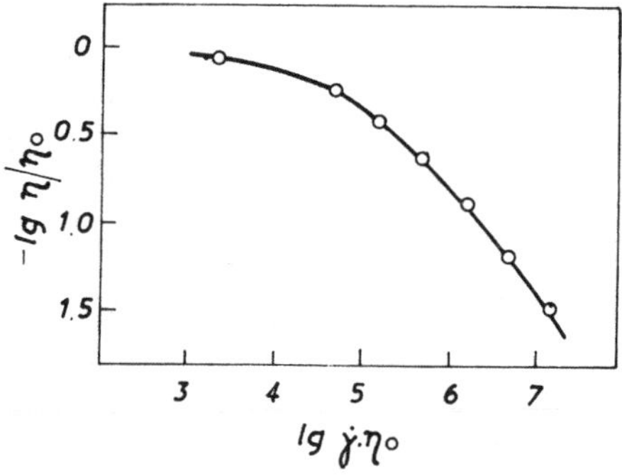

FIG. 30 Temperature-invariant dependence (Vinogradov function) of melt viscosity of IPP. (From Ref. 7.)

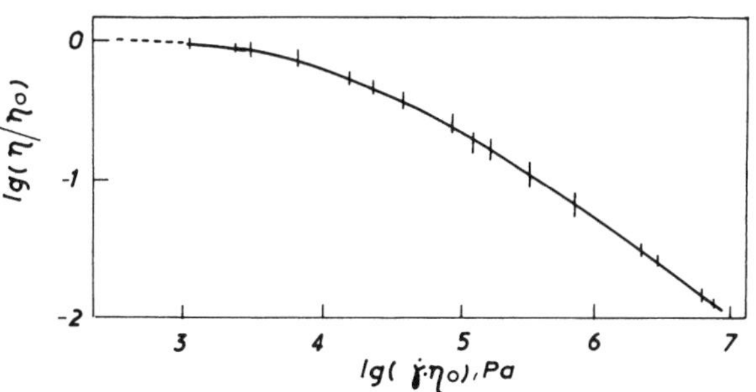

FIG. 31 Temperature-invariant dependence (Vinogradov function) of melt viscosity (η/η_0 vs $\dot{\gamma} \cdot \eta_0$) of PIB at various temperatures from −20°C to 80°C. (From Ref. 22.)

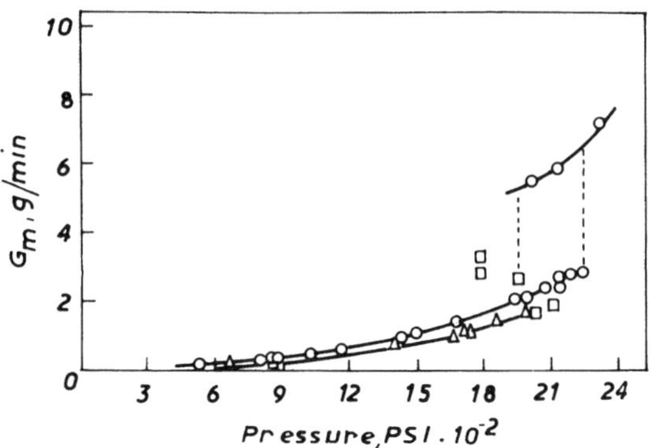

FIG. 32 Mass flow rate of Marlex 6009 HDPE vs. extrusion pressure relations at 186°C with standard die, $L_d/R = 0.0391$, $R = 0.0391$; (○) constant pressure; (△) constant speed; (□) oscillation flow. (From Ref. 72.)

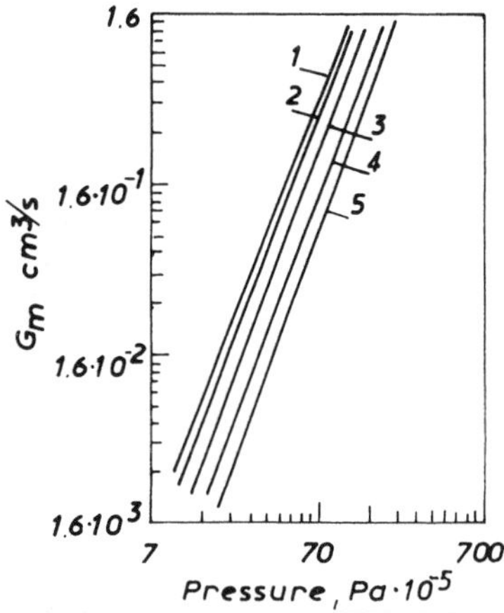

FIG. 33 Mass flow rate of LDPe ($\rho = 0.92\,\text{g cm}^{-3}$, $I_m = 2.1\,\text{g}/10\,\text{min}$) vs extrusion pressure relations. Entrance capillary pressure, Pa × 10^{-5}: 1, 140; 2, 350; 3, 700; 4, 1400; 5, 1750. (From Ref. 16.)

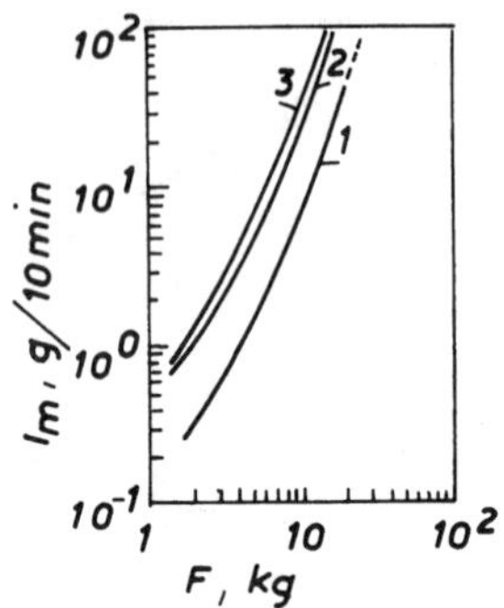

FIG. 34 Dependence of melt index of IPP on loading at various temperatures, °C: 1, 190; 2, 230; 3, 250. (From Ref. 12.)

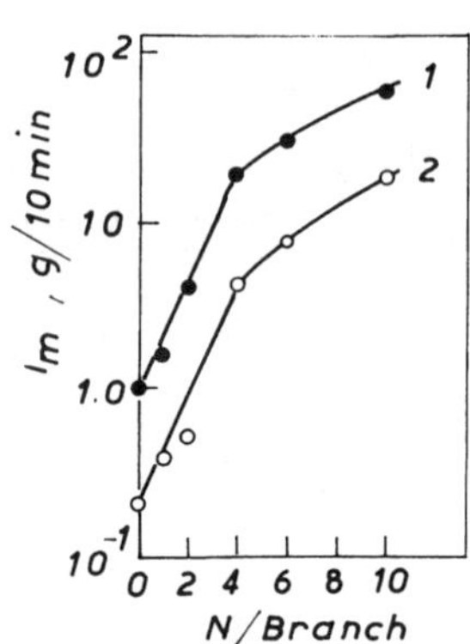

FIG. 36 Effect of length of short-chain on melt index of poly-1-butene: 1, $[\eta] = 1.0$ dL g^{-1}; 2, $[\eta] = 2.0$ dL g^{-1}. (From Ref. 51.)

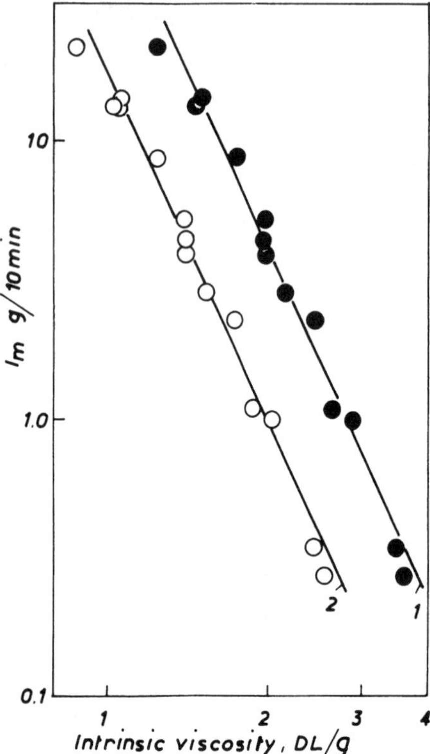

FIG. 35 Melt flow rate as a function of intrinsic viscosity for polypropylene homopolymers: 1, $[\eta]$ decalin; 2 $[\eta]$ trichlorobenzene. (From Ref. 74.)

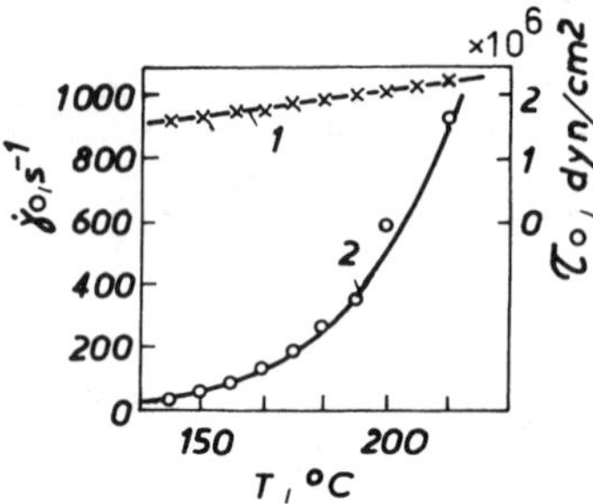

FIG. 37 Effect of melt temperature on onset of elastic turbulence in polyethylene. 1, Critical shear stress; 2, critical shear rate. (From Ref. 79.)

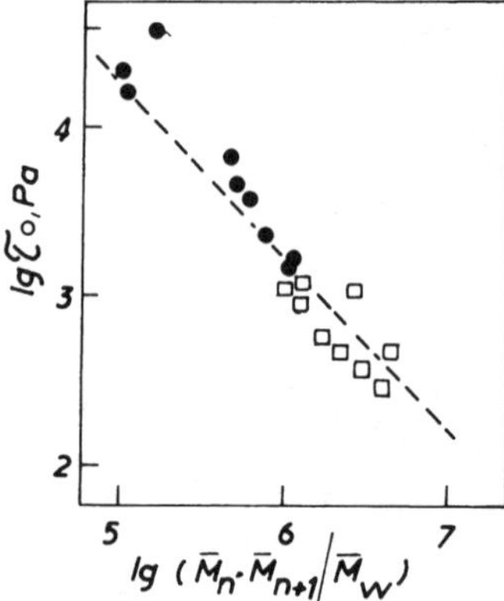

FIG. 38 Dependence of critical shear stress (at which non-Newtonian behavior begins) on molecular weight distribution of PE (●) and PIB (□). (From Ref. 22.)

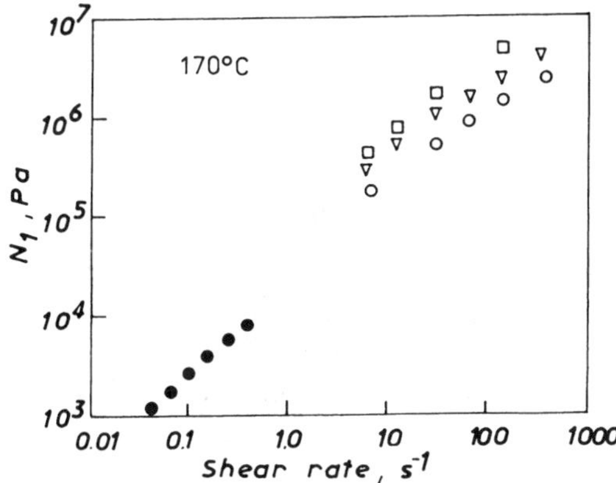

FIG. 39 First normal stress difference as measured and as predicted by Tanner's theory of HDPE ($\rho = 0.962\,\mathrm{g\,cm^{-3}}$; $I_m = 0.72\,\mathrm{dg\,min^{-1}}$; $M_n = 2.1 \times 10^4$; $M_w = 12 \times 10^4$) using swell values evaluated at three different times: (●) experimental; N_1 calculated using Tanner's equation with B given by (□) equilibrium swell, (▽) 20-s swell data, (○) B_o (instantaneous). (From Ref. 57.)

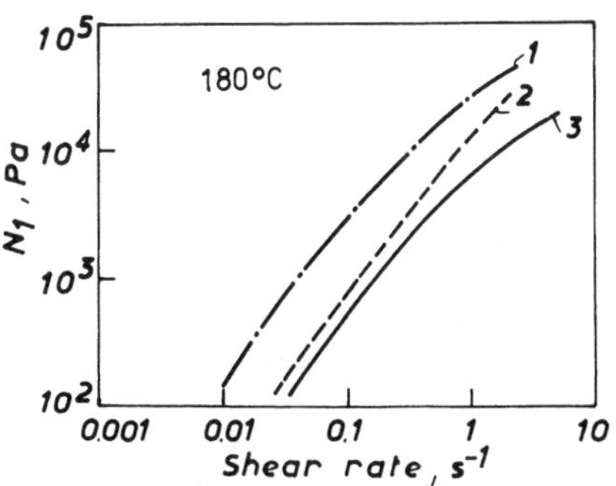

FIG. 41 First normal stress difference vs shear rate of LDPE, ρ, $\mathrm{g\,cm^{-3}}$: 1, 0.915; 2, 3, 0.920. I_m, g/10 min (190°C): 1, 2–2; 3–1, 6. (1) DFDQ 440, moderately low molecular weight, high degree of long-chain branching. (2) Polythene 5600, medium molecular weight, high degree of branching. (3) Sclair 15-11E, narrow molecular weight distribution. (From Ref. 46.)

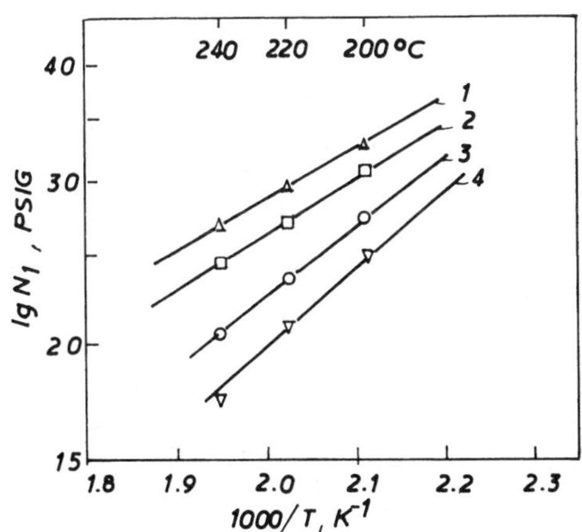

FIG. 40 First normal stress difference vs reciprocal temperature for HDPE at various shear rates, s^{-1}: 1, 700; 2, 500; 3, 300; 4, 200. (From Ref. 80.)

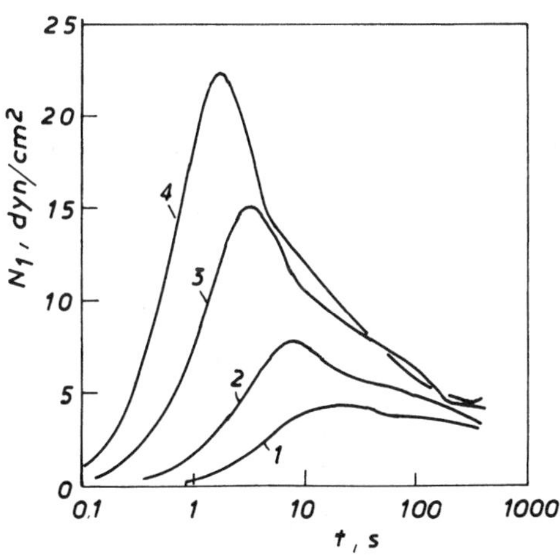

FIG. 42 Transient first normal stress difference N_1 for different constant shear rates ($\dot{\gamma}$, s^{-1}: 1, 1; 2, 2; 3, 5; 4, 10) plotted vs a logarithmic time scale. LDPE; $T = 150$°C. (From Ref. 48.)

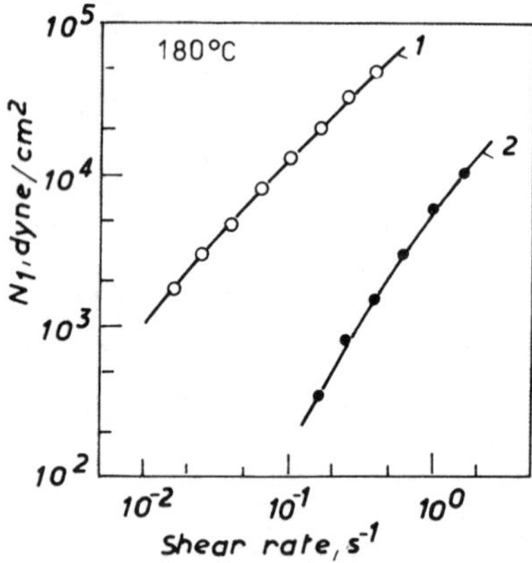

FIG. 43 Principal normal stress difference as a function of shear rate at 180° for IPP. I_m, g/10 min: 1, 3.7; 2, 25.0. $M_w \times 10^{-5}$: 1, 3.39; 2, 1.79. (Data from Ref. 27.)

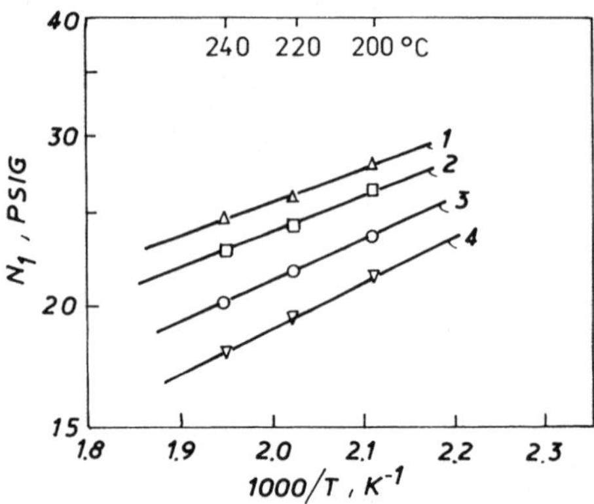

FIG. 44 Plots of log N_1 (first normal stress difference) vs reciprocal temperature for polypropylene. Shear rate $\dot{\gamma}$, s^{-1}: 1, 700; 2, 500; 3, 300; 4, 200. (From Ref. 80.)

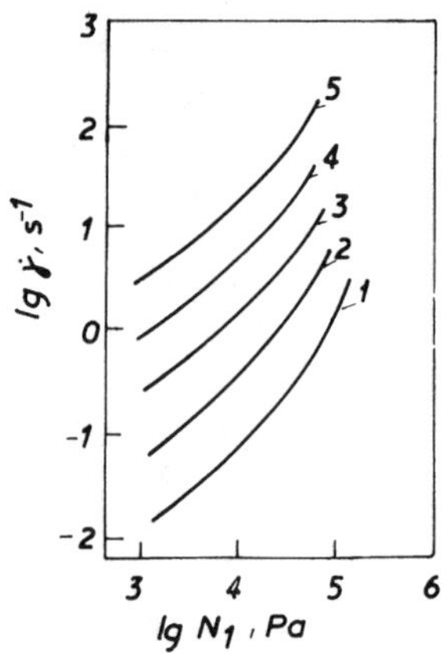

FIG. 45 First normal stress difference N_1 vs shear rate for PIB. Temperature, °C: 1, 22; 2, 40; 3, 60; 4, 80; 5, 100. (Data from Ref. 82.)

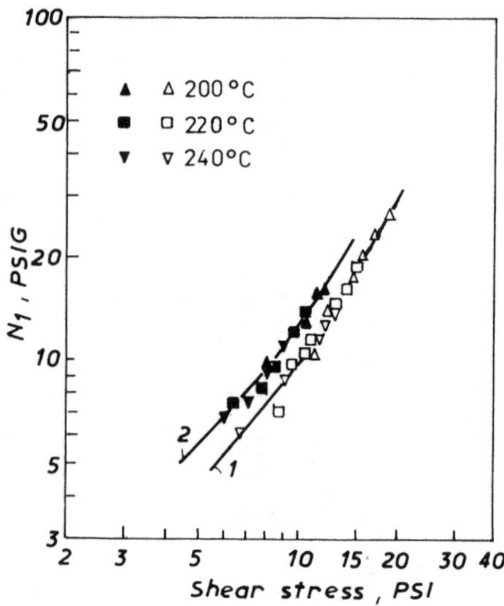

FIG. 46 Plots of log N_1 (first normal stress difference) vs lg τ (shear stress) for HDPE (1) and IPP (2) melts. (From Ref. 80.)

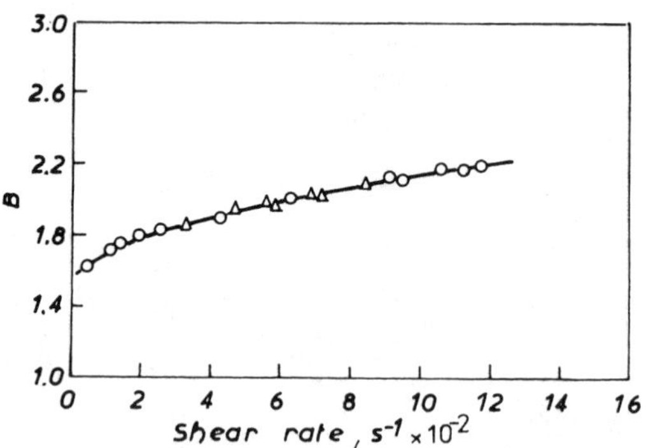

FIG. 47 Die swell ratio (B) of HDPE (Marlex 6009) as a function of apparent shear rate: (○) constant pressure; (△) constant speed. (From Ref. 72.)

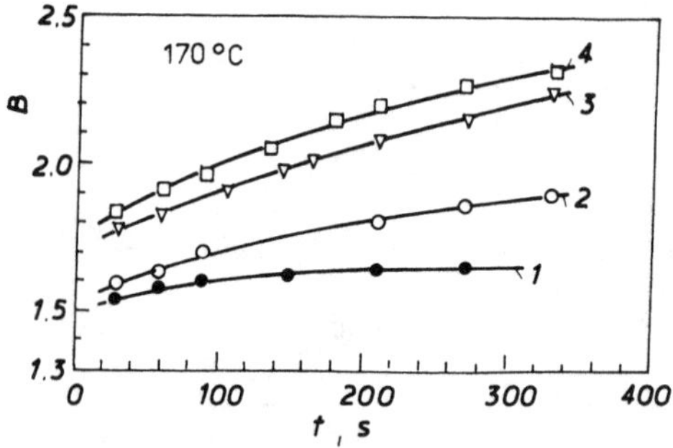

FIG. 48 Die swell of HDPE. Time at four shear rates, $\dot{\gamma}$, s^{-1}: 1, 7; 2, 73; 3, 367; 4, 750. (From Ref. 57.)

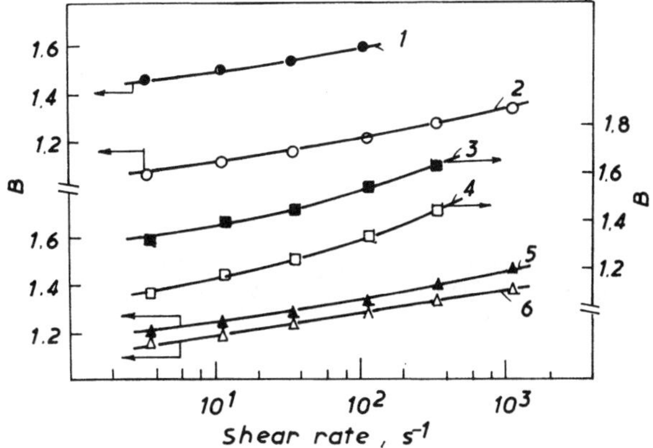

FIG. 49 Die swell behavior of LDPE with different shearing histories: (curves 1, 3, 5) swell ratios of solvent-treated materials measured with a capillary ($L_d/D = 59.83$); (2, 4, 6) swell ratios of full sheared materials measured with a capillary ($L_d/D = 59.83$). Temperature, °C: 1, 2–190; 3, 4–160; 5, 6–130. I_m, g/10 min: 1, 2–0.6; 3, 4–8.1; 5, 6–24. ρ, g cm^{-3}: 1, 2–0.919; 3, 4–0.911; 5, 6–0.928. $\bar{M}_n \times 10^{-5}$: 1, 2–5.31; 3, 4–3.35; 5, 6–0.67. $\bar{M}_w/\bar{M}_n$: 1, 2–17.0; 3, 4–27.7; 5, 6–7.4. (From Ref. 18.)

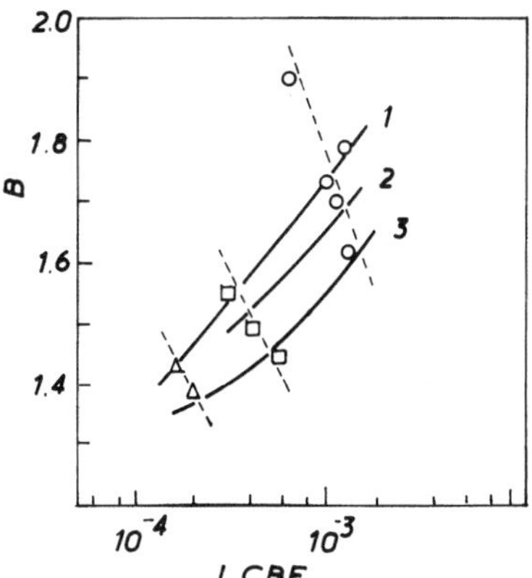

FIG. 51 Relationship between long-chain branching frequency (LCBF) and swell ratios of solvent-treated materials (LDPE): circles, squares, and triangles designate resins of the A, B, and C, family, respectively. I_m, g/10 min: 1, 3–5; 2, 7–8; 3, 22–24. Resin A, $\rho = 0.914 - 0.919$ g cm^{-3}, I_m 0.6–22.6 g/10 min; $\bar{M}_w/\bar{M}_n = 17$–27.7. Resin B, $\rho = 0.924$ g cm^{-3}; I_m 3.1–23.0; $\bar{M}_w/\bar{M}_n$ 9.4–11.5. Resin C, $\rho = 0.928$ g cm^{-3}; I_m 4.5–24.0 g/10 min; $\bar{M}_w/\bar{M}_a$ 7.4–8.2. (From Ref. 84.)

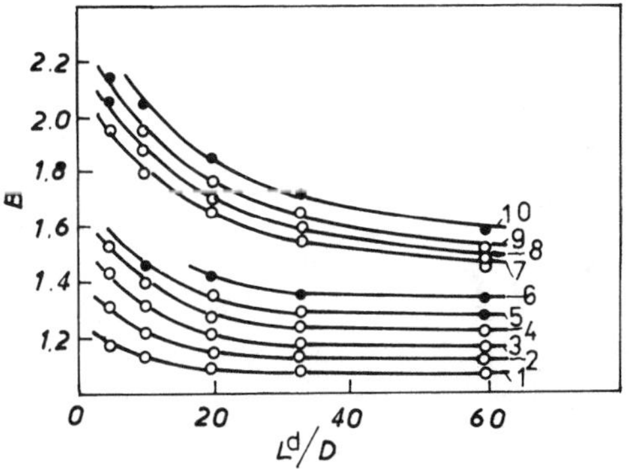

FIG. 50 Die swell behavior of LDPE. Full sheared, curves 1–6; solvent treated, curves 7–10. Shear rate, $\dot{\gamma}$, s^{-1}: 1, 3.5; 2, 11.7; 3, 35.0; 4, 116.7; 5, 350.1; 6, 1167.0; 7, 3.5; 8, 11.7; 9, 35.0; 10, 116.7. (From Ref. 18.)

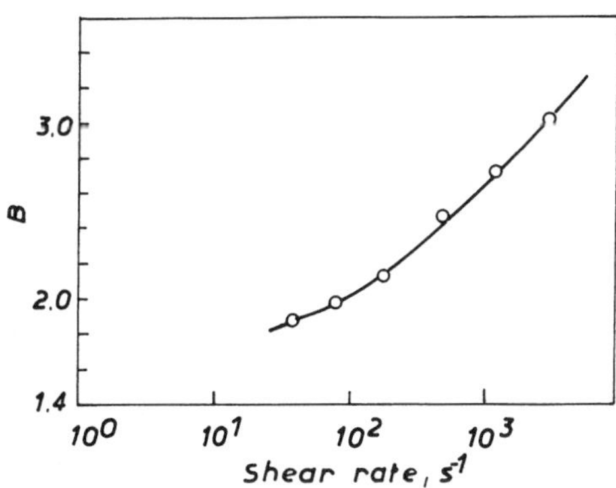

FIG. 52 Equilibrium die swell vs shear rate of IPP. $T = 190$°C. (Data from Ref. 57.)

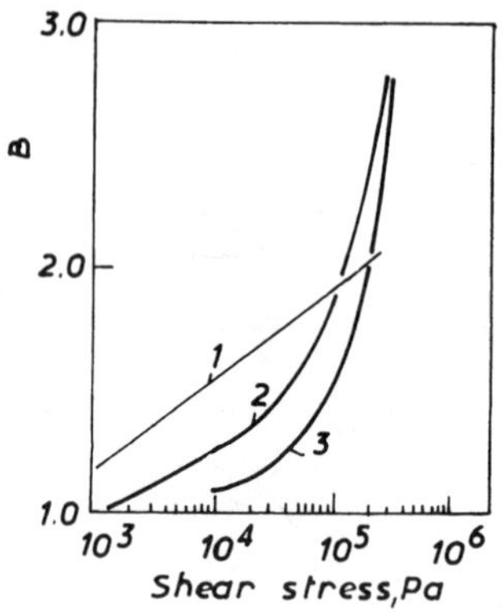

FIG. 53 Die swell vs shear stress of PO. Curve 1, LDPF; curve 2, PIB; curve 3, IPP. (From Ref. 22.)

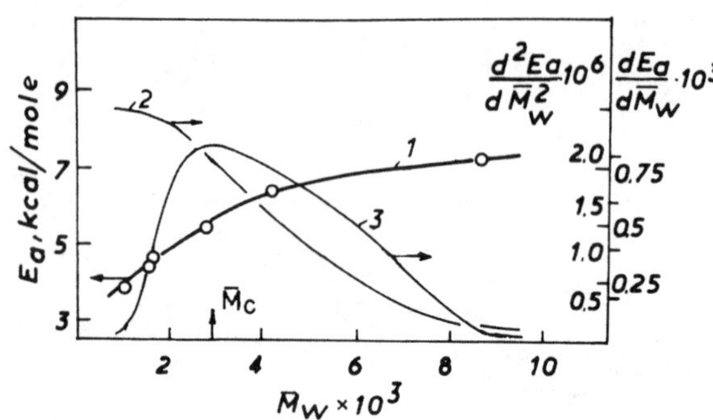

FIG. 56 Effect of molecular weight on energy of activation of melt flow of LDPE. Curve 1, $E_a(\bar{M}_w)$; curve 2, $dE_a/d\bar{M}_w$; curve 3, d^2E_a/dM_w^2. (From Ref. 5.)

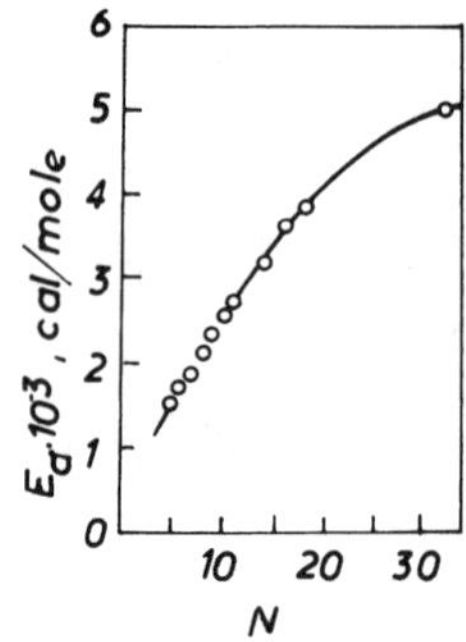

FIG. 54 Activation energy of melt flow of paraffins vs number of C atoms. (Data from Ref. 16.)

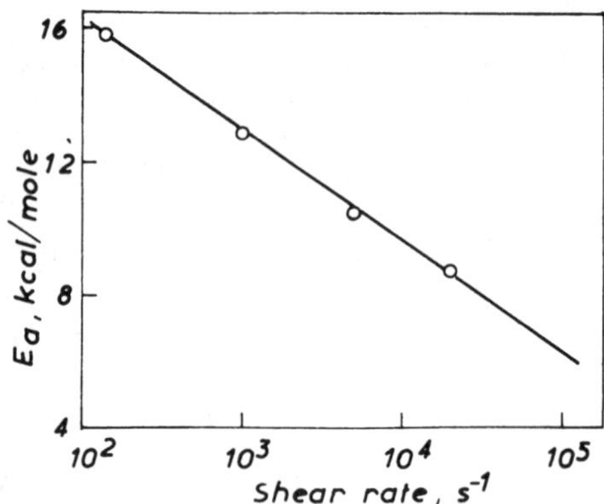

FIG. 57 Effect of shear rate on energy of activation of melt flow of polypropylene. (From Ref. 51.)

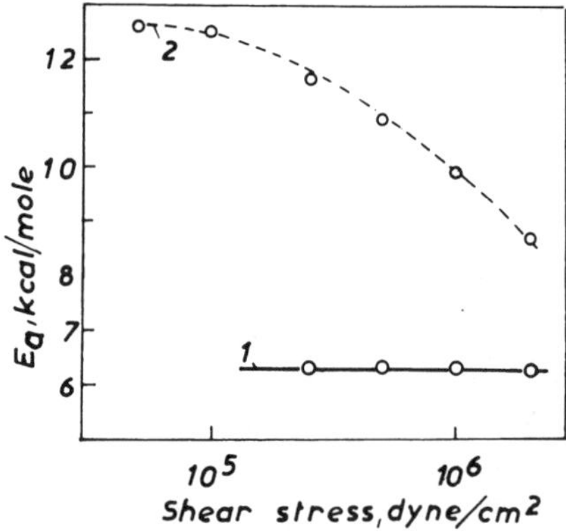

FIG. 55 Activation energy for steady shear viscosity at constant stress of LHDPE (curve 1) vs stress, between 150°C and 190°C (curve 2, branched LDPE). (Data from Ref. 25.)

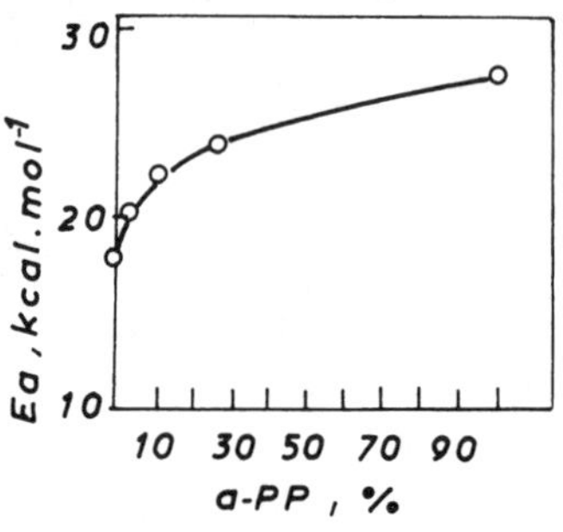

FIG. 58 Effect of content of APP on energy of activation of melt flow of polypropylene. (From Ref. 70.)

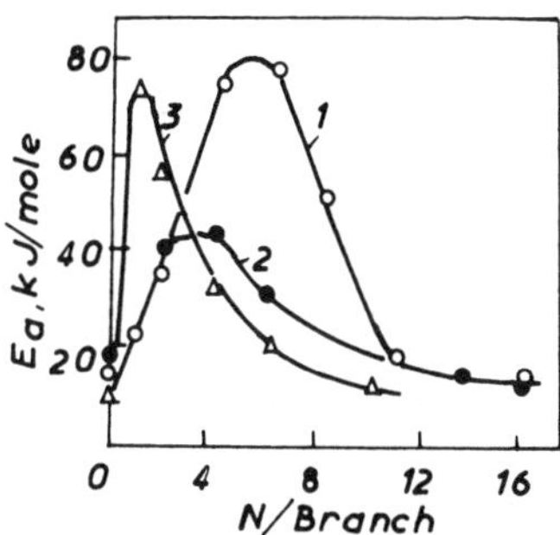

FIG. 59 Dependence of activation energy of melt flow of α-PO on number of C atoms in the side chain. Results obtained from different authors: curve 1, [145]; curve 2, [146]; curve 3, [51]. (From Ref. 22.)

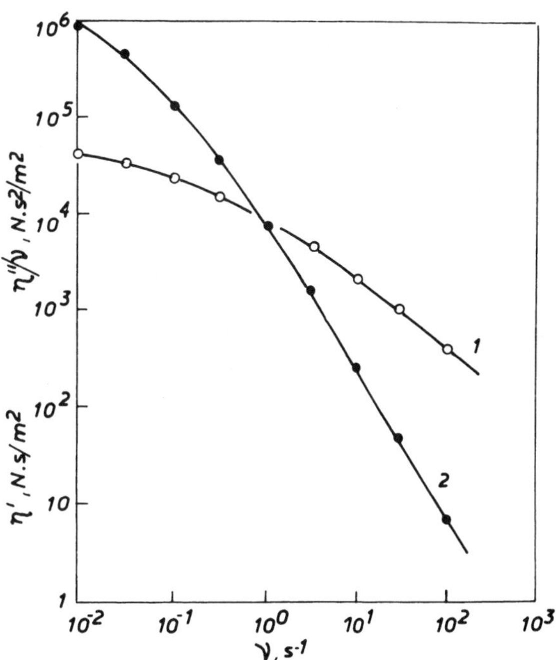

FIG. 61 Dynamic viscosity components η' and η'' as a function of frequency for LDPE. Curve 1, $\eta'(\nu)$; curve 2, $\eta''/\nu(\nu)$ (see Figure 4 of Ref. 88). (From Ref. 24.)

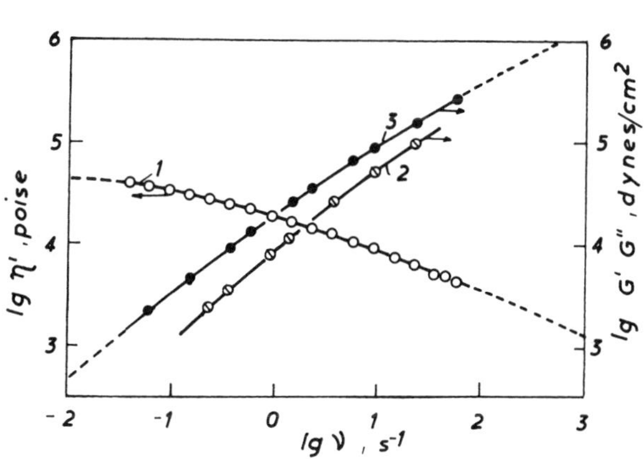

FIG. 60 Dynamic viscosity and dynamic moduli plotted against frequency for HDPE. Curve 1, $\eta'(\nu)$; curve 2, $G'(\nu)$; curve 3, $G''(\nu)$. (From Ref. 41.)

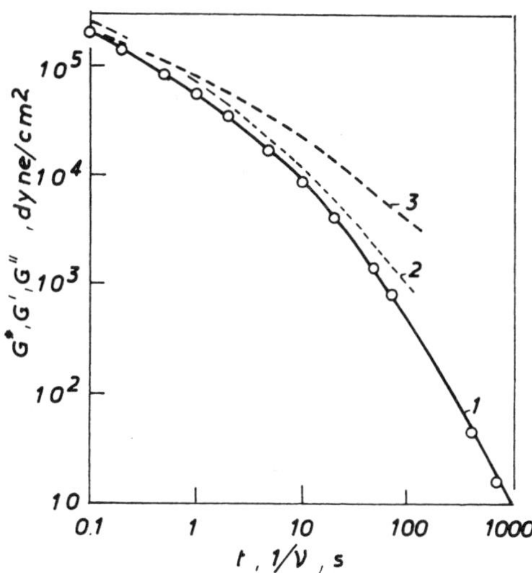

FIG. 62 Shear relaxation modulus $G^*(t)$ (curve 1), $G'(1/\nu)$ (curve 2), and $G''(1/\nu)$ (curve 3) of LDPE vs time (t) ($1/\nu$). Temperature 150°C. (Data from Ref. 48.)

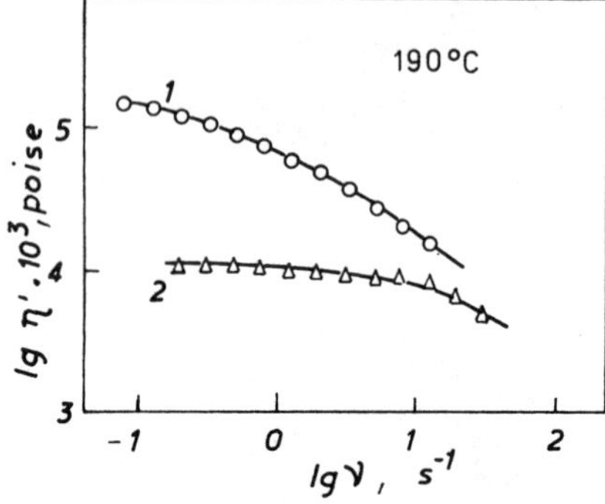

FIG. 63 Dynamic viscosity component η' plotted against frequency for IPP. I_m, g/10 min: 1, 4; 2, 18.5 $\bar{M}_w \times 10^{-3}$: 1, 344; 2, 222; $\bar{M}_w/\bar{M}_n$: 1, 5.9; 2, 3.0. (Data from Ref. 89.)

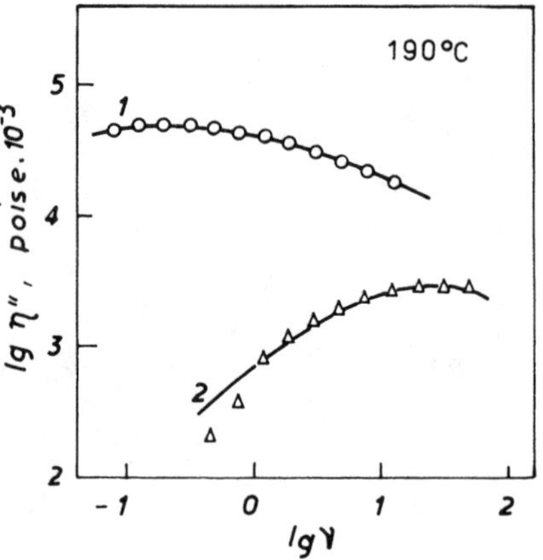

FIG. 64 Dynamic viscosity component η'' plotted against frequency for IPP. I_m, g/10 min: 1, 4; 2, 18.5. $\bar{M}_w \times 10^{-3}$: 1, 344; 2, 222; $\bar{M}_w/\bar{M}_n$; 1, 5.9; 2, 3.0. (Data from Ref. 89.)

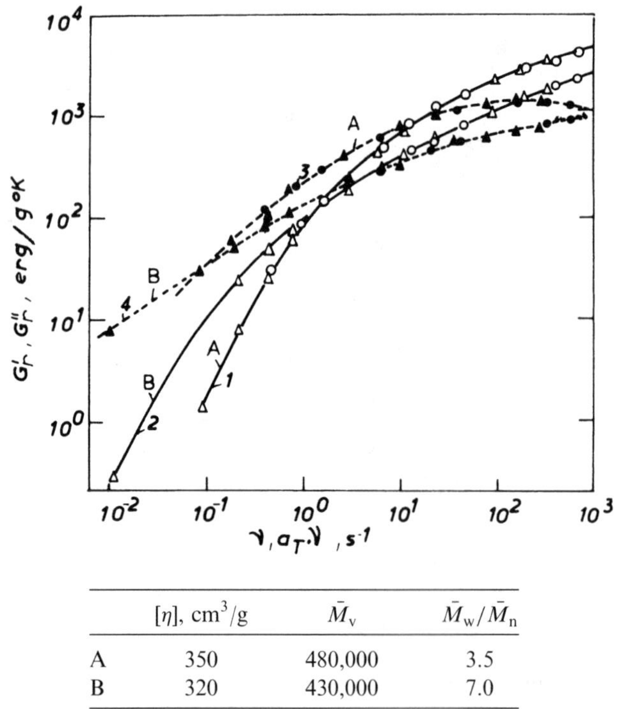

	[η], cm³/g	$\bar{M}_v$	$\bar{M}_w/\bar{M}_n$
A	350	480,000	3.5
B	320	430,000	7.0

FIG. 65 Double logarithmic plot of reduced storage modulus $G_r' = G'/\rho T$ (curves 1, 2) and $G_r'' = G''/\rho T$ (curves 3, 4): (△, ▲) against angular frequency ν (for 210°C) and (○, ●) against $a_T\nu$ (for 190°C) for both samples of IPP A and B. (Data from Ref. 81.)

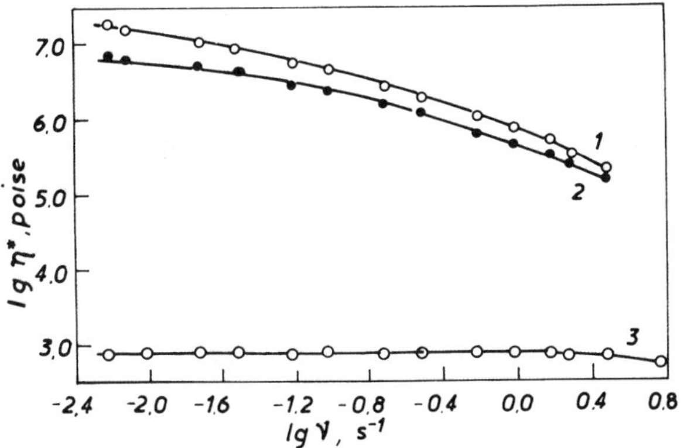

PIB	$\bar{M}_w \times 10^{-5}$	$\bar{M}_w/\bar{M}_n$
1	0.847	2.99
2	8.48	2.54
3	15.0	2.38

FIG. 66 Dynamic viscosity–frequency curves for the polyisobutylenes (1, 2, 3) at 160°C. (From Ref. 30.)

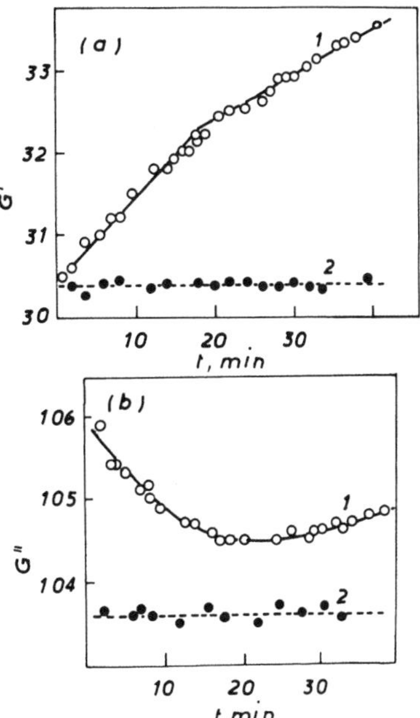

FIG. 67 Storage and loss moduli [G' (a) and G'' (b)] of PE fraction (curves 1) ($\bar{M}_w = 159{,}000$, $\bar{M}_n = 147{,}000$) in arbitrary units as a function of the time of annealing at $T_2 = 160°C$. The melt has been previously annealed during 40 min at $T_1 = 200°C$ and the origin of the time is taken 3 min after the beginning of the cooling from T_1 (when the temperature T_2 is stabilized). The dotted line shows the modulus of the IPP sample (curve 2) ($\bar{M}_w = 500{,}000$; $\bar{M}_n = 11{,}000$) as a function of the annealing time at $T_2 = 190°C$; this time is much smaller than the relaxation time. (From Ref. 90.)

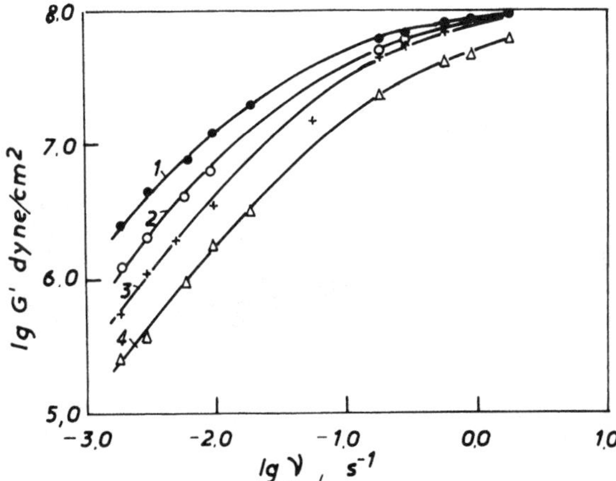

FIG. 68 Plot of G' vs frequency for PIB at various temperatures, °C: 1, 116; 2, 140; 3, 160; 4, 190. (From Ref. 30.)

2. Polarization (χ) of PO and their depencence on type of bond, number of C atoms, etc.
3. Double-refraction (birefringence) (Δn) of isotropic and oriented PO.
4. Haze (*Hz*), gloss (*Gl*), and transparency (*Tr*) of PO and their dependence on the processing conditions.

The optical properties and parameters and their dependence on different factors indicated in Sec. II are exemplified in Tables 13–25 and Figures 69–91.

(Text continues on p 355)

TABLE 12 Optical properties and parameters of PO and their relationship from different factors, [Op. Pr. (F_i)]

N	PO	Refractive index n	Refraction R	Polarization χ
1	2	3	4	5
1.	HDPE, UHMW-HDPE, LHDPE	$n(T)$: [126] p V-18 (Table 13)	$R(T)$: [126] p V-18	
2.	LDPE, LLDPE	$n(T)$: [126] p V-18 (Table 14)	$R(T)$: [126] p V-18	
3.	Different PEs	$n(\alpha_c)$: [127] (Table 15); n (0): [127] (Table 16)		
4.	IPP			
5.	Different PO, PE, PP, α-PO	n: [128] p 131, [126] p III-241 (Table 17)	R (type of main chain link): [128] p 51 (Table 18) R (atom group): [128] p 48	χ (type of bond) [129] (Table 19); $\chi(\alpha_c)$: [129] (Table 20); χ (number of C atoms): [129] (Fig. 69.)

N	Double refraction (birefringence) Δn	Haze, gloss, and transparency Hz, Gl, Tr
1	6	7
1.	Δn (size of crystal): [130] (Fig. 70); $\Delta n(\varepsilon, T)$: [131] (Fig. 71: $\Delta n(\varepsilon, t)$: [131] (Fig. 72)	
2.	$\Delta n(\varepsilon, \alpha_c)$: [130] (Fig. 73); $\Delta_n(\varepsilon_{bl})$: [132]; $\Delta n(T)$: [133] (Fig. 74)	Hz, Gl (Ex.p.fr.): [135] (Table 22); Hz(PI): [136] (Fig. 77); Hz, Gl ($\bar{M}_w\bar{M}_n$): [47] p 21 (Fig. 78); $Hz(B)$: [137] (Fig. 79; Hz(FLH): [137] (Fig. 80); $Hz(G_m, B)$: [138] p 220; HL_d, θ_d): [137] (Fig. 81); Gl(PI): [136] (Fig. 82); Tr(PI): [136]; $Tr(\nu)$: [128] p 125 (fig. 83), [144]; Tr, Hz (t_{bl}): [139] (Fig. 84); Tr, Hz (FLH): [139] (Fig. 85)
3.	$\Delta n(\varepsilon, T)$: [128] p 58; $\Delta n(\varepsilon_0, \alpha_c)$: [128] p 58	$Hz(T_d)$: [12] p 397 (Fig. 86); $Hz(\rho)$: [140] p 106; Hz, Tr, Gl(FLH): [12] p 398 (Fig. 87) Tr, s, Tr, b, Hz: [139]; Tr, s (AH): [139] (Fig. 88) $Tr(\nu)$: [141] p 301; Gl, Hz ($\bar{M}$)): [134] (Table 23)
4.	Δn(, T): [134] (Fig. 75); $\Delta n(\dot{\gamma})$: [81] (Fig. 76)	$Tr(\nu)$: [128] p 126 (Fig. 89)
5.	Δn (type of crystal): [135] p 180 (Table 23)	Hz: [126] p VIII-9 (Table 24); $Tr(z)$: [128] p 125–127 (Table 25): [126] p VIII-9; $Tr(\nu)$: [128] p 126, 127 (Fig. 90, 91)

TABLE 13 Refractive index (n) and specific molar refractivity of high-density polyethylene (Marlex 50)

Temp., °C	V_{Sp} (cm^3 g^{-1})	n	$r = R/M$
130	1.261	1.4327	0.3273
139.3	1.270	1.4297	0.3297
150.6	1.281	1.4261	0.3283

Source: Data from Ref. 126.

TABLE 14 Refractive index (n) and specific molar refractivity of LDPE (Alathon 10)

Temp., °C	V_{Sp} (cm^{3}g^{-1})	n	$r = R/M$
90	1.159	1.4801	0.3293
100	1.178	1.4693	0.3283
108	1.200	1.4575	0.3297
113	1.239	1.4432	0.3286
118	1.250	1.4396	0.3289
124.8	1.256	1.4368	0.3288
			Av. 0.3290

Source: Data from Ref. 126.

TABLE 15 Refractive indices of unoriented linear polyethylene films

Films	Volume fraction crystallinity	Refractive index (n)
LLDPE		
HU-1	0.562	1.5264
HU-2	0.622	1.5306
HU-3	0.550	1.5238
HU-4	0.625	1.5316
HU-5	0.657	1.5359
HU-7	0.698	1.5369
HU-8	0.571	1.5260
HU-10	0.716	1.5371
HU-11	0.602	1.5288
HU-12	0.564	1.5249
HU-13	0.611	1.5321
HU-14	0.710	1.5404
HU-1A1	0.647	1.5317
HU-1A2	0.650	1.5335
HU-1A3	0.656	1.5336
HU-1A4	0.668	1.5346
LHDPE		
Marlex A1	0.829	1.5483
Marlex A2	0.864	1.5512
Marlex A3	0.889	1.5546

Source: Ref. 142.

TABLE 16 Anisotropic refractive index values $n(\theta)$ for the oriented hercules polyethylene samples*

Angle (deg.)	H_1	H_2	H_3	H_4	H_5
0	1.5540	1.5615	1.5623	1.5671	1.5697
10	1.5517	1.5577	1.5614	1.5669	1.5670
20	1.5486	1.5570	1.5583	1.5604	1.5621
30	1.5452	1.5516	1.5522	1.5558	1.5552
40	1.5425	1.5457	1.5461	1.5478	1.5488
45	1.5418	1.5407	1.5441	1.5414	1.5480
50	1.5371	1.5346	1.5354	1.5383	1.5396
60	1.5313	1.5301	1.5298	1.5333	1.5335
70	1.5296	1.5263	1.5262	1.5275	1.5275
80	1.5269	1.5250	1.5199	1.5208	1.5239
90	1.5228	1.5226	1.5181	1.5189	1.5206

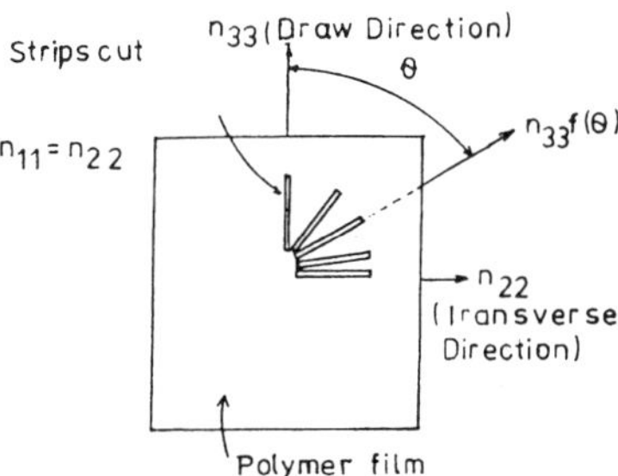

*Schematic representation of the strip-cutting procedure for the examination of the planar optical properties of uniaxially oriented films. The samples H_1–H_5 corresponded to the samples HU-1-HU-5 (Table 15).
Source: Ref. 127.

TABLE 17 Refractive indices *n* of PO

N	PO	*n*	Temp., °C	Ref.
1	2	3	4	5
1.	Polypropylene	1.49	—	[128]
	atactic, density 0.857 g cm^{-3}	1.4735	20	[126]
	density 0.9075 g cm^{-3}	1.5030	20	[126]
2.	Polyisobutene	1.505–1.510	—	[128]
				[126]
3.	Polyethylene	1.51–1.52	25	[128]
	density 0.914 g cm^{-3}	1.51	20	[126]
	density 0.94–0.945	1.52–1.53	20	[126]
	density 0.965	1.545	20	[126]
4.	Poly-4-methyl-l-pentene	1.467	—	[128]
		1.459–1.465		[126]
	n_c	1.464		[128]
	n_f	1.505		[128]
5.	Poly-1-decene	1.4730		[126]
6.	Polybutene (isotactic)	1.5125		[126]
7.	Polyethylene–vinyl acetate (80%–20% vinyl acetate)	1.47–1.50		[126]
8.	Polyethylene–propylene (EPR rubber)	1.4748–1.48		[126]

Source: Data from Ref. 128 and 126.

TABLE 18 Refraction of Repeat Group in PO

Polymer	Repeat group	Refractive index (*n*)	Refraction of group (*R*)	
			Calculated	Experimental
1. Polyisobutylene	—CH—$C(CH_3)_2$—	1.5089	18.39	18.48
2. Polyethylene	—$CH—_2$	1.51	4.56	4.62

Source: Ref. 128.

TABLE 19 Bond polarizability parameters ($Å^3$)*

Bond	*f*	$\chi_{\parallel}$	$\chi_{\perp}$
C—C	0.50	0.303	0.303
C—H	0.70	0.486	0.486

* Required parameters for alkanes, then, are the parallel and perpendicular components $\chi_{\parallel}$ and $\chi_{\perp}$ for C—C and C—H bonds. In addition, the locations of the centers along the bonds are needed. For the C—C bond, this is obviously halfway along the bond. For the C—H bond, the location is expressed as the fractional distance along the bond from C to H, *f*(C—H).
Source: Ref. 129.

TABLE 20 Polarizabilities (χ) (per CH_2) and refractive indices (n) for polyethylene

State	χ_1, $Å^3$	χ_2, $Å^3$	χ_3, $Å^3$	n_1	n_2	n_3	$\langle n \rangle$
Isolated planar zig-zag PE chain* (calcd)	1.48	1.42	3.32				
Crystal† (calcd)‡	2.42	2.40	3.13	1.51	1.50	1.60	1.56
Crystal§ (exptl)				1.54	1.54	1.60	1.56

* Index 1 = transverse in-plane, 2 = out-of-plane, 3 = chain direction.
† Index 1 = a, 2 = b, 3 = c.
‡ The bond polarizabilities determined for alkanes (Table 19) are used to calculate polarizabilities (per CH_2) for the polyethylene crystal.
§ From Ref. 127.
Source: Ref. 129.

TABLE 21 Birefringence of polymer spherulites ($\Delta n_s = n_r - n_t$)*

Type of crystal	Position of index n_3 with respect to the spherulite radius	Sign of the spherulite birefringence
Uniaxial $n_1 = n_2 < n_3$	**Parallel**	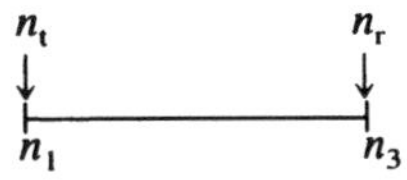 Positive
	Perpendicular	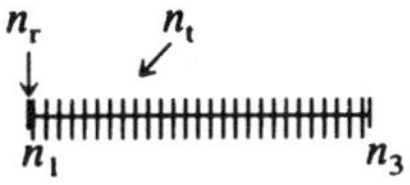 Negative
Biaxial $n_1 < n_2 < n_3$	**Parallel**	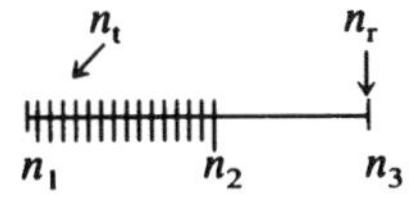 Positive
	Perpendicular	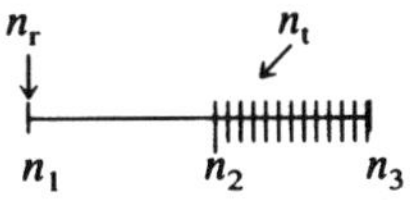 Negative
	Perpendicular	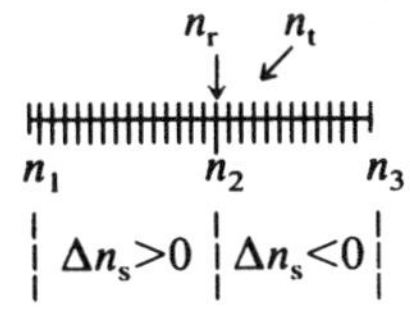 Mixed

$\Delta n_s > 0$ | $\Delta n_s < 0$

* The birefringence of a spherulite Δn_s is defined as $n_r - n_t$, where n_r is the refractive index parallel to the spherulite radius and n_t is the refractive index perpendicular (tangential) to the radial direction. If the amorphous phase is assumed to be isotropic. Δn_s is directly related to the orientation of crystallites with respect to the radial direction in the spherulite: In polymer crystals, the largest refractive index n_3 is along the molecular chain axis, while n_1 and n_2 are the refractive indices along the electric symmetry axes perpendicular to the chain axis.

In uniaxial PE crystals $n_1 = n_2$. If these uniaxial crystals are oriented along a spherulite radius, two types of spherulite birefringence are possible. If the chain axis (n_3) is parallel to the spherulite radius, then $n_r \geqslant n_t$, since under this condition $n_r = n_3$ and $n_t = n_1$. The spherulite then has a positive birefringence. If the chain axis (n_3) is perpendicular to the spherulite radius $n_r = n_1$ and n_t varies from n_1 to n_3. Thus $n_t \geqslant n_r$ and, for this case, the spherulite has a negative birefringence.

In the most general case $n_1 \neq n_2 \neq n_3 (n_3 > n_2 > n_1)$. Such crystals are called biaxial. if the chain axis is parallel to the spherulite radius, then, as for uniaxial crystals, Δn_s will be positive. However, if the chain axis is perpendicular to the apherulite radius, two possibilities can be considered: (1) If n_1 is parallel to the spherulite radius, Δn_s is negative since n_1 is less than both n_2 and n_3, and hence $n_r < n_t$. (2) If n_2 is parallel to the radius, $n_r = n_2$ and n_t varies from n_1 to n_3. So n_r can be either greater or smaller than n_t. Then sign of the spherulite birefringence is a function of crystallite orientation. In such a case, the birefringence is called mixed birefringence.
Source: Ref. 135.

TABLE 22 Optical properties of blown films and extrusion swelling and I_m of fresh and processed materials (LDPE)

Sample	Density, g cm^{-3}	Extruder passing frequency	Haze	Relative haze*	Gloss	Relative gloss*	Swelling ratio	I_m, g/10 min
A	0.921	0	6.2	0.71	10.4	1.13	1.33	0.55
		5	4.4		11.7		1.25	0.67
B	0.919	0	15.2	0.35	4.7	2.02	1.55	0.56
		5	5.3		9.5		1.23	0.87
C	0.924	0	4.0	0.73	12.5	1.06	1.33	1.37
		5	2.9		13.2		1.29	1.49
D	0.921	0	4.6	0.70	12.1	1.07	1.35	1.96
		5	3.2		13.0		1.29	2.06
E	0.924	0	9.1	0.53	8.9	1.37	1.50	2.97
		5	4.8		12.2		1.37	3.25
F	0.924	0	7.5	0.77	10.7	1.15	1.45	7.96
		5	5.7		12.3		1.41	8.23

* The relative haze and relative gloss are defined as the ratio of the haze and the gloss of the fresh material to those of the material extruded for the fifth time.
Source: Ref. 135.

TABLE 23 Optical properties of different samples of LDPE

Resin	$\bar{M}_w$	$\bar{M}_w/\bar{M}_n$	Gloss (%) ASTM D-523	Haze (%) ASTM D-1003
A50a	170,000	6.4	10.1	5.0
A50b	225,000	8.4	8.0	8.7
A50c	193,000	7.4	8.6	8.0
A70a	165,600	6.4	9.3	6.3
B18a	127,200	5.6	10.4	6.0
B18b	150,900	7.0	8.8	7.2
B20	167,000	9.5	9.1	6.7
B30a	456,400	19.9	7.7	9.5
B30b	495,000	25.5	7.9	8.6
PTR 35	101,800	5.1	10.9	5.9
B80a	710,000	35.5	9.4	5.6
B80c	330,400	18.1	10.0	7.0
PTR 43	631,500	30.5	7.9	8.6

Source: Ref. 134.

TABLE 24 Haze, ASTM D-1003 (%)

Material	Low	High
Polypropylene	1.0	3.5
Polyethylene, medium density	2	40
Polyethylene–vinyl acetate	2	40
Polyethylene, low density	4	50
Polymethyl-1-pentene	<5	
Polyethylene, high density	10	50

Source: Data from Ref. 126.

TABLE 25 Optical transmissivity of PO

N	Polymer	Thickness, mm	*Tr*
1.	Polyethylene (LDPE)	0.1	0.9
		1.0	0.78
2.	Polypropylene	0.09	0.9
		1.15	0.81
3.	Poly (4-methylpentene-1)	0.107	0.92
		1.15	0.90
4.	Polyethylene–vinyl acetate (3 wt% vinyl acetate)	0.1	0.90
		1.0	0.72

Source: Data from Ref. 128.

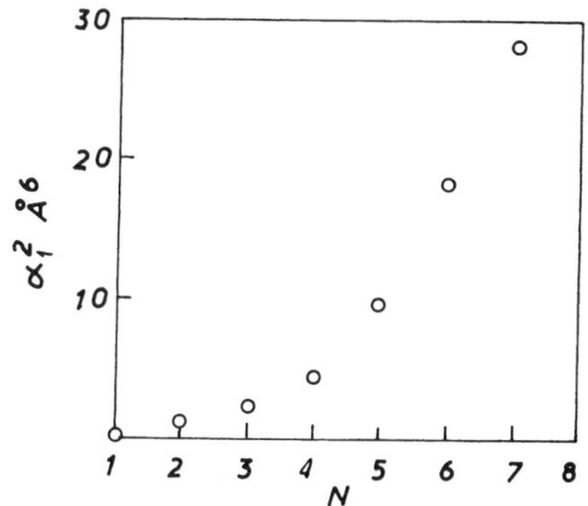

FIG. 69 Squared optical anisotropy α_1^2 vs carbon atom number N for linear alkanes. (Data from Ref. 129.)

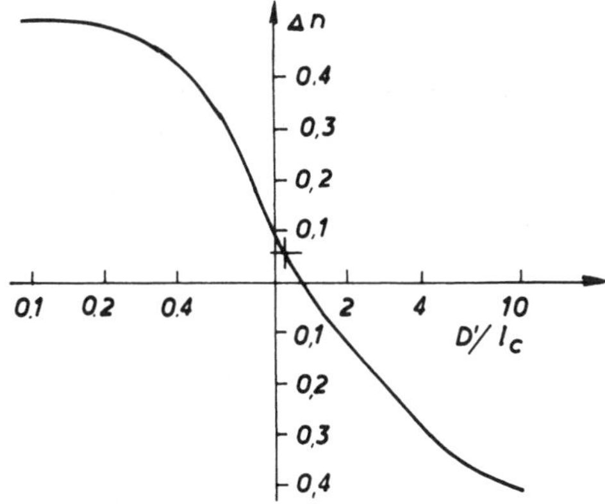

FIG. 70 Birefringence of an isolated PE crystal as a function of the diameter to length ratio of a cylindrical crystal. (From Ref. 130.)

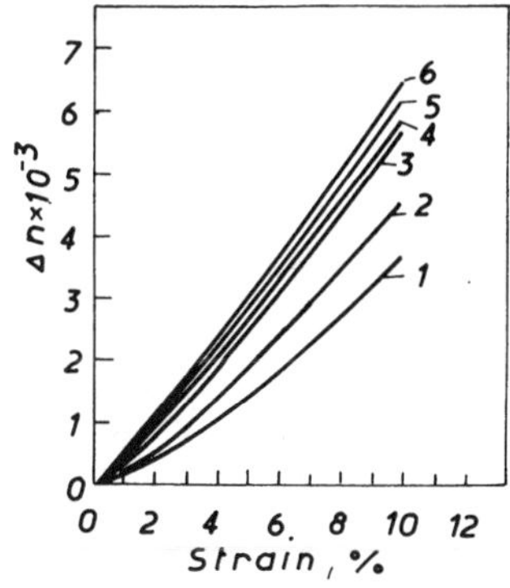

FIG. 71 Variation of birefringence of HDPE with strain at constant rate of strain at various temperatures, °C: 1, 20; 2, 30; 3, 40; 4, 50; 5, 70; 6, 90. (From Ref. 131.)

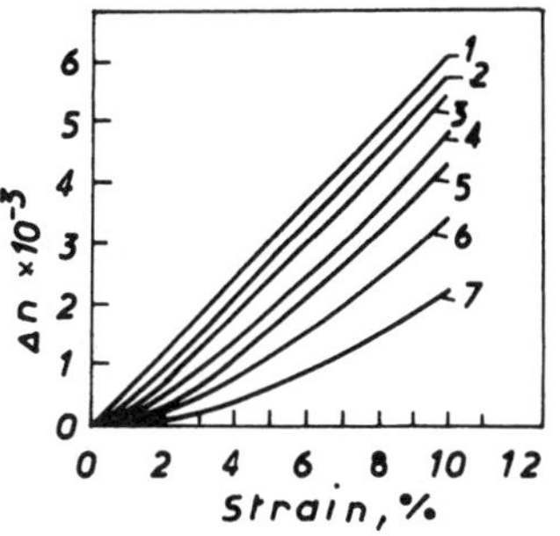

FIG. 72 Theoretical curve of birefringence of HDPE against strain for various relaxation times, s: 1, 0; 2, 1; 3, 3; 4, 10; 5, 30; 6, 50; 7, 100. According to the phenomenological theory of dynamic birefringence proposed by Stein, Onogi, and Keedy, the birefringence Δn shown by a Maxwell element having the single relaxation time τ_o in an experiment at constant rate of strain is given by $\Delta n = V_d[Bt + (A - B)\tau_o = (1 - e^{-\upsilon\tau_o})]$ where V_d is the rate of elongation and A and B are the strain-optical coefficient of the spring and the dashpot of the element, respectively. The following values are used for A, B, V_d, and τ_o: $A = 0$; $B = 0.06$; $V_d = 0.1\%/s$; $\tau_o = 0–100$. (From Ref. 131.)

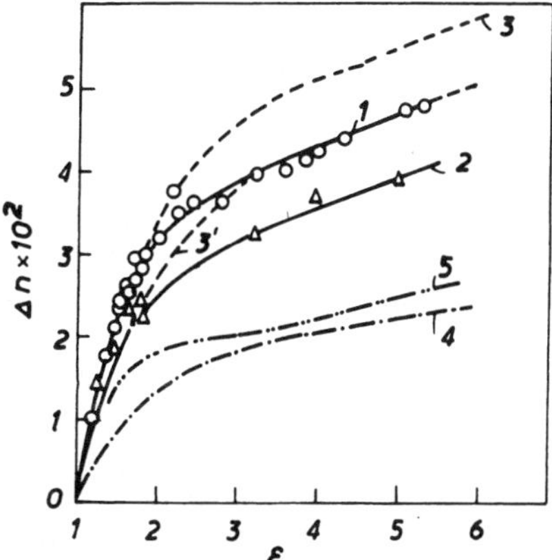

FIG. 73 Birefringence vs draw ratio. (1) LDPE stretched at 33°C, no annealing [143]. (2) LDPE stretched at 65–100°C, annealed 1 h at 103°C [144]. (3) Calculated with Eq. (1): upper curve (3) with $\Delta n_{cl} = 5.93 \times 10^{-2}$, lower curve (3′) with $\Delta n_{cl} = 4.92 \times 10^{-2}$. Using the two-phase representation [Eq. (2)], one arrives at (4) crystal contribution to Δn and (5) amorphous contribution to Δn. $\Delta n = f_c \Delta n_{cl}$ (1) $\Delta n = f_c$ $(v_c \Delta n_c + f_a(1 - v_c)\Delta n_a)$ (2) where v_c is the volume fraction of crystallites, Δn_c and Δn_a are the birefringences of the anisotropic microphases, i.e., the crystallites and the amorphous layers, and Δn_{cl} is the birefringence of the clusters. The orientation parameter (f) is defined by $f = [3(\cos^2\theta) - 1]/2$. (From Ref. 130.)

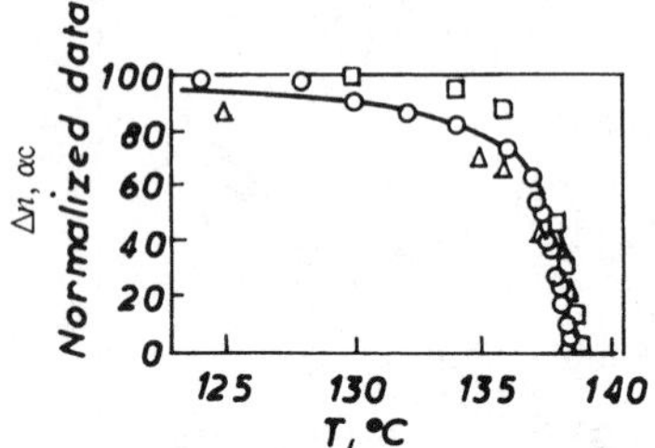

FIG. 74 Normalized birefringence Δn and wt% crystallinity of LHDPE plotted as a function of sample temperatures. The curves are horizontal from about room temperature to about 130°C. The birefringence at 25°C was calculated for a slice of 3 μm thickness. 25°C data: (□) $\Delta n = 0.063$, (△) $\Delta n = 0.048$. (○) $\Delta n = 0.067$: the line drawn gives dilatometric data for a sample of 0.992 weight fraction crystallinity at 25°C. (From Ref. 133.)

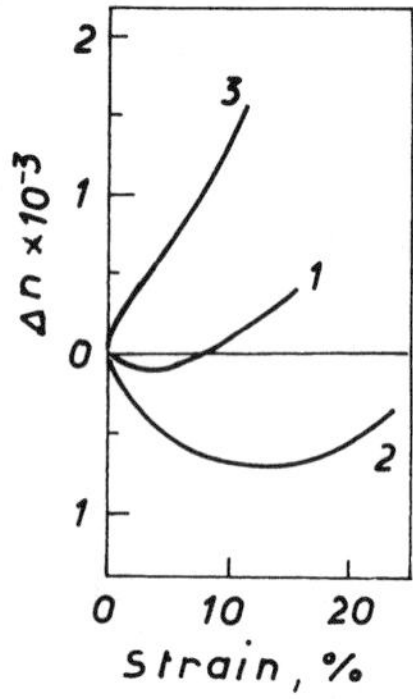

FIG. 75 The birefringence-strain curves for the undrawn (curve 1) and drawn films (curve 2, 3) of IPP, measured at a constant speed of 10%/min at a rom temperature. Curve 2, drawn temperature 150°C, stretched to $\varepsilon_{or} = 1.3$; curve 3, drawn temperature 100°C, stretched to $\varepsilon_{or} = 3.9$. (From Ref. 134.)

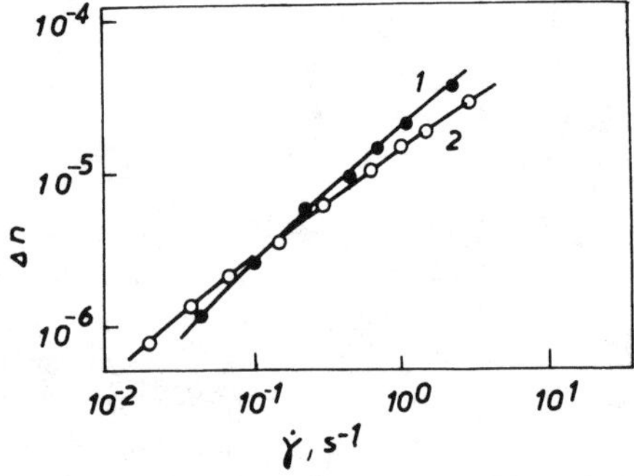

Sample 1: $[\eta] = 350\ \mathrm{cm^3\ g^{-1}}$, $\bar{M}_v = 480{,}000$, $\bar{M}_w/\bar{M}_n = 3.5$
Sample 2: $[\eta] = 320\ \mathrm{cm^3\ g^{-1}}$, $\bar{M}_v = 430{,}000$, $\bar{M}_w/\bar{M}_n = 17$

FIG. 76 Double logarithmic plot of flow birefringence against shear rate for two samples of IPP. Measurement temperature 210°C. (From Ref. 81.)

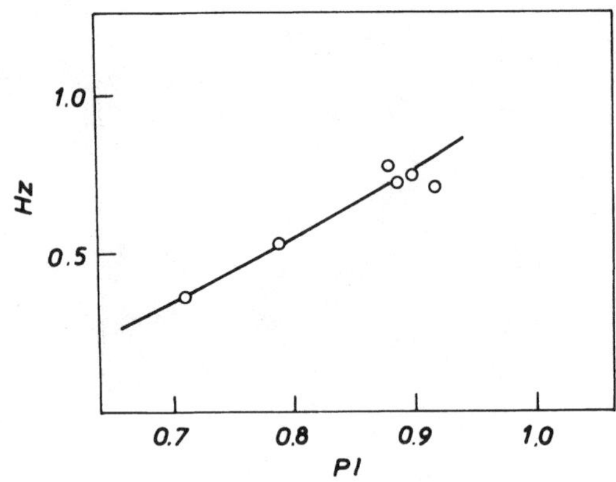

FIG. 77 Relationship between relative haze (Hz) and processing index (PI) of LDPE. $\mathrm{PI} = B_e/B_e$ (where $B_e < B_r$) where B_e is the equilibrium die swell of the extrudate after the shearing and B_r is the swell ratio of the material in the stable state after solvent or heat treatment. When the PI is unity, the viscoelastic properties of the materials are independent of the shear processing; but as the PI becomes smaller than unity, the materials become sensitive to the shearing. (From Ref. 136.)

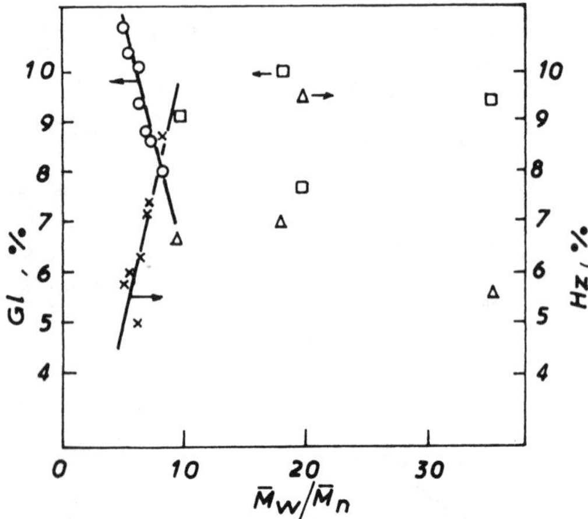

FIG. 78 Gloss and haze of LDPE vs polydispersity ($\bar{M}_w/\bar{M}_r$). Symbols: Type resins II ($\rho = 0.9195$–$0.9219\ \mathrm{g\,cm^{-3}}$; $I_m = 0.32$–3.76)—gloss (○), haze (×). Type resins I($\rho = 0.9147$–$0.9244\ \mathrm{g\,cm^{-3}}$, $I_m = 0.40$–21.50)—gloss (□), haze (△). (From Ref. 47.)

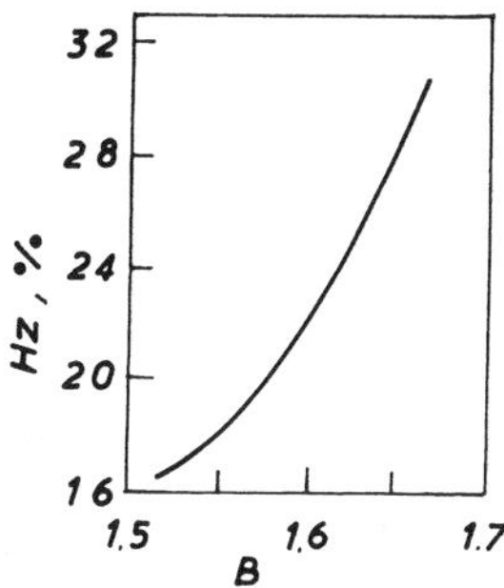

FIG. 79 Relationship between relative haze and swelling ratio of LDPE ($\rho = 0.918$ g cm^{-3}, $I_m = 7$ g/10 min). (From Ref. 137.)

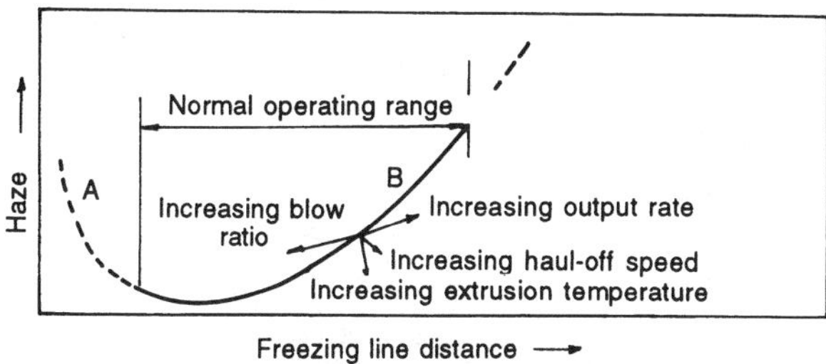

FIG. 80 Effect of freeze line distance and other operating variables on the haze of LDPE film. (From Ref. 137.)

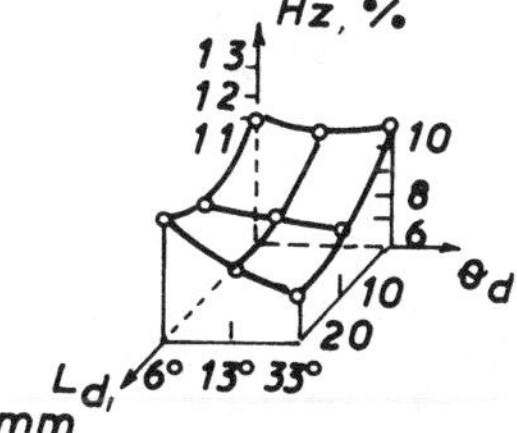

FIG. 81 Effect of die profile on relative haze of foil of LDPE ($\rho = 0.930\,\mathrm{g\,cm^{-3}}$, $I_m = 1.0\,\mathrm{g/10\,min}$) (190°C, 21.6 N). (From Ref. 137.)

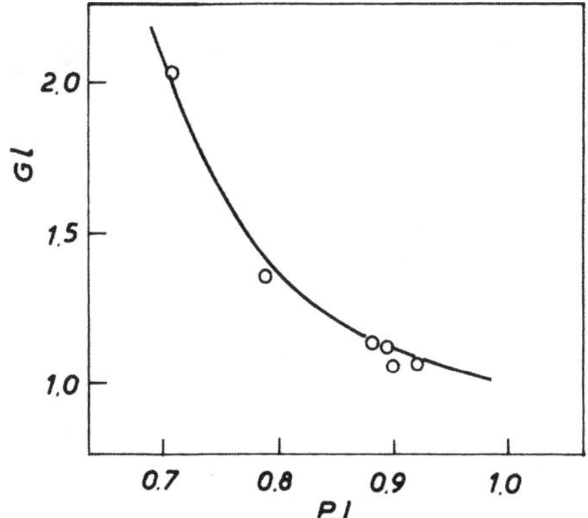

FIG. 82 Relationship between relative gloss and processing index of LDPE. (From Ref. 136.)

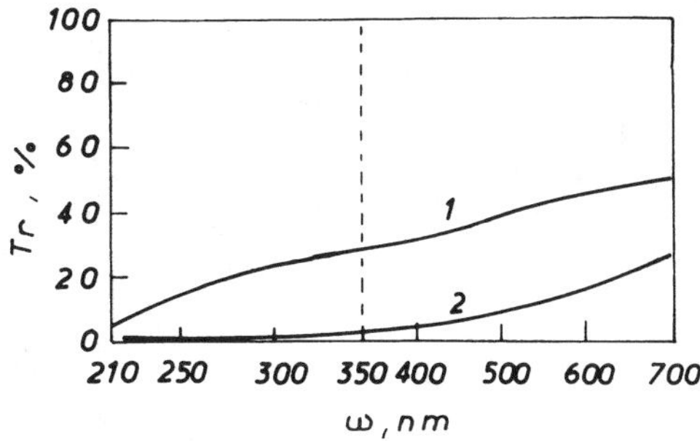

FIG. 83 Relationship between transmission of LDPE and wavelength. Thickness, mm: 1, 0.1; 2, 1. (Data from Ref. 128.)

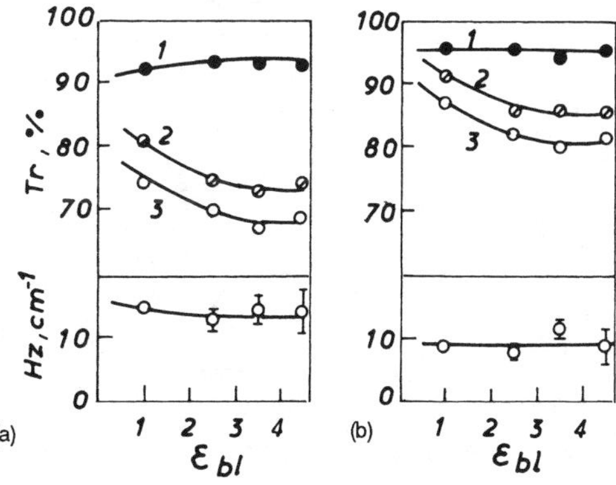

FIG. 84 Transmissivities $Tr(3)$, Tr, s(2), Tr, b(1) (normalized to film thickness of 54 μm) and turbidity as a function of blow-up ratio: (a) LLDPE; (b) LDPE. The direct transmission factor is defined as the ratio of the intensity of the beam transmitted through the sample to the intensity incident on the sample. The total direct transmission factor (Tr) is the product of a "surface phase" direct transmission factor Tr, s and a "bulk phase" direct transmission factor Tr, b, i.e., $Tr = Tr$, s $\cdot Tr$, b. The turbidity (Hz) of the bulk phase was also calculated as Tr, b $= \exp(-z \cdot Hz)$. (From Ref. 139.)

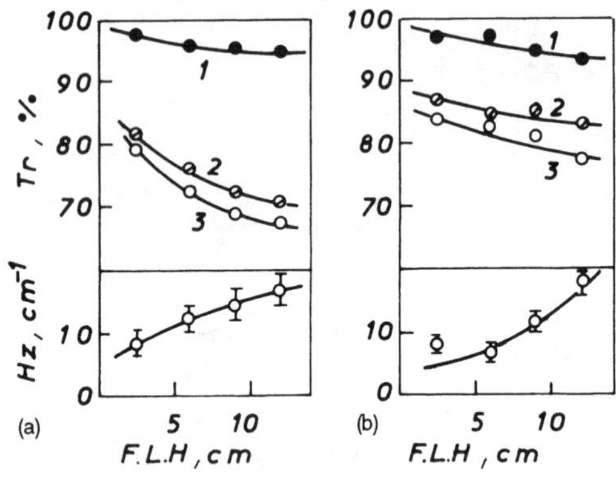

FIG. 85 Transmissivities $Tr(3)$, Tr, s(2), Tr, b(1) (normalized to film thickness of 54 μm) and turbidity as a function of frost line height: (a) LLDPE, (b) LDPE. (From Ref. 139.)

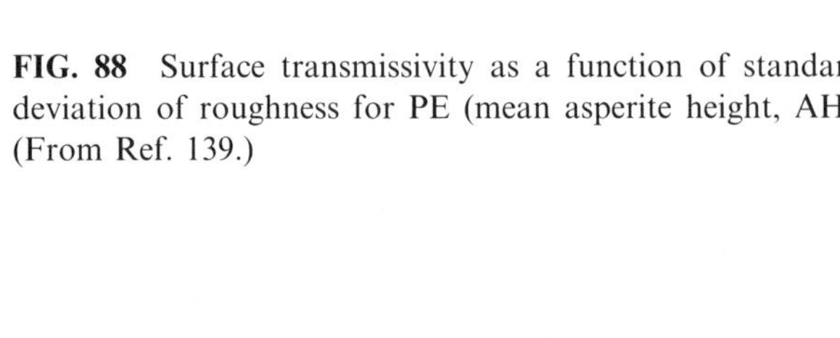

FIG. 88 Surface transmissivity as a function of standard deviation of roughness for PE (mean asperite height, AH). (From Ref. 139.)

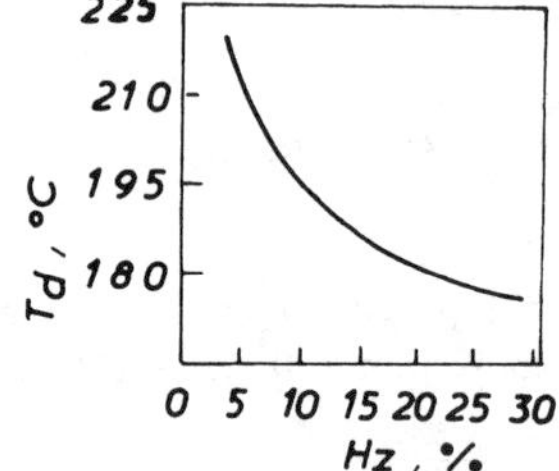

FIG. 86 Effect of die temperature on relative haze of foil of PE. (From Ref. 12.)

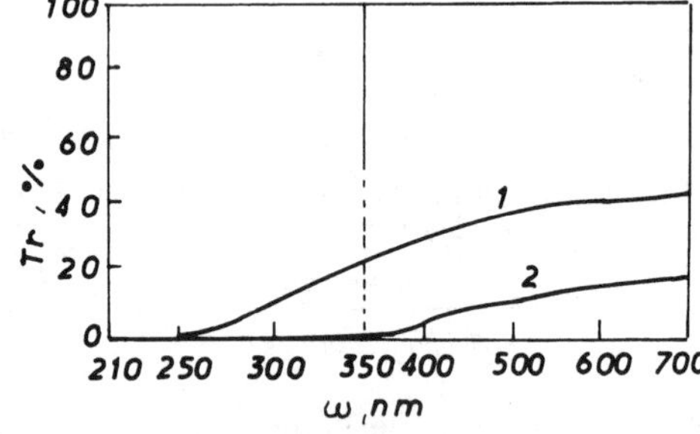

FIG. 89 Relationship between transmission of IPP and wavelength. Thickness, mm: 1, 0.09; 2, 1.15. (From Ref. 128.)

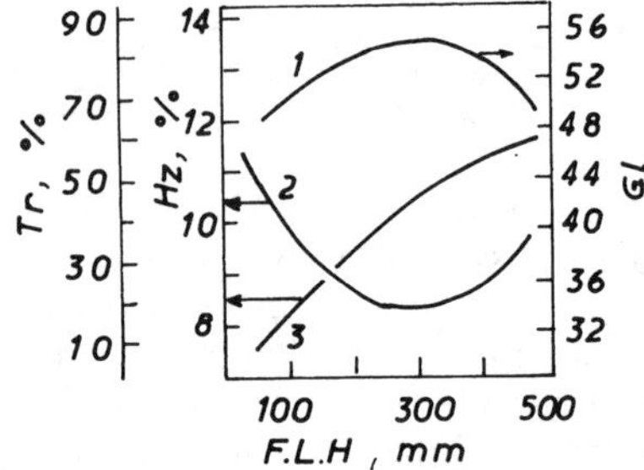

FIG. 87 Effect of freeze-line distance on gloss (curve 1), relative haze (curve 3), and transmission (curve 2) of film of PE. (From Ref. 12.)

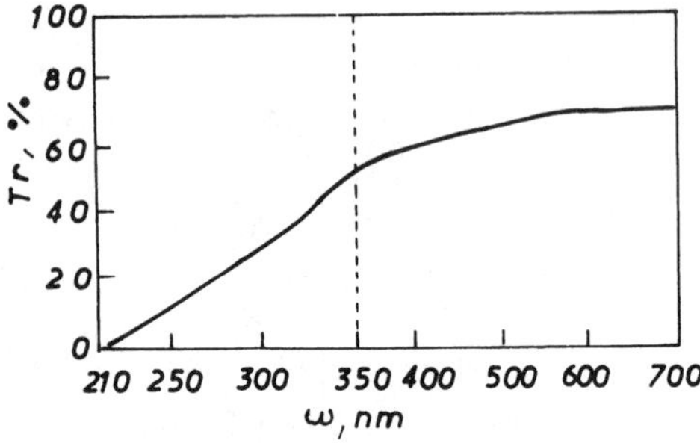

FIG. 90 Relationship between transmission of poly(4-methyl-pentene) and wavelength. Thickness, mm: 1.15. (From Ref. 128.)

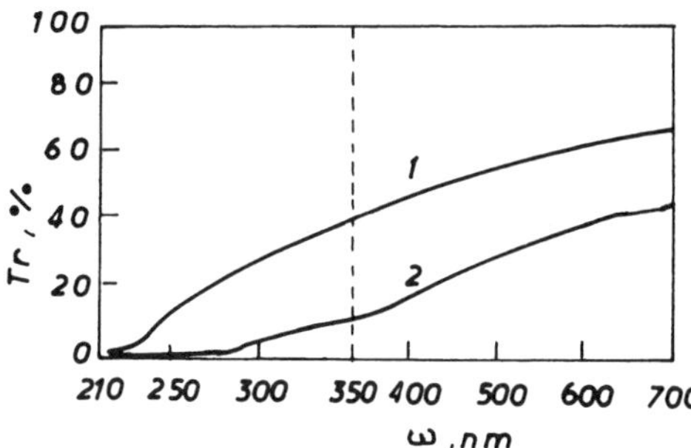

FIG. 91 Relationship between transmission of polyethylene–vinyl acetate copolymer (3 mol% vinyl acetate) and wavelength. Thickness mm: 1, 0.1; 2, 1.0. (Data from Ref. 128.)

SYMBOLS AND ABBREVIATIONS

a_T	empirical shift factor; a_T is obtained by shifting the viscosity data at various temperatures to a reference temperature
B	die swell
BPE	branched PE (LDPE)
E_a	activation energy of melt flow
f, f_c, f_a	orientation parameters
F_i	factor
FLH	frost line height
g	ratio of the intrinsic viscosities of branched and linear molecules at the same molecular weight
G', G''	dynamic moduli
G^*	complex shear relaxation modulus
G_m	melt flow rate
Gl	gloss
HDPE	high-density polyethylene
Hz	haze (turbidity)
I_m	melt index
IPP	isotactic polypropylene
J_e^o	shear compliance (steady-state elastic compliance)
L_d	length of die
LDPE	low-density polyethylene
LHDPE	linear high-density polyethylene
LLDPE	linear low-density polyethylene
LPE	linear PE (LLDPE)
$\bar{M}$	average molecular weight
MDPE	middle-density polyethylene
n	n_D, n_c, n_f refractive indices
$n(\theta)$	refractive index measured at various angles with respect to the film draw direction
n_w	weight-average number of branch points per molecule
Δn	birefringence
Δn_a	birefringenes of the anisotropic amorphous layers
Δn_c	birefringenes of the anisotropic crystalline microphases
Δn_{cl}	birefringence of the clusters
Δn_s	birefringence of the spherolites
N_1	first normal stress difference
p	pressure
PB	polybutene-1
PE	polyethylene
PEP	polyethylene-propylene block copolymer
PIB	polyisobutylene
PO	polyolefins
PP	polypropylene
Pr	properties
R	refraction
r	specific mole refractivity
t	time
T	temperature
T_{or}	orientation temperature
Tr	transparency (transmission)
Tr, b	bulk transmissivity
Tr, s	surface transmissivity
UHMWHDPE	ultrahigh molecular weight high-density polyethylene
v_c	volume fraction of crystallinities
V_d	rate of deformation ($d\varepsilon/dt$); draw rate
V_{sp}	specific volume
$V_{sp;m}$	specific melting volume
X	degree of polymerization
z	thickness
α_1^2	squared optical anisotropy
α_2	amorphous intercrystalline transition lying just below the α_c relaxation
α_c	crystallinity
tan δ	loss (dynamic-mechanical) factor
ε	elongation (draw) ratio; stretching strain
ε_{bl}	blowing ratio
η	apparent non-Newtonian viscosity
η^*	complex dynamic viscosity
η', η''	dynamic viscosity
η_μ	elongation (extensional) viscosity
η_0	zero shear viscosity
θ	angle between tensile axis and the orientation direction

θ_d	angle of the die lips
ν	frequency
ρ	density
τ	shear stress
ω	wavelength
$\dot{\gamma}$	shear rate
$\dot{\gamma}_o$	critical shear rate
χ	polarizability parameter

REFERENCES

1. RM Aliguliev, DM Khiteeva, AH Ivanova, AA Mamedov. Vysokomol. Soedin., 26: 2254, 1984.
2. L Macskasi. Plast. Kautsch. 36: 388, 1989.
3. HC Dae. J Appl Polym Sci 17: 1403, 1973.
4. D Romanini, A Savadori, G Gianotti. Polymer 21:1092, 1980.
5. EE Yakobson, LA Faitelson, LL Sulzhenko, VP Kovtun, NM Domareva, IP Briedis, VV Arefeva. Mech Polymerov 6: 963, 1973.
6. JW Teh, A Rudin, HP Schreiber. Plastic Rubb Proc Appl 4: 149, 1984.
7. HP Frank. Rheol Acta 7: 344, 1968.
8. HP Frank. Rheol Acta 7: 222, 1968.
9. ED Pashin, JuE Polyak, ML Fridman. Dokl Akad Nauk SSSR 281: 392, 1985.
10. ChD Han. Rheology in Polymer Processing, New York: Academic Press, 1976 (in Russian) Khimiya, Moskva, 1979.
11. AI Ivanchenko, VA Pakharenko, VP Privalko, Ef Petrushenko, GI Khmelenko, LA Ivanova. Handbook of the Thermal Properties and Rheological Properties of Polymers (in Russian), Kiev: Naukova Dumka, 1977.
12. Kunststoff-Handbuch. Vol. 4, Polyolefine (R. Vieweg, A. Schley, A. Schwarz, eds.). Munchen: Carl Hanser Verlag, 1969.
13. VI Bukhgalter, EF Meshterova, LP Chizhukova. Plast Massy 1: 33, 1969.
14. DM Best, SL Rosen. Polym Eng Sci 8: 116, 1968.
15. MG Dodin. Int J Polym Mater 11: 115, 1986.
16. RAV Raff, KW Doak, eds. Crystalline Olefin Polymers, Part 2 (in Russian). Moskva: Khimiya, 1970.
17. DV Ivanjukov, MI, Fridman. Polypropylene (Properties and Application) (in Russian), Moskva: Khimiya, 1974.
18. M Rokudai, T Fujiki. J Appl Polym Sci 26: 1343,1981.
19. JW Teh, A Rudin, HP Schreiber. Plastic Rubb Proc Appl 4: 157, 1984.
20. T. Kataoka, S Ueda. J Appl Polym Sci 12: 939, 1968.
21. GC Berry, TG Fox. Adv Polym Sci 5: 261, 1968.
22. GV Vinogradov, AYa Malkin. Rheology of Polymers (in Russian), Moskva: Khimiya, 1977.
23. A. Santamaria. Mater Chem Phys 12: 1, 1985.
24. IP Briedis, LA Faitelson. Mech Polymerov 1: 120, 1976.
25. MS Jacovic, D Pollock, RS Porter. J Appl Polym Sci 23: 517, 1979.
26. J McKelvey. Polymer Processing, New York: Wiley, 1962, p 42.
27. W Minoshima, JL White, JE Spruiell. Polym Eng Sci 20: 1166, 1980.
28. AN Dunlop, HL Williams. J Appl Polym Sci 14: 2753, 1970.
29. VP. Privalko. Handbook of Polymer Physical Chemistry, Vol. 2, Properties of Polymers in Solid State (in Russian), Kiev: Naukova Dumka, 1984.
30. AN Dunlop, HL Williams. J Appl Polym Sci 20: 193, 1976.
31. H Takahashi, T Matsuoka, T Kurauchi. J Appl Polym Sci 30: 4669, 1985.
32. D Romanini, B Barboni. Chemie et Industrie (Italy) 61: 3, 1979.
33. JM Dealy, WMKW Tsang. J Appl Polym Sci 26: 1149, 1981.
34. AV Shenoy, DR Sainie. Rheol Acta 23: 368, 1984.
35. HM Laun. Rheol Acta 21: 464, 1982.
36. BH Bented, JD Slee, CA Richter. J Appl Polym Sci 26: 1001, 1981,
37. D Acierno, FP La Mantia, D Romanini, A Savadori. Rheol Acta 24: 566, 1985.
38. B Anding, S Waese. Plast Kautschuk 29: 245, 1982.
39. HM Laun, H Schuch. J Rheol 33: 119, 1989.
40. HP Schreiber, EB Bagley, D C West. Polymer 4: 355, 1963.
41. RN Shroff M Shida. J Polym Sci A-2 8: 1917, 1970.
42. BH Bersted. J Appl Polym Sci 30: 3751, 1985.
43. IP Briedis, LL Sulzhenko, LA Faiitelson, EE Yakobson. Mech Polymerov 6: 1129, 1973.
44. JJ Labaig, Ph. Monge, I Bednarick. Polymer 14: 384, 1973.
45. CD Han, YJ Kim, HK Chuang, TH Kwack. J Appl Polym Sci 28: 3435, 1983.
46. JM Dealy, A Garcia-Rejon, MR Kamal. Can J Chem Eng 55: 651, 1977.
47. P Starck, JJ Lindberg, Angew Makromol Chem 75: l, 1979.
48. I Meissner. J Appl Polym Sci 16: 2877, 1972.
49. GD Han, HJ Yoo. J Rheol 24: 55, 1980.
50. S Middleman. The Flow of High Polymers, Continuum and Molecular Rheology, New York: Interscience, 1968; Moskva: Mir, 1971.
51. RL Combs, DE Slonaker, HW Coover, Jr. J Appl Polym Sci 13: 519, 1969.
52. V Dobrescu, G Andrei, A Câmpeanu, C Andrei. Rev Roum Chim 33: 399, 1988.
53. JA Brydson. Flow Properties of Polymer Melts, London: Plastics Institute, 1970.
54. LS Bolotnikova, VG Baranov, LM Beder, YaN Panov, SYa Frenkel. Vysokomol Soedin 8-24: 14, 1982.

55. JL White, D Huang. Polym Eng Sci 21: 1101, 1981.
56. FF Meshterova, II Vavilova, RYa Alekseeva. Plast Massy 7: 49, 1978.
57. A Garcia-Rejon, JM Deary, MR Kamal. Can J Chem Eng 59: 76, 1981.
58. HJ Yoo, CD Han. J Rheol 25: 115, 1981.
59. EL Kalinchev, MB Sakovceva. Properties and Rheology of Thermoplast (in Russian), Leningrad: Khimiya, 1983.
60. DA Seano, ed. Electrical Properties of Polymers, New York: Academic Press, 1982.
61. DH Sebastian, JR Dearborn. Polym Eng Sci 23: 572, 1983.
62. Y Ide, JL White. J Appl Polym Sci 22: 1061, 1978.
63. J Meissner. Pure Appl Chem 42: 551, 1975.
64. J Meissner. Rheol Acta 10: 230, 1971.
65. J Meissner. Rheol Acta 8: 78, 1969.
66. J Meissner. Trans Soc Rheol 18: 405, 1972.
67. PK Agrawal, WK Lee, JM Lorntson, CI Richardson, KR Wisjbrun, AB Metzner. Trans Soc Rheol 21: 355, 1977.
68. T Raible, SE Stephenson, J Meissner, MH Wagner. J Non-Newton Fluid Mech 11: 239, 1982.
69. KE Wissbrun. Polym Eng Sci 13: 342, 1973.
70. ML Fridman, GV Vinogradov, AYa Malkin, OM Ponadii. Vysokomol Soedin A-12: 2162, 1970.
71. DW Van Krevelen. Properties of Polymers: Correlations with Chemical Structure, Amsterdam: Elsevier, 1972.
72. A Rudin, RJ Chang. J Appl Polym Sci 22: 781, 1978.
73. SK Bhateja, EH Andrews. Polym Eng Sci 23: 888, 1983.
74. MD Baijal, CL Sturm, J Appl Polym Sci 14: 1651, 1970.
75. BA Krencel, VI Kleiner, LL Stockaya. α-Polyolefins (in Russian), Moskva: Khimiya, 1984.
76. A Weill. J Non-Newton Fluid Mech 7: 303, 1980.
77. JE Guillet, RL Combs, DF Slonaker, DA Weemes, HW Coover, Jr. J Appl Polym Sci 9: 767, 1965.
78. JE Guillet, RL Combs, DE Slonaker, DA Weemes, HW Coover. J Appl Polyml Sci 8: 757, 1965.
79. ER Howells, JJ Benbow. Trans. Plast Inst 30: 242, 1962.
80. CD Han. Polym Eng Sci 11: 205, 1971.
81. JWC Adamse, H Janeschitz-Kriegl, IL Den Otter, JLS Wales. J Polym Sci A-2-6: 871, 1968.
82. GV Vinogradov, AYa Malkin, VF Skumski. Rheol Acta 9: 155, 1970.
83. RA Mendelson, EL Finger. J Appl Polym Sci 17: 797, 1973.
84. M Rokudai. J Appl Polym Sci 26: 1427, 1981.
85. V Verney, R Genillon, J Niviere, JF May, Rheol Acta, 20: 478, 1981.
86. VI Bukhgalter, RI Belova. Plast Massy 2: 75, 1978.
87. JM Torregrosa, A Weill, J Druz. Polym Eng Sci 21: 768, 1981.
88. MS Jacovic, D Pollock, RS Porter. J Appl Polym Sci 23: 517, 1979.
89. V Verney, A Michel. Rheol Acta 24: 627, 1985.
90. B Millaud, J Rault. Macromolecules 17: 340, 1984.
91. LH Tung. J Polym Sci 3: 409, 1960.
92. WL Peticoles, J Watkins. J Am Chem Soc 79: 5083, 1957.
93. WF Busse, R Longworth. J Polym Sci 5: 49, 1962.
94. HP Schreiber, EB Bagley J Polym Sci Polym Lett 1: 365, 1963; WJ Bontinek. Rheol Acta 8: 328, 1969.
95. RA Mendelson, WA. Bowles, EL Finger. J Polym Sci A-2 8: 105, 1970.
96. J. Miltz, A Ram. Polymer 12: 685, 1971.
97. G Locati, L Gargant. Proc 7th Int. Congress Rheol, Gothenburg, August, 1976, p 520.
98. P Briedis, LA Faitel'son. Polym Mech 12: 100, 1976.
99. A Ram. Polym Engl Sci 17: 793, 1977.
100. S Pedersen, A Ram. Polym Engl Sci 18: 990, 1978.
101. K Sakamoto, T. Kotaoka, J Fukusava. J. Soc Mater Sci Jpn 15: 377, 1966.
102. K. Chayashidao. Rheol Acta 2: 261, 1963.
103. S Jamanoyti, K Jasuno. J Chem Soc Jpn A-66: 1468, 1963.
104. R Longworth, ET Pieski. J Polym Sci B-3: 221, 1965.
105. HP Frank. Rheol Acta 5: 89, 1966.
106. JE Guillet, RL Combs, DE Slonaker, DA Weemes, HW Coover, Jr. J Appl Pdym Sci 9: 767, 1965.
107. WW Graessley. Acc Chem Res 10: 332, 1977.
108. H Schott, WS Kagan., J Appl Polym Sci 5: 175, 1961.
109. SL Aggerwal, L Marker, MJ Carrano. J Appl Polym Sci 4: 77, 1960.
110. J. Meissner. In: EH Lee, ed. 4th International Congress on Rheology, 1963, Part 3, p 427, New York: Interscience, 1965
111. RA Mendelson. Trans Soc Rheol 9: 53, 1965.
112. L. Boghetich, RE Kratz. Trans Soc Rheol 9: 225, 1965.
113. RS Porter, J. F Johnson. J Polym Sci C-15: 365, 1966.
114. RS Porter, JR Knox, JF. Johnson. Trans Soc Rheol 12: 409, 1968.
115. V Seminov. Adv Polyml Sci 5: 387, 1968.
116. LL Blyler. Rubber Chem Technol 14: 823, 1969.
117. RN Shroff LV Cancio, M Shida. Trans Soc Rheol 21: 429, 1977.
118. F Deri, A. Piloz, JF May. Angew Makromol Chem 55: 97, 1976.
119. R Genillon, E Deri, JF May. Angew Makromol Chem 65: 71 1977.
120. JS Wang, RS Porter, JR Knox. J Polym Sci B 8: 671, 1970.
121. JD Ferry. Viscoelastic Properties of Polymers, 2nd ed., New York: Wiley, 1970, p 605.
122. I Havlicek, V Vojta, S. Kasmer, E Schlosser. Makromol Chem 179: 2467, 1978.
123. MN Matrosovich, VG Kravchenko, JuA Kostrov, VG Kulichikhin, LP Braverman. Vysokomol Soedin B-22: 357, 1980.

124. AA Miller. Macromolecules 3: 674, 1970.
125. M Hoffmann. Makromol Chem 153: 99, 1972.
126. J Brandrup, EH Immergut, eds. Polymer Handbook, New York: Interscience, 1960.
127. AR Wedgewood, JC Seferis. Polym Eng Sci 24: 328, 1984.
128. TA Speranskaya, LI. Tarutina. Optical Properties of Polymers (in Russian), Leningrad: Khimiya, 1976.
129. RH Boyd, L Kesner. Macromolecules 20: 1802, 1987.
130. M. Pietralla, HG Kilian. J Polym Sci Polym Phys Ed 18: 285, 1980.
131. Y Fukui, T Sate, M Ushirokawa, T Asada, S Onogi. J Polym Sci A-2 8: 1195, 1970.
132. M Gilbert, DA Hemsley, SR Patel. Br Polym J 19: 9, 1987.
133. G Czornyj, B Wunderlich. Makromol Chem 178: 843, 1977.
134. A Tanaka, K Tanai, S Onogi. Bull Inst Chem Res Kyoto Univ 55: 177, 1977.
135. GH Meeten, ed. Optical Properties of Polymers, New York: Elsevier, 1986.
136. M. Rokudai. S. Mihara, T Fujiki. J Appl Polym Sci 23: 3289, 1979.
137. PL Clegg, ND Huck. Plastics (London) 26: 107, 1961.
138. IL Fridman. Processing Technology of Crystalline Polyolefins (in Russian) Moskva: Khimiya, 1977.
139. H. Ashizawa, JE Spruiell, JL White. Polym Eng Sci 24: 1035, 1984.
140. JP Watson. Anal. Proc 22: 105, 1985.
141. LZ Ismail. Polym Test 7: 299, 1987.
142. SN Kolesov, Structural Electrophysics of the Polymeric Dielectrics, Himia Uzbekistan, Tashkent, 1975.
143. H Stahl. Diplomarbeit, Univenitat Ulm, 1978.
144. S. Onogi, A. Tanaka, Y. Ishikawa, T Igarashi. Polym J 4: 467, 1975.
145. RS Porter, I Wang, IR Knox. J Polym Sci B-8: 671, 1970.
146. K Shirayama, T Matsuda, S Kita. Makromol Chem 147: 155, 1971

14

Solution Characterization of Polyolefins

Jimmy W. Mays
University of Alabama at Birmingham, Birmingham, Alabama

Aaron D. Puckett
University of Mississippi Medical Center, Jackson, Mississippi

I. INTRODUCTION

The mechanical and physical properties of polyolefins are influenced, sometimes profoundly, by molecular weight. For example, paraffinic alkanes are incapable of forming fibers or films possessing useful mechanical properties. On the other hand, polyolefin fibers and films constitute a major segment of the polymer industry and find broad utility in everyday life.

The rheological properties of polymer melts are affected by both molecular weight and molecular weight distribution (MWD), as well as by long chain branching. With polyolefins, slight differences in the MWD, especially at the high molecular weight end of the distribution, will strongly impact the processing behavior of the material. This effect is especially true in processes involving elongational flow, such as fiber spinning and blown film formation. Molecular weight and MWD also affect the physical properties of the finished polymeric items.

The determination of molecular weights of synthetic polymers is complicated relative to that of low molecular weight compounds because of polydispersity, i.e., all the chains in a given sample of polymer are never the same length. Consequently, various *average* molecular weights such as the number-average molecular weight ($\bar{M}_n$) and the weight-average molecular weight ($\bar{M}_w$) defined in Eqs. (1) and (2) respectively must be employed:

$$\bar{M}_n = \frac{\sum N_i M_i}{\sum N_i} \tag{1}$$

$$\bar{M}_w = \frac{\sum N_i M_i^2}{\sum N_i M_i} \tag{2}$$

Here N_i is the number of molecules of molecular weight M_i. Methods are well established for measurement of these two average molecular weights and for examining the entire molecular weight distribution (MWD) of macromolecules. These characterization techniques employ dilute polymer solutions, and thus special difficulties are encountered with the many polyolefins that are semicrystalline in the vicinity of room temperature and soluble only at temperatures where thermal degradation can occur.

In this chapter, we describe the major techniques for measuring molecular weights and characterizing MWD of polyolefins. These methods include classical techniques such as light scattering, osmometry, viscometry, and fractionation. Also discussed are newer and evolving techniques such as temperature rising elution fractionation (TREF), supercritical fluid (SCF) separations, and the combination of size exclusion chromatography (SEC) with various molecular-weight-sensitive detectors. Characterization of long chain branching is briefly considered.

II. DETERMINATION OF MOLECULAR PROPERTIES BY LIGHT SCATTERING

A. Theory

A number of experimental techniques are available for the determination of $\bar{M}_w$ of macromolecules. These methods include equilibrium sedimentation and the scattering of neutrons, X-rays, and light. The light scattering method is by far the dominant technique at the present time.

When light passes through a molecular system, even one in which no absorption occurs, transmission is less than 100% because of scattering of a portion of the light in all directions. Lord Rayleigh [1] related the intensity of light scattered from dilute gases to their optical properties via

$$\frac{i(\theta)r^2}{I_0} = \frac{2\pi^2}{\lambda^4 N_A} \frac{(n_0 - 1)^2 M}{c} (1 + \cos^2 \theta) \tag{3}$$

where $i(\theta)$ is the intensity of scattered light per unit volume at a given angle (θ) relative to the incident beam, r is the distance from the scatterer to the detector, I_θ is the intensity of the incident beam of wavelength λ, N_A is Avogadro's number, n_0 is the refractive index of the medium, and M and c are the molecular weight and concentration of the gas. The quantity $i(\theta)r^2/I_0$, which is the fraction of light scattered at a given θ, r, and I_0, is commonly known as the Rayleigh ratio (R_θ).

Subsequent work by Einstein [2], Debye [3], and others [4–6] extended Rayleigh's theory to the treatment of simple liquids and dilute polymer solutions. For polymers, an optical constant K may be defined

$$K = \frac{4\pi^2 n_0^2 (\mathrm{d}n/\mathrm{d}c)^2}{\lambda^4 N_A} \tag{4}$$

where $\mathrm{d}n/\mathrm{d}c$ is the refractive index increment of the polymer–solvent system. Also, to extract molecular parameters of the polymer, scattering by the solvent must be subtracted out; this results in the use of the excess Raleigh factor, $\bar{R}_\theta$, where

$$\bar{R}_\theta = R_{\theta,\mathrm{soln}} - R_{\theta,\mathrm{solv}} \tag{5}$$

Combination of Eqs. (3)–(5) leads to the well known relationship (6) between $\bar{R}_\theta$ and $\bar{M}_w$:

$$\frac{Kc}{\bar{R}_\theta} = \frac{1}{\bar{M}_w P(\theta)} + 2A_2 c \tag{6}$$

where $P(\theta)$ is the particle scattering factor which depends on polymer size and shape. A_2 is the second virial coefficient, a thermodynamic parameter that provides a measure of polymer–solvent interactions.

At very small angles the value of $P(\theta)$ approaches unity (assuming the particles are not too large); this forms the basis of low-angle laser light scattering (LALLS), where a plot of $Kc/\bar{R}_\theta$ versus c yields $\bar{M}_w$ from the intercept and A_2 from the slope:

$$\frac{Kc}{\bar{R}_\theta} = \frac{1}{\bar{M}_w} + 2A_2 c \tag{7}$$

If the macromolecules are not too small relative to the wavelength of light, light scattering can also yield the z-average mean-square radius of gyration, $\langle S^2 \rangle$, from the angular dependence of the scattering intensity. For a random coil $P(\theta)$ is defined by

$$P(\theta)^{-1} = 1 + \frac{16\pi^2 N_0^2}{3\lambda^2} \langle S^2 \rangle \sin^2 \frac{\theta}{2} \tag{8}$$

Incorporation of Eq. (8) into Eq. (6) leads to

$$\frac{Kc}{R_\theta} = \frac{1}{M_w} \left[1 + \frac{16\pi^2 N_0^2}{3\lambda^2} \langle S^2 \rangle \sin^2 \frac{\theta}{2} \right] + 2A_2 c \tag{9}$$

which allows the construction of a double extrapolation procedure or "Zimm Plot" [7] shown in Fig. 1 to be constructed. Here $Kc/\bar{R}_\theta$ is plotted versus $\sin^2(\theta/2) + kc$, where k is a numerical factor chosen to provide appropriate spacing of the data points. The slope of the $C \to 0$ line, i.e., the angular dependence of scattering intensity, yields $\langle S^2 \rangle$, so long as $\langle S^2 \rangle^{1/2} \geq$ ca $\theta/20$. Similarly, A_2 is obtained from the concentration dependence at zero angle, while the mutual intercept is $\bar{M}_w^{-1}$. Light scattering is an especially useful means of polymer characterization,

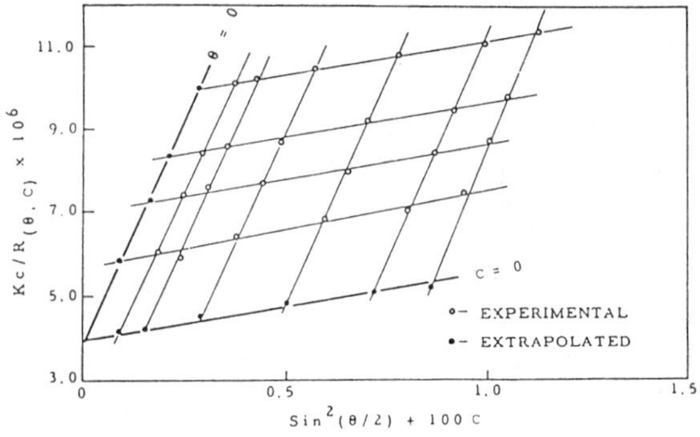

FIG. 1 Zimm plot for fractionated polyethylene sample in 1-chloronaphthalene at 135°C. (Reprinted with permission from R Chiang. J Polym Sci 36: 91, 1959. Copyright 1959, John Wiley & Sons, Inc.)

partly because information concerning molecular mass, molecular size, and thermodynamic interactions can all be obtained from a single experiment.

B. Light Scattering from Polyolefin Solutions

Special difficulties are associated with dilute solution characterization of semicrystalline polyolefins such as polyethylene (PE) and stereoregular polypropylene (PP). These polymers are only soluble in the vicinity of their crystalline melting points, generally $> 100°C$. Thus, all handling and clarification of solutions and measurements must be conducted at temperatures where this class of polymers is susceptible to thermo-oxidative degradation. Temperatures of ca 120–140°C are normally used with high density PE, while for isotactic PP, which has a higher melting point, temperatures of 130–150°C are commonly employed. While it might initially seem that measurements would best be conducted at the lowest possible temperature so as to minimize potential problems with degradation, aggregation is frequently encountered under such conditions, leading to inaccurate molecular weight measurements. Wagner and Dillon [8] have described a satisfactory method for preparing molecular solutions of ultrahigh molecular weight polyethylene (UHMWPE). This polymer is especially slow to dissolve because of its large number of chain entanglements, high degree of crystallinity, and extremely high molecular weight. Wagner and Dillon [8] found tetralin to be unsatisfactory as a solvent, presumably due to the tendency of this solvent to form peroxides. The use of nitrogen sparged, stabilized (0.1% Santonox R) decalin was, however, satisfactory. The UHMWPE was cut into pieces one millimeter or less in size, added to the stabilized solvent, and shaken gently under nitrogen for 1 hour at 180°C. Filtration was also conducted at 180°C through 0.5 μm poly-(tetrafluoroethylene) filters, presumably under nitrogen. Reproducible, low-shear intrinsic viscosity data were obtained using solutions prepared in this manner [8]. This method of solution preparation should also prove useful with easier to dissolve lower molecular weight PEs and with other solvents.

Grinshpun and coworkers [9–11] have studied the dissolution behavior of both PE and isotactic PP in trichlorobenzene and other solvents. For work with isotactic PP they added 0.1% Irgafos D13-168 (2,4-di-*tert*-butylphenylphosphite) and 0.1% Ionol (butylated hydroxytoluene) to the sample solutions. At dissolution temperatures $>145°C$ degradation occurred, even in the presence of stabilizers. Heating of solutions for about 30–50 h in trichlorobenzene at 145°C yielded reliable and reproducible M_w and A_2 values via light scattering experiments [11].

Another solvent that has been used with some success for light scattering measurements is α-chloronapthalene [9, 18]. Both high and low density PE have been dissolved in α-chloronapthalene (stabilized with 0.05% 4,4′thiobis(3-methyl-6-*tert*-butyl phenol) at 135–145°C. This solvent is attractive because it has a large dn/dc value ($dn/dc \cong 0.18$ mL g^{-1}, therefore allowing very low concentrations to be employed when LALLS or refractive index detectors are used. However, some investigators have reported problems with aggregation of high molecular weight species in α-chloronapthalene that could not be broken up by additional thermal treatment up to 160°C [10, 18].

Additional references on preparation of polyolefin solutions should be noted [12, 13]. Most workers now favor a strategy of heating initially at a higher temperature (160–170°C) for a shorter time (1–2 h), followed by additional heating at about 140°C until dissolution is obtained.

Special attention must also be given to solution clarification procedures. Certain preparative centrifuges can be operated at elevated temperatures; alternatively, filtration using porous glass or poly(tetrafluoroethylene) filters can also be readily adapted to high temperature applications. Ideally, the clarification process should be conducted under an inert atmosphere to avoid degradation and at a sufficiently high temperature to avoid partial fractionation of the polymer. Excellent advice concerning light scattering experiments on polyolefins can be found in the review by Chiang [14] and in the book by Kratochvil [15].

Another requirement for molecular weight measurements using light scattering is the concentration dependence of the refractive index. Selected refractive index increments for polyolefins are listed in Table 1. The dn/dc value for a given polymer–solvent pair depends not only on wavelength but also varies weakly with temperature as well. Differential refractometers capable of operating at elevated temperatures are commercially available from several companies.

The rather close agreement between values of dn/dc for PE and PP in 1,2,4-trichlorobenzene at a given wavelength indicates that this would be a particularly well-suited solvent in which to perform studies on ethylene-propylene copolymers, since compositional variations in fractions will have only a small effect on dn/dc.

Amorphous polyolefins such as polyisobutylene or atactic PP are soluble in common organic solvents at

TABLE 1 Selected Refractive Index Increments of Polyolefin Solutions

Polymer	Solvent	λ (nm)	Temp. (°C)	d*n*/d*c* (mL g^{-1})	Ref(s).
HDPE	1-chloronaphthalene	436	135	−0.215	16
		546	135	−0.192	16
		633	135	−0.183	16
		546	135	−0.196	17
		633	145	−0.189	18
		633	135	−0.177	19
	1,2,4-trichlorobenzene	436	135	−0.125	16
		546	135	−0.110	16
		633	135	−0.107	16
		633	145	−0.098	9
		546	135	−0.106	17
		633	145	−0.112	18
		633	135	−0.104	19
	ortho-dichlorobenzene	633	135	−0.056	19
	diphenylmethane	546	142	−0.125	20
	n-decane	436	135	0.117	16
		546	135	0.114	16
		633	135	0.112	16
LDPE	1-chloronaphthalene	546	125	−0.195	21
LDPE	1-chloronaphthalene	633	145	−0.185	18
LLDPE	1-chloronaphthalene	633	145	−0.180	18
LDPE	1,2,4-trichlorobenzene	633	145	−0.108	18
LDPE	1,2,4-trichlorobenzene	633	135	−0.091	19
LLDPE	1,2,4-trichlorobenzene	633	145	−0.106	18
Isotactic PP	1-chloronaphthalene	436	135	−0.205	22
		546	135	−0.184	22
		633	135	−0.173	22
		546	150	−0.191	23
		546	125	−0.189	24
		546	140	−0.188	25
	1,2,4-trichlorobenzene	436	135	−0.121	22
		546	135	−0.107	22
		633	135	−0.102	22
Atactic PP	1-chloronaphthalene	436	135	−0.211	22
		546	135	−0.188	22
		633	135	−0.177	22
	1,2,4-trichlorobenzene	436	135	−0.123	22
		546	135	−0.111	22
		633	135	−0.105	22
	1-chlorobutane	546	25	0.108	26
		633	23	0.088	27
	tetrahydrofuran	633	23	0.079	28
Polyisobutylene	*n*-heptane	436	25	0.142	29
		633	25	0.143	30
	cyclohexane	436	25	0.101	29
		633	25	0.095	30
Poly(1-butene), isotactic	1-chloronaphthalene	436	135	−0.206	22
		546	135	−0.187	22
		633	135	−0.177	22
	1,2,4-trichlorobenzene	436	135	−0.120	22
		546	135	−0.103	22

TABLE 2 Selected References to Light Scattering Studies of Polyolefins

Polymer	Comments	Ref(s).
PE	Studied both low and high density PE via LALLS and SEC-LALLS	19
PE	Study by 6 laboratories on an NBS standard PE by LALLS and SEC–LALLS	9
PE	Classic early studies on light scattering of linear and branched PEs	32–35
PE	Reported the first online (with SEC) determination of radii of gyration and molecular weight of linear PE using a MALLS detector	36
PE	Studied HDPE, LDPE, and LLDPE using SEC–LALLS. They noted shear degradation of high MW PE when small particle size columns and low flow rates is necessary to obtain true molecular weights of such materials	37
PE	Performed a systematic evaluation of the performance of LALLS and MALLS detectors for SEC. It was noted that MALLS detection offers the advantage of determining the radius of gyration at each retention volume.	38
PE	Used SEC–MALLS to evaluate MWD and branching in PE resins. It was shown that differences that are not detected by conventional SEC, but that are important to end-use performance, are detected by SEC–MALLS	39
PE	Studied linear and branched PE using a combination of classical light scattering, SEC, viscometry, and NMR. Used the data to characterize the molecular structure of LDPE	40
PP	An excellent study combining osmometry, SEC, viscometry, and light scattering studies on isotactic PP fractions	41
PP	A recent study combining LALLS and SEC–LALLS studies of isotactic PP. Careful consideration is given to solution preparation	11
PP	Early studies on isotactic, atactic, and syndiotactic PP	25, 42–45
Poly-1-butene	Study of fractionated atactic and isotactic poly(1-butenes) by osmometry, viscometry and light scattering	46

ambient temperatures; selected dn/dc values for these two materials are also listed in Table 1. A much more extensive listing of dn/dc values for polyolefins and some olefinic copolymers is available in the Polymer Handbook [31]. Selected references to light scattering studies on polyolefins are presented in Table 2.

C. Dynamic Light Scattering from Polyolefin Solutions

Only a few papers have appeared on the application of dynamic light scattering (DLS; also known as photon correlation spectroscopy) to solutions of crystalline polyolefins [47–52]. The details of DLS can be found elsewhere [53, 54]. The limited number of papers on DLS of polyolefins is largely due, we believe, to the lack of commercial instrumentation suitable for use at highly elevated temperatures.

Chu *et al.* [47] used DLS to study the diffusion behavior of linear and branched polyethylene in 1,2,4-trichlorobenzene. Diffusion coefficients and hydrodynamic radii were measured. In addition, from the decay time range of photon correlation measurements the molecular weight distributions of moderately broad commercial polyethylene samples could be evaluated by DLS. Thus, DLS can complement SEC as a means for evaluating polydispersity of polyolefins. Helmstedt [49, 50] and Wu [51] have studied the molecular weight dependence of the diffusion coefficient for branched and linear PE, respectively in TCB. Once this relationship is "calibrated" for a particular system, DLS can provide a convenient means of estimating molecular weight. Wu and Lilge [51] confirmed Chu's [47] claim that DLS could provide an estimated of polydispersity, although they noted that polydispersities measured by DLS tended to be narrower than those estimated by SEC. More recently, Wu [52] demonstrated that high temperature DLS is a very sensitive technique for detecting a very small amount of degradation in the high molecular weight tails of PE after extrusion. This small amount of degradation could dramatically affect the viscoelastic properties of the PE but it was hard to detect by SEC.

III. DETERMINATION OF NUMBER-AVERAGE MOLECULAR WEIGHTS BY OSMOMETRY

Number-average molecular weights, $\bar{M}_n$, for polyolefins are most conveniently measured via colligative properties measurements. Especially useful are the complimentary techniques of membrane osmometry (MO) and vapor pressure osmometry (VPO). Cryoscopy and ebulliometry are not nearly so popular, in part because of well-documented experimental difficulties [55, 56].

MO and VPO are complementary techniques, with the former being most useful when $30{,}000 < \bar{M}_n < 500{,}000$ and the latter technique being preferred when $\bar{M}_n < 20{,}000$. Furthermore, commercial MO and VPO instruments, suitable for use at elevated temperatures, are readily available.

A. Membrane Osmometry

In MO a dilute polymer solution is separated from pure solvent (of the same type used to prepare the solution) by a semipermeable membrane, which allows transport of solvent but not solute. Solvent molecules will diffuse from the solvent side into the solution side due to the lower chemical potential of the solvent in the solution. This solvent diffusion ceases when the potential difference is zero. In practice, with "conventional" osmometers this occurs when the hydrostatic pressure of a column of liquid equals the osmotic pressure, π, caused by the difference in chemical potential. An osmometer is depicted in Fig. 2.

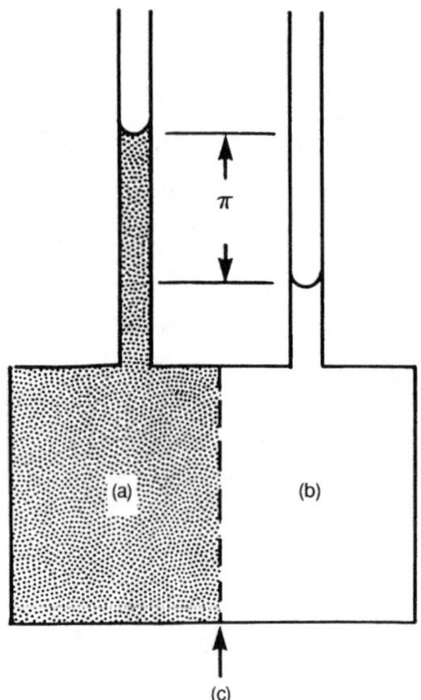

FIG. 2 Schematic illustration of the principles of membrane osmometry. A polymer solution in compartment (a) is separated from pure solvent in compartment (b) by a membrane (c), which will allow solvent (but not solute) to pass through. Solvent diffuses from (b) to (a) until the osmotic pressure is just balanced by the hydrostatic pressure. The difference in heights of the column at equilibrium is the osmotic pressure π.

For ideal solutions, values of π can be related to solute molecular weight M by van't Hoff's law:

$$\pi \approx \frac{cRT}{\bar{M}} \qquad (10)$$

where c is the concentration, R is the gas constant, and T is absolute temperature. Since macromolecular solutions of suitable concentrations for measurement of π rarely behave ideally, an extrapolation to infinite dilution must be carried out. A virial expansion of the form

$$\pi/c = RT(1/\bar{M}_n + A_2c + A_3c^2 + \cdots) \qquad (11)$$

where A_2 and A_3 are virial coefficients, has been shown to allow reliable extrapolation of π/c values to infinite dilution [57–59]. Because of the increasing contribution of the third term in Eq. (11) for high molecular weight polymers, plots of π/c versus c are frequently curved upward at higher concentrations. Consequently, plots of $(\pi/c)^{1/2}$ versus c are preferred in many cases because they are more apt to be linear [57–59].

A number of commercially available membrane osmometers operate satisfactorily at elevated temperatures. The major problem with membrane osmometry continues to be the poor stability of regenerated cellulose at >100°C temperatures [14, 60–62]. Despite the limitation of regenerated cellulose membranes, membranes with improved performance have not been developed. Another limitation of membrane osmometry for polyolefin characterization is the problem of diffusion of low MW polymer components across the membrane because of the very broad MWD typical for polyolefins. This results in apparent $\bar{M}_n$ values which are larger than the true values. Typical solvents suitable for use at high temperatures include stabilized trichlorobenzene, tetralin, decalin and xylene.

Detailed references to membrane osmometry studies on polyolefins are rare. Noteworthy are the studies of Billmeyer and Kokle [62], who compared M_n values determined for polyethylenes by various techniques, and of Atkinson and Dietz [41], who characterized a set of 12 well-fractionated polypropylenes by various methods.

B. Vapor Pressure Osmometry

In VPO, two matched temperature-sensitive thermistors are placed in a carefully thermostated environment, which is saturated with solvent vapor. A drop of pure solvent is placed on one thermistor, and polymer solution is placed on the other. The lower vapor pressure of the solvent in the solution results in condensation of solvent vapor onto the solution thermistor. This creates a temperature imbalance (condensation causes an increase in temperature of the solution thermistor), which is usually detected as a change in resistance (ΔR) across a Wheatstone bridge. ΔR is related to the polymer concentration (c) and molecular weight by the equation

$$\left(\frac{\Delta R}{c}\right)_{c\to 0} = \frac{K}{\bar{M}_n} \tag{12}$$

where K is the calibration constant, determined by performing ΔR measurements on solutions of a substance of known molecular weight.

An extrapolation of a plot of $\Delta R/c$ versus c to infinite dilution yields an intercept of $\bar{M}_n^{-1}$.

VPO avoids the problem with diffusion of low molecular weight components across a membrane (as in MO) and is thus applicable to samples with lower molecular weights. The very broad molecular weight distributions of many polyolefins can lead to problems in applying MO for $\bar{M}_n$ determination because of diffusion of these smaller molecules across the membrane. Unfortunately, VPO suffers from lower inherent sensitivity than MO and is generally most useful when $\bar{M}_n \leq$ ca 20,000. Thus, to date, not many reports of successful application of VPO to crystalline polyolefins have appeared [41, 62].

Recent improvements in the design of vapor pressure osmometers have led to instruments capable of measuring $\bar{M}_n$ values as high as about 10^5 [63, 64]. These instruments have recently been applied to the analysis of both PE and PP [61, 65]. In both studies 1,2,4-trichlorobenzene was used as the solvent at ca 140°C with sucrose octaacetate used as the calibrant. In general, it was found [61, 65] that $\bar{M}_n$ results in accord with those generated by other techniques were obtained for narrow molecular weight distribution materials (standard PEs, and hydrogenated, anionically produced polybutadienes). For very polydisperse commercial polypropylenes, $\bar{M}_n$ via VPO was less than $\bar{M}_n$ determined by size exclusion chromatography (SEC) in many instances. This result could reflect the sensitivity of VPO to low molecular weight components not detected by SEC. These components may be oligomeric or polymeric, or they may be additives or residual solvents. Care must be taken to remove the latter components prior to the VPO experiment if they are present, otherwise $\bar{M}_n$ values measured will be erroneously low. MO would not detect such low molecular weight materials since they would diffuse through the membrane. In an earlier comparative study of various methods for determining $\bar{M}_n$ of polyethylenes, Billmeyer and Kokle [62] found $\bar{M}_n$ by MO to be consistently higher than values determined by other techniques.

IV. INTRINSIC VISCOSITIES OF POLYOLEFINS

The intrinsic viscosity $[\eta]$ is usually obtained using the Huggins equation

$$\frac{\eta_{sp}}{c} = [\eta] + k'[\eta]^2 c \tag{13}$$

where η_{sp} is the specific viscosity taken as $(\eta - \eta_0)/\eta_0$ where η and η_0 are solution and solvent viscosities, respectively, c is concentration, and k' is the Huggins coefficient, which has a value of about $\frac{1}{3}$ for linear flexible polymers in thermodynamically good solvents. Thus, a plot of η_{sp}/c versus c yields $[\eta]$ as the intercept. The intrinsic viscosity is a measure of the average hydrodynamic size of the polymer chains in solution and has units of reciprocal concentration. The value of $[\eta]$ for a given sample will depend on solvent quality, temperature, polymer structure, molecular weight, and molecular weight distribution. In part because $[\eta]$ depends on several factors, there is no theoretical way to relate $[\eta]$ directly to polymer molecular weight. Instead, an empirical Mark–Houwink equation [66, 67] of the form

$$[\eta] = KM^a \tag{14}$$

must be established for each combination of polymer, solvent, and temperature. Since synthetic polymers are never monodisperse, the question arises as to which average molecular weight should be utilized for such purposes. Flory [68] showed long ago that $[\eta]$ reflects the viscosity-average molecular weight ($\bar{M}_v$) of the sample where

$$\bar{M}_v = \left(\frac{\sum N_i M_i^{1+a}}{\sum N_i M_i}\right)^{1/a} \tag{15}$$

where N_i is the number of molecules of molecular weight M_i, and a is the empirical "Mark–Houwink" exponent of Eq. (14). For random coils, $\bar{M}_v$ will fall between $\bar{M}_n$ and $\bar{M}_w$, but will be closer to the latter. Thus, $\bar{M}_w$ values are preferred for establishing the Mark–Houwink parameters K and a. Obviously, more reliable values are also obtained when well-fractionated samples are employed.

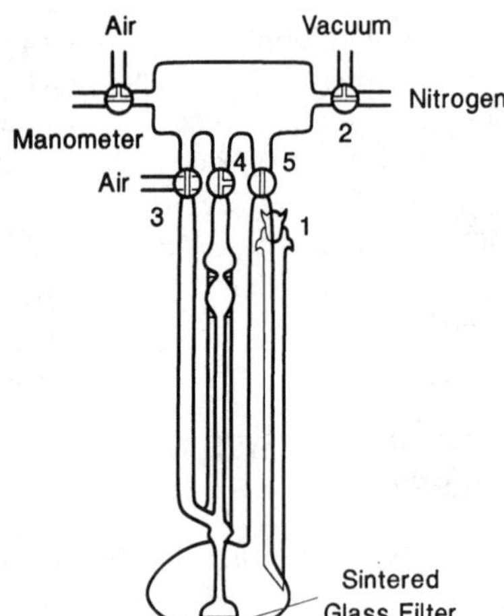

FIG. 3 A viscometer suitable to use at elevated temperatures with polyolefin solutions. Dissolution, filtration, and measurements can all be conducted under an inert atmosphere. Solvent and polymer are introduced through tube 1. The viscometer is then flushed with nitrogen and evacuated through stopcock 2. The viscometer and its contents may then be heated to the desired temperature while the polymer dissolves. Filtration and measurements are conducted by manipulating stopcocks 3, 4, and 5. (Reproduced with permission from Ref. 69. Copyright 1966, John Wiley & Sons, Inc.)

Once established, the Mark–Houwink parameters allow $\bar{M}_v$ to be computed from Eq. (12) based on the relatively straightforward, simple, and accurate (relative to light scattering and osmometry) measurement of solution viscosities. Viscosity determinations can usually be accomplished with a simple capillary viscometer, although with crystalline polyolefins the problems of solution preparation, filtration, and possible chain degradation at elevated temperatures must be considered. Nakajima and coworkers [69] have described how dissolution, filtration, and measurements can all be conducted in a sealed capillary viscometer under an inert atmosphere. A diagram of their viscometer is presented in Fig. 3. At very high molecular weights ($>10^6$), such as with UHMWPE, a substantial dependence of shear rate on measured viscosities is observed. Wagner and Dillon [8] have described the construction and use of a Zimm–Crothers low-shear viscometer that could be operated at 135°C under an inert atmosphere.

Selected Mark–Houwink coefficients for various polyolefins are tabulated in Table 3. For polyethylene, the Mark–Houwink parameters recommended by Wagner [70] are listed. For other polymers, an effort was made to choose the more reliable relationships based on polydispersity of samples, number of

TABLE 3 Mark–Houwink Parameters for Various Polyolefins

Polymer	Solvent	Temp. (°C)	$K \times 10^3$ (ml g^{-1})	a	MW method	Ref(s).
PE	decalin	135	62	0.70	LS	71
PE	1,2,4-trichlorobenzene	135	52.6	0.70	LS/MO	72
PE	1,2,4-trichlorobenzene	135	30.1	0.75	LS	73
PE	1,2,4-trichlorobenzene	130	39.2	0.725	LS/MO	74
PE	1-chloronaphthalene	130	55.5	0.684	LS/MO	74
PE	tetralin	130	51	0.725	MO	75
PE	1,2-dichlorobenzene	138	50.6	0.7	LS	76
Isotactic PP	decalin	135	23.8	0.725	LS/MO	41, 77
Isotactic PP	1,2,4-trichlorobenzene	135	19.0	0.725	LS/MO	41, 77
Isotactic PP	1,2-dichlorobenzene	135	24.2	0.707	LS/MO	41
Isotactic PP	1,2,4-trichlorobenzene	145	15.6	0.76	LS	78
Atactic PP	decalin	135	15.8	0.77	MO	79
Atactic PP	tetralin	135	19.3	0.74	MO	79
Atactic PP	toluene	30	21.8	0.725	MO	79
Atactic PP	tetrahydrofuran	30	46.0	0.66	LS	28
Isotactic poly-1-butene	ethylcyclohexane	70	7.34	0.80	LS/MO	46
Isotactic poly-1-butene	*n*-nonane	80	5.85	0.0	LS/MO	46
Atactic poly-1-butene	cyclohexane	30	18.0	0.721	LS	80
Polyisobutylene	cyclohexane	25	13.5	0.740	LS	30
Polyisobutylene	*n*-heptane	25	15.8	0.697	LS	30
PS	1,2,4-trichlorobenzene	135	12.1	0.707	LS/MO	23

samples, molecular weight range, etc. Relationships established under theta conditions are omitted because of the additional experimental complexities that are introduced under these conditions.

V. DETERMINATION OF MOLECULAR WEIGHT DISTRIBUTION BY FRACTIONATION

The properties of high polymers are both a function of molecular weight and molecular weight distribution. Two polymers having the same weight-average molecular weight may process very differently and exhibit substantial differences in mechanical properties if their distributions are vastly different. The following sections will describe methods to determine MWD from the weights and molecular weights of fractions of a polymer sample.

A. Fractionation Methods

The foundation for classical fractionation was established over fifty years ago and methods have changed very little [81]. Theoretical treatments have been given by Tompa [82], Voorn [83], and Huggins and Okamato [84]. Basically, the method entails the separation of a polymer sample into fractions of different molecular weights and narrower molecular weight distributions based upon solubility. A new "laboratory manual" on polymer fractionation has recently been published [85].

Fractionation methods can be grouped as precipitation or extraction methods based on the fact that the solubility of a polymer molecule decreases with increasing molecular weight. Therefore, the addition of a non-solvent to a solution of a polydisperse polymer progressively precipitates fractions of decreasing molecular weight (fractional precipitation). Alternatively, the solid polymer may be extracted with solvents of increasing power to obtain fractions of increasing molecular weight (fractional extraction).

The power of a solvent can be determined from the Huggins interaction parameter, χ_1, which represents the polymer–solvent interactions. The parameter can be calculated experimentally from vapor pressure or osmotic measurements using Eq. (16):

$$(\ln a_1 - \ln v_1)v_2 - 1 = \bar{X}_n^{-1} + \chi_1 v_2 \tag{16}$$

where a_1 is the solvent activity, $\bar{X}_n$ is the number average degree of polymerization, and v_1 and v_2 are the volume fraction of solvent and polymer, respectively. Values of χ_1 decrease with increasing solvent power. The interaction parameter also exhibits a temperature dependence which can be described by

$$\chi_1 = A + B/T \tag{17}$$

where A and B are empirical constants. Computer simulations of solvent precipitation of polymers have been used to confirm that the Flory–Huggins theory describes the process well [86–89]. Therefore, the solvent power may be adjusted using solvent composition and temperature gradients to obtain phase separation in polymer solutions or preferential solvation of polymer. The separated phase or solvated polymer will contain a concentration gradient of all the different molecular weight species present that is molecular weight dependent. The average molecular weights of the fractions can be determined by the methods discussed previously and used to construct cumulative and differential molecular weight distribution curves. Selected solvent/nonsolvent combinations used for fractionation of polyolefins are given in Table 4 [31].

B. Data Analysis

The simplest method for estimating the molecular weight distribution from fractionation data is to plot the weight of the fractions against the value of the molecular weight of each fraction. The molecular weight of the fractions are determined using osmometry, light scattering, or viscosity measurements. A continuous line is drawn through the points, and the plot is taken as the integral distribution of molecular weights. The slope of this curve, dw/dm, when plotted against molecular weight gives the differential distribution. This method assumes that the fractions are monodisperse.

A better method which does not assume a monodisperse fraction has been introduced by Schulz and Dinglinger [90]. In this method it is assumed that one half of the fraction contains molecular weight species that were less than the average molecular weight of the fraction and the other half of the fraction had a molecular weight greater than the average. The cumulative weight fraction $C(M_i)$ for the ith fraction can be calculated by adding one half the fraction's weight to the sum of the weights of the other fractions. The cumulative weight $C(M_i)$ is plotted against the molecular weight of each fraction and the integral and differential molecular weight distributions can be determined graphically. Both of the methods described only give estimates of the molecular weight distribution. More elaborate methods based upon assumptions of the distributions of species within a fraction have been

TABLE 4 Solvent and/or Solvent/Nonsolvent Combinations for Fractionation of Polyolefins

Polymer	Solvent	Method
PE	2-Ethylhexanol–decalin (85:15)	Precipitation at low temperature
	Tetralin–benzyl alcohol (60:40)	Precipitation by decreasing temperature, 165–105°C
	Toluene–*n*-butanol	Precipitation by decreasing temperature, 115–100°C
	Xylene–triethylene glycol	Precipitation
	Xylene–polyoxyethylene	Precipitation by temperature, 130–175°C
	Tetralin–2-butoxyethanol	Column elution at 126°C
	Tetralin	Column extraction, variable temperature
	Xylene–diethylene glycol–monomethyl ether	Column elution, 126°C
	o-Dichlorobenzene	SEC, 130–138°C
	1,2,4-Trichlorobenzene	SEC, 135°C
Poly-1-butene	Cylcohexane–acetone	Precipitation at 35°C
	1,2,4-Trichlorobenzene	SEC, 80–135°C
Isotactic PP	Cyclohexanone–ethylene glycol or dimethyl phthalate	Precipitation 130–135°C
Atactic PP	Benzene–methanol	Precipitation
	Benzene–acetone	Precipitation
PP	*o*-Dichlorobenzene–diethylene glycol–monomethyl ether	Column elution 168–172°C
	Tetralin	Column extraction, increasing temperature
	o-Dichlorobenzene	SEC, 135°C
	Tetralin	TREF, 20–150°C

described and tend to give a more accurate estimation of the true molecular weight distributions [91–93].

C. Fractionation of Polyolefins

Fractional precipitation of crystalline polymers must be conducted above the melting point of the crystals to eliminate the effects of crystallinity on solubility. In addition, with polypropylene branching and tacticity differences can cause a reversal in the order of fractionation. In most cases polyolefins are fractionated from dilute solutions in poor solvents to eliminate these effects. However, the direct extraction of powdered high density polyethylene with solvents has been reported [94].

Figure 4 illustrates an apparatus used for fractionation of polyethylene [95, 96]. Xylene was used as the solvent and triethylene glycol the nonsolvent. The polymer-rich phase is less dense and will remain at the top of the jacketed vessel, allowing the solvent-rich phase to be transferred to the adjacent vessel. The polymer-rich phase may then be diluted with additional xylene and withdrawn as a fraction. Additional nonsolvent may then be added to the solvent-rich solution to obtain another fraction. The temperature of both containers was maintained constant by a saturated vapor of 2-ethoxyethanol provided through ports connected to the joints at the bottom of the vessels.

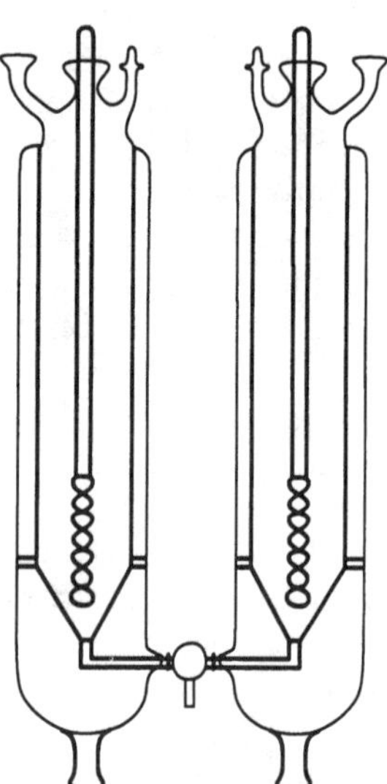

FIG. 4 Schematic illustration of an apparatus for the batch fractionation of polyethylene at elevated temperatures. (Reprinted with permission from *Crystalline Olefin Polymers*, RAV Raff and KW Doak, eds. Wiley-Interscience, New York, 1965, p 555.)

In some cases it may be advantageous to use a temperature gradient instead of, or in addition to, a nonsolvent to obtain fractions. Theta conditions for various polyolefins are summarized in Table 5. Fractionation of polyethylene using temperature

TABLE 5 Theta Solvents for Selected Polyolefins

Polymer	Solvent composition, vol/vol	Temperature (°C)
Isotactic poly(1-butene)	Anisole	89
	Cyclohexane/*n*-propanol (69/31)	35
Polyethylene	Biphenyl	118–125
	Bis(2-ethylhexyl) adipate	145
	Diphenyl ether	161–165
	n-Dodecanol	137–174
	n-Hexane	133
Poly(isobutylene)	*i*-Amyl *n*-valerate	21–22
	Chlorobenzene/propanol	
	(79.7/20.3)	14
	(76.0/24.0)	25
	(67.5/33.5)	49
	Chloroform/*n*-propanol	
	(79.5/20.5)	14
	(77.1/22.9)	25
	(57.9/42.1)	49
	Phenetole	86–87
Isotactic poly(1-pentene)	Phenetole	55.8
	2-Pentanol	62.4
Isotactic polypropylene	Diphenyl	125
	Diphenyl ether	143–146
Syndiotactic polypropylene	*i*-Amyl acetate	42

Source: From Ref. 31.

gradients and nitrobenzene [97] and low molecular weight polyethylene glycol [98, 99] have been reported.

Fractional extraction or solvation of polyolefins has usually employed chromatographic techniques. A block diagram illustrating the method is shown in Fig. 5. In these methods, the polymer is deposited on an inert substrate such as sand and then packed into a column. A varying mixture of solvent and nonsolvent is then passed through the column to attain fractionation. In most cases increasing concentrations of solvent are used to obtain low to high molecular weight fractions. The method was first described by Desreux and Spiegel [100] for the separation of low density polyethylene. Others have investigated fractional chromatography to fractionate high density polyethylene [101, 102] and polypropylene [103–107]. Investigators have studied the effect of solvent/nonsolvent combinations, temperatures and supports on fractionation efficiency. Porter *et al.* [105] demonstrated that atactic polypropylene could be fractionated at 156°C using *p*-xylene/methanol and isotactic polypropylene could be fractionated at 172°C using *o*-dichlorobenzene/methyl carbitol. These investigations reported no appreciable crystallinity effects upon the fractionation, but determined that the initial fractions contained higher syndiotactic contents. It is expected that fractionation methods will allow critical characterization of PP prepared using metallocene chemistry since these initiators allow improved control of M and MWD, tacticity, degree of branching and branch length of PP.

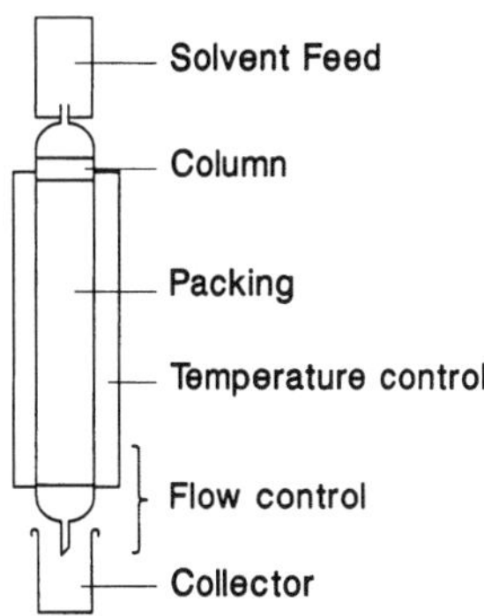

FIG. 5 Block diagram of a column fractionation apparatus. (Reprinted with permission from RM Screaton, in Newer Methods of Polymer Characterization (B Ke ed.), Wiley-Interscience, New York, 1964, p 450.)

D. Supercritical Fluid Fractionation of Polyolefins

Supercritical fluid (SCF) fractionation [108, 109] is a technique that can be used to separate polymers on the basis of molecular weight and composition. In this section, we will briefly discuss the use of SCF methods for fractionating according to molecular weight. Strategies for using SCF techniques for separating on the basic of composition (crystallinity) of ethylene copolymers is given at the end of Sec. VII on temperature rising elution fractionation.

Many of the aspects of SCF fractionation are similar to solvent/nonsolvent fractionation, except that normally a single solvent is used at a constant temperature, and pressure is varied to adjust the thermodynamic solvent quality of the SCF. Isothermal increasing pressure profiling is most commonly employed on a laboratory scale, although isothermal decreasing pressure profiling is also a feasible method. At a given temperature, polymer solubility is enhanced by increasing pressure. In practice, the polymer is often coated onto a porous substrate or packing at a temperature above the polymer melting temperature but at a pressure too low to effect solubilization. This results in a thin polymer coating on the packing. Fractions of varying molecular weight are collected by incrementally increasing pressure while maintaining a flow of SCF through the column. Propane has been used as a SCF solvent for separating polyethylene [110]. Propylene has been used in separating copolymers of ethylene and vinyl acetate [111]. Starting with a high density polyethylene parent material having a M_w/M_n, of 5.89, fourteen fractions having M_w/M_n from 1.30–2.11 were obtained [110].

VI. DETERMINATIONS OF MOLECULAR WEIGHT DISTRIBUTIONS BY SIZE EXCLUSION CHROMATOGRAPHY

A. Theory

Size exclusion chromatography (SEC), also known as gel permeation chromatography, is a method which allows polymer components to be separated based upon hydrodynamic size. SEC has become the more popular term because it is more descriptive of the separation process. The method uses a packed column of porous glass or a crosslinked gel (e.g., polystyrene crosslinked with divinyl benzene) as the separation medium. Separation is based upon the ability of molecules to permeate the microporous volume of the packed column. Small molecules are capable of permeating a larger number of pores and are, therefore, retained on the column longer. Large molecules are excluded from many of the pores and are thus eluted before the smaller molecules. The total volume which may be seen by the polymer molecule is that which is composed of the interstitial volume and the pore volume. All molecules must see the interstitial volume and a portion of the pore volume for separation to occur. If all of the molecules are so small that they see the entire volume or so large that they do not see any pore volume, then no separation will be accomplished.

A typical SEC system consists of a solvent reservoir, pump, injector, packed column or columns, and detectors as illustrated in Fig. 6. The most widely used packing material is polystyrene gel. Detection methods include those based on refractive index ultraviolet, infrared, viscosity and light scattering. The detectors must be very sensitive to small changes because low concentrations (0.1–2 wt%) of polymer are injected to prevent overloading of the columns.

In addition to steric exclusion, other secondary mechanisms of separation may be involved. The major secondary separation processes are adsorption, entrapment, degradation, ionic exclusion, and ionic attraction. However, with proper selection of packing material, solvent and flow rate these effects can usually be overcome.

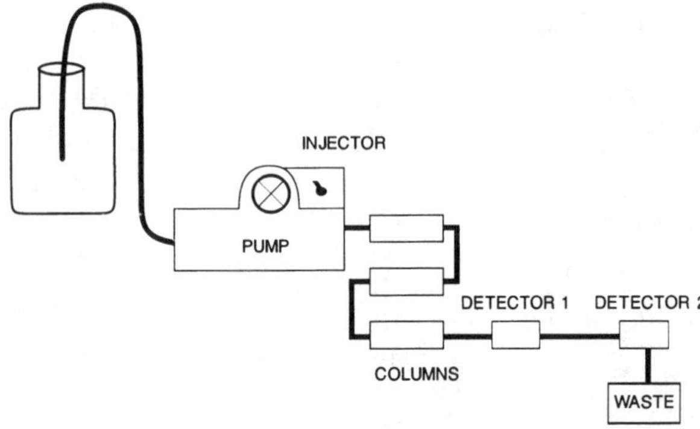

FIG. 6 Schematic diagram of an SEC system.

When the secondary effects are eliminated, a SEC system may be calibrated using standards of known molecular weight. After calibration the system may be used to very accurately determine the molecular weight and molecular weight distribution of a polymer sample. Calibration is always required unless a molecular weight sensitive detector is used to convert measured elution volumes to a molecular weight or mass.

The most widely accepted calibration method for SEC is "universal calibration" proposed by Grubisic and co-workers [112].

These investigators suggested that the product of a polymer's molecular weight and intrinsic viscosity, which is proportional to its hydrodynamic volume, could be used to characterize the separation phenomenon. They showed that a plot of log $(M[\eta])$ versus elution volume was linear and applicable to a variety of polymers. The calibration function may be described by

$$V_e = A + B \log(M[\eta]) \tag{18}$$

where A and B are empirical constants. For quantitative SEC, narrow molecular weight standards are evaluated and a calibration plot is constructed. Narrow distribution polystyrene standards are normally used for calibration but other narrow and broad molecular weight standards have been used.

Using the calibration and SEC data, the number-average and weight-average molecular weights described by Eqs. (1) and (2), respectively, can be calculated. The number of molecules within each fraction N_i is assumed to be proportional to h_i/M_i and is calculated from the concentration detector output and M_i calculated using

$$M_i = [(10^{(V_{ei}-A)/B})/K]^{[1/(1+a)]} \tag{19}$$

where

V_{ei} = elution volume of the ith increment
M_i = molecular weight of the ith increment
K, a = Mark–Houwink values
A, B = universal calibration constants
h_i = the detector output for the ith increment

Therefore a SEC system calibrated using the universal method can be a powerful tool to determine molecular weight and molecular weight distributions when Mark–Houwink values for the polymer in the SEC solvent are known.

The most common detectors used in SEC are based on refractive index and UV absorption. Recently online LALLS detectors have been introduced which allow molecular weight determinations using SEC without calibration. The LALLS detector is placed in series with a concentration detector so that $\bar{R}_\theta$ and C can be measured for each elution volume increment. Because the concentration of the polymer is very small, the scattering Eq. (7) reduces to

$$\frac{Kc}{\bar{R}_\theta} = \frac{1}{\bar{M}_w} \tag{20}$$

where K is the optical constant, c is the concentration, and $\bar{R}_\theta$ is the excess Rayleigh factor. The dn/dc value is assumed to be constant and not a function of molecular weight. Various molecular weight averages can be calculated using the general equation

$$M_x = \frac{\sum C_i M_i^x}{\sum C_i M_i^{x-1}} \tag{21}$$

where C_i is the concentration and M_i is the molecular weight of the ith increment (for $x = 0$, $M_x = \bar{M}_n$; $x = 1$, $M_x = \bar{M}_w$; $x = 2$, $M_x = \bar{M}_z$).

In addition to molecular weight, LALLS detectors in conjunction with SEC may be used to estimate polymer branching. The branching index based upon viscosity [113] may be defined by

$$g' = ([\eta]_{br}/[\eta]_l)_M \tag{22}$$

where $[\eta]_{br}$ and $[\eta]_l$ are the intrinsic viscosities of the branched and linear polymer of equivalent molecular weight. If the universal calibration is valid for the polymers under investigation, the branching index may be determined from a SEC experiment with online LALLS using

$$g' = (M_l/M_{br})_{V_e}^{a+1} \tag{23}$$

where M_l and M_{br} are the molecular weight of the linear and branched polymer eluting at the same V_e. The exponent a is the Mark–Houwink exponent for the linear polymer.

Another recent improvement in detectors is an online viscometery detector. Using the pressure drop across a small capillary and the output of an in-series concentration detector, intrinsic viscosities can be calculated as a function of elution volume. These measurements allow viscosity-average molecular weights to be calculated for each elution volume and the construction of a molecular weight distribution. This detector also may be used to characterize branching.

B. Size Exclusion Chromatography of Polyolefins

As with classical fractionation, SEC of polyolefins must be conducted at elevated temperatures. Typically SEC of polyethylene and polypropylene is carried out at 145–150°C. Common solvents used are TCB and 1,2-dichlorobenzene for polypropylene, and TCB and 1-chloronaphthalene for polyethylene. Some problems with molecular weight determinations have been reported because of the formation of large stable aggregates in polyethylene and polypropylene solutions. These aggregates can be trapped in the SEC columns and lead to large differences in calculated molecular weight averages. However, methods to prepare aggregate free solutions have been described and good agreement between static light scattering and SEC measurements have been obtained [9–11]. In addition the universal calibration method has been shown to be applicable to polypropylene in 1,2-dichlorobenzene at 135°C [41]. In this study polystyrene and fractionated polypropylene samples were shown to fall on the same calibration line. SEC has also been coupled with LALLS and viscosity detectors to characterize branching of polyolefins including ethylene–propylene copolymers [77, 114, 115].

C. Special Problems in SEC of Ultra-High Molecular Weight Polyethylene UHMWPE

There are many problems associated with analysis of the average molecular weights and molecular weight distribution of UHMWPE. These include: (1) degradation during dissolution of this polymer at elevated temperatures; (2) problems with the polymer not dissolving completely due to its high molecular weight and high degree of crystallinity; (3) shear degradation of UHMWPE during SEC analysis; (4) difficulty in establishing a calibration curve that is valid for UHMWPE.

As discussed previously, in Sec. II.B, Wagner and Dillon [8] have described a satisfactory method for preparing solutions of this polymer. More recently, Xu *et al.* [116] dissolved UHMWPE in dichlorobenzene (containing 0.125% w/v BHT) by heating at 170°C for 6–8 h. Solutions were not filtered prior to SEC analysis because of concerns over possible removal of high molecular weight species during the filtration process. These authors also conducted SEC analysis at 170°C, instead of the usual 135–145°C used for conventional PE analysis. This was done because UHMWPE is not completely soluble in chlorinated aromatic solvents even at 145–150°C due to their high degree of crystallinity [116]. Commercial SEC systems are presently available that allow analyses to be conducted at temperatures as high as 210°C. To confirm that SEC analysis of PE could be conducted at 170°C without causing thermal degradation, Xu *et al.* analyzed LLDPE at 135, 160, and 170°C, obtaining identical results for average molecular weights and polydispersities.

As pointed out by deGroot and Hamre, degradation of high molecular weight PE can be a problem in SEC unless large particle size columns with low flow rates are employed [37]. Xu and coworkers used 10 μm particle size packings at a flow rate of 1 mL min^{-1}; these conditions would be expected to lead to shear degradation and falsely low molecular weights based on the findings of de Groot and Hamre [37]. Numerous other authors have noted problems with shear degration during SEC analysis of very high MW polymers. Giddings [117] has published a thorough analysis of fundamental obstacles in SEC of ultrahigh molecular weight polymers. He suggests that field flow fractionation (FFF) techniques offer major advantages for analysis of very high molecular weight macromolecules because of the low shear rates encountered during FFF experiments.

Calibration of the SEC column is a major problem during analysis of UHMWPE. Linear PE standards having very high molecular weights are not available. UHMWPE often has high MW components present with $M > 10^7$. Xu *et al.* [116] extrapolated linearly a PE calibration curve based on standards having molecular weights of 120,000 or less. As they noted [116], such an extrapolation is hazardous because calibration curves are rarely linear at molecular weights much higher than that of 10^6. Some alternative calibration procedures include universal calibration and the use of light scattering detectors. Polystyrene standards are commercially available with molecular weights up to around 2×10^7. However, the use of Mark–Houwink coefficients established over a lower molecular weight range for ultrahigh molecular weight materials requires extrapolation, which can introduce errors. The better approach appears to be the use of direct light scattering measurements of M_w of the eluent at various elution volumes, thus avoiding the need for calibration. A problem exists here when LALLS photometers are used; namely, the particle scattering function ($P(\theta)$) can often *not* be taken as unity at a small but finite angle because of the large radii of UHMWPE molecules [see Eq. (8)]. Thus, the use of a MALLS photometer, which allows extrapolation to infinitely small angle, is preferred for application to UHMWPE.

VII. TEMPERATURE RISING ELUTION FRACTIONATION

A rather new characterization method for complex polymer compositions is temperature rising elution fractionation (TREF). An excellent review of the method and application has been given by Glockner [118]. The basis of TREF is the relationship between structure, crystallinity, and dissolution temperature. Higher degrees of crystallinity require higher dissolution temperatures. The method is especially useful for studying short chain branching (SCB) distributions for polyethylene. Branching is present in many polyethylenes and creates significant problems for determining molecular weight from viscosity or SEC. Branching has been shown to increase the breadth of the molecular weight distributions up to $\bar{M}_w/\bar{M}_n = 20$ [32].

TREF is very similar to the column fractionation technique discussed earlier. A diagram of the basic apparatus is shown in Fig. 7. The support is an inert material such as sand or glass beads. The polymer is dissolved at elevated temperatures and introduced into a heated column at which time the flow is stopped and the temperature decreased at a controlled rate. As the temperature is dropped the polymer is thought to deposit on the column as illustrated in Fig. 8. The potential for crystal formation is largely determined by the SCB content with more highly branched fractions forming crystals at lower temperatures and are the last to be deposited on the support. Solvent is then passed through the column and the temperature

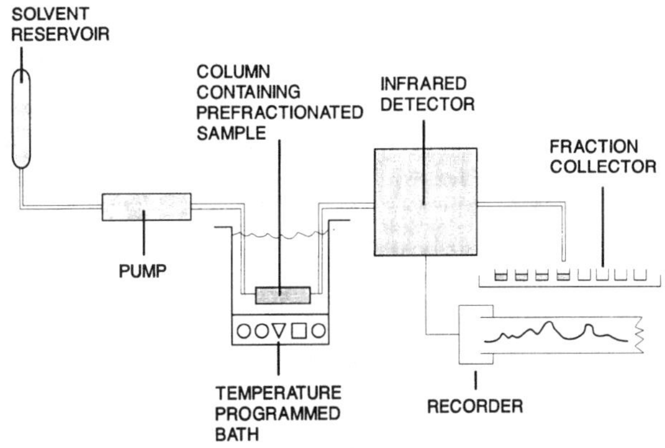

FIG. 7 Schematic diagram of a TREF apparatus. (Reprinted with permission from Ref. 119.)

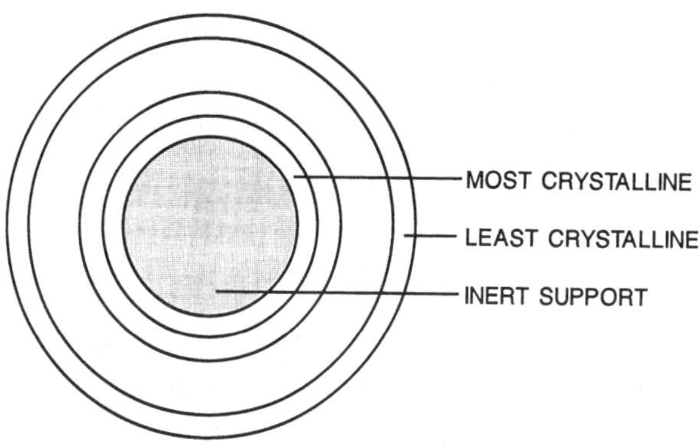

FIG. 8 Conceptual illustration of the layers formed on a spherical support after precipitation or crystallization in a TREF column. (Reprinted with permission from Ref. 119.)

increased at a controlled rate to elute fractions. Fractions are collected as a function of temperature. Usually an IR detector is used monitoring the C-H stretch at 3.41 μm. Fractions are eluted with increasing crystalline contents as a function of time and temperature.

The effects of molecular weight on separation has been shown to be negligible over a wide molecular weight range of linear polyethylenes [119]. The reason the separation is not molecular weight dependent has been ascribed to a very controlled crystallization process from dilute solution. By using a very slow and controlled decrease in temperature (in some cases over the course of days), the crystallization process is more thermodynamically controlled and the kinetic effect of chain length on crystallization is diminished.

The initial use of analytical TREF was to characterize polyethylene fractions [120]. Mirabella and Ford used TREF to fractionate LDPE and LLDPE for further characterization using SEC, X-ray ^{13}C-NMR, DSC and viscosity measurements [121]. These investigators determined that the melting behavior of LLDPE correlated well with a multimodal SCB distribution and the distribution became narrower with increasing molecular weight. Mingozzi and collaborators [122] have applied both analytical and preparative TREF to the analysis of tacticity distribution in polypropylene. The results showed that both methods could be used successfully: analytical TREF gave faster qualitative results on the polymer microstructure, while preparative TREF, with subsequent analysis of the fractions, could yield detailed quantitative tacticity information. Hazlitt has described an automatic TREF instrument to measure SCB distributions for ethylene/α-olefin copolymers [123].

Development of high performance analytical TREF for polyolefin analysis has recently been reported by Wild and Blatz [124]. Their goal was to develop TREF to the point where it is economical, fast, and easy to carry out, making it approach analytical SEC in convenience. These authors:

1. Conducted crystallization in an oil bath containing multiple dilute polymer solutions. The bath is slowly cooled to $< 0°C$.
2. The crystallized polymer was loaded by co-filtering the polymer/solvent mixture directly into the elution column.
3. Eution is conducted in a programmed GC oven capable of operation from below room temperature.
4. Used a matched reference column with embedded temperature probe to monitor actual temperature.
5. Infrared detection was used rather than refractive index detection because of superior baseline stability.
6. Computerized data acquisition and analysis was employed.

Soares and Hamielec recently reviewed TREF of polyolefins [125] and compared analytical and analytical TREF (Table 6). TREF is a powerful tool that can be used to fractionate complex polymer systems according to composition, tacticity, and crystallinity. Once the materials have been fractionated, conventional methods may be used to characterize the molecular weight and molecular weight distribution of the fractions. Recently Dawkins and Montenegro have coupled TREF with SEC to characterize styrene/

TABLE 6 Comparison of Preparative and Analytical TREF

Preparative TREF	Analytical TREF
Fractions are collected at predetermined temperature intervals	Continuous operation
Information about structure is obtained off-line by other analytical techniques	Information about structure is obtained online via a calibration curve
Requires large columns and large sample sizes	Requires small columns and small sample sizes
Time-consuming but can yield detailed information about polymer microstructure	Fast but generates less information about polymer microstructure

butyl methacrylate copolymers [126]. More, recently, Lederer and coworkers [127] have coupled preparative TREF with SEC/LALLS in characterizing the distribution of molar mass and comonomer content (w_0) of medium density polyethylenes. They concluded that these experiments lead to poor resolution of the MWD and w_0. Better results were obtained by molecular weight fractionation (the Holtrup method [94]), followed by TREF fractionation.

Watkins and coworkers [110] recently described a supercritical fluid version of TREF: critical solvent, isobaric temperature rising elution fractionation (CITREF). In this process, a fractionation is performed by passing a SCF through the columns, which contain deposited polymer, at a constant pressure, while incrementally increasing the temperature. Thus, this process separates on the basic of melting point. Watkins *et al.* (110) demonstrated that by combining isothermal increasing pressure SCF fractionation, which separates based on molecular weight, with CITREF, which separates on the basis of crystallinity, it is possible to obtain PE fractions that have narrow distributions of both molecular weight and chain branching [108].

Quite recently, Karoglanian and Harrison [128] noted the similarity of compositional distribution information generated by differential scanning calorimetry (DSC) and TREF. They showed that DSC thermograms could be generated from TREF chromatograms, and vice versa.

VIII. CHARACTERIZATION OF LONG CHAIN BRANCHING

Long chain branching occurs in many polyolefins, e.g., LDPE. The presence of long chain branching affects the solution properties of the polymer, as well as its properties in the melt and in the solid state. Since long chain branching causes decreases in the dimensions of a polymer relative to that of the linear material having the same composition and molecular weight, dilute solution properties may be used to characterize the extent of branching [113, 129, 130]. A parameter that reflects the size difference directly is the "g parameter",

$$g = \left(\frac{R^2_{g'\mathrm{br}}}{R^2_{g'\mathrm{l}}} \right) \tag{24}$$

$R_{g'\mathrm{br}}$ and $R_{g'\mathrm{l}}$ are the radii of branched and linear polymers, respectively, compared at the same molecular weight, M. In general, the greater the degree of branching the lower the value of g. Theoretical expressions are available which allow one to compute the number of branch points per molecule from the measured value of g [113, 129–131]. In practice this method is difficult to apply to crystalline polyolefins for several reasons. Firstly, it is difficult to accurately measure R_g of such materials due to problems in working at elevated temperatures. Secondly, most branched polyolefins have very broad molecular weight distributions, and the measured R_g is a z-average parameter. It is well known [132] that $R_{g,z}/M_w$ is virtually independent of the level of branching in such polymers. Thirdly, our efforts to find reliable relationships between R_g and M for linear PE and PP in the literature were not successful. While some data exist, they either do not yield reliable scaling exponents (υ) in the $R_g \propto M^{\upsilon}$ relationship, they do not correct for polydispersity effects in this relationship, and/or they do not give realistic values of the Flory–Fox parameter, Φ, which is a function of both R_g and $[\eta]$ [133].

The intrinsic viscosity is also reduced by the presence of branching. The parameter g',

$$g' = \left(\frac{[\eta]_{\mathrm{br}}}{[\eta]_{\mathrm{l}}} \right)_M \tag{25}$$

compares $[\eta]$ values for branched polymer ($[\eta]_{\mathrm{br}}$) to linear polymer ($[\eta]_{\mathrm{l}}$) at the same molecular weight. The smaller the g' value, the higher the degree of branching. The g' parameter is much easier to measure experimentally than is g. Furthermore, reliable Mark–Houwink coefficients are available for many polyolefin/solvent systems (Table 3). The only limitation of using the g' parameter to evaluate the number of branch points (λ) present in a molecule is that theoretical relationships g' and λ, are only available for star polymers under theta conditions [130].

The relationship between g' and g is given by

$$g' = g^{\varepsilon} \tag{26}$$

where ε is predicted theoretically to have values between $\frac{1}{2}$ and $\frac{3}{2}$ [130, 134]. Experimentally, most values of E are found to be in the range of 0.5–1 [135, 136]. Once the value of ε is known for a particular system, g can be calculated from g' and then λ can be computed using theory [113].

The use of SEC-based methods for characterizing long chain branching offers advantages of speed, automation, and small sample size requirements. Drott and Mendelson [21], Cote and Shida [137], and Kurata *et al.* [138] have developed methods for determining the degree of branching from the SEC chromatogram and the intrinsic viscosity of the whole polymer using iterative procedures. More common nowadays is the use of light scattering or intrinsic viscosity detectors online for detecting and quantifying long chain branching. If universal calibration is valid for the polymer under investigation, the branching index g' may be determined from an SEC experiment using online light scattering using

$$g' = (M_{\mathrm{l}}/M_{\mathrm{br}})_{V_e}^{a+1} \tag{27}$$

where M_{l} and M_{br} are the molecular weights of the linear and branched polymers eluting at the same V_e. The exponent a is the Mark–Houwink exponent for the linear polymer. If MALLS detection is used online and both the branched polymer and a linear reference material are analyzed, g can be evaluated as a function of molecular weight [139].

REFERENCES

1. Lord Rayleigh. Phil Mag 41: 447, 1871.
2. A Einsteine. Ann Physik 33: 1275, 1910.
3. P Debye. J Appl Phys 15: 338, 1944; J Appl Phys 17: 392, 1946; J Phys Colloid Chem 51: 18, 1947.
4. BH Zimm, PM Doty. J Chem Phys 12: 203, 1944.
5. PM Doty, BH Zimm, H Mark. J Chem Phys 13: 159, 1945.
6. RS Stein, P Doty. J Am Chem Soc 69: 1193, 1947.
7. BH Zimm, J Chem Phys 16: 1099, 1948.
8. HL Wagner, JG Dillon. J Appl Polym Sci 36: 567, 1988.
9. V Grinshpun, KF O'Driscoll, A Rudin. J Appl Polym Sci 29: 1071, 1984.
10. V Grinshpun, KF O'Driscoll, A Rudin. ACS Symp Ser 245: 274, 1984.
11. V Grinshpun, A Rudin. J Appl Polym Sci 30: 2413, 1985.
12. LA Utracki, MM Dumoulin. ACS Symp Ser 245: 91, 1984.
13. R Lew, D Suwanda, ST Balke. J Appl Polym Sci 35: 1049, 1988.
14. R Chiang In B Ke, ed. *Newer Methods of Polymer Characterization*, New York: Interscience, p 471, 1964.
15. P Kratochvil. *Classical Light Scattering from Polymer Solutions*, Amsterdam: Elsevier, 1987.
16. J Horska, J Stejskal, P. Kratochvil. J Appl Polym Sci 24: 1845, 1979.
17. M Hert, C Strazielleo. Makromol Chem 184: 135, 1983.
18. V Grinshpun, KF O'Driscoll, A Rudin. Am Chem Soc Symp Ser 245: 273, 1983.
19. TB MacRury, ML McConnell. J Appl Polym Sci 24: 651, 1979.
20. RK Agarwal, J Horska, J Stejskal, O Quadrat, P Kratochvil, P Hudec. J Appl Polym Sci 28: 3453, 1983.
21. EE Drott, RA Mendelson. J Polym Sci Part A-2 8: 1373, 1970.
22. J Horska, J Stejskal, P Kratochvil. J Appl Polym Sci 28: 3873, 1983.
23. H Coll, DK Gilding. J Polym Sci Part A-2 8: 89, 1970.
24. NE Weston, FW Billmeyer, Jr. J Phys Chem 65: 576, 1961.
25. R Chiang. J Polym Sci 28: 235, 1958.
26. S Florian, D Lath, Z Manasek. Angew Makromol Chem 13: 43, 1970.
27. JW Mays, LJ Fetters. Macromolecules 22: 921, 1989.
28. Z Xu, JW Mays, X Chen, N Hadjichristidis, F Schilling, HE Bair, DS Pearson, LJ Fetters. Macromolecules 18: 2560, 1985.
29. T Matsumato, N Nishioka, H Fujita. J Polym Sci Polym Phys Ed. 10: 23, 1972.
30. LJ Fetters, N Hadjichristidis, JS Lindner, JW Mays, WW Wilson. Macromolecules 24: 3127, 1991.
31. *Polymer Handbook, 3rd Edition.* J Brandrup and E. H. Immergut, eds. New York: Interscience, Section VII, 1989.
32. FW Billmeyer. J Am Chem Soc 75: 6118, 1953.
33. LD Moore. J Polym Sci 20: 137, 1956.
34. QA Trementozzi. J Polym Sci 23: 887, 1957.
35. R Chiang. J Polym Sci 36: 91, 1959.
36. T Housaki, K Satoh. Makromol Chem Rapid Commun 9: 257, 1988.

37. AW deGroot, WJ Hamre. J Chromatogr 648: 33, 1993.
38. L Jeng, ST Balke, TH Mourey, L Wheeler, P Romeo. J Appl Polym Sci 49: 1359, 1993.
39. U Dayal. J Appl Polym Sci 53: 1557, 1994.
40. E Nordmeier, U Lanver, MD Lechner. Macromolecules 23: 1072, 1990.
41. CML Atkinson, R Dietz. Makromol Chem 177: 213, 1976.
42. F Danusso, G Moraglio. Makromol Chem 28: 250, 1958.
43. JB Kinsinger, RE Hughes. J Phys Chem 63: 2002, 1959.
44. P Parrino, F Sebastiano, G Messina. Makromol. Chem. 38: 27, 1960.
45. H Inagaki, T Miyamoto, S Ohta. J Phys Chem 70: 3420, 1966.
46. WR Krigbaum, JE Kurz, P Smith. J Phys Chem 65: 1984, 1961.
47. B Chu, M Onclin, JR Ford. J Phys Chem 88: 6566, 1984.
48. JW Pope, B Chu. Macromolecules 17: 2633, 1984.
49. M Helmstedt. Makromol Chem Macromol Symp 18: 37, 1988.
50. M Helmstedt. Prog Coll Polym Sci 91: 127, 1993.
51. C Wu, D Lilge. J Appl Polym Sci 50: 1753, 1993.
52. C Wu. J Appl Polym Sci 54: 969, 1994.
53. R Pecora. Dynamic Light Scattering, New York: Plenum, 1976.
54. RB Flippen. In: HG Barth, JW Mayes, eds. Modern Methods of Polymer Characterization, New York: John Wiley, 1991.
55. CA Glover. Cryoscopy In: PE Slade, ed. *Polymer Molecular Weights*, New York: Marcel Dekker, Chap. 4, 1975.
56. G Davison. Ebullioscopic methods for molecular weights. In: LS Bark, NS Allen, eds. *Analysis of Polymer Systems*, London: Applied Science, chap. 7, 1982.
57. TG Fox, PJ Flory, AM Bueche. J Am Chem Soc 73: 285, 1951.
58. WR Krigbaum, PJ Flory. J Polym Sci 9: 503, 1952.
59. WR Krigbaum, PJ Flory. J Am Chem Soc 75: 1775, 1953.
60. EH Immergut, S Rollin, A Salkind, H Mark. J Polym Sci 12: 439, 1954.
61. FM Mirabella, J Appl Polym Sci 25: 1775, 1980.
62. FW Billmeyer, V Kokle. J Am Chem Soc 86: 3544, 1964.
63. AH Wachter, W Simon. Anal Chem 24: 90, 1969.
64. DE Burge. J Appl Polym Sci 24: 293, 1979.
65. JW Mays, EG Gregory. J Appl Polym Sci 34: 2619, 1987.
66. H Mark. Der Feste Korper, Leipzig, 1938.
67. R Houwink. J Prakt Chem 157: 15, 1940.
68. PJ Flory. *Principles of Polymer Chemistry*, Ithaca, NY: Cornell University Press, 1953, p 313.
69. A Nakajima, F Hamada, S. Hayashio. J Polym Sci Part C 15: 285, 1966.
70. HL Wagner. J Phys Chem Ref Data 14: 611, 1985.
71. R Chiang. J Phys Chem 69: 1645, 1965.
72. A Peyrouset, R Prechner, R Panaris, H Benoit. J Appl Polym Sci 19: 1363, 1979.
73. TG Scholte, NLJ Meijerink. Br Polym J 133: June, 1997.
74. HL Wagner, CAJ Hoeve. J Polym Sci Polym Phys Ed. 11: 1189, 1973.
75. JL Atkins, LT Muus, CW Smith, ET Pieski. J Am Chem Soc 79: 5089, 1957.
76. HS Kaufman, EK Walsh. J Polym Sci 26: 124, 1957.
77. TG Scholte, NLJ Meijerink, HM Schoffeleers, AMG Brands. J Appl Polym Sci 29: 3763, 1984.
78. V Grinshpun, A Rudin. Makromol Chem Rapid Commun 6: 219, 1985.
79. G Moraglio. Chim Ind Milan 41: 879, 1959.
80. P Hattam, S Gauntlett, JW Mays, N Hadjichristidis, RN Young, LJ Fetters. Macromolecules 24: 6199, 1991.
81. GV Schultz, Z Phys Chem B46: 173, 1940; B47: 155, 1940.
82. H Tompa. Polymer Solutions, London: Butterworths, chap 7, 1956.
83. MJ Voorn. Adv. Polym. Sci. 1: 192, 1959.
84. ML Huggins, H Okamoto. In: MJR Canton, ed. Polymer Fractionation, New York: Academic Press, 1966.
85. F Francuskiewicz. Polymer Fractionation. Berlin: Springer-Verlag, 1994.
86. LH Tung. J Polym Sci 61: 449, 1962.
87. K Kamide. In: LH Tung, ed. Fractionation of Synthetic Polymers, New York: Marcel Dekker, chap 2, 1977.
88. R. Koningsveld, AJ Staverman. J Polym Sci Part A-2 6: 367–383, 1968.
89. K Kamide, Y Migazaki. Polym J 12: 205, 1980.
90. GV Schulz, A Denlingin. Z Phys Chem B 43: 47, 1939.
91. LH Tung. In: MJR Cantow, ed. Polymer Fractionation, New York: Academic Press, chap 8, 1967.
92. G Beall. J Polym Sci 4: 483, 1949.
93. R Koningsveld, CAF Tuijnman. J Polym Sci 39: 445, 1959.
94. W Holtrup. Makromol Chem 178: 2335, 1977.
95. LH Tung. In: JI Kroschwitz, ed. Encyclopedia of Polymer Science and Technology, 2nd ed., New York: Wiley, vol 7, 1987, pp 298–326.
96. RAV Raff, KW Dock eds. Crystalline Olefin Polymers, New York: Wiley-Interscience, p 555, 1965.
97. RB Richards. Trans Faraday Soc 42: 10, 1946.
98. A Nasinic, C Mussa. Makromol Chem. 22: 59, 1957.
99. H Wesslau. Makromol. Chem. 26: 96, 102, 1958.

100. V Desreux, MC Spiegels. Bull Soc Chim Belges 59: 476, 1950.
101. RW Ford, JD Ilavsky. J Appl Polym Sci. 12: 2299, 1968.
102. PS Francis, RC Cooke Jr., JH Elliot. J Polym Sci 31: 453, 1958.
103. TE Davis, RL Tobias. J Polym Sci 50: 227, 1961.
104. RA Mendelson. J Polym Sci Al: 2361, 1963.
105. RS Porter, MJR Cantow, JF Johnson. Makromol. Chem 94: 143, 1966.
106. K Yamaguchi. Makromol. Chem. 132: 143, 1970.
107. T Ogawa, Y Suzuki, T Inaba. J Polym Sci A1 10: 737, 1972.
108. MA McHugh, VJ Krukonis, JA Pratt. Trends Polym Sci 2(9): 301, 1994.
109. MA McHugh, VJ Krukonis. Supercritical Fluid Extraction, Principles and Practice, 2nd ed. Boston: Butterworth-Heinemann, 1994.
110. JJ Watkins, VJ Krukonis, PD Condo Jr., D Pradhan, P Ehrlich. J Supercrit Fluids 4: 24, 1991.
111. B Folie, M Kelchtermans, JR Shutt, H Schonemann, V Krukonis. J Appl Polym Sci 64: 2015, 1997.
112. Z Grubisic, P Rempp, H Benoit. J Polym Sci B5: 753, 1967.
113. BH Zimm, WH Stockmayer. J Chem Phys 17: 1301, 1949.
114. DE Axelson, WC Knapp. J Appl Polym Sci 25: 119, 1980.
115. LI Kulin, WL Meijerink, P Starck. Pure Appl. Chem 60: 9, 1403, 1988.
116. J Xu, P Ji, J Wu, M Ye, L Shi, C Wan. Macromol Rapid Commun 19: 115, 1998.
117. JC Giddings. Adv Chromatogr 20: 217, 1982.
118. G Glockner. J Appl Polym Sci App Polym Symp 45: 1, 1990.
119. FM Mirabella Jr. Proceedings of the International GPC Symposium 87, May 11–13, 1987, Itasca, IL, pp 180–198.
120. L Wild, TR Ryle. Polym. Prep Am Chem Soc Polym Chem Div 18: 182, 1977.
121. FM Mirabella Jr., EA Ford. J Polym Sci Part B: Polym Phys 25: 777, 1987.
122. I Mingozzi, G Cecchin, G Morini. Int J Polym Anal Charact 3: 293, 1997.
123. LG Hazlitt. J Appl Polym Sci Appl Polym Symp 45: 25, 1990.
124. L Wild, C Blatz. Development of high performance TREF for polyolefin analysis. In TC Chung, ed. New Advances in Polyolefins, New York: Plenum: 1993, pp 147–157.
125. JBP Soares, AE Hamielec. Polymer 36: 1639, 1995.
126. JV Dawkins, AMC Montenegro. Br Polym J 21: 31, 1989.
127. N Aust, I Beytollahi-Amtmann, K Lederer. Int J Polym Anal Charac 1: 245, 1995.
128. SA Karoglanian, IR Harrison. Thermochim Acta 288: 239, 1996.
129. WH Stockmayer, M Fixman. Ann NY Acad Sci 57: 334, 1953.
130. BH Zimm, R Kilb. J Polym Sci 37: 19, 1959.
131. EF Casassa, GC Berry. J Polym Sci A-2 4: 881, 1966.
132. W Burchard. Adv Polym Sci 48: 1, 1983.
133. Y Fujii, Y Tamai, T Konishi, H Yamakawa. Macromolecules 24: 1608, 1991.
134. GC Berry. J Polym Sci A-2 9: 687, 1971.
135. JW Mays, N Hadjichristidis. J Appl Polym Sci Appl Polym Symp 51: 55, 1992.
136. J Roovers. Encyclopedia of Polymer Science and Engineering, vol. 2, New York: Wiley, 1985, 487.
137. JA Cote, M Shida. J Polym Sci A-2 9: 421, 1971.
138. M Kurata, M Abe, M Iwama, M Matsushima. Polym. J. 3: 729, 1972.
139. C Jackson, YJ Chen, JW Mays. J Appl Polym Sci 5: 179, 1996.

15

Thermodynamic Properties of Polyolefin Solutions

Luc van Opstal
BASF Antwerpen, Antwerpen, Belgium

Erik Nies
University of Technology, Eindhoven, The Netherlands

Ronald Koningsveld
Max Planck Institute for Polymer Research, Mainz, Germany

I. INTRODUCTION

Polymer solutions and blends play an important role in polymer science and technology and the basic foundations of their properties, such as thermodynamics and rheology, require fundamental as well as applied studies. In this chapter we concentrate on the thermodynamic properties of polyolefin solutions which are important for understanding the state of systems used in bulk or solution polymerization, and also for the design of fractionation procedures with the objective to prepare fractions large enough to allow the study of their properties.

In bulk and solution polymerization it is an advantage for process designers to have a knowledge of the phase separations that may occur during production. Such phenomena are mostly undesired and the best way to find methods to avoid them rests in a thorough study of the thermodynamic properties of the system in hand. This chapter provides the tools for such an approach. It is demonstrated that molecular models exist, capable of supplying a nearly quantitative description of available data on partial miscibility of polyolefin solutions. Such models cannot by themselves predict phase behavior; some measured data will have to be fitted to the model for a given system, after which much experimental time can be saved letting the model's predictions guide further experimentation.

Bulk and solution polymerization are both in industrial use in the synthesis of polyolefins, and lead to products that contain many components. These components may differ on several points among which the chain length is the foremost, and its distribution in polyolefins may vary largely in width. A number of processing and product properties are influenced by the chain-length distribution that therefore needs to be investigated. Fractionation techniques must be applied for the purpose, and are based on the molar mass dependence of physical properties like miscibility or solubility.

Until about 40 years ago, the determination of the molar mass distribution was the prime objective of polymer fractionation by phase separation. Since the advent of size exclusion chromatography the need for such time consuming and essentially ineffective classic fractionation methods has disappeared. However, today chromatographic techniques are still incapable of producing large-size fractions of the order of 100 g and phase separation methods continue to be needed for the preparation of large amounts of narrow-distribution polymers that cannot be obtained by direct synthesis. The most important polyolefins, polyethylene and polypropylene, fall into this category and a discussion of fractionation by distribution between two liquid phases, or between a liquid and a crystalline phase, is appropriate.

Most of the experimental examples cited in this chapter refer to polyethylene because that very important industrial product has received most attention in the literature. The treatments of fractionation and liquid–liquid separation as such, developed here, are completely general and can be applied to other polyolefins if need be.

Fractionation procedures must be amenable to evaluation and criteria for width and location of a molar mass distribution must be defined. Such criteria exist, e.g., in the ratio of weight- and number-average molar masses: M_w/M_n, and the ratio of centrifuge- and weight average molar masses: M_z/M_w. The ratio M_w/M_n can be shown to be related to the variance, σ_n^2, of the distribution of the total amount in moles among the chain lengths present in the sample:

$$\sigma_n^2/M_n = M_w/M_n - 1 = U_n \tag{1}$$

Similarly, the variance of the mass distribution is given by

$$\sigma_w^2/M_w = M_z/M_w - 1 = U_w \tag{2}$$

Schulz introduced the "*Uneinheitlichkeit*", U [1], equal to zero for strictly single-component substances. It is seen from these equations that $M_z \geq M_w \geq M_n$, the equalities referring to strictly monodisperse samples containing a single component only. Large values of the ratio M_w/M_n, such as 10 or more, are not uncommon for polyolefins, but synthesis conditions also exist leading to values of about 2. Even then, there are still many thousands of homologous components in the polymer [2, 3], which jeopardize unambiguous studies of properties as a function of molar mass.

There is an extensive literature on fractionation which we do not attempt to review here. The reader is referred to a number of review articles [1–7].

II. PREPARATIVE FRACTIONATION OF POLYOLEFINS

A. Fractionation Procedures

Strictly one-component samples do not exist in synthetic polymer practice. There are various ways in which the individual components in a polymer may differ. If all the chains are linear, they will show a spectrum of chain lengths, the molar mass distribution (mmd). Other molecular characteristics that add to the multicomponent nature of polymers are: molecular architecture (short- and/or long-chain branching), and co- or termonomer distribution in co- or terpolymers. In this section the emphasis is on the distribution in chain length; the other aspects of polydispersity will be touched upon in the next section.

If solubility or miscibility are the properties used to separate components in a multicomponent polymer, one cannot expect the separations to be very effective. The relative differences between the neighboring chain lengths are extremely small, and so are the differences in the property used. Nevertheless, procedures have been developed that still permit the isolation of reasonably sharp fraction distributions, and are in use, particularly to prepare large-size polyolefin fractions. We discuss these procedures briefly: they are fractionation by liquid–liquid phase separation, and fractionation by crystallization from solution.

B. Liquid–Liquid Phase Separation

Fractionation by liquid–liquid phase separation relies on the molar mass dependence of the distribution coefficient which, in terms of the Flory–Huggins–Staverman theory [8–13] (FHS), reads

$$\phi_{ib}/\phi_{ia} = e^{\sigma m_i} \tag{3}$$

where ϕ_{ia} and ϕ_{ib} are the volume fractions of component i of the polymer in phase a and b, respectively. The separation parameter σ depends on parameters of the FHS model, and one the whole polymer concentrations (volume fractions) ϕ_a and ϕ_b in the two phases, and m_i stands for the length of the macromolecular chain i, relative to the size of the solvent molecules. The theory predicts the two phases to represent one polymer solution (a), which is very dilute and is prefered by the shorter chains in the mmd, and another one (b) which is a concentrated solution (typical ϕ_b values are 0.1–0.3). Hence, the miscibility gap in a polymer solution is asymmetric and located in the solvent-rich part of the phase diagram.

Setting up the material balance for each of the components one obtains x_b, the relative amount of polymer in the concentrated phase b,

$$x_b = \int w_b(M)\,dM = \int w(M)(1 + re^{-\sigma m})^{-1}\,dM \tag{4}$$

where $w(M)$ is the mass distribution, normalized to 1 g of whole polymer, $w_b(M)$ is the mass distribution of the fraction in the concentrated phase. It follows that $0 \leq x_b \leq 1$. The fraction in the dilute phase, x_a, equals $1 - x_b$. The symbol r stands for the ratio of the volumes of the two phases: $r = V_a/V_b$. Since the initial polymer concentrations are usually small, the r values encountered in fractionation experiments are large, the concentrated phase always having the smaller volume.

The FHS model relates the relative chain length, m, to molar mass M by

$$m = v_p M / V_s \tag{5}$$

where v_p is the specific volume of the polymer in solution (assumed independent of M), and V_s is the molar volume of the solvent.

Assuming model distribution functions for $w(M)$ one can calculate the efficiency of fractionation schemes, two of which are being considered here:

(a) *Precipitation fractionation*, in which the separation is effected in such a fashion that x_b is small, say < 0.1. The fraction is the material in the concentrated phase.

(b) *Extraction fractionation*, in which the opposite end of the x_b scale is employed ($x_b > 0.9$). The fraction is now the material present in the dilute phase.

With precipitation the fractionation starts at the high molar mass end of the mmd; with extraction the lower chain-length fractions are isolated first.

We show in Fig. 1 two typical examples of the results such model calculations lead to. The original distribution functions $w(M)$ are asymmetrical, as often found in polyolefins, and we present two M_w/M_n ratios (10 and 2) which are values arising frequency in olefin polymerization. The M_w values are 132 kg mol^{-1}, and the M_z/M_w ratios equal the M_w/M_n values.

We note that precipitation fractionation (low end of the x_b scale) does not lead to favorable results (low M_w/M_n and M_z/M_w ratios), unless extremely low initial polymer concentrations are used, which is most undesirable from a preparative standpoint. At more realistic initial concentrations the even less attractive result prevails that the fraction obtained by precipitation may have a wider distribution than the parent polymer. Such effects are related to the asymmetry of the initial mmd [6, 13].

Extraction, on the other hand, arises from the calculation as the obvious procedure to use in preparative fractionations, the resulting fraction distribution having quite a small width which, as an additional bonus for preparative purposes, does not depend greatly on the initial polymer concentration.

The essentials of these results, obtained by model calculation, have been verified by experiments on the system diphenylether/poly(ethylene) (Fig. 2). Fractionation calculations have been performed by

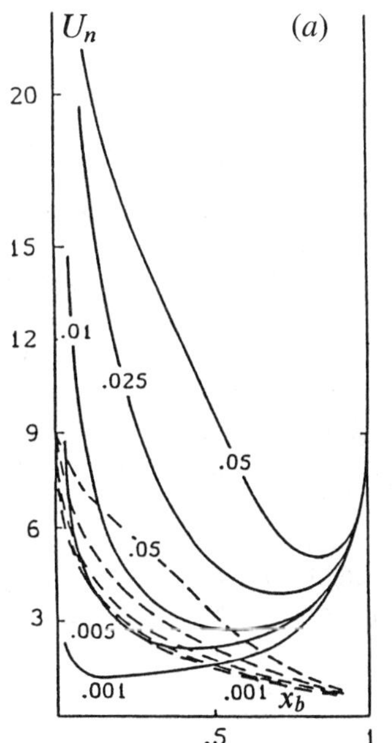

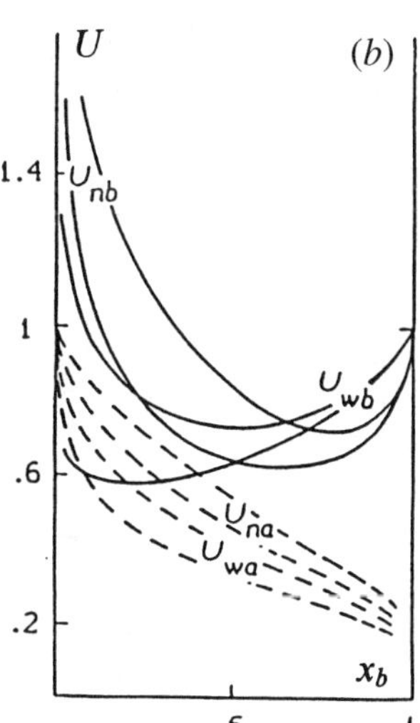

FIG. 1 Calculated $(U_n)(x_b)$ and $(U_w)(x_b)$ curves for two initial distributions: (a) $M_w/M_n = M_z/M_w = 10$, $M_w = 132$ kg mol^{-1}; (b) $M_w/M_n = M_z/M_w = 2$, $M_w = 132$ kg mol^{-1}. ($U_n = M_w/M_n - 1$; $U_w = M_z/M_w - 1$.) Solid curves: fraction in concentrated phase, dashed curves: fraction in dilute phase. Curves in (a) are labeled with the initial polymer concentration. In (b) there are two sets of two curves for U_n and U_w, the top one referring to an initial concentration $\phi = 0.02$, the other to $\phi = 0.01$ [2].

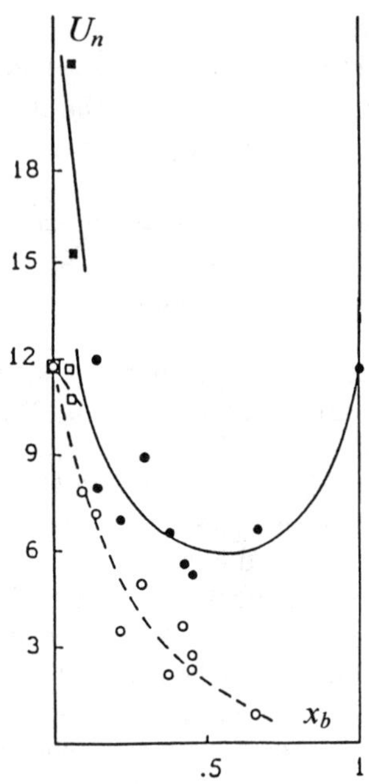

FIG. 2 Experimental $(U_n)(x_b)$ curves for the system diphenylether/polyethylene. Filled symbols: fraction in concentrated phase, open symbols: fraction in dilute phase. Initial concentrations of 1 and 2% by weight (circles and squares, respectively). Initial sample: linear polyethylene: $M_w = 153$ kg mol^{-1}, $M_w/M_n = 13$, $M_z/M_w = 6$ [2].

many authors and led to similar results (see the list of references in [6]).

A decrease of the fraction distribution's width goes with a decrease of its yield, and sufficient quantities of well-fractionated material will need a considerable amount of initial polymer. A 100 g size of the fraction will require several kg of initial polymer to be handled, e.g., in 10% solution, and a pilot plant will soon be needed to carry out such separations. This has been done [14–17] and some typical results are shown in Table 1. The solvent system used was a mixture of xylene and triethylene glycol (good and poor solvent, respectively). With a single solvent, to which Figs. 1 and 2 apply, the controlling variable is the temperature, a decrease of which will increase x_b. It is also possible to work isothermally and let the effect of decreasing solvent power be brought about by an increase of the amount of poor solvent in the solvent mixture.

We see in Table 1 that, though considerable sharpening of the distribution has been obtained in these one-step separations, M_w/M_n and M_z/M_w values close to unity cannot be realized. This was already clear from Fig 1. A one-step extraction is effective but

TABLE 1 Large-Scale Extraction Fractions of Linear Poly-(ethylene) (Average Molar Masses in kg mol^{-1}, Mass in g)

Sample	M_n	M_w	M_z	M_w/M_n	M_z/M_w	Mass
Initial	12	153	900	13	6	
Fraction 1	92	140	330	1.5	2.4	133
Fraction 2	9	55	300	6	5	543

would have to be repeated many times if very narrow fraction distributions are to be obtained.

Countercurrent extraction should be the answer to the problem and has indeed been suggested by a number of authors [18–20]. Calculations by Englert and Tompa [21] have demonstrated that truly impressive results should be obtainable, provided some experimental problems could be solved. The latter relate to the extremely slow phase segregation caused by the high viscosity of the concentrated phase which is a handicap for continuous operation. Further, a binary solvent system in which the solvent power of the two phases is different, though not too much, is not easy to find. It has been a fairly recent accomplishment of Wolf *et al.* [22–24] to overcome these practical problems and realize large-scale fractionation by continuous liquid–liquid extraction. They included poly(isobutylene) in their studies and also succeeded in applying the method succesfully to linear poly(ethylene) [25].

C. Fractional Crystallization from Solution

Though small, the chain-length dependence of the liquid–liquid distribution coefficient is yet large enough to supply a useful principle of fractionation provided the conditions are carefully chosen. This statement is even more relevant for fractionation by crystallization from solution. The melting point of crystallizable polymers depends very little on chain length but the fact that some polyolefins crystallize from dilute solution in folded-chain conformation appears to enhance the sensitivity to chain length needed for fractionation [26]. Pennings found later that rheological effects like elongational flow further increase the selectivity of crystallization according to chain length, with the result that quite effective fractionation of polyethylene and polypropylene from dilute solution could in this fashion be realized [27, 28].

In Pennings' procedure a mildly stirred polyolefin solution is cooled slowly. In the first stages of the process the longest chains deposit in the form of long fibrous crystals which build up a tenuous fabric-like network. The fibrillar crystals contain long thin backbones in which the polymer chains are highly extended. The fibrils adhere to the stirrer and form a substrate for shorter chains to crystallize at lower temperatures in a folded-chain conformation. After a suitably low temperature is reached the polymer has crystallized completely, and we have obtained a structure, well accessible to the solvent, with the low end of the mmd concentrated in layers adjacent to the solvent. In the dissolution procedure that now follows, the polymer is stepwise dissolved in pure solvent at increased temperatures and the successive fractions usually show a monotonous increase in average molar mass, and cover a large range. The technique is very effective, in particular in the low molar mass range, as is illustrated by the data in Table 2. The method has been scaled up to a pilot plant operation and initial amounts of polymer as large as 5 kg could be split up into 12 fractions, the M_w values of which indicated that sizeable shifts in average molar mass can so be obtained [3].

Since Pennings' method is based on differences in crystallization and dissolution temperatures, any structural feature bringing such differences about can be analyzed with the method. For instance, the degree of short-chain branching in polyethylene (linear low density) is known to markedly influence the crystallization temperature from solution. Fractions prepared with Pennings' technique will differ not only in average chain length, but also in degree of short-chain branching, and additional information is needed to interpret the fractionation results. Size exclusion chromatography and differential scanning calorimetry have been used for the purpose [29]. In Pennings' original papers [27, 28] the separation of stereoblock structures from polypropylene and of ethylene–propylene block copolymer fractions from a high-impact polypropylene has been reported. In more recent years a simpler and faster, though less effective variation suitable for routine use, called TREF (temperature rising elution fractionation), has come into widespread use (see [30] and Chapter 14 of this book).

TABLE 2 Crystallization-Dissolution Fractions of Linear Poly(ethylene) (Average Molar Masses in kg mol^{-1}, Mass in g)

Sample	M_n	M_w	M_z	M_w/M_n	M_z/M_w	Mass
Initial	8	75	250	9	3	
Fraction 1	11	11	12	1.0	1.1	33
Fraction 2	8	9	13	1.1	1.4	26
Fraction 3	8	9	11	1.1	1.2	13

III. PARTIAL MISCIBILITY IN POLYOLEFIN SOLUTIONS

A. The System *n*-Alkane/Polyethylene

1. Experimental Data

Until 1960 few polymer solutions were known to exhibit lower critical miscibility behavior (LCM), i.e., to separate into two fluid phases upon heating [31]. LCM behavior was only found in polymer systems in which both constituents are highly polar, e.g., aqueous solutions of poly(ethylene oxide) [32]. The miscibility behavior in these systems is probably related to a change in entropy associated with the formation of hydrogen bonds at low temperature.

In 1960 Freeman and Rowlinson demonstrated that solutions of polyolefins in hydrocarbon solvents may also show LCM [31]. These authors reported minimum solution temperatures (supposed to represent critical temperatures; LCST) for solutions of polyethylene, poly(isobutylene) and polypropylene in *n*-alkanes (*cf.* Table 3). These observations were not altogether unexpected. It was known at the time that hydrocarbon mixtures show LCM in the vicinity of the vapor/liquid critical point of the more volatile component if the

TABLE 3 LCST and UCST Solvents for Polyolefins of Various Average Molar Mass [31] (PIB: Polyisobutylene; PE: Polyethylene; PP: Polypropylene)

Solvent	Polymer	LCST, °C	UCST, °C
Ethane	PIB	<0	36
Propane	PIB	85	103
n-Butane	PIB	miscible	
Isobutane	PIB	114	142
	PIB	swells, but insoluble	
n-Pentane	PE	swells, but insoluble	
	PP	152	202
	PP	105	201
	PP	136	201
	PIB	75	199
Isopentane	PIB	54	189
Neopentane	PIB	insoluble	
Cyclopentane	PIB	71	—
n-Hexane	PE	127	—
	PIB	128	—
2.2-Dimethylbutane	PIB	103	—
2.3-Dimethylbutane	PIB	131	—
Cyclohexane	PE	163	—
	PIB	139	—
n-Heptane	PIB	168	—
n-Octane	PIB	180	—
Benzene	PIB	150–170	—

components differ sufficiently in molecular size (i.e., molar volume) and energy. Ethane, for instance, exhibits LCM with *n*-alkanes ranging from tetracosane to heptatriacontane [31], whereas propane is miscible in all proportions with the *n*-alkanes up to $C_{37}H_{76}$, but only partially miscible with lubricating oils [33].

Since Freeman and Rowlinson's publication [31] the amount of experimental information in the system *n*-alkane/polyethylene has increased considerably. Papers by Nakajima *et al.* [34–36], Kodama and Swinton [37, 38], Koningsveld *et al.* [39], and Kennis *et al.* [40] deal with the molar mass dependence of the LCM in terms of cloud-point curves of linear polyethylene in various alkanes, ranging from *n*-pentane to *n*-tridecane, and in binary mixtures of *n*-alkanes. Galina *et al.* [41] reported a spinodal and a cloud-point curve for a polyethylene sample ($M_w = 177$ kg mol^{-1} in *n*-hexane, measured with Gordon's Pulse Induced Critical Scattering technique [42, 43].

Some of these data are displayed in Figs. 3 and 4. The polyethylene sample used in [39] (Fig. 3) has a very broad molar mass distribution; values for the number-, weight-, and centrifuge-average molar masses are: $M_n = 8$, $M_w = 177$, and $M_z = 1200$ kg mol^{-1}, respectively. The cloud points for high polymer concentration were determined with the phase volume ratio method [44]; those for low polymer concentrations, however, had to be established by visual observation because the phase volume ratio becomes very large when the concentration drops below certain values. Owing to the broad mmd the critical concentration and temperature differ considerably from the precipitation threshold values, i.e., they do not coincide with concentration and temperature of the extremum in the cloud-point curve [39, 44]. Furthermore, *n*-hexane and the polyethylene sample in hand differ in weight-average molar mass by a factor of 2500 and, as a consequence, the miscibility gap is very asymmetrical and confined to the solvent-rich part of the phase diagram.

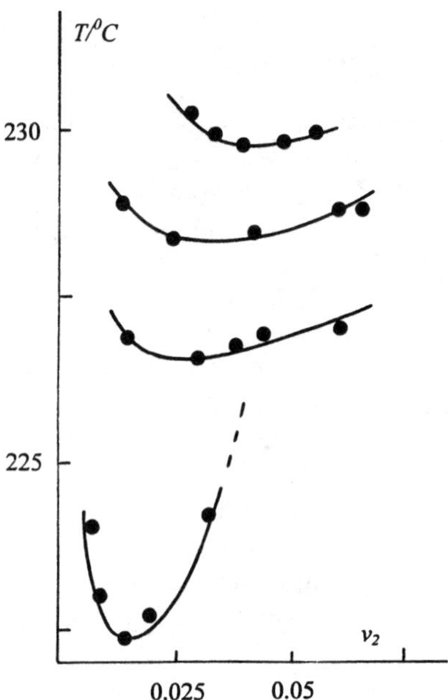

FIG. 4 Cloud-point curves for solutions of linear polyethylene in *n*-octane [36]. Weight-average molar mass values in kg mol^{-1}: 77, 93.5, 125, 205 (top to bottom). Curves drawn by hand. Volume fraction polymer: v_2.

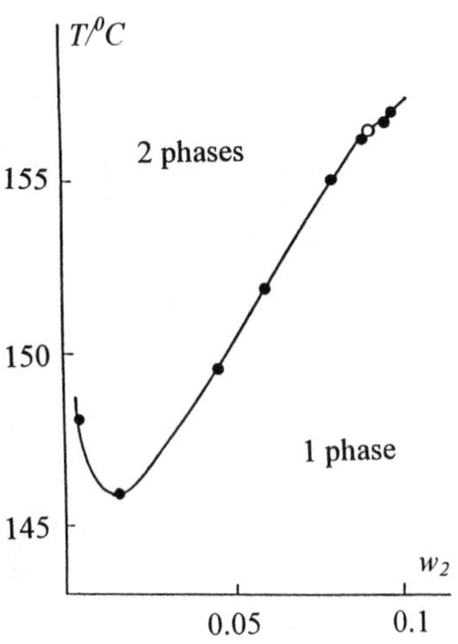

FIG. 3 LCM cloud-point curve for *n*-hexane/linear polyethylene [39]. Cloud points: ●; critical point: ○; polymer samples. $M_w = 177$ kg mol^{-1}; $M_w/M_n = 22$, $M_z/M_w = 7$. Curves drawn by hand. Weight fraction polymer: w_2.

The data in Fig. 4 refer to linear polyethylene samples with less broad distributions since they were obtained by fractionation of a commercial Marlex 50 type polyethylene into 10 fractions. The shape of the curves indicates that some of them would intersect, if extrapolated. This points to considerable residual polymolecularity in the fractions so that identification of minima with critical points is not permitted. Nakajima *et al.* [35, 36] nevertheless analyzed the data assuming the identification to be allowed and, using the Shultz–Flory method [45], derived values for the critical temperatures for solutions of polyethylene of infinite molar mass in *n*-alkanes. This specific temperature is known as the theta temperature, Θ. Nakajima *et al.*'s Θ values, listed in Table 4, agree within 5 K with those

TABLE 4 Theta Temperatures (°C) for Linear Poly(ethylene)

Solvent	Ref. 31	Ref. 37
n-Pentane	About 80	—
n-Hexane	133.3	127.9
n-Heptane	173.9	174.0
n-Octane	210.0	214.1
n-Nonane		249.3
n-Decane		278.4
n-Undecane		303.0
n-Dodecane		327.7
n-Tridecane		356.5

calculated in a similar fashion by Kodama and Swinton [37]. Since the samples of the latter authors were reported also to be nonuniform in molar mass, the discrepancy can probably be ascribed to all systems under investigation not having been close enough approximations of truly binary systems [46, 47]. The Θ value for the system *n*-pentane/poly(ethylene) is based on data for samples of low molar mass ($M_w = 5\text{–}23\ \text{kg mol}^{-1}$) and are therefore approximate.

The experiments that led to the data represented in Figs. 3 and 4 were all performed on systems confined to small glass tubes, sealed under vacuum. Hence, the pressure varies along the cloud-point curve and equals that of the vapor which is in equilibrium with the mixed liquid. The pressure changes involved are small and can be ignored in theoretical calculations.

Kennis *et al.* [40] reported on the influence the addition of nitrogen exerts on the location of the miscibility gap for linear polyethylene in *n*-hexane at pressures up to 7.5 MPa in the temperature range 393–453 K for small concentrations of nitrogen. Liquid–liquid phase boundaries (cloud points) were determined with the aid of a Cailletet-like apparatus in which the phase separation was observed visually. Experimental cloud-point isopleths (p vs T at constant composition) for linear polyethylene in *n*-hexane were reported to be nearly parallel; values for $\mathrm{d}p/\mathrm{d}T$ are positive which indicates upper critical solution pressures (at constant T) and lower critical solution temperatures (at constant p) (Fig. 5). The addition of small amounts of nitrogen shifts these isopleths to higher pressures and lower temperatures.

Data by Wolf [48, 49] should be mentioned in this connection. This author investigated the influence of

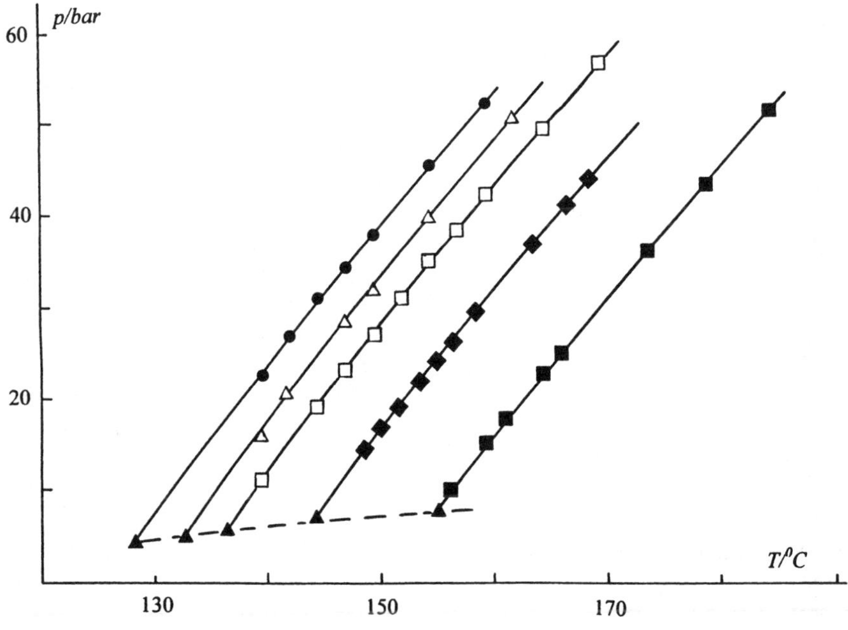

FIG. 5 Experimental liquid–liquid isopleths for the system *n*-hexane/linear polyethylene ($M_w = 177\ \text{kg mol}^{-1}$, $M_w/M_n = 22$, $M_z/M_w - 7$) [40]. Polymer weight fractions: 0.0053 △, 0.0055 (●), 0.0298 (□), 0.0666 (◆), 0.1313 (■). Filled triangles: locus of three-phase equilibrium liquid/liquid/vapor.

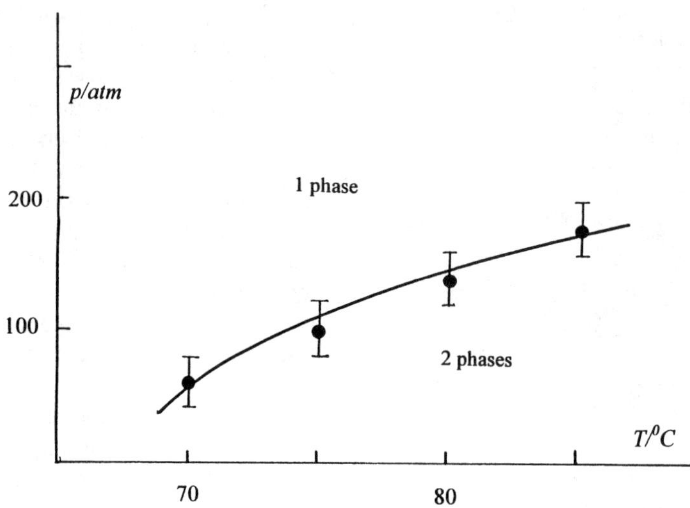

FIG. 6 Critical curve extrapolated to infinite chain length of the polymer for the system *n*-pentane/poly(isobutylene) [48, 49].

pressure on the location of liquid/liquid critical points for solutions of polyolefins in *n*-alkanes. Figure 6 shows the critical curve for *n*-pentane/poly(isobutene) for infinite chain length of the polymer. The data indicate that, at constant temperature, miscibility improves with increasing pressure. According to Wolf, mixtures of polyethylene and long *n*-alkanes (*n*-hexane and longer) behave similarly.

So far, we have discussed the miscibility behavior of solutions of linear polyethylene in long-chain *n*-alkanes at ambient and elevated pressures. It was also mentioned that *n*-pentane dissolves only low molar mass polyethylene. However, it is likely that, at pressures above 200 bar, polyethylene of high molar mass will also be at least partially miscible with *n*-pentane. Ehrlich and Kurpen [50] reported on the influence pressure has on the location of the miscibility gap for solutions of a slightly branched polyethylene sample with a molar mass of 247 $\mathrm{kg\,mol^{-1}}$ in ethane, *n*-propane, and *n*-butane. It was found that the sample dissolves in ethane at pressures as high as about 1200 bar and at temperatures between 120–140°C. This is well above the critical point of the solvent and the term liquid–fluid separation instead of liquid–liquid can appropriately be used here.

Since linear polyethylene may be regarded to represent a *n*-alkane homologue of very large chain length and broad chain-length distribution, solutions of linear polyethylenes in *n*-alkanes might be expected to be amenable to a simple theoretical treatment. In addition, pure *n*-alkanes and their mixtures with linear polyethylene would be likely to provide a suitable starting point for group contribution models since they can be conceived as containing two functional groups only (see Sec. III.A.3).

An explanation for the LCM behavior exhibited by *n*-alkane/polyethylene systems is hard to give, but might be sought in the decreasing configurational energy and the increasing molar volume of the solvent with rising temperature when its vapor/liquid critical point is approached so that it becomes a less good solvent under such conditions. This tentative explanation suggests that theoretical treatment of miscibility data on linear *n*-alkane/polyethylene solutions should include free-volume contributions (see Secs. III.A.3 and III.A.4).

In the past several theoretical studies have been concerned with the mutual solubility of linear polyethylene and *n*-alkanes. In the course of such investigations phase behavior, or *pVT* relations, of pure *n*-alkanes has to be dealt with. In the following, three of such models will be discussed briefly; Flory's Equation of State theory (EoS), the Mean-Field Lattice Gas (MFLG) model, and the Simha–Somcynsky (SS) theory.

2. Flory's Equation of State Approach

Using a procedure elaborated by Prigogine, Flory and coworkers [51, 52] developed an EoS theory for non-spherical molecules, such as *n*-alkanes. This model considers a chain molecule as a linear sequence of terminal and mid-chain segments which are endowed with a hard-sphere potential. Further, the terminal groups are allowed to exert intermolecular forces differing from those of the mid-chain segments. This EoS treatment can easily be extended to cover mixtures of chain–molecule liquids. The theory provides good descriptions of liquid isotherms up to pressures of 1400 bar at temperatures between 30 and 120°C for a series of *n*-alkanes ranging from *n*-hexane to linear poly(ethylene), although deviations between theory and measured liquid densities become quite large above 1000 bar (about 5%).

This mode was subsequently used to predict excess properties (excess volume, excess enthalpy, ΔH^e) of binary mixtures of *n*-alkanes. For instance, the decrease of ΔH^e and the reversal of its sign with a change in the temperature, observed in several systems, are correctly predicted though the magnitude of the effect is underestimated. In addition, LCM behavior was predicted to occur in mixtures of poly(ethylene) and *n*-alkanes, in agreement with experimental findings. Calculated LCST's, however, were only in qualitative agreement with the observed data, e.g., the location of the miscibility gap for *n*-hexane/polyethylene was predicted to be about 60°C too low.

3. Mean-Field Lattice Gas Model

The mean-field lattice gas model (MFLG) represents a pure component by a lattice, the sites of which are randomly occupied by n_1 moles of molecules. The molecules are allowed to occupy m_1 lattice sites each; the volume per lattice site, v_0, is kept constant. The total number of sites equals $n_0 + n_1 m_1 (= N_\phi)$ where n_0 is the amount of vacant sites in moles. Pressure and temperature changes affect the density of the system and can be dealt with by appropriate variations of n_0. In one of the more elaborate versions of the model (see, e.g., ref. [53]), the Helmholtz free energy, ΔA, of mixing n_0 vacant sites with n_1 moles of molecules is

$$\Delta A/N_\phi RT = \phi_0 \ln \phi_0 + (\phi_1/m_1) \ln \phi_1 + \Gamma_{01}\phi_0\phi_1 \tag{6}$$

where ϕ_0 and ϕ_1, the site fractions of vacant and occupied sites, respectively, are related to the molar volume V:

$$\phi_1 = 1 - \phi_0 = n_1 m_1/N_\phi = m_1 v_0/V \tag{6a}$$

In applications of Eq. (6) it has been found useful to write

$$\Gamma_{01} = a_{01} = b_{01}(T)/Q \tag{6b}$$

where $Q = 1 - c_{01}\phi_1$, a_{01} and c_{01} are parameters related to differences in size and shape between the constituents of the mixture, and b_{01} is related to the change in internal energy upon mixing.

Equation (6) can easily be extended to cover quasi-binary systems containing a solvent 1 and a polymer 2 with a chain-length distribution:

$$\Delta A/N_\phi RT = \phi_0 \ln \phi_0 + (\phi_1/m_1) \ln \phi_1 + \sum (\phi_{2_i}/m_{2_i}) \ln \phi_{2_i} + \Gamma_{01}\phi_0\phi_1 + \Gamma_{02}\phi_0\phi_2 + \Gamma_{12}\phi_1\phi_2 \tag{7}$$

where the quantities Γ refer to the three binary interactions, and Q is now given by

$$Q = 1 - c_{01}\phi_1 - c_{02}\phi_2 \tag{7a}$$

Expressions for the EoS, spinodal and critical conditions, and coexisting phase compositions can be derived with the use of standard thermodynamic procedures. General expressions for multicomponent systems can be found in [54].

The MFLG model describes the vapor/liquid critical point ($v \equiv l$), v/l equilibrium data and isotherms of pure components such as *n*-pentane and other *n*-alkanes quite well (Fig. 7) while polymers also fall within the scope of the model. Since linear polyethylene and *n*-alkanes consist of identical repeat units it has been assumed that, in a first approximation, the parameters for *n*-alkane/polyethylene mixtures can be set equal to zero [55]. This assumption proved to be too simplistic since the locations predicted for spinodal curves were found to be only in qualitative agreement with the measured curves and locations of miscibility gaps. However, Fig. 8 illustrates that values for mixture parameters can be found that provide a fair description of the measured LCM behavior and its pressure dependence for the system *n*-alkane/linear polyethylene [56, 57]. The predictive power of the procedure is considerable, as is witnessed by Fig. 9 in which the location of cloud points in pressure–temperature–composition space for *n*-octane/*n*-nonane/linear PE mixtures is predicted remarkably well in terms of the nearby spinodals.

4. The Simha–Somcynski Hole Theory

In the MFLG theory the effects of compressibility are related to the presence of vacancies on the lattice. On the other hand, in the EoS theory of Flory and coworkers a completely filled lattice is assumed and the pVT contributions are due to changes in the volume of the lattice sites or cells. Finally hole theories, which for polymer systems were initiated by Simha and

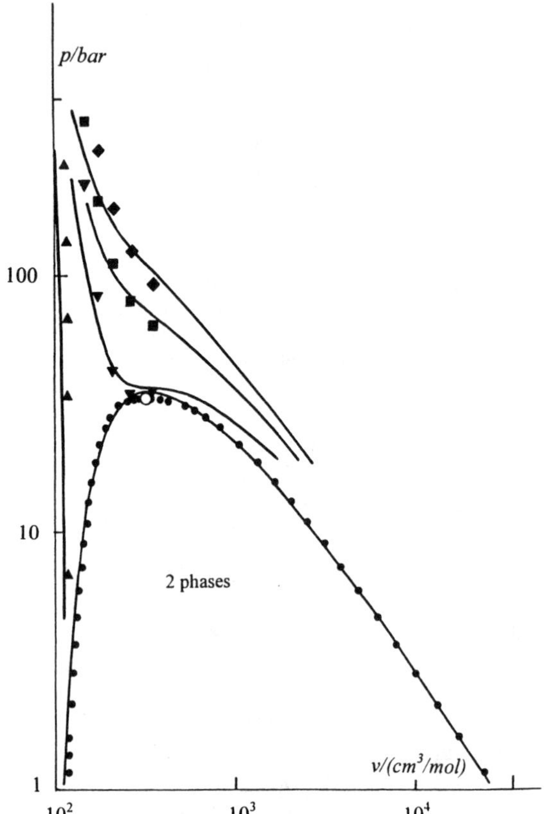

FIG. 7 MFLG description for *n*-pentane [54] of liquid–vapor equilibrium (●), the liquid–vapor critical point (○) and isotherms for some temperatures in °C: 37.8 (▲), 200 (▼), 260 (■), 320 (◆).

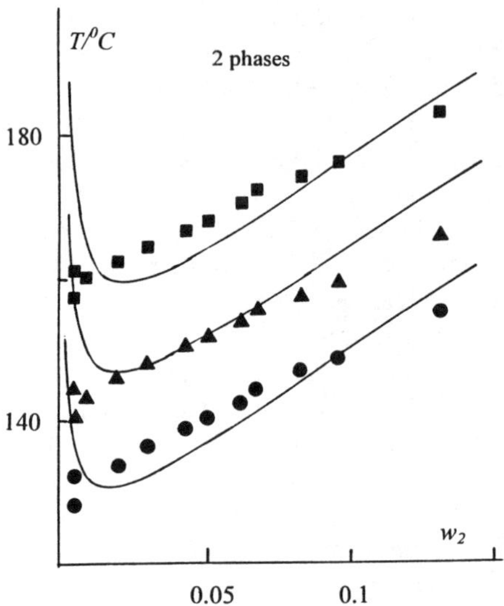

FIG. 8 Comparison of MFLG calculated spinodal curves with experimental cloud points for linear polyethylene in *n*-hexane at indicated pressures in bar [56]. Parameter values in Table 5. Experimental data from [40]. Pressures in bar: 6 (●), 25 (▲), 50 (■).

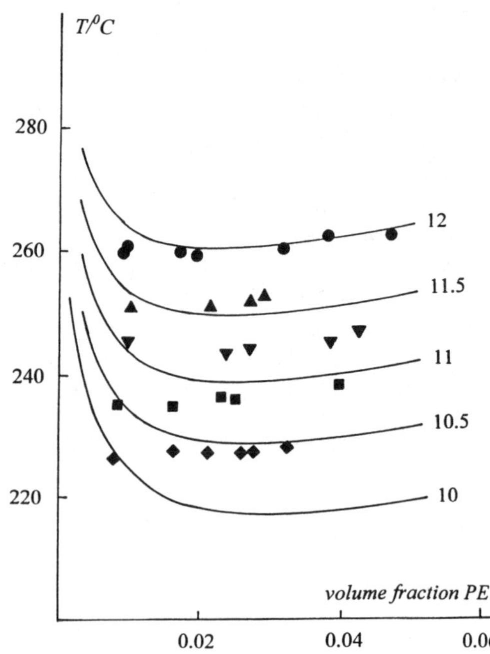

FIG. 9 Comparison of experimental cloud points with predicted spinodal curves for indicated pressures (in atm) for solutions of linear polyethylene ($M_n = 37$ kg mol^{-1}, $M_w = 49$ kg mol^{-1}) in mixtures of *n*-octane and *n*-nonane as a function of the mole fraction x_n of nonane in the mixed solvent, top to bottom: $x_n = 1$, 0.726, 0.473, 0.240, and 0, respectively [56].

Somcynski [58], allow for both vacancies and a variable cell volume.

The first applications of the Simha–Somcynski (SS) hole theory concentrated on the equation of state of low and high molar mass components and mixtures thereof [59–63]. The results proved that the SS theory is quite successful in this respect. In a subsequent application the miscibility behavior of solutions has been considered [64–66] including the system *n*-hexane/PE. Several refinements have been introduced into the SS theory which resulted in a more accurate evaluation of thermodynamic properties without the introduction of additional adjustable parameters [66–69].

In the SS theory a real system of volume V containing N n-mers of molar mass M is replaced by N s-mers occupying s consecutive sites on a lattice. In order to enhance configurational disorder only a fraction $y = Ns/(Ns + N_h)$ of the lattice sites is occupied, leaving N_h lattice sites vacant.

The partition function Z at given N, V, and T is determined by the summation over all possible values of y, viz.

$$Z(N, V, T) = \sum_y Z(N, V, T, y) = \sum_y g(y) l_f(y)^\nu \{\exp(-E_0(y)/kT)\} \quad (8)$$

where g stands for the combinatorial entropy related to the mixing of vacancies and segments; l_f is the segmental free length related to the free-volume contribution characteristics of cell theories; $\nu = 3sNc_s$; $3sc_s$ is the effective number of degrees of freedom of the s-mer, and E_0 is the internal energy of the system which is assumed to be equal to the lattice energy.

The importance of the c parameter was first recognized by Prigogine *et al.* [70]. It is to be taken as a measure of the perturbation of internal rotations of the chain in addition to motions of the chain as a whole, in the dense medium. The Flory approximation [12, 13] is invoked to represent the combinatorial entropy

$$\ln(g/Ns) = -s^{-1}\ln(y) - \{(1-y)/y\}\ln(1-y) \quad (9)$$

The internal energy is given by

$$E_0 = (\varepsilon_c/2) y N \{s(z-2)+2\} \quad (10)$$

where z is the lattice coordination number.

In the above equation ε_0 is the contact energy of a pair of segments, which is assumed to obey a Lennard-Jones 6–12 potential

$$\varepsilon_c = \varepsilon^*(A(\tilde{\omega}^{-4} - 2B\tilde{\omega}^{-2}) \quad (11)$$

where $\tilde{\omega} = yV/(Nsv^*)$ is the reduced cell volume, and ε^* and v^* are the maximum attraction energy and corresponding cell volume, respectively. The constants A and B are set equal to $A = 1.011$ and $B = 1.2045$ [71]. The free length l_f also depends on volume according to

$$l_f = y(\{\tilde{\omega}^{1/3} - 2^{-1/6}) + (1-y)\tilde{\omega}^{1/3}\} v^{*1/3} \quad (12)$$

The Helmholtz free energy A is related to the partition function Z by

$$A = -kT \ln Z \quad (13)$$

where k is Boltzmann's constant. Classic thermodynamic relations allow the calculation of all desired properties, such as EoS, excess volumes, spinodals, binodals, and critical conditions.

The SS theory can be generalized to multicomponent systems [61, 66] by adopting a simplification introduced by Prigogine *et al.* in their theory of mixtures, which is based on the simple cell model [70]. That is, the repulsion volume v^* and the maximum attraction energy ε^* are replaced by an average instead of dealing explicitly with the problem of packing differently sized segments. This approximation requires the restriction to nearly equal-sized segments (not necessarily equal-sized chemical repeat units). For a binary systems of components a and b we have the following relation for the mean interaction parameters

$$\langle\varepsilon^*\rangle\langle v^*\rangle^m = \varepsilon^*_{aa} v^{*m}_{aa} q_a^2 + 2\varepsilon^*_{ab} v^{*m}_{ab} q_a q_b + \varepsilon^*_{bb} v^{*m}_{bb} q_b^2 \qquad m = 2, 4 \quad (14)$$

where $q_i = N_i[s_i(z-2)+2]/\{\sum N_j[s_j(z-2)+2]\}$ stands for the contact fraction of component i.

So far, the only polyolefin solutions treated with the SS theory refer to the system n-hexane/PE, for a wide range of molar masses and involving the assumption that the PE samples were molecularly monodisperse [64]. It was shown that LCM binodals and spinodals at various pressures can be represented by master curves over a reasonable temperature distance from the critical point (Fig. 10). It was furthermore demonstrated that the prediction of LCM phase behavior in

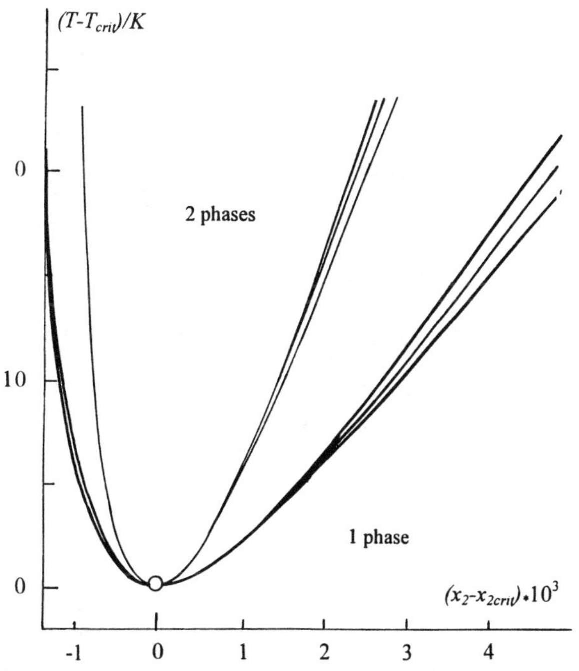

FIG. 10 LCM binary binodals (heavy curves), spinodals (light curves) and critical point (○) in terms of $(T - T_{crit})$ vs. $(x_2 - x_{2_{crit}})$ for 100, 50, and 1 bar (top to bottom). Calculated for n-hexane/polyethylene with the Simha–Somcynski theory [64]. Mole fraction of polymer: x_2.

n-hexane/PE is related to the existence of a critical free-volume difference between the constituents. A comparison of the theoretical predictions with experimental data has not yet been carried out.

5. Recent Theoretical Developments

In the last decade attempts have been made to develop improved models, capable of providing quantitative predictions of thermodynamic properties of polymeric systems. We are not aware of any application so far to polyolefin solutions and therefore merely mention some relevant references [72–74], concentrating in this chapter on the older and still useful approaches.

B. The Effect of Short-Chain Branching

Short-chain branching is being used to modify properties of polyethylene and also plays an important role in liquid–liquid phase equilibrium in polymer solutions. Phase behavior during polymer production by solution (co)polymerization will therefore be affected by the molecular architecture and quantities like separation temperature and/or pressure as a function of, for instance, monomer conversion should preferably be predictable by model calculations.

For the system branched diphenylether/polyethylene it has been shown [75] that the two-phase region may be shifted by more than 10°C compared with that of a linear poly(ethylene) sample of about equal number and weight average molar mass (Fig. 11). This information indicates that chain branching may jeopardize a miscibility based fractionation of poly(ethylene) (see Sec. I). At a certain stage of the fractionation process, the polymer being separated will be a mixture consisting of species with a given molar mass and low degree of branching, as well as species which are more branched but have a compensating higher molar mass.

The FHS model, developed independently by Staverman [8, 9], Huggins [10, 11], and Flory [12, 13], suggests that a modest amount of long-chain branching should not change thermodynamic properties noticeably, the number of branch points per chain being negligible. The effect of a sizeable amount of relatively short side chains should be noticeable, however, and can easily be introduced into the model to obtain a meaningful description of the thermodynamic behavior of solutions of branched chain molecules.

A branched polyethylene chain may be considered to be composed of end, linear middle segments and branched middle segments (cf. Fig. 12). Using the strictly regular approximation [76] and Staverman's concept of contact-surface areas [77], we may define the interaction term g in the well-known FHS expression for ΔG, the free enthalpy of mixing,

$$\Delta G/N_\phi RT = (\phi_1/m_1)\ln\phi_1 + \sum[(\phi_{2_i}/m_{2_i})\ln\phi_{2_i}] + g\phi_1\phi_2 \quad (15)$$

by [78]

$$g = b(1-c')/(1-c'\phi_2) \quad (16)$$

where

$$c' = 1-\rho_2-(\rho_3-\rho_2)\delta-(\rho_1-\rho_2)(\eta\delta+2/m_n) \quad (17)$$

In the latter equation, δ and η represent the number of branched middle segments per macromolecule and the "average effectivity" of the end segments, respectively. The latter parameter reflects overall effects of a side chain. For very short branches the effectivity will be close to zero ($0<\eta<1$); with growing length of the side chain η will go to unity. In the derivation of Eq. (15) it was assumed [78] that the interaction energies between solvent molecules and polymer segments and between polymer segments mutually are all identical. As a result, Eq. (15) contains one parameter b which is related to an average interaction energy. The various specific contact surface areas, s_0, s_1, s_2, and s_3, of solvent molecules, end segments and linear and branched middle segments, respectively, appear in the factors ρ, defined by

$$1-\rho_i = s_i/s_0 \quad (18)$$

To check the viability of Eq. (16) the system diphenylether/branched polyethylene has been used as a test case [78]. Experimental liquid–liquid critical points, measured with the phase volume ratio method [44] for samples varying in degree of branching, may serve the purpose together with some experimental spinodal temperatures, measured with the Pulse-Induced Critical Scattering technique of Gordon *et al.* [41–43]. The data were used to calculate values for the parameters in Eqs. (17) and (18) (details of the fitting procedure can be found in [54]).

Diphenylether is known as a Θ solvent for polyethylene with a Θ temperature well above the melting point of the polymer. The above analysis yields $\Theta = 160.8$°C, in good agreement with values reported in the literature [79]. Calculated spinodals are in quantitative agreement with experiment (Fig. 13). Significant deviations occur only at low polymer concentrations. It has been established earlier [80]

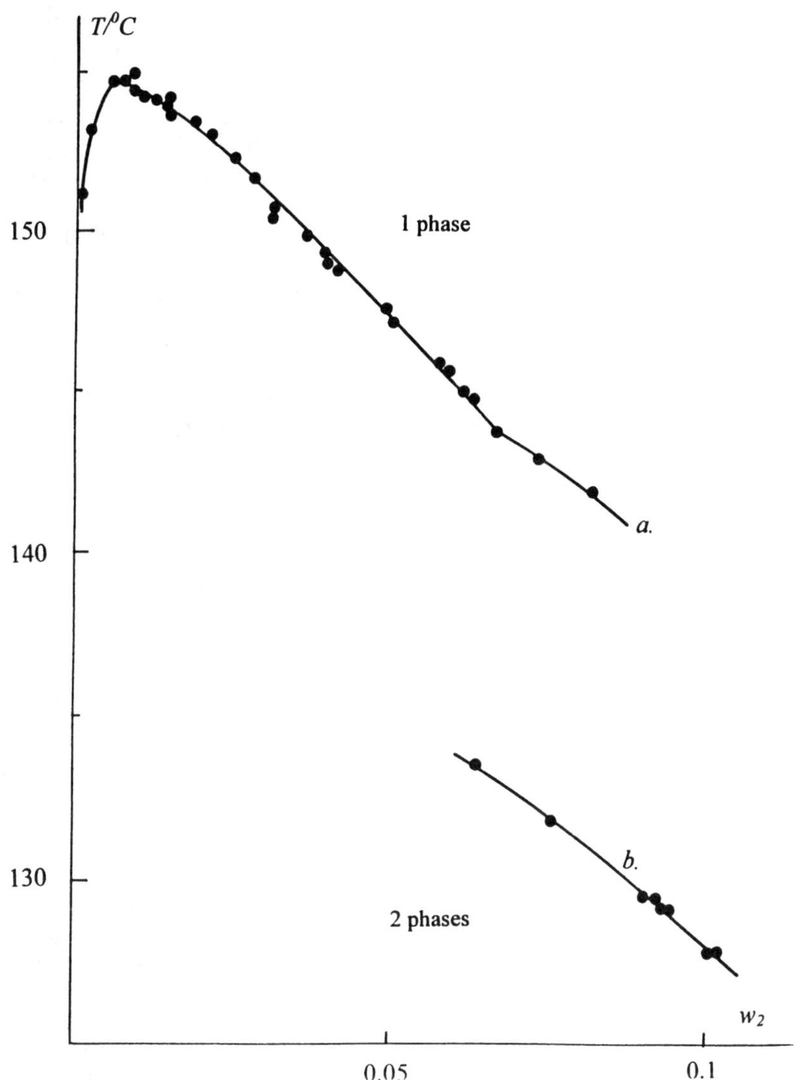

FIG. 11 Cloud-point curves in diphenylether for linear and branched polyethylene at comparable number- and weight-average molar mass [75]. (a) linear: $M_n = 8 \times 10^3$, $M_w = 1.5 \times 10^4$; (b) branched: $M_n = 1.1 \times 10^4$, $M_w = 1.6 \times 10^4$.

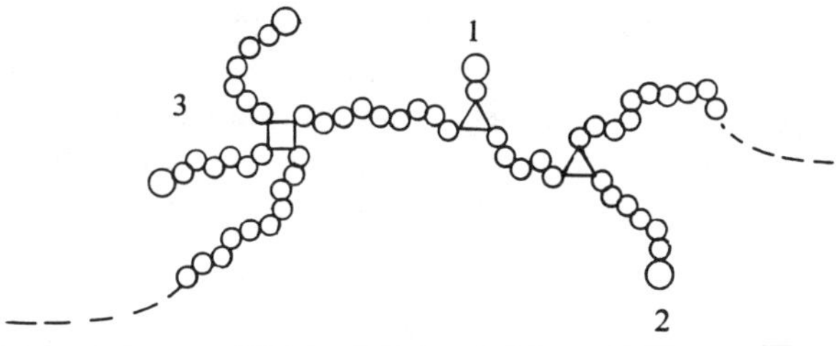

FIG. 12 Side-chain "effectivity" η for short (1: $0 < \eta < 1$) and medium-sized side chains (2: $\eta = 1$; 3 : $\eta = 2$) emanating from tertiary (△) or quaternary (□) branch-bearing segments [75].

that such deviations may be attributed to the nonuniformity of local polymer segment concentration which cannot be ignored at high dilution. At low concentration the polymer coils will be separated by regions of pure solvent and some of the assumptions underlying Eq. (15) are not valid. Reference [80] describes a procedure to deal with the situation to which we refer the reader for details. It is worth noting that the system diphenylether/linear polyethylene obeys the original FHS equation (g independent of concentration and molar mass) to a considerable extent [81].

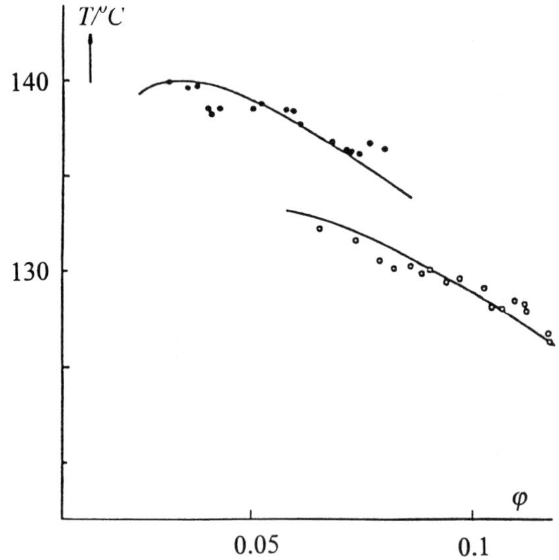

FIG. 13 Comparison of experimental (●, ○) and calculated spinodals (——) for branched poly(ethylene) in diphenylether [75]. Top: $M_w = 230\,\text{kg mol}^{-1}$, 2.3 end groups/100 C, $\eta = 0.84$. Bottom: $M_w = 70\,\text{kg mol}^{-1}$, 2.2 end groups/100 C, $\eta = 0.83$. Volume fraction polymer: ϕ.

Equation (17) suggests that the present treatment, if applied to a polyethylene sample with a small but significant number of very short side chains per chain, should yield a small value for η. One of the polyethylene samples was a hydrogenated polybutadiene with a small amount of 1–2 addition, and should contain a few ethyl branches. This sample indeed yielded the smallest value for η.

C. Solutions of Copolymers

An important parameter in phase behavior of polymer systems is the chemical composition of the chains. For example, in Schmitt *et al.*'s work on the miscibility of poly(methylmethacrylate) and poly(styrene-co-acrylonitrile) [82] it was found that a minute change in the chemical composition of the statistical copolymer splits the single lower-critical miscibility gap into two and adds an upper consolute two-phase range.

The influence of the copolymer's chemical composition on thermodynamic properties can be explored theoretically with a molecular consideration based on the rigid lattice model. Simha and Branson [83] derived the following expression for the interaction function g in the FHS Eq. (15):

$$g = g_{1\alpha}\phi_\alpha + g_{1\beta}\phi_\beta - g_{\alpha\beta}\phi_\alpha\phi_\beta \tag{19}$$

It distinguishes between three different interactions in the system, one for each solvent-homopolymer interaction ($g_{1\alpha}$ and $g_{1\beta}$) and a third for the comonomer α–β interaction ($g_{\alpha\beta}$), but neglects differences in size and shape between the various units. If allowance for such differences is made, the strictly regular treatment leads to the following expression for Λ, which replaces the term $g\phi_1\phi_2$ in Eq. (15) [84]:

$$\begin{aligned}\Lambda = {} & \phi_1 Q^{-1}[g_{1\alpha}\sum \phi_{\alpha i}\phi_i + g_{1\beta}\sum \phi_{\beta i}\phi_i \\ & - (s_\alpha s_\beta/s_1^2)g_{\alpha\beta}\sum(\phi_{\alpha i}\phi_{\beta i}\phi_i/q_i)] \\ & + \tfrac{1}{2}(s_\alpha s_\beta/s_1^2)^2 Q^{-1} g_{\alpha\beta}\sum\sum[d_{ij}^2\phi_1\phi_j(q_i q_j)^{-1}]\end{aligned} \tag{20}$$

where

$$Q = \phi_1 + \sum q_i\phi_i$$

$$q_i = (s_\alpha\phi_{\alpha i} + s_\beta\phi_{\beta i})/s_1$$

and

$$d_{ij} = \phi_{\alpha i} - \phi_{\beta j}$$

where

- ϕ_1, ϕ_i = volume fraction of solvent molecules 1 and copolymer molecules i, each occupying 1 and m_i lattice sites, respectively
- s_1, s_α, s_β = molecular interacting surface area of solvent molecules and comonomer units α and β, respectively.
- $\phi_{\alpha i}, \phi_{\beta i}$ = volume fractions of repeat units α and β in copolymer chains i.

Further, $g_{1\alpha} = s_\alpha \Delta w_{1\alpha}/RT$, $g_{1\beta} = s_\beta\Delta w_{1\beta}/RT$, and $g_{\alpha\beta} = s_1\Delta w_{\alpha\beta}/RT$, in which Δw_{hk} stands for the interchange (free) energy for h–k contacts. Equation (20) accounts for a possible effect of a distribution in chemical composition through the $\phi_{\alpha i}$ and $\phi_{\beta i}$ variables but ignores sequence-length distribution effects.

Several authors studied the influence of the chemical composition on miscibility behavior of copolymer solutions. However, most of these studies were restricted to the measurement of cloud points which complicates theoretical treatment. One investigation [85] reports more complete phase diagrams, in terms of spinodals, cloud points, and critical points, of solutions of poly(ethylene-co-vinylacetate) (EVA) in diphenylether. The EVA samples were commercial samples from different origins possessing broad molar mass distributions and vinyl acetate (VA) contents varying from 2.3 to 12.1 wt%. Samples with higher VA content could not be used because of the

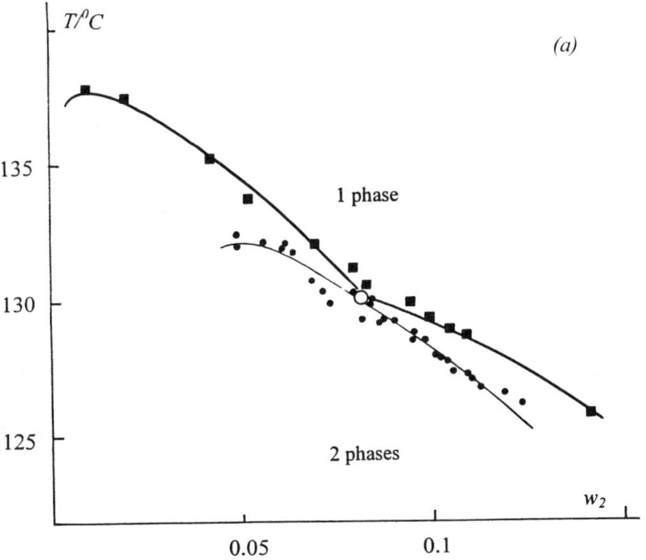

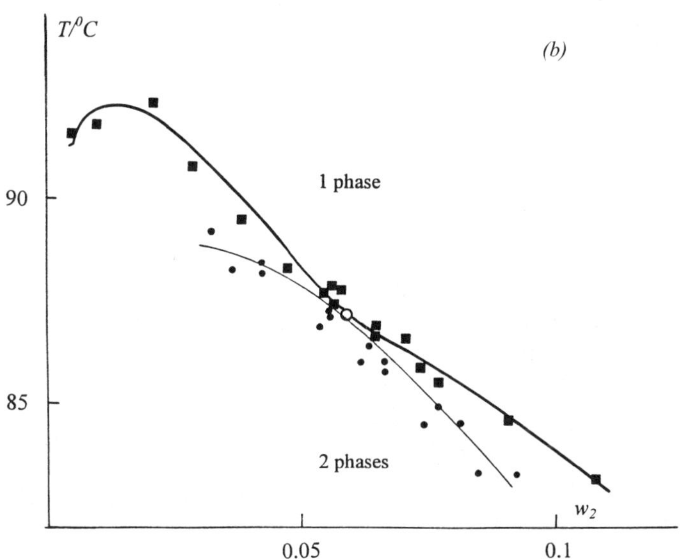

FIG. 14 Measured phase diagrams for the system diphenylether/EVA. Cloud points: ■ [84]; ●: spinodal point; ○: critical point. Curves drawn by hand. (a): Polymer 2.3 wt% VA, $M_w = 465\,\mathrm{kg\,mol^{-1}}$, $M_w/M_n = 9$, $M_z/M_w = 5$; (b) polymer 12.1 wt% VA, $M_w = 300\,\mathrm{kg\,mol^{-1}}$, $M_w/M_n = 4.5$, M_z/M_w (by GPC). Weight fraction EVA: w_2.

interference of polymer crystallization with liquid–liquid phase separation. Included in the study was a branched low-density polyethylene sample. Critical points were measured with the phase volume ratio method [44], spinodal points were determined with the PICS technique [41–43], and cloud points were determined visually. It was found that the miscibility gap in solutions of EVA in diphenylether, which is of the upper-consolute type, shifts to lower temperatures if the VA content of the polymer is increased. Figure 14 gives two examples for VA contents of 2.3 and 12.1% by weight.

EVA copolymers are considered to be classical examples of random copolymers since the reactivity

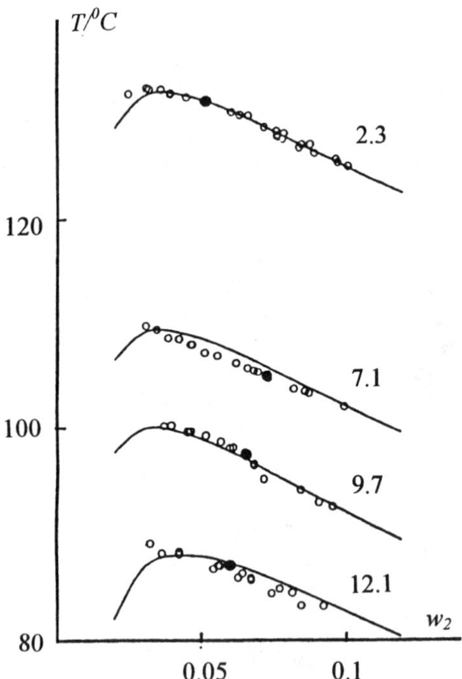

FIG. 15 Comparison of experimental (○) and fitted spinodal points (——) for diphenylether/EVA for indicated VA contents. Experimental and calculated critical points (●) coincide [85]. Weight fraction polymer: w_2.

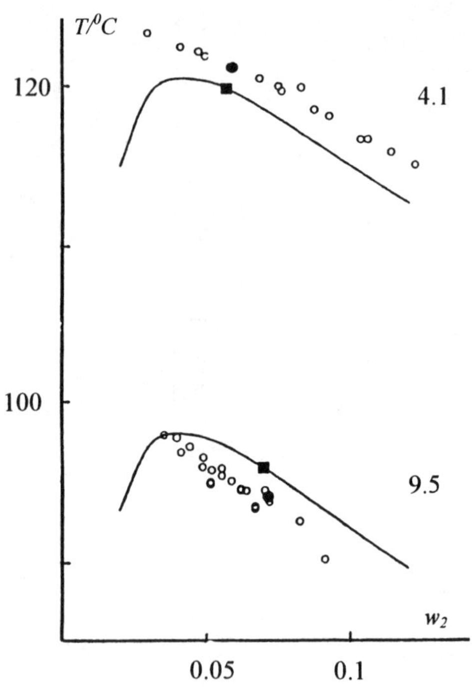

FIG. 16 Prediction of spinodal (——) and critical points (■) for diphenylether/EVA compared to relevant experimental data, not used in the fit of Fig. 15 (experimental, ○: spinodal points; ●: critical points) [85].

ratios of the two monomers are nearly equal. Fractionation of EVA samples has furthermore revealed [86] that the EVA samples do not possess a wide distribution in VA content. Ignoring the latter distribution at all, we may simplify Eq. (20), reducing it to

$$
\begin{aligned}
g &= a + (b_0 + b_1/T)/Q \\
b_1 &= g_{1\alpha}\phi_\alpha + g_{1\beta}\phi_\beta - (s_\alpha s_\beta/s_1^2)(\phi_\alpha\phi_\beta g_{\alpha\beta})
\end{aligned}
\tag{21}
$$

Equation (21) contains the usual entropy correction terms a and b_0 [87]. The interaction parameters in Eq. (21) were adjusted to spinodal and critical points on four EVA samples in diphenylether. Descriptions of the data used were quite acceptable (Fig. 15), but prediction by interpolation for samples with different VA content is far from quantitative (Fig. 16). Several reasons for this failure can be advanced. First of all, chain branching was ignored in the derivation of Eq. (20). However, the EVA samples were synthesized at elevated pressures (up to 1500 bar) and must therefore be expected to contain branched molecules. Secondly, coupled GPC viscosity measurements on the samples indicate that long-chain branching is negligible below a molar mass of 50 kg mol^{-1} [88], which complicates theoretical treatment. Thirdly, the average molar masses were obtained from GPC measurements in tetrahydrofurane at 20°C and polystyrene samples were used as standards. Finally, the treatment ignores dilute solution effects. Little can be done in the latter respect because of lack of literature reports on dilute solution properties of copolymers. In view of the simplifications introduced the present treatment may be considered to supply a quite satisfactory practical approach.

D. The System Ethylene/Polyethylene; "Breathing-Lattice" Model

The synthesis of low-density (long-chain branched) polyethylene (LDPE), which proceeds along a free radical mechanism, is performed in bulk ethylene at temperatures and pressures far above those of the liquid–vapor critical point of the monomer ethylene. The production of LDPE takes place in autoclaves at pressures between 1400 and 3500 bar and temperatures up to 600 K. About 10–35% of the ethylene is converted into LDPE. The separation of LDPE from the reaction mixture is carried out at 150–300 bar. It is interesting to note that polymerization impedes measurement of pVT behavior of ethylene at 473 K, even for pressures below 400 bar [89].

Fluid phase separations occur in the system ethylene/LDPE under synthesis and recovery conditions, and knowledge of the influence of p and T on miscibility behavior for various molar masses is therefore of practical importance. Because of the relatively high pressures and temperatures involved (supercritical conditions) it is meaningful to refer to the phase separations in hand as fluid–fluid phase equilibria.

Various authors reported on such phase equilibria in supercritical solutions of linear PE or LDPE in ethylene (see [90–94]). It is unfortunate, however, that in many cases essential information on the molecular characteristics of the samples used is missing (degree of branching, molar mass distribution) which renders quantitative treatment of the measured phase diagrams impossible.

Chain branching cannot be neglected in the analysis of experimental data which is demonstrated clearly in Figs. 11 and 17. The miscibility gap for the linear sample is seen to be located at about 400 bar above that for a comparable branched sample [95]. Experimental cloud-point curves, e.g., those reported by Swelheim *et al.* [94] are in fact only the top parts of large two-phase regions which, upon a decrease in pressure gradually change in character from fluid–fluid to vapor–liquid equilibrium. This experimental information, the fact that LDPE melts at about 125°C at 1 bar, and the pVT data for pure ethylene, allow the construction of the $p(T)$ projection of the Bakhuis–Roozeboom $p(T, \phi_2)$ diagram for ethylene/LDPE (Fig. 18). Heavy curves refer to $p(T, \phi_2)$ regions actually measured and light curves can be added on the basis of general thermodynamic considerations. A full discussion can be found in [96].

The influence of pressure and temperature on fluid phase relations in the system ethylene/polyethylene being well documented in a qualitative sense [90–94], the effect of molar mass (distribution) is usually left out of consideration. Only De Loos *et al.* [95, 97] studied fluid phase relations in terms of cloud points and critical points in the system ethylene/linear PE for eight well-characterized polymer samples. Pressures ranged up to 2000 bar, temperatures were between 390 and 450 K, and a polymer concentration range of 0 to 30 wt% was studied. The weight-average molar masses varied from 3.7 to 118 kg mol^{-1}. Cloud points were determined visually and critical points were measured with the phase volume ratio method.

De Loos *et al.*'s extensive work presents a very useful set of data to test or develop theoretical models so that quantitative insight may be obtained about the phase separations that occur during polymerization

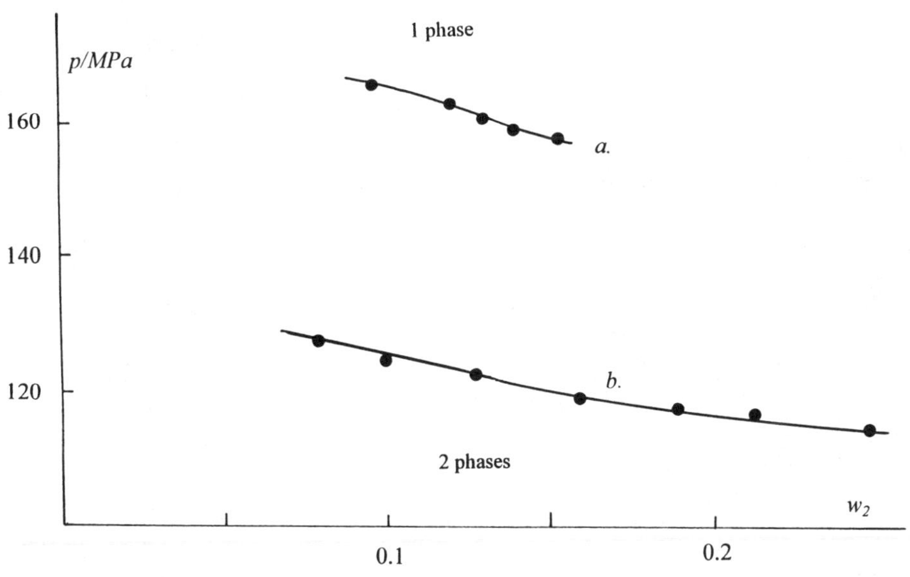

FIG. 17 Experimental cloud-point curves for ethylene/polyethylene showing the influence of chain branching. Temperature: 423.15 K. (a) Linear PE sample, $M_w = 55$ kg mol^{-1}, $M_w/M_n = 6$; (b) branched PE sample, $M_w = 54$ kg mol^{-1}, $M_w/M_n = 8$ [94, 95].

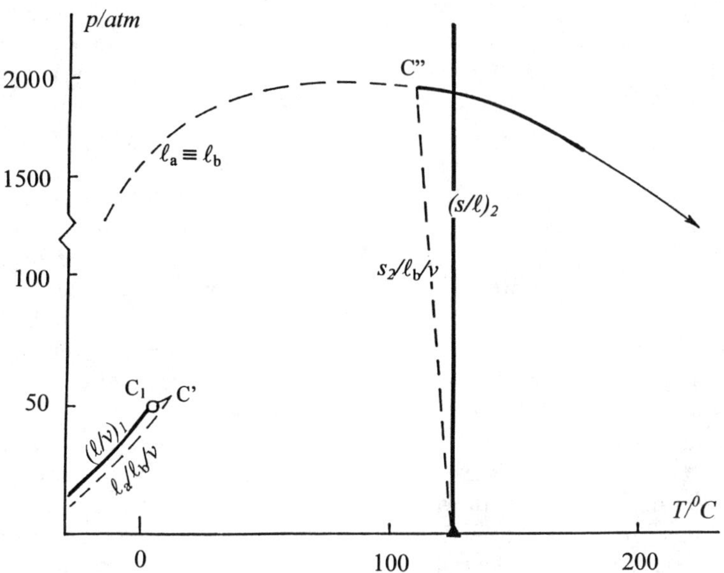

FIG. 18 Bakhuis–Roozeboom $p(T)$ projection for the system ethylene/branched polyethylene. Heavy curves are based on experimental data, light curves are estimations. Triple point polyethylene: ▲; three-phase equilibria: solid PE/liquid b/vapor, $s_2/l/v$; liquid a/liquid b/vapor, $l_a/l_b/v$; C_1: liquid/vapor (l/v) critical point of ethylene; C' and C'': critical end points; critical demixing curve: $l_a \equiv l_b$; pressure dependence of PE melting point: $(s/l)_2$ [96].

and expansion of the reaction mixture. Flory's Equation of State (EoS) theory and the Mean-Field Lattice Gas (MFLG) model supply treatments that introduce free-volume contributions to a lattice-like basic concept. Both approaches have been investigated [54, 95] in attempts to reproduce all critical data simultaneously. It was found that in both models the parameter values calculated for the individual samples differ substantially among each other. For example, one finds in the MFLG treatment that the interaction parameters for the ethylene/PE mixture change monotonously with the average molar mass of the sample [54]. Consequently, a simultaneous description of all samples with the same parameter values is, at the moment, not possible and neither are predictions reaching beyond the scope of the measurements.

An alternative is provided by the so-called Semiphenomenological (SP) treatment in which the simple FHS rigid-lattice expression [Eq. (15)] is extended with the aid of classical thermodynamic relations of general validity [54, 96, 98, 99]. We have

$$\Delta V^e = (\partial \Delta G/\partial p)_{T,\phi_i} \tag{22}$$

$$\alpha = (1/V)(\partial V/\partial T)_{p,\phi_i} \tag{23}$$

$$\beta = -(1/V)(\partial V/\partial p)_{T,\phi_i} \tag{24}$$

The excess volume is referred to as ΔV^e, and α and β stand for the thermal expansion coefficient and the isothermal compressibility.

Integration of Eqs. (23) and (24) for constant α and β leads to an exponential dependence of V on p and T:

$$V = V_0^* \exp[\alpha(T - T_0)] \exp(-\beta p) \tag{25}$$

where V_0^* is the volume at zero pressure and some reference temperature. Assuming α and β to be representative for pure substances as well as mixtures we may express the excess volume as

$$\Delta V^e = \Delta V_0^* \exp[\alpha(T - T_0)] \exp(-\beta p)\phi_1\phi_2 \tag{26}$$

Combining Eqs. (22), (26) and the FHS expression for ΔG [Eq. (15)], followed by integration in p, yields a general form for $g(p, T)$. Expansion of the exponentials and truncation after the quadratic terms finally results in

$$g = (A + BT + CT^2)(D + Ep + Fp^2)/T \tag{27}$$

Any of the coefficients A to F may be expected to depend on concentration and the degree of complexity in which Eq. (27) can be applied will be dictated by the available experimental data. The SP treatment resembles a cell theory (see Sec. III.A.3) since the volume fractions are assumed not to vary with p and T, but the volume of the lattice sites changes with these variables ("breathing lattice").

Analysis of De Loos *et al.*'s data with the SP treatment [54, 95] revealed that a single expression for the interaction function suffices to describe all data simultaneously. Predicted cloud-point curves come quite close to experimental values although deviations may become large at low polymer concentration. Improvement can be obtained on this point if the interaction term g is allowed to depend strongly on concentration [54].

The applicability of the SP treatment for practical purposes thus appears to be well supported by the present analysis of the system ethylene/poly(ethylene). Its value is greater than that, however. The system cyclohexane/poly(styrene) has recently been analyzed with the SP treatment. Here we have a complex phase behavior to deal with, upper as well as lower critical miscibility, and miscibility improving or declining with increasing pressure, depending on the polymer molar mass. Again the SP approach allowed all data to be covered in a nearly quantitative fashion with a single interaction function [54, 98].

If the objective is more limited, and data on a single polymer sample must be analysed, either Flory's EoS or the MFLG model may still serve a useful purpose. The molecular architecture occurring in HDPE (long and short branches) appears to be well covered by the MFLG model if extended in the sense indicated in Sec. III.B. Figure 19 illustrates that the increase in miscibility brought about by chain branching (Fig. 11) can be covered by an appropriate introduction of Eq. (17) into the MFLG model [100].

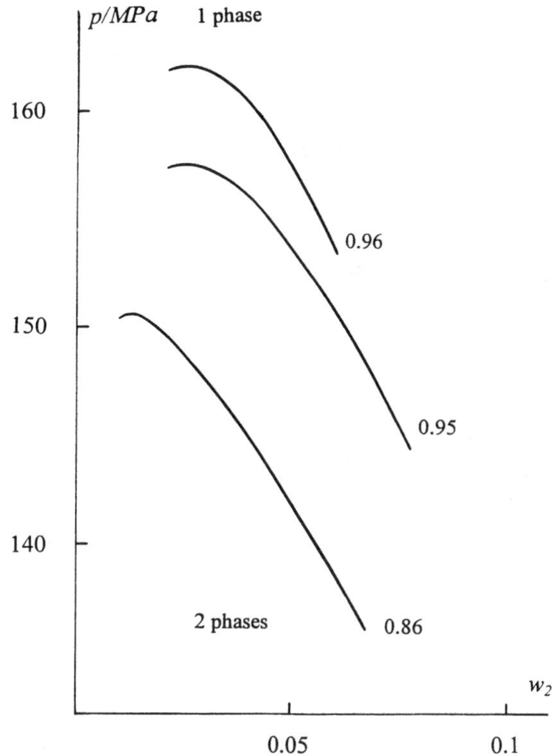

FIG. 19 Spinodals calculated with the MFLG model for the system ethylene/polyethylene showing the solubilizing effect of increased branching [decreasing values of c' in Eq. (16)], as indicated [100].

IV. CONCLUSION

In this chapter the emphasis was on the support equilibrium thermodynamics can supply in dealing with two practical problems met in polyolefin production and characterization. Liquid–liquid phase separations in particular are essential in preparative fractionation and occur frequently during synthesis and/or product recovery. We have demonstrated that simple molecular models exist that are capable of describing known data and sometimes predict behavior that has not yet been observed. Useful recipes for preparative fractionation can be derived from these models that are also helpful in ordering phase separations that may be of practical interest, such as fluid–liquid separations during the bulk polymerization of ethylene, or demixing during solution (co)polymerization of ethylene in hydrocarbons. Although no simple model exists, capable of *ad hoc* prediction of thermodynamic properties of polyolefin solutions in undocumented cases, the treatments discussed here do supply a manageable mathematical framework that greatly reduces the need for extensive experimentation.

SYMBOLS AND ABBREVIATIONS

A	Helmholtz free energy of mixing
A–F	Coefficients
a_{01}, c_{01}, b_{01}	Parameters related to differences in size and shape and respectively change in internal energy of mixing
a, b_0	Entropy correction terms
b	Parameter related to average interaction energy
E_0	Internal energy of the system
EoS	Flory's equation of state
FHS	Flory–Huggins–Staverman theory

g	Interaction function
$g_{1\alpha}, g_{1\beta}$	Solvent homopolymer interaction
$g_{\alpha\beta}$	Comonomer interaction
G	Free enthalpy of mixing
k	Boltzmann constant
l_f	Segmental free energy
LCM	Lower critical solubility
LCST	Lower critical solubility temperature
mmd	Molar mass distribution
m_i	Length of the macromolecular chain i relative to the size of the solvent molecule
m_1	Lattice sites
MFLG	Mean-field Lattice Gas theory
N_s	Number of sites
N_h	Number of holes
n_1	Number of molecules
n_0	Amount of vacant sites in moles
r	V_a/V_b
R	Gas constant
s	Number of mers
s_0, s_1, s_2, s_3	Specific contact surface area of solvent molecules, end segments and linear and branched middle segments
SS	Simha–Somcynski theory
SP	Semiphenomenological treatment
T	Temperature
U_n	M_w/M_v
U_w	M_z/M_w
v_0^*	Volume at zero pressure
V^*	Cell volume
V_s	Molar volume of the solvent
v_p	Specific volume of polymer
v_o	Volume per lattice site
Δv^e	Excess volume
v^*	Repulsion volume
V	Molar volume
V	Volume of system
w	Weight fraction of polymer
$w(M)$	Mass distribution normalized to 1 g polymer
$w_b(M)$	Mass distribution of the fraction in concentrated phase
x_b	The relative amount of the polymer in concentrated phase
Δw_{hk}	Free energy for *h–k* contacts
y	$N_s/(N_s + N_h)$
z	Lattice coordination number
Z	Partition function
α, β	Thermal expansion coefficient and the isothermal compressibility
Γ_{ij}	Three binary interactions
δ	Number of branched middle segments per macromolecule
ε_q	Contact energy of a pair of segments
ε_c	Lennard-Jones 6–12 potential
ε^*	Maximum attraction volume
φ_1	Volume fraction of solvent molecules
ϕ_i	Volume fraction of copolymer molecules
$\phi_{\alpha i}, \phi_{\beta i}$	Volume fractions of repeated units α and β
Φ_{1a}, Φ_{1b}	Volume fraction of component 1 in the phase *a* (very diluted) and *b* (concentrated)
Φ_0, Φ_1	Site fraction of vacant and occupied sites
$\nu = 3sNc_s; 3sc_s$	is the effective number of degrees of freedom
$\tilde{\omega}yV/N_{SV}$	the reduced cell volume
η	Average effectivity of the end segments separation parameter
σ	Separation parameter

REFERENCES

1. GV Schulz. In: HA Stuart, ed. Die Physik der Hochpolymeren. Berlin: Springer, 1953, Vol. II, Chapter 17.
2. R Koningsveid. Adv Polym Sci 7: 1, 1970.
3. R Koningsveld, LA Kleintjens, H Geerissen, P Schutzeichel, BA Wolf. In: G Allen, C Price, C Booth, eds. Comprehensive Polymer Science. Oxford: Pergamon, Vol. I, 1989.
4. LH Cragg, H Hammerslag. Chem Rev 39: 79, 1946.
5. GM Guzman. In: *Progress in High Polymers*. New York: Heywood, 1961, Vol. I, p 113.
6. MJR Cantow, ed. *Polymer Fractionation*. New York: Academic Press, 1967.
7. LH Tung, ed. *Fractionation of Synthetic Polymers*. New York: Marcel Dekker, 1977.
8. AJ Staverman, JH Van Santen. Recl Trav Chim 60: 76, 1941.
9. AJ Staverman. Recl Trav Chim 60: 640, 1941.
10. ML Huggins. J Chem Phys 9: 440, 1941.
11. ML Huggins. Ann NY Acad Sci 43: 1, 1942.
12. PJ Flory. J Chem Phys 9: 660, 1941.
13. PJ Flory. J Chem Phys 10: 51, 1942; 12: 425, 1944.
14. R Koningsveld. Chem Weekbl 57: 129, 1961.
15. J Van Schooten, H Van Hoorn, J Boerma. Polymer 2: 161, 1961.
16. VN Kuznetsov, VB Kogan, LA Venkstern, SD Vogman, TA Usatova, VA Morozova. Vysokomol Soyed A12: 2768, 1970.
17. Z Roszkowski. Polymery (Warsaw) 16: 445, 1971.

18. KE Almin. Acta Chem Scand 11: 1541, 1957; 13: 1263, 1274, 1278, 1287, 1293, 1959.
19. LC Case. Makromol Chem 41: 61, 1960.
20. K Jäckel. *Festschrift Carl Wurster*. Ludwigshafen am Rhein; BASF AG, 1960, p 269.
21. A Englert, H Tompa. Polymer 11: 507, 1970.
22. BA Wolf, H Geerissen, J Roos, P Amareshwar. Patent Application GFR P 32 42 130.3, 1982.
23. H Geerissen, J Roos, BA Wolf. Makromol Chem 186: 735, 1985.
24. H Geerissen, J Roos, P Schützeichel, BA Wolf. J Appl Polym Sci. 34: 271, 287, 1987.
25. H Geerissen, P Schützeichel, BA Wolf. Makromol Chem 191: 659, 1990.
26. R Koningsveld, AJ Pennings. Recl Trav Chim 83: 552, 1964.
27. AJ Pennings. J Polym Sci, Part C 16: 1799, 1967.
28. AJ Pennings. In: D McIntyre, ed. Characterization of Macromolecular Structure, NAS Publ. 1573, 1968, p 214.
29. VBF Mathot, MFJ Pijpers. Polym Bull 11: 297, 1984.
30. L Wild. Adv Polym Sci 98: 1, 1990.
31. PI Freeman, JS Rowlinson. Polymer 1: 20, 1960.
32. GN Malcolm, JS Rowlinson. Trans Faraday Soc 53: 921, 1957.
33. RE Wilson, PC Keith, RE Haylett. Ind Eng Chem 28: 1065, 1936.
34. A Nakajima, F Hamada. Kolloid ZZ Polym 205: 55, 1965.
35. A Nakajima, F Hamada. Rep Progr Polym Phys Japan 9: 41, 1966.
36. F Hamada, K Fujisawa, A Nakajima Polym J 4: 316, 1973.
37. Y Kodama, FL Swinton. Br Polym J 10: 191, 1978.
38. Y Kodama, FL Swinton. Br Polym J 10: 201, 1978.
39. R Koningsveld, LA Kleintjens, E Nies. Croat Chem Acta 60: 53, 1987.
40. HAJ Kennis, ThW De Loos, J De Swaan Arons, R Van der Haegen, LA Kleintjens. Chem Eng Sci. 45: 1875, 1990.
41. H Galina, M Gordon, BW Ready, LA Kleintjens. In: WC Forsman, ed. *Polymers in Solution* New York: Plenum, 1986.
42. KW Derham, J Goldsbrough, M Gordon. Pure Appl Chem 38: 97, 1974.
43. M Gordon, P Irvine. Macromolecules 13: 761, 1980.
44. R Koningsveld, AJ Staverman. J Polym Sci, Polym Symp 61: 199, 1977.
45. AR Shultz, PJ Flory. J Am Chem Soc 75: 3888, 5631, 1953.
46. H Tompa. Polymer Solutions. London: Butterworth, 1956.
47. R Koningsveld, AJ Staverman. J Polym Sci A6: 305, 325, 349, 1968.
48. BA Wolf. Ber Bunsenges Phys Chem 75: 924, 1971.
49. BA Wolf. Adv Polym Sci. 10: 109, 1972.
50. P Ehrlich, JJ Kurpen. J Polym Sci A1: 3217, 1963.
51. PJ Flory, RA Orwoll, A Vrij. J Am Chem Soc 86: 3507, 3515, 1964.
52. RA Orwoll, PJ Flory. J Am Chem Soc 89: 6814, 6822, 1967.
53. LA Kleintjens, R Koningsveld. Sep Sci Techn 17: 215, 1982.
54. L van Opstal. PhD Thesis, University of Antwerp, 1991.
55. LA Kleintjens, R Koningsveld. Coll Polym Sci 258: 711, 1983.
56. R Van der Haegen, LA Kleintjens, L van Opstal, R Koningsveld. Pure Appl Chem 61: 159, 1989.
57. L van Opstal, R Van der Haegen. unpublished results.
58. R Simha, T Somcynski. Macromolecules 2: 341, 1969.
59. A Quach, R Simha. J Appl Phys 42: 4592, 1971.
60. RK Jain, R Simha. J Chem Phys 70: 2792, 1979.
61. RK Jain, R Simha. J Chem Phys 72: 4909, 1980.
62. RK Jain, R Simha R. Macromolecules 13: 1501, 1980.
63. R Simha, RK Jain. Coll Polym Sci 263: 905, 1985.
64. E Nies, A Stroeks, R Simha, RK Jain. Coll Polym Sci 268: 731, 1990.
65. A Stroeks, E Nies. Polym Eng Sci. 28: 1347, 1988.
66. A Stroeks, E Nies. Macromolecules 23: 4092, 1990.
67. E Nies, A Stroeks. Macromolecules 23: 4088, 1990.
68. E Nies, H Xie. Macromolecules 26: 1683, 1993.
69. H Xie, E Nies. Macromolecules 26: 1689, 1993.
70. I Prigogine, N Trappeniers, V Mathot. Discuss Faraday Soc 15: 93, 1953.
71. JE Jones, AE Ingham. Proc Soc London A107: 636, 1925.
72. WG Madden, AI Pesci, KF Freed. Macromolecules 23: 1181, 1990.
73. KF Freed, J Dudowicz. Theor. Chim Acta 82: 357, 1992.
74. KS Schweizer, JG Curro. Adv Polym Sci 116: 319, 1994.
75. LA Kleintjens, HAM Schoffeleers, R Koningsveld. Ber Bunsenges Phys Chem 81: 980, 1977.
76. EA Guggenheim. *Mixtures*. Oxford Clarendon Press, 1952.
77. AJ Staverman. Rec Trav Chim 56: 885, 1937.
78. LA Kleintjens, M Gordon, R Koningsveld. Macromolecules 13: 503, 1980.
79. JE Mark. Physical Properties of Polymers Handbook. New York: American Institute of Physics, 1996.
80. R Koningsveld, WH Stockmayer, LA Kleintjens, JW Kennedy. Macromolecules 7: 73, 1974.
81. S Vereecke. PhD Thesis, Leuven, 1998.
82. BJ Schmitt, RG Kirste, RG Jelenic. Makromol Chem 181: 181, 1980.
83. R Simha, H Branson. J Chem Phys 12: 253, 1944.
84. R Koningsveld, LA Kleintjens. Macromolecules 18: 243, 1985.
85. R Van der Haegen, L van Opstal. Makromol Chem 191: 1871, 1990.

86. J Echarri, JJ Irvin, GM Guzman, J Ansorema. Makromol Chem 180: 2749, 1979.
87. R Koningsveld, LA Kleintjens. Macromolecules 4: 637, 1971.
88. D Lecacheux, J Lesec, C Quiveron, R Prechner, R Panaras, H Benoit. J Appl Polym Sci 29: 1569, 1984.
89. DR Douslin, RH Harrison. J Chem Thermodyn 8: 301, 1976.
90. P Ehrlich. J Polym Sci A3: 131, 1965.
91. R Steiner, K Horle. Chem Ing Techn 44: 1010, 1972.
92. F Nees. PhD Thesis, Karlsruhe, 1978.
93. G Luft, A Lindner. Angew Makromol Chem 56: 99, 1976.
94. T Swelheim, J De Swaan Arons, GAM Diepen. Rec Trav Chim 84: 261, 1965.
95. ThW De Loos. PhD Thesis, Delft, 1981.
96. R Koningsveld, WH Stockmayer, E Nies. *Polymer Phase Diagrams*. Oxford: Oxford University Press, in press.
97. ThW De Loos, W Poot, GAM Diepen. Macromolecules 16: 111, 1983.
98. L van Opstal, R Koningsveld. Polymer 33: 3433, 1992.
99. R Koningsveld. Adv Coll Interf Sci. 2: 151, 1968.
100. LA Kleintjens. PhD Thesis, Essex University, 1979.

16

General Survey of the Properties of Polyolefins

Cornelia Vasile
Romanian Academy, "P. Poni" Institute of Macromolecular Chemistry, Iasi, Romania

I. PHYSICAL CONSTANTS AND CHEMICAL PROPERTIES

Polyolefins have particular properties—synthetically given in Table 1—that permit both their identification and their differentiation from other polymeric materials; they also determine the end use of the items in which they enter.

Their properties depend on a great number of parameters, such as molecular structure, tacticity, composition of copolymers and modified polyolefins, molecular weight and molecular weight distribution, morphology, environment, and so on.

Most of these dependences have been discussed in other chapters, and together with the general table of required properties they lead to a thorough characterization of polyolefin materials.

II. TESTING AND SPECIFICATIONS

Standardization, testing, and characterization of plastics are very important both for the choice of suitable materials and for the prediction of their behavior during service time and also for the selection of special items versus the standard ones.

Hundreds of test methods referring to polyolefin properties are actually employed. They consist of both physical (hardness, stiffness, tensile properties, solubility, viscosity, etc.) and chemical (acetone extractable, carbonyl content, etc.) tests, representing the objects of international, national, or industrial standards and specifications. Some of them, elaborated by ASTM Committee D-20, Subcommittee XII, Polyolefin Plastics, are indicated in Table 1.

They are continuously developing, being subject to repeated revision (at least every five years) [6] due to the progress recorded in instrumentation, synthesis of new materials, attainment of new performance characteristics, and so on.

The quality control of polyolefin finished products will be presented in the following chapters.

(Text continues on p 411)

TABLE 1 General survey of the properties of some polyolefins

Property	LLDPE	LDPE	HDPE	PP
1. Usual appearance Clarity, transparence				
Film	Transparent	Transparent to opaque	Translucent to opaque	Transparent to translucent
Solid products	Hazy to opaque	Hazy to opaque	Hazy to opaque	Hazy to opaque
2. Elastic behavior		Flexible, resilient	Flexible, resilient	Hard
3. Pyrolysis bchavior	Becomes clear, melts, decomposes, vapors are barely visible	Becomes clear, melts, decomposes, vapors are barely visible	Becomes clear, melts, decomposes, vapors are barely visible	Becomes clear, melts, decomposes, vapors are barely visible
4. Ignition behavior	Continues to burn after ignition, yellow flame with blue center, burning droplets fall off Slight paraffin-like odor	Continues to burn after ignition, yellow flame with blue center, burning droplets fall off Slight paraffin-like odor	Continues to burn after ignition, yellow flame with blue center, burning droplets fall off Slight paraffin-like odor	Continues to burn after ignition, yellow flame with blue center, burning droplets fall off Slight paraffin-like odor
5. Chemical resistance ASTM D 543	Attacked by strong acids; unaffected by strong alkalis, diluted alkalis, and diluted acids	Attacked by strong acids; unaffected by strong alkalis, diluted alkalis, and diluted acids	Attacked by strong acids; unaffected by strong alkalis, diluted alkalis, and diluted acids	Attacked by strong acids; unaffected by strong alkalis, diluted alkalis, and diluted acids
6. Solvent resistance ASTM D 543		Soluble some aromatics above 60°C	Unaffected below 80°C	Unaffected below 80°C
7. Odor	None	None	None	None
8. Specific gravity ASTM D 792; D 1505 (g cm^{-3})	0.918-0.925	0.91–0.93 c: 1.0 am: 0.85	0.941–0.965	0.9–0.917 c: 0.932–0.943 am: 0.855–0.858 s: 0.87–0.89
9. Specific volume ASTM D 792 (ml kg^{-1})		1091–1061	1061–1037	1119–1048
10. Impact strength ASTM D 256; D 1709 (J/12.7 mm)	1.3	No break	1.02–8.15	0.27–4.25 2.2–12
10a. Notched Izod impact (J m^{-1})				i: 25–100 g: 670–780 m: 25–650
11. Tensile strength ASTM D 638; 651 (MN m^{-2})	22.7	(4) 6.9–15.9	21.4–38	29–38.6 m: 13.4–36.2
12. Tensile modulus ASTM D 638 (N mm^{-2})		55.1–72	413–1034	1032–1720
13. Elongation in tension ASTM D 638 (%)	600	90–800	(12) 50–900	i: 200–260 (40–50) m: 210 (37)

PB-1	PIB	P4MP	EP	EVA	Fluorinated EP
Hazy to opaque		Transparent clear		Transparent to opaque	Translucent
Flexible, resilient, hard	Leathery or rubbery, soft				
Becomes clear, melts, decomposes, vapors are barely visible	Melts, vaporizes, gases can be ignited	Melts, decomposes, vaporizes, white smoke			
Continues to burn after ignition, yellow flame with blue center, burning droplets fall off		Burns quietly Paraffin-like odor of vapors also rubber			
Slight aromatic odor					
Attacked by strong acids; unaffected by strong alkalis, dilute alkalis, and diluted acids	Attacked by strong acids; unaffected by strong alkalis, diluted alkalis, and diluted acids	Attacked by strong acids; unaffected by strong alkalis, diluted alkalis, and diluted acids	Attacked by strong acids; unaffected by strong alkalis, diluted alkalis, and diluted acids		Unaffected by most chemicals
				Attacked by chlorinated hydrocarbons and aromatics	Unaffected by most solvents
None					None
0.91–0.92	0.91–0.93	0.83	0.90	0.925–0.95	2.14–2.17
				1070	466–459
			16	No break	No break
26–30	20–60(100)	24–28	21–26	10–18	17.2–24.2
180				10–80	
300–380	1000	13–22	400–500	750–900	

(Continued)

TABLE 1 (Continued)

Property	LLDPE	LDPE	HDPE	PP
14. Modulus of elasticity in tension ASTM D 747 (MN $m^{-2}10^{-2}$)		1.18–2.42	5.33–10.4	8.92–13.8 i: 2000–2800 N mm^{-2} m: 1700–3100 N mm^{-2}
15. Flexural strength ASTM D 790 (MN m^{-2})		No break 38–48	13.8–20.3	(11.7) 34.5–55
15a. Flexural modulus (MPa)				i: 1000–2000 s: 380–760 m: 500–2400
16. Compressive strength ASTM D 695 (MN m^{-2})		Excessive cold flow	(16.5) 19–25	58.8–59 38–58
17. Compressive modulus ASTM D 695 (N mm^{-2})				1000–2100
18. Hardness (Rockwell) ASTM D 785; 1706		D 41–D 48 (Shore)	D.55–D 90 (Shore)	R 80–R 110 70–80 (Shore)
19. Abrasion resistance Taber mg/100 cycles		10–15	2–5	m: 30–80
20. Haze ASTM D 1003 (%)		4–50	10–50	i: 1.0–3.5 s: 20–48
20a. Gloss				i: 85–90 m: 116
21. Transmittance ASTM D 1003 (%)		10–80	0–40	55–90
22. Refractive index ASTM D 542 (η D)		1.49–1.506– 1.526	1.52–1.54	1.49 a: 1.4725
23. Dissipation (power) factor ASTM D 150, D 1531; D 669 (10^6 Hz)		<0.0005	<0.0003	<0.0002–0.0003 <0.002
24. Dielectric constant ASTM D 150; D 1531 (10^6 Hz)		2.25–2.35	2.25–2.35	2.2–2.3
25. Dielectric strength ASTM D 149 Short time 0.125″ thick (V mm^{-1} 10^{-2})		181–276	>316 190–200	>316 200–260
26. Volume resistivity ASTM D 257 at 23°C and 50% RH (ohm cm^{-1})		$>10^{16}(1.10^{17}–5.10^{18})$	$> 10^{16}(6.10^{15})$	$>10^{16}–10^{17}$
27. Arc resistance ASTM D 495 (S)		135–160		135–185
28. Thermal expansion ASTM D 696 (mm mm^{-1} °$C^{-1}10^{15}$)		16–18	11–13	11
29. Specific heat (kJ $kg^{-1}K^{-1}$) at 25°C	1.859	2.315	2.22–2.3	1.93 i, c: 1.789 s, c: 1.805 am: 2.349
30. Thermal conductivity ASTM C 177 (W m^{-1} °C^{-1} 10^{-3})		3.37	4.63–5.22	1.38

PB-1	PIB	P4MP	EP	EVA	Fluorinated EP
					3.45–4.8
			10–12	20–26	11.05
					19.7
200					
D 65		167–174	R 80–R 85	D 27–D 36 (Shore)	R 25
		<5		2–40	
i: 1.5125	1.50–1.51	1.459–1.465	1.47–1.48	1.47–1.50	1.34
		0.00007–0.0005		0.02–0.05	0.0003
		2.12	2.3	2.6–3.2	2.1
			320–370	248–312	197–237
	10^{10}–10^{13}	10^{16}	10^{15}	$1.5.10^{8}$	10^{19}
		> 90		8–80	
				16–20	8.3–10.5
i,c: 1.783 a: 2.092	1.948	ic: 1.725		2.8	1.18
				0.3	2.1

(Continued)

TABLE 1 (Continued)

Property	LLDPE	LDPE	HDPE	PP
31. Heat distortion temperature, ASTM D 648 (°C 4.5 MN m^{-1} 10 MN m^{-2})		32–41 40–50	43–60 60–82	54–66 90–100
32. Softening point (Vicat) (°C)		85–100	112–132	150 138–155
33. Glass temperature (°C)		–120	–80→–90	–10; –18; –27
34. Melting range (°C)	136.5	105–120 depends on branching	120–130	i: 160–170 s: 138; 125–148 m: 130–165
35. Crystallization temperature (°C)		107	124–134	176
36. Crystal system	Triclinic	Orthorhombic Monoclinic	Orthorhombic Monoclinic	i: monoclinic i: triclinic i: hexagonal s: orthorhombic
36a. Crystallinity (%)				i: 55 s: 21–29 (39–50)
37. Unit cell parameters	t	t	o m	m t h o
(Å) a:	4.285	4.285	7.4 8.09	6.7 13.4 12.7 14.5
b:	4.82	4.82	4.9 2.53	20.9 6.5 12.7 5.6
c:	2.54	2.54	2.5 4.79	6.5 10.9 6.3 7.4
38. Heat of fusion ASTM D 3417–75 (kJ kg^{-1})		140–280.5		i: 189–250 s: 50.2 (32–71)
39. Entropy of fusion S_u (J $mol^{-1}K^{-1}(S_u)_v$			9.6–9.82 7.42–7.72	
40. Transition and relaxation temperatures associated with amorphous phase (°C) α:		50–95	47–67	30→77
β:		0→22	–20→47	0
γ:		–120→–98	–133→–53	–80
41. Molecular properties: Branches		20–30 ethyl and butyl branches/1000 C	Mainly linear <10 ethyl branches/1000 C	
Number of double bonds/1000 C		<0.3	<0.2	<1
Types of unsaturation (%): ASTM D 3124–72 vinylidene		80	30	Present
vinyl		10	45→100	
transvinylidene		10	20	
CH_3 placement racemic				i: 1.4 s: 91–96
CH_3 placement meso				i: 98 s: 3–8

PB-1	PIB	P4MP	EP	EVA	Fluorinated EP
54–60			51–56	60–64 (34)	
			135	64	285–295
–25	–70→–60	c: 18 am: 29			
125–135 I: 106.5 II: 124 III: 135.5 124	44	above 240			285–295
Three crystalline modifications I: rhombic II: tetragonal III: orthorhombic	Orthorhombic	Tetragonal			
I II III	o	II			
17.7 14.9 12.5	6.94	18.66			
17.7 14.9 8.9	11.96	18.66			
6.5 20.8 7.6	18.6	13.84			
75–130	214	132.5–234			

(Continued)

TABLE 1 (Continued)

Property	LLDPE	LDPE	HDPE	PP
Molecular weight $\bar{M}_w$		50,000–300,000	50,000–300,000	220,000–700,000
$\bar{M}_n$		10,000–40,000	5,000–15,000	38,000–60,000
$\bar{M}_w/\bar{M}_n$	<5	2–50	4–15	i: 5–12; s: 1.7–2.6 m: very narrow MWD
42. Chain conformation N*P/Q		2*1/1	2*1/1	2*3/1 4*2/1
43. Low temperature performance		Good	Good	Fair
44. Brittleness low temperature ASTM D 746°C		–156	–156→–73	25
45. Effect of sunlight		Surface crazing Embrittles	Surface crazing Embrittles	Surface crazing Embrittles
46. Environmental stress crack resistance ASTM D 1693 failure max %		20–50	20–50	
47. Melt index or melt flow rate ASTM 1238–15 T (g/10 min)	5.6	0.46–19.9	0.46–19.9	i: 0.4–22.9 or <35 s: 2.9–8.9 m: 0.2–1500
48. Machining qualities		Fair	Excellent	Excellent
49. Moldability		Excellent	Good	Excellent
50. Mold shrinkage at injection				
$(mm\ mm^{-1})$		0.02–0.035	0.02–0.035	0.015–0.025
(%)		1–5	2–5	1–2.5
51. Water absorption ASTM D 570 (1/8″ thickness % 24 hours)	<0.015	<0.01	<0.01	<0.01
52. Permeability, 10^{-3} $(cm{-}3\ cm\ cm^{-2}\ s^{-1})$ $(cm\ Hg)^{-1})$ at 25°C				
water vapors		90; 420	60; 12	51; 160
oxygen		2.88; 15	0.403	2.3; 4
carbon dioxide		12.6; 55	0.36; 13	9.2; 12
helium		4.9	1.14	38
hydrogen				41
nitrogen		0.969	0.143	0.44
53. Critical surface tension $(mN\ m^{-1} \equiv dyn\ cm^{-1})$		25.5–36	28–34.1	29–34

PB-1	PIB	P4MP	EP	EVA	Fluorinated EP
I: 2*3/1	2*8/5	2*7/2			
II: 2*11/3		2*4/1			
II: 2*40/11					
				Very good	Excellent
			−73		
Surface crazing Embrittles				Embrittles	Unaffected
			3.3		
				Fair	Good
				Good	Good
				0.007–0.012	0.001–0.005
			1–2	0.7–1.1	
0.3–2			0.01	0.05–0.13	0
			12–450		
			0.403	2800	
		32.3	0.36	59.2	
		92.6	1.14–31	132	
		101.0			
		136.0	0.143–4.87		
		7.83			
	27	25	28	37	

(Continued)

TABLE 1 (Continued)

Property	LLDPE	LDPE	HDPE	PP
54. Solvents	Aliphatic cycloaliphatic, and aromatic hydrocarbons; higher aliphatic esters and ketones, halogenated hydrocarbons, xylene, *p*-xylene, tetraline, and decaline	Aliphatic cycloaliphatic, and aromatic hydrocarbons; higher aliphatic esters and ketones, halogenated hydrocarbons, xylene, *p*-xylene, tetraline, and decaline above 60°C	Aliphatic cycloaliphatic, and aromatic hydrocarbons; higher aliphatic esters and ketones, halogenated hydrocarbons, xylene, *p*-xylene, tetraline, and decaline above 80°C	Aliphatic, cycloaliphatic, and aromatic hydrocarbons; higher aliphatic esters and ketones, halogenated hydrocarbons, xylene, *p*-xylene, tetraline, and decaline IPP: above 80°C APP: room temp
54a. Xylene soluble (%)				i: 80–95 m: 90–99
55. Nonsolvents	All common solvents at room temperatures Triethylene glycol; 2-butoxyethanol	All common solvents at room temperatures Triethylene glycol; 2-butoxyethanol	All common solvents at room temperatures Triethylene glycol; 2-butoxyethanol	All common solvents at room temperatures Triethylene glycol; 2-butoxyethanol
56. Solubility parameters $(J\ m^{-3})^{1/2}10^{-3}$	14.3	15.7–17.99	16.8	18.8–19.2
57. Burning rate ASTM D 635 (in min^{-1})		1.02–1.06	1.02–1.06	0.75–0.83

The density and chain structure are frequently used to differentiate the sorts of PE: ASTM D 1248–60 T i: isotactic; (conventional) a: atactic; s: syndiotactic; c: crystalline; am: amorphous; m: metallocene catalysts obtained.
LLFPE, LDPE, HDPE—linear low density, low density, and (respectively) high density poly(ethylene): PP—poly(propylene); PB-1—poly-(butene-1); PIB—poly(isobutylene); P4MP—poly(4-methylpentene-1); EP—ethylene–propylene copolymers; EVA—ethylene–vinylacetate copolymers.
Source: Refs. 1–13.

PB-1	PIB	P4MP	EP	EVA	Fluorinated EP
Aliphatic, cycloaliphatic, and aromatic hydrocarbons; higher aliphatic esters and ketones, halogenated hydrocarbons, xylene, *p*-xylene, tetraline, and decaline	Aliphatic, cycloaliphatic, and aromatic hydrocarbons; higher aliphatic esters and ketones, halogenated hydrocarbons, xylene, *p*-xylene, tetraline, and decaline THF Dioxane anisol CS_2	Aliphatic, cycloaliphatic, and aromatic hydrocarbons; higher aliphatic esters and ketones, halogenated hydrocarbons, xylene, *p*-xylene, tetraline, and decaline Above 80°C	Aliphatic, cycloaliphatic, and aromatic hydrocarbons; higher aliphatic esters and ketones, halogenated hydrocarbons, xylene, *p*-xylene, tetraline, and decaline	Aliphatic, cycloaliphatic, and aromatic hydrocarbons; higher aliphatic esters and ketones, halogenated hydrocarbons, xylene, *p*-xylene, tetraline, and decaline	
All common solvents at room temperatures Triethylene glycol; 2-butoxyethanol	Lower ketones, alcohols, nitromethane	All common solvents at room temperatures	All common solvents at room temperatures	All common solvents at room temperatures	
16.49	14.5–16.47				
					1.02–1.06

REFERENCES

1. D Braun. Simple Methods for Identification of Plastics. München: Hanser, 1986.
2. TR Crompton. Analysis of Polymers. An Introduction. Oxford: Pergamon, 1989, p 6.
3. J Brandrup, EH Immergut, eds. Polymer Handbook, First, Second, and Third Editions. New York: Wiley Interscience, 1964, 1975, 1989.
4. A Le Bris, L Lefebvre. Ann Chim 9(78): 336, 1964.
5. M Rätzsch, M Arnold. Hoch Polymere und Ihre Herstellung. Leipzig: VEB Fachbuchverlag, 1973, p 140.
6. Annual Book of ASTM Standards, Parts 35 and 36: V Shan. Handbook of Plastics Testing Technology, Second Ed. John Wiley & Sons Inc. New York, 1998.
7. N Goldenberg, I Diaconescu, V Butucea, V Dobrescu. Polyolefins, Bucharest: Ed. Tchnicå, 1976, pp 11, 163, 256.
8. E Horowitz. Encycl Polym Sci Technol 12: 771, 1970. M Silberberg, R Supnik Encycl Polym Sci Technol 13: 596, 1970.
9. G Gianotti, A Capizzi. Europ Polym J 6: 743, 1970.
10. EP Moore Jr. Polypropylene Handbook. Munich: Hanser, 1996; The Rebirth of Polypropylene: Supported Catalysts. Munich: Hanser, 1998.
11. AK Kulshreshtha, S Talapatra. Chap. 1 and in Handbook of Polyolefins Ed. C Vasile, Marcel Dekker, Inc., 2000, p 1–69.
12. Y Sakata, AP Unwin, IM Ward. J Mater Sci 30: 5841, 1995.
13. H Uehara, Y Yamazaki, T Kanamoto. Polymer, 37: 57, 1996.

17

Degradation and Decomposition

Cornelia Vasile
Romanian Academy, "P. Poni" Institute of Macromolecular Chemistry, Iasi, Romania

I. INTRODUCTION

The degradation of polymers involves several physical and/or chemical processes accompanied by small structural changes which lead nevertheless to significant deterioration of the quality of the polymeric material (i.e., worsening of its mechanical, electrical, or aesthetic properties) and finally to the loosening of its functionality. Degradation is an irreversible change resembling the phenomenon of metal corrosion. Decomposition defines the chemical processes occurring under the action of heat, oxygen, chemical agents, etc., and results in the formation of nonpolymeric products or in a new structure completely different from the original. It is the advanced stage of degradation, and distinction between the two terms is rarely made [1].

II. TYPES OF DEGRADATION AND DECOMPOSITION

The initiating or degrading agent defines the type of degradation or decomposition as shown in Table 1 [1–11].

The most common types of degradation occur through chemical reactions (chain scission, cross-linking, side group elimination, chemical structure modification, etc., or their various combinations); also, there exist a few important types of degradation, such as environmental stress cracking and thermal embrittlement, resulting from physical changes (disruption of polymer morphology or of secondary bonds, or changes in the conformation of macromolecules).

It is very difficult to study a "pure" type of degradation; actually, this would only be theoretically justified as, under practical conditions, the various agents act simultaneously. For example, biodegradation occurs with measurable rate at 60–70°C; oxidative degradation is faster at $T > 100°C$ (thermo-oxidative degradation) or in the presence of light (photo-oxidation). Also it is difficult to distinguish between thermal and thermochemical degradations because the impurities or additives presented in the polymeric material might react with the matrix if the temperature is high enough.

Environmental (outdoor or indoor) degradation, weathering or aging is induced by all chemical, physical, biological, and mechanical agents (solar light, temperature, oxygen, humidity, microorganisms, rain, wind, harmful atmosphere emission, abrasive action of powders). The main factors affecting the behavior of polymers under degradative agents' action are: oxygen availability, temperature, impurities, residual catalyst, physical form (molten versus solid), morphology (crystallinity), air pollutants, radiation exposure, chemical exposure, part thickness, stress in the material, comonomer content for copolymer, presence of additives, and so on.

III. IMPORTANCE OF DEGRADATION AND DECOMPOSITION STUDIES

Special attention must be paid to the degradation of polymers in all stages of their life: synthesis, processing,

TABLE 1 Degrading agents and types of degradation [1–11]

Degrading agent	Type of degradation or decomposition
Light (ultraviolet, visible)	Photochemical degradation
X, γ rays, fast electrons, etc.	High-energy radiation-induced degradation
Laser light (pulsed mode)	Ablative photodegradation, involving photothermal and/or photochemical processes; laser flash photolysis
Electrical field	Electrical aging
Plasma	Corrosive degradation, etching
Micro-organisms	Biodegradation or biological degradation
Enzymes *in vivo*, complex attack	Bioerosion; hydrolysis, ionization or dissolution
Stress forces	Mechanical degradation, fatigue
Abrasive forces	Physical degradation, physical wear, environmental stress, cracking
Ultrasound	Ultrasonic degradation
Chemicals (acids, alkalis, salts, solvents, water)	Chemical degradation and/or decomposition, reactive gases, etching, solvolysis, hydolysis
Heat	Thermal degradation and/or decomposition
Oxygen, ozone	Oxidation, oxidative degradation and/or decomposition, ozonolysis
Heat and oxygen	Thermo-oxidative degradation and/or decomposition; combustion
Light and oxygen	Photo-oxidation

and service. Thus, even in synthesis reactors or in processing devices, degradation may occur if some overheating appears or if materials are subjected to high temperatures, mechanical forces, or oxygen action during drying, grinding, extrusion, molding, etc. Under inadequate synthesis or processing conditions, a great number of structural defects or impurities may be incorporated into the polymeric material, which will influence all properties and degradation behavior as well.

The level of degradation (degree of deterioration) reached in obtaining the fabricated products has considerable influence on the length of service life. A higher degree of deterioration and a shorter service life will lower product quality.

A good understanding of the physical and chemical aspects of degradative processes and of the decomposition of polymers facilitated the development of important practical applications, such as:

Establishment of the possibility of preventing or retarding this undesired, deleterious, degrading process by improving the synthesis and processing conditions of the polymeric materials, and also by suitable additive use, X-ray cross-linking of polyethylene (PE) applied in pipe fabrication in order to prolong service life.

Surface treatments by laser or plasma irradiation or by use of chemical agents (HNO_3, O_3, $KMnO_4$+ H_2SO_4, inducing better compatibility, wettability, printability, dyeability, adherence, etc. (which also represents a good method for morphological study, as the amorphous regions are preferentially attacked).

Photochemical, mechanical, high-energy initiation frequently used in the synthesis of block and grafted copolymers.

Controlled degradation for the improvement of melt flow index, MWD, and production of waxes.

Production of photosensitive materials or biodegradable polymers with adjustable life.

Materials and energy recovery from polymer wastes by pyrolysis and combustion.

IV. AGING-ENVIRONMENTAL DEGRADATION

Aging of plastics is a very complex process of degradation of special practical interest. It begins, as with the aging of any organism, with its birth (production). When a polymer is subjected to harsh and aggressive environments, many agents can initiate or develop physical and/or chemical processes, that change the material properties, i.e., aging occurs [6, 9, 12, 13]. At least five types of degradation are implied as the simultaneous action of sunlight, oxygen, temperature, and harmful atmospheric emission. As an example, during their use, milk crates are successively or simultaneously subjected to sunlight, low or high temperatures, chemical stress cracking agents (detergents), multiaxial stress during handling and transport, sudden impact under load, etc. [14].

Most forms of environmental attack begin at the polymer surface. An "oxidation skin" (surface layer) having a high oxygen concentration thus appears; it is brittle, and the surface crazing is evident, showing the morphological changes. Sometimes it is obviously colored, leading to aesthetic failure. Then the process advances inside by a diffusion-controlled mechanism.

After a certain period of time—the induction period—it shows a sudden change in all macroscopic properties, often followed by a catastrophic failure (embrittlement and total disintegration). The induction period coincides with the effective end use or service time of polymeric material; it is considered the time of 50% loss in the initial properties.*

V. ENVIRONMENTAL CORROSION OF POLYMERS

Due to the increase of atmospheric pollution a new type of degradation is developing, and its effects are increasingly pronounced [15–19]. Air pollutants such as particulate matter, SO_2, N_xO_y, hydrocarbons, ozone, atomic oxygen, singlet oxygen, etc., cause serious corrosion of polymeric materials, having both a mechanical and an oxidative character.

The action of air pollutants is developed in several stages, namely, loss of gloss by abrasion, crazing and cracking, leaching, loss of reinforcement, and breakup [15].

Electrochemical corrosion is observed in polymer (EVA, PE, butyl rubber) coated metal surfaces [16].

Polyolefins react with N_xO_y or SO_3 by thermally or photochemically initiated reactions leading to nitro, nitrite ester, nitrate ester, peroxyacetyl nitrates, or, in the polymer chain, carbonyl, hydroxyl, sulfones or sulfide groups appearing. These reactions are accompanied by chain scission or cross-linking reactions.

The cumulative actions of many atmospheric pollutants significantly affect polymeric materials. Specifically, the degradative action of ozone, atomic oxygen, and singlet oxygen will be discussed below. However, the concentration of air pollutants is very important in terms of the long-term properties of materials; environmental corrosion is negligible, compared to photochemical, thermal, and mechanical aging.

* See also Controlled lifetime PO materials, Chapter 19.

VI. CHEMICAL DEGRADATION

Stereoregular polyolefins are inert to the attack of most chemical reagents such as alkalis (in all concentrations); inorganic (in diluted solutions) and organic acids (in all concentrations); aqueous solutions of neutral, acid, and basic salts; and even $KMnO_4$, $K_2Cr_2O_7$ and KNO_3. They are only attacked by oxidizing agents such as H_2SO_4 ($c > 80\%$), which cause rapid deterioration of PE mechanical properties; HNO_3 reacts with PE at an appreciable rate even at low concentrations. This is a decomposition process (acidolysis) but other processes of the same type rarely happen in polyolefins.

Liquid or gaseous chlorine or fluorine attack the polyolefins; however, LDPE tubes are used for conveying gaseous chlorine or in water treatment plants. Dry chlorine does not lead to the embrittlement of LDPE, but bromine and iodine are absorbed and diffuse through PE, with a worsening of the mechanical properties.

Certain liquids can weaken the surface layer of the polymers, inducing environmental stress cracking through solvation or swelling [20].

VII. BIODEGRADATION*

A. Introduction

Living organisms can attack synthetic polymers chemically or mechanically. The mechanical mode refers to the attack by certain rodents and insects (especially for plastic insulation of electrical cables placed underground).

Biodegradation is the ability of microorganisms to influence abiotic degradation through physical, chemical, or enzymatic action. Interplay between biodegradation and different factors in the biotic and abiotic environments is very important, whereas the differences between biotic and abiotic degradation may be interpreted as being the result of biodegradation only [21].

The main factors involved are heat, stress, oxygen, pollutants, and water. Water is necessary because the microorganisms' survival depends on it. In polymers, biodegradation occurs preferentially at the surface, where the polymer can come in close contact with the culture medium or microorganisms. Therefore it depends on the particle size and surface area.

Their activity on the polymer is studied by growth tests of microorganisms on solid agar media (fungi or bacteria test) for a definite period of time (~3 weeks); by the soil burial test; by changes in molecular weight, structure, crystallinity, density, weight loss, mechanical, optical or dielectric properties, scintillation reading method consisting of the examination of the evolution of $^{14}CO_2$ from the labeled samples, etc. [3, 22–29]. The burial test is very important for appreciation of the behavior of polymers in the composting operation used for waste disposal.

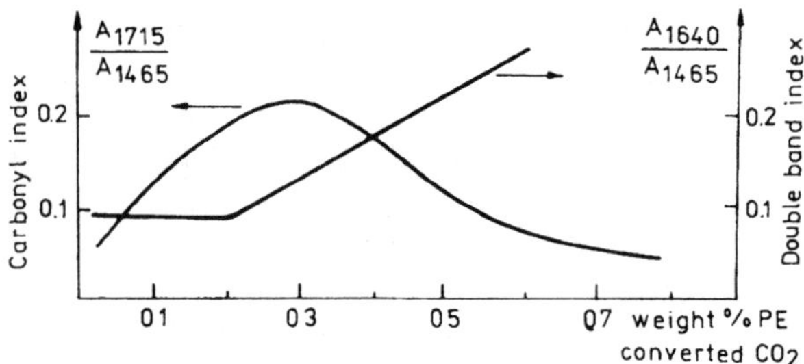

FIG. 1 Carbonyl and double bonds indices vs. the weight of PE converted in carbon dioxide. (Adapted from Refs. 26 and 30b.)

B. Biodegradation of Polyolefins

Frequently microbial attack is first directed to the additives and oligomers rather than the macromolecules themselves.

Polyolefins have a high resistance against microbial attack because the enzymatic action occurs at the chain ends only. Biodegradability was observed only for normal paraffins having a molecular weight ≤ 450. Low molecular weight PE waxes or fractions ($\bar{M} = 600$–800) are partially degraded, while high molecular weight and branched PE waxes are not degraded under the same conditions. It was observed that low molecular weight material, carbonyl, ester double bonds, etc., increase biodegradability [21, 28, 29]. This means that before biodegradation or in a strong synergism with it, another kind of degradation must occur, leading to a reduction of the molecular weight (rate determining step) up to a value at which the microorganisms can attack and biodegradation can begin.

The PE photo-oxidation facilitates the microorganism's attack because carbonyl groups are formed during this process; its consequence is increased hydrophilicity, solubility, diffusion of water, and surface area. Degraded PO in biotic and abiotic environments show an increased polydispersity, the MWD curve is shifted towards the low molecular weight region, structural changes also being important. The degraded PO behave similar with heterogeneous copolymers with increased polydispersity and long chain branching. The effect is pronounced with prolonged degradation [21].

It has been found that, during biodegradation of PE, the curve of variation of the carbonyl index shows a maximum (Fig. 1) while that of the double bonds index is linear.

On the basis of these results, Albertsson [21, 26, 27, 30] proposed a three-stage mechanism for PE biodegradation. The first stage has a constant degradation rate and a certain type of equilibrium with the environment is achieved; CO_2 evolution and oxygen uptake are fast and a rapid change in properties is observed. The second stage is characterized by a parabolic decline of the degradation rate and slow changes in properties and oxygen uptake. Finally, in the third stage, rapid deterioration of structure (high rate), collapse of structure, mineralization of polymer takes place. Because PE is a resistant polymer, few indications of the last stages exist.

This mechanism may be summarized as follows. The previously produced $>C{=}O$ groups are attacked by micro-organisms which degrade the PE chains to shorter segments, ester, acid, etc., groups being formed. These groups undergo enzymatic action, the end product being acetyl coenzyme A, which enters in the citric acid cycle whose ultimate end products are CO_2 and H_2O.

By the consumption of oxidized groups during biodegradation, the total energy of the system decreases, whereas during other oxidative degradation processes the total energy of the system increases (usually oxygen-containing groups appear).

Like each type of degradation, biodegradation also has several undesired effects. Uncontrolled microbial decomposition might occasionally cause pollution of the groundwater and of the surrounding air. In order to avoid pollution, biodegradable polymers and aerobic composting procedures have been developed.

The long-term properties and recycling in nature of polymers depend on the synergetic interaction of physical and chemical abiotic factors and also on the biotic effects of the surroundings.

VIII. MECHANICAL AND ULTRASONIC DEGRADATION

A. Mechanical Degradation

Mechanical degradation comprises both physical fracture phenomena and chemical changes induced by mechanical stresses. It is met in operations such as stirring (in solutions or other states), grinding, milling, processing in extruders, filling, flowing of solutions through capillaries, etc., as well as in products running under continuous or periodic stresses [31, 32]. During these operations heat is generated, and in most cases a thermomechanical degradation thus occurs. The extent to which the mechanical degradation appears depends on the state of the polymeric material (glassy, crystal, viscoelastic, molten, solution), on the means of applying the mechanical stress, and also on the stress value.

Mechanical degradation in a solid polymer is due to the voids and fracture planes, whereas in viscoelastic and molten states it is caused by stretching of parts of the macromolecule.

Mechanical energy can be dissipated (de-excitation) via various relaxation phenomena (conformational changes) in a harmless way, or bond scission can occur. The latter phenomenon is possible when the former is impeded (rigid material).

The capability of absorbing and storing mechanical energy depends on molecular weight and molecular weight distribution, crystallinity, temperature, etc. [20].

Crystalline, cross-linked, oriented, and high molecular weight PE (especially ultrahigh molecular weight polyethylene (UHMWPE)] is less sensible to crack propagation, fatigue, and surface or volume microvoid formation. At a high value of deformation, a preferential increase of the number of crazings, to the detriment of their dimensions, is observed. If a certain stress is exceeded, i.e., stress cracking, then failure occurs.

The brittle fracture of olefin polymers appears under load on a very short timescale. Brittle fracture time characterizes the mechanical behavior of a polymer, as brittleness temperature also does.

At lower stress, polymers can exhibit similar cracking at much longer periods of time, which is influenced by the environment. The environmental stress cracking is essentially a surface physical phenomenon.

Chain stability increases with decreasing chain length, with a limiting or critical degree of polymerization existing beyond which scission is less important [3]. Evidence for this polymerization degree is the detection of free radicals in mechanical degradation of PE, only at chain lengths above 70–100 Å (corresponding to a critical polymerization degree of 80 structural units).

If a gradually increasing stress is applied to a semicrystalline polymer, the initial main chain scission occurs almost exclusively in the amorphous regions: first the shortest tie molecules, then longer tie molecules are broken, and so on.

Initiation of polyolefin mechanical degradation increases with the applied stress [33], yet the effect is pronounced if a limit of stress is exceeded and for a deformation of 5–12%.

Like other types of degradation, mechanical degradation occurs, as has been mentioned, by a free radical chain mechanism [34], but in most cases there is a preference (nonrandom) for breaking in the middle portion of the chain, which induces an important change in molecular weight distribution [35, 36]. Shear degraded products show a significant narrowing of their molecular weight distribution, smaller molecular weight and crystallinity degree, the electrical and mechanical properties being significantly worse and the transition temperatures changed [37–39].

There are many similarities between thermal and mechanical degradation. The activation energies of mechanical degradation are in good agreement with those corresponding to the first stage of thermal degradation, as seen from values in Table 2 [36].

Structural defects, branching points, double bonds, oxygen-containing groups, decrease the thermal and mechanical stability [36]. At high temperature shearing in an extruder, new chemical structures not present prior to shearing corresponding to vinyl, vinylidene, and trans-vinylene chain ends were observed. That should mean that, during polymer shearing, the disproportionation via β scission of the main chain radicals to form olefins is one of the main elementary steps. A strong preference was shown for abstraction of methine protons of the E/P copolymers [36a]. In PE, depending on the temperature and mechanical stress, both chain scission and an increase in molecular weight can occur. Also, a significant increase in the number of branches has been observed. Scission and enlargement reactions (especially branching) are not mutually exclusive but competitive. The long-chain branchings formed predominantly at low temperature will influence, even

TABLE 2 Activation energies of thermal and thermomechanical degradation

Polymer	E_I (kJ mol^{-1}) First stage	E_{II} (kJ mol^{-1}) Second stage	%	U_0 (kJ mol^{-1}) Mechanical degradation
	Thermal degradation			
PE	96.6–105	294.0	15	105
PP	109.2	239.4	12	121.8 (126)

Source: Data from Ref. 36.

in very low amounts, the rheological properties [37], melt flow index (MFI) being much smaller at low temperature as compared with the case of thermal degradation. The long-chain branching formation and crosslinking, over the whole temperature range, proceed almost exclusively in the thermomechanical degradation of a PE containing a relatively high content of vinyl end groups.

Morphological and structural changes are also observed when the applied stress is periodic; after a certain number of cycles, the polymer presents dynamic fatigue as in the case of static fatigue [40].

Large molecules cannot be straightened without rupture. The force necessary for chain rupture is estimated to the value for C—C bond splitting. According to Stock's law, parabolic distribution of MW should be awaited as a result of degradation of macromolecules with nearly equal length. From experiments it has been established that macromolecule scission appears near the center. According to the "yo-yo" model, only the central parts of molecules become straight and the end remains in the form of a coil. As concentration increases and interaction between macromolecules rises it is expected that the rupture site—molecular center—will be gradually lose its locality because of entanglements. Other models have also been proposed [40d].

B. Ultrasonic Degradation

Ultrasonic degradation has, on the one hand, features similar to those of mechanical degradation and, on the other hand, some different ones [41, 42]. The decrease in molecular weight is relatively fast, but it slows down in the latter stages of degradation, finally reaching approximately the same value for different samples; the molecular weight distribution of degraded samples is narrow. The rate of degradation increases with the intensity of the ultrasonic waves. It is more reproducible than mechanical degradation by high-speed stirring or capillary flow.

Cavitation is a prerequisite for mechanical excitation of macromolecules and for ultrasonic degradation; the gas presence is also necessary for the initiation of degradation in solutions, which is also influenced by the solvent nature [43].

Along with the application of mechanical and ultrasonic degradation should be mentioned the reduction of the drag and the turbulent flow of liquids, viscosity index improvement for lubricating oil (important applications for PIB and its oligomers), improvement of behavior in the cycles of freezing and thawing, increase of rubber plasticity by increase in mastication time, mechanochemical synthesis of block and graft copolymers, etc.

IX. HIGH-ENERGY RADIATION-INDUCED DEGRADATION

A. General Remarks

X-rays, γ-rays, high-energy electrons, and so forth have appreciably high kinetic energy, which may cause bond dissociation (especially in the presence of oxygen) by ionization and/or free radical formation, and finally the material's failure.

However, at low doses (0–60 kGy), irradiation is used both for the improvement of long-term properties of some polymers (crosslinking of PE pipes, cables, etc.) and for medical applications. For example, in the sterilization of medical items, a dose of 25 kGy of ^{60}Co γ-radiation has been used to destroy bacteria, fungi, and spores [44–46].

At higher doses, in the presence of oxygen or during long periods of time, degradation occurs. The long-lived trapped or primary radicals represent an excellent means to initiate auto-oxidation (hydroperoxide group formation) in mild conditions, which leads to storage degradation. The result is a less complex product mixture than in the case of other kinds of degradation [47].

B. Aspects of Polyolefin Radiochemical Degradation

Radiochemical degradation occurs by three simultaneous competitively distinct processes [48, 49]: oxidative scission in amorphous phase, crosslinking and crystal destruction and/or chemicrystallization. The increase of polymer density was explained by crystallization in amorphous regions. In fact, chemicrystallization induces only an apparent increase of crystallinity and of density, observed in PE oxidation too, due to the formation of polar groups (—OH, >C=O, —COOH) which would increase intermolecular forces of attraction, resulting in a high density [50].

In anaerobic irradiation at low doses, only crosslinking occurs, whereas scission is more frequent in the oxidative medium, all changes being dependent on the depth of the oxidized layer, absorbed dose, dose rate, thickness of sample, macromolecular orientation, temperature, presence of additives, etc.

In the case of PE irradiation, it has been established that the superficial oxidized layer undergoes essentially chain scission, whereas the behavior of nonoxidized core layers is dominated by crosslinking; therefore, the samples become more and more heterogeneous as the dose increases and a skin core structure appears. If in an inert medium, γ-rays penetrate in all directions of the material; in an oxidative medium, the phenomenon is oxygen diffusion-controlled. A depth distribution will exist in the variation of carbonyl index, crystallinity, melting point, degradation products, and so on.

In most cases of polyolefin irradiation, some volatile products are detected, such as hydrogen, low hydrocarbons (methane to butane), etc. [41, 51]. Unsaturation and branching have also been identified [52]. The chain end ketones are major products of PP scission that is radiochemically induced.

All the above-mentioned changes evidence the type and extent of degradation, which can be appreciated from the ratio of crosslinking to chain scission. Polyolefins are divided according to their behavior in electron beam, in crosslinkable (PE) and degradable (PP, PIB) polymers [44]. In all cases, changes in molecular weight are observed, which can be appreciated by the relation [52]:

$$\bar{M}_n^0/\bar{M}_n = 1 + Y - X \qquad (1)$$

where $\bar{M}_n^0$ and $\bar{M}_n$ are molecular weight before and after irradiation, Y is the number of scissions per mole, and X is the number of crosslinks per mole.

The changes are particular to each polymer and depend on the environment (Fig. 2). PP and P4MP are severely degraded, and in the case of PE the most important effect of radiochemical aging is a loss in ductility [53]. Syndiotactic PP developed by metallocene catalysts is much more stable to γ-radiation during sterilization than PP [61].

UHMWPE is increasingly used in replacement surgery. One of the causes for the degradation of UHMWPE in artificial joints may lie in the sterilization treatment employed before surgery. The oxygen uptake in UHMWPE increases upon irradiation and continues to do so as the material ages. The scission process is especially important to the structural integrity of the polymer over time, as long chains are broken. The resultant shorter chains are thus able to pack together more easily, leading to a lower molecular weight material with a higher crystallinity and density (similar to HDPE after self aging for 5 years). This evolution is further complicated by oxidative degradation which can stiffen the molecular chains and lead to embrittlement and a reduced fatigue and wear resistance. This structural evolution may be responsible for a limited life of the polymer component, currently 15 years or less. The chain scission is the dominant mechanism in the UHMWPE's response to irradiation. The initial absorption of ionizing energy leads to the damage [54].

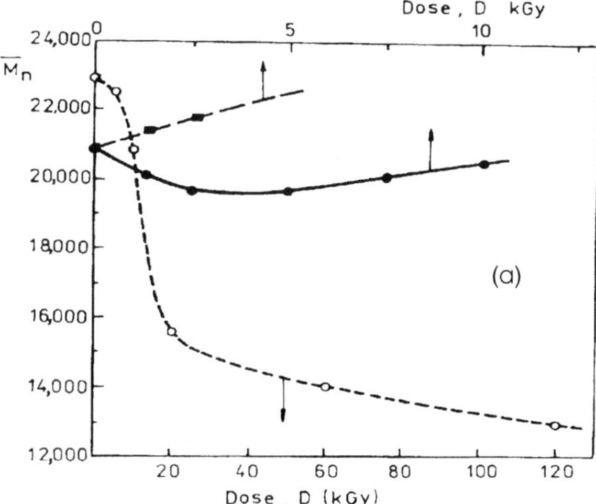

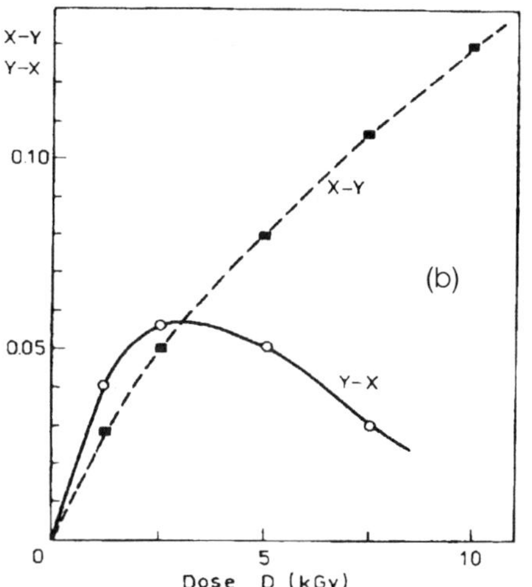

FIG. 2 (a) Molecular weight change vs. irradiation dose for the irradiated PP in air (○) and PE in air (●) and vacuum (■). (b) Number of scissions and cross-linkings for irradiated PE in vacuum (■, - - - -) and in air (○, - - - -). (Adapted from Ref. 52.)

In the EPR loaded with amine type antioxidants, free radicals produced by ionizing radiation on polymer interact with antioxidant with R-NO$^\bullet$ stable radical formation [55]. The oxidation products increase with decreasing dose rates due to the increased time for oxygen diffusion.

P-H $\xrightarrow{\text{X-rays}, \gamma\text{-rays, fast electrons}}$ Active species: radicals, ions, excited groups, and caged macro radicals

→ Cross-linking—prevalent for PE in inert atmosphere, at the beginning of the process, nonoxidized core zone; in crystalline, oriented zone; and also in highly stereospecific PP, EPDM, etc.

→ Scission—prevalent for PP, PIB, and superficial oxidized layer of PE and also in amorphous regions

The mechanism of high-energy-induced degradation in an oxidative medium is a free radical one, similar to the general mechanism of oxidation, with only the specific initiation step [44, 46, 56, 57] (see above).

The kinetic analysis of the process is based on Fick's law, modified by a term taking into account oxygen chemical consumption [53, 58, 59]. The activation energy for thermo-oxidative degradation is 60 kJ mol^{-1} [60] and for oxygen diffusion is 23 kJ mol^{-1} [58].

X. OXIDATIVE DEGRADATION

A. Introduction

In many types of degradation the interference of oxidative attack, which has an accelerating role, happens [56, 62–65], i.e., auto-oxidation. However, in oxidative degradation, the most commonly known phenomena are photo-oxidation and thermal oxidation ($T < 300°C$), the effects of which are discoloration, hardening, and surface cracking or flaking. These two phenomena are mainly differentiated by the initiation step, since other aspects are similar.

B. Initiation of Oxidative Degradation

1. Initiation of Photodegradation and Photo-Oxidation

Photo-oxidation is very closely related to outdoor weathering. The main factors involved in the process are light (sunlight) and oxygen (also its excited form: singlet oxygen and ozone). The main cause of photo-oxidation is nevertheless ultraviolet light ($\lambda = 290$–400 nm) which, although it constitutes only 4–5% of the total radiation, provides enough energy (419–217 kJ $Einstein^{-1}$) to break the chemical bonds. As a consequence of this, free radicals are formed by excitation of absorbing functional groups ("chromophores") in polymer [66].

Polyolefins, having only saturated (C-C and C-H) bonds, are nonabsorbing UV light substances and should be stable during irradiation. However, they degrade rapidly when exposed to light, the effects being evident a few months later for PE and earlier for PP, PB-l, and other polyolefins. Some chromophores are also produced during fabrication, storage, and processing that play a key role in photodegradation initiation. Thus, the following chromophores have been evidenced [67–84]: oxygen–polymer charge transfer complexes (O_2 ... PO), catalyst residues, hydroperoxides (excited with $\lambda_e = 320$ nm), hydroxyl groups ($\lambda_e = 320$ nm), carbonyl groups ($\lambda_e = 270$–360 nm), singlet oxygen, isolated (λ_e up to 250 nm) and conjugated (visible light) double bonds, polynuclear aromatic hydrocarbons (PAH) (naphthalene, phenanthrene, hexahydroxypyrene) absorbed from the environment, and other nonchemically bonded impurities. Some additives such as antioxidants, fillers, flame retardants, etc., and also degradation products of polymers and/or additives are sources of UV-absorbing chromophores.

It is recognized unanimously that photo-initiation is characterized by a large randomness due to the statistical distribution of the chromophores in the material and the fact that all chromophores accelerate photodegradation. Their relative importance in the initiation step is highly debatable. The following orders have been established:

Scott *et al.* [67–70]:
POOH $>\!\!>C{=}O > [C{=}C \ldots O_2]$ or
POOH$>Ti^{4+}$ residue$>[PP^* \ldots O_2]$ for PP initiation and
POOH$>\ >C{=}O > \ >C{=}C< \ >[LDPE \ldots O_2]$ for LDPE photoinitiation.

Carlsson and Wiles [71–73]:
Ti^{4+} residue$\geq$ROOH$>$PAH$\approx$ $>C{=}O$ $>$RO—OR$\gg$ (O_2 ... PP].

Kuroda and Osawa [74–77]:
Extracted precipitates $>C{=}O>$PAH$\cong$—$CH{=}CH_2$. The extracted precipitates have 0.44%O and $\bar{M}_n \approx 1289$, and contain large amount of impurities with —OH, $>C{=}O$, ester groups, double bonds, etc.

Winslow [78, 79] and Lamaire *et al.* [80–83] concluded that hydroperoxides formed at 85–95°C do not show a photoinitiating effect.

Gugumus [84], studying the thermally peroxidized samples, considers that the reactions of the excited charge transfer complexes $[PO\ldots O_2]^*$ are responsible for the initiation of photo-oxidation, since non-initiating reactions occur by the photolysis of hydroperoxides and ketone groups.

Photolysis of various chromophores explains the formation of products.

(a) *Ketone photolysis.* This takes place by the Norrish I and Norrish II mechanisms, according to the following schemes [84–91]:

Norrish I

$$\sim CH_2{-}\overset{\overset{O}{\|}}{C}{-}CH_2{-}CH_2\sim \xrightarrow{h\nu} \sim CH_2{-}\overset{\overset{O}{\|}}{C}^{\cdot} + {}^{\cdot}CH_2{-}CH_2\sim$$

or

$$\sim CH_2^{\cdot} + {}^{\cdot}\overset{\overset{O}{\|}}{C}{-}CH_2{-}CH_2\sim$$

Norrish II occurs via a six-membered cycle intermediate if the ketones possess at least one hydrogen atom in γ-position with respect to $>C{=}O$.

$$\xrightarrow{h\nu} \sim CH_2{-}CH\left(\begin{array}{c} H\cdots{}^{*}O \\ CH_2{-}CH_2 \end{array}\right)C{-}CH_2\sim \longrightarrow \sim CH{=}CH_2 +$$

$$\left[\sim CH_2{-}\underset{\underset{OH}{|}}{C}{=}CH_2\right] \longrightarrow CH_3{-}\underset{\underset{O}{\|}}{C}{-}CH_2\sim \quad \text{or} \quad \sim\overset{\overset{OH}{|}}{\underset{\underset{CH_2}{|}}{C}H}{-}\underset{\underset{CH_2}{|}}{C}H\sim$$

The ratio for the Norrish type I to type II reaction in PO and olefin copolymers is reported to be about 1/50. Since the predominant type II reaction leads to vinyl unsaturation, the infrared absorbance of this group is often use to determine the extent of the reaction [78, 87–91]

(b) *Hydroperoxide decomposition.* In most oxidative reactions it is considered that hydroperoxides decomposition (respectively photolysis) (HPD) is a key step with the discussions being centered on their role as initiating or chain-branching agents. If the role of initiating agents is still under dispute [67–70, 80–84], their role as branching agents is well established:

$$ROOH \xrightarrow{t^{\circ} \text{ or } h\nu} RO^{\cdot} + OH^{\cdot}$$

Until recently they were considered the main photosensitizers, but Lemaire *et al.* [80–83] and Gugumus *et al.* [84] proposed a mechanism including both inter- and intramolecular reactions, which led to acids, saturated ketones, and transvinylene groups. They consider that only tertiary hydroperoxides would have a photosensitizing role.

The macro radicals formed by HPD will be in close proximity and would not be rapidly separated by diffusion, while the cage reaction between them would yield the identified products and confirm some experimental data on the accumulation of functional groups during photo-oxidation (Fig. 3) or thermo-oxidation (Fig. 4).

It is worth mentioning that, while ketone and vinylene group formation presents an induction period, transvinylene groups have a constant rate of appearance for a long period of time. Therefore, the three kinds of functional groups do not originate from the same type of reaction.

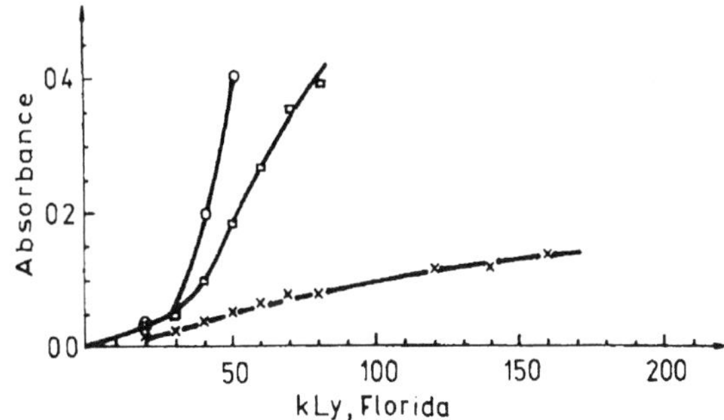

FIG. 3 Variation of various chemical groups during photo-oxidation of HDPE (2 mm plaques) in outdoor exposure conditions in Florida: (○) carbonyl (1748 cm^{-1}), (□) vinyl (909 cm^{-1}), and (×) transvinylene (967 cm^{-1}) groups. (From Ref. 84b.)

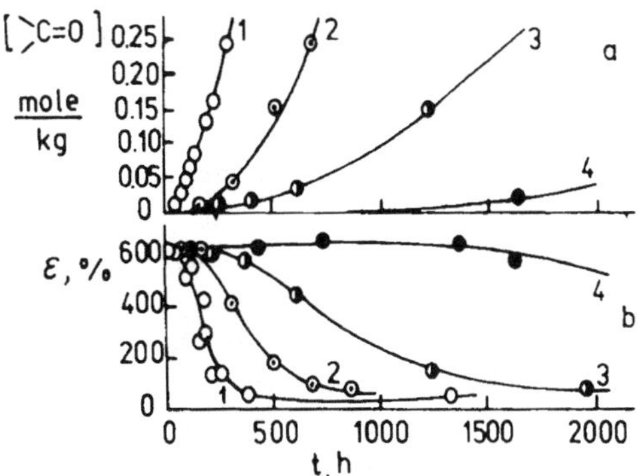

FIG. 4 Kinetic curves of carbonyl groups formation (a) and change in the relative elongation at break (b) during PE oxidation at various temperatures: 1. 100°C; 2. 90°C; 3. 80°C; 4. 60°C. (Adapted from Ref. 92.)

The additional chromophores, hydroperoxides or carbonyl groups are created during chain propagation in all types of HPD mechanisms proposed; they may give rise to the initiation of new chain reactions upon prolonged irradiation and thus to rapid deterioration of polymers.

HPD is dependent on the chemical structure, so it is very difficult to generalize. In PE, the hydroperoxide groups are isolated and lead to crosslinking and scission. In PP and PB-1 very labile blocks of hydroperoxide groups appear, so that β scission is the most important reaction [71–73]. Particularly for P4MP, due to the existence of tertiary carbons both in the backbone and in the side groups, two kinds of hydroperoxides appear [93]. The kinetic parameters are also different, the activation energy being 102.8 and 97 kJ mol^{-1}, respectively [94]. For the HPD of other polyolefins, the activation energy varies between 75 and 86 kJ mol^{-1}, while for the polyolefins containing double bonds this is lower, about 52 kJ mol^{-1} [95]. The HPD will be particularly complex with copolymers (also for LLDPE) and polymer blends, as many more kinds of reactions have to be considered [96–101].

(c) *Charge transfer complex.* The role of the charge transfer complex as the initiating agent in photodegradation has been demonstrated by inter- and intramolecular reactions [84] which would explain the PE crosslinking by combination of two alkyl or peroxy radicals.

(d) *Oxygen singlet and ozone.* $^1O_2$ is formed during photoirradiation by transfer of energy from the excited $>C{=}O$ groups; it may react rapidly with the double bonds by an "ene" type of process or with tertiary carbon atoms giving hydroperoxides [63, 72]:

$$-CH_2-CH{=}CH-CH_2\sim + {}^1O_2 \longrightarrow -\underset{\displaystyle |}{\overset{OOH}{CH}}\!\!-CH{=}CH-$$

The ozone action [92, 102] is very important for the utilization of polymeric materials in various domains, especially in strong electric fields. The peculiarity of the ozone oxidation is short kinetic chain, high radical concentration, greater share of the ozone with intermediate radicals, as well as the localization of the process at the surface and in amorphous regions. For example, the following reactions can take place:

$$\begin{array}{lcl} RO_2^{\cdot} & & RO^{\cdot} + 2O_2 \\ RO^{\cdot} + O_3 & \longrightarrow & RO_2^{\cdot} + O_2 \\ R^{\cdot} & & RO^{\cdot} + O_2 \end{array}$$

Singlet oxygen lifetime in bulk polymer matrices of poly(4-methyl-1-pentene) is $\sim 18 \pm 2$ µs [103].

(e) *Metals and metallic compounds.* Some metals and/or metallic compounds are external impurities (in ppm range) originating from synthesis (catalysts used in polymerization processes such as heterogeneous organo aluminum/titanium complexes in the Ziegler–Natta process, metal oxides of Cr, Mo, V in the Phillips process or metallocene catalysts), processing (processing equipment and containers can introduce traces of Fe, Ni, Co, Al or Cr), storage, contact with metal corrosion (e.g., contact with copper wires in cable insulation), pigments (ZnO, TiO_2, Fe_2O_3), stabilizers and vulcanization promotors (in PO elastomers), atmospheric, contact with sea water, etc. Some of these impurities catalyze or retard the attack of oxygen on polyolefins. It is well known that certain transition metal complexes prevent thermal and photochemical degradation of polymers, as exemplified by Ni and Zn dialkyldithiocarbamate which are widely used as thermal and photostabilizers. On the other side iron dialkyldithiocarbamate is a strong pro-oxidant used in biodegradable compositions. It had been mentioned that $CuCl_2$ and copper stearate, butyrate and acetate have a retarding effect on HDPE photodegradation while $FeCl_3$ markedly accelerates PO photodegradation. Their accelerating role is exemplified by the following reactions [67, 71, 104–107]:

Catalytic decomposition of hydroperoxides:

$$2ROOH \xrightarrow{M^{n+}/M^{(n+1)+}} RO^{\cdot} + ROO^{\cdot} + H_2O$$

Direct reaction with the substrate:

$$2RH + MX_2 \longrightarrow 2R^{\cdot} + M + 2HX$$

Activation of oxygen:

$$M^{n+} + O_2 \leftrightarrows M^{n+}\ldots O_2, \text{ or } M^{(n+1)+} + O_2^{\cdot -}$$

$$M^{n+} + O_2 + RH \longrightarrow M^{n+} + R^{\cdot} + {}^{\cdot}OOH \text{ (or ROOH)}$$

$$O^{\cdot -} + H^{\cdot} \longrightarrow HO_2^{\cdot -}$$

Decomposition of inorganic compounds:

$$MX \xrightarrow{h\nu} M + X^{\cdot}$$

$$X^{\cdot} + RH \longrightarrow R^{\cdot} + HX$$

Photosensitizing reaction by energy transfer to a polymer molecule:

$$MX \xrightarrow{h\nu} M^* + X^{\cdot}$$

$$M^* + RH \longrightarrow M + RH^*$$

$$RH^* \longrightarrow R^{\cdot} + H^{\cdot}$$

Transition metal salts ($CuCl_2$, $FeCl_3$) became very reactive in the presence of water. UV and/or visible light irradiation causes an electron transfer from Cl^- to metal cation. The chlorine radical produced can abstract a hydrogen atom from a polymer molecule: the resulting alkyl radical is very easily oxidized by air accompanied by chain scission. Metallic salts promote the breakdown of hydroperoxide groups to free radicals, accelerating radical reactions. The relative activity of a series of metals should be independent of the substrate, depending only on the redox potential of the metal. In fact the relative activity is unpredictable, being an unsolved problem for photodegradation. The following series were established:

For IPP: $Co^{3+} > Mn^{3+} > Cr^{3+} > Fe^{3+} > Cu^{2+}$. Oxygen uptake of IPP in the presence of Co^{3+} takes place at lower temperatures than in the presence of Co^{2+}. The catalytic effect of metal stearates in IPP oxidation in solid state varies in the following order:

$Co > Cr > Mn > Cu > Fe > V > Ni > Ti \cong Pb \cong Ca \cong Ag \cong Zn > Al > Mg \cong Cd > Ba \gg Sr$. For the oxidation in solution, the order is significantly changed [108].

For EPDM: $Co^{2+} > Fe^{3+} > Cu^{2+} \gg Na^{+} = Ni^{2+} > Ce^{1+} > Pb^{2+} > Sn^{2+} > Zn^{2+}$.

Co ions show the strongest catalytic effect whereas Cu, Fe and Mn, frequently accused of being degradation catalysts, are not always so, and Pb and Zn are not always inert as might be expected. Synergetic and antagonistic effects may appear in commercial polymers which generally contain traces of a variety of metals.

The higher the valency of the metal, the higher is its catalytic activity. In the case of complex combinations, it seems that more ionic character induces a more effective catalyzed degradation [109]. Complex formation decrease the rate of PP oxidation due to the decrease of concentration centers [110]. Much effort will have to be exerted to establish the influence of the ligand. In the retarding (stabilizing) action, the metals and metallic compounds can be scavengers for free radicals, screening or UV absorbers, quenchers of excited state energy or singlet oxygen [65–71]. Special attention has been paid to the copper- or aluminum-catalyzed oxidation, since PP, PE, and PB-1 are of greatest interest as insulation materials for electrical cables. In the polyolefins containing additives, knowledge of metal and metallic compound action is very important (see Chapters 19–22).

Although the additives are widely used to improve properties and processing, very few investigations to establish the degradation resistance have been reported [111–114]. The nucleating agents decrease the radiation stability of polymers, because the radicals generated are present in higher proportion in nucleated polymers. The hydroperoxide groups concentration formed during γ-irradiation exposure decreases as the spherulite size increased. There was a sharp decrease after 3 weeks' exposure; for longer than 6 weeks an increase in mechanical properties has been observed [114].

2. Initiation in Thermo-Oxidative Degradation

Several steps in thermo-oxidative degradation are shared with those described for photodegradation. The oxygen attack is favored by the temperature and sample characteristics. At ambient temperature most polyolefins are quite stable. The formation and decomposition of hydroperoxides constitute the key step, as these are the main branching agents, too. The rate of HPD depends on its structure being higher in the presence of vinyl and ketone groups [115]. The role of various radicals in PO degradation can be appreciated by their stability (see Table 14).

C. Experimental Studies in Oxidative Degradation of Polyolefins

The most commonly applied methods in the study of oxidative degradation are: determination of changes in intrinsic viscosity, molecular weight and molecular weight distribution, number of scissions, oxygen uptake, spectrometric methods, thermal methods, chemiluminescence, determination of electrical and mechanical properties and of gel content, etc., all helping to establish the induction period.

Processing oxidative stability is established by hot-milling at about 170°C. The "sticking point" is determined when the polymer adheres to the cooler of the two walls.

Due to the fact that, especially in the first moments of oxidation, the amounts of oxygen containing groups or products is very low, their identification by the above-mentioned methods is difficult. For detailed studies, Carlsson *et al.* [116] proposed FTIR techniques of great sensitivity, recorded after some derivatization reactions; these have been accomplished by means of phosgene and iodine to convert hydroperoxides to other compounds, diazomethane to convert acids and peracids to methyl esters, sulfur tetrafluoride to convert acid to acid fluorides, or nitric oxides to convert alcohols and hydroperoxides to nitrites and nitrates.

In advanced stages of oxidation, a great number of secondary reactions are developed progressively, the process being diffusion-controlled.

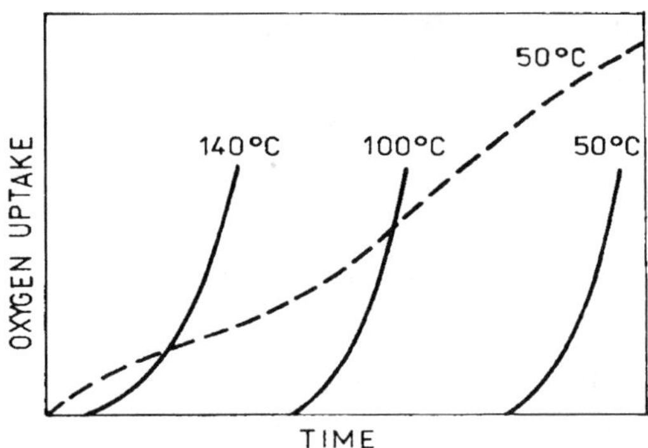

FIG. 5 Oxygen uptake vs. time for photo-oxidation (- - - - -) and thermal oxidation (————) at various temperatures for a branched PE. (Adapted from Ref. 4.)

The oxygen uptake curve for polyolefin degradation has an S shape (similar to some curves for the evolution of products (see Figs. 3 and 4) with a pronounced induction period whose duration increases with the stability of polymer and varies with the reaction conditions [92]. The initiation by heat leads to a longer induction period than in photo-oxidation (Fig. 5) and this induction period decreases with increasing temperature [4, 92, 117].

The general equations of these curves are as follows [92]:

For low degrees of oxidation:

$$v = a(t - \tau)^2 \tag{2a}$$

For various degrees of oxidation:

$$v = Ae^{-kt} \tag{2b}$$

where v is the oxidation rate, τ the induction period, t time, k rate constant, and a and A are constants describing an autoacceleration process promoted by HPD; the induction period is over when the HPD is finished. Other equations have also been proposed [84c].

The relative oxidative stabilities of various polyolefins are dependent on their chemical composition, molecular structure, molecular weight, morphology, etc. The following order has been established [101]: HDPE > LDPE > P4MP > PP.

In theoretical investigations, oxidative studies in solution, liquid phase (melting), and very thin films are preferred, as the secondary reactions can be largely minimized along with the effects of morphology [115].

When exposed to degradation both structure and morphology of semicrystalline polymers may be modified.

1. Structural Effects

However, the oxidative degradation mechanism is approximately the same for many polyolefins, the rate of elementary steps depends on the chemical structure of the polymer chain. The chain length of oxidation varies greatly depending on chemical structure of the polymer unit, the reactivity of peroxy and alkoxy radicals segmental mobility of the polymer chains, the light intensity, the permeability to oxygen and many other factors. PP and other polyolefins with tertiary carbons are vulnerable to oxidative degradation [4, 92, 117, 118]. After 1000 h exposure to oxygen under the same conditions, the cumulative oxygen uptake by the branched PE was five times greater than that of linear polymer (Fig. 6). PIB with methylene groups partially shielded by inert methyl groups is much more stable than PE. However due to the fact that the densities of the amorphous and crystalline regions are approximately the same, the entire volume of polymer is accessible to oxygen and the oxygen uptake is high, with all side groups being converted to hydroperoxides.

APP oxidizes more rapidly than IPP; while APP undergoes a random scission, the oxidation of IPP is nonrandom [115b]. The induction period for IPP oxidation is longer and the maximum hydroperoxide concentration is lower than that of APP.

With the increase in reaction time, in polyolefins there appear a change of molecular weight and a shift of molecular weight distribution peaks towards lower and lower molecular weights, the curves showing a strong narrowing. In PE, increase and decrease of molecular weight are competitive; in other words, there is a competition between crosslinking and scission reactions. In contrast, in PP oxidation the scission starts in high molecular weight fractions [121].

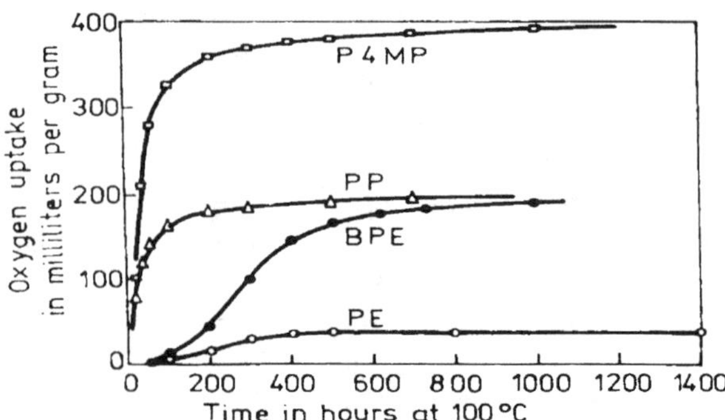

FIG. 6 Effect of the chemical structure on the oxygen uptake-time curves of some polyolefins. (Adapted from Refs. 4, 119, and 120.)

LLDPE (E/B and E/hexene copolymer) artificial and natural aging depends on the comonomer nature [122]. In E/P copolymer thermo-oxidation, the degradation mainly affects polypropylene units.

The polyolefin elastomers behavior depends on their unsaturation. Macromolecules with a reduced unsaturation mainly undergo the reaction of backbone scission while for a high unsaturation degree (>1.3%) concomitantly with the scission reaction, crosslinking takes place. Ethylene–propylene–diene terpolymers are less stable due to diene units and at increased P content, the chemical nature of the diene determines the predominant reaction, e.g., dicyclopentadiene favors the scission while ethylenenorbornene facilitates the crosslinking reactions. It has been found that EPDM containing ethylidene norbornene is significantly more oxidatively resistant than EPDM containing dicyclopentadiene [123].

2. Effects of Morphology

In terms of the influence of morphology, the following characteristics of oxidative processes have been established [52, 119, 120, 124–133]:

1. Oxidation occurs at the surface. Deeper in the polymer mass, the oxygen supply is limited at a certain level by diffusion, therefore the oxygen uptake is dependent on the sample thickness. In articles with higher surface to volume ratios such as films and fibers, the physical properties deteriorate more rapidly upon oxidation.
2. Apparently, the compact crystalline regions are inaccessible to oxygen; it penetrates preferentially in amorphous regions and its accessibility is roughly proportional to volume fraction of amorphous material and density [119, 120, 130].
3. In P4MP, the oxygen is constantly absorbed, the crystallinity decreasing during oxidation [125].
4. The effect of orientation is a drastic reduction in both permeability and solubility and, as a consequence, a significant reduction in oxidative degradation and a prolonged induction time (and corresponding prolonged service life) (Fig. 7) [52, 101, 126) both for y and UV irradiation.
5. Air oxidation leads to the formation of microcracks perpendicular to the drawing direction in IPP films.
6. The pronounced influence of morphology on oxidative behavior results in a high nonhomogeneity, even the kinetic parameters of the elementary steps of reaction having various values corresponding to crystalline and amorphous regions [127, 128].

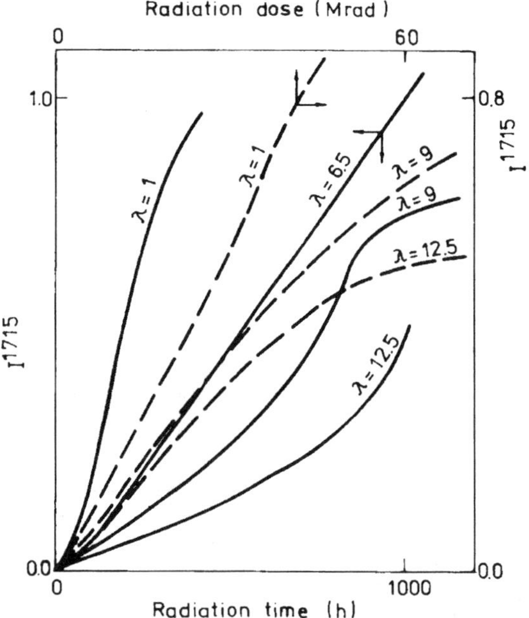

FIG. 7 Variation of carbonyl index (I_5) with radiation dose and radiation time in γ irradiation (- - - -) and UV irradiation (———) at 40°C of HDPE films with various draw ratios λ (final length/initial length); drawing temperature 60°C with a rate of 3 mm min^{-1}. (Adapted from Ref. 52a.)

There is also a nonhomogeneity in the distribution of the reagent, so that the local rates of the chemical reactions can differ markedly from the average [131], especially for copolymers, blends, and composites [132]. At high temperatures, all distributions become narrower.

7. The increase in crystallinity from 39% for unweathered to 55% for weathered LLDPE [9], Fig. 8, is due to the formation of new groups,

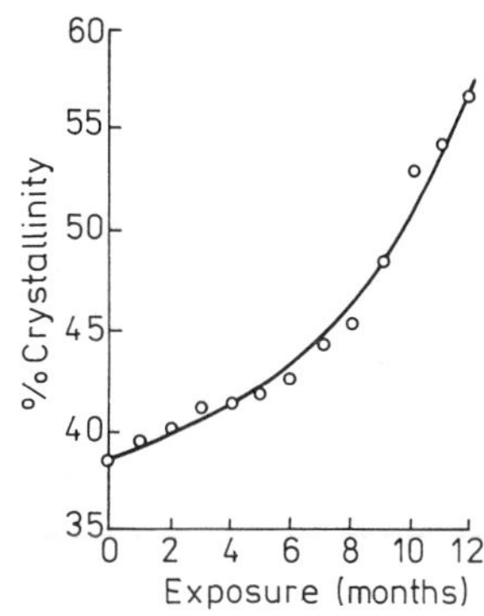

FIG. 8 The variation in crystallinity of LLDPE with exposure time [9a].

chain breaking and crosslinking. PE is a semicrystalline polymer that can be considered to behave like a two-phase system, a well-ordered crystalline phase dispersed in a less rigid amorphous phase. The gradual increase in crystallinity according to many authors is due to oxidative crystallization (interaction of polar groups formed) and scission of constrained chains in the amorphous region [9, 134–136].

In the case of PP, the main effects of photo-oxidation are the reduction of the macromolecular chain and the formation of new chemical groups, especially carbonyl and hydroperoxides—Fig. 9. The reduction in molecular weight and increase in polarity—Fig. 9a—increase the crystallizability while chemical irregularities decrease the crystallizability. As results chemicrystallization appears [137, 138]. The crystallization rate of PP is drastically reduced with exposure time, with the exception of the materials weathered for only 3 weeks which crystallized faster than virgin polymer at the same temperature—Fig. 9c. In the same time a γ-phase is developed during crystallization of highly photo-degraded samples—Fig. 9b. The spherulites showed "normal texture" in the middle and a fibrilar character towards the periphery, because the impurities tend to concentrate at the boundaries of the growing spherulites. The melting thermograms showed double peaks due to both reorganization during heating and the presence of different crystal populations. All these observations have an inherent importance in the recycled degraded plastics. The most important practical consequence of chemi-crystallization is spontaneous formation of surface cracks by contraction of the surface layers, causing deterioration of mechanical properties after short-term exposure. The strained or entangled sections of the molecules can then be relaxed and further crystallization occurs by the rearrangement of the freed molecule segments. Chemi-crystallization occurs near to the surface. The reduction in mechanical properties was greater in nucleated PP despite the observation that the extent of chemical degradation was almost identical. An increase in crystallinity—chemicrystallization—by rearrangement of molecule segments released by chain scission in the noncrystalline regions is also observed, meaning a higher molecular orientation and surface densification.

All types of oxidation give very similar products but in highly varying quantities. Beside carbon dioxide, water, and oxidized polymer, the main products of oxidative degradation are mixture of aldehydes, ketones, alcohols, acids, esters, etc. [110, 115, 121] (see Table 7).

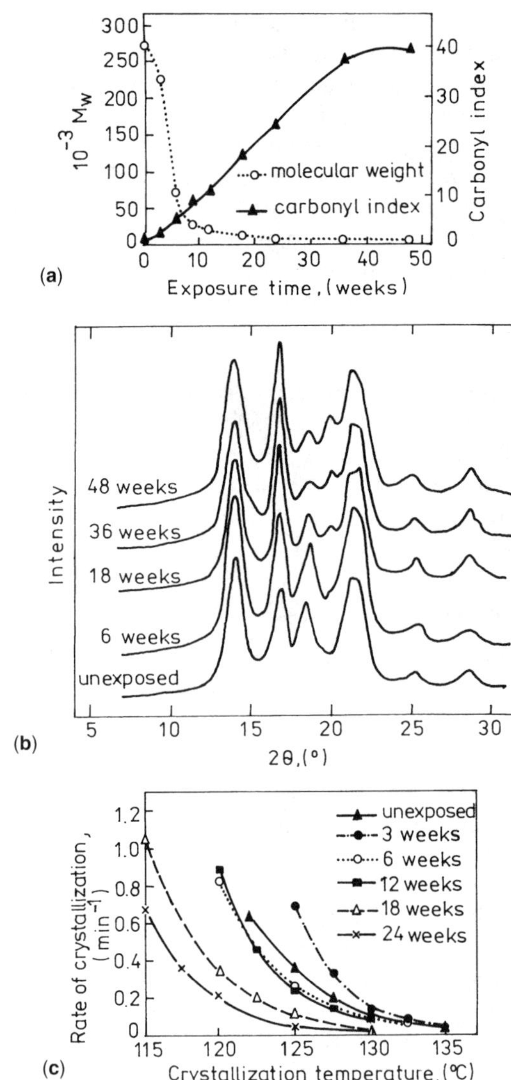

FIG. 9 Variation of MW, carbonyl index (a), crystalline structure (b) and rate of crystallization (c) of PP with exposure time [137].

D. Mechanism of Oxidative Degradation

Hydrocarbon polymers oxidize by a typical cascading free radical chain mechanism with degenerated branching [84c, 121, 139–142], similar to that proposed by Bolland and Gee [143] for volatile hydrocarbon oxidation, which can be presented by the following scheme:

(I) Initiation takes place by homolytic bond dissociation under the action of different kinds of energy:

Reaction rate expression

(1) $RH \xrightarrow{k_i} R^{\cdot}$ $\quad w_1$

(1a) RH containing chromophores

$$+ O_2 \xrightarrow{T, h\nu, M^{n+}} R^{\cdot} + {}^{\cdot}HO_2 + M^{(n-1)+}$$

In spite of the accumulation of a considerable amount of experimental data, the mechanism on this stage is not fully understood.

(II) Propagation:

(2a) $R^{\cdot} + O_2 \xrightarrow{k_{p1}} RO_2^{\cdot}$ $\quad k_{p1}[R^{\cdot}][O_2]$

This reaction is limited both by sample thickness and by the relative diffusion rates of oxygen and of volatile products.

(2a′) $RO_2^{\cdot} + RH \xrightarrow{k_{p2}} ROOH + R^{\cdot}$ $\quad k_{p2}[RO_2^{\cdot}][RH]$

or

$2R^{\cdot} + {}^{\cdot}OOH$

Since this oxidation reactions occur in a polymer cage the products may recombine to form a hydroperoxide

(2b) $RH + H^{\cdot} \longrightarrow H_2 + R^{\cdot}$

(2c) $R^{\cdot} + R'H \longrightarrow RH + R'^{\cdot}$

and the reaction of the small free radical fragments:

(2d) ${}^{\cdot}O_2H + RH \longrightarrow R^{\cdot} + H_2O_2$

(2e) ${}^{\cdot}OH + RH \longrightarrow R^{\cdot} + H_2O$

The hydroperoxide is formed either in the initiation step or in propagation. The homolytic cleavage of hydroperoxide enables an additional radical to be formed. This is a crucial step toward increasing the concentration of radicals in the polymer. It often appears as the most critical set of elementary reactions in photo-oxidation. In HDPE, hydroperoxidation occurs essentially on the secondary carbon atoms to the saturated chain end in the α-position with respect to the vinylidene groups (defects in concentration range from $c = 1.5 \times 10^{-2}$ to $3 \times 10^{-2}\,mol\,kg^{-1}$). Hydrogen-bonded hydroperoxides are unstable at 85°C as opposite to the isolated hydroperoxides formed in the α-position with respect to vinylidene groups, the last being characteristic for thermo-oxidation at low temperatures. In 1-octene LLDPE, the variation in branch points have nonsignificant influence on the photo-oxidation rate [144, 145]. ROOH accumulates in radio-oxidation, ketone groups are prevalent over acidic groups in thermo- and radio-oxidation whereas acid groups dominate in photo-oxidation.

(IIa) Chain transfer:

(2f) $RO_2^{\cdot}, RO^{\cdot}, HO^{\cdot}, R'^{\cdot} + RH \longrightarrow$

$R^{\cdot}$ + molecular compounds

$R'^{\cdot}$, $RO_2^{\cdot}$ and $RO^{\cdot}$ are the main chain carrying radicals in autooxidation. If a peroxy radical is formed, then the next step can be the intramolecular abstraction of hydrogen atoms. The intramolecular hydrogen abstraction is roughly three times more likely than the intermolecular route [146]. The intramolecular abstraction will cause the radical sites to migrate down the polymer chain. During the migration, the free radical abstracts the hydrogen of primary, secondary or tertiary carbons, forming corresponding radicals. A tertiary sterically hindered carbon radical is thermodynamically favored. The hindered nature of these radicals lowers the probability of forming a crosslink between macromolecules. This characteristic differentiates PP from the unbranched type of PE, wherein crosslinking is a very common phenomenon leading to an initial increase in viscosity.

(IIb) Chain branching:

(3a) $ROOH \xrightarrow{k_{p3}} RO^{\cdot} + {}^{\cdot}OH$ $\quad k_{p3}[ROOH]$

(3b) $2ROOH \longrightarrow RO^{\cdot} + RO_2^{\cdot} + H_2O$

(3c) $ROOH + RH \longrightarrow RO^{\cdot} + R^{\cdot} + H_2O$

(3d) ketone groups $\xrightarrow{h\nu, \text{Norrish I}} RCO^{\cdot} + R^{\cdot}$

Reaction products of the radical chain process (carbonyl groups, double bonds) again act as chromophores. At low concentrations, the HPD is a pseudo-unimolecular step. The lifetime of peroxyl radicals is 10^{-2} s, for alkoxy radicals 10^{-8} s; hence the reaction rate depends on the stability of the alkoxy (and other) radicals and varies according to the following sequence [4]:

$$CH_3{-}\underset{CH_3}{\overset{CH_3}{\underset{|}{\overset{|}{C}}}}{-}O^{\cdot} \approx -\underset{CH_3}{\overset{CH_3}{\underset{|}{\overset{|}{C}}}}{-}CH_2{-}O^{\cdot} > CH_3{-}\overset{CH_3}{\overset{|}{C}H}{-}O^{\cdot}$$

$$> CH_3{-}CH_2{-}CH_2{-}O^{\cdot} > CH_3{-}O^{\cdot}$$

(III) Termination:

$$\left.\begin{array}{lll} \text{(4a) } 2R^{\bullet} \xrightarrow{k_{t1}} & \text{inert products,} & k_{t1}[R^{\bullet}]^2 \\ \text{(4b) } 2ROO^{\bullet} \xrightarrow{k_{t2}} & \text{ketones,} & k_{t2}[RO_2^{\bullet}]^2 \\ \text{The reaction of these radicals takes place according to the Russell mechanism resulting in ketones [146a].} & \text{secondary alcohols, carboxylic acids, vinyl groups} & \\ \text{(4c) } R^{\bullet} + ROO^{\bullet} \xrightarrow{k_{t3}} & & k_{t3}[RO_2^{\bullet}][R^{\bullet}] \\ \text{(4d) } R^{\bullet} + RO^{\bullet} \xrightarrow{k_{t4}} ROR & & k_{t4}[R^{\bullet}][RO^{\bullet}] \\ \text{(4e) } RO^{\bullet} + H^{\bullet} \longrightarrow ROH & & \\ \text{(4f) } ROO^{\bullet} + H^{\bullet} \longrightarrow ROOH & & \end{array}\right.$$

As with any termination reaction, this will have the overall effect of stabilizing the polymer because of a decrease in the overall number of radicals. When an unsaturated site is formed it will be more prone to oxidation than the saturated site. All these remarks are important for polymer stabilization.

(IV) Secondary steps such as:

$$\text{(5a) } -CH(OO^{\bullet})-CH_2- \longrightarrow \sim -CH_2-C{=}O + {}^{\bullet}OH$$
$$\searrow -CH-CH- \sim + {}^{\bullet}OH \text{ (epoxide, bridged by O)}$$

(5b) The end vinylidene groups appear by intermolecular hydrogen transfer, involving a six- or five-membered cycle, the intermediate state having an activation energy of 46–71 kJ mol^{-1}.

(V) Chain scission. There are various paths which can lead to chain scission. The most common is a unimolecular β scission of carbon- and oxygen-centered radicals. The resulting products from carbon-centered radicals are olefins and a new carbon radical, which can re-enter the oxidation cycle. From the alkoxy radical, carbonyl-containing molecules and another carbon-centered radical are formed according to the reactions:

$$\sim -CH_2-C(CH_3)(O^{\bullet})-CH_2-CH(CH_3)-CH_2-CH(CH_3)-CH_2- \sim$$
$$\longrightarrow \sim -CH_2-C(=O)-CH_3 + {}^{\bullet}CH_2-CH(CH_3)-CH_2- \sim$$
$$\longrightarrow {}^{\bullet}C(CH_3)(CH_3)-CH_2- \sim$$

Other chain scission reactions would include the Norrish I and Norrish II type of carbonyl "backbiting" mechanism described above. The main effect of the chain scission is the reduction in the molecular weight of the polymer, leading to a change in many of the polymer properties. One of the most detrimental is the loss of toughness of the polymer. In addition, the scission will produce products which will tend to cause an increase in the color of the polymer and the generation of oxygenated compounds with adverse effect to the taste and odor properties of the degraded polymer. Variation both in MW and MWD during oxidation is a proof of these important reactions [147].

Oxidative degradation has a very complex mechanism, occurring by several simultaneous and successive reactions. The individual reactions and even the elementary steps [80, 84] may be affected by chemical structure, morphology of polymers, reaction conditions, and so forth.

Some differences appear among various kinds of oxidation, e.g., the role of hydroperoxides is more important in thermal oxidation, initiation being more complex, the initiation of the photo-oxidation depending on incident light intensity at a certain wavelength since it is primarily a surface reaction and the length of kinetic chain is much shorter, etc.

EVA copolymers oxidize in the β-position with respect to the acetyl groups. In thermal oxidation a polyene chain appears, while in photooxidation cross-linking by C—O—C bonds is also possible [99–101].

E. Kinetics of Oxidative Degradation

The kinetics of oxidative degradation has been developed by Emanuel and Buchachenko [13, 92], Tüdös *et al.* [115, 121, 141, 148], Shlyapnikov *et al.* [95, 127, 149], and others [84, 87–91, 133, 150, 151]. The thermal and photochemical degradation of solid PO has traditionally been studied within a kinetic framework developed for the auto-oxidation of liquid hydrocarbons.

Spectroscopic methods (particularly transmission and attenuation total reflectance IR spectroscopy have produced concentration profiles of oxidation products such as ketones, aldehydes, acids, and alcohols as a function of time [152]. These profiles show an induction period before a rapid increase in concentration to a steady concentration with time which has been interpreted as the limiting oxidation rate of the polymer. Oxygen uptake experiments confirmed the general oxidation profile for PE and PP powders and films. The degenerated branching agent is the polymer hydroperoxide and the limiting oxidation rate is controlled by the rate of the propagation reaction. Such an approach, if applied to the solid oxidizing polymer, would theoretically enable the ultimate service life of the material to be determined from the kinetic curve. A failure criterion is established and the appropriate rate coefficient can be determined accurately enough. Very sensitive analytical methods such as XPS and oxygen uptake using very sensitive pressure transducers and chemiluminiscence are used to determine the rate at the earliest stages of oxidation. The exoenergetic termination reaction of peroxy radicals occurs by the Russell mechanism. The terminating radicals may be either primary or secondary so that a six-membered transition state can be formed by the two peroxy radicals, which would lead to an alcohol singlet oxygen and a triplet excited carbonyl chromophore.

In the case of PP, the chain carrying radical is tertiary and the usual termination reaction of the peroxy radicals involves an intermediate tetraoxide that cannot lead to an emissive carbonyl. This restriction has led to the proposal of a variety of light emitting reactions. PP hydroperoxide prepared by controlled oxidation will produce chemiluminiscence (CL) when heated under nitrogen. Several reaction mechanisms have been suggested to account for this phenomenon and they do not involve peroxy radical intermediates. Up to around 180°C, all of the hydroperoxides may be decomposed.

In the liquid state, the induction time is taken as that time for the total consumption of antioxidants, after which the oxidation proceeds at the uninhibited state. The linear part of the concentration—time curve is considered to be a measure of the steady oxidation rate. The application of this approach to the integral CL curve for PP powder during oxidation at 150°C immediately reveals a difficulty in the definition of both the induction period and the steady rate of oxidation.

The magnitude of the induction period in any oxidation experiment depends on the sensitivity of the method used to measure it. The oxidation of even a single particle of powder of PP is highly heterogeneous and requires a new interpretation of the kinetic data. The physical spreading from an initial center may play an important role.

The reaction mechanism involves numerous elementary steps to which both the short-lived active centers ($R^\bullet$, $RO^\bullet$, $RO_2^\bullet$, $HO^\bullet$, $HO_2^\bullet$) and a few long-lived intermediates (ROOH) participate.

The initiation rate (w_0) is specific for thermal, chemical, irradiation, and photooxidation [56, 84, 150, 151], yet it may be assumed constant for a certain short time interval.

The kinetic analysis takes into account the following conditions [148]: (1) the kinetic chain is high enough, $\nu > 10$; (2) the substrate concentration is 10^2–10^4 times higher than that of the dissolved oxygen; (3) $k_{p1}/k_{p2} \sim 10^{-6}$–$10^{-8}$ and $[R^\bullet] \ll [RO_2^\bullet]$, k_{p1} being extremely high as, in this reaction, a free radical reacts with $^3O_2$ without any activation energy; reaction (4b) dominates among the termination reactions.

With these assumptions the following equation for the rate of autooxidation has been obtained:

$$-\mathrm{d}[O_2]/\mathrm{d}t = -\mathrm{d}[RH]/\mathrm{d}t = k_{p2}[RO_2^\bullet][RH] = k_{p1}[R^\bullet][O_2] \tag{3}$$

For each reacting species, a differential equation of the concentration change may be written. The Bodenstein–Semenov steady-state principle can be applied only for short-lived active species; therefore, either the obtaining system is analytically nonsolved or its complex solution is impracticable.

The practicable kinetic expressions for the main kinetic values are obtained in limiting cases, such as:

(a) in the beginning of the process, $t = 0$, $[ROOH]_0 = 0$, and

$$[ROOH] = w_0 t\left(1 + \frac{k_{p3}\nu_0 t}{2}\right) \tag{4}$$

where:

$$w_0 = k_{p2}[RH](w_1/k_{t2})^{1/2} \tag{5}$$

is the initial rate of the process and ν_0 is the initial kinetic chain length value:

$$\nu_0 = w_0/w_1 \tag{6}$$

The relation indicates the accelerating character in the first moments of the process.

(b) in the advanced stage of the reaction, by integration of the differential equation, at long reaction time t_1,

$$[\text{ROOH}] \longrightarrow \infty$$

$$[\text{ROOH}] = [\text{ROOH}]_\infty(1 - \sqrt{[\text{ROOH}]_1/[\text{ROOH}]_\infty}) \times \exp[-k_{p3}(t - t_1)/2]^2 \quad (7)$$

which is the equation for an S-shaped curve, in agreement with experimental results (see Figs. 3 and 4) having the induction time of

$$\tau = t_1 + \frac{1 - [\text{ROOH}]_t/[\text{ROOH}]_\infty}{k_{p3}} \quad (8)$$

The equations are especially valid for thermo-oxidative degradation, as in photo-oxidation the role of hydroperoxides is less important, their curve of formation showing a maximum; an improvement of the theory for other situations is also required. For the kinetics of hydroperoxide decomposition in a general case no stationary state for radical concentration is postulated. The induction period tends to a finite value when the initial ROOH concentration tends towards zero. The rate of hydroperoxide decomposition in the PE/PP blends is an increased function of the PP content.

Other kinetic treatments consider the diffusion phenomenon of oxygen within a sample in nonstationary conditions or nonhomogeneity of the system [87–91, 149, 150, 153].

Gugumus [154] and Mikheyev [155] elaborated two kinetic models to describe the semicrystalline nonhomogeneous PO oxidation using the following main concepts: the oxidation process represents a superposition of two processes: a homogeneous initiation of reactive chains inside the amorphous domains, and a heterogeneous spreading of the chains from one amorphous domain (reaching a limiting peroxidation level) to another by migration of the low molecular weight initiators. It postulated that the heterogeneous process proceeded from the initially more oxidized sample surfaces towards deeper layers and the rate of the oxidation spreading from one domain to another is significantly lower that the rate of developed homogeneous oxidation inside the amorphous domain. The important condition implied is that the induction period must obey the laws of the homogeneous oxidation mechanisms. The surface layers reaching the ultimate peroxidation level become a source of LMW migrating peroxides. However PP photo-oxidation proceeds irregularly even within the amorphous phase; (1) initiation occurs at high rates in localized zones probably associated with catalyst residues and other defects in the polymer; (2) stabilizers are enabled to inhibit this process but they limit the spreading of the oxidation; (3) during the induction period a branched chain reaction may occur that produces a wide variety of oxidation products including water and CO_2; (4) the induction period is controlled by the physical spreading; (5) the sigmoidal profile represents the statistical increase of the oxidizing fraction and not a kinetic curve corresponding to a homogeneous free radical chain reaction. According to Gugumus's concept in the spreading model, three macroscopic kinetic stages are distinguished: induction period, self acceleration stage, and a stage characterized by the constant oxygen uptake—see Figs. 4 and 5. The self-accelerated step is described by an exponential kinetic law, while Mikheyev considers most suitable a parabolic dependence. The heterogeneous mechanism of oxidation is localized within the amorphous domains (surrounded by crystallites) and not involve the crystalline phase. The rigid crystallites influence the structure of the noncrystalline phase. To take into consideration many of the physical and chemical features of oxidation, Mikheyev *et al.* [155] considered a micellar sponge model of an amorphous supermolecular structure. In this sponge distinct microporous zone exist, so explaining the variation in the HPD rate in various structural zones and the inhomogeneous additive (especially stabilizer) distribution and consequently their local action and "intra-cage" process performed inside micropores. Wide possibilities remain for further improvements and detailed study of the elementary chemical steps. The critical layer thickness, at which the oxidation of the PO is exclusively controlled by chemical kinetics, is 120 μm; beyond this value, the thermo-oxidative process will be affected by diffusion.

A wide range of activation energies for the oxidative reactions of uninhibited polyolefins has been reported, such as 75.6–80.0 kJ mol^{-1} for IPP oxidation at 40–100°C and 138-231 kJ mol^{-1} for 120–170°C temperature range; for HPD, the activation energy is 102–108 kJ mol^{-1} and a pre-exponential factor of $2.3 \times 10^{12} - -3.2 \times 10^{12}$ min^{-1} [156, 157], $n = 1, E = 67$ kJ mol^{-1} for the temperature range of 240–290°C [158]. The activation energy for EPR oxidation is 86.2 kJ mol^{-1} and for EPDM, 61 kJ mol^{-1} for the temperature range of 150–180°C [159].

XI. THERMAL DEGRADATION

A. Thermal and Thermo-Oxidative Stability

Thermal and thermo-oxidative stability, or heat resistance, is the ability of polymers to preserve their properties and composition within certain limits upon increasing temperature in inert or/and oxidative atmospheres. Thermal stability is limited by the strength of the bond that requires the least energy to break and does not necessarily reflect the strength of the dominant bond type. Using the quantum chemical modelling Chamot [160] evaluated activation enthalpies for several bonds in PO (see Table 3).

On the basis of these data, he concluded that both in PP and butene polymers, virtually all backbone bonds have lower enthalpy for dissociation than any of the side chain C—C bonds. The theoretically perfect alternating *i*-butene/*n*-butene copolymer could be more stable than a PP with head-to-head linkages from orientation irregularities.

PO contain thermolabile chemical defects (peroxide, unsaturation, branch points, etc.) that are weaker than those of the main chain. The thermal stability increases in the order: PIB < branched PE < PP < linear PE. The defects reduce the decomposition temperature below ~ 400°C, the value for pure polymethylene [161].

TABLE 3 Activation enthalpies of some bonds

Bond	Range of ΔE (kJ mol^{-1})
Propylene polymers	
Side chain bonds—methyl	244.4–254.1
Backbone bonds—stereoregular/isotactic	216.7–269.2
tacticity reversals	215.5–227.2
Orientation irregularities	193.2–246.9
Butene polymers	
Side bonds	219.7–131.4
Methyl, ethyl	211.3
Backbone bonds	
b-butene/isobutene	206.6–207.1
iso-butane/isobutene	165.5–182.3

Bond	Dissociation energy (kJ mol^{-1})	Bond	Dissociation energy (kJ mol^{-1})
Ordinary C—C	340.2 (311–319.2)	C=C	425
↓			
C—C—C	374 (336)	C≡C	840
↓			
—C—C=C	231	H—H	432
↓			
C—C—C=C	260.4	H—C	370 (424–508.2)
↓			
CH_2=CH—CH_2—*n*—C_4H_9	237.3	H—O	419
Allyl—C_2H_5	300.3	H—F	562
Allyl-*i*—C_3H_7	291.9 (241.5)	H—Cl	428
allyl—CH_2—*n*—C_nH_{2n+1}	241.5–247.8	C—F	436 (457–449)
Allyl—CH_2—CH_2-allyl	155.4	C—CI	293 (327.6)
C_6H_5—CH_2—CH_2—C_6H_5	197.4	C—Br	249.7
C_6H_5—CH_2-*n*—C_nH_{2n+1}	241.5–273	C=O	314 (378)
↓			
C—C—	263	C—N	224
|			
C_6H_5		C=O	727 (730.8)
—C—C—	372	O=O	267.9
| ←			
C_6H_5		N—H	351.7

*Other possible values, for the dissociation energies are indicated in parentheses.
Source: Data from Refs. 94, 160, 167, 289, and 290.

Several criteria for evaluating thermal and thermo-oxidative stability have been proposed [162–165], among which the most important are the temperatures at which the polymer weight decreases up to a certain conversion or during a certain period; the initial change rate of a property at a given temperature; the activation energy for the initial stage of degradation; the temperature at which the oxygen consumption reaches a certain value over a certain period of time (induction period), etc. Each of them has some limits, so that it is best to correlate the results obtained using several criteria [166].

B. Structure–Stability Relationship

Principally, polyolefins possess a higher heat stability than most other, polymers, yet comparison of the stability of high hydrocarbons (e.g., hexadecane is stable up to 390°C) with the PE one shows a lower stability of the latter. It must be concluded that initiation is not due to the scission of C—C or C—H bonds; another much more sensitive structure must be responsible for this.

The parameters that may change polyolefin stability are [4, 167–176]:

a. Structural ones: weak links, oxygen-containing groups, double bonds, allylic C—C bonds, tertiary carbon atoms (branching or side groups), cross-links, head-to-head or tail-to-tail linkages.
b. Molecular weight and molecular weight distribution.
c. Morphological parameters: degree of crystallinity, stereoregularity and orientation, chain rigidity.
d. Others: processing, storage, and running conditions; thermal history; presence of additives and non-bonded impurities (catalyst residue); etc.

The structure of polyolefins very much affects their stability (Fig. 10), branching being responsible for low stability, as tertiary and also allylic position hydrogens are very susceptible to free radical or oxygen attack.

Polymethylene and linear PE exhibit a maximum of the volatilization rate (at about 20–26% conversion), which is absent in branched PE. Thermal and thermo-oxidative stability of metallocene-prepared syndiotactic PP is higher than Ziegler-prepared IPP. That means that the stereoregularity is one of the dominant facts determining the stability of PP [177, 178].

Double bonds and allylic C—C bonds are very reactive with oxygen and ozone, while short crosslinks increase chain stiffness and thermal stability. All structural effects are better evidenced in the thermal studies of olefin copolymers which, with a few exceptions, have lower stabilities than the corresponding homopolymers [170, 175, 176].

In modified polyolefins, the variation of thermal stabilities is dependent on the modification type because other labile structures appear. In chlorinated APP and PE, thermal stability decreases with increased chlorine amount [180–182]. The labile groups are tertiary chlorine, vicinal chlorine, allylic chlorine, unsaturation, etc.

Generally, in polyolefins, molecular weight decreases appreciably before the evolution of volatile products [179]. For example, in the case of PE, molecular weight decreases at 270°C, the evolution of volatiles begins at 370°C, and an enlargement in molecular weight occurs simultaneously [183]: for PP,

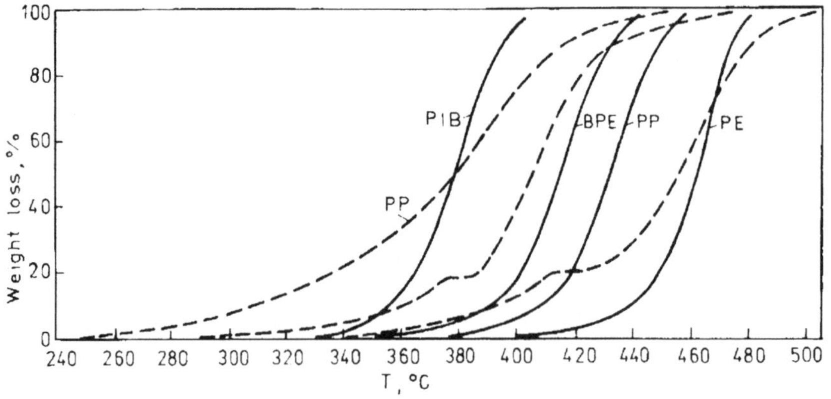

FIG. 10 Weight loss curves for thermal (———) (heating rate 1.7°C min^{-1}, nitrogen) and thermo-oxidative (- - - -) (heating rate 12.4°C min^{-1}, air) decomposition of some polyolefins. (Adapted from Refs. 4, 174, 179, and 310.)

molecular weight drops abruptly at 230–250°C, while the evolution of volatiles begins at up to 300°C. Commonly, the high molecular weight tail is reduced and the molecular weight distribution peak narrowed.

In Table 4 some data concerning thermal and thermo-oxidative stabilities of polyolefins are presented. It can be seen that, with some exceptions, the thermal stabilities, judged by T_d values, roughly correlate with D_{R-R} and T_c values. Thermo-oxidative degradation occurs at lower temperatures than the thermal degradation.

XII. DEGRADATION AND CHEMICAL TRANSFORMATION OF PO DURING PROCESSING

During melt processing, the polymer is exposed to severe conditions. The fabrication involves exposure to heat, oxygen, and mechanical shear, and the residence time in this harsh environment can range from a few seconds to several minutes or even hours. At processing temperatures of 200–300°C the oxidative reaction rates of PO are extremely rapid. Viscosity increases or decreases depending on whether scission or crosslinking predominates [184]. Crosslinking and scission can alter extrusion rates, affect melt strength, melt extensability and orientation, cause melt fracture or create gels (crosslinked network particles dispersed in films). Unsaturation plays an important role in determining the crosslinking versus scission balance as a function of processing temperature. Ziegler HDPE with low vinyl concentration and hydrogenated vinyl-free LLDPE did not cross-link under extrusion conditions whereas Phillips HDPE with high vinyl concentration did [185]. LLDPE vinyl concentration decreased during melt processing and was accompanied by crosslinking. A ^{13}C NMR study of melt processed E/1-hexene medium density copolymer showed that vinyl reduction was accompanied by LCB formation and "T"-type linkages between main chain alkyl radicals and vinyl end groups. Sixty percent of the vinyl decay was attributable to LCB formation [186]. The absence of crosslinking in PP, despite the formation of unsaturated groups, led to the hypothesis that steric hindrance of higher substituted olefins prevented the addition reaction observed with HDPE vinyl groups. Crosslinking was attributed to addition of alkyl radicals to olefin groups. The activation energy for addition to vinyl groups was approximately 18 kJ mol^{-1}, while the activation energy for β cleavage of secondary alkyl radicals to form vinyl groups was approximately 91 kJ mol^{-1}. Crosslinking dominates at lower temperatures and scission dominates at higher temperatures.

In extruder, the output influences viscosity, duration of degradation, rate of mechano-chemical initiation and in some cases, temperature as a result of self-heating. During processing changes appear in MW, carbonyl and hydroperoxide groups. It is reflected not only on the "momental" changes of properties but the lifetime left. Fortunately, a very limited amount of air has access to the melt during the extrusion or molding. Further, a nitrogen purge at the throat of an extruder has an additional marked effect on the maintenance of molecular weight [187]. The influence of mechanical field on the MWD function is specific. Its high molecular weight part is being cut off and the width of MWD is narrowed along with the degradation. Ziegler HDPE does not really change MFI and MWD or the polydispersity index (PI)—Table 5—after multiple processing while Phillips HDPE with a high concentration of vinyl groups decrease MFI and increase distribution width in the same conditions [188].

Degradation in melts is less studied. Chemistry of degradation in melts seemed not to differ much from reactions in concentrated solutions. Unfortunately, attempts to use the data of degradation theory of macromolecules in diluted solution for the prediction of processing results are unsuccessful even in an qualitative level. The situation is complicated by the fact that mechanical degradation is just the beginning of many staged process in presence of oxygen and temperature. It is very hard to choose even on the level of comparatively simple models what is mainly mechanical or thermo-oxidative degradation. An attempt was made to create a general picture in which both classes of chemical reactions are included. Mechanical degradation is a source of alkyl radicals and initial step of thermal oxidation—Scheme 1.

Superposition of mechanical degradation and thermal oxidation is expressed in increased radical concentration as compared with that for pure auto-oxidation.

The main experimental proofs on the mechanism of degradation during processing are: sharp change in MW with temperature increase during mechanical degradation; presence of three temperature ranges of MW change, namely low temperature oxidative degradation, intermediate region of mechanically MW increase, and high temperature mechanical degradation and correspondance of these MW changes with double bonds concentration. In HDPE undergone to multiple repeated extrusions, the increase of MW only

TABLE 4 Thermal and thermo-oxidative stabilities of some polyolefins

Polyolefin	D_{R-R} (kJ mol^{-1})	CE	T_d (°C)	T_c (°C)	T_h (°C) vacuum, $\frac{1}{2}$ h	$T_{10\%}$ (°C) (155)		T_{50} (°C) (155)		E (kJ mol^{-1})	T_0 (°C)
						TG	Py–GC	TG	Py–GC		
1. Polymethylene					415					302.4	
2. PE	< 415.8	1.0	400 (410)	400	406						165
3. Branched PE					404					264.6	
4. Commercial PE Hifax 1400-J						437	428	458	444		
5. PP	< 357		380 328–410	300	387	405	395	436	405	243	130 (120)
6. PIB	< 310.8	1.2	340 288–425	50	348	341	346	376	366	205.8	
7. Poly-4-methyl-1-pentene			291–341								160 (180)
8. Polyvinylcyclohexane					369					205.8	
9. Polycyclohexylethylene			335–391								300
10. EPDM (butadiene norbonene, dicyclopentadiene, 1,4-hexadiene)			325 350–370		455 410–420						
11. EPR											290
12. E/TFE			200–240								

D_{R-R}, dissociation energy into radicals; CE, cohesive energy per 5 Å chain length; EPR, ethylene–propylene rubber; E/TFE, ethylene–tetrafluoroethylene copolymer; T_d, decomposition temperature; T_c, ceiling temperature; T_h, temperature where 50% weight loss occurs upon heating of the polymer; T_{10} and T_{50} are temperatures corresponding to 10% or 50% volatilization, determined by means of thermogravimetry (TG) or pyrolysis–gas chromatography (Py–GC); E, activation energy for thermal degradation; T_0, temperature at which 0.2 mol kg^{-1} of oxygen is consumed in air for 30 min.

Source: Data from Refs. 4, 160–162, 167, and 170.

TABLE 5 Influence of multiple extrusion on fluidity and MWD width of HDPE [187]

T(°C)		220°C			240°C			260°C			280°C		
Number of extrusions		1	3	5	1	3	5	1	3	5	1	3	5
Phillips	MFI	2.0	1.4	1.2	2.3	1.5	1.2	2.8	2.0	1.0	3.4	1.7	1.1
	PI	5.9	9.7	13.2	7.4	11.2	15.7	7.8	11.6	14.2	7.5	10.0	15.4
Ziegler	MFI	4.2	4.0	4.1	4.3	4.1	4.3	3.8	4.0	4.2	3.9	4.6	5.2
	PI	1.2	1.5	1.5	1.0	1.3	1.7	1.3	1.7	2.3	3.6	2.8	1.02

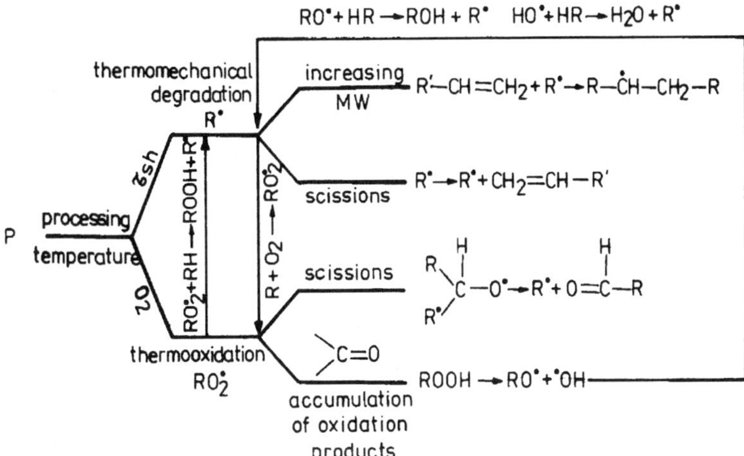

SCHEME 1 Transformation of macromolecules during processing.

for samples containing vinyl double bonds was observed—Table 5. The effect of oxygen in this process appeared from the reactions with oxygen and diffusional limitations on the oxidation rate. For PP, the oxidation rate is 5×10^{-2} mol kg^{-1} s^{-1}, diffusion coefficient 1×10^{-4} cm s^{-1} and $[O_2] = 1 \times 10^{-4}$ mol kg^{-1}.

XIII. CONTROLLED DEGRADATION AND CROSSLINKING

Physical, mechanical and processing properties of PO can be enhanced by controlled degradation or regulated crosslinking.

Controlled rheology is intentionally performed by mechanical means or irradiation for the purpose of improving flow characteristics, raising or lowering viscosity—visbreaking—and narrowing molecular weight distribution. The high temperature extrusion, a common technique in the plastic and rubber industry, is a technology established for many years. As in processing operations, polymers undergo the three types of degradations under high mechanical shear, temperature and oxygen, all occurring by a radical mechanism. Taking into account the changes occurring in the chemical structure, molecular weight, and molecular weight distribution at low degradation which are accompanied by changes in polymer properties, the controlled degradation processes have been developed in order to improve polymer quality (decreased rigidity, increased tensile strength, etc.). A "speciality polymer" is thus obtained from the corresponding "commodity" one.

Thermal bath (250–400°C, 1–24 h) or thermo-oxidative ($T < 240$°C, $t < 1$ h, air flow) and reactive processing in the presence of initiators (dialkylperoxides) are procedures currently used [189–194]. The initiators (as a powder or solution) are introduced in the compounding step, in the feed part of an extruder, in melt, along with the extruder, or in the final injection molding step.

The reactive extrusion of PE [189] at low initiator concentration (< 0.08 wt%) is used for melt flow index

(MFI) modification while at high initiator concentration (of at least 0.2 wt%) it is industrially applied for the production of crosslinked PE [191].

The modification of MFI is larger for LLDPE and HDPE than for LDPE. Two simultaneous competitive reactions (chain scission and extension) occur, the chain extension reactions being favored at low temperatures [189b].

Controlled rheology or controlled flow properties of polypropylene are increasingly applied in fiber production and in injection molding processes. It offers several processing advantages, such as high speed in the fiber spinning and in injection molding, lower processing temperatures, fewer rejects, higher dimensional fidelity (shape retention) in injection molding, thinner molding walls, etc. [195]. The PP leaving the extruder has a higher MFI, lower MW and narrower MWD.

The modification of MFI, molecular weight, and molecular weight distribution (Figs. 11 and 12) depends on peroxide concentration [188, 191]. It can be observed that the high molecular tail disappears while the low molecular weight tail remains unchanged.

Controlled degradation of APP in air at 150–300°C in the presence of di-*tert*-butyl peroxyisopropylbenzene increases the polymer functionality and reduces its isotactic content and branching. The product can be used in the processing of polyolefins and in adhesive

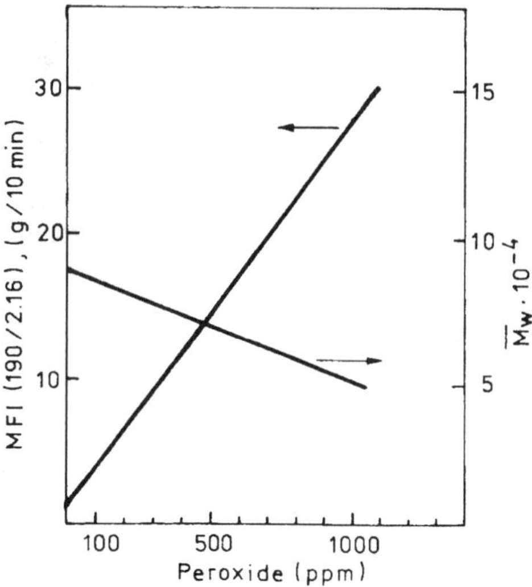

FIG. 11 Dependence of melt flow index (MFI) and $\bar{M}_w$ of PP degradation on peroxide amount. Degradation at 190°C, 5 min, 2,5-dimethylhexane-2,5-di-*tert*-butylperoxide as initiator. (Adapted from Ref. 191.)

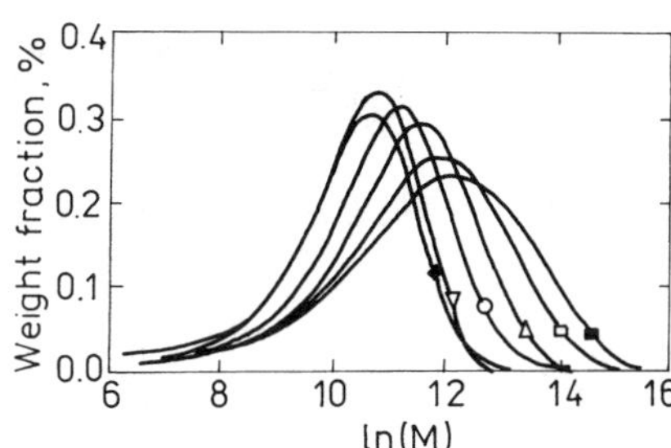

FIG. 12 Modification of molecular weight distribution curves at reactive extrusion of PP at 220°C, screw speed 44 rpm, at various initiator concentrations of ■ 0.000 virgin PP; □ 0.02 peroxide and △ 0.01 wt %; ● 0.3%; ▽ 0.5% and ◆ 0.6. (Adapted from Ref. 202a.)

compositions [196, 197]. The changes in olefin elastomer (EP, EPDM) structure during high temperature extrusion consist of a decrease in molecular weight and double bond formation of vinyl, vinylidene, and vinylene groups that will affect the behavior and stability to the oxidative degradation during storage and application [198]

The modification of PIB with peroxides is employed for the obtainment of a good additive for lubricating oils [199].

An interesting result was reported by Nanbu *et al.* [200], who used a silica–alumina catalyst at 280°C and obtained a highly branched polymer (71 branches per 1000 carbon atoms) from HDPE, while Konar *et al.* [201] eliminated the short branches from LDPE by oxidative degradation, in the presence of a phase transfer catalyst, as permanganate.

The theoretical treatments of controlled degradation are based on its radical character [189, 191, 192, 202] and chemically (peroxide) initiated degradation, or pure chain random scission, where Saito's integral is valid [202a].

The chain scission rate constant and the initiator decomposition efficiency are considered as the most important kinetic parameters of this process, because the chain scission is the main reaction which dominates control of MWD. By the deterministic and stochastic modelling procedures and Monte Carlo simulation it has been established that all kinetic parameters depend on the reaction time, 30 s giving the most reliable estimations [202].

Crosslinking of PE is a well-established technology. It improved form stability, creep resistance, and stress cracking [203]. Radiation, principally γ-radiation, can be used to graft (10–30 kGy dose), crosslink (5–25 kGy) or degrade (100–1500 kGy) LDPE.

The alkyl radicals produced double bonds and allyl radicals which also participated in crosslinking. The cross-linking rate was strongly dependent on temperature and reaction terminated quickly at $T > 180°C$ [204].

XIV. THERMAL AND THERMO-OXIDATIVE DECOMPOSITION AND COMBUSTION

A. Products of Thermal and Thermo-Oxidative Decomposition and Combustion

The qualitative and quantitative determination of the thermal and thermo-oxidative decomposition and combustion products advances closely with the development of high-performance investigation techniques. Early on, only the main products or classes of compounds were studied, whereas recently, special importance has been given to the evidence of products appearing as traces, since these might cause serious problems both for human life (by their toxicity, especially in incomplete combustion) and for the purification of the products obtained from polymer waste pyrolysis (or incineration), applied for their recovery as high-grade materials.

Important contributions in this field have been made by Levin [205]. Chaigneau and Le Moan [206], Tsuchiya and Sumi [207], and many others. Lattimer [208] found two oligomer series in the mass spectra of PO, that of 1-alkenes and that of α, ω-alkadienes. Vinyl groups are formed from secondary (or tertiary) macroradicals β-scission. α, ω-Alkadenes are cleaved from second (or tertiary) vinyl terminated macroradicals. Vinyl groups content increases with increasing pyrolysis temperatures.

Although polyolefins have an apparently simple structure, recent studies have indicated the enormous complexity of the mixtures resulting from their decomposition and combustion; see Tables 6–8.

The decomposition products of polyolefins can be classified into seven groups: hydrogen, *n*-alkanes, isoalkanes, alkenes, isoalkenes, aromatics and high boiling point compounds. In some cases, coke or a carbonaceous residue appears.

The reaction conditions (temperature, heating rate, residence time, pressure, oxygen concentration, etc.) and sample characteristics (structural, morphological. etc.) may markedly change the composition of the resulting mixtures (Table 6) and even the nature of the main reaction products. That is why in many cases the results are in apparent disagreement; the absence of a product indicates either that it was not produced under the experimentally used conditions or that the analytical techniques employed were not suitable.

As temperature increases, the rate of decomposition increases too. The amount of heavier compounds decreases, while that of lighter compounds increases [209] (Fig. 13).

At low temperatures (350–500°C) PE and PP decomposition leads to wax-like products, while at high temperatures (> 600°C) gaseous hydrocarbons are predominant. The number and amounts of thermal decomposition products of PP are greater when they are obtained by distillation-pyrolysis (longer time) than by flash pyrolysis (shorter time) [209].

Several similarities appear in the composition of gaseous pyrolysis products of different polyolefins, but the liquid products have many more peculiarities

(Text continues on p 444)

TABLE 6 Global composition of thermal decomposition products of PO

Polyolefin	Alkanes	Isoalkanes	Olefins	Dienes	Arenes	Monomer
PM	25		50	25		1
PE	58		17.7	11	12.8	0.1
LDPE	56.2	10.6				
Ziegler PE	53.8	4.2				
Phillips PE	54.1	3.9				
PP	8.15		78–75	1–2	1.0	0.17
IPP	9.4		84.5	1.9		
APP	12.0		77.5	1.0		
PIB						20.0

Flash pyrolysis, $T = 550°C$, inert atmosphere.
Source: Data from Refs. 3, 209, and 210–212.

TABLE 7 Products identified as resulting from thermal (T) and thermo-oxidative (TO) decomposition and combustion (C) of some polyolefins

		PM [207c]			PE [205, 213, 214]			PP [209, 215, 216]			PIB [207b, 217, 218]			PB-1 [219]			P3MB [220]			P4MP [221]			PCH [222]		
No.	Compound	T	TO	C	T	TO	C	T	TO	C	T	TO	C	T	TO	C	T	TO	C	T	TO	C	T	TO	C
1.	Acetaldehyde		m	t		m	t		m	t											l				
2.	Acetic acid		m	t		m	t		m	t										7	5–7	t			
3.	Acetone		m	t		m	t		c	t		m						m	t		m	t			
4.	Acetylene	l	l	l		l	l	l	l		l	l		l	l					l	l				
5.	Acenaphthalene	t	t		t	t																			
6.	Acrolein		t	t		c	t		t	t											2–7 c				
7.	Acrylic acid			l			l																		
8.	Anthracene	t	t	t	t	t	t	t	t	t															
9.	Benzene	l	l	t	l	l	t	l	l	t															
10.	Benzanthracene			t			t																		
11.	Benzocoronene	t	t	t	t	t	t																		
12.	2,3-Benzofluoranthene	t	t	t	t	t	t																		
13.	1,2-Dibenzofluorene	t	t	t	t	t	t																		
14.	2,3-Dibenzofluorene	t	t	t	t	t	t																		
15.	Benzo(a)pyrene	t	t	t	t	t	t																		
16.	Benzo(c)pyrene	t	t	t	t	t	t																		
17.	Benzoperylene	t	t		t	t																			
18.	Bianthracene	t	t		t	t																			
19.	1,3- and 1,4-Butadiene				3–6	l																			
20.	Butanol		l	t		l	t																		
21.	*n*-Butane	m	m	t	5–15	m	t	m	m	t															
22.	1-Butene	m	m	t	10–16	m	l	m	m					m						m					
23.	2-Butene, cis and trans	m	m	t	m	m	l	m						m						m					
24.	Butyric acid					l	l																		
25.	Butanal								m																
25a.	Butenal											m													
26.	Butine						t																		
27.	Butyrrolactone					m	t																		
28.	Caproic acid					m	t																		
29.	Carbon dioxide		l	vh		l	vh		l	vh		l	vh		l	vh		l	vh		l	vh		l	vh
30.	Carbon monoxide		l	m		l	m		l	m		l	m		l	m		l	m		l	m		l	m
31.	Chrysene		t	t		t	t																		
32.	Coronene	t	t			t	t																		
33.	Crotonaldehyde		t			t	t																		
34.	Crotonic acid		t	t		t	t											c							
35.	$C_6H_{17}CO$									t													l		
36.	Cyclohexane	l	l	l	l	l	l																l		
37.	Cyclohexene				l	l	l																		
38.	Cyclopropane				l	l	l																		
39.	Decanal					l	l																		
40.	Decane	m	m	t	m	m	t																		
41.	1-Decene	m	m	t	m	m	t	l	t	t															
42.	2,4,4,6,6,8,8,10,10,12-Decamethyl tridecene 1 and 2										m	m													
43.	Dibenzofurane								t																
44.	Dibenzopyrene						t																		
45.	2,4- or 2,2- Dimethylbutene	m	m	t	m	m	t										0–2								
46.	Dimethylbutane										m														
47.	Dimethylheptane							m	m	t															
48.	3,3-Dimethylheptene				m	m	t																		
49.	2,4-Dimethylheptane, 1,5 or 6				l	l	t	c 27–42	c	t															
50.	4,6-Dimethyl-6-heptane-2-one					l	t			t															

TABLE 7 (Continued)

No.	Compound	PM [207c]			PE [205, 213, 214]			PP [209, 215, 216]			PIB [207b, 218, 218]			PB-1 [219]			P3MB [220]			P4MP [221]			PCH [222]		
		T	TO	C	T	TO	C	T	TO	C	T	TO	C	T	TO	C	T	TO	C	T	TO	C	T	TO	C
51.	2,4-Dimethylhexane				l	l	t																		
52.	2,4- and 2,3- Dimethyl-hexene, 1 and 2					m	m	t			t						m								
53.	4,6-Dimethylnonene							0.1–2.3	l	t															
54.	2,4-Dimethyl-7-octene-2-one									t															
55.	Dimethylovalene					t	t																		
56.	2,4- Dimethylpentadiene													m			m	m	t						
57.	2,4- and 2,3-Dimethylpentane				l	l	t				m	m		m			m								
58.	2,4-Dimethylpentene, a and 2							0.6			m	m													
59.	Dimethylpropane										m														
60.	Decanol		l	t		l	t																		
61.	Decanone		l	t		l	t																		
63.	Docosene	l	l	t	l	l	t			t															
64.	Dodecan	m	m	t	m	m	t																		
65.	Dodecanal					t	t																		
65a.	2,4,4,6,6,8,8,10,10,12,12,14-Dodecamethyl-pentadecene, 1 and 2										m														
66.	Dodecene-1	m	m	t	m	m	t		l	l	t				t										
67.	Eicosene	m	m	t	m	m	t			t															
68.	Epoxide					t	t																		
69.	Ethane	m	m	t	4–21	m	t													m	t				
70.	Ethene				4–7	m	t	m		t				m	m					m	m				
71.	Ethanol					l															l				
72.	Ethanal		m	t		l	t		m	t											l				
73.	Ethylbutadiene													m			m								
74.	Ethylbenzene				vl	l	t	vl	l	t															
75.	Ethylcyclopentene				l	l	t																		
76.	Ethylcyclohexane																								m
77.	2-Ethyl-1-hexene				m	m	t								m										
78.	2-Ethyl-1-pentene													c	c										
79.	Ethylstyrene								t																
80.	Fluoranthrene				t	t	t																		
81.	Formaldehyde					t	t		m					m				m	t						
82.	Formic acid					t	t				m														
83.	Furane					t	t																		
84.	Heneicosene-1				m	m	t			t															
85.	Heptacosene				m	l	t			t															
86.	Heptadecane				m	m	t			t															
87.	1-Heptadecene				m	m	t																		
88.	1,3;2,4-; 1,6-Heptadiene				l	l	t																		
89.	2,4,6,8,10,12,14-Hepta-methyl-1-pentadecene							m																	
90.	2,4,4,6,6,8,8-Hepta-methylnonene, 1 and 2										m														
91.	Heptanal					l	t																		
92.	Heptanol					l	t																		
93.	Heptane	c	m	t	c	m	t							c											
94.	2-Heptanone					m	t																		
95.	1; 2; 3-Heptene	c	m	t	c	m	t							c											
96.	Hexacene				m	m	t																		
97.	Hexacosene				m	l	t																		
98.	Hexadecane				m	m	t																		
99.	1-Hexadecene				m	m	t																		
100.	1,3- 1,4- 1,5- Hexadiene				l	l	t	l	l	t															

TABLE 7 (Continued)

No.	Compound	PM [207c]			PE [205, 213, 214]			PP [209, 215, 216]			PIB [207b, 218, 218]			PB-1 [219]			P3MB [220]			P4MP [221]			PCH [222]		
		T	TO	C	T	TO	C	T	TO	C	T	TO	C	T	TO	C	T	TO	C	T	TO	C	T	TO	C
101.	2,4,6,8,10,12-Hexa-methyl-1-tridecene							c																	
102.	2,4,6,8,10,12,-Hexa-methyl-I-pentacene							m																	
103.	2,4,6,8,10,12,14-Hexa-methylheptadecane							m																	
104.	2,4,4,6,6,8-Hexamethyl-nonene, 1 and 2										m														
105.	Hexanal					m	t																		
106.	Hexane	m	m	t	m	m	t																		
107.	2-Hexanone					m	t																		
108.	1- 2- 3-Hexene	c	c	t	c	m	t							c	m	t									
109.	Hydrogen				0.5	m		m	m		m	m		m	m		m	m		m	m				
110.	Hydroperoxides					m			m			m			m			m			m				
111.	1-Hydroxy-2-propanone								m																
112.	Hydroxyvaleric acid					m	t																		
113.	Indenofluoranthene				t		t																		
114.	Isobutanal					m	t														m				
115.	Isobutanol																				t				
116.	Isobutene				m	m	t	2.5–3.3	m		1.7–33.9[a] 64–81[b] c		c	t	l	l	t			c 52–59	c	t			
117.	Isobutenyl cyclohexane									t	t														
118.	Isobutyric acid																				l				
119.	Isodecane				l	l	t																		
120.	Isononane	m	m	t	m	m	t																		
121.	Isooctane				m	m	t																		
122.	Isopropanol					m	t														t				
123.	Isovaleraldehyde					l	t														m				
124.	Isovaleric acid					l	t														l				
125.	Isoxazole					t	t																		
126.	Methane	m	m	t	7.7	m	t	m			m	m	t	m		t				m	m	t			
127.	Methanol					t	t		m													t			
128.	2-Methylbutane	m	m	t	m	m	t							m	m	t				1.5–3					
129.	Methylbutadiene						t																		
130.	2-Methylbutene, 1 or 2				m	m	t	15.5	m	t	m	m	t	c						m					
131.	3-Methylbutene	m	m	t	m	m	t													m					
132.	Methylcyclohexane				l	l	t																l		
133.	1-Methylcyclohexene				l	l	t																l		
134.	Methylcyclopentene				l	l	t																		
135.	Methylcyclopropylketone								m																
136.	2-Methyl-2-propen-1-ol								m																
137.	2-Methyl-4-ethylhexane				l	l	t																		
138.	Methylethylketone					l	t																		
139.	2-Methyl-3-ethylpentane				m	m	t																		
140.	2-Methyl-3-ethylpentene 2				m	m	t	m	m	t															
141.	Methylforminate									t															
142.	2-; 3- and 4-Methylheptane							0.8-1.7	m	t				m	m	t									
143.	4-Methyl-5-heptene-2-one								t																
144.	6-Methyl-1-heptene							2.6	m	t										m	m				
145.	2-Methyl-1-heptene				m	m	t	m																	
146.	3-Methyl-1-heptene				m	m	t																		
147.	4-Methylheptene, 1 or 2							2.6	m	t															
148.	5-Methylheptene, 2 or 3													m											
149.	2-Methylheptene-3				m	m	t			t			t												
150.	5-Methyl-1,3-hexadiene						1.7			t												m			

TABLE 7 (Continued)

		PM [207c]			PE [205, 213, 214]			PP [209, 215, 216]			PIB [207b, 218, 218]			PB-1 [219]			P3MB [220]			P4MP [221]			PCH [222]		
No.	Compound	T	TO	C	T	TO	C	T	TO	C	T	TO	C	T	TO	C	T	TO	C	T	TO	C	T	TO	C
151.	2,3-Methylhexane										m														
152.	3,4-Methylhexene-1				m	m	t				m										m				
153.	4,5-Methythexene-2				m	m	t													m					
154.	2-Methylhexene-3																			m					
155.	3-Methyl-3,5-hexadiene							m																	
156.	Methyl-4,5-methylenephenanthrene				t	t	t													m					
157.	2-Methyl-1,4-pentadiene																								
158.	2-Methylpentane				m	m	t	1.7	m	t				m						m					
159.	2,3,4-Methylpentene-1				m	m	t	c 14–21	m	t										c					
160.	2-Methylpentene-2				m	m	t																		
161.	2-Methylpentene-3				m	m	t																		
162.	Methylphenanthrene				t	t	t																		
163.	2-Methylpropane				m	m	t	m	m	t	m	m	t				2–4	c	t						
164.	2-Methyl-2-propenal					m	t																		
165.	2-Methyl-2-propanal								m									c							
166.	2-Methylpyrene				t	t	t																		
167.	Methylvinylketone					t	t		m																
168.	Nonadecane				m	m	t		t																
169.	Nonadecene-1				m	m	t		t																
170.	1,3-Nonadiene				l	l	t																		
171.	2,4,6,8,10,12,14,16,18-Nonamethyl-1-nonadecene							m																	
172.	2,4,4,6,6,8,8,10,10-Nonamethylhendecene, 1 and 2										m														
173.	2,4,6,8,10,12,14,16,18-Nonamethyl-1-uncosene							m	m																
174.	Nonane				m	m	t																		
175.	1-Nonene				m	m	t	l	l	t															
176.	Nonanone						t																		
177.	$C_9H_{18}CO$								m																
178.	Nononal					t	t																		
179.	Octacosene				m	m	t																		
180.	Octadecene				m	m	t																		
181.	2,4,4,6,6,8,8,10-Octamethyl hendecene, 1 and 2										m														
182.	1-Octadecene				m	m	t			t															
183.	1,3-Octadiene				m	m	t																		
184.	Octonal					t	t																		
185.	Octanol					t	t																		
186.	Octane				c	c	t																		
187.	Octanone						t																		
188.	1-Octene				c	c	t																		
189.	Ovalene				l	l																			
190.	Pentacosene				l	l	t			t															
191.	Pentadecanol					l	t																		
192.	Pentadecane				m	m	t			t															
193.	Pentadecene				l	l	t			t															
194.	1,3-; 1,4-Pentadiene, cis and trans				m	m	t													m					
195.	2,4,6,8,10-Pentamethyl-1-tridecane							m																	
196.	2,4,4,6,6-Pentamethylheptene, 1 and 2										m														

TABLE 7 (Continued)

No.	Compound	PM [207c]			PE [205, 213, 214]			PP [209, 215, 216]			PIB [207b, 218, 218]			PB-1 [219]			P3MB [220]			P4MP [221]			PCH [222]		
		T	TO	C	T	TO	C	T	TO	C	T	TO	C	T	TO	C	T	TO	C	T	TO	C	T	TO	C
197.	2,4,6,8,10-Pentamethyl-1-tridecene							m	m	t															
198.	Pentanal					m	t																		
199.	Pentane				m	m	t	8,5	m	t				c						0.2–0.5					
200.	Pentanol					l	t																		
201.	2,4,4,6,6-Pentamethylheptene, 1 or 2										c	c													
202.	Pentanone					t	t		m																
203.	Pentene-1				m	m	t	m	m	t				c	c	t				m					
204.	Pentene-2, cis and trans				0.5–9	m	t																		
205.	Peropyrene				t	t	t																		
206.	Phenanthrene				t	t	t																		
207.	Phenylacetylene						t																		
208.	Phthalate									t															
209.	Propadiene				l	l				t															
210.	Propanol					t	t																		
211.	Propanal								m																
212.	Propane				18	m	t	1.1	m	t	m	m	t	m						c 28–4.2	m	t			
213.	Propene				24 c	m	t	20 c 0.2–25	m	t											m				
214.	Propine				l	l														t					
215.	Propionic acid					t	t																		
216.	Pyrene				t	t	t																		
217.	Rubicene				t	t																			
218.	Styrene						t																		
219.	Tetracosene				m		t			t															
220.	Tetradecanal						t	t																	
221.	Tetradecane				m	m	t																		
222.	1-Tetradecene				l	l	t		t																
223.	2,4,6,8-Tetramethylhendecane							m																	
224.	2,4,4,6-Tetramethylheptene, 1 and 2										m														
225.	Tetrahydrofuran					t	t																		
226.	2,4,6,8-Tetremethylhendecene							c 8-18	m																
227.	2,4,6,8-Tetramethyl-1-nonene							0.9–1	l	t															
228.	2,2,4,4-Tetramethylpentane										m														
229.	4,6,810-Tetramethyl-8,10-undecene-2-one									t															
230.	4,6,8,10-Tetramethyltridecane							0.3–2	l	t															
231.	4,6,8,10-Tetremethyltridecene							2–6	l	t															
232.	Toluene				t	t	t		t																
233.	Tricosene				m	l	t		t																
234.	Tridecanal					t	t																		
235.	Tridecan				m	m	t		t																
236.	1-Tridecene				m	m	t	2–6	t																
237.	2,4,4,6,6,8,8,10,10,12,12,14,14-Tridecamethylpentadecene, 1 and 2										m														
238.	Trimethylbenzenes							l	m	t															
239.	2,3,3-Trimethylbutene-1				t	t	t																		
240.	Trimethylcyclopentane				t	t	t																		

TABLE 7 (Continued)

		PM [207c]			PE [205, 213, 214]			PP [209, 215, 216]			PIB [207b, 218, 218]			PB-1 [219]			P3MB [220]			P4MP [221]			PCH [222]		
No.	Compound	T	TO	C	T	TO	C	T	TO	C	T	TO	C	T	TO	C	T	TO	C	T	TO	C	T	TO	C
241.	4,6,8-Trimethylhendecane							0.4	0.7																
242.	4,6,8-Trimethyl-hendecene-1							0.6																	
243.	2,4,6-Trimethylheptane							1.3–2	m	t															
244.	2,4,6-Trimethylheptene-1							1.3–20	m	t															
245.	2,2,5-Trimethylhexane				l		t																		
246.	Trimethylhexenes				m	t	t	m	m	t	m	m	t												
247.	2,4,6-Trimethyl-6,8-nonadiene-2-one									t															
248.	2,4,6-Trimethyl-1-nonene, eritro and treo							c 4–10																	
249.	Trimethylnonane							m																	
250.	2,4,4-Trimethylpentene, 1 or 2										m														
251.	2,2,3-Trimethylpentene										m														
252.	2,4,4-Trimethylpentane										m	m													
253.	2,2,4-Trimethylpentane										m														
254.	2,3,4-Trimethylpentene				c	l	t																		
255.	2,3,4-Trimethylpentane										m														
256.	4,6,8-Trimethylundecane-2-one									t															
257.	Triphenylene				t	t	t																		
258.	2,4,6,8,10,12,14,16,18,20-Undecamethyl-1-tricosene							m																	
259.	2,4,4,6,6,8,8,10,10,12,12-Undecamethyltridecene, 1 and 2										m														
260.	Undecanol					t	t																		
261.	Undecanal					t	t																		
262	Undecane				m	l	t																		
263.	Undecene-1				m	m	t		t																
264.	Valerolactone					t	t																		
265.	Water			l	vh		l	vh		l	vh	l	vh	l	vh				l	vh		l	vh	l	vh
266.	Vinylcyclohexane																						m	l	t
267.	Vinylcyclohexene																						m	l	t
268.	The heavier product identified				C_{94}			C_{50}			C_{28}														
269.	Xylenes, other products						t			t															
270.	C_6H_5—$(CH_2)_n$—C_6H_5					t	t																		
271.	$C_{11}H_{24}$																m								
	$C_{11}H_{22}$																m								
	$C_{12}H_{24}$																m								
	$C_{13}H_{28}$																m								
	$C_{17}H_{34}$																m								

[a] Reported to polymer.
[b] Values represent the percentage of compound in volatile products.
m, main product; l, low concentration; t, traces; c, characteristic products; vh, very high concentration.
Source: Data from Refs. 205–209, 212–225.

TABLE 8 Thermal decomposition products of some olefin copolymers [213, 219, 223, 226–228]

Copolymer	Thermal decomposition products
Ethylene–propylene	Increased quantity of *methane*, ethane, ethylene isoalkanes (2 and 3-methylalkanes). Cycloalkanes
Ethylene–1-butene	Great amount of *ethane*, ethylene, n-C_4, i-C_5, n-C_5, $5MC_5$, n-C_6, n-C_7, $3MC_7$, $3EC_6$, n-C_8, n-C_9, $3MC_9$, n-C_{10}, $5EC_9$, $3,7MC_9$, MEC_9, n-C_{11}, i-C_{11}, etc.
Propylene–l-butene	C_2, C_3, n-C_4, i-C_5, n-C_5, $2MC_5$, n-C_6, $2MC_6$, $3MC_6$, n-C_7, $3MC_7$, $2,4MC_7$, $2,4MC_8$, $2M4EC_7$, $3,5MC_8$, $3M5EC_8$, $2,6M4EC_7$, $4,6MC_9$, $2M4EC_8$, $3M5EC_8$, $3,5MC_9$, $3M5EC_9$, etc.
Ethylene–1-hexene	High amount of *n-butane*, i-C_4, i-C_5, $2MC_5$, $3MC_5$, CyC_5, $2MC_6$, MC_6, $2MC_7$, $4MC_7$, $3MC_7$, $3EC_6$, $2MC_8$, $4MC_8$, $ECyC_6$, $3MC_8$, $4MC_9$, $5MC_9$, $4EC_8$, $iPCyC_6$, $BuCyC_5$, $2MC_9$, n-$PCyC_6$, $3MC_9$, $4MC_{10}$, $5MC_{10}$, *sec*BuC_62MC_{10}, $4MC_{10}$, $nBuCyC_6$, $3MC_{10}$, etc.
Ethylene1-1-octene	High amount of *n-hexane*, all decomposition products identified for ethylene–1-hexene copolymer
Ethylene–propylene–diene monomer (DCPD)	*Propylene*, n-C_4, butene, benzene, toluene, traces of styrene, CyC_5, $MCyC_5$, $MCyC_6$, $ECyC_5$, n-$PCyC_5$, $1,2MCyC_6$,
Ethylene–propylene–norhornene	*Methylcyclopentane* $MCyC_5$, $MCyC_6$, $1,2MCyC_6$, norbornene, isoprene, dimethylbutadiene
Ethylene–vinyl acetate	Acetic acid, methane, carbon dioxide, olefins, polyene chain
Ethylene–tetra-fluorethylene	Tetrafluorethylene, ethylene, vinylidene fluoride, fluorhydric acid, 1,3,3,3-tetrafluorpropylene, 3,3,4,4,5,5,5-heptafluorpentene-l, 3,3,3-trifluorpropylene
Ethylene/CO	Cyclopentanone, furane, water, monomers, LMW ketones
Polyolefin/sulfur dioxide	*Sulfur dioxide* and corresponding monomers as butene-l, hexene. heptene-l, octene-1, benzene, toluene, xylenes, indene, ethylene, tetraline, octadecene, octadiene, cyclopetene, cyclypentadiene, bicyclopentene
Chlorinated polyethylene ethylene	*HCl, benzene*, toluene, xylenes, naphthalene, polymer chain
Sulfochlorinated polyethylene	*HCl, sulfur dioxide*, polyene chain

n, *i*- normal and isomer hydrocarbons; M - methyl, E- ethyl; P-propyl, Bu-butyl, Cy-cyclo.

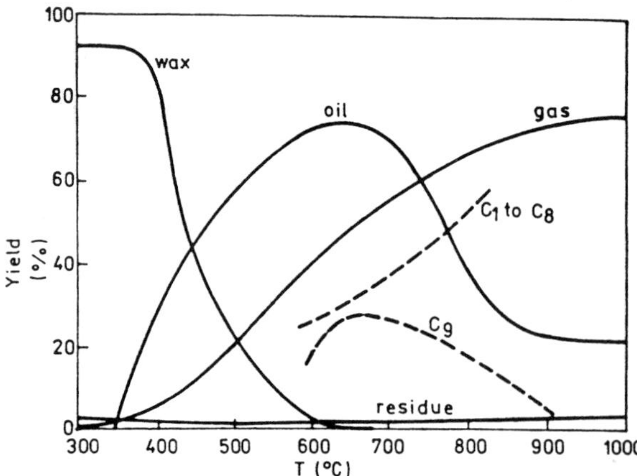

FIG. 13 Formalized curves (————) of variation of yield of global pyrolysis products with temperature of pyrolysis and some examples for certain products (- - - - - - -). (Partially adapted from Ref. 209.)

for each polymer, which permits their identification (see below).

During thermal and thermo-oxidative decomposition, the nonvolatile material formed presents three types of unsaturation; vinyl ($R—CH=CH_2$), trans-vinylene ($R—CH=CH—R'$), and vinylidene [229]

$$(R—CH_2—\underset{\underset{CH_2}{\|}}{C}—CH_2—R')$$

and also oxygen-containing groups, such as $=C=O$, $—OH$, $—COOH$, $—COOR$, $C—O—C$. etc. The branching increase leads to an increased probability of isomer formation.

Most products of thermal decomposition appear in the presence of oxygen (thermo-oxidative decomposition) too; besides, a great number of oxygen containing compounds are present (Fig. 14 and Table 7). Although various aldehydes, ketones, and other oxygenated species are formed at low temperatures, these

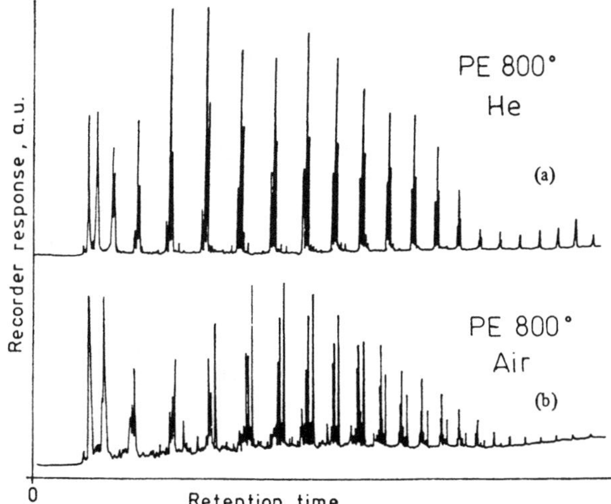

FIG. 14 The pyrograms of PE, obtained at (a) 800°C, 20 s, 25°C min^{-1}, He, 4°C min^{-1} to 280°C; (b) 10 s, air, FID, SE-54 fused silica capillary. (From Ref. 231.)

decompose above 350°C losing CO_2 and forming smaller hydrocarbons fragments [183].

The pyrolyzates from PE (He, 800°C) present a pyrogram with a series of triplets of peaks, corresponding to a homologous series of alkadienes, alkenes, and alkanes. In oxidative atmosphere, each group of peaks contains two additional peaks, assigned to a straight-chain alcohol and aldehyde [230]. The relative quantities of oxygen-containing compounds are [121]:

For PE: acetaldehyde > acetone + propionaldehyde ≫butyraldehyde > methanol > ethanol.

For PP: acetone > acetaldehyde > methylacrolein> methyl ethyl ketone + butyraldehyde > methyl propyl ketone.

The amount of unsaturated and aromatic hydrocarbons is higher in an oxidative atmosphere [215].

In the composition of the combustion products, most products of thermal and thermo-oxidative decomposition appear in very low concentrations. Their composition depends on the mode of operation, e.g., in closed or open space, flameless or flame combustion, oxygen concentration, etc. The major products of combustion are distinguished by higher concentrations of the thermally stable compounds (CO_2, H_2O, CO, low molecular weight compounds, aromatics, etc.) and also by some toxic products. The apparent lethal concentrations (ALC_{50}, 30 min exposure period) of pyrolysis products of polyolefins is about 11.8 mg L^{-1}, which indicates that they are not unusual or extremely toxic [205, 222, 232].

Modified polyolefins (chlorosulphonated PE) [233] and copolymers have a particular mechanism of decomposition; therefore some products of their corresponding homopolymers and some specific ones may appear (Table 7).

B. Structural Information

Degradative methods such as pyrolysis (pulse or programmed mode) and decomposition in various atmospheres are by far the most common techniques applied for polymer studies. Thermally produced small molecules can be further analyzed by means of several specialized techniques [227, 234–242]: evolved gas analysis, pyrolysis–gas chromatography (Py–GC) or –mass spectroscopy (Py–MS), field ionization/field desorption, Py–IR, Py–FTIR, thermogravimetry–MS, and also by high-resolution chromatography [230, 243–245], e.g., capillary column, on-column cryogenic focusing, hydrogenation or other chemical modification of polymers or their pyrolyzates.

Qualitative identification of polyolefins, copolymers, or blends can be achieved by comparison of the "fingerprint" pyrograms, chromatograms (Fig. 15), mass spectra, and IR spectra with some of the data given in various publications [227, 231].

In addition to identifying structural units, degradative techniques can provide information about additives, impurities, residuals, structural defects, etc. However, in all determinations, it is absolutely necessary to maintain identical conditions so that the fragmentation of macromolecules will be reproducible. Pyrolytic methylation is also used for the identification of polymers [246].

Quantitative determination aims at the establishment of the composition of both copolymers and blends, and also of the microstructure of macromolecules, branching, tacticity, structural defects, and so on. In this respect, the surface, height, and other values of the corresponding peak of a particular component (denoted by "c" in Table 7 and italics in Table 8) are determined.

Some well-characterized copolymers can be used as model compounds in these types of studies [210, 223]. To, give some examples, the region between C_{11} and C_{13} is most informative for a PE short-branch chain; $2MC_6$, and $3MC_7$, methyl alkanes are indicative of butyl branches [214, 223]. Seeger and Barrall [247] determined the short branches of PE from the isoalkane peaks. Also very often employed is the ratio of isoalkane or alkene to *n*-alkane peaks [248]. The tacticity of PP or of E/P copolymers can be established by

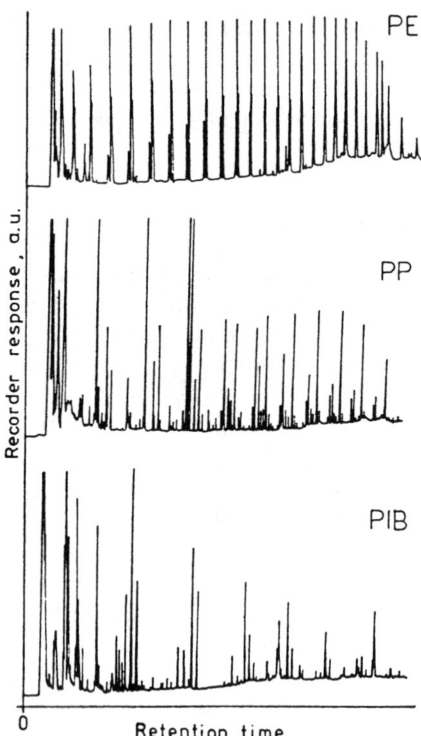

FIG. 15 Capillary gas chromatography of polyolefin pyrolyzates. Model 120 Pyroprobe, pyrolysis temperature 750°C, column 50 × 0.25 mm; SE-54 capillary, initial temperature 500°C for 3 min, heating rate 8°C min^{-1} to 250°C. (From Ref. 231.)

the analysis of tetramer and pentamer fragments [224, 248]; Tsuge *et al.* [249, 250] proposed to differentiate between APP and IPP and establish the structural defects in PE, the chlorinating of polymers, and pyrolysis and identification of the resulted substituted aromatic hydrocarbons [249, 251].

The obtained precision of determinations is very good (<1%) [248], even at ppm concentration levels.

C. Combustion Behavior of Polyolefins

Combustion represents an exothermal reaction of a combustible material with oxygen (carburant) and depends on both. For polymers in general and especially for most polyolefins, combustion is an endo- and exothermal multistage process (heating, melting, decomposition, inflammation, etc.) [252]. When the decomposition temperature is reached, more or less oxidized products will evolve. The volatile products mix with the surrounding air or with other gaseous oxidants and at a certain composition, when the mixture becomes inflammable, a flame appears [253, 254]. Thus burning occurs in the gaseous phase, with combustion products (see Table 4) and heat resulting. A part of that heat is transferred back and the cycle continues, according to a "candle-burning" model [255, 256].

Generally, it is more difficult for oxygen to reach the burning surface during vigorous combustion than under the mild conditions of thermo-oxidative decomposition [253].

The flame of a burning polymer is a self-fed and stable one when the rate at which the inflammable products are produced is equal to the rate of their combustion.

Due to the complexity of the process, fire behavior cannot be predicted, at least for the time being, with any confidence, as no methods for quantifying the properties of the combustible materials exist. However, several parameters were introduced as measures of flammability or fire properties [254–265], such as combustibility; ease of ignition or ignitability; ignition or self-ignition time or temperature, oxygen index; rate of surface spread of flame; rate of heat release or of burning; propensity to produce smoke and toxic gases; heat of combustion; limit concentration of ignition; optical density of fumes or of emitted soot; and so forth [265]. Some of these characteristics, corresponding to polyolefins, are given in Table 9.

Pyrolysis of polyolefins is insignificantly accelerated by oxygen, which explains the poor correlation between the pyrolysis rates and the oxygen indices [241]. The ignition time of HDPE is 1.5 times higher than that of PP. PE drips, but not as extensively as PP. In flames of PE there is a clearly blue region, which is less pronounced for PP. The formation of soot and smoke is more intensive for PP [225].

D. Mechanisms of Thermal and Thermo-Oxidative Decomposition and Combustion

1. Mechanism of Thermal Decomposition [207, 210–212]

For quite a long time polyolefins have been considered generally degrading through a random scission mechanism, PE being the best example [207, 210–212]. The decomposition of a polymer by a purely random scission mechanism implies the formation of a pyrolyzate containing a wide distribution of volatile and/or nonvolatile materials, without preference for any product and without hydrogen elimination. Data

TABLE 9 Some characteristics regarding the combustion behavior of polyolefins

Characteristic	PE	PP	PIB	CPE
General behavior	Extensive charring of surface	Slight charring of surface	Incomplete burning of combustion products	0–50% Cl burns; 50% CI it does not burn
Heat of gasification (kJ g^{-1})	2.4	2.5		
Heat flow (kJ monomer $unit^{-1}$)	84			
Radiant heat flux (kW m^{-1})	19–34	17–42.5		
Ignition time (s)	524–83	372–46		
Ignition temperature (°C)	360–367	332–340		
Oxygen index	15–21	17.7–21	17.7–1.83	
Heat of combustion (kJ kg^{-1})	43,440 (46,650)	46,704	46,858	
Sell-ignition temperature (°C)	850	550		
Rate of burning (mm s^{-1})	3.2–12.9	7.5–17.2		
Surface temperature (°C)	625			

Source: Data from Refs. 255, 256, and 266–270.

presented in Table 7 show some exceptions even in PE decomposition: there is a preferential production of certain products [209, 212, 271] such as series of oligomers (alkanes or alkenes) for PE, PP, and PB-l; a great amount of monomer from PIB [226]; and so on. At high temperatures, a great amount of hydrogen is present, increasing the unsaturation and arene content. Sometimes a carbon-rich residue also appears. In order to explain the experimental results, at least seven reaction pathways must be included, in a general scheme (Scheme 2) of a free radical chain reaction mechanism for polyolefin decomposition [243, 244, 272–277].

(1) *Initiation* by homolytic random or at the chain end scission with radicals generation by breaking of C—C bonds at low temperatures or C—C and C—H bonds at high temperatures. It takes place with a higher probability at weak points as allylic positions with respect to double bonds or at tertiary hydrogens for C—H bonds. Scission tendency increases with increasing $+I_s$ effect of substituents R_1 and R_2, thus PE < PP< PIB < P4MP:

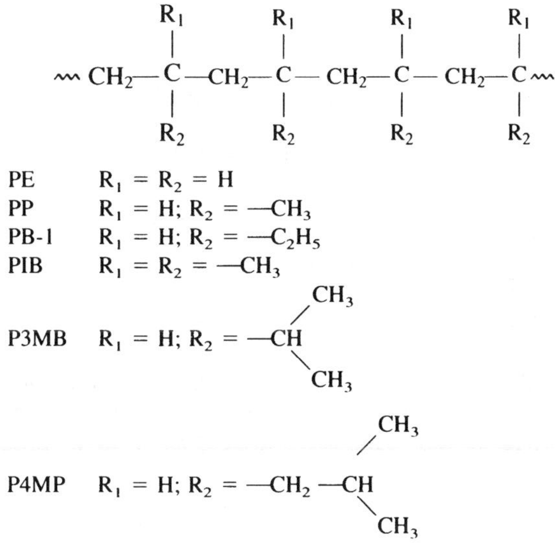

$$\sim CH_2-\underset{R_2}{\overset{R_1}{C}}-CH_2-\underset{R_2}{\overset{R_1}{C}}-CH_2-\underset{R_2}{\overset{R_1}{C}}-CH_2-\underset{R_2}{\overset{R_1}{C}}\sim$$

PE $R_1 = R_2 = H$

PP $R_1 = H; R_2 = -CH_3$

PB-1 $R_1 = H; R_2 = -C_2H_5$

PIB $R_1 = R_2 = -CH_3$

P3MB $R_1 = H; R_2 = -CH(CH_3)_2$

P4MP $R_1 = H; R_2 = -CH_2-CH(CH_3)_2$

SCHEME 2

$$(a)\ \sim(R_1R_2)C—CH_2—\underset{\underset{CH_2}{\|}}{C}—CH_2—(R_1R_2)C\sim \longrightarrow \sim(R_1R_2)\dot{C} + \dot{C}H_2—\underset{\underset{CH_2}{\|}}{C}—(R_1R_2)C\sim$$

$$(b)\ \sim(R_1R_2)C—CH_2—\underset{\underset{R_2}{|}}{C}=CH_2 \longrightarrow \sim(R_1R_2)\dot{C} + \dot{C}H_2—\underset{\underset{R_2}{|}}{C}=CH_2$$

$$(c)\ \sim(R_1R_2)C—CH_2—\underset{\underset{R_1}{|}}{C}=\underset{\underset{R_2}{|}}{C}—CH_2—(R_1R_2)C\sim \longrightarrow \sim(R_1R_2)\dot{C} + \dot{C}H_2—\underset{\underset{R_1}{|}}{C}=\underset{\underset{R_2}{|}}{C}—CH_2\sim$$

$$(d)\ \sim(R_1R_2)C—CH_2—(R_1R_2)C—CH_2—\overset{\overset{H\,(R_1)}{|}}{\underset{\underset{R'(R_2)}{|}}{C}}—CH_2—(R_1R_2)C\sim \longrightarrow$$

R′ — branch

$$\dot{H} \text{ or } \dot{C}H_3, \dot{C}_2H_5, \text{etc.} + \sim(R_1R_2)C—CH_2—(R_1R_2)C—CH_2—\underset{\underset{R'(R_2)}{|}}{\dot{C}}—CH_2—(R_1R_2)C\sim$$

(2) *Depropagation or Depolymerization* producing monomer

$$\sim CH_2—(R_1R_2)C—CH_2—(R1R_2)\dot{C} \longrightarrow \sim CH_2—(R_1R_2)\dot{C} + \underset{\text{monomer}}{CH_2 = (R_1R_2)C}$$

Its probability increases with decomposition (pyrolysis) temperature and for the same temperature in the following order: PIB>P4MP>P3MB>PB-1>PP>PE.

(3) *Nonpreferential or preferential intramolecular hydrogen transfer* (terminal radical → internal radical ≡ radical isomerization) and subsequent β scission. α scission is also considered possibly to occur [94], leading to a wide spectrum of alkanes, alkenes, and dienes; some products as oligomeric series are preferentially formed by a backbiting or coiling process which involves a cyclic intermediate (e.g., one to five transfer leads to a pseudo-six-membered ring):

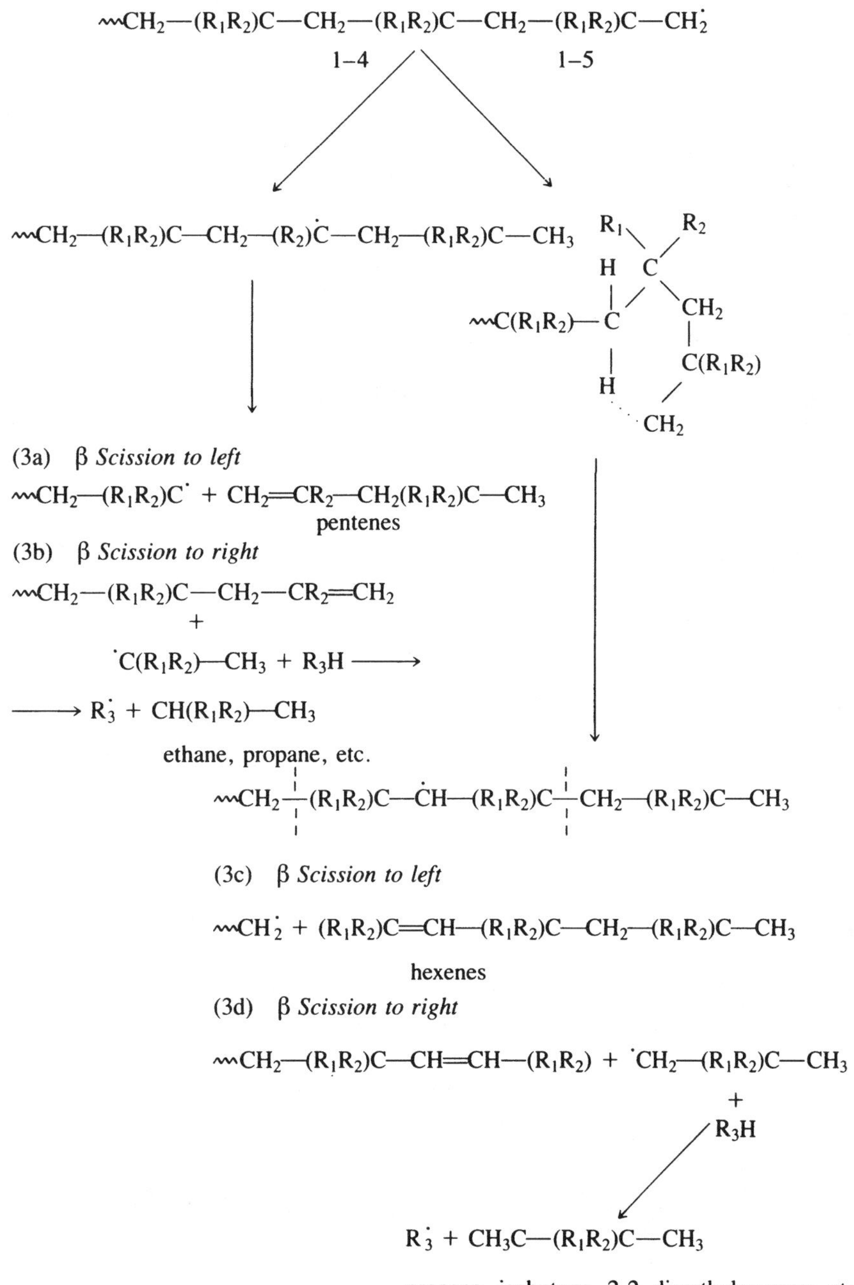

(4) *Intermolecular hydrogen abstraction*, explaining the crosslinking and formation of hydrogen, alkanes, terminal olefins, etc.

Polyolefin	Preferential intramolecular transfer
PE	1–5; 1–9; 1–17; 1–21; 1–23 leading to C_6, C_{10}, C_{14}, C_{18} alkenes and C_3, C_7, C_{11}, C_{15} alkanes
PP	1–3; 1–5; 1–7; 1–9; 1–13 in secondary macroradicals; 1–6; 1–10; 1–12 in primary radicals
PIB	To methyl or methylene group in primary and tertiary radicals, leading to dimers, trimers, and higher oligomers
PB-1	1–5

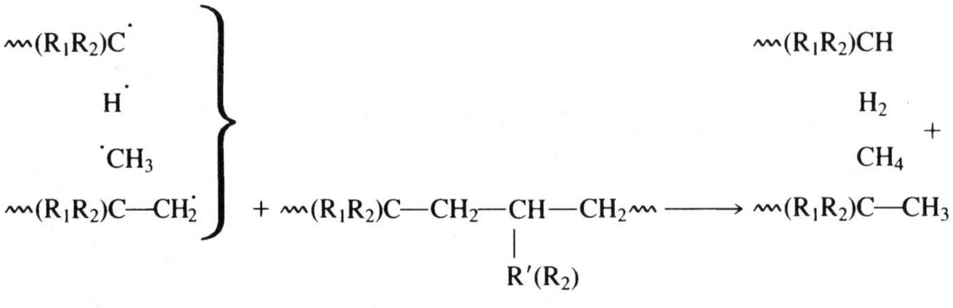

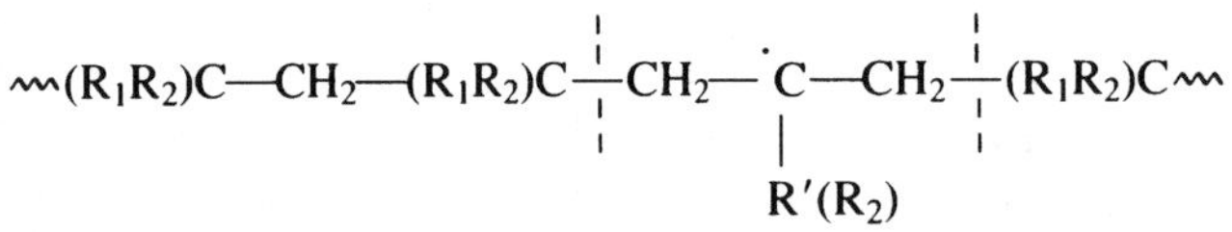

$$\left.\begin{array}{l} \sim(R_1R_2)C^{\cdot} \\ H^{\cdot} \\ {}^{\cdot}CH_3 \\ \sim(R_1R_2)C{-}CH_2^{\cdot} \end{array}\right\} + \sim(R_1R_2)C{-}CH_2{-}\underset{\displaystyle R'(R_2)}{\underset{|}{CH}}{-}CH_2\sim \longrightarrow \begin{array}{l} \sim(R_1R_2)CH \\ H_2 \\ CH_4 \\ \sim(R_1R_2)C{-}CH_3 \end{array} +$$

$$\sim(R_1R_2)C{-}CH_2{-}(R_1R_2)C \,\vdots\, {-}CH_2{-}\underset{\displaystyle R'(R_2)}{\underset{|}{\dot{C}}}{-}CH_2 \,\vdots\, {-}(R_1R_2)C\sim$$

(4a) β *Scission to left*

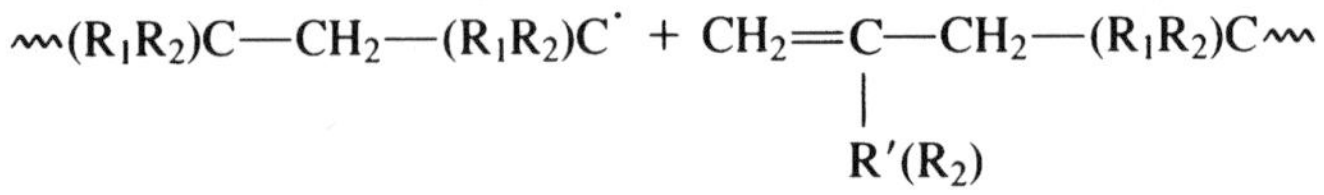

$$\sim(R_1R_2)C{-}CH_2{-}(R_1R_2)C^{\cdot} + CH_2{=}\underset{\displaystyle R'(R_2)}{\underset{|}{C}}{-}CH_2{-}(R_1R_2)C\sim$$

(4b) β *Scission to right*

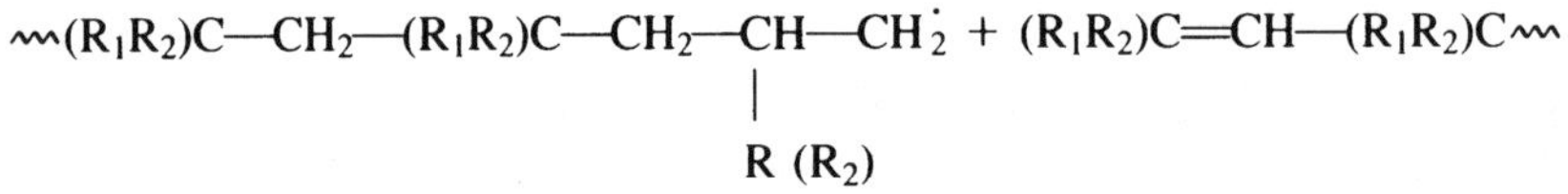

$$\sim(R_1R_2)C{-}CH_2{-}(R_1R_2)C{-}CH_2{-}\underset{\displaystyle R\ (R_2)}{\underset{|}{CH}}{-}\dot{C}H_2 + (R_1R_2)C{=}CH{-}(R_1R_2)C\sim$$

(4c) Side group scission

$$\sim(R_1R_2)C{-}CH_2{-}(R_1R_2)C{-}CH_2{-}\underset{\displaystyle CH_2}{\underset{||}{C}}{-}CH_2{-}(R_1R_2)C\sim + R''^{\cdot}$$

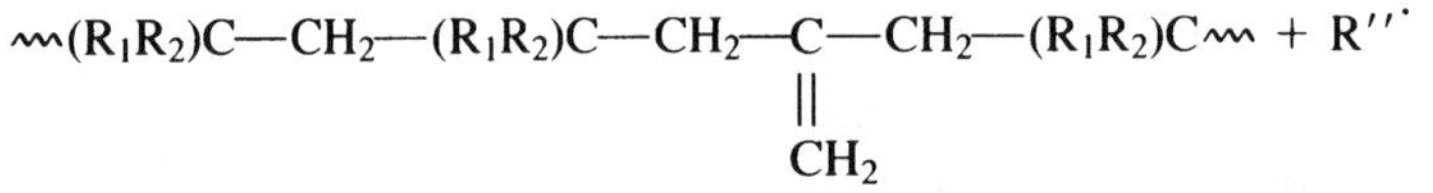

(5) *Isomerization of vinyl group* leads to internal double bonds:

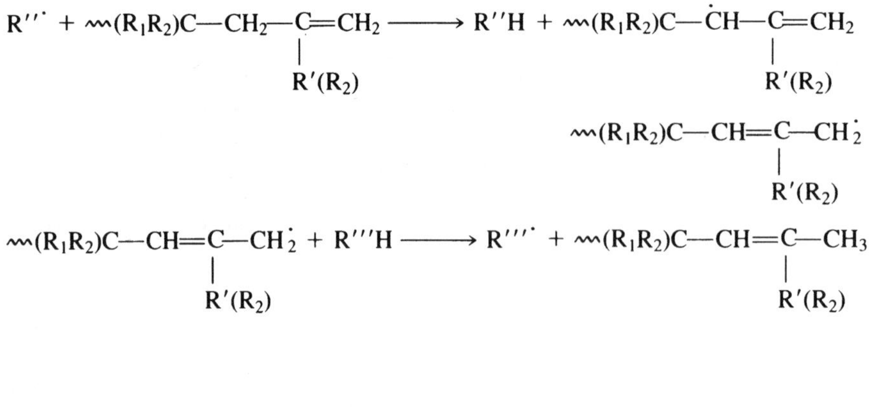

$$R''^{\cdot} + \sim(R_1R_2)C{-}CH_2{-}\underset{\displaystyle R'(R_2)}{\underset{|}{C}}{=}CH_2 \longrightarrow R''H + \sim(R_1R_2)C{-}\dot{C}H{-}\underset{\displaystyle R'(R_2)}{\underset{|}{C}}{=}CH_2$$

$$\sim(R_1R_2)C{-}CH{=}\underset{\displaystyle R'(R_2)}{\underset{|}{C}}{-}\dot{C}H_2$$

$$\sim(R_1R_2)C{-}CH{=}\underset{\displaystyle R'(R_2)}{\underset{|}{C}}{-}\dot{C}H_2 + R'''H \longrightarrow R'''^{\cdot} + \sim(R_1R_2)C{-}CH{=}\underset{\displaystyle R'(R_2)}{\underset{|}{C}}{-}CH_3$$

(6) *Termination by recombination or disproportionation* of macroradicals explains crosslinking, formation of nonvolatile compounds, hydrocarbons with short and long branches and terminal unsaturation.

(6a) *Recombination*

$$\sim(R_1R_2)C{-}\dot{C}H_2 + {}^{\bullet}CH_2{-}(R_1R_2)C\sim \longrightarrow \sim(R_1R_2)C{-}CH_2{-}CH_2{-}(R_1R_2)C\sim$$

$$\sim(R_1R_2)C{-}\dot{C}H_2 + \sim(R_1R_2)C{-}CH_2{-}\dot{C}H{-}CH_2{-}(R_1R_2)C\sim \longrightarrow$$

$$\sim(R_1R_2)C{-}CH_2{-}CH(CH_2{-}C(R_1R_2){-}CH_2{-}){-}CH_2{-}(R_1R_2)C\sim$$

$$H_3C^{\bullet} + {}^{\bullet}H \rightarrow CH_4$$

$$H_3C^{\bullet} + {}^{\bullet}CH_3 \rightarrow CH_3{-}CH_3$$

(6b) Disproportionation, its importance increases at high temperatures

$$\sim(R_1R_2)C{-}\dot{C}H_2 + {}^{\bullet}CH_2{-}(R_1R_2)C\sim \longrightarrow \sim R_2C{=}CH_2 + CH_3{-}(R_1R_2)C\sim$$

$$\sim(R_1R_2)C{-}\dot{C}H_2 + \sim(R_1R_2)C{-}{}^{\bullet}CH{-}CH_2{-}(R_1R_2)C\sim \longrightarrow \sim(R_1R_2)C{-}CH_3 + $$

$$+ \sim(R_1R_2)C{-}CH{=}CH{-}(R_1R_2)C\sim$$

(7) Cyclization, aromatization and polymerization lead to cyclic alkanes and alkenes, mono and polynuclear arenes and coke. It is very important at high temperatures and in oxidative atmosphere, its importance decreases in the order PE > PP > PIB.

$$\xrightarrow{\text{scission and aromatization}} H_2 + \text{benzene, toluene, xylenes, PAH}$$

The prevalance of each reaction pathway depends first on sample characteristics, and second on the heating conditions. The facility of intramolecular chain transfer to different atoms of the chain varies with the chain flexibility and steric factors. The coiling (or backbiting) process involves a pseudocyclization step (five- or six-membered ring) in intramolecular transfer. Thus, in polymethylene decomposition, additional transfer along the chain is more probable; in PE decomposition, an extensive intra- or intermolecular transfer takes place, dependent on branching, while in PP decomposition radical isomerization is the most important process and in PIB decomposition, due to the alternate quarternary carbon atoms, sterically hindered, intramolecular transfer becomes less important, depolymerization to monomer being the major reaction pathway [1, 41, 207b, 225, 276]. Strauss *et al.* [278, 279] established that thermal decomposition of PIB takes place by random initiation with a very small zip length with inter- and intramolecular transfer prevailing soon after the initial break. Sawaguchi and Seno [280], proposed a detailed mechanism and kinetics for PIB thermal decomposition consisting of the three steps, all of them included in the scheme presented above. The difference is that the third step is a diffusion-controlled termination by bimolecular reaction between macroradicals and vaporization of volatile radicals. The rates of the chain end initiation and termination depend on molecular weight by M^{-1} and M^{-n} respectively. The rate of termination increases with decreasing molecular weight. The volatiles consist of the monomeric compounds (10–30 wt%), mainly isobutene monomer and the volatile oligomers (90–70 wt%) ranging from dimers to decamers. Isobutylene is formed by depolymerization (direct β scission) of R_p and R_t radicals, while volatile oligomers are formed by the intramolecular hydrogen abstraction (backbiting) of R_p and R_t and the subsequent β scission at the inner position of the main chain.

The most important products from PP are compounds with $3n$ carbons (oligomers) [212], yet the major products are different as a function of the experimental conditions. Thus, besides 2,4-dimethylpentene, identified as major products were ethane, propene, 2-methylpentene [207c], 2-pentene, propene, 2,4,6-trimethylnonene-1 [212, 279], and propene, tetramethylhendecene, and trimethylnonene [209]. Stereoisomerism also influences the PP decomposition mechanism [280a].

For P3MB-1 [228, 281], P4MP-1 [221, 282], and other polyolefins with bulky side groups, the reactions involving side groups become important:

$$\sim CH_2-\underset{\underset{\underset{\displaystyle CH_3\ \ CH_3}{CH}}{|}}{\underset{CH_2}{\overset{|}{CH}}}\sim \;\longrightarrow\; \sim CH_2-\dot{C}H\sim \;+\; CH_2=C(CH_3)-CH_3 \qquad \text{or} \qquad \sim CH_2-\underset{\underset{CH_3}{|}}{\underset{CH_2}{\overset{|}{CH}}}\sim \;\longrightarrow\; \sim CH_2-\dot{C}H\sim \;+\; CH_2=CH_2$$

The mechanism of nonpolar copolymers (EPR, EPDM, etc.) decomposition generally differs from those of the corresponding homopolymers due to a boundary effect limiting the depropagating step [283]. In the case of polar copolymers (EVA, EVC) or modified polyolefins (CPE), decomposition occurs in two stages. The first involves the elimination of side polar groups with the formation of small molecules (CH_3COOH, HCl, etc.) and of a polyene chain. In the second one, the break of the polyene chain takes place, with aromatic hydrocarbons as main products.

2. Mechanism of Thermo-Oxidative Decomposition

To a great extent, the mechanism of thermo-oxidative decomposition is similar in the initial stages to that for thermal degradation. Higher temperatures lead to shorter fragments by a thermal decomposition mechanism, oxygen having only a catalytic effect. Sometimes oxidation occurs even in the gaseous phase.

3. Mechanism of Combustion

The burning mechanism of polyolefins is similar to that of gaseous hydrocarbons, i.e., a chain radical mechanism with both direct and indirect branching (via aldehyde, hydroperoxide, etc., acting as a molecular intermediate) [284–286], as a function of temperature, which nearly makes impossible any generalization. Moreover, in the gaseous phase, some ionic reactions have been postulated [252], but the molar fraction of these species is only 10^{-7} at atmospheric pressure whereas that of the radical species is 10^{-2}–10^{-4}.

Due to the branching of the kinetic chain and the great exothermicity of the oxidation reactions, the rate of heat release from the system may become smaller than the rate of heat accumulation, so that the temperature rises dramatically. The reaction is continuously accelerated, and self ignition and explosion of the reactants mixture become critical phenomena.

E. Kinetics and Thermodynamics of Polyolefin Decomposition

1. Some Thermodynamic Considerations

Only several general considerations on the thermodynamics of polymer decomposition are known [287]. They are restricted to decomposition in solution where the chain flexibility and number of polymer–solvent contacts are important parameters, changing during decomposition. It has been established that the change of the free energy is favorable to the decomposition process. After decomposition the number of molecules becomes greater, and all components of free energy change with chain length. Generally, polyolefins decompose by reactions which are not favored at room temperature, the free enthalpy of the system having positive values of 2–68.2 kJ mol^{-1} [288].

Above a ceiling temperature, the tendency of decomposition is easy to explain in terms of breaking of the chain into small and more stable fragments (products), resulting in an increase in the entropy of the system.

Therefore, decomposition depends mainly on the dissociation rates of bonds in the molecular structure rather than on thermodynamic properties. The dissociation energies of some bonds were given in Table 3.

2. Kinetics of Thermal Decomposition

Kinetics is the theoretical basis for the understanding of a reaction mechanism. The improvement of instrumentation has been decisive for a complete analysis of the reaction products and much progress has been made in this direction.

Although more powerful computers are now available, the kinetics is not in an advanced phase and able to generalize all experimental results. In order to summarize the development of the kinetic treatment for the thermal decomposition of polyolefins, the following simplified scheme is proposed:

(1) Initiation

(a) Chain end initiation $P \xrightarrow{k_e} R^{\bullet}_{q} + R^{\bullet}_{E}$

(b) Random scission initiation $P_n \xrightarrow{k_s} R^{\bullet}_{m} + R^{\bullet}_{n-m}$

(2a) Depolymerization or depropagation by intramolecular chain transfer

$$R^{\bullet}_{n} \xrightarrow{k_p} R^{\bullet}_{n-j} + P_j \begin{cases} j = 1, \text{depolymerization}, P_j \text{ monomer} \\ j = 2, 3, \ldots, \text{for oligomeric series of} \\ \qquad \text{alkanes or alkenes} \end{cases}$$

(b) Intermolecular chain transfer

$$R^{\bullet}_{n} + P_s \xrightarrow{k_{tr}} P_n{-}CH_2{-}CH_3 + P_r{-}CH{=}CH_2 + R^{\bullet}_{s-r}$$

The competition between the two types of transfer has considerable influence on the reaction mechanism.

(3) Termination

(a) First-order $R^{\bullet}_{n} \xrightarrow{k_{t1}} P_n$

(b) Second-order

disproportionation $R^{\bullet}_{m} + R^{\bullet}_{n} \xrightarrow{k_{t2}} P_m + P_n$

recombination $R^{\bullet}_{m} + R^{\bullet}_{n} \xrightarrow{k_{t2}} P_{m+n}$

the last reaction is negligible at high temperatures. Where P_n is the total number of polymer molecules and $R^{\bullet}_{n}$ is the total number of radicals; n, j, s, r are the polymerization degrees; $k_e, k_s, k_p, k_{tr}, k_{t1}$ and k_{t2} are the rate constants for the corresponding steps of decomposition. Starting from this general reaction mechanism scheme, several mathematical approaches have been developed. Thus Gordon [291] used probability theory and Simha and Wall [289, 292–295], Boyd [296–298], Inaba and Kashiwagi [299, 300], and Chan and Blake [301] solved differential rate equations. Statistical theory and other procedures have also been used [302, 303].

In all kinetic treatments, three important parameters are introduced: $1/\gamma$, average length of zip (e.g., number of elementary steps between initiation and termination or the ratio between the probability of propagation and the sum of the probability of termination and transfer); σ, transfer constant; and L, maximum length of volatile fragments.

Rate equations are formulated for the change in the concentration of P_n or $R^{\bullet}_{j}$, the following assumptions being used: (1) steady state or pseudosteady state is established for the overall radical concentration and for that of each radical species; (2) the rate constants are independent of the polymerization degree of the reacting species; (3) all radicals are sufficiently reactive, so that they will react before vaporization, irrespective of their polymerization degree.

Due to the fact that the kinetic equations are not linear, they cannot be directly integrated. In order to obtain the solutions, one should resort to: (1) direct numerical integration using computers (difficult to use in practice because of the complex dependences obtained); (2) integration using some suitable molecular weight distributions for the initial polymer such as monodisperse [293]; most probable [289, 294-296]; exponential type [291a, 299, 300]; Poisson, Schulz, Zimm [296] which is considered as preserved or not (which corresponds better to real situations) during decomposition; (3) use of a simplified solution for special (in which one feature of the general mechanism is absent) or limiting (zip length longer or smaller than the average polymerization degree) [304] cases; or (4) for the initial stages of decomposition [41, 179, 301, 305–307].

The last procedure is preferred since at high conversions, the process is diffusion controlled and many reactions overlap. In all cited references, the complete table including information on rate equations, changes in molecular weight, activation energies, etc., corresponding to various cases have been presented.

To distinguish among various mechanisms, one has to establish the correlations between experimental values, such as changes in weight loss, molecular weight, molecular weight distribution, and the rate of evolution of volatile compounds as a function of conversion or time.

In Fig. 16, the dependence of weight loss (w) and M on conversion for three types of mechanisms is presented [179, 304].

The changes in molecular weight experimentally obtained—Fig. 17—are characterized by a shift of MWD curves towards the lower molecular weight and a narrowing of the width of the initial distribution. The values M_w, M_n and polydispersity all decreased with increasing temperature and residence time.

The number of chain scissions is only about one or less at temperatures at/or below the ceiling temperature of 300°C. At temperatures higher than the ceiling temperature a significant number of chain scissions were obtained. Polymethylene shows a maximum of the volatilization rate at 25% conversion (pure random scission) (Fig. 18), while for branched PE no maximum in the rate of conversion is evidenced. There is therefore a good agreement between theory and experimental results.

A good agreement between theoretical treatment and experimental results for polyolefins is obtained by introducing two types of bonds—weak and normal—which are to be broken [41, 179, 288, 289, 293–295].

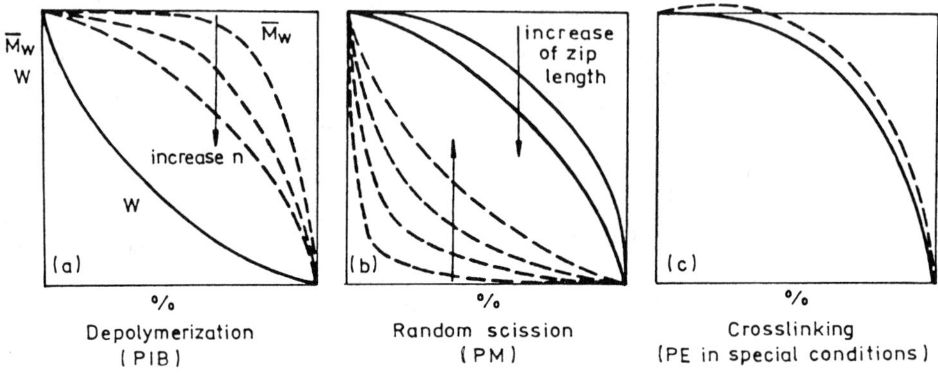

FIG. 16 Dependence of weight loss (w) and $\bar{M}_w$ on conversion for various mechanisms of polyolefin decomposition. In (a) the increase of the initial values of n refers to the fulfillment of the condition: n becomes comparable to the zip length. (Adapted from Refs. 179 and 304.)

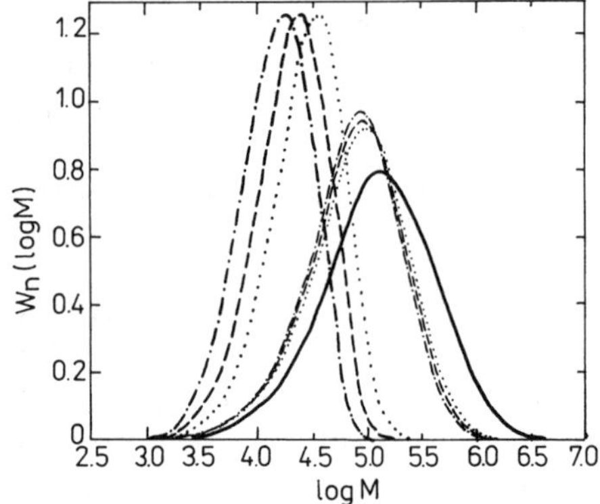

FIG. 17 MWD curves of IPP undegraded (————) and thermally degraded at 275°C and 325°C for 6 (··········), 12 (- - - - - - -), and 24 h(–·–·–·–) (From Ref. 301.)

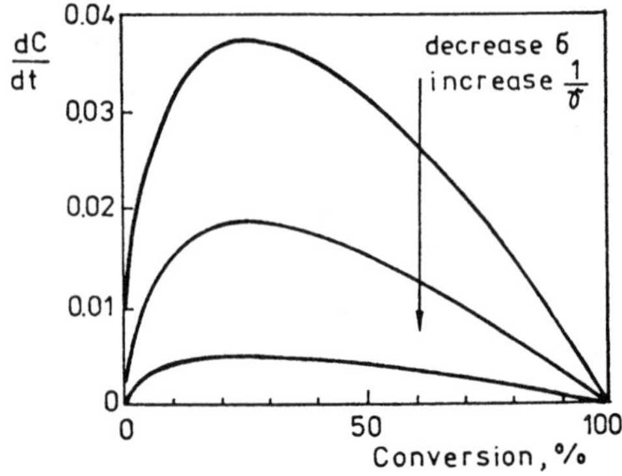

FIG. 18 Rate of conversion vs. conversion for different zip lengths or σ values (Adapted from Refs. 295 and 304.)

For initial stages of decomposition, the rate of volatilization (dC/dt) can be evaluated using the following equation [295]:

$$\frac{dC}{dt} = \frac{L(L-1)k}{N} + \frac{s(s-1)bk}{2N} + \frac{sbk'}{N} \tag{9}$$

while for the variation of the average polymerization degree:

$$\frac{1}{\bar{P}_{n,t}} - \frac{1}{\bar{P}_{n,0}} = \frac{1-e^{-kt}}{1+L(1-e^{-kt})} \tag{10}$$

where N is the number of carbon atoms from macromolecules; s, the length of branches; b, the number of branches per molecule, $s \ll L$; k and k', rate constants for the breaking of normal and weak bonds $k' > k$, $P_{n,0}$ and $P_{n,t}$ are initial and at "τ" time polymerization degrees.

In reality, any decomposition process must be described by a complex system of rate equations, implying many ways of initiation, propagation, and termination, according to the mechanism involved.

3. Kinetics of Thermo-oxidative Decomposition of Burning

If in the first stage of thermo-oxidative decomposition the kinetic treatment for thermal degradation (Sec. E) can be applied, the kinetics of burning is very complex, as kinetics of both thermal and thermooxidative decomposition must be included along with the equations of mass and heat transfer [303].

4. Kinetic Parameters for Polyolefin Decomposition

The kinetic parameters of thermal and thermo-oxidative decomposition vary within very large limits, under both isothermal and dynamic heating conditions, The reasons are multiple as are their explanations. Thus, Flynn [12] considers that the changes in the rate limiting step take place as the reaction temperature and conversion region are changed, while Burnham and Braun [308] introduced an activation energy distribution model—experimentally confirmed [310]—for the complex materials and for the reaction with a complex mechanism. Analysis of the kinetic parameters values given in Table 10 also shows that:

Both thermal and thermo-oxidative decomposition of polyolefins are characterized by a set of kinetic parameters.

There is a slight effect of molecular weight on the E values, while that of branches and other structural defects is much more evident.

The activation energies for thermo-oxidative decomposition are smaller than those for thermal decomposition.

Some values for the elementary steps of thermal decomposition of polyolefins have been also evaluated [41, 290] as follows:

Elementary step	Activation energy (kJ mol^{-1})
Initiation	284.5–296.5; 336
β scission	138.16
Depropagation	21–33.6; 77.4 [301]
Secondary hydrogen abstraction	52.5–79.8
Termination by disproportionation or recombination	4–8.4, 40.6 [301]
Diffusion in molten state	28.14
Diffusion in solid state	40.74

XV. THERMAL TREATMENT OF POLYOLEFIN WASTES

A. Thermochemical Procedures—Feedstock Recycling

1. Degradative, Decomposition Procedures for Feedstock Preparation

The main objective of the degradative, decomposition procedures for feedstock preparation processes is to convert the long-chain polymers into smaller hydrocarbon chains that can be readily processed in existing industrial equipment. The level of metals, halogens (chlorine), and nitrogen are also very important for use as higher value refinery and petrochemical feedstocks. At this time such recycling, of course, depends upon economics, but the driving force is primarily environmental.

Degradative extrusion is a way of preparing commingled plastics scraps for chemical materials recycling, which allows highly diverse mixed plastics to be homogenized, sterilized, compacted, chemically modified, dehalogenated, and converted into a form suitable for transport. So, the processed plastics wastes are safely stored or used. The process may be directed to obtain end products as gasoline, waxes, fuels, etc. [318].

During the degradative extrusion process, the reduction in molecular weight under thermal mechanical work and chemical agents takes place. Chemical agents such as air, oxygen, steam, hydrogen, and degradation promoting additives, metal oxides and other catalysts, are added to the feedstock or fed into the barrel of the extruder to facilitate degradation. As

TABLE 10 Kinetic parameters for thermal and thermo-oxidative decomposition of olefin polymers

Thermal						Thermo-oxidative					
		Kinetic parameters						Kinetic parameters			
Polyolefin	Temperature range (°C)	E (kJ mol^{-1})	A (min^{-1})	n	Ref.	Polyolefin	Temperature range (°C)	E (kJ mol^{-1})	A (min^{-1})	n	Ref.
PM	345–396	286.02	$10^{17.6}$		222	HDPE	500–800	39			303
$L = 72$		301.05	1.9×10^{16}								
PE								59–63			94
$M_w = 11{,}000$	375–436	192.6	$10^{12.64}$		222						
16,000	365–436	220.9	$10^{14.8}$		222	LDPE	203–250	103–157			312
23,000	375–436	276.8	$10^{18.8}$		290	LLDPE	203–250	63–153			312
20,000	360–392.6	263.8		1		XPE		105–121.8			313
$L = 72$		182.7	4×10^{13}		288	PP	TG	52.7–79.1	3.2×10^4	1	314
$= 129$		294	8×10^{19}					78	1.6×10^6		290
$= 154$		336						104			94
Branched PE	350.9–372.6	168–231	Wide spectrum	0.5	309	PIB		197–218.4			222
	$\alpha =$			1	310						315
LDPE	Two 0.35	214.6	3.4×10^{15}	1.45	311a	Copolymers					
	stages 0.65	262.1	1.4×10^{18}	0.85		IB/*p*-fluorostyrene		247.8			315
LLDPE	250–360	235.2	3×10^{10}		311b	IB/*p*-bromostyrene		243.6			315
PP	336–366	242.8				IB/*p*-chlorostyrene		239.4			315
IPP		214–235			277						
	320–341	230.8			222	IB/α-methylstyrene		197.4			315
	250–300	272.2			222	EPR	250–340	63–84		1	317a
$L = 48$		247.8–285.6					360–500		113.4–197.4		317a
APP		234–237	$10^{14.4-16.8}$	1	94	EPDM	175–500	270			316
					277						
PIB	360–326	205.2–256.8	4×10^{18}	1	222	EVA	$T < 370$	87.4			94
			172–218		279						
P3MB-1	300–550				220	P3MB-1 amorphous	50–240	63.84			210
P4MP-1	291–341	224.0			222	Crystalline		84.84			210
PCH	321–336	205.2			222	CPE	$T > 700$°C	39.3			317b
EPDM	175–455–505	270		1	316						
EVA	260–290	180	7×10^{12}	1	94						
E/S	320–380	205–247		0.9	175						

E, global activation energy; A, preexponential factor; *n*, reaction order; α, conversion degree, PM, polymethylene.

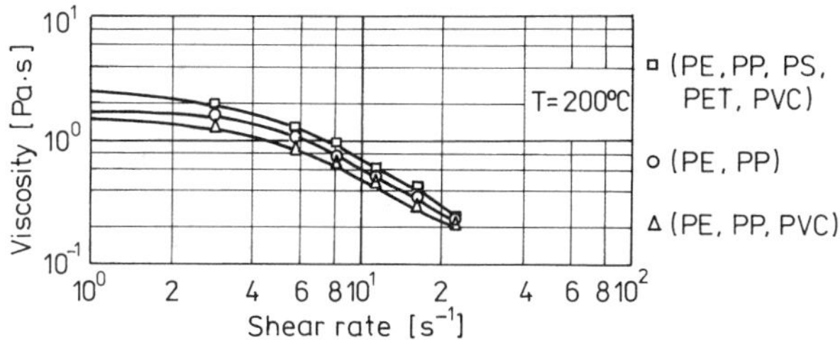

FIG. 19 Viscosity vs. shear rate of mixed plastic waste undergoing degradative extrusion. (From Ref. 319.)

a result homogeneous liquid melt, granulated solid or gas ($T > 500°C$) are obtained. The process is carried out in both cascade extruders or in single line corotating or/and counterrotating twin screw extruders. The halogens are eliminated and recovered. Melt viscosities depend on degradation temperature, shear stress and residence time—Figure 19.

Degraded extrudate can be mixed with crude oil residue in certain proportions, the mixtures showing enough stability in their properties. Pre-treatment costs at the Leuna plant (1993) was of the order of 178 DM/t.

2. Low Temperature Pyrolysis or Cracking

Low temperature pyrolysis or cracking of plastics waste gives high boiling liquids or waxes as potential feedstocks for the steam-cracking or fluid-bed catalytic cracking (FCC) processes currently used to produce olefins for polymerization. PO degradation and degradation products have been described above. In commingled plastics some differences appear. The presence of halogen-containing components can modify the degradation behavior of PO.

The problem of technological utilization of polyolefin wastes is of great importance when considered economically and ecologically. This importance arises from both the valuable products obtained and the enormous quantity of wastes resulting from synthesis, processing, and from domestic and industrial consumption, too. The last two waste types have a high degree of contamination with halogen-containing polymers and compounds of heavy metals which lead to serious difficulties in the application of the process. Dehalogenation is required as a physical or chemical pretreatment of waste to reduce chlorine content below 200 ppm or even 20 ppm.

Chlorine atoms from the decomposition of chloroparaffin fire retardants increase the rate of secondary hydrogen abstraction, accelerating degradation to volatile products but also favoring crosslinking and increasing molecular weight of the remaining polymer. At higher temperatures, halogen-containing components of waste lower thermal stability of crosslinked PE. The HCl evolved can add to terminal double bonds in cracked products of PE. At temperatures below 500–550°C no detectable cyclization occurs and the waxes are linear, being very suitable for subsequent steam cracking to olefins [320].

In PE/PP mixtures, the presence of significant amounts of PP can increase the rate of PE degradation, allowing higher yields of liquids at a lower temperature than in the case of PE alone [321]. The maximum yield of product suitable for recycling as a petrochemical feedstock requires temperatures in the range of 400–600°C together with tailoring the residence time to minimize over cracking and coke production. In these conditions the yield of light gas can be limited to that required for process heat (10–15%). The main known processes are: liquid phase treated bench reactor (Veba in Bottrop, Germany and Fuji, Japan), degradative extrusion (IKV, Aachen), hot tube (Conrad process, American Plastic Council, USA), fluidized bed (British Petrochemical BP's Grangemouth, Fina, DSM, EniChem, Atochem, American Plastic Council).

Waxes obtained from POs waste can be further steam cracked resulting in monomers (ethylene and propylene (20–40%), methane (10–14%), and gasoline and liquids (6–20%) in main products. Steam cracking may be conducted by blending with naphtha, the obtained monomer content being higher than that resulted from naphtha only. FCC of degradation products of PO and representative commingled plastics stream showed that high yields of gasoline are produced. Copyrolysis with coal is also practiced [322].

3. Pyrolysis

Several reviews on the pyrolysis processes [223, 224, 323–325] should be mentioned. Because the polymers are low-conducting materials, various types of pyrolysis processes have been performed such as continuous or discontinuous; in an inert or an oxidative atmosphere; gasification, liquefaction, or carbonization (PTGL or PTGC processes); by direct, indirect, or combined transfer of heat; or by *in situ* generation of heat by burning a part of the pyrolyzing material. The thermochemical processes used to recycle the feedstock from plastics waste are usually those employed in the petrochemical industry such as visbreaking, steam cracking, catalytic cracking, liquid and gas phase hydrogenation, pyrolysis, coking, and gasification. Plastics waste can also serve as reducing agents in blast furnaces. Due to the very similar characteristics and also for smaller capacity necessary in comparison with petrochemical plants it is considered completely uneconomic to build new plants dedicated to the plastic waste stream. The only solution is to use existing petrochemical complexes, adapting them to supplement their usual feedstocks with plastics.

By pyrolysis of polyolefins, waxy, liquid, and gaseous products are obtained; their yields depend on the working conditions such as heating rate, temperature, gas flow, residence time, type of reactor, mode of heat supply, etc. High temperatures favor gas formation, while increased residence time leads to charring [326] (Fig. 13).

Moreover, it is possible to control the properties of each pyrolysis product by changing the reaction conditions. For example, the molecular weight and molecular weight distribution of the waxes [327] can be changed, as can the boiling point of oil [328] or the composition of gases [329–334], etc. Different pyrolysis processes are conducted toward the result of high yields in one of the reaction products (Table 11).

Most pyrolysis processes differ both by the type of pyrolysis reactor employed, such as melting bath reactor (pot or autoclave)*. Poly-Bath Distillation (Sanyo Electric Co., Kawasaki Heavy Industrie, Mitsui Petrochemical Waste Management Institute, Ruhrchemie) [328, 335–338]: reactor with solid heat carrier (Garrett Occidental and Tosco, USA) in rotary drum or in vertical shaft ("P Poni" Institute) kilns (Kobe Steel, Japan) [339–342]; fluidized bed reactor (Hamburg University, CRE Eckelman, Occidental Petroleum Corporation, Union Carbide, Nippon Zeon, Japan Gasoline) [329–334, 343–345]; flash pyrolysis (Procedyne System. US Department of Energy), Tyrolysis uses cross-flow reactor, (Foster Wheeler Power Product Ltd) [346, 347]; extruder type reactor [348–350]: rotary drum reactor (Herbold, Herko-Kiener, GWU, Monsanto-Landgrand), Purator, Australia [325], vacuum [351], etc., as well as by the obtained pyrolysis products.

Polyolefin wastes can be pyrolyzed alone or in mixtures with other materials, especially if found in industrial or municipal refuse. The quality of waxes may be improved by introduction in the pyrolysis reactor of steam, inert gas, air, oxygen, or ozone flow, or by extraction of the low molecular weight products [327, 328, 348, 349]. In this manner the olefin content is reduced and the color and odor are eliminated. They are employed as external or internal lubricants in plastics processing, in the rubber industry, and in coating manufacture.

Pyrolysis oil is a potential source of synthetic crude oil; it can be used in mixtures with diesel fuel, as cracking raw material, or can be introduced in the refining flow in order to obtain valuable chemicals [341a, 352, 353].

The pyrolitic oil may be used directly as a liquid fuel, added to petroleum refinery feedstocks or catalytically upgraded to transport grade fuels, being also a potential source of valuable chemical feedstocks.

The pyrolytic oils from municipal solid waste contain substantial concentration of PAH consisting mainly of naphthalene, fluorene and phenanthrene and their alkylated substituents. Some species were of known carcinogenic or mutagenic activity, e.g., methylfluorenes, phenanthrenes, chrysene, and methylchrysene. The PAH were formed via Diels–Alder and deoxygenation secondary reactions. Increase in reactor temperature and residence time increased PAH concentration [382].

Other authors have shown that polychlorinated benzenes and dioxines are destroyed under pyrolysis with indirect heating at temperature above 700°C [383].

4. Pyrolysis Kinetics

In order to develop the mathematical model, both kinetic and heat transfer phenomena have been considered. The HDPE particles are discharged onto the fluidized bed in the upper part of the reactor. The volatiles evolved from primary reactions underwent secondary cracking reactions [384].

Kinetic laws:

(a) for the overall primary decomposition reaction: zero order kinetics:

$$\mathrm{PE} \xrightarrow{k_1} a\mathrm{G_p} + b\mathrm{A_p}$$

G_p, primary gas: A_p, primary waxes and tars.

(b) Three reactions scheme:

$$\mathrm{P} \begin{cases} \xrightarrow{k_1} a\mathrm{G_1} + b\mathrm{A_1} \\ \xrightarrow{k_2} \mathrm{P^*} \xrightarrow{k_3} a'\mathrm{G_2} + b'\mathrm{A_2} \end{cases} \tag{11}$$

Secondary reaction:

$$b\mathrm{A_p} \begin{cases} \xrightarrow{k_{S1}} \mathrm{G_s} \text{ (secondary gases)} \\ \xrightarrow{k_{S2}} \mathrm{S_s} \text{ (secondary solids or heavy tars)} \end{cases} \tag{12}$$

Simulation programs based on this scheme, evaluates product yields ($Y_{\mathrm{calc.}}$) with the following relation:

$$Y_{\mathrm{calc}} = \sum_{i=1}^{N_1} \Delta \mathrm{G}_{\mathrm{PO}_i} + \left(\frac{k_{\mathrm{S1}}}{k_{\mathrm{S1}} + k_{\mathrm{S2}}} \right) \sum_{j=1}^{N_1} \Delta A_{\mathrm{PO}_j} X_{\mathrm{S}_j}$$

* Brackets indicate which companies designed the pyrolysis units with respective reactor.

TABLE 11 Pyrolysis products and processes used in the recovery of polyolefin wastes

Waste	Pyrolysis conditions	Main product obtained	Ref.
LDPE	350–400°C, 420 min nitrogen	Waxes, $\bar{M}_n = 600$	354
	450°C–525°C, inert gas, sieve bottom reactor	Waxes and gases	355
	130°–150°C in solution, oxygen or air flow	Oxidized waxes	356
	250–450°C, 35–45 min, oxygen or oxygen +air	75% oxidized waxes having 18–21 >C=O/1000C 8–9 OH/1000 C, soft	357
PE or PP	280–450°C, 22.3–26.3 MPa, supercritical water	Monomers for repolymerization, increasing pressure, gas production is reduced	358
LDPE, LMWPE, APP, PP	130–300°C, long time, 250–450°C, short time melting bath	Reduced viscosity waxes, oils, gases	359
LDPE+HDPE	350–385°C, 3 h, steam or nitrogen, melting bath	Waxes, $\bar{M}_n = 800$, penetration 5.6	360
	350–400°C, dry distillation in wax bath	Waxes, heating oil, gases	337
HDPE	Melt extruded (150°C) into pyrolysis reactor, 350–500°C, 10–30 min, steam	Waxes, $\bar{M}_n = 2000$, softening point 50–68°C, dripping point 115°C	361
PE. EVA, EPR	130–180°C, oxygen 120–700 min, in presence of organic peroxides, Hoechst AG	Oxidized waxes, emulsifiable, with small polydispersity	362
PP	Oxidative pyrolysis, 350–450°C, Ruhrchemie	Ketones, 95% liquids or waxes	363
PP and copolymers E/B	Powder samples, turbulent bed, 115–150°C, air flow, 5 min–2 h	Waxes having reduced viscosity and increased solubility	193a
Polyolefins	Molten bath, 119–212°C in oxygen-containing gas flow	Waxes	364
PE,PP,EPR	Extruder reactor, two-stage pyrolysis, 520–615°C distillation at 300–390°C	9–24% Gases, 75–89% heating oil	365
Polyolefins from municipal wastes	Rossi procedure	Combustible oil, gas, and carbon	366
	Extruder-type reactor (I) 160–400°C, (II) 400–600°C; (III) 600–800°C	Liquid and gas, fuels	349
APP, PE, LMWPE	450–520°C, 60–180 min, 2–3.5 atm, helical coil reactor, Procedyne system	Grade fuel oil and gas	346
APP	450–640°C, oxidative pyrolysis in air, fluidized bed	7–36% gas, 50–88% fuel oil with $\bar{M}_n = 430–670$, boiling point 240–310°C	345
60%PE/40%PS	$<$ 440°C	84.1% liquid; 57.1% S; 27.7% α-olefins	367
Polyolefins or mixed wastes	Hydrogasification 7–18% water or partial oxidation, 650–1400°C, 5s	Pipeline gas, ethylene	368
Mixed plastics, elastomers	Fluidized bed, sand 420–450°C, Japan Gasoline	Gases, oils, monomers	343, 369
	600–750°C, Hamburg University, Grimma	Gases, oils	329-333
PP	Hot steam, fixed bed, 650°C, 2.7 s	65% gases with 26% propene	370
Mixed wastes	Molten bath with salts, 420–480°C	Light oil, aromatics, paraffin waxes, and monomers	371
PO	320–400°C, 2 steps, 1 atm	Waxes, paraffins, oils	372
PE, mixed plastics /heavy oil	400–500°C, 0.005–0.4 bar, further processing by hydrogenation and separation 800–900°C; 350–2500 ms	75–95% olefm monomers	373
Mixed plastics	Delayed coker	85% distillates, 14% gases, $<$1% coke	374
Mixed plastics	Fluidized bed, various conditions	Aromatic hydrocarbons, aliphatic oils, waxes, low molecular weight, hydrocarbons, monomers	331, 332

TABLE 11 (Continued)

Waste	Pyrolysis conditions	Main product obtained	Ref.
Mixed plastics	350–390°C, dehydrochlorination	Low viscosity fuel for pumping	375
PP	CS_2 and iron supported catalyst, 380°C	>90% naphta, kerosene,	376
EPDM	350–450°C, 5–30 min, hydrothermal, NaOH super critical conditions, desulfuration, liquefaction	Gas oil, oily materials	377
PO	PARAC Process, Webau Demonstration Plant 20,000 t/yr	Paraffin waxes	378
Mixed plastics	Fluidized bed, gas cleaning section that removes impurities to acceptable levels for downstream refinery or petrochemical plants	Hydrocarbons	379
PO	Hydrous pyrolysis, 450°C, 160 atm		380
Mixed plastics (chlorine-containing)	Stepwise pyrolysis, 330–440°C		381
	Conrad process—American Plastics Council Variable residence time and pyrolysis temperatures, rotary kiln reactor, CaO for HCl capture	Feedstock for steam crackers giving higher amount of monomers, 86% naphta grade products	381a
Mixed plastics automobile shredded 3 wt% chlorine	Energy and Environmental Research Corporation and US EPA, pilot scale, "spouted bed" fluidized regime; ablative gasification, 500–800°C Fe_2O_3 active gasification catalyst and sorbent for HCl	Petrochemical feedstocks	381b

ΔG_{POi} and ΔA_{POj} are the quantities of gases and respectively primary tars generated in the "t" interval per gram of the PE discharged; X_s is the reaction extension.

A good agreement between calculated and experimental yields is obtained for a wide temperature range—Fig. 20.

Activation energies are around 196 and 232 kJ mol^{-1} for reactions 11 and 12 respectively. The values of the kinetic constants at 1000 K are: $k_{S1,1000K} = 9.93–11.175\,s^{-1}$ and $k_{S2,1000K} = 3.09–2.85\,s^{-1}$, depending on the kinetic model used.

Other kinetic treatments were presented [385–387] which tested the kinetics of the liquid hydrocarbon pyrolysis to PO thermal degradation.

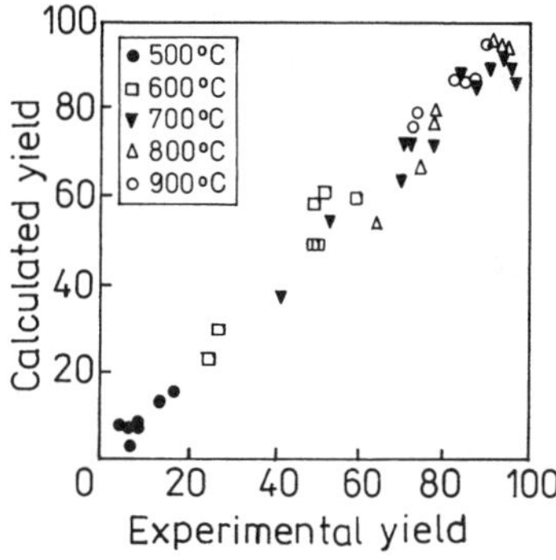

FIG. 20 Calculated vs. experimental yields of total gases. (From Ref. 384.)

A less simplified approach was followed by Darikavis *et al.* [387], who fitted their experimental data by means of a multiple independent parallel reaction kinetics. Their model assumed that volatilization products were produced from a large number of independent parallel first order reactions.

The broad variation in the kinetic parameters of decomposition (see Table 10) is essentially due to two reasons: differences in PO characteristics (MW, presence of weak links, impurities, etc.), and differences in experimental conditions for which kinetic parameters were evaluated. In particular a single step model is not able to cover, using the same kinetic parameters, a wide range of heating rates, temperatures, and conversion levels. It has been established that the reaction order is equal to 1 and decreases with the aging time of the waste [388].

A further and deeper analysis of pyrolysis mechanism allows us to better characterize the product distribution.

B. Hydrogenation

Feedstock recycling processes of plastics waste have similar requirements to those imposed on processes for upgrading cool and heavy crude-oil residues. Hydrocracking permits the recycling of mixed and/or contaminated plastics. In hydrocracking the heat breaks macromolecules into highly reactive free radicals and molecules that are saturated with molecular hydrogen as they form.

The Veba Combi-Cracking process is suitable for converting residue oil and plastics waste.

Hydrogenation itself usually takes place in a series of two or three column reactors [325b, 389] at a variable (60 MPa–200 MPa) hydrogen pressure in the presence of catalysts such as Ni/Mo or Co/Mo on Al_2O_3. In these conditions, unsaturated compounds are hydrogenated, and heterocompounds are removed.

The products obtained from PO by this process are 5–20 wt% gaseous hydrocarbons and 95% gasoline, Diesel and lubricating oil.

To feed the reactor, the plastics waste has previously undergone visbreaking alone or in the mixture with heavy petroleum derived vacuum residues [390–392].

The process can also be conducted in a coal–oil plant as has been demonstrated in Bottrop, Germany. The procedure allows conversion of dangerous polychlorinated aromatics to other harmless compounds.

C. Gasification

Gasification is a partial oxidation in a restricted supply of oxygen at temperatures up to 1600°C and a pressure of 150 bar. The gasification agents are: O_2, air, flue gas, steam, CO_2 and, in exceptional cases, H_2. Gasification yields a synthetic gas (CO and H_2) with various heating values and applications as lean gas and water gas (4600–12,500 kJ N^{-1} m^{-3}); synthesis gas and reduction gas (12,500 kJ N^{-1} m^{-3}), town gas and strong gas (1670–2000 kJ N^{-1} m^{3}) or rich gas and synthetic gas (25,000–37,000 kJ N^{-1} m^{-3}.

The PO wastes mixed with O_2 and steam run through the following steps: preheating and cracking (vaporized by radiant heat from the flame and reactor walls), oxidation, and gasification at ignition temperature.

The known processes of partial oxidation are fixed-bed gasification (Lurgi EcoGas Koppers and Winkler), melt gasification (Sheel, Texaco), adapted for mixtures containing plastics wastes as: Union Carbide; SFW (Saarberg-Fernwärme-GmbH, ZSG, Kiener, Babcock-Rohrback, Eisenmann, Pulse, Voest-Alpine AG (high temperature gasification process), thermoselect SA-Locarna, LAVBAG (pressurized-oil gasification; entrained-flow gasification of Veba Oel Technologie GmbH; Menger and Fischer, etc. [325b].

Synthetic gases are used as feedstocks for many chemicals such as urea, ammonia, alcohols, acids, amides, resins, etc., and also as fuel.

A solid product fraction in the pyrolysis of PE in a fixed-bed reactor was found to be mainly the result of the secondary tar-cracking process. Complete conversion of PE is generally observed when pyrolysis is performed by vacuum or flash pyrolysis [393].

The actual mechanism of formation of the solid fraction in the pyrolysis of PE and the influence of reaction temperature and reactor geometry are still open problems. Their elucidation is important in order to minimize solid formation in gasification and pyrolysis reactions that may cause sooting and fouling operating problems of these processes.

Cozzani found two types of coke: a filamentous one that is believed to be the result of a metacatalyzed coking process that takes place on the reactor walls, and a spherical coke that is mainly formed by a gas-phase mechanism [394].

D. Reduction in Blast Furnaces

Comparisons of plastics waste with the standard coke, coal and heavy oil reductants indicates that its composition makes it fundamentally suitable for recycling in blast furnaces. Pretreated by degradative extrusion or untreated plastics waste as liquid are injected in the blast furnaces, reducing the oil consumption [325].

E. Catalytic Pyrolysis or Thermocatalytic Decomposition

Catalysts are used in pyrolytic procedures both for the decrease (with 150–250°C) of the pyrolysis temperature and for the modification of the reaction products composition [275, 395–400].

Various catalysts are used, such as metals, metallic oxides, salts, zeolites, activated carbon, etc. (Table 12), which act according to the following simplified scheme.

$$\text{Polyolefins} \xrightarrow[\text{}]{\text{Free radical mechanism of decomposition noncatalytic or slightly catalytic}}$$

$$\begin{array}{l}\text{Large fragments}\\ \text{and small quantity}\\ \text{of low molecular}\\ \text{weight compounds}\end{array} \longrightarrow$$

$$\xrightarrow[\text{further cracking in several steps: isomerization dehydrocyclization, aromatization, ring expansion reactions, etc.}]{\text{Catalytic by radical mechanism (activated carbon) or ionic mechanism (electrophilic catalysts) involving}}$$

$$\begin{bmatrix}\text{Aliphatic, alkanes, isoalkanes,}\\ \text{alkenes isoalkenes,}\\ \text{cycloalkanes,}\\ \text{cycloalkenes}\end{bmatrix}$$

$$\text{Dehydrocyclization} \longrightarrow \text{aromatics} \longrightarrow \text{dealkylation} \longrightarrow \text{simple aromatics} \longrightarrow \text{coking} \longrightarrow \text{coke}$$

TABLE 12 Catalytic systems used in polyolefin decomposition

Waste	Catalytic system and reaction conditions	Obtained products	Ref.
Polyolefins	Al_2O_3/SiO_2, 300–450°C. Nichimen, Japan: 400-650°C, CHO Rands, Japan	Heating oil, isobutene	363
	Mesoporous silica KFS-16 with hexagonal pore structure, accelerated decomposition and KSM-5	Liquids, kerosene and diesel oil	363a
PP	Silica–alumina and CaX zeolites,	Isoalkanes, methyl and alkyl-	395
APP	fixed or fluidized bed. 480°C, 380°C	branched	395a
	Silica–alumina, fluidized bed, 500°C	Fuel oil	402
	Zeolites, 280–400°C, fluidized bed		403
	Silica–alumina, mordenite, ZSM-5, fixed bed, two-stage pyrolysis, 400–450°C, 420–460°C	26–37% gases 47-66% liquids with high content of aromatics	397
	H-75M-5,liquid phase pyrolysis, 200–340°C	Low pour point oil, gasoline	404
	H_2/hydrogenation catalyst, 350–500°C Agency Ind. Sci. Techn, Japan	Gasoline of high isooctane content	363
	Cr_2O_3, MnO_2, CuO supported on asbestos or firebrick, two-stage pyrolysis, fixed bed, 420–450°C, 450–500°C	25–35% gases, 75–65% liquids	399 400
PE	3–5% In_2O_3 and 0.5–1%CdO ceramic supported, fluidized bed, 740–780°C, steam or nitrogen	81–93% gaseous olefins with 33–43% ethylene, 8.4% oil	405
Polyolefins	Metal-supported carbon, Me = Pt, Fe, Mo, Mn, Co, Ni, Cu, fixed bed, 526°C, nitrogen	*n*-alkanes and 30–45% aromatics	396
PE,PIB, butyl rubber	H^+ $[MeAlCl_4OH]^-$. Me = Li, Na, K, Mg, Ca, 300–400°C	90–95% C_4 hydrocarbons 45–60% isobutane and 19–25% isobutene	401
PIB and	$NaAlCl_4$. 200°C	Gaseous products, isobutene	406
butyl	NaX-zeolites, 330–350°C		407
rubber	$AlCl_3/CaCl_2 = \frac{1}{2}$–2.5, 1 h, 250–350°C	59–39% gases mainly C_4 41–61% liquid	408
PE	NaCl, $BiCl_3$, 0.7–1 mol, 280°C–330°C	C_4 hydrocarbons	409
PP	$TICl_3$+ MgClEt (organo-magnesium),6 h, 70°C, boiling heptane	PP waxes	410
PE	4% $AICl_3$, supported activated carbon, 200°C	Synthetic oil, gasoline	411
Mixed plastics PE,PP,PS	ZSM-5, liquefaction	Gasoline, kerosene, diesel fuel	404a
PE+PS	Fe_2O_3/S	Cracking and hydrogenation	412
Mixed plastics	Silica–alumina, 380–430°C	Fuel oil, gasoline	413
PE,PP,PVC, PET PS,ABS	semibath and continuous		414
PE,PP,PS	Silica–alumina, H-ZSM-5, sulfonated zirconia	Product distribution depends on catalyst acidity	415
PE/PP	S, $NiMo/Al_2O_3$ ZSM-5, 230–450°C, 500 psig of H_2	Twofold increase of gasoline in presence of S	416
PP	Various zeolites HY zeolites	Narrow product distribution	417
PE/PP	380–430°C acid activated silica–alumina, ZSM-5, nonacidic mesoporous silica catalyst (folded sheet material)	Increased yield of liquids	418
Mixed plastics	410°C silica alumina	Fuel oil	419
PP/PA or PET	Thermocatalytic process		420

As in all heterogeneous catalytic processes, the rate of each step will change with the type of catalyst, contact time, temperature, etc.

Metal-supported activated carbon catalysts are active for aromatics formation [395, 396], Pt having a high catalytic activity for toluene demethylation. Zeolites are active in cracking and ZSM-5 has a selective activity due to the pore structure [334, 397–399], giving a high content of aromatics, while $MeAlCl_4$ catalysts are selective for C_4 hydrocarbon formation [401]. A detailed study of the influence of the PO structure on catalytic decomposition over silica–alumina indicated that LDPE and LLDPE produced only smaller wax-like compounds compared with HDPE and XPE.

The global activation energy for ionic reactions is very low, varying between 48 and 151 kJ mol^{-1}, the reaction rate being high [401].

The long contact times increase the gaseous products yield but at the same time catalysts will be deactivated due to the deposited coke, which reduces the surface area. The acid catalysts, which strongly absorb aromatics, are rapidly deactivated.

Economic efficiency was evaluated at a revenue of 175–200 DM/t waste by combining the pyrolyse with a petrochemical plant.

F. Incineration—Energy Recovery

The incineration of waste is virtually taboo in some countries, in others such as Switzerland and Japan it is the method of choice for waste disposal. However, recently most countries have come to desire as complete an inertization and mineralization of waste as possible because the residues have to be dumped and must not react afterwards, and these residues must be reduced to a minimum due to shortage of landfill space.

There are two reasons for a low degree of acceptance for "energy recovery" from waste, the first one is fear for environmental pollution, and the second one are the corrosion problems. Waste incineration plants are not operated with the aim of producing energy. The main purpose is, and remains, to reduce the volume of waste to a considerable degree by means of incineration. On the other hand, only inert waste is allowed to be landfilled. New technologies for waste incineration must accomplish all demands regarding pollution.

Combustion in pure oxygen at $T_{max} \sim 1600°C$ with chemical conversion of the reaction gas offers most of these requirements (RWT + pilot plant from Aachen design industrietechnik Alsdorf GmbH). Polymer wastes, and especially polyolefin wastes, having a high calorific value (36–46 MJ kg^{-1}) can yield, by incineration with heat recovery, an important contribution to reducing the energetic needs (in plants and municipal heating, for the obtainment of hot water, steam, etc.).

The most suitable processes for the energy recovery from polyolefin wastes are: (l) direct incineration, and (2) thermal and mechanical treatments. The latter procedures are applied for the production of gaseous, liquid, or solid fuels in a complex pyrolysis–gasification–incineration system.

Synthetic material wastes cannot be directly burnt in common incinerators due to their specific combustion behavior [421, 422] as follows.

Burning, melting, and decomposition take place as endothermic processes. Thus the grates cool and hang up.

The complex hydrocarbon mixtures resulted from decomposition from the air explosive mixtures (2–5 m^3 N air kg^{-1} gas). That is why the combustion is violent and the shock waves extinguish the flames. To prevent this phenomenon the oven must be sufficiently long and equipped with an auxiliary burner.

The air quantity needed for burning is very large (15–25 m^3 N kg^{-1} waste). As constituents of municipal wastes, the polymers lead to the increase of caloric value, improving combustion especially when humidity is high. After a certain limit (the optimum concentration is 6–10%), they form a reducing atmosphere inside the incinerator, the air necessary increasing along with temperature, consequently increasing the amount of toxic nitrogen oxides [422]. Hence a limited combustion must be practiced.

The smaller the size of combustible material, the higher are the combustion rates and the shorter the combustion times. For films and foam materials, the burning time is a few seconds, while for the other forms it is much longer.

A very efficient agitation is also necessary to assure a good contact between combustible and carburant, and thus a complete combustion.

Soot has a very low density, is spread at long distance (>1 km), and contains toxic substances. Hence the imposed limits for its concentration in surroundings are very low (<100 mg N m^{-3}). It is also necessary that incinerators be provided with washing and neutralizing equipment [423].

The volume decrease after burning is approximately 95%, the ash amount being very small, so that a large landfill domain will now be available to other utilities, incineration being considered a hygienic procedure for waste disposal.

Now and then plastics wastes are treated separately by a complex system of pyrolysis–gasification–

incineration as special wastes. In this case, the optimum conditions are achieved in incinerators with a rotary drum, coupled with a fixed bed or fluidized bed reactor, followed by a postcombustion chamber [422–427]. In the postcombustion chamber, a water spray device will exist in order to control the temperature and to reduce the risk of explosion. The operating temperature varies between 700° and 1200°C.

Specialized incinerators for plastic wastes are manufactured by Kawasaki Heavy Industries, Tokuma Boiler Co. Ltd., Okumura, Mitsubishi Heavy Industries, Nippon Carbon Co., Katayose Japan, Mitsubishi Petroleum Co., Burnway Society Copenhagen, Sarotti Chocolate Factory a Feriro Company, John Thurley Ltd. from Harrogate, Redman Heeman and Frauds, Spaulding Fibre Co. USA, etc. [428, 429].

Recently, combustion in suspension and in fluidized beds has been developed [428–431], along with small modular incinerator systems [432]. Small incinerators are used especially in the medical field for hospital waste incineration.

Cocombustion is also practicable in common incinerators together with coal, wood, and other combustible materials [428].

Energy recovery from plastic waste can also be achieved in cement kilns (see sec. XV.D) [432].

The price of incineration will rise continuously due to the necessity of automatically controlling combustion, the development of burning in suspension, and the control of the emission of toxic gases and their neutralization. Emission limits in French incinerators are given in Table 13 [433].

TABLE 13 Emission limits in French incinerators [433]

Emission type	Existing units and new units before Feb 1997 ($mg\ N^{-1}\ m^{-3}$)	New units after Feb 1997 ($mg\ N^{-1}\ m^{-3}$)
Dust	30	10
HCl	50	10
HF	2	1
SO_2	300	50
NO_x	—	—
CO	100	50
Cd + Hg	0.2	—
Cd + Ti	—	0.05
Hg	—	0.05
Dioxines/furanes	—	0.1 $ng\ N^{-1}\ m^{-3}$
TOC	20	10

Some people consider that in the future incineration will be replaced by other techniques, such as pyrolysis, which offers such advantages as simpler devices, lower operating temperatures, smaller quantity of gases to clean, greater feasibility, larger capacities of the plants, stockable products or pipeline gases are obtained, the recovered metals (from municipal wastes) are less oxidized, and so forth.

A detailed analysis of the plastic addition in the combustor of municipal waste has proven their positive effect. The ash quality was improved through further reduction of heavy metals content and the polychlorinated dioxines amount was not increased even for a 8–10 wt% PVC content [434].

XVI. CONCLUDING REMARKS

Polyolefin degradation and decomposition constitutes the topic of an enormous number of publications, so that it is practically impossible to cite them all. Most of them refer to polyethylene and polypropylene and some to polyisobutylene and polybutene-l. There are shortcomings in the study of metallocene-polyolefins, copolymers and modified polyolefins.

Both degradation and decomposition are mainly dependent on structural defects, with photodegradation also strongly dependent on nonbonded chemical impurities.

All types of degradation and decomposition occur according to a free radical chain mechanism, leading to a very complex mixture of reaction products; only a few ionic species are present during combustion and high-energy-induced degradation. In most practical situations, there appears a superposition of the different kinds of degradation; oxidative attack especially cannot be prevented, so that any reaction mechanism is additionally intricated with the product mixture, as well.

Each kind of degradation has some distinctive features. For example, biodegradation begins at a certain molecular weight, while mechanical degradation slows down from a certain molecular weight on.

Due to their radical nature, all polymers can be stabilized against rapid deterioration by a suitable additive. Physical–mechanical degradation alone is not influenced by antioxidants or other additives. Controlled degradation is a suitable way to enhance some PO properties.

XVII. FUTURE DEVELOPMENTS

Future developments will depend on the general trends in polymer science and technology. "Specialty" polyolefin polymers obtained by synthesis or processing will be the object of future degradation studies.

It will be necessary to establish some correlations between the reciprocal influences of various degradative factors in order to better understand the aging phenomenon. Intensive studies must be undertaken in polluted atmospheres in order to establish the action of corrosive and pollutant agents whose concentration in the environment increases continuously.

The kinetics and thermodynamics of in degradation, decomposition, and combustion must also be developed in order to generalize and minimize the number of limiting assumptions.

The nature of the participating radicals in degradation/decomposition at high temperatures must be clarified, taking into account the fact that primary radicals are not stable at these temperatures yet are included in all the presented mechanisms (Table 14).

Technologies for optimal large scale liquefaction, gasification, and combustion still need basic research directed towards the thermal efficiency of the systems, operational problems, and pollutant emission control.

TABLE 14 Thermal stability and light sensitivity ranges of radicals formed in some polyolefins [435]

Radical	Method of radical generation	Range of thermal stability (K)	Range of light sensitivity λ (nm)
$-CH_2\dot{C}HCH_2-$	Radiolysis Photolysis Mechanical	<200–300	300
$-CH_2\dot{C}HCH_3$	Photolysis Mechanical	<120–140	230–240
$-CH_2-\dot{C}H_2$	Photolysis Mechanical	<120–140	—
$-CH{=}CH-\dot{C}H-CH_2-$	Radiolysis	<290–340	255
$-(CH{=}CH)_2-\dot{C}H-CH_2-$	Radiolysis	—	285
$-CH(CH_3)-\dot{C}H-CH(CH_3)-$	Photolysis Radiolysis	<250–290	<230
$-CH_2-\dot{C}(CH_3)-CH_2-$	Photolysis Radiolysis Mechanical	—	—
$-CH(CH_3)\dot{C}H-C(CH_3){=}CH-$	Radiolysis	—	255
$-(C(CH_3){=}CH)_2-\dot{C}(CH_3)-CH_2-$	Radiolysis	—	310
$-C(CH_3)_2\dot{C}H-C(CH_3)_2-$	Photolysis Radiolysis Mechanical	<200–220	<230
$-CH_2\dot{C}(CH_3)_2$	Radiolysis Photolysis Mechanical	<183	—
$RO_2^{\bullet}$	Radiolysis	<260–280 (PE)	250–280
	Photolysis	<250–290 (PP)	
	Mechanical	<200–220 (PIB)	

There is a general lack of fundamental information needed to extrapolate the results towards full scale process description. All known models do not take into account the rigorous and exhaustive description of the chemistry of thermal decomposition of polymers and describe the pyrolysis process by means of a simplified reaction pathway, while each single reaction step considered is not entirely representative of a complex network of reaction.

SYMBOLS AND ABBREVIATIONS

a, A, B	Constants
APP	Atactic polypropylene
b	Number of branches per molecule
BPE	Branched polyethylene
CPE	Chlorinated polyethylene
dC/dt	Rate of volatilization
DTA	Differential thermal analysis
E/P	Ethylene–propylene copolymer
EPDM	Ethylene-propylene-diene copolymer
E/S	Ethylene–styrene copolymer
EVA	Ethylene–vinyl acetate copolymer
EVC	Ethylene–vinylchloride copolymer
FTIR	Fourier transform infrared spectroscopy
GC	Gas chromatography
HDPE	High-density polyethylene
HPD	Hydroperoxide decomposition
IB	Isobutylene
IPP	Isotactic polypropylene
IR	Infrared
k	Rate constant
k_e, k_s, k_p, k_{t1}, k_{t2},	Rate constants for initiation at the chain end, or by random scission, propagation, termination for first and second order, respectively
$1/\gamma$	Average length of zip in thermal decomposition
L	Maximum length of volatile fragments
LDPE	Low-density polyethylene
LLDPE	Linear low-density polyethylene
LMWPE	Low molecular weight polyethylene
$\bar{M}_n$	Number average molecular weight
MFI	Melt flow index
P_n or R_n	Polymer or radical molecules with n, j, s, or r polymerization degree
PAH	Polynuclear aromatic hydrocarbons

PB-1	Polybutene-1
PCH	Polycyclohexyl ethylene
PE	Polyethylene
PIB	Polyisobutylene
P3MB	Poly-3-methylbutene
P4MP	Poly-4-methylpentene
PP	Polypropylene
MS	Mass spectrometry
n	Degree of polymerization
N	Number of carbon atoms per molecule
PM	Polymethylene
Py	Pyrolysis
s	Length of branches
t	Time
TG	Thermogravimetry
TOC	Total oxygen consumption
UV	Ultraviolet
UHMWPE	Ultrahigh molecular weight PE
v	Oxidation rate
w_0	Initial rate of oxidative reaction
XPE	crosslinked PE
τ	Induction period
σ	Transfer constant
υ_0	Initial kinetic chain length in oxidative reaction

REFERENCES

1. K Hatada, RB Fox, J Kahovec, E Marechal, I Mita, V Shibaev. Pure Appl Chem 68(12): 2313–2323, 1996.
2. N Grassie. The Chemistry of High Polymer Degradation Processes. London: Butterworth, 1956; Encyclopedia of Polymer Science Technology, Vol. 4. New York: Interscience, 1968, p 647.
3. W Schnabel. Polymer Degradation: Principles and Applications. Berlin: Akademie Verlag, 1981; Polym Eng Sci 20: 688, 1980.
4. WL Hawkins, FH Winslow. Degradation and stabilization. Stability and structure relationship. In Crystalline Olefin Polymers. RAV Raff, KW Doak, eds. London: Interscience, 1964, p 361; 1965, p 819.
5. HA Lueb, ThCF Schoon. J Polym Sci Symp 57: 147, 1976.
6. F Severini, R Gallo, S Ipsale. Arab J Sci Eng 13: 533, 1988.
7. M Inoue. Plas Ind News 34: 183, 1988.
8. MJ Kubey, G Field, M Styzinsky. Materials Forum 13: 1, 1989.
9. FS Qureshi, HS Hamid, AG Maadhah, MB Amin. In: JM Buist, ed. Progress in Rubber and Plastic Technology, Vol. 5. London: Plastics and Rubber and Rapra Technology Limited, 1989, p 1.
   a. SH Hamid, AG Maadhah, MB Amin, In: SH Hamid, MB Amin, AG Maadhah, eds. Handbook of Polymer Degradation, New York: Marcel Dekker Inc, 1993, pp 218–228.
10. ES Qureshi, MB Amin, AG Maadhah, SH Hamid. Polym Plast Technol Eng 28: 663, 1989; 28: 475–492, 1989.
11. AYe Chmel', AM Kondyrev. Visokomol Soedn A-30: 2391, 1988.
12. HJ Flynn. Thermochim Acta 134: 115, 1988; Polym Eng Sci 20: 675, 1980: Proc 2nd European Symposium on Thermal Analysis. London: Heyden, 1981, p 223.
13. NM Emanuel, AL Buchachenko. Physical Chemistry of Aging and Stabilization of Polymers. Moscow: Izd. Nauka, 1982, p 356; Polym Eng Sci 20: 662, 1980.
14 AK Kulshreshtha, VK Kaushik, GC Pandey, S Chakrapani, YM Sharma. J Appl Polym Sci 37: 669, 1989.
15. B Ranby, JP Rabek. Environmental Corrosion of Polymers. Causes and Main Reactions in Effects of Hostile Environments. ACS Symp. Series 229, 1983, Chap 17, p 291.
16. GS Shapoval, AA. Pud, VP Kuhar, AP Tomilov. Dokl Akad Nauk SSSR 305: 395, 1989.
17. RB Seymour Plastics vs Corrosives, New York: John Wiley & Sons, 1982.
18. HHG Jellinek. Reactions of Polymers with Pollutant Gases in Aspects of Degradation and Stabilization of Polymers. Amsterdam: Elsevier, 1978, chap 9, p 432.
19. A Fink-Sontag. Int. Ann. Conf. ICT, 1989; 20th Environ. Test 90's pp 51/1–51/13.
20. JB Howard. Stress cracking. In: RAV Raff, KW Doak, eds. Crystalline Olefin Polymers, Part 2. New York: Interscience, 1964, Chap 2, p 47.
21. AC Albertsson, S Karlsson. Polym Mater Sci Eng, ACS Symp. Ser 58: 1988, p 65; J Appl Polym Sci 35: 1289, 1988; Prog Polym Sci 15: 177, 1990; B Erlandsson, AC Albertsson, S Karlsson. Polym Degrad Stab 57: 17–23, 1997.
22. VM Kestel'man, VI Yarovenko, EJ Melnikova. Int Biodeter Bull 8: 15, 1972.
23. GJL Griffin. J Polym Sci Symp 57: 281, 1976.
24. B Dolezel. Br Plast 40: 105, 1967.
25. B Dolezel, L Adamirova, Z Naprstek, P Vodracek, Biomaterials 10: 96, 1989.
26. AC Albertsson. Thesis, Royal Institute of Technology, Stockholm, 1977; In: AV Patris, ed. Int. Conf. on Advances in the Stabilization and Controlled Degradation of Polymers. Vol. 1. Lancaster: Technomic, 1989, p 115.
27. AC Albertsson. J Appl Polym Sci. 22: 3419, 1978; Combined Biological and Physical Impact on PE Building Material, Report No. 35, Swedish Plastics and Rubber Institute, 1982; 19th Prague Microsymp. on Mechanism of Degradation and Stabilization of Hydrocarbon Polymers, Preprints, Prague: Czechoslovak Akad Sci, 1979.
28. G Collin, JD Cooney, DM Wiles. Int Biodet Bull 12: 67, 1976.

29. G Collin, JD Cooney, DJ Carlsson, DM Wiles. J Appl Polym Sci 26: 509, 1981.
30. a. AC Albertsson, B Ranby. J Appl Polym Symp 35: 4243, 1979.
b. AC Albertsson, SO Andersson, S Karlsson. Polym Degrad Stab 18: 73, 1987.
31. a. MI Knuniat, IYa Dorfman, AN Kriucikov, EY Prut, NS Enikolopian. Dokl Akad Nauk SSSR 293: 1409, 1987.
b. WO Drake, JR Pauquet, RV Todesco, H Zweifel. Angew Makromol Chem 176/177: 215,1990.
32. a A Peterlin. J. Polym Sci. Symp. C-32: 297, 1970.
b. A Peterlin. Int J Fracture 11: 761, 1975.
33. MYa Rapoport, GE Zaikov. Eur Polym J 20: 409, 1984.
34. KL Devries, M Igarashi, E Chao, J Polym Sci Symp 72: 11 1,1985.
35. KW Scott, J Polym Sci Symp 46: 321, 1974.
36. VR Regel', OE Poznyakov, AV Amelin, Mekhanika Polymerov 1: 16, 1975.
a. AC Colbert, JG Didier, L Xu. Macromolecules 29(27): 8591–8597, 1996.
37. GR Rideal, JC Padget. J Polym Sci Symp 57: 1, 1976.
38. EJ Globus, NG Kabanova, NM Kostenko, AM Lobanov, BI Sajin. Plasticeskie Massy 2: 77, 1989.
a. I Xu. Macromolecules 111 189–194, 1997.
39. EP La Mantia, V Citta, A Valenza, S Roccasalvo. Polym Degrad Stab 23: 109, 1989.
40. a. C Vasiliu Oprea, V Bulacovschi, A Constantinescu. Polymers: Structure and Properties. Bucharest: Tehnica Publishing House, 1986, chap 14, p 244, chap 15, p 316.
b. J Runt, M Jacq, JT Yeh, KP Gallangher. Polym Preprints 29: 134, 1988.
c. N Kaiya, A Takahara, T Kajiyama. Polym. Preprints 29: 130, 1988.
d. GE Zaikov, VM Goldberg, Encyclopedia of Fluid Mechanics. Vol 9, Polymer Flow Eng. Gulf Publ Co. NY, 1990 p 403.
41. HHG Jellinek. Degradation of Vinyl Polymers. New York: Academic Press, 1955 p 125, 164, 231, 306; Encyclopedia of Polymer Science and Technology, Vol. 4. New York: John Wiley & Sons, 1966, p 740; J. Polym Sci 5: 264, 1950; 9: 369, 1959.
42. YuYa Nel'kenbaum, IK Prokofev, YuA Sangalov. Visokomol Soedn A-28: 1058, 1986.
43. YuYu Nel'kenbaum, IK Prekofev, YuA Sangalov. Visokomol Soedn 28(5):1180–1186, 1986.
44. L Mascia. Electron beaming of polymers for advanced technologies. In: E Martuscelli, C Marchetta, L Nicholais, eds. Future Trends in Polymer Science and Technology. Lancaster: Technomic: 1988, p 25.
45. a. AS Hoffman. Applications of synthetic polymeric materials in medicine and biotechnology. In 44, p.193; b. JL Williams. Polym Preprints 28: 309, 1987.
46. a. RE Becker, DJ Carlsson, JM Cooke, S Chmela. Polym Degrad Stab 22: 313, 1988.
b. K Schipschack, A Bartl, L Wuckel. Acta Polym 36: 494, 1985.
47. A Tidjani, Y Watanabe. J Appl Polym Sci 60(11), 1839–1845, 1996.
48. VM Belyaev, SI Kogan, MD Pukshanskii, VP Budtov, AP Zemskova. Visokomol. Soedn A-31: 165, 1989.
49. L Audouin-Jirakova, G Papet, J Verdu. Eur. Polym. J. 25: 181, 1988; J Polym Sci Polym Chem Ed A-25: 1205, 1987.
50. J Verdu, L Audouin. Mater Tech, 85: 31–38, 1997.
51. EJ Lawton, AM Bueche, JS Balwit. Nature 172: 76, 1953; Ind Eng Chem 46: 1703, 1954.
52. D Babic. Makromol Chem Macromol Symp 28: 231, 1989.
a. G Akay, T Tincer. Polym Eng Sci 21: 8, 1981.
b. T Cimen, T Tincer, G Akay. Macro'83, Bucharest, Sept. 5–7, 1983, V.-II p 184.
c. EP La Mantia, G Spadaro, D'Acierno. MACRO '83, Bucharest, Vol. 14, Sept. 5–7, 1983, p 192.
53. a. G Papet, L Jirakova-Audouin, J Verdu. Rad Phys Chem 29: 65, 1987; 33: 329, 1989.
b. AM Naggar, LC Lopez, GL Wilkes. J Appl Polym Sci 39: 427, 1990.
54 M Goldman, R Gronsky, R Ranganathan, L Pruitt. Polymer 37(14): 2909–2913, 1996.
55. S Baccaro, B Caccia, S Onori, M Pantaloni. Nucl Instrum Meth Phys Res B-105: 97–99, 1995.
56. SW Shalaby. J Polym Sci Macromol Rev 14: 419, 1979; SL Kuzina, AI Mikhailo. High Energy Chem 31(4): 239–243, 1997.
57. T Zaharescu, S Jipa. Polym Test 16(2): 107–113, 1997.
58. MV Belousova, VD Skrida, OE Zgadzai, AI Maklakov, IV Potapova, BS Romanov, DD Rumyanthev. Acta Polym 36: 557, 1985.
59. SP Fairgrive, JR McCallum. Polym Degrad Stab 11: 281, 1985.
60. AA Dalinkevich, IM Pisarev. Russ Polym News 2(3): 16–21, 1997.
61. J Donohue. Worldwide Metallocene Conference, MetCon'96, June 12–13, Houston and SPO'95, pp 61–73.
62. VYa Shlyapintokh, Photochemical Transformation and Stabilization of Polymers. Moscow: Izdatel'stvo Khimia, 1979, p 344.
a. NM Livanova, GE Zaikov. Int J Polym Mater 36(2): 23–31, 1997.
63. a. B Ranby, JE Rabek, Photodegradation, Photo-oxidation and Photostabilization. London: Interscience, 1975; J Appl Polym Sci 23: 2481, 1979.
b. JE Rabek, J Lucki, B Ranby. Eur Polym J 15: 1089, 1101, 1979.
64. a. JE McKellar. Photosensitized degradation of polymers by dyes and pigments. In: NS Allen, JE McKellar, eds. Photochemistry of Dyed and Pigmented Polymers. London: Applied Science, 1980.
b. NS Allen, J Homer, JE McKellar. J Appl Polym Sci 21: 2261, 3147, 1977.

65. L Reich, SS Stivala. Autooxidation of Hydrocarbons and Polyolefins. New York: Marcel Dekker, 1969.
66. X Hu. Polym Degrad Stab 55(2): 131–171, 1997.
67. MU Amin, G Scott, LMK Tillikeratne. Eur Polym J 11: 85, 1975.
68. E Rasti, G Scott. Eur Polym J 16: 1153, 1980.
69. G Scott, ACS Symp Ser 25: 340, 1976.
70. a. SAl Malaika, G Scott. In: NS Allen, ed. Degradation and Stabilization of Polyolefins. London: Applied Science, 1983.
b. G Scott. Atmospheric Oxidation and Antioxidants. Amsterdam: Elsevier, 1965, p 188.
71. DJ Carlsson, DM Wiles. J Macromol Sci Rev Macromol Chem C-14: 65, 155, 1976.
72. DJ Carlsson, DM Wiles. J Polym Sci Polym Lett Ed 11: 759, 1973.
73. DJ Carlsson, DM Wiles. Macromolecules 2: 587, 597, 1969; 4: 174, 1971.
74. H Kuroda, Z Osawa. Makromol Chem Makromol Symp 27: 97, 1989; J Polym Sci Polym Lett Ed. 20: 577, 1982.
75. Z Osawa. In: HHG Jellinek, ed. Degradation and Stabilization of Polymers, Vol 1. New York: Elsevier, 1983, chap 3.
76. Z Osawa, H Kuroda, Y Kobayashi. J Appl Polym Sci 29: 2834, 1984.
77. Z Osawa, T Takada, Y Kobayashi. Macromolecules 17: 119, 1984.
78. G Gooden, MY Hellman, EH Winslow. J Polym Sci Polym Chem Ed A-24: 3191, 1986.
79. EH Winslow. Pure Appl Chem 49: 495, 1977.
80. J Lemaire, R Arnaud. Polym Photochem 5: 243, 1984.
81. R Arnaud, J Lemaire. In: N Grassie, ed. Development in Polymer Degradation, Vol. 2. 1979, p 159.
82. JM Ginhac, LJ Gardette, R Arnaud, J Lemaire. Makromol Chem 182: 1017, 1981.
83. R Arnaud, JY Moison, J Lemaire. Macromolecules 17: 332, 1984.
84. a. E Gugumus. Makromol Chem Macromol Symp 25: 1, 1989.
b. E Gugumus. Makromol Chem Macromol Symp 27: 25, 1989.
c. E Gugumus. Angew Makromol Chem 176/177: 27, 1990; 182: 85, 111, 1990.
85. JE Guillet, RGW Norrish. Proc R Soc London A-233: 153, 1955.
86. AM Trozzolo, EH Winslow. Macromolecules 1: 98, 1968.
87. R Gooden, MY Hellman, DA Simoff, HE Winslow. In: NS Allen, JE Rabek, eds. New Trends in the Photochemistry of Polymers. London: Elsevier, 1985.
88. E J Golemba, JE Guillet. Macromolecules 5: 63, 212, 1972.
89. SKL Li, JE Guillet. Photochemistry 4: 21, 1984.
90. SKL Li, JE Guillet. J Polym Sci Polym Chem Ed 18: 2221, 1980.
91. GH Hartley, JE Guillet. Macromolecules 1: 165, 413, 1968.
92. NM Emanuel. Papers of IUPAC MACRO '83 Bucharest, Sept. 5–9, Plenary and Invited Lectures, Part 2, 1983, p 369; NM Emanuel, D Gal. Modeling of Oxidation Process. Budapest: Akademiai Kiado, 1986.
93. E Zitomer, AH Diedwardo. J Macromol Sci Chem A-8: 119, 1974.
94. EI Kirilova, ES Shul'ghina. Aging and Stabilization of Polymers. Leningrad: Izd Himia, 1988, pp 11, 123, 174, 197, 201, 215.
95. VP Pleshanov, GS Kiryushkin, SM Berlyant, YuA Shlyapnikov. Visokomol Soedin A-29: 2019, 1987.
96. a. L Lamaire, R Arnaud, J Lacoste. Acta Polym 39: 27, 1988.
b. L Peeva, B Fileva, S Evtimova, V Taveteva. Angewandte Makromol Chem 176/177: 79, 1990.
97. G Geuskens. MS Kabamba, Polym Degrad Stab 4: 69, 1982; 5: 399, 1983; 19: 315, 1987.
98. G Geuskens. Makromol Chem, Macromol Symp 27: 85, 1989. Proc Microsymp. Degradation, Stabilization and Combustion of Polymers, Stara Lesna, June 18–22, 1990, p 148.
99. V Dobrescu, C Andrei. Developments in Chemistry and Technology of Polyolefins. Bucharest: Editura Stiintifica si Enciclopedica, 1987, chap 5, p 129.
100. LI. Lugova, VM. Demidova, EO Pozdniakova, EN Matveeva, AE Lukovnikov. Plasticeskie Massy 8: 61, 1974; 9: 49, 1974.
101. LS Shibryaeva, SG Kiryushkin, GE Zaikov. Visokomol Soedn A-31: 1098, 1989.
102. AA Popov, AV Russak, YeS Popova, NN Komova, GYe Zaikov. Visokomol Soedn A-30: 159, 1988.
103. PR Ogilby, M Kristiansen, DO Martire, RD Scarlock, VI Taylor, RL Clough. In: RL Clough, NC Billingham, KT Gilen, eds. Polymer Durability, Adv Chem Series 249, 1993 pp 113–124.
104. DG Lin. Plasticeskie Massy 12: 52, 1988.
105. Z Osawa. Polym Degrad Stab 20: 203, 1988.
106. M Mlinac, J Rolich, M Bravar. J Polym Sci Symp 57: 161, 1976
107. A Kaczmarek, JF Rabek. Angew Makromol Chem 247, 111–130, 1997; JF Rabek. Polymer Photodegradation. Mechanisms and Experimental Methods. London: Chapman & Hall, 1995.
108. Z Osawa, T Saito. Stabilization and Degradation of Polymers. Adv Chem Ser. 169, 1978, p 159.
109. Z Osawa, K Kobayashi, E Kayano. Polym Degrad Stab 11: 63, 1985.
110. NM Livanova, GE Zaikov. Int J Polym Mater 36: 23–31, 1997; Polym Degrad Stab 57: 1-5, 1997.
111. F Yoshii, G Meligi, T Sasaki, K Makunchi, AM Rabie, S Nashimoto. Polym Degrad Stab, 49: 315, 1995.
112. T Sterzynski, M Thomas. J Macromol Sci Phys B-34: 119, 1995.

113. I Manaf, F Yoshii, K Makouchi. Angew Makromol Chem 227: 111, 1995.
114. MS Rabello, JR White. J Appl Polym Sci 64: 2505–2517, 1997.
a. S Nishimoto, T Kagiwa. In: SH Hamid, MB Amin, AG Maadhah, eds. Handbook of Polymer Degradation. New York: Marcel Dekker, 1993.
115. a. M. Iring, S Laszlo-Hedvig, T. Kelen, E Tiidiis. J Polym Sci Symp 57: 55, 65, 89, 1976.
b. M Iring, S Laszlo-Hedvig, E Tüdös, T. Kelen. Polym Degrad Stab 5: 467, 1983.
116. DJ Carlsson, R Brousseau, C Zhang, DM Wiles. In: JL Benham, JE Kinstle, eds. ACS Symp. Series No. 364, 1988, chap 27, p 376; Polym Degrad Stab 17: 303, 1987.
117. HP Frank. Polym Eng Sci 20: 678, 1980; R Pazur, M Troquet, JL Gardette. J Appl Polym Sci Polym Chem Ed 35(9): 1689–1701, 1997.
118. L Matisova-Rychla, J Rychly. In: RL Clough, NC Billingham, KT Gilen, eds. Adv Chem Series 249, 1993 Polymer Durability pp 175–193.
119. WL Hawkins, W Matreyek, EH Winslow. J Polym Sci 41: 1, 1959.
120. EH. Winslow, JC Aloisio, WL Hawkins, W Matreyek, S Matsuoka. Chem Ind (London) 533; 1465, 1963.
121. E Tüdös, M Iring, Acta Polym 39: 19, 1988; Prog Polym Sci 15: 217, 1990.
122. A Tidjani, R Arnaud, A Dasilova. J Appl Polym Sci 47(2): 211–216, 1993.
123. NL Maecker, DB Priddy. J Appl Polym Sci 41: 21–32, 1991.
124. a. M Mucha. Acta Polymerica 40: 1, 1989; Colloid Polym Sci 264: 1, 1986.
b. M Mucha, M Kryszewski. MACRO '83, Bucharest, Sept. 3–5, 1983, V.-14, p 290.
125. NC Billingham, P Prentice, TPJ Walker. J Polym Sci Symp 57: 287, 1976.
126. TV Monakhova, TPA Bogaevskaya, BA Gromov, YuA Shlyapnikov. Visokomol Soedn B-16: 91, 1974.
127. SG. Kiryushkin, YuA Shlyapnikov. Polym Degrad Stab 23: 85, 1989.
128. VL Maksimov, T G Agnitzeva. Visokomol Soedn B-29: 920, 1987.
129. A Torikai, H Shirakawa, S Nagaya, K Fueki. J Appl Polym Sci 40: 1637, 1990.
130. PJ Luongo. J Polym Sci B-1: 141, 1953.
131. AJ Buchachenko. J Polym Sci 57: 299, 1976.
132. VA Belyi, LS. Koretskaya. Mekh Kompoz Mater (Zinatne) 2: 262, 1987.
133. M Mucha, M Kryszewski. Colloid Polym Sci 258: 743, 1980.
134. AA Popov, NYa Rapoport, GE Zaikov. Oxidation of Oriented and Stressed Polymers. Moscow: Khimya, 1987.
135. HD Hoekstra, JL Spoormaker, J Breen. Angew Makromol Chem 247: 91–110, 1997.
136. L Tong, JR White. Polym Eng Sci 37(2): 321–328, 1997.
137. MR Rabello, JR White. Polymer 38(26): 6379–6387, 1997; 38(26): 6389–6399, 1997; Plastics, Rubbers Comp. Process Appl 25(5): 277, 1996.
138. R Gallo, F Severini, S Ipsale, N Del Finti. Polym Degrad Stab 55(2): 199–207, 1997.
139. P Pagan. Polym Paint Colour J 177(4189): 648, 1987.
140. A Garton. In: NS Allen, ed. Photo-Oxidation Mechanisms of Commercial Polyolefins in Developments in Polymer Photochemistry, Vol. 1. London: Applied Science, 1980, chap 4.
141. E Folders, M Iring, E Tudos. Polym Bull 18: 525, 1987.
142. R Pazur, M Troquet, JL Gardette. J Polym Sci A Polym Chem, 35: 1689–1692, 1997.
143. a. JL Bolland, G Gee. Trans Faraday Soc 42: 236, 1942.
b. JL Bolland. Proc R Soc London 186: 230, 1946.
144. J Lemaire, P Dabin, R Arnaud. In: M Vert, J Feijen, AC Albertsson, G Scott, E Chieline, eds. Mechanism of Abiotic Degradation of Synthetic Polymers. Cambridge: Royal Society of Chemistry, 1992, pp 30–41.
145. S Verdu, J Verdu. Macromolecules, 30(8): 2262–2267, 1997.
146. N Grassie. In: G Scott, ed. Development in Polymer Stabilization, Vol 1. London: Applied Science, 1979, pp 221–229.
a. MB Mattson, B Stenberg. In: RL Clough, NC Billingham, KT Gilen, eds. Polymer Durability. Adv Chem Ser 249, 1993, pp 235–249.
147. S Giros, L Audouin, J Verdu, P Delprat, G Marat. Polym Degrad Stab 51(2): 125–131, 1996.
148. E Tüdös, Z Fodor, M Iring, Angew Makromol Chem 158/159: 15, 1988.
149. a. YuA Shlyapnikov, SP Kiryushkin, AN Marin. Stabilization of Polymers Against Oxidation. Moskva: Izv. Himia, 1986, p 236.
b.YuA Shlyapnikov. Makromol Chem Macromol Symp 27: 121, 1989.
150. NG Billingham. Makromol Chem Macromol Symp 28: 145, 1989.
151. TM Kollmann, DGM Wood. Polym Eng Sci 20: 684, 1980.
152. M Cellina, GA George, NC Billingham. In: RL Clough, NC Billingham, KT Gilen, eds. Polymer Durability, Adv Chem Series 249, 1993, pp 159–174.
153. a. YeT Denisov, AI Vol'pert, VP Filipenko. Visokomol Soedn A-28: 2083, 1986.
b. AV Kargin, YuB Shilov, YeT Denisov, AA Yefimov. Visokomol Soedn A-28: 2236, 1986.
154. F Gugumus. Polym Degrad Stab 52(2) 131–159, 159–168; 53(2): 161–169, 1996.
155. YuA Mikheyev, LN Guseva, GE Zaikov. Int J Polym Mater, 35: 193–272, 1997; Polym Sci Ser B 39(5-6): 241–256, 1997; Visokomol Soedn Ser B Vol 39(6): 1082–1098, 1997.

156. B Catoire, V Verney, A Michel. Makromol Chem Macromol Symp 25: 199, 1989.
157. L Goldfarb, CR Foltz, DC Messersmith. J Polym Sci Polym Chem Ed 10: 3289, 1972.
158. JCW Chien, JKY Kiang, Makromol Chem 181: 47, 1980.
159. T Zaharescu, I Mihalcea. Materiale Plastice (Bucharest) 31(2): 139–142, 1994.
160. E Chamot. Symp on Industrial Applications of Quantum Chemistry, ACS 27th Regional Meeting Akron, May 31–June 2, 1995.
161. RP Chartoff. In: EA Turi, ed. Thermal Characterization of Polymeric Materials, 2nd ed. London: Academic Press, 1997, pp 68-80.
162. ET Eggertsen, EH Stress. J Appl Polym Sci 10: 1171, 1966.
163. S Ciutacu, D Fatu, E Segal. Thermochim Acta 131: 279, 1988.
164. JK Gilham, AE Lewis. J Polym Sci Symp C-6: 125, 1964.
165. a. AA Miroshnichenko, MS Platitsa, VI Volkov, VA Dovbnya. Visokomol Soedn A-30: 2516, 1988.
b. AA Mirosnichenko, MS Platitsa, TP Nikolaeva. Visokomol Soedn A-30: 2523, 1988.
166. C Vasile, EM Călugaru, A Stoleru, M Sabliovschi, E Mihai. Thermal Behaviour of Polymers, Bucharest: Ed. Academiei, 1980, pp 167–190.
167. M Bornengo. Mater. Plast Elastomers 4: 383, 1966.
168. RH Hansen. In: RT Conley, ed. Thermal Stability of Polymers. New York: Marcel Dekker, 1970, p 153.
169. L Reich, SS Stivala Elements of Polymer Degradation. New York: McGraw-Hill, 1971.
170. L Reich, SS Stivala. Polym Eng Sci 20: 645, 1980.
a. MJ Pohlen. In: WJ Bartz, ed. 11th Int Colloq on Industrial and Automotive Lubrication, Jan 13–154, 1998, Vol III p 1979.
171. SM Gabbay, SS Stivala. Polymer 17: 121, 137, 1976.
172. SS Stivala, SM Gabbay. Polymer 18: 807, 1977.
173. a. LA Wall, JH Flynn. Rubber Chem Technol 35: 1157, 1962.
b. LA Wall, JH Flynn. J. Res. NBS 70A: 487, 1966.
174. C Vasile, L Odochian, I Agherghinei. J Polym Sci Polym Chem Ed A-26: 1639, 1988.
175. JR McCallum, K Peterson. Eur Polym J 10: 471, 477, 1974.
176. NA Slovokhotova, MA Margupov, VA Kargin. Visokomol Soedn A-6: 1974, 1964.
177. H Mori, T Hatanaka, M Terano, Macromol Chem Rapid Commun 18(2): 157–161, 1997.
178. Z Osawa, M Kato, M Terano. Macromol Chem Rapid Commun, 18(8): 667–671, 1997.
179. HA Schneider. Thermochim Acta 83: 59, 1985; Survey and critique of thermo-analytical methods and results. In: HHG Jellinek, ed. Degradation and Stabilization of Polymers. Amsterdam: Elsevier, 1983, chap 10, p 506.
180. KS Minsker, AA Berlin, RB Panceshnikova, ED Antonova. Visokomol Soedn B-29: 171, 1987.
181. VN Urazbaev, RB Panceshnikova, KS Minsker. Visokomol Soedn B-29: 445, 1987.
182. L Weishen, S Lianghe, S Devon, L Buoliang, Polym Degrad Stab 12: 375, 1988.
183. A Holmstrom, EM Sorvik. J Appl Polym Sci 18: 761, 779, 3153–3178, 1974; J Polym Sci Symp 57: 33, 1976.
184. WG Oakes, RB Richards. J Chem Soc 2929, 29311, 1949.
185. RT Johnston, EJ Morriston. In: RL Clough, NC Billingham, KT Gilen, eds. Polymer Durability, Adv Chem Series 249. 1993, pp 651–682.
186. JC Randall. In: O Guwen, ed. Cross-linking and Scission in Polymers. Dordrecht: Kluwer, 1990, pp 60–62.
187. VN Gol'dberg, GE Zaikov. Int J Polym Mater 31(1-4): 1–39, 1996.
188. McCoy, J Benjamin, G Madras. AIChEJ 43(3): 802–810, 1997.
189. a. D Suwanda, R Lew, ST Balke. J Appl Polym Sci 35: 1019, 1033, 1988.
b. T Bremner, A Rudin. Plast Rubber Proc Appl 13: 61, 1990.
c. D Suwanda and ST Balke, Am Chem Soc ANTEC '89; 589–592, 1989.
d. McCullough, D James, JF Bradford. US Pat 5, 587,434.
190. ST Balke, D Suwanda, R Lew. J Polym Sci, Polym Lett. Ed. 25: 313, 1987.
191. M Dorn. GAK 35: 608, 1982; Adv Polym Technol 5: 87, 1985.
192. P Huedec, L. Obdrzalek. Angew Makromol Chem 89: 41, 1980.
193. a. Badische Anilin and Soda Fabrik Aktiengesellschaft. German Pat. 2,004,491, 1969.
b. Chemische Werke Huels AG, French Pat. 1,377,951, 1963.
c. Sumimoto. German Pat. 2,311,822, 1969.
d. K Rauer, P Demel. German Pat. DE 3,642,273, 1987.
e. D Spanikova, E Spirk, S Kamenar, JB Reca, P Rosner, Z Smolarova, T Vojtech. Czech. Pat. 243,904, 1985.
194. E Chiang. J Polym Sci 28: 235, 1958.
195. R Rado. Chem Prumysl 12: 209, 1962.
196. VP Nehoroshev, EG Balahonov, DI Davidov, EI Levin, LP Gossen. Plasticeskie Massy 2: 82, 1989.
197. MP Keropian, VI Gorohov, Yu G Riseev, LYa Galishnikova. Plasticeskie Massy 9: 25, 1987.
198. AC Kolbert, JG Didier, L Xu. J Polym Sci Part B Polym Phys 35: 1955–1961, 1997; Macromolecules 29(27): 8591–8598, 1996; Polym Preprints 38(1): 819, 1997.
199. YuA Miheev. Visokomol Soedn B-28: 908, 1986.
200. H Nanbu, Y Ishihara, H Honma, T Takesue, T Ikemura. J Chem Soc Jpn 765, 1987.
201. J Konar, SK Ghosh. Polym Degrad Stab 22: 43, 1988.

202. a. C Tsoganakis, Y Tang, J Vlachopoulos, AE Hamielec. Polym Proc Eng 6: 2960, 1988.
b. J Appl Polym Sci 37: 681, 1989.
c. Polym Plast Technol Eng 28: 319, 1989.
d. C Huang, TC Duever, C Tzoganakis. React Engineering 3(1): 43–63, 1995.
e. 5(1&2): 1–24, 1997.
f. JBP Soares, T Shouli AE Hamielec. Polym Rection Eng 5 (182): 25–44, 1995.
g. VJ Triaca, PE Gloor, S Zhu, AN Hrymak, AE Hamielec. Polym Eng Sci 33 (8): 445–454, 1993.
203. N Akmal, AM Usmani. In: C Vasile and RB Seymour, eds. Handbook of Polyolefins, 1st ed. New York: Marcel Dekker, 1993, pp 553–560.
204. T Yamazakim, T Seguchi. J Polym Sci Part A Polym Chem 35(2): 279–284, 1997.
205. a. M Paabo, BC Levin. Fire Mater 11: 55, 1987.
b. BC Levin. Fire Mater 11: 143, 1987.
c. JA Conesa, R Font, A Marcilla. Energy Fuels 11(1): 126, 1997.
206. M Chaigneau, G Le Moan. Ann Pharm Francaises 28: 259, 417, 1970; 29: 259, 1971.
207. a. Y Tsuchiya, K Sumi. J Polym Sci A-I-6: 415, 1968:
b. Y Tsuchiya, K Sumi. J Polym Sci A-1-7: 813, 1969.
c. Y Tsuchiya, K Sumi. J Polym Sci A-1-7: 1599, 1969.
d. Y Tsuchiya, K Sumi. J Polym Sci Polym Lett Ed. B-6: 357, 1968.
208. RP Lattimer. J Anal Appl Pyrolysis 31: 203–221; 1995; Rubber Chem Technol 68: 785–793, 1996.
209. MT Sousa Pessoa de Amorim, G Comel, P Vermade. J Anal Appl Pyrolysis 4: 73, 1982.
210. a. L Michailov, P Zugenmaier, HJ Cantow. Polymer 12: 70, 1971.
b. L Michailov, P Zugenmaier, HJ Cantow. Polymer 9: 325, 1968.
211. B Kolb, G Kemmner, KH Raiser, EW Cieplinski, LS Ettre. Z Anal Chem 209: 302, 1965.
212. E Kiran, JR Gilham. J Appl Polym Sci. 20: 2045, 1976.
213. a. Y Sugimura, S Tsuge. Macromolecules 12: 512, 1979.
b. E Willmott. J Gas Chromatogr 7: 101, 1969.
214. a. DH Ahlstrom, SA Liebman, KB Abbas. J Polym Sci Polym Chem Ed 14: 2479, 1976.
b. SA Liebman, DH Ahlstrom, WR Starnes, EC Schilling. J Macromol Sci A-17: 935, 1982.
215. J Mitera, J Michal. Fire Mater 9: 111, 1985.
216. GS Kiryushkin, AP Mar'in, Yu A Shlyapnikov. Visokomol Soedn A-22: 1428, 1980.
217. WJ Bailey, EP Ragelis. Polym Preprints 3-2: 403, 1962.
218. R Pazur, M Troquet, JL Gardette. J Polym Sci A Polym Chem Ed 35: 1689–1701, 1997.
219. J Voigt. Kunststoffe 54: 2, 1964.
220. KS Minsker, AG Liakumovich, YuA Sangalov, OD Svirskaia, VN Korobeinikova, AN Gazizov. Visokomol Soedn A-16: 2751, 1974.
221. SM Gabbay, SS Stivala, L Reich, J Appl Polym Sci 19: 2391, 1975.
222a. N Grassie, A Scootney. In: EM Immergut, J Brandrup, eds. Polymer Handbook, 2nd ed., 1975, p II-473.
222b. T Morikawa. Fire Mater 12: 43, 1988.
223. a. J van Schooten, JK Evanhuis. Polymer(London) 6: 343, 561, 1965.
b. M Blazsko. J Anal Appl Pyrolysis 25: 25–35, 1993; 35, 221–235, 1995; 39: 1–25, 1997.
224. a. C Vasile. Chemical processes for recovery of secondary polymeric materials. In: M Rusu, O Sebe, C Vasile, D Staicu, eds. Recovery of Secondary Polymeric Materials. Bucharest: Ed. Tehnica, 1989, chap 4, p 295.
b. CN Cascaval. Study of Polymers by Pyrolysis-Gas Chromatography. Bucharest: Ed. Academiei, 1983, p 66.
225. a. M Seeger, HJ Cantow. Makromol Chem 176: 2059, 1975.
b. W Klusmeier, A Kettrup, KH Ohrbach. J Therm Anal 35: 497, 1989.
226. JL Wuepper. Anal Chem 51: 997, 1979.
227. TR Crompton. Analysis of Polymers: An Introduction. Oxford: Pergamon Press, 1989.
228. O Ciantore, M Larrari, F Ciardelli, SD Vito. Macromolecules, 30(9): 2589–2597, 1997.
229. MR Thompson, C Tzoganakis, GL Rampel. J Polym Sci Part A Polym Chem Ed 35 (14): 3083–3086, 1997.
230. T Wampler, E. Levy. Analyst 111: 1065, 1986; J Anal Appl Pyrolysis 8: 153, 1985.
231. a. CDS Chemical Data Systems. One page Application Note 4 and 18.
b. CG. Smith. J Anal Appl Pyrolysis 15: 209, 1988.
232. DA Kourtides, WJ Gilwee, CJ Hilado. Polym Eng Sci 18: 675, 1978.
233. KS Minsker, AM Steklova, GE Zaikov. Polym Plast Technol Eng 36(2): 325–332, 1997.
234. CG Smith, RA Nyquist, SJ Martin, NH Mahle, PB Smith, AJ Paszter. Anal Chem Appl Rev 61: 214R, 1989.
235. WJ Irwin. J Anal Appl Pyrolysis 1: 89, 1979.
236. a. IC NcNeill. J Polym Sci A-4: 2479, 1966.
b. IC McNeill. Eur Polym J 6: 373, 1970.
c. IC McNeill. Polym Eng Sci 20: 668, 1980.
d. R McGuchan, IC McNeill. Eur Polym J 3: 511, 1967; 4: 115, 1968.
237. PJ Tayler, D Price, GJ Milnes, JH Scrievens, TG Blease. Int J Mass Spectrom Ion Proc. 89: 157, 1989.
238. T Hammond, RS Lehrle. Br Polym J 21: 23, 1989.
239. TP Wampler. J Chem Ed 63: 64, 1986.
240. TP Wampler. J Anal Appl Pyrolysis 15: 187, 1989.
241. SH Hamid, WH Prichard. Polym Plast Technol Eng 27: 303, 1988.
242. J Paulik, H Macskasy, E Paulik, L Erdey. Plaste Kautsch. 8: 588, 1981.
243. TP Wampler, EJ Levy, J Anal Appl Pyrolysis 8: 65, 1985.
244. a. SA Liebman, TP Wampler. Pyrolysis and GC in Polymer Analysis. New York: Marcel Dekker, 1985.

b. SA Liebman, TP Wampler, EJ Levy. J High Resol Chromat. Commun 7: 172, 1984.

245. a. Y Sugimura, T Nagaya, S Tsuge, T Murata, T Takeda, Macromolecules 13: 928, 1980.
b. T Nagaya, Y Sugimura, S Tsuge. Macromolecules 14: 1797, 1981.
c. T Tsuge, Y Sugimura, T Nagaya. J Anal Appl Pyrol 1: 221, 1980.
246. JK Haken, PI Iddamalgeda. J Chromatogr A756 (1-2): 1–2, 1996.
247. M Seeger, EM Barrall. J Polym Sci Polym Chem Ed. A-1-13: 1515, 1975.
248. a. G Audisio, G Bajo. Makromol Chem 176: 199, 1975.
b. D Deur Siftar, V Svob. J Chromatogr 51: 57, 1970; J Gas Chromatogr 5: 72, 1967.
249. S Tsuge, T Okumoto, T Takeuchio. Macromolecules 2: 200, 1969.
250. H Seno, S Tsuge, T. Takeuchi. Makromol Chem 161: 185, 195, 1972.
251. Li Wenguang, J. Xiabin, X. Zhengyuan, F Zhilin. Chin J Appl Chem 3: 20, 1986.
252. J Brossas. Polym Degrad Stab 23: 313, 1989.
253. CE Cullis. J Anal Appl Pyrolysis 11: 451, 1987.
254. DW van Krevelen. Flammability and flame retardancy of organic high polymers. In: Advances in Chemistry: Thermally Stable Polymers. Warsawa, 1977, p 119.
255. a. GWV Stark. Trans Mater Eng 84: 25, 1972.
b. SJ Burge, CFH Tipper. Combust Flame 13: 495, 1969.
256. a. PC Warren. Soc Plast Eng 27: 17, 1971.
b. R Friedman, JB Levy. Combust Flame 7: 195, 1963.
257. HE Thompson, DD Drysdale. Fire Mater 11: 163, 1987.
258. G Camino, L Costa, E Casorati, G Bertelli, R Locatelli. J Appl Polym Sci 35: 1863, 1988.
259. a. G Camino, R Amaud, L Costa, J Lamaire. Angew Makromol Chem 160: 203, 1988.
b. G Bertelli, G Camino, E Marchetti, L Costa, R Locatelli. Angew Makromol Chem 169: 137, 1989.
c. G Camino, L Costa, MP Luda. Papers of Microsymposium Degradation, Stabilization and Combustion of Polymers, June 18–22, Stara Lesna, 1990, p 14.
260. DE Stuetz, AH Di Edwardo, F Zitomer, BP Barnes. J Polym Sci Polym Chem Ed l3: 585, 1975.
261. E Dittmar, R Capron, JM Jouany, M. Guerbet. Polym Degrad Stab 23: 377, 1989.
262. A Tkac. J Polym Sci Symp 57: 109, 1976.
263. JR Richard, C Vovelle, R Delbourgo. 15th Symp. on Combustion, Tokyo, August 25–31, 1974, p 205.
264. RG Gann, R Diert, MJ Drews. Flammability. In: Encyclopedia of Polymer Science and Engineering, 2nd ed., vol. 7, New York: Interscience, 1986, p 154.
265. RE Lyon. Int SAMPE Symp Exhib, 41, 1996, Book 1 pp 687–708, Materials and Process Challenges, Aging Systems, Affordability, Alternative Applications.
266. M Chatain, L Chesne. Rev Gen Caoutch Plast 50: 695, 1973.
267. G Rietz. Plaste Kautschuk 36: 181, 1989.
268. ShA Nasibullin, YuN Hakimullin, IN Zaripov. Zh Prikl Himii 60: 241, 1987.
269. a. AS Maltzeva, GM Ronkin, YuE Frolov, EP Yakushina, AI Rozlovskii, Dokl Akad Nauk SSSR 292: 897, 1987.
b. RE Lyon. 6th Annual Conf on Flame Retardancy Connecticut May 23–25, 1995, pp 1–9.
270. a. BC Levin, MS Maya Paabo, JL Gurman, SE Harris, E Braun. Toxicology 47: 135, 1987.
b. V Babrauskas, BC Levin, RG Gann. ASTM Stand. News 14: 28, 1986.
271. WJ Bailey, LJ Baccei. Polym Preprints 12–2: 313, 1971.
272. MJ Roedel. J Am Chem Soc 75: 6110, 1953.
273. VD Moiseev. Plasticeskie Massy 12: 3, 1963.
274. VD Moiseev, MB Neiman, VI Suskina. Chemical Properties and Modification of Polymers. Moskva: Nauka, 1964, p 86; Visokomol Soedn 9: 1383, 1961.
275. C Vasile, P Onu, V Bărboiu, M Sabliovschi, G Moroi. Acta Polym 36: 543, 1985.
276. SL Madorsky. Thermal Degradation of Organic Polymers. New York: Interscience, 1964, p 93.
277. JRY Kiran, PC Uden, JCW Chien. Polym Degrad Stab 2: 113, 1980.
278. SL Madorsky, S Straus. J Res NBS 53: 361, 1954.
279. D McIntyre, JH O'Mara, S Straus. J Res NBS 68A: 153, 1964.
280. T Sawaguchi, M Seno. Polymer, 37(25): 5607–5617, 1996; Macromol Chem Phys 196: 4139, 1995; l97: 215–221, 1996; Polymer J 28: 392–399, 1996; Polymer 37: 3697–3672, 1996.
a. T Sawaguchi, M Seno, Macromol Chem Phys 197 (12): 3995–4015, 1996.
281. T Shimono, M Tanaka, T Shone. J Anal Appl Pyrolysis 1: 189, 1979.
282. L Reginato. Makromol Chem 132: 125, 1970.
283. Y Shibasaki. J Polym Sci A-1-5: 21, 1967.
284. LR Sochet. La Cinetique des Reactions au Chaines. Paris: Dunod, 1971, p 2.
285. VN Kondratiev. Chain reactions. In: CH Bamford, CEH Tipper, eds. Comprehensive Chemical Kinetics. Amsterdam: Elsevier, 1969, pp 90, 154.
286. IA Schneider, Chemical Kinetics. Bucharest: Ed. Didactica si Pedagogica, 1974, p 177.
287. GE Zaikov, Russ Chem Rev 62(6): 603–620, 1993; Int J Polym Mater 24 (1-4): 1–27, 1994; ACS Publ 208th ACS National Meeting, Washington DC August 21–25, 1994, p 125; EF Vainstein, GE Zaikov. Oxid Commun, 19(3): 32361, 1996.
288. LA Wall. NBS Special Publ. No. 357, 1970, p 47.
289. R Simha, LA Wall, PJ Blatz. J Polym Sci 5: 615, 1950.
290. B Dickens. J Polym Sci Polym Chem Ed 20: 1065, 1170, 1982.

291. a. M Gordon, LR Shenton. J Polym Sci 38: 157, 179, 1959.
b. M Gordon. J Phys Chem 64: l, 1960; Trans Faraday Soc 53: 1662, 1957; Soc Chem Ind (London) Monograph, 13: 163, 1961.
292. R Simha. Trans Faraday Soc 54: 1345, 1958; J Appl Phys 12: 569, 1941.
293. a. R Simha. J Chem Phys 24: 796, 1956.
b. R Simha, LA Wall, J Bram. J Chem Phys 29: 894, 1958.
294. R Simha, LA Wall. J Polym Sci 6: 39, 1951; J Phys Chem 56: 707, 1952.
295. a. LA Wall, S Straus, JH Flynn, D McIntrye, R Simha. J Phys Chem 70: 53, 1966.
b. LA Wall, SL Madorsky, DW Brown, S Strauss, R Simha. J Am Chem Soc 76: 3430, 1954.
c. LA Wall. Effect of Branching on the Thermal Degradation of Polymers, SCI Monograph No. 13, in Thermal Degradation of Polymers, ACS 1961, p 146.
296. RH Boyd. J Chem Phys 31: 321, 1959; J Polym Sci 49: Sl, 1961; J Polym Sci A-1-5: 1573, 1967.
297. RH. Boyd, TP Lin. J Chem Phys 45: 773, 778, 1966.
298. RH Boyd, R Simha. Macromolecules 20: 1439, 1987.
299. A Inaba, T Kashiwagi. Macromolecules 19: 2412, 1986; 20: 1440, 1987; Eur Polym J 23: 871, 1987.
300. T Kashiwagi, A Inaba, A Hamins. Polym Degrad Stab 26: 161, 1989.
301. JH Chan, ST Balke. Polym Degrad Stab 57(2): 113, 127, 135, 1997.
302. R Pathria, VS Nanda. J Chem Phys 30: 1322, 1959.
303. E Williams. Chemical kinetics of pyrolysis. In: Advances in Chemical Engineering, Vol. l. 1974, p 197.
304. RH Boyd. The relationship between the kinetics and mechanism of thermal depolymerization. In: RT Conley, ed. Thermal Stability of Polymers. New York: Marcel Dekker, 1970, chap 3, p 47.
305. T Hammond, RS Lehrle. Pyrolysis GCL. In: Comprehensive Polymer Science Series, Vol. 1, C Booth, C Price, eds. Polymer Characterization. Oxford: Pergamon Press, 1988, chap 27, p 589.
306. RS Lehrle. J Anal Appl Pyrolysis 11: 55, 1987.
307. a. RS Lehrle, JC Robb, JR Suggate. Eur Polym J 18: 443, 1982.
308. AK Burnham, RL Braun. Kinetics of Polymer Decomposition, Informal Report UCID-212 93, Dec. 1987.
b. AK Burnham, RL Braun, RW Taylor, T Coburn. ACS Symp 34(1): 36, 1989.
c. AK Burnham, MS Oh, RW Cawford, AM Samoun. J Energy Fuels 3: 42, 1989.
309. M Mucha. J Polym Sci Symp 57: 25, 1976.
310. C Vasile, E Costea, L. Odochian. Thermochim Acta 184: 305, 1991; Rom Acad Sci Papers Chem Sect IV (XVIII), 125–140 (1995).
311. a. AL Lipskis, AV Kviklys, AM Lipckiene, AN Maciulis. Visokomol Soedn A-18: 426, 1976.
b. LN Raspopov, GP Belov, IN Musaelian, NM Cirkov. Visokomol Soedn B-15: 812, 1973.
312. J Vogel, K Meissner, C Heinze, B Poltersdorf. Plaste Kautsch 36: 13, 1989.
313. a. IM Piroeva, VA Mrdoyan, YuK Kobalyan, SM Ayropetyan. Arm Khim Zh 41: 379, 1988 b. Y Kodera, WS Cha, BJ Mecoy, Prepr Pap ACS Div Fuel Chem 42(4): 1003–1007, 1997.
314. AR Horrocks, JA D Souza, Thermochim Acta 134: 255, 1988.
315. ZA Sadykhov, NA Nechitailo, GP Afanasova. Visokomol Soedn A-15: 637, 1973.
316. AS Deuri, K Bhowmick. J Therm Anal 32: 755, 1987.
317. a. AA Sokolovskii, NN Borisova, LG Angert. Visokomol Soedn A-17: 1107, 1975.
b. MA Serageldin, H Wang. Thermochem Acta 117: 157, 1987.
318. W Michaeli, V Lackner. Angew Makromol Chem 232, 167-185, 1995.
319. G Menges, V Lackner. In: J Brandrup, M Bittner, Michaeli, G Menges, eds. Recycling and Recovery of Plastics. Munich: Hanser, 1996, pp 413–422.
320. J Randall, L Sharp. Fluidized Bed Cracking of Single and Mixed Plastics APC/APME Exchange Meeting, Brussels, July, 1993.
321. C Vasile, M Brebu, V Dorneanu, RD Deanin. Int J Polym Mater 38: 219–247, 1997.
322. P Straka, J Buchtele, J Kovanva. 38th Microsymp "Recycling of Polymers" Prague, July 14–17, 1997, p 22.
323. AG Buekens, JG Schoeters, SPEJ 29: 4l, 1973; Basic Principles of Waste Pyrolysis, Review of European Processes in Thermal Conversion of Solid Wastes and Biomass. ACS Symp. Ser Vol. 30, 1980, p 397.
324. B Baum, C Parker. Solid Waste Disposal, Vol. 2, Reuse, Recycle and Pyrolysis. Michigan: Ann Arbor, 1974, pp 1–79.
325. a. W Bischofberger, R Born. Verfahrens und Umwelttechnische Analyse neuer Thermische Prozesse in der Abfallwirtschaft, Vol. I. Pyrolyse. Munchen: Technische Universitat, 1989, p 89.
b. W Kaminsky, H Sinn. In: J Brandrup, M Bittner, W Michaeli, G Menges, eds. Recycling and Recovery of Plastics. Munich: Hanser, 1996, pp 434–444.
326. M Ishiwatari. J Polym Sci Polym Lett Ed 22: 83, 1984.
327. A Caraculacu, C Vasile, G Caraculacu. Acta Polym 35: 130, 1984.
328. S Tsutsumi. Recycling of Waste Plastics, Sanyo Electric Company Papers, Environmental Division, Conversion of Refuse to Energy, Montreux 3–5, Nov. 1975, Conference Papers IEEE Catalog Number 75 CH 1008-2CRE Zurich: Eiger AG; p 567.
329. W Kaminsky, H Sinn, J Janning. Chem. Ing Tech 46: 579, 1974; Angew Chem 88: 737, 1976.
330. W Kaminsky, H Sinn. Kunststofe 68: 284, 1978; Angew Makromol Chem 232: 151–165, 1995.

331. a. W Kaminsky. J Anal Appl Pyrolysis 8: 439, 1985.
b. Makromol Chem Macromol Symp 57: 145–160, 1992.
c. CM Simon, C Eger, H Kastner, W Kaminsky. Ber Dtsch Wiss Ges Erdoel Erdgas Kohle Tagungsber, 1996, 9603, pp 337–344.
332. W Kaminsky, J Menzel, H Sinn. Int J Recycling Plastics: Conservation and Recycling 1: 91, 1976; J Anal Appl Pyrolysis 38: 75–87, 1996; JS Kim, W Kaminsky, B Schlesselman. J Anal Appl Pyrolysis 40: 365–372, 1997.
333. a. W Kaminsky. Resource Recov Conserv 5: 205, 1980.
b. U Bellman, W Kaminsky. Ulmwelt. 19: 336, 1989.
334. a. W Kaminsky. Entsorg Prax H-9: 392, 1987.
b. H Lechert, V Wolbs, Q Sung, W Kaminsky, H Sinn (ASEA Brown Boveri AG). European Pat. Appl. 321, 807, 1989.
335. H Ito, M Yamada, Y Nozaki, H Kobe. German Pat. 2,210,223, 1972.
336. a. M Endo. Jpn. Plast 7: 29, 1973.
b. Y Saheki. Jpn Plast 6: 22, 1972.
337. E Schaub, S Speth (Ruhrchemie Aktiengeselschaft). German Pat. 2,205,001, 1972; Chem. Ind. Tech. 45: 526–29, 1973.
338. C Vasile, L Odochian, N Hurduc, M Sabliovschi, CN Cascaval, IA Schneider. Rom. Pat. 71,349, 1977.
339. LJ Ricci. Chem Ing News 52(23), 1974; Chem Eng 83(16): 52, 1976.
340. HW Schnecko. Chem Ing Tech 48: 443, 1976.
341. a. C Vasile, M. Rusu. Science Technics (Bucharest) 31: 22, 1980.
b. C Vasile, RD Deanin, M Mihaies, Ch Roy, A Chaala. Int J Polym Mater 37: 173–199, 1996.
c. C Vasile, S Woramoncha, RD Deanin, M Mihaies, Ch Roy, A Chaala. Int J Polym Mater, 38: 263–273, 1997.
342. BJ Flanagan. Conversion of Refuse to Energy, CRE Conf Proc, Montreaux 3–5 Nov. 1975, p 220.
343. Japan Gasoline Co. and Nippon Zeon Co. Papers, Japan, 1974, 1975.
344. GM Molton, EL Compton (Occidental Petroleum Corp.) German Pat. 2,255,484, 1972; British Pat. 1,353,067, 1974.
345. H Nishizaki, M Sakakibara, K Yoshida. Pyrolysis of atactic polypropylene in a fluidized bed. In: Effective Use of Natural Resource and Urban and Industrial Wastes, 1977, p 169.
346. J Bhatia, HK Staffin, Waste Plastic Pyrolysis, Technical Papers presented at Recycling World Congress, New Orleans, April 1982.
347. R Fletcher, H. T Wilson. Resource Recov. Conserv. 5: 333, 1981; Foster Wheeler Power Products Limited, Firm Papers: The Role of Pyrolysis in the Disposal of Waste Tires, 1980.
348. a. M Yoshida, M Watanabe, K Tohma, M Noda. US Pat. 3,984,288, 1976.
b. H Tokushige, A. Kosaki, T Sakai (Japan Steel Works Ltd.). US Pat. 3,959,357, 1976.
349. a. E Lewis. German Pat. 2,737,698, 1978; British Pat. 919,920.
b. Y Tsukagoshi, KA Tsuchiya, T Inoue, TM Ogawa, H Komori, T Ayukawa, MK Murakami (Kabushiki Koisha Niigata Tekkosho, Japan). German Pat. 2,222,267, 1978.
350. EM Calugaru, C Vasile, M Sabliovschi, CN Cascaval, H Darie, C. Zaharia. Rom. Pat. 78,462, 1979.
351. A Chaala, H Darmstadt, C Roy. J Anal Appl Pyrolysis, 39: 79–96, 1997.
352. a. G Collin. Internationaller Recycling Congress, CRE/MER, Verlag fur Umwelttechnik Berlin, 1979, p 700.
b. G Collin, G Grigoleit, GP Bracker. Chem Eng Tech 50: 836, 1978.
353. C Vasile, G Moroi, M Sabliovschi, C Boborodea, A Caraculacu. Acta Polym 37: 419, 1986.
354. SYa Haikin, TG Agnivtzeva, AG. Sirota, MD Pukshanski, Visokomol Soedn B-18: 92, 1976.
355. E Piiroja, H Lippmao. Makromol Chem, Makromol Symp 27: 305, 1989.
356. MB Jerman D Fles Polymery (Zagreb) 9: 55, 1988.
357. EK Piiroya, EA Ebber, EE Yakob, KB Kisler, VM Gal'perin, VS Bugorkova (Tallin Polytechnic Institute), USSR Pat. 1,399,304, 1985.
358. FO Azzam, BS Kochei, S Lee. Proc Ann Int Pittsburg Coal Conf 11th Vol 2, pp 1046–1050, 1994; K Arai, 38th Microsymp "Recycling of Polymers" Prague Academy of Sciences, Czech. R. July 14–17, 1997, ML 6.
359. a. K Oshima. US Pat. 3,956,414, 1976.
b. Y Kitaoka, K Murata, K Hama, M Hashimoto, K Fujiyoshi (Mitsubishi Shipbuilding and Engineering Co. Ltd. and Mitsui Petrochemical Industry Ltd.). US Pat. 3,832,151, 1974.
360. a. Mountray Ltd, Tomkins and Co. British Pat. 1,450,285. 1974.
b. RM Joyce (E. I. du Pont de Nemours and Co.), US Pat. 2,372,001, 1962.
c. F Gude, E Klimpel, W Nagengist. German Pat. 2,419,477, 1974.
361. GA Celemin. Spanish Pat. ES, 539,973, 1986.
362. a. M Ratzsch, G Kotte, KD Ebster (VEE Leuna Werke). German (East) Pat. 128,875, 1977.
b. Thermooxidative breakdown of polyolefins into waxy products. Paper presented at Seminar on Recycling of High Polymeric Wastes, Dresden Sept. 18–23, 1978.
c. M Rätzsch, KD Ebster, C Wild, R Liebich, H Wiegleb, G Kotte. Plaste Kautschuk 28: 306, 1981.
363. a. J Brandrup, Kunststoffe, Verwertung von Abfällen, in Ulman Encyclopedie der Technische Chemie, 4th ed. Vol. 15, Weinheim: Verlag Chemie, 1978, p 411; Kunststofe 65: 881, 1975; Ind Ant 99: 1535, 1977.
b. Y Sakata, MA Uddin, A Muto, Y Kanada, K

Koizumi, K Murata. J Anal Appl Pyrolysis, 43: 15–25, 1997.
364. JD Upadhyaya (Moore and Munger). US Pat. 4,624,993, 1986.
365. A Belestrini. Traitment des matieres plastiques et du caoutchouc par pyrolyse, Paper presented at Seminar on Recycling of High Polymeric Wastes, Dresden, Sept. 17–23, 1978.
366. L Guaglimi. Petrol Int. 28: 50, 1982.
367. WC McCoffrey, MJ Brues, DG Cooper, MR Kamal. J Appl Polym Sci 60(12): 2133–2140, 1996.
368. H Bockhorn M Burckschafte Chem Eng Technol 61: 813, 1989.
369. a. Plast Rubber Week, Feb. 15, 1974, p 5.
b. T Oyamoto, S Ochiai, K Uda, K. Kakigi, K Orio. (Mitsubishi Jukogyo Kabushiki Kaisha). British Pat. 1,423,420, 1973.
370. T Kuroki, T Sawaguchi, T Hoshima. Nippon Kagaku Kaishi 2: 322, 328, 1976.
371. GE Bertolini, J Foutaine. Conserv Recycl 10: 331, 1987.
372. A Tille, K Knebel, HJ Derdulla, M Richter, R Jauch, H Ultrecht. German Offen DE 19512 029, 1996.
373. FS Sodero, FB Berruti, A Leo. Chem Eng Sci 51(11): 2805–2810, 1996; S Lovett, F Berruti, LA Behle. Ind Org Chem Res 36(11): 4436–4444, 1997.
374. H Predel. Ber Dtsch Wiss Ges Erdoel Erdgas Kohle Tagungsber, 1996, 9603, 79–81.
375. U Setzer, K Hedded. Ber Dtsch Wiss Ges Erdoel Erdgas Kohle Tagungsber, 1996, 9603, 329–336.
376. I Nakamura, K Fujimoto. Shigen Kankyo Taisaku 32(16): 1543–1549, 1996; CA 126–132282j.
377. T Tenndi, F Toshinari, H Enomoto. Shigen to Sozai 112(13): 935–946, 1996; CA 126: 132305.
378. M Gebauer. J Utzig Chem Tech(Leipzig) 49(2): 57–62, 1997.
379. S Hardman, DC Wilson. BP Chemicals 38th Microsymp Recycling of Polymers, July 14–17, 1997 Prague SL1.
380. G Audisio. 38th Microsymp Recycling of Polymers, July 14–17, 1997 Prague SL11.
381. a. H Bockhorn, A Hornung, U Hornung. 38th Microsymp Recycling of Polymers, July 14–17, 1997 Prague SL12.
b. MW Meszaros. In: P Rader, SD Baldwin, D Cornell, GD Sadler, RF Stockel, eds. Plastics, Rubber and Paper Recycling. A Pragmatic Approach, ACS Symposium Series 609, chap 15, pp 170–182, 1995.
c. SJ Pearson, GD Kryder, RR Koppang, WR Seeker. In: SD Baldwin, D Cornell, GD Sadler, RF Stockel, eds. Plastics, Rubber and Paper Recycling. A Pragmatic Approach, ACS Symposium Series 609, chap 16, pp 183–193, 1995.
382. PT Williams, S Besler. J Anal Appl Pyrolysis 30: 173, 1994.
383. B Hinz, M Hoffmockel, K Pohlmann, S Schädel, I Schimmel, H Sinn. J Anal Appl Pyrolysis 30: 35–46, 1994.
384. a. JA Conesa, A Marcilla, R Font. J Anal Appl Pyrolysis 30: 101–120, 1994.
b. AES Green, JP Mullin, RT Pearce. Chem Phys Processes Combust 1994, 325–328.
385. RWJ Westerhaut, JAM Kuipers, VPM. van Swanij, Chem Eng Sci 51(10): 2221–2230, 1996.
386. E Ranzi, M Dente, T Faravelli, G Bozzano, S Fabini, R Nava, V Cozzani, L Tognotti. J Anal Appl Pyrolysis 40–41: 305–319, 1997.
387. GS Darivakis, JB Howard, WA Peters. Combust Sci Technol 74: 267, 1990.
388. D Srivastava, AK Nagpal, GN Mathur J Polym Mater 13(4): 289–291, 1996.
389. C Vasile, C Savu, E Voicu. Bosovei Rivista dei combustibili 49(6): 129–136, 181–188, 1995; 51(4): 161–174, 1997.
390. J Burgtorf H Meier. Deutsch Wiss. Ges Erdoel, Erdgas Kohle, Tagungsber, 9603, 281–289, 1996; CA: 126–104911.
391. K Nieman. Deutsch Wiss. Ges Erdoel, Erdgas Kohle, Tagungsber, 9603, 281–289, 1996, pp 35–43.
392. MM Ibrahim, E Hopkins, MS Sochra. Fuel Process Technol 49 (1–3); 65–73, 1996.
393. V. Cozzani. Ind Eng Chem Res 36: 5090–5095, 1997; 36, 342–348, 1997.
394. V Cozzani, C Nicolello, L Petarca, M Rovatti, L Tognotti. Ind Eng Chem Res 34: 2006–2020, 1995.
395. a. Y Uemichi, Y Kashiwaya, M Tsukidato, A Ayama, H. Kanoh. Bull Chem Soc Jpn 56: 2768, 1983.
b. Y Uemichi, A Ayama, T Kanazuka, H Kanoh. J Chem Soc Jpn 735, 1985.
c. Y Sakata, MA Uddin, K Koizumi, K Murata. Chem Lett 245, 1996; Polym Degrad Stab 53(1): 111, 1996.
396. Y Uemichi, Y Makino, T Kanazuka. J Anal Appl Pyrolysis 14: 331, 1989; 16: 229, 1989.
397. C Vasile, P Onu, V Bärboiu, M Sabliovschi, M. Florea/ Acta Polym 39: 306, 1988; MACRO '83 Bucharest, Sept. 4–8, 1983, vol. 6, p 165; Papers of 2nd Symp. on Zeolites in Modern Technologies, IASI, Oct. 30–31, 1987, p 91; Papers of lst Symp. on Zeolites in Modern Technologies, IASI, Oct. 28–29, 1983, pp. 284, 292.
398. a. CrI. Simionescu, C Vasile, P Onu, V Barboiu, M Sabliovschi, G Moroi. Thermochim Acta 134: 301, 1988.
b. Vasile E Buruiana. Polymer '91, Melbourne, Feb. 10–13, 1981, Abstract Book, Paper P45.
399. C Vasile, EM Calugaru, M Sabliovschi, CN Cascaval. Materiale Plastice (Bucharest) 21: 54, 1984.
400. C Vasile, CN Cascaval, EM Calugaru, M Sabliovschi, IA Schneider ("P.Poni" Institute). Rom. Pat. 72,451, 1977; 74,577, 1978.

401. a. EF Gumerova, SR Ivanova, EL Ponomareva, VP Butov, AA Berlin, KS Minsker. Visokomol Soedn B-31: 607, 1989.
b. SR Ivanova, EF Gumerova, KS Minsker, GE Zaikov, AA Berlin. Prog Polym Sci 15: 193, 1990.
402. VC Smith. US Pat. 4,151,216, 1979.
403. D Engelharth, H Pfaff. German Offen 19517096, 1996; CA 126: 6053j.
404. a. T Fukuda, K Saito, S Suzuki, H Sate, T Hirota. US Pat. 4,851,601, 1989.
b. T Hirota, FN Fagan. Makromol Chem Macromol Symp 57: 161–173, 1992.
c. VJ Fernandez, AS Araujo, GJ Fernandez, J Therm Anal 49(1): 255–260, 1997.
405. YuG Egiazarov, BKh Cherches, LL Potapova, LV Gulyakevich (Institute of Physical and Organic Chemistry AN BSSR). USSR Pat. 1,397,422, 1986.
406. AA Berlin, EF Gumerova, SP Ivanova, KS Minsker, MM Karpasas. Vysokomol Soedn B-29: 604, 1987.
407. DV Kazachenko, KS Minsker, SR Ivanova, ED Zhidkova (Bashkir State University). USSR Pat. 1,337,383, 1985.
408. SR Ivanova IYu Ponedel'kina, V Romanko, KS Minsker, AA Berlin, MM Kukovitshii, NM Taimolikin, NV Tolmacheva, RE Kurochkina (Bashkir State University). USSR Pat. 1,351,913, 1985.
409. KS Minsker, RM Masagutov, SR Ivanova, EF Gumerova, KM Vaisberg, EA Kruglov, YuL Sorokina, AI Yunkin, AD Radnev (Bashkir State Univ., Scientific Research Institute of Perochemical Prod.) USSR Pat. 1,366,501, 1988.
410. M Avaro, JC Bailly, P Mangin (Naphtachemie). French Pat. 2,235,954, 1975.
411. L Balog, L Bartha, G Deak, N Nemes, M Kovacs, E Dobest, F Denes, E Parkas, J Kantor. Hung. Pat. 41,826, 1987.
412. J Wann, A Pemg, T Kamo, H Yamaguchi, Y Sato. Preprint Paper Am Chem Soc Div, Fuel Chem, 42(4): 972–977, 1997.
413. a. Y Sakata, MA Uddin, A Muto, K Kaizumi, M Narazaki, K Murata. Meeting of the Society of Chem Eng (Japan) G201, p 73 Kyoto Japan, 1996; The Fifth Asian Conf on Fluidized Bed and Three Phase Reactors, December 16–20, 1996, Hsitou, Taiwan pp 375–380; Polym Degrad Stab 56: 37–44, 1997.
Y Sakata. Proceedings of ISFR '99, Japan, Res. Ass. for Feedstock Recycling of Plastics p 99–102.
b. C Vasile, et al. idem, p 107–1100.
c. M Brebu, et al. idem, p 123–125.
414. K Liu, HLC Meuzelaar. Fuel Processing Technol 49: 1–15, 1996.
415. R Lin, DL Negelein, RL White. Preprint Paper ACS Div Fuel Chem, 42(4): 982–986, 1997.
416. MM Ibrahim. Energy Fuels 11(4): 926–930, 1997.
417. W Zhao, S Hasegawa, J Fujita, S Yoshii, T Sasaki, K Maruuchi, J Sun. Polym Degrad Stab 53: 129–135, 1996.
418. Y Sakata. 38th Microsymp Recycling of Polymers, July 14–17, 1997 Prague SL8.
419. MA Uddin, Y Sakata, A Mato, K Murata. 38th Microsymp Recycling of Polymers, July 14–17, 1997 Prague P-16; MA Uddin, K Koizumi, Y Sakata. 5th World Congress of Chemical Engineering, San Diego, July 14–18, 1996, Vol IV, pp 373–378.
420. CC Elan, RJ Evans, S Czernik. Preprint Paper ACS Div Fuel Chem 42(4): 993–997, 1997.
421. C Vasile, CN Cascaval, EM Calugaru, IA Schneider. Materiale Plastice (Bucharest) 16: 186, 1979; C Vasile, M Brebu, A Spac, H Darie. Annals "Al. I. Cuza" Univ IASI, Chemistry V: 213–218. 1997.
422. Y Saheki. Jpn. Plast. 22, 1972; Plast Ind News 18: S33, 1972.
423. N Dotreppe-Grisard. Tribune de Cebedeau 348: 467, 1972.
424. G Tellier. Rev. Gen. Caoutch. Plast. 50: 157, 1973.
425. K W Stookey, AK Chatterjee (Torrax System Incorporated USA) French Pat. 71,313,96, 1970.
426. S Kurisu, S Ochiai, K Uda, M Kakigi, K Orio (Mitsubishi Heavy Industries Ltd.). British Pat. 1,423,420, 1973.
427. WR Brown, GO Goldenbach, DR Moody, MA O'Hagen, FM Placer. British Pat. 2,027,527, 1978.
428. Refused-Fired Energy Systems in Europe: An Evaluation of Design Practice. An Executive Summary, Report SW-771, US Environmental Protection Agency, 1979.
429. Second (SW-122/1974) and Fourth (SW-600/1977) Report to Congress Resource Recovery and Waste Reduction, US Environmental Protection Agency.
430. VH Marcos. Heat Recov Sci 7: 465, 1987.
431. AJ Minchener, EA Rogers. Production of Energy from Waste. Fluidized Bed Combustion of Low Grade Materials. Contract Nr 188-77 EEUK, Final Report. Commision of the European Communities, Luxemburg, 1980.
432. R Fraunfelker. Small Modular Incinerator Systems with Heat Recovery. A Technical Environmental and Economic Evaluation. Executive Summary, Report SW-797, US Environmental Protection Agency, 1979.
433. HJ Knopf. In: J Brandrup, M Bittner, W Michaeli, G Menges, eds. Recycling and Recovery of Plastics, Munich: Hanser, 1996, pp 779–795.
434. JP Peyrelongue. Waste Management, October 1997, 51.
435. MYa Mel'nikov, EN Seropegina, Int J Polym Mater 31: 41–93, 1996.

18

Biocompatibility and Biodeterioration of Polyolefins

K. Z. Gumargalieva
N. N. Semenov Institute of Chemical Physics of the Russian Academy of Sciences, Moscow, Russia

G. E. Zaikov and A. Ya. Polishchuk
N. M. Emanuel Institute of Biochemical Physics of the Russian Academy of Sciences, Moscow, Russia

A. A. Adamyan and T. I. Vinokurova
A. A. Vishnevskii Institute of Surgery of the Russian Academy of Chemical Sciences, Moscow, Russia

I. INTRODUCTION

Polymer plastics have already found broad application in different areas of industry, and particularly in the medical industry due to their reasonably high resistance to many environmental factors. The solution of two important but opposite problems, the prediction of material durability and material recycling, requires permanent study of the stability of material properties.

II. BIOCOMPATIBILITY: BIOMEDICAL ASPECTS

A. Interaction of Polyolefins with Biomedical Media

Once implanted into the body medical polymers must show specific properties without interaction with surrounding tissues or with the body as a whole. So far there is no recognized definition of biocompatibility since there is no material parameter or biological tests which could be used as a quantitative characteristic of this property of the polymer [1].

Biocompatibility generally means that the polymer can exist in contact with blood and enzymes without undergoing degradation or provoking thrombosis, breakdown of tissues, or harmful, immune, toxicological or allergenic effects.

Higher molecular weight polyethylene usually shows a low toxicity [2]. The degradation of the molecules results in formation of low molecular weight fragments and even of monomers. The reciprocal dependence of the toxicity of polymers of a certain homologous series on their molecular weight has been reported [2].

The rate of degradation of the polymeric implant in the biological medium alters material biocompatibility. Generally, the compatibility of polymeric materials depends either on initial interactions with physiological components or on the stability of implants in the surrounding biological medium.

Degradation of polymers in biological medium is a complex physicochemical process comprising diffusion of the medium components in the polymer and transformation of chemically unstable bonds.

Depending on the ratio between the rates of diffusion and of chemical reaction, the degradation process can be limited by different stages (diffusion or chemical reaction):

The rates of diffusion and of chemical reaction are of the same order of magnitude. Reaction takes place in a specific reaction zone whose size increases with time, finally covering the size of the polymer, i.e., reaction takes place in the internal diffusion-kinetic zone.

The rate of diffusion is one order (or more) of magnitude higher than that of chemical reaction. As solubility of low molecular weight compounds in the polymer is complete, degradation occurs in the whole volume of the polymer, i.e., in the internal kinetic area.

The rate of diffusion is one order (or more) of magnitude lower than that of chemical reaction. In this case degradation occurs in the thin surface layer, i.e., in the external diffusion-kinetic zone.

For the film of thickness l, the changes in thickness are derived from the equation

$$l = l_0 - K^{s}_{eff} C^{s}_{cat} t / \rho \tag{1}$$

where K^{s}_{eff} is the rate constant of degradation on the polymer surface, C^{s}_{cat} is the concentration of catalyst on the polymer surface, and ρ is the polymer density.

The changes in weight are derived from the equation

$$m = m_0 - K^{s}_{eff} C^{s}_{cat} st \tag{2}$$

where s is the polymer surface in contact with the medium.

For a fiber of radius r, the following equations can be written:

$$r = r_0 - K^{s}_{eff} C^{s}_{cat} t / \rho \tag{3}$$

$$m^{1/2} = m_0^{1/2} - K^{s}_{eff} C^{s}_{cat} (\pi l / \rho)^{1/2} t \tag{4}$$

Having a crystalline structure, polyolefins represent the materials which degrade in the external diffusion-kinetic zone [3]. The degradation occurs in the thin reaction layer, the size of which is generally impossible to find due to the absence of K_{eff} and D_{cat}. As generally recognized the thickness of this layer tends to be zero, and the degradation actually occurs on the material surface.

B. Catalysts Derived from Biological Media

An analysis of the literature shows that the following substances should be considered as catalysts: water, salts, and enzymes. The selection of these substances was made partly because they are the most widespread and partly because their catalytic activity is now recognized.

1. Water

The water content in biological media is large. For example, the human body contains approximately 75% water, mainly in the intracellular fluid of the tissues and in the plasma. Thus polymers implanted into any part of the body must be in contact with water.

Abundant information on the sorption and diffusion of water in various polymers is available. The data acceptable for polyolefins application in medicine are shown in Table 1.

The table data show slow water diffusion in PE and PP, thus indicating stability of carbon-chain polyolefins against water.

2. Salts

Salts are found in biological media in large amounts. Table 2 shows the ionic composition of liquids in the human body [4]. The plasma and intracellular liquid contain proteins in anionic form due to the presence of carboxylate groups.

TABLE 1 Water sorption (c^0_w) value and water diffusion coefficient (D) for polyolefins used in medicine

Polymer	T/°C	c^0_w/(g/100 g)	D/(10^{-9} cm^2 s^{-1})
Polypropylene (PP)	25	0.007	2.4
Polyethylene (PE)	25	0.006	2.3
$\rho = 0.923$			

TABLE 2 Ionic composition of the liquids in the human body (mequiv L)

Ion	Plasma	Liquid in tissues	Intracellular liquid
Na^+	138	141	10
K^+	4	4.1	150
Ca^{2+}	4	4.1	40
Mg^{2+}	3	3	40
Cl^-	102	115	15
HCO_3^-	26	29	10
PO_4^{3-}	2	2	100
SO_4^{2-}	1	1.1	20
Organic acids	3	3.4	—

```
                               ⁻O
                                \ /
                                 Y
  O          OH                 / \                O           HO
  ||   H2O   |                  OH        \ /      ||           \ /
 ~C-X <----> ~C-X~  ------------------->   Y  ---> ~C  + HX~  +  Y
             |                            / \      |            / \
             OH                          O   O     OH          ⁻O
                                        /    |
                                       H     H
                                       |    /
                                       |   O   X
                                        \ /
                                       ~C-OH
```

Electrolytes diffuse in hydrophobic polymers with a mechanism similar to the transport of gases and vapors. Therefore, in the case of electrolytes with a high vapor pressure (for example, hydrochloric acid) the c^0 and D values are close to the corresponding values for water in the same polymers. For electrolytes with a low vapor pressure (for example, chlorides and phosphates) we found very low values of c^0 and D, i.e., hydrophobic polymers do not sorb these electrolytes to a significant extent.

The above general scheme can be suggested which shows how salts catalyze processes leading to the degradation of polymers containing carbonyl groups [3]. Thus salts (especially phosphates) have a strong catalytic effect on the degradation process in polymers containing carbonyl groups.

3. Enzymes

Enzymes play an active role in the degradation of polymers implanted in the human body. However, only recently have these effects received experimental justification.

Using the method of quantitative histoenzymology [5] Salthouse studied the activity of various enzymes in a capsule on the surface of materials after implantation for different times into rats (Table 3). An increase in the enzyme activity takes place in all the materials after 14 d, associated with an increased phagocytosis in the implantation region. After the heating of the surgical incision and the formation of a capsule from the connective tissue an equilibrium concentration of enzyme is apparently reached on the polymer surface.

Enzymes diffuse easily in the capsule, and are adsorbed on the surface of polymers in different concentrations. The surface concentration of enzyme catalysts depends either on the number of enzyme species in the bulk of the capsule or on the competitive adsorption of other proteins (provoking no catalysis), lipids, etc. The mechanism of the action of enzymes on polymers is extremely complex because the majority of polymers are not specific substrates for the enzymes.

TABLE 3 Activity of enzymes in the capsule on the surface of various materials

		Activity		
Group of materials	Time after implantation	Acid phosphates	Aminopeptidase	Oxyreductase*
Polypropylene	7	+†	±	+
Polyethylene	14	++	+	+
Polyurethane	28	±	±	±
Polytetrafluoroethylene	42	—	—	±

*Combined activities of succino- and lactohydrogenase and of cytochromexidase.
† List of symbols: − = no activity, ± = very low activity, + = moderately high activity, ++ = substantial activity.

C. Stability of Polyolefins in the Living Body

As reported above, a number of polyolefin parameters allow us to refer these materials to stable polymers.

1. Polyethylene

Polyethylene (PE) has been used in surgery since the 1950s: at that time a weak response of the body to an implant was reported. For example, only formation of so-called granulated tissue around the polymer was found after PE implantation under rat skin for 3 weeks following by vascularization after 6 weeks. After 12 months no inflammation was found in the layer of the fibrous tissue whereas fibrous tissue had become thinner [6]. However, later Calnan [7] reported the hardening of such a sponge and formation of a dozen fibrous tissues that was not acceptable for flexibility of soft tissues. Calnan also found the reaction of inflammation around the implant. The opposite results reported for PE in [6] and [7] are quite natural since so far no reliable quantitative criteria have been developed for biocompatibility of biomedical polymers. This is due to the complicated character of body–implant interactions in the case of a capsulated interface. Collection of data in this area defined the fields of PE application. In particular, polyethylene is not recommended for substitution of the soft tissues whereas it is a reasonable material to substitute for bone tissue (i.e., head of the hip bone and other elements of the pelvic bones). The first application of high density polyethylene (HDPE) for substitution of bone tissue showed large wear of the material [8]. Later, improvement in the synthesis technology allowed one to get polyethylene of different molecular weight (e.g. higher molecular weight polyethylene), thus enabling wide PE application to prosthesizing the pelvis.

High density conditions and application of special catalysts allow one to get ultra high molecular weight polyethylene of molecular weight 4×10^6 g mol^{-1}. Using the method of hot pressure, slabs can be obtained which serve as the initial material for the preparation of the elements of endoprosthesis. Such a treatment results in the excellent properties of polyethylene which basically depend on temperature. Tables 4 and 5 collect PE parameters as they depend on molecular weight whereas Table 5 also shows comparative properties of polyethylene and of bone cement.

During the last 20 years ultra high molecular weight polyethylene (UHMWPE) presents itself in a good light for endoprosthesizing pelvic bones. However, this does not mean that UHMWPE application solves all the problems. Although it shows excellent antifrictional properties and good ability of dry sliding, its wear must be also mentioned. For example, for the pelvis joint the couple metal/PE shows wear of 0.2 mm h^{-1}, and the couple ceramics/PE shows wear of 0.1 mm h^{-1}. The wear provokes formation of particles and a consequent negative response of the living body such as formation of granules around the foreign body. Biological ageing of the material leads to the loss of its positive properties. Table 6 shows positive and negative characteristics of polyethylene used as an endoprosthesis.

2. Polypropylene

Polypropylene (PP) is used in medicine due to its high chemical stability and favorable mechanical properties. In the early 1970s it was applied for lining the valves of artificial hearts, and for ball joint prostheses. It is now

TABLE 4 Properties of polyethylene

Property	Unit	High density polyethylene low density	Low density polyethylene high density	Ultra high molecular weight polyethylene
Molecular weight	g mol^{-1}	5×10^4	2×10^4	4×10^6
Density at 23°C	g cm^{-3}	≤ 0.9200	≥ 0.9200	0.9380
Melting area	°C	105–110	130–135	135–138
Impact resistance at 23°C	mJ mm^{-2}	6	13	140–160
Shape stability at 1.8 H mm^{-2} according to ISO/R7S.A	°C	=35	=45	=95
Wear stability		–	+	++

– = no activity, + = moderately high activity, ++ = substantial activity.

TABLE 5 Properties of polyethylene in comparison with properties of bone cement

Property	Unit	Bone cement, PMMA	Ultra high molecular weight polyethylene
Density at 23°C	g cm^{-3}	1.20–1.25	0.9380
Water uptake	%	about 2	0.01
Temperature limit (short time interval)	°C	90	100
Elastic modulus at 23°C	N mm^{-2}	4400–5200	800–1000
Tensile yield stress:	N mm^{-2}		
in air at			
23°C		38–44	41
40°C		about 38	about 37
after exposure for 300 days in Ringer* solution at 40°C (tested in air at 23°C)		33–36	?
Tensile stress at:	N mm^{-2}		
23°C		—	22
40°C		—	about 16
Elongation at break	%	1	about 450
Compression yield stress:	N mm^{-2}		
in air at 23°C		85–130	20–30
after exposure for 300 days in Ringer solution at 40°C (tested in air at 23°C)		89–90	?
Impact viscosity	J mm^{-2}	1.5–2.0	No break point

* Composition of Ringer solution: NaCl = 9 g; $NaHCO_3$ = 0.2 g; $CaCl_2$ = 0.2 g; KCl = 0.2 g; glucose = 1 g; H_2O = 1 L.

TABLE 6 Properties of ultra high molecular weight polyethylene as an implant material

Characteristics	Properties to be improved
Biocompatibility (implant and products of ageing)	Mechanical stability
Springing ability	Hardening (as it depends on geometry)
Antifrictional ability	Hardening (as it depends on geometry)
Viscosity and plastic characteristics	Wear
	Resistance to biological aging
	Tendency to brittle destruction
	Thermosterilization

commercially available as suture threads from the Ethicon Company (USA).

Results of histology show that PP of "medically pure" grade provokes only a moderate response of the tissues of the living body [5, 9, 10] as compared with PP containing traces of catalysts and stabilisers.

(a) Aging of Polypropylene Fibers in vivo. PP fibers prepared from a complex thread (diameter of a thread unit = 0.033 mm) in the form of a wicker braid have been investigated. The maximal period of tests *in vivo* in the skin cellular tissue of a rabbit was 4 years. The samples exposed *in vivo* were taken from the rabbit at fixed time and washed to remove connective tissue. The samples of original material underwent the same treatment. Then the properties of original and exposed materials were compared.

The alteration of mechanical properties of the fibers by biological aging are illustrated in Fig. 1. After 4 years fibers lose about 60% of their initial durability and about 80% of their initial elongation. There can be noted a linear character of a durability decrease with an average rate of 1.13% per month, whereas a sharp decrease of elongation was observed within half a year (40–50% with a rate of 8.3% per month) following by slow decrease with a rate of 0.8% per month. The weight loss of the samples, being not visible during

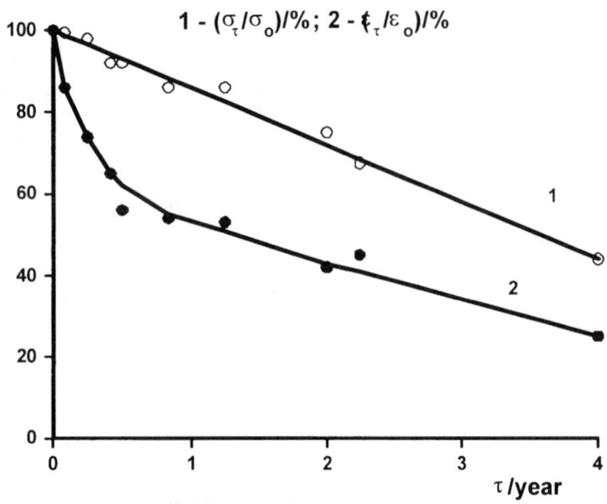

FIG. 1 Stress (1) and elongation (2) at break of PP fibers in the living body.

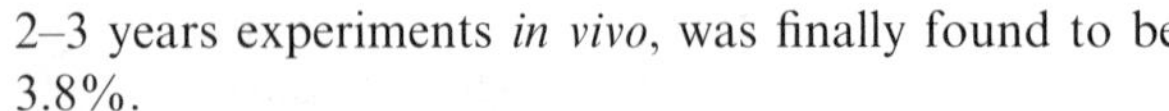

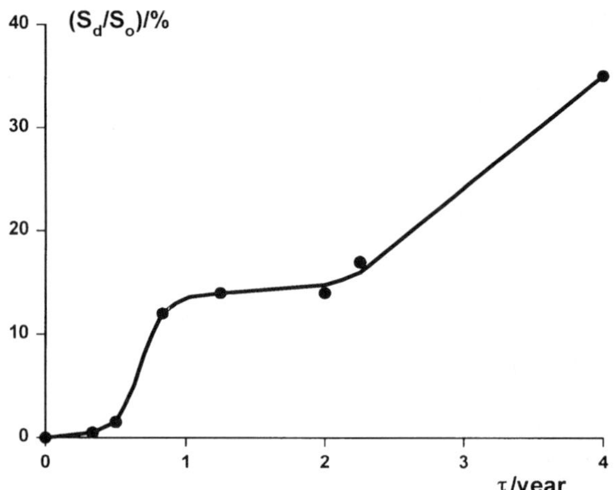

FIG. 2 Alteration of the relative amount of defects on the surface of fibers by implantation.

2–3 years experiments *in vivo*, was finally found to be 3.8%.

Molecular weight (M_η) which was measured of 80,000 using characteristic viscosity of PP solution in decaline decreased slowly against time of implantation. However, after 4 years the decrease in molecular weight was 17% (Table 7).

The surface of fibers was investigated by the methods of light and electronic microscopy in order to clarify the reasons for the loss of mechanical properties of fibers and increase in their brittleness. At an earlier time of implantation fibers morphology remains basically unchanged. Cracks appear after 5 months. The development of cracks accelerates on the surface of fibers during the first year.

During the second year this process is quantitatively stabilized whereas some quantitative changes can be noted. During the first 10 months mainly small and medium sized cracks are observed which further develop into large cracks located perpendicular to the axis of the fibers. The later appearance of the cross cracks lead to the breakdown and the separation of the fiber fragments. The above destruction processes take place in a relatively thin layer of the fibre surface (about 1 mm or 3% of the fiber diameter) and do not affect the molecular weight of the polymer and material weight loss. The weight loss, which was measured at 3.5% after 4 years of implantation, was caused by the partial separation of the fiber surface and destruction of the surface layer. The curves representative of the surface effects are shown in Fig. 2. To avoid the effect of temperature on the analysis of the process of PP ageing in the living body the fibers were studied at 37°C light free in air, in water, and in Ringer solution. Results of the investigation of mechanical properties (Fig. 3), and of measurements of weight and molecular weight of the polymer showed that all these parameters remained unchanged during all the period of exposure (4 years). No changes were also observed in fibers morphology.

TABLE 7 Viscosity of PP solutions after material implantation in the living body

Period of implantation, months	$[\eta]$ (mL g^{-1})
Original material	1.08
Sterilized material	1.08
6	1.02
15	1.01
27	1.02
48	0.9

Therefore, the conclusion can be drawn that the ageing of PP fibers is only affected by the living body. To understand the role of structural changes, which occurred in PP fibers, in the process of fibers destruction, stress relaxation was studied using the tensile method. Figure 4 shows curves representative of stress relaxation of PP fibers (original, and implanted for 2 and 4 years) through the dependence of the ratio between remaining and initial load on time (τ).

There were no visible changes in orientation of supermolecular structures of fibers found in the time

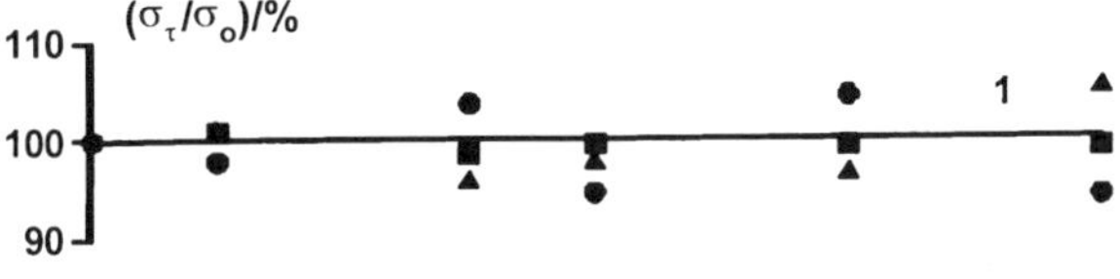

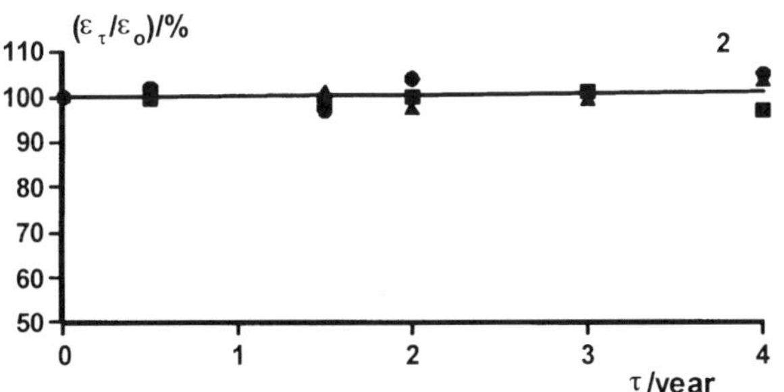

FIG. 3 Stress (1) and relative elongation (2) at break of PP fibers after exposure in model media. Symbols: ● = air, 37°C; ■ = water, 37°C; ▲ = Ringer solution, 37°C.

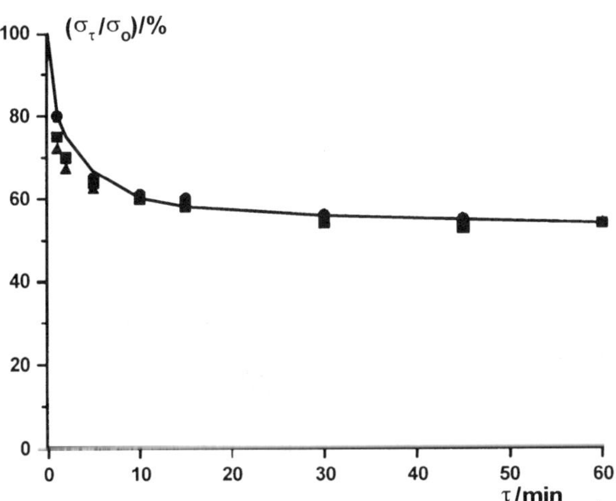

FIG. 4 Alteration of stress relaxation of PP fibers by period of implantation (shown by symbols: ● = 0; ■ = 2 years; ▲ = 4 years).

frame of the experiment; all results were within experimental error (6–10%).

To realize the changes occurred in orientation of the fibers, X-ray analysis of PP fibers (either initial or implanted for 2 and 4 years) was performed. Some reorientation of structural elements of the fibers was detected which referred to the angle of crystallite reorientation (ϕ). The angle ϕ varies from 8° (original fibers) to 13–15° for samples implanted for 4 years.

Therefore, looking at the results of an analysis of PP fibers after their implantation in underskin cellular tissue of rabbits, the following conclusion can be drawn. The ageing process of fibers occurs which is justified by

a certain decrease of polymer molecular weight;
some reorientation of the structural elements of fibers;
loss of mechanical properties (e.g., stress and relative elongation at break);
destruction of the surface layers of fibers, resulting in the development of cracks.

The most typical are changes of such properties as relative elongation (in the first weeks) and surface quality (after 3–5 months of implantation).

(b) Aging of PP Fibers in Contact with Elements of Connective Tissue and Liquid of Tissue. The experiment described in the previous subsection was performed either with Russian or Italian (trade mark "Moplen") PP implants. The results of the investigation of Moplen fibers, which were implanted in underskin cellular tissue of rabbits for 4 years, showed no changes in their properties. The weight of samples, molecular weight and mechanical characteristics remained unchanged in comparison with their initial values.

Therefore, PP and Moplen fibers showed different behaviors in the same medium of implantation. The reason for such a difference is apparently the difference in polymer capsulation. In the PP (Russian) case the sample was germinated by connective tissue following capsulation of separated fibers and fiber groups, whereas the Moplen sample was capsulated as a unit without germination of tissue between threads and fibers.

Since the growth of connective tissue is a well known response of the body to inflammation process provoked by foreign bodies (e.g., PP and Moplen implants), the absence of germination of Moplen samples by connective tissue can be explained by the more inert character of this material than of polypropylene. Since Moplen properties remain unchanged even after 4 years of implantation, the slow interaction of this polymer with living tissue can be assumed. To test this assumption, a more detailed analysis of the effect of connective tissue and tissue liquid on polymer was made.

The following method for the implantation of samples was developed. The first group of PP braid was implanted directly into the underskin tissue of the experimental animals (white rats) whereas the second group was placed inside silicone pipes with a length of 40 mm and internal diameter 5 mm. The germination of the samples of the first group began from the first week of implantation *in vivo*. The samples of the second group were basically in contact with tissue liquid. Although the germination of these samples was not completely prevented, the formation of capsules began after a 3 month delay in comparison with the first group.

Mechanical tests, which were performed after 3 months of implantation, showed (Fig. 5) the decrease of relative elongation at break of 40 and 25% for the first and second groups of samples, respectively. The highest rate of decrease of the relative elongation was observed during the first 2–3 months. For all periods of study, the difference between values of relative elongations remained within the range of 10–15%. The loss of durability was also observed more for the first group, and remaining for both groups of samples within the range of 10–12%.

The microscopic study showed the appearance of cracks perpendicular to the axis of the fiber. The cracks' character was typical for chemical degradation. Figure 6 shows that 3% of the surface was destroyed after 3 months for the first group of samples and after 9 months for the second group.

The difference in mechanical properties and surface quality between the two groups of samples can be explained by the different mechanism of the destruction of the surface layer of the material. For the first group aseptic inflammation and formation of the capsule of connective tissue occurred close to the surface of the fibers whereas for the second group tissue liquid served as a barrier to the inflammation process.

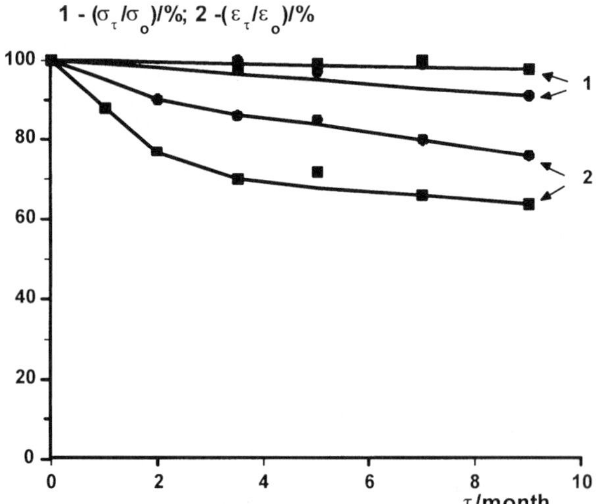

FIG. 5 Stress (1) and relative elongation (2) at break of polypropylene fibers implanted in the living body. Symbols: ● = samples germinated by connective tissue; ■: samples in contact with tissual liquid.

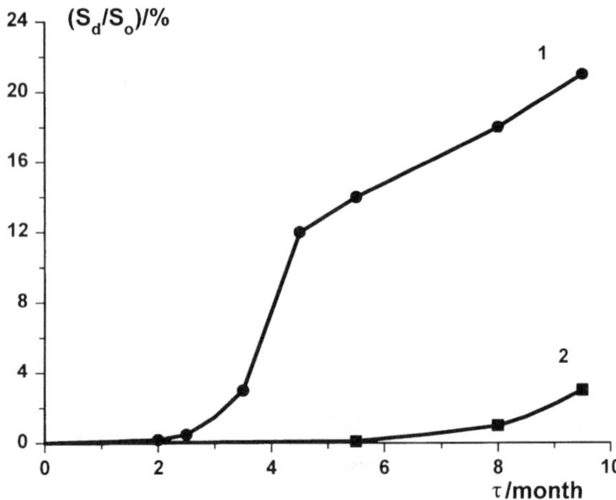

FIG. 6 Destruction of the surface of PP fibers in conditions of their germination by connective tissue (1) and in tissue liquid of the body (2).

Therefore, degradation processes can be assumed in the surface layers of PP fibers which alter mechanical properties of the material, leading to the formation of cracks.

The reasons for the formation of cracks can be the internal structure of the fibers and the anisotropic distribution of stresses in the fibers. The elements of connective tissue contribute to the development of cracks, thus provoking a difference in the properties of PP fibers for the two groups investigated.

D. Aging of Polypropylene Films

To understand the role of PP orientation, the samples of bi-axis oriented films were studied. Films were implanted under the skin of white rats for a period of 1 year and 7 months. The thickness of films (0.03–0.035 mm) was relevant to the diameter of fibers earlier studied.

Either stabilized or nonstabilized film samples of polypropylene were studied. Most PP stabilizers, which are applied to prevent polymer degradation during processing, are toxic, although polypropylene itself is an inert material. Hence, for experiments with animals, a stabilizing system based on Irganox 1010 was chosen which is sufficiently harmless [12].

Prior to experiments, the efficiency of stabilizing systems based on Irganox 1010 and on two similar Russian compounds, phenosan-23 and phenosan-28, was evaluated.

The efficiency of the stabilizing systems was evaluated by estimating the induction period of oxidation of the original PP films and analyzing the changes in material properties after isothermal heating at 150°C [13].

Results showed the same induction period for all three systems (55–60 min) whereas isothermal heating for 6 h at 150°C caused no changes in the mechanical properties of the material. Therefore, the resistance of all three PP formulations to the thermo-oxidation process is substantially the same. Therefore, the same stablizing effect of these additives during PP processing can be assumed.

For *in vivo* studies film stabilized with phenosan-23 and nonstabilised PP film was selected. Results of analysis of the effect of implantation on the mechanical properties of PP films are shown in Fig. 7.

These properties remain unchanged *in vivo* for stabilized films. No changes in material structure nor chemical degradation were detected, whereas for non-stabilized PP films, the stress at rupture reduced by 20% after 19 months of implantation. Probably this parameter is the best for the characterization of PP aging in the living body.

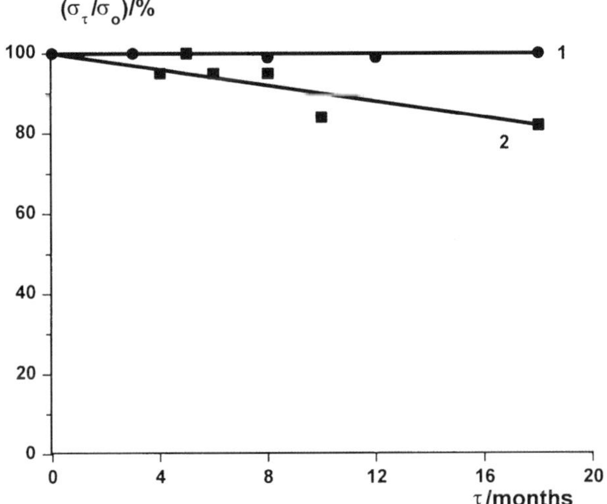

FIG. 7 Stress at rupture of films of stabilized (1) and non-stabilized polypropylene (2) after exposure in the media of the living body.

III. BIODETERIORATION OF POLYETHYLENE UNDER ACTION OF MICRO-ORGANISMS

Biodeterioration (biodamaging) of polyolefin materials proceeds at their contact with living organisms and may lead to a change of the exploitation properties. In general the following processes may cause biodamage:

adsorption of micro-organisms or substances existing in tissues of the living organism on the material surface;

decomposition of the material as a result of specific influence (living organisms use the polymeric material as a food source) or under the influence of metabolism products.

As the first process proceeds, the chemical structure of polyethylene does not change as a rule. The material plays the role of the support on which the adhesion and the growth of colonies of micro-organisms (bio-overgrowth) or the formation of a collagen-like capsule take place. Adhesion of micro-organisms is the initial stage of material bio-overgrowth which defines all further views of bio-overgrowing and biodamaging of polymeric materials. At the stage of bio-overgrowth the determination of the amount of biomass on the surface of the polymeric material is of great interest because it alters surface quality (optical, adhesive, etc.).

The second process leads directly to the aging of polymeric material under the influence of chemically active substances. In this case the "bulk" exploitation properties, such as mechanical, dielectric and others, will be affected.

A. Kinetics of Biomass Growth: Methods and Results

The growth and development of microscopic fungi directly on the solid surfaces of polymers is usually estimated by the growth of the diameter of colonies of a definite type or by selection of microscopic fungi. This method, which follows the five-level Russian standard scale (GOST 9.049-75*), is connected with the difficulties of biomass recognition in amounts of $\mu g\ cm^{-2}$ in the initial stages of growth.

To obtain kinetic parameters of accumulation of biomass we have applied a sensitive radio-isotopic

* GOST 9.049-75 specification: United system of protection against corrosion and aging. Polymeric materials. Laboratory tests on resistance to affect of microscopic fungi.

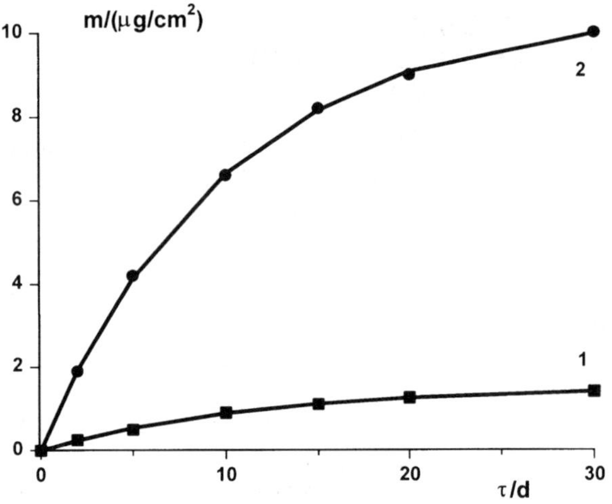

FIG. 8 Kinetic curves of biomass accumulation of the microscopic fungi on PE (1) and cellophane (2) surface.

method [14, 15]. Polymeric films were contaminated by suspension of microscopic fungi of 10^6 cells mL^{-1} in water or in a nourishing medium [6, 7]. The amount of biomass was measured as the difference between the original and degraded samples for each polymer. The intensity of irradiation was measured by a liquid scintillation counter "Mark-388."

The growth of micro-organisms was estimated according to the change in the amount of dry biomass per unit of the sample surface.

Figure 8 shows the kinetic curves of biomass accumulation on the surface of PE and a hydrophilic cellulose derivative (cellophane). In both cases the kinetics of accumulation follows an exponential trend:

$$m/m_{\infty} = 1 - \exp(-kt) \tag{5}$$

where m is biomass accumulated at time t, m_{∞} is biomass at equilibrium, and k is an apparent constant of biomass growth.

Table 8 shows m_{∞} values, initial rate of biomass growth (v_{init}), and k for two types of spore suspensions derived from the logarithmic trend of Eq. (5), in water and in a nourishing medium of Chapeck – Dox. The values of m_{∞} and v_{init} are two times larger than those in water, whereas k values are nearly the same for both media. Larger bio-overgrowth of cells in the case of cellophane links to the polymer's hydrophilicity.

B. Kinetics of Adhesion of Micro-organisms

Biodegradation follows the interaction of micro-organisms with the polymeric surface and begins with the adhesion of micro-organisms. Adhered cells act as aggressive bioagents, precipitating exoferments or other molecular substances. For this reason the quantitative parameters of adhesion are definitive for further stages which are bio-overgrowing (biomass accumulation and biodegradation) [4, 16, 17].

Adhesion strength (F_{adh}), measured by the method of centrifugal detachment, may serve as a macroscopic parameter accessible for quantitative estimation. To estimate F_{adh} a suspension of cells of micro-organisms (10^6 spores mL^{-1}) was attached to the surface of a polymeric film and exposed for a certain time interval at different external conditions (temperature (T, °C) and humidity (ϕ, %)). Then films were fixed on metal plates and centrifuged in the field intensities acting perpendicular to the surface. The number of cells (γ_F), which was detached from the surface into the bulk of the centrifugal glass filled with distilled water at the particular field intensities, was calculated with the help of an optical microscope. Adhesion was estimated according to the force of the spore's detachment from the polymeric surface

$$F_{\text{adh}} = (1/675)\pi^3 r^3 \omega^3 R(\rho_{\text{cell}} - \rho_{\text{l}}) \tag{6}$$

where r is the radius of the spores, ω is the angular rate of rotation, R is the distance to rotor axis, ρ_{cell} is the

TABLE 8 Equilibrium biomass (m_{∞}) on the surface of various polymeric materials and the initial rate of biomass accumulation (v_{init})

	Treatment by mixture of spores in Chapeck–Dox medium			Treatment by mixture of spores in water		
Investigated material	m_{∞} (μg cm^{-2})	v_{init} (μg cm^{-2} d^{-1})	k (10^{-6} s^{-1})	m_{∞} (μg cm^{-2})	v_{init} (μg cm^{-2} d^{-1})	k (10^{-6} s^{-1})
Cellophane	10.5 ± 1	0.60 ± 0.05	1.0	5.80 ± 0.60	0.40 ± 0.080	1.5
Polyethylene	1.5 ± 0.20	0.27 ± 0.02	0.9	0.27 ± 0.02	0.01 ± 0.001	1.3

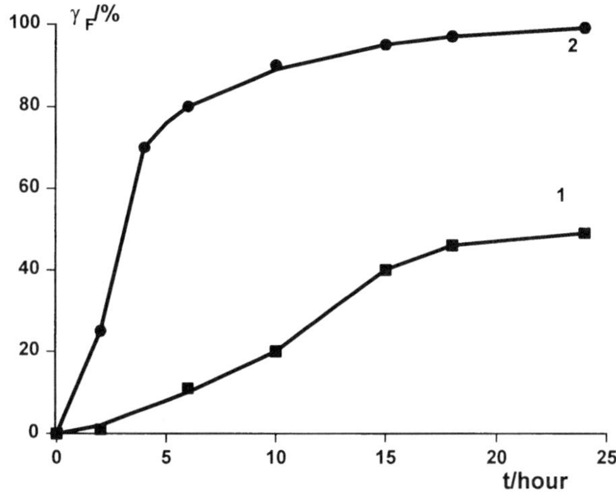

FIG. 9 Kinetic curves of the *Aspergillus niger* spores adhesion to the polymer surface at $\phi = 98\%$ and $T = 22°C$.

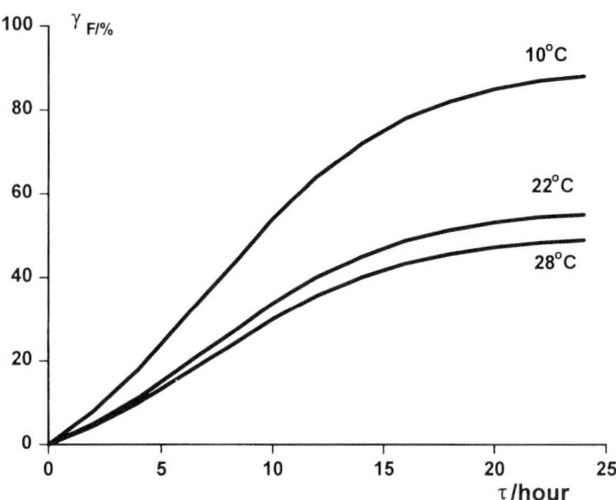

FIG. 10 Kinetic curves of the *Aspergillus niger* spores adhesion to the polyethylene surface at $\phi = 98\%$ and different temperatures.

TABLE 9 Parameter of adhesion of *Aspergillus niger* conidia to the surface of different polymeric materials

Material	k (h^{-1})	F^{50} (dyn $cell^{-1}$)	γ_∞ (%)
Polyethylene	0.06	3.3×10^{-4}	55
Cellophane	0.36	1.6×10^{3}	85

density of spores (cells), and ρ_l is the density of the liquid in which the detachment is performed.

The effect of the material nature on adhesion of spores is shown in Fig. 9, presenting kinetic curves of adhesion on different polymers at certain hygrothermal conditions. Adhesion equilibrium was reached within 24 h. Table 9 shows calculated values of adhesion intensities and of constants of formation of adhesion intensities γ_∞.

Figure 10 shows the dependence of the adhesion index of *Aspergillus niger* spores on time at constant humidity and different temperatures. Table 10 collects the values of parameters calculated from these kinetic curves for two substances representative of two extreme solubilities in water (polyethylene and cellophane). Table 10 shows the same rate constant by the formation of adhesion intensities measured at different temperatures and the increase of adhesion intensity with the decrease of temperature. This tendency is valid for both polymers.

Figure 11 shows kinetic curves of adhesion of *Aspergillus niger* spores to the surface of polyethylene at constant temperature and variable environmental humidity. The constants of adhesion, which were calculated at different humidifies and collected in Table 11, showed remarkable changes in k values as γ_∞ changes were sufficiently small.

The values of adhesion parameters which were calculated from the kinetic curves (Fig. 12) of adhesion of conidia of different microscopic fungi spores are collected in Table 12.

TABLE 10 Adhesive parameters of interaction for polyethylene and cellophane at different temperatures and relative humidity $\phi = 10\%$

	Polyethylene			Cellophane		
T (°C)	k (h^{-1})	F^{50} (dyn $cell^{-1}$)	γ_∞ (%)	k (h^{-1})	F^{50} (dyn $cell^{-1}$)	γ_∞ (%)
10	0.06	5.2×10^{-4}	85	0.36	2.6×10^{-3}	90
22	0.06	3.3×10^{-4}	55	0.36	1.6×10^{-3}	85
38	0.06	1.3×10^{-4}	50	—	—	—

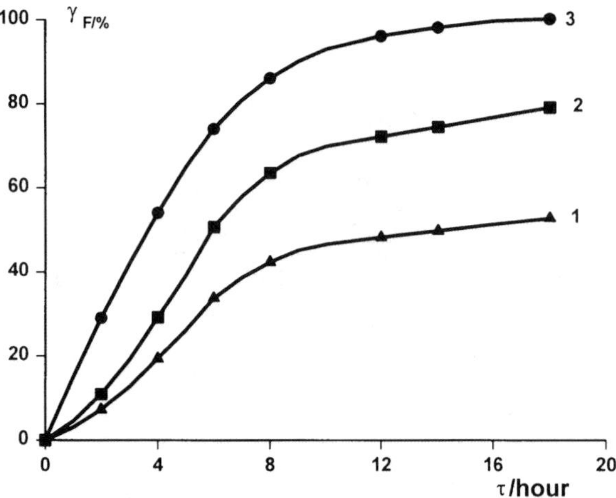

FIG. 11 Kinetic curves of the *Aspergillus niger* spores adhesion to the polyethylene surface at $T = 22°C$ and different humidifies: 1: $\phi = 0\%$, 2: $\phi = 30\%$, 3: $\phi = 98\%$.

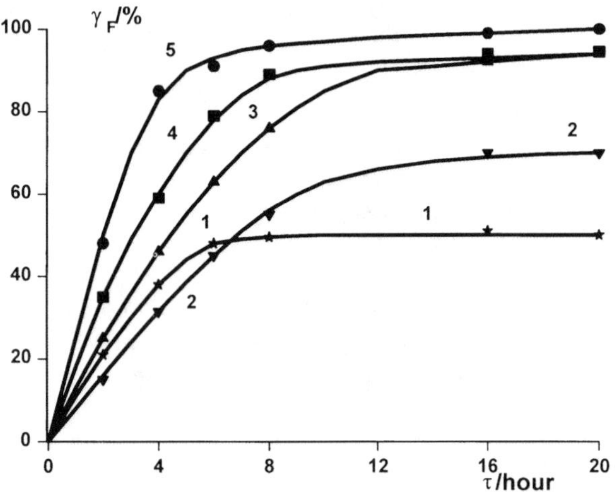

FIG. 12 Kinetic curves of adhesion of different fungi to polyethylene surface. 1: *Aspergillus niger*; 2: *Penicillium cyclopium*; 3: *Paec. varioti*; 4: *Penicillium chrysogenium*; 5: *Aspergillus terreus*.

TABLE 11 Adhesive parameters of conidia for polyethylene at different humidity and constant temperature 10°C

ϕ (%)	k (h^{-1})	F^{50} (dyn cell^{-1})	γ_∞ (%)
0	0.08	3.0×10^{-4}	70
30	0.66	5.2×10^{-4}	85
100	0.56	1.9×10^{-3}	100

The final scope of the investigation of the quantitative regularities of adhesion and bio-overgrowth was obtaining macroscopic kinetic parameters of the biodamaging ability of polymers. The experimental data collected in Table 13 show the link between adhesion and the growth of biomass. Additional study is required to model media to better understand this type of degradation.

Biostability of the material can be predicted with the help of experimental values of microscopic kinetic parameters of the interaction of microorganism cells (spores conidia) with the polymeric surface. In the general case biostability (B) depends on adhesion intensity, the amount of biomass, and apparent rate constants of degradation of accessible bonds:

$$B \sim (F_{\text{adh}} \Delta m_\infty k^{\text{adh}})^{-1}$$

For example, the values of biostability of cellulose calculated from this expression (humidity, $\phi = 98\%$) was equal to 0.5×10^6 s cm^2 dyn^{-1}μg^{-1} whereas this parameter calculated for polyethylene was 5 orders of magnitude higher: 0.4×10^{11} s cm^2 dyn^{-1} μg^{-1}.

Therefore, it was shown that the investigation of kinetic regularities of microscopic processes, such as adhesion, bio-overgrowth, and biodegradation, allows us to make models of the mechanism of the complex processes of biostability and biodecomposition of polymeric materials.

TABLE 12 Parameter of adhesion of microscopic fungi to the surface of polyethylene at $\phi = 10\%$ and $T = 22°C$

Fungus type	k (h^{-1})	F^{50} (dyn cell^{-1})	γ_∞ (%)	r(μm)
Aspergillus niger	0.23	0.26×10^{-1}	80	5.0 ± 0.5
Aspergillus terreus	1.96	0.31×10^{-3}	94	1.0 ± 0.05
Paec. varioti	0.30	0.76×10^{-1}	98	5.5 ± 0.7
Penicillium chrysogenium	0.40	0.45×10^{-3}	94	1.5 ± 0.06
Pinicillium cyclopium	0.50	0.10×10^{-1}	55	2.6 ± 0.4

[2] r = radius of spores for different type of fungi.

TABLE 13 Parameters of biodamaging ability of PE and cellophane

Material	Adhesion parameters*			Growth, m_∞(μm cm^{-2})	Degradation, k_{enzyme}(s^{-1})
	k^{adh} (h^{-1})	F^{50} (dyn cell^{-1})	γ_∞ (%)		
Polyethylene (LDPE)	1.6×10^{-5}	3.3×10^{-4}	55 ± 5	1.5 ± 0.2	1.2×10^{-9}
Cellophane	1.0×10^{-4}	1.6×10^{-3}	85 ± 5	10.5 ± 1.0	0.5×10^{-6}

*k^{adh} = constant of adhesion of *Aspergillus niger* microscopic fungus conidia; F = adhesion intensity of a single cell to polymer surface; γ_∞ = extreme adhesion index; m_∞ = biomass per surface unit; k_{enzyme} = rate constant of degradation affected by enzyme.

IV. BIOMEDICAL, TECHNICAL AND ECOLOGICAL PROBLEMS OF BIODESTRUCTION OF POLYOLEFINS

A. Polypropylene

Polypropylene can undergo only oxidative degradation. Therefore, PP should be stable in the tissues of the living body, decomposing only under the effect of oxidative enzymes and oxygen dissolved in the living medium.

The overall picture of the degradation of PP filaments and films in the body is characterized by the following phenomena: a worsening of mechanical properties (decrease in the breaking load and in the relative elongation) [18]; formation of cracks in the initial stages of the implantation and fragmentation of the polymeric implant at longer times (the fragmentation to be observed quicker in PP samples without antioxidants) [18]; unchanged molecular weight during the implantation; appearance of C=O absorption bands in the IR spectra of the PP samples [19]; an increased activity of oxyreductase and cytochrome oxidase in the capsule surrounding the implant [5].

The above experimental results on PP degradation in the living body show that this polymer undergoes the S type mechanism of degradation under the effect of oxidizing enzymes which cannot penetrate into the polymer matrix [Eqs. (3), and (4)]. It is well known that oxidative degradation accompanies the breakdown of the main chain, thus leading to the decrease in molecular weight. However, since the catalysts are unable to penetrate into the bulk of the polymer, the breakdown of bonds occurs only on the surface, provoking negligible decrease in the molecular weight of the whole sample. The accumulation of carbonyl compounds also takes place on the surface of cracks rather than over the whole volume of the sample.

Special experiments which were performed with PP exposed in a solution of cytochrome oxidase did not show the expected result because of the rapid deactivation of this enzyme in experiments *in vitro*. Therefore, experiments with bacteria, which produce oxidative enzymes able to open C—C bonds, are required to understand the role of these enzymes in PP oxidation.

B. Polyethylene

The biodegradation of PE was detected when the possibility of exploitation of polymeric by-products with the aid of micro-organisms and bacteria was studied. It was found that commercial polyethylenes (especially LDPE) undergo nearly no biodegradation by the radical mechanism [20, 21].

The following features of the biodegradation of PE have been detected:

The main condition for the biodissociation is an increase of the polymer surface (e.g. by using powder) [22].

The dissociation process takes place with a decrease in weight, accelerating during initial stages of exposure and then slowing down.

Samples of high density polyethylene (HDPE) exposed with micro-organisms are found to include carbonyl groups, especially in the presence of fillers sensitive to oxidases and esterases.

PE labelled with ^{14}C was used to study the rate of degradation of slowly decomposing PE [24]. Samples incubated with micro-organisms for different times (up to 3 years) were irradiated with UV light in order to accelerate the degradation, and the weight loss caused by the action of the bacteria and micro-organisms was determined. Extrapolation to zero time of irradiation yields the weight loss due to the biodegradation. The weight loss of the powders was found by this method to be 0.5% in 3 years.

These data indicate the S-type mechanism of PE degradation. The decomposing species can be oxidative enzymes generated by bacteria and micro-organisms, for which PE acts as a carbon feed medium.

Combining the data for PP and PE degradation we can conclude that, although these polymers decompose slowly, the rearrangement of their structure during oxidation can lead to undesirable consequences for the living body.

The main problem, which limits further application of polyethylene in orthopaedics and other areas of implant surgery, is the uncertain effect of the living body on the physicomechanical characteristics of the material. Specific oxidative ferments make the process of biodegradation random.

The experimental finding of the existence of adhered cells and bio-overgrowth on the surface of PE samples allows one to assume the possible biodamage of this polymer.

Alternatively, the ecological problem of polyolefin wasting provokes a need for the rapid decomposition of polyolefins with the aid of micro-organisms for the alteration of polymer structure by irradiation or by other processes initiating the breakdown of structural chemical bonds [25–27].

V. CONCLUDING REMARKS

Being exposed in the medium of living body and affected by micro-organisms the polyolefins undergo chemical modifications due to the breakdown of the main chain and alteration of the supermolecular structure.

The kinetic study of the effect of biofactors on polymers enables one to determine the start and duration of the biodegradation process. The area of application of polyolefins (threads, bone tissue prosthesis, burn coatings, details of commercial products to be stored or used) is very important for this study.

Following the general classification, polyolefins can be referred to as inert polymers with moderate biostability in comparison with other groups of polymers (polyesters, polyamides, etc.), losing only to fluorinated polymers. Either durability or acceleration of biodegradation (an ecological problem) of polyolefins can be governed by macrokinetic parameters in the case of development of polyolefin composites.

NOMENCLATURE

B = biostability
F = adhesion intensity of a single cell to polymer surface
F_{adh} = adhesion strength
ϕ = humidity
γ_F = number of cells
γ_∞ = extreme adhesion index
k = apparent constant of biomass growth.
k^{adh} = constant of adhesion of *Aspergillus niger* microscopic fungus conidia
k_{enzyme} = rate constant of degradation affected by enzyme
m = biomass accumulated at time t
m_∞ = biomass at equilibrium
r = radius of spores for different type of fungi
r = radius of spores
R = distance to rotor axis
ρ_{cell} = density of spores (cells)
ρ_l = density of the liquid in which detachments is performed
T = temperature
v_{init} = initial rate of biomass growth
ω = angular rate of rotation
D = water diffusion coefficient
D_{cat} = catalyst diffusion coefficient
c^0_{cat} = water sorption coefficient
C^s_{cat} = concentration of catalyst on the polymer surface
ε_0 = initial elongation at break
ε_τ = elongation at break at time T
l = film thickness
l_0 = initial film thickness
K_{eff} = rate constant of degradation
K^s_{eff} = rate constant of degradation on the polymer surface
m_0 = initial sample mass
m = sample mass
η = viscosity
t, τ = time
r_0 = initial fibre radius
r = fibre radius
ρ = polymer density
s = polymer surface to be in contact with medium
S_0 = initial surface
S_d = surface with defects
σ_0 = initial stress at break
σ_τ = stress at break at time τ
− = no activity
± = very low activity
+ = moderately high activity
++ = substantial activity

REFERENCES

1. J Bruck. Biomed Mater Res 11: 1–6, 1977.
2. R Lefaux. In: Y Champetier, ed. Chimie et Toxilogie des Matieres Plastiques. Paris: Compagnie Fran d'Etudions, 1964, pp 57–65.

3. YV Moiseev, GE Zaikov. Chemical Stability of Polymers in Aggressive Media (Russian). Moscow: Khimiya, 1977.
4. E Laifut. Transfer Phenomena in the Living Systems (Russian). Moscow: Mir, 1977.
5. TN Salthouse. J Biomed Mater Res 10: 197–201, 1976.
6. Z Neuman. Br J Plast Surg 9: 195–203, 1957.
7. JS Calnan. Br J Plast Surg 16: 1–6, 1963.
8. J Biny. Acta Path. Microbiolog Scand. Suppl 16: 105-109, 1968.
9. TN Salthouse, BF Matlaga. J Surg Res 19: 127–132, 1975.
10. NC Liebert, RP Cherhoff, SL Cosgrofe, R McCuskey. J Biomed Mater Res 10: 939–942, 1976.
11. TI Vinokurova, IB Rozanova, SM Degtyareva. Biodegradation of polyethylene. Proceedings of the 3rd Symposium on Synthetic Polymer Materials for Medicine (Russian), Belgorod-Dnestrovskii (USSR), 1977, pp 32–34.
12. TV Merezhko, TI Vinokurova. In: Actual Problems of Modern Surgery (Russian). Moscow: Meditsina, 1977, pp 142–148.
13. D Lebaner. Medizintechnik, 105: 19–23,1985.
14. M Kinley, R Kelton. Verh. Int. Theor. Limol. 21: 1348-1351, 1981.
15. FF Mazur. Biological Tests of Woodpulp by Radioisotope Methods (Russian). Moscow: Gosstroiizdat, 1959.
16. KC Marshall. Microbial Adhesion and Aggregaton. Berlin: Springer Verlag, 1979.
17. YV Moiseev, GE Zaikov. Chemical Resistance of Polymers in Reactive Media. New York: Plenum Press, 1984.
18. RW Postlethwait. Ann. Surg, 190: 10–13, 1979.
19. J Dolezel. Plasty Kau 13: 257–261, 1976.
20. AC Albertson, OA Svek, K Sigbritt. Polymer Degrad Stabil 18: 73–78, 1987.
21. NB Wykrisz. Plast Polym 42: 195–198, 1974.
22. IP Fisher. Biocompatibility of polymeric materials. Proceedings of the First Biomaterial Congress, Vienna, 1980, pp 20–21.
23. M Minac, I Mynjko. Biodegradation of heterochain polymers. Proceedings of the Second International Symposium on Degradation and Stabilisation of Polymers, Dubrovnik, 1978, pp 8–12.
24. AC Albertson, B Ranby. Polyolefines biodegradation. Proceedings of the Third International Symposium on Biodegradation, Kingston, 1975, pp 743–745.
25. AY Polishchuk, GE Zaikov. Multicomponent Transport in Polymer Systems for Controlled Release. Amsterdam: Gordon and Breach, 1997.
26. GE Zaikov. Polymers for Medicine. Commack: Nova Science Publishers, 1998.
27. KZ Gumargalieva, GE Zaikov. Biodegradation of Polymers. Commack: Nova Science Publishers, 1998.

19

Controlled-Lifetime PO Materials

Cornelia Vasile
Romanian Academy, "P. Poni" Institute of Macromolecular Chemistry, Iasi, Romania

I. INTRODUCTION

A controlled degradable polymer is a polymer which degrades at a predictable time.

The scientific development of new polymeric materials was earlier directed at obtaining the most stable plastics possible. Modern technologies demand instead that polymers be easily adapted to today's strict requirements established for the protection and preservation of our environment. The long-term properties of polymers should be predictable: polymers should stable if needed, or there should be a way to control the degradation time of polymeric materials.

The plastics industry may be entering a new phase in which it will be possible to design polymers with a variety of lifetimes by carefully controlling formulation and fabrication. New polymer materials will offer well-organized structures, bio- and environmentally adaptable materials, use of renewable resources to a larger extent than today, thus forming highly specialized materials with multifunctionality as well as adaptability.

A time-controlled polymeric system which would be "environmentally friendly" should present three phases in the transformation. (a) Very limited variation in useful properties corresponding to the induction period or to very slow chemical transformation. This phase is well controlled by the use of efficient stabilizers; the role of additives in creating tailormade plastics and in generating long-term innovative products is commonly accepted. Stabilizers, sensitizers, pigments, lubricants, processing aids, impact modifiers and other additives (see Chaps. 20–22) enable safe and effective processing and guarantee for example, mechanical, thermal or other properties and also a controlled lifetime. (b) The reactions proceed rapidly, loss of physical properties is observed and transformation into small particles results. (c) The solid fragments should be converted not into CO_2 and water but into bio-assimilable low molecular weight compounds. All polymers, synthetic as well as native, are degradable when exposed to the environment. It is only the rate of degradation that differs greatly from polymers that are nearly inert (e.g., PE, PVC) to those that are highly susceptible to the degradation processes. Aging of polymers might be slow, but it is irreversible and it results in the decrease of the essential properties of the polymers.

An approach which is beginning to reduce the litter and waste problem is to promote plastics with accelerate environmental degradation/decomposition under the action of UV light from the sun, heat, oxygen, moisture and micro-organisms.

These directions are summarized in Table 1.

II. LIFETIME DETERMINATION OR PREDICTION—NATURAL AND ARTIFICIAL AGING OF POLYOLEFINS AND THEIR SERVICE LIFE

There is always a natural concern regarding the durability of polymeric materials, because if the useful time of these materials can be predicted their maintenance

TABLE 1 Time-controlled degradation

Material	Initial mechanism of degradation	Control	Degradation products including biodegradable products*
PO	Oxidation initiated by light, heat, transition metals, etc	Antioxidants metal ions inhibitors UV stabilizers photosensitizers	Carboxylic acids
PO/natural polymer (mainly starch)	Natural polymer degradation	Natural polymer content and/or antioxidant, pro-oxidant, degrading additives content and ratio	Hydrocarbons, alcohols, ketones, keto acids, carboxylic acids and sugars

*See Table 7 Chap. 17 and Table 2, this chapter.

and replacement can be planned [1]. Testing in that environment is the best idea.

Service life can be evaluated from the variation in time of the mechanical, rheological, thermal (TG, DSC), and electrical properties, of carbonyl and vinyl groups absorption (FT–IR), of molecular weight (GPC), and so on, under natural aging conditions [2, 3] of exposure; also, it can be predicated from artificial aging experiments [4–12].

Besides structural, compositional, morphological, and dimensional (especially thickness) sample characteristics, behavior during natural aging also depends on geographic location (latitude, altitude, mountain, sea, desert, etc.), solar radiation intensity, season, (thermal cycling), environment (indoor or outdoor; hot, cold, or seawater; soil type, chemicals), rain, humidity, atmospheric contamination (pollutants, smoke, fog), etc.

Severe exposure conditions significantly reduce service life. Thus, the service life of unprotected PE in the dark at 20°C is 8–10 yr, and of stabilized or crosslinked PE in a moderate climate is 15–20 years [7, 13]. In India, Pakistan, or Saudi Arabia, where the sun intensity is greater (250 kW cm^{-2}), the weathering life is only 2–5 yr, even if the products (pipes, tubes) contain much more antioxidants (2%). For other PE products (films), the lifetime in Saudi Arabia is only 3 months, while in Florida it is twice as long [14–16]. Temperature variations in the environment can also be a potential factor for the degradation of the polymeric material, diminishing the lifetimes of the products. These factors are especially important in climates with large differences between summer and winter temperatures (tests for these environmental changes are found in temperature-cycling experiments) During autumn and winter there was almost no oxygen uptake. The differences between accelerated and outdoor weathering are probably due to a change of the mechanisms leading to oxygen uptake. In accelerated aging most of the oxygen is consumed in a propagation step giving all the expected products, whereas in outdoor weathering most of the oxygen is consumed by an initiation reaction caused by a CTC [17] of oxygen and the polymer and yields water, these complexes having higher stability at low temperatures. Decreasing the temperature or increasing the oxygen pressure should lead to more initiation through the CTC. Under these conditions an accelerated weathering test would probably correlate better with outdoor weathering.

In the future, due to increase in pollution, and consequent depletion of the ozone layer and increased UV radiation, the service life of all plastics will probably be continuously reduced.

Immersion in water or in solutions (for cables, storage vessels, batteries, or marine applications) can reduce the lifetime by more than a factor of 3 [18].

Artificial long-term and short-term aging experiments employ more severe conditions than some of those met with natural aging, in a shorter period of time. The obtained results are extrapolated afterwards in order to appreciate the service life. With this aim in view, two methodologies are currently used [19].

1. The kinetic approach considers that, for example 55 elementary steps are involved in the kinetic scheme of PO photo-oxidation. The extrapolated induction period is obtained on the Arrhenius plot

[20]. In this case, the kinetic parameters for the chemical process are evaluated and on this basis the lifetime is also calculated. Due to the change in the kinetic parameter values with temperature region and other experimental conditions, the evaluation of service life in other conditions is not always possible [21, 22].

2. Accelerated aging based on the simulation methodologies in natural and artificial conditions. Simulation of natural exposure conditions is achieved in different ways such as oven aging; long-term immersion in aerated hot water (used especially for pipes); thermal–electrical aging via application of an a.c. voltage of 60–70 kV for 3–13 h (used for cables, etc.). Frequently used are the weatherometer or xenon tests, which approximate natural sunlight (xenon lamp) and work in light–dark cycles, with water spray at different temperatures. Because not all of the influences involved in natural aging can be reproduced, the results of artificial aging must be carefully interpreted [4, 16, 23]. The long-term (differential thermal analysis) (DTA) is more sensitive than other experiments. Figure 1 indicates the possibility of using oven aging coupled with determination of the variation of different properties in order to appreciate the service life of PE and PB-1 pipes. The similar behavior of PB-1 and XPE in these products upon aging has to be mentioned too. The service life depends on the production conditions [24].

This technique cannot be used in the case of hot water pipes, where long-term hydrostatic pressure tests give better results [7b]. Oxygen is not usually regarded as an experimental variable in the study of polymer weatherability since the oxygen concentration in the environment is essentially constant. However, the oxidation

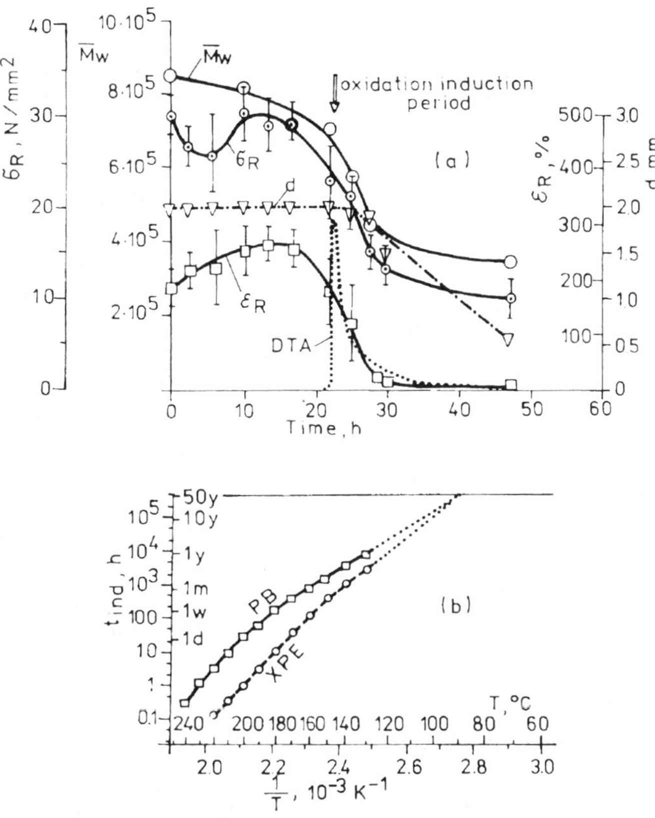

FIG. 1 (a) Effect of oven aging at 200°C on tensile strength (σ_R), elongation at break (ε_R), viscosimetric average molecular weight ($\bar{M}_w$), nonembrittled wall thickness (d), and on long-term DTA curve of a pipe (20 × 20 mm) of PB-1. (b) Logarithms of oxidation induction time vs. reciprocal absolute temperature in the case of oven aging of PB-1 (20 × 2.0 mm) and XPE (20 × 2.5 mm) used in pipe fabrication (d, day; w, week: m, month; y, year). (Adapted from Refs. 5 and 7.)

of materials determines significant changes in all properties.

The oxygen uptake test measures the drop in pressure in a closed system. This test gives errors if gaseous oxidation products are formed. Oxidative induction time is determined according to ASTM procedure D 3895-80 and adapted methods [20]. This test can also be used to assess the antioxidant efficiency.

New methods for evaluation environmental-stress degradation and weathering of plastics have been proposed [25] based on mechanical behavior, acoustic spectroscopy, ultra-acceleration of degradation by plasma irradiation, etc.

Outdoors experiments seems to give the most reliable results. The statistical analysis systems using many models showed that parabolic models have higher reliability, as also appear from Fig. 2.

Hamid and Amin [26] found that the acceleration factor in using an artificial weathering is about three times that of natural weathering (e.g., 5000 h of artificial weathering $\cong$ 14,000 h of natural weathering), while other authors [27] found an acceleration factor of 7.5–8.5.

LDPE films deteriorate after a shorter period of UV aging than HDPE films whereas the films from a blend of two polymers show an intermediate behavior under the same exposure conditions [28]. The lifetime of PB-1 hot water pipes at 68–70°C and 25 N mm^{-2} is about 50 years [24b]. E-CO copolymers degrade rapidly after only 6 weeks [24c].

In severe conditions of exposure, polypropylene (PP) products and films have a lifetime of only 700–1000 h, still decreasing 2–3 times in the presence of metallic impurities [29]. PP fibers are more stable, due to their morphological characteristics.

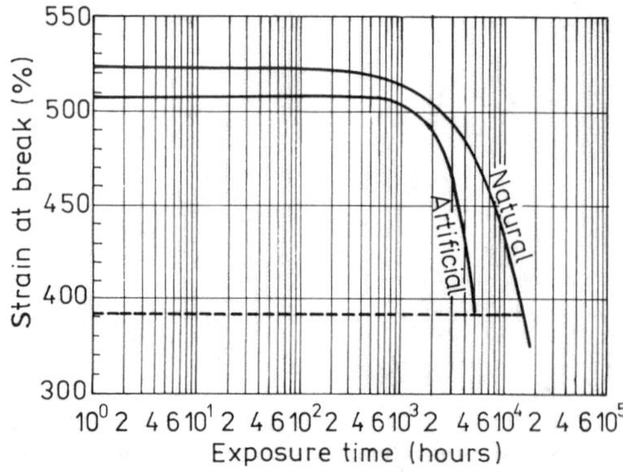

FIG. 2 Change in mechanical properties of PE exposed in natural and artificial environments [26].

Compared to PE, the corresponding ethylene copolymers are much more sensitive to aging processes due to the presence of tertiary carbon atoms along the chain, which leads, as in the case of PP, to increased susceptibility to oxidation.

The following order has been established with respect to stability against aging [30]: HDPE > LDPE > PP > EVA > EPDM.

For all polyolefins, aging resistance is greater after crosslinking [30] and in composites [31–34]. Reprocessed materials have a low aging resistance [35, 36].

The combined actions of two or more agents, particularly electrical, thermal, and mechanical stresses, which are almost unavoidable, lead to an increase by at least 2 times the rate of aging and to a shorter service life [37, 38], especially for cables and other insulating materials.

Knowledge of the mechanisms of degradation, migration and diffusion of additives and/or degradation products in materials allows precise prediction of the useful lifetime of the materials. The degradation of a synthetic polymer or of a synthetic polymer/natural polymer blend is a continuous process and the diffusion and migration of the degradation products are responsible for the interaction of polymers with the environment. The chemical structure of polymers is critically important in determining how degradable the substances are to artificial degradative and/or to biological degradation process.

A correlation between natural and artificial weathering was considered for lifetime prediction in a short exposure time. It was found that the confidence level of predicting time on the basis of artificially accelerated exposure trials is dependent on many parameters which include time, material, equipment, etc.

In products with a large wall thickness, the influence of weathering is often limited to a surface layer, either due to limited oxygen diffusion or to limited UV penetration. The depth of this layer may be small (~0.5 mm) compared with the whole wall thickness (~4 mm) but it can causes brittle fracture. Mechanical behavior depends on the oxidation (degradation) profile, a critical degradation profile accounted for the failure of samples, that is time of failure. Tensile tests were performed on films microtomed from the samples [39]. The density of weathered samples can increase because of chemicrystallization, increase in polar groups, oxygen uptake or the loss in volatile products. The decrease in the nominal strain corresponded to an increase in the vinyl index and in the density. The carbonyl index showed too much scatter.

The deficiencies in artificial weathering that led to poor or inconsistent correlation between such tests and outdoor exposures are due to the fact that the simulation of dynamic factors of natural weather (combination of radiant energy, thermal energy and humidity) is not possible in an artificial device [40].

Gijsman *et al.* [41] did not find any correspondance between outdoor weathering and an accelerated test of PE aging. The oxygen uptake necessary to give the same drop of the elongation at break and to form the same amount of carbonyl groups and unsaturation in outdoor weathering was twice that in the accelerated weathering. A combination of many techniques led to new insights and reduced the errors.

In general, outdoor aging is too slow to be useful in the development of stabilizer formulation or for quality control. FT–IR studies [42] suggested that the lack of correlation may be due to the differences in degradation mechanisms in accelerated and outdoor testing.

Besides these considerations we suppose that the acceleration factor for accelerated weathering to translate them to natural weathering is not yet very well evaluated, especially for each period of natural weathering. Also, only temperature was considered as a factor, not pressure, humidity, etc.

III. DEGRADABLE POLYMERS

Degradation is a change in chemical structure involving a deleterious change in properties. Deterioration is a permanent change in the physical properties of a plastic. Various kinds of degradation can take place in PO materials. Degradation can be promoted by increasing the effects of these environments and is achieved in several ways, creating the following materials [43, 44]:

(a) Photodegradable polymer is a polymer in which degradation results from the action of natural daylight; such as copolymers containing functional groups which are photochemically active—E/CO copolymers; vinyl ketones.
(b) Enhanced photo-oxidability obtained by adding photosensitizers or photodegradants that are additives which promote environmental degradation (organosoluble metal ion carboxylate or sulfur complexed transition metals such as Fe III dithiocarbamate).
(c) Biodegradable polymer is a polymer in which degradation is mediated at least partially by the biological system (such as only starch or starch $+Fe^{3+}$ or Mn^{2+}); Some plastics can be made so that they biodegrade at a predetermined rate in living systems. Biodegradable materials are chemically transformed by the action of biological enzymes or micro-organisms into products which themselves are capable of further biodegradation.
(d) Bioabsorbable polymer is a polymer that can be assimilated by a biological system.
(e) Polymer fragmentation: the polymer molecule is broken up or segmented into lower molecular weight units such as auto-oxidizable polymers, which then biodegrade.
(f) Polymer erosion means dissolution or wearing from a surface.
(g) Hydrolytically degradable plastics, less frequently met in PO materials.
(h) New derivatives of natural polymers and water soluble polymers.

The most convenient route is to use cheap synthetic polymer and add a biodegradable or photo-oxidizable component. A more expensive solution is to change the chemical structure by introducing hydrolyzable or oxidizable groups in the chain of a synthetic polymer.

A. Photodegradable PO

As has already been shown, polyolefins are subjected to photodegradation upon exposure to sunlight. Additives and impurities catalyze the breaking of the polymer chain by a series of UV-initiated free radical reactions. Some efforts focus on the addition of photosensitive species, such as carbonyl groups or metal complexes, to accelerate these processes.

The photodegradability of polymers can be induced by introducing ketone groups into the polymer chains as a result of copolymerization with monomers as CO (2%) (E-CO copolymers marketed by DuPont, Union Carbide and Dow Chemical) or vinyl ketones (Guillet copolymers marketed by Eco-Plastics Ltd., trade name Ecolyte). Incorporation of 2–5% ketone comonomers in PE gives a master batch (MB) that is blended with natural resins in a ratio of 1/9 or 1/20 and then transformed into products. In the presence of UV radiation, these products degrade by Norish type I or type II reactions. The rate of photodegradation can be controlled by adjusting the percentage of CO in the copolymer.

There are still unanswered questions regarding the extent to which it degrades into nonpolymeric products rather than just disintegrating into smaller particles of PE [44].

The lifetime can be programmed to be 60–600 d of outdoor exposure. There are uses for garbage, grocery bags, mulch film, food packaging, etc.

Photosensitized or photodegraded PO films have an extensive photochemical damage that failed to promote subsequent mineralization in soil [45, 46].

Light-catalyzed polymer modifications increase susceptibility to fragmentation of plastics by exposure to light, but significant degradation will not continue when modified plastics are buried in soil or compost. Environmental conditions in compost and soil can prevent the activation cycling of transition metal catalysts, reducing the capacity of modified plastics to degrade.

B. Enhanced Photo-Oxidability: Additives to Increase the Environmental Degradation

PO can be made degradable by means of additives. The types of additives include: aromatic ketones (benzophenone and substituted benzophenones [47], quinone), aromatic amines (trisphenylamine), polycyclic aromatic hydrocarbons (anthracene, certain dyes such as xanthene dyes), or transition metal organic compounds. The transition metal compounds of Fe, Co, Ni, Cr, Mn are widely used. Organo-soluble acetyl acetonates of many transition metals are photo-oxidants and transition metal carboxylates are also thermal pro-oxidants. Co acetylacetonate appears to be an effective catalyst for chemical degradation of PP in the marine environment. The preferred photoactivator system is ferric dibutyldithiocarbamate with a concentration range of 0.01–0.1%. Scott has patented the use of organometallic compounds like iron (ferric) dibutyldithiocarbamate or Ni-dibutyldithiocarbamate [48]. Cerium carboxylate [49] and carbon black are also used in such materials [50].

Other additives which facilitate degradation are vegetable oils (unsaturated lipids), and low molecular weight compounds, because they are inherently more accessible to chemical reactions. Eventually, when the molecular weight of the polymer is reduced to about 5000, biodegradation occurs. One reservation regarding the use of these additives is the potential toxic effects of heavy metal residues.

Plastigone, developed by Oilead of Plastopil Hazorea, Israel, contains a minimum of two metal (especially iron) complex additives known as antioxidant-photoactivators, for example an iron complex used together with at least one Ni or Co complex to act as a metal ion deactivating compound and also as a photosensitizer. The effect of such additives is exemplified in Fig. 3.

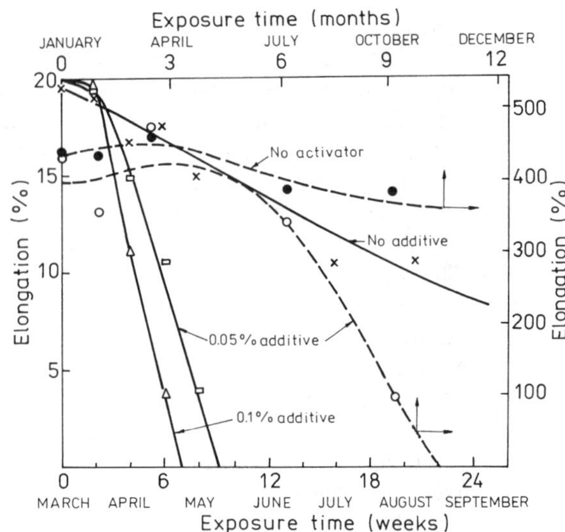

FIG. 3 Elongation at break vs. exposure time of PP tapes (———) in South Africa and LDPE (- - - - -) blow film in machine direction (60 μm) in England [48].

At high concentration these chemicals act as stabilizers to thermal oxidation, but they are consumed in the stabilizing process and their concentration is lowered. They then become effective as catalysts of photo-oxidative degradation. Thus they act as delayed action photosensitizers, giving a built-in induction period [51].

Catalysts used in commercial PO products include complexes of unsaturated fatty acids and stearate salts of a transition metal such as iron or manganese. In the presence of light or at elevated temperatures transition metal fatty acid complexes generate free radicals that carry out chain scission of PO.

Many such PO products contain a proprietary photosensitization additive.

Regeneration of the reduced metal ion by redox reaction during catalysis is essential to continue PO degradation. Fe^{3+} is the thermodynamically favored oxidation state for iron under aerobic and alkaline conditions, whereas Fe^{2+} is favored under anaerobic and acidic conditions. Under most disposal and environmental conditions Mn^{2+} is favored. It is therefore unlikely that a particular metal catalyst will perform equally well in a wide range of disposal situations. Due to environmental pH values it is difficult for such catalysts to be recycled for further free radical generation. Both Fe^{3+} and Fe^{2+} salts can precipitate as insoluble oxides or sulfides under environmental conditions, reducing the polymer degradation potential.

Therefore photodegradation is promoted by irradiation but this is restricted to the surface of the material.

It rarely leads to complete removal of the waste, but the chemically modified fragments produced may biodegrade. The films degrade rapidly when sunlight exposed but remain unchanged when they are buried in compost.

C. PO/Natural Polymer Blends

A solution to the accumulation of polymer waste in the environment is to develop biodegradable polymers which is a blending of the synthetic polymers with natural ones. Starches, wood flour, cellulose, some kinds of lignin, waxes, casein, oils, etc., have all been used. This is an economical and easy way to make an environmentally acceptable material.

All materials of organic origin, including natural polymers have an inherent tendency to decomposition. For many decades this property of the natural polymers has been regarded as a major disadvantage, especially if they were used as a material for products of which long life is required. The efforts were directed towards developing materials of possible infinite durability or at least improving durability of the existing materials. Now, the industry supplies daily many tons of indestructible packaging and other products which have a negative effect on the environment. So, scientists started searching for compromising solutions to reduce the durability of the products to obtain a controlled lifetime. An ideal material would be one that would undergo destruction and absorption by the environment after a certain pre-determined time.

Consequently, the role of the natural polymer was reassigned, whether modified or used in a mixture with synthetic polymers. The properties of such materials are predestined so to make them absorbable after a predictable time by the natural environment without harm to the latter. It is expected that the additive will degrade in the environment and thus lead to a decrease in solid waste pollution.

If the specimen contains biodegradable fragments at the surface once these fragments are removed, the cohesion of material is reduced, the ratio of its surface area/volume is increased, and the material becomes more open to water permeation, micro-organism attack being possible in a deeper and deeper layer. The water accessibility and photo-oxidation of materials favor degradation starting with the hydrophilic component. Diffusion, migration, and leakage of additives and/or degradation products increase the degradation rate and leave the polymer matrix more brittle, the material being much more prone to future disintegration.

In the mixtures, the synthetic polymer retards microbiological attack due to its hydrophobic nature which slow down the enzymatic activity. This effect is maximum with the polymers whose main chain is built of carbon atoms alone, such as PO. The natural component has a speeding-up effect of biodeterioration.

Therefore the incorporation of natural polymers in long-term-use products (building or electric use materials) must be carefully planned. In these application fields free-sugar polymers (such as lignins) have to be used because they favor microorganism growth and change the superficial properties of the products.

The natural polymers derived from various vegetable and natural tissues are highly diversified either in composition or behavior. In its turn, biodegradation is promoted by enzymes and may be aerobic or anaerobic. Operating in all environments: burial in soil, surface exposure, water immersion, *in vivo*, etc., enzymatic attack may lead to the complete removal of the products from the environment.

The main problem, associated with the use of natural polymers in polymer blends (as fillers), is their hydrophilic nature to repulse the generally hydrophobic polymer matrix. Because hydrophilic natural polymers and hydrophobic PO are immiscible at a molecular level, simple mixing produced blends that tend to phase separation. Compatibility between them can be improved by introducing reactive functional groups on each polymer. The functional group is expected to react during blending. Chemical modification is a promising technique for overcoming lignin's frequently observed adverse effects on mechanical properties of solid polymers and on viscosity or cure rate of resins systems—see Table 5.

D. PO/Starch Blends

Currently the most commonly used biopolymer for incorporation into PE is starch.

As corn starch is a native agricultural product, inexpensive (about 10 cents lb^{-1}) and available annually in multimillion ton quantities it could be considered a replacement for petroleum-based plastics as starch–graft copolymers, starch–plastic composites, and starch itself. [52].

PE/starch films were formulated in US by Otey *et al.* [53]. They developed a process for extrusion, compounding and blowing of starch as a thermoplastic film at 5–10% moisture. The concept of using corn starch as filler to accelerate the degradation process was developed by Griffin in the UK in 1973 [54]. Griffin found that the degradation of starch/LDPE film in compost is accelerated by the absorption of unsaturated lipids. Peroxides were generated and consequently auto-oxidation was enhanced. The Coloroll Company offered in Europe shopping bags using this blend of starch/dotriacontane ($C_{32}H_{66}$), as the biodegradable additive has a positive effect on the degradation factors.

Cole *et al.* [55] investigated the degradability of corn starch based PE and modeled the role of biodegradation by a (scalar) percolation theory. Based on these researches Archer Daniels-Midland (ADM) Company developed a technology to make starch-based plastics. These materials were proposed for applications such as agricultural mulch and disposable packaging.

1. Preparation of the Blends

Extrusion is the preferred method for compounding gelatinized starch plastic composite materials, although batch mixing is also effective. The processing temperature is below 230°C, that is the decomposition temperature of the starch. The high shear destroyed the starch granules in minimal time.

The open roller/press method is frequently employed for laboratory scale studies in the preparation of natural polymer/synthetic polymer blends.

To increase the percentage of natural polymer incorporated, compatibilizing agents such as silanes, EVA, maleic anhydride copolymers, dispersing agents as petroleum resins or partial crosslinking were tested with good results.

2. Biodegradation Behavior of Starch/PO Composites

The biodegradation rate of such composites depends on starch content, temperature, moisture, pH, micro-organism type, presence of transition metal salts, film thickness, surface area, and processing procedure.

Addition of corn starch to LDPE has a stabilizing effect on the thermo-oxidation mode while the same additive destabilizes the LDPE to UV irradiation. This property would diminish the risk of degradation during the processing of such materials.

Amylolytic Arthrobacter bacteria have the ability to utilize starch as the sole carbon source in liquid culture media. One of a group of bacteria designed as LD76 degraded up to 80% of starch in starch–PE–ethylene acrylic acid (EAA) plastic films in 60 d in a liquid culture media, but PE and EAA components remained largely nondegraded [56]. *Lactobacillus amylovaras bacterium*, isolated from corn waste fermentation, secreted amylase that rapidly degraded starch granules. The concentration of starch in plastics influences its availability in microbial attack. Almost 40% by weight (30% by volume) is required in formulation to assure interconnectivity of starch domains. At very low concentrations of starch only surface starch would be accessible to direct attack by micro-organisms.

The starch removal from starch–plastic composites by amylose is evidenced in the FT–IR spectrum of starch–EAA–PE treated with Bacillus sp XI amylase for 72 h at room temperature. The carbohydrate -OH and C-O absorbance bands are drastically reduced—Fig. 4.

Pure starch degrades very rapidly in compost after 49 d [57]. Soil and compost tests showed that in PO/starch blends, starch alone is the biodegradable fraction. The PE/starch blends degrade faster by light than by compost conditions (elevated temperature of 50–60°C, micro-organisms). In soil, molecular weight is reduced from 225,000 to 110,000 but CO_2 evolution measurements indicated that only the starch component is degraded.

Removal of starch presumably leaves a porous residue of components that are not normally biodegradable. The plastic residue with increased surface area relatively to solid plastics may speed its degradation by environmental factors. Residual plastics in "biodegradable" starch/PE composites is degraded by auto-oxidizable chemicals added to the formulations. Flexible film bags containing 6% starch degrade

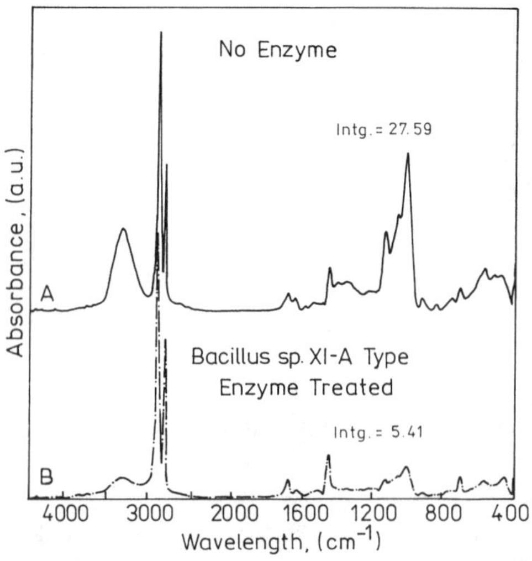

FIG. 4 Removal of starch from starch-EAA-PE-films by Bacillus sp XI amylose [52].

after 3–6 yr. For rigid materials 10–15% starch is recommended.

Patents and applications assigned to Fertec in Italy describe starch plastic composites that are reputed to be totally degradable [58].

The major pathways for degradation in seawater and aquatic media (lakes, rivers, streams, etc.) include: (a) attack by fungi and bacteria, attack by invertibrates; mechanical degradation resulting from disassociation of starch of film matrix, (b) photodegradation, chemical degradation of matrix [59]. The peroxides begin to degrade the polymer chain by an auto-oxidation mechanism in the presence of oxygen in seawater [60]. The second mechanism is tremendously enhanced by the increase in surface area provided by the first mechanism.

PO, mainly PP/starch blends, were developed for marine applications, especially to fabricate fasteners or ties. This was a must because of the presence of significant amounts of thermoplastic debris in the world's oceans that poses a hazard to marine life, including several protected species. For fishing gear or lobster pots and traps remaining under water, possible degradation routes are hydrolysis, chemical degradation, and slow biodegradation. Photodegradation is least likely to occur due to the absence of UV light at depth. Therefore, the use of commercially available enhanced photodegradable PE would not be suitable for such applications.

Starch tends to absorb water. In the absence of any microbial activity starch does not seems to play any role in bringing about the disintegration of the samples. In the presence of high microbial activity like in soft mud or sludge, it plays an important role. The extent of microbial activity depends on location. In the absence of additives, unstabilized PP did not show any significant chemical degradation in the marine environment.

The long term properties of the starch-filled LDPE have been monitored, showing that an initial thermolysis or photolysis is necessary to decrease the induction time of degradation. Starch/PO composites are about 10 times more susceptible to a combination of thermal oxidation and biodegradation than the corresponding pure LDPE [61]. Outdoor weathering conditions are a multiaction mode of degradation where photo-oxidation due to the sun's radiant energy is very important; if instead composting is carried out, thermal oxidation may be the dominant mode of degradation, since temperatures up to 70°C are reached during some periods over the lifetime of a compost.

3. Complex Systems

Complete degradation of blends depends on a combination of oxidative and biological processes to break down the polymer fragments. ECOSTAR is a product obtained from regular corn starch treated with a silane coupling agent to make it compatible with hydrophobic polymer, and dried and mixed with an unsaturated fat or fatty acid auto-oxidant to form a master batch that is added to PE. The auto-oxidant interacts with transition metal complexes present in soil or water to produce peroxides which attack the synthetic polymer. For 6% starch in Ecostar the degradation time is 3–6 yr, for 15% starch it is 6–12 months when buried.

To increase the starch load up to 40–60% in PO a gelatinized starch is used in the films of poly(ethylene-co-acrylic acid) (EAA) or a mixture of EAA/LDPE or LDPE/EMA (ethylene–maleic anhydride). The difficulty with this system is that the high level of filler seriously impairs the mechanical properties, especially in thin films, and the starch is subjected to moisture-absorption problems. LDPE–Mn stearate/starch showed a decrease of MW during thermo-oxidation and an increase of MW during UV irradiation. In the mixtures, it has been established that ethylene–acrylic acid copolymer accelerates LDPE oxidative degradation while plasticized starch inhibits it [62].

To increase the compatibility of starch with PE, maleic anhydride was grafted onto LLDPE [63] or PE/starch/photoactivator coated gelatin that accelerates the photodegradation rate. Gelatin was eliminated by biodegradation [64], while starch was modified by introduction of cholesterol units [65]. The functional groups introduced onto the polymer chain can react with the hydroxyl group of starch or the carboxyl group of modified starch.

Both weight loss and FTIR data indicated that in a river more than 80% of the starch depletion was dramatically reduced with increased EAA content of the plastic—Fig. 5.

Shoren *et al.* [66] suggest that EAA somehow interferes with starch accessibility to micro-organisms and their hydrolytic enzymes. Amylose in starch forms V-type exclusion complexes with EAA that make it highly resistant to starch-hydrolyzing enzymes [66, 67].

This might also explain why some of the starch in plastic films always remains conserved even after prolonged exposure to highly biodegrading conditions. Also, in this regard, PE is implicated in encapsulating starch under certain processing conditions, forming

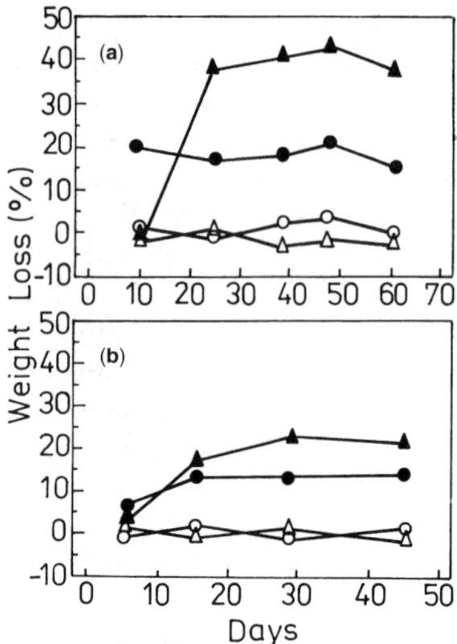

FIG. 5 Weight loss in film exposed in river (a) and laboratory amylolytic bacteria LD76 culture (b) with starch-LDPE-EAA (▲), Starch EAA (●), LDPE (△), EAA (○) [67].

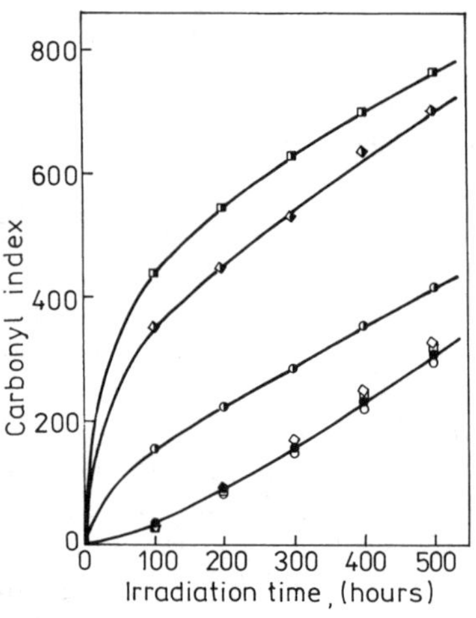

FIG. 6 The carbonyl index as a function of irradiation time: (●) pure LDPE; (○) LDPE 3.9% starch; (◇); 8% starch; (◑) 10% MB; (◇) 15% MB and (◧) 20% MB [73].

an impremeable barrier to hydrolytic enzyme [68, 69]. Eastman Kodak Co. (UK) has developed, under the Byoplastic tradename, a photo- and biodegradable starch–plastic composite [68]. The slower degradation rate in water is explained by a deactivation or leaching out of the pro-oxidant during aging. LDPE/starch did not degrade during 11 weeks at 80°C.

According to the actual trends in additive use in polymers, a system composed of starch and a pro-oxidant such as manganese stearate with a styrene–butadiene copolymer as compatibilizing agent was developed [70, 73].

9–15% modified starch/5% aluminum or manganese stearate/5%Co acetyl acetonate is a master batch (MB) used by many authors [70].

Bioplatese is manufactured by Manzinger Papierwerke GmbH, Munich from LDPE/carbohydrate/fatty acids. Fatty acids react with the salts in soil to release peroxides which destroys the enzyme-terminated molecular chain.

To disintegrate PP/starch blends within 6–9 months in a marine environment it is desirable to bring about chemical degradation of PP by using certain metal catalysts like aluminum stearate/Co(III) acetylacetonate (within the sample or in seawater) and auto-oxidants like corn oil or fatty acids, in addition to the possible biodegradation of the starch filler [71].

Pretreatment of the blends, either thermal (50°C) [72] or photochemical [46, 73], increases the environmental degradation rate—Fig. 6.

In environmental degradation of PE, however, the main degradation agents seem to be UV light and/or oxidizing agents.

Environmental degradation is a complex process in which both abiotic and biotic factors degrade polymers. In the abiotic environment, a continuous increase in the amount of carbonyl compounds with exposure time can be seen, contrary to the observed decrease in the biotically aged samples.

During abiotic and biotic degradation, series of low molecular compounds are released into the surrounding environment. In addition the remaining polymer will be changed in regard to its crystallinity, lamella thickness and overall morphology.

Carboxylic acids as a major product category degrade according to the β oxidation of fatty acids, removing two fragments. The process requires activation by adenosine triphosphate and coenzyme A. The acetyl coenzyme A enters the citric acid cycle and is cycled and used in either the catabolic or the anabolic cycle. Apart from the absence of low molecular weight carboxylic acids in the biotic environment, the lactone and ketone detected during abiotic degradation seem to have been assimilated by the biotic environment. So GC–MS "fingerprinting" of the degradation products is useful to differentiate between abiotic and biotic degradation of various PO materials [74b, 74c].

Mono- and dicarboxylic acids were the major products identified in both environments. Ketoacids were formed in both air and water but hydrocarbons and ketones were only identified after aging in air. Blends without pro-oxidants did not form any degradation products. By a liquid phase (diethyl ether) and a solid phase (nonpolar dimethylsiloxane and polar carbowax fibers) extraction followed by GC–MS with derivatization of products (methanize dicarboxylic acids) the following products were identified—Table 2.

The degradation products formed depend on the type of polymer, degradation mechanism(s), and also the type of additive present in the material [75].

Starch-filled iron catalyzed PE produces more carboxyl, resulting in more rapid microbial growth. After removing the micro-organisms the surface of the oxidized PE was found to be eroded with substantial reduction in sample thickness, while the molar mass in the polymer remained unchanged.

It is shown that the microbial ezo enzymes are able to recognize relatively high molar mass carboxylic acids and remove them from the surface of the polymer under conditions where water is not able to remove them by physical leaching. Bioerosion of the fragmented polymer is evidenced by reduction of the sample thickness.

The oxidation products of oxidized PE are unlikely to present a threat to the environment and by conversion to biomass they contribute to the fertility of the soil. Abiotic iron catalyzed photo- and thermo-oxidative degradation is the rate limiting step in the bioassimilation process.

No water-soluble toxic compounds were produced during the first days of degradation but after a longer period, toxicity related to small oxidative compounds as formaldehyde and acetaldehyde was observed [76].

The prediction of the environmental interaction of polymers need knowledge not only of the toxicity of the degradation products but also of the degree of exposure of the living cells to the degradation products.

Oxygen is usually required for all degradation and disregarding some important exception (e.g. anaerobic putrefaction) oxygen is required in most biodegradation processes. Photo-oxidation increases the amount of low molecular weight material by breaking bonds. The surface area is increased through embrittlement and erosion, improving the possibility of further degradation [77a].

Extracellular bacterial peroxidases and H_2O_2 oxygenases have been evidenced in the biotransformation studies and this process may lead to further carboxylation of the polymer matrix on exposure to soil micro-organisms. Most of the studies in the biodegradation of polymers have primarily been concerned with the

TABLE 2 Identified degradation products formed in LDPE /Master batch (starch, a pro-oxidant as manganese stearate with styrene–butadiene copolymer as compatibilizing agent), Starch-EMA and LDPE during different aging times in water, air and thermal aging [72, 74, 75]

Water

Hydrocarbons: Butane to higher alkanes as *n*-octane, *n*-nonane, *n*-decane, *n*-dodecane, *n*-tridecane, *n*-tetradecane, hexadecane heptadecane, octadecane, nonadecane, eicosane, heneicosane, docosane, tricosane, tetracosane, pentacosane, hexacosane, heptacosane, octacosane, nonacosane; alkenes

Alcohols: ethanol, butanol, 1-pentanol, isopropyl alcohol, l-docosanol

Carboxylic acids: formic acid, acetic acid, propionic acid, butyric acid, valeric acid, caproic acid, butanoic acid, pentanoic acid, hexanoic acid, heptanoic acid, octanoic acid, nonanoic acid, decanoic acid, undecanoic acid, dodecanoic acid, tridecanoic acid, tetradecanoic acid, pentadecanoic acid, hexadecanoic acid, heptadecanoic acid, octadecanoic acid, benzoic acid, lactic acid

Dicarboxylic acids: butanedioic acid, pentanedioic acid, hexanedioic acid, heptanedioic acid, octanedioic acid, nonanedioic acid, decanedioic acid, undecanedioic acid, dodecanedioic acid, tridecanedioic acid, tetradecanedioic acid, pentadecanedioic acid, hexadecanedioic acid

Ketoacids: 4-oxopentanoic acid, 5-oxohexanoic acid, 6-oxoheptanoic acid, 7-oxooctanoic acid, 8-oxononanoic acid, 9-oxodecanoic acid, 10-oxoundecanoic acid, 11-oxododecanoic acid, 12-oxotridecanoic acid, 2-oxopentanedecanoic acid

Aldehydes

Ketones: 2-decanone, 2-undecanone, 2-dodecanone, 2-tridecanone, 2-tetradecanone, 2-pentadecanone, 2-hexadecanone, 2-heptadecanone, 2-octadecanone, 2-nonadecanone, 2-eicosanone, 2-heneicosanone, 2-docosanone, 2-tricosanone, furanone, 2-furancarboxaldehyde, other furanones

Lactones: butirolactone

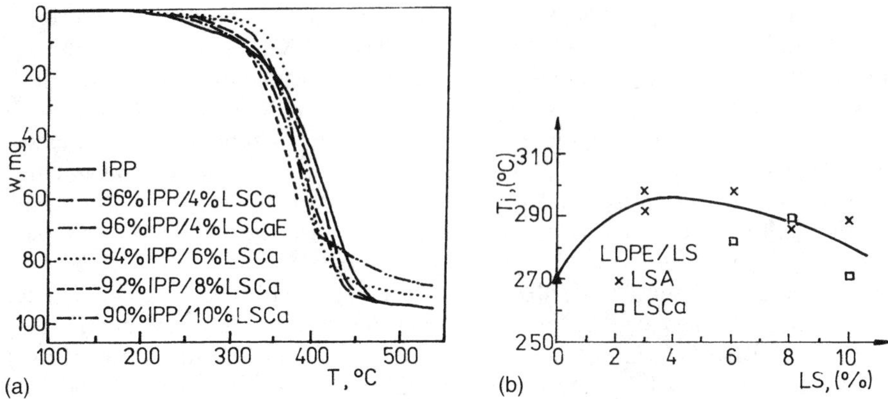

FIG. 7 (a) TG curves of various IPP/calcium (Ca) or ammonium (A) lignosulfonates LS blends; (b) Onset decomposition temperatures vs. LS content [78].

rate of mineralization of the polymer to carbon dioxide and water. However, this is not the most likely or even the most desirable outcome of the environmental polymer waste. As in the case of nature's lignocellulosic litter, the formation of humic materials in the soil, either by composting or by natural assimilation of dead microbial material, provides value-added products for agriculture. The starch appears to play no part in the biodegradation of the PE. Biomass appear to be the major product formed by bioassimilation of PE oxidation products [77b].

E. PO/Lignin Blends

The composite material, wood, consists of approximately 70% fiber component and 30% lignin. The lignin noncarbohydrate, nonfiber polymer matrix component of natural (woody) plants is a phenolic polymer matrix in which cellulose fibers are embedded. It is obviously "designed" as a high impact strength, thermally resistant, thermoset polymer which performs best in combination with highly crystalline cellulosic fibers. Therefore lignin can be a good component in engineering plastics.

The lignin role in PO blends depends on the lignin nature, provenance and content. Due to its phenolpropanoidic structure similar to that of hindered phenols it can act as a stabilizer [78] or initiator of PO degradation [79].

The onset decomposition temperature of PO is shifted to higher temperatures for the blends containing 2–4% lignosulphonates (LS), and the weight loss rate is slower (Fig. 7), the mixtures having a higher thermal stability.

Incorporation of low amounts (3–8%) of lignosulfonates into PO leads to a significant increase of the physicomechanical indices so the tensile strength is two times higher and elongation at break 6 times higher with respect to those of LDPE or IPP—Table 3.

Moreover the LDPE/LSNH$_4$E mixture crosslinked with DDM is very stable to fatigue during mechanical stress, resisting under load for 12 consecutive cycles before breaking.

Glass transition temperatures of the mixtures were higher with respect to those of LDPE. After 6 months

TABLE 3 Physicomechanical indices of PO/lignosulfonates blends [78, 110]

No.	Sample	Tensile Strength (MPa)	Elongation at break (%)
1	LDPE	13	120.5
2	LDPE/3%LSNH$_4$E	26.4	631
3	LDPE/5%LSNH$_4$ Oxidized	11.8	171
4	LDPE/3%LSNH$_4$E/1.5%DDM	24.1	599
5	IPP	15.3	85
6	IPP/2%LSNH$_4$E/1.5% PA	19.1	98
7	IPP/4% LSNH$_4$E/1.5% PA	20.7	636

PA: phthalic anhydride, for other abbreviations, see list at the end of chapter.

degradation by soil burial, all values decrease, probably due to the morphological modification, phase separation, and crystallinity changes due to the apparition of the degradation products.

Lignosulfonates are very hydrophilic polymers absorbing a high quantity of water. Also by its structure the lignin has many chromophore groups able to absorb light and it is very sensitive to oxidation. These are requirements for creating potentially environmental degradable materials, though lignin is recognized by its resistance to biodegradation.

There is no mechanism capable of breaking carbon–carbon bonds, which explains the fact that PO are fairly inert to biodegradation. Nature uses however peroxidases for degradation of the complex lignin structure and these enzymes could also degrade the PO molecules [80].

By combination of the chemical procedures of component modification with the physical ones of surface treatment (photo-oxidation plasma and/or electron bombardment of the surface) [46, 81a] the bio/environmental disintegrable polyolefin material can be obtained by increasing LSR content.

It has also used a low molecular weight lignin chemically modified by an epoxidation reaction, (LER) which increases the functionality and water solubility, PP-g-maleic anhydride as compatibilizer and diaminodiphenylmethane as crosslinking agent. All of these characteristics may be requirements for an increased biodegradability/disintegration rate. Glycidyl methacrylate-grafted PP improved the compatibility of IPP with epoxy modified lignin and it allowed one to incorporate up to 15% LER in IPP under the same conditions of the blend processing as films [81a].

In the case of the commercial PE/starch blends it has been established that the films were much affected by aqueous medium, while PO/lignin blend films undergo various molecular, macromolecular, and morphological changes, especially in soil [81].

For long term biodegradability tests, the films buried in soil cultivated with *Vicia X. Hybrida hort* in greenhouse conditions changed their properties after a period of 7 months, these conditions being aggressive as laboratory-accelerated conditions due to the complex phenomena determined by the action of rhizospheres [81b].

IPP became less stable, while DTG curves of the LS containing films are shifted to high temperatures, probably due to the change of $LSNH_4$ to LSMe with high stability due to the complex formation in the aged films as is demonstrated by Mössbauer spectra of iron lignosulfonates [81c]. The behavior is similar to that of natural porphyrin [82].

It is well known that the nature of the counterion governs the mechanism of thermal decomposition of lignosulfonates the lignosulfonic acid and aminonium lignosulfonates being less stable. However, several ions, such as Cu^{2+}, intensify the oxidative processes, or it was clearly demonstrated that in all stages of lignin biodegradation the key process is oxidative [82]. Enzymes bound the active oxygen species by an electron transport system. The current model considers that oxo-iron or manganese porphyrin cation radical species are effective for the C–C bond cleavage of lignin compounds.

In order to improve the environmental degradability, the photo-oxidation of the film surface was performed, so we can control the initiation step. After UV exposure the films changed their color from transparent light green-yellow to dark yellow. Superficial and deep cracks appear starting with LER particles and the changes in all components of the free surface energy are characteristic (Fig. 8b) for the unstable surfaces in respect to photo-oxidation. The components of the blends have a pronounced oxidation degree both of the chain ends and of aromatic rings placed at the surface. The hydroxyl, carbonyl and carboxyl groups (from peracids) are formed. LER acts as a chromophore for IPP, because UV absorbance increases with exposure time. The weight loss increased (Fig. 8a) and elongation at the break decreased.

Therefore, as a result of surface photo-oxidation the polarity is increased, chromophore groups appeared, and both components have a certain oxidation degree and as a consequence they would be much sensitive to the environmental attack, because the protective PO coating is partially degraded. It can be remarked, increased opacity of the film, the decrease of the film thickness, and for several films the weight loss reached 20%, characteristic for environmentally degradable materials and really they are partially or totally disintegrated and could be assimilated as humic substances.

Due to the ablation by erosion of the superficial layer, the values of the polar components of the free surface energy are smaller than that of photo-oxidized films.

Other proofs on the release in soil of degradation products have been obtained by means of the soil burial test, both by the observation of the plant growth and its composition. Only photo-oxidized samples influence the dynamics of growth, the vegetative mass increased, and blooming was a little affected being retarded; chlorophyll and carotenoidic pigment contents decreased (Table 4).

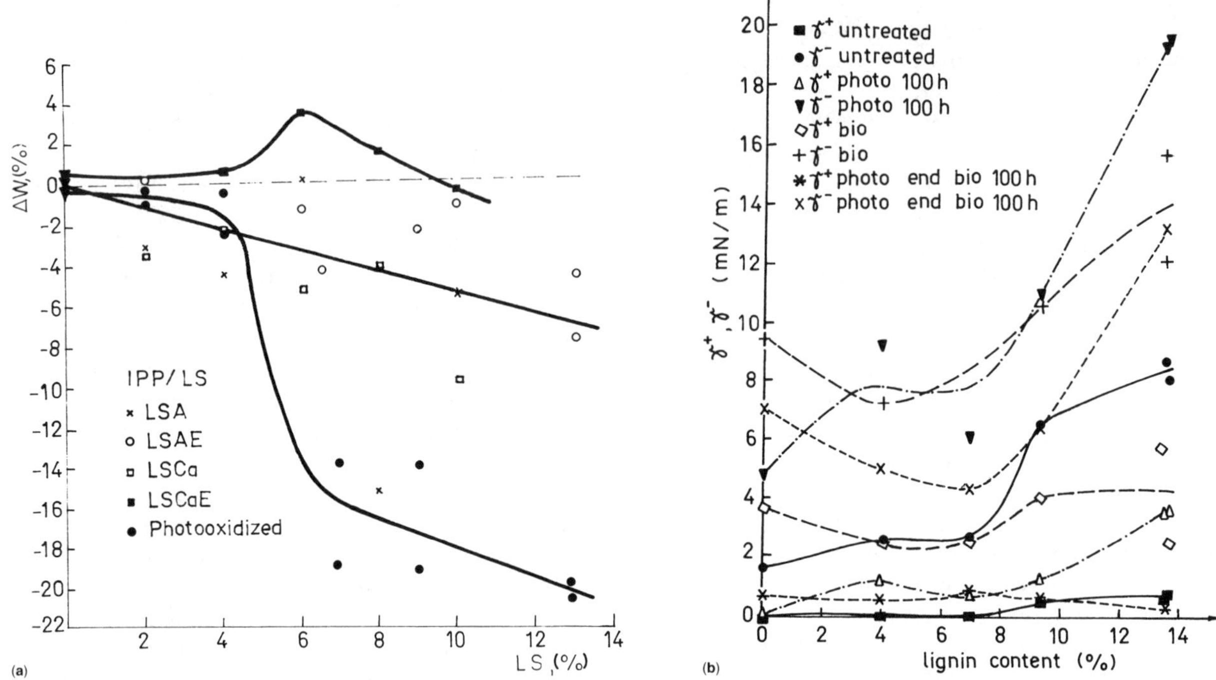

FIG. 8 The variation of the weight loss (a) and of the components of surface free energy (b) with lignosulfonates content [81].

TABLE 4 Mineral content of the *Vicia Hybrida Hort* grown in the presence of IPP/LER bends buried in soil [81]

		Macroelements (mg/100 g dried mass)			Microelements (mg 100 g dried mass)			
Sample	Mineral content (%)	Na	Ca	Mg	Fe	Mn	Cu	Zn
Reference plant (without polymers buried in soil)	10.4	632	278	86	0.3	0.4	0.4	1.7
IPP/9%LER	14.7	1072	675	299	5.5	2.5	0.8	4.5
IPP/13%LER	11.2	1019	527	185	0.4	1.3	0.6	2.3
IPP/9%LER photooxided	8.6	1068	445	101	1.1	1.5	1.2	2.4
IPP/13%LER photo-oxidized	9.6	725	348	114	2.2	1.4	0.8	2.2

The manganese, cooper and iron amounts have been increased with increased LER content. That means that for a short period of time, when the quantity of the released products is higher than 20% the physiological state of plant is perturbed.

From lignin, a great variety of compounds are evolved, including aldehydes, alcohols, carboxylic acids. From polyolefins, up to hundreds of degradation products have been identified, the main ones being esters in biotic media, and carboxylic acids in abiotic media. All these compounds decrease the pH of the environment; some are water soluble and could be bioassimilated, leading to the change in anabolism, the enzymatic activity being perturbed.

In PP/organosolv lignin or oxidized lignin with up to 2–3 wt% lignin an increase in photostability was evidenced (radical scavenger) [79] while at higher lignin content it initiates radical reactions [79b]. Lignin could be used to replace some phenolic antioxidants in food applications. Other effects of lignin incorporation in PO are: increase in conductivity, hydrophilicity, improved printability imparts biodegradability characteristics [79c]. Biodegradation is observed even at 4 wt% prehydrolysis lignin in PP if the Phanerochaete Chrysosporium specialized micro-organism is used. The growth of fungus formed a weblike network over the exterior surface.

Lignin copolymers are also degraded by white rot fungi [83]. It has been assumed that micro-organisms capable of lignin degradation may also be able to attack plastics [84], according to the following explanation.

Generally, lignin is considered a nonbiodegradable material. However, wood contains ~30% lignin and it is naturally transformed in humus. Living organisms consume many vegetables containing lignin and biodegrade them. That means that the lignins are biodegradable materials, but their biodegradation rate is very slow. This situation occurs with native lignin from woody plants, but lignins as by-products from pulp and paper industry, known as industrial or commercial lignins, have different behaviors. They have a lower molecular weight, but lost a part of the functional groups, although it, as a general structure, remains many active sites. The microbial recalcitrance of lignin is due to its high molecular weight, chemical heterogeneity, and absence of regular hydrolyzable intermonomeric linkages [82]. The lignin does not have readily hydrolyzable bonds recurring at periodic intervals along a linear backbone, which is a requirement for enzymatic activity. Instead, lignin is a three-dimensional amorphous, aromatic polymer composed of oxyphenylpropane units, with a seemingly random distribution of stable carbon–carbon and ether linkages between monomer units. The degradation of lignin by the anaerobic environment is limited to the low molecular weight derivatives. Only oligomers and monomers of lignin (< 600 Da) have been shown to be mineralized, while higher molecular weight lignin is not anaerobically degradable.

There are many organisms known to degrade lignin, namely white rot, brown rot, soft fungi, and bacteria. Chief among them are the white rot fungi belonging to the family Basidiomycete. Soil bacteria are undoubtedly active in metabolizing the fragments and probably the lignin macromolecule, because the biological organisms in soil are both abundant and varied.

Biological activity is closely related to surface area and, hence bacteria—having by far the largest surface to volume ratio of all member of the soil community—might be expected to exert a large proportion of the activity in the lignin humus pathway. Lignin and humus into anaerobic environments are protected from microbial attack and along with other biosynthethized organisms are, probably, precursors of coal. The developing of hyphae of the white rot fungus releases an enzyme called ligninase. But this is no ordinary enzyme, it does not bind to a specific substrate that fits snugly into a complementary active site. This lack of specificity is why biochemists took so long to find it. Because the structure of lignin is random, the usual rules of stereochemical intimacy between enzyme and substrate no longer apply.

The ligninase works by "firing" at lignin highly reactive chemical fragments called radical cations, in conjunction with H_2O_2 produced by the fungus.

Even though lignins are very heterogenous in nature, they have one common feature: they all possess alkoxylated benzene rings, which is exactly the object of the attack by ligninases. Ligninases are oxidative, extracellular, and nonspecific.

Humus may be formed as a metabolic by-product of lignolytic soil micro-organisms. Humus is a collection of dark colored organic components of soil and other sedimentary accumulations. It is composed of relatively refractory organic materials resulting from microbial decay of plants and animal tissue. Humus influences the structure and texture of soil, increasing aeration and moisture holding capacity. It is also able to store and gradually release nutrients to the surrounding environment. Humus may participate in a more direct manner in plant nutrition by an actual involvement in plant metabolism and by stimulation of the plant enzyme system.

The general process must be oxidative in nature resulting in an increased carboxyl acidity and a decreased alcoholic hydroxyl content. Increased acidity can be explained only by ring cleavage of some of the aromatic molecules.

The environmental degradation of the PO/natural polymer blends may proceed by one or more mechanisms including microbial degradation in which micro-organisms such as fungi and bacteria consume the material, macro-organism degradation in which insects or other macro-organisms masticate and digest the material, photodegradation in which UV irradiation produces free radical reactions with chain scission, chemical degradation, thermal and thermo-oxidative degradation, etc. [87].

TABLE 5 PO/natural polymer blends [95]

Blend	Characteristics	References
HDPE/corn or rice, potatoes, starch	Improved mechanical and thermal properties	88
HDPE/wood fibers + silanes HDPE/nuts bark	Improved mechanical properties	89
PE/Linocellulosics	Geosynthetic products	90
Thermoplastics/Bagasse	Building materials	91
Cellulose esters/LLDPE	Modification of cellulose fibers through esterification lower high surfce energy of cellulose to that of PO	92
PE, EVA/Hydroxypropyl lignin	EVA produces blends with superior strength properties	52, 93
PP/milled wood/PP-g-MA	Improved mechanical properties	94
25%PE/40%starch/25%EAA or PVA/10% urea	Improved mechanical properties; biodegradable in aquatic media	56
Starch-g-methyl vinyl ketone		72
Starch/oxidized PE		96
Starch/olefin Copolymers as: E-co-acrylic acid	Functional groups of olefin copolymers interact with —OH groups of starch	53, 97
E-co-vinyl alcohol		98
E/P-g-maleic anhydride E/maleic anhydride E/vinyl acetate-maleic anhydride Styrene-maleic anhydride	50–80 wt% starch in blend composition; tensile strengths of blends containing functional groups were superior compared to the blends made from nonfunctionalized polymers; the torques were higher and water absorption increases	99
-PE/polyester or PE/polyester/polylketone		100
PE/rapidly oxidized unsaturated bloc S/isoprene/copolymer		101
PE/various kinds of starch		102
60–70%PP/9–15% modified starch/5% aluminium stearate/5%Co acetyl acetonate	Good mechanical properties	87, 103 59, 72a
Lignin/synthetic polymers	Engineering plastics, adhesives, additives, etc.	104
PP/MA-g-PP/wood fiber composites	Low cost materials Satisfactory mechanical properties	105
Grafted lignin or pieces of wood with alkene monomers without or with plastics	Rapidly and completely degraded under terrestrial conditions by fungi or other flora or fauna	106
Lignin/LDPE/animal or vegetable oil or its oxidized products/metal salts optionally surfactants, starch or oxidized waxes	Biodegradable thermoplastic resin compositions	107
Lignin/PE	Lignin-filled plastics, water degradable plastics	108
PE/lignin Polybutadiene/lignin	Highly filled (up to 64%) lignin PE with good mechanical properties, PE catalytic-grafted on lignin surface improved the adhesion with PE matrix	109
PO/lignosulfonates	Medium term products. High valorization of various kinds of lignin, by-products in pulp and paper industry	110
2–10 wt% Lignin (organosolv, Kraft or beech prehydrolysis)/IPP	Composited with low price, photooxidation, engineering plastics, increased photostability	79

Abiotic oxidation must precede the onset of biotic degradation which is shown to occur readily at $M \approx 40{,}000$. Bioassimilation involves further oxidation catalyzed by transition metal ions and probably by oxo enzymes from the micro-organisms. 80%PCL/20%PE blends lost 85% of their initial elongation only after 18 weeks [85]. However it was demonstrated that starch degrades faster than PCL [86].

Summarizing, the following solutions were applied to obtain PO/natural polymer blends with useful properties and/or biodegradation characteristics—Table 5.

Increased used of polysaccharides and other natural polymers would reduce our dependence on nonrenewable petrochemicals from which the synthetic polymers are derived.

IV. TEST METHODS TO ASSESS THE BIODEGRADABILITY OR RESISTANCE TO MICROBIAL DETERIORATION [111–122]

National and international standards were developed to assess the resistance of plastics in normal use or to predict the preservative effects of added biocides—Scheme 1 and Table 6.

The only standard test generally recognized for testing biodegradability of plastic films is ASTM G21-70, reapproved in 1985. Evidence for biodegradability includes relative rate of growth of fungi on the plastic surface (rated visually) and/or progressive change in physical, optical or electrical properties versus time of incubation. Mixed fungi, cultures of bacteria or other micro-organisms can be used.

ECOLYTE PE passes this test and other tests after photodegradation. It is not possible by this test to determine whether the plastic component of ECOSTAR is biodegradable or not because the micro-organisms attack the starch.

ASTM biometer test method D 5210 and soil burial test methods results are in good accord, some results being computer simulated by percolation theory [113].

The available test methods for plastics may largely be classified as liquid culture, solid medium, soil burial, and weathering tests, some of them being listed below: [114, 115].

1. Liquid culture tests
   tests of the microbial utilization of liquid components;

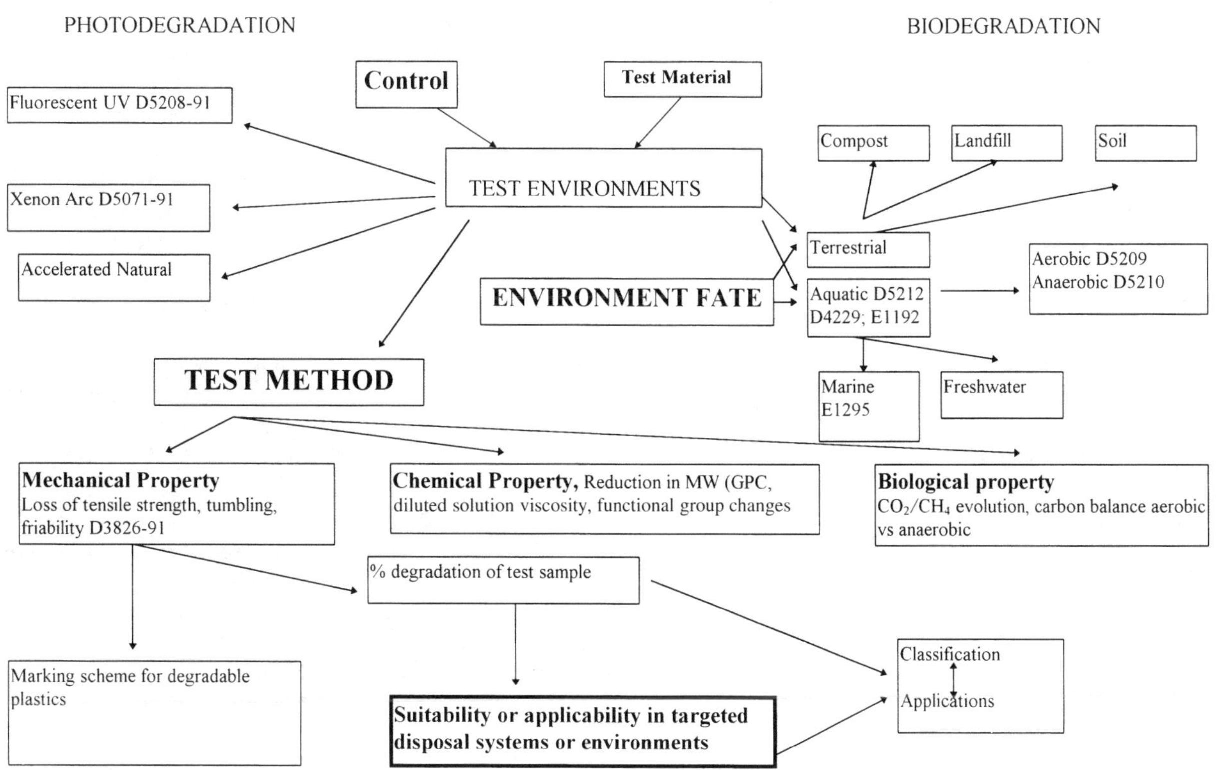

SCHEME 1 Standards development protocol.

TABLE 6 Test standards for evaluating the resistance of plastics to microbial attack [111]

ASTM	G21-90	Standard practice for determining resistance of synthetic polymeric materials to fungi
	G22-76	Standard practice for determining resistance of plastics to bacteria
	G29-90	Standard practice for determining algal resistance of plastic films
ASTM	5209	Aquatic aerobic test, can be used to simulate the biodegradability under real life conditions
	D5210	Anaerobic test; composting test
	D5338-9	Respirometric test [112]
DIN	40046	Climatic and mechanical tests for electronic components and electronic equipment; test *J*; mold growth
	53 739	Testing of plastics; determination of behavior under the action of fungi and bacteria: evaluating by visual examination or measurements of change in mass or physical properties
IEC	68-2-10	Electrical engineering; environmental testing procedures; test *J* and guidance; mold growth.
ISO	846	Plastics; determination of behaviour under the action of fungi and bacteria. Evaluation by visual examination or measurement of change in mass or physical properties
	ISO/CD 14855	Respirometric test; disintegration of the material during fermentation phase, no negative effects on compost quality

tests of the biocidal effectiveness of liquid components;
determination of the mycelial dry weight;
determination of colony counts;
determination of turbidity;
determination of oxygen consumption

2. Solid medium tests
   test of the microbial utilization;
   test of the biocidal effectiveness;
   determination of the colony diameter;
   determination of oxygen consumption or oxygen uptake test
3. Soil burial test [116]
4. Municipal activated sludge test. This very active biological medium is obtained from waste water treatment plant. It is an extremely active mixture of biologically active agents including both bacteria and fungi.
5. Compost
6. Weathering tests
   field tests;
   test in a tropical chamber;
   laboratory tests, moist chamber.

Mainly test methods with fungi and bacteria have been developed as most damage was caused by these organisms, as mold fungi are predominantly isolated from attacked plastic materials—Table 7. But as bacteria, yeasts, and actinomycetes are also known to deteriorate plastics they are also included in some of the test methods. While pure cultures are mainly used for liquid culture and solid medium tests, the soil burial tests and the weathering tests are performed using a natural microbial microflora. The use of pure culture assay, which has the advantage of being reproducible from lab to lab and permit the distinction between degradation due to chemical and photodegradation and degradation due to biological activity.

Soil is a highly variable entity in terms of its composition and biological activity, both varying with location, season, pretreatment, and storage. To date it is impossible to standardize soil as a biodegradation test medium. Most test protocols use these soils with different texture characteristics namely: sandy, loam and clay textured soil, and under near optimal conditions of moisture, aeration, temperature and pH [115].

The test periods of the different tests are quite different. Liquid culture and solid medium tests are conducted for a few days up to several weeks. Soil burial tests are usually carried on for several weeks or months. Weathering tests under field conditions require one or even several years.

The extent of degradation is also determined on the plastic material itself. The following methods are frequently applied and also the tests presented above for lifetime prediction [113–122]:

Weight loss determination is a relatively sensitive method to determine mass change caused by microbial attack; It is a very easily applicable method.

TABLE 7 Standard test fungi [111, 115, 95]

	ASTM G21(1990)	DIN 53739 (1993)	GOST 11326.0 (1978)	IEC68-2-10 (1991)	ISO 846 (1978)	MIL STD-810 (1989)	NF X-41-513-15 (1981)	IBRG ring tests (1996)
A. tenuis								+
Aspergillus amstelodami			+				+	+
A. flavus			+			QM380		+
A. niger	ATCC	DSM1957		ATCC6275	ATCC6275	QM386		+
A. Sydowii	9642							+
A. terreus				PQMD821				
A. versicolor			+			QM432		+
Aurebasidium pullulans				ATCC9348				
Chaetomium globosum			+		ATCC6205	QM459	+	+
Cladosporium herbarum	ATCC9348							+
Gliocladium virens	ATCC6205	DSM1962		IAM 5061	ATCC9645			+
Memnoniella schinata							++	+
Myrothecium verrucaria	ATCC9645						+	+
Paecilomyces varioti		DSM1963	+	IAM 5001	ATCC 10121		+	+
Penicillium brevi-compactum			+				++	+
P. cyclopium			+				+	+
P. funiculosum		DSM1961		IAM7013		QM391	+	+
P. ochro-chloron	ATCC9644			ATCC9112	IAM7013			
Scopulariopsis brevicaulis			+	IAM5146				
Stachybotrys atra							+	+
Sterigmatocystis nigra		DSM1944					+	+
Trichoderma T-1							+	+
T. lignorum			+				+	+

Lignocellulosics degrading microorganisms are: wood destroying and soil fungi white rot fungus Phanerochaete crysosporium, Streptomyces viridosporus T7A, Streptomyces badius [119] *Nocardia* and *Streptomyces*. A variety of Ascomycetes and Fungi Imperfecti are now known to be lignin degraders, as are certain representatives of both the eubacteria and the actinomycetes. Soft rot fungi are Ascomycetes and Fungi Imperfecti. Soft rot fungi *Paecilomyces sp, Allescheria sp and Fusarium. -Aspergillu fumigatus*, and *Fusarium*; Bacteria: *Nocardia, Bacillus, Pseudomonas, Arthrobacter and Azotobacter, Streptomyces badius, Actinomycetae, Xanthomonas Streptomyces flavovireus, Endoconidiophora adiposa* [95]

However some of the weight loss can be masked by the incorporation of oxygen into the material. To make residual weight loss measurements more accurate and convenient, the solvent extraction techniques for the retrieval of polymer incubated in soil have been applied. However, for some polyethylene based materials these techniques give erroneous results because of PE crosslinking (partially insoluble in 1,2,4-trichlorobenzene). Ideally, different measurements for biodegradation should correlate molecular weight determination when applied to the same polymer [120].

The rate of biodegradation is measured by CO_2 production relative to that of compounds of similar chemical structure produced in nature which are known not to accumulate in the environment [117]. Calibration with a known biodegradable material is necessary. Biodegradation rate index (BRI) = ΔCO_2 sample/ΔCO_2 control. As control the following biodegradable materials were selected: sugar (sucrose), cellulose, lauric acid, methyl stearate or cis-polyisoprene [117].

The monitoring of carbon-14 labeled carbon dioxide evolution detected by scintillation counting is a very sensitive method but requires the expensive synthesis of radiolabeled test substrates. It is closely related to the ultimate degradation of the test material. ^{3}H-PE is also used in biodegradation testing [121]. Rate of biodegradation is measured by CO_2 production relative to that of compounds of similar chemical structure produced in nature which are known not to accumulate in the environment [117].

V. PROBLEMS

Several problems appear with the application of degradable materials:

Starch based products should be kept out of the recycling stream as starch tends to clump in the melt, making reprocessing difficult.

The degradation products have not yet been characterized.

Scraps generated during the manufacture of degradable products will tend to be less suitable for reprocessing. Many applications are implemented in agriculture. It is considered that they will be less applicable to solve litter (waste) problems.

VI. CONCLUDING REMARKS

The development of polymers such as PHA, PVA blends, polysaccharides and lignin-based blends or copolymers is continuing in studies designed to enhance the environmental degradability of plastics. Although numerous advantages are proposed for the development and marketing of starch–plastic products numerous challenges remain. Study of the influence of compounding variables on morphology–physical properties and biodegradability can provide a basis for tailoring properties of starch plastics to fit specific applications. Influence of the structure of the natural polymer (starch, lignin, etc.) is not fully understood.

Verification of the degradation of plastics in aqueous systems is a part of the worldwide effort to develop environmentally responsible packaging and delivery material. One of the major issues raised in connection with degradable plastics is the fate and toxicity of degraded products.

Fully degradable PO materials have not yet been developed.

SYMBOLS AND ABBREVIATIONS

BRI	Biodegradation rate index
CTC	Charge transfer complex
DDM	Diaminodiphenylmethane
DSC	Differential scanning calorimetry
DTA	Differential thermal analysis
EAA	Ethylene acrylic acid copolymer
EMA	Ethylene–maleic anhydride copolymer
EVA	Enthylene–vinyl acetate copolymer
EPDM	Ethylene-propylene-diene rubber
FTIR	Fourier transformation–infrared spectroscopy
GC-MS	Gas chromatography–mass spectroscopy
GPC	Gel permeation chromatography
HDPE	High density PE
LDPE	Low density PE
LLDPE	Linear low density PE
LER	Epoxy modified lignosulfonate
LS	Lignosulfonates
LSCa, LSA or $LSNH_4$	Calcium, ammonium lignosulfonates
MB	Master batch
PCL	Polycaprolactone
PE	Polyethylene
PHA	Polyhidroxyalkanoates
PO	Polyolefins
PP	Polypropylene
PVA	Polyvinyl alcohol
PVC	Polyvinylchloride
TG	Thermogravimetric method

REFERENCES

1. KT Gillen, M Celina, RL Clough, J Wise. Trends Polym Sci, 5(8): 250–257, 1997.
2. F Severini, R. Gallo, S. Ipsale. Arab J Sci Eng 13: 533–539, 1988.
3. AK Kulshreshtha, VK. Kaushik, GC Pandey, S Chakrapani, YM Sharma. J Appl Polym Sci 37: 669, 1989.
4. HY Flynn. Thermochim Acta 134: 115, 1988; Polym Eng Sci 20: 675, 1980.
5. E Kramer, I Koppelmann. Polym Eng Sci 27: 945, 1987; Kunststoffe 73: 714, 1983.
6. E Kramer, I Koppelmann. Polym Degrad Stab 16: 261, 1986.
7. a. E Kramer, I Koppelmann, I Dobrowsky. Angew Makromol Chem 158/159: 187, 1988.
   b. E Kramer, I Koppelmann, Guendouz. Angew Makromol Chem 176/177: 55, 1990.
8. E Kramer, J Koppelmann, I Dobrowsky. J Therm Anal 35: 443, 1989.
9. M Mucha, M Kryszewrki. Acta Polym 36: 648, 1985.
10. GC Derringer. J Appl Polym Sci 37: 215, 1989.
11. HJ Fleming. Thermochim Acta 134: 15, 19R8.
12. A Markiewicr, KJ Fleming. J Polym Sci Polym Phys Ed 24: 1713, 1986; 25: 1885, 1987.
13. CK Haywood. Oxidation and aging. In: A Renfrew, P Morgan, eds. Polythene: The Technology and Uses of Ethylene Polymers, 2nd ed. London: Iliffe & Sons Ltd., Interscience, 1960, p 131.
    a. SH Hamid, FS Qureshi, MB Amin, AG Maadhah. Polym Plast Technol Eng 28: 475, 649, 1989.
14. FS Qureshi, HS Hamid, AG Maadhah, MB Amin. In: JM Buist, ed. Progress in Rubber and Plastic Technology, Vol. 5. London: Plastics and Rubber and Rapra Technology Limited, 1989, p 1.
    a. SH Hamid, AG Maadhah, MB Amin. In: SH Hamid, MB Amin, AG Maadhah, eds. Handbook of Polymer Degradation. New York: Marcel Dekker 1993, pp 218–228.

15. AL Andrady, MB Amin, AG Maadhah, SH Hamid, K Fueki, A Torikai. Material Damage in Environmental Effects, Panel Report UNEP, 1489, chap 6, p 55.
16. a. SH Hamid. Twelfth Annual International Conference on Advances in the Stabilization and Controlled Degradation of Polymers, Lucerne, May 21–23, 1990, p 131.
17. F Gugumus. Makromol Chem Macromol Symp 25: 1, 1989; Angew Makromol Chem 176/177: 241, 1990; Polym Degrad Stab 34: 205, 1991.
18. JL Henry, A Garton. Polymer Preprints 30(1): 183, 1989.
19. a. L Lamaire, R Arnaud, J Lacoste. Acta Polym 39: 27, 1988.
    b. L Peeva, B Fileva, S Evtimova, V Taveteva. Angew Makromol Chem 176/177: 79, 1990.
20. L Woo, SY Ding, MTK Ling, SP Westphal. J Therm Anal 49(1): 131–138, 1997; Soc Plast Eng ANTEC'97/506, 3606–3610.
21. WO Drake. J. Polym Sci Symp 57: 153, 1976.
22. AR Horrocks, IA. D'Souza. Thermochim Acta 134: 255, 1988.
23. NM Emanuel, AL Buchachenko. Physical Chemistry of Aging and Stabilization of Polymers. Moskva: Izd. Nauka, 1982, p 356; Polym Eng Sci. 20: 662, 1980.
24. a. H Rainer Sasse, A Boue. Angew Makromol Chem 158/159: 335, 1988.
    b. M Ifwarson, T Traenkner. Kunststoffe 79: 827, 1989.
    c. AL Andrady. J Appl Polym Sci 39: 363, 1990.
25. F Ohishi. Angew Makromol Chem 232: 187–192, 1995.
26. SH Hamid, MB Amin. J Appl Polym Sci 55: 1385–1394, 1995.
27. A Tidjani. J Appl Polym Sci 64 (13): 2497–2503, 1997.
28. RN Gupta, AB Mathur, GN Mathur. IUPAC MACRO '83, Bucharest, Sept 5–9, 1988, vol. 5, p 160.
29. A G Sokolova, GP Tolstov, GA Musaev. Plast Massy 7: 21, 1986.
30. G Kammel, R Wiedenmann. J Polym Sci Symp 57: 73, 1976.
31. a. Yu Kobal'nova, OE Kuzovleva, LM Yarusheva, PV Kozlov, AA Fed. Mekh Kompoz Mater (Zinatne) 6: 1095, 1987.
    b. A Daro, M Trojan, R Jacobs, C David. Eur Polym J 26: 47, 1990.
32. IYu Kobal'nova, OE Kuzovleva, LM Yarusheva, LD Ashkinadze, AM Vinogradov, PV Kozlov, AA Fed. Plast Massy 9: 34, 1987.
33. AS Deuri, K Bhowmick. J Therm Anal 32: 755, 1987.
34. VP Skvortzov, VN Kuleznev, LO Bunina, VI Sergheev, VP Petrova. Plast Massy 5: 39, 1989.
35. A. Valenza, EP La Mantia. Arab J Sci Eng 13: 497, 1988.
36. N Goldenberg, MO Sebe, O Frangu, D Staicu, E Constantinescu. IUPAC MACRO '83, Bucharest, Sept 5–9, 1983, vol 3, p 268.
37. JP Crine, S Haridoss, KC Cole, AT Bulinski, RJ Densley, SS Bamji. IEEE Int Symp Electrical Insulation, Boston, June 5–7, 1988.
38. P Cygan, B Krishnakumar, JR Laghari. IEEE Trans Electr Insul 24: 619, 1989.
39. JCM de Bruiyn. In: RL Clough, NC Billingham, KT Gillen, eds. Polymer Durability. Adv Chem Series 249, Amer Chem Soc Washington, 1996, pp 599–620.
40. J Lemaire, JL Gardette, J Lacoste, P Delprot, D Vailllant. In: RL Clough, NC Billingham, KT Gillen, eds. Polymer Durability. Adv Chem Series 249, Amer Chem Soc Washington, 1996, pp 577–598.
41. P Gijsman, J Hennekens, K Jansser. In: RL Clough, NC Billingham, KT Gillen, eds. Polymer Durability. Adv Chem Series 249, Amer Chem Soc Washington, 1996, pp 621–636.
42. A Titjani, R Arnaud. Polym Degrad Stab 39: 285–292, 1993.
43. AC Albertsson, S. Karlsson. J Macromol Sci Pure. Appl Chem A-33: 1565–1570, 1996.
44. PJ Hocking. J Macromol Sci Rev Macromol Chem Phys C-32(1): 35–59, 1992.
45. AA Yabannavar, R Bartha. Appl Environ Microbiol Oct: 3608–3614, 1994.
46. C Vasile, MC Pascu, MM Macoveanu, C Boghinå. Cell Chem Technol in press; M Pascu, M Gheorghiu, GH Popa, Proceedings of the 14th Inter Symp On Plasma Chem. August 2–6, 1999 Prague, M Hrabovsky, M Konrad, V Kopecky eds. Czech Acad Sci Vol IV p 1877–1882.
47. R Acosta, M Garcia, A Rosales, C Gonzales, L Berlanga, G Arias, NS Allen. Polym Degrad Stab 52: 11–17, 1996.
48. G Scott. Polymer Age 6: 54, 1975; Plast Eng 40: 803–811, 1997; G Scott. Paper presented at the 23rd Aaron Katchalski Katzir Conference, "Polymers and Environment", May 10–16, 1996, Rehovot, Israel.
49. YC Lin. J Appl Polym Sci 63(6): 811–818, 1997.
50. D Gilead, G Scott. British Pat. 1 586 344, 1978.
51. MJ Robey, G Field, M Styzinsky. Materials Forum 13: 1–10, 1089.
52. CL Swanson, RL Shogren, GF Fanta, SH Imam. J Environ Polym Degrad 1(2): 155–166, 1993.
53. FH Otey, RP Westhoff, CR Russell. Ind Eng Chem Prod Res Dev 16: 305–308, 1977; US Pat 4,337,121, 1982; Starch 31: 163–165, 1979; FH Otey, RP Westhof, WM Doane. Ind Eng Chem Prod Res Dev 19(4): 592–595, 1980; SPI Proc Symp on Degradable Plastics, Washington DC, June 10, p 39, 1987.
54. GJL Griffin. UK Pat 1,485,833, 1972; 4 016 117, 1977; Adv Chem Series 134: 159, 1974; Proc. Symp. on Degradable Plastics, SPI, Washington DC June 10, p 47, 1987; GJL Griffin. J Polym Sci Polym Symp 57: 281, 1978; Pure Appl Chem 52: 399, 1970; Int. Pat. Appl. PCT/GB 88/00386, 1988; US Pat 4983651, 1991.

55. RP Wool, MA Cole. Engineering Plastics, Engineering Materials Handbook ASM International Version 3, July 1988; New Orleans.
56. GF Fanta, CL Swanson, RL Shogren. J Appl Polym Sci 44: 2037–2042, 1992.
57. M Vikman, M Itavaara, K Poutanek. J Environ Polym Degrad 3: 23–29, 1995.
58. C Bastioli, V Belotti, R Lombi *et al.* Eur. Pat Appl 0400531, 1990; PCT Int Pat Appl. WO 2023, 02024, 02025, 1991.
59. KE Gonzales, SH Patel, X Chen. J Appl Polym Sci 43: 405–415, 1991.
60. WJ Maddever, GM Chapman. Plast Eng 1989: p 31; WJ Maddever, PD Campbell. Modified starch based environmentally degradable plastics in SA Barenberg, YL Brash, R Harayen, AE Redpath, eds. Degradable Materials: Perspectives, Issues and Opportunities, Boca Raton, CRC Press, 1990.
61. B Erlandsson, S Karlsson, AC Albertsson, Polym Degrad Stab 57: 15–21, 1997; J Environ Polym Degrad 1: 241, 1993.
62. D Bikiaris, J Prinos, C Panayiotou. Polym Degrad Stab 56(1): 1–9, 1997.
63. R Chandra, R Rustgi. Polym Degrad Stab 56(2): 185–202, 1997.
64. BS Yoon, MH Suh, SH Cheung, JE Yie, SH Yoon, SH Lee. J Appl Polym Sci 60(10): 1677–1685, 1996.
65. BG Kang, SH Yoon, SH Lee, JE Yie, BS Yoon, MH Suh. J Appl Polym Sci 60(11): 1977–1984, 1996.
66. RL Shoren, AR Thompson, RV Greene, SH Gordon, GL Cote. J Appl Polym Sci 47: 2279–2286, 1991.
67. a. K Arevalo-Nino, CF Sandoval, LJ Galan, SH Imam, SH Gordon, RV Greene. Biodegradation 7: 231–237, 1996.
b. SH Imam, SH Gordon, AR Thompson, RE Harry-O'Kuru, RV Greene. Biotechnol Tech 7: 791–794, 1993.
68. T Nakashima, M Matsuo. J Macromol Sci Phys B-35 (3&4): 659–679, 1996.
69. SH Imam, SH Gordon, A Burgess-Cassler, RV Greene. J. Environ Polym Degrad 3: 107, 1994.
70. B Erlandsson, S Karlsson, AC Albertsson. Polym Degrad Stab 55: 237–245, 1997.
71. KE Gonzales, SH Patel, X Chen. Polymer Preprints 30(1): 503, 1989.
72. HD Hwu, HM. Lin, SF Jian, JH Chen. ANTEC'97/373, 2026–2030.
72. a. M Hakkarainen, AC Albertsson, S Karlsson. Environ Polym Degrad 5(2): 67–73, 1997; J Appl Polym Sci 66: 959–967, 1997.
73. AC Albertsson, S Karlsson, Macromol Chem Macromol Symp 48/49: 395–402, 1991.
74. M Hakkarainen, AC Albertsson, S Karlsson. J Chromatogr A-271: 251–263, 1996.
b. AC Albertsson, C Barenstadt, S Karlsson, T Lindberg. Polymer 36(16): 3075–3083, 1995.
c. M Hakkarainen, AC Albertsson, S Karlsson. J Appl Polym Sci 66: 959–967, 1997.
75. AC Albertsson, S Karlsson. Macromol Chem Macromol Symp 118: 733–737, 1997; J Environ Polym Degrad 4(1): 51–53, 1996.
76. KE Johnson, AL Pametto, IL Somasundaram, J Coats. J Environ Polym Degrad 1(2): 111, 1993.
77. a. AC Albertsson, S Karlsson. Macromol Chem Macromol Symp 48/49: 395, 1991.
b. J Lemaire, R Arnaud, P Dabin, G Scott, SAl Malaika, S Chohan, A Fauve, A Maaroufi, J Macromol Sci Pure Appl Chem A-32(4): 731, 1995.
78. C Vasile, MM Macovenu, G Cazacu, VI Popa. Natural Polymer Waste Recovery in E Martuseelli, P Musto, G Ragosta eds. Volume of UNESCO Publ 5th Mediterranean School on Polymers for Advanced Technologies, 1997, 4–6 October Capri, pp 185–210.
79. a. B Kosikova, M Kacurakova, V Demianova. Chem Papers 47(2): 132–136, 1993; J Appl Polym Sci 47: 1065–1073, 1993.
b. B Kosikova, K Miklesova, V Demianova. Europ Polym J 29(11): 1495–1497, 1993; B Kosikova, A Revajova, V Demianova. Europ Polym J 29: 1995–1997, 1993; 31(10): 953–956, 1995.
c. B Kosikova, V Demianova. Proc 6th Int Conf Biotechnology in the Pulp and Paper Industry. Advances in Applied and Fundamental Research. E Srebotnik, K Messner, eds. Vienna: Facultas Universitat Verlag, 1996, pp 637–640.
80. AC Albertsson, S Karlsson. Controlled Degradation by Artificial and Biological Processes Chap 44 in: K Hatada, T Kitayama, O Vogl, eds. Macromolecular Design of Polymeric Materials. New York: Marcel Dekker, 1997, pp 793–802.
81. a. C Vasile, M Downey, B Wong, MM Macoveanu, MC Pascu, JI Choi, C Sung, W Baker. Cell Chem Technol 32: 61–88, 1998; Europ Symp on Polymer Blends, Maastrich, May 12–15, 1996, p 389; C Vasile, M Macoveanu, M Pascu, E Profire. Paper presented at the 23rd Aaron Katchalski Katzir Conference, "Polymers and Environment", May 10–16, 1996, Rehovot, Israel; MM Macoveanu, L Constantin, A Manoliu, MC Pascu, L Profire, G Cazacu, C Vasile. Cell Chem Technol in press; MM Macoveanu, N Georgescu Buruntea, MC Pascu, M Căsăriu, A Ioanid, P Vidraşcu, C Vasile. Cell. Chem Technol in press.
b. MM Macoveanu, P. Vidraşcu, MC Pascu, L. Profire, G Cazacu, I Mândreci, H Darie, C Vasile. Bull Botanical Garden of Jassy 6: 487–490, 1997; MC Pascu, MM Macoveanu, P Vidraşcu, I Mândreci, A Ionescu, C Vasile. Bull Botanical Garden of Jassy 6: 493–499, 1997.
c. CrI Simionescu, V Rusan, CI Turta, SA Bobcova, MM Macoveanu, G Cazacu, A Stoleriu. Cell Chem Technol 27: 627–644, 1993.

82. a. M. Shimada, T. Higuchi. Microbial, enzymatic and biomimetic degradation of lignin. In: NS David, Han N Shiraishe, eds. Wood and Cellulose Chemistry. New York: Marcel Dekker, 1991, chap 12.
b. JF Kennedy, GO Phillips, PA Williams, eds. Lignocellulosics. Science, Technology Development and Use. New York: Ellis Horwood, 1992; 8th Int Symp. on Wood and Pulping Chemistry, Helsinki, June 6–9, 1995, vol III.
83. OR Milstein, A Gersonde, M Hutterman, J Chen, JJ Meister. Rotting of thermoplastics made from lignin and styrene by white-rot basidiomycetes. In: RE Hinchee, DB Anderson, FB Metting, GD Sayles, eds. Applied Biotechnology for Site Remediation. Boca Raton LI: Lewis Publisher, 1994.
84. B Filip. In: M Vert, J Feijen, AC Albertsson, G Scott, E Chiellini, eds. Biodegradable Polymers and Plastics. Cambridge: Royal Society of Chemistry, 1992, p 45.
85. L Tilstra, D Johnsonbaugh. J Environ Polym Degrad 1(4): 247, 1993; D Goldberg. J Environ Polym Degrad 3: 61–67, 1995.
86. D Goldberg, JF Rocky, RF Eaton, BS Samuels, KL La Cavao INDA-TEC, Int. Nonwoven Fabrics Conf., Book of Papers, (1990); 55–66 ABIPST, 62 13412, 1992; Y Tokiawa, A Iwamoto, MK Yama. Enzymatic degradation of polymer blends. Proc 3rd Int Scientific Workshop on Biodegradable Plastics and Polymers, Osaka, Japan, 1993 p 36.
87. SM Goheen, RD Wool. J Appl Polym Sci 42: 2691, 1991.
88. GJL Griffin, SA Hashemi. Iran J Polym Sci Technol 1(1): 45–54, 1992.
89. RG Raj, BV Kokta. Ann Tech Conf Soc Plast Eng. 49-th, 1883, 1910, 1993.
90. MM Sain, BV Kokta D Maldas. J Adhesion Sci Technol 7(1): 49–61, 1993; Z Krulis, D Michalkova, J Mikesova, BV Kokta. 38th Microsymp Recycling of Polymers, Prague, July 14–17, 1997, p. 21.
91. SH Hamid, AG Maadhah, AM Usmani. Polym Plast Eng 21(2): 173–208, 1993.
92. P Jandura, BV Kokta. 38th Microsymp Recycling of Polymers, Prague, July 14–17, 1997, p 10
93. SL Clemniecki, WG Glassers. ACS Symp Ser 397: 452–463, 1989.
94. Ann Tech Conf Soc Plast Eng 49th, 1886, 1993.
95. C Vasile, MM Macoveanu, G Gazacu, VI Popa Roum Q Chem Rev 6(2): 85–111, 1998.
96. JL Jane, AW Schwabacher, SN Ramrattan, JA Moor. US Pat 5 1 115 000, 1992.
97. FH Otey, RP Westhoff; WM Doane. Ind Eng Chem Prod Res Dev, 23: 284, 1984; 26: 1659, 1987.
98. ER George, TM Sullivan, EH Park. Polym Eng Sci 34: 17, 1994; D Vega, MA Villar MD Failla EM Valles. Polym Bull 33: 229–235, 1996.
99. M Bhattacharya, UR Vaidya, D Zhang, R Narayan. J Appl Polym Sci 57: 539, 1995; D Ramkumar, UR Vaidya, M Bhattacharya, M Hakkarainen, AC Albertsson, S Karlsson. Eur Polym J 32(8): 999–1010, 1996; UR Vaidya, M Bhattacharya. J Appl Polym Sci 52: 617–628, 1994; Z Yang, M Bhattacharya, UR Vaidya Polymer 37(11): 2137–2150, 1996; UR Vaidya, M Bhattacharya, D Zhang. Polymer 36(6): 1179–1188, 1995.
100. AR Ghaham. PCT Int. Appl. WO 9212187, CA 18:102921y; US Appl., 633 352/1990.
101. M Gottichermann, M Trojan, A Daro, C David. Polym Degrad Stab 39(1): 55–68, 1973.
102. W Funke, W Bregthaller. In: F Meuser, DJ Manners, WB Seibel, eds. Prog Plant Carbohydr Res Second Part Proc Int Symp, 7th, 1992 Publ 1995, pp 161–164.
103. AC Albertsson, S Karlsson. J Appl Polym Sci, 22: 3419–3433, 1978.
104. J Wang, JRS Manley, D Feldmann. Prog Polym Sci, 17(4): 611–646, 1992.
105. M Lu, JR Collier, BJ Collier. ANTEC'95, pp 1433–1437.
106. JJ Meister, JM Chen, US Pat 789 360; 930621.
107. Takenaka Komuten Co, Japan. Jpn Kokai Tokyo Koho, 93-323836.
108. J Kubat, HE Stroemvall, Plast Rubber Process Appl 3(2): 111–118, 1983.
109. S Casenave, A Ait-Kadi, B Brahimi. Soc Plast Eng, ANTEC'95, 1438-1442; B Brahimi, B Riedl, A Ait-Kadi Soc Plast Eng, ANTEC'95, 2576–2580.
110. CrI Simionescu, MM Macoveanu, C Vasile, F Ciobanu, M Esanu, A Ioanid, P Vidrascu, N Georgescu Buruntea. Cell Chem Technol 30: 411–429 (1996); CrI Simionescu, MM Macoveanu, C Vasile, Fl Ciobanu, G Cazacu; High-Technology Composites in Modern Applications. Int. Symp on Advanced Composites, Corfu, Greece, September 18–22, 1995 Symposium Book pp 222–232.
111. a. M Pantke. General Aspects and Test Methods in Microbially Influenced Corrosion of Materials. Berlin: Springer Verlag, 1996, pp 380–391.
b. M. Itavaara, M Vikman. J Environ Polym Degrad 4(1): 29–36, 1996.
c. R Narayan. Development of standards for degradable plastics by ASTM Subcommitee D-20.96 on environmentally degradable plastics. In: M Vert, J Feijen, AC Albertsson, G Scott, E Chiellini, eds. Biodegradable Polymers and Plastics. Cambridge: Royal Society of Chemistry, 1992, p 176.
112. G Bastioli. 38th Microsymp Recycling of Polymers, Prague, July 14–17, 1997, PSL15.
113. RF Wool, D Raghavan, S Billieux, G Wagner. In: M Vert, J Feijen, AC Albertsson, G Scott, E Chiellini, eds. Biodegradable Polymer and Plastics. Cambridge: Royal Society of Chemistry, 1992, p 112.
114. IM Gidman, N Kirkpatrick, PJ Nicholas. Method and developments for determination of degradability of plastics. An overview of research activities at PIRA

International. In: M Vert, J Feijen, AC Albertsson, G Scott, E Chiellini, eds. Biodegradable Polymer and Plastics. Cambridge: Royal Society of Chemistry, 1992, p 249.

115. A Corti, G Vallini, A Pera, F Cioni, R Solaro, E Chiellini. Composting microbial ecosystem for testing the biodegradability of starch-filled PE films. In: M Vert, J Feijen, AC Albertsson, G Scott, E Chiellini, eds. Biodegradable Polymers and Plastics. Cambridge: Royal Society of Cambridge, 1992, p 245.

116. R Bartha, AV Yabannavar. Biodegradation testing of polymers in soil. In: JD Hamilton, R Sutcliffe, eds. Ecological Assessment of Polymers. Strategies for Product Stewardship and Regulatory Programs. New York: Van Nostrand Reinhold, 1996, chap 4, p 53; AV Yabannavar, R Bartha. Soil Biol Biochem, 25(11): 1469–1475, 1993; Applied Environ Microbiol 60: 2717–2722, 1994.

117. GE Guillet, HX Huber, J Scott. Studies of the biodegradation of synthetic plastics. In: M Vert, J Feijen, AC Albertsson, G Scott, E Chiellini eds. Biodegradable Polymers and Plastics. Cambridge: Royal Society of Chemistry, 1992, p 545.

118. M Pantke, KY Seal. Material Organisms 21: 151–164, 1986, 25: 87–98, 241–253, 1990.

119. AL Pometto, KE Johmson, M Kim. J Environ Polym Degrad 1(3): 213, 1993.

120. R Bartha, AV Yabannavar, MA Cole, DJ Hamilton. Chap 9 Plastics. In: JD Hamilton, J Sutcliff, eds. Ecological Assessment of Polymers. Strategies for Products Stewardship and Regulatory Programs. New York: Van Nostrand Reinhold, 1996, p 167–183.

121. JR Coats. J Environ Polym Degrad 5(2): 119, 1997.

122. EL Sharabi, R Bartha. Appl Environ Microbial 59: 1201–1205, 1993.

20

Additives for Polyolefins

Cornelia Vasile
Romanian Academy, "P. Poni" Institute of Macromolecular Chemistry, Iasi, Romania

I. INTRODUCTION

The role of additives in creating tailor-made plastics and in generating innovative products is commonly accepted. The versatility of the new materials was enhanced by new processing technologies and above all by designing special formulations tailored to the different fields of applications. The numerous and diverse applications of polyolefins would not be possible without the development of suitable additives. Virgin PO from commercial processes are very susceptible to air oxidation. If stored unstabilized at ambient temperature, the physical properties deteriorate rapidly over various periods of time (even weeks for PP) depending on polyolefin structure, physical form, temperature, available oxygen, etc. The uncontrolled oxidation is exothermic. It must be prevented even in the production reactor before drying and storage, by so-called "in process" stabilization. This is achieved by the addition of a few parts per million of an antioxidant. In particular PP products could not even exist without the addition of stabilizers.

During the production and utilization of polyolefins, their characteristics may be modified in a such a way that they lose their usefulness due to the various kinds of degradation (see Chap. 17). Processing of polyolefins takes place at relatively high temperatures (> 100°C), which involves, too, high frictional and shearing forces at high speeds, these contributing to the deterioration of the material. Knowledge of the degradation of polyolefins upon processing and service life is the basis for selection of a stabilizer or a stabilization system to fulfill the requirements of a plastic under certain exposure conditions.

Additives are applied for the preservation of some properties (stabilization against the action of heat, oxygen, light, etc.), to facilitate their processing in different items, and/or for the modification of some properties for special purposes (see Chapter 21).

Usually the first two objectives must be concomitantly considered and fulfilled. High effectiveness cannot be obtained one-sidedly by the addition of a single additive. That is why complex systems of additives (antioxidants, light stabilizer mixtures, or complex formulations of several additives) are used.

The distinct description of the role and action mechanism of an additive or of a class of additives (as presented briefly in the following) is justified only theoretically.

Many books and review articles have detailed descriptions of additives [1–11].

II. TYPES OF ADDITIVES

The main groups of additives used for polyolefin stabilization, processing, etc., are schematically presented below along with their tradenames and producers or suppliers, given in Tables 1–4.

1. *Antioxidants (inhibitors or heat stabilizers)* assure protection against thermal and oxidative

(Text continues on p. 522)

TABLE 1 Commercially available antioxidants

Tradename	Manufacturer/supplier (for addresses see Appendix 1)	Chemical class and examples[a]
Advamond TPP	Advance, Germany	Triphenylphosphite (0.5–3)
Advastab 405	Advance, Germany	2,2′-Methylen-bis(6-*tert*-butyl-4-methylphenol)
Advastab 406	Advance, Germany	2,2′-Thiobis(6-*tert*-butyl-4-methylphenol)
Age Rite	RT Vanderbilt USA	K,N
Age Rite 415	Anchor, UK	4,4′-Dialkyldiphenylamine (0.5–3)
Aghidol 1	Russia	2,6-Di-*tert*-butyl-4-methylphenol (0.1–0.6)
Aghidol 42	Russia	2,6-Di-*tert*-butyl-4-metoxymethylphenol (0.01–0.5)
Aghidol 7	Russia	2,2′-Diethylenebis-6-*tert*-butyl-4-ethylphenol (0.25–1.5)
Aghidol 5	Russia	3,3′,5,5′-Tetra-*tert*-butylbiphenyl diol 4,4′ (0.5–2.0)
Alkofen B,BP	Russia	2,6-Di-*tert*-butyl-4-methylphenol (0.1–0.6)
Anox BP	Ente Nazionale Idrocarburi, Italy	D, I, J, K
AN-2	Ethyl, USA	4,4′-Methylene bis (2,6-di-*tert*-butylphenol) (0.2–1.0)
AO	Song Woun Ltd, Korea	I, J
AO 40	Russia	1,3,5-Trimethyl-2,4,6-tri(3,5-di-*tert*-butyl-4-hydroxybenzyl)benzene
AO SD-13	Czech Republic	*N*-(1,3-Dimethylbutyl)-*N*-phenylphenylene diamine 1,4 (0.25–1.0)
Antigene	Sumitomo Chemical Co Ltd, Japan	K (< 0.20)
Antioxidant 4,41	Czech Republic	2,6-Di-*tert*-butyl-4-methylphenol (0.1–0.6)
Antioxidant 762	Ethyl, USA	(2,6-Di-*tert*-butyl-4-metoximethyl phenol (0.01–0.5)
Antioxidant X	Bayer, Germany	2,6-Di-*tert*-butyl-4-metoximethyl phenol (0.01–0.5)
Antioxidant 13	Czech Republic	2,6-Di-*tert*-butyl-4-octadecylpropionyl phenol) (0.5–2.0)
Antioxidant 425	Anchor, UK	2,2′-Methylene bis(6-*tert*-butyl-4-ethylphenol) (0.25–1.5)
Antioxidant	Russia	2,2′-Methylene bis(6-*tert*-butyl-4-methylphenol) (0.5–1.0)
Antioxidant 2246	Anchor, UK, American Cyanamid Co., USA	2,2′-Methylene bis(6-*tert*-butyl-4-methyl phenol) (0.5–1.0)
Antioxidant H	Colorom, Codlea Romania	1,2-Dihidro-2,2,4-trimethylchinoline polymer
Antioxidant 330	Ethyl, USA	2,4,6-Tris-(3,5-di-*tert*-butyl-4-hydroxy benzyl)mesitilen (0.05–5.0)
Antioxidant 420	Naugatuck, USA	*N*-Isopropyl-*N*′-phenyl phenylene diamine 1,4 (up to 1)
Antioxidant S	Czech Republic	*N*-Isopropyl-*N*′-phenyl-phenylene diamine 1,4 (up to 1)
Antioxidant SD	Germany	*N*-Isopropyl-*N*′-phenyl phenylene diamine 1,4 (up to 1)
Antioxidant 6	Czech Republic	Mixture of methylbenzyl phenyl phosphites (0.1–0.3)
Antioxidant TBM-6	Organo-Synthèse, France	4,4′-Thiobis(6-*tert*-butyl-3-methylphenol) (0.5–1.0)
Antioxidant 736	Etyhyl, USA	4,4′-Thiobis(6-*tert*-butyl-2-methylphenol)
AS-5	Catalin, USA	2,2′-Methylene bis(6-*tert*-butyl-4-methylphenol) (0.5–1.0)
ASM PEN	Bayer, Germany	*N*-phenylnaphthyl amine-2 (0.2–1.5)
ASM 4020	Bayer, Germany	*N*-(1,3-Dimethylbutyl)-*N*′-phenylphenylene diamine-1,4 (0.25–1.0)
BHT	Koppers, USA; Monsanto, UK	2,6-Di-*tert*-butyl-4-methyl-phenol (0.1–0.6)
Bisalkofen MZP	Russia	2,2′-Methylene bis(4-methyl-6-(1-methylcyclo-hexyl-1)phenol) (0.001–5.0)
Binox	M Shell, UK	4,4′-Methylene bis(2,6-di-*tert*-butyl phenol) (0.2–1.0)
Bayer 4020	Bayer, Germany	*N*-(1,3-Dimethylbutyl)-*N*′-phenylphenylene diamine-1,4 (0.25–1.0)
BMC	Russia	4,4′-Thiobis(6-*tert*-butyl-3-methylphenol)
CAO	Ashlan Chemical Co., USA	A, B, D
CAO-I,3	Catalin, USA	2,6-Di-*tert*-butyl-4-methyl phenol
CAO 20	Catalin, USA	4,4′-Methylene bis(2,6-di-*tert*-butyl-phenol (0.2–1.0)
CAO-5, CAO-14	Ashland, USA	2,2′-Methylene bis(6-*tert*-butyl-4-methylphenol) (0.5–1.0)
Carstab	Morton Thiokol, USA	J, K
Cyanox	American Cyanamid Co., USA	B, G, L, O
Cyanox 8	American Cyanamid Co., USA	Condensation product of diphenylamine with di-isobutylene (0.5–2.0)
Cyanox 2246	American Cyanamid Co., USA	2,2′-Methylene bis(4-methyl-6-*tert*-butyl phenol)

TABLE 1 (Continued)

Tradename	Manufacturer/supplier (for addresses see Appendix 1)	Chemical class and examples
Cyanox 425	American Cyanamid Co., USA	2,2′-Methylene bis(4-ethyl-di-*tert*-butyl phenol)
Cyanox LDTF	American Cyanamid Co., USA	Dilauroil-thio-dipropionate
Chimer 18	Chimosa	Distearyl-thio-dipropionate
Dalpac	Hercules Powder Co., USA	2,6-Di-*tert*-butyl-*p*-cresol)
DBPC	Koppers, USA	2,6-Di-*tert*-butyl-*p*-cresol) (0.1–0.6)
Diafen	Russia	*N*-(1,3-Dimethylbutyl)-*N*′-phenylphenylene diamine-1,4 (0.25–1.0)
Diafen FP	Russia	*N*-Isopropyl-*N*′-phenylphenylene diamine-1,4 (up to 1)
Diafen 13	Russia	
Difenam DMB	Russia	Condensation product of diamine with diisobutylene (0.5–2)
Ethanox	Ethyl Corp., USA	A, B, G, I
Ethyl Anti-oxidant	Ethyl Corp., USA	3,3′,5,5′-Tetra-*tert*-butyl-diphenyl diole-4,4′ (0.5–2.0)
Fenozam 1	Russia	Methyl ester of 3,5-di-*tert*-butyl-4-hydroxyphenyl propionic acid (0.5–2)
Fenozam 23	Russia	3,5-Di-*tert*-butyl-4-bydroxy phenyl propionic acid and pentaerithrite ester (0.1–1)
Fenozam 30	Russia	Diethyl-3,3′-(2,2′-thio bis-(3,5-di-*tert*-butyl-4-hydroxy phenyl)di-propionate (0.1–0.3)
Flectol	Monsanto Europe SA	K
FAU 13	Russia	2-(3,5-Di-*tert*-butyl-4-hydroxy-phenyl 4-hydroxy-phenylethyl)-2,3-dihydro-1,2,3,4-propionylhydrazino)-5-(3,5-di-*tert*-butyl-oxaphosphodiazole) (0.1–0.3)
FA Systems	Ciba Additives	Hydroxylamine, high molecular weight
GA-80	Sumitomo Chemical Co. Ltd., Japan	G
Garbefix	Soc Francaise D'Organo-Synthèse, France	A, B, L, M
Good-rite	BE Goodrich Co., USA	G, J, K
Good-rite	BE Goodrich Co., USA	4′4′-Dialkyl diphenyl amine (0.5–3.0)
Hostanox	Hoechst AG, Germany	C, L, M
HPM	Soc. Francaise D'Organo Synthèse, France	C
Ionol	Shell Nederland	D
Ionol SR	Shell Nederland	2,6-Di-*tert*-butyl-4-methyl phenol
Ionox	Shell Nederland	G
Ionox 220	Shell Nederland	4,4′-Methylene bis(2,6-di-*tert*-butyl-phenol) (0.2–1.0)
Ionox 330	Shell Nederland	2,4,6-Tris(3,5-di-*tert*-butyl-4-hydroxy, benzyl) mesitilen
Irgafos	Ciba-Geigy, Switzerland	M
Irgafos PEPQ	Ciba-Geigy, Switzerland	Tetrakis-(2,4-di-*tert*-butylphenyl)-4,4′diphenylene propionate
Irgafso 12	Ciba-Geigy, Switzerland	Hydrolytical stable organophosphite
Irgafso 168	Ciba-Geigy, Switzerland	Tris 2,4-ditertbutylphenyl phosphite, aromatic phosphite for color critical applications
Irganox	Ciba-Geigy Corp., USA	A, B, E G, H I, J, K, L, O
Irganox 245	Ciba-Geigy, Switzerland	Triethylenglycol-bis-3-(3-*tert*-butyl-4-hydroxy-5-methyl phenyl propionate)
Irganox 259	Ciba-Geigy, Switzerland	1,6-Hexamethylenebis-3-[(3,5-di-*tert*-butyl-4-hydroxyphenyl)propionate
Irganox 414	Ciba-Geigy Corp., USA	4,4′-Butylidene bis(2-*tert*-butyl-5-methyl phenol) (0.01–0.5)
Irganox 415	Ciba-Geigy Corp., USA	4,4′-thio bis(6-*tert*-butyl-3-methyl phenol (0.5–1.0)
Irganox 1076	Ciba-Geigy Corp., USA	2,6-Di-*tert*-butyl-4-octadecyl propionyl phenol
Irganox 1010	Ciba-Geigy Corp., USA	Pentaerythrityltetrakis-3-(3,5-di-*tert*-butyl-4-hydroxyphenyl) propionate (0.1–1.0)
Irganox 1098	Ciba-Geigy Corp., USA	*N*, *N*′-Hexamethylene-bis-3,5-di-*tert*-butyl-4-hydroxyphenyl) propionamide

TABLE 1 (Continued)

Tradename	Manufacturer/supplier (for addresses see Appendix 1)	Chemical class and examples
Irganox 1330	Ciba-Geigy Corp., Switzerland	1,3,5-Trimethyl-2,4,6-tris(3,5-di-*tert*-butyl-4-hydroxybenzyl)benzene
Irganox 1520	Ciba-Geigy Corp., Switzerland	2-Methyl-4,6-bis(octylthiomethylphenol)
Irganox 3114	Ciba-Geigy Corp., Switzerland	1,3,5-Tris(3,5-di-*tert*-butyl-4-hydroxy benzyl)cyanuric acid
Irganox 3125	Ciba-Geigy Corp., Switzerland	Trinuclear phenol
Irganox B215	Ciba-Geigy Corp., Switzerland	Blend Irgafos 168/Irganox 10102/1
Irganox B225	Ciba-Geigy Corp., Switzerland	Blend Irgafos 168/Irganox 1010 1/1
Irganox B900	Ciba-Geigy Corp., Switzerland	Blend Irgafos 168/Irganox 10764/1
Irganox B1171	Ciba-Geigy Corp., Switzerland	Blend Irgafos 168/Irganox 10981/1
Irganox PS802	Ciba-Geigy Corp., Switzerland	Distearylthiodipropionate
Irganox LO-1	Ciba-Geigy Corp., USA	Condensation product of diphenyl amine with di-isobutylene
Irgastab TPP	Ciba-Geigy Corp., USA	Triphenylphosphite (0.5–3.0)
Irgastab SN-55	Ciba-Geigy Corp., USA	Tri(*n*-propylphenyl-phosphite) (0.1–1.0)
Ipognox 44	Poland	*N*-Isopropyl-*N*′-phenyl phenylene) diamine-1,4′ (up to 1)
Isonox	Schenectady Chemicals Inc., USA	B, D
Keminox	Chemipro Kasei Ltd., Japan	A, B, C, D, G
Lowinox	Chemische Werke GmbH, Germany	A, B, D, E, I, J, K, L, M
LZ-MB-I	Russia	4,4′-Methylene bis(2,6-di-*tert*-butyl-phenol) (0.2–1.0)
LZ-TB-3	Russia	Bis-3,5-di-*tert*-butyl-4-hydroxybenzyl) sulfide (0.5–1.0)
Mark	Argus Chemical SA-NV, Belgium	L, M, O
Methen 44-26s	Korea	Bis(3,5-di-*tert*-butyl-4-hydroxy benzyl) sulfide (0.1–0.5)
Mixxim	Fairmount Chem. Co. Inc., USA	B, M
Naugard	Uniroyal Ltd., UK	B, C, I, J, K, L, M, N, O
Naugard BHT	Uniroyal Inc., UK	2,6-Di-*tert*-butyl-4-methyl phenol (0.1–0.6)
Naugard 431,	Uniroyal Inc., UK	Phenolic antioxidants (low volatility)
Naugard P	Uniroyal Inc., UK	Tri(*n*-nonylphenyl phosphite) (0.5–2.0)
Naugard PAN	Uniroyal Inc., UK	*N*-Phenylnaphthyl amine (0.1–0.5)
Naugard 445	Uniroyal Inc., UK	4,4′-Dialkyl diphenylamine (0.5–3.0)
Naphtam-2	Russia	*N*-Phenyl naphthyl amine-2 (0.2–1.5)
NG 2246	Russia	2,2′-Methylene bis(6-*tert*-butyl-4-methyl phenol)
Naruxol 25	Naugatuck, USA	2,2′-Methylene bis(6-*tert*-butyl-4-ethyl phenol)
Naugawhite	Uniroyal Ltd., USA	B
Neozon A	Russia	*N*-Phenylnaphthyl amine (0.1–0.5)
Neosone A	ACNA, Italy	*N*-Phenylnaphthyl amine (0.1–0.5)
Neosone D	Cyanamid, USA	*N*-phenylnaphthyl amine (0.1–0.5)
Negonox	ICI Ltd., UK	C
Nonox	ICI Americas, USA	B, D, K, O
Nonox AN	ICI Americas, USA	*N*-Phenylnaphthyl amine
Nonox TBC	ICI Americas, USA	2,6-Di-*tert*-butyl-methylphenol
Nonox WSP	ICI Americas, USA	2,2′-Methylene bis(4-methyl-6-(1-methyl cyclohexyl-1) phenol)) (0.01–5.0)
Nonox D	ICI Americas, USA	*N*-Phenylnaphthyl amine-2
Nonox ZA	ICI Americas, USA	*N*-Isopropyl-*N*′-phenylphenylene diamine 1,4
Nonox ZC	ICI Americas, USA	*N*-(1,3-dimethylbutyl)-*N*′-phenylphenylene diamine 1,4
Nonox OD	ICI Americas, USA	Condensation product of diphenylamine with diisobutylene
Nonflex ALBA	Seiko, Japan	2,5-Di-*tert*-butyl hydrochinone
Oxi-chek	Ferro Corp., Chem. Div., USA	B, I
OS	ICI, UK	2,6-Di-*tert*-butyl-4-methyl phenol
Perkadox	Akzo Chemicals, Netherlands	D
Phosclere	Amsterdam, Netherlands	L
Permanax	Vulnax International, France	B, D

TABLE 1 (Continued)

Tradename	Manufacturer/supplier (for addresses see Appendix 1)	Chemical class and examples
Polyguard	Uniroyal Ltd., UK	M
Polyguard	Naugatuck, USA	Tri-(*n*-Nonylphenyl phosphite) (0.1–1.0)
Plastonox 425	Cyanamid, USA	2,2′-Methylene bis(6-*tert*-butyl-4-ethyl phenol)
PDA 8, 10, 14	Benson, USA	4,4′-Dialkyl diphenyl amine (0.5–3.0)
Ronox	Borzesti, Romania	Tri-Nonylphenyl) phosphite
Samilizer	Sumitomo Chemical Co. Ltd., Japan	A, B, I, J, L, M
Samilizer BBM	Sumitomo Chemical Co. Ltd., Japan	4,4′-Butylidene bis(2-*tert*-butyl)5-methyl phenol
Sandostab P-EPQ	Sandoz, Switzerland	Tetrakis(2,4-di-*tert*-butyl-phenyl-4,4′-diphenylene-diphosphite)
Saantonox R,E	Monsanto Europe SA	A, 4,4′-Thio bis(6-*tert*-butyl-3-methyl phenol)
Santowhite	Monsanto Co., USA	A, B, D, K
Santowhite MK	Monstanto Co., USA	4,4′-Thio bis(6-*tert*-butyl-3-methyl phenol) (0.5–1.0)
Santowhite powder	Monsanto Co., USA	4,4′-Butylidene bis(2-*tert*-butyl-5-methyl phenol (0.01–0.5)
Santoflex 13	Monsanto Co., USA	K
Santoflex 36, IP	Monsanto Co., USA	*N*-Isopropyl-*N*′-phenylphenylene diamine-1,4 (up to 1)
Santoflex 13	Monsanto Co., USA	*N*-(1,3-Dimethylbutyl)-*N*′-phenylphenylene diamine-1,4 (0.25–1.0)
Santicizer	Monsanto Co., USA	A, M
Santovar A,O	Monsanto Co., USA	2,5-Di-*tert*-butyl hydrochinone
SE-10	Hoechst-Celanese	Disulfide
Seenox	Shipro Kasei Ltd., Japan	A, B, G
Stabilizer TPP	Germany	Diphenyl phosphite
Tominox	Yoshitomi Pharmaceutical Ind, Japan	I, J
Topanol	ICI, UK	B, D, K
Topanol BHT, O	ICI Americas, USA	2,6-Di-*tert*-butyl-4-methyl phenol
Topanol CA	ICI Americas, USA	1,1,3-tris(5-*tert*-butyl-4-hydroxy-2-methylphenol) butane
TBMd	Organo-Synthèse, France	4,4′-thio-bis(6-*tert*-butyl-3-methyl phenol)
Tioalkofen	Russia	2,2′-thio-bis(4-methyl-4-α-methylbenzyl phenol) (0.1–0.3)
Tioalkofen MBP	Russia	
Tioalkofen BM	Russia	4,4′ -thio-bis(6-*tert*-butyl-3-methylphenol)
Ultranox	Borg-Warner Chemicals, USA	A, B, D, I, K, M
Ultranox 276	General Electric, USA	Octadecyl-3,5-di-*tert*-butyl-4-hydroxyhydrocinnamate
Ultranox 618	General Electric, USA	Aliphatic phosphite
Ultranox 626	General Electric, USA	Bis-(2,4-di-*tert*-butylphenyl) pentaerythrityl diphosphite)
Ultranox 641	GE Speciality Chemicals, USA	Solid phosphite based butylethyl propane, has improved handling characteristics, granted FDA
Virid	Victoria, Romania	Tris-(3,5-Di-*tert*-butyl-4-hydroxybenzyl-izocianurate
Vulcanox	Mobay Corp., USA Bayer AG, Germany	B, D, K, L
Weston	Borg-Warner Chemicals, USA	M, O
Weston 399	General Electric, USA	Aromatic phosphite liquid
Weston 503B	General Electric, USA	Trisnonylphenyl phosphite
Weston 504B	General Electric, USA	
Weston 505B	General Electric, USA	
Wing Stay	Goodyear SA, France, USA	A, D, E, K
Wytox	Olin Corp., USA	B, D, K, M
Yoshinox	Yoshitomi Pharmaceutical, Japan	A, B

[a] In parentheses are given the technologically recommended concentrations.
Source: Refs. 1–10.

degradation during processing and/or environmental exposure under working conditions. More than 500 compounds are commercially used as antioxidants; see Table 1. They can be classified as:

(a) *Chain breaking or primary antioxidants*
  Sterically hindered phenols that are radical scavengers
  secondary aromatic amines
  aminophenols
  aromatic nitro- and nitroso-compounds
(b) *Preventive or secondary antioxidants* [1–8]
  organic phosphites and phosphonates which are hydroperoxides or peroxides decomposers
  sulfur-containing acids and sulfides
  thioethers and thioalcohols
  thioesters of thiopropionic acid with fatty alcohols
  dithiocarbamates (*N*-zinc dibutyldithiocarbamate), dialkyl dithiocarbamates,
  thiodipropionates
  mercaptobenzimidazoles
  amino-methylated indoles
  aminopyrazoles
  metallic chelates
(c) *Mixtures of primary and secondary antioxidants.* More than 380 patents dealing with new compounds acting as antioxidants are recorded each year. (Commercially available antioxidants given in Table 1 correspond to the following classes: A: thiobisphenols; B: alkylidene-bisphenols having molecular weight 300–600; C: alkylidene-bisphenols having molecular weight > 600; D: alkylphenols; E: di-3-*tert*-butyl-4-hydroxy-5-methylphenyldicyclopentadiene; F, G: hydroxybenzyl compounds having molecular weight 300–600 and > 600 respectively; H: acylaminophenols (hydroxyanilides); I: 3-(3,5-di-*tert*-butyl-4-hydroxyphenyl) propionate; J: poly(hydroxyphenylpropionates) of molecular weight >600; K: amines; L: thioethers; M: phosphites and phosphonites; N: zinc dibutyldithiocarbamate; O: blends of secondary and primary antioxidants.)
(d) *UV absorbers* that directly interact with UV light. By absorbing UV light before it has a chance to energize a chromophore, the formation of a radical is prevented. The UV absorbers do not absorb all UV light, so some radicals are formed but are neutralized by the hindered amine, primary antioxidants as secondary stabilizers due to their synergism with hindered amines. UV absorbers alone impart only a moderate UV stability to PO.

2.1. *Antistatic agents* employed for minimizing the accumulation of static electricity; have the following functions: technological—improve processing, fire prevention—safety in product exploration [1, 10] (Table 2). An antistat has a limited solubility in the PO matrix and a hydrophilic character. Over a period of time, the antistat migrates to the surface of the polymer. The hydrophilic portion is excluded from polymer matrix and resides on the surface and static charges may dissipate. Concentration range from 0.1 ppb up to 1.0 ppb are need. The surface active antistats are not permanent.

The antistats can be classified as "oily" or "waxy" surface deposits. Also known are internal (glycerolmonostearates and ethoxylated secondary amines) and external (polyols) antistats. They may pose problems during sealing, printing or adhesion operations. They must accomplish functionality as mold release agents. To control the surface electrical properties, they can be used in conjunction with a conductive agent. The surface resistivity must be below $10^{14}\ \Omega\,cm^{-2}$.

The most frequently used antistatic agents are

(a) Cationic compounds–quaternary ammonium salts; high concentrations are necessary
(b) Anionic compounds–high concentrations are necessary
  alkyl sulfonates, sulfates or phosphates
  dithiocarbamates or carboxylates of alkali metals and of alkaline earth metals
(c) Nonionic compounds
  polyethylene glycol esters or ethers
  fatty acid esters
  ethanol amines
  mono- or diglycerides
  ethoxylated fatty acids

2.2. *Conducting additives* reduce specific resistance to Ω cm.

The following conducting additives can be used:
aluminum flakes coated with coupling agents
steel microfilaments
silvered glass fibers and spheres
nickel surface coated textiles
special carbon black and carbon fibers
organic semiconductors

3. *Biosensitizers and photosensitizers* used for the obtainment of polyolefins with predictable lifetimes [11–13].
   (a) Transition metal (Co, Ni, Cr, *Zn*) · N, N'-diethyl-diselenocarbamates or other metallic complexes. Fe complexes are mostly used as Fe(III)-acetylacetonate, Fe(III)-2-hydroxy-4-methylacetophenone oxime
   (b) Blends with natural macromolecular compounds (starch, cellulose, etc).

4. *Peroxides and crosslinking agents* have the following functions [1, 14, 15]: vulcanization of PO saturated rubbers; crosslinking (especially PE and ethylene copolymers) in order to improve the dimensional stability; control of MFI and MWD of PO (especially PP); chemical modification (grafting, chlorinating, etc.) initiation. Chemical classes are
   (a) Organic peroxides—Table 3
   alkyl peroxides
   hydroperoxides
   peroxyesters
   diacyl peroxides
   peroxy ketals, etc.

   In PP, peroxides promote extrusion degradation. The advantages of the controlled rheology resins lay in their high MFI and narrower MWD, preventing more output and higher orientation in some fine denier fiber operations There are some interactions between peroxides as a radical source and stabilizers, the performance of each of them being adversely affected.
   (b) Silanes. These are frequently used in the presence of coagents, such as ethylene glycol, dimethacrylate, diallylterephthalate, triallylcyanurate, triallylisocyanurate, trimethylolpropanetrimethacrylate, *n*-phenylenedimaleimide, 1,2-cis-butadiene, triazine, etc.
   (c) Photoinitiators for PO (especially PE) crosslinking or other chemical modifications [16–20]; inorganic and organic chlorides and other compounds containing chlorine such as S_2Cl_2, SO_2Cl_2, PCl_3, chlorosilanes, $CHCl_3$, C_2Cl_6, CCl_4, C_2Cl_4, oxalic chloride, 1,2,3,4,-5,8,-hexaehloronaphthalene, chloranhydride; chlorendie anhydride, hexachlorobenzene.
   (d) Aromatic ketones and quinones such as benzophenone, nitrobenzophenone, 9,10-anthraquinone, 1,2-benzanthraquinone, 2-methyl- or ethylchloroanthraquinone, 1,4-naphthaquinone, 1,4-benzoquinone, xanthone, 4-chlorobenzo-phenone+triallylcyanurate, 4,4°-dichlorobenzophenone,
   (e) Other compounds such as di- and triphenylamine, azocompounds, benzyl sulfide, benzyl sulfoxide, phenyl sulfoxide.
   (f) Mixtures of sensitizers such as diphenylamines and tetracyanoethylene, 7,7,8,8-tetracyanoquinodimethane, *p*-chloranil, maleic anhydride, etc. [18].

5. *High-polymeric aciditives for improving impact strength* (see Chapter 22);
   5.1. Elastomeric compounds are incorporated in the thermoplastic ones: EPR, EPDM, etc.
   5.2. Compatibilizers are increasingly used to link different incompatible components in PO blends and composites.

6. *Light stabilizers, photostabilizers* are chemical compounds interfering in the physical and chemical processes of light-induced degradation. There are four main classes of light stabilizers having different modes of action [1, 5, 10].
   6.1. *UV absorbers,* their protection action is based on a filter effect, absorbing harmful radiation. Classes:
   (a) hydroxybenzophenones
   (b) hydroxyphenylbenztriazoles
   (c) cinnamate type
   (d) oxanilide type
   (e) salicilates
   (f) carbon black and other pigments, which act also as antioxidants
   (g) other compounds as resorcinol monobenzoates
   6.2. *Quenchers* confer stability through energy transfer
   (a) benzotriazole type UV absorbers
   (b) benzophenone type UV absorbers
   (c) nickel compounds used almost exclusively for PO
   6.3. *Hydroperoxide decomposers* play an important role in hydroperoxide group decomposition. The most important are metal complexes of sulfur or phosphorous containing compounds.
   6.4. *Free radical scavengers* act by free radical scavenging reactions. By this mechanism is explained the stabilization due to the addition of *n*-butylamine-nickel-2.2′-thiobis(4-*tert*-phenolate), nickel-bis-3.5-di-*tert*-butyl-4-hydrophosphonic acid mono butyl ester, 2-hydroxy-4-dodecyl oxybenzophenone. The hindered amine type light stabilizers (HALS) have the highest effectiveness.

TABLE 2 Commercially available antistatic agents

Tradename	Producer/supplier	Examples
Additive EN, N	ICI, UK	
Advastat	Advance Chem. Corp., USA	
Antistatic Agent	Hexcel Chemical Products Zeeland, USA	
Anti Static S.p.S.	Rhone Poulenc, France	Quaternary ammonium compounds
Antistatin	BASF AG, Germany	
Armostat	Akzo Chemicals Inc., USA	
Armostat 300, 400, 600	Armours Hess Chem., Ltd, UK	Ethoxilated amines
Interstat	Akzo Chemicals Inc., Netherlands	
Atmer	ICI Ind. Kortensburg, Belgium	
Adogen	Sherex Chemical Co., USA	
Barostat	Chem. Werke, Germany	
Catofor 02, 05, 06, 09, 15	Glovers Ltd., UK	Ethoxylated amines; *N*-bis-2-hydrox-ethyl-alkyl amine
Cemulcat Cequartyl	Soc. Francaise d'Organo-Synthèse, France	
Cyastat	American Cyanamid Co., USA	Stearoamides
Dehydrat	Henkel & Cie, Germany	
Ethoguard	Armour Hess Chem. Ltd., UK	Quarternary ammonium compounds
Gafac	GAF Corp., USA	
Hostastat	Hoechst AG, Germany; Hoechst Celanese Corp., USA	
Irgastat	Ciba-Geigy AG, Switzerland Ciba-Geigy Corp., Switzerland	
Ken-Stat KSMZ100	Kenrich Petrochemicals, USA	Aminozirconate and sulphenylzirconate reduce translucence
Lankrostat	Lankro Chem. Ltd., UK	
Larostat	Jordan Chemical Co., USA	
Markstat	Argus Chemical Belgium Div., Witco Corp., USA	
Nopcostat	Diamond Shamrock Chemicals Co., USA	
Sandin	Sandoz Huningue SA, France	
State. Eze	Fine Organics Inc., USA	
Stater SDG, MOG, PK	Pretext Cie, France	
Statexan	Bayer AG, Germany, USA	Alkyl sulfonates
Syntron C-101, 109	Pretext Cie, France	
Tebestat	Th Boehme KG, Germany	
VP-FE-2	Hoechst AG, Germany	Ester of glycerine with fatty acids
Witamol 60	Chemische Werke, Hüls, Germany	Polyglycol ester

Source: Refs. 1, 10.

6.5. Combined formulations of several light stabilizers and antioxidant systems are more effective in photostabilization.

7. *Lubricants* and other processing aids: their addition solves several processing problems. The trade names and suppliers of lubricants are given in Table 4 according to the following classification: A: fatty alcohols and dicarboxylic acid esters; B: fatty acid esters of glycerols and other short-chain alcohols; C: fatty acids; D: fatty acid amides; E: metal salts of fatty acids (metal soaps); E: oligomeric fatty acid esters (fatty acid complex esters); G: fatty alcohol fatty acid esters; H: wax acids, their esters and soaps; J: polar PE waxes and derivatives, waxlike polymers; K: nonpolar PO waxes; L: natural and synthetic paraffin waxes; M: fluoropolymers; N: combined lubricants; O: silicones, silica, silicates, etc. as antiblocking agents; P: other compounds such as $CaCO_3$, etc., which have a similar effect.

(Text continues on p. 528)

TABLE 3 Commercially available organic peroxides and their producers

Tradenames	Producer	Chemical structure
Dicup R,T	Hercules, Inc., USA	Dicumyl peroxide 90–100% solid, 40% on carrier
Dicup 40C	Hercules, Inc., USA	
Dicup 40KE	Hercules, Inc., USA	
Dicup 40-MB	Hercules, Inc., USA	40% master batch
DTBP	US Peroxygen Div., USA	Di-*tert*-butyl-peroxide, 95% liquid
Esperox IOXL	US Peroxygen Div., USA	*Tert*-butyl-peroxy-3,3,5-trimethyl-cyclohexane
Experox KXL	US Pergamon Div., USA	50% on carrier
Hercules D-16	Hercules, Inc., USA	*Tert*-butylcumyl peroxide, 95% liquid
Interox DYBP	Peroxid Chemie GmbH, Germany	2,5-Dimethyl-2,5-di-*tert*-butyl-peroxyhexane-3, 90% liquid
DHBP	Peroxid Chemie GmbH, Germany	2,5-Dimethyl-2,5-di-*tert*-butyl-peroxyhexane 90% liquid
DHBP-45-IC	Peroxid Chemie GmbH, Germany	45% on carrier
DHBP DTBP	Peroxid Chemie GmbH, Germany	Di-*tert*-butyl-peroxide, 95% liquid
DHBP DIPP	Peroxid Chemie GmbH, Germany	Bis-(*tert*-butyl-peroxyisopropyl) benzene, 90–100% solid
DIPP-40-IC	Peroxid Chemie GmbH, Germany	40% on carrier
DIPP-40-G	Peroxid Chemie GmbH, Germany	40% on carrier
DIPP BCUP	Peroxid Chemie GmbH, Germany	*Tert*-butylcumylperoxide, 95% liquid
BCUP-50-IC	Peroxid Chemie GmbH, Germany	40–50% on carrier
DCUP	Peroxid Chemie GmbH, Germany	Dicumyl peroxide, 90–100% solid
DCUP-40-IC	Peroxid Chemie GmbH, Germany	40% on carrier
NBV-40-IC	Peroxid Chemie GmbH, Germany	*n*-Butyl-4,4-di-*tert*-butyl-peroxy-valerate, 40% on carrier
CH-SOAL, FT	Peroxid Chemie GmbH, Germany	1,1-Di-*tert*-butyl-peroxy-cyclohexane, 50% liquid formulation
TBPB-50-IC	Peroxid Chemie GmbH, Germany	*Tert*-butyl-peroxy-benzoate, 50% on carrier
BP-50-P-SI	Peroxid Chemie GmbH, Germany	Dibenzoyl peroxide, 50% paste
DCLBP-50-P-SI	Peroxid Chemie GmbH, Germany	Bis-(di-chloro-benzoyl)peroxide. 50% paste
CLBP-50-P-SI	Peroxid Chemie GmbH, Germany	Bis-(*p*-chlorobenzoyl)peroxide, 50% paste
BU-10-IC	Peroxid Chemie GmbH, Germany	2,2-Di-*tert*-butyl-peroxybutane, 40/50% liquid formulation
Lucidol S-50S	Akzo Chemicals BV, The Netherlands	Dibenzoyl peroxide, 50% paste
Luperco 230XL	Luperox GmbH, Germany	*n*-butyl-4,4-di-*tert*-butylperoxy-valerate, 40% on carrier
331-80B	Luperox GmbH, Germany	1,1-Di-*tert*-butylperoxy-cyclohexane, 50% liquid formulation
231 LMB	Luperox GmbH, Germany	1,1-Di-*tert*-butyl-peroxy-3,3,5-trimethylcyclohexane, 40% master batch
231 XL	Luperox GmbH, Germany	40% on carrier
P-XL	Luperox GmbH, Germany	*Tert*-butylperoxy benzoate, 50% on carrier
AST	Luperox GmbH, Germany	Dibenzoyl peroxide, 50% paste
CST	Luperox GmbH, Germany	Bis-(2,4-dichlorobenzoyl)peroxide, 50% paste
233 XL	Luperox GmbH, Germany	Ethyl-3,3-di-*tert*-butyl peroxy-butyrate 40% on carrier
220	Luperox GmbH, Germany	2,2-Di-*tert*-butyl-peroxy-butane, 40/50% liquid formulation
130	Luperox GmbH, Germany	2,5-Dimethyl-2,5-di-*tert*-butylperoxy-hexyne 3, 90% liquid,
Luperco 130 XL	Luperox GmbH, Germany	45% on carrier
Luoperco 130 EVA-25	Luperox GmbH, Germany	25% master batch
Luperco 101	Luperox GmbH, Germany	2,5-Dimethyl-2,5-di-*tert*-butylperoxy-hexane, 90% liquid
Luperco 101 XL	Luperox GmbH, Germany	45% on carrier
Luperox Di	Luperox GmbH, Germany	Di-*tert*-butyl peroxide
Luperox 802 G	Luperox GmbH, Germany	Bis-(*tert*-butyl peroxyisopropyl)benzene, 80% master batch
Luperox 801	Luperox GmbH, Germany	*Tert*-butylcumyl peroxide, 95% liquid
Luperox 500R, 500T,	Luperox GmbH, Germany	Dicumyl peroxide, 90–100% solid
540C, 540 KE, 540G	Luperox GmbH, Germany	40% on carrier
540 PE, 540 LMB	Luperox GmbH, Germany	40%, master batch
Nypex BS	Nippon Oil and Fats Co., Japan	Dibenzoyl peroxide, 50% paste
CS	Nippon Oil and Fats Co., Japan	Bis-(*p*-chloro-benzoyl)peroxide, 50% paste
Perhexa V	Nippon Oil and Fats Co., Japan	*n*-Butyl-4,4-di-*tert*-butylperoxy-valerate, 40% on carrier
Perhexa 3M	Nippon Oil and Fats Co., Japan	1,1-Di-*tert*-butyl-peroxy-3,3,5-trimethyl cyclohexane, 40% on carrier
Perhexa 22	Nippon Oil and Fats Co., Japan	2,2-Di-*tert*-butyl-peroxy-butane, 40–50% liquid formulation

TABLE 3 (Continued)

Tradenames	Producer	Chemical structure
Perbutyl Z	Nippon Oil and Fats Co., Japan	*Tert*-butyl-peroxy benzoate, 50% on carrier
Perhexyn 25B	Nippon Oil and Fats Co., Japan	2,5-Dimethyl-2,5-di-*tert*-butyl-peroxy-hexyne-3, 90% liquid
Perhexan 25B	Nippon Oil and Fats Co., Japan	2,5-Dimethyl-2,5-di-*tert*-butyl-peroxy-hexane, 90% liquid
Perbutyl D	Nippon Oil and Fats Co., Japan	Di-*tert*-butyl peroxide, 95% liquid
Perbutyl P	Nippon Oil and Fats Co., Japan	Bis-*tert*-butyl-peroxy-isopropyl) benzene, 90–100% solid
Perbutyl P-40	Nippon Oil and Fats Co., Japan	40% on carrier
Perbutyl C	Nippon Oil and Fats Co., Japan	*Tert*-butyl-cumyl-peroxide, 95% liquid
Perbutyl C-40	Nippon Oil and Fats Co., Japan	40–50% on carrier
Percumyl D	Nippon Oil and Fats Co., Japan	Dicumyl peroxide, 90–100% solid
Perkadox 14-90,14S	Akzo Chemicals BV, Netherlands	Bis-*tert*-butyl-peroxyisopropyl) benzene, 90–100% solid
Perkadox 14-40B	Akzo Chemicals BV, Netherlands	40% on carrier
Perkadox 14-40K	Akzo Chemicals BV, Netherlands	
Perkadox 14-40 MB	Akzo Chemicals BV, Netherlands	40% master batch
Perkadox 14-40A	Akzo Chemicals BV, Netherlands	
Perkadox 14-40B	Akzo Chemicals BV, Netherlands	
Perkadox BC/BC	Akzo Chemicals BV, Netherlands	Dicumylperoxide, 90–100% solid
Perkadox BC-40B,	Akzo Chemicals BV, Netherlands	Dicumylperoxide, 40% on carrier
BC-40K	Akzo Chemicals BV, Netherlands	
BC-40MB; BC-40A	Akzo Chemicals BV, Netherlands	Dicumylperoxide, 40% master batch
Peroximon DE	Montefluos SpA, Italy	Di-*tert*-butyl peroxide, 95% liquid
Peroximon F	Montefluos SpA, Italy	Bis-(*tert*-butylperoxy-isopropyl) benzene, 90–100% solid
Peroximon F-40	Montefluos SpA, Italy	40% on carrier
Peroximon F-40MG	Montefluos SpA, Italy	40% master batch
Peroximon DC	Montefluos SpA, Italy	Dicumyl peroxide, 90–100% solid
Peroximon MC 40, 40K	Montefluos SpA, Italy	40% on carrier
Peroximon DC 40MG	Montefluos SpA, Italy	40% master batch
Peroximon S 164	Montefluos SpA, Italy	1,1-Di-*tert*-butyl-peroxy-3,3,5-trimethylcyclohexane, 40% on carrier
Trigonox 145	Akzo Chemicals BV, AE, Netherlands	2,5-Dimethyl-2,5-di-*tert*-butylperoxy hexyne-3, 90% liquid
Trigonox 145B	Akzo Chemicals BV, AE, Netherlands	45% on carrier
Trigonox 101	Akzo Chemicals BV, AE, Netherlands	2,5-Dimethyl-2,5-di-*tert*-butylperoxyhexane, 90% liquid
Trigonox 101-50D	Akzo Chemicals BV, AE, Netherlands	45% on carrier
Trigonox 45B, 40	Akzo Chemicals BV, AE, Netherlands	45% on carrier
Trigonox B	Akzo Chemicals BV, AE, Netherlands	Di-*tert*-butyl-peroxide, 95% liquid
Trigonox B-50E	Akzo Chemicals BV, AE, Netherlands	50–74% liquid formulation
Trigonox T	Akzo Chemie, USA	*Tert*-butylcumyl peroxide, 95% liquid
Trigonox T-50D	Nouiy Chemicals, USA	40–50% on carrier
Trigonox 17-40B	Nouiy Chemicals, USA	*n*-Butyl-4,4-di-*tert*-butylperoxy-valerate, 40% on carrier
Trigonox 17-40, MB	Nouiy Chemicals, USA	40% master batch
Trigonox 22-B-50	Nouiy Chemicals, USA	1,1-Di-*tert*-butyl-peroxy-cyclohexane, 50% liquid formulation
Trigonox 29-40B	Nouiy Chemicals, USA	1,1-Di-*tert*-butyl-peroxy-3,3,5-trimethylcyclohexane, 40% on carrier
Trigonox 29-40MB	Nouiy Chemicals, USA	40% master batch
C	Nouiy Chemicals, USA	*Tert*-butylperoxy-benzoate, 100%
C-50 D	Nouiy Chemicals, USA	50% on carrier
DB-50	Nouiy Chemicals, USA	2,2-Di-*tert*-butyl-peroxy-butane, 40/50% liquid formulation
USP 400P	US Peroxygen Div., USA	1,1-Di-*tert*-butyl-peroxy-cyclohexane, 50% liquid formulation
USP 333XL	US Peroxygen Div., USA	Ethyl-3,3-di-*tert*-butylperoxy-butyrate
USP 333 KXL	US Peroxygen Div., USA	40% on carrier
Vulcup R	Hercules, Inc., USA	Bis-(*tert*-butyl-peroxyisopropyl) benzene, 90–100% solid
Vulcup 40C, 40KE	Hercules, Inc., USA	40% on carrier
Vulcup 40MB	Hercules, Inc., USA	40% on master batch

Source: Adapted from Ref. 1, 2.

TABLE 4 Commercially available lubricants and related (mold release) agents

Tradename	Supplier/producer	Chemical class
Abril Wax	Abril Industrial Waxes Ltd., UK	A, D, G
Advawax	Advance Chem. Corp., USA	Mixture
Aero	American Cyanamid Co., USA	Metallic stearates
Ampacet 100520	Ampacet Dow Corp., USA	5% erucamide
Ampacet 100329	Ampacet Dow Corp., USA	
Ampacet 100342	Ampacet Dow Corp., USA	20% Diatomaceous silica
Armoslip	Akzo Chemical Inc., USA	D
Armid H,T,	Soc. Bezons, France	Fatty amides
Arwax	Rohm and Haas Co., USA	Paraffin waxes
A.C. Polyethylene	Allied Corp. Int. NV-SA, Netherlands	J, K
Barolub	Chemische Werke Germany	A, B, C, D, E, F, G, J, K, L, N
Cab-o-sil	G.L. Cabot Inc., USA	Antiblocking agent silica gel
Carbowax	Union Carbide Corp., USA	Polyethylene glycol and methoxypolyethylene glycol
Carbune	Beacon Co., USA	Synthetic waxes
Chemetron 100	AMC-Corp., USA	Synthetic waxes
Crodacid	Croda Universal Ltd., UK	C
Crodamid	Croda Universal Ltd., UK	D
Dapral L	Akzo Chemicals GmbH, Germany	Unspecified
Dri-Slip	3M Co, USA	Antislipping agent, demolding agent, fluorinated compounds
Elwax	Du Pont de Nemours, USA	EVA copolymers
Epolene Waxes	Eastman Chemical Int., USA	J, K
Estol	Unichema Chemicals Inc., USA	B, G
Fluorolube	Hooke Electrochem Co., USA	Low molecular weight fluorinated polymers for high temperatures
Fransil	Soc. Neo., France	Demolding agents, silicon compounds
Glycmonos	Comiel, Prodotti Chimici Indus, Italy	B
Hoechst Wachs CaF-I, CaF-2	Hoechst AG, Germany	D, H, J, K
Hostalub H-1,	Hoechst Celanese Corp., USA	D, H, K, L, M, N
Hostalub FA 1	Hoechst Celanese Corp., USA	Amide wax
H-II, PH	Hoechst Celanese Corp., USA	
G-2503,	Hoechst Celanese Corp., USA	
2504	Hoechst Celanese Corp., USA	
HaroChem	Harcros Chemicals, Netherlands	
HaroGel	Harcos Chemicals, Netherlands	E
Harowax	Harcos Chemicals, Netherlands	A, B, C, D, H, L
Irgawax	Ciba Geigy GmbH, Germany	A, B, C, D, G, L, N
Interstab L	Akzo Chemicals Inc., Germany	A, B, C, D, E, G, H, J, K, L, N
Lankroplast	Lankro GmbH, Germany	A, C, D, J, N
Lankromark	Lankro GmbH, Germany	E
Ligalub	Peter Greven Fettchemie, Germany	A, B, C, D, E, G, L
Kel-F	3M Co., USA	Low molecular weight polytrichlor-fluorethylene
Listab	Chemson Gesellschaft für Polymer-additive Germany	E
Loxiol	Henkel KGaA, Germany	A, B, C, E, F, G, L, N
Loxamid	Henkel Corp., USA	D
Lubriol	Comiel, Prodotti Chimici Industriali, Italy	A, B, D, F, G, H
Naftolube	Chemson Gesellschaft für Polymer-additive Germany	A, B, D, G, L
Naftozin N	Chemson Gesellschaft für Polymer-additive Germany	C
Pationic 1061	Patco Polymer Additives, USA	Glycerol esters, FDA approved
1083	Patco Polymer Additives, USA	

TABLE 4 (Continued)

Tradename	Supplier/producer	Chemical class
Polywachs	Chemische Werke, Hülls Marl AG, Germany	Polyethylene glycol
Polywax	Petrolite Corp., USA	LMWPE
Pristerene	Unichema Chemie, Germany	C
Prifrac	Unichema Chemie, Germany	C
Priolube	Unichema Chemicals Inc., USA	G
Realube	Reagens Societe per Azioni Industria Chimica, Italy	A, B, C, D, E, F, G, K, L
Sicolub	BASF AG, Germany	D, E, H, J, N
Staniyor	M & T Chimie SA, France	C, D, E, H
Swedlub	Swedstab AB, Sweden	A, G, L
Swedstab	Swedish AB, Sweden	E
Synthewax Comiel	Prodotti Chimici Industriali, Italy	D
TL 127	Liquid Nitrogen Process, USA	Fluorocarbon powder
Tospearl	GE Silicones, USA	Fine particle, silicone fully crosslinked, for thin films
Unislip	Unichema Chemicals Inc., USA	D
Uniwax	Unichema Chemicals Inc., USA	D
Ultrasil	Chem. Fabrik, AG, Germany	Antiblocking agent, silicon compounds
Vestowax	Hülls AG, Germany	J, K, L
Vinlub	Commer, Italy	B, D, G
Vidax 73-U	Du Pont de Nemours, USA	LMWPTFE
Wachs BASF	BASF AG, Germany	H, J, K, N

Source: Refs. 1, 2, 10.

8. *Metal chelators or metal deactivators* retard effectively metal-catalyzed oxidation. Generally they are chelating agents that deactivate transition metal ions by forming inactive or stable complexes: Classes (see also Table 5) are: amides of aliphatic and aromatic mono- and dicarboxylic acids and their *N*-mono-substituted derivatives (*N,N′*-diphenyl oxamide) cyclic amides as barbituric acid; hydrazones and bishydrazones of aliphatic or aromatic aldehydes (such as benzaldehyde, salicylaldehyde) or *o*-hydroxyaryl ketones; hydrazides of aliphatic and aromatic mono- and dicarboxylic acids as well as *N*-acylated derivatives; bisacylated hydrazine derivatives; heterocyclic compounds such as: melamine, benztriazoles 8-oxyquinoline, hydrazone, and acylated derivatives or hydrazinotriazines, aminotriazoles and acylated derivative polyhydrazides; sterically hindered phenols + metal complexing groups, Ni salts of benzyl phosphonic acids; pyridinethiol/Sn-compounds; *tert*-phosphorus acid ester of thiobisphenol; various combinations with antioxidants.

9. *Nucleating and clarifiers agents* determine the increase of the rate of nucleation by athermal and heterogeneous nucleation (see Chapter 7) and increase the rate of nucleation, leading to the decrease of the spherulite dimensions and increased crystalinity. All clarifiers nucleate but not all nucleators clarify well. Commercially available nucleators can be classified into two groups named "melt sensitive" and "melt insensitive" [22]. With the clarifying agents, spherulites become smaller than the wavelength of the visible light so allowing more light to pass through and product will be transparent. A new technology for clarifying PP is claimed by Mulliken Chemical. Milland 3988 grade PP gives increased transparency and is approved in PP formulations by FDA, HPB, Canada, and BGA, Germany.

The *melt sensitive nucleators* have a melting point which is below or near the normal processing temperature for PP-based resins, while the *melt insensitive nucleators* do not melt at normal processing temperature. The nucleation mode differs between the melt sensitive and melt insensitive nucleators. The first ones set up a "physical gelation network" within the matrix that assures a high dispersion and therefore a high degree of nucleation. The melt insensitive

(Text continues on p. 531)

TABLE 5 Commercial metal deactivators (metal chelator) suitable for PO

Source	Structure	Tradename (designation)
American Cyanamid, USA	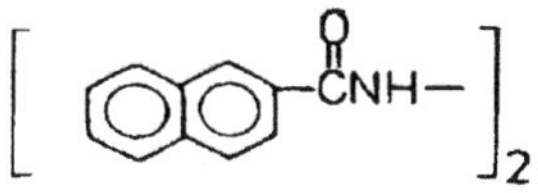 [–CNH–]$_2$ (C=O)	BNH
American Cyanamid, USA	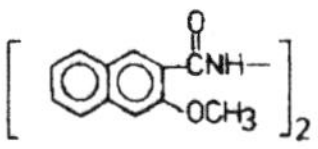 [–CNH–, –OCH_3]$_2$	MBNH
Argus Chemical SA Belgique	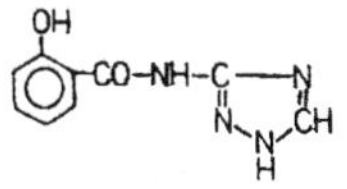 OH, CO–NH–C, N, N, CH, N, H	Mark CDA-I and $(CDA\text{-}I)_2$, and related compounds
Argus Div., USA	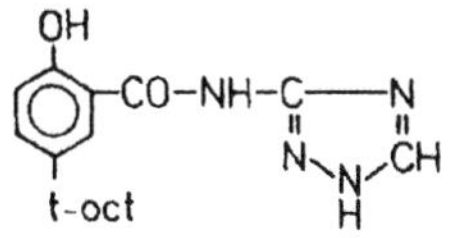 OH, CO–NH–C, N, N, CH, N, H, t-oct	$(CDA\text{-}I)_{oct}$
	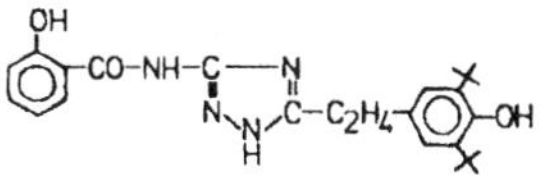 OH, CO–NH–C, N, N, C–C_2H_4–, OH, N, H	(CDA-I)-A
	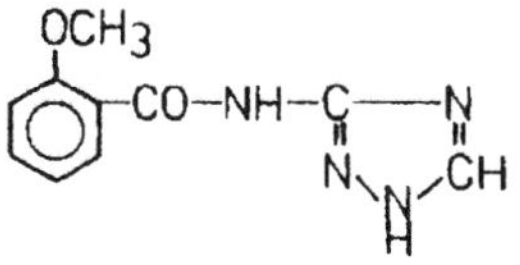 OCH_3, CO–NH–C, N, N, CH, N, H	(CDA-I)-OR
	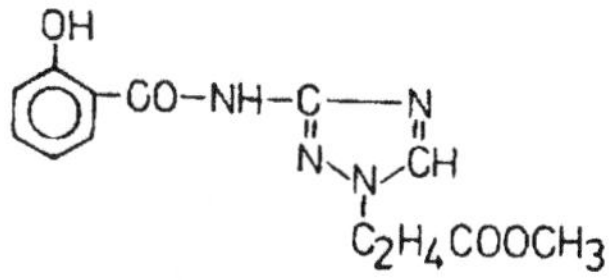 OH, CO–NH–C, N, N, CH, N, $C_2H_4COOCH_3$	(CDA-I)-NR
	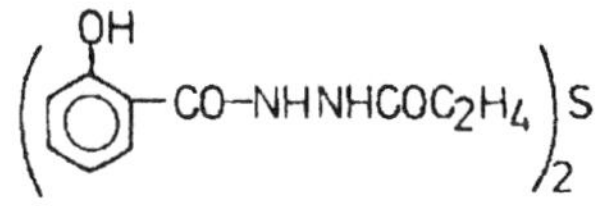 (OH, CO–NHNHCOC_2H_4)$_2$ S	(CDA-3) and $(CDA\text{-}3)_2$
	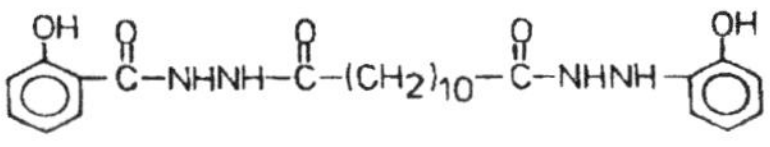 OH, C(=O)–NHNH–C(=O)–$(CH_2)_{10}$–C(=O)–NHNH–, OH	CDA-6

TABLE 5 (Continued)

Source	Structure	Tradename (designation)
Ciba-Geigy AG, Germany American Hoechst Corp., USA	HO–⟨Ph(+)₂⟩–CH_2CH_2–C(=O)–NHNH–C(=O)–CH_2CH_2–⟨Ph(+)₂⟩–OH (+ = tert. butyl group)	Irganox MD 1024
Ciba Geigy Corp., USA	[HO–⟨Ph(+)₂⟩–CH_2–CH_2–COO–CH_2]$_4$≡C	Irganox 1010
Eastman Chemical International, Switzerland Eastman Chemical Products, USA	[⟨Ph⟩–CH=NNHC(=O)–]$_2$	Eastman Inhibitor OABH
Hoechst AG, Germany, USA	[HO–⟨Ph(+)(CH_3)⟩–S–⟨Ph(+)(CH_3)⟩–O–]$_3$P	Hostanox VP OSP
Monsanto Industrial Chemicals Co., USA RT Vanderbilt Co. Inc., USA	[C_2H_5OC(=O)–⟨Ph⟩–NHC(=O)–]$_2$	For HDPE
	[⟨Ph⟩(C=O)₂>NNHC(=O)–]$_2$	
Sherwin William Co.,	Benzotriazole (N=N–NH fused to benzene ring)	
Ube Industrial Ltd.,	CH_3OC(=O)–⟨Ph⟩–C(=O)NHNHC(=O)CH=CH–⟨Ph⟩	
	[⟨Ph⟩–CH=CHC(=O)NH–]$_2$	
	HO–⟨Ph⟩–C(=O)NHNHC(=O)CH=CH–⟨Ph⟩	

TABLE 5 (Continued)

Source	Structure	Tradename (designation)
	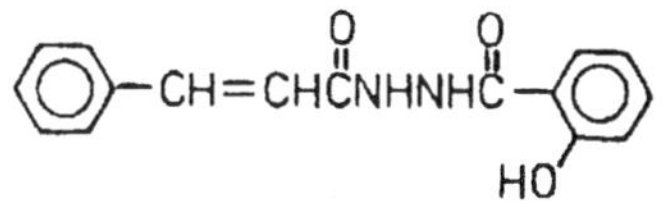 C_6H_5–CH=CHC(O)NHNHC(O)–C_6H_4(HO)	For HDPE
Uniroyal Inc., USA	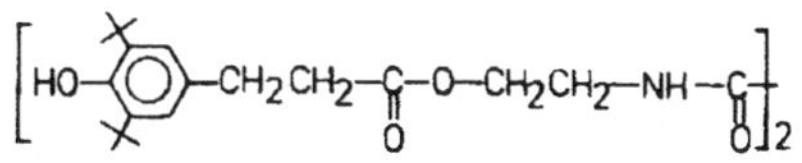 [HO–C_6H_2(t-Bu)$_2$–CH_2CH_2–C(O)–O–CH_2CH_2–NH–C(O)–]$_2$	Naugard XL-1

Source: Refs. 1, 2, 21.

nucleators act as a single point nucleation site within the matrix.

So, the clarity of the compound is effected by the dispersion of the finely divided nucleators determined by the mixing time in a high speed mill.

Melt sensitive nucleators include the sorbitol (bisbenziliden)-based compounds with the following structure:

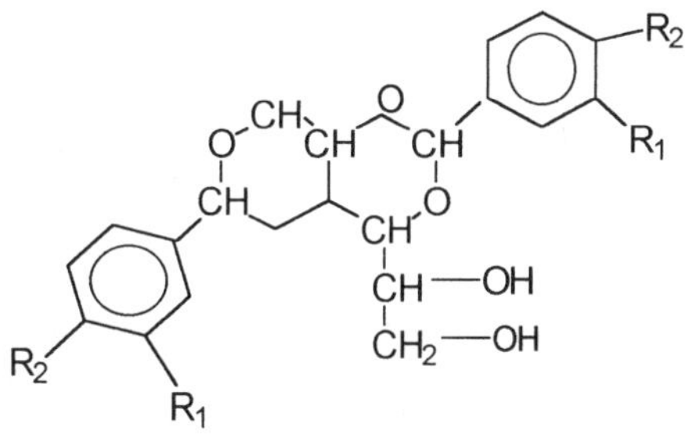

where R_1 is —H or —CH_3 and R_2 is —H, —CH_3 or CH_2—CH_3.

Their decomposition at elevated temperatures causes two problems: the lost of the nucleating property, and the formation of aldehydes as decomposition products. Although the odor of the aldehydes is pleasant it is undesirable in most applications.

The following nucleators are known:

(a) Melt insensitive: dibenzylidene sorbitol (DBS) very active. They are less effective as clarifiers. Alkali metal or aluminum salts of aromatic and aliphatic carboxylic acids (Al, Na, K, Li-benzoate or *tert*-butyl benzoate), which are particularly active in PP.

(b) Inorganic additives such as talc, silica, kaolin, and catalyst residues, which have a poor effect.

(c) Organic compounds such as salts of mono- or polycarboxylic acids: sodium succinate, sodium glutarate, sodium cinnamate, sodium capronate, sodium-4-methyl valerate, Al-phenyl acetate, which has a medium effect; potasium stearate, very effective for HDPE; polymers such as ethylene/acrylic ester copolymers.

10. *Special additives* see Chapter 21:
    blowing agents
    flame retardants
    pigments and colorants
    fillers and reinforcements

11. *Other additives*
    11.1. *Plasticizers* reduce polymer–polymer interactions. They are used 0–50 wt% in sulfochlorinated PE.
    11.2. *Extenders* used in PO rubbers.
    11.3. *Compatibilizing agents* prevent phase separation in multicomponent systems.
    11.4. *Coupling agents* increase the permanency of the effect of an additive:
        silanes
        titanates
        zirconates
        functionalized PO
    11.5. *Adhesives*
    11.6. *Solvents, dilutes*
    11.7. *Antifungal compounds, bactericides, bacteriostats, microbiocides–biostabilizers* prevent biological growth on surface of articles. The following classes [23] are used:
        10,10′-oxybis(phenoxarsin);
        N-(trichloromethylthio)phthalimide and
        N-trifluoromethylthio)phthalimide;

tributyltinoxide and derivatives;
copper γ-hydroxyquinoline;
zinc dimethyldithiocarbamate;
diphenylantimony-2-ethylhexanoate.

11.8. *Additives for control surface properties* modify the frictional and adhesion properties. Daniel Products' Slip Ayd surface conditioner range number 14 is a PE micronized powdered wax giving a range of particle sizes and hardness with improved resistance to blocking and abrasion. For rapid reduction in the coefficient of friction of blow films, there is a highly loaded oleoamide slip concentrate from Colortech with high content (10%) of active ingredients, cost effective as a mold release agent for injection molding of the PO elastomers.

11.9. *Shrinkage modifiers* are commonly inorganic fillers; A new and revolutionary approach to obtain low and zero-shrink is with the use of a combination of Low Profile Additive and $CaCO_3$. In these additives, the $CaCO_3$ particles are coated with the appropriate polymer, resulting in a more efficient low profile action. The products are in the form of beads of 80 µm mean diameter and are free-flowing and less dusty than other powders. During compounding they disperse easily.

11.10. *Additives for the improvement of machinability and adhesion resistance* make the system harder and more easily cut. They also include inorganic fillers such as powdered metals, wood flour, $CaCO_3$, sawdust, clay, talc, etc.

11.11. *Barrier forming additives* improve barrier properties and influence permeability reduction. An aluminosilicate with platelet type particles in the nanometer size range reduces permeability to gases by up to 45 times and is a new mineral filler for PO, EVA films, etc. They are increasingly used in packaging and are of great interest for automobile fuel tanks. This technology is developed by Nanocar, Arlington USA.

11.12. *Degradation additives* are developed due to the problems of recycling, but experience has shown that their practical use is less than was originally estimated. The consumers and legislators began to appreciate that the existing landfills do not meet the conditions for efficient degradation. The interest for these additives decreased for the producers of trash and grocery bags but it still remain for the agricultural and horticultural sector. Ketones and cyclic ketones are useful as degradation agents and for MFI control of PO [24].

11.13. *Antifogging agents* prevent build-up of condensation inside food packaging. Although Ampacet Antifog PE MB has optimum antifog properties and is acceptable for food contact, a new Accurel additive has recently been developed. It is based on LDPE microporous carrier with 25 wt% of a special ester (monooleate). For 1% level it is very effective and has a good stability and imparts clarity to the film. Studies of different sorbitol fatty acid esters used as antifog agents in PE revealed that their mobility depends on the length of the hydrocarbon chain, while the antifog performance is influenced by the ratio of the —OH groups to the hydrocarbon chain length [25].

11.14. *Acid scavengers* neutralize catalyst residues and prevent equipment corrosion. The choice of the acid scavengers can affect the overall acidity/basicity of a resin and influence the reactions of many of the organic additives in the system. Also, the acid scavengers as metallic stearates will impart to the polymer a certain lubricity, mold release or slip properties. As acid scavengers can be used: metallic stearates (of sodium, calcium and zinc, zeolites structures (dihydro talcite, both synthetic and natural) CaO and ZnO and other metallic salts based on lactic or benzoic acid. Environmental Products has introduced PTA-210 and PTA-310 for increased thermal stability of PE, EVA, EMA, EVAA, PP, and ionomers. They are used to neutralize acidic catalyst residues. Concentration level is very low (0.2–0.5%). The hydrocalcite L-SSRH introduced by Reheis has a similar effect.

III. ACTIVITY AND STRUCTURE–EFFICIENCY RELATIONSHIPS

A. Phenolic Antioxidants

The stabilization of the polymers is still undergoing transition from an art to a science as the

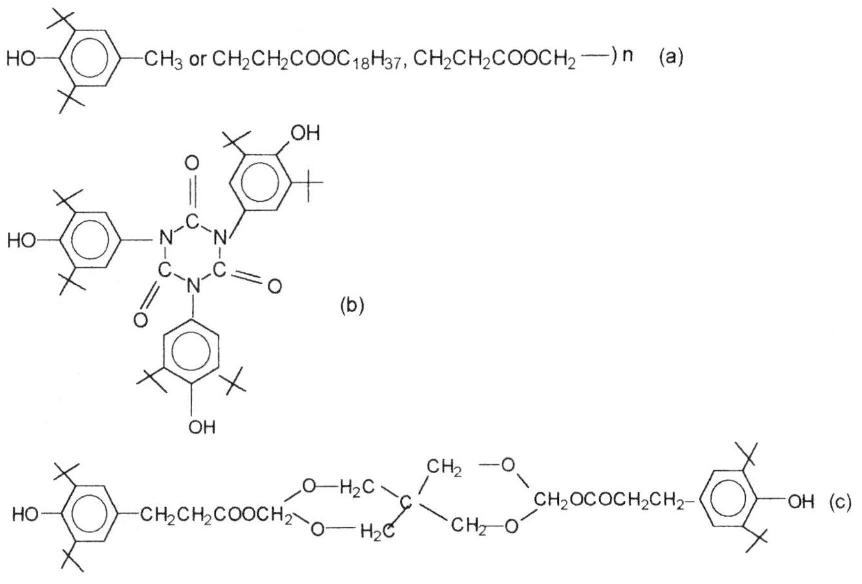

SCHEME 1

mechanism of degradation becomes more fully understood.

Chemical structures of several phenolic antioxidants are presented in Scheme 1. Hindered phenolic antioxidants (AO) are the chemically most diversified group of stabilizers. Phenolic moities also constitute an important part of UV light absorbers and metal chelators.

As has been already mentioned in Chapter 16, the radicals are involved in the chain initiation, propagation, branching, and termination steps of thermally, catalytically, mechanically or radiation induced processes. Two kinds of free radicals are of key importance: carbon centered macroradical $R^{\bullet}$, and oxygen centered alkyl peroxyl $POO^{\bullet}$, acyl $RO^{\bullet}$ and acylperoxyl $RC(O)OO^{\bullet}$. Suitable stabilizers or stabilizer systems can inhibit or delay degradation [26].

$R^{\bullet}$ and $ROO^{\bullet}$ can be scavenged by chain breaking antioxidants by an electron–acceptor process. The electron donors such as aromatic amines scavenge $ROO^{\bullet}$ while some transformation products are able to scavenge both $R^{\bullet}$ and $ROO^{\bullet}$. Hydroperoxides and peroxyacids rank among the most dangerous species formed in polymers in an aerobic environment. Hydroperoxide decomposing (HD), antioxidants (AO), light stabilizers (LS) or metal deactivators (MD) are applied to prevent hydroperoxide homolysis. The reactivity of $R^{\bullet}$ (formed by breaking C—C or C—H bonds during PO processing and in solid polymers degrading in an oxygen deficit atmosphere) is influenced by microenvironment. Crosslinking or chain scission change the properties of the polymer. Therefore, scavenging of the primarily formed $R^{\bullet}$ radicals immediately after their formation is very important in polymer stabilization [27]:

$$RH \xrightarrow[\text{mechanical forces}]{t^{\circ}C,\ M^{n+},\ h\nu,\ \text{radiation}} R^{\bullet}\ (RR'C{=}O^{*}) \xrightarrow{O_2/RH} ROO^{\bullet}\ (\downarrow \text{H Donor}) \xrightarrow{RH} ROOH + R^{\bullet} \xrightarrow{RO^{\bullet}\ \text{scavenger}} RH$$

RH ↑ LS, MD; ROOH + R• → RO• + •OH

R• scavenger ↖ ↗

ROO• scavenger + HD, AO, LS → stable compounds

RO• scavenger

Chain breaking or primary antioxidants interfere in the various steps of degradation as in the chain propagation step either by a chain breaking by a donor mechanism or by a chain breaking acceptor mechanism, deactivating alkyl and alkylperoxy radicals. Phenolic and secondary aromatic amine type antioxidants act according to a chain breaking donor mechanism [28] such as:

$$RO^{\bullet} + AH \xrightarrow{k_{inh}} ROOH + A^{\bullet} \qquad E_a = -32.6\,\text{kJ/mol}$$

$$RO^{\bullet} + A^{\bullet} \longrightarrow ROOA\ \text{nonradical product}$$

$$A^{\bullet} + RH \longrightarrow AH + R^{\bullet} \qquad E_a = 123\,\text{kJ/mol}$$

less probable while by a chain-breaking acceptor mechanism the aromatic nitro and nitroso compounds act, this mechanism also being presented for the transformation products of phenols having a conjugated dienone (quinone methides, stilbene quinones and benzoquinones) structure.

A mechanism of the sterically hindered or semi-hindered phenol action in PO stabilization has been elaborated by Pospisil who explained the additive consumption [29–31].

The main reactions of phenolic antioxidants take place with peroxy radicals, oxygen, singlet oxygen, phenoxy radical disproportionation, etc.

Specifically, the hindered phenols are able to transfer their phenolic hydrogen to the generated radical, a nonradical product being formed. The stable hindered phenoxy radical will not abstract more hydrogen from a polymer matrix:

$$\mathrm{ROOC{-}CH_2{-}CH_2{-}C_6H_4{-}OH} + \mathrm{R^{\bullet}} \rightarrow \mathrm{RH}$$
$$+ \mathrm{ROO^{\bullet}} \rightarrow \mathrm{ROOH}$$
$$+ \mathrm{RO^{\bullet}} \rightarrow \mathrm{ROH}$$

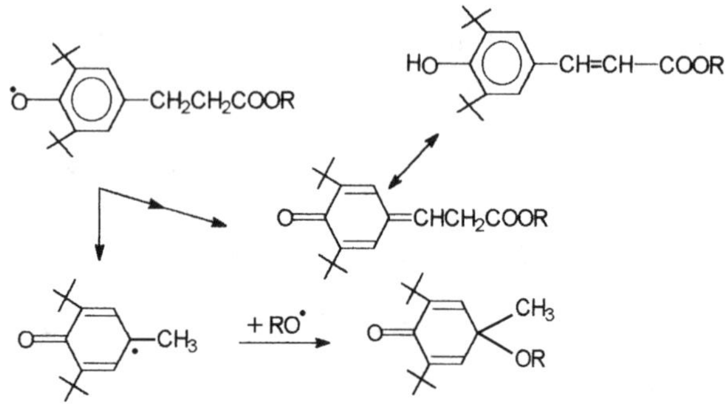

The explanation of the stabilization mechanism of hindered phenols by quinonmethides formation led to improvement in the light stabilization action. So phenols with a suitable substitution can also act as radical scavengers. The quinonmetides by a disproportionation mechanism regenerate the phenolic groups and its activity:

The initial radicals $\mathrm{R^{\bullet}}$ and $\mathrm{RO^{\bullet}}$ are effectively removed from participation in the propagation steps. The effect is not permanent if a newly formed hydroperoxide can undergo homolytic cleavage with the formation of other two radicals (see Eq. (3a), Chapter 16).

The phenoxy oxygen-based radical is easily delocalized to a carbon atom forming a quinone-like structure. Phenolic antioxidants having, in positions 2 or 4, substituents bearing at least one hydrogen atom in the α-carbon atom farm quinone methides as a consequence of mono-electron oxidation. The quinone-methides formed *in situ* from phenolic antioxidants have a considerable contribution to the overall stabilization effect during long-term heat aging of PO by their free-radical scavenging capacity and in combination with phenols. These new chemical structures are formed from the original additives after deactivation of oxidizing species heterogeneously distributed in polymer matrix. Some of these products contributed to the integral stabilization effect and/or are even a necessary condition of effectivity. Stabilizers are preferentially consumed by the oxidizing species and in the surface layers especially during the induction period [32]. Finally, the concentration of the effective form of stabilizer is reduced below a level no longer able to protect the degrading polymer and the failure occurs in this location. Besides this consumption, the stabilizers are depleted by photolysis, radiation, atmospheric oxidants, catalysts, and metallic compounds. In particular, nitrogen oxides and sulfur oxides can react with phenolics at room temperature to form colored (typically yellow) structures. The discoloration process is referred as "gas fading". When gas fading occurs, the hindered phenol used in the stabilization system forms a compound which can turn the products (such as fibers or films) pink, blue or yellow. To avoid, eliminate or at least minimize the gas fading problem, the following solutions have been given:

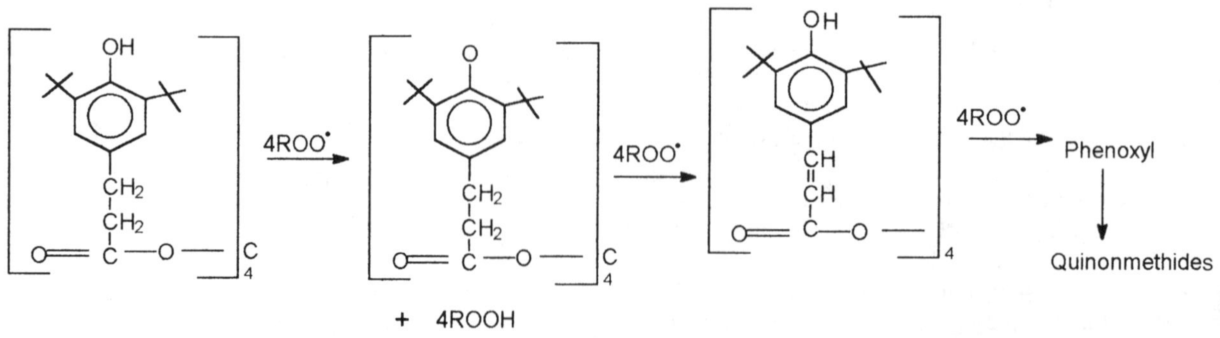

(a) use of hindered phenols resistant to gas fading and structures b and c (Scheme 1);
(b) use stabilization systems such as phosphite or hydroxylamine, the last being much more suitable because it also offers melt stability.
(c) Very recently, benzofuranone derivatives have been described acting basically as C(alkyl) radical scavengers [33]; non-discoloring processing stabilizers.

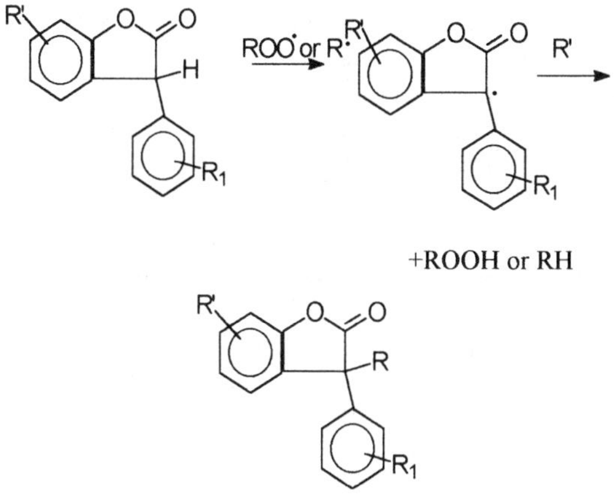

+ROOH or RH

(d) A mixture of heavy hydroaromatics was reported as effective melt and long-term stabilizers, and antioxidants for PO [34]. Besides discoloration of polymers, quinone methides may participate in aged polymers in reactions with free radicals and compounds prone to 1,6-addition. This accounts for aromatization of the conjugated dienoid system and may be connected with regeneration of the phenolic function. Substitution of the 4-position by a tertiary alkyl prevent discoloration due to QM formation.

Based on quinonmethide (QM) chemistry, Pospisil indicted the following structural regularities for structure–stabilizers' action in PO [37]:

(a) If the general structure of hindered phenols antioxidants (HPA) is considered:

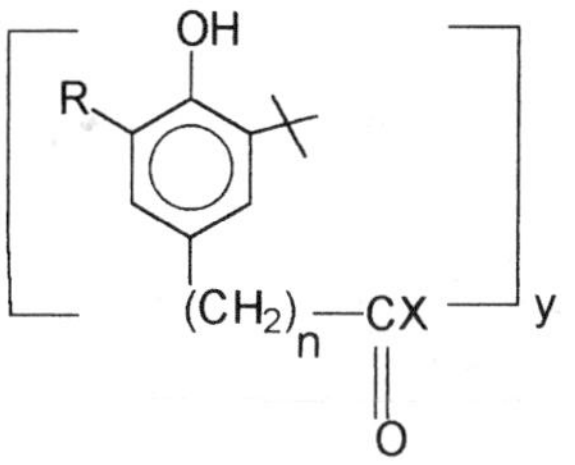

the structural changes in the 4-position change the phenol function. For example esters X = O, y = alkyl, $n = 1 \sim 4$, or X = $O(CH_2)_3$, y = 0, $n = 2$ are processing and long-term heat stabilizers. Sulphur-containing X = $O(CH_2)_2$, y = S, $n = 2$ is an effective antioxidant both for PO and diene rubbers while hydrazide (X = NH, y = 0, $n = 2$) and oxamide (X = $O(CH_2)_2NHCO$, y = 0, $n = 2$) are metal deactivators for PO in contact with Cu. It was also supposed that the lower sterical hinderance of the phenolic moiety increases the synergetic effect via in-cage cooperation with other stabilizers. If the QM is the main end-products of chemical consumption of HPA, it is also responsible for inactive consumption by oxidizing or sequesting metallic impurities, oxidation with environmental NO_n resulting 4-nitroso or nitrocyclohexadienones, photolysis, photo-oxidation and radiolysis.
(b) The oxygen centered radicals are also scavenged by secondary aromatic and planar partially hindered amines. These compounds act as antifatigue agents and antiozonants. Their mechanism of action is relatively complicated. It includes formation of aminyls, nitroxides, nitrones, quinones, imides, and deeply colored condensation and cyclization products. Such a chemical transformation limits the application of amines in light colored polymer products.

In PP stabilization a strong heterosynergism of dihydroacrydine with activated thioethers was observed.
(c) The scavenging of $R^{\bullet}$, $ROO^{\bullet}$ and $RC(O)OO^{\bullet}$ radicals, deactivation of ROOH and peracids have also been included in the mechanism of HPA activity.

The efficiency of the phenolic antioxidants is related to their chemical structure, especially steric hindrance by alkyl substituents in the 2- and 6-position and substitution in the 4-position [35]:

(i) contribution of the fully sterically hindered phenols to stabilization consists of the stoichiometric reaction between the phenol and peroxy radicals;
(ii) partially hindered phenols having a hydrogen atom on the α-carbon give quinonemethide; contribute to stabilization by stepwise reactions resulting in stable transformation products in an "over-stoichiometric way" [36];
(iii) cryptophenols that are not substituted in the 2- or 6-positions and do not contribute to stabilization.

Esters are processing and long-term stabilizers, sulfur containing compounds are antioxidants for PO and diene rubbers, while oxamides are metal chelators.

Hydroperoxides (HP) are the primary molecular photo-oxidation products and the most dangerous chromophores in photodegradation. HP are stable only at ~120°C, decomposing at high temperatures. Therefore their study in PO melt is questionable.

Adjusting the molecular architecture of stabilizers to maximize their activity, physical persistence and compatibility with polymeric matrix is of great importance. Beneficial co-operative effects in stabilizer combinations are also obtained by influencing the physical factors such as migration.

B. Activity Mechanisms of Amines

Hindered amines were initially designed to act as light stabilizers (HALS), because the aminoxy radical is only formed during exposure to light. Today they gain increasing importance as long term thermal stabilizers (HAS).

Cyclic hindered amines are applicable as light stabilizers for most polymers. Aromatic and heterocyclic amines have a crucial importance for stabilization of PO rubbers and coatings. They have antioxidant, antifatigue, antiozonant and photostabilizing activities due mainly to the scavenging activities of R$^{\bullet}$ and ROO$^{\bullet}$ as the key stabilization pathway [38]. Hydroxyl amines also function like primary antioxidants and radical scavengers.

The stabilization mechanism of HAS on photo- and thermo-oxidation was attributed to the HAS as such with a more or less pronounced contribution of their oxidation products [39]. For example, high molecular weight HAS protects from thermal oxidation even in the absence of their oxidation products and at temperatures as high as 135°C. Tensile strength, carbonyl index, and other properties vary with aging time without significant induction period (Fig. 1).

HAS can protect polymers from thermo-oxidative degradation according to the two main mechanisms of stabilization already encountered, i.e., free radical scavenging, and hydroperoxide decomposition. The first type of stabilization reaction is based mainly on oxidation products of the amines, though for some time radical scavinging was considered to also be

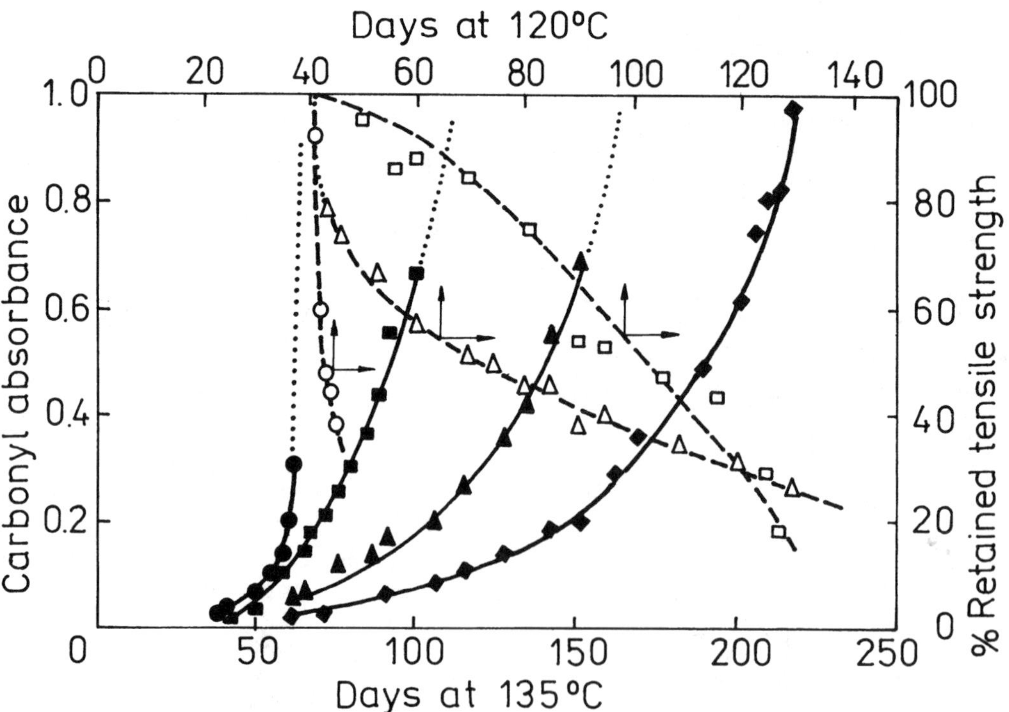

FIG. 1 Carbonyl absorbance and tensile strength of PP stabilized with phenolic (●, ○) and various HAS (■, □) [39].

responsible for the light stabilizing activity of HAS. The reactions partially shown in the following schemes involve on the one hand hydrogen abstraction from hydroxyl amine and hydroxylamine ethers and on the other hand scavenging of alkyl radicals by nitroxyl radicals. Since the hydrogen atom of hydroxyl amine is very labile its abstraction by free radicals is very easy. The rate constants for hydrogen abstraction by peroxy radicals from hydroxylamine and phenolic antioxidants are comparable.

The secondary amino group in the α-position to the aromatic nucleus is the key functionality. Several structures such as substituted diphenylamines, *N*-phenyl-1-naphthyl amine are antioxidants, *N,N′* disubstituted 1,4-phenylenediamines; 6-substituted or oligomeric 2,2,4-trimethyl 1,2-dihydroquinolines, 2,2-disubstituted 1,2-dihydro-3-oxo(phenylimino)-3H indoles, substituted phenol triazines and nitrogen containing heterocyclic compounds act as antioxidant and antifatigue agents. Compounds like 4,4′-bis(α, α'-dimethylbenzyl)diphenyl amine are applicable in homosynergetic combination with HP for PO stabilization. Most authors considered oxidation products of HAS (especially nitroxyl radicals) to be the key to their excellent light stabilization performance. Some recent works showed that the HAS mechanism is quite complex. In addition to the various transformation products the charge transfer complex of the parent amine with polymer can play an important role in PO stabilization [40, 41]. The aromatic amines seem to be more effective antioxidants at similar compositions than phenol derivatives, showing a better resistance against irradiation during PE as tubes processing [42].

The aromatic aminyls $=N^{\bullet}$ are the primary radical species involved in the amine stabilizing mechanism able to scavenge $R^{\bullet}$ and $ROO^{\bullet}$ radicals.

The intermediary N-O-R formed by the reaction with the peroxy radical $ROO^{\bullet}$ is returned to the reactive nitroxyl radical in a regenerating process or "Denisov cycle" [43].

Regeneration of the nitroxide is the main reaction envisaged in the HAS cyclic activity [22, 44]:

R—(piperidine)NH —Oxidation→ ; R• ; R—(piperidine)NO• ⇄ R—(piperidine)NOR ; ROH ; ROO•

An important point about HAS is that the nitroxyl species is regenerative. However, the cyclic regeneration eventually ends. Some of the regenerative products are inefficient radical scavengers and some of the stabilizer is lost during exposure. The high molecular weight compounds have been shown to be effective as thermal stabilizers.

The HAS have a lower effectivity as heat stabilizers than phenols as thioethers. Only their combination with HPA assures a good stabilization. In the mixtures with hindered phenols, HAS have almost no influence

$=N^{\bullet} \xrightarrow{R^{\bullet}} =N\text{-}R \longrightarrow =N\text{-}H + =C{=}C=$

$=N^{\bullet} \xrightarrow{ROO^{\bullet}} =NO \xrightarrow{R^{\bullet}} [=NO + R^{\bullet}] \longrightarrow =NOR$

$[=NO + R^{\bullet}] \longrightarrow =NOH + =C{=}C=,\ \text{etc.}$

$=NOR \longrightarrow =NOH + =C{=}C=$

$=NOR \longrightarrow =NO + ROH$

$=NOR \longrightarrow =NO + ROOR\ \text{or}\ C{=}O,\ ROH$

on the rate of thermo-oxidation at low temperatures. Their nitroxyl derivatives, however, always exhibit synergism, most pronounced when both stabilizers are used in equimolecular ratios. During photo-oxidation phenols lower the efficiency of HAS.

Modern stabilizer systems contain HAS in combination with phosphorus hydroperoxide decomposers. Both aliphatic and aromatic phosphites act synergetically when used together with phenols, they having hydrogen peroxide decomposing capability, but the chain breaking activity is not important. However when HAS is in the presence of compounds which form acids upon irradiation or heat, like thiosynergists, halogenated flame retardants, and crop protection agents, salt formation occurs and a strong antagonistic effect is observed. Blends of HAS with different molecular mass and kinetics of the aminoxy radical formation show synergetic effects compared with the use of a single HAS at the same concentration level.

HAS-phosphites and phosphonites containing amine and phosphorus unit in one molecule are highly effective inhibitors of photo- and thermo-oxidation and exhibit lower critical antioxidant and longer induction periods than phosphite alone. They even exceed the efficiency of phenols in many cases.

Transformation products of phenolic antioxidants act differently and in many cases contrary to under photo- and thermo-oxidative conditions, also influencing the efficiency of stabilizer mixtures in different ways. Phenol–HAS mixtures also behave synergetically during thermo-oxidation, but show antagonistic effects during photo-oxidation.

Ciba Geigy has a patent [45] on the use of 2-hydroxyphenyl-1,3,5-triazines as UV absorbers with low volatility suitable to use at elevated processing temperatures.

Sulfur compounds give an antagonistic effect with secondary and tertiary HAS. All new systems need further investigation to assure a safe use.

Several bifunctional, autosynergist molecules were synthesized as: Phenol/HAS, UV absorbers/HAS, HAS/phosphites, phenol thioethers, such as:

The hindered phenol type structure of Vitamin E imparts an intrinsic antioxidant activity that was found to be 250 times greater than that of BHT and the amount of Vitamin E retained after successive extrusion passes is higher than of Irganox type antioxidants. Grafting of antioxidants on PO chain prevents the migration. High efficiency have antioxidants containing maleate function or two functional groups [46].

Some antioxidants also give protection to polymers against the effect of light both by their inherent structure or by a synergetic effect with light stabilizers [47a] while the sterically hindered amines are quite efficient as antioxidants [47b].

In respect to their photochemical behaviour, the antioxidants are classified as weakly absorbing photochemically inactive (2-hydroxybenzophenone for PP), weakly absorbing, photochemically active (phenols in PO, act as HD and are rapidly consumed by irradiation), and light resistant strongly absorbing (azocompounds) that influence their effect as light stabilizer. Apart from this behavior, compatibility, distribution, mobility and reaction capacity with solid polymer are important characteristics, too. Light protective effect of antioxidants is predominantly determined by light resistance as for UV screeners. Two directions are known for the elaboration of light protective systems: modification of stabilizer structure and selection of synergetic mixtures. In the first case groups with high light resistance (alkyl, ester) or able to regulate antioxidant distribution (use of a grafted antioxidant) are introduced.

The synergetic effects have been explained by the following coupled mechanisms: diffusion of antioxidant from the volume protected by UV into the oxidizing surface layer; quenching of the excited state of antioxidants by light stabilizers; increase of solubility and/or redistribution of stabilizer in the presence of other additives; protection of the light stabilizer by antioxidant and stabilization by regeneration of another additive.

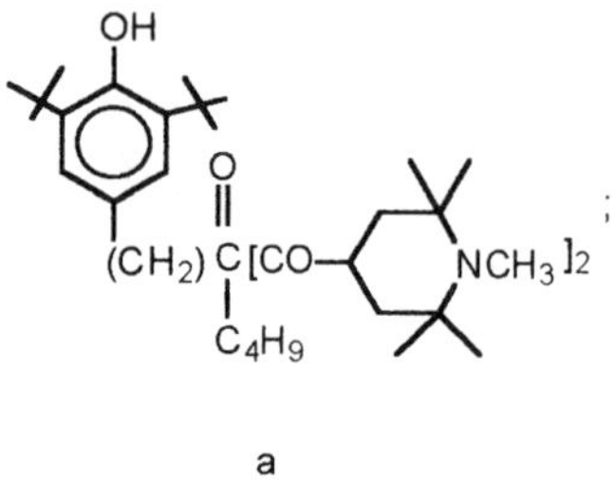

OH; (CH_2) C[CO– ; $NCH_3]_2$; C_4H_9 ; a

P—[O— ; $NR]_2$; b

NH ; NH ; P ; O ; O ; c

OH ; SR ; SR ; d

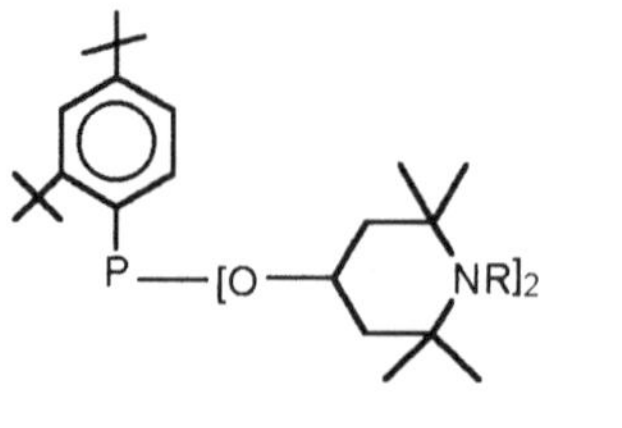

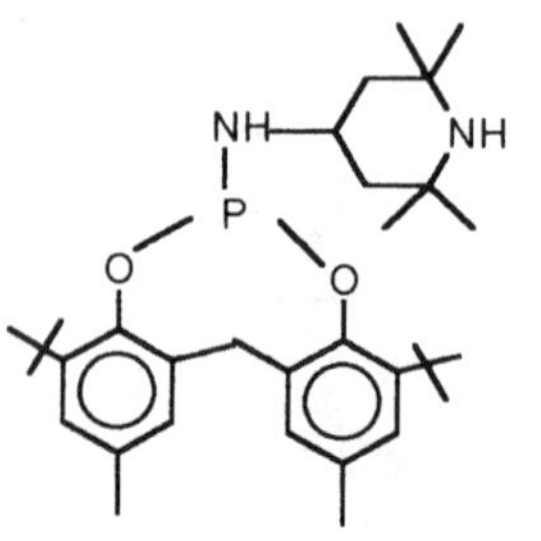

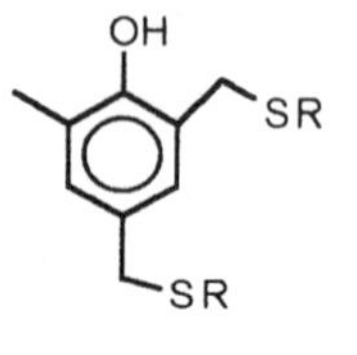

C. Preventive or Secondary Antioxidants and Combined Systems

Preventive and secondary antioxidants decompose hydroperoxides without intermediate formation of free radicals, preventing chain branching [20]. They are termed "secondary" because their best performance is achieved in the presence of primary antioxidants. They also contribute to melt flow and odor stabilization during processing. Aliphatic phosph(on)ites esters act only as secondary HD antioxidants while sterically hindered ortho-*tert*-alkylated aromatic compounds are capable of acting also as a primary radical chain breaking reaction.

Two chemical classes serve as hidroperoxide decomposers antioxidants: activated organic compounds of sulfur (long-term heat stabilizers) and organic compounds of trivalent phosphorus (processing stabilizers). Aliphatic phosphites are more effective processing stabilizers than aromatic phosphites and faster consumed. During their action trivalent phosphorus compounds are transformed in pentavalent compounds.

The antioxidant charge-transfer mechanism of phosphites as processing stabilizers is related to direct scavenging of molecular oxygen:

$$[(RO)_3P\ldots O_2)] \rightarrow [RO)_3P^+OO^-] + (RO)_3P \rightarrow 2(RO)_3P^{\cdot}{=}O$$

High hydrolysis stability for phosphites is necessary to prevent their deactivation and danger of corrosion of processing equipment. The consumption of phenols is reduced by phosphites during melt treatment. In the complex formulation, apart from combined action of original compounds, the mechanism must include the interaction between transformed products.

The stabilizing action of phosph(on)ites is based in three main mechanisms:

1. Oxidation

$$\underset{\text{phosphite}}{POOH + P(OR_1)_3} \rightarrow \underset{\text{alcohol}}{ROH} + \underset{\text{phosphate}}{O{=}P(OR_1)_3}$$

R_1: alkylor aryl radical

2. Substitution
3. Hydrolysis

$$H_2O + {>}P{-}OAr \rightarrow H{-}P{=}O + Ar{-}OH$$

A similar mechanism was proposed for sulfur containing secondary antioxidants:

$$POOH + R'{-}S{-}R'' \rightarrow ROH + R'{-}\overset{\overset{O}{\|}}{S}{-}R''$$

or

$$R'{-}\overset{\overset{O}{\|}}{\underset{\underset{O}{\|}}{S}}{-}R''$$

The products formed by hydrolysis are effective primary and secondary antioxidants. Polyfunctional antioxidants are formed during the application of cyclic arylene phosphites and various HD and chain breaking moieties which autosynergetically enhance their antioxidative activity. In addition to decomposing hydroperoxides, aromatic phosphites can also react with unsaturated (vinyl) groups in the polymer, coordinate with transition metal residues, help to preserve the hindered phenol, and prevent discoloration by reacting with quinoidal compounds.

As a class, phosphites, particularly alkyl or mixed alky–aryl phosphites, have poor hydrolytic stability. Aromatic phosphites provide improved hydrolytic stability but do not equal the antioxidant performance of alkyl or mixed alkyl–aryl phosphite. Hydrolytic stability can be improved either by addition of small amounts of organic bases or incorporation of amine functionality. The fluorophosphites have good thermal and hydrolytic stability [48].

The phosphite efficiency in hydroperoxide reduction decreases in the order: phosphites > alkyl phosphites > aryl phosphites > hindered aryl phosphites, Hindered aryl phosphites can also act as chain breaking primary antioxidants when substituted by alkoxy radicals but their activity is lower than that of hindered phenols but in oxidizing media at high temperatures the hydrolysis of aryl phosph(on)ites takes place and produces hydrogen phosph(on)ites and phenols which are effective chain-breaking antioxidants. Multifunctional stabilizers such as those containing HAS–phosph(on)ites moieties show a superior efficiency due to the intramolecular synergism—Table 6 [49].

A great diversity of sterically hindered phenols is commercially available. Aromatic amines are often more powerful antioxidants than phenols, yet their application is limited to vulcanized elastomers, due to their staining properties. Aromatic phosphites and phosphonites are preferable due to their hydrolytic stability.

It is desirable to combine the effectiveness of primary and secondary antioxidants by mixing or by synthesis of compounds with two structural functions in one and the same molecule.

In recent years the use of phenolic antioxidants in combination with suitable stabilizers has become state

TABLE 6 Induction period and relative oxidation rate after induction period in thermo-oxidative degradation of 0.1 mm PP film at 180°C [49]

Antioxidant (0.02 mol kg)	Induction time (min)	Relative oxidation rate
None	45	1
BHT	410	0.49
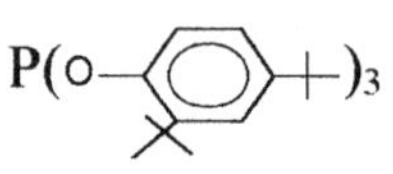 P(O–)3	230	0.5
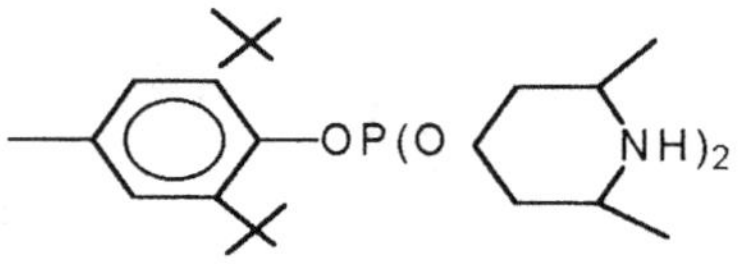 –OP(O NH)2	1020	0.015
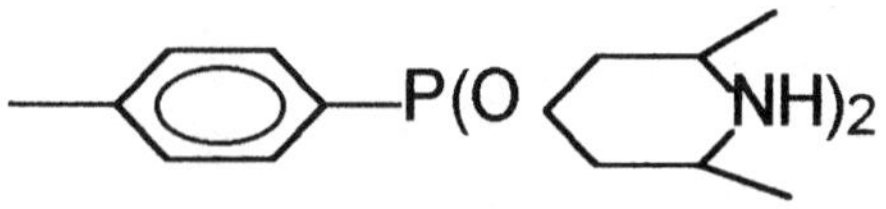 –P(O NH)2	430	0.17

TABLE 7 Systems of stabilizers containing HPA or HAS-phosph(on)ites [50, 51]

System	Advantages
0. Sterically hindered phenols	Do not contribute to melt stabilization
1. 1p sterically hindered phenols/2p phosphite	Protect phenols from excessive consumption during melt processing, improves long-term behavior; improve resistance to gas yellowing
2. Sterically hindered phenols/fluorophosphonites	Improve hydrolytic stability, process color and long term heat aging, excellent heat aging characteristics and under gamma radiation or high temperature moisture exposure [512]
3. HP/HAS	Long term stabilization
4. Sterically hindered phenols/organosulphurcompounds as dilaurylthiopropionate	Long term behavior is much improved
5. Phenolic aryl phosphite/sterically hindered phenols with two different functional groups	Autosynergism [51b]

of the art. Several examples are given in Table 7. Homosynergism is met when stabilizers have the same effective mechanism while the heterosynergism appears in combinations of stabilizers that have different effective mechanisms and complement each other [50].

By combining primary and secondary antioxidants, synergetic effects are frequently observed. Both the phosphites and thioesters are synergetic with hindered phenols. A significant technical example of synergism is observed on combining sterically hindered phenols with phosphites or phosphonites for the stabilization of polyolefin melts. Also, the combination of several stabilizers leads to a higher efficiency up to 200% [11, 19]. Another example is the use of distearyl thiopropionate for long term aging of PP. At an oven temperature of 150°C, a typical test temperature, the use of a thioester by itself fails in a few days. A typical hindered phenol can give lifetimes of 20–35 d, while in a ratio 1 : 4 combination a hindered phenol/thioester yields lifetimes of 80–100 d (source: Montell POs). The pheno/phosphite/epoxy system give good results for the stabilization of mineral-filled PP.

The FS system developed by CIBA Additives seems to be a breakthrough for color-critical PO applications. These new systems do not contain phenolic antioxidants so they provide good color stability and gas fade resistance, long term thermal, and UV light stability. They improve the activity of the HAS. Doverphos HiPure 4, introduced by Dover Chem. Corp, is a high purity processing and heat stabilizer, cost effective, and FDA approved for food contact applications and medical devices.

In order to prevent the loss of antioxidants by migration, extraction by water or organic solvents and volatilization, macromolecular antioxidants have been synthesized. The active antioxidant molecules can form macromolecular substances or polyaminophenols, polyphosphonites and polythiophosphonites, polyepoxides, etc. The active moities can be in the main macromolecular chain (by copolymerization), in side groups (by grafting), at the chain end, or as a bridge between macromolecules, generally by polymer analogous conversion (e.g., electrophilic conversion of olefin and dienes in reaction with phenols and aminophenols [52]). Some natural polymers such as starch or lignin act as antioxidants in PO [53].

D. UV Screeners

The *UV screeners* themselves are subjected to photodegradation during long exposure period, leaving the substrates susceptible to weathering and failure. This phenomenon may limit the practical lifetime of many materials. The degradation of both benzophenone and benzotriazole in PP was attributed to free radical attack at the —OH group or through energy transfer from excited chromophores. All classes of UV stabilizer have high absorption at the wavelengths (290–325 nm) that cause degradation of polymer and they must harmlessly dissipate the energy that they have absorbed [54]. The benzophenone, benzotriazoles, oxanilides, and triazines are photostable because their excited states can undergo a rapid internal hydrogen transfer. The reverse reaction is exothermic, producing heat which is dissipated through the matrix [55]. The mechanism for harmlessly dissipating the absorbed energy by internal hydrogen transfer would be disrupted and other reactions leading to destruction of the chromophore could result. A second pathway for destruction results if free radicals due to photo-oxidation of the matrix abstract the phenolic hydrogen of the UV screener, leading to oxidation of the chromophore. HAS stabilize all UV screeners.

The photochemistry of cyanoacrylates (Uvinal N-35) probably involves a charge-separated species that would be subjected to attack by nucleophiles (water) and the addition of free radicals across the double bonds would result in the loss of the chromophore.

A simple way to evaluate the photostability of a UV screener is to measure the change in absorption upon irradiation due to variation in UV absorption of various screeners and in various matrixes. It is necessary to model the degradation process to find a meaningful way to normalize the rate of loss. If one assumed that the rate of photodegradation of a UV screener is proportional to the light flux, then the rate of loss is given by

$$d[\text{uvs}]/dt = \Phi I[\text{uvs}]$$

where Φ is the quantum yield, I is the light flux, and [uvs] is the concentration of the UV screener (in evaluation, it must take into account nonuniformity of [uvs] through the film cross section and loss of screener). But, according to the Lambert–Beer law, [uvs] is directly proportional to absorbance A:

$$A = \varepsilon[\text{uvs}]L$$

where ε is the absorption coefficient for monochromatic light at the wavelength and L is the film thickness.

The rate of photodegradation depends on the structure of the screener and matrix, concentration, light source, etc. The factors that affect the rate of matrix degradation and free radical production will also affect the stability of the screener. Where HAS are effective in stabilizing the matrix, the lifetime of the screener may be increased as well to give an extra "synergetic" effect. The matrix seems to play the overriding rate in determining the lifetime of most commercially available screeners. Strongly hydrogen bonding media can disturb the internal hydrogen bond that is essential for screener stability while matrices that are subjected to rapid oxidation can produce radicals that react with screeners and lead to degradation.

An apparent zero-order kinetics was established for the loss of the screener [22, 56], that can be explained by the high absorbing, nonstirred nature of the system and by the negligible quantum yield at the wavelengths of the low-absorbing "tail". The photodegradation of currently available UV screeners occurs at such a rate that most of the screener will be depleted from the surface layers of a coating or in the bulk after only 3–5 y of direct sun exposure or after several thousand hours in an accelerated device. An upper limit on the performance of stabilized materials could be supposed.

E. Activity of the Other Additives

PO occupy an important position as insulating materials in both the electrical and electronic fields. Unfortunately a high electrostatic charge may appear on the surface, due to the low surface conductivity. This phenomenon is not desirable, due to dust attraction (heavy contamination of plastic surface), disturbing processing procedure (by possible prevention of the correct winding of calendered films or melt-spun fibers), and production delays; also films may adhere one to another, the printability of the final products being impaired by static charges, while spark discharge can produce serious accidents, danger of fire, or explosion.

Antistatic agents prevent or reduce all these manifestations. They impart a long term antistatic property, while physical treatments (by flame or corona discharge) have only a short-term effect. Long-term antistatic properties can be achieved by (a) surface application of an external antistatic agent from a solution (c ranging in the ppm domain); (b) incorporation of an "internal" antistatic agent (c from 0.05 to 2.5%); (c) incorporation of conducting additives (c from 5 to 10%).

A minimum level of humidity is generally necessary to obtain an antistatic effect; therefore, antistatic agents will be hydrophilic substances (e.g., amino derivatives, polyethylene glycol esters, and ethoxylated fatty amines, which are ideal for PE and PP) that reduce the surface resistivity of plastics by more than $10^5\ \Omega\,cm$.

They have both a hydrophilic part (absorbing water on the surface) and a hydrophobic one, which confers a certain compatibility with polymers. The structure must be varied according to the structure and morphology of PO. For incorporated antistatic agents, the migration of the additive to the plastic surface determines the effectiveness. This diffusion depends on the solubility and concentration of the additive, molecular structure and molecular weight of the antistatic agent and of polymeric matrix, degree of crystallinity, mode of processing and cooling of the finished products.

Conducting additives are of particular importance in the screening of high-frequency emission in electrical and electronic devices.

A large part of the litter problem consists of polymeric substances, among which polyolefins are preponderent due to their wide application as packaging materials. As was already shown, PO exhibit a significant resistance to micro-organisms attack and to sunlight irradiation. Three methods have been developed for the obtainment of bio- and photodegradable materials [12], namely introduction of ketone groups in polymer chains (copolymerization) [3], incorporation (0.01–0.1 phr) of *photoactivators or biosensitizers*, leading to bio- and photodegradation (iron complexes being commercially used), and mixing with natural polymers (biodegradable materials) leading in time to polymer material disintegration (see Chapter 19).

Crosslinking is a technology employed with special types of PO. It produces tough, hardwearing compounds, especially suitable for electrical cable, wire and pipe applications because of the improved heat resistance, dielectric properties, aging, creep rupture and wear/abrasion resistance [56]. Some additives give the same result as radiation crosslinking. Special preparation and processing technologies are required. Silane crosslinking works by first grafting reaction of the groups onto the PE molecule with addition of silanes and very small amounts of peroxide in a one- or two-stage extrusion process. The moisture crosslinking is accelerated by a tin catalyst. Two processes are known: Sioplast (1968) developed by Midland Silicones (Dowcorning), and Monosil developed by BICC and Maillefer (1974), while Huls introduced a silane formulation under the name Dinasylan Silfin. This process requires lower capital investment than peroxide crosslinking for a higher output rate (about 2–5 times). However silane grafting of PP is not yet a commercial product. The following silanes are mainly used: vinyltrimethoxyethoxysilanes, 3-methacryloxypropyltrimethoxysilane, etc. Copolymers of vinyl silane with ethylene or ethylene–butyl acrylate have also been prepared to replace grafted copolymers.

Internal lubricants influence the rheological processes inside the polymers without reducing their properties significantly. Polymer blends and composites (reinforced and filled PO) have a high requirement of lubricants for their processing.

Lubricants are multifunctional, inducing the following types of effects, the corresponding substances being defined according to the effect produced, namely:

I. Effects on rheological properties:

- plasticizing;
- viscosity reduction (internal lubrication action), for example PE and PP waxes enable MFI to be increased, which is too low; reduction of heat dissipation. The higher the viscosity, the more mechanical energy is converted into heat, which may lead to local overheating and degradation of material;

- release effect consisting of the reduction of the adhesive forces between the polymer melt and solid surfaces; friction reduction; slip effect prevention of melt fracture. High viscosity at high shear speeds may lead to the surface phenomenon known as melt fracture.
- The *shark skin effect* can be avoided by using special additives such as fluoropolymers. These coat the machine and the die surfaces; the wall slippage rate is increased, and thus the stick-slip effect is counteracted.

II. Effects on the finished part surfaces by accumulation of additive on their surfaces:

- Mold release effect—the demolding force required to eject a solidified part from the mold is reduced so that the finished article is ejected from the hot mold without deformation. External release agents are based on PE waxes, fluoropolymers or high molecular weight polymers (or a mixture of these) in a solvent solution or in a water emulsion. Many types are noninflammable. They are generally sprayed on the mold surfaces and consist of silicone compounds in traces. They affect adhesion of ink and paint.
- Slip effect imparted by the slip agents. They are of exceptional importance in the production of PO films, assuring good handling properties, particularly on automatic packaging machines. Compounds which have shown usefulness as slip agents include primary amides, secondary amides, and ethylenebisamides with the following structures: stearamide

$$CH_3(CH_2)_{16}—CONH_2$$

stearylstearamide

$$CH_3(CH_2)_{16}—CONH—(CH_2)_{10}—CH_3$$

and *N,N'*-ethylenebis stearamide

$$CH_2(CH_2)_{16}—CON\ CH_2CH_2NCO—(CH_2)_{16}CH_3$$

The main structural characteristics are molecular weight and unsaturation. The unsaturated compounds will migrate faster and therefore yield lower coefficient of friction. The selection of an appropriate slip agent must take into view a balance between the processing conditions and the end use applications. The slip agent must not cause deleterious effects on the clarity of films and must not interfere with corona treatment or printing operation. In PE film production, oleic acid amide is frequently used, while in PP syringe production the erucamide is preferred.

Antiblocking (and antislip effect) agents microscopically roughen the film surface, being used especially for LDPE heavy duty sacks. Blocking is a term describing the polymer film sticking to itself as a result of storage in roll form. Typical antiblocks are silica, talc, diatomaceous earth, and primary amides. Antiblocks are used in the 0.05 ppb to 0.2 ppb range. The oriented films need less antiblock agent with finer particle size than the unoriented films.

III. Other effects

- Dispersing effect. Many lubricants assist the breakdown of agglomerated solids, this effect being widely used in the production of pigment preparations. For PO, PE waxes are generally used due to their better wetting and penetrating properties.
- The water repellence effect is very important for the products used in insulating compounds for electrical equipment as well as in talc reinforced PP for leisure furniture.

The crosslinking reaction takes place by a free radical mechanism, the radicals formed by *organic peroxide* decomposition abstracting a hydrogen atom from the substrate (crosslinkable polymer). Various secondary competitive reactions can also occur.

As a rule, in the presence of other additives [16], the peroxide concentration has to be increased, while sometimes, in order to control the density and yield of the crosslinking reaction, coagents, which also impart better storage stability and processability, are used.

The best photo-crosslinking initiator for PE is 1,2,3,4,5,8-hexachloronaphthalene; XPE obtained by means of 4-chloro-benzophenone and triallylcyanurate has improved homogeneity of crosslinking [17]. The typical conditions are 150–180°C for 10–15 min.

Introduction of ethylene copolymers and other polymers as *high polymeric impact modifiers* in PE or especially in PP is very important for applications in the automotive industry, where high impact strength is required even at very low temperatures (−20°C). The selection of components of the blend is essential for the production of quality products [57]—see also Chapter 22.

Lubricants can play an important role in the improvement of processability in terms of faster more homogeneous mixing/compounding, lower energy consumption, better flow properties in molding or extrusion, higher output, faster set-up time, and quick easy release of the molded part from the mold. All of these can be summed as better processability. A judicious use of modern additives is an essential part of the competitiveness of a processor.

Lubricants influence melt rheology in the desired way, acting also on machine parts and molds, thus resulting in finished articles with smooth, glossy surfaces.

The friction-reducing effect on the finished product is described as a slip effect; the suitable substances used for achieving it are known as *slip agents*. External lubricants are added directly to the metal surfaces of the processing machine or to the mold to prevent sticking of the polymer melt.

- Binding capacity. Lubricants are suitable binders for powdery solids, used for the production of nondusting forms of some additives. Ready-to-use lubricant/stabilizer mixtures are delivered as flakes or fine beads.

The interaction between metals and metallic compounds and polymers is extremely complex, leading either to polymer degradation (accelerators or photosensitizers) or to polymer stabilization (stabilizers, retarders, metal deactivators).

The function of a *metal chelator (deactivator)* is to form an inactive or much less active complex with the catalytically active metal species. The most active catalysts of hydroperoxide decomposition are derivatives of transition metals (Fe, Co, Mn, Cu, V), which are easily oxidized or reduced, having different oxidation states with comparable stability [21]. Special attention must be paid to the inhibition of copper-catalyzed thermo-oxidative degradation of PO used as insulating materials, especially for electrical cables.

PO having a medium crystal growth rate (PP) or rapid homogeneous crystallization (PE) are influenced by foreign substances. In the presence of *nucleating agents*, under the same cooling conditions, the resulting spherulites will be considerably smaller than in PO without nucleating agents. Nucleated polymers have a finer grain structure, which is reflected in their physicochemical characteristics, such as less brittleness, improved optical properties, more transparency or translucency, high degree of crystallinity, involving increased hardness, elasticity modulus, tensile strength and yield point, elongation at break, and impact strength. During processing, short cycle times in the injection molding are achieved by the incorporation of nucleating agents; after-crystallization is also prevented, so that after-shrinkage of finished articles on prolonged storage is diminished. For example, nucleated PE films have lower haze, higher gloss, and improved clarity; the injection molded parts from nucleated PP have better strength characteristics, reduced tendency to delamination, excessive flash formation, sinking, and voiding.

Generally, PO do not require *plasticizers*. These act as lubricating agents, lowering the processing temperature, modifying processing properties and finished product properties, and imparting rubber-like extensability and permanent flexibility.

Biocides or antimicrobial agents inhibit bacterial growth, improving aesthetic and sanitary properties. They are not absolutely necessary in PO composition. However they are incorporated in the formulations of some items (such as packaging). Several examples of such compounds are: 10,16-oxybisphenoxy arsine (OBPA-trade name); 2-*n*-octyl-4-isothiazolin-3-one-1,2-dicarboxyimide (Folpet-trade name).

Many additives have more than one effect on the compound. Carbon black which is widely used as a pigment also functions as a light-shield, an electrically conductive component, and a reinforcement.

Coupling agents are used especially in polymer composites. They enable different phase systems (PO matrix and filler or reinforcements) to be linked together. Active chemicals, they are usually applied as coatings or treatments to additives. Titanates (monoalkoxy, neoalkoxy, chelate, cicloheteroatoms) treated inorganics promote adhesion, improve dispersion, preventing phase separation, and inhibit corrosion.

Zirconate coupling agents provide an alternative, correcting the disadvantages of silanes and titanates. They do not produce color in contact with phenols other than nitrophenols. They do not interact with HAS and they often improve UV stability. Neoalkoxy zirconates provide novel opportunities for adhesion of polymers to metal substrates. They still have high cost.

IV. CONCENTRATION RANGE

Usually, the concentration at which the above-mentioned additives are incorporated in PO is quite low, not exceeding 1 (several) %, as follows—Table 8—yet it must be increased in special/severe conditions of processing/utilization.

The light stabilizer concentration will be a function of the inherent stability of PO to be stabilized, of the

TABLE 8 Concentration range for various additives

Additive	*c* (%) (usually)	*c* (%) (severe conditions)
Antioxidants	0.02–0.5	0.5
Antistatic agents	0.5–2	2–3
Biosensitizers/photosensitizers	1.5–5	
Crosslinking agents	1.5–1.8	4–11 (high level is for EVA)
High polymeric impact modifiers	0–40 thermoplastic phase	40 Elastometric phase
Light stabilizers	0.05–0.5	1–2 extrusion process Tropical climate
Lubricants	0.3–3	
Metal deactivators	0.05–0.2	0.2–0.5
Nucleating agents	ppm range, up to 0.5	

specific applications, and of the presence of other additives in the formulation (e.g., antioxidants, pigments, fillers, etc.)

Satisfactory mechanical properties can be obtained only at lowest dosage levels (especially for peroxides).

V. INCORPORATION*

As already shown, as polyolefin degradation may occur in every stage of the polymer's life, it is advisable to add some kind of stabilization as early as possible, even in the polymerization process, or to add the *antioxidant* before the polymer comes into contact with the air.

During polymerization, the antioxidant is to be added in a form adapted to the manufacturing process: as a liquid (when coadditives such as phosphite or thioether may serve as solvent), an emulsion, etc.

In other stages (drying, pelletizing, processing), the master batches are preferred as powder or as master fluff. In a similar mode, light stabilizers are incorporated, but the addition by master batches is much more important in this case. Master batches now include formulations for electrical conductivity, antistatic properties or other specific performance.

Antistatic agents are applied on the surface from the solution or are incorporated into the polymer.

Commonly, *crosslinking agents including peroxides* are applied in a phlegmatized form, preblend with mineral fillers or extender oil and sometimes with stabilizers. Also, preblends of peroxides in polymers such as EPDM, EVA, PE, etc., are commercially offered. Such mixtures facilitate peroxide handling.

* See also "Compounding" in Chapter 29.

Nucleating agents are incorporated as powder/powder mixtures, as suspensions or solutions, or in the form of master batches.

Other additives are incorporated in one of the ways described earlier into the polymer melt, i.e., at high temperatures when the solubility is much higher, so that one can assume that in general they are molecularly dissolved in the polymer melt.

The nucleating agents are preferably milled down to 5 μm in a roll mill and are mixed with PO powder.

In order to safely handle, readily formulate and compound, two kinds of additive delivery systems [2] have been recently developed, namely the Xantrix system developed by Montell, and the Accurel system developed and patented by Akzo-Nobel. Xantrix named a broad range of additive concentrates (acid neutralizers, antifogging agents, biocides, antioxidants, antistatic agents, fragrances, nucleators, oils, slip agents, UV stabilizers) in the form of dry free-flowing PP, LDPE, and HDPE spheres containing up to 30% of liquid or low melting additives. This system is not influenced by heat history and suffers minimal degradation, and uniform dispersion up to 75/1 (or higher) is offered. Xantrix peroxide concentrates are used in visbreaking of PP and reactive processing. All grades met FDA regulations if required.

The Accurel system has a microporous polymer "sponge" (e.g., of PP, LDPE, HDPE, polyamide) used as carrier for incorporating liquid additives into a thermoplastic melt. The carrier is produced by a "thermal phase separation" process. A microcellular structure with cells of ~5 microns connected by pores of 0.5 microns and internal surface $90\,m^2/g$, absorb up to 2.5 times its own weight of liquid. The key to this solution is the miscibility of a polymer with an additive at processing temperature. By rapid cooling, liquid additive is retained within the polymer structure by capillary forces. The obtained material (powder or granules) is dried and easy to handle and is termed "superconcentrate". The sponge can be reloaded after it has been emptied of additive or it can be used for another application, such as filtration. Additives used in the Accurel system include silicone oils, silanes, peroxides, antistats, lubricants, and antifog additives.

ICT's Atmer super concentrates offer high active loading (0–50 wt%) giving less contamination and permitting like-to-like mixing of resin and additive concentrate pellets. Atmer 8103 is a 50% active concentrate of distilled glycerol monostearate in PE, as a lubricant processing aid or antistatic agent in POs. Atmer 8112 and 8174 are 20% active antifogging agents in PE carrier for LDPE and LLDPE film applications. Hüls reported RIMPLAST technology for modification of PP with silicone IPNs to reduce viscosity, melt pressure, improve processing, and can produce surface and structural changes with high impact strength. Montell's Xantrix 3035 peroxide concentrates is effective in grafting reactive monomers to PP during processing at 0.5–2 wt%, while Xantrix 3055 is for high temperature vis-breaking of PP. Both grades are free-flowing spheres of PP carrier resin and additives with no mineral filler, mainly for production of fine fibers giving enhanced filtration and barrier performance. Xantrix ADS 8015 is a hydrophilic concentrate added at 3–5 wt% during processing, making PP surfaces more resistant and thus useful in medical applications.

VI. REQUIREMENTS

The essential requirements for additives are: quality, cost-effectiveness, food-use acceptance, and safety in handling. Most additives are now offered in dust-free pellets for easy of processing and safety in the workplace. There are also specific requirements for each field of applications such as fibers, films, and rubber. The general requirements for additives may be classified into three groups: physical, chemical, and toxicological [58]. They can be briefly described as:

(1) Permanent, having low volatility under processing and service conditions, nonmigratory, nonleaching, nonbleeding, and nonsweating of compound involving no loss of performance,
(2) Good processing stability;
(3) Good heat, light, hydrolytic, and chemical stability;
(4) Sufficient retention in the polymer substrate;
(5) Good dispersion;
(6) Sufficient mobility during service lifetime;
(7) Without color, free from odor and taste;
(8) Compatible in complex compounds (blends);
(9) Ease to handling;
(10) Nonflammable and nontoxic, safe use for additives and their decomposition products.

Low molecular weight additives (stabilizers, dyes, antistatic agents, lubricants) introduced into polymers may be released into the environment, thus changing the polymer's properties, reducing its service life, and polluting the surrounding medium and materials in contact with polymer [11, 59]. The process includes additive diffusion toward the polymer surface and its subsequent removal by evaporation, washing out, or diffusion into the material contacting the polymer.

If the additive, having emerged on the surface, is not removed, it forms a new phase, the process being known as blooming or sweating (exudation).

The loss of additives depends on their solubility, their molecular weight, the nature and morphology of the polymer and of the environment, on temperature, rate of cooling of the product, etc.

The observation of blooming is very often used to judge compatibility; this appreciation should be effectuated at temperatures as close as possible to the temperature applied to the corresponding products.

In PO, solubility and volatility requirements are of considerable theoretical and practical interest due to their nonpolar character, while most additives are polar compounds.

For example, 50 days to blooming of UV absorber in PE is observed, while for EVA, under the same conditions, more than 330 days before blooming is observed.

HAS with high molecular weight have a reduced migration, although sometimes low molecular weight compounds are preferred for surface protection. *N,N′*-diphenyloxamide, the first commercial metal deactivator for PO, is avoided, due to its volatilization during processing.

The additive loss to the surrounding medium is of great importance for all applications involving the contact of plastic articles with extractive media, by dry cleaning of textiles, and by aqueous media from containers, pipes, tubes, and washing machine parts. Washing out is appreciably faster for aromatic amines than for alkyl phenols. For applications. in contact with food, FDA and BDA regulations recommend liquid antioxidants based on Vitamin E such as BASF Univul.

Low volatility of stabilizers is very important when they are added during polymerization (prestabilization) not to be lost in the final stages of polymer manufacture (devolatilization, isolation, drying/palletizing, etc.).

In order to increase the extraction stability of additives the following methods are available: Synthesis of polymeric additives (especially antioxidants) with molecular weight > 3000. Copolymerization with special monomer containing inhibiting structural units or free radical initiated grafting of antioxidants on polymer chain, physical stabilization by orientation of the polymer (stretching), which lowers the diffusion rate, treatment of the additive with suitable agents (coupling agents, resins, etc.). The first two procedures have not yet led to widespread practical use, mainly for economic reasons.

A certain incompatibility is required for antistatic and nucleating agents. Antistatic agents would exude continuously, due to their surface action. However, they should not influence transparency too much. They should not be too volatile under heat or stress, and they should not produce reactions with the polymer, its degradation products, or other additives.

Additives should contribute as little as possible to polymer discoloration on use. For this reason, aromatic amines are used only exceptionally in plastics; 2,6-di-*tert*-butyl-4-methyl phenol and *n*-octadecyl-3,5-di-*tert*-butyl-4-hydroxy-phenyl-propionate produce discoloration by their oxidation products; therefore they have to be avoided. During polyolefin aging, yellowing can usually be attributed to the additives, to their interactions, or to their oxidation products.

In order to prevent, or at least to minimize the discoloration produced by the phenolic antioxidants, it is recommended to combine them with coadditives such as phosphites and thioethers.

Additives (especially stabilizers) should not be decomposed during different thermal treatments of the polymer. Under processing and service conditions, sufficient thermal stability is needed to avoid additive consumption and to suppress the formation of decomposition products that may be dangerous to health.

Organic peroxides must be safe to handle during transportation, storage, and processing. Their decomposition must be rapid, so that the crosslinking reaction occurs at the desired temperature, without any tendency to premature crosslinking (prevulcanization, scorch) and only crosslinking should occur as chemical modification. Organic peroxides should preserve their reactivity in the presence of other ingredients (fillers, reinforcements, pigments, etc.). Each organic peroxide possesses its own specific hazardous features and requires special precautionary and safety measures. Only relatively stable peroxides can be used for crosslinking reactions, in order to avoid the dangerous situations of explosion and fire, as, generally, they are incorporated at fairly high temperatures. Dicumyl peroxide is considered as a safe product.

It is important to store crosslinking peroxides for longer times than one week at temperatures above 30°C, although after several months at $T < 30°C$ most peroxides exhibit no reduction in their activity.

Possible health hazards may arise during handling and usage of peroxides; thus contact with eyes, skin, and mucous membranes should be avoided.

A nucleating agent should be wetted or absorbed by the polymer. Its melting point should be higher than that of the polymer.

In order to improve storage stability of phosphites prone to hydrolysis, small amounts of bases (such as tri-isopropanol amine) are added or blended with water-repellent waxes or other hydrophobic compounds.

All additives should be homogeneously dispersible into the polymer at corresponding dimensions; for example, in the case of nucleating agents, the particle dimension should be as fine as possible (1–10 μm).

Handling safety is required for all additives with respect to industrial hygiene. They should have low toxicity and should not cause irritation of the skin or mucous membranes.

Static charges represent one of the most dangerous sources of ignition for the explosion of gases or vapor/air mixtures and of dusts. The incorporation/application of antistatic agents is one of the requirements of national safety for plastic article usage.

For food packaging applications, both additives and polymers need specific toxicological regulations from national health authorities. Because of severe requirements concerning safety, the demand for low-dust versions of additives is steadily increasing, to reduce to a minimum any contamination of human beings on handling.

The choice of additives used in products that come into direct contact with food (packaging) or skin (textiles), and that can migrate into them, has to be made according to toxicological laws, toxicological properties being of crucial importance.

The toxicological risk of an additive depends on the dose, the chemobiological mechanism, and the products resulting from its transformation in time during its entire life.

During PO processing at elevated temperatures, highly volatile additives may contaminate the atmosphere; vapors or gases may lead to toxic effects for humans, but appropriate protection measures (local exhaust devices) can prevent this.

Absorption of dusts and vapors by inhalation and through the skin are the main routes of human contamination in the industrial use of additives. That is why additive manufacturers must supply additives in low-dusting forms. Sometimes even a microencapsulation technique is used.

There is always the possibility that incorporated additives, monomers, oligomers, etc., will migrate in small amounts to the products' surface and into the surrounding media, and thus directly or indirectly into the human body. This happens with food packaging materials, children's toys, packaging materials for pharmaceuticals, textiles, etc. Finally a subchronic or chronic exposure of man will result. Additional requirements must be formulated with respect to product specific interactions between the packaging material and packaged goods. These interactions depend on the diffusion properties of additives, the nature of the packaged food, the contact time between the plastic and the food, the storage temperature, the interaction between different additives, etc.

The migrated amount and the toxicity of the additive determine whether it is suitable for use in food packaging.

The manufacturer of plastic packaging has to know in what amounts an additive migrates into food; this amount must be determined by simulating all conditions in which it is used (temperature, contact time, food simulants). The food simulants used are distilled water, aqueous acetic acid, aqueous ethanol, and natural fats or oils; 95% ethanol and ethyl acetate are recommended as fat simulants for PO. Only 10, 50, or 100 ppb (parts per billion) of additives are tolerated in food. Numerous methods of additive analysis have been developed [1, 60].

Additives used in plastics packaging have to be judged by toxicologists who consider both human health and environmental protection. The legislation regarding the use of additives in food packaging in various countries varies under the specific conditions of each country [61].

The requirements for the detection limits of traces of an additive migrated into food have to become more stringent, and specific methods of detection have to be elaborated.

Knowledge of LD_{50} is of special importance for toxicity estimation. In this respect, tests on laboratory animals have to be performed.

Toxicological problems and environmental implication related to chemistry of additives especially stabilizers and peroxides remain an open problem.

VII. TESTING AND SELECTION OF ADDITIVES

The selection of a particular additive or additive system for a resin formulation is based on knowledge of a polymer's weakness and the conditions under which it will be processed and used.

Determination of the relationship chemical structure–*efficiency of stabilizers* and their mechanism and kinetics of action in polymer matrix is important to choose the most perspective ones for practical use and also to predict the characteristics of new potentially active stabilizers. Many methods have been elaborated in this respect.

The effectiveness of *antioxidants or stabilizers* is usually tested from three aspects:

(a) Processing stability by repeated extrusion in plastograph mixing chamber by the melt flow rate (ISO 1133, DIN 53735, ASTM 1238-88) torque (Brabender plastocorder) or solution viscosity (ISO 1628-3) are measured;
(b) Long term stability—after the circulating air oven aging, physicomechanical indices (DIN 53455, ISOR527, DIN 53448) or yellowness index (ASTM D 1925-70) are determined;
(c) Light stability by artificial weathering (in Atlas Weather-Ometer, Xenotest, artificial weathering DIN 53387 (ISO4892) or Suntest UVCON, QUV and SEPAP) or natural outdoor weathering in various climates followed by carbonyl index, surface texture, etc., determination.

The most common methods coupled with the three mentioned tests are those also used for evaluating thermal and thermo-oxidative stability, namely measurement of the oxidation induction period by the oxygen uptake test, manometry, DTA, DSC, TG, EGA and EGD in isothermal and nonisothermal (dynamic) conditions; oven aging in static or circulating air at 50–150°C, chromatography UV, IR, ESR, NMR spectroscopy, chemiluminescence, viscosimetry, mechanical indices determination, etc. Other criteria refer to discoloration, embrittlement, cracking, loss of elongation, etc. Hydroperoxides concentration can be determined by iodometric titration. The "method of inhibitors" can be used as chemiluminescence to determine the rate of the radicals' formation [62].

DSC, TG, and DTA can be used in correlation with the oxidative induction time (OIT) changes in molecular weight and tensile properties to establish both thermal stability and stabilizer efficiency—Figure 2.

The OIT can be determined by DTA, DSC, and TG experiments more accurately in isothermal conditions.

Kharitonov, Zaikov and Gurevich [65] have established a correlation between the structure of the molecules and the mechanism of their oxidation by means of a high sensitive differential manometric device linked to a computer. A detailed study of all kinetic values of the oxidation process and its mechanism has been developed. This allows one to predict the oxidation behavior of compounds, the antioxidative activity and inhibition mechanism. The potential of the beginning of anodic oxidation of phenolic stabilizers in PP is in good accordance with induction period results.

Chemiluminescence coupled with manometric and spectroscopic methods permit estimation of kinetic values which are useful to establish the efficiency of stabilizers with various structures. These have been evaluated [66]: the rate of oxidation at small extent of conversion, induction period of inhibited and non-inhibited oxidation (τ), the inhibition constant k_{inh}, i.e., the rate constant of the reaction of an inhibitor with the peroxy radical, the inhibition coefficient f, which determines the quantity of chains terminated by one inhibitor molecule, etc. The specific feature of the behavior of stabilizers in solid state (polymers) is the dependence of stabilizer reactivity on compatibility, solubility, chemical and physical stability of stabilizer against high temperature of the treatment of polymer, etc. Among the reactions of oxidation of polymers in solid state, the initiated oxidation reaction of PP is the most studied. This reaction may be a test for the investigation of a stabilizer's ability to terminate the oxidation chain. A comparison of hindered phenol efficiency is given in Table 9.

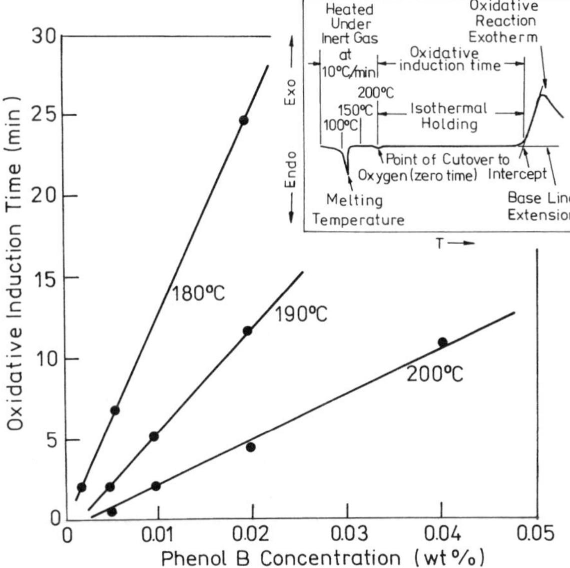

FIG. 2 Oxidative induction time determined from DSC curves versus phenolic antioxidant concentration [63, 64].

Such dependences have also been established for phosphite stabilizers. Similar kinetic studies were made by Denisov [67].

Antistatic agent testing consists of testing the complete coating of the surface of the finished part with active ingredients. The durability of antistatic agents is determined by the labeling technique with isotopes, measurement of the surface tension, flame photometry, thin-layer chromatography, diffusion coefficients, etc.

A long term effect decreases by evaporation or wiping of the antistatic molecules from the surfaces, especially for external antistatics. These phenomena are compensated for over a long period of time by migration from the bulk.

For the samples, carefully and exactly treated (humidity and storage time are very important), the following procedures are employed: determination of chargeability, of the discharge rate, and of dust attraction, and characterization of conductivity (the half-life time being determined) by measuring the surface and bulk resistance. Between the surface resistance and the half-life time there exists a correlation permitting the

TABLE 9 Rate constants for chain termination for phenolic antioxidants in PP at 80°C [66]

No.	Para substituent in phenol	$k_{\text{inh}} \times 10^{-3}$ (kg/mol^{-1} s^{-1})
1	$—OCH_3$	8.1
2	$—OC—(CH_3)_3$	3.9
3	$—CH_3$	3.5
4	$—C_6H_5$	1.8
5	$—C(CH_3)_3$	3.2
6	—Cl	2.7
7	—CHO	0.6
8	$—COCH_3$	0.5
9	$—COC_6H_5$	1
10	—CN	0.6

appreciation of the effectiveness of antistatic agents; surface resistance $< 10^9\,\Omega$ and half-life time 0 s is excellent for an antistatic; surface resistance of 10^9–$10^{11}\,\Omega$ and half-life time 1–10 s is good to moderate; and for surface resistance $> 10^{12}\,\Omega$ and half-life time > 60 s the antistatic agent is not effective.

Efficiency of bio- and photosensitizers is established by the well-known biodeterioration methods of [68] fungal resistance (ASTM G 21-70 or G 22-76) environmental test methods and other biodeterioration test techniques (see also Chapter 19).

Selection of *peroxides* is primarily determined by polymer and processing conditions (residence time and temperature in the processing equipment). The advance in the crosslinking reaction is determined by gel-fraction content using hot xylenes for extraction.

Light stabilizers and metal deactivators are tested by accelerated and outdoor weathering, chemiluminiscence methods, etc. At the current level of understanding there seems to be no way to predict just which type of absorber will be the most suitable in any particular matrix. The rate of UV absorber photodegradation is sufficiently high that the outdoor lifetimes of many coatings and articles could be limited to the absorbers that were added to protect them [69]. Many efforts are underway to develop the understanding of the chemistry and kinetics of UV absorber loss that will be essential to the creation of highly weatherable products.

Testing of lubricants is based on the determination of the flow behavior of plastic melts (capillary viscosimetry, spiral flow test, demolding force, by plastograph, extrusiometer, melt strength, roll-mill tests, etc.) or by finished part testing (gloss and smoothness, slip measurements, determination of static friction, etc.).

Nucleating agents and high polymeric impact modifier efficiency are judged by the modifications in optical (polarizing microscope) and mechanical properties.

Although additives have been treated separately no distinct lines exist between the categories, some of them even having multiple functions.

Physical effects which influence the performance of the small molecular materials in polymers are not well known yet. Experimental results proved that the protection time of primary antioxidants, UV stabilizers, etc., is related to the ratio S^2/D where S denotes the solubility, and D the rate of diffusion. The additives with low solubility and high diffusion rate migrate rapidly towards the surface, so the polymer loses the additive due to volatilization and leaching of additive from the surface. Therefore some applications are where additive migration to the polymer surface has a positive effect such as antistatic and antifogging agents. Migration of antioxidants is a major concern in applications involving polymers in direct contact with food and the human environment [70].

Antioxidants undergo oxidative transformation as a consequence of their antioxidant function. This transformation results in a chemical loss of antioxidants, which can occur during processing, reprocessing and recycling, or during service life.

Physical losses occur because of volatilization, poor solubility, diffusion, leachability, blooming, etc. Efficiency means inherent activity and physically retained by polymers. There are still less information about chemical nature and migration of transformation products of additives. The antioxidants diffuse in PE hot water pipes leading to their consumption and pipe deterioration [71].

Diffusion of benzotriazioles and HAS in PP is mainly influenced by their molecular weight. For HAS the following relation was proposed [71]:

$$E_d = A + B \ln M_w$$

where A and B are constants, and E_d is the activation energy of diffusion. Several studies on the relation between physical parameters and stabilization efficiency showed decreased efficiency of stabilizers with high molecular weight. Increased solubility of stabilizer improves the performance whereas an increase in diffusion rate has the opposite effect. Decreased stabilization efficiency with increased molecular weight is due to a decreased homogeneity of distribution of active stabilizing functionalities throughout the polymer [72]. Polymer bound stabilizers as polymerized acrylic derivatives of HAS offer a favorable combination of these properties. Polymer bound additives are at least partially soluble in the polymer and are present in a metastable alloy. Whereas acrylates are known for the tendency to form homopolymers, maleate derivatives are typically copolymerization monomers and usually do not form homopolymers. In reactive processing of a PO with maleated functionalized stabilizers, usually a single molecule is attached to the polymer. Fumarates and maleimides are also typical derivatives with a low tendency to homopolymerization and give systems with good performance [2, 7]. Chmela and Hrdlovic [73] have synthesized photoreactive oligomeric light stabilizers—terpolymers containing benzophenone or HAS groups with weak links that are light sensitive. These links cleave the oligomeric chain of the additive, and produce shorter fractions with high mobility.

VIII. INTERACTION OF ADDITIVES

In complex formulations, the additives interact with each other, the effects being either advantageous or deleterious. Several examples are significant. Firstly it has to mentioned the use of complex systems for stabilization with excellent results, that were mentioned above. Phenols, phosphites and HAS as well as their transformation products interact during thermo- and photo-oxidation of PO [74]. TiO_2 has a catalytic effect on thermo- or photodegradation. It interacts with hindered piperidine light stabilizers and hindered phenolic antioxidants [75]. Atanase is ineffective while on rutile pigments strong synergism will be observed with antioxidant and HAS. Talc influences the UV exposure stability of PP [76]. A carbon black surface oxidized by treatment with H_2O_2/HNO_3 improves the photostability of LDPE but the effect was not observed in PP. Silica enhances polymer crystallinity acting as nucleating agents. AO and HAS absorbed on silica have reduced activity more important in photodegradation [77].

The brominated flame retardants can cause catastrophic deactivation of the HAS [78]. In formulations containing aromatic brominated flame retardants, combination of UV absorbers and a new class of siloxan with HAS provide the highest level of UV stabilization. Dyes influence also PO photostability [79].

IX. ANALYSIS OF ADDITIVES

The experimental techniques used for the analysis of additives are numerous and varied.

Chromatographic and spectroscopic methods—GC–MS, HPCL, HPCL–UV–VIS, NMR—coupled also with extraction (supercritical fluid extraction [80–83]) are the most useful.

Thermal methods—DSC, DMA, DTA, TG, TMA—are useful especially in the cases when it is difficult if not impossible to quantitatively extract an additive from its polymer matrix prior to chromatographic analysis. If the additive is incompatible with the resin it can be detected and determined by either melting temperature, heat of fusion, changes in heat capacity, glass transition, etc. [84–86]. Conversely, when an additive is soluble in a polymer its concentration can be estimated from shifts in T_m and T_g of the resin. Nonpolymeric additives and fire retardants or plasticizers can be quantitatively vaporized at temperatures below the degradation temperature of the host polymer and identified by IR or MS analysis. The TG method is capable of detecting antioxidants at concentrations above 0.01 wt% (Figure 3) [84].

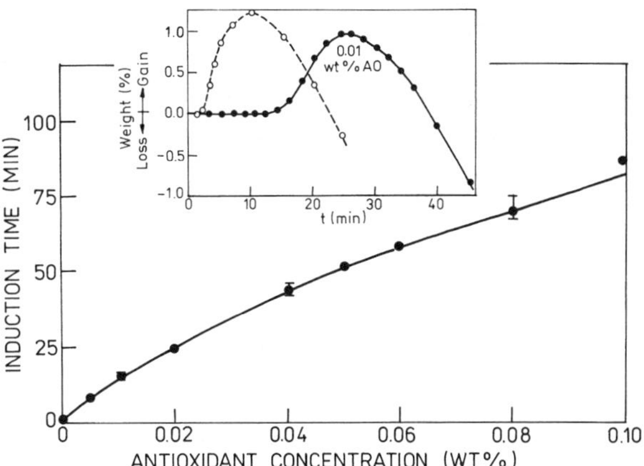

FIG. 3 A calibration curve of the time to oxygen absorption by LDPE versus concentration of 4,4′-thiobis(3-methyl-6-*tert*-butylphenol) (Santonox R) at 200°C [84].

A. Optimal Conditions

For OIT determinations by DTA and DSC analyses: pure oxygen at a flow rate 50 ± 5 mL/min; isothermal measurements between 150–230°C, aluminum pans, 2–8 mg samples as disk, onset temperature 60°C.

For TG: isothermal small samples weighing less than 10 mg are heated at 160°C/min at a fixed temperature (~200°C) in oxygen; dynamic: temperature range 20–50°C, heating rate 0.1–0.4°C/min.

For GC flame ionization detection, BP-5 capillary column, 280–300°C.

For GC-MS: J&W DB-5MS column; 280–300°C, detector 190°C, 70 eV.

X. ADDITIVES FOR RECYCLING OF THE POLYOLEFIN WASTE

A. Introduction

Recycling plastics still suffers from the imputation of being of low value and consequently only acceptable for "secondhand applications". The know-how of the recycling technologist consists in how to reformulate the material to provide it with a new life.

The main problem in post-consumer plastics recycling is due to the chemical damage such as degradation undergone by the polymers during processing steps and by the products during their lifetime and, moreover for the heterogeneous recycling, to the incompatibility among the different phases. New functional groups formed enhance the sensitivity of the recyclate to thermal and photodegradation [87].

Material properties of recyclates are influenced by the history of the polymer synthesis, primary processing and application, and the recovery source. Irreversible chemical transformation arising in degradation result in polymer bound structural heterogenities (such as alcohols, carbonyl, unsaturation, and crosslinking). They are localized irregularly in the polymer bulk. Some of them are considered (photo)-active impurities. As a result, a material made from recycled plastics differs from the relevant virgin plastics in sensitivity to the attacking environment. The contamination increases with degree of oxidation and may even limit the material reuse due to a strong sensitizing effect. Nonpolymeric impurities admixed or generated (salts of transition metals and transformation products formed as a consequence of sacrificial and depleting consumption of stabilizer during the polymer first life) during the polymer first life are of specific interest. Structural inhomogenieities and residual impurities (the last often can be removed by purification) are present in recyclates. Already during the first processing of the virgin plastic and the first life irreversible changes (predamages) of the polymer chain take place induced mechanochemically, chemically or by radiation. Through carbon centered (alkyl) or oxygen centered (alkylperoxyl, alkoxyl, acyl, acylperoxyl) free radicals, oxygen containing structures (hydroperoxides, carbonyl, alcohols, carboxyls, lactones) and olefinic unsaturated groups (vinyl, vinylidene, allylic) are formed. In addition, crosslinked polymer chains can be formed. Disproportionation and depolymerization result in LMW products. The concentration of the new structures (structural inhomogeneities) increases with the previous application field (e.g. outdoor, high temperature or mechanical stress). The oxygenated structures are more sensible or have photosensitive effects so recyclates are more sensible to oxidation; they even initiate oxidation in blends with virgin materials or other polymers.

The prodegradant structural inhomogeneities and nonpolymeric impurities may negatively affect the recyclate lifetime. With the formation of new pro-oxidative functional moieties a substantial part of the stabilizers added for the first lifetime are simultaneously consumed. As a consequence the mechanical properties of commingled plastic recyclates generally do not meet material expectations. Increase in resistance to degradation by proper restabilization and the recovery of physical properties by compatibilizers represents the state of the art in material recycling of plastics for demanding applications.

Physical damage as surface cracks might be reversed by reprocessing while for chemical changes such as crosslinking chain scission, and new formation functional groups are irreversible.

Less-easily degraded components of recyclates are contaminated during reprocessing by photosensitizing prodegradants (such as chromophores) from heavily degraded components. Hence the degradation products are distributed in the whole mass of recyclate and act in the system as oxidation promoters. The accelerating effect of the added aged recyclate on the degradation of virgin polymer has been observed in thermal aging and weathering.

To reduce the negative effects of the recycling steps two main ways can be adapted: (a) restabilization, and (b) adding fillers, modifier agents, compatibilizing agents, etc.

PP is more sensitive to thermomechanical degradation. The molecular weight of PP decreases drastically with the number of reprocessing steps, with a concomitant reduction in mechanical properties, in particular of elongation at break.

Fillers improve some mechanical properties such as modulus and tensile strength but worsen the processibility and the elongation at break and impact strength. In many cases the beneficial effect on the cost of material is also obtained.

The additives can play an important role both in the homogeneous and in the heterogeneous recycling.

Closed loop applications from clearly defined sources, e.g., bottles crates, waste bins, automotive parts, require a careful adjusted restabilization system with sufficient processing and light stabilizers. At least the amount of stabilizers used up in the first life has to be compensated for.

B. Restabilization of Recyclates [88–97]

Restabilization of post-consumer recyclates is necessary for effective protection against the detrimental influence of shear forces and heat during processing, against UV light and oxygen to modify the processing and aging behavior, while for outdoor service, besides this protection, a reduced quantity of stabilizer residues is required. Restabilization allows for closed loop recyclates. Long-term thermal stability that maintains properties allows long-term applications even under heat stress, UV light stability maintains

TABLE 10 Differences between recyclates and virgin materials [88–97]

Predamage	Composition	Contamination
• Chemical changes	• Different polymers	• Other polymers
• Carbonyl group content	• Polymer types/grades	• Metal traces (catalyze auto-oxidation)
• Hydroperoxide content	• Production method (catalyst)	• Contact media
• Double bonds concentration, initiation sites	• Different producers	• Inorganic impurities
• Discoloration	• Phase separation	• Organic impurities and their degradation products (inks, paints, adhesives)
• Crosslinking	• Additives	• *Sources of contamination*
• Molecular weight change	• Predamage	• hydrolysios of catalyst residue
	• Fluctuating composition	• degradation products (e.g., HCl, humic substances)
		• phosphite hydrolysis
		• $(RO)_3P \xrightarrow[H_2O]{HX} H_3PO_3$ (phosphoric acid)
		• over-oxidized thiosynergist (sulfenic or sulfonic acids
		• contact media, e.g., battery cases promote hydrolysis
		• processing stabilizer
		• corrosive/metal activation
		• Metal $+HX \rightarrow Me^{n+}$
		• Metal oxide $+HX \rightarrow Me^{n+}$
		• protonation of hindered amines (HAS) 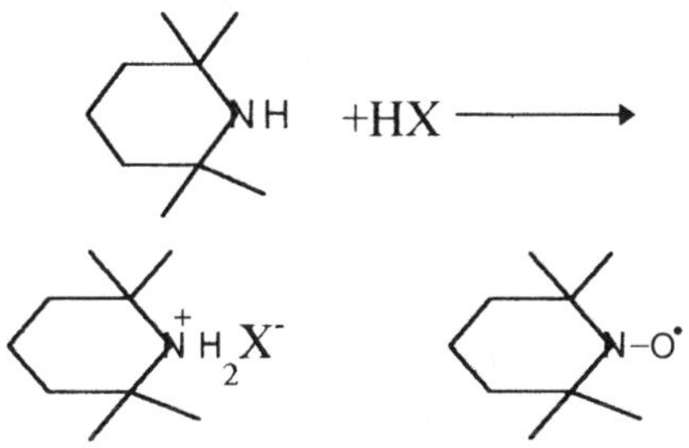
Consequences		
• Branching/crosslinking (gel formation)		
• Modified degradation		
• Change of mechanical properties		
• Change of processing behavior		

properties during outdoor exposure and prolongs products for either second service time, allowing higher value-added products and applications.

The concept of restabilization and repair has been introduced in order to provide recyclates with similar properties to their original use ("closed loop") as a potential substitute of virgin material. Since 1991 Ciba Additives in the framework of many R&D projects established that recyclates behave differently to virgin material as is shown in Table 10.

A successful stabilizer system for recyclates has to take the special requirements discussed above into consideration. Predamage is introduced into the recyclate during processing and the lifetime of the plastic article and the fluctuating composition of recyclates including the presence of contaminants is responsible for a modified degradation and processing behavior. In an aged state macromolecules are attacked and broken-off material is brittle under impact [98]. Preconditions for viable recycling are: separation, additives analysis, and restabilization [87].

As mentioned above, many stabilizers are deactivated to other products during their first service: for example, phenolic antioxidants to quinonemethides, cyclohexadienones or benzoquinones; phosphates are formed from phosphites; various oxidation products of sulphides, including organic acids of sulfur, from sulphidic heat stabilizers; and substituted piridin-4-one and open-chain nitroso and nitro compounds, salts and other transformation products arising from HAS. Stabilizers are also lost by diffusion and migration.

Of these new compounds, considered as nonpolymeric impurities, only quinonmethides and transformation products of sulfides contribute beneficially to the resistance to degradation.

Some of these products can act as fillers or nucleating agents. Consumption of stabilizers generally results in insufficient protection of the recyclates for the second reprocessing and application.

Restabilization is one very effective approach to improve the quality of any recyclate by increasing the substitution factor. As a result the recyclate can substitute for virgin plastic in selected applications with comparable performance.

Restabilization must be achieved as the end-products of processing of the recyclates are in outdoor applications, such as non-pressure pipes, cable ducts, waste bins, crates, bottles, drums, pallets, playground material, plastic lumber, etc.

Ciba has a new organizational unit "Venture Group Additives for Plastic Recycling" that elaborated new solutions, new stabilizer formulations and is developing standards for materials and products from recyclates [88–91]. Tailor-made stabilizer systems under the trademarks RECYCLOSTAB, RECYCLOSSORB, or RECYCLOBLEND are specifically developed for post-consumer recyclates by addressing the degradation profile and contamination. The stabilization system (processing, thermal, light stabilizers and costabilizers) in connection with stabilizer residues has to be reformulated and carefully adjusted according to the degradation history and contamination. Recyclostab developed by Ciba for commingled plastics has a good performance. Resistance to thermal degradation is doubled compared with straight reclaim from painted PP bumpers while mainly ultimate elongation is improved.

Heat stabilizers can also play a role in recycling thermoplastics, when material is again subjected to heat. Some stabilizer systems introduced into the original compound retain their effectiveness through subsequent re-processing, but there may be a need to restabilize waste plastics for more efficient recycling.

The additives for recycling have two major functions, namely to enhance the properties of plastics recovered from waste, and also to facilitate the recycling process providing fast and effective identification of types of plastics-tracer "additives".

In the stabilization/restabilization of post-consumer recyclate plastics the degree of degradation, amount of residual stabilizers and extent of contamination has to be taken into account. Generally, the waste contains many structural defects and impurities, being much more sensitive to degradation.

Synergetic hindered phenols/hydrolysis resistant phosphite offers a long-term stabilization of PP. The HDPE bottle waste is restabilized with 0.2–0.3% of a phenolic antioxidant. Standard HAS grades, such as Ferro AM 806 with MMA–polymer backbone, are used to produce furniture and plastic lumber made of POs. Quantum offers a Spectratech line of colorant and additive concentrates for recycling [96].

PP recyclates are stabilized by 0.2% RECYCLOSTAB 411 stabilizer mixture addition containing antioxidants and costabilizers [99]. PP from battery cases is commercially reclaimed and reused. This requires a stabilization with 0.2% Recyclostab 451. The stabilization of PP/EPDM painted parts (bumpers) from the automotive industry is effective with a T-1 stabilizer mixture comprising an antioxidant, a phosphite, and a thioether that are thiosynergists. Restabilization with a combination of a process stabilizer and a high and a low molecular weight hindered amine, increased by twice the weathering time of the retained properties. Only selected oxirane key compounds are able to eliminate to a great extent the detrimental influence of paint residues in EPDM–bumper material on the mechanical properties and long term properties of material. Although PE gives less problems in restabilization, this depends on the first application. The HDPE recyclates are effectively restabilized with 0.1% Irganox 1010 and 0.1% Irgafos 168, while LDPE film recyclates are best stabilized with a mixture of Irganox 1010/Irgastab 168 1/4 and also with Recyclostab 411. The last stabilizer system is also suitable for the polyolefin fraction of mixed plastics waste. LDPE film recyclate is stabilized with Recyelostab 421-Ciba, specially designed for thin films.

The HAS compounds increase appreciably the lifetime of the PO waste by restabilization against UV light, the best results being obtained with Recyclostab 811 containing antioxidants and light stabilizers.

The sensitizing effect of the aged PO was almost suppressed in the following cases [94] (see Table 11).

Figure 4 gives an example of the effect of these stabilizer systems.

C. Repair

1. Fibers, Compatibilizers and Impact Modifiers

These are the additives mainly used to improve the properties of the products obtained from waste plastics. The Dutch Bennet company found that the addition of its BRC 200 compatibilizer to a mixed

TABLE 11 Re-stabilization of the aged PO and applications of recyclates [94]

Types of Waste	Stabilizer	Application
LDPE film recyclate	Recyclostab 421-Ciba antioxidants + costabilizers	Thin film, substitution of virgin material
HDPE waste bins	0.4% Recyclossorb 550- Ciba, basic stabilization + HAS	Waste bins, fulfill existing standards of original level quality
PP battery cases	0.2% Recyclostab 451-Ciba, antioxidants + costabilizers	Same applications as original
Pigmented PP from crates	HAS-1 + LS-2	
PP/EPDM bumper	Basic stabilization + low and high molecular weight HAS	
PP/EPDM bumper	Selected oxirane compounds	
EPDM from bumpers	P-1/AO-1 1/1 + 0.2% HAS-1 + 0.4% HAS-2 of different molecular weight	
PE/PP bottle fraction	0.2% Recyclostab 451	MFI is constant during multiple processing steps
HDPE body + PP caps +PET and PVC as impurities	Recyclostab 451	
HDPE five year old pigmented bottle crates	0.1% HAS-1 or P-1/AO-1 2/1 + 0.1 HAS 1 + UV absorber of benzotriazole type LS-1	Imparts protection against blending of the pigment
Mixed PO + carbon black	Recyclostab 451(9 antioxidants + costabilizers) HAS + Chimasorb 944 or Tinuvin 783 (blend of antioxidants) Tinuvin 622 Recylostab 550 costabilizers + HAS)	Reduce negative effect of the carbon black on thermal oxidation
PO + 2.5% carbon black	0.2 (AO-1 + P-1) + HAS-3 1/2	
PE/PA film coextruded film from food packaging	HAS + selected antioxidants	Injection moulding and extrusion
LDPE/PA-6	Mixture Irganox 1098/Irganox 1078/Irgafos 168/Chimassorb 944	Processing, long-term stabilization
HDPE, PO, mixed plastics	Recyclostab 411 (antioxidants + costabilizers)	Processing, long-term stabilization
LDPE films	Recyclostab 411(antioxidants + costabilizers	Processing, long-term stabilization
PP, PO blends	Recyclostab 451	Processing, long-term stabilization
Mixed plastics 55–60% PO + 15–20% styrenics + 5–8% PVC + 8% PET	Recyclostab 811 aromatic phosphite P-10.1% + 0.05% AO-1 + 0.1–0.2%(HAS-1 + HAS-2/LS-1+ antioxidants (calcium salts of organic acids and/or synthetic hydrocalcite+ octadecyl-3-(3,5-di-*tert*-butyl-4-hydroxyphenyl)propionate	

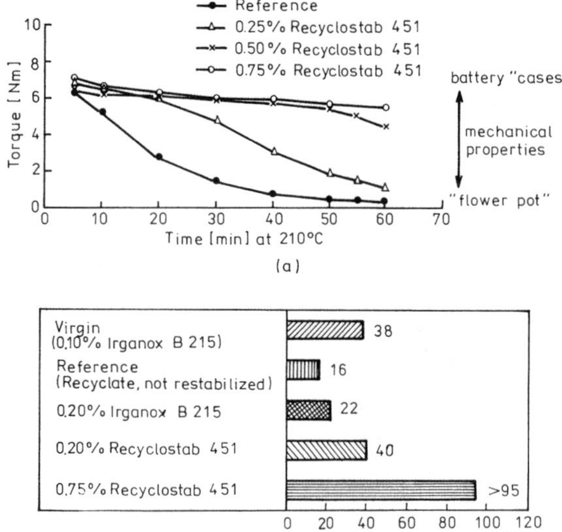

FIG. 4 Processing stability (a) and long-term stability (b) of PP scraps from battery cases [91, 92].

waste stream of equal parts of LDPE, LLDPE, HDPE, VLDPF, and PP homo- and copolymer boosted its impact strength to higher than the single components in injection molded parts. Compatibilizers containing branched polyethylene, EPR, EVA, and elastomers are added to scrap HDPE, LDPE and PP to make drainage pipes.

Du Pont Fusabond compatibilizers, mainly based on maleic anhydride grafted POs, are useful in the manufacture of multilayer film waste that can be reprocessed as video film cassettes or bottles.

Vasile, Roy and Deanin [100–103] proposed a locked loop between the tertiary and secondary recycling of plastic waste. A part of the plastic waste was pyrolyzed in mild conditions. Liquid and waxy pyrolysis products are modified with maleic anhydride. These modified products were tested with good results in processing of LLDPE/PP waste and of light autofluff. They act both as compatibilizers, melt flow improvers, and lubricants.

2. Compatibilizers in Mixed Plastics Recyclates

Commingled post-consumer recyclates containing mixtures of polymers that differ in structure and polarity with limited miscibility have low tensile and impact properties. Both nonreactive and reactive compatibilizers are applied. Nonreactive compatibilizers are diblock or multiblock copolymers. Compatibilizers principally consisting of mixtures of POs are of major interest for recyclates. Random E/P copolymers are effective compatibilizers for LDPE/PP, HDPE/PP or LLDPE/PP blends [94]. Other examples are given in Table 12.

Multicomponent or multifunctional compatibilizers are partially advantageous together with the right conditions of compounding favoring dispersion [97–100].

Reactive compatibilizers consist of functional or reactive additives interacting *in situ* with components of the blends. This allows the phases to be held together by covalent bonds. Some examples are given in Table 13 [104].

Approaches that use specially synthesized reactive compatibilizers and that can account for the formation of compatibilizing crosslinks *in situ* are mostly too expensive for commingled post-consumer waste recyclates.

Reactive blending or extrusion in the presence of a free-radical initiator or reactive monomer have a better chance. Combining shear–temperature effect in the

TABLE 12 Compatibilizers used for processing of mixed plastic recyclates [94]

Mixture	Compatibilizers
LDPE/PP	Random E/P
HDPE/PP	VLMWPE
LLDPE/PP	Poly(S-block (E-co-B-1)-block S) triblock copolymers SEBS
LDPE or HDPE/PS	Graft copolymers PS-g-PE or PS-g-EPDM; block copolymers S-co(E-butadiene diblocks) S-butadiene-b triblocks SEBS Hydrogenated S/isoprene rubber Hydrogenated PS-block-butadiene
PE/PVC	Poly(ethylene-g-vinyl chloride), partially chlorinated PE
PE/PET	EPDM, SEBS
Commingled recyclates	S/butadiene; S/E-co-P or hydrogenated S-B-S triblocks

TABLE 13 Reactive compatibilizers for processing of the mixed plastics recyclates [94, 104]

Mixed plastics waste	Reactive compatibilizer
PE + PP + PS + PET + PA	PE, PP, EPDM, PS, E/P copolymers functionalized (grafted) with maleic anhydride or acrylic acid, succinc anhydride, styrene, trimelitic anhydride
PA/PE; PET/PE films	Epoxy functionality and ionomers such as poly(E-co-MMA-co-butylacrylate) with Zn counterion
PS + PP + PE	5% Kraton D

application of high-stress shearing favoring the formation of carbon centered radicals is a promising approach.

3. Other Additives for Recyclates

Desiccants, such as CaO, eliminate the need for drying, thus avoiding porosity and metal corrosion problems. CaO scavenges up to 32% of its own weight in moisture. Croxton and Garry Company produces a range of Fluorox grades differing mainly in particle size: Garosorb products are powders or pourable pastes, while multidisperse E-CAO-80P is a nondusting polymer bound granular form.

A *peroxide concentrate* has been developed by Polyvel Inc. for melt-flow modification of PP recyclate. A 1% loading can increase the MFI of standard PP to over 150 g/10 min.

The "*tracers and markers*" are developing in waste plastic products reflecting x-ray or infrared signals and triggering separation mechanisms. The UK research agency PIRA in collaboration with Bayer AG are working on this problem under a research grant supported by the EU. Eastman Chemical in US demonstrated the feasibility of an organic filler on the laboratory scale.

XI. NEW DEVELOPMENTS AND FUTURE TRENDS

The main future trends in additives are in response to legislative and commercial pressure. It is necessary to develop new compounds and master batches, which are nondusting, non-toxic, and easier to handle. The elimination of heavy metals and other hazardous ingredients is done by replacing them with alternative systems. Free-dust, free-solvent formulations reduce the risk of explosions. The performance will be improved by surface treatments of additives that create bonding with the matrix and the development of synergetic relationships or multifunctional additives. Efforts are also being made to improve economics, lower densities, higher yield, and lower concentrations. A potentially fertile field is the synergism between components where better performance is vital, as well as properties such as weathering and flammability.

Demand is increasing for additives which will render recycled plastics more easily reprocessable. The problem of identifying different plastics in a waste stream might also be solved by development of suitable "tracer" additives.

Due to the side effects of additives in formulations, current development is aimed at single multifunctional ingredients or hybrid systems or at exploiting synergetic relations between two or more additives. This approach is being used increasingly in the development of technology for UV light stabilization and flame retardance.

New technologies provide commodity PO with modified sensitivity to degradation. Depletion of the ozone layer accounting for a potential increase in the intensity of UV radiation should be anticipated by relevant UV stabilizers. Effective radiation-resistant protection during and after sterilization by energetic radiation is necessary.

Commercially feasible approaches to complex stabilization of polymer blends are required.

SYMBOLS AND ABBREVIATIONS

A	Absorbance
DSC	Differential scanning calorimetry
DMA	Differential mechanical analysis
DTA	Differential thermal analysis
GC–MS	Gas chromatography—mass spectroscopy
HAS	Hindered amine stabilizers
HALS	Hindered amines light stabilizers
HPA	Hindered phenols antioxidants
HDPE	High density polyethylene
HD	Hydroperoxide decomposer
HP	Hydroperoxide

HPLC	High pressure liquid chromatography
LS	Light stabilizer
MD	Metal deactivator
OIT	Oxidative induction time
PO	Polyolefin
PP	Polypropylene
QM	Quinonmethide
TG	Thermogravimetry
T_m	Melting temperature
T_g	Glass transition temperature
TMA	Thermo-mechanical analysis
ε	Absorption coefficient
Φ	Quantum yield

REFERENCES

1. R Gächter, M Müller, and PP Klemchuk, eds. Plastic Additives, Third edn. Munich: Hanser, 1990.
2. J Murphy. Additives for Plastics Handbook, Oxford: Elsevier Advanced Technology, 1996; V Dobrescu, C Andrei, Progress In Chemistry and Technology of Polyolefins. Bucharest: Ed. Stiintfica si Encciclopedica, 1987, chap. 5, p 129.
3. Modern Plastics Encyclopedia. New York: McGraw-Hill Inc.
4. VA Karghin, ed. Encyclopedia Polymerov in Sovetskaia Encyclopedia, Moskva, 1972 and 1977.
5. B Ranby, JF Rabek. Photodegradation, Photo-oxidation and Photostabilization of Polymers. London: Wiley Interscience, 1975.
6. Polymer Symposium "Degradation and Stabilization of Polyolefins". J Polym Sci Symp 1976; 57:73, 161, 171, 181, 191, 197, 267, 319, 329.
7. EI Kirilova, ES Shul'ghina. Aging and Stabilization of Thermoplastics. Leningrad: Izd. Himia, 1988.
8. E Tudos, Z Fodor, M Iring. Angew Makromol Chem 1988; 158/159: 15.
9. G Scott, ed. Developments in Polymer Stabilization. London: Elsevier, 1979, 1981, 1987, 1991.
10. S Horun. Additives for Polymer Processing. Bucharest: Ed. Tehnica, 1978.
11. G. Scott. Makromol Chem Macromol Symp 1989; 28: 59; J Polym Sci, Symp 1976; 57: 357.
12. W Schnabel. Polymer Degradation. Principles and Practical Applications. (Berlin: Akademie Verlag, 1981, p 123.
13. JE Guillet. Polymers with controlled lifetimes. In: JE Guillet, ed. Polymers and Ecological Problems. New York: Plenum, 1973; F Sitek, JE Guillet, M Heskins, J Polym Sci Symp 1976, 57: 343.
14. I Chodak, I Chorvath, I Novak, K Csomorova. Europ Polym 1992; 28: 107.
15. P Kurian, KE George, DJ Francis. Europ Polym 1992, 28: 113.
16. PV Zamotaev. Markomol Chem Macromol Symp 1989; 28: 287.
17. YL Chen, B Ranby. J Polym Sci, Polym Chem 1989; A-27: 4051.
18. B Ranby. In: O Vogl and EI Immergut, eds. Polymer Science in the Next Decade. New York: John Wiley and Sons, 1987, p 121.
    a. F Ding, H Kubota, Y Ogiwara. Europ Polym J 1992; 28: 49.
19. RF Becker, DJ Carlsson, JM Cooke, S Chmela, Polym Degrad Stab 1988; 22: 313.
20. K Schwetlick, J Pionteck, A Winkler, U Huhner, H Kroschwitz, WD Habicher. Polym Degrad Stab 1991; 31: 219.
21. Z Osawa. Polym Degrad Stab 1988; 20: 203.
22. RE Becker, LPJ Burton, SE Amos. Additives. In: EP Moore Jr, ed. Propylene Handbook. Munich: Hanser, 1966, p 176–209.
23. M Pantke. General aspects and test methods. Heitz *et al.* eds., Microbially Influenced Corrosion of Materials. Berlin: Springer Verlag, 1996, pp 380–391.
24. R Torenbeek, J Meijer, AH Hegt, G Bekendam. PCT Int Appl WO 96 03 397, 1997.
25. E Foldes, A Szigeti-Erdey. Soc Plast Eng. ANTEC'97/233, 3024–3028.
26. G Scott. In: G Scott, ed. Atmospheric Oxidation and Antioxidants. Amsterdam: Elsevier, 1993, vol 2, chapter 5.
27. S Hosoda, H Kihara, Y Seki. In: RL Clough, NC Billingham, KT Cilen, eds. Polymer Durability, Adv. Chem Series 249, Am Chem Soc, Washington D.C., pp 195-211, 1996.
28. ON Nikulicheva, VP Fadeeva, VA Logvienko. J Therm Anal 1996; 47: 1629–1638.
29. J Pospisil. Polym Degrad Stab 1988; 29: 181–202.
30. J Pospisil, S Nespurek. Polym Degrad Stab 1991; 34: 85; 1993; 39: 103–120; 1993; 40: 217–220.
31. J Pospisil, S Nespurek. Polym Degrad Stab, 1996; 54: 7–14; 1997; 54: 15–21; Adv Polym Sci 1980; 36: 69; Photooxidation reactions of phenolic antioxidants. In: NS Allen, ed. Developments in Polymer Chemistry Vol 2. London: Applied Science Publication, 1981, p 53.
32. YuA Shlyapnikov, NK Tyuleneva. Intern J Polym Mater 1997; 36: (1,2) 119–129.
33. P Dubs, R Ptüeland. Chimia 1994; 48: 117–120.
34. J Kubo, H Onzura, M Akiba. Polym. Degrad Stab 1994; 45: 27–32.
35. WD Habicher, I Bauer, K Scheim, C Rautenberg, A Lepack, K Yamaguchi. Macromol Symp 1997; 115: 93–125.
36. H Zweifel, Macromol Symp 1997; 115: 181–201.
37. J Pospisil, S Nespurek. Macromol Symp 1997; 115: 143–163.
38. J Pospisil. In: RL Clough, NC Billingham, KT Gillen, eds. Polymer Durability, Adv Chem Series 249, Am Chem Soc, Washington D.C., pp 271–285, 1996.

39. F Gugumus. Polym Degrad Stab 1994; 44: 299–322.
40. ET Denisov. Polym Degrad Stab 1989; 25: 209.
41. PP Klemchuk, ME Gande. Polym Degrad Stab 1988; 22, 241; 1990, 27: 65; F Gugumus. Polym Degrad Stab 1993; 39: 117.
42. M Iring, Z Fodor, M Body, P Baranovics, T Kelen, F Tudos. Angew Makromol Chem 1997; 247: 225–238.
43. YB Shilov, YT Denisov. Vysokomol Soedn 1974; A-16: 2316; Polym Sci USSR 1974; 16/18: 2686.
44. J Sedlar in A Patsis ed. Advances in the Stabilization and Controlled Degradation of Polymers, Technomie Publ Co Inc, Basel 1989, Vol 1 p 227.
45. A Schmitter, A Burdeske, M Slongo, JL Birbaum. US Pat 5 208 788 Feb 22 1994 to Ciba Geigy Ltd.
46. S Al-Malaika, S Issenbuth. In: RL Clough, NC Billingham, KT Gillen, eds. Polymer Durability. Adv Chem Series 249, Am Chem Soc, Washington, D.C., pp 425–439, 1996.
47. a. VB Ivanov, VYa Shlyapintokh. Polym Degrad Stab 1990; 28: 249–273.
    b. YA Shlyapnikov, GS Kiryushkin, AP Marin. Antioxidative Stabilization of Polymers. Chichester; Ellis Horwood, 1996.
48. J Scheirs, J Pospisil, MJ O'Connor, SW Bigger. In: RL Clough, NC Billingham, KT Gillen, eds. Polymer Durability. Adv Chem Series 249, Am Chem Soc, Washington, D.C., 1996, pp 359–374.
49. K Schwetlick, WD Habicher. In: RL Clough, NC Billingham, KT Gillen, eds. Polymer Durability. Adv Chem Series 249, Am Chem Soc, Washington, D.C., 1996, pp 349–357.
50. H Zwiefel. In: RL Clough, NC Billingham, KT Gillen, eds. Polymer Durability. Adv Chem Series 249, Am Chem Soc, Washington, D.C., 1996, pp 375–396.
51. a. GJ Klender. In: RL Clough, NC Billingham, KT Gillen, eds. Polymer Durability, Adv. Chem Ser 249, pp 397–423, 1996.
    b. M Ghasemy, H Ghasemy, Iranian J Polym Sci Technol 1995; 8(1): 4–10.
52. RZ Bigalova, VP Mislinska, GE Zaikov, KS Minsker. Oxidation Comm 1997; 20(1): 139–144; 1995; 14(11): 1705–1721; Int J Polym Mater 1996; 31: 131–152.
53. C Vasile, G Cazacu, MM Macoveanu, VI Popa. Fifth Mediterranean School on Polymer for Advanced Technology E Martuscelli, P Must, G Ragoste eds, Capri, Oct. 4–7, 1997 UNESCO Publ and references therein, pp 185–209.
54. JE Pickett, JE Moore, Polym Degrad Stab 1993; 42: 231–244.
55. JF Rabek. Photostabilization of Polymers. London: Elsevier, 1990, pp 209–230 and references therein.
56. S Al-Malaika, G Scott in N Allen ed Degradation and Stabilization of Polyolefins, Appl Sci. Elsevier, 1983.
57. J. Fortelny, J Kovar, Europ Polym J 1992; 28: 85.
58. J Lemaire. Caoutch Plast 1996; 73(754): 68–72.
59. PA Marin, Yu A Shliapnikov. Polym Degrad Stab 1991; 31: 181.
60. GS Matz, J Chromatogr 1991; 587(2): 205.
61. CIBA-GEIGY AG. Additives for Plastics, Elastomers, Synthetic Fibers for Food Contact Applications, Positive List, Part 1, 1986.
62. YA Gurevieh, IG Arzamanova, GE Zaikov, eds. Quantitive Aspects of Polymer Stabilisation. Commack, New York: Nova Science, 1996, pp 1–27.
63. E Kramer, J Koppelmannn. Polym Eng Sci 1987; 27: 947.
64. T Schwartz, G Steiner, J Koppelmann, J Appl Polym Sci 1989; 38: 1.
65. VV Kharitonov, BL Psikha, GE Zaikov, MI Artisis. Oxidation Comm 1995; 18(3): 256–266; Chem Phys Rep 1997; 16(3): 419–429; VV Kharitonov, BL Psikha. Int J Polym Mater 1996; 31: 171–182.
66. YaA Gurvich, IG Arzamanova, GE Zaikov. Oxidation Comm 1996; 19: 1–24.
67. ET Denisov. Bound Dissociation Energies, Rate Constants, Activation Energies and Enthalpies of Reactions in Handbook of Antioxidants, CRC Press, Boca Raton, FL, 1995, p 605.
68. D. Allsop, K. J. Seal. Introductian to Biodeterioration. London: Edward Arnold 1986, chap. 5, p 95.
69. JE Pickett. Macromol. Symp 1997; 115: 127–141.
70. Plastic Technology, Manufacturing Handbook. New York; Plaspec.
71. J Viebke, UW Gedde. Polym Eng Sci 1997; 37(5): 896–911; V Dudler, C Moinos, In: RL Clough, NC Billingham, KT Gillien, eds. Polymer Durability. Adv Chem Series 249, Am Chem Soc, Washington, D.C., 1996, pp 441–453.
72. J Malik, A Hrivik, DQ Tuan, In: RL Clough, NC Billingham, KT Gillen, eds. Polymer Durability. Adv Chem Series 249, Am Chem Soc, Washington, D.C., 1996, pp 455–471; Polym Degrad Stab 1995; 47: 1–8.
73. S Chmela, P Hrdlovic. In: RL Clough, NC Billingham, KT Gillen, eds. Polymer Durability. Adv Chem Series 249, 1996, pp 473–482.
74. W Habicher, I Bauer, K Scheim, C Rautenberg, A Lossack, K Yamaguchi. Macromol Symp 1997; 115: 93–125.
75. NS Allen, H Katami, In: RL Clough, NC Billingham, KT Gillen, eds. Polymer Durability. Adv Chem Series 249, Am Chem Soc, Washington, D.C., 1996, pp 537–554.
76. MS Rabello, JR White. J Appl Polym Sci 1997; 64 (13): 2505–2527.
77. NS Allen, M Edge, T Corrales, C Liauw, F Catalina, C Peinaudo, A Minihan, Polym Degrad Stab 1997; 56(2): 125–139.
78. RL Gray, RL Lee. Angew Makromol Chem 1997; 247: 61–72.
79. JCVP Moura *et al.* Dyes Pigm 1997; 33: 173–196.

80. GN Foster, In: J Pospisil, PP Klemchuk, eds. Oxidation Inhibition on Organic Materials 3rd edn. Boca Raton, FL: CRC Press, 1990, vol 2, pp 299–347.
81. T Macko, B Furtner, K Lederrer. J Appl Polym Sci 1996; 62(13): 2201–2207.
82. DW Allen, DA Leathard, C Schmith. Chem Ind (London) 1989; 16(2): 38–39.
83. AI Bauer, R Franz, O Piringer. J Polym Eng 1996; 15(1-2): 161–180.
84. HE Bair. Thermal analysis of additives in polymers. In: EA Turi, ed. Thermal Characterization of Polymeric Materials, 2nd edn. London: Academic, 1997, p 2263.
85. A Manivannan, MS Seehra. Preprint Papers ACS Div Fuel Chem 1997; 42(4): 1029–1032.
86. KGH Raemaekers, JCJ Bert. Thermochim Acta 1997; 295 (1–2): 1–58.
87. G Menges. Materials Fundamentals in Plastics Recycling, Chapt 1 in Y Brandrup, M Bittner, W Michaeli, G Menges eds. Recycling and Recovery of Plastics 1996, Hanser, Munich 3–19.
88. H Herbst, G Capocci. TPO's in Automotive, October 28–30, 1996 Novi MI. Automotive Plastics Recycling Closed Loops by Restabilization.
89. R Phaendner, H Herbst, K Hoffmann, 1997 Prague Meetings, 38th Microsymp on Recycling of Polymers. Acad Sci Czech Republic, Prague, July 14–17.
90. H Herbst, R Phaendner, K Hoffmann, R'97 Recovery, Recycling Re-Integration Congress, Geneva, Switzerland, February 4–7, 1997 II16–II21.
91. K. Hoffmann, Glober'96, Environmentally Technologies, Davos, Switzerland, March 18–22, 1996, p 17.1.1.
92. K Hoffrnann, H Herbst, R Phaendner. Addcon'96 Paper 4.
93. H Herbst, P Flanagan. Plastics recycling—added value by restabilization. In: N Hoyle, DR Karsa, eds. Chemical Aspects of Plastics Recycling. Proc Symp on Chemical Aspects of Plastics Recycling, UMIST, Manchester. July 3–4, 1996 Special Publ., No 199. Cambridge: Royal Society of Chemistry, 1997.
94. J Pospisil, S Nespurek, R Pfaendner, H Zweifel. Trends Polym Sci 1997, 294–299.
95. R Phaendner, H Herbst, K Hoffmann, F. Sitek. Angew Makromol Chem 1995; 232: 193–198.
96. J Pospisil, F Sitek, R Phaender. Polym Degrad Stab 1995; 48: 351–358; Ciba Geigy. Recyclostab, Product Information Publ. 28886/e Basel, 1992.
97. R. Pfaendner, H. Herbst, K. Hoffmann, 38th Microsymp on Recycling of Polymers, Prague, July 14–17, 1997, p SL7.
98. A Baldizar, TU Gevert, M Markinger. Durability Build Mater Compon 7th Proc Int Conf, 1996, 1, pp 683–692.
99. H Herhst, K Hoffmann, R Pfaendner, F Sitek. Improving the quality of recyclates and additives (stabilizers). In: J Brandrup, M Bittner, W Michaeli, G Menges, eds. Recycling and Recovery of Plastics. Munich: Hanser, 1996, pp. 297–314.
100. C Vasile, M Brebu, H Darie, RD Deanin, V Dorneanu, DM Pantea, OG Ciochina. Int J Polym Mater 1997; 38: 219–247.
101. C Vasile, RD Deanin, M Mihaies, Ch Roy, A Chaala, W Ma. Int J Polym Mater 1997; 37: 173–199.
102. C Vasile, S Woramongconcha, RD Deanin, A Chaala, Ch Roy. Int J Polym Mater 1997; 38: 263-273.
103. C Vasile, M Mihaes, RD Deanin, M Leanca, Th Lee. Int J Polym Mater 1998; 41: 335–351.
104. G Obieglo, K Romer. Kunststoffe 1996 86(4): 546–547.

21

Special Additives: Blowing Agents, Flame Retardants, Pigments, Fillers, and Reinforcements

Cornelia Vasile
Romanian Academy, "P. Poni" Institute of Macromolecular Chemistry, Iasi, Romania

Mihai Rusu
"Gh. Asachi" Technical University, Iasi, Romania

Important changes in the properties of plastics resulting from the incorporation of special additives permit their use in applications where the polymer alone would have had small chance to meet certain performance specifications.

The types of additives discussed here lead to important physical changes. Thus blowing agent use permits the production of cellular structures; flame retardants improve thermal and combustion characteristics; pigments and colorants change the aesthetic aspect; while physical and mechanical properties are modified by active fillers and reinforcements.

Some of the additives used in polyolefin compositions may play a multiple role; sometimes they can even have a negative effect. For example, many pigments also act as photostabilizers, the protective action against light being fulfilled by the colored pigments while white pigments, such as the anatase form of TiO_2, are photosensitizers. Therefore the choice of a suitable additive for a particular end use is very important, as it involves knowledge of the behavior of each type of additive [1].

I. BLOWING AGENTS

A. Role and Mode of Action

Blowing agents are gaseous, liquid, or solid materials able to produce a foam structure. If the cells (pores) are formed through a change in the physical state of a substance (for example, expansion of a compressed gas, evaporation of a liquid, or dissolution of a liquid), the acknowledged definition is that of a physical blowing agent, while if the cells are formed by evolution of gases as thermal decomposition products of a material, this is known as a chemical blowing agent. The physical phenomenon or the decomposition reaction must take place within a narrow temperature range [1–5].

B. Classification of Blowing Agents

Over the years, many types of blowing agents have been used, but few of them have commercial importance, others being eliminated for various reasons such as low efficiency or versatility, toxicity, difficulties of dispersion, or economics.

As physical blowing agents for polyolefins (PE, PP, EP copolymers, etc.), low boiling halogenated hydrocarbons are used in the presence of nucleating agents (e.g., silica, silicates, sulfides) that facilitate the development of small and uniform cells [3]. Physical blowing agents are efficient and inexpensive, and some of them (such as the above-mentioned ones) are self-extinguishable; yet they have high toxicity.

Chemical blowing agents can be inorganic or organic substances, or exothermal and endothermal according to reaction type [3–5]. As inorganic chemical blowing agents the following have been used: ammonium carbonate, bicarbonate and nitrite, sodium carbonate or bicarbonate, and alkaliboronhydride, among which only sodium bicarbonate has remained in use over the years, although its importance has also been diminished by the introduction of organic chemical blowing agents. In spite of their low price, chemical inorganic blowing agents present some major disadvantages such as difficulty of dispersal and effective use in high pressure expansion processes and high alkalinity of their decomposition residues, which affects other properties of the materials.

Exothermal blowing agents release more energy during decomposition than is needed for the reaction. Once started, decomposition continues spontaneously after energy supply. Typical agents are hydrazides and azo compounds. Due to possible skin irritation precaution should be taken when handling these substances. Azo compounds impart a yellow color to the products. Endothermal blowing agents require energy for decomposition. Shorter cooling periods and molding cycles are needed. The materials are bicarbonate and citric acid.

As an alternative to blowing agents, in 1980 lightweight void-forming pellets or manufactured from ultrafine ceramic foam $\rho = 100$–$200\ kg/m^3$ have been introduced for PP. Teepull from Filtec are stable to 1050°C, do not generate smoke or hazardous gases and resist attack by most chemicals. They are used to form foams.

The following classes of organic compounds are now used as blowing agents for polyolefins [2]:

(a) Azo compounds. Azo compounds are: azodicarbonamide, modified azodicarbonamide, etc. For modification additives ("kickers") or other chemical blowing agents are employed, e.g., from the sulfohydrazide group, which act as nucleating agents, decrease the decomposition temperature, and reduce the toxic gases (cyanuric acid) and plate-out phenomenon or lead to pores with uniform dimensions. Effective kickers are polyols, urea, amines, certain organic acids and bases, metallic compounds such as lead, zinc, or cadmium compounds, and certain fillers and pigments.
(b) Hydrazine derivatives. Hydrazine derivatives are such as 4,4′-oxybis(benzenesulfohydrazine) (which is suitable for the manufacture of articles for food applications), diphenylsulfone-3,3′-disulfohydrazine, diphenylene oxide-4,4′disulfohydrazide, and trihydrazino triazine.
(c) Semicarbazide-*p*-toluene sulfonyl semicarbazide.
(d) Azoisobutironitrile. This decomposes at 95–98°C, giving 136 cm^3/g gas yield, very effective; yet its decomposition products are toxic.
(e) Dinitrosopentamethylenetetramine.

Some of their characteristics are given in Table 1. The following are producers and tradenames of the blowing agents of present interest in polyolefin applications.

Commercial Blowing Agents for PO

Blowing agent	Manufacturer
Azobis	Hiraki Kasei, Osaka, Japan
Azobul	PCUK (Produits Chimiques Ungine Kuhlmann) Paris la Defense, France
Azocel	Fairmount Chemical Co. Inc., Newark, NJ, USA
Blowing agents	Societe Francaise d'Organo Synthèse, Gennevillers, France
Cellmic	Sanyo Kasei Co., Ltd., Osaka, Japan
Celogen	Uniroyal Chemical Division of Uniroyal, Inc., Naugatuck Chem. Div., CN, USA
Cellular	Dover Chem. Corp., USA
EPI cor 972 endothermic concentrated	EPI, TX, USA
EPI cell 700 exothermic concentrated	EPI, TX, USA
Espon	Quimica Heterocicla Mexicana SA. Ecatepec Edo de, Mexico
Expandex	Stepan Chemical Co., National Petrochemical; Olin Corp., Chemicals Group, Stamford, CN, USA.
Ficel	FBC Industrial Chemicals Hauxton, Cambridge, UK
FM 2182H	Quantum Chem Co., OH, USA
Freon	Du Pont de Nemours, Belgium and Switzerland
Genitron	Whiffen & Sons FBC Industrial Chemicals, Hauxton, Cambridge, UK
Hacel	Hubron Rubber Chemicals Ltd., Manchester, UK
Hydroglycerol/ azodicarbamide	BP Chemicals, UK
KEM TEC	Sherwin-Williams Chemicals, Cleveland, OH, USA
Kempore	Olin Corp., Chemicals Group, CT, USA
Lucel ADDA	Lucidol Division, Wallace & Tiiernan
Neocellborn	Eiwa Chemical Ind. Co. Ltd., Kyoto, Japan
Nitropore	Div. Inc. Wilmington MA, USA
Porofor	Bayer AG, Leverkusen, Germany
Unicell	Dong Jim Chemical Ind. Co. Ltd., Seoul, Korea
Unifoam	Otsuka Chemical Co. Ltd., Osaka, Japan
Vinstab	Hebron SA, Barcelona, Spain
Vinyfor	Eiwa Chemical Ind. Co. Ltd., Kyoto, Japan

Source: Refs 1, 5, and 6.

TABLE 1 Some characteristics of the chemical blowing agents used in polyolefin processing

Blowing agent	T_d(°C)	Gas yield ($cm^3\ g^{-1}$)	Range of concentration used (%)	LDS_{50} (mg kg^{-1} body weight)	Uses
Azodicarbonamide	200–215	220	0.1–4	6400	LDPE HDPE PP EVA
Modified azdicarbonamide	155–220 180–200	150 200	0.3–2		PE, PP EVA
4,4′-oxybis-benzene sulfohydrazide	157–160	125 160	0.5–2	>5200	PE, PP EVA. etc., for cable insulation
Diphenyl-3,3′-disulfohydrazide	120–140 155	110		>5000	PE, EVA
Diphenyleneoxide 4,4′-Disulfohydrazide	175–180	120		>2500	
p-Toluenesulfonylsemi carbazide	228–235	140		10,000	PE, PP
Trihydrazino-triazine	265–290 275	190 225			Injection and extrusion of PP and PE
N,*N*′-Dinitrozopentamethylene tetramine	130–190	265		0.5–3	HDPE, PP

Source: Refs. 1 and 5.

C. Concentration Range

The concentration used varies with the kind of manufacturing product, as follows: 0.25–0.5% for films, from 0.5–0.75% for tubes and profiles, and from 0.3–0.75% respectively for injection molded parts. Smaller quantities of blowing agents, varying between 0.05–0.1%, are used for the elimination of sink marks in thick-walled injection molded parts.

D. Requirements

A good blowing agent should meet the following requirements:

Its decomposition temperature should be at least 15°C lower than the processing temperature of the plastics.

It should be incorporated and dispersed easily in the polymeric material.

On decomposition, it should generate a high gas yield (nevertheless, not explosive).

Neither the blowing agent nor its decomposition products (gas or residue) should present health hazards or have corrosive effects; also they should impart no objectionable odor.

The decomposition residue should be compatible with the polymeric material, that is, it should not exudate or cause discoloration.

Finally, it should not adversely affect the thermal, electrical, or mechanical properties of the plastics.

E. Incorporation into the Compound

The following ways of incorporation are practicable: (a) directly dispersable blowing agent powders; (b) master batch with a dispersing agent; (c) the simplest way is the application on the surface of the granules to be processed previously treated with a bonding agent such as paraffin oil or butyl stearate; (d) in an eccentric tumbling mixer or in similar equipment; (e) using automatic metering and mixing devices mounted directly on the injection molding machine or extruder in place of the hopper; (f) liquid blowing agents are injected straight into the barrel of the injection molding machine or the extruder via a measuring and pumping system; (g) physical blowing agents can be introduced under pressure in plastics or may be incorporated into the polymer mass during its manufacture.

F. Testing

As gas yield mainly characterizes the efficiency of a blowing agent, the testing methods are related to gas evolution. In this respect, the ASTM D 1715-60T has been developed. Other methods, such as EGA, EGD, chromatography, DTA, TG, etc., are applied also in order to establish gas yield, mode and temperature of decomposition, qualitative and quantitative determination of all decomposition products, rate of decomposition, explosion risk, toxicity, etc.

G. Uses

Blowing agents are used in high pressure compression molding processes, calendering with subsequent expansion, rotational foaming, mold foaming, sintering and extrusion blow molding, extrusion with subsequent crosslinking, etc., from each processing technology to obtain a large variety of products with special foaming characteristics [6].

H. Developments

Replacement of halogenated foaming agents is one of the recent trends, due to ozone depletion and the greenhouse effect. Some companies, such as BASF, have concentrated on completely halogen-free alternatives. Flammability was also a problem, but it was solved by using high-boiling foaming agent or water, absorbed onto a support such as zeolite, silica gel or activated charcoal, being released at the reaction temperature.

II. FLAME RETARDANTS

A. Role and Mechanism of Action

The flammability of polyolefins is a definite hazard in many of their major markets. Thus it has become very common practice to add flame retardant chemicals to reduce their flammability. Flame retardants strongly reduce the probability of the burning of the combustible materials in the initiating phase of fire (it is generally known that, in strong, long-lasting fires, no material resists), thus increasing the application range of these materials. Flame retardancy is required to protect life or some products' properties.

Flame retardants exert many different modes of action as a function of the chemical nature of the polymer–flame retardant systems and of the interactions between the components. It is considered that inhibition of burning is achieved by modification of either the condensed phase or the dispersed or gaseous phase in a physical and/or chemical mode. In the condensed phase, the following may occur: dilution of the combustible content, dissipation of heat, shielding of surface by charring or forming of a glassy coating or gas or liquid barrier reducing the heat transfer, alteration of the distribution of the condensed pyrolyzates, lowering of the decomposition temperature (due to the endothermic decomposition of the flame retardant), influence on the combustion mechanism both in the condensed and in the gaseous phase (flame retardant interrupts the chain reaction by radical acceptor species), while in dispersed phase, by generation of noncombustible gases (N_2, CO_2, H_2O, HX) that dilute the fuel gas and tend to exclude oxygen from the polymer surface [7, 8]. Commonly, vapor phase active flame retardants or those acting physically are less efficient than condensed phase-active retardants, which directly reduce fuel gas generation.

Up to now, no coherent theory, to be generally applied to all classes of flame retardants and polymers, exists.

B. Classification of Flame Retardants

According to their chemical structure, mode of action and mode of incorporation in the polymer, inorganic and organic flame retardants are usually classified as additives and reactives. Despite the large number of different flame retardant additives, most commercial retardants currently belong to one of the following six chemical classes: alumina trihydrate (ATH) and $Mg(OH)_2$, organochlorine compounds, organobromine compounds, organophosphorus (including halogenated phosphorus) compounds, antimony oxides, and boron compounds.

There are also flame retardants chemically bound to the substrate, such as epoxy intermediates, polycarbonates, polyesters, and flexible or rigid urethane intermediates.

Many inert fillers and additives, such as talc, $CaCO_3$, crushed marbles, and clays, can also act as flame retardants or thermal sinks. At high loadings (>20%wt) they lead to mass dilution and slow heat generation, favoring charring and reducing flammability and smoke generation. Generation of volatile, combustible gases can also be prevented by protecting the surface against the heat of the flame by a thermal barrier such as an intumescent coating. Thus in the case of thermal sink action, the intrinsic flammability of the substrate remains unchanged. The effectiveness of the coating strongly depends on maintaining the physical integrity of the intumescent coating. Intumescent systems (such as melamine–formaldehyde resins), developing an insulating, hard-ablative char outer layer, reduce smoke generation by shielding the underlying combustible material (as in the case of EPDM).

Smoke suppressants are also important additives in fire retardance. Visible smoke may accompany or precede the development of heat and toxic gases. It results from an incomplete combustion. Additives such as antimony oxide, metal (Ba, Ca, Zn) borates, $Al(OH)_3$, $Mg(OH)_2$, $MgCO_3$, and magnesium oxychloride are used as fillers and flame retardants, but they are also good smoke reducers.

C. Flame Retardants for Polyolefins [2, 7–18]

1. Aluminum Trihydrate (ATH)

Aluminum trihydrate contains 34.6% water (0.84 g/cm^3), which it loses above 220°C, absorbing 280 cal/g to evaporate the water and another 190 cal/g to heat the alumina, a strongly endothermic process that slows the rate of pyrolysis. The evolved water dilutes the combustible gases and cools the flame, while alumina residue acts as a thermal shield on the polymer surface, by its physical mode of action. Its primary flame retardant mechanisms are its ability to cool the polymer by evaporation, and to blanket it in water vapor that excludes atmospheric oxygen. ATH is an inefficient additive compared with those producing chemical inhibition of burning, an amount greater than 20 wt% being necessary. However, it is still used in most cases (58–70% from the total amount of flame retardants used for PO). It is widely used as a low-cost filler, primarily as a flame retardant that produces no smoke or toxicity. Pyrogenic silica produces a similar effect with ATH, at less than 20 wt%, but it is much more expensive.

It has two drawbacks: (1) 40–65% by weight must be used in polyolefins, causing stiffening and embrittlement; (2) it decomposes at processing temperatures above 180–230°C, which limits it to polymers that can be processed below those temperatures.

In polyolefins, it is used mainly in LDPE wire and cable insulation. Other uses reported or claimed include EVA wire and cable and foam carpet-backing; EMA wire and cable; EPR and EPDM wire and cable, electrical insulation, roofing, and automotive parts—whitewall tires, hoses, seals, and weather stripping; XLPE wire and cable; chlorosulfonated polyethylene; and HDPE and polybutene. There is considerable interest in its use in PP cabinets (less smoke than polystyrene), carpeting, and wire and cable, but processing temperature and embrittlement both pose problems.

2. Magnesium Hydroxide [18, 19]

Magnesium hydroxide is more stable than aluminum trihydrate, up to 300–340°C, permitting processing up to 275°C. Beyond these temperatures, in a fire it releases 31% by weight of water (0.73 g/cm^3), absorbing more enthalpy than aluminum trihydrate. In polyolefins it gives good flame retardance, smoke suppression, and char formation. High concentrations are needed, 30–75% by weight, depending on the specific polyolefin and the level of flame retardance needed. For maximum effectiveness, 60% $Mg(OH)_2$ by weight is most often recommended for polyethylene, EVA, EEA, EPDM, and PP. Major uses are in wire and cables.

Magnesium hydroxide filler treated with magnesium stearate has improved filler/matrix interaction and maleic anhydride functionalized EPR increased impact strength. The composition is very efficient in fire-retarded PP [20].

3. Borates [21]

Zinc borates (ZB) have multifunctional properties such as smoke suppressant, flame retardant, afterglow suppressant, char promoter, anti-arcing agent, anti-tracking in electrical applications, corrosion inhibitor, fungicide activity, and improves oil resistance. During polymer combustion, the zinc borate promote char formation and prevent burning drips. It accelerates photocure and improves oil resistance. It has synergetic effects in halogen-free applications with $Al(OH)_3$, $Mg(OH)_2$, SiO_2 or silicone polymers, ammonium polyphosphate, red phosphorus, Sb_2O_3, nitrogen compounds, etc. (see synergists below).

Some characteristics of borates are given in Table 2. FIREBRAKE ZB has been used as a fire retardant in PO wire and cables as well as electrical parts. It improves elongation properties of PO, crosslinked EVA-electrical parts, etc. [22].

4. Halogen-Containing Compounds

Due to the release of halogen acid during decomposition, halogen-containing compounds interrupt the chain reaction of combustion by replacing the highly reactive $HO^{\bullet}$ and $H^{\bullet}$ radicals by the less reactive $X^{\bullet}$; for example:

$$HO^{\bullet} + HX = H_2O + X^{\bullet}$$

$$H^{\bullet} + HX = H_2 + X^{\bullet}$$

TABLE 2 Characteristics of borates as fire retardants

Borate	Starting dehydration temperature (°C)
Boric acid	70
Boric oxide	—
Ammonium pentaborate $(NH_4)_2 \cdot 5B_2O_3 \cdot 8H_2O$	120
Barium metaborate $BaO \cdot B_2O_3 \cdot H_2O$	200
Zinc borate FIREBRAKE ZB $2ZnO \cdot 3B_2O_3 \cdot 3.5H_2O$	290
Zinc Borate FIREBRAKE 415 $2ZnO \cdot yB_2O_3 \cdot zH_2O$	415
Dehydrate zinc borate FIREBRAKE 500 $2ZnO \cdot 3B_2O_3$.	—

(a) Organic Chlorine [19]. Chlorinated paraffins with the highest possible chlorine content (~40%) are technically very important. Their main disadvantage is low thermal stability, being applicable only in thermoplastics with processing temperatures below 200°C.

Aliphatic chlorine is more reactive than aromatic and thus more efficient for flame retardance, and its effect generally increases with concentration. Chlorinated paraffins ($\leqslant$ 74% Cl) are low in cost and are widely used in polyethylenes; but since they decompose in processing above 190°C, their use is mainly limited to LDPE. Recent improvements in thermal stability up to 235°C suggest that they may also be useful in PP [18, 23].

Whereas highly chlorinated paraffins are solid fillers that stiffen polyolefins, chlorination of polyethylene itself reduces crystallinity, making the polymer more flexible as well as flame retardant [23].

Cycloaliphatic chlorine is more stable and able to withstand processing temperatures in polyolefins. Of the many structures explored in research and development, the one chosen for most commercial usage is the adduct of 2 mols of hexachlorocyclopentadiene with 1 mol of cyclooctadiene, containing 65% Cl and stable up to its melting point of 350°C [19]. It is generally used with antimony oxide synergist and is effective in LDPE, HDPE, XLPE, and PP [19, 24, 25]. The main problems are embrittlement, due to the high filler loading, and smoke generated by volatilization of chlorine and antimony oxide.

Perchloropentacyclodecane is considerably used in PO compositions because of its improved thermal stability, electrical properties, and resistance to migration. In PP compositions it prevents dripping during burning and produces a carbonaceous coating on extinguishment.

(b) Organic Bromine. Organic bromine is more reactive than chlorine, and it is a more effective flame retardant [19, 26, 27]. Typical estimates suggest that bromine is twice as effective on a weight basis: 20% bromine is generally enough for good flame retardance, and synergism with phosphorus or antimony oxide can lower the requirement to 6–7% [27, 28].

Bromine compounds which are particularly efficient are brominated phenol respectively bisphenol A ether, hexabromocyclododecane, and bromine containing Diels–Alder reaction products based on hexachlorocyclopentadiene [29].

Aromatic–aliphatic bromine compounds, the bis-(dibromo propyl)-ether of tetrabromobisphenol A or bromomethylated as well as aromatic-ring-brominated compounds such as 1,4-bisbromomethyltetrabromobenzene, combine the high heat stability of aromatic-bound bromine with the significant flame retardant effect of aliphatic-bound bromine. They are therefore used mainly as flame retardants for thermoplastic PO and PO fibers.

Brominated diphenyl ethers like pentabromodiphenyl ether (mixture mainly containing pentabromo diphenyl ether, tetra- and hexabromodiphenyl ether), octabromodiphenyl ether (mixture of mainly nonabromodiphenyl ether and octabromodiphenyl ether), and decabromodiphenyl ether are excellent flame retardants for those thermoplastics to be processed at high temperatures.

Oligomeric bromine-containing ethers based on tetrabromoxylydene dihalides and tetrabromobisphenol A (molecular weight between 3000 and 7000) are useful flame retardants for PP. The molar efficiency of halogen compounds as flame retardants decreases in the following order I > Br > Cl > F. In practice, chlorine and bromine compounds are seldom used without a synergist such as Sb_2O_3, which increases effectiveness or reduces the cost (Table 3). These combinations produce a considerable amount of carbonaceous residue when burnt in polyolefins such as PP. The predominant flame retardant action of chlorine is centered in the condensed phase while bromine is active in the gaseous phase.

In early commercial development, the instability of organo-bromine caused serious problems in processing and weathering; but careful selection of stabler structures, most of them aromatic, the rest cycloaliphatic, has permitted rapid market growth, so that current use of 68 million pounds a year in the U.S. makes them the

TABLE 3 Use of Organo-bromine in flame-retarded formulations

LDPE	68%	68%						
HDPE			53%					
XLPE				70%	70%			
EPR						72%		
PP							58%	58%
DBDPO[a]	25%			20%			22%	
EBTBPI[b]		32%	27%		20%	21%		22%
Sb_2O_3	7%	10%	6%	10%	10%	7%	6%	6%
Talc			14%				14%	14%
Clay		24%						
Oxygen index		27	29	26	26	28	26	
UL-94 rating	V-0	V-0	V-0	V-0	V-0	V-0	V-0	V-0

[a]Decabromo diphenyl oxide.
[b]Ethylene bis(tetrabromo phthalimide).
Source: Refs. 19, 21, 22, and 26.

leading class of flame retardants. They are almost always synergized by antimony oxide, making them still more efficient. While organo-bromine is more expensive than organo-chlorine, its higher effectiveness may make it quite competitive; and the use of lower concentrations of additive also reduces the deleterious effects on other properties, such as embrittlement and tinting strength.

Since the technology is still developing, the industry has not yet focused its choice on one or two leading organo-bromine compounds. Rather, many manufacturers and many users of these flame retardants report serious interest in a wide variety of structures. Those most frequently mentioned in the technical literature, in approximate order of importance, are [2, 19, 21, 22, 25, 26], as follows: decabromo diphenyl oxide, ethylene bis(tetrabromo phthalimide), tetradecabromo diphenoxy benzene, ethylene bis(dibromo norbornane dicarboximide), octabromo diphenyl oxide, hexabromo cyclo dodecane, tetrabromo dipentaerythritol. In Europe, the organo-bromine compounds mentioned are quite different, mostly based on ethers of tetrabromo bisphenol A; and while Europeans are more concerned about possible toxicity, paradoxically they are more likely to use polybrominated biphenyls, which would be taboo in the US [2, 28].

5. Antimony Oxide

Antimony trioxide alone is not a flame retardant; but when it is used in combination with organic halogen, their synergistic interaction greatly reduces the amount of organic halogen required to produce flame retardance [2, 19, 27, 30, 31]. A variety of mechanisms have been proposed to explain this synergism, generally involving the intermediate formation of antimony halides and oxyhalides, and most of them involving free-radical inhibition of vapor phase oxidation. The optimum halogen/antimony atomic ratio has been variously estimated from 1/1 to 3/l; in some systems the ratio is fairly critical, in others quite broad. Overall market analysis gives an organohalogen/antimony oxide weight ratio of 2.37/1 [28].

In flame retarding polyolefins, typically, 40% organochlorine can be replaced by 8% organochlorine + 5% antimony oxide, or 20% organobromine can be replaced by 6% organobromine + 3% antimony oxide [27]. Consequently, almost all polyolefin formulations based on organic halogen also contain antimony oxide to make it as effective as possible (Table 4).

6. Mixtures of Flame Retardants—Synergists

Flame retardants can act at different stages of the combustion process; hence in many cases a combination of retardants is employed. Sometimes in such cases a synergetic effect can be observed. For example, antimony oxides (Sb_2O_3 and Sb_2O_5) in concentrations of 3–8% (sometimes 15%) serve as synergists and strengthen the effect of organic halogen compounds (the halogen/Sb ratio is commonly from 1/1 to 5/1), although they themselves are not flame retardants.

In a fire, organic halogen and antimony oxide both produce dense white smoke, which can obscure the victim's view of the exit and keep him from reaching it [30]. Efforts to minimize this problem have often involved replacement of part or all of the antimony oxide by other metal oxides, especially zinc borate [19, 21, 22, 30]. Since $2ZnO{\cdot}3B_2O_3{\cdot}3.5\ H_2O$ loses 14% water at 300–450°C, forming glass and char, it

TABLE 4 Formulations of fire resistant PO insulators

	Examples (parts per weight)						
Components	1	2	3	4	5	6	7
Ethylene-α-olefins	100	100	100	100	100	100	100
ATH	60	100	100	100	150	—	—
Red phosphorus	20	20	20	20	—	30	—
Zinc borate	10	10	30	—	—	—	30
Properties							
Oxygen index (%)	26	37	40	32	26	24	22
Volume resistivity (Ω cm)	1×10^{16}	5×10^{15}	3×10^{15}	1×10^{16}	1×10^{15}	4×10^{16}	1×10^{16}
Dielectric loss (%)	1.2	2.1	2.3	2.0	2.2	1.0	1.1
Tensile strength (kg/mm^2)	1.7	1.5	1.1	1.7	1.4	1.7	1.7
Elongation at break (%)	580	540	500	550	490	600	600

Dicumyl peroxide 2.5; antioxidant 0.5.
Source: Ref. 22.

can also partly replace aluminum trihydrate and magnesium hydroxide [22]. Others that have been mentioned include molybdenum oxide, barium borate [30], ferrocene [30, 31], nickelocene [31], aluminum acetyl acetonate, and calcium carbonate [31]. One study claimed that, while it took 66% of chlorine to flame retard polyethylene foam, 35% chlorine + 5% sulfur could also do it synergetically [27].

Melting and dripping are common problems in flammability of polyolefins, particularly at low concentrations of flame retardant. A frequent practice to minimize this is the addition of fillers, particularly 10–15 phr talc [19, 23, 26] to increase the melt viscosity and probably also the thermal conductivity. One study also reported the benefit of 10% titanium dioxide, probably acting as a heat sink. Such techniques are most often used to raise a flame retardant rating from UL-94 V-2 to V-0.

Other commonly used synergists are boron compounds, titanium oxide, molybdenic oxides, zinc oxides or borates, antimony silicates, phosphorous-halogens, etc. The synergists are used for economic reasons, as decreasing the quantity of flame retardants, being also of special interest for transparent thermoplastics.

A mixture of chlorinated and brominated flame retardants has a synergetic effect that allows retardant levels to be lowered, resulting in improved physical properties and lower cost formulations [32].

In halogen-containing PO, a typical formulation for wire and cable application consists of decabromodiphenyl oxide or Dechlorane Plus, antimony oxide and zinc borate. It is interesting to note that, in presence of Irganox 1010, the zinc borate can significantly improve the physical properties of the wire and cable (aged elongation). The zinc borate is used in conjunction with decabromodiphenyl oxide, Sb_2O_3 and $Mg(OH)_2$ in an ethylene-vinyl acetate copolymer (EVA) and a LDPE blend for wire insulation applications. The use of ZB/Sb_2O_3 in conjunction with ethylene bis-(tetrabromophthalimide) in EVA gives a nonblooming wire insulation. The combination of a chlorine source (Dechlorane Plus) and bromine source is known to display synergy in fire retardancy. Sakamoto of Sumitomo Electric reported the use of tris(tribromoneopentyl) phosphate (replacing decadromodiphenyloxide), the zinc borate and Sb_2O_3 to achieve the VW-1 test with no dripping for LDPE and EVA wire insulation.

In halogen-free PO systems, the zinc borate is recommended to be used in conjunction with ATH and/or $Mg(OH)_2$ to achieve the most cost effective fire test performances. At a total loading of 80–150 phr, the weight ratio of ZB to ATH is normally in the range of 1:10 to 1:1. This combination can not only provide smoke reduction but can also form a porous ceramic residue, that is important thermal insulator for unburned polymer. The benefit of using the ZB and red phosphorous in halogen free PE insulator is illustrated in Table 4.

The synergetic action between pentabromotoluene, chlorinated paraffins and Sb_2O_3 appear to be more efficient than mixtures of any two alone.

7. Phosphorus Compounds [27, 30]

Phosphorus is the most flame retardant element known; typically 5% P in polyolefins gives good flame retardance. Generally organic phosphorus is much better than inorganic, either because it is more miscible or because it can burn to phosphorus oxide glass [27, 30]. Unfortunately, this requires 20% or more of organo-phosphorus compound, and its high polarity is poorly miscible with polyolefins [17, 27, 30]. While most studies have found little success with inorganic phosphorus, there are occasional reports of usefulness with materials such as PCl_3, $POCl_3$, H_3PO_3, H_3PO_4, $PNCl_2$, ammonium phosphate, and dithiopyrophosphate. [19, 23, 25, 27].

Generally organic halogen–phosphorus combinations are synergistic, producing flame retardance at much lower concentrations [25, 27, 30, 31]. Typically where 5% P or 40% Cl are required for good flame retardance, 2.5% P + 9% Cl work equally well; and where 5% P or 20% Br are required, 0.5% P + 7% Br work equally well [27]. In many cases 3–6% by weight of organic bromine–phosphorus compounds is sufficient [27].

Elemental red phosphorus is itself highly combustible; but once it is properly compounded into solid plastics, it also offers good flame retardance. In polyolefins, 8% produced good resistance to burning, and it could be further synergized with polyacrylonitrile or melamine [30].

It seems that, while the phosphorus compounds facilitate polymer decomposition, the phosphoric acid formed reacts with the polymer and produces char layer phosphates as well as water and other noncombustible gases. Quaternary phosphonium compounds are recommended as flame retardants for PO, but brominated phosphorus esters are only infrequently applied in PO compositions due to their poor compatibility.

Polyarylphosphate oligomers offer lower opacity, smoke toxicity and corrosion [33].

8. High Molecular Weight or Polymeric Flame Retardants

Attempts have been made to synthesize higher molecular weight or polymeric flame retardants containing halogen or/and phosphorus or to introduce these elements by copolymerization. Chlorinated PE and chlorosulfonated PE are selected for flame retardant uses largely because their chlorine content is high.

In their turn, polyolefins including chlorinated PE, chlorosulfonated PE, and other polymers can be flame retarded by (1) structural modification by halogenation (for example, elastomers such as EPDM), graft copolymerization with chloroethylvinylphosphate, allyltrichloroacetate, etc., obtaining the halogenated copolymers as chlorotrifluoroethylene or hexafluoropropene–vinylidene fluoride, etc.; (2) utilization of flame retardants such as brominated aromatics, borosiloxanes derivates, red phosphorus/$Al(OH)_3$, antimony/bromo alkyl methyl phosphinic acid, $Mg(OH)_2$, etc; (3) utilization of intumescent systems such as pentaerythritol/phosphate, melanine or cyclic nitrogen compounds, etc. Polymeric flame retardants as well as reactive flame retardants do not lead to problems of leaching, migration, or volatility, which usually accompany additive flame retardants; instead, the problem still remaining is the problem of compatibility with the substrate or their mutual compatibility.

9. Intumescents

Intumescent coatings are formulated so that, when they are exposed to a fire, they turn into a foamed char that covers the plastic, protecting it from heat and atmospheric oxygen and thus smothering the fire. They are typically composed of a carbonific (source of carbon) such as pentaerythritol, an acid catalyst such as an amine phosphate, and a foaming agent such as amine or melamine [19, 28]. These are normally post-coatings, but several recent proprietary formulations have been offered for addition directly to PP at concentrations of 25–30% by weight. As an example maleimine phosphates/trioxide [34] is mentioned. A combination of silicone fluid, magnesium stearate, decabromodiphenyl oxide, and aluminum trihydrate has been reported to give good flame retardance without embrittlement in PP [19].

A list of tradenames of flame retardants for PO is given in Table 5.

D. Requirements

For the most suitable choice of a flame retardant the following requirements have to be first considered.

It should provide a durable effect when only small amounts are added, so that the mechanical, physical, and optical properties of the flame retarded polymer should not be affected.

The manner of incorporation should be easy and cheap.

It should not cause corrosion of the processing equipment, and it should not plate out.

It should not decompose at the processing temperature; neither toxic effluent nor abundant smoke should occur.

In fiber treatments, they should be durable under leaching, soaking, and washing.

(Text continues on p 574)

TABLE 5 Flame retardants for thermosplastics

Tradename	Chemical class	Supplier
Adine	Organic bromine compounds	Ugine Kuhlmann M & T Chimie 92062, Paris, France
Amgard	Phosphoric acid ester	
Amgard CD	Ammonium bromide/phosphate	Albright & Wilson, Ltd. Rockingham Works, Avonmouth, Bristol BS11 0YT, UK
Amsperse	Submicron Ag_2O_3 in plasticizer	Amspec Chemical Corp. NY, USA
AZ-60; AZ-44; AZ-55; AZ-80	As, Mg, Zn, Mo compounds	Anzon 2595 Aramigo Ave., PA, USA
Bromkal	Organic bromine compounds	Chemische Fabrik Kalk, GmbH, Germany
Buan Ex	Permanent FR	PA Corp, Box 840, Valley Forge, PA USA
BYK	Permanent FR	BYK, CT, USA
Caliban	Brominated compounds Antimony oxide	KMZ Chemicals, Ltd.
CDP	Cresyl diphenyl phosphate	Union Carbide Chemicals Co. Div. Carbide Corp.
Celluflex	Tricresyl phosphate	Celanese Chemical Co.,
179C	Phosphorous compounds	Celanese Corp. of
CEF		America, Industrial Chem. Dept., Route 202–206, North Somerville, NJ 08876 USA
Chlorowax	Chlorinated paraffins	Diamond Shamrock Plastics Div., Cleveland, OH 44114, USA
Cerechlor	Chlorinated paraffins	Imperial Chem. Industries, Ltd., Millbank, London SW IP 3JF, UK
CP-40	Chlorinated paraffins 40–60% Cl	Hercules Chem. Corp. Occidental Petroleum Corp.
Dechlorane	Chlorine resp. chlorine and bromine containing cycloaliphatics as perchloropentacyclodecane suitable for PP	Hooker Chemicals, Plastics Corp., Niagara Falls, NY 14302, USA PO Box 728
Disflamoll	Organic phosphates and halogenated organic phosphates	Bayer AG, 5090 Leverkusen, Germany
DPK, TDF,	Tris 2-ethyl hexyl phosphate	Bayer AG, 5090
P, TPP, TKP	Tricresyl phosphate	Leverkusen, Germany
Exoit	Chlorinated paraffins Organic phosphates	Hoechst AG, 6230 Frankfurt 80/M, Germany Postfach 800320 Hoechst Celanese Corp., Industrial Chem. Dept., Route 202–206 North Somerville, NJ 08876, USA

TABLE 5 (Continued)

Tradename	Chemical class	Supplier
Firebrake	Inorganic flame retardants	US Borax & Chem. Corp. Los Angeles, CA 90010, USA
Fireguard	Organic bromine compounds	Teijin Chem. Co., Minatoku 6–21, 1-chome, Nishishinbasi, Tokyo, Japan
	Brominated aromatics 20–80% Br, $T_d = 400°C$	Tokyo Soda, Tokyo, Japan
Firemaster	Brominated aromatics	Great Lakes Chem. Co.,
Fire Fighters	20–80% Br, $T_d = 400°C$	West Lafayette, IN 47904, PO Box 2200, USA
	NI Tribromophenyl/ maleimide brominated polyaromatic 20–80% Br for EPR	IMI Institute R & D Haifa, Israel
Fire Shield	Organic bromine compounds, antimony trioxide	PPG Industrie Inc. AMF O'Hara, PO Box 66251, Chicago, IL 60666, USA
Firex	Organic halogenated compounds with antimony trioxide	Dr Th Böhme KG 8192, Geretsried, Germany
Flexol TDF TCP	Phosphates	Union Carbide Corp., USA
Halowax	Chlorinated paraffins	Koppers Co., Inc.
Kemgard	Mg, Ca, Zn complexes	Sherwin Williams Chem., 1700 W Fourth St., KS, USA
Kronitex TBP MX TDF	Tributylphosphate Cresyldiphenylphosphate	FMC Corp. Org. Chem. Div., 2000 Market St., Philadelphia, PA 19103, USA
Mastertek PP 5601 5602 5603 5604	As_2O_3, chlorine and/or bromine masterbatches	Campine America 3676 Davis Road, PO Box 526 Dover, OH 44622, USA
Onchor 23A	Antimony trioxide	National Chem. Corp.
Phosgard C-22R	Poly (β-chloroethyl-triphosphate)	Monsata Chem. Corp., Org. Chem. Div.
Phosflex	Organic phosphates	Staufier Chem. Co., Specialty Chem. Div., Westport, CT 06880, USA
Proban	Phosphoric acid esters Quaternary phosphonium	Albright & Wilson, Rockingham UK Works Avonmouth, Bristol BS11 0YT, UK PO Box 26229, Richmond, VA 23260, USA
Protenyl BM BSP NP	Unspecified composition for PO fibers	Protex Chemicals, Ltd.

TABLE 5 (Continued)

Tradename	Chemical class	Supplier
Pyro-Check	Organic bromine compounds	Ferro Corp. Chem. Div., Bedford, OH 44146, USA
Pyrovatex	Organic phosphates Halogenated phosphorus compounds	Ciba Geigy AG, 4002 Basel, Switzerland
PX-90	Tricresylphosphate	Pittsburg, Chem. Co.
Reomol 21P TTP	Cresyldiphenylphosphate	Ciba Geigy AG, Basel, Switzerland
Sandoflam	Organic bromine compounds	Sandoz AG, 4002, Basel, Switzerland
Santicizer	Phosphoric acid esters, halogenated organic compounds	Monsanto Co., 800 N Lindbergh Blvd, St Louis, MO 63167, USA
Saytex	Halogenated cycloaliphatics	Ethyl Corp., Bromine Chemicals Div., 451 Florida Blvd, Baton Rouge, LA 70801, USA
Spiriflam	Halogen-free, free-flowing powders, intumescent	Montell, Three Little Falls Center 2801 Centerville Rd., Wilmington, DE, USA
Tardex	Organic bromine compounds	ISC Chemicals, Avonmouth, St. Andrew's Road, Bristol BS11 9HP, UK
Thermogard	Antimony trioxide for PO fibers	Metall & Termit Corp., Harshaw Chem Co., 1945 E. 97th St Cleveland, OH 44106, USA M & T Chem. Inc., PO Box 1104, Ratsway, NJ 07065, USA
T23P	Tris 2,3-dibromopropyl phosphate	Michigan Chem. Corp.
Unichlor	Chlorinated paraffin	Neville Chem. Corp., Neville Island, Pittsburg, PA 15225, USA

Flame retardants and smoke suppressants	Suppliers
Smoke suppressants Molybdenum compounds Zinc borates	AMAX Polymer Additives Group, Suite 320, 900 Victors Way, Ann Arbor MI 48108, USA
Dispersions of phosphates esters, chlorinated paraffins, zinc borates $Al(OH)_3$, Sb_2O_3	Americhem Inc., 2038, Main Str. Cuyahoga Falls OH 44222, USA Ampacet Corp., 250S, Terrace Ave, Mount Vernon, NY 10550, USA Podell Industrie Inc., Entin Rd, Clifton, NJ 67016, USA
Sb_2O_3	AMSPEC Chem. Corp. At the Foot of the Water Street, Gloucester City, NJ 08030, USA Asarco Inc., 180 Maiden Lane, NY 10038, USA Laurel Industrie Inc., 29525 Chagrin Blvd, Suite 206, Pepper Pike, OH 44122, USA Nyacol Inc., Megunco Rol, PO Box 349, Ashland, MA 01721, USA Cookson Ltd., Uttoxeter Road Meir, Stoke-on-Trent ST3 7PX, UK Société Industrielle de Chimique de l'Aisne BP 46F02301, Chauny, France
Sb_2O_3, zinc borate	Claremont Polychemical Corp., 501 Winding Rd, Old Bethpage, NY 11804, USA

TABLE 5 (Continued)

Flame retardants and smoke suppressants	Suppliers
Phosphorus/nitrogen intumescent coatings	Borg-Warner Chemicals, Inc. International Center, Parkersburg, WY 26102, USA
	Montefluor, Via Principe Eugenio 1/5 I-20155, Milano, Italy
Clorinated paraffins	CII, Inc., PO Box 200, Station A, Willowdale, ON M2N 5S8, Canada
Halogenated compounds	ICI, PLC Mond Div., PO Box 13, Runcorn, Cheshire, UK
	Dover Chem. Corp., ICC Ind., Inc., 720 Fifth Ave, NY 10019, USA
	Keil Chem., Div. Ferro Corp., 3000 Sheffield Ave, Hammond, IN 46320, USA
	Pearsall Chem. Div., Witco Chem. Corp., PO Box 437, Houston, TX 77001, USA
	Hüls, AG, Postfach 1320, D-4370 Marl, Germany
	M & T Chemie SA, Alée des Vosges, La Défense 5, Cedex 54, F92062, Paris la Défense, France
	Asahi Glass Co., Ltd., Chiyoda Building 2-1-2, Marunouchi, Chiyoda-ku, Tokyo 100, Japan
	Otsuka Chem. Co., Ltd., 10-Bunge-machi, Higashi-ku, Osaka 540, Japan
Phosphate esters, halogenated phosphorus compounds, chlorinated paraffins, aluminum hydroxide, antimony oxide, zinc borate	Harwick Chem. Corp., 60S Seiberling St, Akron, OH 44305, USA
	Isochem Resin Co., 99 Cook St, Lincoln, RI 02865, USA
	Olin Corp., Chem. Div., 120 Long Ridge Rd, Stamford, CT 06904, USA
	Ware Chem. Corp., 1525 Stratford Ave, Stratford, CT 06497, USA
	Sumitomo Chem. Co., Ltd., New Sumitomo Building, 15–5 chome, Kitahama, Higashi-ku, Osaka, Japan
	Sanko Chem. Co., Ltd., 16 Toori-cho, 8-chome, Kurume City, Fukuoka 830J, Japan
Smoke-reducer based molybdenum	Climax Molybdenum Co., Ltd., London SW1X 7DL, UK; 1600 Huron Pkwy, Ann Arbor, MI 48106, USA
Organic bromine compounds	Ameribrom, Inc., 1250 Broadway, NY 10001, USA
	Dai-Chi Chem. Ind., Osaka, Japan 2-2-1 Higeshi Sakashita, Itbashi-ku, Tokyo 174, Japan
	Dead Sea Bromine Co., Potash House, PO Box 180, Beer-Sheba 84400, Israel
	Dow Chem. Co., Midland, MI 48640, USA
	Eurobrom B. V. Patentlaan 5, NL 2288 AD Rijswijk, Holland
	Hitachi Chem. Co., Ltd., Shinjuku Mitsui Build. 2-1-1 Nishi-Shenjuku, Shenjuku-ku, Tokyo 160, Japan
	Makhteshim Chem. Works, Ltd., PO Box 60, Beer-Sheva 84100, Israel
	Marubeni Co., Fine Chem. Soc., Osaka (A777) CPO Box 1000, Osaka 530–91, Japan
	Nippon Kagaku Co. Ltd., Tokyo Kaijo Building 2-1 Marunouchi-1-chome, Chiyoda-ku 100, Tokyo, Japan
	Riedel-de-Haen AG, Wunstarferstr 40, D-3016, Seelze, Germany
Antimony based compounds	Chemische Fabrik Grünau GmbH, Postfach 120, D-7918, Illertissen, Germany
	Nyacol Products Inc., Ashland, MA 01721, USA
Phosphate esters	Akzo Chemical Division, Speciality Chem. Div., Westport, CT 06880, USA
	Condea Petrochemie GmbH, Mittelweg 13, D-2000, Hamburg 13, Germany
	Daihachi Chem. Ind. Co., Ltd., 3–54, Chodo Higashi, Osaka City, Osaka Pref. 577, Japan
	Kyowa Hakko Kogyo Co., Ltd., Ohtemachi Building 6-1, Ohte-machi-1-chome, Chiyoda-ku, Tokyo 100, Japan
	Mitsui Toatsu Chem., Inc., 2-5 Kasumigaseki-3-chome Chiyoda-ku, Tokyo 100, Japan
	Mobil Chem. Co., Phosphorus Div., PO Box 26683, Richmond, VA 23621, USA
Aluminium hydroxide	Aluminium Comp. of America Alcoa Labs., Alcoa Center, PA 15069, USA
	Kaiser Chemicals Div., Kaiser Aluminum & Chem. Corp., 300 Lakeside Dr, Oakland, CA 94643, USA
	SOLEM Industries, Inc., Subsid. of J. M. Huber Corp., 4940 Peachtree Industr. Blvd, Narcross, GA 30071, USA
	BA Chemicals, Ltd., Chalfont Park, Gerrards Cross, Bucks SL9 OQB, UK
	Martinwerk GmbH, Postfach 1209, D-5010, Bergheim, Germany
	Metallgesellschaft AG, Rauterweg 14, D-6000, Frankfurt 1, Germany
	Vereinigte Aluminum Werke AG, Postfach 2468, D-5300, Bonn, Germany

TABLE 5 (Continued)

Aluminum trioxide	Campine NV Nijverheidsstraat 2, B-2340, Beerse, Belgium
Nitrogen-containing compounds	Süddeutsche Kalkstickstoffwerke AG, Postfach 1150/1160, D-8223, Trostberg, Germany
Halogenated compounds phosphate esters	Asahi Denku Kogyo KK, Furukawa Building 8, 2 chome, Nihonbashi-Murmachin Chuo-ku, Tokyo, Japan
Phosphorus and nitrogen compounds	Chisso Co., 7-3 Kounouchi-2-chome, Chiyoda-ku, Tokyo 100, Japan
Halogenated and nitrogen-containing compounds	Nissan Chem. Ind., Ltd., Kowa, Hitotsubashi Building 7-1, Kanda Nishiki-cho-3-chome, Chiyoda-ku, Tokyo 101, Japan
Molybdenum compounds	Sherwin Williams Co., Chem. Div., PO Box 6506, Cleveland, OH 44101, USA
Zinc borates	US Borax, 3075 Wilshire Blvd., Los Angeles, CA 90010, USA
	Borax Holdings, Ltd., Borax House, Carlisle Place, London SW1P 1HT, UK
Antimony oxide, antimony sulphide, zinc and magnesium oxide	Anzon Ltd., Cockson House, Willington Quay, Wallsend, Tyne & Wear NE28 6UQ, UK
Zinc oxide	Vieille Montagne, 19, rue Richer, F 75442, Paris, Cedex 09, France

Examples of inorganic phosphorus compounds: H_3PO_3, H_3PO_4, PCl_3, $POCl_3$; Examples of organic phosphorus compounds; $ClCH_2POCl_2$; $CH_3Ph_2PO_4$; $(PhCH_2)_3$ PO dialkylphophinic acid; $R_3P{=}CHCOCO{=}PR_3$, R = phenyl, etc.
Examples of phosphorus–halogen systems; tris-2,3-dibromopropyl-phosphates bis(betachloroethyl)vinyl phosphate.
Examples of halogen–metal oxide flame retardants; Sb_2O_3 (1–40%) with chlorinated PP or phthalimide derivative (40–57% chlorine) with aliphatic or aromatic bromine compounds.
Source: Refs 2, 8, 9, 12.

The lower melting point and processing temperatures of polyethylenes permit the use of chlorinated paraffins, whereas PP requires stabler additives [19, 26]. Since PP is less stable than polyethylene, it usually requires higher concentrations of flame retardants. The decomposition temperature of the polymer must be matched to the decomposition temperature of the flame retardant in order to obtain maximum effectiveness; thus polyethylene is better with stabler aromatic bromine, whereas PP is better with less stable aliphatic bromine [18, 19, 26].

Flammability of PP becomes particularly critical in fibers, where solid additives may interfere with spinning and mechanical properties. A wide variety of flame retardants are therefore explored for such applications [35], often as coatings rather than additives. For example, addition of a liquid containing 45% Br + 15% Cl, to polyvinyl acetate or acrylic latex, produces a flame retardant backing for PP fabrics, for applications such as upholstery [18].

In polyethylene and PP, the addition of increasing concentrations of flame retardants has reasonably linear effects. In poly-1-butene, on the other hand, the optimum concentrations of ethylene bis(dibromo norbornane dicarboximide)/antimony oxide peaked sharply at 4–5/2–2.5%, dropping off dramatically below and above this range [26].

Low molecular weight polymers (high melt flow index) melt and drip, often carrying away heat and flame and allowing the sample to self-extinguish, giving UL-94 V-2 ratings and requiring little flame retardant [19]. For V-0 ratings, the polymer requires higher concentration of flame retardant and/or large addition of solid filler to increase its melt viscosity [23]. This makes processing more difficult, mechanical properties worse (impact strength and integral hinge fail), and appearance less glossy [26]. Here is where more powerful flame retardants, finer-particle-size fillers, elastomeric additives, and coupling agents all become more critical [18, 19].

E. Incorporation

Incorporation of flame retardants in polyolefins is difficult, due to their high crystallinity (PE) or high processing temperature (PP) as well as to their incompatibility with most organic compounds.

The following ways of incorporating are used: (a) thermoplastic powders are blended with flame retardants in a fluid or turbine mixer, extruded by means of a twin screw extruder, and pelletized; (b) previously pelletized thermoplastics are mixed with powders of flame retardants in a Banbury mixer; then the mixture is transferred to a roll mill and into a belt granulator;

(c) flame retardant concentrates (50–80% flame retardant) are premixed intensively with a thermoplastic, a good homogenization being obtained during subsequent extrusion or injection molding; (d) during polymerization or flame retardant can be reacted with the polymer, prior to extrusion or applied to the fiber or fabric as an after-treatment.

F. Concentration Range

The amount of flame retardant used varies with its effectiveness, conditions of exposure to fire, and the physical form of the product (molding, laminates, foams, films, etc.).

According to the exposure conditions, severe, moderate, or light, various classes of flame retarded products are designed. The following codes govern flammability behavior in building A, B, C or I, II, III, etc., for low to easy flammability according to ASTM E119 and E84; DIN 4102.

Some typical formulations for flame retarded PO are [2]:

For HDPE, (a) 5% octabromodiphenyl, 1.6% chlorinated paraffin, and 3.5% Sb_2O_3 or (b) 10% decabromodiphenyl and 10% Sb_2O_3.

For LDP, (a) 6% chlorinated paraffin (70% chlorine) and 3–5% Sb_2O_3 or (b) 4.9% octabromodiphenyl, 1.6% chlorinated paraffin (70% chlorine), and 3.6% Sb_2O_3.

For PP (a) 3.5% bis(dibromopropyl ether) of tetrabromobisphenol A and 2.0% Sb_2O_3 or (b) 27% dechlorane plus and 13% Sb_2O_3. Other possible combinations are given in Refs. 14, 15 and 18. The required flame retardant amounts for PO fibers are up to about 30% of the weight of the fibers. Average amounts required for the fire retardant element to render PO self-extinguishable are [16] 40% Cl; 5% Sb_2O_3 + 8% Cl; 20% Br; 3% Sb_2O_3 + 6% Br; 5% P; 0.5% P + 7% Sb_2O_3. Smaller quantities are necessary in synergetic systems.

G. Testing Methods

As has been mentioned already in Chap. 17, burning phenomena are accompanied by an infinitely great number of parameters that influence each other. These include the type, duration, position vs product, and intensity of the ignition source, ventilation, and the shape and thickness of products, the properties of the material and field of application of combustible materials respectively flame retarded polymers. In such a complex system, it is impossible to measure all variables or to separate the main effects of burning from the secondary effects accompanying a fire (smoke development, toxicity, and corrosivity of combustion gases).

The attempts to standardize the testing methods have led to the development of specialized standards for building materials, transportation and storage materials, electrical engineering, furniture and furnishings, etc. Other tests have been proposed for textiles, foams, road vehicles, streetcars, railways, aircraft, mining, and home furnishings.

Unfortunately, these are only a few internationally accepted standards; often each country or even some companies have adopted standards for various applications of plastics [9] or testing flame retardants. As examples, in Table 6 some important standards are listed.

For commercial use, the most common test method is prescribed by Underwriters Laboratories as UL-94. A vertical sample is ignited at its bottom end and burns like an inverted candle. If it burns less than 5 s, it is classified as V-0. If it burns 5–25 s, it is classified as V-1. If it burns and drips, and the burning drops fall on surgical cotton and ignite it, it is classified as V-2. It is commonly observed that burning and dripping can remove both flame and heat from the sample, thus extinguishing the flame; but the burning drops are very likely to ignite the cotton. Thus the V-2 rating is the easiest to achieve, V-0 the most difficult. In some products, V-2 is sufficient; in others, V-0 is absolutely required.

As a results of the rapid increase in the fire model there is a great demand for the fire property and specialized test methods. New test methods have been developed such as [36] cone calorimeter, heat release measurements, etc. With these methods, the following fire properties have been measured for PO: critical heat flux, thermal response parameter surface radiation loss, heat of gasification, flame heat flux limit, yield of products, heat of combustion, corrosion index, flame extinction index, fire propagation index, etc. [37].

Heat release rate is considered the most important fire parameter. Unfortunately, no analytical results for heat release rate in terms of chemical or physical properties of materials are available. Several relationships have been established between ignition temperature, time to ignition, heat of gasification, mass loss rate and heat release rate with thermodynamic and transport properties [38–41] but they are too general.

Smoke corrosivity has been the subject of intense debate in recent years. ASTM D5485 is the most promising method for its determination. All smokes is corrosive, although acid gases tend to be highly

TABLE 6 Standardized methods for flame retardant testing

Standard number	Method
	(I) *Methods Referring to the Burning Phenomenon*
1. ASTM D-1929-77(85) ISO 5657	Standard Test Method for Ignition Properties of Plastics
2. ASTM D 3713-78	Measuring Response of Solid Plastics by a Small Flame
3. ASTM D 2859-76	Standard Test Method for Flammability of Finished Textile Floor Covering Materials (Pill Test)
4. ASTM E84-84	Standard Test Method for Surface Burning
ISO/DP 5658	Steiner Tunnel
DIN 4102	Characteristics of Building Materials. Flame Spread and Smoke Development from Tunnel Test
ISO/DP 5660	Rate of Heat Release Test
5. ASTM E 286-69	Standard Test Method for Surface Flammability of Building Materials Using a 2.44 m Tunnel Furnace
6. ASTM E648-78(86)	Test Method for Critical Radiant Flux of Floor Covering System Using a Radiant Heat Energy Source
7. ASTM D 3675-83	Test Method for Surface Flammability of Flexible Cellular Materials Using a Radiant Heat Energy Source
8. ASTM E 906-83	Standard Test Method for Heat and Visible Smoke Release Rates for Materials and Products
9. ASTM E 662	Standard Test Method for Specific Optical Density of Smoke Generated by Solid Materials
10. ASTM D 2843	Standard Test Method for Density and Smoke from the Burning or Decomposition of Plastics
11. ASTM E 906-83	Standard Test Method for Heat and Visible Smoke Release Rates for Materials and Products
12. ASTM D 2863-77	Standard refers to Oxygen Index Determination
13. ASTM D 2843 ISO DP 8887	Fire Effluents (smoke obscuration, toxicity and corrosivity)
	(II) *Methods Referring to the Classes of Materials*
1. ASTM E 119-83 ISO/DP 5658	Standard Methods for Fire Tests of Building Construction and Materials
2. ASTM E 136-82 ISO 1182	Standard Test Methods for Behavior of Materials in a Vertical Tube Furnace at 750°C
3. ASTM D 635-63(81)	Flammability of Rigid Plastics. Extent of Burning and Burning Time
4. ASTM D 568-77 ISO R 1326	Flammability of Flexible Plastics
5. ASTM D 3801-80	Flammability of Solid Plastics
6. ASTM D 1433-77	Flexible Thin Plastic Sheeting Supported
7. ASTM D 1230-85	Flammability of Fabrics—Clothing Textiles
8. ASTM D 1692-59T ISO 3582	Flammability of Plastics Foams and Sheeting
9. ASTM D 2633-82	Standard Methods of Testing Thermoplastic Insulated and Jacketed Wire and Cable
10. ASTM 1360-58	Test for Fire Retardance of Paints
11. FMV SS 302	For Materials Used for Vehicle Interior
ISO 3795-76	Federal Motor Vehicle Safety Standard
MU 302	Vehicle Safety Standard
12. UFAC	Upholstered Furniture Action Council. Cigarette Test Related to a Voluntary Standard for Upholstered Furniture
13. FF5-74	Children's Sleepwear Standard
14. VDE Standards	For Electrotechnical Products

TABLE 6 (Continued)

Standard number	Method
	(II) *Methods Referring to the Classes of Materials*
15. DIN4102 ISO 1182-79, BS 476	Refer to building materials
16. ASTM D 757 ISO/R 181 DIN 53459	Test for Solid Electro-Insulating Materials

Other test have been proposed for textiles, foams, road vehicles, streetcars, railways, aircrafts, mining and home furnishings.
Source: Refs 8, 11, and 52.

corrosive. Corrosion depends also on smoke amount, humidity, and temperature [42].

H. Uses of Flame-Retarded PO

The flame-retarded PO are required in applications such as wire and cables, fuel tanks, pipe for oil and gas, pipes for industrial and mining use, fibers and filaments, furniture and appliances. Tens of millions of pounds of flame retardants are needed for these applications.

There are a number of uses such as in building, electronics, furniture, and fitting applications where products should have adequate fire resistance, such as:

Cable insulation from LDPE and HDPE especially uses $Al(OH)_3$.

Vehicle construction and vehicle interiors in aircraft and ship construction require special flame-retarded products, while a moderate severity is enough for automotive construction and for automotive interiors.

Ducts of flame-retarded PO are frequently used in the building industry; drain pipes are manufactured in Europe from flame retarded PP. Roof films are obtained from flame retarded PE. Appropriately flame retarded PO are used in the production of furniture.

Household dust bins, various injection-molded parts for electrical installations, life-saving devices, attachment parts, profiles and sheets.

Flame retarded polyolefin elastomers are used for cushioning and flooring, wire and cable coverings, belting insulation, coating fabrics, and coating adhesives and sealants. EPDM is used in single-membrane roofing.

Polyolefin elastomers have a great variety of uses encompassing artificial hearts, earth mover tires, glues, tarpaulins, rubber conveyor belts, to name just a few; many of them should be flame retarded.

Flame-retarded PP fibers are used in the manufacture of indoor–outdoor carpets, in upholstery and apparel or barrier fabrics; monofilaments are used for ropes, cordage, outdoor furniture, webbing, automobile seat covers, filter cloths, etc.

Flame-retarding man-made fibers and their mixtures still poses difficulties for the textile finisher, the process being very expensive. Most commercial flame retardants may alter the aesthetics or performance of the finished product.

I. Problems and Side Effects with the Use of Flame Retardants

Flame retardant additives are reactive, which means that they are less stable than the polymers; they may decompose during normal processing and/or use. This limits the choice of additives that can be used, and it further limits the processing conditions that may be used. These additives are particulate fillers. Large amounts are needed to flame retard polyolefins; they are difficult to disperse uniformly; and they result in serious stiffening and even embrittlement, particularly in PP.

Organic halogens + antimony oxide are the most frequently used types of flame retardants. While they can produce high LOI and achieve UL-94 V-0 ratings, they generally produce large amounts of smoke, which obscures the visibility of safe exits, and they also produce gases that are toxic and corrosive. Tests for smoke are fairly standard, but practical tests for toxicity remain elusive, and corrosion is merely mentioned qualitatively as a secondary consideration. Much current effort is directed toward reducing the amount of smoke generated when flame retardants are used.

High concentrations of solid flame retardants are difficult to disperse in molten polymers, particularly the fluid melt of PP. Poor dispersion can aggravate embrittlement and perhaps even reduce flame retardant effectiveness. Masterbatching all the flame retardant into part of the molten polymer generally improves dispersion [26]. Surface treatment with a coupling agent is recommended and probably widely practiced. For example, coating alumina trihydrate and magnesium hydroxide with 2% organosilane has been reported to reduce embrittlement [19]. Organotitanate treatment of aluminum trihydrate, magnesium hydroxide, and organobromine + antimony oxide has been reported to improve both mechanical properties and flame retardant efficiency.

One of the deleterious effects of halogenated flame retardants on the stabilization occurs in UV-stabilized formulations. Hindered amines being basic, are adversely affected by the presence of halogenated flame retardants. The source of this antagonism is an interaction between the basic amine and the acidic byproducts of the halogenated (typically brominated) flame retardants that can result from excessive processing temperatures.

The inorganic-based flame retardants also affect the ability of hindered phenols and thioesters to yield good long-term heat aging values that are reduced from 26 days heat aging period to only one day in presence of 1/1 ratio hindered phenol antioxidant/ZnO [43]. A key health and environment issue regarding flame retardants is the possible formation of dioxins and related compounds [44]. Brominated dioxins and furans can be produced from brominated diphenyl ether, a highly effective flame retardant for PO under combustion conditions. Brominated dioxin/furans are less toxic than their chlorinated counterparts. Special regulations for flame retardants and flame-retarded materials have been elaborated [45, 46].

Other problems related to the flame retardants in plastics appear because of recycling, marking, ecolabelling and product take-back [47]. Incineration and processing raised questions about formation of toxic products when brominated flame retardants are used, so environmentally preferable products have to be found.

Progress is necessary in the establishment of the ecotoxicological properties of each common or new flame retardant.

J. Recent Developments in Flame Retardants

During the last few years no substantial new classes of efficient flame retardants have been developed, existing developments are mainly based on chemical modification of structure which already exist. Recent developments are related to: replacement of halogenated types due to ecological and economic reasons, reduction of toxic emissions, synergetic effect combinations development, multifunctional systems combining two or more fire control function.

A range of halogen-containing flame retardants are developed in reactive systems in which the polymer chain is modified by grafting flame retardant molecule in the polymerization process as dibromostyrene-g-PP. A melt blendable phosphorus/bromine additive for PP fibre and injection moulding (Rheoflam PB-370 from FMC Corporation) combines the two components in the same molecule allowing them to act synergetically and prevents generation of corrosive HBr during processing.

Zinc hydroxystannate/zinc stannate developed by International Tin Research Institute as replacement for Sb_2O_3 is a smoke suppressant, non-toxic, safe and easy to handle, effective at low addition level.

III. PIGMENTS AND COLORANTS

A. Role and Mode of Action

Most plastic articles are commercialized in colored form, color increasing the attractiveness of the final product. Color change is a selective absorption, wavelength dependent, or/and light scattering (reflectance).

Dyes can only absorb light and not scatter it. Thus the articles will be transparent. The optical effect of pigments is somehow different, so if the refractive index of the pigment appreciably differs from that of the plastic and for a certain specific particle size, reflectance of the light takes place, and thus the plastic becomes white and opaque, while if selective light absorption takes place, a colored and opaque article is obtained.

B. Types of Colorants and Pigments [2, 5, 48, 49]

For changing a plastic's color, carbon, inorganic pigments and organic pigments, dyes, and special pigments are employed, the last group including fluorescent whitening agents, nacreous pigments, pearlescent pigments and metallic pigments. Generally the dyes are soluble in plastics, whereas the pigments are virtually insoluble or partially soluble.

1. Inorganic or Mineral Pigments

Inorganic or mineral pigments are metallic oxides and salts. They have good light, weathering, and thermal stability and do not migrate, yet they cannot be used for the obtainment of transparent products; the colors obtained are not bright and the variety of shades is reduced. The amount of inorganic pigment required for a good coloration is greater (1-2%) than that of organic pigment (0.05–1%).

Sometimes the great quantities of inorganic pigments incorporated into plastics lead to lowering of the mechanical properties; however they are preferred for outdoor uses (e.g., in building materials). These pigments are grouped according to the color achieved: white (TiO_2, ZnS, ZnO, lead oxide, $BaSO_4$, $CaCO_3$ and aluminum, calcium or magnesium silicates in a decreasing order of white shade); yellow (various chromates, hydroxides alone or treated with colloidal silica); red (Fe_2O_3, $PbCrO_4$, $PbMoO_4$, $PbSO_4$, etc.); green (Cr_2O_3, CoO + ZnO), blue and violet (Prusian blue, manganese violet), brown and black (iron oxide brown, carbon black, black As_2S_3), and other pigments, and various combinations of colorants mentioned for the obtainment of intermediate shades.

Besides white pigmentation, white pigments impart to materials white reflectance, increased brightness and modification of other colors (tints). They may also give higher UV-protection and weather resistance.

Carbon black has also useful antistatic properties, can provide electrical conductivity and gives effective UV screening so it can be described as an universal additive. There are about 100 different grades of carbon black today. Depending on production processes the following types of carbon black are known: furnace black (10–100 μm) channel black (10–30 μm), lampblack (60–200 μm), thermal black (100–200 or 300–500 μm), acetylene black (30–40 μm), etc.

Black *antimony sulphide* has a strong black color having also flame retardant properties. Colcolor E60R is a new black pigment under development at Degussa very effective in recycling which requires black overcoloring.

2. Organic Pigments, Dyestuff or Colorants

Organic pigments give a large variety of shades, but their stabilities are inferior to those of inorganic pigments, having, too, a more pronounced migration tendency. Dyes are transparent and give bright colours in light. The main groups of organic pigments are monoazo and diazo pigments (red), indigo and anthraquinone, phthalocyanine (green, blue), perylene, flavanthrone, quinacridones (red), polycyclic, and diazine pigments, diarylene (yellow), orange indathrone (blue), dioxazine (violet), and many other groups of organic colorants [50]. Reddish violet dioxazine pigments have extremely high tinctorial strength. They are used for obtaining various shades of red, blue, green, and yellow.

3. Fluorescent Whitening Agents

Fluorescent whitening agents act as fluorescent dyes, although they are colorless. They improve the initial color of plastics, which often is slightly yellowish, by imparting a brilliant white to the end-use articles. Also they increase the brilliancy of colored and black pigmented articles. This function is achieved by the absorption of UV radiation (360–380 nm), converting it to longer wavelengths and re-emitting it as visible blue or violet light. In a polymer with yellow tint, the emitted light makes up for the absorbed blue light to make the polymer appear whiter and to glow even in normal lighting. The chemical classes of fluorescent whitening agents are benztriazoles, benzoxazoles, pyrene, triazine, coumarins, etc.

Dyes are especially suitable for outdoor use, with very good weatherability. Some FWA exhibit intrinsic light-collecting properties which are exploited, notably in Israel, for selective forcing of vegetables. Orange 240 is for traffic safety, while Red 300 is for laser dyes or colorants for agricultural films or greenhouses. The concentration recommended is 0.05%.

4. Nacreous or Pearlescent Pigments

Nacreous pigments and metallic flake pigments are lamellar pigments having unique optical properties due to their plate-like shape and to the high reflectivity of the pigment particles. They impart a pearly or nacreous luster to objects. The main nacreous pigments are basic lead carbonate, which has the highest pearl brilliance and is used in HDPE, LDPE bottles, and PP articles; natural pearl essence; lead hydrogen arsenate or phosphate, and titanium dioxide coated with muscovite mica, which gives a white pearl effect in blow molded PE containers.

Pearlescent pigments are based on thin platelets of transparent mica coated with TiO_2 or iron oxide, producing interference patterns. They increase lustre, whitness and coverage, and have sparkle effects. The following grades are known: Mearlin, Dynacolor, Tek Pearlite, etc., differing by TiO_2 particle size.

5. Metallic Flake Pigments

Metallic flake pigments are constituted of copper, copper alloys (bronzes) of various compositions (92% Cu, 6% Zn, 2% Al-pale gold bronze to 67.75% Cu, 31% Zn, 0.25% Al-green gold bronze), and aluminum. The appearance of the aluminum pigmented products depends on particle shape and size. Aluminum flakes impart a silver grey color, while small spheres give a sparkling aspect.

Plastics pigmented with aluminum can be recycled and recycled PO may be optically improved by addition of aluminum pigments. Among recent developments Obron Atlantic offers Mastersave an easily dispersed pelletized 80% aluminum pigment concentrate, which usually does not require predispersion. They are available as pastes (a slurry with mineral spirits). The major use of metallic flake pigments is in protective and decorative coatings.

Since these pigments provide a barrier to light, moisture, heat, corrosive gases, etc., such coatings are applied to oil and water tanks, highway bridges, transmission towers, railway signal towers, etc. Metallic powders are also used as pigments.

6. Special Pigments and Applications

(a) Laser Marking. Recently active developments of pigments suitable for laser marking systems have been produced. Pulse transversal excited atmospheric pressure CO_2 lasers at a pulse length of only a few seconds cause carbonization and therefore a change in color of plastics surrounding the pigment. Carbonization and bubbling can occur together, producing a light-grey marking. The limits for the energy density required to mark PE and PP pigmented with 0.5% Iriodin LS810 is 3 J/cm^2 visible marking and 3–7 J/cm^2 for easily legible marking. These pigments are particularly useful in plastics recycling.

(b) Photochromic and Thermochromic Pigments. Photochromic and thermochromic pigments are micro-encapsulated liquid crystal systems giving precise color changes (from colorless to highly colored) of specific temperatures or exposed to UV light. They are particularly interesting in agricultural applications where use in "polytunnels" can protect plants from over exposure to ultraviolet and infrared radiation.

A Chameleon organic pigment concentrate for PO was introduced by Victor International Plastics. Color is activated from −25 to +58°C.

7. Other Pigments

Other pigments serve, too, as anticorrosives (lead oxides, phosphates), flame retardants (metaborates), fungicides, antivegetatives, etc.

Some general information about pigments and dyes for polyolefins is given in Table 7.

A detailed list may be consulted in Pigmente und Fullstofftabellen (M. and O. Lickert Loats, 1987) and in Modern Plastics Encyclopedia (McGraw-Hill, Inc., New York 1983/1984).

C. Requirements

Dyes and pigments must accomplish requirements concerning strong covering power, good thermal (240–280°C) stability and light fastness, good weather and chemical resistance; also they should present resistance to bleeding and to migration. Especially for articles used in food packaging, they should not present any tendency to migration from any kind of preparation, e.g., solvent bleeding, contact bleeding, or blooming.

Three other requirements, namely stability to processing, absence of plate-out during processing, and modification of rheological and mechanical properties according to the product's performance, are dependent on the compatibility of the components of the system and on the possible interaction between colorants and other components of the formulations. Sufficient compatibility is also necessary in order to avoid exudation.

With fibers, in addition wash and rub fastness are required.

Pigmental or tinctorial properties are dependent not only on the chemical structure of the pigment class but also on the size and shape of particles, particle size distribution, crystallinity and crystalline form, dispersability, and finish and surface coating of the pigments as well as on the type of polymer and manufacturing process of the pigment itself and of the plastic product. These characteristics strongly influence color shade, covering power, color strength, etc.

In coloration of polyolefins, especially for PP, due to the high processing temperature, most organic pigments cannot be used because of their pronounced tendency of migration. Among those that could possibly be used are blue or green phthalocyanines, perylene, quinacridones, isoindolines, azocondensation pigments, and a few others, due to their high level of stability. However, they cause problems in the injection molding of large articles, such as milk crates from HDPE, since they accelerate the nucleation and thus induce stresses in the longitudinal and transversal directions, which is accompanied by a dimensional instability. These effects do not appear in PP and its copolymers.

With manganese, copper, and zinc pigments, there is a risk of catalytic decomposition of polyolefins that is possibly combined with odor nuisance. Coloration with manganese flakes is still practised for articles with short lifetimes, but it tends to be replaced by more stable strontium compounds.

For products for outdoor use, such as bottles and shipping crates made from HDPE, the following pigments are recommended; white opac (rutile TiO_2 after-treated), yellow (cadmium yellow, nickel titanium yellow, chrome titanium yellow), red (cadmium red, iron oxide red), blue (cobalt blue, ultramarine blue), green (cobalt green, chrome oxide green), brown (iron oxide brown, chrome iron brown), black (carbon black-furnace).

One should avoid the use of TiO_2 together with organic pigments and UV absorbers due to a considerable degree of degradation and embrittlement after only a few months. By surface treatments such effects are minimized. ZnS offers a good alternative to TiO_2 pigments. It is used for plastics which are nonabrasive, catalytically inactive, dry lubricants, and impart good dielectric properties.

In the case of nacreous pigments of the highest pearl lustre, the following requirements must be accomplished besides complete dispersion: diameter of particles between 15–30 μm, optical thickness about 1.38 μm, high refractive index, smooth surfaces, complete orientation of the individual platelets, and avoidance of fragmentation.

Metallic pigments (such as Al) have a poor resistance to UV light, oxidation, and chemicals.

In countries with hot, damp climates, a shift in the shade of the product results after a relatively short time, due to oxidative attack. Thus ultramarine pigments are less resistant to weathering, and sulfidic cadmium yellow pigments become blue.

Therefore, although inorganic pigments exert a protective action against aging, PO articles used in outdoor applications should be stabilized against the influence of light.

Ultramarine pigments have excellent heat stability, good transparency, are easily dispersable, do not produce wrapping in PO, are nontoxic, nonmigratory and have worldwide approval for food contact and are also safe for use in toys.

D. Incorporation into the Compound

Pigments are supplied on the market in the following forms: single pigment powders, liquid pigment concentrates, solid pigment concentrates (the pigment is fully and finely dispersed in a liquid or, respectively, in a solid carrier), and special pigment mixtures, which are combinations of inorganic pigments and/or fillers with organic pigments of lower dispersability. In mixtures, antioxidants, light stabilizers, and antistatic agents, etc., may also be used.

Combinations of several pigments are necessary in order to achieve an accurate final shade or a specific optical property. Also, pigment concentrates such as granules composed of colorant and processing polymers are preferred in order to avoid dust in polymer processing. They assure an optimum dispersion and are easier to measure.

Dry powder pigments have particles of 0.01–0.1 microns, aggregates of 100 microns, and agglomerates. The fine milling of aggregates and agglomerates is very important in order to obtain a good coloration, because they should be fully dissolved or homogeneously distributed in the finished articles.

The dispersion process of pigments consists in the following steps: breakdown, wetting, distribution, and stabilization.

Breakdown is achieved by means of mills, crushers, high speed mixers, and edge runners.

Dispersion in a liquid, in pasty phase, or in polymer melts takes place on extruders, calendars, kneaders, triple roll mills and attrition mills. Also, ultrasound and thermal energy based procedures are used for the production of a fine pigment powder. Inorganic pigments, due to their ionic lattice, are more readily dispersable than organic ones.

Wetting is accelerated by raising the temperature or sometimes by the addition of low molecular weight additives such as plasticizers or adhesion promoters with the medium; for example, butyl stearate is added to incorporate fluorescent whitening agents; often it is added as a concentrate or as a master blend (10% active whitener blend with chalk or a plasticizer).

To improve the performance of pigments, they should be subjected to an after-treatment of the particles' surface, which increases dispersability and improves the resistance to light, weathering, and chemicals, or may be micronized.

Incorporation may take place into the mass during manufacture or during melt or solvent spinning; by dry blending before forming processes with plastics as powder or pellets.

Perylene pigments are widely used especially for melt coloration of PP, PE, etc.

Plastics are colored during their processing, either at the compounding stage with slow running or high

(Text continues on p 593)

TABLE 7 Some colorants and pigments for polyolefins

No.	Pigment	Producer	Chemical description	Color: Shade	Resistance to: Light	Resistance to: Heat T_d, °C	Resistance to: Migration	Uses
0	1	2	3	4	5	6	7	8
1.	Alon	Godfrey Cabot Inc., USA	Aluminum	White				GU
2.	AmapPast Orange GXP Red DF. Violet PR Blue HB	Color Chem, Intern			E			GU
3.	Blankophor	Bayer AG, Leverkusen, Germany	Benzotriazole phenyl coumarins					FWA
4.	Bayplast yellow GY-5680	Bayer AG		Slight opaque	E	heat stable		GU
5.	Cadmofix	Bayer AG	Cd (S, Se)					GU
6.	Cadmopur yellow	Bayer AG	(Cd, Zn)S	Yellow pure	E	300	N	GU
6a.	Cadmopur red	Bayer AG	(Cd) (S, Se)	Red pure	E	300	N	GU
6b.	Cadmopur orange	Bayer AG	(Cd) (S, Se)	Orange pure, 1/10	E	300	N	GU
7.	Calcocid Calcodur Calcosin Calcozoic	American Cynamid Co., USA	Colorant series					GU
8.	Carbolac lamp black	G. L. Cabot, Inc., USA	Carbon black					used also as filler
9.	Chromolene	SNCI, France						PE, PP
10.	Chromthal yellow 3G	Ciba Geigy Ltd Bau R-1038.5.15 CH-4002 Basel Switzerland	Diazo-condensation compounds	Yellow pure, 1/10	E	260	N	LDPE
	Chromthal DP			Yellow to red bright		270–285		PO
11.	Chromthal 8G, GR, 6G					280	CT	HDPE
12.	Chromthal 2 RLTS	Ciba Geigy	Tetrachloro-isooindolinone	Yellow pure	E VG	280	N CT	LHPE HDPE

13. Chromthal red BRN	Ciba Geigy	Diazo compounds	Red pure	VG	280	N CT	LDPE HDPE
Chromthal GTAP 2RLP	Ciba Geigy	Anthraquinone-based	Replace diarylide and heavy metals pigments, easily disposable				PO fibers
a. Chromthal orange TRP	Ciba Geigy		Blue				PO film PP fibers
b. Chromthal HRP			Yellow brilliant				
14. Chromthal Red A3B	Ciba Geigy	Anthraquinone	Red, pure	E	280	N	HDPE PO
15. Chromthal Scarlet RN	Ciba Geigy	Diazo compds.	Red pure, 1/10	VG	280	N CT	LDPE HDPE
16. Chromthal Bordeaux RN	Ciba Geigy	Tetrachloro-thioindigo	Red pure	E VG	260	LM CT	LDPE HDPE
17. Chromthal Blue	Ciba Geigy	Cu-phthalocyanine	Blue pure, 1/10	E	280	N CT	LDPE HDPE
18. Chromthal blue A3R	Ciba Geigy	Anthraquinone	Blue	E	270	N	PO
19. Chromthal green	Ciba Geigy	Halogenated phthalocyanine	Green	E	300	N CT	PO
20. Chromthal brown 5R	Ciba Geigy	Diazo compds.	Brown pure, 1/10	E	260	N	PO
21. Chrome oxide green	Bayer AG	Chrome oxide	Green pure, 1/4	E	300	N CT	GU
22. Cobalt blue	Degussa	Spinel $CoAl_2O_4$	Blue	E	300	N	GU
23. Cobalt green	Degussa	Spinel $(Co, Ni, Zn)_2(Ti, Al)O_4$	Green pure	E	300	N	GU
24. Colanyl	Hoechst AG, Germany	Inorganic pigment concentrates 50–70%	Colorant series				GU
Combibatch	Reed Spectrum		Colorant series + functional additives				GU
25. Crystex	Kali-Chemi, Stauffer GmbH, Germany		Colorant series				GU
26. Cyandur	American Cynamid Co., USA		Organic Colorant sries				GU
27. Dalmar	Du Pont de Nemours, USA		Pigments				GU
28. Diazo	GAF Corp., USA		Azo compds.				GU
29. Duol	Du Pont de Nemours, USA		Colorant series				GU
30. Duranol	ICI, UK	Dispersion pigments					GU

TABLE 7 (Continued)

No.	Pigment	Producer	Chemical description	Color Shade	Resistance to Light	Resistance to Heat T_d, °C	Resistance to Migration	Uses
0	1	2	3	4	5	6	7	8
31.	Durazol	ICI, UK	Dispersion	Colorant series				GU
32.	Eastman	Eastman, Chem. Kingsport (Tn) USA	Bisbenz-oxazoles					FWA
33.	Euthylen	BASF AG, Germany		Colorant series				GU
34.	Fluolite	ICI, Manchester UK	Pyrene Triazine					FWA
35.	Graphtol yellow RCL	Sandoz, Inc.	Diazodiaryl	Yellow	E	250	LM	LDPE HDPE
36.	Graphtol yellow GRBL	Sandoz, Inc.	Diazodiaryl	Yellow	S-G	240	LM	LDPE
37.	Graphtol flame Red 3RL	Sandoz, Inc.	Sr salt of a monoazo acid	Red	G-P	260	LM	LDPE HDPE
38.	Hansa	GAF Corp.	Colorant series					GU
39.	Helio fast scarlet EB	Bayer AG	Anthanthrone	Red 1/10	G		LM	LDPE
40.	Helio fast yellow GRN	Bayer AG	Diazo, diaryl	Yellow	S-P	240	LM	LDPE
41.	Helio fast HRN	Bayer, AG	Diazo diaryl	Yellow 1/10	P	250		LDPE
42.	Helio fast red violet Er06	Bayer, AG	Tetrachlorothio indigo	Red	E VG	260	LM	PO
43.	Hostalux	Hoechst AG, Frankfurt	Bisbenzoxazole					FWA
44.	Heliogen	BASF AG	Cu-phthalocyanine	Blue	E	280	N	PO
45.	Heliogen green K	BASF AG	Chlorinated phthta-locyanine	Green	E	280 300	N	PO
46.	Heucosin yellow	Heub	Monoclinic, stabilized $Pb(Cr, S)O_4$		S-G	240	N	PO
47.	Heucotron red	Heub	Lead chromate molybdate	Red		250	N	PO
48.	Hombitan	Sachtleben Chem. GmbH	TiO_2					GU
49.	Hostaperm yellow H4G, H3G	Hoechst AG	Monoazobenzimidazole	Orange	E 1/10	260	N	PO

50. Hostaperm scarlet GO	Hoechst AG	Anthanthrone	Red	S		LM	PO
51. Hostaperm pink E	Hoechst AG	Substituted quinacridone	Red pure	VG G	280	N	PO
52. Hostaperm fast red EG	Hoechst AG	Substituted quinacridone	Red pure	VG	280	N	PO
53. Hostaperm red E5B	Hoechst AG	Quinacridone	Violet	VG	280	N	PO
54. Hostaperm violet ER	Hoechst AG	Quinacridone	Violet	VG	280	N	PO
55. Indazin	Castella Farb. werke AG	Colorant series					GU
56. Intramin	Hoechst AG	Naphtol type					GU
57. Irgacete Irgafinex	Ciba Geigy AG	Organic colorant series					GU GU GU
58. Irgalith yellow B3RS	Ciba Geigy	Diazo diaryl	Yellow	E-VG	250	LM	PO
59. Irgalith yellow BAWP	Ciba Geigy	Diazo diaryl	Yellow	S-G	240	LM	PO
Irgalith blue		α-phtalocyanine	Blue	E			GU, transparent appl.
Irgazin TR				E			PO
60. Iron oxide red	Bayer, BASF	Fe_2O_3	Red	E	300	N	GU
		Fe_3O_4	Brown	E	240	M	GU
61. Janus	Bayer AG	Dispersion series					GU
62. Kronos	Titan Gellschaft, GmbH	Ti compounds					GU
63. Leukopur	Sandoz AG, Basel, Switzerland	Naphto-triazole coumarins					FWA
64. Light yellow	Bayer	Nickel, Cr titanium yellow	Yellow pure	E	300	N	GU
Light fast yellow 62R	Bayer			G			GU
65. Light blue	Bayer	Spinel $CoAl_2O_4$	Blue pure	E	300	N	LDPE, HDPE
66. Lithol	BASF AG, Germany	Ba salt of a monoazo color acid	Red				
67. Lithol scarlet K 3700	BASF, Germany	Ba salt of a monoazo color acid	Red pure 1:10	G P	250	LM	LDPE HDPE

TABLE 7 (Continued)

No.	Pigment	Producer	Chemical description	Color: Shade	Resistance to: Light	Resistance to: Heat T_d, °C	Resistance to: Migration	Uses
0	1	2	3	4	5	6	7	8
68.	Lithol scarlet K4160	BASF, Germany	Sr salt of a monoazo acid	Red 1:10	P	260	LM	PO
69.	Lithol scarlet K 4460	BASF	Cd salt of a monoazo color acid	Red pure / 1:10	S / P	240	LM	LDPE / HDPE
70.	Lithol fast scarlet K 4260	BASF	Mn salt of a monoazo color acid	Red pure / 1:10	G / S	240	LM	LDPE / HDPE
71.	Lithol bronze red K 3660	BASF	Ba salt of a monoazo color acid	Red pure / 1:10	S / P	250	LM	LDPE / HDPE
72.	Lithol rubine K 4631	BASF	Ca salt of a monoazo color acid	Red pure / 1:10	S / P	250	LM	LDPE / HDPE
73.	Light green	Bayer	Spinel $(Co, Ni, Zn)_2 (Ti, Al)O_4$	Green 1/4	E	300	N	GU
74.	Lithopone	Bayer AG, Germany	$ZnSO_4 + ZnO$					GU
75.	Luxol	Du Pont de Nemours, USA	Series					GU
76.	Mapicotan	Columbia Carbon Corp., USA	Red iron oxides					GU
77.	Microgels	SNCI Cie, France	Liquid colorants					PO
78.	Midland	Dow Chem. Co., USA	Indigo					GU
79.	Monolite	ICI, UK						GU
80.	Naphtanil	Du Pont de Nemours, USA						GU

81. Novoperm yellow FGL	Hoechst, Germany	Monoazo	Orange pure 1/10	E VG	260	N	LDPE
82. Novoperm yellow HR	Hoechst	Diazo-diaryl	Yellow pure 1/10	E VG	250	LM	PO
83. Novoperm carmine HF4C	Hoechst	Monoazo-benzimidazole	Carmine pure 1/10	E VG	260	N	LDPE HDPE
84. Novoperm marron HFM	Hoechst	Monoazo-benzimidazole	Marron 1/10	VG	240	LM	LDPE HDPE
85. Novoperm red H2BM	Hoechst	Mn salt of monoazo color acid	Red pure 1/10	S	240	LM	LDPE HDPE
86. Novoperm red MR	Hoechst	Tetrachloro-thioindigo	Red pure 1/10	E VG	260	LM	LDPE HDPE
87. Oracet	CIBA Geigy Switzerland	Colorants series					GU
88. Oralith							
89. Orasolo							
90. Orion red	Sherwin Williams Co., USA	Red pigments					GU
91. Oxide	Sherwin Williams Co., USA	ZnO					
92. Paliotol yellow K0961	BASF, Germany	Quinophtha-lone	Yellow pure	E	260	LM	LDPE
93. Paliotol yellow K1841D	BASF, Germany	Isoindoline	Yellow pure	E	250	N	LDPE HDPE PO food contact
94. Paliotol green K9781	BASF, Germany	Fe complex of nitroso-naphtol	Green pure 1/4	P	220	M	LDPE HDPE
95. Paliogen yellow L1870	BASF, Germany	Flavanthrone	Yellow pure	E	280	N	LDPE HDPE
96. Paliogen red violet K4980	BASF, Germany	Tetra-chloro thioindigo	Red 1/10	G	260	LM	LDPE HDPE
97. Paliogen red K3871, 3911	BASF, Germany	Perylene pigment	Red pure	E	300	LM	LDPE HDPE
98. Paliogen violet L5890	BASF, Germany	Dioxazine	Violet pure 1/10	G S	260	LM	LDPE HDPE
99. Paliogen blue I5470	BASF, Germany	Anthra-quinone	Blue pure 1/10	E	270	N	LDPE HDPE

TABLE 7 (Continued)

No.	Pigment	Producer	Color: Chemical description	Color: Shade	Resistance to: Heat Light	Resistance to: T_d, °C	Resistance to: Migration	Uses
0	1	2	3	4	5	6	7	8
100.	Paletinecht	BASF, Germany	Chromium pigments					GU
101.	Permachrom	Sherwin Williams, USA						GU
102.	Permaton	American Cynamid Co.	Organic pigments serie					GU
103.	Permanent yellow H10G	Hoechst AG Germany	Diazodiaryl	Yellow 1/10	G	260	LM	LDPE HDPE
104.	Permanent red HFT	Hoechst AG Germany	Monoazobenz-imidazole	Red pure 1/10	VG	260	N	LDPE HDPE
105.	Permanent red TG	Hoechst AG Germany	Naphthalene-tetracarboxillic acid derivative	Red pure	E	280	N	LDPE HDPE
106.	Permanent red BL	Hoechst AG Germany	Perylene	Red 1/10	G	300	LM	GU
107.	Permazo	Allied Chem. Corp., USA	Azo pigments					GU
108.	Pigment lake red LC	Hoechst	Ba salt of monoazo color acid	Red pure 1/10	S P	250	LM	LDPE HDPE
108a.	Plastblak PE614	Cabot	Master batches	Black White	E			GU
109.	Plasticone red	Sherwin Williams Co.	Red pigments					GU
109a.	Plastwite PE7474	Cabot	White pigments					GU
110.	Polymon	ICI, UK	Organic pigments					GU
111.	Pontachrome	Du Pont de Nemours, USA						GU
111a.	Protint	Hanne		Transparent tint				PP clarified
112.	PV yellow GR	Hoechst AG Germany	Diaryl pigments	Yellow 1/10	P	240	LM	PO
113.	PV fast yellow H2G	Hoechst AG Germany	Monoazo-benzimidazole	Yellow pure 1/10	E VG	260	N	LDPE
114.	PV yellow H10G	Hoechst AG Germany	Diazodiaryl	Yellow pure 1/10	VG G	260	LM	LDPE HDPE

115. PV Carmine HF3C	Hoechst AG Germany	Monoazo-naphthol AS	Carmine pure 1/10	VG S	280	N	LDPE HDPE
116. PV fast marron HFM	Hoechst AG Germany	Monoazo-benzimidazole	Red pure 1/10	VG	240	LM	LDPE HDPE
117. PV fast red HFT	Hoechst AG Germany	Monoazo-benzimidazole	Red pure	VG	260	N	LDPE HDPE
118. PV carmine HF4C	Hoechst AG Germany	Monoazo-benzimidazole	Red pure	VG	260	N	LDPE HDPE
119. PV fast pink E	Hoechst AG Germany	Substituted quin-acridone	Red 1/10	VG	280	LM	LDPE HDPE
120. PV red MR	Hoechst AG Germany	Tetrachlorothio-indigo	Red pure	E VG	260	DM	LDPE HDPE
121. PV fast orange GRL	Hoechst AG Germany	Perinone	Orange pure, 1/10	VG-G	260	M	LDPE HDPE
122. Radglo	Radiant Color Co, USA						FWA
123. Ramapo	Du Pont de Nemours, USA	Phthalocyanine					GU
124. Rapidazol	Naphtol, Offenbach AG, Germany	Naphtol colorants					GU
125. Raven	Columbian Carbon Corp.,	Colorant serie					GU
125a. Reed lite	Reed Spectrum	14 colors	Free of heavy metals				GU
126. Rubanox	Sherwin Williams Co., USA	Rubin colorants					GU
127. Rutinox HD	British Titan Products, Ltd., England	Rutile TiO_2					GU
128. Sandorin blue RL	Sandoz, Inc.	Anthraqui-none	Blue pure 1/10	E	270	N	LDPE HDPE
129. Sanylen	Sandoz, Inc.	Organic and inorganic colorants					PO
130. Sicoplast	Siegle GmbH, Germany						GU
131. Sicotrans red K2915	BASF AG, Germany	Fe_2O_3 transparent	Red pure 1/4	E	300	N	GU
132. Sicotan yellow K1101	BASF AG, Germany	Nickel (Ti, Hi, Sb)O_2 titanium yellow	Yellow 1/4		300	N	GU
133. Sicotan yellow K2011	BASF AG, Germany	Chrome (Ti,Cr,Sb)O_2 titanium yellow	Yellow pure	E	300	N	GU

TABLE 7 (Continued)

No.	Pigment	Producer	Color: Chemical description	Color: Shade	Resistance to: Heat Light	Resistance to: T_d, °C	Resistance to: Migration	Uses
0	1	2	3	4	5	6	7	8
134.	Sicotherm yellow K1301	BASF AG Germany	(Cd, Zn)S	Yellow pure	E	300	N	GU
135.	Sicopal yellow K2395	BASF AG Germany	Zinc iron	Yellow pure	E	300	N	GU
136.	Sicopal brown K2595	BASF AG Germany	Zinc iron	Yellow pure	E	300	N	GU
137.	Sicomin yellow K1422 (chrome yellow)	BASF AG Germany	Monoclinic, stabilized $Pb(Cr, S)O_4$	Yellow pure 1/4	S G	240	N	LDPE HDPE
138.	Sicomin yellow K1630	BAF AG Germany	Monoclinic, high stabilized $Pb(Cr, S)O_4$	pure 1/4	E	250		HDPE LDPE
139.	Sico Fast yellow K1351	BASF AG Germany	Diazodiaryl	Yellow pure 1/4	S-P P	240	LM	LDPE
140.	Sicotherm red K3201	BASF, AG, Germany	Cd(S, Se)	Red pure 1/4	E	300	N	GU
141.	Sicomin red K3030	BASF AG, Germany	Lead chromate molybdate, highly stabilized molybdate	Red pure 1/4	VG	250	N	LDPE HDPE
	L3022		$Pb(Cr, Mo, S)O_4$	Red pure 1/4	P S	240	N	LDPE HDPE
142.	Sicotherm orange K2801	BASF AG, Germany	Cd(S, Se)	Orange pure 1/4	E	300	N	HDPE high weather resistance
143.	Sico fast orange K2850	BASF AG, Germany	Azo	Orange pure 1/10	S P	240	M	LDPE HDPE

144. Sicopal blue K7310	BASF AG, Germany	Spinel $CoAl_2O_4$	Blue pure 1/4	E	300	N	LDPE HDPE
145. Sicopal green 9996	BASF AG, Germany	Cr_2O_3	Green pure 1/4	E	300	N	GU
146. Sicopal green K9710	BASF AG, Germany	Spinel $(Co, Ni, Zn)_2$ (Ti, Al)O_4	Green pure 1/4	E	300	N	GU
147. Sicopal brown K2595	BASF AG, Germany	ZnO iron	Brown pure 1/10	E	300	N	GU
147a. Sicopal FK4237FG	BASF AG, Germany	Bi	Yellow	Bright			HDPE
147b. Sicoton		Vanadate	No dusting	Exact dosing			HDPE
148. Solvoperm	Hoechst AG, Germany	Soluble colorants					GU
149. Solfast	Sherwin Williams Co.	Organic pigments					GU
149a. Spectrotech	Quantum	White pigments	White	E			PO
149b. Ti pure	Du Pont	TiO_2	White				PO
150. Tinopal	Ciba Geigy AG, Basel, Switzerland	Triazole-phenylcoumarine					FWA
150a. Tint Ayd AL499	Danice Products	Iron oxides	Red	E			GU food appl.
150b. Tint Aid FC							GU
151. Tiofine	N. V. Titandi-oxide Fabr.	TiO_2	White				
151a. Tiona	SCM Chemicals	TiO_2 organic surface treated					GU
152. Tioxide	British Titan Products, England	TiO_2 rutile, anatase	White				GU
152a. Tioxide TR27	Tioxide	TiO_2 rutile	White				GU
153. Titanox	Titanium Pigment Co., USA	Titanium compounds					GU
153a. TST/200	Ampacet	TiO_2 concentrate	White				GU, PO
154. Ultramarine blue	BASF AG, Germany	Na, Al silicate, sulfide	Blue pure 1/10	E	300	N	LDPE HDPE
155. Uvitex	Ciba Geigy AG, Basel	Bisbenzoxazoles bis(styryl) biphenyl					FWA GU
155a. VAT Dyes	Keystone Aniline		All colors				GU
156. Violite	Rhode Island Laboratories Inc., England	Luminescent agents, ZnS, (Zn, Cd, Co, Ca)S					FWA

TABLE 7 (Continued)

No.	Pigment	Producer	Color: Chemical description	Shade	Resistance to: Heat Light	T_d, °C	Migration	Uses
0	1	2	3	4	5	6	7	8
157.	Vynamon	ICI, UK	Colorant series					GU
158.	White fluor	Sumitimo Chem. Co. Ltd., Osaka Japan	Styryl bis benzoxazoles					FWA
159.	White PE and MB19	Ampacet	Inorganic master batches	White	E			PE, PP
160.	White 6 or 18	Ampacet	White concentrates					PE, PP
161.	Yellow PO411, 10412	Cerdec Draker field	clean	Greenish	E			PE, PP

CT-careful testing; E-excellent; FWA-fluorescent whitening agents; G-good; GU-general use; LM-low migration; M-migration; N-no migration; P-poor; S-satisfactory; T_d-decomposition temperature; VG-very good. The light fastness appreciation relates to full shade (pure) and reduction 1/10 or 1/4 with a stabilized rutile: HDPE, LDPE, PE, PO, PP-high density, low density polyethylene, polyolefins, and polypropylene respectively.

Source: Refs. 2, 5, and 6.

speed mixers or directly on the plastic processing machines, e.g., extruder, injection molding machine, foaming unit, etc.

Nacreous pigments are applied either as surface coatings or incorporated in plastics in injection molded extrusion and articles. They have to be added toward the end of the mixing cycle after all other components have been blended in order to minimize fragmentation of the platelets.

Inadequate dispersion of pigments can lead to fluctuation in color intensity, deviation in shade, streaking, clogging of the screen packs in the extruder, tearing of monofilaments and film tapes, printing problems due to the inhomogeneities at the surface, drop in mechanical strength (the notch effect), reduced abrasion resistance due to unsatisfactory coating of the pigments, etc.

E. Concentration Range

The amount of pigment to be added depends on the depth of shade required, the layer thickness, and the opacity, varying between 1–10%.

For fluorescent whitening agents the concentration is 100–500 pp; for special application it is 1000 ppm.

F. Testing of Pigments

The approval or rejection of a pigment requires information about bleeding, finess of dispersion, mass color, tinting strength, hiding power (opacity), heat resistance, migration, etc.; some examples of typical tests being the following [49]:

ASTM D 153-54(66)	Specific Gravity of Pigments
ASTM D 279-31(63)	Bleeding of Pigments
ASTM D 387-60	Mass Color and Tinting Strength of Color Pigments
ASTM D 963-65	Standard Specifications for Copper Phthalocyanine Blue
ASTM D 1210-64	Finesse of Dispersion of Pigment Vehicle Systems
British Standard 3483-1962	Oil Absorption of Pigments
British Standard 2661-1961	Hiding Power
British Standard 1006-1961	Fastness to Light Scale for Pigments
ISO R 105	Hiding Power Scale for Pigments
ISOR 105 Part I	Solvent Staining Scale for Pigments
ISOR 105 Part II	Fastness to Light Scale for Pigments

Heat resistance is measured by the change in color of samples introduced into a forced-air oven for a certain time or under processing conditions.

For testing of white pigments, the Ciba-Geigy Plastic White Scale is used.

Toxicological problems occur. Some whitening agents are allergic; other cause mucous membrane iritation in high concentration. Apart from this, no adverse effect can be attributed at present to pigments.

G. Problems with Pigment Use

Although desirable, organic pigments present problems because they are more resistant to breakdown and dispersion with limited heat stability and less interaction with the matrix.

If a pigment which is not light or thermally stable is used in a stable formulation, the overall effect will be to destabilize the material. In this case it is useful to employ additional stabilization or a proper pigment has to be used.

In most cases, pigments are supplied as dry powders of various specific gravities and bulking values. Care is needed in handling because dry powder may constitute a dust health hazard. It is recommended that a safe-to-use, free-of-dust formulation be employed. Pigment master batches are widely used. They come in a granular form in which the concentrated pigment is dispersed in a polymer carrier (PE) which is compatible with the matrix. A good dispersion of pigment in PO master batches is achieved with waxes [51].

PE and PP waxes are used as carriers for such pigments but it has been difficult to ensure complete avoidance of agglomerates with consequent blocking screens.

H. New Developments in Pigments, Dyestuffs and Special Effects

New ionomers color dispersants have been developed by Hüls (Vestowax P930V) and Allied Signal (ACLyn). In the last case there are three types : LMWPE/ionomer based, resin based, and LMWPE modified. They provide the best color dispersion and allow higher pigment concentration under the same processing conditions. They have:

easier forms for use and incorporation into compounds;

color concentrates;
improved dispersability;
better thermal stability;
replacement of heavy metals; and
introduction of rare earths.

A new class of organic pigments introduced by Ciba give bright, transparent colors, have excellent resistance to weathering, heat and light, and could be an alternative to heavy metal colorants. There are many research efforts for cadmium pigment replacements due to the potential pollution of waste landfills. The alternative pigments should exhibit all the characteristics of the replaced pigments and will not significantly increase cost. PP, HDPE, XPE and mixture producers cannot abandon cadmium pigments without a 4–5 times cost increase, significant reduction in color shades offered, and a loss in productivity.

Ciba has brought out a range of yellow pigments based on bismuth vanadate, some of which resist up to 300°C and are mainly used in PO since they do not show any phenomenon of wrapping. These pigments have liveliness and freshness of tint, but heat stability is insufficient relative to the cadmium yellows which they are supposed to replace.

IV. FILLERS AND REINFORCEMENTS

A. Role and Mechanism of Action

Fillers and reinforcements are solid additives that differ from the plastic matrices with respect to their composition and structure. Modern fillers can take on many of the functions of reinforcements. Usually fibers and lamina structures are counted as reinforcements while the ball type additives are counted as fillers. Inert fillers or extender fillers increase the bulk, solve some processing problems, and lower the price; no improvement is seen in mechanical or physical properties compared with unfilled polymer, though by increased thermal conductivity they improve production rates.

Active fillers, enhancers, and reinforcements produce specific improvements of certain mechanical or physical properties, including modulus, tensile and impact strength, dimensional stability, heat resistance, and electrical properties.

The use of extender fillers can result in the following changes in the properties of thermoplastics [2, 52, 53]: increases in density, in modulus of elasticity, in compressive and flexural strength (stiffening), in hardness, in heat deflection temperature, and in lowering the temperature dependence on mechanical properties, improving the surface quality and lowering shrinkage.

Enhancers and reinforcing fillers induce the following improvements in thermoplastics [2]: increase in tensile strength and tensile stress at break and in compressive and shear strength, increase of the modulus of elasticity and stiffness of the composite material, increase of heat deflection temperature and decrease of the temperature dependence of the mechanical values, improving the creep behavior and bend–creep modulus, as well as the partial impact strength, reducing the viscoelastic yield under load and lower shrinkage.

The action of active fillers can be attributed to three causes [53–56], namely (1) chemical bond formation between filler and material is to be reinforced; (2) immobilization of polymer segments attached to the filler surface by secondary or primary valence bonds, leading to a possible structuring of the polymer matrix, an interfacial layer with characteristic properties thus appearing (the increase of T_g values is a proof for this assumption); (3) when the polymer molecules are subjected to stress with energy absorption, they can slide off the filler surface; the impact energy is thus uniformly distributed and the impact strength increased.

For plastics the most important characteristics of the fillers are: particle shape, grain distribution, specific surface and value of surface energy, thermo-oxidative stability, UV stability for outdoor applications, moisture content, and water-soluble compounds. High surface energies produce dispersion problems, reducing mechanical properties. Surface treated fillers become moisture-resistant, have reduced surface energy, give reduction in melt viscosity, improvement in dispersion, improvement in processing characteristics, in stabilization, and improvement of end-product surface.

Detailed theories are presented in Chap. 25 and many other reviews [56–63].

B. Types of Fillers and Reinforcements

Fillers and reinforcements can be differentiated by the ratio between length (or length and width) to thickness (l/t) as follows [52]:

(a) Fillers as irregularly shaped granules having $l/t \geqslant 1$.
(b) Enhancers such as short fibers, e.g., wood flour, milled or chopped glass fibers, wollastonite, whiskers, talc, etc, l/t varying from 10 to $\gg$100.
(c) Reinforcements such as filaments, nonwoven or woven textile products, l/t being very large.

(d) Special active reinforcements used principally in elastomers, such as carbon blacks, pyrogenic highly dispersed silica (Cab-O-Si-Aerosil) or precipitated ultrafine, carboxylated rubber coated $CaCO_3$ (Fortimax).

Fillers and reinforcements suitable for polyolefins are [57–74] mineral fillers as: natural and precipitated calcium carbonates, talc, mica, silica and silicates, metal powders [73], kaolin, carbon black, aluminum trihydrate, wollastonite, wood flour, asbestos, glass spheres, glass fibers, reinforcing fibers, etc. Carbon fibers (graphite), whiskers, etc., are also used.

Their tradenames and some characteristics are given in Table 8 [2, 5], and in Table 9 are indicated some changes in the properties of polyolefin composites as compared with unfilled polymers.

Calcium carbonates can occur naturally in the form of chalk, limestone, marble, etc., from which the filler is obtained by fine milling or synthesized by precipitation. Due to its low price and to the improvement of polymer properties and aging resistance, natural calcium carbonate is the most important filler used in plastics. Small amounts of finely dispersed $CaCO_3$, silica, or various silicates reduce sticking and improve the paperlike feel of PE films.

Compared with natural, ground calcium carbonates, synthetic precipitated calcium carbonate fillers show the following disadvantages [75]. They are more expensive than ground chalk. Due to the larger surface, the shearing forces during processing are appreciably higher; high filler addition is thus not possible. They also have greater absorptive effect on plasticizers, stabilizers, lubricants, etc. The adhesion calcium carbonate–polymer matrix can be improved by a surface treatment, commonly using stearates.

Talc is a natural hydratated magnesium silicate with the formula $3MgO{\cdot}4SiO_2{\cdot}H_2O$. Talc occurs in four particle shapes: fibrous, lamellar, needle-shaped, and modular; however, only the lamellar form is used in commercial applications, which determines its good slipping properties. In PP, talc gives a good balance of rigidity and impact strength. High purity gives very good long-term thermal stability, making compounds ideal for use in packaging, including odor sensitive food contact applications.

Lamellar reinforcement *mica* is obtained from muscovite or phlogopite minerals. The types not suitable for the electrical industry are used in ground form as fillers and reinforcements [2, 63]. The decisive factor as regards the reinforcing action is the ratio of the diameter to the thickness of the lamellae.

Silanization of the filler surface facilitates the incorporation of mica in PO, the aminosilanes used, in particular, resulting in good mechanical properties of the compounds.

Silica and silicates are synthetic and natural. The natural ones (sand, quartz, quartzite, novawite, tripoli, and diatomaceous earth) differ in their particle size, degree of crystallinity, and hardness. β-Quartz is, moreover, the hardest of the common minerals. Quartz has a density of 2.65 g cm^{-3} and a Mohs hardness of 7.

Synthetic silicates are obtained by relatively complicated procedures such as fine spherical primary particles, which can form agglomerates and aggregates. Depending on the manufacturing process, the surface may be very large, attaining values of 50–800 $m^2\ g^{-1}$. Apart from the chemical composition, all synthetic silicas are not crystalline. All silicas and silicates are supplied with various silane coatings.

Metals and metallic powders, consisting of aluminum, bronze, copper, and nickel, are always used in PO, if products with very high thermal or electrical conductivity are required. Heavy metal powders additionally increase the resistance to neutron and gamma rays [66]. Metals in fiber form and metallic oxides are also supplied as fillers and reinforcements for PO.

Kaolins, also known under the names of porcelain earth and china clay, are hydrated aluminum silicates possessing a clearly determinable crystal lattice with a plate-like, hexagonal structure. Kaolins consist of primary and secondary kaolinites. In their lamellar structure, primary kaolins have a ratio of length to thickness of 10/1. Kaolins generally possess a high degree of whiteness and are electrical nonconductors; they are highly resistant to chemicals, even to strong acids.

Kaolins used in PO composites should have fine particles of 0.6–6 μm and a Mohs hardness of 2.5.

The calcinated kaolin form is appreciably harder than natural kaolin; it considerably improves the electrical properties of a polymer. Kaolins are obtainable with various silane coatings that facilitate dispersion of the filler in the plastics.

Carbon black is a special form of carbon obtained through partial combustion of liquid or gaseous hydrocarbons. It can be used as a black pigment (particle dimensions of 15–20 μm), as an improver of electrical conductibility (particle dimension 17, 24, and 90 μm), and as a filler/reinforcement material (particle dimension 23–28 μm) [2, 5, 63].

Aluminum trihydrate is a nonflammable white powder insoluble in water. It is generally used as a

(Text continues on p 606)

TABLE 8 List of trade names and manufacturers of fillers and reinforcements used for polyolefin composites

Trade name	Manufacturer	Characteristics	Observations
1	2	3	4
Alumina trihydrate		Extender, flame retardant, smoke suppressant	
Martinal	Martinswekre GmbH, Bergheim/Erft., Gedr. Giulini Ludwigshafen/Rh., Vereinigte Aluminiumwerke AG, Bonn, Germany	Nonflammable white powder; crystalline, lamellar structure; nonabrasive filler; density 2.4 g/cm^3, Mohs hardness: 2.5 to 3.5	LDPE, PP, increases the resistance to corrosion, acids, heat, aging and mechanical properties, low shrinkage
Asbestos	Asbestos Corp. GmbH, Nordenham Grace GmbH, Worms, Germany	There are six kinds of asbestos from chrysolite and amphibole group, as fiber or prismatic crystals. Color varies from white, green and brown; Mohs hardness 2.5 to 6; fiber diameter of 20 to 90 nm and length up to 500 nm, mean strength; use-temperature limit 1510°C; σ =2500 N/mm^2; elastic modulus = 160×10^3 N/mm^2	
Natural calcium carbonate			
Albacor	Phizer Chas Co., USA	Mean particle diameter 0.5 to 44 μm; specific surface 1 to 15 m^2/g; pH 9 to 9.5; specific gravity at 25°C of 2.65 g/cm^3, specific heat = 0.867/kg K; thermal conductivity 2.4–3.0 W/m K, Mohs hardness 3; aspect ratio length/diameter ~1	LDPE, LLDPE, HDPE, LHDPE, EVA, PP, PB-1. Improves viscosity, hardness, some physico-mechanical and electrical properties, reduces shrinkage at forming
Calibrite	Omya GmbH, Köln, Germany		
Calcit	VHZ, Langelsheim, Germany		
Calcite	Diamond Alkali Corp., USA		
Carborex	Omya GmbH, Köln, Germany		
Durcal	Omya GmbH, Köln, Germany		
Hydrocarb	Omya GmbH, Köln, Germany		
Micromya	Omya GmbH, Köln, Germany		
Millcal	Diamond Alkali Corp., USA		
Marfil	Witco Chem. Co., UK		
Mülicarb	Omya GmbH, Köln, Germany		
Multiflex	Phillips Petroleum Corp., USA		
Non-Fer A-1	Diamond Alkali Corp., USA		
Omya	Omya GmbH, Köln, Germany		
Omyahte	Omya GmbH, Köln, Germany		
	Omya GmbH, Köln, Germany		
Ulmer Weiss	Ulmer Fullstoff Vertrieb GmbH, Ulm, Germany		
	Nordbayn.Farben-u. Mineralwerke, Hof, Germany		

Microsohi	Vereinigte Kreidewerke, Dammann KG, Söhlde, Germany		
Precipitated calcium carbonate			
Calofort U-50	John and Sturge, Ltd., UK	Mean particle diameter 0.004 to 0.07 µm, cube Specific surface 32 to 40 m^2/g	LDPE, LLDPE, HDPE, LHDPE, EVA, PP, PB-1 Particle size of 0.5 µm; good; thermal stability; 0.1 µm, good mixing, smooth surface, good gloss and abrasion resistance; 0.3 µm, good dispersion, high hot resistance
Calmote	Witco Chem. Co., UK		
Nofacal	Nordbayn.Farben-u. Mineralwerke, Hof, Germany		
Socal	Deutsche Solvay-Werke GmbH, Sollnger Obligs, Germany Solvay Cie, Belgique		
Omya BLR	Omya GmbH, Köln, Germany		
Polcarb S	English China Clays, UK		
Sturcal	John Struge, Ltd., UK		
Varon	Croxton-Garry, Ltd., UK		
Winnofil	Imperial Chemical House, Müllbank London, GW 1, UK		
Vical	Gewerkschaft Viktor Chemische Werke, Castrop-Rauxel, Germany Kreidewerke Schäfer, Diez/Lahn, Germany		
Whitetex 2	Columbian Intern. Co., UK		
Carbon Black			
Arogen	JM Huber Corp., USA	0.5 to 2.5% in LDPE, HDPE, EVA, PB-1. Thermal and UV stabilizer, good thermal and aging resistance and also surface resistivity; increases tensile strength; it is used in compositions for cables insulation, sheets, pipes, conduits, and injection molded products	
Aromax	Cabot, France		
Black Pearl	Cabot GmbH, Hanau, Germany		
Condulex	Columbian Carbon Deuschland, Hamburg, Germany		
Corax	Union Carbide Corp., USA		
Dixie	Degussa AG, Germany		
Elfblack	Union Carbide Corp., USA		
Kosmos	Columbia Carbon Corp., USA		
Micromex	Goodfrey Cabot Inc., USA		
Mogol	Degussa AG, Germany		
Printex corax	Cabot Carbon Corp., USA		
Regal 300, 600	Cabot Carbon Corp., USA		
Sterling	Cabot Carbon Corp., USA		
Super-Carboval	Cabot Carbon Corp., USA		
Vulcan	JM Huber Corp., USA		
Wyex E, PC	Union Carbide Co., USA		
WYB or Thornel or VYB			
Carbon graphite fibers			
Fortafil	Great Lakes Carbon Corp., USA		
Kynol	Carborund Co., USA		

TABLE 8 (Continued)

Trade name	Manufacturer	Characteristics	Observations
1	2	3	4
Rhigilor	La Carbone Lorraine, France		
Rigilor	Serafim Co., France		
Thornel	Union Carbide Europa SA Genf, CH		
	Courtaulds, UK		
	Sigri Elektrographit GmbH, Mettingen, Germany		
Glass fibers			
	Bayer AG, Leverkusen, Germany	See also Table 10, Hot resistant; resistant to aggressive chemicals; heat resistant up to 650°C; high tensile strength, good elasticity; very high brittleness	LDPE, HDPE, PP PB-1
	Czech-Petrocarbon Co. Ltd., UK		
	Gevetex Textilglas GmbH, Düsseldorf, Germany		
	Klöckner-Scott Glasfaser, GmbH, Dortmund-Mengede, Germany		
Fibralloy	Fibralloy 300-Monsanto Chem. Co., USA		
	Owens/Coring Fibreglass, Wiesbaden, Germany		
	Saint Gobain Industries, Chambery, France		
	Vetrotex International, Geof, CH		
	Temp R/Tape Polypenco Inc., USA		
Tuyglas, Tygmot	Fothergill and Harvey Ltd., UK		
Glass spheres, filled glass			
Ballotini	Ballotini Europe, France	Solid spheres; diameter 4 to 44 μm; density 2.5 g/cm^3; are used together with coupling agents, as silanes	10–40% in LDPE, PP. Good homogeneity and flow properties; uniform shrinkage; increases modulus and compression resistance
	Potters-Ballotini GmbH, Kirchheim-bolanden, Germany		
	3M, St Paul, MN, USA		
Kaolin			
Argelac	Soc. Rhone Progil, France		20–45%, good abrasion resistance; low water absorption; good electrical properties
Argine GH115	Argigee, Montgujon, France		
China Clay	English China Clay Sale Comp., Ltd St. Austell, Cornwall, UK		
Claytey AM3	Franterre SA, France		
Diapone P	The News Conseil Mines, UK		For LDPE, PP. The two-dimensional oriented PP films with 50% kaolins are opaque like paper
China clay	G.H. Luh GmbH, Walluf, Germany		
China fill	Gesellschaft für Rohstoffveredlung GmbH, Hirschau, Germany		
Gepico	Godfrey Cabot, USA		

Franclay	Franterre SA, France		
Kaolin	Gebr. Dorfner OHG, Hirschau, Germany	Thermal conductivity 1.9 W/m K; specific heat 0.2 J/kg K	
Microsika	SIKA Cie., France		
Speswhite	English China Clays Sales St. Austell, Cornwall, UK		
Supreme	English China Clays, Sales, E. Kick, Schnailtenbach, Germany		
Metals and metallic oxides			
Maglite	Meck Co., USA		HDPE, PP
Magnesium oxide	Lehmann & Voss Co., Hamburg, Germany		MgO and TiO_2 improve hardness; < 70% ZnO in PP lead to good aging resistance
Metallic oxides	C. Schlenk, Nürnberg, Germany	Increase heat and electrical conductivity; increase weathering properties	
Metallic powders	G.M. Langer & Co., Ritterhude, Germany		
	Lieubau Chemie, Hamburg, Germany		
Zinc oxide	VHZ Langelscheim/Harz, Germany Zinkweiss-Handelsgeselschaft, Oberhausen, Germany		
Zinc dust	Lindur GmbH u. Co., K, G, Köln, Germany Stolberger Zincoli GmbH, Stolberg, Germany		
Mica			
Concord mica	Sciama SA, Paris, France	Delaminated $\rho = 2.9$ g/cm^3 mica flakes have thickness up to 1 μm; thermal conductivity 2.5 W/m K; specific heat 0.86 J/kg K	LDPE, LLDPE, PP, excellent electrical and thermal properties, good dimensional stability
English mica	The English Mica Comp., Stamford, CT, USA		
Glimmer			
Micro-Mica	Norwegian Talc, Bergen, Norway		
Silican natural			
Aktisil	R. Hoffmann & Söhne KG, Neuburg, Germany	Specific surface 50 to 800 m^2/g	60% in LDPE, PP improve electrical properties, moisture resistance, minimum shrinkage, good flow properties. Microtonized silica gels are also nucleating agents or antiblocking agents
Coloritquarz	Gesellschaft für Rohstoffveredlung GmbH, Hirschau, Germany		
Cristobalit	Quartzwerke GmbH, Frechen, Germany		
Elmin	Quartzwerke GmbH, Frechen, Germany		
Latexyl	Silice et Kaolin, France		
Extrusil	Füllstoff Gesellschaft, Germany		
Gloxil	Gesellschaft für Neuburger Kieselweiss, Neuberg, Germany		

TABLE 8 (Continued)

Trade name	Manufacturer	Characteristics	Observations
1	2	3	4
Grenette	Sanson S.A., Paris, France		Increases abrasivity of sample surface, improves modulus and resistance at compression; 15–45% quartz reduces crazing and shrinkage
Crystal quartz	Dörentruper Sand- und Thonwerke GmbH, Dörentrup, Germany		
Crystal Quartz sand	Dudinger Glassandwerk Bock u. Co., Dudinger, Germany		
Hi-Sil	Columbia South Chem. Co., USA		
Quartz flour	Quartzwerke, GmbH, Frechen, Germany Gesellschaft für Rohstoffveredlung GmbH, Hirschau, Germany		
Sigrono	Westdeutsche Quartzwereke Dr. Müller GmbH, Dosten, Germany		
Sikron	Quartzwerke GmbH, Frechen, Germany		
Silbone	Quartzwerke GmbH, Frechen, Germany		
WQO-Quartz flour	Westdeutsche Quartzwerke, Dr. Müller GmbH, Dosten, Germany		
Silica synthetic			
Aerosil	Degussa, Frankfurt/M., Germany		LDPE, HDPE, PP. Colloidal silica 0.5–3% increases tixotropy, is antiblocking agent; increases dielectric properties
Cab-O-Sil	Cabot GmbH, Hanau, Germany		
Syloid	Grace GmbH, Worms, Germany		

Talc			
Cascade, Mistron Micro-Talc Micro-Talkum	Chemag, Frankfurt/M., Germany Norwegian Talc, Bergen, Norway J. Scheruhn, Hof, Germany Naintsch Mineralwerke GmbH, Graz, Austria Nordbaryische Farben-und Mineralwerke, Hof, Germany S.A. des Talcs de Luzenac, Luzenac-sur-Ariege, France	Density 2.9 g/cm^3 platelets; 40 to 62% SiO_2; 0.2 to 11% Al_2O_3 and others, labellar structure Thermal conductivity 2.1 W/m K; Specific heat 0.86 J/kg K Various colours	LLDPE, LDPE, LDHDPE, VA, PP, PB-1. Rapid injection cycle, good resistance, Rockwell hardness 88–97; <25% controls the flow
Wollastonite			
Casiflax Tremin Wollastonite 100, 300	Quartzwerke GmbH, Frechen, Germany Quartzwerke GmbH, Frechen, Germany Vereinigte Harzer Zinkoxyde, Langelsheim, Germany	Blade crystal, microfibrous, 4–6 μm in diameter, triclinic crystal system; brilliant white, specific gravity 2.9 g/cm^3, $1/t > 100$, Franklin fibre, US Mohs hardness 4.5; melting point 1540°C; coefficient of thermal expansion 65×10^{-6} mm/mm^1 °C	HDPE, PP
Wood flour			
Vavonite	Vavassewr Co. Ltd., UK	Particle size 60 to 80 μm; bulk density 182 to 285 kg/m^3; specific volume 3.5 to 5.49 m^3/kg limit dimensions of fibers 43.0–4.3/17.1–4.3 182.0–86/34.4–8.6	Improves the surface gloss and moisture resistance

TABLE 9 Physicomechanical properties of some polyolefin composites compared with unfilled polymers

Polymers or composite materials	Density, kg/m^3	Elasticity modulus, MPa	Elongation at break, %	Limit resistance to, MPa			Impact strength kJ/m^2		Shore hardness
				Stretching	Bending	Compression	Without notch	With notch	
1	2	3	4	5	6	7	8	9	10
UHMWHDPE	934		450	45				Without break	40
HDPE	948–960	6200	200–800	23–25			Without break	2–12	
HDPE + a % glass fibers									
$a = 20$	1100	6500	1.5	50.0	70				
$a = 30$	1180	6210	1.9	86.2			239.9	69.3	75
HDPE + a % calcit									
$a = 33–37$	1200		20.0	19.0			35.0		
$a = 40–44$	1400		10.0	19.0			20.0		
$a = 48–52$	1400		6.0	18.0			14.0		
HDPE + a % asbestos									
$a = 0$			277	23.6			158.5		5.7
$a = 5$			23	25.6			42.4		5.9
$a = 10$			12	25.2			26.4		6.0
$a = 15$			5	25.0			21.3		6.75
$a = 20$			5	26.0			15.7		6.43
$a = 25$			3	27.2			6.45		8.6
$a = 30$			0	25.4			7.45		7.7
$a = 35$			0	23.7			8.17		7.4
$a = 40$			0	26.3			13.23		7.2
Croslinked HDPE	944			19.6					2.84
Crosslinked HDPE + a % wood flour									
$a = 5$	951			20.8					3.81
$a = 10$	975			23.4					3.91
$a = 20$	1008			26.2					4.19
$a = 30$	1032			23.0					6.60
LDPE	923		600	11.03					52
LDPE + a weight parts $CaCO_3$									
$a = 25$	1065		200	11.13					55
$a = 50$	1184		50	11.71					57
$a = 75$	1288		20	11.83					59
$a = 100$	1378		10	11.5					61
EVA (5% vinyl acetate)	926		600	14.18					47

EVA (5% vinyl acetate) + a weight parts $CaCO_3$								
a = 25	1066		500	9.45				49
a = 50	1186		170	9.57				49
a = 75	1290		60	10.65				55
a = 100	1380		40	11.43				56
EVA (10% vinyl acetate)	927		700	13.05				45
EVA (10% vinyl acetate) + a weight parts $CaCO_3$								
a = 25	1068		500	9.85				49
a = 50	1187		250	8.64				50
a = 75	1291		100	9.20				52
a = 100	1381		50	10.20				55
EVA (15% vinyl acetate)	937	12.89		700				40
EVA (15% vinyl acetate) + a weight parts $CaCO_3$								
a = 25	1078	11.30		650				42
a = 50	1198	8.56		500				43
a = 75	1302	7.69		350				44
a = 100	1392	7.68		150				46
EVA (15% vinyl acetate) + 0.5 weight parts stearic acid	938	14.22		700				35
EVA (15% vinyl acetate) + 0.5 weight parts stearic acid + a weight parts $CaCO_3$								
a = 25	1078	11.13		700				38
a = 50	1198	8.95		700				40
a = 75	1301	7.09		600				42
a = 100	1392	6.47		600				44
PP	900	19300	24.6	40.0	46.0	100	1.63	
PP + 40% $CaCO_3$ + silane	1200	25000	29.0	30.0		70.0	3.32	
PP + 30% $CaCO_3$ + 10% glass fibers	1230		26.0	35.0	45.0	26.8	4.35	
PP + 15% glass fibers + 25% mica	1200	45000	1.3	42.0	59.0		3.81	
PP + 25% tuff + 15% glass fibers	1220	42200	3.4	60.0	84.0	36.4	6.0	

TABLE 9 (Continued)

Polymers or composite materials	Density, kg/m^3	Elasticity modulus, MPa	Elongation at break, %	Limit resistance to, MPa			Impact strength kJ/m^2		Shore hardness
				Stretching	Bending	Compression	Without notch	With notch	
1	2	3	4	5	6	7	8	9	10
PP + a % glass fibers									
$a = 20$		58000		100			3.1		
$a = 30$	1130	69000	2.5	96.5	124.1	82.4	239.6	101.3	98.0
$a = 40$	1220	77500	1.2	74.5	88.0		24.5	9.8	
PP + a % asbestos									
$a = 30$	1000–1300			39.0	75.0				
$a = 40$				38.0	53.0		13.7		
$a = 50$		37000		32.0			2.1		
PP + 8% carbon fibers	960	45000		30–32					
PP + a % talc									
$a = 20$		2690	Flow limit 39	4.1				25.1	
$a = 30$		2840	31.99	10.5	64.54				
$a = 40$		4 140	37.25	3.1				23.43	
PP + a % $CaCO_3$									
$a = 30$		2930	70.0	30.68	66.74			7.32	
$a = 35$		3040	17.2	30.06	60.74			7.26	
$a = 40$		3250	11.7	24.68	58.95			5.19	
PP + 40 % silicate	1250	65000		32.0	48.0		3.46		
PP + 40 % wood flour				40.0	41.0		3.38		

PP + 40 % wood flour + 1 % dibutyl phthalate		41.6	43.0		3.75
PB-1	290	26.0	21.6		
PB-1 + a % treated kaolin					
$a = 10$	210	22.0	15.0		
$a = 20$	170	20.0			
$a = 30$	120	18.0	8.0		
PB-1 + a % treated chalk					
$a = 10$	20.0	24.0			
$a = 20$	14.0	22.0			
$a = 30$	11.0	14.0			
PB-1 + a % glass fibers (length of fibers 5–10 mm)					
$a = 10$	160.0	28.0			
$a = 20$	110.0	28.0			
$a = 30$	80.0	27.0			
PB-1+a % Kiselgur (specific area 0.6 m^2/g)					
$a = 10$	230	25.0			
$a = 20$	190	23.0			
$a = 30$	170	20.0			

EVA, ethylene vinyl acetate copolymer; UHMWHDPE, ultra high molecular weight HDPE.
Source: Ref. 70.

flame retardant but acts also as a filler, increasing the stiffness of materials and improving flow and electrical properties. It is a nonabrasive filler with a low density (2.4 g/cm^3) and a Mohs hardness ranging between 2.5–3.5.

Wallastonite is a naturally occurring calcium metasilicate available in surface coated forms of various types. New grades under development indicate its strong growth and its potential for the replacement of calcined clay and other mineral used in thermoplastics

Wood flour is obtained by fine milling of soft or hard wood wastes; the flour obtained is then successively subjected to sorting and drying operations, leading to many sorts with the range of characteristics listed in Table 8. Wood flour is used for thermoformable PP sheets.

Asbestos fibers have been used in the past. Its use has ceased following discoveries of its health hazards.

Glass spheres, both solid and hollow, are obtained from sodium borosilicates or silica. As fillers, and also as flame retardants, solid glass spheres have diameters ranging from 4–5000 μm. For plastics the size typically used is 30 μm in diameter and densities of 2.5 g/cm^3 are preferred. All have the same spherical shape and are transparent and compression-resistant. They have controlled granulometry and high thermal stability. They increase flowability and contribute to an improvement in stress distribution. Molded parts filled with glass spheres exhibit isotropic behavior, shrinkage in the reinforced material being the same in all directions, so that the properties sometimes can be previously predicted. Solid glass microspheres (diameter $< 50\,\mu m$) improve modulus, compressive strength, hardness, and surface smoothness. Favorable flow properties permit high filler content. The action of glass spheres is decisively influenced by their wettability; silanes are therefore often used as coupling agents.

Molding compounds containing glass spheres can be extruded or injection molded. Hollow glass spheres displace the same volume of PO as solid spheres, but are lighter in weight. Dimensional stability, lower viscosity, and improved flow are the main advantages. The typical density range is 1.1 g/cm^3.

Expandable microspheres can also be used. They are thermoplastic microspheres encapsulating a gas. They are used for foamed items and for weight reduction. The recent developments in microspheres include Ecosphere hollow glass microballons from Emerson & Cuming for highly filled PP.

Glass fiber. From the melt of naturally occurring silicon, boron, aluminum oxides, etc., textile glass fibers, mainly 10–20 μm in diameter, are produced by means of a mechanical drawing method from molten glass [2, 5, 61]. The fibrous products include strands, filament yarns, staple fibers, and staple yarns. A surface treatment is applied and then they are chopped into specific lengths. The theoretical minimum critical length for a reinforcing fiber has been estimated to be about 50–100 times its diameter [76]. The chopped fiber is available in lengths of 3.2–25 mm. Some applications require the use of milled fiber with typical length of 1.6–0.8 mm. Milled fiber is used in applications where stiffness, dimensional stability, high flow during molding, and more uniformly isotropic mold shrinkage are considered more important than tensile strength or toughness. The surface treatment consists of lubricants, sizings, and coupling agents and it is necessary to reduce strand breakage during processing and provide compatibility with matrix.

Two-dimensional structures are subdivided into glass fibers, fleeces, mats, coverings, fabrics, felts, and sheets. Depending on their composition, many kinds of glass grades with distinct properties and applications may be mentioned; see Table 10 [2, 5, 68].

Alkali-containing (soda-lime–silica, eventually boron, too) grade A glass is used in articles (such as windowpanes and bottles) not excessively subjected to stress and exposed neither to the action of weathering nor to moisture. Grade C glass is a chemical glass suitable for applications that require greater resistance to acids, while dielectric grade D glass is limited to electrical insulating materials. Most commonly used is grade E glass, calcium–aluminum–boron silicate glass developed for electrical application. E glass has a good resistance to heat and water, fair resistance to bases and low resistance to acids. S and R glass are high-strength reinforcements but cost significantly more than E glass, up to seven times more.

For special applications, M and R grades are indicated for products subjected to extremely high stress and respectively for products with high strength and good thermal stability, while S glass with high strength is used for aircraft and rocket construction. These types are very expensive. L glass is used in applications requiring radiation protection.

The glass fiber surface must be protected against physical and chemical stresses and against the influence of moisture during storage. These requirements are achieved by applying a sizing agent (in the case of continuous filament) or a lubricant (in the case of staple fibers) and respectively coupling agents to the glass surface (silanes, titanates, etc.) in order to assure a good bond between the matrix and the glass.

TABLE 10 Composition and some properties of different types of glass fibers

Type of glass fibers Property	A glass soda lime	C glass	D glass	E glass borosilicate	L glass	M glass	R glass	S glass
1	2	3	4	5	6	7	8	9
SiO_2, %	72.5	65.0		54.0		53.5		64.0
$Al_2O_3(Fe_2O_3)$, %	1.5	4.0		15.0		0.5		26.0
B_2O_3, %		5.0		8.0				
MgO, %	3.5	2.0		4.0		9.0		10.0
$Na_2O + K_2O$, %	13.0	8.0		0.8				
BaO, %						8.0		
TiO_2, %						8.0		
CaO + ZnO + LiO, %						8.0		
CaO, %	9.0	14.0		18.0		13.0		
Density, g/cm^3	2.48–2.50	2.49	2.16	2.54	4.30	2.89	2.5	2.49
Strength of monofilament, N/mm^2	2450–2900	3000	2500	3500	4000	3500	4750	4900
Modulus of elasticity, N/mm^2	45,000–60,000	70,000	52,000	73,000	73,000	12,400	83,000	87,000
Softening point, °C	785	749	763	846	708	863	928	815
Mohs' hardness	6			4.5–6.5				
Diameter, μm	2–100			1.3				
Fiber length, mm				0.2–6				
Dielectric constant 10^6, Hz Ω cm				6.5–7			6.0–8.1	
Refractive index, 25°C				1.55–1.566			1.541	

Source: Refs. 2, 5, 6, and 67.

Advantages of glass fibers over other reinforcements include a favorable cost/performance ratio with respect to dimensional stability, corrosion resistance, heat resistance and ease of processing.

As coupling agents for polyolefin composites, the most often used are γ-amino propyltrioxy silane, γ-glucidoxypropyltrimetoxysilane, γ-mercaptopropyltrimetoxy silane, and some polymeric coupling agents such as maleic anhydride–propylene copolymers.

Carbon fibers. There are a wide variety of types of carbon fiber produced for the use in composites, varying for example in degree of graphitization and diameter. Two principal types are available in chopped and sized form, polyacrylonitrile-based and pitch-based. Compared to glass fibers, carbon fibers offer higher strength and modulus, lower density, outstanding thermal and electrical conductivity, but much higher cost. In addition they are very chemically resistant and naturally slippery. The low density and high mechanical properties of carbon fibers allow great flexibility in formulating composites.

Polymeric fibers. All types of chemical fibers known can be employed for polyolefin reinforcement; thus aramid (aromatic polyamide) polyamides, polyesters polyvinyl alcohol fibres, cellulose fibers [2, 5, 52, 70–72, 77], etc. are used; the polyolefin fibers (such as UHMWPE) themselves are applied as reinforcements in other composites.

Synthetic fiber reinforced plastics have significantly lower moduli than glass-reinforced plastics in the direction of the fibre, but there is a higher resistance to damage by distortion. Glass fiber-reinforced thermoplastics combine the good properties of plastics with those of inorganic glass and in their property values approach the level of metals. The long fibres technology is developing. The theory suggests that an improvement of $\sim 50\%$ in mechanical properties should be produced by increasing fiber length from 0.3 mm to 2 mm. In practice, however, it is unlikely that this degree of improvement could be obtained due to complexities associated with fibre orientation and skin/core effects. Long fiber PP is potentially the most interesting due to the relatively low cost of the matrix. A typical range is of the form of 15 mm chips with 20–50% glass content. The properties include high dimensional stability and impact resistance, low wrapping, good surface finish, and elimination of the usual effects of shrinkage.

Also used are metallic fibers, carbon fibers, graphite, "whisker" fibers, fibers obtained from titanates having good pigmentating properties, boron fibers (diameter = 0.1–0.12 mm), ceramic fibers, etc. Ceramic fiber include alumina, boron, silicon carbide, alumina–silica, etc. The physical properties of these reinforcements in matrices compare favorably to glass and other fibers. Ceramic fibers suffer from two significant limitations: cost, and an inherent brittleness.

Generally, glass fibers lose their properties at 300–400°C, while asbestos, boron, and carbon fibers maintain their properties at higher temperatures.

In order to obtain materials with various heat conductivities the following fillers can be used: NaCl, AgCl, quartz, graphite, Al_2O_3 (in PE).

C. Requirements

The main requirements for the properties of fillers and reinforcements needed for the production of composite materials to be used in specific applications are [2] the following.

Low moisture absorption and high bulk density; they should preserve their properties during storage prior to compounding.

Optimum compounding is achieved with fillers and reinforcements with mean particle size, intimate wettability through the polymer matrix, which does not present static charge, no shortening of the reinforcing fibers taking place, which means a good dispersion behavior.

Filler particles should be as round as possible with a small specific surface, low surface energy, and low absorptivity, assuring thus a low viscosity during compounding.

A high compounding speed is obtained with fillers having low specific heat and high thermal conductivity.

During processing, low shrinkage, low internal stresses, no cracking, and rapid demolding are specific to composites with fillers and reinforcements with low specific heat, high thermal conductivity, low thermal expansion, uniform filler distribution in the matrix, and a good adhesion between reinforcement and plastic. Sometimes, optimum adhesion requires the use of sizes or coupling agents.

Fillers having a low degree of hardness, small, round particles, good thermal stability, subjected to surface treatment, do not provoke abrasion in the processing machine.

For a controlled modification of the various properties of the composite materials, certain characteristics of the filler/reinforcement are necessary.

A composite with high tensile strength and elongation is obtained using a filler/reinforcement having a high strength in comparison with the matrix, high length/diameter ratio, and good fiber/matrix adhesion, as well as a good distribution in the matrix while, for high flexural strength, it is very important to obtain, in addition, a smooth surface in the finished article.

Fillers with low compressibility and small round particles are suitable for the production of composites having high compressive strength from crystalline polymers.

Fibrous or lamellar reinforcement with the high length/diameter ratio, high modulus of elasticity in comparison with the matrix, high orientation in the direction of the force profile, and good adhesion are used for high stiffness, high modulus of elasticity composites. It is a high strand integrity product with good fiber feed and handling characteristics in 3.17 and 4.76 mm chopped lengths. The high performance carbon/PEEK types are "pre-pregs" in which continuous filament and matrix have been combined by a form of stretching process. They require only placing in position and heating to fuse the thermoplastic PP matrix. The interfacial bonding is improved. PP sheet moulding compounds and bulk moulding compounds known as Glass Mat Thermoplastics (GMTs) are compounded in granules for injection moulding and extrusion.

Production of high impact strength composites requires long fiber reinforcement with nonperfect adhesion with the polymer matrix.

Good long-term behavior and fatigue and weathering resistance of the composite materials are achieved with filler/reinforcement with permanent polymer/matrix bond, good resistance to heat, light, water, chemicals, etc.

In order to obtain other improvements of the composite properties, correspondence with the filler/reinforcement properties is necessary, so both components should have good thermal, electrical, chemical, and optical properties, low water absorption, high density, etc.

Also, a low cost results when a low cost filler is used with low processing cost and with maximum possible degree of filling; other requirements regarding the properties being, of course, fulfilled.

All these influences/modifications are evidenced by the data presented in Table 9.

D. Concentration Range

The amount of filler or reinforcement used varies within large limits, as a function of the desired properties of the end products (which can sometimes be previously predicted). As can be seen from Table 9, polyolefins cannot incorporate large quantities of filler, which should lead to a significant lowering of the product cost without any worsening of their properties.

The incorporation of a filler into a linear polymer invariably results in the formation near the filler surface of a boundary interphase structurally different from the pure polymer. However, the properties of the boundary layer and therefore the properties of the filled polymer crucially depend on the method of preparation employed [78, 79]. On the basis of the absolute values of the filler content and dimension of the boundary layer an intuitive classification of filled polymers into "low loaded" and "high loaded" can be made.

E. Incorporation into the Polymer

Incorporation of fillers and reinforcements in PO is a complex procedure occurring in various ways, as a function of the characteristics of the materials used (shape, particle dimensions, wetting by polymer, etc.) and of the quantity to be incorporated in the polymer as powder or melting.

Powdered fillers, and also spheres or flakes, are incorporated in the compounding step, while in the case of fibrous reinforcement materials (especially glass fibers), three modes of incorporation are used. In the first procedure, glass fiber strands are impregnated with polyolefin melt, and after solidification the strands are cut into rodlike granules; the fibers incorporated have therefore the same length as the rods, being oriented in the longitudinal direction of the latter. In the second procedure, the polyolefin powder is blended with short fiberglass, the mixture being then extruded and granulated. This procedure assures a uniform distribution of the reinforcement material in the polymer matrix but unfortunately favors the mechanical destruction of the fibers.

In order to avoid destruction, the glass fibers are first cut to the necessary length and then introduced into the polymer melt.

Pre-heating fillers before mixing has many advantages, such as: improving the mixing rate, reducing energy consumption in processing, reducing wear in equipment, and improving product quality. Unheated fillers may have greater levels of condensed and absorbed surface moisture, requiring longer mixing time. Compounding with mineral fillers and reinforcements may present problems but surface treatments and dispersing agents will help lightweight fillers, as hollow ceramic or glass expandable beads are increasingly interesting.

F. Testing Methods

Testing of filled/reinforced materials is made primarily by determination of the change of the properties of the polymeric materials. Thus the following properties are taken into consideration: mechanical (Young's modulus, tensile strength, etc.), thermal and burning properties, electrical properties, etc. [71].

In the analysis and specification of fillers, several properties are of particular importance, as follows: ASTM designation referring to the particle size as D 1366-53T (Reporting Particle Size Characteristics of Pigments); C 92-46 (Sieve Analysis and Water Content of Refractory Materials); E 20-62T (Analysis by Microscopic Methods for Particle Size Distribution of Particulate Substances of Subsieve Sizes); B 293-60 (Subsieve Analysis of Granular Metal Powders by Air Classification); ASTM designation referring to surface area, D 1510-60 (Iodine Absorption Number of Carbon Black); ASTM designation referring to specific gravity, D 153-54 (Specific Gravity of Pigments), ASTM designation referring to bulk density, D 1513-60, (Pour Density of Carbon Black); to pH value, D 1512-60, or oil absorption, D 281-31 (Oil Absorption of Pigments), D 1483-60.

Other tests include the determination of purity, freedom from coarse particles, etc., and other properties that are extensively treated in many books [2, 5, 57, 61, 67, 70–72].

G. Uses and Problems

Some examples of the application of polyolefin composites are given in Table 11. The predominant fibers used for reinforcement are made of glass or carbon (graphite). Polymeric and metal fibers have their use in specialized circumstances. Mineral fibers usage is low and declining because of health concerns and lower performance, although these materials were once of considerable commercial importance. Natural fibers have yet to demonstrate performance levels to justify their usage in any but nondemanding applications. Although they provide an improvement in stiffness and impact resistance, their use is severely limited by their relatively low resistance (strength loss sets in around 124°C and thermal degradation commences around 163°C), so they could impart a dark coloration to the composite, tend to degrade quickly on exposure to sunlight and microbial attack, and absorb water and oils with diminished mechanical and dielectric properties.

Typically the presence of fillers causes a decrease of stabilization performance due to the physical or chemical phenomena. The physical aspects of the decrease is due to the absorption of the antioxidants onto the surface of the inert filler (observed especially for $CaCO_3$, carbon black, and silicates). If the antioxidant is immobilized it is unable to protect the polymer. Usually a modest increase in the amount of stabilizer

TABLE 11 Fields of applications and manufactured products from filled or reinforced polyolefins

Polyolefin	Filler, reinforcement	Field of application and manufactured products
LDPE	Chalk	Films, household products, short-use-time products
	Talc	Panels, profiles for building construction, baths, containers, tanks, feeders, clearing tanks for chemical industry
	Mica/silica	Ventilation systems, tubes, pipes, for chemical industry
	Kaolin/chalk	Vehicle construction, drainage tubes and tubes for the circulation of cooling liquids
	Glass spheres	Vehicles and road construction, marks
	Wood flour	Agriculture and agricultural machines, enclosed type cab for tractors
	Glass fibers	Cable insulation, pipes, tanks, shaped pieces, telecommunication and vehicle construction
HDPE	Calcite	Films, household article, decorative articles, items obtained by injection molding
	Volcanic tuff	Vehicle construction, household articles, objects obtained by injection molding
PP	Talc	Short-use-time goods, furniture pieces, sport articles
	Mica	Fans, decorative articles, vehicle industry
	Asbestos	High-pressure pipes, tanks, coatings for pipes, industrial and urban construction
	Glass fibers	Fans, pieces for electrical, electronic, radio, telephone, large diameter pipes, household articles, gears, school furniture, vehicle construction, sport articles, etc.
	Various high temperature stabilized	Compression moulded sheets, filter elements, extruded sheets, rods, waste water pipe systems, profiles

suffices to overcome this negative influence. Metallic impurities (iron based) in the talc play a major role in the lack of stability in the final formulation because they catalyze the hydroperoxide decomposition, accelerating the branching reaction (see Chap. 17). Several grades of talc are supplied by manufacturers having various effects on stability or it is coated with epoxy resins. Other solutions to prevent the effect on stability is the use of a formulation containing metal deactivator.

H. New Developments

With a special phosphite esters coating, Snowfort 4000K100 has been developed by Croxton and Garry as a reinforced filler in PO with good rheological properties and impact performance. In the field of reinforcements, the new developments consist of long fibre and high performance fibers (such as aramid) for injection moulding products and the improvement of the surface treatments developing new coupling agents.

Recently E-CR glass (Corrosion-Resistant) was developed. It improves the resistance to acidic corrosion, being particularly designed for reinforcement of plastics submitted to an acidic environment. Vetrotex CertainTeed has developed Twintex, a commingled reinforcement of unidirectional fiber and PP filaments. A glass reinforcement offering superior mechanical properties in compounding PP is M Star Stran from Schüller Mats and Reinforcements.

REFERENCES

1. M Smith. Polymer Technology. New York: Reinhold, 1985, p 497.
2. R Gächter, H Müller, eds. Plastics Additives Handbook. Munich: Hanser, 1983; H Hurnik. Chemical blowing agents, p 619; H Jenkner. Flame retardants for thermoplastics, p 535; E Herrmann, W Damm. Colorants for plastics, pp. 471, 508; K Berger. Fluorescent whitening agents, p 585; AW Bosshard, HP Schlumpf. Fillers and reinforcements, p 397.
3. HR Lasman. Encyclopedia of Polymer Science and Technology, vol. 2, New York: Wiley, 1965, p 532.
4. Gh Manea. Cellular Plastics Materials. Bucharest: Ed. Tehnică, 1978, p 125.
5. S Horun. Additives for Polymer Processing. Bucharest: Ed. Tehnică, 1978, pp. 71, 80; S Horun. Applications of Plastics Materials. Bucharest: Ed. Tehnicâ, 1975, p 200.
6. J Murphy. Additives for Plastics Handbook. Oxford: Elsevier Advanced Technology, 1995.
7. HE Mark, SM Atlas, SW Shalaby, EM Pearce. Combustion of polymers and its retardation. In: M Lewin, SM Atlas, EM Pearce, eds. Flame Retardant Polymeric Materials. New York: Plenum Press, 1978, vol 1, p 15.
8. Flammability Now, Conference Proceeding, April 11–12 1989. Manchester: The Textile Institute, papers 4, 7, 8, 10, 1989.
9. J Troitzsch. Kunststoffe 1987, 77(10): 1078; International Plastics Flammability Handbook. Principles, Regulations, Testing, and Approval, 2nd edn. Munich: Hanser, 1990.
10. DE Lawson. Rubber Chem. Technol 1986; 59: 455.
11. RG. Gann, RA Dipert, MJ Drews. In: M Bikales, Overberger, Marges, eds. Encyclopedia by Polymer Science Engineering. 2nd edn, vol 6, pp 154–210, 1986.
12. RR Hindersinn, GM Wagner. In: M Bikales, O Overberger, Marges, eds. Encyclopedia of Polymer Science and Technolgy, vol. 7. New York: Wiley, p 1, 1967.
13. C Vasile, FI Popescu, O Petreus, M Sabliovschi, A Airinei, I Agherghinei. Bull. Inst. Politechnic Iaşi, Sect. II 1985; 31(35): 91–99.
14. MW Ranney. Fire Resistant and Flame Retardants Polymers. Park Ridge, NJ, Noyes Data Corp, 1974, pp 190, 342.
15. Y Yehaskel. Fire Resistant and Flame Retardant Polymers, Recent Developments. Park Ridge, NJ, Noyes Data Corp., 1979, pp 165, 389.
16. WA Reeves, GL Drake, RM. Parkins. Fire Resistant Textile Handbook. Westport, CN: Technomic Publ. Inc., 1974, p 125.
17. Akzo. Technical Bulletins; Alcoa. Technical Bulletins.
18. RD Deanin. In: C Vasile, RB Seymour eds. Handbook of Polyolefins, 1st edn. New York: Marcel Dekker, 1993, pp 696–705.
19. J Green. In: JT Lutz, ed. Thermoplastic Polymer Additives. New York: Marcel Dekker, 1989, Chap 4.
20. J Wang, JF Tung, PR Hornsby. J Appl Polym Sci 1996; 60(9): 1425.
21. KK Shen, DJ Ferm. Borate fire retardants in plastics. Paper presented at Twentieth Int Conf on Fire Safety, San Francisco, California, January 12, 1995.
22. KK Shen, DJ Ferm. Plastics Compounding, September/October 1985, November/December 1988, US Borax Research Corp., 1995 Fire Retardant Chemical Ass. Fall Conference.
23. Dover Chemical Corp. Technical Bulletins; Ethyl Corp. Technical Bulletins.
24. RD Deanin, SB Driscoll, AK Laliwala. SPE ANTEC 1976, 22: 578.
25. R. Markezich. ENE SPE, Oct. 10, 1991.
26. J Green. In: M Lewin, SM Atlas,. EM Pearce, eds. Flame Retardant Polymeric Materials, vol 3. New York: Plenum Press, 1982, chap I.
27. JW Lyons. The Chemistry and Uses of Flame Retardants. New York: Wiley, 1970, pp 21–23, 282–297.
28. AS Wood. Mod. Plastics 1991; 68(9): 54.
29. RD Royer, P Gearlette, I Finberg, G Reznick. Recent Adv Flame Retard Polym Mater. 1996 (publ. 1997); 7: 175–185.

30. CE Cullis, MM Hirschler. The Combustion of Organic Polymers. Oxford: Clarendon, 1981, pp 211–213, 237–240, 276–296, 307–320.
31. CJ. Hilado. Flammability Handbook for Plastics. Technomic Publ, Lanchester 1982, chap 2, 3, 141, 153.
32. RI Markezich, DG Aschbacher. In: GE Nelson, ed. Fire and Polymers II, ACS Symp Series 599, 1995, pp 65–76.
33. RD Deanin, M Ali. In: GL Nelson, ed. Fire and Polymers II, ACS Symp Series 599, 1995, pp 56–65, chap 3.
34. W Zhu, EC Weil, S Mukhapadhyav. J Appl Polym Sci 1996, 62(13): 2267–2280.
35. MM Gauthier, RD Deanin, CJ Pope. Polym Plast Technol Eng 1981; 16(1): l.
36. A Townson, In: GE Nelson, ed. Fire and Polymers II, ACS Symp. Series 599, 1995, pp 450–497.
37. J Wang. In: GE Nelson, ed. Fire and Polymers II, ACS Symp Series 599, 1995, pp 518–535.
38. RE Lyon. Proc Int SAMPE Symp Exhib 1996; 1: 698–707.
39. ML Jenssesn. In: GE Nelson, ed. Fire and Polymers II, ACS Symp Series 599, pp 409–422, 1995.
40. MA Dietenberger. In: GE Nelson, ed. Fire and Polymers II, ACS Symp Series 599, pp 453–449, 1995.
41. PC Warren. in: WL Hawkins, ed. Polymer Stabilization. New York: Wiley, 1972, chap 7; EH Winson. In: WL Hawkins, ed. Polymer Stabilization. New York: Wiley, 1972; chap 3.
42. MM Hirschler. In: GE Nelson, ed. Fire and Polymers II, ACS Symp Series 599, 1995, pp 553–578.
43. E Moore. PP Handbook. Munich: Hanser 1996, p 189.
44. C Vasile, G Neamtu, V Dorneanu, V Gavat, M Medrihan. Roum Chem Q Rev 1997; 3(4): 243–265.
45. L Marcia. Recent Adv Flame Retard Polym Mater 1995; 6: 344-349.
46. D Lenoir, K Kampke-Thiel. In: GE Nelson, ed. Fire and Polymers II, ACS Symp Series 599, 1995, pp 377–392.
47. GE Nelson. In: GE Nelson, ed. Fire and Polymers II, ACS Symp Series 599, 1995, pp 578–592.
48. R Levene, M Lewin. The fluorescent whitening in textiles. In: M Lewin, SB Sello, eds. Handbook of Fiber Science and Technology, vol 1. New York: Marcel Dekker, 1984, p 257.
49. AP Hopmeier. Encycl Polym Sci Technol 1969; 10: 157; LM Greenstein, AE Petro. Encycl Polym Sci Technol 1969; 10: 193.
50. Y Nagao. Prog Org Coat 1997; 31(1-2): 43–49.
51. C Hardt, C Hahn, W Schafer. Kunststoffe Plast Europe 1996; 86(3): 364–365.
52. H Saechtling. International Plastics Handbook for the Technologist, Engineer and User, 2nd edn. Munich: Hanser, 1987, pp 40, 157, 401, 441, BG 24.
53. JA Manson, LH Sperling. Polymer Blends and Composites, New York: Plenum 1978, p 373.
54. HG Elias. Macromolecules. Basel: Huhig and Wepf, 19, 1990, p 802.
55. K Dinges. Kautsch Gummi, Kunstst 1979; 32: 748.
56. LE. Nelson, PE Chen. J Mater Sci 1968; 3: 38.
57. WC. Wake. Fillers for Plastics. London: Butterworth, 1971.
58. PD Ritchie. Plasticizers, Stabilizers and Fillers. London: Illife Book Ltd. 1977, p 253.
59. WV Titow, BJ Lanham. Reinforced Thermoplastics. London: Applied Science Publishers, 1975, p 9.
60. YuS Lipatov. Physical Chemistry of Filled Polymers. Shawbury: Rubber and Plastics Association of Great Britain, 1977.
61. L Mascia. The Role of Additives in Plastics. London: Edward Arnold, 1974, p 86.
62. YuS Lipatov. Interfacial Phenomena in Polymers. Kiev: Naukova Dumka, 1980, p 211.
63. HS Karz, JV Milewski. Handbook of Fillers for Plastics. New York: Van Nostrand Reinhold, 1987.
64. JV Milewski, HS Karz. Handbook of Reinforcements for Plastics. New York: Van Nostrand Reinhold, 1987.
65. VA Paharenko, V Zverlin, EM Kirienko. Filled Thermoplastics-Handbook. Kiev: Technica, 1986, pp 5, 74, 138.
66. SK Bhattacharya. Metal-Filled Polymers. New York: Marcel Dekker, 1986, p 144.
67. PJ Wright. Acicular wollastonite as filler for polyamides and polypropylene. In: A Whelan, JL Craft, eds. Development in Plastics Technology-3. New York: Elsevier, 1986, p 119.
68. DG Matles., Glass fibers. In: G Lubin, ed. Handbook of Fiber Glass and Advanced Plastics Composites. New York: Van Nostrand Reinhold, 1969, p 143.
69. J Weiss, C Bord. Les Materiaux Composites, vol 1. Paris: CEP Edition, 1983, B-3.
70. AR. Bunsell. Fibre Reinforcements for Composite Materials. Amsterdam: Elsevier, 1988.
71. Zh Vsesoiuz. Himiceskie Obscestva Im. DI Mendeleeva 1988; 34(5): 435–566.
72. DH Solomon, DG Hawthorne. Chemistry of Pigments and Fillers. New York: Wiley, 1983, pp 1–178.
73. AK Giri. J Appl Phys 1997; 81(3): 1348–1350.
74. A Savadori, M Scapin, R Walter. Macromol Symp 1996; 108: 183–202.
75. K Mitsushi. Angew Makromol Chem 1997; 248: 73–83.
76. RF Jones. Guide to Short Fiber Reinforced Plastics, Munich: Hanser, 1998, 5–14.
77. P Bataille, L Richard, S Sapieha. Polym Comp 1987; 10(2): 103–108.
78. VP Privalko, VV Novikov. The Science of Heterogeneous Polymers. Structure and Thermophysical Properties. New York: Wiley, 1995.
79. YuS Lipatov, VP Privalko. Visokomol Soed Ser 1984, 26: 257–260.

22

Polyolefin Polyblends

Rudolph D. Deanin, Margaret A. Manion, Cheng-Hsiang Chuang, and Krishnaraj N. Tejeswi
University of Massachusetts at Lowell, Lowell, Massachusetts

I. INTRODUCTION

The individual members of the polyolefin family offer a fairly broad spectrum of structures, properties, and applications. This spectrum can be broadened even further by blending individual polyolefins with other polymers. Furthermore, many other polymers can be improved by adding polyolefins to them. This is an area of major commercial importance, and one in which both theoretical research and practical development are currently very active.

A. Commercial Importance

Major commercial blends of polyolefins with each other include addition of low molecular weight polyethylene to ultra-high molecular weight polyethylene to improve processability, addition of low density polyethylene to linear low density polyethylene to improve processability, addition of ethylene–propylene–diene rubber (EPDM) to polypropylene to improve low temperature impact strength, and addition of PP to EPDM to produce thermoplastic elastomers. Major commercial blends of polyolefins and olefin copolymers with other polymers include nylon in high density polyethylene for impermeability, maleated EPDM in nylon and other engineering thermoplastics for impact strength, chlorinated polyethylene and ethylene–vinyl acetate in polyvinyl chloride as impact modifiers or plasticizers, and the growing use of CPE and maleated polyolefins as compatibilizers in other polymer blends. One intensive survey collected a detailed list which included 60 commercial polyolefin polyblends [1]. This is a demonstration of their large and rapidly growing practical significance.

B. Miscibility and Compatibility

Over four decades, polyblend scientists and engineers used the terms *miscibility and compatibility* loosely and/or interchangeably; many still do, and most of the literature is written in that way. In recent years scientists have begun to recognize a very important distinction between these terms. For absolute clarity and understanding, *thermodynamic miscibility* describes polymer blends which are completely miscible and homogeneous down to the molecular level, and do not show any phase separation at all. In contrast, practical compatibility describes polymer blends which have useful practical properties in commercial practice. It should be emphasized that most of the commercially useful polyblends have practical compatibility even though they do not have thermodynamic miscibility; in fact, they usually form multi-phase morphologies which produce a synergistic advantage in balance of properties, not available from any single polymer.

1. Thermodynamic Miscibility

For two polymers to be completely miscible down to the molecular level, the mixing process must produce a decrease in free energy ΔG [2].

$$\Delta G = \Delta H - T\Delta S$$

Enthalpy ΔH depends on the attraction/repulsion between the two polymers; usually, unlike molecules repel each other, so ΔH is generally positive (unfavorable to mixing). Entropy ΔS results from the randomization which occurs on mixing; small solvent molecules produce large randomization, so most solvents are miscible, whereas large polymer molecules produce very modest randomization on mixing, not enough to overcome the repulsion between unlike molecules ($+\Delta H$). Thus most polymer blends do not have thermodynamic miscibility, so they separate into two or more micro-phases.

Beyond this general principle, there are several factors which may favor thermodynamic miscibility in specific polymer blends. These may be discussed with particular reference to polyolefins.

(a) Polarity. If two polymers have very similar polarities, this reduces the repulsion (ΔH) between them, and permits randomization (ΔS) to favor thermodynamic miscibility [5]. Since polyolefins are aliphatic hydrocarbons of similar polarity [3, 4], this tends to favor thermodynamic miscibility in pairs of polyolefins [5, 6]. Similarly, copolymerization of olefins with more polar comonomers increases polarity, occasionally producing thermodynamic miscibility with more polar polymers [7]. Several examples will be noted in the course of this review.

(b) Molecular Weight. Since randomization on mixing (ΔS) is inverse to molecular weight, it would be expected that lower molecular weight polymers would be more thermodynamically miscible with each other, and therefore tolerate somewhat wider differences in polarity [2, 5, 6].

(c) Specific Group Attraction. Instead of considering the overall polarities of the polymers in a blend, some researchers focus attention on the possibility of attractive forces between specific functional groups in the polymer molecules, which may reduce repulsion or even favor direct attraction (ΔH) between them [7, 8, 44, 45]. In simple polyolefins with only London dispersion forces between them, the degree of attraction may be questionable; but in olefin copolymers, such specific group attractions may provide critical improvement, particularly for practical compatibilization as noted below.

(d) Co-Crystallization. Crystallizable polymers may be thermodynamically miscible, and be truly miscible in the melt; but on cooling, each polymer will separate and form its own unique crystal structure. Occasionally two polymers have such similar isomorphous crystal structures that they can both enter the same crystal lattice and co-crystallize, thus forming a single homogeneous solid product [9]. This has important consequences in several polyolefin polyblends, as will be noted later in this review.

2. Compatibilization

When a polymer blend is thermodynamically immiscible, separates into two or more phases, and gives poor practical properties, it is generally assumed that either (a) the particle size of the dispersed domains is not optimum, or else (b) the immiscibility of the two phases produces weak interfaces between them, which fail too easily under stress. The most popular way of trying to solve these problems is the addition of a third ingredient, or reaction during the blending process to form a third ingredient. If this succeeds, the third ingredient is then called a *compatibilizer*. In some cases the compatibilizer might actually produce complete thermodynamic miscibility. In most cases, it simply acts as a surfactant to reduce the domain size of the dispersed phase, or acts as an interfacial adhesive between the two phases. If it forms an interfacial layer of significant thickness, this is referred to as a transitional *interphase*. Success is judged by improvement in practical properties, producing practical compatibility. Practical compatibilization is most often accomplished by one of the following techniques [10, 11].

(a) Additive Compatibilizers. Block and graft copolymers, whose segments resemble the two polymer phases in structure or polarity, may orient at the interface and thus bond the two phases together more strongly. In basic research, these are best synthesized beforehand, and then added during the blending process. Theoretically, a monomolecular layer at the interface should be sufficient; and experimentally, only a few percent of the additive are needed. Practically, it is often more economical to form the compatibilizer by reactive processing during the blending process, as discussed below (b).

Specific group attractions between the two polymers may overcome the usual enthalpic repulsion. A few groups can provide stronger bonding at the interface. A larger number of groups can create a transitional interphase. Still larger numbers of groups can produce more or less complete miscibility, approaching homogeneous one-phase systems. Such groups can be built into polyolefins during polymerization by copolymerization with more polar monomers, or they can be added after polymerization by grafting or other post-polymerization reactions. Here again, in commercial practice it may be more economical to add them by reactive processing during the blending process.

Mutual solvents can enhance the formation of a transitional interphase. When two polymers are partially miscible, they may form such a transitional interphase directly, producing practical compatibility. Even when they are not sufficiently miscible by themselves, it may be possible to add a third ingredient which is miscible with both of them, and thus can act as a mutual solvent to produce such a transitional interphase and practical compatibility. Such a mutual solvent is usually a polymer or plasticizer which is miscible, or at least compatible, with both phases. It must be used in fairly large amount, so it will also act as a third major ingredient and contribute its own properties to the blend.

Low-modulus elastomers may be added to compatibilize multi-phase polyblends whose weak interfaces suffer from brittle failure. The elastomer forms a soft rubbery interphase which cushions mechanical or thermal stress and thus reduces brittleness.

Reinforcing fibers are sometimes added to incompatible polyblends to increase their resistance to mechanical stress. The fibers are much longer than the dimensions of the incompatible micro-phases, and bridge across the weak interfaces between them, producing much higher strength.

Interpenetrating polymer networks can inter-disperse two immiscible polymers down to a fine scale of phase separation, and usually use crosslinking to stabilize this morphology [13]. This can produce a remarkable synergism of properties.

(b) Reactive Processing. Block and graft copolymers, and specific group attractions, can be formed during polymerization, generally by copolymerizing olefins with functional comonomers; or they can be formed by post-polymerization reactions before the blending process. While this offers precision in basic research, it often adds considerably to the cost in commercial practice. Thus there is growing interest in reactive processing, to produce these compatibilizing structures directly during the polymer blending process. Leading types of reactive processes are the following.

Free-radical reaction between polyolefins can be produced by peroxide, high energy radiation, and thermal and/or mechanical shear [12]. When radicals of the two polymers recombine with each other, this immediately produces block or graft copolymers.

Carboxylic acid groups are most often introduced by grafting maleic anhydride onto polyolefins, before or during the blending process. They may also be introduced by copolymerization or grafting of acrylic or methacrylic acid. For reactive processing during blending, they react readily with amine groups of nylons, with hydroxyl groups of polyesters, or with epoxy or oxazoline groups of other polymers.

Epoxy groups are introduced by grafting glycidyl methacrylate onto polyolefins. They react readily with carboxylic acid, amine, and other active hydrogen groups on other polymers.

Oxazoline groups are introduced by copolymerization or grafting of vinyl oxazoline comonomers. They react readily with carboxylic acid groups on other polymers.

Most of these compatibilization techniques have been applied to polyolefins and olefin copolymers, and will be illustrated throughout this review.

C. Polyolefin Blends in General

Before turning to polyblends of specific polyolefins, it is worth noting several surveys which cover polyblends of the polyolefin family as a whole.

1. Properties

Those which polyolefins generally contribute to polyblends include high melt strength and elasticity, viscosity, and shear sensitivity; low polarity, dielectric constant, and loss; and water repellency [14]. Polyolefin toughness and processability are increased by blending with thermoplastic elastomers such as ethylene–propylene, and styrene–ethylene–butylene–styrene [14]. Polyolefin–polystyrene blends permit easy fibrillation; polyolefin–polyethylene terephthalate blends give self-texturing fabrics [14]. Barrier properties of polyolefins are often improved by combination with ethylene–vinyl alcohol, polyvinylidene chloride, and polyamides [15], either by blending or by laminating; blends of polyolefins with nylons or polycarbonate permit balanced control of permeability and water absorption [14]. Adding 2–4% of polyolefins to engineering resins often improves their processability and impact resistance [16].

2. Compatibilizers

When used for blending polyolefins with other polymers, these include many examples of the general principles listed earlier [10, 17].

By contrast, most studies do specify individual polyolefins, so the remainder of this review is best organized under each of these specific polyolefins.

II. POLYETHYLENE POLYBLENDS

A. Polyethylenes in General

Many reviews and some research reports refer to polyethylenes in general. These will be reviewed first, before turning to individual types of polyethylene.

1. Blends with Other Polymers in General

Addition of polyethylene powder to other polymers has been recommended to increase surface lubricity and abrasion resistance [18].

2. Blends with Ethylene Copolymers

Addition of ethylene copolymers to polyethylenes has been used to improve toughness, impact resistance, and chemical resistance in films and other forms [19]. Blends of polyethylene with ethylene–vinyl acetate formed two continuous phases, which could then be stabilized by crosslinking [20]. Addition of chlorinated polyethylene to polyethylene is helpful in reducing flammability [18].

3. Blends with Polypropylene

These are miscible in the melt [5] but crystallize separately on cooling, nucleating each other as they do so [21]. They may be compatibilized by addition of ethylene–propylene block copolymers [19], or by attaching acid groups to one polymer and basic groups to the other to strengthen the interface between them and thus retain their ductility [22].

4. Thermoplastic Elastomers

These are sometimes improved by adding polyethylene for thermoplastic processability and strength. In ethylene–propylene–diene rubber (EPDM), adding PE produces progressive increase in modulus [23]. Blends of ethylene–ethyl acrylate with PE are used in some commercial thermoplastic elastomers [24]. Allied Chemical ET is a blend of PE with butyl rubber [25, 26]. "Thermoplastic natural rubber" (TPNR) is made by blending NR with PE or PP, then peroxide-crosslinking the NR to improve moldability and resistance to compression set [24]. PE or EVA was blended with vinyl silicone rubber, forming 1–5 μm silicone domains, which were then crosslinked and grafted by mechanical shear and/or 180°C melt processing [24, 25].

5. Blends with Polystyrene

They are immiscible [27], and sometimes take advantage of their extreme immiscibility: extrusion at high shear rates produces fibrils [28], and extrusion and stretching of films produces synthetic paper which is opaque, tough, and waterproof [29, 30]. Research to compatibilize PE + PS blends has most often used styrene–ethylene–butylene–styrene (SEBS) block copolymers to produce adhesive bonding at the interface [10, 19, 31]. In other studies, triallyl isocyanurate + dicumyl peroxide was used to improve tensile properties [32], and PE–acrylic acid + styrene–oxazoline polymers were coreacted to form amido-ester crosslinks [33].

Blends of PE with styrene–maleic anhydride copolymer were coarse and weak. When *t*-*N*-butyl aminoethyl methacrylate was grafted onto the PE, it reacted with the maleic anhydride to form amide block copolymer at the interface, giving finer morphology, and higher tensile and impact strengths [179].

6. Engineering Thermoplastics

These have often been modified by addition of PE. Blends with polyoxymethylene showed complex anomalous behavior due to interfacial phenomena [34]. PE appeared useful as a melt flow promoter in polyphenylene ether [19]. Dispersion of PE in polycarbonate improves melt flow and energy absorption for automotive applications [19], so PC producers offer such grades commercially [35]; fine stable polyethylene domains may be produced by adding PE-PS or SEBS block copolymers [36, 37].

Blends of PE with polyethylene terephthalate were compatibilized by addition of SEBS, better by maleated SEBS, forming an intricate multi-domain morphology and giving synergistic increase in ultimate elongation and toughness [19, 180]; this may be important in recycling. Blends of PE with polybutylene terephthalate were compatibilized by transesterifying PBT with EVA and then adding this to the blend; this reduced particle size and increased interfacial adhesion [181]. Blends of PE with polycaprolactone appeared miscible, giving a single T_g [38].

Blends of PE with nylons are immiscible and incompatible [27], and are generally compatibilized by grafting maleic anhydride onto the polyethylene, or by copolymerizing acrylic or methacrylic acid into the polyethylene [10, 39]. This certainly produces hydrogen bonding, and probably amidification or transamidification to graft copolymers, as indicated by rheology and morphology [182]. Such blends can be used either to toughen the nylon or to increase the organic impermeability of the PE.

7. Cross-Linked Polyethylene

This has been recycled by chopping it finely and blending with virgin PE to produce thermoplastic compositions. This would appear to be analogous to "dynamically vulcanized" thermoplastic elastomers, as described later in this review.

B. Density Blends

Polyethylenes of different densities are generally miscible in the melt [40, 41], but careful mixing is required to cope with differences in melting point and melt viscosity [42]. On cooling, blends of ultra-high molecular weight polyethylene with high density polyethylene or linear low density polyethylene are isomorphous and cocrystallize to a single homogeneous product; this is useful to improve processability of UHMWPE. Similarly, blends of HDPE + LLDPE are isomorphous and cocrystallize to a single homogeneous product [9, 40, 41, 46–51]. Miscibility was greatest when the comonomer in LLDPE was heterogeneously distributed rather than homogeneously [183]; and it was best when blends were prepared by extrusion rather than by milling or solution blending, producing highest crystallinity, melting point, strength, and elongation [184]. Adding HDPE to LLDPE was useful for improving processability and stiffness [52].

Blends of HDPE + low density polyethylene were immiscible even in the molten state [43], and formed separate crystal structures with generally poorer properties [27, 41, 42, 46–48, 50, 53–55]. On the other hand, one study observed that they formed mixed crystals with intermediate melting point [185]. Some studies reported that practical properties were proportional to blend ratio [56–60]; and a few even showed synergistic improvement of properties such as film drawdown and stiffness [56, 57, 61, 62].

LLDPE + LDPE form different and separate crystal structures [27]. Nevertheless most LLDPE in commercial use is blended with up to 20% of LDPE to improve film processing, impact strength, and clarity [52, 61].

Medium density polyethylene is fairly similar to and miscible with LDPE [53]; and ternary blends of HDPE + MDPE + LDPE showed some interesting synergism [60].

C. High Density Polyethylene + Other Polyolefins

1. Blends with Polypropylene

During processing of these blends, heat, electron beam, or dicumyl peroxide crosslink the PE but degrade the PP; thus strength and heat resistance depend on the PE content [63, 64]. Blends of PE and PP were immiscible in either the amorphous or the crystalline phase [65]. The two polymers tend to form mixtures of crystal structures, and each affects the crystallization of the other [66–70]. Studies on mechanical properties gave mixed results. Improvements in modulus, ultimate tensile strength, and heat deflection temperature [71] suggested good binding in the amorphous interphase [72]; but use of HDPE to improve low temperature impact strength and environmental stress–crack resistance required a compatibilizer such as 5% of ethylene–propylene rubber [42].

2. Blends with Ethylene–Propylene Rubber

Toughness of HDPE films was improved by adding EPR; they were not miscible, but formed micro-size morphologies [186]. PE blocks in the rubber were miscible with HDPE [73] and may even have cocrystallized with it [74], while the rest of the rubber remained as a separate phase. This gave homogeneous tensile modulus and yield properties, S-shaped ultimate tensile strength and low temperature modulus curves, and greatly improved impact strength [75]. One market estimate suggested 290,000 tons of HDPE are impact-modified in this way [19].

3. Blends with Other Ethylene Copolymers

One study found that blends of HDPE with ethylene copolymers were partially miscible in both the amorphous and cocrystallized phases [187]. Addition of styrene–ethylene–butylene–styrene block copolymer to HDPE improved its flexibility and impact strength [76]. Blends with ethylene–vinyl acetate copolymers were immiscible [27].

4. Blends with Polybutylenes

Toughness of HDPE was improved by adding polyisobutylene; they were not miscible, but formed micro-size morphologies [186]. In blends of HDPE with butyl rubber, plots of properties vs blend ratio were mostly S-shaped, indicating useful two-phase systems [77]. Melt viscosity and ultimate elongation exhibited unusual bimodal peaks at high and low HDPE/butyl ratios, where the morphology was fibrillar or laminar, whereas fairly equal ratios were simply particulate [188].

Blends of HDPE with poly-1-butene crystallized separately; small amounts of HDPE nucleated and changed the crystallinity of the P-1-B [78].

5. Blends with Ionomers

In one study, addition of commercial ethylene ionomers to HDPE increased melt flow and impact strength, without loss of tensile strength or heat deflection temperature; these latter gave S-shaped curves indicating useful two-phase systems [79]. Blends with another commercial ionomer gave moduli and heat deflection temperatures which were linear functions of polyblend ratio, indicating good practical compatibility; addition of HDPE to ionomer improved impact strength [189].

Addition of sulfonated EPDM, as the zinc salt, plus zinc stearate, to HDPE produced thermoplastic interpenetrating polymer networks with continuous phases, showing synergistic increase of tensile strength, elongation, and impact strength [80].

D. High Density Polyethylene with Other Polymers

1. Useful Blends

(a) Nylons. Polyamides are immiscible with HDPE, and form dispersed domains whose size increases with concentration [27, 81]. Addition of polyethylene containing carboxyl or anhydride groups produces polyethylene–polyamide graft copolymers which act as surfactants or interfacial adhesives, reducing domain size [81, 82]. In extrusion of film and sheet, and in blow molding, the polyamide domains form broad thin lamellae parallel to the surface. This retains the physical properties of HDPE and combines the different impermeabilities of HDPE and polyamide. A major commercial example is the DuPont Selar series, containing 3–20% of nylon 6-66 copolymer [16, 82, 83].

(b) Polyethylene Terephthalate. Soft drink bottles are primarily PET, but many of them are glued to a HDPE base, which may be difficult to remove. Thus recycling often produces contaminated batches of HDPE + PET, with inferior properties [84]. Addition of 5–20% of styrene–ethylene–butylene–styrene block copolymer softened modulus and tensile yield strength, but increased ultimate elongation and impact strength, probably by forming an interfacial adhesive layer.

(c) Polyvinyl Alcohol. HDPE blends with PVOH were compatibilized by adding ethylene ionomers, which complexed with PVOH at the interface, giving good tensile and barrier properties [190].

(d) Butadiene–Acrylonitrile Rubber. HDPE + NBR were compatibilized by adding either maleated polyethylene or phenolic-resin-modified polyethylene. This gave finer morphology and higher tensile strength [191].

(e) Rubber Recycling. Adding ground scrap vulcanized rubber to HDPE increased its flexibility. Adding HDPE to ground scrap vulcanized rubber increased its tensile strength [86].

(f) Polycarbonate. Adding 5% of HDPE to PC increased its notched Izod impact strength tenfold [15, 87]. HDPE + PC were compatibilized by adding phenoxy resin or ethylene–methacrylic acid ionomer. They reacted with the polycarbonate to produce graft copolymers [192].

(g) Lignin. Lignin is a solid alkyl phenol polymer which is a by-product of the papermaking industry. When it was added to HDPE, it acted as an inert filler, without causing the stiffening commonly seen with calcium carbonate or carbon black [88].

(h) Mixed Automotive Plastics. Recycling of junked autos produces a mixture of plastics which is not economically separable. Mastication and high pressure molding produced sufficient compatibilization for useful properties, but melt flow was too low for conventional processing. Addition of 5–20% of very low molecular weight HDPE increased melt processability into the range of commercial processes [85].

2. Other Interesting Blends

Polystyrene was immiscible with HDPE, forming coarse domains. Addition of 9% styrene–ethylene–butylene–styrene block copolymer gave much finer domain structure [19]. This compatibilizing effect was also observed in PS + polyphenylene ether blends with HDPE, where the styrene blocks of SEBS were attracted to the PS and PPE, improving interfacial adhesion and changing crazing dilation into shear yielding [89].

Polyvinyl chloride blends with HDPE had coarse domains and brittle properties. Slurry chlorination of HDPE attacked only the amorphous phase, producing a block copolymer, which then compatibilized the PVC + HDPE blends, reducing domain size and increasing ultimate elongation [31].

When polydimethyl siloxane was grafted onto PE and this graft copolymer was added to HDPE, the silicone chains came to the surface in islands, lowered surface tension, and increased the wetting angle [90].

E. Linear Low Density Polyethylene + Other Polyolefins

1. Different Linear Low Density Polyethylenes

Individual LLDPE manufacturers use different α-olefins to reduce the regularity and crystallinity of linear polyethylene. Even though their products have similar densities and melt flow indices, they are not identical, and not necessarily miscible with each other. LLDPEs made from 1-butene and 1-hexene were miscible at low molecular weight but not at high molecular weight [42].

2. Ultra High Molecular Weight Polyethylene

This was added up to 6% into LLDPE. On stretching, the blends showed strain hardening, which is useful in producing blown film [91].

3. High Density Polyethylene

This may be added (5–20%) to LLDPE to improve film processability and stiffness, and 5–50% to increase molded stiffness [52].

4. Low Density Polyethylene

Most commercial LLDPE is blended, primarily with LDPE [52]; this is the most important polyolefin blend in terms of total tonnage [42]. Addition of up to 20% LDPE [52] increases melt strength in blown film [92] and improves extensional melt flow for film blowing and wire coating [42]. Melt index and activation energy are below proportional prediction for the blends [92]. Overall higher production rates produce significant cost savings [52].

The two polymers are not miscible; they crystallize separately and show their individual melting points [41, 42, 92]. Total crystallinity is below the proportional prediction [92]. LLDPE gives the higher crystallinity and haze [92]; increasing LDPE improves clarity [52]. LLDPE contributes most to physical properties [93], and film strength is above either of the pure polymers [92].

5. Polypropylene

Compatibility with LLDPE depended on matching viscosities; when this was done, blends were more homogeneous and had higher tensile modulus and strength [193, 194]. Compatibility was also improved by adding ethylene–propylene or styrene–ethylene–butylene–styrene block copolymers; this gave faster and finer crystallization [195].

6. Elastomers

Addition of a series of elastomers to LLDPE gave linear or S-shaped change in physical properties. Ethylene–propylene rubber, butyl rubber, and especially SEBS all increased flexibility, tear and impact strengths, and low temperature flexibility [94, 95].

F. Linear Low Density Polyethylene + Other Polymers

1. Natural Rubber

LLDPE + NR was compatibilized by adding ethylene–b-isoprene diblock copolymer, as judged by mechanical properties [196].

2. Rubber Recycling

Adding scrap vulcanized rubber to LLDPE increased its flexibility and ultimate elongation [86].

3. Styrene Block Copolymers

Styrene–butadiene–styrene and styrene–isoprene–styrene block copolymers were added to LLDPE, and increased flexibility, tear and impact strengths, and low temperature flexibility [94, 95]. When SBS and SEBS block copolymers were blended with LLDPE in ratios from 100/0 to 0/100, hardness and modulus generally gave S-shaped curves indicating two-phase behavior; while low-temperature torsional modulus, tensile yield strength, ultimate tensile strength, ultimate elongation, and melt index gave smooth curves or even straight lines, indicating homogeneous behavior [197]. Medium molecular weight SBS gave the most homogeneous behavior. In practice, addition of these elastomers to LLDPE increased ultimate tensile strength and elongation somewhat, and increased flexibility, especially at low temperature.

4. Polyvinyl Alcohol

Blends with LLDPE were compatibilized by ethylene ionomers, which complexed with PVOH at the interface, giving good tensile and barrier properties [190].

5. Polycarbonate + LLDPE

These formed two-phase systems, and domain size increased with PC content. Stretch orientation improved mechanical properties in some cases [96].

6. Liquid Crystal Polyester

Addition of a semiflexible LCP to LLDPE decreased melt viscosity and nucleated crystallization. This increased modulus of the LLDPE [198].

7. Mixed Automotive Plastics

Recycling of junked autos produces a mixture of plastics which is not economically separable. Mastication and high pressure molding produced sufficient compatibilization for useful properties, but melt flow was too low for conventional processing. Addition of 5–20% of very low molecular weight LLDPE increased melt processability into the range of commercial processes [85].

G. Low Density Polyethylene + Other Polyolefins

1. Higher Polyolefins

(a) Polypropylene. In some studies, LDPE + PP mechanical properties followed the rule of mixtures [199]. In other studies, blends were immiscible both in the melt and in the solid state; LDPE decreased spherulite size and ultimate tensile strength of PP [97, 98]. Adding ethylene/propylene 60/40 random copolymer or EPDM provided solid phase dispersant, which decreased melt flow, modulus, and tensile strength, but increased ultimate elongation, impact strength, and dynamic mechanical behavior [99, 100]. Adding ethylene/propylene 8/92 block copolymer gave better tensile strength but less improvement of impact strength. Peroxide compatibilization failed because it degraded the PP.

(b) Poly-1-Butene. The incompatibility of LDPE with P-1-B was used to make easy-opening heat-seal adhesives for packaging. Coextrusion of the two polymers provided the heat-seal adhesive. Under tension, the seal peeled easily because the two incompatible polymers permitted easy cohesive failure. Optimum P-1-B/LDPE ratios were either 92-75/8-25 or 25-8/75-92. The intermediate range was too weak, the pure polymers too strong [101].

2. Elastomeric Tougheners

(a) Ethylene–Propylene Rubber. Five percent of commercial LDPE is toughened by addition of elastomers [19], primarily EPR copolymers. High ethylene/propylene ratios appeared homogeneous in most tests, but miscibility was inverse to propylene content [102]. Plots of modulus, yield, and low-temperature stiffening vs LDPE/EPR ratio were homogeneous, while ultimate tensile strength curves were S-shaped and ultimate elongation actually synergistic [75].

(b) Butyl Rubber. Plots of properties vs LDPE/butyl ratio indicated partial miscibility forming multiphase systems [77]. Tensile yield elongation was L-shaped; tensile modulus, yield strength, and low temperature stiffening were all S-shaped; and ultimate elongation, generally one of the most sensitive properties, was actually U-shaped.

H. Low Density Polyethylene + Ethylene Copolymers

1. Ethylene–Vinyl Acetate

Melt processability of LDPE was improved by blending with EVA copolymers containing 10–28% VA. Softening of modulus was controlled primarily by overall VA content of the blends. Tensile necking was converted to smooth rubbery elongation by adding larger amounts of VA. Transparency was produced by adding EVA containing 10% VA and then stretching the polyblend film [200]. At 28% VA, blends formed fine-structured interpenetrating polymer networks in which T_g suggested miscibility, and tensile strength peaked synergistically at 50/50 ratio in the blends [201].

2. Ethylene/Ionomers

LDPE and ethylene–acrylic acid copolymers containing 4.6 and 6.5% AA were immiscible in thermodynamic analyses [203]. Adding a commercial ethylene ionomer to LDPE increased melt flow, tensile modulus and strength, and especially ultimate elongation [204a]. In another study, modulus and heat deflection temperature were linear functions of polyblend ratio, indicating good practical compatibility; and adding LDPE increased the impact strength of the ionomer [189].

I. Low Density Polyethylene + Other Polymers

1. Natural Rubber

This and LDPE were immiscible [103–105], forming thermoplastic elastomers with two continuous phases or IPNs. These were improved by adding compatibilizers such as epoxidized NR + maleated LDPE or sulfonated ethylene–propylene-diene rubber (EPDM) + maleated LDPE. These increased tensile strength, elongation, and adhesive peel strength.

2. Rubber Recycling

When done by adding LDPE to vulcanized scrap, this increased tensile strength. Adding vulcanized scrap to LDPE increased its flexibility [86].

3. Styrene–Butadiene–Styrene

This block copolymer increased flexibility, ultimate tensile strength and elongation, and low temperature flexibility of LDPE [76].

4. Polystyrene

This is very immiscible with LDPE, lowering crystallinity and most properties [50, 106]. Compatibilization by grafting or by addition of styrene–ethylene–butylene–styrene block copolymer gave finer phase dispersion and improved tensile strength and elongation [19, 31].

5. Polyvinyl Chloride

These blends with LDPE were weak and brittle. Adding chlorinated polyethylene, or PE grafted with 26% methyl methacrylate, greatly reduced domain size and increased ultimate elongation [31, 107]. Adding hydrogenated polybutadiene–b–styrene–acrylonitrile–b-polybutadiene reduced domain size, increased interfacial adhesion, and improved mechanical properties [204b].

6. Nylon 6

These blends with LDPE are sometimes reported as incompatible [110], at other times praised for their benefits. For example, addition of 10% of LDPE more than tripled impact strength [205]; and LDPE has been reported to improve ultimate tensile strength and elongation, impact strength, water resistance, and dimensional stability [108, 109]. In one study, blends were compatibilized by adding a terpolymer of ethylene/vinyl alcohol/vinyl mercaptoacetate, which reduced domain size and improved mechanical properties [202].

7. Silicone Rubber

This provided flame retardance in LDPE and was synergized further by lead, titanium, vanadium, and tin compounds [111].

III. ETHYLENE COPOLYMER POLYBLENDS

A. Ethylene–Propylene Rubber Polyblends

EPR random copolymers, and EPDM diene terpolymers for higher reactivity, have been widely used in polymer blends. Their use as the major ingredient, and as additives to nonpolyolefins, will be discussed here. Their use as additives in other polyolefin systems can be found throughout this entire survey.

1. Thermoplastic Elastomers

EPR random copolymers are inherently thermoplastic and rubbery in nature; but to give them useful levels of strength, creep resistance, heat resistance, and solvent resistance, they are commonly blended, or block or graft copolymerized, with crystalline polyolefins, particularly polypropylene, which on cooling separate as crystalline domains and act as "thermoplastic crosslinks and reinforcing fillers."

(a) Linear EPR + HDPE or LDPE. These gave property vs EPR/PE ratio curves which were homogeneous, S-shaped, or synergistic, depending on the sensitivity of the test [75, 102]. EPR + ethylene ionomer gave synergistic increase of tensile yield elongation [206]. Sulfonated EPDM ionomer blended with HDPE or PP to give thermoplastic IPNs with co-continuous phases, which produced synergistic improvement of tensile and impact properties [80]

Most thermoplastic polyolefin elastomers are blends of PP with EPR or EPDM [10, 18, 24, 112-114]. The blends are two-phase systems whose domain size depends on ethylene/propylene ratio, molecular weight, and process conditions [2, 5, 115, 116]. Various properties vs EPR/PP ratio may be homogeneous, L-shaped, or S-shaped [75, 117].

(b) Dynamic Vulcanization. This is produced by crosslinking the elastomer particles of thermoplastic elastomers during melt mixing [24, 25, 83, 118–121]. The continuous PP matrix still gives thermoplastic processability, while vulcanization of the elastomer particles provides stability of the morphology and improves melt strength, mechanical properties, permanent set, fatigue, hot strength, and hot oil resistance.

2. Ethylene–Propylene Rubber + Other Polymers

(a) Blends with Diene Rubbers. Natural rubber ozone resistance was improved by adding 35–40% of EPDM [122]. The best results obtained using dicyclopentadiene as termonomer, and with finer dispersion of the EPDM domains.

Polybutadiene blends' morphology depended on the relative viscosity of the two elastomers [122]. The lower viscosity polymer formed the continuous matrix. At equal viscosity, and at higher viscosity, domain size became smaller.

Styrene–butadiene general purpose rubber (SBR) ozone resistance was improved by adding EPDM [108]. The blend was improved by grafting [31], and gave a paintable rubber for automotive use [123].

(b) Polystyrene. Its impact strength was improved by adding 5–25% of EPDM, even better by use of 1% $AlCl_3$ to produce grafting and crosslinking [124]. PS + polyphenylene ether blends with sulfonated EPDM ionomer were compatibilized by adding sulfonated PS [10].

(c) Polyesters. For example, polycarbonate, polyethylene terephthalate, and polyhbutylene terephthalate have been impact-modified (toughened) by addition of EPDM [10, 15, 83], generally by carboxylating the EPDM or even grafting styrene-acrylonitrile copolymer onto it.

(d) Thermoplastic Polyurethane Elastomer. This was improved by reaction with EPR which had previously been grafted with maleic anhydride [17].

(e) Polyamide. Its impact strength is often increased greatly by blending with polyolefins [10, 19, 35, 55, 125], probably mainly with EPDM grafted with maleic anhydride, although a number of other ethylene copolymers, grafts, and ionomers are often mentioned in the literature.

B. Ethylene–Vinyl Acetate Copolymer Polyblends

The full range of EVA copolymers provides a wide range from crystalline to amorphous and from nonpolar to polar, so it is not surprising that they have proved interesting and useful in blends with a variety of other polymers.

1. Blends with Polyolefins

(a) Paraffin Wax. This has been flexibilized and strengthened by blending with EVA [126].

(b) Crosslinkable Polyethylene. Together with EVA at polyblend ratios near 50/50, it has formed two continuous phases; crosslinking between them minimized phase separation [20].

(c) LLDPE. Its heat-sealability, toughness, and environmental stress–crack resistance were improved by adding 5–25% of EVA [52].

(d) LDPE. These blends formed fine-phase interpenetrating polymer networks with a single T_g proportional to blend ratio [201], and showed synergistic increase in melt flow and tensile strength [200]. They also showed improvements in puncture resistance, toughness, low temperature flexibility, and environmental stress–crack resistance [61]. Increasing VA content brought softer modulus, uniform stretching instead of necking, and increasing transparency [200].

(e) Polyethylene Ionomer. This formed two-phase systems and gave S-shaped curves for most properties vs polyblend ratio [206]. EVA tended to increase rebound and low temperature flexibility, suggesting that the EVA phase tended to dominate these properties.

(f) EPDM. This was miscible with high ethylene/vinyl acetate ratio copolymers in most tests, but DSC was very sensitive and showed phase separation [102].

(g) Polypropylene. Its blend morphology showed complex dependence on composition [127]. Modulus and yield strength were fairly proportional to blend ratio, while ultimate strength and impact strength gave S-shaped curves typical of two-phase systems [207]. At best, balance of properties was most useful in wire and cable, footwear, and automotive applications.

(h) Poly-1-Butylene. These blends with EVA could be coextruded onto packaging to form heat-seal adhesives; phase separation made it easy to open the seal later by cohesive failure [101].

2. Polystyrene

This and EVA formed two-phase systems [128]; increasing VA content improved interfacial adhesion and impact strength. Styrene–maleic anhydride copolymer was blended with EVA by grafting glycidyl methacrylate onto the EVA in reactive processing [129]. Methyl methacrylate–acrylonitrile–butadiene–styrene (MABS) graft copolymers were incompatible with EVA, showing severe losses of mechanical properties; increasing VA content of the EVA narrowed the difference in polarity, but was not sufficient to produce compatibility [128].

3. Chlorinated Rubber

It had limited miscibility with EVA [126].

4. Polyvinyl Chloride

This and EVA form a series of interesting and useful polyblends. Thermodynamic miscibility increases with increasing VA content and polarity, or by incorporating CO or SO_2 into the EVA [5, 130–132]. At low to medium VA content, the copolymers are rather immiscible with PVC, but the microphases are strongly bonded to each other, and the EVA domains provide high impact strength in the rigid PVC matrix [108, 133–135]; grafting PVC to EVA may also help [30]. At high VA content, or with CO or SO_2 termonomers, miscibility increases, and the EVA becomes a polymeric plasticizer [136], sometimes even forming the continuous matrix phase [134]. These blends also showed promising UV stability [137].

5. Polybutylene Terephthalate

Its compatibility with EVA was improved by maleating the EVA, producing interfacial adhesion and impact strength [208].

6. Other Polymers

Where miscibility with EVA has been observed these include polyvinyl nitrate and cellulose acetate butyrate [132].

C. Chlorinated Polyethylene

Chlorination of polyethylene can be run to produce a wide range of chlorine contents with increasing polarity, and a broad spectrum of random or block copolymers with high to zero crystallinity. These have demonstrated a surprising range of useful compatibility and even miscibility with other polymers.

1. Polyethylene

Its flammability was reduced by blending with CPE, reducing the need for additional flame retardants [18].

2. Epoxidized Natural Rubber

When containing 25–50 mol% epoxy groups it was miscible with amorphous CPE containing 48% Cl [138]. Low chlorine block copolymer was still crystalline and had limited miscibility.

3. Polyvinyl Chloride

The miscibility with CPE and useful properties cover a wide range [19, 114]. Medium chlorine content produces rigid vinyls of high impact strength [139–141], and larger amounts of CPE act as polymeric plasticizers. Low chlorine content is too crystalline, while high chlorine has too much steric/polar stiffening.

4. Polymethyl Methacrylate

The miscibility with CPE has been reported as a function of chlorine content and temperature [19, 38, 46, 55].

5. Nylon 6

The impact strength was increased up to four-fold by addition of CPE containing 36–42% Cl, with good retention of modulus and strength [109, 205].

D. Acid Copolymers and Ionomers

Polyolefins containing carboxylic acid groups, sometimes neutralized to form ionomers, form much stronger intermolecular hydrogen bonding and ionic attractions than simple polyolefins, and can thus contribute greatly to practical compatibility or even molecular miscibility of polyblends, particularly blends with more polar polymers. Occasionally sulfonated polyolefins offer similar benefits. "Carboxylation" of polyolefins has been noted occasionally throughout this survey. In the current section the emphasis is on carboxylic and sulfonic acid copolymers and their ionomers.

1. Low Density Polyethylene

Blends with ethylene/acrylic acid copolymers containing 4.6–6.5% AA were immiscible in thermodynamic analyses [203]. Adding ethylene ionomer to LDPE increased melt flow, tensile modulus, strength, and especially elongation [204].

2. Ethylene–Propylene Rubber

These blends with ionomer were two-phase systems which showed synergistic increase of tensile yield strength; and ionomer increased the rebound of EPR [206].

3. Styrene–Ethylene–Butylene–Styrene

This block copolymer melt flow was improved by adding ethylene ionomer [79]. T_g vs blend ratio was S-shaped, indicating a two-phase system; and tensile modulus and yield showed evidence of synergism [206].

4. Reclaimed Rubber

This was ground and functionalized with amine-reactive liquid rubber, to improve its compatibility with carboxylated LLDPE [142].

5. EPDM and Butyl Rubber

These were both sulfonated and neutralized to zinc ionomers. Tensile strength of blends was homogeneously dependent on polyblend ratio [80].

6. Polystyrene

Its ultimate elongation was increased by addition of ethylene–sodium methacrylate ionomer [79], and one study observed a surprising synergistic increase of impact strength [189].

7. ABS

Its melt flow was increased by addition of ethylene–sodium methacrylate ionomer [79].

8. Polyvinyl Chloride

Its melt flow was increased by addition of ethylene–sodium methacrylate ionomer [79].

9. Polyepichlorohydrin Rubber

Blends with sulfonated EPDM zinc ionomer gave synergistic increase in tensile strength [80].

10. Polybutylene Terephthalate

Its tensile strength was increased by addition of polyethylene ionomer [204].

11. Polyurethane

Its blends with ethylene–sodium methacrylate ionomer showed increases in melt flow, modulus, and low temperature flexibility [79].

12. Polyamides

They have probably benefitted most from addition of ethylene copolymers with acrylic or methacrylic acid and their ionomers [1, 15, 35, 109, 143–147, 204, 205]. Some studies indicate that reaction between carboxylic acid groups on the polyolefin, and amine end groups on the polyamide, or transamidification with amide groups in the polyamide, forms graft copolymers which improve the morphology, interfacial adhesion, and properties. Benefits reported include improvements in melt flow, ultimate elongation, impact strength, and the balarice between modulus vs impact strength.

E. Miscellaneous Ethylene Copolymers

1. Ethylene–Acrylic Ester Copolymers

These are mentioned occasionally in the polyblend literature. For example, when ethylene/methyl acrylate copolymer was blended with polybutylene terephthalate, maleating the copolymer improved interfacial adhesion and impact strength [208]. Blends of ethylene–ethyl acrylate copolymers with PE are mentioned as commercial thermoplastic elastomers [24]. Ethylene copolymers with ethyl acrylate and carbon monoxide, with acrylonitrile, and with dimethyl acrylamide all provided strong hydrogen bonding, producing miscibility with polyvinyl chloride [132]. Ethylene–methyl methacrylate copolymers were fairly miscible with EPDM, but miscibility depended on ethylene content and the sensitivity of the test [102].

2. Styrene–Ethylene–Butylene–Styrene

These block copolymer blends with LLDPE gave fairly homogeneous properties [95].

IV. POLYPROPYLENE POLYBLENDS

The major practical development of polypropylene polyblends has been the addition of ethylene–propylene rubber to improve impact strength; but a wide variety of other polypropylene polyblends have been studied in exploratory research and sometimes mentioned for practical usefulness.

A. Blends with Ethylene–Propylene Rubber

Literature estimates indicate that 10% or more of commercial PP is toughened by addition of 10–20% of EPR [19, 148]. The major improvements are impact strength at room temperature and especially at low temperatures, environmental stress-crack resistance, and filler tolerance [18, 19, 108, 143, 148–151]. The major market is in the auto industry, while other uses include appliances, signs, sports equipment, and tool handles. At the other end of the scale, as noted earlier, addition of PP to EPR produces saturated thermoplastic elastomers.

A number of mechanisms have been studied to understand these improvements. Optimum rubber domain size is variously estimated from 0.4–2 microns, or even bimodal distributions, to produce ductility through crazing and shear yielding [19, 148, 152–154]. PP blocks in the EPR nucleate crystallization of the PP matrix, producing smaller spherulites; while mobility of EPR promotes higher crystallization of the PP matrix [155, 156]. Dynamic mechanical studies showed three separate loss peaks (transitions) at 25, −40, and −120°C [157]. Injection-molded samples also had a PP skin surrounding an EPDM-filled core [158]. In one study, PP + EPR were compatiblized by putting succinic anhydride on the PP and an amine end group on the EPR, then reacting them to graft them together, and using this graft copolymer to improve compatibility, giving finer dispersion and better low temperature impact strength [209].

B. Blends with Other Polyolefins

1. Polypropylenes of Different Molecular Weights

These produced bimodal crystal texture. High MW nucleated low MW, increasing modulus and strength of fibers [159].

2. Polyethylene

Their blends with PP were probably miscible in the melt [5], but nucleated each other's separate crystallization on cooling [21], producing brittle blends. These were compatibilized by adding ethylene–propylene block copolymer [19] or by grafting acid and basic monomers onto the two phases [22] to retain their inherent ductility.

(a) High Density Polyethylene. Blends with PP are immiscible, crystallizing separately and affecting each other's crystallinity in complex ways [65–70]. Most reports indicate that this increases modulus, environmental stress–crack resistance, low temperature impact strength, and heat deflection temperature [42, 57, 66, 71]. Best results were obtained by adding EPR or polyisobutylene as compatibilizers, forming microsize morphologies or an amorphous IPN interphase [42, 72, 186].

(b) Linear Low Density Polyethylene. Its compatibility with PP depended on matching viscosities of the two phases, giving higher homogeneity, modulus, and strength [193, 194]. Compatibilization was also favored by adding ethylene–propylene block copolymers or styrene–ethylene–butylene–styrene block copolymers, which gave faster and finer crystallization [195, 199].

(c) Low Density Polyethylene. Together with PP they were clearly immiscible in rheological studies [97] and incompatible in many property studies [57]. Addition of LDPE improved ultimate elongation and impact strength of PP [160], especially when EPR, EPDM, or natural rubber–PP graft copolymer were added as compatibilizers [99, 100]. While glass fiber reinforcement of PP was degraded in melt processing, addition of LDPE reduced this degradation [161].

(d) Ultra Low Density Polyethylene. This improved low temperature toughness by forming ULDPE domains surrounded by a shell of amorphous PP as interphase [210].

3. Ethylene Copolymers

(a) Ethylene Ionomer. These blends with PP produced a proportional decrease in modulus, thus flexibilizing PP to any desired level [189].

(b) Styrene–Ethylene–Butylene–Styrene. This block copolymer promoted high impact strength through shear yielding and viscoelastic energy dissipation [162–164].

(c) Ethylene–Vinyl Acetate and Ethylene–Ethyl Acrylate. These copolymers were added to PP, improving flexibility, impact strength, environmental stress–crack resistance, and filler tolerance [108, 127]. These found uses in autos, wire and cable, and shoes.

4. Higher Polyolefins

(a) Propylene-α-Olefin. These block copolymers, such as α-hexene, are weak elastomers. Blending such copolymers with PP produced cocrystallization to a single T_g and T_m for each blend, with improved elastomeric properties [74, 113].

(b) Poly-1-Butene. This appeared fairly miscible with PP [126, 131, 165], but gave separate T_g's [38]. PP accelerated the normally slow crystallization of P-1-B [9, 19].

(c) Butyl Rubber. This improves flexibility, impact strength, environmental stress–crack resistance, and filler tolerance of PP [19, 108].

(d) Saturated Polycyclopentadiene. This was miscible only in the amorphous phase of isotactic PP, and did not affect its crystallinity at all [166].

C. Blends with Other Polymers

1. Diene Elastomers

(a) Polybutadiene, Natural Rubber, and Styrene–Butadiene General Purpose Rubber. These were added to PP to improve its toughness, particularly at low temperature [19].

(b) "Thermoplastic Natural Rubber." This was produced by blending with PP and using peroxide to crosslink the NR. This improved moldability and compression set resistance [24].

(c) Rubber Recycling. By adding PP this increased tensile strength. Adding reground vulcanized rubber to PP increased its flexibility [86].

(d) Styrene–Butadiene–Styrene. This block copolymer was added to PP and increased its melt flow, flexibility, tensile strength, ultimate elongation, and impact strength [76]. Styrene–butadiene star-block copolymer was added to PP and increased flexibility, ultimate elongation, and impact strength; it was superior to EPDM, particularly at −60°C [167].

(e) Butadiene–Acrylonitrile Rubber. Blends with PP were compatibilized to form thermoplastic elastomers

with good hot-oil resistance, by adding block or graft copolymers of the two as macromolecular surfactants [118].

2. Styrene Plastics

(a) Polystyrene. The blends with PP were compatibilized by addition of styrene–ethylene–butylene–styrene block copolymer [19].

(b) ABDS This improved impact strength of PP; the effect peaked at 10% ABS [168]. These blends were further improved by addition of LDPE to give lower viscosity and finer dispersion [169].

(c) Polyphenylene Ether. This was compatibilized with PP by adding SEBS block copolymer, increasing ductility and impact strength [211].

3. Polyesters

(a) Polycaprolactone. This and PP amorphous phases were partially miscible [170], improving dyeability of PP fibers with disperse dyes [171].

(b) Polyethylene Terephthalate. This was compatibilized with PP by grafting 6% of acrylic acid onto the PP. This improved extrusion, gave finer domains, and improved modulus, strength, and impact strength. Low permeability suggested that both polymers contributed to crystalline tortuosity [172, 173, 212].

(c) Liquid Crystal Polyester. It increased tensile strength and fatigue lifetime of PP, but not toughness [213].

4. Nylon 6

The blends with PP were incompatible, giving low strength and elongation. Each polymer nucleated separate crystallization of the other [21]. Grafting maleic anhydride (or acrylic acid) onto PP, and then grafting this to nylon 6, decreased crystallinity, improved and stabilized the morphology, and improved interfacial adhesion, strength, elongation, impact strength, thermal stability, water resistance, and dispersion of glass fiber reinforcement [31, 36, 129, 174, 175, 214–216]. Proper sequence of grafting, dry blending, and injection molding produced optimum laminar morphology, with the best tensile, impact, and barrier properties [10, 216].

5. Nylon 11

The blends with PP were compatibilized by grafting acrylic acid onto the PP, which increased interfacial adhesion [217].

6. Nylon 12

The blends with PP were compatibilized by maleated PP, giving finer dispersion and improved mechanical properties [218].

7. Automotive Plastics

When obtained from junked cars, the mixture is not economically separable. Addition of 5–20% of very low molecular weight PP increased melt processability into the range of commercial processes [85].

V. HIGHER POLYOLEFIN POLYBLENDS

A. Poly-1-Butene

1. Polyethylenes

These are immiscible with P-1-B. Low concentrations of HDPE nucleated and changed the crystallinity of P-1-B [78]. Blends with LDPE or EVA were coextruded to form heat-seal adhesives for packaging; these could be reopened easily because the incompatible polymers provided low peel strength [101].

2. Polypropylene

Together with P-1-B it formed miscible blends [126, 131, 165], in which PP accelerated the usually slow crystallization of P-1-B [9, 19]. These showed two separate T_g's but were compatible for practical purposes [38].

3. Chlorinated Polyethylene

This was immiscible with P-1-B, forming two-phase systems [176]. At high P-1-B content, tensile properties were quite good, suggesting good interfacial adhesion.

B. Polyisobutylene and Butyl Rubber

1. Polyethylenes

These generally gave two-phase blends with PIB and butyl rubber. Properties vs polyblend ratio usually indicated that these phases were partially miscible solid solutions [77]. Crosslinked phases increased stability of morphology and properties [26]. Intensive studies of HDPE + Butyl blends gave bimodal peaks for melt viscosity and ultimate elongation vs blend ratio, which were explained by two continuous laminar/fibrillar phases [188].

2. Ethylene Propylene–Diene Rubber (EPDM)

This and butyl rubber were both sulfonated and converted to zinc ionomers. Polyblends gave homogeneous tensile strengths [80].

3. Polypropylene

These blends with PIB gave L-shaped curves for modulus and impact strength vs polyblend ratio [56].

4. Polymethyl Methacrylate

This was blended with 20% of PIB, which increased tensile modulus, strength, and resistance to weathering [177].

5. Polydiethylsiloxane

This was completely miscible with PIB oils, which broadened their useful temperature range and increased their mechanical and thermal stability [178].

C. Poly-4-Methylpentene-1 and Poly-4-Methylhexene-1

Blends of these two higher polyolefins with each other cocrystallized to form isomorphous blends, a relatively rare and interesting morphological phenomenon [9, 126].

REFERENCES

1. LA Utracki. Polymer Alloys and Blends. Munich: Hanser, 1989, pp 256–264.
2. RL Scott. J Chem Phys 17: 279, 1949.
3. AFM Barton. CRC Handbook of Solubility Parameters and Other Cohesion Parameters. Boca Raton: CRC Press, 1983, chap 14.
4. EA Grulke. In: J Brandrup, EH Immergut, eds. Polymer Handbook, 3rd edn. New York: John Wiley, 1989, pp VII-544–557.
5. S Krause. In: DR Paul, S Newman, eds. Polymer Blends. Orlando: Academic Press, 1978, vol 1, chap 2.
6. S Krause. In: J Brandrup, EH Immergut, eds. Polymer Handbook. New York: John Wiley, 1989, pp VI-347–370.
7. RD Deanin, SP McCarthy, HA Kozlowski. SPE ANTEC 34: 1846–1848, 1988.
8. JW Barlow, CH Lai, DR Paul. In: BM Culbertson, ed. Multiphase Macromolecular Systems. New York: Plenum Press, 1989, pp 505–517.
9. LA Utracki. Polymer Alloys and Blends. Munich: Hanser, 1989, p 62.
10. NG Gaylord. J Macromol Sci-Chem A26(8): 1211–1229, 1989.
11. LA Utracki. Polymer Alloys and Blends. Munich: Hanser, 1989, pp 124–129.
12. RJ Ceresa. Block and Graft Copolymers. London: Butterworths, 1962, chap 5.
13. LH Sperling. Interpenetrating Polymer Networks and Related Materials. New York: Plenum Press, 1981.
14. AP Plochocki. In: DR Paul, S Newman, eds. Polymer Blends. New York: Academic Press, 1978, vol 2, pp 320–322.
15. LA Utracki. Polymer Alloys and Blends. Munich: Hanser, 1989, pp 11–13.
16. LA Utracki. Polymer Alloys and Blends. Munich: Hanser, 1989, p 207.
17. K Kreisher. Plastics Tech 35(2): 67–75, 1989.
18. RD Deanin. Encyc Polym Sci Tech. New York: John Wiley, 1977, Suppl vol 2, pp 458–484.
19. DR Paul, JW Barlow, H Keskkula. Encyc Polym Sci Eng. New York: John Wiley, 1988, vol 12, pp 399–461.
20. P Mukhopadhyay, G Chowdhury. Polym-Plast Technol Eng 28(5&6): 517–535, 1989.
21. B Lotz, JC Wittmann. J Polym Sci, Part B: Polym Phys 24: 1559–1575, 1986.
22. O Olabisi, LM Robeson, MT Shaw. Polymer-Polymer Miscibility. New York: Academic Press, 1979, p 350.
23. AK Datta, S Bhattacherjee, CK Das. J Mater Sci Lett 5: 739–740, 1986.
24. JR Wolfe. In: NR Legge, G Holden, HE Schroeder, eds. Thermoplastic Elastomers. Munich: Hanser, 1987, chap 6.
25. EN Kresge. In: DR Paul, S Newman, eds. Polymer Blends. New York: Academic Press, 1978, vol 2, chap 20.
26. AK Datta, S Bhattacherjee, CK Das. J Mater Sci Lett 5: 319–322, 1986.
27. LA Utracki. Polymer Alloys and Blends. Munich: Hanser, 1989, pp 172–173.
28. DC Baird, R Ramanathan. In: BM Culbertson, ed. Multiphase Macromolecular Systems. New York: Plenum Press, 1989, pp 81–82.
29. JA Manson, LH Sperling. Polymer Blends and Composites. New York: Plenum Press, 1976, p 279.
30. O Olabisi, LM Robeson, MT Shaw. Polymer-Polymer Miscibility. New York: Academic Press, 1979, p 343.
31. DR Paul. Polymer Blends. New York: Academic Press, 1978, vol 2, chap 12.
32. PV Ballegooie, A Rudin. Polym Eng Sci 23(21): 1434, 1988.
33. J Curry, P Andersen. SPE ANTEC 36: 1938, 1990.
34. Y Lipatov. J Appl Polym Sci 22: 1895–1910, 1978.
35. LA Utracki. Polymer Alloys and Blends. Munich: Hanser, 1989, pp 267–270.
36. RP Quirk, J-J Ma, CC Chen, K Min, JL White. In BM Culbertson, ed. Multiphase Macromolecular Systems. New York: Plenum Press, 1989, pp 107–137.
37. S Endo, K Min, JL White, T Kyu. Polym Eng Sci 26(1): 45–53, 1986.
38. LA Utracki. Polymer Alloys and Blends. Munich: Hanser, 1989, pp 95–98.
39. LA Utracki. Polymer Alloys and Blends. Munich: Hanser, 1989, p 115.

40. T Kyu, P Vadhar. J Appl Polym Sci 32: 5575–5584, 1986.
41. T Kyu, S-R Hu, RS Stein. J Polym Sci, Part B. Polym Phys 25: 89–103, 1987.
42. LA Utracki. Polymer Alloys and Blends. Munich: Hanser, 1989, pp 201–205.
43. PJ Barham, MJ Hill, A Keller, CCA Rosney. J Mater Sci Lett 7: 1271–1275, 1988.
44. SP Ting, EM Pearce, TK Kwei. J Polym Sci, Polym Lett 18: 201, 1980.
45. TK Kwei, EM Pearce, BY Min. Macromol 18: 2326, 1985.
46. LM Robeson. In: BM Culbertson, ed. Multiphase Macromolecular Systems. New York: Plenum Press, 1989, pp 177–212.
47. NK Datta, AW Birley. Plastics and Rubber Proc Appl 2: 237–245, 1982.
48. MP Farr, IR Harrison. ACS Polym Preprints 31(1): 257–258, 1990.
49. S-R Hu, T Kyu, RS Stein. J Polym Sci, Part B. Polym Phys 25: 71–87, 1987.
50. LA Utracki. Polymer Alloys and Blends. Munich: Hanser, 1989, pp 85–87.
51. P Vadhar, T Kyu. Polym Eng Sci 27(3): 202–210, 1987.
52. LA Hamielec. Polym Eng Sci 26(1): 111–115, 1986.
53. AA Donatelli. J Appl Polym Sci 23: 3071–3076, 1979.
54. HH Song, DQ Wu, B Chu, M Satkowski, M Ree, RS Stein, JC Phillips. Macromol 23: 2380–2384, 1990.
55. LA Utracki. Polymer Alloys and Blends. Munich: Hanser, 1989, pp 29–60.
56. RD Deanin, RR Geoffroy. ACS Org. Coatings Plastics Chem 37(1): 257–262, 1977.
57. RD Deanin, MF Sansone. ACS Polym Preprint 19(1): 211–215, 1978.
58. HB Hopfenberg, DR Paul. In: DR Paul, S Newman, eds. Polymer Blends. New York: Academic Press, 1978, vol 1, chap 10.
59. J Martinez-Salazar, FJ Balta-Calleja. J Mater Sci Lett 4: 324–326, 1985.
60. MR Shishesaz, AA Donatelli. Polym Eng Sci 21(13): 869–872, 1981.
61. P Filzek, W Wicke. Kunststoffe 73(8): 427–431, 1983.
62. A Garcia-Rejon, C Alvarez. Polym Eng Sci 27(9): 641–646, 1987.
63. FP LaMantia, A Valenza, D Acierno. Polym Degrad Stabil 13: 1–9, 1985.
64. C Sawatari, M Matsuo. Polym J 19(12): 1365–1376, 1987.
65. O. Olabisi, LM Robeson, MT Shaw. Polymer-Polymer Miscibility. New York: Academic Press, 1979, pp 163–164.
66. AJ Lovinger, ML Williams. ACS Org Coatings and Plastics Preprints 43: 13–18, 1980.
67. OF Noel, JF Carley. Polym Eng Sci 24(7): 488–492, 1984.
68. F Rybnikar. J Macromol Sci-Phys B27(2&3): 125–144, 1988.
69. J Martinez Salazar, JM Garcia Tijero, FJ Balta Calleja. J Mater Sci 23: 862–866, 1988.
70. LA Utracki. Polymer Alloys and Blends. Munich: Hanser, 1989, pp 108–110.
71. RD Deanin, GE D'Isidoro. ACS Org Coatings and Plastics Preprints 43: 19–22, 1980.
72. NA Erina, LV Kompaniets, AN Kryuchkov, MI Knunyants, EV Prut. MCMAD7:24(1): 454–460, 1988.
73. R Greco, C Mancarella, E Martuscelli, G Ragosta, Y Jinghua. Polymer 28: 1922–1928, 1987.
74. C-K Shih. Polym Eng Sci 27(6): 458–462, 1987.
75. RD Deanin, ST Lim. SPE ANTEC 29: 222–223, 1983.
76. RD Deanin, Y-S Chang. SPE ANTEC 31: 949–950, 1985; J Elast Plastics 18: 35–41, 1985.
77. RD Deanin, RO Normandin, CP Kannankeril. ACS Coatings and Plastics Preprints 35(1): 259–264, 1975.
78. K Kishore, R Vasanthakumari. Polymer 27: 337–343, 1986.
79. RD Deanin, W-F Liu. ACS Polym Mater Sci Eng 53: 815–819, 1985.
80. H-Q Xie, B-Y Ma. In: BM Culbertson, ed. Multiphase Macromolecular Systems. New York: Plenum Press, 1989, pp 601–617.
81. G Serpe, J Jarrin, F Dawans. Polym Eng Sci 30(9): 553–565, 1990.
82. PM Subramanian, V Mehra. Polym Eng Sci 27(9): 663–668, 1987.
83. LA Utracki. Polymer Alloys and Blends. Munich: Hanser, 1989, p 7.
84. TD Traugott, JW Barlow, DR Paul. J Appl Polym Sci 28(9): 2947–2959, 1983.
85. RD Deanin, DM Busby, GJ DeAngelis, AM Kharod, JS Margosiak, BG Porter. ACS Polym Mater Sci Eng 53: 826–829, 1985.
86. RD Deanin, SM Hashemiolya. ACS Polym Mater Sci Eng 57: 212–216, 1987.
87. LA Utracki. Polymer Alloys and Blends. Munich: Hanser, 1989, p 8.
88. RD Deanin, SB Driscoll, RJ Cook, MP Dubreuil, WN Hellmuth, WA Shaker. SPE ANTEC 24: 711, 1978.
89. MC Schwarz, H Keskkula, JW Barlow, DR Paul. J Appl Polym Sci 35: 653–677, 1988.
90. W Li, B Huang. Polym Bull 20(6): 531–535, 1988.
91. LA Utracki. Adv Polym Tech 5(1): 41–53, 1985.
92. H Schule, R Wolff. Kunststoffe 77(8): 744–750, 1987.
93. B Leroy. Plastics and Rubber Proc Appl 8: 37–47, 1987.
94. WE Baker. Polym Eng Sci 24(17): 1348–1353, 1984.
95. RD Deanin, MA Burduroglu. SPE ANTEC 36: 1850, 1990.
96. M Ajji. Polym Eng Sci 29(21): 1544–1550, 1989.
97. LA Utracki. Polymer Alloys and Blends. Munich: Hanser, 1989, pp 205–207.
98. NS Yenikolopyan, AM Khachatryan, NM Styrikovich, VG Nikol'skh, AS Kechek'yan. Polym Sci USSR 27(8): 1891–1896, 1985.

99. S Al-Malaika, EJ Amir. Polym Degrad Stabil 16: 347–359, 1986.
100. W-Y Chiu, S-J Fang. J Appl Polym Sci 30: 1473–1489, 1985.
101. CC Hwo. J Plastic Film Sheet 3: 245–260, 1987.
102. HW Starkweather. J Appl Polym Sci 25: 139–147, 1980.
103. S Akhtar, PP De, SK De. J Mater Sci Lett 5: 399–401, 1986.
104. NR Choudhury, TK Chaki, A Dutta, AK Bhowmick. Polymer 30: 2047–2053, 1989.
105. NR Choudhury, AK Bhowmick. J Appl Polym Sci 38: 1091–1109, 1989.
106. R Wycisk, WM Trochimczuk, J Matys. Eur Polym J 26(5): 535–539, 1990.
107. B Boutevin, Y Pietrasanta, M Traha, T Sarraf. Polym Bull 14: 25–30, 1985.
108. O Olabisi, LM Robeson, MT Shaw. Polymer-Polymer Miscibility. New York: Academic Press, 1979, pp 340–341.
109. RD Deanin, SA Orroth, RI Bhagat. Polym-Plast Technol Eng 29(3): 289–295, 1990.
110. FP LaMantia, A Valenza. Eur Polym J 25(6): 553–556, 1989.
111. M R MacLaury, AL Schroll. J Appl Polym Sci 30: 461–472, 1985.
112. G Holden. In: M Morton, ed. Rubber Technology. New York: Van Nostrand Reinhold, 1987, chap 16.
113. C-K Shih, ACL Su. In: NR Legge, G Holden, HE Schroeder, eds. Thermoplastic Elastomers. Munich: Hanser, 1987, chap 5.
114. O Olabisi, LM Robeson, MT Shaw. Polymer-Polymer Miscibility. New York: Academic Press, 1979, p 322.
115. J Karger-Kocsis. Polym Eng Sci 27(4): 254–262, 1987.
116. DJ Lohse. Polym Eng Sci 26(21): 1500–1509, 1986.
117. S Danesi, RS Porter. Polymer 19: 448–457, 1978.
118. AY Coran. In: NR Legge, G Holden, HE Schroeder, eds. Thermoplastic Elastomers. Munich: Hanser, 1987, chap 7.
119. JGM van Gisbergen, HEH Meijer, PJ Lemstra. Polym 30: 2153–2157, 1989.
120. CS Ha, SC Kim. J Appl Polym Sci 35: 2211–2221, 1988.
121. DH Kim, SC Kim. Polym Bull 1(4): 401–408, 1989.
122. ET McDonel, KC Baranwal, JC Andries. In: DR Paul, S Newman, eds. Polymer Blends. New York: Academic Press, 1978, vol 2, chap 19.
123. O Olabisi, LM Robeson, MT Shaw. Polymer-Polymer Miscibility. New York: Academic Press, 1979, p 348.
124. S Shaw, RP Singh. J Appl Polym Sci 38: 1677–1683, 1989.
125. RJ Gaymans, RJM Borggreve. In: BM Culbertson, ed. Multiphase Macromolecular Systems. New York: Plenum Press, 1989, pp 461–471.
126. O Olabisi, LM Robeson, MT Shaw. Poymer-Polymer Miscibility. New York: Academic Press, 1979, pp 264–267.
127. S Thomas. Mater Lett 5(9): 360–364, 1987.
128. RD Deanin, TJ Pickett, J-C Huang. ACS Polym Mater Sci Eng 61: 950–954, 1989.
129. SM Brown, CM Orlando. Encyc. Polym Sci Eng. New York, John Wiley, 1988, pp 181–184.
130. MM Coleman, EJ Moskala, PC Painter. Polymer 24: 1410–1414, 1983.
131. LA Utracki. Polymer Alloys and Blends. Munich: Hanser, 1989, p 22.
132. O Olabisi, LM Robeson, MT Shaw. Polymer-Polymer Miscibility. New York: Academic Press, 1979: many references throughout the text.
133. RD Deanin, NA Shah. ACS Org Coatings Plastics 45: 290–294, 1981.
134. NA Shah, RD Deanin. SPE NATEC 1982, p 286; J Vinyl Tech 5(4): 167–172, 1983.
135. RD Deanin, SS Rawal, NA Shah, J-C Huang. ACS Polym Mater Sci Eng 57: 796, 1987; ACS Adv Chem Ser 222: 403–413, 1989.
136. O Olabisi, LM Robeson, MT Shaw. Polymer-Polymer Miscibility. New York: Academic Press, 1979, p 325.
137. T Skowronski, JF Rabek, B Ranby. Polym Eng Sci 24(4): 278–286, 1984.
138. AG Margaritis, JK Kallitsis, NK Kalfoglou. Polymer 28: 2121–2129, 1987.
139. RD Deanin, MR Shah. ACS Org Coatings Plastics 44: 102–107, 1981.
140. RD Deanin, W-ZL Chuang. SPE ANTEC 32: 1239, 1986; J Vinyl Tech 9(2): 60, 1987.
141. Y-D Lee, C-M Chen. J Appl Polym Sci 33: 1231–1240, 1987.
142. W Baker, M Duhaime, K Oliphant. IUPAC, Montreal, 7/8-13, 1990.
143. S Newman. Polymer Blends. New York: Academic Press, 1978, vol 2, chap 13.
144. O Olabisi, LM Robeson, MT Shaw. Polymer-Polymer Miscibility. New York: Academic Press, 1979, p 306.
145. WJ MacKnight, RW Lenz, PV Musto, RJ Somani. Polym Eng Sci 25(18): 1124–1134, 1985.
146. RD Deanin, J Jherwar. SPE PMAD RETEC, Newark, NJ, 11/6-7, 1985, pp 24–28.
147. LA Utracki. Polymer Alloys and Blends. Munich: Hanser, 1989, pp 245–246.
148. K-D Rumpler, JFR Jaggard, RA Werner. Kunststoffe 78(7): 602–605, 1988.
149. G Heufer. Kunststoffe 78(7): 599–601, 1988.
150. IR Lloyd. Plastics Rubber Proc Appl 1 (4): 351–355, 1981.
151. G Scott, E Setoudeh. Polym Degrad Stabil 5: 11–22, 1983.
152. FC Stehling, T Huff, CS Speed, G Wissler. J Appl Polym Sci 26(8): 2693–2711, 1981.
153. BZ Jang, DR Uhlmann, JB Vander Sande. J Appl Polym Sci 30: 2485–2504, 1985.
154. J Karger-Kocsis, A Kallo, VN Kuleznev. Polymer 25: 279–286, 1984.
155. Z Bartczak, A Galeski, E Martuscelli, H Janik. Polymer 26: 1843–1848, 1985.

156. R Greco, C Mancarella, E Martuscelli, G Ragosta, Y Jinghua. Polymer 28: 1929–1936, 1987.
157. JM Hodgkinson, A Savadori, JG Williams. J Mater Sci 18: 2319–2336, 1983.
158. J Karger-Kocsis, I Csikai. Polym Eng Sci 27(4): 241–253, 1987.
159. BL Deopura, S Kadam. J Appl Polym Sci 31: 2145–2155, 1986.
160. AGaleski, M Pracella, E Martuscelli. J Polym Sci, Polym Phys 22(4): 739–747, 1984.
161. M Arroyo, F Avalos. Polym Composites 10(2): 117–121, 1989.
162. AK Gupta, SN Purwar. J Appl Polym Sci 30: 1777–1798, 1985.
163. AK Gupta, SN Purwar. J Appl Polym Sci 30: 1799–1814, 1985.
164. AK Gupta, SN Purwar. J Appl Polym Sci 31: 535–551, 1986.
165. LA Utracki. Polymer Alloys and Blends. Munich: Hanser, 1989, p 65.
166. E Martuscelli, M Canetti, A Seves. Polymer 30: 304–310, 1989.
167. J Karger-Kocsis, Z Balajthy, L Kollar. Kunststoffe 74(2): 104–107, 1984.
168. AK Gupta, AK Jain, BK Ratnam, SN Maiti. J Appl Polym Sci 39: 515–530, 1990.
169. AK Gupta, AK Jain, SN Maiti. J Appl Polym Sci 38: 1699–1717, 1989.
170. NK Kalfoglou. J Appl Polym Sci 30: 1989–1997, 1985.
171. JV Koleski. In: DR Paul, S Newman, eds. Polymer Blends. New York: Academic Press, 1978, vol 2, chap 22.
172. P Bataille, S Boisse, HP Schreiber. J Elast Plastics 18: 228–232, 1986.
173. M Xanthos, MW Young, JA Biesenberger. Polym Eng Sci 30(6): 355–365, 1990.
174. F Ide, A Hasegawa. J Appl Polym Sci 18: 163–174, 1974.
175. SS Dagli, M Xanthos, JA Biesenberger. SPE ANTEC 36: 1924, 1990.
176. JK Kallitsis, NK Kalfoglou. J Appl Polym Sci 32: 5261–5272, 1986.
177. EG Kolawole. Eur Polym J 20(6): 629–633, 1984.
178. YA Sangalov, YY Nel'kenbaum, IK Prokof'ev. UDC 621.892.099,6.678.742.4 + 678.84, Plenum Publishing Corp, 1987, pp , 241–244.
179. Z Song, WE Baker. J Appl Polym Sci 44(12): 2167–2177, 1992.
180. TL Carte, A Moet. J Appl Polym Sci 48(4): 611–624, 1993.
181. I Pesneau, MF Llauro, M Gregoire, A Michel. J Appl Polym Sci 65(12): 2457–2469, 1997.
182. BK Kim, SY Park, SJ Park. Eur Polym J 27(4-5): 349–354, 1991.
183. SY Lee, JY Jho, YC Lee. ACS Polym Mater Sci Eng 76: 325–326, 1997.
184. JN Hay, X-Q Zhou. Polymer 34(11): 2282–2288, 1993.
185. DR Rueda, L Malers, A Viksne, FJ Balta Calleja. J Mater Sci 31 (15): 3915–3920, 1996.
186. N Dharmarajan, DR Hazelton, EJ Kaltenbacher. Plastics Eng 46(5): 25–27, 1990.
187. SJ Mahajan, BL Deopura, Y Wang. J Appl Polym Sci 60(10): 1517–1525, 1996.
188. RD Deanin, RK Datta. SPE ANTEC 40: 2436–2438, 1994.
189. RD Deanin, RS Ciulla. ACS Polym Preprints 37(1): 384, 1996.
190. CP Papadopoulou, NK Kalfoglou. Polym 38(16): 4207–4213, 1997.
191. J George, R Joseph, S Thomas, KT Varughese. J Appl Polym Sci 57(4): 449–465, 1995.
192. L Mascia, A Valenza. Adv Polym Technol 14(4): 327–335, 1995.
193. M Bains, ST Balke, D Reck, J Horn. Polym Eng Sci 34(16): 1260–1268, 1994.
194. M Bains, ST Balke, D Reck, J Horn. Polym Eng Sci 34(16): 1260–1268, 1994.
195. V Flaris, W Wenig, ZH Stachurski. Mater Forum 16(2): 181–184, 1992.
196. C Qin, J Yin, B Huang. Polymer 31 (4): 663–667, 1990.
197. RD Deanin, MA Burduroglu. SPE ANTEC 36: 1850, 1990.
198. FP LaMantia, C Geraci, M Vinci, U Pedretti, A Roggero, LI Minkova, PL Magagnini. J Appl Polym Sci 58(5): 911–921, 1995.
199. JZ Liang, CY Tang, HC Man. J Mater Proc Tech 66(1-3): 158–164, 1997.
200. RD Deanin, T-JA Hou. ACS Polym Preprints 34(1): 896–897, 1993.
201. I Ray, D Khastgir. Polymer 34(10): 2030–2037, 1993.
202. EF Silva, BG Soares. J Appl Polym Sci 60(10): 1687–1694, 1996.
203. J Horrion, PK Agarwal. Polym Eng Sci 36(14): 1869–1874, 1996.
204a. RD Deanin, W Chu. ACS Polym Mater Sci Eng 76: 296–297, 1997.
204b. E Kroeze, G ten Brinke, G Hadziioannou. Polymer 38(2): 379–389, 1997.
205. RD Deanin, SA Orroth, RI Bhagat. Polym-Plast Technol Eng 29(3): 289–295, 1990.
206. RD Deanin, SA Orroth, H Bhagat. ACS Polym Mater Sci Eng 78: 156, 1998.
207. RD Deanin, CJ Parikh, V Ghiya. 4th Pacific Polym Conf, Kauai, 12/15, 1995, p 520.
208. T-K Kang, Y Kim, G Kim, W-J Cho, C-S Ha. Polym Eng Sci 37(3): 603–614, 1997.
209. S Datta, DJ Lohse. Macromol 26(8): 2064–2076, 1993.
210. SP Westphal, MTK Ling, L Woo. J Eng Appl Sci Mater Proc 54th Ann Tech Conf, 5/5-10, 1996, vol 2, pp 1629–1633.
211. MK Akkapeddi, B Van Buskirk. Adv Polym Tech Conf Add Mod Polym Blends. 11(4): 263–275, 1992.
212. M Xanthos, MW Young, JA Biesenberger. Polym Eng Sci 30(6): 355–365, 1990.

213. M Kawagoe, W Mizuno, J Qiu, M Morita. Japan J Polym Sci Tech 52(6): 335–348, 1995.
214. J-D Lee, S-M Yang. Polym Eng Sci 35(23): 1821–1833, 1995.
215. SN Sathe, S Devi, GSS Rao, KV Rao. J Appl Polym Sci 61(1): 97–107, 1996.
216. RM Holsti-Miettinen, KP Perttila, JV Seppala, MT Heino. J Appl Polym Sci 58(9): 1551–1560, 1995.
217. Z Liang, HL Williams. J Appl Polym Sci 44(4): 699–717, 1992.
218. Y Long, RA Shanks. J Mater Sci 31(15): 4033–4038, 1996.

23

Compatibilization of Polyolefin Polyblends

Rudolph D. Deanin and Margaret A. Manion
University of Massachusetts at Lowell, Lowell, Massachusetts

I. INTRODUCTION

Polyolefins are frequently blended with each other and with other polymers. When these blends have useful properties, we call them "compatible."

Most polymer blends are immiscible for several reasons [1, 2]:

Polymers of different polarity and/or hydrogen bonding agglomerate with their own kind and reject the other kind, forming separate phases. Even at the interface, they reject each other (positive enthalpy), giving a weak interface which cannot resist stress (low strength, elongation, and impact strength).

Large polymer molecules disentangle with difficulty (slowing the kinetics of mixing), and they cannot mix as completely (low entropy).

Polymers which crystallize immediately form a separate phase, which is rarely miscible with the second polymer.

While such phase separation may benefit certain specialized properties such as melt fluidity, easy-opening packaging, and lubricity, in most products the loss of strength, ductility, and impact strength is a serious handicap, so most polyblends are therefore labelled "incompatible."

Polyblending offers the possibility of combining the best properties of both polymers in the blend, particularly in a two-phase system [1–4]. Thus, when gross phase separation causes incompatibility, it is highly desirable to reduce the size and morphology of phase separation, and in particular to strengthen the interface between the phases, especially by broadening it from a thin sharp change in composition and properties, to a broader gradual transitional interphase which is more able to distribute and withstand the stresses applied in the normal use of the product. When such efforts are successful, the process is called "compatibilization" [1–11].

Compatibilization may be accomplished in one or more ways:

Physical processes such as mixing shear and thermal history can be used to modify domain size and shape.

Physical additives such as plasticizers, reinforcing fibers, and especially block and graft copolymers, can be used to increase the attraction between molecules and/or between phases.

Reactive processing can be used to change the chemical structure of the polymers and thus increase their attraction for each other.

Each of these techniques will be discussed in turn.

II. PHYSICAL PROCESSES

If two polymers have some degree of compatibility, this can often be enhanced by simple physical processes. When polyolefins are blended with each other, their low polarities are similar enough so that physical processes alone may be sufficient to produce compatibility. When they are blended with more polar polymers, even though physical processes alone may not be sufficient, they can often enhance the compatibilizing effect of additives and chemical reactions [12].

Most important, of course, are mixing energy and particularly shear. Also very important are melt viscosities of the two polymers, and particularly their viscosity ratio. Processing temperature has a variety of effects. Crystallization is a primary cause of phase separation, not only in single polymers but also in polymer blends. Stretch orientation has major effects on phase morphology. And demixing/coalescence/annealing effects can sometimes reverse these processes.

A. Mixing Energy

Increasing mechanical energy of mixing generally disperses the minor polymer into smaller and smaller domains, and this often improves mechanical properties. Design and operation of the mixer determine the *intensity* of mixing energy supplied to the blending process. In blending of HDPE + PP, a corotating twin-screw extruder gave the most compatibilization, a Maddock extruder less, and a simple single-screw extruder least, as judged by impact strength of the blends [13]. Twin-screw mixing of PP + polyisoprene was best when the forward and reverse kneading blocks were placed closest together [14]. Solid state shear extrusion of HDPE + LLDPE + PP, applying high pressure and shear directly to the mixed powders, and heavy cooling to prevent frictional fusion, blended them efficiently enough to give mechanical properties proportional to composition, indicating compatibilization [15].

Extensity of mixing is simply the total mixing time. Most studies have found that, as expected, increasing mixing time produced more thorough blending and favored compatibility. In twin-screw blending of HDPE + polystyrene, increasing mixing time gave finer and finer domain structure [16]. A very detailed study of melt mixing of LLDPE + polystyrene in a Haake roller blade mixer observed successive stages of the process—surface melting of large particles, followed by shear-tearing of the softened surface to produce finer particles—until the entire blend consisted of fine domains [17].

B. Shear

Mechanical energy of mixing is applied to the polymers primarily by shear. In the initial mixing/melting process, frictional/thermal heating of the surfaces of the large particles softens them enough that the shearing action between them tears small particles off the surface; this continues until the entire blend reaches a finer domain size [17]. In melt blending of LDPE + polystyrene, the viscous PS applied enough shear to disperse LDPE into finer domains [18]. Generally, as increasing shear produces finer domain size, more and more properties become proportional to composition, indicating practical compatibility.

Furthermore, increasing type and duration of shear produce a succession of morphologies [12], such as encapsulation, ellipsoidal, laminar, and fibrillar domains [19]. Extrusion blending and injection molding of HDPE + polystyrene gave elongated domains which reinforced axial strength [16]. Extrusion/drawing of HDPE + nylon 6 gave ellipsoidal domains which became more elongated with increasing drawdown ratio [20]. Similarly, increasing shear flow of PIB + polystyrene gave increasing elongation until it reached a critical shear rate at which elongated domains dispersed into finer domains [21]. Generally, the sequence sphere–ellipse–lamina–fibril–finer sphere may recur again down to smaller and smaller domain sizes, properties more proportional to composition, and generally greater compatibility.

C. Melt Viscosity

Blending of polymers has a variety of effects on rheology [22]. For miscible polymers the effects are fairly linear, whereas for immiscible polymer melts the effects can be much more complex. Generally a low viscosity ratio of dispersed phase/continuous matrix phase permits the viscous matrix phase to apply more shear forces to the fluid dispersed domains, deforming spheres into ellipsoids, laminae, and even fibers, which then break down into smaller spheres and may even repeat the cycle all over again [19]. In blends of viscous polystyrene with fluid LDPE, this shearing effect produced much finer LDPE dispersed domains [18]. In blends of viscous polyurethane elastomer with fluid PP, melt spinning and drawing elongated the PP into *in situ* reinforcing fibrils which increased the modulus of the polyurethane matrix [23].

More dramatically, the less viscous polymer tends to form the continuous matrix phase; and the more viscous, undeformable dispersed domains remain large and discrete. Such effects have been observed even when the fluid polymer was only a minor fraction of the total volume, such as 25% EVA + 75% polyvinyl chloride [24].

D. Temperature

Blending temperature affects many of these physical factors in various ways. In terms of thermodynamic equilibriuim, most polyblends are less miscible at higher temperature [25]. Increasing temperature generally decreases melt viscosity, favoring kinetic approach to equilibrium. The differential effects of temperature, on the viscosities of the two polymers in a blend, will of course change their viscosity ratio, with profound effects on morphology and properties as noted above. And any non-equilibrium morphology, such as domain fineness, elongation, laminar and fibrous domains, will tend to revert toward larger equilibrium spheres. Thus processing temperature can have many of these interacting effects on polyblend morphology and properties.

E. Crystallization

When a polymer crystallizes, this immediately creates a two-phase system, with profound effects on properties. When one or both polymers in a blend crystallize, this creates additional phases, with much more complex morphology and effects on properties [26]. In many systems, the mere presence of a second polymer can affect the crystallization of the first polymer. For example, synergistic increase of modulus, strength, and heat deflection temperature in blends of PE with PP [27, 28] was probably due to the fact that PE nucleated the crystallization of PP, giving finer morphology [29].

F. Stretching/Orientation

Shear flow can produce ellipsoidal and fibrillar domains, not only in melt processing as noted above, but also in post-process stretching/drawing as well. Extrusion/drawing of PP + liquid crystal polyester produced elongation and fibrillation of the LCP domains, which increased modulus and strength of the composite [30]. Spinning/drawing of PP in polyurethane elastomer led to *in situ* formation of PP fibers, which increased the modulus of the polyurethane [23].

G. Annealing/Coalescence/Demixing

Thus many studies have observed that shear flow and stretching can reduce domain size and can elongate dispersed domains into ellipsoidal, laminar, and fibrous morphology, producing compatibility and reinforcement. Since these morphologies are nonequilibrium in nature, it is not surprising that annealing/relaxation can permit the blends to re-equilibrate back toward more thermodynamically stable morphology, often referred to as coarsening, coalescence, or demixing. In extrusion/molding of HDPE + polystyrene, shear flow in extrusion and injection molding produced elongated domains and increased axial strength; whereas compression molding, involving little shear flow and much more annealing, permitted coarsening/coalescence toward larger domain size and spherical shape, with consequent loss of strength again [16].

Thus physical processes alone can contribute toward practical compatibilization of polyblend systems. It should be remembered, however, that while such solely physical processes may affect the *degree* of compatibility, they do not generally make a *qualitative* difference between incompatible and compatible polyblends. For such qualitative improvement in compatibility, additives and/or reactions are generally required.

III. PHYSICAL ADDITIVES

The most common way of compatibilizing a polymer blend is to add a third ingredient, which does not react chemically, but contributes physical effects which increase miscibility, decrease domain size, change domain structure, increase interfacial bonding, or make other changes which enhance practical properties. Most popular additives are block or graft copolymers, while others frequently mentioned include third polymers in general, or less often solvents, plasticizers, surfactants, adhesives, and reinforcing fibers.

A. Block Copolymers

Theoretically, the optimum compatibilizer for an immiscible blend of polyA + polyB should be a block copolymer of polyA–polyB, which would migrate preferentially to the interface between the two phases, orient at the interface, send block A into phase A and block B into phase B, and thus bond the two phases together, with a primary covalent bond across the interface [6, 7, 9, 10, 31, 32]. Diblock copolymers fit the theoretical model best, whereas multiblock and star-block would have more difficulty conforming at the interface [10]; but some researchers have found experimentally that these more complex structures may work equally well or occasionally even better.

Theoretically, optimum block length is a compromise between two types of considerations [6, 9, 10]. (1) Short molecules should migrate more easily and completely to the interface, and should saturate the interface efficiently without long wasteful tails, and thus contribute more efficiently to decreasing interfacial tension, whereas longer molecules would have more trouble reaching the interface, and their long tails would just be wasted material that could not contribute any more to bonding at the interface. On the other hand, (2) long molecules could penetrate more deeply into each phase, entangle more intimately with the molecules in each phase, and thus resist pulling out under stress, so that they would contribute much more to interfacial adhesion. Theoretical modelling has led to estimates that optimum block length for simply coating the interface might be 10–20 monomer units, whereas experimental measurements of interfacial adhesion and practical properties have suggested that optimum molecular weights may have to be much higher than this [9, 10].

Concentration of block copolymer required to saturate the interface is occasionally theorized or measured in the range of 0.5–4.0% [32–34]. When larger amounts are added, they often separate as independent micelles, which is inefficient and wasteful. On the other hand, model calculation suggests that, when the block copolymer is high molecular weight, and most of the length is "wasted" away from the interface, $MW = 10^4$ would require a concentration of 2% but $MW = 10^5$ would require 20% [10]; and experimental researchers have sometimes reported optimum concentrations as high as that.

Simple theory says that a blend of A + B should be compatibilized by a block copolymer of A–B. Where this is not readily available, theorists generally concede that the blocks of the compatibilizer need not be *identical* with A and B, so long as they are *miscible* or at least *compatible* with A and B [6, 7, 9, 10, 31]. For example, typical estimates conclude that the solubility parameters of the blocks in the compatibilizer must be within 0.2–1.0 units of those in A and B [6, 7]. In practice, most researchers compromise between the two extremes, using a block copolymer A–C to compatibilize a blend of A + B, by choosing block C for compatibility with B.

An interesting variation on the pure block copolymer is the tapered block copolymer, which is formed by copolymerizing two monomers which enter the polymer chain at very different rates. If A enters the chain faster than B, the first part of the chain will be almost polyA. As A is consumed, and the B/A ratio rises, B will enter the chain occasionally, and with increasing frequency. Toward the end of the reaction, with very little A left, the last part of the chain will be almost polyB. While this does not fit the theoretical model nearly as well, experimental results suggest that such tapered block copolymers may actually be extremely effective compatibilizers. The explanation may lie in their ability to form a broader interphase, with a gradual transition/modulation between the two pure phases.

1. Identical Blocks

When researchers want to compatibilize A + B, they rarely have the ideal block copolymer A–B available, so they usually compromise by using a block copolymer A–C in which block C is hopefully compatible with polymer B. Typical examples may be grouped by the type of polyolefin in the blend.

(a) Polyethylene in General. PE + polystyrene was compatibilized by adding styrene–ethylene/butylene, styrene–ethylene/butylene–styrene, or ethylene–propylene block copolymers, which gave finer dispersion, and greater interfacial adhesion and ultimate elongation [7, 32].

PE + polyvinyl chloride was compatibilized by adding chlorinated polyethylene which had a block copolymer structure, containing 36% Cl [10].

(b) High Density Polyethylene. HDPE + polystyrene was compatibilized by adding ethylene–styrene diblock copolymer, which reduced domain size and increased tensile strength, elongation, and impact strength [35]. It was also compatibilized by adding styrene–ethylene/butylene–styrene block copolymer, which reduced domain size [16, 36].

HDPE + polyphenylene ether was compatibilized somewhat by adding ethylene–styrene diblock copolymer [37].

(c) Linear Low Density Polyethylene. LLDPE + polystyrene was compatibilized by adding styrene–ethylene/butylene–styrene or styrene–ethylene/propylene block copolymers, which reduced domain size and increased toughness (area under the stress–strain curve) [17, 33, 38, 39].

LLDPE + polyphenylene ether was compatibilized by adding ethylene–styrene diblock copolymer, which decreased and stabilized domain size, and increased impact strength [37].

(d) Low Density Polyethylene. LDPE + polystyrene was compatibilized by adding styrene–isoprene–styrene block copolymer, which decreased domain size [18], or by adding styrene–ethylene/butylene–styrene block copolymer, which increased tensile strength, elongation, and ductility [40].

(e) Ultra Low Density Polyethylene. ULDPE + polystyrene was compatibilized by adding styrene–ethylene/butylene–styrene block copolymer, which reduced domain size and increased impact strength [38].

(f) Ethylene/Propylene Rubber. EPR + polystyrene was compatibilized by adding styrene–ethylene/butylene–styrene diblock or triblock copolymer, at concentrations up to 15%. This decreased domain size and increased impact strength [41–43].

(g) Chlorinated Polyethylene. CPE + polystyrene was compatibilized by adding styrene–epoxidized butadiene diblock copolymer, as judged by dynamic mechanical analysis, tensile strength and elongation [44].

(h) Polypropylene. PP + polystyrene was compatibilized by adding styrene–ethylene/propylene diblock copolymer. Up to 5% reduced interfacial tension and up to 20% reduced domain size [38, 45].

2. Miscible/Compatible Blocks

The preceding examples show that many of the successful block copolymer compatibilizers are of the A–C type, where block A is identical with phase A, whereas block C is different than phase B but is miscible or at least practically compatible with it. Further extension of this principle leads to the assumption that successful compatibilizers might also include block copolymers of the type C–D, where C is different from phase A but is miscible or at least compatible with it, and where D is different from phase B but is miscible or at least compatible with it. A number of experimental studies have demonstrated that this is often successful and practical.

(a) Polyethylene in General. PE + polyphenylene ether was compatibilized by adding styrene–butadiene diblock or triblock copolymers [46].

PE + polyethylene terephthalate was compatibilized by adding styrene–ethylene/butylene–styrene block copolymer, increasing interfacial adhesion and ultimate elongation [32].

(b) High Density Polyethylene. HDPE + PP was compatibilized by adding styrene–ethylene/butylene–styrene block copolymer, increasing interfacial adhesion and ultimate elongation [32].

HDPE + polyethylene terephthalate was compatibilized by adding styrene–ethylene/butylene–styrene block copolymer, which increased ductility [47].

(c) Linear Low Density Polyethylene. LLDPE + polyvinyl chloride was compatibilized by adding chlorinated polyethylene. Lower molecular weight was more effective. Up to 5% CPE increased tensile strength, while ultimate elongation increased continually with CPE content [48].

LLDPE + polyvinyl chloride was also compatibilized by adding up to 5% of isoprene–4-vinylpyridine diblock copolymer, acting as emulsifier to reduce domain size and interfacial tension. Polystyrene–acrylic acid diblock copolymer was somewhat less effective [49].

LLDPE + polyphenylene ether was compatibilized by adding ethylene–styrene diblock copolymer, which reduced and stabilized domain size and increased impact strength [37].

(d) Low Density Polyethylene. LDPE + polystyrene was compatibilized by ethylene–styrene/butadiene diblock copolymer [50].

(e) Ethylene/Propylene Rubber. EPDM + polyvinyl chloride was compatibilized by adding CPE, giving higher impact strength [51].

(f) Polypropylene. PP + high impact polystyrene was compatibilized by adding styrene–butadiene block copolymers, which reduced domain size. Triblock and pentablock copolymers also increased interfacial adhesion, ultimate elongation, and impact strength [52].

PP + styrene/maleic anhydride was compatibilized by adding styrene–butadiene or styrene–ethylene/butylene block copolymers [53].

PP + polyphenylene ether was compatibilized by adding styrene–ethylene/propylene or styrene-ethylene/butylene block copolymers, which improved dispersion and properties [54].

3. Tapered Block Copolymers

An interesting variation on the pure block copolymer is the tapered block copolymer, which is formed by copolymerizing two monomers which enter the polymer chain at very different rates. If A enters the chain faster than B, the first part of the chain will be almost polyA. As A is consumed, and the B/A ratio rises, B will enter the chain occasionally, and with increasing frequency, forming random copolymer of decreasing A/B ratio. Toward the end of the reaction, with very little A left, the last part of the chain will be almost polyB. While this does not fit the theoretical model nearly as well, experimental results sometimes suggest that such tapered block copolymers may actually be even more effective compatibilizers than pure block copolymers [9]. The explanation may lie in their ability to form a broader interphase, with a gradual transition/modulation between the two pure phases.

For example HDPE + polystyrene was compatibilized by adding up to 1.5% of ethylene–styrene tapered block copolymer, reducing domain size and increasing impact strength and fracture toughness [55, 56]. For another example, EPR + polystyrene was compatibilized by styrene–ethylene/butylene tapered block copolymer; increasing molecular weight or concentration decreased domain size [41].

B. Graft Copolymers

It is easy to visualize how a block copolymer orients at the interface and binds two immiscible polymer phases together to give a strong interface and good practical compatibility. This is why most compatibilization research has concentrated on block copolymers as physical additives. Unfortunately, most block copolymers are difficult and expensive to synthesize, which is a serious liability in industry.

By comparison, graft copolymers are usually easier and more economical to manufacture [9], so they are generally preferred in industry. Theoretically, while grafted side-chains should be readily available at the interface, they would block accessibility to the backbone, so it could not contact and penetrate its desired phase efficiently [6, 9]. Despite this theoretical handicap, graft copolymers have often proved to be very effective compatibilizers in practice. This simply indicates that our theoretical modelling requires greater sophistication.

In practice, most of our major commercial polyblends are compatibilized by graft copolymers: high-impact polypropylene, high-impact polystyrene, ABS, high-impact rigid polyvinyl chloride, and high-impact nylons. Theoretically, some examples from experimental research illustrate the performance of graft copolymers as compatibilizers for polyolefin polyblends. Using the same notation as before, they may be classified according to whether the graft copolymer has both structures identical with the separate phases A and B, or whether its action is based partly on miscibility/compatibility with the major phases.

1. Identical Structures

HDPE + lignin were compatibilized by grafting HDPE onto the lignin. This strengthened the interface and increased the modulus [57].

LLDPE + polystyrene was compatibilized by its graft copolymer, which increased the impact strength [38].

LDPE + polystyrene was compatibilized by its graft copolymer, which reduced domain size and increased interfacial adhesion [10].

2. One Identical Structure, One Miscible/Compatible

LLDPE + polystyrene was compatibilized by adding a graft copolymer of polystyrene on EPDM. This reduced domain size and increased toughness, as measured by the area under the stress–strain curve [39].

LLDPE + polyvinylidene fluoride was compatibilized by a graft copolymer of methyl methacrylate onto LLDPE. This produced mixing across the interface and bonded the two polymers to each other [58].

PP + polyvinyl chloride was compatibilized by a graft copolymer of polyethylene oxide onto PP. This produced mixing across the interface and bonded the two polymers to each other [58].

C. Other Third Polymers

A variety of other third polymers may be added to incompatible polyblends to improve compatibility. Most of them are random copolymer structures with flexible or rubbery properties. Their compatibilizing action may be visualized in three ways:

Individual attractions between atoms or groups in the third polymer and atoms or groups in the two major polymers may favor miscibility or at least practical compatibility.

Solvency of the third polymer may permit it to mix partially with both of the major polymers.

Rubberiness of the third polymer may cushion the weak interface between the two major polymers and thus prevent brittle failure.

These are illustrated by some typical examples from the experimental literature.

HDPE + PP was compatibilized by adding 5% of EPR [22].

HDPE + polystyrene and LDPE + polystyrene were compatibilized by adding ethylene/styrene random copolymers, which were semimiscible in the polyethylene phase. They decreased domain size and increased interfacial adhesion, ultimate elongation, and tensile toughness. Increasing their molecular weight also increased toughness [59].

HDPE + polyethylene terephthalate was compatibilized by adding 5% EVA, which increased ultimate elongation and impact strength [60].

HDPE + butadiene/acrylonitrile rubber was compatibilized by adding maleated HDPE, which reduced domain size, strengthened the interface, and increased tensile and impact strength [61, 62].

HDPE + acrylonitrile/methyl acrylate 75/25-g-butadiene/acrylonitrile rubber was compatibilized by adding polyethylene ionomer [63].

HDPE + nylons were compatibilized by adding polyethylene ionomer. This gave finer morphology, higher tensile strength, and increased the barrier properties of the HDPE [22, 64, 65].

HDPE + polyethylene terephthalate was compatibilized by adding carboxylated styrene–ethylene/butylene–styrene block copolymer, increasing impact strength 10-fold [66].

Whereas block and graft copolymers act primarily at the interface, and are effective at very low concentrations, these random copolymers probably act by much more gross mechanisms, and are generally effective at considerably higher concentrations. At such high concentrations, of course, they are also more likely to shift bulk properties as well, which may be for better or worse.

D. Solvents, Plasticizers, Surfactants, and Adhesives

Some of the concepts and phenomena of monomeric chemistry have been extended to polymer blends as a way of helping to understand compatibility and compatibilization.

1. Solvents

If two immiscible polymers are both soluble in the same solvent system, it may be possible to blend them in solution. Generally at high dilution they may form a homogeneous solution, but with increasing concentration the system will reach a point at which it separates into immiscible phases, each containing both polymers but in different ratios [25]. Complete removal of the solvent, under carefully controlled conditions, may help produce a more compatible blend than simple melt blending.

Many researchers extend this concept to the addition of a third polymer which is miscible with each of the two base polymers. When the third polymer improves compatibility, they refer to the same concept of a mutual solvent.

2. Plasticizers

Monomeric liquid plasticizers are closely analogous to monomeric solvents. If two polymers are immiscible, but are both miscible with the same plasticizer, adding this plasticizer to the polymer blend may improve its compatibility. For example, blends of CPE with polyvinyl chloride were compatibilized by addition of 50% of dioctyl phthalate plasticizer [67]. In commercial practice, polyvinyl chloride + polyurethane are sometimes compatibilized in a similar way.

3. Surfactants

Mixtures of water with organic liquids are emulsified by addition of monomeric surfactants, molecules which have one end attracted to the organic liquid and the other end attracted to water, and which orient accordingly at the interface. This reduces interfacial energy and interfacial tension, giving much finer and stabler droplet size. When two immiscible polymers are compatibilized by addition of a third polymer, particularly a diblock copolymer, this generally reduces interfacial energy and interfacial tension, giving much finer and stabler domain size; so many researchers refer to the effect as emulsification, and to the compatibilizer as an emulsifier or surfactant [7, 9–11, 68].

Some examples in polyolefin blends are:

Polyolefins + polyethers were compatibilized by adding graft terpolymers of polyolefin and polyether on a silicone backbone, giving very low surface tension [69].

HDPE + nylon 6 was compatibilized by adding polyethylene ionomer, which gave finer domain size due to emulsification and lower interfacial tension [70].

LLDPE + polyvinyl chloride was compatibilized by adding up to 5% of isoprene–4-vinyl pyridine diblock copolymer, acting as an emulsifier to reduce phase diameter and interfacial tension. Polystyrene–acrylic acid diblock copolymer was somewhat less effective [49].

LLDPE + PP was compatibilized by adding maleated pyrolyzed mixed polyolefin waste [71]. This would appear to be a surfactant type of effect.

4. Interfacial Adhesives

When immiscible polyblends fail prematurely under stress, electron microscopy frequently shows that their two phases break cleanly away from each other at the weak interface. When a compatibilizer makes the blend more resistant to stress, electron microscopy frequently shows surfaces that have been distorted and torn before failure. This leads many researchers to refer to the compatibilizer as an interfacial adhesive [7, 9–11]. The mechanism is usually explained by the compatibilizer penetrating and entangling in each of the two phases, and thus providing primary covalent bonding across the interface. Since theories of adhesion do vary greatly, some researchers simply refer to the compatibilizer as adhering to the surfaces of both phases; and some researchers have actually studied macroscopic models, placing a thin layer of compatibilizer between films of the two incompatible polymers, and measuring the adhesive bond which forms.

E. Reinforcing Fibers

When polyA and polyB are so immiscible that the interface between them is weak and incompatible, it may often be possible to add reinforcing fibers which bridge across the interface successfully, bond separate domains to each other, and thus make the polyblend much more resistant to stress. This has been particularly successful in recycling of mixed polymer blends.

Some examples in polyolefin blends are:

Mixed polyolefin waste (HDPE + LDPE + PP) was compatibilized by adding recycled paper fibers to increase modulus [72].

PP + polystyrene was compatibilized by adding glass fibers, increasing modulus, weldline strength, and impact strength [73].

More surprisingly, HDPE + polyurethane was compatibilized by adding kaolin clay, as judged by reverse gas chromatography [74]. Adsorption of polymer on the clay surface favored the thermodynamics of compatibility.

IV. REACTIVE PROCESSING

Addition of a physical compatibilizer is obviously the simplest and most straightforward technique for the average plastics processor. However, the compatibilizer must be matched to the polymers in the blend, with segments either identical to the base polymers, or else similar, miscible, or at least compatible with them. Such tailor-made compatibilizers are rarely available commercially, and when they are, they are usually speciality materials, made in small volume and quite expensive. For many polyblend systems, they do not exist at all, and require considerable research and custom synthesis, which are very expensive.

An alternative technique is to synthesize the compatibilizer by a chemical reaction directly in the melt blending process [7–9, 11, 31]. Chemists refer to this as "*in situ*" synthesis, while engineers generally use terms such as reactive processing, reactive extrusion, or reactive blending. Ideally, this should eliminate the extra, often costly, step of making the compatibilizer, and the necessity of adding it during the melt blending process.

As compared with conventional monomeric reactions, analogous polymer reactions are hindered by slow diffusion, low concentration of reactive groups, steric hindrance, and short time in the extruder [8]; and polyolefins in particular have very little natural reactivity available for rapid efficient synthesis of compatibilizer structures. On the other hand, only low concentrations of block or graft copolymer are required to saturate the interface and produce optimum compatibilization. Thus there has been a rapid growth of interest in reactive processing over the past decade, and it is now the most active aspect of compatibilization research.

When a polymer does not have enough natural reactivity for reactive processing, the polymer may be modified ahead of time to add the functional groups that will be needed [8, 9, 11]. This may be done during polymerization by adding a comonomer to supply the functional groups; or it may be done after polymerization by specific post-polymerization reactions. Thus polyolefins are often functionalized during polymerization by copolymerizing them with acrylics, or post-polymerization by grafting maleic anhydride or acrylics onto the polyolefin backbone. Similarly, for blending with a second polymer, the great variety of polymers B permits a great variety of modification reactions (Table 1) [8].

In most polyblend systems, the base polymers A and B are not spontaneously reactive with each other. It is necessary to add one or more additional ingredients, either catalysts or more often coreactants, to produce the compatibilization reaction [8, 11, 31]. Furthermore, while it might be simplest to feed all the ingredients to the back of the extruder screw, theory and/or practice may show that two or more reactions might best be performed in sequence to produce optimal compatibilization. For neatness, these could be carried out at different times, but this would add to the complexity of the process and to the final cost. Thus some processors have learned to feed the first ingredients to the back of the screw, and then inject other ingredients into ports along the screw, to separate and run all the reactions in their proper sequence, so that they become a series of discrete chemical reactions, but are carried out economically as a single extrusion process [11].

As noted earlier for physical compatibilizers, the compatibilizer for polyA + polyB might optimally be a block or graft copolymer of A–B, but often it may be easier, and just as effective, to make a compatibilizer A–C where C is merely compatible with B, or even C–D where C is compatible with A and D is compatible with B [7, 8, 11, 31]. Such compatibility may come from similarity, polarity, hydrogen bonding, or ionic groups. Incidentally, these may actually be superior to primary covalent bonds, because they generally weaken on heating and thus facilitate thermoplastic processability [8, 9].

TABLE 1 Compatibilizing Reactions [8]

Type	Polymer A	Polymer B	Product
Free Radical			
Active hydrogen	R_3CH	HCR'_3	$R_3C{-}CR'_3$
Vinyl	$R^{\cdot}$	$CH_2{=}CHR$	$R{-}(CH_2CHR)_n$.
Addition			
Urethane	$RNCO$	HOR'	$RNHCO_2R'$
Carbodiimide	$RN{=}C{=}NR$	$HOCOR'$	$RNHCONRCOR'$
Condensation			
Esterification	RCO_2H	HOR'	RCO_2R'
	RCO O RCO	HOR'	RCO_2R' RCO_2H
Amidification	RCO_2H	H_2NR'	$RCONHR'$
Imidification	RCO O RCO	H_2NR'	RCO NR' RCO
Dehydrohalogenation	RNH_2	BrR'	$RNHR'$
Interchange			
Alcoholysis	RCO_2R	HOR'	RCO_2R'
Transesterification	RCO_2R	$R'CO_2R'$	RCO_2R'
Aminolysis	$RCONHR$	H_2NR'	$RCONHR'$
Transamidification	$RCONHR$	$R'CONHR'$	$RCONHR'$
Ester–amide Interchange	RCO_2R	$R'CONHR'$	$RCONHR' + R'CO_2R$
Ring opening			
Epoxy + alcohol	O $RCHCH_2$	HOR'	$RCHOHCH_2OR'$
Epoxy + acid	O $RCHCH_2$	$HOCOR'$	$RCHOHCH_2OCOR'$
Epoxy + amine	O $RCHCH_2$	H_2NR'	$RCHOHCH_2NHR'$
Oxazoline + alcohol	N RC O	HOR'	$RCONHCH_2CH_2OR'$
Oxazoline + acid	N RC O	$HOCOR'$	$RCONHCH_2CH_2OCOR'$
Ionic bonding			
Amine + acid	RNH_2	$HOSO_2R'$	$RNH_3^+\ {}^-OSO_2R'$
Ionomer + ionomer	$(RSO_3)_2Zn$	$(R'SO_3)_2Zn$	RSO_3ZnOSO_2R'

Compatibilization reactions for polyolefin polyblends may conveniently be classified as free radical, acid + amine, acid + hydroxyl, and a broad variety of other miscellaneous reactions.

A. Free Radical Reactions

For saturated polyolefin homopolymers, reactivity is limited primarily to the tertiary hydrogen atoms at branch points in the polymer molecule. These can be abstracted by peroxides or radiation, producing free radicals which can then graft to each other, or to unsaturated monomers or polymers in the vicinity.

1. Two Saturated Polyolefins

LLDPE + polystyrene was heated with peroxide, forming mid-chain free radicals which coupled to produce graft copolymers, and thus compatibilized the polyblend [75].

HDPE + LLDPE + PP was compatibilized by solid-state shear extrusion. In addition to simple physical blending, the system appeared to form copolymers between the ingredients, giving good mechanical and thermal properties [15]. It has long been recognized that high mechanical shear can break polymer molecules in the middle of the chain, producing radicals which can then cross-couple with each other, forming block copolymers [76], and that is most probably what occurred in this study.

LDPE + PP were compatibilized by heating with peroxide. LDPE crosslinked, while PP degraded, so the LDPE/PP ratio determined the resulting balance of properties [77, 78].

2. Saturated + Unsaturated Polyolefins

HDPE or PP were compatibilized with EPDM by heating with dicumyl peroxide. This produced PE or PP radicals which added readily across the double bonds in the EPDM, increasing melt viscosity, tensile strength, and ultimate elongation [79].

Ethylene/methyl acrylate copolymer + EPDM (containing ethylidene norbornene as termonomer) were blended at 180°C, producing graft copolymer. The ester group in methyl acrylate greatly activates the tertiary hydrogen atom, which forms free radicals very easily, and these could easily add to the double bond in ethylidene norbornene. In this study, grafting increased modulus and strength [80].

3. Saturated Polyolefins + Unsaturated Monomers

One of the most popular processes in reactive compatibilization is the grafting of maleic anhydride (MAH) onto polyolefins, to make them more reactive in the blending process [81–83]. For example, heating LDPE, PP, or EPR with peroxide produced free radicals which added easily across the double bond of MAH, thus making the polyolefins easily reactive with nylons, polyesters, and other second polymers in polyblends.

Other unsaturated monomers which have been used in the same way include acrylic acid and glycidyl methacrylate [8, 11].

In olefin copolymers, the comonomer can be chosen to introduce a wide variety of functional groups, in measured concentration, thus greatly broadening the range of reactions which can be used to produce compatibilization. Altogether, these offer a number of one-step chemical reactions which can be used to provide reactive compatiblization.

B. Acid + Amine Reactions

Blending of polyolefins with polyamides offers the possibility of combining some of the best properties of each. Thus polyolefins generally offer melt strength, flexibility, lubricity, impact strength, electrical resistance, low dielectric constant and loss, water resistance, and low price; while polyamides offer melt fluidity, rigidity, strength, high temperature performance, and solvent resistance [84]. Thus it is not surprising that research and commercial development in these systems has been very energetic.

The major problem is that polyolefins are nonpolar, while polyamides are polar and hydrogen bonded, so the two families are not only immiscible, but generally incompatible [85]. The problem has been solved primarily by grafting a low concentration of carboxylic acid groups onto the polyolefins, and then, during melt blending, reacting these with the amine end-groups of polyamides, or possibly by interchange with the amide groups in the polyamide chain, to form polyamide–g-polyolefin graft copolymers, which tend to orient at the interface and thus compatibilize the polyolefin and polyamide phases, to combine some of the best properties of each. At high polyolefin/polyamide ratio, the polyamide generally contributes impermeability to gasoline; while at high polyamide/polyolefin ratio, the polyolefin generally contributes impact strength and water resistance. These considerable benefits have made this system the most popular area in recent and current reactive compatibilization research.

1. Polyolefins

Most of this research has concentrated primarily upon the major polyolefins—PP, HDPE, LLDPE, LDPE [54, 65, 81, 83, 86–111]—and to a lesser extent upon the more flexible metallocene polyolefins [112] and upon EPR and EPDM [90, 105, 113–119]. A few more specialized studies have looked at EVOH [120] and bromobutyl rubber [121].

2. Carboxylating Agents

The most popular reagent for carboxylating polyolefins has been maleic anhydride (MAH), generally using a peroxide to strip an unstable hydrogen off of the polyolefin backbone, and thus creating a midchain free radical which can add to the C=C double bond of MAH quite efficiently [122]. Occasionally, in place of MAH, researchers may use acrylic or methacrylic acid in the same way [91, 97]. More surprisingly, they sometimes use succinic acid in a similar way [100]. A more complex and ingenious way to graft succinic acid is by use of its peroxy half-ester, which can both produce a radical on the polyolefin and simultaneously contribute a succinic acid radical to couple with it [101]. To compatibilize EVOH with nylon, it is reacted with succinic acid to produce the half-ester [120].

Aside from maleated polyolefins, MAH is often grafted onto styrene–ethylene/butylene–styrene block copolymer, which is then used as the reactive compatibilizing agent [88–90]. Styrene/maleic anhydride copolymer has also been used to supply the MAH reactivity [93, 115, 121]. Researchers who use polyethylene ionomers as the compatibilizing agent may well be relying on its residual carboxylic acid groups to produce the reactive compatibilization in the same way [65, 89, 111, 118, 124]. For a more unique way of grafting the acid reactivity onto polyolefins, acrylamide-*t*-butyl sulfonic acid has been grafted onto polyolefin and then used to provide the grafting reaction with nylon [125].

3. Amino Polymers

The primary object of all this ingenuity has been to produce a graft copolymer of polyolefin–g–nylon, either to increase the impact strength and water resistance of nylon, or to increase the gasoline impermeability of polyolefin. Basic research has concentrated mainly on nylon 6, because every molecule offers an amine end group, which can react with MAH to form an imide graft point [126]; but commercial practice appears to offer equal numbers of high-impact grades of both nylons 6 and 66 [127], suggesting either that they insure that there are amine end-groups on the nylon 66, or else that the interchange reaction between polyolefin carboxylic acid and nylon internal amide groups is perfectly reliable [128].

Some less common amino polymers which have been reported in the research literature include aminated PP [93] and aminated EPR [93, 100], polyoxypropylene diamine [105], dimethylamino ethanol [121], and liquid crystalline polyester–b–polyamide [122].

C. Acid + Hydroxyl Reactions

Reactive compatibilization of polyolefins with other polymers has frequently been accomplished by esterification. This has most often been practiced on PP [98, 100, 122, 129–134], and occasionally also on polyethylenes [112, 135–137]. The polyolefin has usually been carboxylated by grafting maleic anhydride (MAH) onto it [138], occasionally by grafting acrylic acid [133, 134] or succinic acid [100]; in several studies it has been added as ethylene/acrylic acid copolymer [136], maleated styrene–ethylene/butylene–styrene block copolymer [130, 132], or styrene/maleic anhydride copolymer [139, 140].

The second polymer in these blends has usually been either a hydroxyl-containing polymer, which could esterify directly, or a polyester which could interchange with a polyolefin carboxylic acid group, to produce a polyolefin–g–polyester graft copolymer. Researchers have used clearly hydroxyl-containing polymers such as EVOH [137, 139, 140] or hydroxylated EPR [100], or more often hydroxy-terminated polymers such as polyphenylene ether [131], polyethylene terephthalate [129, 130, 132, 134, 136], polybutylene terephthalate [112, 131], or liquid crystal polyesters [98, 122, 133]. In some studies they have actually demonstrated ester formation, while in others they have either theorized it or else left open the possibility of simple hydrogen bonding between the carboxylic acid and the hydroxyl or ester group.

Practical motivation for these studies targets several fields. A large number of studies aim to facilitate recycling of mixed polyolefin + polyester packaging polymers by compatibilizing them together. Another group of studies aims to use liquid crystal *in situ* fibrils to reinforce PP mechanical properties [98, 122, 133]. Still another goal is to combine the properties and economics of PP with the properties of engineering plastics [129–132, 134]. Several studies have aimed to crosslink EVOH into water-absorbable products [135, 139, 140]. And several have aimed to increase the impact strength of PP [100] and polybutylene terephthalate [112].

D. Miscellaneous Compatibilization Reactions

In addition to the three major families of compatibilization reactions outlined above, organic polymer chemists have applied their ingenuity to a great variety of other reactions which can be used to compatibilize polyolefin polyblends. These may be classified as reactions of certain functional groups: epoxy, carboxylic acid, hydroxyl, amine, oxazoline, and miscellaneous others.

1. Epoxy Reactions

Glycidyl methacrylate is an unusual monomer, combining two independent types of high reactivity: the free-radical reactivity of the methacrylate group and the epoxy reactivity of the glycidyl group. The methacrylate group has usually been grafted onto polyolefins [130, 131, 141–148], or sometimes copolymerized with ethylene [90, 116, 143, 149–151] or with styrene/acrylonitrile copolymer [152, 153], to give them epoxy reactivity for compatibilization reactions. Then reaction of the epoxy group with hydroxyl-containing polymers has been used to produce ether-linked graft copolymers and thus compatibility. These have almost always been polyesters, assuming that they had hydroxyl end-groups [158]. One study reacted the epoxy group with polyphenylene ether, again assuming that it had a hydroxyl end-group [131].

Actually these polyesters might also have had carboxylic acid end-groups, which would react even more readily with epoxy to form ester-linked graft copolymers. In several studies, the epoxy group was clearly reacted with polymers containing carboxylic acid groups, producing compatibility in this way [141, 144, 145, 150, 152, 153].

The epoxy group also reacted very readily with the amine end-groups of polyamides, forming amine-linked graft copolymers for compatibilization with nylons [90, 147, 149].

2. Carboxylic Acid Reactions

Carboxylic acid groups have usually been incorporated into polyolefins by grafting MAH onto them, occasionally by grafting or copolymerization with acrylic or methacrylic acid. For compatibilization of polyblends, they have most often been reacted with amine, hydroxyl, or glycidyl groups in the second polymer, as detailed above. Occasionally they have been reacted with other groups or additives to produce compatibilization: urethane [23], oxazoline [82, 144, 154], ionomer [155], or magnesium hydroxide [156a].

3. Hydroxyl Group Reactions

Hydroxyl end-groups have most often been reacted with carboxylic acids to produce ester-linked graft copolymers, as detailed above. Their second most frequent use, as detailed above, has been reaction with glycidyl groups to produce ether-linked graft copolymers, also detailed above.

4. Amine Group Reactions

Amine end-groups have most often been reacted with carboxylic acids to produce amide-linked graft copolymers, as detailed above. Their second most frequent use, as detailed above, has been reaction with glycidyl groups to produce amine-linked graft copolymers, also detailed above.

5. Oxazoline Group Reactions

Vinyl or isopropenyl oxazoline has been grafted onto polyolefins [144] or copolymerized with styrene [82, 154], and then reacted with carboxylic acid or anhydride polymers to produce compatibilization.

6. Miscellaneous Other Compatibilization Reactions

(a) Bromine + Ammonium. Isobutylene was copolymerized with *p*-methyl styrene, then brominated on the *p*-methyl group. Butadiene/acrylonitrile rubber was terpolymerized with acrylic acid, then neutralized with tetrabutyl ammonium hydroxide. Reaction between these two modified polymers split out tetrabutyl ammonium bromide, forming an ester-linked graft copolymer, which compatibilized the blend of the two elastomers [156b].

(b) Phenolic Resins as Compatibilizers. PP + polyurethane was compatibilized by addition of phenolic resol resin [157]. Cure of the phenolic resin produced compatibilization, probably by reaction with the N—H groups in the polyurethane and formation of an interpenetrating polymer network.

HDPE + butadiene/acrylonitrile rubber was compatibilized by addition of dimethylol phenolic resin, which cured and compatibilized the blend [61, 62]. Cure reactions of diene rubbers with phenolic resins have been observed before [159], and probably formed an interpenetrating polymer network in this study.

(c) Polybutadiene + Sulfur + Magnesium Hydroxide. Mixed polyolefin waste was compatibilized by compounding with maleated liquid polybutadiene, sulfur, and magnesium hydroxide [156a]. Chemical similarities to rubber compounding and vulcanization suggest that this formed an interpenetrating polymer network.

V. COMMERCIAL PRACTICE

Current commercial practice is still quite new and secretive. Most commercial blends name the polymers in the blend, but rarely the technique or additives used to compatibilize them. From the public literature and informal discussions, it would appear that the most common technique is peroxide-initiated grafting of maleic anhydride onto an active hydrogen or a vinyl group in the backbone of polymer A, to attract or react it with a more polar group in polymer B. Most common examples of polymer A are polyethylene, polypropylene, styrene–ethylene/butylene–styrene, ethylene/propylene rubber, and EPDM. Most of the compatiblizers which are sold commercially are probably of this type, and most in-house compatibilization is probably also of this type.

Other fairly common commercial compatibilizers include ethylene multipolymers with glycidyl methacrylate, ethylene/methacrylic acid ionomers, styrene–ethylene/butylene–styrene, and chlorinated polyethylenes.

These are used primarily to combine the impact resistance of polyolefins with the engineering properties of more polar thermoplastics. They are also used to combine the water resistance of polyolefins with the gasoline impermeability of polyamides.

VI. PRESENT AND FUTURE TRENDS

Current commercial practice has already made rapid progress in improving polyblend compatibility and thus expanding the range of materials and properties which they offer to the plastics industry. The precise relationships between polyblend composition and morphology, and their specific effects on balance of properties, are only partly understood, and most current research and development is concerned with more thorough exploration of these relationships. The main drawbacks to progress are our lack of knowledge about these interrelationships, and the high level of industrial secrecy common in such a nascent stage of development.

Future development will see two competing trends. On the one hand, major polymer producers will develop and sell standard grades of compatibilizers for major polyblend markets. On the other hand, with growing understanding and experience in compatibilization, increasing numbers of polyblend processors will develop their own proprietary ingredients and techniques for compatibilizing the polyblends they sell.

Another area that must see serious development progress is increased understanding of the effects of compatibilizers and processes on morphology and final properties of polyblends. Only in this way can the field become more mature and productive.

One other area of serious interest is the recycling of mixed polymers in solid waste. Neither consumers nor recyclers have been able to separate most of the polymers in solid waste, and reuse of these mixed materials has been far from satisfactory. Hopefully, development of more successful compatibilizers and compatibilization techniques should make these mixed plastics more useful, and thus contribute to solving the growing problems of solid waste.

In all these areas, since polyolefins are the largest family of commercial polymers, their role in these developments will be a major factor in successful commercial practice.

REFERENCES

1. DR Paul, S Newman. Polymer Blends. Orlando, FL: Academic Press, 1978.
2. LA Utracki. Polymer Alloys and Blends. Munich: Hanser, 1989.
3. RD Deanin. In: HF Mark, NG Gaylord, NM Bikales, ed. Encyclopedia of Polymer Science and Technology. New York: John Wiley & Sons, 1977, Suppl vol 2, pp 458–484.
4. RD Deanin. In: K Finlayson, ed. Advances in Polymer Blends and Alloys Technology. Lancaster, PA: Technomic Publishing Co., 1992, vol 3, chap 1.
5. S Datta, DJ Lohse. Polymeric Compatibilizers. Munich: Hanser, 1996.
6. NG Gaylord. Chemtech June 1976: 392–395.
7. NG Gaylord. J Macromol Sci - Chem A26 (8):1211–1229, 1989.
8. NC Liu, WE Baker. Adv Polym Tech 11 (4): 249–262, 1992.
9. RL Markham. Adv Polym Tech 10 (3): 231–236, 1990.
10. DR Paul. In: DR Paul, S Newman, ed. Polymer Blends. Orlando, FL: Academic Press, 1978, chap 12.
11. M Xanthos, SS Dagli. Polym Eng Sci 31 (13): 929–935, 1991.
12. HFH Meijer, PJ Lemstra, PHM Elemans. Makromol Chem, Macromol Symp 16: 113-135, 1988.
13. RA Malloy, CB Thorne. SPE ANTEC 43: 3098–3101, 1997.
14. TAM Wozencroft, DI Bigio, FL Magnus, C Kiehl. SPE ANTEC 43: 2606–2613, 1997.
15. K Khait, SH Carr. SPE ANTEC 43: 3086–3089, 1997.
16. MN Bureau, H El Kadi, J Denault, JI Dickson, S Frechinet. SPE ANTEC 42: 2220–2224, 1996.
17. LY Yang, TG Smith, D Bigio. SPE ANTEC 40: 2428–2432, 1994.
18. R Potluri, CG Gogos, MR Libera, SS Dagli. SPE ANTEC 41: 3172–3176, 1995.
19. LA Utracki. Polymer Alloys and Blends. Munich: Hanser, 1989, pp 229–231.
20. J Gaona-Torres, R Gonzalez-Nunez, BD Favis. SPE ANTEC 43: 2597–2601, 1997.
21. ZJ Chen, RJ Wu, MT Shaw, RA Weiss, ML Fernandez, JS Higgins. Polym Eng Sci 35 (1): 92–99, 1995.
22. LA Utracki. Polymer Alloys and Blends. Munich: Hanser, 1989, pp 201–209.

23. ZS Petrovic, S Milledge, I Javni. SPE ANTEC 43: 3479–3482, 1997.
24. RD Deanin, NA Shah. J Vinyl Tech 5 (4): 167–171, 1983.
25. LA Utracki. Polymer Alloys and Blends. Munich: Hanser, 1989, chap 2, especially pp 30–33, 50–51, 88–92.
26. LA Utracki. Polymer Alloys and Blends. Munich: Hanser, 1989, pp 52–64.
27. RD Deanin, MF Sansone. Polym Preprints 19 (1): 211–215, 1978.
28. RD Deanin, GE D'Isidoro. Org Coatings Plastics Chem 43: 19–22, 1980.
29. AJ Lovinger, ML Williams. Org Coatings Plastics Chem 43: 13–18, 1980.
30. YC Liang, AI Isayev. SPE ANTEC 43: 2626–2630, 1997.
31. L Mascia, M Xanthos. Adv Polym Tech 11(4): 237–248, 1992.
32. DR Paul, JW Barlow, H Keskkula. In: HF Mark, NM Bikales, CG Overberger, G Menges, ed. Encyclopedia of Polymer Science and Engineering. New York: John Wiley & Sons, 1988, vol 12, pp 399–461.
33. P Ghodgaonkar, U Sundararaj. Am Chem Soc, Polym Mater Sci Eng 75: 439–440, 1996.
34. LA Utracki. Polymer Alloys and Blends. Munich: Hanser, 1989, p 127.
35. H-F Guo, NV Gvozdic, DJ Meier. Am Chem Soc, Polym Mater Sci Eng 75: 444–445, 1996.
36. D Bourry, BD Favis. SPE ANTEC 41: 2001–2009, 1995.
37. JR Campbell, SY Hobbs, TJ Shea, DJ Smith, VH Watkins, ICW Wang. In: BM Culbertson, ed. Multiphase Macromolecular Systems. New York: Plenum Press, 1989, pp 439–459.
38. ML Arnal, A Covuccia, AJ Muller. SPE ANTEC 43: 3775–3779, 1997.
39. T Li, M Roha, A Hiltner, E Baer. SPE ANTEC 38: 2635–2638, 1992.
40. J Simitzis, C Paitontzis, N Economides. Polym Polym Composites 3 (6): 427–434, 1995.
41. P Cigana, BD Favis. Am Chem Soc, Polym Mater Sci Eng 73: 16–17, 1995.
42. P Cigana, V Benoit, A Tremblay, BD Favis. SPE ANTEC 43: 1527–1531, 1997.
43. M Matos, P Lomellini, BD Favis. SPE ANTEC 40: 1517–1521, 1994.
44. SN Koklas, NK Kalfoglou. Polym 35 (7): 1433–1441, 1994.
45. ME Matos, C Rosales, AJ Muller, BD Favis. SPE ANTEC 43: 2552–2556, 1997.
46. SG Cottis, KM Natarajan, D Elwood, X Xu. SPE ANTEC 39: 492–498, 1993.
47. A Spaak. SPE ANTEC 31: 1158–1163, 1985.
48. P He, H Huang, W Xiao, S Huang, S Cheng. J Appl Polym Sci 64: 2535–2541, 1997.
49. H Liang, BD Favis, Y Yu, A Eisenberg. SPE ANTEC 43:1764–1768, 1997.
50. LA Utracki. Polymer Alloys and Blends. Munich: Hanser, 1989, p 125.
51. Y-D Lee, C-M Chen. J Appl Polym Sci 33: 1231–1240, 1987.
52. Z Horak, V Fort, D Hlavata, F Lednicky, F Vecerka. Polym 37(1): 65–73, 1996.
53. D Chundury. US Pat 5,321,081,1994.
54. D Gilmore, J Kirkpatrick, MJ Modic, SPE ANTEC 36: 1228–1233, 1990.
55. MF Champagne, MM Dumoulin. Am Chem Soc, Polym Mater Sci Eng 75: 443, 1996.
56. MF Champagne, MM Dumoulin. SPE ANTEC 43: 2562–2566, 1997.
57. S Casenave, A Ait-Kadi, B Brahimi, B Riedl. SPE ANTEC 41: 1438–1442, 1995.
58. T Tang, H Chen, X Zhang, L Li, B Huang. Polym 35(19): 4240–4242, 1994.
59. CP Park, GP Clingerman. SPE ANTEC 42: 1887–1891, 1996.
60. A Khelifi, F Lai. SPE ANTEC 34: 1824–1827, 1988.
61. J George, R Joseph, S Thomas, KT Varughese. J Appl Polym Sci 57: 449–465, 1995.
62. J George, L Prasannakumari, P Koshy, KT Varughese, S Thomas. Polym-Plast Technol Eng 34 (4): 561–579, 1995.
63. M Shah, SA Jabarin. SPE ANTEC 41: 1580–1587, 1995.
64. B Fisa, A Bouti, BD Favis, F Lalande. SPE ANTEC 37: 1135–1139, 1991.
65. H Potente, X Gao. SPE ANTEC 40: 3050–3054, 1994.
66. I-M Chen, C-M Shiah. SPE ANTEC 35: 1802–1806, 1989.
67. M Champagne, P Perrin, RE Prud'homme. SPE ANTEC 37: 1037–1040, 1991.
68. CC Chen, JL White. Polym Eng Sci 33 (14): 923–930, 1993.
69. I Yilgor, E Yilgor, J Venzmer, R Spiegler. Am Chem Soc, Polym Mater Sci Eng 75: 283–284, 1996.
70. N Chapleau, BD Favis, PJ Carreau. SPE ANTEC 43: 1695–1699, 1997.
71. C Vasile, RD Deanin, M Mihaies, C Roy, A Chaala, W Ma. Int J Polym Mater 37: 173–199, 1997.
72. Z Krulis, D Michalkova, J Mikesova, BV Kokta. Polyolefin composites reinforced by cellulosic fibers of paper waste. Symposium on Recycling of Polymers, Institute of Macromolecular Chemistry, Prague, 1997, p 21.
73. M Huang, D Done, C Wei-Berk. SPE ANTEC 40: 2494–2498, 1994.
74. VV Shifrin, YS Lipatov, AY Nesterov. Polym Sci USSR 27 (2): 412–417, 1985.
75. V Flaris, WE Baker. SPE ANTEC 41: 3164–3167, 1995.
76. RJ Ceresa. Block and Graft Polymerization. New York: Wiley-Interscience, 1962.
77. DW Yu, CG Gogos, M Xanthos. SPE ANTEC 36: 1917–1920, 1990.
78. DW Yu, M Xanthos, CG Gogos. SPE ANTEC 37: 643–646, 1991.

79. OS Rodriguez-Fernandez, S Sanchez-Valdes, R Ramirez-Vargas, I Yanez Flores, ML Lopez-Quintanilla. SPE ANTEC 43: 2636–2640, 1997.
80. S Mohanty, PK Bhattacharya, GB Nando. SPE ANTEC 42: 2234–2239, 1996.
81. MK Akkapeddi, B Van Buskirk, GJ Dege. SPE ANTEC 40: 1509–1513, 1994.
82. PJ Perron. Achieving higher performance in engineering alloys via reactive compounding. Compalloy 1989, pp 263–273.
83. E Rizzardo, M O'Shea. Plastics News International: Dec 1994, p 25.
84. RD Deanin, Polymer Structure, Properties, and Applications. Boston: Cahners Books, 1972, chap 8.
85. O Olabisi, LM Robeson, MT Shaw. Polymer-Polymer Miscibility. New York: Academic Press, 1979.
86. H Asthana, K Jayaraman. SPE ANTEC 43: 1101–1105, 1997.
87. KA Borden, RP Herrle, AG Lee, CR Manganaro. SPE ANTEC 42: 2982–2987, 1996.
88. K Chandramouli, SA Jabarin. SPE ANTEC 39: 2111–2119, 1993.
89. CC Chen, JL White. Polym Eng Sci 33 (14): 923–930, 1993.
90. D Chundury, BGJ Bitsch. US Pat 5,317,059, 1994.
91. SS Dagli, KM Kamdar, M Xanthos. SPE ANTEC 39: 18–22, 1993.
92. RJ Datta, MB Polk, S Kumar. Polym-Plast Technol Eng 34 (4): 551–560, 1995.
93. N Dharmarajan, S Datta, G Ver Strate, L Ban. Polym 36 (20): 3849–3861, 1995.
94. V Flaris, J Boocock, R Wissmann, K Hausmann. Novel compatibilizers for polyamide/polypropylene blends. DuPont Company, 1996.
95. H Garmabi, MR Kamal. SPE ANTEC 43: 2687–2691, 1997.
96. CG Hagberg, JL Dickerson. SPE ANTEC 42: 288–294, 1996.
97. B Jurkowski, K Kelar, D Ciesielska, R Urbanowicz. Kautschuk Gummi Kunststoffe 47: 642–645, 1994.
98. J Kirjava, T Rundqvist, R Holsti-Miettinen, M Heino, T Vainio. J Appl Polym Sci 55: 1069–1079, 1995.
99. JS Lin, EY Sheu, YHR Jois. J Appl Polym Sci 55: 655–666, 1995.
100. DJ Lohse, S Datta. SPE ANTEC 38: 1036–1038, 1992.
101. SC Manning, RB Moore. SPE ANTEC 43: 2976–2980, 1997.
102. KM McLoughlin, SJ Elliott, EB Townsend IV, MG Elliott. SPE ANTEC 43: 2986–2990, 1997.
103. MJ Modic. SPE ANTEC 39: 205–210, 1993.
104. MJ Modic, LA Pottick. SPE ANTEC 37: 1907–1909, 1991.
105. TTM Phan, AJ DeNicola Jr., LS Schadler. SPE ANTEC 43: 2557–2561, 1997.
106. D Roberts, RC Constable, S Thiruvengada. SPE ANTEC 42: 1831–1838, 1996.
107. S Sanchez-Valdes, I Yanez-Flores, C Guerrero-Salazar, M Lopez-Quintanilla, F Orona-Villareal, R Ramirez-Vargas. SPE ANTEC 43: 1149–1153, 1997.
108. DK Setua, S Lim, JL White. SPE ANTEC 38: 2686–2688, 1992.
109. AN Thakkar, SJ Grossman. SPE ANTEC 37: 647–650, 1991.
110. J-Y Wu, W-C Lee, W-F Kuo, H-C Kao, M-S Lee, J-L Lin. Adv Polym Tech 14 (1): 47–58, 1995.
111. J-T Yeh, C-C Fan-Chiang, S-S Yang. J Appl Polym Sci 64: 1531–1540, 1997.
112. LM Sherman, Plastics Tech June 1996, 23–24.
113. LL Ban, MJ Doyle, MM Disko, G Braun, GR Smith. In: PJ Lemstra, LA Kleintjens, ed. Integration of Fundamental Polymer Science and Technology. New York: Elsevier, 1989, pp 117–122.
114. JCC Cheng, RH Chang, FJ Tsai, SH Chang. SPE ANTEC 35: 1830–1834, 1989.
115. D Ghidoni, GC Fasulo, D Cecchele, M Merlotti, G Sterzi, R Nocci. J Mater Sci 28 (15): 4119–4128, 1993.
116. KM Kamdar, SS Dagli. SPE ANTEC 39: 1249–1254, 1993.
117. R Legras, P Marechal, JM Dekoninck. SPE ANTEC 40: 1504–1508, 1994.
118. C-CM Ma, F-H Chang, D-K Chen, H-C Shen. SPE ANTEC 36: 1866–1873, 1990.
119. G Maglio, R Palumbo. In: M Kryszewski, A Galeski, E Martuscelli, ed. Polymer Blends: Processing, Morphology, and Properties, Vol 2. Proc Second Polish-Italian Joint Seminar on Multicomponent Polymeric Systems, Lodz, 1982, pp 41–56.
120. M Pracella, M Simonetti, P Laurienzo, M Malinconico. A study of nylon 6/modified EVOH blends for recyclable packaging films. Symposium on Recycling of Polymers, Institute of Macromolecular Chemistry, Prague, 1997, p 26.
121. A Tremblay, S Tremblay, BD Favis, A Selmani, G L'Esperance. SPE ANTEC 41: 1559–1563, 1995.
122. HJ O'Donnell, A Datta, DG Baird. SPE ANTEC 38: 2248–2252, 1992.
123. Refs 54, 81, 83, 86, 87, 89, 90, 92, 94–99, 102–110, 112–118, 122.
124. C Lavallee, BD Favis. SPE ANTEC 37: 973–977, 1991.
125. Z Yin, Y Zhang, X Zhang, D Wang, J Yin. Int J Polym Mater 37: 23–32, 1997.
126. Refs 65, 83, 86–92, 95, 97, 98, 102, 104, 107, 108, 110, 114, 115, 117, 118, 120, 124.
127. Plastics Technology Manufacturing Handbook. New York: Plaspec, Annual.
128. Refs 54, 98, 99, 101, 106, 109, 113.
129. M-K Chung, D Chan. Polym Int 43: 281–287, 1997.
130. M Heino, J Kirjava, P Hietaoja, J Seppala. J Appl Polym Sci 65: 241–249, 1997.
131. HK Kotlar, KL Borve. SPE ANTEC 41: 1843–1847, 1995.
132. J-C Lepers, BD Favis, SL Kent. SPE ANTEC 43: 1788–1792, 1997.
133. MM Miller, JMG Cowie, JG Tait, DL Brydon, RR Mather. Polym 36 (16): 3107–3112, 1995.

134. MW Young, M Xanthos, JA Biesenberger, SPE ANTEC 35:1835–1842, 1989.
135. NG Gaylord, H Ender, LR Davis. SPE ANTEC 40: 1646–1648, 1994.
136. S Kim, CE Park, JH An, D Lee, J Kim. Polym J 29 (3): 274–278, 1997.
137. A Prasad, P Jackson. Am Chem Soc, Polym Mater Sci Eng 75: 281–282, 1996.
138. Refs 98, 112, 122, 129, 131, 135, 137.
139. J Curry. SPE ANTEC 41: 1838–1842, 1995.
140. PS Hope, JG Bonner, JE Curry. Pure Appl Chem 68: 1665–1682,1996.
141. WE Baker, B Wong, L Chen. Am Chem Soc, Polym Mater Sci Eng 75: 437–438, 1996.
142. CM Chen, HR Lee, SY Wu, CP Chen, HC Kao. SPE ANTEC 40: 2002–2006, 1994.
143. G-H Hu, Y-J Sun, M Lambla. Polym Eng Sci 36 (5): 676–684, 1996.
144. NC Liu, HQ Xie, WE Baker. Polym 34 (22): 4680–4687, 1993.
145. TM Liu, R Evans, WE Baker. SPE ANTEC 41: 1564–1571, 1995.
146. Y-J Sun, G-H Hu, M Lambla. SPE ANTEC 42: 270–276, 1996.
147. H-H Wang, W-C Lee, D-T Su, W-J Cheng, B-Y Lin. SPE ANTEC 41: 2105–2109, 1995.
148. L Yao, C Beatty. SPE ANTEC 43: 2577–2581, 1997.
149. JW Barlow, G Shaver, DR Paul. Polymer blend compatibilization through chemical reaction. Compalloy, 1989.
150. S Hojabr, EM Lundhild. A novel approach to compatibilization of polyethylene-poly(ethylene terephthalate) blends. DuPont Company, 1996.
151. RM Holsti-Miettinen, MT Heino, JV Seppala. J Appl Polym Sci 57: 573–586, 1995.
152. SC Stinson, Chem Eng News Jan 15, 1996, 20–22.
153. L Yao, C Beatty, SPE ANTEC 43: 2582–2586, 1997.
154. M Saleem, WE Baker. J Appl Polym Sci 39: 655–678, 1990.
155. C-S Ha, Y-W Cho, Y Kim, J-H Go, W-J Cho. Am Chem Soc, Polym Preprints 37 (1): 375–376, 1996.
156a. Z Krulis, D Michalkova, F Lednicky, Z Horak, M Sufcak. Recycling of polyolefin waste using maleinized liquid polybutadiene. Symposium on Recycling of Polymers, Institute of Macromolecular Chemistry, Prague, 1997, p 20.
156b. P Arjunan. US Pat 5,397,837, 1995.
157. W-Y Chiang, B Pukanszky, W-C Wu. SPE ANTEC 42: 2230–2233, 1996.
158. Refs 116, 130, 131, 143, 146, 148, 150, 151.
159. NJL Megson. Phenolic Resin Chemistry. New York: Academic Press, 1958, pp 126–129.

24

Surface Modifications of Polyolefins by Gas-Phase Methods

Mariana Gheorghiu and Gheorghe Popa
"Al. I. Cuza" University, Iasi, Romania

Mihaela Cristina Pascu
"Gr. T. Popa" Medicine and Pharmacy University, Iasi, Romania

I. INTRODUCTION

For most polyolefins, in particular polyethylene (PE) and polypropylene (PP), the outermost surface is hydrophobic and inert due to a preponderence of nonpolar bonds. This is why extensive work to develop practical and economical methods for surface modification of polyolefins has been carried out by many workers.

Over the years, several methods have been developed in order to modify polymer surfaces for improved adhesion, wettability, printability, dye uptake, etc. These methods include mechanical and wet chemical treatments, and exposure to gas phase processes like corona discharge, flame, UV/ozone, glow discharge plasmas, and particle beams.

Mechanical roughening alone has limited effectiveness. Wet chemical treatments with solvents, strong acids or bases, have problems with uniformity and reproductibility. Furthermore, wet chemical treatments are becoming unacceptable due to the environmental and safety considerations. Uniformity and reproductibility are also criticisms against flame treatments.

Modifications of polymer surfaces by plasma treatments, both corona and low pressure glow discharges, present many important advantages and overcome the drawbacks of the other processes mentioned above. For these reasons, plasma processes have been widely accepted over the years in various industrial applications.

This chapter comparatively presents the most used gas phase methods of treatment for improving polyolefin properties (e.g., UV/ozone and laser irradiation, particle beam bombardment, plasma and corona treatment), as well as their main applications. This chapter starts with the presentation of gas phase methods, which involve only one active element. This one can be either UV radiation or laser beam (Section III) with well-defined wavelength and intensity, or charged particle beams (Section IV) with given energy and fluence. Afterwards, the treatments in complex medium like cold plasma (Section V) or corona and silent discharge (Section VI) are presented.

The schematic diagram of the reactor that can be used in order to realize surface modifications of polymers by UV/ozone treatment is presented in Fig. 1a. The film samples are placed in a stainless steel reactor and are irradiated with UV radiation coming from a UV light source, which could be an ozone-generating mercury vapor grid lamp. In order to supplementary generate ozone, an ozone generator, using a silent discharge, could be added.

In order to obtain laser-induced chemical reactions, the substrate, which can be moved, is placed at the focal point of the laser beam in a chemical chamber filled with a reactant gas (Fig. 1b). The laser beam passes through a fixed optical system and is focused by means of a lens.

Electron sources include heated filaments, photoelectric emission, field emission, hollow cathodes and secondary electron emission. Ions can be extracted from a low pressure discharge plasma. After ions and electrons are generated, they are extracted through an

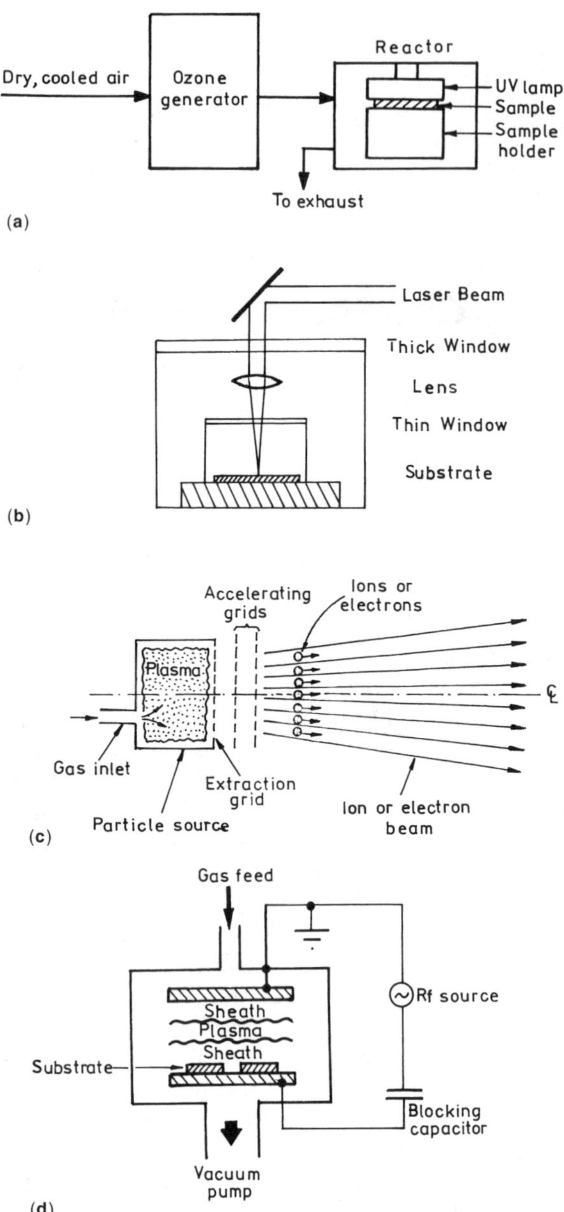

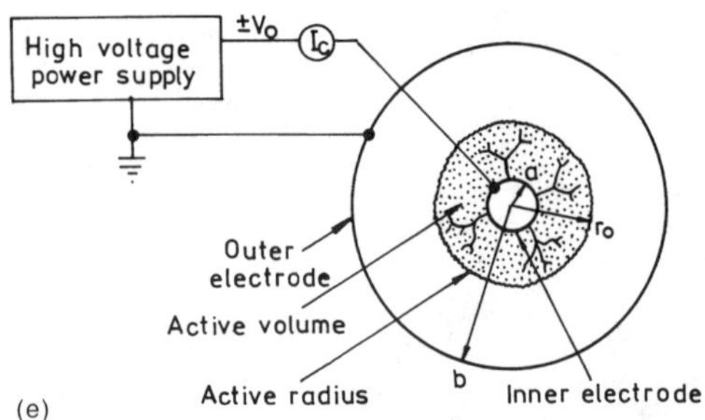

FIG. 1 Simplified schema for: (a) the reactor used for UV/ozone treatment of polymers; (b) the reactor used in order to obtain laser-induced chemical reactions; (c) ion or electron source; (d) an idealized low pressure discharge; (e) the corona discharge.

extraction grid. The accelerating grids allow us to form them into an unidirectional monoenergetic beam (Fig. 1c).

Low pressure discharges are a gas medium in which feedback gases are broken into positive ions, electrons, and chemically reactive species which then flow to the substrate surface and physically or chemically react with it. An idealized discharge in plane parallel geometry, shown in Fig. 1d, consists of a vacuum chamber containing two planar electrodes and driven by a rf or dc power source. The substrates are placed on one electrode, feedstock gases are admitted to flow through the discharge and effluent gases are removed by the vacuum pumps.

Besides ions, electrons, and excited species, radiations with a large spread of wavelength, from bulk plasma, arrive at the substrate surface.

A peculiar electrical discharge is corona, which occurs in regions of high electric field near sharp points or wires, in electrically stressed gases (Fig. 1e).

Corona treatment is, by far, the most widely used plasma technique, because it has the advantage of operating at atmospheric pressure, the reagent gas usually being the ambient air. However, this circumstance is also an important limitation on its chemical effects, other than surface oxidation; its operation using reagent gases other than air is uneconomical and hazardous. Fundamental and applied aspects of corona treatment are discussed in publications by Briggs and coworkers [1, 2], Gerenser and coworkers [3, 4], and Strobel *et al.* [5, 6].

The industrial use of *plasma processing* has been developed mainly by the microelectronics industry since the late 1960s, for the deposition of thin film materials and plasma etching of semiconductors, metals, and polymers such as organic photoresist. The third type of plasma process for surface modification is currently used in areas other than microelectronics, namely in aerospace, automotive, biomaterials, and packaging, to name only a few examples. The potential for obtaining unique surface modifications by plasma treatment is widely recognized [7].

Ozone-generating UV light has been used for many years to clean organic contaminants from various surfaces [8–10], to increase the wettability of different polymer surfaces (like PE, PP, PET and PS) [11–13], and their adhesive properties and dyeability [14–16].

UV treatment of surfaces was also used [17–20] in order to accelerate aging of polymeric materials (in this way the polymers become more environmentally friendly materials).

Investigations in the *high-energy radiation* effects in polymers have been conducted in three main areas [21]: nuclear and high voltage accelerator technology, spacecraft technology, and, more recently, microelectronics technology (for example, ion-beam litography and conductive pattern inscription). While the first two directions mainly concern the effects of ionizing radiation (UV, electrons, and γ rays) on the bulk properties of materials, the third one mainly utilizes charged particles to induce surface modifications.

Significant progress has been made in improving surface-sensitive mechanical, chemical, and physical properties, such as hardness and resistance to wear, abrasion, erosion or chemicals, by ion beam processing [21]. These improvements in properties not only enhance the performance of polymers in current applications, but also open up totally new product areas, such as—in particular, for polymeric materials—in machinery components, where the inherent softness and poor abrasion resistance of polymers have been the major limiting factors.

Ion beam modification of polymers was also very useful in preparing polymer surfaces for metallization [22], ion bombardment of the polymer surface prior to metal film deposition leading to the enhancement of the metal–polymer interaction.

Radiation grafting [22–27] is a very versatile technique by which surface properties of a polymer can be tailored through the choice of different monomers, the most common radiation sources used in this case being high energy electrons, γ radiation, and UV and visible light.

II. UV AND OZONE TREATMENT

A. General Aspects

The most important criterion which has to be fulfilled in order to use the photochemical method for obtaining improved properties of polymeric surfaces is that the polymer substrate becomes photoreactive during irradiation [28].

UV radiation may lead to changes in polymer color and to degradation of their physical properties, especially in polyolefins, ABS, PVC, and PC [29].

As was pointed out in the introduction, ozone-generating UV light is widely used to clean organic contaminants from various surfaces [10, 30–32]. The parameters which influence the UV cleaning procedure are: the contaminants initially present, the precleaning procedure, the wavelengths emitted by the UV source, the atmosphere between the source and the sample, and the time of exposure. The cleaning mechanism [10] seems to be a photosensitized oxidation process, in which the contaminant molecules are excited and/or dissociated by the absorption of short wavelength UV light. Simultaneously, atomic oxygen is generated when molecular oxygen and ozone are dissociated by the absorption of short and long wavelengths of radiation. The products of the excitation of contaminant molecules react with atomic oxygen to form simpler molecules, such as CO_2 and H_2O, which desorb from the surface.

UV/ozone treatment has also been used in order to increase the wettability of different polymer surfaces, such as PE, PP, PET, and PS [11–13].

The individual and combined effects of UV light and ozone on PE films were studied by Peeling and Clark [33] for treatment times of the order of hours in an oxygen atmosphere, these treatments leading to fully oxidized surfaces. In order to characterize the mechanisms of oxidation in weathering situations, Rabek *et al.* [11] studied the effect of ozone and atomic oxygen on PP, in the presence of UV light, for 1–8 h; they found out that extensive oxidation led to the embrittlement of the polymer, indicating some effect on the bulk of the polymer. Dasgupta [34] studied the oxidation of PP and PE in water, through which ozone was passed. Appearance of oxidation groups was obtained by UV light/ozone treatment of polyolefins (PO) [15], this type of treatment also leading to an improvement in wettability, adhesive properties [14-16] and dyeability [15] of PO surfaces.

UV treatment of surfaces is also widely used [17, 18, 35–37] in order to accelerate the aging of polymeric materials, by increasing the speed of the chemical reactions and to increase their crystallinity [38, 39].

The mechanism and extent of oxidative degradation depend [40] on the inherent polymer sensitivity (the chemical structures, i.e., bond strength and chromophores), the morphology (crystalline vs amorphous areas) and the presence of additives (i.e., stabilizers, pigments, fillers, nucleating agents, or flame retardants).

During the plasma treatment of polymers, UV irradiation also plays an important role. It was found, by plasma induced luminescence measurements [36, 41, 42], that one of the main emission mechanisms and, in the meantime, the fastest one is that involving the photoluminescence induced by UV irradiation during the exposure in the plasma. The photoluminescence component decreases quickly, the lifetime being 2 s [41]. Whereas plasma interaction induces strong surface transformations, UV probably acts through more selective processes such as initiation of decomposition of hydroperoxides [44]. Because the UV photons penetrate deeply into the polymer volume, the emission is not really a surface emission, but a signature of the material bulk.

B. UV Irradiation

Novak [16] irradiated 0.10–0.15 mm thick foils of unstabilized isotactic polypropylene (IPP) ($M_W = 2.16 \times 10^{15}$) with UV radiation of 366 nm wavelength. He studied the effect of the treatment time and of the aging time on the free surface energy (γ_s) of modified IPP, for different distances of the surface sample from the discharge lamp (Fig. 2). Free surface energy increases up to 1–2 min of UV treatment in comparison with the virgin surface, γ_s stabilization taking place after almost 2 min of treatment. For a higher distance of the UV source from the polymer surface, the efficiency of the modification decreases, a fact evidenced by a smaller increase in the values of γ_s. The decrease of the free surface energy as a function of the aging time is very small (ca. 4.5%).

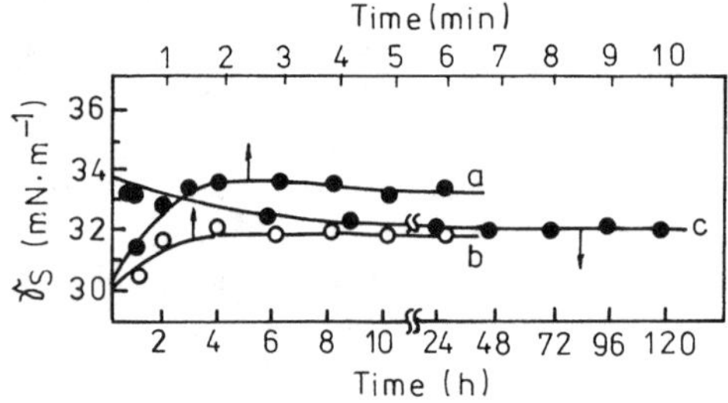

FIG. 2 Variation of the free surface energy (γ_S) of PP, UV irradiated, with: (i) the treatment time, for different distances of the UV source from the polymer surface: (a) $d = 50$ mm (●); (b) $d = 100$ mm (○); (ii) the aging time: (c) ($d = 50$ mm, $t = 2$ min) (●). (Adapted from Ref. 16.)

In order to overcome the pollution problems (especially of soil), problems induced by the accumulation in time of the photo-oxidative resistant polymers as waste, in the composition of these materials there are incorporated very small quantities (~2 wt%) of photosensitizers (photoinitiators) [19, 43–45]. Kosikova *et al.* [20, 46] mentioned that lignin-containing blends at high concentrations (>8–10 wt%) of lignin should present a satisfactory rate of photo- and/or biodegradation. The obtained mixtures could also be exposed to UV or high-energy radiation, this method leading to the initiation of polymer photo-oxidation.

Ternary blends of IPP/epoxidized lignin (LER)/compatibilizing agent as glycidylmethacrylate grafted polypropylene (PP–g–GMA) have been photooxidized [47] using a UV source with intensity of 1.120 kW/m^2, for 50, 100, and 200 h. The composition of IPP/LER blends is presented in Table 1.

The microscopy results showed that, even for smaller exposure times, the irradiated samples presented different behaviors: after a 200 h exposure period, all samples containing LER/PP–g–GMA totally disintegrated. All films changed their color, from transparent green/light yellow to yellow or dark yellow, with increasing lignin content and exposure time.

Incorporation of functionalized (epoxidized) hydrophilic lignin into the hydrophobic isotactic PP changed the superficial properties of the later component of the blends. UV irradiation of the studied mixtures also induced surface modifications in respect with the untreated samples (Fig. 3).

The dispersive component γ_S^{LW} of the surface free energy shows a continuous increase with the exposure time, regardless of the LER content. Differences between the studied films appear in the variation of the acid–base component (γ_S^{ab}) of the surface free energy, as

TABLE 1 Composition of the IPP/epoxy-modified lignin blends

No.	Symbol	IPP (wt%)	Grafted IPP (wt%)	Type of grafted IPP	Lignin (wt%)	DDM (wt%)
1	IPP	100.00	—	—	—	—
2	S6	96.00	—	—	3.96	0.03
3	S5	79.07	13.95	PP-g-GMA-II	6.97	0.102
4	S1	81.71	9.08	PP-g-GMA-I	9.08	0.135
5	S3	73.66	13.05	PP-g-GMA-PP-I	13.06	0.196
6	S4	73.70	13.06	PP-g-GMA-PP-II	13.05	0.195

Source: Ref. 47.

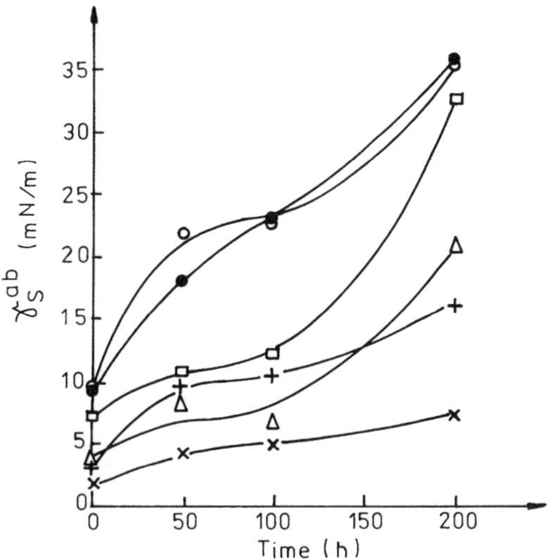

FIG. 3 Variation with the UV irradiation time of the acid–base (γ_S^{ab}) component of the surface free energy for IPP/LER blends. IPP (×), S6 (+), S5 (△), S1 (□), S3 (○), S4 (●). (From Ref. 47.)

well as in the surface free energy ($\gamma_S = \gamma_S^{LW} + \gamma_S^{ab}$) with the UV irradiation period.

The samples with low LER content show constant or slightly larger values than those of the unexposed films, until 100 h of exposure time, a behavior close to that of IPP. For longer UV irradiation times (>100 h), γ_S^{ab} shows a sudden increase. S3 and S4 mixtures, containing higher amounts of LER, have high γ_S^{ab} values for the entire UV exposure time; this behavior is characteristic for the unstable surfaces in respect to photo-oxidation.

The explanation for the increase of the surface free energy due to the oxygen-containing groups formation is proved by the changes in the IR spectra. In the 1600–2000 cm^{-1} range, a decrease of the intensity of the 1650 cm^{-1} absorbtion bands and an increase of the 1720–1800 cm^{-1} band stood out. This behavior has been assigned to the pronounced oxidation, UV-initiated, of the films. LER could play a chromophore role, absorbing UV radiation and thus leading to the excited states with various reactivities present. These states can initiate a new series of oxidation–reduction reactions, resulting in phenolic radicals and electron-excited species. These radicals, together with other active species, initiate oxidation of IPP matrix [46, 48] and an increase of chromophore groups of lignin [49]. The existence of chromophores in the UV-irradiated samples was also proved by UV spectra [47].

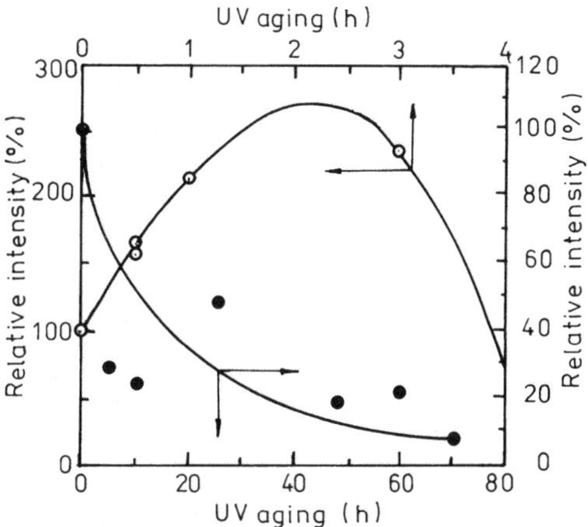

FIG. 4 Electroluminescence intensity for PP films vs. the UV aging time: short irradiation time (○) and long irradiation time (●). (From Ref. 35.)

The UV irradiation of PP-based films could also be used to increase their susceptibility to the attack of physical, chemical and biological factors [50], in this way making them integrable in the environment.

PP films of 18 μm thickness, containing a small quantity of antioxidant (hindered phenol), were exposed to UV radiation (275–375 nm), at 53°C, for times from 0.5–71 h [35]. For these samples, there were studied the electroluminescence (EL) exhibited by the films when applying an ac voltage and the relative molecular weight, for different times of UV aging. The EL intensity as a function of the exposure time shows two distinct behaviors (Fig. 4). For short treatment times (≤ 3 h), the EL intensity increases in respect with the nonaged sample, maybe due to the fact that the antioxidant, initially, protects the polymer. In the case of longer times of UV exposure, the EL intensity decreases with aging time, because the polymer itself starts to degrade and the light intensity rapidly decreases. A nonnegligible decrease in the relative molecular weight was also evidenced [35], as well as in the case of the EL intensity, which also decreased with the aging time.

C. UV/Ozone Treatment

Low density polyethylene (LDPE) and polypropylene (PP) sheets of 0.5 mm thickness were investigated after the UVO treatment (the oxygen flow rate ranging from 0.5–0.7 L/min and the ozone flow rate between 4.5–20.3 mL/min) [15], in order to show the induced modifications.

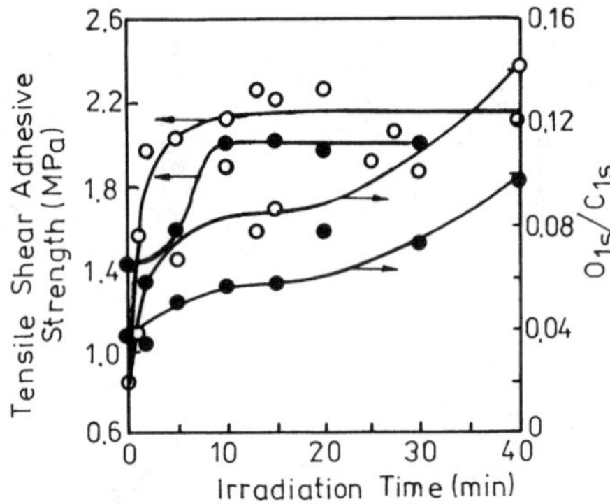

FIG. 5 O 1s/C 1s and the tensile shear adhesive strength as functions of the irradiation time, for PE (○) and PP (●) sheets. The treatment conditions: 40°C temperature, 3 L/min O_2 flow rate. (From Ref. 15.)

Changes of O1s/C1s and of the tensile shear adhesive strength with irradiation time of PE and PP sheets are shown in Fig. 5. The initially faster oxidation of the PE shows that the ratio of O/C can reach a level which changes the surface composition distinctly, after an irradiation time of approximately 1 min. This observation is very important from the practical point of view. It can also be observed that the tensile shear adhesive strength increases rapidly in the first 10 min of UVO treatment, an increase which coincides with the increase of the oxygen functional groups, measured from the ESCA spectra. Furthermore, the bulk tensile strength of modified PE or PP samples was the same or slightly higher than that of pure PE or PP without UV treatment, this fact leading to the conclusion that UVO treatment can not only improve the tensile shear adhesive strength, but also maintain the bulk property of polyolefine samples.

D. Efficiency of Various Combinations of UV Light and Ozone

In order to investigate the efficiency of various combinations of UV light and ozone in modifying the surface chemistry of PP, thermally extruded, biaxially oriented PP films 0.03 mm thick, underwent different treatment regimes [30]: UV/air, UV/air + ozone, and ozone only.

In Fig. 6, the effect of treatment time on the oxygen uptake on PP and on the advancing contact angle of water on the PP surface is presented, for different treatment regimes.

PP was oxidized when treated with UV/air, UV/air + ozone, and ozone alone, relative to the control PP, which contained no detectable oxygen.

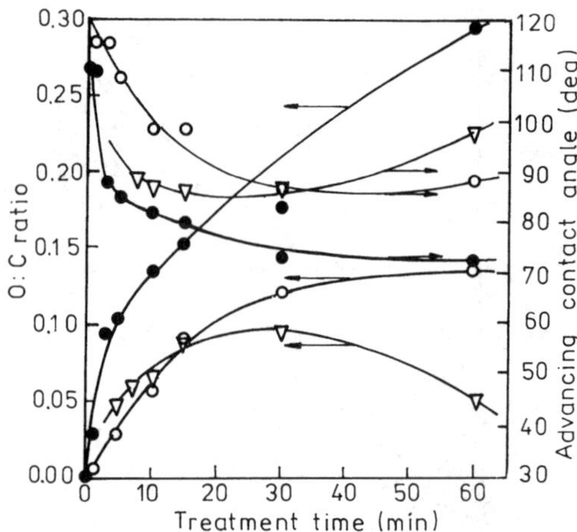

FIG. 6 The oxygen uptake (O : C ratio) on PP and the advancing contact angle of water on PP, for different treatment regimes: (○) ozone, (●) UV/air + ozone, (▽) UV/air. (From Ref. 30.)

With increasing treatment time, the advancing contact angle of water on the PP surface is diminished for all the treatment regimes. In the case of the ozone-only treatment, the contact angle reaches a plateau after 30 min of irradiation. The UV/air + ozone treatment resulted in the fastest initial reaction rate and also in the highest final O : C ratios, while the ozone-only treatment led to a polymer surface with consistently higher advancing contact angle.

The optimal treatment times for each of these regimes, appreciated by a value of the O : C atomic ratio of approximately 0.12, were found to be 15, 10, and 30 min [31] in the case of, respectively, PP–UV/air, UV/air + ozone, and ozone-only treatment.

The obtained results in the case of ozone-only treatment support the reaction mechanisms of chain addition rather than scission, while the UV/air treatment, with the high transmission of UV light and low levels of reactive species, should result in much more chain scission and crosslinking of the polymer [30, 51], as well as oxidation at the surface. Embrittlement of the polymer under these conditions appeared after only 30 min.

Concerning the UV/air + ozone treatment [30, 52], it is clear that PP is modified more quickly and to a greater extent than PP exposed to UV/air or ozone alone. These effects can depend greatly on the UV-absorbing characteristics of polymers [52]. The treated PP surface presented distinct mounds that were removed after water washing, an observation which demonstrates the presence of low molecular weight

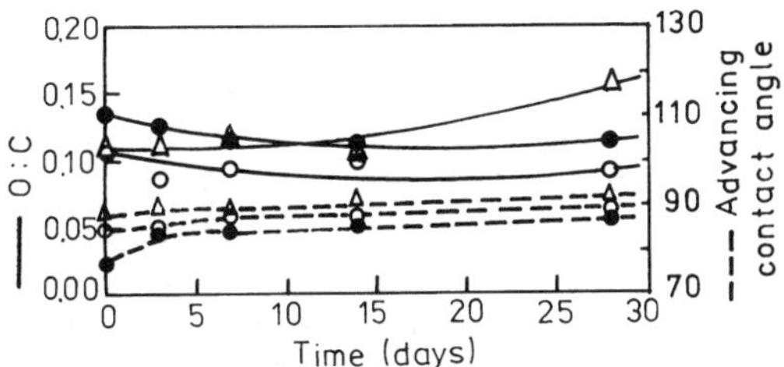

FIG. 7 Effect of aging on the O : C ratio and on the contact angle of water on PP treated: for 15 min with UV/air (○); for 10 min with UV/air + ozone (●); for 30 min with ozone only (△). The contact angle of water on the untreated PP surface is of 110 deg. (Adapted from Ref. 31.)

oxidized material (LMWOM) at the polymer surface. FTIR spectra evidenced a high concentration of carbonyl functionalities in the top 1–2 μm, while XPS analysis showed no gradient of oxygen concentration in a surface layer of 10 nm.

The study concerning the influence of each type of treatment on the PP film surface was continued with experiments on the aging behavior of PP [31]. In Fig. 7, there are presented the variations of the O : C ratio and the advancing contact angle of water on PP surfaces differently treated, versus the aging time. It can be observed that the treated PP is very stable in terms of both O : C ratio and contact angle values, for all three types of treatment. Strobel *et al.* [31] attributed the stability of PP on aging to the high molecular weight of the base PP resin and the lack of interaction between the highly oxidized surface layer and the unmodified bulk, due to the nonpolar character of the PP surface.

III. PHOTOLYTIC SURFACE MODIFICATION WITH LASER RADIATION

A. General Aspects

During the interaction of UV laser radiation with polymers, the absorption of UV photons leads to electronic and/or vibrational excitation of chromophoric groups in organic materials, depending on the laser intensity and material properties. UV laser treatment of a polymer can involve three different interaction phenomena [53–56]:

(a) photochemical changes, below the fluence threshold for ablation F_T, which imply changes in chemical composition, molecular structure, surface energy, reactivity, and electrical potential of the surface; these changes lead to the production of a modified surface morphology and microstructure;
(b) photochemical and/or thermal ablation, for fluences above F_T;
(c) thermal degradation, for fluences above the threshold F_T.

Concerning their reactivity towards UV laser light, polymers may be classified into two categories [53]:

(a) weakly absorbing polymers, like PE, PP, PVF_2, for which a photothermal process (thermal degradation) dominates the interaction; in this case, laser treatment does not produce ablation of these materials, but—under high fluence and high repetition rate—melting can occur;
(b) strongly absorbing polymers, like PVC, PS, where a photochemical process (photoablation) dominates the interaction.

Because the absorption coefficient of the polyolefinic materials is smaller than that of other polymers (e.g., PET) over the whole range of UV laser radiation [57], they cannot be modified directly by laser light, without doping with an UV-absorbing (e.g., benzophenone) molecule. UV absorption by chromophoric impurities (e.g., carbonyl, hydroperoxy, and olefinic groups), obtained by polyolefin processing, acting as dopants and causing heating of the polymeric surface by photochemical secondary reactions cannot be excluded.

From the technological point of view, UV laser treatment makes it possible to alter in a controllable manner the physicochemical structure of polymeric materials, leading to interesting modifications of surface properties, such as better wettability [53, 54, 58], adherence [53–55], and printability [53]. In the meantime, treatment with higher power lasers leads to a significant increase of the surface temperature [59].

The UV-pulsed laser treatment of polymeric materials can induce local changes in the roughness of their surfaces [60]; this modification is of great interest, because the creation of heterogeneities on the polymer surface can enhance specific interactions with living tissues, improving their biocompatibility. Short wavelength (193 nm) UV laser irradiation of biological and polymeric materials [10] has been shown to be capable of etching the materials with great precision, via "ablative photodecomposition" and without significant heating.

In the case of semicrystalline polymers, the interaction with laser beams of higher power density and for short interaction time conditions can be utilized for the production of a controlled surface amorphization [59], without a significant modification of the quality of the film, which remains semicrystalline in bulk. Another application of the photochemical surface modification is as a pretreatment for the subsequent metallization of polymers [54]; this process is very important for getting clean and active surfaces, with good wetting and adhesion for subsequent plating.

The new properties obtained by laser irradiation may be interpreted [61] by the appearance of extended systems of π orbitals.

B. Influence of Laser Treatment on the Ablation, Morphology, and Functional Group Formation on PO Surfaces

PE and PP films were irradiated in air with an argon–fluorine excimer laser ($\lambda = 193\,\text{nm}$) [53]. The 193 nm absorption coefficients, the penetration depths (calculated for 99% of the deposited laser energy), and the "etch depths" (given by the ratio between the thickness of the film and the number of shots before the film is pierced), at different fluences of the laser beam, for PE and PP, are presented in Table 2.

The polymers with low absorption coefficients (like PE and PP) have low ablation rate at high fluence, as long as a low (≤ 3 Hz) repetition rate was used. When the pulse rate is raised to 10 Hz, the number of pulses necessary to pierce the lower absorbing polymers decreases dramatically. In conclusion, it could be said that the piercing of PE and PP is a bulk melting of the polymer and not a real ablation process.

The broadening often observed for the peaks in the XPS core level spectra of laser irradiated polymers could be caused [53] by either a chain crosslinking or a roughening of the surface under this type of treatment.

Murahara *et al.* [58] irradiated PP surfaces with an ArF excimer laser radiation ($\lambda = 193\,\text{nm}$) in open air in tap water. The fluence of the laser beam was $12.5\,\text{mJ/cm}^2$ and the number of shots ranged between 0 and 10,000. In this case, the photodissociated hydrogen atoms effectively dehydrogenated the PP surface, and OH groups, obtained by the photodissociation of H_2O, replaced the hydrogen atoms of the PP surface. In this way, the treated surface was photomodified to a hydrophilic surface only in the photoirradiated areas, a fact evidenced by the gradually decrease of the water contact angle with increasing laser shots. In the meantime, the tensile shear strength was 7 times higher than that of the untreated sample. Because the surface morphology of laser irradiated PP was the same as for the nonirradiated surface, it was concluded that the adhesion enhancement between the PP surface and an epoxy adhesive was not caused by the anchoring effect due to the surface roughness, but results from chemical interaction between the OH groups and the respective adhesive.

A modification of the morphology of PP fibers and films irradiated in a vacuum chamber evacuated to 1 Pa, with a pulsed UV F_2 excimer laser, at a wavelength of $\lambda = 157\,\text{nm}$ and fluences of 50 and $128\,\text{mJ/cm}^2$ [57], was observed for the samples treated with more than 20 pulses at $128\,\text{mJ/cm}^2$. These samples exhibited the characteristic surface morphology found on PET and PA fibers, that is a "roll" or "ripple" structure. It was evidence of the significant dependence of the appearance of the structure on the various parameters of the treatment, especially on the number of applied pulses and on the laser fluence. The mean distance of the "rolls" increases very slightly with the number of pulses, while an increasing depth of the structure yields a coarser appearance. For a given number of pulses, the laser fluence strongly influences the morphology of the irradiated polymer. The fact that the surface modification of the PP fibers is observed only above a certain number of laser pulses could be a sign for an "incubation" effect in the ablation process. A possible reaction might be the exothermic disproportionation

TABLE 2 Absorption coefficients (a), penetration depths of UV and "etch depths" measured by piercing for PE and PP

Polymer	a (cm^{-1})	Penetration depth (μm) (99% energy)	Fluence (mJ/cm^2)	"Etch depth" (E) (μm/pulse) 1 Hz	3 hz	10 Hz
PE	6.5×10^2	70	300	< 0.05	< 0.05	0.6
PP	5.4×10^2	85	300	< 0.05	< 0.05	1.0
			1200	0.15	0.18	> 1.0

Source: Ref. 53.

of the radical chain ends, which are produced by a photoinduced breakage.

Breuer *et al.* [55] irradiated PP foils of 40 μm thick, with an excimer laser with wavelengths of 308, 248, and 193 nm, in different gaseous atmospheres (oxygen, helium or air). The density of the laser energy was varied between 50 and 2000 mJ/cm^2. For these treatment conditions, it was observed that irradiation of PP at high energy density (> 500 mJ/cm^2) leads to ablation, but no chemical changes were detected in the wavelength range used. For energy densities < 500 mJ/cm^2, formation of C=O groups could be detected and a dependence on the laser parameters and on the surrounding atmosphere was found. In this treatment regime no damage of the surface could be observed, but on increasing the irradiation dose (e.g., the number of laser pulses), specific morphological changes such as microcraters, cracks or "ripples" appeared, depending on the experimental conditions. The maximum concentration of C=O bonds was obtained at 193 nm. At 248 nm slightly less chemical changes were found, while for 308 nm these types of changes were significantly reduced. Due to the strongly different photon energies of the laser sources used, this behavior of C=O formation suggests a dominating photochemical mechanism of the oxidation process. Formation of oxygen functional groups, such as C—O and C=O groups, was also shown by Chi-Ming Chan [22] for PP, UV laser irradiated in air, water and ozone. XPS spectra showed that the carbonyl bondings were formed in the uppermost surface layers.

Referring to the atmosphere influence on the PP laser irradiation, carbonyls were not detected at any wavelengths when the irradiation was performed in a helium atmosphere, and, in air, the concentration was less than that in oxygen. This fact showed that heterogeneous reactions with the surrounding gas atmosphere occur. In fact, it is well known [62] that the irradiation of PP with UV laser light in an oxidizing environment results in the oxidation of its surface, the presence of the strongly polar groups on the polymer surface positively influencing the physisorption and, hence, the adhesion strength in the irradiated zones, when appropriate conditions are applied.

IV. TREATMENT WITH PARTICLE BEAMS

A. Ion Beam Interactions

1. General Aspects

The impact of energetic ions on polymer samples results in modifications of their surfaces that can have disturbing and damaging effects, but it can be also used to probe surface properties. The main properties of polymers which can be modified under ion beam interactions are: rheological properties (solubility and molecular weight distribution) [63], electrical conductivity [61, 64–70], optical properties [64, 66], mechanical properties [71], surface texture [22, 72], biocompatibility [73, 74], crystalline state [64, 75, 76], adhesive [66], and surface [72, 73, 77–79] properties. Gas evolution during the irradiation of most polymers, which leads to irreversible changes of the molecular structure, has often been reported [71].

Since the penetration depth of low energy ions in a solid is very small, ions seem to be very important for the modifications in the first few nanometers of the polymer surface [72].

Bombardment with low energy ions ($50\,\text{eV} < E < 600$ eV) may be expected to alter the structure, morphology, and chemical properties of materials, since this energy regime is higher than that required for sputtering. Forward sputtering leads to densification; sputtering followed by ion replacement leads to chemical changes, and sputter removal alters the morphology of the surface [80]. The sputtering action of the ion beam can be used for surface etching, e.g., in order to obtain concentration depth profiles [81].

Apart from the role played by the thermal effect in volatile product emission, it has been shown that the thermal damage may be so severe that changes of the target sample are visible under the microscope [82] and, in some cases, it can produce drastic modifications in viscoelastic properties of polymers [83].

Surface bombardment intrinsically causes a number of additional effects, such as implantation, recoil implantation of surface atoms, mixing, and defect creation, which are disturbing for the goal of a well defined ordered surface [84–87].

It is apparent that there are two basically different processes induced by swift heavy ions passing through a polyolefin [88]:

(a) local intratrack reactions, which lead to the formation of etchable damage;
(b) reactions induced by active species leaving the tracks and diffusing into the surrounding matrix, which may cause a mutual influence of tracks, even at fairly low ion fluences.

The ion bombarded polymers have been shown to undergo a continuous evolution of the primary chemical structure to a "final" carbonaceous material, still keeping a "memory" of their primary stoichiometry and chemical structure [63].

Low molecular mass ions are emitted when a polymer is bombarded with low energy ions in the low fluence regime, these ions being characteristic for the polymer molecular structure [89].

2. Degradation Studies on PO Ion Bombardment

Delcorte *et al.* [89] studied a series of saturated aliphatic polymers, differing only by the pendant group (i.e. polyethylene PE, polypropylene PP, and polyisobutylene PIB), in order to gain a better understanding of the secondary ion formation in these systems. For polymer surface degradation studies, alternated sequences of ToF–SIMS analysis periods (pulsed Ga^+ beam, $D = 10^{12}$ ions/cm^2) and continuous bombardment ($5 \times 10^{12} \leq D \leq 2 \times 10^{14}$ ions/cm^2) were utilized.

The ToF–SIMS intensity variation of the most characteristic peaks of PP and PIB is shown in Fig. 8(a, b), as a function of the cumulated ion dose. For PIB, the decrease with the ion dose of characteristic fragments at 83 and 97 amu is already initiated at the first point of measurement, while for PP, the peak at 69 amu increases first to a maximum value, before decreasing. The direct decrease in the intensity of the most characteristic ions is related to the degradation of the precursor, while the maximum observed for the most important peak of PP shows competition between two different processes: the degradation, and the ion formation which may be induced by the initial degradation. The close relationship between the secondary molecular ion production and the polymer degradation under the ion beam bombardment was pointed out.

Svorcik *et al.* have irradiated 15 μm thick PE foils with 40 keV Ar^+ ions, at room temperature and at fluences ranging from 1×10^{12} to 1×10^{15} ions/cm^2, from both sides, in order to amplify optical changes [73].

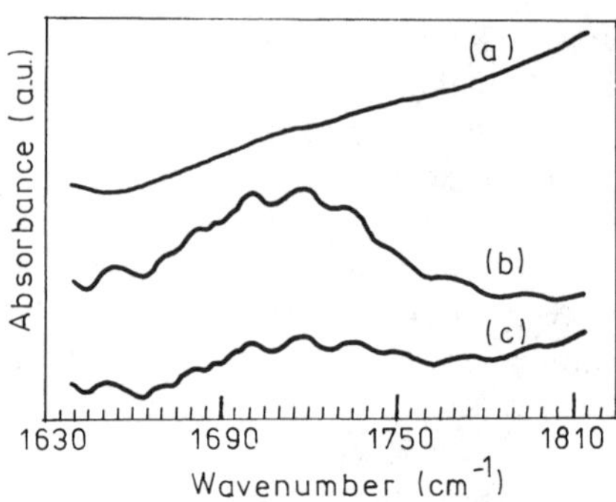

FIG. 9 Part of the IR spectrum of the PE samples irradiated with 40 keV Ar^+ ions at different fluences: (a) 1×10^{13} ions/cm^2; (b) 5×10^{14} ions/cm^2; (c) 1×10^{15} ions/cm^2. (Adapted from Ref. 73.)

The changes in the 1630–1820 cm^{-1} region of IR spectra induced by irradiation at different fluences are presented in Fig. 9. One can see that irradiation at fluences above 1×10^{13} ion/cm^2 leads to an absorbance increase in the region 1710–1765 cm^{-1}, which is related to the presence of oxidized structures such as carboxyl, carbonyl and ester groups. The concentration of these polar structures increases with increasing ion fluence up to a fluence of 5×10^{14} ions/cm^2 and then it declines. This leads to the conclusion that, for an ion fluence of 1×10^{15} ions/cm^2, the oxidized structures created in the early stages of ion irradiation are also degraded, i.e., deoxygenation processes prevail over oxidation processes [73].

In Fig. 10, the valence band spectra of the pristine and Ar^+ 500 eV bombarded PE and PP, at a fluence of about 5×10^{14} ions/cm^2, are presented. The shape of the obtained spectra for ion bombarded polymers

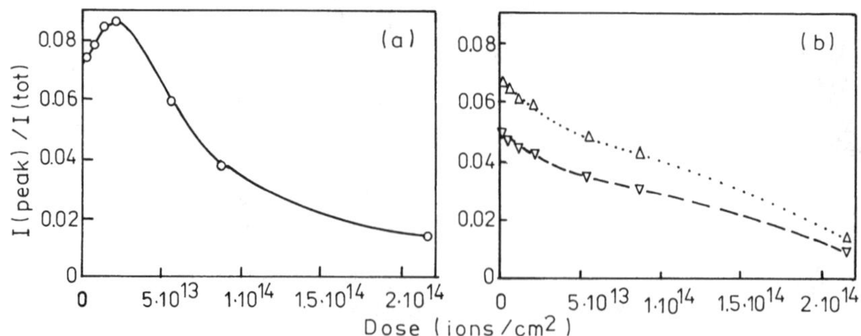

FIG. 8 ToF–SIMS intensity dependence on the cumulated Ga^+ (15 keV) ion dose for the most characteristic peaks of: (a) PP 69 amu (○) and (b) PIB 83 amu (△) and 97 amu (▽). (Adapted from Ref. 89.)

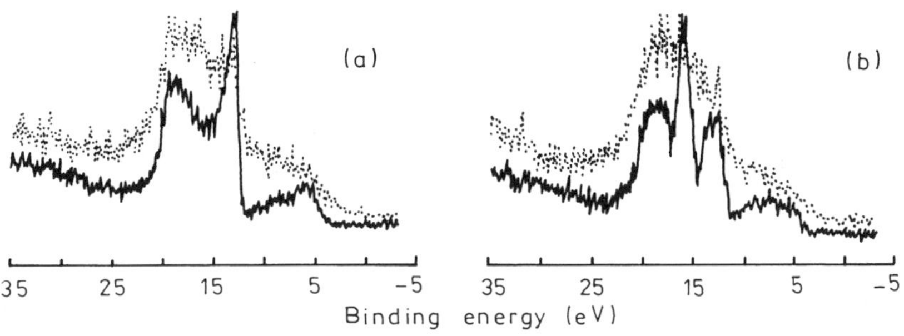

FIG. 10 Valence band spectra of: (a) PE and (b) PP for pristine (——) and Ar^+ 500 eV bombarded samples (········). (Adapted from Ref. 66.)

resembles that of graphite. The disappearance of the strong and sharp peak of PP, from about 16 eV, which comes from the methyl pendant group, means that this well defined pendant group no longer exists in the ion bombarded PP. This behavior should be explained by the incorporation of the methyl groups into the graphitization of the ion-bombarded surface [66].

3. Surface Properties Modification of PO by Ion Bombardment

Generally, improvement of surface energy by ion assisted reactions in a reactive gases environment is mainly due to the polar force [90]. The measured dependence of the polar component of the free surface energy (γ_S^P) on the irradiation fluence, for the 40 keV Ar^+ ions irradiated PE sample [73], is shown in Fig. 11. The polar component of the free surface energy is nearly constant up to a fluence of 5×10^{13} ions/cm^2, while for higher fluences, it increases rapidly, for the fluence of 1×10^{15} ions/cm^2 being about four times higher than for the pristine, untreated PE.

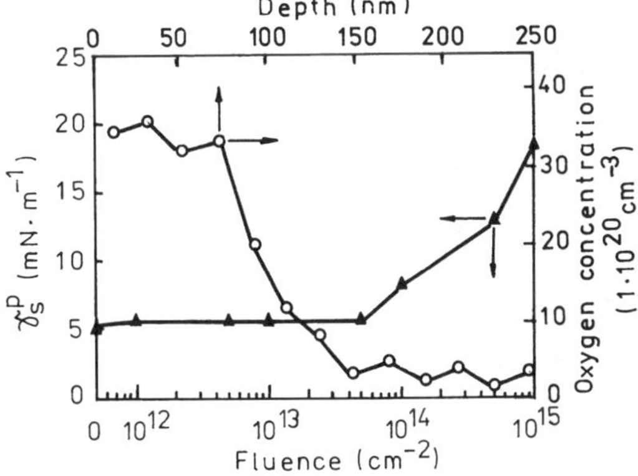

FIG. 11 The variation of the polar component of the surface free energy (γ_S^P) versus the ion beam fluence, for the PE sample, 40 keV Ar^+ ions irradiated and the concentration depth profile of oxygen incorporated in the PE surface, Ar^+ irradiated at a fluence of 5×10^{14} ions/cm^2. (Adapted from Ref. 73.)

From the point of view of the surface properties of the modified PE, the depth of oxygen penetration is also of interest. It is known that oxygen diffuses in the degraded layer from the residual atmosphere in the implanter and creates chemical bonds with the originating radicals at the place of polymer damage, for example in the form of carbonyl groups. The concentration profile of oxygen following the Ar^+ ion implantation, at a fluence of 5×10^{14} ions/cm^2, into PE is also shown in Fig. 11. The oxygen depth profile has a broad concentration plateau from the sample surface to a depth of about 100 nm, which is followed by an sudden concentration decrease.

All these results show that irradiation with 40 keV Ar^+ ions leads to the production of oxidized (i.e., polar) groups in the PE surface layer.

AFM images of PP and PE pristine surfaces and bombarded with an Ar^+ beam of 500 eV and, respectively, with a N^+ beam of 50 keV, for a fluence of 5×10^{14} ions/cm^2, were recorded [66] in order to show the effect of ion beam bombardment on the surface morphology. In the case of PP, the AFM images reveal that the ion bombarded surface gets smoother than the pristine one, this fact proving again that the ion bombarded surface graphitize under the ion bombardment. For higher ion energies and fluences, a different situation could be found. For N^+ 50 keV bombarded PE, dendrite-like morphology was observed. This was also shown in the Ar^+ 50 keV bombarded surface and for the PP sample, too. The dendrite-like morphology could arise from the heat emitted by the discharge of the local accumulated excess charge during the ion bombardment [66], the developed structure being strongly dependent on the type of polymeric material [91].

4. Changes in Electrical Properties of Polyolefins Induced by Ion Beam Bombardment

The ion irradiation of different polymers results in a significant decrease of their sheet resistivity R_S. Such results have also been obtained for PE [65, 73, 77, 83] and PP [68]. The enhanced electrical conductivity depends on ion beam parameters (ion species, energy, dose and beam current) as well as on the constituents and structures of the initial polymers. Large electronic energy deposition by high energy (MeV) ions transforms insulating polymers into highly conductive materials, whereas small electronic energy deposition from low energy (keV) ions forms less conductive materials [70].

It is well known that the electrical conductivity of polymers is related to the concentration of conjugated double bonds [73]. It has been shown that the irradiation of PE with 40 keV Ar^+ ions produces conjugated double bonds, the concentration and conjugation length of which is an increasing function of the ion fluence. The electrical sheet resistance, R_S, decreases, probably due to progressive carbonization of the PE surface [73].

The time dependence of the sheet resistivity R_S for PE samples irradiated with 150 keV Ar^+ ions, at a fluence between 5×10^{14}–1×10^{15} ions/cm^2, was also shown by Svorcik *et al.* [65]. As a result of the ion irradiation at a fluence of 5×10^{14} ions/cm^2, the sheet resistance drops by at least two orders of magnitude in comparison to that of the pristine material. During the first 500 h, R_S increases spontaneously by about 10–15% and after that it remains constant within the measuring errors. The ion irradiation to the fluence of 1×10^{15} ions/cm^2 leads to an additional resistivity drop by about one order of magnitude and the time dependence of R_S is analogous to that of samples irradiated to the lower ion fluence (Fig. 12).

It is supposed that the increased electrical conductivity of the polymers modified by ion irradiation may be affected also by the excess of radicals residing on conjugated polymer chains [65]. Time dependence of the number of unpaired electrons N_e (radicals) for the 150 keV Ar^+ irradiated PE, to different fluences, is also presented in Fig. 12. It is evident that, during the first 500 h, the electron number falls rapidly by 43–55%, after which the decrease in electron number slows down; during the time interval from 500–3300 h, a decrease by only 20–35% is observed. For the irradiation at higher ion fluence, the initial number of radicals increases by about one order of magnitude.

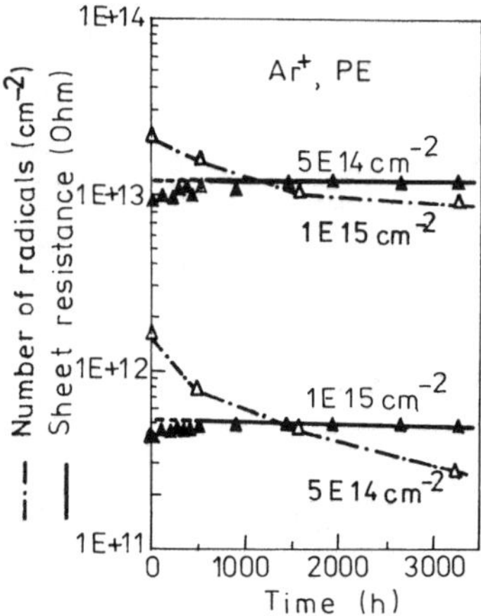

FIG. 12 Time dependence of the sheet resistance (R_S) and of the number of radicals per cm^2 for PE samples irradiated with 150 keV Ar^+ ions to different fluences: 5×10^{14} ions/cm^2; 1×10^{15} ions/cm^2. (Adapted from Ref. 65.)

The experimental findings that:

the very rapid decrease of the number of unpaired electrons during the first 500 h is accompanied by only a slow increase in the sheet resistance

the observed additional decrease of the number of unpaired electrons after 500 h is not accompanied by a corresponding resistivity increase

may be due to the fact that only a part of the radicals created during ion irradiation affects the electrical conductivity of the modified polymers. The number of radicals that may contribute to the enhanced conductivity of the ion beam modified polymers is a nonlinear, decreasing function of the time elapsed from the moment of ion irradiation [65].

B. PO Modifications by Electron Beams

1. General Aspects

Under the influence of radiation producing low densities of energy deposition (such as fast electrons, γ- and X-rays), the energy transferred to the molecules of the medium results in primary "ionizations" and "excitations". Polyatomic molecules, having absorbed energy, will suffer bond cleavages, giving rise to nonsaturated fragments (free radicals), which are responsible for most chemical transformations observed in polymers [92, 93]. It is well known [94] that, upon exposure of polymers to ionizing radiation, new chromophore groups are generated, which efficiently absorb light in the UV, visible and IR region of the spectrum. The main effects of electron beam irradiation are chain scission, oxidation and unsaturation, depending on dose rate, and oxygen content [95].

The dominant initial chemical event in the radiolysis of PE can be considered as being described by the following reaction [92]:

$$-CH_2-CH_2CH_2- \longrightarrow H^{\cdot} + -CH_2-CH^{\cdot}-CH_2 \quad (1)$$

The hydrogen atom generated in the process will abstract a hydrogen from a neighboring molecule:

$$H^{\cdot} + -CH_2-CH_2-CH_2- \longrightarrow H_2 + -CH_2-CH^{\cdot}-CH_2 \quad (2)$$

or from its closest neighboring methylene group within the same molecule, a double bond being formed:

$$H^{\cdot} + -CH_2-CH^{\cdot}-CH_2- \longrightarrow H_2 + -CH_2-CH{=}CH- \quad (3)$$

or the two polymeric fragments formed in (1) and (2) combine to form a crosslink:

$$2\ -CH_2-CH^{\cdot}-CH_2- \longrightarrow \begin{array}{c} -CH_2-CH-CH_2- \\ | \\ -CH_2-CH-CH_2- \end{array} \quad (4)$$

Hydrogen evolution, unsaturation and crosslinking are the three main reactions which account for the major chemical transformations in irradiated PE. In addition, small amounts of low molecular weight hydrocarbons are detected in the evolved gases. Gases are formed as a result of atom or side-chain abstraction and the nature of the gas closely reflects the composition of the macromolecule. From Table 3, it can be seen that the chemical nature of the gases closely corresponds to that of the side groups in the repeating unit.

TABLE 3 Gases formed under the electron irradiation of some polymers

Polymer	Repeating unit in the polymer chain	Gases formed
Polyethylene	$-CH_2-CH_2-$	H_2
Polypropylene	$-CH_2-CH(CH_3)-$	H_2, CH_4
Polyisobutylene	$-CH_2-C(CH_3)_2-$	CH_4

Source: Ref. 92.

It was observed that branching (as a result of the cleavage of side chains) introduces weak points in the polymer structure, the susceptibility towards degradation increasing as branching increases. Thus, polyethylene (PE) is essentially crosslinked by irradiation with very little main-chain scission (crosslinking being the prevailing degradation process at higher irradiation fluences [96]), polypropylene (PP) exhibits both crosslinking and degradation, whereas polyisobutylene (PIB) only suffers chain scission.

The irradiation of polyolefins with 12 MeV electrons resulted in the creation of free radicals and in an increase of free volume fraction as well. The competition of macromolecular splitting and cross-linking was reported for PE irradiated with 8 MeV electrons [96].

Double bonds located at the chain ends are formed as a result of chain scission, in the case of PIB, this fact being shown by the following sequence of reactions [92]:

$$-C(CH_3)_2-CH_2-C(CH_3)_2-CH_2- \rightarrow -C(CH_3)_2-CH_2^{\cdot} + {}^{\cdot}C(CH_3)_2-CH_2- \quad (5)$$

$$\rightarrow -C(CH_3)_2-CH_3 + CH_2{=}C(CH_3)-CH_2- \quad (6)$$

Influence of radiation-induced chemical transformation is also observed in the modification of bulk properties of polymers. The most evident changes for this kind of properties of irradiated polymers are the modifications of their molecular weight, due to the crosslinking or degradation [92, 97–99]. The gradual reduction of the molecular weight of the irradiated polymer weakens it and leads to important losses of its valuable properties when high radiation doses are used [92]. Crosslinking increases the modulus and the hardness of the polymer, an effect that only becomes apparent for high crosslink densities. In the case of partially crystalline polymers, such as PE, moderate crosslinking imparts to the material a nonmelting behavior, this one exhibiting elasticity above its crystalline melting point [92, 100]. Thus, it is possible that, at some dose of the electron irradiation, the concept of crystal melting becomes obsolete; despite localized remnants of pseudo-crystalline ordering, the constituent molecular stems are so highly cross-linked that the "crystal" is unable to melt [97].

2. Polymer Surface Oxidation Induced by Irradiation with Electron Beams

Svorcik *et al.* [96] have irradiated PE foils, 15 μm thick ($M_n = 180{,}000$, $\rho = 0.945\,g/cm^3$) in air and at room temperature, with a 14.89 MeV electron beam. The electron flux was 247 Gy/min and the samples were irradiated to fluences ranging from 57.6 to 576 kGy.

The irradiation of polymers in air is accompanied by significant oxidation, which is a result of a diffusion-controlled reaction of ambient oxygen with reactive products of polymer degradation. The oxidation as a function of the electron fluence can be seen from Fig. 13, where differential IR spectra, obtained as a difference between the IR spectra from irradiated and pristine PE, are shown. The irradiation at fluences above 120 kGy leads to a strong increase of the absorbance in the region of 1700–1750 cm^{-1}, attributed to oxidized structures such as carbonyl (—C=O), carboxyl (—COOH), and ether (—COO—) groups. The concentration of the oxidized structures increases for higher electron fluences, but it does not achieve saturation within the fluence range examined.

For the same samples, the polar component of the surface free energy (γ_S^P) increases linearly with increasing the irradiation fluence, the irradiation to the maximum fluence of 570 kGy leading to an increase of γ_S^P by a factor of 1.4 with respect to the pristine PE.

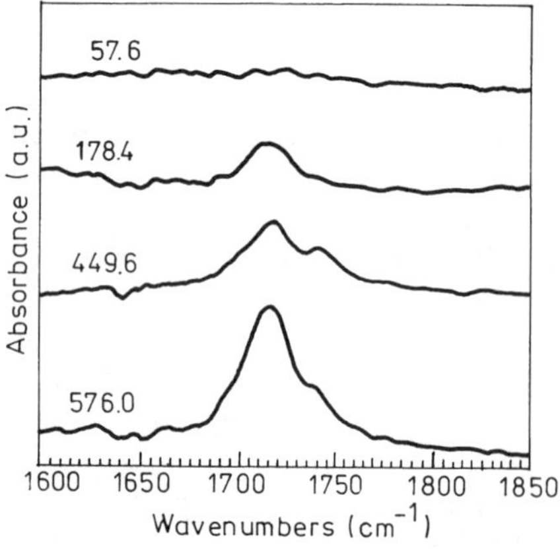

FIG. 13 Differential IR spectra of the PE irradiated to different electron beam fluences (57.6, 178.4, 449.6 and 576.0 kGy). The spectra were obtained as a difference between the IR spectra of irradiated and pristine PE specimens. (Adapted from Ref. 96.)

The irradiation with 15 MeV electrons up to a fluence of 570 kGy did not change the content of crystalline phase and the melting temperature of the investigated PE, a result which was also observed for PP irradiated with 12 MeV electrons.

3. Influence of Electron Beam Irradiation on the Crystallinity of PO

Chemical transformations, such as crosslinks or double bonds, which are radiation induced defects [92, 97, 101, 102], result in a degree of crystallite destruction. Therefore, as the absorbed dose increases, the crystallinity should decrease. In Table 4 [103], there are presented the doses required for crystallographic effects of electron irradiation in various polymers. The dose necessary for the destruction of crystallinity in linear PE is much larger than that corresponding for the other polymers. In fact, it was shown [97] that radiation damage occurs preferentially within the amorphous regions of the material.

Boudet *et al.* [101] studied the degradation of monocrystals of PE by electronic microscopy for acceleration tensions between 1 and 2.5 MV. Observing the electron diffraction pattern, it could be seen that the diffracted beams diminish and then disappear, due to the complete loss of crystallinity for an electron critical dose value D_C received by the sample (D_C—the dose for which the diffraction traces disappear [101]).

Under electron beam irradiation of isotactic polypropylene/lignin-modified epoxy (IPP/LER) blends in an anomalous discharge in argon, at 13.3 Pa and the intensity of the current discharge of 10 mA, for different discharge tensions (600–800 V) and various

TABLE 4 Doses required for crystallographic effects of electron irradiation in various polymers

Polymer	Dose (C/m^2), necessary for the destruction of crystallinity
Linear polyethylene	102 ± 4
Poly(vinylidene fluoride), α phase	31 ± 2
Poly(vinylidene fluoride), β phase	39 ± 2
Poly(vinylidene fluoride), γ phase	36 ± 2
Poly(trifluoroethylene)	14 ± 1

"±" reflects the standard deviation for 7–10 measurements. PE single crystals were grown from a dilute solution in *p*-xylene.
Source: Ref. 103.

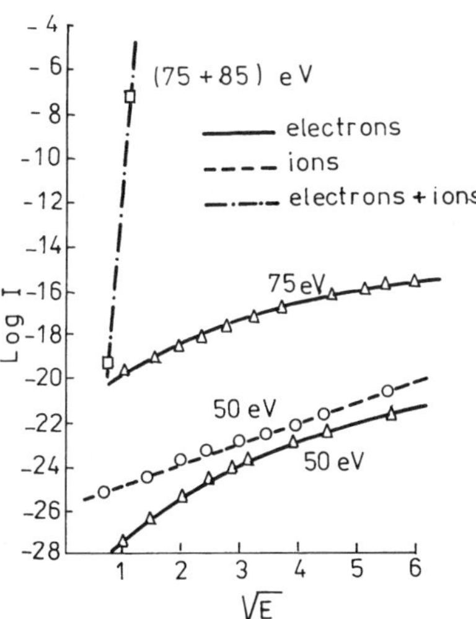

FIG. 14 Dependence of the amplitudes of the fugacity currents on the applied voltage (E) for incident particles with different energies. (Adapted from Ref. 104.)

exposure times (l, 5, 10 min) (M Pascu, personal communication, 10th Conference on Plasma Physics, Applications, Iasi, June 4–7, 1998), the increase of the degree of crystallinity in respect with the untreated samples was observed; this behavior denotes a decrease in the amorphous fraction, compared with the crystalline one. Changes in the refractive index and in the dielectric function values were also evidenced after the applied treatment, changes that were dependent on the exposure time.

4. Efficiency of the Simultaneous Action of Ions and Electrons on PO

De Lima *et al.* [104] studied the consequences of the impact of low energy (50–300 eV) ions and electrons, acting separate or simultaneously, on 8 μm PP films. The amplitudes of the measured fugacity currents versus the energy of the incident particles are presented in Fig. 14. It is clearly evidenced the efficiency of ions and electrons when they are acting together; this efficiency was also shown by electronic paramagnetic resonance, which evidences the presence of free radicals which partially justifies the measured dielectric losses.

V. COLD PLASMA TREATMENTS

A. Generalities

In a low pressure ($\leq$ 133 Pa), high-frequency ($\geq$ 1 MHz) discharge, the heavy particles (gas neutral atoms, molecules or ions) are essentially at ambient temperature ($\approx$ 0.025 eV), while electrons have enough kinetic energy (1–10 eV) to break covalent bonds and to cause further ionization. The chemically reactive species thus created by electron-neutral inelastic collisions can participate in gas-phase (or homogeneous) reactions or in heterogeneous reactions with solid surface in contact with plasma. The efficiency of producing electron-ion pair is greatest at microwave (MW > 100 MHz) than at "low" ($\leq$ 100 MHz) frequencies. This is attributed to a significantly higher fraction of energetic electrons in the tail of the electron energy distribution function (EEDF) and to a higher electron density value [105, 106].

Energetic particles and photons interact strongly with the polymer surface during the plasma treatment. In plasma, which do not give rise to thin film deposition, four major effects are normally observed: surface cleaning (removal of organic contamination), ablation or etching of material from the surface (removal of a weak boundary layer), crosslinking or branching, and modification of the surface-chemical structure either during plasma treatment itself, or upon re-exposure to air. Each of these processes is always present, but one may be favored over others, depending on the gas chemistry, the reactor design and the operation parameters.

Almost all commercial polymer materials contain additives or contaminants such as oligomers, antioxidants, mould release agents, solvents or anti-block agents. Because these materials often have a very similar chemistry to that of the base polymer, they will react with plasma in a similar way to polymer. Oxygen containing plasmas, at normal power levels (typically, a few mW/cm^2), are suitable for removing organic contaminants from polymeric surfaces, but at sufficiently long exposure times (tens of seconds).

Ablation or plasma etching is distinguished from cleaning only by the amount of material that is removed.

CASING (Crosslinking via Activated Species of INert Gases) was one of the earliest recognized plasma treatment effects on polymer surfaces [107]. It occurs in polymer surfaces exposed to noble gas plasmas. Ion bombardment or vacuum ultraviolet photons (λ < 175 nm) can break C—C or C—H bonds and the free radicals resulting can react with other surface radicals or with other chains in chain-transfer reactions. If the polymer chain is flexible or if the radical can migrate along it, this can give rise to recombination, unsaturation, branching or crosslinking. There are also some earlier papers about the crosslinking of polypropylene (PP) in oxygen plasma [108, 109].

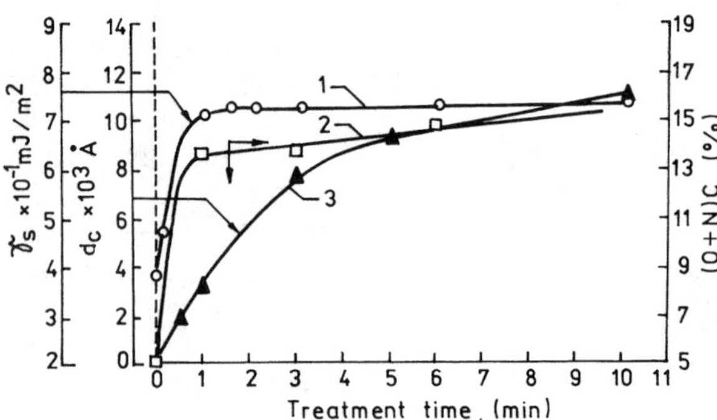

FIG. 15 Surface energy γ_s (curve 1), surface chemical composition (O + N)/C (curve 2), and crosslinking depth (curve 3) of PE vs argon plasma treatment time (60 W, 13.3 Pa).(Adapted from Ref. 112.)

The most reported effect of plasma is the deliberate alteration of the surface region with new chemical functionalities. Reactive plasma is used to add polar functional groups, which can dramatically increase the surface free energy of the polymer. Plasma can also be used for surface fluorination and silylation, surface-chemical changes, which tend to reduce wettability and superficial strength.

Results about functionalization and cross-linking of polyolefins in different gases (inert gases, oxygen, nitrogen, air, etc.) will be presented. Some considerations about the stability in time of these modifications are also given. Experiments in "model situations" in plasma treatment of polyolefins are presented at the end of this section.

B. Surface Functionalization

1. Inert Gases

Gerenser and coworkers have resorted to in situ plasma treatments, in the photoelectron spectrometer preparation chamber, in order to eliminate the problem associated with exposure to the atmosphere [110, 111]. PE treated in low frequency (60 Hz) plasma, in argon, 60 s, shows a somewhat broader the single C 1s peak, a slight decrease in the depth of valley between the various features of the XPS valence band spectrum. These subtle changes may be related to minor surface damage such as chain scission. Also, based on SEM images, no observable differences on the macroscale could be detected between untreated and plasma treated PE. No detectable differences were observed between high (HD) or low density (LD) PE [110].

Plasma treatments done with a small amount of oxygen mixed with argon, in the same conditions mentioned above, introduce reactive functional groups onto the PE surface. The amount of incorporated oxygen approximately tracks the percentage of oxygen in the gas mixture [111].

Other investigations on argon plasma treated PE, but not in situ, show very different results. Therefore, in [112] it is shown that argon RF plasma treatment of PE determines both an increase of surface energy γ_S and (O + N)/C ratio, with almost the same increasing rates (Fig. 15, curves 1, 2). This fact seems to indicate that Ar plasma treatment introduces polar groups just in a very thin layer, similar to that the contact angle measurement can probe [112].

Some results on argon treated PP also show an important increase in the polar component of the surface energy, from 1.1 mN/m for untreated to 36.6 mN/m for 5 min treated PP (treatment conditions: pressure 26.6 Pa, treatment time 5 min, RF power 600 W) [113].

2. Oxygen, Nitrogen, Air

Oxygen and nitrogen plasma treatments in situ incorporate oxygen and nitrogen, respectively, into the PE surface [110, 111]. The oxygen plasma is more reactive than the nitrogen plasma. This is proved by the initial rate of oxygen incorporation (~ 11% O/10 s), by comparison with nitrogen incorporation (~ 4% N/10 s) and the time required to reach saturation (30 s in oxygen, 60 s in nitrogen plasma, respectively). The degree of incorporation, at the saturation level, is approximately 20 at% in each case [110].

The line-shape analysis of the C 1s spectrum for the oxygen plasma treated PE indicates the formation of C—O, C=O (or O—C—O) and HO—C=O (or O=C—O—C) with C—O the most prevalent. The concentration of O—C species is roughly twice that of O=C species, as is determined from the O 1s spectrum [110].

The two peaks found in the C 1s spectrum of nitrogen treated PE, in situ, have been assigned to C—N (286.1 eV) and C=N (287.4 eV). The formation of C—N predominates [110].

Similar results were obtained for oxygen plasma treated PE not in situ. A time-power dependence of the surface concentration of oxygen, which in LDPE ranged from 5.5% at low level treatment (5 s at 10 W) to 18.8% at high level treatment (60 s at 80 W) has been established [114]. However, as opposed to in situ oxygen plasma treatments, where no detectable differences were observed between HD and LDPE [110], lower amounts of oxygen are incorporated in

HDPE [114]. Derivatization of the treated PE samples by trifluoracetic anhydride (TFAA) shows that only a minor fraction of oxygen containing surface groups are hydroxyl groups. (It is well known that reaction between TFAA and hydroxyl groups introduces three fluorine atoms for each hydroxyl). A larger amount of fluorine was observed on HDPE [114].

Some treatments of PE in air and nitrogen plasma, using a very low power plasma (< 1 W), were done with the intention of using such a treatment in commercial processes [115]. XPS investigations were realized, as in Gerenser's works, by directly attaching the plasma cell to the preparation chamber of the Scienta ESCA 300 spectrometer. In Table 5 are given the atomic percentages of different species in air and nitrogen plasma treated PE. Relative intensities of different carbon, oxygen and nitrogen species are also given.

Both plasma treatments, in air and nitrogen, introduced oxygen- and nitrogen-containing species into the PE surface (Table 5). Plasma-treated PE in air shows a significant increase in the amount of O/C functionalities over the first few seconds and a saturation level is attained after 20 s of treatment. Besides the double and single-bonded oxygen, a third unidentified component was apparent at 534.5 eV. Nitrogen plasma treatment introduces amines, imines, and amides. The saturation occurs after the first minute of treatment. The amount of chemically shifted species introduced by the nitrogen plasma after 60 s is about 26% of C 1s spectrum, whereas for the same time of treatment in the air plasma the amount is 21% [115].

There are some differences between the results of Gerenser's group and that obtained by O'Kell *et al.* at very low power levels. The XPS valence band spectra for PE treated at a higher power level (10 W) in nitrogen shows that the intensity of the two peaks at high binding energy decreases significantly and the peaks are considerable broader (Fig. 16, curve 2) as comparing with untreated PE (Fig. 16, curve 1). A new intense feature centered at 6.7 eV is present, which, in opinion of the authors, is due to N 2p lone-pair orbital [110]. It is considered that this extra peak corresponds to the carbon attached as a side group on the polymer chain and it is an evidence for crosslinking. No such signal is present in the valence band spectra of PE treated in air (curve 3, Fig. 16) or nitrogen (curve 4, Fig. 16) plasma at very low power level, which indicates that crosslinking did not occur in these cases [115].

Both treatments, in air and nitrogen plasma, increased the surface energy of the PE film to the same level after 20 s. After one minute of treatment, the surface energy polar component of the nitrogen-treated PE is much higher (12 mN/m) than that of the air plasma-treated PE (7 mN/m). This confirms the XPS results presented above [115].

TABLE 5 The atomic percentages of carbon, oxygen, and nitrogen and relative intensities of different species for treated PE in very low power air and nitrogen plasmas (from Ref. 115)

	Untreated PE	Air-treated PE	Nitrogen-treated PE
C%	99.8	87.2	77.0
O%	2.0	10.7	3.0
N%	0.0	2.1	20.0
<u>C</u>—C		78.7	76.0
<u>C</u>—O		12.8	
<u>C</u>=O		4.4	
HO—<u>C</u>=O		4.1	
<u>C</u>—N			9.0
<u>C</u>=N			8.0
<u>C</u>ONH2			7.0
<u>O</u>—C		49.0	
<u>O</u>=C		43.0	
534.5 eV		8.0	
<u>N</u>—C			48.0
<u>N</u>=C			39.0
401.5 eV			13.0

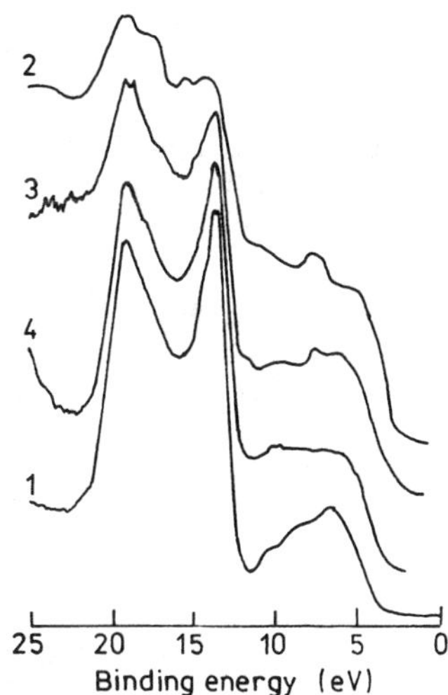

FIG. 16 Valence band spectra of: untreated PE: curve 1, nitrogen-treated PE at higher power level: curve 2 (adapted from Ref. 110), PE treated in air: curve 3 and nitrogen plasma: curve 4 at very low power level. (Adapted from Ref. 115.) Treatment time: 60 s.

3. Other Gases

H_2 plasma is the least effective in surface functionalization, which is proved by both increasing rate and saturation level of surface energy ($\gamma_S = 57\,mN/m$ as compared with 66 mN/m for oxygen-treated HDPE) [112].

Kinetics of the gas-phase *halogenation* of PE was investigated by using XPS data [116]. The initiation step is the breakdown of a halogen molecule into two free radicals. The propagation reactions proceed until the termination steps result in consumption of all the reactive species. Both experimental results about total carbon and chlorine content of the PE films vs chlorination time and the theoretical curve (Fig. 17) suggest a classical first-order kinetic law. XPS data on the concentration of each carbon species allows us to obtain the total kinetic picture. Therefore, for 600 s treatment time, a $CH_2/CHCl/CCl_2$ ratio of 1/1/0.5 is obtained. Based on stoichiometry, these data suggest that adjacent carbon atoms can be monochlorinated, but that dichlorination can only occur on carbon atoms adjacent to unchlorinated carbon atoms. For chlorination times longer than 600 s, the reaction rates are very slow. The bromination kinetics are also first order, but $\sim 10^5$ times slower than that of chlorination. The rate of iodination was found to be slower than that of bromination [116].

The hydrophilic surface modification of CCl_4 plasma-treated PP was investigated [117]. For CCl_4 plasma-treated PP at a discharge current of 50 mA, the dispersive component γ_S^d is higher than the polar one (γ_S^p). For treatments at discharge currents of more than 75 mA γ_S^p is higher than γ_S^d. The elemental composition of CCl_4 plasma-treated PP (Table 6) shows that the chlorination is favorable at a low discharge current of 50 mA, while at a higher discharge current (> 75 mA) the oxidation of PP becomes predominant. The chlorination occurring in the mild CCl_4 plasma is restricted within a surface layer of 36 Å, while the

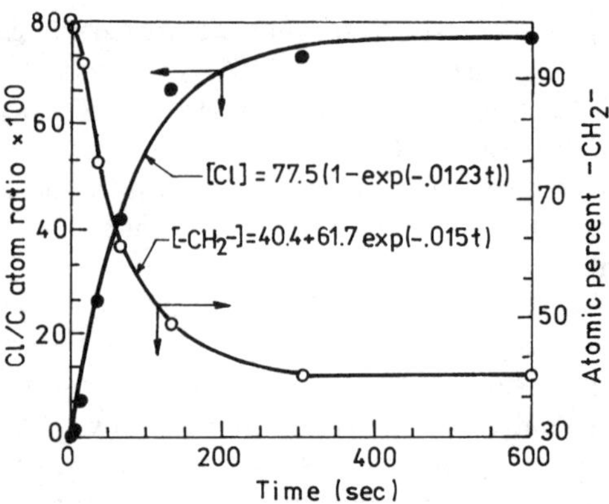

FIG. 17 Unreacted carbon concentration (○: experimental data points; solid line: theoretical curve) and chlorine/carbon atom ratio (●: experimental data points; solid line: theoretical curve) vs. chlorination time. (Adapted from Ref. 116.)

modification in the strong CCl_4 plasma occurs predominantly at the inner layers [117].

Fluorine-containing plasmas can replace hydrogen atoms in polymer molecules with fluorine atoms. This process reduces the surface free energy, reduces the diffusion length of solvent molecules, and produces a barrier layer. The barrier effect in PP occurring after CF_4, SF_6 and SOF_2 plasma modification was investigated by Friedrich *et al.* [118]. Such plasma treatments give a barrier effect of 95%, 93% and 65%, respectively, in *n*-pentane.

In most cases, the surface modifications induced by plasma treatment are well described in terms of chemical characterization, but little attention is focused on the morphologic properties. Some morphologic transformations of isotactic PP treated in different gas plasmas, found by X-ray scattering (transmission mode), FTIR spectroscopy and DSC analysis, are presented in [119]. The thermal effects, evaluated from the gas temperature given by the rotational temperature of molecular nitrogen in C $^3\Pi_4$

TABLE 6 Depth profile of chlorine and oxygen incorporated into CCl_4 plasma-treated PP (from Ref. 117)

	Untreated PP		CCl_4 plasma treated PP at 50 mA discharge current		CCl_4 plasma treated PP at 150 mA discharge current	
Sampling depth (Å)	Cl/C	O/C	Cl/C	O/C	Cl/C	O/C
17	0.02	0.38	0.31	0.29	0.14	0.47
36	<0.01	0.02	0.35	0.20	0.23	0.60
49	<0.01	0.15	0.11	0.71	0.28	0.83

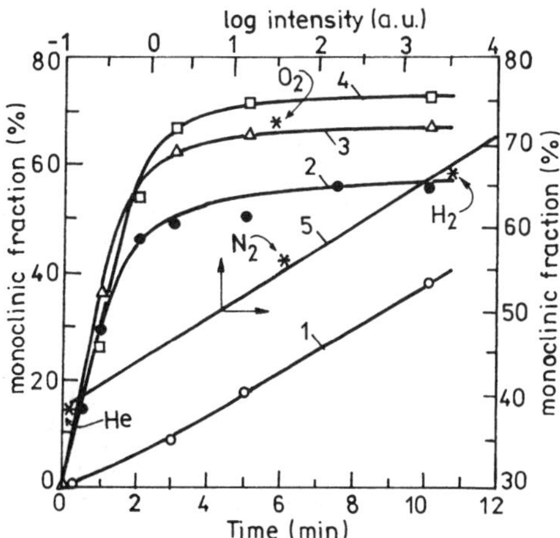

FIG. 18 Dependence of α crystalline phase content of PP vs. treatment time and plasma gas: helium: 1, nitrogen: 2, hydrogen: 3, oxygen: 4. Relationship between the α crystalline phase proportion appearance and the VUV emissions of different plasmas: curve 5 (pressure 30 Pa, RF power 100 W, flow rate 20 sccm). (Adapted from Ref. 119.)

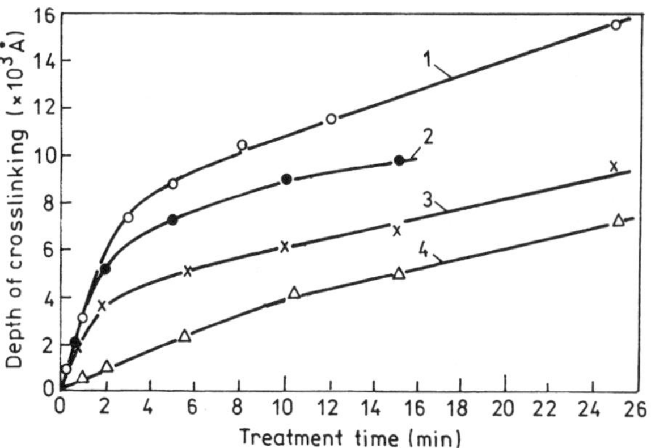

FIG. 19 Crosslinked depth vs. treatment time for different gases: argon: 1, hydrogen: 2, oxygen: 3, nitrogen: 4. Treatment power 60 W, pressure 13.3 Pa. (Adapted from Ref. 112.)

radiative state, affect the polymer crystallinity, but are not the predominant ones. The gas nature directly influences the α crystallization kinetics, as is shown in Fig. 18 (curves 1–4). The α crystalline (or monoclinic) phase proportion appearance is exponentially dependent on the plasma VUV emission for the treatment in He, N_2 and H_2 plasmas (Fig. 18, curve 5). The oxygen plasma leads to a higher α phase content that could be deduced from its VUV emission. This feature was explained by the attachment of oxidized groups that are, in most cases, chromophore groups. Therefore, the possibility that low molecular mass fractions be formed is higher. Such fractions can easily crystallize [119].

C. Crosslinking

For the same treatment conditions the degree of crosslinking (estimated from the remaining gel fraction) is the same, for PE treated in situ, whatever the gas plasma (argon, oxygen or nitrogen) [110].

The crosslinking of HDPE proceeds at two different rates, whatever the plasma gas (Fig. 15, curve 3 for Ar and Fig. 19 for other gases). In the first 2–4 min the increase in crosslinked depth d_C is very fast. This region corresponds to the direct action of the active species (including UV) in the plasma on the HDPE surface. When a dynamic balance between ablation and crosslinking is reached, the growth rate of the crosslinked layer becomes much slower [112].

The control of crosslinking can be achieved firstly by using plasma gases, which strongly emit VUV radiation. From this point of view the efficiency of crosslinking roughly follows the sequence: He > Ne > H_2 > Ar $\approx O_2 \approx N_2 \approx$ air. These results were mentioned in the earlier works of Clark and Dilks [108, 109] and confirmed in a more recent paper [112].

In some recent works the important role of metastable states in surface crosslinking has been shown [120, 121]. As was pointed out in an earlier work of Clark [109], helium seems to be the most efficient for the crosslinking due to the large amount of energy available to transfer toward the polymer surface via ion neutralization, Auger de-excitation, and Penning ionization of the polymer. The crosslinked polymer content depends in a critical manner on the residence time in the discharge of the inert and residual oxygen-containing species and the concentration of these species. The changes in surface work function of the electrons from the polymer, in the earliest stage of the treatment, could involve an increase in the rate constant for the surface de-excitation of the He metastable states [120].

The influence of the discharge power on the crosslinked depth in PE depends on the gas nature (Fig. 20). This dependence is similar to that for crosslinked depth-treatment time (Fig. 19) for argon-treated PE (Fig. 20, curve 1). There is an almost linear relationship for PE treated in hydrogen or nitrogen (Fig. 20, curves 2 and 3) and when PE is treated in oxygen, curve 4 shows three different sections [112].

The depth of the crosslinked layer is not much influenced by the discharge gas pressure in the range 1.3–13.3 Pa [112].

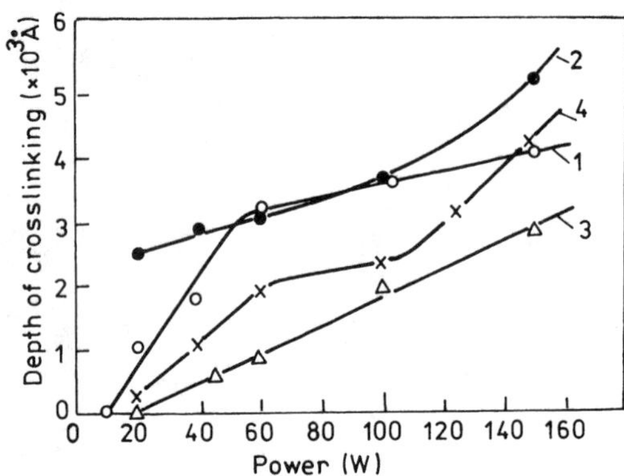

FIG. 20 Crosslinked depth in HDPE vs. treatment power for different gases: 1: argon, 2: hydrogen, 3: nitrogen, 4: oxygen. Treatment time 60 s, pressure 13.3 Pa. (Adapted from Ref. 112.)

More details about the role of crosslinking in different applications will be presented later in this chapter.

D. Stability

The long-term stability of a modified polymer surface is an important factor for further utilization. This stability depends, for the same polymer, on the storage environment as well as the treatment conditions.

Therefore, in situ argon plasma modified PE surfaces were found to be very stable with little or no change after 72 h in the vacuum treatment chamber, whatever the gas plasma treatment [110]. However, exposure of these reactive surfaces to the laboratory atmosphere resulted in drastic changes. The argon in situ plasma treated PE (60 s) exhibits the smallest change: approximately 2% adsorbed oxygen [110].

More recent papers also show that argon plasma treated LDPE and PP, for short times (< 15 s), incorporate almost exclusively oxygen, but at longer treatment times (> 15 s) some nitrogen ($< 2\%$) incorporation was observed. The oxygen incorporation is complete within 5 min upon atmospheric exposure [122].

"Saturation" and "stable" levels were determined for argon plasma treated PE and PP [122]. "Saturation" is the level for which no more oxygen is incorporated by further treatment. The "stable" level is the modification level below which no measurable, by XPS, amounts of material are removed by rinsing with a polymer nonsolvent (i.e. MeOH). Table 7 shows the O/C ratio and percentages of carbon components in one of five environments (hydrocarbon $\underline{C}$—C, $\underline{C}$—H, alcohol/ether $\underline{C}$—O, aldehide/ketone $\underline{C}$=O, carboxylate/ester $\underline{C}$OOH/R and carbonate CO_3^{2-}), for untreated and argon treated polymers, at the "stable" level of treatment, at the "saturated" level, and after rinsing of the saturated surface. A β shift, corresponding to a carbon atom adjacent to carboxylate functionality was also considered.

Stable surfaces are characterized by a high selectivity towards C—O functionalities and low levels of ketonic/carboxylate and carbonate functionalities. Saturated surface exhibits more carboxylate and carbonate functionalities, present as low molecular weight material (LMWM), which is subsequently removed upon rinsing (Table 7). In terms of oxygen incorporation, the "saturation" and "stable" treatment levels are higher for LDPE than for PP. Also the depth of modification is greater in LDPE than in PP. The authors explained the differences in treatment level and depth observed as a consequence of differences between the probabilities of crosslinking G(X) and of chain scission G(S). In PP the high value of G(S) (0.6–1.1) leads to a lower "stable" level with more LMWM. A low depth of modification is observed at a low level of treatment because the surface material is removed in the plasma.

TABLE 7 O/C ratio and C 1s peak fit results of LDPE and PP argon plasma-treated (from Ref. 122)

		Percentage of carbon in each functionality of C 1s				
Sample	O/C	$\underline{C}$—C, $\underline{C}$—H	$\underline{C}$—O	$\underline{C}$=O	$\underline{C}$OOH/R + β	$\underline{C}O_3$
PE untreated		100	0	0	0	0
stable		86.5 ± 1.1	7.9 ± 0.8	3.1 ± 0.4	1.2 ± 0.3	0.6 ± 0.3
saturated	0.27 ± 0.03	80.3 ± 1.1	8.2 ± 0.8	5.1 ± 0.4	2.6 ± 0.3	1.8 ± 0.3
rinsed	0.16	82.8 ± 1.6	8.4 ± 1.1	4.3 ± 0.4	2.4 ± 0.3	0.0 ± 0.3
PP untreated		100	0	0	0	0
stable		84.3 ± 1.2	9.6 ± 0.6	3.9 ± 0.3	1.1 ± 0.3	0.5 ± 0.3
saturated	0.19 ± 0.01	80.6 ± 1.2	8.9 ± 0.6	4.8 ± 0.3	2.1 ± 0.3	2.1 ± 0.3
rinsed	0.11 ± 0.01	88.5 ± 0.7	7.0 ± 0.5	2.8 ± 0.3	1.0 ± 0.2	0.2 ± 0.1

The crosslinking is favorable ($G(X) \approx 0.5$–0.9) at higher treatment levels. In contrast, in PE the crosslinking ($G(X) = 1.7$–3.5) is favored over chain scission ($G(S) \approx 0$–0.88), even at low treatment levels. An increase in the depth of modification arises, because crosslinks formed at the surface provide resistance to etching [122].

The study of the aging processes of isotactic PP treated in NH_3 plasma showed a decrease of the wettability, an increase of about 40% (from the initial, immediately after treatment, value) of the advancing contact angle being apparent (Fig. 21, curve 1, $\Delta\theta/\theta$) [123]. XPS data, at zero electron take-off angle (ETOA), shows that the nitrogen and oxygen uptake on the surface remained constant with the aging time, suggesting the molecular rearrangement takes place through a surface layer thinner than 5 nm. The contact angle of water increases very slowly (less than 10%) in the first 4 days and remained constant for longer aging times, for PP pretreated in He plasma (Fig. 21, curve 2). This was explained by an intense crosslinking induced by the short wavelength photons and the He metastable species into the PP surface. Such a barrier layer prevents the uppermost polar groups being buried inside the polymer [123].

Let us discuss now the problem of the stability of polyolefins surface modifications induced by treatments in the oxygen, nitrogen or air plasmas.

In [114, 115] was shown an important aspect: the oxygen surface contamination of the polymer surfaces treated in oxygen plasma is negligible. The authors to prove this feature, have used two oxygen isotopes O_2^{18} and O_2^{16}, alternatively as plasma medium and aging environments.

The oxygen plasma-treated PE, in situ, lost almost 25% of the incorporated oxygen, adsorbing a small amount of nitrogen (~ 1%) and carbon (~ 5%) [111]. On initial exposure to air, a fairly rapid and significant decrease in incorporated oxygen takes place, whatever the treatment time (Fig. 22). After this initial effect, the surfaces treated at shorter times remain stable with time (curves 1 and 2, Fig. 22). The long-time treated surface (60 s) continues to exhibit a decrease in incorporated oxygen with aging time (curve 3, Fig. 22) [111].

A surface-treated polymer in reactive plasmas is an important class of highly asymmetric surface structure. The dynamics of such a structure is considered, in the last years, to be the key to the investigation of the aging of these surfaces. The highly asymmetric surface structure causes much stress and different stress-induced relaxation mechanisms like outdiffusion of untreated subsurface molecules through the modified layer, short range reorientation of macromolecular segments or side chains, and chemical reactions between introduced groups and the surroundings or the chemical groups of parent polymer [124].

Samples of oxygen plasma-treated PP stored in air, at room temperature, lose their wettability, namely both the advancing and receding angles increase with the aging time [124]. In particular, the advancing angle reaches the value for untreated PP, while the receding angle does not fully recover. No noticeable changes of

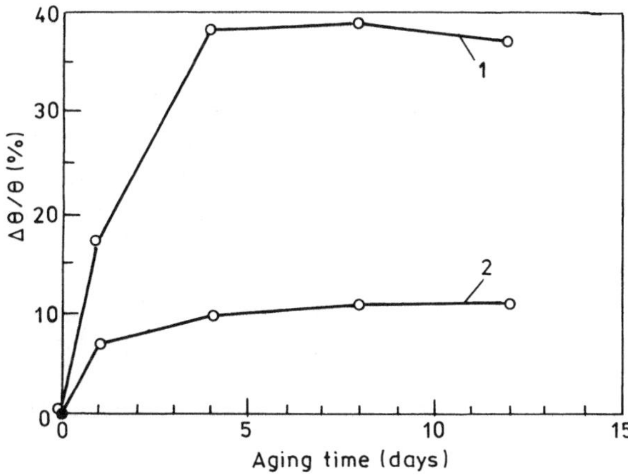

FIG. 21 Dependence of the water contact angle variation with the aging time for: 1: PP treated 1 s in NH_3 (60 W, NH_3 flow rate 150 sccm, 150 Pa); 2: PP pretreated in He (5 W, He flow rate 100 sccm, 100 Pa, 30 s). (Adapted from Ref. 123.)

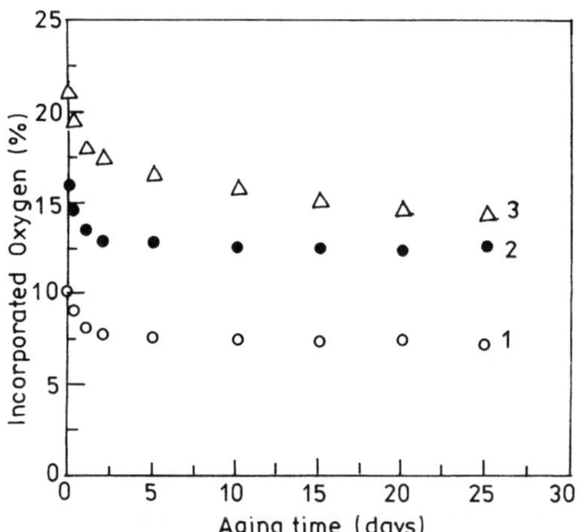

FIG. 22 Incorporated oxygen (%) vs. aging time for PE treated in oxygen plasma for 5 s (curve 1), 15 s (curve 2), 60 s (curve 3). ETOA = 38. (From Ref. 111.)

the XPS surface composition is apparent, which suggests that reorganization occurs in a layer thinner than the XPS sampling depth (< 5–7 nm). These results and those obtained by static secondary ion mass spectrometry (SSIMS) [122] sustain that the hydrophobic recovery of oxygen plasma-treated PP goes on through rearrangement within the modified layer through short-range reorientation of side groups and that the fully recovered surface still contains polar groups [124]. Apparent activation energy for recovery of 58.1 kJ/mole was calculated [125]. Two opposite mechanisms are operating: minimization of the surface free energy tries to build up an homogenous hydrocarbon-like surface, while hydrogen bonding interactions between the introduced surface groups tend to hinder the destruction and dipping into the nonpolar untreated bulk. The value of the fully recovered receding angle tends to increase as the aging temperature increases, which suggests that more hydrogen bonds can be broken, with a greater thermal energy input [126].

In contrast, oxygen plasma-treated HDPE shows a very small hydrophobic recovery and the advancing angle of the fully recovered surface is $\sim 24^\circ$, irrespective of the aging temperature [125]. A possible explanation of this difference between the PP and HDPE arise from their different behavior under oxygen plasma action, as was pointed out above. The extensive crosslinking of HDPE can be associated with low mobility of macromolecules. This is not the only feature which explains the stability of oxygen-treated PE. A major difference exists between the amount of oxygen introduced by plasma treatment in the PP and HDPE. For 20 s treatment time (power: 100 W, gas flow rate: 8 sccm, pressure: 2 Pa) PP reaches an O/C of 0.19, while HDPE 0.31 [127]. Therefore, the opinion of the authors is that a high degree of crosslinking with a large amount of oxygen-containing surface groups interacting by hydrogen bonding, appears necessary in order to greatly reduce recovery.

Gerenser *et al.* showed that nitrogen plasma-treated PP exhibits interesting changes on exposure to atmosphere. Surface composition is changed from 80% C, 20% N and 0% O to 83% C, 12% N and 5% O after 30 s exposure to laboratory conditions [110]. Since the amount of oxygen incorporated is very close to the nitrogen lost and in particular, the imine type nitrogen (N=C) is totally absent, the following hydrolysis reaction of the imine to its parent carbonyl was proposed [110]:

$$R\text{—}CH\text{=}NH + H_2O \rightarrow R\text{—}CH + O + NH_3$$

The replacing nitrogen with oxygen is also apparent for nitrogen-treated PE at very low power levels [115]. However, there is a change in relative intensity of the C 1s signal, implying that an additional oxidation was also taking place. These features persuaded the authors to propose the following reaction, which dominates the aging process in the first minute of air exposure [115]:

$$C\text{—}NH_2 + H_2O \rightarrow C\text{—}OH + NH_3$$

After this initial period of aging, the amount of nitrogen remains relatively constant even after 6 d, whereas the amount of oxygen increases with aging time. The long term process is a gradual oxidation of the main polymer itself [115].

E. Investigations in "Model Situations"

The system polymer–plasma is one of the most complex and usually it is very difficult to establish a relationship between the different plasma species and the most important mechanisms involved in polymer surface modifications. For this reason some experiments in so-called "model situations" were realized. Such a "model situation" can be either a "model polymer" or a "model gas phase".

Several polymers and their corresponding *model molecules* treated in argon or oxygen RF (13.56 MHz) plasmas were investigated by Clouet *et al.* [128–130]. Hexatriacontane ($C_{36}H_{74}$) has been chosen as a model compound of HDPE. The influence of plasma parameters (treatment time, RF power, pressure, gas flow rate) on the degradation rates and the formation and evolution of the degradation products have been investigated. Some conclusions are important [128–130]:

The degradation rates increase linearly with the treatment time and the applied power for both oxygen and argon plasmas.

CO, CO_2, H_2O, and H_2 are the main degradation products in an oxygen plasma and H_2 is the main one detected by mass spectrometry in an argon plasma.

The concentration of CO and CO_2 reaches a maximum faster than those of H_2O and H_2. A steady state is established after 2 min.

Degradation rate increases with oxygen gas flow rate, but the argon flow rate has no effect on degradation.

Pressure has no effect on the degradation rate both in an oxygen and argon plasma. Since the electron energy decreases and its density increases with increasing pressure, the degradation behavior observed is probably a result of these two antagonist effects.

For the Ar plasma the model molecules have a larger O/C increase [$d(O/C) = O/C_{final} - O/C_{initial}$] than their corresponding polymers because of the length of their CH_2 chains. The differences between the d(O/C) of PE and C_{36} treated in oxygen plasma are due to the desorption of short fragment chains into the gas phase.

In a discharge plasma the polymer sample is in a complex flow of energetic species, like radicals, ions, electrons and VUV and, as it was pointed above, it is impossible to study separately the reactions between the surface and each species. A *model gas phase* can be an electron, an ion or a neutral beam (with a well-defined current density and average energy), UV radiation or a remote discharge plasma. Since the interaction of UV and particle beams with polyolefins are presented in other sections of this chapter, here some new results about remote discharge plasmas are pointed out.

By locating the polymer sample outside the main plasma region, one can greatly reduce the number of reactive species. One can selectively eliminate the influence of ions or electrons that exists inside the plasma. A special kind of remote plasma is used in [131], the sample being placed downstream, behind the cathode of a dc glow discharge. Langmuir probe measurements in nitrogen show that extremely weak treatment conditions characterized by a low electron density ($\sim 10^{11}/m^3$, nearly independent of the distance from the cathode) and a concentration of specific cations (mainly N_2^+) dependent on the distance from the cathode are realized. The probe characteristics of oxygen, nitrogen, and hydrogen discharge, at the same distance from the cathode, differ considerably. A very low concentration of cations is present in the case of oxygen, while for hydrogen discharge a high concentration of cations is observed. In both oxygen and hydrogen discharges no electrons are detectable and interpretation of the experimental data in terms of Langmuir theory is impossible.

PE treated in both nitrogen and oxygen shows a sharp increase of the surface chemical heterogeneity in a first reaction stage. The only difference is that in the case of the nitrogen discharge the heterogeneity is mainly caused by electron-donor groups, whereas in the case of oxygen both electron-donor and electron-acceptor groups are detected. In a second mainly reaction stage, the identical values of the surface free energy and its components for oxygen- and nitrogen-treated PE clearly indicate a determining influence of the highly reactive neutral oxygen species, mainly molecular singlet oxygen [131]. The collision rate of oxygen molecules, deriving from air leakage at a basic pressure of 2 Pa, at 300 K, differs only about two orders of magnitude from that of nitrogen molecules at the working pressure of 40 Pa. This difference is compensated by a higher cross section for the reaction of the oxygen plasma species with the polymer surface and the high reactivity toward polymeric alkyl radicals formed during plasma treatment. Similar effects of trace oxygen are not observed for the hydrogen-treated PE. At a lower oxygen content ($< 0.1\%$) the reactive oxygen species seems to be quenched by the hydrogen species and/or oxygen-containing groups formed at the polymer surface are hydrogenated [131].

Downstream from the discharge only relatively long-lived reactive species can exists. Changing gas flow rate or/and applied power can further separate these species. The *remote plasma* of different gases was investigated by Foerch *et al.* [132–136] and Granier *et al.* [137]. After identifying O (^{3}P) and O_2 ($^1\Delta_g$) in the downstream gas, it was shown that atomic oxygen is the species responsible for the modification of polymers. At high atom flux (short distance from the main plasma region) γ_s^p saturates, which is considered to be linked to a degradation. At larger distances γ_s^p increases with oxygen atom fluence. Both O (^{3}P) and O_2 ($^1\Delta_g$) seem to influence γ_s^p.

LDPE treated in remote nitrogen and remote oxygen plasma has similar depths of modification, but treatment in oxygen is considerably faster. Also the stability appears to differ considerably. The total oxygen detected on the surface of remote nitrogen plasma-treated LDPE decreases from 46% to 31%, while oxygen introduced in remote oxygen plasma decreases from 25% to 15% [134].

Studies by Kill *et al.* on the downstream air plasma products and effects produced on PE surface [136] are complementary to those of Granier *et al.* [137]. They were realized in sufficiently similar flow and plasma conditions. Despite a large concentration of nitrogen present in the air plasma, there is no evidence of any nitrogen functionalization of the PE surface. The spectroscopic analyses suggest two plasma regimes that are of particular interest for their effect on the oxidation of the surface of PE. Conditions of low flow rate and high power are favorable in the maximum production of O (^{5}S), the *O (^{5}S) regime*; conditions of high flow and high power optimized the O (^{3}P) concentration, the *O (^{3}P) regime*. The C 1s spectra show the difference in the distribution of oxidation products of PE treated in the O (^{5}S) regime (Fig. 23a) and in the O (^{3}P) regime (Fig. 23b). For the O (^{5}S) regime the analysis of the

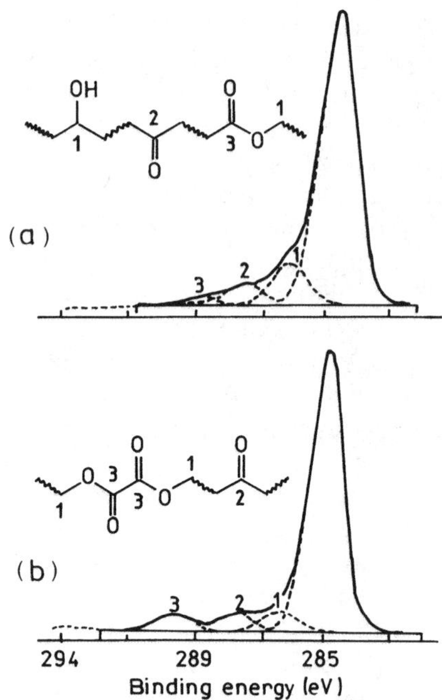

FIG. 23 C 1s spectra of PE treated in: (a) O (^{5}S) regime and (b) O (^{3}P) regime. Exposure time 3 s. (Adapted from Ref. 136.)

type and amount of oxidized carbons in the spectrum shows that alcohol functionalities are preferentially formed on the PE surface (8.7% C—O, 4.4% C=O, 2.2% CO_2). Similar amounts of three forms of oxidized carbon (4.3% C—O, 4.3% C=O, 5.8% CO_2) are apparent in the C 1s spectrum of PE treated in the O (^{3}P) regime. The total oxygen concentrations were 13.1% for PE treated in the O (^{5}S) regime and 14% in the O(^{3}P) regime. Formation of alcohols involves insertion of an oxygen atom of the appropriate symmetry, such as O(^{1}D), into a C—H bond to form alcohol. Since the O(^{5}S) do not have the parity necessary for insertion, the authors supposed that singlet atomic oxygen states are also present under these conditions, but were not detectable with the range of their spectroscopic equipment (500–1500 nm). Oxygen atoms in the O(^{3}P) state do not have appropriate symmetry for insertion, but can initiate oxidation by H atom abstraction [136].

VI. CORONA AND SILENT DISCHARGE

A. Generalities

Two types of atmospheric electrical discharge are used: *corona* and *silent discharge*. It must be pointed out that corona discharge can also be obtained at lower pressures.

The corona discharge is obtained between two electrodes, one of which is of small surface area. Some important parameters are: electrode geometries, electrode–surface distance (usually should not exceed 2–3 mm), the polymeric film speed, the value of the high voltage (~20 kV) with frequencies ranging from 10–20 kHz, and sometimes pressure. The *normalized energy* groups together some of these parameters:

$$E = P/dv \qquad (\mathrm{J/cm^2})$$

where P is the net power (in W), d is the electrode width (in cm), and v is the film velocity (in cm/s).

Many fundamental and applied aspects of corona treatment are discussed in the earlier publications by Briggs and coworkers [1, 138], Evans [139, 140], Owens [141], Gerenser *et al.* [3, 4], Strobel [5, 6], and Amouroux *et al.* [142, 143].

Another type of atmospheric electrical discharge is a parallel-plate dielectric barrier configuration, known as a silent discharge. A dielectric material covers the grounded planar electrode. Within each alternating cycle of high voltage, electrons arriving at the dielectric surface build up sufficient space charge to terminate and prevent complete sparkover to the substrate [144, 145]. Ozone is reported to be the major constituent of a parallel-plate dielectric barrier air discharge, its concentration being more than 100 times greater than electrons or ions and 10 times greater than other excited molecular species. In addition, ozone is by far the longest lived species following a "microdischarge", its lifetime being up to a second, compared to 10^{-4} s for other reactive molecules and 10^{-7} s for positive ions [144].

In the following the main characteristics of corona (at atmospheric and lower pressure) and silent discharge treatments of polyolefins by comparison with treatments in low pressure cold plasmas are presented.

B. Atmospheric Corona Discharge

Corona discharge treatment *in air* proceeds through surface oxidation, which is accompanied by considerable chain scission leading to the formation of water-soluble low molecular weight oxidized material (LMWOM).

The degradation process associated with the formation of so-called nodules were observed for both PE and PP. Such nodules were observed when gaseous atmosphere contains water vapor [6].

The SEM images of PP films at different treatment times show that the morphological aspects depend on the relative humidity under which the corona discharge treatment takes place. At about 60% RH, nodules appears after a few seconds and their size increase with treatment time, reaching a maximum value of about 20 μm [146]. For longer exposure times crystals appear, coexisting with nodules. This last feature suggested that oxidized products forming nodules seem to reorganize themselves in the form of crystals. When the treatment is realized at about 5% RH no nodules were observed, but small crystals (~ 1 μm) appear directly after a few minutes of treatment. The products that constitute the nodules and crystals are monoacid and diacid oligomers, with or without an alcohol or vinyl function. The crystallization of oligomers could be the result of the ozonolysis reactions, leading to the formation of products such as α-β ethylenic acids having a chelate hydroxyl function [146].

These results confirm some earlier ones that roughness is usually generated only at high humidities (> 50% RH) and at normalized energy greater than 1.7 J/cm^2 [5].

Treatments in low pressure plasma, 60 s, do not determine surface roughness of PE, no matter the gas [110]. Even longer exposure times (30 min) just slightly increase the surface roughness of PE as is shown by AFM images [147]. More than this, other works showed that air plasma treatment, 30 s, attenuates the macroroughness of untreated PE and PP [148].

The presence of LMWOM complicates the interpretation of wettability measurements made on treated polymer samples. The liquid should not react with or dissolve the solid surface. In the case of unwashed corona treated films, this constraint is not fulfilled when using polar liquids. For *advancing* water contact angle in air on unwashed corona-treated polymer, dissolution of LMWOM alters the local surface tension of the water at the advancing liquid front. The surface energy of LMWOM itself is likely to be different from the insoluble material. The effect of these two factors makes the interpretation of advancing-angle data on unwashed samples difficult. The *receding* angle on an unwashed surface should measure the wettability of the same surface as the receding angle on washed samples. However, dissolved LMWOM may lower the surface tension of the water at the retreating liquid front, determining the decrease of the measured contact angle rather than the actual value [149]. None of these ambiguities is encountered in interpreting wettability measurements on washed samples. The surface restructuring leads to modifications of the wetting hysteresis defined as [150]:

$$H = \gamma_{lv}(\cos\theta_r - \cos\theta_a)$$

where γ_{lv} is the interfacial free energy of the liquid–vapor interface, θ_r is the receding angle, and θ_a is the advancing angle.

From the point of view of wettability (given by γ_s^p) the efficacity of the corona treatment of LDPE does not increase with injected energy W above a certain critical value W_c (Fig. 24, curve 1), while the wetting hysteresis H has a sharp increase at W_c followed by a slight decrease at higher W (Fig. 24, curve 2) [150]. Therefore the authors established two domains of injected energy. At lower energies ($W < W_c$) the surface roughness is not affected while γ_s^p increases with W. H linearly increases with γ_s^p, which is considered to be due to the surface mobility of polar groups and their reorientation in the presence of water. At higher energies ($W > W_c$) the surface became more heterogeneous and rough because of the nodules' appearance. Washing these nodules leaves behind a surface like an untreated one, but more heterogeneous [150].

Data about corona-treated PP [149] are compared with those reported by Morra *et al.* for oxygen plasma-treated PP (see also Section V.D) [125, 126]. For both oxidation processes the chemical composition of a 2–5 nm layer into the treated polymer surface, analyzed by XPS, is very stable at ambient temperature.

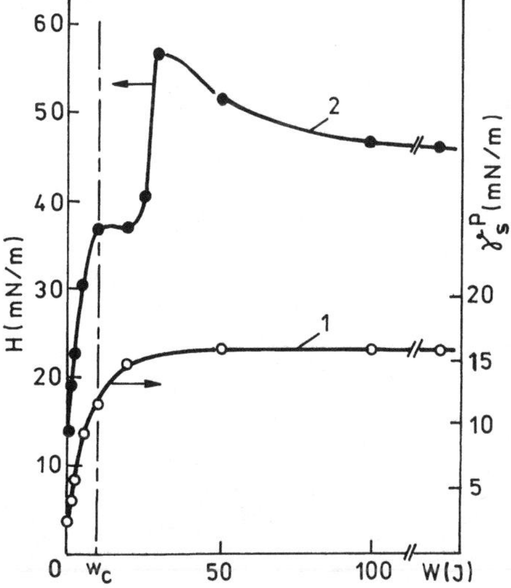

FIG. 24 Variation of γ_s^p (1) and of the wetting hysteresis H for water (2) vs. input energy W (LDPE surface was washed after treatment). (Adapted from Ref. 150.)

The main difference between their results is that for corona-oxidized PP the polar group content diminishes during aging at 100°C while no losses of oxygen plasma-treated PP stored at temperatures up to 135°C is observed. Since the water washing has no effect on the surface properties of oxygen plasma treated PP [126], the difference is readily explained by the fact that no LMWOM is formed during such kind of treatment. Therefore, in the absence of oxidized material of sufficiently low molecular weight to be water-soluble, reorientation is the only significant process that occurs during aging of the PP oxidized surface. When LMWOM is present its migration can cause a loss of oxidized material from the surface region, but only at elevated storage temperatures. There is no possible mechanism for interaction between the highly polar-corona oxidized materials and the highly non-polar bulk PP region. For this reason a significantly barrier to diffusion of LMWOM into the bulk is present at room temperature. At elevated temperatures the LMWOM is able to diffuse away from or evaporate from the surface [149].

A comparative study by static SIMS (SSIMS) of the effects of a variety of ac plasmas with that of air corona on PP is realized in [151]. The SSIMS spectrum of corona-treated PP exhibits a higher yield of mass 43 (Fig. 25b) by comparison with the untreated sample (Fig. 25a). Mass 43 amu could be due to a hydroxyl or carbonyl group on the pendant methyl group or the tertiary carbon atom. The increase of hydrophylicity can be concluded from the presence of water (18 amu). The small yield of 45 amu may be indicative of a carboxyl group. The peaks at higher mass 226/227 are the most intense in the spectra. After rinsing in water, all

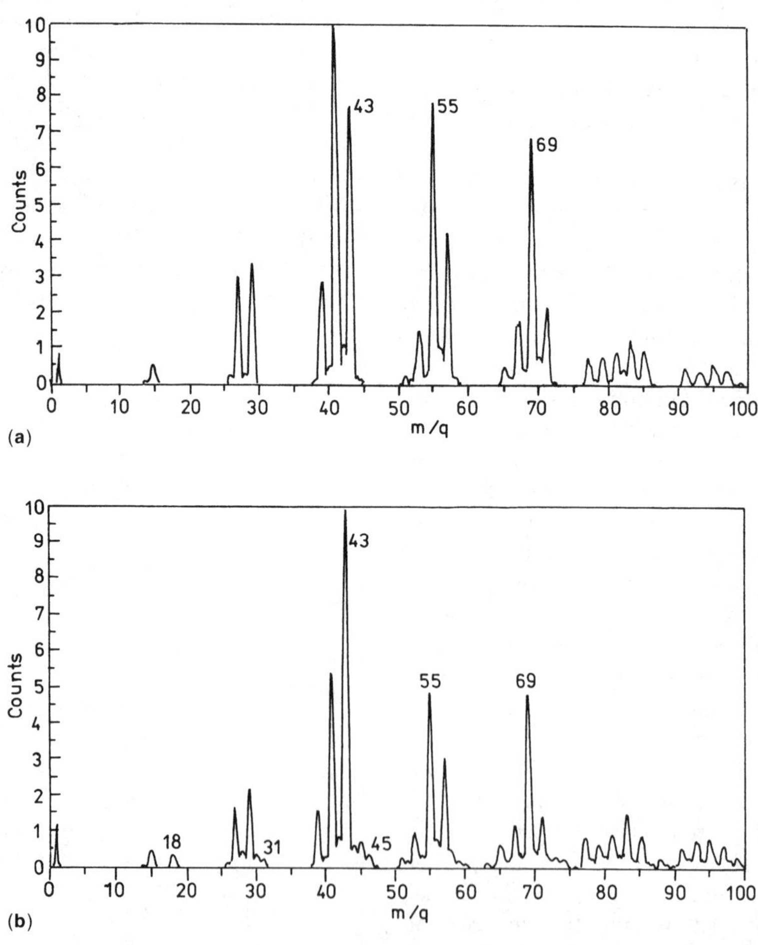

FIG. 25 Positive SIMS spectra of: (a) untreated PP, (b) corona-treated PP, (c) 1 min hydrogen plasma-treated PP, (d) 15 min argon plasma-treated PP, (e) 1 min air plasma-treated PP. Plasma treatment conditions: 820 V, 60 Hz, 13 Pa. Corona treatment conditions: normalized energy 12.7 J/cm^2, 75% RH. (Adapted from Ref. 151.)

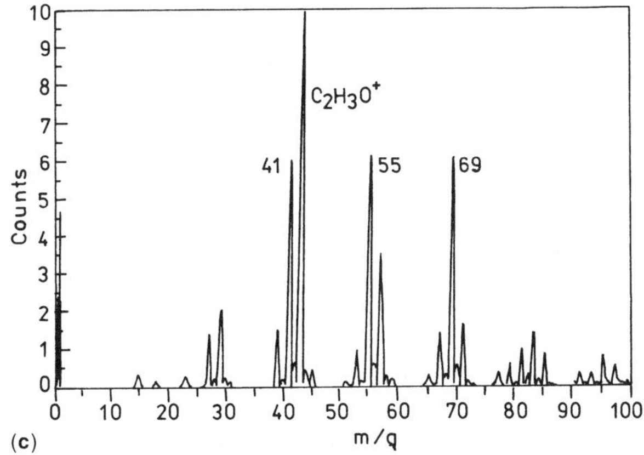

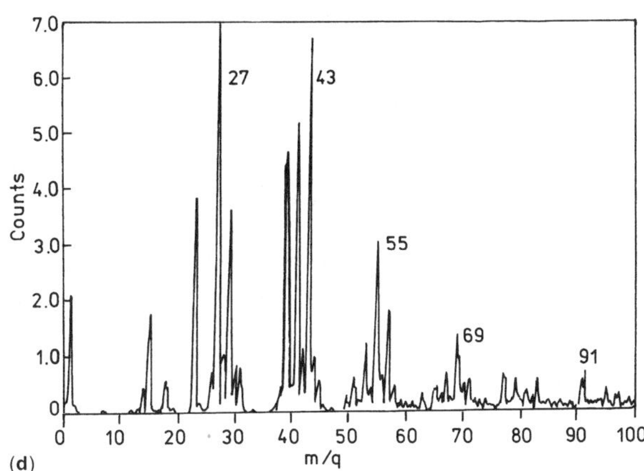

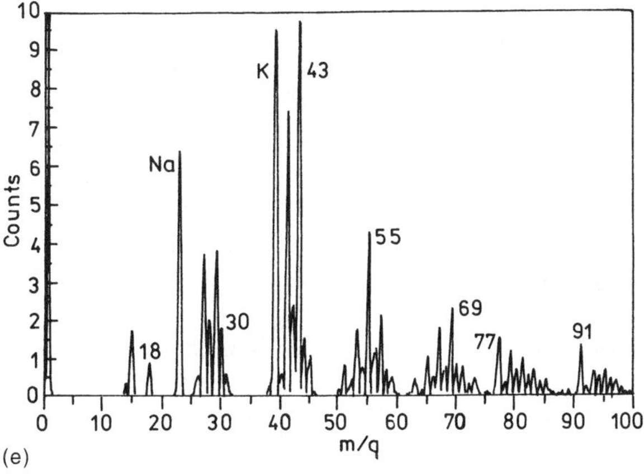

of these new peaks have completely disappeared. Therefore these results also suggest that most of the incorporated oxygen appears to be in a low molecular weight material. The series of peaks that appear after corona treatment of PP is also observed in corona treatment of other polymers (PE, PS, PET), which suggests a rather universal mechanism of low molecular weight material formation. In corona, as well as plasma treatments, a high degree of unsaturation, for which a measure is the ratio 27/29, is found for very short treatment times. For longer exposure times or higher normalized energy the unsaturation decreases. It is possible that many polymers initially form some sort of graphite-like structure, which reacts at a slower rate with oxygen [151]. The spectrum of PP obtained after *hydrogen plasma* treatment is virtually identical to that of corona-treated PP after washing (Fig. 25c). No unsaturation or low molecular weight material is observed which suggests that hydrogen creates reactive sites in polymer which either cross-link or adsorb oxygen upon exposure to the atmosphere. The treatment *in argon* plasma, at short exposure times does not induce strong changes, while at longer treatment times the main effect is a higher degree of unsaturation (Fig. 25d). The spectrum of *air plasma*-treated PP shows high yield of Na^+ and K^+, which are interpreted as indicative of the polymer etching (Fig. 25e). A certain degree of unsaturation is detected and polymer has cross-linked to some extent. The even number peaks starting at 28 amu are indicative of primary and secondary amines. The peaks that are interpreted as low molecular weight compounds were never observed in plasma treatments even after longer exposure times. The spectrum does not change upon rinsing with water [151].

C. Low Pressure Corona Discharge

Publications from Amouroux's group report some important results about polymer treatment in low pressure corona discharge, also including surface treatments of polyolefins in nonpolymerizable gases like nitrogen, and ammonia [123, 142, 152, 153]. Their investigations were realized in 70 kHz RF corona discharge, usually at a pressure in the range 100–200 Pa. Very short treatments (23–400 ms) and short treatments (1–14 s) were carried out.

Their SIMS data are presented by considering an internal standard, the 27 amu peak ($CH{=}CH_2^+$), which is less affected by analysis conditions. For PP treated for very short times in nitrogen corona, the decrease of low mass species (i.e. 41 : $C_3H_5^+$) with the treatment time (Fig. 26, curve 1) is considered to be due to a cleaning effect of the discharge. A low incorporation of oxygen is evaluated from evolution with treatment time of the ratio 43/27 (43 : $COCH_3^+$), shown in Fig. 26, curve 2. The evolution of the even mass peaks (28, 30, 40, etc.) shows that nitrogen

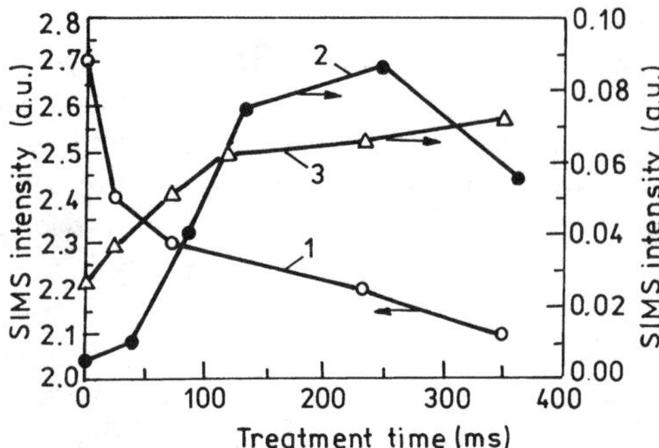

FIG. 26 Evolution of relative intensities of some SIMS peaks with treatment time of PP in nitrogen low pressure corona (200 Pa, 280 sccm, 60 W): 1: 41 amu, 2: 43 amu, 3: 30 amu. (Adapted from Ref. 152.)

incorporation starts for treatment longer than 23 ms and increases with treatment time (Fig. 26, curve 3, $30:CNH_4^+$). No increase in mass peaks > 100 amu has been observed which proves that the discharge has no degradation effect for such short treatment times (< 0.5 s) [152, 153].

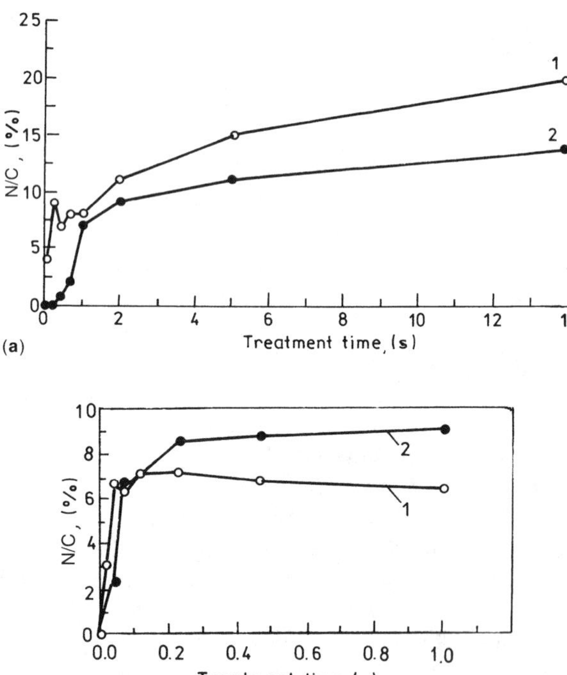

FIG. 27 (a) O/C (curve 1) and N/C (curve 2) vs. treatment time for PP treated in low pressure nitrogen corona. (b) O/C (curve 1) and N/C (curve 2) vs. treatment time for PP treated in low pressure ammonia corona. (Adapted from Ref. 153.)

XPS data, for the entire domain of treatment times investigated, show a saturation trend of surface functionalization, for longer exposure times ($t > 4$s) (Fig. 27a).

Both XPS and surface energy data proved that nitrogen incorporation is faster in ammonia than in nitrogen low pressure corona discharge. 60% of the increase of the surface energy is due to γ_s^p for PP treated in ammonia, while for PP treated in nitrogen only 20% from this increase is due to γ_s^p [152, 153]. For ammonia treated PP N/C ratio sharply increases for very short treatment times ($t < 0.046$ s) and saturates at 10% for times longer than 0.23 s (Fig. 27b, curve 2). N/C ratio has greater value than O/C ration, in opposition with the results for PP treated in nitrogen (Fig. 27) [154].

D. Silent Discharge

As was pointed out above, atmospheric silent discharge has as a main characteristic the higher ozone concentration.

The ozonation rate constants ($mol^{-1}\,s^{-1}$) reported in the literature for some polyolefins are in order [154]:

PP 0.080	PE 0.046	PIB 0.012

The relative degree of oxidation in air silent discharge (Table 8), as compared with untreated surfaces, is entirely consistent with the trend observed for ozonation rates: PP > PE > PIB (polyisobutylene) [145, 146]. A comparison of results reported by Greenwood *et al.* with previous studies, where polymer substrates were exposed to only ozone gas over similar exposure times (30 s), indicates that the surface oxidation is one order of magnitude greater in the case of silent discharge treatment. Saturated hydrocarbon polymers ozonize via a peroxy radical mechanism (as it was shown in Section II.C), a wide variety of oxidized carbon groups being obtained.

For LDPE and PP both low-pressure plasma (LPP) in air and silent discharge (SD) gave similar O/C ratios and comparable concentrations of various types of oxidized carbon functionalities (Table 8) [148]. Both silent discharge and low pressure air plasma treatments gave rise to carbon singly bonded to oxygen. Air plasma and silent discharge attenuates the macroroughness, the effect being most prominent for dielectric barrier treatment. PIB is more susceptible to chemical attack by the silent discharge treatment than by low pressure plasma [148].

TABLE 8 O/C ratios and relative amounts of carbon functionalities for some polyolefins after silent discharge (SD) or low pressure air plasma (LPP) treatments (from Ref. 148)

	PP		PE		PIB	
	LPP	SD	LPP	SD	LPP	SD
O/C	0.29	0.29	0.21	0.21	0.05	0.12
$\underline{C}$—H	70.5	67.1	78.1	76.2	90.8	83.2
—$\underline{C}$—CO_2	5.3	5.9	4.0	4.5	1.1	2.5
≡$\underline{C}$—O—	9.8	11.4	7.7	8.5	5.3	8.6
=$\underline{C}$=O/—O—$\underline{C}$—O—	5.1	7.7	3.9	4.5	1.5	2.9
—O—$\underline{C}$=O	5.3	6.0	4.2	4.5	1.1	2.5
—O—$\underline{C}$O—O—	4.0	1.9	2.3	1.8	0.2	0.3

VII. APPLICATIONS

A. General Considerations

Some applications of polyolefins surface modifications induced by gas phase treatment were already mentioned in previous sections of this chapter.

In many applications polyolefins are not used alone, but they are integrated in different components and composite materials or they are subjected to more or less "aggressive" medium. Two opposite surface characteristics are usually required:

high surface energy, strong wettability and polar surface are used for adhesion with other macromolecules, metals, adhesives, paints and varnishes;

low surface energy, weak wettability and apolar surface are usually applied for some biological and industrial applications.

In the following we will focus on those surface modifications that involve either bonding enhancement or diminishing of polyolefins to other materials. UV/ozone, corona and plasma treatments are already in commercial applications while particle beams are not yet well enough investigated to be used at commercial level, even though they are very promising.

The influence of additives on the changes in polymer properties after gas-phase treatments will be briefly presented at the end of this chapter.

B. Bonding Enhancement

Adhesion is fundamentally a surface property. High technologies such as aeronautics, aerospace, electronics and automotive industries are among the prime users of assembly by adhesive bonding. There are also numerous applications in more traditional sectors such as the wood, building, shoe, and packaging industries, etc.

Adhesion between two substrates is a very complex phenomenon since it implies multidisciplinary knowledge of the physical chemistry of the material surfaces, fracture mechanics, strength of materials, rheology etc.

In practice each assembly is specific, but the principal steps prior to the realization of a bounded joint are the same [154]: treatment of the surfaces to be joined, stress distribution analysis in the joint and design of the assembly, and measurements of the adherence level of the bond formed.

The preparation of the surfaces being together is often decisive in adhesion. Therefore, in the following we will briefly discuss the effects of different gas phase treatments of polyolefins (mainly plasma and corona discharge) on the adhesion for three representative groups: polymer–polymer or polymer adhesive bonding, metal–polymer bonding, and polymer–matrix composites.

Bonding enhancement can be regarded as resulting from the following overlap effects: (i) removal of organic contamination and of weak boundary layers by cleaning, (ii) cohesive strengthening of the polymeric surface by the formation of a thin cross-linked layer that mechanically stabilizes the surface and serves as a barrier against the diffusion of LMW species to the interface, and (iii) creation of chemical groups on the stabilized surface that result in acid–base interactions and in covalent linkages believed to yield to the strongest bond [155].

1. Polymer–Polymer, Polymer–Adhesive

Table 9 gives examples of data about typical bonding improvement for some polyolefins, by comparing conventional surface treatments (chemical, abrasion, and flame) with corona and plasma exposure. It can be seen

TABLE 9 Typical examples of bond strength improvement of some polyolefins after various gas-phase treatments (from Ref. 155)

Material	Substrate	Surface treatment	Load at failure (lb)
HDPE	Self	Ar plasma	353.1
		flame	144.1
		corona	123.9
		chemical	152.1
PP	Self	nitrous oxide plasma	110.4
		oxygen plasma	95.7
		flame	30.7
		corona	11.1
		chemical	309.9

that plasma exposure results in superior bond strength, except for PP chemically treated [155]. Conventional surface treatments have often been found to be inadequate for the complete removal of the contaminants which, even at a concentration level of less than one billionth of a gram per cm^2 surface area, already affect adversely adhesion bond strength.

Some Japanese patents claimed the improvement of adhesion of polyolefins to other polymers. Therefore moldings from a blend of PP, ethylene propylene rubber, and HDPE were treated in gas plasma to impart excellent adhesion to baked polyurethane coatings [156]. Similarly in another Japanese patent [157] PP treated in air microwave plasma has superior peel strength to the coated protective film of polyurethane resin.

Even though such treatments were applied a long time ago only relatively recently have papers presented the direct experimental evidence for the mechanisms involved in bonding between a plasma-generated surface and various adhesives. The changed surface chemistry facilitates reaction of the adhesive with surface species during curing, to form covalent bonds with the plasma-treated interphase [155].

Most experimental data follow the expected relationship that the bond strength will improve with improved wetting. The polymer surface needs to have similar or greater surface energy γ_s than the carrier solvent of the adhesive or paint being applied. This is to allow the liquid to spread the coating evenly over the surface and to "hang" there until bonds are formed between the coating and surface, while the solvent evaporated [154].

However, it is possible to obtain excellent bonding with very poor water wetting. Under optimum conditions for a given polymer-adhesive system, the bond strength can generally be improved to the point where the bond failure is cohesive in the weakest material and not in the bonding line.

For example, in the earliest paper of Schonhorn and Hansen it was shown that the maximum tensile strength of the laminated structure Al–epoxy–He treated PE–epoxy–Al is obtained after 5 s of plasma exposure, which corresponds to a crosslinked surface thickness of about 200–500 Å. For untreated PE this appears to be the upper limit of the weak boundary layer [158].

Remote nitrogen plasma-treated LDPE, containing 15% N, was found to require an average force of 160 g/cm to separate the two films stuck together with an epoxy adhesive, while a force of only 15 g/cm was required for unmodified LDPE. LDPE containing larger amounts of nitrogen indicated a very high joint strength. Even low concentrations of oxygen (15%) improve the peel strength to such an extent that rupture of the polymer strips occurs before separation. In contrast, samples treated in a corona discharge were found to separate as readily as untreated LDPE [154].

Surface printability studies, in the same conditions as above, indicate different trends using water-based inks on nitrogen and oxygen plasma-treated surfaces. Ink adhesion appeared to be improved with increasing percentages of nitrogen [154]. During the remote nitrogen plasma treatment the main species formed are initially amines with smaller amounts of imines and amides, whose relative concentrations increase with longer exposure times (> 10 s) [133]. However, it is not clear whether it is the formation of amides and imines or the increased number of amines available for bonding which causes the apparent ink adhesion [154]. After oxygen remote plasma treatment of LDPE, ink adhesion decreases slightly with higher percentages of oxygen on the surface, a better adhesion being observed for the surface containing only 15% oxygen. Water-based ink adhesion on oxygen-rich surfaces was shown to be a direct consequence of hydrogen bonding between hydroxyl groups on the polymer surface and carbonyl groups in the ink. Since oxygen plasma-treated LDPE has approximately equal concentrations of C—OH (or C—O—C), carbonyl and carboxyl groups, the surface is not particularly rich in C—OH [154]. There is no loss of ink adhesion with time even though the stability study indicates a loss of functionality. The samples containing slip agents show significant decrease in ink adhesion. This proves that significant blooming of slip agents occurs, even the plasma treatment appears to remove them from the surface.

Corona-treated LDPE showed very poor and inconsistent printing properties. Even the wettability of the surface had changed; most of the ink could be removed from the modified surfaces [154].

The results suggest that another type of chemical bonding mechanism might be of importance, namely acid–base interactions.

Interfacial bonding has been studied for the case of PE/PE and PE/PET laminates without adhesives, following treatment in low pressure MW air plasma or in an ambient air corona [159]. The adhesion force was found to exhibit a pronounced maximum for a surface concentration of bonded oxygen between 11–14%, independent of the type of treatment. The maximum adhesion occurs when the concentration of hydroxyl, ether or epoxide groups (basic) is highest and that of carboxyl (acid) groups is lowest (Fig. 28). These results suggest that the highest adhesion force appears when the surface is mechanically stabilized by crosslinking and when the effect of a weak boundary layer due to excessive LMWOM is minimal (low carboxyl concentration) [159]. It has to be established whether the maximum bond strength results in part from direct covalent bonding across the interface between the two laminated surfaces.

For food packaging applications a thin silicon oxide film on PP may be required to ensure O_2 and H_2O barrier properties [160]. Such thin films can be grown using plasma deposition techniques. A pretreatment of PP is necessary to improve adhesion. Both treatments in Ar and Ar–N_2 mixture plasmas enhance the adhesion of the silica layers to PP. This is attributed to crosslinking (detected by in situ UV–visible ellipsometry measurements) and to nitrogen functionalities, which result in acid–base interactions between PP and silica and/or in C—N—Si bonds at the interface [160].

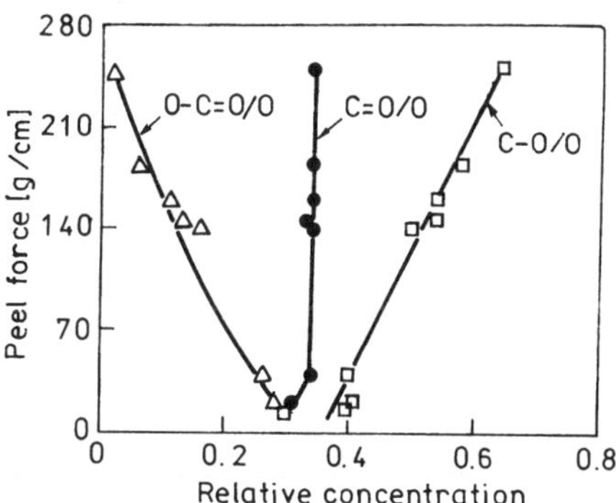

FIG. 28 Peel force of PE/PE laminates vs. relative concentrations of oxygen containing groups. (Adapted from Ref. 160.)

2. Polymer–Metal Adhesion

Buckstrand was among the first to show that evaporated metals react with oxygen-containing polymer surface, which can lead to metal–oxygen–carbon (M—O—C) type linkages [161].

In situ XPS studies revealed the presence of Ag—O—C and Ag—N—C linkages on oxygen and nitrogen plasma-treated PE, respectively [162, 163]. In contrast, no chemical bonding effects are observed on untreated or argon plasma-treated PE. The Ag valence-band structure suggests that the plasma-induced species act as nucleation and chemical bonding sites, resulting in a much smaller average cluster size for vacuum-deposited Ag. Also it was found that metal–PE adhesion is improved according to the following sequence of plasma gases used: $Ar < O < N_2$.

Similar effects were observed in another in situ study of Mg on PP [164]. The highest sticking probabilities for evaporated Mg atoms was found on a PP surface following exposure to a N_2 plasma or following argon ion bombardment at a dose of 5×10^{15} ions/cm^2.

More recently, in situ XPS and X-ray Auger electron spectroscopy investigations were focused on the characterization of chemical bonding of an inert (Cu) and reactive (Al) metal with an inert aliphatic PE before and after oxygen plasma treatment of the polymer surface [165]. When no chemical interaction takes place diffusion of the deposited metal atoms into the polymer can occur. The growth mechanism depends on the balance between the adsorption of the metal on the polymer and the cohesion energy of the deposited metal. On untreated PE no chemical bond is formed with both Cu and Al. The Cu/PE interface is not so sharp as Al/PE because of the diffusion of Cu into the polymer. The smaller atomic radius of Cu (1.28 Å) compared to Al (1.43 Å) could be one reason for this difference. XPS shifts, for such metal/polymer interfaces with no interfacial chemical bond formation, arise mainly from charging effects. Charging in the early stage of metallization is larger for Cu also due to metal atom diffusion into the polymer. In contrast to untreated PE, Al interacts with oxygen plasma-treated PE, though the formation of an Al—O—C complex at low coverage. Due to the chemical interaction between Al and the treated PE, the Al atoms lose their mobility and form small aggregates around the anchoring points or adsorption sites before the formation of a metallic layer. The chemical interaction

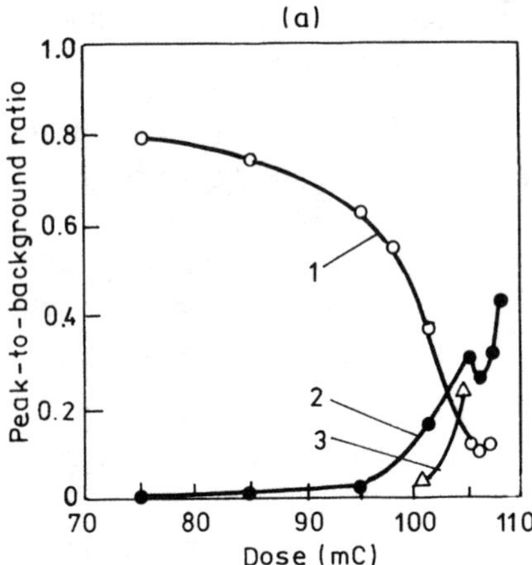

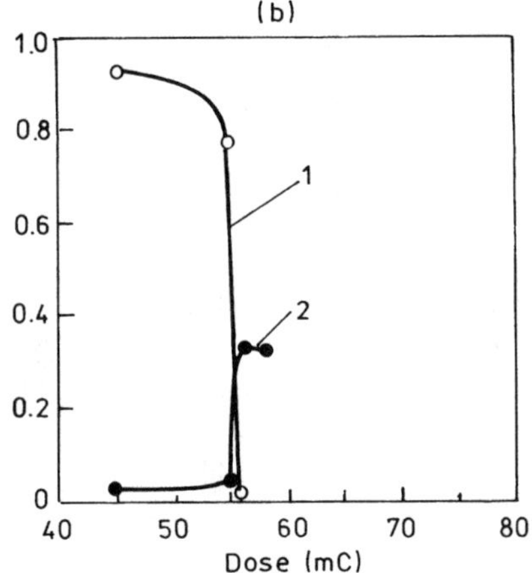

FIG. 29 Auger analysis of the Al–PP interface after: (a) NH_3 plasma treatment, (b) He plasma pretreatment. Al_{metLVV}-1, Al_{oxLVV}-2, C_{KLL}-3. (Adapted from Ref. 123.)

is limited by desorption of the oxygenated functionalities. Cu does not undergo a strong chemical interaction with treated PE and it diffuses into the polymer.

In [123] the difference in the Auger profiles of the Al metallized PP following different low-pressure corona discharge treatments is shown. In the case of NH_3 plasma-treated PP a broader interface is observed (Fig. 29a). The Al_{metLVV} (63.5 eV) peak persists for higher ion doses (Fig. 29a, curve 1) which allow measurement of the carbon (Fig. 29a, curve 3). The interdiffusion of Al in the subsurface layers of NH_3-treated PP is also shown by the Al_{oxLVV} (52 eV) peak evolution with dose (Fig. 29a, curve 2). In contrast, in the case of NH_3 plasma treatment with a He plasma pretreatment, AES shows a sharp interface (Fig. 29b). A drastic decrease of Al_{met} peak is observed and no intermixing of Al with carbon was observed.

3. Polymer–Matrix Composites

In the case of composite materials, the plasma treatments of the filler can be very effective in promoting adhesion because of the large treated surface area. The modifications of interfaces not only enhance the bond strength between the components, but it can also improve the electrical and aging characteristics of the composite by reducing the penetration of water vapor and other contaminants.

Much research has been done over the years on the treatment of particulate filler materials by exposure to selected plasma gases. In a nonpolar polymer host like LDPE the adhesion is promoted by minimizing the acid or base properties of the fillers. For example, in [166] $CaCO_3$, TiO_2, and carbon black (CB) were treated in MW plasma of various gases. The effects of these treatments on γ_s^d, numerical index of acidity/basicity K_i (defined as the difference between the donor DN and acceptor AN numbers DN − AN) are given in Table 10. The Ar treatment increases γ_s^d value for all three particulates, which can be due to the elimination of weakly bonded adsorbates. $CaCO_3$ remains strongly basic, while in the case of TiO_2 it increases the acidity of this surface. The polymerization of methane

TABLE 10 Effect of plasma treatments on $CaCO_3$, TiO_2, and carbon black (from Ref. 166)

Particulate	Treatment	γ_s^d (mN/m)	K_i
$CaCO_3$	None	44.6	7.4
	Ar	47.5	6.7
	CH_4	30.3	0.5
	Ar + CH_4	27.2	0.1
	C_2F_4	26.8	−0.8
	NH_3	45.5	8.3
TiO_2	None	53.0	−2.7
	Ar	55.0	−3.1
	CH_4	37.9	−1.5
	Ar + CH_4	31.5	−1.2
	C_2F_4	26.1	−2.6
	NH_3	50.2	2.4
Carbon black	None	42.8	−3.6
	Ar	44.0	−3.2
	CH_4	40.3	−1.1
	Ar + CH_4	37.6	−0.9
	C_2F_4	38.5	−3.1
	NH_3	41.9	1.1

sharply modifies the acid–base character of $CaCO_3$. In particular, after the combination of Ar and CH_4 treatments, $CaCO_3$ is nearly nonpolar. For TiO_2, plasma polymerization of CH_4 leads to a sharp reduction of the surface energy and to a milder acid interaction potential. Generally for TiO_2, treatments are less successful in producing neutral surfaces than in the case of $CaCO_3$. Plasma treatments in C_2F_4 shift the surface interaction potential to the acidic sites in the case of $CaCO_3$ and has no effect on the K_i value of TiO_2. The NH_3 treatment increases the surface basicity of both CaCO3 and TiO_2 surfaces. The effectiveness of plasma treatment is generally reduced for carbon black. One possible reason for this is the very high surface area of the solid which may lead to stronger interparticle association and, therefore, to less complete coverage by plasma-induced moieties.

The effects of modifying particulate surfaces are illustrated in terms of the mechanical properties of the filled LDPE. In Fig. 30 are given the values of Young's modulus vs $CaCO_3$ concentrations (wt%) for different particulate surface treatments. Addition of unmodified filler raises the modulus (Fig. 30, curve 1) as may be expected. Modifications of the acid–base interaction characteristics and of the surface energies yield major changes in the effect of filler on modulus. Treatments in Ar, CH_4 and both Ar and CH_4 plasma are most effective in increasing the modulus of the compounds (curves 2 and 3, Fig. 30), while the treatments in C_2F_4 (curve 5) and NH3 (curve 4) plasmas lead to opposing results.

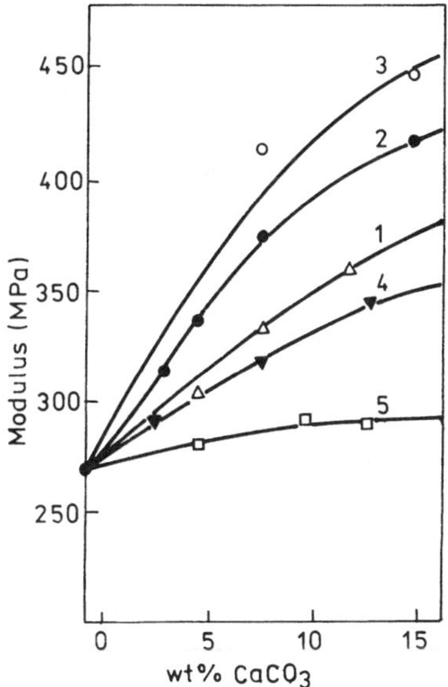

FIG. 30 Effects of plasma-treated $CaCO_3$ filler on the modulus of filled LDPE. 1: untreated, 2: Ar, 3: Ar + CH_4, 4: NH_3, 5: C_2F_4. (Adapted from Ref. 166.)

Ion implantation of polymeric material produces irreversible changes in macromolecular structure and can be used, as was shown in Section IV, to alter near-surface characteristics. This capability has been exploited to improve interfacial bonding characteristics between organic fibers and epoxy matrices in polymer reinforced composite materials. For example, a threefold improvement in interfacial shear strength of ultrahigh molecular weight PE in Epo-828 epoxy matrices was achieved by irradiation with 10^5 cm^{-2} of 100 eV Ti^+ ions without reducing fiber tensile property [167]. Also, implantation with other high-energy ions like Ar^+, N^+ leads to the introduction of polar groups which promote chemical bonding to the matrix resin.

C. Applications in Medicine

Polyolefins are quite widely used in medicine, where their biocompatibility is of great importance.

The energetic characteristics of a polymer, optimum for its biocompatibility, has not been hitherto solved.

It has been assumed that materials that do not adsorb proteins, enzymes, cells, and other elements of biologic liquids show high biocompatibility. For example, it was supposed that a polymer that does not adsorb blood protein should be resistant to thromb formation. The high adhesion of eye liquid components leads to lens dimness.

On the contrary, it was considered that a polymer is biocompatible if it exhibits a high ability to adsorb albumin and a low ability to adsorb other proteins.

In [168], the authors used the geometric method proposed by Kaelbe and Moacanin for analyzing energetic characteristics of solid surfaces. Results of the analysis support the Andrade hypothesis, that materials with zero interfacial tension at their water–solid interface show maximum biocompatibility. PE tubes (cut from subclavian catheters) were treated in air glow discharge plasma for 20 min. Since there is no glow discharge inside the tube, the surface is supposed to be modified due to reaction with long-lived reactive plasma particles. The treatment of PE has been shown to significantly decrease the polymer-water interfacial tension.

A surface modified with polyethylene oxide (PEO) has been shown to exhibit resistance to fouling by protein adsorption and platelet adhesion [169]. This is mainly due to the nonionic hydrophilic characteristics and its high chain mobility in water. Such nonfouling surfaces are important to biological and medical applications such as diagnostic assays, drug-delivery systems, biosensors, bioseparations, implants, and medical devices. A nonfouling polymer surface was prepared using an inert gas discharge treatment of a LDPE surface, which had been precoated with an oleyl PEO surfactant (Brij99), and PEO-PPO-PEO triblock copolymer surfactants (Pluronic) [170]. Adsorption of ^{125}I-labeled baboon fibrinogen and in vitro adhesion of ^{111}In-labeled baboon platelets examined the nonfouling properties of the treated surfaces. The protein adsorption studies support the proposed crosslinking of PEO surfactants to the LDPE via the alkyl segment of Brij99 or via PPO segment in Pluronic.

Several companies offer different surface treatment techniques suitable for device manufacturing with medical-grade plastics [Source: Internet], such as:

(a) *BSI Corp. (Eden Prairie, MN)* uses light-activated surface modification for chemical immobilization of molecules with predefined characteristics to most commonly employed biomaterials.
(b) *Rapra Technology LTD*, having representatives in a lot of countries (Germany, Benelux, Nordic Area, USA, Japan, Korea, Malaysia, Australia/New Zealand, Italy, Brazil, India, South East Europe, Singapore, France, Southern Africa, Chile, Israel) utilizes oxidation of PE for applications in biomaterial domain.
(c) *Spire Corp. (Bedford, MA)* uses ion beam processing, like ion implantation (the most common application being the treatment of artificial joint components made from ultrahigh molecular weight polyethylene (UHMWPE), with benefits limited only by the shallow depth of implantation) and ion beam assisted deposition (IBAD). IBAD applications include deposition of infection-resistant coatings or sealant coatings, production of flexible biosensor circuits on polymer substrates and enhancement of polymethyl methacrylate cement adhesion to UHMWPE orthopedic components.
(d) Talison Research (Sunnyvale, CA) and *Plasma Etch, Inc. (Carson City, Nevada)*, deal with plasma surface engineering. Typical applications for plasma surface engineering include application of hydrophilic coatings to contact lenses, modification of catheter components to enhance adhesion for bonding or coating, treatment of diagnostic devices for chemical functionalization, and preparation of implant components to receive biocompatible coatings. A polyethylene shaft could be argon–oxygen plasma processed along with a poly(ethylene terephthalate) balloon, the bond strength thus increasing over 50%.

Other applications are being developed.

D. The Influence of Additives on the Changes in PO Properties by Gas-Phase Methods

Very low concentrations of additives modify radiation-induced reactions in macromolecules. By reacting with radicals and other reactive species, notably oxygen, such additives can greatly affect the course of radiation-induced reactions [66]. The presence of additives and stabilizers may influence the discoloration level of the irradiated polymers [49, 171]. The addition of chemical compatibilizers and the use of plasma, corona, and ozone treatments that can introduce oxygen-bearing moieties on the surfaces could be employed to increase the interaction and adhesion between materials [14].

Irradiating PE films of 75 μm thick, doped with CB to different concentrations, with 2.6 GeV Au^{24+} ions, to relatively low doses (1×10^8 and 1×10^{11} ions/cm^2), Svorcik *et al.* [85] found the percolation threshold (PT) for the resistivity of the PE + CB mixtures (PT indicating a transition into the region of metallic-like conductivity) for a CB concentration ranging from 5.5 to 6.0 wt%. The strong PT reduction for an irradiation dose of 1×10^{11} ions/cm^2 was attributed to double bonds created by ion irradiation, which form conductive trajectories between CB grains.

In order to show the influence of ionizing treatment on plastic packaging in the context of food safety, PP samples were irradiated with an accelerated electron beam [172]. In the presence of air, alkyl radicals ($P^\bullet$) formed by ionization under vacuum are quickly oxidized into peroxyl radicals ($POO^\bullet$), whose stability varies with their location. After ionization of PP films doped with amine light stabilizers (of the HALS or hydroxyphenylbenzotriazole type), reaction derivatives of the aminoxyl type ($> NO^\bullet$) have been detected. After contact of irradiated films doped with light stabilizers with a liquid food simulating fatty acid esters, $> NO^\bullet$ derivatives were detected in the contact media. The contamination could come either by reaction of the stabilizer in the packaging before migration, or by migration followed by reaction with hydroperoxides present in the contact medium.

UV irradiation of IPP blends containing different concentrations of epoxidized lignin (LER) [49] led to superficial properties modification in respect with the irradiated virgin IPP (changes depending on LER content in the mixtures) and to a pronounced photo-oxidation induced by the LER presence in the studied blends, which could act like a chromophore.

IPP stabilized with some 2′,4′-dihydroxy phenyl ketones [173, 174] oxidizes at smaller rates under UV irradiation, compared with the unstabilized IPP. Changes in structure and properties of LDPE films containing spirooxazine as stabilizers [175] showed that this one acts as an UV absorber and shows a synergetic effect with HALS-type stabilizer.

VIII. NEW TRENDS

One of the more important trends is the utilization of two or more treatment media, either simultaneously or successively.

Also, surface modification by UV or UV/electron beam grafting of monomers of such acrylates is well documented in the literature [25–27]. Photosubstitution at carbonyl sites with alcohols or amines, to form ester or amide groups, respectively, always has an advantage over pure processes of photo-oxidation in improving wettability. These surface modification processes seem to be very beneficial in facilitating the fabrication of composite membranes with hydrophilic barriers on top of hydrophobic support, such as PP, with applications in separation technology and medicine [25].

Since O_2–plasma treatment alone is a rather poor technique for the preparation of a hydroxyl-functionalized PE surface (as was pointed out in Section V.B) the coupling of plasma and wet treatments is a more efficient method of such surface functionalization. For example, the density of surface reactive groups on LDPE and HDPE is greatly increased by aqueous $NaBH_4$ treatment [114].

Rather than use either MW or RF power to sustain the plasma, often two power sources are combined to generate the so-called "mixed" or dual-frequency plasma. While MW excitation generates a high concentration of active species the role of the RF power is to create a negative DC self-bias voltage on the powered, electrically isolated substrate folder. In dual-frequency MW–RF processing, independent control of the RF power allows us to vary the energy bombarding the substrate surface, with values ranging from a few eV to several hundreds of eV, and with fluxes of up to 10^{16} ions/cm^{2}s [155].

The new trend in the investigation of the mechanisms involved in surface modification of PO by treatments in discharge plasmas, is the utilization of so-called model situations, as was shown in Section V.E.

IX. CONCLUDING REMARKS

1. The main gas phase methods for surface modifications of polyolefins are presented. In the main the new trends in experimental methods and results are considered.
2. UV irradiation of polymers may induce changes in their color and degradation of their physical properties (surface as well as bulk properties). Ozone-generating UV light is widely used in order to clean organic contaminants from various surfaces, while the combined effects of UV light and ozone lead to fully oxidized surfaces, increasing—in the meantime—the wettability, adhesive properties, and dyeability of PO surfaces.
3. The impact of energetic ions on polymeric materials is very important for the modifications in the first few nanometers of the polymer surface, this fact having disturbing and damaging effects for the respective surface, but being also sometimes used to probe surface properties. Bombardment with low energy ions alters the structure, morphology, and chemical properties of materials, the main properties which can be modified by this procedure being: rheological properties, electrical conductivity, optical and mechanical properties, surface texture, crystalline state, biocompatibility, adhesive and surface properties.
4. Upon exposure of polymers to ionizing radiation (such as fast electrons, γ- and X-rays), new chromophore groups are generated, these ones efficiently absorbing light in the UV, visible and IR region of the spectrum. The main effects of electron beam irradiation are chain scission, oxidation and unsaturation, depending on dose rate, and oxygen content. Influence of radiation-induced chemical transformations is also observed in the modification of bulk properties of polymers, the most evident changes of this type being the modifications of their molecular weight due to crosslinking or degradation. Crosslinking increases the modulus and the hardness of the polymer and, in the case of partially crystalline polymers, imparts to the material a nonmelting behavior.
5. Results about surface functionalization and crosslinking of PE and PP, in different gas plasmas, are presented. It is shown that there is an important difference if XPS analysis were made "in situ" or not. All the PO surfaces incorporate oxygen after exposure to air, no matter the gas phase. PE and PP differ both from the point of view of incorporated oxygen and from the stability. The extensive

cross-linking of PE and the large amount of oxygen-containing surface groups explain its stability. In contrast, the high probability of chain scission and a lower content of oxygen functionalities can justify the large recovery found for PP.

6. Corona treatments in air of PO proceeds through surface oxidation, which is accompanied by considerable chain scission for both PP and PE.
7. Silent discharge treatments of PE and PP gave a similar O/C ratio with that obtained in low pressure plasma treatment in air.
8. Some applications of the surface modifications of PO induced by gas phase methods are presented.

NOMENCLATURE

a	absorption coefficient
ac	alternating current
AES	Auger electron spectroscopy
AFM	atomic force microscopy
amu	atomic mass unit
AN	acceptor number
Brij 99	oleyl PEO surfactant
CASING	Cross-linking via Activated Species of Inert Gases
CB	carbon black
d	electrode width
D	dose of the ion beam
dc	direct current
d_C	crosslinking depth
D_C	electron critical dose
DN	donor number
DSC	Differential Scanning Calorimetry
E	energy of the incident ions
EEDF	electron energy distribution function
EL	electroluminescence
ETOA	electron take off angle
F_T	fluence threshold for ablation
FTIR Spectroscopy	Fourier Transform IR Spectroscopy
G(X)	cross-link probability
G(S)	chain-scission probability
H	wetting hysteresis
HDPE	high density polyethylene
K_i	numerical index of acidity/basicity
LDPE	low density polyethylene
LMW species	low molecular weight species
LMWOM	low molecular weight oxidized material
LPP	low pressure plasma
M—O—C type linkages	metal–oxygen–carbon type linkages
MW plasma	microwave plasma
N_e	number of unpaired electrons (radicals)
P	net discharge power
PEO	polyethylene oxide
RF power	radio frequency power
RH	relative humidity
R_S	electrical sheet resistance
SD	silent discharge
SEM	Scanning Electron Microscopy
SSIMS	Static Secondary Ion Mass Spectrometry
UVO treatment	ultraviolet/ozone treatment
v	film velocity
VUV radiation	Vacuum Ultraviolet radiation
W	injected energy in the case of corona treatment
W_C	critical value of the injected energy
XPS	X-ray Photoelectron Spectroscopy
θ_a	advancing contact angle
θ_r	receding contact angle
λ	wavelength
γ_{LV}	interfacial free energy of the liquid–vapor interface
γ_S	free surface energy of the solid
γ_S^{LW}	Lifshitz–van der Waals (dispersive) component of the free surface energy of the solid
γ_S^{ab}	acid–base component of the free surface energy of the solid

REFERENCES

1. D Briggs. In: DM Brewis, ed. Surface Analysis and Pretreatment of Plastics and Metals. New York: Macmillian, 1982, pp 199–226.
2. RJ Ashley, D Briggs, KS Ford, RSA Kelly. In: DM Brewis, D Briggs, eds. Industrial Adhesion Problems. New York: John Wiley & Sons, 1985, pp 213–218.
3. LJ Gerenser, JF Elman, MG Mason, JM Pochan. Polym 26: 1162–1166, 1985.
4. JM Pochan, LJ Gerenser, JF Elman. Polym 27: 1058–1062, 1986.
5. M Strobel, C Dunatov, JM Strobel, CS Lyons, SJ Perron, MC Morgen. J Adhes Sci Technol 3: 321–335, 1989.
6. M Strobel, CS Lyons, C Dunatov, SJ Perron. J Adhes Sci Technol 5: 119–130, 1991.
7. SL Kaplan, PW Rose. In: D Satas, ed. Plastics Finishing and Decoration. New York: Van Nostrand Reinhold, 1991, pp 91–100.

8. JR Vig. J Vac Sci Technol A3: 1027–1034, 1985.
9. NS McIntyre, RD Davidson, TL Walzak, R Williston, M Westcott, A Pekarsky. J Vac Sci Technol A9: 1355–1359, 1991.
10. JR Vig. In: KL Mittal, ed. Treatise on Clean Surface Technology. New York: Plenum Press, 1987, pp 20–23.
11. JF Rabek, J Lucky, B Ranby, Y Watanabe, BJ Qu. In: JL Benham, JF Kinstle, eds. Chemical Reactions on Polymers. Washington DC: American Chemical Society, 1988, pp 187–200.
12. J Peeling, MS Jazzar, DT Clark. J Polym Sci, Polym Chem Ed 20: 1797–1805, 1982.
13. P Hedenberg, P Gatenholm. J Appl Polym Sci 60: 2377–2385, 1996.
14. DM Brewis, D Briggs. Polym 22: 7–16, 1981.
15. B Gongjian, W Yunxuan, H Xingzhou. J Appl Polym Sci 60: 2397–2402, 1996.
16. I Novak. Die Angew Makromol Chem 231: 69–77, 1995.
17. L Tong, JR White. Polym Eng Sci 37: 321–328, 1997.
18. R Gallo, F Severini, S Ipsale, N Del Fante. Polym Degrad Stab 55: 199–207, 1997.
19. B Kosikova, A Revajova, V Demianova. Europ Polym J 31: 953–956, 1995.
20. B Kosikova, M Kacurakova, V Demianova. J Appl Polym Sci 47: 1065–1073, 1993.
21. EH Lee. In: MK Ghosh, KL Mittal, eds. Polyimides. Fundamentals and Applications. New York: Marcel Dekker, 1996, pp 471–503.
22. CM Chan. Polymer Surface Modification and Characterization. Cincinnati: Hanser/Gardner Publications, 1994, pp 1–18.
23. E Uchida, Y Ikada. Curr Trends Polym Sci 1: 135–146, 1996.
24. M Pasternak. J Appl Polym Sci 57: 1211–1216, 1995.
25. A Wirsen, AC Albertsson. J Polym Sci A Polym Chem 33: 2039–2047, 1995.
26. A Wirsen, AC Albertsson. J Polym Sci A Polym Chem 33: 2049–2055, 1995.
27. A Wirsen, KT Lindberg, AC Albertsson. Polym 37: 761–769, 1996.
28. RK Wells, JPS Badyal. Macromol 26: 3187–3189, 1993.
29. J Murphy. Additives for Plastics Handbook. Oxford: Elsevier Advanced Technology, 1996, pp 121–137.
30. MJ Walzak, S Flynn, R Foerch, JM Hill, E Karbashewski, A Lin, M Strobel. J Adhes Sci Technol 9: 1229–1248, 1995.
31. JM Hill, E Karbashewski, A Lin, M Strobel, MJ Walzak. J Adhes Sci Technol 9: 1575–1591, 1995.
32. A Hollander, JE Klemberg-Sapieha, MR Wertheimer. J Polym Sci A: Polym Chem 33: 2013–2025, 1995.
33. J Peeling, DT Clark. J Polym Sci 21: 2047–2055, 1983.
34. S Dasgupta. J Appl Polym Sci 41: 233–248, 1990.
35. J Jonsson, B Ranby, C Laurent, C Mayoux. IEEE Trans Diel Electric Insul 3: 148–152, 1996.
36. J Jonsson, B Ranby, F Massines, D Mary, C Laurent. IEEE Trans Diel Electric Insul 3: 859–865, 1996.
37. A Tidjani. J Appl Polym Sci 64: 2497–2503, 1997.
38. MS Rabello, JR White. Chemi-crystallization of PP under UV irradiation. Proc of Annual Technology Conference-Society of Plastics Engineering, 1997, pp 1738–1742.
39. MS Rabello, JR White. J Appl Polym Sci 64: 2505–2517, 1997.
40. H Herbst, P Flanagan. In: W Hoyle, DR Karsa, eds. Chemical Aspects of Plastics Recycling. UK: Bookcraft (Bath) Ltd, 1997, pp 180–196.
41. F Massines, P Tiemblo, G Teyssedre, C Laurent. J Appl Phys 81: 937–943, 1997.
42. G Teyssedre, P Tiemblo, F Massines, C Laurent. J Appl Phys 29: 3137–3146, 1996.
43. CI Simionescu, C Vasile, MM Macoveanu, F Ciobanu, M Esanu, P Vidrascu, M Buruntea. Cell Chem Technol, in press.
44. C Vasile, M Downey, B Wong, MM Macoveanu, M Pascu, JH Choi, C Sung, W Baker. Cell Chem Technol 32: 61–88, 1998.
45. C Vasile, MM Macoveanu, VI Popa, M Pascu, G Cazacu. Roum. Chem Q Rev 6: 85–111, 1998.
46. B Kosikova, M Kacurakova, V Demianova. Chem Papers 47: 132–136, 1993.
47. M Pascu, MM Macoveanu, C Vasile, A Ioanid, RC Oghina. Cell Chem Technol, in press.
48. DJ Carlsson, DM Wiles. J Macromol Sci Rev Macromol Chem C14: 65-106, 1976.
49. EJ Chupka, TM Rykova. In: JF Kennedy, GO Phillips, PA Williams, eds. Lignocellulosic Science, Technology, Development and Use. West Sussex, UK: Ellis Horwood, 1992, pp 565–580.
50. M Pascu, MM Macoveanu, A Ioanid, C Vasile. Cell Chem Technol, in press.
51. M Doytcheva, D Dotcheva, R Stamenova, A Orahovats, C Tsetanov, J Leder. J Appl Polym Sci 64: 2299–2307, 1997.
52. M Strobel, MJ Walzak, JM Hill, A Lin, E Karbashewski, C Lyons. J Adhes Sci Technol 9: 365–383, 1995.
53. Y Novis, R De Meulemeester, M Chtaib, JJ Pireaux, R Caudano. Br Polym J 21: 147–153, 1989.
54. H Frerichs, J Stricker, DA Wesner, EW Kreutz. Appl Surf Sci 86: 405–410, 1996.
55. J Breuer, S Metev, G Sepold. J Adhes Sci Technol 9: 351–363, 1995.
56. W Kesting, D Knittel, E Schollmeyer. Angew Makromol Chem 191: 145–161, 1991.
57. W Kesting, T Buhners, E Schollmeyer. J Polym Sci B : Polym Phys 31: 887–890, 1993.
58. M Murahara, M Okoshi. J Adhes Sci Technol 9: 1593–1599, 1995.
59. S Etienne, J Perez, R Vassoille, P Bourgin. J Phys France 1: 1587–1608, 1991.
60. P Viville, O Thoelen, S Beauvois, R Lazzaroni, G Lambin, JL Bredas, K Kolev, L Laude. Appl Surf Sci 86: 411–416, 1995.
61. J Davenas, XL Xu, G Boiteux, D Sage. Nucl Instrum Meth Phys Res B39: 754–763, 1989.

62. J Breuer, S Metev, G Sepold, OD Hennemann, W Kollek. Appl Surf Sci 46: 336–341, 1990.
63. G Marletta. Nucl Instrum Meth Phys Res B46: 295–305, 1990.
64. A Charlesby. Nucl Instrum Meth Phys Res B105: 217–224, 1995.
65. V Svorcik, R Endrst, V Rybka, V Hantowicz. Mater Lett 28: 441–444, 1996.
66. JW Lee, TH Kim, SH Kim, CY Kim, YH Yoon, JS Lee, JG Han. Nucl Instrum Meth Phys Res B 121: 474–479, 1997.
67. CA Straede. Nucl Instrum Meth Phys Res B68: 380–388, 1992.
68. V Svorcik, V Rybka, V Hnatowicz, I Micek, O Jankovskij, R Ochsner, H Ryssel. J Appl Polym Sci 64: 723–728, 1997.
69. YQ Wang, LB Bridwell, RE Giedd. Desk Ref Funct Polym: 371–86, 1997.
70. YQ Wang, LB Bridwell, RE Giedd. Desk Ref Funct Polym: 387–404, 1997.
71. J Davenas, P Thevenard. Nucl Instrum Meth Phys Res B80/81: 1021–1027, 1993.
72. P Groning, OM Kuttel, M Collaud-Coen, G Dietler, L Schlapbach. Appl Surf Sci 89: 83–91, 1995.
73. V Svorcik, V Rybka, V Hnatowicz, K Smetana Jr. J Mater Sci: Mater Med 8: 435–440, 1997.
74. M Gheorghiu, G Popa, OC Mungiu. J Bioact Compat Polym 6: 164–177, 1991.
75. M Gheorghiu, G Popa, M Pascu, C Vasile. In: KL Mittal, ed. Metallized Plastics. New York: Marcel Dekker, 1997, pp 269–279.
76. M Gheorghiu, M Pascu, C Vasile, G Popa. Int J Polym Mater 40: 257–275, 1997.
77. V Svorcik, V Rybka, O Jankovskij, V Hnatowicz. J Appl Polym Sci 61: 1097–1100, 1996.
78. H Seunghee, L Yeonhee, K Haidong, K Gon-Ho, L Junghye, Y Jung-Hyeon, K Gunwoo. Surf Coat Technol 93: 261–264, 1997.
79. M Gheorghiu, M Pascu, G Popa, C Vasile, V Mazur. Int J Polym Mater 40: 229–256, 1997.
80. UJ Gibson. Nucl Instrum Meth Phys Res B74: 322–325, 1993.
81. E Taglauer. Surf Sci 64: 299–300, 1994.
82. M Cholewa, C Bench, BJ Kirby, GJF Legge. Nucl Instrum Meth Phys Res B54: 101–108, 1991.
83. V Svorcik, I Micek, O Jankovskij, V Rybka, V Hnatowicz, L Wang, N Angert. Polym Degrad Stab 55: 115–121, 1997.
84. E Taglauer. Nucl Instrum Meth Phys Res B98: 392–399, 1995.
85. DG Armour. Nucl Instrum Meth Phys Res B89: 325–331, 1994.
86. JL Sullivan, SO Saied, T Choudhury. Vacuum 43: 89–97, 1992.
87. A Ishitani, K Shoda, H Ishida, T Watanabe, K Yoshida, M Iwaki. Nucl Instrum Meth Phys Res B39: 783–786, 1989.
88. P Apel, A Didyk, A Salina. Nucl Instrum Meth Phys Res B107: 276–280, 1990.
89. A Delcorte, LT Weng, P Bertrand. Nucl Instrum Meth Phys Res B100: 213–216, 1995.
90. SK Koh, SC Choi, WK Choi, HJ Jung, HH Hur. Mater Res Soc Symp Proc 438: 505–510, 1997.
91. H Gunther. K Hartmut. Vide: Sci Tech Appl: 119–121, 1996.
92. A Chapiro. Nucl Instrum Meth Phys Res B32: 111–114, 1988.
93. T Czvikovszky. Radiat Phys Chem 47: 425–430, 1996.
94. VK Milinchuk. Nucl Instrum Meth Phys Res B105: 24–29, 1995.
95. W Burger, K Lunkwitz, G Pompe, A Petr, D Jehnichen. J Appl Polym Sci 48: 1973–1985, 1993.
96. V Svorcik, V Rybka, V Hnatowicz, M Novotna, M Vognar. J Appl Polym Sci 64: 2529–2533, 1997.
97. AS Vaughan, SJ Sutton. Polymer 36: 1549–1554, 1995.
98. HC Kim, AM El-Naggar, GL Wilkes. J Appl Polym Sci 42: 1107–1119, 1991.
99. K Tohyama, T Tokoro, S Mitsumoto, M Nagao, M Kosaki. Conf Electr Insul Dielectr Phenom 1: 146–149, 1997.
100. SK Datta, AK Bhowmick, TK Chaki, AB Majali, RS Despande. Polymer 37: 45–55, 1996.
101. A Boudet, C Roucau. J Phys 46: 1571–1579, 1985.
102. SR Karmakar, HS Singh Col Ann 97–100: 102–106, 1997.
103. AJ Lovinger. Macromolecules 18: 910–918, 1985.
104. PG de Lima, J Lopez, B Despax, C Mayoux. Rev Phys Appl 24: 331–335, 1989.
105. CM Ferreira, J Loureiro. J Phys D: Appl Phys 17: 1175–1188, 1984.
106. M Moisan, C Barbeau, R Claude, CM Ferreira, J Margot, J Paraszczak, AB Sá, G Sauvá, MR Wertheiner. J Vac Sci Technol B9: 8–25, 1991.
107. RH Hansen, H Schomhorn. J Polym Sci Polym Lett Ed B4: 203–209, 1966.
108. DT Clark, A Dilks. J Polym Sci Polym Chem Ed 15: 2321–2345, 1977.
109. DT Clark, A Dilks. J Polym Sci Polym Chem Ed 17: 957–976, 1979.
110. LJ Gerenser. J Adhesion Sci Technol 1: 303–318, 1987.
111. LJ Gerenser. J Adhesion Sci Technol 7: 1019–1040, 1993.
112. Y Yao, X Liu, Y Zhu. J Adhesion Sci Technol 7: 63–75, 1993.
113. M Gheorghiu, D Turcu, P Heinish, V Holban A Mihaescu. Investigation on plasma treatment of polymers. Proc IUPAC Int Symp on Plasma Chem (ISPC-6), Bucharest, 1983, pp 92–95.
114. M Morra, E Occhiello, F Garbassi. J Adhesion Sci Technol 7: 1051–1063, 1993.
115. S O'Kell, T Henshaw, G Fanow, M Aindow, C Jones. Surf Interface Anal 23: 319–327, 1995.
116. JF Ehnan, LJ Gerenser, KE Goppert-Berarchicci, JM Pochan. Macromolecules 23: 3922–3928, 1990.

117. N Inagaki, S Tasaka, M Imai. J Appl Polym Sci 48: 1963–1972, 1993.
118. JF Friedrich, L Wigant, W Unger, A Lippitz, H Wittrich, D Prescher, J Erdmann, HV Gorsler, L Nick. J Adhesion Sci Technol 9: 1165–1180, 1995.
119. F Poncin-Epaillard, JC Brosse, T Fahles. Macromolecules 30: 4415–4420, 1997.
120. M Gheorghiu, F Arefi, J Amouroux, G Placinta, G Popa, M Tatoulian. Plasma Source Sci Technol 6: 8–19, 1997.
121. G Placinta, F Arefi, M Gheorghiu, J Amouroux, G Popa. J Appl Polym Sci 66: 1367–1375, 1997.
122. RM France, RD Short. J Chem Soc Faraday Trans 93: 3173–3178, 1997.
123. M Tatoulian, F Arefi-Khonsari, I Mabille-Rouger, J Amouroux, M Gheorghiu, D Bouchier. J Adhesion Sci Technol 9: 923–934, 1995.
124. M Morra, E Occhiello, F Garbassi. J Colloid Interf Sci 132: 504–508, 1989.
125. E Occhiello, M Morra, F Garbassi, P Humphrey. J Appl Polym Sci 42: 551–559, 1991.
126. M Morra, E Occhiello, F Garbassi. In: JJ Pireaux, P Bertrand, JL Bredas, eds. Polymer Plasma Surface Modifications. Bristol: IOP Publishing, 1992, pp 407–427.
127. M Morra, E Occhiello, L Gila, F Garbassi. J Adhesion 33: 77–94, 1990.
128. F Clouet, MK Shi. J Appl Polym Sci 46: 1955–1966, 1992.
129. F Clouet, MK Shi. J Appl Polym Sci 46: 2063–2074, 1992.
130. R Prat, MK Shi, F Clouet. J Macromol Sci Pure Appl Chem A 34: 471–488, 1997.
131. I Bennish, A Holländer, H. Zimmermann. J Appl Polym Sci 49: 117–124, 1993.
132. R Foerch, NS McIntype, DH Hunter. J Polym Sci Polym Chem Ed 28: 193–204, 1990.
133. R Foerch, J Izawa, NS McIntype, DH Hunter. J Appl Polym Sci Appl Polym Sym 46: 415–437, 1990.
134. R Foerch, J Izawa, G Spears. J Adhesion Sci Technol 5: 549–564, 1991.
135. R Foerch, G Kill, MJ Walzak. J Adhesion Sci Technol 7: 1077–1089, 1993.
136. G Kill, DH Hunter, NS McIntyre. J Polym Sci Polym Chem Ed 34: 2299–2310, 1996.
137. A Granier, D Chereau, K Henda, R Safari, P Leprince. J Appl Phys 75: 104–110, 1994.
138. AR Blythe, D Briggs, CR Kendall, DG Rance, VJI Zichy. Polymer 19: 1273–1288, 1978.
139. JM Evans. J Adhesion 5: 1–9, 1973.
140. CY Kim, JM Evans, DAI Goring. J Appl Polym Sci 15: 1365–1375, 1971.
141. DK Owens. J Appl Polym Sci 19: 265–271, 1975.
142. V André, F Arefi, J Amouroux, Y de Puydt, P Bertrand, G Lorang, M Delmar. Thin Solid Films 181: 451–460, 1989.
143. V André Y de Puydt, F Arefi, J Amouroux, P Bertrand, JF Silvain. In: E Sacher, JJ Pireaux, SP Kowalczyk, eds. Metallization of Polymers. New York: ACS Books, 1990, pp 423–432.
144. B Eliasson, M. Hirth, U Kogelschatz. J Phys D: Appl Phys 20: 1421–1433, 1987.
145. OD Greenwood, S Tasker, JPS Badyal. J Polym Sci Polym Chem 32: 2479–2486, 1994.
146. NF Belkacemi, M Goldman, A Goldman. Surface evolution of PP submitted to corona treatment or aging processes. Proc Int Symp on Plasma Chemistry, Leicestershire, 1993, pp 1210–1215.
147. A Ringenbach, Y Jugnet, TM Duc. J Adhesion Sci Technol 9: 1209–1228, 1995.
148. OD Greenwood, RD Boyd, J Hopkins, JPS Badyal. J Adhesion Sci Technol 9: 311–326, 1995.
149. M Strobel, CS Lyons, JM Strobel, RS Kapaun. J Adhesion Sci Technol 6: 429–443, 1992.
150. N Takahashi, A Aouinti, J Rault A Goldman, M Goldman. Physique des surface et des interfaces. CR Acad Sci Paris 305: 81–84, 1987.
151. WJ van Oij, RS Michael In: E Sacher, JJ Pireaux, SP Kowalczyk, eds. Metalization of Polymers. New York: ACS, 1989, pp 60–87.
152. V André, F Arefi, J Amouroux, G Loravy. Surf Interface Anal 16: 241–245, 1990.
153. J Amouroux. Rom Rep Phys 45: 241–285, 1993.
154. G Fourche. Polym Eng Sci 35: 957–967, 968–975, 1995.
155. EM Liston, L Martinu, MR Wertheimer. J Adesion Sci Technol 7: 1091–1127, 1993.
156. JP 58.208.337 (Dec 5, 1983).
157. JP 58.147.433 (Nov. 5, 1983).
158. H Schonhorn, RH Hansen. J Appl Polym Sci 11: 1467–1483, 1967.
159. L Martinu, JE Klemberg-Sapieha, HP Schreiber, MR Wertheimer. Vide, Suppl 258: 13–20, 1991.
160. S Vallon, A Hofrichter, B Drévillon, JE Klembers-Sapieha, L Martinu, F Poncin-Epaillard. Thin Solid Films 290/291: 68–73, 1996.
161. JM Burkstrand. J Vac Sci Technol 15: 223–226, 1978.
162. LJ Gerenser. J Vac Sci Technol A 6: 1897–2903, 1988.
163. LJ Gerenser. In: E Sacher, JJ Pireaux, SP Kowalczyk, eds. Metallization of Polymers. ACS Symp Ser 406. New York: ACS, 1990, pp 433–452.
164. S Nowak, R Mauron, G Dietler, L Schlapbach. In: KL Mittal, ed. Metallized Plastics 2: Fundamental and Applied Aspects. New York: Plenum Press, 1991, pp 233–244.
165. A Ringenbach, Y Jugnet, TM Duc. J Adhesion Sci Technol 9: 1209–1228, 1995.
166. HP Schreiber, FS Germain. In: KL Mittal, HR Anderson, eds. Specific Interactions and their Effect on the Properties of Filled Polymers. Utrecht: VPS, 1991, pp 273–285.
167. DS Grummon, R Schalek, A Ozzello, J Kalantar, LT Drzal. Nucl Instrum Meth B59/60: 1271–1275, 1991.
168. AY Kuznetsov, VA Bagryansky, AK Petrov. J Appl Polym Sci 47: 1175–1184, 1993.
169. JD Andrade, S Nagaoka, S Cooper, T Okano, SW Kim. Trans ASAIO 33: 75–84, 1987.

170. MS Shen, AS Hoffman, BD Ratner, J Feijen, JM Harris. J Adhesion Sci Technol 4: 1065–1076, 1993.
171. JR Laghari, AN Hammoud. IEEE Trans Nucl Sci 37: 1076–1083, 1990.
172. D Marque, A Feigenbaum, AM Riquet. J Polym Eng 15: 101–115, 1996.
173. A Cascaval, M Esanu, MC Pascu, A Warshawsky, C Vasile. Mem Sci Sect Rom Acad 1:107–115, 1993.
174. C Vasile, H Seliger, A Cascaval, M Esanu, M Casariu, MC Pascu. Romania Pat 1993.
175. A Warshawsky, D Turcu, C Variu, DM Pantea, MC Pascu, C Vasile. Sci Ann "Al. I. Cuza" Univ III: 25–34, 1995.

25

Polyolefin Composites: Interfacial Phenomena and Properties

Béla Pukánszky
Technical University of Budapest, Budapest, and Hungarian Academy of Sciences, Budapest, Hungary

I. INTRODUCTION

The growth rate in the use of particulate filled polymers, including those prepared from polyolefins, is very rapid in all fields of application [1]. Household articles, automotive parts and various other items are equally prepared from them. In the early days mostly particulate fillers were introduced into polymers and the sole reason for their application was to decrease cost. However, as a result of filling, all properties of the polymer change; in fact a new polymer is created. Some characteristics improve, while others deteriorate, so properties must be optimized to utilize all the potentials of modification. Optimization must include all the factors influencing properties from component characteristics through structure to interactions.

The characteristics of polymer composites are determined by the properties of their components, composition, structure, and interactions [2]. These four factors have equal importance and their effects are interconnected. The specific surface area of a filler, for example, determines the size of the contact surface between the filler and the polymer, and thus the amount of the interphase formed. Surface energetics influences structure, but also the effect of composition on properties, as well as the mode of deformation.

This chapter focuses attention on the discussion of the most relevant questions of interfacial adhesion and its modification in particulate filled and fiber reinforced polyolefins. However, this cannot be done without giving a brief account of the four factors mentioned above, which determine the properties of particulate filled polymers. The effect of these factors on the mechanical properties of particulate filled polyolefins is outlined in the following section. Interactions can be divided into two groups; particle/particle and matrix/filler interactions. The first is often neglected although it may deteriorate the properties of the composite; often the only reason for surface modification is to hinder the occurrence of aggregation. Similarly important, but very contrasting, is the question of the formation and properties of an interphase; a separate section will discuss this. The significance of interfacial adhesion is also shown by the numerous efforts to modify them. The most important surface treatment techniques, their effects, mechanism, optimization, and eventual use are discussed in more detail in a later section. Finally, a short section is dedicated to the processing and application of particulate filled polyolefins.

II. FACTORS DETERMINING THE PROPERTIES OF PARTICULATE FILLED POLYMERS

As was mentioned above, four main factors determine the properties of modified polymers: the characteristics of the components, composition, structure, and interaction. All four are equally important and must be adjusted to achieve optimum properties and economics.

A. Component Properties

The properties of the matrix strongly influence the effect of the filler or reinforcement on composite properties; the reinforcing effect increases with decreasing matrix stiffness. In elastomers true reinforcement takes place, and both stiffness and strength increase [3]. This is demonstrated well in Fig. 1, where the tensile yield stress of polyolefin composites containing the same $CaCO_3$ filler is plotted against composition for two different matrices. LDPE is reinforced by the filler, while the yield stress of PP continuously decreases with increasing filler content [4, 5]. For the sake of easier comparison the data were plotted on a relative scale, related to the yield stress of the matrix. The direction of the change in yield stress or strength is determined by the relative load bearing capacity of the components [5, 6]. The filler carries a significant part of the load in weak matrices: it reinforces the polymer. The extent of stress transfer depends on the strength of the adhesion between the components. If this is weak the separation of the interfaces takes place even under the effect of small external load [7, 8]. With increasing matrix stiffness the effect of the interaction becomes dominating, the relative load bearing capacity of the components is determined by this factor. In a stiffer matrix larger stresses develop around the inclusions and the probability of dewetting increases and it becomes the prevailing micromechanical deformation process.

The structure of crystalline polymers, including polyolefins, may be significantly modified by the introduction of fillers. All aspects of the structure change on filling: crystallite and spherulite size, as well as crystallinity, are altered as an effect of nucleation [9]. A typical example is the extremely strong nucleation effect of talc in polypropylene [10, 11], which is also demonstrated in Fig. 2. The nucleating effect is characterized by the peak temperature of crystallization, which increases significantly on the addition of this filler. A $CaCO_3$-modified PP composite is shown as a comparison; the crystallization temperature changes much less in this case. Increasing crystallization temperature leads to the increase of lamella thickness and crystallinity, while the size of the spherulites decreases on nucleation. Increasing nucleation efficiency results in increased stiffness and decreased impact resistance [12].

Numerous filler characteristics influence the properties of particulate filled polymers [13, 14]. Chemical

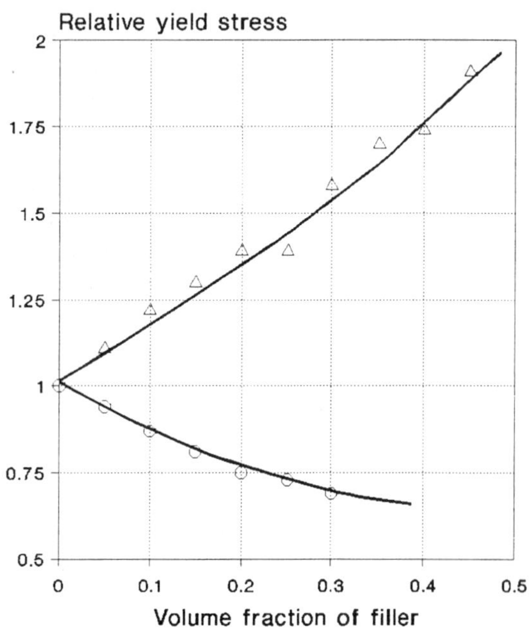

FIG. 1 Effect of matrix properties on the relative tensile yield stress of polyolefin/$CaCO_3$ composites; (○) PP, (△) LDPE.

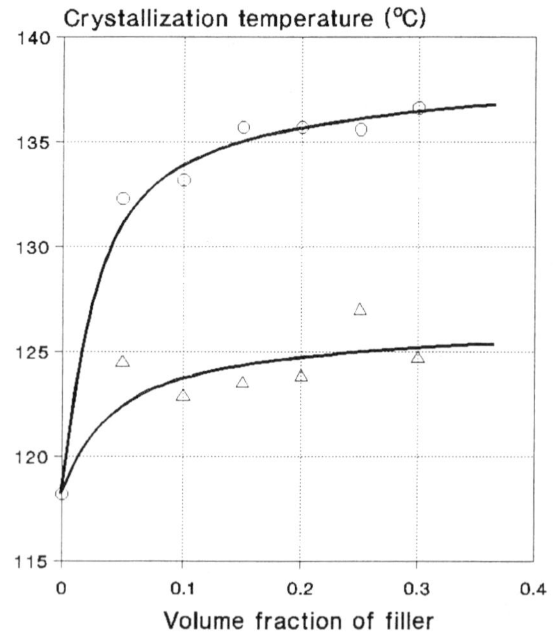

FIG. 2 Nucleation effect of particulate fillers in PP: (○) talc, (△) $CaCO_3$.

composition, and especially the purity, of the filler have both a direct and an indirect effect on its application possibilities and performance. Traces of heavy metal contamination decrease stability; insufficient purity leads to discoloration. High purity $CaCO_3$ has the advantage of white color, while the grey shade of talc-filled composites excludes them from some application fields.

The mechanical properties of polymer composites containing nontreated fillers are determined mainly by the characteristics of the particles, i.e., particle size and particle size distribution. Strength, sometimes also the modulus, increases, while deformability and impact strength usually decrease with decreasing particle size. Particle size in itself, however, is not sufficient for the characterization of any filler; the knowledge of the size distribution is equally important [15]. In the case of large particles the volume in which stress concentration is effective is said to increase with particle size and the strength of the matrix/filler adhesion also depends on it [7, 8]. The other end of the particle size distribution, i.e., the amount of small particles, is equally important. The aggregation tendency of fillers increases with decreasing particle size. Extensive aggregation, however, leads to insufficient homogeneity, rigidity, and lower impact strength. Aggregated filler particles act as crack initiation sites under dynamic loading conditions [16].

The specific surface area of fillers is closely related to their particle size distribution; however, it has a direct impact on composite properties, as well. The adsorption of both small molecular weight additives, but also that of the polymer, is proportional to the size of the matrix/filler interface [14]. The adsorption of additives may change stability, while matrix/filler interaction significantly influences mechanical properties, first of all yield stress, tensile strength, and impact resistance [5, 6].

Fibers and anisotropic particles reinforce polymers, and the effect increases with the anisotropy of the particle. In fact, fillers and reinforcements are very often differentiated by their degree of anisotropy (aspect ratio). Plate-like fillers, like talc and mica, reinforce polymers more than spherical fillers and the influence of glass fibers is even stronger [17]. Anisotropic particles orientate during processing, and the reinforcing effect depends very much also on orientation distribution.

The hardness of the filler has a strong effect on the wear of the processing equipment, which, however, is influenced also by the size and shape of the particles, composition, viscosity, speed of processing, etc. [15].

The surface free energy (surface tension) of fillers determines both matrix/filler and particle/particle interaction. The former affects the mechanical properties, while the latter determines aggregation. Both interactions can be modified by surface treatment.

The thermal properties of fillers differ significantly from those of thermoplastics. This has a beneficial effect on productivity and processing. Decreased heat capacity and increased heat conductivity reduce cooling times [15]. Changing thermal properties also result in alteration of the skin–core morphology of crystalline polymers, and thus the characteristics of injection molded parts change as well. Large differences in the thermal properties of the components, on the other hand, lead to the development of thermal stresses, which influence the performance of composites under the effect of external load.

B. Composition

Composition, i.e. the amount of filler or reinforcement in the composite, may vary over a wide range. The effect of changing composition on properties is clearly seen in Fig. 1. The interrelation of various factors determining composite properties is also demonstrated in the figure. The same property changes in a different, direction depending on the characteristics of the matrix, but interfacial adhesion may have a similar effect as well. The goal of modification is either to decrease cost or to improve properties, e.g., stiffness, dimensional stability, etc. These goals usually require the introduction of the largest possible amount of filler into the polymer, which, however, may lead to the deterioration of other properties. The optimization of properties must always be carried out during the development of composites. Numerous models are available, which describe the composition dependence of the various properties of composites [2, 13, 17]. With their help composite properties may be predicted and the optimization carried out. A somewhat more detailed discussion of this issue is found in Section III of this chapter.

C. Structure

The structure of particulate filled and fiber reinforced polymers seems to be simple: a homogeneous distribution of particles is assumed in most cases. This, however, rarely occurs and often special particle-related structures develop in the composites. The most important of these are aggregation and the orientation of anisotropic filler particles. The first is related to the interactions acting in a particulate filled polymer.

Thus it will be discussed in detail in a subsequent section (see Section IV.A).

D. Interfacial Interactions

Particle/particle interactions induce aggregation, while matrix/filler interactions lead to the development of an interphase with properties different from those of both components. Both influence composite properties significantly. Secondary van der Waals forces play a crucial role in the development of these interactions and they are modified by surface treatment. Occasionally reactive treatment is also used; its importance is smaller in thermoplastics than in thermoset matrices. However, reactive treatment must always be used in fiber reinforced polymers in order to achieve the necessary adhesion between the components. Strong adhesion is a primary condition of good stress transfer, assuring the proper performance of the composite.

III. MECHANICAL PROPERTIES OF FILLED AND REINFORCED POLYMERS

Particulate and fiber reinforced polymers are used in many fields of application, but mostly in areas where they are subjected to the effect of various external loads. Mechanical properties are strongly influenced by the presence of the filler or reinforcement and by all factors mentioned in the previous sections. The study and prediction of mechanical properties is very important in both research and development.

A. Stiffness

Fillers or reinforcements are used mainly to increase the composite modulus; stiffness always increases on the introduction of such components into a polymer matrix [17]. Decreased stiffness can be recorded only as an effect of erroneous modulus measurement: if it is carried out at large elongation, the matrix/filler interface separates, a void forms, and the composite modulus decreases as a result.

The extent of modulus increase depends on many factors. Although sometimes a linear dependence on composition is reported [18], stiffness usually increases exponentially with the amount of filler used. Interfacial interactions have a smaller effect on this property than on others. This weak effect is shown by Fig. 3, where the Young's modulus of PP/$CaCO_3$ composites is plotted against the specific surface area of the fillers [16]. Also the easy dewetting of large filler particles

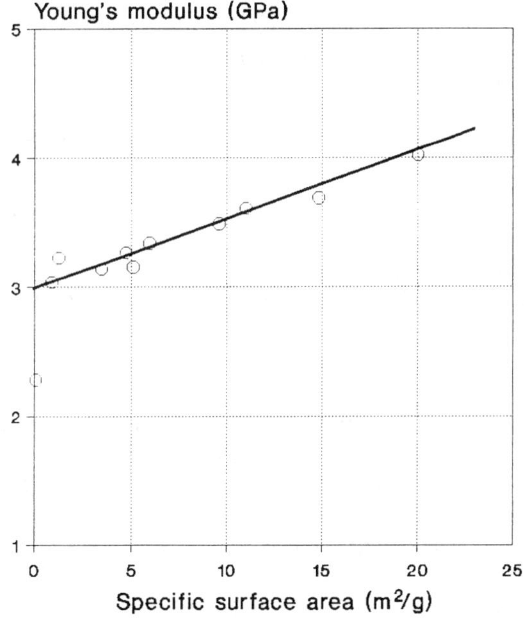

FIG. 3 Weak effect of interfacial interaction on the stiffness of PP/$CaCO_3$ composites.

and the resulting low composite modulus is demonstrated by the value measured at the smallest specific surface area (0.3 g/m^2). The particle size of this filler is 130 μm. Moreover, structure influences stiffness only slightly, since several of the fillers included in the study aggregate, but this does not change the correlation of Fig. 3 at all.

On the other hand, anisotropy and orientation strongly affect composite stiffness. Anisotropy of fillers and short fibers is usually expressed by their aspect ratio, i.e., the ratio between the largest and smallest dimensions of the particle. A close correlation was shown to exist between stiffness and aspect ratio [15], but the orientation of the particles is equally important, and the effect of anisotropy cannot be evaluated without knowing the particle orientation and orientation distribution as well. However, these latter are rarely determined and reinforcement is claimed to be solely the effect of anisotropy, which is misleading and erroneous.

The elastic modulus is the most often modelled property of composites. This can be explained by the large number of data that are available in the literature and by the relative simplicity of modelling caused by the use of the principles of linear elasticity. The available models can be classified into four categories: phenomenological models [19], bounds [20, 21], self-consistent, and semiempirical models [22, 23]. The

Lewis–Nielsen or modified Kerner equation [17] is one of the most successful models:

$$G = G_m \frac{1 + AB\varphi_f}{1 - B\psi\varphi_f} \tag{1}$$

$$A = \frac{7 - 5\nu_m}{8 - 10\nu_m} \tag{2}$$

$$B = \frac{G_f/G_m - 1}{G_f/G_m + A} \tag{3}$$

$$\psi = 1 + \left(\frac{1 - \varphi_f^{\max}}{\varphi_f^{\max^2}}\right)\varphi_f \tag{4}$$

where G, G_m, and G_f are the shear modulus of the composite, the matrix, and the filler, respectively, φ_f is the volume fraction of the filler, and ν_m is the Poisson's ratio of the matrix. The equation contains two structure-related parameters. A is associated with the shape of the filler, i.e., anisotropy, while Ψ is associated with its maximum packing fraction ($\varphi_f^{\max}$). However, the definition of the two parameters is not completely clear. A is defined sometimes as $A = 2k_e$, where k_e is the aspect ratio of the particles [24], but the validity of the correlation is not verified sufficiently. On the other hand, $\varphi_f^{\max}$ is supposed to depend only on geometric factors, i. e. on particle shape, size distribution, and packing [17], but interaction was also shown to influence this quantity [25]. Moreover, McGee and McCullogh [26] derived a correlation for Ψ different from Eq. (4). Although the application of models to predict the composite modulus is useful to obtain approximate values, to reveal the development of special structures, or to determine the effect of unknown factors, they must be used with the utmost care because of the above mentioned contradictions, which are further increased by the uncertainty of input values. For example, Poisson's ratios between 0.25 and 0.30 and moduli between 19.5 and 50 GPa can be found in the literature for $CaCO_3$ [27–29].

B. Yield and Ultimate Properties

Yield stress indicates the upper limit of the load which can be applied during the use of a product, while all kinds of material are extensively characterized by their strength. In the absence of a yield point, composite strength has the same function as yield stress in less rigid materials. Yield strain is used much less frequently for the characterization of particulate filled, and especially fiber reinforced materials, but elongation-at-break is often reported.

The yield stress of numerous polymers decreases with increasing filler content. This behavior is often generalized, but numerous examples are shown when true reinforcement takes place, i.e., yield stress increases with composition. Since yield stress is influenced by many factors and composition dependence may change according to the matrix/filler combination studied, often contradictory statements are published in the literature [5, 6, 30, 31]. Tensile strength shows a very similar composition dependence as yield stress. However, different behavior can be observed at low filler content and high elongation, where changing specimen cross section and the orientation of the matrix polymer also influence tensile strength. Yield strain and elongation-at-break decrease almost invariably with increasing filler content.

The error committed by false generalization has already been shown by Fig. 1. The $CaCO_3$ used decreases the yield stress of PP, indeed, but it has a reinforcing effect in the LDPE matrix, i.e. matrix properties strongly influence the effect of the filler [4]. Also changing interfacial interactions may profoundly affect the composition dependence of composite yield stress. The effect of the interaction is determined by the size of the interface and the strength of the interfacial adhesion. The first depends on the specific surface area of the filler and can be changed over a wide range. In Fig. 4

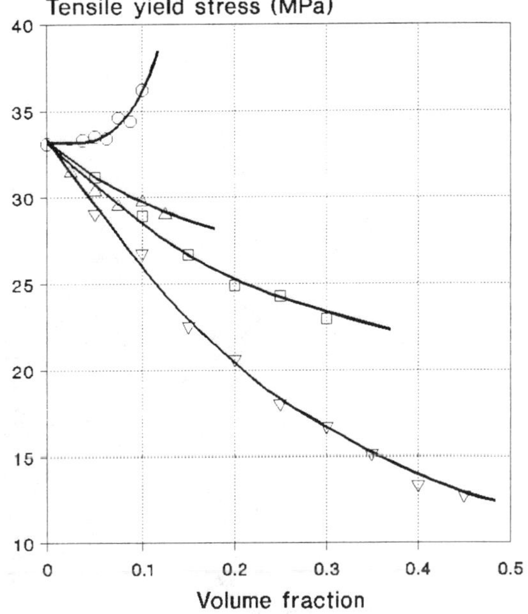

FIG. 4 Effect of the size of the interface on the yield stress of PP/$CaCO_3$ composites: A_f: (▽) 0.5, (□) 3.3, (△) 16.5, (○) 200 m^2/g.

the yield stress of PP/$CaCO_3$ composites is plotted against filler content [5]. Yield stress decreases in most cases, but true reinforcement is achieved at very high specific surface areas. Reinforcement can be brought about also by increasing the strength of interaction through surface treatment as well as with the application of anisotropic fillers.

Filler anisotropy has a similar effect on yield stress and tensile strength as well as on modulus: both increase with increasing aspect ratio and orientation. The two factors are interconnected here too, so both must be known to predict composite properties accurately. The important effect of fiber orientation is shown by Fig. 5, where the strength of glass fiber reinforced PP composites is plotted against the alignment of the fibers [32]. A considerable increase in strength is observed when the fibers have parallel orientation to the direction of the external load (0°), as expected.

The number of models describing the composition dependence of yield stress and strength is much smaller than for the modulus. The reason is clear; the relatively large deformation does not allow the use of the principles of linear elasticity and the strong effect of a large number of factors makes the development of exact models very difficult. Mostly empirical or semiempirical models have been developed up till now. The Nicolais–Narkis [33] model is used frequently, but it often fails, since it assumes zero interaction between the components. Moreover, it predicts zero matrix cross-section at a filler content less than 1, which cannot be true, of course. A more realistic model takes into account also interfacial interactions [5, 34]:

$$\sigma_y = \sigma_{y0} \frac{1 - \varphi_f}{1 + 2.5\varphi_f} \exp(B_y \varphi_f) \tag{5}$$

where σ_y and σ_{ym} are the yield stress of the composite and the matrix, respectively, while B_y is a parameter related to the relative load bearing capacity of the filler. B_y. depends on interaction, both on its strength and on the size of the interface. This can be expressed in the following way [5]:

$$B_y = (1 + A_f \rho_f l) \ln \frac{\sigma_{yi}}{\sigma_{yo}} \tag{6}$$

where A_f and ρ_f are the specific surface area and density of the filler, while l and σ_{yi} the thickness and property of the interphase. Although the correlation could be used successfully in many cases [5, 34], it has several drawbacks. B_y cannot be calculated from the properties of the filler or reinforcement and it is also influenced by structure. Changes in the morphology of the matrix polymer on filling, aggregation or orientation of anisotropic particles, all modify the value of B_y. On the other hand, all these effects can be studied by applying the model to experimental results.

The model was also extended to describe the composition dependence of the tensile strength of heterogeneous polymer systems, including particulate filled and fiber reinforced polymers, as well as blends [6, 35, 36]. During the elongation of the specimen the cross-section of the specimen decreases continuously and orientation leads to strain hardening. These effects have to be taken into account thus the modified correlation takes the form:

$$\sigma_T = \sigma_{T0} \lambda^m \frac{1 - \varphi_f}{1 - 2.5\varphi_f} \exp(B_T \varphi_f) \tag{7}$$

where σ_T and σ_{T0} are the true strength of the composite and the matrix, respectively ($\sigma_T = \sigma\lambda$, $\lambda = L/L_0$, relative elongation). The introduction of true stress accounts for the changing specimen cross-section while λ^m is for strain hardening. Also this correlation was successfully used for many heterogeneous polymer systems, but mostly for polyolefin composites [6].

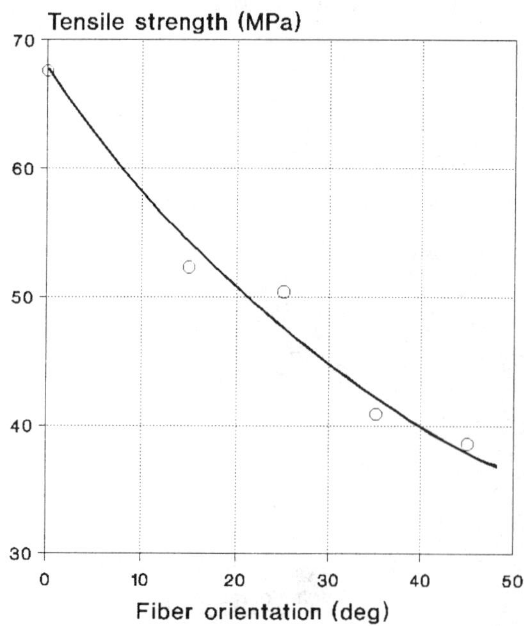

FIG. 5 Correlation of fiber orientation and strength in short glass fiber reinforced PP composites [32].

C. Impact Resistance

Fracture, and especially impact resistance, are crucial properties of all materials used in engineering applications. Similar to yield stress, the impact resistance of particulate filled polymers is also assumed to decrease with filler content, which again is not necessarily true. Fracture and impact resistance often increase or show a maximum as a function of filler content in both thermoplastic and thermoset matrices [37, 38]. Several micromechanical deformation processes take place during the deformation and fracture of heterogeneous polymer systems. The new deformation processes initiated by the heterogeneities always consume energy, which leads to an increase of impact resistance. Energy consumption is different in the various deformation mechanisms, thus also the extent of property change and composition dependence may vary according to the actual processes taking place during deformation and according to their relative weight. Deformation mechanisms leading to the increased plastic deformation of the matrix are the most efficient in improving fracture and impact resistance.

The composition dependence of the critical strain energy release rate of particulate filled PP is presented in Fig. 6 [39]. The maximum in fracture resistance is evident, as well as the different effect of the two fillers, talc and $CaCO_3$. The same factors influence fracture resistance as the other mechanical properties, i.e., composition, structure (aggregation, orientation of anisotropic filler particles), and interaction. Impact resistance was shown to decrease with increasing interaction between the matrix polymer and the filler and this correlation proved to be valid both in thermoplastic and thermoset matrices [39]. The role of micromechanical deformations is more pronounced in fracture than in other deformation and failure modes, shown also in multicomponent PP composites containing a filler and an elastomer at the same time [40]. Contrary to expectations, the presence of an elastomer does not always improve impact resistance on the one hand, while a filler with the appropriate adhesion and particle size may lead to better impact resistance, on the other. The fracture resistance of fiber reinforced composites depends very much on the length and orientation of the fibers.

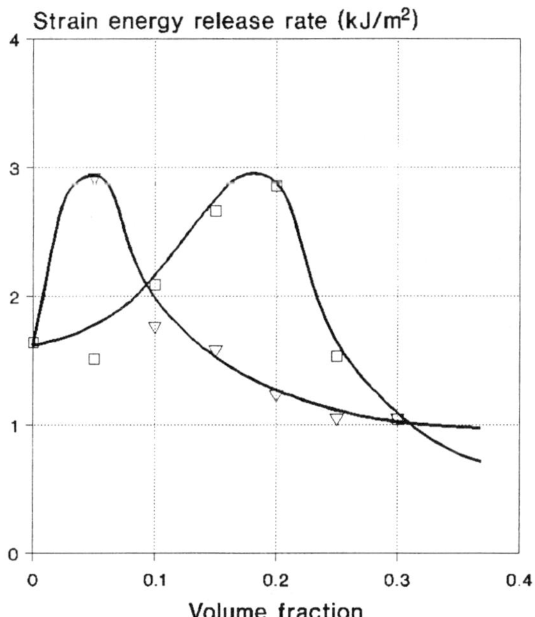

FIG. 6 Maximum in the impact resistance of particulate filled PP composites: (□) $CaCO_3$, (▽) talc.

Because of the effect of a large number of factors influencing fracture resistance and due to the increased role of micromechanical deformation processes, the modelling of this property is even more difficult than that of other composite characteristics. Nevertheless, a relatively large number of models have been published up to now. One of the earliest models [41] is based on the Nicolais–Narkis equation [33], but it proved to be of very limited value, since it neglected a large number of factors, including the effect of interaction. More advanced models take this effect into account formally, but neglect it in practice, based on the argument that the interaction is usually weak [42, 43]. Another approach uses the idea of line strength [44], but it has been shown that this factor is important only in very rigid systems [45]. The model of Evans [46] takes into account the plastic deformation taking place during fracture. Although the model is theoretically sound and rigorous, it is overly complicated for practical use. Jančař [47] introduced bounds and applied the model for the description of the impact resistance of PP composites. Apart from the fact that some doubts may be raised concerning the theoretical background of the approach, bounds are of limited use in practice. The semiempirical model applied for the description of the composition dependence of other mechanical properties [5, 6] can be extended also to fracture and impact resistance [39]. Although the model could be used successfully for a large number of composites both with thermoplastic and thermoset matrices, it has the same limitations as in the case of other properties, i.e., yield stress and tensile strength [5, 6].

The various models presented in this section are very useful for the study of particulate filled and fiber reinforced polymers. They can point out the most important factors influencing the properties of such materials and the comparison of experimental data to the prediction of the models may give valuable information about their structure. However, the majority of the models cannot predict composite properties from component characteristics and composition. As a consequence, experimentation remains very important during the development of filled and reinforced materials.

IV. INTERACTIONS

As was mentioned earlier, two types of interaction must be considered in particulate filled polymers: particle/particle and matrix/filler interactions. The first is often neglected even by compounders, in spite of the fact that it may deteriorate composite properties significantly especially under the effect of dynamic loading conditions [48].

A. Aggregation

Aggregation is a well known phenomenon in particulate filled composites. Experience has shown that the probability of aggregation increases with decreasing particle size of the filler. The occurrence and extent of aggregation is determined by the relative magnitude of the forces which attract the particles, on the one hand, or try to separate them, on the other. Particulate filled polymers are prepared by melt mixing of the components; thus the major attractive and separating forces must be considered under these conditions. Ess and Hornsby [49] carried out a detailed investigation of particle characteristics and processing conditions on the dispersive mixing in particulate filled PA and PP. They listed the principal adhesive forces, which are mechanical interlocking, electrostatic forces, van der Waals forces, liquid bridges, and solid bridging in increasing order of adhesive strength. Particles are separated mainly by shear and elongational forces during homogenization.

Since aggregation is also an important phenomenon in other areas (pigments, paints, powder handling, etc.) numerous studies deal with the interaction of particles [50]. When two bodies enter into contact they are attracted to each other. The strength of adhesion between the particles is determined by their size and surface energy [51, 52], i.e.

$$F_a = \tfrac{3}{2}\pi W_{AB} r_a \tag{8}$$

where F_a is the adhesive force between the particles, W_{AB} is the reversible work of adhesion, and $r_a = r_1 r_2/(r_1 + r_2)$, an effective radius, if the size of the two interacting particles is different.

In the presence of fluids, i.e. in suspensions, but also in the polymer melt during homogenization, further forces act between the particles. Adams and Edmondson [52] specifies two attractive forces, depending on the extent of wetting of the particles. In the case of complete wetting a viscous force acts between the particles which are separated from each other with a constant rate. If the particles are wetted only partially by the fluid (melt), liquid bridges form and capillary forces develop among them. These can be divided into two parts: a hydrostatic component, and one related to surface tension.

Balachandran [53] analyzed the role of electrostatic forces in the adhesion between solid particles and surfaces. According to the available information this role is not completely clear, and the results are contradictory. Electrostatic forces can be significant in the case of polymer and semiconductor particles. These forces have four main types: Coulomb, image charge, space charge, and dipole forces. The magnitude of all four is around 10^{-7}–10^{-8} N, and they are significantly smaller than the other forces acting between the particles. However, if the particles are small, these forces can be larger than their weight: These forces may play an important role during the approach of particles, but they are dissipated on contact and van der Waals or other adhesive forces become dominant. The effect of electrostatic forces is important only in the case of charged particles or surfaces.

The number of forces separating the particles is smaller. Repulsive forces act between particles with the same electrostatic charge. The mixing of fluids leads to the development of shear forces which try to separate the particles. The maximum hydrodynamic force acting on spheres in a uniform shear field can be expressed as [52, 54]

$$F_h = -6.12\pi\eta r^2\dot{\gamma} \tag{9}$$

where η is the viscosity of the melt and $\dot{\gamma}$ is the shear rate.

An analysis and comparison of the adhesive and separating forces acting between the particles shows that the most important ones are the adhesive and hydrodynamic forces. Thus their relative magnitude determines the extent of aggregation during the homogenization of a polymer composite. This is somewhat in contradiction with the statement of Ess and

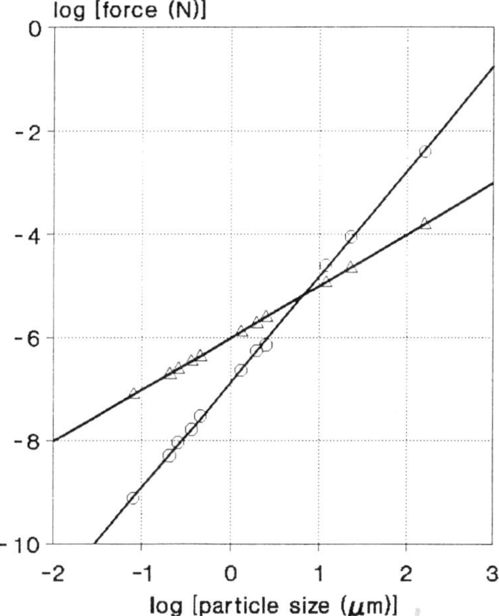

FIG. 7 Adhesion and shear forces acting during the homogenization of PP/$CaCO_3$ composites; effect of filler particle size: (○) shear, (△) adhesion.

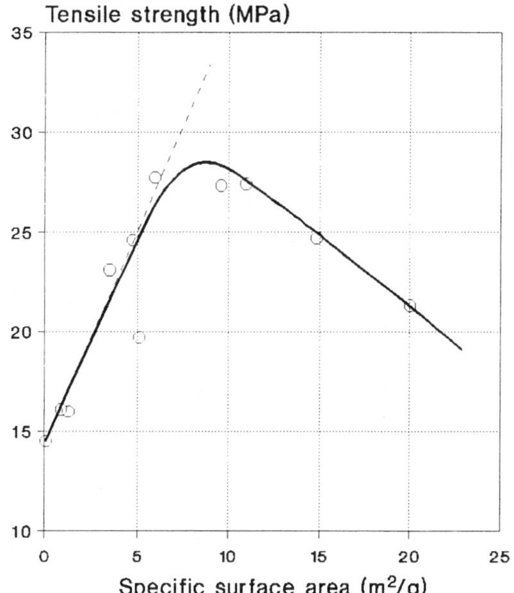

FIG. 8 Effect of aggregation on the strength of PP/$CaCO_3$ composites.

Hornsby [49], who found capillary forces more important than the van der Waals interaction.

Both adhesive and hydrodynamic forces depend on the size of the particles. The two forces were calculated for $CaCO_3$ fillers of various particle sizes introduced into a PP matrix. The results are presented in Fig. 7. Below a certain particle size, adhesion exceeds shear forces, and the aggregation of the particles takes place in the melt. Since commercial fillers have a relatively broad particle size distribution, most fillers show some degree of aggregation and the exact determination of the particle size, or other filler characteristics where aggregation appears is difficult. Experiments carried out with 11 different $CaCO_3$ fillers showed this limit to be around 6 m^2/g specific surface area [16].

Since the relative magnitude of adhesive and shear forces determine the occurrence and extent of aggregation in a composite, the ratio of the two forces may give information about the possibilities of avoiding or decreasing it:

$$\frac{F_a}{F_h} = k\,\frac{W_{AB}}{\eta\dot{\gamma}r} \tag{10}$$

An increase in shear rate and particle size will lead to decreased aggregation. Naturally both can be changed only in a limited range. Excessive shear leads to the degradation of the polymer, while large particles easily separate from the matrix under the effect of external load. According to Eq. (10), also lower reversible work of adhesion results in decreased aggregation tendency. Nonreactive surface treatment invariably leads to the decrease of surface tension and W_{AB}, and thus to decreased aggregation, improved processability, and mechanical properties.

The presence of aggregates is almost always detrimental to the properties of composites. Fig. 8 demonstrates this effect. Initially the strength of PP/$CaCO_3$ composites increases with increasing specific surface area of the filler, but it decreases strongly when aggregation takes place at small particle sizes (large A_f). The effect is even more pronounced in the case of impact properties, the fracture resistance of composites containing aggregated particles decreases drastically with increasing number of aggregates [9, 48].

B. Matrix/Filler Interaction

Both the polymers used as matrices in thermoplastic composites and the fillers have the most diverse physical and chemical structures. Thus a wide variety of interactions may form between them. Two boundary cases of interaction can be distinguished: covalent bonds, which rarely form spontaneously, but can be created by special surface treatments, and zero interaction, which does not exist in reality, since at least secondary van der Waals forces act between the components.

In practice the strength of the interaction is somewhere between the two boundary cases. Interaction between two surfaces in contact with each other can be created by primary or secondary bonds. The most important primary forces are the ionic, covalent and metallic bonds. The bonds formed by these forces are very strong, their strength is between 80–550 kJ/mol for covalent and 600–1200 kJ/mol for ionic bonds [55]. The secondary bonds are created by van der Waals forces, i.e. by dipole–dipole (Keesom), induced dipole (Debye), and dispersion (London) interactions. The strength of these interactions is much lower, between 1–20 kJ/mol [55]. Hydrogen bonds form a transition between the two groups of interactions, both in character and strength (20–40 kJ/mol). Beside the attractive forces created by the above-mentioned secondary forces, repulsive forces may also act between two surfaces due to the interaction of their electron fields. The final distance between the atoms is determined by the equilibrium of the attractive and repulsive forces. As was mentioned before, zero interaction does not exist, one of the forces always acts between surfaces in contact, but beside these forces numerous other factors may influence interaction, i.e., the strength of the adhesive bond.

The number of these factors and their complicated relationship is clearly shown by the numerous theories which attempt to describe the phenomenon of adhesion. These theories represent different approaches to the problem, describe one interaction well, but usually do not offer a general solution [56, 57].

The theory of *mechanical interlocking* explains adhesive bonding by the physical coupling of surface irregularities, roughness. It can be applied for problems emerging in the textile and paper industry, but cannot describe the adhesive interaction of smooth surfaces like glass.

According to the *theory of interdiffusion*, adhesion is caused by the mutual diffusion of the molecules of the interacting surfaces. This theory can be applied for polymer blends, but its use is limited when solid surfaces are in contact (aggregation, filler/matrix interaction).

Derjaugin [58] explains the significant interaction which is sometimes created between polymers and metals by *electrostatic interaction*. According to his reasoning the polymer and the thin metallic film layer correspond to an electric double layer, which forms by charge transfer between the surfaces. The significance of this theory in particulate filled polymers is very limited.

Adhesion is created by primary and secondary forces according to the theory of *adsorption interaction*. This theory is applied most often for the description of interactions in particulate filled or reinforced polymers [59]. The approach is based on the theory of contact wetting and focuses its attention mainly on the influence of secondary forces. Accordingly, the strength of the adhesive bond is assumed to be proportional to the reversible work of adhesion (W_{AB}), which is necessary to separate the two phases with the creation of two new surfaces. The Dupré equation relates W_{AB} to the surface (γ_A and γ_B) and interfacial (γ_{AB}) tension of the components, i.e.,

$$W_{AB} = \gamma_A + \gamma_B - \gamma_{AB} \tag{11}$$

One of the major problems of the application of this theory for solids is the determination of interfacial tension, which cannot be measured directly; it is usually derived from thermodynamic calculations. Good and Girifalco [60] developed the first theory for the calculation of γ_{AB}, which did not gain practical use because it contained an adjustable parameter. The most widely accepted solution was suggested by Fowkes [61, 62]. He assumed that surface tension can be divided into components:

$$\gamma = \gamma^d + \gamma^p + \gamma^i + \gamma^H + \gamma^\pi + \gamma^m + \cdots \tag{12}$$

where the superscripts d, p, i, H, π, and m indicate interactions created by dispersion, dipole-dipole, induced–dipole, hydrogen, π, and metallic bonds. Only dispersion forces act between apolar surfaces and in such a case interfacial tension can be expressed by

$$\gamma_{AB} = \gamma_A + \gamma_B - 2(\gamma_A^d \gamma_B^d)^{1/2} \tag{13}$$

Although it was assumed that Eq. (13) is valid also when an apolar material enters into interaction with a polar one, in practice polar surfaces interact with each other more often. Several attempts were made to generalize the correlation of Fowkes for such cases and the geometric mean approximation gained the widest acceptance. This considers only the dispersion and polar components of surface tension, but the latter includes all polar interactions [63]. Thus interfacial tension can be calculated as

$$\gamma_{AB} = \gamma_A + \gamma_B - 2(\gamma_A^d \gamma_B^d)^{1/2} - 2(\gamma_A^p \gamma_B^p)^{1/2} \tag{14}$$

The surface tension of two thermoplastics and three fillers are listed in Table 1. Large differences can be observed both in the dispersion, but especially in the

TABLE 1 Surface tension of several polymers and fillers (γ_s); dispersion (γ^d) and polar (γ^p) components

	Surface tension (mJ/m^2)		
Material	γ^d	γ^p	γ_s
PP	32.5	0.9	33.4
PMMA	34.3	5.8	40.1
$CaCO_3$	54.5	153.4	207.9
Talc	49.3	90.1	139.4
SiO_2	94.7	163.0	257.7

polar component of surface tension. The majority of polymers has a surface tension in the same range, in fact between that of PP and PMMA. Those listed in the table represent the most important particulate fillers and reinforcements used in practice, since clean glass fibers possess similar surface tensions as SiO_2. Surface treatment lowers the surface tension of fillers significantly (see Section VII.A).

Although Eq. (14) tries to take into account the effect of polar interactions in adhesion, lately the role of acid–base interactions became clear and theories describing them are gaining importance rapidly [64]. The boundary case of such interactions is the formation of covalent bonds between two surfaces, which, however, cannot be described by Eq. (14). As a consequence Fowkes [65] suggested that the reversible work of adhesion should be defined as

$$W_{AB} = W_{AB}^d + W_{AB}^{ab} + W_{AB}^p \tag{15}$$

where W_{AB}^{ab} is created by acid–base interactions. According to Fowkes the polar component can be neglected, i.e., $W_{AB}^p \sim 0$. Thus W_{AB} can be expressed as [65, 66]

$$W_{AB} = 2\sqrt{\gamma_A^d \gamma_B^d} + nf\Delta H^{ab} \tag{16}$$

where ΔH^{ab} is the change in free enthalpy due to acid–base interactions, n is the number of molecules interacting on a unit surface expressed in moles, and f is a conversion factor, which takes into account the difference between free energy and free enthalpy ($f \sim 1$).

The value of ΔH^{ab} can be calculated from the acid–base constants of the interacting phases by using the theory of Drago [67] or Guttman [68]. Drago [67] suggested two constants for each material and divided the compounds into acids and bases. ΔH^{ab} can be calculated as

$$-\Delta H^{ab} = C_A C_B + E_A E_B \tag{17}$$

where subscripts A and B stands for acid and base. C expresses the covalent, while E the electrostatic character of the interaction.

Guttman [68] characterizes materials by a donor (DN) and acceptor (AN) number, which indicate the acidity or basicity of the material. According to this theory

$$-\Delta H^{ab} = \frac{AN \times DN}{100} \tag{18}$$

In spite of the fact that theoretically the approach of Drago is sounder, Guttman's approach, i.e., the use of the AN and DN numbers is more accepted. These numbers can also characterize amphoteric materials and a much larger database exists for this theory. Schreiber [59], for example, lists these parameters for a large number of polymers and other components added to them.

In most cases the strength of the adhesive bond is characterized acceptably by W_{AB} values derived from the above presented theory. Often, especially in apolar systems, a close correlation exists between W_{AB} and the macroscopic properties of the composite in particulate filled polymers, also shown, for example, by the dependence of composite strength on W_{AB} presented in Fig. 9 for PP/$CaCO_3$ composites. However, the bond strengths determined with various mechanical tests are usually much lower than the values derived from

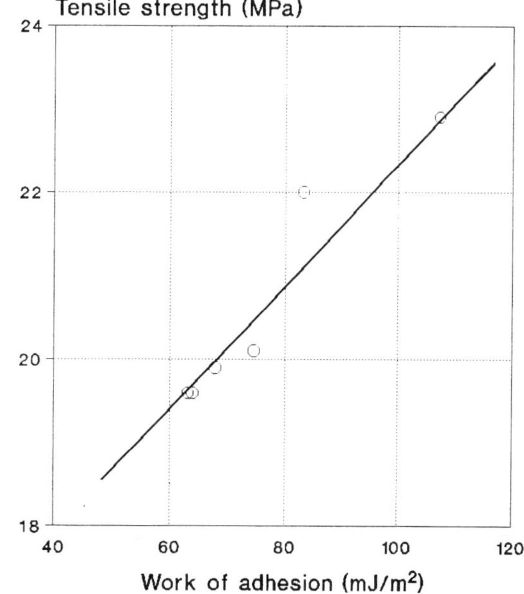

FIG. 9 Effect of interfacial adhesion on the strength of PP/$CaCO_3$ composites: $\varphi_f = 0.2$, $r = 0.9\,\mu m$.

theoretical calculations. This difference can often be attributed to the imperfections of the interface, which lead to the decrease of bond strength. The different elastic properties of the components usually result in the development of stress concentrations, which can cause either the fracture of the interface or lead to the cohesive failure of one of the phases. This observation leads to the development of the theory of *weak boundary layers* [69].

In spite of the imperfections of the adsorption interaction approach, W_{AB} can be used successfully for the characterization of matrix/filler interaction in particulate filled polymers. In these materials debonding is usually the dominating micromechanical deformation process. Stress analysis has shown that the debonding stress (σ^D) depends on the reversible work of adhesion in the following way [8]:

$$\sigma^D = -C_1\sigma^T + \sqrt{\frac{C_2 W_{AB}}{r}} \tag{19}$$

where σ^T is thermal stress, and C_1 and C_2 are factors determined by the geometric conditions of debonding. The prevailing mechanism of deformation depends on the strength of the interaction: in the case of weak adhesion, debonding takes place and yield stress decreases with increasing filler content, while strong adhesion leads to the yielding of the matrix polymer or to other micromechanical deformation processes [4, 70]. Fig. 10 sufficiently supports both the influence of adhesion (W_{AB}) on composite properties and the dominating role of debonding in the studied PP/$CaCO_3$ composites.

Since the effective range of the forces creating the adhesive bond is much smaller than the irregularities of the surface, appropriate wetting is an important condition of interaction. Wetting has thermodynamic and kinetic conditions. The thermodynamic conditions are usually satisfied since polymers having a relatively low surface tension completely wet the high energy surface of fillers (see Table 1) [71]. However, the kinetic condition is more difficult to fulfill: the high viscosity melt cannot always penetrate into small crevices, narrow channels or the space inside an aggregate. Thus surface defects form, which finally lead to the premature failure of the composite.

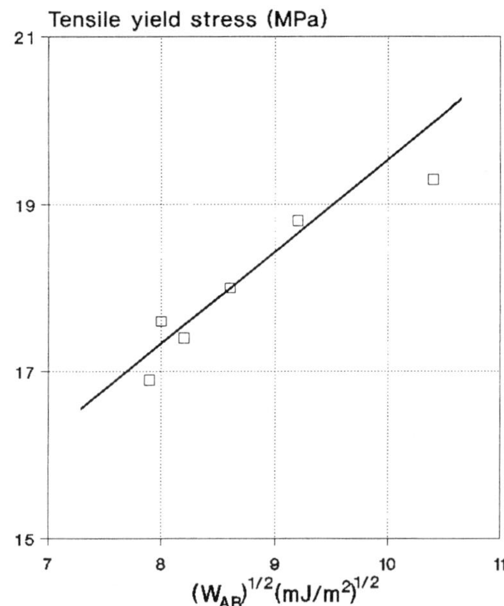

FIG. 10 Debonding in PP/$CaCO_3$ composites; effect of adhesion: $\varphi_f = 0.2, r = 1.8\,\mu m$.

V. INTERFACE AND INTERPHASE

As Table 1 shows, nontreated fillers and reinforcements have high energy surfaces. During the almost exclusively used melt mixing procedure, the forces discussed previously lead to the adsorption of polymer chains onto the active sites of the filler surface. A layer develops as a result which has properties different from those of the matrix polymer [72–76]. Although the character, thickness, and properties of this interlayer or interphase is a much discussed topic, its existence is already an accepted fact.

A. Properties

In semicrystalline polymers the interaction of the matrix and the filler changes both the structure and the crystallinity of the interphase. Interaction may lead to increased nucleation or to the formation of a transcrystalline layer on the surface of anisotropic particles [77]. The structure of the interphase, however, differs drastically from that of the matrix polymer [78, 79]. Because of preferential adsorption of large molecules, the dimensions of the crystalline units may change (usually decrease). Such preferential adsorption has been proved by GPC measurements after separation of adsorbed and nonattached molecules of the matrix [78, 79]. The decreased mobility of the chains also affects the kinetics of crystallization. Kinetic hindrance leads to the development of small, imperfect crystallites, forming a crystalline phase of low heat of fusion [80].

Atomistic simulation of atactic polypropylene/graphite interface has shown that the local structure of the polymer in the vicinity of the surface differs in many ways from that of the corresponding bulk. The density profile of the polymer displays a local maximum near the solid surface and the backbone of the polymer chains develops considerable parallel orientation to the surface [81]. The transcrystallinity observed on the surface of many anisotropic fillers may be caused by this parallel orientation due to adsorption.

The decreased mobility of adsorbed chains has been observed and proved in many cases both in the melt and in the solid state [82, 83]. The overall properties of the interphase, however, are not completely clear. Model calculations indicated the formation of a soft interphase [80], on the one hand, while the increased stiffness of the composite is often explained with the presence of a rigid interphase [84, 85], on the other. The contradiction obviously stems from two opposing effects. Imperfection of crystallites and decreased crystallinity of the interphase should lead to lower modulus and strength, as well as to larger deformability. The hindered mobility of adsorbed polymer chains, on the other hand, decreases deformability and increases the strength of the interlayer.

In many models the properties of the interphase are assumed to be constant, independent of the distance from the filler surface. However, in the case of a spontaneously formed interphase, they must change from the surface of the filler to the homogeneous matrix. Recently a model was developed which assumed an interlayer with continuously changing properties in order to describe the stress–strain behavior of particulate filled polymers [4, 86]. Stress analysis and model calculations indicated that changing interlayer properties significantly influence deformation and stress fields around the particles. A comparison with experimental data showed reasonable agreement between experiments and calculation only when the development of a hard interlayer was assumed. Further considerations proved that, beside elastic properties, the yield stress also changes continuously in the interlayer [4]. The composition dependence of composite yield stress is determined by the relation of debonding stress and the yield stress of the matrix. Strong interaction leads to matrix yielding, while weak adhesion results in debonding as the dominating micromechanical deformation process [4].

B. Thickness

The thickness of the interphase is a similarly intriguing and contradictory question. It depends on the type and strength of the interaction and values from 10 Å to several microns have been reported in the literature for the most diverse systems [25, 76, 78, 81, 87, 88]. Since interphase thickness is calculated or deduced indirectly from some measured quantities, it also depends on the method of determination. Table 2 presents some data for different particulate filled systems. The values indicate that interphase thicknesses determined from mechanical properties are considerably larger than those deduced from theoretical calculations or from extraction experiments [25, 78, 81, 88–91]. The data supply further proof for the adsorption of polymer molecules onto the filler surface and for the decreased mobility of the chains. Thermodynamic considerations and extraction experiments yield thicknesses which are not influenced by the extent of deformation. In mechanical measurements, however, deformation of the material takes place in all cases. The specimen is deformed even during the determination of modulus. With increasing deformation the role and effect of the immobilized chain ends increase and

TABLE 2 Interphase thickness in various particulate filled polymers

Matrix	Filler	Technique	Thickness (μm)	Reference
HDPE	SiO_2	Extraction	0.0036	87
HDPE	SiO_2	Extraction	0.0036	78
PP	SiO_2	Extraction	0.0041	78
PP	Graphite	Model calc.	0.001	81
PS	Mica	Dynamic mechanical measurement	0.06	88
PMMA	Glass	Dynamic mechanical measurement	1.4	88
PU	Polymeric	Dynamic mechanical measurement	0.36–1.45	90, 91
PP	$CaCO_3$	Young's modulus	0.012	89
PP	$CaCO_3$	Yield stress	0.15	89
PP	$CaCO_3$	Tensile strength	0.16	89

the determined interphase thickness increases as well (see Table 2) [89].

The thickness of the interphase must depend on the strength of the interaction. As was pointed out before, the interaction is usually created by secondary van der Waals forces. Although the range of these forces is small, the volume affected by the decreased mobility of the chains attached to the surface is much larger when the material is deformed, also shown by the larger interphase thicknesses determined by indirect, mechanical measurements (see Table 2). This volume and the thickness of the interphase can be estimated by Eq. (5) presented in Section III.B. The thickness of the interphase can be derived from Eq. (6) if the experiments are carried out with at least two fillers of different particle sizes.

Interphase thicknesses are plotted as a function of W_{AB} in Fig. 11 for $CaCO_3$ composites prepared with four different matrices: PVC, plasticized PVC (pPVC), LDPE, and PP. l changes linearly with increasing adhesion. The figure proves several of the points mentioned above. The reversible work of adhesion adequately describes the strength of the interaction, or at least it is proportional to it, the interaction is created mostly by secondary forces and, finally, the thickness of the interphase strongly depends on the strength of the interaction.

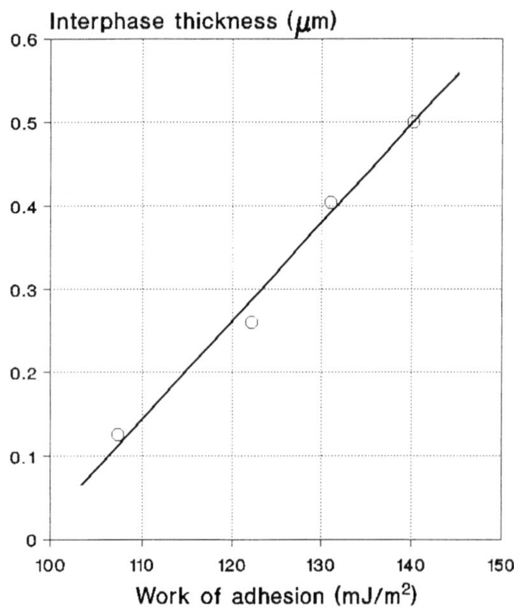

FIG. 11 Effect of interfacial interaction on the thickness of the interphase determined in various polymer/$CaCO_3$ composites (PP, LDPE, pPVC, PVC in increasing order).

C. Effect on Composite Properties

The amount of polymer bonded in the interphase depends on the thickness of the interlayer and on the surface area, where the filler and the polymer are in contact with each other. The size of the interface is more or less proportional to the specific surface area of the filler, which, on the other hand, is inversely proportional to particle size. The effect of immobilized polymer chains depends on the extent of deformation, the modulus shows only a very weak dependence on the specific surface area of the filler (see Fig. 3).

Properties measured at significantly larger deformations, e.g., tensile yield stress or tensile strength, show a much more pronounced dependence on A_f than modulus [92], shown by Fig. 4. As was expressed before the figure shows that yield stresses larger than that of the matrix can also be achieved, i.e., even spherical fillers can reinforce polymers [5, 6]. In the case of strong adhesion, yielding should be initiated at the matrix value, and reinforcing must not be possible. The reinforcing effect of spherical particles can be explained only by the presence of an interphase having properties somewhere between those of the polymer and the filler [5, 6], in accordance with the considerations presented above.

VI. DETERMINATION OF INTERFACIAL INTERACTION

Because of their importance many attempts are made to determine the thickness and properties of the interphase as well as the strength of the interaction. Various techniques are used for this purpose, but a detailed account on them cannot be given here. As a consequence the most often used techniques are briefly reviewed with a somewhat more detailed account of some specific methods having increased importance. A more extensive description about surface characterization techniques can be found in a recent monograph by Rothon [13].

A. Interphase Composition

The composition of spontaneously formed interphases is basically the same as that of the matrix. It is difficult, if not impossible, to detect any changes in it as an effect of the interaction. The thin interlayer cannot be studied by microscopic techniques and spectroscopy cannot characterize it either because of the small volume involved.

Valuable information is obtained about a spontaneously formed interphase by atomistic simulation [81]. Such calculations reveal the density profile in the vicinity of the surface, the orientation of the molecules, and even the properties of the interphase. Unfortunately the comparison of such theoretical calculations to experimental data is difficult, or at least lacking at the moment.

Spectroscopic techniques are extremely useful for the characterization of filler surfaces treated with surfactants or coupling agents. Spectroscopy makes possible the study of the chemical composition of the interlayer, the determination of surface coverage and possible coupling of the filler and the polymer. This latter is especially important in the case of reactive coupling, since, for example, the application of organofunctional silanes usually leads to a complicated polysiloxane layer of chemically and physically bonded molecules [93]. The principles of these techniques can be found elsewhere [13, 94–96]; only their application possibilities are discussed here.

X-ray photoelectron spectroscopy (XPS) gives a spectrum with highly element-specific peaks, allowing direct elemental analysis. The oxidation state of elements can be determined from the spectrum, while chemical shifts give information about the functionality and substitution of organic compounds. Secondary ion mass spectrometry (SIMS) yields charged fragments of molecules covering the surface. Dynamic SIMS is useful for trace element analysis, while static SIMS is used for the structural characterization of surfaces. Although a handbook of standard spectra is available for typical molecules used for surface treatment [97], the evaluation of the spectra may be difficult. Auger electron spectroscopy (AES) is another potential technique for surface analysis, but according to Ashton and Briggs [98] technical problems arise during measurement due to the beam damage of organic materials and charging of highly insulating powders.

Diffuse reflectance infrared spectroscopy (DRIFT) gives valuable information about the chemical and physical changes taking place on a filler surface as an effect of adsorption. The analysis can be carried out on powders, which is a great advantage of the technique. Changes in the characteristic vibrations indicate chemical reactions, while a shift in an absorption band shows physicochemical interactions. The spectrum is recorded in Kubelka–Munk units, which are proportional to concentration. The FTIR spectra of (3-stearyl-oxypropyl) triethoxysilane coupling agent (SPTES) and the DRIFT spectra of a $CaCO_3$ filler covered with the compound are compared in Fig. 12. Polycondensation of the silane takes place on the surface of the filler; the spectra of the adsorbed layer and the original coupling agent differ considerably from each other. A detailed analysis gives information even about the structure of the polysiloxane layer.

FIG. 12 (a) FTIR spectrum of a monomeric silane coupling agent (SPTES), (b) DRIFT difference spectrum of $CaCO_3$ treated with 1.6 wt% of the silane.

Adsorption–desorption techniques are often used for the study of surface treatment or for the modelling of polymer/filler interaction. Adsorption is determined by gravimetric techniques or, recently, more frequently by inverse gas chromatography (IGC). The strength of the interaction is usually deduced from the measured retention time or volume [99, 100].

B. Thermodynamic Parameters

The quantitative determination of the strength of the matrix/filler interaction is difficult: in most cases indirect techniques are used for its estimation. These employ the principles discussed in Section IV.B. The surface tension of the components, interfacial tension or acid–base interaction parameters must be known in order to calculate the reversible work of adhesion. Adsorption or desorption of small molecular weight compounds having an analogous structure to the polymer can be used for the estimation of interfacial interactions.

In flow microcalorimetry a small amount of filler is put into the cell of the calorimeter and the probe molecule passes through it in an appropriate solvent. Adsorption of the probe results in an increase of temperature and the integration of the area under the signal gives the heat of adsorption [98]. This quantity can be used for the calculation of the reversible work of adhesion according to Eq. (16). The capabilities of the technique can be further increased if a HPLC detector is attached to the microcalorimeter. The molar heat of adsorption can also be determined with this setup.

Lately, the most frequently used technique for the determination of thermodynamic properties of a solid surface is inverse gas chromatography [59, 101–103]. In IGC the unknown filler or fiber surface is characterized by compounds, usually solvents, of known thermodynamic properties. IGC measurements can be carried out in two different ways. In the most often applied linear, or ideal, IGC infinite concentrations of n-alkanes are injected into the column containing the filler to be characterized. The net retention volume (V_N) can be calculated by

$$V_N = (t_r - t_0)Fj_0 \tag{20}$$

where t_r is the retention and t_0 the reference time, F the flow rate of the carrier gas and j_0 is a correction factor taking into account the pressure difference between the two ends of the column. The dispersion component of the surface tension of the filler can be calculated from the retention volume of n-alkanes

$$-RT \ln V_n = Na(\gamma_{LV}\gamma_s^d)^{1/2} \tag{21}$$

where V_n is the net retention volume of an alkane, a is the surface area of the adsorbed molecule, γ_{LV} the surface tension of the solvent, and N is the Avogadro number. The left hand side of Eq. (21) is a linear function of a $(\gamma_{LV})^{1/2}$. If the measurements are carried out with polar solvents, the deviation from this straight line is proportional to the acid–base interaction potential of the solid surface [59].

The other approach, nonlinear or finite dilution IGC, consists of the determination of adsorption isotherms. When the surface of a solid is only partially wetted by a liquid, it forms a droplet with a definite contact angle (θ). The interaction of the components is expressed by the Young equation

$$\gamma_{SV} = \gamma_{SL} + \gamma_{LV} \cos\theta \tag{22}$$

where γ_{SV} is the surface tension of the solid in contact with the vapor of the liquid, γ_{LV} the surface tension of the liquid, and γ_{SL} the interfacial tension. Due to wetting, γ_{SV} is smaller than the surface tension of the solid measured in vacuum (γ_{S0}), the difference is the spreading pressure (π_e), i.e.,

$$\gamma_{S0} - \gamma_{SV} = \pi_e \tag{23}$$

The value of π_e is very small for low energy surfaces, but it cannot be neglected for fillers: on the contrary, π_e can be used for the calculation of the thermodynamic characteristics of their surface. The spreading pressure can be derived from the adsorption isotherm [102]

$$\pi_e = RT \int_0^p \Gamma\, d\ln p \tag{24}$$

where p is the vapor pressure and Γ the moles of vapor adsorbed on a unit volume of the filler. If the measurement is carried out with apolar solvents the dispersion components of the surface tension can be determined from the spreading pressure

$$\pi_e = 2(\gamma_S^d\gamma_{LV})^{1/2} - 2\gamma_{LV} \tag{25}$$

Eq. (25) is derived from the Young [Eq. (22)], the Dupre [Eq. (11)], and the Fowkes [Eq. (13)] equations by assuming complete wetting ($\cos\theta = 0$). Measurements with polar solvents give the dispersion component of the surface tension, but also acid–base constants, and W_{AB} can be calculated from them:

$$W_{AB}^{spec} = 2\gamma_{LV,p} + \pi_{e,p} - 2(\gamma_S^d\gamma_{LV,p}^d)^{1/2} \tag{26}$$

where W_{AB}^{spec} is the polar or acid–base component of W_{AB} and the p subscript indicates a polar solvent. The finite dilution IGC is a more tedious technique than linear IGC, but it makes possible the direct determination of γ_s^p or the corresponding acid–base constants.

In order to calculate the polymer/filler interaction, or more exactly the reversible work of adhesion characterizing it, the surface tension of the polymer must also be known. This quantity is usually determined by contact angle measurements, or occasionally the pendant drop method is used. The former method is based on the Young, Dupre and Fowkes equations [Eqs. (22), (11) and (13)], but the result is influenced by the surface quality of the substrate. Moreover, the surface properties (structure, orientation, density) of polymers usually differ from those of the bulk, leading to biased results. The accuracy of the technique may be increased by using two or more liquids for the measurements. The use of the pendant drop method is limited due to technical problems (long time to reach equilibrium, stability of the polymer, evaluation problems, etc.). Recently new techniques like fiber break up and relaxation measurements are used more and more frequently for the direct determination of interfacial tension. Occasionally IGC is used for the characterization of polymers as well [59].

C. Estimation of Interaction

The bond strength of adhesives is frequently measured by the peel test [104]. The results can often be related to the reversible work of adhesion. However, such a measurement cannot be carried out on particulate filled polymers. Even the interfacial shear strength, widely applied for the characterization of matrix/fiber adhesion, cannot be used for particulate filled polymers. The interfacial adhesion of the components is usually deduced indirectly from the mechanical properties of the composites with the help of models describing composition dependence. Naturally, such models must also take into account interfacial interactions.

The number of such models is limited. One of the existing models was developed to describe the composition dependence of the tensile yield stress of particulate filled composites [5] [see Eq. (5)]; with some modification tensile strength and fracture properties can be also analyzed with it. Parameter B of the model is related to stress transfer, i.e., to interfacial adhesion. Since interaction depends on the size of the interface (contact area) and the strength of the interaction, both influence the value of B, which can be determined from the linear plot of the relative yield stress $[\sigma_{yrel} = \sigma_y(1 + 2.5\varphi_f)/\sigma_{y0}(1 - \varphi_f)]$ against filler content. Yield stresses presented in Fig. 4 are plotted in this form in Fig 13. The slope of the straight lines gives parameter B, which is proportional to the interaction (to the contact area between the filler and the matrix, in this case). Changes in the strength of the interaction can also be expressed by B, as will be shown in the following sections (Sections VII.B and VIII.C).

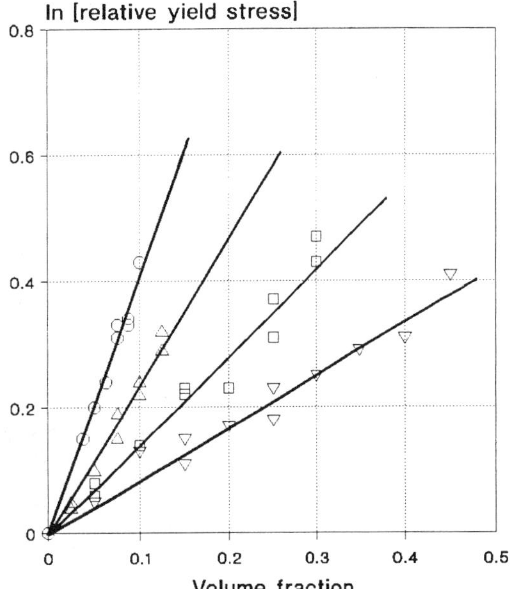

FIG. 13 Quantitative estimation of the effect of interaction, influence of contact surface. Symbols are the same as in Fig. 4.

VII. MODIFICATION

The easiest way to modify interfacial interactions is the surface treatment of fillers. The compound used for the treatment (coupling agent, surfactant, etc.) must be selected according to the characteristics of the components and the goal of the modification. The latter is very important: surface modification is often regarded as magic, which solves all the problems of processing technology and product quality. However, we must be aware of the fact that both particle/particle and matrix/filler interactions take place in particulate filled polymers. Surface treatment modifies both, and the properties of the composite are determined by their combined effect. Beside its type, the amount of the material used for the treatment must also be optimized in order to achieve good performance and economy. Five surface modification techniques are discussed in the following sections.

A. Nonreactive Surface Treatment

The oldest and most often used modification of fillers is the coverage of their surface with a small molecular weight organic compound [25, 105]. Usually amphoteric compounds, surfactants, are used which possess one or more polar groups and a long aliphatic chain. A typical example is the surface treatment of $CaCO_3$ with stearic acid [25, 105–107]. The principle of the treatment is the preferential adsorption of the polar group of the surfactant onto the surface of the filler. The high energy surfaces of inorganic fillers (see Table 1) can often enter into special interactions with the polar group of the surfactant. Preferential adsorption is promoted by the formation of ionic bonds between stearic acid and the surface of $CaCO_3$, but in other cases hydrogen or even covalent bonds may form, too. Surfactants diffuse to the surface of the filler even from the polymer melt, which is further proof for the preferential adsorption [108]. Because of their polarity, reactive coupling agents also adsorb onto the surface of the filler. If the lack of reactive groups does not make chemical coupling with the polymer possible, these exert the same effect on composite properties as their nonreactive counterparts [85, 109, 110].

One of the crucial questions of nonreactive treatment, which, however, is very often neglected, is the amount of surfactant to use. It depends on the type of the interaction, the size of the treating molecule, its alignment to the surface, and on some other factors. The determination of the optimum amount of surfactant is essential for the efficiency of the treatment. Insufficient amounts do not bring about the desired effect, while an excess of surfactant leads to processing problems as well as to the deterioration of the mechanical properties and appearance of the product [102, 111].

The amount of bonded surfactant can be determined by some simple techniques. A dissolution technique proved to be very convenient for the optimization of nonreactive surface treatment and for the characterization of the efficiency of the treating technology as well [102, 112]. First, the surface of the filler is covered with increasing amounts of surfactant, then the nonbonded part is dissolved with a solvent. The technique is demonstrated in Fig. 14, which presents the dissolution of stearic acid from a $CaCO_3$ filler. In practice, surface treatment should be carried out with the amount irreversibly bonded to the surface of the filler (c_{100}). The filler can adsorb more surfactant (c_{max}), but during compounding a part of it can be removed from the surface by dissolution or simply by shear and deteriorating properties.

The specific surface area of the filler is an important factor which must be taken into consideration during surface treatment. The amount of irreversibly bonded surfactant depends linearly on it [102]. ESCA studies carried out on a $CaCO_3$ filler covered with stearic acid have shown that ionic bonds form between the surfactant molecules and the filler surface and that stearic acid molecules are oriented vertically to the surface [102]. These experiments have demonstrated the importance of both the type of the interaction and the alignment of surfactant molecules to the surface. A further proof for the specific character of surface treatment is supplied by the fact that talc and silica adsorb significantly smaller amounts of stearic acid at a unit surface than $CaCO_3$. The lack of specific interaction in the form of ionic bond formation results in a significantly lower amount of irreversibly bonded molecules [102].

The surface free energy of the filler decreases drastically as a result of the treatment [25, 112]. Lower surface tension means decreased interfacial tension and reversible work of adhesion as well. Figure 15

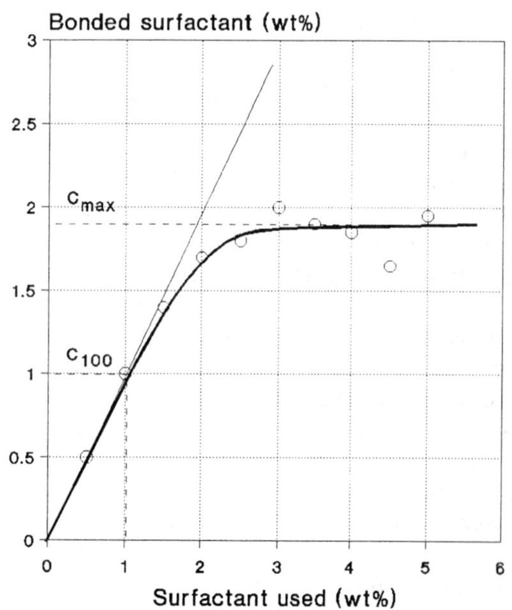

FIG. 14 Dissolution of stearic acid from the surface of a $CaCO_3$ filler.

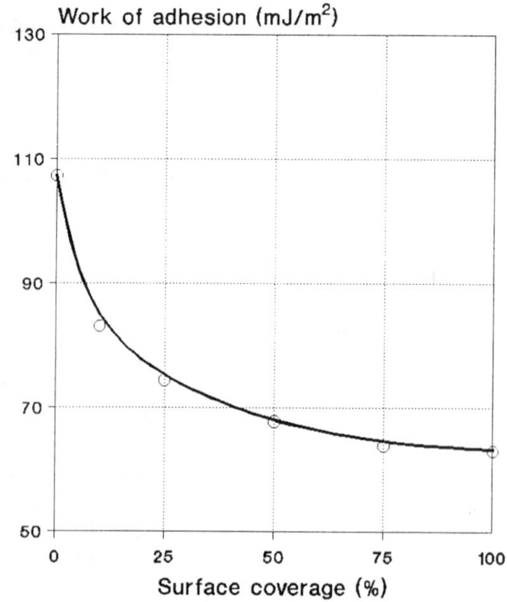

FIG. 15 Decrease of the strength of interaction (W_{AB}) in PP/stearic acid treated $CaCO_3$ composites with increasing surface coverage.

shows the change of W_{AB} plotted against the surface coverage for PP composites containing a stearic acid treated $CaCO_3$ [102]. The decrease of W_{AB} results in weaker particle/particle and matrix/filler interactions. One of the main goals, major reason, and benefit of nonreactive surface treatment is the first effect, i.e., to change the interaction between filler particles. A weaker interaction leads to a considerable decrease in aggregation, improved dispersion and homogeneity, easier processing, better mechanical properties and appearance. The improvement in the mechanical properties, and first of all in impact strength, is often falsely interpreted as the result of improved wetting and interaction of the components. However, wettability (S_{AB}) is defined as

$$S_{AB} = \gamma_A - \gamma_B - \gamma_{AB} \tag{27}$$

where subscript A relates to the wetted surface and $\gamma_A > \gamma_B$, and it decreases with increasing surface coverage of the filler. This fact is demonstrated also by the increasing contact angle of organic. liquids on the surface of treated fillers [112].

As an effect of nonreactive treatment not only particle/particle, but matrix/filler interaction decreases as well. The consequence of this change is decreased yield stress and strength, as well as improved deformability [25, 106]. This is demonstrated by Fig. 16 showing a

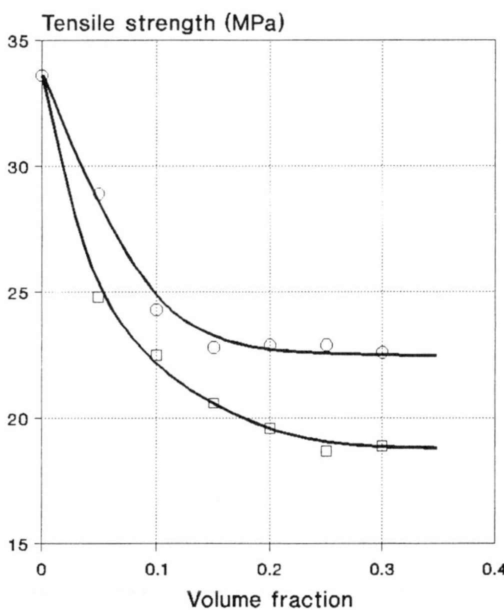

FIG. 16 Effect of the surface treatment of a $CaCO_3$ filler with stearic acid on the strength of its PP composite: (○) nontreated, (□) 100% surface coverage.

decrease in the tensile strength of PP/$CaCO_3$ composites as a result of treatment with stearic acid. Adhesion and strong interaction, however, is not always necessary or advantageous in particulate filled composites; plastic deformation of the matrix is the main energy absorbing process in impact, which decreases with increasing adhesion [39, 113, 114] (see Section III.C).

B. Coupling

Reactive surface treatment assumes the chemical reaction of the coupling agent with both of the components. The considerable success of silanes in glass reinforced thermosets led to their application in other fields; they are used, or at least tried, in all kinds of composites, irrespective of the type, chemical composition or other characteristics of the components. The reactive treatment, however, is even more complicated than the nonreactive. The polymerization of the coupling agent, and the development of chemically bonded and physisorbed layers, render the identification of surface chemistry, the characterization of the interlayer, and the optimization of the treatment very difficult [93]. In spite of these difficulties significant information was collected on silane coupling agents [115]; much less is known about the mechanism and effect of other compounds like titanates, zirconates, etc. [105].

Silane coupling agents are successfully used with fillers and reinforcements which have reactive -OH groups on their surface, e.g. glass fibers, glass flakes and beads, mica and other silicate fillers [72, 116, 117]. The use of silanes with fillers like $CaCO_3$, $Mg(OH)_2$ wood flour, etc., were tried, but proved to be unsuccessful in most cases [30, 118]; sometimes contradictory results were obtained, even with glass and other siliceous fillers as well [119]. Acidic groups are preferable for $CaCO_3$, $Mg(OH)_2$, $Al(OH)_3$, and $BaSO_4$. Talc cannot be treated successfully either with reactive or nonreactive agents because of its inactive surface; only broken surfaces contain a few active —OH groups. The chemistry of silane modification has been extensively studied and described elsewhere [93, 115]. Model experiments have shown that a multilayer film forms on the surface of the filler; the first layer is chemically coupled to the surface, this is covered by crosslinked silane polymer, and the outer layer is physisorbed silane. The matrix polymer may react chemically with the coupling agent, but its interdiffusion with the polysiloxane layer, i.e. physical interaction, also takes place [120, 121].

Recent studies showed that the adsorption of organofunctional silanes is usually accompanied by polycondensation not only on silicates, but also on other fillers. The high energy surface of nontreated mineral fillers bonds water from the atmosphere, which initiates the hydrolysis and polymerization of the coupling agent [122, 123]. FTIR and DRIFT spectra of a $CaCO_3$ filler treated with an aminosilane coupling agent (3-aminopropyl) trimethoxysilane (AMPTES) is presented in Fig. 17. All four spectra are the same, in spite of the different treatment and sample preparation conditions. Spectrum A was recorded on a filler treated from *n*-butanol solution, while spectra B and C were taken from a filler treated by dry-blending. The treated filler was washed with THF (spectrum C) and the similarity of the two spectra (B and C) shows that the polysiloxane layer adsorbs strongly to the surface. Moreover, this spectrum is the same as the one recorded on a polymerized film of the neat aminosilane (spectrum D). The experiments prove that polycondensation takes place, indeed, but also that the polysiloxane layer is coupled strongly to the filler surface even if it does not contain active —OH groups. However, the amount of adsorbed coupling agent, its structure, and properties depend very much on the chemical composition of the organofunctional group of the compound. This is obvious if we compare the spectra presented in Figs. 12 and 17, but also if we study the dissolution curve of the two coupling agents (Fig. 18). The different chemical structure of the silanes leads to considerably different adsorptions. The figure shows also that the simple dissolution technique can also be applied advantageously for the study of reactive coupling agents [122].

Although the chemistry of silane modification of reactive silica fillers is well documented, much less is known about the interaction of silanes with polymers. It is more difficult to create covalent bonds between a coupling agent and a thermoplastic, since the latter rarely contains reactive groups. Reactive treatment is especially difficult in polyolefins, although occasionally successful attempts are reported as is shown in Fig. 19 for a PP/mica composite [116]. The results presented in the figure were obtained by the application of an aminosilane. In PP composites the most often used coupling agent is the Z 6032 product of Dow Corning, which is *N*-β-(*N*-vinylbenzylamino)-ethyl-γ-aminopropyl trimethoxy silane [30, 72]. Although one would expect that coupling occurs through the reaction of free radicals as suggested by Widman *et al.* [124], the

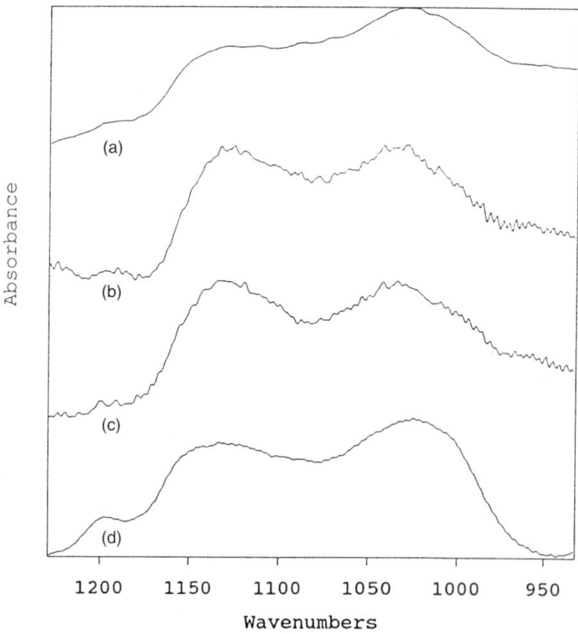

FIG. 17 (a) DRIFT difference spectrum of $CaCO_3$ treated with 0.8 wt% AMPTES from *n*-butanol solution; FTIR difference spectra of $CaCO_3$ treated with 1 wt% AMPTES, (b) before washing with THF, (c) after washing with THF, (d) spectrum of condensed neat AMPTES.

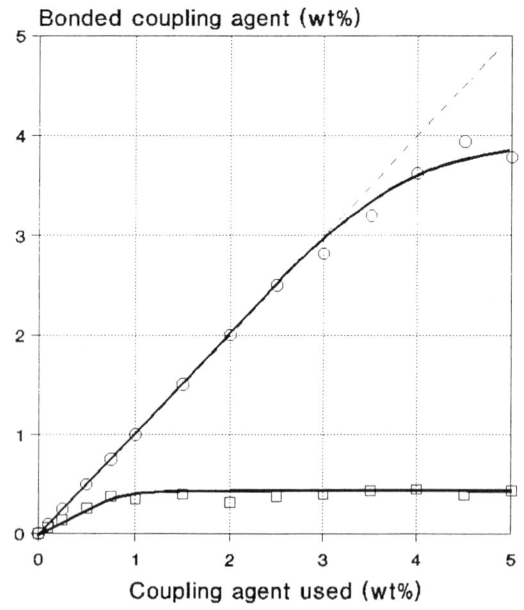

FIG. 18 Adsorption of various silane coupling agents on the surface of a $CaCO_3$ filler: (○) aminosilane (AMPTES), (□) aliphatic silane (SPTES).

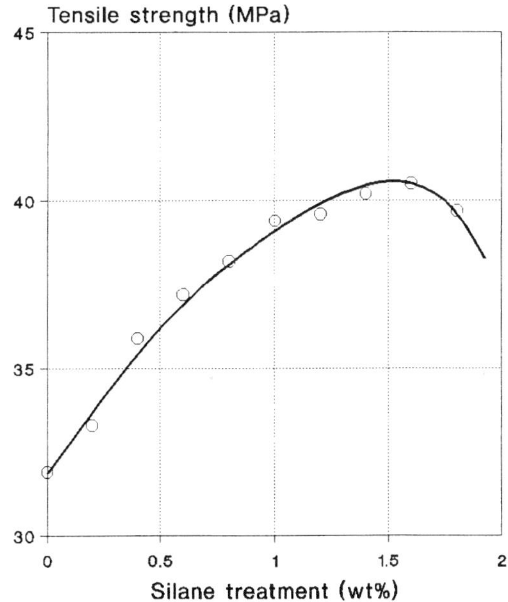

FIG. 19 Reactive coupling of mica to PP by an aminosilane coupling agent [116].

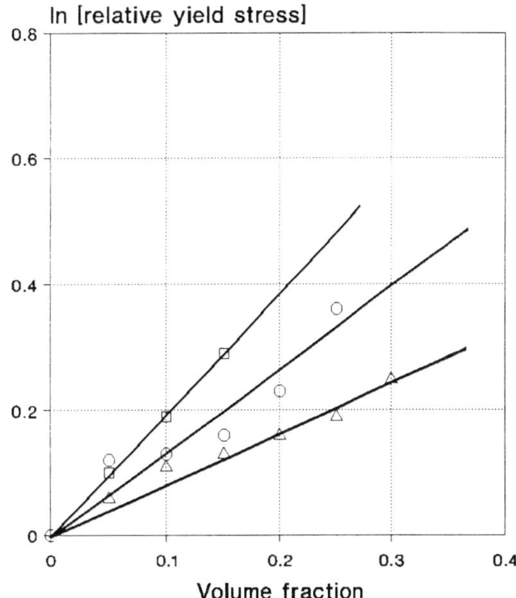

FIG. 20 Effect of surface treatment on the tensile yield stress of PP/$CaCO_3$ composites: (△) stearic acid, (○) no treatment, (□) aminosilane.

latest results show that polypropylene oxidizes during processing even in the presence of stabilizers and the formed acidic groups react with the aminosilane forming covalent bonds [125]. The presence of reactive coupling and the strength of the interaction can be proved by using the model presented in Section III.B. Figure 20 shows the linear plot of relative yield stress against the volume fraction of the filler for three different treatments in PP/$CaCO_3$ composites. It is obvious that stearic acid acts as a surfactant, while the aminosilane is a reactive coupling agent. The strength of the interaction is reflected well by the slope of the straight lines, i.e. by parameter B.

Considering the complexity of the chemistry involved, it is not surprising that surface coverage has an optimum here too, similar to nonreactive surface treatment. This is proved by Trotignon and Verdu [116] finding a strong increase and a maximum in the strength of PP/mica composites as a function of the silane used for the treatment (see Fig. 19). The optimization of the type and amount of coupling agent is also crucial in reactive treatment. Although "proprietary" treatments may lead to some improvement in properties, they might not always be optimal or cost effective. The improper choice of coupling agent may lead to insufficient or even deteriorating effects. Hardly any change in the properties is observed in some cases or the effect can be attributed clearly to the decrease of surface tension due to the coverage of the filler surface by an organic substance, i.e. nonreactive treatment [109, 110].

C. Modified Polymers

The coverage of the surface of a filler with a polymer layer which is capable of interdiffusion with the matrix proved to be very effective both in stress transfer and in forming a thick, diffuse interphase with acceptable deformability. In this treatment the filler is usually covered by a functionalized polymer, preferably by the same polymer as the matrix, which is attached to the surface by secondary, hydrogen, ionic or sometimes even by covalent bonds. The polymer layer interdiffuses with the matrix, entanglements are formed and, thus strong adhesion is created. Because of its increased polarity, in some cases reactivity, usually maleic anhydride or acrylic acid modified polymers are used, which adsorb to the surface of most polar fillers even from the melt. This treatment is frequently used in polyolefin composites, since other treatments often fail in them, on the one hand, and functionalization of these polymers is relatively easy, on the other. Often a very small amount of modified polymer is sufficient to achieve significant improvement in stress transfer [126, 127].

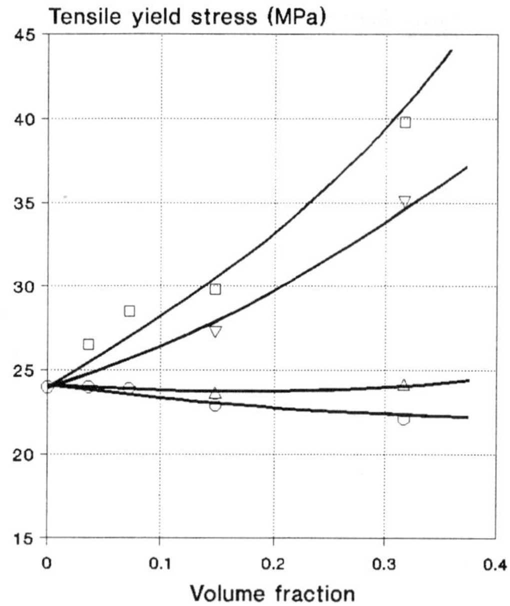

FIG. 21 Surface treatment with a functionalized polymer in PP/cellulose composites. Molecular weight of MA–PP: (○) nontreated, (△) 350, (▽) 4500, (□) 3.9×10^4.

The experiments of Felix and Gatenholm [128] clearly demonstrate the role of interdiffusion in the interphase and the importance of entanglement density. They introduced maleic anhydride modified PP (MA–PP) into PP/cellulose composites and achieved a significant improvement in yield stress. The increase depended on the molecular weight of MA–PP (Fig. 21). Model calculations have shown that the improvement of properties has an upper limit and a plateau is reached at a certain, not too high, molecular weight. The maximum effect of functionalized PP was found with fillers of high energy surfaces [119, 129, 130] or with those capable of specific interactions, e.g., ionic bond formation with $CaCO_3$ [127, 131] or chemical reaction with wood flour, kraft lignin or cellulose [126, 129]. Functionalized polymers are used for the treatment of glass fibers in reinforced PP composites as well. In fact, two surface treatment techniques are combined in this case: reactive coupling and the application of functionalized polymers. The glass fiber is treated with an aminosilane coupling agent which forms covalent bonds with MA–PP or AA–PP. This latter interdiffuses with the matrix polymer as was explained above.

D. Other Reactive Techniques

Although coupling assumes the covalent bonding of the components, many other approaches are tried to increase the adhesion between the filler or reinforcement and the polymer matrix in which chemical reactions are involved. A wide variety of compounds, reactions, and techniques are used in order to achieve this goal, some of them combining the principles presented in previous sections. The main difference between reactive coupling and the techniques presented in this section lies in the introduction of components, usually peroxides, which activate the coupling reaction or the initiation of special reactions, often polymerization. The reactions initiated in this way result in the modification of the chemical and physical structure of the matrix polymer and/or produce composites with special structures and properties.

A frequently used technique is the covering of the filler particles with a reactive monomer and then initiating its polymerization and coupling to the matrix. A recently reported example is the preparation of PP/$CaCO_3$ composites with a filler covered by acrylic acid and dicumil peroxide [132]. Radical reactions lead to a decrease in the molecular weight of PP, but also create a large amount of insoluble polymer. Unfortunately the properties of the product were not disclosed. A similar approach, but with another active component, *m*-phenylene-bis-maleinimide, was used for the improvement of the properties of highly filled polyolefin, PP and PE, composites [133]. Proper selection of processing temperature and the amount of maleinimide led to a significant improvement in tensile strength and impact resistance compared to the unmodified composite. However, processability deteriorated proportionally to the improvement in mechanical properties, indicating that chemical reactions take place during processing; indeed, even in the absence of peroxides. The changes in properties were attributed to radical reactions between the polymer and the active component, but were not proved directly [134].

Another route to modify interfacial interactions combines the coverage of the filler surface with a polymer layer (see Section VII.C) and the subsequent polymerization of a monomer, or chemical reaction with the matrix [135]. A peroxidic oligomer is prepared by the copolymerization of various components, e.g., maleic acid, 2-*t*-butylperoxy-2-methyl-5-hexen-3-in and styrene. The surface of a filler, usually $CaCO_3$, is covered with the oligomer from an aqueous solution. The decomposition of the peroxidic groups in the presence of monomers leads to a composite, where the filler is evenly distributed in the polymer matrix and adhesion of the components is excellent. Up to now,

PMMA and PS composites were prepared by direct polymerization of the corresponding monomers. However, an attempt was made to prepare polyolefin composites by using a filler treated with the reactive peroxidic oligomer [136]. PP degraded, while the crosslinking of polyethylene took place when such a filler was introduced into it. The mechanical properties of PE/$CaCO_3$ composites improved considerably as an effect of this treatment (see Fig. 22). Rheological measurements, as well as a SEM study, indicated that crosslinking took place exclusively in the vicinity of the filler [137]. Although the technique yields composites with interesting properties, its practical utility is in doubt, because of the highly specific character of the reactive oligomer and the tedious preparation of the treated filler.

Recently much attention is paid to the preparation of polyolefin composites by direct polymerization of the monomers in the presence of fillers. The technique was introduced separately by two research groups [138–140] and it offers several advantages. Uniform distribution of the filler can be achieved even at high filler contents, a very high amount of filler (> 90 wt%) can be incorporated into the polymer, the structure and molecular weight of the matrix polymer can be controlled, the wear of the processing equipment can be avoided, and the adhesion of the components is greatly improved [141].

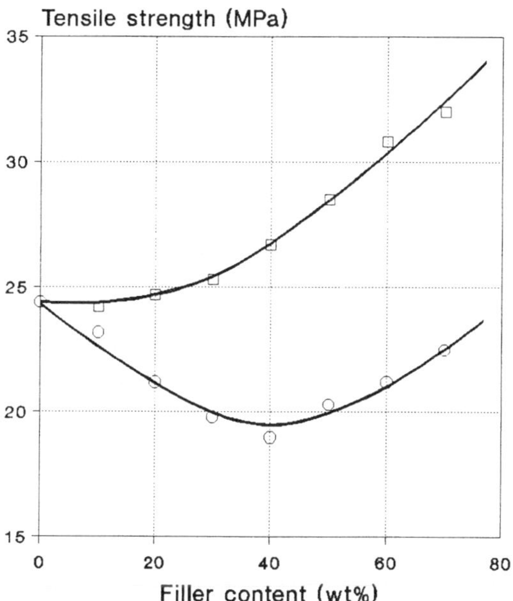

FIG. 22 Effect of chemical modification of the matrix around the filler on the strength of PE/$CaCO_3$ composites: (○) unmodified, (□) modified.

The preparation of such composites usually consists of three steps. First, the filler must be thoroughly dried and all volatiles removed from its surface. In the next step the surface of the filler is activated, it is covered by the polymerization catalyst, then polymerization is carried out with the activated filler. Recently ethylene was polymerized by a Ziegler–Natta catalyst on caolin and barite surfaces. The activity of the rather complicated $(BuO)_4Ti/BuMgOct/EtAlCl_2/Et_3Al$ catalyst depends on the ratio of the components [142, 143]. Optimization of the composition leads to 300 kg PE/g Ti.h yield on the surface of kaolin ($d = 1.4\,\mu m$), but a lower yield (39 kg PE/g Ti.h) was achieved on barite particles ($d = 3.0\,\mu m$). The molecular weight could be regulated by the introduction of H_2 and octene. Purity and purification techniques of the components influenced yield as well as the properties of the product considerably. In this process first the cocatalyst, $AlEt_3$, is attached to the surface of the filler. This step leads to the deagglomeration of the filler particles shown by sedimentation measurements. Then the catalyst is prepared and polymerization is carried out as mentioned above. The structure and properties are claimed to be considerably different from those of composites prepared by the traditional melt blending technique [144].

The same procedure was repeated with a metallocene catalyst as well [144]. The polymerization filling of a large number of fillers (kaolin, glass beads, porous silica, $Mg(OH)_3$, and graphite) has been carried out. SEM analysis showed that the filler is covered first by a thin polymer layer, then fibrils bridge the gaps between the particles, and finally a space-filling process takes place. The molecular weight of the polymer was regulated by the introduction of H_2. In the case of silica and kaolin, elongation and impact resistance increased, compared to the same properties of composites prepared by melt blending, but no improvement was observed with glass beads. However, composite properties were compared at a single filler content and the characteristics of the fillers have not been revealed, thus the advantages of the process cannot be judged at all. The improvement in the physicomechanical properties of the composites has been tentatively explained in terms of interfacial adhesion, homogeneous filler dispersion, and molecular weight of the polyolefin matrix, based on observations obtained by SEM, atomic force microscopy, and image analysis experiments [144]. Although this procedure also shows some potential, its practical application seems to be difficult at the moment. Nevertheless, it might be the only route to the preparation of particulate filled ultra high molecular weight polymers, or composites with extremely high filler contents [141].

E. Soft Interlayer—Elastomer

External loading of a polymer containing hard particles creates stress concentrations around the inclusions, which in turn induce local micromechanical deformation processes. Occasionally these might be advantageous for increasing plastic deformation and impact resistance, but usually they damage the properties of the composite. The encapsulation of the filler particles by an elastomer layer changes the stress distribution around them and modifies the local deformation processes. Encapsulation can take place spontaneously, it can be promoted by the use of functionalized elastomers (see Section VII.3), or the filler can be treated in advance.

The coverage of the filler with an elastomer layer was studied mostly in PP composites. PP has a poor low temperature impact strength, which is frequently improved by the introduction of elastomers [145]. The improvement in impact strength, however, is accompanied by a simultaneous decrease of modulus, which cannot be accepted in certain applications; a filler or reinforcement is added to compensate for the effect. Although most of the papers dealing with these composites agree that the simultaneous introduction of the two different types of material (elastomer and filler) is beneficial, this is practically the sole statement they agree on. A large number of such systems were prepared and investigated, but the observations concerning their structure, the distribution of the components, and their effect on composite properties are rather controversial. In some cases separate distribution of the components and independent effects were observed [146, 147], while in others the encapsulation of the filler by the elastomer was seen [148, 149]. These different structures naturally lead to dissimilar properties as well. The composition dependence of shear modulus agrees well with values predicted by the Lewis–Nielsen model in the case of separate dispersion of the components, while large deviations are observed when the filler particles are embedded into the elastomer [24]. Model calculations have shown that a maximum 70% of the filler particles could be embedded in the elastomer, while even in the cases when separate dispersion of the components dominates, at least 5–10% of the particles are encapsulated. The results indicate that complete encapsulation or separate dispersion cannot be achieved by the mere combination and homogenization of the components.

Extensive investigations [147, 150, 151] have shown that the final structure is determined by the relative magnitude of adhesion and shear forces acting during the homogenization of the composite. Similar principles can be used for the estimation of adhesion as in the case of aggregation, since the same forces are effective in the system. If shear forces are larger than the strength of adhesion, separate distribution of the components occurs, while strong adhesion leads to embedded structure [see Eq. (10)]. The three parameters which influence structure are adhesion (W_{AB}), particle size (r), and shear rate ($\dot{\gamma}$). Fillers usually have a wide particle size distribution, very often in the 0.3–10 μm range. As a consequence, composites will always contain small particles which are encapsulated and large particles which are separately distributed [150].

All the experiments carried out on such systems have shown the primary importance of adhesion in structure formation and in the resulting effect on properties. Numerous attempts have been made to prepare composites with an exclusive structure, i.e., complete coverage or separate distribution. In most cases PP or elastomers modified with maleic anhydride or acrylic acid were used to promote the formation of one of the structures; separate dispersion with MA–PP and embedding with MA–EPDM [129, 130, 151, 152]. Although the desired effect was observed in almost every case, the exclusiveness could not be proved. Similarly contradictory is the effect of structure on properties, especially on impact resistance. In three-component PE composites better impact resistance/modulus ratios could be achieved when an embedded structure was created, but Kelnar [131] has observed lower impact strength on the application of AA–EPDM than with a nonmodified elastomer. Although the relation of structure and impact resistance could not be determined unambiguously up to now, experiments carried out on a series of PP/elastomer/$BaSO_4$ composites proved that a simultaneous increase of impact resistance and stiffness can be achieved with proper selection of the components (Fig. 23). Such composites show a positive deviation from the usual, inverse correlation of stiffness and impact resistance. Composition, adhesion, and particle size of the filler all influence properties through the modification of structure and micromechanical deformation processes [40].

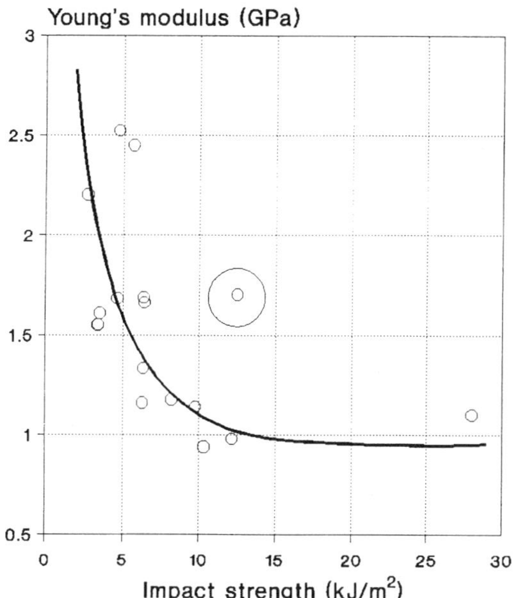

FIG. 23 Effect of composition and structure on the mechanical properties of PP/elastomer/$BaSO_4$ composites.

VIII. PROCESSING AND USE OF PARTICULATE FILLED POLYMERS

The majority of polyolefin composites, including particulate filled, short fiber or glass mat reinforced polymers, are produced by melt processing techniques [153]. Usually the polyolefin is melted in the appropriate equipment, the filler or short fiber is added, and the composite thoroughly homogenized. Structure-related phenomena take place during the homogenization process which require special attention and the proper selection of processing equipment to cope with them. Particulate fillers have a relatively wide particle size distribution, in the range of 0.1–10 μm, depending on the grade in question. As was discussed in detail before (see Section IV.A) small particles tend to form aggregates, which damage the mechanical properties, and especially the impact resistance, of composites. As a consequence the main objective during the homogenization of particulate filled polymers is to destroy the aggregates and to achieve the highest possible homogeneity. The length of reinforcing short fibers decreases during processing [154]. Attrition must be avoided as much as possible since the reinforcing effect of short fibers depends on their length. The conditions of homogenization must be chosen to assure maximum dispersion, but minimum shear must be applied to avoid attrition. Glass mat reinforced composites are produced almost exclusively from PP by the compression of a high temperature, low viscosity melt into the mat to achieve proper wetting [155]. This is done on a special processing line, but the technology is not discussed here.

The homogenization of heterogeneous polymer systems may take place according to two mechanisms: distributive and dispersive mixing [156]. Changes in the spatial distribution of the dispersed component occur during the former, while the second involves the breakdown of aggregates. The efficiency of distributive mixing depends on the total shear deformation of the melt, while that of dispersive mixing on the intensity of shear applied. Usually high shear stresses are needed to separate aggregated particles. The two mechanisms usually take place simultaneously during homogenization, particles from ruptured aggregates are distributed evenly in the melt. As was described earlier, the occurrence and extent of aggregation depends on the relative magnitude of adhesion and shear forces acting in the melt during homogenization [16]. Three factors determine aggregation: the strength of interaction (W_{AB}), particle size (r), and the intensity of shear ($\dot{\gamma}$). Particle size depends on the grade selected, adhesion may be modified by surface treatment, while the level of shear is determined by the processing technology. Shear stress can be changed in a relatively wide range, the upper limit is determined by the thermomechanical degradation of the polymer due to frictional heating and the mechanical rupture of polymer chains.

Break up of aggregated particles may take place by rupture or by erosion. During rupture the aggregates split into a few fragments which are aggregates themselves, while small, individual particles are detached from the aggregates in erosion. Most of the studies related to disaggregation were carried out on carbon black dispersed in elastomers [157, 158]. These studies showed that usually both mechanisms take place simultaneously. The same mechanisms also act in thermoplastics; their relative weight is determined by the structure and adhesion strength of the aggregates and by processing conditions [159–161].

The selection of the equipment used for the homogenization of particulate filled polymers depends on numerous factors. Its availability and installation costs may usually play an important role, but high homogenization efficiency and productivity are required from new machines. Optimum shear conditions are needed to satisfy the first requirement, while continuous operation assures the second.

Several batch machines may be used for the preparation of particulate filled polymers including two roll mills and various internal mixers. Two roll mills have been extensively applied in the rubber industry and gained some importance also in the preparation of blends and composites with thermoplastic matrices. Although the homogenization efficiency of the equipment is good, the batch operation hinders its industrial application. Today it is used mainly in laboratories for the homogenization of all kind of compounds. Internal mixers are applied to a larger extent. The efficiency of distributive mixing is excellent in this apparatus, but batch operation and the difficulty of cleaning represent its main drawbacks. Considerable attempts have been made to overcome these difficulties, and single- and two-stage continuous internal mixers are constructed and used in practice [162, 163].

The acceptable price and the availability of single screw extruders resulted in many attempts to use them also for the production of heterogeneous polymer systems. Because of the low mixed volume the melt is subjected to only moderate shear strains and also the developing shear stresses are limited in this machine, leading to very low efficiency both in distributive and dispersive mixing. The screw is often modified to include distributive and/or dispersive mixing elements in the melt-conveying zone [164], but even these cannot increase mixing efficiency to an acceptable level. Sometimes static mixing elements are applied to improve homogeneity [165]. Although shear stresses and strains can be controlled by such devices, they often cannot achieve the homogeneity required from the product. It can be generally stated that the mixing ability of conventional single screw extruders is limited even when they are supplemented with specially designed mixing devices, and they should not be used when thorough mixing and high product quality are required.

Ko-kneaders possess excellent mixing efficiency achieved by the special operational principle of this equipment. The screw is interrupted at regular intervals and the barrel is equipped with stationary kneading pins positioned in accordance with the interruptions. The screw rotates and reciprocates simultaneously, and thus the material is subjected to high shear strain, while also the shear stress can be controlled by the size of the gap between the kneading pins and the flights of the screw. The apparatus is very flexible and especially useful for the homogenization of heat-sensitive materials, but the relative high installation costs hinder its wider application.

Twin-screw extruders are used most often for the production of particulate filled thermoplastics [166]. A wide variety of machines are available in the market, which differ in the extent of screw intermeshing, in their relative direction of rotation (co- or counterrotating), and in several other technical details. The barrel and screw of modern equipment can be assembled from separate sections and the screw can be designed according to the material to be homogenized. Metering, kneading, and shear elements can be positioned at various places along the screw and components can also be fed in at several stages. The equipment is very flexible and the screw design can be easily changed to fit any requirements. Particulate filled thermoplastics are frequently produced in corotating intermeshing twin-screw extruders. A further advantage of this equipment is that single sections of the barrel and the screw can be replaced if necessary; at certain places considerable wear of equipment occurs inevitably during the production of filled and short fiber reinforced polymers.

Usually several ancillary operations must also be carried out during the preparation of particulate filled composites. Premixing of the components in high-speed fluid mixers increases homogeneity; care must be taken with polyolefins to avoid overheating which might lead to the melting of the polymer. Reliable feeding is crucial for obtaining composites with constant composition and properties. Volumetric and gravimetric feeding devices might be used, the latter being more expensive but much more reliable than volumetric devices. Gravimetric feeding equipment consists of a feeding screw, designed according to the particle characteristics of the raw material, which is attached to a balance. The dosage of liquids is also possible in a similar way, but instead of a screw, a gear pump is used for feeding. The pre-treatment of fillers might be necessary before processing. Fillers have high energy surfaces, thus they adsorb moisture; excessive amounts of volatiles may result in processing problems, so they must be removed. Moreover, the homogenization equipment must be equipped with a venting device, with venting or vacuum zones in twin-screw extruders. Surfactants or coupling agents can be introduced, i.e., surface treatment carried out in high-speed fluid mixers.

Originally particulate fillers were introduced into the polymers to decrease their price. However, price reduction cannot be achieved any more because of the cost of compounding [18]. With the exception of PVC the application of fillers and reinforcements must be justified by some improvement in composite properties for all polymers. Fillers and reinforcements are usually introduced to increase the stiffness and dimensional stability of the composite, but other benefits may also be gained, among others improved productivity, modified crystalline structure or special properties (flame retardancy, conductivity, etc.). $CaCO_3$, talc, and short glass fibers are used in the largest quantities for the modification of thermoplastics. In 1986 about 845,000 tons of $CaCO_3$ was used in Europe, while the application of all other fillers and reinforcements amounted only to 150,000 tons. The majority of $CaCO_3$ was introduced into PVC compounds to prepare both rigid and plasticized, products [18]. Considerable amounts of fillers and reinforcements are also introduced into PP. Somewhat more talc is added to this polymer than $CaCO_3$, but the amount of the two fillers is commensurable. Much less fillers and reinforcing material are used for the modification of PE and other polymers.

As was mentioned in the previous paragraphs PP is frequently modified with fillers and reinforcements. The main goal of modification is the increase of stiffness and dimensional stability, but with the proper selection of the matrix polymer and the introduction of additional components, e.g., elastomers, the properties of modified PP can be changed over a very wide range. This is properly demonstrated by Table 3, where the properties of some commercial modified PP grades are compared to each other. The matrix polymer of the composites is different in most cases; homo, block, and random copolymers are used alike, leading to considerable differences in stiffness, impact resistance, and HDT. The impact modified grades compete with HIPS and ABS, while the performance of fiber reinforced grades approaches that of engineering thermoplastics.

Modified PP is used in many fields of application, but mainly as structural materials. The extensive use of talc can be explained by the platelet-like particle geometry of this filler leading to reinforcement. It also has strong nucleation efficiency in PP, resulting in changing crystalline morphology. Nucleation leads to increased productivity and to a more homogeneous structure of injection molded parts. One of the major user of modified PP is the automotive industry. Instrument panel supports, under-hood elements (air filter covers, timing chain covers, heater boxes, battery boxes, etc.), and other parts are frequently prepared from filled or reinforced PP. Good mechanical properties, lower cost, and recyclability lead to the increased use of PP in this area. Bumpers are also often prepared from modified PP [167, 168]. The requirements are very stringent in this case: high impact resistance and sufficient stiffness must be accompanied by scratch resistance, dimensional stability, low thermal expansion coefficient, etc. Beside the filler, which is usually talc, these compounds frequently also contain an elastomer. The properties of the compound may be optimized by adjusting its structure as was described in Section VII.E. Some large structural parts are produced from glass mat reinforced PP by stamping. A considerable amount of $CaCO_3$-modified PP is used for the production of garden furniture. Smaller amounts of $CaCO_3$ filled PP are utilized in food packaging material and automotive parts. Parts of some household appliances, like washing machine soap dispensers, are

TABLE 3 Properties of some commercial particulate filled and reinforced PP grades

Filler									
Type	Amount (wt%)	Density (g/cm^3)	MFI* (g/10 min)	Young's modulus (GPa)	Yield stress (MPa)	Tensile strength (MPa)	Elongation-at-break (%)	Impact strength† (kJ/m^2)	HDT (°C)
Talc	40	1.25	1.7	3.65	35	25	18	3.8	74
Talc	30	1.13	1.6	2.50	29	22	85	14.0	62
Talc	20	1.04	2.6	2.15	23	19	130	20.0	57
$CaCO_3$	30	1.14	25.0	1.80	26	18	126	6.1	51
$CaCO_3$	40	1.24	4.7	1.90	23	21	330	28.6	56
Glass	10	0.96	3.0	2.10	—	30	10	12.7	109
Glass‡	30	1.14	3.5	5.10	—	85	8	5.3	144
$BaSO_4$	25	1.13	25.0	1.85	26	18	640	4.4	53
$BaSO_4$	50	1.51	20.0	2.10	22	19	56	3.7	58

* 230°C/2.16 kg.
† Notched Charpy impact resistance.
‡ Short glass fiber.

also prepared from talc filled PP. Small amounts of other fillers are also used in PP, e.g., silica is added to bioriented PP films as an antiblocking agent.

Much less filler is used in PE [18]. In several areas fillers cannot be applied at all because of processing problems, like in blow molded bottles, or because of stringent regulations, as in gas pipes. The average level of filler use in PE ranges from a few tenth of a percent to about 20%. TiO_2 is often applied as white pigment, while $CaCO_3$ and china clay are introduced into films to reduce stretch and to give antislip and antiblock, as well as thermal barrier, properties. The incorporation of calcined clay in 0.1–1.0% loadings leads to a good antiblocking effect. Fillers added to film grade polymers improve productivity and printability, as well as surface hardness. The largest amount of filler, mostly $CaCO_3$ and some TiO_2, is used for the production of artificial paper. This product has the advantage of water resistance, high mechanical strength, and good printability. Its application is continuously increasing in many areas. Although fillers are used in increasing amounts in most polyolefins, the most important matrix polymer is PP and it will remain so for the near future.

IX. TRENDS IN POLYOLEFIN COMPOSITES

Fillers and reinforcements are extensively used for the modification of the properties of PP, but they are applied in some other polyolefins as well. The growth rate of composites is faster than that of the base polymers because the introduction of a new product into the market is faster and cheaper by modification than through the development of a new polymerization technology. The improvement of compounding technology, the faster reaction time of compounding companies, as well as their closer relationship with the customer [169], increase the production and use of tailor-made products and the amount of composites applied. This tendency gains further momentum by the significant oversupply of polyolefins forecast for the next 4 years [170]. About one third of all compounds consists of modified polyolefins and this ratio is expected to remain constant or even grow slightly for the near future [169]. Besides the use of increasing amounts of composites, their quality improves as well due to the considerable effort put into research and development in this field.

Producers of traditional fillers, like $CaCO_3$ and talc, constantly improve the characteristics and quality of their products. Recently two new grades of $CaCO_3$ were introduced by one of the largest suppliers of mineral fillers [171]. The first is a master batch used in the production of HDPE bottles, houseware closures, etc., which contains 90% filler and 10% proprietary olefinic binder. The new filler is said to improve both properties and processability. The other new $CaCO_3$ product enhances LLDPE film properties dramatically. The impact and tear strength, as well as the stiffness of the films, improves, which allows considerable (10–20%) downgrading in thickness, resulting in significant savings. Similar to $CaCO_3$, new grades of talc were also commercialized recently [172]. Beside the upgrading of traditional fillers, completely new products are also developed and introduced into the market. Currently, a $Ca(OH)_2$ filler was developed which can be used in a wide variety of polymers including polyolefins [173]. The well known drawback of this filler, that it reacts with atmospheric carbon dioxide to give $CaCO_3$, is inhibited by the formation of a $CaCO_3$ shell around a $Ca(OH)_2$, core. The new product is claimed to offer considerable technical and financial advantages.

Similar trends can also be observed in the field of reinforcements. A large amount of glass fiber reinforced PP is used in structural applications and especially in the automotive industry. A considerable increase in the length of the fibers is observed in new products. Fiber length changes between 10–25 mm even in injection-molded grades [174]. Other technologies, such as compression molding and stamping, allow the use of long fibers without the danger of attrition during the shaping operation. Commingled glass and PP fibers dramatically improve the wetting out of the glass, resulting in better interaction and homogeneity [175]. Continuous fiber reinforcement is achieved with this technique, with the advantage of using the simpler and usually much faster thermoplastic processing technology instead of thermoset processing methods. The increased use of natural fibers is also forecast in polyolefin composites. A wide variety of such fibers, e.g. flax, hemp, jute, sisal, and banana fibers, are already applied in the automotive industry, offering technical advantages such as high impact strength and rigidity accompanied by lighter weight and easier reprocessing [176, 177]. Surface treatment technology is crucial both in particulate filled and fiber reinforced composites, because it assures the necessary dispersity in the first case and efficient stress transfer in the second.

Similar to fillers and reinforcements, the products prepared from modified polyolefins are also under constant change and improvement. Simultaneously increasing stiffness and impact resistance remains one of the major goals of development for products used in structural applications. The automotive industry puts extremely stringent requirements on such compounds. Besides good mechanical properties, they must be characterized by easy processing, low linear expansion coefficient, paintability, etc. [178]. Often the combination of traditional materials appears in new areas of application, like the TiO_2 pigmented HDPE which was recently introduced in the US market for the production of milk bottles [179]. Although further development is expected also in traditional areas, some completely new composites are also expected to appear in the market soon. Master batches and composites based on metallocene polymers are already available or under development at most compounding companies [180]. The combination of the flexibility in the adjustment of properties offered by metallocene technology and the advantageous characteristics of new fillers will lead to compounds with attractive property combinations and advantageous price/performance characteristics. In the last few years the application of polyolefin based thermoplastic olefin (TPO) and thermoplastic elastomer (TPE) compounds increased especially rapidly in the automotive industry and further increases are expected in this area [180]. Most of these products contain metallocene polymers and compete successfully with similar compounds prepared from EPDM and EPR rubbers.

As a result of intensive research a completely new class of materials, the nanocomposites, emerged in the last decade. These materials can be divided into two major groups. Monomers are introduced between the crystal planes of layered silicates, usually, sodium montmorillonites (bentonite, smectite) and polymerized in the first [181, 182], while nanoparticles are precipitated in the polymer matrix by a sol–gel technology in the other [183, 184]. Polyolefin films coated with such a composite prepared from a blend of EVA and poly(ethylene-co-vinyl alcohol) copolymer containing TiO_2 or silica nanoparticles may improve their gas barrier properties tremendously [185]. The intended application of these products is in the packaging industry. Nanoclays, on the other hand, may find their place in structural applications. 2–5% of these fillers improves mechanical properties, decreases gas permeation, and imparts flame retardancy to the composite and the enhanced properties are accompanied by considerable weight savings [176]. Although most of the montmorillonite nano-composites are prepared with PA and epoxy matrices [181, 182], polyolefin composites are expected to appear in the very near future as well [186, 187].

X. CONCLUSIONS

According to some authors interfacial interaction is the determining factor in all heterogeneous polymer systems, including particulate filled and fiber reinforced polymers. Both particle/particle and matrix/filler interactions play an important role in the determination of composite properties. The effect of the former is clearly detrimental, it decreases strength and especially impact resistance of the composite. The occurrence and extent of aggregation is determined by the relative adhesion and shear forces during the homogenization of the components. Due to matrix/filler interactions an interphase forms spontaneously in composites with properties different from that of both components. The strength of adhesion can be acceptably characterized by thermodynamic quantities, mainly by the reversible work of adhesion. The modification of interactions is achieved through the surface treatment of the filler. In particulate filled systems, nonreactive treatment is applied the most often, but lately other techniques have gained in importance as well. The type of surface treatment must be selected according to its goal, the surfactant or coupling agent must be adjusted in accordance with the chemical character of the components, and the amount used must be optimized as well. Proper modification may improve properties significantly; occasionally it can be the condition of the successful application of the composite. The properties of particulate filled polymers depend on component characteristics, composition, structure, and interaction. Although numerous models are available for the prediction of composite properties, they must be applied with great care. Particulate filled polyolefins are prepared almost exclusively with melt blending techniques; twin-screw extruders seem to be the most appropriate for this purpose. Most polyolefins are modified with fillers and reinforcements, but the most important matrix polymer is PP. Particulate filled and fiber reinforced PP composites are used in all fields of applications, from household appliances to automotive parts.

ABBREVIATIONS AND SYMBOLS

a	surface area of an adsorbed molecule
f	conversion factor
j_0	correction factor for pressure difference in IGC
k	constant
k_e	aspect ratio
l	thickness of the interphase

m	strain hardening parameter
n	number of molecules taking part in interaction
p	vapor pressure
r	particle radius
r_a	equivalent radius, $r_a = r_1 r_2/(r_1 + r_2)$
t_0	reference time
t_r	retention time
A	parameter of the modified Kerner equation
A_f	specific surface area of the filler
AN, DN	Guttman's acceptor and donor numbers
B_y, B_T	constant related to stress transfer
C	geometric constant
C_1, C_2	geometric constants in debonding
C_A, C_B	Drago's acid base constants (covalent)
E_A, E_B	Drago's acid base constants (electrostatic)
F	flow rate of carrier gas in IGC
F_a	adhesive force
F_h	hydrodynamic force
G, G_0, G_f	shear modulus of composite, matrix and filler
ΔH^{ab}	free enthalpy change due to interaction
L_0, L	initial and actual length of a specimen
N	Avogadro number
R	universal gas constant
S_{AB}	wettability
T	absolute temperature
V_N	retention volume
W_{AB}	reversible work of adhesion
W^d_{AB}, W^p_{BA}	dispersion and polar component of W_{AB}
W^{spec}_{AB}	acid–base component of W_{AB}
γ	surface tension
γ_A, γ_B	surface tension of the components
γ_{AB}	interfacial tension
γ^d_A, γ^p_A, γ^d_B, γ^p_B	dispersion and polar components of surface tension for components A and B
γ_{LV}	surface tension of liquids
γ_{S0}	surface tension of solids in vacuum
γ_{SV}	surface tension of solids in vapor
γ_{SL}	solid–liquid interfacial tension
$\dot{\gamma}$	shear rate
η	viscosity of the media
λ	relative elongation, $\lambda = L/L_0$
ν_m	Poisson's ratio of the matrix polymer
θ	contact angle
π_e	spreading pressure
ρ_f	density of filler
σ_{yi}	yield stress of the interphase
σ^D	debonding stress
σ^T	thermal stress
σ_{y0}, σ_y	yield stress of the matrix and the composite
σ_{T0}, σ_T	true stress of the matrix and the composite
φ_f	volume fraction of filler
φ_f^{max}	maximum packing fraction
Γ	moles of vapor adsorbed on a unit surface
Ψ	parameter related to φ_f^{max}
ABS	acrylonitrile butadiene styrene terpolymer
AES	Auger electron spectroscopy
AMPTES	(3-aminopropyl) triethoxysilane
DRIFT	Diffuse reflectance infrared spectroscopy
FTIR	Fourier transform infrared spectroscopy
HDPE	high density polyethylene
HDT	heat deflection temperature
HIPS	high impact polystyrene
IGC	inverse gas chromatography
MA–PP	maleinated polypropylene
PP	polypropylene
PVC	poly(vinyl chloride)
pPVC	plasticized PVC
SIMS	Secondary ion mass spectrometry
SPTES	(3-stearyloxypropyl) triethoxysilane
XPS, ESCA	X-ray photoelectron spectroscopy

REFERENCES

1. D Vink. Kunststoffe 80: 842–846, 1990.
2. B Pukánszky. In: J Karger-Kocsis, ed. Polypropylene. Structure, Blends and Composites. London: Chapman and Hall, 1995, vol 3, pp 1–70.
3. A Krysztafkiewicz. Surf Coatings Technol 35: 151–170, 1988.
4. G. Vörös, E. Fekete, B. Pukánszky. J Adhesion 64: 229–250, 1997.
5. B Pukánszky, B Turcsányi, F Tüdös. In: H Ishida, ed. Interfaces in Polymer, Ceramic, and Metal Matrix Composites. New York: Elsevier, pp 467–477, 1988.
6. B Pukánszky. Composites 21: 255–262, 1990.
7. P Vollenberg, D Heikens, HCB Ladan. Polym Compos 9: 382–388, 1988.
8. B Pukánszky, G Vörös. Compos Interfaces 1: 411–427, 1993.

9. AM Riley, CD Paynter, PM McGenity, JM Adams. Plast Rubber Process Appl 14: 85–93, 1990.
10. J Menczel, J Varga. J Thermal Anal 28: 161–174, 1983.
11. M Fujiyama, T Wakino. J Appl Polym Sci 42: 2739–2747, 1991.
12. B Pukánszky, I Mudra, P Staniek. J Vinyl Additive Technol 3: 53–57, 1997.
13. R Rothon, ed. Particulate-filled Polymer Composites. Harlow: Longman, 1995.
14. HP Schlumpf. Chimia 44: 359–360, 1990.
15. HP Schlumpf. Kunststoffe 73: 511–515, 1983.
16. B Pukánszky, E Fekete. Polym Polym Compos 6: 313–322, 1998.
17. LE Nielsen. Mechanical Properties of Polymers and Composites. New York: Marcel Dekker, 1974.
18. M Hanckock. In: R. Rothon, ed. Particulate-filled Polymer Composites. Harlow, Longman, 1995, pp 279–316.
19. RA Dickie. In: DR Paul, S Newman, eds. Polymer Blends. New York: Academic Press, 1978, vol 1, 353–391.
20. JC Halpin, JL Kardos. Polym Eng Sci 16: 344–352, 1976.
21. Z Hashin. J Appl Mech 29: 143–150, 1962.
22. R Hill. J Mech Phys Solids 13: 213–222, 1965.
23. RM Christensen, KH Lo. J Mech Phys Solids 27: 315–330, 1979.
24. J Kolařík, F Lednický, B Pukánszky. In: FL Matthews, NCR Buskell, JM Hodgkinson, J Morton, eds. Proc 6th ICCM/2nd ECCM. London: Elsevier, 1987, vol 1, pp 452–461.
25. B Pukánszky, E Fekete, F Tüdös. Makromol Chem, Macromol Symp 28: 165–186, 1989.
26. S McGee, RL McCullough. Polym Compos 2: 149–161, 1981.
27. PHT Vollenberg. PhD Thesis, Eindhoven University of Technology, Eindhoven, 1987.
28. VP Chacko, FE Karasz, RJ Farris. Polym Eng Sci 15: 968–974, 1982.
29. K Mitsuishi, H Kawasaki, S Kodama. Kobunshi Ronbunshu 41: 665–672, 1984.
30. PHT Vollenberg, D Heikens. J Mater Sci 25: 3089–3095, 1990.
31. RT Woodhams, G Thomas, DK Rodgers. Polym Eng Sci 24: 1166–1171, 1984.
32. RK Mittal, VB Gupta, P Sharma. J Mater Sci 22: 1949–1955, 1987.
33. L Nicolais, M Narkis. Polym Eng Sci 11: 194–199, 1971.
34. B Turcsányi, B Pukánszky, F Tüdös. J Mater Sci Lett 7: 160–162, 1988.
35. B Pukánszky, F Tüdös. Makromol Chem, Macromol Symp 38: 221–231, 1990.
36. E Fekete, B Pukánszky, Z Peredy. Angew Makromol Chem 199: 87–101, 1992.
37. SK Brown, Br Polym J 12: 24–30, 1980.
38. T Vu-Khanh, B Sanschagrin, B Fisa. Polym Compos 6: 249–260, 1985.
39. B Pukánszky, FHJ Maurer. Polymer 36: 1617–1625, 1995.
40. Sz Molnár, B Pukánszky, CO Hammer, FHJ Maurer. Impact fracture study of multicomponent polypropylene composites, accepted in Polymer.
41. CB Bucknall. Adv Polym Sci 27: 121–148, 1978.
42. K Kendall. Br Polym J 10: 35–38, 1978.
43. K Friedrich, UA Karsch. Fibre Sci Technol 18: 37–52, 1983.
44. FF Lange. Phil Mag 22: 983–992, 1970.
45. AG Evans. Phil Mag 26: 1327–1344, 1972.
46. AG Evans, S Williams, PWR Beaumont. J Mater Sci 20: 3668–3674, 1985.
47. J Jančař, AT DiBenedetto, A DiAnselmo. Polym Eng Sci 33: 559–563, 1993.
48. V Svehlova, E Poloucek. Angew Makromol Chem 153: 197–200, 1987.
49. JW Ess, PR Hornsby. Plast Rubber Process Appl 8: 147–156, 1987.
50. MJ Adams, MA Mullier, JPK Seville. In: BJ Briscoe, MJ Adams, eds. Tribology in Particulate Technology. Bristol: Adam Hilger, pp 375–389, 1987.
51. D Tabor. In: BJ Briscoe, MJ Adams, eds. Tribology in Particulate Technology. Bristol: Adam Hilger, pp 206–219, 1987.
52. MJ Adams, B Edmondson. In: BJ Briscoe, MJ Adams, eds. Tribology in Particulate Technology. Bristol: Adam Hilger, pp 154–172, 1987.
53. W Balachandran. In: BJ Briscoe, MJ Adams, eds. Tribology in Particulate Technology. Bristol: Adam Hilger, pp 135–143, 1987.
54. SL Goren. J Colloid Interface Sci 36: 94–96, 1971.
55. KW Allen. Phys Technol 19: 234–240, 1988.
56. S Wu. In: DR Paul, S Newman, eds. Polymer Blends. New York: Academic Press, vol 1, pp 243–293, 1978.
57. KW Allen. J Adhesion 21: 261–277, 1987.
58. BV Derjaugin. Research 8: 70–74, 1955.
59. HP Schreiber. In: G Akovali, ed. The Interfacial Interactions in Polymeric Composites. Amsterdam: Kluwer, pp 21–59, 1993.
60. RF Good. J Colloid Interface Sci 59: 398–419, 1977.
61. FM Fowkes. Ind Eng Chem 56: 40–52, 1964.
62. FM Fowkes. In: FM Fowkes, ed. Hydrophobic Surfaces. Proc Kendall Award Symp, New York: Academic Press p. 151–163, 1969.
63. S Wu. J Macromol Sci, Rev Macromol Chem C10: 1–73, 1974.
64. KL Mittal, HR Anderson. Acid-base Interactions: Relevance to Adhesion Science and Technology. Utrecht: VSP, 1991.
65. FM Fowkes, In: KL Mittal, ed. Physicochemical Aspects of Polymer Surfaces. New York: Plenum, pp 583–602, 1981.
66. FM Fowkes. In: KL Mittal, HR Anderson, eds. Acid-base Interactions: Relevance to Adhesion Science and Technology. Utrecht: VSP, pp 93–110, 1991.
67. RS Drago, GC Vogel, TE Needham. J Am Chem Soc 93: 6014–6026, 1971.

68. V Gutmann. The Donor-Acceptor Approach to Molecular Interactions. New York: Plenum, 1978.
69. KL Mittal. Polym Eng Sci 17: 467–473, 1977.
70. B Pukánszky, G Vörös. Polym Compos 17: 384–392, 1996.
71. HW Fox, EF Hare, WA Zismann. J Phys Chem 59: 1097–1106, 1955.
72. CY Yue, WL Cheung. J Mater Sci 26: 870–880, 1991.
73. PHT Vollenberg, D Heikens. Polymer 30: 1656–1662, 1989.
74. JE Stamhuis, JPA Loppé. Rheol Acta 21: 103–105, 1982.
75. M Sumita, H Tsukihi, K Miyasaka, K Ishikawa. J Appl Polym Sci 29: 1523–1530, 1984.
76. U Zorll. Gummi, Asbest, Kunstst 30: 436–444, 1977.
77. MJ Folkes, WK Wong. Polymer 28: 1309–1314, 1987.
78. G Akay. Polym Eng Sci 30: 1361–1372, 1990.
79. FHJ Maurer, R Kosfeld, T Uhlenbroich. Colloid Polym Sci 263: 624–630, 1985.
80. FHJ Maurer, R Kosfeld, T Uhlenbroich, LG Bosveliev. 27th Int Symp. on Macromolecules, 6–9 July, Strasbourg, France, 1981, pp 1251–1254.
81. KF Mansfield, DN Theodorou. Macromolecules 24: 4295–4309, 1991.
82. S Patel, G Hadziioannou, M Tirrell. In: H Ishida, JL Koenig, eds. Composite Interfaces. New York: Elsevier, pp 65–70, 1986.
83. J Jančař. J Mater Sci 26: 4123–4129, 1991.
84. PHT Vollenberg, D Heikens. In: H Ishida, JL Koenig, eds. Composite Interfaces. New York: Elsevier, pp 171–176, 1986.
85. SN Maiti, PK Mahapatro. J Appl Polym Sci 42: 3101–3110, 1991.
86. G Vörös, B Pukánszky. J Mater Sci 30: 4171–4178, 1995.
87. FHJ Maurer, HM Schoffeleers, R Kosfeld, T Uhlenbroich. In: T Hayashi, K Kawata, S Umekawa, eds. Progress in Science and Engineering of Composites. Tokyo: ICCM-IV, pp 803–809, 1982.
88. K Iisaka, K Shibayama. J Appl Polym Sci 22: 3135–3143, 1978.
89. B Pukánszky, F Tüdös. In: H Ishida, ed. Controlled Interphases in Composite Materials. New York: Elsevier, pp 691–700, 1990.
90. J Kolařík, S Hudeček, F Lednický, Faserforsch Textiltechn 29: 51–56, 1978.
91. J Kolařík, S Hudeček, F Lednický, L Nicolais. J Appl Polym Sci 23: 1553–1564, 1979.
92. B Pukánszky. New Polym Mater 3: 205–217, 1992.
93. H Ishida, JL Koenig. J Polym Sci, Polym Phys 18: 1931–1943, 1980.
94. D Briggs, MP Seah. Practical Surface Analysis, vol 1, Auger and X-ray Photoelectron spectroscopy. Chichester: Wiley, 1990.
95. D Briggs, MP Seah. Practical Surface Analysis, vol 2, Ion and neutral spectrocopy. Chichester: Wiley, 1992.
96. CN Banwell. Fundamentals of Molecular Spectroscopy. New York: McGraw-Hill, 1972.
97. D Briggs, A Brown, JC Vickerman. Handbook of Static Secondary Ion Mass Spectrometry. Chichester: Wiley, 1989.
98. DP Ashton, D Briggs. In: Ref 13, pp 89–121, 1995.
99. NA Eltekova. Adsorption Properties of Modified Surface of Silica and Titania. Proc Eurofillers '95, pp 285–288, 1995.
100. JF Delon, J Yvon, L Michot, F Villieras, JM Dases. Vapor phase adsorption of alkylamines on talc and chlorite. Proc Eruofillers '95, pp 17–20, 1995.
101. H Balard, E Papirer. Prog Surf Coatings 22: 1–17, 1993.
102. E Fekete, B Pukánszky, A Tóth, I Bertóti. J Colloid Interface Sci 135: 200–208, 1989.
103. U Panzer, HP Schreiber. Macromolecules 25: 3633–3637, 1992.
104. C Bonnerup, P Gatenholm. J Adhesion Sci Technol 7: 247–262, 1993.
105. RN Rothon. In: Ref 13, pp 123–163, 1995.
106. J Jančář, J Kučera. Polym Eng Sci 30: 707–713, 1990.
107. J Jančář, J Mater Sci 24: 3947–3955, 1989.
108. G Marosi, G Bertalan, I Rusznák, P Anna, Colloids Surf 23: 185–198, 1987
109. P Bajaj, NK Jha, RK Jha. Polym Eng Sci 29: 557–563, 1989.
110. P Bajaj, NK Jha, RK Jha. Br Polym J 21: 345–355, 1989.
111. RG Raj, BV Kokta, F Dembele, B Sanschagrain. J Appl Polym Sci 38: 1987–1996, 1989.
112. E Papirer, J Schultz, C Turchi. Eur Polym J 12: 1155–1158, 1984.
113. RC Allard, T Vu-Khanh, JP Chalifoux. Polym Compos 10: 62–68, 1989.
114. M Bramuzzo, A Savadori, D Bacci. Polym Compos 6: 1–8, 1985.
115. EP Plueddemann. Silane Coupling Agents. New York: Plenum, 1982.
116. JP Trotignon, J Verdu, R De Boissard, A De Vallois. In: B Sedláček, ed. Polymer Composites. Berlin: Walter de Gruyter, pp 191–198, 1986.
117. LJ Matienzo, TK Shah. Surf Interface Anal 8: 53–59, 1986.
118. RG Raj, BV Kokta, C Daneault. Int J Polym Mater 12: 239–250, 1989.
119. E Mäder, KH Freitag. Composites 21: 397–402, 1990.
120. H Ishida, JD Miller. Macromolecules 17: 1659–1666, 1984.
121. H Ishida. In: H Ishida, G Kumar, eds. Molecular Characterization of Composite Interfaces. New York: Plenum, p 25–50, 1985.
122. Z Demjén, B Pukánszky, E Földes, J Nagy. J Interface Colloid Sci 194: 269–275, 1997.
123. EJ Sadler, AC Vecere. Plast Rubber Process Appl 24: 271–275, 1995.
124. B Widmann, HG Fritz, H Oggermüller. Kunststoffe 82: 1185–1190, 1992.

125. Z Demjén, B Pukánszky, J Nagy Jr. Polymer 40: 1763–1773, 1998.
126. S Takase, N Shiraishi. J Appl Polym Sci 37: 645–659, 1989.
127. J Jančář, M Kummer, J Kolařík. In: H Ishida, ed. Interfaces in Polymer, Ceramic, and Metal Matrix Composites. New York: Elsevier, pp 705–711, 1988.
128. JM Felix, P Gatenholm. J Appl Polym Sci 50: 699–708, 1991.
129. I Kelnar. Angew Makromol Chem 189: 207–218, 1991.
130. WY Chiang, WD Yang. J Appl Polym Sci 35: 807–823, 1988.
131. J Jančář, J Kučera. Polym Eng Sci 30: 714–720, 1990.
132. RA Venables, A Tabtiang. Reactive surface treatment for calcium carbonate filler in polypropylene. Proc Eurofillers 97, Manchester, 1997, 107–110.
133. V Khunová, MM Sain. Angew Makromol Chem 224: 9–20, 1995.
134. V Khunová, MM Sain. Recent developments in the reactive processing of magnesium hydroxide filled polyolefins, Proc Eurofillers '97, Manchester, 1997, pp 199–202.
135. S Voronov, V Tokarev. Design of interphase layers in composites using peroxy macroinitiators. Proc Eurofillers '97, Manchester, 1997, pp 37–40.
136. V Tokarev, U Wagenknecht, S Voronov, K. Grundke, O Bednarska, V Seredyuk. Activation of filler surface with peroxidic oligomeric modificators. Proc Eurofillers '97, Manchester, 1997, pp 61–64.
137. U Wagenknecht, V Tokarev, B Kretzschmar, S Voronov. Design of polyolefin compounds with fillers activated by peroxide oligomers, Proc Eurofillers '97, Manchester, 1997, pp 207–210.
138. LA Kostandov, NS Enikolopov, FS Dyachkovskii, LA Novokshonova, OI Kudinova, YuA Gavrilov, TA Maklakova, Kh-MA Brikenshtein. USSR Pat 763379, 1976.
139. EG Howard, RD Lipscomb, RN MacDonald, BL Glazar, CW Tullock, JW Collette. Ind Eng Chem Prod Res Dev 20: 421–428, 1981.
140. EG Howard, BL Glazar, JW Collette. Ind Eng Chem Prod Res Dev 20: 429–433, 1981.
141. LA Novokshonova, IN Meshkova. New filled polyolefins with special properties obtained by polymerization filling, Proc Eurofillers '97, Manchester, 1997, pp 411–414.
142. F Hindryckx, P Dubois, R Jerome, P Teyssie, M Garcia Marti. J Appl Polym Sci 64: 423–438.
143. F Hindryckx, P Dubois, R Jerome, P Teyssie, M Garcia Marti. J Appl Polym Sci 64: 439–454.
144. P Dubois, M Alexandre, R Jerome, M Garcia Marti. Polymerization-filled composites by supported metallocene based catalyst. Proc Eurofillers '97, Manchester, 1997, pp 403–406.
145. J Karger-Kocsis, A Kallo, VN Kuleznev. Polymer 25: 279–286, 1984.
146. YD Lee, CC Lu. J Chin Inst Chem Eng 13: 1–8, 1982.
147. J Kolařík, F Lednický. In: B Sedláček, ed. Polymer Composites. Berlin: Walter de Gruyter, pp 537–544, 1986.
148. JE Stamhuis. Polym Compos 9: 280–284, 1988.
149. AK Gupta, PK Kumar, BK Ratnam. J Appl Polym Sci 42: 2595–2611, 1991.
150. B Pukánszky, F Tüdös, J Kolařík, F Lednický. Polym Compos 11: 98–104, 1990.
151. WY Chiang, WD Yang, B Pukánszky. Polym Eng Sci 32: 641–648, 1992.
152. J Kolařík, F Lednický, J Jančař, B Pukánszky. Polym Commun 31: 201–204, 1990.
153. I Manas-Zloczower, Z Tadmor. Mixing and Compounding of Polymers; Theory and Practice. Munich: Hanser, 1994.
154. AG Gibson. In: J Karger-Kocsis, ed. Polypropylene. Structure, Blends and Composites. London: Chapman and Hall, vol 3, pp 71–112, 1995.
155. LA Berglund, ML Ericson, In: J Karger-Kocsis, ed. Polypropylene. Structure, Blends and Composites. London: Chapman and Hall, vol 3, pp 202–227, 1995.
156. HEH Meijer, JMH Janssen. in Ref. 153, pp 85–147.
157. SP Rwei, SW Horwatt, I Manas-Zloczower, DL Feke. Int Polym Process 6: 98–102, 1991.
158. SP Rwei, I Manas-Zloczower, DL Feke. Polym Eng Sci 31: 558–562, 1991.
159. SP Rwei, I Manas-Zloczower, DL Feke. Polym Eng Sci 32: 130–135, 1992.
160. YJ Lee, I Manas-Zloczower, DL Feke. Polym Eng Sci 35: 1037–1045, 1995.
161. YJ Lee, DL Feke, I Manas-Zloczower. Chem Eng Sci 48: 3363–3372, 1993.
162. MR Kearney. Plast Eng 23: 377–403, 1991.
163. EL Canedo, LN Valsamis. Int Polym Process 9: 225–232, 1994.
164. GPM Schenkel. Int Polym Process 3: 3–32, 1988.
165. S Middleman, Fundamentals of Polymer Processing. New York: McGraw Hill, 1977.
166. JL White. Twin-screw extrusion - technology and principles. Munich: Hanser, 1989.
167. JP 07118462, Impact-resistant and rigid automobile bumpers made of polyolefin compositions, Mitsubishi Kagaku Kk, Japan; Nissan Motor.
168. EP 580069, Injection molding compositions for automobile bumpers, Sumitomo Chemical Co. Ltd., Japan; Toyota Jidosha Kabushiki Kaisha.
169. P Mapleston. Mod Plast Int 27 (7): 56–59, 1997.
170. MW Shortt. Plast Eng 53 (4): 20–21, 1997.
171. G. Graff. Mod Plast Int 27 (10): 32–33, 1997.
172. G. Graff. Mod Plast Int 28 (5): 30–31, 1998.
173. G. Graff. Mod Plast Int 28 (12): 48–49, 1998.
174. P Mapleston. Mod Plast Int 28 (1): 107, 1998.
175. JA Grande. Mod Plast Intl 28 (8): 34–36, 1998.
176. V Wigotsky. Plast Eng 54 (9): 26–33 (1998).
177. S Moore. Mod Plast Int. 27 (10): 41–42, 1997.
178. RD Leaversuch. Mod Plast Int 28 (1): 42–43, 1998.
179. JH Schut. Mod Plast Int 27 (12): 35–36, 1997.

180. V Wigotsky. Plast Eng 53 (4): 26–31 (1997).
181. DC Lee, LW Jang. J Appl Polym Sci 68: 1997–2005, 1998.
182. JW Gilman, T Kashiwagi, JD Lichtenhan, Sampe J 33: 40–46, 1997.
183. N Juangvanich, KA Mauritz. J Appl Polym Sci 67: 1799–1810, 1998.
184. MA Harmer, WE Farneth, Q Sun. J Am Chem Soc 118: 7708–7715, 1996.
185. S Moore. Mod Plast Int 29 (2): 29–30, 1999.
186. MR Nyden, JW Gilman. J Comp Theor Polym Sci 7: 191–198, 1997.
187. HG Jeong, HT Jung, SW Lee, SD Hudson. Polym Bull 41: 107–113, 1998.

26

Functionalized Polyolefins: Synthesis and Application in Blends and Composites

M. Avella, P. Laurienzo, M. Malinconico, E. Martuscelli, and M. G. Volpe
Istituto di Ricerca e Tecnologia delle Materie Plastiche, CNR, Arco Felice (NA), Italy

I. INTRODUCTION AND HISTORICAL BACKGROUND

Polyolefins (PO) are a very important class of commercial polymers in the world today and are used in a wide range of applications. Despite their versatility, they suffer from certain drawbacks that exert a limiting influence on the range of applications. They are nonpolar and therefore exhibit poor hygroscopicity, printability, and dyeability. In addition they have poor dispersibility with inorganic fillers like talc and mica in composites and poor miscibility in blends, even with themselves, and alloys with polymers like nylons, polyesters, engineering thermoplastics, and so forth. This restricts their use in several new emerging technologies.

One means of overcoming these drawbacks is the introduction of a small amount of polar groups onto the polymer backbone by postpolymerization reactions using the polymer as a substrate. The polar groups impart new properties to the substrate polymer without significantly affecting the backbone properties of the polymer. Chemical modification produces polymers difficult or impossible to obtain by direct polymerization. It enables continuous variation in physical and chemical properties by controlling the extent of the reaction. Properties can be changed in a targeted way to improve biocompatibility, fire retardancy, adhesion, or the ability to blend with other polymers. Chemical modifications are also desirable when changes are required only on the surface of the PO to enhance adhesive properties or water absorption. Some of the functionalized polymers may also serve as reagents (e.g., ion exchange resins, etc.). Attaching a pharmacologically active unit onto a polymer chain is a relatively recent development for the controlled release of drugs and pesticides. Recycling is also possible using this approach.

Graft reactions are generally accompanied by chain scission and crosslinking side reactions. For example, in the case of PE, crosslinking is the predominant side reaction; for iPP and PIB, chain scission is used; and EPR use both crosslinking and chain scission. Such side reactions can be controlled and used for several applications, as, for example, in the peroxide-promoted controlled degradation of PP during melt extrusion, used for the production of controlled rheology (CR) grade resins. CRPP offer unique rheological properties with shortened polymer chain length and narrowed molecular weight distribution, as they exhibit a much higher melt flow index and reduced viscosity and elasticity. In the fibre production field, this leads to easier melt drawing of filaments and faster line speeds. CRPP is also suitable for thin-walled injection-moulding applications. Initiator concentration was found to be the predominant parameter controlling the flow properties of the polymers.

Several types of grafting reactions have been developed; we will briefly consider some of them, with relative examples.

A convenient approach is to bubble halogen into a suspension of the polymer in the presence of an initiator (light or peroxides). In fact, chlorination of PO in aqueous suspension is the most widely used commercial process. Radical initiation is the first step, followed by abstraction of a hydrogen radical from the polymer. The grafted product is formed if the halogen is easily available to react. However, several side reactions, such as chain breaking, coupling and so forth, makes the mechanism of such reactions complicated [1–3].

Solution or bulk functionalization of PO by unsaturated functional molecules is largely studied. In particular, maleic anhydride (MAH)-grafted polymers are widely used commercial products [4–7]. As in the case of halogenation, the mechanism is not clearly understood. Overall, MAH-grafting reactions are complicated. Reactions in an extruder are particularly difficult to predict. Several schemes of reactions and structures of the grafted polymers have been proposed [7–10]. The insertion of a MAH group into the backbone of PO has also been reported. A commercial product from BP Chemicals, Polybond^{TM}, has been reported to be maleated selectively at the chain ends. Ionic crosslinked rubber-like polymers were obtained from the reaction of maleic anhydride-grafted PO with some alkali metal compounds [11].

The functionalization of PP and atactic PP with α-methylstyrene, diethylfumarate and diethylmaleate in solution and in bulk, in the presence of dicumylperoxide has been reported [12]. The degree of functionalization was found to depend on the microstructure of the macromolecule.

Gaylord [13–15] has succeeded in grafting substantial amounts of strictly alternating styrene/maleic anhydride chains onto a PO backbone by injecting styrene/maleic anhydride mixtures. These grafted copolymers are particularly attractive as the succinic anhydride from grafted copolymers is highly reactive and represent a route to the synthesis of new polymers.

Antioxidants and stabilizers of low molecular weight tend to be lost from the polymer by evaporation and degradation during high temperature melt mixing operations and migration during the service life of the product. This can be prevented to some extent if the stabilizers are chemically bound to the polymer. Bhardwaj *et al.* [16] synthesized monomeric UV stabilizers based on benzophenone and grafted them thermomechanically onto PP, LDPE, and PS under melt processing conditions. Ageing studies showed that the graft stabilizers are more efficient than the corresponding physically blended stabilizers. This application represent a cost-effective way of improving the stability of PO during extrusion, with no additional process steps or expensive initiators.

Yamauchi [17, 18] reported the grafting of methyl methacrylate onto PE and PP oxidized with ozone. The active species, identified by ESR spectroscopy as peroxy radicals, were converted into hydroperoxide, which can be broken down by heating, giving rise to hydroxyl species able to initiate the graft copolymerization of methyl methacrylate.

An alternative route to solution and bulk grafting methods has been proposed by Lee and coworkers [19], which used a solid-phase graft copolymerization technique to graft maleic anhydride to polypropylene with the aid of an interfacial agent and a catalyst. The reaction was conducted on powdered polypropylene at temperatures well below the melting point of PP. During the entire reaction the polymer remained a powder. This technique offers the advantage of low temperatures and low operating pressures.

More recently, in order to select and minimize the use of solvents and hazardous reagents, the technology of surface modification has been developed. In addition, it is considered an environmentally friendly technology. Using this method, it might be possible to obtain materials that cannot be obtained by other methods. Further photofunctionalization can be done without affecting bulk degradation or crosslinking [20, 21]. Recently, an easy, cost-effective method to prepare functionalized PE has been reported in which PE is photo-oxidized and then melt-blended with nylon [22]. Carbonyl groups are formed during oxidation, and amine groups from the nylon then react with carbonyls, giving rise to copolymers that stabilize the blend. The authors have shown photo-oxidized PE from waste could be very effective in preparing such PE/nylon blends.

Another interesting example of surface modification is represented by the modification of hydrophobic polypropylene surfaces by carbon dioxide plasma [23] and carbon tetrachloride plasma [24] treatment to convert them into hydrophilic surfaces. This technique has potential applications in coatings and adhesion to other materials such as polymers, metals, and ceramics [24]. Radiation-induced reactions for LDPE/PP blends have also been recently reported [25].

Chemical modification can also be obtained starting from monomers, by the introduction of reactive moieties directly during polymerization.

The Ziegler–Natta catalyst technology is inherently limited for direct functionalization as the catalysts are easily poisoned by acidic or basic impurities. One way to overcome this limit has been proposed by Datta *et al.* [26] which introduced a functional monomer protected by masking the acidic or basic groups with trialkylaluminum. Another interesting example is the introduction of borane groups [27] which are used as intermediates for various functional groups or as initiators for grafting reactions. The synthesis of vinylidene-terminated olefin oligomers [28–31] or of PO containing vinyl unsaturations well removed from the backbone [32, 33] is a valid route for the preparation of graft or block copolymers. Other routes of PO synthesis based on free-radical processes are not very useful in obtaining modified PO, as they are not easily controlled [34–36].

As previously stated, one of the major applications of functionalized PO are in the field of heterogeneous polymer–polymer blends as interfacial agents. In heterogeneous blends, one polymer (A) is dispersed within the other polymer (B), a continuous phase, and they are immiscible. This type of blend has an important potential advantage in that it provides the additivity of the phase properties along with new features derived from a unique and particular morphology. For such engineered properties the minor phase should be evenly dispersed in finely divided domains throughout the continuous phase. Such properties are generally achieved by suitable compatibilizers, which can interlink the phases. The compatibilizers may be a polymer that is miscible with the individual homopolymers or *ad hoc* prepared by functionalizing one or both of the component polymers. PO offer an excellent price/performance ratio and so represent one of the more attractive constituent for the industrial manufacture of blends. In fact, many examples of blends of PO with other polymers of technological interest containing functionalized macromolecules as compatibilizers are reported in the literature. Blends of nylon 6 and PP in the presence of PP–g-MAH obtained by reactive blending in extruders or by melt-mixing in a Brabender-like apparatus have been extensively studied (37–39). PP–PA6 grafted copolymers have been isolated and characterized. The PP–g-MAH copolymers have also been used as interfacial agents in composites of PP with wood cellulose [40–44]. Maleic anhydride and acrylic acid grafted polypropylenes are also reported in ternary composites PP/elastomer/short glass fibre and PP/elastomer/filler ($CaCO_3$) [45]. The use of PO functionalized by diethylmaleate in PO/PVC blends is also reported [46, 47].

The objective of this review is to summarize our recent studies on the functionalization of PO, with particular regard to synthesis and chemical–physical characterization, and on their applications as interfacial agents for polymer blends and composites. This review is subdivided in four sections:

1. Functionalization of ethylene–propylene rubbers.
2. Rubber modification of polybuthyleneterephthalate by reactive blending concurrently with polymerization reaction.
3. Functionalized POic rubbers as interfacial agents in toughened thermosetting matrices.
4. Functionalized polypropylene as interfacial agents in polypropylene matrix composites with vegetable fibre.

II. FUNCTIONALIZATION OF ETHYLENE–PROPYLENE (EPR) RUBBERS

A. Grafting of Monoethylmaleate and Its Cyclization Reaction

As reported in the introduction, grafting of maleic anhydride (MAH) is widely reported [48–52]. The modified polymer can be used as starting materials for the preparation of graft copolymers. Such copolymers have been extensively used as interfacial agents in blends of two incompatible polymers [53–57]. For example, the presence of EPR-g-succinic anhydride (EPR–g-SA) in blends of EPR and polyamide 6 led to a large improvement in the mechanical properties and in the morphological appearance of the blend, mainly attributable to the formation of (EPR–g-SA)–g-PA6 copolymer molecules [56, 58].

HC, H_2C, C=O, O, C=O + H_2N~~~CONH~~~ →

HC, H_2C, C=O, N~~~CONH~~~, C=O

Another method to achieve similar results is to add EPR–g-SA in the course of the hydrolytic polymerization of caprolactam.

The preparation of EPR–g-SA is normally carried out by solution grafting of EPR with maleic anhydride using peroxides as initiators. Such a process is not convenient for industrial preparation, where bulk or aqueous suspension processes are preferred. Unfortunately, maleic anhydride is very corrosive in normal extruders, and in aqueous suspensions this molecule leads to the less reactive maleic acid. A convenient monomer which, in principle, can be grafted in suspension or in bulk is monoethylmaleate (MEM); it can, upon grafting, be converted to the cyclic form of the anhydride by simple heating [59]. Hereafter, we describe the solution grafting of MEM onto EPR.

1. Experimental

(a) Materials. EPR was a random ethylene–propylene copolymer (67 mol% C_2), kindly supplied by Dutral SpA, $M_W = 1.8 \times 10^5$; melt flow index (100°C) = 40 g/min.

Dibenzoylperoxide was recrystallized from absolute ethanol and stored under vacuum.

Maleic anhydride, a Fluka reagent analytical grade, was used without further purification.

All the solvents were of analytical grade and were purified according to standard procedures when necessary. Monoethylmaleate is prepared according to standard procedures [60].

(b) Techniques. Titrations were carried out by a Metrohm AG-CH-P100 Herisau potentiometer equipped with a processor for the analysis of the collected data. An Ag/AgCl electrode in a saturated solution of LiCl in isopropanol was used as reference electrode.

The IR spectra were obtained with a Nicolet 5DXB FTIR apparatus, equipped with a controlled temperature cell which allows scans of temperature from −200–250°C (SPECAC 20-100).

The viscosity measurements were performed at 135°C by a Cannon-Ubbelhode viscosimeter. Concentrations of 100 mg/10 mL in tetrahydronaphtalene were used.

(c) Typical Grafting Reaction. In a flask equipped with a nitrogen inlet and a condenser, 5.0 g of EPR were dissolved in 100 mL of chlorobenzene (CB) at room temperature. When the dissolution was complete, the temperature was allowed to rise to 85°C and a solution containing 8.0 g (5.5×10^{-2} mol) of monoethylmaleate in 10 mL of solvent was added to the system, together with 0.50 g (2.07×10^{-3} mol) of dibenzoylperoxide, previously dissolved in 10 mL of CB. After two hours, the reaction was stopped and the polymeric product was precipitated in methanol, repeatedly washed with acetone, and finally dried in a vacuum oven for 24 h at 50°C. The grafting degree is evaluated by potentiometric titration [60].

(d) Cyclization of EPR–g-MES in o-Dichlorobenzene Solution. 2 g of EPR–g-MES were dissolved in 40 mL of solvent in a flask equipped with a condenser. After the dissolution was complete, the temperature was raised to 160°C while N_2 was allowed to bubble into the solution.

Samples were taken off at chosen intervals of time and film for IR analysis were evaporated directly onto a KBr disk.

2. Results and Discussion

(a) Solution Grafting of Monoethylmaleate onto EPR. The grafting reaction has been studied in solution, using dibenzoylperoxide as radical initiator and anhydrous chlorobenzene as solvent. In all experiments, the temperature has been kept at 85°C because at higher temperatures the free monoethylmaleate is converted to maleic anhydride. Nevertheless, as we can see from the spectrum of Fig. 1, the final product always contains a little amount of grafted succinic anhydride for the establishment of an equilibrium between monoester and anhydride. The presence of a certain amount of grafted anhydride cannot be avoided when the reaction temperature is further decreased (70°C). Actually, it is possible that the cyclization occurs after the grafting onto EPR.

Chlorobenzene is chosen as solvent because it allows higher grafting efficiency compared with other aromatic solvents [61], while dibenzoylperoxide as initiator has a sufficiently high decomposition rate at the reaction temperature.

The reaction has been investigated for two different amounts of peroxide, i.e., 41 and 103 mmol DBPO/100 g EPR. Quantitative results in terms of mmol of MES grafted/100 g EPR vs time for both DBPO concentrations are reported in Fig. 2 for a total reaction time of three hours. The behavior is similar to that of other molecules previously investigated. In fact, at the lower concentration of DBPO a plateau is reached in about one hour of reaction. In similar grafting experiments, maleic anhydride gives a higher final conversion [61]. The higher reactivity of MA can be explained with the enhanced activation of the double bond towards the addition to macroradicals due to a stronger electron-attracting property and higher delocalization ability of the anhydride group with respect to the ester group. According to this explanation, the monoethylmaleate reaches the plateau at a value of conversion intermediate between those reached by MA and a diester group, the dibutylmaleate.

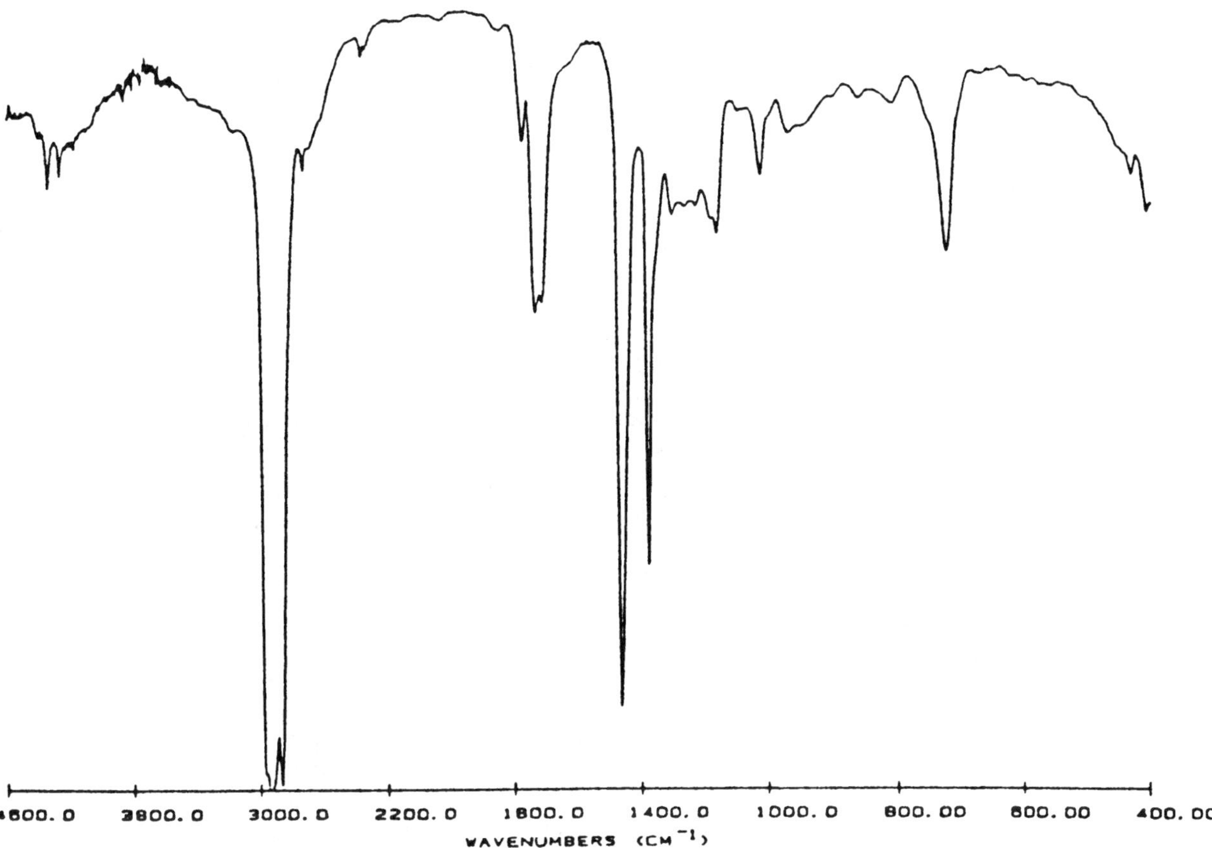

FIG. 1 FTIR spectrum of EPR–g-MES at 6.2% grafting degree.

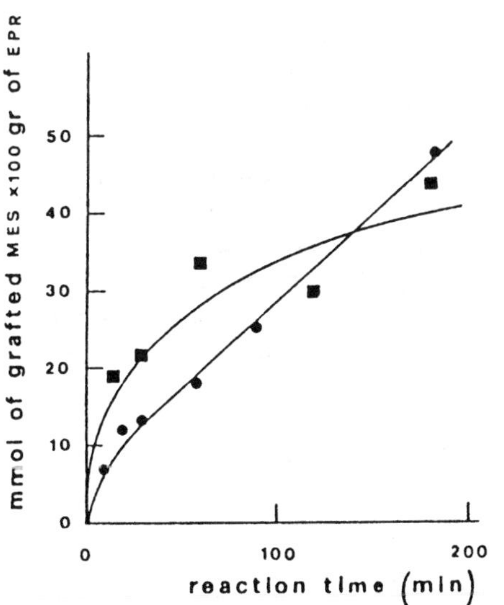

FIG. 2 Dependence of the grafting degree of MEM on the reaction time. Reaction conditions: $T = 85°C$; chlorobenzene = 100 mL; EPR = 5.0 g; MEM = 8.0 g; DBPO (■) = 0.5 g and (●) = 1.25 g.

The curve relative to the higher DBPO concentration is quite different from the previous one also from a qualitative point of view, because a plateau is not reached in the timescale of the experiment. It is believed that, at higher concentrations of peroxide, part of the primary radicals R$^{\bullet}$ is involved in oligomerization of MEM molecules which can, in turn, graft onto EPR molecules [62]. This fact indicates that the grafting of MEM is very sensitive to the concentration of primary radicals, which influences the mechanism of the grafting reaction.

Furthermore, the grafting of unsaturated molecules onto EPR chains leads to a decrease of the reduced viscosity of the POic substrate [61]. Measures of reduced viscosity at different reaction times confirm this trend, as we can see from the plot of Fig. 3. Obviously, a greater concentration of peroxide leads to a greater degradation of EPR chains.

An investigation of grafting degree as a function of the amount of starting peroxide has been performed, and the results are reported in the plot of Fig. 4. The

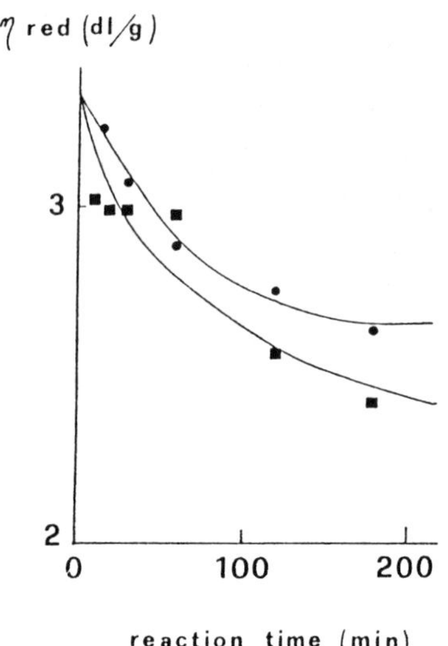

FIG. 3 Dependence of the reduced viscosity of EPR–g-MES on the reaction time. Reaction conditions: $T = 85°C$; chlorobenzene = 100 mL; EPR = 5.0 g; MEM = 8.0 g, DBPO (●) = 0.50 g and (■) = 1.25 g.

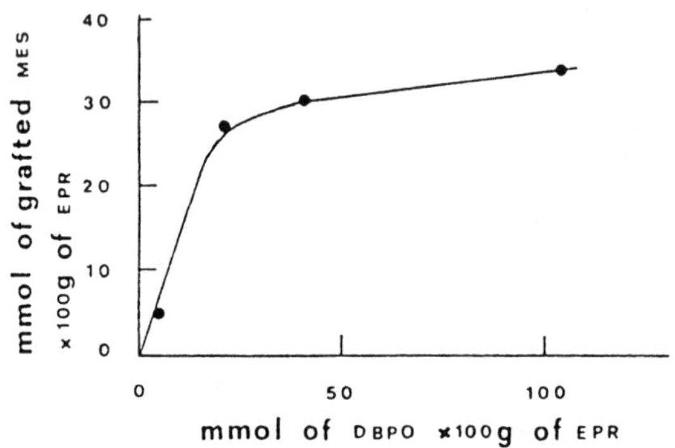

FIG. 4 Influence of the DBPO concentration on the grafting degree. Reaction conditions: $T = 85°C$; $t = 2$ h; solvent = 100 mL; MEM = 8.0 g: EPR = 5.0 g.

curve shows a marked increase in the functionalization degree at increasing DBPO concentrations, up to 30 mmol per 100 g EPR, while it reaches a plateau at higher concentrations. This trend is common to other unsaturated molecules and can be ascribed to three main effects: (a) partial saturation of reactive sites onto EPR; (b) decreasing efficiency of the initiator, due to recombination reactions between primary radicals, and (c) increasing probability of secondary reactions of macroradicals.

(b). Thermal Cyclization of EPR–g-MES to EPR–g-SA. The cyclization of monoethylsuccinate groups grafted onto EPR is an equilibrium reaction [59]:

$$-HC(COOH)-H_2C-C(=O)OCH_2CH_3 \quad \overset{\Delta}{\rightleftharpoons} \quad -HC-C(=O)-O-C(=O)-CH_2 \text{ (cyclic)} \; + \; CH_3CH_2OH$$

Preliminary investigations on an EPR–g-MES at high grafting degree (> 5%) showed that the cyclization occurs at $T > 130°C$ while for $T > 170°C$ the degradation of EPR begins to be evident. Studies on the cyclization in bulk have been performed on films obtained by casting from a xylene solution directly onto a KBr disk at room temperature. The coated KBr disk was put inside the temperature cell which was already at the test temperature, and spectra have been collected at selected times. While collecting spectra, the cell was carefully flushed with a stream of N_2.

Figure 5 shows the spectra of a film of EPR–g-MES, in the spectral range of interest, for a thermal treatment at 150°C. The anhydride band, already present in the starting EPR–g-MES, reaches the same intensity of the ester acid in less than 30 min, and after 3 h the ratio of anhydride/ester inverts completely.

Figure 6 shows the spectra relative to studies performed in a solution of *o*-dichlorobenzene. The behavior in solution at 160°C is quite different; in fact, after many hours the conversion has not yet reached 50%, and only after several days is the anhydride band more intense than the ester band. From a closer inspection of the spectra, it is seen that the principal peak of the anhydride group gradually shifts from 1775 cm^{-1} (in the EPR–g-MES before heating) to 1783 cm^{-1} (after heating). This last frequency is more characteristic of free SA groups [61]. The initial shift towards lower frequencies can be attributed to polar interactions (hydrogen bonding) between the carbonyl moiety of SA and the acidic part of MES:

$$-HC-C(=O\cdots H-O-C(=O)-CH-)-O-C(=O)-CH_2 \text{ (cyclic)} \qquad H_3CH_2CO-C(=O)-CH_2-$$

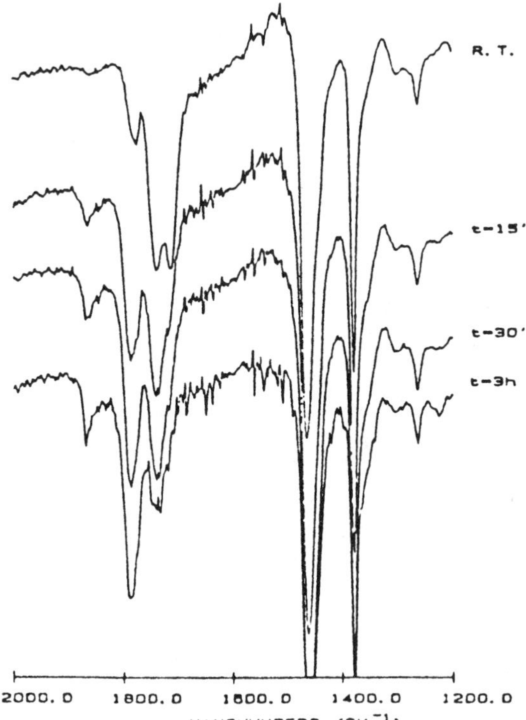

FIG. 5 FTIR spectra of EPR–g-MES at different times during the heating process in bulk at $T = 150°C$ (the poor resolution of the baseline is due to the lower resolution obtainable in the presence of the temperature cell accessory). The first spectrum is referred to the starting EPR–g-MES at room temperature.

Increasing the concentration of SA groups and decreasing that of MES the possibility for such interactions to occur decreases, so the anhydride band shifts to the frequency characteristic of a free five-membered cyclic anhydride.

Anhydride groups could in principle also be formed between a carboxylic and an ester group of two different molecules of MES:

$$\text{—HC(COOH)—H}_2\text{C—COOEt} \; + \; \text{EtOOC—CH—CH}_2\text{—COOH} \longrightarrow$$

$$\text{—HC(H}_2\text{C—COOEt)—C(=O)—O—C(=O)—CH(CH}_2\text{—COOH)—} \; + \; \text{EtOH}$$

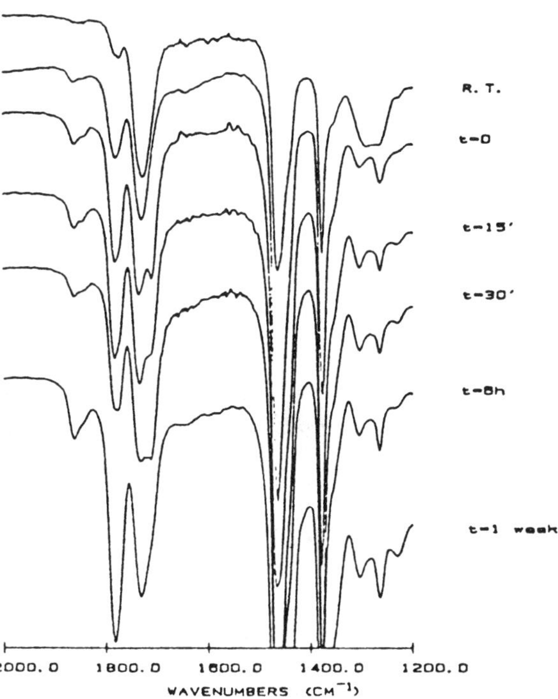

FIG. 6 FTIR spectra of EPR–g-MES at different times during the heating process in solution at $T = 160°C$. The first spectrum is referred to the starting EPR–g-MES at room temperature.

The formation of anhydride linkages between groups grafted onto different EPR chains should lead to cross-linked EPR. To investigate the existence of such cross-links, studies of solubility in xylene of EPR–g-MES after cyclization have been performed.

A sample which had been heated in bulk at 150°C for 3 h (see Fig. 5) was treated with boiling xylene for a long period of time. A gel fraction was actually found whose IR spectrum, obtained on a film compression-molded at 150°C, shows a high content of anhydride. In contrast, the soluble fraction, recovered from xylene, shows an IR spectrum similar to that of the starting material, i.e., before any heat treatment, with a very small anhydride band. We can conclude that the bulk cyclization reaction proceeds via an inter-chain anhydride formation, which is a bimolecular process. A similar treatment with boiling xylene was made on a sample of EPR–g-MES after solution cyclization. Also in this case the sample shows a small gel fraction which presents at IR analysis no differences from the soluble fraction, thus indicating that both mechanisms of inter- and intra-chain anhydride formation can occur in this case, even if the monomolecular mechanism is preferred, due to the dilution of the system.

B. Innovative Elastomeric Networks Based on Functionalized Ethylene–Propylene Rubbers and Hydroxyl Terminated Polybutadiene

Elastomeric networks with thermoreversible linkages are receiving growing interest as it is possible to reprocess them several times by simple reheating. In this field, thermoplastic elastomers (ABS) [63] and ionomers (Surlyn) [64] have been widely developed and applied. Several attempts have been reported to apply the same principles to statistical copolymers of ethylene with maleic anhydride or acrylic and methacrylic acids [65], crosslinked with diols and/or ethoxylamines [66]. The idea which sustained these papers was to take advantage of the thermoreversibility of ester linkage.

New elastomeric networks based on saturated ethylene–propylene rubbers grafted with succinic anhydride groups (EPR–g-SA) crosslinked with a hydroxyl-terminated polybutadiene (HTPB) are hereafter described. Infrared techniques are employed to follow the kinetics of the monoesterification reaction and to assess its potential thermoreversibility, either on the macromolecular system (EPR–g-SA + HTPB), or on a model system, formed by EPR–g-SA and a low molecular weight diol, namely 1,9-nonandiol.

1. Experimental

(a) Materials. The starting EPR was the random copolymer Dutral CO054, previously described. HTPB (hydroxyl terminated polybutadiene), Polysciences, $M_W = 2800$, was dried by vacuum stripping at 50°C and stored under N_2

The hydroxyl number of HTPB was determined by using a standard procedure previously reported [67]. The resulting hydroxyl number, 0.62 meq/g, corresponding to a functionality of HTPB molecules of 1.72, is slightly lower than that reported by the producers (0.8 meq/g).

1,9-Nonandiol and triethylamine (TEA), Fluka reagent grades, were used without further purification.

All the solvents were of analytical grade and were dried according to standard procedures.

An EPR modified by insertion of 1.3 wt% of succinic anhydride groups (13 mmol/100 g of EPR) (EPR–g-SA), was prepared according to a previously reported procedure [68].

(b) Typical Esterification Reaction. In a flask equipped with a nitrogen inlet, 1.00 g of EPR–g-SA (13 mmol of grafted anhydride/100 g of EPR) was dissolved in 50 mL of a suitable solvent (toluene, chlorobenzene). When the dissolution was complete, a solution of the same solvent containing stoichiometric amounts of hydroxyl groups and TEA as catalyst was added under vigorous stirring.

Cast films for IR analysis were obtained by evaporating the solution under vacuum directly onto KBr disks at room temperature.

2. Results and Discussion

Two separate series of experiments were carried out in order to study the reactivity of EPR–g-SA towards low M_W diols with or without the addition of a tertiary amine as catalyst, and the reactivity of EPR–g-SA towards a long-chain hydroxyl-terminated polybutadiene (HTPB). For both of them, the same experimental procedure was followed: to ensure an intimate mixing of the reactants, they were all dissolved in a common solvent at room temperature and in a certain stoichiometric amount (in solution, the kinetics of esterification is very slow, thus the degree of reaction is negligible). Subsequently, the solvent is quickly removed by evaporation under vacuum at room temperature directly onto KBr disks to obtain a film which is used for IR analysis.

This procedure allows us to follow the reaction in bulk, i.e., in the best kinetics condition, and at the same time to overcome all problems arising from poor miscibility of the reactants [66, 69].

The degree of esterification has been followed by FTIR techniques, measuring the variation of the $1785\,cm^{-1}$ band attributable to C=O symmetric stretching of anhydride grafted onto EPR. The carbonyl stretching of the semiester ($1730\,cm^{-1}$) cannot be followed in a quantitative approach, as it is strongly influenced by the presence of the carbonyl stretching of the acidic moiety [70].

It must be pointed out that a straightforward quantitative comparison among different kinetics cannot be done only on the basis of IR spectra, as different thicknesses of the films cannot be avoided and any attempt to normalize different kinetics rating the $1785\,cm^{-1}$ band to a "regularity" band gives quite irregular numbers.

(a) Esterification of EPR–g-SA by 1, 9-Nonandiol. It is well known in the literature [59] that primary alcohols with a hydrocarbon chain longer than C_5 have a lower tendency to react with cyclic anhydride functions than alcohols with short aliphatic chains. The esterification is an exothermic reaction which follows a second-order kinetic.

At room temperature, the equilibrium conditions are attained only after a long time and at a conversion of around 50% of the starting anhydride. It is possible to accelerate the attainment of equilibrium by the addition of a suitable catalyst, such as tertiary amines. The effect of such amines is to coordinate itself to the alcoholic function, as shown by:

$$\text{R}'\text{—HC—C(=O)—O—C(=O)—CH}_2\ (\text{cyclic anhydride}) + \overset{\delta+}{\text{NR}''_3}\cdots\text{H—}\underset{\delta-}{\text{O—R}} \longrightarrow$$

$$\text{R}'\text{—HC—COOR},\quad \text{H}_2\text{C—C(=O)—}\overset{\delta-}{\text{OH}}\cdots\cdots\overset{\delta+}{\text{NR}''_3}$$

where R, R′, and R″ are normally alkyl groups. It is conceivable to assume that such a reaction pattern can occur also when R′ is a POic substrate, as in the case of EPR–g-SA.

Several experiments have been carried out by varying the SA/TEA molar ratio and measuring the $1785\,\text{cm}^{-1}$ absorbance of the anhydride group, while keeping constant the 1/1 ratio of SA/OH. The collected spectra at selected times are reported in Figs. 7(a), 8(a), and 9(a) for three values of the SA/TEA ratio, and the data are summarized in the plots of Figs. 7(b), 8(b), and 9(b).

It clearly emerges that the addition of equimolar amounts of TEA to the SA groups causes a more significant increase in the degree of reaction. Such a result is consistent with the proposed model, in which the amino catalyst remains coordinated to

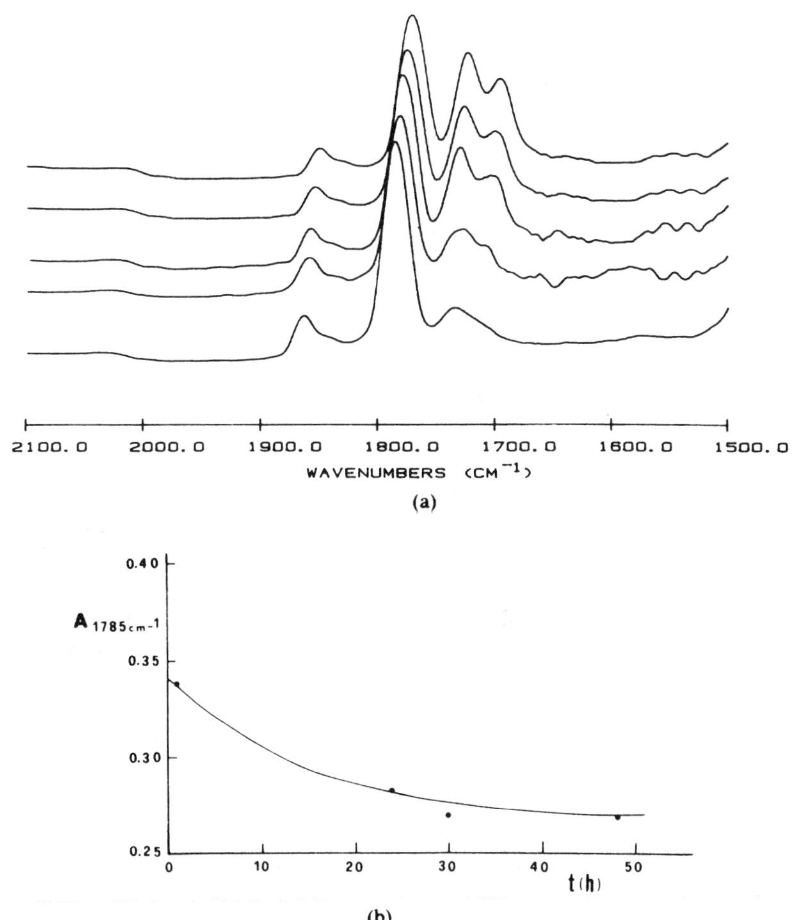

FIG. 7 (a) FTIR spectra of EPR–g-SA/nonandiol at different reaction times. Reaction conditions: molar ratio SA/OH 1/1, molar ratio SA/TEA 1/1; $T = 25°\text{C}$. (b) Absorbance of the $1785\,\text{cm}^{-1}$ band of anhydride versus reaction times. Reaction conditions: molar ratio SA/OH 1/1, molar ratio SA/TEA 1/1; $T = 25°\text{C}$.

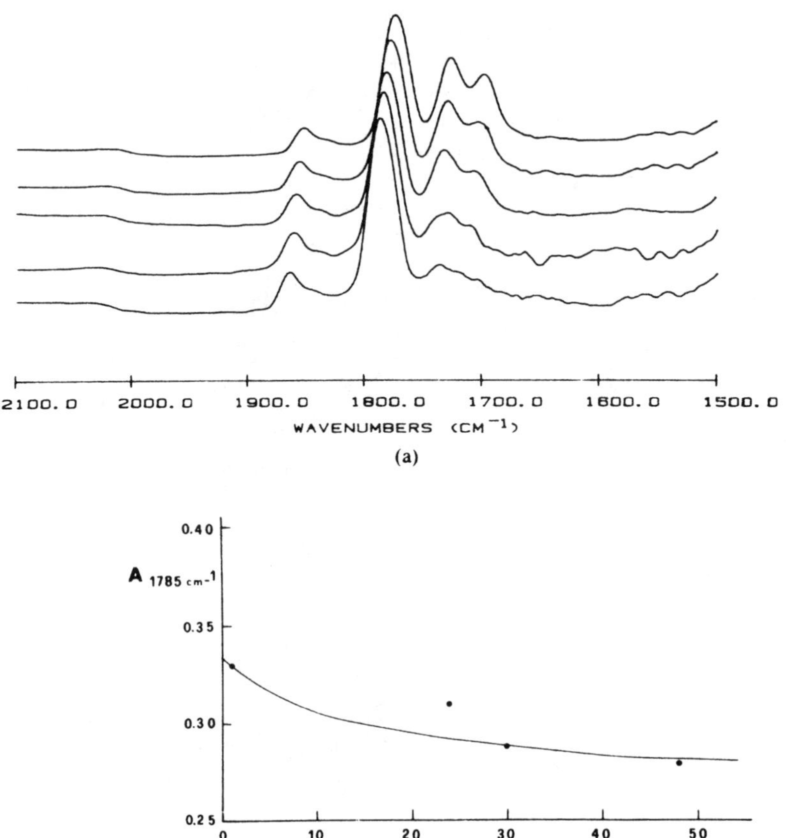

FIG. 8 (a) FTIR spectra of EPR–g-SA/nonandiol at different reaction times. Reaction conditions: molar ratio SA/OH 1/1, molar ratio SA/TEA 2.5/1; $T = 25°C$. (b) Absorbance of the $1785\,cm^{-1}$ band of anhydride versus reaction times. Reaction conditions: molar ratio SA/OH 1/1, molar ratio SA/TEA 2.5/1; $T = 25°C$.

the acidic function. The effect is quite relevant in our case, as the POic backbone creates a medium of low dielectric constant, in which weak dipolar interactions are stabilized.

In principle, a parameter which should affect the equilibrium conversion is an excess of one of the reactive species. We tried to gather evidence of it by adding up to twice the molar amount of diol necessary to react with the anhydride function (i.e., two —OH equivalents per SA equivalent) in the absence of TEA.

It was found that the attainment of equilibrium without catalyst is very slow. Thus it may be concluded that the excess of diol by itself scarcely influences the rate of reaction, leaving moreover the presence of unreacted diol inside the polymeric film, even after a long reaction time. The addition of an excess of diol can be considered only in conjunction with the addition of tertiary amine.

Then, in order to check the influences of both catalyst and excess diol on the kinetics of reaction, in view of the subsequent reaction between EPR–g-SA and HTPB, we performed a final experiment in which the reaction conditions chosen were: ratio SA/TEA 2.5/1 and ratio SA/OH 1/1.5, collecting spectra up to 4 days. The results are shown in Fig. 4. From a comparison with the results obtained under the conditions of Fig. 8(a), it emerges that an excess of diol, in the presence of TEA, positively influences both the kinetic and the equilibrium conversion.

(b) Reaction of EPR–g-SA with HTPB. In the framework of a study on the synthesis of elastomeric interpenetrated networks, the reaction between —OH terminated polybutadienes and EPR–g-SA is of peculiar interest as the monoesterification reaction is said to be thermoreversible [68].

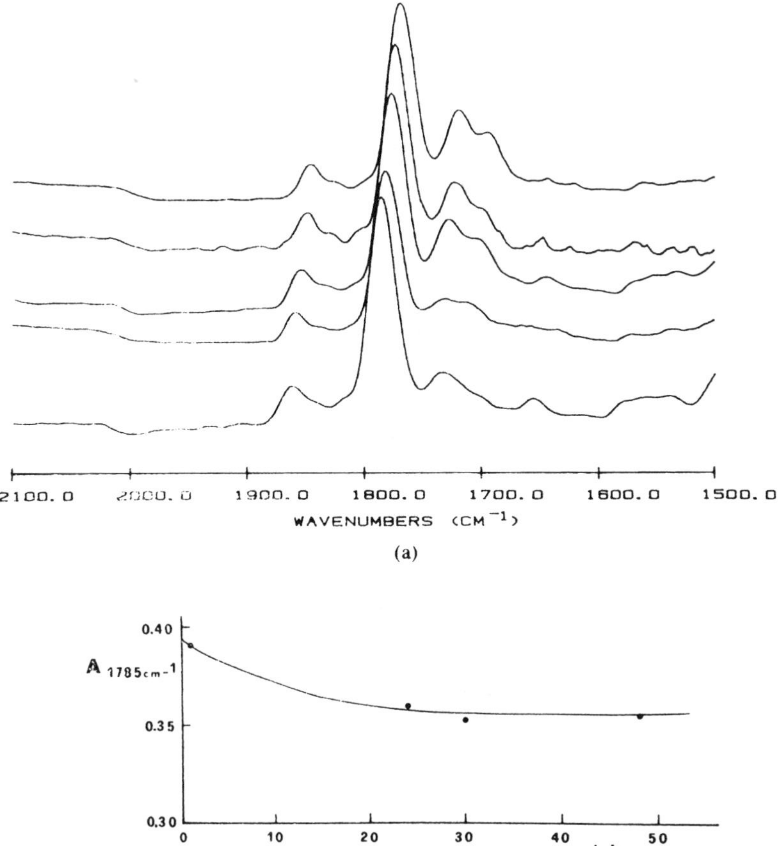

FIG. 9 (a) FTIR spectra of EPR–g-SA/nonandiol at different reaction times. Reaction conditions: molar ratio SA/OH 1/1, molar ratio SA/TEA 10/1; $T = 25^\circ C$. (b) Absorbance of the $1785\,cm^{-1}$ band of anhydride versus reaction times. Reaction conditions: molar ratio SA/OH 1/1, molar ratio SA/TEA 10/1; $T = 25^\circ C$.

Several authors have carried out experiments by using short or long chain diols or alkylamines polyethoxylate [66, 70] but there is some disagreement on the thermoreversibility of the monoesterification reaction carried out on a polymeric substrate. Decroix *et al.* reported that at temperatures higher than 220°C the reaction does not occur at all, independently from the length of diol chain, while at lower temperatures a maximum in the rate of reaction occurs for a certain short length of the chain.

In contrast, Lambla *et al.* [69] reported that, in their system, it is possible to obtain monoesterification at temperatures as high as 230°C; they conclude that the reversible reaction is not predominant even under those conditions.

Initially, we have carried out a kinetics experiment by using HTPB instead of nonandiol under the same conditions as in Fig. 10. The results are reported in Fig. 11, where it is evident that the kinetic behavior is analogous to that previously shown for the short chain diol.

The thermoreversibility of the monoesterification reaction between nonandiol or HTPB and EPR–g-SA has been studied by using a calorimetric cell attached to a FTIR spectrophotometer. The stoichiometric reaction conditions chosen are: ratio SA/OH, 1/1.5. The procedure was as follows: the sample, after casting from solution onto KBr disks at room temperature under vacuum, was rapidly brought to 220°C and left at this temperature for 30 min; such a procedure was chosen because at $T > 200^\circ C$ the esterification reaction does not occur at all [66]; we can consider this situation as the zero reaction time (actually, as shown by the first spectrum of Fig. 12, a certain amount of ester is already present at $T = 220^\circ C$, probably due to the time necessary to cast the film and to bring it to 220°C).

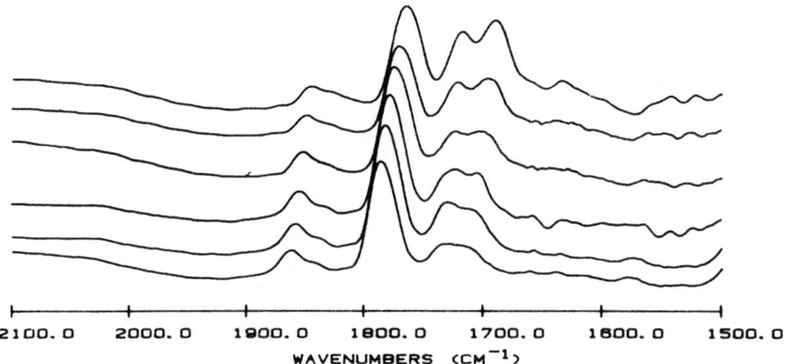

FIG. 10 FTIR spectra of EPR–g-SA/nonandiol at different reaction times. Reaction conditions: molar ratio SA/01—11/1.5, molar ratio SA/TEA 2.5/1; $T = 25°C$.

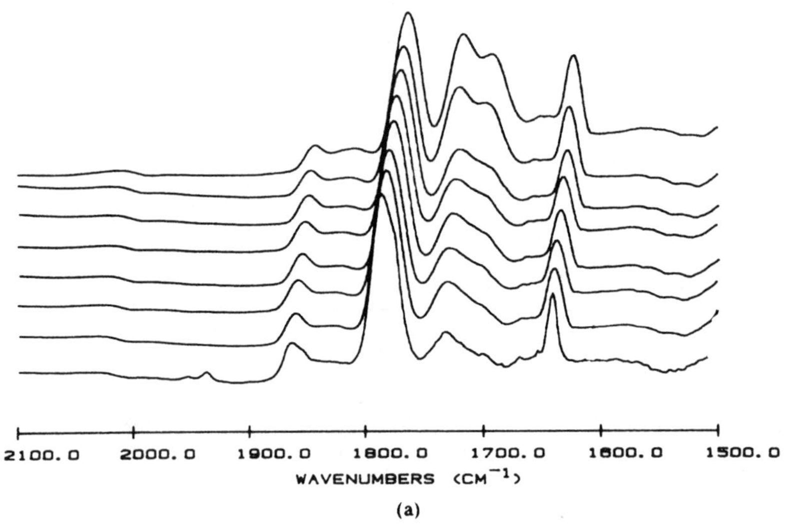

FIG. 11 FTIR spectra of EPR–g-SA/HTPB at different reaction times. Reaction conditions: molar ratio SA/OH 1/1.5, molar ratio SA/TEA 2.5/1; $T = 25°C$.

Successively, the sample was cooled down to 40°C while collecting spectra at $T = 200$, 180, 160, 140, 120, 100, 80, 60, and 40°C (the overall cooling time was ca. 2 h). The trends, reported in Figs. 12 and 13, suggest that the reaction involving nonandiol is much less influenced by thermal treatments than HTPB. In the latter case, there is a clear tendency in favor of esterification as the temperature decreases.

Occasionally, the sample was reheated up to 220°C and maintained at this temperature for 2 h to gather evidence of thermoreversibility; but it did not occur, and we just recorded that the monoesterification reaction did not proceed any longer.

From the above, it is possible to conclude that: (1) di-α,ω-hydroxylated polymers do not react with anhydride functions at high temperatures; (2) the reaction, due to its exothermicity, can be accelerated by cooling the system down to room temperature; (3) once the reaction has occurred, it is substantially irreversible; and (4) shorter chain diols are less influenced by temperature treatment than polymeric diols.

The thermal irreversibility of the reaction between HTPB and EPR–g-SA could be explained by the following: in the case of monoethyl- or monomethyl-succinic derivatives of EPR–g-SA, an increase in temperature favors the cyclization reaction not only from a thermodynamic point of view, but also because

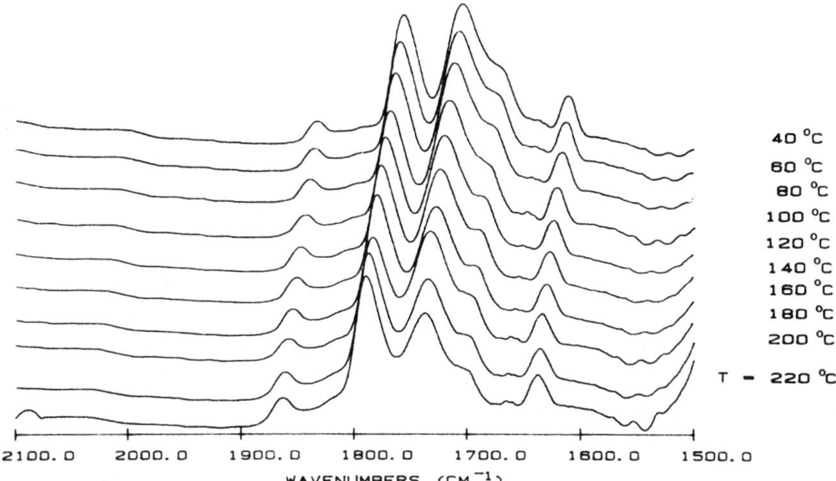

FIG. 12 FTIR spectra of EPR–g-SA/HTPB at different temperatures. Reaction condition: molar ratio SA/OH 1/1.5.

the volatile alcohol which is formed (i.e., methanol or ethanol) is removed from the reactive system. In contrast, by raising the temperature, it is not possible to remove the polymeric diol from the reactive sites, due to the nonvolatile nature of the diol and to the low dielectric constant medium, which, as previously said, stabilizes polar interactions and hydrogen bonding between anhydride and —OH groups.

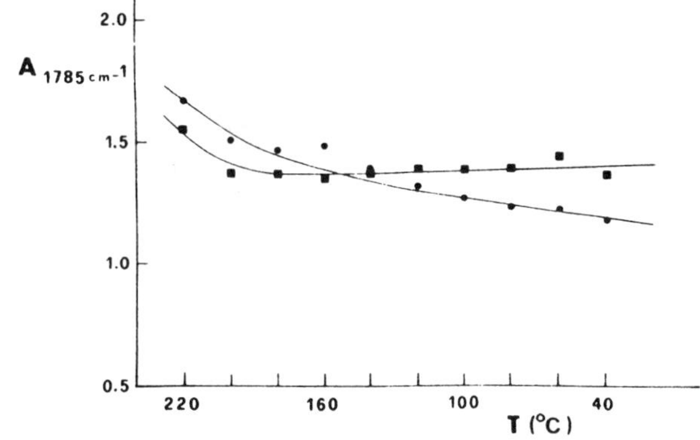

FIG. 13 Absorbance of the $1785\,cm^{-1}$ band of anhydride versus temperatures during the slow cooling: (■) EPR–g-SA/nonandiol, (reaction condition: molar ratio SA/OH 1/1.5). (●) EPR–g-SA/HTPB (reaction condition: molar ratio SA/OH 1/1.5).

III. RUBBER MODIFICATION OF POLYBUTYLENETEREPHTHALATE BY REACTIVE BLENDING CONCURRENTLY WITH POLYMERIZATION REACTION

Polybutyleneterephthalate (PBT) is of growing interest as a material for injection molding. In fact, its rate of crystallization is faster than that of the other widely used linear polyesters such as polyethyleneterephthalate (PET). However, PBT shows a low impact resistance, particularly at low temperatures. For this reason the uses of PBT are limited. The usual method to overcome this limitation is to add a second elastomeric phase to the PBT matrix. Rubber modification of PBT has been realized by melt blending with preformed rubbers such as poly(ethylene–co-vinylacetate) (EVA) and poly(ethylene co-vinylalcohol) (EVOH) [71], or by adding end-capped polymers to produce a second flexible component during PBT polymerization.

A different method is based on the possibility of having block copolymers as a dispersed phase between PBT and elastomeric polyethers. The blends and the copolymers are obtained directly during the polymerization of PBT by dispersing the polyether, end-capped with reactive groups, in the starting monomeric reactive mixture [72, 73].

Our studies on rubber-modified nylons by using EPR–g-SA rubbers gave blends with very high impact properties [56]. Blends were prepared concurrently with the polycondensation reaction of ε-caprolactam [74–76].

The excellent results obtained lead us to follow the same approach to improve the impact resistance of PBT.

In the present section a method of preparing blends with good interfacial adhesion between PBT and EPR opportunely functionalized with alcoholic or ester groups is reported.

A. Experimental

1. Materials

EPR (Dutral COO54) modified by the insertion of succinic anhydride (EPR–g-SA) and dibutylsuccinate (EPR–g-DBS) were prepared, at different grafting degrees, by using a solution method of grafting previously described [61]. Dimethylterephthalate (DMT), Fluka reagent grade, was used without further purification. 1,4-butandiol (BDO), Fluka reagent grade, was purified by vacuum distillation. Titanium tetraisopropylate ($Ti(iP)_4$), an ICN Pharmaceutical product, was purified by distillation at a pressure of 1 mmHg and a temperature of 120°C. All the solvents used were of analytical grade and were purified according to standard procedures.

2. Standard Preparation of EPR–g-Ethanolsuccinimide (EPR–g-ESI)

10 g of EPR–g-SA 1.5% by weight of anhydride (15 mmol SA/100 g of EPR) were dissolved in 200 mL of xylene. 0.37 mL of ethanol–amine (EA), corresponding to an excess of 4/1 moles EA/moles SA, were added to the solution, together with 0.43 mL of triethylamine (TEA) as catalyst, at a temperature of 120°C. After 1 h, the reaction mixture was cooled to room temperature and precipitated in acetone, repeatedly washed, and dried in a vacuum oven at a temperature of 60°C overnight.

3. Typical Procedure for a Binary 90/10 Blend Preparation

A mixture of 89.2 g (0.46 mol) of DMT, 82.8 g (0.92 mol) of BDO, and 5.7 mL of a 0.5% by volume solution of $Ti(iP)_4$ in $CHCl_3$ (corresponding to a molar ratio ($Ti(iP)_4$/DMT of $1 \times 10^{-4}/1$) was charged into a cylindrical glass vial, equipped with a mechanical stirrer and a distillation apparatus. The mixture was heated, using an oil bath, up to 200°C and left to react (transesterification step) for 4 h under N_2 while distilling methanol as a by-product. The temperature was then raised to 255°C within 30 min and the pressure was gradually reduced to 0.5 mmHg. The polycondensation step was continued for 30 min. At this point, 11.2 g of rubber were added to the polymerizing mixture and the polymerization was further continued for 90 min. After cooling, the white reaction product was recovered and finely granulated in a laboratory mill.

TABLE 1 Codes and compositions (wt/wt) of the prepared PBT/functionalized rubber blends

Code of blends	Rubber type	Grafting degree (%wt)	% by weight of rubber
PBTEPR (90/10)	EPR	—	10
PBTDBS2 (90/10)	EPR–g-DBS	2	10
PBTDBS9 (90/10)	EPR–g-DBS	9	10
PBTESI0.5 (90/10)	EPR–g-ESI	0.5	10
PBTEPR (80/20)	EPR	—	20
PBTDBS2 (80/20)	EPR–g-DBS	2	20
PBTESI0.5 (80/20)	EPR–g-ESI	0.5	20

Codes and contents of rubber of the prepared blends are reported in Table 1.

4. Techniques of Characterization

FTIR, viscosity and DSC apparatuses are described in the previous sections.

The morphological characterization of the prepared blends was carried out using a scanning electron microscope (SEM), Philips Model 501. Micrographs of surfaces fractured in liquid N_2 and of microtome-faced surfaces of blends after exposure for 30 min to boiling xylene or *o*-dichlorobenzene vapors were obtained. Before the examination, samples were coated with gold–palladium.

Charpy impact tests were performed by using a Ceast fracture pendulum at different temperatures (from −50°C up to room temperature).

B. Results and Discussion

1. Polycondensation of PBT

PBT has been synthesized according to well known standard procedures [77, 78]. The resulting polymer is characterized by a crystallinity degree of about 40%, with a melting point of 225°C and a T_g of 51°C. The number average molecular weight, as obtained by viscosimetric analysis using a semiempirical relation between the intrinsic viscosity and M_n [79], was 24,000.

2. Preparation of Blend Concurrent with Polymerization of PBT

The addition of a rubbery phase during the polymerization of a monomer of a second polymer to obtain a rubber-modified thermoplastic material has two main advantages towards the melt-mixing process of high

M_W preformed polymers. Firstly, the synthesis of the polymer and rubber modification are done in one step, saving time and reducing the machining of materials, which always produces some degradation. Secondly, the lower viscosity of the polymerization medium, at least when the rubber is added, allows a more favorable mixing with respect to the dispersion of the rubbery phase and makes it possible to obtain a better dispersion of one phase into the other.

Following this procedure, the preparation of blends PBT/functionalized EPR have been carried out by adding the rubbery component during the high temperature polymerization of butandiol and dimethylterephthalate. The functional groups grafted onto EPR must be able to participate in the transesterification equilibrium occurring during the polycondensation. In this respect, the ester functionality of EPR–g-DBS or the alcoholic functionality of EPR–g-ESI are both interesting because they take part in the polycondensation equilibrium with the following reactions:

$$\text{(succinimide)N—}(CH_2)_2OH + HO\!\left[(CH_2)_4—OC(=O)—C_6H_4—C(=O)O\right]_n\!(CH_2)_4—OH \longrightarrow$$

$$\text{(succinimide)N—}(CH_2)_2\!\left[OC(=O)—C_6H_4—C(=O)O\right]_x\!(CH_2)_4—OH + HO—(CH_2)_4—OH$$

EPR-g-DBS

$$\text{(succinate)}(C(=O)OC_4H_9)_2 + HO\!\left[(CH_2)_4—OC(=O)—C_6H_4—C(=O)O\right]_n\!(CH_2)_4—OH \longrightarrow$$

$$\text{(succinate)}C(=O)—O\!\left[(CH_2)_4—OC(=O)—C_6H_4—C(=O)O\right]_x\!(CH_2)_4—OH,\ C(=O)OC_4H_9 + C_4H_9OH$$

The problem which must be faced when preparing similar blends is when, in the polymerization, the second component must be added in order to obtain an optimal state and mode of dispersion of rubber. From our previous studies on the polyamide functionalized rubber blends prepared directly during the caprolactam polymerization, we know that there are four critical factors which influence the attainable degree of dispersion of a functionalized rubber inside a polymerizing matrix: (1) the reactivity of grafted groups; (2) the grafting degree; (3) the length of the polymerizing chains at the moment the rubber is added; and (4) the viscosity of the reaction medium at that time.

The first critical factor will be discussed later. With regard to the third and fourth critical factors, it must be considered that the grafting of a growing PBT chain onto a molecule of functionalized EPR is a heterogenous process which occurs at the interface between the two immiscible components. A PBT molecule, upon grafting, is at least partially segregated from the solution, so it will have less chance to grow. Furthermore, once a growing PBT chain has reacted with the ester or alcoholic groups of the functionalized EPR, it has only one residual alcoholic end group, so it is also intrinsically less able to polymerize via the transesterification reaction. In this respect, the degree of polymerization at the moment in which the rubber is added is very important for the emulsifying efficiency: chains which are too short will be segregated inside the rubber and fail to act as emulsifiers. On the other hand, if there is too high a degree of polymerization, the viscosity of the medium will increase very sharply, and the physical dispersion of the rubber inside the matrix will be very difficult to achieve. A compromise between these two opposing factors led us to choose 30 min for the polycondensation step as a suitable time for the addition of the rubber.

As a matter of fact, kinetic experiments have demonstrated that after 30 min the polymer has reached an M_n of 10,000 while developing a good degree of crystallinity. Moreover, the viscosity of the molten polymer seems to be relatively low.

Some general observations concerning the preparation of some of the blends are discussed below.

(a) Blends with EPR–g-ESI. It was observed that when an EPR–g-ESI having a high grafting degree (6.8% by weight) is used, the resulting PBT/EPR–g-ESI blends show a macroscopic phase separation of the rubber. On the other hand, blends containing an EPR–g-ESI at 0.5% grafting level are easily stirrable and the molecular weight of PBT at the end of the reaction is rather large (21,000). Moreover, the final material is characterized by a very fine distribution of the rubber component. The macroscopic phase separation between rubber and PBT which occurs when an EPR–g-ESI at high grating degree is used can be interpreted on the basis of our previous experience on rubber-toughened polyamide 6 [75]. The alcoholic derivative of succinic anhydride, ESI, causes strong interactions between different grafted EPR chains, due to the occurrence of extended hydrogen bonding. When the outer shell of a rubbery particle has reacted with growing PBT chains, the weak shear forces of our single-blade stirrer are not sufficient to disrupt the entanglements of the rubbery domains, leading to a blend macroscopically phase separated.

(b) Blends with EPR–g-DBS. A different behaviour is shown by EPR–g-DBS. In fact, in this case we obtained good dispersion of the rubber even for a grafting degree of 9%. This difference is probably due to the lack of either hydrogen or polar interactions between the DBS groups and to the extensive degradation of the starting EPR backbone, which occurs during the grafting of DBS molecules [61].

(c) Blend Analysis. Selective extraction of blends with 10% rubber content by trifluoroacetic acid have been made in order to separate and characterize their components. By treatment with this solvent we obtained three phases in all cases: on the bottom, a clear solution containing as solute pure PBT which is completely soluble in the acid; an opaque solution in the middle, mainly containing the copolymer functionalized EPR–g-PBT; and at the top a supernatant condensed rubbery phase.

The three phases have been separately analyzed by FTIR and DSC techniques. Furthermore, the pure PBT, recovered by vacuum stripping of the solvent from the clear solution, was characterized by viscosimetric analysis to determine its molecular weight.

The DSC thermograms of extracted PBTs irrespective of blend composition, all show the same values of T_g and T_m ($T_g = 51°C$, $T_m = 226°C$; see Fig. 14 as an example). In Fig. 15, FTIR spectra related to some of the extracted PBT are reported. Their appearances are identical to commercial grade PBT. However, in the case of PBTDBS9 blend, the existence of a broad band at 3300 cm^{-1}, which has been attributed to a high concentration of —OH terminal groups, confirms the low molecular weight of the extracted PBT.

The copolymeric phases of PBTESI0.5, PBTDBS2, and PBTDBS9 blends have been recovered by centrifugation of the milky solutions. DSC traces relative to these copolymers are reported in Fig. 16. The copolymers extracted from the blends show an endothermic peak at 226°C, close to the melting point of plain PBT. In Fig. 17, FTIR spectra of the copolymers are

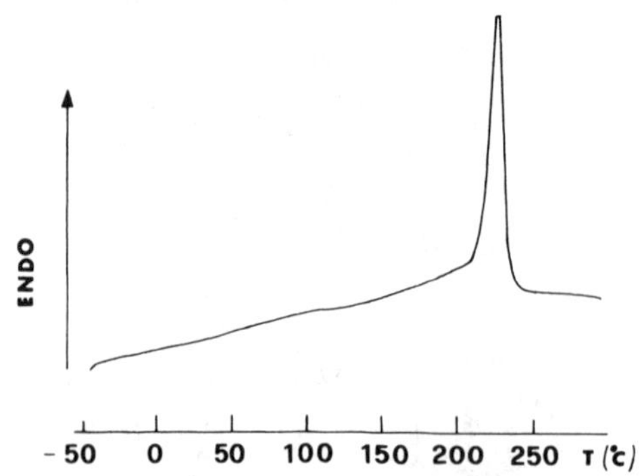

FIG. 14 DSC thermogram of PBT extracted from PBTDBS9 90/10; heating rate = 20°C/min.

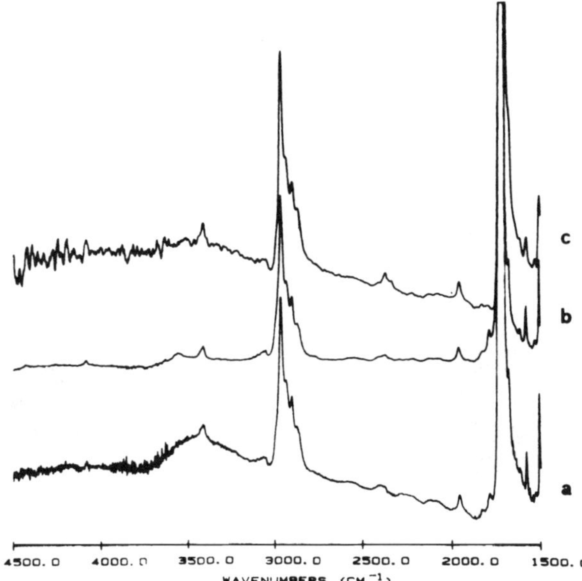

FIG. 15 FTIR spectra of PBTs extracted from (a) PBTDBS9; (b) PBTESI0.5; (c) PBTEPR.

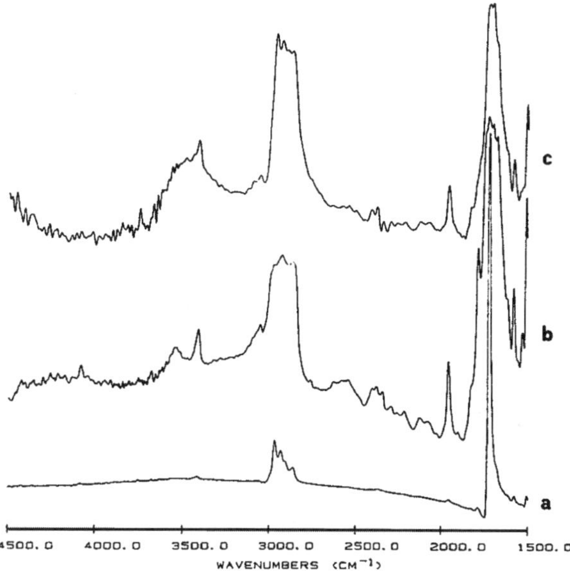

FIG. 17 FTIR spectra of copolymers extracted from (a) PBTESI0.5; (b) PBTDBS2; (c) PBTDBS9.

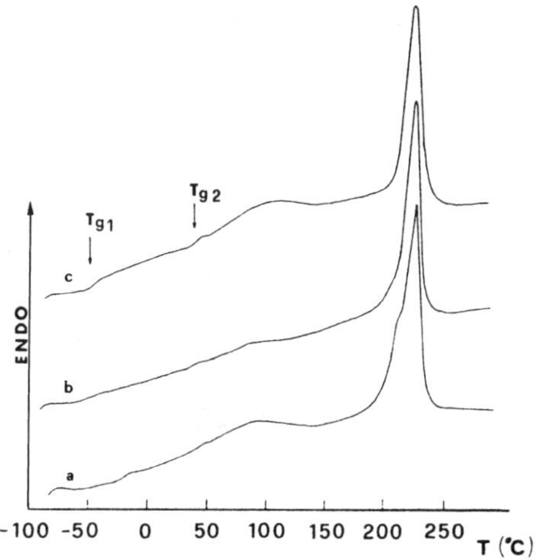

FIG. 16 DSC thermograms of copolymers extracted from (a) PBTESI0.5 (b) PBTDBS2 (c) PBTDBS9, heating rate = 20°C/min.

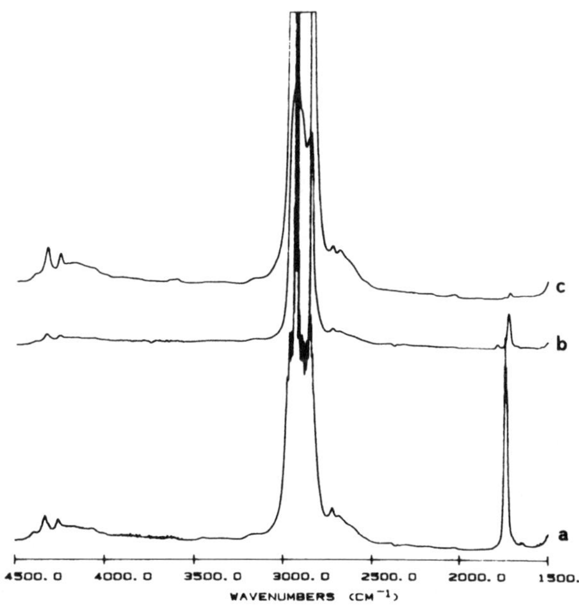

FIG. 18 FTIR spectra of rubbers extracted from (a) PBTDBS9; (b) PBTESI0.5; (c) PBTTEPR.

reported. The spectrum (a) refers to the copolymer (EPR–g-ESI)-g-PBT.

The presence of the rubbery phase is confirmed by the small band at $1784\,cm^{-1}$, corresponding to the secondary peak of the imide group grafted onto the EPR.

The FTIR spectra of the insoluble, in trifluoroacetic acid, supernatant rubbery phases of PBTDBS9, PBTESI0.5, and PBTEPR blends are shown in Fig. 18. The DSC thermogram of the rubbery phase from the PBTESI0.5 blend is shown in Fig. 19, as an example. The DSC shows the T_g typical of EPR's at about −45°C and the small peak of melting at about 45°C, attributed to short regular sequences of ethylene

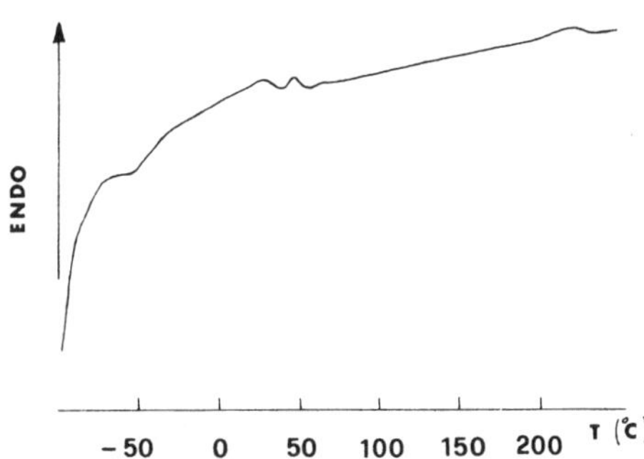

FIG. 19 DSC thermograms of EPR–g-ESI extracted from PBTESI0.5; heating rate = 20°C/min.

in the EPR. No evidence of the PBT phase is observed. It can be seen that both DSC and FTIR analysis confirm that this rubber has the same characteristics of the starting functionalized EPR. Thus it may be concluded that the supernatant phase obtained after extraction of the blends with trifluoroacetic acid is mainly constituted of unreacted rubber.

3. Phase Structure Studies

Scanning electron microscopy has been utilized to investigate the morphology of fracture surfaces and smoothed surfaces after etching of samples of PBT and blends. Samples as obtained from polymerization have been first compression-moulded at 260°C and then fractured in liquid N_2. Smoothed surfaces have been obtained by using a microtome and analyzed after exposure to xylene or *o*-dichlorobenzene (blend PBTESI0.5) vapor.

Micrographs of fracture surfaces of 90/10 and 80/20 blends are shown in Figs. 21 and 22. The micrograph of the fracture surface of samples of plain PBT, for comparison, is shown in Fig. 20.

The PBTEPR 90/10 blend shows a "cheese-like" morphology. The presence of large spherical particles (about 5–10 μm) uniformly distributed in the PBT matrix without evidence of interfacial adhesion is observed (see Fig. 21a). From the SEM examination of the fracture surfaces of PBTDBS2, PBTDBS9, and PBTESI0.5 (90/10) blends it emerges that in all cases a dispersed phase, almost spherical in shape, with dimensions lower than that of the PBTEPR blend is present. Moreover, the average dimensions of such domains, as well as the adhesion to the PBT matrix, are strongly dependent upon the graft degree of the functionalized rubber and upon the chemical nature of the functional group.

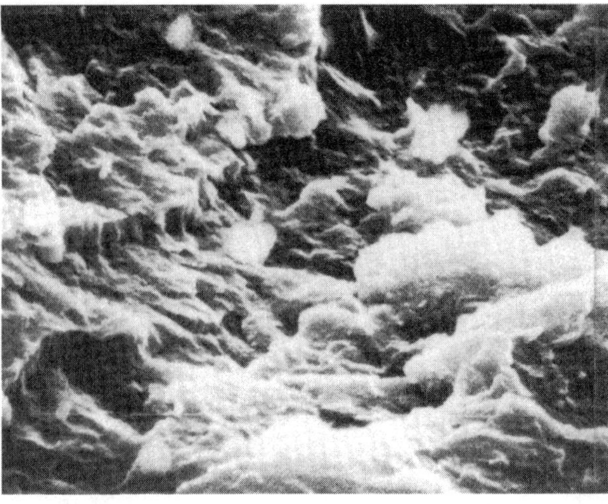

FIG. 20 SEM micrograph of fracture surface of PBT, 2500×.

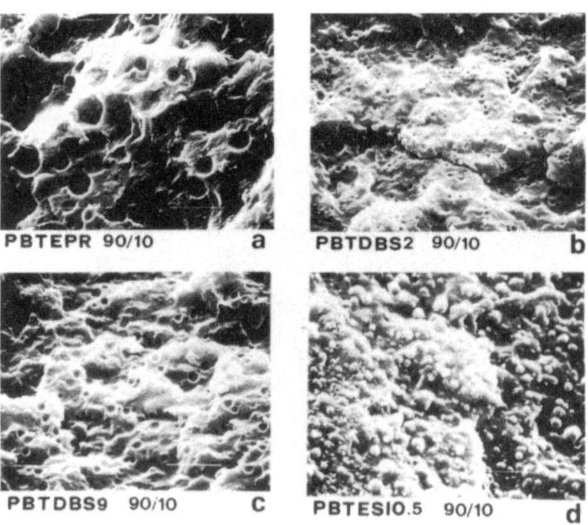

FIG. 21 SEM micrographs of fracture surfaces of 90/10 blends (a) PBTEPR; (b) PBTDBS2; (c) PBTDBS9; (d) PBTESI0.5, 2500×.

In the case of the PBTDBS blends, domains with lower average dimensions are obtained when an EPR–g-DBS rubber with a grafting degree of 2% is used (compare Figs. 21b and 21c). The adhesion of dispersed domains to the PBT matrix seems to be very low both in the case of PBTDBS2 and PBTDBS9 blends. Fracture surfaces of PBTESI0.5 90/10 blend are characterized by the presence of a dispersed phase with a rather large size distribution (0.5–5 μm). Moreover, in such blends the domains show a strong adhesion to the matrix (see Fig. 21d).

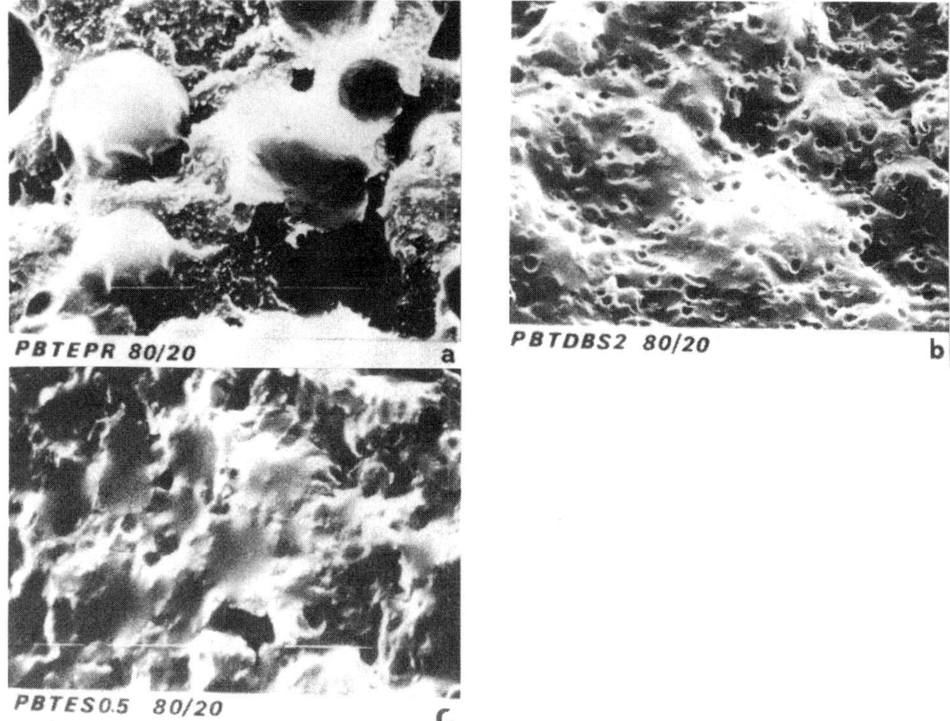

FIG. 22 SEM micrographs of fracture surfaces of 80/20 blends (a) PBTEPR; (b) PBTDBS2; (c) PBTESI0.5, 2500×.

The SEM micrographs in Figs. 22a, b, and c show the mode and state of dispersion of the rubbery phase for PBTEPR, PBTDBS2, and PBTESI0.5 (80/20) blends. Examination of the micrographs clearly shows that a higher rubber concentration in the case of PBTEPR and PBTDBS2 blends only induces an increase in the average dimensions of the dispersed particle. In addition, the PBTESI0.5 (80/20) blend shows larger domains than the corresponding 90/10 blend, but the adhesion of these domains to the matrix seems to be very strong.

No almost free spherical particles emerging from the surface are found. The morphology observed in this blend is accounted for by assuming that the crack propagation front generates the rupture of the dispersed domains without causing any nucleation from the fracture surface. This last result probably indicates that the molecular structure of the graft copolymer PBT-functionalized EPR (i.e., number of grafts per surface area unit, length of PBT grafts, etc.) is influenced by rubber concentration. This is especially evident in the case of PBTESI blends.

The SEM micrographs of the smoothed and etched surfaces of PBTEPR, PBTDBS, and PBTESI (90/10) blends are shown in Figs. 23a, b, c, and d. A comparison with Fig. 21 indicates that a large part of the domains are etched out from the surfaces.

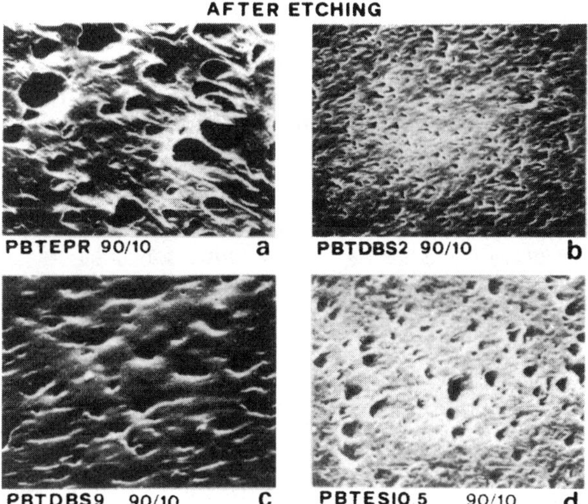

FIG. 23 SEM micrographs of smoothed surfaces after etching of 90/10 blends (a) PBTEPR; (b) PBTDBS2; (c) PBTDBS9; (d) PBTESI0.5, 2500×.

Thus it may be concluded that only a relatively small amount of rubber reacts with PBT preformed

TABLE 2 Results of the thermal analysis by DSC of the blends. Heating rate = 20°C/min; $\Delta H^0 = 31.5\,\text{kJ/mol}$

Code blend	ΔH_f (J/g of PBT)	T_g(°C)	T_m(°C)	X_c of PBT phase (%)
PBT	58	51	225	41
PBTER (90/10)	43	43	225	33
PBTDBS2(90/10)	49	48	226	34
PBTDBS9(90/10)	50	44	227	38
PBTESI0.5(90/10)	49	47	226	36
PBTEPR (80/20)	51	45	225	35
PBTDBS2 (80/20)	51	49	224	36
PBTESI0.5 (80/20)	45	45	224	32

chains, when added during the polymerization, giving rise to the formation of EPR–g-PBT graft copolymers insoluble in xylene and *o*-dichlorobenzene.

4. Thermal Analysis

Results of thermal analysis of samples of pure PBT and samples of as-obtained blends are reported in Table 2. All experiments have been performed at a heating rate of 20°C/min. Only a single T_g was detectable, which is clearly attributable to the homo PBT-rich phase. The T_g of the rubbery phase is not observable, due to the low content of EPR in the blends. The T_g of PBT generally decreases in the blends: this effect can be attributed to some plasticization of the matrix due to the presence of the rubber.

No variations in the values of T_m, are observed, while the crystallinity degree of blends seems to be lower than that of plain PBT. Similar results are consistent with studies on the crystallization behavior of nylon 6/functionalized EPR blends and indicate that the presence in the melt of an elastomeric phase disturbs the crystallization process of the matrix [80].

5. Impact Properties and Fractographic Analysis

The Charpy impact strength values for all the investigated blends are shown as a function of the testing temperature in Fig. 24. The PBT homopolymer is taken as reference material in order to evaluate the impact performance of the blends.

Scanning electron micrographs of fracture surfaces of pure PBT and PBT/rubber blends as obtained by the fracture tests at room temperature are shown in Fig. 25. All the pictures have been taken near the notch tip in the region of crack initiation. As can be seen in Fig. 24,

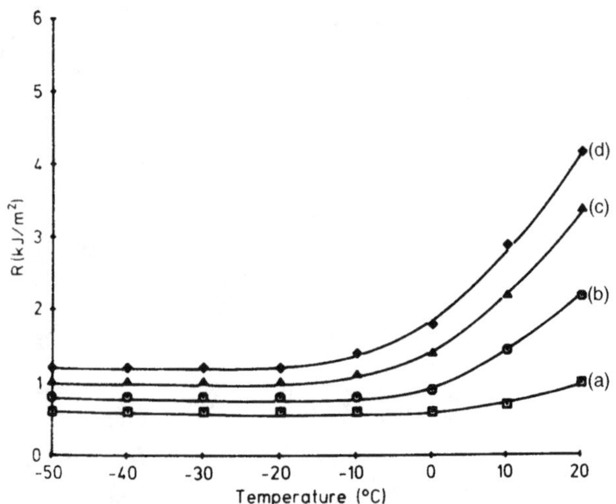

FIG. 24 Impact strength R of PBT homopolymer and of blends as function of the testing temperature. (a) PBT homopolymer; (b) PBT/ISPR blend; (c) PBT/DBS blend; (d) PBT/ESI blend.

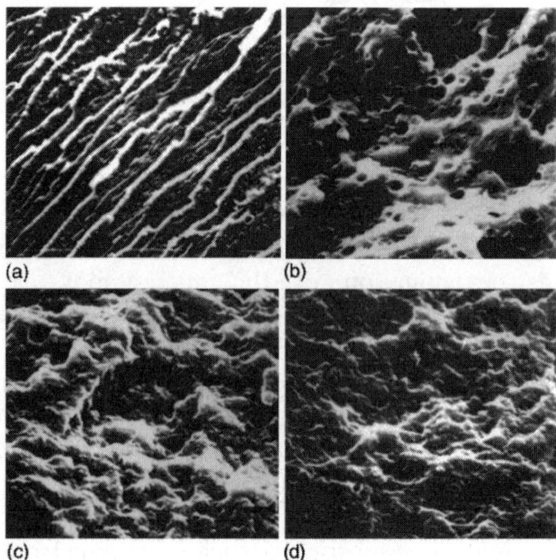

FIG. 25 SEM micrographs of fracture surfaces at room temperature. (a) PBT homopolymer (160×); (b) PBT/EPR blend (640×); (c) PBT/ESI blend (640×); (d) PBT/DBS.

the impact strength R of pure PBT (curve A) remains almost constant at very low values over the whole of the investigated temperature range. Only a very slight enhancement is observed at temperatures higher than 10°C. This is consistent with the fracture surfaces of PBT shown in Fig. 25a, which display features typical of a brittle material, suggesting that the fracture occurs mainly by crack formation.

The addition of unmodified EPR (curve B) or of modified EPRs (curves C and D) produces a small improvement in the R values with respect to the pure PBT in the region of low temperatures (up to about $-10^\circ C$). Beyond this temperature, however, a more substantial rise in the impact strength is achieved. This enhancement is more marked for the blends with modified EPR and is greater for the blend containing EPR–g-ESI as the rubbery phase.

The above results can be interpreted on the basis of the chemistry of grafted groups onto EPR.

Morphological studies on the PBT/EPR, PBT/DBS, and PBT/ESI blends have shown that the mode and state of dispersion of the rubbery phase is strongly affected by the chemical nature of the grafted groups present on the EPR chains. In particular were found: (a) domains of large size (5–10 μm) with no interfacial adhesion for the PBT/EPR blend; (b) domains with lower dimensions but with low degree of adhesion for PBT/DBS blends; (c) tiny domains (~0.5 μm) together with domains of medium size (~5 μm), both strongly adherent to the matrix for PBT/ESI blends. The different morphologies observed in such blends were ascribed to the fact that the ability to form a suitable graft copolymer (EPR–g-PBT) during the processes of blending and polymerization is greater when the rubber is functionalized with alcoholic groups. This copolymer, acting as emulsifier and/or compatibilizing agent, reduces the interfacial energy and improves the interfacial adhesion.

The morphological features mentioned above, together with the fractographic analysis reported in Fig. 25b–d, can be used to explain the given impact properties. In fact, it is widely accepted that, in heterogeneous blend systems, the particles of the disperse phase can act as sites of stress concentrations. Therefore, as soon as the stress level around the particle overcomes the yield stress of the matrix, energy-dissipating mechanisms (crazing and/or shear yielding) become active and an increasing toughening is achieved. The extent of such a phenomenon is related to the dimensions of the rubbery particles and to their adhesion to the matrix. In the case of the blend with unmodified EPR (see Fig. 25b), the fracture surface shows a "cheese-like," morphology and the surface of the holes left by the rubber particles appears to be very smooth.

In addition, during the fracture process, a large number of domains were pulled away from their original positions, indicating a poor adhesion between the rubbery phase and the polyester matrix. In keeping with these findings the capability of such rubber particles to stabilize localised deformations such as crazing in the immediate neighborhood of the notch tip turns out to be very low.

Therefore only craze initiation will eventually occur whereas craze termination will not. As a consequence, only a small amount of fracture energy can be dissipated before the crack development takes place and a slight increase in the impact strength can be obtained (see Fig. 24, curve B).

By replacing EPR with a functionalized EPR the impact properties increase (see Fig. 24, curves C and D) and the morphology of the fractured surfaces change drastically as can be readily seen from Figs. 25c and 25d. In both cases there seems to be evidence of rubbery domains finely dispersed and well embedded in the PBT matrix. As a matter of fact the dispersed phase is barely visible.

The fractographic features, as previously mentioned, can be accounted for by assuming that, during the process of blend preparation, at least part of the modified EPR is involved in the formation of a graft copolymer.

The presence of such a copolymer as shown by us in the case of polyamide 6/EPR blends, improves the mode and state of dispersion of the unreacted EPR, thus favoring the formation of crazes and their termination. Therefore, the material is able to absorb a larger amount of energy before the final fracture occurs, improving the impact strength (see Fig. 24, curve C and D).

Finally a careful inspection of Figs. 25c and 25d reveals also a greater roughness of the fracture surface of the PBT/EPR–g-ESI blend, in accordance with its better impact properties (Fig. 24). Such a behavior may be related to a higher degree of adhesion as well as to the presence of a small number of larger particles which can act as more favorable sites for craze initiation.

C. Polyamide 6/Functionalized Rubber Blends

Although in the present review we have mainly investigated the influence of properly functionalized PO on commodity polyesters, it must be stressed that other thermoplastic matrices (PA6, PA6,6 etc.) have also been modified with a rubbery functionalized component. Generally, two different blending routes, in the case of polyamides, are utilized:

1. Melt mixing of the high molecular weight thermoplastic polymer with an elastomer.
2. Dispersion of the rubbery component during the polymerization of the thermoplastic polymer.

Melt mixing method 1 utilizes well established technologies and is therefore generally preferred to obtain rubber-modified thermoplastics of improved impact properties. As the two components are immiscible, the addition of a third component (compatibilizing agent) is frequently used to increase the adhesion between the elastomeric and the thermoplastic phases and to achieve finer dispersions of the rubbery particles in the thermoplastic matrix.

Modified elastomers bearing functional groups along the chain are frequently used as precursors of compatibilizing agents because they may form graft copolymers reacting with the thermoplastic polymer during the blending process (reactive blending). The functionalized elastomer may also be the only rubber component. The use of such modified rubbers is particularly attractive for condensation polymers which have reactive chain ends and are generally obtained by a step polymerization mechanism. These features greatly enhance the potential use of method 2.

Several examples dealing with PA6 and PA6,6/polyolefin systems have been reported in patents. The materials are generally melt blended and graft copolymers or networks are formed by means of radical initiators or directly by mechanical degradation, with an appreciable improvement of the impact resistance of the polyamides. Authors have studied the PA6/isotactic polypropylene (iPP) system to which an isotactic polypropylene functionalized by insertion of anhydride groups react with PA6 amino end groups to yield an iPP/PA6 graft copolymer. A similar approach has been followed by other authors on PA6/high density polyethylene blends. In both cases the results were very promising.

The previously cited routes 1 and 2 have both been explored for the realization of a EPR-modified PA6 characterized by interesting technological performances [35, 56], particularly resistance to impacts at temperatures below the ambient temperature, where PA6 is normally deficient due to its glass transition well above room temperature. To give an indication, an increase of as much as 40 times in the notched Charpy impact resistance are reported already at 0°C.

IV. FUNCTIONALIZED POLYOLEFINIC RUBBER AS INTERFACIAL AGENTS IN TOUGHENED THERMOSETTING MATRICES

The practical application of thermoset resins, epoxies and polyesters, is often limited because of their brittleness at room temperature. A method which has been adopted to overcome this limitation is based on the addition of reactive low molecular mass rubber to the brittle matrix [81].

Low molecular weight polybutadiene and butadiene–acrylonitrile copolymers terminated with carboxyl, vinyl, amine, epoxy, phenol, and hydroxyl groups have been widely used as toughening agents both for epoxy and polyester resins. Thermally reactive isoprene–acrylonitrile and ethylacrylate–butylacrylate copolymers have also been used [82, 83].

According to Rowe [83] and Nicholas *et al.* [84] in the case of epoxies and polyesters a liquid rubber is suitable for toughening if the following basic conditions are satisfied:

(a) the rubber is compatible with the starting uncured base resin; it phase separates as small domains, during the gelation process, distributed throughout the matrix;
(b) the molecular structure of the rubbers is characterized by the presence of functional end groups which are able to react chemically with the base resin.

The initial compatibility between rubber and uncured resin, their relative chemical reactivity and rate, and nature of phase separation processes, during the cure reactions, mainly determine the mode and state of dispersion of separated rubber domains and consequently the mechanical response of the final cured rubber modified thermoset resin.

More recently Crosbie and Philips [85, 86] investigated the toughening effect of several reactive liquid rubbers (carboxyl terminated butadiene–acrylonitrile, vinyl terminated butadiene–acrylonitrile, hydroxyl terminated polyether, polyepichlorohydrin) and an unspecified experimental reactive liquid rubber developed by Scott Bader Ltd. on two different polyester resins: a flexibilized isophthalic–neopentyl glycol polyester resin, PVC compatible and an epoxy modified polyester resin, which is preaccelerated. The results of these studies are summarized as follows:

1. Three of the four liquid rubbers produced, upon curing, a dispersion of second-phase rubber-rich particles in the polyester resin matrix, whereas the compatible rubber did not produce any detectable dispersion even at relatively high content.
2. The mode and state of dispersion of particles were found to be highly dependent upon both rubber and resin formulation.
3. Toughness increased generally with the presence of dispersed rubber phase. Studies of chemical

reactivity between two samples of low molecular weight polybutadiene differing in the nature of the terminal functional groups with a commercial polyester resin are hereafter reported.

Analytical techniques, including IR and DSC, were used to study the rubber–resin chemical reactivity.

A. Experimental

1. Materials

(a) Base Resin Commercial Samples. These samples of uncured unsaturated polyester resin (PER) were supplied by SNIAL (Colleferro, Roma, Italy) with the tradename of H35. The base resin is usually available as a solution containing about 68 wt% of a prepolymer dissolved in a styrene monomer. The base prepolymer is normally prepared by the reaction of a saturated diol (propylene glycol) with a mixture of phthalic and maleic anhydride (mole ratios respectively 1 : 0.7 : 0.3).

The so-called "conventional unsaturated polyester prepolymer" is characterized by the presence, along the backbone chains, of randomly distributed ester groups and carbon–carbon double bonds. The molecular structure may be schematically represented as follows:

```
HO—C—Φ—C—[O—CH—CH2—O—C—Φ—C—O—CH—
   ||   ||     |          ||   ||    |
   O    O     CH3         O    O    CH3

     CH2—O—C—CH=CH—C—]O—CH—CH2—OH
           ||      ||  n  |
           O       O     CH3
Φ = phenyl ring
```

The chains will be on average hydroxyl and carboxyl terminated together with some minor fraction of di-hydroxyl and di-carboxyl terminated chains. The acid number of the prepolymer (defined as the mg of potassium hydroxide used for the titration of 1.0 g of prepolymer) ranges from 40–45.

The base prepolymer was characterized by using the ^{1}H-NMR technique. The spectrum (see Fig. 26) was obtained in $CDCl_3$ solution using a ^{1}H270 MHz BRUKER apparatus. From this analysis a content of fumarate of about 95 wt% of total maleic anhydride used was calculated (TMS was used as internal standard). Such a result indicates that double bonds along the chains are practically all in a trans configuration.

(b) Low Molecular Mass PO Rubbers.

(a) As starting rubber a sample of hydroxyl terminated polybutadiene (HTPB) $M_n = 1350$, supplied by Polyscience, was used. The molecular structure of HTPB was determined by analysis of 270 MHz ^{1}H-NMR spectrum in $CDCl_3$ solution, using TMS as the internal standard [67]. The spectrum is reported in Fig. 27. It was found that about 70% of the units are arranged in a 1, 2 arrangement and 30% in a 1,4 cis/trans arrangement. The hydroxyl number was determined by using the following procedures: first the hydroxyl end groups were quantitatively converted into isocyanate end groups (see later), then an excess of dibutylamine was added, and finally the amine excess was titrated with 0.033 N HCl solution in *o*-dichlorobenzene/isopropanol 9/1 by volume. The resulting hydroxyl number, 1.2 meq/g, is in good agreement with an average functionality of HTPB molecules slightly lower than 2.

(b) The isocyanate terminated polybutadiene (PBNCO) was obtained by reaction of HTPB with toluene diisocyanate (TDI). The TDI, supplied by Fluka, was a 80/20 mixture of 2,4 and 2,6 isomers; it was used without further purification. 10 g of HTPB were mixed with 1.78 mL of TDI (molar ratio TDI/HTPB 5% exceeding the stoichiometric

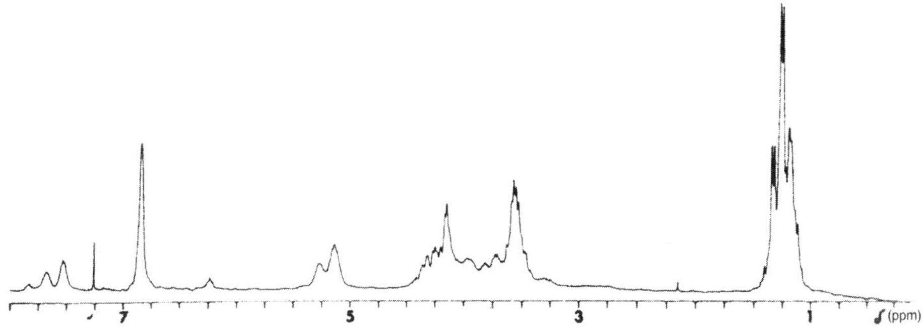

FIG. 26 ^{1}H-NMR spectrum of PER resin.

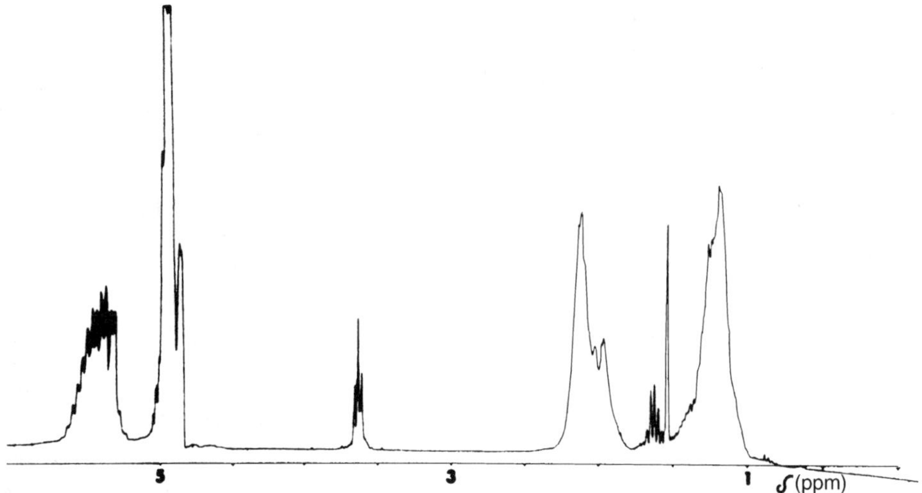

FIG. 27 ^{1}H-NMR spectrum of HTPB rubber.

2 : 1 balance) for 10 min at 70°C under nitrogen atmosphere and mechanical stirring.

(c) Techniques of Characterization. FTIR and DSC apparatuses are described in the previous sections.

Morphological characterizations The morphology of the materials was investigated by optical (OM) and scanning electron microscopy techniques (SEM).

The phase separation was followed by means of a Leitz polarizing optical microscope with a Mettler hot stage.

The mode and state of dispersion of the minor component and its volume fraction were determined by scanning electron microscopy on fractured surfaces after metallization with AuPd alloy. A scanning electron microscope (Philips Model 501) was used throughout.

For a better resolution of the overall morphology an etching technique was also used: fracture surfaces of blends were exposed for 2 min to boiling *n*-heptane vapors to remove preferentially the rubbery component. Such samples were subsequently prepared for SEM examination.

Impact tests. Charpy impact tests were performed by using a Ceast fracture pendulum at two different temperatures (−20°C and room temperature).

(d) Preparation of Rubber-Modified Polyester Resins: Cure Conditions. The procedure for the preparation of cured rubber modified polyester resins (RMPER) was the following: first, 10 g of HTBP or PBNCO rubber were dispersed in the styrene solution containing the polyester prepolymer by mechanical stirring for 30 min at 80°C (a polymerization reaction flask, capacity 250 cm^3, equipped with a mechanical stirrer and vacuum outlet was used); then the mixture was cooled to room temperature and air bubbles were removed under reduced pressure. At this point the initiator was added under stirring; the mixture was poured into a mould consisting of either glass or metal glazing plates, separated by a flexible rubber gasket held by springs, and finally cured at 80°C for three hours.

Methyl ethyl ketone peroxide (Butanox, 1 wt%) and Co-octoate (0.25 wt%) were used as catalyst and accelerator, respectively (the first is available as 50% wt solution in dimethyl–phthalate while the second was already present in the starting formulation, as naphthenate solution).

The code and the composition of RMPER materials investigated are given in Table 3.

TABLE 3 Code and compositions* of rubber modified unsaturated polyester resins investigated

Sample code	Rubber content wt%	Base resin content wt%
PER	0	100
PER/HTPB-10	10	90
PER/HTPB-20	20	80
PER/PBNCO-10	10	90
PER/PBNCO-20	20	80

*The composition (wt%) refers to the base starting (PER) resin (prepolymer + styrene + catalyst).

B. Results and Discussion

1. Preparation of Isocyanate Terminated Polybutadiene (PBNCO)

The isocyanate-end capped polybutadiene (PBNCO) was prepared by a condensation reaction of diisocyanate with the hydroxyl terminal groups of rubber chains. In order to limit chain extension of HTPB through coupling reaction with TDI, a small excess of 5% with respect to the stoichiometric quantities of the latter was used. Larger excess are useless, as with an appropriate choice of the (reaction temperature, only the isocyanate group in para position will react [87]. The reaction was followed by DSC and IR spectroscopy.

The DSC thermograms obtained by heating the reaction mixture from room temperature to about 230°C with a scan rate of 20°C/min show a large exothermal peak starting from room temperature (see Fig. 28). The maximum observed at about 140°C is probably related to the fact that at this temperature the reaction involving mainly the terminal groups of HTPB and the —NCO groups of 2,4 TDI in para position presents the highest rate of conversion.

The small peak at about 160°C is likely to be accounted for by the reaction between the —NCO groups in ortho position of 2,6-TDI and the terminal groups of HTPB. As a matter of fact it is well known from the literature [87] that the reactivity of the isocyanate groups in the para position of 2,4TDI is about eight times larger than the groups of 2,6TDI. To promote the low temperature reaction only, and to avoid coupling, the experiments were effected at temperatures of 70°C. The scheme of the reaction is the following:

2 OCN—R—NCO + HO—PB—OH →

OCN—R—NH—CO—O—PB—O—CO—NH—R—NCO

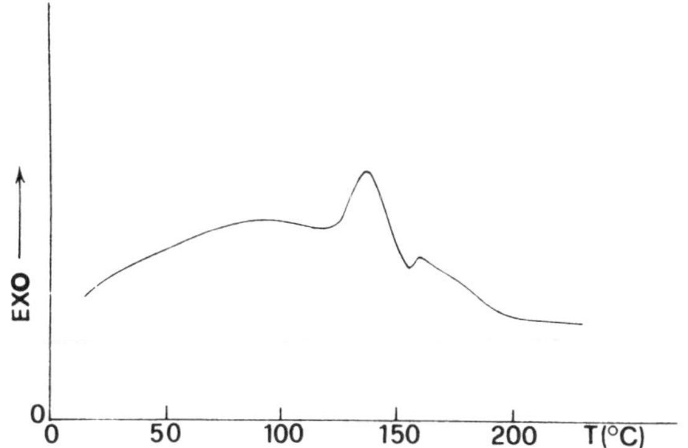

FIG. 28 DSC thermogram of HTPB/TDI mixture.

where R = toluene group and PB = polybutadiene chain.

The IR spectra of pristine TDI and HTPB are compared with that of the reaction product, mainly PBNCO, in Fig. 29. The infrared spectra, after 10 min of reaction at 70°C, show a strong absorption band at 1710–1735 cm^{-1}, corresponding to the stretching of the —C=O belonging to the (—NH—C(=O—O—) urethane groups of the formed PBNCO molecules (see Fig. 29c).

2. Reaction of PBNCO with Unsaturated Polyester Prepolymer

It was found that PBNCO and the unsaturated polyester prepolymer easily react through a condensation reaction that leads to the formation of urethane (—O—C=O—NH—) linkages.

In principle, in the course of the reaction, diblock and triblock copolymers may form according to the following scheme:

OCN—R—NH—CO—O—PB—O—NH—R—NCO

+2HO—PER—COOH ⟶

HOOC—PER—O—CO—NH—R

—NH—CO—O—PB—O—NH

R—HN—CO—O—PER—COOH

(A–B–A TRIBLOCKS COPOLYMER FORMATION)

OCN—R—NH—CO—O—PB—O—CO—NH—R—NCO

+HO—PER—COOH ⟶

OCN—R—NH—CO—O—PB—O

—CO—NH—R—NH—CO

O—PER—COOH

(A–B DIBLOCKS COPOLYMER FORMATION)

The reaction is carried out at 80°C under vigorous stirring adding the styrene solution containing the prepolymer (base PER resin) to the PBNCO. The reaction was followed by DSC and IR. As shown in Fig. 30 the thermogram obtained by heating the reacting mixture (PBNCO–PER resin) from room temperature to 180°C is characterized by the presence of a broad exothermic peak centered at 60°C followed by a sharp peak centered at 145°C, probably due to less-reactive NCO groups). From the area of the first peak the heat generated gave, for the 90/10 PER/PBNCO (wt/wt) composition, about 143 kJ/mol of hydroxyl terminal groups, well in agreement to literature data [88].

On the contrary the DSC thermogram of HTPB/PER blends is flat, indicating that in such a case no

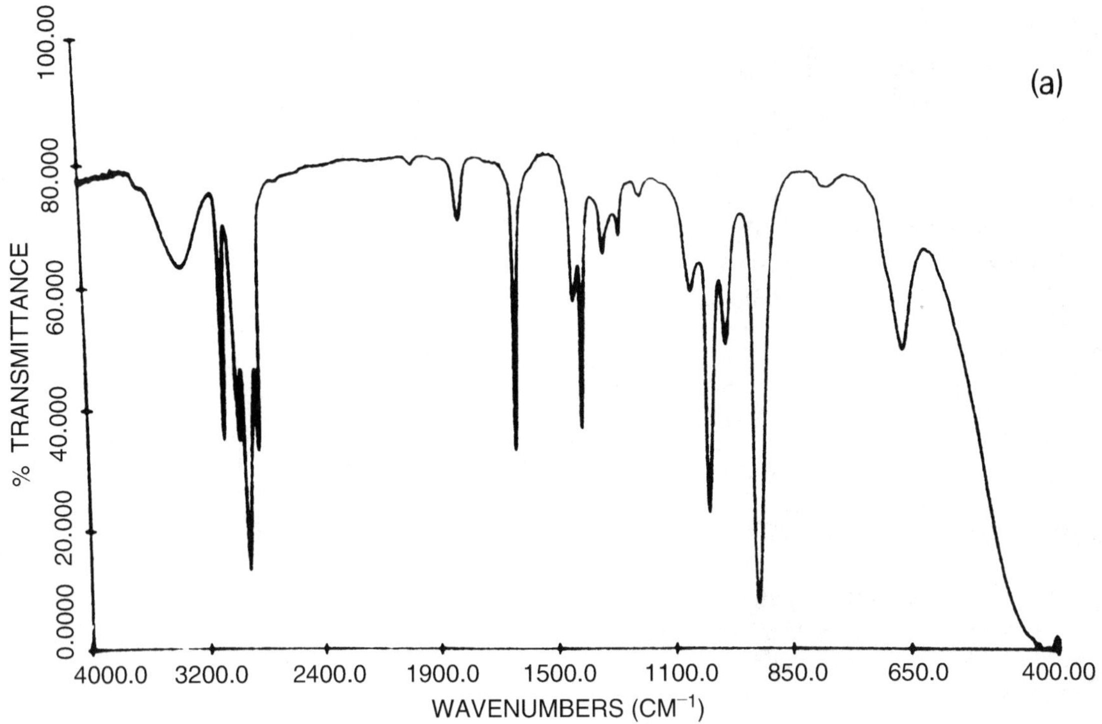

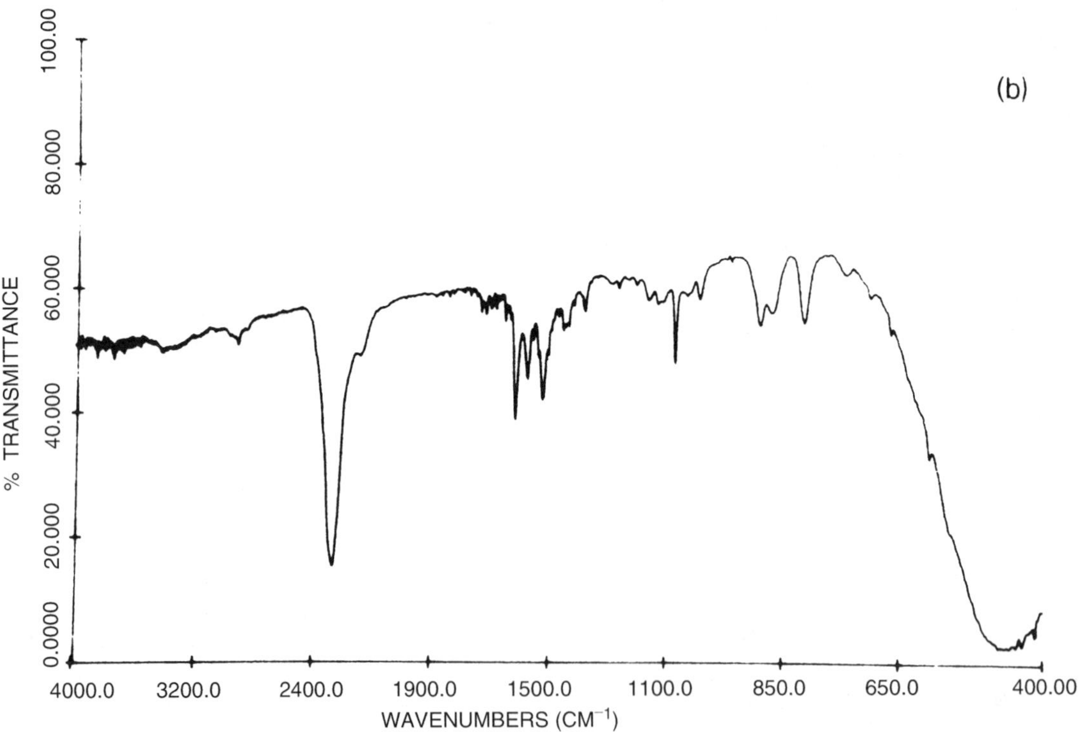

FIG. 29 Infrared spectrum of: (a) HTPB, (b) TDI, (c) HTPB/TDI mixture after reaction to give PBNCO.

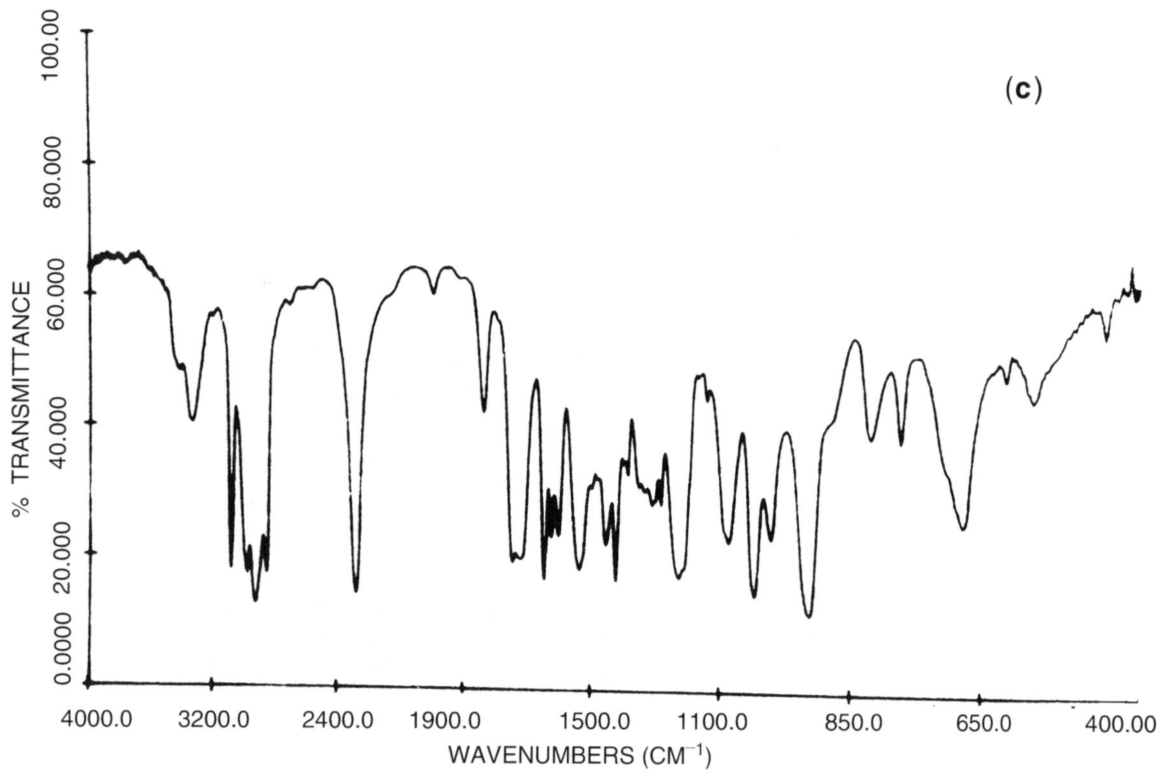

MODIFICATION OF RESINS

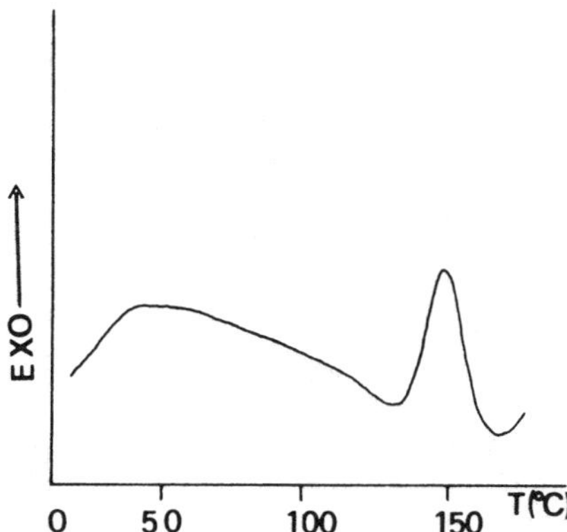

FIG. 30 DSC thermogram of PER/PBNCO (90/10) mixture.

reaction occurs, at least under our experimental conditions.

Infrared spectra of PBNCO/PER 80/20 blends before and after reaction are shown in Fig. 31a and b, respectively. From the analysis of the IR spectra it emerges that the band of —NCO isocyanate group at 2260 cm^{-1} practically disappears after reaction, indicating that by the end of the reaction almost all the isocyanate terminated PB chains have been reacted. Thus it may be concluded that preferable A–B–A triblock type copolymers are formed.

3. Morphology and Phase Separation Studies in PER/PBNCO and PER/HTPB Blends

(a) Phase Structure Before the Curing Process. The phase structure of PER/PBNCO blends was investigated by optical microscopy by squeezing drops of the viscous mixture between two glass plates. The observations were first made after mixing at room temperature, when the rate of conversion related to the reaction of formation of A–B–A copolymer is very low, and then after storing the blend samples at about 80°C for 30 min, that is after completion of the reaction of copolymer formation.

It may be observed that before or at the early stage of copolymer reaction the PER/PBNCO mixture is separated into droplet-like domains (see Fig. 32). In contrast, the mixture at the end of the

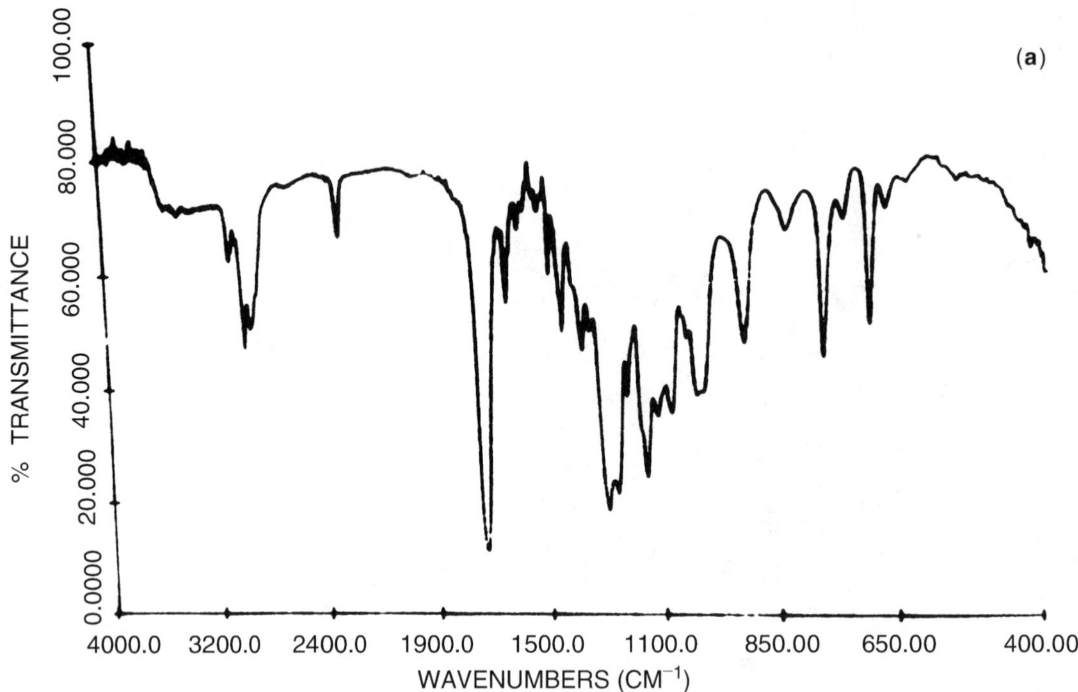

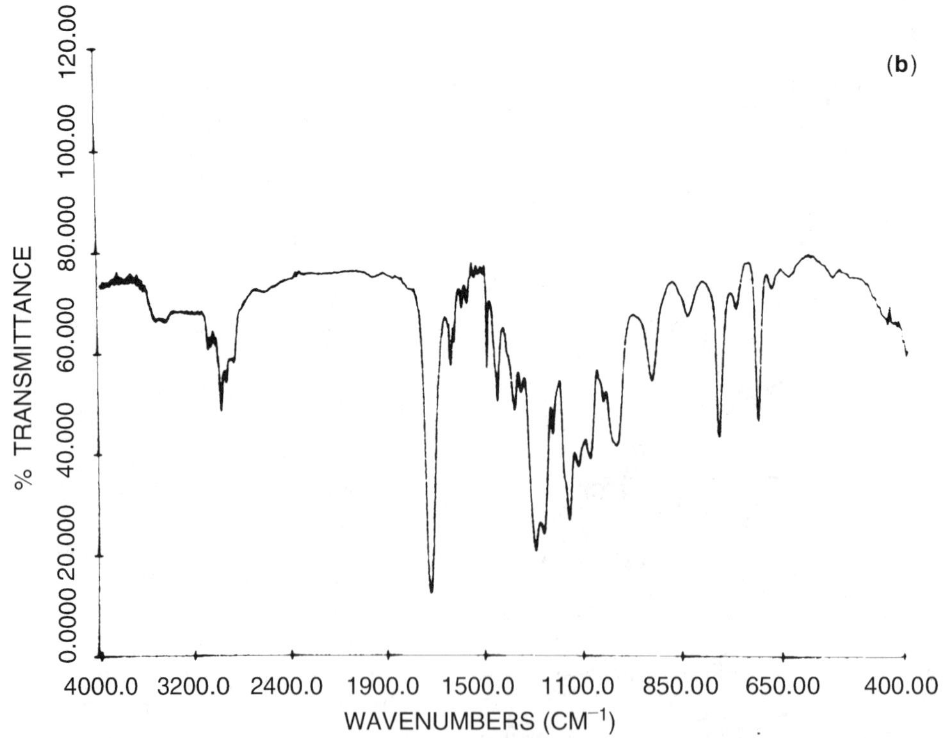

FIG. 31 Infrared spectrum of: (a) PER/PBNCO (80/20) mixture at room temperature before reaction, (b) PEP/PBNCO (80/20) mixture after reaction.

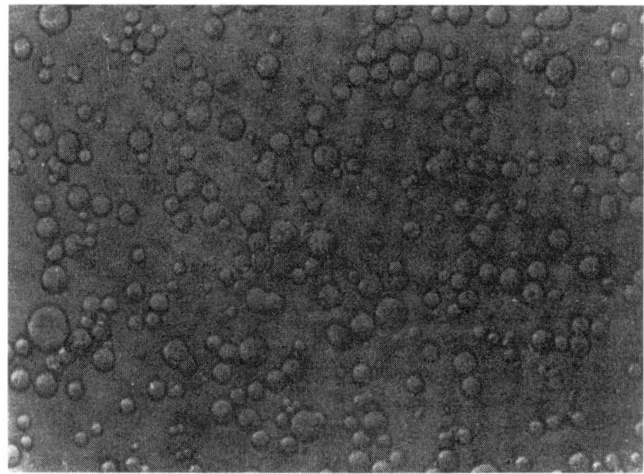

FIG. 32 Optical micrograph of PER/PBNCO (90/10) film before reaction and curing (100×).

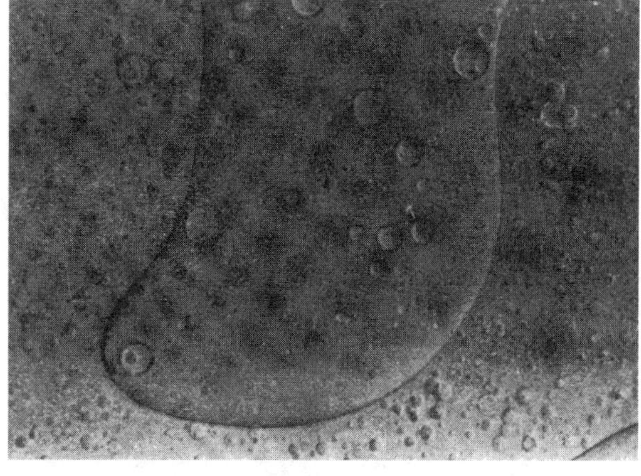

FIG. 34 Optical micrograph of PER/HTPB (80/20) film before curing (100×).

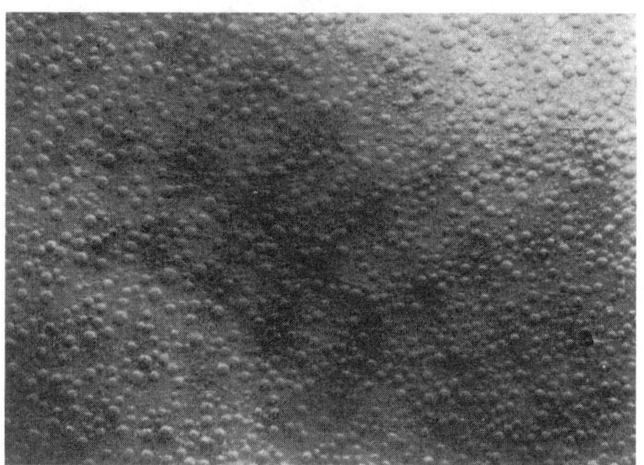

FIG. 33 Optical micrograph of PER/HTPB (90/10) film before curing (100×).

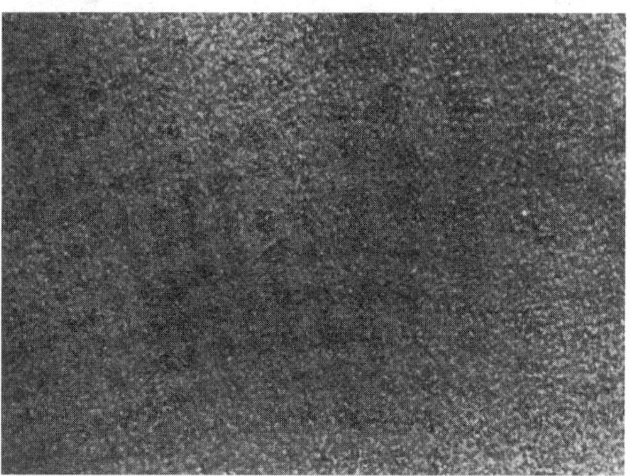

FIG. 35 Optical micrograph of PER/PBNCO (90/10) mixture after reaction and curing process (100×).

reaction appears homogenous under optical microscope. It is interesting to point out that PER/HTPB 90/10 blends before curing show a morphology similar to that observed in PER/PBNCO blends before the formation of A–B–A copolymer (see Figs. 32 and 33), with a clear phenomenon of separation of a rubbery-rich phase.

In contrast, after the thermal treatment (30 min at 80°C) the blend PER/PBNCO becomes transparent under the microscope, while the phase structure of PER/HTPB blends is not stable as coalescence processes induced by time and/or temperature are observed. This is a counter proof of the formation of a ABA block copolymer in the former blend, well emulsified inside the matrix. The mode and state of dispersion of the rubber component in the case of PER/HTPB 80/20 blends appears to be more complicated as domains of rubber with subinclusions of resin are observed (see Fig. 34).

(b) Phase Structure After Cure. The phase structure of PER/PBNCO and PER/HTPB blends 90/10 and 80/20 (wt/wt) in composition, after the cure process was investigated by optical microscopy on thin films and by SEM observation of the fractured surfaces. It can be seen that the start of gelation causes the phase separation of the compatible PER/PBNCO blends, and small domains around 2 μm in size, uniformly distributed all over the sample, strongly adherent to the matrix are visible (see Figs. 35, 36, and 37). It can be seen (see

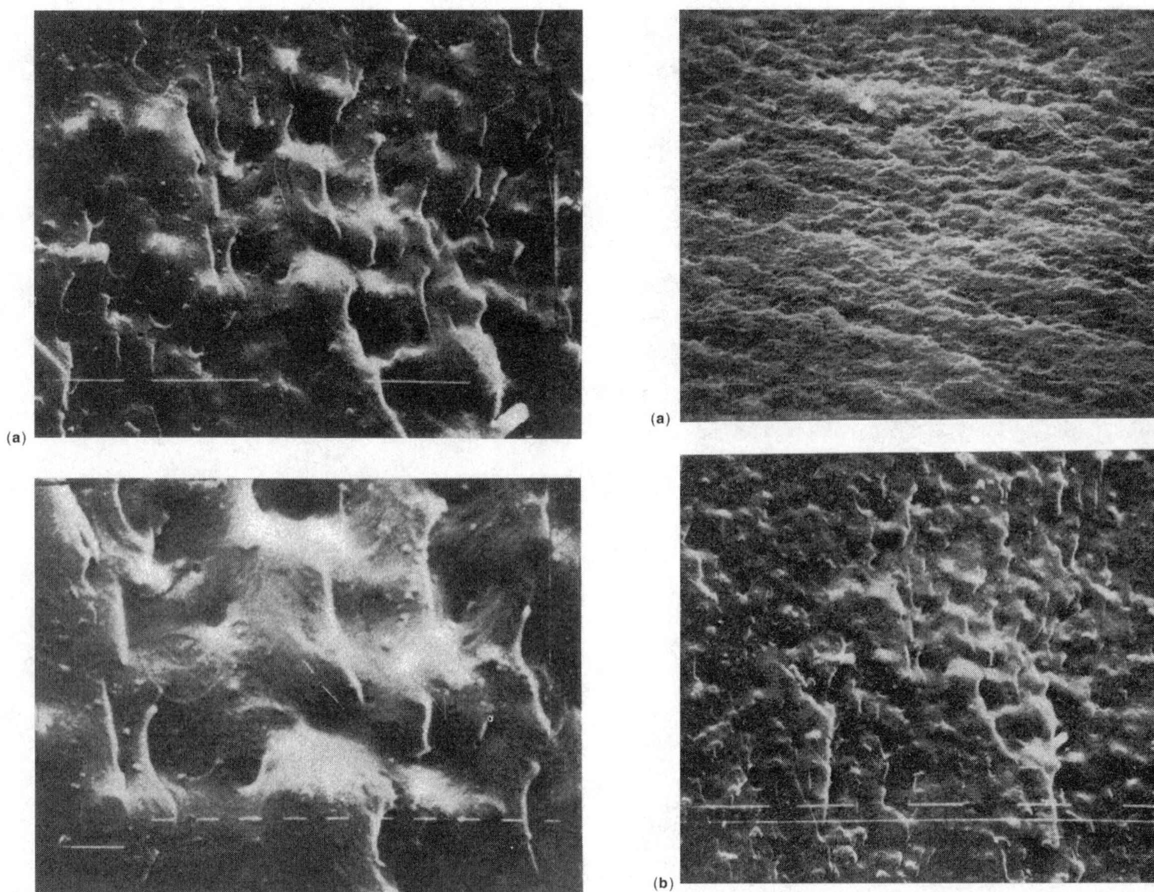

FIG. 36 SEM micrographs of PER/PBNCO 90/10 mixture after rubber–prepolymer reaction and curing process at two different magnifications: (a) 2500×, (b) 5000×.

FIG. 37 SEM micrographs of PER/PBNCO (80/20) mixture after rubber prepolymer reaction and curing process at two magnifications. (a) 640×; (b) 1250×.

Figs. 36a and 37b) that the moving crack front seems to interact with the dispersed particles.

Tails or steps are formed at the border of the inclusions. The phase structure of PER/HTPB blends is much more intricate. As shown by the optical micrographs of Fig. 38 large rubbery domains, probably containing resin subinclusions, coexist with very small domains. The SEM analysis of fractured surfaces show the presence of dispersed domains with a dimension distribution much wider than in the case of PER/PBNCO blends.

Moreover these domains seem to scarcely adhere to the matrix as holes with smooth walls are largely visible on the surface of the samples (see Fig. 39).

It is interesting to observe that the crack front bores out between the domains, whilst still remaining pinned at all positions when it has encountered the particles (see Fig. 39).

4. Glass Transition Studies

The glass transition temperatures (T_g) of pristine HTPB, PBNCO, and PER and PER/HTPB and PER/PBNCO blends are reported in Table 4. From the values of this table it emerges that:

The glass transition value of PBNCO is higher than that of HTPB (−9°C and −16°C respectively), probably related to the formation of urethane-type hydrogen bonding.

In both types of post cured blends, i.e. PER/HTPB and PER/PBNCO, two glass transition are observed, the one at lower temperature is attributable to rubber segments while that at high temperature belongs to the cured resin.

For PER/HTPB modified resins the T_g of rubber remains constant whereas the T_g of resin seems to decrease proportionally to the HTPB content (PER

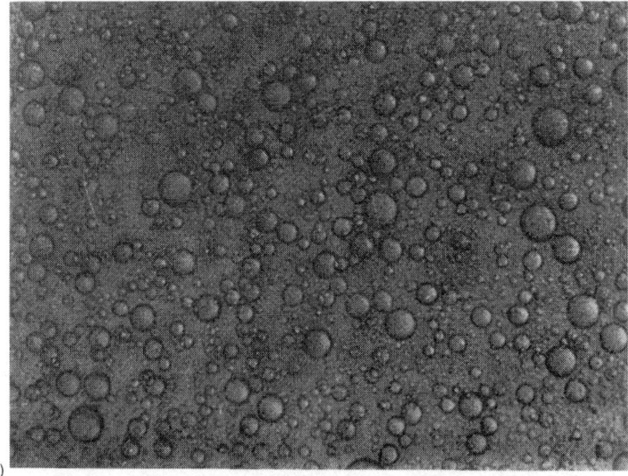

(a)

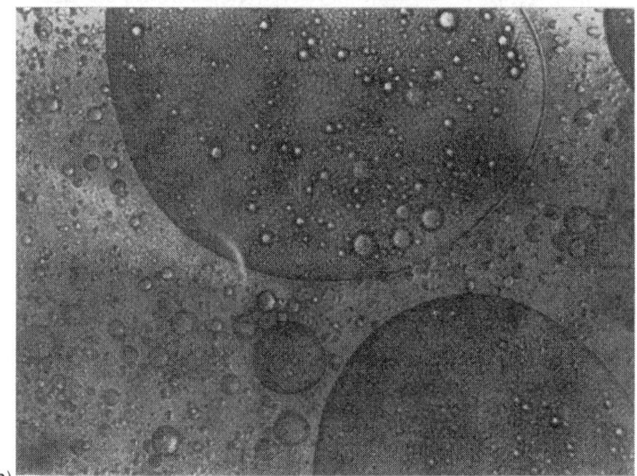

(b)

FIG. 38 Optical micrograph of a thin film of PER/HTPB blends after curing process (100×). (a) (90/10); (b) (80/20).

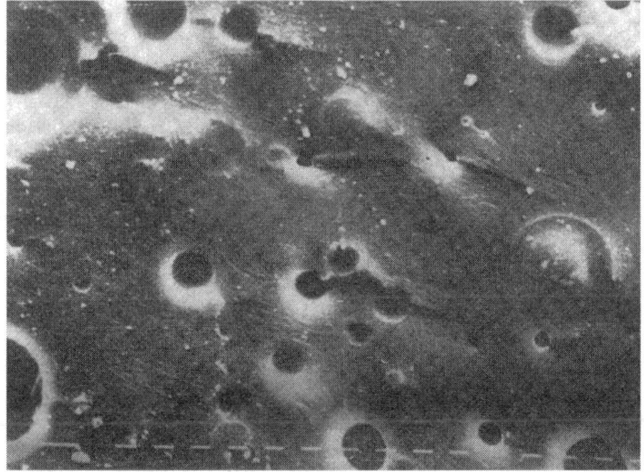

FIG. 39 SEM micrographs of fractured surfaces of PER/HTPB (90/10) blend after curing process 320×.

TABLE 4 Glass transition temperature of starting materials and of rubber modified post cured resin %

Material (Code)	Glass transition temperature (°C)	
HTPB	−16	
PBNCO	−9	
PER		66
PER/HTPB (90/10)	−13	64
PER/HTPB (80/20)	−17	55
PER/PBNCO (90/10)	0	55
PER/PBNCO (80/20)	0	51

(100%): $T_g = 66°C$, PER (90%): $T_g = 64°C$ and PER (80%): $T_g = 55°C$). This effect is likely due to uncompleted phase separation of the rubber inside the polyester matrix.

For PER/PBNCO modified resin the lowering in the T_g of resin seems to be even larger; at the same time a substantial increase in the T_g of rubber is observable.

The overall picture which emerges is that the presence of rubber influences the glass transition temperature of resin. The fact that the T_g of the rubber for PER/PBNCO post cured resin is larger than that of pristine PBNCO may be accounted for by a reduction of free volume related to the formation of the A–B–A copolymer.

5. Fracture Analysis

The fracture behavior of unmodified and rubber-modified polyester resins was examined by using the concepts of the linear elastic fracture mechanics (LEFM). From this approach two parameters can be determined which describe accurately the conditions for the onset of crack growth in brittle materials (89).

One is the stress intensity factor (K), a parameter that determines the distribution of stress ahead of the crack tip. The fracture occurs when K achieves a critical value K_c given by [89]:

$$K_c = \sigma y(a)^{1/2} \tag{1}$$

where σ is the failure stress, a is the initial crack length, and y is a shape factor depending on the specimen geometry. The other parameter often used is the critical strain energy release rate (G_c), which represents the energy necessary to initiate the crack propagation. This can be expressed in terms of fracture energy by means of the equation [89]:

$$G_c = U/BW\Phi \tag{2}$$

where U is the fracture energy corrected by the kinetic energy contribution, B and W are specimen thickness and width, and Φ is a shape factor related to the specimen compliance (C) and to the crack length by the relationship [89]:

$$\Phi = \frac{C}{\mathrm{d}C/\mathrm{d}(a/w)} \tag{3}$$

Φ can be calculated for any specimen geometry experimentally and theoretically [90].

The fracture test used here for K_c and G_c determinations was the Charpy test conducted by means of an instrumented pendulum at an impact speed of 1 m/s. For all the examined materials a set of specimens ($B = 6.0$ mm, $W = 6.0$ mm) with different notch lengths and test span of 48 min were broken at room temperature and at $-20°$C. Therefore loads and energies as function of time were recorded. According to Eqs. (1) and (2) a plot of σ_y vs $1/(a)^{1/2}$ and U vs BWΦ should give a straight line with K_c and G_c as slope. The y values used to calculate K_c were those given by Brown and Srawley [91] and the failure stress σ was determined at the maximum load on load–time curves. For G_c calculation the Φ values were taken by Plati and Williams (92), and the energy data were corrected by the kinetic energy contribution. An example of K_c and G_c determination is reported in Figs. 40 and 41 for the pure polyester resin.

Similar straight lines were obtained for both PER/PBNCO blends. Such result indicated that the approach followed is valid for all materials investigated.

The values of G_c obtained by energy measurements were also compared with those calculated using the

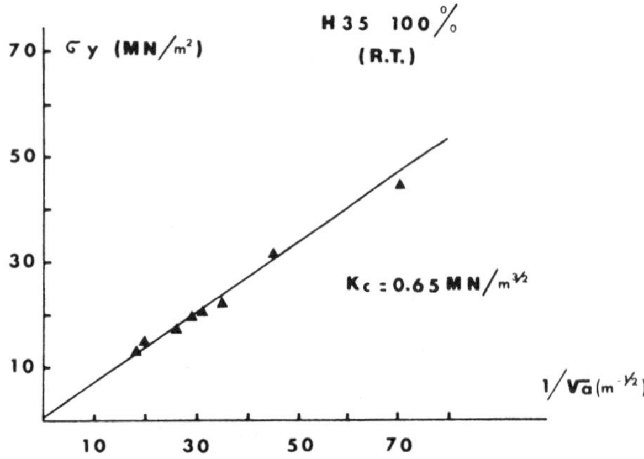

FIG. 40 σ_y as a function of $1/(a)^{1/2}$ for unmodified resin PER at room temperature.

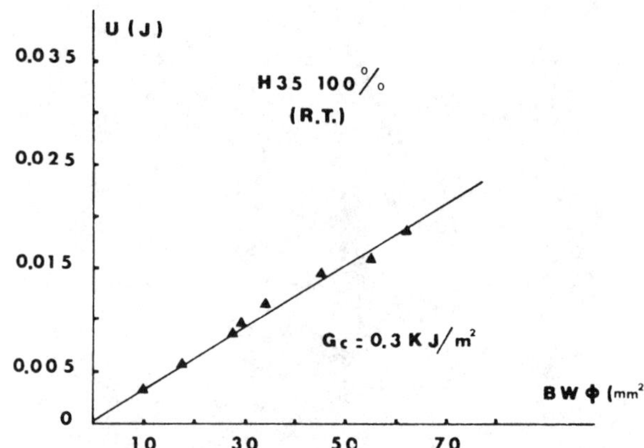

FIG. 41 Impact fracture energy, U, as a function of BWΦ for unmodified resin PER at room temperature.

equation: $G_c = K_c^2/E$ (plane stress condition) where E is the Young's modulus [90].

E values were determined by a rebound test performed on unnotched specimens at low impact speed (0.2–0.3 m/s) using the same instrumented pendulum apparatus.

The fracture toughness parameters and modula for all the materials studied are collected in Table 5.

As can be seen at room temperature the blends containing HTPB rubber show K_c and G_c values slightly lower than those of pure polyester resin, in contrast, a significative improvement in fracture resistance is achieved when the rubber is functionalized with isocyanate groups (PBNCO rubber). In fact the PER/PBNCO (90/10) and PER/PBNCO (80/20) blends show K_c and G_c values about twice and three times higher than the pure resin. Such an increase is accompanied by a slight decrease in the elastic modulus. Furthermore it can be observed that the decrease in the testing temperature ($-20°$C) causes only a slight enhancement in the K_c values, probably due to an increase of the modulus which yields a higher strength whereas the G_c values remain practically unchanged.

6. Morphological Investigation on Fracture Surfaces

Scanning electron micrographs of broken surfaces of all the investigated blends, obtained by the fracture tests at room temperature, are shown in Figs. 42 to 47. The fracture surface of plane H35 samples (see Figure 42) is typical of a brittle polyester resin: the macro-cracks propagate linearly, leading to a rough stepped surface. A comparison of the surface fracture

TABLE 5 Modulus (E), K_c and G_c at room temperature and at −20°C*

	Room temperature				$T = -20°C$			
Materials	K_c MN/m$^{3/2}$	G_c(KJ/m^2) by energy measurement	E MN/m^2	$G_c = K_c^2/E$	K_c	G_c(KJ/m^2) by energy measurement	E (MN/m^2)	$G_c = K_c^2/E$
Pure resin PER	0.65	0.3	2630	0.2	1.0	0.35	3625	0.3
PER/HTPB (90/10)	0.5	0.20	2370	0.1	0.8	0.25	2910	0.22
PER/HTPB (80/20)	0.55	0.25	1780	0.15	0.7	0.28	2430	0.2
PER/PBNCO (90/10)	1.3	0.85	2470	0.7	1.6	0.8	3160	0.8
PER/PBNCO (80/20)	1.2	0.83	1815	0.8	1.5	0.8	2680	0.8

* Charpy impact tests were performed by using a CEAST fracture pendulum at two different temperatures (−20°C and room temperature).

FIG. 42 SEM micrograph of fractured surface of plane PER resin (640×).

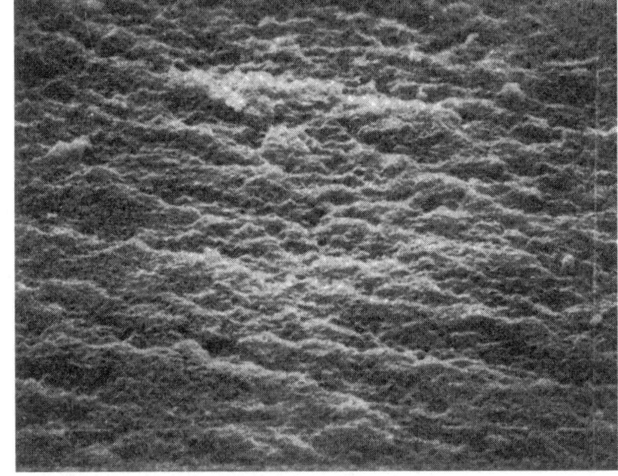

FIG. 43 SEM micrograph of fractured surface of PER/PBNCO 90/10 blend (640×).

of PER/HTPB and PER/PBNCO blends leads to the following observations:

(i) The dimensions of the dispersed rubbery phase in the case of PER/PBNCO blends are much smaller (average diameter 5 μm in 90/10 blends) than those of PER/HTPB blends (average diameter 30 μm in 90/10 blends) (compare Figs. 43 and 46).

(ii) The average dimension of the rubbery particles seems to increase, with increasing the content of rubber in both PER/HTPB and PER/PBNCO blends (see Figs. 44 and 47).

(iii) It can be observed that, on the broken surfaces of both blends (see Figs. 45 and 46), the crack front interacts with the dispersed domains. Tails, or steps, are formed at the rear of inhomogeneities due to the meeting of the two arms of the crack front from different fracture planes. According to Lange (93) such evidence led to a crack-pinning mechanism for the fracture propagation. According to this mechanism as a crack begins to propagate within the material the crack front bows out between the rubbery domains whilst still remaining pinned at all positions where it has encountered the dispersed particles. New

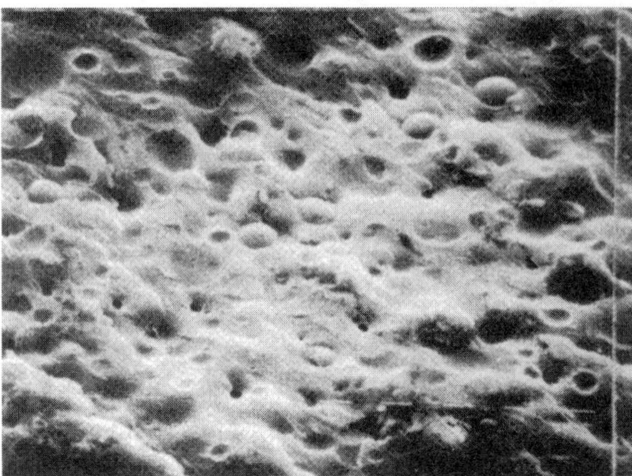

FIG. 44 SEM micrograph of fractured surface of PER/PBNCO 80/20 blend (640×).

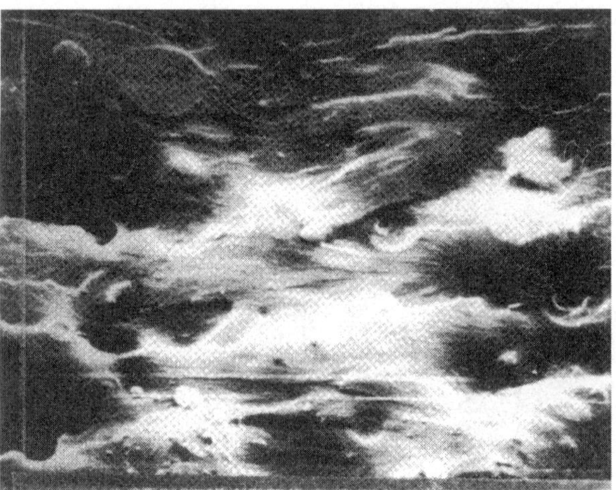

FIG. 45 SEM micrograph of fractured surface of PER PBNCO 90/10 blend 0 (2500×).

fracture surfaces are formed while the length of the crack front is increased.

The overall amount of energy required and then the degree of toughness enhancement of the PER materials will be certainly dependent on factors such as volume fraction, particle size, interparticle distance, interface structure and adhesion between dispersed domains and matrix.

(iv) The volume fraction of dispersed domains, as measured from SEM micrographs, results to be in the case of PER/HTPB larger than the calculated one (see data in Table 6).

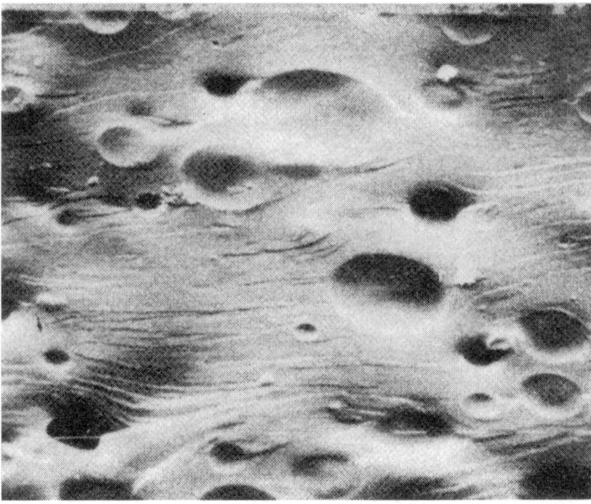

FIG. 46 SEM micrograph of fractured surface of PER/HITB 90/10 blend (640×).

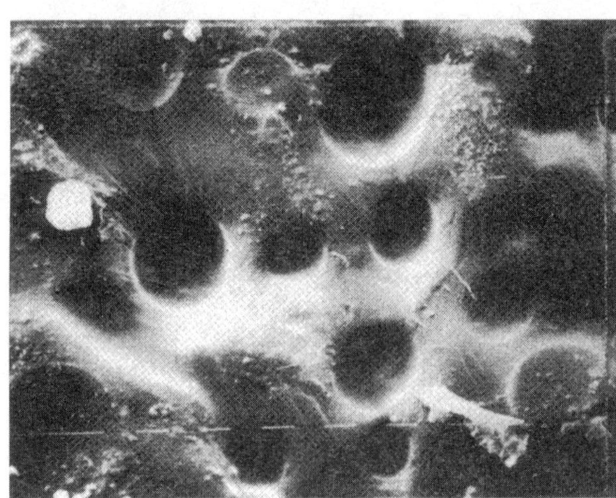

FIG. 47 SEM micrograph of fractured surface of PER/HTPB 80/20 blend (640×).

In order to know more about the nature of the dispersed phase in PER/HTPB and PER/PBNCO blends the broken surfaces were, prior to SEM analysis, etched with *n*-heptane vapor.

It can be observed that in the case of PER/HTPB blends the dispersed domains are only partly etched by the solvent (see Figs. 48 and 49): a sponge-like structure is obtained after etching.

This observation indicates that in such blends the dispersed domains do not contain only rubber but also the same inhomogeneities coming from the fact that

TABLE 6 Calculated and measured volume fraction of rubber domains for PER/HTPB and PER/PBNCO blends. (For the calculation of the volume fractions the following density values were used PER = 1.305 g/cm^3; HTPB = PBNCO = 0.897 g/cm^3)

Blend	Volume fraction calculated	Volume fraction measured
PER/PBNCO (90/10)	0.14	0.14
PER/PBNCO (80/20)	0.27	0.30
PER/HTPB (90/10)	0.14	0.20
PER/HTPB (80/20)	0.27	0.40

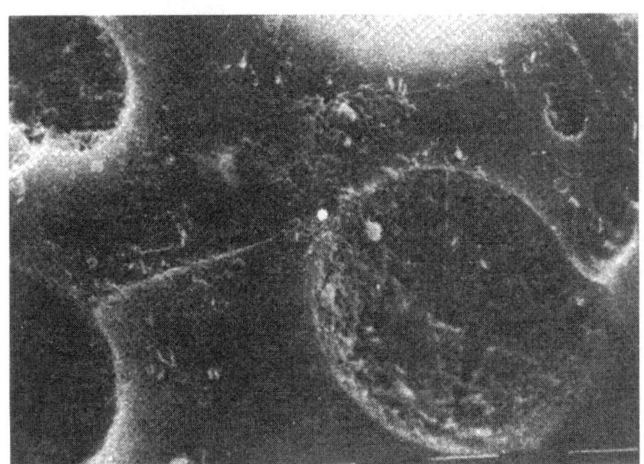

FIG. 48 SEM micrograph of fractured etched surface of PER/HTPB 90/10 blend (1250×).

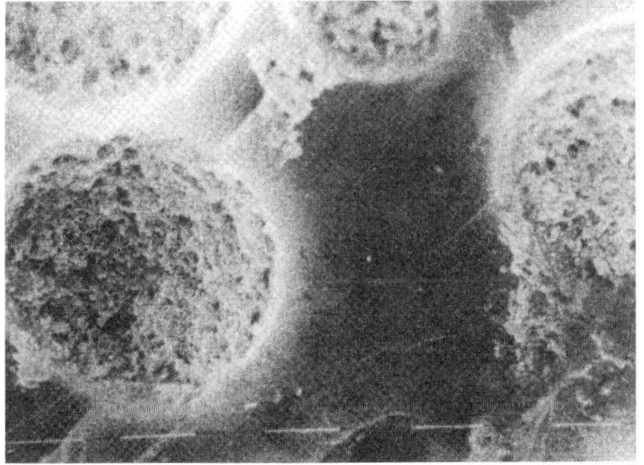

FIG. 49 SEM micrograph of fractured etched surface of PER/HTPB 80/20 blend (1250×).

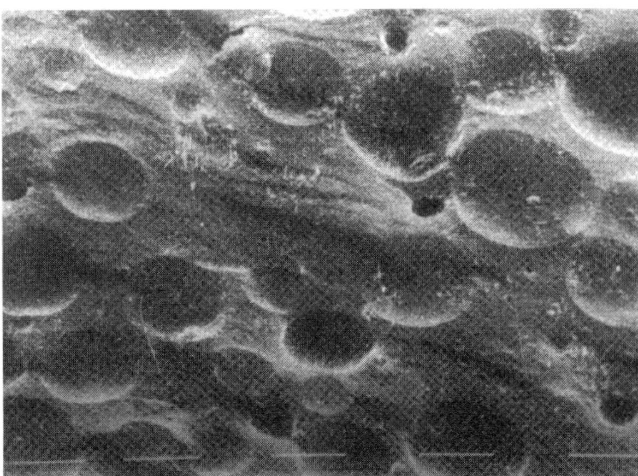

FIG. 50 SEM micrograph or fractured etched of PER/PBNCO 80/20 blend (1250×).

during the blending and before the cure process a certain amount of styrene is likely to occur.

The above hypothesis could explain the larger volume ratio of dispersed domains if measured for PER/HTPB blends. In the case of PER/PBNCO in contrast the segregation of PB between the PER blocks in the ABA copolymers should avoid the occurrence of such a phenomenon. As matter of fact almost no etched out materials is found in the case of PER/PBNCO blends (see Fig. 50). Such finding supports the idea that in those blends the rubbery domains are chemically bonded to the PER matrix.

V. FUNCTIONALIZED POLYPROPYLENE AS INTERFACIAL AGENTS IN POLYPROPYLENE MATRIX COMPOSITES WITH VEGETABLE FIBERS

Vegetable fibers are increasingly finding applications in polymeric reinforcements because of their desirable characteristics such as high specific properties [94, 95]. They are also biodegradable, renewable, and of low cost. The reinforcement potential of plant fibers lies in their cellulose content, fiber size, and hydrogen surface bonds [96, 98]. Some studies have been reported on composites based on a biodegradable polyhydroxybutyrate (PHB) reinforced by the steam-exploded wheat straw fibers [99, 100]. The steam explosion technique is a physical process applied on a wooden structure. It separates the lignocellulosic material into its main components, namely cellulose fibers, amorphous lignin, and hemicellulose. The last two substances are then removed and strong fibers with a high content of cellulose are finally produced. Fibers obtained through this process are known to be more reactive due to increased surface area [101]. However, the steam-explosion process has the disadvantage of producing relatively short fibers. This renders steam-exploded fibers reinforced composites inferior in certain properties compared to those of synthetic fiber-reinforced composites such as glass fiber reinforced thermosets and thermoplastics. In this chapter an attempt has been made to apply the steam-explosion technique to wheat straw fibers and broom from fibers deriving the branches of bottom plants.

A traditional chemical extraction process that uses an alkaline sodium hydroxide solution has also been applied to the same wooden material. Fibers extracted by the two methods have been used as reinforcement for the two commercial polypropylenes: the conventional isotactic polypropylene, and maleic anhydride functionalized isotactic polypropylene (iPPMA). The composites have also been subjected to a water absorption treatment. To assess the reinforcement effect of fibers, composite characterization techniques and determination of mechanical properties have been performed with particular attention to the fiber–matrix interfacial characteristic. The results have been compared with those of short glass reinforced polypropylene composites.

A. Experimental

1. Fiber Extraction

The vegetable fibers were subjected to two different fiber extraction processes: the alkaline treatment process, in which broom branches were heated to 40°C for 20 min in an aqueous solution of 4% (wt) sodium hydroxide. The fibers obtained were washed in water and dried, and the steam-explosion processes was applied whereby broom branches were steam heated at 190°C under a steam pressure of about 20 bar for 2 min using a Delta Lab EC 300.

The extracted fibers were then dried in an oven at 80°C for enough time to remove water absorbed during the steam-explosion process.

2. Straw Fibers/Polypropylene—Specimen Preparation

The fibers were chopped into short-length filaments using an electric grinding mill. Afterwards they were mixed with polypropylene using a Brabender-like apparatus (Rheocord EC, Haake Inc.) operating at 190°C for 10 min.

TABLE 7 Summary of the prepared composites and their composition and code

Materials	Composition (wt%)	Code
iPP	100/0	iPP
iPPMA	100/0	iPPMA
iPP/SEP straw fibers	80/20	iPP80/20
iPP/SEP straw fibers	50/50	iPP50/50
iPPMA/SEP straw fibers	90/10	iPPMA90/10
iPPMA/SEP straw fibers	80/20	iPPMA80/20
iPPMA/SEP straw fibers	70/30	iPPMA70/30
iPPMA/SEP straw fibers	50/50	iPPMA50/50

The prepared composites are listed and coded in Table 7.

Afterwards the material was chopped in the electric grinding mill and dried in a stove for 24 h at 55°C.

The milled material was placed between two teflon sheets in a 1.0 mm (or 3.5 mm for fracture tests) thick steel frame. The whole system was inserted between the plates of a hydraulic press heated at 190°C and kept without any applied pressure for 7 min, allowing complete melting. After this period a pressure of 10 MPa was applied for 3 min.

Previous thermogravimetric analysis had established that under such processing conditions no appreciable thermomechanical degradative effects were detected. To perform tensile and fracture tests the sheets were successively cut by a mill.

Prior to fracture testing, the samples were notched as follows: first a blunt notch was produced by using a machine with a V-shaped tool, and then a sharp notch of 0.2 mm depth was made by a razor blade fixed to a micrometric apparatus. The final value of notch depth was measured after fracture by using an optical microscope.

An attempt to blend iPPMA with untreated straw (not steam-exploded) was also effected for iPPMA80/20, but the resulting material was lacking in compactness, showing that no cohesive forces between the two components were present.

3. Broom Fibers/Polypropylene—Specimen Preparation

The fibers obtained from the two extraction processes (alkaline and steam explosion) were compounded with the polypropylene matrices by melt mixing, using a Brabender-like apparatus (Rheocord EC, Haake Inc.) operating at 190°C for 10 min: for 7 min, without applying any pressure, to allow complete melting of

TABLE 8 Summary of prepared composites and their composition and code

Materials	Composition (wt%)	Code
iPP	100/0	iPP neat
iPPMA	100/0	iPPMA neat
iPP/NaOH extracted fiber	50/50	iPP/broom 50/50
iPP/exploded fiber	50/50	iPP/broom SEP 50/50
iPPMA/NAOH extracted fiber	90/10	IPPMA/broom 90/10
iPPMA/NAOH extracted fiber	70/30	iPPMA/broom 70/30
iPPMAINAOH extracted fiber	50/50	iPPMA/broom 50/50
iPPMA/exploded fiber	90/10	iPPMA/broom SEP 90/10
iPPMA/exploded fiber	70/30	iPPMA/broom SEP 70/30
iPPMA/exploded fiber	50/50	iPPMAlbroom SEP 50/50
iPPMA/iPP/NAOH extracted fiber	5/45/50	iPPMA/iPP/broom 5/45/50
iPPMA/iPP/exploded fiber	5/45/50	iPPMA/iPP/broom SEP 5/45/50 89

the matrix, and subsequently for an additional 3 min at a pressure of 10 MPa. The composites were slowly cooled to room temperature and then released from the mold.

Some composites were also prepared using the fibers obtained from the two extraction processes and polypropylene, to which a small percentage of iPPMA (5%) had been added, to determine if the modified polypropylene can be used as a compatibilizing agent and impact improvement in the mechanical properties of the resulting composites.

All examined samples in Table 8 are listed and coded.

B. Results and Discussion

1. Wheat Straw Fibers/Polypropylene

(a) Thermal Analyses. The nucleating ability of the straw fibers on the crystallization process of iPP was tested by a nonisothermal crystallization experiment carried out by using the DSC technique. All samples were heated from 30°C to 200°C at a scan rate of 20°C/min and kept at this temperature for 10 min in order to destroy any trace of crystallinity (run I); afterwards the samples were cooled to 30°C by using a prefixed scan rate (run II). Finally, a third run (run III), similar to run I, was performed. The crystallinity indices of the samples were calculated by the following relation: $X_c = \Delta H^*/\Delta H^0$ where ΔH^0 is the specific heat of fusion of 100% crystalline iPP, taken as 209 J/g [102]. In Table 9 the thermal data obtained are shown and the following conclusions emerge:

(i) In the first fusion cycle the crystallinity content of iPP is quite the same for the composites both in iPP and in iPPMA; furthermore the melting point of iPP is the same for the composites both in iPP and iPPMA. Differences in the values of melting point and X_c between iPP and iPPMA can be ascribed to the lower crystallinity content of iPPMA with respect to iPP, owing, probably, to the presence of maleic anhydride.

(ii) No nucleating ability of the straw fibers for the iPP matrix was observed in the composites and by using all cooling scan rates.

(iii) Run III, after nonisothermal crystallization, showed no variation of the X_c values with increasing fiber content, as found in run I.

(b) FTIR Analyses. Figure 51 shows the spectrum of iPPMA as received and after activation by heating the film at 180°C for 5–30 min. The nonactivated sample shows a band at 1713 cm^{-1} with a shoulder at 1737 cm^{-1} ascribed to the dimeric form of dicarboxylic acids [103]. The thermal activation induces the appearance of two bands at 1785 and 1861 cm^{-1}, respectively, whose intensity increases as a function of the activation time. These bands were attributed to a cyclic anhydride group [104].

In addition, the intensity of the band at 1713 cm^{-1} decreases, suggesting the transformation from the less reactive hydrolized acid form to the more reactive cyclic anhydride one.

Figure 52 shows the spectra of the iPPMA copolymer (a), of the steam-exploded wheat straw (b), and of the straw/iPP composite (c).

The disappearance of the cyclic anhydride band is observed as well as the occurrence of some bands in the deconvoluted spectrum (see insert in Fig. 52) ranging from 1771 to 1645 cm^{-1}; the bands being not superimposed with those of the wheat straw and attributed [105] to the presence of acid or ester groups. The latter are due to the reaction between the iPPMA copolymer

TABLE 9 Apparent melting point, crystallinity index, and crystallization temperature of all samples*

Parameter	iPPMA neat (iPP neat)	iPPMA/straw 90/10	iPPA/straw 80/20 (iPP/straw 80/20)	iPPMA/straw 70/30	iPPMA/straw 50/50 (iPP/straw 50/50)
T_m(°C)	158 (166)	159	159 (167)	160	159 (165)
X_c(%)	35 (42)	35	36 (39)	35	37 (42)
T_c(°C)	117 (117)	117	117 (120)	118	117 (115)
X_c(%)	34 (41)	37	35 (40)	33	39 (44)
T_m(°C)	153 (162)	156	155 (161)	153	153 (162)
X_c(%)	31 (39)	34	33 (39)	32	32 (42)

* Scan rate = 20°C/min.

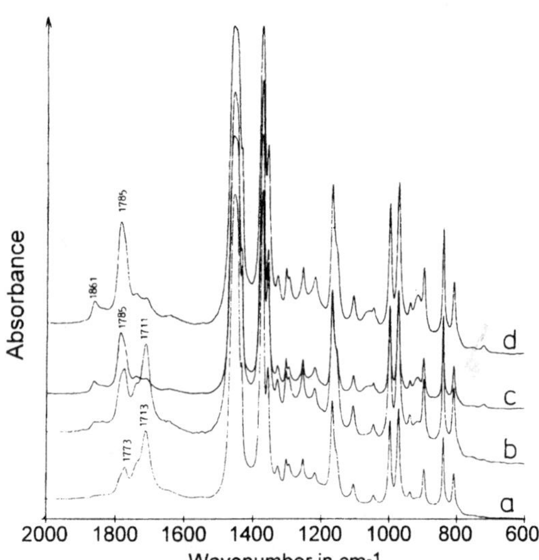

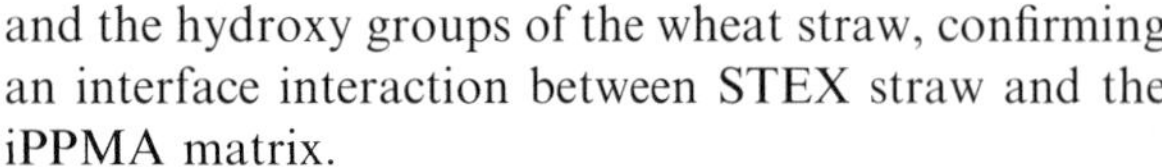

FIG. 51 FTIR spectra of iPPMA; as received (a), and after activation heating at 180°C for 5 (b), 10 (c) and 30 min (d).

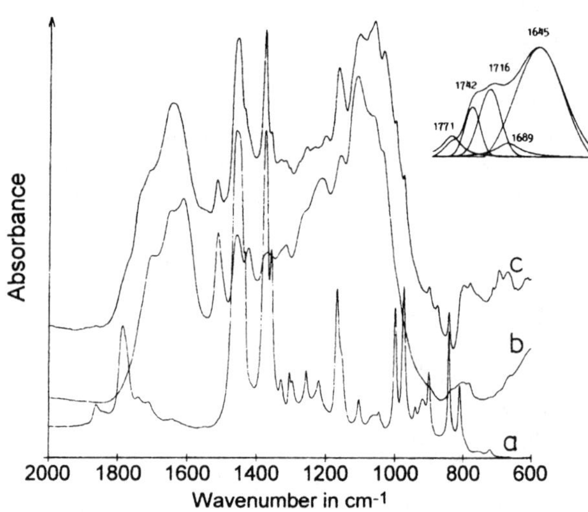

FIG. 52 FTIR spectra of (a) iPPMA activated at 180°C for 30 min; (b) steam-exploded wheat straw; (c) iPPMA 50/50. Insert shows a deconvolution.

and the hydroxy groups of the wheat straw, confirming an interface interaction between STEX straw and the iPPMA matrix.

(c) Impact Tests and Fractographic Analysis. The values of the critical stress intensity factor, K_c for all examined samples, calculated according to the Linear Elastic Fracture Mechanics (LEFM) theory, as a function of straw fiber content are shown in Fig. 53. The values of K_c seem to linearly increase with increasing fiber content for iPPMA-based composites, while the K_c values of iPP-based composites remain almost constant.

The strong performance of fracture parameter, found in the iPPMA composites, is probably due to a stronger chemical interfacial adhesion owing to the presence of reactive radicals of maleic anhydride, but also its nucleating action producing a smaller size of spherulites and consequently a more pronounced presence of amorphous phases allows it to absorb more energy during the fracture test.

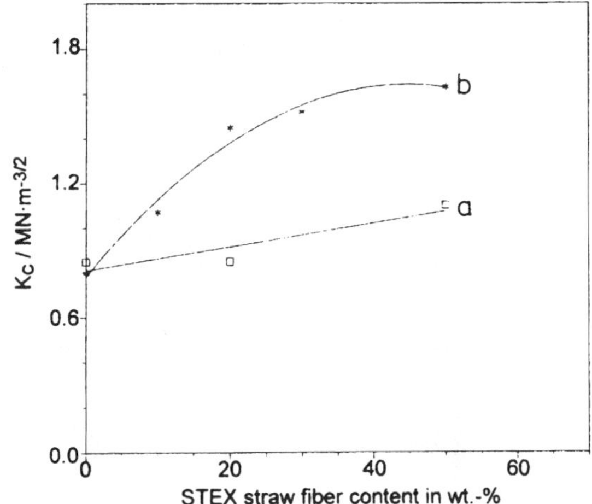

FIG. 53 Critical stress intensity factor, K_C, of (a) iPP and (b) iPPMA-based composites as a function of fiber content (ASTM D256).

(d) Tensile Properties. Tensile properties such as tensile modulus (E), strength at break (σ_b), and elongation at break (ε_b) were evaluated from the stress–strain curves. The results are reported in Table 10. It can be observed that the stiffness of the material (E) slightly increases with the fiber content for the iPPMA-based composites, while it remains almost constant in the case of iPP-based composites.

Concerning the variation of strength at break (σ_b) as a function of straw fiber content, it is observed that the parameter decreases for composites, whereas an opposite trend, in the case of the iPPMA composite, is observed. This decrease found for iPP composites is likely to be due to weak interfacial bonding with the matrix. On the contrary, the high σ_b values for iPPMA-based composites again have to be attributed to the presence of maleic anhydride on the iPP backbone, that is able to graft the STEX fiber terminals.

The low values of elongation at break (ε_b) confirm a brittle behavior of the examined composites. In particular, the iPP-based composites have shown a decrease of elongation with increasing fiber content, whereas the iPPMA-based composites have shown only a slight increase.

In Fig. 54 the SEM micrographs of fractured samples (by tensile tests) for iPP80/20 (a, b) and iPPMA80/20 (c, d), respectively, are shown.

Both the samples showed a brittle behavior, but iPPMA-based samples seem to show a better dispersion of fibers together with a better covering with respect to iPP-based samples. Moreover more regions of plastic deformations seem to appear in the case of iPPMA composites (see Fig. 54c).

(e) Dynamic–Mechanical Tests. Figure 55a shows the curves of storage modulus (E') and loss modulus (tan δ) of the neat iPP matrix, iPP8O/20 and iPP50/50 composites, respectively, as a function of temperature; in Fig. 55b the same curves of iPPMA-based composites are plotted. From the analysis of the figures it can be deduced that:

(i) The modulus E' is increased with increasing fiber content for iPPMA-based composites; in the opposite way the modula are decreased in iPP-based composites. This is found to agree with tensile results.

In all cases the storage modulus, E', drops as the sample goes through the matrix–glass transition. The decrease of E' with temperature is characteristic of semicrystalline polymers and the identical slopes of the curves indicate a similar degree of crystallinity owing to the same thermal treatment.

(ii) Over the temperature range in which the samples have been analyzed, only the glass transition (T_g) is observed. The T_g, taken as the maximum of the

TABLE 10 Tensile properties of iPPMA and iPP-based composites (ASTM D638)

Parameter	iPPMA/straw (iPP neat)	iPPMA/straw 90/10	iPPMA/straw 80/20 (iPP/straw 80/20)	iPPMA/straw 70/30	iPPMA/straw 50/50 (iPP/straw 50/50)
E	1.2	1.3	1.4	1.5	1.6
(GPa)	(1.2)		(1.2)		(1.1)
σ_b	19.7	29.0	27.8	26.2	30.4
(MPa	(15.0)		(11.4)		(8.9)
ε	1.9	2.9	2.7	2.2	2.4
(%)	(1.4)		(1.0)		(0.9)

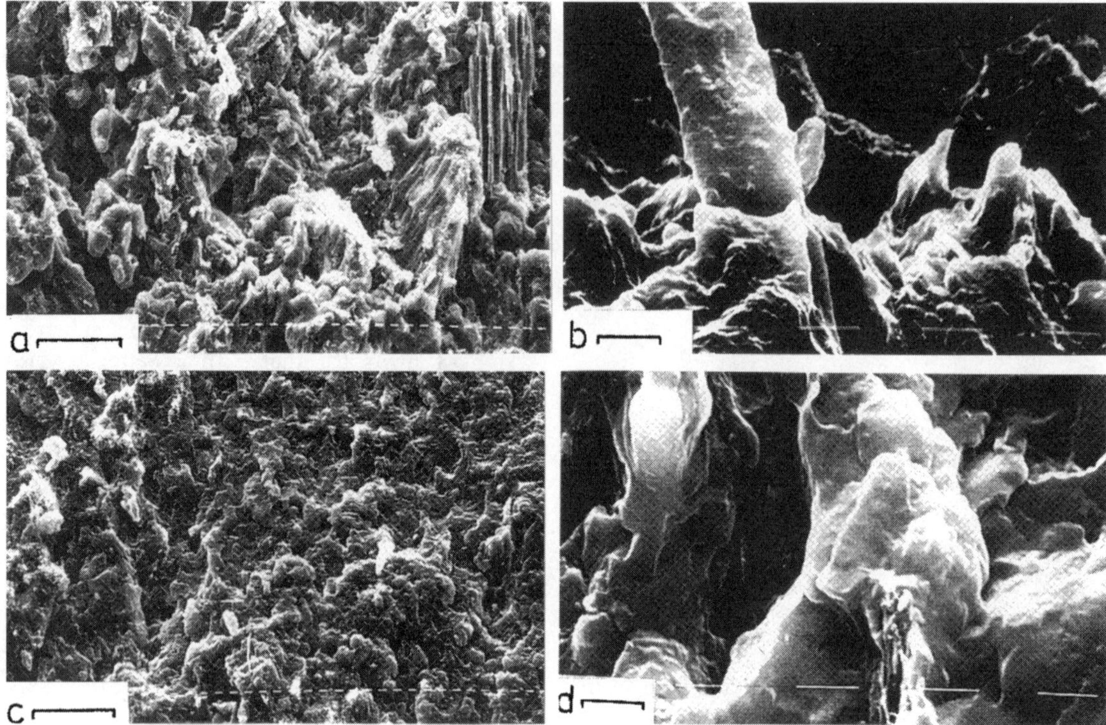

FIG. 54 Scanning electron micrographs of fractured surfaces of iPP and iPPMA based composites; (a) iPP80/20, (b) iPP80/20, (c) iPPMA80/20, (d) iPPMA80/20; length of the bars: (a, c) 160 μm; (b, d) 10 μm.

tan δ peak, is not affected by the presence of filler both in iPP and iPPMA based composites.

The loss modulus tan δ is related to the dissipated energy [106] by the following equation:

$$\tan\delta = E_d/(E_0^2 - E_d^2)^{1/2}$$

where E_0 is the total work done during a cycle of oscillation, and E_d is its irreversibly lost component, i.e., the fraction of the deformation energy dissipated as heat.

(f) Water Absorption. Water absorption is one of the most serious problems that prevents a wide use of natural fiber composites; in fact, in wet conditions, this phenomenon strongly decreases the mechanical performances of composites. This latter can be attributed to the fact that, generally, the lignocellulosic materials are hydrophilic and the plastics are hydrophobic. Consequently, the resulting interface bond between natural fibers and polymer is too weak. Only by creating a stronger interface between matrix and reinforcement material is it possible to reduce the hygroscopicity of lignocellulosic-based materials.

Water absorption tests on our samples were performed by keeping the samples immersed at room temperature for one month and the increase of weight due to the water absorbed was periodically measured. The results of the tests are reported in Fig. 56. The findings show that the two homopolymers (iPP and iPPMA) absorb about the same amount of water. In contrast a lower quantity of water can penetrate into the iPPMA50/50 composite with respect to the iPP-based composite (iPP50/50), owing to a better adhesion of the fibers to the matrix. Tensile tests, after this treatment, were also performed. The results are shown in Table 11 and the following can be pointed out:

(i) A constant decrease of tensile properties (E, σ_b, and ε_b) for all wet samples was observed.

(ii) Only a slight worsening of tensile strength at break was found in the case of iPPMA50/50. The parameter (σ_b) can give an indication of the resulting fiber–matrix adhesion in the composites, confirming, in our case, the positive action of maleic anhydride groups that allow to improve the interface between the two phases, avoiding

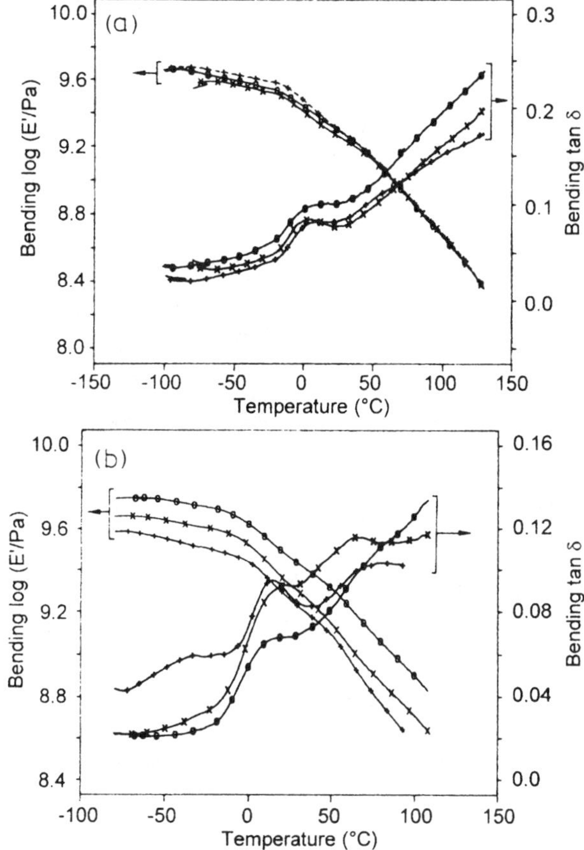

FIG. 55 DMTA storage modulus, E, and loss modulus, $\tan\delta$, vs temperature; (a) (+) neat iPP, (×) iPP80/20, (●) iPP50/50; (b) (+) neat iPPMA, (×) iPPMA80/20, (●) iPPMA50/50.

an easy penetration of water molecules into the composites.

(g) Conclusions. The following conclusions from the results reported can be drawn:

- FTIR analysis shows the presence of ester linkages in the case of iPPMA composites, allowing a superior interfacial adhesion with respect to iPP-based composites.
- The presence of STEX straw fibers in both matrices (iPP and iPPMA) does not produce a worsening of thermal properties of composites (T_m, X_c and T_c).
- The fracture toughness parameters (K_c and G_c) increase with STEX straw fiber content, especially in the case of iPPMA composites.
- The mechanical properties, as shown by tensile tests, for iPPMA composites increase with the straw fiber content; an opposite trend in the case of iPP composites series is observed. These results have also been confirmed by fractographic analyses performed by SEM.
- The presence of maleic groups is able to create a strong interface between polypropylene and lignocellulosic material, limiting the penetration of water molecules under wet conditions.

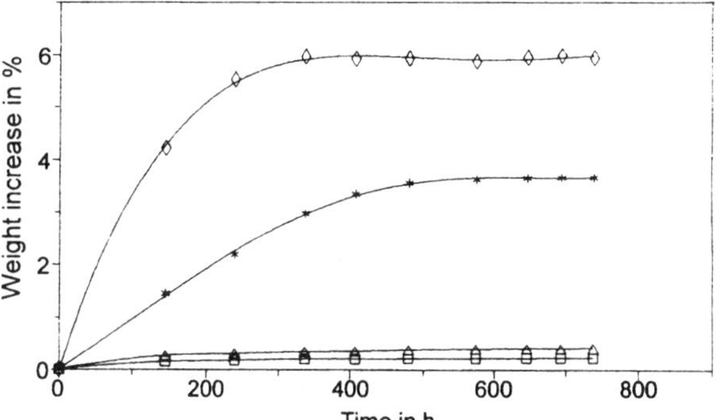

FIG. 56 Water absorption test: increase of weight of the samples vs time; (□) neat iPP, (▲) neat iPPMA, (◇) iPP50/50, (*) iPPMA50/50.

TABLE 11 Tensile properties of iPPMA and iPP-based composites (ASTM D638) after water absorption test

Parameter	iPPMA neat (iPP neat)	iPPMA/straw 50/50 (iPP/straw 50/50)
E (GPa)	1.1 (11.1)	1.3 (1.0)
σ_b MPa)	12.7 (11.3)	28.4 (6.1)
ε_b (%)	1.3 (1.4)	1.7 (0.7)

2. Broom Fibers/Polypropylene

(a) Fiber Characterization. Wide-angle X-ray spectroscopy (WAXS) was used to determine the structure of the fibers obtained by the two extraction procedures. Here, the lignocellulosic material, in provider form, was subjected to a Ni-filtered CuKα radiation using a Siemens 500 D Diffractometer equipped with a 5 cm^{-1} scintillator counter and a linear amplifier.

In Fig. 57 the WAXS diffraction patterns for NaOH-extracted broom fibers, SEP broom fibers, and pure cellulose in the crystallographic form I are reported.

From these graphics the following can be observed: Both extracted fibers are in the crystallographic form I of the cellulose. Both spectra show the 22.6° (2θ) reflection related to the (002) crystallographic plane and the

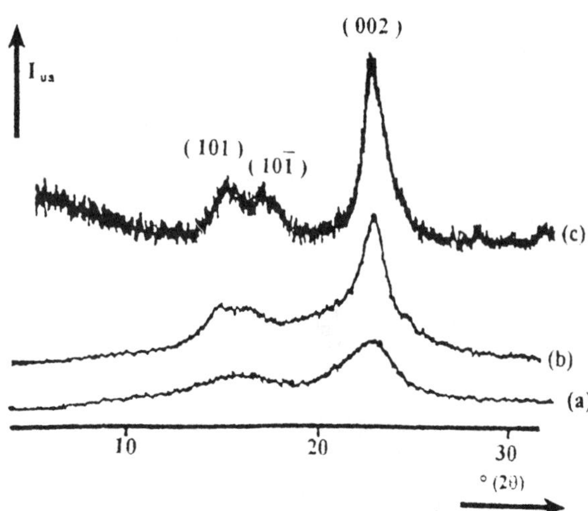

FIG. 57 WAXS diffraction patterns: (a) NaOH-extracted fiber, (b) SEP fiber, (c) pure cellulose.

two broad and unresolved reflections in the range from 13° to 18° (2θ) corresponding to the ($10\bar{1}$) and the crystallographic planes of cellulose I, respectively. The SEP extracted fibers are characterized by a higher degree of crystallinity with respect to the NaOH treated ones. The crystallinity indices, calculated according to the Segal method [105] are $(I_c)_{\text{SEP}} = 0.77$ and $(I_c)_{\text{NaOH}} = 0.60$, respectively. Finally, the sharpness of all the reflections clearly show that, in the case of SEP extracted fiber, the crystal size seems to be higher with respect to that of the NaOH extracted one.

These different crystallization behaviors can be related, in the case of SEP treatment, to a reorganization at high temperatures of amorphous and/or paracrystalline cellulose regions, releasing the strains existing in the native cellulose, mainly due to the interactions with hemicelluloses and lignin

Morphological studies on the fibers were carried out by scanning electron microscopy using a Philips 501 SEM after metallization of the fiber surfaces by means of a Polaron sputtering apparatus with Au–Pd alloy.

In Fig. 58a and 58b the SEM micrographs for NaOH-extracted broom fibers and SEP broom fibers are reported, respectively. From these figures it can clearly be deduced that the exploded fibers present a smooth and clean surface with respect to the alkaline-treated ones, confirming that the steam-explosion process, performed on lignocellulosic materials, produces an effective strong defibrillation giving rise to a high number of single shorter fibers. Moreover, it seems

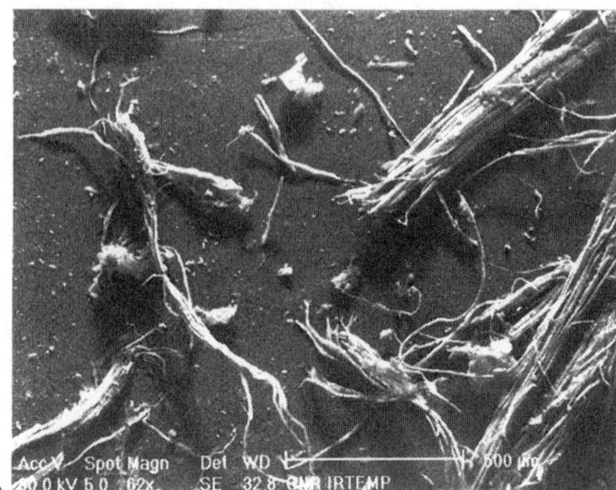

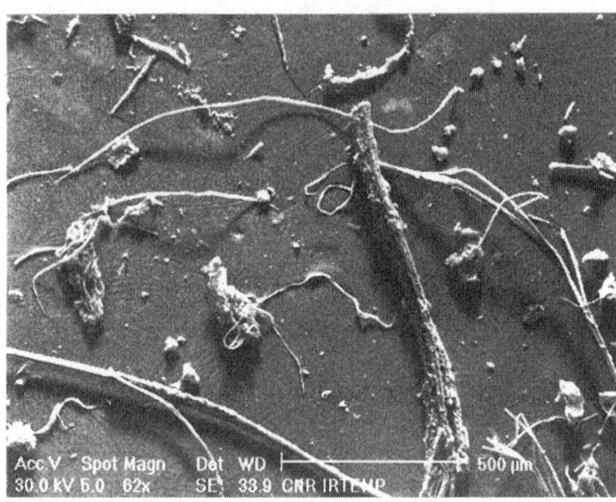

FIG. 58 SEM micrographs: (a) NaOH-extracted fiber, (b) SEP fiber.

that the NaOH-extracted fibers are not single separated, but rather bundled.

It should be noted that in addition to the structural modification of wood, as described above, there are also mechanical effects; this last, owing to the adiabatic expansion of vapour, produce morphological variations able to increase the specific surface. This led to a higher reactivity of SEP and resulted in higher availability of the hydroxyl groups distributed on the surface of cellulose.

(b) Thermal Analysis of the Composites. In Tables 12, 13, and 14 the findings obtained from DSC measurements, for NaOH-extracted fibers and SEP-extracted fibers, iPPMA-based composites, and iPP-based materials, are listed, respectively. The calculated thermal parameters are: the apparent melting temperature measured in the first fusion run (T'_m[I])

TABLE 12 Apparent melting point measured in the first and second fusion run (T'_m [I] and T'_m [II]), crystallinity content (X_c), crystallization temperature (T_c), and glass transition temperature (T_g) of iPPMA-based composites reinforced with NaOH extracted broom fibers

Parameters	iPPMA Neat	iPPMA/Broom 90/10	iPPMAIBroom 70/30	iPPMAIBroom 50/50
T'_m [I] (°C)	160.0	160.9	160.8	161.6
X_c (%)	36.3	37.9	39.7	49.7
T_c (°C)	115.2	117.8	120.7	123.7
X_c (%)	40.0	40.1	40.4	40.7
T'_m [II] (°C)	159.1	159.0	159.1	158.3
X_c (%)	36.7	38.6	38.3	35.9
T_g (°C)	−10.8	−8.3	−5.2	−8.7

TABLE 13 Apparent melting point measured in the first and second fusion run (T'_m [I] and T'_m [II]), crystallinity content (X_c), crystallization temperature (T_g), and glass transition temperature (T_g) of iPPMA-based composites reinforced with exploded broom fibers

Parameters	iPPMA Neat	iPPMA/Broom SEP 90/10	iPPMA/Broom SEP 70/30	iPPMA/Broom SEP 50/50
T'_m [I] (°C)	160.0	160.9	161.7	162.2
X_c (%)	36.3	36.5	40.3	46.7
T_c (°C)	115.2	115.3	119.4	120.3
X_c (%)	40.0	46.8	39.8	44.3
T'_m [II] (°C)	159.1	159.7	157.4	156.0
X_c (%)	36.7	41.0	41.6	41.0
T_g (°C)	−10.8	−9.5	−8.8	−7.3

TABLE 14 Apparent melting point measured in the first and second fusion run (T'_m [I] and T'_m [II]), crystallinity content (X_c), crystallization temperature (T_c), and glass transition temperature (T_g) of iPPA-based materials reinforced both with NaOH extracted fibers, exploded fibers, and of ternary composites

Parameters	iPPMA Neat	iPPMA/Broom 50/50	iPPMA/Broom SEP 50/50	iPPMA/iPP/Broom m 5/45/50	iPPMA/iPP/Broom SEP5/45/50
T'_m [I] (°C)	167.8	167.6	164.0	163.2	163.2
X_c (%)	40.4	50.2	45.7	51.5	51.0
T_c (°C)	117.3	124.3	122.3	125.2	121.5
X_c (%)	43.7	48.3	45.2	49.2	42.5
T'_m [II] (°C)	164.2	166.3	165.5	162.3	160.0
X_c (%)	38.0	42.7	40.3	49.7	42.5
T_g (°C)	−9.2	−7.6	−7.6	−14.7	−10.0

and in the second fusion run (T'_m[II]), the crystallization temperature (T_c), the crystallinity indices for the three processes (X_c), and the glass transition temperature.

For both iPPMA-based composites series one can observe a slight increase in T'_m[I] and X_c in the first fusion run for the composites with respect to neat iPPMA. This fact probably can be attributed to the strong interaction between cellulosic fibers and iPPMA matrix. The fibers may "extract" selectively, the maleic anhydride groups (constitutional defects), from the polypropylenic backbone, leading to composites having a matrix characterized, on average, by a higher degree of constitutional regularity, with respect

FIG. 59 Optical micrograph: iPPMA spherulites growing on a NaOH-extracted fiber.

TABLE 15 Young's elastic modulus (E), tensile strength at break (σ_b), and percent elongation at break (ε_b) of all the samples

Materials	E (Gpa)	σ_b (Mpa)	ε_b (%)
iPPMA neat	1.2	18.3	2.0
iPPMA/broom 90/10	1.4	22.9	2.1
iPPMA/broom 70/30	1.8	27.6	2.2
iPPMA/broom 50/50	2.5	29.2	1.5
iPPMA/broom SEP 90/10	1.2	15.0	1.6
iPPMA/broom SEP 70/30	1.4	15.1	1.3
iPPMA/broom SEP 50/50	1.5	11.2	0.8
iPP neat	1.0	16.5	5.8
iPP/broom 50/50	0.8	9.5	1.6
iPP/broom SEP 50/50	1.3	3.2	2.2
iPPMA/iPP/broom 5/45/50	2.1	16.7	0.9
iPPMA/iPP/broom SEP 5/45/50	1.2	9.5	2.2

to the bulk in the neat iPPMA, and consequently higher T'_m and X_c.

Moreover, the increase of T_c in composites with respect to the neat matrix can be attributed to the nucleating ability of the cellulosic fiber on the polymeric material as was observed by means of optical microscopy. For this purpose, in Fig. 59 an optical micrograph of iPPMA spherulites growing on a NaOH extracted fiber is reported. The figure shows a pronounced nucleating effect of the cellulosic fiber on the polypropylenic matrix. It is possible to note that the radius of the bulk crystallized spherulites and of the fiber nucleated spherulites seem to be comparable in their dimensions, so that only the nucleation and not the growth process is influenced by the presence of cellulosic fiber.

(c) Tensile Tests. The values of tensile properties such as strength at break (σ_b), tensile modulus (E), and percent (ε_b) of the composites are shown in Table 15.

From these parameters, the following considerations can be deduced: (a) iPP based composites show a tensile Young's modulus comparable to that of iPP homopolymer. Conversely, the tensile strength at break and the percentual elongation at break show a sharp decrease compared to the neat matrix. These . findings are probably due to weak fiber/matrix interfacial adhesion. (b) iPPMA-based composites, reinforced with NaOH-extracted fibers, exhibit a strong increase of both Young's elastic modulus and tensile strength at break. These performances can probably be ascribed to a stronger fiber/matrix interface, caused by the interactions between maleic anhydride grafted on polypropylene and —OH groups on cellulosic fibers. (c) The iPPMA-based materials reinforced with SEP-extracted fibers show only a slight decrease in tensile strength at break, while the elastic modulus remains almost constant with respect to the neat iPPMA matrix. (d) The two ternary composites, having a composition of 45% wt of iPP, 5% wt of iPPMA, and 50% wt of fibers, both NaOH treated and SEP extracted, present interesting behavior: their tensile properties are comparable with that of the related iPPMA-based material and also are higher with respect to iPP-based composites.

To explore the possibility for a more diffuse utilization of broom fibers composites in substitution for short glass fiber materials, a comparative analysis of properties of these two kinds of composites was performed, and the results are reported in Fig. 60(a, b). In these graphics Young's modulus (a) and tensile strength at break (b), normalized with respect to the materials density, are reported as a function of fiber content.

The trends of specific Young's modulus for the two types of composites show that the two materials have comparable behaviour and, in particular, that the 50/50 broom fiber composite has a tensile modulus higher than the 70/30 glass fiber one.

Furthermore, the broom fiber composites show, for all the examined compositions, an even higher specific tensile strength at break compared to the glass fibers reinforced materials. Therefore, the broom fibers show a reinforcement capability comparable to that of glass fibers.

(d) Water Absorption Tests. One of the most undesirable properties of vegetable fibers is their

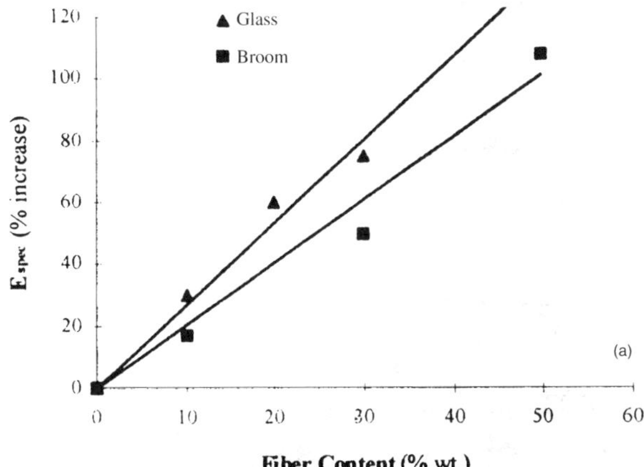

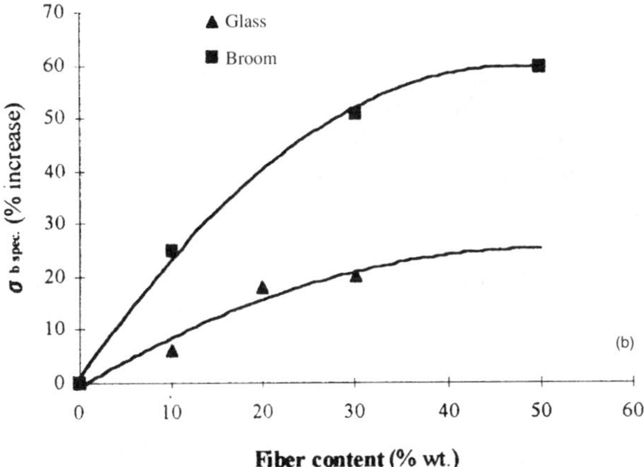

FIG. 60 Tensile test: percent increase in Young's elastic modulus (a) and in tensile strength at break (b) as a function of fiber content for glass reinforced and NaOH-extracted, broom fibers reinforced composites.

dimensional instability due to moisture absorption. This phenomenon is mainly caused by the hydrogen bonding between water molecules and the hydroxyl groups present in the cellulose structure. Clearly, a strong fiber/matrix interfacial adhesion can help to avoid the water penetration reducing the hygroscopicity, and, consequently, the worsening in the mechanical performances of materials.

Water absorption tests on our samples were performed according to the methods described in the experimental section.

The results, shown as percent weight increase as a function of immersion time, are summarized in Fig. 61, from which the following can be deduced: (1) as expected, the homopolymers (iPP and iPPMA) show the lowest water absorption; (2) the two iPP-based materials (iPP/broom 50/50 and iPP/broom SEP 50/50) are characterized by the shortest saturation time: the maximum water absorption was obtained after an immersion time of about 200 h; moreover, for these samples, the highest saturation percentage (about 12% wt) was found; clearly, these findings can be ascribed to the poor fiber/matrix adhesion as already shown by their mechanical behavior; (3) the two iPPMA-based composites reinforced with NaOH-extracted and SEP-treated fibers show different amounts of water absorbed. In the case of SEP-extracted fiber composites, a lower quantity of water penetrates than with the NaOH-extracted fiber materials. This finding is probably due to the lower crystallinity index of these latter fibers with respect to the exploded ones, as demonstrated by WAXS measurements. Moreover, the good fiber/matrix interfacial adhesion, obtained by using SEP-extracted fibers as reinforcements for iPPMA matrix, can also contributes to reduce the water absorption. (d) The two ternary composites reveal, probably, also a strong fiber/matrix interface and have a low percentage of water absorbed at saturation and a long saturation time (for NaOH-extracted fiber composites about 7% of water absorbed and a saturation time of 800 h, for SEP fiber materials 9% and 800 h, respectively); this is in agreement with results of tensile tests.

(e) Conclusions. The following conclusions from the results can be drawn:

1. Broom fibers obtained by steam explosion seem to produce better fiber/matrix interfacial adhesion with respect to those extracted by alkaline treatment, as clearly shown by SEM analysis and water absorption tests.

   Nevertheless, the mechanical behavior of composites reinforced with NaOH-extracted broom fiber is superior with respect to that of the SEP fiber composites; this finding can probably be attributed to the minor structure damage produced by the chemical treatment that allows one to obtain longer and bundled fibers.
2. The utilization of an iPPMA matrix as an alternative to a conventional polypropylene produces composites with good final properties. This occurs because of the presence of maleic groups that are able to create a strong interface between matrix and cellulosic material.
3. The strong affinity of the iPPMA matrix for the lignocellulosic fibers was also demonstrated by the behaviour of the two ternary blends. These were

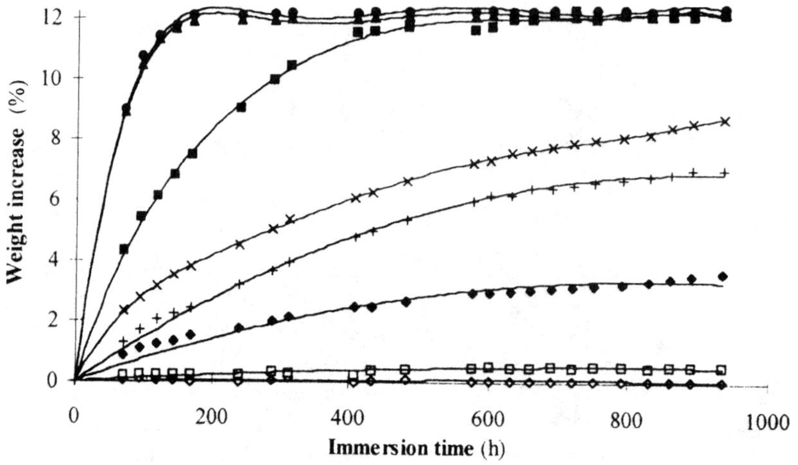

FIG. 61 Water absorption tests: weight increase of the samples as a function of time. (◇ iPP neat; ▲ iPP/Broom; ■ iPPMA/Broom; × iPPMA/iPP/Broom; □ iPPMA neat; ● iPP/Broom SEP; ◆ iPPMA/Broom SEP; + iPPMA/iPP/Broom SEP).

prepared by using a matrix constituted of 90% of iPP and only 10% of iPPMA, and they present properties very similar to those of whole iPPMA-based composites. A possible explanation of this finding can be given by supposing that the cellulosic fibers "capture" the maleated polypropylenic chains by creating an esteric linkage between the cellulosic —OH and the maleic anhydride molecules grafted on the polypropylenic backbone, at the same time excluding the iPP bulk to participate at the interface.

4. The use of SEP fibers as reinforcement for iPPMA-based composites contribute to reduce the penetration of water in the structure with respect to the NaOH-extracted fibers: this effect can be due both to the higher crystallinity of the SEP fibers that reduces the water diffusion through the amorphous region, and to better fiber/matrix interfacial adhesion.
5. The NaOH-extracted fiber iPPMA-based composites present specific mechanical properties comparable to those of analogous short glass fiber reinforced materials; this can lead, in the future, to a real possibility for broom fibers to be used as alternative materials to the standard synthetic reinforcement, also owing to the fact that they are cheap, light, and biodegradable.

VI. COMMERCIAL FUNCTIONALIZED PO AND FUTURE TRENDS

Many companies have added to their standard production grades PO a few modified products. Some products are explicitly defined as "maleic anhydride grafted" polypropylene and polyethylene (AlliedSignals A-C products and Montell Hercoprime products); some other companies refer to them as specialized modified PO (Dupont's Bynel and Fusabond products). Their applications range from resin modifications (toughener, mainly) to adhesive layer for multilayer packagings, to coatings for better printing of PO. At moment, no reliable figures for market demand are available. Also to be fully exploited is the possibility of such functionalized PO to work as compatibilizer for post-consumer mixed plastic scraps recycling. One example is already proposed on the market, recalled as a "universal compatibilizer" (Bennet), with some qualitative data indicating its efficiency. Even though it has not declared its chemical nature, from its characteristics it seems to clearly belong to the wide and steadily growing family of functionalized PO.

ACKNOWLEDGMENTS

The authors are indebted to Dr L Calandrelli for precious support in realizing this work.

SYMBOLS AND ABBREVIATIONS

ABS	acrilonitrile-butadiene-styrene copolymer
BDO	butandiol
CB	chlorobenzene
CRPP	controlled rheology polypropylene
DBPO	dibenzoylperoxide
DBS	dibutylsuccinate

DSC	differential scanning calorimetry
DMT	dimethylterephthalate
EA	ethanolamine
ESI	ethanolsuccinimide
EPR	ethylene–propylene rubber
EVA	ethylene-vinylacetate copolymer
EVOH	ethylene-vinylalcohol copolymer
FTIR	Fourier transform infrared spectroscopy
HTPB	hydroxyl-terminated polybutadience
iPP	isotactic polypropylene
iPPMA	maleic anhydride-grafted isotactic polypropylene
LDPE	low density polyethylene
MAH	maleic anhydride
MEM	monoethylmaleate
MES	monoethylsuccinate
NMR	nuclear magnetic resonance
PA6	polyamide 6
PBNCO	isocyanate-terminated polybutadiene
PBT	polybutyleneterephthalate
PE	polyethylene
PER	uncured unsaturated polyester resin
PET	polyethyleneterephthalate
PHB	polyhydroxybutyrate
PIB	polyisobutylene
PO	polyolefins
PS	polystyrene
PVC	polyvinylchloride
RMPER	cured rubber modified polyester resin
SA	succinc anhydride
SEM	scanning electron microscopy
SEP	steam explosion process
STEX	steam exploded
TDI	toluenediisocyanate
TEA	triethylamine
TGA	thermogravimetric analysis
TMS	tetramethylsilane
WAXS	wide angle X-ray spectroscopy

REFERENCES

1. K Solc. Polymer Compatibility and Incompatibility: Principles and Practices. New York: Harwood, 1982.
2. RC Kowalski. Reactive Extrusion, Principles and Practices. New York: Hanser, 1992.
3. EM Fettes, Chemical Reactions of Polymers. New York: Interscience, 1964, pp 142–293.
4. BK Kim, SY Park, SJ Park. Eur Polym J 27: 349, 1991.
5. A Simmons, WE Baker. Polym Eng Sci 29: 1117, 1989.
6. Z Song, W E Baker. J Appl Polym Sci. 41: 1299, 1990.
7. NC Liu, WE Baker, KE Russell. J Appl Polym Sci 41: 2285, 1990.
8. NG Gaylord, M Metha, V Kumar. Org Coat Appl Polym Sci Proc 46: 87, 1981.
9. NG Gaylord, M Metha. J Polym Sci Polym Lett 20: 481, 1982.
10. KE Russell, EC Kelusky. J Polym Sci Part A, Polym Chem 26: 1189, 1988.
11. Y Minoura, M Ueda, S Mizunuma, M Oba. J Appl Polym Sci 13: 1625, 1969.
12. G Ruggeri, M Aglietto, A Petragnani, F Ciardelli. Eur Polym J 19: 863, 1983.
13. NG Gaylord. J Polym Sci Part C 31: 247, 1970.
14. NG Gaylord, H Antropiusova, BK Patnaik. J Polym Sci Polym Lett Ed. 9: 387, 1971.
15. NG Gaylord, A Takahashi, S Kikuchi, RA Guzzi. J Polym Sci Polym Lett Ed. 10: 95, 1972.
16. YN Sharma, MK Naqvi, PS Gawande, IS Bhardwaj. J Appl Polym Sci 27: 2605, 1982.
17. A Yamaoka, K Ikemoto, T Matsui, J Yamauchi. Nippon Kagaku Kaishi (J Chem Soc Japan Chem Ind Chem) 11: 1919, 1989.
18. J Yamauchi, A Yamaoka, K Ikemoto, T Matsui. J Appl Polym Sci 43: 1197, 1991.
19. R Rengarajan, VR Parameswaran, S Lee, L Rinaldi. Polymer 31: 1703, 1990.
20. JE Wyman, IJ Rangwalla, SV Nablo. Polym Mater Sci Eng 60: 463, 1989.
21. RP Singh. Prog Polym Sci 17: 251, 1992.
22. D Curto, A Valenza, FP La Mantia. J Appl Polym Sci 39: 865, 1990.
23. JGA Terlingen, HFC Gerritsen, AS Hoffman, J Feijen. J Appl Polym Sci 57: 969, 1995.
24. N Inagaki, S Tasaka, M Imai. J Appl Polym Sci 48: 1963, 1993.
25. DW Yu, M Xanthos, CG Gogos. J Appl Polym Sci 52: 99, 1994.
26. S Datta, G Ver Strate, EN Kresge Am Chem Soc Polym Preprints 899, 1992.
27. TC Chung, D Rhubright, GJ Jian. Macromolecules 26: 3467, 1993.
28. T Duschek, R Mulhaupt. Am Chem Soc Polym Preprints 170, 1992.
29. R Mulhaupt, T Duschek, B Rieger. Makromol Chem Macromol Symp 48/49: 317, 1991.
30. T Shiono, H Kurosawa, O Ishida, K Soga, Macromolecules 26: 2085, 1993.
31. RD Leaversugh. Mod Plastics 111, 1994.
32. DJ Lohse, S Datta, EN Kresge. Macromolecules 24: 561, 1991.
33. a. S Datta, DJ Lohse, Macromolecules, 26: 2064, 1993. b. DJ Lohse, LJ Fetters, MJ Doyle, HC Wang, C Kow. Macromolecules 26: 3444, 1993.
34. JA Moore. Reactions on Polymer. Boston: Reidel, 1973, p 73.
35. CE Carraher Jr, JA Moore. Modification of Polymers. New York: Plenum Press, 1983.
36. G Andrews, L Dawson. Encyclopedia of Polymer Science and Engineering, 2nd edn. New York: Wiley, vol 6, p 495.

37. CC Chen, E Fontan, K Min, J White. Polym Eng Sci 28: 69, 1988.
38. HK Chung, CD Han. In: CD Han, ed., Polymer Blends and Composites in Multiphase Systems. Washington, DC: ACS, 195–213, 1984.
39. SJ Park, BK Kim, HM Jeong. Eur Polym J 26: 131, 1990.
40. HE Stromwall. PhD dissertation, Chalmers University of Technology, Gothenburg, Sweden, 1984.
41. H Kishi, M Yoshika, A Yamanoi, N Shiraishi. Mozukai Gakkaishi 34: 133, 1988.
42. S Takase, N Shiraishi. J Appl Polym Sci 37: 645, 1989.
43. RT Woodhams, G Thomas, DK Rogers. Polym Eng Sci 24: 1160, 1984.
44. P Bataille, L Ricard, S Sapieha, Polym Compos 10: 103, 1989.
45. I Kelnar. Angew Makromol Chem 189: 207, 1991.
46. E Benedetti, F Posar, A D'Alessio, P Vergamini, M Aglietto, G Ruggeri, F Ciardelli. Br Polym J 17: 34, 1985.
47. E Benedetti, A D'Alessio, M Aglietto, C Ruggeri, P Vergamini, F Ciardelli. Polym Eng Sci 26: 9, 1986.
48. FP Baldwin, G Ver Strate. Rubber Chem Technol 45(3): 834, 1972.
49. GD Jones. In: EM Fettes, ed. Chemical Reactions of Polymers. New York: Interscience, 1964, p 247.
50. D Braun, U Eisenlhor. Angew Makromol Chem 55: 43, 1976.
51. F Severini. Chim Ind (Milan) 60: 743, 1978.
52. Y Minoura, M Ueda, S Minozuma, M Oba. J Appl Polym Sci 13: 16259, 1969.
53. DR Paul. In: DR Paul, S Newman, eds. Polymer Blends. New York: Academic 2–14, 1978.
54. G Riess. In: E Martuscelli, R Palumbo, M Kryszewski, eds. Polymer Blends: Processing, Morphology and Properties. New York: Plenum, 1980, p 123.
55. G Illing. In: E Martuscelli, R Palumbo, M Kryszewskiy eds. Polymer Blends: Processing, Morphology and Properties. New York: Plenum, 1980, p 167.
56. S Cimmino, L D'Orazio, R Greco, G Maglio, M Malinconico, C Mancarella, E Martuscelli, R Palumbo, G Ragosta. Polym Eng Sci 24: 48, 1984.
57. F Ide, A Hasegawa. J Appl Polym Sci 18: 963, 1974.
58. L Amelino, S Cimmino, R Greco, N Lanzetta, G Maglio, M Malinconico, E Martuscelli, R Palumbo, C Silvestre. Preprints of Plasticon 81-Polymer Blends. Warwick, 1981.
59. TC Trivedi, BM Culberston. Maleic Anhydride. New York: Plenum, 1982.
60. N Lanzetta, P Laurienzo, M Malinconico. Polym Bull, 22: 603, 1987.
61. G De Vito, N Lanzetta, G Maglio, M Malinconico, P Musto, R Palumbo. J Polym Sci Polym Chem Ed. 22: 1335–1347, 1984.
62. B Immirzi, N Lanzetta, P Laurienzo, G Maglio, M Malinconico, E Martuscelli, R Palumbo. Makromol Chem 188: 951–960, 1987.
63. CH Basdekis. ABS Plastics. New York: Reinhold, 1964.
64. A Eisenberg, M King. Ion Containing Polymers. New York : Academic, 1977.
65. EA Bekturov, LA Bimendina. Adv Polym Sci 41: 99, 1981.
66. JC Decroix, JM Bouvier, R Russel, A Nicco, CM Bruneau. J Polym Sci 12: 229, 1975.
67. M Malinconico, E Martuscelli, MG Volpe. Int J Polym Mat 11: 295, 1987.
68. M Avella, R Greco, N Lanzetta, G Maglio, M Malinconico, E Martuscelli, R Palumbo, G Ragosta. In: E Martuscelli, R Palumbo, M Kryszewsky, ed. Polymer Blends. New York, 1980, p 191.
69. M Lambla, A Killis, H Magnin. Eur Polym J. 15: 489, 1977.
70. JM Bouvier, CM Bruneau. Bull Soc Chim France. 11–12: 1093, 1977.
71. F Pilati, G Attalla, Proc Europhysics Conf on Reacting Processing, Naples, 1986, Conference Preprints.
72. RJ Cella. J Polym Sci, Polym Symp 42: 727, 1973.
73. GK Hoeschele. Chimia 28: 544, 1974.
74. S Cimmino, L D'Orazio, R Greco, G Maglio, M Malinconico, C Mancarella, E Martuscelli, R Palumbo, G Ragosta. Polym Eng Sci 25: 1933, 1985.
75. R Greco, N Lanzetta, G Maglio, M Malinconico, E Martuscelli, R Palumbo, G Ragosta, G Scarinzi. Polymer 27: 299, 1986.
76. R Greco, M Malinconico, E Martuscelli, G Ragosta, G Scarinzi. Polymer 28: 1185, 1985.
77. F Pilati, P Manaresi, B Manaresi. IUPAC 26th Symp on Macromolecules, Mainz, 1979, vol 1, p. 231.
78. P Manaresi. In: Giornate di Studio sulla Policondensazione. Naples: Arco Felice, 1981, pp 13–22, Conference Preprints.
79. FH Borman. J Appl Polym Sci 22: 2119, 1978.
80. E Martuscelli, F Riva, C Sellitti, C Silvestre. Polymer 26: 270, 1985.
81. R Burns. In: DE Hudgin ed. Polyester Molding Compounds. New York: Marcel Dekker, 1982.
82. PD Tetlow, JF Mandell, FJ McGarry. 34th Annual Technical Conf on Reinforced Plastics/Composite Institute, Washington, 1979, Section 23-F, p 1.
83. EH Rowe. 34th Annual Technical Conference on Reinforced Plastics/Composite Institute, Washington, 1979, Section 23-B, p.1.
84. CS Nichols, GJ Horning. 38th Annual Technical Conference on Reinforced Plastics/Composite Institute, Washington, 1983, Section 19-B, p.1.
85. GA Crosbie, MG Philips, J Mat Sci 20: 182, 1985.
86. GA Crosbie, MG Philips, J Mat Sci 20: 563, 1985.
87. RG Ferrillo, VD Arendt, AH Granzow. J Appl Polym Sci 28: 2281, 1983.
88. W Hazer, K Hererreiter. Makrom Chem 180: 939, 1979.
89. AS Kinloch, RJ Young. Fracture Behaviour of Polymers. London: Applied Science Publishers, 1983.

90. G Williams. Stress Analysis of Polymers, 2nd edn. New York: John Wiley & Sons, 1980.
91. WF Brown, J Srawley. ASTM, STP. 410, 1966.
92. E Plati, JG Williams. Polym Eng Sci 15: 470, 1975.
93. FF Lange. Phil Mag 22: 983, 1970.
94. P Zadorecki, AJ Michel. Polym Compos 29: 69, 1989.
95. JF Kennedy, GO Phillips, PA Williams. Wood Processing and Utilisations, Chichester: Ellis Horwood, 1989.
96. D Maldas, BV Kokta. J Polym Mat 14: 165, 1990.
97. M Dalvag, C Klason, HE Stromvall. Int J Polym Mat 11: 9, 1985.
98. C Klason, J Kubàt, H E Stromvall. Int J Polym Mat 10: 159, 1984.
99. M Avella, E Martuscelli, B Pascucci, M Raimo, B Focher, A Marzetti. J Appl Polym Sci 49: 2091, 1983.
100. B Focher, A Marzetti, V Crescenzi. Steam Explosion Techniques: Fundamentals and Application, Philadelphia: Gordon and Breach, 1991, p 331.
101. RM Marchessault, S Coulomba, H Morikawa. D Robert. Can J Chem 60: 2372, 1982.
102. S Brandrup, EH Immergut and EA Grulke, eds. Polymer Handbook. New York: Interscience, 1999, vol 5, p 24.
103. JM Felix, P Gatenholm. J Appl Polym Sci 42: 609, 1991.
104. EL Saier, L Petrakis, IR Cousins, WJ Helman, JF Itzel, J Appl Polym Sci 12: 2191, 1968.
105. L. Segal, JJ Creely, AE Martin Jr, CM Conrad Text Res J 29: 786, 1959.

27

Polyolefin Fibers

Aurelia Grigoriu and Vasile Blaşcu
"Gh. Asachi" Technical University, Iasi, Romania

I. INTRODUCTION

PP fibers are the most recent synthetics to achieve commercial importance. They appeared on the fiber market in the early 1970s and represent by now the fourth major class of fibers besides the other traditional synthetic fibers: polyamides (PAs), polyesters (PESs), and polyacrylics. Their starting point was the application of Ziegler and Natta's method for the production of isotactic PPs, which can be processed into films and fibers. Montecatini applied for the patents in 1954 [1, 2].

Four generations of Ziegler–Natta catalysts (ZN catalysts) laid the foundations for the continuous rise of PO fibers.

The rapid progress recorded in the PP fiber domain can be explained by the developments in fiber-forming polymers (such as catalysts and polymerization processes, controlled thermal and/or peroxide reduction of the molecular weight of the polymerization products, and limited molecular weight distribution polymers). To meet the requirements of both quality and economy, the polymer properties should be tailored in an increasingly specialized way, on considering both the individual fiber types and their processing. In time, other methods have been developed, including high and ultrahigh speed methods, split-film technology, spray technology, the spunbond (spinbond) process, and melt-blown systems.

The development of very active stereoselective metallocene catalyst systems, as well as advanced polymerization and pelletizing processes, yielded a large number of highly crystalline PP grades of remarkable purity and homogeneity, with a broad spectrum of interesting rheological properties [3–15]. By various modifications of their fine structures and also by the utilization of additives (stabilization, finishing and coloration), the PO was finally adapted to meet the requirements of many different applications, among which fiber forming is of special importance.

These new catalysts offer PO producers further steps in polymerization reactor outputs and improvements in polymer purity. Metallocene PP (mPP) is a cleaner, clearer, glossier polymer with a more uniform molecular size that could overcome the well known deficiencies of current isotactic PP (inadequate resilience and crush resistance, unsatisfactory dyeability, low sticking and softening temperature, processing difficulties to achieve maximum tensile strength, incompatibility of light stabilizers and flame retardant for PO fibers).

Metallocene catalysts change the well known isotactic PP to a more sophisticated PO polymer which will be able to push aside the nylons and polyesters in floorcovering and in the technical textiles markets [14].

PP has proved to be a highly economical raw material, with the lowest energy content of all synthetic fibers, and a very favorable environmental position. POs are gradually taking their place as major manmade fibers in the international textile industry.

II. POLYOLEFIN FIBER FORMING PROCESSES

A. Background

PO fiber production represents a successful and consistent development of the classical fiber process; a field which has benefitted from all the scientific, technological, and economic progress of the various spinning processes applied to the main synthetic fibers (PA, PES, and PAC).

Thus, for these classes of fibers the classical spinning process was randomly applied for PP in the so-called grid period (1955–1965), beginning with modified spinning methods, which avoided expensive separate steps. Nowadays, fully integrated processes including spinning, drawing, and texturing have been developed and applied [15].

Among the methods for converting POs into fibers and filaments, dry and wet spinning, and melt spinning should be mentioned.

In dry spinning, the polymer is dissolved in a suitable solvent and the solution is extruded through a spinneret into a stream of heated gas to rapidly evaporate the solvent from the polymer and thus from the polymer filaments.

In wet spinning, a polymer solution is extruded through a spinneret directly into a coagulating bath. The spinning speed is slower than in the dry method, and the composition and temperature of the coagulating bath should be carefully regulated to set up the freshly spun filaments and to ensure that the proper amount of solvent is removed from them.

With a few exceptions (e.g., gel-spinning technology), the above-mentioned methods of spinning from polymer solutions have historical importance only. Melt spinning, however, is still used [16].

In melt spinning, the polymer—in powder or pellet for—is heated to high temperature until it becomes molten and then extruded through a spinneret in the shape of filaments, which are reinforced and put into contact with the cooling air surrounding the freshly extruded filaments. This method allows fairly high and ultrahigh spinning speeds, yet it requires considerable extrusion pressure and high melt temperature [17].

Besides these general procedures, applied to the production of PO fibers before 1965, other modern ways have been developed and used for melt spinning. They include high and ultrahigh speed spinning, the split-film method, the spunbond process, and the melt-blown system.

In recent years, as a result of the increases in the speed and throughputs of textile machines of all types, most of the manmade fibers formerly produced in several stages, can now be made in the one-step or high speed process. The simplifcation of multistage processing, coupled with high and ultrahigh speeds, made possible the production of yarn and staple fibers with far fewer operating personnel. In addition, yarn packages became larger and led producers to develop machines that will automatically handle and transport the product, which resulted in further manpower reductions [18]. Nowadays, many modes of single-stage PO fiber processing are in use, draw-texturing, spin-drawing and spin-draw-texturing included.

In draw-texturing, the drawing stage of synthetic yarn manufacture is combined with the texturing process on one machine. Drawing and texturing may take place in separate, usually consecutive, zones of a machine (sequential draw-texturing) or together in the same zone (simultaneous draw-texturing). Spin-drawing or draw-spinning is a process for spinning filaments of partially or fully oriented yarn (POY or FOY) in which more orientation is introduced between the first forwarding device and the takeup; that is, spinning and drawing are integrated sequential stages.

Spin-draw-texturing is a process for making textured yarns in which the spinning, drawing, and texturing stages represent integrated, sequentially steps in a single machine [19].

The tape yarn and film method (initially called the split-film method or fibrillation of plastic films) consists of extruding or casting thin films or tapes of thermoplastic crystalline POs, followed by their stretching to obtain a substantially parallel orientation of the fiber-forming crystallites embedded in an amorphous matrix. Because the interchain bonds are missing, lateral cohesion is low and the films or tapes can be spliced very easily into a network of interconnected fibers (fibrillation).

In the spunbond or spinbond process, fine filaments are extruded at high draw ratios under the impact of a hot airjet and collected to form a sheet (web) of continuous filaments randomly arranged and bonded together in different ways (self-binding, heat sealing with a binder fiber, or external bonding with special agents such as vinyl or acrylic compounds). These special non-woven products are called spunbonded or spunlaid fabrics [20–26].

Compared with the spunbond process, the melt-blown system (spray method) produces microdenier (sprayed) fibers. Molten or dissolved fiber-forming PO polymers are forced with a spray gun or through a multiple-hole extruder to disrupt the filament into a high velocity hot air jet, to form superfine fibers less than 1 denier. These fibers form a very light web with random distribution which can be tightly compressed to sheets resembling paper for different uses [27–29].

B. Procedures

1. Dry Spinning

In dry spinning, large amounts of heat should be applied to the freshly spun filaments to remove the solvent. The expense is considerable, and sometimes thermal damage may occur to the fiber. A dry spinning process for the PE polymer, that does not require expensive equipment and large quantities of heat for solvent removal [30], involves the following steps: *dissolution, filtration, extrusion, water quenching, first air stretching, water stretching, second air stretching, spooling.*

An alkoxyalkyl ester may be used as solvent, to give a PO solution that can be dry spun [4]. Another form of dry spinning, namely flash spinning, may be applied to the formation of PO monofilaments [31]. A homogeneous polymer solution is extruded at lower pressure in a nonsolvent for the polymer, and a multifibrous yarnlike strand is formed. After certain physical treatments (shaking, washing, textile processing), the strand becomes useful for a number of applications (wall coverings, cigarette filters, tarpaulins, etc.).

The recent improvements for flash spinning plexifilamentary film-fibril strands of a fiber forming PO consist of:

Utilization of a C_{4-7}, hydrocarbon/cosolvent spin liquid with a highly reduced ozone depletion hazard as compared to the halocarbon spin liquids currently used for making such strands commercially. The resulting plexifilamentary film-fibril strands have increased tenacity and improved fibrillation compared to strands flash spun from 100% hydrocarbon spin liquids [32].

A fibrillated three-dimensional PP plexifilamentary fibre of an isotactic PP has a microwave birefringence more than 0.07, a M_w/M_n less than 4.3 and a MFR of 3.5–10 [33].

Utilization of a mixture of methylene chloride and carbon dioxide as a solvent [34–38].

Utilization of a spin liquid with a C_{1-4} alcohol (50% by weight of the total spin liquid) for spinning of a crystalline PO at 130–300°C with an electrostatic charging step caried out in an atmosphere comprising one-charge-improving compound [39–42].

2. Wet Spinning

Although wet spinning produces fine denier and high tensile strength, it is slower than melt spinning. Moreover, many of the solvents and precipitating liquids are difficult to work with and may present an explosion hazard. An improved process [43] involves realization of a solution in paraffins or cycloparaffins (boiling points 150–300°C), its extrusion into a coagulating bath, removal of most of the solvent in subsequent extraction baths, and then stretching. Extrusion of filaments is carried out at about 180–220°C, with the paraffins or cycloparaffin hydrocarbons in liquid state. The steps for wet spinning are: *mixing, extrusion, filament forming, solidification (cooling), drawing, drying, spooling.*

The rapid extraction of the solvent prevents proper filament formation and makes the filament crumbly. That is why good results may be obtained by methylene chloride extraction after solidification with water. A wet spinning process with solvent extraction and stretching consists of [43]: *extrusion, filament solidification, first solvent extraction, first water stretching, second solvent extraction, second hot air stretching, takeup.*

Another wet spinning process involves extrusion of a solution of PO in fatty acid into a caustic bath [44].

A modern wet spinning technology, known as solution gel spinning, is now used for ultrahigh molecular weight PE (UHMWPE) (500,000–5000,000), to obtain extended-chain PE fibers.

Gel spinning is a two-stage process for the production of high performance linear polyethylene (HPPE) [45, 46]. The first stage is the formation of a substantially unoriented gel with a very low chain entanglement density.

In the second stage, this lightly entangled network is drawn at an appropriate temperature to yield effective draw ratios of 20. During this process, the microstructure is transformed, from a folded-chain, to an essentially chain-extended morphology of the highly oriented gel-spun filaments. The solvent serves to disentangle the polymer chain during spinning and drawing, with the formation of an extended-chain configuration. This special structure imparts strength and stiffness to the yarn [47]. For orientation in the fiber direction and subsequent crystallization, a one- or multiple-drawing process follows. The yarn may be treated with suitable finishes and dye acceptors before spooling.

The solution-gel spinning technology for UHMWPE involves the following steps: *polymer solution preparation, metering, spinning, coagulation, heatset(drawing), additive treatment, spooling.*

Of special technical interest is the possibility of obtaining PP and PE filament yarns with high strength and high modulus of elasticity, by means of high draw ratios (around 1 : 30), at elevated temperatures or in a nonisothermal multistage drawing process. This is achieved by a high orientation without chain folding, the PE being basically better suited for this process than PP [48–50].

A two-stage drawing process of the paracrystalline as-spun fibers of low orientation yields drawn fibers of very high tenacity (85 cN/tex) with linear density around 0.4 tex.

The first stage, carried out at 60°C, results in oriented paracrystalline fibers which are then transformed during the second stage, at a temperature of 140°C, into highly oriented monoclinic fibers.

To produce high modulus and high strength fibers, a continuous zone-drawing (CZD) method was applied to the isotactic polypropylene fibers. The CZD treatment was caried out five times, at a drawing temperature of 150°C, stepwise, with increasing the applied tension from 14.8 MPa for the first treatment (CZD-1) to 207.0 MPa for the fifth (CZD-5), at a running speed of 500 mm/min. The applied tension increased. The CZD-5 fiber obtained has a birefringence of 0.0348 and a crystallinity of 67.3%. The evidence for the existence of an α-form in the CZD fibers was provided by WAXD and DSC measurements. The CZD-5 fiber has a Young's modulus of 14.7 GPa and a tensile strength of 1.08 GPa [51].

Several methods for producing high-modulus fibers from flexible-chain polymers have been developed. Research work in this field has led to fibers of high modulus, but the technological solutions provided are very expensive and demand the realization of special machinery [52].

3. Melt Spinning

(a) General Considerations. A wide spectrum of fiber properties can be created by varying the process parameters (spinning speed, drawing conditions, thermal treatment, etc.) or by the variation of the characteristics of the polymer feed stocks (average molecular weight, intrinsic viscosity, melting capacity). Such alterations allow a manufacturer to tailor the product to different end uses. The rheological and crystallization properties of PP polymers influence mainly tensile properties. Depending on the crystallization speed specific to every polymer, the spun yarns can be amorphous or crystalline. A correlation may be noticed between the spinning speeds, PP yarn structure, and deforming properties. Thus, the crystalline factor and birefringence tend to reach a maximum value for spinning speeds over 2500 m/min. The noticeable differences of PP yarn orientation are reflected in the behavior of yarns at deformation [53, 54].

PP polymers with a defined molecular structure have the following characteristics: narrow molecular weight distribution (MWD), suitable melt flow index (MFI) and melt flow rate (MFR), low shear sensitivity in the melt, low swelling behavior under the die, low yarn tension, and low crystallization rate in the melt. Thus, the following conclusions can be drawn: processing and fiber properties include lower processing temperatures, higher aperture stresses, higher spinning speeds, finer yarn counts, and greater count precision [1, 55].

Obvious processability advantages are obtained by using narrow MWD polymers, when the filaments' count decreases as speeds increase. Interesting self crimping characteristics may be obtained by making side-by-side heterofilaments incorporating normal and narrow MWD polymers.

When spinning the PP polymer, the resulting fiber material is nondrawn and nonoriented, which means that several processing speeds (drawing, crimping, heat setting, cutting, packaging, etc.) are required before the finished fiber is ready for conversion in the textile industry.

In the conventional two-stage process, the freshly spun subtow is first fed into cans and later processed on the fiber line. In the single-stage compact process, the freshly spun material is directly processed on the fiber line [56–58].

The decision of whether to apply one- or two-step melt spinning processes is rather complex and involves different criteria, such as line fiber capacity, fiber type, building costs, and labor and energy costs.

The compact system is usually chosen for capacities lower than 10,000 tons/yr, while capacities ranging from 20,000 to 50,000 tons/yr are mostly achieved with the conventional process. The operating speed in the conventional process is 10–20 times higher. The spinning speed for the compact process is about 50 m/min; however, to achieve the same throughput, the spinnerets should have a higher number of holes—about 30,000 holes per spinning position.

For bicomponent fibers, a conventional line is generally chosen, while for spun-dyed fibers, compact spinning is more appropriate. A compact spinning plant requires a normal-sized building, while conventional spinning calls for custom facilities.

For lower capacities ($<$ 10,000 tons/yr), labor and energy costs are lower with compact spinning. For capacities higher than 20,000 tons/yr, conventional spinning is more economical, the higher the line capacity.

(b) Conventional Spinning Method. The conventional spinning method (long system) [55] is a two-step process, namely spinning and drawing.

A conventional spinning plant essentially consists of melting and filtration systems, melt distribution pipes for static mixtures, melt dosing pumps, a side color melt extruder, spinning packs, a quenching system, a takeup system, spin-finish baths, and a draw and heat system [29, 30].

The melt spinning steps of high viscosity α-olefin polymers [59] are: *mixing, extrusion, quenching, first stretching, heatset, second stretching, winding-up.*

A diagram for the direct production of helically crimped PP filaments [60] includes the following operations: *melting, heating, quenching, cold drawing, winding, crimping.*

Such installations will be economical only from a certain level of production capacity. For PP fibers to be dyed in the melt, moreover, these plants are inflexible: since the exact shade cannot be seen before the end of the draw, it is imposible to correct the color.

The conventional fiber producing route (long system) is essential for the production of certain types of continuous filament and for the equipment used in association with the spunbonding process.

(c) Short Compact Spinning Method. The short spinning method was developed when PP fibers were introduced, to avoid strong degradation during spinning. The components of a compact spinning plant and those for a conventional spinning plant are basically the same: melting screw, mixer, production pipe, filter, spinnerets, quench box, finish applicator, and takeup system [55].

Low spinning speeds (30–150 m/min) and spinnerets with a high number of holes ($\approx$ 15,000/die) are typical for achieving reasonable output rates and logically lead to a continuous, one stage spin-draw process, since the spinning tows can be guided directly onto the fiber processing line.

The short spinning process offers the following significant advantages for producing spin-dyed PP fibers: color can be checked by inspecting the final fiber after spinning and can be directly modified by adjusting the color metering device; the shape of the spin fiber tows is even; the amount of two spin finishes and the crystalline structure of the fibers do not change, as happens with storage in spinning cans; the finished fiber is available immediately, and therefore elongation and tenacity values can be adjusted without affecting the final fiber titer; finally the blow technique and spinneret size allow a titer range of 1.7–200 dtex (e.g., a compact spinning plant developed by Automatic in cooperation with Fleissner [61]). The output rates per spinneret in the fine and middle titer range correspond to those of conventional plants. In the coarse titer range, the compact spinning plant exceeds by far the conventional process [62–65].

An improvement on the existing short spinning staple line utilizing rectangular spinnerets is the Modern Finetex staple fiber production line [66], developed by Meccaniche Moderne, in Italy. This integrated short spinning line uses circular heads and annular spinerets. The line is composed of a chip feeder and a color dosing unit, plus extruder, filter, distribution manifold, pulling and spin finish applicator, slow drawing unit, steam over-drawing, fast drawing godet, finish application unit, tow collector, crimping machine, thermosetting oven, tow tensioning device, cutter, and baler press.

Fineness (down to 1 denier), and cooling and stretching uniformity are easy to obtain, so that uniformity in the physical and mechanical properties (tenacity, elongation, dyeability, etc.) is guaranteed. All in all, the process represents an economical continuous method of obtaining finer PP filaments.

(d) Stretch Spinning. In stretch spinning, filaments are substantially stretched during a certain stage between spinning (extrusion) and collection. Since the process involves substantial stretch, high tenacity yarns are assumed.

PP fibers and filament yarns with higher tenacity can be achieved by the following combinations: spinning extruders with special mixing devices, spinning tools with melt metering pumps, and spinnerets with optimized geometrical arrangements and spacing (which permit simple separation of the filament yarns); uniform filament cooling conditions (which create a highly drawable morphology) and drafting devices that allow a uniform drawing force transfer without slippage, are also useful in this respect.

A plant developed by Barmag AG (Germany) is representative. This facility, which produces PP fibers of up to 10 cN/dtex tenacity [67–70], has the following characteristics:

- Extrusion technique, by which a high degree of disentanglement in the PP melt is achieved through special homogenizing efficiency.
- Cooling conditions of the filaments, by means of which a highly drawable morphology is set.

- Thermomechanical treatment, assuming a high degree of orientation of the filaments.

(e) High Speed Spinning: Background. An ideal spinning process should yield synthetic fibers with a high degree of orientation and crystallinity. In this respect, a one-step process based on high-speed spinning should be preferred as an industrially effective approach that ensures the manufacture of high modulus, high tenacity fibers.

In a two-step process that has been widely applied for many years, the yarn, having a relatively low degree of orientation and crystallinity, is melt-spun at a speed of 1000–1500 m/min, whereupon the undrawn yarn is subjected to additional treatments, such as drawing and annealing at 400–1000 m/min to increase the degree of orientation and crystallinity. Then, melt-spinning and drawing are combined into a coupled spin-drawing process that contributes to increased productivity and high economic effects. Nevertheless, increases in manufacturing costs (energy, raw material and labor) have made it necessary to study and industrialize high speed spinning as a highly productive, energy-saving process [71–80] that demonstrated a number of technological and qualitative advantages [81, 82], such as:

- Higher molecular orientation of the material, which gives better storage properties.
- Less restrictive requirements regarding air conditioning for spinning.
- Stable package formation.
- Easier handling and less cross-sectional deformation during texturing.
- Higher dyeability and purity of the yarn.

The much higher process economics is due to the higher capacity per spinning unit, to the reduction of energy and labor costs, and to the omission of the separate draw-twisting step (since the yarn has a high enough preorientation to permit postdrawing during texturing). Whereas conventional spinning processes with takeup speeds of less than 2000 m/min require two steps, namely draw-twisting and texturing, pre-oriented yarn (POY) needs only one draw-texturing step.

To obtain high speed spun fibers with properties similar to those of fully oriented conventional yarn, a spinning speed over 5000 m/min is required. Compared with conventional FOYs, the high speed spun yarns have the following characteristics: lower tenacity and lower Young's modulus, greater elongation, less shrinkage, and fewer differences caused by a different internal structure.

As in high speed spinning processes, orientation of the macromolecules takes place within hundredths of a second, and the spinning processes impose severe requirements on the raw materials, such as melt viscosity consistency, material homogeneity, moisture content, and fine dispersion of additives (e.g., dulling agents, dyestuffs, stabilizers). An additional criterion for PPs is molecular weight distribution. The rates of crystallization in PE and PP are extremely high (see Chapter 9), and only the effects of molecular orientation can control the high speed melt spinning in these polymers. The development of fiber structure advances with the progress of molecular orientation and crystallization. Higher spinning temperatures result in a lower rate of crystallization and cause relaxation of the oriented molecules.

The birefringence of PP fibers below 2000 m/min strongly increases with increasing spinning speed; at a takeup exceeding 2000 m/min, birefringence increases linearly and slowly. Correlation of density and shrinkage with spinning speed is similar to that of birefringence. Above 2000 m/min, the molecules among the oriented crystals are stretched along the fiber axis. Such stretched molecules in the amorphous state are responsible for the large increases in Young's modulus and heat shrinkage values. However, the amorphous chains between the lamellar crystals are not stretched very much below 2000 m/min; above 2000 m/min, the number of molecules stretched between lamellar crystals increases, and these molecules reduce recovery, through the destruction of the lamellar structure by large deformation. Elongation decreases with the spinning rate.

PP is one of the polymers difficult to employ in high speed melt spinning: filaments often break, the mechanical characteristics tend to saturate at rates above 1000 m/min, and Young's modulus is relatively low.

Plants adopting high speed spinning will aim at full automation. To detect abnormal conditions, for example, a monitoring system should be composed of on-line measuring devices, collecting process data such as pressure, denier, amount of spin-finish, yarn tension, and winding system parameters. Table 1 lists the correlations of some spinning parameters.

The manufacture of high quality yarn at high takeup speeds obviously requires high speed winders. The extruder and the lower part of the spinning machine are basically the same as for conventional spinning. New processes have created a variety of new flat fibers (monofilaments, fine denier filaments, noncircular cross section filaments, staple fibers,

TABLE 1 Spinning conditions for various polymers

Fiber type	Extrusion temperature (°C)	Takeup speed (m/min)	Quenching temperature (°C)
PP	230	300–6000	220
	258	300–6000	248
	285	300–6000	275
PE	280	200–3000	278

Source: Ref. 81.

conjugated filaments, blended filaments). A typical modern high speed line for filament production [83] comprises : *extrusion, quenching (air cooling, coagulation bath), drawing, finish application, takeup.* A flow sheet for a continuous one-step, high speed process (Neumag, Germany, Fare SpA, Italy or Synthetic Industries, USA) for the production of PP staple fibers with high speed cutting consists of the following steps: *extrusion, quenching, finish application, crimping, cutting, texturing, cooling, takeup.*

(f) High Speed Spinning: BCF Yarn Production. Initially, high speed spinning was intended for the production of flat yarns; later on, an advanced one-step technology involving very high speed spinning was proposed for producing textured, or bulked continuous filament (BCF) yarns.

As a principle, individual continuous filaments are processed in a texturing machine to introduce durable crimps, coils, loops, or other fine distortions along the length of the filaments. Variation of the filament overfeed and machine type permits the application of several texturing procedures (false twist texturing, hot airjet or steam jet texturing, impact texturing, edge or gear crimping, knit–deknit texturing), and thus yarns with different textures and properties can be obtained.

BCF production may be realized in three ways: *discontinuously* (i.e., spinning, drawing, and texturing); *half-continuously* (i.e., spinning and draw-texturing), and *continuously* (i.e., spinning-draw-texturing) [83–86].

From the standpoint of the manufacturing processes, the PP filament yarns are divided into three groups:

1. Spin-dyed BCF, a three-dimensional, crimped yarn produced partly by hot air texturing, mainly in the continuous spin-draw process (a one-step method) and partly in the undrawn yarn (UDY) with subsequent draw-texturing, to produce drawn texturized yarn (DTY).
2. Spin-dyed continuous filament (CF), a fully drawn yarn (FDY), manufactured in an one-step, spin-draw process; it can be converted to air-texturized yarn (ATY) in a separate process.

   The spinning speed and draw ratio are within the conventional limits, which is the reason for the introduction of the term CF, while FDY is now used for fine titer uncrimped PES and PA filament yarns, produced in a fast spin-draw process.
3. Spin-dyed multifilament POY–DTY or POY–ATY spun at high speeds, drawn residually in the simultaneous draw-texturing process, either friction-texturized or cold-air-texturized.

Some examples of BCF yarn processes follow. The German firm Neumag AG has developed economical high speed spinning processes especially for PP filaments yarns for textile applications [56], including:

1. High speed spinning for POY production, further processed to DTY yarns in the simultaneous draw-texturing process.
2. Spin-drawing for CF yarn production (i.e., FDY), either used as a crimped flat yarn or further processed to a structural yarn (BCF yarns) in the air-intermingling process (spin-draw-texturing method).

The Neumag line is equipped with automatic devices for a precise control of temperature, pressure, and yarn tension. For spin-dyed yarns, the metering procedure developed by Neumag assures precise addition, directly into the extruder, of a very highly concentrated master batch (colored chips or powder), to ensure absolute homogenization. The system grants an absolute uniformity in dyeing and avoids repeated thermal degradation, due to repeated melting.

In this process, all spinning positions can only be operated with the same product or the same color. In the case of product changeover and small lots, this means increased waste levels and, consequently, higher conversion costs.

The Maxflex feed system represents an innovative technology aimed at increasing flexibility in the production of man-made fibers [87, 88]. It should be the unit of choice in all those cases where speciality products are called for in an increasing competitive market, products that cannot be achieved economically with conventional technologies.

Maxflex stands for maximum flexibility in feeding additive and/or polymers into a melt stream at the last possible moment. In a spinning system, the last possible moment is immediately ahead of the spin pump that divides the main stream into up to 10 partial streams and transports them to the spin packs. By blending in additives and/or other polymers at this point, in the extreme case, a different product or a different color can be processed on each spinning position that can be selectively used as a pilot one. In addition, in this position, parallel to the production operations, small lots can be produced, differing from the main production by special additive blends.

Especially when the processing products are needed in smaller lots, the spinning line can be operated with markedly higher cost effectiveness by using Maxflex.

Another spin-draw-texturing Mackie line [89] for producing BCF–PP carpet yarns may attain draw speeds of up to 2000 m/min on 2500 denier/120 filament yarns. The texturing process is carried out by a stuffer box/jet entangling technique, in which the jet fluid is hot air. The texturing process must be absolutely uniform when spun-dyed yarns are being handled. Slight variations of color can be overcome in the production of a spun yarn by using a fiber mixture, but since such measures are impossible with BCF yarns, color irregularities may result.

The yarns are automatically doffed and can be used directly on a tufting machine without any intermediate twisting process. Another development for this line may be concerned with intermingling, by a jet technique, of the different filaments in individual yarns: this should make possible novel multicolor effects such a marl, tweed, or even mock-space-dyed items.

The waste reclamation unit is ancillary to the Mackie fiber production equipment and uses any type of PP waste, from fiber production start-up rejects to edge trims from woven or needled fabrics. The waste is deposited in a mixed form on a feed conveyor which transports it to a melting box of vertical configuration. A two-stage, self-regulating wing system is used to feed the melted material to extruders which, in turn, feed coarse strands into a bath of cooling water. The solidified strands are drawn into a regranulating unit capable of producing granules for a variety of purposes, from fiber production to baler strand manufacture.

A short PP staple fiber extrusion Mackie CX unit (developed by Mackie & Sons Ltd, UK) [89] can produce fibers below 1.5 denier, possessing the capacity to produce superbulk fiber through the introduction of a three-dimensional crimp, which represents another method of texturing. The Mackie unit allows different degrees of bulk in a controlled fashion. A typical Mackie CX Superbulk line consists of granule hoppers, melt chambers (each extrusion section consists of two heat chambers employing a gravity-fed dry pump unit, internal and external heaters, multihole interchangeable spinneret plates, and air cooling and drawing rolls), a first oven, a water tank, stretch godets, a second oven, a crimping unit, and a coiler.

The process involves building of differential strains into each fiber, producing a bicomponent effect in a homopolymer item. A thermomechanical route is employed to achieve what is actually a morphological result. The crimp lies latent in the fibers, so that they are readily processed on, for example, the semiworsted spinning system; the bulk form occurs when the yam is subjected to dry heat.

After 1990, a new PP–BCF generation was developed with the Autocrimp process of Extrusion Systems Ltd (UK) [90–92]. During extrusion, a turbulence in the PP polymer, which is in its glass transition phase, is generated. This turbulence is maintained when the filaments pass into crystallized phase after extrusion and it is concentrated towards one side of their cross section (asymmetrical cooling). The reduced stresses into the filaments cause uneven tensions after drawing and relaxing and produce a pronounced three-dimensional helical crimp effect in fibres, similar to that in wool. After heat setting in the relaxed state of the fibres, crimping became permanent and could only be removed by another setting process, at a higher temperature and under tension.

This is a major advance on zigzag crimped fibers produced by a stuffer box, which is only a mechanical deformation of the fibers and can be easily pulled out. In the conventional crimping process the PP filaments were cooled symmetrically after extrusion and they had no propensity to self crimp. With the Barmag STM 16 Speetex (Barmag AG, Germany) [86, 88, 89, 93] multicolor BCF–PP yarns can be produced in an on-line process. The line consists of the main extruder, the side extruder with pump and mixer for color master batches and additives, the spinning unit, and the draw-texturing-winding machine with tangling devices. The same line offers the possibility of manufacturing space-dyed fiber for carpets, (multicolor yarns obtained by the application of various colorants at intervals along a yarn.)

In parallels, efforts have been made to eliminate intermediate steps for technical yarns for BCF–PP filaments, too; as a result, spin-draw-winding and spin-draw-texturing processes are used for PP filament production.

For staple fibers, a slow spinning process, consisting of spin-drawing, and stuffer box-texturing, has been developed [15, 94] for fibers of about 3 dtex. In this case, pleating of the tow into cans is no longer necessary, since the material passes at slow spinning speeds (60–100 m/min) directly to the fiber line, being drawn in the usual way by the stuffer box process. A two-dimensional staple fiber suitable for brilliant and expensive qualities in carpeting is obtained. Another method is a spin-draw-blow-texturing process (high speed spinning process) applied to higher titer fibers. In the high speed blow-texturing process, winding is replaced by combining the yarn to form a tow, which is further cut to staple fiber. In this case, a three-dimensional air-crimped fiber, the appearance of which corresponds to that of bulk continuous filament yarns, may be obtained. The main advantage of this staple fiber, in comparison with the BCF yarn, is that its optimal blendability permits the production of unigoods and velvets without any dyeing problems.

Specialty PP fibers are produced on the compact short spinning process by Plasticisers Ltd (UK) such as Charisma™, Helta™, and Fixset™ [95].

Charisma™ is a blend of high and low shrink fibers. When heat-set in the steam the yarn spun from this blend can develop considerable bulk as the high shrink fiber component contracts and moves towards the centre of the yarn cross section, whilst the low shrink fibers, the majority of the blend, concentrate outwards. Such yarns have a large area of cross section, a soft handle, and produce a much improved cover and appearance in carpets.

Helta™ is a lofty fibre used in high shrink blends, to bring added covering power and soft handle to low and medium weight carpet constructions.

Fixset™ is a heat-sensitive locking fibre which forms in the blended yarn a network of fibres binding the others together after heat-setting.

(g) High Speed Spinning: Bicomponent Fiber Production. Many different processes have been proposed for the production of conjugate or bicomponent synthetic fibers by utilizing:

- different polymers streams;
- a conjugate structure developed chemically;
- a differential melt temperature in the polymer stream;
- a differential degradation and heat stabilization of the polymer streams;
- an asymmetrical filament cross section;
- an asymmetrical cooling of the filaments.

Microfilament flat yarns from bimodal PP blends with two different molecular weights (MFI of 35 and 200, respectively) were processed. The spinning trials proved that an increasingly higher tenacity appears for increasing amounts of high molecular weight PP [96].

A PP conjugate fiber with a crystallinity of 30–50% is composed of two crystalline PP grades with different isotactic ratios (under 0.930 and above 0.945) [97] or with two melt flow rate (MRF) ranges (0.5–3.0 and 60–100 respectively) [98–101].

A composite fiber with an island-in-a-sea cross section comprises at least two different polymers, one of which is a water-insoluble PO and the other is a water-soluble polymer. The islands have an average fineness not greater than 0.5 denier per filament and are uniformly distributed across the cross section of the fiber. A molten blend of highly crystalline PP and linear low density polyethylene (LLDPE) is melt-spun to produce biconstituent fibers having a denier of less than 30 [102]. The improvement is that the LLDPE used has a melt flow rate in the range of 12–120 g/10 min [103].

Hot-melt adhesive conjugate fibers are composed of two different kinds of POs having different melting points, or of other PO blends, such as: PE/PP; PE/polymethylpentene, PP/polymethylpentene [104, 105].

The attempts to produce PE/PP bicomponent fibers stem from a desire to obtain fibers with different, improved properties [106–109]. This spun product uses skin-core bicomponent filaments, the meltable skin being intended for sticking the filaments together in the crossing points during aftertreatment, by calendering or hot bonding. The combination of PP as core (melting temperature 169°C) and PE as skin, with a melting temperature of 125°C, is known.

Melt spinning of a poly(ethylene–octene)-copolymer with elastomeric properties shows that this material is also a melt spinnable polymer. The mechanical properties obtained for the spun fibers are similar to those of an elastane yarn and are only influenced by the variation of the spinning conditions. This however is not acceptable for textile applications, due to the low melting temperature (60°C) of this polymer. It is expected that, by melting the basic polymer with an appropriate PO copolymer, possessing a high degree of crystallinity and a higher melting point, the softening temperature should be raised. The expectation is that in a defined concentration range of the blend components, cocontinuous phase structures occur, which should result in distinctive advantages for the fracture mechanics of the melt spun fibers [110].

A PP-based copolymer was developed by the continuous filament spinning process on a Plantex, one step, draw-texturing unit. The extrusion melt temperature was 230°C, the draw ratios were 3.0 and

3.5 : 1, and the texturing temperature ranged from 150°C. The filament fineness was 17–18 denier and that of the filament textured yarn 1400/80 denier. After twisting (3.5 twists/inch), a subsequent heat-setting at 140°C for a total dwelling time of 60 s was performed. New process opportunities have allowed for increased penetration into product areas such as injection molded parts, air quenched blown films, extruding films and coatings, calendered sheets, thermoformed sheets, and bulked continuous filament extrusion. These copolymers have thermal and physical properties different from those of the traditional PP-based fibre resins [111].

Pigmented and unpigmented biconstituent staple fibers have been produced using Barmag's compact melt spinning line, at melt temperatures between 200 and 240°C. The biconstituent fibers, having a significantly wider range of thermal bonding as well as controllable shrinkage, offer an option for new nonwoven materials of balanced strength and softness.

The poor dyeability of PP fibers was overcome by blending PP with a second polymer (e.g., polyester) at high temperature, in the melt state, followed by its formation into a filament through a melt spinning process. It is possible to obtain light, medium, and dark colors with the new fibers.

The physical modification of the PP fiber to create dye access channels has been shown for a Novatron X-a bicomponent item in a matrix fiber [68]. The dye access channels are incorporated into the fiber at extrusion, the scale or the ripole structure being caused by the rheological differences between the PP and the dye acceptor (Novatron-X additive) at the time of extrusion. Fibers containing Novatron-X are softer and silkier than normal PP items, with improved retention of set after the spun PP yarns have been heat-set. Novatron fiber may be piece-dyed, loose-stock-dyed, hank-dyed (alone or blended with wool), winch-dyed, or printed, when used in a tufted carpet.

A PA–PP mixture was processed into a fiber by using an interfacial compatibilizing agent based on a PP–maleic anhydride copolymer [112].

The ES-Fiber (Chisso Co., Japan) is a modified side-by-side bicomponent item: one side is PP and the other is high density polyethylene (HDPE) [65]. HDPE has a melting point of around 130°C and acts as a thermal bonding agent in lightweight nonwovens. The PP component is unaffected by temperature at this level and retains its original characteristics after bonding has taken place. After heating, the web containing the bicomponent fibers forms a strong bond at fiber cross-over points, and the resulting nonwoven fabric has a bulky and soft texture [112, 113].

The ES-Fiber can be blended with virtually any other synthetic or natural fiber; although the bicomponent part may be as low as 30% of the total value, it still performs its bonding function satisfactorily. The ES-Fiber is used in coverstock, in nonwovens for filtration applications in geotextiles, in oil absorption materials, and as wadding.

A new spinning technology, namely mesh spinneret spinning, has been developed by Teijin, Japan [114]. This technology can produce a new type of mono- or bicomponent filamentary bundles formed by extruding a molten polymer or polymer macroblend through several small openings in the mesh spinneret. According to this process, in which heat is generated by the partitioning members assuring the small openings of the spinneret, heat is instantly supplied to the polymer during its passage through the small openings. As a result, viscosity, temperature a.s.o, can be controlled so that the polymer can be smoothly separated from the extrusion surface and converted into fine streams.

A bicomponent filamentary bundle (e.g., polyethylene terephthalate/PE, 60 : 40) obtained by this method can be easily crimped in a stuffer box and then converted into staple fibers by cutting. The bicomponent fibers develop crimps to a greater extent and are very useful in soft nonwoven production by ordinary dry-laid and thermal bonding methods.

In addition, a bicomponent filamentary bundle can be easily converted into an interconnected parallel web by proper thermal treatment. There are also composite sheets composed of interconnected parallel webs and burst fiber sheets used for a new type of nonwoven.

The blended PP/polyamideurea fibers have better sorption and electric properties (higher sorption of water vapor and dyestuffs, lower electrostatic charge) and changed tensile properties, e.g., lower tenacity [115].

Courtaulds Specialty Fibers have developed a X-ray-detectable yarn which is intended for use in surgical swabs. Micropake is a multifilament PP yarn loaded with barium sulphate during melt spinning, and wrapped with a fine polyester. The loading meets the X-ray opacity test requirements of the British Pharmacopeia. Micropake is appropriate for bleaching and steam sterilization and can be used in woven products [116, 117].

(h) High Speed Spinning: Fibers from Polymer Blends. A technique applied for many years to improve the mechanical properties of the synthetic

fibers is blending of polymers [118, 119]. Fibers' properties may be adjusted to specific end-use requirements by careful control of the blend composition and phase morphology [120].

Blending of thermotropic LCP (liquid crystalline polymer based on aromatic copolyester of 1,4-hydroxybenzoic acid, 2,6 hydroxynaphtoic acid or a copolymer of 2,6-hydroxynaphtoic acid, 4-aminophenol and terephtalic acid, namely Vectra 900 and Vectra 950 from Hoechst Celanese Co. and Rodrun LC 3000 from Unitika Co.) as a minor component with PP (weight proportion of LCP/PP is 10/100) yields, after melt extrusion, fibers consisting of LCP fibrils surrounded by a PP matrix. The use of a compatibilizing agent promoting adhesion between the two phases (PP and LCP) will also increase LCP fibril fragmentation during the drawing process, with consequent impairment of the fibre's mechanical performances [121–126].

By blending the liquid crystal polymers (LCP) with conventional fiber forming polymers it is possible to obtain fibres reinforced at the molecular level. By incorporating a LCP component with high strength and high modulus into fibers with poor thermal stability (such as PO), the thermal properties can be improved.

The fibril–matrix structure of the polyblend fibers can be enhanced by drawing, and the as-spun fibers reinforced with LCP have higher strength and higher modulus than the conventional fibers [127–129].

Graft side-chain liquid crystalline functional copolymers based upon acrylic acid-functionalized polypropylene (PP–AA) were assessed as compatibilizers for polypropylene/liquid crystalline polymer (LCP) polyblend fibers. The compatibilization effect observed was found to depend upon the liquid crystalline phase temperature range of the compatibilizer, the most favorable one being observed in a liquid crystalline state during fiber melt extrusion and hot drawing, thus contributing to a lubricating effect in the blend, and resulting in enhanced physical properties [126].

(i) High Speed Spinning: PO Composites. Composite materials can be tailored for specific applications, with a large variety of reinforcement and matrix systems.

The use of ultra high strength PE fibers in composites is restricted, due to their relative low melting point, poor interfacial bonding to many polymeric matrices, high creep, and low strength perpendicularly to their axes. To overcome these problems, the fibers are modified in various ways: crosslinking, chemical etching, calendering, etc. [130, 131].

These fibers are increasingly used because of their extremely high stiffness and highest strength to weight ratios of all materials [132]. Thermoplastic composites including woven yarns of comingled PP filaments and glass fibers offer a material form suitable for high volume structural automotive applications [133].

Several producers have developed a long fiber PP technology, which is the most interesting, due to the relatively low cost of the matrix material (chips of 15 mm with 20–50% glass content). The properties include high dimensional stability, low warping, good surface finish and elimination of shrinkage, and good stability at elevated temperatures (150°C). The properties of typical long fiber thermoplastic compounds of PP/40% glass fibre are: density 1220 kg/m^3; impact strength 20 kJ/m; flexural modulus 7.5 GPa; tensile strength 110 MPa; heat distorsion temperature 156°C [134].

PP/glass fibers blends can also be used for replacing steel in the chemical industry for vessels for corrosive substances. A polyolefinic matrix with almost 13% addition of glass fibers represents the optimum variant from the viewpoint of the chemical and thermomechanical stresses [135].

For composites obtained from LCF (long carbon fibers)/PP matrix/PANI (polyaniline complex), electrical conductivity was much higher and surface resistivity much lower that in the case of separate LCF/PP and PP/PANI complex composites. By using the PANI complex, the percolation threshold of LCF composites could be moved towards a reduced fiber content. Electrical conductivity measurements over the thickness of the sample and in different directions showed a synergetic effect [136].

PP/flax or /hemp composites were obtained in two steps: mixing of both components in a roll mill at high temperature and extrusion. There are some important problems to be solved in this procedure: the higher sensitivity to temperature of the natural cellulosic fibres versus the synthetic polymer, cutting of fibres into suitable pieces and preparation of their surface to make it more compatible with the aliphatic polymer, etc.

The PP/natural fiber ratios, the procedure parameters (especially temperature and pressure) cause changes in the structure of the components and of the final products, with notable influences on their properties, allowing to predict the applications spectrum [137, 138].

(j) Fiber Processing After Melt Extrusion. In the chemical industry, the main processing steps after extrusion of PP fibers are quenching, drawing, spin-finish treatment, crimping, heat-setting and shrinking, cutting, and packaging [139, 140].

Quenching is the step in which the temperature of the melt-spun filaments is lowered (by water or by air) very rapidly and/or at a controlled rate, soon after extrusion.

For spinning of the textile PP mono- and multifilaments, the best solution is a laminar crossflow air quench duct with a minimum ratio of air quantity to air velocity. For spinning of fiber staple tows, the crossflowing quench air can be applied for rectangular spinnerets, i.e., inside-to-outside radial quenching, or double-direction turbulence quenching [141].

Optimization of melt spinning conditions involves not only quenching—it is just as important to ensure that the auxiliaries (i.e., the supply air conditions, room air conditions, geometry of the building) have to be optimized, as well [142].

Elastomeric mono- and polyfilaments can be produced by melt spinning with subsequent quenching at room temperature and final setting at high temperatures (130–140°C) for several minutes up to 24 h. A specific orientation of crystallites, perpendicular and along the fiber axis can be observed. These filaments have values of elongation at break ranging from 200–650%, tenacity of 1.0–2.5 cN/denier, and a 85–90% elastic recovery upon 25% elongation [26].

Filaments of high and superhigh tenacity (> 15 cN/denier) can be obtained by using PP polymers with the highest isotactic content (≥94%). To avoid orientation, quenching involves no stretching, and final drawing takes place up to 1 : 34 at 135°C [143], or, in three steps, at a total draw ratio of 1 : 12 and at temperatures increasing up to 140–150°C [26]. The elastic modulus of these fibers is 96–110 cN/denier, while elongation at break is 18–24%.

Drawing is performed between two draw stands with godets, the latter running faster than the former; simultaneously, the tow is heated with hot water. Essential for obtaining a good drawing effect are the free slip tow passage and uniform heat transfer for each filament, especially for fiber types with high elongation (e.g., resilient carpet fibers with about 130% elongation or fine titered soft types with elongation of about 350%). All these fibers require excellent heat transfer and low drawing forces.

The spin-finish treatment improves fiber qualities and extends their end uses. Fiber products can also be finished to gain good dyeability or additional resistance to UV and visible light, to flame and thermal aging, and so on.

Crimping is performed on the stuffer box. The crimper serves to impart a sinusoidal or sawtooth-shaped crimp to the tow. Crimping is accomplished by overfeeding the tow into a steam-heated stuffer box with a pair of nip rolls. The tow is still wet with the finish solution, which acts as a lubricant and aids heat transfer in the stuffer box. The over-feed folds the tow in the box, forming bends or crimps in the filaments. The bands are heat-set by steam injection into the box. When the pressure of the tow being forced into the stuffer box exceeds a set pressure applied by a hydraulic cylinder, the box opens and the crimped tow is released.

This crimp is improved through fiber-to-fiber adhesion, but it must also show good ammenability to downstream processing (e.g., carding). This explains the so-called two-dimensional crimp of all staple fibers produced today.

The primary purpose of crimping is to make individual filaments adhere to each other, while secondary crimping ensures cohesion of the tow itself. The shape and number of crimps per centimeter are influenced by the following parameters: roller pressure, stuffer box pressure, design and geometry, and fiber density in the stuffer box. The MW, MWD, and isotactic content of the polymer affects crimp stability, amplitude, and ease of crimping. For optimum adjustment of the crimping parameters, trials must be run under simulated production conditions.

The heat setting of the tow is performed in a duct or in a perforated drum dryer; fiber bonding shrinkage after this process should be less than 1%.

To blend fibers with each other or with cotton and wool, the materials must be cut into staple fibers having an exactly controlled length and a certain crimp to increase frictional adhesion to each other. The cutting equipment consists of a cutting reel equipped at its circumference with cutting blades and a nip roll mounted at a distance of about 5 mm to the cutting edges of the knives. The staple length is determined by the distance between the knives.

After cutting, the staple fibers fall into a funnel and are transported to the balling press. Fully automatic balling presses, operating several small fiber lines and enabling them to run a fully automatic packaging system, are now available.

Several individual steps lead from the continuous, endless filaments or staple fibers to the ultimate goods sold by the textile industry. To arrive at these end products, the textile industry applies both mechanical (spinning, weaving, knitting, tufting, and related techniques) and chemical (dyeing, printing, finishing, etc.) technologies. In such processes, the individual steps are essentially mechanical, electronic, physical or chemical, or a different combination of these.

C. Nonconventional Technologies

1. Film Tape and Film–Fiber Processing

(a) General Remarks. The development of film-to-fiber processing was started in the early 1960s with much enthusiasm. Film processing is relatively simple and inexpensive, making the manufacture of film tapes, split-film yarns, and fibrillated film fibers a reasonable proposal.

It is only polymers with weak intermolecular forces, producing films at high stretch ratios, with pronounced anisotropy, that are favorable choices for film fibrillation, film-slitting, or film-cutting techniques. For these reasons the PO polymers, especially HDPE and isotactic PP, have gained growing importance for such applications.

Film processing by melt extrusion for PO polymers is widely applied [144–146], involving the following steps: polymer melt extrusion (through a flat or an annular die); film solidification (through water quenching, chill-roll, or air cooling); uniaxial stretching; heat relaxation or setting; and final takeup. Cutting into tapes and slicing or splitting into finer fibrous materials may be performed after the film's solidification, during stretching operations, or before the final takeup.

The processing conditions (extrusion temperature, speed, solidification parameters, etc.) have a considerable influence on stretching and therefore on the final mechanical and thermal properties of the resulting products, since uniaxial deformation of the films causes the different structural units forming the spherulitic substructure to be rearranged in the elongation direction. After deformation, the structure depends on the initial structure and on the drawing parameters (drawing ratio, temperature, time, etc.) [147–149].

For PP films with different average molecular weights, at different temperatures and drawing ratios, some physical and mechanical properties are correlated with the structural parameters. Increase of strength and decrease of elongation at break with increase of the drawing ratio were noticed. Tensile strength increases more rapidly, and higher values were recorded for films of higher molecular weight. Elongation at break does not show a specific behavior. A better alignment of macromolecules with increases in the stretching ratio was indicated by a rise in the optical anisotropy, as determined by the general orientation (amorphous and crystalline) [150].

(b) Film Tape Processes. The film tapes developed for HDPE and isotactic PP are widely used. There are many publications dealing with this subject [145, 151–155]. For example, PE monofilaments [151, 152] or PO fibrillated fiber [152, 153] can be obtained by the application of the following steps: extrusion, cooling, air stretching, fibrillation, winding-up.

Two principal methods of producing uniaxially stretched film tapes are known, as follows:

1. Processes in which tape cutting is made before stretching: a tubular or a flat extruded film is cut into primary tapes 1–20 mm wide, then stretched to achieve the desired properties and dimensions. Cutting of an unstretched film normally leads to tapes with a more pronounced monoaxial orientation and a higher strength anisotropy. To avoid expensive creeling and beaming operations, monoaxially oriented films were used directly as a warp on a weaving and knitting machine, with cutting performed on the loom (Chemie Lenzing AG, Austria).
2. Processes in which tape cutting is carried out after stretching of the film has been performed: cutting of a stretched film prevents width shrinkage, introduces some cross-orientation, and reduces the length-splitting tendency of the film.

With an average efficiency rate of 95–98%, extrusion plants for producing drawn film tapes produce wastes whose recycling constitutes an important economical problem.

Traditionally, the material conveyed into a bin is regranulated off-line and can be fed into the extruder of the tape line. During reprocessing of the material, additional degradation takes place, which influences negatively the physical characteristics of the tape. In this procedure, extra costs for energy, personnel, transportation of material, and operation of the regranulating plant are involved.

A much more economic solution is offered by inline recycling (Barmag AG). The wastes are fed directly to a mill and shredded into granules, which are collected pneumatically into a storage silo, from where, via a metering screw and a conveyor, they are introduced together with the fresh granules into the extruder

According to this recycling concept, the film tape line functions waste-free, in an almost fully integrated cycle [156].

(c) Film Fiber and Film Yarn Processes. The fibrillating film is a polymer film in which molecular orientation has been induced by stretching until the material is capable of being converted into the yarn or twin by manipulation (e.g., by twisting under tension, brushing, rubbing, etc.); the result is the formation of a

longitudinally split structure (split fiber) [19]. Owing to their chemical structure, PO polymers have a limited tendency to form strong secondary bonds between the individual molecules with high tensile anisotropy, which makes them very suitable to film-to-fiber technologies. Film-to-fiber technology involves the same general steps applied in the manufacture of film tape. The main features of this method lie in the techniques of film splitting, film slicing, and film cutting, which may be classified into three groups:

1. Random (uncontrolled) mechanical fibrillation performed by brushing, rubbing, twisting, or false twisting, by air jet or sand blast treatment, by special ways of stretching, or by ultrasonic treatments. This approach results in an incomplete separation into filament segments of widely varying widths, lengths, and fiber cross sections. The yarnlike products have filament segments still interconnected in a coarse network-like structure, with varied properties and displaying a wide statistical distribution.
2. Random (uncontrolled) chemomechanical fibrillation, achieved by the introduction of melt additives. Such chemical substances act as weak spots and improve the splitting tendency during stretching and in subsequent mechanical splitting treatments. Examples of these additives are in forming gas compounds, soluble salts (which create a microvoid structure in the film), and incompatible polymers (used to increase film splittability). During drawing, these additives are extended in the draw direction, thus forming well-defined starting points for the film's splitting into relatively fine, network-like fibrous systems (known as split-film fibers or split fibers). Longitudinal splitting is also incomplete, but the cross sections are more regular than those obtained by uncontrolled mechanical splitting.
3. Controlled mechanical film-to-fiber separation, performed by a needle roller equipped with teeth, by embossing or profiling techniques, or by slicing or cutting techniques. In these cases, more or less well-defined separation into regular network-like fibrous structures is achieved. Because of their relatively good uniformity, these processes compete with conventionally synthetic products (coarse multifilament yarns or staple fibers), and one should speak here of slit-film fibers or slit fibers [157].

An example of the application of the embossing method is the Fibrilex yarn (Smith & Nephew Ltd, UK), in which the basic technique is to emboss a pattern of parallel indentations onto the hot melt immediately after extrusion. This results in a series of continuous and uniform fibrils connected by relatively fine membranes extending across the film's width. The film is next divided into tapes of the required denier, and processed through drawing and winding operations. It is then twisted to produce a continuously fibrillated yarn (300–600 denier), with high tenacity (6.5–8.0 cN/ denier) and regular round cross section.

2. The Spunbonded Process

The fascinating idea of producing textiles directly from spinnerets, in one operation, originated in the early 1960s, and was successfully realized in the mid-1980s; the products became known as spunbonded nonwovens or simply spunbonds [158].

Spunbond manufacture involves converting a polymer into fine filaments by a spinning process, distributing these filaments as uniformly as possible over a surface, and then joining them together (by calendering, needle punching, or thick thermal bonding), to produce a continuous textile web.

In the spunbonded process, filament formation can be accomplished with one large spinneret having several thousands holes or with banks of smaller spinnerets containing at least 40 holes. After exiting the spinneret, the molten filaments are quenched by a cross-flow air system, then pulled away from the spinneret and attenuated (drawn) by high pressure air.

Filament denier (diameter) has a significant effect on fabric quality and process efficiency. It is determined by melt viscosity, throughput per spinneret hole, and velocity of the attenuating air. Higher denier filaments give fewer breaks in spinning, finer denier gives desirable fabric properties such as softness, uniform coverage, and strength. Filament diameter is also affected by the distance of the drawing from the spinneret, type of drawing device, quenching conditions, melting strength, polymer rheology, and its quality (e.g. MRF, NMWD).

The properties of the starting polymer (such as MFI and MWD), the spinning conditions (such as draw ratio) and the calendering parameters (e.g., T and P) affect the properties of thermobonded PP nonwoven web.

Web characteristics and variations in bonding can offer a variety of specific products with different fabric properties.

Fabric weight is determined by the throughput per spinneret hole, number of holes, and speed of the forming belt. A *vacuum* is maintained on the underside of the belt. This *vacuum* is very important to uniform fabric formation and to remove the air used in attenuating the filaments.

Air handling and control is critical to the process. The fabric formation area is frequently sealed off from the rest of the line to improve control. Various methods are used to aid fabric formation and improve its uniformity. Examples include stationary or moving deflectors, and the use of static electricity or air turbulence to improve the dispersion of filaments.

To improve fabric integrity, it passes through compression rolls, which can also seal off the forming chamber.

Heat-bonding methods for webs employing the thermal characteristics of PP as a bonding medium became more popular and more compact than those in which part of the process involved removing water (used in binder dispersions).

There are many possibilities of making the fibers to achieve bonding. A process for obtaining fibers from thermoplastic polymers consists of placing melting thermoplastic polymer particles on a heated grid, so that the thermoplastic polymer flows through the grid and is collected in a chamber underneath the grid, followed by injection of an additive into the melted thermoplastic polymer and formation, without mechanical stirring, of a homogeneous mixture made of the additive and melted thermoplastic polymer in the chamber before extruding fibers from the homogeneous mixture [159, 160].

Generally, propylene homopolymers and copolymers formed using metallocene catalysts have a lower melting behavior than the previously mentioned polypropylene polymers. This behavior is of use in the production and utilization of fibers and fabrics which depend either on lower melting behavior in general, or on a differentiated melting point between two fabrics or fibers, to achieve bonding. Spunbonded or melt-blown fabrics show bonding at lower temperatures with fewer pin-holes [161]. The thermotropic polymer, used as bonding fibers for spunbonds, is a blend of one isotactic polymer having an isotacticity of at least 85 and up to 40% by weight of one atactic polymer with an isotacticity of at least 20. The blend may also have a fluorocarbon content between about 0.1–3.0% by weight. The fabric may be produced by spunbonding and melt blowing and has good strength, barrier properties, and softness [162].

Calenders, ultrasonic techniques, short ovens, and suction drum units to bond webs containing PP fiber are economically attractive.

Spunbonded nonwovens are softer and more uniform than common nonwovens; in addition, they have higher opacity and absorption capacity, improved isolation properties, and excellent tensile strength. Microfiber PP spunbonded nonwovens can be easily dyed, printed, flocked, sterilized, cut without linting, and, if requested, given a water-repellent, antistatic or antimicrobial finish. Applications of these products involve carpet backings, packings, interlinings, and so on.

All the known methods for spunbond fabrication with spin extruders process the chips into a melt, which is then guided by means of spinning heads with internal melt piping systems and spinning pumps and extruded through the spinnerets (with different round, rectangular cross sections) to the filaments. Following the filament path, the main takeup ways to obtain the filament bundle are [163, 164]:

1. Hot air flows parallel down the melt, on both sides of the filament rows, drawing the filament down without breaking.
2. Hot compressed air leaves the spinning head close to the melt drops and elongates them to long filaments up to breaking (e.g., the Exxon Melt Blow Process; see the next section).
3. Hot compressed air is delivered as a skin with the melt as a core, elongating the filament up to breaking.
4. A low air pressure jet takes over, cools, and solidifies the filaments without breaking.
5. Compressed air guns draw down, cool, and solidify the filaments.
6. Heated godets takeup and draw down the filaments, whereupon a crossflowing airjet quenches them.

After the filament bundle has been spread in the laying system (e.g., electrostatic system, gas dynamic stream, laying jets and blades, elliptic oscillators, multilaying with cross-moving directions), one obtains a warp to form a fabric with defined properties (space weight, width, edge width, individual titer, and thickness of the fleece, etc.) and therefore defined end uses (covers for baby diapers, heavy geotextiles, etc.).

There are many compact spunbond systems with considerable flexibility in the basic process (e.g., the Reifenhäuser or the Lurgi Docan process).

The Lurgi Docan spunbonded process (Lurgi and Mineralöltechnik GmbH, Germany) [82] is a one-step operation in which a random web is formed by spinning and drawing continuous filaments, which are then laid down on a moving screen belt where the web is built up. The Docan process flow consists of: *feeding. extrusion, spinning, cooling, draw-off and lay-down, web forming, post treatment.*

Thicker needle-punched webs and a broad range of coated and laminated products can be produced on a Docan line equipped with a calender.

An extremely special nonwoven spun product is the mat from individual filament loops, produced by the free fall (2–5 m/min) of the filaments from the melt—that is, by takeup from the spinneret hole through the filament's own weight (e.g., Erolan product, developed by Oltmanns GmbH, Germany). Upon touching, the individual filaments are welded together by the latent heat from the filament, and then they are cooled and solidified by means of water. The thickness of the mat can be adjusted between 4–40 mm.

The dyed PP spunbonds are very important and they can be produced by three methods: mixing of natural chips with master batch chips, mixing of natural chips with dyestuff powder, and melting of the master batch in a second extruder and metering of this melt into the main extruder, where it is mixed with the main melt stream.

Extrusion coating and laminating are ways to add value to the existing product line of spunbonded PP [165].

3. The Melt-Blown Process

The melt-blown process was developed in the 1950s and put to commercial use during the 1970s [163, 164, 166]. The initial patent design of Exxon [167] was executed later by the melt-blown plant system known as Automatic/Fourné [168, 169] and by the Reicofil melt-blown system of Reifenhäuser GmbH (Germany).

Parametric studies of the Reicofil spunbonding process have been developed [170–174]. The variables studied were: polymer throughput rate, polymer melt temperature, primary air temperature, bonding temperature, venturi gap, bonding pressure, web basis weight, and ambient air conditions. Polymer throughput weight and rate has the most significant effect on the final fiber diameter and on the physical properties of the webs.

Bonding temperature has a minimal effect on the final crystallinity of the webs, but it has a noticeable influence on their thermomechanical behavior. Localization of crystallization during spunbonding of PP sets in along the spin line. In contrast to conventional melt blowing, the drawing force on the extruded molten polymer in spunbonding is provided here by air accelerating into a conveying duct. The flow of the melt-blown system includes the following steps [168]: chip feeding, heating (optional), extrusion, melt-blown spinning, cooling (optional), and filament takeup.

In the melt-blown process, the molten polymer moves from the extruder to the special melt blowing die. As the molten filaments exit the die, they are contacted by an airjet of high temperature and high velocity. This airjet is drawn rapidly and, in combination with the quenching air, it solidifies the filaments. The entire filament-forming process takes place within 7 mm of the die. Die design is the key to producing a quality product.

The melt-blown die is made up of a spin die and two air jets. Many filaments are spun .through a multiplicity of orifices, and the hot high speed air is blown along both sides of the die top, which causes a fiber breakage, resulting in fibers with 0.05 dtex. The pressure compensation system for the air jets provides uniformity of the end products.

The fabric is formed by blowing the filaments directly onto a forming wire, about 200–400 mm from the spinnerets. This distance influences fabric uniformity and strength, while the belt speed and throughput determine fabric weight.

Melt blowing requires very high MFR (above 200 g/10 min), narrow MWD, ultraclean, low-smoke PP grades, to obtain the finest possible fibers. Gels and dirt can cause filament breaks, leading to a defect known as shots. These are pinpoint-sized PP beads in the fabric that affect porosity, uniformity, and texture of the fabric, being the principal cause of off quality fabric. Shots can also be caused by excessive melt or process air temperatures, but lower melt or process air temperatures increase filament size, adversely affecting fabric uniformity and barrier properties.

The principal advantage of the melt-blown process is that it can make fine filaments and produce very light weight fabrics with excellent uniformity. The result is a soft fabric with excellent barrier properties, meaning effective filtration characteristics and resistance to penetration by aqueous liquids. This property is vital in medical applications as concerns on blood-borne diseases increase.

The disadvantage of melt-blown fabrics is the lower strength compared with spunbonds. The optimum solution is a widely used composed construction of back spunbonds (for their strength) and melt-blown fabrics (for barrier properties).

The Reicofil spunbonding process is simple, consisting of polymer transport and controlled mixing, extrusion, filament production, cooling and orientation, collecting and distribution of the filaments to form the web, bonding or consolidation (needle punching; impregnation; foam bonding; pressure, chemical, or heat bonding, etc.), edge trimming and winding of the spunbonded fabric. The Reicofil melt-blown line includes computer systems for the automatic line control of speed, temperature, melt flow index, and other parameters. The Reicofil melt-blown system opens up a variety of further applications because of its facility for blowing melt-blown fibers onto air-permeable substrates (spunbonded fabrics, textiles, paper, multilayer composites, etc.). The Reicofil melt-blown line is designed for polymers with very low viscosity (PP and PE, copolymers and polymer blends, addition of pigments, wetting agents, and UV stabilizers) [175].

D. Polyolefin Fibrous Products

1. Fibrous Product Types

The main types of fibrous products obtained by various extrusion methods are presented in Table 2 [89, 155, 176–190].

In the last decade, a considerable and impressive diversification of PO fibers occurred. That is why an important number of studies and patents have been devoted to quite varied aspects, referring to:

short fibers [111, 191–220];
bicomponent and biconstituent fibers [221–232];
technical fibers [233–235];
hollow fibers [236–239];
porous fibers [240–244];
autocrimp and tridimensional crimped fibers [90, 245–252];
microfibers [253–260];

TABLE 2 Correlation between the production method and the fibrous product type

Extrusion method	Fibrous product
Dry spinning	PO mono- and polyfilaments
Wet spinning	PO mono- and polyfilaments
Melt spinning	Regular PO fibers with medium and coarse counts
Short spinning (compact spinning)	Filaments
	special fine filaments
	bicomponent filaments
	spun-dyed filaments
	modified cross section fibers (> 6.7 dtex)
	staple fibers: stuffer box texturing fibers (3 dtex)
	metallocene PO fiber (2; 17; 110 dtex)
High speed spinning	Flat filaments
	fine denier filaments (< 1.7 dtex)
	bicomponent filaments (conjugated and blended filaments)
	spun-dyed filaments
	high tenacity filaments (> 50 cN/tex)
	elastomeric filaments
	textured yarns: UDY–DTY, FDY–ATY, POY–DTY, POY–ATY
	blow texturing staple fiber (higher titer)
	polyfilaments (BCF) (200–1600 dtex)
Spunbond process	Spunbonded nonwovens (spunbonds or spinbonds)
Melt-blown process	Continuous filamentary webs
	Microfibers (0.03 dtex)
Film tape method	Film tapes
Film-fiber method	Split fiber (or split-film fiber)
	Slit fiber (or slit-film fiber)
	Fibrillated film fibers

Source: Refs. 89, 155, 176–190.

flame-retardant fibers [261–269];
antistatics [270–272];
high shrinkage POs [273–277];
fibers from UHMWPE [278–283];
hydrophobe and bioactive POs [284–290];
mono- and polyfilaments [291–297];
BCF polyfilaments [298–302];
surgical threads [303–305];
high strength PO fibers [110, 119, 306–320].

2. Producers of PO Fibrous Products

PO fibers have much to offer to the textile industry in terms of physical properties and growing versatility. Price has been one of their principal advantages, and they remain, potentially, the cheapest synthetic fibers; indeed, the properties of these relatively new synthetic fibers have not yet been fully realized.

Interest in these new fibers grew very fast after 1960 and many companies throughout the world now produce PO products.

Table 3 lists the major producers in North America, Europe, and Asia, together with their tradenames and product types [89, 321–385].

In recent years, many PO fiber producing companies have merged (Table 4).

3. Evolution of Production and Consumption of PO Fibrous Products

Industrial production of PO fibers, especially PP fibers, began in 1960. By 1965, the production amount of PP fibers from all synthetic fibers was 3.9%.

After intensive research and development work between 1970 and 1975 better PP yarns and fibers were produced and accepted on the market. From this time onwards, production and applications expanded, so that today, PP yarns and fibers have a legitimate existence.

PP fibers (including film tapes, monofils and spunbounds) are assumed to have already exceeded acrylic fibers worldwide, and are now in third plane.

During the 1980s, the PO (mainly PP) were considered to be among the most dynamically expanding sectors of the fiber market. During the 1990s, the world production of PP fibers of all types (staple fibers, filament yarns, BCF yarns, monofilaments, film tapes, and spunbonds) increased annually by approx. 200,000 tonnes.

Numerous smaller and medium-sized manufactures have extensive product ranges based on this polymer [175]: staple fibers (including finest titers up to 1 dtex), filament yarns (including BCF yarns), spunbonds (including melt blowns), monofilaments, and film tapes (including fibrillation). World production, including extensive film tape production, surpassed 3.4 million tonnes in 1995 (with the upward trend continuing), the share in the total synthetic fiber production being 16.3%.

According to Fiber Organon estimates, PO production (including PE fibers), came to 3.96 million tonnes worldwide.

According to estimates of the European Textile POs Association (EATP) Brussels, Belgium, production in West Europe during 1995 stagnated at 1.43 million tonnes. The Association's statistics includes 190,000 t spunbonds (including meltblown). Out of the total production in Western Europe, 32% are staple fibers, 27% film fibers, 18% multifilament yarns (including BCF), and 5% tapes.

Leading PP fiber producers in Western Europe are Belgium, Germany, Italy, Great Britain and France.

Contrary to the situation in Western Europe, PP filament yarns (including monofilaments and spunbonds) are much more marketable in the United States than PO staple fibers. In Japan, where PP fiber production is less prevalent, staple fibers and BCF yarns are used.

Table 5 shows the evolution of PP fibers' world production between 1965 and 2001.

In 1992 the production ratios were: USA 35.5% Western Europe 38%, East Asia 5% and, from the view point of the good's range: 36% staple, 23% filament, 31% film fiber, 8% spunbonded and meltblown. Of the first 10 largest PP fiber producers, 5 were from USA: Amoco, Kimberly-Clark, Hercules, Synthetic Industries, and Phillips.

According to estimates of the European Textile POs Association (EATP) Brussels, production in Western Europe in 1994 increased by 10% to 1.44 million tonnes. In USA, the production expanded by more than 12% to 1.146 million tonnes (staple fibers: +14%, filament yarns: +18%, film tapes: +3%). Other important producers of PP fibers are China, Indonesia, Taiwan, Japan, Korea, and India (Table 6).

PP application is hindered by dyeing difficulties, but its cheapness has resulted in very fast growth in the developed world since 1980, especially in carpets, nonwovens, and ropes. Carpet growth is slower in the third world, comparative with disposable nonwovens. However, low cost ropes and nets, and items like geotextiles and agrotextiles are likely to find rapid deployment worldwide.

In Western Europe, PP fiber is expected to take 20% of world consumption in the year 2005.

(Text continues on p 795)

TABLE 3 Polyolefin producers

Tradename	Company (country)	Product type
Achieve	Exxon Chemical (USA)	Fi
Alpha	Amoco Fabrics & Fibers Co. (USA)	Fi, St
Alpha Carpet	La Lokeroise (B)	Fi
Amco	American Manufacturing Co. (USA)	Fi, St.
Asota	Asota GmbH (A)	Fi, St
Asota M10, M11, M12	Asota GmbH (A)	Fi (UV stabilized)
Asota M30	Asota GmbH (A)	Fi (multilobal and hollow structure)
Asota FR-1034	Asota GmbH (A)	Fi (flame-retardant)
Asota AM Plus	Asota GmbH (A)	St (antimicrobial, antidust)
Asota AM Sanitized	Asota GmbH (A)	St (antimicrobial)
Asota FH	Asota GmbH (A)	St (high tenacity)
Asota FM	Asota GmbH (A)	St (medium tenacity)
Asota FL	Asota GmbH (A)	St (low tenacity)
Atrex	Arova Schaffhausen AG (CH)	Fi
Austrofil	Starlinger Co.	Fi
Base	Neuberger SpA (I)	SB
Beklon	NV Beklon Fibers (B)	St
Betelon	TTB Tech Tex GmbH (D)	Fi
Bluebell	Belfast Ropework PLC (GB)	Fi, St
Bonafil	Bonar Textiles Ltd. (GB)	Fi
Charisma	Plastics Ltd. (GB)	St
Chisso Polypro	Chisso Fiber Co. (J)	Fi, St
Corovin	Corovin GmbH (D)	SB
Crackstop	Chemifiber A/S, Varde (DK)	St
Dal-Tex	Don & Low Nonwovens (GB)	SB
Danaklon	Danaklon A/S (DK)	St
Danaklon ES; ES-HB; EA, Soft 61-300	Chisso Fiber Co. (J) under license by Danaklon A/S (DK)	St
Danaklon ES/C; ES-E	Danaklon A/S (DK)	St (PE/PP bicomnponent fibers)
Danaklon 6830A; 6831A; 6835A	Danaklon A/S (DK)	Fi (bicomponent PP/PE fibers)
Danaklon HY-Color HY-Dry, HY-Medical	Danaklon A/S (DK)	Fi, St
Danaklon HY Speed	Danaklon A/S (DK)	Fi (high performance fiber)
Danaklon HY-Strength	Danaklon A/S (DK)	Fi (high performance fiber)
Danaklon HY Comfort	Danaklon A/S (DK)	St
Danaklon Star	Danaklon A/S (DK)	Fi (modified cross sections)
Delebion	Filatura di Delebio SpA (I)	Fi
Delta Carpet	La Lokeroise (B)	Fi
Donfil	Don & Low PLC (GB)	Fi, St
Downspun	PFE Ltd. (GB)	St
Drake PFE	Moplefan SpA (I)	Fi (antibacterial)
Duron	Plastics Ltd (GB)	Fi
ECOfil	L.Kapell GmbH, Duren (D)	Fi

TABLE 3 (Continued)

Tradename	Company (country)	Product type
Ekor	Chemosvit	Fi
Elustra	Hercules Inc. (USA)	St
Embryon	Neste (Finland)	Fi (antistaining)
Essera	Amoco Fabrics & Fibers Co.(USA)	Fi, St
Evotex	Lenzing AG (A)	Tf
Fibertex	Crown Zellerbach (USA)	SB
Fibrilex	Smith & Nephew Ltd. (GB)	Fi, Ff
Fibrite C	Plastic LTD. (GB)	Fi, Ff
Finesse	Amoco Deutschland GmbH (D)	St
Fixset TM	Plasticisers Ltd (UK)	Fi (shrinkable)
Gymlene	F.Drake (Fibers) Ltd. (GB)	St
Gymlene Auto	F.Drake (Fibers) Ltd. (GB)	Fi, St (UV stabilized)
Helta TM	Plasticisers Ltd (UK)	Fi (shrinkable)
Hercules T-190, T-196	Hercules Inc. (USA)	Fi, St
Herculon	Hercules Inc. (USA)	Fi, St
Herculon Type 123	Ullmann & Bamforth (USA)	Fi
Hitex	Lurgi SpA (I)	Fi
Holmestra	HJR Fiberweb AB (S)	SB
Hostacen TM	Hoechst AG (D)	Fi
Hyfib	Synthetic Fabrics Ltd. (GB)	Tf
IHS	Danaklon A/S (DK)	St
Innova	Amoco Fabrics & Fibers Co. (USA)	Fi
Istrona	Chemicke Zavody (Slovakia)	Fi, St
Jacob	Chemicke Zavody (Slovakia)	Fi, St
Kim Cloth	Kimberly Clark (USA)	SB
Krenit	Danaklon A/S (DK) Chemfiber A/S, Varde (DK)	St, Sc
Lanora	Polyolefin Fibers & Engineering PLC (GB)	St
Lanora	Montell Polyolefins Inc., Wilmington (GB)	St
Leolene	F.Drake & Co. (GB)	Fi
Linz PP	Chemie Linz AG (A)	St, Fi
Lutrasil	Lutravil (D)	SB
Lyondell	Polymers (USA)	St, Fi
Macrolene	Sergel SpA (I)	St
Marquessa Lana	Amoco Fabrics (USA)	Fi
Marvess	Phillips Fiber Co. (USA)	St, Fi
Meraklon	Moplefan SpA (I)	Fi, St
Meraklon BCF	Neofil SpA (I)	Fi
Meraklon CI	Montedison Ltd. (I)	Fi, St (UV stabilized)
Meraklon CM	Montedison Ltd. (I)	Fi, St
Meraklon DL	Himont Ltd.(I)	St, Fi (anionic printable fiber)
Meraklon DO	Himont Ltd.(I)	St, Fi (anionic dyeable fiber)
Meraklon HS	Montedison Ltd. (I)	Fi
Meraklon PPL	Montedison Ltd. (I)	Fi

TABLE 3 (Continued)

Tradename	Company (country)	Product type
Meraklon SR-AB	Himont Ltd. (I)	Fi (bactericidal and fungicial filaments)
Meraklon SR-LAN	Himont Ltd. (I)	Fi (filaments with woolike appearance)
Meraklon SR-SH	Himont Ltd. (I)	Fi (heat-set filaments)
Metocen	Hoechst AG (D)	Fi
Midilene S701, S900, S1000	ICIT Fibresin (Romania)	Fi (technical yarns)
Nifil	McClear & L'Amie Ltd (GB)	Fi
Novachrome	PFE Ltd. (GB)	St
Novatron	PFE Ltd. (GB)	St
Oplene	Jeil Synthetic Fiber Co (Korea)	Fi, St
Opor	Chemosvit (Czechia)	Fi
Pavlen ZH 12; ZK 35	Chemosvit (Czechia)	St
Polital	Adolff AG (D)	Tf
Polifin	Polifin Ltd., (South Africa)	Fi
Polybond	Wayn-Tex Inc. (USA)	SB
Polyfelt	Polyfelt GmbH (A)	SB
Polyklon	Polymekon SpA (I)	St
Polyloom	Chevron Research Co. (USA)	Fi
Polyost PP	Polyost NV (B)	St
Polyprim	Godfreys of Dundee Ltd. (GB)	Wb
Polyprop	Syn-Pro (USA)	Fi
Polyren	Westfalia Spin. SG (D)	Tf
Polysec	Godfreys of Dundee Ltd. (GB)	Wb
Polysteen	Steen & Co. (D)	St
Polyunion	Polyunion Kunst. (D)	Fi
Popril	Magyar Viscos. (H)	St, Fi
PP-F	Raj.Petro.Synth. (India)	Fi
PP-BCF	Tolaram Pvt. Ltd. (Indonesia)	Fi
Primaflor	Billermann KG (D)	Fi, St
Prolene	Chemosvit (Czechia)	Fi
Proplon	Neomer Ltd. (India)	St
Protel	Amoco Fabrics (USA)	Fi, St
Pylen	Mitsubishi Rayon Co. (J)	Fi, St
Quintana BCF	Mopleton SpA (I)	Fi
Radilene	Deufil GmbH (D)	Fi
Reileen PP	Reinhold KG (D)	St
Rifil	Rifil Ltd. (I)	Fi
Riga	Chemosvit (Czech)	F
Sanitized MB P 87-21	Asota GmbH (A)	St (antimicrobial)
Sanitized MB P 91-40	Asota GmbH (A)	St (antimicrobial)
Scott Fab	H&A Scott Ltd. (GB)	Ff
Serlene	Sergel SpA (I)	St
Sodoca	Sodoca Ltd. (F)	SB
Sodospun	Sodoca Ltd. (F)	SB

TABLE 3 (Continued)

Tradename	Company (country)	Product type
Sophista	Kuraray (J)	F, St
Spectra 900, 1000	Allied Chemical (USA)	Fi, St (high modulus fiber)
Spleitex Hanf	Hanfwerke Fusen AG (D)	Tf
Super	Chemosvit (Czechia)	Fi
Supersoft GHT	PFE LTD + Extrusion Inc. (UK)	St
Superbulk C	Danaklon (DK)	St
Synbac	Synthetic Int. Ltd. (USA)	Tf
Taybac	Wallace Works Ltd. (GB)	Tf
Tenfor	Snia Fiber (I)	Fi
Tenfor	Phillips (USA)	Fi
Thiobac	TTC Polyolefins Br. (NL)	Wb
Thiolan	TTC Polyolefins Br. (NL)	Fy
Thiolon	TTC Polyolefins Br. (NL)	Fi
Tippfil	Tiszai Vegyi Kombinat (H)	St
Toplene	Rifil SpA (I)	Fi
Toyobo Pylen	Toyobo Co. (J)	St
Trace	Amoco Fabrics Co. (USA)	Fi
Typar	DuPont (USA)	SB
Type XPN	Low Bros & Co. (GB)	Tf
Tyvek C, E	DuPont (USA)	PE-SB
Vegon	Faserwerk Bottrop GmbH (D)	St
Westflex	Esbjerg Tovvaerk. (DK)	St

Source: Refs. 89, 321–385.

TABLE 4 Company mergers in the PO fiber field

Countries/Region	Companies	New name	PE or PP own process
Europe	BASF + Hoechst	Targor	Novolen
	BASF + Shell	Elenac	
	Neste + Statoil + PCD	Borealis	Borstar
	Shell + Himont	Montell	Spherilene
	Union Carbide + Atochem	Aspell	Unipol
	Union Carbide + ENI	Polimeri	Unipol
Japan	Mitsubishi Chemical + Tonen	Japan Polychem	Mitsubishi Slurry
	Mitsui P.C. + Mitsui Toatsu	Mitsui Chemicals	Evolue
	Mitsui P.C. + Ube	Grand Polymer	Hypol
	Showa Denko + Nippon P.C.	Japan Polyolefins	
USA	BP + Dow	Technologie Allianz	Innovene
	Huntsman + Rexene	Huntsman	El Paso
	Lyondell + Millenium	Equistar	
	Phillips + Sumitono	Sumika	Sumitomo

Source: Refs. 382, 383, 386.

TABLE 5 PP filament and fiber production

Year	PP quantity (1000 t)	Share in the synthetic fiber production %
1965	80	3.9
1970	338	6.7
1975	716	8.8
1980	1048	9.0
1985	1518	10.7
1990	2227	12.7
1993	2960	15.1
1995	3459	16.3
1997	3580	35.0
1998	3640	38.5
2001	3850	40.4

Source: Refs. 387–389.

Such perspectives indicate that PP will represent, in 2050, 8% of the world's conventional fibers, and will undoubtedly continue to a higher share later [392].

Table 7 shows the production of PP-based raw materials type using PP, in 1995 and 1997, in Western Europe.

E. Modification of the PO Fiber-Forming Polymers

1. Generalities

Modification of PO fibers, aiming at either improving their properties or at the realization of new ones, is one of the priorities in the strategy of fiber development and application. It is known that modifications of the PO fibers may be superficial or deep.

TABLE 6 World production of PO fibers (1000 t) in 1993 and 1994

Region/Country	1993	1994	±%	Region/Country	1993	1994	±%
Western Europe	1149	1203	+5	*Latin America*	189	292	+7
Benelux	216	230		Brazil	68	73	
Germany	198	207		Argentina	43	43	
Italy	176	181		Mexico	29	30	
Great Britain	157	165		Chile	16	19	
France	85	88		Colombia	11	12	
Spain	68	72		Peru	8	10	
Turkey	55	59		Venezuela	8	9	
Denmark	44	44		others	6	6	
Portugal	39	41					
Austria	34	36		*Asia*	848	906	+7
Greece	24	25		China	330	350	
Switzerland	19	20		Indonesia	122	133	
Finland	15	17		Taiwan	108	103	
Sweden	13	12		Japan	80	81	
Norway	6	6		Korea	64	71	
				India	55	67	
Eastern Europe	168	177	+5	Philippines	43	47	
CIS	45	47		Saudi Arabia	20	26	
Hungary	35	40		Israel	14	15	
Poland	29	30		Iran	12	13	
Czech Rep.	20	18					
Slovakia	12	14		*Africa*	33	34	+3
Yugoslavia	11	11		Egypt	17	18	
Bulgaria	9	10		South Africa	16	16	
Romania	7	7					
				Australia/ New Zealand	38	39	+3
North America	1021	1146	+12				
USA	959	1082					
Canada	62	64					

Source: Refs. 390, 391.

TABLE 7 Production of PP fibrous type products in Western Europe (1000 t)

Type	1995	1997
Short fibers	450	470
Split films	390	260
Multifilaments	260	270
Spunbonded and meltblown	190	250
Tapes	75	222
Monofilaments	50	50
Others	10	20
Total	1425	1542

Source: Refs. 393, 394.

As nonpolar fibers, possessing a very low superficial energy, PO fibers are wholly hydrophobic. For some applications, this hydrophobic character constitutes an advantage, while, for others, it appears as a severe shortcoming. That is why different methods have been applied for PO fiber hydrophilization, one of them involving the introduction of some hydrophilic modifiers in the polymer [395–402].

When incorporated—by extrusion—into a PO matrix, the hydrophilic modifiers do not offer any appreciable wettability, but when they are used in conjunction with polar substrates, PO wettability improves significantly. These polymers provide improved dyeability with a broad range of disperse dyes. The final shade of the dyed product is brighter, deeper, and sharper than prior to PP fibers, with improved washfastness and crockfastness.

Blending of fibers based on i-PP, a polar group material (e.g. ethylene copolymer, acrylic acid or maleic anhydride), a hydrophilic modifier (0.1–2% monoglyceride), and a salt of a linear alkyl phosphate, either in the form of nonionic, anionic or cationic, highly wettable, spinnable fibers are provided.

Another characteristic of a PO material is its spreading factor, which may be defined as the ratio of a liquid drop length on a surface as a function of time. It is possible to achieve wetting of PP by incorporating both modification through grafting and alloying techniques utilizing polar additives and hydrophilic modifiers. Either of them by itself will not impart wettability. This PP/polar group material hydrophilic modifier material combination may be used for various applications wherein reasonable or high wettability is desired, such as diapers, pads, filters, tea bags, or battery separators, formed of woven or nonwoven fibers.

A hydrophilic or wettable polymer provides highly desirable material features, such as: permanence, wickability, and extra comfort. These attributes are highly desired in product applications such as diapers, adult incontinence products, and sanitary napkins, where a nonwoven web comes into contact with the body or entry point for any fluid penetration.

From a commercial standpoint, durability could have numerous applications in both absorbent and nonabsorbent products.

Several attempts and successes may be recorded on the modification of PO fibers, so that a possible classification of theirs should consider several criteria. Among them, the (physical or chemical) method applied, and also the scope for which the modification is made (e.g., improving tinctorial capacity) are most recommended.

2. Physical Methods

Physical methods are based on treatment with radiation (UV, γ), corona, or plasma and are mainly aimed at inducing surface modifications.

Thus, through UV irradiation, grafting of 2-hydroxyethyl methacrylate on PO fibers [403–405] was attained, while the utilization of γ radiation permitted the activation of PO fibers prior to their grafting with various vinylic monomers [406–414], and the corona treatment was applied for the activation of PO fiber surfaces, for various ends—increasing hydrophilicity especially [415–418].

The cold-plasma treatment improved the fibers' superficial properties, such as adhesion of the epoxydic resin for the realization of composite materials [419–423] and improved wettability and rewettability [424–426].

3. Chemical Methods

The modifications induced through chemical methods follow—unlike the physical ones—changes inside the fiber, as well, with significant consequences in wetting, plastifying, comfort, dyeing, etc.

Through radical and/or chemical initiation, polar vinylic monomers (acrylic acid, maleic anhydride) have been grafted onto the PO fibers [427].

With a view to improving their hydrophilicity properties, the PO fibers have been treated with solutions of sodium hypochlorite [428, 429] and mixtures of organic solvents [430]. Such treatments also influence—in a positive manner—the fibers' mechanical characteristics, such as: absolute strength, breaking elongation, and initial modulus.

4. Coloration of PO Fibers

In spite of intensive work, the problem of obtaining a simple dyeable PO fiber comparable to the other synthetic fibers remains unsolved. Economical and ecological problems, as well as increased requirements concerning the stability of dyeing and brightness of shades, will maintain and increase the industry's interest in the melt-dyeing methods of PO fibers.

PO fibers with a paraffinic hydrocarbon structure contain no reactive or polar functional groups, which would anchor the dyestuff within the polymer to achieve good fastness to light and good resistance to rubbing, laundering, or dry-cleaning.

POs are hydrophobic and difficult to dye in that they lack dye sites to which dye molecules may become attached. One approach to color PO fibers has been to add colored inorganic salts to fiber spinning.

Generally, and particularly for PP fibers, the dyeing problem is closely connected with humidification, and fiber hygroscopicity. The wetting phenomenon is further extended to the dyeability future. Wettable and dyeable PO technology and applications, with all the diverse aspects involved, are presented in the literature of the field.

Nonvolatile acids or bases, or materials such as polyethylene oxides or metal salts, have been added to polymers prior to fiber formation to increase the affinity of the fiber for disperse, cationic, acid, or mordant dyes. To improve dyeability PO fibers may be chemically grafted, with appropiate monomers, after fiber formation.

The disperse dye allows for the cost-effective production of fibers that preferably have good light fastness and, in at least some instances, good wash fastness, and good crocking (bleeding) properties. Generally, the dye will have a comparatively high mass to polarity ratio and will be only slightly polar. The literature [321–330, 431–473] puts forward several ideas with respect to coloration of PO fibers.

(a) Conventional Dyeing Methods of Unmodified Polymer. These cannot provide a large range of shades with acceptable fastness properties. Nevertheless, besides mass coloration to limited shades, coloration of PP, even after modification, has remained a problem for decades.

For example, disperse dyes give only tings to unmodified polypropylene fibers. Application of disperse dyes of different structures and properties on unmodified PP fibers has been reported [427]. However, in most cases, it was found that the saturation values were very low on unmodified PP. A new Kromalon technology was developed, with disperse dyes, very good color consistency and color fastness, as well as all the benefits and values of the PO fiber: low specific gravity, low static generation, mildew, stain moisture, wear, and bleach resistance. Especially developed for the floor covering industry, the Kromalon technology can be used as an accent yarn in space dyed carpets as well as a face fiber for printed and piece-dyed carpets [50, 454].

Quite interesting is the observation on the high temperature dyeing of PP fabrics with alkylaminoanthraquinone dyes containing alkyl substituents of varying chain length [474]. Experiments shoved that dyeings with good fastness properties (crockfastness and washfastness) can be attained if at least one octyl substituent is present in the side chain.

Classification of disperse dyes used for PP fibers can be made according to their performance [475], as follows:

low and medium performances
monoazo: Yellow 62, 151, 191, 183, Red 48 : 2, 175, 187, 57 : 1, 48 : 3, Brown 25;
disazo Yellow 180;
phtalocyanine: Blue 15:1;
high performance:
azo condensation: Yellow 93, 95, 155; Red 144, 166, 214, 242;
monoazo: Yellow 181;
monoazo heterocyclic: Yellow 182;
isoindolinone: Orange 61;
quinacridone: Red 122, 202, Violet 19;
perylene: Red 149;
anthraquinone: Red 177;
dioxazine: Violet 23, 37;
indanthrone: Blue 60;
Cu-phtalocyanine: Green 7, Blue 15:3;
heterocyclic Ni complex: Red 257;

(b) Polymer Modification Before Extrusion or Onto the Fiber. This was by attaching dye-receptive groups (e.g., copolymerization, grafting, sulfonation, halogenation, amination, phosphorylation). In the last decade, polymeric dye receptors for disperse-dyeable polypropylene fibers have been used.

Good results were obtained for polypropylene produced by melt blending, containing different quantities of 2-vinylpyridine/styrene copolymer, dyed subsequently with disperse dyes. It is found that industrially acceptable shades can be attained by selecting the amount and type of copolymer, and the disperse dye.

Light fastness of the modified dyed polypropylene fibers appears to be excellent.

Another method applied lately, i.e., dyeing of fibers based on a PP composition, involves the formation in the fiber of a composition containing PP and an ethylene copolymer, and also the fiber's exposure to a disperse dye.

Dyeing of PP fibers with some azo disperse dyes of basic character has been intensively studied via the chlorination route. The results of dye uptake and fastness properties are explained as based on the inductive effect, electrostatic interaction, and steric effect of these dyes.

The development work to make PP dyeable for textiles, e.g., by copolymerization (in either chain or side groups), had not been satisfactory enough for these modifications to be ready for sale. These complex, expensive procedures have a limited practical application.

(c) Addition of Dye-Receptor to the Polymer Before Extrusion. These were, for example, organometallic compounds (of nickel, aluminum, chromium, zinc) and nitrogen-containing polymer systems (of vinyl pyridines, pyrolidone derivates). The nickel compounds bind selected dyes with formation of very stable metal complexes having good wash fastness and somewhat good light fastness but poor rub fastness. This dyeing method has many disadvantages, including: the nickel-modified fibers have a self-shade that prevents the production of clear shades; the number of suitable disperse dyes is limited; the use of dye acceptors raises the price of the fiber. Out of the different modifications in the area of dyeable PP types, however, the so-called Ni-modified products are almost the only ones marketed today, particularly in the BCF sector. With special metal-complex dyes, exceptional light fastness properties are achieved.

(d) Production of Bicomponent and Biconstituent Fibers. This was done from the association of PP with PES, PA, and PUR polymers, as well as ethylene copolymers containing tertiary amine groups. These approaches permit a broader range of dyestuffs, including disperse, acid, premetallized, vat, and cationic dyes, applied in common dyehouse procedures. For attaining good fastness properties, dyes must be carefully selected.

III. CHARACTERIZATION OF POLYOLEFIN FIBROUS PRODUCTS

Because of their very low moisture retention, PO fibers tend to accumulate static charges. However, soiling performance is kept at an extremely low level owing to the polymer used in the extrusion of the fiber as well as to a judicious selection of the spin finishes.

Generally, PO fibers have low thermal conductivity, very sensitive to crystallinity. A typical example is isotropic polyethylene, the thermal conductivity of which varies at room temperature from 0.15–0.67 W/m K, while the crystallinity ratio varies from 0.4–0.9. The thermal conductivity of highly oriented PE fibers may attain values higher than 30 W/m K above 200 K. Consequently, this is the highest value ever measured on a polymer, although very recently a value of 13 W/m K was reported for an oriented polyacetylene film [476, 477]. Table 8 summarizes the average properties of PO fibers [27, 478–540], and Table 9 presents comparatively the properties of PP versus other fibers, the obvious advantages being observed for PP.

PP has excellent extensibility, good resistance to wear through repeated bending (flex abrasion), and a high knot resistance. PP yarns have a high shear strength and are resistant to rot and mildew.

The general mechanical model introduced by Grubb [541] for semicrystalline high modulus fibers is also used for linear POs. This model assumes an interruption of crystal continuity in the mechanically active part of the fiber. Lowering the temperature or increasing the strain rate would cause a fibrillar transformation in the amorphous phase. On the other hand, increasing the temperature or lowering the strain rate is expected to activate the crystalline phase, thus introducing a significant viscoelastic contribution to deformation. This model is confirmed by reports on the slow recovery of fibers after unloading [542–545].

Out of the physical properties, the filaments' fineness is of special importance for subsequent textile processings. The fineness of PP filaments obtained through different spinning processes varies over a large domain (Table 10).

PP threads and fibers are normally round or otherwise correspond to the cross section of the spinneret hole, e.g. trilobal or Y-shaped for BCF yarns.

PP fibers and yarns have good textile and technical properties, low weight, and good processability and are not expensive.

The tenacity in the standard production can be described as sufficient and, in the case of high tenacity yarns, as very good. High bulk elasticity and dimensional stability make PP fibers very suitable for different fields of applications (Tables 11 and 12). Light and weather stability are only sufficient with corresponding stabilization. PP fibers are hygienic, easy-care and

TABLE 8 Comparison of some basic properties of different PO polymers used to produce staple fibers, continuous filaments and oriented tapes for possible use in textiles

Properties	Polyolefin type		
	LDPE	HDPE	PP
Molecular weight, M	$(2\text{–}4) \times 10^4$	$> 3 \times 10^4$	$> 10^5$
Specific gravity, kg/m^3	920–940	940–970	900–910
Crystallinity, %	40–55	60–80	60–70
Water absorption, %	0	0	0.03
Moisture regain at 20°C, 65RH, %	0	0	0.5
Absorption rate, %	0	0	0.5
Dry tenacity, cN/denier	1–3	4–8	6.5–9.0
Wet tenacity, % of dry	98–100	100	100
Elongation at break (dry and wet), %	20–60	10–30	18–60
Melting point, °C	115–122	135–140	162–173
Softening point, °C	100	124–132	149–154
Glass transition temperature, °C	−100	−100	−15 to 10
Shrinkage in boiling water after 20 min, %	10–15	5–10	0–3
Dry air, 130°C			2–4%
Elastic recovery on 5% elongation, %			90–100
Elastic modulus on 10% extension, cN/denier			20–90
Crimping level at texturing, crimps/10 cm			30–200
Heat setting			125–130°C
Deterioration			280°C
Tensile strength (dry), GN/m^2	0.08–0.25	0.35–0.60	0.6–0.8
Breaking strain (dry), %	20–80	10–45	10–40
Tensile strength (wet), GN/m^2	0.08–0.25	0.35–0.60	0.4–0.6
Breaking strain (wet), %	20–80	10–45	10–40
Initial modulus (dry), GN/m^2	2–4	3–6	5–8
Tenacity, cN/tex	filaments normal 15–60	filaments high tenacity 70–90	fiber normal 13–30
E modulus, cN/tex	300–900		100-450

Source: Refs. 27, 478–540.

TABLE 9 Comparative properties of PP versus other fibers

Fiber	Energy content, tonnes oil equiv/tons	Specific gravity g/cm^3	Relative diameters for the same denier	Relative energy content per unit volume	Moisture regain 65% RH, 20°C	Thermal conductivity relative to air = 1.0
PP	2.8	0.91	1.0	1.0	<0.1	6.0
Polyacrylic	6.9	1.17	0.88	3.2	1–2	—
Polyamide	5.5	1.14	0.81	2.5	4	10.5
Polyester	3.9	1.38	—	2.1	0.4	8.8
Cellulosic	2.7	1.52	—	1.6	12–14	11.0
Cotton	0.7	1.54	0.83	0.4	7–8	17.5
Wool	1.4	1.32	—	0.7	16–18	7.3

Source: Refs. 95, 506.

TABLE 10 Fineness of PP filaments obtained through different spinning processes

Spinning process	Finished individual titer or average range
Conventional LOY	1–40 dtex
Conventional POY	1–10 dtex
Very slowly downwards, with air cooling	3–150 dtex
Downwards, with water cooling	25–300 dtex
Upwards, with air cooling	3–300 dtex
Blown fiber	0.01–10 dtex
Spunbond	2–15 dtex
Film yarn	
from flat film	10–200 dtex
from blown film	900–3500 dtex
Upwards through wire mesh with air cooling	<50,000 dtex

Source: Refs. 520, 524, 546.

TABLE 11 Staple fiber line: fiber properties and end uses

Application	Title (dtex)	Tenacity (cN/dtex)	Elongation (%)	Draw tension (N/dtex)	Hole counts
Coverstock	1.5–2.5	2.0–2.5	200	0.6–0.8	150,000
High tenacity fiber (geotextile)	3.5–4.5	5.5–6.5	40–60	2.0–2.2	80,000
Medium tenacity fiber (geotextile)	3.5–6.0	3.5–4.0	50–70	1.5–1.7	80,000
Textile apparel and cotton blends	3.0–5.0	5.0–5.5	60–80	1.8–2.0	80,000
Home textiles	4.0–6.0	3.5–4.0	60–80	1.2–1.4	80,000
Carpets and automotive yarn	6.0–17 13–24	2.5–3.0	150–180	0.5–0.6	45,600 20,180
Needle punch products	15–60	2.5–3.0	150–180	0.5–0.6	
Cigarette filter tow	4.5–5.5	3.0–3.5	60–80	1.4–1.6	~8,000

Source: Refs. 522, 527, 547–549.

odorless. Spin-dyed products have high staining fastness.

PP waste products can be easily and ecologically regenerated or changed into precipitable gases (e.g. *vacuum* pyrolysis).

The resistance of PO fibers to temperature is good only up to values of 70°C (Table 12).

In recent times TROL-type PO fibers—showing good behavior up to 125°C—have also been obtained, in low amounts [50]. PO fibers' pilling effect takes average values, while wrinkling depends on the fibers' fineness (being generally quite pronounced).

The PO fibers' chemical properties are generally good (Tables 13 and 14). Resistance to mineral acids is very good, although they are liable to the attack of concentrated nitric and sulphuric acids. Resistance to alkali is also very good, although they may be attacked by concentrated hot lye.

The identification of PO fibers may be realized neither through solubilization in trichloroethylene, tetrachloroethylene, *n*-heptane and decaline nor through tests of coloration in neocarmin WW (no coloration occurring), neocarmin MS (pale to rich pink), neocarmin TA (yellow), and in carminazorol (pale blue).

Consequently, PO fibers' properties may be structured as follows:

- good and very good:
  1. good tensile strength;
  2. high abrasion resistance;
  3. low specific gravity = high covering power;
  4. chemically inert;

TABLE 12 PP fiber properties in correlation with the application field

Characteristic \ Application	Apparel	Furniture cloths	Tapistry	Carpets	Hygienic and medical products	Interior automotive articles	Technical articles	Geotextiles	Base for synthetic leather
Small specific gravity	*	*	*	*	*	*	*	*	*
Nonhygroscopicity	*	—	—	*	*/o	—	*	*	—
Low thermic conductivity	*	—	*	*	—	—	—	—	—
Low chemical affinity	*	*	*	*	—	*	—	—	—
Color fastness	*	*	*	*	—	*/o	—	—	—
High abrasion fastness	*	*	*	*	—	*/o	—	—	—
Static electrical charge	—	—	—	*/o	—	—	—	—	—
Good light fastness	—	*	*	*	—	*/o	—	—	—
Chemical resistance	*	—	—	—	—	—	o	*	—
Thermal stability	—	—	*	—	o	*	*	—	—
Nonirritating skin	*/o	—	—	—	*/o	—	—	—	—

* important; o - essential.
Source: Refs. 533–536, 550.

TABLE 13 PO comparative behavior

Polymer	PP		PE	
Chemical	A	B	A	B
dilute acid	3	3	3	3
concentrated acid	3	2	3	2
dilute alkali	3	3	3	3
concentrated alkali	3	3	3	3
salt (brine)	3	3	3	3
mineral oil	2	1	2	1
UV	1	0	1	0
UV (stabilized)	3	2	3	2
heat (dry) 100°C	3	2	1	0
steam 100°C	1	0	3	3

A, short term use; B, long term use; 0, no resistance; 1 moderate resistance; 2 passable resistance; 3 good resistance.
Source: Refs. 529–533.

5. nonwater absorbent = nonstaining;
6. can be thermally bonded and thermoformed;
7. comfortable next to the skin;
8. can be produced in microdenier structures.

- lower properties or problems to solve:

1. process problems with achieving maximum tensile strength;

TABLE 14 PO fibers' chemical properties

	Residual tenacity after, %	
Treatment at 70°C	6 h	21 h
H_2SO_4 (55%)	100	95
CH_3COOH glass	100	95
KOH (40%)	100	95
NaOH (40%)	100	95
NaClO (5% active chlorine)	100	95

Source: Refs. 505, 510–513.

2. inadequate resilience/crush resistance in some types of floor coverings (relatively to nylon 6 and 66);
3. absence of satisfactory dyeability (relative to other textile fibers);
4. a low sticking/softening temperature (relatively to other apparel textile fibers);
5. difficulties to adequately make flame retardant fibers and filaments.

The still defficient properties of PO fibers may be improved by the production of PO, through metallocene catalysis. In these cases, polydispersity decreases from 4.5 to 2.5, the MFI index is improved, as well as tenacity and elongation to break.

IV. QUALITY CONTROL

The structure and properties of PO products are influenced by their manufacturing history. Thus, knowledge of PO product properties is important to both producers and users. Indeed, variations in the properties may lead to undesired variations in the final product.

For these reasons, all characteristics should be controlled along the entire production process; in this respect, many methods for the investigation of structure and properties will attest to product quality. There is an increasing tendency in PO production to use central microprocessors, which can check all parameters, saving time, minimizing errors, and facilitating a high level of quality assurance. Many groups and firms have built highly automated lines incorporating advanced robots with the most modern electronic components and artificial intelligence versatility applied to yarn handling, sampling for analysis, sorting, storage, weighing, and commercial packaging of PO fibers and products. Examples include Filteco SpA, Trico SpA, and Nov Engineering SRL (Italy); Barmag AG, Rieter AG, Seydel AG, and Schlumberger AG (Germany) [443, 444].

A highly mechanized production operation facilitates the precise computerization of plants. The main feature of the innovations in textile testing is full digitalization, bringing about numerous advantages such as efficient computer control.

A microprocessor for an extrusion plant assures detection and compensation of variations in the product parameters (temperature, pressure, speed, winding width, tension, etc.), facilitates setting of plant parameters according to input programs, and furnishes a production protocol. Some of the devices for quality control that have been developed are given in the following:

1. The IROS-100 infrared on-line system of Automatic AG (Germany) for controlling molten polymers by evaluating the percentage of additives, identifying release agents, monitoring the degree of polymerization, determining thermal and oxidative degradation, and so on.
2. The RMD-87 system of Karl Fischer AG (Germany) for residual moisture determination in the spinning raw material.
3. The fiber speeder of Enka Tecnica (Italy) for contactless measurement of thread speeds between 50 and 10,000 m/min.
4. Enka Tecnica's Fraytec, for optoelectronic registration of filament breaks in either high speed spinning or in spin-drawing.
5. The IR-equipped finish analyzer of Duratec Inc. (USA) and Enka Tecnica's finish tester, for determining the finish content on spinning bobbins.
6. The DYC control computer system of Dienes Apparatenbau GmbH (Germany) for combining the control of temperature and speed in the spin-draw process.
7. The Vibroscop, Vibrodyn, and Vibrotex models of Lenzing AG (Austria) for determining linear density, tensile force-elongation behavior, and crimping properties.

The most important testing methods for PO fibers are the following: wide- and small-angle X-ray analysis (to determine the degree of crystallinity), the density gradient method, optical methods (to determine birefringence and fiber morphology), differential scanning calorimetry (DSC, for the study of thermal properties), and tensile analysis (for the study of mechanical properties).

Table 15 lists some current standards for PO product testing [445, 446].

V. APPLICATIONS OF PO FIBROUS PRODUCTS

A. General Utilization

PO fibrous products have much to offer the textile and associated industries in terms of properties and growing versatility. Table 16 gives the correlation between the starting products and their end uses [321, 322, 567–585]. In recent years, diversification of PO products has been observed, along with the development of new types, tailored for each end product.

The staple fiber properties of PP (fineness, length, crimping, finish, color fastness, etc.) determine its web properties. A high denier fiber will result in a fabric of high bulk and resilience, while a fiber with a lower denier will produce a softer web [586]. Fiber length influences web elasticity and its tensile strength, while the cross section plays an important role in light reflectance and fiber friction behavior. Type and length of crimp influence tensile properties as well as the coherence of the fibers in the web. Tensile strength, stretching, and shrinkage are essentially determined by drawing, a process that influences the macromolecular orientation and finally the thermobonding capacity [587].

TABLE 15 Standards for PO product testing

Product	Test	Standard
Polymer	Melt flow index	BS 2782/PT7 Method 720A/1979; ISO 1133; ASTM D 2654/1989 A
Fibers	Cross	BS 2043/1968 and 1989; ASTM D 2130/1990; IWTO 8/1966(E); ISO 137; IWSTM 24/1980
	Mass and length	BS 5182/1975 and 1987; IWTO 16/196(E); ISO 2646
	Moisture regain	ASTM D 1576/1984; ASTM D 2654/1989; IWTO 33/1988(E); IWTO 34/1985(E)
	Linear density	ASTM D 1577/1979; DIN 53812; DIN 53830 T_2
	Tensile properties	ISO 5079; ASTM D 3822/1982; AFNOR 607008; DIN 53815; BS 3411/1971 and 1984; BS 1610/1985; DIN 51221; DIN 53834 T_1, DIN 53840 T_2
	Tensile strength of a bundle	BS 5116/1974 and 1986; ASTM D 1445/1990; ASTM D 2534/1985; ISO 3060; IWTO 32/1982 (E)
	Thickness	BS 2043/1968 and 1987; ASTM D 2130/1990; IWTO 8/1966 (E); IWSTM 24/1980
Yarns	Blended yarn analysis	BS 2793/1980
	Coefficient of friction	ASTM D 3108/1989; ASTM D 3412/1986
	Linear density	BS 2010/1983 and 198 6; ASTM D 1907/1989; ASTM D 2260/1987; IWTO 22/1982 (E); ISO 2060
	Single-end yarn strength	ASTM D 2256/1988; ISO 2062
	Twist testing	ASTM D 1422/1985; IWTO 25/1970 (E)
	Yarn appearance	ASTM D 2255/1990
	Yarn crimping	BS 6663/1986
Carpets	Retention of appearance after pile disturbance	BS 6659/PT 1&2/1986
	Carpet flammability	BS 4790/1987
	Carpet pile height	IWSTM 20/1980
	Carpet thickness	BS 4051/1987; BS 5808/1979; ISO 1765
	Drum testing (resistance on walking)	DIN 54323 PT 1&2
	Dynamic loading	BS 4052/1987; ISO 2094; IWSTM 123/1980; BS 4939/1987; ISO 3416
	Roller chair test	DIN 54324
	Tuft withdrawal	BS 5229/1975 and 1982; ISO 4919; IWSTM 127/1980; IWSTM 198/1974
	Wearing resistance to threading wheels	DIN 54322; DIN 53866 T_3
Interior automotive textiles	Hot-light tests	DIN 75202
	Test for inflammability	DIN 22103
	Fog effect	DIN 75201
	Conductivity tests	ASTM D 257
	Light fastness	ASTM 2859-76, BS 4790, BS 476, ASTM E 136-73, DIN 4102, AS 1530, SABS 5
	Flame spread	BS 476, 1968 Part 7

Source: Refs. 445, 446, 551–566.

TABLE 16 Uses of PO Fibrous Products

Fibrous product	End use
PE filaments (> 0.7 dtex)	Filters, ropes, cables fishnets, awnings, floor coverings, clothing material, tarpaulins, protective clothing against wet and cold, lightweight ballistic armour, carpet backing, electrical cable sheating, insulation nets
	PP staple
Coarse counts (60–120 dtex)	Tufted and woven carpets, nonwoven fabrics
Medium counts (3–60 dtex)	Carpet industry, upholstery fabrics, blankets, wool and acrylic fiber blends
Fine counts (< 3 dtex)	Nonwovens, clothing and household sector, filters, cotton and rayon blends, backings for pile fabrics, face masks, protective clothes, oil- and water-absorbing clothes, medical and hygiene sector, underwear, cigarette filters, synthetic leather
	PP filaments
Textured yarns (including BCF yarns)	Narrow fabrics, webbings (strapping), industrial applications, upholstery, workwear, footwear, knitwear, sportswear, women's and men's outerwear, fashionable women's stockings and socks, car seat fabrics, furniture fabrics, wall covering fabrics, blankets
Flat continuous filaments (100–4000 dtex)	Women's hosiery, knit sweaters, socks, underwear, and outerwear; laundry bags, automotive upholstery; rope and fishnets; backings for pile fabrics; tufted carpets; carpet backing, bag sewing thread; floor covering sector
PP film items 2.5 mm wide Film tapes, 3.5 mm thick Strappings, 0.3–0.8 mm thick Slit tapes, 0.05 mm thick Fibrillated film fibers and split film yarns of 5–1200 dtex	Artificial grass, woven carpets, tufted carpet backings (primary or secondary backings), sewing thread, filtration fabrics, scrims, narrow fabrics, webbings, packaging materials (wool pack fabrics, coal bags, nail bags), technical fabrics, household textiles, roofing materials, geotextiles, sunshades
Melt-blown products (< 0.03 dtex) and spunbonded products	Fine filters, filtration piping oil and water absorbers, geotextiles, technical sheeting, container engineering, automotive packaging, felts, floor coverings, medical goods, coating substrates, blankets, furnishing fabrics, laminated products, water-, wind and weather resistant clothing

Source: Refs. 321, 322, 567–585.

The steady growth of polypropylene's share in the nonwoven industry is explained through PP properties such as chemical inertia, hydrophobicity, resistance to bacteria, low electrostatic charge, insulating capacity, low density and melting range, good volumetric yield, suitable deniers and staple lengths, low conversion costs, and the possibility of further simplifying the fiber-spinning and fleece production processes.

In recent years, PO thermoplastic fibers have been used both in homocomponents and heterocomponents [99, 588–668], as bonding agents in the production of nonwovens. There are many reasons for this interest in thermal bonding: the economics of adhesive bonding, the production of novel thermally bonded structures through the use of bicomponent fibers, and the introduction of new or improved fiber types specifically designed for use as binder fibers.

Composite applications have been so far largely confined to those involving low temperatures, below 50–70°C, and either low or transitory loading, owing to their marked susceptibility to creep and associated phenomena.

Within the current range of applications, there may be further long-term problems associated with fatigue, particularly in ropes, nets, and flexible composites, such as diaphragms and membranes [100, 133, 137, 165, 669–708].

Monofilaments are combined to make ropes and twines. BCF are used mainly in carpet face yarns and upholstery fabrics. Higher isotacticity PP (i.e., metallocene i-PP) BCFs are more soil resistant and easier to clean, a benefit in upholstery fabrics and carpets. The additive package should not contribute to gas fading and thermal yellowing. In outdoor carpet or furniture, additional UV and heat stabilizers are usually incorporated for longer service life.

The technology to cut PP or HDPE films into tapes dates back to the mid-sixties. Such tapes were used for direct processing on existing looms or stranders as substitutes of natural hard fibers, e.g., jute, sisal, or hemp. Today, *ca.* 2000 lines are in operation throughout the world, producing annually about 1.5 million tons of stretched tapes. Regardless of application, they are woven, knitted, twisted, or stranded. With woven tapes of PP, large surface cover can be achieved with minimal raw material requirements. Tufted tapes find application in sports stadiums and for outdoor carpeting. Baler twines and ropes made from tapes are stretched to such a degree that, in the longitudinal direction, they splice open into a netting of fibers.

B. Textile Applications

1. Floor Coverings

By far the most important end uses of PO fibers at the moment are for carpets and floor coverings. The high growth in the consumption of PP fibers, mainly in the production of floor coverings (tufted, needlepunched, and woven carpets), has occurred partly because almost simultaneously with the production of PP fibers, the needling technique was found to be ideal for the production of carpets and floor coverings in which properties such as volume stability, low density and moisture absorption, resistance to abrasion and soiling, and a luster similar to wool are of prime importance.

Carpets containing PP fibers have a high resistance to abrasion and excellent dimensional stability. The PP fiber does not absorb or react with most staining agents, and it offers exceptional resistance to soiling (easy-care properties). PP carpets will not impart electrical charges, and they have good antistatic properties. They are available in dope-dyed, multicolored with space-dyeing or differential dyeing effects, triangular cross section, and outdoor-stabilized styles [709, 710]. PP staple fibers are used as 100% PP yarns, but frequently they are mixed with PES and wool for Berber or Saxony qualities.

Today, more than 90% of tufted carpets use PP as the primary backing (mostly woven) while, in the case of secondary backing, jute is still the dominant material [711]. The synthetic fabrics offer a combination of strength, dimensional stability, and uniformity that makes them truly all-purpose carpet backing materials.

Needlepunch carpets are used especially in contact rather than domestic situations. Products are used in schools and universities, in sports halls for tennis courts and bowling greens, as well as for outdoor carpeting.

Specialty fibers which enhance the ability of PP to offer improved bulk cover and appearance in the finished carpet and thereby support the future expansion of PP in this market are now produced. The most usual fiber is BCF–PP. For this field, fibers with a large fineness range (6–110 denier), modest tenacity (3–4 g/denier) and high elongation at break (80–100%) have been achieved to give the required high tensile strength to the fiber. Polymer grades with intermediate melt flow indexes and molecular weights are preferred. Using a trilobal fiber, a shrinkable fiber type is made from a solid shade carpet, from plain velours to saxony and plush qualities. Also twists and frizes and heather-blends woven carpets are made. The main advantage is the elegant, silky luster with the bulk and further on, the superb pile appearance [95, 712–732].

The general trend over the last 5 years is very positive. The share of PP versus other raw materials in the development of tufted carpets, which represents 77% of the total surface of carpets and floor coverings consumed in Europe, can be discussed.

Comparing 1990 with 1997, one can observe a decrease for nylon, wool, acrylic, and polyesteric fibers. The only fibre that is increasing is PP, from 25% to 35% of the total (Table 17).

In the last 25 years, PP proved, by its qualities, to be the best fiber for synthetic sport surfaces, due to its low price, player comfort, ball–surface interaction and durability, response to UV and visible light, frictional properties, environmental aspects, and playing characteristics [733–742].

Attempts have been made to reduce the adverse effect of friction, without creating a dangerously slippery surface. These include the use of slippery

TABLE 17 Evolution of fibers consumption for floor coverings (%)

Raw material	1990	1992	1997
Nylon	62.3	61.6	58
Acrylics	1.6	1.4	0.5
PP	25.1	28.1	35
Polyesters	2.5	2.5	1.5
Wool	8.5	6.4	5

Source: Refs. 731, 732.

copolymers and the change of fiber and yarn profiles. PP takes up a negligible proportion of water, even at 100 RH and, more importantly, it maintains its dimensions, such as strength and abrasion resistance. A structured PP needlepunch ensures a surface with consistent ball bounce and controlled player/surface interaction.

Cylindrical fibers with 100 denier have been used in a fairly open loop construction, together with a PP tape of rectangular cross section with a cut end. PP offers a smaller friction surface.

A woven surface using knit–deknit PP yarn allows contact mainly with the sides of the tufts, thus reducing the scratchiness of the other cut yarn ends.

Sport surfaces must resist temperatures between −35°C (cricket games in Canada) up to +60°C (bowling green in Spain).

2. Apparel and Household Textiles

To satisfy the household textile and apparel markets, PP offers a variety of very useful chemical and physical apparel properties which, along with its cost-effectiveness, make it a most suitable and attractive choice for the designer of new products.

Versatile and reliable technology for coloration, and production of fibers with a highest level of quality and service enables a fashion-color market to be satisfied.

For upholstery, PP slowly replaces viscose because it has a lower price, a better covering power through its smaller density, enhanced abrasion resistance and bright clear colours from the dyeing solution, and even PP short fibers in blend with others fibers (e.g., wool) can be used to obtain yarns suitable for such purposes. To this end, PP needs to be fireproofed.

Texturized PP yams can be easily knitted and will have a market in sportswear and underwear. For these applications, PP is not a cheaper substitute for other materials; it simply offers better performance. Moisture transport through the fabric is a comfort factor, and here PP yarn performs much better than cotton or any other fiber because water is transported within the PP yarn, not stored there, so that the yarn (and the fabric) remain absolutely dry and the garment is therefore very comfortable to wear. The protruding loops confer to the fabric a soft and pleasant handle, a spunlike appearance, and a very high wearer comfort [743–747].

Recently, PP fibers have penetrated more and more into the clothing industry, mainly in the form of the so-called integrated multilayered knit goods for underwear and sports and work clothing—a new generation of textiles containing PP.

The growing use of PP fibers in functional garments (sportswear) is overcoming prejudices about the physiological discomfort of these fibers. In special original two-layer sports underwear, the inner hydrophobic PP layer has the ability to transport moisture onto its surface and away from the skin [748–750]. This function creates a dry microclimate, which avoids heat leakage and allows the best possible insulation as a result of the dry air close to the skin. As an example, Stomatex PE has a much lower density value, making it a better insulator and also exceptionally buoyant and lightweight. This means it is excellent for cold climate protective clothing such as liners for survival suits, sailing suits, and mountaineering suits. These Stomatex suits can also maintain their thermal insulative properties when wet or under water [751].

The physical characteristics of PP fibers can be very useful in clothing, for example a high bulk-to-weight ratio is valuable for garments. The inherent thermal-insulative and moisture-wicking performance of PP, generally, is much better than that of other synthetic fibers. PP is affected by some solvents commonly used in dry-cleaning, however, and its resistance to heat is not particularly high.

3. Medical and Hygienic Textiles

These articles can be applied on a very large area, from disposable baby nappies to feminine hygiene products, from medical pads and dressings to gowns and drapes.

Until 1970, viscose was the main product in this field. Today, PP accounts for 90% of these products because of its easy thermal bonding, price, and simple availability. According to a report of the European Association for Textile POs, consumption of non-woven medical and sanitary products has risen by over 160% between 1980 and 1987. Further large increases in the future are foreseen.

For these articles, carded and spunlaid products have been used [752–760]. The success is explained by the technological progress recorded, such as: high speed of carding, over 200 m/min, and production of fine fibers, 2.2 dtex or even lower, and then their use in the area of personal hygiene, such as coverstock items for disposable diapers and incontinence pads.

To enhance the comfort characteristics, a coverstock utilizing essentially hydrophobic PO fibers is made temporarily hydrophilic, and thereby it maintains the ability to allow aqueous fluids to pass through. The composition containing polar materials and hydrophilic modifiers as additives imparts semi-permanent wetting characteristics (e.g., the contact angle is lower, below 80°) [761–763]. PO fibers are more and more largely used in medical fields, in different areas and under different textile forms, such as: nonwoven laminates with controlled porosity for medical sterile application (having low air permeability, high water vapor transmission, low liquid water permeability, and barrier properties against bacteria), nonwovens for hygienic articles, surgical gowns and drapes, masks, dressings, wipes and pads, spunbonded PE sheets (resistant to penetration by microorganisms but permeable to sterilizing gas), mesh for surgical use, bulkable porous nonwoven fabric suitable for use in diapers and sanitary towels, or ligament prostheses [117, 764–766].

Special PO fibres may be obtained by incorporating additives (like spundyes, pigments, special finishes) during the fibre production. This technique makes them permanently antimicrobial. The 100% pure products or those in blends with natural fibres or other synthetics, are used in all areas in which protection against bacteria and fungi makes sense: socks, underwear, caps, gloves, shoe inlays, slippers, wipe-cloths, carpets, bath-maths, mattress and bed tickings, blankets, bed-linen, etc.

The practice of blending PP with wool, rayon, polyamide, and acrylic fibers has increased because of PP's good balance of properties, coupled with its low specific gravity.

Wool blends are used in carpets, furnishing fabrics, curtains, wall coverings, and outerwear applications (school jumpers, women's sweaters). Beneficial effects on such properties as tensile strength, abrasion resistance, dimensional stability, resistance to pilling and bagging, a wool-like look and handle could be achieved by blending PP with wool.

PP and cotton spun yarns are used in various types of underwear, for track suits, men's socks, schoolwear, hand-knitting yarns (thermoknit yarns), work clothes, blankets, needlepunched fabrics for slippers and other footwear, and associated outlets.

C. Technical Applications

Market prospects for technical textiles are generally encouraging but substantial growth in the use of high performance materials will depend on the level of polymer and fibre prices.

Well established polymers such as polyesters, POs, polyamides, and cellulose continue to account for the vast bulk of technical usage.

Current developments in the properties and performance of these materials often receive far less attention than newer fibers, although they are likely to have a more profound impact upon the capabilities and growth of the technical textiles industry as a whole.

Recent progress in the technology of fibre extrusion has important implications for the optimum scale of production and, ultimately, for the structure and organization of the fibres supply industry.

Price–performance ratios, such as fibres tenacity/price and fibres modulus/price, are very favorable for PP fibres.

PO materials look increasingly set to enter some important new markets just as the number of established end-uses has begun to mature. The technologies for the production of high tenacity yarns and the high speed extrusion of relatively fine PP fibres and filaments, as well as nonwoven processes such as melt blowing, create opportunities for further development and exploitation of the many valuable properties of these and related classes of materials [39, 115, 131, 767–777].

For technical textiles end use, 20% of the PO fibres production is involved.

1. Agro- and Geotextiles

Since the 1960s, agro- and geotextiles have matured into well-established and understood materials. Agro- and geotextiles have great prospects of development because of the environmental laws of the developed countries, and also as they become an essential part of civil engineering, being used in landfill/waste disposal situations, but also in areas such as road and rail construction, erosion control, and the development of water resources [778–800].

Geotextiles must have mechanical and chemical resistance, resistance to mildew, rot, and the effects of soil and weather.

The significant requirements for a suitable geotextile dynamic filter in road pavements are: permeability, pore size, thickness, structure incompressibility, tensile strength, and resistance to deformations. PP meets these requirements, so 65–70% of geotextiles are made of it.

Geotextile fabrics are based on woven, nonwoven and composite fabric structures whose properties are determined primarily by the fiber/yarn components present in terms of intrinsic tensile properties and chemical durability. Most geotextiles are expected to have lifetimes well in excess of 20 years and to provide varying degrees of mechanical reinforcement and containment to the civil engineering structures in which they are incorporated.

Most of PP geotextiles are spunbonded, but also made of staple fibers which generate exceptionally strong needlepunched fabrics, resistant in all areas, piercing including. Such products are used as protection for membranes in landfill, but also in farms for lagoons and slurry pits.

Due to the strict long term durability requirements (chemical resistance against highly concentrated liquid media) the fabric for geotextiles can be made of HDPE tape yarn. The geotextile packing material is a double woven spacer fabric with a uniform cross section, based on a defined spacer length between the two fabric layers.

2. Packaging Products

The European packaging industry is under heavy pressure, partly due to the cheap (dumped) imports from China, Indonesia, and Thailand and, more recently, from some Eastern European countries. The rapid decrease of the 50 kg PP woven sacks' (or light bags) consumption is one of the most striking developments of the last years. This phenomenon can be observed in all countries and is mainly due to the competition from palletized PO sacks, of 500 kg bags, and from bulk transportation. On the other hand, a spectacular increase in the consumption of Flexible Intermediate Bulk Containers (FIBC) up to 2000 kg has been noticed. Evolution of the packaging products in Europe denotes the increase of bags and FIBC [719, 801–803] (Table 18).

3. Ropes and Twines

The total use of PO in the consumption of ropes and twines has increased from 98,000 tonnes in 1992 to 127,000 tonnes in 1997. This is mainly due to the reduced activity in the fishing industry which traditionally has a high consumption of ropes [719, 804].

The evolution of the principal groups of fibres destined for ropes and twines, from 1990 to 1997 (Table 19), is interesting. Increases of PP and decreases of PA and PES may be observed.

TABLE 18 Consumption of PP packaging products in Europe (1000 t)

Packaging product	1990	1992	1997
FIBC	20	35	74
Bags (500 kg)	10	17	34
Sacks	50	25	23

Source: Ref. 719, 801–803.

TABLE 19 Fibers destined to ropes and twines (1000 t)

Fiber	1990	1992	1997
PP	75	79	108
PE	8	9	9
PA, PES	15	15	10

Source: Refs. 719, 804.

4. Automotive Interior Textiles

This area is a great consumer of PP textiles, which have been used not only in floors and seats, but also in parcel shelves, boot linens, interior door panels, and headliners [131, 133, 769, 773, 805, 806].

Products used in this field must have great light and UV stability, especially those with exposure to the outside. In this case, PP fibers are competing with PA and PES.

It is possible that PP fibers will further replace polyamides and polyesters in automotive applications, since recent developments have shown that it is feasible to stabilize the PP fiber to levels far exceeding those for PA and PES ones. The specific gravity advantage of PP is particularly useful in the manufacture of car carpets; its excellent cover-to-weight ratio was attractive to the automotive industry, in which reduction of weight currently equates with savings in fuel.

Advantages of using PP consist of its ability to meet the color fastness and UV stability requirements, its natural resistance to staining, and its color consistency after solution dyeing. Also, there is the advantage of 100% recycling PP compounds. PP fibers with 17–33 dtex are used both for tufted and needlepunched products, as well as spun PP and BCF–PP in tufted carpets.

PP fibers of 6.7–11 dtex, with high UV stability are used for parcel shelves. For headliners, fibers with 3.3 dtex are used.

5. Filtering Media

POs have applications in filtering and the industrial wipes market due to some of their properties, namely: small density compared with other fibers, chemical resistance (both to acid and alkali media), and easy cleaning without absorbing dust and mud.

PP are used in filtration both as needlepunched products and spunbonded melt-blown sandwich structures. The outer spunbonded layer gives the product strength, whilst the inner microfiber confers exceptional filtering properties [807–821].

Hydrophilized porous membranes with large pore diameter are obtained from hollow PP fibers having a stocked, multicellular structure that gives excellent filtration media.

PP filtering materials are used both in dry (for dust control and removal of particles including automative filtration, cabin air filters, vacuum cleaner bags, masks) and wet media (such as filters for paintings, water, chemicals, alcoholic drinks).

PP is used for oil wipes and for absorbing oil spillages. To this ends, filters with hydrophobized fine fibers are used to enhance oil absorbtion via stitchbonded route.

VI. FUTURE TRENDS

The developmental trends pointing towards the year 2020 are clear: more specialty custom-made fibers for specific applications and manufactured in flexible production plants.

Feedstocks for making both PE and PP are expected to be in plentiful supply, thus ensuring that POs will continue to be the lowest cost polymers for fiber making.

In addition, complete recycling systems can operate with a limited number of raw materials, thus affording a relative degree of purity to the recycled product. Together with his customers, the fiber manufacturer will look for environmentally safe solutions. In many cases, the use of PO fibers will in fact be the solution, and these fibers will probably gain an even increased importance in the future [822]. The following developments are to be expected [16, 144, 823–826]:

1. Use of controlled rheology to improve polymer uniformity and consistency, representing important factors in the production of fibers, and better quality BCF yarns for carpets and for the production of nonwoven materials.
2. New polymers obtained by new technologies (e.g. new catalysts—metallocenes), such as mPE and mPP, with different properties as well as all kinds of blends and even copolymers.
3. Polymers are one part of the possibilities of new PO fibers and others are the wide range of additives which will offer nearly infinite possibilities.
4. Development of new stabilizers for PP against light degradation, thermo-oxidation and against high temperature processing.
5. Development of color concentrate lines for textile applications. The color concentrate is melted in a small lateral extruder and metered into the mixer, where excellent mixing with the polymer is achieved.
6. Higher number of filaments per position in all spinning processes and longer winding heads with more packages per mandrel.
7. Higher package weights (from 1–2 kg in 1950 to 35–50 kg in 1990 and to 1000–2000 kg in 1997–2000).
8. Increase of efficiency through the prevalence of medium capacity spinning lines and reduction of energy costs by the introduction of continuous processes into the field of staple fiber manufacture.
9. Increasing computerization and automation in overall spinning processes, winding operations included; takeup speeds will be established around 6000 and 3000 m/min for textile filaments and BCF yarns, respectively; for classical staple fibers, the limit is 2000 m/min, while for high tenacity filament production, ultrahigh spinning speed (8000–10,000 m/min) can be expected.
10. New developments for a high speed texturing process designed to achieve spin-draw-texturing of filaments.

The use of PP in fiber production will continue to grow, due to its secure raw materials (monomer) situation, easy processing, and its environmentally favorable character.

As the economy picks up in Europe, the construction industry should offer interesting growth opportunities for PP in carpets as well as woven and nonwoven geotextiles. In the medical industry, PP will get a boost, as radiation sterilizable grades become available.

Nonwovens will eventually benefit from the research going on to produce PP types that will yield down elastic fibres, PP with higher softening points and dyeability, which will also extend the use range of the material.

Around the turn of the century, preference will certainly be given to integrated processes, the more so as continuous computerization and automation come closer to fulfilling the essential prerequisite for such processes, namely the elimination of inhomogeneities and defects. PP fibers will gain further ground in specific sectors, because they can be used in more types of textile and textile-related structures than any other major polymer.

Finally, we may expect the new industrial nations, such as China, Hungary, Czech Republic, Slovakia, and Romania to begin to develop technologies and processes that will bring prosperity.

ABBREVIATIONS

γ	gamma ray
A	Austria
AFNOR	French standard
AS	Austrian standard
ASTM	American Society for Testing Materials
ATY	air texturized yarn
B	Belgium
BCF	bulk continuous filament
BS	British standard
CF	continuous filament
CH	Switzerland
CIS	Community of Independent States
CZD	continuous zone - drawing
D	Germany
DIN	German standard
DK	Denmark
DSC	differential scanning calirometry
DTY	drawn texturized yarn
EATP	European Textile Polyolefins Association
F	France
FDY	fully oriented yarn
Ff	fibrillated fabric
Fi	filament
FIBC	Flexible Intermediate Bulk Container
FOY	fully oriented yarn
Fy	fibrillated yarn
GB	Great Britain
H	Hungary
HDPE	high density polyethylene
I	Italy
i-PP	isotactic polypropylene
IR	infrared
ISO	International Standard
IWSTM	International Wool Society for Testing of Materials
IWTO	International Wool Textile Organization
J	Japan
LCF	long carbon fiber
LCP	liquid crystalline polymer
LDPE	low density polyethylene
LLDPE	linear low density polyethylene
MFI	melt flow index
MFR	melt flow rate
mPE	metallocene polyethylene
mPP	metallocene polypropylene
MW	molecular weight
MWD	molecular weight distribution
PA	polyamide
PACs	polyacrylics
PANI	polyaniline
PE	polyethylene
PES	polyester
PO	polyolefin
POY–ATY	partially oriented yarn–air textured yarn
POY–DTY	partially oriented yarn–drawn textured yarn
POY	partially oriented yarn
PP	polypropylene
PUR	polyurethane
RH	relative humidity
S	Sweden
SABS	Swiss standard
SB	spunbound or meltblown
Sc	spun cable
St	staple fibre
Tf	tape fabric
TROL	thermic resistant olefin
UDY	undrawn yarn
UHMWPE	ultrahigh molecular weight polyethylene
USA	United States of America
UV	ultraviolet
WAXD	wide angle X-ray diffraction
Wb	woven backing
ZN catalysts	Ziegler–Natta catalysts

REFERENCES

1. K. Schäfer. Man-Made Fiber Year Book (Chemiefasern/Textilind) 1988: 36–42.
2. ***Chemiefasern/Textilind 35/87: 658–663, 1985.
3. H Sinn, W Kaminsky. Advances in Organometallic Chemistry, vol 18, New York: Academic Press, 1980, p 99.
4. HH Birtzinger, D Fischer, R Mülhaupt, R Roger, R Waymonth. Angew Chem 107: 1225–1231, 1995.
5. F Köller, S Bornemann, M. Arnold. Polyolefin-copolymere und unterschieldlich strukturierten α-olefinen. Proc Polymer Materials, Merseburg, 1998, pp 435–442.
6. *** Res Discl 354(10): 13–19, 1993.
7. *** Nonwoven Rep Int 299: 5–6, 1996.
8. *** Chem Fiber Int 46: 98–103, 1996.
9. W Kaminsky, K Kulper, HH Brintzinger, FR Wild. Angew Chem Int Ed Engl 24: 507–512, 1985.
10. G Gleixner, A Wolmar. Fibers of Metallocene PO. Proc 37th Man-Made Fibers Congress, Dornbirn, 1998, pp 1–8.

*** Note: These references have no authors.

11. R Muhlhaupt. Gummi, Fasern, Kunststoffe 49: 394–399, 1996.
12. JP Peckstadt. Market Perspectives of the European PO Textile Industry, Proc 37th Man-Made Fibers Congress, Dornbirn, 1998, pp 1–8.
13. GA Harpell, S Kavesh. (to Allied Co) US Patent 4,455,273 (1984).
14. ND Scott. Metallocenes and Polypropylene - New Solutions to Old Problems? Proc World Textile Congress on Polypropylene in Textiles, Huddersfield, 1996, pp 19–26.
15. D Blechschmidt, H Fuchs, A Vollmar, M Siemon. Proc World Textile Congress on Polypropylene in Textiles, Huddersfield, 1996, pp 308–323.
16. L Riehl. Man-Made Fiber Year Book (Chemiefasern/Textilind) 1987: 34–39.
17. PR Cox. (to Monsanto Chemical Co.) US Patent 3,093,612 (1963).
18. ***Chemiefasern/Textilind 39/91: 1083–1088, 1989.
19. S Davies. Text Horiz 7: 20–28, 1989.
20. SR Beech, CA Farnfield, P Whorton, JA Wilkins. Textile Terms and Definitions, 8th edn Manchester: The Textile Institute, 1988, pp 235–238.
21. OL Shealy. Text Res J 35: 322–327, 1965.
22. H Jorder. Chemiefasern/Textiltechnik 17: 730–738, 1967.
23. E Guandique, E Parrish, M Katz. (to E.I. du Pont de Nemours and Co) US Patent 3,117,055 (1964).
24. GA Kinney. (to E.I. du Pont de Nemours and Co) US Patent 3,314,344 (1967).
25. CL Nottebohm. (to Carl Freundenberg) US Patent 3,035,943 (1962).
26. JE McIntire. (to Imperial Chemical Industries) US Patent 3,304,220 (1967).
27. HF Mark and NG Gaylord eds. Encyclopedia of Polymer Science and Technology, Plastics, Resins, Rubbers, Fibers, vol 9. New York: Wiley-Interscience, 1968, pp 113–128.
28. WA Wente, EL Boom. Ind Eng Chem 48: 1342–1347, 1955.
29. H Shin. (to E.I. du Pont de Nemours and Co) European Patent 0517693 A-1 (1992), B-1 (1997).
30. De Till, CR Shaliman. (to American Viscose Co) US Patent 2,810,426 (1957).
31. WH Howard. (to Monsanto Co) US Patent 3,210,452 (1965).
32. SL Samuels, H Shin. (to E.I. du Pont de Nemours and Co) European Patent 0572570 A-1 (1993), B-1 (1995).
33. K Shimura, J Nakayama. (to Asahi Kasei Kogyo Kabushiki Kaisha) US Patent 5,436,074 (1995).
34. DM Coates, GS Huvard, H Shin. (to E.I. du Pont de Nemours and Co) US Patent 5,202,376 (1993).
35. S Cloutier, LM Manuel, VG Zigoril. (to E.I. du Pont de Nemours and Co) US Patent 5,415,818 (1995).
36. FJ Tsai, VA Topolkaraev. (to Kimberly-Clark Worldwide Inc) US Patent 5,762,840 (1998).
37. H Shin. (to E.I. du Pont de Nemours and Co) US Patent 5,250,237 (1993).
38. H Shin, SL Samuels. (to E.I. du Pont de Nemours and Co) US Patent 5,147,586 (1992).
39. D McGinty; ET Powers, H Shin, RK Siemionko, DM Taylor. (to E.I. du Pont de Nemours and Co) US Patent 5,643,525 (1997).
40. H Shin, ET Powers. (to E.I. du Pont de Nemours and Co) European Patent 0690935 (1996).
41. K Shimura, Y Nakayama. (to Asahi Kasei Kogyo Kabushiki Kaisha) US Patent 5,512,357 (1996).
42. DP Zafiroglu. (to E.I. du Pont de Nemours and Co) US Patent 5,268,218 (1993).
43. W Jurgeleit. (to Vereinigte Glanzstoff Fabriken) US Patent 3,084,465 (1962).
44. M Levine. (to Hercules Powder Co) US Patent 3,017,238 (1962).
45. M von Dingenen. J. Tech Text Int 1(5): 24–29, 1992.
46. A Sengonul, MA Wilding. J Text Inst 85(1): 1–10, 1994.
47. IC Petrea, M Leancă, GE Grigoriu, A Grigoriu. The Structure of Oriented PP films (I). Ann Bucharest Univ Phys 32: 49–54, 1983.
48. H. Zimmermann. Text Horiz 6: 24–29, 1981.
49. Y Ohto, T Kuroki, Y Oie. (to Toyo Boseki Kabushiki Kaisha) US Patent 5,547,626 (1996).
50. IC Wang, MG Dobb, JG Tomka. J Text Inst 86(3): 383–392, 1995.
51. A Suzuki, T Sugimura, T Kunugi. Kobunshi Ronbunshu 54/5: 351–358, 1997.
52. DG Svetec. Tekstilec 40/5-6: 119-124, 1997.
53. GE Grigoriu, A Grigoriu, M Leancă, IC Petrea. The Structure of Oriented PP films (II). Ann Bucharest Univ Phys 32: 63–68, 1983.
54. YL Lee, RS Bretzlaff, RP Wool. J Polym Sci Polym Phys Ed 22: 681–683, 1984.
55. A Schweitzer. Man-Made Fiber Year Book (Chemiefasern/Textilind) 1986: 59–61.
56. H Enneking. Man-Made Fiber Year Book (Chemiefasern/Textilind.), 1986: 66–72.
57. L Riehl, H Reiter. Chemiefasem/Textilind 33/85: 22–29, 1983.
58. D Maragliano, F Derri. (to Monsanto Co.) US Patent 2,947,598 (1960).
59. E Baratti. (to Montecatini) US Patent 3,092,891 (1963).
60. B E Martin. (to E.I. du Pont de Nemours and Co) US Patent 3,093,444 (1963).
61. K Riggert. Man-Made Fiber Year Book (Chemiefasern/Textilind) 1988: 26–34.
62. M Jambrich, I Diacik, E Kamitra. Chemiefasern/Textilind 35/87: 31–39, 1985.
63. A Schweitzer. Chemiefasern/Textilind 36/88: 671–679, 1986.
64. F Hensen. Chemiefasem/Textilind 34/86: 178–184, 1984.
65. M Bussmann. Chemiefasern/Textilind 35/87: 668–675, 1985.
66. ***Man-Made Fiber Year Book (Chemiefasern/Textilind) 1987: 78–85.

67. K Schäfer. Man-Made Fiber Year Book (Chemiefasen Textilind) 1986: 48–54.
68. ***Chemiefasern/Textilind 35/87: 75–88, 1985.
69. Barmag Information Service, Remscheid-Lennep. No 31, 1991, pp 1–105.
70. Barmag Spinning Machines, Remscheid-Lennep. No 56, 1991, pp 1–32.
71. J Shimizu, K Toriumi, Y Imai. Sen-i Gakkaishi 33: T255–T260, 1977.
72. J Shimizu, N Okui, and Y Imai. Sen-i Gakkaishi, 35: T405–T409, 1979.
73. J Shimizu, N Okui, Y Imai. Sen-i Gakkaishi 36: T166–T173, 1980.
74. CA Garber. ES Clark. J Macromol Sci Phys B4: 499–505, 1970.
75. CA Garber. ES Clark. Int J Polym Mater 1: 31–39, 1971.
76. R Hoffmeister. Fiber Prod 4: 5–13, 1976.
77. R Hoffmeister. Text Prax Int 33: 1479–1486, 1978.
78. R Hoffmeister. Fiber Prod 6: 7–131, 1978.
79. R Hoffmeister. Textiltechnik 30: 558–564, 1980.
80. R Hoffmeister. Chemiefasern/Textilind 31/83: 915 (1981).
81. A Ziabicki, H Kowas eds. High Speed Fiber Spinning. New York: John Wiley & Sons, 1985, pp 118–187.
82. Lurgi Express Information, Firma Papers, Fiber Technology Division, Frankfurt am Main, No 1305,1979, pp 1–10.
83. M Mayer. Man-Made Fiber Year Book (Chemiefasern/Textilind) 1986: 33–39.
84. *** Int Fiber J 12(4): 26–28, 1997.
85. H Schellenberg. Man-Made Fiber Year Book (Chemiefasern/Textilind) 1986: 96–101.
86. D Ahrendt. Man-Made Fiber Year Book (Chemiefasern/Textilind) 1988: 56–52.
87. K Schaefer, G Stausberg. Maxflex - increased flexibility and economy of spinning lines as a result of a new additive feed system, Proc 37th Man-Made Fiber Congress, Dornbirn, 1998, pp 1–10.
88. D Ahrendt. Man-Made Fiber Year Book (Chemiefasern/Textilind) 1987: 104–110.
89. D Ward. Text Month 10: 37–42, 1981.
90. PT Slack. Autocrimp - the new fiber extrusion process for PP Hi-loft fibres. Proc World Textile Congress on Polypropylene in Textiles, Huddersfield, 1996, pp 335–347.
91. PT Slack. Tech Text Int 6: 3–5, 1997.
92. ***Chem Fiber Int 47: 148–149, 1997.
93. Barmag Speetex STM 16 - Spin draw texturing line. Remscheid-Lennep. Ex 61/2, 1991, pp 1–24.
94. A Schweitzer. Herstellung von feinen PP-fasern. Proceedings of the 37th Man-Made Fibers Congress, Dornbirn, 1998, pp 1–14.
95. DG Ellis. Novel technique for the production of PO floorcoverings. Proc World Textile Congress on Polypropylene in Textiles, Huddersfield, 1996, pp 289–297.
96. R Gutmann. Chemiefasern/Textilind 44/96: 675–682, 1994.
97. M Taniguchi, Y Fujiyama. (to Chisso Co) US Patent 5,451,462 (1995).
98. T Asanuma, T Shiomura, S Kimura, N Uchikawa, Y Kawai, K Suchiro, S Fukushina. (to Mitsui Toatsu Chemicals Inc) US Patent 5,624,621 (1997).
99. RK Gupta, JE Mallory, K Takeuchi. (to Hercules Inc) US Patent 5,629,080 (1997).
100. A Winter, A Vollmar, F Kloos, B Bachmann. (to Hoechst AG) US Patent 5637666 (1997).
101. TJ Stockes, AE Wright, K Ofosu. (to Kimberly-Clark Co) European Patent 0685579 A-2 (1995), A-3 (1996), B-1 (1998).
102. JS Dugan. (to BASF Co) US Patent 5,405,698 (1995).
103. I Kotter, K Jung, W Grellmann, S Seidler, T Koch, J Filbig, M Gohleitner. Zähigkeit und Morphologie von PP-Copolymeren. Proc "Polymerwerkstoffe '98", Merseburg, 1998, pp 467–474.
104 Z Jezic, GP Young. (to The Dow Chemical Co) US Patent 5,133,917 (1992).
105. H Nishio, T Noma. (to Chisso Co) US Patent 5,130,196 (1992).
106. AS Gaikhman, YP Gaiuza, SI Mesin, VG Andreev. Vysokomol Soedin A, 35: 547–553, 1993.
107. ZP Jezic. Chemiefasern/Textilind 39/9: 1074–1082, 1989.
108. ZP Jezic, GP Young. (to The Dow Chemical Co) US Patent 4,839,228 (1989).
109. F Hensen. Adv Polym Technol 3: 348–354, 1983.
110. R Vogel, G Schauer, G Schmack. Chem Fiber Int 46/6: 421–423, 1996.
111. MR Conboy. Polypropylene-based copolymers for BCF. Proc World Textile Congress on Polypropylene in Textiles, Huddersfield, 1996, pp 27–42.
112. T Graf, O Durcova, L Vasko. Intercolor'91, Budapest, 1991, pp 71–78.
113. *** Text Mon 10: 10–18, 1983.
114. S Novota. Man-Made Fiber Year Book (Chemiefasern/Textilind.), 1988: 62–74.
115. M Kristofic, A Marcinin, A Ujhelyiova, V Prchal. Vlakna Textil 4(1): 14–17, 1997.
116. Courtaulds Specialty Fibers. Med Text 4: 2–10, 1997.
117. Courtaulds Specialty Fibers. High Perform Text 4: 3–6, 1997.
118. Y Qin, MM Miller, DL Brydon, JMG Cowie, RR Mather, RH Wardman. Fibres blended from PP and LC polymers. Proc World Textile Congress on Polypropylene in Textiles, Huddersfield, 1996, pp 60–68.
119. Y Qin, MM Miller, DL Brydon. Text Asia 27(12): 54–56, 1996.
120. MC Manning, RB Moore. J Vinyl Additive Tech 3(2): 184–189, 1997.
121. Y Qin, DL Brydon, RP Mather, RH Wardman. Polymer 34: 1196–1199, 1993.
122. Y Qin, DL Brydon, RP Mather, RH Wardman. Polymer 34: 1202–1204, 1993.
123. Y Qin, DL Brydon, RP Mather, RH Wardman. Polymer 34: 3597–3599, 1993.

124. MM Miller, JMG Cowie, JG Tait, DL Brydon, RP Mather. Polymer 36: 3107–3111, 1995.
125. MM Miller, JMG Cowie, DL Brydon, RP Mather. Macromol Rapid Commun 15: 857–862, 1994.
126. MM Miller, JMG Cowie, DL Brydon, RP Mather. Polymer 38/7: 1565–1568, 1997.
127. Y Qin. Technische Textilien 39(4): 182–189, 1996.
128. Y Qin. Tech Text 54: 735–739, 1994.
129. Y Qin. Tech Text 54: 873–879, 1994.
130. AG Andreopoulus, PA Tarautili. Tech Text Int 5/10: 26–28, 1996/1997.
131. *** High Perform Text 11: 12–19, 1996.
132. HH Kausch. L Berger, CJ Plummer, A Bals. Structure and deformation of UHMPE fibres and of the composite obtained thereform. Proc 35th International Man-Made Fibers Congress, Dornbirn, 1996, pp 1–10.
133. MD Wakeman, TA Cain, CD Rudd, AC Long. The compression moulding of woven yarns of comingled glass and PP for use in high volume structural automative applications. Proc World Textile Congress on Polypropylene in Textiles, Huddersfield, 1996, pp 265–275.
134. J Murphy. Additives for Plastics Handbook. Oxford: Elsevier Advanced Technology, 1996, pp 121–165.
135. B Langer, C Bierögel, W Grellmann, G Aumayr, J Fiebig. Deformations- und Bruchverhalten von PP/GF - Werkstoffen für medial-thermische Beanspruchung. Proc "Polymerwerkstoffe'98", Merseburg, 1998, pp 164–173.
136. R Taipalus, T Harmia, K Friedrich. Studies of the synergy effect between PANI-complex and carbon fibers in PP regarding its electrical conductivity. Proc "Polymerwerkstoffe'98", Merseburg, 1998, pp 184–192.
137. D Paukszta, J Garbarczyk, R Kozlowski. PP/Flax or /Hemp Compositions. Proc World Textile Congress on Polypropylene in Textiles, Huddersfield, 1996, pp 357–365.
138. T Sterzyoski, B Triki, S Zelazny. Polymer 40. 448–453, 1995.
139. VB Valkai. Synthetic Fibers. Weinhein: Verlag Chemie, 1981, pp 123–135.
140. M Ahmed. Polypropylene Fibers-Science and Technology: Amsterdam Elsevier, 1982, pp 154–176.
141. T Tekao. Man-Made Fiber Year Book (Chemiefasern/Textilind) 1987: 52–59.
142. F Fourné. Man-Made Fiber Year Book (Chemiefasern/Textilind) 1988: 53–62.
143. WC Sheehaw, TB Cole. J Appl Polym Sci 8: 2358–2364, 1964.
144. HF Mark. Fiber development in the 20th century. Proc Fiber Producers Conference, Greenville, 1988, Session 1&2, pp 1–10.
145. HH Krässig, J Lenz, HF Mark. Fiber Technology, vol 4. New York: Marcel Dekker, 1984, pp 232–267.
146. D Ward. Text Mon 12: 27–37, 1983.
147. M Leanca, GE Grigoriu, A Grigoriu, C Ciocan. Study of the influence of drawing parameters on the structure and properties of PP films. Proc 26th IUPAC International Symposium on Macromolecules, Florence, 1980, vol. 3. Preprints, pp 81–84.
148. A Grigoriu, GE Grigoriu, M Leanca, C Ciocan. Structural modifications at the spinning of PP yarns. Proc 26th IUPAC International Symposium on Macro-molecules, Florence, 1980, vol. 3. Preprints, pp 77–80.
149. RJ Samuels. Makromol Chem 4: 241–248, 1981.
150. GE Grigoriu, A Grigoriu, M Leanca, IC Petrea. The structure and properties of PP films obtained from homologue polymers mixtures. Proc IUPAC Macro'83, vol IV, Section VI, Bucharest, 1983 pp 34–37.
151. JR White. (to E.I. du Pont de Nemours and Co) US Patent 2,920,349 (1960).
152. RL Rush. (to Phillips Petroleum Co) US Patent 3,112,160 (1963).
153. O Rasmussen. (to Phillips Petroleum Co) US Patent 3,165,563 (1965).
154. VV Yurkevich AB Pakshver. Technology of the Production of Synthetic Fibers. Moscow: Khimiya, 1987, pp 45–89.
155. N Asandei A Grigoriu. Chemistry and Structure of Fibers (in Romanian). Bucharest: Academic Publishers, 1983, pp 180–192.
156. G Strausberg. Technische Textilien 40(2): 64–69, 1997.
157. FK Gilhaus. Man-Made Fiber Year Book (Chemiefasern/Textilind) 1987: 87–92.
158. L Gerking. Man-Made Fiber Year Book (Chemiefasern/Textilind) 1987: 121–129.
159. D Bechter, A Roth, G Schaut. Melliand Textilberichte 78(3): 164–167, 1997.
160. RL Lilly, RH Blackwell. (to BASF Co) US Patent 5,614,142 (1997).
161. GA Stahl, JJ McAlpin. (to Exxon Chemical Patents Inc) European Patent 0760744 (1997).
162. JL McManus, PM Kobyliver, CJ Albertelli. (to Kimberly-Clark Worldwide Inc) US Patent 2,985,029 (1994).
163. F. Fourné. Man-Made Fiber Year Book (Chemiefasern/Textilind) 1988: 66–72.
164. M Wehmann. Chemiefasern/Textilind 40/92: 756–762, 1990.
165. GW Anderson. Value added approach to expand markets of spunbonded PP. Proc INDA'TEC'96, Crystal City, North Carolina, 1996, pp 110–116.
166. *** Chemiefasern/Textilind 40/92: 275–283, 1990.
167. F Fourné. Chemiefasern/Textilind 28/81: 441–449, 1979.
168. F Fourné. Chemiefasern/Textilind 38/90: 690–694, 1988.
169. F Fourné, B Cramer. Chemiefasern/Textilind. 40/92: 757–782, 1990.
170. S Molkan, L Wadsworth, C Davey. Int Nonwoven J 62(2): 5–9, 1994.

171. S Molkan, L Wadsworth, C Davey. Int Nonwoven J 6(2): 10–13, 1994.
172. S Molkan, L Wadsworth, C Davey. Int Nonwoven J 6(2): 14–18, 1994.
173. S Molkan, L Wadsworth, C Davey. Int Nonwoven J 6(2): 19–24, 1994.
174. AC Smith, WW Roberts. Int Nonwoven J 6(1): 18–23, 1994.
175. *** Textil Prax Int 45: 11–18, 1990.
176. A Koszka. Magyar Textiltech 33: 117–120, 1980.
177. N J M Jakobs. TUT Text Usages Tech 6: 27–28, 1992.
178. AA Turetskii, SN Chvahm YaA Zubov, NF Bakaev. Vysokomol Soed A 35(5): 529–535, 1993.
179. JP Penning, AA De Vries, AJ Penning. Polymer Bull 31 (2): 243–248, 1993.
180. E Van Gorp, J Van Dingenen. Tech Text Int 3(2): 12–16, 1994.
181. *** Chemiefasern/Textilind 44/96 (6): 400–404, 1994.
182. R Vogel, G Schauer, R Beyreuth. Chem Fiber Int 45(4): 268–269, 1995.
183. Hercules Inc. High Perform Text 3: 2, 1994.
184. V Caldas, GR Brown, RS Nohr, JG MacDonald, LE Raboin. Polymer 35(5): 899–907, 1994.
185. J. Hudak, S Kohut. Vlakna Text 2(1): 88–91, 1994.
186. M Navone. Chem Fiber Int 45(1): 60–64, 1995.
187. *** Chem Fiber Int 46(2): 110–117, 1996.
188. *** Chem Fiber Int/Man-Made Fiber Year Book (Chemiefasern/Textilind) 1996: 80–86.
189. J Breuk. Chem Fiber Int/Man-Made Fiber Year Book (Chemiefasern/Textilind) 1997: 74–76.
190. L Decraemer. Tech Text Int 7/4: 15–16, 1998.
191. SP Krupp, JO Bieser, EN Knickerbocker. (to Dow Chemical Co) US Patent 5,254,299 (1993).
192. J Kamo, T Hirai, H Takahashi, K Kondo. (to Mitsubishi Rayon Co) US Patent 5,294,338 (1994).
193. F De Candia, G Romano, AO Baranov, EV Prut. J Appl Polym Sci 46(10): 1799–1806, 1992.
194. R Janarthanan, SN Garg, A Misra. J Appl Polym Sci 51(7): 1175–1182, 1994.
195. E Andreassen, OJ Myhre. J Appl Polym Sci 50(10): 1715–1721, 1993.
196. WF Wong, RY Young. J Mater Sci 29(2): 520–526, 1994.
197. RF Saraf. Polymer 35(7): 1359–1368, 1994.
198. YC Bhuvanesh, VB Gupta. Polymer 35(10): 2226–2228, 1994.
199. MS Silverstein, O Breuer. Polymer 34(16): 3421–3427, 1993.
200. ME Orman. J Geotech Eng-ASCE 120(4): 758–761, 1994.
201. A Bauer, D Hofmann, E Schulz. Acta Polym 43(1): 27–29, 1992.
202. AJ Peacock. (to Exxon Chemical Inc) US Patent 5,272,003 (1993).
203. FJ Wortmann, KV Schulz. Polymer 35(10): 2108–2116, 1994.
204. GA Stahl, JJ McAlpin. (to Exxon Chemical Inc) European Patent 0700464 (1996).
205. JA Mahood. (to General Electric Co) European Patent 0702054 (1996).
206. J Ebel, G Gleixner. (to Asota GmbH) European Patent 0629724 (1994).
207. AF Galambos. (to Himont Inc) European Patent 0634305 (1995).
208. Extrusion Systems Ltd. High Perform Text 3: 9–12, 1996.
209. *** Chem Fiber Int 45(2): 116–122, 1995.
210. J Beswick. (to Play-Rite Ltd) GB Patent 2,287,478A (1995).
211. D Butterfass, J Schuhmacher, MW Mueller. (to BASF AG) US Patent 5,466,411 (1995).
212. R Mathis. Properties and development of spin finishes for fine PP for textile applications. Proc World Textile Congress on Polypropylene in Textiles, Huddersfield, 1996, pp 210–222.
213. M Forero. (to Pavco SA) US Patent 5,336,562 (1994).
214. JJ McAlpin, GA Stahl. (to Montell North America Inc) US Patent 5,571,619 (1996).
215. D Butterfass, HJ Schuhmacher, MW Mueller. (to BASF AG) US Patent 5,565,269 (1996).
216. K Ito, I Ueno. (to Asahi Kasei Kagyo KK) US Patent 5,607,636 (1997).
217. *** Chem Fiber Int 47(2): 148–149, 1997.
218. GA Stahl, JJ McAlpin. (to Exxon Chemical Patents Inc) US Patent 5,736,465 (1998).
219. KM Kirkhland, CP Weber. (to Allied-Signal Inc) European Patent 0 520 023 (1992).
220. RS Nohr, JG MacDonald, PM Kobylivker. (to Kimberly-Clark Worldwide Inc) US Patent 5,744,548 (1998).
221. GP Young. (to Dow Chemical Co) US Patent 5,401,331 (1997).
222. S Kavesh. (to Allied-Signal Inc) US Patent 5,248,471 (1993).
223. K Yagi, H Takeda. (to Mitsui Petrochemical Industries Ltd) US Patent 5,246,657 (1993).
224. Danaklon A/S. High Perform Text, 4: 4, 1993.
225. HO Cedarlad, JD Seppala. (to Leucadia Inc) European Patent 0 627 986 (1994).
226. Y Takai. (to Daiwabo Create Co) US Patent 5,356,572 (1994).
227. SL Samuels, VG Zboril. (to Du Pont Canada) European Patent 0501689 (1992).
228. BG Bertelli. (to Montell North America Inc) US Patent 5,536,672 (1996).
229. RB Cook. (to Berkley Inc) US Patent 5,540,990 (1996).
230. Toyo Boseki Kabushiki Kaisha. High Perform Text 10: 2, 1997.
231. RF Tietz. (to E.I Du Pont de Nemours) US Patent 5,130,069 (1992).
232. O Osawa, RS Porter. Polymer 35(3): 540–544, 1994.
233. FM Lu, JE Spruiell. J Appl Polym Sci 49(4): 623–631, 1993.
234. *** Deutsche Seiler-Zeitung 113(4): 43–46, 1994.
235. RB Cook. (to RB Cook) US Patent 5,749,214 (1998).

236. ER Kafchinski, TS Chung. (to Hoechst Celanese Co) US Patent 5,213,689 (1993).
237. J Sestak. Chemicke Vlakna 42 (3/4): 108–116, 1992.
238. Hercules Inc. Nonwoven Rep Int 265: 10–11, 1993.
239. Danaklon A/S, Vliesstoff Nonwoven Int 8(5): 180, 1993.
240. RE Howard, J Young. (to Entek Manufacturing Inc) US Patent 5,230,843 (1993).
241. I Takahashi, S Hayashi, Y Iida. (to Ube-Nitto Kasei Co) European Patent 0565720 (1993).
242. J Kamo, T Hirai, H Takahashi, K Kondon. (to Mitsubishi Rayon Co) US Patent 5,238,642 (1993).
243. *** High Perform Text 6: 3–4, 1994.
244. E Kamei, Y Shimomura. (to Ube Industries Ltd) US Patent 5,435,955 (1995).
245. J Kato, T Yoneyama, K Shimura, Y Nakayama, K Kanekiyo. (to Asahi Kasei Kogyo KK) US Patent 5,286,422 (1994).
246. CP Schobesberger. Recyclable PP fabrics for awnings made of asota F 13. Proc World Textile Congress on Polypropylene in Textiles, Huddersfield, 1996, pp 223–230.
247. ML Delucia, RL Hudson. (to Kimberly-Clark Worldwide Inc) European Patent 0805886 (1997).
248. D Heeren, E Winzenholer. (to BASF AG) European Patent 0 804 370 (1997).
249. PT Slack. High Perform Text 12: 3–4, 1995.
250. M Taniguchi, H Ito, Y Tsujiyama. (to Chisso Co) European Patent 0 809 722 (1997).
251. RK Gupta, JH Harrington. (to Hercules Inc) US Patent 5,763,334 (1998).
252. RK Gupta, JE Mallory, K Takeuchi. (to Hercules Inc) US Patent 5,733,646 (1998).
253. M Konya, K Yagi, A Fukui, M Ogawa. (to Mitsui Petrochemical Industries Ltd) European Patent 0518316 (1992).
254. Institute of Chemistry, Beijing. High Perform Text 3: 2–3, 1994.
255. Israel Fibre Institute. Nonwoven Rep Int 258: 6, 1992.
256. Ampacet Co. Fire and Flammability Bull 1: 6, 1994.
257. Eastern Color & Chemicals Co, Fire and Flammability Bull 11: 7, 1993.
258. *** Res Discl 353: 593, 1993.
259. CR Davy, TC Erderly, AK Mehta, CS Speed. (to Exxon Chemical Patents Inc) US Patent 5,322,728 (1994).
260. RE Kozulla. (to Hercules Inc) US Patent 5,318,735 (1994).
261. RM Broughton, DM Hall. J Appl Polym Sci 48(9): 1501–1513, 1993.
262. Don & Low (Holdings) Ltd. Nonwoven Rep Int 271: 5–6, 1993.
263. P Bott. Vliesstoff Nonwoven Int 7(6/7): 153, 1992.
264. G Landoni, C Neri. (to Enichem Synthesis SpA) US Patent 5,447,991 (1995).
265. G Braca, G Bertelli, L Spagnoli. (to Himont Inc) European Patent 0638671 (1995).
266. G Braca, G Bertelli, L Spagnoli. (to Montell North America) US Patent 5,494,951 (1996).
267. F Drake (Fibres) Ltd. Nonwoven Rep Int 303: 5, 1996.
268. EJ Termine, RW Atwell, HA Hodgen, NA Favstritsky. (to Great Lakes Chemical Co) US Patent 5,380,802 (1995).
269. A Scarfe. (to JWI Ltd) European Patent 0641401 (1995).
270. JH Harrington. (to Hercules Inc) European Patent 0557024 (1993).
271. L Spagnoli, G Braca, L Pinoca. (to Himont Inc) European Patent 0629720 (1994).
272. JH Harrington. (to Hercules Inc) US Patent 5,547,481 (1996).
273. L Clementini, A Galambos, L Giuseppe, K Ogale, L Spagnoli, ME Starsinic. (to Himont Inc) European Patent 0552810 (1993).
274. MM Arvedson, GE Wissler. (to Exxon Chemical Patents Inc) US Patent 5,171,628 (1992).
275. Tama Plastic Industry. High Perform Text 8: 12–13, 1994.
276. G Barsotti. (to Moplefan SpA) European Patent 0663965 (1995).
277. *** High Perform Text 6: 2, 1995.
278. Y Kouno, Y Itoh, K Yagi. (to Mitsui Petrochemical Industries Ltd) US Patent 5,302,453 (1994).
279. S Cloutier, LM Manuel, VG Zboril. (to Du Pont Canada Ltd) European Patent 0597658 (1994).
280. S Cloutier, LM Manuel, VG Zboril. (to Du Pont Canada Ltd) European Patent 0598536 (1994).
281. E.I. Du Pont de Nemours & Co High Perform Text 4: 5, 1994.
282. PV Zamotaev, I Chodak. Angew Makromol Chem 210: 119–128, 1993.
283. AD Bartle. GB Patent 2268756, 1993.
284. S Sakazume, T Miyamoto, H Shimizu. (to Nippon Petrochemicals Co) European Patent 0678607 (1995).
285. RW Johnson, TW Theyson. (to Hercules Inc) US Patent 5,403,426 (1994).
286. W Billner. (to Rieter Ingostadt Spinnereimashinenbau AG) US Patent 5,367,868 (1994).
287. RK Gupta, RJ Legare. (to Hercules Inc) European Patent 0619393 (1995).
288. L Pinoca, T Narni. (to Montell North America Inc) European Patent 0743380 (1996).
289. A Marcincin, A Ujhelyiova, J Legen. Vlakna Textil 4(2): 38–43, 1997.
290. RH Frey, L Sellin, HP Brehm. (to RH Frey, L Sellin, HP Brehm) US Patent 5,681,657 (1997).
291. H Uda, T Takahashi, R Kamei, T Sano. (to Showa Denko KK) US Patent 5,283,025 (1994).
292. T Amornsakchai, DLM Cansfield, SA Janrad, G Pollard, IM Ward. J Mater Sci 28(6): 1689–1698, 1993.
293. U Hitoshi, T Tetsnya, K Ryosuke, S Takeshi. (to Showa Denko KK) European Patent 0551131 (1993).

294. WF Wong, RJ Young. J Mater Sci 29(2): 510–519, 1994.
295. M Russel. Tech Text Int 7: 12–13, 1992.
296. VI Vlasenko, NP Bereznenko. Textil Chemia 22(4): 102–107, 1992.
297. WH Tung, LR Simonsen, YV Vinod, F Werny. (to EI Du Pont de Nemours & Co) US Patent 5,284,009 (1994).
298. I Beswick. Tech Text Int 7: 14, 1992.
299. RS Nohr, YG MacDonald, PM Kylivker. (to Kimberly-Clark Worldwide Inc) US Patent 5,744,54 (1998).
300. G Vita, G Ajroldi, M Miami. (to Ausimont SpA) US Patent 5,460,882 (1995).
301. E Gomez. Rev Indust Text 320: 44–51, 1994.
302. J Bandenbender, E Gartner. (to Bayer Faser GmbH) European Patent 0784107 (1997).
303. JC Brewer, CK Liu. (to United States Surgical Co) European Patent 0526759 (1993).
304. CK Liu, JC Brewer. (to United States Surgical Co) US Patent 5,217,485 (1993).
305. CK Liu. (to United States Surgical Co) European Patent 0585814 (1994).
306. OT Turunen, JFors, E Thaels. (to Borealis AS) US Patent 5,474,845 (1995).
307. Y Ohta, T Kuroki, Y Oie. (to Toyo Boseki Kabushiki Kaisha) US Patent 5,547,626 (1996).
308. J Kamo, T Hirai. (to Mitsubishi Rayon Co) US Patent 5,547,756 (1997).
309. JJ Dunbar, S Kavesh. (to Allied-Signal Inc) US Patent 5,578,374 (1997).
310. A Dockery. Am Text Int 26(11): 74, 1997.
311. Textile Mon 3: 32, 1998.
312. CR Davy, TC Erderly, AK Mehta, CS Speed. (to Exxon Chemical Patents Inc) European Patent 0670918 (1995).
313. E Kapell. TROL- Temperature Resistant Olefin. Proc World Textile Congress on Polypropylene in Textiles, Huddersfield, 1996, pp 43–49.
314. *** Res Discl 363: 384, 1994.
315. *** Res Discl 363: 380, 1994.
316. JS Duggan. (to BASF Co) European Patent 0618316 (1994).
317. S Kavesh, DC Prevorsek. (to Allied-Signal Inc) US Patent 5,736,244 (1998).
318. L Pinoco, R Africano, G Braca. (to Montell North America Inc) US Patent 5,747,160 (1998).
319. D Leucks, E DeWergifosse, RF Venneman. (to Fina Research SA) US Patent 5,753,762 (1998).
320. NS Berke, HK Fololiard. (to WR Grace and Co). US Patent 5,753,368 (1998).
321. *** Am Text Int 8: 63–65, 1989.
322. T Palpeyman. Text Month 12: 29–33, 1981.
323. *** Chemiefasern/Textilind 35/87: 632, 1985.
324. PA Koch. Chemiefasern/Textilind 39/91: 1063–1070, 1989.
325. F Haider. Chemiefasern/Textilind 40/92: 640–644, 1990.
326. ***Chemiefasern/Textilind 40/92: 1024–1028, 1990.
327. BT Hoevel. Chemiefasern/Textilind 40/92: 1029–1033.
328. M Zdenek. Magyar Text 35: 141–144, 1982.
329. ***Chemiefasern/Text 34/86: 6–9, 1984.
330. G Druzsbaczky. Magyar Textitech 35: 36–41, 1982.
331. ***Chemiefasern/Textilind 34/86: 6–8, 1984.
332. Am Text Int 5: 8–13, 1990.
333. S Davies. Text Horiz 8: 23–26, 1989.
334. *** Text Month 8: 27–30, 1981.
335. D Ward. Text Month 6: 20–23, 1981.
336. D Ward. Text Month 10: 47–48, 1980.
337. *** Text Month 8: 23–25, 1983.
338. C Nathan. Text Mon 10: 49–50, 1980.
339. *** Chem Fiber Int 45(2): 74–75, 1995.
340. *** Textile Industries Dye of Southern Africa 12(6): 195–197, 1993.
341. N Butler. Tech Text Int 3(3): 109–111, 1994.
342. P Lennox-Kerr. Pakistan Textile J 42(5): 403–406, 1993.
343. WF Rauschenberger. Recursive modelling of the fibre spinning process. Proc World Textile Congress on Polypropylene in Textiles, Huddersfield, 1996, pp 50–59.
344. *** L'ind Text 1264: 49–50, 1995.
345. *** Man-Made Fiber Year Book (Chemiefasern/Textilind) 1996: 20.
346. FG Rewald. Nonwovens Industry 25(4): 30–31, 1994.
347. D Harison. Nonwovens Industry 24(6): 36–41, 1993.
348. M Jambrich, D Budzok, A Stupak, P Jambrich. Vlakna Textil 2(2): 83–89, 1995.
349. EE Stone. J China Text Inst 5(1): 1–7, 1995.
350. Extrusion Systems Ltd. High Perform Text 4: 14–15, 1993.
351. JP Peckstadt. Chem Fiber Int 45(5): 338–340, 1995.
352. Fina Inc. Nonwoven Rep Int 273: 11, 1993.
353. Committe International de la Rayonne & des Fibres Synthetiques. Nonwoven Rep Int 271: 10–11, 1993.
354. J & J Haigh. Nonwoven Rep Int 265: 8, 1993.
355. WAB Davidson. Am Text Int 3: 58–60, 1994.
356. *** Nonwoven Rep Int 293: 14, 1995.
357. *** Nonwoven Rep Int, 299: 7, 1996.
358. *** Nonwoven Rep Int, 302: 2, 1996.
359. R Shishoo. Int Text Bull. Nonwovens/Industrial Textiles 42(2): 5–6, 1996.
360. P Bottcher. Int Text Bull. Nonwovens/Industrial Textiles 42(2): 8–10, 1996.
361. Don & Low Nonwovens Ltd. Nonwoven Rep Int 296: 8–9, 1995.
362. JG Tomka. Polypropylene - the world's strongest fibre. Proc World Textile Congress on Polypropylene in Textiles, Huddersfield, 1996, pp 1–11.
363. DSM High Performance Fibers BV. Deutsche Seiler-Z 115(3): 212, 1996.
364. JP Peckstadt. Chem Fiber Int 45(6): 214–222, 1995.
365. Jacob Holm & Sons A/S. Med Text 7: 3–4, 1995.
366. Danaklon A/S. Nonwoven Rep Int 283: 1, 1994.
367. Kobe Steel Ltd. Nonwoven Rep Int 283: 8–10, 1994.
368. *** Nonwoven Rep Int 285: 2, 1994.

369. O Tickell. New Scientist 147: 19, 1995.
370. M Prat Ponsa. Revista de Quimica Textil 128: 70–78, 1996.
371. M Jambrich, A Stupak, P Jambrich. Fibres Text East Europe 5(1): 14–20, 1997.
372. D Harrison. Nonwovens Ind 28(4): 22–23, 1997.
373. *** Nonwovens Rep Ind 313: 12–13, 1997.
374. Asota GmbH. Melliand Textilberichte 78(4): 196–198, 1997.
375. J Rupp. Int Text Bull. Nonwovens/Industrial Text 43(2): 44–47, 1997.
376. *** Chem Fiber Int 47(2): 137, 1997.
377. R Paul, SR Naik, GS Shaukarling. Synthetic Fibres 26(1): 27–33, 1997.
378. *** Chem Fiber Int 47(3): 171, 1997.
379. JP Peckstadt. TUT Textiles Usages Tech 25: 14–15, 1997.
380. JP Peckstadt. Asian Text J 618: 65–71, 1997.
381. A Dockery. Am Text Int 26(9): 114, 1997.
382. JW McCurry. Textile World 148(1): 67–69, 1998.
383. ***Nonwoven Rep Int 323: 50, 1998.
384. A Dockery. Am Text Int 27(1): 78, 1998.
385. JD Langley, BS Hinkle. US Patent 5,728,451 (1998).
386. M Ratzsch. Polyolefins, potentials and chances. Proc Polymer Materials, Merseburg, 1998, pp 420–424.
387. ***Man-Made Fiber Year Book (Chemiefasern/Textilind) 1994: 26.
388. ***Man-Made Fiber Year Book (Chemiefasern/Textilind) 1996: 16.
389. ***Text Asia 8: 97, 1993.
390. ***Man-Made Fiber Year Book (Chemiefasern/Textilind) 1993: 6.
391. ***Man-Made Fiber Year Book (Chemiefasern/Textilind) 1995: 6.
392. TFN Johnson. Chem Fiber Int 46: 280–284, 1996.
393. ***Man-Made Fiber Year Book (Chemiefasern/Textilind) 1996: 37.
394. DSM High Performance Fibers. Res Discl 407: 14–16, 1998.
395. E Nezbedova, J Kucera, A Zahradnickova. Long time brittle failure in HDPE. Proc Polymer Materials, Merseburg, 1998, pp 457–466.
396. G Teteris. Der Einfluss der Recyclingmethode auf den Polyolefinabbau. Proc Polymer Materials, Merseburg, 1998, pp 494–501.
397. PJ Sheth. (to Lyondell Petrochemical Co) US Patent 5,614,574 (1997).
398. AE Garovaglia, JG MacDonald. (to Kimberly-Clark Co) US Patent 5,620,788 (1997).
399. PJ Sheth. (to Lyondell Petrochemical Co) European Patent 0770152 (1997).
400. M Kristofic, A Ujhelyiova. Fibres Text East Europe 5(2): 51–53, 1997.
401. DL Davis. (to National Aeronautics and Space Administration) US Patent 5,683,813 (1997).
402. WJ Freeman, RK Gupta. (to Hercules Inc) US Patent 5,683,809 (1997).
403. SR Shukla, AR Athalye. J Appl Polym Sci 51(9): 1567–1574, 1994.
404. SR Shukla, AR Athalye. J Appl Polym Sci 49(11): 2019–2024, 1993.
405. N Kabay, A Katakai, T Sugo, H Egawa. J Appl Polym Sci 49(4): 599–607, 1993.
406. K Kaji, Y Abe, M Murai, N Nishioka, K Kosai. J Appl Polym Sci 47(8): 1427–1438, 1993.
407. K Kaji, I Yashizawa, C Kohara, K Komai, M Hatada. J Appl Polym Sci 51(5): 841–853, 1994.
408. V Koul, S Guha, V Choudhary. Polymer Int 30(3): 411–415, 1993.
409. LA Lebedeva, TV Druzhinina. Zhurnal Prikladnoi Khimii 65(11): 2565–2569, 1992.
410. LH Sawyer, MA White, GW Knight. (to Dow Chemical Co) US Patent 5,185,199 (1993).
411. VV Pilyngin, DA Kritskaya, AN Ponomarev. Vysokomol Soed B 35(1): 34–38, 1993.
412. VV Pilyngin, DA Kritskaya, AN Ponomarev. Vysokomol Soed B 35(1): 30–33, 1993.
413. MM Miller, JMG Cowie, DL Brydon, RR Mather. Polymer 38(7): 1565–1568, 1997.
414. YT Ambrose, CE Bolian. (to Kimberly-Clark Worldwide Inc) US Patent 5,683,795 (1997).
415. H Chtourou, B Riedl, BV Kokta, A Adnot, S Kaliaguine. J Appl Polym Sci 49(2): 361–373, 1993.
416. H Chtourou, B Riedl, BV Kokta. J Colloid Interface Sci 158(1): 96–104, 1993.
417. K Ando, G Kondo. Kobunshi Ronbunshu 51 (12): 712–718, 1994.
418. S Gao, Y Zang. J Appl Polym Sci 47(11): 2065–2071, 1993.
419. S Gao, Y Zang. J Appl Polym Sci 47(11): 2093–2101, 1993.
420. DA Biro. J Appl Polym Sci 47(5): 883–894, 1993.
421. MS Silverstein, O Breuer. J Mater Sci 28(17): 4718–4724, 1993.
422. P Spence, L Wadsworth, T Chihani. J Mater Sci 31(6): 221–229, 1996.
423. S Tokino. Sen-i Gakkaishi 51(4): 232–237, 1995.
424. S Tokino, T Wakida, Y Sato, T Kimura, H Uchiyama. Sen-i Gakkaishi 51(4): 186–191, 1995.
425. J Kaur, R Barsola, BN Misra. J Appl Polym Sci 51(2): 329–336, 1994.
426. AJ DeNicola, RC Sams. (to Himont Inc) European Patent 0525710 (1993).
427. AK Samanta, DN Sharma. Indian J Fibre Text Res 20: 206–210, 1995.
428. RF Tietz. (to E.I. Du Pont de Nemours and Co) US Patent 5,130,069 (1992).
429. N Morgan. Tech Text Int 2(6): 16–18, 1993.
430. D Dordevic, D Petrov. Tekst Indust 44(10-12): 24–28, 1996.
431. F Fordenwalt. Am Dyest Rep 54: 34–36, 1965.
432. VL Erlich. Mod Text Mag 46: 23–25, 1965.
433. VL Erlich. Mod Text Mag 46: 36–39, 1965.
434. MN Molocea, A Grigoriu. Bull LAIC 3: 45–50, 1982.
435. DD Gagliardi. Am Dyest Rep 54: 41–44, 1965.

436. E Eddington, FL Slevenpiper. Am Dyest Rep 52: 31–36, 1963.
437. E Eddington, FL Slevenpiper. Am Dyest Rep 52: 9–13, 1963.
438. E Eddington, FL Slevenpiper. Am Dyest Rep 52: 51–56, 1963.
439. RG Curtis, DD Dellis, GM Bryant. Am Dyest Rep 55: 44–47, 1964.
440. JJ Press. Am Dyest Rep 55: 34–37, 1964.
441. B Biehler. Chemiefasern/Textilind 29/81: 848–849, 1979.
442. *** Man-Made Fiber Year Book (Chemiefasern/Textilind) 1986: 70.
443. *** Am Text Int 8: 30–31, 1991.
444. ITMA'91, Hannover, 1991, Papers of the Exhibition.
445. International Wool Textile Organization Specifications, Cape Town, 1988.
446. 1989 Annual Books of ASTM Standards, Section 7, Textiles. Philadelphia: American Society for Testing and Materials.
447. K Ando, G Kondo. Kobunshi Ronbunshu 51(12): 795–800, 1994.
448. AF Johnson, NWR Brown, SW Tsui. (to University of Bradford) GB Patent 2,315,275A (1998).
449. H Studer. Chem Fiber Int 47(5): 373–374, 1997.
450. RK Datta, K Sen, AK Sengupta, RS Gandhi. Influence of mass colouration on structure and properties of polypropylene feeder and draw textured yarns. Proc World Textile Congress on Polypropylene in Textiles, Huddersfield, 1996, pp 231–248.
451. C Ripke. Pigments and their influence on PP fibre production and quality. Proc World Textile Congress on Polypropylene in Textiles, Huddersfield, 1996, pp 193–209.
452. V Schrenk, HC Porth. Vestowax P 930 - the novel pigment carrier for colouring polypropylene fibres and filaments. Proc World Textile Congress on Polypropylene in Textiles, Huddersfield, 1996, pp 140–153.
453. PJ Sheth. Wettable and dyeable polyolefin - technology and applications. Proc World Textile Congress on Polypropylene in Textiles, Huddersfield, 1996, pp 109–120.
454. PJ Sheth, V Chandrashekar, RR Kolm. (to Lyondell Petrochemical Co) US Patent 5,550,192 (1996).
455. D Blechschmidt, W Kittelmann, H Halke. Chem Fiber Int 46(5): 352–355, 1996.
456. J Falguera Xifra, R Sacrest Villegas. European Patent 0738799 (1996).
457. B Kaul, C Ripke, M Sandri. Chem Fiber Int(Man-Made Fiber Year Book): 72–75, 1996.
458. A Marcincin, A Ujhelyiova, K Marcincin, T Marcincinova. Vlakna a Textil 3(3): 92–99, 1996.
459. H Schmidt, K Twarowska-Schmidt. Fibres Text East Europe 5(1): 51–52, 1997.
460. T Kuroki, Y Ota. (to Toyo Boseki Kabushiki Kaisha) US Patent 5,613,987 (1996).
461. V Prchal, M Kristofic, L Lapcik Jr, B Havlinova. Fibres Text East Europe 5(1): 48–50, 1997.
462. PJ Sheth, V Chandrashekar, RR Kolm. (to Lyondell Petrochemical) European Patent 0770156 (1997).
463. F Bleses. (to Borealis NV) European Patent 0780509 (1997).
464. V Schrenk, K Assmann. Chem Fiber Int 47(5): 375, 1997.
465. L Ruys. Chem Fiber Int 47(5): 376–384, 1997.
466. I Rumsey. GB Patent 2,314,349A (1997).
467. SR Karmakar, HS Singh. Colourage Ann: 97–100, 1997.
468. VS Gupta. Indian J Fibre Text Res 22(4): 236–245, 1997.
469. J Broda, E Sarna, A Wlochowicz. Fibres Text East Europe 6(1): 55–58, 1998.
470. Centexbel. Melliand Textilberichte 78(10): E141, 1997.
471. E Sarna, A Wlochowicz, J Broda. Przeglad Wlokienniczy 52(4): 8–11, 1998.
472. E Bach, E Cleve, E Schollmeyer. DWI Reps 121: 5–11, 1998.
473. *** Res Discl 409: 591, 1998.
474. W Oppermann, H Herlinger, D Fiebig, O Staudenmayer. Melliand Textilberichte 77(9): 588–592, 1996.
475. B Karil, C Ripke, M Saudri. Man-Made Fiber Year Book (Chemiefasern/Textilind) 1996: 72.
476. L Piraux, L Kinany. Solid State Commun 70: 427–435, 1989.
477. B Poulaert, JC Chielenst, C Vandenhende, JP Issi, R Legras. Polym Commun 31: 149–154, 1990.
478. M Lewin, B Sello (eds). Handbook of Fiber Science and Technology, vol 1, Chemical Processing of Fibres and Fabrics, Fundamentals and Preparation, part B. New York: Marcel Dekker, 1984.
479. A Koszka. Magyar Textilech 33: 325–328, 1980.
480. K Meyer. Chemiefasern. Leipzig: VEB Fachbuch Verlag, 1981.
481. VL Erlich, EM Honneycutt. Man-Made Fibers. New York: Wiley-Interscience, 1968.
482. Yu I Mitchenko, RF Tsiperman, LD Rudneva, AN D'yachkov, SA Gribanov. Vysokomol Soed B 34(10): 9–14, 1992.
483. AM Stolin, VI Irzhak. Vysokomol Soed A 35(7): 902–904, 1993.
484. GK El'yashevich, EA Karpov, VK Lavrent'ev, VI Poddubnyi, MA Genina, Yu F Zabashta. Vysokomol Soed A 35(6): 681–685, 1993.
485. VA Mariknin, LP Mysnikova, MD Uspenskii. Vysokomol Soed A 35(6): 686–692, 1993.
486. Y Ohta, H Sugiyama, H Yasuda. J Polym Sci B 32(2): 261–269, 1994.
487. LI Kuzub, AI Efremova, EN Razpopova, OS Svechnikova, NI Shut, VI Irzhak. Vysokomol Soed B 35(6): 308–311, 1993.
488. K Ando, M Takahashi, R Togashi, Y Okumura. Sen-i Gakkaishi 49(6): 323–330, 1993.
489. N Ogata, M Miyagashi, T Ogihara, K Yoshida. Sen-i Gakkaishi 51(6): 248–256, 1995.

490. FA Ruiz, TAPPI J 79(5): 139–141, 1996.
491. PJ Sheth. (to Lyondell Petrochemical Co) US Patent 5,464,687 (1995).
492. E Andreassen, EL Hiurichsen, K Grostad, OJ Myhre, MD Braathen. Polymer 36(6): 1189–1198, 1995.
493. FJ Wortmann, KV Schulz. Polymer 36(12): 2363–2369, 1995.
494. S Kaufmann, A Bossmann. Vlakna Textil 2(1): 2–6, 1995.
495. H Kusanagi. Kobunshi Ronbunshu 53(5): 311–316, 1996.
496. R Kwiatkowski, A Wlochowicz, S Rabiej. Przeglad Wlokienniczy 49(8): 8–10, 1995.
497. T Takahashi, A Konda, Y Shimizu. Sen-i Gakkaishi 51(7): 303–312, 1995.
498. G Kolay, C Ogbouna, PS Allan, MJ Bevis. Chem Eng Res Des 73(A7): 798–809, 1995.
499. V Prchal, B Havlinova, P Hodul, A Marcincin. Vlakna Textil 2(1): 71–74, 1994.
500. FJ Wortmann, KV Schulz. Polymer 36(2): 315–321, 1995.
501. MIA ElMaaty, DC Bassett, RH Olley, MG Dobb, JG Tomka, IC Wang. Polymer 37(2): 213–218, 1996.
502. LE Govaert, T Peijs. Polymer 36(23): 4425–4431, 1995.
503. M Sugimoto, M Ishikawa, K Hatada. Polymer 36(19): 3675–682, 1995.
504. D Wang, AAK Klaassen, GE Janssen, E DeBoer, RJ Meier. Polymer 36(22): 4193–4196, 1995.
505. E Kapell. TUT Textiles Usages Tech 19: 35–36, 1996.
506. L Hes, M De Aranjo, R Storova. Thermal comfort properties of socks containing. POP filaments. Proc World Textile Congress on Polypropylene in Textiles, Huddersfield, 1996, pp 133–139.
507. K Kosegaki, K Iwase, H Sano, T Miwa. Kobunshi Ronbunshu 52(8): 491–496, 1995.
508. M Jambrich, A Murarova, A Stupak, P Jambrich, J Lalik, J Klatic. Vlakna Textil 4(1): 196–200, 1994.
509. M Ishikawa, M Sugimoto, K Hatada, T Tanaka. Kobunshi Ronbunshu 52(3): 149–154, 1995.
510. S Nagai, M Ishikawa. Kobunshi Ronbunshu 52(3): 125–133, 1995.
511. M Ishikawa, M Sugimoto, K Hatada, T Tanaka. Kobunshi Ronbunshu 52(3): 134–140, 1995.
512. A Marcincin, J Legen, E Zemanova, J Stupak, P Jambrich. Vlakna Textil 2(1): 7–14, 1995.
513. J Tino, M Klimova, I Chodak, M Jacobs. Polymer Int 39(3): 231–234, 1996.
514. N Marangone, L Bertoli, G Colomberotto. (to Savio Macchine Tessili SRL) European Patent 0620845 (1994).
515. DG Svetec, SM Kveder. Fibres Text East Europe 4(1): 50–52, 1996.
516. K Kitao, H Turuta. Kobunshi Ronbunshu 52(8): 497–503, 1995.
517. *** Res Discl 382: 110, 1996.
518. *** Res Discl 386: 384, 1996.
519. R Vogel, G Schauer, R Beyreuther. Chem Fiber Int 45(4): 268–269, 1995.
520. VB Gupta, YC Bhuvanesh. Indian J Fibre Text Res 21(3): 167–178, 1995.
521. *** Nonwoven Rep Int 306: 6–7, 1996.
522. D Blechschmidt, H Fuchs, A Vollmar, M Siemon. Chem Fiber Int 46(5): 332–336, 1996.
523. DL Dorset. Polymer 38(2): 247–253, 1997.
524. A Galeski, T Kowalewski, M Przygoda. Fibres Text East Europe 4(3): 95–99, 1996.
525. N Wilson. J Electrostat 28(3): 313–316, 1992.
526. J Revilakova, M Jambrich. Textil Chemia 22(4): 86–96, 1992.
527. MS Silverstein, O Breuer. J Mater Sci 28(15): 4153–4158, 1993.
528. RS Nohr, JG MacDonald. (to Kimberly Clark Co) US Patent 5,283,023 (1994).
529. RW Johnson, TW Theyson. (to Hercules Inc) European Patent 0516412 (1992).
530. J Cao, T Hashimoto. Polymer 34(13): 2707–2710, 1993.
531. S Rastogi, JA Odell. Polymer 34(7): 1523–1527, 1993.
532. YMT Engelen, CWM Bastiaansen, PJ Lemstra. Polymer 35(4): 729–733, 1994.
533. LE Govaert, CWM Bastiaansen, PJR Leblans. Polymer 34(3): 534–540, 1993.
534. TM Zelazny, T Wierzbowska, JR Koscianonski. Przeglad Wlokieriniczy 51(3): 19–23, 1997.
535. E Mielicka, H Kaczmarska, M Baczynska, Fibres Text East Europe 5(1): 58–62, 1997.
536. A Murarova, M Jambrich, I Vyskocil. Fibres Textiles East Europe 5(1): 53–54, 1997.
537. A Ujhelyiova, A Marciuncin, E Zemanova. Vlakna Textil 3(4): 134–140, 1996.
538. E Szucht. Przeglad Wlokienniczy 51(4): 19–22, 1997.
539. ND Scott. Chem Fiber Int (Man-Made Fiber Year Book): 8–9, 1997.
540. N Tamura, T Sakai, T Hashimoto. Sen-i Gakkaishi 53(11): 489–493, 1997.
541. DT Grubb. Polym Sci Polym Phys Ed 21: 165–172, 1983.
542. W Hoogsteen, H Kormelink, G Eshuis, G Tenbrinke, AJ Penning. J Mater Sci 23: 3467–3474, 1988.
543. JM Rossignol, R Sequela, F Rietsch. Polymer 29: 43–45, 1988.
544. WP Leung, CL Choy, C Xu, Z Qi, R Wu. J Appl Polym Sci 36: 130–143, 1988.
545. JM Rossignol, R Sequela, F Rietsch. Polymer 31: 1449–1452, 1988.
546. F Fourne. Man-Made Fiber Year Book (Chemiefasern/Textilind) 1994: 26–28.
547. ***. Man-Made Fiber Year Book (Chemiefasern/Textilind) 1994: 69.
548. F Schultze-Gebhardt. Technische Textilien/Technical Textiles 36: E 135–136, 1993.
549. F Fourne. Man-Made Fiber Year Book (Chemiefasern/Textilind) 1994: 33–36.
550. P Olivieri. Tinctoria 2: 53–55, 1993.
551. H Eichinger. Tech Text Int 11: 22–25, 1992.
552. British Textile Bonding. High Perform Text 9: 2–3, 1992.

553. CD Desper, SH Cohen, AO King. J Appl Polym Sci 47(7): 1129–1142, 1993.
554. OV Nikitina, LI Kuzub, VI Irzhak. Vysokomol Soed A 35(5): 554–558, 1993.
555. Geo Fabrics Ltd. Nonwoven Rep Int 275: 9–10, 1994.
556. KK Leonas. INDA J Nonwovens Res 5(2): 22–26, 1993.
557. Dow Plastics. Med Text 4: 2–3, 1993.
558. Amoco Fabrics. High Perform Text 4: 4–5, 1993.
559. Geofabrics Ltd. High Perform Text 2: 5, 1994.
560. N Butler. Tech Text Int 3(3): 26–27, 1994.
561. Du Pont de Nemours International SA. Tech Text Int 3(3): 10, 1994.
562. Reemay Inc. Tech Text Int 2: 10, 1994.
563. Allied Signal Inc. Tech Text Int 2(6): 7, 1993.
564. Institut Textile de France. L'ind Text 1263(3): 13–14, 1995.
565. B Laszkiewicz. Przeglad Wlokienmiczy 49(9): 3–9, 1995.
566. S Anselment. High lightfast staple and endless yarns of PP for the automotive industry. Proc World Textile Congress on Polypropylene in Textiles, Huddersfield, 1996, pp 366–378.
567. B Lazic, R Jovanovic, P Skundric. Hemijska Vlakna 33(1/4): 26–42, 1993.
568. A Takahashi, K Tatebe, M Onishi, Y Seita, K Takahara. Kobunshi Ronbunshu 50(6): 507–513, 1993.
569. E Leflaive. TUT Text Usages Techniques 10: 24–25, 1993.
570. MW Miligan, F Lu, LC Wadsworth, RR Buntin. INDA J Nonwovens Res 5(3): 8–12, 1993.
571. X Jiu, H Wu, J Huang. J China Text Univ 18(3): 9–17, 1992.
572. W Bendkowska. Przeglad Wlokienniczy 47(5): 111–112, 1993.
573. T Daras, G Reyed. TUT Textiles Usages Techniques 7: 34–35, 1993.
574. A Takahashi, K Tatebe, M Omishi, Y Seito, K Takahara. Kobunshi Ronbunshu 50(6): 515–521, 1993.
575. S Ghosh, M Dever, H Thomas, C Tewksbury. Indian J Fibre Textile Res 19(3): 203–208, 1994.
576. J Kaluzka; C Nowicka, P Wcislo. Przeglad Wlokienniczy 50(4): 7–10, 1996.
577. J Kaluzka, M Lebiedowski. Przeglad Wlokienniczy 51(3): 23–26, 1997.
578. P Wcislo, C Nowicka. Fibres Text East Europe 5(1): 63–65, 1997.
579. J Kaluzka, M Lebiedowski. Fibres Text East Europe 5(1): 21–25, 1997.
580. E Szucht. Fibres Text East Europe 5(1): 38–41, 1997.
581. E Vonwiller. Am Text Int 26(6): 48–52, 1997.
582. GE Katz. High Perform Text 8: 12, 1997.
583. W Gador, J Grzybowska-Pietras, E Jankowska. Przeglad Wlokienniczy 51(7): 16–18, 1997.
584. Maniffatura Tessile Fruilana SpA. High Perform Text 6: 7–8, 1997.
585. S Lukie, P Jovanic, Z Blagojevic, M Asanin. Tekstil Industrija 46(1-2): 29–31, 1998.
586. A Watzl. Man-Made Fiber Year Book (Chemiefasern/Textilind) 1988: 64–65.
587. W Albrecht. Textilveredlung 23: 47–52, 1988.
588. Reemay Inc. Disposables Nonwovens 22(5): 2–3, 1993.
589. Du Pont de Nemours (Luxembourg) SA. Nonwoven Rep Int 274: 4, 1994.
590. Y Seito, S Nagaki, K Tatebe, K Kido. (to Terumo KK) US Patent 5,139,529 (1992).
591. EG Gammelgaard. Nonwoven Rep Int Nonwovens Rep Yearbook 269: 12–14, 1993.
592. M Dever. TAPPI J 76(4): 181–189, 1993.
593. RE Howard, J Young. (to Entek Manufacturing Inc) US Patent 5,230,949 (1993).
594. A Winter, A Vollmar, F Kloss, B Bachmann. (to Hoechst AG) European Patent 0 600 461 (1994).
595. Du Pont de Nemours International SA. Nonwoven Rep Int 276: 6, 1994.
596. Exxon Chemical. Vliesstoff Nonwoven Int 8(3): 88, 1993.
597. J Engelhardt. Nonwoven Rep Int. Nonwovens Rep Yearbook 269: 14, 1993.
598. European Association of Textile Polyolefins. Nonwoven Rep Int 260: 4–5, 1992.
599. Fibertex APS. Nonwoven Rep Int 263: 11, 1993.
600. Vitaweb. Nonwoven Rep Int 263: 12, 1993.
601. HS Lim, H Shin. (to E.I. Du Pont de Nemours and Co) US Patent 5,308,691 (1994).
602. Du Pont de Nemours International SA. Du Pont Mag, Europ Ed 87(1): 19–21, 1993.
603. RS Nohr, JG MacDonald. (to Kimberly-Clark Co) US Patent 5,300,167 (1994).
604. JF Reed, M Swan. (to Minnesota Mining & Manufacturing Co) US Patent 5,324,576 (1994).
605. K Yamamoto, S Ogata. (to Chisso Co) European Patent 0557889 (1993).
606. GA Harpell, S Kavesh, I Palley, DC Prevorsek. (to Allied-Signal Inc) US Patent 5,135,804 (1992).
607. *** Res Discl 342: 785–786, 1992.
608. J van Dingenan, A Verlinde. Tech Text Int 5(1): 10–13, 1996.
609. Exxon Chemical Co. Med Text 3: 2, 1995.
610. RE Kozulla. (to Hercules Inc) US Patent 5,487,943 (1996).
611. *** Disposables Nonwovens 23(4): 1–2, 1994.
612. *** Disposables Nonwovens 23(5): 14–16, 1994.
613. K Ando, G Kondo. Sen-i Gakkaishi 51(4): 164–168, 1995.
614. CJ Molnar, JR Molnar. (to CJ Molnar, JR Molnar) US Patent 5,490,351 (1996).
615. G Barsotti. (to Moplefan SpA) US Patent 5,514,751 (1996).
616. JD Langley. (to Kappler Safety Group) US Patent 5,409,761 (1995).
617. SP Anjur, MF Kalmon, AJ Wisneski. (to Kimberly-Clark Co) GB Patent 2,296,511A (1996).

618. AL McCormack, LJ Garrett, KL English. (to Kimberly-Clark Co) European Patent 0691203 (1996).
619. I Ikkanzaka, K Ikeda, Y Takai. (to Kanai Juyo Kogyo Co; to Daiwabo Create Co) US Patent 5,487,944 (1996).
620. JW Suominen, S Makipirtti, H Bergholm. European Patent 0667406 (1995).
621. G Lesca, V Giannella. (to Himont Inc) US Patent 5,368,927 (1994).
622. DC Strack, LA Connor, SW Gwaltney, AL McCormack, SE Shawver, JS Schultz. (to Kimberly-Clark Co) US Patent 5,336,552 (1994).
623. DJ McDowall, LH Sawyer, DC Strack, TK Timmous. (to Kimberly-Clark Co) European Patent 0674035 (1995).
624. DS Everhart, RE Meirowitz. (to Kimberly-Clark Co) US Patent 5,439,734 (1995).
625. K Niitsuma, T Kawamata, S Hayakawa. (to Fuji Photo Film Co) US Patent 5,370,917 (1994).
626. DC Strack, TN Wilson, DV Willittis. (to Kimberly-Clark Co) European Patent 0604736 (1994).
627. CY Cheng, CR Davey. Int Nonwoven J 6(3): 38–42, 1994.
628. K Sasaki, H Ohara, N Imaki. (to Mitsubishi Chemical Co) US Patent 5,484,651 (1996).
629. DC Strack, TN Wilson, DV Willits. (to Kimberly-Clark Co) US Patent 5,482,772 (1996).
630. RS Nohr, JG MacDonald, ED Gadsby, DS Everhart. (to Kimberly-Clark Co) US Patent 5,525,415 (1996).
631. O Turunen, J Fors, E Thaels. (to Borealis Holding A/S) European Patent 0642605 (1995).
632. T Mouzen, M Watanabe. (to Kitz Co) European Patent 0659468 (1995).
633. DS Everhart, RE Meirowitz. (to Kimberly-Clark) European Patent 2,282,817A (1995).
634. R Stentiford. Nonwoven Rep Int 288: 27, 1995.
635. *** Nonwoven Rep Int 289: 8–9, 1995.
636. J Kaluzka, C Nowicka, P Wcislo. Przeglad Wlokienniczy 50(4): 7–10, 1996.
637. GJ Anderson, ET Arcilla, CI Eyberg, AS Mundt. (to Minnesota Mining & Manufacturing) US Patent 5,418,022 (1995).
638. CM Vogt, B Cohen, CJ Ellis. (to Kimberly-Clark Worldwide Inc) US Patent 5,656,361 (1997).
639. RK Gupta, JE Mallory, K Takeushi. (to Hercules Inc) US Patent 5,654,088 (1997).
640. *** Nonwoven Rep Int 306: 6–7, 1996.
641. *** Nonwoven Rep Int 305: 12, 1996.
642. JA Austin, DD Newkirk. (to Fiberweb North America Inc) US Patent 5,543,206 (1996).
643. LLH Van der Loo, RC Van der Burg. (to DSM NV) US Patent 5,569,528 (1996).
644. *** Nonwoven Rep Int 307: 8, 27, 1996.
645. JH Harrington. (to Hercules Inc) US Patent 5,582,904 (1996).
646. *** Textile Mon 9: 47–48, 1996.
647. *** Nonwoven Rep Int 309: 29, 1996.
648. G Barsotti. (to Moplefan SpA) US Patent 5,585,172 (1996).
649. EM Gillyns, VL Paqnay, JA Raush. (to E.I. Du Pont de Nemours) European Patent 0755331 (1997).
650. Filament Fiber Technology Inc. Med Text 2: 2, 1997.
651. European Association for Textile Polyolefins. Med Text 1: 9–10, 1997.
652. MD Powers. (to Kimberly-Clark Co) US Patent 5,597,647 (1997).
653. Hercules Inc. Med Text 3: 3, 1997.
654. AR Oleszczuk, SL Gessner. (to Fiberweb North America Inc) US Patent 5,616,408 (1997).
655. JL McManus, PM Kobylivker, CJ Albertelli. (to Kimberly-Clark Co) GB Patent 2,307,488A (1997).
656. L Pinoca, R Africano, L Spagnoli. (to Montell North America Inc) US Patent 5,631,083 (1997).
657. M Suzuki. (to Unitika Ltd) GB Patent 2,312,447 (1997).
658. S Mukaiada, K Iguchi, K Tanaka. (to Sanyo Chemical Industries Ltd) US Patent 5,676,660 (1997).
659. S Mukaiada, K Iguchi, K Tanaka. (to Sanyo Chemical Industries Ltd) US Patent 5,676,664 (1997).
660. RD Pike, KL Brown, PW Shipp. (to Kimberly-Clark Worldwide Inc) European Patent 0789612 (1997).
661. TC Erderly, BC Trudell, WR Braudenburger. (to Exxon Chemical Patents Inc) European Patent 0795053 (1997).
662. SK Ofosu, PM Kobylivker, ML DeLucia, RL Hudson, JJ Sayovitz. (to Kimberly-Clark Worldwide Inc) US Patent 5,681,646 (1997).
663. SE Shawver, PW Estey, LA Conner. (to Kimberly-Clark Worldwide Inc) European Patent 0812371 (1997).
664. *** Nonwoven Rep Int 319: 45, 1997.
665. AL McCormack. (to Kimberly-Clark Worldwide Inc) US Patent 5,695,868 (1997).
666. SE Shawver, LW Collier. (to Kimberly-Clark Worldwide Inc) US Patent 5,695,849 (1997).
667. RS Nohr, JG MacDonald. (to Kimberly-Clark Worldwide Inc) US Patent 5,696,191 (1997).
668. LM Delucia, RL Hudson. (to Kimberly-Clark Worldwide Inc) US Patent 5,714,256 (1998).
669. SA Young. Safety Protective Fabrics 1(1): 10–11, 1992.
670. DJ Bettge, G Hinrichsen. Comp Sci Technol 47(2): 131–136, 1993.
671. K Nishino, S Sasagawa, H Katsurayama, T Igane, T Kido. (to Mitsui Petrochemical Industries Ltd) US Patent 5,188,895 (1993).
672. C Hart. (to Kappler Safety Group) European Patent 0505027 (1992).
673. AS Hansen. (to Danaklon AS) US Patent 5,330,827 (1994).
674. MB Fitzgerald. GB Patent 2,268,759A (1993).
675. MW Groves, MD Smith, SR Smith. (to Tryam Trading Inc) GB Patent 2,265,628A (1993).
676. PJ Hine, IM Ward, RH Olley, DC Bassett. J Mater Sci 28(2): 316–324, 1993.

677. CL Choy, I Fei, TG Xi. J Polym Sci B 31(3): 365–370, 1993.
678. TT Zufle. (to TT Zufle) European Patent 0507942 (1992).
679. Y Takai. (to Daiwabo Create Co) European Patent 0535373 (1993).
680. J Luo, D Xu, Y Liu, X Li, B Li, H Zhang. (to Academia Sinica Institute of Chemistry) GB Patent 2,258,869 (1993).
681. K Yagi, H Takeda. (to Mitsui Petrochemical Industries Ltd) US Patent 5,143,977 (1992).
682. ZF Li, DT Grubb. J Mater Sci 29(1): 189–202, 1994.
683. DT Grubb, ZF Li. J Mater Sci 29(2): 13–26, 1994.
684. AG Andreopoulos, K Liolios, A Patrikis. J Mater Sci 28(18): 5002–5006, 1993.
685. RH Olley, DC Bassett, PJ Hine, IM Ward. J Mater Sci 28(4): 1107–1112, 1993.
686. DW Woods, PJ Hine, IM Ward. Comp Sci Technol 52(3): 397–405, 1994.
687. I Sei, S Masayasu, T Hirokazu. (to Chisso Co) European Patent 0696655 (1996).
688. JD Dulaney, RK Gupta, RE Kozulla, RJ Legare, RG MacLellan. (to Hercules Inc) European Patent 0696654 (1996)
689. AL McCormack. (to Kimberly-Clark Co) GB Patent 2,285,408A (1995).
690. GA Romanek, RE Moon, ML Marienfield, SS Guram. (to Amoco Co) US Patent 5,538,356 (1994).
691. *** Chem Fiber Int 45(2): 135, 1995.
692. H Kimura, K Yamada, Y Hirato. (to Nippon Steel Co) European Patent 0628674 (1994).
693. O Darras, RA Duckett, PJ Hine, IM Ward. Comp Sci Technol 55(2): 131–138, 1995.
694. AP Slater, R Turner. (to Courtaulds Aerospace Ltd) GB Patent 2,276,935A (1994).
695. TA Medwell. (to Courtaulds Aerospace Ltd) GB Patent 2,276,934A (1994).
696. R Turner. (to Courtaulds Aerospacce Ltd) GB Patent 2,276,933A (1994).
697. SD Long, PJ Hine, PD Coates, AF Johnson, RA Duckett, IM Ward. Plastics, Rubber Comp. Process Appl 24(5): 277–283, 1995.
698. DP Zafiroglu. (to E.I. Du Pont de Nemours and Co) European Patent 0686213 (1995).
699. Y Katon. (to Aisin Seiki KK) US Patent 5,503,093 (1995).
700. S Backman, S Erikson, A Miller. (to S Backman) European Patent 0648303 (1995).
701. GS Boyce. (to Euro-Projects (LTTC) Ltd) GB Patent 2,310,822A (1997).
702. Dimotecch Ltd. High Perform Text 11: 10–12, 1996.
703. M Shaker. Polym Adv Tech 8(6): 327–334, 1997.
704. JS Nelson, AL Price. (to Safariland Ltd) US Patent 5,619,748 (1997).
705. EMPA St. Gallen. Deutscher Seiler-Z 116: 309, 1997.
706. SL Gessner, LE Trimble. (to Fiberweb North America Inc) US Patent 5,733,822 (1998).
707. C Nowicka. Przeglad Wlokienniczy 52(3): 13–16, 1998.
708. SW Gerry. (to K 2 Inc) US Patent 5,714,224 (1998).
709. B Ridgway. Carpet Rev Wkly 11: 42–46, 1979.
710. W Dirschke. Man-Made Text India 27: 7–10, 1984.
711. *** Chemiefasern/Textilind 35/87: 75–82, 1985.
712. WH Kinkel. (to WH Kinkel) European Patent 0569955 (1993).
713. LLH Van de Loo. TUT Text Usages Tech 9: 28–30, 1993.
714. M Jacobs, J Van Dingenen. Tech Text Int 2(5): 10–13, 1993.
715. DSM High Perform Fibers. Nonwoven Rep Int 282: 5, 1994.
716. DHP Schuster, GA Fels, H Spors. (to Akzo Nobel Faser AG) European Patent 0675220 (1995).
717. WH Tung, LR Simonsen, YV Vinod, F Werny. (to E.I. Du Pont de Nemours and Co) European Patent 0694092 (1996).
718. *** Du Pont Mag 90(1): 14–16, 1996.
719. H Wunsch, BJ Launchbury. Novel techniques for the production of polyolefin floorcoverings. Proc World Textile Congress on Polypropylene in Textiles, Huddersfield, 1996, pp 298–307.
720. SW Shalaby, M Deng. (to Smith & Nephew Richards Inc) European Patent 0714460 (1996).
721. DK Lickfield, MHS Berman, RF Hyslop, AR Oleszczuk, SL Gessner, JA Austin. (to Fiberweb North America Inc) US Patent 5,484,645 (1996).
722. SIMA Promatech. Carpet Rug Ind 22(11): 48–50, 1995.
723. M Gosten. Carpet Rug Ind 22(12): 44–48, 1994.
724. *** Du Pont Mag 89(1): 28, 1995.
725. A Hoyle. Carpet Rug Ind 23(7): 39–46, 1995.
726. PJ Degen, JY Lee. (to Pall Co) GB Patent 2,285,638A (1995).
727. EM Gillyns, DR Stochmel, EM Ebers. (to E.I. Du Pont de Nemours and Co) European Patent 0619849 (1994).
728. *** Int Carpet Bull 284: 6–7, 1997.
729. *** Int Carpet Year Book 1997: 8.
730. H Wunsch, BJ Launchbury. Int Carpet Year Book 1997: 33–37.
731. JM Long, KA Snyder. (to Synthetic Industries Inc) US Patent 5,604,009 (1997).
732. G Fischer. Int Carpet Year Book 1998: 14–15.
733. *** World Sports Activewear 2(2): 34–35, 1996.
734. J Beswick. Polypropylene on textile surfaces for sports and recreation. Proc World Textile Congress on Polypropylene in Textiles, Huddersfield, 1996, pp 331–334.
735. *** World Sports Activewear 2(1): 20–21, 1996.
736. DJ Herliky. (to Aquatic Design Inc) European Patent 0686088 (1995).
737. V Armond, K Bailey. Tech Text Int 4(1): 18–20, 1995.
738. *** World Sports Activewear 2(4): 22–23, 1996.
739. J Beswick. World Sports Activewear 2(4): 24–29, 1996.
740. J Beswick. Nonwoven Rep Int 310: 26–28, 1997.
741. D Buirski. World Sports Activewear 3(2): 42–43, 1997.
742. DJ Herlihy. (to Aquatic Design Inc) US Patent 5,631,074 (1997).

743. ME Hain, CK Liu. (to United States Surgical Co) US Patent 5,292,328 (1994).
744. M Jacobs, T ter Beek. Tech Text Int 5(3): 3, 1996.
745. *** Tech Text Int 5(3): 3, 1996.
746. J Ford. Text Mag 24(3): 11–15, 1995.
747. YC Bhuvanesh, VB Gupta. Polymer 36(19): 3669–3674, 1995.
748. HJ Newmann. US Patent 5,210,877 (1993).
749. Allied Signal Inc. High Perform Text 6: 12–13, 1995.
750. E Witczak, S Tarkowska. Fibres Text East Europe 4(3): 139–142, 1996.
751. A Smith. Text Horiz 16(2): 40–42, 1996.
752. United States Surgical Co. Med Text 1: 4–5, 1995.
753. RH Pearce, RJ Hulme. (to Smith & Nephew PLC) US Patent 5,403,267 (1995).
754. AJ Peacock. (to Exxon Chemical Patents Inc) European Patent 0674726 (1995).
755. 3 M Co. Med Text 3: 7–8, 1996.
756. WD Hanrahan, AS Patil. Text Tech Int: 173–175, 1996.
757. United States Surgical Co. Med Text 7: 2, 1994.
758. *** Nonwoven Rep Int 317: 32–33, 1997.
759. AI Kobashigawa, JD Roe. (to Kerr Co) GB Patent, 2,310,438A (1997).
760. DJ Lennard. (to Ethicon Inc) European Patent 0726078 (1996).
761. RB Quincy, RS Nohr. (to Kimberly-Clark Co) European Patent 0736118 (1995).
762. Coville Inc. High Perform Text 11: 3–4, 1996.
763. Courtaulds Specialty Fibres. Med Text 6(2): 3, 1997.
764. P Hodul, I Vykocil, M Jambrich, Z Kolvekova. Fibres Text East Europe 5(1): 55–57, 1997.
765. *** Res Discl 406: 101, 1998.
766. EG Gammelgaard. Med Text 12: 10–15, 1997.
767. Reemay Inc. Disposables Nonwovens 25(2): 2, 1996.
768. RL Levy, CE Bolian, MT Morman, LE Preston. (to Kimberly-Clark Co) US Patent 5,492,753 (1996).
769. Du Pont. Du Pont Mag 89(4): 16–18, 1995.
770. RS Nohr, JG MacDonald. (to Kimberly-Clarke Co) US Patent 5,567,372 (1996).
771. Sorbent Products Co Inc. Tech Text Int 5(10): 11, 1996.
772. Du Pont de Nemours (Luxembourg) SA. High Perform Text 3: 8–9, 1997.
773. VS Borland. Am Text Int 26(3): 2–6, 1997.
774. MS Nataraj, KL McManis. Geosynth Int 4(1): 65–69, 1997.
775. JD Geimann, RK Gupta. (to Hercules Inc) US Patent 5,712,209 (1998).
776. GA Stahl, JJ McAlpin. (to Exxon Chemical Patents Inc) US Patent 5,723,217 (1998).
777. VK Kothari, A Das. Geotext Geomem 12(2): 179–191, 1993.
778. JS Dugan. (to BASF Co) US Patent 5,736,083 (1998).
779. RR Berg, R Bonaparte. Geotext Geomem 12(4): 287–306, 1993.
780. M Kharchafi, M Dysli. Geotext Geomem 12(4): 307–325, 1993.
781. Don & Low Ltd. High Perform Text 1: 3, 1993.
782. M Kaushke. TUT Text Usages Tech 6: 43–44, 1992.
783. CJ Sprague. Geotech Fabrics Rep 11(1): 14–20, 1993.
784. GV Rao, RK Gupta, MPS Pradhan. Geotext Geomem 12(1): 73–87, 1993.
785. JW Cowland, SCK Wong. Geotext Geomem 12(8): 687–705, 1993.
786. EJG Klobbie, RH Lok. (to Lankhorst Touwfabrieken BV) European Patent 0559252 (1993).
787. K Sen, P Kumar. Polypropylene geotextiles: long term stability and prediction of service life. Proc World Textile Congress on Polypropylene in Textiles, Huddersfield, 1996, pp 275–288.
788. AR Horrocks. Polypropylene geotextiles. Proc World Textile Congress on Polypropylene in Textiles, Huddersfield, 1996, pp 96–108.
789. A Comer, M Kube, K Sauyer. Geotext Geomem 14(5–6): 313–325, 1996.
790. MH Wayne, KW Petrasic, E Wilcosky, TJ Rafter. Geotech Fabrics Rep 14(7): 26–29, 1996.
791. C Kershner. Geotech Fabrics Rep 14(7): 37–38, 1996.
792. AK Ashmawy, PL Bourdeau. Geosynth Int 3(4): 493–515, 1996.
793. *** Water Eng Manage 143(7): 17, 1996.
794. JP Modrak. (to Hercules Inc) US Patent 5,564,856 (1996).
795. A Chodynski, A Jachniak, M Rygiel. Fibres Text East Europe 4(3): 105–107, 1996.
796. MA Lichtwardt, AI Corner. Geotech Fabrics Rep 15(3): 24–28, 1997.
797. R Stetinford. Polypropylene fibre – a sales and marketing perspective. Proc World Textile Congress on Polypropylene in Textiles, Huddersfield, 1996, pp 12–18.
798. JC Stormont, KS Henry, TM Evans. Geosynth Int 4(6): 661–672, 1997.
799. WC Smith. Text World 147(12): 62–66, 1997.
800. T Baker. Geotech Fabrics Rep : 48–52, 1998.
801. HU Xiaomiu, Li Chunbo, Duan Qifu, Diao Jinlun. Filtr Sep 32(4): 307–311, 1995.
802. J Barnes. Filtr Sep 32(3): 199–202, 1995.
803. Minnesota Mining and Manufacturing Co. High Perform Text 4: 10, 1996.
804. *** Tech Text Int 6(8): 10, 1997.
805. B Lauder. Mineral filled geotextiles - the ultimate protection system for synthetic liners in safe management. Proc World Textile Congress on Polypropylene in Textiles, Huddersfield, 1996, pp 325–331.
806. U Genter, O Hotz. PP in the automotive and sportswear sectors. Proc World Textile Congress on Polypropylene in Textiles, Huddersfield, 1996, pp 249–264.
807. P Lofts. (to Scapa Group PLC) GB Patent 2,289,694A (1995).
808. C Freudenberg, KH Morweiser, H Stini, K Veeser. European Patent 0674933 (1995).
809. Reemay Inc. Med Text 6: 6, 1995.
810. ME Jones, AD Rousseau. (to Minnesota Mining and Manufacturing Co) European Patent 0616831 (1994).

811. Lydall Inc. Med Text 4: 3, 1995.
812. Y Ogaki, L Bergmann. Tech Text Int 4(10): 16–18, 1995.
813. Ube-Nitto Kasei Co. High Perform Text 4: 2, 1995.
814. Reemay Inc. Nonwoven Rep Int 301: 3, 1996.
815. JF Dhennim. TUT Text Usages Tech 21: 40–42, 1996.
816. Reemay Inc. Disposables Nonwovens 25(5): 3, 1996.
817. CL Weimer. US Patent 5,593,398 (1997).
818. P Wcislo, J Sojka-Ledakowicz, W Machnowski. Przeglad Wlokienniczy 51(3): 31–35, 1997.
819. T Wierzbowska. Fibres Text East Europe 5(1): 66–69, 1997.
820. M Dever, WT Davis, AA Arrage. TAPPI J 80(3): 157–167, 1997.
821. P Wcislo, JS Ledakowicz, W Machnowski. Nonwoven Rep Int 313: 32–34, 1997.
822. A Bauer, LO Madsen. Chemiefasern/Textilind 41/93: 1189–1194, 1991.
823. *** Am Text Int 3: 5, 1990.
824. D Ahrendt. Magyar Textiltech 35: 19–24,1985.
825. M Jambrich, I Diacik. Magyar Textiltech 35: 273–284, 1985.
826. IM Ward. Composites 22: 341–346, 1991.

28

Technological Aspects of Additive Use for Polyolefin Fiber Production

Aurelia Grigoriu
"Gh. Asachi" Technical University, Iasi, Romania

George Ervant Grigoriu
Romanian Academy, "P. Poni" Institute of Macromolecular Chemistry, Iasi, Romania

I. GENERAL REMARKS

The necessity of tailormade additives for PO fibers is related to fiber manufacturing process conditions and to application diversity. Major developments in this field have been recorded in the last 20 years. For a number of PO fiber applications, the existing disadvantages (e.g., low light stability, low softening point, tendency to creep, and the impossibility of bath dyeing) either are not an obstacle or can be surmounted [1].

Antioxidants, antistats, fillers, flame retardants, UV screeners, slip agents, metal deactivators, whiteners, pigments, and dye receptors are added to polyolefin powder, chips, or granules prior to fiber manufacture. These additives help to overcome fiber sensitivity to heat, UV radiation, and flame and to improve some other properties. In the melt-in-melt system, PP fiber-forming polymers can be mixed with these additives to achieve specific formulations.

PO fibers have great versatility: they may be UV-stabilized to a required degree; the normal minimal static buildup may be reduced by means of conductive materials; and the burn performance of PP (comparable to that of most synthetic fibers) can be improved by the incorporation of flame-retardant protection via halogen- or phosphorus-containing additives.

The basic stabilization of PP (i.e., antioxidants and processing additions) is usually carried out by the polymer producer, immediately after polymerization, before separation, drying, and storage to minimize degradation of the polymer in the molten state at temperatures between 200 and 300°C. Finally PO needs a convenient stabilization against thermal and photooxidative degradation for the application in mind.

There are three directions for incorporation of additives into PO fiber, as follows: polymer stabilization in the chemical industry; coloration of PO polymers through polymer modification or dye-receptor addition before extrusion, or by the melt-dyeing method applied in many variants; films, fibers, and film tapes are colored almost exclusively with concentrated granules, whose amounts may vary between 1 and 10%, in correlation with the depth of shade required, layer thickness, and opacity, and fastness to light, rubbing, and washing; and PO fiber finishing (surface coating) for the improvement of textile processability, handling easy care properties, soiling properties, antistatic properties, slip and stick behavior, luster, whiteness degree, shrinkage, splitting tendency, flammability, and mechanical and chemical properties.

Extraneous substances such as stabilizers, metal deactivators, dye receptors, whiteners, nucleating agents, and oxidants for controlled rheology are applied, to a certain extent, by producers of polymers, while others such as stabilizers, antistats, flame retarders, and master batches of pigments are employed by fiber producers; also, some lubricanting, antistatic, and softening agents as well as preparations for mechanical processing, are applied even during textile manufacturing.

The surface coating of different PP textile products (e.g., stample, filaments, yarns, woven and nonwoven structures, knitting goods, etc.) permits both the passage through the main processing steps in the chemical industry (see Chapter 27) and the transformation to end products in the textile industry (spinning, weaving, knitting, tufting and related techniques, dyeing, printing, etc.). Two methods are usually applied to incorporate additives in a fiber-forming polymer [2, 3], namely *preparation of master batches*, i.e., the polymer and the additive are blended and fused; a polymer tow is made from the melt, which is further processed into granules and then into yarns; and *mechanical blending* of the polymer and the additive, followed by direct spinning into fibers.

The former method produces an extremely homogenous blend, but it has the disadvantage of a considerable amount of polymer degradation due to the thermal stress manifested twice during processing. The latter method is used particularly when the extruder is equipped with both static and dynamic mixers, which permit the production of a very homogenous blend.

II. STABILIZATION OF PO FIBER FORMING POLYMERS

A. Light and Ultraviolet Stabilization

The oxidative tendency of PP initiated via its labile tertiary hydrogen atoms sensitized by both heat and UV light is largely presented elsewhere (see Chapter 17).

The development of nondiscoloring, nonextractable, non-gas-fading antioxidants is the solution for a good thermal stabilization.

Some antioxidants may function as UV stabilizers and vice versa, and typical examples are HAS compounds.

The choice of antioxidant type and the combination of additives influences several fiber properties. PP grades can be tailored to suit minimal degradation where mechanical properties (tenacity, elongation at break, etc.) are important or increased, but they have limited degradation where thermal bonding is a greater consideration (nonwoven webs). Color change due to unsuitable conditions of processing or storage needs to be considered, particularly where color specifications are critical. Significant yellowing would influence to a certain degree the final fiber color.

Long term heat ageing data are necessary for applications where mechanical strength over a long period of time is important (e.g., staple fibers for carpets), but it is not as significant for disposable applications are concerned [4].

One aspect of stabilizers' effectiveness is their resistance to leaching and chemical attack. Modern additives are usually large molecular species, often polymeric, difficult to remove from the polymer [5].

The combination of high molecular mass phenolic antioxidants with phosphites or phosphonites confer adequate protection. They are now the state of the art for PP, HDPE, and LLDPE stabilization. Phenolic antioxidants can confer sufficient processing stability to Ziegler-type high density polyethylene (HDPE, Ti-based catalysts) used for tape manufacturing. However, with HDPE, as with PP fibers, the combinations of phenolic antioxidants with phosphites or phosphonites yield improved processing stability. These combinations are characterized by good extraction resistance, low volatility, and positive influence on substrate color [6, 7].

The high molecular mass of phenolic antioxidants confers substantial protection against thermo-oxidative damage. If required, long-term heat resistance of polyolefin fibers may be increased by addition of an extra amount of phenolic antioxidant, the performance in this respect increasing with the square root of antioxidant concentration. Improved long-term heat resistance of polyolefin fibers can also be achieved with polymeric HAS [8–10].

For PP films with different thickness values, the performance of various phenolic antioxidants increases with specimen thickness. However, the data obtained for one form cannot be used-to predict behavior in another form (i.e., from thick plaques to thin tapes). The best antioxidant for PP fibers is calcium bis-[(3,5-di-*tert*-butyl-4-hydroxy-phenyl) phosphonic acid monoethyl ester].

Controlled rheology PP (known as CR-PP) is a chemically degraded PP with lower average molecular weight and narrower molecular weight distribution that can be processed at considerably reduced temperatures. The peroxide used for the controlled degradation has a slight interference with the stabilizing system; the lower stabilizer's concentration (0.05–0.2%) should be sufficient for most applications. The chemical degradation does not significantly influence processing or long-term stability of CR-PP fibers [8, 11–16].

High concentrations of additive would be uneconomic and technically limited, while many applications (such as polyolefins) are in very thin sections, such as films and fibers.

For PP fibers and tapes, new oligomeric and polymeric hindered amine light stabilizers (HAS) show outstanding performances as light stabilizers. Commercially, they represent most important light stabilizers, followed by UV absorbers such as benzophenones and benzotriazoles. Benzophenones are good general-purpose UV absorbers for clear polyolefin systems, and can also be used in pigmented compounds. Concentrations are usually about 0.25–1.0% [17].

HAS confer by far the best UV stability to polyolefin fibers. In the absence of chemically new classes of UV stabilizers for polyolefins—HAS, introduced commercially almost 20 years ago, can still be considered as the newest class in this sense—the improvements in UV stability are found in the use of the stabilizer combinations. For polyolefin fibers, these combinations do not involve HAS with completely different UV stabilizer types such as UV absorbers or Ni stabilizers. They are based on different stabilizers of the HAS type [18].

On the one hand, combinations of low molecular mass HAS with some polymeric HAS can show synergistic effects with respect to UV stability of PP fibers. The effectiveness of low molecular weight stabilizers, with a large migration tendency, is diminished by special thermal treatments applied to fibers (e.g., crimping, latexing, setting, etc.) [16, 19].

On the other hand, combinations of two high molecular mass HAS can also yield marked synergism in PP fibers, tapes, and multifilaments. The quantitative laws deduced from experimental data, i.e., performance of HAS in PP increasing linearly with HAS concentration and performance in PE directly proportional with the square root of HAS concentration, are also valid with HAS combinations [20–22].

The polymeric HAS can be used without exception in both natural and pigmented PP fibers (staple fibers, multifilaments, tape fibers, etc.) because they have low volatility, high superior compatibility, sufficient migration and extraction resistance, good UV stability after laundering and dry cleaning, good resistance to gas fading, and considerable contribution to heat stability.

Blends of long chain *N,N′*-diallylhydroxylamines, selected phosphites, and selected hindered amines are effective in providing processing, long term heat and light stability performance, and especially gas fade resistance to PP fibers in the absence of a traditionally used phenolic antioxidant [23, 24].

An improved processing stabilizer consists of a phosphite and metal salt of a lactic acid [25, 26].

Novel pyperidine–triazine co-oligomers are useful as light and heat stabilizers for organic materials, in particular as light stabilizers for polypropylene fibers [27].

New pyrazole- and piperidine-containing compounds and their metal complexes are used as light stabilizers for PO polymers. These compounds are highly compatible with polymers, thermostable, and capable of maintaining their photostability under conditions of continuous and extensive irradiation [28].

A mixture of a hindered phenolic compound and a hindered piperidine one can be used as a light stabilizer for PO fibers [29].

Polypropylene discolors and loses its properties on exposure to nonionizing radiation such as gamma rays. It has been found that the synergistic additive package of Ultranox-626, Chimassorb-944, and Milliken XA-200 in polypropylene is more effective than any binary or single combination or single one of these additives [30].

A method of stabilizing a polyolefin nonwoven web against actinic radiation involves two additives to the melt before spinning of the polyolefin fibers: benzotriaziolyl - containing polydialkyl siloxane and polyalkyl piperidyl - containing polydialkyl siloxane [31].

A polyolefin fiber may comprise a mixture of a phosphite, to provide thermal and ultraviolet light stabilization, and an additive selected from a hindered phenolic antioxidant or hindered amine light stabilizer [32].

Pigments interfere with HAS in PP fibers; several studies have been devoted to such aspects [8, 16, 33, 34], large differences being found for different series of stabilizers and pigments, while synergetic effects may appear or not. Pigments can increase or decrease the destructive action of sunlight. The addition of light and heat stabilizers prevents or at least reduces these negative effects.

Additives contribute to the stability of polymers but, under certain circumstances, some of them do not behave in a neutral manner. They can have indirect effects (via interaction with other components of the sophisticated formulations) or direct detrimental ones (by a radical attack of the polymer) on PP stability. For example, humid ingredients cause hydrolysis of aliphatic phosphites and lead consequently to phosphorus acid and other phosphoric acid derivatives [35].

Under severe conditions, some piperidinyl derivatives manifest a degradative potential on PP; this effect on PP processability is more or less pronounced in relation with the number of sterically free piperidyl nitrogens.

The unsuitable HAS in PP fibers may affect: process security (degraded PP leads to fiber breaks), productivity (damaged PP allows smaller extrusion speed), stability of the finished fibers (the lower the pre-degraded state of the PP, the more the long term stability is), dosage rate (fibers with high amounts of critical HAS might not be processable) [36].

A new stabilizing system was proposed, i.e., a blend of HAS (as long-term thermal stabilizer), a dialkylhydroxylamine (nonphenolic radical scavenger), and/or a phosphite (for preventing chain branching) [37]. The relative ratios can be adapted to optimize the performance of the stabilizing system to fit the particular needs of the final fiber applications [38].

Besides light stabilizers and pigments, other factors (for example polymer quality, sample preparation, thermal treatments and latexing, intrinsic PP stability, chemical nature, and amount of additives and spin finishes, climatic zones of exposure, etc.) do influence PP fiber stability.

For example, the reason for the negative influence of various pigments on the PP light stability is not fully understood yet, but there seems to be a relationship between light fastness and light stability. Although many pigments are neutral in this respect, there are some (e.g., the CI Pigment Reds 48 and 57) that impair the effect of the stabilizer or, in some cases, actually promote its effect (such as CI Pigment Red 214, CI Pigment Black 7) [39, 40].

Various nitrogen-containing chromophores (azo, nitroso and aromatic amino compounds) decompose to free radicals when excited by UV light, hence causing initiation of the degradation process. In contrast, other pigments may block some of the damaging rays, resulting in the shown degradation of PO polymers [41]. Even efficient light stabilizers (e.g., HAS) cannot always compensate for the negative influence of the pigments [33].

If the interference of pigments with light stabilizers has not been known before, it is advisable to use a HAS–benzotriazole-type UV absorber (0.1–0.2%) combination, which protects both the PP polymer and the pigment against fading.

B. Flame Retardance

The technologies used in flame-retardant (FR) textile materials evolved in parallels with the introduction of new types of fibers and fabrics and of new flame retardants, to various polymers as well as to the end products made from them. For PO fiber forming polymers, both flame - retardant (FR) finishes, applied to the end product, e.g., woven, knitted, or nonwoven fabrics, as well as additives added to the polymer before or during fiber formation and before textile processing took place, are used. The finishes and additives are usually extraneous materials and, due to their relatively high concentration in the textile material (10-30%), they may change its nature and behavior [42–44].

Flame retardation of polypropylene can be done by the application of chlorine and bromine derivatives with and without Sb_2O_3 as a synergist and phosphorus compounds as well as inert fibers. Bromine compounds are much more effective for PP than chlorine compounds and aliphatic compounds are more effective than aromatic ones.

As a rule, the FR agents are added to the melt before spinning. The processing temperatures of PP are high (250–300°C) and any FR agent added should remain stable in this temperature range. Aromatic bromine compounds such as decabromodiphenyl oxide (DBDPO), which decomposes at 340°C, and alicyclic chlorine derivatives such as Dechlorane Plus, which decomposes above 350°C, fulfill this condition and can be used in conjunction with Sb_2O_3 as additives. The quantities are, however, high and change the physical properties markedly.

A typical composition that was stated to be suitable for application on PP fibers and films consists of 2% Sb_2O_3 and 4% highly effective bromine-containing alicyclic compound. This composition would increase the LOI of PP from 17.4 to at least 25.2 which is adequate for self extinguishing behavior [43].

Red phosphorus in amounts up to 8% from the weight of the PP polymer was found to be effective [45, 46]. In combination with bromine compounds, it was found as capable of replacing Sb_2O_3 [45, 47] in a typical formulation [45]. 1.38% red phosphorous along with 5% 1-(2,4,5-tribromophenyl)-1-(2,4,5-tribromophenoxy)-ethane (TBPE) were applied to PP. Formulations including red phosphorus and polyacrylonitrile or melamine were found to be highly effective [48]. The mechanism in this case is believed to be based on the effect of the P_2O_5 produced by the oxidation of the red phosphorus on the surface of the burning P-containing acrylonitrile additive. The P_2O_5 enhances the dehydration of the acrylonitrile and thus an insulating char layer is formed on the surface.

When PP fibers are used in carpets, they are FR treated with PVC or chlorinated paraffins together with Sb_2O_3 [49]. This additive is added to the melt and, in order to obtain optimal distribution, a preliminary mix extrusion is performed. The melt viscosity decreases; therefore, the temperature of the extrusion can be decreased to below 260°C. Only a slight matting effect is said to be noticed on the fibers and the mechanical properties are allegedly practically unchanged [50].

The amounts of FR agent needed are determined by the fineness of the fibers and the nature of the textile structure, and range from 1.5–6.2%. The flame retardant was found to be compatible with UV stabilizers and pigments usually included in PP fibers [50–52]. Flame retardance can be obtained also by incorporation in PO spun melts: halogen-containing products (5-15%), antimony oxide (2–8%), and a grafted PP copolymer with halogenated monomeric units or halogenated bisphenol derivatives as grafted chains [53–56].

Halogen-containing flame retardants have a significant negative influence on PP fibers stabilized with HAS [57]. The possible solutions include development of new pigments and halogen-free retardants.

PP filaments for knitting were treated with flame retardants (tetrabrom-dipentaerythritol and antimony oxide). The addition had a positive effect of flame stabilization, while the physicomechanical properties ranged within normal limits [58].

C. Other Additives

Fluoropolymer-based additives are very useful, offering low temperature processability, chemical resistance, weatherability, self-extinguishing, and flexibility. Available as powders, pellets, dispersions or master batch concentrate (at additions of 250–1000 ppm), they are designed for use in PO film extrusion to eliminate melt fracture or problems arising from die build-up when running PO polymer with fillers and pigments [17, 59].

Compounds containing clarifying agents (e.g., pimelic acid derivatives, dibenzylidene sorbitol, etc.) can be processed on sheet and film extrusion lines up to 270°C but, due to the rapid crystallization rate, they will process differently from nonnucleated PP [60]. It is recommended that normal chill-roll temperatures should be used. Warmer cooling surfaces with faster resin extrusion speeds will give a clearer more glossy stiffer film, as well as a higher output [17, 61, 62].

Modern fillers can take on many of the functions of classical reinforcements. There is no standard for judging a filler against a reinforcement. As well as fibers, silicates in layers with lamina structure are counted as reinforcements. Ball-glass-type additives are counted as fillers.

Fiber-type reinforcements such as glass [63], rubber [64], and asbestos (and, recently, calcium silicates or calcium carbonate [65]) show high rigidity and stability, associated with stiffness. The polymer matrix must transfer the load to the fibers and, if the stress is applied in the same direction as the main orientation of the fiber, the capacity of reinforcement is maximized [66, 67].

The additives of the PO fibers affect their susceptibility to biological deterioration. Antioxidants and UV stabilizers retard the changes often necessary for microbial growth initiation. In the case of certain metallic catalysts, an active resistance to microbial attack is observed. The lubricants used during the weaving process of PP (e.g., vegetable oils) promote the development of fungal colonies which, although not affecting the tensile strength of the fabric, may cause extensive staining.

A antimicrobial treatment is useful for some applications of PP fibers: clothing for work and sport wears, floor coverings, geotextiles, fishing nets, and ropes.

The lubricants used in the spinning of PO are highly exposed to microbial and fungal attack with negative effects.

PP fibers can be antimicrobially finished by adding a master batch of the sanitized products to the other adequate additives such as: UV and light stabilizers, pigments, antistatic and flame retardant agents, etc. [68].

Chemical types of preservatives or biocides required in PO production are reflected in Table 1.

In general, polymeric materials are engineered to resist environmental degradation. But with increased social awareness of the large amounts of plastics entering the waste stream, there is a renewed interest in

TABLE 1 Chemical types of biocides for PO fiber products

Inorganic compounds
Salts of mercury, copper, zinc, chrome
Organic compounds
Phenolics (phenol and homologues, chlorophenols, chlorocresols, nitrophenols)
Nitrogen-containing compounds (quaternary ammonium salts, triazine and derivatives, oxazolidines, imidazolines)
Sulphur-containing compounds (chlorophenylsulphides, chlorodimethylsulphones)

Source: Refs. 69, 70.

producing highly degradable polymers. The addition to the process of small amounts of peroxides is an effective method of modifying the PP rheology, improving its processability and degree of recyclability [71].

With appropriate selection of additives (stabilizers or sensitizers) one can influence the end uses, and the lifetime of the products (from disposable articles, such as diapers, to long life geotextiles). Traditional polymers can be modified to be sensitive to the weathering process which can include one or any combination of actions by living organisms, light, heat, oxygen, water, ionizing radiation, or chemical reagents.

Photosensitive additives to initiate or change the PO degradation include: metallic organic and inorganic compounds containing iron, cobalt, copper, and manganese, monomers containing functional groups which absorb UV radiations (carbonyl and ketone pendant groups), and biodegradable additives (e.g. corn starch) [41].

The introduction, before spinning, of hydrophobic (e.g., polysiloxane) or hydrophilic agents (alkoxylated fatty amines, primary fatty acid amines) confer to the PO fibers and to the articles made of them the desired properties [72–76].

High performance PO fibers are produced by including in the dilute solution of an ultrahigh molecular weight PE or PP, some polymeric additives such as: lower molecular weight polyolefins, oxidized polyolefins, olefin copolymers, polyolefin graft copolymers, and polyoxymethylenes. The PO fibers so prepared have melting points above 140°C and exhibit improved adherence to matrices and resistance to fibrillation [77, 78].

Melt-dyeing methods (spin-dyeing, pigmentation, mass-dyeing, dope-dyeing) are the predominant procedures for PP fibers at present, along with many variants, such as dry coloring (blending of dry pigments with the polymer pellets before extrusion), melt dispersion (compounding of pigment and polymer before the final extrusion), and concentrate pigmentation (dry blending or blending and compounding of master batches with grey polymer chips for final extrusion).

Pigmentation techniques are now becoming more sophisticated. So far, dyeing systems for PP have been very expensive; nevertheless modest successes have been attained with nickel and metallic dyes despite their considerable cost. At present, spin-dyeing remains the only reasonably priced method. When spun-dyed fibers are used, the colors do not fade. Sometimes two to four colors are obtained in each yarn, creating a subtle effect in upholstery fabrics or carpets. Today, spin-dyeing fulfills almost all coloring requirements with respect to both color range and fastness properties.

With an appropriate selection of pigments, textile fastness properties, such as those relating to laundering, perspiration, rubbing, and light/weather, are of the highest level.

Pigments used for PO fibers must have some technical characteristics, such as: efficient thermal stability; light fastness adequate for the intended uses of the finished articles; no migration (blooming nor contact bleed); compatibility with other additives (stabilizers, etc.) and lack of photodestructive effect on polymer; and lack of negative influence on the fiber's mechanical properties.

The pigments exhibit a fixed particle size, however, and a tendency to form agglomerations. For finetitered PP fibers, there is a limit in the production of deep shades and, at the same time, the compression life of the spinning filters is shortened. The smaller the amount of pigment to be incorporated into the total melt, the more difficult it is to disperse the pigments homogeneously. As a rule, the minimum amount should not be below 2%. There have been used mostly organic pigments because of their light and bright shades and good overall fastness and less anorganic pigments. Anorganic pigments must have particle sizes smaller than 1 μm, in order to obtain colors and required fineness for PO [79].

During the last few years there have been interesting innovations in PO coloration, especially by EMS-Inventa AG, which can produce dye concentrates in up to 90 different colors shades [80].

The PP fibers, being nonpolar matrices, cannot bind polar dyes. The additives may be diluted, prior to spinning, by blending with the raw material and then extruded and pelleted. Manufacturers supply such blended pigments as master batches. The most well-known pigment preparations (with low molecular size PE binder as granules or powders that can be metered readily) are Lufilen and Luprofil (BASF AG, Germany), Bayplast (Bayer AG, Germany), Remafin and PV-Echt (Hoechst AG, Germany), Filofin (Ciba-Geigy AG, Switzerland), and Sanylene F, Sandorin, Graphtol, Sanduror (Sandoz AG, Switzerland) [81, 82].

Modern spinning plants with integrated metering systems facilitate economical production even on a small scale with rapid, accurate color changes. The requirements of shade uniformity differ, according to the end use of the spun fiber. Thus, continuous filaments call for the best possible color consistency, whereas slight unevenness in shade does not become apparent with staple fibers. The important factors in this respect are accuracy of metering, the mixing efficiency of the extruder itself and of the screw, and the ratios of additives to raw materials.

Good results have been obtained with metering extruders in which the additives are melted separately before being incorporated into the melt. To improve the mixing effect of the extruder, a mixing cam can be installed at the screw front section, as well as a static mixing element between the extruder exit and the spinning head, or several layers of coarse wire filters with different mesh arrangements. The 3DD mixer has been designed and developed for the precise and reproductible mixing of different polymers, adding colors, antistatics, stabilizers, water repellants, antibacterial additives, and others. The size of the mixer matches the throughput of the main extruder. The mixer is available either as an extension to the main extruder or as a separately driven mixing component [83].

Spin dyeing can be carried out with the following methods: addition of master batch granulates to the naturally white spinning quantity, admixture of the granulates, addition of dyestuff powder in the extruder entrance, injection of the master batch melt into the last third of the compression zone and/or into the entrance of the metering zone or before 3DD mixer [84].

Melt dyeing with pigments results in excellent fastness and has advantages over bath dyeing with regard to costs, energy consumption, and pollution.

The pigments used for coloration of PP fibers are hot soluble in the polymer but they are in a crystalline form. Being a foreign substance in the polymer matrix, they can act as a kind of filler, having influences on the physical properties of the fibers.

Polyolefins in the melting state have no aggressive action against pigments.

Also, dye association with PO-specific additives can enhance tinctorial capacity of PO fibers [85].

Some of the commercial HAS available today cause, in some instances, a reduction in color yield, especially with organic pigments. For improving coloring fastness of textile articles, PO incorporated in a melt of blue or green phtalocyanine pigments or red azo or disazo ones in combination with two primary stabilizers, substantially free of secondary stabillizers. For example, Chimassorb 119 results in the best hot-light stability for the investigated samples [86]. It is well established that pigmentation (e.g., type and amount of pigment) can have a marked effect on the resulting light stability of PP fibers [87].

Special results were obtained by Holliday Pigments Ltd. (England) with ultramarine blue pigment (Premier F), which allows dyeing with 3% concentration of 4 dtex PP filaments [88]. The physicomechanical filaments' characteristics are comparable with those obtained with TiO_2 and with other organic pigments. PE and PP waxes are used as carriers for pigments, but it has been difficult to ensure complete avoidance of agglomerates, with consequent blocking of melt screens (as used in extrusion spinning of fibers). There are new dispersions developed by Hüls (Vestowax P 930 V, Tegowax P 121) that improve this situation in fiber spinning. It also produces increased color strength in injection moulding and extrusion, allowing reduction in the pigment concentration [89]. Spun-dyed products have a high staining fastness.

In spite of intensive work, the problem of obtaining a simple dyeable PO fiber comparable to the other synthetic fibers remains unsolved. Economical and ecological problems, as well as increased requirements concerning the stability of dyeing and brightness of shades, will maintain and increase the industry's interest in the melt-dyeing methods of PO fibers.

III. SPIN FINISHES FOR PO FIBER PROCESSING

The generation of static electricity on fibers leads to a variety of problems both during manufacturing and consumer use. For instance, in manufacturing, static generation may cause difficulties in opening, extruding, or carding fibers, in positioning layers of textile materials, and in keeping fabrics lint- and soil-free during processing.

Among all common synthetic fibers, PP exhibits the highest friction. The use of various internal antistatics, which are mixed with the PP powder before extrusion (0.1–1%), have not yet offered the best solution, because of the incompatibility between the nonpolar PP and the mostly strong polar antistatic and also because of the lack of thermal stability in the melt. The application of external antistatic agents (0.5–2%) after fibers' extrusion of the fibers is a most convenient way.

The task of lubricating fine PP fibers is complicated furthermore by the fact that the classic best performing lubricants on the other major fibers, like low viscosity mineral oils and fatty esters, cannot be used. These types of products wet PP very well, but tend to migrate into it and swell it in an unacceptable way because of their chemical similarity with the PP molecule.

In the production of synthetic filaments it has been customary for years to use auxiliary agents known as spin finishes which, applied in a ratio of 0.2–1.0%, confer gliding, antistatic, and hence lubricating properties onto the synthetic filaments [90–95]. Spin finishes are usually mixtures of antistatic agents (Table 2), lubricants (e.g., mineral oils) and emulsifiers. Most of the known antistatic agents that function by improving the electric conductivity of textiles are hygroscopic compounds that contribute to conductivity by ensuring the presence of water and disolved ions.

The antistatic agents recommended for PO fibers are quaternary ammonium salts of fatty acids, alkanolamines, alkanolamides, polyglycolesters, alkoxylated triglicerides, and ethoxylated tertiary amines (Table 2). The quaternary ammonium compounds and nonionic surfactants dominate this field because of their higher oil solubility and hygroscopicity.

Quaternary ammonium derivatives of fatty compounds are probably the most widely used, nondurable antistats for textile materials because of their combination of very high antistatic effectiveness and softening action on textiles. Quaternary ammonium compounds usually impart very good static protection at levels as low as 0.25% of the weight of the fabric.

In some applications, the low water solubility of these fatty quaternary ammonium salts is a problem. The substitution of poly(oxyethylene)groups for one or more of the alkyl groups present on the nitrogen atom improves their solubility. Also, perquaternized alkylene-diamine derivatives are reported to offer improved water solubility.

During processing, the influence of the spin finishes is mainly observed on the frictional and antistatic behavior. The ability of a spin finish to spread quickly and evenly over the surface of a textile filament or fiber is becoming increasingly recognized as one of the key requirements of high performance products.

Enhanced surface coverage (ESC) spin finish systems are the new formula applied to PP fibers [101]. The incorporation of wetting agents in a spin finish formulation allows aqueous diluted finish in solutions, emulsions or dispersion form to wet evenly on an untreated fiber surface.

The function of an ESC spin finish is to provide a higher degree of uniform surface spreading of the spin finish in the water-free state applied either neat or from aqueous solution. The resulting benefit of ESC translates into subsequent improved performance of the key functional parameters of the applied spin finish (improved friction, lubrication and electrostatic properties of the finished fiber). Typical application levels of ESC spin finishes on polypropylene fibers will be in the region of 0.2–0.5%, depending upon the processing route and end use—around half the level used typically for traditional finishes. The quantity of spin finish distributed along the surface of the fibers will be of the order of molecular thickness.

The distribution of the ESC spin finish indicates a more even coverage compared with the broken and discontinuous bands of the standard finish. The practical effects and benefits to be seen in fiber, yarn, and fabric processing obtained by employing the ESC technology finishes are presented in the following.

In *worsted spinning*, the spin finishes fulfill the following targets [102, 103]:

even drafting during all drawing stages, influencing sliver and yarn evenness;
sliver cohesion and correct can deposit;
good antistatic properties;
elimination of sliver breakage and loop formation;
elimination of thermic damages, in particular at the ring traveller;
high yarn strength.

Surface coating greatly influences subsequent mechanical processing of PO staple fibers. The purpose of PO

TABLE 2 Properties of antistatic finishes

Product	Concentration, %	Efficiency	Compatibility with other additives
Quaternary compounds	0.5–1.5	Very good	Good
Amines of ethoxylated fatty acids	0.5–2.0	From good to very good	Good up to 1.0%
Alkylamides of fatty acids	1.0–2.0	From moderate to good	Good up to 1.0%
Special alkylamines	1.0–3.0	From moderate to good	Good up to 0.5%
Polyetherglycols	2.0–3.0	From weak to moderate	Good up to 2.0%

Source: Refs. 96–100.

coating is to create more favorable conditions for mechanical carding and spinning. As a result of this process, the fiber breakage rate in carding is reduced, as well as static electricity formation and fluff formation, while the amount of wastes produced in carding and spinning is considerably reduced. At the same time, cohesion, flexibility, and elasticity of fibers are improved.

The improved frictional and antistatic properties provided by ESC finishes have allowed benefits to be gained in the production and processing of staple fibers for both ring and Dref spinning. Better coverage has resulted in lower fiber/metal friction even at much reduced finish application rates. Advantages found in practice include:

spin finish usage at 0.3–0.4% (approximatively 50% of normal);
higher spinning speeds on ring frames;
better yarn uniformity;
better static protection provided without addition of ionic agents.

Nonwoven processing lubricants have the following requirements: easy fiber separation, good web formation and coherence, good antistatic properties, and fiber damage reduction due to low fiber-to-metal friction.

Improved evenness of finish coverage and better friction performance have given benefits in several nonwoven applications:

excellent surface coverage by a hydrophilic finish has provided improved re-wetting of fabrics in end-use applications;
the ability of organic-based finishes to spread well has allowed the replacement of some environmentally questionable finishes where such impact is important;
reduced finish application level, whilst maintaining the necessary finish properties, has led to fewer application and downstream problems.

Before high-speed winding, filaments have been lubricated with the aim of improving their antistatic properties and soiling behavior, or of reducing the number of end breaks during texturing [for example, PEG (polyethyleneglycols) as additives up to 5%]. Special types of spin finishes that can be mixed into the water are applied for air jet texturing yarns [104, 105].

For the *filament sector*, conventional finishes consist traditionally of mineral oil and emulsifier mixtures (e.g., 70–50% mineral oil; 30–50% emulsifiers). During heating and setting processes, these spin finishes make the control of cracking residues interfering with other chemical agents or representing air pollutants difficult.

The increased speed requirements of modern textile processing methods led to modern finishes based on water-soluble or self emulsifying mixtures of polyalkylene glycols and/or ester or ether derivatives. The introduction of polyalkylene glycols in spin finishes has several advantages, such as good thermohydrolysis behavior, substantially lower cracking residues in heating and setting processes, and tailoring of the products to the filaments, in terms of viscosity and molecular weight [92].

The increased affinity between yarn and finish arising from the ESC technology, along with the better frictional properties, has allowed improved spinning and processing of POY yarns for air jet texturing. The practical advantages demonstrated to date include:

reduced sling during application of neat finishing even at high finish levels;
lower required spin finish level;
improved texturing performance;
better woven fabric uniformity.

Phosphates are essential components of many spin finishes, softening agents, etc. They have good heat and chemical stability, and they impart special friction and antistatic properties to the fiber. Their concentrations in the finishing bath before drawing range between 0.7-0.8% [104]. The adhesion properties of PP filaments are influenced, too, by the different chemical composition of the phosphates. For textile processing of *PO staple fibers*, the finer the denier is, the higher become the demands of the spin finishes. In general, hydrophilic, essentially water-soluble, lubricants must be selected as core components in spin finishes for PP multifilaments and staple fibers.

Typical examples are fatty acid polyglycol esters and their homologs, alkoxylated long chain alcohols, alkoxylated triglycerides, and the salts of phosphated alcohols or alkoxylated alcohols [103]. Spin finishes for low dtex PP sample fibers tend to exhibit low emulsion viscosities, medium static and dynamic surface tensions, while overall low surface tensions are more desirable for coarse fibers [105–107].

Table 3 lists some special additives for PO fiber processing.

For the hygienic sector, 1.5–2.5 denier PP staple fibers with a yarn oil level of 0.6% are used. For PP polymers, gas-fading free spin finishes (e.g., Silastol GFE; DG; GF 18) are necessary. For highly stabilized

TABLE 3 Special additives for PO fiber processing

Tradename	Company	Remarks
ABM 101, 107	Riverdale Color, Brooklyn, NY 11205	Nucleating agents for PO
AST 141, 142	Riverdale Color, Brooklyn, NY 11205	Antistatic agents for PO
Alfetal	Zschimmer u. Schwarz Chem. Fabrik, D-5420 Lahnstein, Germany	Emulsifier for melt spinning (0.1–0.2%)
Amgard NL	Albright & Wilson Co., PO Box 48, Aurora, NC 27806, USA	Flame retardant (20–30%)
Antelim PM	Zschimmer u. Schwarz Chem. Fabrik, D-5420 Lahnstein, Germany	Spin finish for PP monofilaments (0.5–1.0%)
BK 2011, 2035, 2083	Henkel KgaA, D-4000 Dusseldorf-13, Germany	Spin finish for PP staple fiber (0.4–6%)
Chimassorb 119, 944	Ciba-Geigy Ltd., Basel, Switzerland	Light stabilizers (HAS) for PO fibers
Chimassorb 81	Ciba-Geigy Ltd., Basel, Switzerland	UV absorber for PO fibers
Delion	Takemoto Oil and Fat Co. Ltd., Minatomachi, Gamogori-city, Japan	Spin finish oil for PP filament and staple fibers (1.0–1.5%)
Duron	Duron Ontario Ltd., Mississaga, Ontario, Canada	Spin finishes with low fogging effects
Dynamar PPA, FX 5920 A, TM	3M, St. Paul, USA	Fluoropolymer-based additive as processing aid
Dynasylan Silfin 14, 51	Hüls AG, D-45764 Marl, PO Box 1320, Germany	Blend stabilizer for PO composite
Dyneon LLC	Hoechst AG, D-6230 Frankfurt am Main, Germany	Fluoropolymer-based additive as processing aid
Elactiv	Zschimmer u. Schwarz Chem. Fabrik, D-5320 Lagnstein, Germany	Antistatic agent for melt spinning (0.5–1.0%)
Esterol BA	Chem. Fabrik Stockhausen GmbH, D-4150 Krefeld-1, Germany	Spin finish for PP filaments (1.0–1.5%)
Fasavin AF/TL	Zschimmer u. Schwarz Chem. Fabrik, D-5420 Lahnstein, Germany	Spin finish for BCF PP yarns (1.0–1.5%)
Fasavin DE/PE	Zzchimmer u. Schwarz Chem. Fabrik, D-5420 Lahnstein, Germany	Spin finish for POY filaments (1.0–1.5%)
Fasavin RC/TL	Zschimmer u. Schwarz Chem. Fabrik, D-5420 Lahnstein, Germany	Spin finish for technical PP Filaments (2.0%)
Favorol SF	Chem. Fabrik Stockhausen GmbH, D-4150 Krefeld-1, Germany	Spin finish for PP staple fibers (0.4–0.6%)
Flerolan	Zschimmer u. Schwarz Chem. Fabrik, D-5420 Lahnstein, Germany	Bactericide for melt spinning (0.1–0.5%)
FS	Ciba- Geigy Ltd., Basel, Switzerland	Antioxidants acting synergistically with HAS, UV stabilizers for PP films and fibers
Genapol	Hoechst Celanese Co., Bridgewater, USA	Polyethyleneglycolethers as spin finishes for PO fibers
Hostalub	Hoechst AG, D-6230 Frankfurt am Main, Germany	Slip and antistatic agent for PP fibers
Hostanox	Hoechst AG, D-6230 Frankfurt am Main, Germany	Antioxidant for PP fibers
Hostatat	Hoechst AG, D-6230 Frankfurt am Main, Germany	Antistatic for PP fibers
Hostavin	Hoechst AG, D-6230 Frankfurt am Main, Germany	Light stabilizers for melt spinning
Irganox 1010, 3114	Ciba- Geigy Ltd., Basel, Switzerland	Antioxidants for PO tapes and fibers
Lamberti Lub K/78	Fratelli Lamberti SpA, 21041-Albizzate, Italy	Spin finish for continuous filament and PP staple fibers (0.8–1.1%)

TABLE 3 (Continued)

Tradename	Company	Remark
LAO 304	Riverdale Color, Brooklyn, NY, 11205	Antioxidant for PO
LBA 30, 38, 39	Riverdale Color, Brooklyn, NY, 11205	Liquid blowing Agents for PO
LST 251, 252, 253	Riverdale Color, Brooklyn, NY. 11205	Light stabilizers for PO
Lubestat	Milliken Chem., B-9000, Gent, Belgium	Staple fiber overspray
Lubricit	Zschimmer u. Schwarz Chem. Fabrik, D-5420 Lahnstein, Germany	Emulsifier for melt spinning
Luchem HA-R 1000	Elf Atochem, USA	Reactive HAS system for long-term stabilization of polyolefins
Millard 3988	Milliken Chem., Gent, B-9000, Belgium	Clarifying agent for PP fibers
Morbisol SC	Fratelli Lamberti SpA, 21041-Albizzate, Italy	Spin finish for PP staple fibers (0.5–0.8%)
Morbisol KV, NF	Fratelli Lamberti SpA, 21041-Albizzate, Italy	Spin finish for PP multifilament yarn (0.8–1.2%)
Polyfix	Schill u. Seilacher, 22113 Hamburg, Germany	Durable antistatic agent for PP fibers
Sanduvor PR-31	Clariant Huningue SA, Basel, Switzerland	HAS for stabilization of pigmented and unpigmented PO matrix polymer
Sinparol PF, VF	Chem. Fabrik Stockhausen GmbH, D-4150 Krefeld-1, Germany	Spin finish for PP filaments (1.0–1.5%)
Santized MBP	Santized AG, Burgdorf, Switzerland	PO antimicrobial finish
SL 36, 18	Riverdale Color, Brookly, NY. 11205	PO fiber lubricants
Stantex 56051, FDA, BGA	Henkel KgaA D-4000 Düsseldorf, Germany	PO spin finish (1.0%)
Syn Fac	Milliken Chem., Gent, B-9000, Belgium	Surfactants for PO fibers
Syn Lube	Milliken Chem., Gent, B-9000, Belgium	PO fiber lubricants
Syn Stat	Milliken Chem., Gent, B-9000, Belgium	Antistats for PO fibers
Synthesin	Dr. Th. Boehme, Chem. Fabrik GmbH, D-8192 Geretsried-1, Germany	Spin finish for PP staple fibers (sanitary and carpet type) (0.4–0.8%) and for PP BCF yarns (0.8–1.2%)
Tallpol	Chem. Fabrik, Stockhausen GmbH, D-4150 Krefeld-1, Germany	Softening agent for PP staple fibers
Tinuvin 326, 327, 328	Ciba-Geigy Ltd., Basel, Switzerland	UV absorber for melt spinning of PE (0.2–0.4%) or PP (0.3–0.6%)
Tinuvin 622, 765, 783, 791	Ciba-Geigy Ltd., Basel, Switzerland	Light stabilizers in PP films and fibers
Ultranox 626	GE Specialty Chem., Parkersburg, USA	Antioxidant for melt spinning (0.1%)
Univul Products	BASF, D-67056 Ludwigshafen am Rhein, Germany	Light stabilizers (HAS) and UV absorbers for PP and HDPE films and fibers
Uvitex OB	Ciba-Geigy Ltd., Basle, Switzerland	Optical brightener for PO powder and granulates or for blow films (0.0005–0.001%)
Westomax P930	Hüls AG, D-45764 Marl, PO Box 1320, Germany	Pigment carriers for PO fibers and filaments
Weston	GE Specialty Chem., Parkesburg, USA	Liquid phosphite stabilizer

Source: Refs. 17, 95–110.

PP polymers, spin finishes will be applied with Polyfix RNEF, Silastol F 36, or Silastol MPF [108].

In the case of 6.0–30 denier PP staple fibers for needle felts, the following requirements have to be met for spin and end finishes: adjusted fiber-to-fiber friction for drawing; good antistatic properties; protection against mechanical damage with needlepunch processes; one-way spin finish, i.e., the same spin finish for all positions.

In this last case, the following spin finish systems are employed: from spinning to the fiber line, one-way finishes, such as Silastol 390 or 308 and Limanol LB 24 or LB 25; in the needlefelt plant, an additional overspray in the mixing chamber to guarantee optimal antistatic protection against mechanical damage at the needlepunch process. The oversprays employed are Limanol G 112, N 43 or N 46.

These higher PP staple fiber deniers require yarn oil levels of 0.6–1.0%.

The conventional processing auxiliaries based on a mixture of fatty acid polyglycolesters, mineral oils, alcylbenzol, and antistatics show undesired fogging behavior. Usually, components of low fogging processing aids are polyethyleneglycol and polyethylene–polypropyleneglycol (e.g., products under the trade name of Duron) [97].

A method for assuring fluid permeability to PP nonwovens consists of applying a fluid-permeable mixture to the PO spun fiber at a rate of 0.1–0.5%. This mixture contains 0–45% surfactant and 55–100% polyoxyalkylene modified silicone [109]. Another solution utilizes 70–95% aliphatic diethanol amide and 30–5% surfactants, including one or more, selected from the following: nonionic surfactants, alkyl phosphate salts, quaternary ammonium salts, alkylimidazolinium salts [110].

IV. REQUIREMENTS FOR PO FIBER ADDITIVES

The special conditions of manufacture and use of PP fibers have a marked impact on the ancillary properties required for stabilizers. High processing temperatures (240–300°C) call for low volatility and high thermal stability of the additives. Subsequent treatments of the fibers, such as crimping, setting and latexing, necessitate, too, low volatility and, in addition, good extraction resistance (also required for textile goods, if considering particularly washing and dry cleaning).

Another requirement specific to PP fiber additives is the gas fading or gas yellowing free effect, i.e., discoloration of fibers by industrial waste gases.

For PP fibers stabilization, specific requirements were identified. The fibers' high surface to volume ratio required stabilizers of very low volatility. For example, inferior light stabilizers show a delayed interaction in the gas-heated drying channel, which results in a red and yellow color of the fibers. Applications such as filters, geotextiles, or active wears necessitate high extraction resistant stabilizers. In colored carpet fibers and hygienic articles (diaper covers), color and gas fading resistance are of critical importance. For combinations of a high molecular sterically hindered phenol and a hydrosterically stable phosphite, discoloration is the main problem which is related to the phenolic nature of the antioxidants [111, 112].

With all basic additives (antistatic agents, HAS, dihydrotalcites, etc.) significant gas fading can be observed, which can be explained either by a color shift in the alkaline environment [113] or by the oxidation of the phenolic antioxidants catalyzed by alkalies [114].

The above considerations suggest that phenol-free stabilizers should be introduced to exclude gas fading in specific applications.

The spin finish properties (wetting time, special adhesive content) must be correlated with the customer's machinery availability to guarantee optimum application. The preparations for softening and antistatic protection of the PP staple fibers must be compatible with the stabilizers, and should be preferably liquid and have suitable fogging and ecological behavior [115]. The most important properties of nondurable antistats include low volatility, low toxicity, heat resistance, oil solubility, nonyellowing characteristics, and low flammability; they should also be noncorrosive, especially if the treated material will come into contact with metallic processing equipment.

A standard requirement for automotive textiles (fabrics, carpets, nonwovens) is the low fogging effect, i.e., the condensation of volatile components of the spin-finished treated material inside the car windows (the so-called window fogging effect) [97].

Pigmentation interferes somewhat with tenacity but, with reasonably low amounts of pigment, good physical properties of the fibers are attainable.

Pigments should not have negative influences, but they must have a positive influence on mechanical properties of fibers, especially after light exposure.

Pigment selection is not justified only through esthetic reasons, but also through their influence on the lifetime of finishing articles.

ABBREVIATIONS

CR-PP	controlled rheology polypropylene
DBDPO	decabromodiphenyl oxide
ESC	enhanced surface coverage
FR	flame retardant
HAS	hindered amine light stabilizer
HDPE	high density polyethylene
LLDPE	low linear density polyethylene
PE	polyethylene
PEG	polyethyleneglycol
PO	polyolefin
PP	polypropylene
PVC	polyvinylchloride
TBPE	tribromophenyloxyethane
UV	ultraviolet

REFERENCES

1. H Zimmermann. Text Horiz 6: 24–28, 1981.
2. H Herlinger, R Gutmann, J Spindler. Man Made Fiber Year Book 1989; (suppl): 42–47.
3. JC Leininger. (to Phillips Petroleum Co.) US Patent 4,965,301 (1990).
4. C Mendonca. Study of the effect of various anti-oxidant systems in spinning of PP fibres. Proc World Textile Congress on Polypropylene in Textiles, Huddersfield, 1996, pp 183–192.
5. AR Horrocks. Polypropylene geotextiles. Proc World Textile Congress on Polypropylene in Textiles, Huddersfield, 1996, pp 96–108.
6. F Gugumus. Polym Degrad Stab 44: 273–297, 1994.
7. JP Galbo, R Seltzer, R Ravichandran, AR Patel. (to Ciba-Geigy Co) US Patent 5,096,950 (1992).
8. SL Gessner. (Fiberweb North America Inc.) US Patent 5,108,827 (1992).
9. RV Todesco, R Diemunsch, T Franz. New developments in the stabilization of polypropylene fibers for automotive application. Paper presented at 32nd Int Man-Made Fibers Congress, 17th Intercarpet, Dornbirn, 1993.
10. RL Hudson. (to Kimberly-Clark Co.) European Patent 0698137 (1996).
11. D A Gordon. Adv Chem Ser 85: 22–28, 1968.
12. H Gysling. Adv Chem Ser 85: 239–245, 1968.
13. G Scott. Developments in Polymer Stabilization, vol 8, London: Elsevier, 1987, pp 239–289.
14. G Phahler, K Loetzsch. Kunststoffe 78: 142–149, 1988.
15. P Hudec, L Obdrzalek. Angew Makromol Chem 89: 41–49, 1980.
16. G Kletecka. (to BF Goodrich Co.) US Patent 5,190,710 (1991).
17. J Murphy. Additives for Plastics Handbook. Oxford: Elsevier, 1996, pp 121–165.
18. G Reinert. (to Ciba- Geigy Co.) US Patent 5,057,562 (1991).
19. A Steinlin, W Saar. Melliand Textil 61: 94–102, 1980.
20. F Gugumus. In: R Gächter, H Mueller, eds. Plastic Additives, 3rd edn. Munich: Hanser, 1990, pp 1–104.
21. F Gugumus. Angew Makromol Chem 176/177: 241–252, 1990.
22. F Gugumus. Polym Degrad Stab 40: 167–176, 1993.
23. DW Horse. GB Patent 2,292,944 (1996).
24. Ciba-Geigy AG. Research Disclosure 345: 32–34, 1993.
25. WP Enlow, RW Avokian. (to General Electric Co.) European Patent 0618314(1994).
26. WP Enlaw. (to General Electric Co.), European Patent 0618315 (1994).
27. V Barsatto. GB Patent 2266531 (1993).
28. IY Kritko, J Azran. (to UV Stab Ltd.) US Patent 5,610,305 (1994).
29. T Ishii, S Yachigo. (to Sumimoto Chem. Co. Ltd). US Patent 5,246,777 (1993).
30. *** Research Disclosure 354: 28–35, 1993.
31. RS Nahr, JG Mac Donald. (to Kimberly-Clark Co.) US Patent 5,244,947 (1993).
32. JA Mahood. (to General Electric Co.) US Patent 5,500,467 (1996).
33. EJ Termine, RW Aatwell, HA Hodgen, NA Favstritsky. (to Great Lakes Chemical Co.) US Patent 5,380,802 (1995).
34. PJ Sheth. Wettable and dyeable polyolefin: technology & applications. Proc World Textile Congress on Polypropylene in Textiles, Huddersfield, 1996, pp 109–120.
35. R Gächter, H Müller. Plastic Additives Handbook. 3rd edn. Vienna: Hanser, 1990, pp 33–34.
36. J Koehler, J Bayer. Impact of some piperidyl derivatives on processability of polypropylene. Proc World Textile Congress on Polypropylene in Textiles, Huddersfield, 1996, pp 69–76.
37. A Dobcek, W Schmidt. PP circular-knit panneling textiles for the automotive interiors. Paper presented at the 34th Int Man-Made Fibers Congress, Dornbirn, 1995.
38. JR Pauquet, C Kröhnke, RV Todesco, J Zingg. Stabilization of polypropylene fibres: a new era has begun. Proc World Textile Congress on Polypropylene in Textiles, Huddersfield, 1996, pp 171–182.
39. C Ripke. Pigments and their influence on PP-fibre production and quality. Proc World Textile Congress on Polypropylene in Textiles, Huddersfield, 1996, pp 193–209.
40. F Gugumus. In: P Klemchuk, J Pospisil, eds. Inhibition of Oxidation Processes in Organic Materials, vol 2. Boca Raton, FL: CRC Press, 1989, pp 29–162.
41. KC Leonas. Text Chem Col 24(10): 13–23, 1992.

*** Note: These references have no authors.

42. M Lewin, S Sello. Handbook of Fiber Science and Technology, vol II Functional Finishes. New York: Marcel Dekker, 1983, pp 75–78; 121–122.
43. J Green. In: M Lewin, SM Atlas, EM Pearce, eds. Flame Retardant Polymeric Materials. New York: Plenum Press, 1982, pp 37–142.
44. HD Metzemacher, R Seeling. (to Lonza Ltd.) US Patent 5,139,875 (1992).
45. GN Davis, HWC Haynes. (to West Point Peperell) US Patent 5,102,701 (1992).
46. M Branchesi, L Spagnoli, G Braca. (Montell North America Inc.) US Patent 5,529,845 (1996).
47. P Barnes. J Vinyl Additive Tech 3(1): 70–77, 1997.
48. D Dickmann, W Nyberg, D Lopez, P Barnes. J Vinyl Additive Tech 2(1): 57–63, 1996.
49. SK Dey, P Natarajan, M Xanthos, MD Braathen. J Vinyl Additive Tech 2(4): 339–345, 1996.
50. KW Van Every, MJ Elder. J Vinyl Additive Tech 2(3): 224–229, 1996.
51. *** Text Inst Ind 4: 124–132, 1981.
52. K Edi. TROL-temperature resistant olefin. Proc World Textile Congress on Polypropylene in Textiles, Huddersfield, 1996, pp 43–44.
53. BL Cline. (Techlon Fibers Co.) US Patent 4,774,044 (1988).
54. Y Nakajima, M Taniguchi. (to Chisso Co.) US Patent 5,618,623 (1997).
55. Y Nakajima, M Taniguchi. (to Chisso Co.) US Patent 5,567,517 (1996).
56. EJ Termine, RW Atwell, HA Hodgen, NA Favstritsky. (to Great Lakes Chemical Co.) US Patent 5,380,802 (1995).
57. RL Gray, RE Lee, BM Sanders. J Vinyl Additive Tech 2(1): 63–69, 1996.
58. Y Charit, M Shimoshani, L Utevsky, P Georgette. Chem Textil 42/94: 790–797, 1992.
59. DE Priester. J Vinyl Additive Tech 3(4): 305–309, 1997.
60. JX Li, WL Cheung. J. Vinyl Additive Tech 3(2): 151–157, 1997.
61. ACY Wong, WWM Leung. J Vinyl Additive Tech 3(1): 64–70, 1997.
62. P Pukanszky, I Mudra, P Staniek. J Vinyl Additive Tech 3(1): 53–58, 1997.
63. P Dave, D Chundury, G Baumer, L Overley. J Vinyl Additive Tech 2(3): 253–258, 1996.
64. C Wang, K Kox, GA Campbell. J Vinyl Additive Tech 2(2): 167–170, 1996.
65. R Gendron, D Binet. J Vinyl Additive Tech 4(1): 54–60, 1998.
66. SC Tiong, RKY Li. J Vinyl Additive Tech 3(1): 89–92, 1997.
67. J Spano, W Steen. J Vinyl Additive Tech 3(3): 237–242, 1997.
68. B Mebes. Tech Text 35(3): 134–139, 1992.
69. TA Oxley, S Barry. Biodeteoration. Chichester: John Wiley and Sons, 1983, pp 28–54.
70. A Guttag. US Patent 5,725,735 (1998).
71. H Wunsch, BJ Launchbury. Int Carpet Yearbook 1997; (suppl 1): 33–37.
72. A Schmalz. (to Fiberco Inc.). US Patent 5,721,048, 1998.
73. KR Gupta, JH Harrington. (to Hercules Inc.) US Patent 5,763,334 (1998).
74. D Urban, W Ladner, A Paul. (to BASF AG) US Patent 5,102,799 (1992).
75. JH Harrington. (to Hercules Inc.) US Patent 5,582,904 (1996).
76. RS Nahr, JG MacDonald. (to Kimberly - Clarke Co.) US Patent 5,567,372 (1994).
77. GA Harpell. (to Allied Co.) US Patent 4,455,273 (1984).
78. GA Harpell. (to Allied Co.) US Patent 4,584,347 (1986).
79. *** Chemical Fiber International 47: 98–102, 1997.
80. *** Man Made Fiber Year Book 1992; (suppl): 58–65.
81. AC Schmalz. (to Fiberco Inc.) US Patent 5,721,048 (1998).
82. *** Am Text Int 8: 30–38, 1989.
83. AK Samanta, DN Sharma, Indian J Fiber Text Res 20: 206–210, 1995.
84. F Fourné. Man Made Fiber Year Book 1994; (suppl): 26–32.
85. RV Todesco, R Dienmusch, F Frantz. Man Made Fiber Year Book 1994; (suppl): 83–92.
86. *** Man Made Fiber Year Book 1994; (suppl): 94–98.
87. F Steinlin, W Saar. Melliand Textil 61: 941–945, 1980.
88. *** Chem Fiber Int 46: 350–358, 1996.
89. V Schrenk, HC Porth. Westowax P 930- the novel pigment carrier for colouring polypropylene fibres and filaments. Proc World Textile Congress on Polypropylene in Textiles, Huddersfield, 1996, pp 140–152.
90. C Ripke, Chem Textil 30/82: 30–38, 1980.
91. R Kleber. Man Made Fiber Year Book 1988; (suppl): 76–82.
92. HJ Harrington. (to Hercules Inc.) US Patent 5,545,481 (1996).
93. HJ Harrington. (to Hercules Inc.) US Patent 5,540,953 (1996).
94. RD Neal, S Bagrodia, LC Trent, MA Pollock: (to Eastman Chem Co.) US Patent 5,677,058 (1997).
95. MS Williams. ICI surfactants. Paper presented at the 37th Int Man-Made Fibers Congress, Dornbirn, 1998.
96. WF Werhast. Kunststoffe 66: 701 (1976).
97. A Riethimayer. Gummi, Asbest, Kunststoffe 26: 76; 182; 298; 419; 506 (1973).
98. R Mathis. Properties and development of spin finishes for fine PP fibre for textile applications. Proc World Textile Congress on Polypropylene in Textiles, Huddersfield, 1996, pp 210–222.
99. A Speidel. Melliand Textil 62: 912–918, 1981.
100. CP Schobesberger. Tech Textil 37: T59–65, 1994.
101. R Niestagge. Year Book (Spinning/Twisting/Winding) 1989; (suppl): 29–38.

102. A Schulberger. Melliand Textil 67: 223–229, 1986.
103. M Acar. Man Made Fiber Year Book 1988; (suppl): 38–45.
104. TM Buck. Fine denier polyolefin fibres for apparel. Paper presented at the Int Man-Made Congress, Dornbirn, 1992.
105. G Barsotti. (to Mopletaan SpA) European Patent 0576896, 1994.
106. R Mathis. Quality assurance and environment protection as practiced by a fibre finish supplier. Paper presented at the Int Man-Made Fibers Congress, Dornbirn, 1994.
107. F Fourné. Chem Textil 43/95: 811–822, 1993.
108. G Mutschler. Chem Textil 37/89: 968–975, 1987.
109. T Kato, Y Takasu, M Minafuji. (to Takemoto Yushi Kabushiki Kaisha) US Patent 5,258,129 (1993).
110. T Kato, Y Takasu, M Minafuji. (to Takemoto Yushi Kabushiki Kaisha) US Patent 4,988,449 (1991).
111. P Klemchuk. Oxidation Inhibition in Organic Materials, vol I. New York: CRC Press, 1990, pp 18–36.
112. RV Todesco, R Diemunsch, T Franz. Tech Textil 36: E135–E142, 1993.
113. P Klemchuk, P Horng. Polym Degrad Stab 34: 333–346, 1991.
114. RS Macomber. J Org Chem 47: 2481–2487, 1982.
115. R Weitenhansl. Chem Textil 35/87: 664–669, 1985.

29

Technological Aspects of Additives Use for Thermoplastic and Elastomeric Polyolefins

Mihai Rusu
"Gh. Asachi" Technical University, Iasi, Romania

Gheorghe Ivan
Tofan Group, Bucharest, Romania

I. THERMOPLASTIC POLYOLEFINS

A. General Aspects

The main groups of additives for TPO composites, their meaning, and mode of action will be treated in Chapters 19 and 20. The aim of the present chapter is to present some practical aspects concerning the use of these various additives (antioxidants, metal de-activators, light stabilizers, antistatic agents, conductive agents, mold release agents, lubricants, antiblocking agents, crosslinking agents, nucleating agents, etc.).

1. Use of Antioxidants

These are used to prevent heat-induced oxidation of TPO, especially during processing but also during the service period of final TPO products [1].

Antioxidants constitute indispensable additives for all TPO composites; they can be incorporated either during the manufacturing process of the polymer or in the compounding step of the mixtures.

Polypropylene is quite thermostable in the absence of oxygen. However, it is highly sensitive to oxidation and cannot be used without adequate stabilization. Usually, PP is protected immediately after polymerization, before separation, drying, and storage. The complete stabilizer package is normally introduced during pelletization [2].

Processing of PP is usually performed at temperatures between 200 and 280°C, exceptionally even up to 300°C [2, 3]. With insufficient stabilization, molecular weight and melt viscosity of PP decrease while the melt flow increases [2].

Generally, the phenolic antioxidants used for long-term heat stability are also effective processing stabilizers for PP. To that purpose they may be used without any other stabilizer, if the performance requirements are not too high [2, 4, 5].

2,6-Di-*tert*-butyl-*p*-cresol was used extensively to provide or to improve processing stability of PP. Combination of 2,6-di-*tert*-butyl-*p*-cresol and high molecular weight antioxidants are still used in practice. However, they are increasingly replaced by combinations of high molecular weight phenolic antioxidants and phosphites or phosphonites [6–8]. In comparison with 2,6-di-*tert*-butyl-*p*-cresol, these phosphorus compounds are characterized by good extraction resistance and low volatility; in addition they improve substrate color.

Efficiency of phenolic antioxidants depends on their structure. It has been found that most weakly hindered phenols used without a costabilizer provide better melt stabilization than highly hindered phenols [9].

However, they confer also more color to the polymer than highly hindered phenols. There are even differences in performance between phenolic antioxidants showing the same steric hindrance [2]. The differences are often less pronounced when these antioxidants are combined with high performance phosphites or phosphonites.

The same structure–efficiency relationship is also observed in the phosphorus-containing antioxidant group. Thus it has been established that stabilization systems with tris(nonylphenyl) phosphite and distearyl pentaerythritol diphosphite are significantly inferior to systems containing tris(2,6-di-*tert*-butylphenyl) phosphite and tetra-kis(2,4-di-*tert*-butylphenyl)-4,4′-diphenylene-bisphosphite [10].

As the choice of the high molecular weight phenolic antioxidant is dictated by the application planned for PP processing stability often needs to be adjusted to the specific requirements.

The most important long-term heat stabilizers for PP are phenols of medium (300–600) and especially high (600–1200) molecular weight [e.g., 4-methyl-2,6-di (α-methyl-benzyl)phenol; 2,2′-methylen-bis(6-*tert*-butyl-4-ethylphenol); 2,2′-methylen-bis-[4-methyl-6(1-methyl-cyclohexyl)phenol], etc.]. They are frequently used together with thioesters as synergists, e.g., dilauryl thiodipropionate, distearyl thiodipropionate, or dioctadecyl thiodipropionate.

The commercial antioxidants mostly used in practice are:

2,6-Di-*tert*-butyl-4-octadecylpropylphenol;
Pentaerythritol tetra-kis-3-(3,5-di-*tert*-butyl-4-hydroxyphenyl)propionate;
1,3,5-Tris(3,5-di-*tert*-butyl-4-hydroxybenzyl)mesitylene;
1,3,5-Tris(3,5-di-*tert*-butyl-4-hydroxybenzyl)isocyanurate;
1,3,5-Tris(3,5-di-*tert*-butyl-4-hydroxyphenyl)butane;
2,2′-Methylen-bis(3,5-di-*tert*-butyl-4-hydroxyphenol).

Apart from the dependence of structure, the efficiency of antioxidants depends also on concentration and of the sample thickness.

Hindered amine light stabilizers may be successfully used for the thermo-oxidative stabilization of PP. Low molecular weight HALS have a less reduced efficiency than high molecular weight ones.

The addition of fillers such as talcum to PP is often accompanied by a considerable reduction of antioxidant efficiency. Carbon black and other pigments have a similar effect, especially at high temperature.

The above-mentioned antioxidants are also suitable for the stabilization of PP thin films. However, the synergetic effects of phenolic antioxidants with phosphorus-containing antioxidants [e.g., pentaerytritol-tetra-kis-3-(3,5-di-*tert*-butyl-4-hydroxyphenyl)propionate and tris(2,3-di-*tert*-butylphenyl)phosphate] was not observed. On the contrary, typical high molecular stabilizers such as polymeric HALS have a pronounced contribution to the stabilization of PP thin films.

As HDPE is less sensitive to oxidation than PP, lower levels of stabilizers can be generally employed. As with PP the additives can be added during a suitable manufacturing step or on pelletizing. The requirements concerning color, volatility, extraction resistance, etc., are more or less the same as with PP. However, compatibility of additives in PE is generally poorer than in PP [2].

Processing stabilization of HDPE can be successfully achieved with the same phenolic antioxidants as in PP stabilization, their combination with phosphorus-containing antioxidants leading to a considerable increase of stabilization efficiency in this case, too. High molecular weight phenolic antioxidants are especially suitable for HDPE.

In comparison with PP sulphur-containing synergists are used only occasionally for the stabilization of HDPE. Preference is given to dilauryl thiodipropionate over distearyl thiodipropionate because the latter is less compatible with HDPE [2]. Long-term heat stability of HDPE can be improved considerably by the addition of high molecular weight HALS [6–8].

Phenolic antioxidants used for PP and HDPE do not have a satisfactory stabilization effect for HMWHDPE. Only a combination of phenolic antioxidants and phosphites or phosphonites provides the necessary processing stability in this latter case.

In principle, the phenolic antioxidants used for PP and HDPE can also be used for LDPE. However, as stressed before, compatibility of additives with LDPE is generally poorer than with other TPO. To avoid blooming, most phenolic antioxidants should not be used at concentrations above 0.1%. Phenolic antioxidants confer already excellent thermo-oxidant stability to LDPE [2]. Long-term stability of LDPE can again be improved considerably by the addition of high molecular weight HALS [6–8].

The primary and secondary antioxidants for LLDPE are the same as for HDPE. However, the compatibility problem is more important in the case of LLDPE than for HDPE. As in the case of other TPO types, the stabilization efficiency depends on the chemical structure and the concentration of antioxidants, which can be varied according to the desired properties. Phosphorus-containing antioxidants do not assure better stability of LLDPE, while polymeric HALS contribute markedly to the increase of the stability of LLDPE.

Stabilization of crosslinked PE is assured by the same antioxidants used for noncrosslinked PE. The selection of the stabilization system with respect to compatibility performance and extraction resistance is a function of the PE to be stabilized (LDPE, LLDPE, HDPE, HMWHDPE) as well as of the crosslinking procedure and intended application. If there is direct contact between copper and polymer, addition of a metal deactivator is mandatory.

The main question in PB-1 stabilization (used exclusively for pipe production) is the extraction resistance of antioxidants with cold and hot water. That is why the high molecular weight phenolic antioxidants are the most suitable.

After this brief survey of the technical problems of TPO stabilization with antioxidants, some practical aspects will be mentioned in the following.

*(a) Pentaerytritol-tetra-kis-[3-(3,5-di-*tert*-butyl-4-hydroxyphenyl) propionate].* This is a highly effective, nondiscoloring stabilizer for PP, HDPE, LDPE, LLDPE, and EVA against thermo-oxidative degradation, even under extreme conditions. It prevents premature deterioration of the physical properties of the TPO during processing and service.

Unlike most commercial antioxidants it is highly effective even without thioesters. For this reason the unpleasant odor and taste frequently imparted to the TPO through the use of thioesters can be avoided. It is because of this particular property that the antioxidant is widely used in the stabilization of TPO and has proved very successful as a stabilizer in materials that come into contact with food, for example, packaging films, various domestic appliances, and other consumer goods. Its high resistance to extraction allows it to be used in TPO articles exposed to hot water solutions (e.g., washing machine components).

This antioxidant can be used for the stabilization of TPO composites, especially for PP with filler (talcum and asbestos) or reinforcements (glass fibers). If it is used for LDPE film stabilization, it offers the following advantages: it permits the production of highly transparent nondiscoloring films; it retards yellowing; it does not cause discoloration on storage in the dark; it has no adverse effect on printability; and it imparts no taste to packaged foods.

It is extremely effective in HDPE. Injection molded articles, containers, extruded and blow films, slit films, and packaging films have provided successful fields in application. As it is virtually nonvolatile even at temperatures around 300°C, it is also very suitable for use in rotational molding.

It gives to TPO used for cable insulation effective protection against aging. As long as the insulation does not come into direct contact with the conductor, satisfactory stabilization can be obtained with it alone. But if there is any contact with the copper wire core, the use of a suitable metal deactivator is recommended.

Its concentration in TPO depends on the end use of the parts and on the protection required. In PP the usual concentration ranges between 0.05 and 0.2%, although in special cases, such as for reinforced PP, it may go up to 0.5%. In PE, 0.05–0.2% antioxidant is used.

*(b) 2,2′-Thiodiethyl-bis-[3-(3,5-di-*tert*-butyl-4-hydroxyphenyl)propionate].* This is a sulfur-containing phenolic antioxidant that effectively stabilizes PP, HDPE, and LDPE against thermal oxidation. It has good compatibility with these materials and does not affect their color. It imparts exceptional processing and end-use stability to TPO. It is particularly suitable for use in carbon black (channel black and furnace black) loaded TPO formulations and LDPE wire and cable insulation. It can also be used in chemically crosslinked PE.

*(c) Octadecyl-3-(3,5-di-*tert*-butyl-4-hydroxyphenyl)-propionate.* This is a versatile general antioxidant for TPO; it can be used alone or in combination with thiodipropionic acid esters. In either case, it gives effective protection against thermo-oxidative degradation during processing and service, thus preventing premature deterioration in the mechanical properties of TPO. It does not affect the initial color of TPO and reduces discoloration usually associated with aging. It also improves the TPO light stability. These properties make it particularly suitable for protecting articles that must have good light stability, such as films.

It has also been proved very successful as a stabilizer in the case of TPO that come into contact with food, such as packaging films and household utensils. For this application it should be used without addition of thiodipropionic acid esters, to prevent the taste and odor of the food from being affected.

Because of its chemical structure, it has good compatibility with TPO, low volatility, and high resistance to extraction.

For stabilizing PP films, it is best used in conjunction with one or more light stabilizers.

It is an effective processing stabilizer for PP. Under extreme processing conditions, it may sometimes be used in combination with 2,6-di-*tert*-butyl-*p*-cresol.

It imparts excellent protection to both LDPE and HDPE, where it offers the following advantages: it gives highly transparent nondiscolored films; suppresses gelling; does not cause discoloration on storage in the dark; and has no adverse effect on printability. Its concentration depends on the end use of the TPO and on the protection required. In PP the usual concentration is 0.1–0.5%, while in PE it is of 0.01–0.2%.

(d) 1,3,5 - Trismethyl - 2,4 - tris(3,5-di - tert *- butyl - 4 - hydroxybenzyl)benzene.* This protects TPO from thermo-oxidative degradation during processing and service.

A wide range of tests has shown that it is most efficient during processing when used alone or in combination with phosphites.

The addition of thioesters causes a significant improvement in long-term stability with little influence on the processing stabilization. For high performance light stable, grade addition of a sterically HALS or combinations of HALS with UV absorber are recommended.

Because of its excellent dielectric properties it is widely used in PE cable insulation and in PE and PP condenser parts. As long as insulation does not come in direct contact with the conductor, it can be used without a metal deactivator.

Its concentration depends on the end use of the TPO and on the protection required. In PP the usual concentration is 0.1–0.5%, while in PE it is 0.02–0.2%.

*(e) 4,4′-Thiobis(3-methyl-6-*tert*-butylphenol).* This is an alkylated thiobisphenol that hinders the TPO degradation under heat and oxygen action. It has a slight discoloration tendency and a good compatibility with TPO.

In small amounts, it assures good protection against thermo-oxidative degradation of PP, HDPE, LDPE, and PB-1. Its action is amplified by the combination with dilauryl thiodipropionate or distearyl thiodipropionate.

Due to its bisphenol structure, it is very sensitive in the presence of carbon black, being very advantageously used for the stabilization of reinforced TPO and crosslinked PE.

The recommended concentration varies within large limits (from 0.05–2%), according to the desired stabilization.

*(f) Tris(2,4-di-*tert*-butyl)phosphite.* This is used as a costabilizer of multinuclear phenol antioxidants with thio compounds. Of prime importance is its action as processing stabilizer; at the same time, it improves color fastness. The recommended concentration: 0.05–0.2%.

(g) Calcium (3,5 - di - tert *- butyl - 4 - hydroxybenzyl - monoethyl)phosphate.* This is a hindered phenol antioxidant for TPO, recommended for applications requiring extraction resistance, low volatility, excellent color and color stability, and superior gas-fading resistance.

It imparts processing and good long-term stability to TPO. It is particularly suitable for use in PP films. It can be used alone or in combination with other additives such as costabilizers (thioesters, phosphites), light stabilizers, and antistatic agents.

*(h) Butyric acid 3,3-bis(3-*tert*-butyl-4-hydroxyphenyl)-ethylene ester.* This is a polynuclear phenolic antioxidant with high molecular weight. Its most important applications are in the field of TPO, especially for PP and HDPE. This antioxidant is used particularly for special stabilization of articles exposed to high mechanical and/or thermal stress (e.g., extremely heat-resistant articles, articles coming into contact with extractive media or metal ions, and filled or reinforced TPO). If it is used for the stabilization of PP its maximum effectiveness is reached either alone or in combination with sulfur synergists and/or phosphites.

In combinations with distearyl thiodipropionate, processing and long-term stability of carbon black or talcum-filled PP can be significantly improved. The long-term stability can be further improved by incorporation of a metal deactivator or a solid epoxide.

The addition of montan waxes has also proved very effective in increasing the extruder output and improving the surface finish of articles manufactured from filled PP. The complexing action of lubricants on metal ions in the talcum and the reduction in shear forces also help to improve processing and long-term stability.

Due to its high efficacity as a processing stabilizer, it provides the required stabilization of HDPE even in very low concentrations. In addition to its use as sole stabilizer, it can be combined with sulphur synergists and/or phosphites. Such systems have proved extremely successful, particularly for very severe processing conditions.

Such antioxidants are used only for some special applications where a superior stabilizer system is needed, e.g., for long-term stabilization of cable and pipe cladding from LDPE or LLDPE. The amount of antioxidant required depends largely on the type of polymer, the field of application, and the desired protection. For PP the amount normally ranges between 0.05–0.2%, while for special applications it can be up to 0.5%. For PE and its copolymers, small amounts are normally used (as a guideline, 0.02–0.1%).

(i) Dilauryl Thiodipropionate. The main application of this compound is as additive to PP at a concentration level of 0.2–0.5% in combination with hindered phenols and other stabilizers. It is also recommended for use in PE, EVA, and PB-1.

(j) Ditridecyl Thiodipropionate. This is recommended for TPO.

(k) Dimyristyl Thiodipropionate. This can be used in the same way as dilauryl thiodipropionate. In some cases, it is preferred for particular applications in PP.

(l) Distearyl Thiodipropionate. This is especially recommended for use in film applications. It can be used in the same application as dilauryl thiodipropionate and is generally preferred when a higher melting point and lower volatility are required.

Long chain alkyl esters of thiodipropionic acid have proved to be excellent antioxidants for the protection of plastic materials such as PE and PP. An important feature in the application and performance of thiodipropionate acid esters is their powerful synergistic action in combined use with phenolic antioxidants.

Hindered phenols, in fact, act as chain autoxidation inhibitors; thiodipropionic acid esters, on the other hand, act as preventive antioxidants. This difference in the mechanism of the stabilizers permits them to act independently and synergistically, providing greater protection than would be predicted by the sum of their individual, but separate, effects.

The normal concentration is about 0.01–1%.

The high efficiency of the antioxidant mixtures is due to their synergistic effect. Table 1 lists several types of synergistic mixtures of antioxidants [11, 12].

2. Use of Metal Deactivators

It is well known that transition metals (Fe, Co, Mn, Cu, V, etc.) accelerate the thermo-oxidative degradation of TPO [5]. This process can be retarded by metal deactivators [13].

In practice, the inhibition of the copper-catalyzed degradation of TPO is of highest importance because of the steadily increasing use of TPO insulation over copper conductors. Here it is mandatory to combine a metal deactivator with an antioxidant if the former does not contain moieties with radical scavenging function.

The more appropriate antioxidants to be combined with metal deactivators in order to stabilize TPO compounds that are in contact with copper are frequently selected from among the above-mentioned ones. They are added in amounts varying from 0.05–0.5% depending on the polymer, the nature of the insulation (solid, cellular, crosslinked), whether the cable is petrolatum filled, and on service conditions.

Metal deactivators in actual use are essentially the following:

N,*N*′-bis[3-(3′,5′-di-*tert*-butyl-4-hydroxyphenyl)propionate] hydrazine;

TABLE 1 Synergetic antioxidant mixtures

Component 1	Component 2	Used mainly for
2,6-Di-*tert*-butylphenol	Dilauryl thiodipropionate	PP
2,6-Di-*tert*-butyl-4-methoxyphenol	Dilauryl thiodipropionate	PP
2,6-Di-*tert*-butyl-4-phenylphenol	Dilauryl thiodipropionate	PP
4-Alcoxy-2,6-diphenylphenol	Distearyl thiodipropionate	PP
2,2′-Methylen-bis(2-methyl-6-methylcylclohexylphenol)	Dilauryl thiodipropionate	PP
1,1,3-Tris(2-methyl-4-hydroxy-6-*tert*-butylphenyl) butane	Distearyl thiodipropionate or dilauryl thiodipropionate	PP
1,1,4,4-(2,6-di-*tert*-butylphenyl) butane	Distearyl thiodipropionate	PP
Octadecyl-3-(3,5-di-*tert*-butyl-4-hydroxyphenyl) propionate	Tetra-kis(2,4-di-*tert*-butylphenyl)-4,4′-diphenylene-diphosphite	LDPE
Pentaerytritol-tetra-kis 3(3,5-di-*tert*-butyl-4-hydroxyphenyl)propionate	Tetra-kis(2,4-di-*tert*-butylphenyl)-4,4′-diphenylene-diphosphate	LDPE HDPE

Source: Ref. 11.

2,2′-oxamidobis-ethyl-3-(3′,5′-di-*tert*-butyl-4-hydroxyphenyl)propionate;
3-Salicoylamido-1,2,4-triazole;
N,N ′-Dibenzoxalylhydrazide.

The latter requires predispersion in a master batch because it is insoluble in TPO.

Treatment with petrolatum in order to improve waterproof resistance of PE insulating cables, or the production of insulations with cellular structure having superior dielectric properties, influences negatively their stabilization. The negative influence of foamed insulation is probably not so much caused by the residues of the blowing agent as by faster diffusion of oxygen to the PE/copper interphase [14].

In the case of filled cables, the migration of additives and the rate of oxygen diffusion are important parameters, because insulation is swollen by the filler to a certain extent.

Stabilization of crosslinked PE can be successfully achieved by using a combination of aminic antioxidants and metal deactivators.

Filled (with talcum, asbestos, calcium carbonate) and/or reinforced TPO have lower thermo-oxidative stability than the basic resin, which is explained by the content of iron or manganese from the filler or by the reinforcement materials that accelerate PE degradation.

Therefore, for a satisfactory stabilization of these composites, the incorporation of metal deactivators is necessary.

The reduced thermal stability of filled/reinforced TPO is certainly due not only to the conceivable negative effects on metal impurities, as found in naturally occurring minerals. The possible sorption of polar additives on the polar surfaces of fillers has to be kept in mind, too. This phenomenon would result in a substantial amount of polymer remaining unprotected against the attack of oxygen, because part of the stabilizer may be firmly held by the filler and would be hence only partially available for the intended function as stabilizer [14].

Some recommendations for the metal deactivators for TPO are necessary.

*(a) N,N′-Bis[3(3′,5′-di-*tert*-butyl-4-hydroxyphenyl)-propionyl] hydrazine.* This a highly efficient metal deactivator that can be used either alone or preferably in combination with conventional antioxidants (sterically hindered phenols). The product is primarily recommended for the stabilization of TPO insulated wire and cable. It has demonstrated effectiveness in LDPE and PP homo- and copolymer resin. It should also prove effective in HDPE insulation.

As a metal deactivator, it is equally effective in talcum-filled TPO.

Best results are obtained when is used in association with pentaerytritol-tetra-kis-[3-(3,5-di-*tert*-butyl-4-hydroxyphenyl)propionate].

Recommended concentration:

Polypropylene and propylene copolymers 0.1–0.5%;
High density polyethylene 0.1–0.2% [in both cases, preferably in conjunction with pentaetrytritol tetra-kis-[3(3,5-di-*tert*-butyl-4-hydroxyphenyl)propionate]];
Low density polyethylene 0.1% [preferably in combination with 0.05–0.1% pentaerytritol-tetra-kis-[3(3,5-di-*tert*-butyl-4-hydroxyphenyl)proprionate]].

*(b) Tris[2-*tert*-butyl-4-thio(2′-methyl-4′-hydroxy-5-*tert*-butyl) phenyl-5-methyl phenyl]phosphite.* This is recommended as metal deactivator for TPO products that come into direct contact with copper or other metals (such as cables, fittings resistant to hot water, etc.). Also, it may be utilized in TPO compositions incorporating metal-containing filling materials (such as talcum or asbestos).

Increased effects are obtained when associated with phenolic antioxidants and sulfur-containing costabilizers. The stabilization effect may be increased by the addition of calcium stearate.

The recommended concentrations range between 0.1–1.0% in the case of PP and 0.1–0.3% for PE.

3. Use of Light Stabilizers

In the following, only some technical aspects concerning the use of light stabilizers will be presented (see also photostabilization).

As compared to other TPO, polypropylene is especially sensitive toward UV radiation [5]. The low stability of unstabilized PP exposed outdoors manifests itself by loss of gloss, formation of surface crazes, chalking, and breakdown of its mechanical properties. The same phenomena are observed in light stabilized PP too, but only after a more or less pronounced time lag [15, 16].

The light stability of PP articles also depends, more or less, on parameters other than the effectiveness of added light stabilizers, e.g., sample thickness, degree of orientation and crystallinity [17], pigmentation, and the presence of fillers and reinforcing agents and of other additives such as phenolic antioxidants and co-stabilizers [18].

For light stabilization of PP representatives of the following light stabilizer classes are mainly used at present: 2-(2′-hydroxyphenyl)benztriazoles; 2-hydroxy-4-alkoxybenzophenones; nickel-containing light stabilizers; 3,5-di-*tert*-butyl-4-hydroxybenzoates as well as sterically hindered amines. Nickel-containing light stabilizers are used exclusively in thin sections such as films, whereas all other classes may be used in thin and thick sections. Combination of low molecular weight HALS with high molecular weight HALS are often synergetic with respect to light stability. They combine the advantages of low molecular weight HALS, e.g., high mobility, with the resistance to migration and extraction of high molecular weight HALS.

For light stabilization of PP thick sections, the specific requirements of the polymer must be taken into account as much as with films. It is quite obvious that preservation of the mechanical properties of plastics is of considerable importance. Moreover, the initial appearance of the articles should be maintained for as long as possible.

Pigments, fillers, and other additives from PP composites influence either positively or negatively the photostabilizing efficiency, depending mainly on the composite color [15].

In recent years, such PP compositions are used as garden furniture. In these applications, UV stabilization is mandatory. Two major problems may arise in this connection. The first one is related to impurities possibly contained in the filler, that may act as sensitizers for photo-oxidation. This is particularly true for transition metals. The second problem is the deactivation of the light stabilizers by the filler. This is caused by the absorption of the polar stabilizers on the polar filler surface. This physical effect can be reduced considerably by the addition of certain amounts of polar substances. Some epoxy compounds have shown excellent performance in this respect [15].

Soon after HALS became available, it was found that sulfur-containing additives such as dilauryl thiodipropionate or distearyl thiodipropionate may have a negative influence on the performance of sterically hindered amines [19]. Therefore it is advisable to avoid using sulfur-containing antioxidants or thiosynergists in combination with HALS or to check their possible negative influence. Even inorganic sulfur-containing compounds can be detrimental to UV stability of HALS-stabilized TPO [20].

Related to the mechanism of the negative action of the sulfur-containing compounds on PP photostabilization with HALS, many hypotheses have been made, which are summarized in Ref. 15.

High density polyethylene photostabilization induces the same problems as in PP light stabilization, hence the same light stabilizers are used.

In this case, the experimental data of artificially accelerated ageing revealed that HALS show much better performance in unpigmented tapes than 2-hydroxy-4-*n*-octylbenzophenone, despite the much higher concentration of the latter. Comparison among the hindered amines favors the poly{[6-[(1,1,3,3-tetramethylbutyl)-imino]-1,3,5 triazine-2,4-dibutyl],[2,2,6,6-tetramethyl-piperidyl)-amino)-hexamethylene-[4-(2,2,6,6-tetramethyl-piperidyl)imido]} [15].

The effectiveness of the HALS in unpigmented HDPE plaques can be further improved by combinations of UV absorber, such as 2-(2-hydroxy-3′-*tert*-butyl-5′-methylphenyl)-5-chlorobenzophenone.

Among the numerous commercial light stabilizers, only a few are suitable for LDPE. The reason is that most light stabilizers are not sufficiently compatible at the concentration necessary for the required protection, meaning that they bloom more or less rapidly [15].

Initially, UV absorbers of the benzophenone and benztriazole type were used to protect LDPE resin [4, 5, 21]. With the development of nickel quenchers, significant improvement of the light stability of LDPE films has been achieved [15]. For economic reasons, combinations with UV absorbers are mainly used. The service life of such films may be increased by raising the concentration of the light stabilizers.

With the development of sterically hindered amines [16], a further step forward could be expected in the increasing UV stability of LDPE, similar to that already achieved in PP and HDPE. However, the compatibility between LDPE and the sterically hindered amines available in the early years was insufficient, so that relatively poor performance was achieved, especially on outdoor weathering. It is only with the development of polymeric HALS that these difficulties have been overcome [15].

Efficiency of various types of light stabilizers in LDPE films depends on the stabilizer structure and less on the film thickness. Combinations of light stabilizers are in all cases superior in efficiency compared with single stabilizers. The most efficient are the combinations of nickel-containing light stabilizers and HALS with UV absorbers (e.g., 2-hydroxy-4-*n*-octylbenzophenone).

Some application fields require the incorporation in the LDPE films of some fillers, e.g., China clay, talcum, or chalk. These fillers have a pronounced antagonistic effect on the light stability of films, which is attributed mainly to transition metals present in the fillers as impurities [22]. In this case, in order to reduce this effect, the light stabilizer concentration is increased.

Ethylene–vinylacetate copolymer filled films exhibit the same behavior. They require high concentration of light stabilizer, even higher than that used in LDPE films.

Low density polyethylene thick sections are stabilized with the same light stabilizers as the LDPE film, with an exception for Ni compounds.

Photostabilization of LLDPE films is achieved in a similar way as for LDPE films; in the first case, the photostability is lower when the films are thinner, so the light stabilizer concentrations have to be increased.

Some of the most important light stabilizers for TPO are described below.

*(a) 2-Hydroxy-4-*n*-octylbenzophenone.* This efficiently protects both LDPE and HDPE. A percentage of 0.2–0.5% of it is recommended for the protection of PE monofilaments and films. For molded items and blow-molded containers, a percentage of 0.15–0.3% is suggested.

In agriculture films, it gives a very good performance, with results that double or even triple the lifetime of the materials. For this application 0.5% is recommended.

It is effective both in natural and in pigmented PP. It is used in doses varying from 0.25–0.5% relative to the polymer weight. The same role given above is applicable, namely, decreasing thickness of the finished products requires increased UV absorber concentration. For PP films, an amount of 0.5% is recommended, whereas for molded articles the content should be 0.2–0.3%.

The lifetime of industrial sacks can be doubled or even tripled using an amount of 0.25–0.5%. An additional protective effect on TPO is obtained by synergetic mixtures with nickel-containing stabilizers.

*(b) 2-(2′-Hydroxy-3′,5′-di-*tert*-butylphenyl)-5-chlorobenztriazole.* This is a light stabilizer for use mainly in PP and HDPE.

The following properties deserve special attention: high thermal stability and low volatility; excellent light stability; and chemical inertness (it is particularly not susceptible to the formation of colored complexes with heavy metal salts).

It is especially recommended for articles with thickness a tenth of a millimeter or more, since efficient UV light absorbtion is dependent on the thickness of the specimen. The concentration recommended is 0.1–0.5%.

Combined with HALS, it frequently results in synergetic effects.

*(c) 2-(2′-Hydroxy-3′-*tert*-butyl-5′-methylphenyl)-5-chlorobenztriazole.* This is a very appropriate UV absorber, especially for incorporation into TPO. Its maximum absorption occurs at 353 μm. It is not sensitive to heavy metal ions and alkaline media; normally it does not give colored reaction products.

To obtain adequate protection of TPO against UV light, the following concentration should be added to the base resin: PP 0.3–0.6%, PE 0.2–0.4%.

It shows low volatility at elevated temperature and high resistance to thermal degradation.

(d) Bis(2,2,6,6-tetramethyl-4-piperidyl)sebacate. This is a light stabilizer belonging to the class of low molecular weight HALS, characterized by effectiveness surpassing that of classic products such as UV absorbers and nickel-containing stabilizers. Thus, applications demanding particularly high light stability for such polymers as PP and HDPE were possible.

In contrast to UV absorbers, its effectiveness does not depend on the polymer thickness. For this reason, its use is particularly attractive in articles with high specific surface such as films and membranes. The product imparts extraordinary light stability also to molding with large cross sections.

It can be combined with UV absorbers. By means of such combinations, further improvement of TPO light stability can frequently be achieved as a result of synergetic effects. Sulfur-containing stabilizers such as thiodipropionic acid esters, or sulfur-containing phenolic antioxidants have sometimes been found to have a negative influence on its effectiveness.

(e) Poly{[6-[(1,1,3,3-tetramethylbutyl)-imino]-1,3,5-triazine-2,4-diyl] [2-(2,2,6,6-tetramethylpiperidyl)-amino]-hexamethylene-[4-(2,2,6,6-tetramethylpiperidyl)-imino]}. This is an oligomeric light stabilizer from the class of sterically hindered amines. Its areas of application include polyolefins, olefin copolymers such as EVA, and PP-elastomer blends. The suitability of the product in crosslinked PE is particularly noteworthy.

It is considerably more effective than UV absorber or nickel-containing stabilizers. Because of its oligomeric structure, with an average molecular weight of about 3000, it is the material chosen for the stabilization of thin articles—particularly films.

The following particularly valuable properties of it are noteworthy: it is a colorless compound, hence color and transparency of TPO are not affected; it excels with extremely low volatility and high thermostability;

it is highly compatible with TPO; and it is easily dispersed and very extraction resistant.

Synergetic effects can be achieved by using it in combination with UV absorbers or other sterically hindered amines.

A master batch in PP is available, as well as a few master batches in LDPE containing some combinations with UV absorbers.

The effect of sulfur-containing stabilizers such as thiodipropionic acid esters or sulfur-containing phenolic antioxidants may be deleterious. It is also possible that contact with sulfur or/and halogen-containing pesticides, as can occur with agricultural films, may impair effectiveness.

(f) Butanedioic Acid Polymer with 4-Hydroxy-2,2,6,6-tetramethyl-1-piperidine-ethanol. This is a polymer of the hindered amine class. It offers superior high light protection of TPO substrates against the harmful effects of UV radiation and has the following advantages: good compatibility, very low volatility, low migration tendency, and excellent thermal stability at the usual processing temperature.

It is particularly recommended for light stabilization of TPO in applications where food approval is mandatory. In some instances, the combination with other light stabilizers such as UV absorbers is advisable. For LDPE agricultural films, it is advantageously combined with other polymeric sterically hindered amines.

Sulfur-containing additives such as thiodipropionic acid esters or other sulfur-containing stabilizers can have a negative influence on its effectiveness; also, certain pigments can influence its effectiveness.

(g) Dimethyl Succinate Polymer with 4-Hydroxy-2,2,6,6-tetramethyl-1-piperidine-ethanol. This is a polymeric light stabilizer of the hindered amine class. Besides offering superior light protection in TPO, it offers the following advantages over the existing light stabilizers: good compatibility with all TPO; very low volatility; low migration tendency; and no negative influence on the polymer color.

*(h) Nickel 2,2′-Thiobis(4-*tert*-octylphenolate)n-butyl.* This is a very effective light stabilizer for TPO. It does not behave as normal UV absorbers, since its absorbtion in the 200–400 μm region is quite poor, but as a quencher, i.e., a compound able to deactivate the excited states, by dissipating the energy that has been absorbed as heat or in some other form that is not dangerous for the polymer.

It is thermally stable, highly compatible with TPO, and resistant to extraction. It also presents antioxidant properties.

It is a very efficient light stabilizer for thin TPO articles, such as LDPE films, PP films, and fibers.

Its use is strongly recommended for inclusion in a mixture with UV absorbers of the benzophenone type, with which it gives a synergistic effect. This combination also improves the cost/performance ratio.

The usable concentration of it in LDPE films varies, depending on the service life required and on the thickness of the film. For example, with 0.25% [2,2′-thiobis(4-*tert*-octylphenolate)]*n*-butyl-nickel and 0.25% 2-hydroxy-4-octyloxybenzophenone in 150–200 μm thick films, an outdoor lifetime of two years is obtained in a temperate climate, whereas a larger lifetime may be achieved with a higher concentration of stabilizers.

It is highly recommended for the light protection of PP fibers.

*(i) Nickel Bis(3,5-di-*tert*-butyl-4-hydroxybenzyl)-monoethyl-phosphate.* This has proved effective for use in PP, PE, and other olefin homopolymers and copolymers. Being a sterically hindered phenol, it acts as a light stabilizer and antioxidant.

It gives polymers, especially TPO, such as PP and PE, dye affinity.

It can be used satisfactorily in combination with other stabilizers, for example other light stabilizers, phosphites, phenolic antioxidants, and their combinations. Its use in combination with a UV absorber such as 2(2′-hydroxy-3′,5′-di-*tert*-butylphenyl)-5-chlorobenztriazole induces increased stabilization with the same total concentration of stabilizer.

It has a good extraction resistance. In the presence of sulfur-containing additives (e.g., thiodipropionic acid esters or cadmium sulphide pigments) it may cause discoloration at high processing temperatures.

The more efficient some antioxidant (component 1) and photostabilizer (component 2) combinations present synergetic effects (Table 2) [11].

4. Use of Antistatic Agents/Conductive Agents

Antistatic agents act by reducing/eliminating the tendency of plastics to retain electrostatic charges, by providing a surface layer (often activated by atmospheric moisture) or by establishing a conductive network within the plastic compounds. They may be used both in an external and internal way. *External antistatic agents* (i.e., for external applications) are applied on surfaces as aqueous or alcoholic solutions. All surfactants are more or less effective, which is the case for many other hygroscopic substances such as glycerin, polyols, or polyglycols, which do not show surfactant activity. Antistatic agents for external

TABLE 2 Synergetic combinations of antioxidants and photostabilizers

Component 1	Component 2	Used mainly for
Octadecyl-3(3,5-di-*tert*-butyl-4-hydroxyphenyl) propionate	2-Hydroxy-4-*n*-octoxybenzophenone	LDPE
Octadecyl-3(3,5-di-*tert*-butyl-4-hydroxyphenyl) propionate	2-(2′-hydroxy-5′-methylphenyl) benzotriazole	PP
Octadecyl-3(3,5-di-*tert*-butyl-4-hydroxyphenyl) propionate	*n*-Butylamine-Ni-2,2′-thio-bis(4-di-*tert*-octylphenolate)	PP
Octadecyl-3(3,5-di-*tert*-butyl-4-hydroxyphenyl) propionate	Zinc-diethyldithiocarbamate	LDPE
Octadecyl-3(3,5-di-*tert*-butyl-4-hydroxyphenyl) propionate	Nickel-diethyldithiocarbamate	LDPE
2,2′-Methylen bis(6-*tert*-butyl-4-methylphenol)	2-(2′-hydroxy-3′-*tert*-butyl-5-methylphenyl)-5-chlorobenzotriazole	PP
Tris(nonylphenyl)phosphite	*p*-Octylphenyl salicilate	HDPE
Tris(nonylphenyl)phosphite	2-Hydroxy-4′-*n*-octoxybenzophenone	HDPE
Tris(3,5-di-*tert*-4-hydroxyphenyl) phosphite	2-(2′-Hydroxy-3,5′-di-*tert*-butylphenyl)-5-chlorobenzotriazole	PP
Dilauryl thiodipropionate	4-Dodecyloxy-2-hydroxybenzophenone	LDPE
Zinc diethyldithiocarbamate	2-Hydroxy-4-*n*-octoxybenzophenone	LDPE
Nickel diethyldithiocarbamate	2-Hydroxy-4-*n*-octoxybenzophenone	LDPE
Zinc dibutyldithiocarbamate	2-Hydroxy-4-*n*-octoxybenzophenone	LDPE

Source: Ref. 11.

applications are used in the ppm range. Utilization of antistatic agents in the mentioned concentrations may decrease the surface resistance from 10^{14}–10^{16} to 10^{8}–10^{10} Ω.

The *internal antistatic agents* are incorporated in polymer compounds. For these antistatic agents, the migration of the additives to the TPO surface determines effectiveness.

In the case of TPO, migration of internal antistatic agents depends on their structure. Thus, in LDPE it starts quite soon after processing and quickly reaches equilibrium, HDPE showing a certain delay while PP develops much more slowly.

The development of effectiveness also depends on the type of processing. While an effect can be observed after processing for compression-molded or injection-molded articles of PE, this will take a much longer time for blow or stretched films, because the rate of diffusion decreases with increasing orientation. In the case of stretched PP films, the achieved effect is inversely proportional to the stretch ratio. This is because of the increased orientation and because the surface/volume ratio increases with increased stretching and more substances will be required for coating the surface. For smaller concentrations, a corresponding longer period of time is needed until the surface is sufficiently coated [23, 24].

If several additives are present concomitantly in a polymer, their interaction should be taken into account. Thus, small concentrations of polyethylene glycols or glycerin result in a 2–20 times increase of the diffusion rate of sodium alkylsulphonates. For TPO, a combination of glycerin monostearate with ethoxylated alkylamines has been shown to yield a particularly quick and long-lasting effect in final products with large surfaces such as oriented films.

On the other hand, migration, and thus effectiveness of antistatic agents, can be lowered considerably by other additives such as carbon black or titanium dioxide, talcum, or wood flour. That may be explained by the fact that additives with large specific surfaces (e.g., fillers and pigments) absorb the antistatic agents and hinder their migration to the products surface. The most widely employed antistatic agents for TPO are ethoxylated fatty amines and fatty acid esters.

The influence played by antistatic agents on the specific surface resistivity, as well as on the physico-mechanical characteristics of TPO, is presented in Tables 3–7 [25].

The usual concentrations for the most important TPO are 0.10–0.20% for ethoxylated alkylamines and 1.0–2.0% for fatty acid esters.

TABLE 3 Effect of anionic antistatic agents on the specific surface resistivity ρ_s of PE

Antistatic agents	Amount, %	ρ_s, Ω LDPE	ρ_s, Ω HDPE
Without		$4 \cdot 10^{15}$	$4 \cdot 10^{15}$
Potassium diizopropylthiophosphite	2	$4.6 \cdot 10^{14}$	
	4	$8.0 \cdot 10^{11}$	
Potassium dioctyldithiophosphate	1	$8.5 \cdot 10^{10}$	$8.0 \cdot 10^{13}$
Potassium didecyldithiophosphate	2	$6.0 \cdot 10^{9}$	$5.8 \cdot 10^{12}$
Potassium (*tert*-octylphenyl) dithiophosphate	2	$6.2 \cdot 10^{10}$	

Source: Ref. 25.

TABLE 4 Effect of cationic antistatic agents on the specific surface resistivity ρ_s of PE

Antistatic agent	Amount, %	ρ_s, Ω LDPE	ρ_s, Ω HDPE
Without		$4 \cdot 10^{15}$	$4 \cdot 10^{15}$
Trimethylethylammonium chloride	1	$6.1 \cdot 10^{13}$	$8.5 \cdot 10^{13}$
Trimethyl (methylferocianyl) ammonium metasulfate	4	$2.1 \cdot 10^{14}$	$1.4 \cdot 10^{14}$
Trimethyl (methyl ferocianyl) ammonium nitrate	4	$2.9 \cdot 10^{12}$	$2.9 \cdot 10^{13}$
Trimethyl (methylferocianyl) ammonium dimetaphosphate	1	$1.4 \cdot 10^{10}$	$6.3 \cdot 10^{10}$
Dimethylethylalkyl (C_{12}-C_{18}) ammonium chloride	1	$1.9 \cdot 10^{13}$	$7.6 \cdot 10^{13}$
Dimethylethyloctadecyl ammonium chloride	1	$5.5 \cdot 10^{9}$	$1.9 \cdot 10^{10}$
Dimethyloctyl (cetylcarboxymethyl) ammonium chloride	1	$4.6 \cdot 10^{9}$	$3.8 \cdot 10^{9}$

Source: Ref. 25.

TABLE 5 Effect of ethoxylated fatty amines on the specific surface resistivity ρ_s of PE

Antistatic agent	Content of ethoxylated group, %	ρ_s, Ω LDPE	ρ_s, Ω HDPE
Without		$1 \cdot 10^{15}$	$1 \cdot 10^{15}$
Ethoxylated laurylamine	4	$3.4 \cdot 10^{11}$	$4.6 \cdot 10^{12}$
	10	$3.8 \cdot 10^{11}$	$9.6 \cdot 10^{12}$
	20	$8.1 \cdot 10^{13}$	$9.1 \cdot 10^{13}$
Ethoxylated alkyl (C_{14}-C_{21}) amine	8	$9.3 \cdot 10^{12}$	$9.8 \cdot 10^{13}$
	30	$8.6 \cdot 10^{14}$	$9.4 \cdot 10^{14}$

Source: Ref. 25.

Normally, antistatic agents are incorporated together with other additives and pigments in mixing equipment commonly used in plastics processing.

Due to their partial incompatibility, which determines effectiveness, a certain degree of slip effect, which may cause feed difficulties during extrusion, has to be taken into account when processing TPO with antistatic agents. The addition of small amounts of friction-increasing fillers (e.g., SiO_2) may be of help. If lubricants are used for the processing of TPO to improve rheological behavior, the lubricant portion may be reduced when applying internal antistatic agents. Liquid antistatic agents can also be charged with feed pumps directly into the metering zone of the processing machines [23].

Polymeric materials are used to an increasing extent in technical fields where the surface resistivity

TABLE 6 Effect of internal agents on the physico-mechanical properties of PE

Antistatic agent	Content of antistatic agent, %	Tensile strength, MPa	Elongation at break, %	Brinell Hardness	Thermal stability, °C
Without		12.9	520	14.2	310
Potassium di-*n*-decyldithiophosphate	2	11.0	520	13.0	220
Calcium di-(*tert*-octylphenyl) dithiophosphate	2	10.9	500	14.0	210
Barium di-(*tert*-octylphenyl) dithiophosphate	2	12.5	480	13.8	180
Monoglyceride ester of capronic acid	1	12.1	684	12.9	170
Monogliceryde ester of miristic acid	1	14.4	666	11.7	132
Diglyceride ester of capronic acid	1	13.5	648	12.3	184
Triglyceride ester of capronic acid	1	12.8	546		
	4	12.2	542		
Trimethyloctodecyl ammonium acetate	1	11.1	580	12.6	170
Trimethyl (cetylcarboxymethyl) ammonium chloride	2	13.0	570	13.5	125
N-methylpyridinium metaphosphate	4	11.8	550		40
N,*N*-ethylcetylmorpholyn bromide	4	13.5	560		76

Source: Ref. 25.

TABLE 7 Effect of internal antistatic agents on the physico-mechanical properties of PP

Antistatic agent	Content of antistatic agent, %	Tensile strength, MPa	Elongation at break, %	Brinell Hardness	Thermal stability, °C
Without		32.2	810	60	98
Ethanolamide of undecilic acid	2	30.0	806	59	90
	5	30.9	800	54	85
Ethanolamide of miristic acid	2	32.1	778	50	87
Ethanolamide of stearic acid	2	31.3	792	61	89
	5	30.7	840	57	85
Salt of methanolamine with	2	30.6	860	57	88
undecilic acid	5	21.8	830	57	86
Salt of triethanolamine with miristic acid	5	28.0	758	55	100
Salt of triethanolamine with	2	27.2	820	53	88
stearic acid	5	29.8	830	57	87

Source: Ref. 25.

must not exceed 10^6 Ω. Such a value cannot be achieved with mechanisms of ionic conductivity as shown by incorporated or external antistatic agents; *electron conductive additives* must be used instead [26–28].

The agents for such conductivity are mainly carbon black and metals. In the case of TPO, a volume resistivity of 10^2 Ω can be achieved with common carbon black types in concentrations of 10–20%. This conductivity level can be reached with two special carbon black types and a nonstatistic filler distribution.

Better conductivity than with carbon black can be achieved with metal particles. Common are copper or aluminum in the form of powder or flakes, as well as brass, carbon, and stainless steel fibers. To increase the effective surface, electrically neutral filler particles are coated with metal, e.g., nickel-plated glass fibers or small spheres, silver or nickel-coated mica or silicates [23, 27, 29].

5. Use of Lubricants, Mold Release Agents, Slip, and Antiblocking Agents

These are added to a compound to migrate to the surface during processing or used to form a separating layer, to prevent sticking to mold surface or other materials.

Polyolefins have inherently good processing properties and do not generally require a lubricant. However, calcium, zinc, or magnesium stearate used as an acid acceptor in HDPE and PP and the stearic acid liberated from them influence flow and demolding behavior [4, 30].

Low molecular weight PE waxes (molecular weight about 2000) are sometimes used as internal lubricants in PE, in order to increase the melt flow index of the polymer or to standardize regrind materials when they do not conform to specifications. Concentrations up to 5% are normally used. Studies have shown that the addition of 2–3% PE wax does not increase the susceptibility to environmental stress cracking but can in fact even reduce it in large injection-molded parts.

Polyethylene waxes are not fully compatible in PP and can therefore lead to overlubrication. If necessary, PP wax can be used instead [30].

In reinforced TPO, nonpolar polyolefin waxes act as internal lubricants, improving processing and, by giving better wetting to fillers, increasing the mechanical strength of the end products. The quantity to be used in TPO varies between 0.2–1.0%.

In TPO, *amide wax* is used as an external lubricant. It is mainly used in flat sheet extrusion, for instance in HDPE films. Amide wax improves the rheological properties of polymers and facilitates further processing of the film in the tape. The normal application concentration is in the range 0.3–1%.

The release effect of the LDPE film used for backing self adhesive coatings can be considerably enhanced by the addition of 2–3% amide wax. It is also available as an antiblocking system in granular form. This antiblocking system is also used successfully in the production of film and sheeting of injection-molded articles from EVA copolymers.

Glycerol monostearate, as well as *ethoxylated fatty amines* with chain lengths of C_{12}–C_{18} not only act an antistatic agents but also provide an external lubricant effect in films and injection molding [28]. The concentration ranges from 0.1–0.5%. Very marked lubricating effects at very low concentration (0.05–0.15%) are displayed by *sodium alkane sulphonates* with chain lengths of C_{10}–C_{18} [30].

Fluoropolymers are also effective lubricants in LDPE and HDPE, but it is seldom necessary to use them in these polymers. They are especially used in LLDPE in order to obtain processing characteristics similar to those of LDPE. In this way, excessive pressure buildup and associated irregularities in the film surface can be avoided, even at high output. This effect is due to coating of the machine and die surface with fluoropolymers, which are incompatible with LLDPE.

In PP external lubricants are sometimes used as *mold release agents*. *Erucamide* is very suitable for this purpose. The use of *partially saponified montan waxes* and sometimes also amide wax and fatty acid esters in talcum-reinforced PP is usually at 0.5% based on the total compound. The optimum effect is obtained if the filler is preheated with the wax in a fluid mixer.

Due to their good dispersing effect, lubricants in reinforced PE help to form smoother finished part surfaces. At the same time, the lubricating effect partially compensates for the poorer flow of talcum reinforced PP melts.

The wetting and dispersing effect of lubricants is also used in the incorporation of fine metal fibers or powder for production of conductive TPO.

For certain applications, it is desirable for finished part surfaces to have good *slip* properties. Lubricating solids such as *fluoropolymer powders* or *graphite* are sometimes incorporated into material for gear wheels, bearings, etc. The fatty acid esters or hydrocarbon waxes commonly used as lubricants for the melt state are gradually extended if used in very high concentrations, thus yielding permanent lubricating effects. Certain additives migrate so strongly to the surface, even during cooling, that a uniform invisible thin coating is formed. These slip agents are exceptionally important in the production of TPO films [30]. They ensure good handling properties, particularly on automatic packaging machines [31].

In PE, oleic acid amide is normally used, while in PP erucamide is preferred. Slip agents are added mainly to film grades at a concentration of 0.1%. The additive is incorporated either by the polymer manufacturer in a pelletizing extruder or by the processor in the form of a master batch. In the latter instance, it is possible to combine the slip agent in any desired ratio with an antiblocking agent and to adjust specifically the slip effect of the finished products.

The effect of *antiblocking agents* is very similar to that of the slip agents and the two groups of products are often used in combination [31]. Blocking of smooth surface hinders destacking and takes off film, especially flexible film, from rolls [32]. Antiblocking agents roughen the film surface. Projected solid particles or particles lying between films make it possible for air to flow in as film layers are separated. The effect is the same as that previously achieved by powdering smooth or sticky surfaces [30].

Colloidal silica from various origins or *chalk* or other fine-particle fillers are normally added. If the particles of antiblocking agent incorporated are too large, they can impede surface sliding. This effect is used for *antislip modification.* The products used here are high molecular weight thermoplastics or elastomers [33]; predispersed in a suitable carrier, they form evenly distributed particles on the film surface [34].

6. Use of Crosslinking and Nucleating Agents

Crosslinking agents are usually added in TPO composites in order to provide a tridimensional structure. The crosslinking process induces an increase of the heat resistance (to 120–150°C) and improves dielectric properties, ageing, creep–rupture, and wear/abrasion resistance.

There are two additive methods:

Crosslinking with organic peroxide;
Silane crosslinking.

Organic peroxides were the first crosslinking method for PE and can also be used for covulcanization of the two types of polymer together. The polymer is crosslinked in the presence of peroxides at elevated temperature, the peroxide acting as an activator and decomposing at 130°C approximately. It subsequently forms free radicals on the polymer chain and two such radical points then crosslink.

Crosslinking with organic peroxides is widely applicable in the processing of thermoplastic TPO. It is achieved for one of the following purposes [35]:

Wire and cable insulation and cable sheathing (e.g., made from PE, CPE);
Profiles continuously crosslinked in salt bath or by steam (PE);
Shoe soles (formed and crosslinked EVA, PE);
Rotational molding (PE, EVA);
Formed articles with closed-cell structure (PE, EVA, CPE).

Selection of peroxides is primarily determined by the TPO to be crosslinked and its processing conditions, such as residence time and temperature as well as by the requirements that have to be met by the finished article, e.g., freedom from odor. A further factor is that the degree of crosslinking is a function of the peroxide quantity, whereas the rate of the crosslinking reaction is determined by temperature.

Among the great variety of peroxides used in polymer crosslinking, those frequently used for TPO crosslinking are given in Table 8.

The addition level for PE ranges, according to the type, from 0.8–7.5 per 100 polymer parts to 1.5–5.2 per 100 parts for EVA. The concrete amount of peroxide required depends on the structure of the peroxide, on the polymer to be crosslinked, and on the presence and quantity of other additives. Satisfactory mechanical

TABLE 8 Peroxides used as crosslinking agents for TPO

Peroxide as crosslinking agent	Decomposition range, °C	Used mainly for
n-Butyl-3,3-bis(*tert*-butyl peroxy) butirate		LDPE, HDPE
1,1-Bis(*tert*-butylperoxi)-3,3,5-trimethylcyclohexane	aprox. 150	LDPE, HDPE
tert-Butylperoxybenzoate	approx. 140	LDPE, HDPE, EVA (ethylene vinyl acetate) at low crosslinking temperature
Dicumylperoxide	approx. 170	LDPE, HDPE, EVA
tert-Butylcumylperoxide	approx. 180	LDPE, HDPE, EVA
Bis(*tert*-butylperoxy izopropyl) benzene	approx. 180	LDPE, HDPE, EVA
Di-*tert*-butylperoxide	approx. 190	LDPE, HDPE, EVA, direct addition to cable insulation and sheathing compounds
Ethyl-3,3-di(*tert*-butylperoxy) butirate		LDPE, HDPE
tert-Butyl-per-2-ethylhexanoate		LDPE, HDPE, EVA

Source: Ref. 36.

properties can already be obtained with a low quantity of peroxides. However, permanent set (foamed articles) does not improve until fairly high quantities of peroxide are used. The mechanical properties generally deteriorate when the maximum quantities are exceeded. Higher quantities than the minimum are likewise necessary if the compounds contain substances that can scavenge free radicals, e.g., antioxidants or extender oils. The negative influence of these types of additives can be partially compensated for by the use of coagents [35]. As coagents these can be used: ethylene glycol dimethacrylate, diallyl therephthalate, trialkyl izocianurate, trimethylol propane trimethacrylate, or *m*-phenylene-dimaleimide.

Acidic fillers (talcum, silicates) have a negative influence on the crosslinking with peroxides of TPO due to the catalytic effect of the materials on the ionic decomposition of peroxides, and also due to the absorbtion of the crosslinking agents on their surface. These facts can be remedied by increasing the peroxide quantity and/or by the addition of coagents.

A suitable choice of the combinations of blowing agents, kickers, and crosslinking agents must not lead to adverse influence on the thermal stability of peroxides or on their effectiveness during crosslinking. It is possible to produce articles with a homogeneous, fine-cell, and closed-cell structure.

Incorporation of the crosslinking agents in PE is applied together with other additives, by means of the classical equipment used in the processing of plastic materials. Peroxide master batches, in pellet form, which instead of an inorganic substance as a carrier contain a polymer compatible with the final mixture, have recently been gaining in importance [33].

Silane crosslinking works by first grafting reactive groups onto the PE molecule by addition of silanes and very small amounts of peroxide, in a one- or two-stage extrusion process. The crosslinking is effected by moisture or water and is accelerated by a tin catalyst [36].

The Sioplas process was developed in 1968 by Midland Silicones (Dow Corning), using peroxide-activated grafting of vinyl silane onto the polymer chain, usually in compounding units or twin-screw extruders. The liquid silane/peroxide mixture is added to the polymer melt by membrane or piston pumps, producing a grafted granular product which, after a catalyst has been added, can be processed to cable, pipe or other finished parts.

Ready-to-use master batches of graft polymers mixed with the appropriate catalyst are supplied by some compounders.

The Monosil process, developed by BICC and Maillefer in 1974, combines the two steps of the Sioplas process, bringing together grafting with catalyst introduction and extrusion. The extruder itself is of particular importance here, using special screw geometry and an L:D ratio of 24–30. Single-screw extruders with specially adapted screws can be used, or one-step silane crosslinking can also be carried out by twin-screw machines (Table 9).

The mixture of vinyl silane, peroxide and catalyst is introduced either in the feed hopper or directly to the polymer melt. Moisture crosslinking is similar to Sioplas, in a water or steam bath and new types of polymers also permit ambient curing without additional water treatment.

This process requires lower capital investment than for peroxide crosslinking, but offers an output rate about 2.5 times higher. The capital investment is higher than for Sioplas, but elimination of one stage can reduce manufacturing costs.

Copolymers of vinyl silane VTMO can also be bonded to polymer chains by copolymerization, to produce vinyl silane–ethylene or vinyl silane–ethylene–butyl acrylate. These can be processed in granular form together with the appropriate catalyst master batch in conventional extruders. They can thus be regarded as alternatives to the grafted versions above but, due to their production process, they are only available as high pressure polyethylene grades.

A silane formulation for crosslinking low cost unstabilized standard polyethylene for cables and pipes has been introduced by Hüls under the name Dynasylan Silfm. In addition to cost advantages there are processing advantages: higher stability reduces scorch effect allowing higher start-up temperatures for the extruder (giving faster incorporation of the silane mixture and higher output rate). Optimum matching of components is vital, since increasing the amount of stabilizer impairs the grafting reaction of the silane.

In recent years, a marked increase of the use of crosslinked PE, in particular for cable insulation, has been observed. Other fields of application that are becoming increasingly important are hot water pipes as well as foamed and crosslinked PE for use as insulating and packaging materials. Although by β radiation crosslinking it is possible to obtain articles free from the products of peroxide decomposition, this process requires a high level of capital expenditure and can only be used to a certain polymer thickness. PE crosslinking is thus predominantly carried on with the aid of peroxides. This is also applied to

TABLE 9 Silane products for sioplas and monosil crosslinking processes

Classification	Description	Main applications
Vinylsilanes	Vinyltrimethoxysilanes	Sioplas process
	Vinyltriethoxysilane, vinyltrismethoxyethoxysilane	Sioplas process
Mixtures of VTMO	– with 7.5% dicumylperoxide	Sioplas process
	– with 8.5% dicumylperoxide	Sioplas process
	– with 6.6% dicamylperoxide, antioxidants and copper stabilizer	Sioplas process: cable applications
	– silicone-modified	Sioplas process: reduced screw contamination for improved ejection rate
	– mixture with crosslinking catalyst and peroxide	Monosil process: low voltage cables
	– mixture with crosslinking catalyst andscorch-resistant peroxide	Monosil process: low voltage cables
	– mixture with crosslinking catalyst and peroxide	Monosil process: medium voltage cables
	– stabilized mixture with crosslinking catalyst and peroxide for unstabilized polymers (contains antioxidants and copper stabilizer)	Monosil process: low voltage cables
Methacrylylsilane	3-methacryloxy-propyltrimethoxysilane (MEMO)	Sioplas process

Source: Ref. 36.

EVA copolymers used for cable and wire insulation and for shoe soles, as well as in polymer blends [35].

Polypropylene normally crystallizes slowly into relatively large crystals known as spherulites. These are larger than the wavelength of visible light and reflect light, reducing clarity and increasing haze in the material. However, when *nucleating (or clarifying) agents* are used, the crystal initiation rate is increased throughout the polymer, meaning that more crystals are growing in the same space; due to this reason, their final sizes will be smaller. So, nucleating agents considerably increase crystal initiation, generating crystals smaller than the wavelength of visible light, which allow more light to pass through [37, 38].

A normal nucleating agent is dibenzylidene sorbitol. Geniset is claimed to accelerate crystal formation and improve transparency of polypropylene.

New technology for clarifying polypropylene is claimed by Milliken Chemical for its Millad 3988. It gives increased transparency, resin throughput, productivity, and enhanced physical properties. It is also approved in PP formulations for food contact applications by agencies worldwide, including FDA (US Food and Drug Administration), HPB (Canadian Health Protection Board), and BGA (Bundesgesundheitsamt, Germany).

The nucleating agent is usually added at the compounding stage. Resins clarified with Millad 3988 can be processed over a wide temperature range:

For injection moulding: Normal melt temperatures of 200–270°C are used. Maximum clarity is achieved with mold temperatures of 20–40°C with a smooth and polished surface (SPI 1-2 finish is recommended; SPI 3 finish is normally acceptable).

For blow molding: many types of injection and injection-stretch blow molding are available for blow molding PP bottles with mold clarity, orienting the material. Recommended processing conditions depend on the type of equipment and specific resin, so the additive supplier should be consulted.

For extrusion: compounds containing Millad 3988 can be processed on sheet and film extrusion lines at up to 270°C but, due to the rapid crystallization rate, they will process differently from non-nucleated PP. If the resin crystallizes too fast, smooth film surfaces will not be achieved and there may be stresses in the product. Generally it is recommended that normal chill–roll temperatures are used. Warmer cooling surfaces with faster resin extrusion speeds will give a clearer more glossy stiffer film, as well as higher output. The exact chilling surface and nominal extruder temperatures depend on resin, film thickness and equipment and optimization trails may be advisable.

In "uncontaminated" HDPE, a nucleating agent such as 1% potassium stearate seemingly has a comparatively large effect, the size of the spherulites being reduced in this way from 73 μm (unnucleated) to 13 μm (nucleated). The addition of nucleated PE or for a higher TPO is believed to have a similar effect in LDPE. In the injection molding of HDPE, during which pronounced low orientation generally occurs in the flow direction, certain organic pigments can give rise to nucleation, resulting in the buildup of high interval stress and undesired distortion phenomena [36].

Addition of adipic or *p*-aminobenzoic acid raises the crystallization temperature by some 15°C, compared with unnucleated PB-l.

In long periods of time, most additives used in small quantities in TPO composites are incorporated during polymer manufacture or in the pelletization phase. Recently, additive manufacturers recommend the master batch. The amounts of additives in these masterbatches vary in large limits (5–40%) depending on their nature and on that of the polymer. In the processing industry, master batches offer the following technical and economic advantages [10]:

Straightforward and very homogeneous dispersion of very small quantities of active substance in plastics, on the usual, processing machines, and in the same way as with normal pigment concentrates.
Cold premixing with slow-running mixers or direct addition via the usual metering units is sufficient.
The usual additions of only 1–3% can be metered with sufficient accuracy using these units.
The good dispersability and metering properties of master batches permit optimal utilization of very small quantities of expensive additives. It is unnecessary to add excess amounts to provide a safety margin as in the case of single additives.
Economic bulk purchase of standard grades of plastic in tankers. Simplified stock keeping.
Fast delivery of standard grade and hence less tied-up capital.
Greater flexibility and rapid response to special customer requirements in terms of quality and also price.
No unnecessary stocking of expensive special granules if the finished article for which they are used does not meet with market success.
More economical use of old extruders through additional polymer stabilization.

Table 10 presents several examples (from Hoechst AG, Germany) for the use of master batch systems for TPO, the application field of additivated TPO, together with the maximum limit of the amount of additive (international regulation) required for the product use in contact with food [10].

Up to now, there are no international standards for identification of the main characteristics of TPO supplied by different firms. That is why each firm has its own system of codes, using figures and/or letters, which precede or follow the trade name of product. In general, these codes provide some information about product characteristics (chemical structure, melt flow index, etc.) and application fields. Recently, some firms added into the codes information about the additive groups incorporated into the TPO.

For example, for the LDPE supplied by BASF (Germany) as LUPOLEN, figures and letters accompanying the tradename have the following significance [39]:

The first two digits in the four-digit LUPOLEN type numeric represent the second and third figures behind the decimal point for the value of the density of the basic material.
The third digit is an internal code.
The fourth digit gives indications on any present additive (0 = no additive; 2 = stabilizer; 4 and 5 = slip agent, according to proportion; 7 = flame retardant).
The first letter following the four-digit numeric indicates the melt flow index (MFI 190°C/2.16 kg, g/(10 min)): D = 0.2; E = 0.5; F = 0.8; G = 1.0; J = 2.7; K = 4; M = 7; S = 20; T = 36). Further letters indicate: P = powder; SK = for high voltage cable; X = development product.
The letter preceding the four-digit numeric indicates that the product concerned is a copolymer or speciality (A = ethyleneacrylic copolymer; V = ethylene–vinyl acetate copolymer; XL = contains peroxide crosslinker).

For PP supplied by the same firm, with the trade name of NOVOLEN, the figures and letters indicate:

The first digit in the four-digit NOVOLEN type numeric gives an indication of the type of polymer (1 = homopolymer; 2 = block copolymer; 3 = polymer blend).
The second digit gives an indication of the stiffness (the greater this digit, the higher the stiffness).
The third digit gives an indication of any present additive (0 = base stabilization; 1 = high temperature stabilization; 2 = lubricant and antiblocking agent;

TABLE 10 Additive master batch systems use according to Hoechst AG technical data

Product	Carrier	Recommended addition, phr	Approval rating under						
			BGA[d] PE(II)	BGA[d] PP(VII)	FDA[e]	HDPE			
						blown film	hollow articles	pipe, profiles	injection moldings
Hostanox system VN 25514	PP	1–2[a]	–	4.5	+				
Hostanox system O 1961	LDPE	1–2[a]	6	7.5	+	+	+	+	+
Hostastat system E 1902	LDPE	2–3	3	3	X	X	X	X	+
Hostastat system E 1903	HDPE	2–3	3	3	+	+	+	+	+
Hostastat system E 1905	PP	3–5	3	3	+				
Hostastat system VN 26427	LDPE	1–5	5	5	+		X		
Hostalub system B 1601	LDPE	1–2[b]	4	4	+				
Hostalub system B 1604	PP	1–2	5	5					
Hostalub system B 1605	LDPE	1–2	4	2	+				
Hostavin system VN 28721	HDPE	1–2	–	–	–	+	+	+	+
Hostavin system VN 25890	HDPE	1–2	–	–	–	+	+	+	+
Hostavin system VN 30373	PP	1–3	–	–	–				
Hostavin system L 1922	LDPE	1–3	–	–	+	X	X	X	X
Hostavin system VN 23285	LDPE	1–3	–	–	–				
Hostavin system VN 30880	LDPE	2–5	–	–	–				
Hostavin system VN 25027	PP	1–2	–	–	–				
Hostaflam system I 1912	LDPE	13–20	–	–	–				
Hostatron system P 1940	LDPE	1.5–3	7.5	7.5	+	+		+	+
Hostatron system P 1941	LDPE	1–2	5	5	+	+	X	+	+
Hostatron system P 1931	LDPE	1–2	5	5	+	+	X	+	+
Hostatron system VN 22621	LDPE	1–2	3	3	+				
Hostatron system P 9937	PE	1–2	2	2	+				
Hostatron system P 9947	PE	1–2	2	2	+	+	X	+	+
Antiblocking system B1615	LDPE	1–2	2	0.4	+				
Antiblocking system B1616	LDPE	2–3[c]	4	4	+				
Antislip system B1613	LDPE	1–2	unlimited	unlimited	+				

+ suitable, X possible, – unstable.

[a] When used in remelting frozen material in an extruder: 4–6%.

[b] For injection molding: 2–5%.

[c] For injection molding EVA: 2–10%.

[d] Bundesgesundheitsamt (German Federal office of Health).

[e] Food and Drug Administration (USA).

Source: Hoechst technical data (1985).

3 = antistatic and antiblocking agent; 4 = nucleating agent; 6 = light stabilizer; 7 = flame retardant; 8 = antistatic agent). The special combination between the third and the fourth digit denotes: 10 = high temperature stabilization; 11 = alkaline washing resistant; 12 = thermal stabilization and stabilization in contact with copper; 42 = nucleating agent, alkaline washing resistant, and antistatic agent.

The first letter from the four-figure groups gives indicators on flow melt index of the polymer (MFI 230°C/2.16 kg, g/(10 min)): E = 0.5; H = 1.8; J = 2.6; K = 4; L = 5; M = 8; N = 11; P = 15; T = 37; Z = 1.5 for MFI (190°C/2.16 kg, g/(10 min)).

The second letter along with the four-figure group indicates that the polymer contains a filler or a reinforcement, the first letter being that of the name (T for talcum) and the following figures indicating the amount (%wt).

B. Compounding

1. General Remarks

In principle there are four possible methods of incorporating additives into polymers, namely [40]:

TABLE 10 (Continued)

Use of the systems in											
LDPE				EVA			PP				
blown film	hollow articles	pipe, profiles	injection moldings	blown film	pipe, profiles	injection moldings	flat film	blown film	hollow articles	pipe, profiles	injection moldings
							+	+	+	+	+
+	+	+	+	+	+	+					
+	+	+	+	+	+	+					
							+	+	+	+	+
	X						+	+	+	+	+
+		+	+	+	X	+					
							+	+	X	X	+
+			X	+		X					
–	–	–	–	–	–	–	X	X	X	X	X
–	–	–	–	–	–	–					
							+	+	+	+	+
+	+	+	+	+	+	+	X	X	X	X	X
+	X	X	+	+	X	X					
+	X	X	+	+	X	X					
							X	+	X	+	+
+	+	+	+	+	+	+					
+	X	X	+	+	X	X	+	+	X	+	+
X	X	X	+	+	X	+	+	+	X	+	+
+	X	X	X	+	X	X	+	+	X	+	+
X	X	X		X	X		+	+	X	+	+
X	X	X		X	X						
X	X	X	+	X	X	+	+	+	X	+	+
+			+	+		+					
+			+	+		+					
+											

Additives are added to the monomer before or during the production of the polymer.

Additives are added during the work-up of the polymer.

Additives are added in a separate manufacturing step (polymer compounding).

Additives are incorporated directly before or during molding.

From these methods, excepting the cases when the antioxidants are directly added in the fabrication flow of polymers in order to avoid the oxidative degradation of polymers in the separation and drying operations, only the last two methods are used for the incorporation of various additives in TPO.

As already mentioned, the additives may be gaseous (some blowing agents), fluid or viscous liquids (blowing agents, pigment pastes), waxy, powdery solids (stabilizers, colorants, pigments, antiblocking agents, blowing agents, fillers, etc.), spheres, fibers, monocrystals, granules, etc.

Gaseous additives are incorporated into polymer exclusively in processing, while liquid additives and those which become liquids in the compounding step due to their good solubility are easily incorporated. That is why compounding mainly deals with the incorporation of solid additives, especially with colorants, fillers, and reinforcements.

In order to get a uniform dispersion of the small quantities of powdery additives in the whole mass of the polymer, commonly master batches are used, obtained by premixing of an additive with a corresponding polymer, thus forming concentrates.

The term "compounding" is occasionally used referring only to the stage of the dilution of the master batches with the polymer.

This technique facilitates the control of small quantities of additives (as colorants) in large scale operations, which aspect is essential when the finest dispersion of additives is required, as for example when using carbon black as a protection agent against UV degradation. The use of master batches reduces the dust in the processing plant.

When compounding TPO with rubbers, it is desirable to prepare a master batch containing approximately equal parts of each component at high processing temperature [41] because the melt remains viscous at high temperature and more heat is generated. Master batches are essential in continuous compounding processes, because of the difficulty of obtaining uniform distribution and comminution of the small amount of additives required in the final compound.

Master batches are obtained as dry powders or granules from polymers with stabilizers and/or colorants, pigments, carbon black, antiblocking agents, flame retardants, etc., the additive concentration ranging from 2–20%, while in the PE/carbon black mixtures it may be 40% and in PE/TiO_2 even 50%.

According to other techniques, the additives are either granulated by using paraffins or low molecular weight TPO, or they are dissolved or dispersed in adequate solvents, obtaining solutions, dispersions, or respectively pastes.

If higher quantities of powdery additives are needed, in order to prevent agglomeration of the fine particles, their surfaces are treated with lubricants [42, 43], but in this case, due to the difficulties of compounding, suitable techniques of feeding must be employed. Also, TPO compounds containing high amounts of powdery additive tend to absorb the liquid additives so that their proportion in the system is apparently reduced. For avoiding such a phenomenon, the powdery additive is added only after the homogenization of all other mixture components.

An adequate procedure for the incorporation of fibrous, lamellar, or crystalline fillers and reinforcements must preserve their form and dimensions [44–46].

2. General Operations in the Compounding of Thermoplastic Polyolefins

However, there are a large variety of possibilities for TPO compounding with additives; most of them can be included in a general scheme such as that presented in Figure 1.

The process includes the following steps [40]:

Storage of starting materials (TPO and additives);
Metering of starting materials;

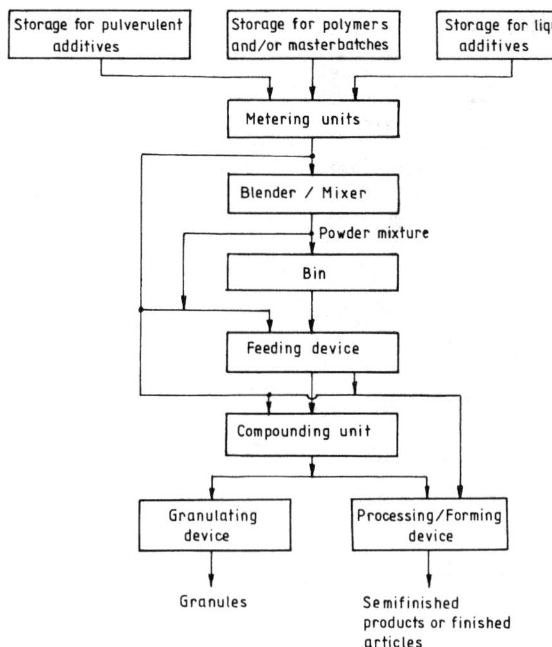

FIG. 1 Flow diagram for the compounding of thermoplastics. (Adapted from Ref. 40.)

Mixing of components;
Feeding in compounding or forming unit;
Compounding or molding of compound;
Storage in bins and/or packaging.

(a) Storage of Starting Materials. Polyolefins as melts, powders or granules, as well as some of the additives employed in the compounding step (mainly fillers, reinforcements, and master batches) are stored in outdoor silos or bins, these materials coming either from the synthesis process or from transport vehicles to processing units. For other additives (such as powdery stabilizers, lubricants, pigments, etc.), a better storage is made in use-bins or directly in transport containers from which they are fed in metering devices.

This last procedure is economical and at the same time advisable from the point of view of safety, since it avoids transfer of materials between containers and the risk of dust formation.

(b) Metering of Starting Materials. Metering of the starting materials in the compounding step can be achieved continuously or batchwise, as gravimetric or volumetric metering. The choice between batchwise and continuous metering depends on the construction and the mode of operation of the downstream units; the required precision of metering determines whether volumetric or gravimetric metering is preferable.

For metering, suitable devices are used, such as various types of balances: containers or weigh hoppers, belt weigh types, etc., for gravimetric weighing or travelling band conveyer or screw conveyer for solid materials and rotameters, piston pumps, etc., for liquid components, in the case of volumetric metering [44].

After metering, the components of the mixture pass (with or without an intermediate storage step) into receiving bins, either at dry blending or directly at a compounding or forming unit.

(c) Mixing of Components. Mixing of TPO with various additives, as a preliminary step of the compounding or forming process, aims at a better homogenization of the blend. The theoretical basis of thermoplastic mixing is extensively treated in many papers [47–53].

Materials introduced into the mixing unit (mixer or blender) are shifted in many ways: by means of a fluidizing gas (fluidized bed mixer or pneumatic mixer), by mixer rotation (tumble blenders) or by the movement of some internal mixing mechanisms (ribbon, Z-blades, etc.) of the mixer (mechanical mixer).

Pneumatic mixers are suitable for the homogenization of powdery materials having components with small differences of granulosity and density, as well as for granulated materials mixing [53–56], but they cannot be used for mixing electrostatically charged materials. Their advantages consist of easy construction and great productivity.

In tumble mixers—Fig. 2—by the rotation of the partially filled drum, the material moves by translation and rotation, and thus homogenization results.

This type of mixer has a reduced efficiency, but it can be improved by changing the drum shape or by equipping it with baffles [53, 55]. There are several types of constructive solutions for mechanical mixers, such as intensive nonfluxing mixers, ribbon blenders, double arm mixers, plug mixers, etc. [42, 55–57]. They assure mixing both at room temperature and at high temperature; moreover, in some of them, the obtained mixture can be cooled [42]; these make them suitable for various forms of the mixing components [42, 55].

The mixtures obtained by dry blending can be passed into the compounding or directly into the forming unit, optionally depending on the composition of the mixtures, the shape of the components, the processing technique, and the type and characteristics of the end products.

Master batches are always passed to a compounding unit and then granulated or comminuted. The granulated mixtures (obtained by dilution of master batches with TPO granules) and those from TPO granules with powdery additives frequently pass directly into the processing process (injection or extrusion). This is the case in the fabrication of thick-walled products, being less used for thin-walled product fabrication. In the latter case, the compounding step is required before processing.

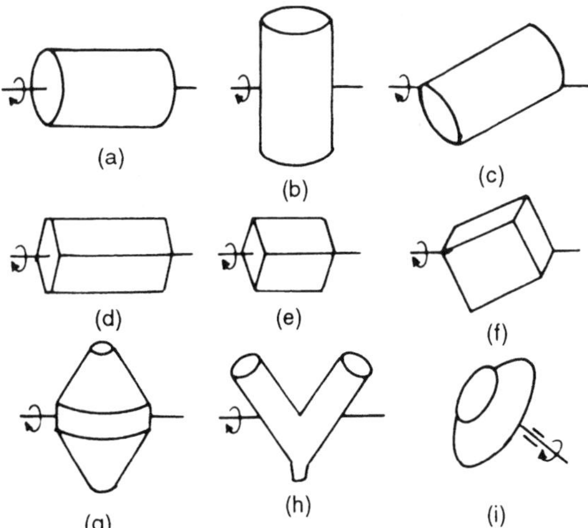

FIG. 2 Different types of tumble mixers. (a), (d), (e) Centric rotary tumble mixers. (b) Tumble mixer. (c), (f) Eccentrically rotary tumble mixers. (g) Double cone mixer. (h) V-shape or twin shell mixer. (i) Mushroom type mixer. (From Ref. 55.)

Powdery mixtures can be introduced either in the compounding–granulating operations or in processing. However, the second way presents some advantages (smaller installation costs and consumption of water, air, supply power, nonintensive mechanical degradation of polymer, etc.). It is still less used for TPO additivation due to the small volumetric density, low flow capacity of the mixtures, high electrostatic charge of the powders, differences in the melting behavior in processing devices, etc. [43, 58, 59]. This is why only the compounding–granulating operation is described in detail in the following.

(d) Feeding in Compounding or Forming Units. Individual components or mixtures obtained by dry blending can be concomitantly or consequently fed into the compounding unit, either by free flow or by forced flow, depending on their characteristics and shape, and also on those of the compounded mixture [60].

Granulated TPO (pure polymers and/or master batches) or granules with powdery additives are fed into compounding units by free flow. Forced flow is especially applied for powdered components or mixtures when consequent feeding of some additives is needed, for example after the plastification of the components previously fed into the compounding unit (internal mixer or extruder).

(e) Compounding. Compounding is the process of incorporating additives into polymers, thus achieving uniformity on a scale appropriate to the quality of the articles subsequently made from the compound. It is also known as hot blending or melt blending, better homogenization being explained as follows: the molten material is subjected to shear in such a way that the interfaces of the phase to be mixed intersect lines of flow produced by the shear. This produces an increase in interfacial area between the two phases until the mixing is considered adequate. Since the increase in interfacial area is proportional to the shear to which the mass is subjected, when the interfaces intersect flow lines, the quality of mixing is proportional to the amount of shear or to the shear stress multiplied by time [48–51, 54, 56, 57, 61–63]. A very important requirement in any practical problem is that the shear stress between the lines of flow intersected by a particle must be great enough to deform and break up the particle. If one component has a much higher viscosity than the other, the shear will all occur in the low viscosity component and there will be no increase in interfacial area. In practical cases, the mixer is designed to produce extremely high rates of shear across small distances, so that the high viscosity particles will become isolated in the high shear area and so be disrupted.

Two main types of equipment are used for hot mixing of thermoplastics, internal mixers, and extruders [44, 64–68].

On a laboratory scale, the small two-roll mill is the most convenient compounding equipment; it was previously commonly used in industry, too.

With an LDPE of MFI = 2 g/(10 min), the front roll is usually controlled at a temperature of about 130°C and the back roll of about 90°C. The PE sheet adheres to the front roll from which it is repeatedly removed, using a doctor knife, and returned to the nip of the mill for an overall time of about 10 min. The additives or the additive master batches can be added by prior tumble blending or directly to the mill after the PE has melted. A typical speed of operation is 24–30 rpm. For simple mixing, an open nip can be used, but for dispersing additives a tight nip is more effective [41, 44].

The internal mixer consists of a heavy steel body enclosing a chamber shaped like two parallel intersecting cylinders. In each cylinder there is a rotor of teardrop shape, with the point coming very close to the cylinder wall—Fig. 3.

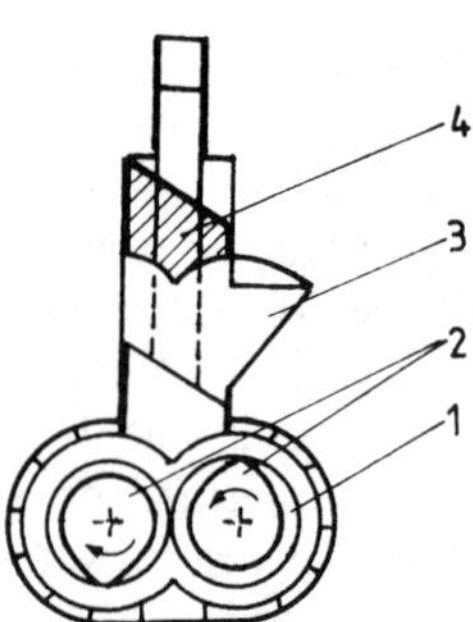

FIG. 3 General arrangement of an internal mixer. (1) Mixing chamber. (2) Rotors. (3) Hopper. (4) Ram.

Material is introduced through an opening along the intersection of the two cylinders and is held in the chamber with a hydraulic ram. Rotor rotation produces an exceedingly high shear on the material between the point of the rotor and the cylinder wall. The rotors are designed and operated so as to throw the mass being mixed back and forth between the two rotors, and also in an irregular pattern back and forth along the cylinder. This assures that all parts of the mass eventually get into the region of high shear where the intensive mixing occurs. The high shear rate produces a great deal of frictional heat. Body and rotor are generally water-cooled to remove as much of this as possible; nevertheless, in a short time, the temperature of the mass rises to the point where the drop in melt viscosity due to increased temperature reduces the effectiveness of the mixing action [42, 44, 48, 69, 70].

Compounding of TPO with additives is performed in internal mixers whose mixing chambers have a free volume of up to 240 L, rotor speeds between 35–70 rpm and a ram pressure of 0.4–0.6 MPa. The compounding temperature ranges between 110–125°C and the mixing time is between 12–15 min.

There are many companies (Mecaniche Moderne, Italy; Farrel Conn. Div. USA; Werner & Pfleiderer, Germany; Bolşevic, Russia, etc.) that produce internal mixers in an extremely large variety of types and dimensions; the characteristics of some of these are presented in Table 11.

Compounding of thermoplastic polymers with different additives may be carried out also with screw, disc, and screw and disc extruders; of these the most used are the screw extruders.

The processes occurring in screw extruders are extensively analyzed in the literature [48–57, 69, 71,

TABLE 11 Characteristics of some types of internal mixers

Mixer type	Producer	Volume of chamber $10^{-3} \cdot m^3$		Speed of rotor, rpm		Pressure of the ram, MPa	Supply power of motor drive, kW
		Total	Useful	Anterior	Posterior		
RS-14M	Bolşevic,	71	45	30	34	0.16–0.22	125
RSDV-140–20	Russia	250	140	17	30	0.5–0.6	320
RSDV-140–30		250	140	25	30	0.5–0.6	630
RSDV 140–40		250	140	33	40	0.5–0.6	700
GK-160V	Werner	240		18	21	0.18–2.0	400
GK-160Vu	& Pfleiderer,	240		18	21	0.2–0.5	550
GK-160VuH	Germany	240		36	42	0.2–0.8	1600
GK-230UK			230	20–60	20–60		650–2000
Banbury	Ferrel		0.45	336			5.5
Medget	Conn. Div.,						
BC	USA		1.1	77	116		6.5
OOC			1.8	62	125		11
1			11	60	120		37
3A and 3D			45	30	70		110
9D			120	21	43		147
11 and 11D			153	20	40		220

Source: Ref. 70.

72]. In the following, only some functional characteristics of the most commonly used types of screw extruders—Fig. 4—for polymer compounding will be mentioned.

Mixing in the single-screw extruder of a simple construction is due to both frictional forces (appearing between the material and the active part of the extruder) and intensive shearing processes in the clearance and to the flowing with different speeds of the material streams. However, simple single-screw extruders (Fig. 4a), even with large L/D ratio [47, 73–78], do not assure good conditions for blend compounding. First, because there are some zones where the material is not mixed, in both longitudinal and transversal flow in the screw channel; secondly, because the exchange between the material layers displaced along the screw channel is rather reduced. This is why screw extruders with intensive mixing zones or single-screw extruders of special design are preferred for the compounding of thermoplastic polymers.

There are numerous types of single-screw extruders with intensive mixing zones—Fig. 4b—these zones being placed in different positions along the screw—Fig. 5a–g—or at its tip—Fig. 5h–j. Irrespective of the construction form or their place, the intensive mixing zones repeatedly divide the processed material and intensify the shearing process, which assure a better dispersion of the different additives in the polymer mass and an improvement of blend compounding [51].

Among the single-screw extruders of special design, the most commonly used are the Ko-kneder and Transfermix types [42, 44, 48, 49].

Compounding Ko-kneder aggregates (Fig. 4e) are single-screws extruders, which present two distinct motions: the normal rotational motion and a reciprocatory motion along the axis. This type of extruder presents a rather complicated construction, due to the teeth that are located on three spiral generatrixes, on the internal surface of the cylinder. The cylinder teeth and the corresponding pairs of screw kneading elements associated with them act like a mill, by imposing supplementary shear forces on the material.

The two motions of the screw are synchronized: the reciprocating one starts only when the slots from the screw are on the direction of the cylinder teeth.

The special design of the screw and the cylinder, together with the combined motion, rotational and axial, of the former, make the shearing and frictional

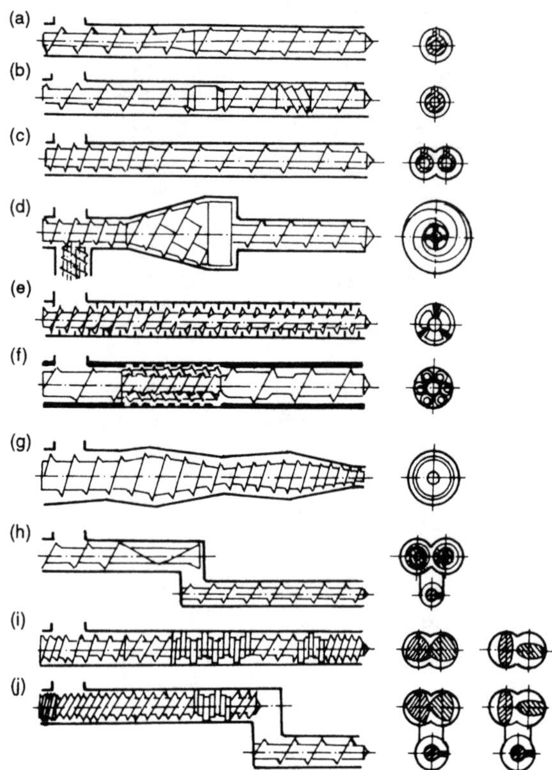

FIG. 4 Compounding extruders. (a) Simple single screw extruder. (b) Single screw extruder with intensive mixing zone. (c) Simple twin screw extruder. (d) Extruder with shear cone screw. (e) Ko-kneder extruder. (f) Planetary gear extruder. (g) Transfermix extruder. (h), (i), (j) Special twin screw extruder.

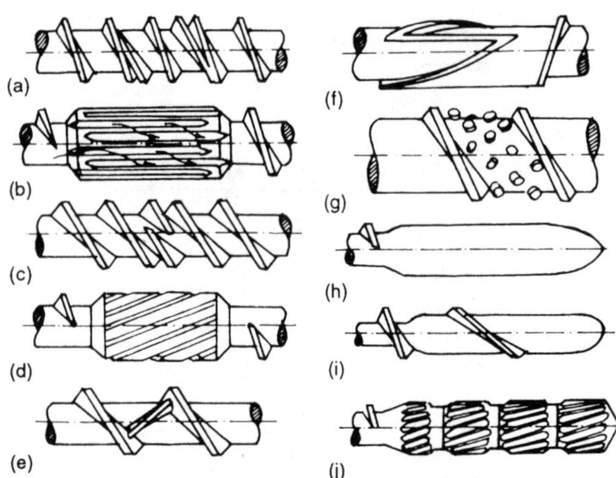

FIG. 5 Types of intensive mixing zones of a screw. (From Ref. 55.)

processes occurring in the Ko-kneder extruder more intensive, allowing a better blend compounding.

Ko-kneder extruders, made for the first time by Swiss Buss Company, are also produced today by the following companies: H. Linden and Werner & Pfleiderer, Germany; Japan Steel Works, Japan; Bolşevic, Russia, etc.; the characteristics of those from Russia are given in Table 12.

Transfermix mixing equipments are single-screw extruders (Fig. 4g) provided with an inner flighted cylinder with a helix contrary to that of the screw. Both channels, from the cylinder and from the screw, have variable depth (about a sine function).

The mixing process of the Transfermix extruder is similar to that of simple single-screw extruders, with the difference that it takes place simultaneously, both in the cylinder and in the screw channel. Due to their variable depths, the material is continuously transferred from one channel to another. At each transfer from one channel to another, the material passes over the clearance between the two flights, where the shear stresses are maximum.

Transfermix extruders are made by Werner & Pfleiderer of Germany; their characteristics are given in Table 13.

The reduced mixing effectiveness of the simple single-screw extruders and the complex design of the special types makes the multiple-screw extruders to be preferred for thermoplastic polymer compounding, the twin-screw extruder being the most commonly used.

Twin-screw extruders (Fig. 4c) are made by many manufacturers (Baker Perkins Inc., Davis Standard Div., Farrel Conn. Div., Gloucester Eng. Co., USA; Bellaplast Mechinenbaum GmbH, Werner & Pfleiderer GmbH, Hermann Berstorff, Breyer GmbH, Reifenhäuser GmbH, Germany; Japan Steel Works, Japan; etc.), the screw diameter varying between 15 mm (Hermann Berstoff) and 380 mm (Werner & Pfleiderer) or even more. Among these, the ZSK type extruders have a special place. They are made by Werner & Pfleiderer. Their uses, performance, and characteristics are given in Tables 14 and 15.

The compounding process in twin- and multiple-screw extruders is complex and depends on the number and construction characteristics of the screws, the rotation way, the degree of gearing, the construction characteristics of the equipment, etc. [53, 71–73, 79, 80].

Disc extruders and those with both disc and screw, although they may be used as compounding equipment for thermoplastic polymer blends, are little used for TPO compounding.

TABLE 12 Characteristics of Ko-kneder extruders made in Russia

Characteristics	Ko-kneder Type			
	ČOS-90	ČOS-200	ČOS-300	ČOS-400
Diameter of moving screw, mm	90	200	300	400
Relative length, *L/D*	7–15	7–11	7–15	7–11
Maximum rotation speed of screw, rpm	120	120	120	120
Supply of power of motor drive for the screw, kW	70	200	400	600
Production, $Kg \cdot h^{-1}$	140	200–1000	260–1250	3000–5000

Source: Ref. 47.

TABLE 13 Characteristics of transfermix type extruders made by Werner & Pfleiderer

Extruder type	Screw diameter, mm			Production, $Kg \cdot h^{-1}$	
	Feeding zone	Metering zone	With degasing	Without degasing	For PE/TiO_2 concentrate (50%)
2/1.2	50	30	18	80	45
3.5/1.7	89	43	90	440	220
4.5/2.2	114	56	200	900	450
6/2.8	152	71	450	2200	1100
8/3.4	203	84	110	5400	2700
10/4	254	102	2200	11000	5500

Source: Ref. 47.

Parameters for TPO compounding in extruders vary within very large limits and depend on the extruder type, on the nature, proportion, shape and characteristics of the components of the blends, on final characteristics of the compounded blend, etc.

(f) Storage in Bins and/or Packaging. The granules resulting from the compounding process may be stored in silos or bins, packaged in sacks or containers, or, more recently, loaded directly into specialized transport devices that are delivered to users.

(g) Units for Thermoplastic Polyolefin Compounding. The thermoplastic polyolefins to be compounded may have different forms: as melts (LDPE, EVA), powders (HDPE, UHMWPE, PP, etc.) and granules (LDPE, LLDPE, HDPE, PP, etc). Mixing TPO with various additives may take place either in the dry or in the melt state.

Dry blending of powdery TPO is carried out in tumble blenders or in mechanical blenders. This is the case when incorporating into TPO of stabilizers, colorants, pigments, flame retardants, antiblocking agents and some reinforcements (as fibers, spheres, single crystals) and fillers, especially if these are heavily dispersed in the polymer mass, their particles being coarse and not treated with lubricants. Depending on the additive concentration, the obtained granules can be either master batches or even be handled for subsequent processing, if the additive ratio corresponds to that of the final mixtures.

A scheme of a simple installation for compounding dry powder mixtures is presented in Fig. 6. The installation is provided with a nonfluxing mixer and a compounding extruder.

The equipment assures the incorporation of fillers (chalk, talc, kaolin, etc.) and other powdery additives into the powdery PP; it can also be used for mixing TPO with textiles or glass fibers previously cut at the required length.

Dry blending of granulated TPO with powdery additives (frequently colorants, pigments, flame retardants, and antiblocking agents) often takes place

TABLE 14 Some examples of ZSK type twin-screw extruders for TPO compounding

ZSK type twin screw extruder		25	30	40	53 57 58
Material (shape of material)	Procedure	Production, kg · h^{-1}, in function of the product			
HDPE, LLDPE (powder)	Compounding, coloration		max 30[a]		60–180
		max. 30[b]		40–100	max. 30
HDPE, LLDPE (pastes)	Concentrates from 15% to 500 ppm, mixtures of additives	max. 15		40–60	60–150
PP (powder)	Compounding, degasing, coloration		max. 30		60–180
		max. 30		40–100	max. 300
	Degradation		max. 30		60–180
		max. 30		40–100	max. 300
	Blending with rubber		max. 25		60–180
		max. 30		40–80	max. 270
LDPE (melt)	Degasing, mixing with additives, homogenization	max. 25		60–100	150–270
LDPE (granules)	Mixing with additives, coloration, homogenization		max. 30		100–180
		max. 30		60–100	max. 280
LDPE, HDPE, PP	Obtainment of master batch		max. 20		30–180
		max. 20		15–100	60–300
	Reinforcement with glass fiber		max. 25		80–150
		max. 25		40–90	130–250
	Incorporation of fillers (clay, chalk, etc.)		max. 30		100–250
		max. 30		50–120	150–350

[a] standard operation.
[b] high power operation.
Source: Werner & Pfleiderer technical data (1989).

also in tumble mixers. During mixing, the powdery additives disperse through granules or adhere to the granule surface; thus a sufficiently good homogeneity for the mixture is achieved. The quality of the mixture principally depends on the shape and dimensions of the granules. Small granules retain more easily the powdery additives than larger granules; this is due to the differences between the specific surfaces of the two types of granules. Granules having irregular surfaces retain the powdery additives much better than do granules with smooth surfaces.

In some cases, in order to facilitate the mixing of TPO granules with powdery additives, mineral oil or other viscous liquids are added to wet the granule surfaces, thus assuring good adhesion between the granules and the powders. This tends to reduce dusting and gives better additive distribution in a smaller time. However, care must be taken to mix for a minimum possible time, because liquids tend to cause agglomeration or lumping of the additive particles on longer mixing.

In this case, the obtained mixture undergoes a compounding–granulation process or passes directly to processing—Fig. 7.

Installations equipped with tumble mixers are also used in the dry mixing of various types of granulated

TABLE 14 (Continued)

70	83 90 92	120 130 133	161 170 177	220 240	280 300	380
characteristics, formulations, drive power, and machine assembling						
	300–800	1000–2500	3000–5000	5000–9000	10000–17000	
200–500	max. 1000	max. 4000	max. 8000	max. 14000	max. 30000	45000
200–300	300–500	1000–1500	1500–2500	4000–6000	7000–10000	
	300–800	1000–1500	3000–5000	5000–9000	10000–17000	
250–500	max. 1000	max. 4000	max. 8000	max. 14000	max. 30000	max. 45000
	300–800	1000–2500	3000–5000	5000–9000	10000–17000	
200–400	max. 900	max. 3500	max. 7000	max. 12000	max. 22000	max. 40000
	300–800	1000–2500	3000–5000	5000–9000	10000–17000	
200–400	max. 800	max. 3000	max. 6000	max. 10000	max. 25000	
300–400	500–800	2000–3000	4000–6000	6000–10000	12000–20000	
	400–800	1600–2800	4200–5700	7000–9500	12000–16000	
300–500	max. 1000	max. 3200	max. 6500			
	150–800	400–2200				
150–600	300–700	800–3400				
	350–700	1000–2000				
300–500	600–1100	1500–3000				
	400–800	1000–2200				
400–700	800–1400	2000–4000				

TPO (LDPE + HDPE, LDPE + LLDPE, PE + PP, etc.) or TPO with additives as granulated master batches. Dry mixing as a preliminary step of the compounding–granulation process is economically justified only in special cases. This is why modern installations for incorporation of additives into TPO employ especially melt mixing by means of all types of equipment, among these the most frequently used being internal mixers and extruders. The process takes place in a single step.

Melt mixing in internal mixers is used for all types of mixtures and especially for compounding TPO with elastomers, such as EPDM, PIB, IIR, etc., in a similar unit to that presented in Fig. 8. The components are added simultaneously or successively. When compounding powdery TPO with elastomers and other additives (solids and liquids) in internal mixers, the last components can be mixed beforehand with thermoplastic TPO on a nonfluxing mixer or on a tumble mixer. At the end of the compounding process in an internal mixer, the blend is discharged onto a roller, converted into a band, cooled, and granulated. Before processing, the master batches must be subjected to a new mixing–compounding–granulation cycle (on internal mixers or on extruders), while the mixtures with corresponding composition for processing can be directly introduced into extrusion or injection processes.

Installations for melt compounding using extruders differ with respect to the nature, characteristics, shape

TABLE 15 Technical characteristics of ZSK type twin-screw extruders from Werner & Pfleiderer

ZSK type twin screw extruders	25	30	40	53	57	58	70	83	90
Standard operation M. . . .		9		50		50		240	
High power operation M. . . .	8.2		32.5	90		96	175	350	
Supply power of motor driver, kW		5.6[a]		32		32		150	
	5.2[b]		20	55		60	108	215	
Rotation speed of screw, rpm	300	300	300	300		300	300	300	
Diameter of screw, mm	25	30	40	53	57	58	70	83	90
Channel depth, mm	4.15	4.7	7.1	5.5	9.5	10.3	12.5	7.5	13.5
Relative length, *L/D*	42	42	48	48		48	48	48	
Length of extruder[c] mm	2500	2600	2600	3800		3400	4000	5300	
Width of extruder[c], mm	780	780	720	550		700	700	700	
Height of extruder[c], mm	1400	1400	1400	1300		1400	1550	1400	
Total weight[c], kg	950[d]	950[d]	850	3000		3000	4900	7000	

[a]Standard operation.
[b]High power operation.
[c]Without other auxilliary devices of *L/D* = 24, with motor.
[d]With installation of operation.
Source: Werner & Pfleiderer technical data (1989).

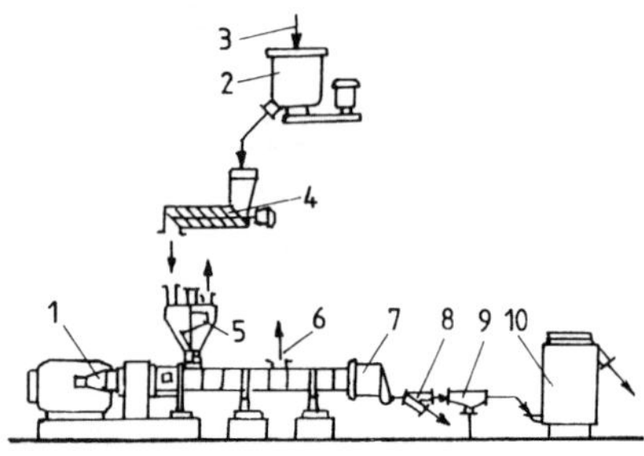

FIG. 6 Flow sheet of a compounding unit of PP with fillers (clay, calcium carbonate, kaolin, etc.) and other additives. (1) Compounding extruder. (2) Nonfluxing mixer. (3) PP powder. (4) Metering screw. (5) Hopper. (6) Degasing. (7) Granulating device. (8) Switcher. (9) Water separator. (10) Dryer for granules. (From Ref. 47.)

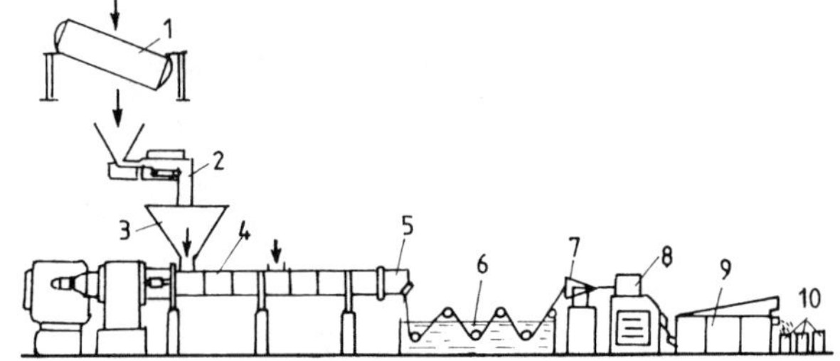

FIG. 7 Flow sheet of a compounding unit of TPO with pigments and/or other additives. (1) Tumble mixer. (2) Belt weight. (3) Hopper. (4) Compounding extruder. (5) Extruder head. (6) Cooling bath. (7) Take-off device. (8) Granulating device. (9) Dryer for granules. (10) Packaging bags. (From Ref. 47.)

and ratio of the mixing components. Figure 9 presents schematically a compounding unit for TPO melts (as LDPE resulted from the demonomerization step) with master batches of additives (stabilizers and antiblocking agents).

The compounded mixture is granulated; the granules are then cooled with water, dried, and sent to packaging.

When solid polymers are used in single-step compounding with extruders, the metering of components have to be done dynamically, directly in the extruder, without storage. The additives are introduced either in the solid polymer, in the feeder hopper, or in the extruder after the melting of the polymers [43]–Fig. 10.

The first way is used for all types of components, the latter solution being preferred when the form and

92	120	130	133	160	170	177	220	240	280	300	380
240	650		650	1750			4400		8800		
385	1050		1160	2500		2750	7000		13000		25000
150	400		400	1080			1805		2890		
240	646		728	1540		1730	2875		4270		9000
300	300		300	300		300	200		160		175
92	120	129	133	160	172	177	222	238	280	300	385
16.3	10.5	19.5	23.5	14.5	26.5	31.5	18	24	20.75	40.75	37.5
48	48		48	42		42	36		36		30
6000	7900		7500	9500		10000	13000		15000		20000
700	1600		1700	2000		2000	2900		3600		4000
1450	1200		1500	1500		1850	1600		1900		2700
7300	15000		20000	26000		30000	50000		70000		12000

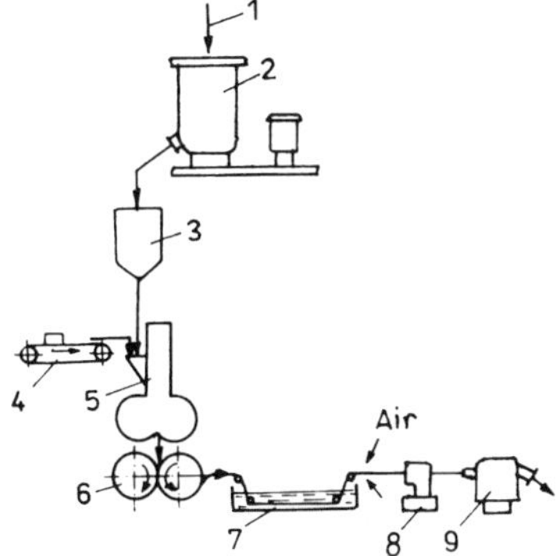

FIG. 8 Flow sheet of the internal mixer compounding line. (1) TPO and other powdery additives. (2) Nonfluxing mixer. (3) Receiving bin. (4) Metering belt for rubber. (5) Internal mixer. (6) Roll mill. (7) Cooling bath. (8) Take-off device. (9) Granulating device.

dimensions of fillers and reinforcements have to be maintained—Figs. 11 and 12.

For the incorporation of glass fibers, these are processed as roving and then cut at the necessary dimensions. This operation is done before feeding into the extruder or directly by the screw of the extruder.

The reasons that some additives (fillers, reinforcements, pigments, etc.) are fed into the extruder after the melting of the blend are mentioned in various papers [47, 54, 81].

Recent developments are related to the intensification of compounding processes in one of the following ways [82]:

Improvement of compounding equipment;
Use of technological aids;
Optimization of the technologies and installations.

As recently twin- and multiple-screw extruders are preferred, the new developments include modifications of screw geometry (number of entrances, depth of channels, places of intensive mixing zones, etc.), increase of the rotation speeds, increase of the torsion moment, reduction of specific energy consumption: all these in order to increase the production rate, to obtain high grades of mixing, etc.

Among the main companies involved in developing and improving compounding extruders, Werner & Pfleiderer and Hermann Berstorff, Germany; Kobe Steel and Japan Steel Works, Japan; and Farrel Conn. Div., USA should be mentioned.

As technological aids, fatty acids and their salts and esters, heavy amines, fatty alcohols, aliphatic hydrocarbons, LMWPE ($\bar{M}_n = 1000$–5000), APP, PIB, PE waxes, silicone oligomers, polyurethane oligomers, etc., are used, as well as some macromolecular compounds such as polyethylene oxide, LDPE, HDPE, EVA, EPDM, PS, etc. [82]. All these products assure a controlled viscosity of compounds, prevent degradative processes, increase production rates, diminish

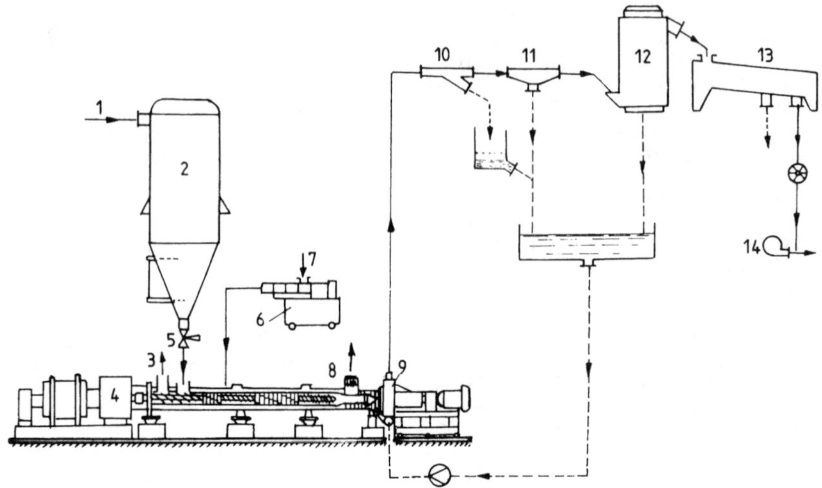

FIG. 9 Flow sheet of a compounding unit for LDPE melt. (1) LDPE melt from polymerization reactor. (2) Melt storage bin. (3), (8) Degasing. (4) Compounding extruder. (5) Valve. (6) Master batch compounding extruder. (7) Master batch components. (9) Granulating device. (10) Switcher. (11) Water separator. (12) Dryer for granules. (13) Vibrating screen. (14) Fan. (From Werner & Pfleiderer company papers, 1989.)

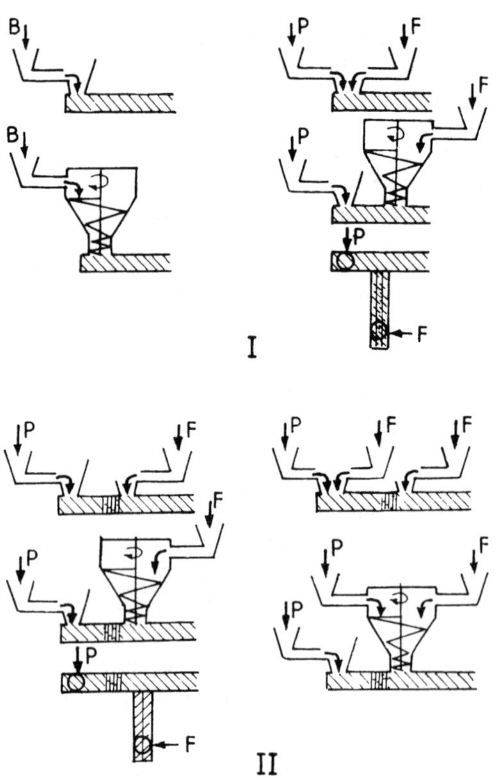

FIG. 10 Possibilities for the feeding of the materials in the compounding extruder (PP, fillers). (I) Filler introduced with solid polymer. (II) Filler introduced in polymer melt (B-blend; P-polymer, F-filler).

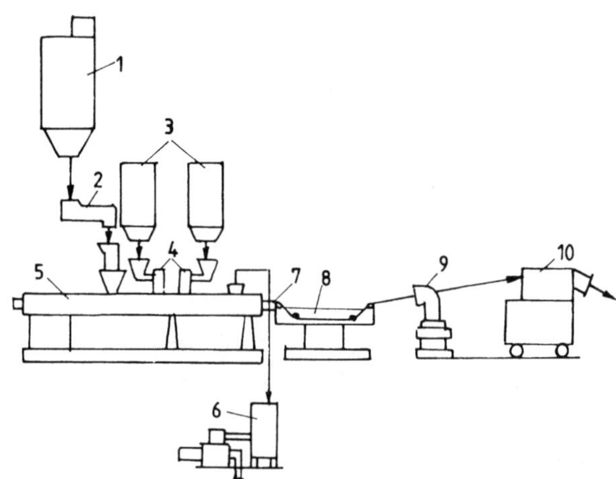

FIG. 11 Flow sheet of the TPO compounding unit by feeding of the pigments in the melt zone of the extruder. (1) TPO use bin. (2) Metering belt. (3) Pigment bins. (4) Gravimetric metering belt. (5) Compounding extruder. (6) Vacuum aggregate. (7) Extruder head. (8) Cooling bath. (9) Take-off device. (10) Granulating device. (From Ref. 54.)

friction between the active parts of the equipment, etc., all these improvements leading to better compounding.

The optimization of compounding technologies and installations aim now to establish the optimum blend composition, depending on the required blend and components characteristics, to choose the most suitable compounding equipment and its operating

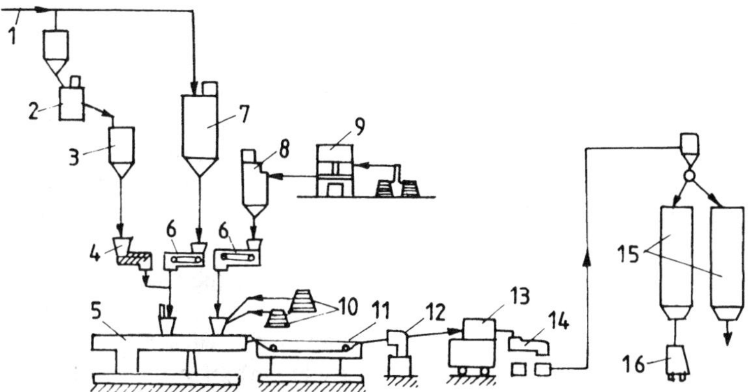

FIG. 12 Flow sheet for the glass fibers reinforced TPO compounding line. (1) Blend components without glass fibers. (2) Nonfluxing mixer. (3) Receiving bin. (4) Metering screw. (5) Compounding extruder. (6) Metering belts. (7), (8) Use bin. (9) Cutting device for glass fibers. (10) Roving. (11) Cooling bath. (12) Take-off device. (13) Granulating device. (14) Vibrating screen. (15) Bin for granules. (16) Packaging device. (From Ref. 54.)

parameters, and also to choose the appropriate technological scheme for the compounding unit [82].

II. ELASTOMERIC POLYOLEFINS

The impressive range of possible applications of elastomers comes about, at least in part, because they can be compounded with other polymers and additives. The nature of the base polymer is of great importance for basic properties of the products manufactured from it, but these properties may be significantly altered by nature and/or the amount of the compounding ingredients used in formulations. On the other hand, the ingredients influence the mixing and processability of rubber compounds, making their crosslinking possible. In designing the formulations, the technical reasons for items that have to meet exacting specifications are of highest importance, while for rubber articles with less demanding service requirements, a lowering of the manufacturing price is possible.

The ingredients used in elastomer formulations may include one or more of the following groups: curing agents, curing accelerators, retarders, plasticizers and extenders, reinforcing fillers and nonreinforcing fillers (extenders), compounding resins (tackifiers, plasticizers), protecting agents (against thermal, oxidative, ozonant, photochemical, and mechanical degradation), pigments and colorants, and other special-purpose additives. As a general rule, the formulations based on classical (crosslinkable) rubbers are more complex (including the curing group) and are realized by manufacturers or by specialized suppliers of tailored compounds. Compounds based on thermoplastic rubbers do not require a curing group, are simpler, and they are often supplied ready for molding (forming).

A skilled compounder should be able to make use of the opportunities available in the attempt to meet the specifications of a given compound, which often include conflicting requirements that must be balanced to obtain a product with optimal cost-to-performance ratio.

A. Additives for Thermoplastic Elastomeric Polyolefins [83–86]

The thermoplastic elastomers are, by definition, materials that can be processed (at high temperatures) in the same way as the thermoplastic materials (i.e., not requiring curing) with the difference that, in the service temperature range, they exhibit the high elasticity characteristic of vulcanized rubbers. Frequently, the suppliers of TEPO claim that their products are ready for use and offer a range of materials tailored for specified physical characteristics and/or specific applications. Nevertheless, sometimes compounding of TEPO may be necessary for technical reasons and often it is cost-effective.

1. Modification with Thermoplastics and/or Extenders

The processability and elastic properties of TEPO can be modified by addition of other more or less compatible polymers or by extension with process oils (paraffinic grades are recommended). Some commercial grades of TEPO contain LDPE or HDPE in partial substitution of PP. Polyethylene tends to form a pseudohomogeneous system with the EP(D)M phase, and its main function is to improve the elasticity at room temperature, to diminish the hardness and stiffness; the improvement of the processability is inferior to that imparted by polypropylene. The extender oils act as softeners of EP(D)M, phase being fully compatible.

2. Fillers

At our present state of knowledge, the elastic performance and other main physical characteristics of TEPO are in no way improved by fillers. Carbon blacks, active fillers acting as true reinforcing agents of traditional (crosslinkable) rubbers, do not produce TEPO modification apart from a general stiffening of the material. The use of talc may provide some advantages in reducing mold shrinkage and surface hardening.

3. Antidegradants

As a general feature, TEPO based on EP(D)M + PP blends exhibit higher stability than other elastomers to oxidative reactions occurring during processing and service. In respect to the thermo-oxidative degradation, both polymers of the system (EP(D)M and PP) behave in a similar manner and the same antioxidants (0.1–0.2% by weight) are suitable. Amine-derivative antidegradants (frequently used in formulations of curing rubbers) are not usual because they are discoloring and/or staining. Hindered phenol derivatives with typical structure (**I**), where R_1 and R_2 are voluminous, usually identical substituents (e.g., *t*-butyl), *n* is usually 2, and R_3 is a long chain (fatty) alcohol (C_{12}–C_{18}), a di- or tetrafunctional alcohol, are successfully used.

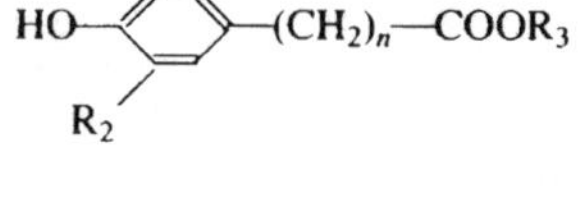

$$HO-C_6H_2(R_1)(R_2)-(CH_2)_n-COOR_3$$

(I)

$$R_1(OH)(R_3)C_6H_2-X-C_6H_2(OH)(R_2)(R_3)$$

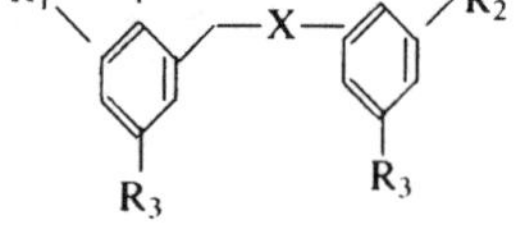

(II)

The long chain in the para position assists dispersion and prevents staining; a higher molecular weight diminishes the volatility and is advantageous for long-term aging, particularly at high temperature. An increase of the molecular weight is possible also by structure (**II**), where R_1, and R_2 are usually *t*-butyl or cycloalkyl, R_3 is $-CH_3$ or another alkyl, and X is usually $-CH_2-$ or $CH_3-CH_2-CH<$. Dialkyldithiopropionates (containing long aliphatic chains for compatibility) are not used alone but in synergistic associations with hindered phenols in order to improve long-term aging behavior of TEPO; they may adversely affect the action of photostabilizers and are not to be used in combination with such antidegradants.

The products intended for use exposed to sunlight may be formulated with photochemical stabilizers (UV stabilizers); whenever it is possible, 1–2% of furnace grade carbon black (with particle diameter of 20–25 nm) represent the least expensive and most efficient stabilizer. In light colored formulations, benzophenone and benztriazole derivatives have been the most popular photostabilizers over the past period; at present, the most powerful class of such stabilizers is represented by HALS, which has a typical structure (**III**):

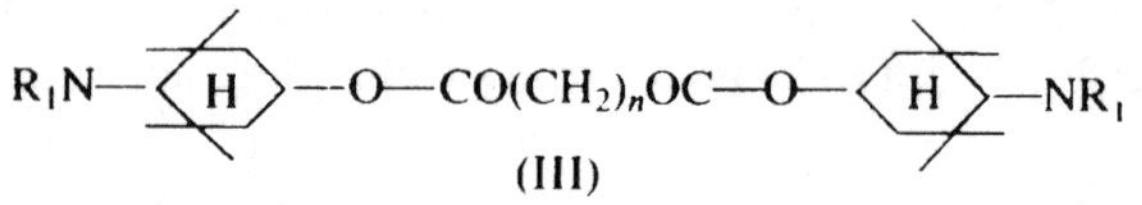

$$R_1N-\langle H \rangle-O-CO(CH_2)_nOC-O-\langle H \rangle-NR_1$$

(III)

where R_1 is H or CH_3 and *n* is 3 or 4.

4. Antiagglomerants

Due to the fact that TEPO are customarily supplied in granular form, separating (antimassing) agents are necessary, particularly for soft (low PP content) granules. When stacked under load for long periods, the agglomeration may be prevented by coating the granules with a thin layer (using 0.2–1% by weight) of silicon-based liquids or fine powders formed from calcium and zinc stearates or HDPE.

5. Crosslinking Agents

Thermoplastic elastomeric polyolefins with superior physical performances are obtained by dynamic vulcanization of the elastomer phase of the system; for these grades, a curing system is used to obtain TEPO.

B. Additives for Butyl Rubbers [86–92]

Butyl rubbers are copolymers of isobutylene with a small amount of isoprene, synthesized by low temperature cationic polymerization. Commercial grades of IIR are distinguished by unsaturation content (0.6–2.2 mole %) and by molecular weight (viscosity).

1. Blends with Other Polymers

IIR is not miscible with general-purpose rubbers because of its dissimilar structure and quite different cure rate, so that blends with these rubbers are not of practical interest. IIR can be blended in any ratio with PIB and with halogenated butyl rubbers (having quite similar structure). Compounds of IIR with polyolefins (LDPE), and particularly with PIB, are encountered in some applications (sheetings, corrosion resistant tank linings). EP(D)M can also be blended with IIR; blends of IIR + EPDM are recommended in the production of inner tubes.

2. Curing Systems [93, 94]

Sulfur curing is widely used in butyl rubber-based compounds. Butyl rubbers require more highly active accelerator systems and lower sulfur loading in order to achieve a suitable complex of physical properties associated with an acceptable high cure rate. Combinations of thiuram and thiazole accelerators are commonly used; dithiocarbamate accelerators give good aging properties and faster cures, but adding some amount of MBTS may be useful to provide processing safety. In order to ensure a very safe processing associated with superior aging properties and low set, sulfur donor blends (3 TMTD/2 DTDM) are used.

Vulcanization with curing resins is a rather slow process requiring high temperatures (for a recent review of the subject see Ref. 95); resulting vulcanizates are characterized by outstanding resistance to high temperature air and steam and also superheated water; the major application is in curing bags and bladders [96].

Curing systems using quinoid derivatives (GMF or dibenzo GMF with oxidizing agents) give very fast cures, and they are used on a large scale in the CV wire insulation process; aging behavior of these vulcanizates is superior to those obtained by sulfur or sulfur-donor systems.

The panorama of vulcanization rates and aging characteristics is illustrated by Table 16.

3. Fillers [97]

Like many other amorphous elastomers, IIR require reinforcing fillers in order to achieve high mechanical characteristics. The types and dosages of fillers used and the effect on processing and on final properties are similar to those of other hydrocarbon elastomers. A maximal response to the highly reinforcing blacks is obtained by heat treatment, when IIR + carbon black is subjected to high temperature mixing (160–200°C); lower temperature is practicable if a chemical promoter is added. Heat treatment gives vulcanizates with improved performances (higher modulus and tensile strength, better resilience and dielectric properties).

Mineral, light-colored fillers are also used, fine particle precipitated silicas being the most effective type of nonblack filler. Careful selection of filler is necessary, particularly when associated with nonsulfur curing systems, because significant undesired interferences are possible.

Reinforcement promoters are sometimes used in black loaded IIR compounds, preferring those that promotes polymer–polymer linkages predominantly, e.g., 1,4-dinitrosobenzene (Polyac) or 1,4-bis(4-nitrosophenyl)piperazine (Promoter 127). A level of 0.2–0.4 phr is adequate and a mixing cycle in which the promoter is completely reacted is essential (needs 3 min at more than 160°C) to avoid undesirable effects on processing safety.

4. Plasticizers and Processing Aids

In order to improve processing, to reduce hardness and modulus, and to improve resilience and set properties, appropriate plasticizers are recommended. A rather high dosage of plasticizers is required when good low temperature behavior is specified. Mineral oil plasticizers are usual in most applications; paraffinic and naphthenic oils are preferred because IIR is characterized by a low solubility parameter; in order to meet, eventually, severe low temperature requirements, ester plasticizers may be useful. Plasticizers (or other additives) incorporating olefinic unsaturation may interfere with crosslinking reactions and are not recommended.

Stearic acid is a usual processing aid in butyl rubber formulations, acting as lubricant and minimizing mill sticking; apart from this action, it also plays the role of activator in the curing system. Hydrocarbon or phenolic tackifying resins are sometimes provided in compounds to assist the adhesion in confection or splicing.

Heat treating agents (discussed in connection with fillers as reinforcement promoters) are sometimes regarded as process aids.

5. Antidegradants

As a general rule, IIR with appropriate unsaturation should be selected for applications with unusual aging potential. Wax blends designed for ozone protection can be added, but the loading must be 8 phr or more because it is more soluble in IIR than in general-purpose rubbers. The ozone resistance of IIR-based

TABLE 16 Curing systems for butyl rubber[a]

Curing system	1	2	3	4	5
MBT	0.5				
MBTS		1			
TMTD	1		3	4	
TEDC		1.5			
Sulfur	2	1.5			
DTDM			2		
GMF				1.5	
Red lead oxide (Pb_3O_4)				5	
CR (Neoprene W)					5
Curing resin (Amberol ST-137)					10
Properties of unvulcanized compounds					
Mooney viscosity ML_{1+4} (100°C)	93	89	87	104	82
Mooney scorch (125°C) t5, min	15	12	>30	2.5	>30
Monsanto rheometer (160°C) t90, min	25	28	65	7	45 (180°C)
Properties of vulcanized compounds (Initial/aged[b])					
Cure temperature, °C	160	160	170	150	190
Cure time, min	25	30	35	11	30
Modulus at 100% elongation, MPa	2.6/0.6	2.3/0.7	1.5/0.8	2.0/1.4	1.9/0.9
Tensile strength, MPa	16.3/0.6	15.4/1.5	17.9/0.5	15.7/5.5	15.8/13.4
Elongation at break, %	480/518	460/620	630/250	560/380	590/350
Hardness, Shore A	65/42	64/50	55/46	64/58	64/56
Compression set (ASTM B) 70 hours at 100°C, %	70	65	76	60	12

[a]Basic recipe (mass parts): 100 butyl rubber (1.6 mol % unsat); 50 N330 (HAF) carbon black; 5 zinc oxide; 1 stearic acid.
[b]Air aged, 48 h at 160°C.
Source: Ref. 91.

vulcanizates can be improved by adding 20–35 parts of EPDM. Chemical antiozonants of the substituted *p*-phenylene diamine type increase the threshold strain but do not brake the growth of cracks once initiated; they are not customarily used in IIR compounds. Nickel dibutyl dithiocarbamate is a useful antiozonant in IIR-based formulations.

6. Examples of Applications

The most important single consumer product is probably the inner tube [98]. The low air permeability associated with good heat, tear, and flex resistance of IIR-based vulcanizates results in long lasting inner tubes that maintain correct inflation pressures longer, increase tire life, and improve driving safety. These attributes have long been recognized and are the reason for the use of IIR in almost all inner tube production. An example of inner tube formulation is presented in Table 17; it is a low cost, easy processing compound based on medium unsaturated IIR that ensures a rather fast cure rate, accommodates high black and oil loading, and provides excellent aging and air retention.

Curing bladders undergo the most severe operating conditions of all compounds produced in any tire factory, and the quality of the bladders can have a significant influence on the efficiency of the whole factory. Over the last decades, IIR bladder technology remained largely unchanged even though over the same period tire technology recorded considerable progress in design, performance, and processing conditions. A typical bladder recipe is presented in Table 18.

Other nontire automotive applications for IIR may include radiator hoses, heater hoses, weather strips and body mounts.

Superior resistance to weathering and aging, good resistance to moisture, and good electrical properties represent the reasons that IIR can be used as insulating

TABLE 17 Inner tube formulation based on butyl rubber

Components	Mass parts	
Butyl rubber 1.6 mole % unsaturation (Polysar Butyl 301)	100	
N660 (GPF) Carbon black	70	
Paraffinic oil	27.5	
Stearic acid	1	
Zinc oxide	5	
MBT	0.5	
TMTD	1.0	
Sulfur	1.5	
Properties of unvulcanized compound		
Mooney viscosity ML_{1+4} (100°C)	45	
Garvey extrusion		
rate, cm/min	64	
die swell, %	60	
appearance	B10	
Properties of vulcanized compound, press cured 10 min at 171°C	**Initial**	**Aged[a]**
Hardness, Shore A	48	55
Modulus at 100% elongation, MPa	1.4	2.0
Modulus at 300% elongation, MPa	4.8	6.9
Tensile strength, MPa	11.9	9.5
Elongation at break, %	640	460
Tear strength (ASTM Die C), kN/m	38	

[a]Air aged, 70 h at 125°C.
Source: Ref. 91.

materials in high and low voltage cables, cable jackets, moisture resistant gaskets, and splicing and terminating. Examples of jacket compounds are presented in Table 19.

Industrial rubber products including conveyor belts, roll covers, gaskets, and sheeting are obtained from IIR-based compounds. The building and construction industries are other important consumers for roofing, flashing, pad lining, and waterproofing. The chemical industry uses a high tonnage of IIR compounds for chemical resistant linings.

C. Additives for Halobutyl Rubbers [99–101]

Halobutyl rubbers are obtained by reacting chlorine or bromine with IIR in solution, in a continuous process. Virtually all of the halogenation takes place at the isoprenyl portions of the chains; steric hindrance of the double bond favors substitution rather than addition, and most of the original unsaturation is retained, although it is largely isomerized (only an isoprenic unit is presented here):

$$-CH_2-\overset{CH_3}{\overset{|}{C}}=CH-CH_2- \xrightarrow{X_2} -CH_2-\overset{CH_3}{\overset{|}{\underset{\oplus}{C}}}-\underset{X}{\underset{|}{CH}}-CH_2- \rightleftarrows \quad X^{\ominus}$$

$$-CH_2-\overset{CH_3}{\overset{|}{\underset{\oplus}{C}}}-CH-CH_2- \longrightarrow -CH=\overset{CH_3}{\overset{|}{C}}-\underset{X}{\underset{|}{CH}}-CH_2- \qquad X^{\ominus}\ X \qquad HX$$

Based on the same virtually saturated PIB backbone, bromobutyl and chlorobutyl rubbers possess the same low permeability toward air, gases and moisture vapor of regular butyl rubbers. Their vulcanizates exhibit similar elastic features and higher vibration dampening in comparison with general-purpose (dienic) elastomers. The vulcanizates of CIIR and BIIR obtained with many curing systems

TABLE 18 Curing bladder recipe based on butyl rubber

Components		Mass parts	
Butyl rubber 1.6 mole % unsaturation (Exxon Butyl 268)		100	
CR (Neoprene WRT)		5	
N339 (HAF) Carbon black		50	
Paraffinic oil		5	
Zinc oxide		5	
Curing resin		7.5	
Properties of unvulcanized compound			
Mooney viscosity ML_{1+8} (100°C)		70	
Mooney scorch (135°C) t5, min		98	
Properties of vulcanized compound, press cured 30 min at 195°C	Initial	Air aged[a]	Steam aged
Hardness, Shore A	48	71	59
Modulus at 100% elongation, MPa	1.2		1.7
Modulus at 300% elongation, MPa	4.4	3.6	6.4
Tensile strength, MPa	12.5	4.4	12.3
Elongation at break, %	680	380	550
Tension set after 300% elongation, %	8	broken	12
Tear strength (ASTM Die C), kN/m	34.7	18.8	34.8

[a]48 h at 180°C.
Source: Ref. 96.

will exhibit similar vulcanizate stabilities, higher than sulfur vulcanizates of IIR at elevated temperature. The main difference between CIIR and BIIR consists in the energy of the carbon–halogen bond (79 kcal/mol for C—Cl and 66 kcal/mol for C—Br), allowing us to expect that BIIR would be inherently more reactive.

1. Blends with Other Elastomers [99, 101]

The thermodynamics of mixing predict that, as a general rule, the polymers will form a two-phase system, and experimental (microscopic) observations support this prediction. The problem in blend curing is to obtain a degree of interfacial crosslinking between dissimilar phases.

In CIIR + BIIR blends, both polymers have identical backbones with similar reactive functionalities; they are miscible at molecular scale, representing an ideal case for covulcanization. CIIR and BIIR can be used interchangeably without significant effect on the state of cure, but BIIR will increase the cure rate.

In IIR + CIIR or IIR + BIIR blends, the elastomer structures are essentially similar, but the different reactive functionalities provide completely different vulcanization behavior. In this situation it is recommended to avoid accelerator systems that will over-cure the halobutyl rubber phase. For BIIR + IIR compounds, as an example, the recommended cure systems may contain, in addition to zinc oxide and stearic acid, one of the following two associations: (i) sulfur 1 phr, MBTS 1.5 phr, TMTD 0.5 phr; (ii) sulfur 1 phr, TBBS 1 phr, TMTD 1 phr.

Halobutyl rubbers are compatible with EP(D)M in all proportions and can be covulcanized using zinc oxide–sulfur cure systems. The properties of the resulting blends are intermediate between those of the participating elastomers, taking into account their ratio. The addition of halobutyl rubbers is of special benefit in blends of EPDM and general-purpose rubbers. In such compounds, EPDM imparts ozone resistance, but it is not sufficiently compatible with dienic elastomers to yield satisfactory dynamic properties. The addition of halobutyl rubbers produces a ternary blend with greatly improved dynamic performance, ozone resistance, and flex resistance, while maintaining other desirable properties of these blends.

Although halobutyl rubbers and natural rubber have quite dissimilar structures and degrees of functionality (and different crosslinking chemistry, too), they can be blended in all proportions to attain desired

TABLE 19 General purpose insulation recipe based on butyl rubber

Components		Mass parts	
Butyl rubber 0.7 mole % unsaturation (Polysar Butyl 100)		100	
Polyethylene (Allied 617-AC)		5	
Zinc oxide		15	
Stearic acid		0.5	
Antioxidant BPH		1.5	
Hydrous magnesium silicate (Mistron Vapor)		25	
Calcined hard clay		100	
Microcrystalline wax		5	
DTDM sulfur donor		1.5	
Accelerator TMTD		1	
Accelerator MBT		1	
Accelerator ZMDC		1	
Properties of unvulcanized compound			
Mooney viscosity ML_{1+4} (100°C)		53	
Mooney scorch (125°C) t5, min		17	
Properties of vulcanized compound, press cured 20 min at 165°C	Initial	Air aged[a]	Bomb aged[b]
Hardness, Shore A	64	66	67
Modulus at 300% elongation, MPa	2.3	3.0[c]	3.0[c]
Tensile strength, MPa	7.0	4.5	5.7
Elongation at break, %	690	490	660
Static ozone resistance (ASTM D1352, 0.03 vol.% ozone, room temperature)	No damage		

[a]14 days in air circulation oven at 121°C.
[b]40 hours in air bomb at 127°C and 0.55 MPa.
[c]Change evaluated from variation of the modulus at 100% elongation.
Source: Ref. 88.

compound properties. The addition of NR to a halobutyl compound is beneficial in (i) increasing green strength and building tack (proportionally with the amount of NR); (ii) improving cured adhesion to high unsaturation compounds; (iii) raising physical properties (modulus, tensile strength) depending on the cure system used.

Styrene–butadiene rubbers can be blended also in all proportions with halobutyl rubbers. Usually, halobutyl rubbers comprise a major part of these blends to ensure maximum resistance to heat and weathering. Styrene–butadiene rubbers do not impart a sufficiently high level of building tack; for that reason some NR may be necessary for blends intended for manufacturing items.

2. Curing Systems

Halobutyl rubbers contain two types of functionalities that enable them to vulcanize with zinc oxide alone. Three parts of zinc oxide are sufficient for cure, but five parts are sometimes recommended and it may be effective in enhancing heat resistance (Fig. 13). Current theory explains that zinc oxide produces C—C crosslinks, ensuring excellent heat stability; most of the zinc oxide forms, during cure, zinc halide, which may be the actual curative and ingredients inhibiting the conversion of zinc oxide to zinc halogenide delay the crosslinking reactions.

Magnesium oxide can function as an activator, scorch retarder, and cure modifier. As an acid acceptor, magnesium oxide activates amine cure and improves heat resistance of thiuram–thiazole cures. It is effective as a scorch retarder with all cure systems (except amine systems). In carbon black filled compounds, magnesium oxide has a marked effect on scorch safety even at low level (0.25 phr), lengthening the available processability time. In mineral filled compounds, magnesium oxide can function in a role

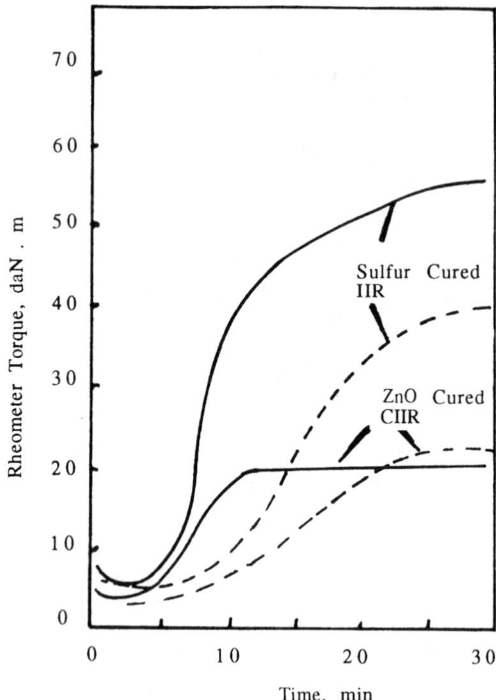

FIG. 13 Comparison of the vulcanization behavior of butyl rubber (Exxon Butyl 268) with sulfur + accelerator and chlorobutyl rubber (Exxon Chlorobutyl 1066) with zinc oxide: (- - - -) at 149°C; (-) at 166°C. Cure system for butyl rubber: MBT, 0.5 phr; TMTD, 1.0 phr; sulfur, 2.0 phr. Cure system for chlorobutyl: ZnO, 5.0 phr; stearic acid, 2.0 phr.

similar to that provided by zinc oxide, particularly with the more active fillers such as hydrated silicas.

Lead oxides (litharge, red lead) should be used in compounds intended for high resistance to hot water and steam.

Curing halobutyl rubbers in blends or in contact with other compounds requires curing systems compatible with the other elastomers and in this case sulfur is customarily used. As little as 0.5 phr of sulfur alone produces, e.g., a significantly high state of cure in BIIR (approximately 0.35 phr of sulfur would be required to realize stoichiometric crosslinking by monosulfidic linkages). Use of sulfur as the only curative results in compounds that undergo reversion; the cured structure can be stabilized by adding zinc oxide. Calcium hydroxide may provide a similar effect, but the unsatisfactory dynamic properties of the resulting vulcanizates prevent the use of such systems.

The association of elemental sulfur with zinc oxide into the curing system of a halobutyl rubber lowers the heat resistance of the vulcanizate, but it provides an easy means of controlling the modulus: the increase of sulfur dosage leads to the increase of the modulus. Benzothiazole and sulfenamides are the most widely used primary accelerators, and the obtained vulcanizates display a good balance of properties. Thiuram di- and polysulfides are used extensively because they give compounds with good physical properties at a rather low cost. At low levels, particularly in halobutyl rubbers + NR blends, thiurams are scorch retarders, but at increased dosages, their retarding effect is less pronounced. Therefore they are used in association with other curatives (elemental sulfur, DTDM) ensuring higher states of cure and good processing safety. Zinc dithiocarbamates, at rather low dosages to avoid scorchiness, are used in applications requiring high crosslinking density, ensuring a low compression set. Good vulcanizate properties are obtained with alkylphenol disulfides (Vultac series), but their relatively high reactivity can diminish processing safety.

Halobutyl rubbers can be crosslinked with phenolic resins by the same mechanism postulated for regular butyl rubbers. The usual association with zinc oxide ensures a high state of cure and the vulcanizates have excellent dry heat resistance. However, they are inferior to resin cured regular butyl rubber in resistance to steam and superheated water. Since halogen is present in the polymer itself, another halogen donor is not necessary in the resin cure of halobutyl rubbers.

Regular butyl rubbers undergo molecular mass reduction under the action of peroxides, but halobutyl rubbers (particularly BIIR) are capable of being crosslinked with peroxides (1–2 phr). In order to attempt an optimum cure and optimum properties, a suitable coagent is required (metaphenylene bismaleimide, triallyl cyanurate, polyfunctional methacrylates).

Amines with enough reactivity are able to react with the halogen present in halobutyl rubbers. By this means, grafted antioxidants can be obtained, and with difunctional amines the crosslinking is possible [102, 103]. The presence of zinc oxide may be favorable, but some special derivatives (like Diak No. 1, Du Pont de Nemours) are able to vulcanize alone, allowing us to obtain zinc-free vulcanizates.

3. Fillers

The response of halobutyl rubbers to carbon blacks is very similar to that of regular butyl rubbers and other synthetic rubbers. To summarize the effect on the main processing and vulcanizate characteristics, the following chart has been developed [99].

	Optimum black property		
	Particle	Structure	Surface activity
Processing operations			
mixing	Large	High	Little effect
calendering	Small	High	Little effect
extruding	Small	High	Little effect
Vulcanizate properties			
reinforcement	Small	Low*	High
high modulus	Small	High	Little effect
abrasion resistance	Small	Low	High
tear strength	Small	Normal	High/normal

* With medium or large particle size, the effect of structure is not important.

A wide range of available mineral fillers may be used with halobutyl rubbers. Silicas give maximum reinforcement among mineral fillers, but they should be added at moderate levels to avoid compound stiffening and interference in vulcanization by adsorption of the components of the cure system; additional dosage of curatives or adsorption competitors (like polyethylene glycols) may be useful. Clays (soft or hard) can be provided in formulations, bearing in mind that they vary in surface pH and therefore may affect the crosslinking process; high clay dosage may cause mill sticking, and require the use of release agents. Talc has a low reinforcing effect (like soft clays), but it does not interfere in curing. Calcium carbonate (whiting) is mainly used as an extender for reducing compound cost and eventually lowering the compound viscosity; it is neutral toward halobutyl cure. Surface-treated grades must be avoided, because the usual modifiers may influence the cure rate. In order to enhance polymer–filler interactions with mineral fillers, the silanes, mercapto silane $HS(CH_2)Si(OCH_3)_3$ and aminosilane $H_2N(CH_2)$—$Si(OCH_3)_3$ are useful; the mercapto or amino groups react with the halogenated unit of the polymer, while alkoxy groups hydrolize and react with the filler surface [101]. Silane pretreated fillers are also on the market.

Combinations of two (seldom three) mineral fillers can be used to obtain a balance of properties and cure rate effects.

4. Plasticizers, Tackifiers, and Processing Aids

Petroleum derived processing oils are customarily used as plasticizers; aliphatic and naphthenic oils are preferred, while aromatic oils with higher solubility parameters, more polar in nature, are poorer solvating agents and can result in the alteration of low temperature flexibility. The unsaturated function present in aromatic oils can also interfere in sulfur vulcanization. For low temperature flexibility, ester-type plasticizers (adipates, sebacates, pelargonates) can be employed.

Among tackifiers recommended for use in halobutyl rubber compounds are nonreactive alkylphenol formaldehyde resins (Amberol ST 149, Rohm & Haas; SP-1045, Schenectady), wood rosin, rosin esters, hydrogenated rosin derivatives, hydrocarbon resins (Escorez, Exxon), polyisobutylene (Vistanex, Exxon), and polybutene.

Paraffin waxes, their mixtures with petrolatum, and low molecular mass polyethylenes are often used as processing aids; Struktol 4OMS (Schill & Seilacher) and mineral rubber also improve processing behavior and can enhance adhesion between halobutyl and unsaturated rubbers.

5. Antidegradants

Halobutyl rubbers with almost completely saturated backbones usually do not require extra antioxidants or antiozonants; the rubber producer adds traces of stabilizers. For severe service circumstances, particularly in blends with unsaturated rubbers, the addition of some antidegradant may be necessary. In the selection of any antidegradant care should be exercized, particularly regarding amine derivatives that may interfere in the crosslinking process, but with proper technology vulcanizates with enhanced stability may be obtained [103]. Ozone resistance and weathering behavior of halobutyl rubber vulcanizates are also dependent on the curing system, on the nature and amount of fillers and plasticizers.

6. Examples of Applications

The tire industry is probably the most important consumer of halobutyl rubbers for inner liners, tubes, tire sidewall compounds, and tire curing members. Other applications include heat resistant conveyor belts, steam hoses, gaskets, pharmaceutical stoppers, chemically resistant tank linings, adhesives, and sealants.

Halobutyl rubbers possessing an extremely low air permeability and compatibility with general-purpose rubbers are suited to inner liner formulations, minimizing air and moisture diffusion, allowing better retention of the proper inflation pressure inside the tire, and minimizing intracarcass pressure and degradation [101].

Inner linear type	Relative inflation retention, %	Intracarcass pressure kPa	Wheel durability km
100 General-purpose rubbers	100	147	32,600
60/40 CIIR/general-purpose rubbers	136	119	56,100
100 CIIR	254	42	72,400

An example of an inner liner compound based on BIIR and NR is presented in Table 20. In practice, blends of halobutyl rubbers with NR or BIIR + CIIR are used. A medium reinforcing carbon black such as GPF (providing adequate green strength and flow properties) associated with paraffinic or naphthenic oils is preferred; a simple curative system provides satisfactory curing results, and MgO is suggested as retarder.

In formulation of the compounds for inner tubes, halobutyl rubbers offer improved resistance to heat softening and growth, associated with desired butyl rubbers properties of excellent air retention, particularly important in severe service conditions (high speed, heavy loading). In Table 21 is presented a practical recipe using Zn0 as curing agent for maximum heat resistance; although initial tensile strength is slightly below the usual level for butyl rubber tubes, properties after aging under severe conditions are superior.

Halobutyl rubber-based compounds are attractive for use in conveyor belts, particularly for conveying hot materials or passing through high temperature zones, because they exhibit good heat, flex, tear, and

TABLE 20 High performance tire inner liner compound based on bromobutyl rubber

Components	Mass parts	
Bromobutyl (2030, Polysar)	80	
Natural rubber, masticated	20	
N660 (GPF) Carbon black	60	
Stearic acid	1	
Tackifying resin (Pentalin A, Hercules)	4	
Paraffinic oil (Sunpar 2280, Sun Oil)	15	
Accelerator MBTS	1.5	
Zinc oxide	3	
Sulfur	0.5	
Properties of unvulcanized compound		
Mooney viscosity ML_{1+4} (100°C)	41	
Mooney scorch (125°C) t5, min	24	
Monsanto Rheometer (165°C)		
ML/MH, daN · m	4.9/15.5	
Optimum cure t90, min	9.7	
Garvey die extrusion (104°)		
Rate, cm/min	82	
Die swell, %	70	
Tel-tak true tack, kPa	22	
Properties of vulcanized compound press cured 10 min at 165°C	**Initial**	**Aged[a]**
Modulus at 100% elongation, MPa	0.7	3.1
Modulus at 300% elongation, MPa	3.5	5.0
Tensile strength, MPa	9.5	5.1
Elongation at break, %	580	310
Tear strength (ASTM Die C), kN/m	28	
Peel adhesion (23°C), kN/m	27.1	
DeMattia flex for 600% cut growth, kc		116

[a]Aged in air, 168 h at 120°C.

Source: Ref. 101.

TABLE 21 Heat resistant inner tube based on chlorobutyl rubber

Components	Mass parts	
Chlorobutyl rubber (1068, Exxon)	100	
N660 (GPF) Carbon black	60	
Paraffinic oil (Flexon 840, Exxon)	20	
Stearic acid	2	
Zinc oxide	10	
Properties of vulcanized compound press cured 8 min at 166°C	**Initial**	**Aged[a]**
Modulus at 300% elongation, MPa	3.7	
Tensile strength, MPa	8.6	3.0
Elongation at break, %	590	370
Hardness, Shore A	43	63

[a]Air aged, 24 h at 163°C.
Source: Ref. 102.

abrasion resistance. Formulations recommended for heat resistance conveyor belts are given in Table 22.

Halobutyl rubbers are widely used in pharmaceutical closures due to the following characteristics: chemical and biological inertness; low extractibility; resistance to heat (sterilization), ozone, and light; low permeability to gases and vapors; and adequate self-sealing and low fragmentation (during needle penetration).

TABLE 22 Heat resistant conveyor belts based on chlorobutyl rubber

	Mass parts	
Components	Carcass	Cover
Chlorobutyl rubber (1066, Exxon)	100	98
Natural rubber, masticated		2
N550 (FEF) Carbon black	25	
EPC Carbon black	20	
N110 (SAF) Carbon black		50
Paraffinic oil (Flexon 765, Exxon)	10	5
Stearic acid	1	1
Phenolic adhesion resin (Amberol ST-149, Rohm & Haas)	3	3
Antioxidant 2246	1	1
Magnesium oxide (Maglite D, Merck)	1	1
Zinc oxide	5	5
Accelerator TMTD	1	1
Accelerator MBTS	2	2
Properties of vulcanized compounds (Initial/aged[a]) press cured 40 min at 153°C		
Modulus at 100% elongation, MPa	4.4/2.2	8.3/3.5
Tensile strength, MPa	14.1/3.5	19.1/4.7
Elongation at break, %	740/175	580/175
Hardness, Shore A	55/64	69/80
Adhesion, MPa		
to cotton	0.24	
to rayon	0.22	
to polyester	0.10	

[a]Air aged, 16 h at 193°C.
Source: Ref. 99.

TABLE 23 Pharmaceutical stopper compounds based on bromobutyl rubber

Components	Mass Parts	
	Sulfur free	Zinc/sulfur free
Bromobutyl rubber (2244, Exxon)	100	100
Calcined clay (Whitetex, Freeport Kaolin)	90	90
Polyethylene AC-617A (Allied Chemicals)	3	3
Zinc oxide	3	
Curing resin (Amberol ST-137, Rohm & Haas)	1	
Hexamethylene diamine carbamate (Diak No. 1, Du Pont de Nemours)		0.5
Properties of vulcanized compounds press cured at 180°C		
Modulus at 100% elongation, MPa	0.9	0.8
Modulus at 300% elongation, MPa	1.8	1.8
Tensile strength	6.2	5.8
Elongation at break, %	810	740
Compression set (22 h at 70°C), %	34	46
Hardness, Shore A	42	39
Extraction results (DIN 58367)		
pH change	−0.5	−1.0
Reducing substances, ml $KMnO_4$ 0.01N/10 ml	0.02	0.04
Metal traces, ppm		
zinc	0.4	0.04
lead	<0.2	<0.2
cadmium	<0.01	<0.01
calcium	0.2	<0.1
magnesium	<0.05	<0.05
iron	<0.02	<0.02

Source: Ref. 101.

Two typical compounds are given in Table 23: one is resin cured (avoiding sulfur) and the other eliminates sulfur as well as zinc oxide using a diamine curative. Calcined clay is preferred as filler; silica (Ultrasil VN3, Degussa) can be used in addition to improve hot tear resistance, but it should be limited because of its retarding effect in this recipe.

Butyl and halobutyl rubbers are widely used in adhesive formulations. Formulations can be adapted for various requirements, e.g., for easy removal from

TABLE 24 Compounds based on chlorobutyl rubber for pressure sensitive adhesives

Components	Mass parts	
	For transparent film	Pigmented, for paper tape
Chlorobutyl rubber (1066, Exxon)	50	50
SBR 1011	50	50
Whiting		100
Magnesium oxide (Maglite K, Merck)	1	1
Tackifying resin (SP-567, Schenectady)	30	20
Chlorinated hydrocarbon (Arochlor 1254, Monsanto)	20	
Tackifier (Indopol H-1900, Amoco)		40

Source: Ref. 99.

glass, for high adhesion to polyolefin plastics, etc. Halobutyl rubbers can be blended with general-purpose rubbers to obtain the desired performances. Typical CIIR + SBR blend formulations are presented in Table 24.

D. Additives for Ethylene–Propylene and Ethylene–Propylene–Diene Rubbers [104–108]

Amorphous and curable materials can be obtained by copolymerizing ethylene and propylene with certain catalysts of the Ziegler–Natta type. The resulting ethylene–propylene rubbers do not contain unsaturation and can only be crosslinked with peroxides. If we add a third monomer during copolymerization of ethylene and propylene, the resulting ethylene–propylene–diene rubbers will have unsaturation, meaning that it became possible to vulcanize them with sulfur and other crosslinking agents. The majority of commercial EPM grades contain 45–85 mole% (40–80 mass%) of ethylene and the most important grades about 50–70 mole%. Many compounds are quoted in the literature as termonomers, but in commercial rubbers only three nonconjugated dienes are used: trans-1,4-hexadiene, ethylidene nonbornene, and dicyclopentadiene.

In resulting EPDM the double bond resides inside groups of the polymer chain. During EPM and EPDM production, the following main parameters can be adjusted to provide the different commercial grades that are currently available: (i) concentration ratio of ethylene and propylene (amorphous or segmented grades); (ii) co- or terpolymerization; (iii) type and concentration of termonomer (vulcanization behavior, physical properties of vulcanizates); (iv) solution or suspension polymerization; (v) molecular weight and molecular weight distribution (viscosity and processability); and (vi) oil extension.

1. Blends with Other Polymers

Etylene–propylene rubbers and low unsaturation grades of etylene–propylene–diene rubbers are used in blends with saturated polymers, most usually with polyolefins (eventually precrosslinked) in order to produce thermoplastic elastomers or elastomer-modified plastomers. Sequential EPDM grades are preponderantly used in such applications. In compounds with IIR, EPDM can improve some manufacturing characteristics and reduce the permanent set of vulcanizates. Blending diene polymers and copolymers (IR, BR, SBR, NBR) with EPDM improves their ozone and weathering resistance. Adding EPDM, about 30% by weight (selecting ultrafast ENB-containing grades), is usual in such applications in order to solve the problem of covulcanization and aging stability at the same time.

2. Curing Systems [109]

Etylene–propylene rubbers can only be crosslinked with peroxides. The presence of the tertiary carbon in polypropylene prevents the crosslinking process and instead leads to chain scission; the copolymers of ethylene and propylene can be crosslinked with peroxides, providing the ethylene concentration is high enough (at least 50 mole%) and rather well distributed along the polymer chain. Terpolymers are increasingly crosslinked with peroxides and they are more efficiently cured than copolymers. DCP containing EPDM give higher cure rates with peroxide in comparison with EPM and ENB containing EPDM. Bearing in mind the curing temperature and the desired physical characteristics of the vulcanizates, a careful selection of peroxide and of coagent is necessary. At the same time, some general rules, as presented in Table 25, should be followed in the formulation of the recipes for peroxide vulcanization of EPDM. The main

TABLE 25 Guide for formation of EPDM compounds for peroxixde vulcanization

Components	Mass parts
EPDM	100
Zinc oxide (for better heat stability)	5–15
Zinc stearate	0.5–1.5
Carbon black (channel grades to be avoided)	variable
Mineral filler (untreated clays and fillers with acid pH to be avoided)	variable
Paraffinic oil	0–90
Antioxidant (e.g., TMQ)	1–2
Peroxide (e.g., 2,5-dimethyl-2,5-di(*t*-butylperoxy)hexane 50% on inert filler)	3–10
Coagent (e.g., *m*-phenylene-bismaleimide)	1–2

Source: Ref. 110.

advantages in peroxide curing are fast curing without reversion, best retention of physical properties after heat aging, excellent compression set, better electrical properties (and avoidance of sulfur in electrical isolation of copper leads) and good color stability during cure and weathering. At the same time, the following disadvantages should be known: higher cost, cure impossible in air, slightly lower physical properties (much poorer tear strength).

The limited amount of cure sites and reactivity in EPDM requires higher levels of curatives than do the diene polymers; ENB grades require 0.5–2 phr sulfur or equivalent amounts of sulfur donors (e.g., DTDM). They can be cured by sulfur and thiazoles/sulfenamides, but with these systems the cure rate may be unsatisfactory (except for fast curing grades). Frequently, thiurams and dithiocarbamates are used as accelerators, and in order to avoid undesired blooming of some accelerators (or their reaction products) and to increase cure rate, mixtures of synergistic accelerators are recommended; optimized preblend accelerator systems are on the market. Reducing the free sulfur (to 0.5 phr or less) and replacing it with a sulfur donor results in an improvement in the heat aging; 5 phr or more zinc oxide are also required for additional activation of sulfur curing systems. Typical sulfur curing systems for EPDM are illustrated in Table 26. In order to obtain the desired fast cure and a good balance of physical characteristics, a proper selection of the EPDM grade is particularly important. In pressureless continuous curing techniques a high viscosity polymer should be used associated with an

TABLE 26 Typical sulfur cure systems for EPDM

	Low cost	Fast cure, nonbloom		Improved heat resistance and compression set	
Components of the cure system	1	2	3	4	5
Sulfur	2.0	2.0	2.0	0.5	0.5
DTDM				1.0	
MBT	1.5	1.5	1.0		
TMTM	1.5				1.5
TMTD		0.8	0.5		
DPMT		0.8			0.5
ZDBC			2.0	3.0	4.0
TETD				3.0	
TDEC		0.8	0.5		
ZDMC				3.0	
Zinc oxide	5.0	5.0	5.0	5.0	5.0
Properties of uncured compounds					
Mooney viscosity MS_{1+4} (121°C)	10	10	11	9	10
Mooney scorch (121°C) t5, min	50	18	21	50	50
Properties of vulcanized compounds (Initial/aged[a]) press cured 20 min at 166°C					
Tensile strength, MPa	15.7/18.7	16.4/22.3	15.4/23.5	15.2/17.8	13.9/17.2
Elongation at break, %	410/130	390/130	370/140	520/230	660/300
Hardness, Shore A	64/76	64/73	62/73	61/69	59/69
Tear strength (ASTM Die C) kN/m	29	27	27	31	37
Compression set (ASTM Method B)[b]					
after 22 h at 70°C, %	18	19	19	18	24
after 70 h at 149°C, %	85	80	70	51	65

[a]Air aged, 7 days at 149°C.
[b]Pellets cured 25 min at 166°C.
Source: Ref. 110.

effective dessicant; use of a vacuum extruder is recommendable. The compounds designed for microwave curing should contain 80 phr or more HAF black to improve the microwave receptivity.

Ethylene–propylene–diene rubbers can also be cured using reactive phenolic resins [111].

3. Fillers

In the case of EP(D)M, all types of fillers can be used. These rubbers, particularly the ones of high molecular weight, accept large amounts of fillers (associated with plasticizers), allowing us to obtain final products with usable properties.

A careful selection of carbon black type and amount allow us to obtain maximum economic benefits while maintaining also the desired processability and product quality. Amorphous and low viscosity polymers require filling with about 70 phr of N330 (HAF) or N550 (FEF) carbon blacks. In the case of crystalline polymers and high molecular weight polymers and crystalline polymers, large amounts of N650 (GPF) or N762 (SFR) carbon blacks can be used for reinforcement. At comparable viscosities, EPDM accept higher loading than EPM. Figure 14 presents a general picture of the effect of carbon blacks in EPDM.

Calcined clays and fine particle calcium carbonates are used to lower the cost and improve the processability of light compounds or to lower the cost of

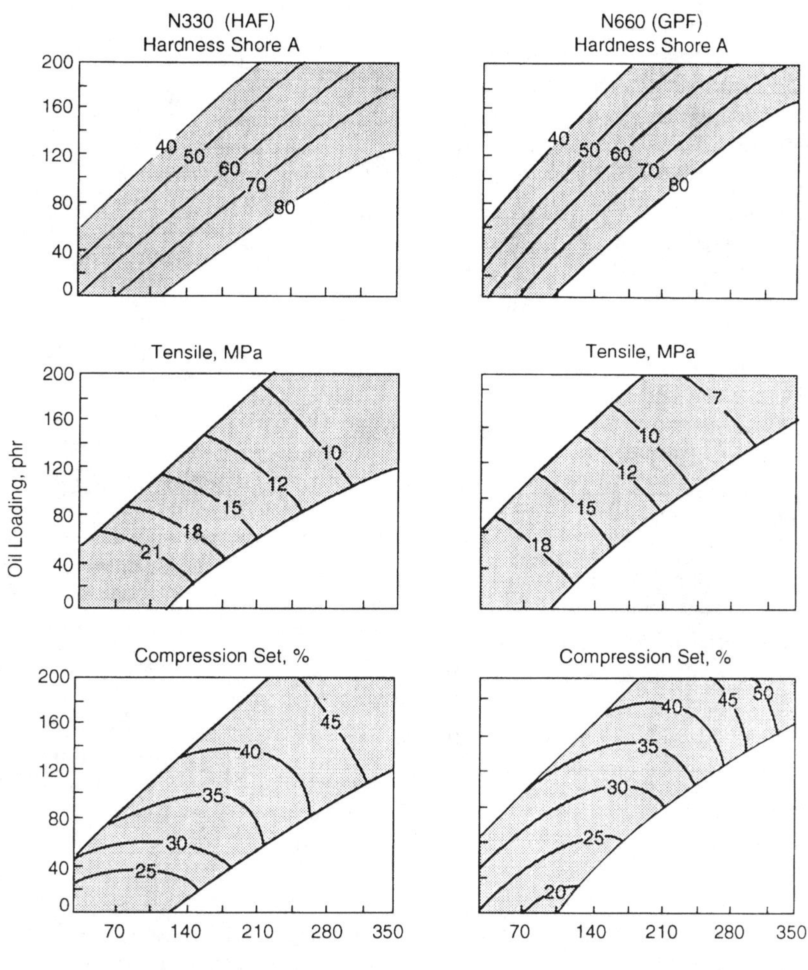

FIG. 14 Influence of carbon black and oil loading in EPDM. Test recipe, mass parts: EPDM (very fast curing, medium viscosity grade), 100 phr; carbon black, varied; paraffinic oil, varied; zinc oxide, 5 phr; stearic acid, 1.0 phr; accelerator MBT, 0.25 phr (increasing at the rate of 0.125 phr per 100 phr oil); accelerator TMTD, 1.00 phr (increasing at the rate of 0.375 phr per 100 phr oil); sulfur, 1.00 phr. Carbon blacks (Cabot): N330, HAF (Vulcan 3); N660, GPF (Sterling V). (From Ref. 112.)

black compounds; calcined clays also reduce the water absorption of vulcanizates, a desired characteristic for cable jackets. Hard clays give good reinforcement but their use is limited because they may retard the vulcanization. A good cost-to-performance compromise can be achieved using a blend of hard and calcined clays. Nonreinforcing fillers like soft clays and ordinary whitings are customarily used only to reduce compound cost; they are often used in sponge compound formulations.

4. Plasticizers, Tackifiers, and Processing Aids

All petroleum oils are compatible with EP(D)M; the most widely used are naphthenic oils. Paraffinic oils of low volatility are recommended for peroxide cure or for products intended for high temperature use. At low temperature, the paraffinic oils (added in large quantity) exude from EPDM vulcanizates or from high ethylene EPDM; from this cause it is advisable to replace them partially by naphthenic oils. Aromatic oils reduce mechanical properties of vulcanizates and interfere in peroxide vulcanization; limited amounts (20 phr) may be useful to improve feeding in extrusion stocks. Synthetic plasticizers like saturated organic esters may be used in order to improve the low temperature behavior, but the dosage must be limited bearing in mind their limited compatibility. Chlorinated plasticizers, useful in improving flame resistance, must be used in limited amounts (no more than 30 phr) for the same reason.

Ethylene–propylene and ethylene–propylene–diene rubber compounds have poor building tack, so, when such a property is required (e.g., for assembling operations), the use of tackifers (cumarone resins, xylene formaldehyde resins, etc.) is indispensable.

Processing aids like stearic acid, zinc or calcium soaps of fatty acids and heavy alcohols are particularly useful in improving filler dispersion; calcium soaps are good rheomodifiers, improving extrusion and injection flow.

5. Antidegradants

The use of protective agents is not necessary in EPM compounds. For EPDM compounds, antioxidants are provided, particularly if they are intended for high temperature (150°C or more) service. In black compounds (or when staining is tolerated) aromatic amines (especially phenylene diamine derivatives) offer the best protection; in peroxide cured compounds, TMQ use is recommended. In nonblack compounds, nonstaining antioxidants should be provided; addition of 5 phr chlorosulfonated polyethylene (Hypalon, Du Pont de Nemours) will improve the state of cure and retention of characteristics on aging. EP(D)M vulcanizates possess an intrinsic resistance to ozone, and the use of antiozonants is not necessary.

6. Examples of Applications

Ethylene–propylene and ethylene–propylene–diene rubbers serve a wide range of markets: the tire industry uses EPDM in compounds (with other rubbers) for sidewalls and tubes, while the nontire field absorbs the most important tonnage for hoses, auto parts, hot material conveyor belts, a large variety of sheetings, profiles and roofings for the building industry, cable insulations, and jacketing. Ethylene–propylene and ethylene–propylene–diene rubbers have many uses outside the conventional rubber industry such as viscosity modifiers in lubricating oils and modifiers (for improvement of low temperature behavior) of polyolefins, but the most important amounts are absorbed in the production of thermoplastic elastomers.

Ethylene–propylene and ethylene–propylene–diene rubbers can be used in compounds for the production of many rubber goods for domestic use and for technical purposes. Thus it is used in garden hoses, where low cost, highly extended black compounds are used for the tube and tough black or light compounds are used for the cover. In automotive radiator hose, EPDM-based compounds, sulfur cured, give products with good heat resistance and with the necessary set of mechanical properties, offering a practical, economical solution, as presented in Table 27. EPDM can also be compounded for use in hoses intended for air, water, special fluids (phosphate and ketone), and high-pressure steam (up to 260°C).

EPDM is used for a variety of automotive parts (weather strips, seals, boots, sponge gaskets, etc.). As a blending polymer in white/black sidewall, EPDM confers ozone and flex resistance; addition of EPDM in butyl rubber-based compounds for air tubes imparts specific properties in manufacture and reduces growth in service.

Conveyor belts represent a large tonnage production, and the outstanding characteristics of EPDM regarding ageing, weathering, and ozone resistance, associated with good mechanical performances recommend its use in this field as exemplified in Table 28.

The modern building industry is using large quantities of rubbers (profiles, sheetings) for sealing, insulation, and roofing. Rubber-based sheets for roof insulation represent a high tonnage application where

TABLE 27 EPDM formulations for tube/cover of radiator hose[a]

Components		Mass parts	
EPDM (Vistalon 3708, Exxon)		50	
EPDM (Vistalon 4608, Exxon)		50	
N650 (GPF-HS) Carbon black		150	
Paraffinic oil (Flexxon 876, Exxon)		80	
Zinc oxide		5	
Stearic acid		1	
Accelerator MBT		0.5	
Accelerator TMTD		3.0	
Sulfur		0.5	
Properties of vulcanized compound steam cured 25 min at 170°C	**Initial**	**Air aged[b]**	**Coolant aged[c]**
Tensile strength, MPa	10.4	12.0	11.5
Elongation at break, %	360	260	330
Hardness, IRHD	70	70	66
Volume swell, %			+5.8

[a]Satisfying Specification A of General Motors.
[b]72 h at 125°C.
[c]7 days at 100°C in coolant 50W/50G.
Source: Ref. 113.

TABLE 28 EPDM formulations for heat resistant conveyor belting

Components	Mass parts	
EPDM (Polysar EPDM 345)	100	
EPDM (Polysar EPDM 965)		100
CR (Neoprene W, Du Pont de Nemours)	5	5
N347 (HAF) Carbon black	80	80
Paraffinic oil (Sunpar 2280, Sun Oil)	40	40
Zinc oxide	10	10
Stearic acid	1	1
Antioxidant TMQ	0.5	0.5
Accelerator ZMBT	2.0	2.0
Accelerator TMTD	0.8	0.8
Accelerator MBTS	3.0	3.0
ZDBC Accelerator	1.5	1.5
Sulfur donor DTDM	0.8	0.8
Sulfur	0.5	0.5
Properties of unvulcanized compounds		
Mooney viscosity ML_{1+4} (100°C)	45	95
Mooney scorch (125°C) t5, min	14	7.5
Monsanto Rheometer (160°C)		
ML/MH, daN·m	6.5/41	11/47.5
Optimum cure t90, min	10.1	8.2
Properties of vulcanized compounds (Initial/aged[a]) press cured for t90 at 160°C		
Hardness, Shore A	74/89	73/89
Modulus at 100% elongation, MPa	2.5/9.5	3.1/10.2
Tensile strength, MPa	15.1/15.6	18.8/17.7
Tear strength (ASTM Die B), kN/m	45.5/37.0	46.0/26.0
Abrasion resistance (DIN), mm^3	170/	160/

[a]Air aging, 96 h at 150°C.
Source: Ref. 114.

EPDM successfully competes with other rubbers (butyl, chloroprene, and chlorosulfonated polyethylene) and other materials (PVC, bitumen compounds, and composites), offering quite satisfactory ageing/weathering characteristics associated with good/excellent mechanical properties at a rather low cost. Table 29 illustrates such an application.

The excellent electrical properties associated with water resistance, heat resistance and toughness recommend EP(D)M for a large number of electrical applications. They are used in the insulation of power cables and ignition wiring because of their excellent response to high voltage and heat resistance, and in cable cover where abrasion and weathering (but not oil) resistance are important; good water resistance (including preservation of the electrical properties in moist atmosphere or after immersion) make possible the use of EP(D)M in underground wires and connectors. Table 30 contains two examples of formulations: for medium/high voltage insulation and for low voltage insulation, which offers a cost/performance ratio competitive with classical NR/SBR compounds.

E. Additives for Chlorinated Polyethylenes [116–118]

Various chlorinated polyethylene grades are produced by chlorination of HDPE in solution or in suspension. The differences between commercial products concern mainly the chlorination degree (25–42 wt% of chlorine), viscosity and crystallinity.

1. Blends with Other Polymers

Various grades of CM can be blended with one another to obtain a compromise of the properties. CM can also be blended with other rubbers: blending with EPDM improves low temperature flexibility and addition of NBR imparts a better resistance to

TABLE 29 EPDM formulations for roofings

Components	Mass parts	
	Low cost	General purpose
EPDM (Vistalon 2555, Exxon)	100	100
N550 (FEF) Carbon black	140	90
Whiting (Omya BL, Plüss-Staufer)	50	
Paraffinic oil (Flexon 815, Exxon)	60	40
Zinc oxide	10	10
Stearic acid	2	1
Tackifying resin	5	5
Antioxidant TMQ		0.5
Accelerator TMTD	1.8	1.0
Accelerator MBTS		1.0
Accelerator DPTD		0.7
Sulfur donor DTDM	1.8	
Sulfur		0.8
Properties of unvulcanized compounds		
Monsanto Rheometer (180°C, arc 5°)		
ML/MH, daN·m	16.5/54	15.0/68
optimum cure t90, min	19	7
Properties of vulcanized compounds (Initial/aged[a]) press cured, 25 min at 160°C		
Hardness, Shore A	78/81	68/69
Modulus at 100% elongation, MPa	3.1/	2.1/4.1
Tensile strength, MPa	7.9/7.1	13.6/13.1
Elongation at break, %	410/305	600/450
Tear strength (DIN 53507A), kN/m	15/	14.3/8.5

[a]Air aging, 7 days at 100°C.
Source: Ref. 113.

TABLE 30 EP(D)M formulations for electrical applications

Components	(A) Medium/high voltage insulation	(B) Low cost low voltage insulation
EPM (Vistalon 504, Exxon)	100	
EPDM (Vistalon 7000, Exxon)		100
Silane treated calcined clay	70	
Soft clay		240
Paraffin wax	10	
Paraffinic oil		75
Zinc oxide	5	10
Stearic acid		1
Antioxidant TMQ	1.5	
Antioxidant MBI	0.5	
Antioxidant ODPA		0.5
Silane A 172 (Dow Corning)	1.0	
Dicumyl peroxide (DiCup 40, Hercules)	7.5	
Accelerator MBTS		1.0
Accelerator TMTD		3.0
Coagent TAC	1.0	
Sulfur		1.5
Properties of unvulcanized compounds		
Monsanto Rheometer (160°C, arc 5°)		
ML/MH, daN·m	12.0/100	7.5/84
optimum cure t90, min	20	10
Properties of vulcanized compounds (Initial/Aged[a])		
Hardness, Shore A	70/	56/
Modulus at 100% elongation, MPa	4.0/	1.9/
Tensile strength, MPa	9.0/7.8	8.7/7.8
Elongation at break, %	260/250	570/560
Volume resistivity, ohm·cm × 10^{-15}	17/2[b]	6.8/
Dielectric constant	2.8/	
Power factor	0.004/	

[a]Compound (A) was aged 7 days at 135°C; compound (B) was aged 10 days at 70°C.
[b]After aging for 24 h in water at 90°C.
Source: Ref. 115.

petroleum derivatives (oils, fuels). Limited amounts of SBR (about 5 phr) are added by some processors as a coagent in peroxide vulcanization.

2. Curing Systems

Chlorinated polyethylene rubbers present saturated polymer chain, which may be crosslinked with peroxides or high energy radiation. The choice of peroxide depends on formulation and vulcanization technology (mainly on temperature obtainable). The following peroxides (curing temperature; phr for commercial, desensitized products) have proved suitable for compounds loaded with carbon blacks or neutral/basic mineral fillers: 1,1-di-*tert*-butylperoxy-3,3,5-trimethylcyclohexane (140–150°C; 6–12 phr); 4,4-di-*tert*-butylperoxy-*n*-butyl valerate (150–160°C; 7–14 phr); dicumyl peroxide (160–170°C; 6–2 phr); bis(*tert*-butylperoxyisopropyl)benzene (165–180°C; 5–8 phr). Good results can be obtained with *tert*-butylperbenzoate (4–6 phr), when an acidic mineral filler is contained. Optimal physical properties are obtained using peroxides in conjunction with coagents like triallyl cyanurate, triallyl trimellithate, trimethylol propane trimethacrylate, and diallyl phthalate.

More recently, curing systems using amine and thiadiazole [119] or triazine thiols [120] have been developed in order to replace the peroxide curing systems.

3. Fillers

Like other polyolefinic rubbers, chlorinated polyethylene rubbers accept large amounts of filler (30–200 phr). Furnace and thermal blacks are customarily used [121]. Among mineral fillers, those with pH ~ 7 are preferable (reinforcing silicas, aluminium silicates, clays, or whitings); using acidic fillers, careful selection of the crosslinking system must be provided. At the same time, it must be remembered that carbon blacks impart a better aging stability in comparison with light-colored fillers.

4. Plasticizers and Processing Aids [122]

The selection of an appropriate plasticizer is governed by two requirements: compatibility with CM and absence, as far as possible, of interference in the crosslinking reaction. Aromatic and naphthenic oils have only limited usefulness because they disturb peroxide curing; paraffinic oils have only limited compatibility. Among the synthetic plasticizers, trialkyl trimellithate imparts good heat resistance, dioctyl sebacate has proven suitable for low temperature flexibility, and dinonyl phthalate realizes a good efficiency-to-price ratio.

In common practice, CM compounds do not require processing aids; however, if necessary, the flow properties of highly loaded compounds can be improved using calcium soaps of saturated fatty acids [123].

5. Antidegradants

Similarly with other chlorine-containing polymers, chlorinated polyethylene rubbers are susceptible to HCl splitting, at high processing/service temperatures. A good stability is achieved using magnesium oxide (10–15 phr) and lead compounds (5–10 phr tribasic lead sulfate, dibasic lead phthalate, lead oxide). Association with an epoxy compound (epoxy resins or a plasticizer like epoxidized soya oil) enhances the effect of lead compounds. Conventional antioxidants are not generally required except in cases of particularly high demands for heat resistance, when 0.1–0.3 phr of TMQ or ADPA can be used.

Zinc oxide and other zinc derivatives (even at low concentrations) reduce the stability of CM and drastically decrease the resistance of the vulcanizates to aging and should therefore not be used under any conditions.

6. Examples of Applications

Easy processing, cost competitiveness and colorability make CM attractive for flexible cords (up to 600 V) and ignition wire jacketing; an example is offered in Table 31. Another major use for CM is in automotive hose covering. While in the cable industry peroxide vulcanization is common practice, for hose compounds both peroxide and amine + thiadiazole curing systems can be used.

Significant quantities of CM are used in the production of sheet goods and in plastics modification.

F. Additives for Chlorosulfonated Polyethylenes [124–128]

Available chlorosulfonated polyethylene commercial grades contain 25–43% by weight chlorine and 0.8–1.5% by weight sulfur, randomly distributed along the saturated polymer chain. The main variable is the chlorine content and it affects various properties, as presented in Fig. 15. The characteristics of the polyethylene used as raw material (molecular weight and molecular weight distribution, branching, crystallinity) are reflected in the properties of the resulting CSM.

1. Blends with Other Polymers

Addition of other polymers to CSM is not a common practice; compounding of CSM with other rubbers and plastics has been investigated in order to improve some performances of the latter; it may be also used as a component of resin curing systems [95].

2. Curing Systems [129, 130]

Curing chlorosulfonated polyethylene rubbers involves a chemistry different from that of general-purpose rubbers. Containing chlorine atoms on the chain, it remembers CM behavior but the presence of more reactive chlorosulfonyl groups introduces some specific features. Metal oxides such as magnesium and lead oxides react in the presence of small amounts of water or weak acids (stearic, abietic) and of accelerators or sulfur donors (TMTD, DPTT, MBT). Magnesium oxide is a good acid acceptor, able to control the pH value in the mixture and is primarily used in general-purpose compounds; dibasic lead phthalate is used for compounds with lower water swell. Polyols are added to improve the solubility of metal compounds in rubber matrixes. In special cases, polyols can be used as curatives associated with bases in order to generate ester bridges [93].

TABLE 31 Compounds based on chlorinated polyethylene for insulation of auto ignition wire

Components	Mass parts		
Chlorinated polyethylene (CP-R-6158, Du Pont de Nemours)	100		
N550 (FEF) Carbon black	30		
Whiting, atomized	125		
Diisodecyl phthalate	30		
Magnesium oxide (Maglite D, Merck)	5		
Triallyl cyanurate	2		
Bis(tert – butylperoxyisopropyl)benzene	4		
Properties of unvulcanized compound			
Mooney viscosity ML_{1+10} (100°C)	75		
Monsanto Rheometer (177°C, arc 3°)			
ML/MH, daN·m	12/130		
optimum cure t90, min	5		
Properties of vulcanized compound press cured 20 min at 160°C	**Initial**	**Air aged[a]**	**Oil aged[b]**
Modulus at 100% elongation, MPa	3.8	7.3	4.9
Tensile strength, MPa	11.4	11.4	35.3
Elongation at break, %	295	295	280
Hardness, Shore A	78	89	74
Tear strength (ASTM Die C), kN/m	2.8		
Brittle point, °C	−30		
Moisture absorption (7 days at 70°C), mg/cm^2	22		

[a]Aged in air, 70 h at 136°C.
[b]Aged in ASTM Oil No. 2, 18 h at 121°C.
Source: Ref. 118.

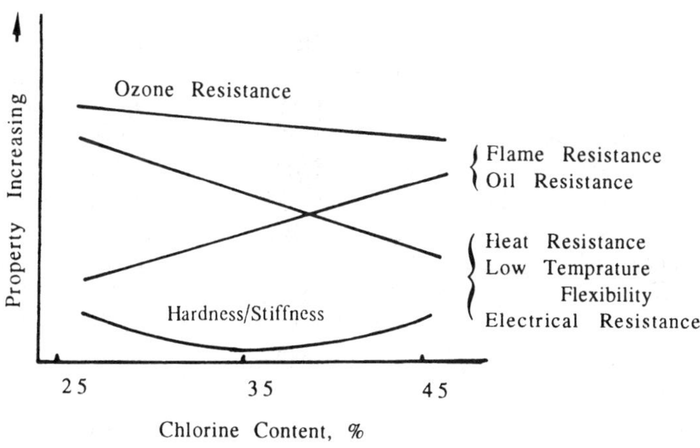

FIG. 15 Effect of chlorine content on properties of chlorosulfonated polyethylene.

With special aliphatic diamines a rapid cure takes place; aromatic diamines, with lower basicity, require higher curing temperatures.

Chlorosulfonated polyethylene rubbers may also be vulcanized with peroxides, which results in final products with better heat stability; small amounts of EP(D)M can play a positive role in these formulations. Peroxide vulcanization may be further improved by adding coagents like methylene bismaleimide, triallyl cyanurate, or triallyl trimelythate.

TABLE 32 Vulcanizable compounds based on chlorosulfonated polyethylene (mass parts)

Compound Type	Components A	B	C
Chlorosulfonated polyethylene	100	100	100
Whiting, micronized	64	120	80
Titanium dioxide	35		
Plasticizer		30	
Magnesium oxide	4	4	10
Pentaerithritol, fine dispersed		3	
Processing aid			
Accelerator TMTD	1.5	1.5	1.5
Sulfur		2	
Dicumyl peroxide		1	

Source: Ref. 128.

Table 32 may be used to adopt a starting formulation in the development of a curing system for CSM compounds. In the case of compounds of type A, crosslinking is realized by divalent metal oxide (MgO) assisted by moisture (atmospheric); it proceeds rather slowly at ambient temperature (as is the case with roofings and geomembranes), resulting in ionic bridges that impart high moduli at the final stage. The compounds of type B are provided with TMTD associated with MgO to ensure development of covalent crosslinks by sulfur and sulfonyl groups; such compounds can be cured at temperatures above 150°C with a variety of other ingredients and achieving high mechanical performances. The compounds of type C are vulcanized with peroxide and form covalent C—C bridges (without involvement of chlorosulfonyl groups). The vulcanizates obtained are characterized by very good heat resistance and excellent compression set, but the other components of the practical formulations should be carefully selected to avoid their interference in the vulcanization chemistry (as presented for EPDM and CM).

3. Fillers and Plasticizers

For compounds based on CSM, the selection of fillers and plasticizers is quite similar to that for CM.

4. Antidegradants

Chlorosulfonated polyethylene rubbers exhibit intrinsic resistance to the attack of oxygen and ozone; properly compounded products have excellent weathering resistance. For very severe service conditions (150–180°C), CSM vulcanized with peroxides can be further improved by adding a synergistic combination of bis(3,5-di-*tert*-butyl-4-hydroxyl)-hydrocinnamate and dilauryl-thiodipropionate [123].

5. Examples of Applications

The key properties obtainable with CSM are [128]:

(i) superior resistance to oxygen and ozone, which can result also (by proper formulations of compounds) in excellent weathering;

(ii) high mechanical performances (including abrasion resistance) without the use of reinforcing fillers;

(iii) good behavior at low temperature (brittle temperature as low as −60°C) by an appropriate formulation;

(iv) excellent stability to corrosive chemicals and good resistance to oils and greases;

(v) excellent electrical properties;

(vi) good heat resistance, corresponding to an "EE" rating by ASTM D2000, is obtainable; properly compounded vulcanizates can also exhibit good flame resistance.

SYMBOLS AND ABBREVIATIONS

ADPA	diphenylamine–acetone condensation product
APP	atactic polypropylene
BDA	Bundesgesundheitsamt, Germany
BIIR	brominated butyl rubber
BPH	2,2′-methylene-bis-(4-methyl-6-*tert*-butylphenol)
BR	butadiene rubber
CIIR	chlorinated butyl rubber
CM	chlorinated polyethylene rubber
CPE	chlorinated polyethylene
CR	chloroprene rubber
CSM	chlorosulfonated polyethylene rubber
DBS	dibenzytidene sorbitol
DCP	dicyclopentadiene
DPMT	dipentamethylene thiuram disulfide
DPTT	dipentamethylene thiuram tetrasulfide
DTDM	dithiodimorpholine
ENB	ethylydene norbornene
EPC	easy processing channel carbon black
EPDM	ethylene–propylene–diene rubber
EP(D)M	either ethylene–propylene or ethylene–propylene–diene rubber
EPM	ethylene–propylene rubber
EVA	ethylene–vinyl acetate
FDA	US Food and Drug Administration
FEF	fast extrusion furnace carbon black
GMF	*p*-quinone dioxime

GPF	general-purpose furnace carbon black
HAF	high abrasion furnace carbon black
HALS	hindered amine light stabilizer
HD	1,4-hexadiene
HDPE	high density polyethylene
HMWHDPE	high molecular weight high density polyethylene
HPB	Canadian Health Protection Board
IIR	butyl rubber (isobutylene–isoprene rubber)
IR	isoprene rubber
LDPE	low density polyethylene
LLDPE	linear low density polyethylene
LMWPE	low molecular weight polyethylene
MBI	2-mercaptbenzimidazole
MBT	2-mercaptbenzothiazole
MBTS	benzothiazyl disulfide
MEMO	3-methacryloxy-propyltrimethoxisilane
NBR	nitrile rubber (acrylonitrile–butadiene rubber)
NR	natural rubber
ODPA	octylated diphenyl amine
PB-1	polybutene-1
PE	polyethylene
PHR	weight parts per 100 weight parts polymer
PIB	polyisobutylene
PO	polyolefin
PP	polypropylene
PS	polystyrene
PVC	polyvinylchloride
SBR	styrene–butadiene rubber
TAC	triallyl cyanurate
TBBS	benzthiazyl-2-*tert*-butylsulfenamide
TEDC	tellurium-diethyl dithiocarbamate
TETD	tetraethyl thiuram disulfide
TMQ	2,2,4-trimethyl-1,2-dihydroquinoline
TMTD	tetramethyl thiuram disulfide
TMTM	tetramethyl thiuram monosulfide
TEPO	thermoplastic elastomer polyolefin
TPO	thermoplastic polyolephin
UV	ultraviolete
VTMO	vinyltrimethoxisilane
VTMOEO	vinyltrimethoxietoxysilane
ZDBC	zinc-dibutyl dithiocarbamate
ZMBT	zinc-2-mercaptobenzthiazole
ZDMC	zinc-dimethyl dithiocarbamate

REFERENCES

1. J Murphy. Additives for Plastics Handbook. Oxford: Elsevier, 1996, p 114.
2. F Gugumus. Antioxidants. In: R Gächter, H Müller, eds. Plastics Additives Handbook, 3rd edn. Munich: Hanser, 1990, p 1.
3. Plastics Handbook, 2nd edn, Munich: Hanser, 1987, p 37.
4. S Horun. Additives for Polymers Processing. Bucharest: Editura Technicá, 1978, p 211.
5. S Horun, O Sebe. Degradation and Stabilization of Polymers. Bucharest: Editura Technicá, 1983, p 296.
6. F Gugumus. Kunststoffe 77: 1070, 1987.
7. F Gugumus. SPE Tech Pap 34: 1447, 1988.
8. F Gugumus. Polym Degrad Stabil 24: 289, 1989.
9. KW Bartzo. SPE Tech Pap, Polyolefines IV, 1984.
10. Höechst Technical Papers, 1973.
11. V Dobrescu, C Andrei. Advances in Polyolefins Chemistry and Technology. Bucharest: Editura Stiintificó şi Enciclopediă, 1987, p 129.
12. C Andrei, P Ioana. Rev Chim (Bucharest) 34: 681, 1983.
13. H Müller. Metal deactivators. In R Gächter, H. Müller, eds. Plastics Additives Handbook, 3rd edn Munich: Hanser, 1990, p 105.
14. ER Wright. J Cell Plast 12: 317, 1976.
15. F Gugumus. Light stabilizers. In: R. Gächter, H Müller, eds. Plastics Additives Handbook, 3rd edn. Munich: Hanser, 1990, p. 129.
16. C Andrei, I Drăgutan, AT Balaban. Photostabilization of Polymers with Sterically Hindered Amines. Bucharest: Ed Academiei Române, 1990, pp 9, 203.
17. EH McTigue, M Blumberg. Appl Polym Symp 4: 175, 1967.
18. J Masek, P Caucik, J Halcik, M Karvas, J Durmis. Plaste Kautsch 24: 404, 1977.
19. F Gugumus. Kunststoffe 77: 1065, 1987.
20. AB Shapiro, LP Lebedeva, VI Suskina, GN Antipina, LN Smirnov, PI Levin. Vysokomol Soedin 15: 3034, 1973.
21. L Mascia. The Role of Additives in Plastics. London: Edward Arnold, 1974, p 131.
22. F Gugumus. Kunststoffe 74: 620, 1984.
23. G. Pfahler. Antistatic agents. In: R Gächter, H Müller, eds. Plastics Additives Handbook, 3rd edn. Munich: Hanser, 1990, p 749.
24. K Rombusch, G Maahs. Angew Makromol Chem 34: 50, 1973.
25. ON Şeverdâev. Antistatic Polymer Materials. Moscow: Himiâ, 1983, p 63.
26. KH Mobius. Kunststoffe 78: 53, 1988.
27. DM Bigg. Electrical properties of metal-filled polymer composites. In: SK Bhattacharya, ed. Metal-Filled Polymers. Properties and Applications. New York: Marcel Dekker, 1986, p 165.
28. M Rusu, M Dărângă, N Sofian, DL Rusu. Materiale Plastice (Bucharest) 35: 15, 1998.
29. R Leaversuch. Mod Plast Int 17: 59, 1987.
30. T Riedel. Lubricants and related additives. In: R Gächter, H Müller, eds. Plastics Additives Handbook, 3rd edn. Munich: Hanser, 1990, p 421.
31. R Keller. Kunststoffe 76: 586, 1986.

32. H Hurnik. Chemical blowing agents. In: R Gächter, H Müller, eds. Plastics Additives Handbook, 3rd edn. Munich: Hanser, 1990, p 811.
33. N Asandei, G Adamescu, M Rusu, S Petrovan, M Nicu. Bull Polyt Inst Jassy XXV (XXX), Section II, Fasc. 1-2, 1980, p 53.
34. K Stange. Kunststoffe 74: 674, 1984.
35. AL Berg., Organic peroxides as cross-linking agents. In: R Gächter, H Müller, eds. Plastics Additives Handbook, 3rd edn Munich: Hanser, 1990, p 833.
36. J Murphy. Additives for Plastics Handbook. Oxford: Elsevier, 1996, p 198.
37. J Janssen. Nucleating agents for partially crystalline polymers. In: R Gächter, H Müller, eds. Plastics Additives Handbook, 3rd edn., Munich: Hanser, 1990, p 862.
38. CJ Kuhre. SPE J 20: I 1 13, 1964.
39. BASF Technical Papers, 1991.
40. G Menzel. Compounds development and compounding of thermoplastics. In: R Gächter, H Müller, eds. Munich: Hanser Publishers, 1983, p 715.
41. A Kennaway, WA Rigby. Compounding. In: A Renfrew, P Morgan, eds. Polythene. The Technology and Uses of Ethylene Polymers. New York: Interscience, 1960, p 409.
42. G Matthews. Polymer Mixing Technology. New York: Applied Science, 1982, pp 50, 92, 135.
43. KM Hess. Kunststoffe 73: 783, 1983.
44. M Rusu, DL Rusu. Polymer Processing Technology, Iasi: Ed Dosoftei, 1995, pp 181, 348.
45. RF Jones. Guide to Short Fiber Reinforced Plastics. Munich: Hanser, 1998, p 15.
46. TOJ Kresser. The processing of crystalline olefin polymers. In: RAV Raff, KW Doak, eds. Crystalline Olefin Polymers. New York: Interscience, 1964, p 399.
47. VA Paharenko, VG Zverlin, EM Kirienko. Filled Thermoplastics-Handbook, Kiev: Technica, 1986, pp 5, 74, 138.
48. VV Bogdanov, RV Torner, VN Krasovskii, EO Reger. Polymer Mixing. Leningrad: Himiâ, 1972, pp 48, 83, 120.
49. VV Bogdanov, VI Mitelkin, SG Savataev. Technological Bases of Polymer Mixing. Leningrad: Izdatel'stvo Leningradskova Universiteta, 1984, p 5.
50. RV Torner. Technological Bases of Polymer Processing. Moscow: Himiâ, 1977, p 202.
51. N McKelvy. Polymer Processing. New York: Wiley, 1962, p 299.
52. Z Tadmor, CG Gagos. Principles of Polymer Processing. New York: Wiley, 1978, p 196.
53. L Basova, V Broâ. Technics of Plastics Processing. Moscow: Himiâ, 1985, p 79.
54. VS Kim, VV Skacikov. Dispersing and Mixing in the Manufacture and Processing of Plastics. Moscow: Himiâ, 1988, pp 42, 68.
55. RZ Tudose. Processes and Devices in the Processing Industry of Macromolecular Compounds, Bucharest: Editura Tehnică, 1976, p 219.
56. D Râbinin, UE Lucac. Mixing Machines for Plastics Processing and Rubber Blends. Moscow: Mashinostroenie, 1972, pp 122, 114.
57. VK Zavgorednego. Devices for Plastics Processing. Moscow: Mashinostroenie, 1976.
58. JJM Cormont, HFH Meijer, H Herrmann, H Herres. Kunststoffe 73: 599, 1983.
59. MO Sebe, N Goldenberg-Serban, A Variu, E Constantinescu, D Stanciu, Gh Manea. Materiale Plastice (Bucharest) 21: 105, 1984.
60. DH Wilson. Feeding Technology for Plastics Processing. Munich: Hanser, 1998, p 105.
61. NP Ceremesinoff. Guide to Mixing and Compounding Practices. Prentice Hall: PRT, 1994, p 17.
62. RH Wildi, C Maier. Understanding Compounding. Munich: Hanser, 1998, p 4.
63. C Ranwendaal. Polymer Mixing: A Self-Study Guide, Munich: Hanser Publishers, 1998, p 11.
64. DB Told. Plastics Compounding. Munich: Hanser, 1998, p 13, 71, 125, 159.
65. C Ranwendaal. Mixing in single-screw extruders. In: I Manas-Zloczwer, Z Tadmor, eds. Mixing and Compounding Polymers. Theory and Practice. Munich: Hanser, 1994, p 251.
66. PG Anderson. Mixing practice in corotating twin-screw extruders. In: I Manas-Zloczwer, Z Tadmor eds. Mixing and Compounding Polymers. Theory and Practice. Munich: Hanser, 1994, p 679.
67. C Ranwendaal. Mixing in reciprocal extruders. In: I Manas-Zloczwer, Z Tadmor, eds. Mixing and Compounding Polymers. Theory and Practice. Munich: Hanser, 1994, p 735.
68. K Inoue. Internal batch mixer In: I Manas-Zloczwer, Z Tadmor, eds. Mixing and Compounding Polymers. Theory and Practice. Munich: Hanser, 1994, p 251.
69. C Bernhardt. The Processing of Thermoplastic Materials. New York: Reinhold, 1959, p 154.
70. G Vostroknutov, MI Novikov, VI Novikov, IV Prozorovskaâ. Processing of Rubber and Rubber Blends. Moscow: Himiâ, 1980, p 148.
71. LPBM Janssen. Twin-Screw Extrusion. Amsterdam: Elsevier, 1978, pp 28, 53, 84, 95.
72. G Schenkel. Plastic Extrusion Technology and Theory. London: Iliffe Books, 1966.
73. C Ranwendaal. Mixing in Polymer Processing. New York: Marcel Dekker, 1991.
74. HS Karz, JV Milewski. Handbook of Fillers for Plastics. New York: Van Nostrand Reinhold, 1987.
75. JV Milewski, HS Karz. Handbook of Reinforcements for Plastics. New York: Van Nostrand Reinhold, 1987.
76. SK Bhattacharya ed., Metal-Filled Polymers. New York: Marcel Dekker, 1986, p 144.
77. PJ Wright. Acicular wollastonite as filler for polyamides and polypropylene. In: A Whelan, JL Craft, eds. oxford: Elsevier, 1986, p 119.
78. Matles. Glass fibers. In: G Lubin, ed. Handbook of Fiberglass and Advanced Plastics Composites. New York: Van Nostrand Reinhold, 1969, p 143.

79. Z Tadmor, I Klein. Engineering Principles of Plasticating in Extrusion. New York: Van Nostrand Reinhold, 1970.
80. R Mihail, A Stefan. Simulation of Processing Processes of Polymers. Bucharest: Editura Tehnică 1989, p 181.
81. Polymer Coloration (translated from Einfurben von Kunststoffen, VDI-Verlag, Dusseldorf), Leningrad: Himiâ, 1980, p 251.
82. VS Vasilenko, GD Eremenko, AS Demina, IB Karacunski, VB Uzdeski, S Sadrina. Future Trends for Intensifying the Compounding Process of Polyolefin Composite Materials. Moscow: Obzornaâ Informatia, Seriâ Polimerizationîe Plastmassî, 1989.
83. BM Walker, ed. Handbook of Thermoplastic Elastomers. New York: Van Nostrand Reinhold, 1979.
84. R Ranalli. Ethylene-propylene rubber-polypropylene blends. In: A Whelan, KS Lee, eds. Developments in Rubber Technology. Thermoplastic Rubbers. London: Applied Science, 1982, chap 2.
85. AA Kanauzova, MA Yumashev, AA Dontsov. Poluchenie termoplasticheskikh rezin metodom dinamicheskoy vulkanizatsii i ikh svoystva, Moscow: TsNIITEneftekhim, 1985.
86. IA Arutyunov, BS Kulberg, TS Fedenyuk. Olefinovye termoelastoplasty. Proizvodstvo i potreblenie. Moscow: TsNIITEneftekhim, 1986.
87. WS Penn. Synthetic Rubber Technology, vol. I. London: Maclaren, 1960, p 101.
88. HE Rooney, ed. Polysar Butyl Handbook, Sarnia: (Canada) Polysar Ltd., 1966.
89. GM Ronkin. Svoistva i primenenie butilkauchuka. Moscow: TsNIITEneftekhim, 1969.
90. GJ van der Bie, JM Rellage, V Vervloet. In: CM Blow, ed. Butyl rubber. Rubber Technology and Manufacture. London: Butterworths, 1971.
91. Polysar Butyl Handbook. Sarnia (Canada): Polysar Ltd., 1977.
92. WD Gunter. Butyl and halogenated butyl rubbers. In: A Whelan, KS Lee, eds. Developments in Rubber Technology. 2. Synthetic Rubbers. London: Applied Science.
93. W Hofmann. Vulkanisation und Vulkanisationshilfsmittel. Leverkusen: Bayer AG, 1965; English translation, London: Maclaren, and New York: Palmerton, 1967.
94. GA Blokh. Organicheskie uskoriteli vulkanizatsii i vulkanizuyushchie sistemy dlya elastomerov. Moscow: Khimiya, 1978; Organic Accelerators and Curing Systems for Elastomers. Shawbury: RAPRA, 1981.
95. G Ivan. Rubber vulcanization by reactive resin systems. Paper presented at the Symposium "Rubber Vulcanization", Prahatice, Czechoslovakia, 1990.
96. C Moore. Bag-o-Matic bladders. Compounding and Processing. Tech Rep 78ET B209, Esso Europe Inc., Machelem, 1978.
97. G Kraus, ed. Reinforcement of Elastomers. New York: Interscience, 1965.
98. HA Cook. Butyl innertube technology. Tech. Rep 77ET-B0270, Esso Europe Inc., Machelem, 1977.
99. Chlorobutyl Rubber. Compounding and Applications. Houston: Exxon Chemical Co., 1973.
100. Polysar Bromobutyl Handbook. Sarnia (Canada): Polysar Ltd., 1976.
101. Bromobutyl Rubber. Compounding and Applications. Houston: Exxon Chemical Co., 1973.
102. DC Edwards. A high performance curing system for halobutyl elastomers. Paper presented at ACS Rubber Div. Meeting, New York, 1986.
103. G Ivan, E Tavaru, M Giurginca, DN Nicolescu. Rev Roum Chim 34: 1017, 1989; G Ivan, E Tavaru, M Giurginca. Acta Polymerica 42: 507, 1991.
104. EP Baldwin, G Ver Strate. Rubber Chem Technol 45: 709, 1972.
105. TM Sukhotkina, NN Borisova. Svoistva etilen-propilenovykh kauchukov i rezin na ikh osnove. Moscow: TsNIITEneftekhim, 1973.
106. L Corbelli. Ethylene-propylene rubbers. In: A Whelan, KS Lee, eds. Developments in Rubber Technology. 2. Synthetic Rubbers. London: Applied Science, 1981.
107. AP Buna. Leitfaden fur den Praktiker. Marl: Huls AG, 1986.
108. OA Govorova, AE Frolov, GA Sorokin. Svoistva rezin na osnove etilen-propilenovykh kauchukov. Moscow: TsNIITEneftekhim, 1986.
109. W Hofmann. Prog Rubber Plast. Technol. 1: 18, 1985; Kaut Gummi Kunstst 40: 308, 1987.
110. JA Riedel, R van der Laan. Ethylene Propylene Rubbers. In: Vanderbilt Co., The Uanderbilt Rubber Handbook, 13th edn. Norwalk: Vanderbilt Co., 1990, p 123.
111. G Ivan, E Bugaru, P Bujenita. Resin curing of EPDM rubber. Activation System Influence on Crosslinking Density. Paper presented at PRI International Conference "Various Aspects of Ethylene-Propylene Based Polymers", Leuven, 1991.
112. DC Novakoski. Profile of Cabot Carbon Blacks in EPDM Rubber, Tech Rep RG-135.
113. Vistalon. Selected Compounds for Automotive, Building and Mechanical Goods. Machelem: Exxon Chemical Co.
114. Polysar EPR. Application Formulary. Fribourg: Polysar International SA, 1991.
115. Vistalon. Selected Compounds for Electrical Applications. Machelem: Exxon Chemical Co.
116. Guidelines for the Formulation of Bayer CM Compounds, No. 006. Leverkusen: Bayer AG, 1978.
117. E Rohde. Kaut Gummi Kunst 32: 304, 1979; 35: 478 1982.
118. AJ Maldonado. Chlorinated polyethylene. In: RF Ohm, ed. The Uanderbilt Rubber Handbook, 13th edn. Norwalk: Vanderbilt Co., 1990, p 190.

119. C Barnes, RT Sylvest. Gummi Asbest Kunstst 36: 150, 290; 1983; WH Davis Jr., JH Flynn. Paper 24, ACS Rubber Div. Meeting, Los Angeles, 1985.
120. K Mori, Y Nakamura. Rubber Chem Technol 57: 34, 1984.
121. G Matenar. Russe in Bayer CM 3630, No. 012, Leverkusen, Bayer AG, 1979.
122. G Matenar, E Rohde. Plasticizers in Bayer CM 3630, No. 013, Leverkusen, Bayer AG, 1979.
123. W Hofmann. Rubber Technology Handbook. Munich: Hanser, 1989, pp 105, 107.
124. R Schlicht. Kaut Gummi 10: WT66, 1957.
125. RH van der Laan. Hypalon Bull No. 3A. Delaware: Du Pont de Nemours, 1969.
126. I Soos, L Horvath. Gumiipari kozlemenyek. Hypalon kautcsuk általános jellemzese es vizsgálata, Budapest: Taurus Research Center, 1970.
127. GM Ronkin. Khlorosulfirovanyi poletilen. Moscow: TsNIITEneftekhim, 1977.
128. GA Baseden. Chlorosulfonated polyethylene. In: RE Ohm, ed. The Uanderbilt Rubber Handbook, 13th edn. Norwalk: Vanderbilt Co., 1990, p 183.
129. JT Maynard, PR Johnson. Rubber Chem. Technol. 36: 963, 1963.
130. IC Dupuis. Basic Compounding of Hypalon. Selecting a curing system. Publication HP-320.1: Delaware: Du Pont de Nemours, 1970.

30

Quality Control of Processed Thermoplastic Polyolefins

Mihaela-Emilia Chiriac
SC INCERPLAST SA–Research Institute for Plastics Processing, Bucharest, Romania

Processed polyolefin tests are based on standard specifications for molded and extruded low and high density polyethylene (LDPE and HDPE), polypropylene (PP), and polybutylene (PB) [1–3].

These specifications provide information to aid in the identification of materials according to types (referring to density), categories (referring to nominal flow rate), grades (referring to tensile strength, elongation, brittleness, dielectric constant, environmental stress-crack resistance), and classes (referring to color and filling). The classifications are made in a manner that allows the seller and the purchaser to agree on what constitutes an optimum commercial batch.

All test methods for processed polyolefins refer to international standards, including those of the United States [American Society for Testing and Materials (ASTM) and Underwriters Laboratories (UL)], Great Britain [British Standards (BS)], Germany [Deutsche Instit für Normung (DIN)], Japan [Japanese Industrial Standards (JIS)], the European Standard (EN), and the International Organization for Standardization (ISO).

I. TESTING OF PIPES, TUBINGS AND FITTINGS [1–31]*

There are many applications for polyolefin compounds in the processing of pipes, tubings, and fittings from polyethylene (PE), crosslinked polyethylene (XPE), biaxially oriented polyethylene (PEO), polypropylene, and polybutylene.

According to the conditions of service, pipes, tubing, and fittings are divided into three main categories:

High pressure pipes, tubings and fittings:
- cold and hot water distribution systems
- gas distribution systems
- liquid transport for industrial processes
- irrigation

Low pressure pipes and fittings:
- municipal sewage
- wall sewer and drainage

Large-diameter pipes in both pressure and nonpressure systems

From the constructive point of view (dimensions, smooth walled, corrugated, etc.), diversity of pipes and fittings is also generated by the conditions of

*The following international standards apply in Section I: ASTM D 2104; ASTM D 2662; ASTM D 2666; ASTM D 2873; ASTM D 3000; ASTM D 3035; ASTM D 3287; ASTM F 667; ASTM F 845; ASTM F 878; ASTM F 892; ISO 3212; ISO 3213; ISO 3549; ISO 3478; ISO 3480; ISO 3503; ISO 4440; ISO 4451; ISO 7279; BS 1972; BS 3284; ISO 4991; ISO 6572; ISO 6730; DIN 4728; DIN 4729; DIN 8075; DIN 8078; DIN 16889; DIN 19535; DIN 19537; DIN 19658; EN 712; EN 728; EN 917; EN 1636-5; EN 1680; EN 1704; EN 1716; EN 12095; EN 12099; EN 12100; EN 12106; EN 12107; EN 12117; EN 12118; EN 12119; ISO 161-1; ISO 1167; ISO 3213; ISO 9323; ISO 9393; ISO 10146; ISO 11922-1, 2; ISO 11420; ISO EN 13479; ISO 12230; ISO 13761; ISO EN 13478.

service (in air, in building walls, underground, in corrosive media, etc.).

The international standards for technical terms used in piping industry are ASTM F 412, ISO 3, and ISO 497, which give a definition of each term. The most important terms for a complete understanding of the notions in this chapter are as follows:

Hoop stress: the tensile stress in a pipe in circumferential orientation, due to internal hydrostatic pressure.

Hydrostatic design stress: the recommended maximum hoop stress that can be applied continuously with a high degree of certainty that the pipe will not fail.

ISO equation: an equation showing the relation between stress pressure and the pipe dimensions:

$$S - \frac{p(D_i + h)}{2h} \quad \text{or} \quad S = \frac{p(D_o - h)}{2h} \tag{1}$$

where

S = hoop stress, MPa
P = hydrostatic pressure, MPa
D_i = average inside diameter mm
D_o = average outside diameter, mm
h = minimum wall thickness, mm

Long-term hydrostatic strength (LTHS): the hoopes stress that, when applied continuously, will cause failure of the pipe at 100,000 h.

Pressure rating: the estimated maximum pressure that the median in the pipe can exert continuously with a high degree of certainty that the pipe will not fail.

Standard dimension ratio (SDR): a specific ratio of the average specified outer diameter to the minimum specified wall thickness D_o/h.

The most important (primary) properties for PE, PP, and PB pipes, tubings, and fittings are presented within cell classification limits in Tables 1–3, to facilitate identification of such materials.

For example, PE 233425 A has the following characteristics:

Density, 0.926–0.940 g/cm^3
Melt index, 0.4–0.15 g/10 min
Flexural modulus, 276–552 MPa
Tensile strength at yield, 21–24 MPa
Environmental stress-crack resistance in 24 h, 50% failure
Hydrostatic design basis, 11.3 MPa
Natural (uncolored) and unfilled.

A. High Pressure Pipes, Tubing and Fittings

For high pressure pipes, tubing, and fittings of all kinds, four properties should be added to those mentioned above.

In the case of black polyolefin pipes, fittings, and compounds, test method ISO 11420 describes two procedures for assessment of carbon black particle and agglomerate size and dispersion for content of less than 3%. The method is applicable also to raw material in pellet form.

For the colored polyolefin piping and fittings, ISO 13949 specifies a test method for assessment of degree of pigment dispersion.

1. *Workmanship*: the pipes, tubing, and fittings shall be homogeneous throughout, having essentially uniform color, opacity, density, and so on. The pipe wall shall be free of cracks, holes, blisters, voids, foreign inclusions, or other visible defects that may affect the wall integrity.
2. *Dimensions and tolerance* (D_i, D_o, h, SDR, SIDR): measured in accordance with ASTM D 2122, ISO 3126 or EN 1636-5. The International Standards ISO 11922 with Part 1-metric series and part 2-inch based series specified tolerance grades for outside diameter, out-of roundness, and wall thickness of polyolefin pipes for conveyance of fluids. They are manufactures with nominal diameters and nominal pressure in accordance with ISO 161-1 and nominal wall thickness in accordance with ISO 4065.
3. *Long-term hydrostatic pressure strength*: determined in accordance with ASTM D 1598, BS 6437, ISO 1167, or DIN 8073.
4. *Short-term hydrostatic pressure*: determined in accordance with ASTM D 11599, BS 6437, ISO 1167, or DIN 2075.

Depending on the diversity of anticipated service, other basic properties are completed in the case described next (Sections 1–3).

1. Pipes, Tubing and Fittings for Hot and Cold Water

Other characteristics that need to be considered in connection with service in hot and cold water are as follows:

Long-term hydrostatic pressure strength at different temperatures (23 and 82°C) measured in accordance with 7.5 ASTM F 877.

TABLE 1 Primary properties for polyethylene

Property	Test method	Cell classification limits					
		1	2	3	4	5	6
Density, g/cm^3	ASTM D 1505 BS 3412 ISO 4056	0.910–0.925	0.926–0.940	0.914–0.955	>0.955		
Melt index	ASTM D 1238 BS 2782; 720A DIN 53753	>1.0	1.0–0.4	0.4–0.15	<0.15		
Flexural modulus, MPa	ASTM D 790 BS 2782; 335 ISO 178 DIN 53457	<138	138–276	276–552	552–758	758–1103	>1103
Tensile strength at yield, MPa	ASTM D 638 ISO 527	<15	15–18	18–21	21–24	24–28	>28
Environmental stress-cracking resistance	ASTM D 1693						
test condition[a]		A	B	C			
test duration, h		48	24	192			
failure, max, %		50	50	20			
Hydrostatic design basis at 23°C, MPa	ASTM D 2837	5.52	6.89	8.62	11.3		

[a]A, natural; B, colored; C, black (2% carbon black); D, natural with ultraviolet stabilizer; E, colored with ultraviolet stabilizer.
Source: Adapted from ASTM D 3350 (1984).

TABLE 2 Primary properties for polypropylene

Property	Test method	Cell classification limits							
		1	2	3	4	5	6	7	8
Tensile strength, MPa	ASTM D 638 ISO 527	5	10	15	20	25	30	35	40
Secant (1% strain) minimum modulus of elasticity	ASTM D 790	100	250	500	750	1000	1250	1500	1750
Minimum Izod impact, J/m	ASTM D 256 BS 2782-335 ISO 180	10	50	100	200	300	400	500	700
Deflection temperature at. 4.64 kgf:/cm^2 min, °C	ASTM D 48 ISO 75 DIN 53461	50	60	70	80	90	100	110	120
Flow rate (condition L), g/10 min	ASTM D 1238 BS 2782; 720 A DIN 53753	≤0.3	0.3–1.0	1.0–3.0	3.0–10	10–20	20–40	40–100	>100

Thermocycling: 1000 cycles between 16 and 82°C while the pipe is subjected to nominal pressure when tested in accordance with 7.5 ASTM F 877.

Heat reversion: test method BS 2782-met. 1102 B.

Longitudinal reversion: Test method ISO 2507.

Stabilizer migration resistance: tested in accordance with 7.10 ASTM F 876.

Moisture content by coulometry: test method EN 12118.

Volatile components: test method EN 12099.

TABLE 3 Primary properties of polybutylene: cell classification limits[a]

Property	Test method	Types					
		I			II		
Density, g/cm^3	ASTM D 1505 BS 3412 ISO 4056	0.905–0.909			0.910–0.920		
Minimum tensile strength, MPa	ASTM D 638 ISO 527	20.7			20.7		
Minimum yield strength, MPa	ASTM D 638 ISO 527	10.3			13.8		
Minimum elongation at break, %	ASTM D 638 ISO 527	300			300		
Flow rate, g/10 min	ASTM D 1238 BS 2782: 720 A ISO 1133 DIN 53753	Categories					
		0	1	2	3	4	5
		<0.25	0.26–0.75	0.76–2.5	2.6–10	10–25	>25

[a]Classes: A, general-purpose and dielectric unpigmented; B, general-purpose and dielectric, in colors (including black and white); C, weather-resistant (black) containing not less than 2% carbon black.
Source: Adapted from ASTM D 2581 (1987).

Determination of oxidation inducing time: test method EN 728.

Resistance to rapid crack propagation (RPC) full-scale test (FST): test method EN ISO 13478.

Resistance to crack propagation for slow crack growth on notched pipes of wall thickness greater than 5 mm: test method EN ISO 13479. It is expressed in terms of time to failure in a hydrostatic pressure test on a pipe with machined longitudinal notches in the outside surface.

Pressure reduction factors for use at temperature above 20°C: test method ISO 13761.

For example, Tables 4 and 5 give the values of two basic properties for two types of polyolefin pipes: crosslinked polyethylene XPE SDR 9 (degree of crosslinking, 64–89%) and polybutylene PB 21-SDR 11.

At the same SDR, under the same conditions, PB pipes maintain greater pressure than PE pipes. For example, the values of minimum sustained pressure at the same SDR are:

PE 3408: 2.19 MPa
PE 2306, 2406, 3406: 1.79 MPa
PB 2110: 2.79 MPa

TABLE 4 Mechanical properties for crosslinked polyethylene (XPE-SDR 9) and polybutylene (PB21-SDR 9) pipes

Property	Test method	Pipes	
		XPE-SDR 9	PB21-SDR9
Sustained water pressure (1000 h), MPa	ASTM D 1598		
23°C		3.62	3.3
37.8°C			3.1
82°C		1.72	
93°C		1.45	
Burst pressure, MPa	ASTM D 1599		
23°C		4.27	3.8
37.8°C			3.6
82°C		1.90	
93°C		1.62	

Source: Adapted from ASTM D 876 (1985) and ASTM D 2666 (1983).

TABLE 5 Physical and mechanical properties of polyethylene gas pressure pipes

Property	Test method	Cell classification limits[a]					
		1	2	3	4	5	6
Flow rate, g/10 min	ASTM D 1238	>1.5	1.5–1.8	1.0–0.4	0.4–0.1	<4.0	<1.5
	BS 2782-720 A						
	DIN 53753	E	E	E	E	F	U
Yield stress (MPa)	ASTM D 1599 or ASTM D 2290-B	17.2–20.7	20.7–24.1	24.1–27.6	>27.6		
1000 h sustained pressure 23°C (MPa)	ASTM D 1598	9.2	9.8	10.3	10.8	11.4	12

Source: Adapted from ASTM F 678 (1982). E, temp. 190°C; total load including piston, 2.16 kg; F, temp. 190°C; total load including piston, 21.60 kg; U, temp. 310°C; total load including piston, 12.50 kg.

PB pipes maintain the same pressure as acrylonitrile–butadiene–styrene (ABS) pipes at the same SDR.

For PP pipes, ISO 3213 lays down the minimum values for expected strength as a function of time and temperature in the form of reference lines for use in calculation of SDR. The reference lines for PP have been agreed upon by a group of experts after considering experimental data and have been accepted by the relevant technical committees in ISO.

The same reference lines are recommended for PB pipes by ISO 12230 and, for crosslinked polyethylene PE-X pipes, by ISO 10146.

2. Pipes, Tubing and Fittings for Fuel Gas Service

When these components are intended for service with fuel gas, one must use the following values:

Yield stress: test method ASTM D 1599 or ASTM D 2290 met B.

Flattening: determined in accordance with 6.4 ASTM D 678.

Thermal stability: tested in accordance with ASTM D 3350.

Chemical resistance (mineral oil and methanol): determined in accordance with 6.3 ASTM F 678.

Elevated temperature service (hydrostatic design basis determined at the specified higher temperature): test method ASTM D 2837.

Outdoor storage stability: determined in accordance with 5.3.6 ASTM F 678.

Table 5 gives the physicomechanical properties and cell classification limits for PE gas pressure pipes, tubing, and fittings.

3. Pipes, Tubing and Fittings for Chemical Fluids

Environmental stress-crack resistance (tested in accordance with 7.8 ASTM F 876) and *chemical resistance* in specific media (test method ISO 4433 or BS 1973) are significant in such service.

B. Low Pressure, Pipes, Tubing and Fittings for Drainage and Waste Disposal Absorption Fields

Smooth-walled pipes are suitable for soil drainage and waste disposal system absorption fields.

Corrugated tubing also can serve in underground applications when soil support is given to flexible walls.

The basic properties of products of these types are as follows.

Workmanship: the same as for high pressure pipes, tubing, and fittings. The surfaces shall be free of excessive bloom, but holes deliberately placed in perforable pipes are acceptable.

Dimensions: given for smooth-walled polyethylene pipes in ASTM F 810 and for corrugated polyethylene tubing in ASTM F 405.

Physicomechanical properties: see Tables 6 and 7.

For polyethylene large-diameter profile wall sewer and drain pipes, ASTM F 894 includes the requirements and test methods for materials, workmanship, dimensions, stiffness, flattening, impact strength, and joint system.

Large-diameter profile wall sewer and drain pipes are used in low pressure and gravity flow applications, while EN 1636-5 includes specifications for pipes and fittings of polyethylene piping system for nonpressure drainage.

TABLE 6 Physical and mechanical properties for smooth-wall polyethylene drainage pipes

Property	Test method	Value for nominal size of 76.2–152.4 mm
Flattening	ASTM F 810	No splitting, cracking, or breaking
Impact resistance	ASTM D 2444	54–95 J
Stiffness at 5% deflection	ASTM D 2412	0.131–0.055 MPa
Environmental stress cracking	ASTM D 1693 ASTM F 810	No cracking or splitting

Source: Adapted from ASTM F 810 (1985).

TABLE 7 Physical properties for corrugated polyethylene tubing

Property	Standing tubing	Heavy-duty tubing
Maximum elongation, %	10	5
Stiffness at 5% deflection, MPa	0.17	0.21
Stiffness at 10% deflection, MPa	0.13	0.175

Source: Adapted from ASTM F 405 (1985); all test methods from this standard.

C. Large-Diameter Pipes

An intermediate type of pipe is the large-diameter polyolefin pipe. Such piping is intended for new constructions and replacement of old piping systems used to transport water, municipal sewage, industrial process liquids, effluents, slurries and so on in both pressure and nonpressure systems.

Large-diameter polybutylene pipe can be used as commercial and industrial process piping at temperatures up to 82°C.

ASTM F 809 M presents the outside diameters and wall thickness dimensions. Mechanical properties are given in Table 8.

D. Joints, Valves and Seals

ASTM D 2609, ASTM D 2657, ASTM D 2683, and ASTM 3261 cover all types of joining for pipes and fittings—socket fusion, butt fusion, and saddle fusion. For all types of fusion, it is necessary to specify four basic properties: workmanship, dimension, short-term rupture strength (not less than the minimum short-term strength of the pipe, and sustained pressure (the fittings and fused pipes shall not fail for a period of 1000 h at the required pressure).

For the characterization of polyolefin fusion joint pipes fittings and valves, one or more of the following tests should also be considered.

Impact resistance: 5.3 ASTM F 905.
Torsion resistance: 5.4 ASTM F 905.
Inspection of joints by ultrasonic testing: ASTM F 600.
Tightness under internal pressure: test method ISO 3458.
Pressure drop in mechanical pipe joining: test method ISO 4059.
Compression: test method BS 5114 or 864.
Resistance to pull-out under constant longitudinal force: test method EN 712 for piping system- end- load

TABLE 8 Mechanical properties for large-diameter polybutylene pipes

	SDR 11: 32.5		
Property	Test method	23°C	82°C
Sustained pressure (1000 h), MPa	ASTM D 1598	2.76–0.86	1.38–0.41
Burst pressure, MPa	ASTM D 1599	3.03–0.93	1.52–0.45

Source: Adapted from ASTM F 809 M (1983).

bearing mechanical joints between pressure pipes and fittings.

Determination of the long-term hydrostatic strength of fittings, valves, and ancillary equipment: test method EN 12107.

Determination of gaseous flow rate/pressure drop relationship for fittings, valves, and ancillaries: test method EN 12117.

Resistance to bending between support for PE valves: test method EN 12100.

Test for leaktightness under and after bending applied to the operating mechanism for PE valves: test method EN 1680.

Resistance to internal pressure and leaktightness for valves: test method EN 917.

Resistance of valves after temperature cycling under bending: test method EN 1704.

For valves and valves with PE spigot ends having a nominal outside diameter greater than 63 mm and intended for the transport of fluids, EN 12119 specifies a test method for the resistance to thermal cycling between −20°C and +60°C.

Impact resistance of an assembled topping tee: test method EN 1716.

For the brackets for rainwater piping system test method EN 12095 for bracket strength is recommended.

In the case of PE/metal and PP/metal adaptor fittings for pipes for fluids under pressure, ISO 9623 specifies the design lengths and size of threads.

It also specifies the dimensions at the socket and spigot ends of the fitting. The adaptor fittings have one plain socket or spigot for fusion jointing a PE socket or spigot to a PE pipe or a PP socket or spigot to a PP pipe. The other component of the fitting is metal and is threaded to enable connection to metal pipes, fittings, valves, and/or apparatus with pipe threads. The pressure-tight joint is achieved by the compression of a gasket.

Metal adaptor fittings are intended for use in pipelines for water supply.

Laboratory research on the testing of pipe, tubing, fitting, and joint systems involves their behavior in service with respect to quality, establishment of the test conditions for determining service life, yield tests, detection of failure, and test for joining and welding.

Examination of internal surfaces by optical microscopy is a rapid quality control method that permits the measurement of particle dimensions (3–5 μm) and the detection of the specific crystalline microstructures that appear during extrusion [4].

The service life may be estimated by extrapolation from long-time pressure diagrams for different conditions of temperature and environment [5].

Another point of interest is the determination of various stress rate test conditions and the influence of surface irregularities and impurities on compression, tensile strength, flexion, and bending, when related to chemical attack at the surface [6].

In each moment of pipe rupture, cyclic fatigue and compression tests allow the estimation of the energetic loss from hysteresis curves generated by consumed energy in fractures and dissipated thermal energy [7].

Yield tests have shown, on the one hand, the mechanism of surface degradation caused by the chemical attack or internal pressure and, on the other hand, the "sweating" defect that appears before failure [8, 9].

Resilience testing of failed samples showed that the failures are oriented parallel to the direction of extrusion, being propagated in a normal direction: fragile, quasifragile, and ductile [10].

Robertson tests have determined the strength and temperature for the translation of fragile failure to ductile failure [11].

Most tests refer to weld and joints in pipe and fitting systems. Quality control involves ultrasonic, X-ray, and holographic tests[12].

Ultrasonic control—the "shadow" technique—may reveal the presence of a hot or a cold weld, discontinuities, and site of failure [13, 14].

Thus, the holographic test evidences the effects of axial tension induced by welding conditions (temperature, pressure, and time), as compared with interferograms. Quality control of welding by holographic test using a He–Ne laser allows estimation of residual tensions increasing with the glass transition temperature as a result of growth in modulus of elasticity [15].

Determination of behavior with different types of mechanical stress permits one to establish the regression equations by which to calculate the maximum effort that will be supported in security conditions by pipes and joining systems [16].

At the same time, the investigator can establish the optimal parameters for welding and for reducing the number of defects [17].

Neutron irradiation tests show that joints behave similarly to the basic material, for all types of mechanical stress.

II. TESTING OF ELECTRICAL INSULATION [32–40]*

Polyolefin and olefin copolymer insulation and jackets for electrical wires and cables are used in the following applications:

Insulating tubing: flexible, semirigid, or rigid, and heat shrinkable for 105 and 125°C;
Electrical conduits and fittings;
Insulation and jackets for electric cables;
Insulation and sheaths for telecommunication equipment and for radiofrequency electrical wires and cables;
Insulating tapes.

A. Flexible, Semirigid and Rigid Insulating Tubing for 105 and 125°C

Insulating tubing is intended for use in air, in dry and damp locations, as part of the internal wiring of electrical devices and equipment.

Requirements for 105° and 125°C flexible, semirigid, and rigid heat-shrinkable polyolefin tubing are presented in Table 9; they refer to shrinkable and crosslinked polyolefin insulating tubing.

B. Electrical Conduits and Fittings

The 90°C HDPE conduit class is for aboveground uses when encasement is not less than 50 mm of concrete and for underground use by burial or encasement in concrete.

Requirements for the 90°C HDPE conduit and fitting class are given in Table 10.

For the colored cable coatings, ISO 13949 specifies a method for the assessment of the degree of pigment dispersion.

C. Insulation and Jackets for Electrical Cables

Beside the compounds in which the characteristic constituents are polyolefins, other products also may be employed for insulation and jackets. These include crosslinked compounds and constituted from chlorosulfonated polyethylene (CP: insulation and jackets for wires and cables used for deep wall submersible pumps subjected to 60, 90, and 105°C); chlorinated polyethylene (CPE: Insulation and jackets for wires subjected to 90°C); ethylene–propylene–diene monomer (EDPM) acceptable for wires subjected to 60–70°C); EP (acceptable for cables subjected to 75 and 90°C as insulation under CP jacket and to 105°C under EPDM jacket); a covulcanizate EP + PE (acceptable for wires without any covering over insulation subjected to 75 and 90°C).

All these kinds of insulation and jacketing are presented in UL 1581. Table 11 gives the mechanical properties for PE insulation class 30.75°C.

D. Insulation and Sheaths for Telecommunication and Radiofrequency Electrical Wires and Cables

For telecommunication and rf wiring/cabling situations, the insulation may be either solid, cellular or foamskin.

Sheaths can be natural or colored. They must consist of a compound based on one of the following materials, compounded and processed to meet the requirements: ethylene homopolymer, mixture of ethylene polymers, and copolymers containing up to 10% of the nonethylene monomer. Compounds and products have a density greater than 0.910 g/cm^3 and the melt index flow value is between 0.05–3.0 g/10 min.

For example, the requirements for PE insulation and sheathing for telecommunications (types 03 and 2), radiofrequency cables (type 03 and 2), and electricity supply cables (type TE_1, TS_1 and TS_2) are given in Tables 12–14. All telecommunication and radiofrequency types of polyethylene insulation and sheathing are discussed in BS 6234.

The tests are made selectively, depending on the application field of insulation and sheath.

Permittivity and the loss tangent determination are measured at a frequency of 1–20 MHz in accordance with either BS 2067 or any other suitable method (e.g., ASTM D 1531).

E. Insulating Tapes

PE insulation tapes are used both at 80°C and at low temperatures to cover joints and welds in wires and cables (with no additional protection).

Requirements for PE insulation tape are given in Table 15.

If polyolefin insulation is intended to serve in a corrosive medium, the stress-cracking test must be included [32].

Research devoted to the flammability of polyolefin insulating tape showed that thermo-oxidative aging

* The following international standards apply in Section II: DIN VDE 0207 and JIS 2380.

TABLE 9 Requirements for rigid, semirigid, and flexible heat-shrinkable polyolefin tubing

Property	Test method (UL 224 paragraph)	Class	
		105°C	125°C
Temperature rating, °C		105	125
Minimum tensile strength, MPa	8	104	104
Minimum ultimate elongation, %	8	200	200
Dielectric withstand and breakdown voltage at 2500 V	11	60	60
Aging in a circulating air oven	8.9	168 h 136°C 60 d, 113°C	168 h 158°C 60 d, 134°C
flexibility	10	Flexible with no cracking or permanent deformation	
variation of initial minimum tensile strength, %	8.14	30	30
variation of minimum ultimate elongation, %	8.14	0	0
dielectric withstand and breakdown voltage	11	At least half unaged specimen but not less than 2500 V	
Deformation	12	1 h, 121°C ≥50% decrease in wall thickness	1 h, 125°C
Heat shock	13	No cracks	
Cold bend	14	No cracks	
Flammability	15 and 16		
Minimum volume resistivity, Ω cm	17	10^{14}	10^{14}
Minimum longitudinal change, %	20	±3	±3
Maximum secant modulus, MPa	18	173	173

Source: Adapted from UL 224 (1981).

TABLE 10 Requirements for high density polyethylene conduits and fittings in the 90°C Class

Property	Test method (UL 651 A paragraph)	Value
Tensile strength	6 (or ASTM D 638)	27.5 MPa
Deflection temperature under heat and load	7	62°C
Low temperature handling or conduct at −20°C[a]	9	No shattering, chips, or cracks
Low temperature handling of chips at −20°C	10	No shattering, chips, or cracks
Water absorption	11 (or ASTM D 577)	1.0001
Crushing resistance	12 (or ASTM D 2412)	70–75%
Resistance to impact	14	No cracks or tears
Dimensional stability and crushing strength of conduit after 168 h at 113°C	16	5%
Bending test of joints	17	Joints not damaged
Axial-pull test on joints in conduit	19	Joints not damaged or separated

[a]For height of 1525 mm.
Source: Adapted from UL 651 A (1981).

TABLE 11 Mechanical properties of polyethylene insulation in the 30.75° class

Property	Value
Minimum tensile strength	9.64 MPa
Minimum ultimate elongation	350%
Aging in a full draft circulating oven for 48 h at 100°C	
restraint of tensile strength	75% of the result with unaged specimens
restraint of ultimate elongation	

Source: Adapted from UL 1581 (1984); all test methods from paragraph 470 of this standard.

TABLE 12 Requirements for insulation and sheathing for telecommunication cables

Property	Test method (BS 6469 paragraph)[a]	Insulation type, 03 and 2			Sheathing	
		Solid	Cellular	Foamskin	Natural	Black
Tensile strength, MPa	2.2	10.0	3.5	3.5	8.0	8.0
Elongation at break, %	2.2	300	150	150	300	300
Carbon black content, %	5.3					2.5 ± 05
Filling compound absorption, %	A	15	15	15	15	15
Change of tensile strength after absorption:						
min. value, MPa	2.2	7.5	3.0	3.0	7.0	7.0
max. variation, %		–25	–25	–25	–15	–15
Resistance to oxidation time, h	B	1000	1000	1000		
Environmental stress-crack resistance	D					
max. failures, compound					2	2
max. failures, sheath					0	0

[a]Requirements only for insulation and sheathing in direct contact with hydrocarbon-based filling compounds only (e.g., petroleum jelly): A, appendix E-BS 6234; B, appendix H-BS 6234; C, appendix J-BS 6234; D, appendix F-BS 6234.
Source: Adapted from BS 6234 (1987).

TABLE 13 Requirements for insulation and sheathing for radiofrequency cables

Property	Test Method (BS 6469 paragraph)	Insulation: types 03 and 2, solid	Sheathing	
			Natural	Black
Tensile strength, MPa	2.2	10.0	10.0	10.0
Elongation at break, %	2.2		300	300
Carbon black content, %	5.3			2.5 ± 0.5

Source: Adapted from BS 6234 (1987).

TABLE 14 Requirements for insulation and sheathing for electricity supply cables

Property	Test method (BS 6469 paragraph)	Insulation: TE_1	Sheathing (over cables)	
			TS_1: 80°C	TS_2: 90°C
Melt flow index, g/10 min	5.1	0.4	0.4	—
Tensile strength, MPa	2.2	10.0	10.0	12.5
Elongation at break, %	2.2	300	300	300
Carbon black content, %	5.3		2.5 ± 0.5	2.5 ± 0.5
Carbon black dispersion, %	A		5	5
Property after aging in air				
temperature, °C		100	100	110
duration, days		10	10	14
elongation at break, %	2.2	300	300	300
Pressure at high temperature				
temperature	4.2			115
max. deformation (%)	2.6			50
Water absorption				
temperature, °C	2.6	85		
duration, days		14		
max. variation of mass, mg/cm^2		1		
Shrinkage test	2.5			
temperature, °C ± 2°C		100		
duration time, h		6		
max. permissible shrinkage, %		4		

Source: Adapted from BS 6234 (1987).

causes a surface migration of low molecular weight components (chlorinated paraffins), a phenomenon that decreases the flame resistance of the product) [33]. Most investigations thus have been concentrated on the continuous detection and testing of the insulating surface by submillimetric wave laser or by scanning computerized images to detect pores, defects, or contamination of the surface [34, 35].

Cable breakdown is due mainly to the distribution of irregularities and contaminations.

When the cable surface is explored continuously, concomitantly with the extrusion process, polyolefin insulations will permit the statistical detection—by absorbed IR energy—of pores, pinholes, or any type of defect or surface contamination (with dimensions about 200 μm) [36].

The distribution of defects as a function of dimensions gives a good image of the compound, extrusion process, and parameters used for a statistical number of cables.

For special applications, such as deep-sea submersible cables, tests for breakdown resistance and performed at hydrostatic pressure, at high and low temperatures between 0 and 100.

TABLE 15 Requirements for polyethylene insulating tape

Property	Test method (UN 510 paragraph)	Value
Thickness	5	0.15 min
Minimum tensile strength	6	10.3 MPa
Minimum elongation	6	60%
Weatherability at 100 h	4	
tensile strength retained		65%
elongation retained		65%
Adhesion strength	8	175 N/m
Dielectric breakdown	7	39.37 kV/mm
Moisture absorption		
retain of breakdown (%)	23°C × 96%UR × 96h	90%
Exposure to heat, 60 days at 87°C	10	Not cracking
Exposure to cold, 2 h at −10°C	11	Not cracking
Deformation	12	40%
Storage	13	175 N/m

Source: Adapted from UL 510 (1982).

TABLE 16 Basic properties for polyethylene film and sheeting

		Type designation			
Property	Test method	1	2	3	4
Density, g/cm^3	ASTM D 1505	0.910–0.925	0.926–0.940	0.941–0.966	
Impact strength, g/min	ASTM D 1709	<40	40–70	>70	
Coefficient of friction	ASTM D 1894	<0.20	0.20–0.40	0.41–0.70	>0.70
Haze, %	ASTM D 1003	<5.0	5.0–9.0	>9.0	
Luminous transmittance, %	ASTM D 1003	<1.0	1.0–2.5	26–50	51–100

Source: Adapted from ASTM D 2103 (1986).

TABLE 17 Basic properties for polypropylene films

		Cell classification system			
Property	Test method	1	2	3	4
Haze, %	ASTM D 1003	<0.1	1.0–3.0	>3.0	
Gloss 45°, %×10^4	ASTM D 1003	>90	80–90	70–79	<70
Coefficient of friction	ASTM D 1894	<0.2	0.2–0.5	0.6–0.8	>0.8
1% secant modulus (machine and transverse directions)	ASTM D 882	40,000–70,000	71,000–100,000	101,000–130,000	
Elongation at break (machine and transverse direction), %	ASTM D 882	>600	400–600	<400	

Source: Adapted from ASTM D 2530 (1981).

TABLE 18 Classification of low density polyethylene films

Property	Test method	Classifications				
		Type	1	2	3	
Minimum drop dart impact resistance, g	ASTM D 1709		40–165	75–255	105–315	
		Class:	1	2	3	4
Gloss units	ASTM D 2457		<30	30–50	50–70	>70
Haze, %	ASTM D 1003		>25	10–25	5–10	0–5
		Surface:	1	2	3	
Coefficient of friction	ASTM D 1894		>0.5	0.2–0.5	>0.2	
Wetting tension, $\times 10^{-3}$ N/m	ASTM D 2578	Finish:	1	2	3	4
		32–34	35–37	38–40	>41	

Source: Adapted from ASTM D 4635 (1986).

III. TESTING OF FILMS AND SHEETS [41–45]*

Another type of product with various applications includes films, foils, and sheets for general use, such as packaging, construction, industrial, and agricultural utilizations.

These films are made of polypropylene, polyethylene homopolymers, ethylene copolymers, and also blends of homopolymers and copolymers including ethylene–vinyl acetate (EVA).

The basic properties of polyethylene and polypropylene films are given in Tables 16 and 17. The five basic properties listed in each table are to be evaluated for each kind of film with specific tests for obtaining a complete image of the film's intrinsic quality.

A. Polyethylene Films

1. Low Density Polyethylene Films

LDPE films are made of low density polyethylene, linear low density polyethylene (LLDPE), and ethylene–vinyl acetate copolymer, EVA/LDPE blends.

Blends of ethylene–vinyl acetate with low density polyethylene have density values up to 0.929 g/cm^3. These films are unpigmented and unsupported.

Classification of low density polyethylene films defines the types, classes, surfaces, and finishes and considers the following properties: impact strength, coefficient of friction, optical properties, and surface treatment (Table 18).

Additional requirements are as follows.

Workmanship (appearance): the films must have the qualities required from a commercial point of view. The customer and the supplier will establish acceptable limits with regard to gels, wrinkles, scratchers, undispersed raw materials, holes, tears, and blisters.

Dimensions: thickness, 0.025–0.100 mm, and width, 1250–3000 mm (both measured in accordance with ASTM D 2103); yield tolerance, ±3 to ±10% (determined in accordance with ASTM D 4321).

Intrinsic quality requirements (including tensile strength, heat sealability, and odor): per Table 19.

2. Medium Density Polyethylene Films

The classification of medium density polyethylene films (Table 20) considers the following properties: stiffness, optical properties, coefficient of friction, and surface treatment.

Dimensions are as follows: thickness 0.025–0.100 mm, and width, 375–3000 mm (both measured in accordance with ASTM D 2103); yield tolerance ±3 to ±10% (determined in accordance with ASTM D 4321).

Physicomechanical requirements are given in Table 19.

B. Nonoriented Polypropylene Films

Films may contain colorants, stabilizers, or other additives (for improvement of surface properties).

In addition to the five basic properties (Table 18), the following dimensional tests are required:

*The following international standards apply in Section III: ASTMD 4635; ASTM D 3981; ASTM D 2673; ASTM D 3015; DIN 16995; DIN 55543; DIN 55545; BS 3012; JIS 6781.

TABLE 19 Physicomechanical requirements for low and medium density polyethylene films

Property	Test method	PE film: Low density	PE film: Medium density
Tensile strength, MPa	ASTM D 882		
longitudinal		11.7	14.1
transversal		8.3	10.5
Elongation, %	ASTM D 882		
longitudinal		225	100
transversal		350	300
Heat sealability ratio	ASTM F 88	0.60–.75	0.60–0.75
Odor	ASTM E 462	≤ 3.5	≤3.5

Source: Adapted from ASTM D 4635 (1986) and ASTM D 3981 (1982).

TABLE 20 Classification of medium density polyethylene films

Property	Test method	Classifications				
1% secant modulus, MPa	ASTM D 882	Type:	0	1	2	3
			<170	170–240	240–345	>345
		Class:	0	1	2	3
Clarity				Low	Moderate	High
Gloss units	ASTM D 2475		<30	30–50	50–70	>70
Haze, %	ASTM D 1003		>25	10–25	5–10	0–5
		Surface:	0	1	2	3
Coefficient of friction	ASTM D 1894		>0.7	0.4–0.7	0.2–0.4	≤0.2
Wetting tension, $\times 10^{-3}$ N/m	ASTM D 2578	Finish:	1	2	3	4
			32–34	35–37	38–40	>41

Source: Adapted from ASTM D 3981 (1982).

Thickness, 0.018–0.254 mm (measured in accordance with ASTM D 374);

Yield tolerance, ±3 to ±10% (determined in accordance with ASTM D 2530).

C. Oriented Polypropylene Films

Films may contain colorants, stabilizers, or other additives, and may be coated for the improvement of performance properties (heat sealability, gas permeability, etc.). Films may be annealed (heat-set) to reduce the unstrained linear shrinkage and shrink tension upon exposure to heat.

Oriented polypropylene (OPP) films may be:

Uniaxially oriented (OPP-U) with maximum tensile strength of 103 MPa in the direction of orientation;

Biaxially oriented (OPP-X), such that both machine direction and transverse direction tensile strength exceed 103 MPa;

Balance oriented (OPP-B), such that machine direction and transverse direction tensile strength exceed 103 MPa but do not differ by more than 55 MPa.

We note the following general requirements:

Appearance: the same as for the polyethylene film (Section III.A.1);

Blocking: the films shall not be blocked excessively (ASTM D 3354);

Thickness: 0.010–0.050 mm (ASTM D 374);

TABLE 21 Cell classifications for oriented polypropylene films

Property	Test method	Cell classification limits 1	2	3	4
Coefficient of friction	ASTM D 1894	<0.2	0.2–0.35	0.36–0.50	>0.50
Linear thermal shrinkage, unrestrained at 120°C, %	ASTM D 2732	0–3	4–6	7–20	>20
Wetting tension, kg/m	ASTM D 2578	<0.035	≥0.35		
Heat sealability, kg/m	ASTM F 88	4–10	11–25	>25	
Minimum gloss	ASTM D 2457	50	70	85	
Maximum haze, %	ASTM D 1003	15	7	3	

Source: Adapted from ASTM D 2673 (1984).

Package yield: ±3 to ±10% (ASTM D 4321)

Films intended for the packaging of food, drugs, and cosmetics must comply with the requirements of the US Food, Drug and Cosmetic Act, as amended.

Specific requirements include the characterization of commercial OPP films. Cell classifications are based on six main properties, as defined in Table 21.

D. Polyethylene Sheets

PE sheets have major applications in the building industry, in agriculture, and in orthopedic surgery. Such sheeting is made of polypropylene, polyethylene, or modified polyethylene, such as ethylene copolymer, consisting of a major portion of ethylene in combination with a minor portion of other monomers or a mixture of polyethylene with a lesser amount of the other polymer.

PE sheeting comes in a variety of colors, opacities, translucencies, and dimensions.

1. Sheets up to 0.250 mm Thickness

Thin PE sheets are intended for use in agricultural and industrial applications. The following properties are to be determined.

Appearance: commercially, the sheets must be as free as possible of gels, streaks, pinholes, particles of foreign matter, undispersed raw material, and other visible defects; they may be natural, colored, translucent, or opaque.

Dimensions: nominal thickness, 0.025–0.250 mm, (ASTM D 374); width and weight measured in accordance with ASTM D 4397.

The physicomechanical properties are presented in Table 22.

TABLE 22 Requirements for polyethylene sheeting 0.025–0.250 mm thick

Property	Test method	Value
Dart drop impact resistance, g	ASTM D 1709 (A)	
for 0.025 mm thickness		40
for 0.250 mm thickness		475
Tensile strength, MPa	ASTM D 882	
longitudinal direction		11.7
crosswise direction		8.3
Elongation, %	ASTM D 882	
longitudinal direction		225
crosswise direction		350
Luminous transmittance	ASTM D 2103	1%
Reflectance, 45°; 0°	ASTM E 97	≥70%
Water vapor transmission rate, g/24 h/m^2	ASTM E 96	
for 0.025 mm thickness		22
for 0.250 mm thickness		2.2

Source: Adapted from ASTM D 4397 (1984).

TABLE 23 Requirements for low and high density polyethylene sheeting, 1.4–19.1 mm

Property	Test method	Sheeting LDPE	Sheeting HDPE
Tensile strength, MPa	BS 2782–301 F	11.2–9.8	19.0
Elongation at break, %	BS 2782–301 F	350	150
Heat reversion for nominal thickness > 1.4 mm, %	BS 4646 4 and 7	3	3

Source: Adapted from BS 3012 (1970) and BS 4646 (1970).

2. Sheets Between 1.4 and 19.1 mm Thickness

Thicker PE sheets (from low and high density polyethylene) are intended for general use, especially for orthopedic and surgical applications.

PE sheets 1.4–19.1 mm thick are supplied in squares or rectangles with clean-cut edges, free from defects, surface blemishes and contamination, air bubbles, blisters, and voids, and substantially uniform in color (natural or black).

The physicomechanical requirements are presented in Table 23.

Research on film foil, and sheet refers especially to such mechanical and functional properties as impact resistance, toughness, chemical resistance, and appearance (pigment and carbon black dispersion).

One of the most important properties—impact resistance—has been tested in various conditions of acceleration of the impact weight (acceleration being recorded on an oscilloscope). In this way, the falling rate, impact strength, and kinetic energy transmitted to the film can be calculated by the Gardner test [41, 42].

Another method consists of measuring the radioluminescent elastic modulus of oriented polyethylene and polypropylene films [43].

Welding of film and foil can be examined by ultrasonic or X-ray tests [44], to determine the effects of welding irradiation (for ≤ 5 Mrad). A significant increase in tensile strength was noted.

The most important tests for toughness consist of filled PE bags subjected to horizontal and vertical vibrations, repetitive free falls from different heights at different temperatures, and storage compression.

Research into the kinetic coefficient of friction on horizontal and inclined surfaces showed that addition of pigments and fillers tends to increase the value of a material [45].

The real functionality of film has been demonstrated by contraction tests in a tension-active medium at high temperatures [46, 47].

Polyethylene film porosity is determined by transfer phenomena manifested in LiCl on an electrical field, which can be observed photometrically or visually. Film porosity decreases with increasing thickness; above 0.100 mm, LiCl transfer ceases film permeability to CO_2, O_2, and N_2 can be measured by the gas chromatography method [48].

Various tests have been performed to determine the dispersion of pigments or dyes. This property depends on the dimensions of the pigment particles, the coloration technique, and the processing technology applied (to the compound).

For rigorous control of film thickness and surface quality (concomitantly with the extrusion process) electro-optical and IR methods may be employed [49].

IV. TESTING OF CONTAINERS, CISTERNS AND BOTTLES [50–65]*

Blow-molded polyolefin containers, cisterns, and bottles can be made of polypropylene, low and high density polyethylene, and their copolymer. These vessels are used for the packaging, storage, and transport of fluid, viscous, and powdered products such as chemicals, intermediates for synthesis, oil products, detergents, paints, water, food products, drugs, cosmetics, and so on.

On considering the wide range of applications and the diversity of substances that contact these containers and bottles, we see that two distinct types of test are to be applied, namely:

Tests concerning geometrical shape and capacity;
Tests for physicomechanical properties.

* The following international standards apply in Section IV: ASTM D 2463; ASTM D 2561; ISO 2248; ISO 2872; ISO 2244; DIN 55441; DIN 55457; BS 4838.

A. Containers

Dimensional measurements are made in accordance with BS 4839 (Parts 1–3) and refer to

Container wall thickness
Container height to neck face
container overall height
Container diameter
Internal neck diameter
External neck and thread diameters
Neck height

Containers are divided according to capacity, as follows:

≤ 5 L
5–60 L
61–210 L

Capacity measurements, made according to BS 4839 (Parts 1–3) refer to:

Net capacity
Minimum brimful capacity
Minimum gross capacity

Performance requirements are presented in Table 24.

In the case of polyolefin containers for petroleum products with a nominal capacity of 26.5 L (7 gal) or less, additional properties, tested in accordance with ASTM D 3435, are as follows.

Resistance to internal pressure: the container shall show no evidence of rupture, crack, or leakage.

Aging: the material shall retain at least 70% of its original tensile strength after 1000 h (ASTM D 2565) or 700 h (ASTM G 23).

Permeability: the filled container (70% isooctane and 30% toluene) shall not have a weight loss greater than 3%.

Petroleum resistance: the material from the container shall evidence no pitting, crazing, softening, bubbling, cracking, tackiness, or decomposition.

Elevated temperature resistance: the container shall not leak.

B. Cold Water Storage Cisterns

Cisterns designed to store cold water are between 16 and 100 L in capacity.

Performance requirements are categorized according to type (Table 25) and routine tests (Table 26).

C. Bottles

Dimension tests for bottles assess height, outer diameter, and wall thickness, determined in accordance with ASTM D 2911 or JIS Z 1703.

Volumetric capacity and tightness tests are determined in accordance with JIS Z 1703.

Physicomechanical properties for polyethylene bottles are presented in Table 27.

Additionally, the susceptibility to soot accumulation may be determined, in accordance with ASTM D 2741.

Various morphological tests and research results concerning wall failure, as a function of the medium, allow one to identify the mechanism by which oxidative degradation occurs.

Internal failure of walled containers or bottles can be evaluated by scanning electron microphotography or by differential scanning calorimetry thermograms, which indicate the endothermal bimodal maximum caused by the diffusion mechanism of low molecular weight components [55].

For pharmaceutical purposes, containers and bottles are tested for permeability by estimating the

TABLE 24 Requirements for polyolefin containers

Property	Performance
Product compatibility	Complete compatibility between container and product to be packed
Environmental stress-crack resistance	No evidence of leakage, cracking, or crazing
Drop impact strength	No sign of rupture or leakage
Top load resistance	No deformation
Adhesion of ink to printed containers	No ink shall be removed
Resistance of product to printing on container	No ink shall be removed
Handle strength	Handle shall remain interact and undamaged
Closure	Resistance to product, and corresponding to type and form of container

Source: Adapted from BS 4839 Part 1 (1972), part 2 (1974), and Part 3 (1977); all test methods from this standard.

TABLE 25 Requirements for polyolefin cisterns

Property	Value
Deformation	2%
Resistance to deflection	≤32 mm
Fatigue resistance	No cracking
Weathering test	In accordance with 8.9 BS 4213
Hot water test	In accordance with 8.10 BS 4213

Source: Adapted from BS 4213 (1986); all test methods from this standard.

transfer rate of the substance through the wall from penetration–time diagrams [56, 57].

In the case of packaging for very aggressive chemicals, mechanical and stress-cracking tests are made in the same aggressive medium [58].

Internal tensions induced during the manufacturing process definitely influence the properties of polyolefin containers and bottles. Conditioning of container specimens is very important because at temperatures higher than the glass transition temperature T_g, the elastic area increases as a result of the partial elimination of these tensions.

A most efficient test method for stress-crack resistance of bottles was established by Mitsui Petrochemical Industries Ltd. The MPC-H_2-A 312-2/709 method consists of deforming filled bottles and immersing them in a tensio-active medium.

All testing procedures for containers and bottles with capacity exceeding 3.78 L (1 gal) must satisfy the security requirements included in the US Hazardous Materials Transportation Act of 1989 and the requirements of Office of Operation and Enforcement, Material Transportation Bureau, as given in the *Federal Register* (49 CFR 178–24.a) [59].

V. TESTING OF COATINGS [66–73: DIN 53283]

Coatings of polyethylene powder are applied by the fluidized-bed method, the flame spraying method, or the dispersion method. Tests of coatings depend on the shape, dimension, and final application of the coated metals.

Appearance test: the product is inspected visually for color, gloss surface, roughness, cracks, and other

TABLE 26 Routine requirements for polyolefin cisterns

Property	Value
Impact resistance	The impact shall neither split nor puncture the cistern.
Tensile strength	7.6 MPa
Elongation at break	
longitudinal	50%
transversal	100%
Resistance to reversion[a]	
for PE with density	
≤0.94 g/cm^3	5.0
for PP with density	3.0
>0.94 g/cm^3	2.0
Sprue strain[a]	max. 0.76 mm
Delamination[a]	Peeling from specimen ends only

[a]Applicable to injection-molded cisterns only.

TABLE 27 Requirements for polyethylene bottles

Property	Test method	Performance
Leak test	JIS K 1703	No leakage shall occur
Drop test	JIS Z 0202	No damage to the stopper shall occur, and airtightness of the stopper shall not be diminished
Boiling test	JIS Z 1703	Variation of internal capacity shall be within 6%
Stress cracking	JIS Z 1703	Shall show at least the value agreed between the parties concerned
Test of damage to stopper	JIS Z 1703	No damage shall occur

Source: Adapted from JIS Z 1703 (1976).

defects that may be detrimental in the intended application.

Thickness: one of three independent tests can be used:

Electromagnetic gaging (JIS K 6766);
Length gaging with micrometer (JIS B 7502);
Dial gage (JIS B 7503), and vernier caliper (JIS B 7507);
Microscopic test (JIS K 6766).

Pinhole test: tester coils scan the coating surface with a high diffusion voltage (JIS K 6766).

D.c. high voltage test: 5 mA and 10 kV (JIS K 6766).

Chemical resistance test: made in the specific medium with which the coating is in contact (22 and 94 h at 23°C and 60 days at 60°C).

The chemicals recommended by JIS K 6766 are presented in Table 28.

Studies and tests of polyolefin coatings measure various stresses appearing during service and refer to surface quality thickness, thermal impact resistance, chemical resistance and adhesivity (efficiency and durability).

For polyethylene coatings on paper, the control of surface quality can be investigated by a multiple-sensor computerized control system [66]. The thickness of very thin coatings must be determined by radiometric recording of neutron absorption [67].

For very large domains of temperatures, the coating generally retains its properties of resistance at 40 cycles between −40 and 80°C (30 min each cycle) [68].

The adhesivity of a coating can be determined with ultrasonic microscopy by calculating the coefficient of reflectivity at the metal–coating interface under 25–27°C. The values of this coefficient show the difference between perfect and imperfect adhesion [69].

Polyethylene coating efficiency on steel is measured by the coefficient of expansion in H_2O, 10% AcOH, and 10% HNO_3, and by capacitance and electrical resistance [70].

TABLE 28 Specific media for imparting chemical resistance to coatings

Chemical	Concentration
Sodium chloride solution	10% aqueous solution
Sulfuric acid	30%
Nitric acid	40%
Sodium hydroxide solution	40% aqueous solution
Ethyl alcohol	95% v/v

Source: Adapted from JIS K 6766 (1977).

The protective effect of PE coatings on concrete is determined by assessing the penetrability of long-term chemical action (5–25% HNO_3, H_2SO_4, and HCl). In this way, it is possible to estimate coating life (5–38 yr for HNO_3, 38–80 yr for H_2SO_4, and 20–50 yr for HCl) [71, 72].

Plasma treatment of polypropylene coatings stimulates adhesion to metals.

Logarithmic mass–time diagrams for chemical resistance permit one to study the increase in solution absorption until penetration of the chemical medium and corrosion of the substrate have occurred [73, 74].

VI. TESTING OF RODS [74, 75]*

Polyethylene rods are used by themselves or as part of other devices in the chemical and building industries, and so on. They are manufactured by extrusion from polyethylenes of low, medium, and high density.

Pieces manufactured from PE rods have great chemical resistance, sometimes replacing pieces made of polytetrafluorethylene (PTFE).

The PE rods may be black (for general use); or natural (when the rod will not be exposed to direct sunlight). Black is essential when the rod is to be exposed to direct sunlight.

In appearance the surface must be free from contamination grooving, air bubbles, blisters, and voids, in accordance with BS 2917 for LDPE and MDPE rods and BS 4645 for HDPE rods.

Physicomechanical requirements are given in Table 29.

Depending on the service medium and the conditions of application, rods and manufactured pieces are additionally tested with regard to impact strength (ASTM D 256 or BS 2782–met 350) and variation of weight in NaOH, HCl, and NaCl (ISO 175).

Internal defects are defected by X-ray or ultrasonic methods for high precision manufactured pieces such as retention balls for atomic reactors.

Similar research is aimed at determining the elastic bending modulus without plastic deformation [74].

Aging in corrosive media can be estimated by ultrasonic wave rate propagation as a function of the stationary time in the medium; this approach takes advantage of the dependence of the speed of propagation on the volatility of the medium [75].

* The following German standards apply in Section VI: DIN 16980, DIN 16774, and DIN 16776.

TABLE 29 Requirements for polyolefin rods

Property	Test method	LDPE and MDPE	HDPE
Dimensions	BS 2919 and BS 4645		
diameter, mm		2.3 + 0.5–152.4 + 1.5	2.3 + 0.5–152.4 + 1.5
length, m		15; 30; 60 and over	15; 30; 60 and over
Tensile strength, MPa	BS 2782–301 F	9.6–10.9	18.5
Elongation at break, %	BS 2782–301 F	350	150
Heat reversion, %	BS 2919	3	

Source: Adapted from Bs 2919 (1982) and BS 4645 (1982).

VII. TEST METHOD FOR BIODEGRADABLE POLYOLEFINS [76–82]

Ecological aspects becomes more and more important for any industrial production and we all should care for the environment in order to preserve the living basis for the coming generation. Biodegradable materials can help us in doing so.

Biodegradable polyolefins are a chance for us to contribute to the reduction of the waste problem.

Today's intelligent products and production processes should respond to the question: "Why should a product which is used possibly only for a few minutes be made of exhaustible raw materials instead of renewable ones?" Biodegradable polyethylene film is available which is in accordance with environmental demands. All over the world there are already composting facilities which can be used to "recycle bioplastics".

Since there are various products for which it is claimed that they are biodegradable, it is necessary to find a generally acceptable testing standard to test all products carefully under the same conditions. Manufactures from Japan, Europe, and USA are testing their products according to ASTM D 5338-1992, the best recognized testing procedure for bioplastics. ASTM D 5338-1992 "Controlled composting test" determines the released carbon dioxid during an aerobic degradation. This method is in accordance with the CEN draft "Evaluation of the ultimate aerobic biodegradability and desintegration of packing material under controlled composting conditions". Only if a material does disintegrate within a standard compost cycle of 10–15 weeks can it be called compostable.

Since the compostability is the Unique Selling Purpose (USP) and main marketing idea for this new product category it has to be determined that the products are accepted by composting companies. This is even more important because this new group of materials called bioplastics are compostable but look like ordinary plastic, which makes it difficult to see the difference between the products.

Therefore, a label for biodegradable products is necessary. Some countries have introduced official labels for biodegradable products. For the German and Benelux market, the latest legislation demands only a fully biodegradable material.

The testing procedure is slightly different the "Controlled Composting Test". DIN 54900 (in press) defines "Biodegradable polymers" and measures the ecotoxicity and heavy metal content. The method will be very similar to the CEN standard.

REFERENCES

1. ASTM D 1248-84.
2. ASTM D 4101-82.
3. ASTM D 2581-80 (87).
4. EA Koslikoskaia, VE Bakhierev. Plast. Massy, 9: 22, 1980.
5. M Ifwarsson. Kunstoffe, 79: 525–529, 1989.
6. A Kettman. Gas-Erdgas, 118: 277–284, 1977.
7. N Kaiy. Jap Polym J, 2117: 523–531, 1989.
8. A Dorsch. Z Werkstofftech, 15: 1172–1776, 1984.
9. N Kaia, A Takahara. Polym J (Tokyo), 21: 523, 1989.
10. P Fluer, DR Roberto. ASTM Spec Tech Publ, 736, Philadelphia: American Society for Testing and Materials, 1981.
11. R Vancrombrugge. Mater Teck, 67: 143–148, 1979.
12. GK Kaigorodov. Svar Proizvod, 1: 13–14, 1982.
13. D Badgerow. Oper Sect Proc Am Gas Assoc D 433–466, 1983.
14. RM Genussov. Int I Fract, 38: R9–R12, 1988.
15. LM Lobanov, VA Pivtorâk. Avtom Svaska, 1: 29–32, 1986.
16. LK Uspenskii, DI Bogdanov. Zavod Lab, 46: 66–67, 1980.
17. VS Loginov. Primer Polim Mater Stroti, 18: 21, 1983.
18. JF Schneider, L Haty. Energy Res Abstr, 13: 10, 1988.
19. K Jung. Plaste Kautsch, 34: 168, 1987.

20. J Polak, L Boubela. 7th Proc Int Conf Internal and External Pipes, Prague, Czechoslovakia, 1987, pp 73–80.
21. LM Moore, GP Marshall. Polym Degrad Stab, 25: 161–180, 1989.
22. D Saint Toyre. J Appl Polym Sci, 38: 147–162, 1989.
23. PS Leevers, JG Williams. J Phys C, 3: 231–236, 1968.
24. VP Tarnogradskii, SA Sergienko. Teknol Oborud Svaski Skluvanya Plastmassov Akad Nauk Kiev URSS, 1987.
25. K Jung. Plaste Kautsch, 34: 168–170, 1987.
26. El Barbari, N Michel, I Menges. Kunststoffberarter, 31: 63–66, 1986.
27. GN Karov, EA Mineev. Avtom Svaska, 10: 67–71, 1985.
28. J Crissman. J Test Eval, 11: 273–278 1983.
29. I Harvath. ZIS Mitt, 21: 723–732, 1979.
30. HM Elder, M Tawashi. J Appl Polym Sic, 22: 1443–1449, 1978.
31. AV Sheney, DR Saini. Polym. Test, 6: 37–45, 1986.
32. LN Chapkina, VI Fvarov, VA Dadyko. Plast Massy, 5: 55–56, 1981.
33. LP Razumovskim, E Aseeva. Polym Degrad Stab, 26: 113–123, 1989.
34. A Cantor. IEEE J Quantum Electron QE-17: 447–489, 1981.
35. WJ Nelson. Report EPRI-EL, 1120: 54 Richmond, VA: Reynolds Met Co., 1979.
36. JN Groegh. IEEE Trans Electr Insul, EI-19: 250–253, 1984.
37. GT Hoang, P Guerin. Proc Int Conf Conduct Breakdown Solid Dielectric, Piscataway, NJ: IEEE, 1983, pp 390–394.
38. KE Gonslaves, SH Patel. Polym Prep, 503–504, 1989.
39. M Saure, H Franke. Report 1980, BMFT-FB-T-80-044, Available from U.S. National Technical Information Service Division of Government Reports, Announce Index, 81(20): 4303, 1981.
40. JP Caine, SC Haridoss. Polym Eng Sci 28: 1145–1149, 1988.
41. P Zoller. Polym Test, 3: 197–208, 1983.
42. NI Williams. Plast Rubber Process Appl, 1: 209–276, 1981.
43. VA Antov. Visokomol Soedin. Ser B, 30: 594–597, 1988.
44. B Kloeckner. Kunststofferarbeiter, 33: 60–63, 1988.
45. K Thompson. TAPPI J, 71: 157–161, 1988.
46. JP Watson. Anal Proc London, 22: 105–109, 1985.
47. HJ Serif, H Wolmann. Pharmazie, 37: 711–714, 1982.
48. P Kuobloch. Plast. Kauc, 21: 134–136, 1984.
49. P Ciclo, M Lamoutague. Adv Instrum, 42: 1601–1614, 1987.
50. A Stainov, K Petkov, V Kreslev. Polym. Degrad Stab, 23: 99–107, 1989.
51. VM Aslanyan, AN Felekyan. Plast Massy, 4: 34–35, 1989.
52. QB Bucknall. Pure Appl Chem, 58: 985–998, 1986.
53. MG Dodur. Plast Massy, 9: 16–17, 1975.
54. M Cotovan, E Obloja. Mater Plast, 11: 410–413, 1974.
55. AK Kulshrestha. J Appl Polym Sci, 37: 669–679, 1989.
56. R Skalotzki, H Wollmann. Pharmazie, 38: 392–395, 1983.
57. N Kallus. Brauwelt, 117: 826–827, 1977.
58. JD Barnes. Soc Plast Eng Tech Pap, 24: 824–826, 1978.
59. Wyle Laboratories, Inc. Report 1984. Available from U.S. National Technical Information Service Division of Government Reports, Announce Index, 85: 175, 1985.
60. BD Bauman. Air Products and Chemical Inc. US Patent 4,552,847 (cl. 136-5; GOIN 15/08), November 12, (1985) Appl. 536, 602, September 28, (1983).
61. J Vogel. Plast. Kautsch, 36: 13–15, 1989.
62. P Eyerer, T Wurster. Plastverarbeiter, 35: 30–34, 1984.
63. F Herrman, H Hoefler. Kunstoffe, 73: 83–87, 1983.
64. MJ Gowood, TJ Sheerman. Polym Test, 1: 191–199, 1980.
65. IA Popov, VA Yurzhenko. Plast Massy, 3: 36–37, 1979.
66. LB Kilman. TAPPI J, 72: 113–116, 1989.
67. GS Pakarkii. Defektoskopya, 8: 1078, 1980.
68. IB Savatinova. Zavod Lab, 46: 173–174, 1980.
69. Y Tsukahara, K Ohire. Ultrasonics, 37: 3–7, 1989.
70. S Okude. Jap Org Coat, 5: 255–271, 1983.
71. J Fabrug, K Binder. Chem Kunstst Aktuell (Sondernumer; Nat Kuenstliche Alterung Kunstst) pp 49–57, 1977.
72. MK Frolova, EN Shevehenko. Plast Massy, 4: 55–57, 1983.
73. YA Usovich, OE Reinvald. Lakokras Mater Ikh Primen, 3: 434, 1983.
74. JM Charrier, AN Gent. Polym Eng Sci, 24: 1172–1173, 1984.
75. B Kaupe. Z Werkstofftech, 15: 157–172, 1984.
76. Degradable Plastics: Standards, Research and Development. U.S. General Accounting Office. Report No. GAO/RCE D-88-208, Gaithersburg MD: US GAO, 1988.
77. JM Gould, SH Gordon, LB Dexter, CL Swanson. Biodegradation of starch-containing plastics. In: JE Glass, G Swift, ed., Biodegradability and Utilization, Washington DC: Am Chem Soc, 1990, pp 65–76.
78. GJL Griffin. US Patent 4.016.117 (1977).
79. RP Wool, D Raghavan, S Billieux, G Wagner. Statics and Dynamics of Biodegradation in Polymer-Starch Blends. In: M Vert, J Feijen, A Albersson, G Scott, E Chiellini, eds. Biodegradable Polymers and Polymers and Plastics. Cambridge: Royal Society of Chemistry, 1991, pp 112–122.
80. JS Peanaski, JM Long, RP Wool. Polyethylene-Starch Blends, J Polym Sic Part B, Polym Phys Ed, 29: 565, 1991.
81. SM Goheen, RP Wool. Degradation of Polyethylene-starch Blends in Soil, J Applied Polym Sic, 42: 2691, 1991.
82. G Vallini, A Pero, F Cecchi, M Briglia, F Perghem. Waste Manag Res, 7: 277, 1989; RG Sinclair, ES Lipinsky, JD Browning, D Bigg, TA Rogens. US Patent 5,444,133-A (1995).

31

Balance of Processability, Performance, and Marketability in Metallocene Polyolefins

Anand Kumar Kulshreshtha
Indian Petrochemicals Corporation Limited, Vadodara, Gujarat, India

Cornelia Vasile
Romanian Academy, "P. Poni" Institute of Macromolecular Chemistry, Iasi, Romania

I. INTRODUCTION

Metallocene-catalyzed polyolefins represent the latest wave of developments in the history of the polyolefin industry. These catalyst systems have been applied to a number of key product families across the chemical industry including: polyethylene, polypropylene, cyclic olefins, styrenics, and specialty chemicals. Activity in the various metallocene technologies is proceeding on a global basis, with many participants already involved in alliance arrangements.

Analysis of the fundamental factors underlying market penetration support the belief that metallocene-catalyzed polyolefins will become a key technology of the 21st century. These considerations include: (1) applicability to current infrastructure and technology (catalyst, process) configuration, (2) performance attributes, (3) cost, (4) development status, (5) strength of the participants, and (6) overlap with existing products/markets. The normal market and technological barriers that hinder all new innovations exist with metallocenes, and it is critical that these barriers be eliminated/lowered over time if maximum commercial success is to be achieved.

This chapter discusses the metallocene polyolefins from their processability point of view. This was the constraint of m-PO resins in the past but has been overcome by advances in technology. Metallocenes compete with conventional POs, and engineering thermoplastics and have no peer. Nothing is going to stop them from ruling and from bringing out novel products and applications.

II. IMPROVEMENTS IN THE PROCESSABILITY OF METALLOCENE POLYOLEFINS

Some remarkable data on the effect of polydispersity on melt index ratio for polyolefins synthesized using heterogeneous titanium-based Ziegler–Natta catalysts and produced with constrained geometry catalysts was obtained. It is generally accepted that Ziegler–Natta catalysts have multiple active center types and consequently produce polyolefins with broad molecular weight distributions. Shear thinning, as expected, increases as the molecular weight distribution broadens for polyolefins produced with these catalysts. On the other hand, polyolefins synthesized with constrained geometry catalysts have narrow molecular weight distributions, with polydispersities near the theoretical value of two for the single-site type catalyst. However, the melt index ratio can be increased at almost constant polydispersity by increasing the long chain branching frequency. It has been shown how to synthesize polyolefins with narrow molecular weight

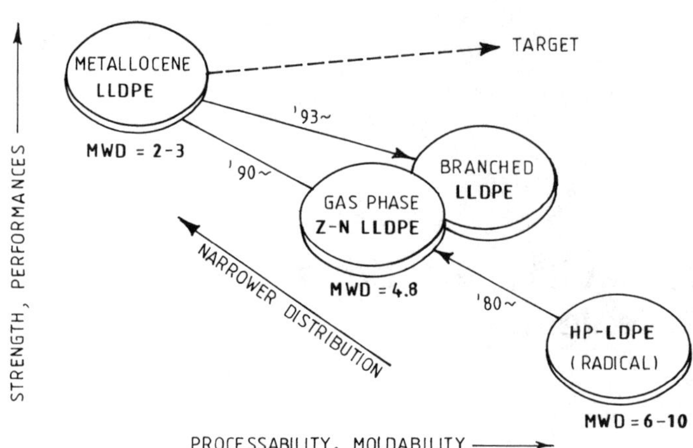

FIG. 1 Balance of strength versus processability and moldability for polyolefins synthesized using different polymerization processes [1].

distribution and sufficient degree of long chain branching that combines the excellent mechanical properties of polyolefins with narrow molecular weight distributions (impact properties, tear resistance, environmental stress cracking resistance, and tensile properties) with the good shear thinning of linear polyolefins with broad molecular weight distribution. Polyolefins with narrow molecular weight distributions and containing no long chain branches generally have poor rheological properties.

An interesting summary of property balances and the position of metallocene copolymers of ethylene α-olefins in the balance of strength versus processability and moldability (Fig. 1) are shown.

A suitable cocatalyst is tris(pentafluorophenyl) borane. There is no evidence in the literature that methylaluminoxane cocatalysts are suitable for the synthesis of polyolefins containing long chain branches. It can be speculated that the presence of methylaluminoxane will promote transfer to aluminum and therefore produce dead polymer chains with saturated chain-ends which are unavailable for long chain branch formation.

Although there is not much information available in the literature concerning optimal solvents for long chain branch formation, it appears that paraffinic solvents are preferable to aromatics. The use of aliphatic solvents for long chain branch synthesis has been recommended. This choice has also been supported for the catalytic system bis(cyclopentadienyl) zirconium dimethyl/perfluorotriphenyl boron. It was found that this catalyst system gave high levels of vinyl ends during polymerization in hexane and few or no vinyl ends in toluene.

It seems that the target for polymers made with these catalytic system is to produce polymer chains having narrow molecular weight distributions but with high levels of long chain branching to further improve processability.

Figure 2 gives a schematic mapping of the processability – toughness profile of various PE blown film resins.

III. APPROACHES TO ENHANCE THE PROCESSABILITY OF METALLOCENE RESINS

A. Dual Catalyst Systems

The normally narrow polydisperse metallocene polyolefins have a significant processing problem that could be solved by considering broadening the molecular weight distribution through blending different metallocene polyolefins, through the use of specific metallocenes that could yield a small amount of long-chain branching, through the use of mixed metallocenes that could form bimodal distributed polymers, or through the variation of the transition metal ratios of the mixed metallocene. The use of Cp_2ZrCl_2 and Cp_2HfCl_2 metallocene at 60°C resulted in monomodal molecular weight distribution polyethylene having a polydispersity index of 2.5. For the following mixed metallocenes Cp_2ZrMe_2, Cp_2TiPh_2, $(Me_5Cp)_2ZrMe_2Me_2$, and Cp_2TiPh_2, simply varying

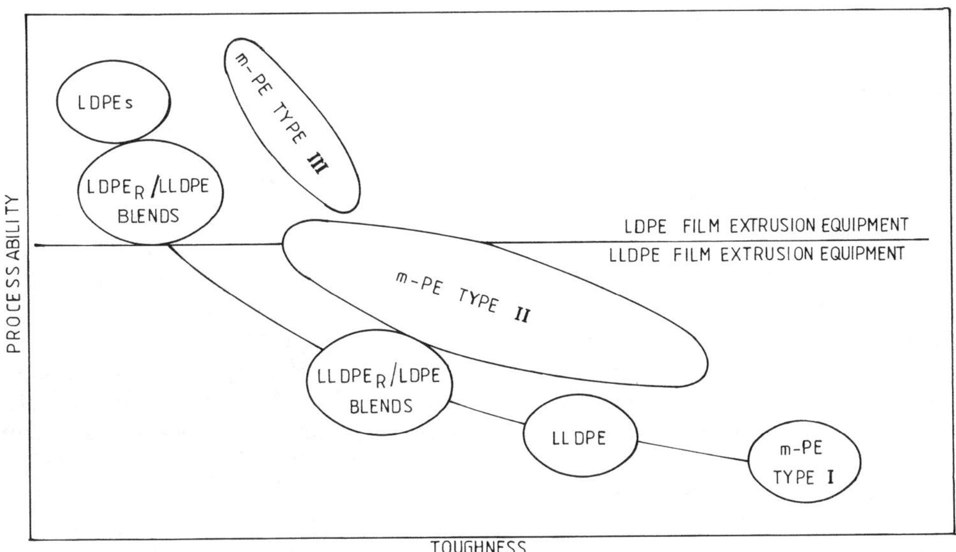

FIG. 2 Schematic map of processability—toughness domains of EXXPOL/UNIPOL m-PEs.

the Ti : Zr ratio results in the variation of the number average molecular weight between 16×10^3 and 63×10^3.

The use of catalyst mixtures consisting of C_{2v}-, C_2- and/or C_3-symmetric metallocenes provides some insight into the understanding of metallocene catalyst activity, comonomer incorporation capability, and the corresponding polymer microstructure. Some of the early works in this area involved the following: (i) Cp_2ZrCl_2 and $En(Ind)_2ZrCl_2$; (ii) Ind_2ZrCl_2 and $En(Ind)_2HfCl_2$; (iii) Cp_2HfCl_2 and $En(Ind)_2ZrCl_2$; (iv) rac-$En(Ind)_2ZrCl_2$ and rac-$En(Ind)_2HfCl_2$; and (v) the chloride-free mixed metallocene $(Cp)_2ZrMe_2$ and $En(Cp)_2ZrMe_2$. These mixed-metallocenes, differing in terms of transition metal, π-carbocyclic ligand, and the σ-homoleptic alkyl/aryl ligand, are characterized by different propagation and termination rate constants. In use, the most significant effects relate to the broadening of the molecular weight distribution through the achievement of bimodality of polydispersity index.

B. Manufacture of Bimodal Polyethylene by Varying Ethylene Pressure

It has been shown that by varying the ethylene pressure during polymerization the MWD of polyethylenes made with a two-site-type silica supported metallocene catalyst can be controlled from broad to narrow and from uni- to bimodal.

The observed MWD bimodality of polyethylene made with the combined catalyst is found to be due to the presence of two different metallocenes. The MWD of polyethylene made with the combined catalyst is slightly narrower than the one expected by the direct combination of the MWDs produced with the individual metallocenes. This might indicate that the two metallocenes interact in the combined catalyst or, more likely, that the supporting conditions for the individual and combined metallocenes were slightly different, resulting in supported catalysts with somewhat distinct behavior.

The MW averages of polyethylene made on $Et[Ind]_2ZrCl_2$ are not very sensitive to ethylene pressure, but that the ones for polyethylene made with Cp_2HfCl_2 increase at higher ethylene pressures. This indicates that the peak separation of the combined MWD will vary with ethylene pressure, which is a very convenient way of controlling the shape of the combined MWD.

The MWD varies from broad and unimodal at the lower pressure (5 psig), to broad and bimodal at intermediate pressures (50–150 psig), and finally to narrow and unimodal at the higher pressure (200 psig). For Cp_2HfCl_2, the MWD shifts to higher molecular weight averages when monomer pressure is increased, which indicates that β-hydride elimination and eventually transfer to MAO (no chain transfer agent was used) is the controlling transfer step. The relative amount of polyethylene made on Cp_2HfCl_2 sites is inversely proportional to ethylene pressure. The MW peak associated with Cp_2HfCl_2 sites becomes increasingly smaller as ethylene pressure increases and at 250 psig the MWD becomes unimodal.

C. Dual Metallocene m-PPs and m-BOPP

One of the main characteristics of resins made with single-site catalysts is narrow molecular weight distribution and uniform composition. In the case of PP, this leads to correspondingly narrow melt temperature ranges and at times dramatic processing limitations. Thus, early commercial uses of single-site PP were in applications where its narrow melt temperature range was not a problem, such as cast film, or in compounds.

Exxon has found a new level of molecular control that can solve all of these limitations. Exxon has developed unique isotactic PP homopolymers and copolymers made in one reactor using dual metallocent catalysts on a single support. The copolymer is a bimodal and probably the first true bimodal resin made in a single reactor. (Fig. 56, Chapter 1).

Bimodal composition distribution, which applies only to copolymers, can produce resins having different concentrations of copolymer, i.e., different melting points in that are the same length.

Exxon describes a dual-metallocene catalyst on one support, referred to as Catalysts 1 and 3, making two different narrow MWD polymers. Catalyst 1 makes long isotactic polymer chains with a 150°C melting point; Catalyst 3 makes shorter isotactic polymer chains with a 135°C melting point. The combination, however, is not a predictable linear blend. Catalyst 3 has an unexpected secondary effect on the long-chain population made by Catalyst 1, reducing them, broadening the MWD, and creating a higher concentration of short chains. This secondary effect also broadens the melt temperature range. The shorter polymer chains from Catalyst 3 also have lower isotacticity or crystallinity.

Exxon said the copolymers are more remarkable than the homopolymers. When ethylene is added, the initial chain-breaking reaction reverses and longer chains form again. The copolymers combine all three "tailored molecular distributions"—molecular weight, composition and tacticity—whereas the broad MWD homopolymers display control of only molecular weight, and tacticity distribution. These dual-metallocene PPs are a very recent development.

The new three-way modality control can broaden melt temperature ranges and make single-site PP that may be used in high speed biaxially oriented film lines.

Biaxial orientation is commercially by far the most important PP film making process. It starts by forming a thick PP sheet, then stretching it in perpendicular directions at close to its melting point. BOPP resins need a low softening point and broad range of stretch temperatures. If the processing temperature window is too narrow, stretching is uneven and the web breaks.

The new bimodal copolymers can make good BOPP films at as much as 10°C below the processing temperatures of current Ziegler PPs. This suggests the possibility of a resin with 'step-out' processability, allowing higher line speed and improved manufacturing economics to deliver levels of shrink performance previously unattainable. The melt temperature window on Exxon's first single-site PPs, commercialized as Achieve resins in 1995, for example, was only a 5°C window (152–157°C), too narrow for BOPP lines. Ziegler-made PPs have a 10°C window. The new bimodal copolymers can make good BOPP films at as much as 15°C below the processing temperatures of current Ziegler PPs.

D. Using Constrained Geometry Catalysts

Dow Chemical constrained geometry catalysts are the most suitable catalysts because they have an "open" metal active center. The active center of these catalysts is based on group IV transition metals that are covalently bonded to a monocyclopentadienyl ring and bridged with a heteroatom, forming a constrained cyclic structure with the titanium center. The bond angle between the monocyclopentadienyl ring, the titanium atom center, and the heteroatom is less than 115°. Strong Lewis acid systems are used to activate the catalyst to a highly effective cationic form. This geometry allows the titanium center to be more "open" to the addition of ethylene and higher α-olefins, but also for the addition of vinyl-terminated polymer molecules. A second very important requirement for the efficient production of polyolefins containing long-chain branches by these catalytic systems is that a high level of dead polymer chains with terminal unsaturation be produced continuously during the polymerization.

E. By Blending

Processing properties of syndiotactic polypropylene have been improved by blending with isotactic polypropylene. It has allowed a syndiotactic polypropylene-based polymer to process well with conventional processing machines in conventional operating conditions. The transparent articles with moderate rigidity can be injection-molded without the sticking problem of polymer to the mold cavity. Film and sheet are also cast with blown film of unusual transparency achieved by quenching with water or air.

F. EXXPOL/UNIPOL Product Technology—Improved Processing Using m-PE Film Resins

EXXPOL metallocene catalysts has been customized for use with the UNIPOL process to demonstrate a range of m-PEs with narrow to broad molecular weight distribution, narrow comonomer distribution, and linear to long-chain branched structures. Prototype, improved processing m-PEs for blown film applications span a wide performance domain of processability and properties. Developmental products are produced which target the processability regime of LDPE/LLDPE blends, are useable on both LLDPE and LDPE equipment, and deliver improvements in film toughness. Technical capability has also been demonstrated to produce m-PEs gas phase with rheology essentially matching that of LDPE.

IV. BREAKTHROUGH METALLOCENE RESINS OFFERING UNIQUE COMBINATIONS OF PERFORMANCE FOR PROCESSORS

A. Dow's ELITE Resins

For unique combinations of stiffness, toughness, stretch, sealability, and other properties not available from traditional polyolefin resins, try ELITE enhanced polyethylene resins from Dow Plastics.

ELITE resins are produced via INSITE Technology, a constrained geometry catalyst and process technology developed by The Dow Chemical Company that provides extraordinary control over polymer architecture.

This design flexibility, along with advanced modeling techniques, has unleashed distinctly improved molecular structures and performance choices that put ELITE resins in a class of their own (see Fig. 3).

Best of all, ELITE resins are molecularly designed to meet or beat performance requirements in a variety of applications, (as shown in Table 1). ELITE resins can even help you differentiate your products in the marketplace and expand your product mix by providing end-users with tougher, more reliable products.

ELITE resins:

Set new performance standards;
Process efficiently on conventional equipment;
Can be tailored to specific end-use or Application requirements;
Are developed faster than ever before for speed to market.

Polymer processors will require more than the performance combinations which will be described. ELITE resins provide a full range of mechanical and processing performance properties, in addition to these significant benefits:

Lower film blocking tendencies for further improved machinability;

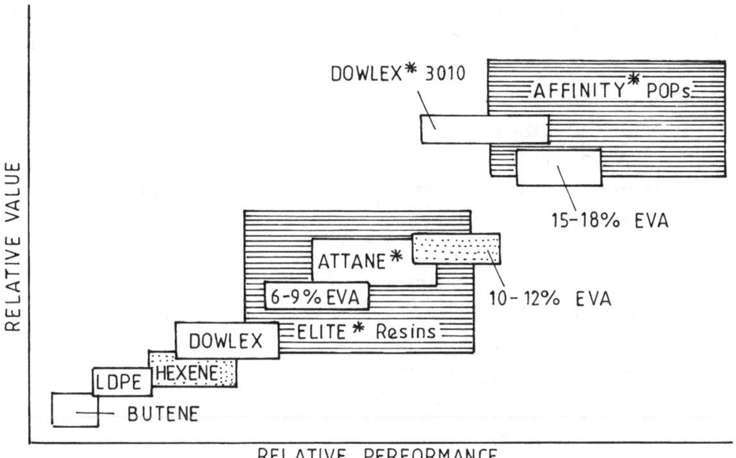

FIG. 3 Positioning of ELITE resins. ELITE enhanced polyethylene resins offer a special balance and value that fills the gap between LLDPE resins and AFFINITY polyolefin plastomers. (*Trademark of the Dow Chemical Company.)

TABLE 1 Markets and Applications for ELITE Resins

Converter markets	Application examples
Consumer and industrial thin- and thick- gauge film.	Heavy duty shipping sacks for chemical and resin packaging, lawn and garden materials, building and construction films, films and backsheets for health and hygiene articles, agricultural films, shopping bags, refuse sacks
Stretch	Blown and cast stretch films
Food and specialty films and laminates	Flowable liquid and dry goods pouches for VFFS* applications, produce and poultry packaging, bag-in-box liners, collation packaging converter films for laminations
Injection molded articles	Thin-wall refrigerated and frozen food containers cups, crates, cases
Rotomolded articles for packaging and durable goods	Toys, tanks, recreational equipment, lawn furniture, construction barricades
Extruded profiles and sheet	Wire and cable jacketing, industrial sheeting, pipe, automotive liners

* VFFS = Vertical form–fill–seal.

Improved hot tack strength for enhanced package integrity in VFFS equipment;
Low extractables for improved taste–odor performance;
Excellent ultimate tensile strength for load bearing capability;
Low temperature impact resistance for cold temperature durability;
High on-pallet load retention for fewer package failures.

1. Stiffness and Impact Strength for Downgauging

For applications such as consumer and industrial films, high dart impact and high modulus are an important property combination. These properties not only ensure abuse resistance and the opportunity to *downgauge* your product without sacrificing performance, but they also offer improved convertibility.

Usually, an increase in modulus (by increasing density) will sacrifice impact strength. Dow Plastics can molecularly design ELITE enhanced polyethylene resins to increase the modulus by 20–25% over a higher α-olefin 0.920 density LLDPE, while increasing the dart impact strength by 40%. Or, *dart impact can be tripled or even quadrupled while maintaining modulus*, providing super-tough films that can accommodate higher levels of recycle or trim and scrap.

2. High Stretch and High Puncture

Depending on the application, blown and cast stretch films ideally provide high stretchability for increased product yields, strong holding force to secure pallets during transportation, and high puncture resistance to prevent failure on corners and irregularly shaped loads.

To provide converters with more versatile products, ELITE enhanced polyethylene resins are designed to outperform LLDPEs, significantly improving on-pallet puncture resistance at moderate prestretch levels for wrapping irregularly shaped loads, while maintaining or increasing ultimate stretch and on-pallet elongation for higher product yields on regular loads. ELITE resins offer this combination of properties while providing a level of processability that is equal to or better than that of LLDPEs.

3. Impact Strength and Processability

Applications such as thin-wall injection molded containers require processability for easier flow and faster cycle times, and impact strength for durability and crack resistance—especially at low temperatures for refrigerated and frozen food applications. Yet for typical HDPE injection molding resins, when the processing index is decreased for improved mold flow and extrudability, dart impact also decreases.

ELITE enhanced polyethylene resins can eliminate that trade-off by maintaining excellent processability while doubling the impact strength of current state-of-the-art HDPEs.

4. Sealability and Stiffness

Although LLDPE resins brought new levels of toughness to packaging films, they sealed at higher temperatures. To achieve lower heat seal initiation temperatures and high hot tack strengths, converters often blend or coextrude with a lower melting point material, such as POP, EVA, or ULDPE. This, however, increases costs and sacrifices stiffness.

With ELITE enhanced polyethylene resins, you can overcome this problem, realise faster packaging line speeds and fewer leakers on vertical form–fill–seal equipment, and achieve better handling characteristics. The lower seal initiation temperature also creates a broader sealing window, while providing extra flexibility in various bag-making processes.

ELITE resins can provide the heat seal initiation of EVA and ULDPEs with a modulus that compares to that of an LDPE. This translates to *an increase in modulus of up to 100%* versus straight EVA or ULDPE.

5. Sealability and Heat Resistance

Some food and specialty packaging applications require heat resistance for hot fill and cook-in capabilities, in addition to low seal initiation temperatures. This combination of properties is often compromised in conventional EVA and ULDPE.

ELITE resins deliver the heat seal initiation temperature of an EVA or ULDPE, and a Vicat softening point comparable to or better than that of LLDPE.

6. Fabrication Compatibility and High Output

ELITE enhanced polyethylene resins are compatible with existing commercial fabrication equipment, including injection molding, blown and cast film extrusion, extrusion coating, rotomolding, profile and sheet extrusion, and thermoforming equipment.

ELITE enhanced polyethylene resins also have excellent compatibility with other polyolefins, allowing efficient blending and coextrusion for cost and performance flexibility.

In addition, ELITE resins are fully stabilized for high temperature extrusion and for trim and scrap reuse.

B. Innovene Technology with INSITE Resins [3]

Products that are not only stronger (making downgauging possible) but also more processable than LLDPE are at a significant advantage over conventional Ziegler and high pressure LDPE technology based products (Fig. 4).

This substantially improves processability of film products, even high strength products with a narrow molecular weight distribution (Fig. 5). Normally this is a difficult combination to achieve due to the trade-off between strength and processability. The reason behind the novel polymer architecture is the constrained geometry within the catalyst itself.

Most importantly, there is no output penalty using existing extrusion equipment. This is key since metallocene products are going to be widely accepted in the volume markets. Innovene with INSITE technology is clearly unique in this respect.

A further benefit of the new products is their outstanding sealability in packaging film applications. The

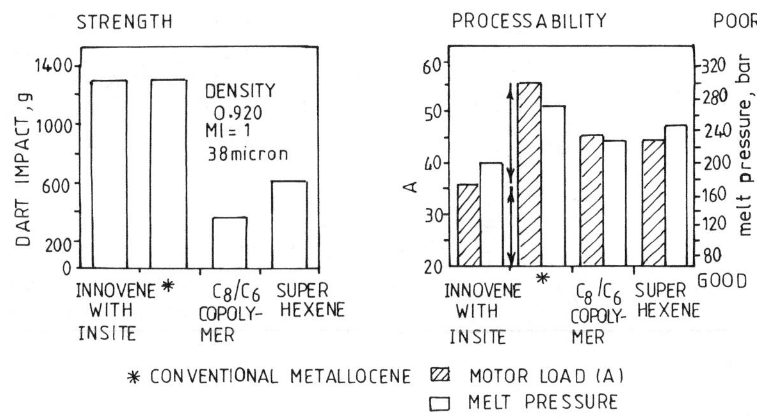

FIG. 4 High performance blown film.

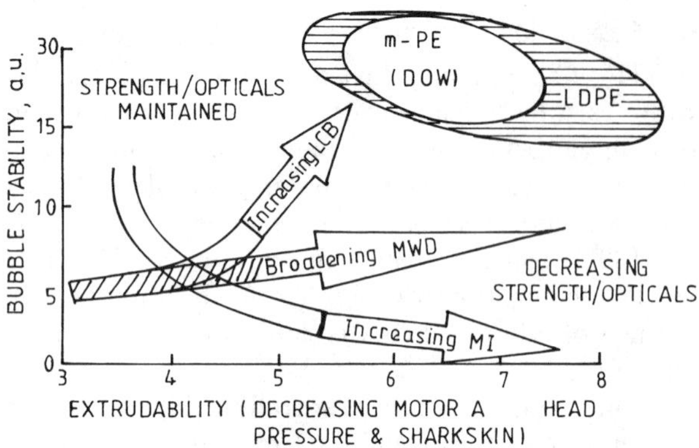

FIG. 5 Product processability—polymer design.

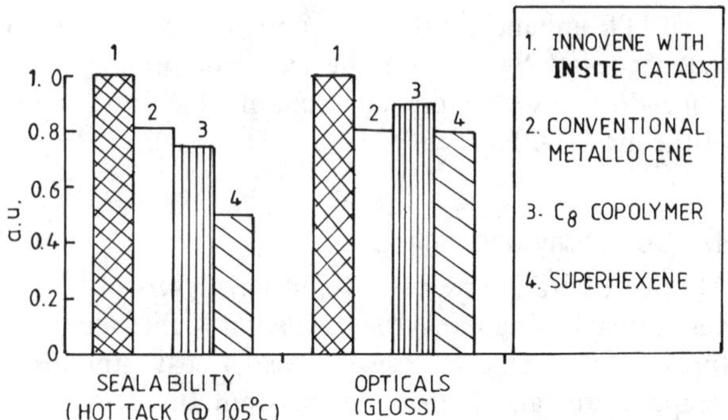

FIG. 6 High performance blown film. @ Trademark of the Dow Chemical Company.

seal starts to form at lower temperature and the hot tack strength is greater than LDPE and even the best-LDPE. This is once again a result of regular distribution of branching in the polymer backbone. The benefits will be faster packaging line speeds, e.g., in form–fill–seal applications.

Long chain branching (LCB) is needed to achieve both improved extrudability and melt strength. Simply broadening molecular weight distribution has little effect on melt strength and increasing melt index improves extrudability but at the expense of melt strength (Fig. 5).

Apart from high performance film, other products can be designed having the processing characteristics of LDPE while retaining the advantages in mechanical properties, heat sealing, and blocking. Optical properties are also good (Fig. 6) and will enable broad application in many packaging markets.

V. GRADES OF METALLOCENE RESINS FOR PROCESSING

Because metallocene catalysis applies to different polymerization techniques, and because the resulting resins can be performance tailored, their processing characteristics differ by grade manufacturer. Current commercial metallocenes appear to be categorized as follows:

(i) The first category targets the commodity resin market, designed to replace traditional LLDPE with a typical density of 0.915 g/cm^3, referred to as mLLDPE, where m is metallocene catalyzed.
(ii) The second and third categories are lower in density (0.865–0.915 g/cm^3) and the resins can be classified as plastomers because of their elastic behavior; plastomers are divided at the 20% comonomer level, the principal monomer being ethylene; polyalkene (polyolefin) plastomers (POPs) are 0.895–0.915 density and contain 20% comonomer; polyalkene (polyolefin) elastomers (POEs) are 0.986–0.895 g/cm^3 density and contain 20% comonomer.
(iii) The fourth type is metallocene catalyzed polypropylene and primarily aimed at the biaxially oriented polypropylene market.

Applications of the new generation resin and their place in the market are given [4] in Table 2.

Table 3 attempts to simplify the correlation between resin properties and processing.

Table 4 shows typical worst case torque requirements for four extruder sizes. The range of screw speed represents plastomers at the low end and LDPE at the high end. The specific torque value can be simply multiplied by the maximum screw speed to determine the recommended drive size. The transition from LDPE to plastomers is the most severe. When running LDPE and plastomers on the same machine, field weakening must be used to satisfy the increased screw speed and lower torque requirements of LDPE. Processing performance of narrow MWD, branched, unimodal and bimodal metallocene polymers is illustrated (Fig. 55, Chapter 1).

A. Single Reactor Gas Phase Innovene Process BP (BP Chemicals) with INSITE (Dow Catalysts) Technology

In August 1996, technical cooperation of BP Chemicals with The Dow Chemical Company was announced in the field of single-site catalyst technology to meet new customer requirements for even stronger

TABLE 2 New Generation Resins in the Marketplace

Resin	Density, g/cm^3	Application	Replacement for
Metallocene polyethylene (mPE) plastomers.	0.900–0.915	Coextrusion, heat seal layers	Ethylene–vinyl acetate (EVA) resins with 9% vinyl acetate, ultra low density polyethylene (ULDPE), very low density polyethylene (VLDPE)
mPE elastomers	0.855–0.900	Stretch–cling films	Ethylene methylacrylate
Metallocene linear low density polyethylene (mLLDPE)	0.916–0.925	Stretch films, industrial sacks, shopping bags, form/fill/seal applications, frozen food packaging	Butene copolymer C4 LLDPE, LDPE–LLDPE blends, EVA, ULDPE, ionomer layers
Metallocene medium density polyethylene (mMDPE)	0.932	Good optics food packaging films, personal care, agriculture	Low density polyethylene (LDPE)
High performance, high clarity LLDPE	0.909–0.912	Linear stretch film, lamination, highly transparent food packaging	Higher α-alkene octene C8 LLDPE, LDPE-LLDPE blends
Ethylene–propylene resins, random heterophasic polypropylene (PP) copolymer	0.890	Coextruded form/fill/seal film, hot fill, heavy duty sacks, medical and personal care films, blends with other PP or PE grades	Soft poly(vinyl chloride), nonpolymer packaging, LDPE, LLDPE

TABLE 3 Impact of Metallocene Polymer Properties on Processing

Polymer property	Interpretation	Implication	Process impact
Narrow molecular weight (MW) distribution with little or no long-chain branching	Fewer low MW 'lubricants'	Higher process pressures, higher torque	Higher melt temperature change, motor size or screw speed
	Fewer low MW 'extractables'	Low film blocking, less chill roll plateout	Easier winding/unwinding reduced, reduced housekeeping.
	Fewer high MW 'stiffeners'	Lower melt strength	Decreased bubble stability, easier drawdown
	Fewer entanglements	Higher clarity, faster melt relaxation	Easier drawdown.
Increased long-chain branching	More chain entanglement	Increased melt tension	Increased bubble stability, less draw resonance
		Increased shear sensitivity	Less torque increase, less die pressure increase
Lower density	Lower softening point	Soft, tacky pellets	Decreased specific rate in groved feed machines, increased specific rate in smooth bore machines.
		Soft, tacky film	Increased collapser friction, increased wrinkling.
	More elastic behavior	Sensitive to tension variation	Harder to wind

TABLE 4 Typical Extruder Torque Requirements

Extruder	Typical screw speed, rev/min		LDPE, kW Cast/(rpm)	LLDPE, kW/(rpm)	mLLDPE kW/(rpm)	Plastomer, kW/(rpm)
	Blown	Cast				
65 mm × 24 : 1	105–125	125–140	0.181	0.260	0.286	0.307
90 mm × 24 : 1	90–115	110–125	0.487	0.753	0.828	0.902
115 mm × 24 : 1	75–105	105–115	1.029	1.595	1.754	1.913
150 mm × 24 : 1	60–85	90–100	2.443	3.781	4.159	4.537

film products. The partnership with Dow has led to the development of unique composition of the matter patents based on the performance of INSITE catalysts in the Innovene process.

VI. CONCLUSIONS

1. Long-chain branching control in constrained geometry metallocene catalyst permits enhanced polyolefin processability on conventional processing equipment. m-PEs type III have processability equal to LDPE and have better performance and properties. Dow and Exxon both offer metallocenes to achieve this goal.
2. Blending of metallocene catalysts (reactor blending) or of resins to give bimodal MWD is not a practical approach compared to the above.
3. The single reactor gas phase process used in conjunction with metallocene catalyst is the best available technology now for improving PO processability and for obtaining high performance PO grades for demanding applications in the market.

SYMBOLS AND ABBREVIATIONS

BOPP	Biaxially oriented polypropylene
Cp	cyclopentadienyl
En	ethenyl
Et	ethyl
EVA	ethylene vinyl acetate copolymer
Ind	indenyl
LDPE	low density polyethylene
LLDPE	linear low density polyethylene
MOA	methylaluminoxane
MWD	molecular weight distribution
mLDPE	metallocene low density polyethylene
mPE	metallocene polyethylene
mPP	metallocene polypropylene
PE	polyethylene
PP	polypropylene
ULDPE	ultra low density polyethylene
VLDPE	very low density polyethylene

REFERENCES

1. AE Hamielec, JBP Soares. Prog Polym Sci 21: 651, 1996.
2. JBP Soares, AE Hamielec. Report, Molecular Structure of Metallocene Polyolefins: The Key to Understanding the Patent Literature, (Personal Communication) 1995.
3. GLC Mackay. Distinctive PE technology. Lecture given at IPCL, Nagothane, India, October, 1998.
4. K Williams. Plast Rubber Proc Appl 27(1): 30, 1998.

Appendix 1

Suppliers and Manufacturers of Polyolefins and Polyolefin Products

This list does not claim to be complete. It is compiled from many sources and the author disclaims responsibility as to the reliability of this information since it is impossible to verify directly.

Supplier/manufacturer	Tradename	Polyolefin type and product
A & E Lindenberg GmbH & Co. KG 5060 Bergisch Gladbach 2, Germany	Lindolen	PO, PP with wood flour for thermoforming and welding
A. Schulman Inc., Akron, OH 44313-1710, USA	Polyflam	PE, PP with inorganic fillers
	Polyfort	PP with inorganic filler
	Polyman	Thermoplastic elastomer
	Polytrope	Propylene copolymer, liquid crystals; molding and extrusion compounds
A. Schulman Inc., Akron, OH 44309-1710, USA	Superohm	Ethylene copolymer, molding and extrusion compounds for electrical use
AB Akerlund & Pausing, Sweden	Repak	PO molding and extrusion compounds regenerated material
Adell Plastics Inc., Baltimore, MD 21227, USA	Adell	HDPE, PP with inorganic filler molding and extrusion compounds
Aeci (Pty) Ltd., Johannesburg, 2000, South Africa	Aecithene	LLDPE molding and extrusion compounds
Airex AG, 5643 Sins, Switzerland	Herex	Olefin copolymer; cellular plastics of high density
Airofoam AG, 4852 Rothrist, Switzerland	Airofom	PE foam, for particular applications
Albis Corp. Rosenberg TX Albis Plastics GmbH, Hamburg, Germany		Reinforced PO
Aldrich Chemical Co., Inc., 4 W St. Paul Avenue, Milwaukee, WI 53233, USA		PE, CPE, PP, CPP, PB, PIB, chlorosulfonated PE, oxidized PE, P4MP
Afelder Korkwaren-Fabrik, 220 Alfeld, Germany	ALKOzell	LDPE, foam

Supplier/manufacturer	Tradename	Polyolefin type and product
Alkudia Empresa para la Industria, Madrid 20, Spain	Alkathermic	LDPE agricultural and greenhouse films
Allied Engineered Plastics Corp., Morristown, NJ 07960, USA AlliedSignal	Aclyn	Ethylene copolymer, raw material
	Paxon	MDPE, HDPE for molding and extrusion compounds, reinforced PO
	Spectra 900	PE fibers
Alveo AG, Lucerne, Switzerland	Alveolux	Ethylene copolymer, crosslinked PE foams, foamed sheets
American Hoechst Corporation, Plastic Division, 289 N Main Street, Leominster, MA 01453, USA		PE
Amoco Chemical Corp., 200 East Randolph Dr. Chicago, IL 60601, USA	Amoco	PO, PE, PP, new types
and Geel, Belgium		PP
Chocolate Bayou, TX, USA		
Cedar, Bayou, TX, USA		PP
Amoco Polymers Alpharetta, GA, USA		Reinforced PO
Ampacet Corp., Mount Vernon, NY 10550, USA	Ampacet	PE, olefin copolymers, PP preproducts, and molding and extrusion compounds
Anchor Plastics Co., New Hyde Park, NY 11040, USA	Aerotuf	PP profiles and pipes
	Aeroflex	LDPE profiles and pipes
AOE Plastic GmbH, 8090 Wasserburg/Inn, Germany	Elpeflex	PE packaging tubes and films; composite films for particular applications
	Trikoron	PE/polyamide composite films
Appryl, Lavera and Gonfreville, France		PP
Arco Chemical Co., Philadelphia, PA 19101, USA	Arcel	Ethylene copolymers foamed
	Arpro	PP foamed
	Durethene	HDPE packaging tubes and films
	Dylan	PE molding and extrusion compounds
	Dypro	Propylene copolymer molding and extrusion compounds
	Super Dylan	HDPE molding and extrusion compounds
Arak Petrochemical, Sandor Khomeni, Iran		PP
Arjomari-Prious SA, 7500 Paris, France	Arjomix	PP with inorganic fillers; glass mats or textiles
Armstrong Products Co., Warsaw, IN 46580, USA	Vibro-Flo	PP products for flame spraying and centrifugal sintering powder
Aristech, Neal WV, USA Laporte TX, USA		PP
Asahi Chemical Industries Co., Ltd., Tokyo, Japan	Copolene	Ethylene copolymers, raw material
	Suntec	HDPE, raw materials, molding and extrusion compounds

Supplier/manufacturer	Tradename	Polyolefin type and product
Asea Kabel, 12612 Stockholm, Sweden	ET-semicon	Thermoplastic elastomers for molding and extrusion compounds
ATO Chem, 92091 Paris, La Defense 5, France	FIT	PO, PP with inorganic fillers
	Lacqtene	PE, ethylene copolymers, PP for molding and extrusion compounds
	Drevac	Ethylene copolymers, raw materials for adhesive and glue
	Plathen	PE, melt adhesives, glue
	Platilon	PP packaging and electrical insulating films, webs, technical films and sheets
Bamberger Polymers, Inc., New Hyde Park, NY 11042, USA	Bapolan	PE, LLDPE, MDPE, PP for molding and extrusion compounds
BASF Aktiengesellschaft, 6700 Ludwigshafen, Germany	Basopor	PE foams
Ropzenburg, The Netherlands	Elastamid GM 261	PE/dicyandiamide resins for molding and extrusion, PP
Wilton, UK	Kuroplast	Olefin copolymers for adhesive and glue, PP
Tarragona, Spain		
	Lucalen	Ethylene copolymer for adhesives
	Lucobit	Ethylene copolymer + bitumen casting and laminating resins
	Lupolen	PE, ethylene copolymer raw materials and for molding and extrusion compounds
	Lutrigen	Chlorinated PE raw materials
	Neopolen	PE foam
	Novolen	PP raw materials and for molding and extrusion
	Oppanol-B	PIB raw materials
Bayer AG, 5090 Leverkusen, Germany	Acralen	Ethylene copolymer aqueous dispersions
	Bayer CM	Chlorinated PE raw materials
	Baylon	LDPE, ethylene copolymer raw materials and for molding and extrusion compounds ethylene copolymer ancillary products for plastics processing
	Baymod	
	Levaflex EP	Thermoplastic elastomers, molding and extrusion compounds
	Levapren	Ethylene copolymer raw materials for adhesives and glue
	Levasint	Ethylene copolymer flame spraying and centrifugal sintering powders
Bergmann GmbH Br Co. KG, 7560 Gaggenau 2, Germany	Bergaprop	PP raw materials
Berndt Rasmussen, Birkerod, Denmark	Reedex-F	PP, webs, technical films and sheetings
Borealis A/S, Beringen, Belgium and Lyngby, Denmark		PP Reinforced PO

Supplier/manufacturer	Tradename	Polyolefin type and product
Bulgaria	Buplen	PP propylene copolymers, raw materials
	Bulen	PE, molding and extrusion compounds
Billerud AB Abt. Nya Produkter Siiffle, Sweden	Monarfol	LDPE, webs, technical films and sheetings
Borden Chemical, Leominster, MA 01453, USA	Proponite	PP composite films
BP Chemicals International Ltd., London SWIW OSU, UK	Breox	Ethylene copolymer raw materials
	Hyvis	PIB raw materials in separated classification
	BP EXP839	XL Olefinelastomer + Me hydrate
	BP Polycure 798	XL polyethylene copolymer + CL add.
	Innovex	LLDPE raw materials
	Rigidex	HDPE for molding and extrusion com pounds
Braspol, Duque De Caxias RJ, Brazil		PP
Braas and Co., 637 Oberursel, Germany	Rhepanol	PIB construction sealing webs, roof coverings
British Cellophane Ltd., Bridgewater Somerset, TA6 4PA, UK	Bricling	LDPE, LLDPE packaging films and tubes
	Duraphene	PE agricultural and greenhouse films
	Propophane	PP packaging films and tubes
	Tensiltarpe	PE webs, technical films and sheeting
Büttikofer AG, 5728 Gontenschwil, Switzerland	Flexathen	PE construction profiles foamed
BXL Plastics Ltd., London, SWIW OSU, UK	Aerowrap	HDPE foamed sheets
	Alkathene	LDPE, MDPE raw materials
	BXL	PE raw material
	Evazote	Ethylene copolymer, foamed
	Hy-Bar	PE, LDPE, LLDPE, ethylene copolymer, PP composite films
	Plastazote	PE, foamed
Cabot Corp., Billerica, MA 01821 USA	Cabelec	HDPE, PP, ethylene copolymer, molding compounds for electrical applications
Caffaro SpA, Milan, Italy	Cloparin	PO
	Solpolac	Chlorinated PE for lacquers
Canadian Industries Ltd., Montreal, Quebec, Canada	Milrol	PE construction site protective webs, agricultural and greenhouse films
	Miltite	PE packaging films and tubes
Carl Freudenberg, 6940 Weinheim/Bergstrasse, Germany	Frelen	Crosslinked or crosslinkable foamed PE
Sparte Spinnvliesstoffe, Lutravie, 6750, Kaiserslautern, Germany	Lutrasil	PP fleece
Carmel Olefin Ltd., P.O. Box 1468 Haifa, 31000, Israel		PE, PP

Supplier/manufacturer	Tradename	Polyolefin type and product
Carlon, Cleveland OH, USA	Carlex	PE pipes
CdF Chemie RT, 92080 Paris, La Defense 2, France	Lotader	Olefin copolymers, raw materials
	Lotrene	LLDPE, ethylene copolymer raw materials, packaging, and electrical insulating films
	Lotrex	LLDPE, molding and extrusion compounds
	Norsoflex	LLDPE, olefin copolymer construction sealing materials without filler
Celanese Engineering Resins, Chatham, NJ 07928, USA	Celstran	PP with inorganic filler
	Vari-cut	PP with glass fibers
Celanese Polymer Specialty Co., Louisville, KY, USA	Coathylene	LDPE, MDPE, HDPE, ethylene copolymer products for particular applications, aqueous dispersions
Cellu Products Co., Patterson, NC 28661, USA	Cellair	PE foam
Cestidur Industries, 69608 Villeurbanne, Cedex, France	Cestidur	HDPE pipes, rods, technical profiles, laminated products
	Cestilene	HDPE pipes, rods, technical profiles, laminated products
	Cestilite	HDPE pipes, rods, technical profiles, laminated products
Chemepetrol, Livinov, Czech Republic		PP
Chemie Export-Import, Berlin, Germany	Scona	Thermoplastic elastomers for processing
Chemie Linz AG 4020 Linz Austria	Asota	PP fibers for particular applications, nonflammable
	Daplen	PE, molding and extrusion compounds
Chempol, Prague 10, Czech Republic	Mikroten	HDPE packaging and electrical insulating films
	Tatrafan	PP packaging films and tubes
Chemisches Kombinat Tisza, 3581 Leninvaros, Hungary	Biafol	PP packaging and electrical insulating films
	Hexcue	MDPE, HDPE molding and extrusion compounds
	Tipolen	LDPE molding and extrusion compounds
Chem. Service, Inc., PO. Box 3108, West Chester, PA 19381-3108, USA		PIB
Chim Import-Export, Bucharest, Romania	Copolarg	Ethylene copolymer molding and extrusion compounds
Chevron Chemical Co., Houston, TX 77253, USA	Poly-Eth	LDPE, LLDPE, ethylene copolymer molding and extrusion compounds
Chiba PP, Chiba, Japan		PP

Supplier/manufacturer	Tradename	Polyolefin type and product
Chisso Petro, Chiba Goi, Japan and Tokyo, Japan		PP Reinforced PO
Chicago Molded Plastic Corp., Wheeling, IL 60090, USA	Campco	PE, PP sheets and other unreinforced semifinished products
China, various prefectures		PP
Clopay Corp. Plastics Products Div., Cincinnati, OH 45202, USA	Clopane	PO packaging and electrical insulating films, other coverings
	Fabtex	PE films and synthetic leather, other coverings
	Microflex	PE flexible films and sheetings, webs and cuttings
	Satinflex	PE flexible films and sheetings, webs and cuttings
	Taff-a-flex	PO, PP flexible films and sheetings, webs and cuttings
	Velva-flex	PE films and synthetic leather
Cobon Plastics Corp., Newark, NJ 07102, USA	Cobothane	Ethylene copolymer tubes
Cole Plastics Ltd., Milton Keynes, UK	Playrite	PE molding and extrusion compounds
ComAlloy International (subsidiary of A Schulman Inc.) Nashville TN, USA		Reinforced PO
Commercial Plastics and Supply Corp., Cornwells Heights, PA 19020, USA,	Comco	PE webs, technical films, and sheetings for automatic thermoforming
Commercial Plastics Ltd., Cramlington, Northumberland, UK	Fablon	PE flexible films and sheeting, webs and cuttings
Cooper Associates, Irvine, CA, USA	Replastic	PP molding and extrusion compounds regenerated materials
Corder Ltd., Billinghurst, West Sussex, UK	Correx	PP with plastic foam core
"Ceroplast" Fritz Müller KG, 5600 Wuppertal 2, Germany	Ceroplast	PE profiles and pipes, films
	Corothene	PE webs, flexible films, sheetings, and cuttings
Courtaulds, Plastics Group, Spondon, Derby DE2 7BP, UK	Courthene	LDPE flexible films and sheetings, sheets and other unreinforced semi finished products
	Hilex	HDPE flexible films and sheetings, sheets and other unreinforced semifinished products
	Propylex	PP webs and sheets for automatic thermoforming and welding for technical purposes
Czech Republic	Frizeta	PP raw materials, molding and extrusion compounds
Cryovac Div., W. R. Grace & Co., Duncan, SC 29334, USA	Cryovac	PE flexible films and sheetings, webs and cuttings, packaging films and tubes, composite films

Supplier/manufacturer	Tradename	Polyolefin type and product
Crystal X-Corp., Darby, PA 19023, USA	Crystalene	PE flexible films and sheetings, webs and cuttings
Cyrolit, USA	Evalon	Ethylene copolymer roof covering
Dealim, Yeochon Complex, Korea		PP
Degussa AG, 6000 Frankfurt/Main, Germany	Colcolor	HDPE, ethylene copolymer molding and extrusion compounds for electrical applications
Deutsche Kapillar-Plastik GmbH & Co., KG, 3563 Dautphetal Germany	Dekalen H	HDPE pipes
	Dekaprop	PP, propylene copolymer nonflammable pipes
DLW AG, 7120 Bietigheim, Germany	Delifol	Chlorinated PE, ethylene copolymer + bitumen roof covering with carrier web
Dr. F Diehl & Co., 7758 Daisendorf, Germany	Diwit	HDPE molding and extrusion compounds, forms and sheetings for special applications
Doeflex Industries, Ltd., Redhill, Surrey, UK	Doeflex	PP webs and sheets for automatic thermoforming, compounds with fillers (chalk and talcum)
Dow Chemical Co., 2020 Dow Center Midland MI 48640 or 48674, USA	Dowlex	LLDPE molding and extrusion compounds and metallocene obtained PO
	DOW 5435 30-11	HDPE + CPE elastomer blend
	DOW 5348-40-1	CPE + fillers
	Primacor	Ethylene copolymer, LLDPE/PS molding and extrusion compounds
	Saranex	PE + polyvinylidene chloride and copolymer composite films
	Srin	Chlorinated PE, molding and extrusion compounds
	Zetabon	Ethylene copolymer, plastic-lined metal
	Zetafax	Ethylene copolymer raw materials
DPP Inc., Kashima and Mizushima, Japan		PP
Draka Plast GmbH, 1000 Berlin 51, Germany	Beroplast	PE, PP pipes, pipes for electrical applications
Drake (Fibres), Huddersfield, UK	Gymlene	PP fibers
Drakopoulos SA, Athens, Greece	Cartonplast	PP sheets and other semifinished goods
DRG Flexible Packaging Pty, Melbourne 3205, Australia	Ventflex	HDPE composite films
DSM, 6160 AP Geelen, The Netherlands NL/Toyobo Co. Ltd., Osaka, Japan	Dyneema	PE fibers for special applications

Supplier/manufacturer	Tradename	Polyolefin type and product
DSM, 6160 AP Heerlen, Netherlands	Kelrinal	Chlorinated PE/thermoplastic elastomers, raw materials
	Keltan	Thermoplastic elastomers, raw materials
DSM, 6160 Heerlen, Netherlands	Stamylan	PE ancillary products for processing molding and extrusion compounds
	Stamylan P	PP ancillary products for processing molding and extrusion compounds
	Stamylex	LLDPE ancillary products for processing molding and extrusion compounds
DSM Resins, 8022 AW Zwolle, Netherlands	Keltaflex	PO ancillary products for processing molding and extrusion compounds
Du Pont Canada Inc., Toronto, Ontario, M5K 1B6, Canada	Sclair	PE, LLDPE, MDPE molding and extrusion compounds
	Sclairfilm	PE packaging tubes and films
	Sclairlink	PE products for particular applications
	Sclairpipe	PE pipes
Du Pont-Mitsui, Polychemical Co. Ltd., Tokyo, Japan	Evaflex	Ethylene copolymers, molding and extrusion compounds
Dynamit Nobel AG, 5210 Troisdorf, Germany	Dyflor	Propylene copolymer/olefin diene copolymer raw materials
	Trocellen	PE, foamed
	Trolen VP35	PP/EPDM sheets and other nonreinforced products
	Trovidur	PE, PP sheets and other nonreinforced products
E.I. du Pont de Nemours & Co. Inc., 1007 Market Street, Wilmington, DE 19898, USA	Alathon	PE, LLDPE raw materials or with glass fibers
	Alcryn	PIB/thermoplastic elastomers, molding and extrusion compounds
	Aldyl A	PE/thermoplastic elastomers, molding and extrusion compounds, pipes
	Bynell	PO for special applications
	Elvaloy	PE/PVC molding and extrusion compounds
	Elvax	Ethylene copolymer for processing
	Keldax	Ethylene copolymer with inorganic fillers, not fibrous
	Microfoam	PP foamed sheets
	Nordel	Thermoplastic elastomers EPDM molding and extrusion compounds
	Nucrel	PE, ethylene copolymer/polyacrylic and polymethacrylic raw materials molding or extrusion compounds
	Selar	Olefin copolymer wood fibers for special applications
	Surlyn	Ethylene copolymer coating materials, flexible films, sheeting, webs and cuttings

Supplier/manufacturer	Tradename	Polyolefin type and product
	Tefiel	Ethylene copolymer/PTFE and copolymer molding and extrusion compounds
	Typar	PP fleece
	Tyvek	PE fleece chlorosulfonated PE
Eastman Chemical Products, PO Box 431 Kingsport, TN 37662, USA	Epolene	PE, PP for special applications
	Tenite	PE, LLDPE, PP propylene copolymer raw materials, for paper and textile refinement, flame spray and centrifugal sintering powder, compounds with glass fibers, structural foams chlorinated polyolefin PE wax
Egeplast, Werner Strumann GmbH & Co., 4407 Emsdetten/Germany	Egelen	LDPE, HDPE pipes
El Paso Chemical Co., PO. Box 3986, Odessa, TX 79760, USA	Rexene	LDPE, LLDPE, ethylene copolymer, PP, propylene copolymer molding and extrusion compounds
Emil Keller AG 9220, BischofszellTG, Switzerland	Kalen	PE pipes
	Kaliten	HDPE pipes
EniChem SpA, 20097 San Donate, Milan, Italy	Eraclear	LLDPE molding and extrusion compounds
	Riblene	LDPE, ethylene copolymers molding and extrusion compounds
	Vitron	LDPE, ethylene copolymer molding and extrusion compounds
Enka Industrial Products Inc., Schaumburg, IL 60195, USA	Accurel	LDPE products for plastic processing
Elastogran Kunststoff Technik GmbH, D-2844 Lemfoirde, Germany		PP sheets, PP filled with wood dust, thermoplastics reinforced with glass fiber mats
Enron Chemical Co., Rolling Meadows, IL 60008, USA	Norchem	LDPE, ethylene copolymer, MDPE, LLDPE, PP molding and extrusion compounds
	Nortuff	PP/HDPE, with inorganic fillers
	Plexar	HDPE, ethylene copolymer raw materials for adhesive and glue
Epsilon Products, Marcus Hook, PA, USA		PP
Erez Thermoplastic, Products Kibbutz, Erez M.P. Ashkelon 79150, Israel	Reztex	PE webs, technical films, and sheetings
Erta N V, 8880 Tielt, BE/P Amold, Northampton, UK	Ertalene	HDPE, pipes, rods, technical profiles, sheets, blocks, and blanks for machining
Etimex GmbH, 7000 Stuttgart I, Germany	Vistalux	PP/polyvinylidene chloride and copolymer packaging films and tubes, composite films

Supplier/manufacturer	Tradename	Polyolefin type and product
Ets. G. Convert, 01100 Oyonnax, France	Naltene	LDPE sheets
	Naprene	PP sheets
	Supernaltene	HDPE sheets
Europlast Rohrwerk GmbH, P.O. Box 130160/Bruchstrasse, D-4200 Oberhausen Il, Germany		PE semifinished products
Eval Corp. of America, Omaha, NE 68103, USA	Eval	Ethylene copolymer packaging tubes and films, composite films
Evode Plastics Ltd., Leicester LE7 8PD, UK	Evoprene	Thermoplastic elastomer molding and extrusion compounds
Ewald Dörken AG, 5804 Herdecke/Ruhr, Germany	Deltaplan	PE construction site protective webs
Exxon Chemical America, 13501 Katy Freeway, Houston, TX 77079, USA	Escomer	PE for processing
	Escor	Ethylene copolymer for processing
	Escorene	LDPE, HDPE, ethylene copolymer, preproducts, coating materials,
	Escorene Alpha	LLDPE molding and extrusion compounds
	Extraflex	PO packaging and electrical insulating films
	Extrel	PP propylene copolymer packaging films and tubes
	Parapol	PIB raw material, PP
	Vistaflex	Thermoplastic elastomers
	Vistanex	PIb raw materials
Exxon, N de Gravenchon, France		PP
Exxon, Baytown, TX, USA		PP
	Exxon EX FR-100	EVA polyolefin + ATH filler
Fatra Napajedla,	Porofol	LDPE agricultural and greenhouse films
Farbwerke, Hoechst, Knapsack, Germany		PP
Favorite Plastics Corp., Brooklyn, NY 11234, USA	Favorite	Ethylene copolymer, PE packaging films and tubes, webs, technical films, and sheetings
Femso-Werk Franz Miiller & Sohn, 6370 Oberursel 1, Germany	Femso	PO, ethylene copolymer, thermoplastic elastomers, profiles and pipes
Ferro Corp., Evansville, IN 47711, USA	Ferrene	Olefin copolymer with inorganic fillers
	Ferrex	Propylene copolymer Plastics Division with inorganic fillers
	Ferro Lene	PP with inorganic fillers
Ferro Eurostar, 95470 Fosses, France	Star Flam	PE, PP with inorganic fillers, nonflammable
	Starglas	HDPE with inorganic fillers
	Starpylen	PP with talcum, with asbestos fibers, with glass fibers

Supplier/manufacturer	Tradename	Polyolefin type and product
Fiberfil Inc. (AKZO), Evansville, IN 47732, USA	Electrafil	PO, with carbon fibers for electrotechnical applications
	Ethofil	HDPE with glass fibers
	Methafil	Polymethylpentene molding and extrusion compounds
Fillite USA Inc., Huntington, WV 25702, USA	Fillene	PP molding and extrusion compounds
Fina Oil & Chemical Co., Cosden Chem Div.,		PP
8350 N. Central Expressway, Dallas TX 75221, La Porte, TX, USA		PP
Fina Chemicals SA, A division of Petrofina, Rue de l'Industrie 52,	Finaprop	PP molding and extrusion compounds
International Polymer Dept., B-1040, Brussels, Belgium	Finathene	LDPE, MDPE, HDPE for blown film, molding and extrusion compounds
Flachglas AG, Bereich Kunststoffe 8480, Weiden, Germany	Thermodet	PP sheets and other nonreinforced semi-finished goods
Flo-pak GmbH, 4040 Neuss, Germany	Deli-Pak	PE foamed sheets
Flexible Reinforcements Ltd., Chorley Clitheroe, Lancashire, UK	Wavelene	LDPE, technical materials with fabric and fleece/wood fibers
Flexmer (Amerplast), Nastola, Finland	Flexipack	PE composite films
Flex-O-Glass, Inc., Chicago, IL 60051, USA	Flex-O-Film	Ethylene copolymer, packaging films and tubes, webs and sheets for automatic thermoforming
	Sur-Flex	Ethylene copolymer, packaging and electrical insulating films, particular applications
Formosa Plastics, Point Comfort, TX, USA		PP
Fränkische Rohrwerke, 8729, Königsberg, Germany	FF	PE, crosslinkable PO, pipes, foamed
	FF-Kabuflex	HDPE, pipes for electrical applications
	FF-Therm	Crosslinked PE, PB-I pipes
Friedrichsfeld GmbH, 6800 Mannheim 71, Germany	Friatherm	Polymethylpentene pipes
Gallard Schlesinger Industries, Inc., 584 Mineola Ave., Carle Place, NY 11514, USA		PE, PP, PIB
Geberit GmbH, 7798 Pfullendorf; Germany	Geberit	HDPE pipes
General Electric Co., Schenectady, NY 12345, USA	Irrathene	Crosslinked or crosslinkable PE flexible films and sheetings, webs and cuttings

Supplier/manufacturer	Tradename	Polyolefin type and product
General Tire Br Rubber Co., Plastics Div., Marion, IN 46952, USA	Boltaron	PE, rigid reinforced
Gerodur AG, Benken im Lithgebiet, Switzerland	Gerodur	PE pipes
Gilman Brothers Co., Gilman, CT 06336, USA	Gilco	PE packaging and electrical insulating films and sheeting, webs and cuttings
Golan Plastic Products, Jordan Valley, IL 15145, USA	Pexgol	Crosslinked or crosslinkable PE, pipes
Gurit-Worbla AG, 3063 Ittigen-Bern, Switzerland	Berlene	PO webs, pipes, technical films and sheetings
	Granulit	PO cleaning material
	Oekolex-G	PO packaging films and tubes, sheets for thermoforming and technical purposes
	Oekolon-G	PO for thermoforming and welding for technical purposes and for automatic thermoforming
Hagusta GmbH, 7592 Renchen, Germany	Hagulen	HDPE pipes
Haka, 9202 Gossau SG, Switzerland	Hakathen	HDPE, PP PB, pipes
Hammond Plastics, Inc., 1001 Southbridge St., Worcester, MA 01610, USA		PE, PP
Hanna Engineered Materials, (formerly CTI Texapol), Norcross, GA, USA		Reinforced PO
Haren KG, Röchling, 4472 Haren/Ems I, Germany	Polystone	PE, PP, rods, profiles, sheets, blocks and blanks for thermoforming and welding
Hegler Plastik GmbH, 8735 Oerlenbach, Germany	Heglerplast	HDPE pipes for special applications
	Hekaplast	HDPE pipes for special applications
	Siroplast	HDPE pipes for special applications
Hemijska Ind., 0dazaci, Serbia		PP
Henkel Corporation 7900 W 78th St., Minneapolis, MN 55435, USA		PE waxes
Hercules Inc., Wilmington, DE 19894, USA	Herculon	PE packaging and electrical insulating films
	Liteplate S	PP plastic-lined metal special applications, low blush copolymers
	Metton	Polydicyclopentadiene, RIM system products
Himont Italia, 20121 Milan, Italy and	Dutral	Thermoplastic elastomers, propylene copolymers, molding and extrusion compounds, liquid crystal polymers
	Dutralene	
Himont, Wilmington, DE 19894, USA	Hi-Fax	PE raw material, special applications
	Hifax 7135	PO products, especially PP products

Supplier/manufacturer	Tradename	Polyolefin type and product
	Hifax CAS3A Hifax CA138A Hifax CA162A	from Catalloy process for injection molding; low blush copolymers. high rubber copolymers, high porosity
	Moplan	PP molding and extrusion compounds
	Moplefan	PP packaging films and tubes, composite films, electrical insulating films
	Moplen	PP propylene copolymer raw material, molding and extrusion compounds
	Moplen-EP	Ethylene copolymer raw material, molding and extrusion compounds
	Mopier	PP raw material, molding and extrusion compounds for special applications
	Pro-fax	PP propylene copolymer raw material, molding and extrusion compounds
	Himont EXP 127-32-6	Intumescent PP
HMC Polymers, Map Ta Phut, Tailand		PP
Hoechst AG, 6230 Frankfurt am Main 80, Germany and Hoechst AG Werk, Gersthofen 8906, Germany	Hostalub	PO for plastics processing
	Hostaphan	PE packaging films and tubes composite films, electrical insulating films
	Hostapren	Chlorinated PE/thermoplastic elastomers
	Hoechst-Wachse	PO for processing
	Hostalen	PE with glass fibers
	Hostalen PP	PP propylene copolymer, molding and extrusion compounds, with talcum, glass fibers, and/or glass balls
Hoechst Australia, Altona, Victoria, Australia		PP
Hoechst, France, Lillebonne, France		PP
Hoechst Iberica, Tarragone, Spain		PP
Honnan Petro, Yeochon, Korea		PP
Honnan Oil, Yeochon, Korea		PP
Huber & Suhner AG, 9100 Herisau, Switzerland	Sucorad	Crosslinked PE or crosslinkable PO pipes
Hüls AG, 4370 Marl, Germany	Vestolen A	HDPE raw material, molding and extrusion compounds, with glass fibers
	Vestolen EM	PP derivatives, molding and extrusion compounds
	Vestolen P	PP derivatives, molding and extrusion compounds with glass fibers
	Vestoplast	PO products for particular applications
	Vestopren	thermoplastic elastomers, molding and extrusion compounds
	Vestowax	PE for processing
Hunstman Chemical Co.,	Polycom	PP, with fillers as chalk or talcum
Salt Lake City, UT 84144, USA		PP
Woodbury, NJ; Longview, TX, USA		PP

Supplier/manufacturer	Tradename	Polyolefin type and product
Hyundai, Daesan, Korea		PP
ICI, Australia, Botany Bay, NSW, Australia		PP
ICI PLC "Visqueen" Products, Stockton-on-Tees,	Politarp	PE technical lining and sealing materials
Cleveland, UK	Visqueen	LDPE, LLDPE, ethylene copolymer, packaging films and tubes, webs, technical films, and sheetings
ICI PLC Welwyn, Garden City, Hertfordshire	Procom	PE, propylene copolymer molding and extrusion compounds
A471HD, UK	Propafilm	PP packaging films and tubes, composite films
	Propafoil	PP films for special applications
	Propaply	PP packaging films and tubes
	Propathene	PP, propylene copolymer raw material and molding and extrusion compounds
Idemitsu Petrochemical Co. Ltd., Tokyo, Japan Amegasaki, Chiba, Japan	X-sheet	PP with glass fibers mats or textiles for special applications
Igoplast Faigle AG,	Igoform	PO molding and extrusion compounds
9434 Au, Switzerland	Igopas	PP, HDPE, pipes, rods, sheets, etc.
INA-Organsko Kemijska, Industrial (INA-OKI), 4100 Zagreb, Croatia	Okiten	LDPE molding and extrusion compounds
India Petrochem Corp.,		PP
Baroda Koyali,		PP
Nagothane Mahar,		PP
Vadodara,		
Indian Petrochemicals Corp. Ltd. Gujarat, India	Koylene	PP molding and extrusion compounds
Ind. Petroquimicas Koppers, Argentinas SA, Buenos Aires, Argentina	Ardylan	PE molding and extrusion compounds
Indelpro, Tampico, Mexico		PP
Isobelec SA, Sclessin/Liège, Belgium	Warlbne	PE pipes, rods, webs, technical films and sheetings, semifinished goods
Israel Petrochemical, Enterprises Ltd, Haifa 31000, Israel	Ipethene	LDPE molding and extrusion compounds
J. H. Benecke GmbH, 3000 Hannover 1, Germany	Acella	PO self-adhesive films, laminating films, fleece
	Corovin	PP fleece
J.T. Ryerson & Son, Inc., Chicago, IL 60608, USA	Mono-Line	HDPE pipes
Kalenborn Schmelzbasaltwerk 5467 Vettelschoss, Germany	Kalen	HDPE sheets and other semifinished products, (pipes)

Supplier/manufacturer	Tradename	Polyolefin type and product
Kalle, Niederlassung der Hoechst AG 6200 Wiesbaden Biebrich, Germany	Suprathen	LDPE packaging films and tubes, webs, technical films and sheetings
	Trespalen	HDPE packaging films and tubes, sheets
	Trespaphan	PP packaging films and tubes, sheets, electrical applications
Kanegafuchi Chemical Ind., Co., Ltd, Osaka Japan	Eperan	PE, foamed
Karl Dickel & Co, 4100 Duisburg, Germany	Cast-Film	PP packaging films and tubes
	Sarafan	PE, LLDPE, PP packaging films and tubes
Kaysersberg, 68240 Kaysersberg, France	Acylux	Propylene copolymer composite films
Kibbutz Ginegar, Israel	Intrasol	LDPE agricultural films and greenhouse films
King Plastic Corp., Venice, FL 34284, USA	Hy-Pact	PE packaging films and tubes, webs, and technical sheets for automatic thermoforming
Koch Chemical Co., Muskegon Specialty Division, 1725 Warner St., Whitehall, MI 49461, USA		PE chromatographic
Korea Petrochemical Ind, Ulsan, Korea		PP
Koro Corp., Hudson, MA 01749, USA	Polykor	PE, PP with inorganic fillers
Kunststoffwerk Höhn GmbH, 5439 Höhn, Germany	Brandalen	PE, PP, PB pipes
Kunstoplast-Chemie GmbH, 6370 Oberursel, Germany	Kunstolen	PE, PP molding and extrusion compounds
LATI SpA, 21 040 Vedano Olona, Varase, Italy	Latene	HDPE, PP, propylene copolymer molding and extrusion compounds
LATIUSA Inc. Mt Pleasant, SC, USA		Reinforced PO
Lenzig AG, 4860 Lenzing, Austria	Lenzing-s-band	PP foamed sheets
	Lenzingtex	PO packaging films and tubes, construction site protective webs, agricultural and greenhouse films, roof coverings
	Lenzingtex Alu	PE composite films, webs, technical films and sheetings
Leschuplast Kunststoffabrik, GmbH, 5630 Remscheid 11, Germany	Leschuplast	LDPE, ethylene copolymer technical lining and sealing materials, construction sealing webs, roof coverings
Liquid Plastics Ltd., Preston PR14 AJ, UK	Soladex	Chlorinated PE coating materials
LNP Corp., Malvern, PA 19355, USA	Stat-Ken	PP raw material, with carbon fibers for electrotechnical applications
Engineering Plastics Inc., (subsidiary of Kawasaki Steel Corp), Exton, PA, USA		Reinforced PO

Supplier/manufacturer	Tradename	Polyolefin type and product
Lonza Werke GmbH, 7858 Well am Rhein, Germany	Ultralen	PP packaging and electrical insulating films
Lyondell Petrochem Bayport, TX, USA		PP
Mayer Enterprises Ltd., Coating Dept, Tel Aviv, Israel	Plastopil	PE agricultural and greenhouse films
Makoto Murata, Tsurusaki Oita and Mizushima, Japan		PP
Mecano-Bundy GmbH, 6900 Heidelberg 1, Germany	Mecanyl-Rohr	LDPE, HDPE, PP pipes, tubes
Mercia Polythene, Kidderminster, UK	Merpol	LDPE packaging and electrical insulating films for special applications
Mikuni Lite Co., Osaka, Japan	Papia	PP, with finely divided organic fillers plastic construction sheets
Mitsubishi Chemical, Industries Ltd., Tokyo, Japan	Novatec	LDPE, LLDPE, HDPE, PP, raw materials, molding and extrusion compounds
Mitsubishi Petrochemical. Co. Ltd., Tokyo, Japan	Beauron	Thermoplastic elastomers, molding and extrusion compounds
	Ecolo F	Propylene copolymer with inorganic fillers
	Hi-Zex	MDPE, HDPE, molding and extrusion compounds
	Milastomer	Propylene copolymer, molding and extrusion compounds
	Neo-Zex	LDPE, LLDPE, MDPE, flame spray and centrifugal sintering powder
	Noblen	PP thermoplastic elastomers molding and extrusion compounds
	Tafmer	Thermoplastic elastomers, molding and extrusion compounds
	TPX	Polymethylpentene with or without fillers, transparent
	Ultzex	LLDPE, molding and extrusion compounds
	Yukalon	LDPE, LLDPE, MDPE, ethylene copolymer raw materials, flexible films and sheeting, webs and cuttings
Ichibara Chiba,		PP
Mitsubishi Kagaku, Mizushima Kashima		PP
Ibaraki, and Yokkaichi, Japan		PP
Mitsubishi Rayon Co Ltd., Tokyo, Japan		Reinforced PO
Mitsui Polychemical, Co., Ltd Tokyo, Japan	Dumilan	Ethylene copolymer products for particular applications
	Mirason	LLDPE, LDPE, MDPE raw materials, molding and extrusion compounds
Mitsui Toatsu, Hiroshima, Japan		PP
Mobil Chemical Co., Films Dept Macedon, NY 14502, USA	Bicor	PP packaging and electrical insulating films, composite films

Supplier/manufacturer	Tradename	Polyolefin type and product
Modified Plastics Inc., Santo Ana, CA, USA		Reinforced PO
Monmouth Plastics Inc., Asbury Park, NY 07712, USA	Empee	HDPE, PP molding and extrusion compounds for special applications
Monomer-Polvmer and Dajac Labs, Inc., 36 Terry Drive, Trevose, PA 19047, USA		PE, CPE, PP, CPP. PE-latex, PIB. chlorosulfonated PE, P4MP, EVA copolymer
Monsanto Co., St. Louis, MO 63167, USA	Santoprene	PO mixtures, raw materials
Montedison SpA, 20121 Milan, Italy	Vinavil	Ethylene copolymer aqueous dispersion
Montefina, Feluy, Belgium		PP
Montell Polyolefins (from	Adstif	PP
Royal Dutch Shell + Himont),	Adflex	PP
Berre l'Etang, France	Adsyd	PP
Koln, Germany		
Brindisi, Ferrara, Terni, Italy		
Pernis, The Netherlands		
Carrington, UK		PP
Varennes, Quebec, Canada		PP
Sarnia, Ontario, Canada		PP
Bayport, TX, Lake Charles, LA, USA		PP
Clyde NSW, Gellan, Victoria, Australia		PP
Montell Advanced Materials Lansing, MI, USA		Reinforced PO
Montepolimeri SpA, 20121 Milan, Italy	Ferlosa	HDPE, synthetic paper and cardboard
Moore & Munger Marketing, Inc., Fairfield Office Center, 140S Hermann Street, Fairfield, CT 06430, USA		Atactic PP, PE waxes
Moplephan UK, Brantham, UK	Bexphane	PP flexible films and sheeting, webs and cuttings
MPD Technology Ltd., London SWI P4QF, UK	Caprez DPP	PP with carbon fibers
Multibase Inc, Copley, OH, USA		Reinforced PO
Murtfeldt GmbH & Co., KG 4600 Dortmund 12, Germany	Werkstoff S	HDPE, sheets and other nonreinforced semifinished products
N Lundbergs Fabriks AB, Fristad, Sweden	Lubonyl	Crosslinked or crosslinkable PO pipes
Naftochim, Burgas, Bulgaria		PP
Nan Ya Plastics Corp., Taipei, Taiwan		Reinforced PO

Supplier/manufacturer	Tradename	Polyolefin type and product
Neste Chemicals GmbH, 300930 Mörsenbroieher Weg 200, D-4000 Düsseldorf 30, Germany		PO fibers, films, and nonwovens
and Neste Polyeten AB, PO box 44, S-44401, Stenungsund, Sweden		LDPE, MDPE, LLUPE, HDPE
Neste OY, SF-06850 Kuloo, Finland		
Nippon Petrochemicals Co. Ltd., Tokyo, Japan	Rexlon	LDPE molding and extrusion compounds
	Softlen	Ethylene copolymer raw materials
Nippon Synthetic Chemical Ind., Co Ltd.,	Soablen	Ethylene copolymer, raw materials, molding and extrusion compounds
Osaka, Japan	Soarlex	Ethylene copolymer, raw materials, molding and extrusion compounds
	Soarnol	Ethylene copolymer/polyvinyl alcohol raw materials for special applications
Nippon Unicar Co. Ltd.,	Uniclene	PE, raw materials
Tokyo, Japan	Uniset	PE raw materials, molding and extrusion compounds
Niederberg-Chemie GmbH, 4133 Neukirchen Vluyn, Germany	Carbofol	Ethylene copolymer + bitumen technical lining and sealing materials
Nisseki Plastic Chemical Co. Ltd., Tokyo, Japan	Staflene	HDPE, ethylene copolymer molding and extrusion compounds
Norddeutsche Seekabelwerke AG, 2890 Nordenham, Germany	Poly-Net	PO flexible films and sheeting, webs and cuttings
Northern Petrochemical Co., Omaha, NE 68102, USA	Norprop	PP packaging and electrical insulating material
Northern Petrochemical Co., Morris, IL, USA	Norlin	LLDPE molding and extrusion compounds
North Sea Petro, Antwerp, Belgium		PP
Norton Co., Specialty Plastics Div., Akron OH 44309, USA	Boronol	Propylene copolymer with inorganic fillers
Novacor Chemicals Ltd., Calgary, Alberta, Canada	Novapol LL	LDPE, LLDPE molding and extrusion compounds for special applications
Nuova Italresina,	Nipren	PP pipes
20027 Rescaldina Milan, Italy	Nirlene	LDPE pipes
Odenwald-Chemie GmbH, 6901 Schöhau, Germany	Cuticulan	LDPE webs, technical films and sheetings
	OC-Plan 2000	Ethylene copolymer + bitumen technical technical films and sheeting
OMV Deutschland, Burghausen, Germany		PP
Osaka Soda Co. Ltd., Osaka, Japan	Daisolac	Chlorinated PE raw material
Ostermann & Scheiwe, GmbH & Co 4400 Münster, Germany	OSMO plast	PO construction profiles

Supplier/manufacturer	Tradename	Polyolefin type and product
Oy Uponor AB, Nastola, Finland	Upolar	HDPE packaging films and tubes
	Upoten	LDPE, packaging films and tubes
4P Folie Forchheim GmbH, 8550 Forchheim, Germany	Plastin	HDPE packaging films and tubes
	Plastotrans	LDPE packaging films and tubes
Paja Kunststoffe Jaeschke OHG, 5064 Rösrath 1, Germany	Pajalen	LDPE packaging films and tubes
Papeteries de Belgique, Brussels, Belgium	Helioflex	PE packaging films and tubes, composite films
	Helioplast	PP packaging films and tubes, composite films
PCD Polymers, Schwechat St Peter Str 25, A-2323, Austria	Daplen PP	PP, Chemical plant construction
	Daplen HDPE	HDPE, gas pipes, drinking water pipe, landfill, ducting pipe, semi-finished products
Pemex, Morelos, Mexico		PP
Pennekamp + Huesker KG 4426 Vreden 1, Germany	Ceram P	HDPE with inorganic fillers
Pertamina, Palembang, Indonesia		PP
Permanite Ltd., Waltham Abbey EN9 1AY, UK	Permabit	Chlorinated PE roof coverings
Petkim-Petro, Aliaga-Izmir, Turkey		PP
Petrarch Systems Inc., Bristol, PA 19007, USA	Rimplast	PO, RIM system products, special applications
Petrimex, Bratislava, Slovakia 82602	Bralen	LDPE, molding and extrusion compounds
	Liten	HDPE raw materials
	Mosten	PP raw materials
	Taboren	PP with glass fibers
	Tatren	PP raw materials
Petroquimica Cuyo, Lujen de Cuyo, Argentina		PP
Petrochemical Works, Ploesti 2000 Brazi,	Ropol	LDPE molding and extrusion compounds
Pitesti 0300, Ploesti-Teleajen, 2000, Romania	Ropoten	MDPE molding and extrusion compounds HDPE molding and extrusion compounds
		LDPE, PP molding and extrusion compounds
Petrochemie Danubia, 4021, Linz, Austria	Multifil	PP fibers for special applications, nonflammable
	Polyfelt	PP fleece
	Strapan	PE composite materials
Petroken, Ensenada, Argentina		PP
Phillips Petroleum Co., Brussels, Belgium	Marlex	LDPE, HDPE, ethylene copolymer, PP raw materials, molding and extrusion compounds

Supplier/manufacturer	Tradename	Polyolefin type and product
	Marlex 130	LLDPE molding and extrusion compounds
Phillips Petroleum Co., Overijsse, Netherlands	Solprene	Thermoplastic elastomer raw materials
Phillips Sumika PP Co, Pasadena TX, USA		PP, new grades
Pierson Ind., Inc., Palmer, MA 01069, USA	Pierson	Ethylene copolymer packaging films and tubes
Plast Labor SA, 1630 Bulle, Switzerland	Multilay	Thermoplastic elastomer products for particular applications
Plastic Suppliers, Blackwood, NJ 08012, USA	Polyflex	Pp packaging and electric insulating films
The Plastic Group (formerly Polyfil, Inc), Woonsocket, RI, USA		Reinforced PO
PMC (division of Clariant Corp), Milford, DE, USA		Reinforced PO
Pol Cekop/MZRIP, Plock, Poland		PP
Polybrasil SA Maua Sao Paolo and Camacari Bahia, Brazil		PP
Polipropileno SA, Sao Paolo, Brazil	Prolen	PP, MDPE molding and extrusion compounds
Polycom Huntsman Inc., Washington, PA, USA		Reinforced PO
Polydress Plastic GmbH, 6120 Michelstadt, Germany	Griffolyn	PE packaging films and tubes with inserts, agricultural and greenhouse films
Polychim, Dunkerque, Belgium		PP
Polymer Composites, Inc., Minneapolis, MN 55416, USA	Fiberod	PP molding and extrusion compounds for special applications
Winona, MN, USA		Reinforced PO
Polymer Corp. Ltd., Reading, PA 19603, USA	Polypenco	HDPE profiles and pipes, sheets and nonreinforced semifinished products
	Ultra Wear	HDPE pipes, rods, technical profiles, sheets, blocks, and blanks for machining
Polipro De Venez, El Tablazo, Venezuela		PP
Polyolefin Co., Pulau Ayer Merbau Is, Singapore.		PP
Polyon-Barkai Kibbutz, Barkai M.P Menashe 37 860, Israel	Thermofilm	Ethylene copolymer packaging films and electrical insulating materials
Polypropylene Malaysia, Kuantan, Malaysia		PP
Polysciences, Inc., 400 Valley Rd., Warmington, PA 18976, USA		Polynorbornediene, P4MP, CPE, chlorosulfonated PE, ethylene copolymers

Supplier/manufacturer	Tradename	Polyolefin type and product
Porvair PLC, King's Lynn, UK	Porvent	HDPE webs, technical films and sheetings poromerics
	Vyon	HDPE sheets, poros
PP India Ltd, Mathora, Uttar, India		PP
PPG Industries Inc., Pittsburgh, PA 15272, USA	Azdel	PP with glass fibers mats or textiles
PPH Rio Grande Do Sul, Brazil		PP
Premix Oy, Rajamiiki, Finland	Pre-flee	LDPE, HDPE, PP molding and extrusion compounds for electrical applications
PQ Colombiana, Cartogena, Colombia		PP
Prolastomer Inc., Waterbury, CT 06708, USA	Prolastic	Thermoplastic elastomer molding and compounds
Protective Lining Corp. Brooklyn, NY 11232, USA	Protectolite	LDPE flexible films and sheetings, webs and cuttings, covers
Quantum Chemical Morris, IL, USA		PP
	Quantum Petrothene XL7403 XL	polyolefin copolymer + ATH filler
	Quantum Petrothene YR 19535 XL	polyolefin copolymer + ATH filler
	Quantum Petrothene YR19543	EVA copolymer + mineral filler
	Quantum Ultrathene UE631	EVA polyolefin copolymer
Rehau AG & Co., 8673 Rehau, Germany	Rau	PO profiles, pipes, sheets, etc.
Reichold Chemicals Inc., White Plains, NY 10603, USA	Blanex	Crosslinked PO, molding and extrusion compounds
	Blapol	PE molding and extrusion compounds
Ren Plastics (Ciba Geigy), East Lansing, MI 8823, USA	Ren-Flex	Thermoplastic elastomer molding and extrusion compounds
Repsol Quimica, Puertollano, Spain		PP
Rexene, Odessa, TX, USA		PP
Rheinische Kunststoff-werke GmbH, 6520, Worms am Rhein, Germany	Renolen	PO packaging films and tubes
Rhône-Poulenc Films, 69398, Lyon Cedex, 3 France	Terthene	PE + polyethyleneterephtalates, composite films
	Vipathene	HDPE + PVC composite films
Rhône-Poulenc Films, 92080 Paris La Defense, Cedex 6, France	Clarylene	PE + polyethyleneterephtalates, composite films for special applications
	Lamithene	PE + regenerated cellulose, composite films
	Nylane	PE + polyamides, composite films
	Suprane	HDPE electrical insulating materials

Supplier/manufacturer	Tradename	Polyolefin type and product
Rilling & Pohl KG, 7000 Stutgart 30, Germany	Ripolit	PE sheets and other nonreinforced – semifinished goods
Roga KG Dr. Loose GmbH & Co., 5047 Wesseling bei Köln, Germany	Ergeplast	PE pipes, profiles with various cross sections
Röchling Sustaplast KG, 5420 Lahnstein, Germany	Sustylen	PE, PP pipes and profiles, flexible films, sheets
Röhrig & Co., 3000 Hannover 97, Germany	Röcothene	PE tubes, rods, profiles with various cross sections, technical profiles, tubes, nonflammable
Romchim, Midia and Ploiesti, Romania,		PP
Ross & Roberts, Inc., Stratford, CT 06497, USA	Arnar	PE films and sheets
Rotrotron Corp., Bohemia, NY 11716, USA	Rotothene	LDPE, LLDPE, MDPE products for particular applications
	Rotothon	PP, flame spraying and centrifugal sintering powder
ROW Co., Wesseling, Germany		PP
RTP Co., Winona, MN 55987, USA	RTP	PO with inorganic fillers for electrical applications, Reinforced PO
Rumianca SpA, 20161 Milan, Italy	Rumiten	PE, LLDPE, HDPE, raw materials, molding and extrusion compounds
Sabic, Riyaah 11422, Saudi Arabia	Ladene	LLDPE molding and extrusion compounds
Sasol, Secunda, Republic of South Africa		PP
Samsung Gen Chem, Daeson, Korea		PO
Sapcofill GmbH, 4030 Ratingen, Germany	Ecopol	PP with inorganic fillers
Saudi European PC, Ibn Zahr, Saudi Arabia		PP
Sea Land Chemical Co., 14820 Detroit Avenue, Cleveland, OH 44107, USA		Atactic PP
Seadrift Polypro Co, Seadrift, TX, USA		PP
Seaboku Poly, Takasago, Japan		PP
Seitetsu Kagaku Co. Ltd., Osaka, Japan	Flo-Thene	PE molding and extrusion compounds
Sekisui Chemical Co. Ltd., Osaka 530, Japan	Lightlon	LDPE electrical insulating films, foamed plastics
	Microlen	PE, foamed
Shell International Chemical Co. Ltd. London SE1 7PG, UK	Carlona	LDPE, HDPE, PP, propylene copolymer, molding and extrusion compounds
	Duraflex	PB pipes

Supplier/manufacturer	Tradename	Polyolefin type and product
Shell Chemical Co., One Shell Plaza, Houston, TX 77002, and Norco LA, USA		PP, PB
Showa Denko K.K., Tokyo, 105, Japan	ACS	Chlorinated PE/PMMA molding and extrusion compounds
	Bifan	PP packaging films and tubes, composite films
	Elaslen	Chlorinated PE raw material
	Sho-Allomer	PP molding and extrusion compounds
	Sholex	LDPE, HDPE, molding and extrusion compounds
Shrink Tubes & Plastics Ltd., Redhill, Surrey RH1 2LH, UK	Canusaloc	Ethylene copolymer pipes for special applications
SIMONA GmbH, 6570 Kirn/Nahe, Germany	Rhiamer	PP, HdPE, pipes
	Rhiatherm	Crosslinked or crosslinkable PO, propylene copolymer pipes
	SIMONA	HDPE technical lining and sealing materials and construction sealing webs
		HDPE, PP pipes, rods, and sheets
S.I.R., 20161 Milan, Italy	Klartene	LDPE packaging and electrical insulating materials
	Plartoon	PE, foamed
Slovchemia, Bratislava, Slovakia		PP
Soltex Polymer Corp., Houston, TX 77098, USA	Fortifier	LDPE, chlorinated PE, HDPE, flame spray, and centrifugal sintering powder, molding and extrusion compounds
	Fortilene	PP, propylene copolymer, molding and extrustion compounds
	Lextar	HDPE synthetic paper or cardboard
	Soltex	HDPE, PP molding and extrusion compounds
Solvay & Cie, 1050 Brussels, Belgium	Alkorflex	Chlorinated PE roof coverings
	Clarene	Olefin copolymer/polyvinyl alcohol copolymer preproducts
	Eltex	LLDPE, HDPE, olefin copolymer raw materials, molding and extrusion compounds
	Eltex P	PP, propylene copolymer raw materials, molding and extrusion compounds
	Wood-Stock	PP with sawdust
Antwerp, Lille, Belgium		PP
Sarralbe, France		PP
Solvay America, Deer Park, TX, USA		PP
Sonobat SA, Battice, Belgium	Tubiflex	PE pipes
South Africans Poly, Sasolburg, Safrpol (Republic of South Africa)		PP

Supplier/manufacturer	Tradename	Polyolefin type and product
Southern Chemical Products Co., PO Box 205, Macon, GA 31297, USA		Crosslinkable PE
Stanley Smith & Co., Isleworth, Middlesex TV7 7AU, UK	Lockite	PE technical profiles
	Vitradur	HDPE sheets and other semifinished products
	Vitralene	PP sheets and other semifinished products
	Vitrapad	PE, PP sheets and other semifinished products
	Vitrathene	PE profiles and pipes, semifinished goods
Statoil Petrochemical Co., Norway	Statoil TPE	Thermoplastic elastomer molding and extrusion compounds
St. Regis Corp., New York, NY 10017, USA	Trip-L-Ply	PE/PP composite films
Steuler Industriewerke GmbH, 5410 Höhr-Grenzhausen, Germany	Bekaplast	PE, PP sheets for thermoforming and welding for corrosion protection
Sumitomo Chemical Co. Ltd., Tokyo 103, Japan; Ichihara Chiba, Japan	Bondfast	Propylene copolymers melt adhesives, glues, and cements
	Esall	PP molding and extrusion compounds
	SumikaFlex	Ethylene copolymer, thermoplastic elastomers raw materials, molding and extrusion compounds
	Sumikagel	Olefin copolymer flexible films, sheetings, webs and cuttings
	Sumikathene	LDPE, LLDPE, MDPE, HDPE, ethylene copolymer molding and extrusion compounds
Sumitomo Chemical Co. Ltd., Osaka, Japan	Evatate	Ethylene copolymer molding and extrusion compounds
Syfan BOOP Films M.V., Haneger, 85140, Israel	Syfan	PP packaging films and tubes
Symalit AG, 600 Lenzburg I, Switzerland	Symalen	PE pipes
	Symalit	PP with glass fibers, mats, or textiles
Taiwan PolyPro Co., Kaosung, Taiwan		PP
Targor (Hoechst + BASF) GmbH		PP
Teknor Plastics Ltd. Orpington, Kent BR6 6BH, UK	Telcon	PO, profiles and pipes, flexible films and sheetings, webs and cuttings, semifinished products
Tetrafluor Inc., El-Segundo, CA 90245, USA	Tetralene	LDPE molding and extrusion compounds for special applications
Thai Polypropylene, Map Ta Phut, and Thai Petchem Ind. Co., Rayong, Thailand		PP
Thermofil Inc, Brighton, MI, USA (subsidiary of Nippon Steel Corp.)		Reinforced PO

Supplier/manufacturer	Tradename	Polyolefin type and product
Thyssen Edelstahlwerke AG, Magnetfabrik, 4600 Dortmund, Germany	Siperm	PE sheets and other semifinished products, porous with glass fibers, mats, and textiles
Titan Himont, Pasir Gudang, Johor, Malaysia		PP
Tiszai Vegyi Komb, Tiszaujvaros, Hungary		PP
Tokuyama, Tokuyama City, Japan,		PP
Tonyang Nylon, Ulsan, Korea		PP
Tonen Chem, Kawasaki, Japan		PP
Tonen Sekiyukagaku K.K., Tokyo, Japan	Tonen	PP with carbon fibers
Toyo Rubber Co., Japan		Newer PO structural foams
Toyo Soda Mfg. Co. Ltd., Tokyo, Japan	Nipolon	LDPE raw material, molding and extrusion compounds
TP Composites Inc., Aston, PA, USA		Reinforced PO
Transfordora De Propilene, Tarragona, Spain		PP
Ube Industries Ltd., Tokyo, Japan	Ubec	LDPE, LLDPE, molding and extrusion compounds with inorganic fillers for electrical applications
Sakai, Osaka, Japan		PP
Ube Polypropylene, Ube City, Japan		PP
UCB NV Filmsector, 9000 Gent, Belgium	Cellothene	PE + regenerated cellulose composite films
Ukichima Polypro, Kawasaki, Japan		PP
Unifos, SE;	News	LLDPE raw material
Unifos Kemi AB, 44 401, Stenungsund, Finland	Visico	Crosslinked or crosslinkable PO/silicones for electrical applications
Union Carbide Corp.,	Unipol PP	PP
Polyolefins Div. Danbury, CT 06817, USA	UCAR	LDPE molding and extrusion compounds for electrical applications
	Glad	PE packaging films and tubes
	UCarb Unigard RE DFDA 1735 NT	Polyolefin Copolymer + mineral filler
	UCarb Unigard RE HFDA 1393 NT XL	Polyolefin copolymer + mineral filler
	UCarb Ucarbsil FR 7920 NT	Polyolefin + mineral filler
	UCarb Unigard HP HFDA 6522NT	Polyolefin copolymer + Br additives
Union ERTS. A, Madrid I, Spain	Ertileno	PE molding and extrusion compounds

Supplier/manufacturer	Tradename	Polyolefin type and product
Unitecta Oberflachenschutz GmbH, 4630 Bochum-Gerthe, Germany	Organat	Ethylene copolymer + bitumen roof coverings
USI Industrial Chemicals, Cincinnati, OH 45249, USA	Microthene	LDPE, LLDPE, ethylene copolymer, raw materials, flame spray and centrifugal sintering powder
	Multifilm	PE packaging films and tubes, composite films
	Petrothene	LDPE, LLDPE, MDPE, HDPE molding and extrusion compounds, with chalk, or carbon fibers
U.S. Industrial Chemical Co., New York, NY 10016, USA	Flamolin	PE molding and extrusion compounds for electrical applications, nonflammable
	Ultrathene	Ethylene copolymer molding and extrusion compounds
	Vynathene	Ethylene copolymer molding and extrusion compounds
Valota, Carobbio degli Angeli, Italy	Valotene	PO agricultural and greenhouse films
Van Leer Plastics Inc., Houston, TX 77240, USA	Valeron	HDPE packaging films and tubes, composite films
VEB Orbitaplast, Weissandt-Gölzau, Germany	Gölzathen	PE pipes, packaging films and tubes, webs, technical films and sheetings, semifinished products
VEB Leuna Werke, Leuna bei Merseburg, Germany	Leunapor	PO foamed sheets/foamed artificial leather
	Le-Wachs	PE for plastic processing
	Mirathen	PE molding and extrusion compounds
	Miravithen	Ethylene copolymer molding and extrusion compounds
VEB Chemische Buna, Schopau, Germany		PP
VEB Eilenburger Chemie Werk, 7280 Eilenburg, Germany	Saxolen	PE sheets, blocks, and blanks for machining, for thermoforming, webs
Veloflex Carsten Thormählen, GmbH & Co., Köln-Reisiek, Germany	Veloflex	PE book binding materials
Vestolen GmbH, Gelsenkirchen, Marl, Germany		PP
Vinora AG, Folienwerk, 8640 Rapperswil, Jona, Switzerland	Vinopren	PP packaging films and tubes
	Vipafin	HDPE packaging films and tubes
Viskoza, Po Folije, 1 Ambalaza 15 300 Loznica, Bosnia/Herzegovina	Lofolen	PP packaging films and tubes
Voltek Inc., Lawrence, MA 01843, USA	Volara	PE foamed sheets
	Volasta	PE structural foams
	Volon	PE pipes, foamed
	Volton	Crosslinked and crosslinkable PO foam

Supplier/manufacturer	Tradename	Polyolefin type and product
Wacker Chemie GmbH, 8000 Münich 22, Germany	Vinnapas	Ethylene copolymer raw materials, aqueous dispersions, solutions
	Wacker VAE	Ethylene copolymer raw materials
Warri Ref & PC Ekpan, Nigeria		PP
Wefapress-Werkstoffe, Beck & Co. GmbH, 4426 Vreden, Germany	Wefapress	HDPE rods, profiles with various cross sections, technical profiles, sheets, blocks, blanks for machining
Western Plastics Corp., Tacoma, WA 98402, USA	Klearcor	PE pipes
Westlake Plastics Co., Lenni, PA 19052, USA	Ethylux	PE profiles and pipes, flexible films and sheetings, webs and cuttings semifinished products
Winzen International Inc., Minneapolis, MN 55420, USA	Winlon	HDPE plastic films and artificial leather
Wolff Walsrode AG, 3030 Walsrode 1, Germany	Waloplast	PE packaging films and tubes, laminating films, composite films
	Waloplast-matt	PE plastic films and artificial leather for medical purposes, laminating films, adhesive carrier films, technical materials with fabric or fleece
	Walothen,	PP packaging films and tubes, laminating films, adhesive carrier films
	Walotherm	PE packaging films and tubes
Worlee-Chemie GmbH, 2058 Lauenburg, Germany	Adekaprene	Chlorinated PP raw material for lacquers
Yhtyneet Paperiteh-taat Oy Valke, Finland	Walkicel	PE composite films with cellulose esters
	Walkicomp	PE/polyamide/PE composite films
	Walkiter	Polyethyleneterephthalate/PE composite films
	Walkivac	Polyamide/PE composite films
Ylopan Folien GmbH, 3590 Bad Wildungen, Germany	Ylopan	LDPE packaging films and tubes, construction site protective webs, agricultural and greenhouse films, laminating films
Yokkaichi Polypro, Yokkaichi, Japan		PP
Yokon, Ulsan, Korea		PP
Yung Chia Chem, Linyuan, Taiwan		PP
Yorkshire Imperial Plastics Ltd., Leeds, UK	Polyorc	PE pipes

Abbreviations: EDPM, ethylene–diene–propylene monomer rubber; HDPE, high density polyethylene; LDPE, low density polyethylene; LLDPE, linear low density polyethylene; MDPE, medium density polyethylene; PB-1, polybutene-1, PE, polyethylene; PIB, polyisobutylene; PP, polypropylene; PS, polystyrene; PTFE, polytetrafluoroethylene; PVC, polyvinyl chloride; RIM, reaction injection molding; CPE-chlorinated PE; CPP, chlorinated PP; P4MP, poly-4-methylpentene- 1.

See also tables in Chapters 1, 2, 18–20, 25, 27.

Appendix 2

Suppliers and Manufacturers of Additives for Polyolefins

Supplier/manufacturer	Additives
Abril Industrial Waxes, Sturmi Way, Village Farm Industrial Estate, Pyle, Mid-Glamorgan CF33 6NU, UK tel: +44-1656 744362 fax: +44-1656 742 471	Fatty amide waxes, polycarboxylic acid waxes, flame retardants
Accrapak Systems Ltd, Burtonwood Industrial Centre, Warrington WA5 4HX, UK	
Acmos Chemische Fabrik Tietjen & Co, Potsfach 10 1069 D-28010 Bremen I, Germany tel: +49-421 51681 fax: +49-421 511415	
Acrol Ltd, Everite Road, Ditton, Widnes WA8 8PT, UK tel: +44-151 424 1341 fax: +44-151 495 1853	Flame retardants, organic pigments
Additive Polymers Ltd, Unit 4 Kiln Road, Burrfields Road, Portsmouth PO3 5LP, UK tel: +44-1705 678 575 fax: +44-1705 678 564	
Air Products plc, Hersham Place, Molesey Road, Walton-on-Thames KT12 4RZ, UK tel: +44-1932 249 273 fax: +44-1932 249 786	
Air Products and Chemicals, Inc., (prev. Akzo) 7201 Hamilton Boulevard, Allentown PA 18195, USA tel: +1-215 481 8935 fax: +1-215 481 8504	Surface modified UHMWPE, adhesion promoters
Akcros Chemicals Ltd, Headquarters, Eccles Site, PO Box 1, Eccles, Manchester M30 0BH, UK tel: +44-161 785 1111 fax: +44-161 788 7886	

Supplier/manufacturer	Additives
Akcros Chemicals America, 500 Jersey Avenue, PO Box 638, New Brunswick, NJ 08903, USA tel: +1-908 247 2202 fax: +1-908 247 2287	
Akcros Chemicals (Asia-Pacific) Ltd, 7500A Beach Road, Z 15-309 The Plaza, Singapore 199591 tel: +65 292 1996 fax: +65 292 9665	
Akro-Plastik GmbH Industriegebiet Scheid D-5665 1 Niederzissen Postfach 67, Germany tel: +49-2636 97 42-0 fax: +49-2636 97 42 31	
Akzo-Nobel Polymer Chemicals PO Box 247 Akzo Chemicals Inc., Amersfoort AE Amersfoort AZ, Netherlands 3800 tel: +31-33 467 767 fax: +31-33 467 6151	Antistatic agents, organic peroxides
Akzo Nobel Faser AG, Accurel Systems, D-63784 Oldenburg, Germany tel: +49-6022 81-478 fax: +49-6022 81-823	Slip/antiblocking agents, antistatic agent, impact modifiers, flame retardants
Akzo Chemical Co., Specialty Chemicals Div. Westport, CT 06880, USA	Flame retardants, carbon fibres
Akzo Chemicals Inc., New Brunswick, NJ 08903, USA	Antistatic agents
Albright & Wilson UK Ltd. Flame Retardants Plastics PO Box 3, 210-222 Hagley Road West Oldbury, B68, 0NN, UK tel: +44-121 420 5312 fax: +44-121 420 5111	Flame retardants
Alcan Chemicals Europe, Chalfont park Gerrards Cross, S19 0QH, UK tel: +44-1592 411 000 fax: +43-1753 233 444	Flame retardants
Alchemie Ltd Brookhampton Lane Kineton CV35 0JA, UK tel: +44-1926 641 600 fax: +44-1926 641 698	Fillers, release agents
Albis Plastic GmbH P.O. Box 280340 D-2000, Hamburg 28, Germany	Antiblocking agents, antioxidants, antistats, blowing agents, color concentrates, corrosion protection pigments fillers, flame retardants, lubricants, masterbatches, mold release agents, paint auxiliaries, thixotropic agents, UV stabilizers
Albright & Wilson Inc., PO Box 26229 Richmond, VA 23260, USA	Flame retardants, phosphorus compounds

Supplier/manufacturer	Additives
Allied Signal Europe NV Performance Additives Haasrode Research Park, B-300 1 Heverlee, Belgium tel: +32-16 391 210 fax: +32-16 391 371	PE copolymers modifiers, ionomer pigments dispersants, micronised PE waxes
Allied Signal Inc Performance Additives, PO Box 1039, 101 Columbia Road, Morristown NJ 07962-1039, USA tel: +1-201 455 2145 fax: +1-201 455 6154	Lubricants, mold reslease agents
Alpha Calcit Fullstoff GmbH & Co, KG, 500 Köln 50 Postfach 1106, Germany tel: +49-2236-89-14-0 fax: +49-2236-40-644	Small particles reinforcing fillers, semi-reinforcing fillers, $CaCO_3$
Alcoa International Inc., 61 Av. d'Ouchy, CH-1006, Lausanne, Switzerland	
Aluminum Company of America, Alcoa Labs, Alcoa Center, PA 15069, USA	Aluminum hydroxide, flame retardants
AmeriBrom Inc 52 Vanderbilt Ave, New York NY 10017, USA tel: +1-212-286 4000 fax: +1-212-286 4475	Brominated compounds, flame retardants
American Cyanamid Co., Wayne, NJ 07470, USA Bridgewater, NJ 08807, USA	Antioxidants, antistats
Americhem Inc., 225 Broadway E, Cuyahoga Falls, OH 44221, USA tel: +1-216 929 4213 fax: +1-216 929 4144	Flame retardants: phosphate esters, chlorinated paraffins, zinc borate aluminum hydroxide, antimony oxide
American Hoechst Corp., Somerville, NY 08876, USA	Metal deactivators
AMAX Polymer Additives Group, Suite 320, 900 Victors Way, Ann Arbor, MI 48108, USA	Flame retardants, smoke suppressants zinc borates, molybdenum compounds
Amoco Chemicals, Mail Code 7802, 200 East Randolph Drive, Chicago, IL 60601-7125, USA tel: +1-312 856 3092 fax: +1-312 856 4151	
Amoco Chemical (Europe) SA, 15 Rue Rothschild, CH-1211 Geneva 21, Switzerland tel: +41-22 715 0701 fax: +41-22 738 8037	
Ampacet Corp., 250 S. Terrace Avenue, Mount Vernon, NY 105501, USA	Flame retardants: chlorinated paraffins, antimony oxide
Ampacet Corp, 660 White Plains Rd, Tarrytown NY 10591, USA tel: +1-914 631 6600 fax: +1-914 631 7278	
Ampacet Europe SA, Rue d'Ampacet 1, Messancy B-6780, Belgium tel: +32-63 38 13 00 fax: +32-63 38 13 93	

Supplier/manufacturer	Additives
Ampacet Europe SA, Rue des Scillas 45, L-2529 Howald, Luxembourg tel: +352-29 20 99-1 fax: +352-29 20 99-595	
Amspec Chem. Corp., Foot of Water Street, Gloucester City, NJ 08030, USA tel: +1-609 456 3930 fax: +1-609 456 6704	Antimony trioxide
Andersons, PO Box 119, Maumee, OH 43537, USA tel: +1-419 891 6545 fax: +1-419 891 6539	
Angus Chemical Co., 1500 E Lake Cook Rd, Buffalo Grove, IL 60089, USA tel: +1-708 215 8600 fax: +1-708 215 8626	
Anzon Inc., 2545 Aramingo Ave., Philadelphia, PA 19125, USA tel: +1-215 427 3000 fax: +1-215 427 6955	Antimony trioxide, ammonium phosphate
Anzon Ltd., Cookson House, Willington Quay, Wallsend, Tyne & Wear NE26 6UQ, UK tel: +44-191 262 2211 fax: +44-191 263 4491	Antimony compounds, zinc and magnesium oxides
API SpA, Via Dante Alighieri 27, I-36065 Mussolente (Vi), Italy tel: +39-424 579 511 fax: +39-424 579 800	Pigment concentrates
APV Baker Ltd, Speedwell Road, Parkhouse East, Newcastle-under-Lyme ST5 7RG, UK tel: +44-1782 565 656 fax: +44-1782 565 800	
Argus Chemical Co, 520 Madison Avenue, New York NY 10022, USA, NY 11231, USA	Antioxidants, metal deactivators
Argus Chemical SA, NV, Bruxelles, and 1629 Drogenbos, Belgium Witco Corp., New York, NY, USA tel: +1-212 605 3600	
Adeka-Argus Chem. Co. Ltd., Urawa City, Saitama Pref. 336, Japan	
Aristech Chemical Corporation, PO Box 2219, 600 Grant St., Pittsburgh, PA 15219, USA tel: +1-412 433 7800 fax: 1-412 433 7721	
Asahi Chemical Industry Hibixa-Mitsui Building 1-2 Yurakucho 1-Chome, Chiyoda-ku Tokyo, Japan tel: +81-3 2 33 12 71	
Asahi Denka Kogyo KK, Furukawa Building, 3-14, 2-chome, Nihonbashi-Murmachin, Chuo-ku, Japan Asahi Glass Co. Ltd, Chiyoda Building, 2-1-2 Marunouchi, Chiyoda-ku, Tokyo 100, Japan	Flame retardants: halogenated compounds, phosphate esters

Supplier/manufacturer	Additives
Asahi Denka Kogyo KK, Furukawa Building 3-14, Nihonbashi-Muromachi 2-chrome, Chuo-ku, Tokyo 103, Japan tel: +81-3-5255 9017 fax: +81-3 3270 2463	
A/S Norwegian Talc, P.O. Box 744 N-5001 Bergen/Norway	Extenders
Asarco Inc., 180 Maiden Lane, New York, NY 10038, USA	Antimony oxide
Ashlands Chemical Co., Columbus, OH 43216, Box 2219, USA tel: +1-614 889 4191 fax: +1-614 889 3735	Antioxidants
Aspanger Geschaftsbereich, Jungbunzlauer GmbH, A-2870 Aspang, Postfach 32, Austria tel: +43-2642 52355 fax: +43-2642 52673	
Astab, c/o Croxton & Garry-Ltd, Curtis Road, Dorking RH4 1XA, UK tel: +44-1306 886 688 fax: +44-1306 887 780	
Aston Industries Ltd, 38 Nine Mile Point Ind. Estate, Crosskeys NP1 7HZ, UK tel: +44-1495 200 666 fax: +44-1495 200 616	
Astor Wax Corp, 200 Piedmont Court, Doraville GA 30340, USA tel: +1-401 348 8083	
Asua Products SA, Ctra Sangroniz 20, E-48150 Soldica (Bilbao), Spain tel: +34-4 453 16 50 fax: +34-4 453 35 66	
Atlas SFTS BV Baumstrasse 39, D-47198 Duisburg, Germany tel: +49-2066 560 34 fax: +49-2066 106 19	
Axel Plastics Research Laboratories Inc., Box 77 0855 Woodside NY 11377, USA tel: +1-718 672 8300 fax: +1-718 565 7447	Mold release agents
BA Chemicals Ltd, Chalfont Park, Gerrards Cross Bucks, SL9 0QB, UK tel: +44-1735 887373 fax +44-1753 889602	Aluminum hydroxide
Bärlocher GmbH, Riestrasse 16, D-80992 München, Germany tel: +49-89 14 37 30 fax: +49-89 14 37 33 12	
Bärlocher USA, West Davis St, PO Box 545, Dover, OH 44622, USA tel: +1-216 364 4000 fax: +1-216 343 7025	Additive systems
Barmag AG, Leverkuser Strasse 65, D-42897 Remscheid, Germany tel: +49-21 91 6 70 fax: +49-21 91 67 17 38	

Supplier/manufacturer	Additives
BASF AG, EPM/KU Unternehmensbereich Spezialchemikalien, D-67056 Ludwigshafen, Germany tel: +49-621 60-0 fax: +49-621 60 723 48 BASF Corp., Chim. Div., Parsippany, NJ 07054, USA	Lubricants, mold release agents
BASF Lacke + Farben AG, Max-Winkelmann-Strasse 80, Postfach 61 23 D-4400 Münster, Germany tel: +49-2501 14-1 fax: +49-2501 14 33 73	Light stabilizers, UV absorbers
BASF Corp Colorants and Additives for Plastics, 300 Continental Drive North, Mount Olive, NJ 07828, USA tel: +1-201 426 2600	Organic pigments
Bayer AG Geschaftsbereich KU, D-51368 Leverkusen, Germany tel: +49-214 30-1 fax: +49-214 30-7407	Blowing agents, flame retardants, impact modifiers, processing aids, stabilizers, UV absorbers, antistatic agents, pigments
Bayer Corporation, 100 Bayer Road, Pittsburgh, PA 15205, USA tel: +1-412 777 2000	
Battenfeld Gloeneo Extrusion Systems, Berry Hill Industrial Estate, Droitwich, Worcs. WR9 9RB, UK tel: +44-1905-775611 fax: +44-1905-776716	
BBU, Bleiberger Bergwerks Union AG, Radetzkystrasse 2, Postfaeh 95, A-9010 Kalgenfurt, Austria tel: +43-42 22 555 25	
Berga Kunststoffproduktion GmbH & Co KG, Thyssen Strasse 19-21 D-1000 Berlin 51, Germany tel: +49-330 414 3030 fax: +49-330 414 4095	
D Bergmann, Th. GmbH & Co. Kunststoffwerk KG, Ringstrasse 1-3, D-7560 Gaggenau 12, Germany tel: +49-7225 10 41-44 fax: +49-7225 68 02-10	
Bl Chemicals Inc Henley Division, 50 Chestnut Ridge Road Montvale, NJ 07645, USA tel: +1-201 307 0422 fax: +1-201 301 0424	
Biesterfeld Plastic GmbH, Ferdinandstrasse 41, D-20095 Hamburg, Germany tel: +49-40-3 20 08-0 fax: +49-40-3 20 08-442	
BIP Chemicals Ltd., PO Box 6, Popes Lane, Oldbury, Warley, W Midlands B69 4PD, UK tel: +44-121 551 1551 fax: +44-121 552 4267	
Blancs Mineraux de Paris (BMP), 40 rue des Vignobles, F-78402 Chatou Cedex, France tel: +33-1 30 52 32 63 fax: +33-1 30 71 46 83	

Supplier/manufacturer	Additives
Beck, Otto, Kunststoff GmbH & Co, Industriestrasse Postfach 12 60, D-3408 Duderstadt, Germany tel: +49-5527 8480	
Boehringer Ingelheim KG, Geschöftsgebeit Chemikalien, D-55216 Ingelheim, Germany tel: +49-6132-77 313 73 fax: +49-6132-77 46 17	Exothermic blowing agents
Borax Holdings Ltd, Borax House, Carlisle Place, London SW1P 1HT, UK	Zinc borate
Borg-Warner Chemicals Inc., International Center, Parkersburg, WV 26102, USA	Flame retardants: phosphorus/nitrogen intumescent compounds, antioxidants
BP Chemicals International, 6th Floor, Britannic House, 1 Finsbury Circus, London EC2M 7BA, UK tel: +44-171-496 4867 fax: +44-171-496 4898	
Brabender Technologic KG, Kulturstrasse 55-74, PO Box 35 01 38, D-47032 Duisburg, Germany tel: +49-203 9884-0 fax: +49-203-9884-174	
Branco Industria e Commercio Ltd., Rua Manoel Pinto de Carbvalho 229, CEP 02712-120 São Paulo SP, Brazil tel: +55-11 265 8666 fax: +55-11 872 3735	Additives, color concentrates
Brockhues AG, D-65396 Walluf, Mühlstrasse 118, Germany tel: +49-6123 797 403 fax: +49-6123 797 418	
L Brüggemann Salzstrasse 123-131, D-74076 Heilbronn, PO Box 1461, Germany tel: +49-71 31 15 75-0 fax: +49-71 31 15 75-65	Flame retardants, heat stabilizers, weathering agents, impact modifiers, mould release agents
Buhler Kunststoffe, Farben u Additive GmbH, Ritzenschattenbalb 1, D-87480 Weitnau, Germany tel: +49-8375 920 10 fax: +49-8375 920 130	
Burgess Pigments Inc, PO Box 349, Sandersville, GA 31082, USA tel: +1-912 552 2544 fax: +1-912 552 1772	Calcined aluminium silicates, calcined days
Bush Boake Allen, Terpene Products, Dans Road, Widnes WA8 0RF, UK tel: +44-151 423 3131 fax: +44-151 424 3268	
Busing & Fasch GmbH & Co, Abt. Kunststoffe Cloppenburger Strasse 138-140, Postfach 25 61, D-26015 Oldenburg, Germany tel: +49-441 43077 fax: +49-441 340 23 50	
Buss AG, Basel, CH-4133 Pratteln I, Hohenrainstrasse 10, Switzerland tel: +41-61-8256-111 fax: +41-61-8256-699	

Supplier/manufacturer	Additives
Byk-Chemie GmbH Abelstrasse 14, Postfach 10 02 45, D-46483 Wesel, PO Box 245, Germany tel: +49-281 6 70-0 fax: +49-281 6 82 45	Additives, antifoaming agents dispersing agents, leveling, wetting and dispersion agents
Byk-Chemie USA, 524 South Cherry St, PO Box 5670, Wallingford, CT 06492, USA tel: +1-203 254 2086 fax: +1-203 284 9158	
C-Tech Corporation, 5-B, 5th Floor, Himgiri 1277 Hatiskar Marg, Mumbai 400 025, India tel: +91-22 422 5939 fax: +91-22 430 9295	
Cabot Corporation Special Blacks Division, 157 Concord Rd, Billerica MA 01821, USA tel: +1-508 670 7042	Carbon black powder and granules
Cabot Plastics International, rue E Vandervelde 131, B-4431 Ans/Loncin, Belgium tel: +32-41 46 82 11 fax: +32-41 46 54 99	Colored masterbatches, white masterbatches, electrically conductive masterbatches
Cabot Plastics International, Interleucenlaan 5, 3001 Leuven, Belgium tel: +32-16 3901 11 fax: +32-16 4012 53	
Cairn Chemicals Ltd, Cairn House, Elgiva Lane, Chesham HP5 2JD, UK tel: +44-1494 786 066 fax: +44-1494 791 816	
Campine America Inc., 3676 Davis Road, PO Box 526, Dover, OH 44622, USA tel: +1-216 364 8533 fax: +1-216 364 1579	
Campine NV, Nijverheidsstraat 2, B-2340 Beerse, Belgium	Antimony trioxide
Etn. Gcbroeders Cappelle NV, Kortrijksrraat 11–5, B-8930 Menen, Belgium tel: +32-56 531 200 fax: +32-56 521 262	
Cerdec Corporation, Drakenfeld Products, West Wylie Avenue, PO Box 519, Washington, PA 15301, USA tel: +1-412 223 5900 fax: 1-412 228 3170	
Certainteed Corporation - FRD, PO Box 860, Valley Forge, PA 19482, USA tel: +1-215 341 7770 fax: +1-215 293 1765	
Chemax Inc., Box 6067, Greenville, SC 29606, USA tel: +1-803 277 7000 fax: +1-803 277 7807	
Chemie-Export-Import Storkower Strasse 133, D-1055 Berlin, Germany	Azocarbonamide, calcium carbonates, carbon blacks, colorants and pigments, emulsifiers, hardeners, lubricants, peroxides, printing inks, silicic acid, solvents, stabilizers, waxes

Supplier/manufacturer	Additives
Chemiehandel SE AG, PO Box, CH-8834, Schindellegi Switzerland tel: +41-1-785 04 44 fax: +41-1 785 01 43	
Chemische Werke Lowi GmbH, Teplitzer Strasse, Postfach 16 60, D-8264 Waldkraiburg, Germany tel: +49-86 38 4011 fax: +49-86 38 819 42	Antioxidants
Chemische Werke Munchen, Otto Börlocher GmbH Riesstrasse 6, D-8000 Munich 50, Germany	Lubricants, stabilizers, antistats
Chemische Fabrik Grunau GmbH, PO Box 120, D-7918 Illertissen, Germany	Antimony compounds
Chemson GmbH, Reuterweg 14, Metalgesell. u Cookson Group, D-60271 Frankfurt am Main, Germany tel: +49-69 159 1590 fax: +49-69 159 2236	Heat/light stabilizers, lubricants, metal soaps
Chemson Polymer-Additive GmbH, Gailitz 195 A-9601 Arnoldstein, Austria tel: +43 4255 2226 fax: +43 4255 2435	
Chevron Chemical Co., Olefins and Derivatives Division, PO Box 3766, Huston, TX 77253, USA tel: +1-713 754 2000	Compatibilizers, impact modifiers
Chinghall Ltd, Ward Road, Bletchley MK1 1JA, UK tel: +44-1908 76227	Additive dispersions
Chrostiki, PO Box 22, 19400 Koropi, Attikis Greece tel: +30-1 6624 692 fax: +30-1 6623 873	Additives, antioxidants, antistats, color concentrates, colorants and pigments, fillers, master batches
Chisso Co, 7-3 Maunouchi 2-chome., Chiyoda-ku, Tokyo 100, Japan	Flame retardants: phosphorus and nitrogen compounds
Chemipro Kasei Ltd., Chuo-Ka Kobe, Japan	Antioxidants
Ciba Geigy GmbH, D-7867 Wehr/Baden, Germany CH-4002 Basel, Switzerland	Additives, pigments, flame retardants, phosphate esters, halogenated phosphorus compounds
Ciba-Geigy Corp., Div., Hawthorne, NY 10532, USA Ciba-Geigy AG, Div. Additive, Basel Ciba Geigy Marienberg GmbH, Bensheim, Germany; Ciba-Geigy Corp., Additives Div., Hawthorne NY 10532, USA	Antioxidants, antistatic agents, lubricants, metal deactivators, optical brightening agents, light/heat stabilizers, processing stabilizers, pigments

Supplier/manufacturer	Additives
Ciba-Geigy AG 8900-Gersthofen, Germany	
Ciba Additive GmbH, PO Box 1640, D-68619 Lampertheim, Germany tel: +49-6206 50 20 fax: +49-6206 50 21 368 Ciba Additives Hulley Road, Macclesfield SK10 2NX, UK tel: +44-1625 665 000 fax: +44-1625 502 674	
Ciba-Geigy AG, Division KA, PO Box CH-4002 Basel, Switzerland tel: +41-61 697 1111 fax: +41-61 697 3974	
Ciba-Geigy Corporation Additives Division, 540 White Plains Road Tarrytown, NY 10591-9005, USA tel: +1-914 785 2000 fax: +1-914 347 5687	
Ciba-Geigy Corporation, Chemicals Division, 410 Swing Road, Greensborough, NC 27409-2080, USA tel: +1-919 632 6000 fax: +1-919 632 7008	
Ciba-Geigy Corp Pigments Division, 335 Water Street, Newport, DE 19804, USA tel: +1-302 633 2000 Ciba-Geigy (Japan) Ltd, 10-66 Miyuk-cho, Takarazuka-shi, Hyogo 665, Japan tel: +81-797 74 2472 fax: +81-797 74 2472	
C.I.I. Inc, PO Box 200 Station A, Willowdale, ON, M2N 5S8, Canada	Chlorinated paraffins
Claremont Polychemical Corp., 501 Winding Road, Old Bethpage, NY 11804, USA	Antimony oxide, zinc borate
Clariant Huningue SA, Rothaustrasse 61, 4132 Muttenz 1, Switzerland tel: +41-61-469-6433 fax: +41-61-469-655	UV stabilizers
Clariant Corp, 4000 Monroe Road, Charlotte, NC 28205, USA tel: +1-704 331 7029 fax: +1-704 331 7112	
Climax Molybdenum Co., 23, 75 Swallow Hill Road Building 900, Pittsburgh, PA 15220-1672, USA tel: +1-412 279 4200 fax: +1-412 279 4710	
Climax Molybdenum Co., 1600 Huron Parkway, Ann Arbor, MI 48106, USA	Molybdenum compounds
Climax Performance Materials Corp., Polymer Additives Group, Amax Center, Greenwich, CT 06836-1700, USA tel: +1-203 629 6000; fax: +1-203 629 6608	
Coatex, 35 rue Ampere, ZI Lyon Nord, F-69727, Crenay Cedex, France tel: +33-72 08 20 00 fax: +33-72 08 20 40	Phosphoric acid dispersing agents

Supplier/manufacturer	Additives
COIM SpA, I-20019 Settimo Milanese (MI), Via A Manzoni 28, Italy tel: +39-2 33505-1 fax: +39-2 33505 249	
Colloids Ltd, Dennis Road, Tanhouse Industrial Estate, Widnes, Cheshire WA8 0SL, UK tel: +44-151 424 7424 fax: +44-151 423 3553	Masterbatches, purging compounds liquid pigment dispersions
Colorant GmbH, Justus-Staudt-Strasse 1, D-6250 Limburg-Offheim, Germany tel: +49-64 31 53391	
Color-Chem International Corp., 8601 Dunwoody Place, Bldg 334, Atlanta, Ga 30550, USA tel: +1-770 993 5500 fax: +1-770 993 4780	
Color-Plastics-Chemie, Postfach 12 05 10, D-5630 Remscheid 11, Germany tel: +49-2191 530 05-9 fax: +49-2191 516 95	Color concentrates, color pastes, colorants and pigments masterbatches
Color-Service GmbH, Offenbacher Landstrasse 107-109, Offenbacher Landstrasse 109, D-63512 Hainburg/Hess, Germany tel: +49-6182 40 34 37 fax: +49-6182 6 68 86	Special universal masterbatch, color concentrates
Color System SpA, Via S. Quasimodo 9, I-20025 Legnano (Mi), Italy tel: +39-331 577 607 fax: +39-331 464 248	
Colores Hispania, Josep Pla 149, Barcelona 08019, Spain tel: +34-3307 13 50 fax: +34-3303 25 05	Lead-free decorative pigments small particles zinc phosphate
Colores y Compuestos Plasticos SA, C/Fedanci, 8 al 10 bajos 2ª Pare Empresarial, 08 190 Sant Cugat del Valle (Barcelona), Spain tel: +34-93 589 26 02 fax: +34-93 589 43 84	
Colortech Inc, 5712 Commerce Boulevard, Morristown, TN 37814, USA tel: +1-905 792 0333 fax: +1-905 792 8118	
Colortronic (UK) Ltd, Matilda House, Carrwood Road, Chesterfield Industrial Estate, Chesterfield S41 9QB, UK tel: +44-1246 260 222 fax: +44-1246 455 420	
Columbian Chemicals Co., 1600 Parkwood Circle, Suite 400, Atlanta, GA 30339, USA tel: +1-404-951-5700	
Comiel SpA, Via Bessarione 1, I-20139 Milano, Italy tel: +39-2569 341 fax: +39-2569 181	Lubricants
Compounding Technology AG, Baumgartenweg 4, CH-4106 Therwil (Basel), Switzerland tel: +41-61 722 0526 fax: +41-661 722 0529	

Supplier/manufacturer	Additives
Constab Polymer-Chemie GmbH & Co, Möhnetal 16 Postfach 1127, D-59602 Rüthen/Möhne, Germany tel: +49-2952 8 19-0 fax: +49 2952 31 40	Additive masterbatches, pigment/dye masterbatches, agricultural film masterbatches, degradable additive masterbatches
Condea Petrochemie GmbH, Mittleweg 13, D-2000 Hamburg 13, Germany	Flame retardants, phosphate esters
Cookson Pigments Inc, 56 Vanderpool Street, Newark, NJ 07114, USA tel: +1-201 242 1800 fax: +1-201 242 7274	
Cookson Matthey, Liverpool Road, Cookson Ltd., East Kisgrove, Stoke on Trent, ST4 3AA, UK tel: +44-1782 794400	Antimony trioxide
Cookson Specialty Additives, 1000 Wayside Road, Cleveland, OH 44110, USA tel: +1-216 531 6010 fax: +1-216 486 6638	
Corduplast GmbH, Postfach 1227, D-4417 Altenberge, Germany tel: +39-2505 2186	Additives, antioxidants, antiblocking agents antiplate-out agents, antistats, blowing agents, flame retardants, lubricants, master batches, metal deactivators, UV absorbers, UV stabilizers
Costenoble GmbH, Postfach 5205, D-6236 Eschborn, Germany tel: +49-6196 44020 fax: +49-6196 481283	
Cp-Polymer-Technik, Berliner Strasse 3-5, D-2863 Ritterhude, Germany tel: +49-4929 1034	
Croda Oleochemicals, Cowick Hall, Snaith, Goole DN14 9AA, UK tel: +44-1405 860 551 fax: +44-1405 860 205	
Croda Universal Inc., 4014 Walnut Pond Drive, Houston TX 77059, USA and North Humberside London tel: +1-713 282 0022 fax: +1-713 282 0024	Lubricants
Crowley Chemical Co., Inc., 261 Madison Ave, New York, NY 10016, USA tel: +1-212 682 1200 fax: +1-212 953 3487	
Croxton & Garry Ltd, Curtis Road, Dorking RH4 1XA, UK tel: +44-1306 886 688 fax: +44-1306 887 780	Calcium carbonate
Cynamid International, One Cyanamid Plaza, Wayne, NJ 07470, USA	Carbon fibers, crosslinking agents, flame retardants
Chimopar SA, Th. Palade 50, 74585 Bucharest, Romania	Stabilizers
Comiel, Prodotti Chimici Industriall SpA, Mailand, Italy	Lubricants

Supplier/manufacturer	Additives
Commercial Society "Synthesis" Oradea, 35 Borş Str., 3700 Oradea, Romania	Fillers, stabilizers
Colorom SA Codlea 27 Lunga, Str. 2252, Codlea, Braşov, Romania	Colorants, pigments
Daihachi Chemical Industry Co. Ltd., 3-54 Chodo Higashi Osaka City Osaka Pref. 577, Japan	Borates, phosphorous-based flame retardants
Dai-ichi Chemical Industries Ltd., 2-2-1, Higashi Sakashita, Itabashi-ku, Tokyo 174, Japan	Brominated flame retardants
Dai-ichi Kogyo Seiyaku Co. Ltd., New Kyoto Center Building, 614 Higashishiokoji-cho, Shimukyo-ku, Kyoto 600, Japan tel: +81-75 343 1181 fax: +81-75 343 1421	
Daniel Products Co Inc, 400 Claremont Ave, Jersey City, NJ 07303, USA tel: +1-201 432 0800 fax: +1-201 432 0266	
Danisco Ingredients, Edwin Rahrsvej 38, Brabrabd DK-8220, Denmark tel: +45-89 43 50 00 fax: +45-86 25 10 77	High concentration antistatics, anti-fog agents, polyglycerol esters
Datacolor International, 6 St Georges Court, Dairyhouse Lane, Altringham WA14 5UA, UK tel: +44-161 929 9441 fax: +43-161 929 9059	
Davis-Standard Compounding Systems, 1 Extrusion Drive, Pawcatuck, CT 06379, USA tel: +1-203 599 1010 fax: +1-203 599 6258	
Davis Standard, Asia Room, 1211 Peninsula Centre, 67 Mody Road Tsimshatsui Kowloon, Hong Kong tel: +852 2 723 1787 fax: +852 2 723 4263	
Davis-Standard Europe, Tricorn House, Cainscross, Stroud GL5 4LF, UK tel: +44-1453 765 111 fax: +44-1453 750 819	
Henry Day & Sons Ltd, Saville Bridge Mills, Dewsbury WF12 9AF, UK tel: +44-1924 464 351 fax: +44-1924 459 211	
Dead Sea Bromine Compounds, PO Box 180, Beer Sheva 85101, Israel tel: +972-7 297 265 fax: +972-7 280 444	Brominated flame retardants
Degussa AG, GB Anorganische, Chemiprodukte (AC-KP) D-60287 Frankfurt am Main, Germany tel: +49-69 2 18-01 fax: +49-69 2 18 32 18	Antiblocking agents, auxiliary rubber agents, bonding agents, calcium silicates, carbon blacks, catalysts, colorants and pigments, fillers, ink soot, paint auxiliaries, silicic acid, soot chips, soot dispersions, soot pastes, soot plastic concentrates, soot wettings, UV absorbers

Supplier/manufacturer	Additives
Dennert Schaumglass GmbH, 8439 Heng, Germany	Microcellular glass spheres, microbiocides
Georg Deifel KG, Mainberger Strasse 10, D-8720 Schweinfurth, Germany tel: +49-9721 1774 fax: +49-9721 185 091	
Diadema Industrias Quimicas I, Ltd., Av Fagundes de Oliveira 190, Piraporinha Diadema, SP 09950-907, Brazil tel: +55-11 7647 1133 fax: +55-11 7617 1155	
Diversey Ltd., Watson Favell Centre, Northampton NN3 4PD, UK tel: +44-1604 405 311	Mold release agents
DJ Enterprises Inc., Box 31366, Cleveland, OH 44131-0366, USA tel: +1-216 524 3879	
Diamond Shamrock Chemicals Co., Morristown, NJ 07960, USA	Antistatic agents
Gebr. Dorfner GmbH & Co, D-92242 Hirschau, Germany tel: +49-9622 82-0 fax: +49-9622 82-69	
Dover Chemical Corp., Dover Chemical Corp sub. ICC Industries Inc., 720 Fifth Avenue, New York, NY 10019, USA ICC Industries Inc., 3676 Davis Rd NW, PO Box 40, Dover, OH 44622, USA tel: +1-216 343 7711 fax: +1-216 364 1579	Flame retardants: chlorinated paraffins, halogenated compounds
Dow Corning Europe, Rue General de Gaulle 62, B-1310 La Hulpe, Belgium tel: +32-2 655 21 11 fax: +32-2 655 20 01	
Dow Corning GmbH, Schossburgstrasse 24, D-65201 Wiesbaden, Germany tel: +49-611-928630 fax: +49-611-24628	
DSM Resins, PO Box 615, Zwolle, NL 8000, Netherlands tel: +31-38 284 911 fax: +31-38 2214 284	Conductive compounds
DuPont CRP 711, Room 221, Wilmington, DE 19880-0711, USA	Barrier-forming additives
DuPont Chemicals, 1007 Market Street, Wilimington, DE 19898, USA tel: +1-302 774 2099 fax: +1-302 773 4181	
DuPont de Nemours European Technical Centre, Antoon Spinoystraat 6, B-2800 Mechelen, Belgium tel: +32-15 44 13 54 fax: +32-15 44 15 10	

Supplier/manufacturer	Additives
DuPont de Nemours International SA, PO Box 50, CH-1218 Le Grand-Saconnex, Switzerland tel: +41-22 717 5111 fax: +41-22 717 6150 Du Pont Asia Pacific Ltd., 1122 New World Office Building, East Wing, Salisbury Road, Kowloon, Hong Kong tel: +852-734 5345 fax: +852-724 4458 DuPont Far East Inc., Kowa, Building No 2, 11-39 Akasaka I-chome Minato-ku, Tokyo 107, Japan tel: +81-5 85 55 11	
DuPont Fibers and Composites, Development Center, Chestnut Run Plaza, Magnolia Run, Wilmington, DE 19880-0702, USA	
Duslo a.s., Sala 927 03, Slovakia tel: +42-706 75 4512 fax: +42-706 75 3018	Magnesium hydroxide
Dynamit Nobel AG, Sparte Chemikalien, D-5210 Troisdorf, Germany tel: +49-22 41 850 fax: +49-22 41 85 27 93	Bonding agents, catalysts, silanes
Eastman Chemical Co., Polymer Additives and Specialty Monomers Business Unit, PO Box 431 Kingsport, TN 37662, USA tel: +1-615 229 2000 fax: +1-615 224 0618	Metal deactivators
ECC International Ltd., John Keay House, St Austell, Cornwall PL25 4DJ, UK tel: +44-1726 74482 fax: +44-1726 623019 ECC International SA, 2 rue du Canal, B-4_5.S 1 Lioxhe, Belgium tel: +32-41 79 98 11 fax: +32-41 79 82 79	High-whiteness marble, ball and calcined clays, anti-blocking agents, coated fine/ultrafine calcium carbonate, chalk whitings, alkaline high purity china clay
ECC Japan Ltd., 10th Central Building, 10-3 Ginza 4-chome, Chuo-ku, Tokyo 104, Japan tel: +81-3 3456 8250 fax: +81-3 3456 8255	
E & P Würtz GmbH & Co., D-6530 Bingen am Rhein 17, Germany	Additives, lubricants, mold release agents
E. Merk, Frankfurter Strasse 250, D-6100 Darmstadt, Germany	Colorants, iriodin pearl luster pigments, mica, pigments
Elf Atochem SA, 4 cours Michelet, La Defense 10, F-92091 Paris la Defense Cedex 42, France tel: +33-14900-8080 fax: +33-14900-8396 Elf Atochem North America, 2000 Market Street, Philadelphia, PA 19103, USA tel: +1-215 419 7000 fax: +1-215 419 7413	Processing aids, modifiers
Ellis & Everard Pie, 119 Guildford St., Chertsey, Surrey KT16 9AL, UK tel: +44-1932 566033; fax: +44 1932 560363	

Supplier/manufacturer	Additives
EM Industries Inc., 5 Skyline Drive, Hawthorne, NY 10532, USA tel: +1-914 592 4660 fax: +1-914 592 9369	Laser-marking additives
Emerson & Cuming Composite Materials Inc., 77 Dragon Court, Woburn, MA 01888, USA tel: +1-617 938 8630 fax: +1-617 933 4318	
EMS-Chemie AG Selnaustrasse 16, CH-8039 Zurich, Switzerland tel: +41-1 201 54 11 fax: +41-1 201 01 55	
Endex Polymer Additives Inc., 2198 Ogden Ave, Suite 131, Aurora, IL 60504, USA	
Engelhard Corp, 101 Wood Ave., Iselin, NJ 08830-0770, USA tel: +1-908 205 5000 fax: +1-908 321 0250	
EniChem Polimeri SR, Via Rossellini 15-17, I-20124, Italy tel: +39-263 331	
Ente Nazionale Indocarburi, 00144 Rome, Italy	Antioxidants
EPI Environmental Products Inc., 103 Longview Drive, Conroe, TX 77301, USA tel: +1-409 788 2998 fax: +1-409 788 2968	
Epolin Inc., 358-364 Adams St, Newark, NJ 07105, USA tel: +1-201 465 9495 fax: +1-201 465 5353	
C H Erbslöh, Kaistrasse 5, Postfach 29 26, D-2000 Hamburg IL, Germany tel: +49-211 390 01-61 fax: +49-211 390 01-21	Plasticizers, flame retardants, light stabilisers
C H Erbslöh, Dusseldorfer Strasse 103, D-47809 Krefeld, Germany tel: +49-2151 525 00 fax: +49-2151 525 152	
Esseti Plast Srl, Via IV Novembre 98, I-21058 Olona (Va), Italy tel: +39-331 641 159 fax: +39-331 375 182	
Ethyl Corp., Bromine Chemicals Division, 451 Florida Boulevard, Baton Rouge, LA 70801, USA Ethyl SA Chemicals Group, London Road, Bracknell RG12 2UW, UK tel: +44-1344 780378 fax: +44-1344 773860	Brominated compounds
Eurobrom B. V., Patentlaan 5, NL-2288 AD Rijswiyr, The Netherlands	Brominated compounds
European Owens-Corning Fiberglass Glass Fibers, 178 Chausee de la Hulpe, B-1170 Brussels, Belgium	

Supplier/manufacturer	Additives
Europol plc, Fauld Industrial Estate, Tutbury, Burton-on-Trent, Staffs DE13 9HR, UK tel: +44-1283 815611 fax: +44-1283 813139	
Everlight Chemical Industrial Corp., 6 Floor, Chung Ting Bldg No 77 Sec 2 Tun Hua S Road Taipei, RC tel: +886-2 706 6006 fax: +886-2 708 1254	
Evode Plastics Ltd., Wanlip Road, Syston, Leicester LE7 8PD, UK tel: +44-1533 696752 fax: +44-1533 692960	
Expancel, Box 13000, 850-13 Sundsvall, Sweden tel: +46-60 13 40 00 fax: +46-60 56 95 18	Thermoplastic microspheres
Exxon Chemical Europe, Mechelesteenweg 363, B-1950 Kraainem, Belgium tel: +32-27 69 31 11 fax: +32-27 69 32 25 Exxon Chemical Europe Inc., Vorstlaan 280, Bld du Souverain, B-1160 Brussels, Belgium tel: +32-2 674 4111 fax: +32-2 674 4129	Polyolefinic modifiers
Fairmount Chemical Co Inc., 117 Blanchard Street, Newark, NJ 07105, USA tel: +1-201 344 5790 fax: +1-201 690 5290	
Faxe Kalk Frederiksholms, Kanal 16, PO Box 2183, DK-1017 Copenhagen K, Denmark tel: +45-33 13 75 00 fax: +45-33 12 78 73	Calcined chalks
Ferro Chemicals Group, 1000 Lakeside Ave, Cleveland, OH 44114, USA tel: +1 216 641 8580 Ferro Plastic Europe, v. Helmenstraat 20, NL-3029 AB Rotterdam, Netherlands tel: +31-10 478 49 11 fax: +31-10 477 82 02 England (Chirk, Wales), France (Ferro) Eurostar, Fosses,	Inorganic pigments
Spain (Castellon), Portugal (Lisbon) 7050 Krick Road, Bedford, OH 44146, USA	Color concentrates, colorants and pigments, flame retardants (brominated compounds), UV stabilizers, antioxidants
Fibrolux GmbH, Oberlindau 23, D-6000 Frankfurt/Main, Germany tel: +49-69 72 8903 fax: +49-69 721212	
Filtec Ltd., Constance House, Waterloo Road, Widnes, Cheshire WA8 0QR, UK tel: +44-151 495 1988 fax: +44-151 420 1407	Ceramic microspheres
Karl Finke GmbH & Co. KG, Harzfelder Strasse 174-176, D-42281 Wuppertal, Germany tel: +49-202 709 06-0 fax: +49-202 70 39 29	

Supplier/manufacturer	Additives
FMC Corp., Chemical Products Group, 1735 Market Street, Philadelphia, PA 19103, USA tel: +1-215 299 6000 fax: +1-215 299 5999 FMC Flame Retardants Global HQ,	Phosphorus-containing flame retardants, phosphate esters
FMC Corporation (UK) Ltd., Tenax Road, Trafford Park, Manchester M17 1WT, UK tel: +44 161 872 3191 fax: +44-161 875 3177	
FMC Corporation NV, Avenue Louise 480 B9, Brussels 1050, Belgium tel: +32-2 645 5511 fax: +32-2 630 6350	
FMC International SA, 4th Floor, Interbank Building, 111 Passeo de Roxas Makati, Metro Manila, Phillippines tel: +63-2817 5546 fax: +63-2818 1485	
Franklin Industrial Minerals, 612 Tenth Avenue North, Nashville, TN 37203, USA tel: +1-615 259 4222 fax: +1-615 726 2693	
Frekote Inc., 164 Folly Mill Road, Seabrook, NH 03874, USA tel: +1-603 474 5541	Mold release agents
Frekote Mold Release Products, Dexter Corp., One Dexter Drive, Seabrook, NH 03874-4018, USA tel: +1-603 474 5541 fax: +1-603 474 5545	
Freudenberg Carl, Postfach 180, D-6940 Weinheim, Germany tel: +49-6201 801 fax: +49-6201 693 00	
Frisetta GmbH, Postfach 49, D-7869 Schonau/Schwarzwald, Germany tel: +49-76 73 10014	
Gabriel Chemie GmbH, Industriestrasse 1, A-2352 Gumpoldskirchen, Austria tel: +43-2252 636 30 fax: +43-2252 636 60	Additives, color concentrates, colorants, masterbatches pigments, printing inks
Gaypa Srl, Via Tribollo 6/8, Minticello Conte Otto (Vi), Italy tel: +39-444 946 088 fax: +39-444 946 027	
GE Specialty Chemicals, PO Box 1868, 501 Avery Street, Parkersburg, WV 26102-1868, USA tel: +1-304-424-5411 fax: +1-304-424-587	Silicone antiblocking agents
GE Speciality Chemicals, General Electric Plastics BV, Plasticslaan 1, PO Box 117, NL-4600 AC Bergen op Zoom, Netherlands tel: +31-1640 32911 fax: +31-1640 32940	Purging compound
Gebrüder Dorfner OHG, Kaolin umb. Quart-sand Werke, PO Box 1120, D-8452 Hirschau, Germany	Fillers

Supplier/manufacturer	Additives
Georgia Marble Co., 1201 Roberts Blvd, Bldg 100, Kennesaw, Ga 30144-3619, USA tel: +1-404 421 6500 fax: +1-404 421 6507	
Gevetex Texilglas GmbH, Postfach 1160, D-5120 Herogenrath, Germany tel: +49-2406 810 fax: +49-2406 792 86	
GFl (UK) Ltd, Marle Place, Brenchley, Tonbridge, UK tel: +44-1892 724 534 fax: +44-1892 724 4099	
Gharda Chemicals Ltd, 48 Hill Road, Bandra (W), Bombay 400 050, India tel: +91-22 642 6852 fax: +91-22 640 4224	
Givaudan SA, CH-1214 Vernier, Switzerland	Antioxidants, light stabilizers, biocides, stabilizers, UV absorbers UV stabilizers
Glaswerk Schuller, Faserweg 1, D-6980 Wertheim, Germany	Glass fiber mats, glass fibers
GlassTech Scandinavia, AB Box 1013, Angelholm 262-21, Sweden tel: +46-431 250 75 fax: +46-431 250 76	Chopped glass fiber for PP
Goldmann GmbH & Co KG, Postfach 3405, Bielefeld 1, Germany tel: +49-521 350 46-48 fax: +49-521 322 452	
Th Goldschrnidt AG, Postfach 10 14 61, D-45116 Essen, Germany tel: +49-201 173 2365 fax: +49-201 173 1838	
Goldschmidt Chemical Corp., 11 West 17th Street, Suite G3, Costa Mesa, CA 926027, USA tel: +1-714 6502161 fax: +1-714 642 9478 Goldschmidt Industrial Chemical Corp., Pitt Metals and Chemicals Div., 941 Robinson Highway, PO Box 279, McDonald, PA 15057-2079, USA tel: +1-412 796 1511 fax: +1-412 922 6657	Release agents, chemical blowing agents, flame retardants, semi-conducting tin dioxide, release coating, color pastes, glycerol fatty acid esters, deaerators, processing auxiliaries, wetting/ softening agents
B F Goodrich Specialty Chemicals, 9911 Brecksville Road, Cleveland, OH 44141-3247, USA tel: +1-216 447 5000 fax: +1-216 447 5750	Antioxidants
B F Goodrich Chemical, Goerlitzer Strasse 1, D-4040 Neuss, Germany tel: +49-21 01 1 80 50 fax +49-21 01 18 00	
Goodyear SA, Rueil Malmaison, France Goodyear, Akron, OH 44316, USA Goodyear Chemical, Av des Tropiques, ZA Courtaboef 2, F-91955 Les Ulis Cedex, France tel: +33-216 796 8295	Antioxidants

Supplier/manufacturer	Additives
Great Lakes Chemical Corporation, One Great Lakes Blvd, West Lafayette, IN 47906, USA tel: +1-317 497 6100 fax: +1-317 497 6123	HAS light/UV stabilizers, phenolic antioxidants, benzophenone stabilizers, antioxidants, phosphite antioxidants, bis(tribomophenoxy)ethane
Great Lakes Chemical (Europe) Ltd., PO Box 44, Oil Sites Road, Ellesmere Port L65 4GD, UK tel: +44-151 356 8489 fax: +44-151 356 8490	
Great Lakes Chemical (Deutschland) GmbH, Overather Strasse 50-52, D-51429 Bergisch Gladbach, Germany tel: +49-2204 9 54 30 fax: +49-2204 5 68 62	
Great Lakes Chemicals, PO Box 2200, West Lafayette, IN 47906, USA	Flame retardants, brominated compounds
Grinit Plast Srl, 22076 Mozzate (Co), Strada Statale 223, 1126, Italy tel: +39-331 831 625 fax: +39-331 833 731	Pellet masterbatches, micronised masterbatches, microballoon masterbatches, non-dusting powder masterbatches
G E Habich's Sohne, Postfach 67; D-3512 Reinhardshagen, Germany tel: +49-55 44 1071 fax: +49-55 44 7 91-72	
Haagen Chemie BV, PO Box 44, Roermond, The Netherlands	Metal soaps, stabilizers, stearates
Hager & Kassner GmbH, Postfach 449, D-4730 Ahlen 1, Germany tel: +49-21 40 90 74 fax: +49-23 82 5151	
Hampton Colours Ltd., Toadsmoor Mills, Brimscombe, Stroud GL5 2UH, UK tel: +44-1453 731 555 fax: +44-1 45 3 73 1 234	
Hans W Barbe, Chemische Erzeugnisse GmbH, D-6200 Weisbaden 13, Germany	Mold release agents
M A Hanna Company, Suite 36-5000, 200 Public Square, Cleveland, OH 44114-2304, USA tel: +1-216 589 4000 fax: +1-216 589 4200	
Hanse Chemie GmbH, Charlottenburger Strasse 9, D-21502 Geesthacht, Germany tel: +49-41 52 80 920 fax: +49-41 52 7 9156	Modified silicone toughness improver, epoxy-silicone additives
Harcros Chemicals BV, Haagen House, Roermond, Netherlands	Lubricants
Harshaw Chemical Co., 1945 East 97th Street, Cleveland, OH 44106, USA	Antimony oxide
Harwick Chemical Corp., 60S Seiberling Street, Akron, OH 44305, USA	Aluminum hydroxide, antimony oxide, flame retardants: (phosphate esters, halogenated phosphorus compounds, chlorinated paraffins), zinc borate
Hartmann Drukfarben GmbH, Postfach 60 03 49, D-6000 Franfurt/Main, Germany tel: +49-69 40001 fax: +49-69 400 02 86	

Supplier/manufacturer	Additives
W Hawley & Son Ltd, Colour Works, Duffield, Belper DE56 4FG, UK tel: +44-1332 840 291 fax: +44-1332 842 570	Color concentrates
Henkel Corp., Ambler, PA 19002, USA	Lubricants
Henkel KGaA Plastics Technology, D-40191 Dusseldorf, Germany tel: +49-211 797-0 fax: +49-211 798-9638	Lubricants
Henkel & Cie, 4000 Düsseldorf, Germany Henkel Corp., 11501 Northlake Drive, Cincinnati, OH 45249, USA tel: +1-513 530 7415 fax: +1-513 530 7581	Antistats, antifogging agents, plasticizers, lubricants, slip agents, dispersing aids, low temperature plasticizers
Herberts GmbH, Christbusch 25, Postfach 2002 44, D-5600 Wuppertal 2, Germany tel: +49-202 894-1 fax: +49-202 89 95 73	Microfine thermoplastic powders
Herberts Polymer Powders SA, PO Box 140, CH-1630 Bulle, Switzerland tel: +41-26 913 5111 fax: +41-26 912 7989	
Hercules Aerospace Co., Composite Products Group, PO Box 98, Magna, UT 84044, USA tel: +1-801 251 5372	Graphite fibers
High Polymer Laboratories Vishal Bhawan, 95 Nehru Place, New Delhi 110 019, India tel: +91-11 643 1522 fax: +91-11 647 4350	Additives
Hitachi Chemical Co. Ltd., Shinjuku Mitsui Building 2-1-1, Nishi-Shinjuku, Shinjuku-ku, Tokyo 160, Japan	Flame retardants, brominated compounds
Hitox Corp. of America, PO Box 2544, Corpus Christi, TX 78403, USA tel: +1-512 882 5175 fax: +1-512 882 6948	
Hoechst Celanese Corp., Industrial Chemicals Dept., Route 202-206, North Sommerville, NJ 08876, USA	Flame retardants, phosphorus compounds, antioxidants, antistatic agents, nucleating agents, dispersing agents, stabilizers, metal deactivators, compatibilizers, HAS, benzophenone stabilizers
Hoechst AG Business Unit-Additive, Gersthofen Postfach 101567, D-86005 Augsburg 1, Germany tel: +49 821 479-0 fax: +49 821 479 28 90	
Hoechst AG, PO Box 800320, D-65926 Frankfurt/Main 80, Germany tel: +49-69 305-0 fax: +49-69 30 36 65	Chlorinated paraffins, phosphorus compounds
Hoffmann Mineral, Postfach 14 60, D-86619 Neuburg (Donau), Germany tel: +49-84 31 53-0 fax: +49-84 31 53-3 30	Functional fillers

Supplier/manufacturer	Additives
Holland Colors America Inc., 1501 Progress Drive, Richmond, IN 47374, USA tel: +1-317 935 0329 fax: +1-317 966 3376	
Holland Colours Apeldoorn BV, Halvemannweg I, Postbus 720, 7300 AS Apeldoorn, Netherlands tel: +31-55 366 3143 fax: +31-55 366 2981	
Holliday Chemical Holdings plc, Birkby Grange, 85 Birkby Hall Road, Birkby, Huddersfield HD2 2YA, UK tel: +44-1484 828 200 fax: +44-1484 828 230	Photochromic dyes
Holliday Pigments Ltd., Morely Street, Kingston-upon-Hull HU8 8DN, UK tel: +44-1482 329 875 fax: +44-1482 223 114	Ultramarine pigments
J M Huber Corp., Solem Division, 4940 Peachtree Industrial Blvd, Suite 340, Norcross, GA 30071, USA tel: +1-770 441 1301 fax: +1-770 368 9908	
Hubron Manufacturing Div. Ltd., Albion Street, Failsworth, Manchester M35 0FF, UK tel: +44-161 681 2691 fax: +44-161 683 4658	
Hüls AG, Paul-Bauman-Str 1, Postfach 1320, D-45764 Marl, Germany tel: +49-23 6549-1 fax: +49-23 49-41 79 Hüls America Inc., 80 Centennial Ave, PO Box 456, Piscatawny, NJ 08855-0356, USA tel: +1-908 980 6800 fax: +1-908 981 5497	Accelerators, additives, additives for electrostatic spray paints, auxiliary rubber bonding agents, crosslinking agents, dispersing agents, emulsifiers, extenders, flame retardants (chlorinated paraffins), hardeners, mold release agents, paint auxiliaries, peroxides, processing aids, solvents, waxes
ICI Specialty Chemicals; 45 Everslaan, B-3078 Kortenberg, Belgium ICI Chlor-Chemicals, PO Box 14, Runcorn WA7 4QG, UK tel: +44-1928 514 444 fax: +44-1928 580 742	Antiblocking agents, antifoaming agents, antioxidants, antistats, demixing agents dispersing agents, emulsifiers, flame retardants, foam stabilizers, lubricants masterbatches, mold release agents, stabilizers, UV stabilizers nucleating agents
ICI Polymer Additives, Concord Pike and New Murphy Road, Wilmington, DE 19897, USA ICI Americas, Wilmington, DE 19897, USA	Antioxidants, antistats
Imperial Chemical Industries PLC, Mond Division, PO Box 13, Runcorn, Cheshire, UK	Chlorinated paraffins
Inbra Industrias Quimicas Ltd., AV Fagundes de Oliveira 190, Piraporinha Diadema, SP 09950-907, Brazil tel: +55-11 745 4133 fax: +55-11 746 2011	Epoxidized soyabean oil, azooliearbonamide blowing agents

Supplier/manufacturer	Additives
Incemin AG, Schachen 82, CH-5113 Holderbank, Switzerland tel: +41-62 893 56 27 fax: +41-62 893 20 17	Flame retardants, aluminium hydroxide, magnesium hydroxide, nucleating/anti blocking agents, synthetic wollastonite
Inducolor SA, De Bavaylei 66, B-1800 Vilvoorde, Belgium tel: +32-2 251 26 04 fax: +32-2 252 45 67	Color and pigment concentrates
Industrie Generali SpA, Via Milano 201, PO Box 28, I-21017 Samarate (Va), Italy tel: +39-331 220537 fax: +39-331 222492	
ISC Chemicals; Ltd., St Andrew's Road, Avonmouth, Bristol BS11 9HP, UK	Brominated compounds
Isochem Resins Co., 99 Cook Street, Lincoln, RI 02865, USA	Flame retardants: phosphate esters, halogenated phosphorus compounds, zinc borates
Italmaster Srl, Via Somma 72, 1-20012 Cuggiono (Milano), Italy tel: +39-2 9724 1328 fax: +39-2 9724 1333	
Jayant Oil Mills, 13 Sitafalwadi, Dr Mascarenhas Road, Mazgaon, Bombay 400 010, India tel: +91-22 373 8810 fax: +91-22 373 8107	Stabilisers, lubricants/antifogging agents, coupling agents, slip/antiblocking agents
Jayell Plastics (Mktg) Ltd, 4 Hawthornden Manor, Bramshall Road, Uttoxeter ST14 7PH, UK tel: +44-1889 566 647 fax: +44-1889 566 638	
Johnson Matthey plc, Precious Metals Division, Orchard Road, Royston SG8 5HE, UK tel: +44-1763 253385 fax: +44-1763 253419	
Jordan Chemical Co., Folcroft, PA 19032, USA	Antistatic agents
Jotun Polymer AS, PO Box 2061, Sandefjord N-320, Norway tel: +47-334 570-00 fax: +47-334 64614	Gelcoats, fillers
Kaiser Chemicals Div., Raiser Aluminum & Chemical Corp., 300 Lakeside Dr., Oakland, CA 94643, USA	Aluminum hydroxyde
Kali-Chemie AG Hans-Bockler-Allee 20, Postfach 220, D-3000 Hannover 1, Germany tel: +49-511 857-1 fax: +49-511 282126	Blowing agents, catalysts, flame retardants, fluorochloro hydrocarbons, solvents
Kaneka Belgium NV, Nijverheidsstraat 16, B-2260 Westerlo-Oevel, Belgium tel: +32 14 21 49 4 fax: +32 14 21 62 23	
Kaneka Corp 3-12, 1-Chome, Motoakasaka, Minato-ku, Tokyo, Japan tel: +81 3 3479 9647 fax: +81 3 3479 9699	Weather resistant impact modifiers, impact modifiers
Kaneka Texas Corp., 17 South Briar Hollow, Houston, TX 77027, USA tel: +1-713 840 1751 fax: +1-713 552 0133	
Karle Finke Farbstoffe, Hatzfelder Strasse 174-176, D-56000 Wuppertal 2, Germany tel: +49-202 70 40 51	

Supplier/manufacturer	Additives
Kautschuk GmbH, Reutenveg 14, D-6000 Frankfurt, Germany tel: +49-69 159-0 fax: +49-69 1592 125	
Kay Agencies, 259 Ramsmruti, Ramdaspeth, Nagpur 440 010, India tel: +91-712 524 418 fax: +91-712 540 520	
Keil Chemical Div., Ferro Corp., 3000 Shelfield Ave, Hammond, IN 46320, USA	Chlorinated compounds
Kemira Pigments Oy, Speciality Products, FIN-28840 Pori, Finland tel: +358-39 341 000 fax: +358-39 341 919	
Kemira Pigments SA, Ave Einstein 11, R-1300 Wavre, Belgium tel: +32-10 232 711 fax: +32-10 229 892	
Kemira Polymers, Station Road, Birch Vale, Cheshire SK12 5BR, UK tel: +43-1663 746518 fax: +44-1663 736605	Polymeric plasticizers
Kenrich Petrochemicals Inc, 140 East 22nd Street, PO Box 32, Bayonne, NJ 07002-0032, USA tel: +1-201 823 9000 fax: +1-201 823 0691	Powder coupling agents
KMG Minerals Div. of Franklin Industrial Minerals, 1469 South Battleground Avenue, Kings Mountain, NC 28086, USA tel: +1-704 739 3616 fax: +1-704 739 7888	Calcium carbonate, mica, wet-ground mica
Korlin Concentrates, 501 Eric Street, Stratford, Ontario N5A 2N7, Canada tel: +1-519 271 1680	
Krahn Chemie GmbH, Grimm 10, D-52000 Hamburg 11, Germany tel: +49-40 32 92-0 fax: +49-40 32 92-322	
Kronos International Inc., Peschstrasse 5, D-1573 Leverkusen, Germany tel: +49-214 3 56-0 fax: +49-214 4 21 50	Titanium dioxide
Kronos Ltd, Barons Court, Manchester Road, Wilmslow SK9 1BQ, UK tel: +44-1625 547 200 fax: +44-1625 533 123	
Kvaerner A/S, PO Box 100, Skoyen N-0212 Oslo, Norway tel: +47-22 96 70 00 fax: +47-22 52 01 22	
Kyowa Hakko Kogyo Co. Ltd., Ohtemachi Building 6-1 Ohte-machi 1-chome, Chiyoda-Ku, Tokyo 100, Japan	Flame retardants: phosphate esters
Lankro Chem. Ltd., Manchester M30 0BH, UK	Antistatic agents
Lari Industria Thermoplastica, Via Baracca 7, I-21040 Vedano, Olona, Italy tel: +39 332 409 111	

Supplier/manufacturer	Additives
Latha Chemical Co, 7 Kumarappa Maistry St., Madras 600 001, India tel: +91-518 653	Carbon black, activated carbon
Laurel Industries Inc., 30195 Chagrin Blvd., Cleveland, OH 44124-5794, USA tel: +1-216 831 5747 fax: +1-216 831 8379	Antimony trioxide, low smoke flame retardant, polyester antimony catalyst
Lehmann & Voss & Co, Alsterufer 19, D-2000 Hamburg 36, Germany tel: +49-40 44197-1 fax: +49-40 44197 219	Auxiliary rubber agents, azodicarbon-amide, carbon blacks, catalysts, cross-linking agents, master batches, paint auxiliaries, thixotropic agents
Llewellyn Ryland Ltd., Haden Street, Birmingham B12 9DB, UK tel: +44-121 440 2284 fax: +44-121 440 028 1	
Gebrüder Lödige Maschinenbau GmbH, Postfach 2050, D-33050 Paderborn, Germany tel: +49-5251 3 09-0 fax: +49-5251 309-123	
Lohmann GmbH & Co KG, Irlicher Strasse 55, Postfach 1201 10, D-5450 Neuwied 12, Germany tel: +49-2631 7860 fax: +49-2631 78 64 67	
Lonza AG, Postfach 5633, Sins, Switzerland tel: +41-42-660111 fax: +41-42-662316 Lonza-Werke GmbH, Postfach 15XO, D-7858 Weil am Rhein, Germany tel: +49-7621 703-0 fax: +49-7621 703-255	Fillers, graphite, lubricants, bisamide wax lubricants, release agents, slip agents, dispersants, antistats antifogging agents, arometic amine chain extenders, sorbitan ester antistat/antifog agents, magnesium hydroxide, aluminium trihydrate
Lonza Inc., 17-17 Route 208, Fair Lawn, NJ 07310, USA tel: +1-201 794 2400 fax: +1-201 793 2515 Lonza SpA, Via Vittor Pisani 31, I-20124 Milan, Italy tel: +39-2 669991 fax: +39-2 669 876 30	
Luzenac, 131 Ave Charles de Gaulle, F-92200 Neuilly, France tel: +33-1 47 45 90 40 fax: +33-1 47 47 58 05	
Lucky Ltd, Twin Towers Bldg 20, Yoido-Dong, Yongdeung Po-ku, PO Box 672, Seoul, S Korea tel: +82-2 789 7407 fax: +82-2 787 7766	
Luperox GmbH, Denzinger Strasse 7, Postfach 1354, D-8870 Gunzburg/Do, Germany tel: +49-8221 980 fax: +49-8221 9 81 66	Organic peroxides
3M Specialty Fluoropolymers Department, 3M Center, Building 220-IOE-10, St. Paul MN 55144-1000, USA tel: +1-612-733-6760 fax: +1-612-737-7686	

Supplier/manufacturer	Additives
3M European Chemical Business Center, 3M (Antwerp) Belgium SA/NV, Canadastraat, B-2070 Zwijndrecht, Belgium tel: +32-2 252 0711	
3M Nederland BV, Industrieweg 24, NL-2382 NW Zoeterwoude, Netherlands tel: +31-715 540 450 fax: +31-715 540 212	
M & T Chemicals, Inc., PO Box 1104, Rahway, NJ 07065, USA	Flame retardants: antimony oxide, halogenated compounds
M & T Chemie SA, Allee des Vosges; la Defense 5, Cedex 54, F-92062, Paris la Defense, France	Flame retardants: halogenated compounds
Magnesia GmbH, PO Box 2168, D-21311 Lüneburg, Germany tel: +49-4131 520 11 fax: +49-4131 530 50	Magnesium oxide
Mallinckrodt Specialty Chemicals 16305 Swingley Ridge Drive, Chesterfield, MO 63017, USA tel: +1-314 895 2000 fax: +1 314 530 2562	
Mallickrodt Specialty Chemicals Europe, Postfach 1268, Industriestrasse 19-2, 1, D-6110 Dieburg, Germany tel: +49-6071 200 40 fax: +49-6071 200 444	
Makhteshim Chemical Works Ltd., PO Box 60, Beer Sheba 84100, Israel	Brominated compounds
Marine Magnesium Co., 995 Beaver Grade Road, Coraopolis, PA 15108, USA tel: +1-412 264 0200 fax: +1-412 264 9020	
R J Marshall Co, 26776 W Twelve Mile Road, Southfield, MT 48034-7807, USA tel: +1-313 353 4100 fax: +1-313 948 6460	Flame retardants, fillers, granite effect fillers, calcium sulphate
Martin Marietta Magnesia Specialties Inc., PO Box 15470, Baltimore, MD 21220-0470, USA tel: +1-510 780 5500 Martin Marietta Magnesia Specialties Inc., Executive Plaza II, Hunt Valley, MD 21030, USA tel: +1-301 667 0200	Flame retardants
Martinswerk GmbH, PO Box 1209, D-5010 Bergheim, Germany	Aluminum hydroxide, fillers, flame retardants
Marubeni Co. Fine Chemicals Sec, Osaka (A 777); Osaka 530-91, Japan	Brominated compounds
Mayzo Inc., 6577 Peachtree Industrial Blvd., Norcross, GA 30092, USA tel: +1-770 449 9066 fax: +1-770 449 9070	
Mearl Corp., PO Box 3030, 320 Old Briarcliff Road, Briarcliff Manor, NY 10510, USA tel: +1-914 923 9500 fax: +1-914 923 9594	Luster pigments

Supplier/manufacturer	Additives
Mearl Corporation, 41 East 42nd Street, New York, NY 10017, USA tel: +1-212 573 8500 fax: +1-212 557 0742	
Mearl International BV, Emrikweg 1X, 2031 BT Haarlem, Netherlands tel: +31-23 318 058 fax: +31-23 351 365	
Merator Inc., 560 Sylvan Avenue, Englewood Cliffs, NJ 07632, USA tel: +1-201 560 3533 fax: +1-201 569 7368	
Mercury Plastius Ltd, Vale Industrial Park, Tolpits Lane, Watford, Herts WD1 8QP, UK tel: +44-1923 778155 fax: +44-1923 771958	
Merck K GaA, Frankfurter Strasse 250, Postfach 4119, D-6427 I Darmstadt, Germany tel: +49-6151 726211 fax: +49-6151 771958	
Metallgesellschaft AG, Reuterweg 14, D-6000 Frankfurt 1, Germany	Aluminum hydroxide, master batches
Microbial Systems International, Gothic House, Barker Gate, The Lace Market, Nottingham NG1 1JU, UK tel: +44-115 952 1181 fax: +44-115 952 1281	
20 Microns Ltd, Talc Division, 307-308 Arundeep Complex, 3rd Floor, Race Course (South), Baroda 380 015, India tel: +91-265 330 714 fax: +91-265 333 755	Calcined china clay, talc, barytes, calcium carbonates
Micropol Ltd., Bayley Street, Stalybridge, Cheshire SK15 1QQ, UK tel: +44-161 330 5570 fax: +44-161 330 7687	Silicone lubricant masterbatch
Miles Inc., Mobay Road, Pittsburgh, PA 15205-9741, USA tel: +1-412 777 2000	
Milliken Chemical Div., Milliken & Co, M-400 PO Box 1927, Spartanburg, SC 29304-1927, USA tel: +1-864 503 2200 fax: +1-864 503 2430 Milliken Chemical, 18 Ham, B-9000 Gent, Belgium tel: +32-9 265 10 84 fax: +32-9 265 11 95	Clarifying agents for PP, reactive colorants
Mitsui & Co Deutschland GmbH, Konigsallee 63-65, D-40215 Düsseldorf, Germany tel: +49-211 93 86 347 fax: +49-211 93 86 348	UV absorber, silicone carbide fiber, graft copolymer compatibilizer, elastomer modifier for PO, aramid fiber
Mitsui Toatsu Chemicals Inc., 2-5 Kasumigaseki 3-chome, Chiyoda-ku, Tokyo 100, Japan	Flame retardants: phosphate esters, brominated compounds, alumina fiber, fused silica filler
Mobay Chemical Corp, Penn Lincoln Parkway West, Pittsburg, PA 15205, USA	Flame retardants: phosphate esters, antioxidants, antistatic agents
Mobil Chemical Co., Phosphorous Div., PO Box 26683, Richmond, VA 23261, USA	Flame retardants: phosphate eaters, halogenated phosphorus compounds

Supplier/manufacturer	Additives
Monsanto Chemicals SA, Av de Tervuren 270-272, B-1150 Brussels, Belgium tel: +32-2 761 41 11 fax: +32-2 761 40 40	Antioxidants, asbestos substitutes, impact modifiers, benzyl phthalate plasticizers
Monsanto Corp., 800 North Lindberg Boulevard, St. Louis, MO 63167, USA	Flame retardants: phosphate esters, chlorinated compounds, halogenated phosphorus compounds
Monsanto Co., Akron, OH-44314, USA	Antioxidants
Montell Polyolefins Functional Chemicals, Three Little Falls Centre, 2801 Centerville Road, PO Box 15439, Wilmington, DE 19850-5981, USA	Halogen-free flame retardants additive delivery systems
Montefluos, via Principe Eugenio 115, 1-20155 Milan, Italy	Phosphorus/nitrogen intumescent compounds
Morton Thiokol, Cincinnati, OH 45215, USA	Antioxidants
Morton International Inc., Morton Plastics Additives 2000 West Street, Cincinnati, OH 45215-3431, USA tel: +1-513 733 2100 fax: +1-513 733 2133	Multifunctional lubricating stabilizers
Multibase Inc., 3835 Copley Road, Copley, OH 44321, USA tel: +1-216 867 5124 fax: +1-216 666 7419	
Multibase SA, ZI du Guiers, F-38380 Saint-Laurent-du-Pont, France tel: +33-76 67 12 12 fax: +33-76 67 12 92	
Murtfeldt GmbH & Co KG, Postfach 120161, D-4600 Dortmund 12, Germany tel: +49-231 251012 fax: +49-231 251021	
Nemitz Kunststoff-Additive GmbH, PO Box 12271, D-48338 Altenberge, Germany tel: +49-2505 674 fax: +49-2505 3042	Chemical blowing agents, antioxidants, UV stabilizers, antistats, slip/anti-blocking agents
Neu Engineering Ltd, Planet House, Guildford Road, Woking GU22 7RL, UK tel: +44-1483-756565 fax: +44-1483-755177	
Nevicolor SpA, Via Maso 27, I-42045 Luzzara (Regio Emilia), Italy tel: +39-522 97 64 21 fax: +39-522 97 65 69	Self-extinguishing masterbatches, color/additive masterbatches, anti-sticking agents
Neville Chemical Co., Neville Island, Pittsburgh, PA 15225, USA	Chlorinated compounds
Nihon Kakaku Sangyo Co. Ltd., 20-5 shitaya 2-chome, Taito-ku, Tokyo 110, Japan tel: +81-3 3876 3131 fax: +81-3 3876 3278	
Nippon Chemical Industrial Co. Ltd., 15-1 Kameido 9-chome, Koto-ku, Tokyo 136, Japan	Phosphorus compounds
Nippon Oil and Fats Co., Chiyoda-ku, Tokyo, Japan	Organic peroxides

Supplier/manufacturer	Additives
Nippon Kayaku Co. Ltd., Tokyo Kaijo Building, 2-1, Marunouchi, I-chome, Chiyoda-ku, Tokyo 100, Japan	Flame retardants: brominated compounds, phosphorus compounds
Nissan Chemical Industries Ltd., Kowa-Hitotsubashi Building, 7-1, Kanda Nishiki-cho, 3-chome, Chiyoda-ku, Tokyo 101, Japan	Halogenated compounds, nitrogen-containing compounds
NOF Corporation, Yurakucho Bldg 2-8 Toranomon 1-chome, Chiyoda-ku, Tokyo 100, Japan tel: +81-3 3283 7295 fax: +81-3 3283 7178	
Norman Rassmann GmbH & Co., Kajen 2, D-2000 Hamburg 11, Germany tel: +49-40 36870 fax: +49-40 3687249	
Novalis Fibres, 129 rue Servient, BP 3052, F-69398 Lyon Cedex 03, France tel: +33-78 62 80 47	
Nouiy Chemicals, Chicago, IL 60606 USA	Organic peroxides
Npco Minerals Inc., 123 Mountain View Drive, Willsboro, NY 12996-0368, USA tel: +1-518 963 4262 fax: +1-518 963 4187	
Nyco Minerals Inc., Europe, Ordrupvej 24, PO Box 88, DK-2920 Charlottenlund, Denmark tel: +3164 33 70 fax: +31 64 37 10	Wollastonite (untreated), ultrafine wollastonite, surface modified wollastonite
Nyacol Inc., Megunco Road, PO Box 349, Ashland, MA 01721, USA	Antimony oxide
Obron Atlantic Corporation, PO Box 747, Painsville, OH 44077, USA tel: +1-216 354 0400 fax: +1-216 351 6224	Aluminium pigment concentrates
Occidental Chemical Corp Technology Center, 2801 Long Road, Grand Island, NY 14072, USA tel: +1-716 773 8100	
Occidental Chemical Corp., Occidental Tower, LBJ Freeway, Dallas, TX 75244, USA tel: +1-214 404 3800 fax: +1-214 404 2333	
Occidental Chemical Europe SA, Holidaystraat 5, B-1831 Diegem, Belgium tel: +32-2-715 6600 fax: +32-2-725 4676	
Occidental Chemical Corp., PO Box 728, Niagara Falls, NY 14302, USA	Chlorinated compounds
Olin Corporation, 120 Long Ridge Road, PO Box 1355, Stamford, CT 06904, USA tel: +1-203 356 3036 fax: +1-203 356 3273	Flame retardants: phosphate esters, halogenated phosphorus compounds, antioxidants
Olin Japan Inc., Shiozaki Bldg, 7-1 Hirakawa-cho 2-chome, Chiyoda-ku, Tokyo 102, Japan tel: +81-3 263 4615 fax: +81-3 023 24031	

Supplier/manufacturer	Additives
OM Group Inc., 2301 Scranton Road, Cleveland, OH 44113, USA tel: +1-216 781 8383 fax: +1-216 781 5919	Mixed metal heat stabilizers
Omya GmbH, Brohler Strasse 11a, D-50968 Köln, Germany tel: +49-221 37 75-0 fax: +49-221 37 18 64	Calcium carbonate, dispersing agents ultra-fine calcium carbonate
The Ore and Chemical Corp., 520 Madison Ave, 27th Floor, New York, NY 10022, USA tel: +1-212 715 5237 fax: +1-212 486 2742	
Osi Specialties Inc., 39 Old Ridgebury Rd., Danbury, CT 06810-5124, USA tel: +1-203 794 3402 fax: +1-203 794 1088	
Otsuka Chemical Co., 2-27 Ote-dori 3-chome, Chuo-ku, Osaka 540, Japan tel: +81-6 943 7711 fax: +81-6 946 0860	
Otsuka Chemical Co. Ltd., 10 Bungo-machi, Higashi-ku, Osaka 540, Japan	Halogenated compounds
Overbeck Handelsgesellschaft GmbH & Co, Breitenweg 29-33, D-28195 Bremen, Germany tel: +49-421 30 92 204 fax: +49-421 30 92 337	Steel wool (fibers)
Owens Coming World Headquarters, Fiberglas Tower, Toledo, OH 43659, USA tel: +1-419 248 8000 fax: +1-419 248 8592 T	Twin-form glass fiber, calcium carbonate, glass fibers
Owens Coming Fiberglass Europe, Chaussee de la Hulpe, B-1170 Brussels, Belgium tel: +32-2 674 82 11 178 fax: +32-2 660 85 72	
Papenmeier GmbH Mischtechnik, Postfach 2140, D-32828 Augustdorf, Germany tel: +49-5237 69-0 fax: +49-5237 6 92 31	
Patco Polymer Additives American Ingrediants Co., 3947 Broadway, Kansas City, MO 64111, USA tel: +1-816 561 9050 fax: +1-816 561 0422	Internal lubricants
Pearsall Chemical Div., Witco Chemical Corp., PO Box 437, Houston, TX 77001, USA	Flame retardants: chlorinated and brominated compounds, chlorinated paraffins
PEC GmbH, Koningstrasse 37-43, D-5202 Hennel 1, Germany tel: +49-2242 5091 fax: +49-2242 1244	
Penn-White Ltd., Radnor Park Trading Estate, Congleton, CN12 4XJ, UK tel: +44-1260 279 631 fax: +44-1260 278 263	
Pergan GmbH, Schlavenhorst 71, D-46395 Bocholt, Germany tel: +49-287 1 99 02-0 fax: +49-2871 99 02 50	Crosslinking agents, initiators, peroxides
Pergan GmbH, D-4190 Kleve, Germany	

Supplier/manufacturer	Additives
Periclase Mishor Rotem DN, Arova 8680, Israel	Magnesium hydroxide
Peroxid-Chemie GmbH, D-8023 Hollriegelskreuth, Germany	Crosslinking agents, hydrogen peroxide inhibitors, peroxides, inhibitors, peroxides, persulfate
Peroxid-Chemie GmbH, Dr-Gustav-Adolph-Strasse 3, D-82034 Pullach, Germany tel: +49-89-74422-0 fax: +49-89 74422-203	
Pierce & Stevens Corporation, 710 Ohio Street (14072), Box 10192, Buffalo, NY 14240, USA tel: +1-716 856 4910 fax: +1-716 856 7530	Polymeric microspheres
Plalloy MTD BV Postbox 3035, 6460 HA Kerkrade, Netherlands tel: +31-45 546 46 53 fax: +31-45 536 25 23	
Plastichemix Industries, 6 Kirti Towers, Tilak Road, Vadodara 390 001, India tel: +91 265 421 844 fax: +91 265 422 129	
Plasticolors Inc., 2600 Michigan Avenue, Ashtabula, OH 44004, USA tel: +1-216 997 5137 fax: +1-216 992 3613	Pigment dispersions, thermoplastic elastomer additives, thickeners additive concentrates
Podell Industries Inc., 3 Entin Road, Clifton, NJ 07014, USA	Flame retardants, dispersions of phosphate esters, chlorinated paraffins, antimony oxide
Polar Minerals, 5060 North Royal Atlanta Drive, Suite 22, Tucker, GA 30084, USA tel: +1-404 934 4411 fax: +1-404 934 4376	
Polycolour Plastics Ltd., Unit 12, Stafford Park 12, Telford TF3 3BJ, UK tel: +44-1952 290 382 fax: +44-1952 290 386	
Polyrex Additives, Rte de Lausanne 23, CH-1400 Yverdon-les-Bains, Switzerland tel: +41-24 22 45 63 fax: +41-24 22 45 64	
Polyvel Inc., 120 N White Horse Pike, Hammonton, NJ 08037, USA tel: +1-609 567 0080 fax: +1-609 567 9522	
Potters-Ballotini, Postfach 12 03 33, D-40603 Düsseldorf, Germany tel: +49-211 29608-0 fax: +49-211 2960818	Conductive additives
Potters Industries Inc., Southpoint Corporate Headquarters, PO Box 840, Valley Forge, PA 19482-0840, USA tel: +1-610-651 4700 fax: +1-610-408 9723	
Potters-Ballotini GmbH, Morschheimer Strasse 9, PO Box 1226, D-6719 Kirchheimbolanden, Germany	Glass globe armoring, marble, hollow glass spheres, solid glass spheres

Supplier/manufacturer	Additives
PPG Industries Inc., PO Box 66251, Chicago, IL 60666, USA	Antimony oxide, brominated compounds, reinforcing silicas for rubbers, continuous strand mat
PP Composites Ltd., 39c Llondow Industrial Estate, Cowbridge, S. Glamorgan CF7 7PB, UK tel: +44-1446 775885 fax: +44-1446 775822	
PPG Industries Inc., One PPG Place, Pittsburgh, PA 15272, USA tel: +1-412 434 2261 fax: +1-412 434 2821	
PPG Industries Fiber Glass BV, PO Box 50, NL-9600 AB Hoogezand, Netherlands tel: +31-5980 13 911 fax: +31-5980 99 649	
PQ Corporation, Box X40, Valley Forge, PA 19482, USA tel: +1-610 651 4200 fax: +1-610 251 9124	Flame retardants
Prayon-Rupel SA Société Chimique, Gansbroekstraat 31, B-2870 Puurs (Ruisbroek-Antwerp), Belgium tel: +32-3 860 9200 fax: +32-3 886 3038	
Premix Oy, PO Box 12 05201 Rajamöki, Finland tel: +358-0-290 1066 fax: +358-0-290 3135	Additives
Provencale SA, 29 av Frederick Miatral, 83174 Brignoles Cedex, France tel: +33-4 94 72 83 00 fax: +33-4 94 59 04 55	Calcium carbonate, barium sulphate
Quantum Chemical Co., 11500 Northlake Drive, PO Box 429550, Cincinnati, OH 45249, USA tel: +1-513 530 6500 fax: +1-513 530 6313	
Quartz & Silice, BP 103, F-77793 Nemours Cedex, France tel: +33-1 64 45 4600 fax: +33-1 64 28 4511	Fused quartz yarn
Quimica Roveri Commercial Ltd., Rua Alvaro Fragoso 344, Vila Carioca, Srio Paolo/SP 04223-000, Brazil tel: +55-11 274 3466 fax: +55-11 63 5659	
Raschig AG, Postfach 211128, D-6700 Ludwigshafen, Germany tel: +49-621 56180 fax: +49-621 58 28 85	
Reagens SpA. Ind. Chimica, I-40016 San Giorgio di Piano, Bologna, Italy tel: +39-51 897 157 fax: +39-51 897 561	Antioxidants, lubricants, stabilizers
Redlands Minerals, PO Box 2, Retford Road, Worksop S81 7QQ, UK tel: +44-1909 537 800 fax: +44 1909 537 801	Doloamite, perlite functional fillers, calcium/magnesium carbonate fillers, white calcite
ReedSpectrum, Holden Industrial Park, Holden, MA 10520, USA tel: +1-508 529 6321	

Supplier/manufacturer	Additives
Reedy International Corp., 25 East Front St, Suite 200, Keyport, NJ 07735, USA tel: +1-908 264 1777 fax: +1-908 264 1189	Chemical blowing agents
Reheis Inc., PO Box 609, 235 Snyder Ave., Berkeley Heights, NJ 07922, USA tel: +1-908 464 1500 fax: +1-908 464 7726	
Reichold Chemicals Inc., Research Triangle Park, NC 27709, USA tel: +1-919 990 7500	Cure initiator
Reinhold Chemie AG, CH-5212 Hausen bei Brugg, Switzerland tel: +44-56 482222 fax: +41-56 482376	
Renger GmbH & Co. KG, D-8613 Strullendorf, Germany	Lacquers and varnishes
Repi SpA, Via del Vecchia Stazione 104-106, I-21050 Gorla Maggiore (Va), Italy tel: +39-331 614 001 fax: +39-331 619 537	Blowing agents, color pastes, phosphate flame retardants, color pastes (silicone rubber)
Resart GmbH, Postfach 3440, D-6500 Mainz, Germany tel: +49-6131 631-0 fax: +49-6131 631142	
Rhein Chemie Rheingau GmbH, Postfach 81 04 09, D-68204 Mannheim, Germany tel: +49-621 89 07-0 fax: +49-621 89 07-555	Release agents
Rheochem Manufacturing Co., 775 Mountain Boulevard Ste 214, Watchung, NJ 07060, USA tel: +1-908 757 0300 fax: +1-908 757 0607	
Rhône-Poulenc Chemicals, Secteur Spécialités Chimiques, Cedex No 29, F-92097 Paris-La Defense, France tel: +33-1 47 68 12 34 fax: +33-1 47 68 13 31	Antistatic agents
Rhône-Poulenc Chemicals, Poleacre Lane, Woodley, Stockport SK6 1PQ, UK tel: +44-161 910 1416 fax: +44-161 910, 1526	
Rhône-Poulenc Inc., CN 7500 Cranbury, NJ 08512-7500, USA tel: +1-609 860 4000	
Riedel-de-Haen AG, Wunstorferstrasse 40, D-3016 Seelze, Germany	Brominated compounds
James Robinson Ltd, PO Box 83, Hillhouse Lane, Huddersfield HD1 6BU, UK tel: +44-1484-435577 fax: +44-1484-435580	
Rohm & Haas Co., Independence Mall West, Philadelphia, PA 19105, USA tel: +1-215 592 3000	

Supplier/manufacturer	Additives
Rohm and Haas Deutschland GmbH, In der Kron 4, D-1000 Frankfurt/Main 90, Germany tel: +49-69 78 99 60 fax: +49-69 789 53 56	Impact modifiers, acrylic impact modifiers
RTP Company, PO Box 439, 580 East Front Street, Winona, MN 55987, USA	Color concentrates, propellants/blowing agents, purging compounds
RT Vanderbilt Co., Norwalk, CT 06855, USA	Antioxidants
Ruhr-Chemie, Hugenottenstrasse 105, Postfach 1429, D-6832 Friedrichsdorf, Germany tel: +49-6172 733283 fax: +49-6172 733141	
RV Chemicals Ltd, Widness, Cheshire WA8 0TE, UK tel: +44-151 424 6101 fax: +44-151 420 4330	
Sachtleben Chemie GmbH, Pestalozzistrasse 4, PO Box 17 04 54, D-4100 Duisburg 17 (Homberg), Germany tel: +49-2136 22-0 fax: +49-2136 226 50	Titanium dioxide pigments, zinc sulphide pigments
Sandoz AG, CH-4002, Basel, Switzerland	Phosphorus compounds
Sandoz Huningue SA 68330, Huningue, France	Antistatic agents
Sanko Chemical Co. Ltd., 16 Toori-cho 8-chome, Kurume City, Fukuoka 830, Japan	Phosphorus compounds
Sanyo Chemical Industries ltd., 11-1 Ikkyo Nomoto-cho, Higashiyama-ku, Kyoto 605, Japan tel: +81-75 541 4311 fax: +81-75 551 2557	
Sarma, Via Lainate 26, I-20010 Pogliano Milanese (MI), Italy tel: +39-2 932 55 363	Additive/color masterbatches, optical brightness, processing aids, antimicrobial agents, calcium carbonate masterbatches, purging agents, flame retardants, antifogging masterbatch, anti-slip masterbatch, antidegradation masterbatches, blowing agents masterbatches, heat/light stabilizers, radiation barriers/UV stabilizers, plasticizers
Sartomer Co., Oaklands Corporate Center, 468 Thomas Jones Way, Exton, PA 19341, USA tel: +1-215 363 41 17 fax: +1-215 363 4140	
Schering AG, Waldstrasse 14, D-4709 Bergkamen, Germany tel: +49-2307 651 fax: +49-2307 69805	
Scherun Talkum-Bergbau, PO Box 1329, D-6870 Hof/Saale, Germany	Talcum

Supplier/manufacturer	Additives
Schenectady Chemicals Inc., Schenectady, NY 12301, USA	Antioxidants
Sherex Chemical Co., Dublin, OH 43017, USA	Antistats
Schuller Mats and Reinforcements, PO Box 517, Toledo, OH 43697-0517, USA tel: +1-419 878 1504	Glass reinforcements
SCM Chemicals - Asia/Pacific, Lot 4, Old Coast Road, Australind WA 6230, Australia tel: +61-97 251 261 fax: +61-67 252 504	Titanium dioxide
SCM Chemicals - Europe, PO Box 26, Grimsby South Humberside DN37 8DP, UK tel: +44-1469 571 000 fax: +44-1469 571 234	
SCM Chemicals World HQ, 7 St Paul St Suite 1010, Baltimore, MD 21202, USA tel: +1-410 783 1120 fax: +1-410 783 1087	
Sekisui Chemical Co Ltd, 2-4-4, Nishi-Tenma, Kita-ku, Osaka 520, Japan tel: +81-6365 4122 fax: +81-6365 4370	
Shell International Chemical Co., Shell Centre, London SE1 7NA, UK tel: +44-171 934 1234 fax: +44-171 934 8060	
Sherwin-Williams Co., Chemical Div. PO Box 6506, Cleveland, OH 44101, USA	Molybdenum compounds
Sherwin-William Chemicals, 1700 W Fourth Street, Coffeyville, KS 67337, USA tel: +1-316 251 7276	
Silberline Ltd, Ranbeath Road, Leven, Fife KY8 5HD, UK tel: +44-1333-424 734 fax: +44-1333-421 369	
Sinloihi Co. Ltd., Kita-Yaesu Bldg, 3F 3-2-11 Nihonbashi, Chuo-ku Tokyo, Japan tel: +81-3 3281 1255 fax: +81-3 3281 1881	
Sintimid Hochleistungs, Kunstoffe GmbH, Postfach 138, A-6600 Reutte/Tirol, Austria tel: +43-5672 702625 fax: +43-5672 70501	
Sitra SpA, Via A Falcone 294, 1-80127 Napoli, Italy tel: +39-81 560 6565 fax: +39-81 560 5655	
SKW Trostberg AG, Postfach 12 62, D-83303 Trostberg, Germany tel: +49-8621 86-0 fax: +49-8621 29 11	Curing agent, powder coatings

Supplier/manufacturer	Additives
Slide Products Inc., PO Box 156, Wheeling, IL 60090, USA tel: +1-708 541 7220	
SNCI, F-74490 Saint-Jeoire-en-Faucigny, France tel: +33-50 35 95 20 fax: +33-50 35 97 01	
SN2A, BP 98, F-13133 Berre I'Etang, France tel: +33-42 10 22 42 fax: +33-42 74 10 15	High conductive acetylene black
Societe Francaise D'Organo Synthese, Neuilly-sur-Seine, France	Antioxidants
Societe Industrielle et Chimique de l'Aisne, PO Box 46, F-02301 Chauny, France	Antimony oxide
SOLEM Industries Inc., sub. J M Huber Corp., 4940 Peachtree Industrial Boulevard Norcross, GA 30071, USA tel: 1-770 441 1301 fax: 1-770 368 9908	Aluminum hydroxide
Solem Europe, PO Box 2 11, NL-9640 AE Veendam, Netherlands tel: +31-5987 51390 fax: +31-5987 51393	Flame retardants, magnesium hydroxide
Solvay et Cie, Rue du Prince Albert 33, B-1050 Brussels, Belgium tel: +32-2 509 61 11 fax: +32-2 509 51393	
Spencer Chemie, Postbus 234, 3130 AE Vlaardingen, Netherlands tel: +31-1 434 06 23 fax: +31-1460 32 38	
Song Wound Ltd., Suweon, 25, Korea	Antioxidants
Stalo-Chemicals GmbH, Postfach 1606, D-49383 Lohne, Germany tel: +49-4442 943-0 fax: +49-4442 943-200	
Struktol Co of America, 201 E Steels Corners Road, Box 1649, Stow, OH 44224-0649, USA tel: +1-216 928 5188 fax: +1-216 928 8726	
Subhash Chemical Industries, S Block Shed W-2 MIDC, Bhosari Poona, 411026 Maharashtra State, India tel: +91-212 790 632 fax: +91-212 792 554	
Suddeutsche Kalkstickstoffwerke AG, PO Box 1150/1160, D-8223 Trostberg, Germany	Nitrogen-containing compounds
Sudwest-Chemie GmbH, Postfach 2120, D-7910 Neu-Ulm, Germany tel: +49-731 70141 fax: +49-731 7070764	

Supplier/manufacturer	Additives
Sumitomo Chemical Co., Ltd., New Sumitomo Building, 155-chome, Kitahama, Higashi-ku, 541 Osaka, Japan 4-5-33 Kitahama; Chuo-ku, Osaka 541, Japan tel: +81-6 220 3272 fax: +81-6 220 3345	Flame retardants: antimony oxide, phosphorus compounds, halogenated compounds, antoxidants
Suzhou Plastic Auxiliaries Factory, Seal Sands Chemicals Ltd, Seal Sands, Middlesborough TS2 1UB, UK tel: +44-1642 546546 fax: +44-1642 546068	
Sun Chemical Corporation, 222 Bridge Plaza South, Fort Lee, NJ 07024, USA tel: +1-201 224 4600 fax: +1-201 223 4392	
Synco Srl, Via G Rossini 21, 1-50041 Calenzano (Fi), Italy tel: +39-55 8878 201	
Syncoglas NV, Drukkerijstraat 9, B-9240 Zele, Belgium tel: +32-52 457 611 fax: +32-52 449 502	Glass reinforcements
Synthecolor SA, 7 rue des Oziers, ZI du Vert Galant, F-95310 Saint-Ouen L'Aumone, France tel: +33-1 34 40 39 50 fax: +33-1 34 64 05 15	
Talcs de Luzenac, PO Box 1162, 31036 Toulouse Cedex, France	Fillers, talcum
Talksteinwerke, P Reithofer, A-8223 Stubenberg 152, Austria	Talcum
TBA Industrial Products Ltd., PO Box 40, Rochdale, Lancs OL12 7EQ, UK tel: +44-1706 47422 fax: +44-1706 46170	
Techno-Polymer, Postfach 1250, D-5982 Neuenrade, Germany tel: +49-2392 6355 fax: +49-2392 6400	
Teijin Chemicals Co. Ltd. 6-21,1-chome Nishishinbashi, Minato-ku, Tokyo, Japan	Brominated compounds
Teknor Apex International 505 Central Avenue, Pawtucket, RI 02861, USA tel: 1-401 725 8000	
Tekuma Kunststoff GmbH, Mollner Landstrasse 75, D-2000 Oststeinbeck, Germany tel: +49-40 7 12 0057	

Supplier/manufacturer	Additives
Tenax Fibers GmbH Co KG, D-42097 Wuppertal, Germany tel: +49-202 32 2338 fax: +49-202 32 2360	
Ter Hell Pjastic GmbH, Postfach 11648, D-3690 Herne, Germany tel: +49-2323 496 10	
Thai Oleochemicals Co. Ltd., 87 Yotha Koad, Taladnoi, Sampantawong, Bangkok 101001, Thailand tel: +66-2 235 9565 fax: +66-2 235 8448	
Thomas Swan & Co Ltd., Crookhall, Consett DH8 7ND, UK tel: +44-1207 505 131 fax: +44-1207 590 467	
Tintometer Co, Waterloo Road, Salisbury SP1 2JY, UK tel: +44-1722 327 242 fax: +44-1722 412 322	
Tioxide International, Lincoln House, 137-143 Hammersmith Rd., London W14 0QL, UK tel: +44-171 331 7746 fax: +44-171 331 7711	
Toho Chemical Industry Co. Ltd., No 1-1-5 Shintomi; Nihonbashi, Chuo-ku, Tokyo 104, Japan tel: +81-3 3555 3731 fax: +81-3 3555 3755	
Townsend Chemicals Pty Ltd., 1 14-126 Dandenong-Frankston Road, Dandenong, Vic 3175, Australia tel: +61-3 9793 6000 fax: +61-3 974 0723	Polymeric plasticizers
Tosoh Corporation, 7-7 Akasaka I-chome, Minato-ku, Tokyo 107, Japan tel: +81-3 3585 9891 fax: +81-3 3582 8120	
Tramaco GmbH, Siemensstrasse 1-3, D-25421 Pinneburg, Germany tel: +49-4101 706-02 fax: +49-4101 706-200	Chemical blowing agents
Transpek Industry, Ltd., Kalali Road, Atladra, Vadorara 390 012, India tel: +91-265 335 444 fax: +91-265 334 141	Sodium hydrosulphite, zinc formaldehyde sulphonate
Th. Boehme KG, 8192, Geretsried, Germany	Antistatic agents
Ube Industries Inc, 31.5 E Eisenhower Str 213, Ann Arbor, MI 48108, USA tel: +1-313 760 573 1	
UFV Ulmer Füllstoff Vertrieb GmbH, PO Box 1126, D-7906 Blaustein, Germany	Calcium carbonates, fillers, silicic acid

Supplier/manufacturer	Additives
Ulmer Fullstoff Vertrieb GmbH, Postfach 1240, D-7906 Blaustein, Germany tel: +49-7304 8171 fax: +49-7304 8176	
Unichema Germany, Postfach 100963, D-46429 Emmerich, Germany tel: +49-2822 720 fax: +49-2822 72276	
Unichema North America 4650 South Racine Ave, Chicago, IL 60609, USA tel: +1-312 376 9000 fax: +1-312 376 0095	Polyol additives, dimer and modifying agents, slip/antiblocking agents, lubricants
Union Carbide Corporation, Old Ridgebury Rd., Danbury, CT 06817, USA tel: +1-203 794 2382 fax: 1-203 794 3088	
Uniroyal Chemical Co. Inc., World Headquarters, Benson Road, Middlebury, CT 06749, USA tel: +1-203 573 2000 fax: +1-203 573 2489	
United Carbon India Ltd., NKM International House, Babubhai M Chinai Marg, Bombay 400 020, India tel: +91-2021 914 fax: +91-2850 406	
United Mineral and Chemical Corp., 1100 Valley Brook Ave., Lyndhurst, NJ 07071-3608, USA tel: +1-201 507 3300 fax: +1-201 507 1506	
Uniroyal Inc., Middlebury, CT 06749, USA	Antioxidants, metal deactivators
Uniroyal Ltd, Bromsgrove, Wores, UK	
US Borax, 26877 Tourney Road, Velancia, CA 91355-1847, USA tel: +1-805 287 5464 fax: +1-805 287 5545	Zinc borates
US Borax, 3075 Wilshire Boulevard, Los Angeles, CA 90010, USA	Zinc borate
US Peroxygen Div., Witco, Richmond, CA 94804, USA	Organic peroxide
US Gypsum Co., 125 S Franklin Street Chicago, IL 60606, USA tel: +1-312 606 4018 fax: +1-312 606 4519	
Vanetti Colori, Viale Kennedy 856, I-21050 Marnate (Va), Italy tel: +39-331 365 267 fax: +39-331 365 297	

Supplier/manufacturer	Additives
Vereinigte Aluminum Werke AG, PO Box 2468, D-5300 Bonn, Germany	Aluminum hydroxide
Verotex CertainTeed Corp Fiber Glass Reinforcements 750E Swedesford Road, PO Box 860 Valley Forge PA 19482 USA tel: +1-215 331 7000 fax: +1-215 293 1765	Glass fibers
Vetrotex International 767 Quai des Allobroges BP 929 F-73009 Chambery Cedex, France tel: +33-79 75 53 00 fax: +33-79 75 54 05	
Vianova Kunstharz AG Altmannsdorfer Strasse 104 A-1120, Vienna, Austria	Color pastes
Viba; Via dei Giovi 6 I-20032 Carmano - Milan, Italy tel: +39-2 615 0541 fax: +39-2 615 1723	Range of additives
Vielle Montagne 19, rue Richer F-75442 Paris, Cedex 09, France	Zinc oxide
Vinnolit Kunststoff GmbH Carl-Zeiss-Ring 25; D-85737 Ismaning, Germany tel: +49-89 96 103-0 fax: +49-89 96 103-103	
Vulnax International 922/3 Saint-Cloud, France	Antioxidants
Wacker Chemie GmbH, Hanns-Seidel-Platz D-8000 Munich, PO Box 22, Germany tel: +49-89 62 79 01 fax: +49-82 62 79 1770	Fillers, mold-release agents, silanes, silicic acid,solvents, thickeners, thixotropic agents
Ware Chemical Corp 1525 Stratford Avenue Stratford, CT 06497, USA	Flame retardants: phosphate esters, antimony oxide, barium metaborate, zinc borate
Werner & Pfleiderer GmbH Theodorstrasse 10; D-70469 Stuttgart, Germany tel: +49-711 897-0 fax: +49-711 897 3999	
Westdeutsche Quartzwerke Dr. Muller GmbH; Postfach 680 D-4270 Dorsten, Germany tel: +49-2362 200 50 fax: +49-2363 200599	
Wilson Color SA 2 rue Melville Wilson, B-5330 Assesse, Belgium tel: +32-83 660 211 fax: +32-83 660 363	
Witco Corporation, One American Lane, Greenwich, CT 0683 1-2550, USA tel: +1-203 552 3294 fax: +1-203 552 2886	Dicumylperoxide, MEK peroxide, peroxyester

Supplier/manufacturer	Additives
WIZ Chemicals, Via G Deledda 11, I-20025 Legnano (Mi) Italy tel: +39-331-545 654 fax: +39-331-546 900	
Worlee Chemie GmbH, Grusonstrasse 22, D-22113 Hamburg, Germany tel: +49-30 733 33-0 fax: +49-40 733 33-296	
Yoshitomi Pharmaceutical Ind. Ltd., Osaka 541, Japan	Antioxidants
Zepperling Kessler & Co., PO Box 1464, D-22904 Ahrensburg, Germany tel: +49-4102 515 10 fax: +49-4102 487 169	Antistats, color concentrates, flame retardants, master batches, concentrated conductive agents, conductive polymer
Zoltek, Carbon and Graphite Div., 3101 McKelvey Road, St Louis, MO 63044, USA tel: +1-314 291 5110 fax: +1-314 291 8536	
Zychem Ltd. 72 Grasmere Road, Gatley SK8 4RS, UK tel: +44-161 428 9102 fax: +44-161 428 3686	Release agents

Commercial light stabilizers for polyolefins of the 2-hydroxybenzophenone class

Trade name	Chemical structure	Melting point, °C	Application field
Advastab 48	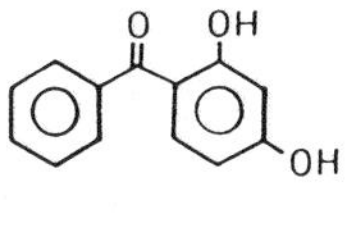	145	Polyolefins
Inhibitor DHBP		143–144	
Chemasorb 22	O OH	144–147	
Uvinul 400		136–140	
Uvistat 12	OH		
Rylex H			
Sumisorb 100			
Cyasorb UV-9	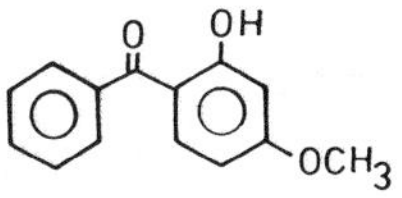	63–64.5	Polyolefin fibers
Uvinul M-40	O OH	62	
Advastab 45		63	
Uvistat 24	OCH_3	62.5–65	
Uvistat 90		63.5–64.5	
Chemasorb 90		62–65	
Cyasorb UV-24	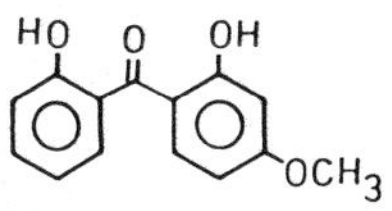 HO O OH	68–70	Polyolefins
Advastab 47	OCH_3		
Uvinul 490	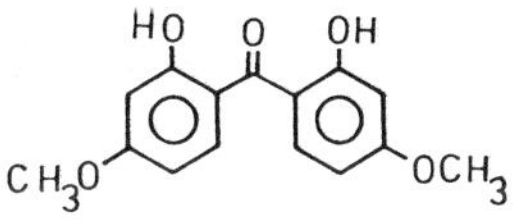 HO O OH CH_3O OCH_3	80	Polyethylene

Trade name	Chemical structure	Melting point, °C	Application field
Cyasorb UV-21	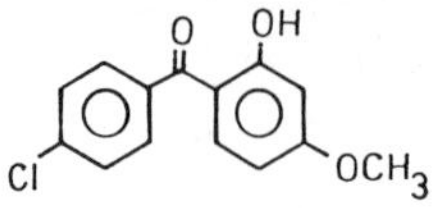 O OH Cl OCH_3	112–114	Polyethylene, polypropylene
Cyasorb UV-39	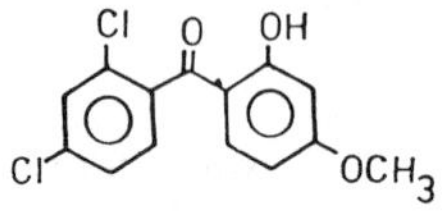 Cl O OH Cl OCH_3	68.5–69.5	Polyethylene
Uvistat 247 Chemasorb 71 Hoechst	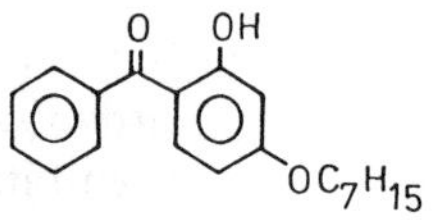 O OH OC_7H_{15}	62.5–65	Polyethylene, polypropylene
Ongrostab HOB	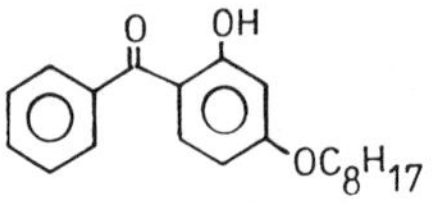 O OH OC_8H_{17}	45–49	Low density and high density polyethylene, polypropylene
Uvinul 408			
Advastab 46			
Cyasorb UV-531		48	
Chemasorb 81		45–49	
Chimassorb 81			
Carstab 700, 705		47	
Viosorb 130			
Sumisorb 130/Seesorb 102			
Chemasorb 125 DOBP Seesorb 103	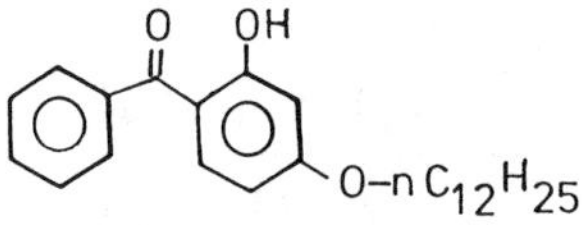 O OH O–n $C_{12}H_{25}$	44–46	Polypropylene, polyethylene
Dastib 242	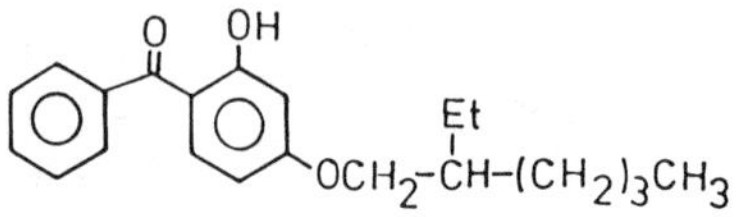 O OH Et OCH_2–CH–$(CH_2)_3CH_3$	Viscous oil	Polyethylene, polypropylene
Dastib 263	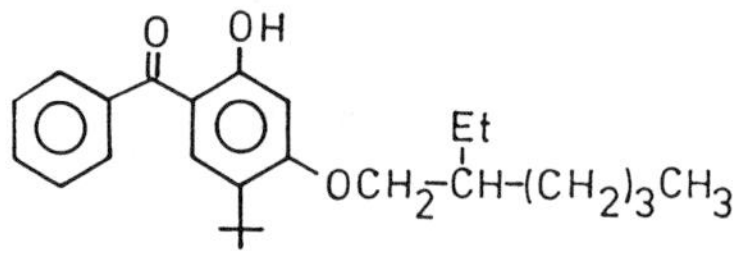 O OH Et OCH_2–CH–$(CH_2)_3CH_3$		Polyethylene, polypropylene
Permasorb A	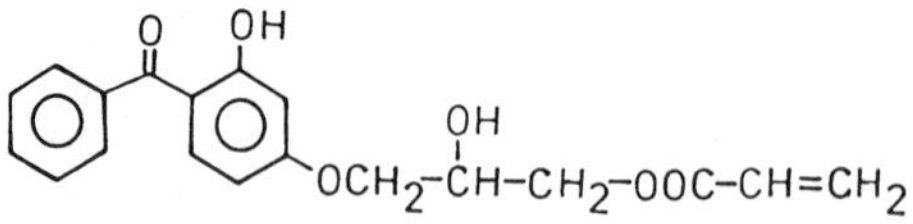 O OH OH OCH_2–CH–CH_2–OOC–CH=CH_2		Polyethylene, polypropylene
Permasorb MA	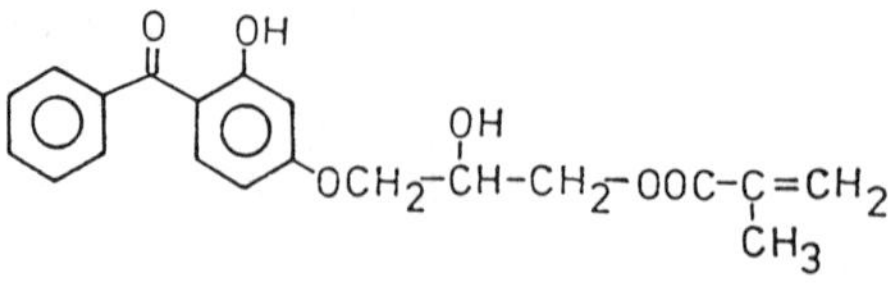 O OH OH OCH_2–CH–CH_2–OOC–C(=CH_2)–CH_3		Polyethylene, polypropylene
UV-CHEK-AM-340 Ferro Corp.	Hydroxybenzoate		PO

Commercial light stabilizers for polyolefins of the 2-hydroxyphenyl-benzotriazole class

Trade name	Chemical structure	Melting point, °C	Application field
LA-34 Asahi Denka Tinuvin PS	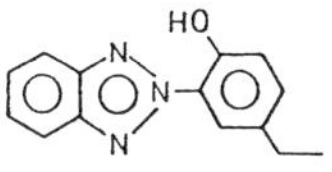		Polyolefins
Tinuvin 326 Usolvin VSE	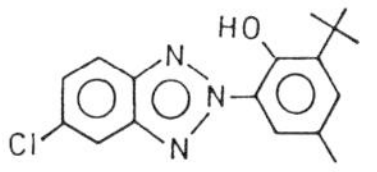	139–141 140	Polypropylene, polyethylene, polybutylene
Tinuvin 327 Viosorb 580 CIBA	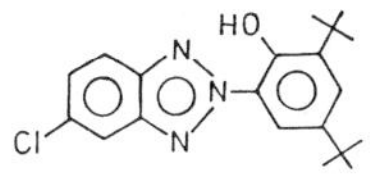	157 154–158	
Tinuvin 328	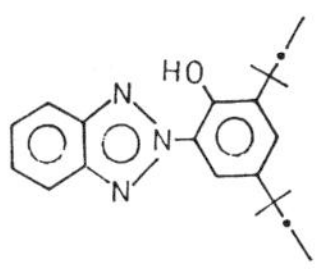	80–83	Polypropylene, polyethylene

Commercial phenylsalicylates as stabilizers for polyolefins

Trade name	Chemical structure	Melting point, °C
Salol UV-Absorber NL/1	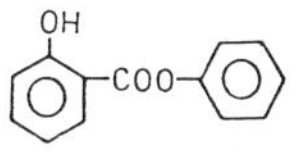	41–43
Eastman Inhibitor OPS	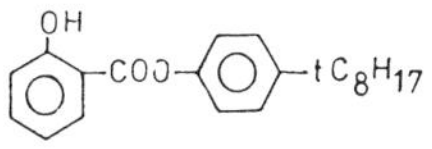	72–74

Some commercial nickel chelates used as light stabilizers for polyolefins

Trade name	Chemical structure	Melting point, °C	Recommended use/ photostabilizing mechanism
Ferro AM 101	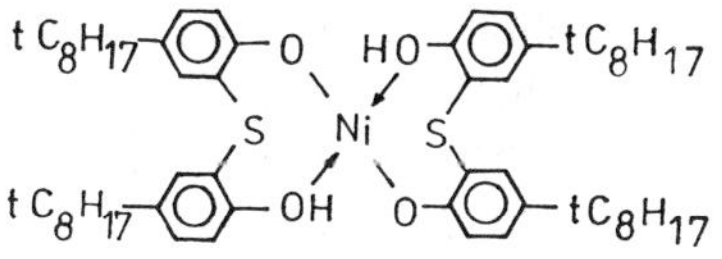		Isotactic polyolefins Excited state quenching; UV absorption
Cyasorb UV 1084 Chimassorb N-705	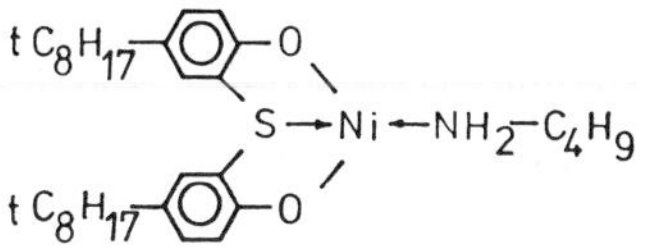	258–261	Polyethylene; polypropylene Excited state quenching; scavenging of free radicals; UV absorption

Some commercial nickel chelates used as light stabilizers for polyolefins

Trade name	Chemical structure	Melting point, °C	Recommended use/ photostabilizing mechanism
Cyasorb UV-2409	tC8H17, O, C6H11, S→Ni←N–(CH2)2OH, tC8H17, O, (CH2)2OH	370 (decomp)	Polyolefins Excited state quenching; scavenging of free radicals; UV absorption
Irgastab 2002	EtO, O, O, OEt, HO–CH2–P, Ni, P–CH2–OH, O, O	180–200	Polyolefins Scavenging of free radicals; metal deactivator
Sanduvor NPU	C9H19, C9H19, H3C–, C=O, Ni, O=C, –CH3, N, N, N, O, O, N, C6H5, C6H5	140–160	LDPE, PP fibers Synergism with Tinuvin 328 and Cyasorb 531

Commercial hindered amine light stabilizers for polyolefins

Trade name	Chemical structure	Physical properties[a]	Recommended applications
Tinuvin 144[b]	n-Bu, COO, N–, C, CH2, COO, N–, HO	White powder, m.p. 146–150°C	Polypropylene (PP)
Tinuvin 622[b]	[O–N-(CH2)2–OOC(CH2)2COO-(CH2)2–N–OOC(CH2)2–CO]n	White powder, melting range 130–145°C; $\bar{M}_n > 2000$; soluble in aromatic and chlorinated hydrocarbons	PP LDPE
Tinuvin 770[b]	OOC-(CH2)8–COO, N, H, N, H	White powder; m.p. 81–86°C; soluble in aromatic and chlorinated hydrocarbons, methanol, acetone	PP HDPE
Sanol LS 744[c]	OOC, N, H	White powder, m.p. 94–95°C; solubility; same as Tinuvin 770	PP HDPE

Commercial hindered amine light stabilizers for polyolefins

Trade name	Chemical structure	Physical properties[a]	Recommended applications
Sanol[c] LS 2626	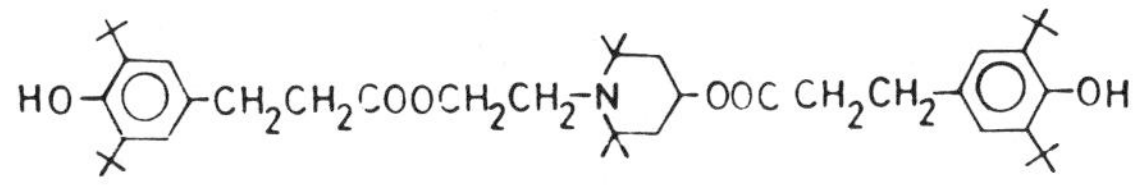 HO–⌬–$CH_2CH_2COOCH_2CH_2$–N⟨⟩–OOC CH_2CH_2–⌬–OH		Polyolefins
Spinuvex A 36[d]	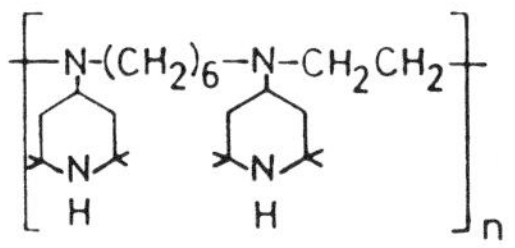 [–N–$(CH_2)_6$–N–CH_2CH_2–]$_n$ (N–H piperidine rings)	Off-white, nondusting, granular; melting range 90–100°C ($\bar{M}_n$ = 1800–2200; soluble in chloroform, benzene; practically insoluble in acetone and ethyl acetate	PP HDPE
Chimassorb 944[b]	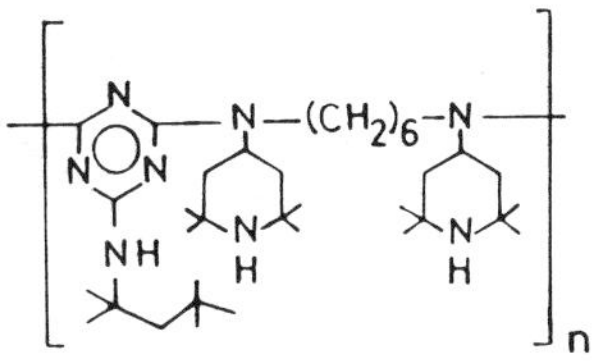 [triazine–N–$(CH_2)_6$–N–]$_n$ (NH; N–H piperidine rings)	Off-white powder; softening range 110–150°C; $\bar{M}_n$ > 2500; soluble in chloroform and hydrocarbons; slightly soluble in low alcohols	PP fibers PP tapes LDPE film Ethylene–vinyl acetate film
Cyasorb 3346[e]	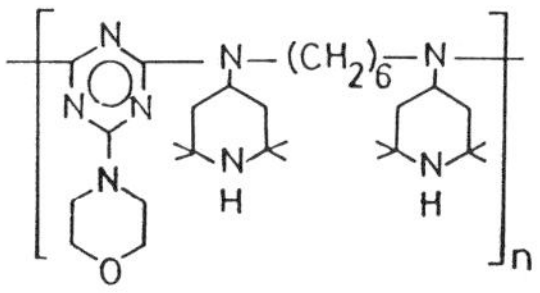 [triazine–N–$(CH_2)_6$–N–]$_n$ (morpholine; N–H piperidine rings)		
Hostavin TM-N 20	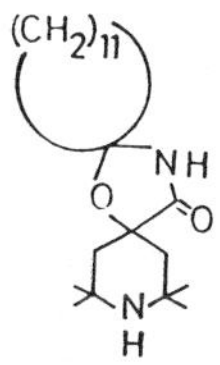 $(CH_2)_{11}$; NH; O; =O; N–H	m.p. 221°C (from ethanol)	Polyolefins PP and HDPE for special applications
Good-rite UV 3034[g]	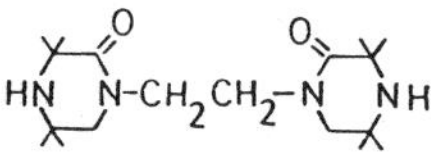 HN⟨⟩N–CH_2CH_2–N⟨⟩NH (=O)	Crystalline white powder, m.p. 134–136°C; very soluble in ethanol, dichloromethane; soluble in toluene	PP stretched tape PP film HDPE
Good-rite 3150[g]	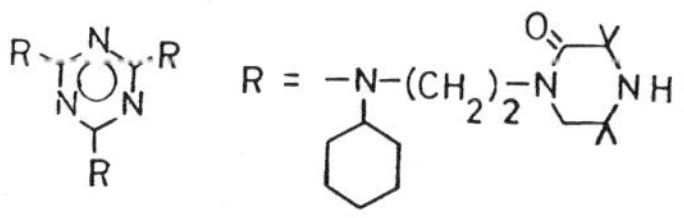 triazine R, R, R; R = –N–$(CH_2)_2$–N⟨⟩NH (cyclohexyl, =O)		PP fibers PP textiles (automotive, marine) Outdoor fabrics and carpets

[a] All commercial HALS are practically insoluble in water. [b] Product of Ciba-Geigy AG (Switzerland). [c] Product of Sankyo Company (Japan). [d] Product of Montell (Italy). [e] Product of American Cyanamid Company (U.S.A.). [f] Product of Hoechst AG (Germany). [g] Product of BF Goodrich Company (U.S.A.)

Index